FOODS & NUTRITION ENCYCLOPEDIA

2nd EDITION

Volume 1
A-H

Audrey H. Ensminger
M. E. Ensminger
James E. Konlande
John R. K. Robson

CRC Press is an imprint of the
Taylor & Francis Group, an **informa** business

Dedicated to —

improved food and nutrition for all people
in a world broken by unshared bread

CRC Press
Taylor & Francis Group
6000 Broken Sound Parkway NW, Suite 300 Boca
Raton, FL 33487-2742

First issued in paperback 2019

ISBN-13: 978-0-367-44965-0 (pbk)
ISBN-13: 978-0-8493-8981-8 (hbk)

Visit the Taylor & Francis Web site at
http://www.taylorandfrancis.com

and the CRC Press Web site at
http://www.crcpress.com

Library of Congress Card Number 93-36692

Library of Congress Cataloging-in-Publication Data

Food & nutrition encyclopedia / Audrey H. Ensminger ...[et al.] —
2nd ed.
p. cm.
Includes bibliographical references and index.
ISBN 0-8493-8980-1 (set). — ISBN 0-8493-8981-X (v.1). — ISBN
0-8493-8982-8 (v. 2)
1. Nutrition — Encyclopedias. 2. Food — Encyclopedias.
I. Ensminger, Audrey. II. Title: Foods and nutrition encyclopedia.
TX349.F575 1993
641′.03—dc20 93-36692
CIP

ABOUT THE AUTHORS

Audrey H. Ensminger, whose expertise is human nutrition, is Adjunct Professor, California State University-Fresno. She (1) completed the B.S. degree in Home Economics, University of Manitoba, Canada, and the M.S. degree in Home Economics, Washington State University; (2) taught at the University of Manitoba, the University of Minnesota, and Washington State University; and (3) served as dietitian for the U.S. Air Force, at Washington State University, during World War II. Audrey Ensminger has lectured throughout the world. She is the senior author of the two widely used human nutrition books, *Foods & Nutrition Encyclopedia* and *Food for Health.*

M. E. Ensminger, whose field is nutrition and biochemistry, is President, Agriservices Foundation, a nonprofit foundation serving world agriculture. Dr. Ensminger (1) completed B.S. and M.S. degrees at the University of Missouri, and the Ph.D. at the University of Minnesota; (2) served on the staffs of the University of Massachusetts, the University of Minnesota, and Washington State University; and (3) served as Consultant, General Electric Company, Nucleonics Department (Atomic Energy Commission). Dr. Ensminger is the author or co-author of 21 widely used books that are in several languages and used throughout the world. Dr. Ensminger is Adjunct Professor, California State University-Fresno; Adjunct Professor, the University of Arizona-Tucson; and Distinguished Professor, the University of Wisconsin-River Falls.

James E. Konlande completed the B.A. degree at Brooklyn College, Brooklyn, NY; and the M.S. and Ph.D. degrees at Rutgers University, New Brunswick, NJ, with a major in physiology, biochemistry, and nutrition. Prior to co-authoring FOODS & NUTRITION ENCYCLOPEDIA, he served as (1) Assistant Professor, Nutrition, the University of Michigan School of Public Health, Ann Arbor, Michigan; and (2) Head, Foods and Nutrition, Winthrop College (The University of South Carolina), Rock Hill, South Carolina.

John R. K. Robson, M.D. was educated in England. He completed (1) the Bachelor of Medicine (M.B.) and Bachelor of Surgery (B.S.) degrees at Durham University, Kings College Medical School; (2) the Diploma in Tropical Medicine and Hygiene (D.T.M.&H. Edin.), Edinburgh University Medical School; (3) Diploma in Public Health—Special subject Nutrition (D.P.H. London), London University; and (4) the Doctor of Medicine (M.D.) at the University of Newcastle on Tyne, Medical School, and he is a Certified Specialist in Clinical Nutrition, American Board of Nutrition. Dr. Robson has done human nutrition work throughout the world. Also, he served as Professor of nutrition and Director of Nutrition Program, School of Public Health, University of Michigan; and as Professor of Nutrition and Medicine, Medical University of South Carolina, Charleston; and Executive Editor, *Ecology of Food and Nutrition*—an international journal. He is a Fellow of the Royal Society of Tropical Medicine and Hygiene; and a member of the New York Academy of Sciences, the Nutrition Society of Great Britain, the Scientific Council Population Food Fund, and the American Society of Parenteral and Enteral Nutrition.

PREFACE

We never had it so good! For most Americans, the 20th century brought substantial improvements in their standard of living. Greater human productivity and technological advances reduced the six-day, 70-hour work week of the turn of the century to a five-day, 40-hour week. Additionally, most people are able to afford comfortable living quarters, a television set, an automobile, an annual vacation, and educational opportunities for their children. Yet, an uneasy concern hangs over most everyone. They're concerned that their current life-style may be jeopardizing their health. Additionally, the media abound with news about low-level radiation, polluted air and water, and cancer-causing chemicals in our food.

THE GOOD OLD DAYS. The Good Old Days were not so good! Most Americans believe that people were healthier and happier before the advent of nuclear power, pesticides, food additives, and computers. True enough, earlier generations were not concerned about the hazards of technology, but they had a more immediate concern—all-too-frequent early death. In 1900, U.S. life expectancy at birth was only 47.3 years.

Nutritional deficiency diseases were common at the turn of the century. Pellagra, scurvy, beriberi, goiter, rickets, and anemia were major causes of disease, morbidity, and death. Today, these diseases are almost nonexistent, thanks to a better understanding of human nutritional requirements, increased variety of foods, and fortification of foods with minerals and vitamins.

A CENTURY OF PROGRESS. We have been blessed with a century of nutritional progress. The cause and cure of nutritional diseases was not understood until early in the 1900s when the biological approach in experiments—the use of laboratory animals (largely white rats and mice, guinea pigs, chickens, pigeons, and dogs)—was ushered in. Their diets were made up of relatively pure nutrients (proteins, carbohydrates, fats, and minerals)—using casein or albumen, lard, and a pure carbohydrate such as dextrin. Deficiencies followed. Then, it was discovered that dramatic cures resulted when minute amounts of a certain mineral(s) or vitamin(s) were added.

Using the biological approach, the mineral era was ushered in during the 20th century. Phosphorus was found to be essential for people and animals in 1918; copper in 1925; magnesium, manganese, and molybdenum in 1931; zinc in 1934; and cobalt in 1935. But the essentiality of selenium was not discovered until 1957, and of chromium until 1959. Then, as recently as 1972, it was found that fluorine and silicon are essential. Also, using the biological approach, the vitamin era was ushered in during the 20th century. In 1912, Casimir Funk, a Polish scientist working in London, discovered thiamin. The discovery of vitamin A followed in 1913. Since that time, the growth of the vitamin family, the isolation of many vitamins, the partial solution of the puzzle of vitamin functions in the body, the discovery of the amazing therapeutic value of minute quantities of these vitamins in the cure of deficiency diseases, the numerous determinations of food composition with respect to vitamins, and the synthesis of most of them, have had a profound effect in nutrition.

THE BIOTECHNOLOGY ERA. On May 23, 1977, scientists at the University of California-San Francisco reported a major breakthrough as a result of altering genes—turning ordinary bacteria into factories capable of producing insulin, so essential to the survival of diabetics. This feat gave rise to a major scientific revolution, called *biotechnology*. While some aspects of biotechnology are decades away from commercial production, others are near, and still others are here now. The entire food team—producers, processors, and consumers—will benefit from biotechnology, from the greater abundance of high-quality products produced more efficiently.

PUBLIC HEALTH MEASURES. Public health measures, more than any other factor, were responsible for reducing the death rate from infectious diseases—the major killers at the turn of the century. Filtering systems, followed by chlorination, gradually eliminated waterborne bacteria that caused such serious diseases as typhoid fever and cholera. Also, pasteurization of milk reduced the transmission of tuberculosis.

HEALTH MEASUREMENTS. The health of a population is usually measured in the following four ways:

1. **Longevity.** The most common measure of life expectancy is the average number of years that the newborn is expected to live. In 1900, the life expectancy in the United States was 47.3 years. In 1989, 89 years later, it was 75.2 years.
2. **Death rate.** The most dramatic indicator of America's improved health is the declining death rate. It fell from 1,800 deaths per 100,000 people per year in 1900 to 535 deaths per 100,000 people per year in 1988.
3. **Infant mortality.** At the turn of the century, about 100 of every 1,000 infants born in the United States died before their first birthday. In 1988, there were only 10 infant deaths per 1,000 births.
4. **Tooth decay.** At the turn of the century, scientists were just beginning to understand that tooth decay is caused by bacteria-fermenting nutrients in the mouth. Today, we have the methods available virtually to eliminate tooth decay. Yet, dental caries and periodontal disease continue to affect a large proportion of Americans and cause substantial pain, restriction of activity, work loss, and cost.

LEADING CAUSES OF DEATH. The leading causes of death in the United States have changed considerably since 1900.

In 1900, the 10 leading causes of death in descending order were: tuberculosis (accounting for 11% of deaths), pneumonia (accounting for 10%), diarrhea and enteritis, heart diseases, nephritis, accidents and violence, cerebrovascular disease, cancer, bronchitis, and diphtheria.

In 1989, the 10 leading causes of death in descending order were: heart disease, cancer, cerebrovascular disease, accidents and violence, lung diseases, pneumonia and influenza, diabetes mellitus, suicide, liver disease, and nephritis.

It is noteworthy that in 1989 (1) only one of the 10 leading causes of death was an infectious disease (namely, pneumonia and influenza); (2) the leading causes of death were chronic diseases of middle- and old-age; (3) heart disease, stroke, and related disorders accounted for 43% of all deaths; (4) cancer ranked second as cause of death; (5) accidents ranked fourth; and (6) lung diseases (chronic bronchitis, emphysema, and asthma) ranked fifth as cause of death, with cigarette smoking accounting for 80 to 90% of deaths due to lung diseases.

GRAYER BUT HEALTHIER. In 1920, only 4.7% of Americans lived beyond age 65. In 1989, approximately 12.5% lived past 65 and 1.2% lived to be 85 or more.

Although 45% of the elderly must limit their activities somewhat due to heart conditions, arthritis, hearing loss, or visual impairment, the vast majority retain their independence—only 6% of those over 65 are in nursing homes.

PREVENTING PREVENTABLE DEATHS. Each person has the power to reduce significantly his/her chances of dying from such feared killers as heart disease and stroke, cancer, or lung diseases. Here is how:

1. **Preventing heart disease and stroke.** Heredity, male gender, and increasing age are risk factors for heart disease and stroke that cannot be changed. However, the following major risk factors can be controlled: cigarette smoking, high blood pressure, and elevated blood cholesterol.

Cigarette smoking is the major preventable risk factor for heart disease—approximately 30% of all coronary heart disease deaths are attributable to smoking. When people quit smoking, the risk of dying from cardiovascular disease declines rapidly. Ten years after quitting, it is almost the same as that of a lifetime nonsmoker.

High blood pressure (hypertension), which afflicts 15 to 25% of Americans, increases the risk of heart disease (heart attack, congestive heart failure, or stroke) and kidney failure. Fortunately blood pressure may be controlled by proper diet, weight reduction, not smoking, restricted salt intake, exercise, and/or medication.

Blood cholesterol levels are affected by age, gender, genetics, and diet. The ratio of low-density lipoprotein (LDL, or "bad") cholesterol to high-density lipoprotein (HDL, or "good") cholesterol in the blood is also an important determinant of heart disease. The Food and Nutrition Board's Committee on Diet and Health recommends that the fat content of the U.S. diet not exceed 30% of the caloric intake, that less than 10% of the calories should be provided from saturated fatty acids, and that dietary cholesterol should be less than 300 mg/day. When the blood cholesterol is extremely high or diet fails to bring it down sufficiently, medication may be necessary.

Additional measures that may help reduce the risk of heart disease include maintaining a reasonable body weight, eating a nutritionally balanced diet, exercising regularly, and reducing stress.

2. **Preventing cancer.** There are two ways to reduce the risk of dying from cancer:

a. Avoid factors known to cause cancer. This includes (1) avoiding smoking cigarettes, which are responsible for about 30% of the deaths from cancer; (2) avoiding unprotected exposure to the sun, which is responsible for almost all of the skin cancer that occurs in the United States; and (3) avoiding excessive consumption of any one food (eat a variety of foods), because diet may be responsible for as much as 35% of cancers.

b. Have cancer check-up as part of the regular physical examination, so as to detect cancer early when there is a greater chance of cure.

3. **Preventing lung diseases.** Cigarette smoking is the major cause of lung diseases in the United States among both men and women; it accounts for 80 to 90% of all lung disease deaths.

WE STILL EAT THE WRONG FOODS. Americans are jogging, quitting smoking, buckling seat belts, and installing smoke detectors—all of which will likely help them live longer. But they are risking disease and early death by eating too much of the wrong foods. Cholesterol intake appears to have decreased since the 1980s; but efforts to limit sugar and salt, get enough vitamins, and control weight have slipped. Such trends cannot be ignored because dietary factors are associated with 5 of the 10 leading causes of death, including heart disease, some cancers, and stroke. Next to stopping cigarette smoking and alcohol abuse, choosing the proper diet is the most important action people can take to improve their health and lessen their chances of contracting disease. A recent study also revealed that only 19% of adults maintain their proper weight for their height, age, and sex.

CONCERN ABOUT NUTRITION AT ALL-TIME HIGH. A 1992 survey revealed that 64% of all consumers were concerned about the nutritional content of their food purchases.

Also, this same survey revealed that half of these consumers fretted about the fat in their food—it was their single largest worry.

A NEW FOOD GUIDE PYRAMID. For almost 50 years, the four food groups were arranged on a wheel, which was hung in classrooms all over America. In 1992, the U.S. Department of Agriculture released the *Food Guide Pyramid*, featuring six food groups, designed to reflect the changing eating habits of consumers and giving the department's official recommendations of what is good for you. The *Food Guide Pyramid* and the six food groups are featured in the "F" Section of *Foods & Nutrition Encyclopedia.*

THE PAST IS PROLOGUE. Despite the many advances that typify nutrition, there are many unknowns. So, the search goes on. But the past is prologue! The future promises foods that are tastier, more nutritious, more abundant, and biologically altered. During the next two decades, America's grocery stores will experience the most dramatic transformation since the first supermarket opened in La Habra, California, in 1915 (See *Foods & Nutrition Encyclopedia*, p. 789, Fig. F-40, for a picture of the first supermarket in America). The transformation will not be in new products introduced by food manufacturers, but in the biotechnology wizardry that produced them.

In the decades ahead, consumers will be able to eat cholesterol-free butter, cheese, and eggs; vegetables produced without the use of pesticides and herbicides; disease-resistant crops; foods that prevent cancer; meats in which saturated fats have been converted into unsaturated fats; wheat bread that contains adequate lysine and methionine; food products with improved mineral, vitamin, and fiber content, along with better flavor and longer shelf life; exotic fruits like kiwi, mango, and papaya available the year-round; and a great array of new products designed for microwave cooking. The pace of the biotechnology revolution will depend on two factors: (1) scientific progress, and (2) public acceptance.

BRIDGE THE GAP AND SPEED THE PROCESS. *Foods & Nutrition Encyclopedia* represents the joint effort of four well-known professional nutritionists. It's for consumers all—for everyone who seeks good health, and for those who counsel with them—physicians, dentists, nutritionists, health experts, and others in allied fields. It's for teachers and students. It's for those who produce, process, and market foods. It's for those who wish to know the "why" of foods and nutrition—for those who wish to be educated rather than indoctrinated. It's for those who want the facts—both the pros and cons—on which they may base a judgment. It's for those who wish to know from whence their food comes—from field to table. It's for those who are concerned with getting the most nutrition from their food dollars. It's for those who desire authoritative and pleasurable reading about foods and nutrition, and their relationship to health.

Foods & Nutrition Encyclopedia is not intended as a source of home remedies. Rather, it is the fervent desire of the authors that the enlightened person will institute a health and disease prevention program, call upon the family doctor when ill health strikes, and be better qualified to assist the M.D. in carrying out the prescribed treatment. Most of all, the authors hope that *Foods & Nutrition Encyclopedia* will bridge the gap between awareness and the application of nutrition, and speed the process of buoyant good health.

ACKNOWLEDGMENTS. For Dr. and Mrs. M. E. Ensminger, *Foods and Nutrition Encyclopedia* was their greatest challenge. They, along with a staff of 6 to 8, devoted more than 7 years to the preparation of the first edition; and they have been revising and updating it ever since. It was a team approach; Dr. and Mrs. Ensminger gratefully acknowledge the contributions of all those who assisted them in the preparation of *Foods & Nutrition Encyclopedia*, Second Edition, without whose dedicated efforts the mountainous task could not have been completed. Special appreciation is expressed to the following staff members for their commitment to excellence and a rigid schedule: Joan Wright and Janetta Shumway who deciphered the authors' hieroglyphics and put them through a typewriter; Randall and Susan Rapp, Rapp Publishing & Typographic Services, who typeset the many changes for the second edition; Margo Williams who prepared the new art that enhances the second edition, Jean Nelson who proofread the copy, and Ran Guang Liang who prepared the superb cover design. Also, we shall be ever grateful to Robin Spencer Palmisano, who at the time of preparing the first edition, was Systems Dietitian, University Hospitals, The Ohio State University, Columbus, Ohio, a very special person and dedicated professional, who contributed so much to Table F-36 Food Compositions. (Presently, Robin Spencer Palmisano is a lawyer and a member of the firm of McGlinchey, Stafford, Cellini & Lang, New Orleans, Louisiana.) Further, we are grateful to Ron Bruce, President, Unisoft Systems Associates, 1340 Dublin Road, Columbus, Ohio, 43215, for permission to continue to use Table F-36 Food Compositions. At appropriate places, due acknowledgement and sincere appreciation is expressed to those who responded so liberally to our call for information and pictures. In the preparation of this preface, authoritative information and statistics were secured from *America's Health*, published by American Council of Science and Health, New York. The unnumbered line drawings in *Foods & Nutrition Encyclopedia* were created by Dynamic Graphics, Inc. If *Foods & Nutrition Encyclopedia* will result in people living healthier, happier, and longer—and enjoying buoyant good health—the authors will feel amply rewarded.

A. H. Ensminger
M. E. Ensminger
J. E. Konlande
J. R. K. Robson

Clovis, California (January, 1994)

INTRODUCTION

(This is an overall view of *Foods & Nutrition Encyclopedia*, and how to use it.)

True to its title, *Foods & Nutrition Encyclopedia*, covers the whole gamut of the three-pronged subject, foods–nutrition–health. A simple definition of each of these terms follows:

A food is any material that is taken or absorbed into the body of an organism for the purpose of satisfying hunger, growth, maintenance, tissue repair, reproduction, work, or pleasure.

Nutrition is the science of food and its nutrients and their relation to health.

Health is the state of complete physical, mental, and social well-being.

Like other sciences, foods and nutrition do not stand alone. They draw heavily on the basic findings of chemistry, biochemistry, physics, microbiology, physiology, medicine, genetics, mathematics, endocrinology, and, most recently, behavior, cellular biology, and genetic engineering. It is noteworthy, too, that much of our knowledge of human foods and nutrition came the animal route. Human nutritionists often use animal subjects for studies, because people usually prove too time-consuming, too costly, too difficult to control, and perhaps undesirable from a medical standpoint.

An overall view of *Foods & Nutrition Encyclopedia* along with instructions on how to use it, is contained in the sections that follow.

AUTHORITY. The authoritativeness of a work is evidenced by the credentials of those who created it. *Foods & Nutrition Encyclopedia* represents the team efforts of four well-known professionals in the field of foods-nutrition-health.

GOALS. At the outset, the authors stated as their objective: To produce the most complete and in-depth foods and nutrition source ever. If they haven't lived up to their objective, it's their fault; and they are willing to let the readers of the encyclopedia make judgments.

(Also see PREFACE.)

SELECTION OF TOPICS. The selection of the topics that are covered in *Foods & Nutrition Encyclopedia* was based on the author's professional experiences and perceptions relative to the needs of consumers, with application to many related fields, including medicine, dentistry, nursing, dietetics, teaching, nutritional science, public health, athletics, homemaking, and food production, processing, distribution and marketing. As the authors selected and treated each of the subjects, they were guided by the simple question: "Is it helpful?" If the answer was in the affirmative, they preceded with the philosophy of "let the chips fall where they may," no matter how complex or how controversial the subject. Also, the authors were ever aware that some contemporary interests are transient; so, they strived to achieve a balance between the timely and timeless.

COMPREHENSIVENESS. *Foods & Nutrition Encyclopedia* covers all aspects of foods–nutrition–health, with adequate historical and interpretive context. Each article includes all relevant aspects of the topic; thus, a few dictionary-type definitions were considered sufficient. Also, the entries reflect the whole gamut of foods–nutrition–health; there are precious few items in any other book on foods and nutrition not found as a subject entry and/or an index entry in *Foods & Nutrition Encyclopedia*.

ILLUSTRATIONS. *Foods & Nutrition Encyclopedia* is profusely illustrated with more than 1,600 pictures and figures, including 16 pages of colored pictures, all of which enhance the work. Each illustration is placed in close proximity to that portion of the subject matter which it clarifies, supplements, or complements. Most of the line drawings were created exclusively for *Foods & Nutrition Encyclopedia* by staff artists.

READABILITY. Writings are like food—both must be used to be of value. In recognition of this fact, the authors made a sincere effort to adhere to the following "musts of good writing":

1. Write as if they were speaking to a college professor, but so that an eavesdropping eighth grader would, quickly and easily, understand what they're saying and get the point.
2. Present human interest and historical events related to the subject.
3. Make events come alive; write with vigor and energy that jumps at the reader.
4. Use a simple, direct style that is easily comprehended.
5. Use words that are easily understood, but use technical terms where needed, with such words defined immediately in the article, thus ensuring understanding.
6. Present both U.S. Customary and Metric Systems of weights and measures.

OBJECTIVITY. To the end that this encyclopedia would be reliable, the authors resolved that the facts and inferences must be accurate and reflect current scholarship. Scholars may and do differ among themselves. Under these circumstances, the authors presented the opposing points of view—the pros and cons, then editorialized.

EASE OF USE. *Foods & Nutrition Encyclopedia* is organized so that readers may quickly and easily find the information that they are seeking. This is achieved through—

1. All topics being arranged alphabetically, using the word-by-word system—the same system that is used in a dictionary or in a library card catalog.

2. Cross referencing to related articles; usually at the end of an article, but within the text of an article when it makes for greater convenience.

3. A comprehensive index, which makes it possible for the reader, easily and quickly, to make a systematic survey of all parts and locations of *Foods & Nutrition Encyclopedia* directly or indirectly pertaining to a given subject. Additionally, where more than one reference page is listed in the index for a given subject, the main section is listed in bold numbers.

ALTERNATE NAMES. Alternate names are used wherever appropriate, with each name indexed, so that the article can be located under any of the alternate names; for example—

Vitamin C (Ascorbic acid)

FOOTNOTING THE NEW AND THE CONTROVERSIAL. The day and age of complete documentation of the literature used as background material for a work of this kind is past—it would be too voluminous. So, *Foods & Nutrition Encyclopedia* is an interpretation, correlation, and application of research findings, based on an extensive review of the literature by the authors. However, literature pertaining to new or controversial material has been documented in footnotes wherever possible.

FOOD COMPOSITION TABLE. Table F-36 Food Compositions is complete; and the edges of the pages are black, thus making for easy and speedy finding. It's without a peer.

(See FOOD COMPOSITIONS.)

PHYSICAL FORMAT. The format of *Foods & Nutrition Encyclopedia* is designed for quick and easy use, including—

1. **Outline.** An outline at the beginning of each major article which gives the reader an overall view of the article and shows the interrelationship of units. For example, four of the selected outline heading of corn follow:

CORN

PROCESSING CORN.

Corn Milling.

WET MILLING.

2. **Dot (bullet) system.** Within a given section, a "dot system" (or "bullet system") is used when it is appropriate, and when it will make for ease of use; for example, under the section of Kinds of Corn, each of the six kinds of corn is made to stand out like dent corn as follows:

- **Dent corn**—This is the most common variety.

3. **Tables.** Tabular presentation, which provides quick answers in concise form, is used wherever possible; for example—

TABLE C-45
CORN PRODUCTS AND USES

Product	Description	Uses	Comments

ABERNETHY BISCUIT

A hard biscuit containing caraway seeds, named after an English surgeon who, in the early 1800s, treated maladies with diet.

ABSINTHE

A yellowish-green alcoholic beverage made from oils of wormwood and anise. The use of absinthe can destroy the nerve centers of the brain. Production of absinthe is banned in many countries, including the United States.

ABSORPTION

• The transfer of a substance through a membrane or the taking in of nutrients or other substances from an outside source; e.g., the passage of substances into the blood and/or lymph system from the digestive tract, through the skin, or by way of the lungs.

• The uptake of water, fat, or other substances by foods.
(Also see DIGESTION AND ABSORPTION.)

ABSORPTION METER

An instrument used to measure the absorption of light, by which a quantitative measure of a colored substance in a solution may be obtained. Many substances, such as minerals, vitamins, and amino acids, will react with a particular reagent to form a colored complex. Since the color developed is proportional to the amount of the substance present, its quantity may be measured by an absorption meter.

ACCLIMATIZE

The process of becoming adjusted to a new environment, especially temperature, altitude, or climate. In individuals acclimatized to high temperatures, the loss of sodium in sweat is decreased, and, to some extent, this loss is compensated for by decreased urinary sodium excretion.

ACEROLA

This cherrylike fruit, which is native to the Americas, grows in tropical and semitropical regions. The edible portion is the richest known source of ascorbic acid (vitamin C). It contains from 1,000 to 4,000 mg of ascorbic acid per 3½ oz (*100 g*) serving. However, most Americans who consume acerola take it in the form of vitamin tablets which supply both vitamin C and vitaminlike substances called bioflavonoids.

(Also see ASCORBIC ACID; FRUIT[S], Table F-47 Fruits of the World; and BIOFLAVONOIDS.)

ACETATE

A salt of acetic acid. The name also refers to the metabolic product called acetylcoenzyme A.

(Also see METABOLISM.)

ACETIC ACID

An organic acid which is produced by (1) the metabolism of nutrients, and (2) vinegar fermentation. It reacts with alkalis to form acetates.

(Also see ADDITIVES, Table A-3; and VINEGAR.)

ACETOACETIC ACID

This ketone acid is formed when the quantities of fat which are burned for energy greatly exceed the amounts of carbohydrate which undergo metabolism. An excessive accumulation of acetoacetic in the body results in the condition called ketosis.

(Also see ACID-BASE BALANCE.)

ACETOBACTER *Acetobacteriaceae*

A genus of bacteria used aerobically (with air) to convert alcohol to acetic acid. An acetobacter, *Acetobacter pasteurianus*, is used in the making of vinegar.

ACETO-GLYCERIDES (PARTIAL GLYCERIDE ESTERS)

These differ from triglycerides in that either one (or sometimes two) of the long chain fatty acids attached to the glycerol molecule is replaced by acetic acid. They are nongreasy and have lower melting points than the corresponding triglycerides.

Aceto-glycerides are used in shortenings and spreads, as films for coating foods, and as plasticisers for hard fats.

ACETOIN

Acetyl methyl carbinol, $C_4H_8O_2$, precursor of diacetyl, which imparts the flavor of butter. Acetoin is a product of fermentation, produced by bacteria during the ripening of cream for churning and by the action of yeast on diacetyl.

ACETONE

A substance that gives the breath a fruity odor. It may accumulate in the blood, breath, and urine when there is an abnormal metabolism of fats.

(Also see KETONE BODIES.)

ACETONEMIA

A buildup of acetone in the blood. This condition occurs only when there is an impairment in breathing. Under normal circumstances most acetone is removed from the body via the breath.

ACETONURIA

The passing of abnormally large amounts of acetone in the urine. This may be a sign of ketosis, which may be brought on by (1) diets high in fat, and low in carbohydrate; (2) starvation; or (3) diabetes.

(Also see DIABETES MELLITUS; KETONE BODIES; and STARVATION.)

ACETYLCHOLINE

A substance released from the ending of certain nerves which (1) stimulates the digestive functions, and (2) slows the heart rate and lowers the blood pressure. It is noteworthy that digestive disorders characterized by hypersecretion of digestive juices and spasms of the alimentary tract are often treated with medications like atropine and belladonna, which counteract acetylcholine and its effects.

(Also see ATROPINE; BELLADONNA DRUGS; and DIGESTION AND ABSORPTION.)

ACETYLCOENZYME A

A key substance formed during the metabolism of carbohydrates, fats, and proteins. It plays important roles in (1) the production of energy, carbon dioxide, and water from the intermediate products of metabolism, and (2) the synthesis of fatty acids, ketone bodies, acetylcholine, cholesterol, and related compounds.

(Also see METABOLISM.)

ACHALASIA

A malfunctioning of the muscular coat around the esophagus. There is less than the normal amount of peristalsis and the lower sphincter fails to relax. Hence, solid food tends to remain in the esophagus because it cannot pass into the stomach. The condition may be corrected by surgery, or by stretching the lower sphincter with a dilator.

(Also see DIGESTION AND ABSORPTION.)

ACH INDEX (ARM, CHEST, HIP INDEX)

This refers to the arm girth, chest diameter, and hip width. It is sometimes used as a method of assessing the state of nutrition.

ACHLORHYDRIA

A condition in which there is a lack of hydrochloric acid in the gastric juices. Achlorhydria often afflicts patients who have pernicious anemia, and sometimes occurs in elderly people. Reports indicate that achlorhydria exists in from 9.8 to 35.4% of the people over 60 years of age. It is suspected that some people develop this disorder because their blood contains antibodies which attack the acid-secreting cells of the stomach and cause them to atrophy and cease functioning. (The development of antibodies against one's own tissues is called autoimmunity.) Lack of stomach acid may also be due to, or aggravated by, deficiencies of iron, niacin, or vitamin B-6. A severe shortage of acid may reduce the absorption of such essential mineral nutrients as calcium, magnesium, copper, iron, and zinc. Some cases of achlorhydria may be corrected by the administration of mineral and vitamin supplements.

(Also see AUTOIMMUNITY; DIGESTION AND ABSORPTION; and DISEASES.)

ACHOLIC

An abnormal condition in which there is a lack of bile secretion from the liver. Acholic people may have poor absorption of fats and fat-soluble vitamins, unless they are given pills or tablets which contain bile preparations.

(Also see DIGESTION AND ABSORPTION.)

ACHROMOTRICHIA

A lack of pigment or graying of the hair. It occurs in rats as a result of pantothenic acid deficiency. The discovery of this condition led to the designation of pantothenic acid as the anti-gray hair vitamin. However, there is no evidence that it prevents the graying of hair in people.

(Also see PANTOTHENIC ACID.)

ACID

A substance which has a pH of 6.9 or lower and is capable of turning litmus indicators red. It is responsible for the sour taste of foods such as lemons, pickles, tomatoes, vinegar, etc. Many acids occur naturally in foods, while others are added for flavoring, or to inhibit the growth of certain microorganisms associated with food spoilage. Acids also occur naturally in the body, such as the hydrochloric acid of the stomach.

(Also see ACID-BASE BALANCE; ACID FOODS AND ALKALINE FOODS; ADDITIVES; and PRESERVATIVES.)

ACID-ASH RESIDUE

A mineral residue which is left after (1) the other nutrients in a food have been metabolized, or (2) a food has been burned to an ash in a laboratory. Ashes of foods give an acid reaction when the predominant chemical elements are chlorine, phosphorus or sulfur, because these elements generally form acids. Foods which are most likely to have an acid residue are breads and cereal products, eggs, fish, meats, and poultry. Milk has an alkaline residue.

(Also see ACID FOODS AND ALKALINE FOODS.)

ACID-ASH RESIDUE DIETS

These diets leave an acid residue in the body because they contain foods rich in chlorine, phosphorus and/or sulfur—such as eggs, meat, fish, poultry, bread, and cereal products. They are often prescribed for patients with kidney stones (calculi) that are believed to be formed under alkaline conditions. Sometimes, these stones dissolve in acid. Therefore, acid-ash diets are used to keep the urine acid.

(Also see ACID FOODS AND ALKALINE FOODS; and DISEASES.)

ACID-BASE BALANCE

An acid is a chemical that can release hydrogen ions, whereas a base, or alkali, is a chemical that can accept hydrogen ions. For the pH, or hydrogen ion concentration, of extracellular fluid to remain normal, a balance between acids and bases must be maintained. This equilibrium is known as the acid-base balance; it refers to the hydrogen ion concentration in the body fluids. When the hydrogen ion concentration is high, the fluids are acidic—the condition is acidosis; when the hydrogen ion concentration is low, the fluids are basic—the condition is alkalosis. Since the chemical reactions of the cells depend very greatly on the hydrogen ion concentration, the acid-base balance must be regulated very precisely.

The degree of acidity is expressed in terms of pH. A pH of 7 is the neutral point between an acid and an alkaline (base). Substances with a lower pH than 7 are acid, while substances with a pH above 7 are alkaline. The normal pH of the extracellular fluids of the body is 7.4, with a range of 7.35 to 7.45. Maintenance of the pH within this narrow range is necessary to sustain the life of cells. The extremes between which life is possible are 7.0 to 7.8.

BUFFERS. The word *buffer* comes from a Middle English root meaning "to protect from blows." In the 17th century, it referred to a coat of armor. In chemistry, a buffer is a mixture of acidic and alkaline components, which protects a solution against wide variations in the pH, even when strong bases or acids are added to it. A solution containing such a protective mixture is called a buffer solution. A buffer protects the acid-base balance of a solution by rapidly offsetting changes in its ionized hydrogen concentration. It works by protecting against either added acid or base.

ACID-FORMING AND BASE-FORMING FOODS. The potential acidity or alkalinity of the foods ingested covers a wide range and depends on the minerals present. On combustion, certain foods, such as most vegetables and fruits, leave an ash in which the basic elements (sodium, potassium, calcium, and magnesium) predominate; hence, they are known as base-forming foods. Other foods, such as cereals, meat, and fish, leave an ash in which the acid-forming elements (chlorine, phosphorus, and sulfur) predominate; these are known as acid-forming foods. Although sulfur is present in foods mainly in neutral form in the sulfur-containing amino acids (methionine, cystine, cysteine), it is oxidized in the body to sulfuric acid; hence, it is an acid-forming mineral. Therefore, foods containing a large amount of protein are generally acid-forming. Contrary to popular belief, citrus fruits are not acid-forming. They do contain citric acid and acid potassium citrate, but the citrate radicals are completely metabolized in the body, leaving only potassium. Thus, many "acid" fruits are really base-forming foods.

(Also see ACID FOODS AND ALKALINE FOODS.)

REGULATION OF ACID-BASE BALANCE. Regulation of the acid-base balance refers to the control of the hydrogen ion concentration in the body fluids. It is very important that the pH of body fluids be maintained within the narrow, slightly alkaline range of 7.35 and 7.45, because variance from this range leads to disruption of normal body processes and activity. The acid-base balance is regulated by chemical buffers, by the lungs (respiratory), and by the kidneys (renal), as follows:

1. **Chemical Buffers.** All of the body fluids contain acid-base buffers. These are chemicals that can combine readily with any acid or base in such a way that they keep the acid or base from changing the pH of the fluids greatly. The three most important chemical buffers are the bicarbonate buffer, phosphate buffers, and protein buffers.

a. **Bicarbonate Buffer.** This buffer, which is present in all body fluids, is a mixture of carbonic acid (H_2CO_3) and bicarbonate ion (HCO_3^-). When a strong acid is added to this mixture, it combines immediately with the bicarbonate ion to form carbonic acid—an extremely weak acid. Thus, this buffer system changes a strong acid to a weak acid and keeps the fluids from becoming strongly acid. However, when a strong base is added to this mixture, the base immediately combines with the carbonic acid to form water and neutral bicarbonate salt.

Loss of the weak acid and the addition of the neutral salt scarcely affect the hydrogen ion concentration in the body fluids. Thus, the carbonic acid-bicarbonate buffer system protects the body fluids from becoming either too acidic or too basic.

b. **Phosphate buffers.** These chemical buffers are especially important for maintaining normal hydrogen ion concentration in the intracellular fluids, because their concentration inside the cells is many times as great as the concentration of the bicarbonate buffer.

c. **Protein buffers.** Like phosphate buffers, protein buffers, including hemoglobin, are especially important within the cells.

In essence, the chemical buffers of the body fluids are the first line of defense against changes in hydrogen ion concentration, for any acid or base added to the fluids immediately reacts with these buffers to prevent marked changes in the acid-base balance.

2. **Lungs (respiratory).** Carbon dioxide combines with water and electrolytes in the extracellular fluid to form carbonic acid in accordance with the following reaction:

$$CO_2 + H_2O \rightarrow H_2CO_3$$

Ultimately, the lungs control the body's supply of carbonic acid. This is so because, normally, respiration removes carbon dioxide at the same rate that it is formed by all cells of the body as one of the end products of metabolism. However, if respiration decreases below normal, carbon dioxide will not be excreted normally; instead, it will accumulate in the body fluids, causing an increase in the concentration of carbonic acid. As a result, the hydrogen ion concentration rises. On the other hand, if the respiration rate rises above normal, the opposite effect occurs; carbon dioxide is blown off at a more rapid rate than it is formed, thereby decreasing the carbon dioxide and carbonic acid concentrations. It is noteworthy that complete lack of breathing for a minute will reduce the pH of the extracellular fluid from the normal of 7.4 down to about 7.1, while over-breathing can increase it to about 7.7 in a minute. Thus, the acid-base balance of the body can be changed greatly by under- or over-ventilation of the lungs.

In the preceding paragraph, the effect of changing the rate of breathing on the acid-base balance is detailed. In this paragraph, the opposite effect of the acid-base balance on respiration is discussed. A high hydrogen ion concentration stimulates the respiratory center in the medulla of the brain, greatly enhancing the rate of ventilation. Conversely, a low hydrogen ion concentration depresses the rate of ventilation. So, the effect of the hydrogen ion concentration on the activity of the respiratory center affords an automatic mechanism for maintaining a fairly constant pH of the body fluids. That is, an increase in hydrogen ion concentration increases the rate of ventilation, which in turn removes carbonic acid from the fluids. Loss of the carbonic acid decreases the hydrogen ion concentration back toward normal. Conversely, diminished hydrogen ion concentration depresses the ventilation, and the hydrogen ion concentration rises back toward normal. This respiratory mechanism for regulating acid-base balance reacts almost immediately when the extracellular fluids become either too acidic or too basic. Thus, acidosis greatly increases both the depth and rate of respiration, while alkalosis lessens the depth and rate of respiration. This respiratory mechanism is so effective in regulating the acid-base balance that it usually returns the pH of the body fluids to normal within a few minutes after an acid or alkali has been administered.

3. **Kidneys (renal).** In addition to carbonic acid, a number of other acids are continually being formed by the metabolic process of the cells, including phosphoric, sulfuric, uric, and keto acids. On entering the extracellular fluids, all of these can cause acidosis. Normally, the kidneys rid the body of these excess acids as rapidly as they are formed, preventing an excessive build-up of hydrogen ions.

Occasionally, too many basic compounds enter the body fluids, rather than too many acidic compounds. This may occur when basic compounds are injected intravenously or when large quantities of alkaline food or drugs are consumed.

The kidneys regulate acid-base balance by (a) excreting hydrogen ions into the urine when the extracellular fluids are too acidic, and (b) excreting basic substances, particularly sodium bicarbonate, into the urine when the extracellular fluids become too alkaline.

The kidneys also conserve base by eliminating extra hydrogen ions through the production and excretion of ammonia (NH_4):

$$NH_3 \text{ (from deamination of amino acids)} + H^+ \rightarrow NH_4$$

If the normal amounts of buffers are present in the blood, and if the lungs and kidneys are normal, one can recover promptly from the effects of severe muscular exercise, intake of acid or base, unbalanced diets (so far as acids and bases are concerned), short periods of starvation, short bouts of vomiting, and other adverse conditions. But there are limits to the adjustments that the body can make; if the capacity is exceeded, the pH changes and the body cells are prevented from performing their functions normally.

ABNORMALITIES OF ACID-BASE BALANCE. Many disorders of the respiratory system, the kidneys, or the metabolic system for forming acids and bases can cause serious derangement of the acid-base balance. Some of the effects of these conditions, compared with the normal pH of the blood of 7.4, are shown in Fig. A-1; and a brief discussion of each condition follows:

Acidosis generally causes depressed mental activity, and, if unchecked, it may culminate in coma and death. Usually the afflicted person will pass into a coma when the pH of the extracellular fluid falls below 6.9.

Alkalosis causes overexcitability, resulting in excessive initiation of impulses, muscle contraction, and even convulsions.

In acidosis, the ionized hydrogen concentration is above normal. In alkalosis, ionized hydrogen concentration is below normal. Either of these abnormal states initiates compensatory responses of the chemical buffers, lungs, and kidneys, which cause body fluids to accept, to release, or to excrete ionized hydrogen. Increases and decreases in ionized hydrogen concentration are changed so that the pH is not significantly changed from its normal range of 7.35 to 7.45. Failure of either the lungs or the kidneys to carry out this function results in acidosis or alkalosis. If the failure is largely related to the lungs (respiratory), it is called respiratory acidosis or respiratory alkalosis. If the failure is mainly related to the kidneys, it is called metabolic acidosis or metabolic alkalosis.

Examples of diseases that affect the lungs and contribute to the development of respiratory acidosis are: pneumonia, emphysema, asthma, pulmonary edema, barbiturate poisoning, morphine poisoning, and congestive heart failure. Common causes of respiratory alkalosis are: extreme emotion, hysteria, or anxiety (causing hyperventilation); labored breathing in response to hot weather, high altitude, or fever (causing hyperpnea); excessive breathing forced upon a patient by a poorly adjusted mechanical respirator; or overstimulation of the respiratory center in the brain, which may be brought about by aspirin, poisoning, meningitis, or encephalitis.

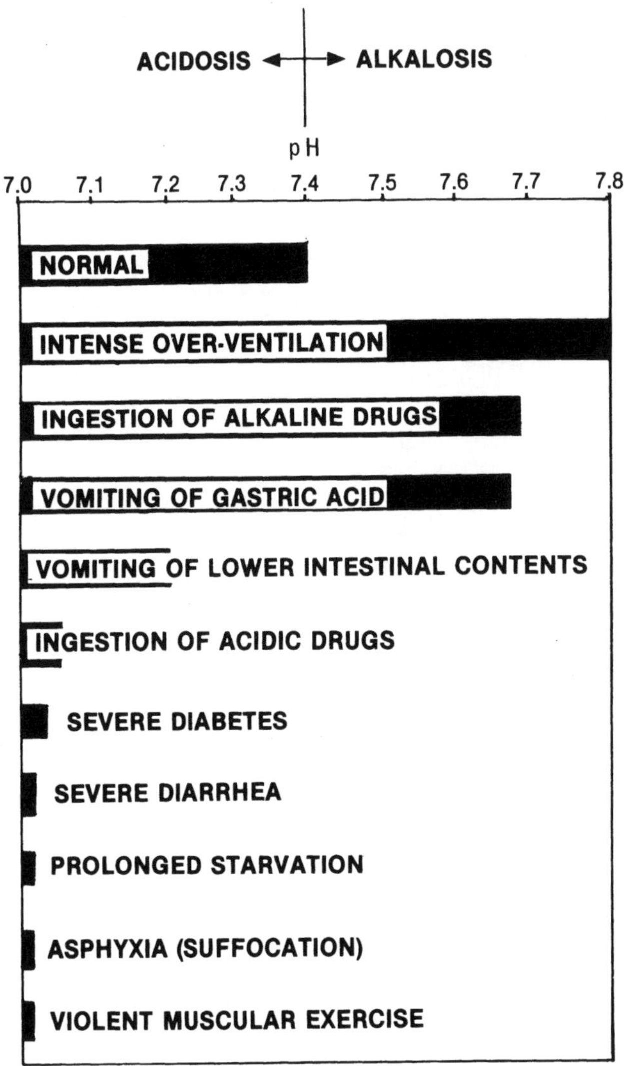

Fig. A-1. pH of the body fluids in various acid-base disorders compared with normal pH.

Examples of metabolic acidosis are: diabetic acidosis, brought about because the body cannot metabolize blood glucose properly and turns for its energy to the catabolism of protein and fat; starvation, when the body turns to its own stores of protein and fat to supply its needs; greatly accelerated metabolism, such as thyrotoxicosis, resulting in the depletion of carbohydrate stores and the burning of protein stores, followed by ketosis; gastrointestinal problems, such as may be caused by prolonged vomiting and the inability to eat, resulting in decreased carbohydrate intake, glycogen depletion, burning of body protein and fat, and finally ketosis; severe diarrhea, induced because large amounts of bicarbonate (HCO_3) and sodium (Na) are swept away with the intestinal contents; and chronic or acute kidney disease, as the kidneys become unable to cope with excess ionized hydrogen concentrations.

Examples of metabolic alkalosis include: initial vomiting, resulting in loss of ionized hydrogen and choline; potassium depletion (caused by insufficient potassium intake, gastrointestinal loss of potassium, or ACTH therapy), inducing alkalosis as ionized hydrogen and sodium move into the cells to replace the lost potassium; and excess intake of alkali powders or sodium bicarbonate, as in long-term ulcer therapy.

In all cases of acid-base inbalance, three basic rules should be followed: (1) rely on a medical doctor for diagnosis and prescribed treatment, (2) treat the primary cause of the acidosis or alkalosis, and (3) make every effort to aid the various compensatory responses of the lungs and kidneys.

ACID-BASE REACTION OF FOODS

This refers to the potential acidity or alkalinity of food, which depends upon the reaction they yield after being broken down (metabolized) in the body, thus releasing their mineral elements. These minerals function in maintaining the acid-base balance in the body. Acid-forming elements are chlorine, phosphorus, and sulfur, while base (alkaline)-forming elements are calcium, sodium, potassium, and magnesium. Foods can be classified as acid foods, alkaline (basic) foods, or neutral foods according to their mineral content.

(Also see ACID-BASE BALANCE; ACID FOODS AND ALKALINE FOODS; and WATER AND ELECTROLYTES, section headed "Acid-Base Balance.")

ACID DETERGENT FIBER (ADF)

The fiber which is extracted from plant foods with acidic detergent, a technique employed to determine indigestible matter. The extract is composed of cellulose and lignin, cell-wall substances which are not digested by man.

(Also see ANALYSIS OF FOODS; CARBOHYDRATES, UNAVAILABLE; and FIBER.)

ACID FOODS AND ALKALINE FOODS

The metabolism of many foods results in a small amount of mineral residue, or ash (so-called because it is similar to the ash remaining after burning material). Only highly refined foods, consisting chiefly of fats, sugars, or starches do not yield an ash. The processing of such foods has resulted in the removal of the mineral elements; the carbon, hydrogen, and oxygen which remain are metabolized to water, carbon dioxide, and energy.

A solution of the mineral residue of a food will, upon testing, give an acid, alkaline (basic), or neutral reaction, depending upon the relative proportions of acid-forming elements (chlorine, phosphorus, and sulfur) and of alkali-forming elements (potassium, sodium, calcium, and magnesium). The type of reaction of the food ash in water is important because it gives an indication of the contribution of the food to the acidity, alkalinity, or neutrality of the body fluids, and, ultimately, to the urine. The kidneys help to maintain the neutrality of the body fluids by excreting the excess acid or alkali in the urine.

FOODS YIELDING AN ACID RESIDUE. The acid-forming elements predominate over the alkali-forming mineral elements in foods containing moderate to large amounts of protein, with the exception of milk and some of the other dairy products which contain sufficient calcium to give an alkaline reaction. Whole grains give an acid reaction disproportionate to their protein content due to the extra phosphorus present in the form of phytates. Although most fruits have an alkaline ash, others like prunes, plums, and cranberries make a net contribution of acid to the body since they contain organic acids that are not metabolized by the body, but which pass unchanged into the urine.

FOODS YIELDING AN ALKALINE RESIDUE. Fruits and vegetables generally contain higher proportions of alkali-forming mineral elements than the acid-forming elements since their protein content is usually low. Corn and lentils, however, are acid forming. Surprising as it may seem, because of their pronounced acid taste, an alkaline residue is formed from tomatoes, citrus fruit, and rhubarb, due to their organic acids (citric, ascorbic, oxalic, and others) being completely metabolized in the body to carbon dioxide, water, and energy. Some nuts (coconuts, almonds, chestnuts) yield an alkaline ash, while others (peanuts, walnuts) yield an acid.

Table A-1 lists foods which yield acid, alkaline, or neutral ashes.

GAPS IN INFORMATION CONCERNING ACID FOODS AND ALKALINE FOODS. The classification of a food ash as acid or alkaline has metabolic significance only when there is certainty as to the proportion of the food minerals which are digested and absorbed. This is not the case with dairy products (a variable amount of dietary calcium is excreted in the stool) or with the phosphorus present as phytates in whole grains. Phytates further complicate the picture by binding with alkali-forming minerals and carrying them into the stool. Similar effects might be expected to occur when diets high in fiber are consumed. Certain additives in food mixtures, such as sodium aluminum phosphate used as an emulsifying agent in pasteurized, processed American cheese, may have an unpredictable effect on the food

TABLE A-1
ACID, ALKALINE, OR NEUTRAL FOODS[1]

Acid-Ash Foods	Alkaline (Basic)-Ash Foods	Neutral-Ash Foods
Bread:	**Cheese, American[2]**	**Arrowroot starch**
White	**Cream**	**Butter**
Whole wheat	**Fruit:**	**Candy, plain**
Rye	Apple	**Coffee**
Cake, plain	Apricots, raw	**Cornstarch**
Cereal:	Apricots, dried	**Lard**
Cornflakes	Banana	**Margarine**
Farina	Blackberries, raw	**Oil, vegetable**
Macaroni	Blueberries, raw	**Postum**
Oatmeal	Cantaloupe	**Sugar, white**
Puffed wheat	Cherries, fresh	**Syrup**
Puffed rice	Currants, fresh	**Tapioca**
Rice	Dates, dried	**Tea**
Shredded wheat	Figs, dried	
Fat, mayonnaise	Gooseberries	
Fruit:	Grapefruit	
Cranberries	Grapes	
Plums	Lemon	
Prunes	Lime	
Meat:	Loganberries	
Bacon	Mango	
Beef, roast	Nectarines	
Cheese, cheddar	Olives, green and ripe	
Cheese, cottage	Orange	
Chicken	Peach, raw	
Eggs	Pear, raw	
Fish, halibut	Persimmon	
Ham	Pineapple, raw	
Lamb	Pineapple juice	
Pork	Raisins	
Veal	Raspberries, black and red	
Nuts:	Strawberries	
Brazil nuts	Tangerine	
Peanut butter	Watermelon	
Peanuts	**Ice Cream**	
Walnuts, English	**Jam**	
Vegetables:	**Milk**	
Corn	**Nuts:**	
Lentils, dried	Almonds	
	Chestnuts	
	Coconut, fresh	
	Sweets	
	Molasses, medium	
	Vegetables:	
	Asparagus	
	Beans, baked	
	Beans, lima	
	Beans, navy	
	Beans, snap	
	Beets	
	Beet greens	
	Broccoli	
	Cabbage, cooked	
	Carrots	
	Cauliflower	
	Celery	
	Chard, Swiss	
	Cucumber	
	Dandelion greens	
	Eggplant	
	Endive, curly	
	Kale	
	Kohlrabi	
	Lettuce	
	Mushrooms	
	Okra	
	Onions	
	Parsnips	
	Peas	
	Peppers	
	Potato, white	
	Potato, baked	
	Potato, mashed	
	Pumpkin	
	Radish	
	Rutabagas	
	Salsify	
	Sauerkraut	
	Squash, summer	
	Squash, winter	
	Sweet potato	
	Tomatoes or juice	
	Turnip greens	
	Turnip	
	Water cress	

[1]Adapted by the authors from data presented in *Mayo Clinic Diet Manual*, 3rd ed., W. B. Saunders Company, Philadelphia, 1961, pp. 182–184.

[2]Ash of this type of processed cheese is alkaline, due to additives such as sodium aluminum phosphate.

residue, depending upon the net absorption and metabolism of the food and additive combination, which cannot be accurately predicted from a knowledge of the separate components.

Nonetheless, tables of acidic, alkaline, or neutral foods provide, for the dietitian, a means of planning menus to accomplish certain metabolic alterations.

DIETARY USES OF ACID FOODS AND ALKALINE FOODS. Under normal circumstances, there is no reason to be concerned about dietary excesses of either acid or alkaline foods, since the neutrality of the blood is maintained by the kidneys, aided somewhat by the lungs. When, however, there are kidney stones or impairment of kidney function, it may be necessary to select foods to obtain an acid, alkaline, or neutral urine.

Kidney stones composed of calcium and magnesium phosphates, carbonates, and oxalates are formed under alkaline conditions; since these salts are insoluble in alkaline solutions. It is, therefore, believed that a diet resulting in an acid urine will help to reduce the formation of such stones. Physicians sometimes recommend that the urine be kept acid by giving the patient cranberry juice (which contains a nonmetabolizable organic acid) several times a day.

Uric acid and cystine stones are formed under acid conditions; hence, a therapeutic diet should, in this case, make the urine alkaline.

Milk is usually allowed in either acid-ash or alkaline-ash diets since much of the calcium is unabsorbed, and it is, therefore, debatable whether this food has an acidic or alkaline reaction.

CAUTION: It is not advisable for people to undertake on their own, without a physician's advice, to change the relative acidity or alkalinity of their body fluids and urine. Overemphasis of certain foods and avoidance of others may result in a deficient or imbalanced diet.

(Also see ACID-BASE BALANCE; and DIETS.)

ACIDOPHILUS MILK

Milk inoculated with *Lactobacillus acidophilus* bacteria for the purpose of establishing this organism in the intestines, where it may have beneficial effects.

(Also see LACTOBACILLUS ACIDOPHILUS; and MILK.)

ACIDOSIS

A disorder characterized by an abnormally high level of acid in the blood, or by a decrease in the blood bicarbonate. The excess acids may consist of ketone bodies, phosphoric acid, sulfuric acid, hydrochloric acid, lactic acid, or carbonic acid.

(Also see ACID-BASE BALANCE)

ACID TIDE

A short-lived increase in the acidity of the blood and the urine which may occur after eating a protein-rich meal that stimulates a heavy secretion of alkaline juice from the pancreas.

(Also see DIGESTION AND ABSORPTION.)

ACINI

Groups of secretory cells in glands such as the salivary glands, the pancreas, and the liver. "Acini" means shaped like a cluster of grapes, which is indicative of the shape of these organized clusters of cells. Their secretions of enzymes and bile feed into ducts that empty into the digestive tract.

ACKEE

The fruit of a tree (*Blighia satida*) which grows in the West Indies and South America. People who eat the unripe or bruised fruit may suffer severe low blood sugar which is accompanied by extreme nausea, convulsions, or even death.

(Also see POISONOUS PLANTS.)

ACNE VULGARIS

An outbreak of pimples or similar eruptions on the face, back, or chest. At one time it was thought that certain foods caused outbreaks of acne, but now this does not seem likely unless the foods cause allergic reactions.

(Also see ALLERGIES; and CHILDHOOD AND ADOLESCENT NUTRITION.)

ACONITASE

This is an enzyme occurring in many animal and plant tissues that accelerates the conversion of citric acids (1) into aconitic acid, and (2) then into isocitric acid.

ACRODYNIA (PINK DISEASE)

A disease, also called pink disease, which occurs in young children, usually between 4 months and 3 years of age. It is characterized by painful red swollen hands and feet, muscular pains that make movement difficult, loss of energy, and general mental and physical sluggishness. Acrodynia may be caused by chronic mercury intoxication. Also, there are other, as yet unknown, causes of the disease. A similar condition occurs in rats on a diet deficient in pyridoxine (vitamin B-6).

ACROLEIN

A bitter substance produced by overheating fats. If consumed in sufficient quantity, it may irritate the stomach and the intestines.

ACROMEGALY

A disorder resulting from the oversecretion of growth hormones by the pituitary after most of the bones have stopped growing. The excess of hormone causes overgrowth of the bones in the face, hands, and feet. It may also lead to a diabetic condition and various other abnormalities.

(Also see ENDOCRINE GLANDS.)

ACTH

An abbreviation for adrenocorticotropic hormone, which is secreted by the pituitary. Under certain stressful conditions, it stimulates (1) the secretion of certain hormones by the adrenal cortex, and (2) a release of free fatty acids from various fatty tissues.

(Also see ENDOCRINE GLANDS.)

ACTIVATE

To convert a substance, such as a provitamin or enzyme, into an active form or different substance.

ACTIVATION OF ENZYMES

In order to function, some enzymes require a metal (such as calcium), or a nonmetal, or both, attached to them, either loosely or tightly; others function independently. For example, calcium activates a number of enzymes including pancreatic lipase, adenosine triphosphatase, and some proteolytic enzymes.

ACTIVATOR

An agent that initiates metabolic activity by another substance. For example, certain trace minerals combine with vitamins to activate some important enzymes. The enzymes may not function unless activated.

(Also see ENZYMES; MINERAL[S]; and VITAMIN[S].)

ACTIVE TRANSPORT

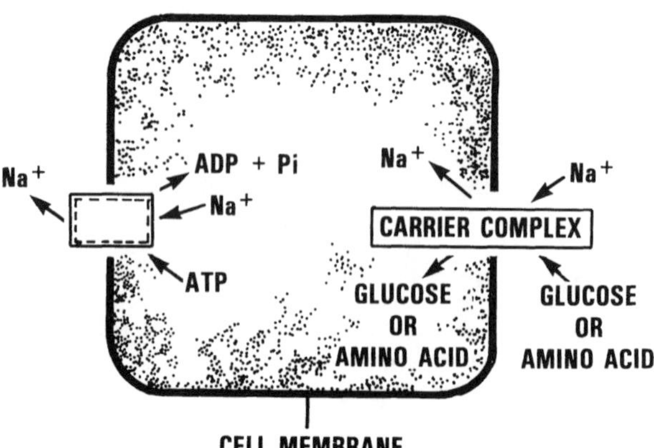

Fig. A-2. Active transport occurs when there is movement against a concentration gradient, thus necessitating the expenditure of energy. A carrier system utilizing Na+ has been implicated in the active transport of glucose and some amino acids.

The process by which one type of molecule is carried through a membrane by another molecule. Active transport requires energy that is derived from the metabolism of nutrients.

(Also see DIGESTION AND ABSORPTION.)

ACUTE CONDITION

A disease or disorder having a sudden beginning, and lasting a short time. Often, the symptoms are severe.

ADDISON'S DISEASE

A chronic condition resulting from an undersecretion of the adrenal cortical hormones. Usually, there is an inability to adapt to such stresses as low blood sugar (hypoglycemia), infection, dehydration, and chilling.

(Also see ENDOCRINE GLANDS.)

ADDITIVES

Contents *Page*

History 9
Need For Additives 9
Yearly Consumption 10
Fear of Food Additives 10
Fear of Mass Poisoning from Food Additives 12
Fear of Cancer from Food Additives 12
Common Food Additives 12
Controls Governing Additives 18
Risk Vs Benefit 19

Contrary to popular belief, the word "additive" refers to a use and *not* a class of substances. The Food and Drug Administration (FDA) defines food additives as follows:

> "...the intended use of which results or may reasonably be expected to result, directly or indirectly, either in their becoming a component of food or otherwise affecting the characteristics of food. A material used in the production of containers and packages is subject to the definition if it may reasonably be expected to become a component, or to affect the characteristics, directly or indirectly, of food packed in the container. 'Affecting the characteristics of food' does not include such physical effects, as protecting contents of packages, preserving shape, and preventing moisture loss. If there is no migration of a packaging component from the package to the food, it does not become a component of the food and thus is not a food additive. A substance that does not become a component of food, but that is used, for example, in preparing an ingredient of the food to give a different flavor, texture, or other characteristic in the food, may be a food additive."[1]

[1] *Code of Federal Regulations*, Title 21, Section 170.3, Revised as of April 1, 1978.

Sometimes additives are divided into two categories: (1) intentional or direct, and (2) incidental or indirect.

1. **Intentional or direct.** These are the additives of known composition that have been purposely added to foods to achieve specific effects during production or processing or to impart or retain desired characteristics.

2. **Incidental or indirect.** These are chemicals which have no planned function in food, but become part of it during some phase of processing, packaging, or storing.

This article deals primarily with intentional additives.

For the most part, additives perform a necessary function in food. The amount of most additives consumed on a yearly basis is relatively small; hence, additives pose a minor health risk. There are over 2,000 additives, many are common chemicals, and all are continually scrutinized for any adverse effects on the health of people. But, to enjoy the benefits from any technology carries a certain degree of risk.

HISTORY. Before the Civil War, people raised most of what they ate and processed it themselves, and food additives were limited mostly to home-grown colorings and substances needed for preservation in storage—vegetable and fruit juices, salt, spices, smoke.

Our system of food supply changed after the Civil War. Thousands of rural people flocked to the cities to work in factories; they now needed food grown and preserved by someone else. Manufacturers of food products sprang up almost everywhere.

Food purity as such was not a major consideration. Cheap and handy methods of preserving foods were important to profits, and scientific knowledge of food chemistry was practically nonexistent.

Dangerous adulteration of foods was commonplace. Chemicals to keep products looking good until they reached the consumer or just to hide the smell and look of spoilage were used without much restraint.

The problem of food additives became acute. Supplying the rapidly growing urban population required constantly expanding facilities, and speed of production took precedence over both quality and safety. For example, copper sulphate, a powerful emetic also known as blue vitriol, was added to canned vegetables to give them that fresh, green look; and salicylic acid, borax and formaldehyde were used generously—and carelessly.

Food and drug protection became an operating function of the Federal Government under the strong leadership of one man. Dr. Harvey Washington Wiley, chief chemist for the U.S. Department of Agriculture in Washington, D.C., announced publicly that the American people were being steadily poisoned by the dangerous chemicals that were being added to food with reckless abandon.

To dramatize the problem, and to learn more about the reactions of the human body to ingestion of these chemicals, he formed, in 1902, what became known as "Dr. Wiley's Poison Squad." Twelve young healthy men, recruited from the Department of Agriculture, pledged to eat nothing except what Dr. Wiley prescribed.

Over a period of 5 years, Poison Squad members were fed measured doses of many kinds of commonly-used food additives. Dr. Wiley was not only concerned about determining the effects of these additives, he was also interested in stirring up the public about the need for a pure food law.

The efforts of Dr. Wiley and the publication of the book, *The Jungle*, by Upton Sinclair, which told of the filth accompanying the production of meat and meat products, were powerful, moving forces that helped persuade Congress to pass the Food and Drug Act of 1906 as well as the Meat Inspection Act of the same year.

In 1927, the Food, Drug, and Insecticide Administration, later to be named the Food and Drug Administration (FDA), was created. Then in 1938, the Federal Food, Drug, and Cosmetic Act of 1938 was signed, broadening the scope of the Food and Drug Act of 1906. Still, the FDA's problem with chemicals in foods became unmanageable. Literally thousands of new compounds were being used in crop production, food processing and packaging. Some were so acutely toxic as to require emergency action by FDA. Such cases were relatively simple, since it was no great problem to prove in court that they were harmful. The big problem was the host of substances which had not been sufficiently tested to show that they were safe. The total situation so-concerned FDA's Commissioner Charles W. Crawford that he decided to discuss it with Congressman Frank B. Keefe of Wisconsin, a strong advocate of pure foods.

On May 9, 1949, Mr. Keefe introduced a resolution calling for a special committee to investigate the use of chemicals in foods. Death ended the career of Congressman Keefe, and Representative James J. Delaney of New York was named chairman of the investigating committee. The "Delaney Hearings" became the platform for a long parade of scientific, legal, agricultural and medical experts who gave their advice on how the "chemicals-in-food" problem should be handled.

Early in the deliberations it became clear that "chemicals in foods" would have to be regulated in at least two separate categories—pesticides, which were used on raw agricultural commodities, and "food additives," which were substances added to improve food products or to facilitate processing or packaging.

The Pesticide Chemicals Act became law in 1954, the Food Additives Amendment was approved in 1958, and the Color Additive Amendments in 1960. In all these laws, the intention of Congress was not to ban the use of food chemicals, but to ensure their safety when properly used. Furthermore, these laws put upon industry the responsibility to prove by scientific research acceptable to the FDA that the substances would be safe as used.

(Also see DELANEY CLAUSE.)

NEED FOR ADDITIVES. Processing food without additives would be to go back to the good old days of baking freshness—good today, stale tomorrow—the days when cottage cheese separated, cookies dried up in 2 days, any food with fat or oil in it became rancid, canned vegetables and fruits were soft and mushy, and marshmallows got too hard to toast. Furthermore, quantities available would be less due to spoilage, and convenience foods would be nonexistent.

But even in the good old days there were additives! The ancient Egyptians used food coloring made from vegetables and insects, while the Romans preserved fruit in honey; and salting foods was common in the Middle Ages. Marco Polo searched for food additives—herbs and spices—during his travels. Also, in many homes, additives are still employed when food is made from "scratch." For example, baking soda (sodium bicarbonate) is added to bread dough, and pectin and sugar are added to make a jam. Many home recipes include the additive acetic acid—vinegar.

The food industry uses additives for one or more of the following four purposes:

1. **To maintain or improve nutritional value**—Many foods are fortified with vitamins and minerals that might otherwise be lacking in a person's diet or that have been destroyed or lost in processing. Common nutritional additives include vitamin D in milk, vitamin A in margarine, and iodine in table salt. Breads and cereals are enriched with B vitamins and iron, lost or destroyed during the milling and processing of grains. Such fortification has helped eradicate once prevalent deficiency diseases, such as rickets, beriberi, pellagra, and goiter.

(Also see ENRICHMENT.)

2. **To maintain freshness**—Foods last as long as they do on the shelf or in the refrigerator because of additives that retard spoilage, preserve natural color and flavor, and keep fats and oils from turning rancid. Preservatives such as the nitrates and nitrites, for example, protect cured meats, fish, and poultry from contamination by the bacterially produced toxin responsible for botulism, a food poisoning illness. Ascorbic acid (vitamin C) keeps uncooked peaches from turning brown. Antioxidants, such as BHA (butylated hydroxyanisole) or BHT (butylated hydroxytoluene), help prevent changes in color, flavor, or texture that occur when foods are exposed to air.

3. **To help in processing or preparation**—A wide variety of compounds are used to give body and texture to foods, evenly distribute particles of one liquid in another, affect cooking or baking results, control acidity or alkalinity, retain moisture, and prevent caking or lumping. Chemicals called emulsifiers give such products as peanut butter and mayonnaise a consistent texture and prevent them from separating into an oily layer at the top of the jar and a dry layer at the bottom. Thickeners create smoothness and prevent ice crystals from forming in frozen foods such as ice cream. Humectants are used to keep moisture in foods such as shredded coconut. Leavening agents (yeasts and baking powder) are essential to make baked goods rise.

4. **To make food more appealing**—The most widely used additives are those intended to make food look and taste better. These include coloring agents, natural and synthetic flavors, flavor enhancers such as MSG (monosodium glutamate), and sweeteners. The characteristic flavor of strawberry ice cream, for example, may come from real strawberries, or it may come from a chemical flavoring. Because consumers associate strawberries with a reddish color, strawberry ice cream is tinted pink. Perhaps, these uses are the most controversial. Industry says the public prefers foods that are colored and flavored, whereas many consumers and consumer advocates believe too many colors and flavors are used. Industry manufactures what consumers buy—successful products stay on the market.

On the other hand, the following are considered inappropriate uses of food additives:

1. To disguise faulty or inferior processes.
2. To conceal damaged, inferior or spoiled foods.
3. To gain some functional property at the expense of nutritional quality.
4. To replace economical, well-recognized manufacturing processes and practices.
5. To use in excess of the minimum required to achieve the intended effect(s).

When used properly, food additives serve a variety of specific functions. Terms used to describe these functions and their definitions are listed in Table A-2.

Table A-2 is a long list. Often for discussion purposes the functions of food additives are classified into the following five key groups:

1. **Nutrients**—These are substances such as vitamins, minerals, and protein added to maintain or improve the nutritional quality of a food.
2. **Preservatives**—This group includes antimicrobials which prevent food spoilage from bacteria, molds, fungi, and yeast, and antioxidants which delay and/or prevent rancidity or enzymatic browning.
3. **Texturizers**—These are substances used to stabilize, thicken, retain moisture, and prevent caking. Texturizers aid processing or preparation, and consumer acceptance.
4. **Colors**—These impart to foods their desired, appetizing or characteristic color.
5. **Flavors**—Spices and natural and synthetic flavors are employed to complement, magnify, or modify the taste and/or aroma of food.

YEARLY CONSUMPTION. By far, the most consumed additives are sugar, salt, and corn syrup. These three, plus such other substances as citric acid, baking soda, vegetable colors, mustard, and pepper, account for more than 98%, by weight, of all food additives used in this country.

Fig. A-3 shows the yearly per capita consumption of food additives. The majority of additives are consumed at a level of about 1 lb *(0.45 kg)* per year—0.04 oz *(1.12g)* per day.

FEAR OF FOOD ADDITIVES. Clearly, with someone putting chemicals in the foods, harmful

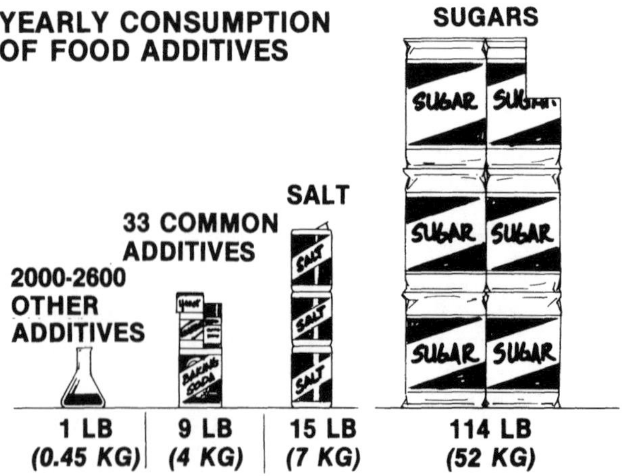

Fig. A-3. Per capita yearly consumption of food additives.

TABLE A-2
TERMS USED TO DESCRIBE THE FUNCTIONS OF FOOD ADDITIVES[1]

Term	Function
Anticaking agents and free-flow agents	Substances added to finely powdered or crystalline food products to prevent caking, lumping, or agglomeration.
Antimicrobial agents	Substances used to preserve food by preventing growth of microorganisms and subsequent spoilage, including fungistats, mold and rope inhibitors.
Antioxidants	Substances used to preserve food by retarding deterioration, rancidity, or discoloration due to oxidation.
Colors and coloring adjuncts	Substances used to impart, preserve, or enhance the color or shading of a food, including color stabilizers, color fixatives, color-retention agents, etc.
Curing and pickling agents	Substances imparting a unique flavor and/or color to a food, usually producing an increase in shelf life stability.
Dough strengtheners	Substances used to modify starch and gluten, thereby producing a more stable dough.
Drying agents	Substances with moisture-absorbing ability, used to maintain an environment of low moisture.
Emulsifiers and emulsifier salts	Substances which modify surface tension in the component phase of an emulsion to establish a uniform dispersion or emulsion.
Enzymes	Substances used to improve food processing and the quality of the finished food.
Firming agents	Substances added to precipitate residual pectin, thus strengthening the supporting tissue and preventing its collapse during processing.
Flavor enhancers	Substances added to supplement, enhance, or modify the original taste and/or aroma of a food, without imparting a characteristic taste or aroma of its own.
Flavoring agents and adjuvants	Substances added to impart or help impart a taste or aroma in food.
Flour treating agents	Substances added to milled flour, at the mill, to improve its color and/or baking qualities, including bleaching and maturing agents.
Formulation aids	Substances used to promote or produce a desired physical state or texture in food, including carriers, binders, fillers, plasticizers, film-formers, and tableting aids, etc.
Fumigants	Volatile substances used for controlling insects or pests.
Humectants	Hygroscopic substances incorporated in food to promote retention of moisture, including moisture-retention agents and antidusting agents.
Leavening agents	Substances used to produce or stimulate production of carbon dioxide in baked goods to impart a light texture, including yeast, yeast foods, and calcium salts.
Lubricants and release agents	Substances added to food contact surfaces to prevent ingredients and finished products from sticking to them.
Nonnutritive sweeteners	Substances having less than 2% of the caloric value of sucrose per equivalent unit of sweetening capacity.
Nutrient supplements	Substances which are necessary for the body's nutritional and metabolic processes.
Nutritive sweeteners	Substances having greater than 2% equivalent unit of sweetening capacity.
Oxidizing and reducing agents	Substances which chemically oxidize or reduce another food ingredient, thereby producing a more stable product.
pH control agents	Substances added to change or maintain active acidity or alkalinity, including buffers, acids, alkalis, and neutralizing agents.
Processing aids	Substances used as manufacturing aids to enhance the appeal or utility of a food or food component, including clarifying agents, clouding agents, catalysts, flocculents, filter aids, and crystallization inhibitors, etc.
Propellants, aerating agents, and gases	Gases used to supply force to expel a product or used to reduce the amount of oxygen in contact with the food in packaging.
Sequestrants	Substances which combine with polyvalent metal ions to form a soluble metal complex, to improve the quality and stability of products.
Solvents and vehicles	Substances used to extract or dissolve another substance.
Stabilizers and thickeners	Substances used to produce viscous solutions or dispersions, to impart body, improve consistency, or stabilize emulsions, including suspending and bodying agents, setting agents, jellying agents, and bulking agents, etc.
Surface-active agents	Substances used to modify surface properties of liquid food components for a variety of effects, other than emulsifiers, but including solubilizing agents, dispersants, detergents, wetting agents, rehydration enhancers, whipping agents, foaming agents, and defoaming agents, etc.
Surface-finishing agents	Substances used to increase palatability, preserve gloss, and inhibit discoloration of foods, including glazes, polishes, waxes, and protective coatings.
Synergists	Substances used to act or react with another food ingredient to produce a total effect different or greater than the sum of the effects produced by the individual ingredients.
Texturizers	Substances which affect the appearance or feel of the food.

[1]*A Comprehensive Survey of Industry on the Use of Food Chemicals Generally Recognized as Safe*, National Academy of Sciences / National Research Council.

substances could be introduced into the diet of numerous people, and consumers could be duped as to the quality of their food. Indeed, there are those who plead for the good old days when foods were simpler. Despite nostalgia for the past, eating was not especially safe at the turn of the century. At that time, it was difficult to protect foods from spoiling; manufacturers freely used pigments containing lead, copper, and arsenic to color candy, pickles, and other foods; jams were created with grass seed, coloring and flavor; and black pepper was extended with charcoal.

While the actual hazards from food additives are small, the fear of them is large—fear fanned by the press, legislation, and consumer groups.

From the standpoint of food safety, additives deserve the least concern. The two major food hazards, in order of their importance, are: (1) microbiological contamination leading to such diseases as food poisoning and botulism, and (2) suboptimal intakes of key nutrients due to ignorance, indifference, or misinformation. And the three minor food hazards, in order of their importance, are: (1) environmental contaminants such as mercury in fish, (2) toxicants that occur naturally in foods, and (3) food additives.

Fear of mass poisoning from additives. Probably due to publicity, many Americans feel that there are poisonous chemicals in their foods. Furthermore, the term "chemicals," has a bad connotation. To say that foods contain chemicals is almost a condemnation. The current reasoning seems to be that foods that contain additives also contain chemicals, and chemicals are bad. But the truth is, foods, regardless of their origin, are chemicals.

Public alarm is now easily generated at the mention of compounds that are unfamiliar or bear exotic names. To illustrate, if it were proposed that acetone, acetal dehyde, methylpropyl ketone, methyl isovalerianate, isoamyl isovalerianate, methanol, o-methylpropylbenzene, or B-phenylethylcaproate be added to fruit as a preservative or to enhance the flavor or aroma, a measure of public opposition could be expected, when in fact, all of these chemicals and many more occur naturally in ripe strawberries.

Fig. A-4 illustrates how a product of "Mother Nature," could be labeled and sound harmful. Yet, the major ingredient, hydrogen oxide, is water.

Everything is made up of chemicals. Chemicals in food are chemicals regardless of how they were put into the food. Even "Mother Nature" can put harmful chemicals into foods. For example, potatoes contain solanine, a poison. In consuming an average of 119 lb *(54 kg)* of potatoes per year, the average American ingests 9,700 mg of solanine — enough to kill a horse. Yet no one dies from eating potatoes, since no one eats their yearly allotment of potatoes in one sitting.

Many plants that are commonly eaten as human food, and are not regarded as poisonous, contain many chemicals, each of which would have detrimental effects in excess. Furthermore, where or how a specific chemical compound is formed is irrelevant as far as the properties and effects of the compound are concerned. Today, many of the simpler organic substances produced by living organisms can be synthesized in the laboratory by chemists. When the synthetic version of a naturally occurring organic chemical is produced in a laboratory or factory, the resulting chemical is identical to the one produced naturally by plants or animals.

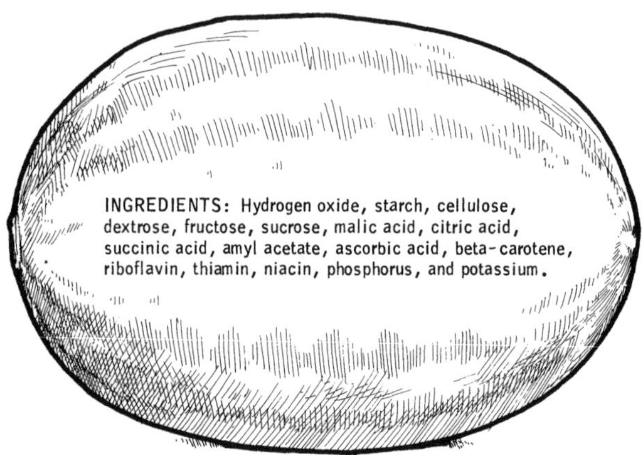

INGREDIENTS: Hydrogen oxide, starch, cellulose, dextrose, fructose, sucrose, malic acid, citric acid, succinic acid, amyl acetate, ascorbic acid, beta-carotene, riboflavin, thiamin, niacin, phosphorus, and potassium.

Fig. A-4. A melon labeled with its ingredients—a product of "Mother Nature."

(Also see BACTERIA IN FOOD; and POISONS.)

Fear of Cancer from Food Additives. Another widespread fear of food additives is that additives will cause, or are causing, cancer. Many food additives have been evaluated for their ability to cause cancer in animals. This has created a growing list of suspected carcinogens. Some are synthetic chemicals while others are naturally occurring chemicals that have been extracted from plants and used in food processing. Some notable examples of food additives which are suspected carcinogens based on animal studies include: oil of calamus; dulcin (*p* phenetylurea); red dye No. 2; butter yellow (*p* dimethylaminoazobenzene); and safrole. Recently, a synthetic flavoring, Cinnamyl anthranilate, used for 40 years, has been cited for causing cancer in rats. It has never been proven that a chemical that causes cancer in a test animal will also cause human cancer.

Despite the suggestion that some food additives cause cancer, there is very little evidence to indicate that anyone has had food additive-related cancer. In fact, a recent large study of cancer risks in the United States indicated that fewer than 1% of all cancer deaths could be related to food additives. Moreover, some food additives such as antioxidants possibly provide a protective effect.[2]

(Also see CANCER.)

COMMON FOOD ADDITIVES. Since there are nearly 3,000 food additives, an attempt to list all of them with their function and use in foods is not within the scope of this book. However, Table A-3 provides a listing and description of the uses of some of the more common additives.

[2]*Chemical and Engineering News,* August 17, 1981, p. 32.

TABLE A–3
COMMON FOOD ADDITIVES

Name	Chemical Formula[1]	Function[2]	Food Use and Comments[3]
Acetic acid	$C_2H_4O_2$	pH control; preservative	Acid of vinegar is acetic acid. Miscellaneous and / or general purpose; many food uses; GRAS additive.
Adipic acid	$C_6H_{10}O_4$	pH control	Buffer and neutralizing agent; use in confectionery; GRAS additive.
Ammonium alginate	Polysaccharide	Stabilizer and thickener; texturizer	Extracted from seaweed. Widespread food use; GRAS additive.
Annatto	Bixin plus several compounds similar to carotene	Color	Extracted from seeds of *Bixa orellana.* Butter, cheese, margarine, shortening, and sausage casings; coloring foods in general.
Arabinogalactan	Polysaccharide	Stabilizer and thickener; texturizer	Extracted from Western larch. Widespread food use; bodying agent in essential oils, nonnutritive sweeteners, flavor bases, non-standardized dressings and pudding mixes.
Ascorbic acid (vitamin C)	$C_6H_8O_6$	Nutrient; antioxidant; preservative	Widespread use in foods to prevent rancidity, browning; used in meat curing; GRAS additive.
Aspartame	A combination of two amino acids; aspartic acid and a methyl ester of phenylalanine	Sweetener, low calorie	Soft drinks, chewing gum, powdered beverages, whipping toppings, puddings, gelatin, tabletop sweetener.
Azodicarbonamide	$C_3H_4O_2N_3$	Flour treating agent	Aging and bleaching ingredient in cereal flour.
Benzoic acid	$C_7H_6O_2$	Preservative	Occurs in nature in free and combined forms. Widespread food use; GRAS additive.
Benzoyl peroxide	$C_{14}H_{10}O_4$	Flour treating agent	Bleaching agent in flour; may be used in some cheeses.
Beta-apo-8′ carotenal		Color	Natural food color. General use not to exceed 30 mg per lb or pt of food.
BHA (butylated hydroxyanisole)	$C_{11}H_{16}O_2$	Antioxidant; preservative	Fats, oils, dry yeast, beverages, breakfast cereals, dry mixes, shortening, potato flakes, chewing gum, sausage; often used in combination with BHT; GRAS additive.
BHT (butylated hydroxytoluene)	$C_{15}H_{24}O$	Antioxidant; preservative	Rice, fats, oils, potato granules, breakfast cereals, potato flakes, shortening, chewing gum, sausage; often used in combination with BHA; GRAS additive.
Biotin	$C_{10}H_{16}N_2O_3S$	Nutrient	Rich natural sources are liver, kidney, pancreas, yeast, milk; vitamin supplement; GRAS additive.
Calcium alginate	Polysaccharide	Stabilizer and thickener; texturizer	Extracted from seaweeds. Widespread food use; GRAS additive.
Calcium carbonate	$CaCO_3$	Nutrient	Mineral supplement; general purpose additive; GRAS additive.
Calcium lactate	$C_6H_{10}CaO_6$	Preservative	General purpose and / or miscellaneous use; GRAS additive.
Calcium phosphate	$CaHO_4P$ $CaH_4O_8P_2$ $Ca_3O_8P_2$	Leavening agent, sequestrant, nutrient	General purpose and / or miscellaneous use; mineral supplement; GRAS additive.
Calcium propionate	$C_6H_{10}CaO_4$	Preservative	Bakery products, alone or with sodium propionate; inhibits mold and other microorgansims; GRAS additive.
Calcium silicate	$CaSiO_3$ Ca_2SiO_4 Ca_3SiO_5	Anticaking agent	Used in baking powder, salt, and food; GRAS for use in baking powder and salt.
Canthaxanthin	$C_{40}H_{52}O_2$	Color	Widely distributed in nature. Color for foods; more red than carotene.
Caramel		Color	Miscellaneous and / or general purpose use in foods for color GRAS additive.

Footnotes at end of table.

(Continued)

TABLE A-3 (*Continued*)

Name	Chemical Formula[1]	Function[2]	Food Use and Comments[3]
Carob bean gum	Polysaccharide	Stabilizer and thickener	Extracted from bean of carob tree (Locust bean). Numerous foods like confections, syrups, cheese spreads, frozen desserts, and salad dressings; GRAS additive.
Carrageenan	Polysaccharide	Emulsifier; stabilizer and thickener	Extracted from seaweed. A variety of foods, primarily those with a water or milk basis.
Cellulose	$(C_6H_{10}O_5)_n$	Emulsifier; stabilizer and thickener	Component of all plants. Inert bulking agent in foods; may be used to reduce caloric content of food; used in foods which are liquid and foam systems.
Citric acid	$C_6H_8O_7$	Preservative; antioxidant: pH control agent; sequestrant.	Widely distributed in nature in both plants and animals. Miscellaneous and/or general purpose food use; used in lard, shortening, sausage, margarine, chili con carne, cured meats, and freeze dried meats; GRAS additive.
Citrus Red No. 2		Color	Coloring skins of oranges.
Cochineal		Color	Derived from the dried female insect, *Coccus cacti,* raised in West Indies, Canary Islands, southern Spain and Algiers; 70,000 insects to 1 lb. Provides red color for such foods as meat products and beverages.
Corn endosperm oil		Color	Source of xanthophyll for yellow color. Used in chicken feed to color yolks of eggs and chicken skin.
Cornstarch	$(C_6H_{10}O_5)_n$	Anticaking agent; drying agent; formulation aid; processing aid; surface-finishing agent	Digestible polysaccharide used in many foods often in a modified form; example foods include baking powder, baby foods, soups, sauces, pie fillings, imitation jellies, custards, candies, etc.
Corn syrup	Mixture of carbohydrates	Flavoring agent; humectant; nutritive sweetener; preservative	Derived from hydrolysis of cornstarch. Employed in numerous foods, e.g. baby foods, bakery products, toppings, meat products, beverages, condiments and confections; GRAS additive.
Dextrose (glucose)	$C_6H_{12}O_6$	Flavoring agent; humectant; nutritive sweetener; synergist	Derived from cornstarch. Major users of dextrose are confection, wine, and canning industries; used to flavor meat products; used in production of caramel; variety of other uses.
Diglycerides	Glycerol plus two fatty acids	Emulsifiers	Uses include frozen desserts, lard, shortening, and margarine; GRAS additive.
Dioctyl sodium sulfosuccinate	$C_{20}H_{37}NaO_7S$	Emulsifier; processing aid; surface active agent	Employed in gelatin dessert, dry beverages, fruit juice drinks, and noncarbonated beverages with cocoa fat; used in production of cane sugar and in canning.
Disodium guanylate	$Na_2C_{10}H_{12}N_5O_8P$	Flavor enhancer	Derived from dried fish or seaweed. Variety of uses.
Disodium inosinate	$C_{10}H_{11}N_4O_8PNa_2$	Flavor adjuvant	Derived from seaweed or dried fish; sodium guanylate a by-product. Variety of uses.
EDTA (ethylenediaminetetraacetic acid)	$C_{10}H_{16}N_2O_8$	Antioxidant; sequestrant	Calcium disodium and disodium salt of EDTA employed in a variety of foods including: soft drinks, alcoholic beverages, dressings, canned vegetables, margarine, pickles, sandwich spreads, and sausage.
FD&C colors: **Blue No. 1** **Red No. 40** **Yellow No. 5**		Color	Coloring foods in general including dietary supplements.

Footnotes at end of table.

(Continued)

TABLE A-3 (*Continued*)

Name	Chemical Formula[1]	Function[2]	Food Use and Comments[3]
Gelatin	Protein	Stabilizer and thickener; texturizer	Derived from collagen by boiling skin, tendons, ligaments, bones, etc. with water. Employed in many foods including: confectionery, jellies, and ice cream; GRAS additive.
Glycerine (glycerol)	$C_3H_8O_3$	Humectant	Miscellaneous and general purpose additive; GRAS additive.
Grape skin extract		Color	Colorings for carbonated drinks, beverage bases, ades, and alcoholic beverages.
Guar gum	Polysaccharide	Stabilizer and thickener; texturizer	Extracted from seeds of the guar plant of India and Pakistan. Employed in such foods as cheese, salad dressings, ice cream, and soups.
Gum arabic	Polysaccharide	Stabilizer and thickener; texturizer	Gummy exudate of Acacia plants. Used in a variety of foods; GRAS additive.
Gum ghatti	Polysaccharide	Stabilizer and thickener; texturizer	Gummy exudate of plant growing in India and Ceylon. A variety of food uses; GRAS additive.
Hydrogen peroxide	H_2O_2	Bleaching agent	Modification of starch, and bleaching tripe; GRAS bleaching agent.
Hydrolyzed vegetable (plant) protein	Protein	Flavor enhancer	To flavor various meat products.
Invert sugar	Equal mixture fructose and glucose	Humectant; nutritive sweetener.	Main use in confectionery and brewing industry.
Iron	Fe	Nutrient	Dietary supplements and foods; GRAS additive.
Iron-Ammonium citrate		Anticaking agent	Used in salt.
Karaya gum	Polysaccharide	Stabilizer and thickener	Derived from dried extract of *Sterculia urens* found primarily in India. Variety of food uses; a substitute for tragacanth gum; GRAS additive.
Lactic acid	$C_3H_6O_3$	Preservative; pH control	Normal product of human metabolism. Numerous uses in foods and beverages; a miscellaneous general purpose additive; GRAS additive.
Lecithin (phosphatidylcholine)	Varies with source	Emulsifier; surface active agent	Normal tissue component of the body; edible and digestible additive naturally occurring in eggs; commercially derived from soybeans. Margarine, chocolate and wide variety of other food uses; GRAS additive.
Mannitol	$C_6H_{14}O_6$	Anticaking; nutritive sweetener; stabilizer and thickener; texturizer	Special dietary foods; GRAS additive; supplies ½ the calories of glucose; classified as a sugar alcohol or polyol.
Methylparaben	$C_8H_8O_3$	Preservative	Food and beverages; GRAS additive.
Modified food starch	Polysaccharide	Drying agent; formulation aid; processing aid; surface finishing agent	Digestible polysaccharide used in many foods and stages of food processing; examples include baking powder, puddings, pie fillings, baby foods, soups, sauces, candies, etc.
Monoglycerides	Glycerol plus one fatty acid	Emulsifiers	Widely used in foods such as frozen desserts, lard, shortening and margarine; GRAS additive.
MSG (monosodium glutamate)	$C_5H_8NNaO_4$	Flavor enhancer	To enhance the flavor of a variety of foods including various meat products; consumers may purchase and use in home cooking; possible association with the Chinese restaurant syndrome.

Footnotes at end of table.

(Continued)

TABLE A-3 (*Continued*)

Name	Chemical Formula[1]	Function[2]	Food Use and Comments[3]
Papain	Protein	Texturizer	Miscellaneous and/or general purpose additive; GRAS; achieves results through enzymatic action; used as meat tenderizer.
Paprika		Color; flavoring agent	To provide coloring and/or flavor to foods; GRAS additive.
Pectin	Polysaccharide	Stabilizer and thickener; texturizer	Richest source of pectin is lemon and orange rind; present in cell walls of all plant tissues. To prepare jellies and similar foods; GRAS additive.
Phosphoric acid	H_3PO_4	pH control	Miscellaneous and/or general purpose additive; GRAS additive; used to increase effectiveness of antioxidants in lard and shortening.
Polyphosphates	Long chains of phosphates	Nutrient; flavor improver; sequestrant; pH control	Numerous food uses; most polyphosphates and their sodium, calcium, potassium, and ammonium salts considered GRAS additives.
Polysorbates	Combination of ethylene oxide and a fatty acid	Emulsifiers; surface active agent	Polysorbates designated by numbers such as 60, 65, and 80; variety of food uses including baking mixes, frozen custards, pickles, sherbets, ice creams, and shortening.
Potassium alginate	Polysaccharide	Stabilizer and thickener; texturizer	Extracted from seaweed. Wide usage; GRAS additive.
Potassium bromate	$BrKO_3$	Flour treating agent	Employed in flour, whole wheat flour, fermented malt beverages, and to treat malt.
Potassium iodide	KI	Nutrient	Added to table salt or used in mineral preparations as a source of dietary iodine.
Potassium nitrite	KNO_2	Curing and pickling agent	To fix color in cured products such as meats.
Potassium sorbate	$C_6H_7KO_2$	Preservative	Inhibits mold and yeast growth in foods such as wines, sausage casings, and margarine; GRAS additive.
Propionic acid	$C_3H_6O_2$	Preservative	Mold inhibitor in breads and general fungicide; GRAS additive; used in manufacture of fruit flavors.
Propyl gallate	$C_{10}H_{12}O_5$	Antioxidant; preservative	Used in products containing oil or fat; employed in chewing gum; used to retard rancidity in frozen fresh pork sausage.
Propylene glycol	$C_3H_8O_2$	Emulsifier; humectant; stabilizer and thickener; texturizer	Miscellaneous and/or general purpose additive; uses include salad dressings, ice cream, ice milk, custards, and variety of other foods; GRAS additive.
Propylparaben	$C_{10}H_{12}O_3$	Preservative	Fungicide; controls mold in sausage casings; GRAS additive.
Saccharin	$C_7H_5NO_3S$	Nonnutritive sweetener	Special dietary foods and a variety of beverages; baked products; tabletop sweeteners.
Saffron		Color; flavoring agent	Derived from plant of western Asia and southern Europe. All foods except those where standards forbid; to color sausage casings, margarine, or product branding inks.
Silicon dioxide	SiO_2	Anticaking agent	Used in feed or feed components, beer production, production of special dietary foods, ink diluent for marking fruits and vegetables.
Sodium acetate	$C_2H_3NaO_2$	pH control; preservative	Miscellaneous and/or general purpose use; meat preservation; GRAS additive.
Sodium alginate	Polysaccharide	Stabilizer and thickener; texturizer	Extracted from seaweed. Widespread food use; GRAS additive.
Sodium aluminum sulfate	$AlNa(SO_4)_2$	Leavening agent	Baking powders, confectionery; sugar refining.

Footnotes at end of table.

(Continued)

TABLE A-3 (*Continued*)

Name	Chemical Formula[1]	Function[2]	Food Use and Comments[3]
Sodium benzoate	$C_7H_5NaO_2$	Preservative	Variety of food products; margarine to retard flavor reversion; GRAS additive.
Sodium bicarbonate	$NaHCO_3$	Leavening agent; pH control	Miscellaneous and/or general purpose uses; separation of fatty acids and glycerol in rendered fats; neutralize excess and clean vegetables in rendered fats, soups and curing pickles; GRAS additive.
Sodium chloride (salt)	NaCl	Flavor enhancer; formulation acid; preservation	Widespread use of salt in many foods; GRAS additive.
Sodium citrate	$C_6H_5Na_3O_7$	pH control; curing and pickling agent; sequestrant	Evaporated milk; miscellaneous and/or general purpose food use; accelerate color fixing in cured meats; GRAS additive.
Sodium diacetate	$CH_3COONaCH_3COOH$	Preservative; sequestrant	An inhibitor of molds and rope forming bacteria in baking products; GRAS additive.
Sodium nitrate (Chile saltpeter)	$NaNO_3$	Curing and pickling agent; preservative	Used with or without sodium nitrite in smoked, cured fish; cured meat products.
Sodium nitrite	$NaNO_2$	Curing and pickling agent; preservative	May be used with sodium nitrate in smoked, cured fish, cured meat products, and pet foods.
Sodium propionate	$C_3H_5NaO_2$	Preservative	A fungicide and mold preventative in bakery products, alone or with calcium propionate; GRAS additive.
Sorbic acid	$C_6H_8O_2$	Preservative	Fungistatic agent for foods, especially cheeses; other uses include baked goods, beverages, dried fruits, fish, jams, jellies, meats, pickled products, and wines; GRAS additive.
Sorbitan monostearate	Ester of sorbitol with the fatty acid stearic acid	Emulsifier; stabilizer and thickener	Widespread food usage such as whipped toppings, cakes, cake mixes, confectionery, icings, and shortenings; also many nonfood uses.
Sorbitol	$C_6H_{14}O_6$	Humectant; nutritive sweetener; stabilizer and thickener, sequestrant	Occurs naturally in berries, cherries, plums, pears, and apples; a sugar alcohol or polyol. Examples of use include chewing gum, meat products, icings, dairy products, beverages, and pet foods.
Sucrose (table sugar)	$C_{12}H_{22}O_{11}$	Nutritive sweetener; preservative	Sugar occurs naturally in some fruits and vegetables. The most widely used additive; used in beverages, baked goods, candies, jams and jellies—an endless list including meat products.
Tagetes (Aztec marigold)		Color	Source is flower petals of Aztec marigold. To enhance yellow color of chicken skin and eggs, incorporated in chicken feed.
Tartaric acid	$C_4H_6O_6$	pH control	Occurs free in many fruits, free or combined with calcium, magnesium, or potassium. In the soft drink industry, confectionery products, bakery products, and gelatin desserts.
Titanium dioxide	TiO_2	Color	For coloring foods generally, except standardized foods; used for coloring ingested and applied drugs.
Tocopherols (vitamin E)	$C_{26}H_{44}O_2$ $C_{28}H_{48}O_2$ $C_{27}H_{46}O_2$ $C_{28}H_{42}O_2$ $C_{29}H_{44}O_2$	Antioxidant; nutrient	To retard rancidity in foods containing fat; used in dietary supplements; GRAS additive.
Tragacanth gum	Polysaccharide	Stabilizer and thickener; texturizer	Derived from the plant *Astragalus gummifier* or other Asiatic species of *Astragalus*. General purpose additive with broad application or potential application.

Footnotes at end of table.

(Continued)

TABLE A-3 *(Continued)*

Name	Chemical Formula[1]	Function[2]	Food Use and Comments[3]
Turmeric		Color	Derived from rhizome of *Curcuma longa.* Food use in general, except standardized foods; to color sausage casings, margarine or shortening; ink for branding or marking products.
Vanilla		Flavoring agent	Used in various bakery products, confectionery and beverages; natural flavoring extracted from cured, full grown unripe fruit of *Vanilla panifolia*; GRAS additive.
Vanillin	$C_8H_8O_3$	Flavoring agent and adjuvant	Widespread confectionery, beverage and food use; synthetic form of vanilla; GRAS additive.
Yellow prussiate of soda	$Na_4Fe(CN)_6 \cdot 10H_2O$	Anticaking agent	Employed in salt.

[1]In some cases, the chemical formula is too complex to include.

[2]Function refers to those defined in Table A-2.

[3]GRAS = GENERALLY RECOGNIZED AS SAFE by the Food and Drug Administration.

CONTROLS GOVERNING ADDITIVES. In the United States, the basic food law is the Federal Food, Drug and Cosmetic Act of 1938. This act gives the Food and Drug Administration (FDA) primary responsibility for the safety and wholesomeness of our food supply. The three important amendments which follow strengthened the act:

1. **The Miller Pesticide Amendment** of 1954 provides for the establishment of safe tolerances (permissible amounts) for pesticide residues on raw agricultural commodities.

2. **The Food Additives Amendment** of 1958 requires premarketing clearance for substances intended to be added to food and for substances occurring in food during processing, storage, or packaging. This amendment includes the Delaney Clause which states that no chemical can be added to food if, in any amount, it produces cancer when ingested by man or animal.
(Also see DELANEY CLAUSE.)

3. **The Color Additive Amendment** of 1960 regulates the listing and certification of color additives. About 1,700 samples are examined annually by FDA scientists.

Prior to the 1958 Food Additives Amendment, the FDA had to prove that an additive was potentially harmful before it could obtain a court order banning its use. Today, the manufacturer must prove to the FDA that an additive is safe under conditions of intended use before the permission to use it is given. The FDA requires extensive testing on at least two species of laboratory animals but not on humans. The manufacturer must present a petition stating results of studies on organ and tissue function including the liver, brain, kidneys, blood forming organs, excretory and reproductive systems. FDA scientists review the petition. If they judge the proposed additive safe and effective, they issue a regulation permitting its use.

• **Animal Testing**—To overcome the prohibitive cost of exposing large numbers of animals to low doses of a test chemical, animals are usually given the highest daily dose of the chemical that they can tolerate without ill effects. The reasoning behind this method is that since only about 500 animals are used to test a chemical, the test could easily miss potential carcinogens unless some method is employed to increase the frequency of cancers in the animals. A high daily dose is currently the method of choice. Once a test demonstrates that a chemical is capable of causing cancer in animals, the next question is whether the chemical will cause cancer in people. No one really knows. The best answer seems to be "possibly." Also, the reverse is often true; that is, human carcinogens will generally cause cancer in animals—but not always. Thus, there are no easy answers, and all testing and interpretations are subject to human error. For example, nitrites, chemicals used in the preservation of meats, were the subject of a study conducted in 1978 at Massachusetts Institute of Technology (MIT). The study reported that nitrite alone fed to rats increased the incidence of cancers of the lymphatic system. Immediately, and before the study was properly reviewed, the USDA and the FDA announced that they would soon ban nitrites. Tempers flared and pork producers lost money due to the implication of bacon containing a cancer-causing substance. The MIT study has now been reviewed by an independent group, and the research has been shown to be in error.

• **Generally Recognized as Safe (GRAS)**—When the Food Additives Amendment of 1958 was adopted, about 200 substances were exempted from the testing requirement because they were judged by experts to be "Generally Recognized as Safe" (GRAS) under conditions of their use in foods at the time. The GRAS list is now much larger. About half of the products are natural flavorings or their derivatives. Vitamins, minerals, and other dietary supplements make up another large group.

In the time since the GRAS list was formulated, better testing methods have been developed. Also, use of some GRAS list items has increased far beyond their use at the time the list was compiled. As a result, several of the items, including cyclamates, were found to be of questionable safety and taken off the market.

RISK Vs BENEFIT. Obviously, as far as human life is concerned no risk is acceptable. On the other hand, those criticizing the use of food additives wish to be assured of absolute safety—an impossible wish. No doubt the furor surrounding the use of additives will continue. But before an outright condemnation of additives the following salient points should be considered:

1. A blanket condemnation of additives overlooks the fact that enrichment of foods with minerals and vitamins has and can do much to eliminate deficiency diseases such as anemia, beriberi, scurvy, rickets, and goiter.
2. The yearly consumption of additives other than sugar and salt is very low.
3. Food products are well labeled and American consumers can choose what they eat from probably the largest selection in the world. This large selection is partly the result of the use of additives.
4. Many allowable food additives are derived directly from food itself. For example, lecithin is found in all living organisms, both plant and animal. It is obtained primarily from soybeans and is used mostly as an emulsifier to keep ingredients in a processed food from separating. Also, a large number of laboratory-created additives are found naturally in foods. Calcium and sodium propionate, for example, are produced during fermentation in the production of Swiss cheese. The propionates are mold inhibitors, used primarily in baked goods.
5. The scientist's ability to detect additives (chemicals) in food, whether intentional or incidental, exceeds his ability to interpret the meaning in terms of human health. Technological advances now allow for the determination of miniscule amounts—billionths and even trillionths—of chemicals in a food. For comparison purposes, one part per billion is one second out of the life of a 32-year-old individual and one part per trillion is one second out of 320 centuries.
6. Additives save money by preventing spoilage and maintaining freshness. It has been estimated that the removal of additives from bread would cost the consumer $1.1 billion per year; from margarine $600 million per year; from meats $600 million per year; and from processed cheese $32 million per year.[3]

All chemicals should not be regarded as harmful. Everything is chemical—foods and additives. Many chemicals posing a hazard in foods have been eliminated, and some additives have been responsible for the elimination of some food hazards.

While it is true that some food additives could be removed from foods, consumers would have to make certain trade-offs. Many additives are in foods primarily because of our life-style where people spend less time preparing, let alone growing foods, thereby devoting more time to other activities. Also, without additives consumers may have to purchase from a limited variety of products with less appeal and at higher prices. Still, there are those who prefer the elimination of all additives even with the loss of convenience, variety, and increased cost. Rather than banishing all additives, or discouraging the production of new ones, *obvious* hazards should be eliminated based on the exercise of *sound* judgment of facts.

(Also see BACTERIA IN FOOD; COLORING OF FOOD; ENRICHMENT; FLAVORS, SYNTHETIC; GUMS; MEAT, sections headed "Meat Curing," "Sausage Making," and "Meat Preservation"; NITRATES AND NITRITES; ORGANICALLY GROWN FOOD; and POISONS.)

[3]Clydesdale, F. M. and F. J. Francis, *Food, Nutrition, and You*, Prentice-Hall, Inc., 1977, p. 101.

ADENINE AND ADENOSINE

These related substances have metabolic roles as (1) components of chemical messengers that help to regulate metabolic processes, (2) parts of nucleic acids, and (3) components of special compounds that store and release chemical energy.

(Also see ENERGY-UTILIZATION BY THE BODY; METABOLISM; and NUCLEIC ACIDS.)

ADENOHYPHYSIS

The frontal lobe of the pituitary gland. It secretes hormones that stimulate the secretion of the other endocrine glands.

(Also see ENDOCRINE GLANDS.)

ADIPOSE TISSUE

A special connective tissue in which fat is stored for later use to provide energy. It also provides body protection against injury by trauma, insulation against excessive loss of heat from the body, and helps to keep certain organs in place.

(Also see TISSUES OF THE BODY; and FATS AND OTHER LIPIDS.)

ADIPOSITY

Obesity or the excessive accumulation of body fat. (Also see OBESITY.)

ADIPSIA

A lack of thirst when water is needed by the body. (Also see APPETITE.)

AD LIBITUM (AD LIB)

This latin term means to take food freely without restriction, in contrast to restrictive feeding where only fixed amounts of food are allowed to be taken.

(Also see APPETITE.)

ADRENAL CORTEX INSUFFICIENCY (ADDISON'S DISEASE)

The cortex or outside of the adrenal gland secretes several hormones which are very important in controlling metabolism. These hormones are broadly called the mineralocorticoids (mineral metabolism) and the glucocorticoids (glucose metabolism). Diseases such as tuberculosis or cancer can cause a destruction of the cortex of the adrenal gland, thereby preventing the proper secretion of its hormones. The resulting condition is characterized by marked pigmentation of the skin, decreased heart size, low blood pressure, and inability to withstand stress.

(Also see ENDOCRINE GLANDS.)

ADRENAL GLAND

One of the endocrine (ductless) glands of the body, located near the kidney, which secretes hormones needed to utilize nutrients.

(Also see ENDOCRINE GLANDS.)

ADRENALIN

A manufacturer's trade name for the hormone adrenaline (epinephrine).

(Also see ADRENALINE; and ENDOCRINE GLANDS.)

ADRENALINE

The common name for the hormone epinephrine, which is secreted by the adrenal glands. Adrenaline stimulates the nervous system and various other tissues—producing effects such as a racing of the heart, elevation of both the blood pressure and blood sugar, and breakdown of the glycogen in muscle to glucose.

(Also see ENDOCRINE GLANDS.)

ADRENERGIC

A drug, nerve transmitter, or similar substance which has the same effects as noradrenaline (norepinephrine).

(Also see ENDOCRINE GLANDS.)

ADRENOCORTICAL HORMONES

The hormones secreted by the adrenal cortex. These hormones help the body to adjust to stresses like low blood sugar, chilling, emotional excitement, and starvation.

(Also see ENDOCRINE GLANDS.)

ADULTERATION

The fraudulent addition of substances to food so as to gain an economic advantage.

(Also see U.S. FOOD AND DRUG ADMINISTRATION.)

Fig. A-5. Look Before You Eat, and see if you can discover any unadulterated food. (19th century cartoon by Opper. Courtesy, The Bettmann Archive, Inc., New York, N.Y.)

ADULT NUTRITION

Contents *Page*

Changes in the Body During Adulthood 21
Health Status of American Adults 22
Self Assessments of Health Status 22
Major Risk Factors for Degenerative Diseases 23
Nutritional Allowances of Adults 24
Dietary Guidelines 24
Planning of Meals 25
Modified Diets and Dietetic Foods 25
Summary 26

Thanks to medical research and the high standard of living enjoyed in the United States, most babies grow up to become adults and many adults retain their vitality well into their later years, as is shown in Fig. A-6.

Many adults are literally "cut down" in the prime of life by one or more degenerative diseases such as atherosclerosis, cancer, coronary heart disease, diabetes, or high blood pressure. Many health professionals and researchers believe that these dreaded conditions result mainly from a combination of hereditary susceptibility and poor health practices such as lack of exercise and unwholesome diets. Others lay some of the blame on the stresses of modern life, since the typical American adult often appears to be caught "in the middle" of the conflicts between the younger and older generations, demands of an employer and his or her family, and putting into prac-

Fig. A-6. Young people who practice good health habits are more likely to retain their health and vitality as they grow older.

tice the principles espoused by leading personalities in health, politics, and religion. A detailed exploration of the interrelationships between these factors is beyond the scope of this article, since many of the relevant topics are covered elsewhere in this book. Instead, this article will be limited mainly to some dietary principles that are believed to be helpful in the maintenance of good health during the adult years.

(Also see CANCER; HEART DISEASE; HIGH BLOOD PRESSURE; OBESITY; and STRESS.)

CHANGES IN THE BODY DURING ADULTHOOD. Most young adults enjoy moderately good health and think little of the often-unnoticed degenerative changes that usually occur gradually, but steadily, as the years pass. These changes are worth considering because they may be slowed by a good diet and regular exercise. One of the most notable changes that occur during the adult years is that many people who were fit and trim in their late teens and early twenties become fat and flabby in middle age. Women may blame some of their increase of body fat on the natural tendency for the female body to lay down stores for pregnancy and lactation, since the estrogen hormones promote the deposition of body fat in females. However, males cannot use this excuse. One indication of the prevalence and seriousness of this problem is the number of adults who consider themselves to be overweight, as shown in Fig. A-7.

Even the men in the U.S. Armed Forces are prone to get fatter as they grow older, as evidenced by the data shown in Fig. A-8.

Fig. A-9 shows that only a few inches of gain around the waist represent quite a few pounds of extra fat.

The onset of these and other important changes generally starts at maturity (following the growth period), then continues ceaselessly until death. But, how fast and how far they develop is determined by a host of factors, including diet, heredity, exercise, stress, etc.

Generally speaking, the following body changes and cautions apply more to young adults than to the aged: (1) peak weight is often attained at middle age, after which there may be a gradual decrease in weight; (2) young adults with a family history of cardiovascular disease should be checked regularly for symptoms; (3) if the blood levels of insulin and glucose are not normal after an overnight fast, adult-onset diabetes should be suspected; (4) reduction in the secretion of thyroid hormones usually occurs later in life; and (5) the marked reactions after menopause—accelerated rate of mineral loss from the bones; thinner, less elastic skin; and hot flashes—usually begin in adult women in their 40s.

The changes that begin in the adult body and become evident to various degrees, and which are usually speeded up and accentuated in the aged, are summarized and detailed in tabular form in the section on

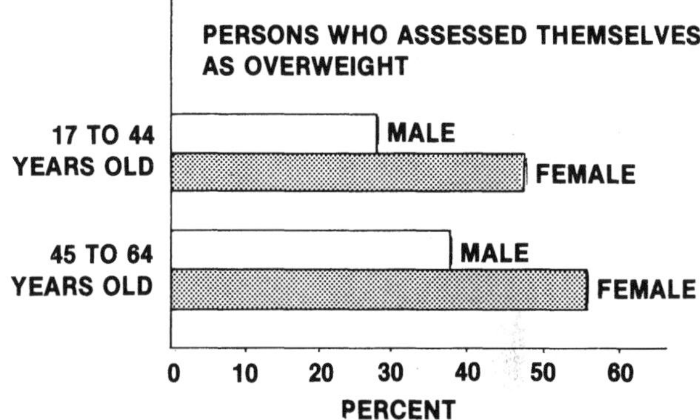

Fig. A-7. Percentages of younger and older adults who consider themselves to be overweight. (Adapted from *Social Indicators III*, U.S. Dept. of Commerce, Chart 2/23)

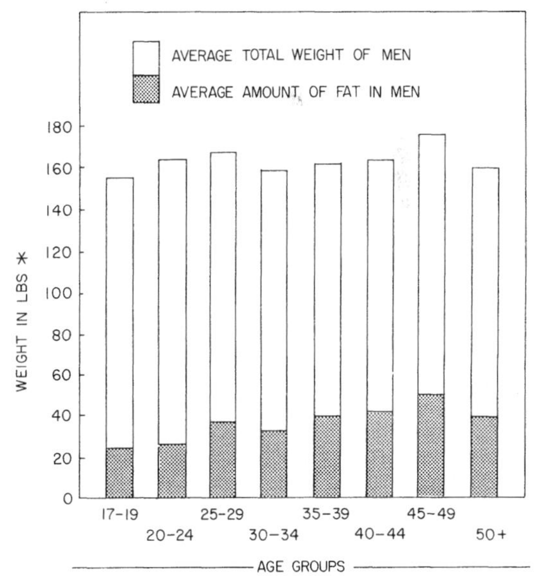

Fig. A-8. Average body fat contents of eight age groups of army men, as found in a study of 112 soldiers. (The bar graph was prepared from data in Krzywicki, H. J., *et al.*, "Alterations in Exercise and Body Composition with Age," *Proceedings of the Eighth International Congress On Nutrition*, Excerpta Medica Congress Series No. 213)

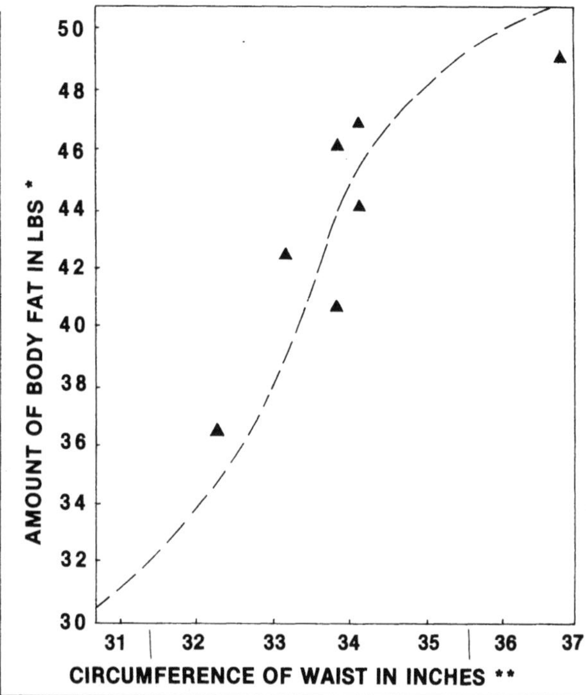

* TO OBTAIN KG, DIVIDE LB BY 2.2

** TO OBTAIN CM, MULTIPLY IN. BY 2.54

Fig. A-9. Correspondence of body fat with waistline measurements in adult males. (The graph was prepared by the authors from data in Krzywicki, H. J., *et al.*, "Alterations in Exercise and Body Composition with Age," *Proceedings of the Eighth International Congress On Nutrition,* Excerpta Medica Congress Series No. 213)

Gerontology and Geriatric Nutrition, Table G-3, Changes That May Occur in the Body During Aging.

HEALTH STATUS OF AMERICAN ADULTS. It is difficult to estimate the extent to which poor nutrition plays a role in the health problems of American adults because (1) few medical examinations are performed for diet-related disorders, (2) there is a lack of agreement among doctors regarding the importance of nutrition in disease prevention, and (3) doctor's offices and other medical facilities would be overwhelmed by crowds of people if every adult obtained a comprehensive physical examination at least once a year. Nevertheless, it is necessary to start somewhere, so that people who are likely to develop disabling conditions can be treated in time to prevent these problems. Therefore, some surveys have been conducted in which adults in various parts of the United States were asked about their health status.

Self Assessments of Health Status. The responses of younger and older adults to questioning about their overall health status are presented in the form of bar graphs in Fig. A-10.

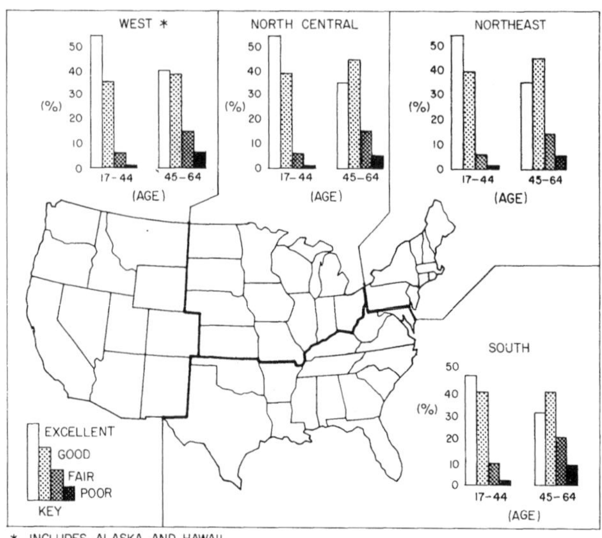

Fig. A-10. Self assessment of health status by adults in the four major regions of the United States in 1973. (Based upon data in *Health, United States, 1975,* U.S. Dept. HEW, pp. 437 and 439, Tables CD.III.1 and CD.III.2)

It may be noted from Fig. A-10 that the majority of adults of ages 17 to 44 considered themselves to be in excellent health, whereas only about 40% of those of ages 45 to 64 considered their health to be excellent. Furthermore, over 90% of the younger adults declared that they were in either excellent or good health, whereas only 77% of the older adults thought so. Hence, there appears to be a moderate decline in health status during adulthood. The prevalence of some of the most common health impairments is given in Fig. A-11.

Fig. A-11 shows that chronic health-impairing conditions are more prevalent in older adults, which suggests that various tissues deteriorate as adults grow older. However, the causes of the these types of deterioration

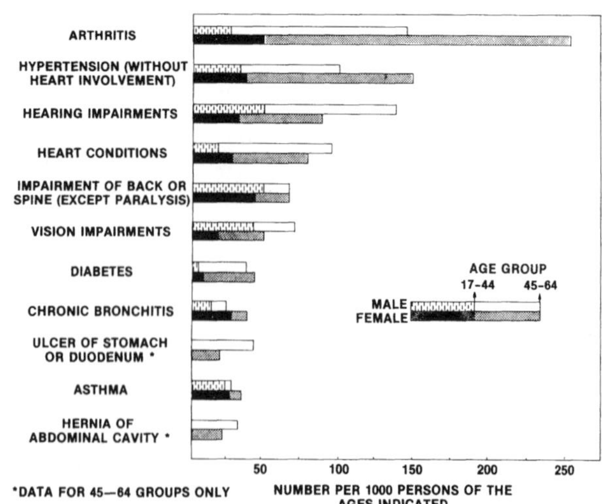

Fig. A-11. Prevalences of selected chronic conditions reported by American adults in health interviews. (Based upon data in *Health, United States, 1975,* U.S. Dept. HEW, pp. 481 and 487, Tables CD.III.23 and CD.III.26)

are poorly understood; so it is difficult to determine the roles of various dietary factors. Hence, health workers have attempted to find statistical correlations between degenerative diseases and certain types of dietary patterns. This approach has led to the identification of what are commonly called "risk factors."

Major Risk Factors for Degenerative Diseases. The leading killer diseases in the United States are heart diseases, cancer, and strokes, which together account for about two-thirds of all deaths. It is thought that many early deaths from these diseases may be prevented by (1) identifying people who have one or more risk factors, and (2) modifying the risk conditions as much as is feasible under the circumstances. Therefore, the major risk factors follow.

- **Abnormal electrocardiogram**—This condition is closely associated with the risk of a heart attack because it usually indicates a defective conduction of the heartbeat through the heart muscle. Most of the sudden deaths of middle-aged people from heart attacks are believed to be associated with conductivity defects. Sometimes, the cause of the abnormality may be identified and corrected. For example, deficiencies of magnesium and/or potassium may be responsible for electrolyte imbalances that can be treated by oral or intravenous administration of the appropriate mineral salts.

- **Alcoholism**—Two of the most direct effects of this vice are impairment of the mental processes and damage to the liver. The former effect contributes greatly to accidents, homicides, and unplanned pregnancies, whereas the latter is believed to be the main cause of cirrhosis of the liver. It is noteworthy that some chronic alcoholics suffer from multiple nutritional deficiencies because of failure to consume adequate diets. Unfortunately, alcoholism is our nation's number one social problem.

- **Family history of cancer, diabetes, heart disease, high blood pressure, and/or obesity**—Certain diseases run in families which appear to have hereditary susceptibilities to various metabolic abnormalities. Hence, people whose close relatives have had one or more of these conditions early in life should go to a doctor for regular physical examinations that include the appropriate laboratory tests.

- **High blood cholesterol and lipoproteins**—Total blood cholesterol is an indicator of a heart attack. Even better indicators are the level of the lipoproteins which carry cholesterol in the blood, specifically high-density lipoproteins (HDL) and low-density lipoproteins (LDL). HDL has a protective role since it transports cholesterol back to the liver, while LDL seems to deposit cholesterol in cells, including blood vessels. Also, evidence indicates that high HDL decreases the risk of heart attack while high LDL increases the risk. (See Fig. A-12.)

- **High blood pressure (hypertension)**—This condition is defined as a systolic pressure of 140 or greater and/or a diastolic pressure of 90 or greater. (See Fig. A-13.)

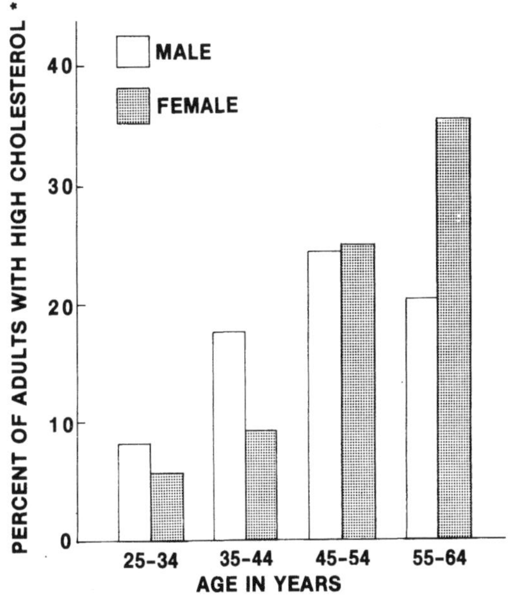

Fig. A-12. Percentages of adults with high blood cholesterol. (Adapted from *Social Indicators III*, U.S. Dept. of Commerce)

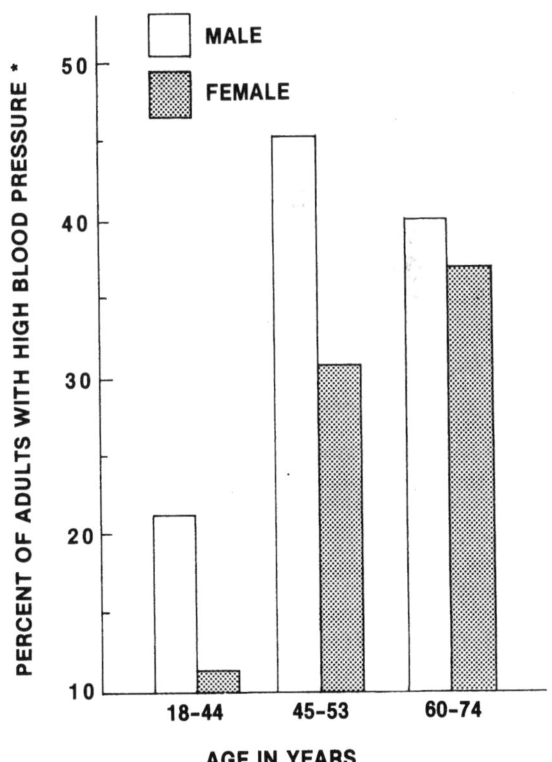

Fig. A-13. Percentages of adults with high blood pressure. (Based upon data from *Health, United States*, U.S. Dept. HEW)

High blood pressure is a dangerous condition because it hastens the development of atherosclerosis and increases the likelihood of an aneurism, heart attack, kidney disease, and stroke. Hence, it has been called the "silent killer." Prompt treatment by means of dietary salt restriction, drugs, and reduction of overweight is believed to decrease the risks.

• **Impaired glucose tolerance (diabeticlike metabolism of carbohydrates)**—This term designates the failure of the blood sugar level to come down to normal within a specified time after the consumption or injection of a test dose of sugar (glucose). Prolonged elevation of the blood sugar after meals increases the risk of diabetic complications such as arteriosclerosis, atherosclerosis, coronary thrombosis, injuries to the eyes and the kidneys, and stroke. Adults who have no previous history of diabetic symptoms are usually helped considerably by reduction of overweight to normal, since this condition occurs much more frequently in the obese than in people whose weights are normal. It is noteworthy that an impaired glucose tolerance sometimes results from the depletion of the body's supply of the essential mineral chromium. The latter circumstance is confirmed by the improvement of glucose tolerance when either inorganic chromium or the glucose tolerance factor (an organic complex that contains chromium) is administered orally or by injection.

• **Lack of exercise**—About 45% of American adults of age 22 and over fail to engage in physical exercise on a regular basis.[4] Hence, it is not surprising that the rates of various diseases increase steadily with age, since lack of exercise is closely associated with the development of obesity and its dire consequences. Furthermore, regular exercise increases the chances of surviving a heart attack since it enlarges the small blood vessels that may take over some of the task of blood circulation when the major coronary arteries are plugged.

• **Obesity**—Life insurance statisticians have found that notable obesity results in significantly increased risks of coronary heart disease, diabetes, gallbladder disease, high blood pressure, kidney disease, osteoarthritis, and stroke. However, people who are only moderately overweight by life insurance tables may have little or no increased risk if they are otherwise in good health. Hence, the decision to undertake a stringent program for weight reduction should be made only after consulting a doctor.

• **Poor dietary practices**—Generally, this designation applies to diets that are high in calories from refined carbohydrates, fats, and starches, but low in proteins, minerals, and vitamins. However, even certain reasonably well-balanced diets may be unsuitable for people with greater than normal tendencies to develop conditions such as arteriosclerosis, atherosclerosis, bowel problems, cancer, congestive heart failure, coronary heart disease, diabetes, high blood fats, high blood pressure, kidney disease, and various types of food intolerances. In such cases, normal diets may be potentially harmful —except those specially modified to meet nutritional needs while avoiding factors that may aggravate the degenerative disorder(s). For example, people prone to congestive heart failure, excessive fluid buildup in the tissues, and high blood pressure may have to go to considerable trouble in avoiding salt-containing foods that supply too much sodium for their conditions.

• **Rapid and/or irregular heartbeat**—Racing of the heart while at rest may be a sign of one or more disorders such as anemia, congestive heart failure, hyperthyroidism, or an infectious disease. Hence, a doctor should be consulted promptly whenever the pulse remains at 90 beats per minute or greater for prolonged periods of time. Also, an irregular pulse may be a sign of a disorder such as a reaction to a drug or other medication, or a severe deficiency of potassium, magnesium, or thiamin. It is noteworthy that the normal resting pulse rates of trained athletes may be as low as 45 to 60 beats per minute.

• **Smoking**—At least seven potential carcinogens are present in cigarette smoke, but the level of smoking needed to cause lung cancer is uncertain, and may even depend upon the overall resistance to injury of the smoker's lungs. Smoking is also believed to be a major contributor to cardiovascular disease because (1) it reduces the supply of vital oxygen to the tissues, (2) it speeds up the heart rate, and (3) it causes an inflammatory disease of the blood vessels.

• **Stress**—Although brief exposures to certain emotional or physical stresses may be stimulating to the body, prolonged severe stresses may cause irreversible damage to tissues and organs. Therefore, people who find it difficult to cope with day-to-day affairs should seek some help from a competent person such as a clergyman, doctor, friend, or psychologist.

NUTRITIONAL ALLOWANCES OF ADULTS. The Recommended Dietary Allowances for adults established by the National Research Council, National Academy of Sciences are given in Table A-4. A more complete presentation of nutrient allowances for different age groups, including adults, along with nutrient functions and best food sources of each nutrient, is given in this book in the section on NUTRIENTS: Requirements, Allowances, Functions, Sources, and its accompanying Table N-6; hence, the reader is referred thereto.

DIETARY GUIDELINES. The first dietary guideline is: *Eat enough of proper foods consistently.* To ensure nutritional adequacy, adults (and those responsible for their diets) are admonished to—

1. Read and follow the section on NUTRIENTS: Requirements, Allowances, Functions, Sources, including Table N-6.
2. Eat a daily diet which includes definite amounts of foods from each of the Six Food groups as detailed in the section on Food Groups, Table F-37.

But, lots of folks don't know what constitutes a good diet; and, worse yet, altogether too many people neglect or ignore the rules even if they know them. In either case, the net result is always the same; we shortchange ourselves on the right amounts of good nutritious foods. As a result, more and more doctors and nutritionists are recommending judicious mineral and vitamin supplementation.

[4]*Health, United States, 1975,* U.S. Dept. HEW, p. 453, Table CD.III.9.

TABLE A-4
MEAN HEIGHTS, WEIGHTS, AND RECOMMENDED DAILY NUTRIENT ALLOWANCES FOR ADULT MALES AND FEMALES[1]

Category	Males			Females[2]		
	19–24 yrs.	25–50 yrs.	51+ yrs.	19–24 yrs.	25–50 yrs.	51+ yrs.
Weight .... lb (kg)	160 (72)	174 (79)	170 (77)	128 (58)	138 (63)	143 (65)
Height .... in. (cm)	70 (177)	70 (176)	86 (173)	65 (164)	64 (163)	63 (160)
Energy .... kcal	2900	2900	2300	2200	2200	1900
Protein .... g	70	70	68	65	64	63
Macrominerals						
Calcium ....mg	1200	800	800	1200	800	800
Phosphorus ....mg	1200	800	800	1200	800	800
Sodium[3] ....mg	500	500	500	500	500	500
Chloride[3] ....mg	750	750	750	750	750	750
Magnesium ....mg	350	350	350	280	280	280
Potassium[3] ....mg	2000	2000	2000	2000	2000	2000
Microminerals						
Chromium[3,4] ....mcg	50–200	50–200	50–200	50–200	50–200	50–200
Copper[3,4] ....mg	1.5–3.0	1.5–3.0	1.5–3.0	1.5–3.0	1.5–3.0	1.5–3.0
Fluoride[3,4] ....mg	1.5–4.0	1.5–4.0	1.5–4.0	1.5–4.0	1.5–4.0	1.5–4.0
Iodine ....mcg	150	150	150	150	150	150
Iron ....mg	10	10	10	15	15	10
Manganese[3,4] ....mg	2.5–5.0	2.5–5.0	2.5–5.0	2.5–5.0	2.5–5.0	2.5–5.0
Molybdenum[3,4] ....mcg	75–250	75–250	75–250	75–250	75–250	75–250
Selenium ....mcg	70	70	70	55	55	55
Zinc ....mg	15	15	15	12	12	12
Vitamins, fat-soluble						
Vitamin A[5] ....mcg RE	1000	1000	1000	800	800	800
Vitamin D[6] ....mcg	10	5	5	10	5	5
Vitamin E[7] .... mg α-TE	10	10	10	8	8	8
Vitamin K[3] ....mcg	70	80	80	60	65	65
Vitamins, water-soluble						
Biotin[3] ....mg	30–100	30–100	30–100	30–100	30–100	30–100
Folate ....mcg	400	400	400	400	400	400
Niacin ....mg	19	19	15	15	15	13
Riboflavin ....mg	1.7	1.7	1.4	1.3	1.3	1.2
Thiamin ....mg	1.5	1.5	1.2	1.1	1.5	1.0
Vitamin B-6 ....mg	2.0	2.0	2.0	1.6	1.6	1.6
Vitamin B-12 ....mcg	2.0	2.0	2.0	2.0	2.0	2.0
Vitamin C ....mg	60	60	60	60	60	60

[1]*Recommended Dietary Allowances*, 10th rev. ed., 1989, NRC–National Academy of Sciences.

[2]The recommendations in this table are for nonpregnant, nonlactating females. Those for pregnant or lactating females are given in NUTRIENTS: Requirements, Allowances, Functions, Sources, Table N-6. (Also see PREGNANCY AND LACTATION NUTRITION.)

[3]These figures are given in the form of ranges of recommended intakes, because there is less information on which to base allowances.

[4]The upper values of the range should not be habitually exceeded, since the toxic levels for many trace elements may be only several times usual intakes.

[5]Retinol equivalents; 1 retinol equivalent = 1 mcg of retinol or 6 mcg beta-carotene.

[6]As cholecalciferol; 10 mcg cholecalciferol = 400 IU of vitamin D.

[7]Alpha-tocopherol equivalents; 1 mg d-α-tocopherol = 1 α-TE.

It is noteworthy, too, that those whose diets are restricted for the treatment of certain disorders may need to use supplements as sources of the nutrients that would ordinarily be provided by foods which are not allowed in their diets. For example, people who cannot drink milk or eat cheese may have to take a calcium supplement that provides this mineral.

Planning of Meals. A convenient way of planning meals is to use a food guide that translates the technical language of nutrients into terms of everyday eating. The Six Food Groups does this. It (1) groups foods in categories which reflect their similarities as good sources of specific nutrients, and (2) gives recommended amounts of each.

(Also see FOOD GROUPS, Table F-37).

Modified Diets and Dietetic Foods. Some people may need to follow diets that restrict certain foods as part of the overall therapy for one or more conditions such as atherosclerosis, cancer, diabetes, gall bladder disease, gout, heart disease, high blood fats, high blood pressure, and obesity. The percentages of adults on the most common types of modified diets are shown in Fig. A-14. However, one should *not* undertake to follow such a diet without the advice of a doctor or a dietician.

It may be noted from Fig. A-14 that the percentages of people on modified diets are much higher for older adults than for younger adults. The high rates of high blood pressure, high blood cholesterol, and obesity indicate the need for greater use of modified diets by people of all ages.

Similarly, many new dietetic foods which have been developed for use by people who are required to follow modified diets may *not* be the best buys for people who do not need dietary restrictions. For example, healthy

adults of normal weight may be wasting their money by purchasing dietetic canned fruits or vegetables.
(Also see MODIFIED DIETS.)

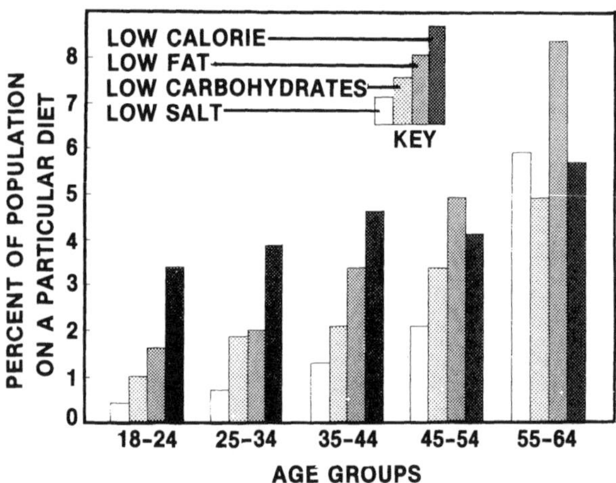

Fig. A-14. Adults of selected age groups who are commonly-prescribed modified diets. (Based upon data in *Social Indicators, III,* U.S. Dept. of Commerce, 1980, p. 106, Table 2/24)

SUMMARY. It is generally recognized that there are changes in appearance, mental outlook, and body functions as individuals pass from youth into middle age and then into old age. Behind the outer changes in people as they grow older are changes in the body and its workings. Some of these changes are not due to aging as such, but to impaired nutrition of body cells. So, adult nutrition involves the past, the present, and the future. An adult's nutritional state at any specific age reflects all his previous dietary history as well as his current food practices. Consequently, a good diet early in life will make for vigorous maturity; and continuance of good adult nutrition will extend the years of usefulness and delay—and, in some instances, even prevent—the appearance of old age. With good nutrition throughout life, it is the privilege of adults to extend their youthful vigor into middle age and beyond and to enjoy buoyant good health.
(Also see NUTRIENTS: REQUIREMENTS, ALLOWANCES, FUNCTIONS, SOURCES.)

ADZUKI BEAN *Phaseolus angularis*

This legume, which ranks next to soybeans in importance among the Chinese and Japanese, recently became known in the United States as a staple food in the Zen Macrobiotic diets. Fig. A-15 shows adzuki beans along with other legume seeds.

ORIGIN AND HISTORY. The adzuki bean is believed to have originated in Japan and China, where it has been cultivated for many centuries. In recent times, the bean was introduced into parts of Africa and the Americas.

The rise in popularity of the adzuki bean among American natural food enthusiasts stems almost entirely from the promotion of Zen Macrobiotic dietary practices by George Ohsawa, a Japanese who adopted certain ideas from Zen Buddhism. The Zen Macrobiotic movement reached its peak in the late 1960s, then declined after there were reports of severe malnutrition and a few deaths attributed to the overzealous practice of the dietary restrictions. However, some Americans discovered that adzuki beans could be made into various tasty dishes, even though the Macrobiotic ideas were of little interest to them.

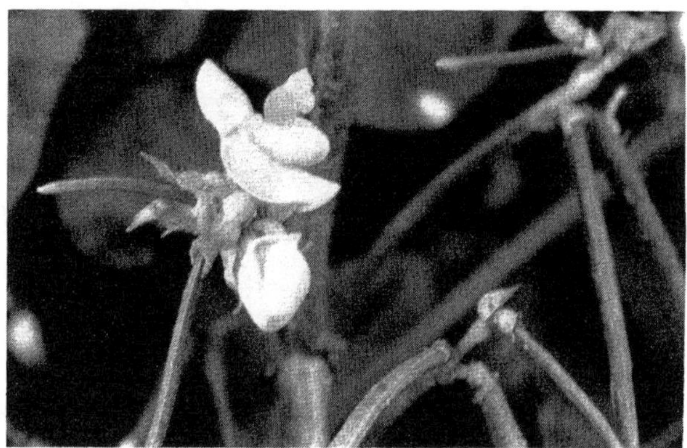

Fig. A-15. Adzuki bean, which is highly esteemed by the peoples of China and Japan. (Courtesy, Minnesota Agricultural Experiment Station, University of Minnesota)

PRODUCTION. Adzuki beans grow well in temperate climates with hot, humid summers. Japanese farmers often grow the adzuki in rotation with rice. From 3 to 5 months are required from planting to full maturation of the beans. Yields between 400 lb *(0.18 metric tons)* and 1,000 lb *(0.45 metric tons)* per acre are obtained.

PROCESSING. Much of the crop is marketed as whole, dried beans. However, some of the beans are ground finely for utilization as a flour or meal in cakes, candies, soups, and nondairy substitutes for milk such as *Kokoh* (a mixture of brown rice, glutinous brown rice, adzuki beans, and sesame seeds).

SELECTION AND PREPARATION. About the only places where adzuki beans are sold in the United States are oriental food shops and natural food stores. The latter are likely to be found in cities and towns which have large colleges and universities.

The beans should be soaked in water overnight, then cooked in a pressure cooker or a saucepan. Pressure cooking saves fuel because it takes only 30 minutes of heating (an additional 30 minutes should be allowed for the pressure to drop while the closed cooker remains undisturbed), whereas the saucepan method requires about 1½ hours of heating (plus another 20 minutes of standing after the heating has been discontinued). Cooked adzuki beans go well with most cereal grains and other vegetables. (Those who consume these beans on a regular basis are usually vegetarians, so they do not mix the beans with meats and poultry. Hence, recipes for such mixtures are not readily available.) However, there doesn't seem to be any culinary reason for not preparing bean and meat mixtures.

Cakes, candies, and cookies may be made from cooked beans that have been mashed and sweetened.

CAUTION: Almost all types of mature beans, including adzuki beans, tend to give people gas; hence, those who are bothered greatly by this condition should eat only green beans (the immature pods and seeds).

Other potentially harmful substances in beans are trypsin inhibitors (agents which impair the digestion of protein) and red blood cell clumping agents. Both of these undesirable components may be rendered harmless by soaking the beans overnight, then cooking them thoroughly. (This admonition is given because some people might be tempted to grind the uncooked dry beans into flour to be used in the preparation of cakes, candies, cookies, and other desserts. However, the baking of products made from raw bean flour is not always sufficient to render them harmless.)

NUTRITIONAL VALUE. The nutrient composition of adzuki beans is given in Food Composition Table F-36.

Some noteworthy observations about the nutrient composition of adzuki bean products follow:

1. Dried adzuki beans have about the same nutritional value as dried common beans. However, the boiled and sweetened canned product sold in specialty shops contains approximately twice as many calories, one-half as much protein, and about 50% more carbohydrates than cooked kidney or navy beans. (Canned adzuki beans are intended mainly as an ingredient of cakes, candies, cookies, and other dessert items.)
2. If it is assumed that unsweetened, cooked adzuki beans are similar in composition to cooked navy beans, then a 3½ oz *(100 g)* portion of the former would supply about the same amount of protein (about 7 g), but twice as many calories (118 kcal), as 1 oz *(28 g)* of cooked lean meat.
3. The calcium content is only about one-fourth of the phosphorus content. Hence, foods that are rich in calcium, but poor in phosphorus (dairy products and green leafy vegetables) should be consumed with adzuki beans to ensure optimal utilization of the dietary calcium.
4. Adzuki beans are rich in iron and potassium.
5. These beans, like most other dried legumes and cereal grains, are almost totally lacking in vitamins A and C. Hence, people eating diets based mainly upon these foods should be certain to (a) eat plenty of fresh fruits and vegetables and/or (b) take vitamin supplements.

Protein Quantity and Quality. People who eat adzuki beans regularly, whether they be Asians or Americans adhering to Macrobiotic diets, are not likely to consume much animal protein. Therefore, the following characteristics of the bean protein merit consideration:

1. From 3 to 4 oz *(98 g)* of cooked beans are required to supply an equivalent amount of protein to 1 to 2 oz *(42 g)* of such animal foods as cheese, eggs, fish, meat, and poultry. Also, on an equivalent weight basis, beans supply from 2 to 4 times as many calories as the animal products.
2. The quality of the protein in the beans is lower than that in most animal foods because of deficiencies of the amino acids methionine and cystine. However, part of these deficiencies may be compensated for by the consumption of products made from corn, rice, wheat, and other cereals which contain ample amounts of the amino acids lacking in the beans.

Zen Macrobiotic Diets. The Macrobiotic diets were developed by George Ohsawa from the practices of Zen Buddhist monks, who sought to still their passions in order to live a more perfect contemplative life. The diets consist of several regimens, ranging from the most severe, which provides only brown rice and water, to the most liberal, which includes cereal grains, vegetables, soups, animal foods, fruits, desserts, and water. Adzuki beans are important components of most of these diets because the bean protein complements the protein in the cereals that are so heavily emphasized.

Unfortunately, some of the young Americans who chose the Macrobiotic way of life were overzealous and restricted themselves to severe regimens which resulted in malnutrition. The saddest cases were those of young growing children, whose nutritional requirements are higher than those for most other groups of people. For example, one infant female was given adzuki bean powder and other ingredients in the form of a mixture called Kokoh, a milk substitute somewhat similar to the emergency supplements utilized in famines, except that in this case the formula was diluted excessively with water.[5] As a result, the baby girl developed protein-energy malnutrition and had to be hospitalized.

The lessons learned from the American experiences with Macrobiotic diets were: (1) radical new diets that contain unfamiliar foods should *not* be adopted unless someone with a sound knowledge of the scientific principles of nutrition is consulted, and (2) regimens which may be harmless when followed by healthy adult males for brief periods of time may be hazardous when applied to children and adult females for extended periods.

(Also see BEAN[S], Table B-10 Beans of the World; LEGUMES, Table L-2 Legumes of the World; and ZEN MACROBIOTIC DIETS.)

AEQUUM

This is the food required for body maintenance under usual or specified activity.

AEROBIC

In the presence of air. The term usually applied to microorganisms that require oxygen to live and reproduce.

(Also see METABOLISM.)

[5]Robson, J. R. K., *et al.*, "Zen Macrobiotic Dietary Problems in Infancy," *Pediatrics*, Vol. 53, March 1974, p. 326.

AEROPHAGIA

The swallowing of air, a habit commonly found in nervous people.

AESCULIN (Esculin)

This is a glycoside found in chestnuts.

AFLATOXIN

Aflatoxins are the most feared of the mycotoxins—toxin-producing molds. There are four main aflatoxins: B_1, B_2, G_1, and G_2, of which B_1 is the most common and the most toxic. A high incidence of cancer, especially liver cancer, is associated with aflatoxin.

(Also see POISONS, Table P-11 Mycotoxins.)

AGAR-AGAR

This material is extracted from red algae *(Rhodophyceae)* and is known to most students as a medium for growing microorganisms, but it has long been used by Asians for making jellies and other food preparations. Although many people use the term agar-agar, food technologists simply call the substance agar. Hence, the latter name will be used throughout this article.

HISTORY OF AGAR. Use of this jelling agent originated in Japan where it is called *Kanten.* Chinese settlers introduced its use to the Dutch East Indies (this area is now known as Indonesia), where it was given the name Agar-Agar. Thence the Dutch took the material to Europe. The idea of using agar as a culture medium for microorganisms was suggested by a Dutch housewife, Frau Fanny Hesse, who had used it to make jellies. She made the suggestion to her husband, Dr. Walter Hesse, who passed it on to Dr. Robert Koch, the famous microbiologist.

PRODUCTION OF AGAR. The modern day production of agar, although modified somewhat from ancient procedures, still is based mainly upon the principles underlying the traditional methods of production.

Harvesting Algae from the Sea. Some of the agar-containing seaweed is collected from rocks exposed at low tide along the sea coast of Japan, but the best material is collected under water by specially trained female divers, called *Ama.* Wearing only trunks, they dive all day long into the sea and bring the algae up to floating tubs or rafts. (These women are unique in having exceptionally large lung capacities and long endurance in the cold water.) Male divers, equipped with diving apparatus, make the collections at depths greater than 60 feet. The algae is sold to commercial processors after it has been dried and partly bleached on the beach.

Extraction and Purification of Agar. The seaweed is extracted by boiling in water in open pots or by autoclaving. The extract is then filtered and further bleached, followed by reextracting in order to obtain the maximum yield of agar. After jelling, freezing, and thawing, the gel is dried and pulverized.

SPECIAL PROPERTIES OF AGAR. The most unique property of this substance is the heat resistance of the gel that it forms (its solutions start to gel at 86°F *[30°C]*, but melting of these gels requires temperatures over 185° *[85°C]*). Also, agar forms strong gels at concentrations of 0.5% (about 1 teaspoon in a quart of water). Finally, it is a nonnutritive carbohydrate; hence, it does not yield food energy.

USES OF AGAR IN FOODS. Although agar costs more than synthetic and other natural jelling agents, it is usually superior to such products because its gels have greater transparency, strength, stability over a range of acidity and alkalinity, and reversability (gels may be formed and melted without decomposition of the agar). Some of the most common food uses follow.

- **Confections**—Jelly-type candies are still made with agar (it is used at concentrations from 0.3 to 1.8% by weight), although starch and pectin are used whenever a transparency and other characteristics of agar gel are not required. A typical candy might be prepared by the following procedure:[6]

1. The agar is soaked for 2 or more hours in water.
2. After all the agar is dissolved, the solution is cooked with sugar to 222°F *(105°C)*, sweetener (corn syrup and/or invert sugar) is added with thorough mixing; and finally, color, flavor, and acid are added.
3. When all of the ingredients are dissolved, the mixture is poured out to cool.

- **Fish, meat, and poultry products**—Canning of soft animal products, such as various meats, in agar gel prevents them from being broken into pieces during shipment. Other jelling agents are not as satisfactory since they are more likely to melt.

- **Fruit and vegetable gels**—Although other jelling agents have replaced agar in most of these products, it is still used in dietetic and vegetarian preparations (it is noncaloric and may be used in place of gelatin).

USE OF AGAR IN HOME RECIPES. Because of its cost, agar is not marketed through the normal food retail channels. It is usually sold only by Chinese and Japanese groceries and by health food stores.

Agar may be used in place of gelatin, but much smaller quantities are needed to produce the same jelling effects (1 to 3 teaspoons *[5-15 ml]* of dry, powdered agar per quart *[0.95 liter]* of water or other liquid will suffice, depending upon the consistency desired). If, by mistake, too much agar is used and the gel is too stiff, it may be remelted with extra water in the top half of a double boiler (boiling water is placed in the bottom half).

(Also see ADDITIVES.)

---

[6]Detrano, V., "Jellies and Gums," *Candy Industry Confectionary Journal*, Vol. 117, 1961, p. 49.

AGEUSIA

The lack of normal tasting ability which may be due to zinc deficiency or other causes.

(Also see TASTE; and ZINC.)

AGING

The processes by which certain body functions decline gradually with the passing of time until there may be great impairment(s). The rate of aging depends upon dietary, genetic, and environmental factors.

(Also see GERONTOLOGY AND GERIATRIC NUTRITION; and LONGEVITY.)

AGING MEAT

The process through which meat is made tender and flavor develops. The tenderizing process starts soon after slaughter. Aging is usually accomplished by holding the cuts at a temperature of 38° to 40°F (*3° to 4°C*) and a relative humidity of 85 to 90%, with a gradual flow of air to provide fresh atmosphere.

ALANINE

One of the nonessential amino acids.

(Also see AMINO ACID[S].)

ALAR (DAMINOZIDE)

Alar is the brand name of the chemical *daminozide*, an organic acid. It is a growth regulator which enables growers to provide consumers top quality apples that maintain firmness and resist bruising, early decay, and cracking. Alar was formerly applied as a spray on some varieties of apples, including Red Delicious, McIntosh, and Staymen, about 60 days before harvest.

On February 26, 1989, the CBS program *60 Minutes* reported that Alar is a potent cancer-causing agent. Apple prices and sales plummeted. The Apple Institute reported that the economic damage cost growers $250 million. Although scientific facts judged Alar to be safe, this verdict was overruled by the court of public opinion.

ALBEDO

The white, fibrous material which lies between the segments and the peel of citrus fruits. It constitutes from 20 to 60% of the whole fruit, and is used in the production of pectin. The albedo is rich in fiber and bioflavonoids (vitaminlike substances which have beneficial effects when consumed along with vitamin C).

(Also see BIOFLAVONOIDS; CITRUS FRUITS; FIBER; and PECTIN.)

ALBINISM

A lack of pigment in the eyes, hair, and skin due to an inborn error of metabolism. People who have this trait are called albinos. Their hair and skin are white, while the corneas of their eyes are pink.

(Also see INBORN ERRORS OF METABOLISM.)

ALBUMEN

The watery material which comprises the white of eggs and which consists mainly of the protein albumin.

(Also see ALBUMIN.)

ALBUMIN

A type of protein which is easily digested and which contains a good proportion of the essential amino acids. It is present in the tissues and body fluid of animals and humans as plasma albumin. It is also in milk, as lactalbumin, and in egg white, as egg albumin. Albumin is soluble in water, and is coagulated and solidified by heat. It is the principal type of protein in the blood, where it plays a major role in regulating the fluid flow between the blood and tissues. However, lactalbumin and egg albumin provoke allergies in some people.

(Also see ALLERGIES; and PROTEIN[S].)

ALBUMIN INDEX

A measure of egg quality and freshness; the ratio of the height of the albumin to the width when an egg is broken onto a flat surface. As the egg deteriorates, the albumin index decreases, i.e., the egg white spreads.

ALBUMINOIDS

Fibrous proteins that have supporting or protective function in the animal, of which there are three types: (1) collagens in the skin, tendons and bones, which are resistant to pepsin and trypsin, and are converted to water-soluble gelatin by boiling with water; (2) elastins in the tendons and arteries, which are not converted to gelatin; and (3) ceratins which comprise the horns, hoofs, feathers, scales, and nails, and which are insoluble in dilute acids and alkalis and are not attacked by any animal digestive enzymes.

ALBUMINURIA

A condition in which some of the albumin in the blood escapes through the kidney's filtering membrane into the urine. Albuminuria often accompanies kidney disease.

(Also see DISEASES.)

ALCAPTONURIA (ALKAPTONURIA)

A rare inborn (recessive) abnormality of metabolism in humans marked by the inability to complete the degradation of tyrosine and phenylalanine; their metabolism ceases at homogentisic acid, which is excreted in the urine. The homogentisic acid oxidizes to black melanoid pigment; hence, the urine of alcaptonurics slowly turns black. The defect appears to be harmless.

ALCOHOL, DENATURED

This refers to ethyl alcohol, which is made *unfit* for drinking, though its usefulness for other purposes is not affected. The following are some of the most commonly used denaturants: methanol, camphor, amyl alcohol, gasoline, isopropanol, terpineol, benzene, castor oil, acetone, nicotine, aniline dyes, ether, pyridine, cadmium iodine, sulfuric acid, kerosene, and diethyl phthalate. These may be used either alone or in combination. One of the prime reasons for denaturing alcohol is taxation purposes.

ALCOHOLIC BEVERAGES

Alcoholic beverages can be divided into three classifications as follows:

1. Beers
2. Distilled liquors
3. Wine

(Also see BEERS AND BREWING; DISTILLED LIQUORS; and WINE.)

ALCOHOLISM

Contents *Page*

Historical Background and Current Trends 30
Metabolism of Alcohol 31
Factors that Affect the Blood Level of Alcohol 32
Beneficial Effects of the Moderate Use of Alcohol 33
Harmful Effects of Alcohol Abuse 33
Occasional Drinking to Excess 33
Habitual Intoxication 33
Theories About the Cause of Alcoholism 36
Physiological Theories 36
Psychological Theories 36
Sociological Theories 36
Signs and Symptoms of Alcoholism 37
Treatment of Alcoholism 38
Correction of Medical, Nutritional, and Psychological Disorders 38
Alleviation or Curing of Dependence on Alcohol 38
Suggestions for Preventing Overconsumption of Alcoholic Beverages 39
Summary 39

Alcoholism is an addiction, a physical dependence on the habitual use of alcohol. It is noteworthy that 12 to 14% of regular drinkers become heavy or problem drinkers, and at least 5% become alcoholics.

In 1990, U.S. companies produced 8 billion gallons of alcoholic beverages; 75% of which was beer. That same year (1990), alcohol cost American society an estimated $136 billion and more than 65,000 lives, 22,000 of them on highways.[7] Years earlier, the seriousness of alcoholism as a public health problem was recognized by the U.S. congress when it established the National Institute on Alcohol Abuse and Alcoholism (NIAAA) in 1971 and gave the newly formed institute the mission of coping with alcoholism through a variety of programs in research, training, prevention, treatment, and rehabilitation. In 1973, the NIAAA founded the National Center for Alcohol Education, which published the book that served as the source of much of the information presented in this article.[8]

HISTORICAL BACKGROUND AND CURRENT TRENDS. Anthropologists believe that alcoholic beverages were produced by almost every prehistoric society that had access to fermentable substances such as fruits, honey, or starchy vegetables. (It appears that only the Eskimos lacked the means to produce these drinks.) The processes for making the beverages were easy to discover because (1) fruits and beverages that contained honey fermented spontaneously, (2) some starchy grains (notably barley, rye, and wheat) contained enzymes that converted starch to sugar during the process of sprouting (germination), and (3) human saliva supplied a starch splitting enzyme that was made to work on starchy foods by chewing them briefly prior to fermentation.

The first widespread use of alcohol by people in ancient societies appears to have been for religious ceremonies and social rites of passage such as those marking birth, puberty, marriage, and death. Many of the writings from the ancient civilizations contain admonitions against drunkeness, while tolerating or even encouraging the drinking of alcoholic beverages. For example, religious Jews have long used wine as a ceremonial drink, yet the rate of alcoholism for these people is about the lowest for any ethnic or religious group that consumes this type of drink. However, other cultural groups have accepted drinking to intoxication as a normal occurrence on certain occasions, as evidenced by the literature of Greece and Rome, and the observations anthropologists have made of certain Latin American Indian tribes.

Abstinence from alcoholic beverages later became a requirement for being a faithful member of non-Christian religions such as Buddhism and Islam; and of Christian denominations such as Baptists, Christian Scientists, Jehovah's Witnesses, Methodists, Mormons, Quakers, and Seventh-day Adventists. The members of these denominations played important roles in reducing the heavy drinking that characterized the early days of America. For example, the settlers of New England imported rum from the West Indies, and sailors and laborers often received liquor for part of their wages. However, the settlement of the Bible Belt (a term coined by the American writer H. C. Mencken to denote the areas of the South and Midwest where religious fundamentalism prevailed) during the period of westward and southward expansion in the early 19th century was followed by a drop in alcohol consumption from about 7 to 2 gal (*27 to 8 liter*) per capita per year. After that, alcohol consumption in the United States remained at about 2 gal (*8 liter*) per capita until the enactment of Prohibition in 1919. (Alcohol consumption increased steadily after the repeal of Prohibition in 1933.)

One of the recent trends which is a cause for great concern is the growing incidence of alcohol abuse among American teenagers. According to a report by the U.S. Department of Health and Human Services, more than 3 million 14 to 17-year-olds had drinking problems.[9] Some of the other current trends in the use of alcohol follow:

1. Heavy drinking is most frequent in young and

[7]Statistics adapted from *National Geographic*, Vol. 181, No. 2, February, 1992, pp. 2-35.

[8]The National Center for Alcohol Education, *The Community Health Nurse and Alcohol-Related Problems*, U.S. Department of HEW.

[9]Millman, R. B., "Alcohol: The Friendly Foe," *Science Year 1982*, World Book-Childcraft International, Inc., Chicago, Ill.

middle-aged males, whereas the proportion of heavy drinkers among females remains about the same for all ages.

2. Almost 40% of adult males are moderate or heavy drinkers, but only 16% of the females are moderate or heavy drinkers.

3. Half of all adult females, but less than one-third of all males, are abstainers or infrequent drinkers of alcoholic beverages.

4. Residents of the Pacific region (West Coast states, Alaska, and Hawaii) and of New England consume the greatest amount of alcohol, whereas residents of the East South Central region (heart of the Bible Belt) consume the least.

5. People who do not have church affiliations have higher rates of heavy drinking and drinking problems. While religion apparently provides a satisfying alternative to alcohol for some, it obviously is not an effective prevention method for everyone.

6. The use of alcohol increases with education and social status. For example, 60% of abstainers have less than an eighth grade education, and the rate of heavy drinkers rises from 6% of those with a grammar school education to 13% of high school graduates and reaches 15% among college graduates. Socioeconomic status shows a similar relationship with proportionately more people at lower levels reporting abstinence and a rise in alcohol-related problems and heavy drinking accompanying rise in social class.

METABOLISM OF ALCOHOL. Although there are several kinds of alcohol, such as methyl (wood) and isopropyl (rubbing) alcohols, *ethyl alcohol* or *ethanol* is the substance contained in all alcoholic beverages. The concentration of ethanol varies from one kind of drink to another. For example, the ethanol (absolute alcohol) content in American distilled spirits varies from 40 to 50%, while table wines range from 10 to 14%, and most American beers contain about 4%.

The term *proof* as applied to distilled spirits refers to ethanol concentration. Standards vary from country to country. In the United States, proof is twice the ethanol concentration. Therefore, an 80-proof whiskey or gin contains 40% alcohol.

Because of its chemical properties, alcohol—like sugar—is classified as a simple, incomplete food with limited nutritional value. It provides calories, yet it lacks proteins, minerals, and vitamins. Hence, people who regularly drink alcoholic beverages but fail to eat properly may develop nutritional deficiencies.

The major aspects of alcohol utilization by the body are outlined in Fig. A-16.

Details of the processes outlined in Fig. A-16 follow.

• **Absorption**—Alcohol enters the body through the mouth and goes immediately to the stomach where it is diluted by the stomach juices. If food is present in the stomach—especially a fatty food—absorption into the bloodstream will be somewhat delayed. The greater the fat content in the stomach the slower the absorption of the alcohol. However, if a drink is diluted with a carbonated beverage, the alcohol will be absorbed more quickly. The carbon dioxide in carbonated beverages can cause the valve between the stomach and the small intestine to open. With the entrance to the small intestine open, the alcohol absorption rate is increased. In addition, the emotional state of the drinker may affect the rate of absorption. Often the effect a drinker expects has some effect on what really happens.

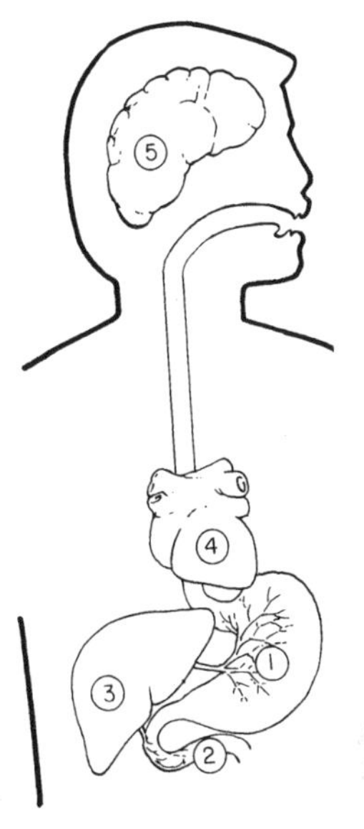

Fig. A-16. Utilization of alcohol by the body. The alcohol is absorbed by the stomach (1) and the small intestine (2) and travels via the bloodstream to the liver (3) where much of it is metabolized. However, some alcohol gets to the heart (4) and is pumped to other tissues such as the brain (5).

• **Transportation and distribution within the body**—Shortly after consumption, alcohol enters the bloodstream from the stomach and the small intestine and begins to travel with the blood to each organ's tissues and cells. Some organs absorb more alcohol from the bloodstream than others. Alcohol is concentrated more heavily in the blood and the brain than in fat or muscle tissue. Since there is such a large volume of blood, the alcohol is diluted and only a very small amount enters the cells at any one time—not enough, under ordinary circumstances, to cause any physical damage or noticeable effects in most organs of the body.

• **Oxidation**—The chemical combination of alcohol with oxygen in the tissues begins in the liver where alcohol is changed at a constant rate to a toxic chemical called acetaldehyde. The acetaldehyde is immediately converted to acetate in the liver and in many cells of the body. The acetate, in turn, is oxidized to carbon dioxide and water, producing heat and energy. Other organs are capable of metabolizing only small amounts of alcohol. When a person drinks alcohol at a faster rate than the body can oxidize it (about ½ oz [*15ml*] of ethanol per hour), intoxication occurs.

• **Effects of alcohol**—The brain is very sensitive to alcohol. Its normal functions of judgment, reasoning, and muscle control are affected by the amount of alcohol contained in 1 or 2 drinks, taken rapidly. At a low concentration in the blood, alcohol may act as a stimulant to the central nervous system, but as the concentration increases, there is a depressant effect. (If drinking is not continued, the brain returns to its normal functioning and the effects of a drink or two will disappear in about an hour when the blood alcohol level returns to normal.) Some typical effects of various blood levels of alcohol follow:

1. **Below a blood alcohol level of 0.03 to 0.05%**, there are generally no effects, or only mild alterations of feelings. *At 0.05%*, impairment of reaction time and coordination may be observed.

2. **At a blood alcohol concentration between 0.05 and 0.10%**, there is usually an increase in feelings of relaxation, sedation, and/or euphoria. Because alcohol can release inhibitions and camouflage fatigue, many people erroneously consider it a stimulant. Alcohol acts like a stimulant only at very moderate blood alcohol levels.

3. **At a blood alcohol level between 0.10 and 0.20%**, most drinkers show visible signs of intoxication: physical and mental impairment affecting perception and performance. Motor coordination deteriorates, judgment is impaired, and reaction time increases. Visual and auditory discrimination are affected; speech is sometimes slurred. When intoxicated, one can be silly and noisy, or gloomy and depressed, or belligerent and aggressive. Some become overly confident in their abilities and take greater risks than they would normally. Some feel romantic, some antisocial, some withdraw quietly into a private fog. One person can manifest all these behaviors in succession during a drinking session.

4. **At a blood alcohol level over 0.20%**, most people are quite sedated. They have trouble perceiving what is going on around them, trouble standing up, and trouble staying awake.

5. **At a blood alcohol level of 0.40%**, a person is in a coma. At 0.50%, one is in danger of dying from alcohol overdose, often by accidental choking. At high blood alcohol levels, absorption of alcohol still in the stomach continues at a constant rate. Death will occur when the respiratory and circulatory centers are anesthetized and cease to function.

Factors that Affect the Blood Level of Alcohol. The leading factors which determine whether or not a drinker becomes intoxicated are as follows:

• **Number of drinks and the concentration of alcohol consumed**—A man who weighs 150 lb (*68 kg*) may avoid becoming intoxicated if he refrains from consuming more than ½ oz (*15 ml*) of pure alcohol per hour. This means limiting drinking to only a single drink such as (1) a 1-oz (*30 ml*) shot glass of spirits, (2) a 4-oz (*120 ml*) glass of table wine, or (3) a 12-oz (*360 ml*) can or glass of beer. Table A-5 gives the alcohol content of some of the most popular beverages.

It is noteworthy that the carbonation of champagne and other sparkling wines increases the rate at which alcohol is absorbed, whereas the nonalcoholic substances (water, sugar, salts, and amino acids) slow the rate of absorption.

• **Speed of consumption**—The body breaks down alcohol and inactivates (oxidizes) it at a fairly steady pace. When a person drinks a lot in a short time, alcohol is absorbed into the bloodstream faster than the body can oxidize it, so it accumulates and the blood alcohol concentration increases. When a person sips a drink, the body has more time to oxidize the alcohol.

• **Setting for drinking**—Hot, noisy, crowded parties are more conducive to intoxication than intimate, relaxed, comfortable settings.

• **Amount of food in stomach**—Alcohol is absorbed faster if the stomach is empty. Food in the digestive tract slows the absorption rate. Hence, drinking alcoholic beverages during or after a meal reduces the likelihood of intoxication.

• **Emotional and physical states**—Fear, stress, anger, and fatigue may alter a person's rate of absorption in various ways.

• **Body weight**—The same amount of alcohol will be in a greater concentration in the bloodstream of a small, light person, than in a large, heavy person.

• **Previous drinking experience**—Prolonged drinking experience results in what is called tolerance or adaptation so that more alcohol is required to achieve the same effect. Psychological adaptation, sometimes connected to tissue adaptation, may also result in a learned ability to control one's self and to behave normally despite a high concentration of alcohol in the blood.

TABLE A-5
ALCOHOL IN SELECTED DRINKS[1]

Drink — 100g (3.5 oz) Portion		Alcohol
		(g)
BEER, 4.5% alcohol by volume		3.6
WINE,		
TABLE, 12.2% alcohol by volume		9.9
DESSERT, 18.8% alcohol by volume		15.3
GIN	80-proof	33.4
	86-proof	36.0
	90-proof	37.9
	94-proof	39.7
	100-proof	42.5
RUM	80-proof	33.4
	86-proof	36.0
	90-proof	37.9
	94-proof	39.7
	100-proof	42.5
WHISKEY	80-proof	33.4
	86-proof	36.0
	90-proof	37.9
	94-proof	39.7
	100-proof	42.5
VODKA	80-proof	33.4
	86-proof	36.0
	90-proof	37.9
	94-proof	39.7
	100-proof	42.5

[1]The authors gratefully acknowledge that these drink compositions were obtained from the HVH-CWRU Nutrient Data Base developed by the Division of Nutrition, Highland View Hospital, and the Departments of Biometry and Nutrition, School of Medicine, Case Western Reserve University, Cleveland, Ohio.

• **General physical health**—State of health is especially important if the liver is impaired in its ability to oxidize the alcohol. The presence of other drugs in the body may alter the effects of the alcohol on body tissues.

BENEFICIAL EFFECTS OF THE MODERATE USE OF ALCOHOL. Alcoholic beverages have been consumed for thousands of years in order to obtain the benefits enumerated below:

1. Lessened feelings of tiredness and fatigue
2. Reduction of tension, anxiety, and pressure; and increased feelings of relaxation and optimism
3. A sense of euphoria
4. Elimination of self-consciousness, and promotion of self-esteem
5. A drink may be a unifying symbol, encouraging a feeling of solidarity among a group of drinkers.
6. Talking and laughing may seem easier, as may singing and storytelling.
7. Social relations may seem less difficult, more honest, healthy, and open. Drinking companions often put their arms around one another, relaxing the "taboo on touching" which usually prevails in American culture.
8. Release of inhibitions, even though there may be some undesirable consequences, may be comforting, even exhilarating.
9. Stimulation of the flow of digestive juices and arousal of the appetite
10. Dilation of the peripheral blood vessels, relieving minor pains (particularly in the elderly) and causing a feeling of warmth
11. Action as a mild sedative, facilitating falling asleep
12. Reduced acuity of the senses, making the drinker less bothered by little annoying sounds and changes of light.

HARMFUL EFFECTS OF ALCOHOL ABUSE. These effects vary according to the circumstances under which drinking occurs, and the extent to which intoxication has become a regular event in a person's life. Therefore, the problems of someone who occasionally becomes drunk at home differ from those of the person who goes out on drinking binges at regular intervals.

Occasional Drinking to Excess. Certain social groups tolerate drunkeness among their members on certain occasions such as special holidays. Hence, the consequences may range from mild annoyances to potentially hazardous behavior, as enumerated below:

1. The drinker may make a nuisance of himself, affecting the way other people respond to him or her, not only during the nuisance period, but later as well.
2. Physical conflicts may be provoked by the intoxicated person's becoming belligerent and insulting.
3. Nausea and vomiting are common occurrences after overdrinking, particularly in inexperienced drinkers.
4. A hangover the morning after overdrinking usually impairs efficient work or enjoyment of life.
5. The possibility of injury in recreational and other activities increases with intoxication.
6. The most serious behavioral problem of the overdrinker is driving while intoxicated. The chance of being involved in an accident increases greatly with the increase in blood alcohol concentration.

Habitual Intoxication. Chronic, excessive drinking often has dire consequences that result from the alteration of normal physiological functions by alcohol, and from the disruptive effects of alcohol behavior on personal functioning and interpersonal relationships. For example, alcoholics have a mortality rate 2½ times greater than average, a suicide rate 2½ times greater than average, and an accident rate 7 times greater than average. Examples of the major physiological, emotional, and behavioral disorders associated with chronic alcoholism follow.

• **Cardiovascular problems**—Alcohol affects the cardiovascular system by (1) causing vasodilation of the peripheral vessels, producing flushing, heat loss, and a sense of warmth; and (2) promoting vasoconstriction of the central blood vessels, producing resistance to the flow of blood, and increasing the work load on the heart. Drinking alcohol is contraindicated when participating in strenuous activity and is ineffective as a means of staying warm.

A disorder known as alcoholic cardiomyopathy has been identified in recent years and occurs most often in men with a long history of alcohol misuse. Thought to be caused by the effects of alcohol on the myocardium, it is characterized by slow or sudden onset of left- and right-sided congestive heart failure, with distended neck veins, enlarged heart, elevated diastolic blood pressure, and fluid collection in the tissues (edema). With abstinence and bed rest, the tissues return to normal and the individual recovers. The earlier the condition is diagnosed, the better.

• **Central nervous system disorders**—Long-term excessive use of alcohol can result in premature aging of the brain. Whereas some difficulties with cognitive function clear up after a period of abstinence, residual problems can remain, usually due to a combination of nutritional deficiencies, repeated alcohol-related convulsions, head injuries, and degeneration of nerve and tissue cells.

One well-known, but not universal, effect of excessive alcohol use is the blackout. Unlike fainting or "passing out," in which a person loses consciousness, an individual experiencing a blackout walks, talks, and acts normally, and appears to be aware of what is happening. Yet later, he or she has no recollection of events during the blackout. Not all persons with alcohol problems experience blackouts. There is no way to predict who will have them and who will not, nor to know when the first one will occur. Blackouts have been reported after ingestion of only a few ounces of alcohol and can occur in alcohol-dependent persons who had not been drinking at the time.

• **Decreased immunity to infections and impaired healing of injuries**—Alcohol-dependent persons frequently have a lowered resistance to infections of all types because of the depressing effect of alcohol on the immunological system, especially white blood cells. Upper respiratory infections and pneumonia are common. Injuries, including surgical wounds, require longer to heal.

• **Development of an addiction to alcohol**—Consuming large quantities of alcohol over extended periods of time results in a decreased sensitivity of the brain to the effects of the alcohol. As a person continues to drink, an increased intake of alcohol is needed to achieve a desired effect. Many alcohol-dependent people do not seem to be intoxicated after drinking large amounts. Their bodies can tolerate higher continuous blood alcohol concentrations without noticeable signs or symptoms of intoxication. The risk of greatly increased tolerance and development of addiction seems greater in those who consume an average of about six or more drinks per day. The increase in tolerance and the development of addiction are gradual processes that usually progress over a period of many years. However, some individuals may experience these developments over a period of only a few years or months.

Usually, withdrawal reaction to the cessation of drinking is not seen unless an individual has ingested the equivalent of a pint of distilled spirits for at least 10 consecutive days. The overall reaction may be mild to severe, and the person may go through only one or all of the successive stages, *depending* on (1) how addicted the person is, (2) whether alcohol is used alone or with other drugs, (3) the amount and duration of the current drug intake, and (4) the particular substances in question. The signs and symptoms of withdrawal are characterized by a state of hyperexcitability, a reaction to the chronic depressive effects of the alcohol. The commonly seen early features of withdrawal include anxiety, anorexia, insomnia, and tremor. Vital signs are elevated with a pulse rate frequently as high as 120 to 140 beats per minute. The patient may complain of feeling shaky inside and appear irritable and hypersensitive to sudden sound and movement. This stage develops a few hours after alcohol intake is stopped, peaks in 24 to 36 hours, and may end abruptly with no further problems.

An additional problem to the ones described for a mild withdrawal reaction is hallucinosis. The person remains oriented and rational. Signs of disorientation beyond brief confusion about time should be observed carefully as an indication of progression of the withdrawal reaction to delirium tremens. An increase in pulse rate may also signal progression of the reaction.

In addition to the problems of mild to moderate withdrawal reaction, delirium tremens, the most severe reaction, is characterized by disorientation, paranoia, and outbursts of irrational behavior with danger of self-inflicted injury or harm to others. Racing of the heart is most common and may be accompanied by fever, rapid breathing, sweating, vomiting, and diarrhea. Without alert and vigorous treatment during this state, death may result directly from shock, malignant elevation of the body temperature (hyperthermia), or secondarily to complicating illness, infection, or injury. The delirium state peaks usually the third day after cessation of drinking and persists 2 to 3 days, ending abruptly and often dramatically. The incidence of delirium tremens among hospitalized alcoholic individuals averages about 5% and may be less in the total alcoholic population.

• **Emotional disturbances**—Alcohol decreases cognitive functions and allows the emotions to predominate. Anger or rage, sadness, and euphoria are commonly experienced during heavy drinking and often displayed in an exaggerated manner. Hence, the person who drinks may become argumentative, hostile, or intent on fighting; tearful and maudlin; or the "life of the party." In contrast, some people drink to dull or escape their feelings. Drinking to lessen the pain of guilt, rage, or sorrow usually provides only temporary relief, and when the effects of the alcohol are gone, the painful feelings return, often with increased intensity. In certain situations, with mental judgment diminished by alchohol and normal fears dampened, a person may take unaccustomed risks. Accidents, homicide, and suicide are serious consequences of alcohol's effect on emotions.

• **Fetal addiction and the fetal alcohol syndrome (FAS)**—When the mother is an active alcoholic, her child may be born addicted and experience withdrawal symptoms within the first week of life.

Heavy drinking during pregnancy can also cause the fetal alcohol syndrome. Most common in the FAS are prenatal growth deficiencies, usually more severe with respect to birth length than birth weight, and postnatal growth deficiencies persisting through early childhood. Failure to thrive is reported despite adequate calorie intake and excellent foster care in the first year of life. Often, the infant has an abnormally small head. Mental deficiency or developmental delay are almost always present. Fine motor dysfunction including tremulousness, weak grasp, and poor eye-hand coordination are present in most children with FAS.

Craniofacial malformation are common features of FAS, which impairs the normal development of the skull, jaws, teeth, nose, eyes, and facial skin. Sometimes variable anomalies of limbs and joints are present, including congenital hip dislocations, abnormalities of the toes, and inability to extend completely the elbows or metacarpal phalangeal joints. Cardiac malformations syndrome encompass an atrial septal defect, a patent ductus arteriosis, and cardiac murmurs representing ventricular septal defects. Anomalies of external genitalia have also been noted.

Recent studies indicate that a direct correlation exists between maternal blood ethanol levels and growth deficiency, especially of the brain, and that these deficiencies are due to the ethanol intake, not caloric insufficiency. A high blood alcohol level during a critical period of embryonic development probably is necessary to produce the FAS. The average alcohol consumption may not be as important as the maximum concentrations obtained during heavy drinking at critical periods. Malformations probably are produced by high alcohol concentrations during the first trimester when embryonic development is taking place. Growth may be the most vulnerable to heavy drinking during the third trimester. Behavioral disturbances, related to the chemical and structural disruption of the central nervous system, may differ depending on ethanol levels at other developmental stages.

Studies indicate that when women who drink heavily significantly reduce their alcohol intake or abstain entirely even as late as the third trimester of pregnancy, the incidence of FAS is significantly reduced. The correlation between FAS and child abuse is being studied currently. Studies have already shown that children who are retarded, malformed, or hyperactive tend to be abused more often than other children.

There are still a number of unanswered questions

about fetal alcohol syndrome, such as the risk of moderate drinking, whether the risk to the fetus is greater at particular times during pregnancy, and the influence of other factors such as smoking and and nutrition. While research continues on these and other issues, caution is advised. The recommended guideline for alcohol consumption during pregnancy is no more than 1 oz *(28g)* of ethanol per day (two drinks).

• **Gastrointestinal disorders**—Alcohol is an irritating chemical and in strong solution can damage the mucosa and result in esophagitis and gastritis. Increased capillary fragility can result in gastric bleeding. Erosion and ulceration of the mucosa can occur, and gastric or duodenal ulcers are a frequent result. The severity of these problems is related to the extent and duration of alcohol misuse.

• **Increased risk of cancer**—An excessive intake of alcohol increases the risk of cancer, particularly of the mouth, pharynx, larynx, and esophagus. In addition, alcoholic patients with cancer show lower survival rates and higher rates of developing another primary tumor when compared to other patients with the same cancer. The highest risk occurs when heavy alcohol intake is combined with heavy smoking, especially for esophageal cancer. One study reported the risk of esophageal cancer as 44 times higher for individuals who took more than six drinks per day and smoked a pack or more of cigarettes per day than for individuals who used little or none of either substance.

• **Liver disease**—Enlargement of the liver occurs with prolonged heavy alcohol use, probably as a result of accumulation of triglycerides in the hepatic cells. This fatty liver condition is usually reversible with abstinence from alcohol and a nutritional diet. Alcohol-induced hepatitis and cirrhosis are more serious disease processes. Each of these ailments is thought by some to be due largely to the direct effects of alcohol on the liver tissue and may occur even in the presence of adequate diet. With cirrhosis, death usually results from hepatic failure, hemorrhage from esophageal varices, portal vein thrombosis, or infection. Approximately 85 to 90% of deaths from cirrhosis are alcohol related.

• **Nutritional deficiencies**—A number of alcohol-related neurological disorders are due to nutritional deficiencies—primarily the B vitamins, including thiamin. These deficiencies result from decreased taste for food, decreased appetite (alcohol is high in calories and suppresses the appetite), and malabsorption of nutrients due to the irritated lining of the stomach and small intestine. A common nutritional deficiency disorder, peripheral polyneuropathy, is characterized by weakness, numbness, partial paralysis of extremities, pain in the legs, and impaired sensory reaction and motor reflexes. This condition is reversible with adequate diet and supplemental thiamin and other B vitamins.

If the polyneuropathy is left untreated, it may progress to Wernicke's encephalopathy. This more serious disorder is also reversible. It is characterized by ophthalmoplegia, nystagmus, ataxia, apathy, drowsiness, confusion, and inability to concentrate. Without treatment, it can be fatal. Another disease, often manifested after improvement from Wernicke's encephalopathy is Korsakoff's psychosis. This condition is characterized by disorientation and memory defect, usually with filling of the gaps with fictitious details, and often polyneuropathy. Many of those who develop this disorder show only limited improvement with treatment and generally require placement in psychiatric institutions or nursing homes. For obvious reasons, these persons require close supervision and assistance with activities of daily living.

• **Pancreatitis**—Acute pancreatitis, which may develop from prolonged alcohol misuse, exhibits symptoms ranging in severity from a gastritislike sensation to severe pain with nausea and vomiting and rigidity of the abdominal musculature. The most serious consequences of this condition include necrosis and hemorrhaging of the organ. Usually all that is required to alleviate the problem is abstinence from alcohol, maintenance of nutrition and fluid balance, and treatment of symptoms.

• **Personality changes**—Alcohol, when long misused, alters the personality. The particular changes caused by alcohol misuse are related to the person's basic personality structure and his individual response to the long-term effects of alcohol. For example, a fun-loving, outgoing person who enjoys life and the company of others may, with problem drinking, become irritable, belligerent, defiant, hostile, isolated, rigid, or stubborn. The preteen or teen-ager may misuse alcohol to avoid the painful aspects of adolescence, a critical period of psychosocial growth and development. Bypassing the "growing up" process leaves the person handicapped as an adult. This is one possible explanation for the emotional immaturity and juvenile behavior often seen in the alcohol-dependent adult.

• **Sensory impairments**—For some people,alcoholic beverages, such as wine, may serve to enhance the flavor of some foods. For others, alcohol reduces the sensitivity to taste and odors, making food less appealing. Tactile response is not affected, but sensitivity to pain is decreased. This is one factor in the increased incidence of burns, cuts, scrapes, and bruises among problem drinkers. At high doses, vision is impaired in terms of decreased resistance to glare (i.e., the eyes take longer to readjust after exposure to bright lights), and there is a narrowing of the visual field (tunnel vision). These effects are particularly significant when a person attempts to drive while under the influence of alcohol. Normal function returns when alcohol is no longer present in the body.

• **Sexual difficulties**—It is a popular notion that alcohol acts as a sexual stimulant. While it may assist in overcoming guilt and lack of self-confidence, promote a feeling of sexiness or amorousness, and release inhibitions, actual performance is impaired. Chronic heavy use of alcohol can result in sexual frigidity or impotency. Generally, the disturbance in sexual function disappears with abstinence from alcohol over a period of several months or more.

• **Sleep disturbances**—While low doses of alcohol induce relaxation and sleepiness, large doses produce sleep disturbances. Experienced as restlessness and early awakening with difficulty returning to sleep, the disturbance is caused by a shortened period of rapid eye movement (REM) or dream sleep. Results include fatigue, irritability, impairment of concentration and memory, and a variety of physical discomforts.

• **Violent and destructive behavior (pathological alcohol reaction)**—After drinking only a small amount of alcohol, some people lose contact with reality, get out of control, and become violent and physically destructive. Such persons may smash windows, throw things, break furniture, or assault another person. The episode may last several hours. The victim usually collapses in a state of exhaustion and awakens with no memory of what occurred. Pathological alcohol reaction is differentiated from blackout because of the extremely violent behavior which characterizes it. There is no way of predicting who will develop this disorder or when. A person in this state may attempt homicide. The mechanism of pathological alcohol reaction, a rare condition, is unknown.

THEORIES ABOUT THE CAUSES OF ALCOHOLISM. Research, to date, has failed to reveal any clear-cut, unequivocal evidence as to the basic causation of either heavy drinking or alcoholism. A number of researchers have approached this question largely from the viewpoints of their own disciplines. An overview of the major theories of the causes of alcoholism follows:

Physiological Theories. The proponents of these theories believe that there is something about the body chemistry of some people that makes them particularly vulnerable to alcohol. They argue that alcohol affects the intricate chemical body functions of these people in an unusual manner quite different from the way it affects other people. If this is true, then it would follow that some people, because of their physical make-up, are destined to become alcoholics if they ever use alcohol.

• **Allergy theory**—The supporters of this theory contend that ingestion of alcohol brings about some sort of allergic type of response, creating a need for more alcohol.

• **Endocrine theory**—The proponents of this theory believe that some endocrine deficiency or imbalance, such as hypothyroidism, hypopituitarism, hypoadrenalism, or even hypogonadaism, results in a predisposition to alcohol misuse and alcoholism.

• **Genetic theory**—According to this theory, predisposition to develop alcoholism is inherited. One method employed in studies designated to test this theory was to compare the rate of alcohol problems among siblings born to an alcoholic parent; some of the siblings were raised by the alcoholic parent and others were raised in nonalcoholic households. While statistics indicate that the rate of alcoholic dependence is markedly greater among relatives of alcohol-dependent persons than in the general population, there is no conclusive evidence to support a genetic theory. In families where the disorder is common, the psychological and socioculture factors also are thought to be of significance. In addition, no evidence is available to confirm that genetic or hereditary factors are responsible for a tendency to ward off alcoholic dependence.

• **Metabolic theory**—The proponents of this theory feel that alcoholism stems from an unidentified metabolic disturbance that causes a craving for alcohol.

• **Nutrition theory**—Those who subscribe to this theory contend that alcoholism stems from some sort of dietary or nutritional deficiency that is supplemented by alcohol.

Psychological Theories. These theories attempt to explain alcoholic behavior in terms of the attempts by alcoholics to meet their psychological needs. Two of the leading theories follow:

• **Alcoholic personality theory**—An alcoholic personality is one which exhibits low frustration tolerance; weak ego strength; exaggerated sensitivities; sexual and emotional immaturity; basic feelings of inferiority, inadequacy, and powerlessness; excessive dependence or conflict between dependency and independency needs; and difficulty in postponing gratification. Although those who develop alcohol dependence commonly do exhibit some or all of these personality traits, many others who also exhibit the same traits do not develop the syndrome. A cause-effect relationship has not been established. Some hold the opinion that persons afflicted with alcoholism have an unconscious "death wish" and that the chronic misuse of alcohol is a way of realizing the wish (i.e., slow suicide). Perhaps this is true in some individuals, but the statement cannot be applied generally. There is no known personality type unique to the problem of alcoholism.

• **Learning theory**—Some researchers have suggested that repeated heavy use of alcohol is a learned behavior pattern. For example, if the individual finds that alcohol relieves anxiety and fear, the emotional reward from reduction or increased tolerance of these feelings provides reinforcement each time the drug is used for that purpose. A conditioned response of drinking thereby becomes strengthened. This theory implies that the behavior leading to alcohol problems can be unlearned.

Sociological Theories. Alcoholism cannot be explained entirely on the individual's personal reaction to alcohol, either physical or psychological. Personal reaction to alcohol does not account for the high rate of alcoholism among certain cultural groups and the low rate among others. Therefore, sociologists have made studies of the drinking patterns and customs of ethnic groups, religious groups, socioeconomic groups, and urban and rural people in order to try to identify the factors that seem to promote alcoholism among certain groups and the control of drinking in others.

• **Cultural theory**—The role that culture plays in the causation of alcoholism is a puzzling one. Reasons for this include varying definitions of alcoholism, varying methods of recording the statistics of alcoholism, and varying expectancies regarding acceptable behavior. For example, Italy is known as a country that has low rates of alcoholism but high rates of drinking. Furthermore, Italian-Americans have lower rates of alcoholism than a number of other ethnic groups in this country. Ireland, on the other hand, has high rates of alcoholism, as do the Irish-Americans in the United States. France, at one time, was known as a country with low rates of alcoholism; more recently, it appears that heavy addiction to drinking is a problem in France. In general, research has led sociologists to suggest that alcoholism will be at its highest level in ethnic groups where tension is high and drinking habits are not subjected to consistent social control.

• **Religious affiliation theory**—Faithful members of religious groups, such as Baptists, Buddhists, Christian Scientists, Jehovah's Witnesses, Methodists, Mormons, Muslims, Quakers, and Seventh-day Adventists, either control their consumption of alcohol strictly or abstain completely from alcoholic beverages. Furthermore, the Jews in the United States and other parts of the world have extremely low rates of alcoholism, even though they have high rates of drinking. Some social scientists speculate that the reason for the low rates of alcoholism among Jews is closely associated with the fact that wine is used as part of a religious ritual and that children associate religion and wine from their early years. Hence, it appears that religious affiliation may exert a strong influence against the development of alcoholism. However, some sociologists believe that the low rates of alcoholism in these groups result in part from other factors such as an atmosphere of security and sense of belonging, family cohesiveness, group pride, and a strong adherence to the basic concepts of law.

• **Socioeconomic theory**—Clues from the evidence on different kinds of drinking and attitudes toward drinking found in the various social classes suggest that the lower-class drinker, in keeping with other aspects of the low-income culture, may be more apt to be an *"impulsive* drinker" who drinks because it is enjoyable and because alcohol is available (somewhat as he engages impulsively in other forms of behavior). He may also drink partly in response to the emphasis on virility and physical strength in the low-income culture; i.e., heavy drinking is often associated with "manliness," and one of the few available ways of expressing virility for the lower-class male is to get drunk.

On the other hand, it is possible that the middle-class alcoholic may be more a *compulsive* drinker who drinks partially to allay what is generally termed "middle-class anxiety," an anxiety over fears of failure, a drive for achievement, a sense of guilt over basic impulses, and discrepancy between the real and the ideal self. The compulsive drinker would seem more apt to be led to further drinking, after he has been inebriated, because of his sense of guilt over this behavior. It may be that the compulsive middle-class drinker has a more basic problem and a more complex one in recovering from alcoholism than the lower-class heavy drinker with his more impulsive life orientation. In fact, it seems possible that lower-class membership may be less conducive to true alcoholism than middle-class membership, in that true alcoholism involves a compulsive need to drink.

While definitive data, as has already been noted, are seriously lacking, it is suggested that an artificially large number of alcoholics might be found in lower socioeconomic groups, partly because their alcoholism brought them to this lower status, whereas their background might actually have been middle-class.

• **Urban environment theory**—There are data which seem clearly to indicate that drinking and alcoholism are more common in cities than in rural areas. Whether or not the differences are as large as the data would lead one to think would seem to depend on more accurate gathering of facts about the actual incidence of these behaviors. Even if it is found that there is a higher rate of drinking and alcoholism in urban areas, this does not necessarily mean that conditions of urban life create alcoholism—or even contribute to it importantly in all cases. One reason for this statement is that alcoholics, like other deviants, may have a tendency to migrate to larger metropolitan areas. This tendency is also probably responsible, to a considerable extent, for the finding that alcoholics are often centered more in one part of the city than another.

Some lay groups, particularly, have tended to think that the availability of alcohol in the environment tends to increase alcoholism and drinking. Several surveys have shown, however, that in cities in which there are a large number of places which sell alcohol, including social centers such as taverns, drinking and alcoholism rates are not necessarily higher.

SIGNS AND SYMPTOMS OF ALCOHOLISM.

Early indications that a person is leading down the path towards alcoholism are often overlooked by his or her relatives and friends, which is unfortunate because about 7% of all drinkers become alcoholics, and it is much easier to treat the disorder in its early stages than after it has become well established. Therefore, some noteworthy signs and symptoms of alcoholism follow:

• **Drinking enough to be moderately high several times a week**—This pattern includes the "martinis for lunch bunch" as well as those who sip continuously at home most afternoons, have extended cocktail hour every night, or settle down for almost daily TV and six-pack sessions.

• **Drinking enough to get really drunk once a week**—The weekend heavy drinkers are in this category: those who think parties are a reason to get drunk and find a party once a week; and workers to whom every Friday means happy hour. The latter are prone to reason: "I can really get drunk tonight because I can sleep late tomorrow."

• **Drinking to oblivion once a month**—Generally referred to as spree drinking, this pattern is often followed by the heavy drinkers who try to cram all the drinks they didn't have on other days into one long drinking session.

• **A tolerance to alcohol is developed**—Greater amounts must be consumed to obtain the desired effects. Often, the increased tolerance to alcohol is accompanied by increased tolerances to drugs such as barbiturates.

• **Occasional periods of temporary amnesia or "blackouts" drinking**—During these periods, the drinker remains conscious but does not remember what happened during the blackout. (Nondrinkers may also have blackouts, but for different reasons such as fatigue, reduced blood flow to the brain, and emotional stress.)

The drinker who regularly drinks to the point of intoxication or sickness has already become an alcoholic, because other people who can control their drinking behavior will usually be able to stop before these undesirable effects occur.

TREATMENT OF ALCOHOLISM. The two major phases of the treatment of alcoholism are (1) correcting the medical, nutritional, and psychological problems brought on by this disorder; and (2) alleviating or curing the dependency on alcohol. These treatments are in many respects interdependent in that total recovery requires that the alcoholic be helped to function as normally as possible. However, they are covered separately in order to simplify the various aspects of the treatment.

Correction of Medical, Nutritional, and Psychological disorders. Doctors and other health professionals are likely to be called upon to assist alcoholics in various stages of debilitation and/or intoxication. Naturally, the highest priority should be given to the life-threatening aspects of the disorder before attending the less critical problems of the patient. Brief discussions of the most commonly encountered conditions follow:

- **Mild to moderate intoxication**—Although most patients can recover from this condition without outside intervention, prompt treatment ensures that the intoxicated persons will not inflict injuries on themselves or other people. Usually, a cold shower, strong coffee, forced walking, and/or induced vomiting will help to speed the return to full consciousness; although, in some cases a doctor may have to administer a stimulant drug. (None of these measures hasten the oxidation of alcohol by the body. Rather, they act as stimulants to the central nervous system, which has been depressed by alcohol.)

- **Alcoholic stupor**—The need for intervention depends upon whether or not the vital signs (body temperature, heart rate, and rate of breathing) are normal since alcohol may depress the centers of the brain that control these functions. When the vital signs are normal, most people can be allowed to return to normal gradually, although someone should be present to check the vital signs at regular intervals.

- **Violent and aggressive behavior**—This condition is sometimes referred to as "pathological intoxication." It may be necessary for the doctor to quiet the patient with a sedative such as one of the barbiturates.

- **Coma**—Sometimes, comatose patients may appear to be asleep, but in fact do not respond to external stimuli in normal ways. The treatment depends upon the depth of the coma and the status of the vital functions.

- **Abstinence or withdrawal disorders**—Alcoholics who have interrupted their pattern of drinking may have the "shakes," hallucinations, convulsive seizures, and/or delirium tremens. Treatment requires medical detoxification and close observation until this critical period has passed. Sometimes, patients may be in shock and require intravenous infusions of saline solutions. Also, these disorders may be aggravated by severe nutritional deficiencies that damage the nervous system. Hence, some doctors may inject large doses of the B vitamins.

- **Nutritional deficiencies**—Many alcoholics eat poorly and are likely to have multiple nutritional deficiencies. However, the most appropriate nutritional therapy can only be chosen after the patient's medical status has been determined by means of close observations and laboratory tests. For example, an alcoholic fatty liver may be treated with a normal-fat, normal-protein, vitamin-enriched diet, whereas more severe liver disease may require restriction of the protein intake to prevent a coma from being induced. Other medical problems such as salt and water retention in the tissue, kidney failure, central nervous system disorders, anemias, and alcoholic heart disease may also require special medical *and* nutritional therapies.

- **Psychological disturbances**—Depending upon the type of disorder, the patient may require institutionalization or may be treated on an outpatient basis. For example, some alcoholics suffer from paranoia and are potentially dangerous to themselves and others. Nevertheless, great strides have been made since 1948 (when the treatment of psychiatric patients with drugs on an outpatient basis was initiated at Rockland State Hospital in New York) in the use of medications to control psychological disorders.

Alleviation or Curing of Dependence on Alcohol. In the United States, the aim of therapists seeking to treat alcoholics in recent years has been the achievement of complete abstinence. Before the emergence of Alcoholics Anonymous, it was generally conceded that alcoholics would continue to drink, and the alcoholic was therefore usually advised to drink in moderation. This treatment was characteristically unsuccessful. After the A.A. experience, the prevailing contemporary view has generally been that the alcoholic must not only abstain from the use of alcohol but must also alter his self-concept, changing his image from that of a drinker to an abstainer. The demand for complete abstinence which the American culture tends to impose upon the arrested alcoholic shapes the prevailing treatment philosophy and is generally reflected in the orientation of the institutions providing this treatment. (Therapists in certain other countries such as Japan do not attempt to achieve complete abstinence.)

A recently developed alcoholism therapy is known as the "total push." It subjects the patient to a variety of treatments such as Alcoholics Anonymous meetings, correction of nutritional deficiencies, drug therapy, group sessions, hypnosis, individual psychological and religious counseling, lectures, and physiotherapies in order to identify those which best meet the patient's needs. Then, the patient is encouraged to continue those deemed most suitable on a long-term outpatient basis. However, there is little objective evidence on the effectiveness of the various therapies. Therefore, many of the treatment programs around the country still rely on drug therapy and various types of group sessions.

• **Drug therapy**—Antabuse (trade name for disulfiram) is a drug widely recommended by a number of clinicians. Its effect is to make the patient violently nauseated whenever he drinks alcohol. It has been found that it works much more effectively if patients are asked to give it to themselves. The drug alone is claimed not to be effective, and it is generally recommended that it be combined with psychotherapy. It is also observed that patients who come to psychotherapy in the first place, who stay with it, and are willing to take antabuse, constitute a highly motivated group. Their high level of motivation is viewed as an important factor in treatment success.

A recent report on the treatment of alcoholism in Japan questions the generally accepted view that alcoholics must be totally abstinent if they are to control their drinking. An experimental plan was used which allowed the alcoholic patients to consume up to 200 cc's of *sake*, the Japanese national wine, daily without experiencing an unpleasant reaction. A cyanide compound, in conjunction with the *sake*, was used to create a feeling of illness when the patients consumed more than this amount. The method effected a reported cure in 76% of the 250 alcoholics treated—an impressive rate. In a modification of this method, the drug is placed in the patient's food without his or her knowledge, and he or she experiences an unpleasant reaction if more than a few ounces of *sake* are consumed.

• **Group session (Alcoholics Anonymous)**—Various programs have been developed in which the psychotherapy is oriented towards (1) the alchoholic and his or her family; or (2) groups of alcoholics who share their trials and tribulations. So far, the most successful programs have been those of Alcoholics Anonymous and the separate organizations that were founded along similar lines.

Alcoholics Anonymous (AA), founded in 1935 in Akron, Ohio, by two desperate alcoholics, a stock broker and a surgeon, now claims 2 million participants in 136 countries. AA is a remarkable fellowship of mutual and spiritual support that has endured in simplicity. It has no dues, no bureaucracy, no minutes; the only requirement for membership is a desire to stop drinking. The cornerstone of the program is *abstinence*.

The success of Alcoholics Anonymous is explained in various ways by social and behavioral scientists. Some see the organization as forming a subculture which offers a definite set of norms and a sure sense of belonging to its members. Its meetings also provide the important sense of social acceptability that many alcoholics apparently need. It is frequently claimed that only an alcoholic can understand another alcoholic. Within Alcoholics Anonymous, a sense of group understanding and support may be achieved. Some social scientists have suggested that attendance at Alcoholics Anonymous meetings in many ways affords the type of relief from other pressures (such as family strains) that drinking once did.

Two of the organizations that operate on principles similar to those of Alcoholics Anonymous are (1) Al-Anon Family Groups, which are fellowships of wives, husbands, relatives, and friends of problem drinkers; and (2) Alateens, which is strictly for teen-age children of alcoholics.

SUGGESTIONS FOR PREVENTING OVER-CONSUMPTION OF ALCOHOLIC BEVERAGES.

In addition to prevention approaches that focus on the alcohol-related behavior of the individual, reduction in alcohol misuse will necessitate changes in the environment surrounding alcohol use, particularly in the laws that regulate sale and consumption of alcoholic beverages. For example, raising the minimum age for purchasing alcoholic beverages, reducing the ethanol content of alcoholic beverages, and establishing stiff penalties for drinking-related motor vehicle offenses are believed to offer potential for discouraging some problem drinking. Moreover, it may be necessary to modify the requirements of some states that restrict drinking to certain bars and restaurants and thus give alcohol a special allure and effect it might not have if available in social settings where the emphasis is on recreation rather than specifically on drinking. A recent study indicates that alcohol consumption and the incidence of alcoholism are usually low when the price of beverage alcohol is high relative to average disposable income. This finding suggests that increasing the price of alcoholic beverages might have some preventive effects. While it is doubtful that this approach would reduce the current number of alcoholics, the rate of increase in new cases might be affected. An educational program integrated with rising prices and other legislative measures most likely would enhance the effectiveness of each strategy.

Establishing standards and guidelines for advertising of alcoholic beverages is another environmentally focused primary prevention strategy. Myths of alcohol enhancing manliness, sexiness, friendships, and so on are often perpetuated in today's beverage alcohol advertising. More women and young people are being depicted in alcohol advertising and new campaigns are being developed to entice these groups to drink more. The following guidelines have been developed to encourage responsible advertising:

1. Reduce the emotionalism surrounding alcohol use, and reduce social pressures to drink by cutting through the myths and meanings alcohol has held.
2. Discourage drinking for its own sake and encourage alcohol use in relation to meals and other leisure activities.
3. Consciously avoid using alcohol as a sedative or coping agent for life stresses.
4. Emphasize responsibility and moderation in the serving and consumption of alcoholic beverages.

SUMMARY. Alcoholism, a compulsive addiction to heavy and frequent drinking, is a behavior over which the afflicted individual has little, if any, control. Untreated alcoholism may cause extensive and severe physical and mental damage and may eventually result in psychosis and/or death.

Although various treatment methods have been recommended, and although there has been considerable activity in some states to develop a variety of treatment and training facilities, there is very little adequate research to indicate that any of these various intervention strategies significantly reduce alcoholism in groups of alcoholics, over time periods of a year or more. Individual cases of cures are reported, especially by members of

Alcoholics Anonymous. However, scientifically precise studies of the impact of this organization have not been carried out, partly because of the principle of the anonymity of its members.

Therefore, prevention of alcohol misuse before it occurs would seem to be one of the best ways to reduce the incidence of alcoholism in the United States. Some examples of alcoholism prevention programs are education of people in schools, churches, and other community settings; guidelines for responsible advertising of alcoholic beverages; public service messages on alcoholism prevention and treatment in the mass media; and legislation to control the availability of alcoholic beverages.

ALDEHYDE

Any of a large group of organic compounds containing the grouping -CHO and holding an intermediate position between the alcohols and acids; for example, formaldehyde, acetaldehyde, and benzaldehyde.

ALDOHEXOSE

A 6-carbon sugar (such as glucose or mannose) containing a -CHO grouping.

ALDOSTERONE

A hormone secreted by the adrenal cortex. It acts upon the kidney tubules to bring about the retention of sodium and water. Sometimes, high blood pressure is brought on by the excessive secretion of this hormone.

(Also see ENDOCRINE GLANDS; and HIGH BLOOD PRESSURE.)

ALEURONE

The protein-rich layer(s) of cells which lie(s) just inside the branny seed coats of cereal grains. These cells contain enzymes that become active when the grains sprout. The activated enzymes digest starch and protein which is stored in the seed so that there is an ample supply of sugars and amino acids for the growth of the embryonic plant. Hence, the enzyme actions in the aleurone layer(s) of sprouted grains are highly useful for the production of alcoholic beverages, because the enzymatically-liberated sugars may be fermented by yeast to produce alcohol.

(Also see BEERS AND BREWING; CEREAL GRAINS; and DISTILLED LIQUORS.)

ALFALFA (LUCERNE) *Medicago sativa*

Alfalfa leaves are often dried, ground, and put into tablet form for use as a human vitamin and unidentified factor supplement. Alfalfa leaf tablets are rich in protein, calcium and trace minerals, carotene, vitamins E and K, all of the water-soluble vitamins, the unidentified factors, and in vitamin D if they are sun cured. (See Food Composition Table F-36.)

Fig. A-17. Alfalfa—nutrition for animals and man. (Courtesy, USDA)

Alfalfa, or lucerne as it is known in many parts of the world, belongs to the pea family, *leguminosae*. The Turkistan and common alfalfas are varieties of the genus, *Medicago*, species *M. sativa*. Alfalfa is a perennial plant (that is, it grows from year to year without replanting) with deep roots, trifoliate leaves, bluish-purple flowers, and kidney-shaped seeds.

Man has grown alfalfa for forage for livestock longer than any other plant. It was likely planted in southwestern Asia long before recorded history. The Persians took it to Greece when they invaded that country in 490 B.C. Historians believe that alfalfa was taken from Greece to Italy during the Ist century A. D., from whence it spread to other parts of Europe. The Spanish explorers took alfalfa to South America during the early 1500s. European colonists introduced it into the United States in 1736.

Today, alfalfa grows throughout the world under widely different conditions, but it thrives best in deep, loamy, well-drained, neutral to slightly alkaline soils, and under irrigation. Like other legumes, alfalfa has the ability to enrich the soil with nitrogen, provided it has the help of nitrogen-fixing bacteria that live in small nodules (lumplike growths) on the roots. The bacteria take nitrogen from the air and change it into a form that plants can use for food.

(Also see LEGUMES, Table L-2 Legumes of the World.)

ALGAE

The algae are primitive plants. As plants they contain chlorophyll and are capable of photosynthesis—converting the sun's energy into food. As part of the food chain, many provide food for fish. Algae vary in size and shape. Some consist of individual microscopic cells while others form flat sheets of narrow filaments or stem-like structures that may be more than 100 ft (*30 m*) long. Algae are able to live almost anywhere—salt water, fresh water, hot springs, polar snow, soil, trees, and rocks. Probably the best known algae are the seaweeds, specifically kelp. Algae are classified as green, brown, or red:

• **Green algae or Chorophyta**—Characteristically, green algae are found in fresh water, though some may dominate tropical marine waters, and some are occasionally found on land. Green algae form an early link in the food chain as food for fish. Furthermore, some species of green algae grown on water or lagoons of waste material provide a potential source of animal and human food. The natives of Lake Chad, Africa, and of Mexico have harvested and eaten algae for years. Grown as a crop, 4 to 16 tons of dried algae can be harvested from an acre each year, but the eating of algae is not widely accepted. The nutrient composition tends to be rather variable, with protein ranging from 8 to 75%, carbohydrate from 4 to 40%, lipids from 1 to 86%, and ash 4 to 45%.

(Also see SINGLE-CELL PROTEIN.)

• **Brown algae or Phaeophyta**—Most brown algae are commonly known as seaweed. These range from low-growing species on rocks, to species with tangled filaments about ⅓ in. (*7.5 mm*) long, to the massive rockweeds and kelps. Their brown color is due to the carotenoid pigment, phaeophytin. Brown algae extract and concentrate a number of chemicals from the sea; consequently, they have been used as fertilizer by coastal farmers. Moreover, they are used for human and animal food in the Orient and Europe. Kelp contains the carbohydrates, mannitol and alginic acid; minerals, notably iodine; and a significant amount of protein. Alginates extracted from kelp are important in textiles, printing inks, medicinals, rubber, ice cream, and salad dressings, as emulsifiers.

(Also see ADDITIVES; and SEAWEED.)

• **Red algae or Rhodophyta**—Most red algae range from a few inches to a foot or two in size. These are chiefly found in the oceans. In Asiatic countries, some red algae are eaten. An important red algae, Irish moss, grows off the rocky coasts of Europe and North America. It is harvested for carrageenan, a food gum, used for gelling, thickening, and stabilizing food mixtures. Another red algae is the source of agar-agar employed by researchers as a bacteriological culture medium, and also used as a thickening agent in foods, as well as numerous industrial uses such as dyeing, printing, sizing, and adhesives.

(Also see ADDITIVES; AGAR-AGAR; CARRAGEENAN; SEAWEED; and SINGLE-CELL PROTEIN.)

ALGINATES

Salts of alginic acid, found in many seaweeds (kelp). They hold large amounts of water and are used (1) as thickeners, stabilizers, and emulsifiers in various foods, and (2) to provide bulk and smoothness in ice cream, synthetic cream, and evaporated milk.

(Also see ADDITIVES.)

ALGINIC ACID

An acid $(C_6H_8O_6)_N$ obtained from seaweeds, used by the food industry as a thickener and emulsifier.

ALIMENTARY TRACT (DIGESTIVE TRACT)

The tubular, and in part sacculated, passage that serves the functions of digestion, absorption of food, and elimination of residual waste products.

(Also see DIGESTION and ABSORPTION.)

ALKALI

A substance which neutralizes acids. It usually refers to a soluble salt or a mixture of soluble salts present in some of the residues of food which remain after complete burning or metabolism of the organic matter.

(Also see ACID FOODS AND ALKALINE FOODS.)

ALKALINE-ASH RESIDUE

A mineral residue which is left after (1) the other nutrients in a food have been metabolized, or (2) a food has been burned to an ash in a laboratory. Ashes of foods give an alkaline reaction when the predominant chemical elements in the ash are sodium, potassium, calcium, and/or magnesium; because these elements generally form alkali. Milk has an alkaline-ash residue because of its high calcium content. Fruits and vegetables are most likely to have an alkaline-ash, whereas breads, cereals, eggs, fish, meats, and poultry are most likely to have an acid-ash residue.

(Also see ACID FOODS AND ALKALINE FOODS.)

ALKALINE TIDE

A temporary increase in the alkalinity of the blood and urine that may occur after the ingestion of a meal which stimulates the secretion of large amounts of hydrochloric acid into the stomach.

(Also see DIGESTION AND ABSORPTION.)

ALKALI RESERVE

This term designates the substances in the blood and body fluid which will neutralize acids. The main alkali in the blood is the bicarbonate ion. The alkali reserve protects the body against the damaging effects of the acids which arise from metabolism. The most significant of these acids are the ketone bodies, phosphoric acid, sulfuric acid, hydrochloric acid, and lactic acid.

(Also see ACID-BASE BALANCE.)

ALKALOIDS

Alkaline, druglike substances found in certain plants. All alkaloids contain nitrogen in a nonprotein form. Some of the important alkaloids used in medicine are atropine,

cocaine, morphine, quinine, and strychnine.
(Also see ATROPINE; and BELLADONNA DRUGS.)

ALKALOSIS

A disorder characterized by a rise in the alkaline content of the blood, or a fall in the blood acid. Alkalosis may result from vomiting, overuse of alkalizing preparations such as sodium bicarbonate, or excessively deep breathing.
(Also see ACID-BASE BALANCE.)

ALKANET

A dye extracted from the root of *Alkanona tinctoria*, which grows in Asia, Hungary, Greece, and the Mediterranean region. It is used to color wines, cosmetics and confectionery. In the United States, alkanet is approved for coloring sausage casings, margarine, shortening, and for marking or branding products.

ALKAPTONURIA

An inherited metabolic disorder in which the urine turns dark upon exposure to air. The abnormality is due to the presence in the urine of abnormally large amounts of homogentisic acid, a compound resulting from the incomplete metabolism of certain amino acids. Usually, the disorder is treated with a low protein diet in order to reduce the amount of homogentisic acid which is formed.
(Also see INBORN ERRORS OF METABOLISM.)

ALLERGEN

Any of a wide variety of substances, or environmental conditions which may provoke an allergic reaction such as asthma, hives, or runny nose, when they come in contact with certain tissues. Some of the most common allergens are beverages, foods, spices, pollen, dust, cosmetics, drugs, vaccines, exposure to cold or heat, sunlight, and various emotional states.
(Also see ALLERGIES.)

ALLERGIES

Contents *Page*

Prevalence and Cost 43
Historical 43
Causes 43
Foods 44
Stress 44
Heredity 44
Inhalants (Hay Fever) 45
Insects 45
Drugs 45
Contactants 45
Physical Agents 45
Bacteria and Fungi 45
Symptoms 45
Clinical Manifestations 46
How Food and Inhalent Allergies Differ 46
Allergic Symptoms in Body Systems 46
Diagnosis of Food Allergies 46
Dietary History 46
Skin Test (Patch Test) 47
Provocative Food Test 47
Elimination Diets; Rowe Elimination Diets 47
The Pulse Test 48
Treatment of Food Allergies 48
Elimination of the Causative Food 48
Denaturation of the Protein 48
Drug Treatments 49
Desensitization (Reduced Sensitivity) 49
Keep Healthy 49
Avoid Stress 49
Maintain Good Nutrition 49
Know Botanical Families 49
Read Labels 49
Treatment of Food Allergies in Children 50
Allergy-producing Foods 50
Milk Allergy 50
Milk-allergic Infants 50
Milk Foods to Avoid 51
Milk Replacements 51
Read the Label 51
Egg Allergy 51
Egg Foods to Avoid 52
Egg Replacements 52
Read the Label 52
Wheat Allergy 52
Wheat Flours to Avoid 52
Wheat Flour Replacements 53
Read the Label 53
Corn Allergy 54
Corn Foods to Avoid 54
Read the Label 54
Legume Allergy 54
Legume Foods to Avoid 54
Read the Label 55
Nut Allergy 55
Seafood Allergy 55
Other Allergy-producing Foods 55
Standards of Identity 55
Allergy-free Products List 55
Allergy-free Recipes 55

An allergy may be defined as any unusual or exaggerated response to a particular substance, called an allergen, in a person sensitive to that substance. Formerly, the condition was called hypersensitivity, but allergy is now the preferred term. Allergies are the result of reactions of the body's immunological processes to "foreign" substances (chemical substances in such items as foods, drugs, and insect venom) or to physical conditions.

Every person has some protection from infections and diseases. This is called immunity. The body develops immunity by forming antibodies to overcome viruses, bacteria, or other substances that are not normally found in the tissues. Likewise, most people have little or no trouble with allergies because antibodies or special mediator cells respond normally to protein-type substances to which they are exposed. However, the allergy-prone person has a protective mechanism which overreacts (it gets

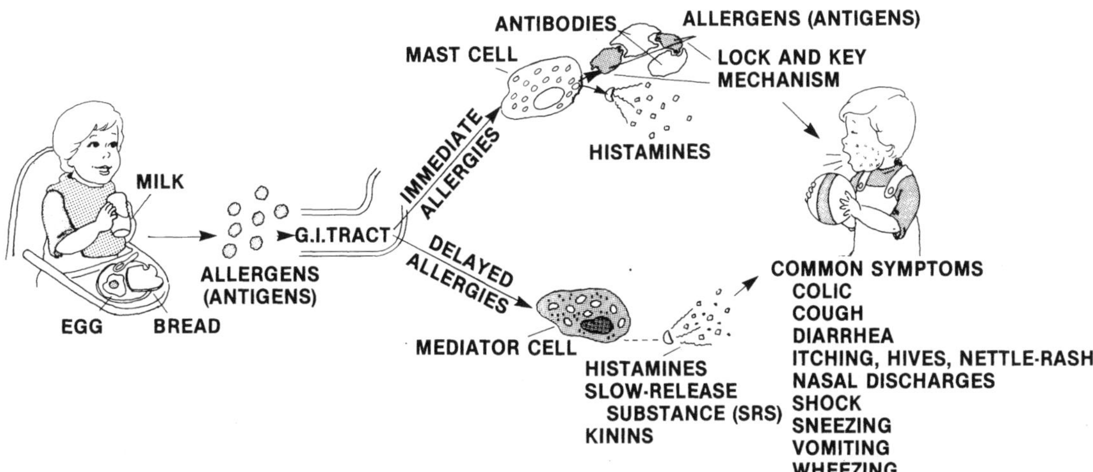

Fig. A-18. The cause and mechanisms of food allergies. Note that different mechanisms are involved in (1) immediate allergies and (2) delayed allergies. In immediate allergies, antibodies "lock" onto an allergen (antigen), effectively immobilizing and neutralizing the allergen, thereby preventing it from damaging body tissues. In delayed allergies, there are no antibodies; instead, allergens are mediated by special cells.

out of control) to certain substances; resulting in allergy symptoms.

Allergy specialists classify allergies into two broad categories: (1) immediate, and (2) delayed.

Immediate allergies, as implied by the name, are those that occur almost immediately (within minutes) after a person is exposed to the allergen (allergy-producing agent). Examples of immediate allergies are: some food allergies, hay fever, bronchial asthma, hives, many drug allergies, and sensitivity to insect venom.

Delayed allergies occur hours (at least 4 hours) or days after exposure to the allergen. Examples of delayed allergies are: some food allergies, poison ivy, cosmetics, metals, soaps, solvents, and rejection of skin grafts and organ transplants.

PREVALENCE AND COST. Allergic diseases are among the most common and most costly health problems, afflicting an estimated 40 million Americans.[10] These are prevalence figures of patients actually afflicted at the time; they do not include those who have had the condition and recovered. About half of these 40 million people suffer from hay fever. The other half suffer from food sensitivities which have been linked to virtually every food in the American diet; but especially to cow's milk (lactose), eggs, gluten, peanuts, soybeans, tree nuts, fish, shrimp and other crustaceans, shellfish, and sulfite. Lactose intolerance in varying forms of severity may affect all segments of the population; it occurs in up to 90% of some ethnic groups, such as Greeks, Arabs, Jews, Black Americans, Japanese, Thai, Formosans, and Filipinos. Celiac disease (an adverse reaction to the protein gluten, which is naturally present in wheat, rye, barley, and, to a lesser extent, oats), afflicts additional people. Another noteworthy statistic is that approximately 9% of all patients seeking medical care at a physician's office do so for one of the allergic diseases.[11]

[10]Estimates by the authors, based on a number of sources.

HISTORICAL. The earliest mention of allergy was by Hippocrates (460 B.C. to 377 B.C.), Greek physician, called the "father of medicine." He commented on the appearance of a skin rash, which no doubt was hives, as a result of the ingestion of milk. In the 1st century B.C., Lucretius, the Roman philosophical poet, recognized that some people reacted adversely to certain foods. His comment on the problem, which has been widely quoted, was: *"Quod ali cibus est aliis fuat acre venenum."* (What is food to one may be fierce poison to others.) Early in the 17th century, Beaumont and Fletcher, English playwrights, echoed the same thought: "What's one man's poison is another's meat and drink."

In 1855, Overton reported (*Southern Journal of Medical and Physical Sciences,* Vol. 3, 1855) on a man whose asthma and other symptoms were caused by wheat.

In 1890, Koch, a German physician who won the 1905 Nobel Prize in physiology and medicine for his work on tuberculosis, found that people and animals developed a sensitivity to tuberculin, the product of the tuberculosis germ. Today, the most certain and practical method of testing domestic animals for tuberculosis is the tuberculin test.

At the close of the 19th century and the beginning of the 20th century, a number of observers began to correlate clinical experience with laboratory findings to show that what had been called idiosyncrasy was an immunologic phenomenon.

CAUSES. Allergies are caused by contact between a foreign substance (an allergen) and sensitive body tissues. But the mechanisms back of immediate allergies and delayed allergies differ.

[11]*Asthma and the Other Allergic Diseases,* NIAID Task Force Report, U.S. Department of Health, Education, and Welfare, Public Health Service, National Institute of Health, NIH Publication No. 79-387, May 1979.

In immediate allergies, the patient has antibodies circulating in his blood and attached to specific cells. So, when a foreign substance enters the body, the body reacts just as it does when attacked by viruses or bacteria. It protects itself by setting up antibodies to neutralize any further attack. These antibodies, which the body has produced, attack the tissue surfaces, and when the allergen, or allergic-producing agent, again enters the body, it is believed that the antibodies separate from the tissue and attack the invading substance. A very slight tissue damage results which apparently causes the release of histamines, chemical substances which are carried by the circulation to the skin and mucous membranes, with subsequent allergy symptoms. The type of symptoms produced depends entirely on the selective sensitivity of the particular part of the body where the antigen-antibody reaction occurs.

In delayed allergies, no antibodies to specific allergens are found circulating in the blood; hence, there is no reaction between an antibody and an allergen. Rather, in delayed allergies, allergens (antigens) are mediated by a special type of cell.

So in immediate allergies, antibodies serve as the body defense to allergens (antigens); in delayed allergies, allergens are mediated by special types of cells.

From the above, it may be concluded that two conditions are necessary to produce an allergic response: (1) the person must first become exposed and sensitized to a particular substance; and (2) a second exposure to the substance must occur, which then provokes the allergic symptoms. The reason why a sensitivity to a particular substance exists in one person and not in another is very complex and not fully understood. Perhaps an individual structural or functional weakness is involved, a theory supported by the fact that, in some cases, allergies may be inherited. Perhaps, too, adrenal glands contribute to allergic responses, and emotional attitudes also seem to be associated.

Allergies can be caused by an innumerable variety of substances, including foods, pollens, dust, molds, animal dander, drugs, cosmetics, toilet articles, dyes, chemicals, fabrics, and poison ivy. Thus, allergies may be caused by substances eaten or drunk (foods), inhaled (dust, pollen, and animal dander), touched (soap, wood, and cosmetics), and injected (insect bites). Sunlight and heat or cold may aggravate allergic response, as may bacteria or parasites in the intestine.

Foods. Foods are frequent causes of allergic reactions. Some are more likely to produce allergic reactions than others, but practically all foods can produce an allergic reaction in some people. The only exceptions to which scientists have yet to find anyone who is allergic are most fats (although not all), refined sugar, and salt. Also, people may be sensitive to single or multiple foods. Protein is especially likely to cause an allergy, and even a minute amount may be enough to cause difficulty. The most common food allergens—that is, the foods which cause allergic reactions most frequently—are milk, eggs, wheat, corn, legumes, nuts, and seafoods. Also, many people are allergic to strawberries, citrus fruits, tomatoes, chocolate, and berries.

Foods are seldom the only factors affecting the allergic patient. Inhalent allergy is almost always present, and contact and drug allergies are likely.

Stress. There is substantial evidence that psychological factors contribute to allergic reactions in hypersensitive individuals, and that both physical and emotional stress increase the sensitivity of a person through allergic attacks. Among the types of emotions that increase the probability of an allergic response are: anger, fear, resentment, worry, and lack of self-confidence. Records indicate that certain individuals have had allergic reactions to a particular allergen when under stressful conditions, but that they did not experience any reaction when the stressful conditions were removed. So, emotions may trigger an allergic attack, but the allergy must already exist.

It is also recognized that when an asthmatic child is exposed to great tension and contention by constant quibbling and screaming at home, treatment will not be effective until something is done to correct the environmental and emotional situation.

Heredity. For many years, scientists have recognized that certain allergies tend to be hereditary—that they are handed down from parent to child. In 1916, Dr. Robert A. Cooke and Dr. Albert Vander Veer, Jr., Cornell Medical College, reported that if both parents have hay fever or another allergy the chances of their offspring developing an allergy are greater than if only one parent is allergic. Also, the child with "double inheritance" (both parents being allergic), is likely to develop allergies at an earlier age.

In 1924, Dr. Cooke and Dr. Will C. Spain carried out a more detailed study on the hereditary role. They found that, if both parents were allergic, an average of 72% of their children developed an allergy. If only one parent had an allergy, 56% of the children became allergic.

It is suggested that parents with a history of allergies in their families exercise the following precautions to reduce their child's risk, although these recommendations are not based on rock-solid evidence: breast-feed the infant as long as possible; avoid cow's milk and cow's milk products for at least 6 months; delay giving eggs, fish, and citrus juices; don't keep pets; don't have wool fabrics or feathers in the house; minimize dust and mold; and don't smoke. And good luck!

Some years ago, the word "atopy" was coined to refer to a group of allergic diseases that occur only in man and are the result of genetic abnormality. Scientists are convinced that a number of major allergic disorders, such as hay fever and bronchial asthma, are strongly influenced by heredity. They also know that some allergies regarded as atopic occur in man and in other animals. Not only that, atopy does not include physical allergy, serum sickness, contact dermatitis, and many forms of food allergy and drug allergy, simply because these disorders are not known to be transmitted genetically. So, there is disagreement about what the word really means, and the original definition is no longer completely valid. Nevertheless, the term remains popular and is widely used.

Inhalants (Hay Fever). This refers to allergic symptoms—sneezing, itching, and weeping—caused from substances inhaled through breathing, such as dust, pollen, and animal dander. The fact that they are introduced through inhalation does not necessarily indicate that the symptoms will be only respiratory; for example, inhaled substances can produce hives, and conversely, foods may cause asthma.

Hay fever is caused by the action of histamine that has been released from mast cells. Antihistamines are drugs that prevent the released histamine from causing symptoms. They are the main treatment for hay fever and help more than two-thirds of the people who try them. For best results, antihistamines should be taken around the clock—usually four times a day. The main problem with antihistamines is their common side effect, sleepiness, made worse when alcohol is also taken.

Immunotherapy (previously called "desensitization shots") has proved its worth in the treatment of hay fever. In a few cases it produces cures; and in many more cases it markedly reduces the severity of symptoms. In immunotherapy, gradually increasing doses of an allergen—beginning with very small amounts—are injected into a patient's skin.

Insects. Insect allergy affects four out of every 1,000 Americans. This refers to the allergic reactions suffered by most everyone from the sting of bees, wasps, hornets, yellow jackets, mosquitoes, bedbugs, fire ants, and other insects. Some people, however, react more violently to the venom of such insects than others. A few acutely allergic individuals go into a state of anaphylactic shock and may die within minutes. People who have a violent allergic reaction to insect stings should consider protecting themselves with a course of venom immunotherapy; in any case, they should keep "sting kits" at hand so that epinephrine can be injected promptly in the event of a sting reaction.

Drugs. It is estimated from 1 to 4% of persons receiving drugs experience an allergic reaction. The drugs that produce the most frequent allergic reactions are penicillin and aspirin. One study showed that penicillin causes an allergic response once in about every 1,900 doses, and that aspirin causes an allergic reaction once in about every 34,000 doses.

Among the many drugs that can produce allergic reactions are sulfa drugs, antibiotics, aspirin, tranquilizers, diuretics, antituberculosis drugs, anticonvulsants, phenobarbital, insulin, and quinine. Also, some people are allergic to the horse serum present in certain antitoxins and other immunity-producing agents.

It is noteworthy that drug allergies greatly resemble food allergies. Note, too, that drug allergies are especially common in patients with multiple food allergies.

Contactants. This refers to materials which can cause allergic skin reactions in persons who come in contact with them by touching or by external physical contact. Among contactant allergens are such everyday substances as cosmetics, perfumes, soaps, shampoos, face lotions, shaving lotions, deodorants, mouthwashes, fabrics (wool and silk, but not cotton and synthetic fibers), and solvents. Also, included in this group are such plants as poison ivy, poison oak, and poison sumac; about 70% of the population develops some sort of skin reaction, such as inflammation and blistering, from contact with the causative agents in these plants.

In a highly sensitive person, contactants can cause inflammation, itching, swelling, and blistering.

Physical Agents. These include heat, cold, and ultraviolet light. In some people, these agents cause a pronounced, localized rash on the exposed area.

Bacteria and Fungi. Bacteria and fungi are known to be sources of allergies in some people.

SYMPTOMS. A wide variety of symptoms are caused by allergies; among them, nasal congestion, cough, chest/throat mucus, wheezing, constipation, diarrhea, stomachache, bloating, poor appetite, eczema, hives, headache, tension, fatigue, breath odor, and sweating. Many of these symptoms are also characteristic of other disorders. For these reasons, in treating an allergy, it is first necessary that the diagnosis definitely establish the cause.

Allergic manifestations may occur within a few minutes of the patient coming in contact with the allergen, or they may be delayed for many hours or even for several days. With food allergies, the time lapse between consumption and the appearance of symptoms is determined by the rapidity with which the food is absorbed from the gastrointestinal tract. The quicker the absorption, the more rapidly the symptoms will develop after eating.

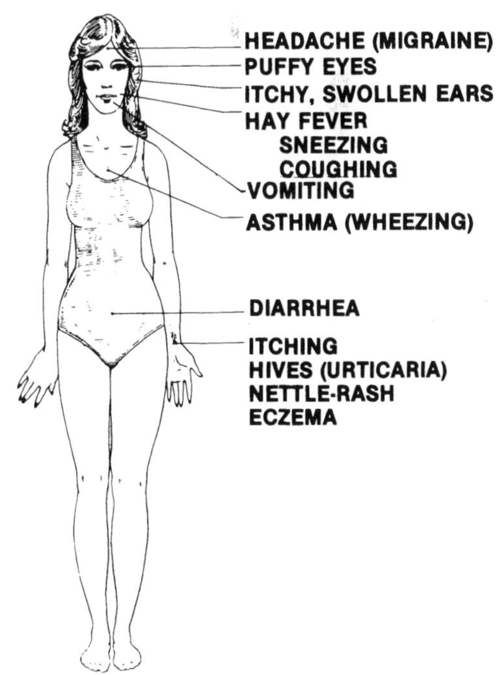

Fig. A-19. Common areas and symptoms of allergic reactions.

Allergic reactions may be trivial and local, e.g., a mild nettle-rash or hives (urticaria), or they may be general and so serious as to result in sudden death. Serious reactions are known as anaphylactic shock. Anaphylaxis can be caused by drugs (penicillin and others), horse serum, insect stings, foods (shellfish, nuts, eggs), pollen extracts, and various other agents.

Clinical Manifestations. The clinical manifestations may be classed as follows:

1. **Skin.** A nettle rash or hives (urticaria) is a common allergic reaction among all ages of people. Eczema is an allergic reaction usually due to sensitivity to an agent applied directly to the skin. But it often occurs as a food allergen in infants and young children and occasionally in adults. A less common but more severe giant form of hives is known as angioneurotic edema.

2. **Respiratory System.** Asthma is an allergic disorder, usually resulting from inhaling the allergen. But often in children, and sometimes in adults, the symptoms are due to a food allergen.

3. **Gastrointestinal System.** Dyspepsia, abdominal discomfort, vomiting, and diarrhea may all be the result of an allergic reaction. There are many causes of these common symptoms. When they persist or recur repeatedly for no obvious reason, a food allergy should be suspected.

4. **Circulatory System.** Anaphylactic shock is due to release of large amounts of histamine which causes widespread vasodilation and circulatory collapse, which is often fatal. It seldom occurs except when an allergen has been injected intravenously. Although anaphylactic shock remains a potential hazard of intravenous feeding, it is very rare with preparations now available.

How Food and Inhalent Allergies Differ. In order to identify and manage food allergies, it is first necessary to recognize the differences between food allergies and inhalant allergies. These differences may be summarized as follows:

1. Food allergies may be present at birth or begin in early childhood, whereas inhalant allergies are unusual under the age of 2 years and rare under the age of 1 year.
2. Reactions to food allergies are usually delayed for several hours and last for several days, whereas reactions to inhalants tend to develop within minutes and to disappear with an hour or so if exposure is not continued.
3. Food allergies tend to cause a wide variety of manifestations, often constituting a systemic disturbance, whereas inhalant allergies tend to be limited in scope, affecting principally (but not exclusively) the respiratory tract.
4. Food allergies usually cause gastrointestinal disturbances, whereas inhalant allergies seldom do.
5. Except in severe cases (and not always then), food allergies may not be revealed by skin tests or other objective methods, whereas these procedures are usually accurate in inhalant allergies.
6. Desensitization (reduced sensitivity), a common method of treating inhalant allergies, is ineffective in food allergies.
7. Detection of most food allergies depends on clinical methods, chiefly, elimination and challenge, which are not adapted to inhalant allergies.

Allergic Symptoms in Body Systems. Table A-6 presents in summary form the symptoms exhibited in various body systems resulting from allergic reactions.

TABLE A-6
SYMPTOMS EXHIBITED IN VARIOUS BODY SYSTEMS RESULTING FROM ALLERGIC REACTIONS

Body System	Allergic Reaction
Cardiovascular system	Anaphylaxis Circulatory collapse Shock
Cutaneous (skin) system	Angioneurotic edema (giant hives of the skin, mucous membranes, and sometimes the internal organs). Eczema Purpura (bleeding into the skin, mucous membranes, and/or internal organs). Urticaria (hives, nettle-rash).
Gastrointestinal system	Abdominal distention Abdominal pain Bad breath Constipation Diarrhea Dyspepsia Nausea Vomiting
Genito-urinary system	Hematuria (blood in the urine)
Locomotor system (organs concerned with movement)	Schonlein's purpura (small hemorrhages in the skin).
Nervous system	Headache (migraine)
Ocular (eye) system	Irritation Redness
Respiratory system	Bronchial asthma Hay fever Head colds Irritation of nasal passages.

DIAGNOSIS OF FOOD ALLERGIES. First, the physician determines by the symptoms whether the patient has an allergic disease. If the presence of an allergy is confirmed, he then proceeds to identify the allergen(s) causing the trouble.

Detection of food allergies may involve one or more of the following diagnostic procedures.

Dietary History. The physician will usually take a food history and perhaps ask the patient to complete a food diary.

In some food allergies, the symptoms develop so rapidly and dramatically, almost immediately after eating the offending food, that the patient is able to make his own diagnosis. But it is not always easy to associate symptoms with any particular food, especially if there is a delay of some hours between the eating of the suspected food and the onset of symptoms. A particular food may be wrongly suspected by the patient to be the cause of his allergy because it happened to be eaten immediately before an attack, whereas the actual causative food may have been eaten much earlier. In cases of doubt, the physician will usually give the patient a diary in which to record with great care and detail over a period of many days or even weeks all the food which he eats and the time of his meals. He should record in the diary any disturbances which he believes may be due to food allergy, noting the nature and intensity of the symptoms and the time of the occurrence. If a study of the diary suggests a relationship between the intake of a certain food and the onset of allergic manifestations, the doctor will usually carry out additional tests.

Skin Test (Patch Test).

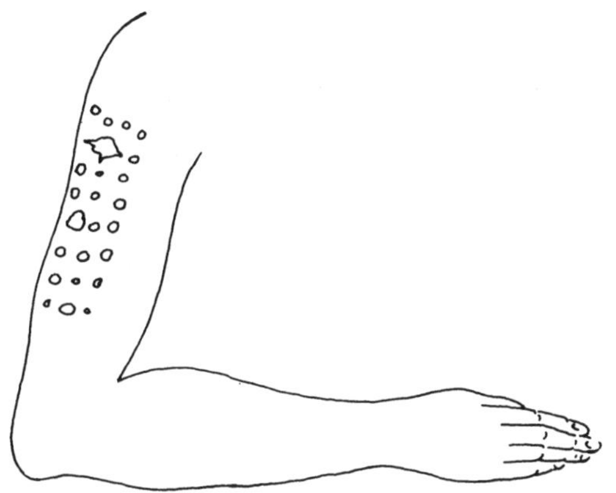

Fig. A-20. The skin test used on a patient to detect the allergic substance and the degree of hypersensitivity. Graded dilutions of the suspected substance were injected into the skin, following which the temporary hivelike wheals of varying sizes appeared, indicating an allergy to the substance tested.

This test has been used much in the past. A minute quantity of an extract of the suspected food (or other substance) is injected into the skin or rubbed into a scratch. If temporary hivelike wheals of varying sizes appear, it is indicative of sensitivity to the food in question. However, because a patient reacts to the skin test should not be accepted as confirmatory evidence for two reasons: (1) it does not necessarily follow that a patient is sensitive to the same protein when taken by mouth; and (2) the skin test may be negative to a protein which is producing allergic reactions elsewhere, e.g., in the gut. Nevertheless, the skin test is a valuable diagnostic tool when used by an allergist. If there is a well-marked reaction to one single article of food, while tests of other food extracts are negative, the skin test may be of value in confirming a suspicious history.

For diagnosing sensitivity to contactants, such as nail polish or poison ivy, the patch test is the test of choice. In the patch test, small squares of absorbent gauze or cotton are dipped in various solutions containing suspected allergens and taped against the patient's skin, usually the forearm or back.

Provocative Food Test. This consists in giving the patient a small quantity of the suspected food camouflaged in a dish in such manner that he is unaware that he is eating the food to which he thinks he may be allergic. If, at the appropriate interval after the meal, the patient records in his diary the typical symptoms, this strongly suggests that the food is responsible. Because of the variability of allergic responses, however, the test should be repeated three times before either a positive or negative result is accepted. If a negative result is obtained on three occasions, it can be concluded that the symptoms are not due to the suspected food. On the other hand, if the tests are positive, then appropriate treatment should be undertaken.

Elimination Diets; Rowe Elimination Diets. If the dietary history, skin tests, and provocative food tests fail to detect the causative food, elimination diets may be helpful. With infants and young children, it is easy to administer strict elimination diets. Thus, it is possible to eliminate from a childs' diet separately, and in order, the common causes of children's food allergies: milk, eggs, and wheat. In adults, the procedure is much more difficult because of the number of foods to which sensitivity may develop. Also, patients may have to be fed on such diets for many weeks before the offending article of food is discovered. And, as the diets are complicated, they are normally only suitable for the investigation of patients in the hospital with a dietetic department and a physician experienced in the subject of allergies. Because such elimination tests are tedious and time-consuming for both patient and doctor, they should be used as a last resort only when symptoms believed to be due to food sensitivity cannot be detected by other tests and are seriously inconveniencing the patient.

Various types of elimination diets have been devised with the object of discovering the offending article. The most widely used elimination diets are those formulated by Rowe and presented in Table A-7.

The elimination diets given in Table A-7 contain food items that have proven least likely to produce allergic response. They should be used as follows in diagnosis:

1. The hypersensitive person is confined to one of the first three diets for 1 to 3 weeks.
2. At the end of this period, the other diets are tried one at a time in the same manner.
3. After all the diets have been tried, if no adverse reactions have occurred, all the foods on the combined diets are acceptable for consumption, and it may be concluded that the allergy is due to a food item not included in any of the diets, or is of nonfood origin.
4. Additional food items not included in any of the diets can then be tried one at a time for periods of 1 to 3 weeks and eliminated from or accepted into the diet on the basis of response or nonresponse. Normally, milk, eggs, and wheat should be tried last as they are the food items most often found to induce an allergic reaction.

Each item should be tried at least three times over a period of several weeks before accepting or rejecting it as a part of the diet.

TABLE A-7
ROWE ELIMINATION DIETS[1]

Diet 1	Diet 2	Diet 3	Diet 4
Lamb	Chicken (no hens) Bacon	Beef Bacon	Milk Plain cottage cheese Cream
Rice Rice biscuit Rice bread Tapioca	Corn Corn pone Corn-rye muffins Rye bread Ry-krisp	*Breads made of any combination of:* Soy, lima, potato, or tapioca. Tapioca	Tapioca
Carrot Lettuce Spinach Sweet potato Swiss chard	Artichokes Asparagus Beets Squash	Carrot Lima beans Peas String beans Tomato	
Grapefruit Lemon Pears	Apricot Peach Pineapple Prune	Apricot Grapefruit Lemon Peach	
Cane sugar Maple syrup or syrup made with cane sugar and flavored with maple. Salt	Cane or beet sugar Karo corn syrup Salt	Cane sugar Maple syrup or sugar made with cane sugar and flavored with maple. Salt	Cane sugar
Sesame oil Olive oil	Sesame oil Corn oil	Sesame oil Soybean oil	
Gelatin-plain or flavored with lime or lemon.	Gelatin-plain or flavored with pineapple.	Gelatin-plain or flavored with lime or lemon.	
Royal baking powder Baking soda Cream of tartar	Royal baking powder Baking soda Cream of tartar	Royal baking powder Baking soda Cream of tartar	
Vanilla extract Lemon extract	Vanilla extract	Vanilla extract Lemon extract	

[1]Adapted by the authors from: Rowe, A. H., *Elimination Diets and the Patients Allergies*, 2nd ed., Lea and Febiger, Philadelphia, Pa., 1944, p. 150.

Table A-8 shows typical daily menus based on each of the first three of the Rowe Elimination Diets listed in Table A-7.

The Pulse Test. Some physicians report that the pulse speeds up following the consumption of foods to which the patient is allergic. The pulse-testing technique is conducted as follows: For one week, the patient should (1) record everything that he eats at every meal; and (2) take his pulse (a) before rising each morning, (b) just before each meal, (c) three times (at 30-minute intervals) after each meal, and (d) just before going to bed. This is a total of 14 pulse counts daily for a week.

If the highest pulse count everyday for a week is not over 84, and if it is the same each day, the person may not be allergic to any food. On any given day, the range from the lowest to the highest count will probably not be more than 16 beats. A count greater than 84, points to a food allergy. Also, if the highest count varies more than two beats from day to day the person may be allergic; for example, if on Monday the count is 72, on Tuesday 78, Wednesday 76, and Thursday 71.

TABLE A-8
SAMPLE MENUS USING ROWE ELIMINATION DIETS 1, 2, AND 3

Diet 1	Diet 2	Diet 3
Breakfast: Half grapefruit Lamb liver sauteed in sesame oil. Rice cakes Lemonade	**Breakfast:** Stewed prunes Chicken livers and bacon Corn-rye muffins Apricot jam	**Breakfast:** Grapefruit Beef bacon Bread made with soy-lima, potato, and tapioca. Peach jam
Lunch: Lamb chop Steamed brown rice Carrots Spinach with lemon Rice biscuit Grapefruit juice	**Lunch:** Baked squash stuffed with chopped chicken and corn. Pineapple upside-down cake made with corn-rye flours.	**Lunch:** Hamburger patty or meatballs String beans Sliced tomato Pie made with soy, lima, potato, and tapioca flours; filling of apricot and tapioca.
Dinner: Roast leg of lamb Baked sweet potato Swiss chard Gelatin with fresh pear. Puffed rice balls made with maple syrup.	**Dinner:** Whole chicken rotisseried or baked. Asparagus Beets Rye bread Fresh peach with sugar	**Dinner:** Swiss steak or beef stew Baked potato Peas Fruit jello with apricot, grapefruit, and peach. Cookies made with allowed flours, oil, and flavoring extract.

TREATMENT OF FOOD ALLERGIES.

Basically, treatment consists of determining the allergen, then avoiding it as much as possible, but, at the same time, following a superior nutritional program.

Elimination of the Causative Food. If the food can be identified and eliminated from the diet, the symptoms will not recur. If the responsible item of food is one which is not consumed regularly (e.g., strawberry, walnut, chocolate, or shrimp), it can easily be avoided. But such avoidance is far more difficult in the case of milk, eggs, and wheat, which are present in so many foods.

Avoidance of offending food may not need to be absolute. Thus, tolerance level is important in the management of food allergy.

Denaturation of the Protein. Sometimes a protein ceases to act as an allergen if it is denatured by heat. Thus, a patient sensitive to raw milk may be able to take with impunity boiled milk and milk in cooked products.

Also, a patient sensitive to lightly boiled eggs may be able to take eggs which have been boiled for at least 10 minutes; and patients who are sensitive to eggs may be able to eat the yoke, especially if well cooked, although the whites may continue to cause symptoms.

Drug Treatments. Drug treatments had best be by prescription of a physician. Such drugs as antihistamines, adrenalin, and ephedrine, and some newer drugs, are often effective in relieving allergic symptoms, but they do not cure the condition. In severe cases the hormonal substances ACTH and cortisone may be used, but their adverse effects limit their long-term use, especially in children. Tranquilizers and sedatives are often indicated because of the close connection between emotion and allergy.

Desensitization (Reduced Sensitivity). Many attempts have been made to desensitize patients who suffer from food allergy by repeatedly injecting or giving by mouth increasing quantities of the allergen. Generally speaking, this method is regarded as of little or no value in food allergies.

Keep Healthy. Physicians have long known that good health will lessen allergies. So, persons subject to allergies should eat a superior nutritional diet, practice good hygiene, get enough rest and sleep, and avoid extreme fatigue.

Avoid Stress. Allergies often appear to involve emotional factors, as well as organic factors; and conversely the allergy may induce tensions which did not originally exist. So, the doctor must determine the place of these emotional factors in the patient's condition. Sometimes, all treatments for the condition fail until the psychological problems are determined and alleviated.

Maintain Good Nutrition. Persons suffering from allergies should eat a balanced diet, containing adequate minerals and vitamins, to insure maximum good health. They should consult the family doctor or a nutrition specialist if they feel they are lacking in nutrients.

Unfortunately, many intelligent allergic patients, knowing something of the nature of their disease, go to extremes to avoid any foods that may aggravate it. Some avoid so many kinds of food that they become severely undernourished or malnourished. This is particularly true in the case of growing children. So, it is emphasized that allergic patients should not be subjected to dietary restrictions without good evidence that their symptoms are due to a food allergen.

Eat a variety of foods rather than concentrating on a limited number. Eat a moderate amount and try to avoid food jags. Eat regularly without overeating at a particular meal or time of day.

Know Botanical Families. There is evidence that all foods of the same botanical family are likely to produce allergic reactions; that is, if one member of the family produces such a response, all of them will likely do so. Since many food items not normally associated are members of the same botanical family, such as asparagaus and onions, a careful study of the botanical family of an identified allergen is advised. Table A-9 presents some common food items classified in the selected botanical families.

TABLE A-9
BOTANICAL CLASSIFICATION OF SOME FOOD ITEMS

Cabbage and mustard family . . .	Broccoli Brussels sprouts Cabbage Chinese cabbage Cauliflower Collards	Horseradish Kale Mustard Radish Turnip Watercress
Cereal family	Barley Cane sugar Corn Malt Millet Oats	Popcorn Rice Rye Sorghum Wheat Wild rice
Gourd family	Cantaloupe Cucumber Pumpkin	Squash Watermelon
Lily family	Asparagus Chives Garlic	Leek Onion Shallot
Nightshade family	Bell pepper Cayenne pepper Eggplant	Potato Tomato
Palm	Coconut Date Olive palm	
Parsley	Anise Caraway seed Carrot	Celery Parsley Parsnip
Pea	Black-eyed pea Chickpea Common bean (navy, kidney, pinto, string) Jack bean	Lentil Licorice Pea Peanut Soybean
Plum	Almond Apricot Cherry	Peach Plum
Sunflower	Dandelion Endive Jerusalem artichoke	Lettuce Safflower Sunflower seed

Read Labels. Labels tell what is in the food package. So, those suffering from allergies should become inveterate label readers. They should read labels continually to make sure that the products which they purchase do not contain ingredients to which they are allergic. Just because a label says "pure, 100% organically grown," "natural" or "organic," does not assure that the product is safe for allergy patients. Such foods may still contain a variety of ingredients, one of which may induce a reaction. Since there are presently no laws requiring definition of certain terms such as "natural" or "organic," consumers must read the labels carefully to evaluate all products. If they are unsure of the ingredients, they shouldn't buy the product. Consideration should be given, however, to the fact that refinement or extensive processing may modify food enough so that it can be tolerated, especially if the allergy is a mild one.

Treatment of Food Allergies in Children. Food allergies in children are of particular concern to parents. If, upon being given a new food, a child responds with wheezing, coughing, running of the nose, colic, vomiting, diarrhea, constipation, or eczema, the new food should be avoided for at least a week or two, then it may be tried again.

Fortunately, children tend to outgrow their sensitivity to food allergens. Foods known to have caused reactions in infancy may be tried months or years later when perhaps they can be taken with impunity.

ALLERGY-PRODUCING FOODS. In this section, we turn to the most important aspect of food allergies—the foods themselves. Because of space limitations, the discussion will be limited to the most common offenders; namely, milk, eggs, wheat, corn, legumes, nuts, and seafoods. It is unfortunate that these foods are allergy-producing because most of them are also essential foods; hence, their avoidance becomes a serious problem in undernourished people and in children or infants.

People may be sensitive to a single food or to multiple foods. When a patient is sensitive to one or two foods only, it is not difficult for him to realize it himself. The difficulty arises, however, when he is allergic to many foods.

Also, allergy to a food may be of various degrees of severity. Symptoms of asthma or hives may not arise if only one food is eaten, provided the sensitivity is mild. However, if a number of such foods are ingested at one time, the symptoms may be pronounced. Another factor which is important in connection with the production of symptoms is the readiness with which a food is absorbed from the gastrointestinal tract. The more quickly the absorption, the more rapid will the symptoms develop after eating.

In addition to the foods themselves, some food additives may cause allergic reactions. An example is tartrazine, an azo dye, which is added to many foods and soft drinks, and which may cause urticaria and asthma. The food preservatives sulphur dioxide and sodium benzoate, present in many orange drinks, have also been responsible for attacks of asthma.

MILK ALLERGY. Although the word "milk" ordinarily means cow's milk, it also applies to milk from any mammal, including human milk and goat's milk. Also, in various parts of the world, milk is obtained from ewes, buffalos, mares, yaks, camels, and reindeers.

Cow's milk is the most common food allergen. Studies show that, among patients known to have food allergies, 65% are allergic to milk. It is not only the most common in all age groups, but it also causes an unusually large variety of symptoms, including nasal stuffiness, hives, eczema, migraine headache, vomiting, diarrhea, or a serious attack of asthma. Also, there are some reports that bed-wetting may be an allergic reaction to milk. Since it ordinarily does not cause severe reactions, at least in patients over 1 year of age, milk allergy is easily overlooked. But the physician or nutritionist who watches constantly for a milk allergy will find that it is amazingly common.

People who must avoid milk do so for different reasons; some have an inherent inability to digest milk properly, known as a lactose intolerance, others have a milk protein allergy. (For information relative to lactose intolerance, see MILK AND MILK PRODUCTS, section headed "Lactose Intolerance.")

- **Protein Allergy**—Milk contains two proteins, casein and lactalbumin, either or both of which can cause an allergic reaction. If a person is allergic only to the casein protein, milk from animals other than cows, such as goats, may sometimes be substituted. Also, heat treatment of cow's milk (as well as dried and evaporated milk preparations) is sometimes helpful in decreasing allergic symptoms, but may not be completely satisfactory because of some cross-reactive milk proteins.

Milk-allergic Infants. Milk is the most nearly perfect food; it's nourishing, delicious, and a valuable source of protein, vitamins, and minerals. Unfortunately, however, it is also a common allergen, capable of producing any of the reactions known to be caused by food allergies.

Mother's breast milk is usually considered the best milk for infants, particularly in families where there is a tendency to develop allergies. However, some allergies occur as a result of foods eaten by the mother which are excreted in the milk; then produce symptoms in the infant. An infant may also develop allergies from other sources.

An infant with a milk allergy may need specially processed milk or milk substitutes. Dried, boiled, or evaporated milk may work because milk protein in these forms has been exposed to a heat process. If these are not suitable because of a more extensive allergy, the infant may require modified milks, such as soybean milk, meat preparations, or synthetic milk (see Table A-10).

TABLE A-10
SUBSTITUTE FORMULAE FOR MILK ALLERGIC INFANTS[1]
(See your physician; he will recommend the proper substitute)

Product	About the Formulae
Isomil (Ross)	Soy protein isolate in a milk-free formula.
Mull-Soy (Borden)	Add water to this milk-free formula which uses soy flour.
Neo-Mull-Soy (Borden)	No corn oil. Uses soy protein isolate.
Nutramigen (Mead Johnson)	Formula in which existing milk proteins have been hydrolyzed enzymatically to the less reactive form (predigested).
Pro-Sobee (Mead Johnson)	Milk-free formula with protein supplied by soy protein isolate.
Soyalac (Loma Linda)	Soy protein and soy oil. Added iron; may or may not be vitamin enriched.

[1]Brand names are used for educational purposes only and do not imply endorsement or discrimination by the authors.

There may be additional complications in the milk-allergenic infant. He may have an allergy to some feed or drug which has been given to the cow from which the milk was taken. Not infrequently, he may also be allergic to a common milk substitute such as soybean products. In instances of double allergy, the physician may recommend a meat formula to supply needed protein. A commonly-used meat formula is MBF (meat base formula), a product of Gerber Company, which contains strained beef hearts, sucrose, oil, thickener, calcium, ascorbate, and vitamins A and D. For the infant allergic to beef, Gerber's manufactures Lambase, similar to MBF, but with lamb as the meat foundation.

Parents who have an infant or child who is allergic to milk should learn to recognize the many products which contain milk. Several foods "hide" milk or milk products in their ingredients. Check the labels carefully and follow the doctor's recommendations.

Milk Foods to Avoid. Milk-sensitive patients should not drink milk as a beverage and they should avoid the foods listed in Table A-11. After a period of abstinence, they may return to eating limited quantities of foods which contain the allergy-inducing substance. For example, a person on a milk-free diet may have limited amounts of bakery products containing milk. This should only be done with a physician's approval, however.

TABLE A-11
MILK-CONTAINING FOODS TO AVOID

Foods	Pertinent Facts
Au gratin, scalloped dishes	Foods with sauces or gravies often include butter or cheese in main ingredients.
Baby foods	Check label carefully.
Beverages, milk-based	Malted milk, hot chocolate, cocoa prepared with milk; and liquid diet preparations.
Breads	Bakery products, such as pies, breads and cakes, containing small amounts of cooked milk may be tolerated. Avoid biscuits and biscuit mixes, muffins and muffin mixes including refrigerated products.
Butter, oleomargarine	Usually permitted in modest amounts.
Cakes	Includes home mixes.
Candy	Chocolate as well as many opaque candies.
Cheese	All types of cheese, including spreads and cheese foods.
Cookies	Avoid cookies containing milk.
Doughnuts, fritters	Includes doughnut mixes.
Dressings	Boiled salad dressings.
Eggs	Scrambled or omelets.
Fish	Canned fish balls.
Ice cream, sherbets	Includes sodas.
Macaroni products	Includes noodles, spaghetti.
Malted milk, hot chocolate, or cocoa	If prepared with milk.
Meats	Dried milk used as filler and binder in meat loaf and processed meats such as hot dogs, luncheon meats and sausage.
Milk, skim milk, buttermilk, cream	Avoid milk, buttermilk, and cream as such; and in prepared foods, including ice cream, sodas, milk sherbet, Bavarian cream mousses, custards, gravies, cream sauces, soups, and chowders. Check the label on packaged foods for evidence of milk or milk products.
Milk, other than whole	Evaporated, powdered or condensed form, either as beverage or ingredient in packaged foods.
Seasonings	Flavored seasonings may contain milk solids.

Milk Replacements. Those who are allergic to milk, need substitutes. Table A-12 presents some substitutes.

TABLE A-12
MILK REPLACEMENTS

Foods	Replacement
Butter	Milk-free margarines.
Coffee creamers	Any of the brands on the market, but check the label.
Cream	Nondairy foods without milk protein.
Meats, luncheon	Unless distinctly labeled as containing a milk product(s).
Milk	To replace calcium found in milk, substitute eggs, legumes, whole grain cereals, salmon, mustard or turnip greens, broccoli, and collards. Also, it may be wise to take a mineral supplement.

Read the Label. The following terms indicate the presence of milk or milk derivatives:

Calcium caseinate
Casein
Caseinate
Casein hydrolysate
Creamed foods
DMS (dried milk solids)
Lactalbumin
Lactate solids
Milk solid pastes
Sweetened condensed milk
Whey or whey solids

EGG ALLERGY. The hen's egg (and eggs from other birds) is an excellent food; it is nourishing, delicious, and a valuable source of protein, vitamins, and iron. Also, it has many uses in cookery. Unfortunately, however, it is also a common allergen, capable of producing any of the reactions known to be caused by food allergies. Also, it is often said that egg allergy is always severe.

Like milk, egg has several different protein fractions which may induce an allergic reaction. They are vitellin, livetin, ovalbumin and other conjugated proteins. The egg white contains the larger amount of protein and is usually a greater offender than the egg yolk. Since the yolk is rich in iron and fat-soluble vitamins, sometimes it may be fed to infants allergic to whole eggs. Sometimes, heating destroys the allergy causative factor(s) of the egg white by coagulating it, with the result that some people who are allergic to eggs may be able to eat hard-boiled eggs without developing symptoms.

One of the great dangers of egg allergy is that its victims, no matter how slight their degree of sensitivity, tend to have violent reactions to certain vaccines grown on chicken embryos (influenza, spotted fever, yellow fever). Some of these vaccines seem to be entirely free of egg protein contamination, whereas others, notably influenza vaccine, are not to be trusted.

Egg Foods to Avoid. Those who are allergic to eggs would do well to avoid the foods listed in Table A-13.

TABLE A-13
LIKELY EGG-CONTAINING FOODS TO AVOID

Foods	Remarks
Baking powder	Read the label; some contain egg white.
Beverages	Prepared drinks for insomnia or underweight may contain egg or egg powder.
Breaded foods	May use egg mixture as part of the coating.
Breads	Baked with glazed crust. Prepared flour mixes for home cooking.
Cakes, cookies	Avoid unless labeled egg-free.
Candies	Cream candies, fondants, marshmallows.
Doughnuts	Includes mixes.
Dressings	Hollandaise, mayonnaise, egg sauces, salad dressings unless known to be egg-free.
Eggs	Fresh, frozen, powdered, cooked in any form. Do not substitute other fowl eggs.
Egg substitutes	Some substitutes on the market with limited cholesterol levels do contain egg white or egg protein and should not be used.
French toast	Avoid unless made with an egg substitute.
Ice cream, sherbets	Unless known to be egg-free.
Meats	Egg used as binder in sausages, croquettes, meat cakes.
Pancakes, waffles	Includes mixes.
Pies	Custard and cream pies.
Poultry	Especially chicken if fricasseed or in broth.
Puddings	Custards, Bavarian cream.
Soups	Broths made with egg.

Egg Replacements. Some replacements for eggs are listed in Table A-14.

TABLE A-14
EGG REPLACEMENTS

Foods	Replacement
Eating or cooking whole eggs	Jolly Joan Egg Replacer (Ener-G Foods, Seattle, Washington)
Cooking—coating for breaded and batter-dipped foods	2 tablespoons flour + ½ teaspoon shortening + ½ teaspoon baking powder + 2 tablespoons liquid.
Leavening agent in recipes	Add ½ teaspoon baking powder per egg deleted.

Read the Label. The following terms indicate the presence of eggs or egg derivatives:

Albumin	Globulin
Dried egg solids	Ovomucin
Egg solids	Vitellin

WHEAT ALLERGY. Wheat is not the only cereal grain to produce allergies. But, because it is found in such a large variety of foods, it is the most burdensome to its victims and to those who prepare their meals. Corn, oats, rye, barley, and rice may also cause an allergic reaction. Buckwheat, which is not actually a cereal or grain, but a member of the rhubarb family, may also cause an allergic reaction. The chief causative factor in wheat is gluten protein.

• **Gluten allergy**—Wheat gluten is extremely versatile because it provides the strong foundation for the leavening action of yeast. For this reason, wheat gluten or wheat flour is frequently used in the baked products made from other grains, including soy flour. Anyone having a wheat allergy should be extremely suspicious of a label that includes "other flours," for they will likely include at least some wheat flour or wheat flour derivatives.

• **Celiac disease** *(Celiac sprue)*—Celiac disease is due to a genetically-inherited intolerance to wheat gluten, a main constituent of wheat flour, which is also present to a small extent in rye, barley, and oats, but not in rice. Such persons develop lesions of the small intestine which lead to diarrhea and malabsorption. It is not known why gluten causes intestinal damage. Symptoms usually arise during the first 3 years of life, but it may affect adults. Patients are relieved if gluten is excluded completely from the diet.

Wheat Flours to Avoid. Patients allergic to wheat gluten can cook and bake with a wide variety of other cereals. But their substitutes should be varied. They may develop an allergy to corn, for example, if they use it as the sole replacement for wheat flour. Rice, potato, and soy flours are less likely to result in increased sensitivity to cereals.

Since not all flours are the same, the proportions of ingredients in recipes will need to be altered. A combination of flours often produces a more palatable product than a single flour. Try several flours until a single substitute or a combination of substitutes is found that fits the taste of the patient. Products made with the substitutes may lack some volume, be a bit crustier than wheat flour products, and be more inclined to crumble. They may also be somewhat higher in fat content, so the shortening in the recipe may be lessened.

A list of likely wheat-containing foods to avoid is given in Table A-15.

TABLE A-15
LIKELY WHEAT-CONTAINING FOODS TO AVOID[1]

Foods	Remarks
Baby foods	Check labels carefully.
Beans, baked	Avoid if wheat products are included.
Beverages	Malted milk, ale, beer, some wines, coffee substitutes, gin, whiskey.
Breads	Bread, bread crumbs, breadings. Biscuits, muffins, popovers, pastries, pretzels; including mixes.
Breakfast cereals	Dry or cooked, if they contain wheat or bran.
Crackers	Crackers and cracker meal.
Cakes, cookies, and candies	Includes cake mixes and cake flour.
Cheese	Spreads and sauces.
Corn bread	Unless known to be wheat-free.
Doughnuts	Includes doughnut mixes.
Dressings	Thickener
Dumplings, fritters	Unless made with other flours than wheat.
Fish	Breaded or floured; canned.
Ice cream, sherbets	Sometimes. Includes ice cream cones.
Jams	If wheat products are used as a thickener.
Macaroni products	Noodles, macaroni, ravioli, spaghetti, vermicelli.
Meats	Avoid floured meats such as swiss steak, most meat loaf, chili con carne, croquettes, and stuffed poultry. Check the label and make certain that there is no wheat flour in processed meats, such as luncheon meats, hot dogs, premolded hamburgers, and sausage.
Pancakes, waffles	Includes mixes.
Pies	Thickener in pie fillings. Includes pastry.
Potato and rice mixes	Read the label.
Puddings	Thickener
Rye bread	Unless known to be wheat-free.
Sauces	Gravies, scalloped dishes. Thickener in commercial cream sauces.
Snack foods	Pretzels
Seasoning mixes	Avoid if it contains wheat products.
Soups	Creamed soups, chowders, bisques. Some soup mixes. Canned varieties with noodles, some bouillon cubes and extracts. Dry soup mixes.
Soy bread	Unless known to be wheat-free.
Vegetables	Commercial thickener in creamed vegetables or vegetable sauces.
Yeast	Avoid yeast that is made from an extract of wheat, or that is combined with a base of a wheat product.

footnote at end of table *(Continued)*

TABLE A-15 (*Continued*)

[1]These foods may contain wheat and/or gluten if purchased in a commercial form. The offending allergen can be removed and an acceptable substitute added, providing the food is made at home.

Wheat Flour Replacements. The following flours can be substituted for wheat flours in gravies, cream sauces, and other dishes requiring thickening: arrowroot, potato starch, rice, rye, soy, or tapioca. In baking, however, no really satisfactory substitute for wheat flour is available; nevertheless, the replacement substitutes shown in Table A-16 are acceptable.

TABLE A-16
WHEAT FLOUR REPLACEMENTS

Foods	
Substitute the following for one cup wheat flour:	
Barley flour	½ cup
Corn flour	1 cup
Cornmeal	¾ cup
Potato starch flour	⅝ cup
Rice flour	⅞ cup
Rolled oats	1⅓ cup
Rye flour	1¼ cup
Rye meal	1 cup
Soya flour	¾ cup
Combination substitutes for 1 cup of wheat flour:	
{ Rye flour	½ cup
{ Potato flour	½ cup
{ Rye flour	⅔ cup
{ Potato flour	⅓ cup
{ Rice flour (10 tablespoons)	⅝ cup
{ Rye flour	⅓ cup
{ Soy flour	1 cup
{ Potato starch flour	¾ cup

Read the Label. The following terms indicate the presence of cereal or cereal derivatives:

- Enriched flour
- Flour
- Hydrolyzed flour
- Modified food starch
- Monosodium glutamate (MSG)
- Self-rising flour
- Sodium glutamate (if from wheat gluten)

CORN ALLERGY. Corn is another common allergy-inducing food. Since it is extremely versatile, it is found in cereals, oils, sweeteners, and starches. It is also used as a filler, extender, and thickener. For these reasons, some form of corn is found in many foods.

Patients vary as to the type of corn to which they are allergic. Some cannot take it in any form but pure corn sugar (glucose). Others have no trouble with immature corn; that is, corn-on-the-cob or canned (either kernel or cream-style) corn. Some patients can eat popcorn and other whole corn products, but they cannot consume corn syrup. For unknown reasons, corn syrup is the most potent offender; corn oil the least.

Corn Foods to Avoid. The use of corn or corn derivatives will vary from brand to brand. Some problems can be eliminated either by preparing the product at home or by purchasing those brands recognized as corn-free (see Table A-17).

TABLE A-17
LIKELY CORN-CONTAINING FOODS TO AVOID

Foods	Remarks
Baby foods	Check labels carefully.
Baking powder	Check label.
Beverages	Carbonated beverages, instant coffee, liquors, instant tea, and liquor mixes.
Breads	Breads, pastries. Baking mixes, including biscuit and doughnut mixes.
Cakes, candies, cookies	Marshmallows, marshmallow creme.
Cereals	Check labels.
Cheese	Possibly trial and error is the only way to determine any corn content.
Crackers, graham	Check label.
Desserts	Cream puffs, including rennet tablets and instant pudding mixes.
Doughnuts	Includes mixes.
Dressings	French dressing, other salad dressings.
Fried foods	Deep fat frying mixtures or breadings; frying fats.
Frostings, icings	Unless home-cooked.
Fruits	Some canned and frozen fruits may contain corn syrup. Read the label.
Gelatin	Capsules
Grits	Avoid corn grits.
Ice creams, sherbets, ices	Ask the manufacturer if not labeled.
Jams	Some jellies and preserves.
Margarine	If made from corn.
Meats	Cooked meats with gravies. Cured or tenderized ham, lunch ham. Cooked sausages, frankfurters, weiners. Bacon. Bologna. Chili preparations, chop suey. Canned meats.
Nuts	Some peanut butter.
Oils	Corn oils and frying oils.
Pancakes, waffles	Includes mixes.
Pies	Cream pies. Pie crusts. Canned pie fillings, containing corn syrup.
Potato and rice mixes	Read the label.
Puddings	Blanc mange, custards.
Sauces	Sauces for sundaes, meats, fish, vegetables. Gravies.
Seasonings	Monosodium glutamate. Salt cellars in restaurants. Seasoning mixes.
Snack foods	Pretzels and corn chips.
Soups	Creamed, thickened and vegetable soups.
Soybean milks	Read the label.
Sugar	Confectioner's or powdered sugar.
Syrups	Commercially prepared syrups.
Tortillas	Unless made with wheat flour.
Vegetables	Harvard beets. Canned peas. Canned or frozen string beans.
Vitamins	Capsules

Read the Label. The following terms indicate the presence of corn or corn derivatives:

Corn solids
Corn starch
Corn syrup
Vegetable starch

LEGUME ALLERGY. Legumes include such foods as peas, beans, and peanuts. A patient with an allergy to legumes, should be extremely careful in selecting food since the legume family includes the following foods:

Acacia
Alfalfa
Arabic
Beans (green, kidney, lima, mung, navy, wax. Soybean, including soya flour and oil)
Carob
Cassia
Karaya
Lentils
Licorice
Locust bean gum
Peas (black-eye, chick, green, split)
Peanuts (includes oil)
Tragacanth

Patients allergic to nuts are often allergic to legumes, also. Yet, some patients tolerate legumes, such as peanuts, despite being allergic to most nuts.

- **Soybean allergy**—An allergy to soybeans is becoming increasingly difficult to live with, since soybeans have become so versatile. They provide oil, flour, and protein. Spun soybean protein concentrate is being used as extenders and replacements for some meats. Soybean products also find extensive use as binders, fillers, and emulsifiers.

Legume Foods to Avoid. Among the legume foods to avoid are those listed in Table A-18.

TABLE A-18
LIKELY LEGUME-CONTAINING FOODS TO AVOID

Foods	Remarks
Baby foods	Check label carefully.
Breads, bakery products	When certain products of the *Leguminosae* family are used as antistalents in breads and baking.
Candies	Check label.
Cereals	Some breakfast cereal preparations and breakfast pastry preparation.
Cheese spreads	Avoid cheese spreads containing legume product.
Crackers	Avoid crackers containing any legume product.
Dressings	Salad dressings, salad oils, imitation salad dressings.
Ice cream	Avoid legume-containing ice cream, such as peanut ice cream.
Margarine	Unless labeled otherwise.
Meats	Sausages, cold cuts, commercial one-dish meals, *if* distinctly labeled as containing a legume.
Macaroni products	Noodles, spaghetti.
Oils	Shortenings, salad oils.
Peanut butter	And peanut spreads.
Potato mixes	Read the label.
Sauces	Steak sauces, seasoning sauces.
Seasonings	Mixes
Snack foods	Flavored potato chips.
Soups	Read the label.
Substitutes	Whipped "nondairy" toppings; substitute milks. Baking chocolate substitutes.

Read the Label. The following terms indicate the presence of legume or legume derivatives:

- Hydrolyzed vegetable protein
- Soya flour
- Soy concentrate
- TVP (textured vegetable protein)
- Vegetable protein concentrate

NUT ALLERGY. The term "nut" is applied to a variety of seeds, not all of which are true nuts. Those most likely to act as allergens are peanut (actually a legume), walnut, pecan, Brazil nut, hazelnut (filbert), and coconut. Less common offenders are almond, cashew, pistachio, macadamia nut, and pinyon nut.

Patients on a nut-poor diet, should avoid nuts of all types: nut crumbs on cookies, cake icings, and ice cream; candies containing nuts; and oils from nuts used in salad oils, lard substitutes, and margarines. Patients allergic to nuts may be allergic to cottonseed meal and oil, also. (Olive oil is permitted.)

A variety of symptoms may result from the eating of nuts, including asthma, nose allergy, hives, or headaches.

SEAFOOD ALLERGY. Fish is a strong antigen, which usually produces severe symptoms. Some patients develop asthma if they are merely exposed to the odor of fish that is being cooked. Hence, no fish-sensitive patient is unaware of his problem.

Patients on a seafood-free diet should avoid fish and shellfish—fresh, canned, smoked, or pickled; fish liver oils and concentrates in vitamin preparations; fish and shellfish stews, bisques, broths, soups, salads, hors d'oeuvres, and caviar. Also, they should avoid licking labels, which may contain a fish glue adhesive; and they should avoid injections of fish origin in the treatment of varicose veins.

Those allergic to fish may be able to eat oysters, shrimp, and lobster. Also, patients unable to eat one type of fish may be able to eat others.

OTHER ALLERGY-PRODUCING FOODS. Many other foods produce allergies in some people; among them, certain spices—especially mustard and oil of wintergreen, cottonseed, castor bean, strawberries, chocolate, and sulfites.

STANDARDS OF IDENTITY. Table A-19, Standard of Identity, lists foods which contain a particular ingredient that may cause an allergic reaction in certain individuals, but which may lack a label listing ingredients.

ALLERGY-FREE PRODUCTS LIST. The following companies will send lists of their products which are free of particular allergy-producing ingredients:

Akins Special Foods, Inc.
P.O. Box 2747
6947 East 13
Tulsa, OK 74119

Chicago Dietetic Supply, Inc.
405 E. Shawmut Avenue
La Grange, IL 60525

Ener-G Foods, Inc.
1526 Utah Avenue South
Seattle, WA 98134

Frito-Lay, Inc.
P.O. Box 660634
Dallas, TX 75266-0634

General Foods Kitchen
250 N. Street
White Plains, NY 10625

General Mills
Nutrition Department
Minneapolis, MN 55440

ALLERGY-FREE RECIPES. Sources of recipes for allergic diets follow. No claim is made relative to the completeness of this list; rather, these are sources known to the authors. Your physician may be able to augment this list. Prices are subject to change; hence, anyone interested should contact the companies for current prices.

Allergy Cooking
Conrad, M.L.
Thomas Y. Crowell Publisher
10 E. 53rd Street
New York, NY 10022

(Continued)

TABLE A-19
STANDARD OF IDENTITY

Food	Milk	Eggs	Wheat	Corn	Soybean
Beverages	Milk, cream		Whiskey	Bourbon whiskey	
Breads	Raisin, whole wheat, rolls	Raisin, whole wheat, rolls	Raisin, whole wheat, rolls	White breads, raisin, whole wheat, rolls.	Raisin, whole wheat, rolls
Candies	Fudge, caramel, chocolate	Nougat, kisses, frappes, marshmallows.			Nougat, kisses, frappes, marshmallows.
Cereal products			Farina	Grits	
Cheese	Standards allow some cheeses to be made from goat's or cow's milk				
Dressings		Salad dressings, mayonnaise	Salad dressings	Salad dressings, mayonnaise	
Fish			Breaded	Breaded	Tuna packed in certain medium.
Flours			Phosphated wheat	Corn, cornmeals	
Fruits				Canned and juices with syrup from corn syrup.	
Ice cream, sherbets	Ice creams, sherbets, ice milk	French-style ice creams, other ice creams, sherbets.		Includes corn syrup solids.	
Jams				Jellies, jams using corn syrup.	
Macaroni products	Macaroni products	Macaroni products	Semolina		
Sauces, condiments, spreads	Margarines			Catsup	Peanut butter, margarines
Vegetables	Vegetables in sauces		Canned green peas, beans, corn (not all brands)		Canned green peas, beans, corn
Miscellaneous	Chocolate products				Lecithin (maybe milk or egg sauce)

Allergy-Free Recipes *(Continued)*

Allergy Cooking Made Easy (for wheat-free, milk-free and egg-free diets)

El Molino Mills
3060 West Valley Boulevard
Alhambra, CA 91803

Allergy Diets

Ener-G Foods, Inc.
1526 Utah Avenue South
Seattle, WA 98132

Ralston Purina Company
Nutrition Service
Checkerboard Square
St. Louis, MO 63199

Allergy Recipes

The American Dietetic Association
620 North Michigan Avenue
Chicago, IL 60611

Baking for People with Food Allergies H.G Bull. No. 147

Superintendent of Documents
U.S. Government Printing Office
Washington, DC 20402

Cooking with Imagination for Special Diets

Grocery Store Products Company
West Chester, PA 19380

Creative Cooking Without Wheat, Milk and Eggs

Shattuck, R.R.
Leisure Dynamics Publishing Group
Cranbury, NJ 08512

Diet Food Finder, The

Casale, J.T.
R.R. Bowker Company
1180 Avenue of the Americas
New York, NY 10036

Easy, Appealing Recipes (for milk-free diets)

Mead Johnson Laboratories
Department 852
Evansville, IN 47721

Eggless Cookbook, The

Sattler, H.R.
Leisure Dynamics Publishing Group
Cranbury, NJ 08512

(Continued)

Allergy-Free Recipes *(Continued)*

Elimination Diets, The

Rowe, A., Jr. and C.E. Sinclair
Lea & Febiger
600 Washington Square
Philadelphia, PA 19106

Good Eating for the Milk-Sensitive Person

Ross Laboratories
625 Cleveland Avenue
Columbus, OH 43216

Good Recipes to Brighten the Allergy Diet

Superintendent of Documents
U.S. Government Printing Office
Washington, D.C. 20402

Gourmet Food on a Wheat-free Diet

Wood, M., and C. C. Thomas
Bannerstone House
301-327 Lawrence Avenue
Springfield, IL 62703

Low Gluten Diet with Tested Recipes

Arthur B. French
Clinical Research Unit
University Hospital
Ann Arbor, MI 48104 (fee charged)

Milk-Free Cookbook, The

Syntex Laboratories, Inc.
Nutritional Products Division
Palo Alto, CA 94304

125 Great Recipes for Allergy Diets

Good Housekeeping
959 Fifth Avenue
New York, NY 10011

Recipes for Goat's Milk (Goat milk can be substituted in any recipe calling for milk. The product will be somewhat richer since the cream in goat milk has not been separated out.)

California Goat Dairyman's Association
P.O. Box 934
Turlock, CA 95380

Recipes to Brighten the Allergy Diet

Best Foods
P.O. Box 307
Coventry, CT 06238

Wheat, Milk and Eggfree Recipes from Mary Alden

Quaker Oats Company
Merchandise Mark Plaza
Chicago, IL 60654

ALLICIN

An antibacterial substance extracted from garlic, *Allium sativum.*

ALLIGATOR PEAR *Persea americana*

Another name for the avocado.
(Also see AVOCADO; and FRUIT[S], Table F-47 Fruits of the World—"Avocado".)

ALLIGATOR SKIN

A roughening of the skin which may be the result of nutritional deficiencies (particularly the lack of vitamin A), poor liver function, poor hygiene, and/or a hot, dry, environment.

ALLOTRIOPHAGY (PICA)

Depraved, or abnormal, appetite characterized by the desire to eat such substances as dirt, chalk, pencils, etc. This condition is usually caused by nutritional deficiencies.
(Also see GEOPHAGIA.)

ALMONDS *Prunus amygdalus*

The almond is a small deciduous, nut-bearing tree which is closely related to the peach, apricot, and cherry tree—all belong to the genus *Prunus,* and bear stone fruits. Edible almonds are also called sweet almonds because there is a variety which is inedible and called bitter almonds. The fruit of the almond is classified botanically as a drupe, like that of the peach but the outer fleshy layer of the almond is astringent, tough, and becomes dry at maturity. This fleshy layer—the hull—splits and releases the nut.

Fig. A-21. Almonds are grown for their fragrant blossoms as well as for their nuts. A leathery hull encases the woody shell of the nut, and splits open when the kernel is ripe. The somewhat flat, oval, brown kernel is one of the most popular nuts.

Almond trees are small trees, growing to 9 to 22 ft (*3 to 7 m*) high. Their leaves are long, pointed and slightly curled. In the early spring, long before leaves appear, almond trees bloom with showy, fragrant pink or white flowers, each about 1½ in. (*3.8 cm*) in diameter. Because of this early flowering habit, the commercial culture of the almond is limited to areas with mild winters and few late spring frosts, though the trees are grown as ornamentals in other regions.

ORIGIN AND HISTORY. The almond probably originated in western Asia and North Africa, but it is now widely grown in the countries that border the Mediterranean Sea. Some almonds were introduced into the United States from Mexico and Spain when missions were established in California. However, these died after the missions were abandoned. Then in 1840, some trees were imported from Europe and planted in New England. However, the climate in New England and down the length of the Atlantic seaboard and into the Gulf States ruled out almonds as a profitable crop. In 1843 some almond trees from the East Coast were planted in California, and now California is the only important almond-producing state.

PRODUCTION. Almonds grow best where there are long hot summers and mild, but not warm, winters. Nut harvest is largest when rainfall during the season is low but the trees are properly cultivated and irrigated. Worldwide, the yearly production of almonds averages 1.3 million metric tons. The five leading almond-producing countries, by rank, are the United States, Spain, China, Italy, and Iran, though almonds are produced in many countries of Africa, Asia, and Europe.

In the United States, almonds are grown on a commercial basis only in California, where an average of 658 million pounds (*299,400 metric tons*) are produced each year. In a recent year the California crop was valued at an estimated $592 million.

Propagation and Growing. Almond trees are propagated primarily by budding, using almond (sweet or bitter) or peach seedlings as rootstalk. Young almond trees are planted 25 to 30 ft (*7.6 to 9.2 m*) apart in orchards, and they produce some almonds after 4 years. Full production is reached by the seventh year after planting. With good management—pruning, fertilizing, pest control, and irrigation—1 acre of almonds can produce 1,500 to 2,200 lb (*680 to 998 kg*) of almond meat (kernels).

There are about 100 varieties of almonds grown in California. In orchards, several varieties that will bloom at the same time must be planted together, since cross pollination is necessary. Honeybee colonies set in almond orchards ensure cross pollination. Almonds grow throughout the summer and then ripen in the early fall.

Harvesting. In many areas around the Mediterranean where almonds are grown, their harvest depends mostly upon hand labor. In California harvesting is quite mechanized.

When most of the hulls have split, the fruits, or nuts, are harvested. The trees are jarred with rubber mallets or a power shaker to shake the nuts down. Then in many California orchards, mechanical sweepers windrow the nuts on a smooth soil surface. Another machine is used to pick up the windrow. As picked up from the ground, there is a mixture of hulls and nuts. Still another machine is used to remove the hulls. In some orchards, large ground cloths spread under the tree are used to collect the nuts as they are jarred from the tree. After hulling, almonds are dried and delivered to a processing plant.

PROCESSING. Almonds are sold unshelled or shelled. They are shelled mechanically, and then the kernels are electronically sorted for defects, graded for size, and packaged. Shelled almonds may also be salted and roasted, blanched, ground into a meal or made into a paste.

• **Almond Oil**—Oil of almond is obtained by expression or distillation of the kernels of sweet (edible) or bitter almonds; primarily, bitter almonds are used for this purpose. The prussic acid (hydrocyanic acid) contained in bitter almonds is eliminated during the process. Almond oil is used in flavoring extracts.

SELECTION. Almonds in shells should be free from splits, cracks, stains, or holes. Moldy nuts may not be safe to eat. Nutmeat should be plump and fairly uniform in color and size. Limp, rubbery, dark, or shriveled kernels are likely stale. If antioxidants are added to delay the onset of rancidity, thus extending the shelf life of packaged nutmeat, they are listed on the package.

PREPARATION. Bitter almonds contain prussic acid and should not be eaten; but sweet almonds are popular eaten alone when dried, or when roasted and salted! Almonds are also used in a variety of products, including candies and rich pastries. Additionally, other imaginative uses for almonds may be found, such as waffles, biscuits, muffins, vegetable salads, topping for baked goods, or as additions to meats, poultry, or seafood salads.

Because of their oil content, almonds need protection from air, and high temperatures. Whole almonds become rancid less quickly than nut pieces and unroasted almonds keep better than roasted ones. For prolonged storage, almonds should be kept cool and dry. Shelled almonds will stay fresh for several months stored in tightly closed containers in the refrigerator. For longer storage, almonds can be frozen in tightly closed freezer containers at 0°F (*−18°C*) or lower.

NUTRITIONAL VALUE. Almonds are very nutritious. Dried almonds contain only about 5% water and each 100 g (about 3½ oz) are packed with 598 Calories (kcal) of energy. Almonds are high in calories primarily because they are 54% fat (oil) which contains 2¼ times more calories per unit of weight than does protein or carbohydrate. Even with the fat content, each 100 g of almonds also contains 19 g of protein, 20 g of carbohydrate, 773 mg of potassium, 5 mg of iron, 3 mg of zinc, and 28 mg of vitamin E. Unsalted almonds have only 4 mg of sodium per 100 g. More complete information regarding the nutritional value of dried or roasted and salted almonds, almond meal, and almond paste is provided in Food Composition Table F-36.

(Also see NUT[S], Table N-8 Nuts of the World.)

ALOPECIA

Loss of hair. Sometimes, the condition is due to a nutritional deficiency. Also, there are various other causes of hair loss such as disorders of the scalp, reaction to certain drugs or chemicals, venereal disease, and infection. Loss of hair also occurs in thyroid deficiency (myxedema) and in certain types of pituitary disorders.

(Also see BALDNESS.)

ALPHA-CHOLESTEROL

This term is a layman's designation for blood cholesterol which is carried on alpha-lipoproteins, in contrast to that carried on beta-lipoproteins. However, the cholesterol molecule itself does not vary, only the lipoprotein which carries it in the blood.

(Also see ALPHA-LIPOPROTEINS; BETA-LIPOPROTEINS; CHOLESTEROL; and HEART DISEASE.)

ALPHA-KETOGLUTARIC ACID

One of the intermediary products in the Krebs cycle; also the product of oxidative deamination of glutamic acid.

(Also see METABOLISM, Fig. M-78 The Krebs cycle.)

ALPHA-LIPOPROTEINS

Special fat-carrying proteins which are produced in the liver. Alpha-lipoproteins are also called high density lipoproteins (HDL). Normally, they contain from 5 to 8% triglycerides, 17 to 30% cholesterol, 21 to 29% phospholipid, and 33 to 57% protein.

(Also see BETA-LIPOPROTEINS; CHOLESTEROL; and HEART DISEASE.)

ALUMINUM (Al)

Aluminum, which is the most modern of common metals, is widely used in buildings and wrappings. Additionally, it is used in food additives, aluminum-containing antacids, and aluminum cooking utensils. Currently, not much is known about the amount of aluminum consumed by people or of its effects, if any. More experimental work is needed.

(Also see MINERAL[S]; and POISONS.)

ALVEOGRAPH

An instrument used to measure the stretching quality of dough; it indicates the stability, extensibility, and strength of doughs.

ALZHEIMER'S DISEASE *Senile Dementia*

This affliction was named after Alois Alzheimer, a German physician, who first described the disease in one of his patients in 1906. About 4 million Americans are estimated to have Alzheimer's disease.

The disease is characterized by a decreased mental acuity. There is increased forgetfulness, failing judgment, a decreased capacity to do simple calculations, and sometimes a disorientation in time and place. The ailment is more prevalent in middle-aged women than in men, but may occur occasionally at an earlier age. In many cases it does not appear until the individual reaches the 70s.

Currently, there is no cure. However, some work has indicated that niacin may be helpful in stimulating blood circulation to the brain. Also, during the last five years, a significant neurochemical abnormality has been found in senile patients. There is a definite deficiency of acetylcholine, which is known to be involved in learning and memory. The precursor of the neurotransmitter, acetylcholine, is choline, one of the vitamin B complex. Some reports have shown that dietary choline may improve learning and behavior.

As in the case of most degenerative diseases, prevention, if at all possible, is better than cure. Thus, a conclusion may be drawn that consumption of choline may be helpful. In 1992, researchers at the Massachusetts Institute of Technology reported that choline levels in the brains of Alzheimer's victims are much lower than normal, thereby lending credence to the theory that choline may be effective in preventing and treating the disease. It is also suggested that adequate trace elements and vitamins be included in the diet throughout life.

Along with a superior diet, it has been found that remaining active and involved, with a positive rather than a negative attitude to life, can be helpful in delaying any onset of senile dementia, whether it develops prematurely or in old age.

AMARANTH *Amaranthus*

Amaranth, a relative of pigweed, was originally grown by the Aztecs in what is now Mexico, where it was used as a basic food as well as in religious ceremonies. It is a tall, bushy plant with broad leaves and a showy flower or seed head. Amaranth is grown for the leaves as a green vegetable and for the seed as a grain. It is a high-protein grain, with 15 to 18% protein which is high in the amino acids lysine and methionine. Amaranth is available in many health food stores.

Fig. A-21a. Amaranth. (Courtesy, Minnesota Agricultural Experiment Station, University of Minnesota)

AMARANTH

Amaranth, also known as Red No. 2, is a dye. It was banned by the Food and Drug Administration for use in foods, drugs, and cosmetics.

AMEBIASIS

Infestation with harmful amoebas, particularly *Entamoeba histolytica.* Cysts of the amoeba may be ingested in contaminated food or water. Also, flies and infested animals may carry the organisms to foods.

(Also see DISEASES.)

AMELOBLASTS

Special epithelial cells surrounding tooth buds in gum tissue, which form cup-shaped organs for producing and depositing the enamel on the surfaces of developing teeth. Vitamin A deficiency results in faulty production of ameloblasts and makes for unsound teeth.

AMERICAN DIETETIC ASSOCIATION (ADA)

This is a professional organization of more than 40,000 dietitians. Its goals are (1) to improve the nutrition of human beings, (2) to advance the science of dietetics and nutrition, and (3) to promote education in these allied areas. The ADA was founded in 1917 during World War I. In 1925, the ADA began publication of the *Journal of the American Dietetic Association,* which it continues today. Furthermore, the ADA publishes other educational material for nurses, doctors, teachers, and other related professionals.

AMERICAN INSTITUTE OF NUTRITION (AIN)

This is a professional society for nutrition scientists, which publishes the *Journal of Nutrition.* The AIN was founded in 1928 to (1) develop and extend nutrition knowledge, and (2) promote personal contact between researchers in nutrition and related fields. Only individuals who have published research and who are engaged in the field of nutrition may be elected to membership.

AMERICAN SOCIETY FOR CLINICAL NUTRITION (ASCN)

This society is a division of the American Institute of Nutrition (AIN), which publishes the *American Journal of Clinical Nutrition.* The aims of the ASCN are (1) to promote education about human nutrition in health and disease, (2) to promote the presentation and discussion of research in human nutrition, and (3) to publish a journal devoted to experimental and clinical nutrition. Members of the American Institute of Nutrition with publications in the field of clinical nutrition may become members of ASCN.

(Also see AMERICAN INSTITUTE OF NUTRITION [AIN].)

AMIDASE

The enzyme which removes amino groups from amino acids and related compounds.

AMINATION

The taking up of an amine group (NH_2) by an amino acid resulting in the amine form of that acid. Amination is one means by which toxic nitrogenous products such as ammonia (NH_3) are removed from the system and made ready for excretion.

AMINE

A substance derived from an amino acid by (1) the action of certain intestinal bacteria, or (2) enzymatic action within the cells of the body.

(Also see AMINO ACID[S].)

AMINO ACID(S)

These are the structural units of protein. The term "amino" indicates the presence of the NH_2 group—a base—while the term "acid" indicates the presence of the COOH groups or carboxyl group—an acid. Since all amino acids possess this unique chemical feature of containing both an acid and a base, they are capable of both acid and base reactions in the body. Thus, they are said to be amphoteric. There are 22 amino acids, the chemical structures of which are shown in Fig. A-22.

PEPTIDE BONDS. In the formation of protein from amino acids, amino acids are linked to each other by peptide bonds. These peptide bonds are formed by the combination of the HN_2 group of one amino acid, and the COOH groups of another, with the elimination of water. Since all amino acids contain both NH_2 and COOH groups (see Fig. A-22), long chains of amino acids may be formed. These long chains are called peptides or proteins. Therefore, digestion of protein involves breaking protein into individual amino acids by enzymatic hydrolysis of the peptide bonds. Amino acids are then absorbed and distributed by the bloodstream to the cells of the body. Whereas, protein synthesis involves the systematic linking of individual amino acids to form body proteins. The nature of protein is determined by the types of amino acids in the protein, and also by the order in which they are joined.

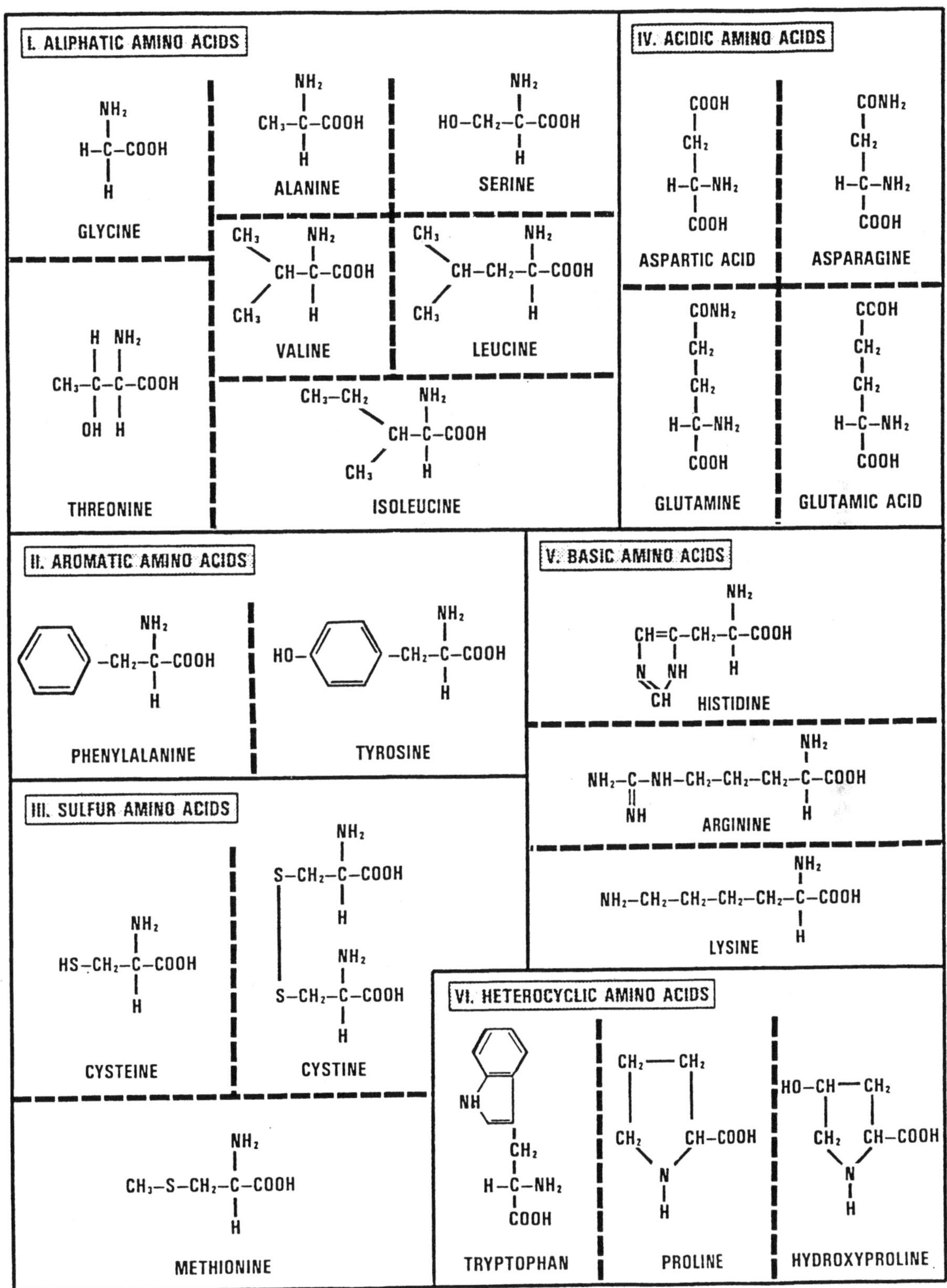

Fig. A-22. Structures of the amino acids.

ESSENTIAL AND NONESSENTIAL. In order for a protein to be synthesized in the body, all of its constituent amino acids must be available. Some of the amino acids can be synthesized within the body. These are called nonessential or dispensable amino acids. If the body cannot synthesize an amino acid from materials normally available, at a speed that will meet the physiological needs of the body for normal growth and development, and it must be supplied in the diet, it is referred to as an essential or indispensable amino acid. The essential and nonessential amino acids are:

Essential (indispensable)	**Nonessential** (dispensable)
Histidine	Alanine
Isoleucine	Arginine
Leucine	Asparagine
Lysine	Aspartic acid
Methionine (some used for the synthesis of cysteine)	Cysteine
Phenylalanine (some used for the synthesis of tyrosine)	Cystine
Threonine	Glutamic acid
Tryptophan	Glutamine
Valine	Glycine
	Hydroxyproline
	Proline
	Serine
	Tyrosine

NOTE WELL: Arginine is not regarded as essential for humans, whereas it is for animals; in contrast to human infants, most young mammals cannot synthesize it in sufficient amounts to meet their needs for growth.

An amino acid is nonessential (dispensable from the diet) if its carbon skeleton can be formed in the body, and if an amino group can be transferred to it from some donor compound available, a process called transamination. Fig. A-23 illustrates the formation of the amino acid alanine by the process of transamination. The body can make alanine readily in any amount needed because the carbon chain is a common metabolic product to which the amino group from another common nonessential amino acid, glutamic acid, can be added.

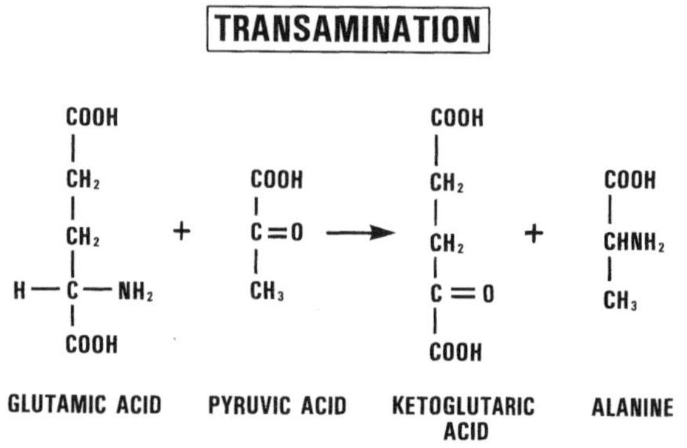

TRANSAMINATION

$$\underset{\text{GLUTAMIC ACID}}{\begin{array}{c}COOH\\ |\\ CH_2\\ |\\ CH_2\\ |\\ H-C-NH_2\\ |\\ COOH\end{array}} + \underset{\text{PYRUVIC ACID}}{\begin{array}{c}COOH\\ |\\ C=O\\ |\\ CH_3\end{array}} \longrightarrow \underset{\text{KETOGLUTARIC ACID}}{\begin{array}{c}COOH\\ |\\ CH_2\\ |\\ CH_2\\ |\\ C=O\\ |\\ COOH\end{array}} + \underset{\text{ALANINE}}{\begin{array}{c}COOH\\ |\\ CHNH_2\\ |\\ CH_3\end{array}}$$

Fig. A-23. The process of transamination.

SPECIFIC FUNCTIONS. Aside from protein formation, some of the amino acids participate in other specific metabolic reactions and demonstrate some interesting characteristics. Hence, a few points deserve mention.

• **Arginine**—This amino acid participates in the formation of the final metabolic product of nitrogen metabolism—urea—in the liver. The process is known as the urea cycle.

• **Cysteine and methionine**—These are the principal sources of sulfur in the diet (see Fig. A-22). Sulfur is necessary for the formation of coenzyme A and taurine in the body. Methionine can convert to cysteine, but not vice versa. Methionine also converts to S-adenosylmethionine an active methylating (CH_3) agent in the formation of compounds such as epinephrine, acetylcholine, and creatine.

• **Glutamic acid**—This amino acid easily loses an amino (NH_2) group and thus participates in the transamination process shown in Fig. A-23. Also glutamic acid is the precursor of gamma-aminobutyric acid, a chemical associated with nervous system function. In passing, it is noteworthy that the sodium salt of glutamic acid, or monosodium glutamate (MSG), is well-known for its flavor enhancing properties.

• **Glycine**—Glycine is the simplest of the amino acids, and it can be derived by the loss of one carbon from serine. Glycine participates in the synthesis of purines, porphyrins, creatine, and glyoxylic acid. Glyoxylic acid is of interest because its oxidation yields oxalic acid of which there is increased formation in the genetic disorder oxaluria. Furthermore, glycine conjugates with a variety of substances thus allowing their excretion in the bile or urine.

• **Histidine**—Histamine is formed by decarboxylation—removal of the COOH groups—of histidine by the enzyme histidine decarboxylase. Histamine is a powerful blood vessel dilator, and it is involved in allergic reactions, and inflammation. Also histamine stimulates the secretion of both pepsin and acid by the stomach.

• **Lysine**—Lysine provides structural components for the synthesis of carnitine which stimulates fatty acid synthesis within the cell. Lysine is apt to be deficient in vegetarian diets.

• **Phenylalanine and tyrosine**—The body can convert phenylalanine to tyrosine, but the reverse reaction does not occur. In normal individuals almost all of the phenylalanine not used in protein synthesis is converted to tyrosine. Tyrosine becomes the parent compound for the manufacture of the hormones norepinephrine and epinephrine by the adrenal medulla (the inner part of the adrenal glands, the small triangular glands lying in front and on top of each of the two kidneys), and of the hormones thyroxine and triiodothyronine by the thyroid gland (which lies in the base of the neck). Also, the pigment melanin which occurs in the skin and retina of the eye forms from the enzymatic conversion of tyrosine. Those individuals not able to convert phenylalanine to tyrosine, due to the hereditary lack of phenlalanine hydroxylase, suffer from the inborn error of metabolism known as phenylketonuria.

• **Tryptophan**—In the body an important neurotransmitter of the brain, serotonin, forms from tryptophan. Serotonin counteracts the effects of epinephrine and norepinephrine, and may possibly improve the duration of sleep. Furthermore, animals with reduced brain levels of serotonin demonstrate behavioral abnormalities, including insomnia. Also, serotonin is a powerful constrictor of blood vessels present in many tissues including blood platelets, and cells of the intestinal mucosa. Following injury, release of serotonin contributes to stopping the blood flow. Some niacin can be manufactured from tryptophan, but not sufficient to meet the total niacin requirements of the body.

REQUIREMENTS. Proteins differ in nutritive value mainly due to their amino acid composition. If one essential amino acid is missing from the diet, a certain protein or proteins will not be formed, and an adult will enter a state of negative nitrogen balance while a child or infant will cease to grow. These two facts provide the basis for experimentally determining amino acid requirements. The estimated essential amino acid requirements shown in Table A-20 may serve as a guide for the selection of dietary protein sources.

TABLE A–20
ESTIMATED AMINO ACID REQUIREMENTS[1]

Amino Acid	Requirement, mg/kg Body Weight/Day			
	Infants	Children		Adults
	(3–4 mo)	(–2 yr)	(10–12 yr)	
Histidine	28	?	?	8–12
Isoleucine	70	31	28	10
Leucine	161	73	42	14
Lysine	103	64	44	12
Methionine plus cystine	58	27	22	13
Phenylalanine plus tyrosine	125	69	22	14
Threonine	87	37	28	7
Tryptophan	17	12.5	3.3	3.5
Valine	93	38	25	10

[1]*Recommended Dietary Allowances*, 10th ed., 1989, NRC–National Academy of Sciences, p. 57.

These requirements are adequate only (1) if the body cells have the ability to synthesize the necessary carbon skeletons (alpha-keto acids) to which the amino nitrogen can be attached; and (2) if the diet provides enough nitrogen for the synthesis of the nonessential amino acids so that essential amino acids will not be used to supply amino groups for the nonessential amino acids.

Even in infants the essential amino acids make up only about 35% of the total need for protein. In adults, essential amino acids account for less than 20% of the total protein requirement. Most proteins contain plenty of dispensable amino acids; usually the concern is to meet the essential amino acid needs, particularly of infants and children.

It should be noted from Table A-20 that on a weight basis infants and children require larger amounts of essential amino acids by virtue of their higher rate of protein synthesis. There is no information on amino acid requirements for pregnancy and lactation.

All proteins are not created equal. Plant foods often contain insufficient quantities of lysine, methionine and cystine, tryptophan, and/or threonine. When a protein source is low in some essential amino acid, that amino acid is termed the limiting amino acid. Lysine is the limiting amino acid of many cereals, while methionine is the limiting amino acid of beans (legumes). In general, the proteins of animal origin—eggs, dairy products, and meats—provide mixtures of amino acids that are well suited for human requirements of maintenance and growth. The egg provides all of the essential amino acids in sufficient quantities and balance to meet the body's requirements without excess. Frequently, the essential amino acid pattern of the egg is used to evaluate the amino acid pattern of other foods by assigning a chemical score or amino acid score. A score of 100 indicates that the food has the same amino acid pattern as eggs, while a score of 60 indicates that the most limiting amino acid is 60% of the amount contained in an egg. Often proteins in the diet which are low in some amino acid can be complemented by the addition of protein from another source. That is, proteins having opposite strengths and weaknesses are mixed. For example, many cereals are low in lysine, but high in methionine and cystine. On the other hand, soybeans, lima beans, and kidney beans, are high in lysine but low in methionine and cystine. When eaten together the deficiencies are corrected. Rather than eating more of a protein low in some amino acid(s) it is much better to supplement with a protein that complements the deficiency. This avoids the possibility of creating an amino acid imbalance which may reduce the utilization of or increase the need for other amino acids. Complementary protein combinations are found in almost all cultures. In the Middle East, bread and cheese are eaten together. Mexicans eat beans and corn, Indians eat wheat and pulses (legumes). Americans eat breakfast cereals with milk. This kind of supplementation works only when the deficient and complementary proteins are ingested together or within a few hours of each other. For those who are interested, the protein and amino acid content for a variety of foods are given in Proteins and Amino Acids in Selected Foods, Table P-37.

(Also see CHEMICAL SCORE; DIGESTION AND ABSORPTION; INBORN ERRORS OF METABOLISM; METABOLISM; NITROGEN BALANCE; PROTEINS; and REFERENCE PROTEIN.)

AMINO ACID ANTAGONISM

Interference with the utilization of certain amino acids by others which are chemically similar.

AMINO ACID IMBALANCE

A condition in which the dietary supply of amino acids is poorly utilized in meeting the body's requirements. An amino acid imbalance usually occurs when the total protein intake is low, and there are excesses of certain amino acids while others are deficient.

(Also see AMINO ACID[S].)

AMINO ACID REFERENCE PATTERN

The amounts of each of the essential amino acids which in combination are believed to meet the body's protein needs in the most efficient manner. Patterns are usually based upon the minimum requirements for infants and young children; or they may be based upon the proportions in which the amino acids occur in such well utilized foods as eggs and human breast milk.

(Also see AMINO ACID[S].)

AMINOACIDURIA

A condition in which abnormally large amounts of amino acids are excreted in the urine. This disorder is usually due to one or more defects in the processes by which the kidneys prevent such urinary loss.

AMINOPEPTIDASE

An enzyme, produced by the intestinal glands, which digests peptides, especially polypeptides, by splitting off the amino acids containing free amino groups.

(Also see DIGESTION AND ABSORPTION.)

AMINOPTERIN

This is a yellow crystalline compound used clinically as an antagonist to folic acid in the treatment of certain leukemias.

AMINO SUGAR

A sugar which contains an amino group. Amino sugars are important constituents of compounds in connective tissues.

(Also see CARBOHYDRATE[S].)

AMMONIA

A potentially toxic alkaline gas which is formed from amino acids or urea by (1) intestinal bacteria, or (2) metabolic activities of cells. Normally, the accumulation of ammonia in the body is prevented by enzymes which convert the ammonia to safer compounds such as urea or amino acids. However, ammonia intoxication is a common occurrence in certain liver diseases.

(Also see DIGESTION AND ABSORPTION; METABOLISM; and PROTEIN[S].)

AMMONIUM

A singly charged positive ion formed during protein metabolism. Under alkaline conditions, the ammonium ion may be converted to ammonia gas; under acid conditions, it forms salts. Also, the ammonium ion is a nontoxic means by which the kidney excretes excess acid.

(Also see METABOLISM; and PROTEIN[S].)

AMPHETAMINES

A group of drugs used as both stimulants and appetite depressants. They are habit forming and create a false sense of well being, which makes it easy for users to increase the dosage and to become dependent on these substances. Also, people with various ailments may suffer severe side effects which are sometimes fatal. Therefore, amphetamines should be used only under a doctor's close supervision.

(Also see APPETITE; and OBESITY.)

AMPHOTERIC

A compound having properties of both an acid and a base and therefore able to function as either. Amino acids and proteins have this dual chemical nature—they contain both an acid group (*carboxyl, COOH*) and a basic group (*amino,* NH_2).

AMYDON

This starch preparation was used for many centuries to thicken liquids. It was made by soaking wheat in water, then sun-drying the liquid, which left the amydon.

AMYLASE

Any one of several enzymes which convert starch to maltose, such as pancreatic amylase (*amylopsin*) and salivary amylase (*ptyalin*).

(Also see DIGESTION AND ABSORPTION.)

AMYLODYSPEPSIA

Inability to digest starch.

AMYLOPECTIN

A complex carbohydrate molecule made up of glucose units linked together in a branched chain. Amylopectin is the major component of starches from corn, rice, and barley in which it often comprises the outer layer of the starch granules. These starches make excellent thickeners for cream soups, gravies, puddings, and white sauces.

(Also see STARCH.)

AMYLOPSIN

A digestive enzyme secreted by the pancreas. It helps to break down (digest) starch to the sugar maltose. Another name for this enzyme is pancreatic amylase.

(Also see DIGESTION AND ABSORPTION.)

AMYLOSE

A complex carbohydrate molecule made up of glucose units linked together in straight chains. It occurs together with amylopectin in many food starches.

(Also see AMYLOPECTIN; CARBOHYDRATE[S]; and STARCH.)

ANABOLISM

A process involving the conversion of simple substances into more complex substances of living cells (constructive metabolism).

(Also see METABOLISM.)

ANACIDITY

A lack of hydrochloric acid in the stomach.

(Also see DIGESTION AND ABSORPTION.)

ANAEROBIC

A type of metabolism which occurs in the absence of oxygen.

(Also see METABOLISM.)

ANALOG

Anything that is analogous or similar to something else. (Also spelled analogue.)

ANALYSIS OF FOODS

Contents *Page*

History 65
Analytical Procedures 66
Chemical Analyses 66
Proximate Analysis 67
Moisture 67
Ash 67
Crude Protein 67
Ether Extract (Fat) 68
Crude Fiber 68
Available Carbohydrate 68
Usefulness of the Proximate Analysis 68
Van Soest Analysis for Fiber 69
Bomb Calorimetry 69
Chromatography 69
Colorimetry and Spectrophotometry 70
Protein and Amino Acid Analyses 71
Biological Analyses 71
Digestibility Trials 72
Nutrient-Deficient Animals 72
Microbiological Assays 72
Microscopic Analyses for Filth and other Adulterants or Contaminants 72
Physical Methods of Analysis 72
Use of Food Composition Data for Planning Menus and Diets 73

Nutritional science owes much of its rapid growth during the 20th century to the continual improvement of biological, chemical, and physical analyses of foods.

Analyses such as the one shown in Fig. A-24 have long provided the data used by human and animal nutritionists in the evaluation of foods and feeds. These analyses are now more important than ever as our leaders strive to find means of feeding an ever-growing world population.

Fig. A-24. In the past, the Kjeldahl procedure shown in the photo was considered to be the most efficient way to measure the protein content of food. In recent years, several other means of protein determination have been developed, most of which are designed for a certain degree of automation.

HISTORY. The French chemist, Lavoisier, is credited with having founded the science of nutrition in the 1770s when he demonstrated that the metabolism of foodborne carbon within the body yielded approximately the same amounts of carbon dioxide and heat as the burning of carbon outside of the body. However, he and other leading scientists of the late 18th century analyzed foods for their hydrogen and carbon contents by combustion. Hence, the early attempts to evaluate the nutritional values of foods was centered on the elemental composition rather than the nutrient composition. Nevertheless, many workers used terms such as albuminous, farinaceous, gelatinous, glutinous, mucilagenous, and oily to describe some of the substances they extracted from foods by various chemical and physical procedures.

Food analysis was advanced significantly by the isolation and purification of certain sugars, fats, and amino acids in the latter part of the 18th century and the early part of the 19th century. During this period, the Swedish chemist Scheele isolated glycerine from a soap (1783), Proust of France crystallized pure grape sugar—glucose (1804), the French chemists Thenard and Gay-Lussac determined the chemical composition of sugar by combustion (1811), Kirchhoff of Russia converted starch into grape sugar (1812), the French chemist Chevreul described many of the components and properties of fats (1811 through the 1820s), and his contemporary the French naturalist Braconnet used acid hydrolysis to convert cellulose-rich plant materials to glucose, and to convert proteins to their constituent amino acids. The growing interest of the era in the composition of foods prompted the German chemist Liebig (who is considered to be the father of organic chemistry) and his co-workers to improve the methods of foods analysis and to suggest the designations of carbohydrate, fat, and protein for the major organic nutrients.

Liebig's prominence as the world's leading agricultural, food, and physiological chemist in the 1840s helped to establish the priority of this type of chemical research for the newly established agricultural experiment stations of Scotland, England, and Germany, and encouraged a movement to establish similar research centers in the United States. Therefore, it is noteworthy that starting in the late 1850s, Henneberg and Stohmann of the Weende Agricultural Experiment Station in Germany developed the system of proximate analysis which is still used in the analysis of foods.

The latter part of the 19th century saw the development and improvement of various types of calorimeters as nutritionists and physiologists strove to fill the gaps in knowledge that resulted from a dependence upon proximate analysis. In 1866, the English chemist Frankland measured the heat content of various foods that he burned in a crude calorimeter. Studies of the caloric expenditure of people and animals had been measured before that time, but little had been done to determine the energy content of foods. Later (1884), the German physiologist Rubner utilized data from dietary studies of the renowned German scientist Voit (Liebig was the mentor of Voit, who in turn taught Rubner) and calculated the heat of combustion of carbohydrates, fats, and proteins. He corrected for the incomplete metabolism of protein in the body by subtracting the heat of combustion of the urinary product urea. Then, in the 1890s, the American nutritionist, Atwater, who had worked for a while with Voit and Rubner, refined the caloric conversion factors of Rubner by correcting them for the percent losses that occurred in digestion. This required many feeding trials with people and animals in which both the foods consumed and the excretory products were analyzed.

Another notable development of the latter part of the 19th century was the establishment in the United States of what was then called the Official Methods of Food Analysis, which were those that had been adopted by the fledgling American Association of Official Agricultural Chemists (now known as the Association of Official Analytical Chemists or simply the AOAC). Atwater and other leading agricultural chemists were on the AOAC committee which screened the various analytical methods that were then utilized. The essential principles of the Weende system of proximate analysis were adopted by the committee, but the procedures were modified. After reviewing the accumulated data of almost 100 years of food analyses, the committee came to the conclusion that the official methods of analysis left much to be desired and that a considerable amount of biological and chemical research was needed to narrow the gap between the data obtained from laboratory analysis and the results observed in feeding trials.

The 20th century ushered in the era of extensive experimentation with purified diets that were tested on people and animals. These procedures led to the discovery of many micronutrients such as essential amino acids, minerals, and vitamins. For example, American farmers had long maintained that yellow corn was nutritionally superior to white corn, but they were rebuked by the agricultural chemists who maintained that proximate analysis showed that the two types of corn were equal in value. Eventually, the farmers had the last laugh when it was found that yellow corn contained much more carotene (vitamin A) than white corn. New methods of biological, chemical, and physical analysis were developed as the nutrition experiments suggested the existence of unknown essential factors.

Today, many good methods of food analysis are available, but the costs of equipment, labor, and supplies limit the use of certain ones to well funded research projects. Hence, the current trend is towards the development of procedures that may be automated and carried out on many samples at a time. Similarly, bioassays that utilize animals are very expensive and are gradually being replaced by those employing microorganisms whenever it is feasible to do so. Yet, much needs to be done to fill the present gaps in our knowledge regarding the composition of our common foods and of those foods consumed by people in the developing countries.

ANALYTICAL PROCEDURES. Foods are analyzed by chemical, biological, and/or physical procedures. Physical procedures include both the older types of examination, which were mainly visual and organoleptic, and the newer types of nondestructive testing that utilize such means as electrical conductance or resistance, light reflectance, and ultrasonics. Each type of procedure has its advantages and disadvantages.

Chemical Analyses. Today, many foods are being analyzed routinely by highly sophisticated chemical procedures. The laboratories in government agencies, universities, and large food manufacturing concerns are often equipped to perform these analyses when needed to determine nutrient compositions in order to (1) enforce or comply with the laws on nutrient labelling; or (2) provide data for nutritional research.

PROXIMATE ANALYSIS. This group of chemical analyses, which has been utilized for over 100 years, still forms the basis for much of our nutrient composition data. A summary of the analytical scheme is given in Table A-21.

TABLE A-21
THE FRACTIONS OF PROXIMATE ANALYSIS

Fraction	Major Components	Procedure[1]
Moisture (dry matter by difference)	Water and any volatile compounds (100% − H_2O = % dry matter).	Heat chopped or pulverized sample to constant weight at a temperature just above boiling point of water. Loss in weight equals water.
Ash (mineral matter)	Mineral elements.	Burn at 930° to 1,110°F *(500° to 600°C)* for 2 hr.
Crude protein (protein averages 16% N; hence, N x 6.25 = crude protein)	Proteins, amino acids, nonprotein nitrogen.	Determine nitrogen by Kjeldahl sulphuric acid digestion.
Ether extract (fat)	Fats, oils, waxes, fat-soluble vitamins, coloring matter.	Extraction with ether.
Crude fiber[2]	Cellulose, hemicellulose, lignin.	Organic portion[3] of the residue after boiling a dried, defatted sample in weak acid and weak alkali.
Available carbohydrate Also known as nitrogen-free extract (NFE)	Starch, sugars, some cellulose, hemicellulose, pectins, and lignin.	Remainder; i.e. 100 minus sum of the other fractions.

[1]Each procedure is applied to a separate, chopped or pulverized, sample of standard weight of the food to be analyzed.
[2]Total carbohydrate = crude fiber + available carbohydrate (NFE).
[3]Organic portion is measured by the loss in weight of the residue after ashing.

Details of the procedures used in proximate analysis follow.

Moisture. Certain food products that are sold by weight may vary considerably in their water contents. Furthermore, the presence of extra water in foods lowers the levels of the other components of the foods. Hence, the U.S. Food and Drug Administration (FDA) and the U.S. Department of Agriculture (USDA) have established standards of identity of some of these items in order to promote fair trade practices.

The determination of moisture in foods requires the use of different techniques. The appropriate method is selected by the analyst based on such characteristics as: (1) presence of volatile matter, (2) possibility of browning of some ingredients, (3) need for low temperatures and vacuum, and (4) the presence of some compounds which might be chemically altered during drying—for example, sugars.

Moisture is now being determined in five ways:

1. **Oven drying.** According to the Association of Official Analytical Chemists (AOAC), samples are heated at 275°F (*135°C*) to a constant weight. The loss in weight represents the amount of water contained in the food. This method is not always suitable for the determination of water content, because the short chain fatty acids and organic acids in certain foods are volatilized and lost in addition to the evaporation of water.
2. **Vacuum drying.** The boiling point of water is lowered in a vacuum. Therefore, samples may be dried at lower temperatures in vacuum ovens so that the losses of compounds other than water are minimized.
3. **Distillation with toluene.** The dry matter of samples having large quantities of volatile acids and bases can be determined by distillation with toluene.
4. **Freeze-drying.** The freeze drier basically consists of heated shelves surrounded by a refrigerated condenser. Samples are frozen prior to freeze-drying. After the frozen sample is placed in the freeze drier, the apparatus is evacuated to an extremely low pressure. Under these conditions, the frozen water crystals within the sample pass directly into a vapor phase without first becoming a liquid. This technique prevents the loss of many of the volatile organic compounds in the sample.
5. **Rapid moisture testing equipment.** Quite often, a processor needs to know the moisture content of a food immediately in order to avoid wasting fuel in operations such as drying. One type of rapid moisture analyzer is an apparatus that comes in two parts—a drier and a small scale. The processor weighs out a portion of the food as specified by the operating directions. The drier consists of a heating element and a fan. The sample is then placed on a screen which allows hot air to pass through the feed sample and dry it until a constant weight is attained (generally about 5 minutes). After drying, the sample is removed from the drier and immediately weighed to determine the loss of water. Although this method of determining moisture content is not extremely accurate, it does permit the processor to get a rapid and reasonably good indication of the dry matter content of the food. A second type of moisture analyzer is an apparatus that measures the conductance or resistance of the food, and the value derived therefrom is compared to a calibration chart. When this type of apparatus is used, it should be periodically checked and calibrated against a standard of which the exact moisture content is known.

Ash. The ash fraction of the proximate analysis represents most of the mineral constituents of the food. Some of the minerals are volatilized and lost from the sample at the high temperature of ashing. Samples in porcelain crucibles are placed in a muffle furnace and ignited at temperatures in excess of 1,100°F (*600°C*). The residue that is left after burning is termed ash. In plants, mineral composition is highly variable since soil conditions determine the makeup of this fraction.

Crude Protein. Protein, on the average, contains about

16% nitrogen. Theoretically, if we know the nitrogen content of a food, we can obtain an estimate of the amount of protein that it contains. When we know the amount of nitrogen, we can estimate crude protein by multiplying the nitrogen content of the food by 6.25 $\left[\frac{100\%}{16\%}\right]$.

The commonly used procedure for determining the nitrogen content of foods is called the "Kjeldahl determination." The organic matter of the sample is destroyed by digestion with sulfuric acid. The nitrogen is then converted to an ammonia compound which is quantitatively released as ammonia during alkaline distillation. The precise amount of ammonia is titrated against a standard solution; and the value obtained is converted to nitrogen content, and finally to protein. A recent development in the use of this procedure has been the automation of the analysis of the digested sample. Fig. A-25 shows a typical automated version of this analysis.

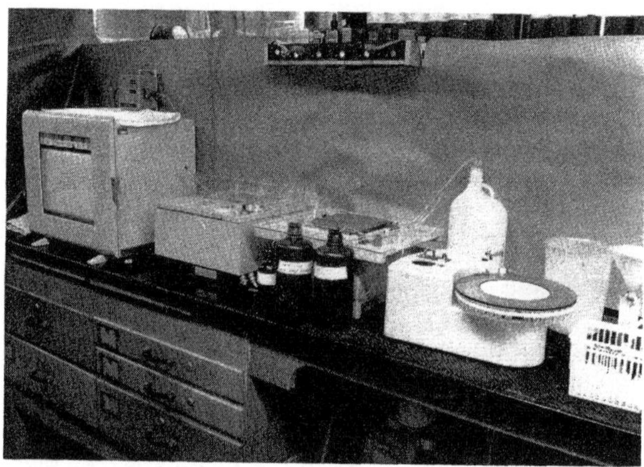

Fig. A-25. An automated analysis of digested samples for percent nitrogen or crude protein. (Courtesy, C.W. Gehrke, University of Missouri)

It should be remembered that the figure derived from the analysis represents only an approximation of the protein content. This procedure involves the following two basic assumptions which are generally applicable to foods:

1. **Proteins contain approximately 16% nitrogen.** In reality, this merely represents an average. Some foods contain proteins that average more than 16% nitrogen, whereas others contain proteins with lower contents of nitrogen.
2. **All nitrogen is in the form of proteins.** For some foods, this may be true. But there are a number of other compounds in certain foods that contain nitrogen; for example, yeasts contain large amounts of nitrogen in nucleic acids.

Ether Extract (Fat). Many people refer to this fraction as the fat content of the sample. This tends to be an oversimplification as the ether extract also contains organic acids, oils, pigments, alcohols, and the fat-soluble vitamins. Many of the complex lipids, such as phospholipids, are not completely extracted by this procedure.

The procedure is exactly what the name implies. The sample is continuously extracted with ether, using a specially designed apparatus. After the extraction is completed, the ether solvent is evaporated and the residue that remains constitutes the ether extract.

Crude Fiber. This fraction is an approximate measure of the undigestible and unabsorbable carbohydrates in foods. Recently, there has been considerable interest in the fiber content of foods, because it is thought to exert protective actions against the development of disorders such as cancer of the colon, diabetes mellitus, diverticular disease, gallstone formation, and ischemic (decreased blood flow) heart disease. Therefore, many foods are now recommended on the basis of their crude fiber content, which is often stated on the label.

To determine crude fiber by proximate analysis, the food sample is first boiled in dilute acid and then in dilute alkali to simulate the digestive action of gastric and pancreatic secretions. The residue of the sample that remains undigested after these boiling procedures is then weighed and ashed. The difference between the weight of the initial residue and that of the ash indicates the amount of fiber present in the sample.

The crude fiber fraction consists of cellulose, hemicellulose, and lignin. However, the digestion procedures remove from 20 to 50% of the cellulose, about 80% of the hemicellulose, and from 50 to 90% of the lignin.[12] Hence, the traditional analytical procedure fails to account for a large part of the true fiber content. In recent years, a system developed by Van Soest, of Cornell, for evaluating digestible fractions of feeds and foods has been receiving considerable attention. This procedure will be discussed later in this article.

Available Carbohydrate. This fraction, which in agricultural chemistry is known as the nitrogen-free extract (NFE), is calculated by the following formula: NFE = 100 − (% moisture + % crude fiber + % ash + % ether extract + % crude protein).

This fraction, represents a catchall for the organic material for which there is no specific analysis. The vast majority of components in this fraction are carbohydrates (starch, fructans, pectins, cellulose, hemicellulose, and lignin), but other substances, such as pigments, organic acids, and water-soluble vitamins, are also present.

Usefulness of the Proximate Analysis. As with all analytical techniques, there are advantages and disadvantages in the use of the proximate analysis for the evaluation of foods.

The *advantages* of the system should not be minimized. They are:

1. **Most laboratories are equipped to run this type of analysis.** Expensive and sophisticated equipment is not needed.
2. **It provides a good general evaluation of the food.** A food that is high in crude fiber will probably be inferior in nutritional value to one that is very low in crude fiber.

[12]Robertson, J. B., "The Detergent System of Fiber Analysis," *Topics in Dietary Fiber Research*, Plenum Press, 1978, p. 6.

Likewise, a food with a high percentage of ether extract is likely to be high in calories.

3. **Much of the data available on food composition is reported in terms of proximate analysis.**

Some of the *disadvantages* of the proximate analysis are:

1. **The system does not define the individual nutrients of the food.** Rather, the fractions represent mixtures of the various nutrients.

2. **It is not accurate.** Crude protein, crude fiber, and available carbohydrates are rough estimates of their respective fractions.

3. **The procedure is time-consuming.** There is little possibility for automation in the proximate analysis. Many of the fractionations involve several weighings of samples and other procedures which must be done by the laboratory technician.

4. **It does not tell how much indigestible material there is in a food.** Unfortunately, the acid-alkali treatment dissolves much of the crude fiber of plant products, making it impossible to predict accurately how much indigestible matter there is in the food. The method overestimates the nutritive value of some foods, underestimates that of others, and fails to indicate how the constituents of the indigestible residue are related to each other or the function(s) some of them perform in digestion.

5. **It does not go far enough.** Proximate analysis does not provide any information relative to palatability, texture, toxicity, digestive disturbances, or nutritional availability. Thus, further steps need to be taken to evaluate a food.

It is likely that the use of proximate analysis will decline in the future as new techniques are utilized, but, for the present, it offers a means of estimating the water, ash, protein, fat, fiber, and carbohydrate contents of food.

VAN SOEST ANALYIS FOR FIBER. This procedure separates and classifies the digestible and undigestible parts of plant cells.

The food sample is initially boiled in a neutral detergent solution to separate the neutral detergent soluble fraction (cell contents) and the neutral detergent insoluble fraction (cell walls). The cell contents are highly digestible (about 98%) and include various sugars, starches, pectins, proteins, lipids, nitrogenous compounds, soluble carbohydrates, and water-soluble minerals and vitamins. Cell walls can be further fractionated by boiling in an acid detergent solution. Hemicellulose is solubilized during this procedure, while the lignocellulose fraction of the food remains insoluble. Cellulose is then separated from lignin by the addition of sulfuric acid. Only lignin and acid-insoluble ash remain upon the completion of this step. This residue is then ashed, and the difference of the weights before and after ashing yields the amount of lignin present in the sample.

BOMB CALORIMETRY. When compounds are burned completely in the presence of oxygen, the resulting heat is referred to as gross energy or the heat of combustion. The bomb calorimeter is used to determine the gross energy of foods, waste products from the body (for example, feces and urine), and tissues.

The unit of measurement of food energy is the kilogram-calorie (commonly designated as the Calorie or kilocalorie) which is defined *as the amount of heat required* to raise the temperature of 2.2 lb (*1 kg*) of water 1 °C (precisely from 14.5° to 15.5°C). With this fact in mind, we can readily see how the bomb calorimeter works.

Briefly stated, the procedure is as follows: An electric fuse wire is attached to the sample being tested, so that it can be ignited by remote control; 2,000 g of water are poured around the bomb; 25 to 30 atmospheres of oxygen are added to the bomb; the material is ignited; the heat given off from the burned material warms the water; and a thermometer registers the change in temperature of the water.

However, some of the heat liberated from the food is absorbed by the metal parts of the calorimeter and is not available for heating the water. This error is corrected for by burning a sample of a benzoic acid standard which will emit a known amount of heat. Other corrections are made for the burning of the fuse wire and the acids produced during the combustion.

CHROMATOGRAPHY. In 1903, Tswett, a Russian botanist, first described his attempts to separate colored substances; hence, the origin of the term chromatography. Today, many of the compounds that are separated and identified by chromatographic techniques are colorless; but new refinements in these techniques enable the food analyst to measure extremely minute amounts of many compounds.

Chromatography separates compounds through the use of two phases—a stationary or fixed phase and a mobile phase. The differences between the various chromatographic techniques lie in the variation of the materials used in these phases. The stationary phase can be either solid or liquid material, while the mobile phase or "carrier" can be either gas or liquid in nature. The various types of chromatography are listed in Table A-22.

TABLE A-22
TYPES OF CHROMATOGRAPHY

Stationary Phase	Mobile Phase	Chromatographic Procedure
Solid	Liquid	Thin-layer chromatography. Ion exchange chromatography. Gel chromatography. Absorption-column chromatography.
Liquid	Gas	Gas-liquid chromatography. Capillary-column chromatography.
Solid	Gas	Gas-solid chromatography.
Liquid	Liquid	Paper chromatography. Partition chromatography. Zone electrophoresis.

Numerous materials from foods, such as proteins, amino acids, sugars, fatty acids, minerals, and many other components, are routinely identified and measured by this type of analytical procedure. In addition to nutrient analysis, chromatography can be adapted to the detec-

Fig. A-26. Preparation of thin-layer chromatographic plates for food analysis. (Courtesy, University of Illinois at Urbana-Champaign)

tion of drug residues, hormones, pesticides, and other food contaminants.

The *advantages* of chromatography are numerous; among them, the following:

1. **Extreme sensitivity.** Many compounds can be detected and measured when present in amounts as small as parts per billion (mcg/kg of sample).
2. **Inexpensive.** Many chromatographic techniques can be adapted by almost any laboratory. Relatively little equipment is needed for many chromatographic procedures.
3. **Rapid.** In techniques such as thin-layer chromatography, a large number of samples can be done simultaneously in a relatively short period of time.
4. **Readily adaptable.** Chromatography can be adapted readily to almost any chemical compound.

On the other hand, there are several *disadvantages*, such as:

1. **Complexity of operation.** Many of the newer techniques involve sophisticated equipment which require the operator to be familiar with the theory and the "art" of chromatography.
2. **Sample preparation.** Most samples require some sort of preparation before they can be chromatographed. This may involve such procedures as extraction, hydrolysis, and/or evaporation.
3. **Sample size.** Samples must be small in order to be chromatographed. This means that the sampling procedure must be carefully planned if the results are to be valid.

(Also see CHROMATOGRAPHY.)

COLORIMETRY AND SPECTROPHOTOMETRY. These procedures are chemical analyses in which light is passed through solutions to yield information about the concentration of certain compounds. A particular wavelength of light is passed through the samples, and the amount of light absorbed by the sample gives an indication of the concentration of the compound being tested. Colorimetry differs from spectrophotometry in that colorimetry is useful for measuring wavelengths in the visible region of the light spectrum whereas spectrophotometry utilizes wave lengths in the ultraviolet, visible, and infrared regions of the spectrum.

Many nutrients are analyzed by chemical procedures which involve either colorimetry or spectrophotometry. Vitamin A analysis is a good example of a colorimetric procedure. The standard assay for vitamin A determination involves the treatment of the sample with antimony trichloride. A deep blue-colored solution is produced, the intensity of which is dependent on the amount of vitamin A in the sample. The solution of unknown concentration is measured in the colorimeter and compared to a series of standards of known concentrations. Spectrophotometric assays are essentially the same as colorimetric assays except the researcher has a more versatile machine with which to work.

The atomic absorption spectrophotometer is one of the most widely used instruments for mineral analysis, having the ability to detect many minerals at concentrations less than 1 part per billion (1 mcg/kg of sample). In addition to its high sensitivity, this machine is readily adaptable to automation, thus presenting the chemist with a rapid, accurate method of food analysis. The atomic absorption spectophotometer works on a slightly different principle than the regular spectrophotometer. The main principle behind this machine is that when certain compounds (for example, minerals) are volatilized, they emit light of a characteristic wavelength. The machine is calibrated to detect this light.

Mass spectrometry is one of the most recently developed procedures for the analysis of foods. Fig. A-27 shows the instruments used in this type of analysis.

Fig. A-27. A mass spectrometer and its accessories being used to make food analyses. (Courtesy, University of Illinois at Urbana-Champaign)

In this procedure, the substance to be analyzed in the spectrometer is usually subjected to various preliminary treatments such as solvent extraction and gas-liquid chromatography in order to isolate the constituents for identification and measurement. Then, the sample is introduced into the spectrometer, where it is ionized by bombardment with electrons, or by other means such as chemical ionization. The ions are passed through electrical and magnetic fields which separate them into a spectrum of light to heavy masses. Electrical instruments then detect and record the ion masses, which are literally "fingerprints" of the substances analyzed.

Mass spectrometry offers a means of certain identification of substances whose identity might otherwise be uncertain when assayed by other analytical procedures. (Many of the procedures of colorimetry, spectrophotometry, and gas-liquid chromatography identify only general classes of compounds.)

PROTEIN AND AMINO ACID ANALYSES. In the past, the Kjeldahl procedure was considered to be the most efficient way to measure the protein content in food. In recent years, several other means of protein determination have been developed.

When colorimetric techniques for the determination of total protein nitrogen are used, a certain degree of automation can be designed, thereby offering distinct advantages over the Kjeldahl procedure. (See Fig. A-24.) However, the various procedures which measure the total nitrogen content give only a rough indication of the total protein content because many foods contain nonprotein nitrogen compounds. Therefore, nutritionists and biochemists have developed techniques that yield estimates of the total true protein. A few of these procedures follow:

1. **Biuret assay.** This calorimetric assay compares light absorbence of the unknown protein solution to the absorbences of standard protein concentrations. A biuret reagent is reacted with the protein to produce a colored solution.
2. **Lowry assay.** This colorimetric assay is based on the presence of tyrosine and tryptophan in protein.
3. **Turbidity measurements.** This assay utilizes turbidity measurements after precipitation has been accomplished in a controlled manner.
4. **Peptide bond method.** This spectrophotometric assay is based on the fact that peptide bonds absorb light in the 195 to 225 nanometer region of the spectrum.
5. **Warburg-Christian assay.** This spectrophotometric assay utilizes ultraviolet absorption analysis after all nonprotein material has been removed by fractionation or dialysis.

In planning certain diets, the amino acid composition of the foods must be considered. While most procedures for determining the amino acid patterns of foods are automated to a considerable degree, much time and labor is still involved in sample preparation. For example, the protein within the food must be completely hydrolyzed into its constituent amino acids. In many cases 6N HC1 is used. Unfortunately, this and many other hydrolysis procedures make for the following problems:

1. Some amino acids are destroyed in the process.
2. Some amino acids are chemically altered during hydrolysis.
3. Some of the peptides remain incompletely hydrolyzed, thereby tying up amino acids which should be measured. Following the hydrolysis of the protein, the resulting amino acids are commonly separated by ion-exchange chromatography and subsequently measured by colorimetric techniques or by gas-liquid chromatography.

This procedure may be simplified somewhat if the levels of only the few limiting amino acid(s) in the foods have to be measured. Then, the sample is hydrolyzed and portions of the hydrolyzate are analyzed by procedures that are specific for the amino acids sought. For example, it is most useful to analyze grains for lysine, which is almost always the limiting amino acid in this type of food. Fig. A-28 shows a typical analysis.

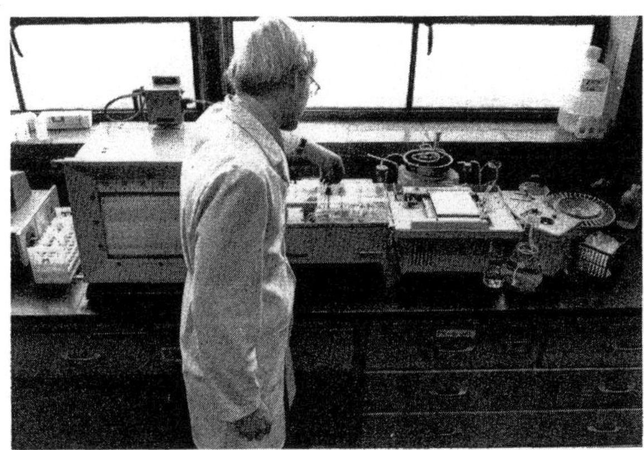

Fig. A-28. Automated analysis of lysine in grain hydrolysates. (Courtesy, C. W. Gehrke, University of Missouri)

Biological Analyses. Chemical analyses do not always provide information about the biological availability of nutrients. For example, the amino acid analysis of hair indicates that it contains sufficient quantities of certain essential amino acids to make it a high-quality protein. Yet, it is not digestible by people, pigs, or poultry. Similar circumstances may limit the utilization of other nutrients such as minerals and vitamins, which may be present in poorly utilized forms in certain foods. Therefore, it is sometimes necessary to utilize biological analyses to fill the gaps in the information provided by chemical analyses.

Biological assays tend to be laborious and time-consuming. Large numbers of samples are needed to produce statistically reliable results, and quite often data obtained from these assays is highly variable. For example, some people do not digest cellulose at all, while others apparently harbor microorganisms in their colon which convert some of it into other substances. In the latter cases, only 60 to 80% of an oral dose of cellulose is recovered in the feces. Hence, digestibility data provides only a crude approximation of the percent absorption of certain complex carbohydrates. The assay utilizing nutrient-deficient animals—rats, mice, guinea pigs, pigeons, or chicks—is particularly cumbersome because (1) the animals should be of approximately the same age, sex, and weight, and (2) time is required to induce deficient conditions in these animals.

(Also see VITAMIN[S], section headed "Biological [Animal] Assay.")

DIGESTIBILITY TRIALS. A digestibility trial is made by determining the percentage of each nutrient in the food through chemical analysis; giving the food to human subjects or test animals for a preliminary period, so that all residues of former foods will pass out of the digestive tract; giving weighed amounts of the food during the test periods; collecting, weighing, and analyzing the feces; determining the difference between the amount of the nutrient fed and the amount found in the feces; and computing the percentages of each nutrient digested. The latter figures are known as the *coefficients of digestibility* for those nutrients.

NUTRIENT-DEFICIENT ANIMALS. In this assay, animals, such as the rat or the chick, are fed diets deficient in a specific nutrient. Growth response curves are developed by the feeding of known amounts of the nutrient to some of the deficient animals. Other deficient animals are given the product to be assayed, and their responses are compared to the growth curves. In addition, the evaluator can observe changes in specific tissues as various levels of the specific nutrients are supplied.

An excellent example of this type of assay is the bioassay of vitamin A. Young female rats are initially fed vitamin A-deficient diets to deplete their reserves. The feed to be assayed and graded levels of known concentrations of vitamin A are fed to the vitamin A-deficient rats. Three parameters are then measured to determine the vitamin A content of the feed:

1. Growth response.
2. Concentration of vitamin A in the liver of the rats from the various treatments.
3. Examination of the degree of cornification of the vaginal epithelium. In a vitamin A deficiency, the lining of the vagina undergoes cornification.

Another example of this type of assay is the bioassay of vitamin D. Rats and chicks are used as test animals; rats respond equally well to D_2 and D_3, whereas chicks respond only to D_3. The assays measure the alleviation (curative test) or the development (prophylactic test) of vitamin D deficiency in terms of the degree of rickets produced.

A bioassay method, known as the *line test*, uses stained longitudinal sections of the distal end of radius bones to evaluate calcification. Usually the test animal is the rat, although the chick must be used if the vitamin D activity of a sample intended for poultry nutrition is to be determined. Young rats from mothers having a deficient supply of vitamin D are kept on a rachitogenic diet so that no calcification occurs in the ends of the long bones. When a test material is fed to these vitamin D-deficient rats, its value as a source of vitamin D is measured by the amount that must be fed for 7 to 10 days to produce a good calcium line (line test) in the ends of the long bones. Standard cod-liver oil is fed to a similar group of animals and is used as a basis of comparison.

(Also see VITAMIN[S], section headed "Biological [Animal] Assay"; VITAMIN D, section headed "Measurement/Assay.")

MICROBIOLOGICAL ASSAYS. In this type of assay, a micoorganism is selected that is known to require the nutrient in question. Therefore, if the nutrient is unavailable, the selected microorganism will not grow. The growth medium is prepared so that it is nutritionally complete except for the nutrient to be tested. Graded levels of the nutrient are then added to the media and a growth response curve is prepared. The sample to be assayed can then be tested and compared to the growth response curve to determine the concentration of the nutrient. Many of the micronutrients, such as the B complex vitamins, are assayed in this manner.

(Also see VITAMIN[S], section headed "Microbiological Assay.")

Microscopic Analyses for Filth and Other Adulterants or Contaminants. "Filth" in foods is considered by the U.S. Food and Drug Administration to consist mainly of insect fragments, and rodent feces and hairs, which are present at higher levels than are consistent with good manufacturing practices. Therefore, special analyses have been developed for the detection of these extraneous materials.

The basic procedure for the isolation of insect and rodent filth from foods consists of (1) boiling a sample of the food in dilute hydrochloric acid for 10 minutes, (2) cooling, (3) neutralizing the acid with sodium hydroxide solution and a phosphate buffer solution, (4) adding a pancreatin suspension (a source of digestive enzymes), and (5) incubating the resulting mixture overnight at 99°F (37°C). Then, the digested mixture is mixed gently with kerosene (in a separatory funnel), which is allowed to separate out from the water and form an upper layer that contains the filth. The water layer is run out through the bottom of the funnel, and the kerosene layer is filtered through a ruled filter paper disk in a Buchner funnel so that the filth is isolated on the filter paper. Then, the filter paper is examined under a microscope and the amount of filth is estimated by counting the particles between the lines on the paper. The basic procedure has to be modified for certain types of foods. For example, fatty foods are usually defatted with a solvent prior to the isolation of the filth.

Microscopic analyses are also used to detect adulterants and/or contaminants such as husks, particles of shell, dirt, and sand in dried herbs and spices. Usually, the herbs or spices are ground finely, and placed in a microscope slide with a few drops of water and glycerine. It is usually necessary for the analyst conducting the microscopic examination to refer to authentic samples, photographs, and/or drawings of the herbs or spices and of the most commonly occurring extraneous materials in order to be certain of making correct judgments about the presence or absence of adulteration or contamination. The presence of large amounts of dirt and sand may be confirmed by ashing the sample and examining the ash.

Physical Methods of Analyses. The great expenditures of time and money that are required for making chemical analyses has spurred a search for more rapid and less expensive means of analysis. Certain physical methods of analysis appear to be the most promising in this regard. For example, it was related earlier in this article (in section headed "Moisture") that water content may be determined by measuring the electrical conductance or resistance of food samples. Descriptions of some other promising procedures follow.

• **Infrared Reflectance Spectroscopy (IRS)**—Collaborative research conducted jointly by the U.S. Department of Agriculture and researchers at Pennsylvania State University has produced a new rapid procedure for the analysis of nutrients in foods and feeds.[13]

In this procedure, ground dried samples of feeds or foods are placed in a special lead-sulfide infrared detector and the reflection of infrared radiation is measured at various wavelengths. The values obtained are then used to calculate the concentrations of constituents such as residual moisture, crude protein, fiber, and fat. It is noteworthy that the calculation requires the use of a special computer program that corrects for the interactions of the infrared reflectance spectrums of the various constituents.

• **Ultrasonics**—These special sound waves have been used to measure the thickness of fat in livestock prior to slaughter. They have also been used on pregnant women to determine certain characteristics of the fetus. It seems likely that we will soon be hearing more about this type of testing since ultrasonic waves are a useful means of harmlessly determining the densities of different material (and therefore the bone, fat, muscle, and water in human and animal tissues).

USE OF FOOD COMPOSITION DATA FOR PLANNING MENUS AND DIETS. Most of the procedures used by dietitians and home economists in their planning of meals and modified diets are based, at least in part, on the nutrient data obtained by food analysis. Basic food exchange lists (based upon the values for protein, carbohydrate, fat, and food energy) have been developed by a joint committee of the American Dietetic Association, the American Diabetes Association, and the United States Public Health Service. These lists give portion sizes of foods within each of several categories of foods (milk, vegetables, fruits, bread, cereals, meats, and fats). The portion sizes have been calculated so as to be approximately equal in nutritive value (with respect to the major nutrients, but not in regard to mineral and vitamin contents). For example, one slice (3" x 2" x 1/8") of meat or poultry is approximately equivalent to one all-meat frankfurter (when there are 8 or 9 per pound).

An intelligent lay person, however, may refer directly to a table of food compositions to obtain information for selecting the foods most appropriate for his individual preferences and health needs. Examples of such use of a food composition table are given under the items which follow:

• **Example #1, menus high and low in fat content**—A typical American dinner menu might consist of the following items: Portions of approximately 100 g (3½ oz) each of tossed salad made with raw vegetables and salad dressing (1 Tbsp), meat with condiment, cooked starchy vegetable (with 2 pats of butter, margarine, or equivalent sauce), cooked green or yellow vegetable (plus butter, margarine, or sauce), dessert (cake, cookies, ice cream, gelatin dessert, pie, or pudding); beverage(s) (alcoholic, carbonated, chocolate, cocoa, coffee, fruit drink or juice, milk, and/or tea); and two thick slices of bread or its equivalent in biscuits, crackers, muffins, and/or rolls (plus 2 pats of butter, margarine, or other spread).

The problem consists of staying within the pattern outlined above and choosing food items in order to make up low-fat and high-fat menus.

• **Procedure**—Select low- or high-fat items from Food Composition Table F-36 in the following sequence:

1. Major courses of salad, meat, vegetables, dessert.
2. Dressings, spreads and sauces for the major courses.
3. Beverages to complement major courses.
4. Breads and appropriate spreads.

When serving size of an item differs from 100 g, convert the values given in the table to the appropriate values by using the following formula:

$$\text{Nutrient content of serving} = \text{nutrient content of 100 g} \times \frac{\text{serving size (g)}}{\text{100 g}}$$

Table A-23 shows sample low-fat and high-fat menus selected to fit a typical American dinner pattern.

The high-fat dinner menu shown in Table A-23 contains 1,855 Calories (Kcal), which is close to the Recommended Daily Dietary Allowance[14] for energy for adult females 51+ years of age (1,900 Kcal), while the low-fat menu contains a little more than half of the recommended allowance (see Table A-4). Furthermore, the latter menu is only moderately low in fat and the percent of energy contributed by fat is in excess of the recent recommendations of nutritionists who wish to prevent heart disease (they advise a restriction of dietary fat to not more than 30% of the total dietary energy).

• **Example #2, substitution of mixed dishes for individual food items**—Perhaps a person who plans menus might wish to substitute a double serving of meat balls and spaghetti for single servings of sliced beef (from a pot roast), mashed potatoes, and gravy. The problem is to determine the effects of this substitution on the nutritive value of the meal.

• **Procedure**—Look up the composition for each of the foods under consideration (see Food Composition Table

[13]Shenk, J. S., *et al.*, "Application of Infrared Reflectance Analysis to Feedstuff Evaluation," *Proceedings, First International Symposium Feed Composition, Animal Nutrient Requirements, and Computerization of Diets*, International Feedstuffs Institute, Logan, Utah, 1977, pp. 242-248.

[14]*Recommended Dietary Allowances*, 10th ed., NRC—National Academy of Sciences, 1989.

TABLE A-23
LOW-FAT AND HIGH-FAT MENUS SELECTED ACCORDING TO A TYPICAL AMERICAN PATTERN OF FOODS FOR DINNER

Type of Food	Low-Fat Menu Food Item	Fat	Energy	High-Fat Menu Food Item	Fat	Energy
	(100 g or specified amount)	(g)	(kcal)	*(100 g* or specified amount)	(g)	(kcal)
Tossed salad (raw vegetables)	Endive (2/3) and cucumber slices (1/3)	.1	12	Lettuce (2/3) and tomato (1/3)	.3	19
Salad dressing	French, 1 Tbsp *(15g)*	5.9	62	Italian, 1 Tbsp *(15 g)*	9.0	78
Meat	Roast chicken, light meat	3.4	166	Porterhouse steak	42.2	465
Relish or sauce for meat or fish	Cranberry sauce, 2 Tbsp *(30 g)*	.1	44	Barbecue sauce, 1 Tbsp *(15 g)*	1.0	14
Starchy vegetable	Boiled potato	.1	65	Baked potato	.1	93
Spread or sauce for starchy vegetable	Meat gravy, 1 Tbsp *(18 g)*	3.5	41	Butter, 2 pats *(14 g)*	12.1	100
Green or yellow vegetable	Cooked carrots	.2	31	Kale	.7	28
Spread or sauce for green or yellow vegetable	Butter, 2 pats *(14 g)*	12.1	100	Cheese sauce, 1 Tbsp *(19 g)*	2.4	33
Bread or other baked product	Rye toast, 2 slices *(40 g)*	.3	56	Baking powder biscuits, 2 medium *(40 g)*	6.8	145
Spread for bread	Apple butter, *2 Tbsp (30 g)*	.2	56	Butter, 2 pats *(14 g)*	*12.1*	*100*
Beverage, first	Chocolate drink, 1 glass *(240 g)*	5.5	183	Ginger ale, 1 glass *(240 g)*	0	74
Beverage, second	Hot tea, 1 c. *(240 g)*	0	5	Hot cocoa, 1 c. *(240 g)*	11.1	233
Dessert	Plain cake with white icing	10.9	276	Pound cake	29.5	473
	Totals:	42.3	1,097	Totals:	127.3	1,855
	Percent of energy contributed by fat—34.8			Percent of energy contributed by fat—61.8		

F-36). The nutritional effects of making this and other substitutions are summarized in Table A-24.

Table A-24 illustrates the following important points which need to be considered when mixed dishes (either homemade or commercial mixtures) are to be substituted for separate food items.

1. It is usually appropriate to serve a double (200 g) or triple (300 g) portion of a mixed dish (the latter portion size for items such as beef and vegetable stews), since these preparations usually contain two or more major components.

2. The meat or fish content of mixed dishes is, in most cases, less than 50% of the composition; so the protein content of a double portion will often be less than that of single portions of meat or fish by themselves. A protein-containing garnish, sauce, or additional food item should, therefore, be added, if it is desired that the protein content be approximately equal to that of a meal containing separate meat or fish items (this addition may not be necessary if the other foods, during the day, supply sufficient protein).

3. Mixed dishes are normally higher in carbohydrate content, but may be lower in fat content than individual meat items.

No attempt was made to determine the mineral and vitamin contents of the menus in the examples which were just cited because it would have made the procedure too time-consuming. Fortunately, the dietitians in many of the larger hospitals have access to computers for menu planning.

Fig. A-29. Computer analyses of the content of foods. (Courtesy, University of Illinois at Urbana-Champaign)

TABLE A-24
EFFECTS OF THE SUBSTITUTION OF MEAT OR FISH CONTAINING MIXTURES FOR MEAT OR FISH PLUS SEPARATE VEGETABLE ITEMS

	Nutrient Composition, Food Mixtures						Nutrient Composition, Single Foods				
Single Food Items	**Energy**	**Protein**	**Fat**	**Carbo-hydrate**	**Fiber**	**Food Items from Mixtures**	**Energy**	**Protein**	**Fat**	**Carbo-hydrate**	**Fiber**
(*100 g* or specified amount)	(kcal)	(g)	(g)	(g)	(g)	(*100 g* or specified amount)	(g)	(g)	(g)	(g)	(g)
Pot roast of beef, sliced	327	26.0	23.9	0	0	Spaghetti with meat-balls in tomato sauce (home recipe), double portion (7 oz, or *200g*)	268	15	9.4	31.2	.6
Mashed potatoes, (milk added)	65	2.1	.7	13.0	.4						
Meat gravy, 1 Tbsp (*18 g*)	41	.3	3.5	2.0	.4	Grated parmesan cheese (1 oz, or *28.4 g*)	100	9.1	6.6	.7	0
Total nutrients	433	28.4	28.1	15.0	.8	Total nutrients	368	24.1	16.0	31.9	0.6
Corned beef, sliced	372	22.9	30.4	0	0	Corned beef hash (with potato) canned, double portion (7 oz, or *200 g*)	362	17.6	22.6	21.4	1.0
Mustard, yellow, 1 Tbsp (*16 g*)	12	.8	.7	1.1	.2						
Potatoes, Irish, boiled (pared)	65	1.9	.1	14.5	.5	Poached egg, 1 medium (*50 g*)	82	6.3	5.8	.4	0
Butter, 2 pats (*15 g*)	100	.1	12.1	.1	0	Tomato catsup, 2 Tbsp (*33 g*)	35	.7	.1	8.5	.2
Cabbage, boiled	20	1.1	.2	4.3	.8	Cole slaw, made with mayonnaise	144	1.3	14.0	4.8	.7
Total nutrients	569	26.8	43.5	20.0	1.5	Total nutrients	623	25.9	42.5	35.1	1.9
Broiled cod	170	28.5	5.3	0	0	Fish cakes, frozen, fried, reheated, double portion (7 oz, or *200 g*)	540	18.4	35.8	34.4	0
Tartar sauce, 2 Tbsp (*33 g*)	177	.5	19.3	1.4	.1						
Potatoes, Irish, french fried	274	4.3	13.2	36.0	1.0	Cheese sauce, 3 Tbsp (*50 g*)	87	4.0	6.5	3.2	trace
Total nutrients	621	33.3	37.8	37.4	1.1	Total nutrients	627	22.4	42.3	37.6	trace

It is noteworthy that the storage of nutrient composition data on computer tapes has made it possible for more thorough analyses of diets and menus to be made by busy dietitians. Hence, it is hoped that more attention will be paid to the mineral and vitamin contents of hospital menus when computer printouts of nutrient composition are readily available to all dietitians.

ANAPHYLAXIS

A severe allergic reaction which often leads to shock and sometimes to death. The extreme reaction is due to a massive release of histamine which causes the blood vessels to dilate and the blood pressure to fall.

(Also see ALLERGIES.)

ANCHOVY PEAR *Grias cauliflora*

A West Indian tree related to the Brazil nut which bears a fruit resembling the mango. This tree and the fruit are referred to as the anchovy pear. It is an oval fleshy fruit that grows 2 to 3 in. *(5 to 8 cm)* long with eight grooves on its brown skin. Anchovy pears are usually eaten pickled.

ANDROGENS

Hormones secreted by the testes whioh promote the development of male sexual characteristics. They also promote the buildup of tissues such as muscie and bone. The secretion of these hormones from the testes is responsible in large part for the larger body mass of males compared to females.

(Also see ENDOCRINE GLANDS.)

ANDROSTERONE

One of the hormones secreted by the testes.

(Also see ENDOCRINE GLANDS.)

ANEMIA

Contents — Page

Production of Red Blood Cells (Erythropoiesis) 76
Functions of Red Blood Cells 77
The Role of Iron in Red Blood Cells 77
The Major Causes of Anemia 77
Types of Anemia 77
Groups of Persons Susceptible to Nutritional Anemias 77
Diagnosis of Nutritional Anemias 81
Signs and Symptoms 81
Laboratory Tests 81
Interpretation of Signs, Symptoms, and Laboratory Tests 82
Treatment and Prevention of Iron-Deficiency Anemia 82
Oral Administration of Iron 82
Tablets 83
Tonics 83
Parenteral Administration of Iron (Injections) 83
Dangers of Therapeutic Administration of Iron 83
Good Food Sources of Iron 83
Factors Affecting the Utilization of Iron in Foods 84
Enrichment and Fortification of Foods with Iron 84
Enrichment 84
Fortification 84

One of the most common disorders which saps the vitality of many persons around the world is a condition where the blood is deficient in either the quantity and/or quality of red cells (erythrocytes). The overall effect of any type of anemia is a reduced supply of oxygen to the tissues of the body. Symptoms and clinical signs of this disorder are paleness of skin and mucous membranes, weakness, frequent tiredness, dizziness, sensitivity to cold, shortness of breath after exercise, loss of appetite, dyspepsia, tingling sensations in the extremities, rapid heartbeat, brittleness and dryness of the nails, soreness and cracks at the corners of the mouth, and atrophy of the tongue papillae, giving it a glossy appearance.

Many studies have shown that the highest prevalence of anemia is in infants and women of reproductive age. A high prevalence is also found in preschool children and adolescents, particularly those from low-income families.

PRODUCTION OF RED BLOOD CELLS (Erythropoiesis). Fig. A-30 outlines the major processes in the life cycle of erythrocytes.

Insufficient oxygen in tissues and blood (hypoxia) is the most important of the various factors which stimulate the production of red cells (other factors are alkalosis which develops at high altitudes, cobalt salts, adrenal cortical and sex hormones, and thyroxine). The process begins with the synthesis and secretion by the liver of a *protein substrate* (PS) in response to a subnormal level of oxygen in the blood circulating in the organ. Similarly, the kidneys produce an enzymelike substance (renal erythropoietic factor or REF) which converts PS to *erythrocyte stimulating factor (ESF)*. Formation of ESF is increased in anemia and after hemorrhage, and decreased when the blood contains a sufficient amount of red cells (regulation of ESF production is altered in disorders such as polycythemia in which there is overproduction of red cells).

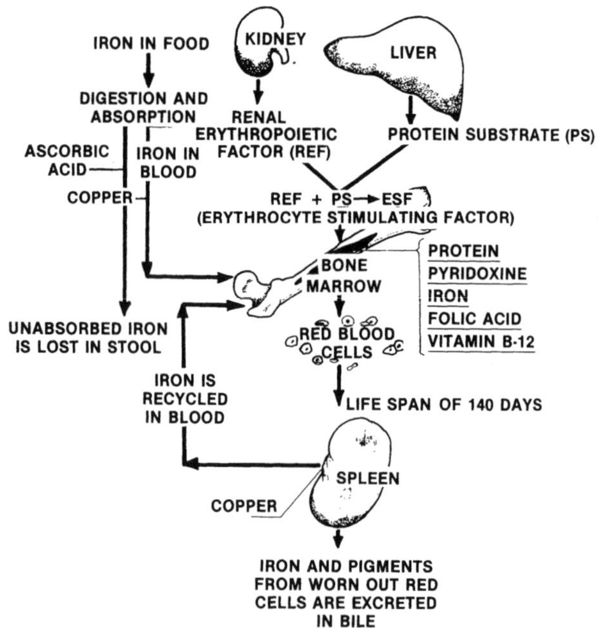

NUTRIENTS NEEDED FOR ERYTHROPOIESIS

I. MINERALS

Iron: Core of hemoglobin molecule

Copper: Part of enzyme ferroxidase which converts iron to ferric form for release from tissues into blood

II. VITAMINS

Ascorbic acid: Reduces iron to ferrous form for absorption

Pyridoxine: Cofactor in synthesis of hemoglobin

Folic acid and Vitamin B-12: Control division, growth, and maturation of red cells

III. PROTEIN

Raw material for hemoglobin and red cells

Fig. A-30. Formation of red blood cells (erythropoiesis).

Stem cells (embryonic blood cells) in the bone marrow are stimulated by ESF to differentiate into erythrocytes. There are several stages in the transformation of stem cells into red blood cells where nutrient deficiencies may lead to abnormalities in the quality and quantity of mature erythrocytes. Lack of protein, iron, or pyridoxine (vitamin B-6) limits the synthesis of hemoglobin, while copper has an indirect, but similar, effect (as part of an enzyme which converts iron to a form suitable for transport in the blood). Availability of iron for hemoglobin synthesis is also affected by dietary ascorbic acid (vitamin C), which helps to promote the absorption of dietary iron. Limitation of hemoglobin synthesis leads to the formation of abnormally small, or microcytic, cells. Cell division requires folic acid and vitamin B-12. Lack of either nutrient results in increased numbers of *megaloblasts*—immature blood cells enlarged in size and containing segmented nuclei.

Mature red cells lose their nuclei within a day or two

of their release into the blood from the bone marrow. The average life span of these cells is about 140 days, but nutrient deficiencies may shorten this period. There is an increased rate of red cell breakdown by hemolysis (liberation of hemoglobin from red blood cells) when there is a deficiency of vitamin E, which protects the cell membrane against substances which cause it to rupture. Similarly, riboflavin (vitamin B-2) deficiency is accompanied by increased hemolysis due to lack of a protective compound generated by an enzyme containing riboflavin.

Iron from the fragments of red cells is stored in the spleen, liver, and bone marrow (reticuloendothelial system). Recycling of this iron requires the action of a copper-containing enzyme.

Maintenance of an adequate supply of erythrocytes, therefore, depends not only on the supply and utilization of iron, but also, to a lesser degree, on ascorbic acid, copper, folic acid, protein, pyridoxine, riboflavin, vitamin B-12, and vitamin E.

FUNCTIONS OF RED BLOOD CELLS. The toroidal (donutlike) shape of erythrocytes helps to ensure their survival, since their travel through the bloodstream might be compared to the passage of a rubber inner tube through a river full of rapids. This shape also provides a large surface area for picking up and releasing oxygen and carbon dioxide by the hemoglobin present in these cells. Red cells also contribute most of the buffering capacity of the blood. *(Buffering is the process by which a substance, present in a solution, prevents excessive changes in the acidity or alkalinity of the solution.)*

The Role of Iron in Red Blood Cells. Transport of oxygen by erythrocytes depends upon an atom of iron attached within the core of each molecule of hemoglobin, and the protective shell provided by the protein part of the molecule. Thus, iron in hemoglobin may loosely bind oxygen whereas other forms of iron may bind oxygen so that it cannot be released readily. The presence of iron-containing hemoglobin in the red cells allows the blood to carry approximately 40 times as much oxygen as could be carried solely by a simple solution of oxygen in the fluid part of the blood. Only 0.5 ml of oxygen can be carried by a simple solution in 100 ml of blood, while approximately 19.5 ml of oxygen can be transported by the red cells which are normally present in the same volume of blood. Likewise, only about 3 ml of carbon dioxide dissolves in 100 ml of blood, but a total of 56 ml of carbon dioxide is actually transported by this volume of blood. Approximately 70% of the carbon dioxide transported in the blood is in the form of bicarbonate ions which are synthesized and broken down within the red cells through the action of an enzyme. About 20% of the carbon dioxide is bound to hemoglobin. The remainder is dissolved in water solution as carbonic acid.

Buffering action is required in the blood because carbon dioxide, formed during energy metabolism in tissues, reacts with water to form carbonic acid. Most of this action is contributed by red cells because oxygenated hemoglobin releases acid in the form of hydrogen ions and deoxygenated hemoglobin picks them up. Thus, the oxygenated hemoglobin furnishes hydrogen ions to compensate for the loss of acid when carbon dioxide is expired from the lungs, while the deoxygenated hemoglobin binds the additional acid formed by metabolism in the tissues. The two-part system of buffering by red blood cells is outlined in the equations which follow.

1. **Buffering in the lungs.**

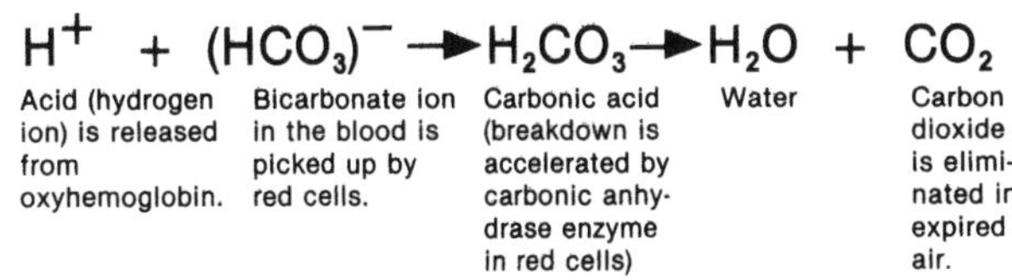
$H^+ + (HCO_3)^- \rightarrow H_2CO_3 \rightarrow H_2O + CO_2$
Acid (hydrogen ion) is released from oxyhemoglobin.
Bicarbonate ion in the blood is picked up by red cells.
Carbonic acid (breakdown is accelerated by carbonic anhydrase enzyme in red cells)
Water
Carbon dioxide is eliminated in expired air.

2. **Buffering in other tissues.**

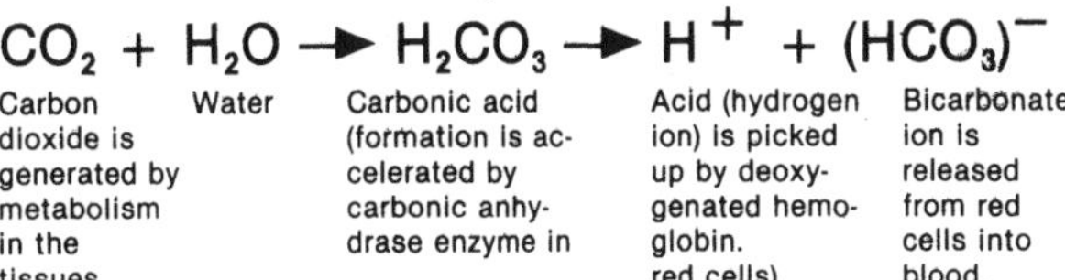
$CO_2 + H_2O \rightarrow H_2CO_3 \rightarrow H^+ + (HCO_3)^-$
Carbon dioxide is generated by metabolism in the tissues.
Water
Carbonic acid (formation is accelerated by carbonic anhydrase enzyme in red cells)
Acid (hydrogen ion) is picked up by deoxygenated hemoglobin.
Bicarbonate ion is released from red cells into blood.

THE MAJOR CAUSES OF ANEMIA. Factors which reduce the quality or quantity of hemoglobin are: (1) loss of blood from the body as in menstruation, or within the body as in internal hemorrhages; (2) increased destruction of red blood cells due to genetic defects, deficiencies of protective nutrients, toxic agents, or Hodgkin's disease (cancer of the lymph nodes); (3) reduced production of hemoglobin and/or red cells in disorders affecting iron utilization (deficiency of gastric acid or disease of the bone marrow), or as a result of deficiencies of nutrients which are needed for red blood cell production; and (4) abnormal hemoglobins or red cells (Mediterranean or Cooley's anemia, and sickle cell anemia).

Types of Anemia. Identification of the correct cause of an anemia is important because indiscriminate use of iron therapy may make some conditions worse (such as hemolytic anemia due to vitamin E deficiency, or anemia due to infectious organisms which thrive on iron). In order to help with such identification, the various types of anemia are described in Table A-25.

GROUPS OF PERSONS SUSCEPTIBLE TO NUTRITIONAL ANEMIAS. Most of the anemias are likely to be more severe in population groups that have high metabolic requirements, or that are subjected to various stresses. A discussion of these groups follows:

• **Infants**—Both human milk and cow's milk are very poor sources of iron. The full-term newborn infant enters life with a store of iron and red cells which should prevent for several months the development of anemia on a diet containing only milk. However, the premature infant is likely to have a much smaller store of iron, and is much more susceptible to hemolytic anemia. Even healthy infants, however, may be depleted of iron (due to small, daily losses in the stool and urine) after several months on a low-iron diet. Also, pasteurized milk has been found to induce, in some infants and children, a small but

TABLE A-25
TYPES OF ANEMIAS

Type	Description	Effects on the Body	Possible Cures
A. *Nutritional in origin:*			
Hemolytic anemia resulting from vitamin E deficiency	Red blood cells have an abnormal membrane (results from a deficiency of vitamin E) which makes them extra sensitive to hemolysis. Most often found in premature infants given milk formulas containing polyunsaturated fats without adequate vitamin E. Also found in adults with a deficiency of the vitamin.	An increased rate of destruction of red cells, resulting in anemia. Edema and some of its consequences (swollen legs, noisy breathing, and puffy eyelids). Severe deficiency of vitamin E in infants may also be accompanied by encephalomalacia (softening of the brain).	Administration of 25 IU of vitamin E daily. (Water soluble forms of the vitamin are more effective than fat-soluble forms.) Avoid administration of iron (interferes with the utilization of vitamin E).[1]
Iron-deficiency anemia	The most common form of anemia. Red cells are reduced in size (microcytic) and contain a subnormal amount of hemoglobin (hypochromic). Also, the cell count is subnormal, due to a decreased production of hemoglobin. Most prevalent in infants, children, and pregnant women.	Deficiency of hemoglobin and its consequences (paleness of skin and mucous membranes, fatigue, dizziness, sensitivity to cold, shortness of breath, rapid heartbeat, and tingling in fingers and toes).	Therapeutic oral doses of iron (60–180 mg / day). Injections of iron dextran or similar compounds may be necessary in severe cases. Care should be taken to avoid excessive doses which might result in toxicity (hemochromatosis).
Megaloblastic anemia (Also see MEGALOBLASTIC ANEMIA.)	Abnormal maturation of red cells resulting in enlarged cells (megaloblasts or macrocytes) with normal concentration of hemoglobin (normochromic). Subnormal cell count. Caused by defiency of folic acid and/or vitamin B-12.	Megaloblastic reaction in the bone marrow (abnormalities of red cells, white cells, and platelets). Macrocytic anemia. Vitamin B-12 deficiency may lead to gastrointestinal and nervous disorders. (Also see effects given below under pernicious anemia.)	First, test for the possibility of pernicious anemia. If there is no deficiency of vitamin B-12, then therapeutic oral doses of 0.1–1.0 mg of folic acid per day are usually effective. Daily injections of 0.01 mg of folic acid per day in cases due to malabsorption.
Pernicious anemia[2] (Also see PERNICIOUS ANEMIA.)	Same as above description of megaloblastic anemia. Additionally, there are gastrointestinal disorders (glossitis, achlorhydria, and lack of intrinsic factor for vitamin B-12 absorption) and neurologic damage (shown by abnormal electroencephalogram).[2] Caused much more frequently by autoimmunity (antibodies to intrinsic factor for vitamin B-12) than by dietary deficiency of the vitamin.	Megaloblastic anemia. Abnormalities of cells lining the gastrointestinal tract. Atrophy of gastric parietal cells and achlorhydria. Synthesis of abnormal fatty acids and their deposition in nerve tissue. Mental dysfunction.	Administration of 100 mcg per day of vitamin B-12 by injection (may be given orally *only* in those cases where absorption is known to be good). Lifelong maintenance injections (100 mcg per month) may be necessary for patients lacking intrinsic factor. *Caution:* Megadoses (1,000 mcg per day) have sometimes resulted in blood clots after the first week of therapy.[3]
Pica associated with iron-deficiency anemia	Abnormal craving for nonfood items leads to the eating of clay, dirt, plaster, paint chips, and ice. Iron deficiency is believed to be the cause of this disorder.	Certain nonfood items (some types of clays) interfere with iron absorption and make the condition worse.	Treat iron deficiency (as described above for iron deficiency anemia). The possibility of chronic blood loss should also be investigated.
Pregnancy anemia	The added requirements of a developing fetus produce in most cases an iron-deficiency anemia, although there may also be a folic acid deficiency.	Effects are the same as those of iron deficiency, except that the anemia may develop more rapidly in the pregnant woman, making her very vulnerable to the effect of blood loss during childbirth.	Therapeutic doses of iron, and folic acid if required (the latter should be considered in the case of women who were taking oral contraceptives prior to pregnancy and who may have had folic acid depletion as a result of the drug) (Also, see cure given above for iron-deficiency anemia.)

Footnotes at end of table.

(Continued)

TABLE A-25 (*Continued*)

Type	Description	Effects on the Body	Possible Cures
Nutritional in origin, (continued)			
Siderotic anemia due to deficiency of pyridoxine (vitamin B-6)	Microcytic, hypochromic anemia similar to that caused by iron deficiency; except that serum iron is normal or at an elevated level. Vitamin B-6 deficiency impairs synthesis of hemoglobin.	(See effects given above for iron-deficiency anemia.)	Oral doses of 50–200 mg of vitamin B-6 per day. *Note:* The disorder may have nonnutritional causes (factors that interfere with synthesis of hemoglobin).
B. *Nonnutritional in origin:*			
Aplastic anemia	Cessation of blood cell production in the bone marrow due to toxic agents, reaction to drugs, and unknown causes.	Great decrease in all blood cells produced in the bone marrow (red cells, white cells, and platelets). All the complications of anemia plus hemorrhages (due to lack of platelets needed for clotting.)	No cures are available, but there may be gradual recovery if the causative factor is eliminated. Patients may be maintained on blood transfusions. There have been some successful bone marrow transplants from close relatives.
Blood loss	Restoration of the full complement of red cells after blood loss is slower than repletion of other constituents of blood.	Effects are the same as those of iron deficiency, but may be more severe, depending upon extent of blood loss.	Same as for iron deficiency, except in severe cases, where transfusion of packed red cells may be necessary.
Familial hemolytic jaundice (spherocytic anemia)	A hereditary disorder in which the red cells are shaped like spheres instead of being of toroidal (donutlike) shape. Jaundice results from excessive destruction of the abnormal cells by the spleen.	Yellowish color of the skin and whites of the eyes (jaundice). Reduction in the number of circulating red cells.	There is no cure for the hereditary disorder, but surgical removal of the spleen is a cure of the effects of jaundice and excessive red cell destruction.
Hemolytic anemia due to deficiency of G6PD enzyme	Increased hemolysis of red cells due to the effects of drugs, toxic agents, and compounds in foods such as fava beans. This disorder is due to an increased susceptibility of red cells in persons who have hereditary deficiencies of Glucose-6-phosphate dehydrogenase (G6PD).	Anemia and jaundice (due to excessive hemolysis). Men are more likely to have the effects since the trait is sex-linked.	Avoidance of the substance(s) which causes hemolysis. Sometimes treatment with hormones such as ACTH and cortisone is helpful. This disease might be confused with spherocytic anemia, which has similar symptoms.
Hemolytic anemia of the newborn due to Rh factor incompatability	Rh-negative mothers develop antibodies against Rh-positive blood towards the end of pregnancy with an Rh-positive fetus (due to some transfer of blood from the fetus). The first child is not usually affected (unless the mother was previously sensitized by abortion or blood transfusion), but there are likely to be blood problems (caused by the transfer of antibodies from the mother) in the infants from subsequent pregnancies.	Destruction of red cells in the newborn infant. In severe cases, there may be almost complete destruction of the infant's red cells and damage to the brain by the accumulation of bilirubin (pigment resulting from the breakdown of hemoglobin).	A severely affected infant may require a total exchange of blood by transfusion. The antibody production by the mother may be blocked by a substance called RhoGam™,[4] which should be administered to Rh-negative women when they first show signs of sensitization (usually after first pregnancies and abortions).
Hookworm or tapeworm infestation	Infestation of the gastrointestinal tract by parasitic worms which feed on blood (hookworm) or nutrients (tapeworm). Anemia may be due to blood loss; deficiencies of folic acid, iron, and/or vitamin B-12.	Anemia, fatigue, irritability, fever, abdominal discomfort, nausea, or vomiting.	Treatment with an antiworm drug to eliminate the parasite. Nutritional therapy (iron in the case of the hookworm; iron, folic acid, and/or vitamin B-12 in the case of the tapeworm).

Footnotes at end of table.

(Continued)

TABLE A-25 (*Continued*)

Type	Description	Effects on the Body	Possible Cures
Nonnutritional in origin, (continued)			
Infection	The production of red cells by the marrow is sometimes inhibited by toxins from an infectious agent.	Anemia and weakness.	Elimination of the infection. Not helped by extra folic acid, iron, or vitamin B-12.
Leukemia	A form of cancer in which there is over-production by the body of white cells. Normal production of other blood cells (red cells, platelets, normal white cells) is prevented by the overgrowth of abnormal leukocytes in the bone marrow and other blood-forming organs (liver, spleen, and lymphatic tissues).	Grayish-white color of blood with a large excess of leukocytes. Death often results from the acute form of the disease and from the chronic form when it is not treated.	Drugs which prevent the growth of the abnormal cells.
Mediterranean anemia (also called thassalemia and Cooley's anemia)	A hereditary disease which is most prevalent in persons whose ancestors came from the Mediterranean basin (Italy, Sicily, Sardinia, Greece, Crete, Cyprus, Syria, and Turkey). Red cells are fragile and contain abnormal hemoglobin.	An increased rate of destruction of red cells. Bone abnormalities, enlargement of the spleen, leg ulcers, and jaundice.	Blood transfusions provide temporary relief, but eventually there is a toxic accumulation of iron (due to increased breakdown of red cells).
Sickle cell anemia	A hereditary disease in which the red cells have a sickle shape due to the presence of an abnormal hemoglobin. Sickle cells cannot carry as much oxygen as normal red cells, and they have a shorter than normal lifetime.	Anemia. Pain in the joints and extremities. Limited ability to perform strenuous exercise. Death may occur if sickle cells clump together and clog blood flow in tissues, such as the brain.	There is no known cure for the hereditary disorder. Transfusions and/or iron and folic acid therapy may be helpful in cases of severe anemia. Pregnant women with the disease are at high risk due to blood loss during childbirth.

[1]"Vitamin E Therapy in Premature Babies," *Nutrition Reviews*, Vol. 33, 1975, p. 206.
[2]"Pernicious Anemia and Mental Dysfunction," *Nutrition Reviews*, Vol. 34, 1976, p. 264.
[3]"A Qualitative Platelet Defect in Severe Vitamin B-12 Deficiency," *Nutrition Reviews*, Vol. 32, 1974, p. 202.
[4]RhoGam™ is a product recently developed by Ortho Pharmaceutical Corp., Raritan, N.J., which acts by transferring protective antibodies (passive immunity) to the sensitized woman, blocking her production of antibodies against Rh-positive blood.

regular loss of blood from the gastrointestinal tract; evaporated milk does not seem to have this effect (heating during evaporation alters the protein).

• **Children**—Low income, lack of education, or other problems of parents may result in a poor diet for some growing children. There may also be stress factors, such as infestation with worms, chronic infections, diarrheal disease, or milk-induced loss of blood from the gastrointestinal tract.

• **Adolescents**—Adolescent females are more likely to be deficient in iron than adolescent males. A survey by the U.S. Department of Agriculture showed that, on the average, intakes of iron were low for this group (about 50% of Recommended Dietary Allowance).[15] The reasons for poor diets among this group include erratic eating patterns, skipping meals to lose weight, and low socioeconomic status. Male and female energy allowances are 3,000 Calories (kcal) for males but only 2,200 Calories (kcal) for females; making it more difficult for the latter to obtain sufficient iron without consuming excess calories. Furthermore, menstrual blood losses plus the possibility of early marriage and pregnancy for females make their iron nutrition critical.

[15]*Dietary Levels of Households in the United States, Spring, 1965*, USDA Ag. Res. Serv. Bull. No. 62, 1969, p. 17.

• **Athletes and others engaged in strenuous activity**—A sedentary person may not experience symptoms of a mild anemia, while a person who attempts hard work or vigorous exercise may feel handicapped by such a condition (due to the increased requirement for oxygen during such activities).

• **Nonpregnant women**—It has been established that there are many healthy, nonanemic, young women who have negligible amounts of iron stores (the liver, bone marrow, and spleen may contain some unused iron which may be drawn upon to meet physiological requirements). These women have no reserves to meet increased needs due to such events as blood loss, pregnancy, and other stresses. Furthermore, some women who use the contraceptive pill may become depleted of folic acid and, therefore, have increased susceptibility to megaloblastic anemia (hormones present in oral contraceptives reduce the utilization of dietary folic acid).

• **Pregnant women**—During the course of pregnancy, there is, on the average, about a 50% expansion of the blood volume compared to that existing prior to the pregnancy. This means that for the average woman who has a blood volume of 9 to 10 pints (about *4 to 5 liter*), the blood volume expands to 13 or 14 pints (about *6 or 7 liter* of blood). The hemoglobin will then be diluted as it becomes distributed in a larger volume of blood. There is, in pregnancy, some increase in synthesis of red blood cells by the body, but the amount of increase in red blood cell production is proportionately less than the expansion of the total blood volume. In almost all pregnant women, the hemoglobin concentration, or amount of hemoglobin found in a given volume of blood, will, therefore, drop below the level found prior to the start of pregnancy. It has been estimated that the total amount of additional iron required in pregnancy is about 540 mg (or, about 2 mg of extra iron per day). While this does not seem like a difficult requirement to meet, it must be noted that there might have to be a daily dietary increase of 20 mg or more of iron due to the low efficiency of iron absorption by the body (10% or less of dietary iron may be absorbed, except when there is a severe deficiency and percent absorption is increased). These women also run increased risk of developing megaloblastic anemia due to folic acid deficiency since the growing fetus also needs this vitamin. Those who become pregnant right after stopping longterm use of oral contraceptives may have had a folic acid deficiency from the start of pregnancy. This makes for concern, since some congenital malformations in infants have been traced to deficiencies of folic acid in the mother.

• **Aged persons**—The elderly (age 65 and older) have higher incidences of iron-deficiency anemia and pernicious anemia. First of all, atrophy of gastric secretory cells occurs more frequently, and, as a result, they are more likely to have deficiencies of hydrochloric acid (aids in the absorption of iron) and of intrinsic factor (required for vitamin B-12 absorption). Second, reduced incomes and decreased energy requirements make it more difficult to select an adequate diet. Finally, many elderly persons live alone and are, therefore, not as likely to prepare well-balanced meals.

DIAGNOSIS OF NUTRITIONAL ANEMIAS.

Advertisements in the mass media suggest to their audiences that a "tired, run-down feeling" is a sign of anemia. Although such a diagnosis may apply to some persons, the symptom of fatigue may have other causes. It is, therefore, necessary to have greater accuracy in the diagnosis of anemia. The two major diagnostic approaches are observation of signs and symptoms and laboratory tests of the blood.

Signs and Symptoms. Anemic persons often have pale skin and mucous membranes along with feelings of tiredness. There usually is a low tolerance to strenuous exercise, and the anemic person quickly becomes "out-of-breath." Often the heart works harder to deliver more blood to the tissues (shown by rapid pulse). These characteristics may indicate, however, any one of several different anemias (including some with nonnutritional causes). It is, therefore, often necessary to confirm diagnosis of signs and symptoms with laboratory tests.

Laboratory Tests. Usually, determination of the quantity and quality (size, color, shape) of red blood cells is sufficient to confirm the general diagnosis of anemia and to identify the specific type of anemia. Occasionally, it is necessary to perform tests on other body tissues or fluids, particularly when anemias other than that due to iron deficiency is suspected. Some of the most commonly used tests follow:

• **Hematocrit (packed cell volume)**—This value is also called *packed cell volume*. It is one of the easiest to obtain. Blood taken from an arm vein or a finger prick is put into a tube of uniform diameter (closed at one end) and centrifuged. Both the total volume of blood (TBV) and the volume of red cells which pack down in the tube (packed cell volume or PCV) are measured.

Then, hematocrit (%) $= \frac{PCV}{TBV} \times 100.$

Subnormal values may indicate anemia due to deficiency of iron, or megaloblastic anemia (results from deficiency of folic acid or vitamin B-12).

• **Hemoglobin**—This measurement, like hematocrit, confirms a diagnosis of anemia, but leaves some uncertainty as to the type. It is expressed as grams (g) of hemoglobin per 100 ml of blood.

• **Mean corpuscular hemoglobin concentration (MCHC)**—This is a derived value, obtained by calculating the ratio of hemoglobin to hematocrit.

Thus, MCHC (%) $= \frac{\text{hemoglobin (g/100 ml)} \times 100.}{\text{hematocrit (\%)}}$

It gives an estimate of the average amount of hemoglobin per red cell. Generally, MCHC is subnormal for iron deficiency anemia, but normal for megaloblastic anemia.

• **Mean corpuscular volume (MCV)**—This measurement requires a count of red blood cells or RBC (see below).

The formula for its computation is:

MCV (cubic microns) $= \frac{PCV}{RBC} \times 100.$

This value tells whether red cells are normal in size (normocytic), reduced in size (microcytic), or enlarged in size (macrocytic). Iron deficiency produces a microcytic anemia, while megaloblastic anemias are macrocytic.

• **Red blood cell count (RBC)**—This measurement is obtained by placing a small drop of blood on a specially calibrated microscope slide, placing the prepared slide on the stage of a microscope, and counting the number of red cells in random areas of the slide (ruled with microscopic squares). The value obtained (millions of cells per cubic millimeter of blood) is useful in distinguishing between the types of anemias (RBC is more subnormal in megaloblastic anemia than in iron deficiency).

• **Schilling Test**—Patients are given a measured oral dose of radioactive vitamin B-12, followed by analyzing the urine and stool for content of the vitamin (by measurement of radioactivity). Subnormal urinary excretion and an elevated excretion in the stool of a test dose of the vitamin indicates poor absorption, and in the case of a megaloblastic anemia, proof that the anemia is due to deficiency of vitamin B-12 rather than folic acid. This distinction is very important, since folic acid will eliminate the symptoms of a megaloblastic anemia due to lack of either vitamin, but in the case of pernicious anemia (vitamin B-12 deficiency) nerve damage will not be prevented by folic acid. (Most severe cases of pernicious anemia are due to a failure of vitamin B-12 absorption rather than to a lack of the vitamin in the diet.)

• **Serum folate**—Measurement of the blood content of folate (the name for folic acid and related compounds with similar vitamin activity) may provide an indication of a deficiency prior to the development of anemia, or help the investigator to decide on the cause of an anemia.

• **Serum vitamin B-12**—The blood level of the vitamin is usually subnormal well in advance of the development of pernicious anemia. It is important to obtain early evidence of a deficiency of vitamin B-12 in order to prevent irreversible damage to the nervous system.

Interpretation of Signs, Symptoms, and Laboratory Tests. Most cases of anemia are due to iron deficiency, although there are several other causes for anemia. The other types of anemia constitute only a small minority of cases. Nevertheless, they need to be considered when persons with anemia do not respond to iron therapy, or when the symptoms of their anemia differ from those normally found in iron deficiency. One of the problems in the diagnosis of iron-deficiency anemia is the disagreement among both scientists and health practitioners as to the levels of hemoglobin which indicate the presence of the disorder.

The World Health Organization considers anemia to be present when blood levels of hemoglobin are below the following values (g/100 ml of venous blood): children aged 6 months to 6 years, 11; children aged 6 to 14 years, 12; adult males, 13; adult females, nonpregnant, 12; adult females, pregnant, 11[16].

Anemia, like other diseases, is accompanied by specific symptoms (see above); and yet a number of investigators have reported that there does not seem to be a direct correlation between these symptoms and marginal blood levels of hemoglobin. Some tentative explanations for this lack of correlation follow:

1. **There is a wide range of oxygen requirements** for persons engaged in different types of activities. It is believed that those with mild anemia may imperceptibly reduce the amount of their activity.
2. **Blood volume is not always in proportion to body size** since persons who engage in strenuous activity on a regular basis have larger blood volumes than sedentary persons of similar body size. For example the average 154-lb *(70 kg)* male has about 10 pt *(5 liter)* of blood, while a well-trained participant in running contests often has about 14 pt *(7 liter)* of blood. This difference is due to the opening up of capillaries in muscle in response to the greater metabolic requirements. The athlete described above might have a hemoglobin value about equal to the sedentary male, but the former actually has approximately 40 percent more hemoglobin in circulation due to his larger volume of blood. Sometimes there is a "training anemia" observed in athletes when the rate of expansion of blood volume outstrips the rate of hemoglobin production. A similar effect is observed in pregnant women who often have a 40 to 50% expansion of their volume of blood, usually with a decrease in hemoglobin concentration.

   It is, therefore, important for clinical examiners to consider the possibility of hemodilution due to expansion of blood volume before making the diagnosis of anemia. Unfortunately, the measurement of blood volume cannot be done on a routine basis in every clinic. (The measurement of blood volume requires the use of a radioisotope of chromium which is injected into the blood; the dilution of this isotope provides a basis for calculating total blood volume.)
3. **Climate may be a factor**, since persons living in the tropics have lower hemoglobin levels. Perhaps, it would be wise to consider whether inhabitants of areas with hot climates require as much oxygen for their metabolism, since it has been found that they have, on the average, a 10% reduction in their basal metabolic rate, compared to persons living in areas with temperate climates.
4. **To a limited extent, the body regulates red cell production** according to needs of the various tissues. For example, a shortage of oxygen stimulates the liver and kidneys to release into the blood factors which step up the production of red blood cells by the bone marrow. However, the ability of the body to respond to urgent needs for hemoglobin depends mainly upon the extent of tissue stores of iron (not easily measured by an examiner).

Diagnosis, then, is difficult for borderline cases of iron-deficiency anemia; but serious cases are easily detected. Some new procedures are needed, however, for routine assessment of the blood status of women likely to be stressed by factors such as the contraceptive pill and pregnancy.

TREATMENT AND PREVENTION OF IRON-DEFICIENCY ANEMIA. Once anemia has been diagnosed, the patient should be given some form of supplemental iron since treatment consisting only of additional dietary iron takes a long time to cure anemia.

Oral Administration of Iron. Iron-containing tablets are a more stable form of ferrous iron than tonics because iron in solution might become oxidized to the ferric form, the absorption of which is very poor.

[16]*Nutritional Anemia*, WHO: Technical Report Series No. 405, Geneva, 1968.

TABLETS. Ferrous sulfate is the best-utilized form of iron available. Ferrous iron in the form of other salts (fumarate, gluconate, and succinate) is equal in effectiveness to ferrous sulfate, but they are more expensive. There is some evidence of an increased effectiveness of so-called "chelated iron" (iron in chemical combination with well-absorbed organic compounds such as amino acids) compared to ferrous sulfate. The usual dose of the latter compound is 200 mg in tablets taken three times a day. Although most forms of oral iron produce at least a mild irritation of the gastrointestinal tract, enteric-coated tablets are not a good solution to the problem since the iron from these tablets is released farther down in the intestine than that from uncoated tablets; consequently, absorption is reduced. Also, tablets may not disintegrate soon enough in the intestines of persons who have diarrhea or abnormally rapid motions of the intestines.

TONICS. These preparations may be preferable to tablets when there are problems concerning the disintegration of the latter. Some of these tonics contain alcohol, but there is disagreement as to its effect on the absorption of iron (some think it produces an improved absorption).

Parenteral Administration of Iron (Injections). This form of medication is generally used for persons unable to benefit from orally administered iron due to disorders of the digestive tract or an intolerance to oral iron. Occasionally, there are persons who do not have any problems with oral iron, but cannot be relied upon to take their medication regularly. Three commonly used preparations for injection are iron dextran, iron sorbitol, and dextriferron.

Dangers of Therapeutic Administration of Iron. Accumulation of excess iron in tissues may result from chronic ingestion or injection of therapeutic doses of iron, particularly when there are abnormal conditions of iron absorption and utilization such as the following: (1) an increased rate of destruction of red cells, (2) transfusion of a large amount of blood, and (3) an abnormally high rate of absorption such as occurs in chronic alcoholism and chronic disease of the liver or pancreas. An increase in nontoxic tissue storage of iron is called *hemosiderosis*, while that resulting in tissue damage is called *hemochromatosis*. Tissue damage from excess iron is most frequently found in the liver (produces fibrosis and cirrhosis), pancreas (results in "bronze diabetes") and the heart muscle.

Some clinicians are concerned over the possibility that routine administration of therapeutic doses of iron to pregnant women might lead to the accumulation of an iron overload in the tissues of the mothers and/or the fetuses. This situation was found to exist in Bantu women who drank large quantities of a beer very rich in iron. Autopsies of Bantu infants showed an excess of iron in their tissues. The infants, however, had died from other causes. Very few studies have been made of the metabolism of therapeutic iron administered during pregnancy.

Good Food Sources of Iron. The uncertainties associated with the diagnosis and treatment of anemia make it desirable to make special efforts to prevent development of the disorder. There is a wide variety of both plant and animal foods supplying iron in quantities sufficient to prevent anemia. Some of the best food sources of iron are listed in Table A-26.

TABLE A-26
GOOD FOOD SOURCES OF IRON[1]

Source	Iron	Energy
	(mg/100 g)	(kcal/100 g)
Liver, hog, fried in margarine	29.1	241
Wheat-soy blend (WSB)/bulgur flour or straight grade wheat flour	21.0	365
Molasses, cane, blackstrap	16.1	230
Wheat bran, crude commercially milled	14.9	353
Liver, calf, fried	14.2	261
Kidneys, beef, braised	13.1	252
Soybean flour, defatted	11.1	326
Liver, chicken broiler/fryer, simmered	8.5	165
Oyster, fried	8.1	239
Eggs, raw yolk, fresh	5.5	369
Apricots, dehydrated, sulfured, uncooked	5.3	332
Sardines, Pacific, canned in brine or mustard, solids/liquid	5.2	186
Prunes, dehydrated, uncooked	4.4	344
Peaches, dried, sulfured, uncooked	3.9	340
Beef, all cuts[2]	3.8	300
Nuts, mixed, dry roasted	3.7	590
Pork, all cuts[2]	3.2	325
Beans, lima, mature, seeds, dry, cooked	3.1	138
Rice, white, enriched, raw	2.9	363
Wheat flour, all-purpose or family, enriched	2.9	365
Raisins, natural, uncooked	2.8	289

[1]These listings are based on the data in Food Composition Table F-36. Some good food sources may have been overlooked since some of the foods in Table F-36 lack values for iron.

Whenever possible, foods are on an "as used" basis, without regard to moisture content; hence, certain high-moisture foods may be disadvantaged when ranked on the basis of iron content per 100 g (approximately 3½ oz) without regard to moisture content.

[2]Values for different cuts range from 3.9 to 2.5 mg/100 g.

It should be noted that unrefined foods usually contain more iron than unenriched, highly processed foods. For example, unenriched white flour contains only about ¼ as much iron as whole wheat flour, and unenriched white rice only about ½ as much iron as brown rice (enrichment of the refined grains with iron restores the amount lost in processing). Much has been said about the iron-binding effects of phytates in whole grains, but it should be noted that these compounds are broken down by a phytase enzyme in yeast when it is used as a leavening agent in baking. Even the choice of a sweetening agent affects the iron content of the diet. For example, the third extraction (or blackstrap molasses) contains 4 times as much iron as the first extraction (or light molasses), 4 ½ times as much as brown sugar, 12 times as much as maple sugar, 32 times as much as honey, and 160 times as much as white sugar.

Factors Affecting the Utilization of Iron in Foods. One of the limitations of tables giving iron content of various foods is that there are differences in the percent utilization of the iron present in these foods. Iron present as heme (in blood-containing foods such as liver and muscle meats) is probably the most efficiently used form of iron from foods. In general, iron from plant foods does not seem to be as well utilized as iron from animal foods. Other food ingredients may aid or hinder the absorption of iron, but only a few such effects are well understood.

Substances which promote the absorption of iron are: (1) vitamin C (helps to keep iron in the ferrous or reduced state); (2) hydrochloric acid from the stomach (neutralizes some of the alkalinity in the small intestine and thereby increases the solubility of iron); (3) lactose or milk sugar (ferments to lactic acid, which acts similarly to gastric acid in the small intestine); (4) iron-containing foods from animal sources (reasons for this effect are not known); and (5) alcohol (a major ingredient in several proprietary remedies for iron-deficiency anemia).

Iron absorption is inhibited by the following: phytates (whole grains, bran), oxalates (spinach and rhubarb), raw soybeans, large excesses of calcium in the diet, protein from egg yolks, and antacids.

The cooking of acid-containing foods in iron pots has been shown to add significant amounts of iron to the food. For example, it was found that spaghetti sauce, cooked for 3 hours in an iron skillet, contained 87.5 mg of iron per 100 g of sauce, compared to 3.0 mg of iron in an equal amount of sauce cooked in a glass dish.[17] In most developed countries, however, pots and pans made of iron have been replaced by those made from aluminum, stainless steel, and other materials.

Enrichment and Fortification of Foods with Iron. A Ten State Nutrition Survey (data were collected from low-income groups in California, Kentucky, Louisiana, Massachusetts, Michigan, New York, South Carolina, Texas, Washington, West Virginia, and New York City) produced evidence that many persons of both sexes had at least borderline anemia. The occurrence of anemia appeared to be correlated with the consumption of foods low in iron. Food consumption patterns were, in general, found to be consistent with those reported from the more comprehensive survey by the U.S. Department of Agriculture (cited earlier in this article).

Preliminary publication of the findings of the Ten State Nutrition Survey led to urging by some prominent health professionals that the Food and Drug Administration (FDA) promulgate new standards for the fortification of flour and grain products with iron. (Federal agencies have regulatory authority over products shipped across state lines.) Present practices, however, should be examined before new measures are considered.

NOTE WELL: Originally, the FDA differentiated between enrichment and fortification, but now the two terms are used interchangeably.

ENRICHMENT. *This term refers to the restoration of some of the nutrients lost during the processing of foods.* At the present time, about ¾ of the states in the United States have laws requiring the enrichment of flour and certain grain products with iron, thiamin, riboflavin, and niacin according to the standards of the FDA (requires that 12 mg of iron be added to each pound of flour). Not all states, however, require that enriched flour be used for the production of all baked goods. Also, some ethnic groups (notably, Blacks from the South) use rice as a staple food (although required by law in but a few states, about ½ of the rice available in retail markets is enriched). It should be noted that the highest incidence of anemia is among Black females (according to the Ten State Nutrition Survey). One of the problems with the enrichment of flour and baked goods is that the form of iron used for this purpose is not well absorbed (iron sodium pyrophosphate is presently being used since other forms discolor white flour products). However, recent research has shown that the bioavailability of certain iron salts can be more than doubled by reducing the particle size of the iron compound to about ¼ of that commonly used.[18]

FORTIFICATION. *This means the addition to food of nutrients in such amounts that their final levels in the food are greater than those that were naturally present.* Some of the common foods which are presently fortified with iron are commercial infant formulas, infant cereals, and breakfast cereals.

In 1971, the FDA proposed that the enrichment standards for flour and grain products be revised so that iron might be added at a level of fortification (40 mg of iron per pound of flour). The proposal met with considerable opposition from some physicians who argued that the level of iron (designed to protect most women against anemia) might have a toxic effect on men (who require *less* iron than women, but eat more food and, therefore, consume *more* iron). Hemochromatosis is found mainly in alcoholic men, but it is not known what might be the effect of long-term ingestion by normal men of increased amounts of iron. Finally, the FDA withdrew the proposal in 1977 because there was a lack of evidence that the benefits of the iron fortification would outweigh the risks.

Although there have been suggestions that white sugar be enriched or fortified with iron, it seems to be more appropriate to add iron to flour and grain products for the following reasons: (1) Many nutritionists are trying to persuade people to cut down on their intake of sugar; and (2) there is a substantial and predictable consumption of flour and grain products by low-income groups.

There has also been a great amount of discussion relative to providing iron-fortified milk formulas for infants. However, these formulas may be too expensive for low-income families. Also, pediatricians and nutritionists usually advise parents to start feeding meats, cereals, and other iron-containing foods to infants by the time they are 3 months of age (when the iron stores from birth

---

[17] Moore, C. V., "Iron," *Modern Nutrition in Health and Disease*, 5th ed., Lea and Febiger, 1973, p. 300, Table 6C-2.

[18] "Physical Acceptability and Bioavailability of Iron Fortified Foods," *Nutrition Reviews*, Vol. 34, 1976, p. 298.

may be depleted). Like the iron used to enrich flour and bread, the iron used to fortify cereals has a low bioavailability. Meats provide the most available form of iron, but their cost sometimes limits their use. Perhaps some of the newly engineered foods, like textured vegetable protein, will provide suitable vehicles for iron fortification.

(Also see ANEMIA, MEGALOBLASTIC; ANEMIA, PERNICIOUS; DEFICIENCY DISEASES, Table D-1 Major Dietary Deficiency Diseases; and IRON.)

ANEMIA, MEGALOBLASTIC

Contents *Page*

Causes of Folate-Deficiency Megaloblastic Anemia .. 85
Deficient Diets 85
Factors Interfering with Absorption 85
Reduced Utilization 85
Increased Requirement 85
Diagnosis of Folate-Deficiency Megaloblastic Anemia 85
Signs and Symptoms 85
Laboratory Tests 86
Treatment and Prevention of Folate-Deficiency Megaloblastic Anemia 86
Oral Administration of Folic Acid 86
Parenteral Administration of Folic Acid 86
Good Food Sources of Folic Acid 86

This disorder is characterized by enlarged and immature red cells (megaloblasts) which are irregular in shape and size but contain the normal concentration of hemoglobin. However, the total amount of hemoglobin circulating in the blood is subnormal, since there are fewer red cells. White blood cells and platelets are often abnormal, since there is a general megaloblastosis of all of the blood cells formed in the bone marrow. Although nowhere near as prevalent as anemia due to iron deficiency, the effects of megaloblastic anemia are as debilitating and, until recently, frequently fatal.

There are two major types of nutritional megaloblastic anemia; one is due to a deficiency of folic acid, and the other is due to a deficiency of vitamin B-12.

• **Anemia resulting from deficiency of folic acid**—Formerly known as *tropical macrocytic anemia*, this disorder is much more common than anemia due to deficiency of vitamin B-12. The main identifying characteristic is the megaloblastosis in the bone marrow and blood.

• **Anemia resulting from deficiency of vitamin B-12**—Known as *pernicious anemia*, this fairly uncommon disorder needs to be considered in the diagnosis whenever there is megaloblastic anemia since, if not treated promptly and correctly, it results in irreversible damage to the nervous system. Also, this disease is likely to be accompanied by lack of both stomach acid (achlorhydria) and intrinsic factor. (The blood of affected persons often contains antibodies which attack both gastric acid secretory cells and intrinsic factor). However, this unique disorder will be discussed in a separate article.

(Also see ANEMIA, PERNICIOUS.)

CAUSES OF FOLATE-DEFICIENCY MEGALOBLASTIC ANEMIA. Deficiencies of folic acid may result from limitations in either the dietary supply, absorption, or utilization of the vitamin. These problems are discussed in the sections which follow.

Deficient Diets. Diets consisting mainly of grains, such as rice, are usually deficient in folic acid. (The major food sources are organ meats, vegetables, yeast, and mushrooms.) Moreover, the vitamin is destroyed by heating, so overcooking of food sources of folate may result in a deficiency. Depletion of body stores of folate occurs after 3 to 6 months on a severely deficient diet.

Factors Interfering with Absorption. The main site of the absorption of the vitamin is the upper one-third of the small intestine (where functional or structural abnormalities will result in malabsorption). Tropical sprue and gluten-induced enteropathy (celiac disease) are, therefore, common causes of folic acid deficiency. Hookworm infestation and blind-loop syndrome (colonization with folate-consuming bacteria) reduce the amount of vitamin available for absorption. Impairment of folate absorption also results from the use of anticonvulsant drugs (such as diphenylhydantoin), barbiturates, oral contraceptives, and ethyl alcohol.

Reduced Utilization. Vitamin B-12 deficiency reduces the utilization of folate, since the former is required for the interconversion of various metabolic forms of the latter. Folic acid antagonists (sometimes used in the treatment of leukemia because rapidly growing cancer cells are much more sensitive to the deficiency than normal cells) also reduce utilization of the vitamin. Breakdown of folate in the liver may be accelerated as a result of increased activity of enzymes stimulated by anticonvulsant drugs and oral contraceptives.

Increased Requirement. Pregnant women have about double the requirement of other persons for folic acid due to the needs of the developing fetus. Women who use oral contraceptives just prior to pregnancy may be depleted of the vitamin. Some of the adverse outcomes of pregnancy have been attributed to folic acid deficiency.

DIAGNOSIS OF FOLATE-DEFICIENCY MEGALOBLASTIC ANEMIA. Special care must be taken in the diagnosis of nutritional anemias so that the correct treatment can be started as soon as possible. Also, the administration of folic acid will correct the megaloblastosis, but not the nervous disorder of pernicious anemia (which, if not treated, leads to irreversible damage).

Signs and Symptoms. Most of the common complaints from victims of this disorder are similar to those of iron-deficiency anemia (tiredness, tingling in the fingers and toes, and smooth, glossy tongue). However, diarrhea is more frequently found in folate deficiency than in the other deficiencies leading to anemia. There is also increased susceptibility to infection, but this problem might also result from deficiencies of other nutrients.

NOTE WELL: Any signs of nervous disorders should lead the examiner to investigate the possibility of pernicious anemia.

Laboratory Tests. The lack of exclusive and specific signs and symptoms makes it necessary for the diagnostician to rely on laboratory test of blood, urine, and bone marrow. A controlled study of the course of development of megaloblastic anemia has shown that there is a sequence of biochemical and hematological changes which are indicative of the various stages of pathology.[19] The developments during the course of a prolonged deficiency of folate are as follows:

1. **Low serum folate** (below 3 nanograms per ml. 1 nanogram = 1 billionth of a g) is the earliest biochemical indicator of folate deficiency except when the patient has eaten a food rich in folate just prior to (within 6 hours) giving a blood specimen. Other tests are not so affected if the body stores of folate are depleted.
2. **Hypersegmentation of neutrophils** (a type of white blood cell having, on the average, three nuclear lobes). An average of more than 3.35 lobes per neutrophil indicates hypersegmentation and is the earliest hematological sign of megaloblastosis. However, this test is not reliable in the case of pregnant women, because there is a tendency for hyposegmentation in pregnancy. Thus, a pregnant woman would most likely have to have a moderate to severe folate deficiency before hypersegmentation might be observed. This abnormality is also present in pernicious anemia.
3. **Elevated urinary excretion** (more than 50 mg per 12 hours) of formiminoglutamic acid (FIGLU) after administration of a test dose of histidine. (Folic acid is required for the complete metabolism of histidine; otherwise, there will be formation of abnormally high amounts of FIGLU.) However, FIGLU excretion is also elevated in vitamin B-12 deficiency.
4. **Low red cell folate** (less than 20 nanograms per ml. 1 nanogram = 1 billionth of a g) indicates depletion of body folate stores (which immediately precedes the development of megalocytosis).
5. **Megalocytosis of red cells** (the mean corpuscular volume of the red cells is increased).

Mean corpuscular volume (MCV) =

$$\frac{10 \times \text{hematocrit (ml red cells per 100 ml blood)}}{\text{Red cell count (millions/mm}^3\text{ of blood)}}$$

The normal value for MCV is 87 ± 5 cubic microns. Values of 98 or more indicate megalocytosis. This abnormal condition also occurs in pernicious anemia, liver disease, and other disorders where there are reductions in the proportions of immature to mature red cells (blood loss, hemolytic and aplastic anemias).

6. **Megaloblastic bone marrow** can be detected by needle biopsy. This test may be necessary when there are coexistent deficiencies of iron and folic acid. However, this finding is also present in pernicious anemia.
7. **Macrocytic and normochromic anemia** serves to differentiate this disorder from iron deficiency anemia (microcytic and hypochromic anemia), but not from pernicious anemia which is essentially the same in hematological characteristics.

It may, therefore, be seen that the diagnostic tests discussed above may be relied on only when the possibility of pernicious anemia has been excluded, although low folate levels in both serum and red cells are strong evidence for a folate deficiency. Fortunately, there are some specific tests for pernicious anemia.

TREATMENT AND PREVENTION OF FOLATE-DEFICIENCY MEGALOBLASTIC ANEMIA. Therapeutic doses of folic acid should be given only after it is certain that the anemia is the result of deficiency of folic acid, rather than deficiency of vitamin B-12.

Oral Administration of Folic Acid. In cases of uncomplicated dietary deficiency, a suitable daily dose of folic acid is 0.1 mg. Larger doses (0.5 to 1.0 mg) should be given when there are complications such as alcoholism, use of certain drugs, hemolytic anemia, pregnancy, malabsorption, and nonnutritional factors or diseases which interfere with the production of blood cells. There does not appear to be any extra benefit from doses greater than 1.0 mg daily. When long-term maintenance therapy is needed, the dose should be 0.1 to 0.5 mg per day, depending upon the nature of the deficiency. It is desirable for the patient to eat at least one fresh fruit or vegetable every day by the time that therapy is discontinued.

Parenteral Administration of Folic Acid (Injections). Parenteral administration should be used in lieu of oral doses when there are serious conditions interfering with absorption, such as sprue, celiac disease, parasites, blind-loop syndrome, and chronic use of interfering drugs.

NOTE: Either oral or parenteral administration of folic acid may cause a rate of red cell production that rapidly depletes the iron stores of the body (indicated by difficulty in obtaining an increase in the hemoglobin level beyond 10 to 11 g per 100 ml of blood). Thus, there may be a need for concurrent therapy with iron (60 to 180 mg daily).

Good Food Sources of Folic Acid. The term folic acid is derived from the word foliage, since dark, green, leafy vegetables are among the best sources of the vitamin. Other good sources are nuts, beans, and whole wheat. Liver and kidneys are good sources, but muscle meats are poor sources. Citrus fruits are much better sources than other fruits.

There may be losses of up to 50% of the vitamin during the cooking and storage of foods. Therefore, it is wise for diets to include at least one fresh, uncooked vegetable or fruit each day. Frozen orange juice contains almost as much of the vitamin as the fresh juice.

(Also see ANEMIA; and FOLIC ACID.)

---

[19] Herbert, V., "Biochemical and Hematologic Lesions in Folic Acid Deficiency," *The American Journal of Clinical Nutrition*, Vol. 20, 1967, p. 562.

ANEMIA, PERNICIOUS

Contents *Page*

Causes of Pernicious Anemia 87
Dietary Deficiency 87
Lack of Intrinsic Factor 87
Other Factors Interfering with Absorption of Vitamin B-12 87
Increased Requirements 87
Diagnosis of Pernicious Anemia 87
Signs and Symptoms 88
Laboratory Tests 88
Treatment and Prevention of Pernicious Anemia 88
Oral Administration of Vitamin B-12 88
Parenteral Administration of Vitamin B-12 (Injections) 89
Good Food Sources of Vitamin B-12 89
Environmental Sources of Vitamin B-12 89

The characteristics of this disease are as follows: (1) megaloblastic anemia (enlarged and immature red cells, and a reduced number of normal blood cells); (2) disorders of the bone marrow (impaired formation of red cells, platelets, and white cells); (3) prolonged bleeding time (resulting from deficiency and abnormality of platelets); (4) inflammation and atrophy of stomach lining (lack of gastric acid and intrinsic factor may result from destruction of stomach lining cells by antibodies); (5) glossitis (smooth, shiny, and inflamed tongue) and bowel disorders; (6) impairment of fatty acid metabolism (accumulation of abnormal fatty acids in the myelin sheath of nerves, and elevated urinary excretion of propionic and methylmalonic acids); and (7) degeneration of the spinal cord and cerebral damage (indicated by abnormalities in an electroencephalogram). Although the blood disorders in pernicious anemia are very similar to those in megaloblastic anemia due to folic acid deficiency, the abnormalities of the nerves, tongue, and stomach are unique in pernicious anemia.

Most cases of this disease are found in persons over 40 years of age. If not promptly treated, death may result in about 3 years after diagnosis of the disease.

(Also see ANEMIA, MEGALOBLASTIC.)

CAUSES OF PERNICIOUS ANEMIA. Severe deficiency of vitamin B-12 results more often from a failure to absorb the vitamin (Addisonian pernicious anemia due to lack of intrinsic factor) than from any other cause. (Very little of the vitamin is absorbed without *intrinsic factor*, a protein secreted by stomach lining cells, which attaches itself to the vitamin and thereby facilitates its absorption in the intestine.) However, an understanding of all possible causes of the disease is important in order that suitable measures may be taken to prevent its deadly consequences.

Dietary Deficiency. Some cases of the disease have been reported in vegans (persons who abstain from eating any type of animal food). Plant foods do not normally contain even trace amounts of the vitamin unless they are contaminated with insects or microorganisms. (This aspect will be discussed in the section of this article—Good Food Sources of Vitamin B-12.) It is not understood, however, why there are not many more cases of the disease among vegans. Vegetarians who eat dairy products and/or eggs are not likely to have a dietary deficiency of the vitamin.

Lack of Intrinsic Factor. This condition, which is the main cause of pernicious anemia, is believed to be the result of an autoimmunity (formation by the body of antibodies against some of its own tissue) of unknown origin. Most victims of this disorder have antibodies to both intrinsic factor and gastric parietal cells (source of hydrochloric acid and intrinsic factor) circulating in their blood. These antibodies attack the parietal cells and intrinsic factor, and interfere with their functions.

Other Factors Interfering with Absorption of Vitamin B-12. Surgical removal of the stomach eliminates the source of intrinsic factor. Infestation of the intestine with fish tapeworm and blind-loop syndrome (colonization by vitamin B-12-eating bacteria) results in competition for the vitamin by the parasitic organisms. Disorders of the ileum (site of absorption) such as ileitis, gluten-induced enteropathy (celiac disease), and sprue may also contribute to the development of a deficiency of the vitamin.

Increased Requirements. The development of vitamin B-12 deficiency disorders may be accelerated under circumstances where there is an increased metabolism (as in hyperthyroidism) or stepped up red cell production. It is believed, for example, that therapeutic doses of folic acid given without vitamin B-12 to victims of pernicious anemia may hasten the depletion of the dwindling supply of the latter vitamin (folate-stimulated production of red cells may increase the rate of utilization of vitamin B-12).

DIAGNOSIS OF PERNICIOUS ANEMIA. Early diagnosis and treatment of this disease is necessary in order to prevent irreversible damage to the nervous system. Although the disease is nowhere near as common as the anemias of iron and folic acid deficiencies, it has been found to occur more often in persons who have some of the characteristics which follow:

- **A family history of autoimmune disorders**—There appears to be genetic tendencies to such disorders (pernicious anemia, Addison's disease, Hashiomoto's thyroiditis).

- **Stomach surgery (gastrectomy)**—Persons who have had part or all of their stomachs removed (total gastrectomy patients are very susceptible). It may take, however, 3 to 5 years (after surgery) for the development of a severe deficiency of vitamin B-12.

- **Chronic achlorhydria and/or atrophic gastritis**—Lack of acid secretion in response to stimulating agents such as histamine or pentagastrin is evidence of the atrophy of gastric secretory cells.

• **Disorders or surgery of the ileum**—Sufferers from regional ileitis (Crohn's disease), or persons who have had the terminal part of the ileum removed are very susceptible to deficiency of vitamin B-12 (the vitamin is absorbed in the ileal region). Blind loop syndrome, celiac disease, or infestation with fish tapeworm (found in persons who eat raw fish) also result in increased susceptibility to the deficiency disease.

• **Neurological or psychiatric disorders**—Although there are several different vitamin deficiencies known to be responsible for some of these disorders, the severe and ultimately fatal consequences of prolonged deficiency of vitamin B-12 make it worthwhile to use diagnostic procedures to rule out the possibility of this deficiency (there has often been dramatic improvement with prompt administration of the vitamin).

• **Strict vegetarianism (veganism)**—Persons who do not eat any animal foods (meat, eggs, dairy products, poultry, fish, insects) may be susceptible to neuropathy, but not show anemia (due to prevention of blood disorders by dietary folic acid).

Some of the criteria used for diagnosis follow.

Signs and Symptoms. Many of the features of this disease may lead the examiner to a diagnosis of iron-deficiency anemia (breathlessness, tiredness, dizziness, pallor of the skin and mucous membranes, and tingling sensations in the fingers and toes). However, there may also be signs of gastrointestinal disorders (sore, raw tongue, loss of appetite, abdominal pain, constipation, and vomiting), and if the deficiency is severe, signs of neuropathy (unsteadiness in walking, loss of memory, and psychiatric disturbances). Laboratory tests may be necessary in cases which are difficult to diagnose.

Laboratory Tests. It is not usually necessary to use all of the tests which follow, but it is worthwhile to have alternative tests for use in the event that there are limitations as to those which may be conveniently performed.

• **Electroencephalogram**—This procedure (measures "brain wave" activity) is more useful in measuring the extent of neuropathy than it is in diagnosis of the blood or gastrointestinal disorders. It is also valuable in measuring the progress achieved during treatment (disappearance of the megaloblastic anemia might be due only to adequacy of folic acid).

• **Gastric secretion**—Gastric contents are sampled for acid at regular intervals after the injection of histamine or pentagastrin. Failure of acid production under such conditions indicates achlorhydria and the possibility of atrophic gastritis which may result in lack of intrinsic factor and, ultimately, in pernicious anemia.

• **Mean corpuscular volume**—An elevated value for this measurement indicates abnormal enlargement of red cells and some type of megaloblastic anemia, but not necessarily pernicious anemia.

• **Schilling Test**—This is the most definitive test for failure of vitamin B-12 absorption. Urinary radioactivity (indicates absorption and subsequent excretion) is measured after the administration of a test dose of radioactive vitamin B-12 (an extremely small, harmless amount is used). Less than 8% excretion of radioactivity in the urine in 24 hours indicates lack of absorption (most likely due to lack of intrinsic factor).

• **Serum vitamin B-12**—Values of less than 150 picograms (l picogram = 1 trillionth of a g) of the vitamin per milliliter indicate a deficiency. Such values show a need for vitamin B-12 supplementation, even though the patient may be free of symptoms of pernicious anemia.

• **Urinary excretion of methylmalonate**—An elevated excretion in the urine (more than 5 mg of methylmalonate daily) indicates a deficiency of vitamin B-12. Administration of the vitamin usually produces a prompt drop in the excretion of this metabolite (unless there has been depletion of body stores of the vitamin). Some patients with vitamin E deficiency have elevated urinary methylmalonate *and* elevated succinate excretion (the latter does not occur in pernicious anemia since succinate formation is subnormal).[20]

TREATMENT AND PREVENTION OF PERNICIOUS ANEMIA.

Megadoses of vitamin B-12 (frequent administration of great multiples of the physiological requirement) are not advisable, and may in some cases be dangerous.[21] (They have resulted in the formation of blood clots.) The best results have been obtained with moderate doses, since excesses are excreted in the urine.

Oral Administration of Vitamin B-12. This mode of treatment should be reserved for persons who have mild disorders due to dietary deficiencies (such as vegans).

Although there has been some success in the treatment of pernicious anemia by oral administration of preparations containing vitamin B-12 and hog intrinsic factor, many of those treated in this way improved for a while, but eventually had relapes of the disease (some of these patients had antibodies to hog intrinsic factor, which is not surprising in view of the fact that many victims of the disease have antibodies to their own intrinsic factor).

Another type of oral administration consists of massive doses of the vitamin (milligram quantities instead of the more physiological microgram amounts). Occasionally, this approach has been successful, but it is unpredictable (some persons apparently absorb minute fractions of massive doses even when intrinsic factor is missing).

An oral maintenance dose for persons whose absorption is good is 1 mcg per day in tablet form.

---

[20] Barness, L. A., "Vitamin B-12 Deficiency with Emphasis on Methylmalonic Acid as a Diagnostic Aid," *The American Journal of Clinical Nutrition*, Vol. 20, 1967, p. 573.

[21] "A Qualitative Platelet Defect in Severe Vitamin B-12 Deficiency," *Nutrition Reviews*, Vol. 32, 1974, p. 202.

Parenteral Administration of Vitamin B-12 (Injections). Most persons suffering from pernicious anemia are treated by injection of vitamin B-12 because of the uncertainty as to the amount of the vitamin which may be absorbed (the disease is usually caused by a disorder of absorption). The usual dose is 100 mcg per day until the signs of the disease disappear. Unless the disorder of absorption is curable (such as infestation with tapeworm or blind-loop syndrome), lifelong maintenance injections of 100 mcg per month will be required.

Good Food Sources of Vitamin B-12. The ultimate source of the vitamin is microbial fermentation. Most people, however, obtain their supply from animal foods in their diets. (The vitamin has to be fed to nonruminant animals, such as poultry and swine; but grazing animals, such as cows and sheep, harbor microorganisms in their rumen which synthesize the vitamin from dietary cobalt and other nutrients.)

Liver and kidney are the best sources of the vitamin, followed by muscle meat, fish, and milk. Surprisingly, traces of the vitamin have been found in such foods as beets, oats, pears, and wheat due to these foods being contaminated by microorganisms in the soil.

An unusual plant source of the vitamin is *tempeh*, a fermented food prepared in Indonesia and other areas of Asia by inoculating soybean cake with Rhizopus mold. Although the vitamin content of the original soybean cake is negligible (about 0.015 mcg per 100 g), the fermentation process results in a more than a 30-fold increase (about 0.5 mcg of vitamin B-12 per 100 g of tempeh).[22] Frequent use of tempeh (dipped in salt brine and fried, or cut in pieces and used in soup) by these people probably protects them against a deficiency of vitamin B-12.

Environmental Sources of Vitamin B-12. It is ironic that the vitamin is synthesized by *E. coli* in the human colon where it is not absorbed. (Some physicians claim that a few persons have an occasional reversal of flow in their intestine and may thereby benefit from the vitamin synthesized in the colon.)

Soil microorganisms such as *actinomyces* and *streptomyces* also synthesize vitamin B-12. Thus, some of the richest enviromental sources of the vitamin are activated sewage sludge, manure, and dried estuarine mud.

Evidence for human benefit from environmental sources of the vitamin consists of the observation that vegans in India rarely show signs of pernicious anemia, whereas those who have migrated to England are more likely to develop the disease (because better sanitation in England reduces the amount of contamination of water and food with microorganisms which synthesize vitamin B-12).

(Also see ANEMIA; DEFICIENCY DISEASES, Table D-1 Major Dietary Deficiency Diseases; and VITAMIN B-12.)

[22]Van Veen, A. G., and K. H. Steinkraus, "Nutritive Value and Wholesomeness of Fermented Foods," *Journal of Agricultural and Feed Chemistry*, Vol. 18, 1970, p. 576.

ANEURINE

The name given to vitamin B-1 (called thiamin in the U.S.) by the British.

(Also see THIAMIN.)

ANEURYSM

A bulging out of (1) a small segment of an artery such as the aorta or one of the cerebral arteries, or (2) the wall of the heart. Little is known about the cause(s) of aneurysms, but nutritional deficiency is suspicioned. The rupture of an aneurysm often leads to sudden death from a massive hemorrhage. Rupture may be prevented by surgery in which one or more dacron patches are sewn onto the artery or the heart.

ANGINA PECTORIS

A sharp pain felt on the breastbone over the heart which may spread to the left arm and to the fingers. It may occur after strenuous exercise, emotional excitement, or other factors that cause the heart to work harder.

(Also see HEART DISEASE.)

ANGIOGRAM

A means of visualizing certain blood vessels through the injection of a medium which may be seen on x rays.

ANGIOTENSIN

A substance formed in the blood from angiotensinogen, a protein which is synthesized in the liver. Angiotensin constricts blood vessels and raises the blood pressure.

(Also see ANGIOTENSINOGEN; and HIGH BLOOD PRESSURE.)

ANGIOTENSINOGEN

A protein that is synthesized by the liver and circulates in the blood. Under certain conditions, it is converted to angiotensin which causes blood vessels to constrict and the blood pressure to rise.

(Also see ANGIOTENSIN; and HIGH BLOOD PRESSURE.)

ANGULAR STOMATITIS

An infection of the skin at the angles of the mouth, characterized by the epithelium protruding into ridges, giving the appearance of fissures. This is a symptom of riboflavin deficiency and of other diseases. But it can also be produced by poorly fitting dentures.

ANIMAL FATS

These are the fats that are isolated from animal tissues and animal sources. Primarily, animal fats include lard, tallow, and butter fat. Overall, animal fats are saturated. That is, the carbon atoms of fatty acids joined to the glycerol molecules in their triglycerides contain all possible hydrogen atoms. The iodine value, a measure of the degree of unsaturation, is low (under 80) for animal fats. The most common long-chain saturated fatty acids of animal tissues are palmitic acid and stearic acid. The saturated short-chain fatty acids—butyric, caproic, and caprylic acid—are common to butter fat. Since animal fats are high in saturated fatty acids they are usually solid, and have fairly high melting temperatures. Certain oils from fishes and marine animals remain liquid at low temperatures. Also, like all fats, they contain about two and one-fourth times as much energy as do proteins or carbohydrates. Each 100 g of animal fat contains about 850 to 900 Calories (kcal) of energy. All of the following are considered animal fats and may exist in a separate form: bacon fat, beef tallow, butter fat, chicken fat, duck fat, goose fat, mutton tallow, turkey fat, lard, pork backfat, and suet. Of course, the meat of animals contains fats similar to these extracted fats.

Two important commercial animal fats are:

1. **Beef fat or tallow.** Tallow is quite hard. It melts at 104 to 117°F (*40 to 47°C*), and has an iodine value of 40 to 48. The term *tallow* is also applied to sheep fat.

2. **Lard.** Lard is rendered pork fat. Before the advent of hydrogenated vegetable shortenings, lard was a major shortening agent. The melting point of lard ranges between 91 and 115°F (*33 to 46°C*), and the iodine value is 57 to 77. The fat of a hog can be influenced by the type of fat in its diet.

(Also see FATS AND OTHER LIPIDS; LARD; and MEAT[S].)

ANIMAL PRODUCTS

Meat, milk, eggs, and other products derived from animals, including four-footed animals, poultry, and fish.

ANIMAL PROTEIN FACTOR (APF)

The term formerly used to refer to an unidentified growth factor essential for poultry and swine and present in protein feeds of animal origin. It is now known to be the same as vitamin B-12.

(Also see VITAMIN B-12.)

ANIMAL STARCH

This name has been given erroneously to glycogen which is a storage form of carbohydrate found in animal tissues such as liver and muscle. It is not starch, however, since true starch is found only in plant materials.

Also see CARBOHYDRATE[S]; and STARCH.)

ANION

A negatively charged ion. Examples are: hydroxyl, OH^-; carbonate, CO_3^-; phosphate, PO_4^-

(Also see WATER AND ELECTROLYTES.)

ANORECTIC (ANOREXIGENIC) DRUGS

Drugs that depress the appetite and are used as an aid in weight reduction. Amphetamine is the most common anorectic drug.

(Also see AMPHETAMINES.)

ANOREXIA

A lack or loss of appetite for food.

(Also see ANOREXIA NERVOSA.)

ANOREXIA NERVOSA

A condition most frequently found in adolescent girls or young adult single women, characterized by pitiful emaciation resulting from self-inflicted voluntary starvation. The term implies that the condition is a neurosis, which is true in the majority of cases. Occasionally, however, it is caused by endocrine imbalances, usually resulting from disorders within the pituitary gland.

Patients are usually of high intelligence, introverted, perfectionistic, compulsive, and overly sensitive. Often it follows a weight-reduction program. To anorexics, food is repulsive and disgusting; and vomiting usually follows any forced feeding. There is concern that repeated vomiting, which causes loss of hydrochloric acid from the stomach and leads to compensatory loss of intracellular potassium, may cause cardiac arrest (Luther, J. Carter, NIH Shaken by Death of Research Volunteer, *Science*, American Association for the Advancement of Science, 25 July 1980, Vol. 209, No. 4455, pp. 475-477).

Normally, the condition is an attempt to solve unconscious emotional conflicts such as: A preceding history of overeating and obesity; with shame at being fat; hostility in parental relations at home, especially with the mother and in sibling rivalry and jealousy; fantasies, such as a fear that the food is poison, or a fear of becoming pregnant from the food; or it may stem from a suicidal death wish or result from feeling of guilt or personal unworthiness with the denial of food as a means of self-punishment. The patient may believe that she has no stomach to hold or digest food; or that she is eating various body parts; or there may be hallucinations, such as hearing voices commanding no food intake; or preoccupation, depression, or withdrawal states may influence the patient's reaction to food and make it impossible for her to eat; or she may refuse to accept a normal adult sex role.

In extreme cases, the loss of weight may be so great that death may ensue if the malady is not corrected in time. Both psychological and medical treatment are necessary, accompanied by dietary regulation and diet therapy. Psychotherapy is necessary in order to enable the patient to realize the problem and to recognize the need for food consumption; hence, it should be continued until recovery is complete. Medical treatment may include small doses of insulin before meals to stimulate need and desire for food. Diet therapy may consist of a nurse giving support to the patient by offering intimate personal attention at meals, such as feeding the patient in an unhurried and acceptable manner and gradually encouraging self-feeding. However, some patients will require tube feeding. If tube feeding must be used as a last life-saving resort, the procedure should be done in a therapeutic manner—never in a punitive manner. Since the nursing aim is to help the patient assume responsibility for her own eating, there should be a consistent approach involving patience and understanding. The patient should be involved in the care plan, given the reasons for it, and helped to see that she is accepted and cared for as a person. Her refusal to eat must be recognized as her inability to do otherwise, not as a perverse unwillingness to do so. As a relationship of confidence and trust builds, the patient may be led to become more actively involved in self-feeding and eventually in group eating.

(Also see ANOREXIA.)

ANOREXIGENIC DRUG

A medication which takes away the appetite. Such drugs are useful in programs designed to reduce the body weight.

(Also see APPETITE.)

ANOSMIA

This is the loss or impairment of the sense of smell, which may be permanent or temporary, depending on whether or not the olfactory nerves are damaged or destroyed completely beyond hope of healing. Such a condition makes it quite difficult for afflicted persons to derive enjoyment from eating.

Loss of the sense of smell may be the result of a mental state, as in hysteria. In some hallucinations, the person imagines that he smells certain odors that are not actually present. Treatment of anosmia due to mental causes is difficult.

Defects of the sense of smell may be caused by dryness of the mucous membranes of the nose, by infection, injury, obstruction, or deterioration of the nasal tissue; or by action of drugs. Also, certain diseases of the brain, a brain injury, or a brain tumor may produce anosmia.

Simple tests to determine the presence of a sense of smell may be made by releasing certain odors and noting the responses.

Adjustment to the loss of sense of smell is usually not too difficult.

ANOXIA

Lack of oxygen, or hypoxia, in the blood or tissues. This condition may result from various types of anemia, reduction in the flow of blood to tissues, or lack of oxygen in the air at high altitudes.

(Also see ANEMIA.)

ANTABUSE

This is a drug used in the treatment of alcoholism, for the purpose of producing a distaste for alcohol. When antabuse is administered to an alcoholic, extreme discomfort, severe nausea, vomiting, and flushing develops, with intolerance to alcohol.

Antabuse should never be given without the full knowledge and consent of the person; and it should never be given to a person who is intoxicated. The drug is best used along with psychotherapy.

(Also see ALCOHOLISM.)

ANTACIDS

These compounds are usually nonabsorbable basic substances or buffers used to treat indigestion and to soothe discomfort associated with this problem. The wide range of disorders referred to under the general term "indigestion" are generally believed by the public to be related, in one way or another, to gastric hyperacidity; hence, there is very wide usage of antacid preparation. Among such disorders are peptic ulcer, gastritis, gastric hyperacidity, heartburn, excessive gas, stomach irritation (from corticosteroid drugs, food intolerance, excessive smoking, or alcoholic stimulation), pylorospasm (spasm of the sphincter between the stomach and the small intestine), nervous dyspepsia, bleeding ulcers, colitis, hiccups, hypermotility (for which an antacid may be used with an antispasmodic drug), diverticulitis, after gallbladder surgery, before Caesarean sections, and anxiety (for which an antacid may be used with a sedative drug). More than likely, anyone watching television for 2 hours or more at a time will see an antacid commercial promising relief from indigestion due to nervous irritation and/or overeating.

TYPES OF ANTACID COMPOUNDS. Although there are dozens of proprietary antacid mixtures available in most drug stores, the formulations are usually based on one or more of about a dozen major antacid compounds (usually alkaline salts or buffers). A discussion of some of the most commonly used substances follows:

- **Aluminum hydroxide**—This is a slow-acting, mild alkalizer, which is usually suspended in water as a gel, and which coats and protects the stomach lining. It is a constitutent of many "over-the-counter preparations" (no prescription is required), and it may be given to patients with kidney failure. It forms an insoluble compound with phosphate, thereby preventing absorption and toxic accumulation of the phosphate ion. The problem with the latter use is that such patients are usually given diets low in phosphate, so considerable care must be exercised to avoid giving an excess of aluminum. Aluminum may be absorbed and deposited in the bones and brain when there is not enough phosphate or similar ions to form insoluble compounds. Also, it is constipating.

• **Aluminum oxide (alumina)**—This compound, also known as alumina, may be mixed with aluminum hydroxide in various preparations to obtain a greater amount of demulcent gel (coating agent). Some clinicians believe that this compound is of little value as an antacid. (Because of its insolubility, it does not neutralize much acid.)

• **Calcium carbonate**—It is obtained from limestone or similar minerals, and is one of the most potent substances in terms of acid neutralizing power. Also, it is useful as a source of dietary calcium when there is sufficient acid to form a soluble calcium salt. There is some danger, however, in absorption of excess calcium. It may be deposited in soft tissues, such as the kidney, and impair their functions. Also, calcium stimulates gastric acid secretion in some people. Therefore, patients receiving this compound should be monitored by blood tests for hypercalcemia. This salt may be constipating when taken chronically.

• **Calcium-magnesium carbonate (dolomite)**—This compound, also known as dolomite, is the major component of a well-known brand of antacid tablets. Magnesium overcomes the constipating property of calcium carbonate. Like calcium, some magnesium may be absorbed; and it may be useful if the magnesium content of foods is low, but excess magnesium causes sedation.

• **Dihydroxy-aluminum aminoacetate**—This salt of aluminum hydroxide (see above) and the amino acid glycine provides mild buffering.

• **Glycine**—It is an amino acid with a good buffering action.

• **Magnesium carbonate**—This compound is similar to calcium carbonate in potency as an antacid, except that it is not constipating.

• **Magnesium hydroxide (milk of magnesia)**—This compound, also known as milk of magnesia, is a mild antacid with a laxative action. It is often combined with aluminum hydroxide.

• **Magnesium phosphate (tribasic)**—This salt may be used in combination with other antacids (salts of aluminum, calcium, or magnesium) to obtain a strong alkalizing effect for treating the acute stage of a stomach ulcer.

• **Magnesium trisilicate**—This is a slow-acting antacid used in mixtures with other similar compounds. There have been a few reports of the formation of silicate stones in persons using preparations containing this compound for long periods of time.

• **Sodium bicarbonate (baking soda)**—This salt, also called baking soda, is one of the few *highly absorbable* antacids. For this reason, it should never be used in large doses,as it might cause alkalosis. Four teaspoons is the maximum safe daily dose. It is best used in combination with weak acids, such as citric (see above) to obtain an effervescent action and a buffer salt, such as sodium citrate.

• **Sodium citrate**—It is a mild buffer salt.

• **Urea (carbamide)**—This compound, also known as carbamide, is a weak alkali used in a proprietary mixture.

OTHER DRUGS USED WITH ANTACIDS. Digestive disorders may result from a combination of hyperacidity plus other conditions such as hypermotility (excessively rapid movement of the stomach and other digestive organs) and anxiety (which further aggravates the first two conditions). Drug combinations are designed, therefore, to correct these other conditions which lead to indigestion. Two types of drugs which are often used along with antacids for this purpose follow:

ANTISPASMODIC DRUGS are those which slow up the motility of the digestive tract. In some cases, they are also anticholinergic—drugs that block the transmission of impulses from the vagus nerve, thereby reducing the secretion of stomach acid and the digestive enzyme, pepsin.

DEPRESSANT DRUGS, such as barbiturates, are used to calm persons whose overanxiety may lead to hypersecretion and hypermotility.

CONDITIONS FOR WHICH ANTACID THERAPY IS USED. Antacids are used in treating a number of clinical conditions of the upper digestive tract. In all of these conditions the use of antacids is based on the assumption that they decrease the acidity of the stomach content. Some of the most common conditions warranting the use of these remedies follow.

• **Esophageal reflux, or esophagitis**—Reducing reflux of stomach acid into the esophagus is the aim of therapy in esophageal reflux. This is partially achieved by antacid therapy, but, for the most part, it is brought about by avoidance of constricting garments, treatment of obesity, reducing the size of meals, avoiding lying down after meals, elevation of the head of the bed, and the use of anticholinergic or antispasmodic drugs that tighten the lower esophageal sphincter and reduce gastric motility.

For esophagitis, liquid antacids are preferred to tablets because they will run down into the esophagus, whereas tablets may not dissolve in the mouth but remain solid until they reach the stomach.

• **Gastric ulcer**—The use of antacids relieves ulcer pain. Patients in the acute stage of gastric ulcer may require antacid every hour until there is evidence that the ulcer has healed. Complete healing usually requires longer than x rays indicate; therefore, hourly feedings of antacids should continue for at least 2 weeks beyond apparent healing as indicated by x rays. Hospitalization is recommended for large ulcers, particularly those in the central portion of the stomach where malignancies are more likely to occur.

• **Duodenal ulcer**—The therapy of duodenal and gastric ulcers is about the same. Some clinicians recommend hourly doses of antacids for about 2 months for duodenal ulcer patients since about 50% of these ulcers heal in about 6 weeks. These patients are advised to use antacids whenever meals are delayed, when they have hunger sensations, or when they are under stresses which they associate with ulcer symptoms.

• **Stress ulcers or erosive gastritis**—This is a serious condition (due to risk of hemorrhage and/or perforation of the eroded area) since mortality ranges from 20 to 70% once bleeding begins. Most clinicians believe that strong antacids should be used to reduce the gastric acidity, particularly in patients who have suffered from shock (due to blood loss) or sepsis (pathogenic organisms in the wound).

• **Prior to Caesarean sections**—It has been found that the acidity of the gastric contents is the main cause of pneumonitis resulting from the aspiration of stomach contents in anesthetized patients. These patients suffer an "asthmalike" reaction characterized by cyanosis, tachycardia, dysphea, and wheezing on exhalation. Much less severe symptoms are experienced by patients after the aspiration of food or liquid low in acid. The gastric emptying time is often prolonged during labor. It is not uncommon for obstetric patients to vomit food eaten 24 to 48 hours previously. Therefore, it seems advisable to give patients awaiting Caesarean sections antacids to prevent the development of aspiration pneumonitis.

• **Renal stones**—Aluminum hydroxide is useful in this condition because aluminum combines with phosphorus in the small intestine and forms insoluble aluminum phosphate, thereby reducing the absorption of phosphorus. Many types of kidney stones contain phosphorus—calcium phosphate, magnesium phosphate, magnesium-ammonium phosphate, or calcium-magnesium-ammonium phosphate.

• **Kidney failure**—Bone disease is a well-known complication of chronic kidney failure. The causes and results of calcium disturbance in kidney disease are complex. Vitamin D is activated in the kidneys. It is now thought that intracellular phosphorus may be a major determinant in the synthesis of the active form of vitamin D. In the kidney failure, blood levels of phosphate are elevated, and this results in a decrease in the activation of vitamin D, and ultimately a reduction in calcium absorption. The antacid therapy, therefore, consists of two parts: (1) aluminum antacids to combine with and precipitate phosphate, which is normally present in the diet; and (2) calcium in some form to correct the hypocalcemia of kidney failure.

DANGERS AND LIMITATIONS OF ANTACID PREPARATIONS. Some of the common problems which result from the use of antacids are:

1. **Diarrhea or constipation** may be expected if any effective antacid is taken in more than a few doses. Aluminum and calcium compounds have a constipating effect, while magnesium compounds have a laxative effect. Therefore, mixtures of constipating and laxative antacids are popular because the effects are balanced to some degree.

2. **Systemic alkalosis** is most likely to occur with excessive use of sodium bicarbonate, since it is the only one of the commonly used antacids which is completely soluble and absorbable.

3. **Sodium overload** is a risk when large doses of antacids are given to patients with sodium retention. Some popular antacids contain from 12 to 269 mg of sodium per ounce. Sodium may be furnished, in some cases, by the use of sodium carboxymethylcellulose as an emulsifying and suspending agent, by the use of sodium compounds in the production of aluminum hydroxide, or by sodium compounds used as preservatives.

4. **Stimulation of gastric secretion** may counteract effects of antacids. Recent studies have shown that calcium carbonate given to decrease gastric acid actually stimulates gastric secretions. At first it neutralizes acid, but later there is an increase in gastric acid secretion.

5. **Calcium deposits** sometimes result from the absorption of the mineral from calcium-containing antacids. It has been found to constitute a serious danger, particularly in the case of patients on a program of frequent doses of calcium carbonate (2 g per hour).

6. **Phosphate depletion** may occur with the use of aluminum antacids. Although the formation of insoluble aluminum phosphate is helpful in treating patients with renal failure, it may cause hypophosphatemia in normal persons. Signs of this disorder are loss of appetite, muscle weakness, and progressive debility. Low blood levels of phosphate permit the development of hypercalcemia which may result in urinary calculi, and which may even cause osteomalacia and osteoporosis due to lack of phosphate to maintain mineralization of bone.

7. **Kidney stones** result when bloodborne minerals form insoluble salts. Aluminum hydroxide may prevent phosphate stones but, in so doing enhance the probability of calcium stones being formed, due to the reciprocal relationship between blood levels of phosphate and calcium ions. Calcium salts directly increase the probability of formation of calcium stones. Silicate stones may result from the use of magnesium trisilicate.

8. **Magnesium retention in kidney failure** may result in elevated blood levels of magnesium (hypermagnesia). Hence, maganesium antacids should not be given to uremic patients. Although normal persons rapidly excrete any excess of magnesium, persons with impaired kidney functions may develop a dangerous toxicity. Excess magnesium causes sedation of the nervous system.

9. **Aluminum retention in kidney failure** usually results from either of two abnormal conditions—(a) absorption of aluminum, or (b) failure of the kidneys to excrete the mineral. Patients on low-phosphate diets and taking large quantities of aluminum antacids may absorb sufficient aluminum to develop deposits in their bones and brain.[23] Aluminum is not absorbed when there is sufficient phosphate in the digestive tract to form an insoluble aluminum phosphate.

10. **Intestinal obstruction** occasionally occurs in patients given antacids. This condition appears in sufferers from hyperacidity who take a combination of antacids and drugs which reduce gastrointestinal motility. Sometimes, cases of blockage of the intestine may require surgery.

[23]Alfrey, A.C., *et al.*, "The Dialysis Encephalopathy Syndrome," *New England Journal of Medicine,* Vol. 294, 1976, p. 184.

11. **Alteration of drug effects** may occur if antacids change the normal patterns of the absorption and excretion of these medications. Antacids usually make the urine alkaline and thereby increase the excretion by the kidneys of acid drugs such as aspirin. Little is known, however, of the possible effects and interactions of antacids and drugs.

12. **Depletion of essential minerals** results when there has been chronic alteration of the normal acidity and alkalinity of the stomach and small intestine. Overalkalinization of the digestive tract may result in greatly reduced absorption of trace minerals such as zinc, copper, iron, and other metallic elements because most of them are more soluble in acid than in alkaline solution.

CAUTION: It is obvious from the above that antacids are not to be chewed like candy, even though they are often flavored with sugar and peppermint. Also, there is need for a person suffering from chronic indigestion to obtain a correct diagnosis of his disorder so that there is neither hyperacidity nor hypoacidity, both of which may give rise to similar discomfort. There is also the possibility that a person with chronic distress may be suffering from some such condition as hiatus hernia.

(ALSO see HIATUS HERNIA.)

Often, nonantacid demulcent preparations (usually mucilaginous substances such as mucin or psyllium seed gel) might be used in place of part of the dose of antacid since these substances have coating and soothing effects. The role of food proteins as buffers of stomach acid are well known, but these substances also stimulate acid secretion by an unpredictable amount.

Finally, persons who have indigestion due to anxiety and other psychological problems might well seek treatment of these underlying disorders.

(Also see DYSPEPSIA; HEARTBURN; and ULCERS, PEPTIC.)

ANTAGONIST

In nutrition, this term refers to a substance which counteracts the beneficial effects of one or more vital nutrients. For example, the substance avidin from raw egg white binds the vitamin biotin. Sometimes, cooking eliminates or reduces antagonistic effects.

ANTHOCYANINS

These are violet, red, and blue flavonoid pigments contained in many fruits, vegetables, flours, and leaves, such as delphinin, pelargonidin, and cyanidin. They are pH sensitive and water soluble, and they contain glucose and anthocyanidins. They become more violet in alkali mediums and more red in acidic mediums. Due to the fact that they are water soluble, they leak out of cut fruits and vegetables during processing and cooking. Anthocyanins will react with metals and become more blue or violet. Therefore, cans must be lined if the natural color of the canned fruit and vegetable containing anthocyanins is to be maintained. Anthocyanins are found in grapes, berries, cherries, beets, and eggplants.

ANTHROPOMETRIC MEASUREMENTS

Simple measurements of the various parts of the human body. They include weight, height, shoulder width, arm circumference, leg circumference, head circumference, etc. These measurements are useful in assessing nutritonal status and weight condition.

(Also see OBESITY.)

ANTI-

A prefix meaning against or opposing; e.g., antiscorbutic means preventing scurvy.

ANTIBIOTICS

Contents *Page*

History 95
Mode of Action of Antibiotics 95
Antibiotics in Agriculture 95
Antibiotics in Foods 95
Some Antibiotic Concerns 96
Allergic Reactions 96
Indirect Additives 96
Vitamin K Deficiency 96
Side Effects 96
Resistance to Bacteria 96
Future of Antibiotics 98

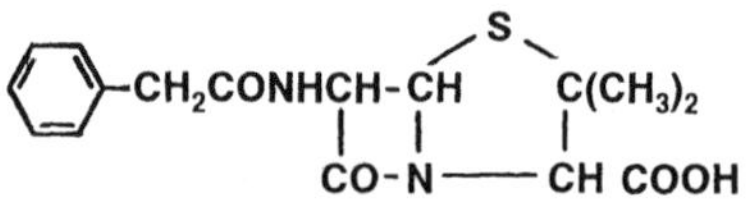

PENICILLIN

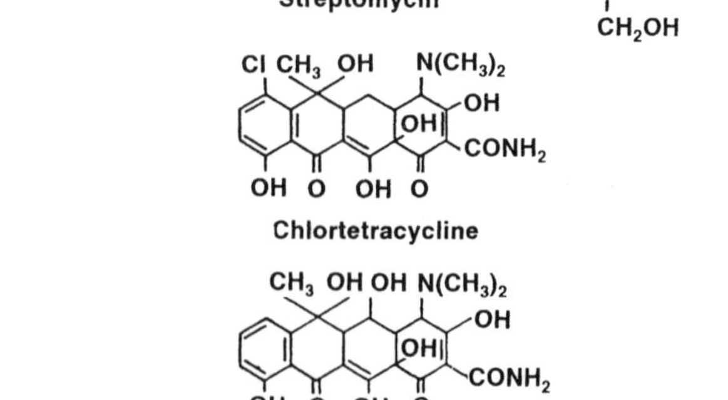

Streptomycin

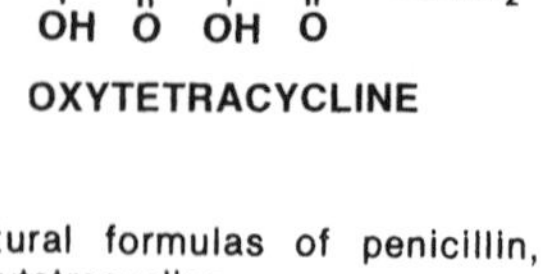

Chlortetracycline

OXYTETRACYCLINE

Fig. A-31. Structural formulas of penicillin, streptomycin, chlortetracycline, oxytetracycline.

Antibiotics are chemical substances produced by living organisms which inhibit the growth of or kill other organisms. Commercially produced antibiotics are substances produced by moldlike organisms grown in nutrient solutions in large steel tanks. Chemists purify these substances and prepare them as concentrates. Then they are tested in animal experiments to determine their potency and toxicity before they can be employed for human use.

As a group, antibiotics exhibit selective antimicrobial activity. Some are active against many gram-positive bacteria, others predominantly against gramnegatives, and a few, the "broad-spectrums," are inhibitors of members of both groups. Others are antifungal only.

The search for effective and nontoxic antibiotics is unceasing. Since the discovery of penicillin in 1928, thousands of antibiotics have been isolated and studied. But relatively few are in active use. The following are among the better known and more widely used antibiotics: bacitracin, chlortetracycline (Aureomycin), erythromycin, neomycin, nisin, nystatin, oxytetracycline (Terramycin), penicillin, pimaricin, streptomycin, and tylosin.

HISTORY. The newer knowledge of antibiotics dates from the discovery of penicillin by Dr. Alexander Fleming, a British scientist, in 1928.[24] Quite by accident, a stray mold spore floated in on the breeze, and landed on a culture plate of bacteria with which Dr. Fleming was working. It inhibited the growth of bacteria. Dr. Fleming correctly interpreted his observation: the possible value of the mold in the treatment of disease, thus ushering in the antibiotic era. However, penicillin did not come into prominence until 10 years later; and it was not until 1944 that streptomycin, the second most widely known of the antibiotics, was discovered by Wakesman, a soil microbiologist, and his colleagues at the New Jersey Station.

Penicillin, streptomycin, and other antibiotics were rushed into use against a long list of human ailments. They were hailed as the "wonder drugs" of our time.

MODE OF ACTION OF ANTIBIOTICS. Numerous theories, each with convincing support, have been hypothesized with respect to the mode of action of antibiotics. No one theory has proven to be the total answer. Rather, the true mechanism is most likely the result of the additive effects from the various proposed mechanisms.

One fact is well substantiated. Antibiotics are effective in controlling certain environmental stresses, as evidenced by the fact that animals raised under germfree conditions exhibit no improved responses to them.

The proven or probable modes of action of antibiotics follow:

1. They may spare certain nutrients. Studies have indicated that antibiotics can replace inadequate intakes of certain vitamins and amino acids.

2. They may selectively inhibit growth of nutrient-destroying organisms while promoting growth of nutrient-producing organisms.

3. They may increase food and/or water intake.

4. They may inhibit growth of organisms which produce toxic waste products or toxins.

5. They may kill or inhibit pathogenic organisms (a) in foods (when used as preservatives), (b) within the gastrointestinal tract, or (c) systemically. Each antibiotic has a distinct mode of disrupting cellular life.

6. They may improve the digestion and subsequent absorption of certain nutrients.

ANTIBIOTICS IN AGRICULTURE. In 1949, McGinnis of Washington State University, and Jukes of Lederle Laboratory, discovered that antibiotics were something new to be added to livestock feeds. Quite by accident, while conducting nutrition studies with poultry, they obtained startling growth responses from feeding a residue from Aureomycin production. Later experiments revealed that the supplement used by Jukes and McGinnis—the residue from Aureomycin production—supplied the antibiotic chlortetracycline. Such was the birth of feeding antibiotics to livestock. A new era in livestock feeding was ushered in—comparable to the vitamin era that was born in 1912.

In addition to their use as growth stimulators, antibiotics are used as nutritional stimulants to promote better feed efficiency in ruminants and swine, and to increase egg production, hatchability, and shell quality in poultry. They are also added to feeds in substantially higher quantities to remedy pathological problems. In the United States, virtually all chickens, turkeys, and milk-fed calves, and an estimated 90% of the swine, are fed antibiotics. A lesser number of finishing cattle and finishing lambs are fed antibiotics. An estimated 40% of all the antibiotics used in this country is fed to livestock.

Many crop sprays contain antibiotics.

When properly used in agriculture in any of the above ways, antibiotics have not been shown to constitute a hazard to human health.

ANTIBIOTICS IN FOODS. Antibiotics may occur in foods in the following three ways:

1. **Naturally.** Nisin, for example, is a naturally-occurring antibiotic, which is sometimes found in milk. Thus, it is noteworthy that gassy fermentations in raw cheese made with milk containing clostridia are controlled by the use of nisin-producing starter cultures.

2. **Through direct addition.** Antibiotics may occur in foods through direct addition as a food additive to aid

[24]Actually, the presence of antibiotics was known much earlier than the discovery of penicillin, but no commercial use was made of them.

in keeping quality of the food. Nisin, for example, is permitted as a direct food additive in many countries.

3. **As an unintentional food additive.** This can result from the carryover of the antibiotic in milk, meat, or eggs as a result of the addition of antibiotics to the feed of animals for growth promotion, or from animal medication for the prevention or treatment of animal diseases. In the case of all antibiotics used in animals, either a zero tolerance or a "negligible residue" in the products has been established. Also, in granting approval for the use of antibiotics in animal feeds and animal medication, the FDA establishes withdrawal periods—the required legal withdrawal period for the antibiotic between the time last received by the animal and the animal's slaughter or product use. Also, where antibiotics are used for intramammary application in mastitis control, the amount of antibiotic that may be used is limited per infusion and the milk must be discarded for a period of 48 to 96 hours following treatment, depending upon the excretion pattern of the antibiotic in the vehicle employed.

The direct addition of antibiotics to foods is not permitted in the United States. In the past, the limited use of tetracyclines (chlortetracyline and oxytetracycline) to extend the shelf life of poultry and fish was permitted. However, questions about development of antibiotic-resistant pathogenic and other microorganims in consumers who were repeatedly exposed to low concentrations of antibiotics led to withdrawal by the U.S. Food and Drug Administration of even this limited usage of these antibiotics as preservatives in poultry and fish.

But many countries favor the use of antibiotics as food preservatives. The FAO/WHO Expert Committee on Food Additives favors the use of nisin as a direct additive to foods. In the United Kingdom, the use of nisin is approved for inclusion in cheese, clotted cream, and any canned food. In New South Wales, Australia, nisin may be added to cheese, canned tomato pulp, canned tomato juice, canned tomato paste, tomato puree, and canned fruit, provided the pH of these items is below 4.5; and the addition to canned soups is permitted, provided the product is subjected to heat treatment sufficient to destroy *Clostridium botulinum.* The use of nisin in cheese is also approved (at levels stipulated by the respective countries) in Belgium, Czechoslovakia, Finland, Italy, Uruguay, and India. It is also approved in France, Israel (except for soft white cheese), Jamaica, Mexico, New Zealand (for export only), Norway (for export only), South Africa, and Spain. Sweden permits nisin in sterile condensed milk and cream, and in cheese, with no limits. In Italy, nisin is approved for use in canned vegetables and in creamed desserts. In Lebanon, Mozambique, Nigeria, Zambia, and the Sudan, nisin is acceptable, although there are no specific laws covering food additives in these countries.

SOME ANTIBIOTIC CONCERNS. When properly used, antibiotics enhance human health; when improperly used, they may be a health hazard. Among the concerns are: (1) allergic reactions, (2) indirect additives, (3) vitamin K deficiency, (4) side effects, and (5) resistance to bacteria.

Allergic Reactions. About 10% of the population of the United States is sensitive to various drugs, with antibiotics heading the list. For this reason, the addition of antibiotics to foods must be carefully controlled; and only small amounts of antibiotics may be employed. It is noteworthy, however, that antibiotic residues in foods are destroyed in cooking.

Indirect Additives. Indirect antibiotic additives may be contained in foods derived from animals treated with or fed antibiotics. The use of antibiotics (1) for the prevention or treatment of diseases like mastitis in lactating cows or blackhead and coccidiosis in poultry, and (2) as growth promotants for animals, involves the risk that residues may be transmitted to foods (meat, milk, and eggs) derived from animal sources. For this reason, regulations for the use of drugs take into consideration the safety of residues of the drugs in food for man. Where relevant, withdrawal periods are required on meat animals prior to slaughter and on milk cows and poultry prior to the food use of milk and eggs, respectively.

Vitamin K Deficiency. The prolonged use of antibiotics may adversely affect the normal bacterial flora of the intestine, with the result that vitamin K deficiency occurs.

Side Effects. In some instances, antibiotics have produced such undesirable side effects as diarrhea, nausea, vomiting, abdominal cramps, and damage to the kidneys or other organs.

Resistance to Bacteria. There has been concern lest the indiscriminate use of antibiotics may lead to increased resistance of certain strains of bacteria which could ultimately render antibiotics ineffective for the treatment and control of diseases.

Some doctors believe that antibiotics should not be used (1) to treat such minor conditions as colds and sore throats, which are readily controlled by simple remedies; or (2) in the continued feeding at low levels to livestock. Those subscribing to this view believe that people's response to antibiotics should not be endangered or their immunity to disease decreased by the use of antibiotics for treatment of minor ailments or by the use of antibiotics as growth promotants in livestock feeds. The debate on this question was intensified in 1977 with the announcement of the FDA Director of his intention to prohibit the use of tetracyclines and penicillin in feeds. His contention was that the ultimate benefit-to-risk ratio is too small to allow the widespread use of antibiotics in livestock feeds. But, under heavy pressure from those who favored wide use of antibiotics, Congress simply ordered more research.

The drug manufacturers and many animal scientists have taken issue with the proposed FDA ban. They point out that this area of animal nutrition has been extensively and intensively examined with the conclusion that antibiotics pose no health hazard to either humans or animals.

The controversy is centered around the ability of certain bacteria to develop resistance to antibiotics and to spread this resistance to other bacteria, a phenomenon known as *episome.* This well-documented ability involves what are termed resistance factors (RF). *An episome is an independent genetic element occurring in addition to the normal cell genetic makeup, which can be transmitted to other bacteria, and which can replicate either as an incorporated piece of the host's genome (chromasomal material) or as a separate unit within the bacteria.* This means that an episome can be fused with the chromosomal material of the host bacteria or can be a separate piece of genetic material in the cytoplasm.

Plasmids are episomes which never become part of the host chromosome; rather, they become a self-replicating unit. Plasmids are responsible for the synthesis of resistance transfer factors (RTF).

Resistance factors are transferred among bacteria via three mechanisms—transformation, transduction, and conjugation.

• **Transformation**—In transformation, the resistance to antibiotics is transferred and incorporataed directly to the genome of the host organism. As shown in Fig. A-32, the recipient cell incorporates a piece of double stranded DNA which contains the genetic information resistance. Once inside the cell the transforming DNA becomes single-stranded with the other strand being broken down and discarded. A break in the host's chromosome then occurs and the transforming fragment then replaces a piece of the host's DNA, thereby producing a hybrid organism. When the cell replicates and divides, one daughter cell has the parent genotype and the other has the newly formed genotype.

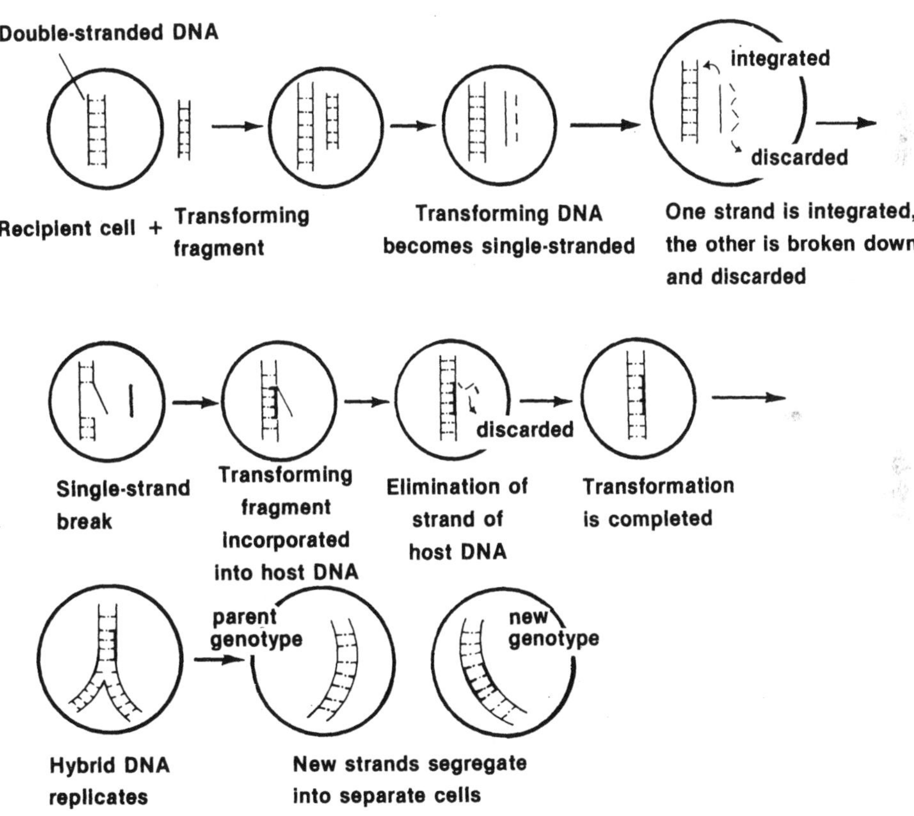

Fig. A-32. Transformation mechanism of genetic recombination.

• **Transduction**—Transduction, as illustrated in Fig. A-33, involves the formation of transducing particles in the donor cell. When the cell lyses (breaks open), these particles are released and eventually come into contact with other bacteria, whereupon the material inside the particle can either be incorporated into the chromosomes of the host bacteria or stimulate lyses of the newly involved cell.

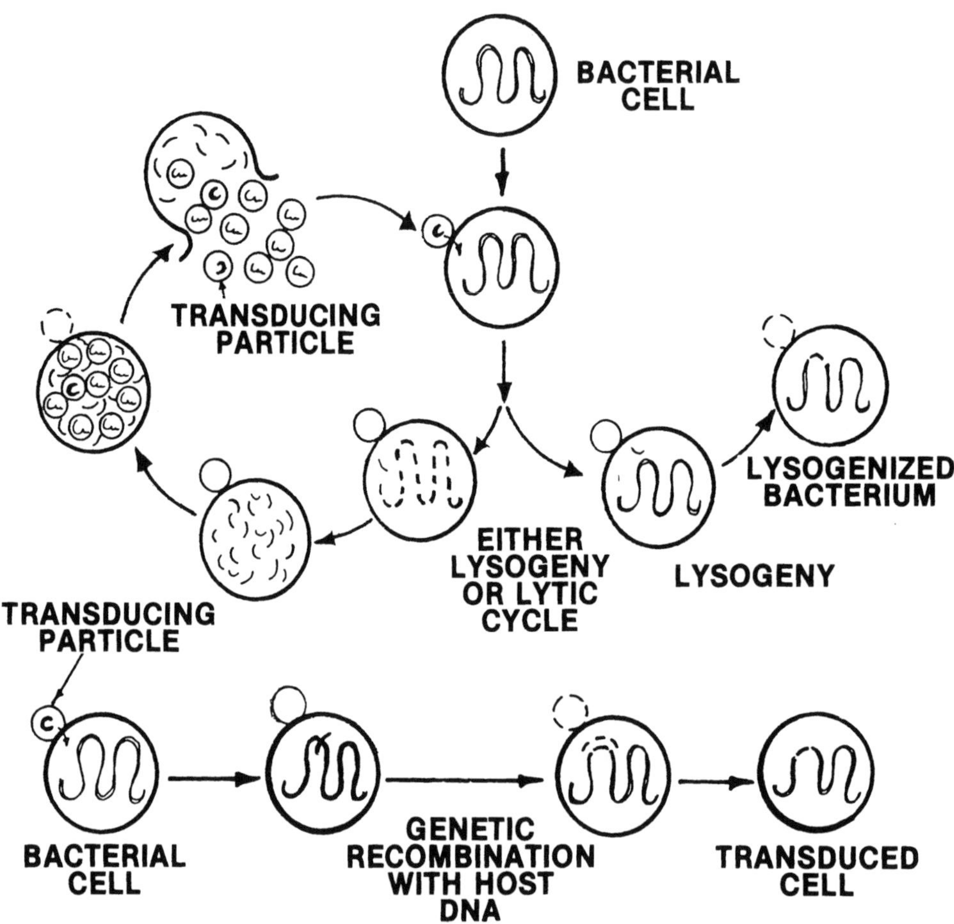

Fig. A-33. Mechanism of transduction.

• **Conjugation**—Conjugation (Fig. A-34) can be described as being similar to the act of mating in higher animals. An extension, referred to as a pilus (plural, pili) joins the donor bacteria to the recipient. Genetic material is then passed from the donor to the recipient. Resistance transfer factors are transferred by this mechanism.

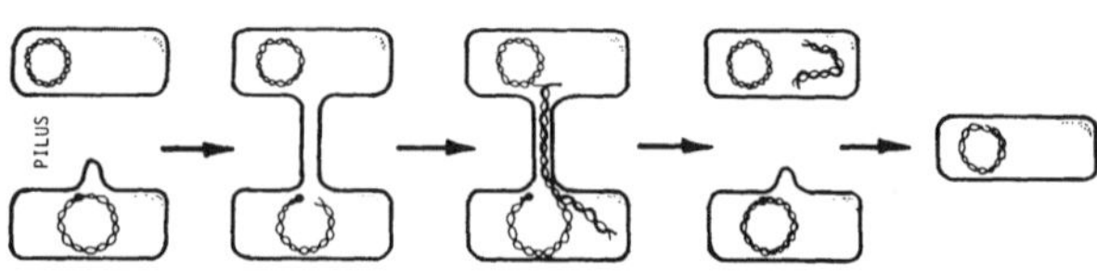

Fig. A-34. Conjugation of bacteria.

FUTURE OF ANTIBIOTICS. Today, more and more antibiotics are less and less effective in the treatment of infections, due to more resistant bacteria. When first used, penicillin was nearly 100% effective against the most prevalent *Staphylococci aureus* that spread hospital-related infection among patients; presently, the drug is far less effective. Both tetracycline and penicillin, once used to cure gonorrhea, now have a failure rate of more than 20% against certain strains. A growing body of evidence indicates that the overuse of antibiotics is helping to make the miracle drugs less effective in treating infections, and that there may come a time when many infections are resistant to all known antibiotics. Before such a disaster strikes, some groups urge setting international standards for prescription, distribution, and advertising.

Also, pressure from consumer groups and many medical people may result in banning many of the antibiotics that are primarily used for medicinal purposes in humans from the list of approved production promoters for animals. However, in the future, an increasing number of antibiotics will likely be developed specifically for the purpose of improving livestock performance. One example is that of bambermycins. This antibiotic was developed solely for use as a production promoter, serving to increase rate of gain and feed efficiency in chickens and swine; it has no medical application. Bambermycins has no toxic effect; does not result in tissue residue—so, no withdrawal period is required; and has a molecular structure completely different from other antibiotics—hence, no cross-resistance to other antibiotics has been found. Strains of bacteria resistant to bambermycins are susceptible to other antibiotics.

ANTIBODIES

Special proteins produced by the body as a defense against invasion by foreign substances called antigens—which may be (1) components of certain drugs or food; (2) infectious microorganisms or parasites; (3) substances from the environment, such as chemicals, dusts, pollen grains, etc; or (4) tissues of the body itself. Normally, antibodies react with antigens and render them harmless. However, certain antibodies may attack body tissue. This abnormal condition is called autoimmunity. Generally, repeated exposure to a specific type of antigen increases the rate at which antibodies against the substance are produced. Antibody production may be impaired under such conditions as (1) advanced aging; (2) inherited inability to produce certain antibodies; (3) oversecretion of stress hormones; or (4) malnutrition.

(Also see ALLERGEN; ALLERGIES; ANTIGEN; AUTOIMMUNITY; DISEASES; and MALNUTRITION.)

ANTICOAGULANT

A substance that interferes with one or more of the processes in the blood clotting mechanism. Such agents may be useful in preventing the formation of unwanted clots, and in the breaking up of such clots once they have formed.

(Also see BLOOD; and VITAMIN K.)

ANTIDIURETIC HORMONE (ADH)

A hormone secreted by the pituitary gland. It acts to reduce the excretion of water in the urine so that the supply of body water is conserved.

(Also see ENDOCRINE GLANDS.)

ANTIDOTE

A substance that counteracts another substance which is poisonous. Sometimes the antidote is as toxic as the poison it counteracts, but the two agents may counteract each other. For example, selenium in tuna fish greatly reduces the toxicity of any mercury which may be present in the fish.

ANTIFOAMING AGENTS (DEFOAMING AGENTS)

These substances suppress or inhibit the formation of foam in systems where it may interfere with processing. Foaming occurs due to the presence of proteins, gases, or nitrogenous materials. Examples of antifoaming agents include octanol, sulfonated oils, organic phosphates, or silicone fluids.

ANTIGEN

Any substance capable of stimulating the production of antibodies against itself. The most common antigens are animal danders, chemicals, dusts, foods, microorganisms, and pollens. Tissue or proteins from the body itself, which are normally enclosed within cells or other structures, may become antigens if they escape into the blood as a result of disease or injury.

(Also see ANTIBODIES.)

ANTI-GRAY HAIR VITAMIN

This designation was once given to pantothenic acid, one of the members of the vitamin B complex. In 1940, when pantothenic acid was first synthesized, it received widespread attention as a possible preventive for gray hair, since it had been observed that the black hair of a rat would turn gray when the animal was deprived of the vitamin, but subsequent studies did not reveal any such benefit to accrue to humans.

(Also see PANTOTHENIC ACID.)

ANTIHEMORRHAGIC

Preventing hemorrhage. A term often applied to vitamin K.

(Also see VITAMIN K.)

ANTIHISTAMINES

Any of various compounds used to treat certain allergic reactions in the body.

ANTIKETOGENESIS

The prevention of ketosis by similating the tricarboxylic acid cycle and thereby bringing about oxidation of the ketone bodies. Fatty acids and ketogenic amino acids are precursors of ketone bodies. The antiketogenic factors are precursors of glucose; these include the carbohydrates, glucogenic amino acids, and the glycerol portion of fat.

ANTIMETABOLITE

A substance bearing a close structural resemblance to one required for normal physiological functioning, which exerts its effect by replacing or interfering with the utilization of the essential metabolite.

ANTIMYCOTICS

Substances that are added to foods to inhibit mold growth, such as sodium and calcium propionate, methyl hydroxybenzoate, quaternary ammonium chloride, sodium benzoate, and sorbic acid.

ANTINEURITIC

A substance used for preventing or treating neuritis.

ANTIOXIDANT

A substance which prevents the reaction of various food constituents with oxygen. This protective effect is desirable because many substances become discolored or spoil when oxidation takes place. A wide variety of synthetic antioxidants is now available. Some of the better known natural antioxidants are vitamins C and E, and the bioflavonoids.

ANTISIALAGOGUES

Substances that stop the flow of saliva.

ANTISTALING AGENTS

These agents slow or prevent the loss of softness, flavor, and consistency of baking products.

(Also see STALING.)

ANTITOXIN

A substance produced within the body which has the power to neutralize a toxin. Sometimes antitoxins produced by cattle, horses, or rabbits are injected into humans to combat fast-acting, potent toxins.

(Also see DISEASES, section on "Vaccination.")

ANTIVITAMIN

A substance which interferes with the action of a vitamin.

(Also see VITAMIN[S], section on "Factors Influencing the Utilization of Vitamins.")

ANURIA

Lack of or defective excretion of urine.

ANUS

The opening at the posterior end of the digestive tract.

(Also see DIGESTION AND ABSORPTION.)

AORTA

The major blood vessel which carries blood away from the heart and down to the lower parts of the body. There are various branches from the aorta which connect with the arteries that carry blood to the arms and to the head.

APASTIA

Refusal to take food.

(Also see ANOREXIA NERVOSA.)

APATHY

Indifference; lack of interest or concern. A type of behavior sometimes found in a deficiency disease.

APHAGOSIS

The inability to consume food.

APHRODISIACS

Substances or foods used to promote sexual desire. Through the years, the following foods have been consumed for this purpose: anise, artichokes, avocados, beans, caraway seed, carrots, chocolate, clams, cloves, cola drinks, fish, garlic, honey, hot sauces, various mushrooms and cheeses, mutton, nutmeg, olives, oysters, peas, peppermint, pistachio nuts, radishes, saffron, shellfish, thyme, tomatoes, vanilla, along with such exotic items as hyena eyes, eel's eggs, bird's-nest soup, shark fin soup, and truffles. However, there is no scientific evidence that any food item possesses such powers.

APOENZYME

The protein part of an enzyme to which the prosthetic group or coenzyme is attached. The coenzyme may be a vitamin.

APOFERRITIN

The protein base in intestinal mucosa cells, which binds with iron (from food) to form ferritin, the storage form of iron.

(Also see IRON, section headed "Absorption, Metabolism, Excretion," Fig. I-17.)

APOLLINARIS WATER

A highly aerated alkaline water, containing sodium chloride and calcium, sodium, and magnesium carbonates. It is obtained from a spring in the valley of the Ahr (Prussia).

APOPLEXY

Another name for a stroke, which may be either a hemorrhage in the brain (cerebral hemorrhage) or a blockage of the flow of blood to the brain (cerebral infarct). In either case, the effect is that certain vital parts of the brain are deprived of a supply of blood.

(Also see CEREBRAL HEMORRHAGE; CEREBRAL INFARCT; and HIGH BLOOD PRESSURE.)

APOSIA

Absence of the sensation of thirst.

APOSITIA

Lack of desire for food.

APPENDICITIS

An inflammation of the appendix, which is often accompanied by (1) pain in the lower right side of the abdomen, (2) a low fever, (3) an elevated white blood cell count, (4) nausea, and (5) tightening of the muscles of the abdominal wall. Appendicitis almost always requires prompt surgery, because the appendix may rupture and the lining of the abdominal cavity become inflamed, producing the condition called peritonitis. The latter condition is very dangerous because it may result in shock, systemic poisoning, and other complications.

It is risky to administer a laxative or an enema to a person who complains of abdominal pain suggestive of appendicitis because these treaments may cause an inflamed appendix to rupture. However, such pain may also be due to constipation, ileitis or ileocolitis, infection of the kidneys, inflamed pouches on the wall of the colon (the condition called diverticulitis), or certain pelvic disorders in women.

(Also see APPENDIX; COLITIS; DIGESTION AND ABSORPTION; and ILEITIS.)

APPENDIX

A narrow, wormlike tube of tissue, closed at one end, with its open end attached to the cecum. Fig. A-35 shows the location of the appendix.

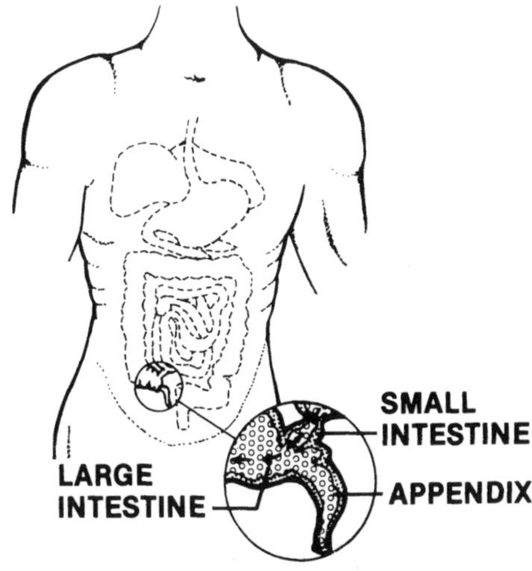

Fig. A-35. The appendix, a nonfunctional tissue which may be very troublesome if it becomes inflamed and appendicitis results.

(Also see APPENDICITIS.)

APPESTAT

A nerve center in the hypothalamus (a small structure at the base of the brain) believed to regulate appetite.

(Also see DIGESTION AND ABSORPTION, section headed "Hunger and Appetite.")

APPETITE

This is the desire for food. A good appetite, which is a sign of good health, is characterized by eating with relish.

Fig. A-36. Children often eat too much or too little. If the condition persists, parents should consult a doctor. (Courtesy, The Connecticut Ag. Exp. Sta., New Haven, Conn.)

A distinction needs to be made between appetite and hunger. Hunger is a physiological desire for food following a period of fasting. Hunger pangs result from contractions of an empty stomach and are rarely felt by anyone who regularly eats adequate amounts of food (especially protein). Appetite, on the other hand, is a learned or habitual response, which arises with the customary intervals of eating and may be influenced by numerous external and internal phenomena. It can become excessive or subject to distorted impulses.

• **Hypothalamic control of appetite**—The hypothalamus (derived from the terms "hypo" meaning below, and thalamus, a region of the brain)—a structure in the ventral region of the diencephalon—has been implicated as one of the major control centers of appetite regulation. Within the hypothalamus, certain areas can be differentiated. Two of these are of particular importance in the regulation of appetite. The first area is that of the lateral hypothalamus. It is commonly called the "feeding center," because, upon stimulation of this region, the person commences to eat whether or not he (or she) is hungry. If this area is damaged, the person loses all desire to eat. The ventro-medial area of the hypothalamus functions as the "satiety center." Stimulation of this region will depress appetite. If the ventro-medial nuclei are destroyed, there is no inhibition of feed intake, and the person will have an uncontrollable appetite. It is believed that there is a chronic activity in the lateral hypothalamus which is kept in check by the inhibitory influence of the ventro-medial area.

Several theories have been advanced as to the exact physiological mechanism which triggers the hypothalamus to tell the person when to eat. While each theory has its merits, there is no conclusive proof in favor of any one of them. In the long run, it seems probable that a combination of factors will provide the answer.

The two theories concerning the hypothalamic control of appetite that have received the most attention are (1) the chemostatic hypothesis, and (2) the thermostatic hypothesis. The chemostatic hypothesis reasons that the hypothalamus is sensitive to circulating blood nutrient levels, such as sugar or lipid. When these levels become too low, the hypothalamus sends signals to begin feeding. Once the blood nutrient level is elevated, stimuli from the feeding center are inhibited and the person feels full. The second theory of appetite control, the thermostatic hypothesis, theorizes that the hypothalamus plays an important role in heat regulation within the body, and that a decrease in hypothalamic temperature will induce feeding.

• **Gastric influences on appetite regulation**—When the gastrointestinal tract is distended, there is normally a cessation of eating. Thus, the actual physical limit of the digestive system has a direct influence on appetite. If a person eats a bulky salad, he may become satiated, even though it does not fulfill his energy requirements.

• **Sociological and psychological factors affecting appetite**—Appetite is affected by such sociological factors as habits, family customs, education, cultural background, environment, and social customs; and by such psychological factors as emotional need, stress, and anxiety.

It is known that the amount of physical activity indirectly influences the appetite. Likewise, it is believed that idleness and boredom stimulate the appetite and produce snacking as a means of breaking the monotony.

It is believed that psychological attitudes can either promote or suppress appetite. Various associations with past pleasures of food consumption and the desirable appearance, odor, and flavor of foods are known to stimulate the appetite. Children are apt to develop cravings for food without realizing the cause. Eating may be a form of compensation for a major loss or disappointment; or a means of securing attention from adults who otherwise might not notice them.

• **Physiological factors**—Various physiological factors, such as mild thiamin deficiency, illness, and infection have been associated with anorexia, or loss of appetite. Occasionally people suffering from specific disorders, such as diabetes, gastric ulcers, or chronic gastritis, develop an appetite out of proportion to their needs. In pregnancy or hysteria, unusual and specialized cravings may develop for particular kinds of food, or even for injurious substances.

Persons with duodenal ulcer seem to have a special form of appetite. Their pain is apt to rise and fall with the stomach's desire for food and content of hydrochloric acid. In such cases, eating tends to relieve the pain, at least temporarily.

Diminution or loss of appetite accompanies many disordered conditions. It is usually one of the symptoms of tuberculosis and anemia. Also, loss of appetite and refusal to eat anything characterizes *anorexia nervosa*, a disease ordinarily related to some form of emotional instability and observed more frequently in women than in men.

The loss of sensitivity to flavors and odors, which often accompanies aging, appears to be related to a concommitant reduction in appetite.

• **Drugs, alcohol, tobacco, and empty calories**—Appetite can be artificially decreased by taking such drugs as Benzedrine and Dexedrine. These drugs are sometimes prescribed for weight reduction, but they should be taken only on the advice of a doctor.

Small amounts of an alcoholic drink will increase appetite. On the other hand, overconsumption of alcohol usually results in loss of appetite. Excessive use of tobacco appears to reduce the appetite; for this reason, addicted smokers often gain weight when they break the habit.

The consumption of empty calories, such as candy and soft drinks, inhibits the appetite.

APPETIZER

A substance that stimulates the appetite.

APPLE *Malus pumila*

Contents Page

Types of Apples 103
Origin and History 104
Production 104
Propagation and Growing 105
Harvesting 105
Storage of Apples 105
Processing 106
Selection 107
Preparation 107
Nutritional Value 108

The apple is a pome—a fruit consisting of firm, juicy, flesh surrounding a core that contains several seeds. It is a close relative of the pear and the quince. With about 7,500 varieties grown worldwide and 2,500 varieties in the United States, size, sweetness, aroma, and crispness vary greatly from one variety to another, as does the color

Fig. A-37. Apple—the king of the fruits. (Courtesy, Agricultural Chemicals Division of Mobay Chemical Corporation, Kansas City, Mo.)

which ranges from shades of red to yellow or green.

Apple trees grow everywhere except in the very coldest and the very hottest climates of the world. They do not thrive in the tropics because they need a period of cold and dormancy to grow properly. Apple trees are a medium-sized tree, with unpruned trees on good soil reaching a height of 30 to 40 ft *(9.2 to 12.2 m)*. Their trunks may be 2 ft *(0.6 m)* or more in diameter. Branches of the apple tree are twisted and spreading, and the tree is about as broad as it is high.

TYPES OF APPLES. In Canada and the United States, 18 to 25 varieties comprise the majority of the commercial apple crop, though many other varieties may be found in home gardens. Nearly all of the major varieties grown today originated as chance seedlings (mutations) before 1900. Table A-27 compares the characteristics of some of the favorite apple varieties. The leading variety of apple in the United States is the Delicious, which originated from a mutation on a farm in Peru, Iowa about 1881.

TABLE A-27
CHARACTERISTICS OF SOME COMMON APPLE VARIETIES

Variety	Origin	Areas Where Grown	Harvest Season	Fruit Color	Use	Flavor
Baldwin	Mass., 1740	New Eng., N.Y., Mich.	October	Medium red	Culinary	Slightly tart
Cortland	N.Y., 1898	N.Y., New Eng.	September	Medium red	Dessert and culinary	High flavor
Delicious	Iowa, 1881	All areas except South. Popular in Northwest, Midwest, Virginia, New Eng.	September	Striped red to full red	Dessert only	Sweet
Golden Yellow Delicious	W. Va., 1890	Northwest, Midwest, Virginia, N.Y., Calif.	September to October	Yellow	Dessert and culinary	Milder than delicious
Gravenstein	Germany imported to U.S. by 1824	Calif.	July to August	Red stripes	Culinary, early dessert, and canning	Mildly acid
Jonathan	N.Y., approx. 1880	Northwest, Midwest, Virginia, Calif.	September	Bright red all over	Dessert and culinary	Pleasantly tart
McIntosh	Ontario, Canada, approx. 1870	N.Y., New Eng., Mich., B.C., Ontario, Canada	August to September	Medium red	Home use and local market	High flavor
Northern Spy	N.Y., approx. 1800	N.Y., New Eng., Mich.	October to November	Bright striped red	Dessert and culinary	Slightly acid
Red Astrachan	Russia. Intro. to U.S. in 1835	Calif.	July	Green to entirely red	Home use and local market	
Rhode Island Greening	R.I., approx. 1740	N.Y., New Eng., Mich.	September to October	Green	Drying and culinary	Refreshing, brisk, acid
Rome Beauty	Ohio, 1848	Midwest, New Eng., Northwest, Virginia, Calif.	October to November	Medium red stripes	Dessert and culinary, baking	Tart
Stayman Winesap	Kansas, 1866	Midwest, Virginia	October to November	Medium red	Dessert and culinary	Tart
Wealthy	Minn., approx. 1860	N.Y., New Eng., Mich., Minn.	August to September	Medium red	Dessert and culinary	Mildly acid

(Continued)

TABLE A-27 (*Continued*)

Variety	Origin	Areas Where Grown	Harvest Season	Fruit Color	Use	Flavor
White Astrachan	Russia. Intro. to U.S. about 1820	Calif.	July to August	Greenish white	Culinary and local market	
White Pearmain	Unknown, date unknown	Calif.	October	Yellow	Home use and local market	
Winesap	Unknown, dates back to colonial period	Northwest, Midwest, Calif.	October to November	Entirely red	Dessert and culinary	Slightly tart
Yellow Newtown (Newtown pippin)	N.Y., early 18th century	Northwest, Oregon, Calif., Virginia	September to October	Green to yellow	Best for drying and culinary	Winelike

ORIGIN AND HISTORY. The commercial-type apple is native to western Asia and/or eastern Europe. Apples were grown by the Greeks as early as the 4th century B.C.

Mention of the apple in mythology also suggests man's long relationship with the apple. In Greek mythology, Hercules traveled to the end of the world to bring back the golden apples of the Hesperides. An apple caused a quarrel between the gods that led to the Trojan War. A Norse myth tells of magic apples that kept people young forever. Dipping for apples, a Halloween game, began among the Celtic people of Britain, as a means of fortune telling. Also, some people consider the apple to be the forbidden fruit of the Garden of Eden, though this is probably attributed to the position of apples in general in mythology.

Supposedly, the Romans took cultivated apples with them when they conquered England, and apple growing became common in England and other parts of Europe. During the 1620s, colonists introduced the apple to North America. Then, as the frontier moved westward, apple trees followed. Some followed with the help of a man named John Chapman (1774-1845), better known to most people as Johnny Appleseed. He carried apple seeds and sprouts with him wherever he traveled and planted them in newly settled areas of the country. Gradually the apple was spread by explorers, Indians, and pioneers. American growers developed new and improved varieties, and apples became an important part of the economy.

Seven kinds of wild apples are native to the United States. Most wild apples are crab apples with small, sour, hard fruit. These are not cultivated to any extent.

Fig. A-38. Gathering apples in the early 1800s. (Photo of the Bettmann Archive, Inc., New York, N.Y., drawn by E. Forbes)

PRODUCTION. Worldwide, the average yearly production of apples is 40 million metric tons. The leading countries, by rank, are the U.S.S.R., China, the United States, and Germany. Other leading countries and their yearly production are indicated in Fig. A-39.

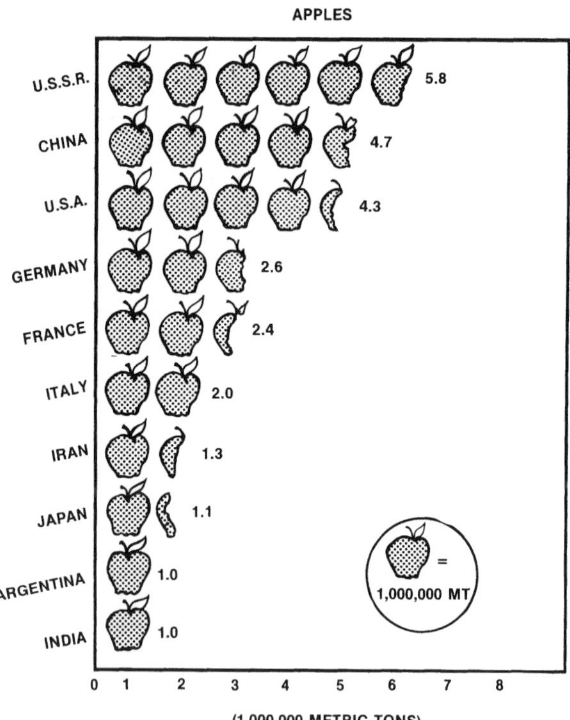

Fig. A-39. Leading apple-producing countries of the world. (Based on data from the *FAO Production Yearbook*, 1990, FAO/UN, Rome, Italy, Vol. 44, pp. 160, 161, Table 69)

Thirty-five states, and parts of Canada and Mexico, produce apples on a commercial basis in North America. In the United States, an average of 4.8 million tons (*4.4 million metric tons*) are produced each year—a crop valued at over $1.4 billion. Among the states, Washington leads in yearly production by growing 49.5% of the U.S. crop. Other leading states are indicated by their yearly apple production in Fig. A-40.

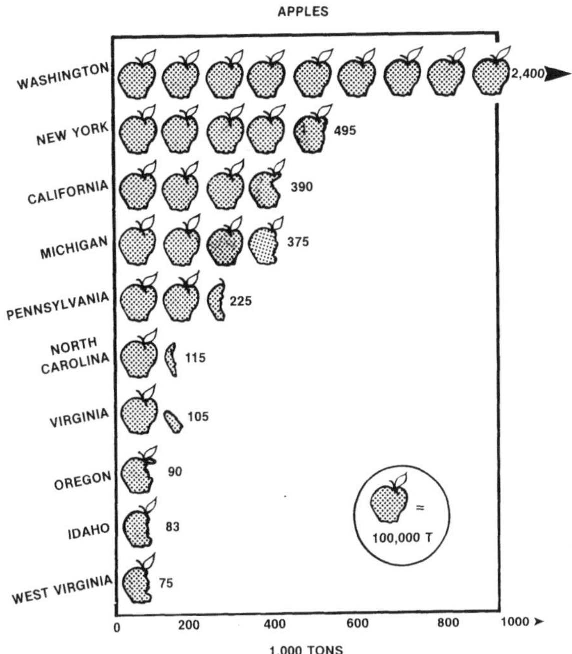

Fig. A-40. Leading apple-producing states. (Based on data from *Agricultural Statistics*, 1991, USDA, Table 255)

Propagation and Growing. Apples are propagated by budding or grafting onto seedling rootstocks, or onto clonal rootstocks of (1) native apples of Europe, such as the French crab, (2) the standard varieties like the Golden Delicious. A dwarf apple tree is produced by grafting or budding onto the rootstocks of the East Malling, from East Malling, England. Dwarf apple trees may reach a height of about 6 ft (*1.8 m*), or some intermediates reach the height of a peach tree.

Apples are grown on a wide range of soils, but they grow best in deep, well-aerated loam soil with abundant water supplies from rain or irrigation. Nursery trees are planted 20 to 30 ft (*6 to 9 m*) apart for standard-sized trees and 6 to 12 ft (*1.8 to 3.6 m*) apart for dwarf trees.

Apple trees generally do not bear fruit until they are 5 to 8 years old, and they reach their peak production around 20 years. Commercial orchard trees are generally replaced after 35 to 40 years.

Apple trees blossom in the early spring. Most apple cultivars (varieties) are self-unfruitful. They require another cultivar for cross-pollination in order to set a commercial crop. During the bloom period, many orchardists make it a regular practice to rent colonies of bees for pollination purposes.

To produce a good crop, orchard management involves pruning, fertilizing, spraying to control pests and diseases, and possible thinning of unripe fruit to ensure a consistent year-to-year crop. Some common diseases and pests of apples which attack the tree and the fruit include: the fungi, apple scab, and cedar rust; the bacterial disease, fire blight; and the Jose scale, leaf roller, aphids, and codling moth (apple worm).

Harvesting. Apples destined to the fresh market, or those which are not processed immediately are hand picked. Apple pickers start with the lower branches and then use long ladders to reach tops of trees. Pickers must be very careful not to bruise apples or they will spoil. Picking apples with their stems is a good pratice, because of higher grade and longer storage.

Fig. A-41. Close-up of golden delicious apples. (Courtesy, USDA)

Some apples may be harvested with a mechanical shaker and catchers when they are destined to immediate processing into juice, cider or other products. However, bruises and cuts turn brown within a few hours, decreasing the value of the fruit.

Following picking, apples continue to respire and the greater the rate of respiration the shorter the storage life. Storage methods seek to slow down the rate of respiration. Low temperatures and/or decreased oxygen slow the rate of respiration. Prior to storage, however, the field heat of apples is often rapidly removed by precooling apples in rooms or chambers where there is rapid cold air circulation. Precooling slows down life processes of the apple, thereby prolonging storage life.

Storage of Apples. There are three main types of storage used for long-term storage of apples: (1) common or nonrefrigerated, (2) refrigerated, and (3) controlled atmosphere. The objectives of all three are the same: to slow respiration, to prevent shriveling and shrinkage, and to inhibit the development of rot organisms.

1. **Common or nonrefrigerated.** This system makes use of the difference between night and day temperatures. Apples are stored in an insulated building and, depending upon the outside temperature, cool air is sucked in from the outside and warm air inside is forced outside until inside and outside temperatures are the same. This system is cheap and efficient, but the first of the crop harvested in the fall when temperatures are warmer may not reach a low enough temperature.

2. **Refrigerated.** Large rooms are kept at 31° to 34°F (*−1 to 1°C*) by refrigeration. Apples are held in these rooms until they are marketed. For some varieties this can be 6 months.

3. **Controlled atmosphere.** This is an airtight room which may or may not be refrigerated. After apples are placed in the room it is sealed, the oxygen content is slowly reduced, and the percentage of carbon dioxide increased due to the respiration of the fruit. Oxygen is maintained around 5% and carbon dioxide at 2 to 5%. The combination of controlled atmosphere and low temperature can keep some varieties in good eating condition for up to a year.

Apple packages sealed with a material of limited permeability, such as polyethylene, can create a situation similar to the controlled atmosphere room—only on a smaller scale. In sealed packages, the carbon dioxide increases as oxygen decreases as a result of normal respiration of the fruit, thus reaching a level where respiration slows. For example, sealed packages have been found to contain 10% oxygen and 5% carbon dioxide, thus in combination with refrigeration they are an efficient storage method for some apple varieties.

PROCESSING. More than half (55%) of the apples in the United States are marketed on a fresh basis. The remaining 45% is processed as canned, dried, or frozen apples, or as vinegar, cider, and juice. Only 16% is canned; 3% is dried; and 2% is frozen. Twenty-three percent is crushed and used to make vinegar, cider, and apple juice.

Apples are canned, sliced or cubed, as pie filling or as applesauce. Slices are usually packed without syrup. However, applesauce is the most important canned apple product. Fig. A-42 outlines the steps involved in canning apples.

Dried apples are trimmed, peeled, cored and sliced, and sulfured prior to drying. They are generally rings or quarters, and they are dried to 25 to 33% moisture. Some apples are dehydrated to 2 to 3% moisture.

Frozen apples are sliced and packed with dry sugar or in a sugar syrup. The major problem with frozen apples is the prevention of enzymatic browning during storage and especially after thawing.

• **Apple juice and cider**—These products are gaining in popularity. Apple juice refers to the product preserved by heat (pasteurization) and packed in hermetically sealed containers. Cider is the unfermented fresh juice from apples. Originally fresh cider was sold at roadside markets, but now it can be purchased in supermarkets. Fresh cider rapidly spoils without the addition of some chemical preservative such as potassium sorbate, or ex-

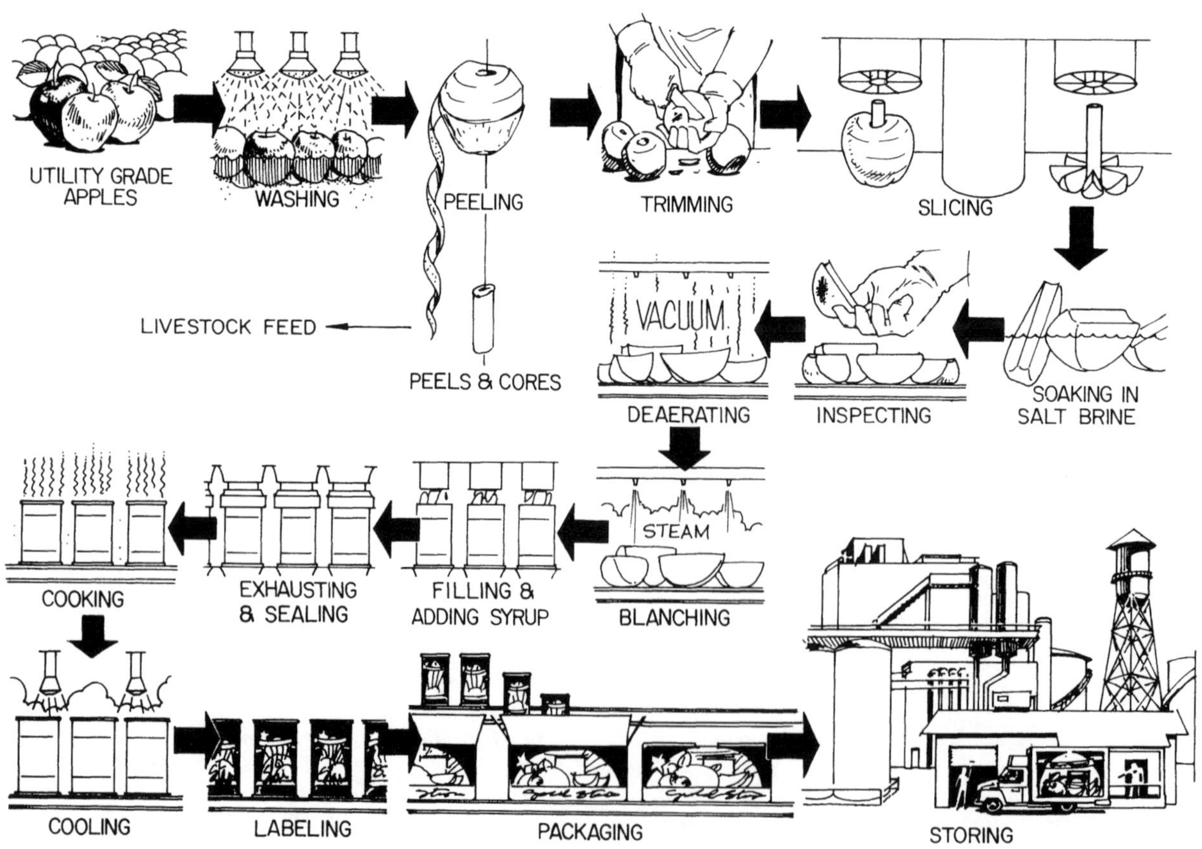

Fig. A-42. Canning apples from the fresh fruit to cans ready for the consumer.

posure to ultraviolet radiation, though refrigeration slows spoilage extending the storage life of fresh cider to 1 or 2 weeks.

Fig. A-43 outlines the commercial production of apple juice.

Other products from apples include cider vinegar, jelly, and apple butter. In Europe, much of the apple crop pressed for cider is fermented to make brandy or wine.

(Also see PRESERVATION OF FOOD; and VINEGAR.)

SELECTION. The many varieties of apples differ widely in appearance, flesh characteristics, seasonal availability, and suitability for different uses (see Table A-27). For good eating as fresh fruit, the commonly available varieties are: Delicious, McIntosh, Stayman, Golden Delicious, Jonathan, and Winesap. Tart or slightly acid varieties such as Gravenstein, Grimes Golden, Jonathan, and Newtown make good pies and applesauce. For baking, the firmer-fleshed varieties—Rome Beauty, Northern Spy, Rhode Island Greening, Winesap and York Imperial—are widely used.

Fresh apples should be firm, crisp, and well-colored. Flavor varies in apples and depends on the stage of maturity at the time the fruit is picked. Apples must be mature when picked to have a good flavor, texture, and storing ability. Immature apples lack color and are usually poor in flavor. They may have a shriveled appearance after being held in storage. Overripe apples are indicated by yielding to a slight pressure on the skin and soft meally flesh. Apples frozen at some time show internal breakdown and bruised areas. Irregular shaped tan or brown areas—scald—may not seriously affect the eating quality of apples.

Most apples are marketed by grade, and many consumer packages show the variety, the grade, and the size. U.S. grades for apples are U.S. Extra Fancy, U.S. Fancy, U.S. No. 1, and combinations of these grades. U.S. No. 2 is a less desirable grade. Apples from the far western states are usually marketed under state grades which are similar to federal grades. The qualities of color, maturity and lack of defects—appearance in general— determine the grade.

PREPARATION. Many apples are enjoyed fresh, eaten out of the hand as a snack. Also apples are enjoyed with other fruits in fruit salads or on fruit plates. Baked apples are an old-time favorite. Apples can also be fried (apple fritters) or stewed. Fresh, canned, or frozen apples are used to make the all American favorite—apple pie. Dried apples are a good snack, or they can be cooked or used in baked goods. Apple jelly and apple butter are all-time bread spread favorites.

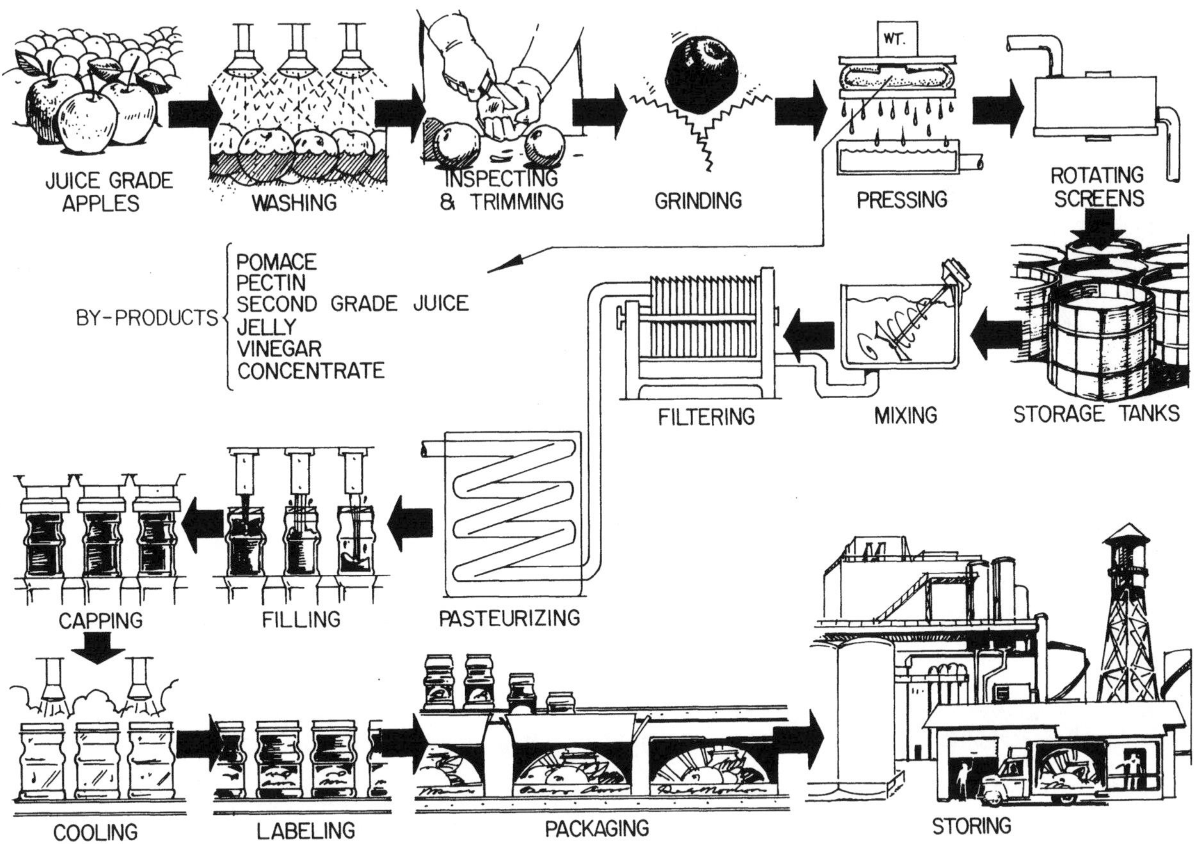

Fig. A-43. The production of apple juice from fresh fruit to a consumer-ready product. Important by-products of the apple juicing process include pectin, jelly, vinegar, and apple concentrate.

NUTRITIONAL VALUE. Fresh apples contain 85% water. Each 100 g (about 3½ oz) contains about 60 Calories (kcal) of energy. One average apple weighs about 150 g. Calories in the fresh apple are derived from the naturally-occurring sugars (carbohydrate) which give the apple its sweet taste. Their taste is also due to small amounts of organic acids. Overall, apples possess minor amounts of the minerals and vitamins. Dried apples contain only 33% water; hence, concentrating the nutrients, including calories. Each 100 g of dried apples contains 239 Calories (kcal) of energy, 3 g of fiber, 405 mg of potassium, and 1.7 mg of iron. More complete information regarding the nutritional value of fresh, baked, canned, and dried apples, apple juice, cider, and applesauce is presented in Food Composition Table F-36.

(Also see FRUIT[S], Table F-47 Fruits of the World.)

APPLE BUTTER

A spread for bread which is made by the prolonged cooking of a mixture of apple pulp and sugar until it turns to a dark brown color and has a caramel flavor. The rural people of Colonial America sometimes added apple cider, orange peels, and quinces to their apple butter. About 2½ lb (*1.1 kg*) of cored, peeled apples (which is equivalent to 3 lb [*1.35 kg*] of fresh apples) are required to make 1 lb (*0.45 kg*) of apple butter.

The nutrient composition of apple butter is given in Food Composition Table F-36.

Some noteworthy observations regarding the nutrient composition of apple butter follow:

Compared to canned, sweetened applesauce, apple butter has (1) a much higher solids content, (2) twice as many calories (186 kcal per 100 g), and (3) four times the potassium content.

(Also see APPLE.)

APPLE CIDER

In the United States, this term usually refers to freshly prepared apple juice that is bottled without pasteurization. Products labeled "apple juice" have usually been pasteurized during bottling. Sometimes, pasteurized apple juice is marketed as cider during the fall, when the demand for cider is great. Unpasteurized cider will ferment within a few days unless preservatives such as sodium benzoate or potassium sorbate are added to it. In Europe, "cider" means fermented apple juice, whereas Americans call fermented cider "hard cider."

It is noteworthy that 100 lb (*45 kg*) of apples yield 65 to 70 lb (*31 kg*) of apple cider or apple juice.

The residue of apple pulp which remains after the juice has been extracted is called pomace and is used as (1) a material for the production of pectin, and (2) an animal feed.

Hard cider contains less alcohol (the content ranges from 0.5 to 8.0%) than wine. It may be plain or sparkling (sparkling cider is made by allowing some fermentation to take place after bottling).

The nutrient composition of unfermented apple cider is given in Food Composition Table F-36.

Some noteworthy observations regarding the nutrient composition of apple cider follow:

A 1-cup serving of apple cider supplies 124 Calories (kcal) and 34 g of carbohydrate. It is a good source of potassium and iron, but it is a poor source of vitamin C, unless extra vitamin C has been added by the manufacturer.

(Also see APPLE.)

APPLE, DRIED

Pieces of apple (usually slices from peeled and cored fruit) which have been dried by artificial means to a moisture content of 24% or less. Discoloration during drying is prevented by treating the pieces with sulfur dioxide. Compared to fresh apples, dried pieces of apple contain about 5 times the level of calories, carbohydrates, and potassium. About 5 lb (*2.3 kg*) of peeled, cored apples (equivalent to between 5½ and 6 lb of fresh apples) are required to make 1 lb (*0.45 kg*) of dried apple pieces. Certain other dried apple products may contain added sugar or syrup.

The nutrient compositions of various forms of dried apple are given in Food Composition Table F-36.

Some noteworthy observations regarding the nutrient composition of dried apples follow:

1. The uncooked dried fruit is rich in calories (275 kcal per 100 g) and carbohydrates (72%). It is also rich in fiber and potassium, but it is low in vitamin C.
2. Cooked, sweetened dried apples have only 40% of the caloric and carbohydrate content of the uncooked fruit.
3. Cooked, unsweetened dried apples have only 28% of the caloric and carbohydrate content of the uncooked fruit.

(Also see APPLE.)

APPLE JACK

This is the American term for apple brandy. It is called "Calvados" in Europe. This brandy, which contains from 55 to 65% alcohol by volume, is produced by distilling fermented apple cider. Most of the world's supply of apple brandy comes from France, where the annual rate of production is usually several times that of the United States.

(Also see DISTILLED LIQUORS.)

APRICOT *Prunus armeniaca*

The apricot, fruit of the apricot tree, is classified as a drupe—a fleshy, one-seeded fruit not splitting open by itself and containing a seed enclosed in a stony endocarp. As the genus name suggests, the apricot is a close relative of the almond, cherry, peach, and plumb. Ripe apricots are orange or yellow, round, and about 1 ¼ to 2 in. (*3 to 5 cm*) in diameter. Their skin is smoother than that of a peach and their edible flesh is sweet, but dry compared to other fruits.

Fig. A-44. Harvesting apricots—an intermediate between the peach and plum. (Courtesy, USDA)

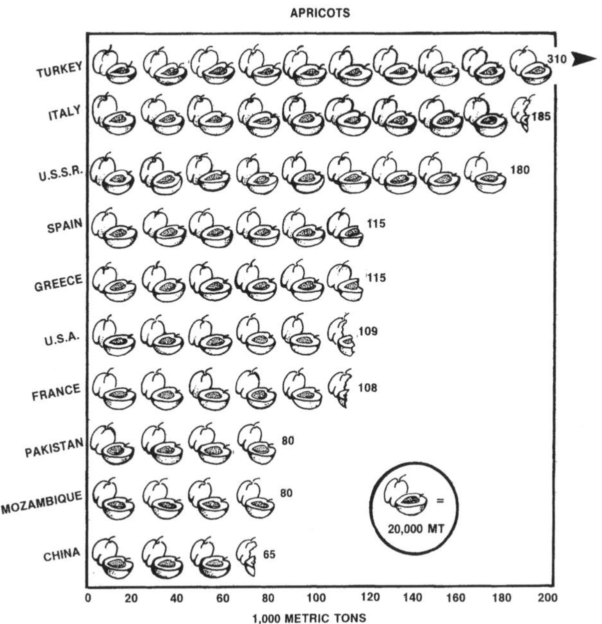

Fig. A-45. Leading apricot-producing countries of the world. (Based on data from the *FAO Production Yearbook,* 1990, FAO/UN, Rome, Italy, Vol. 44, pp. 165-166, Table 72)

Apricot trees require a mild climate. They blossom early in the spring, so they cannot be grown where spring frosts are common. However, a winter chilling is necessary to induce dormancy. Mature trees reach a height of about 30 ft (*9 m*). Delicate pink blossoms cover the trees in the spring. Most commercial apricot varieties are self-fruitful, though they do benefit from bee pollination.

Two other species are also considered apricots but they are not as important: (1) the Japanese apricot, *Prunus mume*; and (2) the black apricot, *Prunus dasycarpa*.

Sometimes the Japanese apricot is grown as an ornamental. The Japanese grow it, and pick the fruits green for pickling in a salt solution. The black apricot is cultivated in some areas of Asia and Europe. It resembles a plum, but in the United States it is considered unattractive and of poor quality.

ORIGIN AND HISTORY. Apricots originated in China (eastern Asia). Alexander the Great is reported to have brought the apricot to Greece. Pliny, about the time of Christ, mentions the fruit as a Roman fruit. Gradually, the cultivation spread northward in Europe, and by 1592 English records mention the "cot." By 1720, the apricot reached Virginia in the New World, and about 1792 it was recorded as growing in the mission orchards of California. Apricot trees grow in many areas throughout the United States, but large-scale commercial production is limited to a few areas.

PRODUCTION. Worldwide, an average of 2.1 million metric tons are produced each year. Leading apricot-producing countries are Turkey, Italy, U.S.S.R., Spain, and Greece. Other leading countries and their production are indicated in Fig. A-45.

In the United States, the commercial crop of apricots comes from California, Utah, and Washington—a yearly crop averaging 122.5 thousand tons (*110 thousand metric tons*). California produces 94% of this crop, which is valued at around $41 million. Principal commercial varieties are Royal, Blenheim, Tilton, Moorpark, and Goldcot.

Propagation and Growing. Apricots are propagated by grafting onto the roots of plum, peach, or other apricots. In orchards the trees are planted from 24 to 30 ft (*7 to 9 m*) apart. Apricot trees bear a light crop usually after 4 to 6 years, and the commercial life of a good orchard is 15 to 20 years. Apricot trees are usually not grown where winter temperatures are lower than -10° to -15°F (*-23° to -26°C*) or where there are fewer than about 700 hr of temperature below 45°F (*7°C*). Apricots bloom early in the spring, thus they may be damaged by spring frosts. During fruit growth, apricots require warm weather to mature, though hot weather can damage the fruit. In humid climates, fungus disease of apricots, which is difficult to control, may develop. Thus, most apricots are grown in dry climates under irrigation. With proper orchard management techniques such as fertilizing, pruning, and pest control, yields average 350 to 400 bu per acre. Larger fruits are produced by thinning the unripe fruit from the tree. Harvesting begins in the early summer.

Harvesting. Ripe apricots are harvested by hand, or with tree shakers which shake the fruit onto catching frames or the ground. The picking season for a given variety usually lasts for about 2 or 3 weeks, and the fruit is harvested in 2 or 3 pickings. Following harvesting, apricots are not stored as fresh produce for any length of time, though they may keep for 1 or 2 weeks at 31° to 32°F (*-0.6° to 0°C*). When apricots are harvested at a firmness permitting storage and shipping, the fruit lacks flavor and dessert quality after ripening. However, this type of fruit is suitable for canning.

PROCESSING. Apricots are marketed as fresh, canned, dried, or frozen fruit. Utilization of the apricot crop in the United States is as follows: 8% fresh, 58% canned, 26% dried, and 7% frozen.

Storage time, and the susceptibility to bruising limit the consumption of fresh apricots. To reach markets, fresh apricots need protective packaging to reduce handling damage.

Apricots may be canned as unpeeled halves, whole peeled or whole unpeeled fruits and canned in water pack, juice pack, or light or heavy syrup. Also some apricots are made into apricot juice, blended with other juices, apricot puree, and apricot concentrate. Canned apricots are first sized, graded, and pitted.

For drying purposes, apricots are cut by machine or hand, exposed to the fumes of burning sulfur in a "sulfur house," and then spread on trays to dry in the sun or placed in a large dehydrator. Dehydrators remove water under controlled conditions of air flow, temperature, and humidity. Sulfuring inactivates enzymes to achieve product quality and storage stability. Blanching is also sometimes used for this purpose. To make apricot leather, whole fruits are washed, blanched, macerated, and then evaporated before being formed into sheets or a thin belt for dehydration. For dried fruit, about 5.5 lb (*2.5 kg*) of fresh fruit yields 1 lb (*0.45 kg*) of dried fruit.

In 1945, 17% of the total U.S. apricot crop was frozen, but since that time the percentage has declined. Due to their high susceptibility to brown and oxidation on exposure to air, apricots are more difficult to freeze successfully than other fruits. Apricots are frozen covered with sugar solution containing 0.1% ascorbic acid (vitamin C).

SELECTION. Fresh apricots are primarily marketed in June through August, but a limited supply of imported apricots is available in larger cities during December and January. Apricots should be plump and juicy in appearance, with a uniform, golden-orange color. Ripe apricots yield to a gentle pressure on the skin. Dull-looking, soft or mushy fruit and very firm pale yellow or greenish-yellow fruit should be avoided since these conditions are indicative of overmaturity or immaturity, respectively. Apricots develop their flavor and natural sweetness on the tree. They should be mature—but firm—at the time of picking.

PREPARATION. Apricots can be eaten fresh out of the hand or prepared in fresh fruit dishes. Canned apricots are eaten as a dessert dish or also used to add a piquant flavor to sauces, salads, or baked goods. Both fresh and dried apricots are cooked to make jam, pies, and puddings. Dried apricots are also a good snack. Frozen apricots are used in confections and baked goods.

Some apricots have sweet kernels inside their stone that can be eaten like almonds, but most of the varieties grown in the United States have bitter pits (kernels). The bitter taste is amygdalin which is poisonous when consumed in large enough quantities.

NUTRITIONAL VALUE. Fresh apricots contain 85% water. Additionally, each 100 g (about 3 ½ oz) provides 51 Calories (kcal) of energy, 319 mg potassium, 2,700 IU of vitamin A and only 1 mg of sodium. The calories in apricots are derived primarily from sugars (carbohydrate) which give them their sweet taste. When canned in syrup, apricots contain more calories due to the addition of sugar. Dried apricots contain only 32% water; hence, many of the nutrients are more concentrated, including the calories. In dried form, they contain 236 Calories (kcal), 1,422 mg potassium, 4.8 mg of iron, and 7,147 IU of vitamin A in each 100 g. Frozen apricots are similar to fresh, with slightly more calories due to the sugar added before freezing. More complete information regarding the nutritional value of fresh, canned, dried and frozen apricots is presented in Food Composition Table F-36.

(Also see FRUIT[S], Table F-47 Fruits of the World.)

APRICOT LEATHER

A chewy sheet of dried apricot pulp that is made by: (1) deskinning and pureeing the fruit, (2) adding a little sweetening and spices, (3) boiling the mixture for about 3 minutes, and (4) drying the puree on a sheet of plastic wrap in a drying tray. The leather makes a nutritious, easily carried, nonperishable snack for between meals, and for camping and hiking trips.

(Also see APRICOT.)

AQUEOUS EMULSION

A stable mixture of two or more immiscible liquids held in suspension by an emulsifier is called an emulsion. An oil and water emulsion is known as an aqueous emulsion, e.g., milk.

ARABINOSE

A five-carbon sugar found mainly in certain root vegetables. It has no known function in man.

(Also see CARBOHYDRATE[S].)

ARACHIDONIC ACID

A 20-carbon unsaturated fatty acid with four double bonds, which occurs in most animal fats and is considered essential in the nutrition of man.

(Also see FATS AND OTHER LIPIDS.)

ARACHIS OIL

Arachis is the genus name for peanut; hence, arachis oil is sometimes used to designate peanut oil.

(Also see PEANUTS.)

ARBUTUS BERRY (STRAWBERRY TREE)
Arbutus unedo

The fruit of an ornamental shrub of the family *Ericaceae* which grows in North America, Mexico, Europe, and Africa.

Although the berries are not very palatable when fresh, they may be made into confections, distilled liquors, liqueurs, and wine. The alcoholic beverages are produced in Algeria, France, Italy, and Spain.

Fig. A-46. The arbutus berry, a fruit that the French use in making confections and liqueurs.

ARGINASE

An enzyme present in most animal cells, which splits arginine to urea and ornithine, the last state of urea synthesis from the amino groups of the amino acids. A deficiency of arginase produces elevated arginine levels in the blood and an accumulation of ammonia in the body, which may result in spastic diplegia (Little's disease), epileptic seizures, and severe mental retardation.

ARGININE

An amino acid which participates in the formation of the waste product urea and the muscle constituent creatine. In certain species such as chickens, pigs, and rats, arginine appears to be essential for growth. It is noteworthy that this amino acid is a potent stimulator for the secretion of growth hormones.

(Also see AMINO ACID[S].)

ARIBOFLAVINOSIS

The term given to a set of symptoms produced by a deficiency of riboflavin (vitamin B-2) or by the presence of a factor which interferes with the absorption and utilization of riboflavin over a period of several months. Ariboflavinosis is characterized by swollen, cracked, bright red lips (cheilosis); enlarged, tender, purplish or bright red tongue (glossitis); cracking at the corners of the mouth (angular stomatitis); congestion of the blood vessels of the conjunctiva; flaking and irritation of the skin in the folds around the nose and in the groin area (seborrheic dermatitis); and ocular disorders, such as vascularization of the cornea, irritation and sensitivity of the eyes, inflammation of the eyelids, watering and mattering of the eyes, abnormal pigmentation of the iris and intense photobia. But, since none of these symptoms are unique to ariboflavinosis, diagnosis is quite difficult. In general, however, ariboflavinosis is associated with lower plasma and urinary riboflavin levels. Less than 14 mcg of riboflavin per 100 ml of plasma and/or less than 200 mcg of riboflavin per gram of creatinine in the urine are indicative of a deficiency condition.

Deficiency symptoms often occur during periods of physiological stress, such as pregnancy, lactation, and rapid growth. Also, ariboflavinosis is often associated with such diseases as pellagra, rheumatic fever, tuberculosis, subacute bacterial enteritis, chronic congestive heart failure, hyperthyroidism, malignancy, chronic fever conditions, and cirrhosis of the liver. Abnormalities of riboflavin metabolism result from all types of severe injuries; and degradation of riboflavin coenzymes appear to be related to shock. Surgical removal of sections of the gastrointestinal tract and chronic diarrhea have both produced malabsorption of riboflavin, followed by ariboflavinosis.

Ariboflavinosis is best prevented dietarily by an adequate diet containing riboflavin rich foods, such as milk, liver, meat, eggs, and certain green leafy and yellow vegetables. Milk is the best common source of riboflavin; one quart will more than fulfill the daily riboflavin requirement. Seeds and whole-grain cereals are not good sources of riboflavin, unless they are enriched and fortified. Enrichment of these products has done a great deal in preventing ariboflavinosis in the United States.

Ariboflavinosis is treated with an adequate diet and administration of 5 mg of riboflavin 3 to 4 times daily. The riboflavin supplement is usually administered orally. However, it is sometimes given intramuscularly, particularly when problems of absorption exist.

(Also see RIBOFLAVIN.)

ARRHYTHMIA

A nonrhythmic, irregular beating of the heart which is sometimes an early warning of a heart attack. However, some people have a slightly nonrhythmic heartbeat for most of their lives. When an arrhythmia causes a serious disturbance of the heart rhythm, the heart is generally inefficient in pumping blood.

(Also see HEART DISEASE.)

ARROWROOT

There are two sources of arrowroot:

1. *C. angustifolia* is found in the Himalayan region and has long been used by the Natives of India. It has a slight yellow color, but it is a good substitute for the West Indian arrowroot.
2. *Maranta arundinacea* is grown in South America, the West Indies, Mexico and Florida. The American Indians used it to heal wounds made by poisoned arrows, which probably accounts for its name.

The starch made from the root is used in foods prepared for children and invalids. Other similar starches which can be substituted for arrowroot are obtained from several other plants of the same genus, *Maranta*, or other genera such as *Zamia*, *Curcuma*, *Tacca*, *Canna*, and *Musa*.

(Also see FLOURS, Table F-26 Special Flours.)

ARSENIC (As)

An element. Most forms are highly toxic when ingested or inhaled in sufficient quantities.

(Also see POISONS, Table P-11.)

ARTERIOLE

A blood vessel which delivers blood from an artery to the capillaries. Arterioles are smaller in diameter than the arteries, yet bigger than the capillaries.

ARTERIOSCLEROSIS

A hardening and thickening of the normally elastic walls of arteries so that they no longer respond readily to changes in the blood pressure. Often, there are calcium deposits in the walls of the arteries which make them brittle. Arteriosclerosis is often accompanied by atherosclerosis.

(Also see HEART DISEASE.)

ARTERY

A large blood vessel carrying blood from the heart to various parts of the body.

ARTHRITIS

Contents *Page*

Osteoarthritis (degenerative joint disease, or hypertrophic arthritis) 112
Causes 113
Symptoms (diagnosis) 113
Treatment 113
Prevention 114
Rheumatoid Arthritis (arthritis deformans, or atrophic arthritis) 114
Causes 114
Symptoms (diagnosis) 114
Treatment 114
Prevention 115
Gout (gouty arthritis) 115
Causes 115
Symptoms (diagnosis) 115
Treatment 116
Prevention 116
Advice to All Arthritics 116

Arthritis is an inflammation of the joints of the body. The term refers to all the conditions that cause stiffness, swelling, soreness, or pain in the joints.

Arthritis is among the oldest diseases known to affect human beings. Evidence of its occurrence has been found in the mummies and excavations of ancient civilizations. Also, Hippocrates (460-377 B.C.), the Greek physician called the "Father of Medicine," described arthritis graphically.

Arthritis is a dreaded and feared crippling disease. The Arthritis Foundation estimates that more than 13 million Americans are afflicted by some form of this condition, and that a quarter of a million more become victims of this disease every year. The human suffering from arthritis is awesome; and the cost in dollars is almost impossible to comprehend. Figures show that arthritis sufferers lose an estimated 115 million days a year from work, which is equivalent to 470,000 persons on sick leave for an entire year. This is more days lost from work than is caused by any other malady except nervous and mental diseases.

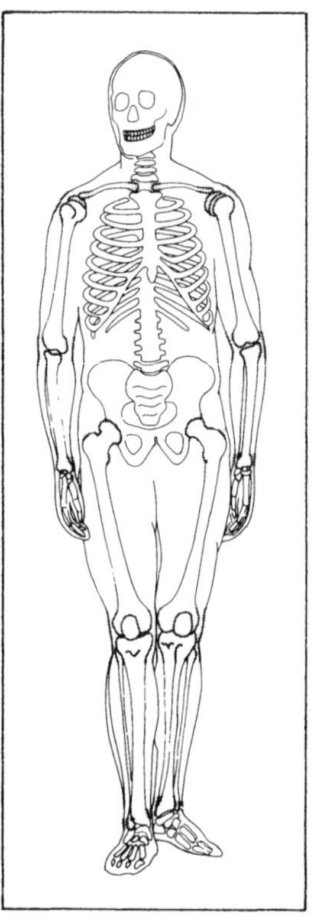

Fig. A-47. Arthritis. The shaded areas show the joints which are most subjected to the wearing away of the linings of the cartilages which reduce friction and serve as "bearings" for the movable parts of the body.

There are many different types of arthritis. Some are caused by infections, some by injury, some by aging, and still others by entirely unknown causes. Infectious arthritis may follow influenza, typhoid fever, tuberculosis, syphilis, or gonorrhea. Arthritis of unknown cause is common, the worst form of which is rheumatoid arthritis.

There are three main kinds of arthritis: (1) osteoarthritis, (2) rheumatoid arthritis, and (3) gout.

OSTEOARTHRITIS (DEGENERATIVE JOINT DISEASE, OR HYPERTROPHIC ARTHRITIS). Osteoarthritis, which is the most common type of arthritis, results from the wear and tear on the joints. Most persons over 50 years of age have osteoarthritis to some degree; for this reason it is often called "old-age rheumatism." However, it often follows injuries, infections and/or diseases afflicting the joints, and it can occur in relatively young people.

Causes. Osteoarthritis is usually produced by a constant strain on a joint which produces a wearing away of the articular cartilage within the joint, especially the weight-bearing joints—the knees, hips, ankles, and spine. Overweight and stress can aggravate the disease.

Symptoms (diagnosis). Osteoarthritis is the most common cause of painful knees, backs, and fingers. But it seldom causes deformity of the joints or crippling. Affected joints may "creak" and grate on movement. Typically, pain is increased by exercise and relieved by rest. Unlike rheumatoid arthritis, osteoarthritis does not produce any nodes under the skin and is not accompanied by constitutional symptoms such as fever and weight loss.

Treatment. The treatment of osteoarthritis consists of weight reduction if overweight, moderation of activity, use of heat, taking aspirin, and, in severe cases, the physician may resort to the injection of cortisone or ACTH into the painful joints and the use of braces to relieve strain. People with osteoarthritis seem to feel better in warm dry climates, like Arizona and New Mexico; wet or damp environments may aggravate the condition.

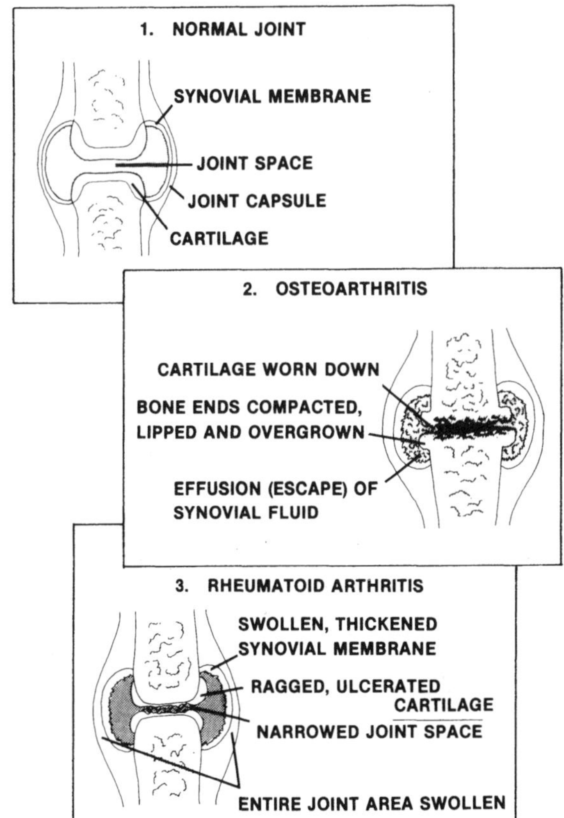

Fig. A-48. Schematic drawings of: (1) A normal joint. (2) Osteoarthritis joint, showing the cartilage worn down; the bone ends compacted, lipped, and overgrown; and effusion (escape) of synovial fluid. (3) Rheumatoid arthritis joint, showing swollen, thickened synovial membrane; ragged, ulcerative cartilage; and narrowed joint space. (See Fig. A-50 for gout.)

Diet may be significant. Osteoarthritis is more common among those who are past middle age—when there is a tendency to put on extra weight, to deposit less calcium in the bones, and sometimes to develop a mild form of anemia. Although these conditions are not caused by the osteoarthritis as such, they can be improved by sound nutrition. So, the doctor will probably advise the person suffering from osteoarthritis to eat more foods rich in calcium, such as milk and milk products, as well as foods containing iron and vitamins; and he will advise the patient to lose weight without deprivation of any of the essential nutrients.

Obesity is to be avoided because extra weight places additional stress on the weight-bearing joints. Patients lose weight more happily on a high-protein, low-carbohydrate, modest-fat, low-caloric diet. So, the following foods should be included and emphasized in the daily diet:

Food	Amount
Milk *(preferably nonfat)*—	3 cups *(0.72 liter)*
Meat—	9 oz *(252 g)*: Each of the following equals 1 oz meat— Meat and poultry, 1 slice (3" × 2" × 1/8") Cold cuts, 1 slice (4 ½" sq × 1/8" thick) Frankfurter, 1 (8-9 per lb) Codfish, mackerel, 1 slice (2" × 2" × 1") Salmon, tuna, crab, ¼ cup Oysters, shrimp, clams, 5 small Sardines, 3 medium Cheese, cheddar, 1 slice (3 ½" × 1 ½" × ¼") Cheese, cottage, ¼ cup Eggs, 1 Peanut butter, 2 Tbsp (limit to one portion per day unless carbohydrate is allowed.
Vegetables(raw or cooked)—	4 servings: 1-2 servings leafy or yellow 1 small serving of potato Other vegetables
Fruit (unsweetened)—	2 servings: 1 citrus 1 other fruit
Cereals and breads (whole grains)—	2 servings
Butter or margarine (fortified)—	5 teaspoons *(25 ml)*

Prevention. An essential element of prevention of osteoarthritis is the avoidance of excessive weight gain, which adds to the wear and tear of weight-bearing joints. Secondary preventive measures are directed at reducing the strain on affected joints by use of a back brace, abdominal support for a sagging abdomen (to take the strain off the back), a neck brace, and adequate rest.

RHEUMATOID ARTHRITIS (ARTHRITIS DEFORMANS, OR ATROPHIC ARTHRITIS).

Rheumatoid arthritis, often called the great crippler, is a progressive, debilitating, and chronic disease that manifests itself primarily by joint pain and inflammation, usually leading to deformities. Studies show that about one out of three arthritic patients have this type. Although this form of arthritis can develop in people of any age, including children just a few months old, it most commonly attacks middle-aged persons. Three times as many women as men get rheumatoid arthritis. Patients with rheumatoid arthritis are usually underweight and undernourished. Rheumatoid arthritis is not related to rheumatic fever, with which it may be confused due to the similarity of names.

People afflicted with rheumatoid arthritis normally have negative nitrogen and calcium balances and reduced tolerances to carbohydrates (starches and sugars). Reduced carbohydrate tolerances are believed to result from the chronic inflammatory state, rather than from any defect in carbohydrate metabolism.

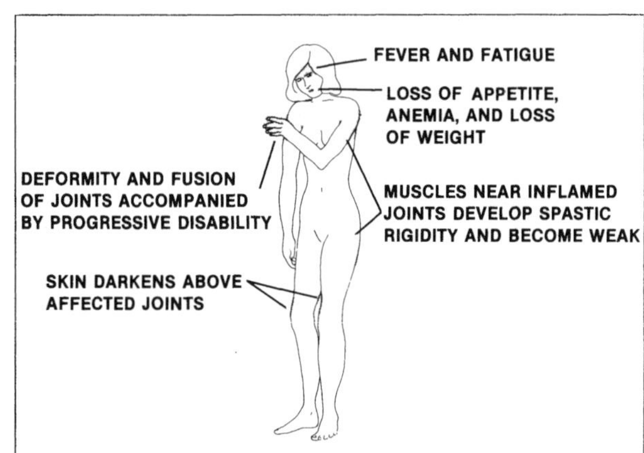

Fig. A-49. Characteristics of rheumatoid arthritis in later stages.

Causes. Although medical science is learning more and more about rheumatoid arthritis, no known cause has been identified and documented. Some experts believe that it results from a generalized infection and that pain, swelling and inflammation in the joints is only one of its manifestations. Others believe that an emotional component influences the course of the disorder. Other possible causes that have been suggested include a cold, damp climate; a bacterium or virus; injury, accident, fatigue, strain to a joint (for example, "housemaid's knee"), shock, allergy, heredity, metabolic disorders, and/or metabolic deficiencies.

Symptoms (diagnosis). Rheumatoid arthritis is progressive and involves many joints. The most common symptoms are swelling and pain of the joints of the fingers, wrists, knees, ankles, or toes, alone or in combination, although other joints may be involved. One hallmark of the disease is that usually both sides of the body are affected; both hands or both ankles, for example. Eventually, the joints stiffen in deformed positions, producing crippling. Other common symptoms are: fever, fatigue, stiffness of joints, weakness of muscles, anemia, anorexia (loss of appetite), and loss of weight. In about 10% of the cases, nodules may appear under the skin, especially around the elbows, wrists, fingers, and occasionally the ankles. Diagnosis in the early stage of the disease may involve x-ray studies; confirmation consists of characteristic changes of bones and joints.

About ¼ of the people who get rheumatoid arthritis recover from the first attack without serious symptoms; ½ suffer only minor discomfort; while the remaining ¼ develop chronic, progressive, disabling arthritis.

Treatment. The treatment of rheumatoid arthritis is difficult and prolonged. But, in most cases, deformity can be prevented. Helpful treatments include a well-balanced diet, adequate rest along with enough physical exercise to prevent stiffening, the application of heat on the joints, maintaining inflamed joints in nondeforming positions (this may require the use of splints and braces), and taking aspirin and other pain-killing drugs as recommended by the physician. Injection of gold salts (gold chloride and gold with sodium thiosulfate) and the use of ACTH or cortisone drugs over long periods of time have helped certain persons through this chronic disease with little or no crippling. Also, intravenous injection of antimalarial drugs is sometimes of benefit in lessening the symptoms in younger persons. In some cases, replacement of patients' joints with plastic or rubber joints enables them to move without pain. There is no cure for rheumatoid arthritis, but more than half of the patients recover eventually.

Due to the chronic disability and severe pain produced by rheumatoid arthritis, afflicted people often fall prey to food faddists and charlatans who offer quick and easy cures. Numerous diets have been devised and promoted as a treatment for rheumatoid and other forms of arthritis; based on low carbohydrate levels, high protein levels, low protein levels, acid-ash, alkaline-ash, low fat levels, high levels of B-complex vitamins, high levels of vitamin C, high levels of sulfur, and/or high levels of vitamin A. Claims are even made that a bowl of cherries will help some gouty patients. In the past, sometimes massive doses of vitamin D were prescribed and used, but this treatment is seldom recommended today, because excess vitamin D has been demonstrated to produce severe, and sometimes even fatal, calcification of the kidneys, through disruption of normal calcium and phosphorus metabolism.

Although food will neither cause nor cure rheumatoid arthritis, good nutrition can affect the way the patient feels, how well he responds to treatment, and how successfully he can resist the ravages of inflammation.

The diet of rheumatoid arthritis patients should be rich in proteins, minerals, and vitamins—all vital to tissue resistance and repair. This calls for plenty of meat, fish, and dairy products, fruits, and vegetables. Liberal amounts of milk and other sources of calcium are recommended. The extra supply of dietary calcium helps avoid the adverse effects of the disease on (1) tissue repair and growth, and (2) bone metabolism; and such a diet helps fortify the body against the general debilitation and possible anemia that may accompany rheumatoid arthritis.

Since rheumatoid arthritis is a chronic, progressive, inflammatory tissue disorder, many patients lose weight and become undernourished. Thus, a high-protein, high-calorie diet, with adequate minerals and vitamins, is recommended until a normal nutritional state is achieved. Also, since anorexia, or loss of appetite, normally accompanies rheumatoid arthritis, it is important that attractive meals be served.

The following daily intake pattern, along with the foods to emphasize, is suggested:

Food	Amount
Milk—	4 cups (1 quart or *0.96 liter)*
Meat, fish, or poultry—	6 oz (168 g) or more (cooked weight)
Eggs, cheese, or nuts—	As main dish or in other foods
Vegetables— (one raw)	4 servings: 1 serving green or yellow 1 or 2 servings of potato or other starchy vegetable 1 or 2 servings of other vegetables
Fruits—	2 servings: 1 serving citrus 1 serving of other fruit
Cereals and breads (whole grains)	4 slices of bread 1 serving of other cereal
Butter or margarine (fortified)—	2 Tbsp (*30 ml*)

Additional foods to meet caloric need. These may be selected from above list of foods.

If the patient is unable to eat this amount of food, small and frequent feedings are indicated. Also, nonfat milk may be used as reinforcing agent to increase the protein content of foods.

Prevention. There is no known preventive for rheumatoid arthritis. But deformities should be avoided if possible.

GOUT (GOUTY ARTHRITIS). To many people mention of gout conjures up a stereotyped picture of a fat old aristocrat sitting with his flannel-wrapped foot propped up on a stool, crippled because of his chronic indulgence in alcohol and food, particularly beef.

In reality, gout is an inherited disorder of metabolism in which uric acid in excessive amounts appears in blood and tissues and produces a painful swelling of the joints of the hands or feet, especially the big toe.

Causes. Gout is caused by a defect in the body's natural process of breaking down *purines* (nucleoproteins, which are compounds of one or more proteins and nucleic acid). This results in the production of too much uric acid, which accumulates in the blood, where it combines with sodium to produce sodium urate. Crystals of uric acid are deposited in tissues around the joints. These deposits cause sudden attacks of swelling, most commonly in the feet.

It is not known why the excess of uric acid appears in the blood, why the excess is not destroyed, or why urates are deposited in the tissue.

Gout tends to be familial; about 25% of the relatives of patients suffering from gout have the disease. Also, it is noteworthy that there is a high prevalence of gout in certain populations, including the Maoris of the Pacific area, and the Blackfoot and Pima Indians of the United States.

Attacks can be brought on by excesses in eating or drinking, or by certain acute medical or surgical problems.

Symptoms (diagnosis).

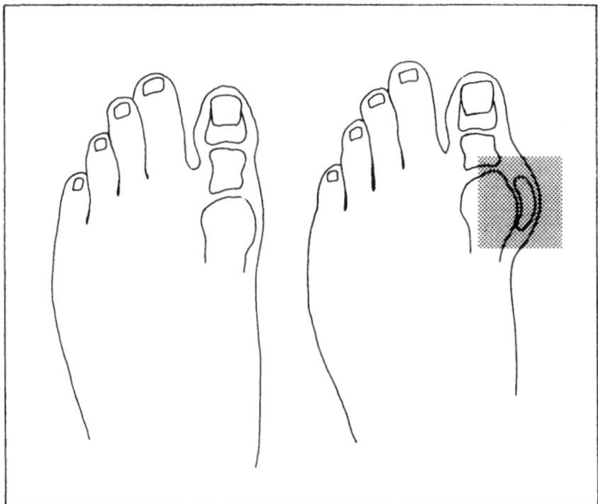

Fig. A-50. Gout, showing (1) normal toe (left) vs (2) swollen big toe (bunion) where the big toe joins the foot.

The major symptoms of gout are those of severe arthritis, a singularly dramatic and painful type of joint inflammation which comes on without warning in acute and recurrent attacks. Ordinarily gout first appears in men in middle or early old age, but it sometimes occurs in young people, also. It affects women infrequently, only about one case in 50. Hippocrates observed the age-sex relationship of gout, and noted that eunuchs were not affected. Typically, an attack of gout begins very suddenly, often at night, with the onset of excruciating pain in a single joint. Most frequently it involves the "bunion joint"—where the big toe joins the foot, but the ankle, the instep, the knee, or virtually any other joint may occasionally be the site of the attack. Usually only one joint or tendon or bursa is affected at a time. If the disease continues uncontrolled, multiple joints may be affected. Bumps (crystalline deposits of urates, known as tophi) on the cartilage of the ear appear in about half of all gout victims. Gout may also begin with the sudden appearance of kidney stones.

Unlike the dull deep-seated ache of rheumatoid arthritis or osteoarthritis, the pain of gout tends to be sharp and agonizing, accompanied by sudden swelling of the affected joint and marked redness of the overlying skin. The pain may be so severe that the victim cannot even bear the touch of light bedclothes on the affected area. If untreated, the attack may last for three or more days before it gradually subsides, followed by the skin of the affected joint peeling and a completely symptom-free period. Typically, early in the disease, an attack may occur once every 4 or 5 months or even at longer intervals, usually affecting just one joint. Later, the attacks become more frequent and multiple joints may be involved. Gout finally becomes chronic.

People in sedentary work are more likely to have gout than manual laborers.

Patients with gout often have increased amounts of uric acid (as urates) in their blood, but such increases occur under a number of other circumstances. Thus, a test that shows an elevated concentration of uric acid in the blood does not necessarily indicate the presence of gout.

Treatment. Treatment consists of limiting the amount of purines in the diet, and of taking drugs under the supervision of a physician. Among the drugs used in the treatment of gout are aspirin, salicylic acid, cortisone, ACTH, Butazolidin, Benemid, uricosuric agents, and colchicine (the alkaloid from the autumn crocus, *Colchicum atumnale, L,* used to treat gout since the 5th century).

It is recognized that dietary measures are effective in reducing the incidence and severity of acute gout attacks. The diet should largely exclude foods that are rich in the white crystalline substances known as *purines,* which includes liver, sweetbreads, kidneys, brains, and anchovies. Also, alcohol and fats should be avoided. Milk, eggs, cheese, cereals, fruits, green vegetables, cocoa, tea, coffee, sweets, and nuts are relatively low in purines.

Gout is not curable, but it may be controlled by proper treatment.

Prevention. In a family where there is gout, men should routinely be given a blood test for uric acid. A diet may be suggested to delay the onset of gouty symptoms, since there are foods which are thought to predispose the formation of uric acid. However, it has not been established that this approach is effective.

ADVICE TO ALL ARTHRITICS. Evidence exists that different forms of arthritis may result from overwork, allergy, exposure to extreme cold conditions, accidents, injuries, infections, diseases, malnutrition and/or severe mental stress. However, none of these causes has been fully documented. Therefore, the specific cause of most forms of arthritis is presently unknown. It is also known that administration of calcium, vitamins A, C, D, E, K, and/or the B-complex vitamins does not uniformly alter arthritic symptoms or the course of the disorder.

Contrary to popular belief, there is no such thing as an "arthritis diet"—a diet that will cure arthritis. However, certain adjustments in the diet of arthritics may be desirable, depending on the kind of arthritis, the treatment and medicines used, and the weight and overall state of health. There is overwhelming evidence that nutritionally balanced meals eaten regularly benefit anyone's overall health, muscle tone, and, in the case of arthritis, build ability to resist the wear and tear of the disease. Although some minor adjustments in specific items may be required, in general the good diet for anyone, whether they have arthritis or not, is based on a selection from four food groups—the milk group, the meat group, the vegetable and fruit group, and the bread and cereal group.

Doctors lament the money that is spent on useless and extravagantly advertised preparations. If some of this money were spent on nutritionally adequate and pleasurable meals, it would help fortify the patient's body and emotions against the pain and crippling of the disease.

The following suggestions provide the soundest advice to arthritics:

1. Consume a nutritious, well-balanced diet.
2. Avoid everyday tensions and get plenty of rest.
3. Maintain proper weight; do not overeat or undereat.
4. Pay no attention to the "free advice" from well-meaning but unqualified friends, relatives, and self-styled food experts.
5. Do not be taken in by the false claims and fancy cures of food faddists, makers of quack medicines, or unscrupulous drug purveyors; they benefit (profit) only those who produce and market them.
6. Consult the family physician or an established arthritis clinic, and follow the advice obtained.
7. Use corticosteroid therapy and/or other medication only under the direction of a reliable physician.

ARTICHOKE, GLOBE *Compositae*

This vegetable, which is considered to be a delicacy, is a thistlelike plant. The edible flower bud is characterized by green leaflike scales (bracts) that enclose it. Both the bracts and the base of the flower (receptacle), which is called the "heart," are eaten. In some places, the tender young leaf stalk may also be consumed. Fig. A-51 depicts a typical globe artichoke.

Fig. A-51 shows both an immature flower head and a mature flower head. It is noteworthy that only the flower bud is edible because the bracts become too tough when the flower reaches maturity and opens.

Fig. A-51. Globe artichoke.

ORIGIN AND HISTORY. The globe artichoke appears to have originated somewhere between North Africa and the eastern end of the Mediterranean, where it has been cultivated for thousands of years. It is believed that both the globe artichoke and the cardoon were domesticated from the same wild plants because they are very similar in their characteristics. The ancient Greeks and Romans were very fond of the artichoke and devoted considerable land to its cultivation. Later, this plant was brought to the Americas by both the French and Spanish explorers.

PRODUCTION. Most of the world's artichoke crop is produced by the countries bordering on the Mediterranean Sea, with over 80% coming from Italy, Spain, and France. Fig. A-52 shows where most of the globe artichokes are grown.

Nearly all of the globe artichoke grown in the United States is produced in the area of Castroville, California, the "Artichoke Capital of the World."

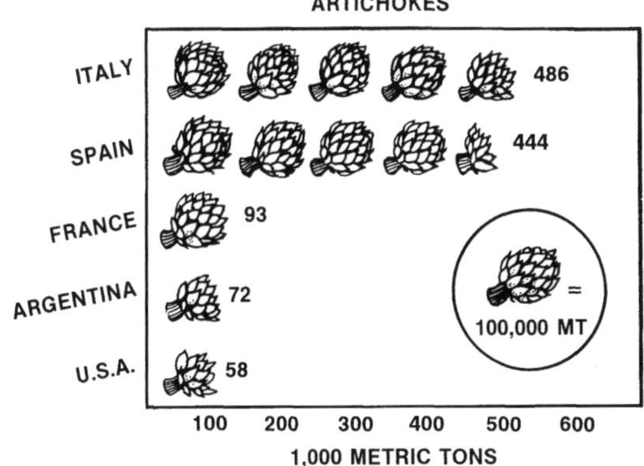

Fig. A-52. Leading artichoke-producing countries of the world, and the tonnage produced by each. (Based upon data from *FAO Production Yearbook*, 1990, FAO/UN, Rome, Italy, Vol. 44, p. 130, Table 51.

Propagation and Growing. Although the aboveground portion dies down each year, the plant continues to grow year after year because new shoots emerge from the root crown (the uppermost portion of the roots, which grows near the surface of the soil). These new shoots or "suckers" are also used to establish new plants. However, the globe artichoke is also propagated from pieces of the root crown itself, which contains more stored food than the suckers.

The globe artichoke is well adapted to the Mediterranean climate, which is characterized by warm winters and cool summers. Frost may kill the plant; and excess heat forces early flowering, which makes the bracts too tough for eating. A rich soil is also required for the production of edible buds. Each plant may grow to a height of 4 ft (*1.2 m*) or more and bear several buds. In California, the crop is irrigated 3 to 5 times during the growing period. Maximum yields are usually obtained during the second and third years after planting. The average yields in California are 6,750 lb per acre (*7,560 kg/ha*).

Harvesting, Grading, Packaging, and Shipping. Artichoke buds are harvested by cutting them off as soon as they reach maturity, but before they start to open, because the edible tissue become increasingly fibrous and tough as the buds open. Fig. A-53 shows artichokes that are ready to be harvested.

After harvesting, artichokes are graded for size, because the larger ones bring higher prices. Then, they are graded for quality so that the best ones may be sold fresh, while the rest may be processed. Each grade of artichoke is packed in separate paper-lined boxes for shipment. Refrigerated railway cars or trucks are used for long distance shipping. About 60% of the crop is marketed fresh, and the rest is processed.

SELECTION AND PREPARATION. The most desirable globe artichokes are compact, plump, heavy in relation to size, somewhat globular, and with large, fresh, fleshy, tight-clinging green leaf scales. Freshness is indicated by the green color, which with age or injury becomes partially brown. Overmature artichokes have hard-tipped leaf scales that are opening or spreading;

Fig. A-53. California artichokes. (Courtesy, Agricultural Chemicals Division of Mobay Chemical Corporation, Kansas City, Mo.)

also, the center formation may be fuzzy and dark pink or purple in color. Leaf scales on such overmature specimens are tough and woody when cooked and may be undesirably strong in flavor.

Seriously discolored artichokes are usually bruised, or lacking in freshness. Bruises appear as dark-colored areas at the point of injury. They may also show mold growth. Bruised or seriously discolored artichokes usually turn grayish-black or black when cooked.

Some of the most common ways of preparing artichokes are: (1) boiling the whole vegetable; (2) stuffing parboiled artichokes with ground meat, bread crumbs, bacon bits, and/or anchovies, and baking; (3) breading and deep frying the parboiled vegetable; and (4) cooking artichoke hearts together with fish, meat, or poultry in a cream sauce.

The vegetable may be served hot or cold with a well-seasoned sauce. Often, artichoke sections are held by the fingers and dipped into the sauce. The tender inner portion is eaten and the remainder is discarded.

NUTRITIONAL VALUE. The nutrient composition of the globe artichoke is given in Food Composition Table F-36.

Globe artichokes may contain some nonnutritive substances which exert medicinal effects, because they have been thought useful in (1) stimulating the flow of bile and other digestive juices, (2) promoting the urinary loss of excessive water from the body, and (3) lowering the blood sugar. (*Note:* The authors make no claim that any of the beliefs regarding the medicinal effects of globe artichoke are accurate; rather, it is hoped that mentioning these claims will encourage further research relative thereto.)

(Also see VEGETABLE[S], Table V-6, Vegetables of the World.)

ARTICHOKE, JERUSALEM

The Jerusalem artichoke, which is native to America, is a close relative of sunflower. It is cultivated for the edible tubers, which grow underground and look like ginger root or misshapen potatoes. Jerusalem artichoke is often recommended for diabetics, as it has no starch; the carbohydrate of fresh tubers is in the form of inulin rather than sugar. But if they are stored for a long period of time, the inulin converts to sugar.

ARTIFICIALLY DRIED

Dried, or dehydrated, by other than natural means, usually by heat.

(Also see PRESERVATION OF FOOD.)

ARTIFICIAL SWEETENERS

Numerous naturally-occurring chemicals have a sweet taste. People, too, can create chemicals which have a sweet taste. Examples of artificial sweeteners include saccharin and aspartame—products of the laboratory. Generally, artificial sweeteners impart sweetness without adding calories.

(Also see SWEETENING AGENTS.)

ASCARIDS

Roundworms, ranging from 8 to 12 in. (*20 to 30 cm*), which may infest the human intestine. The infection enters the body through the mouth, by the ingestion of unclean or contaminated fruits or vegetables and impure drinking water containing eggs of the parasite. A female ascarid lays about 200,000 eggs per day, most of which pass into the feces. Ascarid eggs may remain alive for several years on moist earth in a warm climate such as that in the southeastern United States. In the body, sometimes the worms migrate to the gallbladder, throat, lungs, liver, heart, or nose.

(Also see DISEASES.)

ASCITES

An accumulation of fluid in the abdomen. This disorder may be due to high blood pressure, sodium retention, heart failure, or a severe protein deficiency.

ASCORBIC ACID (VITAMIN C)

Ascorbic acid is another name for vitamin C, the very important substance, first found in citrus fruits, which prevents scurvy, one of the oldest scourges of mankind.

(Also see VITAMIN C.)

ASCORBIN STEARATE

This is a fat-soluble ester of ascorbic acid (vitamin C) and stearic acid. It is often added to food as an antioxidant at concentrations of about 0.1%.

ASCORBYL PALMITATE ($C_{22}H_{38}O_7$)

A derivative of ascorbic acid (vitamin C) and palmitic acid. It is a white or yellowish white powder with a citruslike odor, which is used as an antioxidant for fats and oils and as a source of vitamin C. In amounts of 0.1 to 0.4% by weight of the flour, it serves as an antistaling agent in bakery products. It will retard staling for 2 to 4 days.

-ASE

A suffix used in naming an enzyme; for example, peptidase.

ASH

The minerals in a food. Also, the residue that remains after complete burning or metabolism of the organic matter.

(Also see ANALYSIS OF FOODS.)

ASPARAGINE

A derivative of aspartic acid, a nonessential amino acid. Asparagine is formed in protein metabolism and is found both in tissues and in the blood.

(Also see AMINO ACID[S].)

ASPARAGUS *Asparagus officinalis*

The spears (young shoots) of asparagus, which is a member of the Lily (*Liliaceae*) family, are an expensive, but popular, vegetable. Hence, where the climatic conditions are suitable, many home gardeners grow asparagus. Fig. A-54 shows asparagus spears.

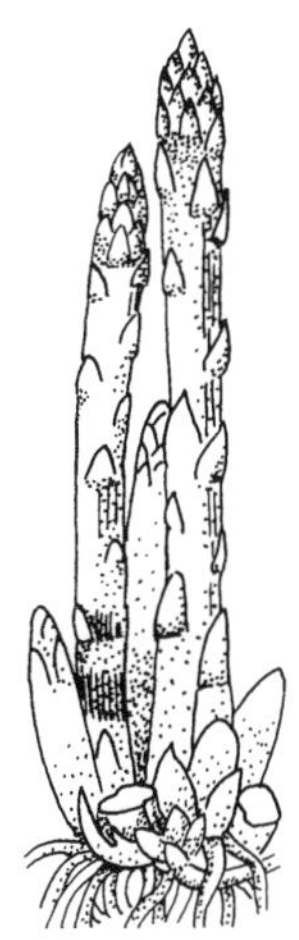

Fig. A-54. Shoots of asparagus.

ORIGIN AND HISTORY. It was native to the eastern Mediterranean, from which it spread and became naturalized over much of the world. However, certain wild varieties have also been found in such widely separated areas as North Africa and South Africa. Also, archaeologists have found evidence that the ancient Egyptians cultivated asparagus. By 200 B.C., the Romans were enthusiastically growing the vegetable and boasted of shoots that weighed as much as 3 lb (*1.3kg*).

During the Middle Ages, the plant was neglected in Europe, but the Arabians kept it in cultivation. However, it was repopularized by Louis XIV, the Sun King, in the 18th century. Since then, it has been developed into varieties with thick, tender spears that are in marked contrast to the slender, fibrous spears of the wild plant.

PRODUCTION. Much of the world's asparagus is produced in the United States, with the rest coming from Europe, Mexico, and Taiwan. Fig. A-55 shows the amounts produced in the leading states during 1990.

The U.S. crop is marketed as follows: fresh, 59%; canned and frozen, 41%.

Propagation and Growing. Asparagus is a perennial; the root crowns send up new shoots each year. Hence, the plant is propagated commercially by (1) grow-

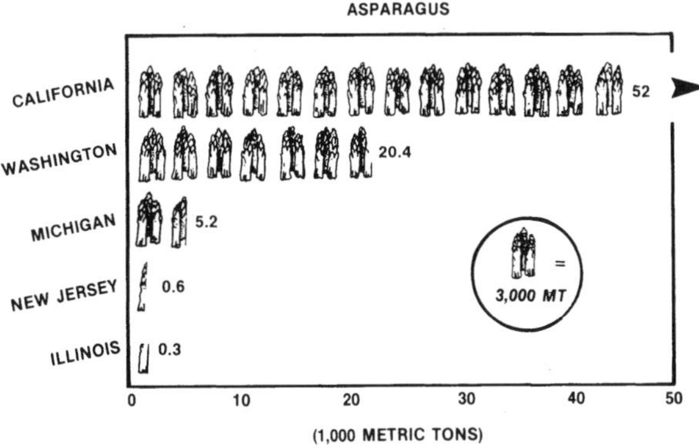

Fig. A-55. The leading asparagus-producing states of the United States, and the tonnage produced in each. (*Agricultural Statistics*, 1991, USDA, Table 202)

ing root crowns from carefully selected seeds, and (2) transplanting the young crowns into a field where the crop may be grown continuously for 15 to 20 years. Each year, the emerging spears are cut daily for 2 to 3 months in the spring. After that, those which emerge are allowed to grow to provide nourishment for the underground parts of the plant. It is noteworthy that asparagus will not grow well unless it has a rest period of from 3 to 5 months, which is usually provided by (1) the cessation of growth after the onset of cold weather in the northerly areas, or (2) withholding irrigation during the mild winters in southwestern United States. It is necessary to protect the root crowns by mulching (applying leaves, straw, or other protective material) in the areas subject to hard freezes.

In the past, it was customary to blanch (whiten) the young shoots by hilling up soil around them, or by other means. Now, green shoots are well accepted and there is no need to incur high labor costs by growing white ones.

Harvesting. Shoots are harvested by cutting or snapping them off, taking care to avoid damage to the root crowns. After harvesting, the shoots are usually washed, graded, tied in bunches, and packed in boxes. In 1990, an average yield of 2,600 lb per acre (*2,919 kg/ha*) was obtained in the United States, with yields of 2,900 lb (*3,256 kg*) and 3,400 lb (*3,817 kg*) in California and Washington, respectively.

PROCESSING. About two-thirds of the U.S. asparagus crop is processed, of which about twice as much is canned as is frozen. Both processes utilize fresh spears that have been washed, graded, cut to uniform lengths, and blanched in boiling water or steam. (The tough pieces cut from the bases of the stems are used to make asparagus soups and other items which are cooked thoroughly so that tenderness results.)

Recently, food technologists developed a freeze-drying process which converts a pound of fresh asparagus to about an ounce of dried product. The freeze-dried vegetable may soon be utilized in soup mixes or in convenience products for backpackers and hikers.

SELECTION AND PREPARATION. To be of best quality, asparagus should be fresh and firm with closed compact tips and the entire green portion tender. Asparagus ages rapidly after cutting; tips become partially open, spread, or wilted, and stalks become tough and fibrous. Tender asparagus is brittle and easily punctured. Slightly wilted stalks may sometimes freshen in cold water, but are usually undesirable. Angular or flat stalks are apt to be tough and woody.

Asparagus costs too much to waste; hence, it should be prepared with a great deal of care. It may be (1) steamed briefly; (2) sliced diagonally and stir fried with fish, meat, poultry, and/or other vegetables; or (3) chopped and cooked in omelets or in soups (usually reserved for the tougher spears).

The vegetable is often served with butter or a sauce such as hollandaise. Asparagus tips au gratin with sliced hard-boiled egg on toast is appealing when served as an hor d'oeuvre with an accompanying dip or sauce, or when tossed in a salad with grated cheese, oil, and/or a well seasoned dressing. Fig. A-56 shows a typical asparagus salad.

Fig. A-56. Iceberg asparagus salad. An appetizing combination of asparagus, iceberg lettuce, and a dressing which contains mayonnaise, tarragon, chervil (dill may be substituted), and a touch of garlic. (Courtesy, California Iceberg Lettuce Commission, San Rafael, Calif.)

NUTRITIONAL VALUE. The nutrient composition of asparagus is given in Food Composition Table F-36.

Asparagus is low in calories and carbohydrates, but 1 cup (*240 ml*) contains more protein (3 to 6 g) than a cup of cooked cornmeal (*2.6 g*) or a cup of cooked rice (*3 to 5 g*). The green stalks are much better sources of vitamin A than the white ones. Also, the vegetable is a fair source of the vitamins thiamin, riboflavin, and niacin. It is noteworthy that canned asparagus contains sufficient sodium to be avoided by people on sodium-restricted diets, who should use either the fresh or frozen vegetable, or else buy a low sodium dietary pack of canned asparagus.

This vegetable was used for medicinal purposes before it was consumed as a food, because it was long believed to be beneficial for conditions such as anemia, arthritis, excessive water retention in the body (dropsy), neuritis, obesity, and rheumatism.

NOTE: The authors make no medicinal claims for this vegetable; rather, these beliefs are reported for informational purposes only.

(Also see VEGETABLE[S], Table V-6 Vegetables of the World.)

ASPARTAME

A low calorie artificial sweetener which was finally approved for human consumption by the Food and Drug Administration (FDA) in mid-August 1981. It is manufactured by G. D. Searle and Company of Skokie, Illinois, and sold under a different brand name. Chemically, aspartame is the combination of the two amino acids, aspartic acid and phenylalanine—a dipeptide. It is about 180 times sweeter than table sugar. Therefore, it provides the same sweetness as sugar with fewer calories. Furthermore, it does not promote tooth decay, nor does it have an after taste.

In 1965, aspartame was serendipitously discovered by James Schlatter of G. D. Searle and Company, who was researching possible new drugs for ulcer therapy. Schlatter synthesized the dipeptide, aspartame, during part of the ulcer research program, but during the purification steps he spilled some on his hands. Shortly thereafter Schlatter licked his finger to facilitate lifting some paper. His finger tasted remarkably sweet. Realizing the potential of a new sweetening agent, G.D. Searle and Company launched an intensive effort, beginning about 1969, to commercialize aspartame. After extensive studies on the safety of aspartame, the FDA approved the use of the product in July 1974. But approval was rescinded in December 1975. The rescinding was due to doubts raised about Searle's testing procedures, and the actions of Dr. John Olney, a psychiatrist at Washington University School of Medicine in St. Louis, and James Turner, a Washington attorney and consumer advocate who challenged the safety of aspartame based on Olney's studies with glutamate. Even after all data was reviewed and the original safety studies of Searle were validated, the FDA was slow to make a decision. Finally, Searle sued the FDA to force a decision. The FDA decided in August 1981 that aspartame is safe and that few compounds have withstood such detailed and repeated testing. Thus, 13 years after intensive research was begun on aspartame, it became a commercially available low calorie sugar substitute. It was available in France, in Luxembourg, and Belgium earlier.

Aspartame is sold to food processors for use in cold cereals, drink mixes, gelatine, puddings, dairy products, and toppings. Consumers can purchase aspartame in tablets or powder for use on or in their foods. Aspartame loses its sweetness during long periods of storage. Also, aspartame is not suitable for baking, since heat causes the loss of sweetness.

NOTE WELL: Aspartame is safe for diabetics, but the phenylalanine released during its metabolism may affect individuals with phenylketonuria. Foods manufactured with aspartame are required to carry a warning for phenylketonurics.

(Also see SWEETENING AGENTS.)

ASPARTIC ACID

A nonessential amino acid which provides amino groups that help to conserve the body's supply of protein. Also, it is involved in the formation of urea and other compounds in the body.

(Also see AMINO ACID[S].)

ASPERGILLUS

A genus of fungi that are commonly-occurring food spoilage organisms. (The name is derived from *aspergillum,* the device used for sprinkling holy water in Roman Catholic rituals, because the head and stalk of the fungi are shaped like the device.) These fungi grow readily on breads and other grain products, legumes, and meats. Certain species even infest such parts of the body as the feet, lungs, and the external opening of the ear. Fig. A-57 shows a typical member of this genus.

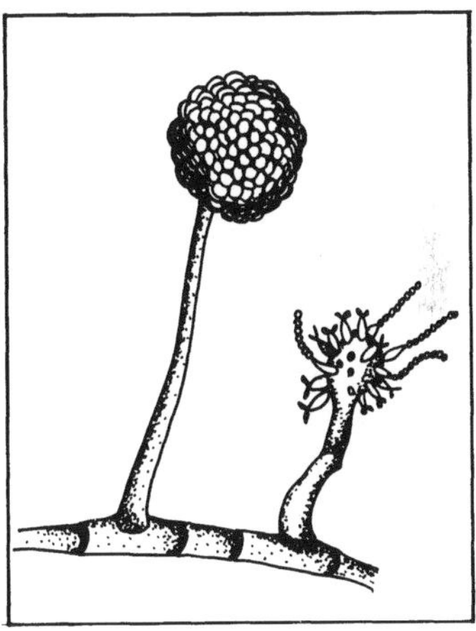

Fig. A-57. *Aspergillus.* A common food spoilage organism.

However, man has learned to utilize certain species of *Aspergillus* for such beneficial purposes as (1) sources of enzymes which convert starches or proteins to sugars or amino acids, respectively; and (2) production of antibiotics, and of organic acids used as food additives. Conversely, *Aspergillus flavus* produces the deadly aflatoxins, which cause liver cancer and other disorders.

(Also see AFLATOXIN; POISONS, Table P-11 Mycotoxins.)

ASSAY

Determination of (1) the purity or potency of a substance, or (2) the amount of any particular constituent of a mixture.

ASSIMILATION

A physiological term referring to a group of processes by which the nutrients in food are made available to and used by the body; including digestion, absorption, distribution, and metabolism.

(Also see DIGESTION AND ABSORPTION; and METABOLISM.)

ASSOCIATION OF OFFICIAL AGRICULTURAL CHEMISTS (AOAC)

The AOAC is a voluntary organization of chemists that sponsors the development and testing of methods for analyzing nutrients, foods, food and color additives, animal feeds, liquors, beverages, drugs, cosmetics, pesticides, and many other commodities. Periodically, they publish methods which are acceptable for these analyses.

ASTHMA

A condition characterized by difficulty in breathing due to obstruction of the bronchial tubes. Usually, it is due to an allergic reaction.

(Also see ALLERGIES.)

ASYMPTOMATIC

Without symptoms.

ATAXIA

A general term which means a lack of coordination of musclar movement., It may be due to various disorders such as damage to certain parts of the nervous system, nutritional deficiencies, intoxication, or a congential disorder.

ATHEROMA

A deposit of fat on or in the walls of arteries. (Also see HEART DISEASE.)

ATHEROSCLEROSIS

A degenerative disease, characterized by the buildup of abnormal patches (plaques) on the walls of arteries, thought to be the main cause of heart attacks.

(Also see HEART DISEASE, Section headed "Atherosclerosis.")

ATHLETICS AND NUTRITION

Contents — *Page*

History 122
Health Benefits from Participation in Sports 124
Metabolism During Athletics 125
Nutrient Requirements of Athletes
During Training and Competition 126
Water and Salt Requirements 126
Meals Taken Before and During Competition 127
Foods and Nutrients Which Are Believed
to Help Performance 127
Steroid Alternatives 128

The terms athletics and sports are generally considered to be synonymous in that they usually refer to various types of recreational physical activities that are often performed in accordance with sets of rules and scoring systems for determing the "winners." Sometimes, games that require little exercising on the part of the participants are called sports. However, in this article, the term refers to athletic contests or recreational activities that require physical activity.

During the past decade, there has been an increasing awareness of the role of nutrition in promoting optimum performance of athletes. What an athlete eats is now recognized to be as important a factor as training, conditioning, and good genes. The nutritional needs of athletes depend on age, sex, body size and composition, activity, and level of training. Recently, researchers have studied the metabolism of nutrients during athletic competition in order to develop certain well founded dietary recommendations, which will be presented in this article.

Fig. A-58. A basketball player needs strength and endurance to play competitively. Training and balanced nutrition help ensure a high level of performance.

HISTORY. Primitive peoples realized that courage and physical fitness were required for success in hunting wild game and fighting neighboring groups that threatened their security. Hence, they sometimes prepared themselves for these occasions by (1) participating in religious ceremonies to win the favors of their tribal gods, (2) eating the hearts and other parts of animals that were thought to confer courage and strength, and (3) performing feats of ability and endurance. (It is noteworthy that certain modifications of these practices have been passed down through the ages and are still utilized by some modern athletes.) Furthermore, some of the long-standing controversies over the effects of diet on performance, such as the one regarding meat consumption vs vegetarianism, continue to be the subjects of heated debates. However, one should consider the historical circumstances under which the various diets were consumed before jumping to conclusions about the merits and demerits of various foods.

The diets of American Indians who engaged in sporting events usually were based upon the foods that were readily available from their environments. For example, the Iroquois Indians, who played lacrosse, ate a varied diet of wild game, fish, vegetables, nuts, berries, and other wild plants because they lived in the bounteous Eastern Woodlands of North America. However, the Maya

Fig. A-59. Iroquois Indians playing lacrosse.

Indians of Central America, who played a primitive form of basketball, ate a predominatly vegetarian diet of beans, corn, squash, and other crops because few animal foods were available to feed the large population. Finally, the Tarahumara Indians in the mountains of Northern Mexico, who have long been champion runners with extraordinary endurance, subsist on a bean and corn diet that provides less than 2,000 Calories (kcal) per day. The diet of these Indians is so inadequate that 80% of their infants and children die before age five. Hence, only the exceptionally strong Tarahumarans survive to adulthood. Therefore, it appears that factors other than diet alone may be responsible for high levels of physical performance by certain groups of primitive people.

On the other side of the world, sports contests became important events in the ancient civilizations of Greece and Rome. The first renowned Greek athletes who participated in the running competitions of the early Olympic Games (which are believed to date back to the 13th century B.C.) ate mainly fruits, grains, legumes, and vegetables like the rest of the people, because the hilly, rocky soil supported only limited crops and small numbers of goats and sheep. Only one province in the northern part of the Greek peninsula had land suitable for raising cattle.

Later, the sports-loving Greeks introduced wrestling and boxing, but there were no weight classes. Hence, the heavier athletes had an advantage over the lighter ones. At that time, some of the champions were hefty-built meat eaters who consumed rather large amounts of food. This led to criticism by philosophers who claimed that an athlete was a slave to his jaw and belly. Recent research findings suggest that the consumption of large amounts of meats and other high protein foods enhanced the appetite so that extra food may be eaten.

A lesser known aspect of the ancient Greek concern for health and physical fitness was the establishment of temples for sick people (the prototypes of modern health resorts and spas) at about 1200 B.C. The patients were treated with rituals that included sleep, diet, exercises, and baths. Later, the Greek father of medicine, Hippocrates (460 to 364 A.D.), built his science upon these principles, and rejected the idea that sickness was a punishment from the gods. Another innovation of the ancient Greeks was the *gymnasium,* which was a sports facility that had a playing area, cloak rooms (Greek athletes performed in the nude), training rooms, baths, and massage rooms. Furthermore, athletic training played an important role in the education of the young men of Greece.

After conquering the Greeks, the Romans continued the tradition of Olympic Games and built many new gymnasiums and public baths where sports such as handball were invented. However, the original spirit of competition gave way to one of entertainment of large audiences and the games were discontinued within a few centuries. The rise of Christianity in Rome was also a factor in the decline of athletics since the early Christians reacted against the carnal excesses of the Romans and substituted literary studies for physical education. The Christian period in Rome began in the 4th century A.D.

Little is known about the history of athletics and nutrition in the period from the 4th to the 18th centuries, since athletic contests were limited to those conducted between people of local towns and were not publicized outside of these areas. Also, the sciences of medicine and dietetics were for many centuries rooted in the writings of Galen (130 to 200 A.D.), a highly-renowned Greek physician who relied mainly on Hippocratic concepts in his practice in ancient Rome. He treated the emperor Marcus Aurelius [121 to 180 S.F.] and the gladiators. Finally, the economic conditions after the fall of Rome in the 5th century A.D. were so bad for so long that most of the people had few choices of foods and little time for sports.

In the latter part of the 18th century, the French chemist, Antoine Lavoisier (1743 to 1794), founded the science of nutrition with his experiments in calorimetry. His work was followed by the research studies of many others, the most notable of which were the investigations of the German chemist, Justus von Liebig (1803 to 1873), the German physiologist Carl von Voit (1831 to 1908) and Max Rubner (1854 to 1932), and the American nutritionist Wilbur Atwater (1844 to 1907). These men developed the principles and procedures for (1) measuring the caloric expenditures during various types of physical activities, and (2) determining the contributions of carbohydrates, fats, and proteins to energy metabolism. However, their recommendations of high protein intakes for strenuous exercise were influenced strongly by (1) Liebig's opinion that a high protein diet was required for heavy muscular work, and (2) the observations of Voit that German laborers consumed about 3,000 Calories (kcal) and 118 g of protein per day. Voit recommended that working men consume 145 g of protein per day, whereas Rubner and Atwater recommended 150 g. It is ironic that these concepts persisted long after Voit himself disproved the idea that extra protein was needed for heavy muscular work and the Swiss investigators Fick and Wislicenus showed in 1866 that mountain climbers required mainly food sources of calories, but that no extra protein was required.

In the early 1900s, several investigators challenged the belief that a high protein intake was required for vigorous physical activity. First, the American physiologist Russell Chittenden (1856 to 1943) demonstrated that his health was improved (he had been troubled by bilious attacks, rheumatism, and sick headaches) and his ability to engage in strenuous exercise (rowing a boat 6 to 10 miles) was enhanced greatly by a reduced intake of calories and protein. The publication of Chittenden's observations was followed by several investigations of the physical performances of meat eaters vs vegetarians. Each of the studies showed that vegetarians had superior endurance. These results were interpreted by certain leading nutritionists to indicate that (1) bacterial action on undigested protein in the bowels of the meat eaters produced toxins which were absorbed into the blood; (2) the vegetarians had no actual physicial superiority, but were more highly motivated for endurance as a result of their dietary convictions; or (3) biases were introduced into the experiments by the selection of vegetarians who were stronger than the meat eaters. In spite of the widespread publicity given to these experiments, little subsequent work was done on this subject.

The first 3 decades of the 20th century also ushered in the eras in which vitamins and hormones were found to be helpful in certain cases of debilitation that were due mainly to deficiencies of these vital substances. However, many athletes and other lay people were often misled by profit hungry promoters of tonics and other nostrums into believing that vitamins and hormones were "metabolic stimulants" that could improve one's performance in strenuous activities.

Practices of questionable merit have persisted right up to the present time, as evidenced by recent reports of wholesale utilization of nutritional supplements and drugs by American athletes. Correction of these practices will be difficult without the widespread and whole-hearted cooperation of athletic trainers and coaches, since the expectations that the taking of certain substances might confer a margin of superiority in competition may often overwhelm the prudent fear of harmful side effects. Fortunately, there has been considerable recent research on metabolism during athletics. Hopefully, the findings of these investigations will be published widely so that some of the myths regarding athletics and nutrition may be laid to rest.

HEALTH BENEFITS FROM PARTICIPATION IN SPORTS. Most of the public believes that athletes are healthier than nonathletes. However, this is *not* always the case, since certain types of regular exercise together with optimal diets are required to achieve and maintain health and physical fitness. For example professional baseball and football players who are moderately obese my have greater risks of developing a degenerative disease than men of normal weight who lead sedentary lives. Furthermore, some athletes consume fad diets or have other bad habits which may be harmful to the body. Therefore, it is necessary that athletes consume prudent diets and practice other good health habits in order to receive the benefits that follow:

• **Development of a collateral circulation in the heart muscle**—It appears that the regular performance of strenuous exercise enlarges the smaller blood vessels which connect the various branches of the coronary arteries. This effect provides a substitute means of supplying blood to the heart muscle in the event that one or more portions of the larger arteries become blocked with atherosclerotic deposits and/or blood clots.

• **Greater endurance for physical activity**—The muscles of trained athletes may be used longer before fatigue develops because physical conditioning for competition speeds up the elimination of lactic acid, a product of energy metabolism which accumulates during strenuous exercise and contributes to a feeling of fatigue.

• **Higher output of blood by the heart**—Vigorous athletic activity increases the amount of blood pumped per minute (the quantity known as the cardiac output) from a resting value of about 6 qt (*5.9 liter*) to 25 qt (*24 liter*). This increase in circulation results from increases in both the heart rate and the amount of blood pumped with each contraction. The latter quantity is called the stroke volume. Athletes who are trained for endurance types of sports usually have the capacities for much higher cardiac outputs than untrained people.

• **Increased aerobic capacity**—The rate at which oxygen can be taken up by the blood is called the aerobic capacity and is usually measured while the subject exercises on a treadmill or a specially equipped exercise bicycle called an ergometer. The latter type of measurement is shown in Fig. A-60.

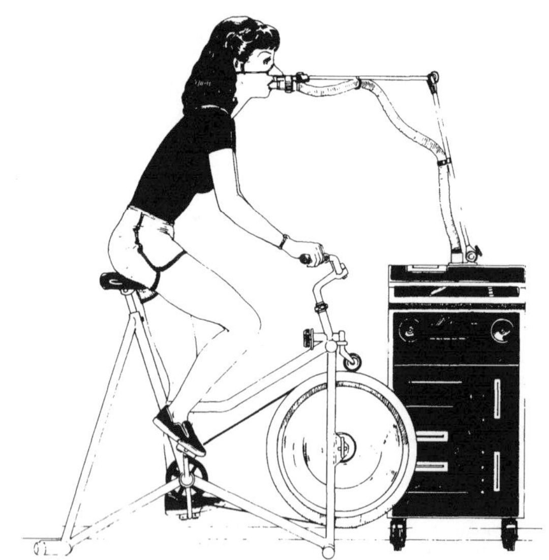

Fig. A-60. Testing for aerobic capacity on a bicycle ergometer.

• **Loss of excess body fat**—Strenuous exercise performed several times a week helps to speed the loss of unwanted body fat, provided that the caloric intake is held constant or moderately reduced. For example, the expenditure of an additional 500 Calories (kcal) per day, 3 times a week in sports such as basketball, handball, jogging, racketball, swimming, or tennis will result in the loss of about 2 ib (*0.9 kg*) per month. If, in addition to exercising, the dietary caloric intake is cut by 500 Calories (kcal) per day, there will be an additional loss of about 4 lb (*1.8 kg*) per month. It is noteworthy that athletes sometimes gain weight during training, but this gain is likely to be almost entirely muscle protein and water. Each pound of body protein is associated with approximately 3 lb (*1.4 kg*) of water.

• **Lowering of the resting heart rate**—Athletic training increases the efficiency of the heart muscle so that more blood is pumped with each contraction. The stroke volume is increased. Hence, fewer beats per minute are required to deliver a fixed quantity of blood to the tissues. It is noteworthy that some well trained athletes have resting pulses which range from 45 to 60 beats per minute, whereas the normal resting pulse ranges from 72 to 80 beats per minute. The advantage of a lowered resting heartrate is that the heart muscle has longer resting periods between contractions.

• **Moderate reductions in diastolic and systolic blood pressures**. Dynamic (isotonic) exercises—those which are characterized by vigorous movements of various parts of the body (bicycling, running, swimming etc.)—may produce moderate reductions (as much as 10 to 15 points) in both the diastolic and systolic blood pressures as a result of the opening up (dilation) of blood vessels within the muscles. This effect occurs because dilated blood vessels offer less resistance to the pumping of blood by the heart. However, static (isometric) exercises—those that involve mainly the development of muscle tension with little or no movement—may cause elevations in the blood pressures.

NOTE: People with cardiovascular disease should *not* perform static exercises without a doctor's approval.

• **Normalization of carbohydrate utilization**—Physical exercise increases the rate at which the sugar in the blood is utilized. For example, the blood sugar levels of well-trained males were found to be about one-third lower after an oral dose of glucose than those in untrained males. Hence, athletes may safely consume greater amounts of dietary carbohydrates than nonathletes. Furthermore, diabetic athletes do not usually require as much insulin as sedentary diabetics.

• **Strengthening of the muscles**—The ability to perform muscular work is increased by the types of athletic training which produce (1) thickening of muscle fibers and (2) increased rates of extraction of oxygen and nutrients from the blood.

The specific benefits from participation in some of the more common sports are summarized in Table A-28.

TABLE A-28
RATINGS OF SELECTED SPORTS THAT PROMOTE PHYSICAL FITNESS AND WELL-BEING[1]

Sport	Physical Fitness					General Well-Being				Total Score
	Stamina	Muscular Endurance	Muscular Strength	Flexibility	Balance	Weight Control	Muscle Definition	Digestion	Sleep	
Jogging	21	20	17	9	17	21	14	13	16	148
Bicycling	19	18	16	9	18	20	15	12	15	142
Swimming	21	20	14	15	12	15	14	13	16	140
Ice or roller skating	18	17	15	13	20	17	14	11	15	140
Handball/squash	19	18	15	16	17	19	11	13	12	140
Skiing—Nordic	19	19	15	14	16	17	12	12	15	139
Skiing—Alpine	16	18	15	14	21	15	14	9	12	134
Basketball	19	17	15	13	16	19	13	10	12	134
Tennis	16	16	14	14	16	16	13	12	11	128
Calisthenics	10	13	16	19	15	12	18	11	12	126
Walking	13	14	11	7	8	13	11	11	14	102
Golf (use of cart or caddy)	8	8	9	8	8	6	6	7	6	66
Softball	6	8	7	9	7	7	5	8	7	64
Bowling	5	5	5	7	6	5	5	7	6	51

[1]Adapted by the authors from *How Different Sports Rate in Promoting Physical Fitness,* U.S. Dept HEW, 1978, pp. 4-5. The ratings are the sums of those from seven American experts who rated each benefit on a scale from 0 to 3. Hence, a rating of 21 indicates that each of the experts thought that the maximum benefit was conferred.

METABOLISM DURING ATHLETICS. The utilization of food energy for muscular work involves (1) the production of energy-rich adenosine triphosphate (ATP) from carbohydrates, fats, and proteins by means of a series of tightly-linked oxidative (aerobic) processes, and (2) the contraction of the muscle fibers which is induced by ATP. High rates of muscular work which exceed the aerobic capacity, such as those performed in sprinting, may also be sustained in part by nonoxidative (anaerobic) processes that produce ATP and lactic acid. However, the accumulation of lactic acid often results in heavy breathing and fatigue which persist for a while after the work has been completed. This condition is called an "oxygen debt" because the elimination of the lactic acid buildup requires extra oxygen.

It is noteworthy that athletic training speeds up the metabolism of lactic acid and retards the development of fatigue. Furthermore, the feeding of salts of lactic acid (lactates) during training apparently stimulates the processes that reduce the lactate accumulation. Therefore, some trainers have conjectured that it might be helpful for athletes in training to consume liberal amounts of fermented milk products such as buttermilk and yogurt which are rich in lactic acid.

(Also see ENERGY UTILIZATION BY THE BODY; and METABOLISM.)

NUTRIENT REQUIREMENTS OF ATHLETES DURING TRAINING AND COMPETITION. One of the most convenient means of estimating the nutritional needs of an athlete is to consider the needs to be the sums of basic requirements for nonathletic people of the same age, sex, height, and body build; plus, the extra nutrients needed to cover the cost of the work performed in playing the sport. The basic recommended nutrient intakes for the major age and sex groups are given elsewhere.

(Also see NUTRIENTS: REQUIREMENTS, ALLOWANCES, FUNCTIONS, SOURCES; and RECOMMENDED DIETARY ALLOWANCES.)

The most important nutritive requirements of athletes are (1) the extra calories to cover the energy cost of the physical activity (unless he or she is trying to lose some body fat by means of caloric restriction), and (2) water to replace that which is lost in sweat. Data on the energy expenditures for various sports are given in Table A-29.

TABLE A-29
ENERGY EXPENDITURES FOR SELECTED ATHLETIC ACTIVITIES[1]

Energy Expenditure	Activity[3]
(kcal per hour)[2]	
170 to 240	Bicycling slowly (6 m.p.h.), table tennis, volleyball, walking slowly (2 m.p.h.).
250 to 350	Bowling, golf, swimming slowly (25 yd per minute), tennis doubles, walking moderately fast (3 m.p.h.).
360 to 500	Bicycling moderately fast (12 m.p.h.), dancing, football, handball, hockey, ice skating, roller skating, skiing (downhill), squash, swimming moderately fast (50 yd per minute), tennis singles, and walking fast (4½ m.p.h.).
510 to 750	Cross-country skiing, jogging (5½ m.p.h.), jumping rope, and running in place.
760 to 1,000	Jogging (7 m.p.h.), and running (8 m.p.h.).

[1]Adapted by the authors from *Exercise and Your Heart,* U.S. Dept. of Health and Human Services, 1981, p. 7; and *Food,* Home and Garden Bull. No. 228, USDA, p. 15.
[2]A range of caloric expenditures is given for each level of activity because the values for individual participants vary according to the intensity of play, body build, and skill of playing. The low end of the range is probably a better estimate for small persons and most women, whereas, the upper end of the range is most applicable to large persons and most men.
[3]To convert to metric see WEIGHTS AND MEASURES.

The extra energy that is required for most sports may be provided by one or more nutritious snacks such as those shown in Table A-30.

TABLE A-30
CALORIC VALUES OF SELECTED POPULAR SNACK FOODS[1]

Food	Calories (kcal)
1 small, 2⅝ in. *(5.9 cm),* navel orange	45
2 medium graham crackers	55
2 Tbsp *(30 ml)* fruit-nut snack[2]	70
2 Tbsp *(30 ml)* peanuts	105
1 cup *(240 ml)* plain low-fat yogurt	145
1 cup *(240 ml)* split-pea soup	195
1 cup *(240 ml)* fruit-flavored yogurt	225
½ cup *(120 ml)* "granola" cereal[3]	280
1 hamburger, 3 oz *(85 g)* patty, on bun	365
12 oz *(360 ml)* chocolate milkshake	430

[1]Adapted by the authors from *Food,* Home and Garden Bull. No. 228, USDA, p. 14.
[2]A mixture of salted Spanish peanuts, raisins, and chopped dates.
[3]A slow-baked mixture of quick-cooking rolled oats, unsweetened wheat germ, unsweetened shredded coconut, coarsely chopped nuts, raisins, vegetable oil, honey, and almond or vanilla flavoring.

It should be noted that the snacks in Table A-30 provide minerals and vitamins in addition to calories. For example, the orange furnishes potassium (a mineral that is very important to the heart muscle and other tissues) and vitamin C (which may improve performance). The other items are good sources of additional minerals and vitamins. Therefore, a variety of nutritious snacks should be consumed, in addition to the well-balanced meals that are desirable for all people. Generally, it is better to obtain nutrients from wholesome foods than to consume *junk foods* that furnish mainly *empty calories* because the former practice minimizes the likelihood that nutritional supplements will be needed.

(Also see NUTRIENTS: REQUIREMENTS, ALLOWANCES, FUNCTIONS, SOURCES; and RECOMMENDED DIETARY ALLOWANCES.)

Water and Salt Requirements. Excessive losses of water (dehydration) and mineral salts (electrolyte depletion) by athletes exercising strenuously in hot environments are believed to be mainly responsible for the heat stroke deaths that occur among high school and college football players. Players who begin practice during the hot, humid "dog days" of August are highly susceptible to dehydration and heat injury because (1) they are often out of condition when they start their training, (2) the protective gear and uniform prevents the escape of heat from the body, and (3) they may fail to consume sufficient water and mineral salts. Runners may also be at risk due to their becoming accustomed to dehydration during practice sessions and failing to drink enough fluid. Finally, wrestlers may deliberately dehydrate themselves in order to lose sufficient weight to qualify for a lower weight class. Often, there is a period as long as 5 hours between weighing in and the beginning of the match. Hence, the wrestlers expect to rehydrate themselves during that interval.

Therefore, players, coaches, and trainers should be aware that the excessive loss of water from the body may have consequences such as fatigue, fever, heat stroke, impairment of sweating, an increased heartrate, and permanent damage to the physiological mechanisms which regulate the body temperature. These consequences may be prevented by the procedures which follow:

1. Athletes should be encouraged to consume from 8 to 10 8-oz glasses *(1.9 to 2.4 liter)* of fluids per day. This includes milk, soups, stews, etc. The liberal consumption of liquids has been shown to boost athletic performances by conferring benefits such as a lower heart rate, increased sweating, and a lower rectal temperature (a measure of overheating in the body).

2. Water should be available during practice sessions and matches that last for two or more hours and for

shorter ones that involve intense activity in a warm environment.

3. Athletes, coaches, and trainers should monitor the loss of body water during exercise by weighing participants before and after practice sessions or matches. Each pound of weight loss is equivalent to 1 pint *(0.47 liter)* of water.

4. Salt tablets may not be needed if athletes salt their food liberally. However, it may be better to use the tablets when needed rather than to salt food heavily since the habitual use of excessive amounts of this seasoning may lead to high blood pressure in certain susceptible people. A good rule of thumb is for athletes to use one 7 grain salt tablet for each pound of weight loss *after* 5 to 6 lb have been lost. Furthermore, each tablet should be taken with at least 1 pint *(0.47 liter)* of water. Taking salt without water aggravates dehydration and may cause nausea.

5. Sport drinks which contain sugar, salt, and other mineral salts (electrolytes) may be suitable for use during practice sessions and contests, provided that overconcentrated products are diluted with water. Coaches should obtain the compositions of these products from the manufacturers or other authoritative sources.

MEALS TAKEN BEFORE AND DURING COMPETITION. Some athletic competitors place great value on pregame meals and on the types of refreshments consumed during competition. For example, the eating of meatball sandwiches before games became a tradition for the New York Giants football team when they won an important game after eating the sandwiches.

Fig. A-61. The pregame meal may make the difference between winning and losing in highly competitive sports such as football.

Other athletes have consumed liquid formula diets before and during athletic contests. The latter practice was popularized by long distance swimmers. However, many athletes are tense prior to the start of competition and may become nauseous if the wrong types of foods are eaten. Therefore, the meal should be cleared rapidly from the stomach and be low in undigestible material such as plant fiber. Furthermore, carbohydrates are the most rapidly utilized and promote the best short-term performances from athletes, whereas fats and proteins require much longer for digestion and are best suited for slower paced events that are conducted over longer periods of time.

A typical pregame meal for an event lasting only an hour or two might consist of cereal, milk, and sugar with toast, butter, and/or jam. Another meal might be an instant breakfast powder mixed with milk. Generally, coaches tell their players to avoid condiments, fatty foods, gravies, and pies.

FOODS AND NUTRIENTS WHICH ARE BELIEVED TO HELP PERFORMANCES. Food fads are prevalent among athletes who are ever in search of an item that will give them an edge in competition. Nevertheless, some foods appear to be better boosters of energy (ergogenic aids) than others. Therefore, details on some of the more popular items follow:

• **Alcoholic beverages**—Beer and champagne have long been consumed by athletes in Europe. In America, the drinking of beer is common after practice sessions and informal competitions. Even the more serious athletes may take beer to relax and sleep well during the night before competition and occasionally on the day of the event to remove inhibitions over extending themselves fully. However, performances that require finely tuned perceptual and motor skills may be impaired somewhat by alcohol.

• **Caffeine-containing beverages**—Coffee, tea, and cola beverages are widely consumed by athletes around the world. Nevertheless, it is still uncertain whether these drinks impair or improve athletic performances. Furthermore, caffeine-containing beverages increase the loss of water in urination (caffeine is a diuretic) and may contribute to dehydration if excessive amounts are consumed.

• **Carbohydrate-rich diets**—Recent studies have shown that the consumption of high carbohydrate diets for 3 or 4 days prior to an athletic event improves performances in sports that require considerable aerobic capacity. The effect of this type of diet is greatest when it is preceded by 3 days of strenuous exercise while consuming a high fat, high protein, low carbohydrate diet that empties the muscles of the stored carbohydrate (glycogen). This practice, which is called "glycogen loading," may not be advisable for all athletes on all occasions, since a middle-aged competitor developed cardiac irregularities while overzealously adhering to the dietary procedure.

• **Dextrose (glucose)**—The ingestion of sugar prior to or during contests improves performances provided that it is taken with sufficient water to prevent irritation of the digestive tract.

• **Honey**—This form of sugar does not appear to offer any advantage over other forms.

• **Iron**—Women athletes may need iron supplementation, particularly when they have heavy losses of blood during their menstrual periods.

• **Orange juice**—The Russians claim that the outstanding performances of their athletes during the 1960 Olympics in Rome were due in part to the consumption of liberal amounts of orange juice, which is rich in potassium (an essential electrolyte) and vitamin C. They won more gold medals than any other country. However, it is still uncertain whether orange juice is an ergogenic food.

• **Protein-rich diets**—These diets help to build muscle during training for events such as weight lifting and wrestling that require great strength. However, they do *not* help performance of trained athletes, and may even impair endurance in certain types of sports.

• **Thiamin**—Where a deficiency of thiamin exists, a thiamin supplement may be advisable for athletes who consume highly refined carbohydrates or sugar such as dextrose and honey to obtain extra calories.

• **Vitamin C**—Present findings indicate that massive doses of vitamin C are not only ineffective in improving the performance of athletes, but may have a negative effect on athletic performance by disturbing the equilibrium between oxygen transport and oxygen utilization.

• **Vitamin E**—Unfortunately, many of the studies of the use of this vitamin by athletes were poorly controlled. Nevertheless, it has been helpful in the treatment of some cases of muscle cramping.

• **Wheat germ oil**—This food might have an effect that is unrelated to the vitamin E it contains, the criteria that were used in testing may not be valid.

STEROID ALTERNATIVES. Steroids, which increase muscle mass, and which have been used in both people and animals, are based on the hormones secreted by the testes, ovaries, and adrenal glands, and are known as *anabolic steroids*. But the use of anabolic steroids on racehorses may reduce fertility in both stallions and mares. Also, most U.S. racetracks forbid the use of steroids within a certain period prior to a race (usually within 48 hours of a race). In human athletic events, the use of anabolic steroids is being curbed by both legislative and educational efforts. But many manufacturers of dietary supplements are now promoting products as substitutes of anabolic steroids, or as *steroid alternatives*. The list of dietary steroid alternatives is extensive; among them, branch-chain amino acids, chromium picolinate, gamma oryzanol and other plant sterols, inosine, and L-carnitine.

Some high school and college athletes are trying to achieve the results attributed to anabolic steroids by using steroid alternatives. Unfortunately, at this time little is known about the steroid alternatives—we do not know whether or how these products affect an athlete's health, or even what many of them contain. More creditable research on each of these products is needed.

Dietary supplements are not illegal; regulatory agencies only become involved in supplements if manufacturers appear to make false claims. However, supplements that are purported to affect the body's structure and function are subject to the drug regulations of the Food and Drug Administration (FDA).

Before taking dietary supplementation of anabolic alternatives, an athlete should seek the counsel of a health professional—a physician, a physical therapist, or nutritionist—even if they advise not using these products until more is known about them.

(Also see the section on HEALTH FOODS, Table H-1, Commonly Considered Health Foods, pp. 1118–1132, for an alphabetical listing and information about these steroid alternatives, along with other supplements. *NOTE:* The authors express their grateful appreciation to Paul Purviance, Physical Therapist, CMS Work-Able, Fresno, California, for his counsel and literature in the preparation of this article.)

ATONIC CONSTIPATION

A condition characterized by retention of material in the digestive tract, particularly in the bowel.

(Also see CONSTIPATION.)

ATONY

Lack of tone or tension. The term is usually applied to muscles which have become flabby and weak.

ATRIA

The upper chambers of the heart which receive blood from the veins and transfer it to the lower chambers (ventricles). Atria are also referred to as auricles.

ATROPHY

Failure of growth, or a wasting of one or more tissues of the body due to such causes as nutritional deficiencies, aging, disease, or injury.

ATROPINE

One of several similar substances, collectively known as the belladonna alkaloids, which are extracted from the deadly nightshade plant (*Atropa belladonna*). Atropine is used mainly to treat disorders of the digestive tract.

(Also see BELLADONNA DRUGS; and DIGESTION AND ABSORPTION.)

ATWATER VALUES (or ATWATER FACTORS)

These are average physiologic fuel values of carbohydrates, proteins, and fats, based on experiments conducted by W. O. Atwater, at Wesleyan University, Middletown, Connecticut. Professor Atwater, who is thought of as the father of American nutrition, published the first extensive table of food values in 1896. Atwater values reflect the number of kilocalories of energy physiologically available from a gram of food. They were derived by Atwater from the heat of combustion and corrected for energy losses from unabsorbed nutrients in the feces and urine. Atwater approximated that on a typical American diet each gram of carbohydrate, fat, and protein will yield 4, 9, and 4 Calories (kcal), respectively. The Atwater values are used extensively in dietary calculations and food analysis.

AURICLES

Another name for the two upper chambers of the heart, called atria.

(Also see ATRIA.)

AUTOIMMUNITY

The presence in the blood of antibodies that attack one or more tissues of the body. One of the first autoimmune disorders to be identified (by Hashimoto in 1912) was Hashimoto's thyroiditis, in which there is an enlargement and deterioration of the thyroid gland. The formation of antibodies against the thyroid is believed to be stimulated by the protein thyroglobin, when it escapes into the blood from its normal confinement within the cells of the gland.

Therapeutic measures for autoimmune disorders include (1) anti-immune drugs like ACTH and cortisone, (2) replacement of hormones and other secretions that are lacking, and (3) optimal nutrition to slow the rate of tissue deterioration.

(Also see ACHLORHYDRIA; ADDISON'S DISEASE; and ANEMIA, PERNICIOUS.)

AVAILABLE NUTRIENT

A nutrient that is available and can be absorbed from the intestine into the blood. Not all nutrients are available or fully available; e.g., when calcium combines with phytic acid to form calcium phytate, the calcium cannot be absorbed.

AVIDIN

A proteinaceous substance in raw egg white which ties up the vitamin biotin so that it is unavailable. Cooking egg white inactivates avidin.

(Also see BIOTIN.)

AVITAMINOSIS

Literally, this means without vitamins. The condition may be due to inadequate intake of vitamins, deficient absorption, increased body requirements, or injection of antivitamins. When referring to a specific vitamin deficiency, avitaminosis is used as a prefix; for example, avitaminosis A.

AVOCADO (ALLIGATOR PEAR) *Persea americana*

The avocado tree is an evergreen, subtropical tree which bears oval or round, green or black fruits containing a single large seed. The fruit is called an avocado, and it may range in weight from ½ to 3 lb (*0.2 to 1.4 kg*). Three varieties of avocados are recognized, primarily by the type of skin: (1) the Mexican with thin skin, (2) the Guatemalan with thick skin, and (3) the West Indian with leathery skin. These three types also differ in their susceptibility to cold injury and in fruit size and quality. The edible portion is the yellow-green pulp which is soft like butter and has a rather nutty flavor.

Fig. A-62. Avocados on the tree. Avocados are the "fat fruit." The buttery flesh contains 16 to 25% fat. (Courtesy, USDA)

ORIGIN AND HISTORY. Avocados are native to Central America. Some early Spanish explorers recorded its cultivation from Mexico to Peru.

PRODUCTION. Each year, about 1.5 million metric tons of avocados are produced worldwide. Avocados are still an important food crop in Central America with Mexico producing about one-fifth of the world production.

U.S. commercial production of avocados is limited to the milder sections of California and Florida, with California producing about 95% of the crop, which in recent years has been worth around $212 million. Some avocados are grown in areas of Texas and in Hawaii. Furthermore, the United States also imports 4,500 metric tons each year.

Propagation and Growing. Avocados can be easily grown from seed, but propagation is primarily by grafting or budding the desired type onto seedling stocks. Avocados are planted 20 to 40 ft (*6 to 12 m*) apart depending upon the variety and whether they are upright or spreading growth.

Seedling avocados begin bearing after 5 to 6 years, while grafted or budded avocados bear in 2 to 3 years.

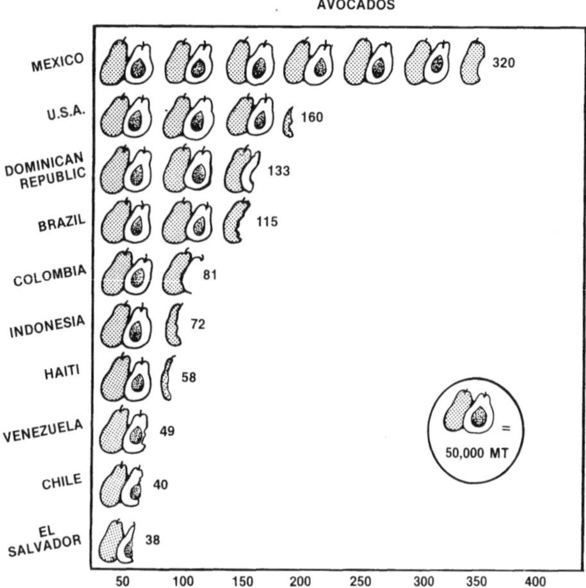

Fig. A-63. Leading avocado-producing countries of the world. (Based on data from *FAO Production Yearbook*, 1990, FAO/UN, Rome, Italy, Vol. 44, pp. 166-167, Table 72)

During growth there is some pruning to train the trees and to cut out dead wood. Wood of the avocado is soft and brittle and liable to wind damage, thus windbreaks are often provided. Some types reach a height of 65 ft *(20 m)*, but budded trees are generally shorter. Well cared for trees are productive for many years.

Pollination by bees is necessary. The pistil of a flower is receptive before the pollen on that flower is mature. Flowers open one day with a receptive pistil, then close and open a second day when the pollen is shed but the pistil is no longer receptive. Pollen-bearing bees bring pollen to pistils while they are receptive. Depending upon the types, the fruits ripen 6 to 15 months after flowering.

Harvesting. Mature fruits may remain on the tree for some time before harvesting, but they are gathered by hand when nearly ripened. Mature fruits will ripen in 3 to 6 days at 80° F *(27° C)*, and in 30 to 40 days at 40° F *4° C)*. Once ripe, avocados are very perishable. Immature fruits will not ripen properly if picked.

Yields vary from 100 to 500 or more fruits per tree, and good Californian orchards yield 6,000 to 12,000 lb per acre *(6,720 to 13,440 kglha)* per year.

PROCESSING AND PREPARATION. Most of the crop is consumed fresh; however, the pulp may be frozen. The avocado does not cook satisfactorily.

Avocados are peeled and seeded, and then used in a variety of fruit dishes, cocktails, dips, salads, salad dressings, and sandwiches. Bacon and avocado sandwiches are a favorite. To avoid the brownish color of avocado flesh that develops when exposed to the air, peeled fruit should be placed immediately in lemon juice until used. Avocados may also be eaten alone, served as halves with lemon juice, vinegar, Worchestershire sauce, and salt and pepper.

• **Guacamole**—Guacamole is a popular Mexican spread made with chopped hot peppers, onions, tomatoes, and mashed avocados.

SELECTION. Ripened fruits are slightly soft, yielding to a gentle pressure on the skin. Avocados that do not yield to the squeeze test may be ripened during 3 to 5 days at room temperature. Ripening can be slowed by refrigeration. Irregular light brown markings sometimes found on the skin have no effect on the flesh. The signs of decay in avocados include dark sunken spots in irregular patches or cracked or broken surfaces.

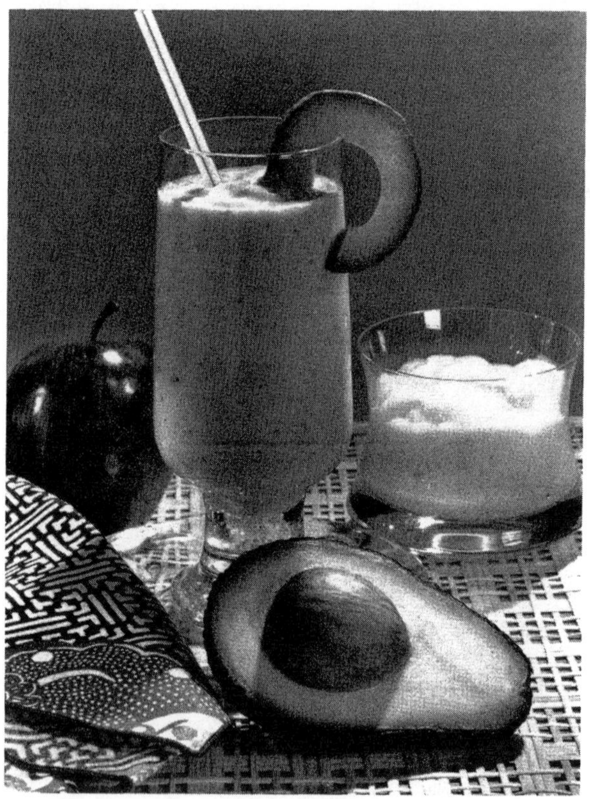

Fig. A-64. Avocados are good with other fruits. The avocado-yogurt "blenderable" breakfast contains lemon-flavored yogurt, avocado, ice, and apple, blended until smooth. (Courtesy, Harshe-Rotman & Druck, Inc., Public Relations, Los Angeles, Calif.)

NUTRITIONAL VALUE. Nutritional data on several varieties of fresh avocados is given in Food Composition Table F-36.

The fat content of avocados ranges from 11 to 25%. Thus, avocados contain 128 to 233 Calories (kcal) per 100 g (about 3½ oz), depending primarily on their fat content. No other fruit has an energy value as high as avocados. Additionally, each 100 g of avocados provides about 1 to 2 g of protein, 500 to 700 mg of potassium, and 300 to 400 IU of vitamin A.

(Also see FRUIT[S], Table F-47 Fruits of the World.)

AVOIRDUPOIS WEIGHTS AND MEASURES

Avoirdupois is a French word, meaning "to weigh." The old English system of weights and measures is referred to as the Avoirdupois System, or U.S. Customary Weights and Measures, to differentiate it from the Metric System.

(Also see WEIGHTS AND MEASURES.)

AZAROLE (NEOPOLITAN MEDLAR) *Crataegus azarolus*

The crab apple-like fruit of a tree (of the family *Rosaceae*) that grows in Algeria, Spain, France, and Italy.

Fig. A-65. The azarole, a relative of the crab apple that grows in the Mediterranean region.

Azaroles are usually yellow-orange to red in color and have a red flesh with a sweet, applelike flavor when they're ripe. However, only the fruit grown in the warmer areas of the countries bordering the Mediterranean ripens fully. In other areas, azaroles may reach maturity, but not ripeness. Unripe fruit must undergo a slight deterioration called "bletting" to become palatable. Azaroles are usually made into jam, jelly, a liqueur, or marmalade.

AZO DYES

This refers to a group of synthetic dyes which have the molecular structure R – N = N – R, a basic structure of two atoms of nitrogen (N) and two radicals (R) of the benzene or naphthalene series. Also, azo dyes are formed from amino compounds by the process of diazotization and coupling. Examples of azo dyes include Chrysoidine Y, Bismark Brown 2R, and Direct Green.

AZOTEMIA

An accumulation in the body of excessive amounts of urea and other nitrogenous wastes due to failure of the kidneys to excrete these products in the urine. The condition is characterized by an abnormally high level of urea in the blood.

(Also see KIDNEYS.)

AZOTOBACTER

A bacterium of the genus *azotobacter*, which is capable of synthesizing protein from atmospheric nitrogen.

Avocado yogurt soup—an easy-to-prepare cold soup for a hot summer day. It consists of avocados, lemon juice, chicken broth, yogurt, onion and celery salt, whirred in a blender and chilled. (Courtesy, California Avocado Commission, Newport Beach, Calif.)

The Golden Rule of Eating: The greater the variety of foods in the diet; the greater the chances of getting everything that the body requires. (Courtesy, California Foods Research Institute for Cling Peach Advisory Board, San Francisco, Calif.)

BABASSU OIL

The babassu is a type of palm which grows all over Brazil. A nondrying edible oil similar to coconut oil may be expressed from the kernels. It is usable in foods and soapmaking, but export from Brazil is small and supplies are limited.

BABCOCK TEST

This is a simple, rapid, accurate, and inexpensive test for determining the fat content of milk and cream, developed by Dr. S. M. Babcock of the University of Wisconsin in 1890. Prior to 1890, milk and cream were sold by volume—regardless of composition—even with added water at times. But it was long recognized that there was dire need for a more equitable basis of payment. Finally, Dr. Babcock developed the butterfat test that bears his name. It became, and still is, an official test in the United States and many other countries. Today, the Babcock Test is being replaced by faster and less variable instrumental methods, but even in these the Babcock Test is normally employed in determining the fat content of samples used to calibrate instruments. Also, the dairy industry is now shifting its attention away from milk fat to milk protein.

• **About Stephen Moulton Babcock (1843-1931)**—Dr. S. M. Babcock was an American agricultural chemist, who served as professor at the University of Wisconsin from 1887 until his death. It is noteworthy that Babcock recognized that he could have patented his test, which would have made him and his heirs wealthy. But he also knew how much dairymen needed the test, therefore, he forfeited a potential fortune and gave the test free to the dairy industry; he put need and service to society ahead of self.

• **Principle of the Babcock Test**—The Babcock Test is made possible by the chemical action of sulfuric acid and the application of centrifugal force. When sulfuric acid and milk are mixed, the acid dissolves the solids-not-fat and leaves the butterfat free to rise. The heat of the reaction liquifies the butterfat and facilitates further separation. The acid also increases the difference between the weight of the butterfat and the solution. The butterfat would never rise completely without the application of force, however. Centrifugal force, applied by whirling the bottles, involves the principle that when a fluid made up of parts having different weights is whirled, the heavier parts are thrown to the outside and the lighter parts are drawn toward the center of force. Because butterfat is lighter than the other constituents, it is brought into the neck of the test bottle, when it is whirled in the centrifuge. The fat content is then determined in the precisely calibrated neck of the test bottle.

• **Why is a fat test needed?**—The fat test is needed in order to establish the market value of milk and cream. But fat tests are important for numerous other purposes. For example, in Cheddar cheese 50% of the dry matter must be fat; hence, fat tests provide a means of standardizing milk for the production of such cheese. Similarly, fat content must be known for the production of nonfat dry milk from skim milk, butter from cream, ice cream from numerous possible sources of fat, and evaporated milk from milk.

BABY BEEF

Meat from heavy calves that are fat enough for slaughter at weaning time, weighing 400 to 700 lb (*180 to 315 kg*) on foot.

(Also see BEEF AND VEAL.)

BABY FOODS

Contents Page

History 134
Importance of Supplementing Milk or Formula at an Early Age 135
Types of Commercial Baby Foods Sold in the United States 136
Special Infant Foods That are Used in the Developing Countries 138
Preparation of Homemade Strained Baby Foods 139
Guidelines for Introducing Baby Foods During the First Year 139
Uses Other Than Infant Feeding 141
Future Prospects 141

Fig. B-1. Commercial infant formulas are carefully produced, and designed to be as close to human milk as is technologically possible. (Courtesy, Gerber Products Company)

The term *baby foods* usually refers to foods other than milk or formula that are fed to infants during the first year of life. There is no specific word for these items in the English language, but in German they are collectively called "beikost." Baby foods are generally used during the period between the time that an infant first requires supplementation of milk or formula and the age when he or she is ready to eat the foods served to the rest of the family. However, some of these items may also be useful in the diets of older children and adults.

HISTORY. It was long thought that infants in prehistoric societies received only breast milk for the first year or two of life. However, there is now evidence from studies of primitive peoples that supplementation of breast-feeding was often started early in infancy. For example, anthropologists have found (1) that Eskimos traditionally gave their babies bits of raw fish and meat; (2) that Australian aborigines supplemented breast-feeding with honey, turtle eggs, fish, meat, fruits, and vegetables; and (3) that Masai tribes in Africa fed blood drawn from their cattle.

The spread of subsistence agriculture and the decline of nomadic hunting, fishing, gathering, and the herding of livestock eventually led to the weaning of infants on diluted gruels made from grains, starchy roots and tubers, and occasionally, legumes. Diets limited to these plant products have often been responsible for the development of protein-energy malnutrition in children. Hence, the settling of people in farming societies may have resulted in an impoverishment in the health of children, unless sufficient amounts of animal products were produced. Therefore, it is noteworthy that even in recent times the agriculturists who fed their infants dairy products, eggs, and meat from cattle, goats, pigs, poultry, and sheep had much healthier children than those who utilized mainly cultivated grain and root crops.[1]

Unfortunately, many of the babies whose mothers were unable to breast feed them did poorly on cow's milk and other foods because little was known about (1) how the nutrient composition of human milk differed from the composition of cow's milk and the various supplemental foods, and (2) sanitary measures which would help to protect foods from contamination by harmful microorganisms. These gaps in knowledge were filled by about the middle of the 19th century when the renowned German chemist Justus von Liebig published his analyses of human milk and cow's milk, along with those of various other foods, and the French scientist Louis Pasteur demonstrated that microorganisms which caused the spoilage of foods could be killed by heating the foods to moderately high temperatures.

In the early part of the 20th century, the research of Sir Frederick G. Hopkins of Cambridge University in England and of Elmer V. McCollum at the University of Wisconsin showed that certain foods contained minute amounts of essential nutrients that were shortly thereafter to be designated as vitamins. These developments helped to create an awareness of the need for feeding infants and children a variety of both animal and plant food. About the same time, studies at the Massachusetts Institute of Technology and the National Canner's Association identified the conditions needed to destroy the harmful microorganisms during canning. Hence, the 1920s ushered in a period of rapid growth in food processing which, ultimately, benefitted infants around the world. Furthermore, it was during this decade that doctors began to recommend the feeding of solid foods before the age of 1 year.

The baby food industry was born in 1928 in response to a challenge of Mrs. Dan Gerber to her husband, whose family owned the Fremont Canning Company in Fremont, Michigan. She asked him if strained food for babies could be produced at the cannery so that many mothers might be spared the tedious chore of home preparation. He investigated the possibility, and soon tins of strained carrots, peas, prunes, spinach, and vegetable soup were being produced and promoted by salesmen who toured the country in Austin automobiles. (These cars were probably the first subcompacts to be sold in the United States.) Other companies soon entered the competition for a share of the growing market and new products were developed.

Commercial baby foods came under fire in the late 1960s and early 1970s after it was learned in 1968 that certain products contained the flavor enhancer

[1]Price, W. A., *Nutrition and Physical Degeneration*, Paul B. Hoeber, Inc.

monosodium glutamate (MSG) which could produce dizziness, headache, numbness and other symptoms in susceptible people. The symptoms were collectively described as the "Chinese Restaurant Syndrome" because they occurred most often in people who had just eaten in Chinese restaurants. (Many Chinese cooks add liberal amounts of MSG to food.) Then, in 1969 several researchers produced brain damage in mice by feeding or

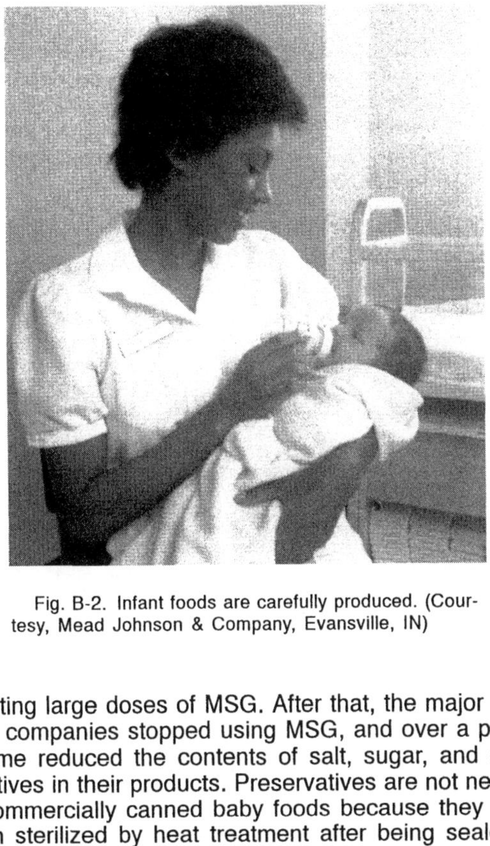

Fig. B-2. Infant foods are carefully produced. (Courtesy, Mead Johnson & Company, Evansville, IN)

injecting large doses of MSG. After that, the major baby food companies stopped using MSG, and over a period of time reduced the contents of salt, sugar, and other additives in their products. Preservatives are not needed in commercially canned baby foods because they have been sterilized by heat treatment after being sealed in the jar. Hence, these products now contain mainly basic food ingredients plus only small amounts of certain seasonings.

IMPORTANCE OF SUPPLEMENTING MILK OR FORMULAS AT AN EARLY AGE. Baby foods, which were once viewed as luxury items, are now seen in a new perspective as a result of the discovery that mothers in primitive societies often fed their babies supplemental foods (some of which were prechewed by the mother) soon after birth.[2] Furthermore, health professionals working with people in the developing countries have begun to recognize that there is usually need to supplement mother's milk when the infant reaches 3 months of age. In the United States, many pediatricians recommend that solid foods may be introduced at 4 to 6 months of age, although many mothers do so much earlier. It appears that mothers in Europe start supplementary foods at 6 to 10 weeks of age. Some of the major reasons for supplementing breast milk or evaporated milk preparations may be deduced from comparisons of the nutrient composition data of these items with the Recommended Daily Allowances (RDAs) for infants which are shown in Fig. B-3.

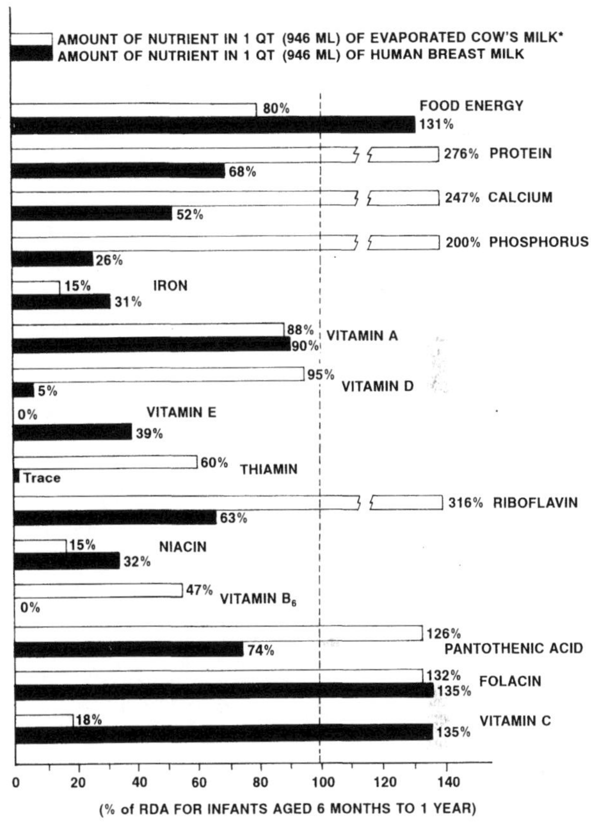

Fig. B-3. Percentages of the RDAs for infants which are supplied by 1 qt (*0.95 liter*) of either human milk or evaporated milk diluted 1:1 with water. (Based upon data from *Recommended Dietary Allowances*, 10th rev. ed., National Academy of Sciences, 1989; and Table F-36, Food Compositions, p. 802 of this book)

It may be seen from Fig. B-3 that for infants at age 6 months, a quart (*0.95 liter*) of diluted evaporated milk provides very high levels of protein, calcium, and phosphorus. Some nutritionists and specialists in infant growth believe that large excesses of these nutrients may be harmful because (1) the infant's kidneys may be taxed in ridding the body of the metabolic waste products, and (2) an excessively rapid and stressful rate of growth may be stimulated since calves that are fed cow's milk dou-

[2]Raphael, D., "Margaret Mead—A Tribute," *The Lactation Review*, 1979, Vol. 4:1, p.2.

ble their birth weight in 5 to 10 weeks, whereas infants require about 21 weeks. Furthermore, the evaporated milk falls far short of providing the allowances for iron, vitamin E, and vitamin C; and is moderately low in calories, thiamin (vitamin B-1) and vitamin B-6 (pyridoxine). Therefore, greater dilution of evaporated milk to lower the levels of protein, calcium, and phosphorus would make the deficiencies of the other nutrients even worse.

Human breast milk provides far too little phosphorus, iron, vitamin D, vitamin E, thiamin, and vitamin B-6; and is low in calories, protein, calcium, riboflavin, niacin, and pantothenic acid. Hence, reliance solely on this food for the feeding of infants aged 6 months and older is likely to result in slower than optimal rates of growth that may be accompanied by various nutritional deficiencies such as anemia.

Although commercially produced infant formulas now provide the proper amounts and proportions of nutrients for infants up to 1 year of age, it is better to introduce solid foods early than to rely solely on the formulas, because some nutritionists, pediatricians, and psychologists fear that failure to give infants solid foods shortly after they have learned to chew may lead to great difficulty in getting them to accept solids later. (Psychologists believe that certain types of learning and behavioral advances are most likely to occur during the periods when the body undergoes the developmental changes that allow the new learning or behavior to be practiced. It is noteworthy that chewing strengthens an infant's throat muscles for the development of speech.) Also, the feeding of sweetened milk or formulas in nursing bottles to infants after some of the teeth have erupted (this stage usually starts at about 6 months of age) may lead to "nursing bottle syndrome," which is tooth decay initiated by prolonged contact of the sweetened mixture with the teeth. This undesirable condition is most likely to result from habitually giving a bottle to an infant at bedtime when the rate of sucking is very slow and the flow of saliva is minimal.

TYPES OF COMMERCIAL BABY FOODS SOLD IN THE UNITED STATES. The items that are most likely to be found in local markets are described in Table B-1.

TABLE B-1
MAJOR TYPES OF BABY FOOD PRODUCTS SOLD IN THE UNITED STATES[1]

Type of Product	Description and Typical Ingredients[2]	Nutritional Value[3]	Comments
Baked foods	Cookies, pretzels, teething biscuits, and toast made from enriched wheat flour, sugar and/or syrup, dairy whey, vegetable shortening, leavening (baking powder or yeast), and various flavorings.	About the same as ordinary breads and cookies. Animal-shaped cookies have added B vitamins.	Useful mainly as teething aids and finger foods, starting at about 5 to 7 months of age.
Cereals, dry, ready-to serve (boxed)	Precooked, tiny flakes that contain cereal flour(s) such as barley, corn, malt, oat, rice, and wheat; plus calcium-phosphate carbonate, iron, thiamin, riboflavin, and niacin. Some items contain fruit(s) and/or soy flour.	A ½ oz (*14 g*) portion provides 22% of the U.S. RDA[4] for iron, thiamin, and riboflavin; and 11% of that for niacin. The high-protein cereal has about double the protein content of the other cereals.	The iron content is utilized better when the cereal is prepared with fruit juice rather than with milk. Some pediatricians recommend that iron-fortified cereals be used up to age 2 or beyond in order to prevent anemia.
Cereals, moist (in jar)	Similar in composition to the dry cereals, except that the water content is about 80% instead of 7%. Some items contain added vitamin C and vitamin B-6.	A ½-jar portion *(64 g)* is approximately equivalent in nutritional value to a ½ oz *(14 g)* portion of a dry cereal.	These items have high water contents and only about ⅕ the solids content of the dry cereals (on a weight basis).
Cottage cheese with pineapple	Cheese food made from nonfat milk, cottage cheese, dry curd cottage cheese, pineapple juice, water, pineapple, rice flour, orange juice, and citric acid.	A 4½ oz *(128g)* jar provides 150 kcal and 45% of the U.S. RDA for protein. Hence, it is similar in nutritive value to the high meat dinners.	Contains almost 50% more calories, but only ½ the protein, ¾ the calcium, and ⅗ the phosphorus of ordinary cottage cheese.
Desserts	Custard or fruit puddings containing water, nonfat milk, milk, fruit and/or fruit juice, sugar, rice flour, corn starch, egg yolks, flavorings, and/or citric acid.	Provide mainly calories (100 to 120 kcal per jar) from the high carbohydrate contents. Some fruit puddings are good sources of vitamin C.	Older children and adults probably enjoy these items as much or more than the baby does. These items may be fed to mature and active infants whose major nutrient needs have already been met by milk, cereals, fruits, vegetables, and meats.

(Continued)

TABLE B-1 *(Continued)*

Type of Product	Description and Typical Ingredients[2]	Nutritional Value[3]	Comments
Egg yolk	Cooked egg yolks and water sufficient for preparation. There are approximately three average egg yolks per jar.	A 4½ oz (*128 g*) jar provides 180 kcal and 100% of the U.S. RDA for protein, 50% of that for riboflavin and vitamin B-6, and 25% of that for vitamin A.	This item is processed so that it is less likely to be allergenic than egg yolks prepared at home.
Fruits	Purees containing fruit, water, and vitamin C. Some products also contain tapioca starch and citric acid.	Most items contain 45 to 300% of the U.S. RDA for vitamin C per 4½ oz (*128 g*) jar.	No sugar is added to these products. Hence, some items have a tart flavor.
Fruit juices	Full strength fruit juice with added vitamin C. Some products also contain added citric acid.	A 4.2 oz (*124 ml*) serving provides 120% of the U.S. RDA for vitamin C.	The orange juice has been specially prepared to minimize the contents of peel and seed, which sometimes provoke allergies.
Meats	Purees of meat or poultry and water. Lamb contains added lemon juice concentrate (to maintain color).	A 3½ oz (*99 g*) jar contains 100% of the U.S. RDA for protein, 50 to 300% of that for riboflavin, 25 to 150% of that for niacin, and 30 to 90% of that for vitamin B-6. (Liver is very rich in vitamin A.)	Beef liver has by far the highest nutrient levels of these products. Hence, ½ jar (*50 g*) would be an adequate serving of this item.
Meat and poultry sticks	Bite size, sausage shaped sticks made from meat and/or poultry, water, calcium reduced skim milk, salt, sugar, onion powder, and garlic.	The contents of a 2½ oz (*71 g*) jar provide 70% of the U.S. RDA for protein.	These items are good "finger foods" for infants that are 7 months or older, providing that they are able to chew foods.
Meat dinners, high	Meat and vegetable purees made from water, meat or poultry, vegetables, rice flour, wheat flour, potatoes, and onion powder. Beef with vegetables also contains added torula yeast.	A 4½ oz (*128 g*) jar supplies from 15 to 45% of the U.S. RDA for protein, and from 15 to 85% of that for vitamin B-6.	Each item contains at least 30% meat or poultry.
Vegetable and meat combinations (soups and dinners)	Purees made from water, vegetables, meat or poultry, rice flour, enriched macaroni or noodles, wheat flour, potato solids, and onion. (Some items also contain dry milk, soy protein and/or yeast.)	A 4½ oz (*128 g*) jar provides 2 to 15% of the U.S. RDA for protein, 2 to 100% of that for vitamin A, and 6 to 25% of that for vitamin B-6.	These items have several times as much protein as vegetables alone, but only about ¼ as much as high meat dinners. The bacon used in vegetables and bacon is cured without added nitrites or nitrates.
Vegetables, creamed	Purees containing water, corn, or spinach, dry milk and rice flour. Creamed spinach also contains oat flour and onion.	4½ oz (128 g) of creamed corn supplies 10% of RDA for protein and 15% of that for vitamin B-6. 4½ oz (128 g) of creamed spinach provides 27% of RDA for protein, 120% of that for vitaminA, 20% of those for calcium and riboflavin, and 15% of that for vitamin B-6.	On a weight basis, strained creamed spinach is about twice as rich in vitamin A as the junior food type of creamed spinach. The levels of the other nutrients are about the same in both products.
Vegetable, plain	Purees made from vegetables and water. Mixed vegetables also contain wheat flour, oat flour, potato solids and onion.	A 4½ oz (128 g) jar contains from 4 to 15% of the RDA for protein, from 2 to 200% of that for vitamin A, from 6 to 40% of that for vitamin C, from 2 to 20% of that for thiamin, from 4 to 20% of that for riboflavin, from 2 to 15% of that for vitamin B-6.	An infant should be given at least one dark green or deep yellow vegetable each day. This requirement may be met by carrots, garden vegetables, mixed vegetables, squash, or sweet potatoes.

[1]The information presented in this table is based upon that obtained from the baby food manufacturers, Table F-36 of this book, and *Recommended Dietary Allowances, 10th edition, National Research Council.*

[2]Ingredients may vary considerably among the different items within each product category. Hence, the purchaser should check the label of the item that is to be purchased.

[3]Nutritional values vary according to the ingredients used. Both types of information are given on the labels of the individual products.

[4]The U.S. RDAs are very liberal nutrient allowances established by the Food and Drug Administration for the standardization of nutritional labeling. In certain cases, they are significantly greater than the RDAs established by the National Research Council. Hence, some foods may provide greater percentages of the nutrient allowances for infants than is indicated by the percentages of the U.S. RDAs given on the product labels.

SPECIAL INFANT FOODS THAT ARE USED IN THE DEVELOPING COUNTRIES. Americans can benefit from a consideration of the problems encountered in feeding infants elsewhere in the world because (1) there are still unexploited opportunities for utilizing our resources and technical expertise to help stave off world hunger, (2) unusual native foods that are used abroad may also be useful in the United States, and (3) the lessons that are learned could be applied to helping impoverished Americans feed their infants.

Although infants in the developing countries have essentially the same nutritional requirements as those in the developed countries, the food resources that are available in each of these situations differ considerably. Too often, the people in the former countries have little opportunity to obtain the animal foods that are readily available in the latter countries and are forced to rely heavily on corn, rice, or wheat products plus starchy fruits and vegetables such as bananas, cassava, Irish potatoes, sweet potatoes, and yams. Unfortunately, the feeding of only grains and vegetables to infants may produce protein-energy malnutrition because (1) they are bulky and are likely to fill the infant before sufficient food is consumed to meet his or her calorie requirements, and (2) the quantity and quality of the dietary protein is often inadequate.

While breast feeding infants for 1 or 2 years may meet some nutrient requirements, the quantity and quality of human milk is usually inadequate to serve as the sole baby food after the age of 6 months. Hence, special infant foods may often be needed to offset the deficiencies of breast milk and the foods that are available locally. Furthermore, the supplementary foods must be good sources of high-quality protein, low in cost, resistant to contamination and spoilage by pathogenic microorganisms, and easy to prepare under primitive conditions. Cereals fortified with protein, minerals, and vitamins come close to meeting all of these criteria. Some examples are given in Table B-2.

It is noteworthy that most of the infant cereals described in Table B-2 are fortified with minerals and vitamins so that in emergency situations these items may be mixed with water to serve as the sole foods. However, it is better if the cereals are supplemented with small amounts of cooked, mashed, and strained local items such as eggs, fish, meats, poultry, and seafood; and edi-

TABLE B-2
PROTEIN-FORTIFIED INFANT CEREALS FOR DEVELOPING COUNTRIES

Product	Description	Comments
Bal ahar	Coarsely ground mixture of bulgur wheat, peanut flour, skim milk powder, minerals and vitamins. Contains 22% protein.	Produced by the government of India. The protein quality is on a par with that of milk.
CSM	Partially cooked mixture of cornmeal, defatted soy flour, nonfat dry milk, soybean oil, minerals, and vitamins. Contains 19% protein.	Developed by the American Corn Millers Federation in accordance with USDA guidelines. Tested and found to be highly satisfactory and acceptable in Latin America and North Africa.
Incaparina	Mixture of corn flour, cottonseed flour, torula yeast, minerals, and vitamins. It has a protein content of about 25%.	Developed by the Institute of Nutrition of Central America and Panama (INCAP), with the aid of funds from Central American governments, Kellogg Foundation, Rockefeller Foundation, and several other foundations. Test marketed in Nicaragua, El Salvador, Guatemala, and Colombia.
Laubina	Finely ground mixture of cooked bulgur wheat, cooked chickpeas, skim milk powder, sugar, minerals, and vitamins.	Developed at the American University of Beirut in Lebanon. Promoted prompt recovery of hospitalized infants suffering from severe protein-energy malnutrition, but was not effective in curing anemia.
Leche alim	Mixture of toasted wheat flour, fish protein concentrate (FPC), sunflower meal, and skim milk powder. Contains 27% protein.	Produced by the Pediatrics Laboratory of the University of Chile in Santiago.
Pronutro	A finely ground mixture of corn, skim milk powder, peanuts, soybeans, wheat germ, sugar, salt, minerals, vitamins, and flavoring(s). It has a protein content of about 22%.	Developed without any governmental or international assistance by Hind Brothers, a private food firm in Natal, South Africa. Effective as skim milk powder in the rehabilitation of infants suffering from severe protein-energy malnutrition (kwashiorkor).
Protex	Defatted mixture of rice bran, germ, and polish. Contains between 17 and 21% protein, plus minerals and vitamins present in the outer layers of the rice grain.	The process for producing Protex commercially was developed by Riviana Foods, Inc. at their Abbeville, La. plant.
Superamine	Finely ground mixture of wheat flour, chickpea flour, lentil flour, skim milk powder, sugar, minerals, vitamins, and flavorings. Contains 21% protein.	Developed as a cooperative project between FAO, WHO, UNICEF, and the Algerian government.

ble leaves, roots, tubers, fruits, stalks, and flowers so that the infant receives essential trace nutrients not supplied by the cereals, and learns to eat a wide variety of foods.

PREPARATION OF HOMEMADE STRAINED BABY FOODS. Home preparation of baby foods instead of using commercially prepared products may be an uneconomical use of time and energy unless the infant is fed foods that have already been prepared for the rest of the family. Then, small portions of the prepared items may be mashed, strained, and fed with a minimum of extra work. (The preparation of large amounts of purees in a blender is likely to tempt the mother to store the food for too long in a refrigerator. Babies are very susceptible to digestive problems caused by spoiled food. Hence, baby food should not be kept in a refrigerator for more than a day or two.) A handy device for preparing small amounts of purees is shown in Fig. B-4.

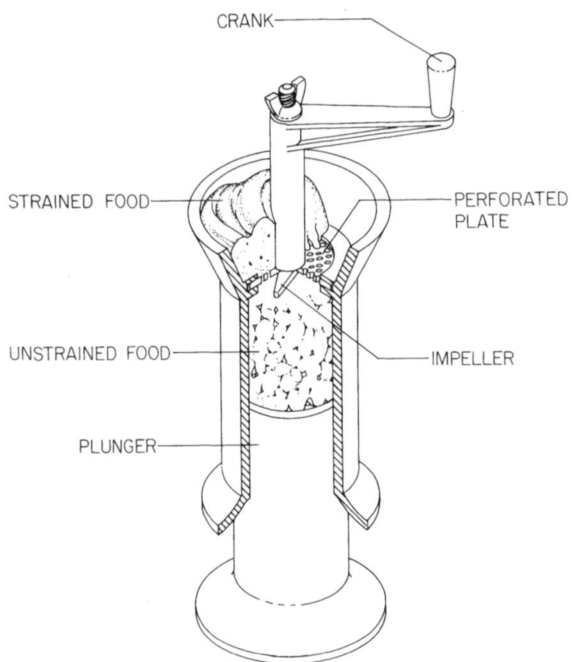

Fig. B-4. A plastic pureeing device for making homemade baby foods. Small pieces of the cooked, unstrained food are placed in the chamber below the perforated metal plate. The plunger is inserted and the device is set upright on a flat, hard surface. Then, the crank is turned with one hand while the outer jacket is pressed down on the plunger with the other hand. The infant may be fed directly from the upper bowl-like receptacle where the strained food is collected.

Instructions for preparing various foods follow.

• **Fruit juices**—Fresh juice should be strained to remove the pulp. To feed fruit juice, put a small amount of juice in a sterilized bottle and add cool boiled water. Each day decrease the amount of water and increase the juice until the baby is getting ¼ cup (*60 ml*) of pure juice daily. Give it cool fruit juice. Heating the juice destroys vitamin C. Do not add sugar to juices. This only adds energy (calories)—not nutrients, and it increases an infant's taste for sweets.

• **Cereals**—Cooked cereals should be strained and made "soupy" with milk. Gradually increase the amount of cereal and thicken it as the infant learns to swallow. Do not add cereal to a baby's bottle. Spoon feeding is important to the development of good eating habits.

• **Fruits**—Either soft canned, cooked fresh, or stewed dried fruits such as peeled apples, peaches, pears, and apricots can be used if mashed thoroughly. Neither sugar nor syrup from the canned fruit should be used. Do not use raw fruit other than ripe bananas.

• **Vegetables**—Fresh, canned, or frozen vegetables may be used. Cook and mash vegetables thoroughly. At first, you may want to strain the vegetables. Do not add salt, spices, fats, or bacon to the vegetables.

• **Fish, meats, poultry, and legumes**—"Table" meats such as ground beef, chicken, liver, and fish may be used. Ground beef should be boiled and mashed as it cooks. Fish, chicken, and liver should be finely mashed. Moisten the meat with broth or milk. Do not fry meat. Canned chicken or fish may be finely mashed and used. Remove all bones, fat, and skin.

Dried beans and peas, thoroughly cooked and mashed, may be used in place of part of the meat. Be sure to include milk or a small amount of meat with the beans. Cook beans without bacon, salt pork, lard, or other fats.

GUIDELINES FOR INTRODUCING BABY FOODS DURING THE FIRST YEAR. While still meeting individual needs, the baby's first year should be a gradual transition to the following *broad* pattern—a working toward including in the daily diet suitable portions of foods from *each* of the following groups:

Milk group, including milk products, cheese, and ice cream.

Meat group, including beef, veal, pork, lamb, liver, poultry, fish, eggs (occasionally dry beans, peas and nuts for older children and adults).

Vegetable-fruit group, including a dark-green or deep-yellow vegetable (for vitamin A value) and a citrus fruit or other good vitamin C source, daily.

Bread-cereal group, emphasizing whole grain, fortified or enriched varieties.

Babies grow and develop at different rates. Some babies need more nutrients at an earlier time than breast milk or a formula can provide. The age for introduction will be variable but most doctors choose sometime within the baby's first month to start fruit juice, and by the third or fourth month to start infant cereals and/or strained foods.

It is noteworthy that some pediatricians consider that an infant is ready for semisolid foods when (1) his or her weight is 11 to 13 lb (*5 to 6 kg*), or double the birth weight; (2) more than a quart of milk or formula is consumed daily; (3) breast feeding fails to meet his or her demands for at least 3 hours after each feeding, and (4) he or she seems to be chronically hungry or dissatisfied. Usually, one or more of these signs are present before there's a possibility of nutritional deficiency. Therefore, parents can offer the first supplementary beverages and foods in a relaxed frame of mind, not worrying whether they're accepted the first day, the first week or even the first month they're offered.

Suggestions for introducing these items are given in Table B-3.

TABLE B-3
SUGGESTIONS FOR INTRODUCING SUPPLEMENTARY BEVERAGES AND FOODS DURING THE FIRST YEAR

Beverage or Food	Stage at Which Supplementary Item is Introduced	Comments
Fruit juices	Fruit juices are usually started *between 2 and 4 weeks of age.* Use juices that are naturally high in vitamin C or those that have vitamin C added. Use fresh, canned, or frozen orange juice or a non-citrus fruit juice fortified with vitamin C. Fresh juice should be strained to remove the pulp.	At first, juices are usually given in small amounts in a nursing bottle. They are prepared by mixing with cool water that was boiled to kill germs. The amount of water may be decreased each day until undiluted juice is fed (usually within a week or so of starting the item). Once weaning has been started, the juices may be given in a cup.
Cereals	Cereals with iron are usually started *between 2½ and 3 months of age.* Iron is needed to prevent iron-deficiency anemia. Read the label. Look for "iron-fortified" cereals. Special baby cereal (in the box) has extra iron added. Quick or instant cream of wheat has extra iron added. Grits and cornmeal are not good sources of iron unless enriched with iron. Use an iron-fortified cereal each day. Cereal is usually offered at the morning and evening feeding. Try rice cereal first as there is less chance of baby having an allergic reaction. Then try oat or barley cereals. Wheat cereals may then be added. Mixed cereals should not be started until all of the single-grain cereals have been accepted by baby without any allergic reaction.	These items are usually the first solid foods fed to a baby. They should be fed by spoon rather than putting them into the baby's bottle because the swallowing of semisolids is an important function to be developed at 2½ to 3 months of age. (At first, it is natural for infants to reject solids by thrusting the tongue out, which is an inborn reflex that protects against choking. However, the tongue thrust reflex disappears as the baby learns to swallow solids.) Swallowing is made easier by holding an infant with its head tipped back slightly and by placing the food near the back of the tongue. The mother should hold the baby in a relaxed manner and refrain from "force feeding" since eating should be a pleasant experience.
Vegetables	Vegetables are usually started *between 3 and 4 months of age.* They provide vitamins and minerals needed for growing. Dark yellow and leafy green vegetables provide vitamin A which helps keep the eyes and skin healthy. Vegetables also add color and new flavors to baby's meals. Bulk from vegetables helps promote regular bowel movements.	Start with a teaspoon *(5 ml)* or so of a mild-tasting vegetable such as beets, carrots, green beans, green peas, or squash. Then, increase the amount gradually to about ½ of a small jar daily, but reduce the amount if the baby develops loose stools. It is desirable for infants to receive at least two different vegetables each day. Therefore, vegetables should have priority over fruits when the baby's appetite is small.
Fruits	Fruits are usually started *between 3 and 4 months.* This is a new "sweet taste" for baby. Fruits supply vitamins and minerals. They also provide bulk to prevent constipation. Include a serving of fruit each day.	Begin with 1 tsp *(5 ml)* and gradually increase until the baby is getting 2 to 3 Tbsp *(30 to 45 ml).* Do not use raw fruit except ripe bananas until baby is older. If the infant's stools become soft and "runny," stop the fruit for a few days; then add a small amount gradually.
Cottage cheese	Strained cottage cheese and fruit mixtures may be introduced *between 3 and 4 months of age,* provided that the infant has already been fed iron-fortified cereals and other semisolid foods on a regular basis. Ordinary cottage cheese may be fed after the baby has been given junior foods.	The nutritive value of cottage cheese is similar to that of milk or formula. Hence, it should be fed in combination with other types of foods so that an excessive emphasis on dairy products is avoided. Only very fresh cottage cheese should be fed to infants since they are very susceptible to diarrhea caused by spoilage microorganisms.
Yogurt	Yogurt may be introduced *between 3 and 4 months of age,* provided that the infant has already been fed iron-fortified cereals, fruits, and vegetables on a regular basis. Plain yogurt mixed with a little strained fruit or juice is better than the sweetened, flavored varieties.	Yogurt has essentially the same nutritive value as the milk (whole or skim) from which it is made. Hence, the feeding of milk or formula plus yogurt without other foods places too much emphasis on dairy foods. Some doctors have treated infant diarrhea successfully by substituting yogurt for milk or formula.

(Continued)

TABLE B-3 (*Continued*)

Beverage or Food	Stage at Which Supplementary Item is Introduced	Comments
Meats	Meats are usually started *by 6 months of age.* They supply protein needed to build and repair muscles and other body tissues, and iron that is needed to prevent the baby from getting anemic. Meat also supplies minerals and vitamins needed to keep the body healthy. Begin with 1 tsp *(5 ml)* of meat and gradually increase to 2 or 3 Tbsp *(30 to 45 ml)* daily. Start with one meat and feed it several days before adding a new flavor. This gives the baby a chance to learn to like the taste. It also gives the mother a chance to see if her baby is allergic to the meat.	Plain meats are best for babies with small appetites because they are more concentrated sources of protein than the meat and vegetable mixtures. However, the latter products may be more suitable for hearty eaters since the feeding of excessive amounts of protein should be avoided. Furthermore, the mixtures teach babies to like blends of flavors that will be encountered later in stews and other mixed dishes.
Egg yolks	Egg yolks are usually started *between 4 and 6 months of age.* Egg yolks (yellows) give an infant iron needed to make red blood. They also give the baby a new taste and texture. Begin with 1 tsp *(5 ml)* and gradually increase until he or she is getting one yolk each day. If a rash develops, see your doctor before giving egg yolk again.	Egg whites should not be used until recommended by your doctor. They may cause the baby to have an allergic reaction.
Baked goods	Dry toast is a good food to develop chewing at *5 to 7 months of age.* To prepare, cut small strips of bread and put in a low (200°F or *93°C*) oven for 1 hour. Store this hard toast in an airtight container. Commercially made teething biscuits are especially shaped for easy grasping. They're hard-baked for biting and to prevent crumbling. The latter is a safety factor since crumbs can choke a baby. That's why babies shouldn't be allowed to eat breads or other dry foods while lying down or unattended. Animal-shaped cookies are also hard-baked and have a thin glaze fortified with several B vitamins.	These products are best suited for use as snacks or desserts after the major food items have been consumed. Therefore, the amounts fed should be limited in order to prevent spoiling of the baby's appetite for other items. Although the first baked goods given infants are usually made from enriched white flour, items made from whole wheat flour should be introduced by age 1, so that the growing child learns to eat more nutritious products.

USES OTHER THAN INFANT FEEDING. Pureed baby foods have often been used by people with various digestive disorders since these items are well suited for bland, low-residue diets. Hence, the purees are thought by many to be suitable mainly for semi-invalids, rather than for normal healthy adults. However, the growing use of electric blenders has created a new interest in pureed items in various recipes. (It is noteworthy that French and Italian cooks have long used purees in their cooking.) Baby foods in jars provide suitable quantities for experimental cookery on a small scale, so that wastage is minimized if the preparations fail to meet expectations. A blender and a strainer may be used to prepare larger quantities once the recipe has been tested and has passed the taste test. Therefore, some recipe suggestions for baby foods are given in Table B-4.

FUTURE PROSPECTS. Food technologists have already developed products which, with a few slight modifications such as reduction of the additive and salt contents, may be used as low cost foods for infants and the rest of the family. Furthermore, there is a need for some prepared infant foods in the developing countries, because preparation of these items by each mother for

Fig. B-5. Supplementing mother's milk or formulas at an early age is important. (Courtsy, Gerber Products Company)

TABLE B-4
PREPARATIONS WHICH UTILIZE BABY FOODS

Preparations	Types of Baby Foods Utilized	Comments
Aspics and gelatin salads	Fruits, meats, and vegetables.	Season liberally with herbs and spices.
Biscuits, muffins, pancakes, and waffles	Cereals, fruits, meats, and vegetables.	Add to batter or dough. Allow for water contents of purees.
Cookies, fruit bars, and spice cakes	Desserts and fruits.	Allow for water content of purees. Use flavorings and spices liberally.
Fruit drinks	Fruit juices and/ or pureed fruits.	Mix 1 part of puree with 2 parts water. Add lemon juice plus honey for extra flavor.
Gravies and sauces	Meats and vegetables.	Substitute purees for starchy thickeners.
Milk shakes	Desserts and fruits.	Add a little cinnamon, cloves, ginger, or nutmeg.
Salad dressings	Meats, fruits, and vegetables.	Mix purees with cottage cheese dressing(s), and/or yogurt.
Sandwich fillings	Fruits, meats, and vegetables.	Mix with cottage cheese, mayonnaise, peanut butter (fruits), salad dressing(s) and/or yogurt.
Souffles	Meats and vegetables.	Allow for water content of purees.
Soups	Dinners, meats, and vegetables.	Use purees for enriching and thickening.
Toppings for cakes, fruits, and ice cream	Desserts and fruits.	These toppings are likely to be lower in calories than most other types.
Toppings for cooked cereals	Cereals (dry, ready-to-eat).	Add a little cinnamon, honey, and milk.

her child is inefficient in terms of the fuel and time that is expended. Some examples follow:

• **Cereals**—The cooking of grain mixtures with water often produces thick, pasty masses that are difficult for infants to swallow and digest. Also, the dilution of these mixtures with extra water to make them more palatable lowers the nutritive value significantly. However, the malting and/or fermentation of cereal products renders them less viscous and more digestible by converting starch and proteins to sugars and amino acids. Furthermore, fermented products are lower in carbohydrates and higher in protein than the unfermented items because some of the carbohydrates are converted to alcohol, which may then be removed by distillation.

• **Dried vegetable powders**—These powders are currently used mainly by food processors, but they also have the potential to reduce the wasteful use of (1) fuel required to ship vegetables with high water contents, and (2) glass jars that are usually discarded after the contents have been used. Many of the dried products may be rehydrated readily with cool water.

• **High-protein breads**—It may be a good idea for infants to learn to eat the slightly heavier and considerably more nutritious breads that are made by adding high-protein supplements such as dry milk and/or soy flour to the conventional types of bread doughs.

• **Indigenous foods**—Certain native foods that may be grown readily without fertilizers in the developing countries have nutritional values that complement, equal, or exceed those of corn, rice, and wheat. For example, flours made from the grainlike seeds of weedy plants (mainly *Amaranthus* and *Chenopodium* species) have been incorporated in experimental foods produced in Peru.

(Also see BREAST FEEDING; and INFANT DIET AND NUTRITION.)

BACILLARY DYSENTERY

An acute infection of the bowel with *Shigella bacilli.* It is characterized by fever; abdominal pain; vomiting; and diarrhea which contains fluid and mucus, blood and pus, if the organisms have ulcerated the intestines. Often the victim becomes dehydrated and loses weight. There is a high death rate in infants from bacillary dysentery. Even when treated with antibiotics, the condition may continue for a long time.

Due to the loss of nutrients, mineral salts, and water, it is essential that nutrients and fluids be consumed by the afflicted person. Frequent small meals with an adequate caloric content, but which are high in protein and low in fat and carbohydrate, are usually most successful. It may be necessary to give fluids intravenously to combat the dehydration.

The source of infection is infected human feces, which is a potential source of contamination in food or water. It is noteworthy that flies may carry the organism to areas where foods are prepared.

(Also see BACTERIA IN FOOD; and FOODBORNE DISEASE.)

BACILLUS

A rod-shaped form of bacteria. These organisms are commonly found in soil, milk, and the intestine of man and animals. Most of the bacilli found in the colon are harmless, but a few species such as *Bacillus cereus* and *Shigella* may cause illness.

(Also see BACTERIA IN FOOD; and FOODBORNE DISEASE.)

BACON

Cured and smoked sides of pork carcasses.

(Also see MEAT.)

BACON, FLITCH OF

A side of cured and smoked bacon.

BACTERIA

Microscopic, single-cell plants, found in most environments; some are beneficial, others are capable of causing disease.

(Also see BACTERIA IN FOOD.)

BACTERIA IN FOOD

Contents *Page*

Bacteria May Be Helpful or Harmful 143
Helpful Food Bacteria 143
Harmful Food Bacteria 144
Food Infections (bacterial) 144
Food Poisonings (bacterial toxins) 148
Bacterial Control Methods 149
Prevention of Bacterial Infection and Food Poisoning 149

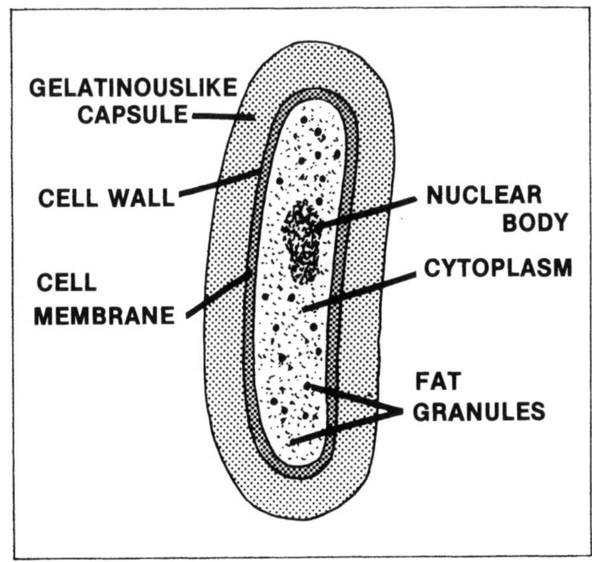

Fig. B-6. Diagram of typical cell structure of bacteria. Although some bacteria have certain special structures, almost all of them have the basic parts shown above.

With the invention of the microscope, scientists of the 1600s and 1700s discovered a whole new world of organisms never previously suspected, a world of bacteria so small that they could not be seen with the naked eye. Then little more than a hundred years ago, Louis Pasteur (1822-1895), the French scientist, demonstrated that certain kinds of these organisms were responsible for the fermentation of grape juice into wine and for the souring of milk.

Bacteria—microscopic, single-celled plants—are very small. Most of them are of the order of one to a few microns in cell length and somewhat smaller than this in diameter. (A micron is 1/25,000 of an inch or 1/1000 of a millimeter.) They can penetrate the smallest opening, many of them can even pass through the pores of an eggshell once the natural bloom of the shell is worn or washed away.

Bacteria may enter food during production, handling, processing, or serving. Since they require nutrients, moisture, and temperatures conducive to their growth, not all bacteria will grow in food; and one species may thrive where another will not. Some bacteria require oxygen from the air—they are known as *aerobic*; others grow in the absence of oxygen—they are *anaerobic*. Some bacteria like temperatures below freezing, others higher temperatures. The minimum temperature for most bacteria is 10°F (*−12°C*), the maximum 140° to 190°F (*60° to 88°C*).

Each year, more than 24 million Americans suffer from botulism, salmonellosis, and other foodborne illnesses. When unreported, untreated, and misdiagnosed cases are added, the number may exceed 70 million. Victims often experience little more than an upset stomach, but nearly 9,000 cases prove fatal.

BACTERIA MAY BE HELPFUL OR HARMFUL.

Most bacteria are helpful—they're benefactors of mankind; without them, no life on earth could exist. But some bacteria are harmful—they're agents of disease and death. This fact offers a convenient means of presenting the subject in the sections that follow.

Helpful Food Bacteria. Fortunately, many bacteria benefit foods and are widely used in biological oxidation reactions and in fermentation processes. Butter and cheese, for example, depend on bacteria for the development of their characteristic flavors. Bacteria are indispensable for making pickles, sauerkraut, olives, yogurt, buttermilk, beer, and wine. They are also employed in the manufacture of corn beef and rye bread. Bacteria are used to enhance the flavor of coffee, being introduced just prior to the roast. Also, a bacterium of the genus *Acetobacter* is used in making vinegar.

In recent years, bacteria have been employed to produce protein. Grown on petroleum-based raw materials—heavy oils, long chain paraffins, etc.—these bacteria, which contain 45 to 85% protein, may eventually have an important role in food production. Good progress has been made in petroprotein production for animals. Product purity, nutritive quality, and gastrointestinal tolerance stand in the way, at present, of the use of petroprotein as human food.

In addition to their role in foods, other helpful bacteria are found in the gastrointestinal tracts of ruminants (cattle, sheep, etc.) and pseudoruminants (horses, rabbits, etc.) which synthesize most, if not all, of the B vitamins and vitamin K; the nitrogen-fixing bacteria which grow in nodules (small bumps) on the roots of legume plants and convert nitrogen gas in the air into chemical forms that can be used by plants and animals; the bacteria that are used in sewage treatment plants to purify water; and the bacteria that are important in the cycling of chemicals in nature—for example, after a plant or animal dies, bacteria help break down the dead organisms into simple molecules that can be used by living plants and animals.

Harmful Food Bacteria. Many bacteria are harmful—they may cause food spoilage or foodborne diseases.

Bacteria are involved in many, but not all, types of food spoilage. The stale, putrid odor of spoiled meat, the foul odor of spoiled egg, and the sour odor of spoiled milk are examples of bacterial food spoilage. Both fresh and canned foods are subject to food spoilage. Some types of food spoilage are caused by enzymes, and others are caused by chemical reactions—for example, by nonenzymatic browning. The control of bacterial spoilage has received a tremendous amount of research attention since the days of Louis Pasteur, who pioneered in the study of bacteria.

Foodborne diseases are of two types: (1) food infections (bacterial), and (2) food poisonings (bacterial toxins). The most common diseases in each of these categories will be discussed briefly in the sections that follow.

FOOD INFECTIONS (bacterial). Food infections result from the ingestion of food with large amounts of viable bacteria, which multiply inside the host (man) and cause infectious diseases. Each specific disease is caused by a specific organism. Because incubation and multiplication of the bacteria take time, symptoms of food infection develop relatively slowly, usually 12 to 24 hours or more after the infected food has been eaten.

Foodborne infections are especially prevalent in poor urban communities with inadequate facilities for storing foods and unsanitary water supplies and lavatories. Under such circumstances, food-related infections are common and contribute to a high sickness and death rate, especially in children.

The usual diet therapy for a bacterial food infection includes: providing small amounts of cracked ice, barley water, and tea until the gastrointestinal disturbance stabilizes. Then, custards, gelatin, eggs, milk, toast, broth and/or cream soups can be given. Solid foods can usually be tolerated on the second or third day and the usual diet can be resumed.

The common bacterial food infections are summarized in Table B-5.

TABLE B-5
FOOD INFECTIONS (BACTERIAL)

Disease	Disease Organism	Signs and Symptoms	How Disease Is Transmitted	Prevention and Treatment	Remarks
Bacillus cereus	*Bacillus cereus*, a saprophytic (bacteria that obtain food by absorbing dissolved organic material) spore-bearing organism widely found in nature.	Severe vomiting 1 hour after ingestion, or diarrhea later. Recovery is rapid.	Usually, the disease occurs among those who eat in Chinese restaurants. The spore may survive normal cooking and produce vegetative forms during cooling.	Do not allow cooked rice to cool over a long period without refrigerating.	*Bacillus cereus* occurs in many samples of uncooked rice.
Brucellosis (Bang's disease, undulant fever, Malta fever)	*Brucella abortus* *Brucella melitensis* *Brucella suis* The genus *Brucella* is named after Sir David Bruce, a British army doctor, who, working in Malta in 1887, first isolated the organism from goats and from soldiers who had been drinking goat's milk.	A recurrent or undulating fever, which may rise to 104° to 105°F *(40° to 41°C)* in the evening, then drop to normal each morning; sweating; fatigue; muscle ache; constipation. If untreated, symptoms may persist for months.	From consuming infected milk or milk products, or from infected animals.	Pasteurizing milk; test of animals, and slaughter of reactors. Treatment with sulfonamides or antibiotics and with heat may be helpful.	The disease affects cattle, sheep (rarely), goats, swine, and man.

(Continued)

TABLE B-5 (*Continued*)

Disease	Disease Organism	Signs and Symptoms	How Disease Is Transmitted	Prevention and Treatment	Remarks
Cholera	*Vibrio comma,* a comma-shaped bacillus.	Fever, severe diarrhea, abdominal cramps, vomiting, intense thirst, followed by collapse. If untreated, a 50% death rate is common. Cholera is one of the most acute and violent infections known to man.	From infected persons and fecal-contaminated food and water.	Isolation of cholera patients, destruction by fire of material passed by patient, well cooked food, chlorinated water, and vaccination. Treatment consists in replacing body fluids and electrolytes.	Epidemics of cholera are rather common in India, Pakistan, and Southeast Asia.
Clostridium perfringens food infection	*Clostridium perfringens* (also known as *Clostridium welchii* or the gangrene organism), a spore-forming anaerobe widely distributed in soil, sewage, and unsanitary food processing plants.	Diarrhea, which is often accompanied by abdominal pain; and headache. Vomiting and fever are uncommon. The incubation period of the disease is 8 to 24 hours. Most patients recover in 24 hours, or at the most within a few days. Rarely fatal.	Infected foods, especially meats and gravies that have been allowed to cool slowly for several hours after cooking. The spores are resistant to heat and will survive boiling for as long as 5 hours.	Cook meats adequately. Do not allow foods, especially meats and gravies, to cool slowly after cooking—refrigerate them promptly.	This is the bacterium that causes gas gangrene when it infects deep wounds.
Escherichia coli food infections	*Escherichia Coli* *Escherichia* was named for the German physician, T. Escherichia, who first studied it.	Acute gastroenteritis in infants, and traveler's disease; characterized by severe watery diarrhea. In these instances, the victims may not have built up an immunity to a particular serotype (a group of closely related microorganisms distinguished by possessing a common set of antigens).	*E. coli* are one of the predominate bacterial flora of the gut. They are excreted in human feces (and urine to some extent). Thence infection is spread to foods and food utensils by flies and human hands. Baby formulas prepared under unsanitary conditions are the usual route of infection.	Avoid contaminated food and water. Avoid overindulgence with food and alcohol. Treatment with sulfa drugs and antibiotics supplemented by kaolin or pectin preparations will often afford speedy relief. Lomotil, a synthetic drug.	This disease occurs worldwide. Sometimes it is known as "tourist diarrhea" or "traveler's dysentery."
Leptospirosis	*Leptospira,* so named because it is the smallest and most delicately formed of the spirochetes (spiral-shaped bacteria).	High fever and intense, hemorrhagic jaundice and hepatitis.	By polluted water—from drinking it or swimming in it, or through cuts or scratches on the skin. From exposure to animals or people with leptospirosis. By consuming food or water that has been polluted, usually by rats.	Prevention consists of avoiding polluted food or water, along with control of rats.	Infected dogs, cattle, horses, swine, rats, and wild animals may pollute water. Leptospirosis is most common among persons who work in foul, watery places such as sewers, rice fields, wartime trenches, and poorly built mines.

(*Continued*)

TABLE B-5 (*Continued*)

Disease	Disease Organism	Signs and Symptoms	How Disease Is Transmitted	Prevention and Treatment	Remarks
Salmonellosis It is named for the American bacteriologist and veterinarian, Daniel E. Salmon, who first isolated the organism in 1885. This is one of the most common foodborne infections in the U.S.; sometimes single outbreaks involve thousands of people.	*Salmonella typhimurium* is the organism most commonly responsible. But there are more than 1,600 types of the genus *Salmonella.*	Diarrhea, abdominal cramps, and vomiting, which usually lasts for 2 to 3 days. The incubation period is 12 to 36 hours. Salmonellosis is rarely fatal except in elderly people and infants.	*Salmonella* bacteria grow rapidly in such cooked foods as meats, eggs, custards, and salads, which have been left unrefrigerated several hours. It may also be transmitted by sewage-polluted water. The organisms may be eliminated 2 to 3 days after the symptoms subside, thereby providing a continuing source of contamination for others.	Refrigerating foods at temperatures below 40°F *(4°C);* hand washing by food handlers; scrupulous cleaning of food processing equipment; and avoiding use of cracked eggs unless thoroughly cooked. *Salmonella* in food is destroyed by a temperature of 140°F *(60°C)* for 20 minutes or 149°F *(65°C)* for 3 minutes. Prevent flies, cockroaches, and rodents from coming in contact with food.	An accurate bacteriological diagnosis of the cause of an outbreak is essential to establish its source.
Shigellosis (bacillary dysentery)	*Shigella* bacteria *The Shigella* genus is named for the Japanese physician, Kiyoshi Shiga, who was the first to discover a species of the organism during an epidemic in Japan in 1898.	Fever, loss of appetite, vomiting, severe abdominal cramps, and massive diarrhea. Young children and frail adults may become dehydrated; hence, care must be taken to maintain their balance of mineral salts.	Spread by fecal contamination of food, water, clothing, and household objects by infected individuals. House flies are also an active agent in its spread.	Good public sanitation and personal hygiene; control of flies; boiling food and water and pasteurizing milk; washing hands before handling foods or eating; and isolation of patients and carriers, especially if they are handling foods. Sulfa drugs or antibiotics are used in treatment.	Epidemics of the disease most commonly occur where large groups are crowded together without adequate sanitation, such as refugee camps.

(Continued)

Determining the microbiological flora in foods. (Courtesy, University of Illinois, Urbana-Champaign, Ill.)

TABLE B-5 (*Continued*)

Disease	Disease Organism	Signs and Symptoms	How Disease Is Transmitted	Prevention and Treatment	Remarks
Tuberculosis (TB)	*Mycobacterium bovis* (tubercule bacillus) Robert Koch identified the causative organism in 1892.	Chronic coughing, usually fever and night sweats, extreme fatigue, loss of appetite, and, eventually, coughing up blood. Enlargement of the cervical and mesenteric lymph nodes.	The bacteria are spread by particles of dust or droplets expelled by a tubercular patient, especially when coughing or sneezing; or introduced into the digestive tract by contaminated foods—such as milk from tuberculin cows or objects placed in the mouth.	Avoid contact with infected people and foods; testing, followed by slaughter of tubercular animals; pasteurization of milk; adequate and nutritious diet, comfortable living quarters; and sufficient daily rest. Successful treatment involves early detection by x-rays or by the tuberculin test—a skin test. Once the diagnosis is established, the physician will likely prescribe an antibiotic. In 1943, Dr. Selman Waksman discovered streptomycin which would kill or slow down the tubercle bacilli, for which discovery he later received the Nobel Prize.	Studies of human mummies have shown that the disease was active in the earliest Egyptian civilizations.
Tularemia (rabbit fever) It was first described in Tulare, California in 1911 (from whence came the name).	*Francisella tularensis*	An ulcerlike sore at the point where the germs enter the skin, followed by headache, aching muscles and joints, weakness, chills, and fever.	About 90% of the reported cases can be traced to handling infected wild rabbits. But the disease has been found in almost every type of small wild animal. Cats and sheep have also been known to be infected.	Prevention consists of wearing protective rubber gloves when handling wild rabbits, and thoroughly cooking meat from wild rabbits at 133°F *(56°C).* Broad-spectrum antibiotics are effective treatment.	Those who hunt rabbits should remember that a slow-running rabbit is probably a sick rabbit, and had best be ignored.
Vibrio parahaemolyticus food infection	*Vibrio parahaemolyticus,* an organism related to the cholera vibrio which grows in sea water.	Profuse diarrhea and dehydration.	Consumption of raw or undercooked sea foods.	Avoid contaminated foods. Cook foods well.	
Versinia enterocolitica (formerly known as *Pasteurella pseudotuberculosis)* food infection		Gastroenteritis	The first human cases reported in 1963 were contracted from infected chinchillas.		

FOOD POISONINGS (bacterial toxins). Food poisoning is caused by the ingestion of bacterial toxins that have been produced in the food by the growth of specific kinds of bacteria before the food is eaten. The powerful toxins are ingested directly, and symptoms of food poisoning develop rapidly, usually within 1 to 6 hours after the food is eaten.

The common bacterial food poisons are summarized in Table B-6.

TABLE B-6
FOOD POISONINGS (BACTERIAL TOXINS)

Disease	Disease Organism	Signs and Symptoms	How Disease Is Transmitted	Prevention and Treatment	Remarks
Botulism This severe form of food poisoning was first described in Germany in 1817 as "sausage poisoning" (the Latin word for sausage is *botulus*), the first food in which it was found.	*Clostridium botulinum* A saprophyte (bacteria that obtain food by absorbing dissolved organic material) which is widespread and found in soils. It forms heat-resistant spores. If the latter are not destroyed by heat in cooking, vegetative forms may grow anaerobically and produce one of the most powerful toxins known. According to one authority on poisons, botulism type A—the most lethal—is 10,000 times as deadly as cobra venom and millions of times more potent than strychnine or cyanide.	Weakness of the eye muscles and difficulty in swallowing, followed by paralysis of the muscles of respiration and death. Symptoms usually begin 18 to 36 hours after the food is eaten. Formerly, about 65% of all instances of botulism were fatal. Today, with early diagnosis, improved emergency hospital care, and the advent of antitoxins, the death rate is much lower. In 1988, 84 cases of botulism were reported in the U.S.	Primarily by eating inadequately cooked home-canned meat and nonacid vegetables (beans, asparagus, corn, and peas).	Adequate cooking. The toxin is inactivated in 10 minutes by heat at 176°F *(80°C)*, but the spore is not destroyed. The food industry uses nitrites as a preservative to prevent the anaerobic growth of *C. botulinum*. Do not use food that shows gas production or change in color or consistency. Burn such foods; otherwise, animals may be poisoned by eating them. Discard canned food that shows bulging in one end of the can. The only known treatment is an antitoxin, which must be of the right type.	The toxin blocks transmission of the neuromuscular junctions. Today, botulism is very rare in the United States. Botulinus-infected foods do not necessarily taste or smell spoiled.
Staphylococcal food poisoning This is by far the most common form of food poisoning observed in the U.S. It is caused by a toxin formed in the food before ingestion.	*Staphylococcus aureus,* primarily.	Vomiting and diarrhea, which may be severe and accompanied by collapse due to dehydration. Ingestion of contaminated food may be followed by symptoms within minutes to 6 hours. Usually, the illness lasts only 1 to 3 days. Mortality is low.	Ingestion of food or water containing the enterotoxin. Many healthy people are carriers of staphylococcal infections, specifically *Staphylococcus aureus.* Foods are readily contaminated by carriers and may, under suitable conditions, provide a good culture medium for growth of the organism. A wide variety of foods have been implicated, but the most common ones are ham, poultry, cream, and custard-filled baked products.	Prevent carriers from contaminating food. Prompt refrigeration of foods at 40°F *(4°C)* or below. Eliminate flies. *Staphylococcus* can be killed by heating to boiling temperature, but toxins may not be destroyed by boiling.	Some strains of *Staphylococcus* produce a powerful enterotoxin which is resistant to heat.

BACTERIAL CONTROL METHODS. From time immemorial, one of man's major food problems has been that of controlling the bacteria of food. Fortunately, bacteria are extremely sensitive to the conditions under which they live; hence, they are relatively susceptible to complete control. Various methods have been practiced through the ages, the most common of which are: (1) chemicals, (2) antibiotics, (3) phages, and (4) physical methods.

1. **Chemicals.** Chemicals are the chief source of bacterial control agents. The list of such agents begins with salt, its use as a preservative for fish and meat dating far back into antiquity. Sugar in high concentrations inhibits the growth of bacteria (and also yeasts and molds). Sugar in solution apparently dehydrates the bacteria as a result of osmosis. Acetic acid acts as a preservative, although it is seldom used alone as is true of other acids. Benzoates and nitrites have long been used as preservatives; their effect is bacteriostatic.
2. **Antibiotics.** The use of antibiotics in foods, and their application in food technology, has been known and studied for many years. Antibiotics may occur in foods (a) naturally, (b) through direct addition as a food additive to aid in production and keeping quality of the food, or (c) as an unintentional food additive resulting from the carry over of the antibiotic in milk, meat, or eggs, as a result of the addition of antibiotics to the feed of animals for growth promotion or medication in the treatment of diseases.

(Also see ANTIBIOTICS, subsection headed Antibiotics in Foods.)

3. **Phages.** This refers to a group of organisms that attack bacteria. They are capable of destroying disease-producing bacteria and living organisms, and they have specific effects on bacteria and food—sometimes good, sometimes bad. Bacteriophages appear to be viruses composed of nucleoproteins, with capacity to multiply. They pass through bacterial filters readily.
4. **Physical methods.** Physical means comprise another category of bacterial control. Heat, of course, is the chief physical control method. Freezing retards proliferation. Also, the preservation of food by radiation has received much study, the goal being to sterilize foods and preserve them indefinitely.

Both high and low temperatures usually inhibit bacterial growth and multiplication. However, some bacteria, referred to as thermophils, grow at relatively high temperatures and produce what is called "flat sour" in canned foods. Other bacteria produce spores, which are quite resistant to heat, as well as other sterilizing agents. Nevertheless, all pathogens may be destroyed by heat; thus, food which has been properly cooked and handled is safe. But, in cooking, the heat may not penetrate the food sufficiently, particularly a large cut of meat (use a meat thermometer to ensure that the desired internal temperature is reached in large cuts of meat); and undercooked foods are dangerous. Furthermore, foods may be contaminated after cooking. Meat, milk, and eggs are excellent growth media for bacteria; hence, foods which have been cooked, improperly stored, then warmed up are especially dangerous. A very important group of bacteria is referred to as psychrophils, which grow and reproduce at the usual refrigeration temperatures. These bacteria are responsible for most of the fresh food spoilage that occurs. Very little bacterial growth occurs in frozen foods, and freezing temperatures bring about a very slow and usually incomplete death to most bacteria. Since the death of bacteria in frozen foods is incomplete, it must be remembered not to allow frozen foods to stand at room temperature after thawing. Fig. B-7 illustrates the temperature ranges of bacterial growth.

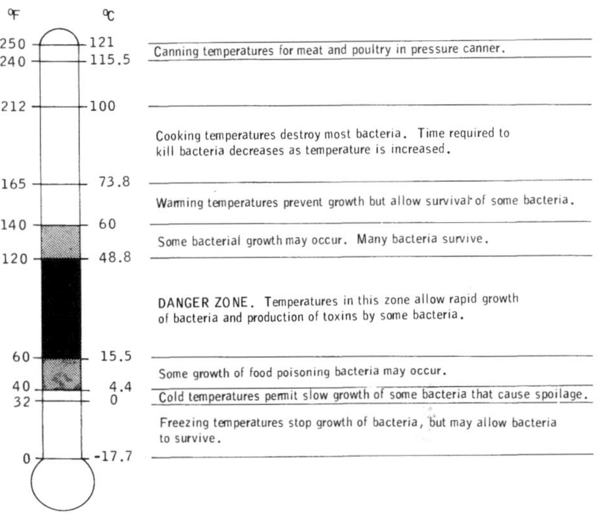

Fig. B-7. Temperature for control of bacteria in meat and poultry. Note that the figure lists the temperatures which retard bacterial growth as well as the temperatures which allow bacteria to multiply.

Bacteria multiply by cell division at a very rapid rate. Within the temperature range of 50° to 140°F (*10° to 60°C*), the growth rate increases tenfold for each 18°F (*–8°C*) increase in temperature, and under favorable conditions bacteria can double their numbers every 30 minutes. Therefore, bacterial numbers can become astronomical in foods left at room temperature for a period of 3 to 4 hours. In fact, it has been estimated that if a single *Escherichia coli* cell were given enough food under optimum conditions, it could produce a mass of bacteria larger than the earth in a relatively short period of time. Of course, it would be impossible to supply enough food to keep the bacteria multiplying at the maximum rate. Fig. B-8 illustrates the rapid bacterial growth which occurs in milk held at room temperature.

PREVENTION OF BACTERIAL INFECTION AND FOOD POISONING. Prevention of infections spread by food and the fecal-oral route depends on scupulous attention to cleanliness along the whole food chain—meat packers, slaughter houses, food manufacturers, retail shops, caterers, restaurants, and kitchens. In all these places care is required to prevent rodents and flies from getting access to food. All food handlers should be scrupulously clean in their personal habits and be provided with clean lavatories and opportunities for washing their hands. Some persons harbor a pathogen and ex-

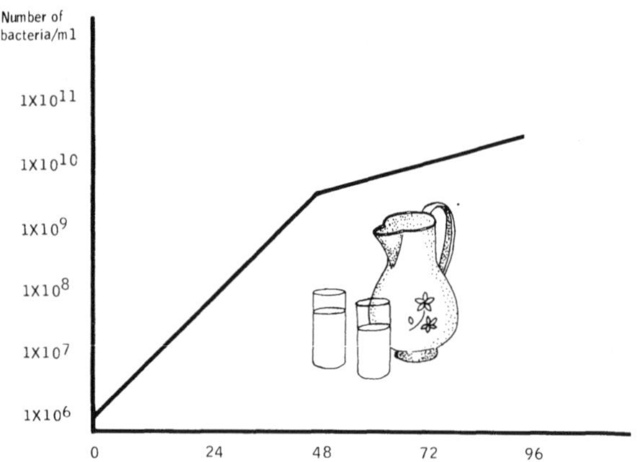

Fig. B-8. Changes in bacterial numbers of milk held at room temperature.

crete it continuously in their feces and/or urine without having any symptoms of disease. Such carriers are particularly dangerous and have been responsible for many outbreaks of disease. Public health authorities are responsible for inspecting commercial food establishments, and they have the legal right to close any premises that are not up to acceptable standards of hygiene.

A general program for preventing bacterial food infections and food poisonings follows:

1. Keep food handling areas spotlessly clean.
2. Wash hands often with soap and water, especially after using the restroom, dressing a wound, and before handling any food; this will help get rid of transient bacteria which can cause disease.
3. Wash raw fruits and vegetable thoroughly in clean water.
4. Don't handle food with hands that have cuts, bruises, or sores on them, *unless* covered by gloves.
5. Don't sneeze or cough on food or in areas where food is being prepared; wear a mask if a nose, throat, or sinus infection is present.
6. Keep body and clothes clean.
7. Wear a hairnet or a cap when handling food.
8. Keep rats, cockroaches, flies and other insects out of areas where food is processed, stored, prepared, or served.
9. Don't use wooden cutting boards—they can't be cleaned properly.
10. Sanitize cutting tools used on raw food before using them on cooked food.
11. Don't smoke in areas where food is processed, stored, prepared, or served.
12. Cook meats to the following minimum internal temperatures: fresh beef, 140°F (*60°C*); fresh veal, pork, and lamb, 170° (*77°C*); turkey, 180°F (*82°C*) (Use a meat thermometer when cooking large cuts of meat.)
13. Keep cooking and eating equipment clean.
14. Keep perishable foods in refrigerator when not being prepared, cooked, or consumed.
15. Avoid consumption of contaminated or partially deteriorated foods.
16. Burn completely all canned foods with abnormal color or composition, putrid odor, or in bulged cans.

BACTERIAL COUNT (PLATE COUNT)

Two methods are commonly employed in determining the degree of bacterial contamination of a sample of material: (1) the plate method, and (2) the direct microscopic count.

• **Plate count method**—In the plate method, a diluted sample (such as milk) is poured onto a standard agar (which is obtained from the stems of certain seaweeds) in a Petri dish where each bacterial cell or group of cells multiplies to produce a colony which is visible to the naked eye. A count of the number of colonies gives the number of bacteria in that portion of the sample that was taken. This is the method most frequently used to estimate the bacterial count of milk. The bacterial count plays a major role in the sanitary quality of milk upon which grades are largely based. Grade A raw milk should not exceed 100,000 bacteria per milliliter; and the bacterial count of Grade A pasteurized milk should not exceed 20,000 bacteria per milliliter.

• **Direct microscopic count method**—In the direct microscopic count method for determining milk quality, 0.01 ml of milk is deposited in 1 sq cm of a glass slide, which is then dried and stained. The number of bacteria is determined by using a calibrated microscope. In addition to providing a bacterial count, this method permits recognition of the type of bacteria present and serves as a general guide to the way the milk was handled prior to sampling. Since the leucocytes and other somatic cells present will also be stained, the direct microscopic count also provides information about possible udder troubles among cows in the herd. Because dead bacteria are stained and may be counted in this method, and because in the plate count each colony is counted as if it developed from a single organism (which isn't always true), direct microscopic counts are usually 3 to 4 times greater than the figures obtained by the plate count method.

BACTERIAL ENDOTOXIN

Poisonous substance present inside bacteria which separates from the cell upon its disintegration.

BACTERIAL EXOTOXIN

Poisonous substance that is liberated during the growth of certain bacteria. Botulism is caused by a bacterial exotoxin.

BACTERIAL SPOILAGE OF FOOD

Certain types of bacteria are involved in many, but not all, types of food spoilage. The stale odor of spoiled meat, the foul odor of a spoiled egg, the souring of milk, and the rotting of fruits and vegetables are familiar examples of bacterial spoilage. Also, the spoilage of canned foods is usually traced to bacterial causes.

(Also see BACTERIA IN FOOD.)

BACTERICIDE

A product that destroys bacteria.

BACTERIOPHAGE

The word *bacteriophage* means bacteria-eater. It refers to a substance that makes living bacteria disintegrate or dissolve. Most scientists believe that phages are viruses; but some believe that phages may be self-reproducing proteins, bacterial enzymes, or hereditary factors in bacteria. Whatever they may be, phages dissolve many kinds of bacteria, particularly those in the intestines. Also, it is noteworthy that bacteriophages exhibit a considerable degree of specificity with respect to the kind of bacteria attacked.

Bacteriophages cause considerable trouble in milk starter cultures. (A culture [starter] is a controlled bacterial population that is added to milk or milk products to produce acid and/or flavorful substances which characterize cultured milk products.) For this reason, commercial suppliers of cultures recognize the need for bacteriophage inhibitors in starter media.

(Also see BACTERIA IN FOOD, section headed "Bacterial Control Methods".)

BAKING POWDER AND BAKING SODA

These convenience leavening powders have brought about many advances in the baking industry, because they have made it possible to bake certain doughs right after mixing, instead of having to wait for the action of yeast to take place. Hence, the products made with baking powder, baking soda, or similar leavening agents are commonly called "quick breads." Also, the use of these powders in baking is referred to as "chemical leavening" in order to distinguish it from the biological leavening obtained with sourdough starters or yeast.

In addition to saving time, baking powder or baking soda allows for the use of a wide variety of ingredients in baked goods. For example, softer, weaker flours may be used in quick breads; whereas, strong flours are usually required for good yeast breads.

HISTORY OF CHEMICAL LEAVENING. It appears that the first use of a crude baking powder occurred in the United States in the 1790s. A substance called "pearlash" (potassium carbonate), which was obtained from wood ashes, was used in baking because it gave off carbon dioxide when heated. The bitter taste of this ingredient spurred the search for a more palatable product.

Around 1835, the first tartrate baking powder was developed. It was a mixture of (1) baking soda (sodium bicarbonate) and (2) cream of tartar (potassium acid tartrate), a substance that was obtained from the residue in wine barrels. A commercial version of this mixture was marketed in 1850.

Later, towards the end of the 19th century, the acid salts monocalcium phosphate and sodium aluminum sulfate were introduced as replacements for cream of tartar.

With the opening of the 20th century, there was a trend toward the increased use of preleavened cornmeal, flour, and special baking mixes. However, most homemakers used baking powder or baking soda until the beginning of World War II. The war was a major factor in the improvement of preleavened mixes because the military services needed standardized products for feeding their men as efficiently as possible. Since the war, the baking mixes have captured most of the market for leavening agents, and now sales of baking powder account for only about 25% of the market.

THE BASIC PRINCIPLES OF CHEMICAL LEAVENING. The actual leavening agent in most baked goods is carbon dioxide gas, whether it be generated from baking powder, sourdough, or yeast. However, chemical leavening agents such as baking powder and baking soda produce carbon dioxide by means of the chemical breakdown of bicarbonate salts rather than by biological means. Some of the basic facts about bicarbonates follow:

1. Baking soda is the common name for sodium bicarbonate, a white powder which yields carbon dioxide when it is (a) dissolved in water and heated, or (b) reacted with an acid. The latter means of leavening is more desirable because reaction with an acid may convert all of the bicarbonate to carbon dioxide and water, whereas heating alone converts most bicarbonates to carbonates, carbon dioxide, and water. Residues of carbonates in baked goods are undesirable because of their bitter and soaplike taste.
2. Household baking powders usually contain a mixture of baking soda and one or more acidic salts, which may be (a) fast-acting, (b) slow-acting, or (c) a combination of (a) and (b). The leavening reaction takes place only when the powders are wetted with water or a liquid containing water. Hence, the old admonition, "keep your powder dry," applies also to the storage of baking powder.
3. Some recipes call for baking soda plus an acidic soured dairy product, rather than for baking powder.
4. Baking ammonia (ammonium bicarbonate) is sometimes used by commercial bakers in the preparation of cookies or crackers. This salt is totally decomposed into ammonia and carbon dioxide by heating. Table B-7 gives the uses, ingredients, and characteristics of the commonly used chemical leavening agents.

TABLE B-7
USES, INGREDIENTS, AND CHARACTERISTICS OF SOME COMMON CHEMICAL LEAVENING AGENTS AND PRELEAVENED PRODUCTS

Type of Leavening Agent or Preleavened Product	Uses	Major Ingredients		Comments
		Source(s) of Acid	Source of Carbon Dioxide	
Baking ammonia	Cookies and crackers made by commercial bakers.	None	Ammonium bicarbonate	Baking causes the decomposition of ammonium bicarbonate and the release of carbon dioxide and ammonia, which act as leavening gases. Wet doughs retain ammonia and have a bitter taste. Hence, this leavening agent is used mainly for flat, dry items such as cookies and crackers, from which the ammonia escapes readily.
Baking soda	Batters or doughs made with acidic ingredients.	Acidic ingredients in the batters or doughs, such as buttermilk, sour milk, or sour cream; fruits or fruit juices; molasses; phosphated flour; and vinegar.	Sodium bicarbonate	It is generally advisable to use at least a slight excess of acid so that the reaction with baking soda is complete. An incomplete reaction results in a residue of sodium carbonate, which has a bitter and soaplike taste.
Phosphate baking powder, double-acting	All purpose household baking powders.	Acidic salts such as monocalcium phosphate, monohydrate (fast-acting), sodium aluminum phosphate (slow-acting), and/or sodium aluminum sulfate (slow-acting).	Sodium bicarbonate	Most household recipes can be prepared with this type of baking powder because (1) the fast-acting acid salt serves well for fast-rising angel food cakes, cookies, crackers, doughnuts, pancakes, and waffles; and (2) the slow-acting acid salt(s) reacts only when heated. Hence, the latter type of salt is very useful in batters or doughs which are held in a freezer or refrigerator for a considerable time prior to baking.
Phosphate baking powder, fast-acting	Angel food cakes, cookies, crackers, doughnuts, pancakes, pizza doughs, and waffles.	Monocalcium phosphate, monohydrate.	Sodium bicarbonate	This type of baking powder releases carbon dioxide immediately upon wetting. Hence, mixing should be completed as rapidly as possible in order to avoid an excessive loss of carbon dioxide.
Phosphate baking powder, slow-acting	Cakes, frozen doughs, or refrigerated batters or doughs.	Monocalcium phosphate, anhydrous; sodium aluminum phosphate; or sodium aluminum sulfate.	Sodium bicarbonate	There is almost no production of carbon dioxide by this type of powder at room temperature, since heating is required to bring about the reaction of the acid salt(s). Hence, batters or doughs made with slow-acting baking power may be kept refrigerated overnight prior to baking.
Phosphated flour	Recipes which utilize baking soda and soured milk products for leavening.	Phosphated flour, which contains monocalcium phosphate or a similar acidic salt(s).	Sodium bicarbonate, which is added as baking soda.	The acidic phosphate is added to the flour so that the batters or doughs will contain sufficient acid to react completely with the soda, since the lactic acid content of soured dairy products is variable.

(Continued)

TABLE B-7 *(Continued)*

Type of Leavening Agent or Pre-leavened Product	Uses	Major Ingredients: Source(s) of Acid	Major Ingredients: Source of Carbon Dioxide	Comments
Self-rising cornmeal	Cornbread	Self-rising cornmeal containing monocalcium phosphate, or a similar acidic salt.	Sodium bicarbonate in the cornmeal.	This type of mix also contains salt. Hence, only eggs plus milk or water need be added. It is used mainly in southeastern United States, because elsewhere homemakers are more likely to use special corn muffin mixes.
Self-rising flour	Biscuits, muffins, pancakes, and waffles.	Self-rising flour, which contains monocalcium phosphate, and/or sodium aluminum phosphate.	Sodium bicarbonate in the flour.	The use of this type of flour has declined, because special mixes for biscuits, pancakes, and waffles have captured much of the market. However, self-rising flour is still used by many cooks in southeastern United States.
Tartrate baking powder	Cakes which contain (1) egg white foams (angel food, chiffon, pound, and sponge), or (2) fruit ingredients like lemon and orange juice or peel.	Cream of tartar (potassium acid tartrate) and tartaric acid.	Sodium bicarbonate	Tartrate baking powders are not used very much because (1) they are more expensive than phosphate baking powders, and (2) they are too fast-acting for most types of baked items. However, the flavor they impart to baked goods is much more desirable than those resulting from the use of other baking powders. Also, tartrates are useful for (1) increasing the volume of egg foams, and (2) enhancing fruit flavors in baked goods, providing sufficient sweetener is used to offset the tart flavor of the tartrates.

NUTRITIONAL VALUES OF CHEMICAL LEAVENING AGENTS. Although most batters and doughs contain only small amounts of baking powder (from 1 to 2%, on the average), this type of leavening agent contributes significant amounts of calcium, phosphorus, sodium, and/or potassium to baked products. Table B-8 gives the amounts of these minerals which are present in some commonly used baking powders and preleavened mixes.

It may be seen from Table B-8 that many of the products, when used in the amounts indicated, contribute from ¼ to ½ of the Recommended Daily Allowances (RDAs) of calcium and phosphorus (the RDA for adults for each of these minerals is 800 mg). This contribution is very important because many people may drink coffee, tea, beer, or other nondairy beverages at their meals. Hence, quick breads may make up for some of the lack of dietary calcium. Nevertheless, it pays for the consumer to read the labels on these products, because there is a wide variation in their calcium contents. For example, tartrate baking powders contain no calcium.

Some people may have to use a special, low-sodium baking powder because they have been advised to restrict dietary sodium. Fortunately, baking powders are available in which the sodium salts have been replaced by potassium salts that serve similar functions.

Most baking powders contain cornstarch. Thus, if an individual suffers from an allergy to corn, a homemade baking powder can be made by sifting together several times: 9 Tbsp (*135 ml*) of cream tartar and 4 Tbsp (*60 ml*) of baking soda.

Other Nutritive Benefits Which May Be Obtained From the Use of Baking Powders. The fast leavening action of most baking powders eliminates some of the need for a highly elastic dough to trap carbon dioxide, because there is less time for the gas to escape prior to baking, which then stiffens the dough. Therefore, flours used for quick breads may be weaker than those that are required for good yeast breads.

Baking powder breads are often strengthened by the addition of eggs because the protein of the egg white acts as a binder. (Egg protein does not work as well in yeast breads because it loses some of its effect during the long time required for the dough to rise.) Strengthening of quick breads allows greater amounts of highly

TABLE B-8
MAJOR CONTENT OF SOME COMMON BAKING POWDERS AND PRELEAVENED PRODUCTS[1]

Leavening Agent or Preleavened Product	Quantity		Minerals			
	Measure	Weight	Calcium	Phosphorus	Sodium	Potassium
		(g)	(mg)	(mg)	(mg)	(mg)
Baking powders, classified by ingredients:[2]						
Sodium aluminum sulfate (SAS), and monocalcium phosphate, monohydrate (MCP)	1 tsp *(5 ml)*	3.0	58	87	329	5
SAS, MCP, and calcium carbonate	1 tsp *(5 ml)*	3.0	173	44	349	0
SAS, MCP, and calcium sulfate	1 tsp *(5 ml)*	3.0	190	47	300	0
MCP, anhydrous (coated)[3]	1 tsp *(5 ml)*	3.0	188	283	246	5
Cream of tartar, and tartaric acid	1 tsp *(5 ml)*	3.0	0	0	219	114
Low-sodium preparation[4]	1 tsp *(5 ml)*	3.0	145	219	0	328
Cornmeal, self-rising	1 cup *(240 ml)*	145.0	1,631	760	2,001	164
Flour, self-rising	1 cup *(240 ml)*	114.0	302	531	1,230	103

[1]Data from Food Composition Table F-36 of this book.
[2]All of the powders also contain sodium bicarbonate, except for the low-sodium product, which contains potassium bicarbonate.
[3]This product is commonly referred to as a "straight phosphate" baking powder.
[4]Sodium-containing salts have been replaced by potassium salts that serve the same functions.

nutritious ingredients to be used without producing a heavy baked product. For example, the maximum amount of liquid skim milk which may be used in yeast breads is about ⅓ cup (*80 ml*) per cup of flour, whereas standard recipes for various baking powder (or baking soda) breads call for greater amounts of milk (from ½ cup [*120 ml*] to 1 cup [*240 ml*] per cup of flour). Similarly, greater amounts of such ingredients as bran, cheese, chopped dates, cornmeal, nuts, raisins, rolled oats, shortening, soy flour, and wheat germ may be added to quick breads.

SUGGESTIONS FOR UTILIZING BAKING POWDER, BAKING SODA, AND PRELEAVENED PRODUCTS IN HOME RECIPES. There is a lot more flexibility in the amounts and types of ingredients which may be used for quick breads than there is for making yeast breads. Hence, the home baker should experiment with various combinations of leavening agents, flours, preleavened mixes, and optional ingredients. However, certain basic principles should be heeded so as to avoid unnecessary waste. Some suggestions follow:

1. All shortenings (butter, margarine, semisolid vegetable shortening, and vegetable oil) may be considered to be interchangeable when measured as liquids at room temperature. The solid fats should be melted, then cooled before measuring.
2. When buttermilk (or any other acid food) is used with baking soda to leaven a dough, the soda should be mixed with the dry ingredients, and the acid added separately at the end of mixing, just before the dough is put into the oven. Otherwise, there may be an excessive loss of carbon dioxide (when the acid reacts prematurely with the baking soda).
3. Baking soda may be dissolved in either plain water or sweet milk prior to its addition, but it should *never* be dissolved in buttermilk or any other acidic fluid.
4. Each cup of buttermilk requires ½ tsp (*2.5 ml*) of baking soda to neutralize the acid completely. This combination is approximately equivalent to 2 tsp (*10 ml*) of baking powder. If naturally soured milk is not availabale, a substitute may be produced by mixing (a) 1 Tbsp (*15 ml*) vinegar or lemon juice, or (b) 1¾ tsp (*8.75 ml*) cream of tartar with sufficient sweet milk to make 1 cup (*240 ml*) of soured product. If the acidity of a soured dairy product is uncertain, it is wise to reduce the baking soda to ¼ tsp (*1.25 ml*) per cup (*240 ml*) of the product, and count the mixture as 1 tsp (*5 ml*) of baking powder. (A slight excess of acid is better tolerated in a baked product than an excess of baking soda, which may impart a bitter, soaplike taste.)
5. Two teaspoons (*10 ml*) of baking powder are also equivalent to (a) 1 cup (*240ml*) of molasses and ½ tsp (*2.5 ml*) of baking soda, or (b) 1¼ tsp (*6.25 ml*) of cream of tartar and ½ tsp (*2.5 ml*) of baking soda.
6. A little extra baking powder or baking soda may be added to strong-flavored mixtures such as spice cake batter if the baker wishes to obtain a lighter product.
7. As much as ¼ tsp (*1.25 ml*) of baking soda per cup (*240 ml*) of flour may be added to speed up the rising of yeast-leavened breads, providing that sufficient time is allowed for yeast fermentation to generate enough acid to neutralize the soda. Another way to combine yeast leavening with chemical leavening is to (a) grow dried yeast overnight in an acidic fruit juice at room temperature, and (b) add the yeast mixture to a batter or dough which contains baking soda. Then the dough or

batter may be (a) baked immediately, since the overnight growth of yeast is usually sufficient to produce a mixture that will impart a yeasty aroma and contribute nutrients; or (b) allowed to rise in a warm place for about 20 minutes prior to baking.

8. Each cup (*240 ml*) of self-rising flour is approximately equal to 1 cup (*240 ml*) of unleavened flour plus 1½ tsp (*7.5 ml*) of baking powder and ½ tsp (*2.5 ml*) of salt.

9. Mildly sweet, cakelike biscuits or soft cookies may be prepared by mixing equal amounts of preleavened biscuit mix and preleavened cake mix. Then the directions on the package of cake mix should be followed (the amounts of other ingredients should be in proportion to the total amount of dry mix), except that the baking temperature and time should be (a) similar to biscuits or muffins, if the batter is baked in small portions, or (b) similar to cake, if the batter is placed in a single baking pan.

10. Lighter baked goods may be prepared without increasing the amount of leavening(s) by (a) separating the egg yolks from the egg whites, (b) adding the egg yolks with the other liquid ingredients and mixing well, and (c) beating the egg whites into a foam, then folding them gently into the batter just before it is placed in the oven.

11. If desired, baking powder can be made in the home kitchen by sifting together the following: 9 Tablespoons (*135 ml*) of cream of tartar and 4 Tablespoons (*60 ml*) of baking soda.

BALANCED DIET

One which provides a person with the proper proportions and amounts of all the required nutrients for a period of 24 hours. In practical use, the term *diet* refers to the foods eaten by a person or a group of people without limitation to the time in which they are consumed.

BALANCE STUDY

A determination of the amount of a nutrient retained by the body under specified conditions. Measurements are made of (1) the amount of the nutrient consumed in food, and (2) amounts lost in the feces, urine, and/or sweat. The body is said to be in positive balance with respect to the nutrient when the intake is greater than the sum of the excretion products. It is in negative balance when the excretion products exceed the intake. At equilibrium, nutrient intake equals excretion.

BALDNESS

This condition, which is known in medical circles as alopecia, is a loss of hair on the skull to the extent that formerly hairy areas have become bare. The ancient Greeks observed that castrated males (eunuchs) and women rarely grew bald. This observation was the basis of the widely held beliefs that (1) baldness is somehow associated with the male sexual hormones, and (2) bald headed men are more virile than men with a full head of hair. However, the only certain connection between male sexuality and baldness is that the sex hormones stimulate the sweat glands to produce an oily secretion called sebum, which is the main constituent of the dandruff that is associated with rapid hair loss.

The tendency for males to become bald at an early age appears to run in certain families, and in certain races. Generally, the hair loss starts at the top of the head and/or at the temples, and progresses until little hair is left, except for a small fringe around the sides and back of the head. This type of hair loss is called "male pattern baldness." Fig. B-9 shows a typical case of this condition.

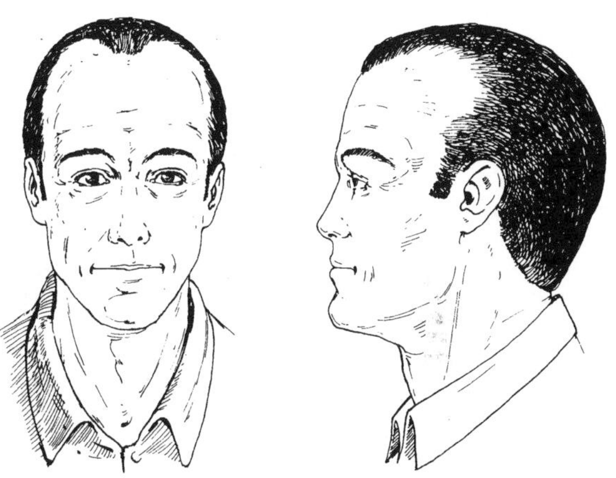

Fig. B-9. Male pattern baldness, a benign condition affecting many middle-aged men.

Some of the other causes of hair loss which may result in baldness are: (1) contact dermatitis, an inflammation of the scalp caused by chemicals, harsh shampoos and soaps, hair dyes and lotions, or poison ivy; (2) damage to the hair follicles by x rays or other therapeutic agents (like those used in cancer chemotherapy); (3) disorders of metabolism associated with conditions such as diabetes, obesity, and thyroid hormone deficiency; (4) emotional stress; (5) feverish illnesses, particularly typhoid and scarlet fever; (6) infections of the scalp such as boils, impetigo, infectious dandruff (sebarrheic dermatitis), and ringworm; (7) malnutrition due to deficiences of amino acids, essential fatty acids, proteins, and various minerals and vitamins; and (8) tightness of the skin on the scalp, because it may impede the delivery of sufficient blood and nutrients to the hair follicles.

The medical specialty that is concerned with the prevention and treatment of hair and skin disorders is dermatology. However, doctors in this field do not offer much hope for the prevention of male pattern baldness, as evidenced by the large number of bald heads which may be observed in any gathering of dermatologists. Nevertheless, other causes of baldness may be countered by appropriate hygienic measures, such as the following:

1. Prompt treatment of abnormal scalp conditions by a doctor.

2. Avoidance of irritation of the scalp by excessive massage, overly stiff hair brushes, and harsh shampoos or soaps.

3. Consumption of a diet containing optimal amounts of the essential nutrients, but avoidance of "shotgun" doses of nutritional supplements, because excesses of certain nutrients interfere with the utilization of others.

It is noteworthy that research on various species of hairy animals has established that the nutrients which might be helpful in the prevention of excessive hair loss are: amino acids, particularly those which contain sulfur (cystine and methionine); the essential, polyunsaturated fatty acids; iodine and zinc; many members of the vitamin B complex; and vitamins A, C, and E. The best way to obtain these nutrients is from a wide variety of nutritious foods such as dairy products, eggs, fish and seafood, fruits and vegetables, meats and poultry, vegetable oils, and whole-grain breads and cereal products.

(Also see TISSUES OF THE BODY.)

BALSAM APPLES *Momordica balsamina*

The orange-red, egg-shaped fruit of a vine (of the family *Cucurbitaceae*) that originated in Africa or Asia, and was brought to the American tropics in the latter part of the 16th century. It is closely related to the balsam pear (*M. charantia*), which is shaped more like a banana. The unripe fruits are usually preferred to the ripe ones, because they grow more bitter with age. Although the very young fruits are sufficiently palatable to eat raw, balsam apples are usually made into pickles and curries, or used as flavoring for fish and meat dishes.

(Also see BALSAM PEAR.)

BALSAM PEAR (BITTER MELON) *Momordica charantia*

The orange-red, furrowed, banana-shaped fruit of a vine (of the family *Cucurbitaceae*) that originated somewhere in Africa or Asia, and was taken to the West Indies and Brazil during the early days of slave trading.

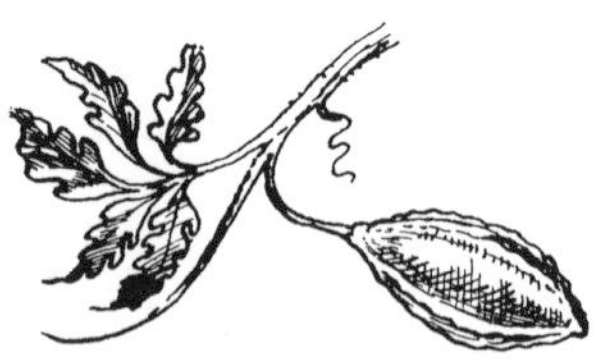

Fig. B-10. Balsam pear.

Usually the bitter-flavored unripe fruits, which range from 2 to 10 in. (*5 to 25 cm*) long, are used as a vegetable, or to make curries and pickles. However, the Chinese add the fruit to fish and meat dishes, and the people of India use the seeds of the ripe fruit as a seasoning. The tender shoots and leaves of the plant may also be utilized as vegetables.

(Also see BALSAM APPLE.)

BANANA *Musa paradisiaca*

Fig. B-11. Bananas grow in bunches, weighing 60 to 100 lb (*27 to 45 kg*). Each bunch is comprised of 9 to 16 clusters of fruit called "hands," containing 12 to 20 separate bananas or "fingers." (Courtesy, Castle & Cooke, Inc.)

Although it looks like a tree, and it is often called a tree, the banana plant is not a real tree because there is no wood in the stem rising above the ground. The stem is actually comprised of leafstalks growing one inside the other; hence, the stem is not really a stem but a pseudostem (false stem). The slightly curved, yellow or red fruit with firm creamy flesh—the banana—is familiar to many people. It represents one of the world's leading fruit crops, and the number one seller in the produce department of American supermarkets. Americans are most familiar with the large yellow, smooth skinned banana known as *Gros Michel* ("Big Mike") or *Martinique*, and the Cavendish varieties. A smaller red-skinned variety is known as the *Red Jamaica* or *Baracoa* banana. Plantains are also a type of banana which are used more like a vegetable than a fruit. Herein, bananas, and plantains are referred to separately.

ORIGIN AND HISTORY. Bananas are believed to have originated primarily in Malaysia about 4,000 years ago, thence they spread over an area from India to the Philippines and New Guinea. People probably used bananas for food long before recorded history. The armies of Alexander the Great found the banana growing in India in 327 B.C. Arabian traders introduced the banana plant into Africa at a very early date, since Portuguese explorers who discovered the Guinea Coast of Africa found bananas growing there in 1482 A.D. Then, soon after the discovery of the New World, explorers took bananas from Africa to tropical America. Thus, the banana plant traveled more than half way around the world to reach the areas of tropical America where, today, about two-thirds of the world's bananas are produced. It was not until the latter part of the 19th century that bananas were brought into the United States in quantities for sale in stores. Even then only those people who lived near seacoast cities where banana schooners docked tasted or saw bananas. Bananas are more perishable than some other fruits; so, specialized, rapid transport needed to be developed before the use of bananas became widespread.

Producing bananas and getting them to market in temperate areas of the world stimulated the creation of large banana plantations. The first big name in the banana export business was the United Fruit Company, formed in 1899, which in 1970 merged into United Brands. Standard Fruit and Steamship Company was established in 1923 as a competitor to the United Fruit Company. In the early days of the banana industry, the large landholdings and one-crop economies made these multinational companies important influences in some Latin American countries.

PRODUCTION. Bananas grow best where the soil is deep and rich, and where the climate is warm and moist. they are raised in the tropics of both the Eastern and Western Hemispheres. However, the most important commercial banana-producing region is Latin America. As shown in Fig. B-12, India is the world's leading banana producer, and Brazil ranks second. Small quantities of bananas are produced in the United States, in Florida along the Gulf of Mexico, and in Hawaii which produces about 6,000 short tons.

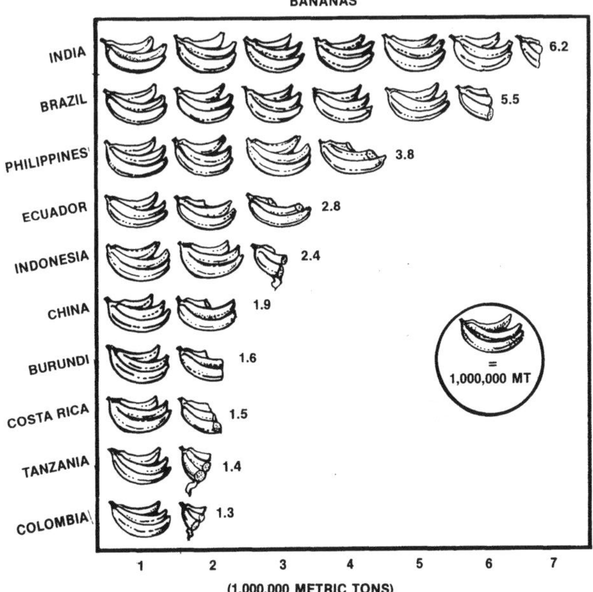

Fig. B-12. The leading banana-producing countries of the world. (Based on data from the *FAO Production Yearbook*, 1990, FAO/UN, Rome, Italy, Vol. 44, p. 169, Table 74)

Although not an important crop for consumption in North America, plantains are an extremely important food crop in some areas of the world. Figure B-13 shows the countries of the world which lead in the production of the large green plantains.

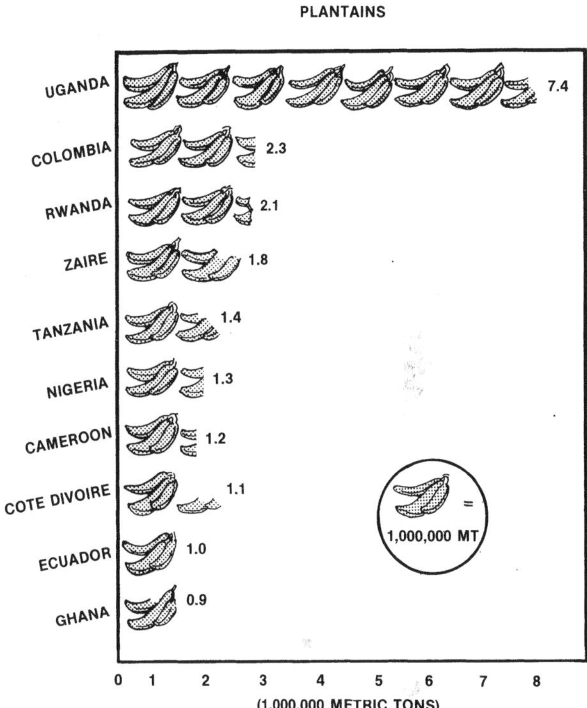

Fig. B-13. The leading plantain-producing countries of the world. (Based on data from the *FAO Production Yearbook*, 1990, FAO/UN, Rome, Italy, Vol 44, p. 169, Table 74)

Propagation and Growing. To grow bananas in a new area, pieces of the fleshy rhizome (roots) from mature banana plants are planted in holes about 1 ft *(30 cm)* deep and 11 to 18 ft *(3.4 to 5.5 m)* apart. In about 3 or 4 weeks, green shoots appear above the ground. Then in less than a year, a cluster of leaves 6 to 10 ft *(2 to 3 m)* long and 1 to 2 ft *(30 to 61 cm)* wide and resembling a huge feather, appear at the top of the banana plant. After 12 to 15 months the plant reaches full size, about 15 to 30 ft *(4.6 to 9.2 m)*, and produces a huge bud at the end of a strong thick stalk. The bud contains many small purple leaves that hold flowers—bracts. As the stalk grows longer, these bracts roll back, revealing rows of small flowers which turn into small green bananas, since the fruit sets without pollination. As the bananas grow and the bracts are shed, the stalk becomes heavier and droops toward the ground, requiring 75 to 150 days to mature to the harvesting stage.

Under favorable growing conditions, bananas may yield 10 or more tons of fruit per acre each year. However, each banana plant bears only one bunch of fruit, and when this is harvested the entire plant is cut down, and another plant which has been growing from the same root stalk takes its place—a cycle that can continue for years.

Harvesting. Bananas are harvested green, a whole bunch at a time by a manual operation. One worker with a knife on a long pole lops the stem of the banana bunch, setting the bunch onto the shoulder of a second man who carries it to a mule, railcar or canoe for transportation to a central gathering point. Nowadays most fruit is boxed for transport and held at 57°F *(14°C)* in a ship's hold to prevent premature ripening. Just before delivery to retail stores, bananas are ripened by warming to 70°F *(21°C)* and the use of ethylene gas in special chambers. Ripened bananas turn from green to the familiar yellow, and starch hydrolizes to sugars, causing the banana to become sweet. Within a period of 10 to 20 days bananas must be harvested, shipped several thousand miles, ripened and sent to retail stores.

PROCESSING. Almost all bananas are consumed fresh—after removing the skin—nature's unique packaging. In most areas of the United States, they can be purchased year round. Only a small amount is dried into banana flakes or chips. More wide use is, however, being made of banana flour from the dried ground unripened fruit. In the tropics where bananas are grown, some are fermented to make beer.

SELECTION. Fresh bananas are at their best when the peel is solid yellow and speckled with brown and still quite firm. Bananas with green tips, or with practically no yellow color, have not developed their full flavor potential. When purchased, bananas should be firm, bright in appearance, and free from bruises or other injury. They will continue to ripen at room temperature. Bananas may be refrigerated, and they will remain good for 3 to 5 days, though the peel will turn dark. Bananas which are bruised, discolored, or dull, grayish and aged in appearance should not be purchased, because they are in various stages of deterioration.

PREPARATION. Ripe bananas are peeled and then eaten whole, sliced, mashed, baked, or dipped in batter and fried. Starchy green bananas and plantains may be made digestible by cooking—baked, boiled or fried.

Bananas may also be used in combination with other foods in fresh fruit dishes, ice cream, pies, gelatin desserts, salads, cookies and cakes.

NUTRITIONAL VALUE. Fresh bananas contain about 70 g of water, 1 g of protein, and 25 g of carbohydrate per 100 g (3½ oz; about one banana), and contribute about 85 to 100 Calories (kcal) of energy. They are good sources of potassium and vitamin A. Dried banana flakes contain only 3% water; hence, many of the nutrients, including calories, are more concentrated. Each 100 g of banana flakes contains 340 Calories (kcal) of energy, 2.8 mg of iron, and 760 IU of vitamin A. More complete information regarding the nutritional value of several varieties of bananas, banana flakes, banana powder and plantains is presented in Food Composition Table F-36.

(Also see FRUIT[S], Table F-47 Fruits of the World.)

BANANA FIGS

Peeled bananas that have been halved lengthwise and dried without sulfuring. (The latter treatment usually prevents darkening of the fruit.) Hence, the product is darkly colored and sticky like figs.

(Also see BANANA[S].)

BANNOCKS

These are flat, round cakes of oat, rye or barley meal, baked on a griddle or in the oven.

BAOBAB (MONKEY BREAD) *Andansonia digitata*

Fruit of the largest tree in tropical Africa, which belongs to the family *Bombaceae* and is one of the longest lived trees in the world.

The fruit pulp is made into refreshing drinks and is also used as a condiment. It is high in calories (290 kcal per 100 g) and carbohydrates (77%), and very rich in fiber, iron, and vitamin C. Surprisingly, the fruit pulp contains more calcium (284 mg per 100 g) than milk, and the calcium to phosphorus ratio is 2.4:1.

BARBADOS GOOSEBERRY *Pereskia aculeata*

A very acidic tropical fruit (of the family *Cactaceae*) that is usually cooked and sweetened before it is consumed.

The Barbados gooseberry has a very high water content (97%) and is low in calories (11 kcal per 100 g) and carbohydrates (2%). It is a good source of vitamin A, but a poor source of vitamin C.

BARBECUE

• To roast or broil red meat, poultry, or fish on a rack or revolving spit before, over or under a source of cooking heat, usually basting with a highly seasoned sauce.

• Red meat, poultry, or fish cooked in or served with a barbecue sauce.

BARBECUE, PIT

A pit (trench) in which wood is burned to make a bed of hot coals over which meat is barbecued. Large cuts of meat are wrapped in paper and burlap, dipped in water, placed over a bed of hot coals, covered with damp soil, and cooked for 12 to 24 hours. Actually the meat is steamed.

BARBITURATES

Sedative drugs which are often used as sleeping pills or anesthetics. The chronic use of barbiturates leads to stepping up of the liver's metabolism of drugs, with the result that greater amounts of drugs are needed to produce a given effect. The need to increase the dosage of a regularly used drug is said to constitute a "tolerance" to the drug.

BARCELONA NUTS

This is another name which is sometimes used for filberts or hazelnuts. It refers to Barcelona, Spain, since many hazelnuts are grown in Spain.

(Also see NUTS, Table N-8 Nuts—"Filbert.")

BARDING

The process of tying a thin sheet of bacon over lean meat.

BARFOED'S TEST

A test for all monosaccharides which gives a red precipitate of cuprous oxide when mixed with Barfoed's solution (copper acetate in acetic acid) and heated in a boiling water bath.

BARIUM MEAL

A mixture of buttermilk or malted milk with a small amount of barium sulfate. The fluid is given orally or as an enema after an overnight fast. The parts of the digestive system containing the barium sulfate can then be seen on a fluoroscope.

(Also see DIGESTION AND ABSORPTION.)

BARLEY *Hordeum vulgare*

Contents *Page*

Origin and History 159
World and U.S. Production 160
The Barley Plant 160
Barley Culture 161
Kinds of Barley 161
  New and Improved Barley 161
Federal Grades of Barley 161
Processing Barley 162
  Milling 162
  Malting (Sprouting) 163
Nutritional Value of Barley 163
Barley Products and Uses 163
Prospects for Barley in the Future 163

Barley has about the same worldwide distribution as wheat, but it is less important in the warmer countries, and it never enters international trade to the same extent. It ranks fourth as a cereal grain, both globally and in the United States.

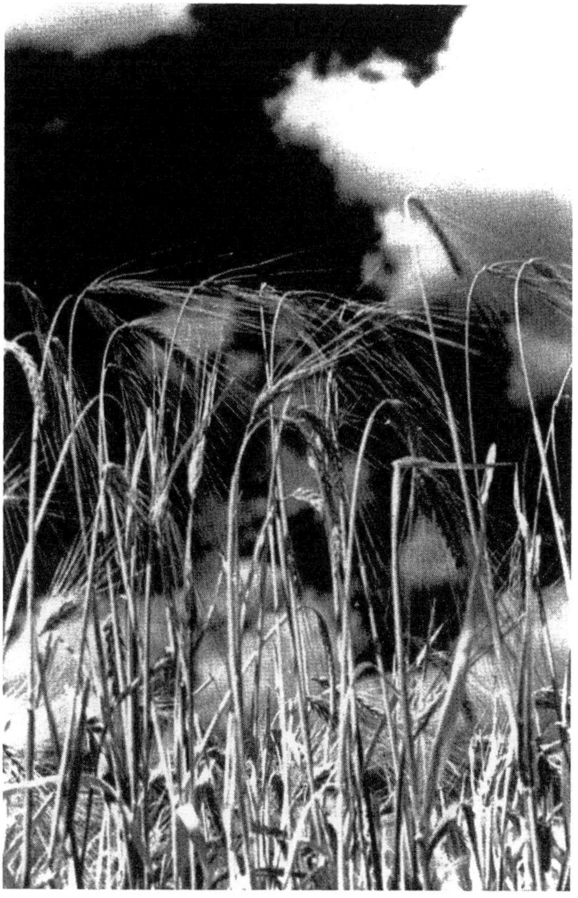

Fig. B-14. A closeup look at a field of barley.

Barley has long been used as a human food in the Asian countries, although its use for this purpose is declining as preferred grains become more plentiful. It produces the most satisfactory malt of all cereals, and is the basis of the best beers and much whiskey in many countries. Also, barley is widely used as a livestock feed.

Foods and beverages made from barley have been used for thousands of years. However, through the years, there have been many new developments in the growing, harvesting, and processing of the grain, and newer knowledge of nutrition has made for increased use of barley by-products for feeding livestock.

(Also see CEREAL GRAINS.)

ORIGIN AND HISTORY. Barley was formerly believed to have originated as early as 7000 B.C. in the dry lands of southwestern Asia, where wild strains of the grain may still be found. Apparently, the primitive people of this area used barley as food for man and beast, and fermented it to make an alcoholic beverage, as evidenced by the oldest known recipe for barley-wine found engraved on a Babylonian brick dated about 2800 B.C.

More recent research points to two primary centers of origin: (1) the highlands of Ethiopia, and (2) southeastern Asia. In any case, barley is one of the oldest cultivated grains, as old as agriculture itself.

The ancient Chinese viewed barley as a symbol of male potency because the heads of grain had heavy beards and contained many seeds. In India, the Hindu god, Indra, was supposed to ripen the barley. Barley is mentioned in the Book of Exodus in connection with the ten plagues.

Ancient Greek athletes, who believed that their prowess in sports depended to a large extent on their diet, ate a barley-mush while in training, because it was easily digested. Greek coins even had designs which showed heads and grains of barley. Later, Roman gladiators were called *hordearii*, meaning eaters of barley.

During the middle ages, the peasants of Europe ate a heavy, coarse bread made from a mixture of barley and rye flours, since the limited availability of wheat at that time restricted its use to the nobility.

The Spanish introduced barley to South America in the mid-1500s. Later, their colonists took the grain to the area which subsequently became southwestern United States. In the early 1600s, English and Dutch settlers took European varieties of barley to eastern United States.

WORLD AND U.S. PRODUCTION. Barley is grown throughout the world. It is the most dependable cereal crop where drought, summer frost, and alkali soils are encountered. Barley is the fourth ranking cereal food crop of the world, being exceeded only by wheat, rice, and corn. World production in 1990 totaled 180 million metric tons. The ten leading barley producing countries by rank were: U.S.S.R., Germany (United in 1990), Canada, France, Spain, U.S.A., U.K., Turkey, Denmark, and Czechoslovakia.

The highest yields in 1990, which each country produced were: United Arab Emirates, 151 bu/a *(8,125 kg/ha)*; France, 121 bu/a *(6,499 kg/ha)*; Switzerland, 106 bu/a *(5,703 kg/ha)*; Zimbabwe, 105 bu/a *(5,660 kg/ha)*; and Denmark, 104 bu/a *(5,620 kg/ha)*. Globally, barley yielded an average of only 46.9 bu/a *(2,251 kg/ha)*.

The United States produced 9,047,290 metric tons (418,856,000 bu) in 1990. The ten leading states in 1990, by rank were: North Dakota, Montana, idaho, Minnestota, South Dakota, Washington, Colorado, California, Wyoming, and Oregon (see Fig. B-16). In yields, the leading states and the number of bushels that each of them produced per acre in 1990, were: Arizona, 105; Utah, 81; Colorado, 80; Nevada, 75; Wyoming, 74; Idaho, 72; Oregon, 70; and Delaware, 70. Nationwide, an average yield of 55.9 bu of barley per acre was obtained in 1990.

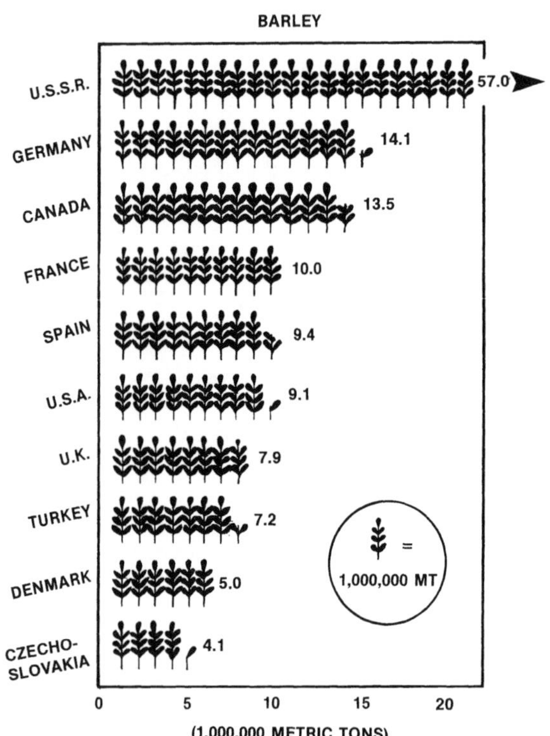

Fig. B-15. Leading barley-producing countries of the world in 1990. (Based on data from *FAO Production Yearbook*, 1990, FAO/UN, Rome, Italy, Vol. 44, pp. 77-78, Table 19)

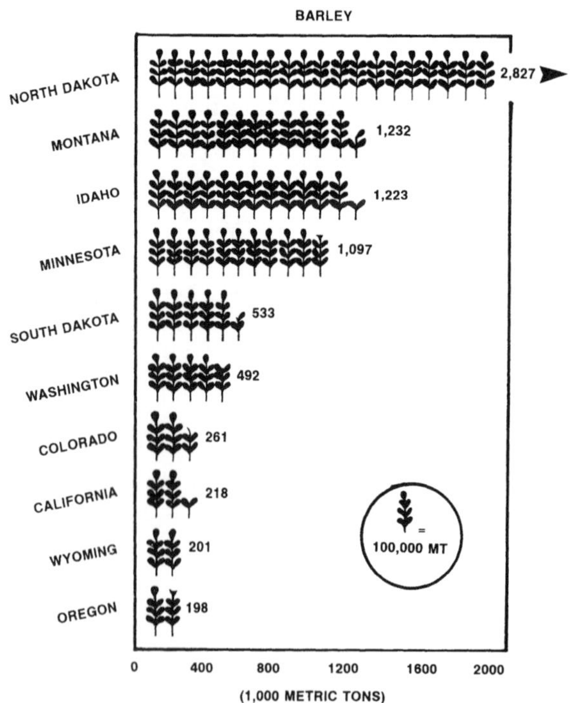

Fig. B-16. Leading barley-producing states of the United States in 1990. (One bushel equals 48 lb, or *21. 8 kg*.) (Data from: *Agricultural Statistics*, 1991, USDA, p. 45, Table 58)

THE BARLEY PLANT. This important cereal grain belongs to the grass family, *Gramineae*. It is classed as genus *Hordeum*, species *H. vulgare*.

The barley plant resembles wheat, both are annual grasses. The stems vary in length from 1 to 4 ft *(30 to 120 cm)*, depending on variety and growing conditions.

The shape and size of leaves vary with variety, growing conditions, and position on the plant. Barley seeds grow in spikes (heads) at the tips of the stems. Most varieties are six-rowed—the single grains grow in three rows on each side of the spike. Although less common, some varieties are two-rowed. At maturity, most varieties of barley have heavy beards or hairlike extensions from the husks of each grain.

BARLEY CULTURE. Barley *(Hordeum)*, of which there are about 20 species, is grown extensively in the temperate zone and its bordering areas throughout the world. It thrives from near the Arctic Circle in Norway to near the equator in the high mountain ranges of Ethiopia. The growing season for barley is short, ranging from 60 to 180 days, depending upon the variety and the climate where it is grown. Thus, it lends itself well to planting in the spring and harvesting before the summer. Like wheat, barley does not do well in hot, humid weather.

The crop is well suited for rotation with other crops in arid areas, such as Arizona and California, where it may be sown in the winter when its water requirements are highest. With such winter-spring production, it does not compete for water with crops that are seeded later in the season, such as cotton, alfalfa, and sorghum.

The drier growing conditions of the northern plains states produce a harder grain which is higher in protein than the softer varieties grown in the midwestern states or on the Pacific Coast of the United States. The higher protein barley is more likely to be used by distillers for making whiskey, whereas the softer barley, with its higher starch content, is preferred by brewers for malting.

Barley is seeded at the rate of about 2 bu *(117 kg/ha)* per acre, by drilling or broadcasting, in well-drained, moderately fertilized soil. Heavy nitrogen fertilization may result in the plant falling over (lodging) and poor development of the grain. Also, overproduction of protein in the grain due to overfertilization makes it less suitable for malting by brewers.

Although high-moisture barley, containing from 25 to 40% moisture, is harvested and used as livestock feed, to prevent spoilage it must be stored in an airtight silo or treated with an organic acid. Therefore, the grain is usually harvested when its moisture content is between 8 and 16%; since the grain may be cracked during harvesting when its moisture content is below 8%, and it may spoil when it contains more than 16%. In the developed countries, most of the barley is harvested with a combine, while in the developing countries many peasants still harvest and thresh the grain by hand. Grain that is to be used for malting must be threshed with care since damaged grains will not germinate properly and are more likely to become moldy during malting.

KINDS OF BARLEY. There are many kinds and types of barley, and there are many different bases for classification; among them, the following:

1. **Hulled vs naked.** In the hulled types, the hull or husk is adherent to the kernel after threshing, whereas in the naked types the hull is loose and easily removed during threshing. Hull-less or naked types of barley are extensively cultivated in southeast Asia, where they are used for food.

2. **Two-row vs six-row.** Based on the arrangement of the grains in the ear, hulled barleys are classed as two-row or six-row. The two-row types predominate in Europe, in parts of Australia, and in western United States. The six-row types are more adaptable to varied environmental conditions, or higher yielding, and are preferred by malters. Six-row barleys are also most common in India and the Middle East.

3. **Spring vs winter.** Spring barley varieties are planted in the spring and mature in the summer of the same year. Winter barley is planted in the fall and harvested the next summer. Spring barley predominates in the United States.

4. **Other classes.** In addition to the above types, barley may have: awned, awnless, or hooded lemmas; and black, purple, or white kernels.

Presently, there are more than 150 varieties of barley grown commercially in the United States and Canada alone, and many additional strains are grown in other parts of the world. The barley collection maintained by the U.S. Department of Agriculture includes more than 1,700 individual strains.

It is noteworthy that malting barley can be used as livestock feed, but feed barley varieties are not satisfactory for malting purposes.

New and Improved Barley. Barley, like other cereal grains, has nutritional deficiencies which make it unsuitable as an unsupplemented staple food for infants and growing children. It has a marginal content of protein, and the level of lysine is disproportionately low in relation to its other essential amino acids.

In 1967, researchers at the Swedish Feed Association in Svalof, Sweden, screened 15,000 barley seed samples, which the U.S. Department of Agriculture supplied from the World Barley Collection, for lysine and protein content. A mutant, naked variety from Ethiopia, which was named Hiproly, was found to contain from 20 to 30% more lysine in its protein than the commonly cultivated barleys.[3] However, the Hiproly seeds are smaller and appear shriveled when compared to the plump seeds of common varieties which have been selectively bred for good malting characteristics. Attempts are now underway to crossbreed Hiproly with other varieties in order to obtain improved quality of seed.

FEDERAL GRADES OF BARLEY. The United States Standards for Barley, established by the U.S. Department of Agriculture, divide it into three classes: barley, western barley, and mixed barley; and into three subclasses: malting barley, blue malting barley, and barley.

There are three official grades of malting barley: U.S. No. 1, U.S. No. 2, and U.S. No. 3. For each grade there are standards for test weight per bushel, sound barley, damaged kernels, foreign material, skinned and broken kernels, thin barley, black barley, and other grains.

There are six grades of barley, other than malt barley: U.S. No. 1, U.S. No. 2, U.S. No. 3, U.S. No. 4, U.S. No. 5,

[3]Munck, L., *et al*, "Gene for Improved Nutritional Value in Barley Seed Protein," *Science*, Vol. 168, p. 985.

and U.S. Sample Grade; with requirements for each grade relative to test weight per bushel, sound barley, total damaged kernels, heat damaged kernels, foreign material, broken kernels, thin barley, and black barley.

PROCESSING BARLEY. The most important uses of barley are for human food, beverages, and livestock feed. In 1990, 47% of the U.S. barley supply was used for food, alcohol, and seed, and 53% was used for feed.

Although freshly harvested and threshed barley does not require additional refinement for feeding to animals, few humans would relish eating unhulled, whole grain or ground grain containing particles of husk. However, man has devised processes for preparing appetizing food and drink from the grain. Livestock also benefit from by-products of these processes. Fig. B-17 outlines the production of some of the common items made from barley. A discussion of milling and malting follows:

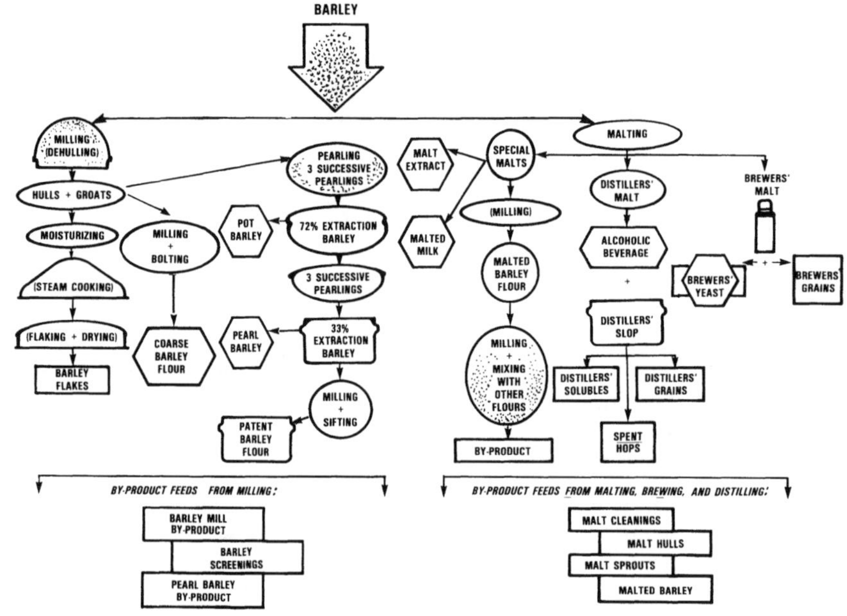

Fig. B-17. Production of beverages, feeds, and foods from barley.

Milling. Like oats and rye, barley grains are covered with hulls or husks, while wheat, rye, and corn have naked seeds. The structure of the dehusked barley grain is similar to that of wheat—both grains have an outer seed coat covering an aleurone layer (the bran layer), a large starchy endosperm, and an oil-containing germ.

Most of the barley used for human food in the United States, and in some other parts of the world, is in the form of whole kernels from which the outer hull or husk and part of the aleurone layer (the bran layer) have been removed. Different types of machines provide scouring or abrasive actions, called pearling, to remove the indigestible hull and all or part of the bran layer. After three successive pearlings, all of the hull and most of the bran are removed; the remaining kernel part is known as pot barley. Two or three additional pearlings, followed by sizing with a grading wheel, produce pearl barley—small, round, white grains of uniform size, from which most of the embryo has been removed. From 100 lb *(45 kg)* of barley, about 65 lb *(29 kg)* of pot, or 35 lb *(16 kg)* of pearl, barley are produced. It is estimated that six pearlings remove 74% of the protein, 85% of the fat, 97% of the fiber, and 88% of the mineral of the original barley.

A variety of products is produced by the milling of whole barley grains to different percentages of extraction. These are listed and described in Table B-9.

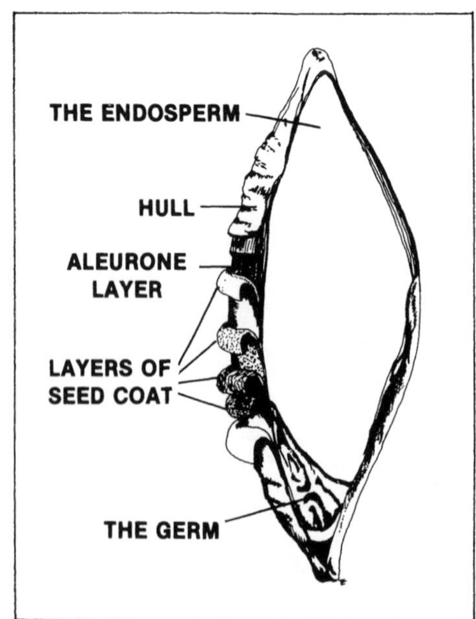

Fig. B-18. Cross section of a barley grain.

Malting (Sprouting). Barley occupies a unique position in malting and brewing, stemming from the fact that the aleurone layer is 3 to 4 cells thick in barley, while most other cereals have only a single layer of aleurone cells. These cells contain the protein that is converted to enzymes during germination of the grain. It follows that barley is better suited for malting than the other cereal grains.

The malting process involves the germination (sprouting) of barley in a controlled state, followed by drying under conditions that retain enzymatic activity.

Malting depends upon development of various enzymes within the aleurone layer of the intact grain of barley when it is germinated under suitable conditions. Failure of grains to germinate results in loss of profit to the malter, so care is taken to select whole, undamaged grains. Also, freshly harvested barley is usually stored for at least 3 weeks prior to malting because there is more likely to be a high rate of germination of barley which has been stored for a short time.

During malting, enzymatic action converts part of the starch to maltose (a disaccharide sugar like lactose and sucrose) and some of the insoluble proteins to soluble proteins. The germination process is stopped by careful heating and drying of the grain (so as to retain enzyme activity) before an excess amount of sprouting has taken place, following which the sprouts are removed from the malted grain by passing the dried malt through revolving sets of wire screens.

It is noteworthy that the term malt may be applied to any grain (rye, wheat, etc.) which has been treated as described above. However, it is generally understood that when the term malt is used alone (without indicating the grain source) it refers to malt from barley.

Malted barley is usually subjected to further processing steps which yield the products listed and described in Table B-9.

NUTRITIONAL VALUE OF BARLEY.

In comparison with corn, barley contains approximately the same percentage of carbohydrates, about 3% more protein, and slightly less fat. Because of the hull, it contains 5% less digestible material.

Malting and milling of barley yield a variety of products which differ widely in their nutrient compositions. Food Composition Table F-36 gives nutrient data for selected barley products.

Pearling barley is a refining process. It removes protein, fiber, minerals, and vitamins. Thus, the decreases in nutrient content which result from the milling of barley are similar to those resulting from the milling of rice, wheat, corn, and other grains.

During the malting of barley, there is an increase in the levels of the vitamin B complex and a decrease in the starch content. Thus, there are more nutrients per calorie in the malted grain. Also, part of the carbohydrate and protein of the original barley grain is digested by the action of enzymes produced during malting. So, the malted product has both increased digestibility and greater palatability due to the roasted malt flavor.

The production of beer and whiskey from barley malt yields such by-products as brewers' grains, brewers' yeast, distillers' grains, and distillers' solubles, all of which are good sources of the B vitamins and various other nutrients.

(Also see BEERS AND BREWING; CEREAL GRAINS, Table C-18 Cereal Grains of the World; DISTILLATION; FERMENTATION; MALT; and UNIDENTIFIED FACTORS.)

BARLEY PRODUCTS AND USES.

Broadly speaking, barley is used in two ways: (1) commercially in foods, beverages, and feeds; and (2) in home recipes.

The production of foods and beverages from barley is almost always linked to the production of animal feeds since the foods and beverages consumed by man are usually produced by milling the whole grain to remove the hull, seed coat, aleurone layer, and embryo; or by malting the grain to make beer or whiskey.

Barley finds little use for human food in North America and Europe, but it is widely used for this purpose in Asia. In the developed countries, the predominant use of barley for human food is the production of beer from malted grain. In the developing countries, however, a considerable amount of the grain is used as a cereal and for bread making. It is inevitable that world food shortages and increased population will force the developing nations to divert even greater amounts of barley and other grains to human use, rather than share precious grain with animals.

Table B-9 presents in summary form the story of barley products and uses.

(Also see BREADS AND BAKING; CEREAL GRAINS; and FLOURS, Table F-25 Major Flours.)

PROSPECTS FOR BARLEY IN THE FUTURE.

Although barley was the first cereal grain to be cultivated by man, other grains have partially taken its place in the production of feeds and foods. However, in the future, barley may regain or surpass its past importance as a result of the following circumstances:

1. Compared to corn, barley is 90 to 95% as valuable as a feed grain. Yet it requires much less water and fertilizer, and it has higher tolerances for cold weather and salty water. These characteristics may be of increasing importance in the future as attempts are made to step up agricultural production in cold or dry climates. For example, the availability of water for irrigation has become such a problem in the southwestern United States and nearby Mexico that it appears that water which has become salty from its use on crops grown on saline soils may soon have to be reused on other crops such as barley. This has prompted the University of California to experiment with the salt water irrigation of barley, since it tolerates salinity better than other grains.
2. Reduction in the cost of synthetic amino acids may make it economically feasible to improve the nutritive value of barley for both man and animals by fortifying its cereal products with lysine and threonine.
3. New and improved varieties, such as Hiproly, have significantly higher levels of protein and the amino acid lysine than the commonly used varieties of barley.
4. Innovations in baking technology have made it possible to add small amounts of dough improvers, like glyceryl monostearate and xanthan gum, to such low gluten flours as barley, in order to produce breads comparable to those made entirely from high-quality wheat.
5. The commercial production of enzymes which digest cellulose (cellulases) may make it possible to convert some of the fiber of barley hulls to sugars which can be digested by chickens, hogs, and man.

TABLE B-9
BARLEY PRODUCTS AND USES

Product	Description	Uses	Comments
		HUMAN FOODS	
Barley 1. Pot barley (Scotch barley)	The remaining kernel following three pearlings, with all of the hull and most of the bran removed.	Commonly used in mushroom and barley soup (a favorite of Jewish and Russian peoples), Scotch broth (a hearty soup), stews, vegetable dishes, and dressings.	*Caution:* It is hazardous to cook barley in a pressure cooker unless at least a small amount of oil or fat is added to the barley and water mixture prior to cooking. Barley forms a starchy foam which may clog the safety valve of the pressure cooker. This may cause a buildup of pressure until the counterweight is blown from the vent. This hazard may be alleviated by first obtaining drippings of fat from either lamb shanks or short ribs of beef, by browning them in the open pressure cooker, then cooking the ingredients together (the meat is placed on a rack over the mixture of barley, drippings, and water) under pressure for 15 min., after which the heating is stopped and the pressure is allowed to drop by itself until it is safe to open the cooker (about 20 min. after the cessation of heating).
2. Pearl barley	The remaining kernel after 2 to 3 additional pearlings beyond the pot barley stage, followed by sizing—small, round, white grains of uniform size, from which most of the embryo has been removed.	Immigrants to the U.S. from Northern Europe and the Middle East brought with them many recipes for tasty dishes which contained refined barley. Some suggetions for preparing barley dishes follow. Barley, which has a flavor that is stronger than polished rice but not as strong as brown rice, may be substituted for rice in many recipes such as casseroles, porridges, puddings, and soups. Also, when cooked in water (use 4 cups *[960 ml]* of water per cup *[240 ml]* of barley), its cooking time is about the same as that for brown rice—about 45 min. The flavor of barley blends well with the flavors of such legumes as soybeans, split peas, or lentils; and with the flavors of other vegetables such as cabbage, carrots, onions, or mushrooms. A typical mixture might contain 1 cup *(240 ml)* of legumes to 3 cups *(720 ml)* of barley. The starchy cooking water from barley has long been used as a broth in invalid diets, to stop diarrhea, and to relieve indigestion.	
Barley flakes	Barley groats that have been soaked, steamed, and rolled.	Cereal in Europe, but not in the U.S.	
Barley flour (Also see MALT FLOUR; WHEAT MALT FLOUR.)	There are two kinds of barley flour: 1. Coarse barley flour milled from barley groats. 2. Patent barley flour, a highly refined flour milled from pearled barley.	Quick-rising doughs (leavened with baking powder) to make biscuits, muffins, and pancakes. Baby foods and food specialties. Small amounts of diastatic malt flours (or the equivalent in malt syrups) may be used by bakers because the enzymes present in these malt products convert the starch of dough to sugars. Patent barley flour is used to make items which are well tolerated by persons with certain digestive disorders.	Doughs made from barley flour lack sufficient elasticity (because of their low-gluten content) for making yeast-leavened breads. However, barley flour may be substituted for wheat flour in recipes for muffins, biscuits, and pancakes which are normally leavened with baking powder. Also, the use of eggs in baked products helps to make up for the lack of gluten in barley flour. These products are useful where an allergy to wheat is encountered. (Also see BREADS AND BAKING.)
Barley groats	Kernels from which the outer hulls and seed coats have been removed.	Porridges in the developing countries.	Usually the kernels are crushed for porridges. (Also see BREAKFAST CEREALS.)

(Continued)

TABLE B-9 *(Continued)*

Product	Description	Uses	Comments
HUMAN FOODS — *(Continued)*			
Infant cereals (Also see INFANT DIET AND NUTRITION.)	Usually made from coarse barley flour, produced by milling the dehulled, whole grain, with added ingredients such as yeast, tricalcium phosphate, salt, iron, niacin, and thiamin.	For human infants.	Many of these items are more nutritious than the ready-to-eat breakfast cereals eaten by older children and adults. (Also see BREAKFAST CEREALS.)
Malted barley cereals	Cooked and ready-to-eat cereals which contain malted barley that has been coarsely ground and then toasted to bring out its full flavor.	Breakfast cereals	Malted barley cereal may also be used to make muffins, pancakes, and other quick breads. (Also see BREADS AND BAKING.)
FERMENTATION PRODUCTS (Also see BEERS AND BREWING; DISTILLED LIQUORS; and FERMENTATION.)			
Beer	The product resulting from a two-step process— 1. Malting 2. Brewing	Alcoholic beverage	Barley was one of the first foods used for the production of beers and whiskeys, because it was readily available, easily malted, and rapidly fermented. Even today, when other materials have replaced part of the barley in these uses, it still has a major role in the production of malts for brewers and distillers. The manufacture of beer from barley comprises two major processes: 1. Malting, a controlled germination process which produces enzymes that are able to convert cereal starch to fermentable sugars. 2. Brewing, which is the process of converting the starch to an alcoholic solution, first by transforming the starch to sugar, then by fermenting the sugar to alcohol by means of yeast.
Brewers' yeast	The yeast that is produced in the brewing of beer. (One lb *[0.45 kg]* of yeast increases to 4 lb *[1.8 kg]* during fermentation.)	Nutrient supplement which provides proteins, nucleic acids, selenium, and the B complex vitamins (exclusive of vitamin B-12).	Brewers' yeast was once distributed by public health nurses in southeastern U.S. to prevent pellagra. It was also given to malnourished refugees after World War II. Presently, there is considerable interest in the selenium content of brewers' yeast, because it is one of the best food sources of this essential mineral. (Also see BREWERS' YEAST; SELENIUM.)
Gin	Distilled spirit flavored with juniper or some other flavoring.	Dutch gin is made by distilling a mash of which at least ⅓ is barley.	
Malt	The product secured by germinating (sprouting) barley in a controlled state, followed by (1) drying under conditions that retain enzymatic activity, and (2) removing of the sprouts.	Malts for brewers and distillers, special breads, extracts, flours, and malted milk.	Starchy materials that are cheaper than malt are sometimes used as brewing "adjuncts" to replace a part of the malt. Among them, unmalted barley flakes, unmalted corn and wheat starches, rice grits, wheat flour, etc. Malt contains various enzymes.

(Continued)

TABLE B-9 (*Continued*)

Product	Description	Uses	Comments
	FERMENTATION PRODUCTS — (*Continued*)		
Whiskey	The product resulting from the distillation of fermented mash cereal grain, followed by aging in wooden barrels from 2 to 8 years. Scotch pot-still or malt whiskey is made only from malted barley. Scotch patent-still or grain whiskey is made from barley and other unmalted cereal grains. Irish whiskey is made from malted barley alone or from malted barley with an admixture of unmalted barley, wheat, rye, or oats. Canadian whiskey is produced from barley, corn, wheat, or rye.	Alcoholic beverage	It is noteworthy that an early American nickname for whiskey was "John Barleycorn." The distinctive smokey flavor of Scotch whisky is due to the peat used for firing the malt kilns, and to the characteristics of the water used.
	LIVESTOCK FEEDS		
Barley, whole or processed	The whole kernel.	All classes of livestock.	The grain is usually processed (by grinding, rolling, or other methods), except for sheep. Barley and barley by-products are excellent feed for livestock. However, care must be taken to ensure that rations based on these feedstuffs are balanced with respect to all the required nutrients, and that such supplements as necessary are used.
Barley distillers' condensed solubles	The product obtained after the removal of ethyl alcohol by distillation from the yeast fermentation of barley or a barley mixture by condensing the thin stillage fraction to a semisolid.	As a source of B vitamins and unidentified factors for swine and poultry.	
Barley distillers' dried grains	The product obtained after the removal of ethyl alcohol by distillation from the yeast fermentation of barley or a barley mixture by separating the resultant coarse grain fraction of the whole stillage and drying it.	Fed chiefly to dairy cattle as a source of protein. Sometimes used in beef cattle and sheep rations.	
Barley distillers' dried grains with solubles	The product obtained after removal of the ethyl alcohol by distillation from the yeast fermentation of barley or a barley mixture by condensing and drying at least ¾ of the solids of the resultant whole stillage.	As a source of B vitamins and unidentified factors for swine and poultry.	
Barley distillers' dried solubles	The product obtained after the removal of ethyl alcohol by distillation from the yeast fermentation of a barley or a barley mixture by condensing the thin stillage fraction and drying it.	As a source of B vitamins and unidentified factors for swine and poultry.	
Barley hulls	The outer covering of barley.	Dairy cows	

(Continued)

TABLE B-9 (*Continued*)

Product	Description	Uses	Comments
	LIVESTOCK FEEDS — (*Continued*)		
Barley mill by-product	The entire residue from the milling of barley flour from clean barley, consisting of barley hulls and barley middlings.	For dairy cattle and swine.	
Barley screenings	Small, broken, or shrunken kernels of barley, chaff, and weed seeds.	Usually finely ground to destroy noxious weed seeds and incorporated in mixed feeds.	
Brewers' dried grains (Brewers' wet grains)	The dried extracted residue of barley malt alone or in mixture with other cereal grain or grain products resulting from the manufacture of wort or beer and may contain pulverized dried spent hops in an amount not to exceed 3%.	Fed chiefly to dairy cattle, at not to exceed ⅓ of the concentrate ration. Not very palatable if fed alone.	Brewers' wet grains are usually dried and sold as brewers' dried grains.
Brewers' dried yeast (Brewers' yeast)	The yeast that develops in the fermentation process.	Primarily in swine and poultry rations, as a source of B vitamins (except B-12) and unidentified growth factors.	
Dried spent hops	The product obtained by drying the material filtered from hopped wort.	Mostly in mixed feeds for dairy cattle.	
Malt cleanings	The product obtained from cleaning malted barley or from the recleaning of malt which does not meet the minimum crude protein standard of malt sprouts.	Mostly in mixed feeds for dairy cattle.	
Malt hulls	The hulls obtained in the cleaning of malted barley.	Mostly in mixed feeds for dairy cattle.	
Malt sprouts	The product obtained from malted barley by the removal of the sprouts. It must contain not less than 24% crude protein.	Mostly in mixed feeds for dairy cattle. Not palatable if fed alone.	Large amounts of malt sprouts in dairy feeds may impart off-flavors to milk.
Malted barley	Barley grain that has been germinated (sprouted) under controlled conditions, followed by drying and removing the sprouts.	Malted barley is occasionally used in calf meals or as an appetizer in fitting stock for show.	
Pearl barley by-product	The entire by-product resulting from the manufacture of pearl barley from clean barley, consisting of the hulls and the outer coats of the kernels.	For dairy cattle and swine.	
Forages:			
Hay	Mature barley plants.	For cattle.	Cut when the grain is in the milk to soft dough stage.
Pasture	Young barley plants.	As winter pasture for cattle, sheep, and swine.	
Silage	Fermented chopped barley plants.	For cattle.	Cut when grain is in the milk to soft dough stage.
Straw	The stalks following threshing.	Bedding for livestock. As a coarse roughage for cattle.	The fibrous straw is suitable for cattle, because, being ruminants, they are able to digest such fiber.

BARM

Yeast formed during the fermentation of alcoholic beverages.

BASAL DIET

A diet common to all groups of experimental subjects to which the experimental substance(s) is added. This type of diet is often used in research studies.

BASAL METABOLIC RATE (BMR)

The heat produced by a person during complete rest (but not sleeping) following fasting, when using just enough energy to maintain vital cellular activity, respiration, and circulation, the measured value of which is called the basal metabolic rate (BMR). Basal conditions include thermoneutral environment, resting, postabsorptive state (digestive processes are quiescent), consciousness, quiescence, and sexual repose. It is determined 14 to 18 hours after eating and when at absolute rest. It is measured by means of a calorimeter and is expressed in calories per square meter of body surface.

(Also see CALORIC EXPENDITURE.)

BASAL METABOLISM

The energy expended by the body when (1) the subject is awake, but there is complete physical and mental rest; and (2) from 14 to 18 hours have elapsed after the last meal. Also, the environmental temperature must be in the comfort zone (usually 70 to 80°F [*21 to 27°C*]) so that no extra work is being done by the body either to heat or to cool itself. The energy expenditure which is measured under these conditions is taken to represent the minimal energy required to sustain the life of a healthy person. The extra energy expended over and above that used for basal metabolism is called the activity increment.

(Also see CALORIC [ENERGY] EXPENDITURE.)

BASE

Any of a large class of compounds with one or more of the following properties: bitter taste, slippery feeling in solution, ability to turn litmus blue and cause other indicators to take on characteristic colors, ability to react with (neutralize) acids to form salts. Bases include both hydroxides and oxides of metals.

BASE BICARBONATE

In the term "base bicarbonate," the word "base" refers to any base that might be combined with bicarbonate. In the main buffer system of the human body, this base is sodium bicarbonate ($NaH\ CO_3$).

BASEDOW'S DISEASE

This disorder—which is also called exophthalmic goiter, Graves' disease, hyperthyroidism, or toxic goiter—is characterized by protrusion of the eyeballs (exophthalmos), swelling of the thyroid gland, breathlessness, an abnormally rapid heartrate, loss of weight despite an increased appetite and hearty eating, high blood pressue, an oversensitivity to heat, increased sweating, nervousness, muscle weakness, and fatigue. Basedow's disease results from an overproduction of hormones by the thyroid gland, and is more likely to occur in females than males.

Treatment consists of such measures as (1) surgical removal of part of the thyroid gland; (2) administration of radioactive iodine, which destroys some of the tissue of the gland; and/or (3) drugs which block the production of thyroid hormones.

It is noteworthy that Basedow's disease may be produced by the administration of iodine supplements to people living in iodine-poor areas. This phenomenon is called the Jod-Basedow effect.

(Also see GOITER.)

BASES

Another name for substances known as alkalis. They may be used to neutralize acids. Some common household bases are ammonia, baking soda, and lye.

(Also see ALKALI.)

BATAAN EXPERIMENT

A study conducted in 1947 on the Bataan Peninsula of the Philippines which established that enrichment of rice and the substitution of enriched rice for highly-polished rice effectively prevents beriberi and similar disorders.

(Also see ENRICHMENT, section headed "The Bataan Experiment"; and THIAMIN.)

BDELYGMIA

The severe dislike for food.

BEACH PLUM *Prunus maritima*

A small fruit of a native American bush (of the family *Rosaceae*) that grows wild on the beaches of the Atlantic Coast from New Brunswick to Virginia.

Fig. B-19. Beach plum, a wild fruit of eastern United States.

Beach plums are about ½ in. (*1 to 1.5 cm*) in diameter and vary in color from dark red to dull purple when they ripen in the late summer or early fall. The fruit is too sour for eating raw, but it makes good jam and jelly. In most cases the plums may be picked freely by visitors to the beaches, but in the resort areas of Cape Cod, Martha's Vineyard, and Nantucket they are harvested by local residents and sold to the tourists.

(Also see PLUMS and PRUNES.)

BEAN(S)

The name *bean* denotes several related plants of the legume family, *Leguminosae*. Both the edible pods and the seeds of these plants are called beans. They are among the most nourishing vegetables eaten by man. Some kinds of beans produce valuable feed for livestock, others yield the raw materials for many kinds of manufactured articles. Being legumes, beans also enrich the soil with nitrogen that their bacteria take from the air.

Some bean plants are low and bushy, others are climbing vines. They have compound leaves, each of which is made up of three leaflets. Their flowers may be white, yellow, red, or purple. Their smooth seeds, may be white, yellow, black, red, tan, cream-colored, or mottled, and their weight may range from 125 to over 700 mg per seed. Their seeds grow in pods divided lengthwise into two halves. When the beans are ripe, the pods split open at the edges. Climbing beans twine their main stalks around poles, strings, or the stalks and branches of other plants.

Fig. B-20. Blackeye beans. (Courtesy, California Dry Bean Advisory Board, Dinuba, Calif.)

• **World and U.S. Production**—Ranked in order, India, Brazil, China, United States, and Mexico are the largest producers. In the United States, Michigan, North Dakota, Nebraska, Colorado, Idaho, and California lead in U.S. production of dry edible beans, while Wisconsin, Oregon, and Michigan lead in U.S. production of snap (green) beans. About 20% of the U.S. acreage devoted to the production of edible beans is used for the commercial production of snap beans and green lima beans. But about 45% of the total value of all edible beans is accounted for by snap beans and green limas; reflecting their higher value than dry beans.

• **Kinds of beans**—Some 100 or more species of beans are cultivated throughout the world.

In the United States and Canada, the most important kinds, exclusive of the soybean, are varieties of the common bean. They were first cultivated by the Indians of South and Central America. Among the varieties of the common bean which are usually eaten after they become fully ripe, when they are called *dry beans*, are kidney beans, navy beans, and pinto beans. Other varieties of the common bean are called *green shell beans*; they are picked when fully grown, but before they have ripened and turned hard. Still other varieties of the common bean, which are picked at an even younger stage, are called *string beans*, or *snap beans*. Some string beans have green pods, and are called green beans; others have yellow pods, and are called wax beans. Both the half-formed seeds and the pods of the string or snap beans are eaten.

Common beans of all varieties have much greater food value than most other kinds of vegetables. The dry mature seeds are rather similar in composition, although they differ widely in eating quality.

• **Production**—Beans are warm-season annuals, sensitive to temperature extremes and requiring a modest amount of moisture. But they are quite tolerant of soil types. Beans should be planted in the spring after the danger of frost is past and when the soil temperature has reached 65°F (*18°C*). Over much of the world, beans are planted, tended, and harvested by hand. In the United States, commercial crops are generally mechanically handled.

• **Utilization**—Beans are consumed as food in several forms. Lima beans and snap beans are used as fresh vegetables, or they may be processed by canning or freezing. Limas are also used as a dry bean. Mung beans are utilized as sprouts.

The use of dry beans for food is dependent upon the seed size, shape, color, and flavor characteristics, and is often associated with particular social or ethnic groups. Popular uses include soups, mixed-bean salads, pork and beans, beans boiled with meat or other vegetables or cereals, baked beans, precooked beans and powder; and, in the Peruvian Andes, parched or roasted beans. In 1990, the per capita consumption of beans in the United States was: dry edible beans, 9.5 lb (*4.3 kg*); and snap beans, 5.6 lb (*2.5 kg*).

Most dry beans contain from 23 to 25% protein, 61 to 63% carbohydrate, and about 1.5% fat. They may be eaten as a substitute for meat. Green shell beans are rich in both proteins and vitamins. Stringless or snap beans are a fair source of energy, but a rich source of vitamins A, B, and C.

Many foods contain components which are harmful at certain stages if they are not processed properly; and uncooked, mature beans are no exception. Fortunately, soaking until the seeds begin to swell, followed by thorough cooking, renders most of these factors harmless. However, more research needs to be done on how to lessen the gas-producing factors in beans. It is known that various types of fermentation and sprouting alleviate much of the gas problem by breaking down the indigestible carbohydrates. But those who suffer discomfort from excess gas may be well advised to eat only immature (green) beans.

(Also see ADZUKI BEAN; BEAN, COMMON; BROAD BEAN; CHICKPEA; COWPEA; HYACINTH BEAN; JACK BEAN; LIMA BEAN; MUNG BEAN; SCARLET RUNNER BEAN; SOYBEANS; TEPARY BEAN; and YARD-LONG BEAN.)

BEANS OF THE WORLD. Only about a dozen species of beans are utilized in the United States. These types are summarized in Table B-10.

Fig. B-20a. Fava beans. (Courtesy, Minnesota Agricultural Experiment Station, University of Minnesota)

Fig. B-20b. Mung beans. (Courtesy, Minnesota Agricultural Experiment Station, University of Minnesota)

TABLE B-10
BEANS OF THE WORLD

Name of Bean	Geographical Areas of Production[1]	Comments
Adzuki bean *(Phaseolus angularis)* (Also see ADZUKI BEAN.)	China, Japan, Korea, and Manchuria.	This bean, which ranks next to soybeans in importance among Chinese and Japanese, recently became known in the United States as a staple food in the Zen Macrobiotic diets.
Broad bean; Fava bean *(Vicia faba)* (Also see BROAD BEAN.)	China produces over 70% of the world's crop. Most of the remainder is grown in Egypt, Italy, United Kingdom, Morocco, Spain, Denmark, and Brazil.	Worldwide, broad beans are an important food crop. However, in the United States it is a minor crop and used almost exclusively as livestock feed. Broad beans contain a substance that triggers favism, an inborn error of metabolism, that afflicts up to 35% of some Mediterranean populations and 10% of American negroes. In sensitive people, the symptoms are brought on by ingesting the beans, or even by inhaling the pollen of the flowering plant.
Chickpea; Garbanzo bean *(Cicer arietinum)* (Also see CHICKPEA.)	India and Pakistan produce over 87% of the world's crop. Most of the remainder is grown in Mexico, Turkey, and Ethiopia.	Some U.S. housewives know this legume as the garbanzo bean. It is among the top dozen vegetable crops of the world, and is most popular in the area extending from the Mediterranean to India.
Common bean; French bean; Kidney bean; Navy bean, Pea bean, Pinto bean, Snap bean *(Phaseolus vulgaris)* (Also see BEAN, COMMON.)	China, India, and Brazil produce about 66% of the world's crop. Most of the remainder is grown in other countries in Africa, the Americas, Asia, and Europe.	In the United States and Canada, the common bean is the most important kind of bean, exclusive of the soybean. Also, it is very versatile; the immature pods and seeds may be picked and used as snap, string, or green beans, or they may be left on the plant until fully mature.
Cowpea; Black-Eyed pea *(Vigna sinensis; V. unguiculata)* (Also see COWPEA.)	Africa, India, China, West Indies, and southeastern U.S.	Despite its name, the cowpea is more akin to the bean than to the pea. Traditionally, black-eyed peas are eaten on New Year's Day in southeastern United States.
Hyacinth bean; Lablab bean *(Dolichos lablab)* (Also see HYACINTH BEAN.)	Southern and eastern Asia and parts of Africa.	This bean is an important crop in India and other dry tropical areas, due to its resistance to drought. There are both field and garden varieties of the hyacinth bean; the field variety is grown mainly for mature seeds and livestock forage, while the garden variety is harvested as immature pods and seeds for human consumption. Also, because of its purple flowers and attractive purplish-red seed pods, the hyacinth bean is sometimes used for ornamental purposes.
Jack bean; Horse bean *(Canavalia ensiformis)* (Also see JACK BEAN.)	Tropical regions around the world, including the southern U.S.	This legume is grown primarily for green manure (for plowing under) or for livestock forage. However, in areas and times of food scarcity, the immature pods and seeds, or the dried mature seeds, are sometimes used for human food.
Lima bean; Butter bean *(Phaseolus lunatus)* (Also see LIMA BEAN.)	Temperate, subtropical, and tropical areas around the world.	This is usually a white, flat bean, which may be eaten either dried or green. There are various small-seeded and large-seeded varieties. The latter types are often called "giant limas." Most of the U.S. production of limas is in California. Various colored varieties are grown in the Caribbean region, but some of these types may be harmful.
Mung bean; Golden gram; Green Gram (Phaseolus vulgaris) (Also see MUNG BEAN.)	India and Pakistan produce much of the world's crop, but some is also grown in tropical and subtropical areas of Africa, the Americas, and Asia.	This is an important bean of Asia. The sprouted seeds of mung beans are eaten as a cooked vegetable or raw in salads.

Footnote at end of table.

(Continued)

TABLE B-10 (*Continued*)

Name of Bean	Geographical Areas of Production[1]	Comments
Scarlet runner bean *(Phaseolus coccineus)* (Also see SCARLET RUNNER BEAN.)	Temperate areas of Europe, Asia, Africa, and the Americas.	This is a popular food bean in cool climates, which it tolerates better than the common bean. Also, its large, bright-red flowers and seeds speckled with red-and-black spots make it an attractive ornamental plant.
Soybean *(Glycine max)* (Also see SOYBEAN.)	The U.S., China, Brazil, and Argentina produce about 88%of the world's crop. Most of the remainder is grown in the temperate areas of Asia and South America.	This bean is the world's leading leguminous crop. It has been grown for over 3,000 years in China, where fermentation and sprouting of the beans have long been utilized to produce highly palatable and nutritious products. Also, soybean oil is one of the world's leading cooking oils. (Even the bean cake left after oil extraction is a highly nutritious base for human foods and livestock feeds.)
Tepary Bean *(Phaseolus acutifolius)* (Also see TEPARY BEAN.)	Western Mexico and southwestern U.S.	This bean is well adapted to hot, dry climates. Most of the world production of the tepary bean is in Mexico and Arizona. Recently, it was introduced into certain drought-ridden parts of Africa.
Yard-Long beans; Asparagus beans *(Vigna unguiculata; V. sesquipedalis; Dolichos sesquipedalis)* (Also see YARD-LONG BEAN.)	Tropical parts of Asia and the Americas.	The name of this bean denotes the unusually long bean pods. Nevertheless, the Latin term *sesquipedalis,* which means a foot and a half long, is a much more accurate description of the length of the bean pods. Yard-long beans are utilized like snap beans.

[1]The principal areas of production according to *FAO Production Yearbook,* Vol. 32, FAO/UN, 1979.

BEAN, COMMON (FRENCH BEAN; KIDNEY BEAN; NAVY BEAN; PEA BEAN; PINTO BEAN; SNAP BEAN; STRINGLESS BEAN; OR GREEN BEAN) *Phaseolus vulgaris*

Contents *Page*

Origin and History 172
Production 173
Propagation and Growing 174
Harvesting 175
Processing 176
Dried Beans 176
Snap Beans 177
Selection and Preparation 177
Dried Beans 177
Snap Beans 179
Nutritional Value 179
Protein Quantity and Quality 179

The common bean is very versatile; the immature pods and seeds may be picked and used as snap beans (also called string beans or green beans), or they may be left on the plant until fully mature for use as dried beans.

Fig. B-21. Beans. (Courtesy, Minnesota Agricultural Experiment Station, University of Minnesota)

ORIGIN AND HISTORY. About 7,000 years ago, this vegetable was grown by the Indian tribes in the Tehuacan Valley of Mexico, and in Callejon de Huaylas, Peru. Apparently, the beans were domesticated independently in each area from a common wild ancestor. However, the separate domestications led to the development of light colored, small seeded varieties in Mexico, and of dark colored, large seeded varieties in Peru. The cultivation of the common bean was spread gradually throughout the Americas by migrating bands of Indians. As a result, the Spanish explorers of the 15th and 16th centuries found the bean crops growing throughout Latin America, and the 17th century British colonists of North

Fig. B-22. Sunday morning in a Boston bakery. Waiting in line for the ever popular baked beans. (Courtesy, The Bettmann Archive, Inc., New York, N.Y.)

America found the Indians growing beans in both New England and Virginia. The basic recipes for Boston baked beans and succotash were derived from those used by the Indians in New England.

Common beans were spread around the world by the Portuguese and Spanish explorers and traders, and they became popular crops in Africa, Asia, and Europe by the early 17th century. Since then, plant breeders have selected plants for large seeds and tender stringless pods. Also, many varieties which grew well only under tropical conditions have now become adapted to the temperate zones.

PRODUCTION. Common beans and lima beans together account for an annual worldwide dry bean production of about 16.3 million metric tons. The U.S. production statistics do not differentiate between the two species of beans. The corresponding production of green beans (mainly common beans) is about 3.1 million metric tons.

Fig. B-23 shows the leading dry bean-producing countries of the world; and Fig. B-24 shows the leading green (snap) bean-producing countries of the world.

In the United States, the production of dry beans is over twelve times the production of snap beans. Fig. B-25 shows the leading dry bean-producing states of the

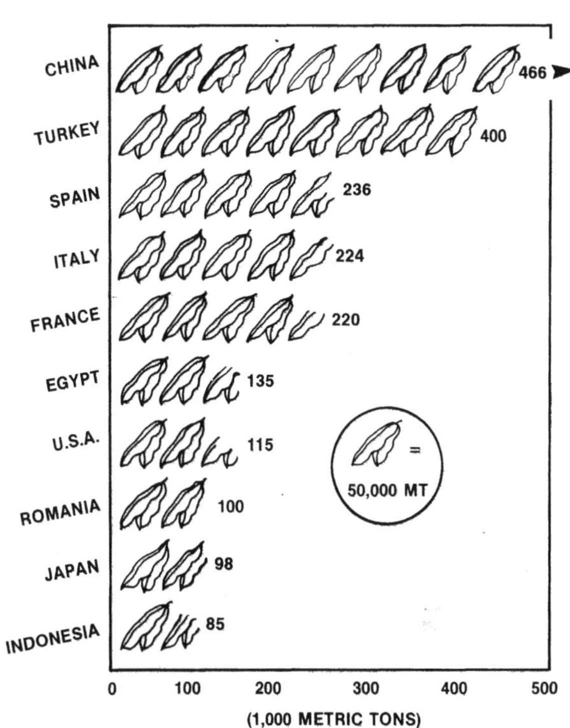

Fig. B-24. Leading green bean-producing countries of the world. (Based on data from *FAO Production Yearbook*, 1990 FAO/UN, Rome, Italy, Vol. 44, p. 144, Table 60)

United States, whereas, Fig. B-26 shows the leading varieties of dry beans produced in the United States.

Fig. B-27 shows the quantities/types of green beans frozen/commercial pack in the United States. Fig. B-28 shows the leading states in the United States in snap bean production for processing.

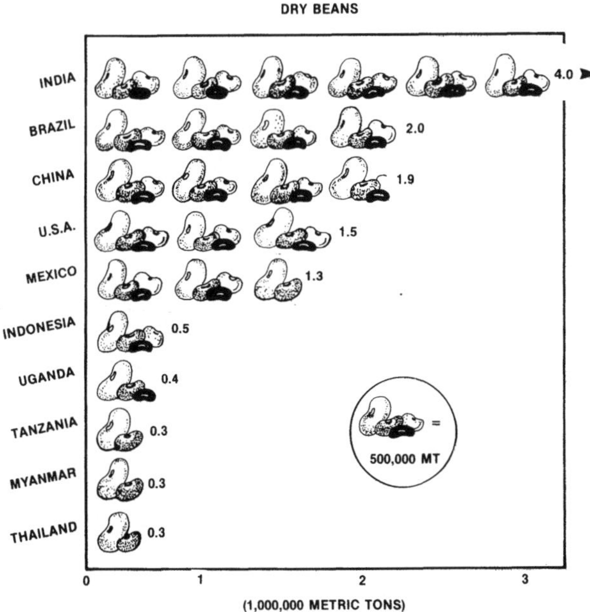

Fig. B-23. Leading dry bean-producing countries of the world. (Based on data from *FAO Production Yearbook*, 1990, FAO/UN, Rome, Italy, Vol 44, p. 100, Table 32)

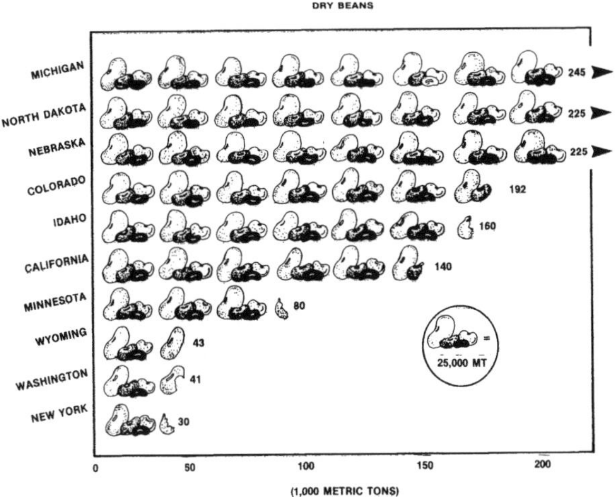

Fig. B-25. The leading dry bean-producing states of the United States. (Based on data from *Agricultural Statistics*, 1991, USDA, p. 241, Table 375)

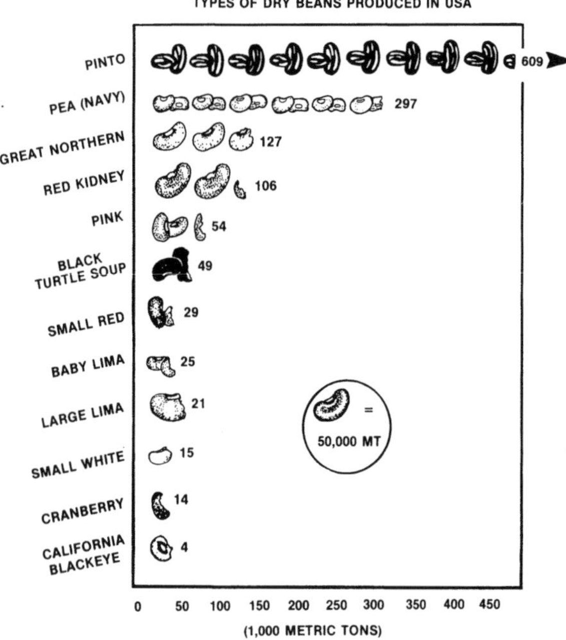

Fig. B-26. The leading varieties of dry beans produced in the United States. (Based on data from *Agricultural Statistics*, 1991, USDA, p. 240, Table 374)

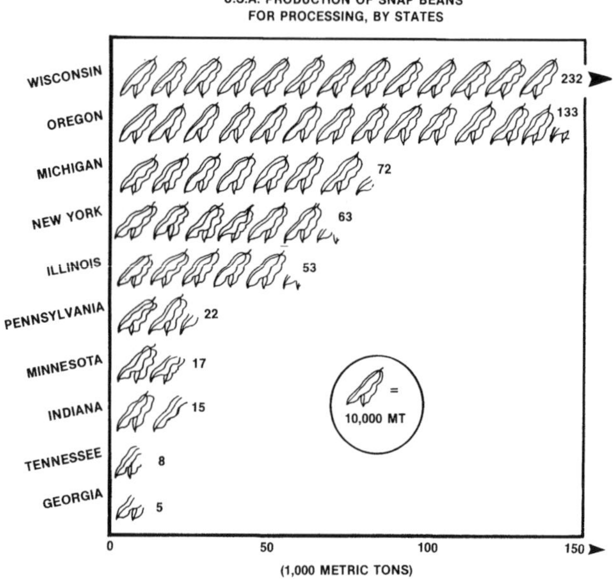

Fig. B-28. The leading states of the United States in snap bean production for processing. (Based on data from *Agricultural Statistics*, 1991, USDA, p. 147, Table 204)

It is noteworthy that a little over one-sixth of the snap bean crop is marketed fresh, while the rest is processed.

Propagation and Growing. The planting of beans in the spring must await the warming up of the soil to about 65°F *(18°C)*; otherwise, the seeds will rot rather than germinate. Fig. B-29 shows bean plants in the early stages of growth.

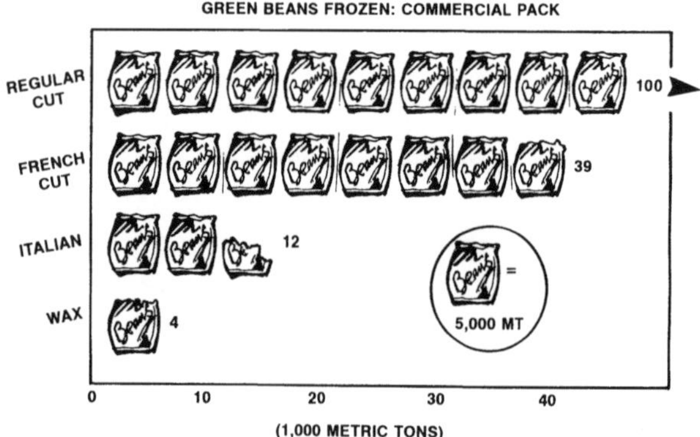

Fig. B-27. Frozen green beans. (Data from *Agricultural Statistics*, 1991, USDA, p. 172, Table 248)

Fig. B-29. Young bean plants. (Courtesy, Union Pacific Railroad Company, Omaha, Neb.)

Even though nodules on the roots of the beans fix atmospheric nitrogen, nitrogenous fertilizers are usually used, because the requirement for this element is very high. This is particularly true for the pole varieties, since they produce much higher yields than the bush varieties. Pole beans are shown in Fig. B-30.

Fig. B-30. Pole beans strung to crossed poles in Florida. (Photo by J. C. Allen & Son, Inc., West Lafayette, Ind.)

However, bush beans are the most commonly grown commercial types in most parts of the United States, except California, because (1) in humid climates climbing beans have to be kept out of contact with the soil, or they may become damaged by disease, (2) providing poles and trellises for pole varieties represents a considerable investment, and (3) the climbers have to be harvested over a longer period of time than the bush beans. Pole beans are grown without supports in California because the dry climate reduces the chance of damage when pods lie on the ground. Fig. B-31 shows a bean plant in the pod stage.

Fig. B-31. A close-up view of green bush bean plants with their seed pods. (Photo by J. C. Allen & Son, Inc., West Lafayette, Ind.)

Three to four months are required for most varieties of beans to mature sufficiently for conversion into dry beans, whereas snap beans (immature pods and seeds) may be produced in about half this time. Consequently, production of the former is most suitable for areas with sufficiently long periods of mild weather, and production of the latter may be the best choice for northerly regions with short growing seasons. However, snap beans are also profitable for the warmer areas such as Florida, because several crops can be grown throughout the year. As a result, fresh snap beans are available continuously in the United States.

Harvesting. Mature beans that are to be dried are harvested after a majority of the pods have ripened, but before they have become dry enough to shatter and spill their seeds. Fig. B-32 shows fully mature bean plants.

Fig. B-32. Navy beans ready for harvesting in Michigan. (Photo by J. C. Allen & Son, Inc., West Lafayette, Ind.)

Usually, the plants are cut by machine and piled in windrows to dry. After drying, they are threshed with a grain combine or a bean harvester.

Snap beans are harvested before the seeds become sufficiently large to cause the pod to bulge. Hand picking is often used for the first harvesting of these beans, because machines usually strip off most of the leaves so that the growth and production of the plants ceases. Hence, the machines are used only for the last picking, as shown in Fig. B-33.

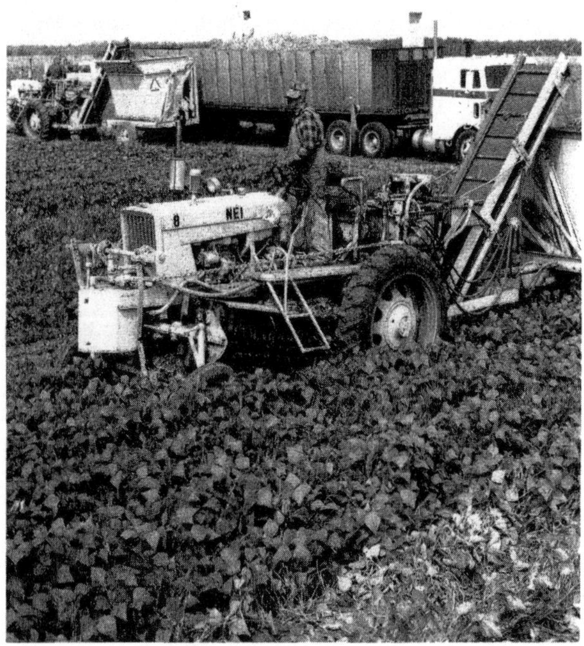

Fig. B-33. Harvesting green beans for canning companies. (Photo by J. C. Allen & Son, Inc., West Lafayette, Ind.)

The average yield of dry beans in the United States is about 1554 lb (*706 kg*) per acre, with 13 states producing well above the average yield.

Average yields of snap beans in the United States are about 3.23 tons per acre, but yields as high as 5.8 tons per acre are obtained in Oregon.

One might wonder why farmers bother to grow dry beans at all, when from 2 to 3 times as many snap beans may be produced on the same amount of land, and both types of beans bring similar prices on the wholesale market. However, dry beans are less expensive to harvest, since the large farms are highly mechanized. Also, dry beans are very resistant to spoilage and may be stored at a low cost, whereas snap beans are highly perishable and must either be kept under refrigeration and marketed within a short time after harvesting, or be canned or frozen. Therefore, the lower yields of dry beans are compensated for by much lower costs of harvesting, processing, storage, shipping, and marketing.

PROCESSING. The demand for convenience foods made from dried beans and snap beans has increased steadily while the use of fresh, unprocessed bean products has declined, because many homemakers go to work and have only limited time for the preparation of meals at home. Hence, some of the recent developments in the processing of these crops are noteworthy.

Dried Beans. Much of this crop is stored and utilized throughout the year for the production of canned items such as beans and pork in tomato sauce, Boston style baked beans, chili con carne and similar products, soups, and dips.

• **Beans with reduced amounts of gas-producing substances**—Some people avoid eating beans because of the gas that they produce.

The effect of beans on gas (flatus) formation in the GI tract has been confirmed in experiments, with the following results:

1. Beans are powerful producers of gas in the GI tract.
2. The CO_2 in the gas is 3 to 6 times higher on bean diets than on a basal diet.
3. Different kinds of beans differ in their capacity to produce gas.
4. When an antibacterial preparation is administered to the subjects, gas production is suppressed, thus showing the influence of bacterial action on gas production.
5. The clostridial group of anaerobes is responsible for flatus formation.
6. Oligosaccharides (carbohydrates containing from 2 to about 8 simple sugars linked together; beyond 8 they are called polysaccharides; raffinose contains 3 simple sugars linked together; stachyose contains 4) are largely responsible for the flatulent properties of beans. Raffinose is a trisaccharide containing 1 molecule each of galactose, glucose, and fructose; whereas stachyose is a tetrasaccharide containing 1 molecule of glucose, 1 molecule of fructose, and 2 molecules of galactose.

• **Beans without gas**—The excess gas (flatulence) following consumption of beans is due to the presence of raffinose and stachyose in beans. The intestinal tract does not produce enzymes capable of splitting these oligosaccharides (raffinose and stachyose), with the result that they pass to the colon where they are fermented by anaerobic microflora and produce gaseous products.

Note: Recently a liquid enzyme product, marketed under the trade name BEANO, came upon the market. It is designed to reduce intestinal gas produced from foods like beans. The enzyme, alphagalactosidase, breaks down indigestible sugars such as raffinose found in the gas-producing beans.[4] This product merits further study.

The gassiness of beans can be greatly lessened by either of the following treatments:

1. **Acid treatment, followed by incubation.** When beans (whole, crushed, or ground) are mixed with water,

[4]*Foods and Nutrition Newsletter*, Cooperative Extension Service, University of Illinois–Champaign, Newsletter No. 17, August, 1991.

the mixture will have a natural pH of 6.0 to 6.5. To activate the enzymes in the seeds, the pH of the mixture is reduced below 6.0, preferably to a pH of 5.0 to 5.5. The pH adjustment is made by adding appropriate amounts of food-grade hydrochloric, sulfuric, phosphoric, or acetic acid.

2. **Eliminating the sugars by soaking-discarding soaking water; cooking-discarding cooking water.** This consists in eliminating most of the tri- and tetra-sugars by (a) soaking the beans at least 3 hours, then discarding the soaking water; (b) covering the beans with boiling water and cooking for 30 minutes, then discarding the water again, followed by rinsing the beans well; and (3) adding fresh water and continuing the cooking process. Do not add any other ingredients to the beans during the first two treatment steps. The raffinose and stachyose are supposed to be absorbed in the soaking and cooking water and removed, at least in sufficient quantity that the beans will not cause the usual degree of gassiness.

• **Flours**—Flours are made from locally grown beans in many countries of the world. Many of these products are being utilized as protein supplements to grain mixtures, in which they augment both the quantity and quality of protein. New types of food additives and processing operations now make it possible to use bean flours in yeast-leavened breads without having undesirable effects on the taste and texture of the baked products.

• **Flour fractions (high-protein and high-starch)**—Air classification, which was originally developed for the processing of wheat flour, is now being used experimentally to separate bean flour into (1) a high-protein fraction containing from 29 to 67% protein, 2 to 11% fat, 2 to 3% fiber, and up to 30% starch; and (2) a high-starch fraction containing from 12 to 16% protein, 1 to 6% fat, 3 to 11% fiber, and 52 to 68% starch. The high-protein fraction may be used as a stabilizer in whipped toppings or foams, whereas the high-starch fraction appears to be useful in baked goods and other items which are presently made mainly from cereal flours. The nutrient composition of the high-starch fraction is similar to those of cereal flours, except that the amino acid pattern of the bean product complements those of the cereal products.

• **Precooked beans**—Many busy homemakers are reluctant to prepare beans because of the long cooking time. Precooked products, which have usually been soaked in a tenderizing solution and steamed or pressure cooked, require only about 35 minutes of cooking time in an open pot, or only a few minutes in a pressure cooker.

• **Protein concentrates**—These products may contain from 40 to 80% protein, and only small amounts of carbohydrates and fats. They are produced by (1) soaking and dehulling dried beans, (2) drying the dehulled beans and grinding them to a fine powder, (3) suspending the powder in a salt solution and coagulating the protein with steam, (4) separating the protein cake from the starchy liquor by centrifugation, and (5) drying and grinding the protein cake into a meal or a flour. Bean protein concentrates may be used as a means of protein fortification for infant formulas, soft drinks, puddings, cereal flours and meals, macaroni and noodle products, baked goods, and other high-carbohydrate foods.

• **Snack foods**—Crispy bean chips, flakes, or wafers are made from soaked, dehulled beans by (1) blending them with a thickening agent, salt, and vegetable oil, (2) pressure cooking the bean dough, and (3) frying slices of the cooked bean dough in deep fat.

Snap Beans. About 16% of this crop is marketed fresh, 64% is canned, and 20% is frozen.

Commercially canned beans are much safer than those canned at home because (1) this vegetable is susceptible to the growth of *Clostridium botulinum*, a species of bacteria which produces a highly lethal toxin; and (2) commercial canning procedures are more certain to destroy spoilage organisms.

Fig. B-34. Canned string beans being removed from water bath. (Courtesy, University of Maryland, College Park)

SELECTION AND PREPARATION. Dry beans are often used in a main dish, whereas snap beans are usually served as a vegetable. Hence, the selection and preparation of these items will be treated separately.

Dried Beans. Money may be saved by purchasing dried beans and cooking them, instead of using canned beans, because (1) each cup of dried beans yields about three times as much cooked beans, and (2) the prices per pound of the various types of uncooked dried beans are close to those of the corresponding canned products. Furthermore, cooking beans at home allows the choice of sauces and seasonings, which are very important in enhancing the appeal of bean dishes. Finally, the soaking and cooking times may be shortened considerably by utilizing the procedures that follow:

• **One-hour hot soak**—This method may be substituted for 15 hours of cold soaking. A pound of dry beans is placed in a saucepan with 8 cups (*1.9 liter*) of water, then heated to boiling and boiled for 2 minutes, then allowed to soak in the water for about an hour. The initial boiling softens the skins of the beans so that water penetrates them much more readily than when they are placed in cold water. However, the water should be soft; otherwise, the bean skins may be toughened to the extent that longer soaking is required. If only hard water is available, it must be softened by adding not more than $^1/_8$ to ¼ tsp (*0.625 to 1.25 ml*) of baking soda per cup (*240 ml*) of dry beans.

• **Pressure cooking**—Although beans cooked with water in a pressure cooker may produce enough foam to clog the vent, foaming may be reduced to a minimum by the addition of a tablespoon (*15 ml*) of oil or melted fat per cup (*240 ml*) of dried beans. Most beans may be cooked within 30 minutes in a pressure cooker, but it is advisable to consult the directions supplied with the cooker which is used.

The most nutritious dishes are those which are mixtures of beans with cheeses, grains, fish or seafood, meats, poultry, and/or other vegetables. Figs. B-35, B-36, and B-37 show some typical items.

Fig. B-35. Chili beans con carne. This dish, which is popular in the California-Arizona-Mexico border areas, is made from dried pink beans, tomato sauce, onion, melted fat or oil, garlic, salt, chili powder, and cumin seed or powder. (Courtesy, The California Dry Bean Advisory Board, Dinuba, Calif.)

Fig. B-36. Savory bean mix. A versatile filling for sandwiches which may be served hot or cold. It contains light red kidney beans, Cheddar cheese, ham, onion, catsup, butter or margarine, garlic salt, mustard, vinegar, and sugar. (Courtesy, The California Dry Bean Advisory Board, Dinuba, Calif.)

Fig. B-37. Beef-and-bean pot. This hearty main dish, which may be served with a tossed green salad and hard, crusty bread, contains short ribs of beef, small white beans, onions, celery, and garlic. (Courtesy, The California Dry Bean Advisory Board, Dinuba, Calif.)

CAUTION: Many foods contain components which are harmful at certain stages if they are not processed properly; and beans are no exception. *Uncooked, mature* common beans contain the following antinutritional and toxic factors: antivitamin E, inhibitors of trypsin, and red blood cell clumping agents (hemagglutinins). Soaking until the seeds begin to swell, followed by cooking (steaming, pressure cooking, or extrusion cooking) until the beans are tender, renders these factors harmless. Also, these components in bean flour are inactivated by pressure cooking prior to grinding. It is noteworthy, too, that various types of fermentation and sprouting, which have been utilized by Asian peoples for centuries, are an effective means of detoxifying beans.

Mature beans also contain gas-producing carbohydrates. But heat treatment is ineffective in lessening this annoyance of beans, caused by certain indigestible carbohydrates being acted upon by gas-producing bacteria in the lower intestine. However, (1) incubation of the raw dried beans in a mildly acid solution at 113°F (*45°C*) to 149°F (*65°C*) for 1 to 2 days, or (2) various types of fermentation and sprouting alleviate much of the gas problem by breaking down the indigestible carbohydrates. But those who suffer discomfort from excess gas may be well advised to eat only immature (green) beans.

Snap Beans. There are many varieties of both green and wax (yellow pod) snap beans. Some are flat, while others are oval or round. Most varieties of snap beans now grown for market are stringless when fully mature, but some develop stringiness as they pass the best stage for harvest. Overmature pods may become fibrous and tough whether stringless or not.

To be of the best quality, snap beans should be fresh in appearance, clean, firm but tender, crisp, free from scars, and reasonably well shaped. All beans in a selected lot should be at approximately the same stage of maturity so that they will cook uniformly. Firm, crisp, tender beans will snap readily when broken. Pods in which the seeds are very immature are the most desirable. Length is generally unimportant if the beans meet the requirements for quality. If seeds are half-grown or larger, pods are likely to be tough or fibrous. Stringiness can be detected by breaking the bean and gently separating the two parts. A dull, lifeless, or wilted appearance indicates that beans have been held too long after picking. Decay appears as a soft, watery, or moldy condition.

The older methods of cooking snap beans are no longer applicable because the modern varieties have been selected for tenderness. Hence, a pound of crosscut pieces will be thoroughly cooked after 13 to 15 minutes in boiling water. Some other ways of preparing snap beans are: (1) tossing in hot oil or fat with minced onion or garlic and/or bits of bacon, (2) boiling French style pieces (thin strips prepared by cutting the beans lengthwise or diagonally) for only 7 to 10 minutes, and (3) cooking them in casseroles, soups, and stews. (The beans should be added near the end of the cooking period when incorporated in dishes requiring much longer cooking.)

NUTRITIONAL VALUE. The nutritive composition of common beans is given in Food Composition Table F-36.

Some noteworthy observations regarding the nutritive value of common beans follow:

1. Cooked dried beans contain about three times as much calories and protein per cup (*240 ml*) as cooked snap beans, because the former have three times the solid content of the latter (snap beans contain more than 90% water). Hence, people trying to lose weight should be encouraged to eat plenty of snap beans, which are filling but not fattening.
2. One cup (*240 ml*) of cooked dried beans contains about the same amount of protein (14 g) as 2 oz (*56 g*) of cooked lean meat, but the beans contain from 50 to 100% more calories.
3. Dried beans contain about one-third as much calcium as phosphorus, but snap beans contain more calcium than phosphorus. It is suspected that an excessively low ratio of dietary calcium to phosphorus might limit the utilization of calcium by the body.
4. Both dried and snap beans provide ample amounts of iron and potassium per calorie.
5. Snap beans are a much better source of vitamins A and C than dried beans.
6. Commercial bean products which contain various meat ingredients are usually much higher in calories, but only slightly higher in protein than similar products without meat. The disproportionate increase in calories is due to the use of high-fat meats.

Protein Quantity and Quality. People are likely to substitute beans for meat in their diets when the latter becomes scarce or expensive. Therefore, certain facts concerning the quantity and quality of the protein in cooked dried beans follow:

1. Cooked beans furnish less protein per gram of food and per calorie than low and medium fat types of cheeses, eggs, fish, meats, milk, and poultry. For example, 3½ oz (*100 g*) of cooked common beans supplying 118 Calories (kcal) are required to provide the same amount of protein (about 7 g) as 1¾ oz (*50 g*) of cottage cheese (53 kcal) or one large egg (82 kcal). Also, the protein quality of the legumes is lower than that of the animal foods.

2. Cooked beans contain from 2 to 4 times the protein of most cooked cereals. For example, a cup of cooked dried beans (*approx. 185 g*) supplies from 14 to 15 g of protein, whereas a cup of cooked rice (*130 g to 205 g*) provides from 3 to 5 g of protein.

3. The protein in beans is moderately low in the sulfur-containing amino acids methionine and cystine. However, the importance of this deficiency has been exaggerated, because the tests of protein quality have been conducted with rats, which have a higher requirement for these amino acids than people. The feeding of a little extra protein by serving ample portions of beans usually ensures that adequate amounts of the deficient amino acids are provided.

4. Mixtures of beans and cereals have a protein quality which comes close to that of meat, milk, and other animal proteins. The highest protein quality is usually achieved in mixtures comprised of 50% bean protein and

50% cereal protein because the amino acid patterns of the two types of foods complement each other. Some examples of food combinations utilizing this principle are corn tortillas and refried beans, baked beans and brown bread, and rice and beans.

5. Beans may be used to upgrade the protein quantity and quality of diets based mainly on cereals and/or starchy foods such as bananas, cassava, and sweet potatoes. The latter foods are the mainstays of people in the developing countries in the tropics. For example, a protein-enriched flour or meal may contain about ⅓ bean flour or meal, and about ⅔ cereal flour made from corn, rice, or wheat. Highly starchy diets may be upgraded considerably by mixing bean flour with the starchy food. For example, impoverished people in the tropical areas of Latin America might well be encouraged to prepare mixtures of bean flour and bananas.

6. Animal protein foods may be extended by the use of legumes. Some approximate "equations" of protein value follow:

(a) 1 frankfurter + ½ cup (*120 ml*) cooked beans = 2 frankfurters.

(b) 10% bean flour + 5% skim milk powder + 85% cereal flour = 10% skim milk powder + 90% cereal flour.

Both examples show that the requirement for expensive animal protein may be halved by the judicious use of legume products. However, it should be noted that the substitution of legumes for animal foods may raise the caloric value of the diet.

(Also see BEAN[S], Table B-10 Beans of the World; LEGUMES, Table L-2 Legumes of the World; and VEGETABLE[S], Table V-6 Vegetables of the World.)

BEAN FLOUR

Lima beans and soybeans are the two most common beans from which flour is made, but flour can be made from any bean.

(Also see FLOURS.)

BEATEN BISCUIT OR SOUTHERN BEATEN BISCUIT

This is an unleavened bread made with flour, shortening, water and milk. The stiff dough is beaten with a rolling pin and folded over many times, then rolled out and cut with a biscuit cutter. When baked, it should be light, even textured, and crack at the edges like crackers.

BEECHWOOD SUGAR (XYLOSE)

Wood sugar is the pentose—five-carbon—sugar which is widely distributed in plant material, especially wood.

(Also see CARBOHYDRATE[S]; and XYLOSE.)

BEEF AND VEAL

Content *Page*

Qualities in Beef Desired by Consumers........... 181
Federal Grades of Beef and Veal................. 181
Beef and Veal Cuts and How to Cook Them....... 181
U.S. Beef Production............................ 184
Per Capita Beef and Veal Consumption........... 184
Beef as a Food.................................. 184

Fig. B-38. King Charles II was so impressed with a platter of beef that was served at one of his feasts, that—according to an old English legend—he ceremoniously arose, touched his sword to the steaming platter and said, "A noble joint it shall have a title. Loin, I dub thee Knight—henceforth thou shall be Sir Loin." (Courtesy, Picture Post Library, London, England)

Beef is the flesh of adult cattle, whereas veal is the flesh of calves.

Beef has been a favorite food of man since Biblical times. The Bible tells of the fatted calf prepared to welcome the prodigal son (St. Luke 15:23).

QUALITIES IN BEEF DESIRED BY CONSUMERS.

Consumers desire the following qualities in beef:

1. **Palatability.** First and foremost, people eat beef because they like it. Palatability is influenced by the tenderness, juiciness, and flavor of the fat and lean.
2. **Attractiveness.** The general attractiveness is an important factor in selling beef to the housewife. The color of the lean, the degree of fatness, and the marbling are leading factors in determining buyer appeal. Most consumers prefer a white fat and a light or medium red color in the lean.
3. **Moderate amount of fat.** Consumers discriminate against too much fat, especially when it must be trimmed heavily.
4. **Tenderness.** Consumers want fine-grained, tender beef in contrast to coarse-grained, less tender meat.
5. **Small cuts.** Most purchasers prefer to buy cuts of beef that are of a proper size to meet the needs of their respective families. Because the American family has decreased in size, this has meant smaller cuts. In turn, this has had a profound influence on the type of animals and on market age and weight.
6. **Repeatability.** The housewife wants a cut of beef just like the one that she purchased last time, which calls for repeatability.
7. **Ease of preparation.** In general, the housewife prefers to select those cuts of beef that will give her the greatest amount of leisure time. Steaks of beef can be prepared with greater ease and in less time than can roasts or stews. Hamburger and sausage are also easy to prepare.

FEDERAL GRADES OF BEEF AND VEAL.

The quality and yield grades of beef and veal are summarized in Table B-11.

TABLE B-11
FEDERAL QUALITY AND YIELD GRADES OF BEEF AND VEAL[1]

Beef		Calf and Veal
Quality Grades	Cutability (Yield) Grades	(Quality Grades Only)
1. Prime[2]	1. Yield Grade 1	1. Prime
2. Choice	2. Yield Grade 2	2. Choice
3. Select	3. Yield Grade 3	3. Select
4. Standard	4. Yield Grade 4	4. Standard
5. Commercial	5. Yield Grade 5	5. Utility
6. Utility		
7. Cutter		
8. Canner		

[1]In rolling meat, the letters "USDA" are included in a shield with each Federal Grade name. This is important, as only government-graded meat can be so marked. For convenience, however, the letters "USDA" are not used in this table or in the discussion which follows.

[2]Cow beef is not eligible for the prime grade. the quality grade designations for bullock carcasses are Prime, Choice, Select, Standard, and Utility.

Note that there are both quality grades and yield grades. Pertinent facts about each type of grade follow:

1. **Quality grade.** Quality refers to the palatability-indicating characteristics of the lean and is evaluated by considering the marbling (flecks of fat within the lean) and firmness of the lean as observed in a cut surface in relation to the apparent maturity of the animal from which the carcass was produced. The maturity of the carcass is determined by evaluating the size, shape, and ossification of the bones and cartilages—especially the split chine bones—and the color and texture of the lean flesh. Superior quality implies firm, well-marbled lean that is fine in texture and has a light red, youthful color.

2. **Yield grade.** It is recognized that variations in conformation (shape) due to differences in muscling do affect yields of lean—and carcass value. This is reflected by the yield grades.

The yield grade of a beef carcass is determined by considering four characteristics: (1) the amount of external fat; (2) the amount of kidney, pelvic, and heart fat; (3) the area of the ribeye muscles; and (4) the carcass weight.

There are five USDA yield grades numbered 1 through 5. Yield Grade 1 carcasses have the highest yields of retail cuts; yield Grade 5 the lowest. A carcass which is typical of its yield grade would be expected to yield about 4.6% more in retail cuts than the next lower yield grade, when USDA cutting and trimming methods are followed.

Until 1989, meat packers could choose to grade or not to grade beef. But, if they graded, they were *required* to grade for both yield and quality. In 1989, the law was changed, separating quality and yield grades of beef; and allowing packers to choose whether beef carcasses are graded for quality, for yield, or for both quality and yield. But packers can continue to choose between *to grade or not to grade.*

BEEF AND VEAL CUTS AND HOW TO COOK THEM.

In order to buy and/or process beef and veal wisely, and to make the best use of each part of the carcass, the consumer should be familiar with the types of cuts and how each should be processed.

Every grade and cut of meat can be made tender and palatable provided it is cooked by the proper method. Also, it is important that meat be cooked at the proper temperature.

Fig. B-39 shows the retail cuts of beef and gives the recommended method(s) of cooking each. Fig. B-40 presents similar information for veal.

(Also see MEAT, section headed "Meat Cooking.")

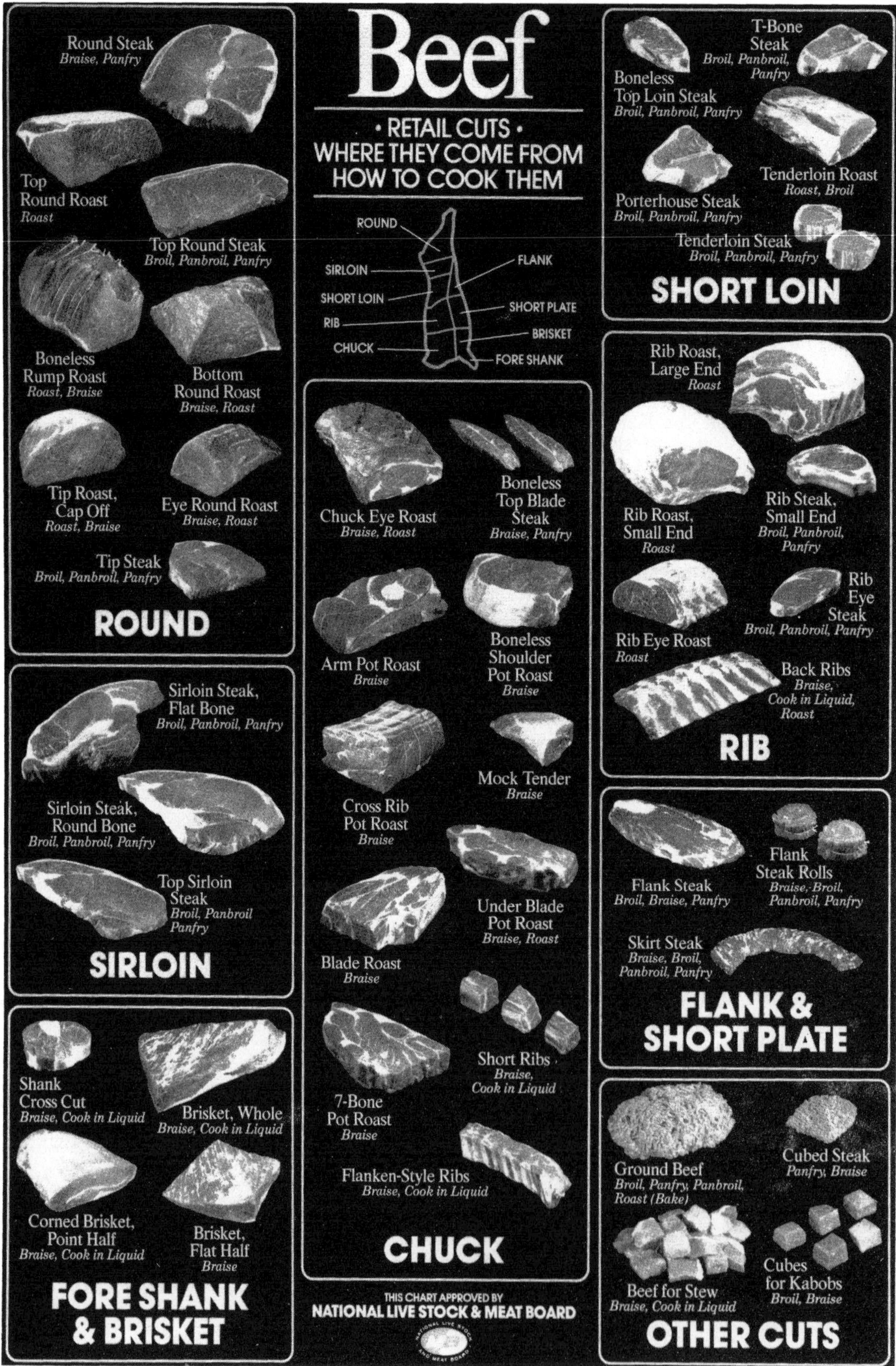

Fig. B-39. Beef. The retail cuts of beef; where they come from, and how to cook them. (Courtesy, National Live Stock and Meat Board, Chicago, Ill.)

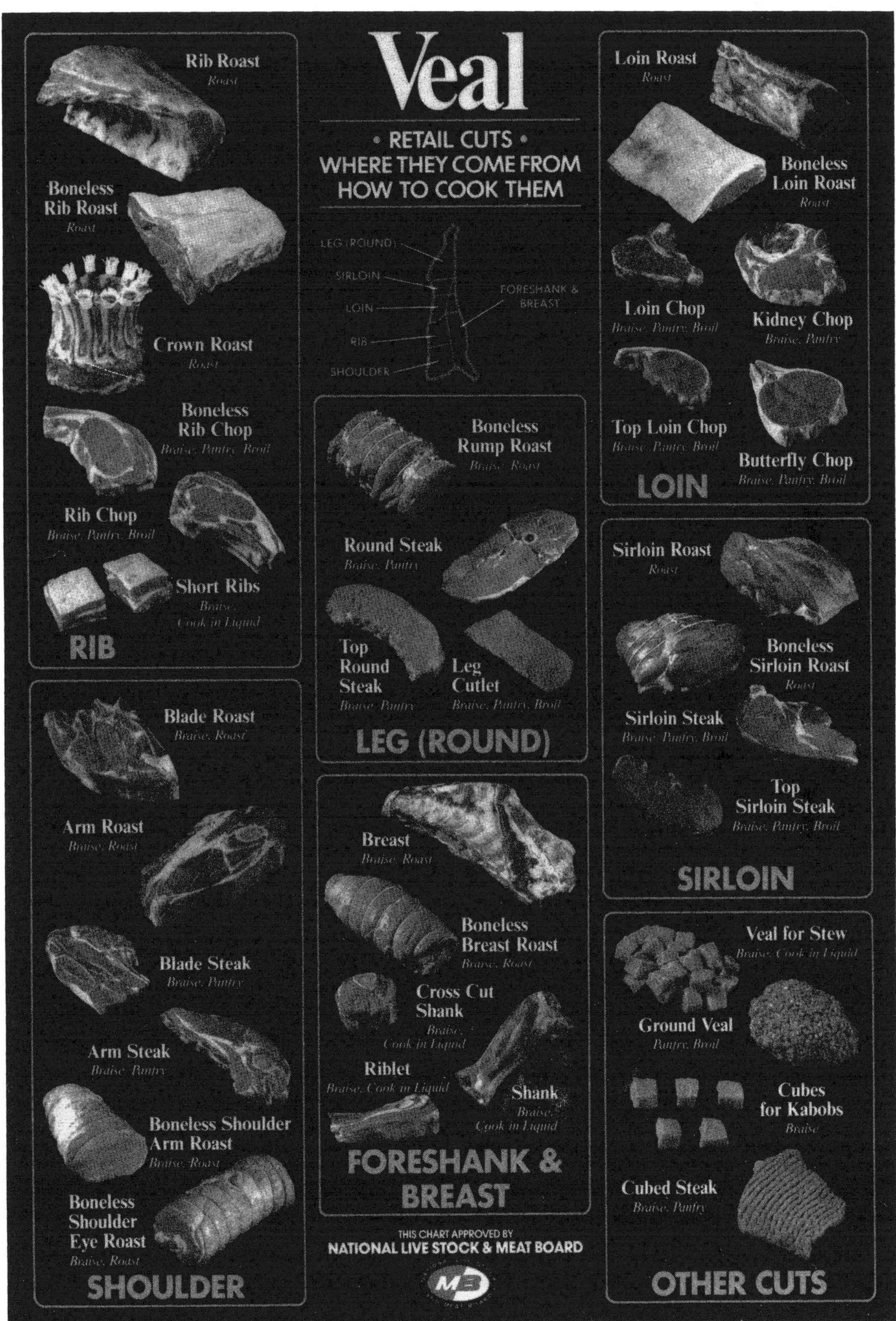

Fig. B-40. Veal. The retail cuts of veal; where they come from, and how to cook them. (Courtesy, National Live Stock and Meat Board, Chicago, Ill.)

U.S. BEEF PRODUCTION. The dominant position of beef production in the United States is attested to by the following statistics:

1. It has 4.9% of the world's human population.
2. It has 8% of the world's cattle population.
3. It produces 22.2 % of the world's beef and veal.
4. It consumes 16.6% of the world's beef and veal.[5]

Thus, 8% of the world's cattle population produces 22.2% of the world's beef. This points up the tremendous efficiency of the U.S. cattle industry.

In 1989, 23.4 billion pounds (*10.7 billion kg*) of beef and veal were produced in the United States; and on January 1, 1991, there were 99.4 million cattle and calves in the United States.

The present and future importance of beef cattle in the agriculture of the United States rests chiefly on their ability to convert coarse forage, grass, and by-product feeds, along with a minimum of grain, into palatable and nutritious food for human consumption.

PER CAPITA BEEF AND VEAL CONSUMPTION. In 1990, the U.S. per capita consumption of beef (carcass basis) was 64 lb (*29 kg*), and the per capita consumption of veal was 0.9 lb (*0.4 kg*).

BEEF AS A FOOD.

Fig. B-41. Beef blade roast. (Courtesy, National Live Stock & Meat Board, Chicago, Ill.)

Beef is a favorite American food; it's good eating, and it's good nutritionally, too.

Beef provides high-quality protein; it contains all the amino acids, and the proportion of the amino acids almost exactly parallels that in human protein. Beef is a good source of several minerals, but it is an especially rich source of iron, phosphorus, copper, and zinc. Beef is also a good source of vitamin A, and of the B vitamins—vitamin B-12, vitamin B-6, biotin, niacin, pantothenic acid, and thiamin. The calories in meat are dependent largely upon the amount of fat it contains.

Special beef products for infants provide important nutrients for growth, development, and resistance to disease and infections early in life. Lean beef is especially good for those who are ill or have undergone surgery; in restoring blood losses, repairing body tissues, and healing.

(Also see MEAT[S].)

BEEF BREAD

The pancreas of a mature animal used for food.

BEEF EXTRACT (MEAT EXTRACT)

An extract of the soluble constituents of beef, made by cooking beef.

BEEF TAPEWORM (MEASLES)

A parasitic infestation in cattle and in the human intestine which may be present without symptoms, or it may be accompanied by anemia, nausea, and diarrhea alternating with constipation. This problem is confined largely to the four states which border Mexico—Texas, New Mexico, Arizona, and California. Preventative measures consist of (1) the use of proper toilet facilities by the help on cattle farms; (2) protection of cattle, feeds,

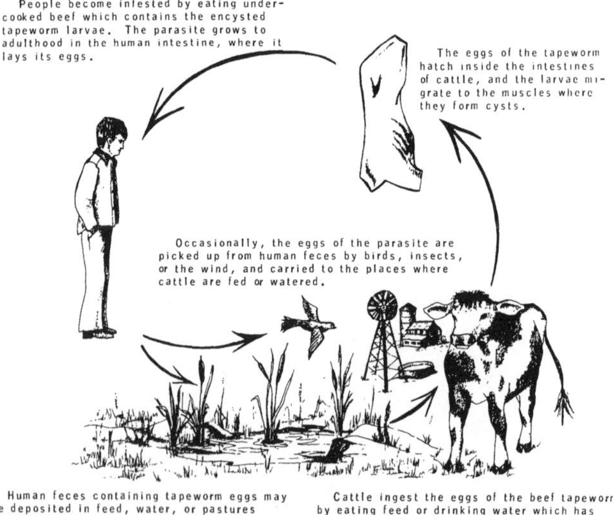

Fig. B-42. The cycle by which the beef tapeworm passes between cattle and man.

[5]Computations made by the authors.

pastures, ponds, and watering troughs from contamination by human feces or sewage; and (3) thorough cooking of beef in areas where infestation is likely to occur.

(Also see DISEASES; and PARASITE INFECTIONS.)

BEEF TEA

A beverage made by extracting finely cut lean beef with hot water. Also, there are commercial beef extracts and cubes that can be used for beef tea. Beef tea, bouillon, and consomme, are different designations for the same beverage.

BEERS AND BREWING

Contents *Page*

History of Brewing 185
Basic Principles of Brewing Beer and Similar Beverages 187
Malting 188
Mashing 188
Hopping 188
Fermentation 188
Finishing 188
By-products of Malting and Brewing 188
Types of Beers and Related Drinks 189
Unusual Beers from Various Parts of the World ... 190
Nutritive Value of Ordinary Beer 192
Mental and Physical Effects of Beer Drinking 193
Beneficial Effects 193
Harmful Effects 194
Planning Meals and Snacks to Accompany Beer 194

The word "beer" is derived from *baere*, the German word for barley, which is the major grain used in the brewing of beer. Brewers make beers and similar alcoholic beverages by fermenting sugary extracts of cereal grains. Usually, the alcoholic content of these beverages ranges from 2 to 8%. Sometimes, other drinks made from such diverse items as fruits, roots and tubers, and herbs are called beers. However, this article is limited to beers made from grains.

Feelings about the drinking of beer range from wholehearted enthusiasm to stern disapproval because of much speculation over the mental and physical effects of the alcoholic content. For example, the British nutritionist Platt stated that the brewing of beer from cereal grains was a type of biological ennoblement.[6] *(Biological ennoblement refers to the improvements in the nutritive values of foods which may be produced by sprouting, fermentation, and similar biological processes.)* On the other hand, the poet Goethe feared that excessive beer drinking and smoking among Germans might lead to an intellectual decline within a few generations.[7] Therefore, it would be wise to postpone judgment on the merits and demerits of beer until after the relevant facts have been considered.

[6]Platt, B.S., "Biological Ennoblement: Improvement of the Nutritive Value of Foods and Dietary Regimens by Biological Agents," *Food Technology, Vol. 18, 1964, pp. 68-76.*

[7]Weiner, M. A., *The Taster's Guide to Beer*, Macmillan Publishing Company, Inc., New York, N.Y., 1977, p. 58.

Fig. B-43. Beer wagon, with barreled beer, drawn by a magnificent eight-horse hitch, reminiscent of days of old. (Courtesy, Anheuser-Busch, Inc., St. Louis, Mo.)

HISTORY OF BREWING. The brewing of cereal beers is probably about as old as grain farming, as evidenced by fermented grains found in clay pots unearthed at the sites of ancient agricultural settlements. Also, it seems likely that various types of brewing were developed independently by native peoples in many places around the world because fermentation may occur wherever there are (1) food-fermenting microorganisms in the air, (2) fermentable foods exposed to the microorganisms, or (3) temperatures warm enough to encourage fermentation. Hence, a wide variety of substances and methods of brewing were used to make primitive beers.

Many historians believe that the first grain beers were brewed before 5000 B.C. by the ancient Chaldeans (the people who settled at the southern ends of the Tigris and Euphrates rivers in what is now Iraq) and the Egyptians. These first beers, made from barley—a grain uniquely suited for brewing, were the forerunners of those now produced in all of the developed countries of the West. (Sprouted grains of barley surpass all other grains in their production of enzymes that convert starches to fermentable sugars.) Gradually, the making of beer grew into a commercial enterprise. It is noteworthy that the Babylonian king Hammurabi (19th century B.C.) passed a law regulating the price of beer so that it might be available to all. By 1300 B.C., the Egyptians exported dried, malted barley cakes to other regions. Beer could be made by merely soaking the barley cakes in water and allowing fermentation to occur.

It is noteworthy that the Greeks and Romans, who depended mainly upon wine to lighten their spirits, initially considered Egyptian beer to be an inferior beverage and ridiculed the people who drank it. They were unaware that the Egyptian laborers who built the pyramids subsisted on a daily ration of three or four loaves of bread, some onions, and two jugs of beer. Nevertheless, they adopted the Egyptians' brewing techniques and promoted them throughout the lands they controlled. However, many of the so-called "barbarian tribes" of western and northern Europe brewed native beers before their conquest by Greece or Rome. In such cases, the influence of the Greeks or Romans brought about greater uniformity in raw materials and brewing procedures, and higher quality in the beers which resulted.

The church came to play an important role in the production of beer, when certain orders of monks improved the older methods of brewing, and invented new ones. Also, monks built the first commercial breweries in many areas of northern Europe during the Middle Ages.

Throughout history, local rulers often established laws governing the quality of beer, because bad beer was considered to be both a fraud and a hazard to health. Germany has long had stringent regulations governing brewing, enforced in the 15th and 16th centuries by such extreme measures as punishment by beatings, banishment, or death for those who made or sold bad beer. Even today, the Germans restrict the name of beer to beverages made from barley, hops, yeast, and water. Regulation of the quality of beer often made this beverage safer than many of the local sources of drinking water. Hence, many people drank beer instead of water.

Meanwhile, the Indians of the New World had developed their own methods of brewing beer from corn, as Columbus learned when he discovered America in 1492. First, they chewed kernels of corn to produce a pulpy mass which was well mixed with their saliva. Then, they took the mass from their mouth and placed it in a large vessel to ferment. Fermentation occurred because the enzyme diastase in the saliva converted some of the corn starch to sugar which was acted upon by the wild yeasts from the air. The Indians of Latin America still use this method to make beers from starchy plants like cassava.

Beer influenced early American history, since the fondness of the early colonists for this beverage was responsible for some of the circumstances under which the United States was founded. First of all, the pilgrims had to land at Plymouth Rock rather than somewhere further south because the supply of beer on the Mayflower had run out. Taverns were soon established which later became the bane of the British, as they were meeting places of the revolutionaries. Also, many of the brewers and innkeepers were leaders of the Revolution. Finally, Washington gave his famous farewell address at France's Tavern in New York.

The beer made in colonial America was actually an unhopped ale, because hops did not grow well in New England. Also, many of the colonial households and taverns made their own brews, as commercial breweries were few and far between. Later, the settlers of Pennsylvania found that both barley and hops grew well there. Hence, the Quaker leader Penn established a large brewery at Pennsburg in that state. Lager beers were introduced by the German settlers who started to migrate to America in large numbers around 1840. Soon, the production of beer moved westward with the Germans, and breweries were set up in almost every sizable town. By 1880, there were a total of 2,272 breweries in the United States.[8]

Even children drank beer with their meals until the westward migration began in the 1840s. About that time, many lay and religious leaders became concerned over widespread alcoholism and the Temperance movement began. Shortly thereafter, total abstinence from alcoholic beverages became a tenet of certain Christian denominations. The Temperance movement culminated in Prohibition, which lasted from 1920 to 1933, when it was repealed. Many breweries closed and never reopened. Prohibition and other measures which provided stringent legal restriction of beer and other alcoholic beverages stemmed partly from the belief that these beverages were essentially alcohol diluted with water, and that they had only harmful effects. It required the development of the science of nutrition to correct this overly simplified point of view.

Many notable discoveries in nutrition were made at the end of the 19th century and the early part of the 20th century. The American public health scientist Goldberger strived throughout the 1920s and early 1930s to convince people that the devastating disease pellagra was due to one or more nutritional deficiencies rather than to an infectious agent or to genetic inferiority. As a result of his work, public health nurses distributed brewers' yeast (a highly nutritious substance produced during the brewing of beer) throughout the southeastern United States. However, it was not often realized that beer itself provided ample amounts of the antipellagra vitamin niacin.

The latest chapter in the story of how brewing has contributed to human welfare began in Germany right after World War II when medical scientists studied the nutritional effects of brewers' yeast. They found that experimental diets based on the European strains of yeast produced liver disorders in laboratory rats. Then, in 1951 the German medical scientist Schwarz came to the United States to work as a visiting scientist at the National Institutes of Health, where he and his coworker Foltz discovered that the American type of brewers' yeast prevented the liver disorder observed in Germany. This work led to their discovery in 1957 that the American yeast contained a complex compound of the element selenium, which they showed to be an essential nutrient. (Perhaps, the nutritional superiority of the American brewers' yeast is due to the growing of grains used for brewing on soils which have a moderately high selenium content.) Two years later, Schwarz and Mertz (Mertz was a German medical scientist who came to the United States) discovered that chromium was an essential nutrient and that the most biologically active form of this element was the glucose tolerance factor from brewers' yeast.

It is ironic that brewers' yeast, which is the most nutritious by-product of brewing, has long been fed to farm animals, and that people benefit indirectly when they eat animal products. It makes one suspect that people might derive more direct nutritional benefits from beers if they were to be made with less filtration and clarification, so that they would contain more yeast particles. The need for the refinement of beer was prompted by the replacement of the opaque beer mug with the transparent beer glass. The beer glass, which allowed drinkers to see what they were consuming, led to a preference for a clearer beer.

[8] *Ibid.*, p. 184.

BASIC PRINCIPLES OF BREWING AND SIMILAR BEVERAGES. The art of brewing has evolved over the centuries into a highly scientific profession, which provides us with a variety of beers and similar products. Brewers also help to put meat, milk, and eggs on the table because the by-products of brewing are fed to livestock. Therefore, the major processes which constitute brewing are noteworthy. Fig. B-44 outlines the production of beer and the various by-products.

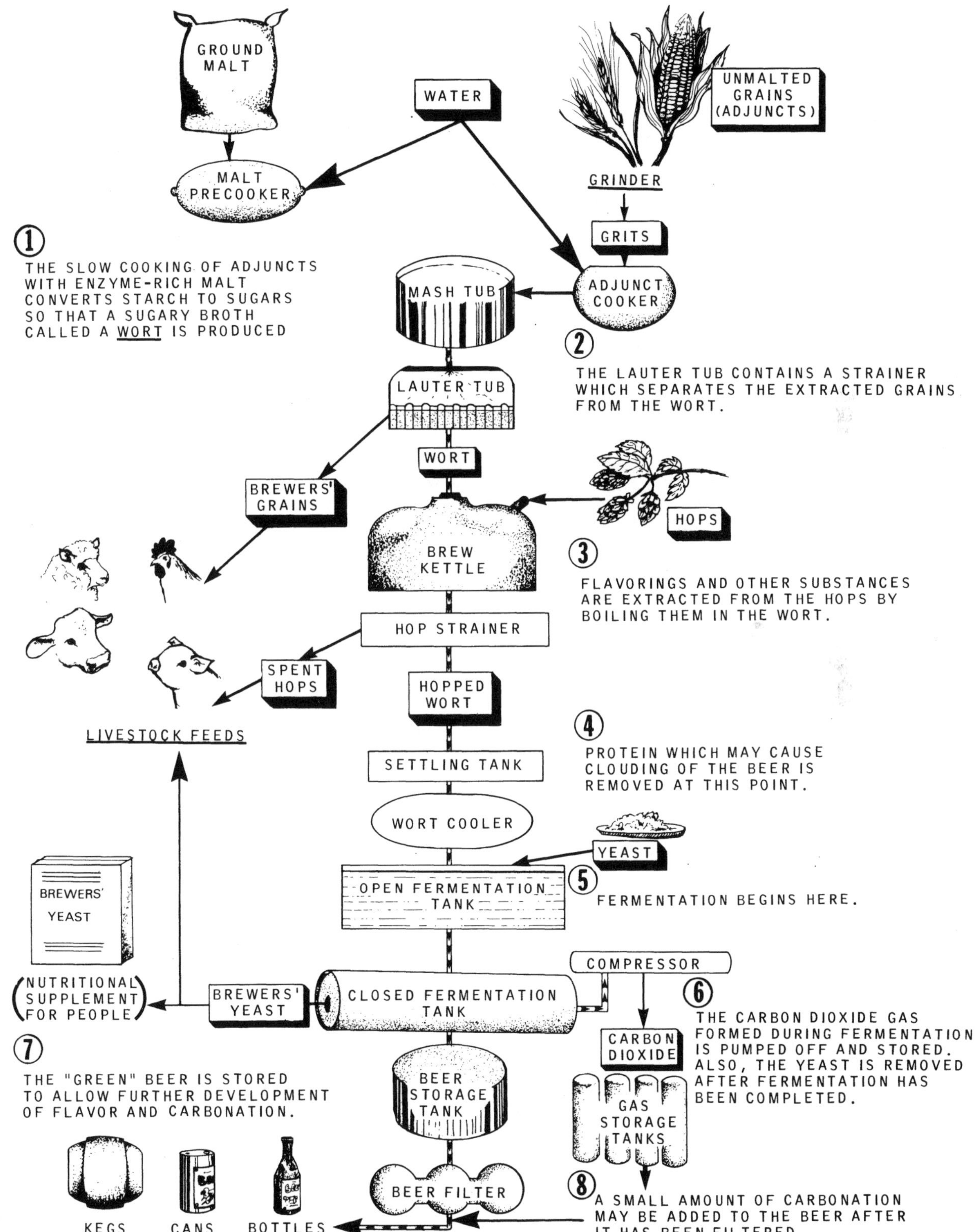

GROUND MALT
WATER
UNMALTED GRAINS (ADJUNCTS)
MALT PRECOOKER
GRINDER
GRITS
① THE SLOW COOKING OF ADJUNCTS WITH ENZYME-RICH MALT CONVERTS STARCH TO SUGARS SO THAT A SUGARY BROTH CALLED A WORT IS PRODUCED
MASH TUB
ADJUNCT COOKER
② THE LAUTER TUB CONTAINS A STRAINER WHICH SEPARATES THE EXTRACTED GRAINS FROM THE WORT.
LAUTER TUB
WORT
BREWERS' GRAINS
HOPS
BREW KETTLE
③ FLAVORINGS AND OTHER SUBSTANCES ARE EXTRACTED FROM THE HOPS BY BOILING THEM IN THE WORT.
HOP STRAINER
SPENT HOPS
HOPPED WORT
LIVESTOCK FEEDS
④ PROTEIN WHICH MAY CAUSE CLOUDING OF THE BEER IS REMOVED AT THIS POINT.
SETTLING TANK
WORT COOLER
YEAST
OPEN FERMENTATION TANK
⑤ FERMENTATION BEGINS HERE.
BREWERS' YEAST
COMPRESSOR
NUTRITIONAL SUPPLEMENT FOR PEOPLE
BREWERS' YEAST
CLOSED FERMENTATION TANK
⑥ THE CARBON DIOXIDE GAS FORMED DURING FERMENTATION IS PUMPED OFF AND STORED. ALSO, THE YEAST IS REMOVED AFTER FERMENTATION HAS BEEN COMPLETED.
CARBON DIOXIDE
⑦ THE "GREEN" BEER IS STORED TO ALLOW FURTHER DEVELOPMENT OF FLAVOR AND CARBONATION.
BEER STORAGE TANK
GAS STORAGE TANKS
BEER FILTER
⑧ A SMALL AMOUNT OF CARBONATION MAY BE ADDED TO THE BEER AFTER IT HAS BEEN FILTERED.
KEGS
CANS
BOTTLES

Fig. B-44. The major operations in the brewing of beer.

Details of the major brewing operations follow.

Malting. Yeast fermentation produces alcohol from sugars, but not from starch. Hence, the most common method of making beer involves the conversion of the starch in cereal grains to sugars by the enzymes produced in the operation called malting.

During malting, whole grain barley, or a similar cereal, is allowed to sprout (germinate) under controlled conditions so that certain enzymes in the grain are activated. The sprouting is stopped at the desired point by careful heating which allows the enzyme activity to be retained. Then, the sprouts are removed from the malted grain by passage through revolving sets of wire screens. Usually, the malt is ground before it is used by brewers.

(Also see MALT.)

Mashing. This process is used to prepare the sugary broth, or wort, which is fermented by yeast. Certain brewers around the world rely solely upon malted barley to yield the sugars in mashing, whereas many others use malt plus other unmalted grains called *adjuncts*, which supply starch and cost less than malt.

Mashing begins with the precooking of the malt and adjuncts in separate vessels to form watery porridges. Then, the cooked malt is combined with the cooked adjuncts in a mash tub, where the mixture is heated slowly so that the enzymes in the mash convert almost all of the starch to sugars. The sugary extract which results is called a *wort*. After mashing has been completed, the mixture of spent grains (brewers' grains) and wort is piped to a lauter tub where the grains settle to the bottom. The hot wort is then filtered through the brewers' grains into the brew kettle. Hot water, used to rinse the grains, is also added to the brew kettle. The rinsed brewers' grains are used as ingredients in livestock feeds.

Hopping. Hops (dried flowers of the hop vine) are added to the wort in the brew kettle to impart to beer its traditional odor and bitter taste. The mixture is then boiled to extract the aromatic components of the hops, and to sterilize and concentrate the wort prior to fermentation. After boiling, the hops are removed by passing the mixture through a strainer. The hopped wort is held in a tank where the excess protein settles out. Tannins from the hops react with the protein so that it comes out of solution. Otherwise, it might cause clouding of the beer. The spent hops are saved for use as feed ingredients. After the protein has been removed, the wort is piped first to a cooler, then to an open fermentation tank.

Fig. B-45. Hops are grown for the making of beer. (Courtesy, Union Pacific Railroad Company, Omaha, Neb.)

Fermentation. This process is started by the addition of yeast to the hopped wort in the open fermentation tank. Once fermentation is underway, the mixture is transferred to a closed fermentation tank, where it is kept for a week or more. The carbon dioxide produced in the fermentation is pumped off and stored in pressurized tanks. Next, the newly formed beer is pumped to a storage tank where it is aged for weeks or months, depending upon the brewery's procedure. The yeast residue, which is known as brewers' yeast, is used as an ingredient of feeds or as a nutritional supplement for people.

Finishing. Some unfermented wort may be added to the beer in storage to produce a mild fermentation which gives the beer a tangy quality. This process is called *krausening*. When the storage period is over, the beer is filtered to remove all residue, and a small amount of carbonation is added.

By-Products of Malting and Brewing. Each barrel of beer (31 gal [*117.8 liters*]) which is produced in the United States is made from the following raw materials: 30 lb (*13.5 kg*) of barley, 10 lb (*4.5 kg*) of corn, 5 lb (*2.25 kg*) of rice, ½ lb (*225 g*) of hops, and ¼ lb (*113 g*) of yeast.[9] However, beer contains from 85 to 90% water, so it may be readily seen that much of the solids in the grains used as raw materials end up in the by-products which remain after the fermentable matter has been extracted. Each of the by-products helps to lower the cost of producing valuable, high-protein animal foods for people, because livestock thrive on such feed ingredients. Therefore, brief descriptions of the by-products follow:

• **Malt sprouts and hulls**—These materials are obtained when newly malted barley is cleaned by passage between revolving wire screens. They are high in fiber, so they are best utilized in feeds for cattle and sheep.

• **Brewers' grains**—This is the grain residue which remains after the malted barley and the adjuncts have been extracted to make a wort. Hence, it is higher in protein and fiber, but lower in energy (due to extraction of starch) than the raw materials. The protein in brewers' grains is utilized more efficiently by cattle and sheep than the protein from soybeans.[10] Also, this feed ingredient contains some unidentified factors which stimulate growth in farm animals.

---

[9]Couch, J. R., "Review of Nutrition Papers from Brewers Feed Conference," *Feedstuffs*, December 5, 1977, p. 8.

[10]*Ibid.*

• **Spent hops**—Hops have long been thought to contain substances that are beneficial to human health. For example, the drinking of hopped beer is thought to help the digestive system cope with a heavy meal. Similarly, the addition of spent hops (those extracted in the brewing of a hopped wort) appear to stimulate the appetites of cattle to consume heavy rations when they might otherwise eat less.[11] It is desirable for cattle to consume heavy rations during (1) fattening for market, or (2) milk production.

• **Brewers' yeast**—The ¼ lb (*112 g*) of yeast used to brew each barrel of beer grows to 1 lb (*454 g*) during fermentation.[12] Hence, there is a net gain of ¾ lb (*338 g*) of yeast per barrel of beer produced, for a total of 123 million lb (*55 mil. kg*) of yeast gained in the U.S. annual beer production of 164 million barrels. The surplus brewers' yeast, which is high in both protein and the vitamin B complex, is used mainly as a nutritional supplement for both man and animals. Perhaps, this yeast also contains some unidentified, nutritionally beneficial factors. For example, research conducted at Rockefeller University in the 1940s showed that it protected laboratory animals against the carcinogenic effects of butter yellow, a synthetic dye which is no longer allowed in foods.

TYPES OF BEERS AND RELATED DRINKS.

The name beer has long been given to a wide variety of undistilled, alcoholic beverages made from fermented extracts of mashed grains. For example, the name itself was derived from the German word for barley, but beers have also been brewed from (1) wheat by the Germans, (2) rice by the Chinese and the Japanese, (3) sorghums and millets by African tribes, (4) rye by the Russians, and (5) corn by the American Indians. Even when beer is made mainly from malted barley plus hops, yeast, and water,the flavor, aroma, and color of the product depend upon other factors such as (1) the type of malt used (various roasting treatments are used after malting to produce the various types); (2) amounts and types of adjuncts, if any; (3) the temperatures and the lengths of time used for mashing, hopping, fermentation, and storage; and (4) the strains of yeast used for fermentation (the identities of many of these strains are well-kept company secrets). Descriptions of some of the most common types of beers and related beverages follow:

• **Ale**—The pilgrims brought ale, rather than beer, to America. Both beverages are made from malted barley, adjuncts, hops, yeast, and water, but ale is stronger and contains more alcohol than beer. Another difference is that production of ale utilizes strains of yeast which rise to the top of the fermentation tank. Hence, it is said to be a "top-fermented" beverage.

• **Beer**—What is commonly called beer in the United States is known as lager in Europe. The process for brewing lager beers was brought to America by the German immigrants who arrived in the 1840s. These beers are made with yeasts which drop to the bottom of the fermentation tank; so, they are said to be "bottom-fermented." Lagers are lighter, lower in alcohol, and contain less hop extract than ales.

• **Bitter**—This British draft beer is one of the driest and most heavily hopped. Generally, the British use from 2 to 3 times more hops to brew their beers than the Americans. When bottled, the beverage is called light ale or pale ale.

• **Bock beer**—At one time, this beer was sold in the spring, when, for unknown reasons, it was associated with the symbol of the goat. Darker, heavier, and sweeter than most other beers, it has long been thought to be more nutritious. The dark color comes from the use of a highly toasted, dark malt.

• **Lager**—The name of this type of beer is derived from the German word *lagern*, which means "to store." It seems that when monks operated many of the breweries in Germany they stored their newly made beer in cool caves during the summer to keep it from spoiling. A lager is a light-colored, mild-tasting beer which was made so popular by the German immigrant brewers that it now claims most of the market for beer in the United States.

• **Light ale or pale ale**—When made in Great Britain, this product is one of the driest and most heavily hopped. The British give it the name "bitter" when it is dispensed as a draft beer.

• **Malt liquor**—American labeling regulations provide that brewed beverages which contain more than 5% alcohol cannot be called beers, but must be designated as malt liquors, ales, porters, or stouts. Usually, malt liquors are higher alcoholic versions of the lager types of beer.

• **Pilsner**—Once, this name applied only to a highly regarded lager beer from Pilsen, Czechoslovakia. Now, it merely signifies that a certain product may bear a similarity to the premium beer from Pilsen.

• **Porter**—This alelike beverage got its name because it was favored by London market porters. It was designed by an enterprising brewer to embody the desirable qualities of several different brews, because customers often asked bartenders to mix them together in a single glass. The name porterhouse steak was later given to the special cut of beef which was served in establishments which dispensed porter.

• **Shandy**—This British drink is a mixture of beer and ginger beer, or a mixture of beer and lemonade—called lemon shandy.

• **Stout**—Some time ago, a demand arose in Great Britain for an "extrastout" porter; hence, this very dark and slightly bitter ale came to be brewed from a dark malt.

---

[11]Ensminger, M. E., J. E. Oldfield, and W. W. Heinemann, *Feeds & Nutrition*, Second Edition, The Ensminger Publishing Company, Clovis, Calif., 1990, p. 462.

[12]Couch, J. R., "Review of Nutrition Papers from Brewers Feed Conference," *Feedstuffs*, December 5, 1977, p. 8.

A burnt taste may be evident when the brewer has used carmelized sugar.

• **Weisse**—The Germans make this cloudy beer from wheat, barley malt, hops, yeast, and water. It is fermented in the bottle, which makes the beer cloudy because some of the particles of yeast remain suspended. The presence of yeast in the beer no doubt raises its nutritional value above other beers, particularly with respect to the vitamin B complex.

Unusual Beers From Various Parts of the World. A study of the primitive beers of tribal Africa prompted the late, renowned British nutritionist Platt to describe the Africans' fermentation of cereal grains as a type of "biological ennoblement."[13] He found that certain African tribes, who consumed about 40% of their corn supply in the form of a crude beer, received substantially greater amounts of the vitamin B complex than they would have received had they consumed all of their corn in an unfermented form. The vitamins synthesized during yeast fermentation apparently made the difference.

On the other hand, certain native beers may be hazardous to health because primitive peoples often use unhygienic brewing practices, and their raw materials may be of poor quality, or even hazardous. Hence, there might be unsuspected toxic contamination of some of these beverages. For example, the Bantu peoples of Africa, who prepare and drink large amounts of homemade grain beers, have high rates of liver cancer and iron poisoning. Among their potentially dangerous brewing practices are (1) the conversion of the grain starch to fermentable sugars by the fungus *Aspergillus oryzae*, which itself is quite safe, except that this beneficial species may sometimes be accompanied by harmful species of *Aspergillus* such as those which produce highly carcinogenic aflatoxins; (2) brewing the beers in iron pots from which the acid beer dissolves considerable iron; and (3) the occasional addition of old lead batteries to the brewing pot, in the mistaken belief that it speeds up fermentation.[14]

The examples just cited show that it is unwise to judge the merits of native beers without a careful consideration of the brewing practices, ingredients, and the quality of the product. It may often be necessary to conduct laboratory studies on the beverages before such judgments can be made. Nevertheless, it is of interest to note how some of these beers are prepared in various parts of the world. Fig. B-46 shows a primitive means by which certain African tribes prepare their grains for brewing.

Descriptions of selected brewing practices follow:

• **African sprouted grain beers**—Although the brewing procedures used by the various tribal peoples of Africa resemble those used elsewhere to make beer from malted barley, they differ in that much of the conversion of the grain starches to sugars is due to the action of enzymes from wild strains of the fungus *Aspergillus oryzae*, in addition to the conversion which results from sprouting alone. Furthermore, there are many different native beers made by various tribes in Africa. Hence, only a few of the more common brewing practices can be described here.

Fig. B-46. African woman grinding grain to be used for brewing beer.

In the mid-continent regions of Africa, corn, millet, or sorghum are sprouted alone or in a mixture by first soaking in water; then, the grain is covered with leaves or soil for a day or so. The sprouted grain may be (1) dried and ground into a flour, (2) cooked with water to make a gruel, or (3) put directly into a vessel containing water, and fermented. Some tribes mix a portion of sprouted grain flour with water to form a paste. Then, they make a yeasty froth from the paste by exposing it to the wild yeasts that are in the air and/or on the surface of the brewing vessel. Other tribes initiate fermentation by adding a soured sweet potato to a watery gruel made from malted grain. Sometimes, cassava (manioc) paste or crushed sugar cane may be added to the fermentation mixture. Fermentation is usually completed within a few days, after which the beer may be boiled, sieved, or consumed.

• **Sake (Japanese rice beer)**—Many people call sake "rice wine" because it has an alcohol content of 14 to 16%, which is more comparable to wine than most beers. However, it is truly a cereal beer because the starches in the rice must be converted to sugars before fermentation can take place. It is noteworthy that the fungus *Aspergillus oryzae*, which is used by the Japanese to convert rice starch to sugars, is the same species which acts on the sprouted grains used to make African native beers. This is no coincidence, since *Aspergillus* fungi are common spoilage organisms that attack grains and legumes around the world. Fig. B-47 outlines the unusual operations by which sake is produced.

[13]Platt, B. S., "Biological Ennoblement: Improvement of the Nutritive Value of Foods and Dietary Regimens by Biological Agents," *Food Technology*, Vol. 18, 1964, pp. 68-76.

[14]Ackerman, L. V., "Some Thoughts on Food and Cancer," *Nutrition Today*, Vol. 7, January/February 1972, p. 3.

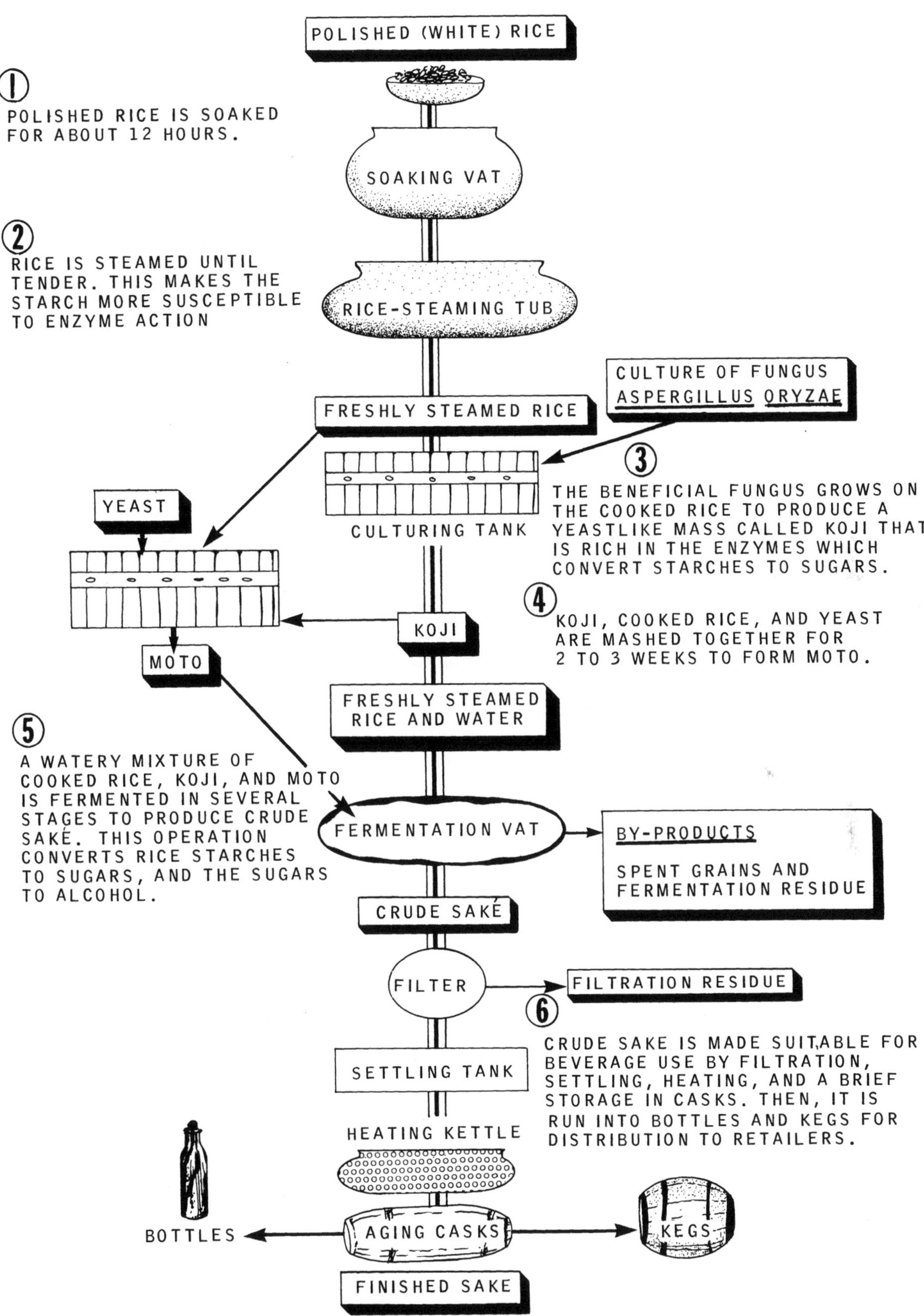

POLISHED (WHITE) RICE
1
POLISHED RICE IS SOAKED FOR ABOUT 12 HOURS.
SOAKING VAT
2
RICE IS STEAMED UNTIL TENDER. THIS MAKES THE STARCH MORE SUSCEPTIBLE TO ENZYME ACTION
RICE-STEAMING TUB
CULTURE OF FUNGUS ASPERGILLUS ORYZAE
FRESHLY STEAMED RICE
3
THE BENEFICIAL FUNGUS GROWS ON THE COOKED RICE TO PRODUCE A YEASTLIKE MASS CALLED KOJI THAT IS RICH IN THE ENZYMES WHICH CONVERT STARCHES TO SUGARS.
YEAST
CULTURING TANK
4
KOJI, COOKED RICE, AND YEAST ARE MASHED TOGETHER FOR 2 TO 3 WEEKS TO FORM MOTO.
KOJI
MOTO
FRESHLY STEAMED RICE AND WATER
5
A WATERY MIXTURE OF COOKED RICE, KOJI, AND MOTO IS FERMENTED IN SEVERAL STAGES TO PRODUCE CRUDE SAKÉ. THIS OPERATION CONVERTS RICE STARCHES TO SUGARS, AND THE SUGARS TO ALCOHOL.
FERMENTATION VAT
BY-PRODUCTS
SPENT GRAINS AND FERMENTATION RESIDUE
CRUDE SAKÉ
FILTER
FILTRATION RESIDUE
6
CRUDE SAKE IS MADE SUITABLE FOR BEVERAGE USE BY FILTRATION, SETTLING, HEATING, AND A BRIEF STORAGE IN CASKS. THEN, IT IS RUN INTO BOTTLES AND KEGS FOR DISTRIBUTION TO RETAILERS.
SETTLING TANK
HEATING KETTLE
BOTTLES
AGING CASKS
KEGS
FINISHED SAKE

Fig. B-47. The production of sake from rice.

Sake differs from most beers in (1) containing from 2 to 3 times as much alcohol, (2) being almost colorless, (3) lacking carbonation, and (4) being more desirable when served warm, whereas other beers are better when served cold.

• **Latin American chewed corn beer (chicha)**—It is not known where the practice of using chewed corn to make beer originated, but it appears to have been spread throughout the Americas by the Indians because Columbus observed it in the Caribbean, and it is still found in the more primitive regions of South America. Certain groups of Indians even have professional chewers who prepare the corn for fermentation into chicha.

• **Russian rye bread beer (kvass or quass)**—This refreshing beverage, which contains not more than 0.7% alcohol, is made by (1) pressure cooking a brown rye bread, (2) treating the cooked bread mash with rye malt, (3) fermenting the malt-treated mash with a combination of yeast and *Bacillus lactis*, and (4) filtering the beer. The drink is usually dispensed cold from special tank trucks which tour the cities during the summer.

NUTRITIVE VALUE OF ORDINARY BEER. The nutritive composition of beer and ale is given in Food Composition Table F-36. The modern preference for a highly clarified beer which is unclouded by protein hazes or yeast sediment appears to have been responsible for a reduction in the nutritive value of this beverage compared to more primitive types of beer. For example, various studies have shown that the African native beers may make significant nutritive contributions to the diets of tribes who rely mainly upon cereal grains and starchy roots and tubers.[15] However, these beers are subjected to only minimal filtration because they are usually consumed from opaque drinking vessels made from clay.

One of the few unclarified beers produced in the developed countries is the German beverage *weisse*, which ranks about the highest in nutritive value among the commercial beers because it is fermented in the bottle and the yeast is left in the product. However, even the ordinary beers are nutritionally superior to nonalcoholic, carbonated drinks. The extent to which some people regard beer as nutritive is evidenced by the fact that females who normally abstain from alcoholic beverages may consume it after childbirth in the belief that it helps to stimulate the flow of breast milk. Hence, the nutrient composition of beer presented in Table B-12 is noteworthy.

It may be seen from Table B-12 that beer supplies liberal amounts of phosphorus, magnesium, riboflavin, vitamin B-6, and niacin.

Although 12 oz *(360 ml)* of beer furnishes only about 9% of the estimated allowance for chromium, few foods other than brewers' yeast supply this mineral in such a readily utilized form, since all of the chromium in beer is alcohol extractable. Recent research has shown that the biological activity of chromium is associated with an alcohol soluble factor which appears to promote better utilization of carbohydrates by the body. Hence, certain chromium-deficient people with diabeticlike tendencies might benefit from drinking beer.

(Also see CHROMIUM.)

The thiamin (vitamin B-1) and pantothenic acid content of beer is low in proportion to its caloric content. Hence, deficiencies of these vitamins might be aggravated by heavy beer drinking, unless other foods rich in the vitamin B complex are consumed.

There is only a small amount of sodium in beer. Thus, it does not seem to be responsible for the water logging effect attributed to it. Rather, the excessive water retention which afflicts some heavy beer drinkers might be due to the salty foods often consumed while drinking. Generally, the consumption of alcoholic beverages is likely to cause the loss of water from the body since alcohol promotes copious urination.

Ounce for ounce, beer contains about the same number of calories as the nonalcoholic, sweetened, carbonated beverages. Hence, anyone who consumes large quantities of any of these beverages might expect to gain excess weight since each 8-oz *(240 ml)* glass supplies about 100 Calories (kcal).

TABLE B–12
NUTRIENT COMPOSITION OF BEER[1]

Nutrient and Unit of Measurement	Amount in 12 oz *(360 ml)* of Beer	RDA For Adult Male[2]	Percent of RDA Furnished By 12 oz *(360 ml)* of Beer
Major constituents:			
Food energy, Calories .....(kcal)	151	2,900	5.2
Protein .... g	1.1	63	.02
Fat .... g	0	—[3]	—
Carbohydrates ..... g	13.7	—	—
Water .... ml	332	2,900	11.4
Alcohol (4.5% by volume) ..... ml	16.2	—	—
Minerals:			
Calcium .... mg	18	800	2.2
Phosphorus ..... mg	108	800	13.5
Sodium .... mg	25	500	5
Magnesium ..... mg	36	350	10.3
Potassium ...... mg	90	2,000	0.5
Vitamins:			
Folate (folic acid) mcg	21.6	200	10.8
Niacin ...... mg NE	2.2	19.0	11.6
Pantothenic acid (B-3) .... mg	0.29	4–7	5.3
Riboflavin (B-2) .. mg	0.11	1.7	6.5
Vitamin B-6 (pyridoxine) .... mg	0.21	2	10.5

[1]Data from Food Composition Table F-36.
[2]The RDAs are from *Recommended Dietary Allowances*, 10th rev. ed. NRC—National Academy of Sciences, 1989.
[3]Blank means that no allowance has been set.

[15]Jones, I. D., "Effects of Processing by Fermentation on Nutrients," *Nutritional Evaluation of Food Processing*, 2nd ed., edited by R. S. Harris and E. Karmas, The Avi Publishing Company, Inc., 1975, p. 346.

MENTAL AND PHYSICAL EFFECTS OF BEER DRINKING. It is difficult to ascribe specific effects to the drinking of specified quantities of beer because (1) the alcohol of the common beers and ales varies widely, (2) higher blood levels of alcohol are reached when beer is consumed on an empty stomach than when it is taken with a meal, and (3) habitual heavy drinkers who are otherwise well nourished metabolize alcohol much more rapidly than light-to-moderate drinkers. Nevertheless, it is important to have some guidelines for distinguishing between safe and dangerous amounts of beer consumption. Therefore, some estimates of these amounts are given in Table B-13.

TABLE B-13
LEVELS OF BLOOD ALCOHOL AND THE EFFECTS WHICH RESULT FROM DRINKING VARIOUS AMOUNTS OF BEER

Level of Alcohol in the Blood	Amount of Ordinary Beer Required to Produce Level of Blood Alcohol[1]	Effects Which are Common at the Designated Blood Alcohol Level	Comments
(%)			
.01-.03	1–3 glasses (8 oz [*240 ml*] ea) or 1–2 cans or bottles (12 oz [*360 ml*] ea) taken on an empty stomach. 1–6 glasses or 1–4 cans when food is taken.	Relaxation of tensions, increased sociability, stimulation of appetite, relief of minor aches and pains, and promotion of sleep. Some people may have a slight impairment of mental and physical functions when their blood alcohol is as low as 0.03%. Also, their responses in automobile driving may be slowed slightly.	Lower levels of blood alcohol and diminished effects result when food is consumed along with beer than when it is taken on an empty stomach.
.04-.06	4-6 glasses or 3-4 cans or bottles consumed without food.	Most people have slowed responses while driving, plus mild reductions in their mental and physical functions, and an increased amount of urination. They may also experience a laxative effect from the beer.	Often, these amounts of beer are so filling that little food is taken. However, habitual heavy drinkers of beer are usually able to consume sufficient food to slow the absorption of alcohol and thereby diminish its effects.
.07-.09	7-9 glasses or 5-6 cans or bottles.	Significant slowing of responses while driving, plus sufficient confusion in interpreting road signs and signals to risk arrest for dangerous driving or "driving under the influence."	Some people may not be able to keep this much beer down if it is consumed in a short period of time (1-2 hours).
.10-.13	10-13 glasses or 7-8 cans or bottles.	Many people may become intoxicated. Almost everyone is likely to have marked mental and physical impairments.	Habitual heavy drinkers may have lower levels of blood alcohol and less impairment.

[1]Calculated for a 154 lb *(70 kg)* adult from data given in McDonald, J. B., "Not by Alcohol Alone," *Nutrition Today*, Vol. 14, January/February 1979, p. 15, Fig. 1. (An alcoholic content of 3.6% by weight was assumed.)

Table B-13 shows that the drinking of a few glasses of beer with a meal results mainly in beneficial effects; whereas the consumption of much larger quantities may be followed by embarrassing, painful, or even disastrous consequences. Furthermore, a little beer may help to calm certain jittery people, but it may make normally peaceful people become more aroused and aggressive. Therefore, some explanations of these effects are appropriate at this point.

Beneficial Effects. The consumption of a few glasses of beer may calm jittery people because small amounts of alcohol slow the activity of the areas of the brain which govern the thought processes, while the areas associated with the other functions are unaffected. Therefore, it is noteworthy that overexcitable persons may be able to improve their performance of both mental and physical tasks after having a few beers. Psychologists have found that optimal performance of mental activities depends upon just the right amount of arousal. Both overarousal and underarousal reduce concentration and performance. Hence, the counteraction of overarousal by small amounts of alcohol or other substances may help to promote a more effective utilization of one's mental abilities.

A few beers may also help to promote sociability by calming feelings of apprehension toward others. The business luncheon is a prime example of how food and drink are employed to bring about a spirit of mutual understanding.

Finally, beer and other alcoholic beverages appear to promote health and well being in ways which are not well understood. For example, a study of over 7,000 Japanese men living in Hawaii showed that those who drank one or two alcoholic drinks per day had only one-half the rate of cardiovascular disorders as those who abstained from these beverages.[16] These findings attest to the wisdom of doctors who prescribe a daily drink for patients suffering from clogged blood vessels (atherosclerosis or arteriosclerosis). Apparently, alcohol relaxes the walls of blood vessels so that their openings are enlarged (it is a vasodilator) and the flow of blood is increased.

Harmful Effects. A few people may fly into violent rages after only a few drinks. It is suspected that they may suffer from a combination of a personality disorder and a tendency to develop low blood sugar.[17] Perhaps, some of these people harbor strong antisocial feelings which they have great difficulty in suppressing. Hence, the diminishing of inhibitions by alcohol causes their violent dispositions to emerge under the slightest provocation. Needless to say, those with such antisocial tendencies would be better off abstaining from all alcoholic drinks.

Fortunately, it is much more difficult to drink too much alcohol in the form of beer than by imbibing wine or distilled liquors which contain much more alcohol. For example, four 8-oz *(240 ml)* glasses of beer provide a quart of fluid, which is about as much as most people may feel comfortable drinking in a short period of time. The alcohol in this quantity of beer will produce only the mildest impairment of mental and physical functions. However, drinking too much beer, or any other beverage for that matter may interfere with the normal nutritive processes by (1) causing feelings of fullness before sufficient food is consumed, (2) provoking diarrhea or vomiting, or (3) causing excessive urination which flushes vitamins and minerals out of the body.

Finally, drinking to the point of intoxication is a very dangerous thing because (1) it may become habitual, (2) control of behavior is lost, and (3) the drinker becomes highly prone to accidents.

(Also see ALCOHOLIC BEVERAGES and ALCOHOLISM.)

PLANNING MEALS AND SNACKS TO ACCOMPANY BEER. It makes good sense to eat food while drinking beer because (1) food reduces the effects of the beer since it slows the rate at which the stomach empties its contents into the small intestine, where most of the alcohol in the beer is absorbed; (2) the tendency to drink too much beer may be curtailed somewhat by slowing the rate at which the stomach empties; and (3) food may provide the protein, minerals, and vitamins lacking in beer so that better nutrition is achieved.

Most beers contain about the same number of calories as the popular brands of sweetened, carbonated soft drinks, although a few special "light" beers may be significantly lower in calories. Hence, the best foods to eat with beer are those rich in protein, minerals, and vitamins, but low in carbohydrates and fats. Also, it is better *not* to eat salty, spicy, or sugary foods with beer because they may provoke excessive thirst and overconsumption of beer. Crafty owners of bars have long supplied such snacks free of charge in order to encourage a greater consumption of their beverages. Some good accompaniments to beer are: (1) cheeses and other dairy products; (2) eggs, fish, meats, poultry, and seafood; (3) fruits, vegetables, and salads; (4) Pretzels, whole grain breads, and other baked goods; and (5) beans, peas, nuts, and other seeds.

Unfortunately, it is a common sight to see outdoor laborers, and people on picnics, washing down large bites of sandwiches or other starchy foods with beer. This practice should be discouraged because it may lead to digestive problems. Foods rich in starch require chewing so that there is ample opportunity for mixing with the digestive juices. It would be better to chew food thoroughly and swallow it before taking a sip of beer. Furthermore, washing food down may promote overeating.

Last, but not least, the practice of spiking beer with various distilled liquors is deplorable because it may prevent the drinker from gauging the amount of alcohol which is consumed, so that intoxication is much more likely to occur than when either type of beverage is consumed alone.

BEET *Beta vulgaris*

This popular vegetable belongs to the same family *(Chenopodiaceae)* as chard and spinach. Both the root and the leaves are eaten. Fig. B-48 shows a typical beet root with some leaves attached.

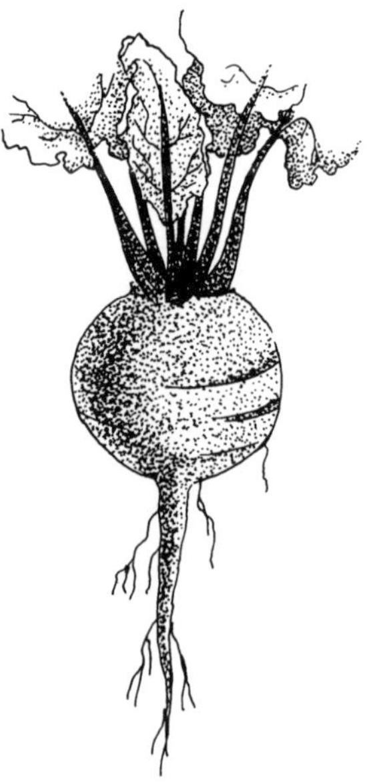

Fig. B-48. Beet.

ORIGIN AND HISTORY. It appears that the greens were gathered from wild beets long before any attempts were made to cultivate the beet for its root.

[16]Yano, K., G. G. Rhoads, and A. Kagan, "Coffee, Alcohol and Risk of Coronary Heart Disease Among Japanese Men Living in Hawaii," *The New England Journal of Medicine*, Vol. 297, 1977, p. 405.

[17]Keller, M., "Alcohol Consumption," *Encyclopedia Britannica*, 15th ed., Encyclopedia Britannica, Inc., Chicago, Ill., 1974, Vol. 1 of Macropaedia, p. 440.

(Beet greens, chard, lambsquarters, and spinach are green leafy vegetables which appear to have a common ancestor.) The wild beet grows along seashores throughout Asia and Europe. Sometime around the early Christian era, the Romans began to cultivate both red and white beets for their roots, while other peoples developed the wild plant for its greens. Later, the barbarian invaders of Rome brought beets with large roots to northern Europe, where they were first used for feeding animals. By the 16th century, the red beet was widely used for food by the British and the Germans, while the white beet was used for fodder. Later, the white beet was found to have a high sugar content.

The first factory for extracting sugar from white beets was built in Silesia in 1801. Ten years later, Napoleon decreed that the sugar beet be utilized as a source of sugar to replace that derived from sugar cane, the supply of which had been cut off by a British naval blockade. In the 19th century, both the red beet and the white beet were brought to the United States, where both crops now thrive.

PRODUCTION. Red beet production in the United States has averaged about 200,000 metric tons per year over the past 20 years. Fig. B-49 shows the leading U.S. markets for beets.

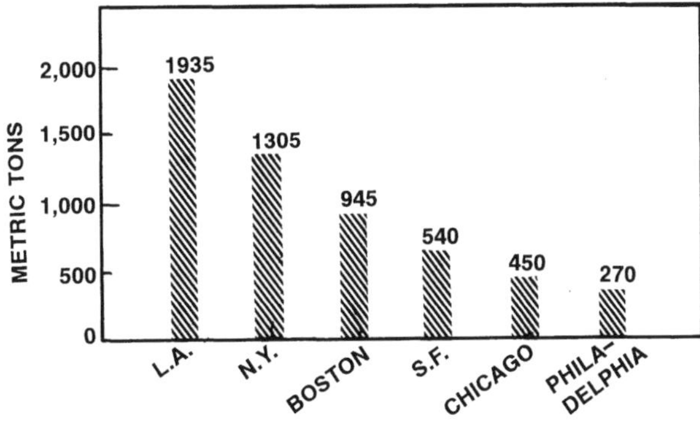

Fig. B-49. Beets shipped to specified markets in 1990. (Data from *Agricultural Statistics*, 1991, pp. 167, 168, Tables 242, 243)

Propagation and Growing. Beets are grown from seeds as an annual crop. The root develops best in cool weather. Hence, the crop is grown in the winter in southern United States, and in the summer in the northern part of the country. A well-drained, moist, and highly fertile soil is required. Ample amounts of fertilizers containing nitrogen, phosphorus, and potassium are often applied by the grower, together with carefully measured amounts of boron. (Some of the leading beet-producing states have soils deficient in boron.) Beet seeds come in clusters called seedballs that usually contain from one to six seeds. Therefore, the young plants are likely to grow too close together, which causes the roots to be small. Also, most of the roots grow close to the surface, so cultivation and weeding is usually done carefully by hand. Thinning of the plants may be delayed until many of the young roots meet the minimum size for marketing. (Actually, small beets are more tender and bring a higher price per unit of weight than the larger ones. Hence, beets grown for canning may not be thinned.)

Harvesting. All of the beets are harvested when most of the roots are between 1½ and 2 in. (*3.75 to 5 cm*) in diameter, since processors prefer beets under 2 in. (*5 cm*) in diameter.

PROCESSING. About 95% of the beets produced in the United States are canned, and only 5% are marketed fresh. Also, beet greens may be marketed fresh in areas near to where they are produced. Beets that are to be canned must be washed, graded for quality and size, and peeled. Peeling is no longer done by hand, but is accomplished by passing the beets through cylindrical vessels in which they are exposed to steam under pressure. This treatment causes the skins to expand and break up so that they may be washed away with high velocity jets of water. The peeled beets may be canned whole, sliced, or cut in narrow strips. Some of the crop is pickled in vinegar.

The freezing of beets is presently in the experimental stage because frozen beets tend to be mushy when thawed. However, it has been found that an ultraslow freezing process (one that requires about 24 hours, compared to the normal freezing time of 2 to 3 hours) will produce only a negligible reduction in firmness, whereas a drastic decrease in firmness occurs in beets frozen within 2½ hours.

SELECTION AND PREPARATION. Good quality beets are relatively smooth and firm. Soft, flabby, rough, or shriveled beets may be tough or woody, or involve excessive waste in preparation. The condition of the beet greens alone does not indicate the quality of the roots, and any defects of the former may be disregarded when only the latter are to be utilized. Decay in beet roots usually appears as soft, wet areas. Beets showing any decay should be avoided.

Good quality beets greens are young, clean, fresh, and tender. Slightly flabby or wilted tops can usually be restored to freshness in cold water if wilting is only in the initial stage. Beet greens showing any indication of a slimy or soft watery condition should be avoided.

Fresh beet greens may be kept for 3 to 5 days, or more, in a refrigerator, but the beet roots with the tops removed may be stored for 2 to 4 weeks without losing their quality. (It is advisable to leave from 2 to 3 in. [*5 to 7.5 cm*] of the leaf stems on the beet roots; otherwise, they will likely "bleed" in storage.) Country people have long stored beet roots over the winter in cold cellars. In such storage, the beet usually dries out somewhat and becomes a little tougher than those stored for shorter periods of time.

Whole beets require about 40 minutes of cooking in boiling water, whereas cut pieces (cubes, slices, or strips) can be cooked in about one-third of that time. Washed beet greens should be served raw in a salad, or after about 10 minutes of cooking.

Beets can be utilized in a variety of vegetable dishes because they are mild flavored and very colorful. Most recipes may be adapted for either cooked fresh beets or canned beets. The cooked vegetable may be added to casseroles, salads, and soups. One of the most popular dishes in eastern Europe is *borscht*, a thick beet soup that is served cold or hot with a dollop of sour cream. Some other suggestions for tasty beet dishes follow:

1. Pureed beets may be used to thicken and add color to various soups and sauces. The time used to prepare a puree may be saved by substituting pureed beets sold as baby food. (It is noteworthy that commercial spaghetti sauce powders may contain dried beet powder which is used as an extender.)

2. Pickled beets may be substituted for cucumber pickles in relish trays, salad plates, and sandwiches. The juice from the beets could be used in gravies, salad dressings, and similar preparations. (Some people prepare low-calorie salad dressings by blending pickled beets with unflavored yogurt, cottage cheese, and various seasonings.)

3. Raw beets and other fresh vegetables may be processed in a blender or a juicer to make a juice drink.

NUTRITIONAL VALUE. The nutrient composition of beets is given in Food Composition Table F-36.

Some noteworthy observations regarding the nutrient composition of beets follow:

1. Cooked beet roots contain about 90% water. Hence, they are low in solids, calories, and other nutrients. However, much of the solid content is readily utilized carbohydrates, such as starches and sugar. Hence, the root part of the vegetable contains about twice as many calories as the leaves.

2. Beet greens are much richer in calcium, iron, vitamin A, and vitamin C than beet roots. Therefore, the two major parts of the plant should be eaten together in order to derive maximum nutritional benefits.

Beet roots have been used medicinally for several centuries, although the chemical nature of potentially beneficial components remain uncertain. There have been some recent controversial reports on the use of beet preparations as preventives against cancer, diabetes, and yellow jaundice. *NOTE:* The authors make no claims for the authenticity of these reports. Rather, they are mentioned for two purposes: [1] informational, and [2] stimulation of research.

People who have eaten large quantities of beets sometimes pass a red-colored urine, due to the red pigments present in the vegetable. However, this red coloration is believed to be harmless.

(Also see VEGETABLE[S], Table V-6 Vegetables of the World.)

BEET SUGAR

Table sugar—sucrose—extracted from the sugar beet. (Also see SUGAR, section headed "Sugar Beet.")

BEETURIA

A harmless condition characterized by red pigmented urine which may occur after the consumption of beets.

BEHAVIOR AND DIET

Contents Page

Behavioral Problems 196
Dietary Causes and Cures 197
Feingold or K-P Diet 197
Sugar Elimination 198
Food Allergies 198
Megavitamin Therapy 198
Brain Transmitters and Diet 198
Malnutrition 199
Overlooked Factors 199

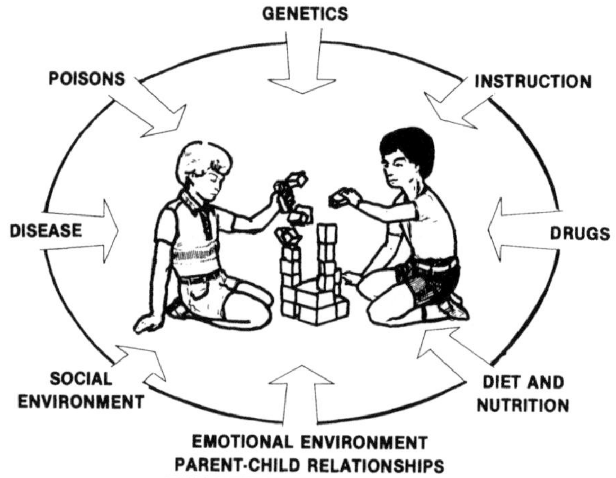

Fig. B-50. Behavior results from a complex interaction of many factors.

Behavior is the manner in which individuals handle or manage themselves in relation to other individuals and their surroundings; or an individual's response to specific conditions. Normal and abnormal behavior are subjective classifications. What is normal or abnormal may be merely a matter of degree and the manner in which a person's behavior is perceived by family, friends, and professionals. There are, however, some general classifications of abnormal behavior, including disruptive behavior, failure to develop learning skills and motor skills, and behavior which breaks moral and social values and laws of the land. Diet—all nutrients entering the body through the mouth,and nutrition—the body processes nutrients undergo after ingestion,have been implicated in forms of abnormal behavior.

BEHAVIORAL PROBLEMS. There are numerous forms of specific behavioral problems, but primarily hyperactivity (hyperkinesis), learning disabilities, and juvenile delinquency and crime have in some way been linked to diet, whether correctly or incorrectly. In some cases, diet is suggested as both the cause and the cure, while in other cases dietary modifications are suggested as cures for abnormal behavior of unknown origin.

• **Hyperactivity**—"Hyperactive," or in medical terminology "hyperkinetic," is a label that has been applied to between 5 and 20% of all school-age children in the

United States. Unfortunately, hyperactivity is a vogue term and some ill-mannered children have been conveniently labeled as hyperactive. Hyperactive children are described as being significantly more restless, easily distracted, inattentive and emotionally labile relative to their peers. Attention is drawn to hyperactive children in situations which require sedentary and attentive behavior; thus, these children usually come under medical and psychological evaluation because of behavioral problems and learning difficulties at school. Furthermore, the hyperactive child may be physically sound but have difficulty with motor skills and coordination.

Many theories on the cause of hyperactivity have been presented, including genetics, brain damage, crowded homes, environmental pollutants, parent-child interactions, fluorescent lights, and diet.

A variety of treatments have been tried with mixed results. These have included stimulants, elimination of fluorescent lighting in the classrooms, determination of allergies, restriction of diet, and, recently, behavioral modification programs. Popular treatments are diet oriented.

• **Learning disabilities**—Life is a learning process involving motor skills, speech, reading, writing, and social skills. The inability to learn may appear at various stages of life, and, due to the many things to be learned, the disabilities are given various labels. Many causes of learning disabilities have genetic roots; others involve social and environmental deprivation, or injury and disease, while still others have nutritional causes. Regardless of the cause, some dietary treatments have been proposed.

• **Juvenile delinquency and crime**—Recently, there have been attempts to lay a large share of the blame for juvenile delinquency and crime upon poor diet. While diet no doubt has a role in emotional well-being, so do many other factors. Moreover, moral bvehavior and respect for the law are taught rather than fed.

DIETARY CAUSES AND CURES. Possible causes and cures are almost as numerous as specific behavioral problems. However, some are more important than others due to their popularity or validity. Unfortunately, popularity and validity do not often coincide in most theories as to the causes and cures of undesirable behavior.

Feingold Or K-P Diet. In 1973, Dr. Ben Feingold, an allergist from the Kaiser-Permanente (K-P) Medical Center in San Francisco, California, introduced his solution to hyperactivity. Dr. Feingold suggested that the elimination of certain chemical compounds in the diet would prevent and treat hyperactivity. Originally, Dr. Feingold believed that salicylates (chemicals related to aspirin) naturally-occurring substances present in many foods—including many fruits, caused hyperactivity. Gradually, Dr. Feingold added (1) artificial food colors and flavors with a chemical structure similar to salicylates, (2) all artificial colors and flavors, and (3) preservatives such as butylated hydroxytoluene (BHT) and butylated hydroxyanisole (BHA). The Feingold or K-P diet is an elimination diet which seeks to remove these chemicals from the diet. Since most all commercially processed and prepared foods contain artificial colors, flavors or preservatives, and since salicylates are present naturally in many fruits and vegetables, it becomes nearly impossible to eliminate them totally from the diet.

Dr. Feingold presents his ideas in a couple of popular books wherein he recommends that many common foods be excluded. Essentially, all manufactured baked goods, luncheon meats, ice cream, powdered puddings, candies, soft drinks, and punches are eliminated. Teas, coffee, margarine, butter, and many commercially produced condiments are excluded. Additionally, many nonfood items such as mouthwash, toothpaste, cough drops, perfume, and some over-the-counter and prescription drugs are taboo. Obviously, eating in restaurants and school cafeterias becomes impossible for a child adhering to the Feingold plan. Foods for the Feingold diet must be prepared from "scratch" in the home. Therefore, Dr. Feingold recommends that the entire family become involved in following the diet plan thereby lessening temptation to stray and preventing the child from feeling different. Following the diet becomes a family affair.

According to Dr. Feingold, 48 to 50% of the children with behavioral problems who strictly adhere to his diet showed marked improvement within 3 to 21 days. Parents of hyperactive children welcomed the Feingold plan, and there are many convincing observations and testimonials on the effectiveness of the Feingold diet. However, hard scientific data is lacking.

Numerous scientists have tested and evaluated Dr. Feingold's hypothesis, and have voiced the following criticisms:

1. Properly conducted scientific trials fail to demonstrate that salicylates and artificial colors and flavors have the dramatic impact on hyperactive behavior that Dr. Feingold predicts. Only a fraction of a percent of hyperactive children may experience adverse reactions to several of the artificial colors and flavors in our food supply.
2. The Feingold diet with its recommendations causes important changes in the family social patterns and life-style that could also affect the hyperactive child's behavior, since hyperactive behavior in some children may be associated with family problems and lack of attention to the child.
3. Embracing the Feingold diet has a placebo effect. That is, there is a high positive attitude toward the program with major changes in family dynamics which augment the expectations of both the family and child. Moreover, there is a desire of the child's parents to believe that the hyperactive behavior is a product of something outside the home.
4. Since the Feingold diet seeks to remove *all* artificial flavors, artificial colors, and preservatives, numerous, and quite chemically unrelated compounds, are condemned. The Feingold diet does not specifically remove an identifiable single substance or group of substances.
5. The long-term effects of indicating to a child that what he or she eats or drinks is responsible for their behavior should be carefully considered.
6. For those who insist on following the Feingold diet plan there is little nutritional risks unless the proposed diet is the older version which eliminated salicylate-

containing fruits, thus was low in vitamin C. Also, following the plan as outlined in his books may prevent some families from seeking needed professional assistance.

Note: Following several years of application, Dr. Feingold revised his recommended diet as follows: After 4 to 6 weeks, the salicylate-containing foods can be returned to the diet of the hyperactive child if no adverse reactions occur. So, today, the artificial additives are the major emphasis in the Feingold diet and the role of salicylates in hyperkinesis has been minimized.

(Also see FEINGOLD DIET.)

Sugar Elimination. Some popular theories suggest that the consumption of refined sugar causes hyperactivity, mental illness, juvenile delinquency, and crime.

Supporters of the "sugar-undesirable behavior theory" explain that consumption of excess sugar leads to the development of reactive or function hypoglycemia—low blood sugar. Then, undesirable behavior results from the mental changes induced by the low blood sugar, or in the case of hyperactive children, their behavior results from the release of hormones from the adrenal gland which attempt to increase blood sugar. Blaming sugar *per se* for behavioral problems is based primarily on anecdotes and is championed by the popular press—not scientific journalism.

When considering the elimination of sugar from the diet as a cure for a certain illness, individuals should examine the following points:

1. Reactive hypoglycemia is not widespread. Moreover, children diagnosed as hyperactive have not shown reactive hypoglycemia when tested.

2. Theories blaming sugar for behavior disturbances present the metabolic machinery of the body in an entirely too simplistic scheme. The crux of physiology is that an individual strives to maintain a constant internal environment—homeostasis—through a multitude of normal changes.

3. Studies which report behavior improvement upon elimination of refined sugar from the diet are troubled with the placebo effect. Behavior changes on the basis of hope and expectations, since observed behavior changes are noted by the individuals and their families—all of those who are aware of the dietary change.

4. If only behavioral problems such as mental illness, juvenile delinquency and crime increased at the rate of increased sugar consumption, there would be less. Consumption of refined sugar has hovered around 100 lb per person per year for the past 70 years, while the crime rate tripled just since 1960.

5. Behavior, like disease, results from a multitude of complex interactions—seldom one cause.

(Also see DENTAL HEALTH; NUTRITION AND DIET; HYPOGLYCEMIA; METABOLISM; and SUGAR.)

Food Allergies. A food allergy is an immunologically mediated adverse reaction to food. Some suggestion has been made that a food allergy may be responsible for behavioral disturbances ranging from hyperactivity to a feeling of tiredness to criminal behavior.

The most common foods causing allergic reactions are milk, eggs, wheat, corn, legumes, nuts, seafood, strawberries, citrus fruits, tomatoes, chocolate, and berries.

Symptoms of an allergy may include nasal congestion, cough, mucus, wheezing, constipation, diarrhea, stomachache, bloating, poor appetite, eczema, hives, headache, tension, fatigue, and sweating. Indeed, many of the symptoms could alter anyone's behavior by making them uncomfortable.

The direct connection between a food allergy and some undesirable behavior remains obscure, and for the time being, hypothetical. Some children with minimal brain dysfunction have been said to have a greater than average frequency of food allergy. But other studies have shown that hyperactive children do not have food allergies, nor has the success of the Feingold diet been related to a food allergy. Overall, a direct connection between a food allergy and behavior receives most of its support from a few individuals who have popularized the theory. Scientists, however, maintain an open mind and continue to conduct well-controlled studies further to elucidate the connections between food allergies and behavior.

(Also see ALLERGIES.)

Megavitamin Therapy. Another popular idea is that massive doses of vitamins may cure hyperactivity and other behavioral abnormalities. Advocates of megavitamin therapy, or orthomolecular psychiatry, propose that optimum molecular concentrations of vitamins are essential for proper mental functioning. No one will argue that dietary deficiencies of such vitamins as thiamin, riboflavin, niacin, panthothenic acid, vitamin B-6 and vitamin B-12 will modify sensory functions (nervous functions), motor ability, and personality. However, orthomolecular psychiatrists base their recommendations on scientific studies which fail to meet the requirements of proper scientific protocol. Moreover, most of the reported efficacy of megavitamin therapy results from anecdotal observations. Studies investigating the benefits of megavitamins with proper scientific protocol have failed to demonstrate improved behavior with megavitamins.

Nevertheless, hope should not be abandoned. There are certain genetic disorders for which megavitamin therapy can be justified and future research may find its justification in other disorders. For example, nine studies reported in the world literature show that megadoses of vitamin B-6 are helpful in treating autism, though more experimental work is needed.[18]

Currently, the American Academy of Pediatrics and the American Psychiatric Association feel that megavitamin therapy is unproved in terms of safety and efficacy.

(Also see VITAMIN[S].)

Brain Transmitters And Diet. Brain function, hence behavior, relies upon the ability of the cells of the brain to synthesize and release normal amounts of transmitters which control brain function by modifying

---

[18]Rimland, Bernard, Ph.D., Letters to the editor, *Science News*, Vol. 119, No. 16, April 18, 1981, p. 243.

the conduction and transmission of nervous impulses. An exciting area of new research is that the possibility exists that the availability of certain nutrients in the food can raise or lower the formation of brain transmitters thereby modifying brain function. The transmitters subject to dietary control include: serotonin, norepinephrine, and acetylcholine.

Serotonin is derived from dietary tryptophan, an essential amino acid. A high carbohydrate (simple sugar) diet increases the serotonin level in the brain of rats and humans. However, there is no direct evidence that an increased concentration of serotonin in the brain affects behavior. Neurologists claim that cells in the brain that release serotonin influence sleep. Tryptophan supplements have been used to treat various sleep disorders, and to improve the mood in depressed patients.

Norepinephrine is also derived from an amino acid—tyrosine. The administration of tyrosine to rats with high blood pressure dramatically reduces blood pressure. This effect seems to be caused by the stimulation of norepinephrine synthesis in the brain. The future may see the use of tyrosine for the treatment of behavioral disorders such as depression.

Choline is the dietary precursor of the neurotransmitter *acetylcholine*. Variations in the daily intake of choline also modify the brain levels of choline and acetylcholine. Some preliminary reports have shown that dietary choline may improve learning and behavior. Recently, clinical trials have shown choline to be successful in the treatment of *tardive dyskinesia*, a disorder of the central nervous system.

The story of the effect of diet on brain transmitters is still incomplete, but it is an area of exciting research—where sound scientific protocol has been followed. Important discoveries ahead may help in understanding the relationship between diet, brain chemistry, and behavior.

(Also see AMINO ACID[S]; and CHOLINE.)

Malnutrition. In humans, it is difficult to say whether nutrition *per se* contributes to behavioral abnormalities, since conditions of malnutrition and a deprived social and emotional environment often coexist. Studies have demonstrated the importance of the mother-child relationship to learning and behavior, and malnutrition may adversely affect this interaction. While the effects of malnutrition are uncertain, the following points deserve consideration:

1. The brain and spinal cord grow rapidly during the last of pregnancy, and continue rapid growth during infancy.
2. Malnutrition during the early life of animals and humans alters the number of brain cells, brain cell size and/or other biochemical parameters, depending upon the time of onset and the duration of the malnutrition.
3. Malnourished animals demonstrate impaired learning abilities.
4. Teachers have observed that hungry, undernourished children are apathetic and lethargic.
5. Childen suffering from kwashiorkor are apathetic, listless, and withdrawn; and they seldom either (a) resist examination or (b) wander off when left alone.
6. Children who recover from marasmus have a reduced head circumference—brain size—even after 5 years of rehabilitation.
7. Numerous studies suggest that infants subjected to acute malnutrition during the first years of life develop irreversible gaps in mental development, and thus never attain their full genetic potential.

(Also see INFANT DIET AND NUTRITION; MALNUTRITION; and MALNUTRITION, PROTEIN-ENERGY.)

OVERLOOKED FACTORS. A host of factors are involved in determining behavior. Too often popularized theories are more the result of "witch hunts," and caution should be exercised when entertaining them. To blame only the diet and nutrition for undesirable behavior is to take an attitude of "the devil made me do it." Human behavior is far too complex even to hint that only one condition in the environment determines it. Rather, individuals would be well advised to consider the effects of such things as (1) smoking and alcohol use during pregnancy, (2) drug abuse during pregnancy and lactation, (3) stress and nutrition during pregnancy, (4) genetic background and genetic errors, (5) family interactions, and (6) social environment.

Unfortunately, blaming the diet for behavior is attractive since it is more palatable to believe undesirable behavior results from conditions outside the individual rather than inherent problems, lack of instruction, or social interactions.

(Also see MINERAL[S]; NERVOUS SYSTEM DISORDERS; PREGNANCY AND LACTATION NUTRITION, section headed "Factors That May Warrant Special Counseling of Pregnant Woman"; and VITAMIN[S].)

BELCHING

The bringing up of gas from the stomach. Excessive belching may be due to (1) swallowing of large amounts of air as a result of nervousness, (2) the drinking of carbonated beverages, (3) a heavy consumption of caffeine containing beverages such as coffee and tea, or (4) force of habit. In certain countries, belching after a meal is considered to be a sign to the host that the food was greatly appreciated. However, frequent and uncomfortable belching may mean that a person has a digestive disorder which requires treatment by a doctor.

BELLADONNA DRUGS

This group of drugs, which is extracted from plants of the Nightshade family, is so named because at one time they were used by ladies to dilate the pupils of their eyes, since ladies with large pupils were considered to be more glamorous. (*Bella donna* in Italian means beautiful lady.) These drugs are now used mainly to reduce the amount of gastrointestinal secretions and to calm the digestive tract in the treatment of such conditions as peptic ulcers, spasms of the gallbladder, diarrhea, stomach and bowel irritation, ulcerative colitis, and heartburn. However, people given these drugs may have such side effects as convulsions, drowsiness, dryness of the mouth, constipation, and blurring of the vision. Hence, they should only be used under a doctor's supervision, even though certain preparations which contain these drugs may be obtained without a prescription.

(Also see ATROPINE.)

BENEDICT'S TEST

A chemical test for determining the presence of reducing sugars; for example, glucose and maltose. Heating with Benedict's reagent—copper sulfate, sodium citrate, and sodium carbonate—gives a green, yellow, or orange-red precipitate, depending upon the amount of reducing sugar present. Benedict's test is used to test for the presence of sugar in the urine.

BENGAL QUINCE (BEL FRUIT; GOLDEN APPLE) *Aegle marmelos*

This fruit grows on a thorny tree (of the family *Rutaceae*) that is native to India and is now found throughout Southeast Asia and the East Indies. The yellow fruit ranges from 2 to 7 in. (*5 to 17.5 cm*) in diameter and has a hard skin and a soft flesh. It is usually eaten fresh, dried, or made into an ice or a sherbet.

Bengal quince is moderately high in calories (133 kcal per 100 g) and carbohydrates (35%). It is a good to excellent source of fiber and potassium, but it is a poor source of vitamin C.

BENIGN

- An illness that is mild in nature.

- A neoplasm (new or abnormal, uncontrolled growth) that is not malignant, not liable to recur.

BENIGN DISORDERS

Mild illnesses or other abnormal conditions which do not generally endanger life. Benign is also used to describe a tumorous growth that is not cancerous.

BENZEDRINE

A member of the amphetamine family of drugs which are stimulants and appetite depressants.

(Also see AMPHETAMINES; and APPETITE.)

BERGAMOT ORANGE *Citrus bergamia*

A small citrus fruit (of the family *Rutaceae*) that is believed to have arisen from a natural crossbreeding (hybridization) of the sour orange (*Citrus aurantium*) with another species of citrus fruit.

Fig. B-51. The bergamot orange.

The flesh of the bergamot is highly acid and of little commercial importance, but the oil from the peel has a distinctive aromatic odor and taste which suits it well for use in candies, medicines, and perfumes. Sometimes, the whole peel is candied. It is also noteworthy that bergamot peel oil is the aromatic base for eau de cologne. Most of the crop is grown in Italy.

(Also see CITRUS FRUITS.)

BERIBERI

This ancient nutritional disease results from a severe deficiency of thiamin (vitamin B-1). It is usually found in areas of the world where diets are high in carbohydrate, but low in thiamin. However, chronic abuse of alcohol may precipitate the disease in persons who might otherwise have access to a balanced diet. A typical victim of beriberi is shown in Fig. B-52.

Fig. B-52. Beriberi, a deficiency disease due to a lack of vitamin B-1 (thiamin) in the diet. Note the cracked skin and swollen legs (edema). In the Far East, beriberi is largely due to the almost exclusive use of polished rice. (Courtesy, FAO, Rome, Italy)

HISTORY OF BERIBERI. Mention of beriberi has been found in ancient Chinese medical writings, but the first European physician to take note of the disease was Bontius, in 1642, when he was a medical officer in the Dutch East Indies, now called Indonesia. Later, other European physicians observed significant declines in the physical activity of the victims of the disease. In 1873, Van Leent, a Dutch naval doctor, observed that the European crew members had significantly fewer cases of beriberi than sailors recruited from the East Indies. He decreased the amount of rice in the Indians' diets and thereby reduced the incidence of beriberi.

A similar result was obtained by Takaki, a Japanese medical officer, who in 1883, showed that the substitution of meat, fish, and vegetables for some of the rice in the sailors' rations resulted in less beriberi than that occurring in the control group of sailors who were fed mainly the rice rations. He attributed the reduction in the incidence of beriberi to the amount of protein in the diet of the sailors given substitute foods for part of their rice rations.

During this same period of time, other physicians observed the high incidence of the disease in congested areas, such as prisons, plantations, and ships, but they looked for an infectious cause. Scientists were then aware of toxic agents, such as *ergotism* and *lathyrism*, and thought that a toxic factor might be associated with either refining or storage of rice. Eijkman, a Dutch surgeon in the Dutch East Indies, tested for an infectious agent by injecting fowl with blood and urine from victims of beriberi. No disease was produced by such measures, but he later noted that the fowl developed the disease when they were fed leftover white rice. He then confirmed his findings with a controlled experiment in which each of two groups of fowl were fed a different type of rice; one group the polished rice, the other unpolished rice. Vorderman, an inspector of the Civil Medical Service in Java and a friend of Eijkman, observed, in 1896, that prisoners who crudely milled their own rice did not have as much beriberi as did those who were given commercially polished rice.

Eijkman's work was continued in 1900 by Grijns, another Dutch surgeon, who conducted extensive laboratory studies on the different parts of the rice grain and later isolated a crude antiberiberi factor from rice polishings.

Several investigators then set to work to try to isolate, in purified form, the substance responsible for the prevention of beriberi. In 1911, Funk, a Polish scientist working in England, believed that he had isolated the pure substance and coined the name "vitamine" for it. However, it was later shown by Elvehjem and his coworkers at the University of Wisconsin that Funk had actually isolated the antipellagra vitamin rather than the so-called antineuritic factor.

The first pure crystalline form of thiamin was isolated from rice polishings in 1926, by Jansen and Donath, Dutch scientists working in Java. A few years later, other workers isolated the vitamin from yeast, wheat germ, and rice bran. Immediate treatment of severe beriberi required a concentrated form of the vitamin, since providing foods which contained the vitamin was not sufficient to stop progress of the disease. The purified form of the vitamin isolated by Jansen was much too expensive to be used for this purpose, so extracts which were crude preparations from rice, yeast, and wheat germ were used. Finally, in 1936, Williams, while working at Merck, Inc., in Rahway, New Jersey, synthesized thiamin. Subsequently, it was possible to synthesize thiamin commercially for treatment of severe thiamin deficiency.

CAUSES OF BERIBERI. The disease is regularly found in populations whose diets contain less than 0.3 mg of thiamin per 1,000 nonfat Calories (kcal). Diets which may cause beriberi are those in which the staple foods are either highly refined, such as white rice, white flour, or degerminated cornmeal, or where there are starchy items like cassava. It is noteworthy that the Irish, who depended upon potatoes, did not develop beriberi, because cooked potatoes contain about 1.0 mg of thiamin per 1,000 Calories (kcal). However, some of the Irish might also have benefited from the drinking of beer and similar fermented beverages which supply about 0.7 mg of thiamin per 1,000 Calories (kcal).

Occasionally, beriberi may be precipitated by stress factors such as pregnancy, lactation, strenuous exercise, or restoration of growth after protein-energy deficiencies. Thus, pregnancy may be debilitating, or even fatal, to a thiamin-deficient woman. Even if the woman herself escapes such morbid consequences, her child may develop beriberi as a result of a low level of thiamin in her milk. Thiamin-deficient men may also be prone to beriberi if they engage in hard labor, since a heightened energy metabolism is likely to hasten the depletion of the body's limited supply of thiamin. (This supply consists of the small amount of thiamin which is present in the body fluids.)

Chronic alcoholism may also precipitate beriberi since alcohol cannot be converted to fat in the body, but must be metabolized by the same thiamin-dependant processes as carbohydrate and protein. Furthermore, alcohol increases urinary excretion of water (diuretic effect) and water-soluble vitamins. A similar, but lesser effect may result from the heavy consumption of caffeine (in beverages such as coffee, colas, and tea) coupled with a diet chronically high in refined foods which are rich in carbohydrates. (Like alcohol, caffeine is a diuretic. Also, it stimulates the utilization of stored carbohydrate for energy.)

FORMS OF BERIBERI. The disease may appear in any one of several forms which may resemble other diseases. Also, there may be a sudden change from one form to another. Therefore, the features of each form are discussed in the sections which follow.

Dry Beriberi. This form is characterized by signs of multiple nervous disorders such as sensations of pins and needles on the legs, muscle pain, delayed pain response, and foot drop, which may indicate the beginning of a brain disorder known as *Wernicke's encephalopathy* (typical signs are double vision, squint-eyedness, disorientation, delusions, and loss of memory). The psychotic effects of the brain disorder are collectively called *Korsakoff's psychosis*, which is particularly distinguished by confused thinking and the making up of stories to fill in gaps of memory (confabulation). This aspect of the disease may eventually result in a stupefied state. These effects may commonly be found in chronic alcoholics, diabetics whose disease has been out of control, and semistarved persons such as prisoners of war.

Wet Beriberi. The presence of edema distinguishes this form from dry beriberi. Other features are extreme loss of appetite, breathlessness, and disorders of the heart which range from palpitation and rapid heart rate to dilation of the heart muscle (myocardium) and congestive heart failure. When the cardiac disorders are severe, the disease may be called *cardiac beriberi*, or *Shoshin beriberi* (a name used in Asia). The physiological processes involved in the development of the cardiac effects are shown in Fig. B-53.

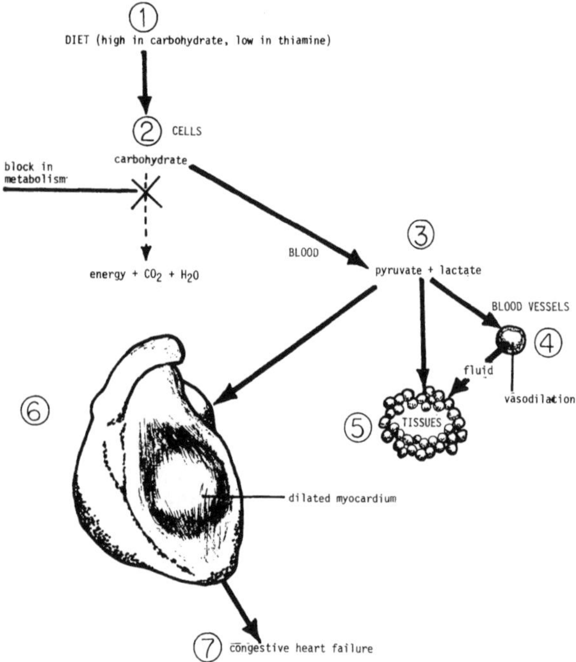

Development of Cardiac Disorders in Beriberi

1. **Dietary imbalance**: Insufficient dietary thiamin for the metabolism of carbohydrate (needed to meet the body's requirement for energy.
2. **Incomplete metabolism in cells**: Oxidation of pyruvate and lactate is blocked by lack of thiamin.
3. **Accumulation of metabolites**: Partial metabolism of carbohydrates continues to produce pyruvate and lactate which accumulate in blood and tissues causes edema.
4. **Dilation of blood vessels**: Effect of elevated blood levels of pyruvate and lactate
5. **Tissue effects**: Fluid from dilated blood vessels collects in tissues.
6. **Effects in the heart**: Cardiac workload is increased (in order to maintain blood flow), myocardium becomes dilated (result of workload plus edema).
7. **Congestive heart failure**: Myocardium becomes exhausted (lack of thiamin for energy production) and fails to pump blood.

Fig. B-53. Physiological processes in cardiac beriberi.

Infantile Beriberi. This form is usually found in infants living in areas of endemic beriberi who have been breast-fed by mothers whose milk is deficient in thiamin. Signs of the disorder are weakness of voice during bawling (complete lack of sound in severe cases), lack of appetite, vomiting, diarrhea, rapid pulse, and cyanosis (in severe cases). The disease results in a high death rate for infants from 2 to 5 months of age. There may be a rapid course of the disease from mild disorders to death.

Juvenile Beriberi. The feeding of extra protein without extra thiamin to protein-deficient children who are stunted in their growth may bring on beriberi because (1) growth resumes and the need for thiamin is increased, and (2) part of the dietary protein may be converted in the body to carbohydrates and/or other substances which require thiamin for their metabolism.

TREATMENT AND PREVENTION OF BERIBERI. Intramuscular injections of from 5 mg to 50 mg of thiamin per day are more effective during the first few days of the therapy for severe beriberi than oral doses of the vitamin since gastrointestinal absorption may be less efficient. Such therapy is usually followed by prompt remission of the cardiac disorders, but the neurological disorders may persist for a while. Oral doses of from 20 mg to 30 mg of thiamin may be given in place of the injections as soon as the patient shows signs of improvement. The last stage of therapy consists of supplying foods or special supplements which are rich in thiamin, such as dried brewers' yeast, dried torula yeast, liver, liver extract, rice polishings, wheat germ, or legumes such as beans and lentils.

In Newfoundland, the enrichment of all white flour with thiamin has eliminated beriberi. It is more difficult, however, to obtain widespread enrichment of rice which requires impregnation of the individual grains with vitamin mix; the enriched rice takes on the yellow color of riboflavin which is objectionable to some rice eaters.

The age old practice in some rice eating areas, such as certain parts of India, of parboiling rice before milling is at least partly effective in preventing beriberi. Parboiling is now done on a commercial basis by first soaking the unhulled grains of rice in water, then parboiling them until their hulls split open. This process causes such water-soluble vitamins as thiamin to migrate deeper into the grain from the outer layers. Hence, these vitamins are retained within the grain when the outer layers are removed by milling.

(Also see DEFICIENCY DISEASES, Table D-1 Major Dietary Deficiency Diseases; RICE; THIAMIN; WHEAT, Section on "Enriched Flour"; and VITAMIN[S], Table V-9.)

BERRY

A type of fleshy fruit in which the seeds are enclosed within the pulp. Some of the most common berries are blackberries, blueberries, currants, cranberries, gooseberries, grapes, raspberries, and strawberries. However, the term is often applied to various fruits which are not strictly berries from a botanical point of view.

(Also see FRUIT[S], Table F-47 Fruits of the World.)

BETA

Second letter in the Greek alphabet, corresponding to the letter "B" in our alphabet. In chemistry and nutrition, the prefix beta is used with a hyphen to indicate the second member of a closely related series of substances.

BETA-CAROTENE

A yellow-orange plant pigment which may be converted in the body to vitamin A. The best food sources of beta-carotene are the yellow, orange, or red fruits and vegetables, or the green leafy vegetables. In the latter, the intense green color of chlorophyll masks the color of the underlying carotene.

(Also see VITAMIN A.)

BETA CELLS OF THE PANCREAS

The cells which synthesize, store, and secrete the hormone insulin.

(Also see ENDOCRINE GLANDS.)

BETA-CHOLESTEROL

This term is a layman's designation for blood cholesterol which is carried on beta-lipoproteins, in contrast to that carried on alpha-lipoproteins. However, the cholesterol molecule itself does not vary, only the lipoprotein which carries it in the blood.

It is important to differentiate between the ways in which cholesterol is conveyed in the blood, because that carried on beta-lipoproteins is much more likely to be deposited in the walls of blood vessels than that borne by alpha-lipoproteins.

(Also see ALPHA-LIPOPROTEINS; BETA-LIPOPROTEINS; CHOLESTEROL; and HEART DISEASE.)

BETA-HYDROXYBUTYRIC ACID

One of the ketone bodies which accumulates in the blood and is passed in the urine when fat is incompletely metabolized.

(Also see ACIDOSIS; and DIABETES MELLITUS.)

BETAINE

A substance closely related to choline. It may serve as a raw material from which the body synthesizes choline. Betaine has been shown in animal experiments to prevent fatty livers when diets high in fat, cholesterol, and/or sugar are consumed. The best sources of betaine are plant foods, particularly beets.

(Also see CHOLINE.)

BETA-LIPOPROTEINS

Special fat-carrying proteins which are produced in the liver. Beta-lipoproteins are also called low density lipoproteins (LDL). Normally, they contain from 10 to 50% triglycerides, 22 to 46% cholesterol, 18 to 22% phospholipid, and 9 to 21% proteins. It is not desirable for one to have a high blood level of beta-lipoproteins because the cholesterol they carry may be deposited in the blood vessels. However, alpha-lipoproteins (HDL) convey cholesterol in a much more stable complex, so that there is little risk of cholesterol deposits when the blood levels of these proteins are elevated. It appears that the levels of beta-lipoproteins may be reduced, and those of alpha-lipoproteins raised by (1) loss of excess body weight, (2) strenuous physical exercise, (3) a diet low in animal fats and cholesterol, (4) drugs such as clofibrate and niacin (this vitamin must be administered in large doses to achieve the desired effects), and (5) moderate amounts of alcohol.

NOTE: Measures designed to alter the levels of blood fats should be undertaken under the supervision of a physician.

(Also see ALPHA-LIPOPROTEINS; CHOLESTEROL; and HEART DISEASE.)

BETA-OXIDATION

A major means by which fatty acids are metabolized in the body.

(Also see FAT[S] AND OTHER LIPIDS; and METABOLISM.)

BETEL LEAVES

Leaves from a creeping plant (*Piper betel*) which are chewed in some parts of the world for a stimulating effect.

(Also see BETEL NUTS.)

BETEL NUTS

The nuts produced by the areta palm (*Areta catethu*), which are chewed in order to obtain stimulant effects.

(Also see BETEL LEAVES.)

BEVERAGES

Any liquid used or prepared for drinking is a beverage. Numerous beverages have been invented, but most of them can be included in one of the following categories:

1. Aromatic and stimulating infusions such as herbal tea, tea, coffee, and roasted grain drinks.
2. Fruit juices, including lemonade and fruit-flavored drinks.
3. Fermented beverages such as beer, wine, and cider.
4. Distilled liquors.
5. Soft drinks or carbonated beverages.
6. Water.

Some noteworthy statistics relative to the consumption of soft drinks in the United States follow:

1. Per capita consumption of soft drinks is estimated at more than 40 gal.
2. Yearly consumption of soft drinks surpasses all other beverages including milk, beer, coffee, or water (22% for water vs 23.8% for soft drinks).

Fig. B-54. Beverages are an important part of entertaining. (Created by Dynamic Graphics, Inc.)

3. Total annual wholesale value of soft drinks is more than $20 billion.

For information about specific beverages, consult the individual articles, and Food Composition Table F-36 of this book.

(Also see ALCOHOLIC BEVERAGES; BEERS AND BREWING; COFFEE; DISTILLED LIQUORS; SOFT DRINKS; WATER; and WINE.)

BEZOAR

A hard, gummy ball which forms in the stomach and which may block the intestines. A bezoar may form when foods rich in gums are swallowed without adequate chewing.

BIBULOUS

Refers to a person who is fond of alcohol beverages.

BICARBONATE

Name used by lay people to designate antacids that contain baking soda or sodium bicarbonate. The chemical definition of bicarbonate is a salt produced by the reaction of carbonic acid with an alkali. Sometimes, the term bicarbonate is used to designate the major alkali present in the blood. However, the correct term for the so-called alkali reserve is bicarbonate ion.

(Also see ACID-BASE BALANCE; ALKALI RESERVE; BAKING POWDER AND BAKING SODA; and BICARBONATE ION.)

BICARBONATE ION

A negatively charged alkali ion derived from bicarbonate salts, or from carbonic acid. It helps to regulate acid-base balance.

(Also see ACID-BASE BALANCE; ALKALI RESERVE; BAKING POWDER AND BAKING SODA; and CARBONIC ACID.)

BICARBONATE OF SODA

Another name for baking soda or sodium bicarbonate. (Also see BAKING POWDER AND BAKING SODA.)

BICYCLE ERGOMETER

A machine, similar to an ordinary bicycle, but fixed to a stand, which is equipped to measure the work performed during pedaling. It is often used in metabolic studies, and in tests of cardiovascular efficiency.

BIFIDUS FACTOR

A substance found in human breast milk which promotes the growth of *Lactobacillus bifidus* in the intestines of infants. *L. bifidus* protects infants against various types of bacterial infections in the digestive tract.

(Also see INFANT DIET AND NUTRITION; and MILK AND MILK PRODUCTS.)

BIGUANIDES

Drugs which are given by mouth for the treatment of diabetes. They are considered to be effective only in the treatment of adult onset diabetes, which usually occurs in people who are overweight. However, there is presently some concern as to whether these drugs are effective, and whether they may even be hazardous to health.

(Also see DIABETES MELLITUS.)

BILBERRY *Vaccinium myrtillus*

The fruit of a wild shrubby plant (of the family *Ericaceae*) that grows wild in the northern parts of Europe and Asia. This berry is closely related to the North American varieties of blueberries. Furthermore, bilberries

Fig. B-55. The bilberry, a hardy wild fruit that resembles its relative the blueberry.

are sometimes referred to as blaeberries (in Iceland), huckleberries, whortleberries, or windberries.

The fruit has a tart flavor which is improved by sweetening. Also, it makes good jams and pies.

BILE

A yellowish-green fluid produced by the liver and stored in the gallbladder, which empties through the bile duct into the duodenum of the small intestine, where it participates in the digestion of fats. It is considered to be one of the digestive juices of the intestinal tract.

(Also see DIGESTION AND ABSORPTION; and GALLBLADDER.)

BILE ACIDS

A group of several similar compounds present in bile, a digestive fluid secreted by the liver, and stored in the gallbladder. Bile acids are (1) produced by the breakdown of cholesterol in the liver, and (2) converted into bile salts by the alkaline constituents of the bile.

(Also see BILE; BILE SALTS; DIGESTION AND ABSORPTION; and GALLSTONES.)

BILE DUCTS

This designation includes (1) the hepatic ducts which collect the bile formed in the liver, (2) the common bile duct connecting the hepatic ducts with the small intestine, and (3) the cystic duct which connects the gallbladder and the common bile duct.

(Also see DIGESTION AND ABSORPTION; and GALLSTONES.)

BILE PIGMENTS

Breakdown products of red blood cells, which are excreted in the bile and the feces. Bile pigments may accumulate in the blood when various liver disorders prevent their excretion in the bile. In such cases, the skin may turn yellow, a condition known as jaundice.

(Also see DIGESTION AND ABSORPTION; and GALLSTONES.)

BILE SALTS

These important components of bile are formed by the reactions between the bile acids and the alkaline substances present in the bile. After the various constituents of bile have been mixed and secreted by the liver, the bile is stored in the gallbladder until food is consumed. Then, the passage of food from the stomach into the small intestine triggers the release of bile from the gallbladder into the duodenum. There, the bile salts help to emulsify the fatty materials in the diet so that they may be digested and absorbed.

Almost all of the bile salts are reabsorbed from the intestine and returned to the liver via the blood. However, certain undigestible carbohydrates, collectively called fiber, bind bile salts so that greater amounts are lost in the stool. This effect tends to reduce the body's content of cholesterol because increased amounts of the sterol are converted to bile salts to replace those which are excreted.

(Also see BILE; CHOLESTEROL; DIGESTION AND ABSORPTION; FIBER; and GALLSTONES.)

BILHARZIASIS

This condition, often called *schistomiasis* at the present time, is an infestation with small flat worms (flukes). It is most common in Asia, Africa, and in parts of Europe where sanitation is poor. Usually the larvae of the flukes penetrate the skin when people wade into contaminated water. However, the flukes may also be present in food.

BILIARY

Pertaining to the bile or to the gallbladder.

(Also see BILE; and DIGESTION AND ABSORPTION.)

BILIARY CALCULI

A disorder in which stones are formed in various parts of the biliary system (the liver, bile ducts, gallbladder, and small intestine.) Another name for this condition is gallstones. The condition is usually more common in people who manifest one or more of the four Fs: female, fat, forty, and fair complexion.

(Also see GALLSTONES.)

BILIARY CIRRHOSIS

A disorder of the liver which is due to blockage of the bile duct(s) and accumulation of bile in the liver and the blood. The condition is characterized by fever, abdominal pain, jaundice, gas, nausea, itching, and an elevated count of white cells in the blood. Prompt surgery is needed to correct the condition.

(Also see BILIARY OBSTRUCTION; and GALLSTONES.)

BILIARY COLIC

A spasm of the muscle surrounding the gallbladder due to (1) blockage of a bile duct by a gallstone; or (2) contraction of the gallbladder against a closed sphincter. Biliary colic is usually accompanied by an attack of severe cramping pain which occurs in the right side of the abdomen just below and in front of the ribs. A doctor should be called at once, when this type of pain is present.

(Also see BILIARY DYSKINESIA; and GALLSTONES.)

BILIARY DYSKINESIA

A malfunctioning of the gallbladder in which (1) it contracts but fails to empty properly because the spincter does not relax and allow the bile to pass; or (2) the muscles around the gallbladder which cause it to contract lack tone and, therefore, fail to do their job.

(Also see BILIARY COLIC; and GALLSTONES.)

BILIARY OBSTRUCTION

A blockage somewhere in the system of bile ducts that (1) prevents bile from passing into the small intestine, and (2) causes it to accumulate in the liver and in the blood.

(Also see BILE DUCTS; and GALLSTONES.)

BILIMBE (CUCUMBER TREE) *Averrhoa bilimbe*

The green, cucumberlike fruit of a tree (of the family *Oxalidaceae*) that is native to Malaysia and now grows in many tropical areas of the world. Bilimbe fruits are usually between 2 and 3 in. (*5 to 7.5 cm*) long and contain a sour, seedy pulp that suits them well for curries, pickles, jams, jellies, and sweetened juices.

The fruit is high in water content (94%) and low in calories (20 kcal per 100 g) and carbohydrates (5%). It is a good source of iron and a fair source of fiber and vitamin C.

Fig. B-56. The bilimbe or cucumber tree which is found in tropical areas.

BILIOUSNESS

A digestive disorder which is usually attributed to some type of malfunctioning of the liver. It is characterized by headache, nausea and vomiting, constipation, lack of appetite, furred tongue, and occasionally slight jaundice.

(Also see DIGESTION AND ABSORPTION.)

BILIRUBIN

One of the bile pigments which result from the breakdown of hemoglobin. Normally, it is excreted into the intestine with the bile. However, blockage of the biliary tract leads to accumulation of bile in the liver and blood. The buildup of pigment often imparts a yellow color to the skin (a condition called jaundice). Newborn babies are particularly prone to brain damage by elevated blood levels of bilirubin.

(Also see DIGESTION AND ABSORPTION.)

BILIRUBINURIA

The presence of the bile pigment bilirubin in the urine. This pigment colors the urine dark. Bilirubinuria is often accompanied by jaundice.

(Also see BILIRUBIN.)

BILIVERDIN

A green pigment resulting from the breakdown of hemoglobin. The pigment is converted to bilirubin in the liver.

(Also see BILE PIGMENTS; and BILIRUBIN.)

BIO-

A prefix denoting life.

BIOASSAY

Determination of the relative effective strength of a substance (as a vitamin, hormone, or drug) by comparing its effect on a test organism with that of a standard preparation.

BIOCHEMISTRY

The study of chemical processes which take place in living organisms.

BIOFLAVONOIDS (VITAMIN P)

Contents Page

History 207
Chemistry, Metabolism, Properties 207
Measurement 208
Functions 208
Disorders Treated with Bioflavonoids 209
Deficiency Symptoms 209
Recommended Daily Allowance of Bioflavonoids . . . 209
Toxicity 209
Bioflavonoid Losses During Processing, Cooking, and Storage 209
Sources of Bioflavonoids 209
Top Food Sources of Bioflavonoids 209

At this time, no evidence exists that bioflavonoids serve any useful role in human nutrition or in the prevention or treatment of disease in humans; hence, this presentation is for two purposes: (1) informational, and (2) stimulation of research.

Bioflavonoids, also known as vitamin P, are a group of natural pigments in vegetables, fruits, flowers, and grains. They appear as companions of natural vitamin C, but they are not present in synthetic vitamin C. Because of the basic yellow color of most of them, they are called flavonoids, after the Latin, *flavus*, for yellow. (Some of these substances are naturally occurring dyes.) To date, about 800 different flavonoids have been identified, of which more than 30 are in the genus *Citrus* alone. Three of the better known bioflavonoids are *hesperidin*, *naringin*, and *rutin*.

Hesperidin is found in the blossom, in the small unripe fruit, and in the peel of the mature sweet orange. Also, it is found in lemons, mandarins, bitter oranges, and citrons. Each mature sweet orange contains almost 1 g of hesperidinlike material of which approximately half can be recovered by commercial procedures. Hesperidin is used extensively as a therapeutic agent in the pharmaceutical industry.

Naringin is the predominant flavonoid in grapefruit. It is distinguished readily from hesperidin by its extreme bitterness, which it sometimes imparts to grapefruit products. Naringin is used in the preparation of beverages and to enhance the piquant flavor of high class confections.

Rutin was first prepared in 1842 by a German pharmacist-chemist, who obtained it from garden rue, *Ruta graveolens*; hence, the name *rutin*. Since then, chemists have found rutin in a number of plants, including tobacco and buckwheat (in the green or dehydrated leaves). Many pharmaceutical laboratories are now marketing dosage forms of rutin, principally in tablets, for the treatment of capillary fragility—a condition in which the smallest blood vessels (1) become abnormally fragile and rupture, so that small hemorrhages occur, or (2) become abnormally permeable, so that they allow substances to pass from the blood into the tissue spaces in larger quantities or of different kinds than normally filter through the capillary walls. Where either of these faults is present, there is increased danger of retinal hemorrhage or apoplexy, particularly in people with high blood pressure. Claims are made that rutin, taken by mouth, will correct these capillary faults in a large proportion of cases. Diabetics frequently have complications of this sort, with loss of vision and even blindness. Some reports indicate that the use of rutin results in arresting progress in the loss of sight, and sometimes improvement of vision, especially in young patients. Rutin has also been used in treating certain types of glaucoma, in treating injury from x-ray irradiation, in treating cold injury (frostbite), and in lessening the severity of the symptoms in cases of hemophilia—the hereditary disease characterized by failure of the blood to clot normally after injury.

HISTORY. In the mid 1930s, Szent-Gyorgyi, the Hungarian scientist, who subsequently (in 1937) won a Nobel Prize in medicine for his work with vitamin C, isolated a material with citrus rind that he called *citrin*, and which he showed consisted of a mixture of flavonoids. His initial test of the new substance with scorbutic guinea pigs, reported in 1936, seemed to indicate that, in combination with ascorbic acid, it was effective in strengthening the body's smallest blood vessels, the capillaries, and in curing scurvy. From this and subsequent work arose the concept of vitamin P, a substance or a group of substances of a flavonoid nature involved in regulation of permeability and maintenance of capillary integrity.

Soon, the race was on! Research was stimulated; studies involving vitamin C and vitamin P were conducted on both humans and animals all over the world; and journal articles attested to clinical successes. Vitamin P (usually in combination with vitamin C) was rushed in to treat a whole host of disorders believed to be related to faulty capillary function; among them, habitual and threatened abortion, postpartum bleeding, nosebleed, skin disorders, diabetes retinitis, bleeding gums, heavy menstrual bleeding, hemorrhoids, and many others.

But the earlier hopes held for vitamin P did not materialize. In 1938, Szent-Gyorgyi reported that subsequent tests failed to confirm the results of his earlier experiments. Similar work in other laboratories verified his conclusions. Although the observation that flavonoids display a synergistic action towards ascorbic acid was demonstrated, the initial claim that some or all of the flavonoids are indispensable food components equivalent to vitamins was not substantiated. As a result, in 1950 the Joint Committee of Biochemical Nomenclature of the American Society of Biological Chemists and the American Institute of Nutrition recommended that the term "vitamin P" be dropped. Following this, the name *bioflavonoids* came into use except in France and the U.S.S.R., where the term *vitamin P* persisted. Then, in the late 1960s, the U.S. Food and Drug Administration (FDA) concluded that, not only were the bioflavonoids not a vitamin, they were without any nutritional value whatsoever.

In the 2 decades following Szent-Gyorgyi's original work, numerous researchers studied the effects of the flavonoids on capillary fragility, infections, the common cold, hypertension, and various hemorrhagic disorders. But no therapeutic value was demonstrated. Also, for flavonoids to be classified as vitamins requires proof (1) that they are essential and indispensable food constituents, and (2) that deficiency syndromes are known which can be cured specifically by their administration. But neither of these prerequisites was fulfilled. So, two schools of thought evolved; those who believed, and those who doubted—with few undecided.

There is a logical explanation, according to the believers, why little therapeutic or nutritive value of bioflavonoids has been demonstrated. They point out that most fruits (especially citrus fruits) and vegetables are rich in bioflavonoids, and that almost everyone gets sufficient of these foods to prevent vitamin P deficiency. It is not that vitamin P is not required, they argue; it's already in the diet.

CHEMISTRY, METABOLISM, PROPERTIES.

• **Chemistry**—The chemical structure of two important bioflavonoids is given in Fig. B-57.

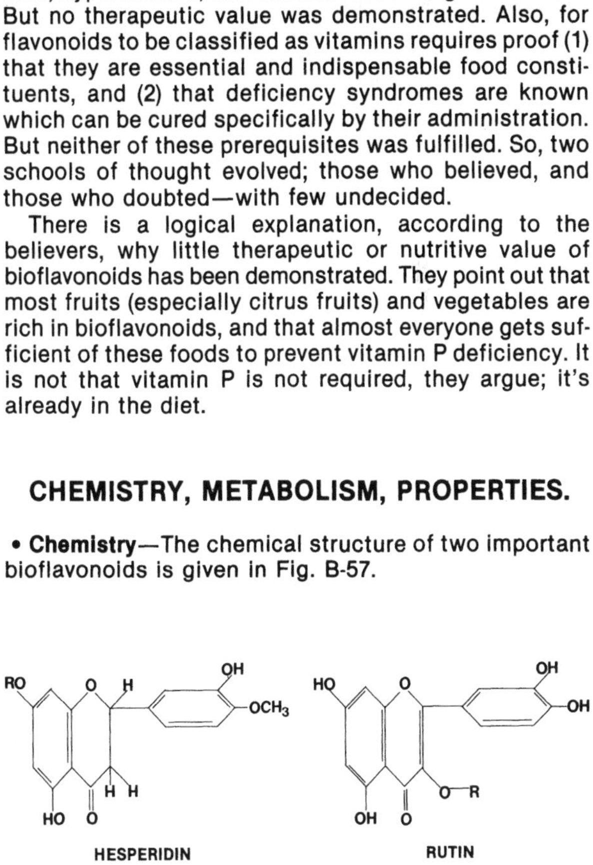

Fig. B-57. Structure of two bioflavonoids. *Hesperidin* is found in citrus fruit; *rutin* is found in buckwheat leaves.

• **Metabolism**—The absorption, storage, and excretion of the bioflavonoids are very similar to vitamin C. They are readily absorbed into the bloodstream from the upper part of the small intestine. Excessive amounts are excreted primarily in the urine.

• **Properties**—Bioflavonoids are brightly colored water-soluble substances. They are relatively stable compounds, resistant to heat, oxygen, dryness, and moderate degrees of acidity, but they are rather quickly destroyed by light.

MEASUREMENT. Dosages of bioflavonoids are given in milligrams (mg).

FUNCTIONS. Bioflavonoids are promoted chiefly because of their function in capillary fragility and permeability. Thus, a knowledge of the physiology of the capillary system is requisite to an understanding of this particular role.

Since all body cells depend on the capillaries to bring them everything they need and to take away wastes, they are important.

The entire cardiovascular system—heart, arteries, and veins—is dependent upon the capillaries. These minute-sized vessels, averaging about 1/2,000 of an inch in diameter, are a part of the microcirculation system—the connecting link between the smallest branches of the arteries and the connecting veins. The segments of this small vessel system in order from the arterial to the venous side are: (1) arteriole, (2) terminal arteriole, (3) metarteriole, (4) capillary, and (5) venule (see Fig. B-58).

The capillaries (1) receive from the bloodstream the oxygen, nutrients, hormones, and antibodies; and (2) take up the wastes. All the rest of the circulatory vessels (the arteries that carry oxygenated blood from the heart to all parts of the body, and the veins that carry the used blood back to the heart for reoxygenation in the lungs) are impermeable. It is only at the capillary level (the network that links the tiniest arteries [arterioles] and the tiniest veins [venules]) that fluid from the bloodstream seeps out of this otherwise closed system and mingles with the fluid that surrounds all the body cells, then seeps back again. For this seepage to take place, the walls of the capillaries must be permeable—but not too permeable. When capillaries are too fragile and break, or become too permeable, blood passes out into the intercellular fluid.

Bleeding into the skin (evidenced by red spots under the skin) and bruises are a sign of capillary breakage and an indication that the capillaries are fragile. Edema (the accumulation of fluid in the tissue) can also result from weak capillaries that are too permeable—capillaries that allow the escape of blood proteins that are needed for retaining the proper osmotic pressure to draw accumulated intercellular fluid back into the bloodstream.

The mechanism by which bioflavonoids exert their claimed influence on capillary fragility and permeability is not fully understood. It is known that capillary breakage is characteristic of the vitamin C deficiency disease, scurvy; and that vitamin C has a vital role in maintaining capillary health. Since bioflavonoids and vitamin C are found together in nature, it is conjectured that they function together in increasing the strength of the capillaries and regulating their permeability. These actions help prevent hemorrhages and ruptures in the capillaries and connective tissues and build a protective barrier against infection.

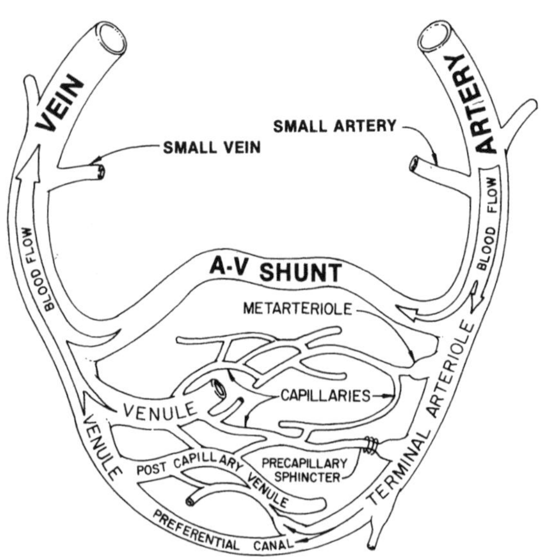

Fig. B-58. Schematic diagram of terminal vascular bed. Arrows indicate direction of flow.

Although flavonoids are not classified as vitamins for higher animal life because of lack of proof that they are essential food constituents, and that deficiency symptoms can be cured by their administration, there are strong indications that they are essential for lower stages of animal life—butterflies, silkworm larvae, crickets, and some species of beetles. Moreover, there are indications that similar essential effects may exist in higher animals and man, too. For example, rutin has been shown to exert a growth-promoting action on young rats and on bacteria. Some authorities believe that a vitaminlike effect of flavonoids may manifest itself only under stress conditions. Moreover, food flavonoids have been shown to exhibit in mammals a great number of specific effects contributing to buoyant good health. Further research on the essentiability of flavonoids as an essential food factor is needed.

In addition to their reputed effect on capillary fragility and permeability and on health, it is claimed that food flavonoids function as follows:

1. **They are active antioxidant compounds in food,** ranking second only to the fat-soluble tocopherols in this regard. This antioxidant effect protects the flavonoid-containing vegetable and fruit foodstuffs from oxidative deterioration, prolongs their shelf life and keeping quality, and improves taste, acceptability, and wholesomeness of mixed dishes by inhibiting the oxidation of the accompanying animal lipids.
2. **They possess a metal-chelating capacity;** and they affect the activity of enzymes and membranes.

3. **They have a synergistic effect on ascorbic acid,** and they appear to stabilize ascorbic acid in human tissues.

4. **They possess a bacteriostatic and/or antibiotic effect,** which is sufficiently high to account for their measurable anti-infectious properties in normal daily food.

5. **They possess anticarcinogenic activity in two ways;** a cytostatic effect against malignant cells (stopping or inhibiting the growth of cells), and a biochemical protection of the cell from damage by carcinogenic substances.

DISORDERS TREATED WITH BIOFLAVONOIDS. Despite the controversial status of bioflavonoids and the lack of experimental evidence of their value, they are being used in the prevention and treatment of the following:

1. Capillary fragility and bleeding.
2. Bleeding gums.
3. Bleeding into the retina of the eye.
4. Certain types of glaucoma.
5. Hemorrhage into the brain.
6. Bleeding kidneys.
7. Female problems, including heavy menstrual bleeding.
8. Varicose veins.
9. Hemorrhoids.
10. Ulcers.
11. Habitual and threatened miscarriages.
12. Bruises that occur in contact sports, like football.
13. Injury from x-ray irradiation.
14. Frostbite.
15. Diabetes and diabetic retinopathy.
16. Thrombosis (a blood clot which may form in the veins of the leg, and subsequently block a major blood vessel—and even cause death). It appears that the bioflavnoids are natural and helpful antithrombosic agents.

No claims are made that the bioflavonoids are the total answer in treating the above disorders or that they should replace established treatments. Rather, they are being used as supplemental preventives and treatments where capillary fragility and permeability are involved.

DEFICIENCY SYMPTOMS. Symptoms of bioflavonoid deficiency are closely related to those of a vitamin C deficiency. The tendency to bleed (hemorrhage) or bruise easily are especially noted.

RECOMMENDED DAILY ALLOWANCE OF BIOFLAVONOIDS. There are no NRC-National Academy of Sciences recommended daily allowances (RDA) of bioflavonoids, similar to vitamin C.

Since synthetic vitamin C does not contain the bioflavonoids, and since bioflavonoids occur with vitamin C in natural food sources, a daily allowance would seem to be indicated. Moreover, most researchers who have worked with the bioflavonoids have found (1) that they are most beneficial when taken with vitamin C; and (2) that in some cases where vitamin C alone is ineffective, the combination of the two may be helpful. Thus, the label on a rather typical vitamin C-bioflavonoid supplement reads as follows:

Each tablet contains:

	mg
Rose hips powder	500
Vitamin C (ascorbic acid)	500
Lemon bioflavonoid*	500
Rutin (buckwheat)*	50

**NOTE:* Need in human nutrition not established.

• **Bioflavonoid intake in average U.S. diet**—It is noteworthy that the average daily intake of flavonoids in the American diet amounts to about 1 g, of which one-half (*0.5 g*) is absorbed from the gut.

TOXICITY. Bioflavonoids are nontoxic.

BIOFLAVONOID LOSSES DURING PROCESSING, COOKING, AND STORAGE. Bioflavonoids are not greatly damaged by food processing or by food preparation in the kitchen unless these are done in strong light. Likewise, losses in storage are minimal, provided they are not expossed to strong light.

SOURCES OF BIOFLAVONOIDS. Bioflavonoids were first discovered in citrus peel. The white pulp of citrus fruits is also a rich source. Citrus fruits contain the bioflavonoids hesperidin and naringin.

Tangerine juice is a very rich source of tangeretin and nobiletin. But frozen orange juice is a poor source of bioflavonoids; the bioflavonoids impart an off-taste to the juice, so squeezing of the pulp is carefully controlled.

Citrus bioflavonoids, made by extracting the pulp which remains after juicing oranges and lemons, are availabale in concentrated form. Sometimes, rose hips, obtained from roses, and rutin, obtained from buckwheat leaves, are included with the citrus bioflavonoids.

TOP FOOD SOURCES OF BIOFLAVONOIDS. Animals are unable to synthesize flavonoids. Moreover, they are quickly metabolized in the body of higher animals. Hence, vegetables constitute the top food sources of the bioflavonoids given in the boxed list.

It is noteworthy that a great deal of flavonoids enter the human body by means of beverages and drinks. Tea, coffee, cocoa, wine (particularly red wine), beer, and even vinegar are important flavonoid sources, accounting for at least 25 to 30% of the total flavonoid intake.

(Also see VITAMIN[S], Table V-9.)

TOP SOURCES OF BIOFLAVONOIDS

Although no standard food composition tables are available showing the bioflavonoid content of foods, the following foods are generally recognized as the richest sources:

- Rose hips
- Buckwheat leaves
- Oranges and lemons (skin and pulp, especially the little white core that runs down the middle)
- Tangerine (juice)
- Grapefruit
- Apricots
- Blackberries
- Black currants
- Broccoli
- Canteloupe
- Cherries
- Grapes
- Green peppers
- Onions with colored skins
- Papaya
- Plums
- Tea, coffee, cocoa, red wine
- Tomatoes

BIOLOGICAL VALUE (BV) OF PROTEINS.

The percentage of the protein of a food or feed mixture which is usable as a protein by a growing child, or an animal. It can be determined by a balance experiment in which a measured intake of nitrogen (N) in food (or feed) is compared to the measured excretion of N in the feces and urine. Then, BV is calculated by the formula which follows:

$$\text{Biological value} = \frac{\text{N intake} - (\text{fecal N} + \text{urinary N}) \times 100}{\text{N intake - fecal N}}$$

The biological value of a protein is a reflection of the kinds and amounts of amino acids available to the animal after digestion. A protein which has a high biological value is said to be of *good quality.*

(Also see PROTEIN[S]; and QUALITY OF PROTEIN.)

BIOLOGY

The study of living organisms which includes (1) plants under the branch of botany, and (2) animals under the branch of zoology.

BIOPSY

The removal of a small piece of tissue for purposes of diagnosis.

BIOS

The designation given to a factor which stimulates the growth of yeast. Later, it was shown that the factor consists of two fractions, bios 1 (inositol) and bios 2 (biotin).

(Also see BIOTIN; and INOSITOL.)

BIOTIN

Contents *Page*

History 210
Chemistry, Metabolism, Properties 211
Measurement/Assay 211
Functions 211
Deficiency Symptoms 212
Biotin Dependency 212
Recommended Daily Allowance of Biotin 212
Toxicity 213
Biotin Losses During Processing, Cooking, and Storage 213
Sources of Biotin 213
Top Food Sources of Biotin 213

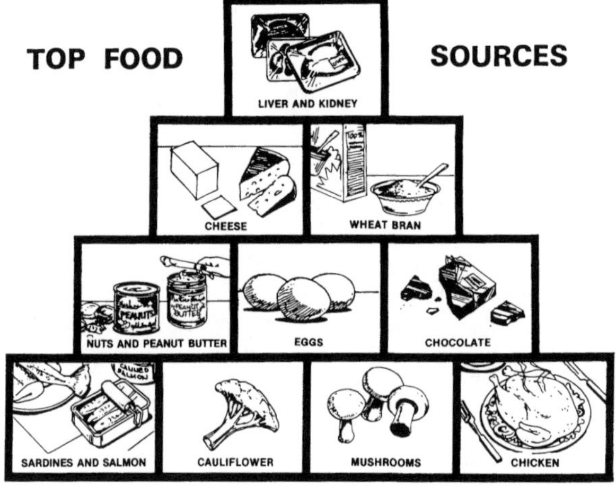

Fig. B-59. Good food sources of biotin.

Biotin, another member of the vitamin B complex, is a water-soluble, sulfur-containing vitamin that is widely distributed in nature and essential for the health of many animal species, including man. It plays an important role in the metabolism of carbohydrates, fats, and proteins.

HISTORY. The history of biotin is the history of the merging of investigations, a chronological record of which follows:

In 1901, Wildiers expressed the belief that yeast required for its nutrition an organic substance which he called "bios."

In 1916, Bateman reported that raw egg white had a detrimental effect on rats, but that the raw egg white was rendered "innocuous" by heat coagulation.

In 1927, Boas in England reported that feeding egg white to rats produced a dermatitis.

In 1933, Allison and co-workers isolated a nitrogen-fixing bacteria in legume nodules, which they named coenzyme R.

In 1936, Kogl and Tonnis of Germany isolated a crystalline substance from boiled yolks of duck eggs, which they called "biotin," since they believed it to be identical to the "bios" factor needed for yeast growth.

In 1937, Gyorgy, the Hungarian scientist, found that

a substance which he named vitamin H, would prevent the pathological condition that resulted from feeding rats and chicks raw egg white.

In 1940, Gyorgy and associates obtained conclusive experimental evidence that coenzyme R, biotin, and vitamin H were the same substance.

In 1942, du Vigneaud and associates, at Cornell, suggested the correct structural formula for biotin based on a study of its degradation products.

In 1943, Harris and his co-workers of Merck and Company synthesized biotin.

Looking back, it is now known that the deficiency (once called egg-white injury) occurs when the biotin in food combines with a factor in the protein of uncooked egg white (called avidin, because of its avidity for biotin); and that when egg white is cooked, avidin is inactivated. This also explains why, in early studies, liver and yeast offered protection against egg-white injury. (Both of them contain sufficiently large amounts of biotin to saturate the avidin completely and leave a surplus of biotin available to meet the needs of experimental animals.)

CHEMISTRY, METABOLISM, PROPERTIES.

• **Chemistry**—Biotin, like thiamin, is a sulfur-containing vitamin. It is a cyclic derivative of urea with an attached thiophene ring. Its structure is given in Fig. B-60.

BIOTIN

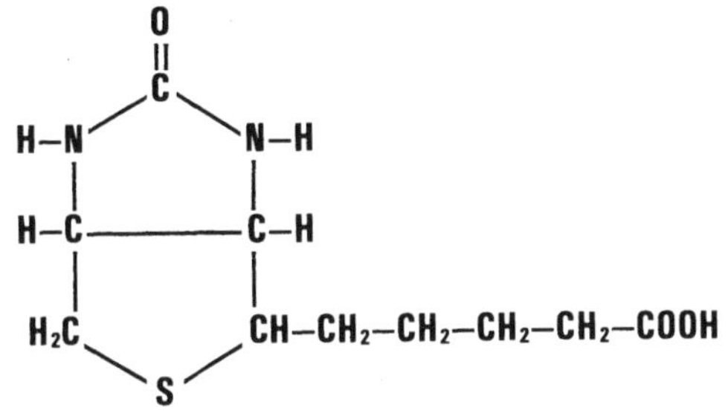

```
            O
            ‖
            C
          /   \
      H–N       N–H
        |       |
      H–C ——————C–H
        |       |
      H₂C       CH–CH₂–CH₂–CH₂–CH₂–COOH
          \   /
            S
```

Fig. B-60. Structure of biotin.

• **Metabolism**—Biotin is absorbed primarily from the upper part of the small intestine. However, avidin, a protein found in raw egg white, binds biotin and prevents its absorption from the intestinal tract. Fortunately, cooking inactivates avidin so that it no longer has the ability to bind biotin.

A considerable amount of biotin is synthesized by human intestinal bacteria, as evidenced by the fact that 3 to 6 times more biotin is excreted in the urine and feces than is ingested. But synthesis in the gut may occur too late in the intestinal passage to be absorbed well and play much of a direct role as a biotin source. Also, several variables affect the microbial synthesis in the intestines, including the carbohydrate source of the diet (starch, glucose, sucrose, etc.), the presence of other B vitamins, and the presence or absence of antimicrobial drugs and antibiotics.

Following absorption, biotin enters the portal circulation. It is stored primarily in the liver and kidneys, although all cells contain some biotin.

Excretion is mainly in the urine. Only traces of biotin are secreted in milk.

Determination of biotin levels in the blood and urinary excretion levels of biotin both provide evidence of biotin status in human beings.

• **Properties**—Biotin is a colorless, odorless, crystalline substance. It is readily soluble in hot water, but only slightly soluble in cold water; and it is stable to heat. It is destroyed by strong acids and alkalis and by oxidizing agents. Also, it is gradually destoyed by ultraviolet light.

MEASUREMENT/ASSAY. No International Units have been defined for the biological activity of biotin.

Analytical results are generally expressed in terms of weight units of pure d-biotin. Purified, high potency solutions of biotin may be assayed by a photometric method based on splitting of an avidin dye complex by biotin. Also, gas-liquid chromatography is sometimes used. However, the microbiological assay is the method of choice for biotin assay of foods. The biotin content of foods can also be determined by using the rat or the chick. Biological assays with rats or chicks are more reliable than the other methods because of determining availability; for example, it has been shown that the biotin in milo, oats, and wheat is less available to the chick than the biotin in corn.

FUNCTIONS. Biotin is required for many reactions in the metabolism of carbohydrates, fats, and proteins. It functions as a coenzyme mainly in decarboxylation-carboxylation and in deamination.

Biotin serves as a coenzyme for transferring CO_2 from one compound to another (for decarboxylation—the removal of carbon dioxide; and for carboxylation—the addition of carbon dioxide), as shown in Fig. B-61.

Numerous decarboxylation and carboxylation reactions are involved in carbohydrate, fat, and protein metabolism; among them, the following:

1. Interconversion of pyruvate and oxaloacetate. The formation of oxaloacetate is important because it is the

CO_2 TRANSFER TO BIOTIN

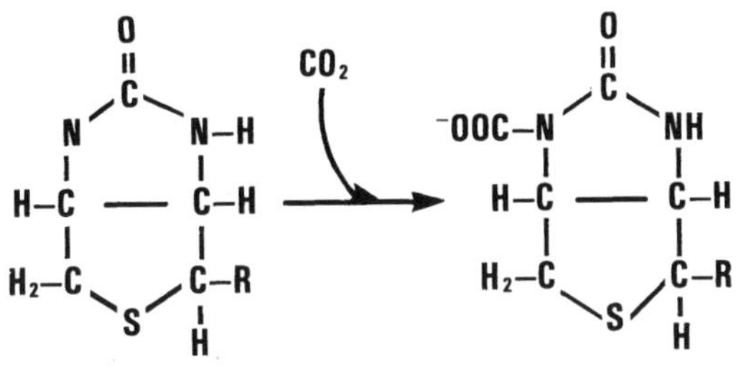

```
       O                                  O
       ‖            CO₂                   ‖
       C                                  C
     /   \           \                  /   \
    N     N–H         \         ⁻OOC–N       NH
    |     |            \             |       |
  H–C  —  C–H   ————————→          H–C   —   C–H
    |     |                          |       |
 H₂–C     C–R                     H₂–C       C–R
     \   / |                          \     / |
       S   H                            S     H
```

Fig. B-61. Transfer of CO_2 by biotin.

starting point of the tricarboxylic acid cycle (TCA), known also as the Krebs cycle, in which the potential energy of nutrients (ATP) is released for use by the body.

2. Interconversion of succinate and propionate.
3. Conversion of malate to pyruvate.
4. Conversion of acetyl CoA to malonyl CoA, the first step in the formation of long chain fatty acids (fat synthesis).
5. Formation of purines, essential part of DNA and RNA, and for protein synthesis.
6. Conversion of ornithine to citrulline, an important reaction in the formation of urea.

Biotin also serves as a coenzyme for deamination (removal of -NH_2) reactions that are necessary for the production of energy from certain amino acids (at least aspartic acid, serine, and threonine); for amino acids to be used as a source of energy, they must first be deaminated—the amino group must be split off.

Biotin is closely related metabolically to folacin, pantothenic acid, and vitamin B-12.

DEFICIENCY SYMPTOMS.

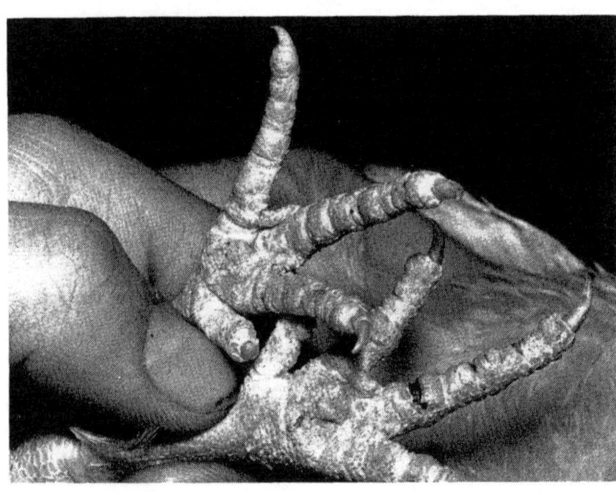

Fig. B-62. Biotin deficiency in a chicken. Note the severe lesions on the bottom of the feet. (Courtesy, H. R. Bird, Department of Poultry Science, University of Wisconsin)

Deficiency symptoms in humans (and in some animals) can be produced (1) by feeding large amounts of raw egg white, containing the biotin-binding glycoprotein known as *avidin;* or (2) by feeding a biotin-free diet in conjunction with a sulfa drug. *Avidin* renders the vitamin nutritionally unavailable, whereas the sulfa drug prevents intestinal synthesis. Deficiency symptoms in man include: a dry scaly dermatitis, loss of appetite, nausea, vomiting, muscle pains, glossitis (inflammation of the tongue), pallor of skin, mental depression, a decrease in hemoglobin and red-blood cell levels, a high cholesterol level, and low excretion of biotin; all of which respond to biotin administration.

There is now substantial evidence that *seborrheic dermatitis* (an abnormally oily skin, which results in chronic scaly inflammation) of infants under 6 months of age is due to nutritional biotin deficiency. In such cases, blood levels and urinary excretion of the vitamin are depressed. Prompt improvement occurs with therapeutic doses of the vitamin, about 5 mg/day, given intravenously or intramuscularly.

Biotin Dependency. A rare inherited disease called "biotin dependency" is known. In people with the disease, the body's use of biotin, the B vitamin necessary for certain metabolic processes, is somehow disrupted. Symptoms include loss of hair, lethargy, coma, and susceptibility to infections. The only treatment is daily doses of biotin. In 1981, medical researchers at the University of California in San Francisco reported successfully (1) diagnosing a biotin deficiency of an unborn baby by examining the amniotic fluid (the fluid was extracted from the womb by a procedure called amniocentesis, then cells from the fluid were grown in various nutrients and compared with normal cells); and (2) giving the mother massive doses of biotin, enough of which passed through the placenta so that the baby was born healthy.

NOTE WELL: Treating an unborn baby via the mother is an important new concept that could be applicable and very beneficial in treating other genetic diseases.

RECOMMENDED DAILY ALLOWANCE OF BIOTIN.

It is difficult to obtain a quantitative requirement for biotin, for the reason that intestinal microflora make a significant contribution to the body pool of available biotin; often humans excrete via the feces and urine considerably more biotin than they have ingested. However, the estimated safe and adequate intakes of biotin are given in Table B-14.

TABLE B–14
ESTIMATED SAFE AND ADEQUATE DAILY DIETARY INTAKES OF BIOTIN[1]

Group	Age	RDA Biotin
	(years)	(mcg)
Infants	0.0–0.5	10
	0.5–1.0	15
Children and Adolescents ...	1.0–3.0	20
	4.0–6.0	25
	7.0–10.0	30
	11+	30–100
Adults		30–100

[1]*Recommended Dietary Allowances*, 10th ed., 1989, NRC–National Academy of Sciences, p. 284.

In general, the combined urinary and fecal excretion of biotin exceeds the dietary intake. It seems likely that the fecal excretion of biotin is an indication of intestinal synthesis, whereas urinary excretion is a reflection of the dietary intake. Published reports of normal values of biotin in blood vary too widely for diagnostic use without carefully controlled observations.

The biotin content of human milk varies widely, but averages about 10mcg/1,000 kcal. Most formulas meet the recommended daily allowances (RDA) given in Table B-14.

Note that Table B-14 shows the National Research

Council (NRC) RDA of biotin for (1) infants, (2) children and adolescents, and (3) adults.

• **Biotin intake in average U.S. diet**—Mixed American diets are thought to provide a biotin intake of 100 to 300 mcg/day for adults. In western Europe, the dietary intake of biotin has been calculated to be between 50 and 100 mcg/day.

TOXICITY. There are no known toxic effects from biotin.

BIOTIN LOSSES DURING PROCESSING, COOKING, AND STORAGE. Considerable biotin is lost in the milling of cereal grains; hence, whole grains are a good source of the vitamin, whereas refined cereal products are a poor source.

Since biotin is stable to heat, cooking losses are not great.

SOURCES OF BIOTIN. Biotin is widely distributed in foods of both plant and animal origin; and it occurs in both the free state and in a form bound to protein. It occurs in the free state in fruits, vegetables, milk, and rice bran; and it occurs partly in a form bound to protein in meats, egg yolk, plant seeds, and yeast. Wide differences exist in the availability of biotin from food sources; for example, the biotin of corn and soybean meals is completely available to test animals, whereas the biotin of wheat is almost completely unavailable. Much needs to be learned about biotin availability. Sources of biotin follow:

• **Rich sources**—Cheese (processed), kidney, liver, soybean flour

• **Good sources**—Cauliflower, chocolate, eggs, mushrooms, nuts, peanut butter, sardines and salmon, wheat bran

• **Fair sources**—Cheese (natural), chicken, oysters, pork, spinach, sweet corn, whole wheat flour

• **Negligible sources**—Refined cereal products; most fruits and root crops

• **Supplemental sources**—Synthetic biotin, yeast (brewers', torula), alfalfa leaf meal (dehydrated)

For additional sources and more precise values of biotin, see Food Composition Table F-36 of this book.

It is noteworthy that the amount of avidin in raw egg white exceeds the amount of biotin in the whole egg. But, since avidin is destroyed by cooking, the usual diet includes little of the biotin-interfering substance.

In addition to food sources, biotin is available in pure synthetic form.

Also, considerable biotin is synthesized by the microorganisms in the intestinal tract, as evidenced by the fact that 3 to 6 times more biotin is excreted in the urine and feces than is ingested.

TOP FOOD SOURCES OF BIOTIN. The top biotin sources are listed in Table B-15.

NOTE WELL: This table lists (1) the top sources without regard to the amount normally eaten (left column), and (2) the top food sources (right column); and the caloric content of each food.

(Also see ADDITIVES, Table A-3; and VITAMIN[S], Table V-9.)

TABLE B-15
TOP BIOTIN SOURCES[1]

Top Sources	Biotin	Energy	Top Food Sources	Biotin	Energy
	(mcg/100 g)	(kcal/100 g)		(mcg/100 g)	(kcal/100 g)
Brewers' yeast	200	283	Chicken livers, simmered	170	165
Chicken livers, simmered	170	165	Beef liver	96	222
Torula yeast	100	277	Cheese, pasteurized process, American	82	375
Beef liver	96	222	Calf liver, fried	53	261
Cheese, pasteurized process, American	82	375	Peanut butter	39	585
Soybean flour, low, high, or full fat	70	386	Walnuts, Persian or English, shelled, chopped	37	694
Rice bran	60	276	Peanuts, roasted and salted	34	585
Wheat bran	57	353	Chocolate, bitter or baking	32	505
Rice polish	57	265	Sardines, Pacific, canned in brine or mustard, solids/liquid	24	186
Calf liver, fried	53	261	Eggs, chicken, fried	20	210
Eggs, chicken, raw, yolk, fresh	52	369	Almonds, shelled	18	638
Lamb liver, fried	41	261	Cauliflower, raw	17	27
Peanut butter	39	585	Mushrooms *(Agaricus campestris)*, raw	16	28
Lamb kidney, raw	37	105	Salmon, Atlantic, canned, solids/liquid	15	124
Walnuts, Persian or English	37	694	Chicken, broiled	11	136
Peanuts, roasted and salted	34	585	Oysters, canned, solids/liquids	9	73
Alfalfa leaf meal, dehydrated	33	215	Wheat flour, whole (from hard wheats)	9	361
Chocolate, bitter or baking	32	505	Spinach, frozen, chopped or leaf	7	25
Hog kidney, raw	32	106	Cheese, natural, camembert (domestic)	6	299
Hog liver, fried in margarine	27	241	Corn, raw, sweet, white/yellow	6	96

[1]These listings are based on the data in Food Composition Table F-36. Some top or rich sources may have been overlooked since some of the foods in Table F-36 lack values for biotin.

Whenever possible, foods are on an "as used" basis, without regard to moisture content; hence, certain high-moisture foods may be disadvantaged when ranked on the basis of biotin content per 100 g (approximately 3½ oz) without regard to moisture content.

BIRCH BEER

This is not really a beer, but an old-fashioned soft drink like ginger beer, lemon beer, and root beer. It is flavored with the oil of wintergreen, or oil of sweet birch and oil of sassafras. These old-fashioned soft drinks were fermented briefly with yeast, bottled, and stored. The fermentation served only to provide the carbonation.

(Also see SOFT DRINKS.)

BITOT'S SPOTS

Shiny, foamlike plaques on the white of the eye. They are usually grayish white and may be scraped off, but they return within a short time. Bitot's spots were originally believed to result from vitamin A deficiency, since they were frequently noted in children whose diets were grossly deficient in vitamin A. However, they have been observed when vitamin A deficiencies were absent, indicating that other nutritional deficiencies may be responsible for this condition.

(Also see DEFICIENCY DISEASES, Table D-2 Minor Dietary Deficiency Disorders; and VITAMIN A.)

BITTERS

• An alcoholic beverage flavored with extracts of bitter substances like quinine.

• A tonic containing bitter matters from such substances as aloes, bitter orange peel, cinchona bark, gentian, or quassia.

BIURET TEST

A chemical test for determining the presence and amount of protein in a solution. The biuret reagent containing copper sulfate is a bright blue. When the biuret reagent is added to a protein in a strong alkali solution, a blue-violet color is formed due to the reaction of the biuret reagent with the peptide bonds. The intensity depends upon the amount of protein.

(Also see PROTEIN[S].)

BLACKBERRY *Rubus spp.*

These species of fruits (of the family *Rosaceae)* usually grow on thorny bushes or trailing vines. Blackberries are similar to raspberries, except that in the case of the former the core of the fruit comes off with the berry when it is picked, whereas raspberry cores (receptacles) remain on the plant when the berry is picked. Both fruits are classified as "aggregate fruits" because each berrry is made up of many tiny drupes or druplets, each of which is considered to be a fruit.

ORIGIN AND HISTORY. Wild blackberries are widely distributed throughout the Northern Hemisphere and have been used as food by man since prehistoric times. Hundreds of varieties have evolved as a result of accidental and intentional crossbreeding (hybridization).

Fig. B-63. A cluster of blackberries. (Photo by J. C. Allen & Son, West Lafayette, Ind.)

The former occurred when the plants or seeds of certain varieties were (1) brought to new locations by animals or man, and (2) allowed to cross with native varieties. Also, the clearing of forests allowed established populations of the plants to spread into the territories of other varieties.

There does not seem to have been any major attempts to cultivate blackberries until the 19th century, which at first seems unusual in view of the very long use of the fruits as a food. The fruit was mentioned in the writings of Aescylus, a Greek, who lived in the 5th century B.C. It seems likely that there was little need to plant the berries since the wild fruit grew abundantly and even became a nuisance in certain areas. Nevertheless, the efficient harvesting of the fruit from wild plants is often difficult when they grow in dense, thorny thickets. Hence, European farmers and gardeners eventually developed cultural methods for this much esteemed crop.

It was not until the middle of the 19th century that cultivated varieties were introduced into the United States, where they soon "escaped" from cultivation and hybridized with wild American species. The fruit became very popular by the late 1800s—when it was much used in fresh fruit dishes, ice cream, pie, and puddings.

Recently, blackberry breeders have sought to develop a thornless plant, but have met with only limited success in this regard. However, the efforts to develop greater disease and pest resistance have been more successful.

PRODUCTION. Statistics on the world production of blackberries are not readily available. However, most of the commercial crop comes from the European countries and the United States. In recent years, the annual U.S. production has averaged around 20,000 metric tons, much of which is produced in Oregon. Smaller amounts are produced in Texas, California, Washington, Michigan, Arkansas, Oklahoma, Alabama, and North Carolina.

Blackberries are not as hardy as raspberries, but they are more tolerant of hot weather. Hence, they are grown mainly in areas that have milder climates than the raspberry-producing regions. The most favorable conditions are found along the Atlantic Coast, in the South Central states, near the Great Lakes, and along the Pacific Coast. It is noteworthy that temperatures below 0°F (*-18°C*) are often injurious to these plants. Blackberries, like raspberries, have biennial stems which grow vegetatively the first year, and bear fruit the second year. After the second year, the stems (canes) no longer bear fruit and are removed. However, the roots of the bush types are perennial and send up new suckers (shoots) every year. The roots of the viny types do not grow suckers unless they have been injured.

The bush types of blackberries are propagated by tip layering. The latter method consists of bending the upper tips of the stems and inserting them in the soil to take root, after which the rooted stem is cut from the parent plant. Many fruit growers start with young plants that are raised in a nursery and certified as disease-free, because blackberries are susceptible to many viruses and other disease organisms. The rooted cuttings are usually transplanted in the early spring, or in November in Texas. A fertile, well-drained, and slightly acid soil is required. Commercial growers usually add nitrogen, phosphorus, and potassium fertilizers because large amounts of soil nutrients are required by the growing plants.

Blackberry vines are often supported on trellises because (1) more fruit-bearing lateral branches are produced when the main stems are kept vertical, and (2) it is easier to cultivate around trellised vines than around those that are left sprawling on the ground. The bushy types also profit from support by trellises because when supported they may be allowed to grow taller without fear of them being damaged by strong winds.

It is very important that the lanes between the rows of blackberry plants be kept free of weeds, grasses, and blackberry suckers which draw on the soil nutrients needed by the crop and sometimes harbor diseases and pests. Irrigation which wets only the soil around the roots is much more desirable than that which wets the leaves and berries and encourages the spread of disease.

Blackberries are harvested after ripening, but before they have become soft and mushy. It is very unwise to allow the berries to become overripe, because they are then very susceptible to mold. Even one or two moldy berries in a box may cause the others to be ruined in a very short time. Bush blackberries may be harvested by a machine which shakes them onto catcher frames, but the berries that grow on vines are still harvested manually. These fruits spoil rapidly if wetted, hence they are washed very briefly just prior to eating or processing.

PROCESSING. About 98% of the U.S. blackberry crop is processed. Most of the processed berries are frozen, but some are canned, and the rest are made into brandy, cakes, desserts, jam, jelly, juice, liqueurs, pie, syrup, and wine.

SELECTION AND PREPARATION. High quality, ripened blackberries are moderately firm and have a purplish-black color. Red-colored blackberries are usually picked before they are ripened fully.

Fresh blackberries are hardly ever available, unless one lives near the berry farms because there is too great a loss of quality during shipment of the fresh fruit over long distances. When they are available, they make an excellent addition to compotes and cold, ready-to-eat breakfast cereals. The frozen fruit is usually best when allowed to thaw completely prior to use, but partial thawing may be sufficient when the berries are to be baked in a cake, pie, or other dessert. Canned berries are very soft and become too mushy when cooked. Hence, it is best to use them in various unheated dishes. The packing medium of canned blackberries usually picks up much of the color and flavor of the fruit, and may be used as a flavoring or syrup.

NUTRITIONAL VALUE. The nutritient compositions of various blackberry products are given in Food Composition Table F-36.

Some noteworthy observations regarding the nutrient compositions of the more common blackberry products follow:

1. The raw fruit is moderately high in calories (58 kcal per 100 g) and carbohydrates (13%). It is an excellent source of fiber (4.1%), a good source of potassium and bioflavonoids (vitaminlike substances with reputed beneficial effects when consumed along with vitamin C), and a fair to good source of vitamin A, vitamin C, and folic acid.
2. Canned blackberries packed in extra heavy or heavy syrup contain about twice the level of calories and carbohydrates present in the raw fruit. However, the caloric content of the fruit itself might be reduced by draining off the packing medium and rinsing the berries in water. It is noteworthy that a few processors pack blackberries in water or juice so that the caloric values of their products are close to those of the raw fruit.
3. Blackberry juice is low in calories (37 kcal per 100 g) and carbohydrates (8%). It is a good source of potassium, but only a fair source of vitamin C. Information regarding the effects of processing on the bioflavonoids is not readily available.
4. Blackberry pie is very rich in calories (243 kcal per 100 g) and carbohydrates (34%). It is a fair to good source of fiber and potassium, but only a poor source of vitamin C.
5. The purplish-black color of blackberries is due mainly to the anthocyanin pigment cyanidin-3-monoglocoside, which is a type of bioflavonoid.

(Also see FRUIT[S], Table F-47 Fruits of the World.)

BLACK BREAD

A dark-colored bread made with rye, which is popular in central and northern Europe.

BLACKSTRAP MOLASSES

In the extraction of sugarcane, blackstrap molasses is the third and final extraction. It has more of the minerals than light or medium molasses. Blackstrap has a strong flavor, so usually only small amounts are eaten. The nutrient composition of blackstrap molasses is given in Food Composition Table F-36.

(Also see SUGAR, section headed "Sugarcane.")

BLACK TONGUE

A nutritional disorder in dogs which is similar to human pellagra, a disease due to a deficiency of the vitamin niacin. Dogs with black tongue were used in the experiments which identified niacin as the cure for this condition.

(Also see NIACIN.)

BLADDER

A hollow saclike organ in which the urine is stored before it is voided. The bladder is surrounded with a muscular coat which contracts when it is stretched. Hence, a full bladder is likely to be emptied by this muscular reflex. However, people can train themselves to delay the emptying of the bladder.

BLAND DIETS

Diets which have long been prescribed for people with disorders such as colitis, hiatus hernia, ileitis, inflammation of the esophagus, and ulcers—illnesses that are commonly characterized by (1) irritation and excessively rapid motions of one or more digestive organs; and (2) oversecretion of digestive juices such as stomach acid and bile. Hence, in an attempt to avoid, or lessen, gas and/or diarrhea, the doctor or dietitian is likely to recommend foods which (1) are soft in consistency and nonirritating, (2) do not overstimulate the flow of gastric acid or other digestive juices, and (3) are relatively low in fermentable carbohydrates and undigestible matter.

A typical dietary prescription for a bland diet might include instructions such as those which follow:

1. The foods which are allowed should be apportioned between six meals—breakfast, a midmorning snack around 10 a.m., lunch, a midafternoon snack around 3 p.m., supper, and an evening snack after 8 p.m.
2. Use only bland foods such as (a) decaffeinated coffee, weak tea, milk, and fruit juices (citris products are to be taken at the end of a meal); (b) dairy products; (c) soft cooked eggs; (d) white bread and other refined grain products; (e) fish, meats, and poultry that are baked or broiled, but *not* fried; (f) baked or boiled Irish potatoes; (g) butter or margarine; (h) soft cooked vegetables; (i) fruits with seeds and skins removed (only one serving of a citrus fruit taken at the end of a meal); (j) mildly seasoned cream soups or sauces *without* bouillon or other meat extractives; and (k) desserts such as plain cakes, cookies, flavored gelatin, puddings, custards, ice cream, and sherbets.
3. Avoid (a) alcoholic beverages, and caffeine-containing drinks such as coffee, cola types of soft drinks, and strong tea; (b) whole-grain breads and cereals; (c) fried items; (d) fibrous fruits and vegetables; (e) raw vegetables and those of the cabbage family; and (f) pungent spices, herbs, and seasonings.

Bland diets may be nutritionally deficient and lead to constipation, so their use should be limited to the acute stages of digestive disorders.

(Also see MODIFIED DIETS.)

BLEEDING TIME

An approximate measure of the blood clotting ability. In this test, either the end of the finger or the earlobe is pricked so that it bleeds, and a piece of filter paper is touched to the wound every 20 seconds. The measure of clotting time is taken to be the length of time after which the filter paper no longer becomes stained with blood. This may be a very rough indicator of nutritional status, and a test for other disorders such as impairment of blood clotting. For example, blood clotting takes an abnormally long time in vitamin K deficiency. However, there are much more sensitive tests of blood clotting time which may be performed when suitable laboratory facilities are available.

(Also see VITAMIN K.)

BLENDED

Combined or mixed so as to render the constituent parts indistinguishable from one another, such as when two or more food ingredients are mixed.

BLENDER TUBE FEEDINGS

(See TUBE FEEDING.)

BLEPHARITIS

A cracking at the outer corners of the eyes that is often associated with deficiencies of riboflavin, pyridoxine, and other B vitamins.

(Also see DEFICIENCY DISEASES; and VITAMIN B COMPLEX.)

BLEPHAROCONJUNCTIVITIS, ANGULAR

A condition in which the lining of the eyelid is inflammed. It is believed to be due to a combination of (1) infection of *Hemophilun duplex* bacteria and (2) a pyridoxine deficiency. Often, the condition can be cured solely by the administration of pyridoxine.

(Also see VITAMIN B-6.)

BLETTING

The overripening and softening of certain fruits in storage.

BLIND LOOP SYNDROME.

A disorder characterized by (1) distention of a section of the small intestine so that a blind loop is formed; (2) stagnation of the flow of digesta in the loop; and (3) an overgrowth of certain bacteria in the loop. Blind loop syndrome may develop after certain types of intestinal surgery or when there are multiple diverticula (outpouchings of the intestinal wall). The syndrome is characterized by abdominal pain (especially after meals), gas, weight loss, diarrhea, constipation, nausea, vomiting, and malabsorption of vitamin B-12 and other essential nutrients. Usually, surgery is needed to correct the condition.

(Also see DIGESTION AND ABSORPTION; DIVERTICULITIS; and VITAMIN B-12)

BLINDNESS DUE TO VITAMIN A DEFICIENCY (XEROPHTHALMIA)

On a worldwide basis, blindness due to vitamin A deficiency is the most prevalent type of blindness. This condition is very common in some parts of the developing countries of Asia, the Middle East, Africa, and Latin America, where diets contain practically no vegetable or animal sources of vitamin A. Fetuses, infants, and preschool children are the most affected due to a high requirement for the vitamin during the development of the tissues of the eye.

HISTORY OF BLINDNESS DUE TO VITAMIN A DEFICIENCY. Night blindness was known in ancient Egypt and Greece, at which time the treatment consisted of feeding roasted ox liver and liver of roosters. Hippocrates prescribed ox liver dipped in honey. It was found that animals suffering from this deficiency stumbled in the dark.

In 1816, Magendie, a French physiologist, noted that ulcers formed on the corneas of dogs' eyes when they were fed unsupplemented diets consisting of wheat gluten, starch, sugar, or olive oil as their sole food.

Dr. Livingston, a British medical missionary, reported that his African bearers suffered from xerophthalmia, which was probably due to their limited diets of coffee, cassava, and meal (corn).

Slaves on Brazilian plantations suffering from night blindness were observed by Gouvea (in 1883) to have better vision at sunrise than at sunset; although it was darker when they went to work in the morning than when they returned from work in the evening. He suspected that exposure to sunlight aggravated the night blindness, and that resting their eyes at night led to their recoveries. We now know, from recent research findings, that strong sunlight rapidly destroys the vitamin A associated with the light sensitive visual pigment in the eyes.

Mori reported, in 1904, that keratomalacia and xerophthalmia were prevalent in Japanese children who ate diets consisting mainly of grains and beans, but that the diseases could be cured by giving cod-liver oil, chicken livers, or eel fat. He, therefore, attributed the diseases to deficiencies of diet fat.

In 1917, Block noted the high incidence of xerophthalmia in Danish children who lived in an institution where they were fed skim milk (the cream removed from milk was used to produce butter for export to Germany) and unfortified margarine. He then conducted a controlled study and reported, in 1919, that milk fat contained a factor which protected children fed whole milk against the disease. About the same time (1917), McCollum and his co-workers at the University of Wisconsin reproduced in rats the observations which had been previously made in humans, and explained that the eye disorders were the result of a fat-soluble factor which they named vitamin A.

Steenbock and Gross reported, in 1919, that their animal studies at the University of Wisconsin showed a vitamin A activity in yellow-colored vegetables, such as carrots and sweet potatoes, while there was practically no activity in white vegetables, such as Irish potatoes and parsnips. Also, in 1919, Palmer and Kempster at the University of Missiori showed that the presence of a yellow pigment was not necessary for vitamin A activity, since chickens fed pigment-free pork liver did not develop a deficiency. The findings of these two groups of investigators were explained in 1929 by Moore Cambridge who showed that rats fed highly purified carotene (the pigment from yellow vegetables) had livers rich in vitamin A.

In 1931, the Swiss scientist, Karrer, determined the structure of the vitamin. Wald of Harvard, who in 1935 explained the role of vitamin A in vision, was awarded the Nobel Prize for his work in 1967.

THE ROLE OF VITAMIN A DEFICIENCY IN THE DEVELOPMENT OF BLINDNESS. Vitamin A is required for the synthesis of the light-sensitive pigment in the retina, and for the normal functioning of the mucous-secreting epithelial cells in the eye. Deficiency of the vitamin leads to *keratinization* (drying and hardening) of these cells. There may be total failure of eye development in the fetus, due to a severe deficiency in the pregnant mother. The vitamin is stored in the liver, so it is less likely that adults will develop a severe deficiency if they have previously had diets containing at least the minimum requirements for the vitamin or its precursors (carotenoid compounds, found in green and yellow vegetables, are converted in the body to vitamin A).

Mild deficiencies, such as *night blindness,* might develop in persons who eat adequate diets, but who are under conditions where destruction of the vitamin is increased within the body, such as in the case of exposure of the eyes to intense sunlight. It should be noted that some cases of vitamin A deficiency may be due to disorders of fat absorption (celiac disease, cystic fibrosis, sprue, liver disease) and transport of the vitamin within the body. In protein deficiency, there may be a shortage of the protein which carries vitamin A in the blood, and blood levels of the vitamin may be low, though liver stores are nearly normal. Some eye disorders have developed when children have been weaned from human breast milk and put on forms of milk low in the vitamin, e.g., sweetened condensed or nonfat milk.

Vitamin A Deficiency and Protein-energy Malnutrition. Xerophthalmia often develops during the rehabilitation of children who have suffered from kwashiorkor, a form of protein-energy malnutrition. A tenative explanation for this observation is as follows: (1) Children develop kwashiorkor on a diet deficient in protein and, in many cases, also deficient in vitamin A; (2) in some cases, the development of xerophthalmia is delayed by a reduction in the amount of vitamin A used by the body, for growth ceases in severe protein deficiency (a greater proportion of the limited body supply of vitamin A is therefore available to protect the eyes); and (3) treatment of protein-deficient children with unfortified skim milk results in the resumption of growth, an increased need for vitamin A to support growth, and the development of xerophthalmia due to further depletion of the body stores of vitamin A.[19]

Special attention should, therefore, be given to providing vitamin A in diets used for the rehabilitation of other nutrient deficiencies.

SIGNS OF VITAMIN A DEFICIENCY IN THE EYE. *Night blindness* is one of the earliest signs of vitamin A deficiency. Another early sign is dryness of the conjunctival membrane of the eye. *Bitot's spot* is a foamy patch alongside the cornea, which might indicate vitamin A deficiency, although it has been found in persons whose supply of the vitamin is known to be adequate. *Xerophthalmia* refers to the effect of the deficiency on the cornea (keratinization and cloudiness), although it has been suggested that the term be used to indicate eye disorders in general which result from vitamin A deficiency. *Keratomalacia* is the softening of the cornea which follows the development of xerophthalmia. If there is not prompt treatment at this stage, there may be perforation of the cornea, extrusion of the lens from the eye followed by infection, and perhaps destruction of the contents of the eyeball.

TREATMENT AND PREVENTION OF EYE DISORDERS DUE TO VITAMIN A DEFICIENCY. Vitamin A should be administered immediately to persons with xerophthalmia in order to arrest the disease. A suitable therapeutic dose is 100,000 IU daily for 3 days. The most rapid therapeutic effects are obtained with intramuscular injections of a water-solubilized form of vitamin A, which is more effective than oral or injected doses of the fat-soluble forms of the vitamin.[20] Then, during convalescence, 30,000 IU of Vitamin A may be given per day in the form of cod- or halibut-liver oil.

Night blindness may be treated by a few daily doses of 30,000 IU of the vitamin.

Vitamin A deficiency may be prevented by providing food sources of the vitamin such as dark green and deep yellow fruits and vegetables, liver, butter or fortified margarine; or if none of these foods are available, fish-liver oil preparations which have been protected against loss of potency by storage in a cool, dark place. Vitamin A is destroyed by heat, light, and oxygen.

NOTE: Extra protein is therapeutic for children whose eye problems are partially due to insufficient protein for the transport of vitamin A in the blood. (It was mentioned earlier in this article that in such cases the liver stores of the vitamin may be normal.)

(Also see DEFICIENCY DISEASES, Table D-1 Major Dietary Deficiency Diseases; and VITAMIN A.)

BLOATERS

Herrings (fish) that have been salted less and smoked for a shorter period of time than red herring. (The latter are well salted and smoked for about 10 days.)

[19]Arroyave, G., "Interrelations Between Protein and Vitamin A and Metabolism," *The American Journal of Clinical Nutrition,* Vol. 22, 1969, p. 1119.

[20]Pereira, S. M., *et al.,* "Vitamin A Therapy in Children with Kwashiorkor," *The American Journal of Clinical Nutrition,* Vol. 20, 1967, p. 297.

BLOOD

The fluid that circulates in the vascular system of vertebrate animals, carrying nourishment and oxygen to all parts of the body and taking away waste products for excretion. Blood consists of liquid plasma containing dissolved nutrients, waste products, and other substances and suspended red blood cells, leukocytes, and blood platelets.

BLOOD CELLS

Three classes of blood cells (corpuscles) are recognized—red cells, white cells, and platelets. All these cells are suspended in the fluid called plasma.

- **Red blood cells**—These cells, which are also called *erythrocytes*, contain a red pigment hemoglobin which binds oxygen. Hence, an insufficient number of these cells (anemia) limits the ability of the blood to carry oxygen to the cells of the body where it is needed.

(Also see ANEMIA.)

- **White blood cells**—These cells, which are also called *leucocytes*, help the body fight against infectious organisms. Certain types of white cells, called phagocytes, are able to ingest and destroy microorganisms.

(Also see SELENIUM.)

- **Platelets (Thrombocytes)**—These tiny circular or oval discs in the blood, numbering about 250,000 per cubic millimeter, are concerned with blood coagulation and contraction of the clot.

BLOOD CHOLESTEROL LEVEL

This measurement is often used for assessing the risk of heart disease. However, it is sometimes misinterpreted because cholesterol does not travel by itself in the blood, but is carried by complex molecules called lipoproteins. For example, beta-lipoproteins, when elevated in the blood, are associated with increased risks of heart disease and stroke; whereas, a high level of alpha-lipoproteins is apparently a good sign. Hence, the measurement of blood cholesterol alone is at best a very crude measurement of the status of cholesterol metabolism in the body.

(Also see ALPHA-LIPOPROTEINS; BETA-LIPOPROTEINS; CHOLESTEROL; and HEART DISEASE.)

BLOOD FATS (LIPIDS)

The type of fatty substances which are most commonly found in the blood are: triglycerides, fatty acids, cholesterol, phospholipids, lipoproteins, and the fat-soluble vitamins A, D, E, and K.

(Also see FATS AND OTHER LIPIDS.)

BLOOD PLATELETS (THROMBOCYTES)

These cells, which are also called thrombocytes, are involved in blood clotting. There may be too few platelets when certain nutritional deficiencies result in anemia. Hence, the blood clotting time may be prolonged in these blood disorders.

BLOOD PRESSURE

The force which the blood exerts against the wall of the blood vessels, due to the pumping action of the heart. When the blood pressure is measured, two readings are generally taken. The higher reading, which is called *systolic blood pressure,* is that which occurs when the heart exerts its maximum force of contraction. The lower reading, which is called the *diastolic blood pressure,* occurs when the heart rests between contractions. Normal blood pressure values center around readings of 120 systolic, and about 80 diastolic. Generally, blood pressure increases a few points with age, but a large increase is indicative of *high blood pressure,* a condition which greatly increases the risk of kidney disease and stroke. High blood pressure probably has multiple causes such as (1) the loss of elasticity in the blood vessel walls, making it necessary for the heart to pump harder to force the blood through these vessels; (2) retention of sodium and water, which increases the blood volume; and (3) clogging of blood vessels so that the openings are smaller. *Low blood pressure may be a sign of heart weakness and it may make the person susceptible to fainting.*

(Also see HIGH BLOOD PRESSURE.)

BLOOD SAUSAGE

A variety of cooked sausage containing a large proportion of blood, usually formulated with pork (skins, cured ham fat, and/or pork snouts and lips and stuffed in a casing).

BLOOD SUGAR

Although several different sugars circulate in the blood, the main blood sugar is glucose. Hence, the blood sugar level is usually considered to mean the same thing as the blood glucose level. In healthy people, the blood sugar is maintained within a narrow range. In diabetes, the blood sugar may be abnormally high; in the condition called low blood sugar (hypoglycemia), the blood sugar is too low.

(Also see DIABETES MELLITUS; and HYPOGLYCEMIA.)

BLOOD SUGAR TEST

Various tests of the blood sugar may be given in order to diagnose certain metabolic disorders such as diabetes and low blood sugar (hypoglycemia). One of the simplest tests is called a fasting blood sugar which is usually made on a sample of blood taken after an overnight fast. Another commonly used test is the glucose tolerance test in which a fasting blood sugar is measured; then a test dose of glucose is given by mouth or by vein and the blood sugar is meaured at regular intervals. Diagnoses are not always based upon blood sugar tests alone because there are many causes of blood sugar abnormalities. Hence, many doctors measure other factors along with the blood sugar, such as the level of insulin in the blood.

(Also see DIABETES MELLITUS; and HYPOGLYCEMIA.)

BLOOD VASCULAR SYSTEM

The vessels that transport blood throughout the body.

BLOOM

This is a mottled, whitish discoloration of chocolate that occurs when part of the fat melts, migrates to the surface, and then resolidifies on the surface as light colored blotches with an accompanying overall loss of gloss. Bloom is an undesirable change that may occur during storage. Additives—emulsifiers—such as, sorbitan monostearate and polysorbate 60, when added to chocolate and chocolate coatings prevent bloom.

(Also see ADDITIVES; and COCOA AND CHOCOLATE.)

BLUEBERRY *Vaccinium* spp.

These much esteemed fruits from bushy and shrubby plants of North America are members of the family *Ericaceae*, which also includes bilberries, cranberries,

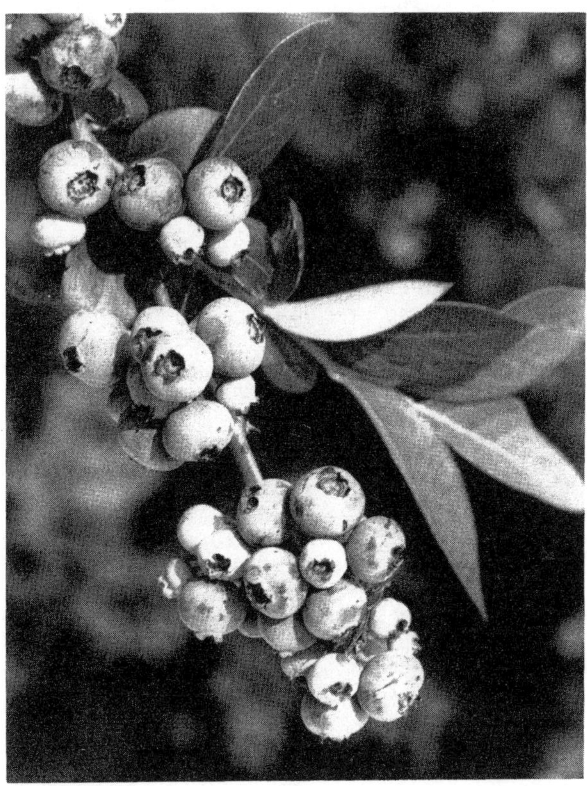

Fig. B-64. Clusters of blueberries. (Photo by J. C. Allen & Son, West Lafayette, Ind.)

huckleberries, and certain nonedible plants such as azaleas, heathers, mountain laurel, and rhododendrons.

Blueberries bear many-fruited clusters, whereas the closely related bilberries bear their fruit singly. Sometimes, blueberries are confused with huckleberries (*Gaylussacia* spp.). However, the former fruits have small, soft seeds, whereas the latter have large, hard seeds.

The major types of blueberries found in North America follow:

- **Dryland, or low blueberry** (*V. pallidum*)—A lowbush-type of berry that grows wild and is most commonly found on hills and ridges from West Virginia and Maryland to Alabama and Georgia.

- **Evergreen or box blueberry** (*V. ovatum*)—This berry is gathered from shrubby plants that grow wild in Washington, Oregon, and northern California.

- **Highbush blueberry** (*V. australe, V. corymbosum*)—The first blueberries to be cultivated commercially were of this type, which grows wild from Maine to southern Michigan, and south to Florida.

- **Lowbush blueberry** (*V. augustifolium*)—This type is the most cold hardy of the blueberries, and grows wild in the northeastern United States and Canada. The spread of this species is encouraged by the clearing of forests and the burning of brush.

- **Mountain blueberry** (*V. membranaceum*)—These excellent-flavored berries are maroon to black in color and are borne by a very drought-resistant species of shrub that grows wild in the Sierra and Cascade mountains from Oregon to British Columbia.

- **Rabbiteye blueberry** (*V. askei*)—A small black blueberry that grows wild and is cultivated in the southeastern United States.

ORIGIN AND HISTORY. The blueberry is native to North America, whereas the closely related *bilberry* originated in Europe. This fruit was used by the Indians to make *pemmican*, which is prepared by mixing sun- or wind-dried strips of meat with melted fat and various types of berries. Little or no efforts were made by the early colonists to bring any of the wild types of blueberries under cultivation because (1) an abundant supply of the wild plants grew wild throughout the eastern United States, and (2) wild blueberries were tart flavored, and the sugary materials suitable for sweetening them were in short supply until the mid-1800s.

Americans began to consume greater amounts of blueberries after the supply of sugar increased sharply in the latter part of the 18th century, when the fruit became a popular ingredient of jams, jellies, pies and tarts. (It is noteworthy that the U.S. per capita consumption of sugar tripled in the period between 1875 and 1935.)

The cultivation of blueberries in the United States was initiated in 1906 by Dr. F. V. Coville, a botanist in the U.S. Department of Agriculture, who selected wild highbush berry plants from New Hampshire and New Jersey for his experiments. He noted that the wild plants thrived in swamps, but did not grow as prolifically in other environments. Hence, he concluded that an acid soil was required. By 1909, he was ready to begin crossbreeding different varieties in order to obtain higher yields of larger and sweeter berries.

Reports of Dr. Coville's work were read by Miss Elizabeth White of Whitesbog, New Jersey (a small town located in eastern Burlington county a few miles south of Fort Dix, that was so named because the White family had extensive cranberry plantings in the local bogs.) Miss White became interested in the cultivation of blueberries because she had often observed the wild bushes growing on her family's land. She began a long collaboration with Dr. Coville, which resulted in her conducting experiments of her own, and her family establishing the first commercial blueberry plantings in America.

The first commercial shipments of blueberries were made in 1916, and the production and utilization of this fruit has increased steadily since then.

PRODUCTION. Most of the world's blueberry crop comes from North America. About three-fourths of this production comes from the United States. Fig. B-65 shows the amounts shipped to the leading markets.

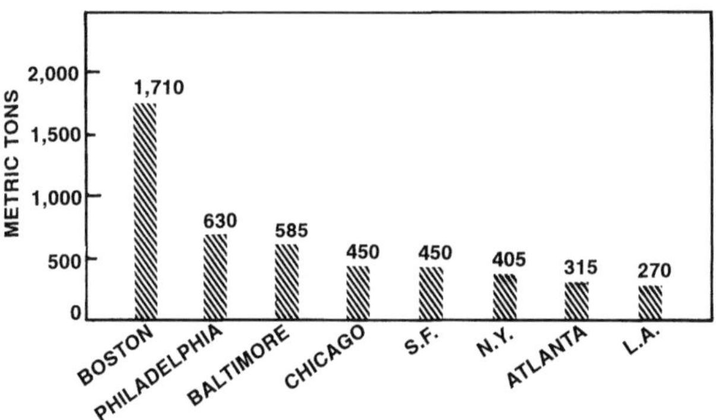

Fig. B-65. Fresh blueberries shipped to 8 specified markets. (Data from *Agricultural Statistics*, 1991, pp. 212, 213, Tables 327, 328)

Much of the U.S. blueberry crop is gathered from wild plants that have become established by natural means such as (1) the dissemination of seeds in the droppings of birds and animals, and (2) vegetative spreading of the established wild plants by rhizomes and suckers. However, special cultural practices are often utilized in order to propagate and/or nurture the highbush, lowbush, and rabbiteye types of plants.

Both the highbush and rabbiteye types are propagated by cuttings. Additionally, the highbush blueberry is propagated by mound layering and grafting; and the rabbiteye blueberry is propagated by suckers. The lowbush type is propagated by vegetative spreading; attempts to propagate this type by cuttings or seeds have not been successful. Blueberries prefer well-drained acid soil. Most blueberries are picked by hand.

PROCESSING. Over two-thirds of the U.S. blueberry crop is processed. More than half of the processed berries are frozen and the rest are canned. Frozen blueberries are packed with or without a 50% sugar syrup, whereas the canned fruit is usually packed in water or in a syrup. Canned blueberry pie fillings, in which the berries are cooked, sweetened, and packed in a thickened liquid, are rapidly gaining popularity in the United States and Great Britain. It is noteworthy that only a small part of the processed berry production is marketed in retail size packages and containers. Most of it is packed in wholesale drums and cans and is utilized by bakeries and the manufacturers of ice cream, jams, jellies, and pies.

SELECTION. High-quality blueberries are plump, of fresh appearance, clean, dry, free from leaves and trash, fairly uniform in size, and of a deep, full color throughout the lot. Ripeness is indicated by the color which may be blue, black, bluish-black, or purplish. The berries may be covered with more or less bloom (a grayish waxy deposit on the skin) depending on the variety.

Overripe fruit has a dull, lifeless appearance and is often soft and watery. Berries held long after picking have a similar appearance and may be more or less shriveled. Freedom from moisture is essential to good-quality berries. Moisture may be caused by natural breakdown, decay, or some form of mechanical injury. Decay is usually indicated by the presence of molds.

PREPARATION. Fresh ripe blueberries may spoil if left out at room temperature for a day or more, but they will keep for 2 or 3 days if stored without washing in a covered container in the refrigerator. The skin of the unwashed fruit is protected by the bloom. Hence, the berries should not be washed until just before they are to be used.

Cultivated berries are only mildly tart and are sweet enough to be eaten without added sugar, whereas the wild berries have a more tangy flavor and are usually best when sweetened a little. Fresh blueberries are good when served with cream in a bowl (with or without added sweetener), or when added to cereals, fruit compotes and fruit salads, ice cream, plain cake, or vanilla pudding. It is noteworthy that the wild berries make better jams, jellies, muffins, pies, and tarts than the cultivated berries.

Frozen and canned blueberries may be substituted for the fresh fruit in most dishes, except that the processed berries are softer and should be given little or no cooking. Therefore, it might be a good idea to use unthawed frozen blueberries in preparations that require a moderately long cooking period. The packing syrup from canned or frozen berries usually has picked up much of the color and flavor from the fruit. Hence, it may be used as a flavoring or a syrup for ice-cream sodas and sundaes, milk shakes, mixed drinks, pancakes, and waffles.

NUTRITIONAL VALUE. The nutrient composition of various blueberry products is given in Food Composition Table F-36.

Some noteworthy observations regarding the nutrient compositions of the more commonly available blueberry products follow:

1. The raw fruit and the unsweetened frozen fruit supply moderate amounts of calories (about 60 kcal per 100 g) and carbohydrates (14 to 15%). They are good sources of fiber, iron, and bioflavonoids (vitaminlike substances with reputed beneficial effects when consumed along with vitamin C), and fair to good sources of potassium and vitamin C.
2. Sweetened frozen blueberries contain nearly double the levels of calories and carbohydrates present in the raw and the unsweetened frozen berries. Furthermore, a given weight of the sweetened product furnishes less of the other nutrients than equal amounts of the unsweetened products.
3. Canned blueberries packed in water are low in calories and carbohydrates because they contain only about two-thirds the levels of the nutrients that are supplied by the raw fruit.
4. Canned blueberries packed in heavy syrup contain about two and one-half times the calories and carbohydrates that are present in canned berries packed in water alone.
5. Blueberry pie is rich in calories (242 kcal per 100 g) and carbohydrates (35%). It is a fair source of potassium and iron.
6. Blueberry turnovers are very rich in calories (405 kcal per 100g), carbohydrates (41%), and fats (25%). They are a good source of iron, but a poor source of potassium and vitamin C.
7. Apple blueberry fruit puree (commonly sold as a baby food) supplies moderate amounts of calories (68 kcal per 100 g) and carbohydrates (16%). It is a good source of vitamin C and a fair course of potassium.
8. The color of blueberries is due mainly to anthocyanin pigments, which are also classified as bioflavonoids.

(Also see FRUIT[S], Table F-47 Fruits of the World.)

BLUE CHEESE

A type of cheese characterized by visible blue-green veins of mold throughout the cheese and by a sharp, piquant flavor. In the family of blue-mold cheese, there are varietal names, but the products differ appreciably only where the milk source varies. For example, roquefort is made only in the Roquefort area of France from sheep's milk; its cow's milk counterpart is known as bleu cheese in other areas of France, blue in the United States, stilton in England, and gorgonzola in Italy.

BLUE VALUE

• When referring to vitamin A, it is a color indicator of the vitamin A; the depth of the transient blue color produced by reaction of the substance with antimony trichloride is proportional to the amount of vitamin A present.

• When referring to starch, it is an index of the free soluble starch present; for example, the amylose present in cornstarch.

BOBOLINK (REEDBIRD; RICEBIRD)
Dolichonyx oryzivorus

A North American songbird related to the blackbird and oriole, named for the sounds of its lovely song—*bob-o-lee, bob-o-link*. It is a serious pest in rice-growing areas in southern United States through which it passes in its fall migration southward toward its winter range south of the Amazon river. The bobolink was formerly regarded as a table delicacy.

BOBWHITE (QUAIL)

This game bird is classified as *Colinus virginianus*, family *Phasianidae* (which includes pheasants and partridges). The bobwhite gets its name from its whistling call, which sounds like *ah bob White*. It is the only kind of quail native to the area east of the Mississippi River. It usually lives in the region of southern Ontario to the Gulf states.

The bobwhite is called a quail in northern and eastern United States and in Canada; and in the South it is called a partridge.

The flesh of the bobwhite makes for good eating.

BODY BUILD

(See BODY TYPES.)

BODY COMPOSITION

Nutrition encompasses the various chemical and physiological reactions which change food elements into body elements. It follows that knowledge of body composition is useful in understanding the individual's response to nutrition.

Based on studies that have been made, there is a wide range in body composition according to age and nutritional state (degree of fatness). Fig. B-66 shows the changes in body composition from infancy to adulthood.

The following conclusions can be made from these figures:

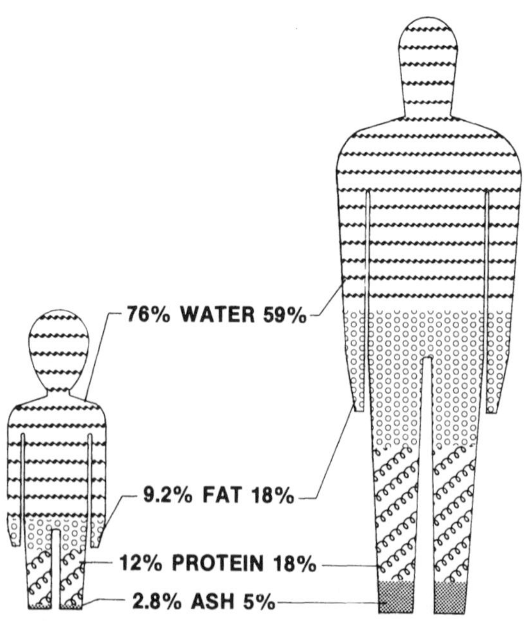

Fig. B-66. This shows the changes in body composition from the infant on the left to the adult on the right.

1. **Water.** On a percentage basis, the water content shows a marked decrease with advancing age and maturity.
2. **Fat.** The percentage of fat normally increases with growth and age. There is considerable difference in the amount of fat in men as compared to women. Men average 12% fat, but women may have as high as 29 to 33% fat.
3. **Fat and water.** As the percentage of fat increases, the percentage of water decreases.
4. **Protein.** The percentage of protein increases slightly during growth, but may decrease if the individual puts on excess weight.
5. **Ash.** The percentage of ash shows a slight increase with age.

The chemical composition of the body varies widely between organs and tissues and is more localized according to function. Thus, water is an essential of every part of the body, but the percentage composition varies greatly in different body parts; blood plasma contains 90 to 92% water, muscle 72 to 78%, bone 45%, and the enamel of the teeth only 5%. Proteins are the principal constituents, other than water, of muscles, tendons, and connective tissues. Most of the fat is localized under the skin, near the kidneys, and around the intestines. But it is also present in the muscles, bones, and elsewhere.

There is a very small amount of carbohydrates (mostly glucose and glycogen) present in the body, found principally in the liver, muscles, and blood.

BODY FLUID(S)

These are contained in two major compartments within the body. The extracellular fluid, the "internal sea" that bathes the cells, is comprised of the blood plasma and the *interstitial* (between the cells) *fluid*. A larger part of the body fluid is contained in the second compartment, the *intracellular* (inside the cell) compartment. Fluids in each compartment have a different chemical composition. Lymph, cerebrospinal, pericardial, pleural and peritoneal fluids are specialized interstitial fluids. Fig. B-67 illustrates the amounts of body fluid contained in

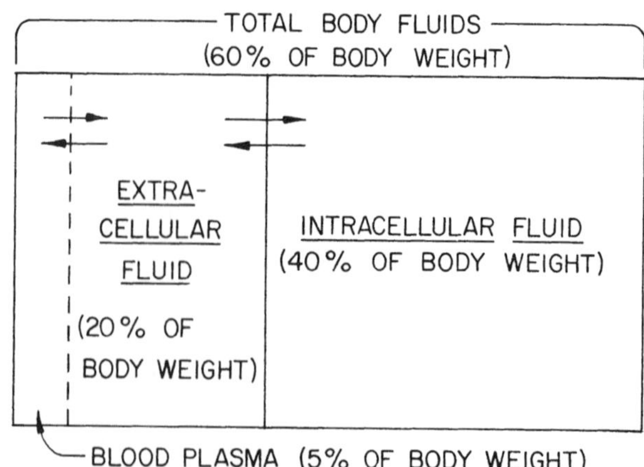

Fig. B-67. Amount of fluid contained in the two major compartments of the body.

each compartment. The digestive system, kidneys, skin, and lungs are routes of fluid input and output. Furthermore, there is a constant exchange of fluids between compartments.

(Also see WATER AND ELECTROLYTES.)

BODY SURFACE AREA

The area covered by the exterior of the body, which determines the heat lost from the body, and therefore basal metabolism.

The body surface area is estimated by plotting a person's height and weight on a standard chart developed by DuBois, based on the following formula:

Surface (cm^2) = weight (kg) .425 × height (cm) .725 × 71.84

BODY TEMPERATURE

The heat produced by metabolism is usually distributed throughout the body. Hence, a good indication of the body temperature may be secured by taking either the oral or the rectal temperature. Body temperature is normally maintained within a narrow range; hence, the normal temperature is usually within a degree or so of 98.6°F (*37°C*). An elevated body temperature—more than 2°F (*1°C*) above normal—suggests that a fever may be present. However, strenuous exercise may raise the body temperature by that amount or more. A subnormal body temperature may mean that there is a circulatory disorder or a metabolic problem such as malnutrition.

BODY TISSUE

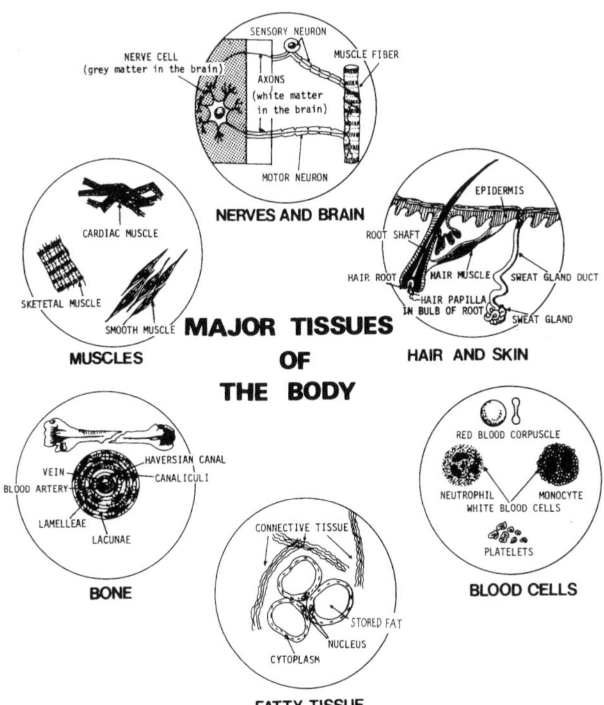

Fig. B-68. Major body tissues.

An aggregate of cells of a particular kind or kinds together with their intercellular substance that form one of the structural materials out of which the body is built, such as connective tissue, epithelium tissue, muscle tissue, and nerve tissue.

(Also see TISSUES OF THE BODY.)

BODY TYPES

This much used system of body typing was devised by Dr. W. H. Sheldon, an American physical anthropologist, in 1940. He recognized three major types of physiques, which he described as follows:

- **Ectomorph**—This type of body is tall and slender, with a delicate bone structure. The trunk is short in proportion to the arms and legs. Furthermore, the fingers, hands, and toes are disproportionately long. Finally, the muscles are wiry, rather than bulky.

  Life insurance companies have found that people with this physique live the longest, so their tables of ideal weights for various heights favor ectomorphs over the two other body types (endomorphs and mesomorphs).

- **Endomorph**—People with this body build are often described as "chubby" or "pleasingly plump," because they appear to be round and soft. Also, they usually have large abdomens, long trunks, and short, but heavily fleshed arms and legs. Lastly, their body measurements from front to back are likely to be greater than their dimensions from side to side.

- **Mesomorph**—This so-called "masculine" physique is characterized by (1) bulky muscles; (2) heavy bones with large joints; (3) and a large chest and broad shoulders, which are notably more prominent than the abdomen or hips. Most people consider the proportions between the trunk and the arms and legs to be ideal. It is noteworthy that some women may have this body build, yet be truly feminine in every respect.

(Also see OBESITY.)

BODY WATER

This term generally designates the total water content of the body. However, the distribution of water within the human body is not uniform. It is distributed between two major compartments: extracellular water, which is outside of the cells; and intracellular water, which is inside the cells. The extracellular water is present in the fluid which bathes the cells (interstitial fluid), blood plasma, cerebrospinal fluid, synovial fluid, and lymph. Normally, the water content in the bodies of adults averages about 60% of the body weight. However, human infants may contain up to 77% water right after birth. Also, as the percentage of fat increases, the percentage of water decreases.

(Also see BODY FLUID[S]; WATER AND ELECTROLYTES; and WATER BALANCE.)

BOILING POINT

The temperature at which the vapor pressure of a liquid equals the atmospheric pressure.

BOLOGNA

A large moist sausage, usually made of beef, veal, and pork, that is chopped fine, seasoned, enclosed in a casing, boiled and smoked.

BOLTING OF FOOD

People who eat too rapidly or greedily are said to be bolting their food.

BOLUS

The name given to the mass of food that is swallowed after chewing.

(Also see DIGESTION AND ABSORPTION.)

BOMB CALORIMETER

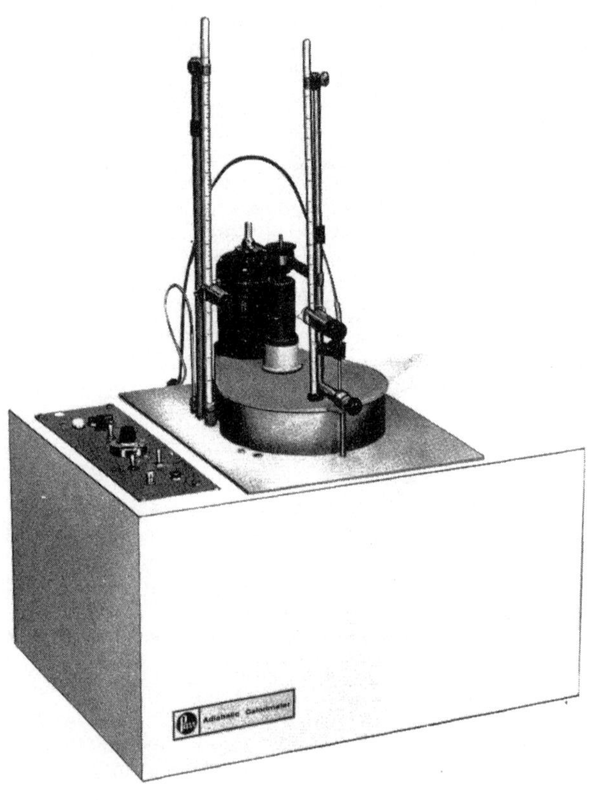

Fig. B-69. Bomb calorimeter for the determination of gross energy. (Courtesy, Parr Instrument Company, Moline, Ill.)

The bomb calorimeter is an instrument for measuring the energy value of food and other materials. It consists of an inner chamber that contains the food sample surrounded by a double-walled insulated jacket that holds water. It works as follows: An electric wire is attached to the material being tested, so that it can be ignited by remote control; 2,000 g of water are poured around the bomb; 25 to 30 atomospheres of oxygen are added to the bomb; the material is ignited; the heat given off from the burned material warms the water; and a thermometer registers the change in temperature of the water.

A *calorie* is the amount of heat required to raise the temperature of 1 g of water 1 °C (precisely from 14.5° to 15.5 °C).

The heat liberated by burning a food in the bomb calorimeter is referred to as the gross energy or heat of combustion. It will coincide with the metabolizable energy in that food only if it can be completely metabolized. For example, proteins liberate 5.65 kcal/g in the bomb calorimeter, but only 4.4 kcal/g in the body where the nitrogen is excreted as urea and uric acid (containing 1.25 kcal/g).

BONE

Contents — *Page*

Structure and Physical Properties of Bone 224
Chemical Composition of Bone 225
Bone Growth and Metabolism 226
Bone Disorders 226

There are 206 separate bones in the human body. They form the skeleton, the framework of hard structures which supports and protects the soft tissues.

STRUCTURE AND PHYSICAL PROPERTIES OF BONE. Bone is very hard and resistant to pressure. Its compressive strength is about 20,000 lb per square inch, and its tensile strength averages 15,000 lb per square inch—considerably higher than white oak.

In the fresh state, it is a pinkish white color externally, and a deep red within. It is composed of two kinds of tissue; an external shell of dense compact substance, within which is the more loosely arranged spongy substance. In typical long bones the shaft is hollowed to form the medullary cavity.

• **The compact bone (cortex)**—The compact tissue, which is always on the exterior of the bone, differs greatly in thickness in various situations, in conformity with the strains and stresses to which the bone is subjected. In the long bones, it is thickest in or near the middle part of the shaft and thins out toward the extremities. At the ends, the layer is very thin, and is especially dense and smooth at joint surfaces.

Fig. B-70. "Skeleton reading." (Courtesy, The Bettmann Archive, Inc., New York, N.Y.)

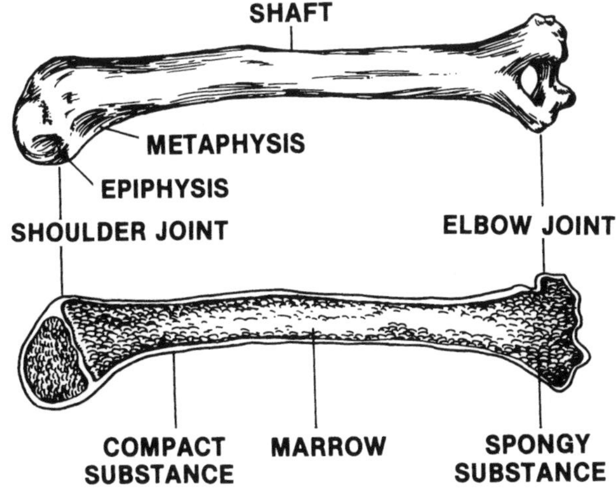

Fig. B-71. The humerus bone—the long bone of the upper arm. *Top:* Humerus bone, showing the three parts: (1) the shaft—the long part of the bone; (2) the metaphysis—the end of the shaft, or diaphysis, where it joins the epiphysis, and (3) the epiphysis—the rounded end.
Bottom: The humerus bone split open to show the inside.

• **The spongy bone (trabeculae)**—The spongy tissue, which forms the bulk of short bones and of the extremities of long bones, consists of delicate bony plates and spicules (small needle-shaped bodies) which run in various directions and intercross—resembling latticework.

• **The periosteum**—This is the membrane which surrounds the outer surface of the bone, except where it is covered with cartilage. It consists of an outer protective fibrous layer, and an inner cellular osteogenic layer. The fibrous layer varies much in thickness, generally being thickest in exposed situations. The osteogenic layer is well developed during active growth, but later it becomes much reduced. The periosteum serves as a place for the branching of blood vessels previous to their entering the bone.

• **The marrow**—The marrow fills the cylindrical cavity in the shafts of the long bones and occupies the spaces of the spongy substance. There are two kinds of marrow in the adult—red and yellow. In the young, there is only red marrow, but later this is replaced in the medullary cavity by yellow marrow. The red marrow contains several types of characteristic cells and is a blood-forming substance, while the yellow marrow is essentially ordinary adipose tissue.

• **Blood vessels of bone**—The blood vessels of bone are very numerous. Those of the compact tissue are derived via minute openings from a close and dense network of vessels in the periosteum. Other branches of the blood vessels enter the extremities of the long bones and supply the spongy bone and the marrow in them.

• **Nerves**—The nerves are distributed freely to the periosteum and accompany the nutrient arteries into the interior of the bone.

CHEMICAL COMPOSITION OF BONE. Chemically, bone consists of organic and inorganic matter in the approximate ratio of 1:2.

The organic matter, which accounts for about one-third of the weight of bone and gives toughness and elasticity, consists largely of collagen in a gel of cementing substance. It may be separated out by immersing a bone for considerable time in dilute mineral acid, following which it comes out exactly the same shape as before, but so flexible that a long bone (one of the ribs, for example) can be easily tied into a knot.

The mineral part, which gives bone its hardness and rigidity, may be obtained by heating bone to a high temperature, completely burning out the organic part and reducing its weight by about one-third. Five-sixths of the mineral part is calcium phosphate, the remainder consists of calcium carbonate, calcium fluoride, calcium chloride, and magnesium phosphate, with small amounts of sodium chloride and sulphate.

BONE GROWTH AND METABOLISM. Growth in length of the long bones normally occurs at the band of epiphyseal cartilage (growth band) lying between the epiphysis and the diaphysis (shaft). The epiphyseal cartilage is a temporary formation which grows by the multiplication of its own cells at the epiphysis end, while at the diaphysis end it degenerates and is replaced by calcified bone. When the epiphyseal cartilage ceases to regenerate and is entirely replaced by bone, the epiphysis unites with the diaphysis and growth ceases. This is referred to as the closing of the epiphysis, or ossification. In the process of ossification, cartilage is replaced by osteoid, which is then calcified (see Fig. B-72).

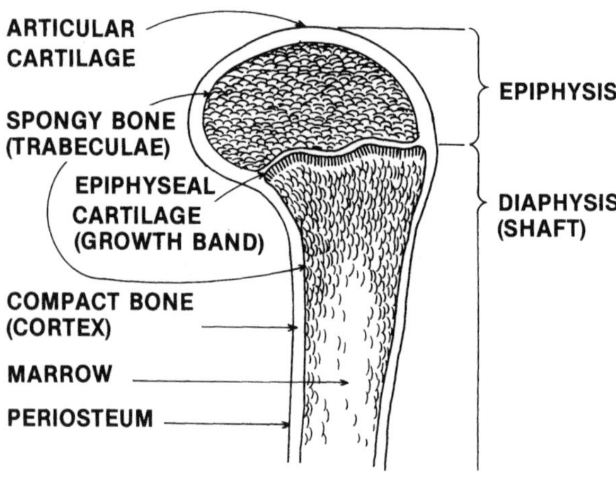

Fig. B-72. Diagram of the longitudinal section of a long growing bone.

In addition to serving as structural material, bones serve as a storehouse for calcium and phosphorus which may be mobilized when the assimilation of these minerals is inadequate to meet body needs. Thus, if the dietary intake of calcium does not meet the requirements, the body can draw upon its bone reserves, which are found in the spongy bone. The exact mechanism involved in the mobilization of bone calcium during periods of stress, i.e., during formation of milk or of fetal bones, is not fully understood. But, it is known that, in some manner, it is under the influence of the parathyroid hormone, and perhaps it is also affected by vitamin D. The phosphorus which is released simultaneously with the calcium does not appear to be used to meet nutritional phosphorus requirements, but is promptly excreted in the urine.

BONE DISORDERS. Bones are subject to numerous disorders, including fractures, congenital defects, infections, tumors, and nutritional diseases. Only the latter will be discussed herein.

• **Osteoporosis**—This is a bone disease of adult people characterized by (1) a descreased amount of bone (called osteopenia) and (2) porous bones. In old age, such bones are brittle and slow to heal when they are broken.

Osteoporosis is due to a failure of normal bone metabolism in the adult. The condition is prevalent in people past the age of 50, particularly women. While there are various other causes, including a defective synthesis of the protein matrix, there is evidence that, in some cases, a long-continued low intake of calcium in early adult life, or during pregnancy and lactation in women, plays a role in its development and that the resulting symptoms can be relieved, if not corrected, by a higher intake of calcium than is normally recommended for adults—by daily intakes of 1,000 to 1,200 mg of calcium per day.

(Also see NUTRITIONAL DEFICIENCY DISEASES; and OSTEOPOROSIS.)

• **Osteomalacia**—This is the adult counterpart of rickets. The symptoms and signs are: pain ranging from dull to severe; bone tenderness on pressure; muscular weakness, with the result that the patient may find it difficult to climb stairs or get out of a chair; a waddling gait; and fractures.

Osteomalacia is caused by a deficiency of calcium, phosphorus, or vitamin D; or an incorrect ratio of the two minerals. The most common contributing factors are: (1) failure to absorb calcium, with excess fecal excretion of calcium phosphate; and (2) dietary deficiency of calcium and vitamin D during the stress demands of pregnancy and lactation.

Treatment consists in rectifying the cause; taking adequate calcium, phosphorus, and vitamin D; and taking the proper ratio of the two minerals (preferably 1:1, although there is evidence that man can tolerate a Ca:P ration between 2:1 and 1:2). Once major deformities are established, they cannot be corrected by diet or drugs; only an orthopedic surgeon can help.

(Also see NUTRITIONAL DEFICIENCY DISEASES; and OSTEOMALACIA.)

• **Rickets**—The word *rickets* is derived from the Anglo-Saxon *wrikken*, meaning to twist. It is a disease of children in which the bones are softened and deformed. Fully developed cases of rickets present the following characteristics: pigeon chest; enlarged wrists and ankles, bowed legs, and knock-knees; enlargement or beading of the ribs, frequently called "the rachitic rosary"; and stunted growth.

The major factors involved in the prevention of rickets are: adequate calcium, phosphorus, and vitamin D, and the proper ratio of the two minerals (about 1:1).

If the disease has not advanced too far, treatment may be successful by supplying adequate amounts of vitamin D, calcium, and phosphorus, and/or adjusting the ratio of calcium to phosphorus.

(Also see DEFICIENCY DISEASES; OSTEOMALACIA; OSTEOPOROSIS; and RICKETS.)

BONE CHARCOAL

The residue from charring bones in a closed retort; sometimes referred to as bone black. It has excellent absorbant properties; therefore, it is used to remove impurities from food items such as crude sugar solutions and vodka.

BONE, COMPACT

A dense type of bone, which forms the outer layer of the long bones.

(Also see BONE, section headed "Structure and Physical Properties of Bone.")

BONE COMPARTMENT

The total sketetal tissue in the body. For example, the bone compartment contains 99% of the body's total metabolic calcium pool.

BONE DENSITY

The mineral compactness of bone, which gives it hardness and rigidity.

BONE MATRIX

The protein "network" (fibers of collagen) of bone in which minerals are deposited.

BONE MEAL

Packinghouse bones are degreased and processed into valuable products.

Bone meal, which is nature's own formulation, is fresh, degreased, ground animal bone. It is used as a mineral supplement for both humans and animals, and as a source of calcium, phosphorus, and needed trace elements. For human use, it is available in both powder and tablet form. Table B-16 gives the complete mineral composition of steamed bone meal on a moisture-free basis.

TABLE B-16
MACRO AND MICROMINERALS OF BONE MEAL, MOISTURE-FREE BASIS[1]

	Unit	
Macrominerals:		
Calcium	(%)	30.71
Phosphorus	(%)	12.86
Sodium	(%)	5.69
Magnesium	(%)	.33
Potassium	(%)	.19
Sulfur	(%)	2.51
Microminerals:		
Copper	(ppm or mg/kg)	11.000
Iodine	(ppm or mg/kg)	34.188
Iron	(%)	.088
Manganese	(ppm or mg/kg)	23.000
Zinc	(ppm or mg/kg)	130.000

[1]Ensminger, M. E., and C. G. Olentine, Jr., *Feeds & Nutrition—complete*, The Ensminger Publishing Company, Clovis, Calif., 1978, pp. 1,366–1,367, Table V–1F.

It is noteworthy that (1) bone meal never contains toxic levels of the essential trace elements, whereas these are sometimes exceeded in man-made mineral formulations; and (2) bone meal trace elements rate high in biological availability, physical and chemical compatability, and storage life.

• **Bone charcoal (bone black)**—This is the product that is obtained as residue from the carbonization of degreased bones (which may be partly degelatinized as well). The bones are carbonized in retorts, following which the residual charcoal is moved into closed iron coolers and protected from exposure to air until cold. It is then crushed and graded. Bone charcoal is employed as a decolorizing and refining medium, chiefly in the sugar industry, which consumes enormous quantities. Fortunately, the char can be revivified by washing and reburning (out of contact with air) for a life of about 2 years.

• **Spent bone charcoal (spent bone black)**—When bone charcoal is exhausted as a decolorizing powder, it is known as spent bone charcoal and is discarded by the sugar manufacturer. But it finds a ready market as a source of phosphate.

• **Bone ash**—This product, which is also obtained by burning fresh bones, is used to make cupels (a small shallow porous refractory cup) for assaying. Also, it is an important constituent of the paste used for English bone china. When treated with sulfuric and phosphoric acids, bone ash yields a substitute for cream of tartar in baking powders.

BONE, SPONGY

A porous type of bone, which forms most short bones and the ends of long bones.

(Also see BONE, section headed "Structure and Physical Properties of Bone.")

BONITO (THE ATLANTIC BONITO: *Sarda sarda*; THE PACIFIC BONITO: *Sarda chiliensis*)

An open-sea fish closely related to the larger tuna and the smaller mackerel. They are found worldwide in warm and tropical waters, but are more coastal in distribution than the tunas. Bonitos have torpedo-shaped bodies, with narrow dark blue lines running from the head toward the back and a silvery underside. Most bonitos weigh about 6 lb (*2.7 kg*), although some weigh as much as 40 lb (*18 kg*) and are over 3 ft (*91 cm*) long. They travel in large schools.

Bonitos are important game and commercial fish and are considered good eating. They are usually canned.

BOOKS

Without claiming that either all or the best nutrition books are listed, the following books are recommended as the kind that will provide valuable reference material on the subject of foods and human nutrition.

Subject Matter	Title of Book	Author(s)	Publisher & Date
General Nutrition	*About Nutrition*	Seventh-day Adventist Dietetic Assn.	Southern Pub. Associates, Nashville, Tenn., 1971
	Applied Nutrition	R. Rajalkshmi	Oxford & IBH Pub. Co., Delhi, India, 1974
	Applied Nutrition	T.J. Wayler R.S. Klein	The Macmillan Co., New York, NY, 1965
	Basic Nutrition	E.W. McHenry G. Beaton	J.B. Lippincott Co., Philadelphia, Pa., 1963
	Body Composition In Animals And Man	Symposium	National Academy of Sciences, Washington, D.C., 1967
	Bogert's Nutrition And Physical Fitness	G.M. Briggs O.H. Calloway	Saunders College Pub., Philadelphia, PA, 1979
	California Nutrition Book, The	P. Saltman J. Gurin I. Mothner	Little Brown and Company, Boston, Mass., 1987
	Consumers All—The Yearbook of Agriculture 1965	U.S.D.A.	Government Printing Office, Washington, D. C., 1965
	Contemporary Nutrition Controversies	T.P. Labuza A.E. Sloan	West Pub. Co., St. Paul, Minn., 1979
	Controversies In Clinical Nutrition	*Ed.:* J.J. Cunningham	George F. Stickley Co., Philadelphia, PA, 1980
	Diet and Health	National Research Council	National Academy Press, Washington, D.C., 1989
	Diet for Living, A	J. Mayer	David McKay Co., Inc, Orangeburg, NY, 1975
	Diets Are For People	C.W. Shearman	Appleton-Century-Crofts, New York, NY, 1963
	Dr. Fishbein's Popular Illustrated Medical Encyclopedia	M. Fishbein, M.D.	Doubleday & Co., Garden City, NY, 1979
	Eating For Good Health	F. J. Stare, M.D.	Doubleday & Co., New York, NY 1964
	Encyclopedia of Food, Agriculture, & Nutrition	*Ed.:* N. Lapedes	McGraw-Hill, New York, NY, 1977
	Facts of Food, The	A.E. Bender	Oxford University Press, Fairlawn, NJ, 1975
	Family Guide To Better Food And Better Health, The	R.M. Deutsch	Meredith Corp., Des Moines, Iowa, 1971
	Family Fare—a guide to good nutrition	U.S.D.A.	Government Printing Office, Washington D.C., 1974
	Focus on Food	L.B. Peck, *et al.*	McGraw-Hill Book Company, New York, NY, 1974
	Food And Health Of Western Man, The	J.L. Mount	John Wiley & Sons, Inc., New York, NY, 1975
	Food And Man	M.E. Lowenberg, *et al.*	John Wiley & Sons, Inc., New York, NY, 1968
	Food And Nutrition	W.H. Sebrell, Jr. J.J. Haggerty	Time-Life Books, New York, NY, 1967
	Food And Nutrition Research	Report of Ag. Research Council/ Medical Council Committee, London	Pendragon House, Inc., Palo Alto, CA, 1974
	Food And The Consumer: Fact vs. Fancy For Today's Concerned Student And Consumer	A. Kramer	AVI Pub. Co., Westport, Conn., 1980
	Food—and your well-being	T.P. Labuza	West Publishing Co., St. Paul, Minn., 1977
	Food Becomes You	R.M. Leverton	Iowa State University Press, Ames, Iowa, 1973
	Foodbook, The	J. Trager	Grossman Pub., New York, NY, 1970
	Food Cultism and Nutrition Quackery	Almquist Wiksell	The Swedish Nutrition Foundation, Stockholm, Sweden, 1970
	Food For Life	F.E. Deatherage	Plenum Pub. Corp., New York, NY, 1975
	Food For Modern Living	I.E. McDermott, *et al.*	J.B. Lippincott Co., Philadelphia, PA, 1967

(Continued)

Subject Matter	Title of Book	Author(s)	Publisher & Date
General Nutrition (continued)	*Food For Sport*	N.J. Smith	Bull Pub. Co., Palo Alto, CA, 1976
	Food For Thought	T.P. Labuza A.E. Sloan	AVI Pub. Co., Inc., Westport, Conn., 1977
	Food For Us All—The Yearbook of Agriculture 1969	U.S.D.A.	Government Printing Office, Washington, D.C., 1969
	Food Fundamentals	M. McWilliams	John Wiley & Sons, Inc., New York, NY, 1966
	Food Science and Nutrition	*Ed.:* F. Clydesdale	Prentice-Hall, Inc., Englewood Cliffs, NJ, 1979
	Food Trends	B. Senauer, *et al.*	Eagan Press, St. Paul, Minn., 1991
	Foods, Diet and Nutrition	K.L. Jones, *et al.*	Canfield Press, Div. of Harper and Row, New York, NY, 1970
	Foods Without Fads	E.W. McHenry	J.B. Lippincott Co., Philadelphia, PA, 1960
	Foundations Of Nutrition	C.M. Taylor O.F. Pye	The Macmillan Pub. Co., New York, NY, 1966
	Fundamentals of Normal Nutrition	C.H. Robinson	The Macmillan Pub. Co., New York, NY, 1978
	Fundamentals Of Nutrition	L.E. Lloyd, *et al.*	W.H. Freeman & Co., San Francisco, Ca, 1978
	Handbook Series in Nutrition and Food	*Ed.:* M. Rechcigl, Jr.	CRC Press, Boca Raton, Fla., 1977
	Health	J. Mayer	Van Nostrand Reinhold Co., New York, NY, 1972
	Health And Human Values: An Ecological Approach	L.A. Jefcoat	John Wiley & Sons, New York, NY, 1972
	Health Robbers, The	*Ed.:* S. Barrett	George F. Stickley Co., Philadelphia, PA, 1980
	Heinz Handbook Of Nutrition, The	B.T. Burton	McGraw-Hill Book Co., New York, NY, 1965
	How To Enjoy Eating Without Committing Suicide	C.D. Ewart	Simon & Schuster, Inc., New York, NY, 1973
	How To Feed Your Family To Keep Them Fit and Happy—No Matter What	"Betty Crocker"	Golden Press, New York, NY, 1972
	Human Nutrition	B.T. Burton	McGraw-Hill Book Co., Manchester, MO, 1976
	Human Nutrition	J. Mayer	C.C. Thomas Pub., Springfield, Ill., 1972
	Human Nutrition—a comprehensive treatise	*Editors:*	Plenum Press, New York, NY
	Vol. 1. *Pre- and Postnatal Development*	M. Winick	
	Vol. 2. *Nutrition and Growth*	D.B. Jelliffe E.F.P. Jelliffe	1979
	Vol. 3A. *Nutrition And The Adult: Macronutrients*	R.B. Alfin-Slater D. Kritchevsky	1980
	Vol 3B. *Nutrition And The Adult: Micronutrients*	R.B. Alfin-Slater D. Kritchevsky	1980
	Vol. 4. *Nutrition: Metabolic And Clinical Applications*	R.E. Hodges	
	Introduction To Nutrition	H. Fleck	The Macmillan Co., New York, NY, 1971
	Introductory Foods	M. Bennion O. Hughes	The Macmillan Co., New York, NY, 1975
	Introductory Nutrition	H.A. Guthrie	Times Mirror/Mosby College Publishing, St. Louis, MO, 1989
	Jane Brody's Nutrition Book	J. Brody	W.W. Norton & Co., New York, NY, 1981
	Let's Talk About Food	American Medical Association	Publishing Sciences Group, Acton, Mass., 1974
	Living Nutrition	F.J. Stare, M.D. M. McWilliams	John Wiley & Sons, New York, NY, 1973
	Man, Food And Nutrition	M. Rechcigl, Jr.	CRC Press, Boca Raton, Fla., 1973
	Meal Management	F. Kinder	The Macmillan Co., New York, NY, 1973
	Meaning of Human Nutrition, The	M.W. Lamb M.L Harden	Pergamon Press, New York, NY, 1973
	Metabolics	L.E. Lamb	Harper & Row, Pub., New York, NY, 1974

(Continued)

Subject Matter	Title of Book	Author(s)	Publisher & Date
General Nutrition (continued)	*Nutrition*	M.S. Chaney, *et al.*	Houghton Mifflin Co., Boston, Mass., 1979
	Nutrition and Behavior	R.B. Kanarek, *et al.*	Van Nostrand Reinhold, New York, NY, 1991
	Nutrition And Family Health Service	L. Anderson J.H. Browe	W.B. Saunders Co., Philadelphia, PA, 1960
	Nutrition And Health Encyclopedia, The	D.F. Tver P. Russell	Van Nostrand Reinhold, New York, NY, 1989
	Nutrition and Physical Fitness	G.M Briggs D.H. Calloway	Holt, Rinehart and Winston, New York, NY, 1984
	Nutrition: An Integrated Approach	R. Pike M. Brown	John Wiley & Sons, New York, NY, 1975
	Nutrition Crisis, A Reader, The	*Ed.:* T. Labuza	West Publishing Co., St. Paul, Minn., 1975
	Nutrition Factor, The	A. Berg	The Brookings Institution, Washington, D.C., 1973
	Nutrition For Good Health	F.J. Stare, M.D. M. McWilliams	Plycon Press, Fullerton, Calif., 1974
	Nutrition For Today	R. Alfin-Slater L. Aftergood	William C. Brown Co., Dubuque, Iowa, 1973
	Nutrition For Today	T.J. Runyan	Harper & Row, New York, NY, 1976
	Nutrition For Your Family's Health	I. Rossman	Emily Post Institute, New York, NY, 1963
	Nutrition in Action	E.A. Martin	Rinehart and Winston, New York, NY, 1971
	Nutrition In Preventive Dentistry: Science And Practice	A.E. Nizel	W.B. Saunders Co., Philadelphia, PA, 1972
	Nutrition Science And You	O. Mickelsen	Scholastic Book Services, New York, NY, 1964
	Nutrition Update	J. Weininger G.M. Briggs	John Wiley & Sons, New York, NY, 1983
	Nuts Among the Berries, The	R.M. Deutsch	Ballantine Books, New York, NY, 1962
	Planning For Health: Development And Application of Social Change Theory	H.L. Blum	Behavioral Publications, New York, NY, 1974
	Principles of Nutrition	E.D. Wilson, *et al.*	John Wiley & Sons, New York, NY, 1979
	Progress In Human Nutrition	*Ed.:* S. Margen, M.D.	AVI Pub. Co., Westport, Conn., 1971
	Recommended Dietary Allowances, 10th edition	Comm. on Dietary Allowances	National Academy of Sciences, Washington, D.C., 1989
	Science Of Nutrition, The	A.M. Thompson	The Macmillan Co., New York, NY, 1972
	Scope Manual On Nutrition	M.C. Latham, M.D., *et al.*	Upjohn Co., Kalamazoo, Mich, 1972
	Understanding Food	L.H. Kotschevar M. McWilliams	John Wiley & Sons, New York, NY, 1969
	Value Of Food, The	P. Fisher A.E. Bender	Oxford Unversity Press, Fair Lawn, NY, 1975
	Whole Foods For You	L. Fryer D. Simmons	Mason and Lipscomb, New York, NY, 1974
	World's Best Food, For Health And Long Life, The	M. Bateman, *et al.*	Houghton Mifflin Co., Boston, Mass., 1981
	World Review Of Nutrition and Dietetics, Vol. 20	*Ed.:* G.H. Bourne	Albert J. Phiebig, White Plains, NY, 1975
	World Review Of Nutrition and Dietetics, Vol. 21	*Ed.:* G.H. Bourne	S. Karger, White Plains, NY, 1975
	You And Your Food	R.B. White	Prentice-Hall, Englewood Cliffs, NJ, 1976
	Your Diet Digestion And Health	E.S. Nasset	Barnes and Noble, New York, NY, 1962
	Your Food and Your Health	R. Carter	Harper & Row, New York, NY, 1964
	Your Good Health	W.I. Bennett, M.D., *et al.*	Harvard Medical School, Cambridge Mass., 1987
Nutrition For The Early Ages	*Baby's Early Years. A Record Book*	L. Moragne R. Moragne	Len Champs, Washington, D.C., 1975
	Clinics in Perinatology—Symposium On Nutrition	*Eds.:* L.A. Barness R.M. Pitkin	W.B. Saunders, Philadelphia, Penn., 1975
	Dietary Lipids and Postnatal Development	C. Galli, *et al.*	Raven Press, New York, NY, 1975

(Continued)

Subject Matter	Title of Book	Author(s)	Publisher & Date
Nutrition For The Early Ages (continued)	*Feed Me I'm Yours*	V. Lansky, and friends	Meadowbrook Press, Wayzata, Minn., 1974
	Fetal And Postnatal Cellular Growth—Hormones And Nutrition	D.B.Cheek	John Wiley & Sons, New York, NY, 1975
	Infant Nutrition	S.J.Fomon, M.D.	W.B.Saunders Co., Philadelphia, PA, 1974
	Nursing Your Baby	K.Pryor	Pocket Books, New York, NY, 1973
	Nutrition And Fetal Development	*Ed.:* M.Winick	John Wiley & Sons, New York, NY, 1974
	Nutrition Programmes For Pre-school Children	*Eds.:* D.B.Jelliffe E.F.P.Jelliffe	Institute of Public Health of Croatia, Zagreb, Yugoslavia, 1974
	Paediatric Priorities In The Developing World	D. Morley	Butterworth & Co., London, United Kingdom, 1973
	Physiology Of Human Pregnancy, The	F.E.Hytten I.Leitch	Blackwell Scientific Pub., London, United Kingdom, 1964
	Real Food For Your Baby	F. Saville	Simon & Schuster, New York, NY, 1973
	Simplified Recipes For Day Care Centers	P.D.Asmussen	Cahners Books, Boston, Mass., 1974
Nutrition For Youths	*Adolescent Nutrition And Growth*	F.P.Heald, M.D.	Appleton-Century-Crofts, New York, NY, 1969
	Down To Earth Cookbook, The	A. Borghese	Charles Scribner's Sons, New York, NY, 1973
	First Foods	M.L.Cronan J.C.Atwood	Charles A. Bennett Co., Peoria, Ill., 1971
	Food Facts For Teen-agers	M.B.Salmon	C.C.Thomas, Pub., Springfield, Ill., 1965
	Food Facts For Young People	P.Arnold P.White	Holiday House, Inc., New York, NY, 1968
	Growth And Development Of The Brain; Nutritional, Genetic, And Environment Factors	*Ed.:* M.A.B.Brazier	Raven Press, New York, NY, 1975
	Growth And Maturation: An Introduction To Physical Development	M.J.Baer	Howard A. Doyle Pub., Cambridge, Mass., 1973
	Health In Elementary Schools	H.D.Cornacchia W.M.Staton L.W.Irwin	The C.V.Mosby Co., St. Louis, MO, 1970
	Help! My Child Won't Eat Right	A.K.Hatfield P.S.Stanton	Acropolis Books, Washington, D.C., 1973
	Human Development	G.Craig	Prentice-Hall, Englewood Cliffs, NJ, 1976
	Malnutrition And Brain Development	M. Winick	Oxford University Press, Fair Lawn, NJ, 1976
	Malnutrition, Learning, and Behavior	*Eds.:* N.S.Scrimshaw J.E.Gordon	The M.I.T. Press, Cambridge, Mass., 1968
	Nutrient Requirement In Adolescence	*Eds.:* J.I.McKigney H.N.Munro	The M.I.T. Press, Cambridge, Mass., 1976
	Nutrition For The Growing Years	M. McWilliams	John Wiley & Sons, New York, NY, 1975
	Protein-Calorie Malnutrition Of Early Childhood: Two Decades Of Malnutrition	E.F.P.Jelliffe	Commonwealth Bureau of Nutrition, United Kingdom, 1975
	Teenage Pregnant Girl, The	*Eds.:* J.Zackler W.Brandstadt	C.C.Thomas, Pub., Springfield, MO, 1975
	Teenage Nutrition And Physique	R.L.Huenemann M.C.Hampton A.R.Behnke L.R.Shapiro B.W.Mitchell	C.C.Thomas, Pub., Springfield, MO, 1974
	Teen-ager's Guide To Diet And Health, The	R.S.Goodhart	Prentice Hall, Englewood Cliffs, NJ, 1964
Nutrition For The Older Ages	*After-50 Cookbook, The*	D. Hamilton	The Swallow Press, Chicago, Ill., 1974
	Aging In American Society	J.D.Manney, Jr.	Univ. of Michigan-Wayne State Univ., Ann Arbor, Mich., 1975
	A Cookbook For The Leisure Years	P. Macdonald	Doubleday & Co., Garden City, NY, 1967
	Mealtime Manual For The Aged And Handicapped	Inst. of Rehabilitation Medicine	Simon & Schuster, New York, NY, 1970
	Nutrition And Aging	*Ed.:* M. Winick	John Wiley & Sons, New York, NY, 1976

(Continued)

Subject Matter	Title of Book	Author(s)	Publisher & Date
Nutrition For The Older Ages (continued)	*Research Planning And Action For The Elderly; The Power And Potential Of Social Sciences*	*Eds.:* D.P.Kent R.Kastenbaum S.Sherwood	Behavioral Publications, New York, NY, 1972
	Understanding Aging: A Multidisciplinary Approach	*Eds.:* M.G.Spender C.J.Dorr	Appleton-Century-Crofts, Englewood Cliffs, NJ, 1975
Allergies	*Allergy Cookbook, The*	C.J. Emerling E.O.Jonckers	Doubleday & Company, New York, NY, 1969
	Caring And Cooking For The Allergic Child	L.L.Thomas	Drake Publishers, New York, NY, 1974
	Coping With Food Allergy	C.A.Frazier	Quandrangle Books, New York, NY, 1974
	Elimination Diets And The Patient's Allergies	A.H.Rowe, M.D.	Lea & Febiger, Philadelphia, PA, 1944
	Food Allergy	F.Speer, M.D.	PSG Pub. Co., Littleton, Mass., 1978
	Questions And Answers About Allergies And Your Child	D.J.Rapp	Drake Publishers, New York, NY, 1974
Cookbooks & Diet Planners For Special Diets	*Bland But Grand*	E.M.Pelz	Doubleday & Co., Garden City, NY, 1969
	Cookbook That Tells You How, The Retirement Food And Nutrition Manual, The	H.E.Zaccarelli	Cahners Books, Boston, Mass., 1975
	Diabetic Gourmet, The	A. Bowen	Harper & Row, New York, NY, 1974
	Dietary Management Of Renal Disease	J.S.Cost	Charles B. Slack, Thorofare, NJ, 1975
	Diet Food Finder, The	J.T.Casale	R.R.Bowker Co., New York, NY, 1975
	Diet In Health And Disease: Rationale And Practice	R.S.Dickie	C.C.Thomas, Pub., Springfield, Ill., 1975
	Fat And Sodium Control Cookbook, The	A.S.Payne D.Callahan	Little Brown & Co., Boston, Mass., 1965
	Feasts For Two	P. Rubinstein	The Macmillan Co., New York, NY, 1973
	Good Sense Family Cook Book, The	G. Maddox	M. Evans & Co., New York, NY, 1966
	Introductory Foods: A Laboratory Manual Of Food Preparation And Evaluation	M.L.Marr T.F.Irmiter	The Macmillan Co., New York, NY, 1975
	Manual Of Clinical Dietetics	Chicago Dietetic Assn.	South Suburban Dietetic Assn., Oak Lawn, Ill. 1975
	Mayo Clinic Renal Diet Cookbook, The	J.D.Margie C.F.Anderson R.A.Nelson J.C.Hunt	Golden Press, New York, NY, 1975
	Prudent Diet, The	I. Bennett M. Simon	David White, New York, NY, 1973
	Recipes For A Small Planet	E.B.Ewald	Ballantine Books, New York, NY, 1973
	Recipes For Long Life	Nutrition Staff Loma Linda Foods	Loma Linda Foods, Riverside, Calif., 1976
	Simplified Diet Manual—With Meal Patterns	Nutrition Section Iowa State Dept. of Health	Iowa State Univ. Press, Ames, Iowa, 1975
	Tasty Cooking For Ulcer Diets	O. Aagaard	Crown Publishers, New York, NY, 1964
	Ulcer Diet Cook Book, The	H. Rubin	J.B.Lippincott Co., New York, NY, 1963
	Why Wild Edibles? The Joys Of Finding, Fixing, And Tasting	R. Mohney	Pacific Search Books, Seattle, Wash., 1975
Disease & Diet Therapy	*Basic Nutrition And Diet Therapy*	C.H. Robinson	The Macmillan Co., New York, NY, 1975
	Clinical Nutrition	V.F. Thiele	The C.V.Mosby Co., St. Louis, MO, 1976
	Controlling Diabetes With Diet	A. Gormican	C.C.Thomas, Springfield, Ill., 1971
	Cooper's Nutrition In Health And Disease	H.S.Mitchell H.J.Rynbergen L.Anderson M.V.Dibble	J.B.Lippincott Co., Philadelphia, PA, 1968
	Counseling And Rehabilitating The Diabetic	*Eds.:* J.G.Cull R.E.Hardy	C.C.Thomas, Springfield, Ill., 1974
	Diabetes And Pregnancy: A Guide For The Prospective Mother With Diabetes	A.L.Graber, *et al*	Vanderbilt Univ. Press, Nashville, Tenn., 1973
	Diabetic Care In Pictures	H. Rosenthal J. Rosenthal	J.B.Lippincott Co., Philadelphia, PA, 1968

(Continued)

Subject Matter	Title of Book	Author(s)	Publisher & Date
Disease & Diet Therapy (Continued)	*Diet Manual, Commonwealth of Pennsylvania, 1974*	Institution Food Research and Services Project	Pennsylvania State Univ., University Park, PA, 1974
	Duncan's Diseases of Metabolism	*Eds.:* P.K.Bondy, M.D. L.E.Rosenberg, M.D.	W.B.Saunders Co., Philadelphia, PA, 1974
	Eating Hints (Recipes And Tips For Better Nutrition During Cancer Treatment)	U.S. Dept. of Health & Human Services	Government Printing Office, Washington, D.C.
	Family Care, A guide ...	*Ed.:* N. Baumslad	The Williams & Wilkins Co., Baltimore, MD, 1973
	Food And Arthritis	G. Maddox	Taplinger Pub. Co., New York, NY, 1969
	Food, Nutrition And Diet Therapy	M. V. Krause M. A. Hunscher	W.B. Saunders Co., Philadelphia, PA, 1972
	Gastrointestinal And Hepatic Pathology Decennial	*Ed.:* S.C. Sommers	Appleton-Century-Crofts, New York, NY, 1975
	Gourmet Food On A Wheat-Free Diet	M.N. Wood	C.C.Thomas, Springfield, Ill., 1967
	Handbook Series In Nutrition And Food Vol. I. *Nutritional Disorders* Vol. II. *Nutritional Disorders*	M. Rechcigl, Jr.	CRC Press, Boca Raton, Fla., 1978
	Handbook Of Clinical Dietetics	American Dietetic Association	Yale Univ. Press, New Haven, Conn., 1981
	How To Live With Your High Blood Pressure	W.A. Brams	Arco Pub. Co., New York, NY, 1974
	Human Nutrition And Dietetics	Sir S. Davidson R. Passmore J.F. Brock A.S. Truswell	Churchill Livingstone, New York, NY, 1979
	Mental And Elemental Nutrients	C.C. Pfeiffer	Keats Pub., Inc., New Canaan, Conn., 1975
	Modern Nutrition In Health And Disease	R.S. Goodhart M.E. Shils	Lea & Febiger, Philadelphia, PA, 1978
	Mowry's Basic Nutrition And Diet Therapy	S.R. Williams	C.V. Mosby Co., St. Louis, MO, 1975
	Normal And Therapeutic Nutrition	C.H. Robinson M.R. Lawler	Macmillan Pub. Co., Inc., New York, NY, 1977
	Nutrition And Dietetic Foods	A.E. Bender	Chemical Pub. Co., Inc., New York, NY, 1973
	Nutrition And Diet Therapy—Reference Dictionary	R.T. Lagua V.S. Claudio V.F. Thiele	The C.V. Mosby Co., St. Louis, MO, 1974
	Nutrition and Diet Therapy	S. R. Williams	Times Mirror/Mosby College Publishing, St. Lous, MO, 1985
	Nutrition and Mental Functions	*Ed.:* G. Serban	Plenum Press, New York, NY, 1975
	Nutrition In Health And Disease	L. Anderson, *et al.*	J.B. Lippincott Co., Philadelphia, PA, 1982
	Nutrition In Nursing	L.S. Boykin	Medical Examination Pub. Co., Flushing, NY, 1975
	Picture Of Health, The	E.D. Eckholm	W.W. Norton & Co., Inc., New York, NY, 1977
	Refined Carbohydrate Foods And Disease — Some Implications Of Dietary Fibre	*Eds.:* D.P. Burkitt H.C. Trowell	Academic Press, New York, NY, 1975
	Review Of Nutrition And Diet Therapy	S.R. Williams	The C.V. Mosby Co., St. Louis, MO., 1973
	Total Parenteral Nutrition. Premises And Promises	*Ed.:* H. Ghadimi	John Wiley & Sons, Inc., New York, NY, 1975
	World Review Of Nutrition and Dietetics, Vol. 23	*Ed.:* G.H. Bourne	Albert J. Phiebig, White Plains, NY, 1975
Education	*Applied Communication In Developing Countries: Ideas And Observations*	A. Fugelsang	Dag Hammarskjold Foundation, Uppsala, Sweden, 1973

(Continued)

Subject Matter	Title of Book	Author(s)	Publisher & Date
Education (continued)	*Focus On Food*	L.B. Peck L. Moragne M.S. Sickley E.O. Washington	McGraw-Hill Book Co., New York, NY, 1974
	Food You Eat, The	J.S. Marr	Evans & Co., Inc., New York, NY, 1973
	International Bibliography Of Nutrition Education	C.M. Taylor K.P. Riddle	Teachers College Press, New York, NY, 1971
	Practical Nurse Nutrition Education	A.D. Shackelton	W.B. Saunders Co., Philadelphia, PA, 1972
Food Composition	*Amino-Acid Content Of Foods*	FAO	United Nations, 1972
	Chemical Analysis Of Foods, The	D. Pearson	Chemical Pub. Co., Inc., New York, NY, 1971
	Composition Of Foods; Agriculture Handbooks *No. 8* *No. 8-1—Dairy And Egg Products* *No. 8-2—Spices And Herbs* *No. 8-3—Baby Foods* *No. 8-4—Fats And Oils* *No. 8-5—Poultry Products* *No. 8-6—Soups, Sauces, And Gravies* *No. 8-7—Sausages And Luncheon Meats*	U.S.D.A.	Government Printing Office, Washington, D.C.
	Food Composition Table For Use In Africa	U.S. Dept. of Health & Human Services and FAO	United Nations, 1968
	Food Composition Table For Use In East Asia	U.S. Dept. of Health & Human Services and FAO	United Nations, 1972
	Food Composition Table For Use In Latin America	National Institutes of Health	N.I.H., Bethesda, MD, 1961
	Food Composition Tables For Use In The Middle East	P.L. Pellett S. Shadarevian	Amer. Univ. of Beirut, Beirut, Lebanon, 1970
	Food Values Of Portions Commonly Used	C.F. Church H.N. Church	J.B. Lippincott Co., Philadelphia, PA, 1975
	McCance And Widdowson's—The Composition Of Foods	A.A. Paul D.A.T. Southgate	Her Majesty's Stationery Office, London, England, 1978
	Nutrients In Foods	G.A.Leveille, *et al*	The Nutrition Guild, Cambridge, Mass., 1982
	Nutritional Qualities Of Fresh Fruits And Vegetables	*Eds.:* P.L. White N. Selvey	Futura Pub. Co., Mt. Kisco, NY, 1974
Food Preparation	*Benevolent Bean, The*	A. Keys M. Keys	Doubleday & Co., Inc., Garden City, NY, 1967
	Cooking Without A Grain Of Salt	E.W. Bagg	Doubleday & Co., Inc., New York, NY, 1964
	Food Fundamentals	M. McWilliams	John Wiley & Sons, Inc., New York, NY, 1974
	Foundations Of Food Preparation	G.O. Peckham	The Macmillan Co., New York, NY, 1974
	Freezing And Canning Cookbook	*Ed.:* N.B.Nichols	Doubleday & Co., Inc., New York, NY, 1973
	Handbook Of Food Preparation, No. 1411	American Home Ec. Assoc.	1971
	Meal Management	F. Kinder	The Macmillan Co., New York, NY, 1973
	Nutrition For Athletes—A Handbook For Coaches	Amer. Assoc. for Health, P.E., and Recreation	Washington, D.C., 1971
	Practical Low Protein Cookery	L.P. Krawitt E.K. Weinberger	C.C. Thomas, Springfield, Ill., 1971
	Putting Food By	R. Hertzberg B. Vaughn J. Greene	Stephen Greene Press, Brattleboro, VT, 1973
	Shopper's Guide. 1974 Yearbook Of Agriculture	U.S.D.A.	Government Printing Office, Washington, D.C., 1974

(Continued)

Subject Matter	Title of Book	Author(s)	Publisher & Date
Food Production & Technology	*Almanac Of Canning, Freezing, Preserving Industries, The*		Edward E. Judge & Son, Inc., Westminster, MD, 1975
	Approved Practices In Fruit And Wine Production	A.H. Scheer E.M. Juergenson	The Interstate Printers & Publishers, Danville, Ill., 1976
	Bakery Products—yeast leavened	D.J. De Renzo	Noyes Data Corp., Park Ridge, NJ, 1975
	Biochemistry: A Case-Oriented Approach	R. Montgomery R.L. Dryer T.W. Conway A.A. Spector	The C.V. Mosby Co., St. Louis, MO, 1974
	Chemicals We Eat, The	M.A. Benarde	American Heritage Press, New York, NY, 1971
	Chemistry And Physiology Of Flavors	H.W. Schultz E.A. Day L.M. Libbey	AVI Pub. Co., Westport, Conn., 1967
	Commercial Processing Of Vegetables	L.P. Hanson	Noyes Data Corp., Park Ridge, NJ, 1975
	Complete Book Of Fruits & Vegetables, The	F. Bianchine F. Corbetta	Crown Publishers, Inc., New York, NY, 1973
	Crops & Man	J.R. Harlan	Amer. Society of Agronomy & Crop Science Society of Amer., Madison, Wisc., 1975
	Designing Foods	National Research Council	National Academy Press, Washington, D.C., 1988
	Developments In Meat Science—2	*Ed.:* R. Lawrie	Applied Science Pub. Englewood, NJ, 1981
	Economics, Marketing, And Technology Of Fish Protein Concentrate, The	S.R. Tannebaum B.R. Stellings N.S. Scrimshaw	MIT Press, Cambridge, Mass., 1974
	Edible Nuts Of The World	E.A. Menninger	Horticultural Books, Inc., Stuart, FL, 1977
	Encyclopedia of Food Technology	A.H. Johnson M.S. Peterson	AVI Pub. Co., Inc., Westport, Conn., 1974
	Encyclopedia of Labeling Meat and Poultry Products	J.C. deHoll, DVM	Meat Plant Magazine, St. Louis, MO, 1978
	Encyclopedia Of The Plant Kingdom, The	*Ed.:* A. Huxley	Chartwell Books, Inc. New York, NY, 1977
	Fabricated Foods	*Ed.:* G.E. Inglett	AVI Pub. Co., Inc., Westport, Conn., 1975
	Food	W. Root	Simon & Schuster, New York, NY, 1980
	Food Additive Tables	*Eds:* E.J. Bigwood, *et al.*	Elsevier Scientific Pub. Co., New York, NY, 1975
	Food Additives To Extend Shelf Life	N.D. Pintauro	Noyes Data Corp., Park Ridge, NJ, 1974
	Food Analysis: Theory And Practice	Y. Pomeranz C.E. Meloan	AVI Pub. Co., Westport, Conn., 1971
	Food Chemistry	L.W. Aurard H.E. Woods	AVI Pub. Co., Westport, Conn., 1973
	Food Dehydration; Practices And Application, Vol. II	*Eds.:* W.B. Van Arsdel M.J. Copley A.I. Morgan	AVI Pub. Co., Westport, Conn., 1973
	Food Law Handbook	H.W. Schultz	AVI Pub. Co., Inc., Westport, Conn., 1981
	Food Sanitation	R.K. Guthrie	AVI Pub. Co., Inc., Westport, Conn., 1972
	Food Science	N.N. Potter	AVI Pub. Co., Inc., Westport, Conn., 1973
	Foods From Animals: Quantity, Quality and Safety	CAST	Ames, Iowa, 1980
	Food Technology—the world over, Vols. 1 & 2	*Eds.:* M.S. Peterson D.K. Tressler	AVI Pub. Co., Inc., Westport, Conn., 1963
	Food Yields Summarized By Different Stages Of Preparation Agriculture Handbook, No. 102	R.H. Matthews Y.J. Garrison	Government Printing Office, Washington, D.C., 1975
	Guide To The Selection, Combination And Cooking Of Foods, Vols. 1 & 2, A	C.A. Rietz J.J. Wanderstock	AVI Pub. Co., Inc., Westport, Conn., 1965

(Continued)

Subject Matter	Title of Book	Author(s)	Publisher & Date
Food Production & Technology (Continued)	*Handbook Of Food Additives*	*Ed.:* T.F. Furia	CRC Press, Boca Raton, Fla., 1975
	Health Plants Of The World	F. Bianchini F. Corbetta	Newsweek Books, New York, NY, 1977
	High-Quality Protein Maize		John Wiley & Sons, New York, NY, 1975
	How To Grow Your Own Vegetables	M. Kressy	Creative Home Library, Des Moines, IA, 1973
	Innards And Other Variety Meats	J. Allen M. Gin	101 Productions, San Francisco, CA, 1974
	Introduction To Food Science And Technology	G.F. Stewart M.A. Amerine	Academic Press, New York, NY, 1973
	Keeping Foods Safe	H. Bradley C. Sundberg	Doubleday & Co., Garden City, NY, 1975
	Leafy Salad Vegetables	E.J. Ryder	AVI Pub. Co., Westport, Conn., 1979
	Lessons On Meat	Staff	National Live Stock and Meat Board, Chicago, Ill., 1972
	Manuals Of Food Quality Control	FAO	United Nations, Rome, Italy, 1980
	Meat	D.J.A.Cole R.A.Lawrie	AVI Pub. Co., Westport, Conn., 1975
	Meat Board Meat Book, The	Barbara Bloch	McGraw-Hill Book Co., New York, NY, 1977
	Meatbook, The	T.M.Evans D. Greene	Charles Scribner's Sons, New York, NY, 1973
	Meat Handbook	A. Levie	AVI Pub. Co., Westport, Conn., 1979
	Meat, Poultry, And Seafood Technology	R.L.Henrickson	Prentice-Hall, Englewood Cliffs, NJ, 1978
	Meat We Eat, The	J.R.Romans P.T.Ziegler	The Interstate Printers & Pub., Danville, Ill., 1977
	Molecular Structure And Function Of Food Carbohydrate	*Eds.:* G.G.Birch L.F.Green	Halsted Press, John Wiley & Sons, New York, NY, 1974
	Natural Poisons	*Ed.:* W. Horwitz	Association of Official Analytical Chemists, Washington, D.C., 1975
	Nutrients In Processed Foods, Vol. 1, Vitamins, Minerals	American Medical Assoc.	Publishing Sciences Group, Acton, Mass., 1974
	Nutritional Evaluation Of Food Processing	R.S.Harris E. Karmas	AVI Pub. Co., Westport, Conn., 1975
	Nutritional Quality Index Of Foods	R.G.Hansen B.W.Wyse A.W.Sorenson	AVI Pub. Co., Westport, Conn., 1979
	Nutrition Technology Of Processed Foods	N.D.Pintauro	Noyes Data Corp., Park Ridge, NJ, 1975
	Our Polluted Food: A Survey Of The Risks	J. Lucas	John Wiley & Sons, New York, NY, 1974
	Oxford Book Of Food Plants, The	G.B.Masefield M. Wallis S.G.Harrison B.E.Nicholson	Oxford Univ. Press, London, England, 1975
	Panic In The Pantry—Food Facts, Fads And Fallacies	E.M.Whelan F.J.Stare, M.D.	The Book Press, Brattleboro, VT, 1975
	Plants For Man	R.W. Schery	Prentice-Hall, Englewood Cliffs, NJ, 1972
	Plant Science	J. Janick R.W. Schery F.W. Woods V.W. Ruttan	W.H. Freeman and Co., San Francisco, Calif., 1974
	Practical Food Microbiology And Technology	H.H. Weiser G.J. Mountney W.A. Gould	AVI Pub. Co., Inc., Westport, Conn., 1971
	Prepared Snack Foods	M. Gutcho	Noyes Data Corp., Park Ridge, NJ, 1973
	Prevention of Microbial And Parasitic Hazards Associated With Processed Foods	Committee on Food Protection, Food and Nutrition Council	National Research Council, Washington, D.C., 1975
	Processed Foods And The Consumer	V.S. Packard, Jr.	Univ. of Minn. Press, Minneapolis, Minn., 1976

(Continued)

Subject Matter	Title of Book	Author(s)	Publisher & Date
Food Production & Technology (Continued)	*Processed Meats*	W.E. Kramlich A.M. Pearson F.W. Tauber	AVI Pub. Co., Westport, Conn., 1973
	Quality Of Horticultural Products	V.C. Arthey	John Wiley & Sons, New York, NY, 1975
	Role Of Sucrose In Foods, The	A. Lachman	AVI Pub. Co., Westport, Conn., 1973
	Science Of Meat And Meat Products, The	*Eds.:* J.F.Price B.S.Schweigert	W.H. Freeman & Co., San Francisco, CA, 1971
	Shortenings, Margarines, And Food Oils	M.T. Gillies	Noyes Data Corp., Park Ridge, NJ, 1974
	Small Fruit Culture	J.S. Shoemaker	AVI Pub. Co., Westport, Conn., 1978
	Snack Food Technology	S.A. Matz	AVI Pub. Co., Westport, Conn., 1976
	Storage, Processing, And Nutritional Quality Of Fruits And Vegetables	*Ed.:* D.K. Salunkhe	CRC Press, Cleveland, Ohio, 1974
	Sunflower, The	C.B. Heiser, Jr.	Univ. of Oklahoma Press, Norman, Okla., 1976
	Sunflower Science And Technology	*Ed.:* J.F. Carter	ASA, CSSA, SSSA, Madison, Wisc., 1978
	Synthetic Foods	M. Pyke	St. Martin's Press, New York, NY, 1971
	Technology Of Food Preservation, The	N.W. Desrosier	AVI Pub. Co., Westport, Conn., 1970
	Technology Of Fortification Of Foods	National Academy of Sciences	National Academy of Sciences, Washington, D.C., 1975
	Temperate-Zone Pomology	M.N. Westwood	W.H. Freeman & Co., San Francisco, Calif., 1978
	Toxicants Occurring Naturally In Foods	Committee on Food Protection, Food and Nutrition Board	National Academy of Sciences, Washington, D.C., 1973
	Toxicity Of Pure Foods	*Ed.:* C.E. Boyd, M.D.	CRC Press, Boca Raton, Fla., 1973
	Tree Fruit Production	B.J.E. Teskey J.S. Shoemaker	AVI Pub. Co., Westport, Conn., 1978
	Tropical Crops—Dicotyledons	J.W. Purseglove	Longman Group Limited, London, England, 1979
	Tropical Crops—Monocotyledons 1 & 2	J.W. Purseglove	John Wiley & Sons, New York, NY, 1972
	Uniform Retail Meat Industry Standards	Industry-wide Coop. Meat Id. Standards Comm.	National Live Stock & Meat Board, Chicago, Ill., 1973
	Use Of Fungi As Food And In Food Processing, The	W.D. Gray	CRC Press, Boca Raton, Fla., 1970 & 1973
	Vegetable Crops	H.C. Thompson W.C. Kelly	McGraw-Hill Book Co., New York, NY, 1957
	Vegetable Growing Handbook	W.E. Splittstoesser	AVI Pub. Co., Westport, Conn., 1979
	World Atlas Of Foods, The	*Ed.:* G. Hale	Simon & Schuster, New York, NY, 1974
Heart	*American Heart Association Heartbook, The*		E.P. Dutton, New York, NY, 1980
	American Heart Association Cookbook, The	R. Eshleman M. Winston	David McKay Co., New York, NY, 1975
	Cooking For Your Heart And Health	M. Waldo	G.P. Putnam's Sons, New York, NY, 1961
	Diet And Atherosclerosis—Vol. 60: Advances in Experimental Medicine and Biology	*Eds.:* C. Sitori G. Ricci S. Gorini	Plenum Press, New York, NY, 1975
	Diet For A Happy Heart	J. Jones	101 Productions, San Francisco, Calif., 1975
	Eat, Drink And Lower Your Cholesterol	F.T. Zugibe	McGraw-Hill Book Co., New York, NY, 1963
	Eat Well And Stay Well	A. Keys M. Keys	Doubleday & Co., Garden City, NY, 1963
	Live High On Low Fat	S. Rosenthal	J.B. Lippincott Co., Philadelphia, PA, 1963
	Low Cholesterol, Lower Calorie Desserts	S. Leinwoll	Charles Scribner's Sons, New York, NY, 1974
	Low Cholesterol Cookbook: Doctor Approved	M. Skinner	The Naylor Co., San Antonio, Tex., 1975

(Continued)

Subject Matter	Title of Book	Author(s)	Publisher & Date
Heart (continued)	*Low Fat-Low Cholesterol Diet, The*	C.B.Y. Bond V. Dobbin H.F. Goffman H.C. Jones L. Lyon	Doubleday & Co., Garden City, NY, 1971
Historical	*Consuming Passions—the anthropology of eating*	P. Fark G. Armelagos	Houghton Mifflin Co., Boston, Mass., 1980
	Food In History	R. Tannahill	Stein & Day, New York, NY, 1973
	History of Nutrition, A	F.V. McCollum	The Riverside Press, Cambridge, Mass., 1957
	Kitchen In History, The	M. Harrison	Charles Scribner's Sons, New York, NY, 1973
	Milestones In Nutrition	S.A. Goldblith M.A. Joslyn	AVI Pub. Co., Westport, Conn., 1964
	Nutrition, Behavior And Change	H.H. Gifft M.B. Washbon G.G. Harrison	Prentice-Hall, Englewood Cliffs, NJ, 1972
	Seed To Civilization: The Story Of Man's Food	C.B. Heiser, Jr.	W.H. Freeman & Co., San Francisco, Calif., 1973
	Story Of Food, The	I.D. Garard	AVI Pub. Co., Westport, Conn., 1974
	That We May Eat, The Yearbook Of Agriculture 1975	*Ed.:* J. Hayes	Government Printing Office, Washington, D.C., 1975
Specific Nutrients	*An Assessment Of Mercury In The Environment*		National Academy of Sciences, Washington, D.C., 1978
	Calcium And Phosphorus Metabolism	J.T. Irving	Academic Press, New York, NY, 1973
	Clinical Applications Of Zinc Metabolism	*Eds.:* W.J. Pories W.H. Strain J.M. Hsu R.L. Woosley	C.C.Thomas, Springfield, Ill., 1974
	Dairy Lipids And Lipid Metabolism	*Eds.:* M.F. Brink D. Kritchevsky	AVI Pub. Co., Westport, Conn., 1968
	Fluorosis	E.J. Largent	Ohio State Univ. Press, Columbus, Ohio, 1961
	Great Vitamin Hoax, The	M.D. Tatkon	The Macmillan Co., New York, NY, 1968
	Improvement Of Protein Nutriture	Comm. on Amino Acids, Food & Nutrition Board	National Research Council, Washington, D.C., 1974
	Influence Of Teeth And Diet And Habits On The Human Face, The	D.M. Davies	Crane, Russak & Co., New York, NY, 1973
	Methods In Carbohydrate Chemistry Vols. I-VII	*Eds.:* R.L. Whistler J.N. Besmiller	Academic Press, New York, NY, 1976
	Newer Methods Of Nutritional Biochemistry, Vols. I-IV	*Ed.:* A.A. Albanese	Academic Press, New York, NY, circa 1970-1972
	Newer Trace Elements In Nutrition	*Eds.:* W. Mertz W.E. Cornatzer	Marcel Dekker, New York, NY, 1971
	New Protein Foods, Vols. 1, 2, 3 & 4	*Ed.:* A.M. Altschul	Academic Press, New York, NY, 1973-1981
	Nutrients In Processed Foods—Fats, Carbohydrates	P.L. White D.C. Fletcher M. Ellis	Publishing Sciences Group, Acton, Mass., 1975
	Nutrients In Processed Foods—Proteins	P.L. White D.C. Fletcher	Publishing Sciences Group, Acton, Mass., 1974
	Nutrients In Processed Foods—Vitamins • Minerals	American Medical Association	Publishing Sciences Group, Acton, Mass., 1974

(Continued)

Subject Matter	Title of Book	Author(s)	Publisher & Date
Specific Nutrients (Continued)	*Nutrition—* *Vol. I—Macronutrients And Nutrient Elements* *Vol. II—Vitamins, Nutrient Requirements, And Food Selection* *Vol. III—Nutritional Status: Assessment And Application*	G.H. Beaton E.W. McHenry	Academic Press, New York, NY 1964-1966
	Nutritive Value Of Triticale Protein	J.H. Hulse E.M. Laing	International Development Research Centre, Ottowa, Canada, 1974
	Protein Nutritional Quality Of Foods and Feeds	*Ed.:* M. Friedman	Marcel Dekker, New York, NY, 1975
	Proteins As Human Food	*Ed.:* R.A. Laurie	AVI Pub. Co., Westport, Conn., 1970
	Second Conference On Vitamin C	*Eds.:* C.G. Burns J.J. Burns	New York Academy of Sciences, New York, NY, 1975
	Symposium: Seed Proteins	G.F. Inglett	AVI Pub. Co., Westport, Conn., 1972
	Symposium: Selenium In Biomedicine	*Eds:* O.H. Muth, DVM J.E. Oldfield P.H. Weswig	AVI Pub. Co., Westport, Conn., 1967
	Single-Cell Protein II	*Eds.:* S.R. Tannenbaum D.I.C. Wang	The MIT Press, Cambridge, Mass., 1975
	Soy Protein And Human Nutrition	*Eds.:* H.L. Wilcke D.T. Hopkins D.H. Waggle	Academic Press, New York, NY, 1979
	Starch And Its Components	W. Banks C.T. Greenwood	John Wiley & Sons, New York, NY, 1975
	Symposium: Sulphur In Nutrition	O.H. Muth, DVM J.E. Oldfield	AVI Pub. Co., Westport, Conn., 1970
	Trace Elements In Human And Animal Nutrition	E.J. Underwood	Academic Press, New York, NY, 1977
	Trace Substances In Environmental Health —VII	*Ed.:* D.R. Hemphill	Environmental Trace Substance Center, Columbia, MO, 1973
	Truth About Fiber In Your Food, The	L. Galton	Crown Pub., New York, NY, 1976
	Vitamins, Vols. I - VII, *The*		Academic Press, New York, NY
	Vitamins And Hormones Many volumes starting in 1942	Many Contributors	Academic Press, New York, NY, 1942
	Vitamins: Their Use And Abuse	J.V. Levy P. Bach-y-Rita	Liveright Pub. Corp., New York, NY, 1976
Obesity (Overweight)	*Brand-Name Calorie Counter, The*	C. Netzer	Dell Pub. Co., New York, NY, 1969
	"But I Don't Eat That Much" A Diet Specialist Answers His Patients' Questions On Once-And-For-All Reducing	M.B. Glenn	E.P. Dutton & Co., New York, NY, 1974
	Carbo-Calorie Diet, The	D.S. Mart	Dolphin Books, Doubleday & Co., New York, NY, 1973
	Childhood Obesity	V. Antonetti	M. Evans & Co., New York, NY, 1973
	Eat And Stay Slim	*Eds:* Better Homes and Gardens	Meredith Press, New York, NY and Des Moines, IA, 1968
	Energetics: Your Key To Weight Control	G. Gwinup	Sherbourne Press, Los Angeles, Calif., 1970
	Exercise Equivalents Of Foods	F. Konishi	Southern Ill. Univ. Press, Carbondale, Ill., 1974
	How The Doctors Diet	P. Wyden B. Wyden	Trident Press Div., Simon & Schuster, New York, NY, 1968
	Low Calorie And Special Dietary Foods	*Ed.:* B.K. Dwivedi	CRC Press, Boca Raton, Fla., 1978
	Nutrition, Physiology, And Obesity	R. Schemmel	CRC Press, Boca Raton, Fla., 1980
	Obesity In Perspective Vol. 2, Parts 1 and 2	*Ed.:* G.A. Bray	Government Printing Office, Washington, D.C., 1975
	Overweight And Obesity: Causes, Fallacies, Treatment	B.Q. Hafen	Brigham Young Univ. Press, Provo, Utah, 1975
	Overweight Society, The	P. Wyden	Wm. Morrow & Co., New York, NY, 1965

(Continued)

Subject Matter	Title of Book	Author(s)	Publisher & Date
Obesity (Continued)	*Pain Of Obesity, The*	A.J. Stunkard	Bull Pub. Co., Palo Alto, CA, 1976
	Psychology Of Obesity: Dynamics And Treatment, The	*Ed.:* N. Kiell	C.C.Thomas, Springfield, Ill., 1973
	Reduce And Stay Reduced On The Prudent Diet	N. Jolliffe	Simon & Schuster Pub., New York, NY, 1964
	Safe And Sure Way To Reduce, The	G. Maddox	Random House, New York, NY, 1960
	Slim Down, Shape Up Diets For Teenagers	G. Maddox	Avon Book Div., The Hearst Corp., New York, NY, 1963
	Stay Slim For Life	I.J. Kain M.B. Gibson	Doubleday & Co., Garden City, NY, 1966
	Thin From Within	J.D. Osman	Hart Pub. Co., New York, NY, 1976
	Think Thin	M.J. Siegel D. Van Keuren	Paul S. Eriksson, New York, NY, 1971
	Your Overweight Child	M.I. Levine J.H. Seligmann	The World Pub. Co., Cleveland, Ohio, 1970
	Your Weight And How To Control It	*Ed.:* M. Fishbein	Doubleday & Co., New York, NY, 1963
Surveys & Proceedings	*Assessment Of Nutritional Status And Food Consumption Surveys* Vol. 20	J.C. Somogyi A. Szczygiel	Albert J. Phiebig, White Plains, NY, 1974
	Chemicals And Health	Report of the Panel on Chem. & Health, Pres. Science Adv. Comm.	Superintendent of Documents, Washington, D.C., 1974
	Clinical Nutrition	*Ed.:* J.C. Somogyi	Albert J. Phiebig, White Plains, NY, 1973
	Comparative Studies Of Food And Environmental Contamination	FAO, IAEA, and WHO	Unipub, New York, NY, 1974
	Dietary Fats And Thrombosis	*Eds.:* S. Renaud A. Nordoy-Basel	Albert J. Phiebig, White Plains, NY, 1974
	Environmental Quality And Food Supply	*Eds.:* P.L.White D. Robbins	Futura Pub. Co., Mount Kisco, NY, 1974
	Laboratory Tests For The Assessment Of Nutritional Status	H.E. Sauberlich J.H. Skala R.P. Dowdy	CRC Press, Cleveland, Ohio, 1974
	Modification Of Lipid Metabolism	*Eds.:* E.G. Perkins L.A. Witting	Academic Press, New York, NY, 1975
	Nutritional Deficiencies In Modern Society	*Eds.:* A.N. Howard M. Baird	Newman Books, Ltd., London, W.1, U.K., 1973
	Nutritional Problems In A Changing World	*Eds:* D. Hollingsworth M. Russel	A. Halsted Press, Div. of John Wiley & Sons, New York, NY, 1973
	Nutrition & Malnutrition: Identification & Measurement	*Eds.:* A.F. Roche F. Falkner	Plenum Press, New York, NY, 1974
	Nutrition Canada, 1975		
	Nutrition, Development and Social Behavior DHEW Pub. No. 73-242	*Ed.:* D.J. Kallen	Government Printing Office, Washington, D.C., 1973
	Nutrition, National Development, And Planning		The MIT Press, Cambridge, Mass., 1973
	Progress In Human Nutrition, Vol. 1	*Eds.:* S. Margen N.L. Wilson	AVI Pub. Co., Westport, Conn., 1974
	Report on the Proceedings of the Workshop on Diet, Nutrition, and Periodontal Disease	Amer. Soc. for Preventive Dentistry	N. Michigan Avenue, Chicago, Ill., 1975
	Sugars In Nutrition	*Eds.:* H.L. Sipple K.W. McNutt	Academic Press, New York, NY, 1974
	Sweeteners: Issues And Uncertainties	Academy Forum	National Academy of Sciences, Washington, D.C., 1975
	Total Parenteral Nutrition	*Eds.:* P.L. White M.E. Nagy	Pub. Sciences Group, Acton, Mass., 1974
World Food	*Biology And The Future Of Man*	*Ed.:* P. Handler	Oxford Univ. Press, New York, NY, 1970
	Bread Of The Oppressed, The	F.B. Floore	Exposition Press, Hicksville, NY, 1975

(Continued)

Subject Matter	Title of Book	Author(s)	Publisher & Date
World Food (Continued)	*By Bread Alone*	L.R. Brown E.P. Eckholm	Praeger Pub., New York, NY, 1974
	Checkmate To Underdevelopment	F. Monckeberg	Press Office, Embassy of Chile, Washington, D.C., 1975
	Conquest Of World Hunger And Poverty	D. Ensminger P. Bomani	Iowa State Univ. Press, Ames, Iowa, 1980
	Diet And Tradition In An African Culture	M. Gelfand	Williams & Wilkins Co., Baltimore, MD, 1971
	Diet For A Small Planet	F.M. Lappe	Ballantine Books, New York, NY, 1971
	Ecology, Food, And Civilization	A.H. Walters	Charles Knight & Co., London, England, 1973
	Feeding The World Of The Future	H. Hellman	M. Evans & Co., New York, NY
	Food And Animals—a global perspective	M.E. Ensminger	Agriservices Foundation Inc., Clovis, CA, 1976
	Food & Man	M.E. Lowenberg E.N. Todhunter E.D. Wilson J.R. Savage J.L. Lubawski	John Wiley & Sons, New York, NY, 1974
	Food And People Dilemma, The	G. Borgstrom	Duxbury Press, Div. of Wadsworth Pub. Co., Belmont, Calif., 1973
	Food, Nutrition, And Health No. 16 in a Series World Review of Nutrition and Dietetics	*Ed.:* G.H. Bourne	S. Karger, White Plains, NY, 1973
	Food—Readings From Scientific American	J.E. Hoff J. Janick	W.H. Freeman & Co., San Francisco, Calif., 1973
	Malnutrition And Food Habits	*Eds.:* A. Burgess R.F.A. Dean	Tavistock Pub., London, E.C.4, 1962
	Man And His Foods: Studies In The Ethnobotany Of Nutrition—Contemporary, Primitive, And Prehistoric Non-European Diets	*Ed.:* C.E. Smith, Jr.	The Univ. of Ala. Press, University, Ala., 1973
	Man, Food, And Nutrition	*Ed.:* M. Rechcigl, Jr.	CRC Press, Boca Raton, Fla., 1973
	No Need For Hunger	R.R. Spitzer	The Interstate Printers & Pub., Danville, Ill., 1981
	Nutrition And Our Overpopulated Planet	S.L. Manocha	C.C.Thomas, Springfield, Ill., 1975
	Nutrition Factor, The	A. Berg	The Brookings Institution, Washington, D.C., 1973
	Nutrition In The Community	*Ed.:* D.S. McLaren	John Wiley & Sons, New York, NY, 1976
	Nutrition, National Development And Planning	*Eds.:* A. Berg N.S. Scrimshaw D.L. Call	The MIT Press, Cambridge, Mass., 1973
	Nutrition—Proceedings Of The Ninth International Congress Of Nutrition, Mexico 1972 *Vol. 1—Review Of Basic Knowledge* *Vol. 2—Prognosis For The Undernourished Surviving Child* *Vol. 3—Foods For The Expanding World* *Vol. 4—Approaches To Practical Solutions*	*Eds.:* A. Chavez H. Bourges S. Basta	S. Karger, White Plains, NY, 1975
	Protein-Calorie Malnutrition	*Ed.:* R.E. Olson	Academic Press, New York, NY, 1975
	This Hungry World	E.S. Helfman	Lathrop Lee & Shepard Co., New York, NY, 1970
	This Hungry World	R. Vicker	Charles Scribner's Sons, New York, NY, 1975
	Too Many: The Biological Limitations Of Our Earth	G. Borgstrom	The MacMillan Co., New York, NY, 1969

BORAX (SODIUM BORATE, SODIUM TETRABORATE)

This is a colorless crystalline substance (chemical formula: $Na_2B_4O_7 \cdot 10\ H_2O$) found in major quantity in the salt deposits of California, Chile, Tibet, Peru, and Canada. Sodium borate was once extensively used as a chemical food preservative, but in many countries, including the United States, its use for this purpose was discarded when it was shown to be potentially toxic in the amounts used. Today, borax is used as follows: In medicines, for ointments and eye washes; in washing powders, water softeners, and soaps; in mixes with clay and other substances to make enamel glazes for sinks, stoves, refrigerators, and metal tiles; in pottery, to make a hard glaze for dishes; in glassmaking, mixed with sand so that it will melt easily and give strong, brilliant glass; in glass cooking utensils and thermometers; and in tanning leather and making paper.

BORIC ACID (BORACIC ACID)

In the United States, boric acid is prepared by treating borax ($Na_2B_4O_7 \cdot 10\ H_2O$) with hydrochloric or sulfuric acid. It forms colorless, odorless, crystals or a white powder that dissolves in water. Its chemical formula is H_3BO_3.

Boric acid is a weak, inorganic acid best known as a mild antiseptic, used in eye lotions and in salves and bandages for burns and wounds. It is also used in insecticides, leather finishing, paints, soaps, wood preserving, and in ceramics and glass manufacturing. Because it is potentially poisonous when taken internally (a lethal dose for normal adults is about ½ oz), its use as a preservative in foods is not permitted in the United States.

BORON

Boron is a biologically dynamic element that affects numerous metabolic processes in higher animals including humans. Further research is needed relative to the nutritional, biochemical, and physiological effects of boron.

BOSTON BROWN BREAD

This is a quick bread which was formerly very popular in New England and the standard fare on Saturday evenings, along with baked beans. It is made with whole wheat flour, rye flour, and cornmeal, plus molasses, milk, and baking soda, and raisins may be added. Traditionally, it is steamed.

BOTULISM

The deadliest of the common types of food poisoning. It is caused by the toxin from *Clostridium botulinus* bacteria. The most common source of botulism is improperly canned food. Most cases of this poisoning are fatal unless immediate medical attention is obtained.

(Also see BACTERIA IN FOOD.)

BOUT

A term used to designate an episode of an acute illness.

BOWEL

Another name for the small and large intestines.

(Also see DIGESTION AND ABSORPTION.)

BOYD-ORR, LORD

John Boyd-Orr was born in Aryshire, Scotland in 1880. He established himself as a British nutritionist, agricultural scientist, writer, and teacher. As a writer, he wrote his best-known book, *Food, Health, and Income,* in 1936. It made important contributions to the science of nutrition and the management of world food supply problems. During 1940, Lord Boyd-Orr served as consultant on food policy to the British government. Then between 1945 and 1948, he served as the first director general of the United Nations Food and Agricultural Organization (FAO). As director general of the FAO, he formulated rationing programs to deal with post-World War II famine, for which he received the Nobel Peace Prize in 1949. Lord Boyd-Orr died in 1971 at the age of 90.

(Also see FOOD AND AGRICULTURAL ORGANIZATION OF THE UNITED NATIONS [FAO].)

BOYSENBERRY *Rubas loganobaccus*

A trailing blackberry that is closely related to the loganberry. (These fruits belong to the family *Rosaceae.*) It was named by Walter Knott, founder of Knott's Berry Farm, after Rudolph Boysen who discovered it in California in 1923. (Genetic studies have shown that it resulted from a spontaneous crossing (hybridization) of a raspberry with a blackberry.)

Fig. B-73. Boysenberry.

The berry has a purplish-black color and is larger than many of the other blackberries. It is grown mainly in California and Oregon. About 95% of the crop is processed by canning, freezing, and baking into pies.

The nutrient composition of various forms of boysenberries is given in Food Composition Table F-36.

Some noteworthy observations regarding the nutrient composition of boysenberries follow:

1. Unsweetened forms of the fruit (fresh, canned, or frozen) are low in calories (35 to 50 kcal per 100 g) and carbohydrates (9 to 11%). However, the sweetened forms contain about twice as many calories and carbohydrates.

2. Boysenberries are a good source of fiber and iron, but only a fair source of Vitamin C. However, the latter shortcoming is offset somewhat by the ample content of bioflavonoids (substances that have beneficial effects when consumed along with vitamin C).

(Also see BLACKBERRY; and FRUIT[S], Table F-47 Fruits of the World.)

BRADYCARDIA

A slower than normal resting heart rate. Bradycardia may be a sign of physical training, because, following training, the heart is able to deliver more blood with each beat. Hence, it does not have to beat as rapidly as it did before training. However, bradycardia may also be due to a disorder of the heart muscle, or other abnormalities.

BRADYPHAGIA

A term meaning abnormally slow eating.

BRAINS

A very tender and delicately flavored meat by-product. The covering membrane should be removed. Brains may be sauteed, creamed, breaded, used in salads, or pan fried.

BRAN

The outer coarse coat (pericarp) of grain which is separated in milling. Wheat, oat, rice, and corn bran are most common. Bran contains a large part of the valuable nutrients found in grains. It is the most popular food for the purpose of increasing bulk in the diet. Most nutritionists recommend that healthy adults get 20 to 35 g of fiber per day.

(Also see CEREAL GRAINS; FIBER; and FLOURS.)

BRAWN

A British term for a pork dish made from the chopped, cooked, and molded edible parts of pig's head, feet, legs, and sometimes tongue. Mock brawn differs in that other meat by-products are used.

BRAZILIAN CHERRY (PITANGA; SURINAM CHERRY) *Eugenia uniflora*

A small ribbed fruit of a shrub or a small tree (of the family *Myrtaceae*) that is eaten fresh or made into jams and pies. Brazilian cherries are red to almost black in color and ¾ to 1 ½ in (*2 to 4 cm*) in diameter, and contain one or two seeds. They are now grown throughout the tropics and subtropics, including Florida. They may also be grown as hedges.

The nutrient composition of the Brazilian cherry (Pitanga) is given in Food Composition Table F-36.

Some noteworthy observations regarding the nutrient composition of the Brazilian cherry follow:

1. The raw fruit has a low to moderately high content of calories (51 kcal per 100 g) and carbohydrates (12 ½%).

2. Brazilian cherries are an excellent source of vitamin A and a good source of vitamin C.

BREADNUT

The nut of a tree (*Brosimum alicastrum*) of Jamaica and Mexico that is roasted and ground into a flour from which bread is made.

BREADROOT *Psoralea esculenta*

The root of a densely hairy plant of western United States used for food.

BREADS AND BAKING

Contents *Page*

History of Baking 244
Basic Principles of Baking 246
Types of Breads and Baked Doughs 246
Unleavened Breads 246
Breads Leavened with Yeast and Other Microorganisms 247
Quick Breads Made with Baking Powder or Baking Soda 248
Products Leavened with Air or Steam 249
Nutritive Values of Breads and Related Items 249
Special Bread Products for Modified Diets 254
Improving the Nutritive Values of Breads and Baked Goods 256
Cornell Bread 256
Suggestions for Adding Highly Nutritious Ingredients to Baked Goods 257
Special Nutritional Supplements to Baked Goods 257
Common Foods That May be Added to Baked Goods 257
Recent Developments in Baking Technology 259

To most people, breads are doughs that have been baked in the shape of loaves. However, the term may also refer to any type of baked product, either leavened or unleavened, that is prepared from a flour or a meal mixed with water and other ingredients. Baking means the cooking of a food by dry heat, whether in an oven or on a flat, heated surface such as a frying pan or a griddle. Occasionally, breads may also be cooked by other processes, such as steaming or frying in deep fat. Fig. B-74 shows squash bread and cranberry tea.

Fig. B-74. Squash bread and cranberry tea—a nutritious break. (Courtesy, USDA)

Bread is a favorite food all over the world. For taste, texture, and social significance, it is unsurpassed. "Breaking bread together" is a well known figure of speech, indicative of people sharing a meal; and the foundation of the Christian church is the Last Supper, when Christ broke bread with his disciples. Hence, it is of more than passing historical interest to determine how ingenious bakers and cooks transformed cereal flours and meals, plus a few other simple ingredients, into tasty products such as bagels, biscuits, breads, cakes, chapaties, cookies, crackers, danish pastry, doughnuts, dumplings, muffins, pancakes, pie pastry, pizza, popovers, puff pastry, rolls, tortillas, turnovers, and waffles.

HISTORY OF BAKING. The first cultivation of wheat and barley appears to have occurred at several places in the Middle East, around 8000 B.C. Most of the earliest agricultural settlements in this region lay within an area called the Fertile Crescent—a strip of land that curved northward and eastward from what is now Israel, to the mountainous border between Iraq and Iran. The early grain farmers used stones for grinding wheat and barley into flours or meals that were mixed with water and baked on flat, heated stones. These primitive breads were dense and chewy because the first bakers did not know how to make doughs rise. Nevertheless, in the Middle East bread was considered "the staff of life"; it was both a daily staple and an integral part of festival occasions.

It may seem surprising that the Egyptians were the first people to become experts in the leavening of bread, because the people of Mesopotamia (an ancient civilization which developed around the Tigris and Euphrates Rivers in what is now eastern Iraq) appear to have had a considerable head start in the utilization of wheat. However, the Mesopotamians started irrigating their crops around 5500 B.C., and their soils subsequently became too salty for the growing of wheat. As a result, they were forced to rely solely upon barley, which tolerates salty soils, for making their breads. There was little opportunity for them to learn how to make breads rise, because accidental leavening by airborne yeasts would have made only a slight improvement in their heavy barley breads. However, they became very proficient in brewing beer from fermented barley.

On the other hand, there was no such difficulty in growing wheat in Egypt, because the soils were irrigated by the salt-free waters of the Nile River. Legend has it that leavening was discovered by an Egyptian baker who set some wheaten dough aside in a warm place. The dough, having been contaminated with wild yeasts and/or bacteria, rose significantly as the microorganism grew and multiplied. No doubt, the person who sampled the accidentally leavened dough after baking was favorably impressed with its light airy texture. Hence, the Egyptians became the first professional bakers. They also invented ovens shaped like beehives, with both a fuel compartment and a compartment for baking.

The type of leavening first used by the bakers of Egypt was a sourdough, made by exposing fresh dough to the wild yeasts and bacteria in the air. Later, they cultivated yeasts in order to obtain greater uniformity in their baked products. It appears that the ancient Hebrews learned to leaven breads from the Egyptians before the 13th century B.C., when they were led out of Egypt by Moses. The flight of the Jews from Egypt is commemorated in the Jewish Passover by the use of an unleavened, crackerlike bread called a matzo, because there had not been time for the use of leavening during the exodus.

Eventually, the ancient Romans adopted the baking practices of the Egyptians and spread them throughout their empire during the early Christian Era. Apparently, Roman bakers were the first to use a mechanical means of mixing dough; which consisted of a large stone basin that held the dough, and wooden paddles that were driven by a horse or a donkey walking around the mixer. They also were the first to use water-driven mills. By the end of the 2nd century A.D., the Roman bakers had organized themselves into guilds that governed their profession. As a result, certain practices, such as the use of the yeasty froth from brewing vats for the leavening of breads, became standardized throughout their empire. The counterparts of our modern-day white breads, which were made for the nobility, required the tedious sifting of whole grain flour through various types of bolting cloths. However, the common people ate only the heavier whole grain breads, which were easier to prepare.

Fig. B-75. Selling bread in Pompeii. (Photo by The Bettmann Archive, Inc., New York, N.Y.)

After the fall of Rome, the peoples of northern Europe went back to their more primitive ways of making breads. For example, bakers in Normandy sometimes allowed dough to rise under the warm blankets of a bed from which the occupants had just arisen. Also, wheat breads were a luxury in much of northern Europe because rye grew better than wheat in the cold, damp climate. Hence, the working class ate a heavy, moist rye bread most of the time. As a result, there were times when people suffered from "Saint Anthony's Fire," which is another name for *ergotism—a fungus disease of grain in which the kernels of grain are replaced by the poisonous fungus.* Rye is much more susceptible to this disease than wheat. People who had unknowingly consumed the fungus in bread suffered from burning sensations in the hands, feet, and chest; some even had convulsions and died. It is noteworthy that there was an outbreak of this disease as late as 1951, in the Rhone Valley of France.

Even though rye grew well in northern Europe, the supplies of this grain sometimes became scarce because of bad weather or blights during the growing season. Hence, it was often necessary to extend bread flour with other flours made from starchy items such as acorns, beans, peas, and potatoes. Some of these breads might have had higher nutritive values than those made wholly from grain flour, because legumes generally contain more protein than grains. Also, blood from animals, especially from oxen, horses, and reindeer, was sometimes mixed into breads.

Throughout the Middle Ages, a feudalistic class system determined the types of foods eaten by people in the various social groups. The more privileged ate their hearty meals on trenchers, which were square slabs of thick, tough bread that served as plates. Following the meal, most of the trenchers were left, and were given to the servants, beggars who came to the door, and the dogs. The term trencherman, meaning a hearty eater, originally meant one who had the good fortune to eat the food served on the trenchers.

The bakers' guilds, which had lost their influence after the fall of Rome, were reorganized in the Middle Ages, and both bakers and millers had important roles under feudalism. Laws were passed which prohibited persons other than professional millers from owning and using mills to grind grain. It was not necessary for baking to be so restricted, because few of the lower class homes had ovens or hearths. Part of the money received by the bakers and millers had to be paid as taxes to the lords and the church. Therefore, small private mills were confiscated whenever they were found. The mills of the professional millers became larger after the first windmill was invented in England in the 11th century. By about the middle of the 14th century, the tall, tower mill had been invented. The latter type of mill had a movable top which carried the sails, and which could be turned to face the wind.

In colonial America, the early settlers built water-driven mills and windmills shortly after their arrival. Without doubt, they used these mills to grind corn for the making of cornbread. It is noteworthy that two of the most revered American presidents—Washington and Lincoln—had at various times been engaged in the milling of grain. Westward migration in America opened up vast new lands that were planted to wheat and corn, and the Great Plains of the United States and Canada soon became the granary for the rest of the world.

The use of a crude type of baking soda had its beginning in the United States in the 1790s, when it was discovered that potassium carbonate prepared from wood ashes gave off carbon dioxide gas when heated. Thus, it became possible to shorten the time required for making breads, because the new powder acted much more rapidly than yeast. Improvements were made in this early type of baking powder, so that, by the 1850s, it was used by bakers around the world.

A steam-driven roller mill for crushing grain was first tested in Switzerland in 1834. But it took almost 50 years for it to be developed for commercial use because the iron rollers wore out too soon. Finally, it was discovered that porcelain rollers did a better job and lasted much longer. As a result of this invention, the baking industry was provided with a large supply of white flour. Roller mills rapidly replaced all other mills because the rollers broke the grain so that the white, starchy part could be separated more easily from the branny seedcoats and the oily germ.

Today, much of the baking industry is mechanized so that only a few people are needed to operate machines that turn out thousands of loaves daily. However, there are still many small bakeries which cater to local needs and turn out special breads that are not produced by the large baking companies. Furthermore, food technologists are ever busy searching for ways to improve breads and to develop new breads for the future.

Fig. B-76. An early advertisement for a Brooklyn baker, suggesting that the quality of his product was comparable to homemade bread. (Photo by The Bettmann Archive, Inc., New York, N.Y.)

BASIC PRINCIPLES OF BAKING. Bread making has always been the subject of awe and reverence. Christ called his disciples the "new leaven." A leaven or a leavening agent is a substance used to make doughs rise. Even today, wheat and bread are considered by the Turkish people to be sacred. Children are taught to kiss bread that has fallen to the floor or ground, and to touch it to their forehead as an apology to God.[21] Nevertheless, the basic operations of baking appear at first glance to be very simple. They consist of the following:

1. Mixing the flour with a liquid, a leavening agent, and other ingredients such as salt, sugar, fat, eggs, raisins, nuts, etc.

[21]Leach, M., and J. Leach, "Meydiha's *Kisir*: a Wheat Dish from Southern Turkey, "*The Anthropologists' Cookbook*, edited by J. Kuper, Universe Books, New York, N.Y., 1977, pp. 64-65.

2. Allowing the dough to rise if yeast or a sourdough starter has been used as the leavening agent. *Yeast* and other microorganisms leaven bread by fermenting sugars in the dough to carbon dioxide gas and water. Hence, carbon dioxide gas is the actual leavening agent. Breads made with baking powder or baking soda do not usually require any rising time before baking because rapid leavening occurs when the doughs are heated.

3. Baking the dough in a pan placed in an oven. Some doughs may be cooked on hot griddles, fried in hot fat, or steamed. Sometimes the baker opens the oven door and is dismayed to find a product that is notably inferior to what is usually expected. For example, the dough may have started to rise, then collapsed as the leavening gas escaped. Fallen doughs are too heavy and too moist. Therefore, doughs must have sufficient elasticity to expand, and they must be strong enough to prevent the escape of the leavening gases. This means that the flour used in baking should contain enough of the protein gluten, which imparts elasticity and strength, or else other measures should be taken to make up for lack of this substance in a flour. Wheat contains the most gluten of all the grains, while rye is a poor second in this respect. All of the other flours are very low in gluten. One of the ways to compensate for the lack of gluten is to add eggs to a dough, because the protein of the egg white acts as a binder. Another way is to use a high baking temperature so that a tough crust is formed rapidly enough to prevent the escape of the leavening gas. Details of the application of these and other principals are given in the section which follows.

TYPES OF BREADS AND BAKED DOUGHS. Over the centuries, many types of baked products were developed by various peoples who attempted to make the best uses of the cereal grains and the baking equipment available to them. No attempt will be made herein to cover all of the diverse items prepared around the world. Instead, some of the more popular breads will be discussed briefly.

Unleavened Breads. These items were made about 10,000 years ago by the first grain farmers. Yet, certain similar products are still popular today. Generally, the most favored ones are made from thin layers of unleavened dough. It is noteworthy that there is often a moderate amount of leavening action in these breads made by the steam generated from the water in the doughs.

• **Chapaties**—These circles of whole wheat dough are eaten throughout India, Pakistan, and Iran. They are prepared by (1), mixing *atta* (the Indian name for whole wheat flour) with salt, *ghee* (an Indian type of clarified butter), and water; forming the dough into thin circles of about 6 in. diameter, and (2) cooking the circles on a griddle until they blister and start to brown.

• **Matzo**—The Jews eat this thin crackerlike, unleavened bread during Passover. Also, matzo meal (crumbs from broken matzos) is used for breading, and for making the chopped fish patties called gefilte fish. Matzos are made from wheat flour and water according to the way prescribed by the Jewish *Talmud*, the authoritative body of Jewish law and tradition.

• **Plain pastry**—This baked dough, which is used for pie crusts and pastry shells, is made from flour, salt, fat, and water. The fat should be semisolid at room temperature so that it produces flakiness by remaining in thin layers or flakes within the dough.

• **Tortillas**—Today, this item is usually prepared by mixing corn or wheat flour with salt and water, then forming the dough into thin circles ranging from 4 to 14 in. in diameter, and baking it on a hot griddle. However, the method of preparation still used by certain Mexican Indians consists of soaking whole kernels of corn in limewater, rubbing the skins off the kernels, crushing the skinned kernels with water to form a dough, and baking circles of the dough on a heated stone or on an iron griddle.

Breads Leavened with Yeast and other Microorganisms. Although baking powders have been available since the 1850s, yeast has continued to be so highly regarded for bread making that during World War II the British Ministry of Food issued instructions to bakers on the ancient art of preparing a yeasty foam (also called a barm) suitable for both leavening and brewing.[22]

Some of the special nutritional benefits that result from leavening doughs with yeast are (1) the bread contains extra vitamins and other nutrients synthesized by the yeast; (2) certain essential minerals, such as chromium, are converted into forms that are well utilized by the body; and (3) phytates (phosphorus compounds found in whole grains and certain legumes, which bind minerals and impede their absorption) are broken down during fermentation.

A typical yeast-leavened white bread may be made by (1) stirring a package of dry, granular yeast into warm milk which contains small amounts of salt and butter, and allowing the mixture to cool; (2) mixing several cups of flour with the milk mixture to make a dough; (3) kneading the dough thoroughly; (4) setting the kneaded dough aside in a warm place to rise until it doubles in bulk; (5) punching down the risen dough; (6) setting the dough aside to rise a second time; and (7) turning the twice-risen dough into greased pans for baking. Most types of plain rolls are made by essentially the same procedure as white bread, except that they may also contain eggs plus a little more fat and sugar. Brief descriptions of some other yeast- and bacteria-leavened products follow:

• **Bagels**—These doughnut-shaped rolls are made from a mixture of flour, salt, sugar, yeast, butter, milk, and egg whites. After rising in a warm place, the bagels are boiled for a short time in water, then removed, and allowed to cool. After cooling, they are brushed with a mixture of water and egg yolk, and baked in an oven.

• **Danish pastry**—This term is sometimes misused by people who think that it refers to various types of sugary buns. However, it refers specificlly to items made from a yeast-leavened dough that is flaky because it contains very thin layers of fat. Prior to rising, layers of the dough are interspersed with sheets of butter or some other type of semisolid fat. The layered dough is repeatedly chilled, folded over on itself, and rolled out so as to make the layers of dough and fat as thin as possible. Then, it is usually coiled or folded around sweetened mixtures of fruit, nuts, or other fillings, after which it is allowed to rise and is baked.

• **Jewish challah bread**—Bakeries which serve Jewish neighborhoods in northeastern United States bake this egg bread on Thursdays so that it may be served on the Jewish Sabbath, which lasts from sundown Friday to sundown Saturday. It is made from dough that is allowed to rise, then is kneaded and made into a braided loaf. The loaf is then allowed to rise again before baking.

• **Malt breads**—These items are sweeter, stickier, and heavier than most yeast breads because malt contains enzymes which (1) convert the starch of the flour into sugars, and (2) digest some of the gluten. Hence, the use of malt results in a weaker gluten structure in the bread, which does not rise as high as ordinary yeast breads. Sometimes, bakers produce imitations of malt breads by using molasses and an inactive malt derivative instead of one that contains active enzymes.

• **Pita, or "pocket" bread**—Although this Middle Eastern bread, or certain variations of it, may be known by different names in different places, it is always characterized by the large hollow center that is formed by baking circles of twice-risen dough for a short time in a hot oven (about 500° or *260°C*). Usually, the baked bread is partly sliced lengthwise, and the pocket is filled with tasty items such as broiled lamb kebab, chopped onions and green peppers, sliced tomatoes, cooked chick peas (garbanzos), or shredded cheese or lettuce.

• **Pizza**—Sometimes, these thin, circular slabs of leavened bread may be almost 2 ft in diameter. Hence, it is easy to see why a famous comedian once referred to pizza as a "baked manhole cover." Before baking, the dough is usually spread with generous amounts of a spicy tomato sauce and shredded cheese, plus other optional toppings like bits of sausage, slices of salami or cooked bacon, fillets of anchovies, sliced mushrooms, chopped olives or green peppers, or various herbs and spices.

• **Raised doughnuts**—These "fried cakes" are made with yeast and a bread flour, and they are allowed to rise before frying in deep fat; whereas cake doughnuts are made with baking powder and a cake flour. Raised doughnuts are sometimes baked in an oven, instead of being fried. The baked items contain fewer calories than the fried items.

• **Salt-rising bread**—Although the procedures used to make this product resemble those used for sourdough breads, the leavening organisms differ from the usual sourdough organisms in that (1) they appear to tolerate both a higher level of salt and a higher temperature, (2) the aroma of the fermentation products is different (more like that of sharp cheddar cheese than sourdough), and

[22]Sheppard, R., and E. Newton, *The Story of Bread*, Routledge & Kegan Paul, Ltd., London, England, 1957, p. 45.

(3) they require a certain amount of cornmeal in the starter, which is a watery dough used as a culture medium for the organism(s).

• **Sourdough bread**—This type of bread is leavened by a starter which often contains beneficial strains of both yeast and bacteria, with the result that two types of fermentation occur. For example, San Francisco's famed sourdough bread is made from a starter containing a bread yeast and a strain of lactic acid bacteria called *Lactobacillus san francisco.* The lactic acid produced during the bacterial fermentation gives the bread its sour taste. Recently, there has been an increased production of sourdough breads. Furthermore, pancakes, biscuits, and rye breads have long been made with sourdough starters.

• **Whole wheat bread**—Doughs made from whole wheat flours are generally kneaded less than those made from more refined flours because the sharp edges of the whole grain seedcoats may cut the strands of gluten in the dough.

Quick Breads made with Baking Powder or Baking Soda. The development of various types of chemical leavening agents, called baking powders, make it possible to prepare breads and other baked items in a much shorter time than is required for yeast-leavened breads. Hence, products leavened with these powders are often called "quick breads" because there is no need for allowing the dough to rise before baking. Some of the basic facts regarding chemical leavening follow:

1. Baking soda is the common name for sodium bicarbonate, a white powder which decomposes and releases carbon dioxide gas when it is dissolved in water and heated, or when it reacts with an acid. About twice as much carbon dioxide is released when baking soda reacts with an acid as when it decomposes in solution. Furthermore, a bitter residue of sodium carbonate remains in the dough when baking soda is used without an acid. Hence, it is almost always used with some type of acidic substance.

2. Baking powders usually contain a mixture of baking soda and one or more acidic salts which react with

TABLE B-17
THE BASICS OF QUICK

Baked Product	Liquid	Fat	Eggs	Sugar	Salt	Baking Powder (P) or Baking Soda (S)	Baking Conditions Time	Temperature	
	◀ --------- (*Amounts of ingredients per c of all-purpose flour*) --------- ▶						(min)	(°F)	(°C)
Biscuits	⅓ to ½ c milk	2 to 4 Tbsp	—	—	½ tsp	1¼ to 2 tsp P	7–10	450	*232*
Cookies	—	¼ c	½	½ c	¼ tsp	1 tsp P	8–12	375	*191*
Cornbread	1 c milk	¼ c oil	1	¼ c	½ tsp	3 tsp P	20–25	425	*218*
Crackers	¼ c milk	1 to 4 Tbsp	—	—	¼ tsp	0 to ¾ tsp S	8–15	350	*177*
Doughnuts	¼ c milk	1½ tsp	½	¼ c	¼ tsp	1 to 2 tsp P	—	—	—
Dumplings	⅓ c of broth from meat, poultry, or fish	—	1	—	½ tsp	2½ tsp P	—	—	—
Gingerbread ...	¼ cup milk	2 Tbsp	—	¼ c	½ tsp	1 tsp S	90	350	*177*
Muffins	½ c milk	1 to 3 Tbsp	½	1 to 2 Tbsp	½ tsp	1¼ to 2 tsp P	20–25	425	*218*
Pancakes	¾ to ⅞ c milk	1 Tbsp	½	0 to 1 Tbsp	½ tsp	1½ to 2 tsp P	—	—	—
Plain cake	¼ to ½ c milk or buttermilk	2 to 4 Tbsp	½ to 1	½ to ¾ c	⅛ to ¼ tsp	1 to 2 tsp P	30–45	350	*177*
Soda bread	⅓ to ½ c buttermilk	2 Tbsp	—	4 tsp	¼ tsp	⅓ tsp S and ¼ tsp P	45–50	350	*177*
Waffles	¼ to 1 c milk	1 to 3 Tbsp	1 to 2	—	½ tsp	1¼ to 2 tsp P	—	—	—

the soda only when they are wetted with water or a similar liquid. Hence, the ingredients do not react when the powder is kept dry.

3. Certain recipes call for baking soda alone, rather than baking powder, because one of the other ingredients is acidic—such as fruit juice or pulp, molasses, sourdough starter, a soured dairy product, or even vinegar.

4. A few baking powders, which are used mainly for cookies and crackers, contain ammonium bicarbonate or ammonium carbonate rather than sodium bicarbonate because the ammonium salts leave no solid residue in the baked products.

The type of leavening agent is only one of the factors responsible for the characteristics of various baked doughs. Additional important factors are the amounts and types of other ingredients such as the flour, liquid, fat, eggs, sugar, and salt; and, the time and temperature used in baking. Table B-17 lists (1) the proportions of the major ingredients, and (2) the baking conditions for some of the common types of quick breads.

(Also see BAKING POWDER AND BAKING SODA.)

BREADS

Other Ingredients, and Instructions
Use minimum liquid for rolled biscuits, maximum liquid for drop biscuits.
Plus ½ tsp vanilla extract or other flavoring(s).
Plus 1 c cornmeal. May also be cooked in a greased skillet.
Extra salt and/or sesame seeds may be sprinkled on the tops of the crackers.
Plus ½ tsp vanilla extract or other flavoring(s). Fry in hot fat until browned on top and bottom.
Plus ⅓ c of nonfat dry milk. Cook 15 min by dropping balls of dough on the top of simmering soup or stew.
Plus 1 tsp ground ginger and 2 Tbsp molasses.
Plus ½ c nuts, bran, blueberries, or raisins, if desired. Half cornmeal and half wheat flour may be used instead of all wheat flour.
Pour batter onto hot griddle and cook until the edges start to brown. Then, turn and cook on the other side.
Plus ½ tsp vanilla or other flavoring(s).
Plus ¼ c raisins, currants, or chopped candied fruits, if desired.
Bake in preheated waffle iron until golden brown.

Products Leavened with Air or Steam. A few baked products are leavened mainly with air or steam, and little, if any, baking powder. The success of such leavening is uncertain unless the batters or doughs are mixed and baked in ways which ensure the retention of the gases. Brief descriptions of some long-time favorites follow:

- **Angel food cake**—The main ingredients of this white, spongelike cake are egg whites, sugar, and flour. Small amounts of salt, flavoring, and cream of tartar are also used. Leavening is provided by the air trapped in the egg white foam, plus steam from the water in the foam.

- **Chiffon cakes**—This light type of cake is similar to angel food in that its leavening depends partly on an egg white foam. However, it differs in that it also contains baking powder, vegetable oil, and egg yolks which help to emulsify the oil.

- **Cream puffs**—Steam is the major leavening agent for these round, hollow shells of puff pastry which are rich in fat and eggs. The mixing of the ingredients is unusual because the fat and water are first boiled together in a sauce pan. Then, the flour is mixed in and the eggs are beaten in, one at a time. Mounds of the batter are dropped onto a baking sheet and baked in a hot oven (400°F or *205°C*) so that (1) steam is generated, and (2) a crust is formed rapidly, trapping the steam within the puffs.

- **Popovers**—These items, like cream puffs, are baked in a hot oven so as to trap steam which produces large cavities. They differ from creampuffs in (1) that they contain less fat and fewer eggs, and (2) their method of preparation.

- **Sponge cake**—This type of cake is similar to angel food in that it depends mainly upon an egg white foam for leavening. It also resembles the chiffon type because it contains egg yolks plus a little baking powder.

NUTRITIVE VALUES OF BREADS AND RELATED ITEMS. Bread has often been called the "staff of life." This implies that it supplies most of the required nutrients. While this characterization might have had some basis when it was applied to the ancient types of coarse, whole grain breads that sometimes contained malted or sprouted grains, it does not seem to be very applicable to the modern types of breads made from highly refined flours. Figs. B-77, B-78, and B-79 show that the levels of selected nutrients found in various baked products may vary considerably.

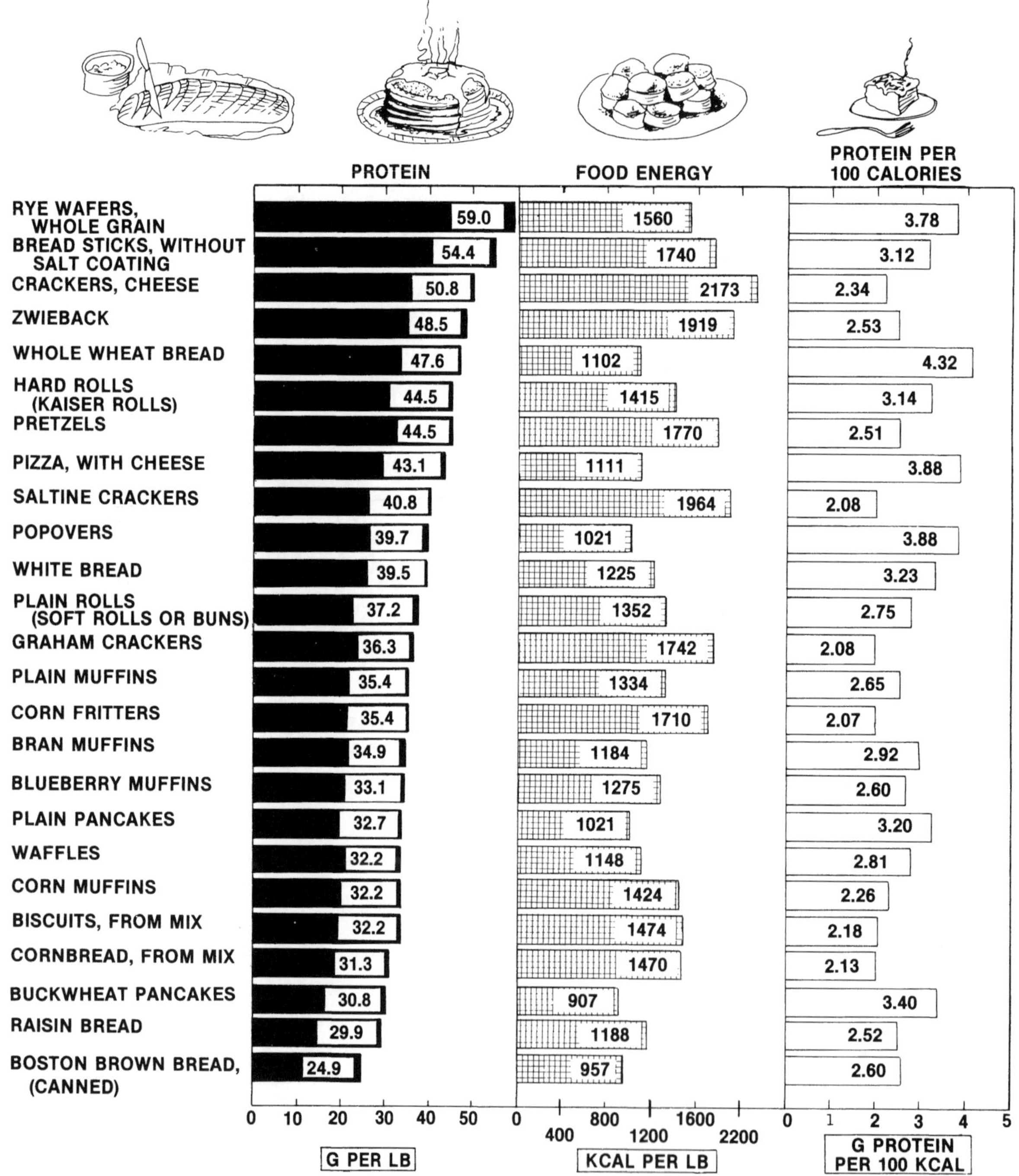

Fig. B-77. Protein and caloric contents of selected breads and baked products. (Based upon data from Food Composition Table F-36 of this book.)

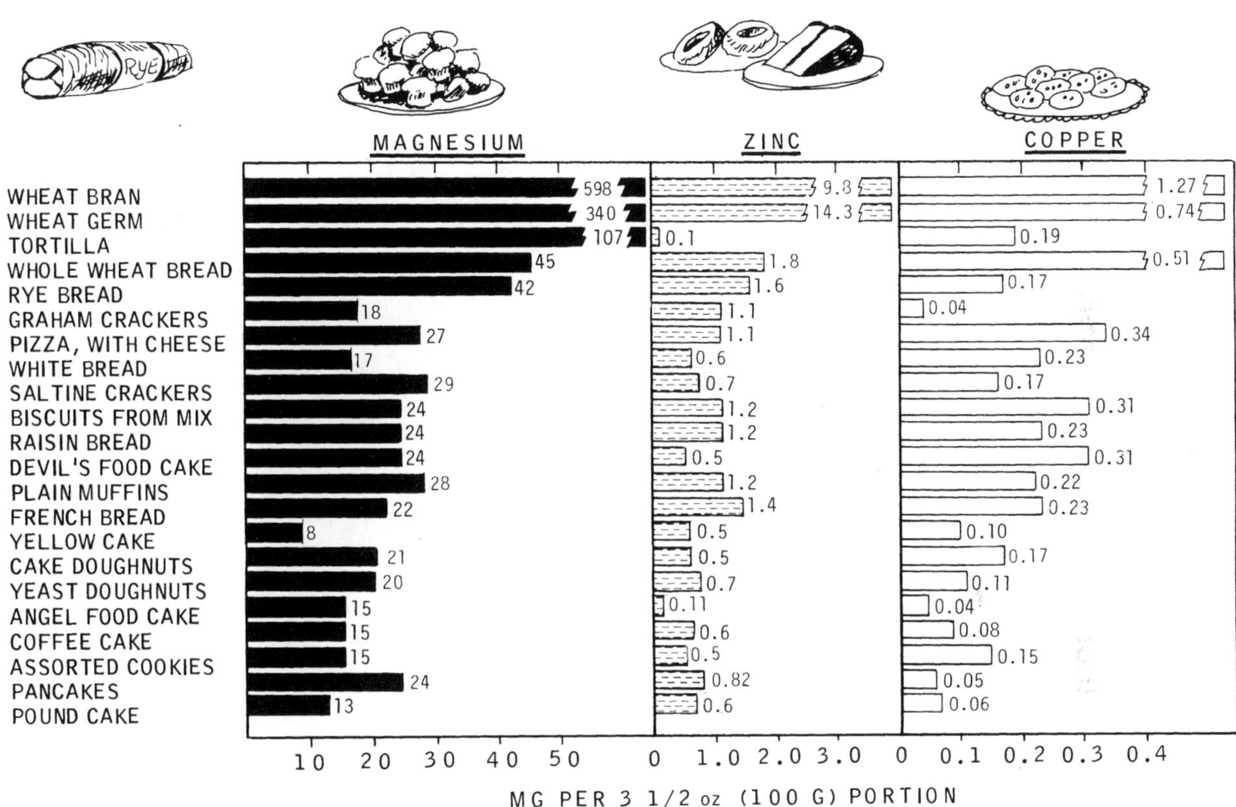

Fig. B-78. Magnesium, Zinc, and Copper contents of selected breads and baked products. (Based upon data from Food Composition Table F-36 of this book)

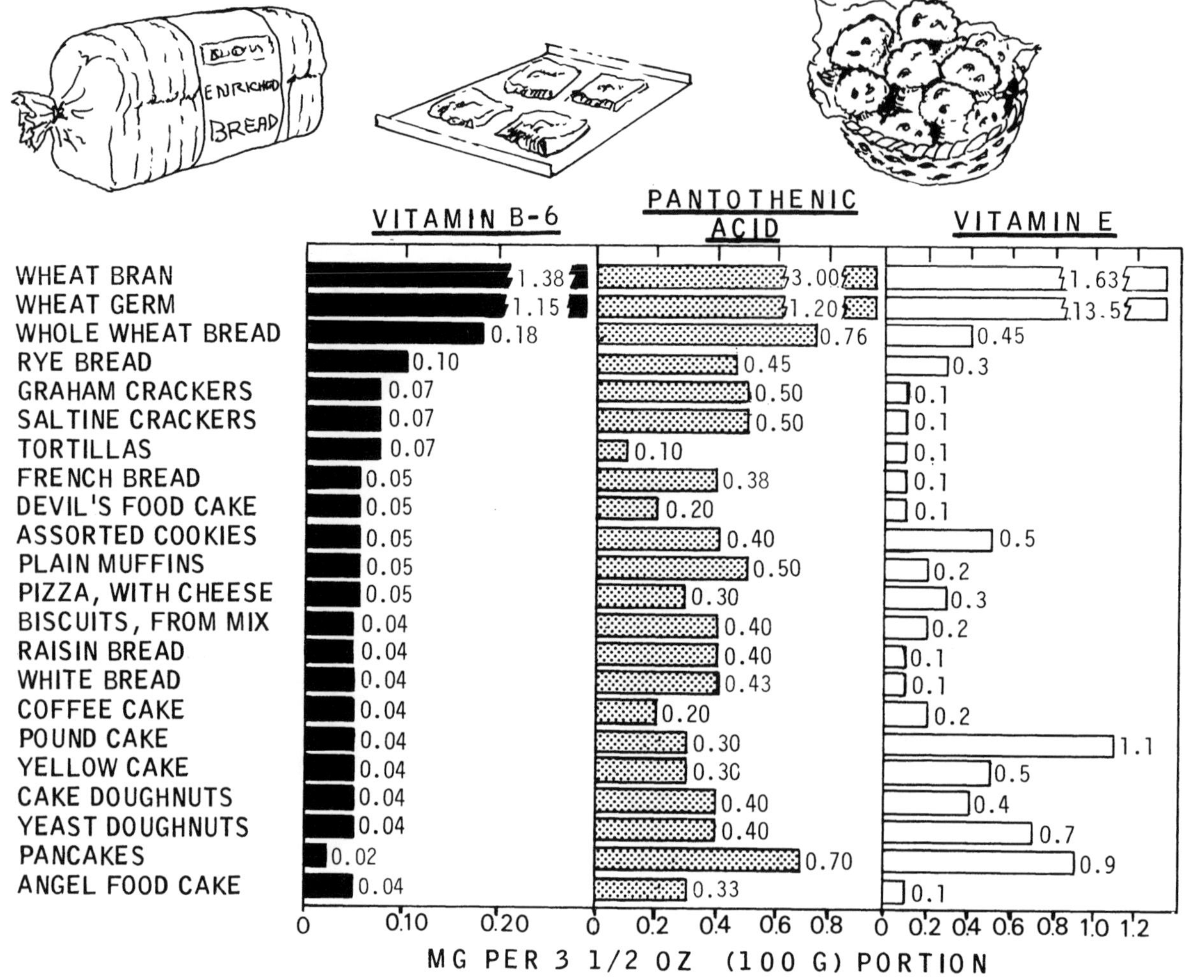

Fig. B-79. Vitamin B-6, Pantothenic acid, and vitamin E contents of selected breads and baked products. (Based upon data from Food Composition Table F-36 of this book)

Comments on Figs. B-77, B-78, and B-79 follow:

1. The items which are highest in protein—both on a per pound basis, and a per calorie basis—are those made from whole grains, such as rye wafers and whole wheat bread. However, products made from highly refined flours may be equally high in protein if they contain milk, eggs, or cheese. For example, Fig. B-77 shows that buckwheat pancakes, popovers, and cheese pizza furnish almost as much protein per calorie as rye wafers.

2. Low moisture items, such as crackers and zwieback (a type of toast), may be good snacks for hikers to carry because they furnish about twice as many calories per pound as a high-moisture item such as Boston brown bread. On the other hand, the latter item may be more filling to people trying to lose weight.

3. Whole grain products generally contain the greatest amounts of minerals and vitamins, although certain items made with white flour plus bran or wheat germ may be close rivals because the latter ingredients are exceptionally rich in these nutrients.

4. The baked products which rank near the bottom of the mineral and vitamin rankings are those made from white flour plus large amounts of fat and sugar. They are rich in calories, but poor in other nutrients. For example, in comparison with whole wheat bread, pound cake contains only ⅓ as much magnesium and ¼ as much vitamin B-6.

There are other pitfalls in relying upon breads to supply much of one's nutritive needs, since even whole grain products may be very deficient in certain minerals and vitamins. Fig. B-80 shows the nutritional merits and demerits of whole wheat bread.

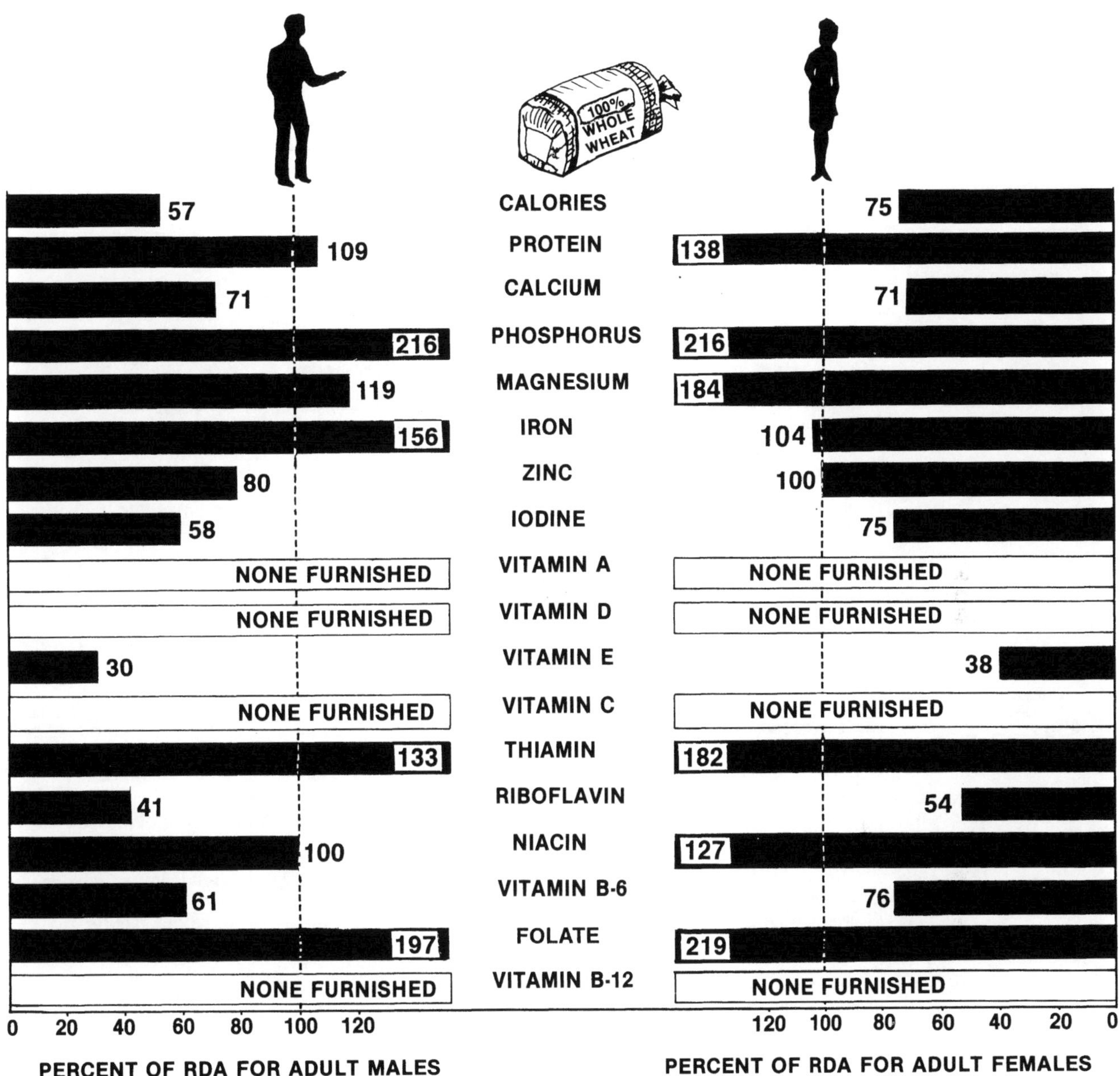

Fig. B-80. The percentages of the Recommended Dietary Allowances (RDAs) for adult (25-50 yr) males and females furnished by 1½ lb (*680 g*) of whole wheat bread. (Based upon RDAs given in *Recommended Dietary Allowances*, 10th ed., NRC–National Academy of Sciences, 1989; and upon nutrient data in Food Composition Table F-36 of this book)

It may be seen from Fig. B-80 that whole wheat bread contains none of the vitamins A, D, C, or B-12. Furthermore, it is a poor source of vitamin E. While a few men might be able to meet their needs for most of the other essential nutrients by eating 2 lb (*0.9 kg*) or more of whole wheat bread per day, most women would gain weight from eating so much bread. This amount of bread would furnish 2,482 Calories (kcal), whereas the energy requirement of most adult females is 2,000 Calories (kcal).

Therefore, it would seem preferable that both men and women (1) limit their consumption of bread and cereals, and (2) obtain a substantial portion of their daily food energy allowance from milk and dairy products, fruits and vegetables, and meats, poultry, fish, eggs, and legumes. In this manner, they would be more likely to meet their needs for the nutrients that are deficient in the breads and cereals.

Special Bread Products for Modified Diets. Many people have been advised by their doctor to modify their diets in order to treat or prevent certain diseases. Furthermore, they may be accustomed to eating plenty of bread with their meals. Hence, it might be beneficial for them to eat modified breads and bakery products which have been prepared in accordance with their dietary prescriptions because these items might make it easier for them to avoid other foods that are not allowed in their diets. Table B-18 lists some of the more common dietary modifications, and the special products that are in accordance with them.

NOTE: The data on special dietetic products are not to be considered as recommendations for people with various health problems to use these products for self treatment. It is the responsibility of the patient's doctor or dietitian to decide which foods best fill a dietary prescription. Therefore, the details which follow are given so that the patient might be better informed in discussions with his or her professional health counselors.

TABLE B-18
SPECIAL BAKERY PRODUCTS

Dietary Prescription	Reasons for the Dietary Prescription	Ingredients in Special Dietetic Bakery Products	Comments
Carbohydrate restriction	Sometimes, disorders such as diabetes, dumping syndrome, high blood triglycerides, low blood sugar, and obesity are improved if the foods which raise the level of blood sugar are restricted.	The carbohydrate content is reduced by replacing some of the flour with nonstarchy ingredients such as bran, cellulose, Jerusalem artichoke flour, ground seedcoats from edible seeds, and nondigestible vegetable gums.	Often, restriction of dietary sugars alone works as well or better than the restriction of carbohydrate in general.
Cholesterol and saturated fat restriction	People who are extra susceptible to atherosclerosis, coronary heart disease, diabetes, or high blood levels of fats such as cholesterol and triglycerides might benefit from this diet.	Products which would normally contain whole eggs are made with egg white, lecithin, and vegetable gums. Also, solid animal fats are replaced with liquid vegetable oils, but not with coconut oil.	Imitation, yolk-free egg products are now sold in most supermarkets. Liquid oils might be unsatisfactory for items such as pastries and pie crusts.
Egg free	It is necessary to eliminate all forms of egg from the diets of those who are allergic to this food.	Special "egg replacers," which are mixtures of various starches, may be used instead of eggs in certain recipes.	Egg replacers are sold by health food stores and dietetic supply companies.
Electrolyte restriction	Patients with high blood levels of certain mineral salts (electrolytes) due to kidney failure require diets low in the salts and low in protein, since some of the products of protein metabolism are electrolytes.	These items are essentially starch cakes made from wheat starch, yeast, mineral-free water, sugar, unsalted margarine with the nonfat solids removed, and a cellulose derivative or a vegetable gum. Sometimes, artificial flavorings and colorings are added.	This type of product provides calories, but no protein, minerals, or vitamins. It is used only during the critical stage of kidney failure.
Energy restriction	Reduction of excess body weight may be aided by the consumption of foods high in bulk and low in calories.	Poorly digested fibrous materials such as bran, cellulose, and seedcoats are added to give bulk and to reduce the caloric content. Vegetable gums, gluten, or gluten flour may be added to strengthen the dough.	The taste and texture of these items may be unappealing when the nondigestible fiber content exceeds 30%. These items may also have a laxative effect.
Fat restriction	A few people lack the means for removal of fats from the blood, so there is a need for a drastic restriction of dietary fat in order to prevent the buildup of dangerously high levels of blood fat.	Most breads and other bakery products may be prepared from modified home recipes which eliminate fats. Commercial mixes for biscuits, muffins, cornbread, rolls, and cakes usually contain too much fat for this diet.	Sometimes, small amounts of special fats called medium-chain triglycerides may be allowed, because they do not accumulate in the blood.
Fiber enrichment	Diets high in fiber might help to prevent constipation, diverticulitis, and other bowel disorders. Sometimes, they may even help to lower elevated blood levels of fats and sugars.	High fiber items are usually made with whole grain flours plus extra fiber in the form of bran, cellulose, and/or seedcoats. They are also used in carbohydrate restricted diets and in energy restricted diets.	Plenty of water should be taken with high-fiber products in order to prevent the formation of dried, hard masses in the lower bowel.

(Continued)

TABLE B-18 *(Continued)*

Dietary Prescription	Reasons for the Dietary Prescription	Ingredients in Special Dietetic Bakery Products	Comments
Fiber restriction	Very bland, low-fiber diets are recommended in order to avoid irritating lesions of the digestive tract such as those which occur in bleeding ulcers, after gastrointestinal surgery, in the acute stages of gastritis or colitis, and when hemorrhoids are either bleeding or severely inflamed.	All baked goods must be made from white flours rather than whole grain flours, and there should be no other fibrous, irritating, or spicy ingredients.	These diets are now used only when critical conditions are present because people with healed lesions appear to tolerate fiber well.
Gluten free	Dietary sources of gluten—a protein abundantly present in wheat, rye, oats, and barley—must be eliminated from the diets of people who cannot tolerate it. This condition sometimes occurs after gastrointestinal surgery, or severe diarrhea.	Only products containing flours made from corn, rice, starchy vegetables like potatoes, or legumes such as soybeans and lima beans may be used. Egg whites and special vegetable gums may be used in place of gluten to strengthen doughs.	People who cannot tolerate gluten must use a minimum gluten diet for the rest of their lives.
Hypoallergenic	In certain allergic conditions, such as asthma, which may be due to a sensitivity to one or more foods, it is best to eliminate the most common sensitizing foods until the actual offenders are identified.	Hypoallergenic breads are generally made without any cereal flours, milk, eggs, or yeast. Generally, the ingredients of these products consist of bean or potato flour, baking powder, sugar, fat, salt, and water. Even vegetable gums may provoke allergies in some people.	Sometimes, cornmeal, corn flour, cornstarch, potato flour, or rice flour may be used to make hypoallergenic items, because only a few people are allergic to these ingredients.
Iron enrichment	Children and women who have iron-deficiency anemia might be better off eating iron-rich foods than taking iron supplements in pill form because the latter may cause digestive upsets and interfere with dietary vitamin E.	Iron-rich ingredients for baked goods include whole grain flours, enriched white flours, soybean flour, wheat germ, molasses, nuts, raisins, and egg yolks.	Whole grain flours, bran, and soybean flour contain phytates which interfere with iron absorption. Yeast fermentation breaks down some of the phytates in whole grain breads.
Milk free	Milk must be eliminated from the diets of people who have milk allergies.	Many types of baked items may be made from water as well as from milk, but it may be better to use imitation milks made from soybeans because they are rich in protein and other nutrients.	Nondairy creamers and coffee whiteners sometimes contain protein derived from milk, so they may aggravate allergies.
Phosphorus restriction	Dietary phosphorus must be kept low for people whose blood level of the mineral is too high. Usually, this condition results from the failure of the kidneys to excrete excesses of phosphorus via the urine.	Only refined flours may be used in baking because whole grains and bran are too rich in phosphorus. Yeast or baking soda may be used for leavening, but baking powders which contain phosphate salts may not be used.	The overall diet should provide enough calories, protein, and calcium to prevent soft tissues and bones from breaking down and releasing phosphorus into the blood.
Potassium restriction	There is a danger of the heart stopping if the kidneys fail to excrete excesses of potassium via the urine. Protein and sodium are also restricted in kidney failure.	Breads low in potassium, sodium, and protein are made from wheat starch, yeast, mineral-free water, sugar, unsalted margarine with the nonfat solids removed, and a cellulose derivative or a vegetable gum.	Tissue breakdown releases potassium into the blood, so it should be minimized by providing adequate calories.
Protein enrichment	Extra dietary protein is needed by patients who are recovering from extensive surgery, severe burns, malnutrition, loss of blood, prolonged illnesses and broken bones.	Even whole grain flours contain only moderate amounts of protein, so baked products may be fortified with high-protein ingredients such as soy flour, bean flours, eggs, nonfat dry milk or evaporated milk, and debittered brewers' yeast powder.	Gluten, gluten flour, vegetable gums, or other additives may be used to strengthen the dough so that it rises properly. Gluten-rich products also provide extra protein.

(Continued)

TABLE B-18 (*Continued*)

Dietary Prescription	Reasons for the Dietary Prescription	Ingredients in Special Dietetic Bakery Products	Comments
Protein restriction	Even moderate amounts of dietary protein may be toxic when (1) the liver loses its ability to detoxify the ammonia generated in protein metabolism, or (2) the kidneys fail to excrete nitrogenous wastes from protein metabolism.	A low protein, energy-rich bread may be made from wheat starch, yeast, water, sugar, margarine or butter, and a vegetable gum or other type of nonprotein additive to serve as a binder in place of gluten.	The diet is high in calories in order to minimize the breakdown of body protein. Sometimes, eggs are used because their metabolism produces little nitrogenous waste when low protein diets are consumed.
Purine restriction	High blood levels of urates, or even gout, may result from (1) the abnormal metabolism of dietary substances called purines, and (2) diets which produce ketosis.	All of the common ingredients except wheat germ and bean flours may be used in low-purine breads. Excessive growth of yeast in dough (2 or 3 prolonged periods of rising) may raise the purine content, because yeast is rich in these substances.	Diets high in protein and fat, but low in carbohydrates, may cause ketosis and elevation of the blood level of uric acid.
Sodium restriction	People who lack the ability to get rid of excess sodium in the urine may develop disorders such as water retention in their tissues, high blood pressure, and congestive heart failure.	No salt or any other sodium compound should be added to items allowed in low-sodium diets. Therefore, yeast should be used for leavening unless a special sodium-free baking powder is available.	A pharmacist may prepare a sodium-free baking powder from certain potassium salts.
Sugar restriction	The limiting of rapidly utilized dietary sugars such as fructose (fruit sugar), glucose (grape sugar), lactose (milk sugar), maltose, and sucrose (cane or beet sugar) may be of some benefit to people with disorders such as diabetes, dumping syndrome, high blood fats, low blood sugar, and obesity.	Breads should be made without any sugary ingredients such as honey, molasses, sugar, syrups, malt, etc. However, items like angel food cake and certain types of cookies cannot be made without sugars, because the textures of these products depend upon the effects of such ingredients.	Certain artificial sweeteners should be used cautiously as sugar substitutes because (1) they do not have the effects on texture that sugars do, and (2) an undesirable flavor may result if their bitter taste is not masked by flavorings and spices such as lemon, orange, cinnamon, and strongly flavored cocoa or chocolate.
Wheat free	Even small amounts of wheat flour may provoke severe allergic reactions in susceptible people.	Flours from any materials other than wheat may be used in baking.	Bakers sometimes develop allergies to wheat as a result of inhaling flour in the air of their bakeries.

Many of the special dietetic items listed in Table B-18 are intended only for people who are on restricted diets which have been ordered by their doctors. It would not be wise for normal, healthy people to eat them because they might develop nutritional deficiencies unless they take pains to obtain the missing nutrients from other foods. Therefore, some suggestions for preparing more nutritious products are given in the sections which follow.

(Also see MODIFIED DIETS.)

IMPROVING THE NUTRITIVE VALUES OF BREADS AND BAKED GOODS. Breads and baked goods have long been major items in the diets of the working classes. For example, the laborers who built the pyramids consumed bread, beer, and a few vegetables. However, they ate a coarse, whole grain bread, which was quite different from the soft, white bread that became readily available to the common people of the Western world following the development of the roller mill in the 19th century. Since then, many attempts have been made to produce special fortified breads which may be both acceptable to most consumers and as nutritious as they are practical.

Cornell Bread. One of the notable achievements in the improvement of white bread was the development of a special bread by the renowned nutritionist Dr. McCay and his colleagues at Cornell University. The bread was made by the Cornell Triple Rich Formula—which was so named because it contained soy flour, skim milk powder, and wheat germ.[23] Table B-19 shows how the nutritive value of breads made according to this formula compare with those made from standard recipes.

[23]Schafer, C., and V. Schafer, *Breadcraft*, Yerba Buena Press, San Francisco, Calif., 1974, p. 58.

TABLE B–19
SELECTED NUTRIENTS FURNISHED BY 1-LB LOAVES OF BREADS MADE FROM VARIOUS INGREDIENTS[1]

Nutrient and Unit of Measurement	White Bread		Whole Wheat Bread	
	Standard Recipe[2]	Cornell Bread[3]	Standard Recipe[4]	with Cornell Ingredients[5]
Calories (kcal)	1065	1050	1097	1082
Protein (g)	32	42	38	48
Protein per unit of food energy (g per 100 kcal)	3.0	4.0	3.5	4.4
Minerals:				
Calcium (mg)	60	305	122	367
Magnesium (mg)	84	167	300	383
Iron (mg)	8.9	9.8	10.5	11.4
Vitamins:				
Vitamin B-6 (mg)	0.6	0.7	2.5	2.6
Pantothenic acid (mg)	2.2	2.7	3.7	4.2
Folic acid (mg)	0.2	0.2	0.2	0.3

[1]The nutrient contents of the baked items were calculated by summing up the contents of the individual ingredients. it was assumed that each of the finished breads contained 35% moisture. Nutrient data for the ingredients were obtained from various USDA sources.

[2]Made from enriched white flour, yeast, sugar, butter, salt, and water.

[3]Same ingredients as standard recipe, plus soy flour, nonfat dry milk, and wheat germ in the proportions developed by Dr. McCay and his associates at Cornell University.

[4]Made from whole wheat flour, yeast, sugar, butter, salt, and water.

[5]Same ingredients as standard whole wheat bread, plus soy flour, nonfat dry milk, and wheat germ in the same proportions as the ingredients used in the Cornell bread.

It may be seen from Table B-19 that the addition of soy flour, skim milk, and wheat germ to standard recipe white bread raises the protein, calcium, and magnesium contents by significant amounts. However, the greatest improvement is made by substituting whole wheat flour for the white flour, *and* adding the special ingredients of the Cornell formula. Then, the fortified whole wheat bread produced by this means contains about half again as much protein, over six times the calcium, over 4½ times the magnesium and vitamin B-6, 1⅓ times the iron, and almost double the pantothenic acid and folic acid as the unfortified standard recipe white bread. White bread has been used for comparison in order to dramatize the differences between the popular breads which contain mainly white flour and water, and those made with more nutritious ingredients. it is noteworthy that the American types of sliced, white breads are usually made with milk plus additives that offset the tendency of milk to make breads heavy. Additional details on this subject are given in the section which follows.

Suggestions for Adding Highly Nutritious Ingredients to Baked Goods. One might question whether it is worth the effort to fortify baked goods with special ingredients when all that is necessary to offset the nutritional shortcomings of unfortified items is to eat sufficient amounts of the other foods which supply the missing nutrients. One answer is that most breads and similar products are more convenient than other foods because they usually keep well for 2 or more days without refrigeration. Most other foods cost more, require refrigeration, and often need cooking. Sandwiches filled with nutritious items may often be a meal in themselves, although fillings may leak out or spoil under certain conditions. Breads, crackers, and cakes which are specially fortified so as to be nearly complete foods may also comprise a meal that needs to be accompanied only by water or another beverage. Therefore, some suggestions for adding various types of ingredients to home baked items follow. (Nutrient composition data for the suggested additives are given in Food Composition Table F-36.)

SPECIAL NUTRITIONAL SUPPLEMENTS TO BAKED GOODS. A little of each of these products goes a long way because they are concentrated sources of nutrients. For example, only ½ cup (*120 ml*) of soy flour, ¾ cup of skim milk powder, and 3 tablespoons (*45 ml*) of wheat germ added to 6 cups of white flour make Cornell Bread much more nutritious than ordinary white bread. (See Table B-19 for details.) Suggestions for adding other nutritional supplements follow:

1. From ¼ to ¾ cup each of bland-flavored items such as alfalfa (leaf) powder, bran, rice polishings, and/or dried whey may be used for every 6 cups of flour, provided that the total amount of these ingredients combined is not more than 1 cup. Larger quantities may make the baked product too heavy and too moist.
2. Ingredients such as blackstrap molasses, bone meal, brewers' yeast, carob powder, dolomite, lecithin granules, and liver powder have unusual tastes and textures. Therefore, it is unwise to use more than 1 teaspoon (*5 ml*) of each of them per cup of flour.
3. Some of the special supplements may interfere with the development of gluten strands in yeast-leavened breads; so, it is best to mix them gently into the dough after it has risen at least once and has been kneaded.
4. The supplements may be mixed right in with the other dry ingredients when breads are leavened with baking powder or baking soda.

COMMON FOODS THAT MAY BE ADDED TO BAKED GOODS. People who eat a lot of bread may appreciate some variety in taste and texture, as well as the nutritional gains which result from fortifying breads with nutritious ingredients. Such variety may be achieved by mixing some of the common foods into bread doughs.

• **Beverages and other liquids**—Fluids such as beer, fruit juices, and wine may be used instead of milk or water in doughs. It was reported that a company of American servicemen stationed in Korea used beer as a leavening agent for pancakes when they ran out of baking powder. Baking soda may be used instead of baking powder when the added fluids are acid. However, this type of leavening takes place as soon as the acid fluid comes into contact with the baking soda, so the oven should be preheated and ready for baking right after the ingredients are mixed, in order to avoid excessive loss of the leavening gas from the batter or dough.

• **Cereals and grain products**—Infant cereals, instant cooking hot cereals, and ready-to-eat breakfast cereals may be added directly to the other bread ingredients, whereas noninstant hot cereals should be cooked first. Extra nutritional value is provided if the hot cereals are cooked in milk instead of water. Cooked whole grains add a nutlike taste and texture, but they also tend to make baked products more moist and heavier. About 1 cup (*240 ml*) of most cereals or grain products may be added for each 6 cups of flour. However, as much as 2 cups of flaked or puffed breakfast cereals may be added.

• **Cheeses**—The flavor of bread is enhanced by almost all of the cheeses, except those that have been mold-ripened (blue, Camembert, and Roquefort); the latter types may contribute moldy flavors. Usually, 2 cups of grated cheese per 6 cups of flour is plenty, except that only 1 cup of cottage cheese should be used, because it may make the baked product too moist and heavy. Cream cheese is so high in fat that it is suitable only for items like cheese cakes and pound cakes.

• **Dairy products**—All of these products except those high in fat (butter and sweet or sour cream) contribute significant amounts of protein, minerals, and vitamins to bread, while also making it heavier and more moist. Therefore, dairy products should *not* be used to replace more than ⅓ to ½ of the liquid required in the recipe. Soured dairy products provide acid so that baking soda may be used instead of baking powder.

• **Eggs**—Egg white foams hold lots of air which helps to leaven baked products. Also, weak doughs are strengthened by the proteins in egg white, whereas strong doughs (those with a high gluten content) may become too tough if too much white is added. Other ingredients, such as sugar and fats, tenderize doughs by interfering with gluten development, as in the case of angel food cake, which is rich in both egg white and sugar. Egg yolks help to disperse the added fat uniformly throughout the product. Finally, egg protein markedly improves the nutritive value of the proteins in the various types of flours.

• **Fruits and fruit products**—These mineral and vitamin rich items may be added to doughs in such forms as (1) whole fruits, like raisins and currants; (2) chopped pieces or slices of fruit; (3) mashed fruits like bananas and avocados; (4) fruit purees; or (5) fruit juices. However, large solid forms of fruit may sink to the bottom of thin batters. Hence, small pieces may be more evenly distributed in the baked product. Pureed fruits add mainly color and flavor since they usually blend into the dough. Those using purees for the first time might wish to start with only small amounts, such as are furnished in small jars of baby food. Fruit juices or purees may often be used instead of the water called for in the recipe. If this is done, the fruit products will usually provide enough acid to warrant the use of baking soda rather than baking powder in the various quick breads.

• **Meats, poultry, and fish**—High protein animal foods may be added to baked goods to increase the quantity and quality of protein. Also, their rich flavors blend well with the milder flavors of most breads. Nevertheless, the person preparing such dishes should be aware that the temperatures that are usually produced in doughs during baking are not high enough to cook raw flesh foods adequately, unless they have been (1) finely ground, (2) sliced very thin, or (3) covered with only a thin batter. Frozen items may remain raw inside even when the outside is well done, so it is best either to cook them slowly for a long time, or to thaw them before cooking. Therefore, precooked meats, poultry, and fish are safer to use in baking. Finally, one might add these ingredients in the form of purees like those added as baby foods.

• **Nuts and seeds**—Combinations of certain nuts and seeds with grain flours furnish higher quality proteins than either type of food alone. Also, nuts and seeds are often good sources of minerals and vitamins which are present in only meager amounts in refined flours. However, large hidden pieces of nuts or seeds in doughs may be unsavory or even dangerous to some people, since they may cause choking if they are accidently swallowed whole. Hence, it is best to chop nuts and seeds into small pieces before using them in baking. About 1 cup (*240 ml*) of the small pieces per 6 cups of flour is about right.

• **Sprouted seeds**—Sprouts supply many of the nutrients found in both seeds and vegetables because extra fiber, vitamins, and other substances are synthesized by seeds during sprouting. Also, freshly sprouted barley and wheat contain active enzymes which digest the starch and protein in flours so that extra sugars and amino acids are made available for the growth of yeast. Hence, the leavening of doughs is speeded up. These effects are most likely to be obtained when sprouting is stopped by drying the grains just after the emerging sprouts are about as long as the kernels of grain. Then, the sprouted grain may be ground into meal or flour so that it may be more conveniently used for baked products. Not more than 1 cup (*240 ml*) of ground sprouts should be used for each 2 to 3 cups of flour.

• **Vegetables**—The practice of adding various vegetables to baked goods probably dates back more than a thousand years to times when poor harvests of wheat and rye forced the bakers of northern Europe to mix their grain flours with substitute flours made from acorns, beans, peas, and potatoes. Therefore, it is not surprising to find that some of the current bakery products may contain carrots, Irish potatoes, pumpkin, sauerkraut, sweet potatoes, and tomato soup. Legumes furnish extra protein, while the yellow and orange vegetables contribute vitamin A. Acid vegetables allow the use of baking soda instead of baking powder. About 1 cup (*240 ml*) of mashed pureed vegetable is about right for 6 cups of flour.

RECENT DEVELOPMENTS IN BAKING TECHNOLOGY. Bakers and food technologists have long experimented with new ingredients and procedures in order to produce the high quality baked products expected by consumers. In their quest for quality, they have worked towards objectives such as (1) reduction of staling and spoilage (bread is usually taken out of supermarkets, or sold at reduced prices, if it has not been sold within 2 to 3 days after delivery); (2) modification of recipes so as to improve nutritive values or meet special dietetic requirements; (3) overcoming of the problems inherent in the mass production of standard products from ingredients which may vary in their baking characteristics; and (4) development of convenience items for consumers who wish to do their own baking. Some recent applications of food technology to the baking industry follow:

- **Air classification of flours into high-protein and low-protein fractions**—This process uses swirling streams of air to separate the finer particles of the high-protein fractions from flour which has been ground extra fine. The high-protein fractions may be used for special breads or for fortifying weaker, low-protein flours made from soft wheats. Low-protein fractions from this process are used alone for producing soft-textured items such as cakes, cookies, and pastries.

- **Artificial sweeteners**—The long-used noncaloric substitutes for sugar have some major shortcomings such as (1) bitter aftertastes, which require masking by certain flavorings; and (2) failure to duplicate the effects of sugar on the textures of baked goods. For example, cakes made with sugar substitutes may have only about one-half the normal volume. However, some promising new products have been developed.

- **Bulking agents**—Certain vegetable fibers may be (1) used in their natural forms (such as bran and psyllium seed); or (2) extracted with acid, fat solvents, or water, after which they may be purified and chemically treated to make them more suitable for use in baking. Products which contain bulking agents are usually (1) lower in calories than those without agents; and (2) more filling and satisfying than low fiber foods. However, baked goods may lose their appeal to most consumers when their fiber content exceeds 30%.

- **Chemical combinations of fats and sugars**—These compounds, which are called glycolipids, occur naturally in wheat flour and appear to be essential for leavened doughs to rise to their maximum volume. However, similar synthetic compounds have been developed for use as additives to special doughs such as those heavily fortified with soy flour or other high-protein ingredients.

- **Emulsifying agents**—Chemically modified fatty acids, such as glyceryl monostearate and calcium-stearyl lactylate, improve the textures and volumes of items made by substituting starchy ingredients like cassava flour and wheat starch for much of the wheat flour. These effects are important because (1) they may help bakers to meet the great demands for breads in the countries of Latin America where wheat is scarce, but cassava (a starchy tuber) is plentiful (a little soy flour may be added to these breads to increase their protein content); and (2) people on low-protein or low-gluten diets may be permitted to eat breads made without wheat flour. These emulsifying agents also make it possible to keep ordinary breads longer before they become stale.

- **Enzymes**—Flours normally contain small amounts of enzymes which digest starch to sugars and proteins to amino acids. These enzyme actions provide nutrients for the growth of yeast and help tenderize the dough. Sometimes, certain flours lack these enzymes, with the result that they must be supplied by other ingredients or added in purified form. European bakers have long depended upon malts (sugary preparations made from germinated barley or wheat) to supply them, but American bakers are now beginning to use pure enzyme preparations derived from various microorganisms.

- **Imitation egg products**—The development of yolk-free products was prompted by the need for some people to cut down on the cholesterol in their diets. These products consist mainly of egg white—which is responsible for the binding effects of eggs on doughs and a leavening effect due to the air trapped in egg foams; but the emulsifying effect of the missing yolk is usually contributed by substances such as lecithin and other emulsifying agents.

  Some total egg-free products made from various starches and/or vegetable gums are now available. These may be used in products for people who are allergic to eggs.

- **Jerusalem artichoke flour**—This product is made from the dried, ground tubers of the Jerusalem artichoke, which is a species of sunflower rather than an artichoke. Although the white, powdery material resembles starch, it is very low in calories because its major constituent is inulin—a complex carbohydrate that is not digested by humans. Furthermore, it has long been used as an ingredient of dietetic pastas for diabetics on carbohydrate-restricted diets. Therefore, replacement of part of the wheat flour with Jerusalem artichoke flour results in a baked product that is lower in both utilizable carbohydrate and calories than similar items where no such substitution was made.

- **Microwave ovens**—These appliances appear to be more energy efficient for baking small masses of dough than conventional gas or electric ovens because (the microwave heat energy is used mainly to heat the food, rather than the air in the oven; and (2) conventional ovens release so much heat into the kitchen that it is often necessary to keep an air conditioner running. However, most of the recipes used in baking have been developed for conventional ovens; hence, it is necessary to develop new ones for microwave ovens. At the present time, bakers are using the latter mainly for operations such as (1) warming yeast-leavened doughs, (2) thawing frozen doughs, and (3) baking items prepackaged in paper or cardboard containers.

- **Mixes for quick, yeast-leavened breads**—The amount of time required for the rising of dough deters many people from preparing yeast-leavened breads at home. This problem has been solved in part by special mixes that shorten the preparation time for yeast breads because they contain ingredients such as (1) malt or other sources of the enzymes which digest the proteins and starch in flour to nutrients which speed the growth of yeast cells, (2) activating agents that hasten the development of gluten strands, (3) gluten maturing agents, or (4) baking powder to augment the leavening action of the yeast.

• **Preservatives**—Some of the common food preservatives cannot be used for yeast breads because they inhibit the growth of yeast. This is not the case with propionates and acetates, which are safe and useful for protecting baked goods against molds and a bacterial growth called "rope" (ropy bread contains sticky, yellowish patches which may be pulled into ropelike threads). Propionates are more desirable than acetates because they contribute a cheeselike taste, whereas acetates make doughs too vinegary.

• **Seedcoat flours that are rich in fiber**—The seedcoats or skins of grains and legumes are high in fiber, so there have been attempts to replace some of the grain flours used in baking with various seedcoat flours. Unfortunately, yeast-leavened breads made with large amounts of these fibrous materials tend to be too heavy and have poor flavors, odors, and textures. Researchers have recently discovered that the seedcoat flours may be improved by treating the seedcoats with acid before grinding them into flour. The addition of egg yolk plus a lecithin derivative also improves these breads.

• **Vegetable gums**—Various types of natural and chemically modified vegetable gums are used by commercial bakers. For example, they are often added to icings, toppings, meringues, fillings, and jellies to bind water so that it is not lost by leakage or by drying out. Likewise, doughs are softened and their textures improved by gums like carrageenan, locust bean gum, gum arabic, and certain chemical derivaties of cellulose. Breads made with gums may be kept longer before they become stale. Finally, certain gums may be used to make special dietetic breads which contain no wheat flour or gluten.

• **Xanthan gum**—This product was developed by USDA scientists who were seeking new uses for corn products. They found that the gum was produced during the fermentation of glucose (a sugar derived from corn starch) by *Xantholmonas campestris* bacteria. It is used to improve the texture and volume of breads lacking gluten, such as those containing large amounts of soy flour.

There seems to be an almost endless variety of baked goods which may be produced by varying the amounts and types of ingredients. Hence, just about everyone should be able to enjoy some of these products, since they may be tailored to fit many types of modified diets. This wide selection of items has been made possible by the persistent efforts of researchers who have sought to improve on what nature has to offer, and who have often succeeded in their quest.

BREAD SPREADS

A slice of fresh whole wheat bread with butter and a sweet spread make a wonderful nutritious addition to any meal, between-meal, or bedtime snack. Down through the ages, breads have been the foundation of most western diets. In ancient days, some of the breads were unleavened; and they still are in some parts of the undeveloped countries. But whether unleavened or leavened, breads have always been eaten with a sweet spread made from the fruits available during the summer months. The various types of spreads for bread are listed in Table B-20, Bread Spreads.

TABLE B-20
BREAD SPREADS

Spread	Ingredients Used	Description
Conserves	All fruits Nuts, optional	Conserves are similar to jams or marmalades, but they usually combine several fruits, and sometimes they contain nuts. They are prepared the same way as jams, and the nuts, if used, are added at the last.
Fruit butters	Apples Grapes Peaches Pears Plums	Fruit butters are the most wholesome of bread spreads because they contain the least amount of sugar to fruit. They are prepared by cooking the fruit and pressing through a sieve. Sugar and spice are added and cooked until thick.
Lemon butter or **Lemon curd**	Lemon, sugar, eggs, and butter.	A favorite spread of the British Isles. It can be bought in jars, or very simply made at home.
Jams	All fruits	The fruit is mashed and cooked before adding the sugar. No more than ¾ lb (*341 g*) sugar should be used for each 1 lb (*454 g*) of fruit.
Jellies	Apples Crabapples Currants Grapes Raspberries	Fruits need to have both acid and pectin. The fruit is cooked until tender and poured into a jelly bag to drip through. Except for very sour fruits, ⅔ as much sugar as juice is the best proportion. Boil the juice rapidly, add the sugar, and pour into jelly glasses when jelly sheets from spoon.
Marmalades	Grapefruit Lemon Orange	Fruits with both pectin and acid are used. Thin slices of fruit and rind are used and the mixture should be clear, bright, and sparkling in color.
Preserves	Most fruits	Preserves contain the whole, or large pieces, of fruit which have absorbed a heavy sugar syrup until they are filled. A good preserved fruit is plump, tender, bright in color, and filled with sweetness.

BREAKFAST CEREALS

Contents *Page*

Effects of a Good Breakfast on Mental and Physical Performance ... 261
History of Cereals ... 261
Production of Breakfast Cereals ... 262
Hot Cereals ... 262
Ready-to-Eat Breakfast Cereals ... 262
Infant Cereals ... 263
Selecting the Most Nutritious Cereals at the Lowest Cost ... 263
Variations in the Ways of Preparing Breakfast Cereals ... 264
The Role of Cereal Mixtures in the Treatment and Prevention of Malnutrition ... 266
The Sugared Breakfast-Cereal Binge ... 266

This term covers a wide variety of grain products that are usually cooked or processed to improve their texture, flavor, and digestibility. Although these products are often made from flours, they differ from breads and other baked goods in that they are not usually leavened; instead, they may be toasted to crispiness like crackers. In the United States and Great Britain, cereals are considered suitable for breakfast primarily, even though certain items like cooked rice may be the basis of every meal in Southeast Asia.

Recently, there has been a lot of controversy in the United States over the nutritional merits and demerits of various breakfast cereals. Therefore, some background information on these products is provided so that consumers might be better informed regarding the types of cereals which are best suited for their particular circumstances.

EFFECTS OF A GOOD BREAKFAST ON MENTAL AND PHYSICAL PERFORMANCE. Many children and adults go to school or work after having had little or no food for breakfast. Some of these people may take only coffee or tea. However, the well-known research studies conducted at the University of Iowa showed that the eating of at least a light breakfast increased the speed of mental response and improved the perfomance of physical work during the late morning hours.[24] Furthermore, the skipping of breakfast sometimes resulted in adverse reactions such as muscle tremors, fatigue, dizziness, nausea, and vomiting when strenuous physical activity was undertaken in the morning.

Most of the researchers who have studied the effects of eating a good breakfast have suggested that the benefits are due mainly to the protein that is eaten at this meal, because high-protein breakfasts were found to be better than low-protein breakfasts in helping to maintain a normal blood sugar level between midmorning and lunch.[25] It is noteworthy that breakfasts consisting chiefly of vegetable protein foods (breads, cereals, peanut butter, and soybean milk) were found to be as effective as those consisting chiefly of animal protein foods (eggs, meat, and milk) in keeping the blood sugar at normal levels.[26] The level of protein eaten at breakfast may depend partly upon the amount supplied by a cereal, because ready-to-eat breakfast cereals are included in about one-fourth of all the breakfasts eaten by Americans.[27] Hot cereals also may make a contribution, particularly in areas such as the southern United States, where hot grits (cornmeal) are still served at many breakfasts. Unfortunately, the degermed meal used to make grits is nutritionally inferior to the whole grains first used by ancient peoples.

HISTORY OF CEREALS. Grains boiled in water—the primitive counterparts of our modern hot cereals—have been consumed by people all over the world for thousands of years. Archaeologists believe that grains were gathered by nomadic bands for many millennia before the first grain farming began about 10,000 years ago. It appears that grains and other foods were boiled in crude bags made from the stomachs and hides of animals. Sometimes, the shells from large shellfish and turtles were used for this purpose. Also, certain cultures boiled their grains to make it easier to remove the tough outer husk. For example, the practice of parboiling rice before grinding off the hulls appears to have originated in India, where the rice eaters have usually been less prone to beriberi than the Southeast Asians who mill the raw paddy rice. Parboiling causes the water-soluble vitamins, such as thiamin, to migrate from the outer layers of the grain to the inner layers. Hence, the loss of vitamins due to milling is reduced by this practice.

The practice of grinding grains before cooking appears to be about as old as grain farming, since grinding stones have been found at the excavated sites of some of the earliest agricultural settlements. No doubt these early farmers soon learned that grain in the form of a meal could be cooked much more quickly than whole grains. In many places around the world, low income people still use the ancient methods of cooking grains.

Another primitive means of processing grains was by roasting, which also made it easier to remove tight-fitting hulls. Often, roasted grains were eaten without additional cooking. These items bore at least a slight resemblance to some of our modern day, toasted ready-to-eat breakfast cereals.

The modern types of ready-to-eat breakfast cereals—which are only about 100 years old—owe their development to Seventh-day Adventists, an American religious sect. This religion was officially founded in 1863 by a small group of people which included an Adventist preacher's wife who later came to be known as Mother White. In 1866, this zealous lady established the Western Health Reform Institute (later known as the Battle Creek Sanitarium, or simply the "San") at Battle Creek, Michigan. People who came to the sanitarium for a health

[24]Tuttle, W. W., M. Wilson, and K. Daum, "Effect of Altered Breakfast Habits on Physiologic Response," *Journal of Applied Physiology*, Vol. I, 1949, p. 558.

[25]Coleman, M. C., W. W. Tuttle, and K. Daum, "Effect of Protein Source on Maintaining Blood Sugar Levels After Breakfast," *Journal of the American Dietetic Association*, Vol. 29, 1953, p. 239.

[26]*Ibid.*, p. 243.

[27]Hayden, E. B., "Breakfast and Today's Lifestyles," *The Journal of School Health*, Vol. XLV, 1975, p. 84.

cure were given austere vegetarian diets based mainly upon minimally refined grain products and fresh fruits and vegetables, because this was in keeping with the dietary principles taught by Mother White. Subsequently, *Dr. J. H. Kellogg*—who had been hired by Mother White to manage the "San"—invented a granolalike, ready-to-eat breakfast cereal for the Adventists who shunned the traditional American breakfast of ham and eggs. Later, Dr. Kellogg and his brother, W. K. Kellogg, founded the cereal company which still bears their name.

Another pioneer of the breakfast cereal industry was *C. W. Post,* who had been a patient at the Battle Creek Sanitarium. Among his inventions was (1) a cereal-based, caffeine-free beverage called Postum for the use by the Adventists, who avoided stimulants such as coffee and tea; and (2) a granular cereal called Grape-Nuts.

The estabalishment of the breakfast cereal industry came about the time that the United States began a transition from a predominantly agricultural society to a more industrialized, urban culture. A strong selling point for the ready-to-eat cereals was—and still is—their convenience. The favorable market for these products inspired such inventions as those of H. Perky, who, at the end of the 19th century, developed and patented the equipment for an entire shredded wheat factory. About the same time, A. P. Anderson invented the process for puffing rice and wheat.

By the beginning of World War II, cereal companies had started to enrich their products with thiamin, riboflavin, niacin, and iron. *Enrichment means the restoration of some of the nutrients that are removed during the processing of a food.* Later, in about 1955, fortification of cereals was started. *Fortification means the addition of certain nutrients to foods in order to provide higher levels of such nutrients than are normally present in the natural, unprocessed foods.*

The steady trend towards the fortification of more and more foods with more and more nutrients spurred the U.S. Food and Drug Administration to make several attempts during the 1960s and the early 1970s to regulate the amounts and types of nutrients that were added to foods. At about the same time, the consumer movement got underway, and some of the leading activists began to attack certain cereals which they classified as vitamin- and mineral-fortified confections. One of the reactions to these controversies was the production of so-called "granolas" or "natural cereals" which in many ways were similar to the first product developed by Dr. Kellogg.

There seems to be every indication that there will be a steady increase in the consumption of commercial cereal products around the world as the developing countries attempt to feed their burgeoning populations. It is noteworthy, too, that certain American cereal and milling companies have assisted in the development of marketing of nutritive cereal mixtures for needy peoples.

PRODUCTION OF BREAKFAST CEREALS.

Few people in the world today eat whole, raw grains because (1) they are too chewy; (2) the raw starch which comprises the greater part of the grains is not very tasty; and (3) they are poorly digested. It is noteworthy that the starch which is present in grains and many other plant foods is often in the form of hard granules that are very resistant to the digestive juices. Cooking starchy foods in water causes the granules to swell and form a pasty mass which is more readily digested. Dry heating or roasting, which alters starch granules by breaking them down into dextrins, enhances the taste of the food. Therefore, some type of cooking or processing is needed to give cereal products the characteristics which are most beneficial and desirable to man.

Hot Cereals. In comparison with ready-to-eat cereals, these products are less processed in the factory, but they require more preparation time in the home, The grains or grainlike seeds which have husks (rice, barley, oats, and buckwheat) are dehulled to produce groats. Naked or huskless grains (corn, wheat, and rye) do not require any special processing prior to cooking, although the cooking time is much longer when they are left whole than when they are broken up into smaller particles. However, many of the products currently on the market contain small pieces of grains which have had the branny layers removed. Quick-cooking cereals are made by precooking the grains and/or cutting them into fine pieces or flakes. Flake-type hot cereals are made by steaming the grain, then passing it between rollers.

Ready-to-Eat Breakfast Cereals. These items are designed to be eaten without cooking; hence, they are cooked at one of the stages in their processing. Usually, grains in the form of a meal or a flour are cooked with added flavorings and other ingredients such as malt, syrups, and heat-stable vitamins and minerals. Then, the cooked cereals are further processed by flaking, puffing, shredding, or other operations. Vitamins and other nutrients that may be altered by heating are sprayed on the cereals after the hot processing operations have been completed. Some typical products follow:

- **Flakes**—This type of product is made by squeezing cooked particles of grain between very heavy rollers that convert each piece into a single flake. Sometimes, flakes are made from tiny pellets of dough prepared from various flour mixtures. Usually, the newly-formed flakes are toasted to a golden brown, after which they may be enriched or fortified with nutrients by spraying.

- **Granules**—. Crunchy bits of cereal are prepared by first baking thoroughly a large breadlike loaf made from a yeast-leavened dough, then breaking it into pieces, which are dried and ground into small particles.

- **Puffed cereals**—Cooked grains or pellets of dough are placed in a closed chamber, subjected to heating and high pressure, then released suddenly by opening the chamber, which is literally equivalent to shooting the pieces from a gun. The blowing up of the cereal into light, porous particles is caused by the sudden expansion of the hot, pressurized liquid which is trapped within each piece. A similar action occurs when popcorn is popped. The cereals may be enriched or fortified after puffing.

• **Shredded grain biscuits**—Long, thin strands or sheets of shreds are produced by forcing cooked grains between tight-fitting, grooved rolls. Then, the strands or sheets are formed into biscuits or bite-sized squares which are baked in an oven. Enrichment or fortification may be done after baking.

• **Special shapes**—These items are usually formed by pushing dough through special dies, then cutting the strands of shaped dough into pieces of the desired size. Like other types of cereals, the specially-shaped pieces are toasted or baked prior to enrichment or fortification.

• **Sugarcoated cereals**—Ready-to-eat cereals are sugarcoated by either spraying on a special syrup, or by tumbling the pieces of cereal in a revolving drum filled with the syrup. The sugarcoated cereal is dried by passing it on a moving belt through a dryer.

Infant Cereals. These items have a bland taste because they are not toasted or flavored like the ready-to-eat breakfast cereals. However, they are easy to prepare by mixing with infant formula or milk and they generally cost less per serving than the other pre-cooked cereals.

Infant cereals are produced by mixing bolted flours (*bolting is the removal of seed coat particles from a flour by passing it through a cloth or a fine sieve*) with water and other heat-stable ingredients to form a paste. The paste is cooked, then spread on a drying drum. Flakelike particles of dried cereal are scraped from the drum as it revolves. Vitamins and other ingredients which might be altered by heating are then added to the dried cereal.

SELECTING THE MOST NUTRITIOUS CEREALS AT THE LOWEST COST. Most of the packaged cereals sold in supermarkets have nutritional labeling, but it may not be easy for the average consumer to evaluate rapidly all of the data on the label while shopping in a crowded store. Furthermore, the various cereal companies appear to be competing with each other to see who can come up with the most heavily fortified ready-to-eat cereals. As a result, these items may supply substantial amounts of many of the essential nutrients. Table B-21 shows the percentages of the Recommended Dietary Allowances (RDAs) for elementary schoolchildren that are furnished by a bowl of cereal and milk, while Fig. B-81 illustrates the distribution of nutrients contributed by the cereal and the milk.

TABLE B–21
AMOUNTS OF SELECTED NUTRIENTS FURNISHED BY 1 OZ OF A TYPICAL READY-TO-EAT CEREAL PLUS 4 OZ OF MILK

Nutrient	Amount in 1 Oz Cereal plus 4 Oz Milk	Percent of RDA[1] for 7- to 10-Year-Old
Calories	192 kcal	10
Protein	6.2 g	22
Calcium	157.4 mg	20
Phosphorus	167.6 mg	21
Magnesium	38.3 mg	23
Iron	3.3 mg	33
Vitamin A	451 mcg RE	64
Vitamin D	5 mcg	50
Niacin	3.4 mg	26
Riboflavin	0.61 mg	51
Thiamin	0.37 mg	37
Vitamin B-6	0.72 mg	51
Vitamin B-12	2.16 mcg	154
Vitamin C	11 mg	24

[1]From *Recommended Dietary Allowances*, 10th ed., NRC–National Academy of Sciences, 1989, pp. 33, 284.

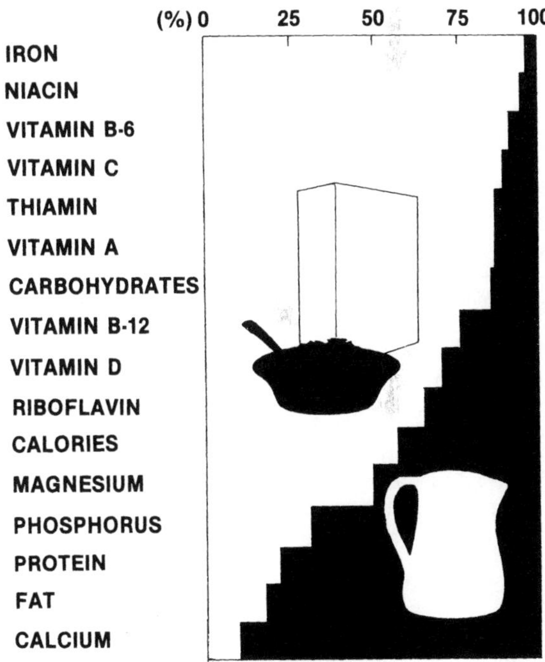

Fig. B-81. The distribution of nutrients contributed by 1 oz of a typical ready-to-eat cereal (shown by white) and 4 oz of milk (shown by black).

As shown in Table B-21, the combination of the typical ready-to-eat cereal and milk supplies ¼ or more of the RDAs for many essential nutrients, but it furnishes only about 1/6 of the allowance for protein. This means that some additional protein should be provided at breakfast, from foods such as extra cereal and milk, eggs, meat, and/or bread.

In the cases of low income people or those on special diets, it might be easier and more economical to obtain extra protein by selecting one of the breakfast cereals that has a higher protein content than the one shown in Fig. B-8l, because protein foods are usually the most expensive items in the diet. If necessary, supplemental vitamins and minerals may be supplied by a multiple vitamin and mineral tablet which costs only a few cents. Therefore, Fig. 82 ranks the major types of breakfast cereals by their protein contents, without regard for their vitamin and mineral contents. (For additional information regarding the nutritive values of these cereals, see Food Composition Table F-36 of this encyclopedia.)

Discussions of each of the types of cereal shown in Fig. B-82 follow:

• **Hot cereals**—The oat and wheat cereals supply the most protein per calorie, while the rice and corn cereals supply the least. Cooked grains of rice furnish more calories and protein per cup than the hot corn cereals, because the water content of the cooked rice is less than that of the cooked corn. In general, hot cereals cost less than either ready-to-eat items or infant cereals because they require less commercial processing.

• **Nutritious toppings for cereals**—Bran and wheat germ are natural accompaniments to cereals because they (or the similar parts of other grains) are usually removed during milling. It is noteworthy that these by-products of milling provide more protein per calorie than most of the grain products, which is a boon for people trying to lose weight. Some hearty people even eat bowls of bran or wheat germ alone with milk.

The cereals which are high in bulk, but low in nutrients—such as farina, and the unsugared puffed cereals—might be good items for filling up heavy eaters who need to watch their weight. Also, the dry, ready-to-eat cereals make good snacks which are less fattening than salted nuts or some types of crackers.

• **Infant cereals**—Except for the rice cereal, these products equal or top the hot cereals and the ready-to-eat cereals in protein content. However, babies may not receive very much of these cereals because mothers usually mix from 2 to 6 tablespoons (*30 to 90 ml*) (equivalent to from 1/6 to ½ oz) of cereal with approximately 1 to 3 oz (*30 to 90 ml*) of formula or milk.[28] Even so, older children and adults may consume larger quantities of these cereals if they do not mind the bland taste.

• **Ready-to-eat cereals**—Certain items in this category are unusually high in protein because they have been fortified with protein-rich ingredients such as wheat gluten, casein (derived from milk), wheat germ, and malt. It is noteworthy that two of these cereals—rice with wheat gluten and casein, and wheat and malted barley granules—are concentrated sources of both protein and energy, so that servings of ½ cup (*120 ml*) or less equal or exceed the protein and caloric values of 1-cup servings of the other cereals.

The best protein sources among the cereals not fortified with this nutrient are the wheat and oat products, since the unfortified rice and corn cereals are low in protein. The presweetened or sugarcoated cereals provide significantly less protein per unit of food energy than their nonpresweetened counterparts. Furthermore, it is much cheaper for people to add their own sugar than it is to pay for it as part of the weight of the cereal, which costs much more per pound than sugar. Finally, the puffed cereals contain only about half as much protein and energy per cup as the nonpuffed products.

VARIATIONS IN THE WAYS OF PREPARING BREAKFAST CEREALS. Generally, most of the cereals eaten at breakfast are convenience-type products that require only a few minutes for their preparation because many Americans allow little time for either the fixing or the eating of this meal. Nevertheless, there are some people who would like to improvise in the preparation of cereals by using ingredients that are more of their own choosing because (1) they prefer to eat whole grain products, and (2) many of the commercial "natural cereals" contain far more sugar and/or other sweetener(s) than they care to eat. Therefore, some variations in the preparation of cereals follow:

1. Save time by cooking brown rice or whole wheat grains in a pressure cooker the night before, and reheating the cereal for a few minutes at breakfast time. Brown rice requires 2 cups (*480 ml*) of water per cup of rice; and from 20 minutes (for firm grains) to 40 minutes (for tender grains) under about 15 lb (*6.75 kg*) pressure. The rice may be made fluffy by allowing it to stand for 20 minutes in the sealed pressure cooker after the heating has been stopped. Each cup of whole wheat grains requires from 2 to 3 cups of water, depending upon whether or not the grain has been soaked for at least 8 hours before cooking. If soaked, about 45 minutes of pressure cooking is required; whereas, at least 1 hour is required if it was not soaked.
2. Give hot cereals a nutlike taste by toasting the uncooked grains slightly, or by browning them in a little hot fat before cooking them in the usual manner.
3. Use about 1 teaspoon (*5 ml*) of honey instead of sugar to sweeten a bowl of hot cereal.
4. Cook buckwheat groats (also known as kasha) by heating carefully for about 20 minutes in 2 cups (*480 ml*) of milk plus ½ cup of raisins for each cup of groats; then allow the cereal to stand in the covered pot for another 10 minutes after turning off the heat. A flavor and texture like that of creamy rice pudding may be produced by the addition of ½ teaspoon of (*2.5 ml*) of cinnamon or nutmeg, 1 tablespoon (*15 ml*) of honey, and ½ cup of evaporated milk.
5. Add wheat germ or bran to hot cereals to get (1) more nutrients per serving, and (2) a stronger flavor which contrasts with the bland taste of most of these items.

[28]Anderson, T. A., and S. J. Fomon, "Beikost," *Infant Nutrition*, 2nd ed., edited by S. J. Fomon, W. B. Saunders Company, Philadelphia, Pa., 1974, p. 416.

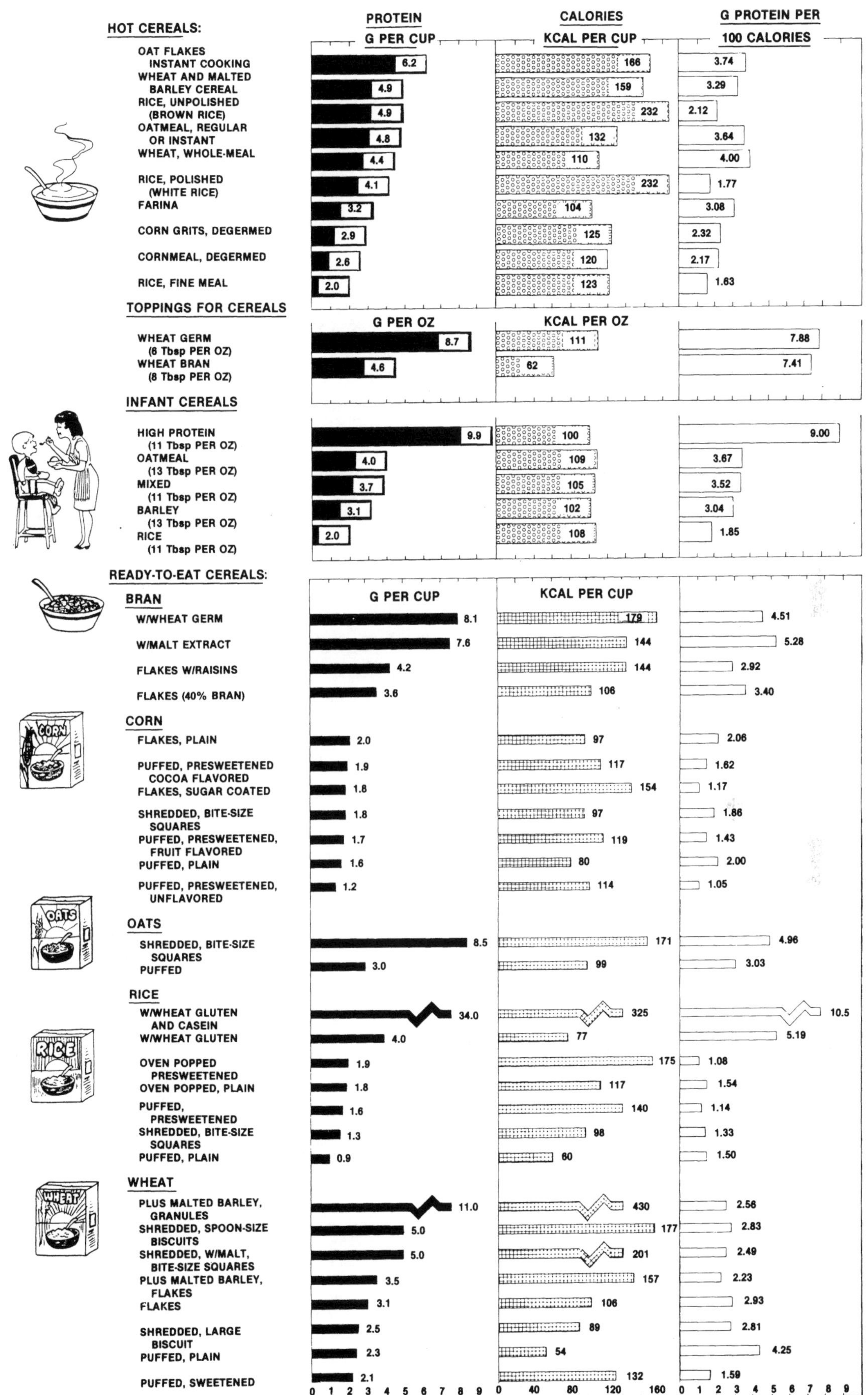

Fig. B-82. Rankings of the major types of breakfast cereals by their protein contents. (Based upon data from Food Composition Table F-36 of this book.)

6. Prepare fried mush from congealed strips or flat cakes of cooked cornmeal by coating them with a mixture of egg plus flour or bread crumbs, and frying them in hot fat until golden brown. Serve with molasses, honey, syrup, or jams; but use these sweeteners sparingly. They may be thinned a little by adding water to them.

7. Do as many people with delicate digestive systems have done—crumble toast, crackers, dry breads, biscuits, or other baked goods into a bowl of milk and eat the mixture with a spoon.

8. Make a ready-to-eat cereal by mixing together grain products such as flours, oats, cornmeal, wheat germ, bran, and/or rice polishings so as to obtain from 3 to 6 cups of cereal; which then should be mixed with 1 cup (*240 ml*)of water; 1 tablespoon (*15 ml*) of honey or other sweetener(s); 1 teaspoon (*5 ml*) of salt; 1/3 cup of vegetable oil, melted butter, or margarine; and other ingredients like cinnamon, shredded coconut, chopped nuts, and/or raisins. The mixture should be baked in a thin layer on a cookie sheet at 350°F (*177°C*) for about 25 minutes, or until dry and slightly browned. If the cereal starts to brown before it is dry, reduce the heat to 250°F (*121°C*). Store the baked cereal in a tightly-closed plastic bag, because sweeteners such as honey cause the mixture to absorb moisture from the air.

9. Pop popcorn with oil plus a little salt. Then, serve it in a bowl with milk.

10. Serve ready-to-eat cereals with "carob milk"; which is prepared by mixing 2 teaspoons of carob powder (sold in health foods stores) with a cup of milk. It is easier to mix the powder if a little milk is added to it and mixed so as to form a smooth paste. Then, the rest of the milk should be added slowly, while stirring.

Fig. B-83. Lunchtime for school children in a rural village in the West African nation of Togo. The cereal mixture was provided by a program operated jointly by the government of Togo and the Food and Agriculture Organization of the United Nations. (Courtesy, FAO, Rome, Italy)

THE ROLE OF CEREAL MIXTURES IN THE TREATMENT AND PREVENTION OF MALNUTRITION. Relief operations in areas of widespread malnutrition often distribute cereals, because these foods are among the most convenient for such operations, and most people will eat them. Sometimes cereals are mixed with other ingredients that supply the nutrients lacking in the grains. Such relief measures often result in at least a temporary improvement in the nutritional status of the people they serve, which shows that rather simple cereal-based diets may be quite adequate. Fig. B-83 depicts a typical feeding operation.

Another example of the use of cereal mixtures for the improvement of nutritional status was a 24-week feeding trial in Tanzania, in which groups of schoolchildren who normally ate only one meal per day (in the late afternoon) were given various types of midday snacks that contained 25 g of protein.[29] The groups of children who received the special snacks in school had greater gains in height, weight, and hemoglobin than a group which received no snack. Apparently, a mixture of cornmeal and nonfat dry milk was as effective as meat powder in promoting growth, except that the latter stimulated a greater increase in hemoglobin.

[29]Latham, M. C., and J. R. K. Robson, "A Trial to Evaluate the Benefit of Different Protein Rich Foods to African School Children," *Nutritio et Dieta*, Vol. 7, 1965, pp. 28-36.

At the present time, a variety of cereal-based, high-protein food products are being tested throughout the world to determine whether they might help to prevent malnutrition. The products contain grain derivatives such as wheat flour, corn flour, cornmeal, and rice flour; plus other ingredients such as nonfat dry milk, fish protein concentrate, soybean derivatives, lentils, chick-peas, yeast, sugar, starch, minerals, and vitamins. Some of the products were developed with the assistance of American cereal companies. Perhaps soon, there will be little difference between the breakfast cereals eaten by Americans, and the special foods given to people in the developing countries of the world. Certainly, cereal grains are one of the least expensive ways of furnishing both protein and calories.

(Also see CEREAL GRAINS; and GREEN REVOLUTION.)

THE SUGARED-BREAKFAST-CEREAL BINGE. The ready-to-eat breakfast cereals of the United States were developed by health food advocates of whom it may be said: They meant to do good, and they did well (financially). In the spirit of the founders, enrichment (restoration of nutrients) followed at the time of World War II, and fortification (the addition of certain nutrients) was introduced in about 1955. Soon the race was on! In an attempt to capture more and more of the market, the sugar content of many breakfast cereals went up and up, ushering in breakfast cereals which have been referred to as "empty calories."

In a study reported in 1979, the U.S. Department of Agriculture revealed that only 3 out of 62 ready-to-eat breakfast cereals contained less than 1% sugar, and that two of the cereals tested contained more than 50% sugar. The USDA analyses, which examined sugars from five different sources, showed that manufacturers used far more table sugar than any other kind. The cereals studied, along with the total percentage of sugar found in each, are given in Table B-22. The second figure (in parenthesis) is the percentage of table sugar (sucrose) in the cereal. The difference between the two figures is the percentage of four other sugars (fructose, glucose, lactose, and maltose) in the cereal.

TABLE B-22
SUGAR CONTENT FOUND IN READY-TO-EAT BREAKFAST CEREALS[1]

Breakfast Cereal	Total Sugar	Table Sugar (Sucrose)
	(%)	(%)
Kellogg's Sugar Smacks	56	43
Kellogg's Apple Jacks	54.6	54
Kellogg's Froot Loops	48	48
Kellogg's Sugar Corn Pops	46	39
Gen. Foods Super Sugar Crisp	46	36
Gen. Mills Crazy Cow (choc.)	45.6	42
Kellogg's Corny Snaps	45.5	45
General Mills Frankenberry	43.7	38
Ral.-Purina Cookie Crisp (van.)	43.5	43
Quak. Oat Cap'n Crunch (berry)	43.3	42
Kellogg's Cocoa Krispies	43	41
General Foods Cocoa Pebbles	42.6	42
General Foods Fruity Pebbles	42.5	42
General Mills Lucky Charms	42.2	36
Ral.-Purina Cookie Crisp (choc.)	41	40
Kellogg's Sugar Frosted Flakes	41	39
Quaker Oats Quisp	40.7	40
Gen. Mills Crazy Cow (Straw.)	40.1	38
Ral.-Purina Cookie Crisp (oat.)	40.1	39
Quaker Oats Cap'n Crunch	40	40
General Mills Count Chocula	39.5	35
General Foods Alpha Bits	38	38
General Foods Honey Comb	37.2	37
Kellogg's Frosted Rice	37	35
General Mills Trix	35.9	33
General Mills Cocoa Puffs	33.3	32
Quak. Oat Cap'n Crunch (p.b.)	32.2	31
General Foods Raisin Bran[2]	30.4	15
General Mills Golden Grahams	30	27
Kellogg's Cracklin' Bran	29	27
Kellogg's Raisin Bran[2]	29	11
General Foods C. W. Post (raisin)[2]	29	18
General Foods C. W. Post	28.7	20
Kellogg's Frosted Mini Wheats	26	26
General Foods Country Crisp	22	18

Footnotes at end of table (Continued)

TABLE B-22 (*Continued*)

Breakfast Cereal	Total Sugar	Table Sugar (Sucrose)
Quaker Oats Life (cinnamon)	21	21
Nabisco 100% Bran	21	19
Kellogg's All Bran	19	16
Gen. Foods Fort. Oat Flakes	18.5	18
Quaker Oats Life	16	16
Nabisco Team	14.1	12
General Foods 40% Bran	13	10
Gen. Foods Grape Nuts Flakes	13.3	7
General Mills Buckwheat	12.2	10
Kellogg's Concentrate	9.3	9
General Mills Total	8.3	7
General Mills Wheaties	8.2	7
Kellogg's Rice Krispies	7.8	7
General Foods Grape Nuts	7	None
Kellogg's Special K	5.4	5
Kellogg's Corn Flakes	5.3	3
General Foods Post Toasties	5	3
General Mills Kix	4.8	4
Ralston-Purina Rice Chex	4.4	4
Ralston-Purina Corn Chex	4	4
Ralston-Purina Wheat Chex	3.5	2
General Mills Cheerios	3	3
Nabisco Shredded Wheat	.6	.6
Quaker Oats Puffed Wheat	.5	.5
Quaker Oats Puffed Rice	.1	.1

[1]Personal communication from Dr. Betty W. Li dated August 20, 1979. Paper entitled "Gas-Liquid Chromatographic Analysis of Sugars in Ready-to-Eat Breakfast Cereals," by Li, B. W., and P. J. Schuhmann, accepted for publication in *Journal of Food Science.* Presentation of the data does not constitute endorsement of any product by the researcher, nor is any discrimination intended with regard to products not tested.

[2]Products containing raisins had fructose and glucose at considerably higher levels than products without raisins.

BREAST FEEDING

Contents *Page*

The Mammary Glands 268
The Breasts 268
Milk Synthesis 269
Milk Ejection or "Let-Down" 269
Care of the Breasts 270
Breast Feeding by Mammals Adapted to Needs 270
The Decline and Resurgence of Breast Feeding 270
Advantages and Disadvantages of Breast Feeding .. 270
Colostrum 272
Feeding Technique 272
Intervals and Adequacy of Feeding 272
Mother's Lactation Diet 272
Weaning the Baby 273

There is abundant evidence that nutrition during pregnancy and early infancy is of great importance in the baby's development and later adjustment to the world in which it must live. For example, the entire period of brain growth begins during the third month of pregnancy and ends around the second or third year of life. Protein, folic acid, and iron intake are especially important, as they are crucial to normal brain develpoment. Inadequate supplies of protein or folic acid will result in fewer brain cells and a small brain. Iron deficiency is a prime consideration, too, because children with anemia show learning disabilities and lack of concentration at school.

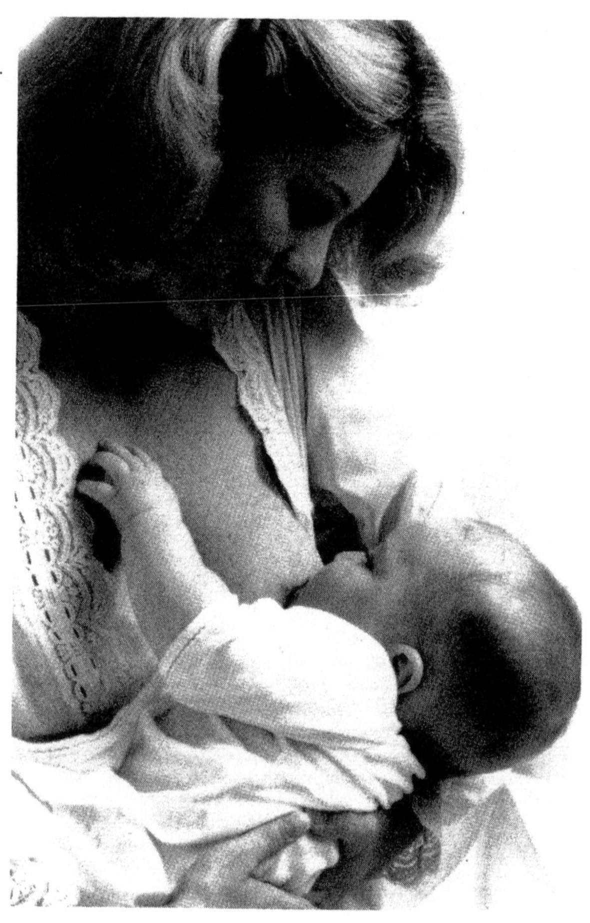

Fig. B-84. Breast feeding. (Courtesy, Gerber Products Company)

Also, growth during the first year is greater than during any other period—birth weight triples by 1 year of age; hence, nutritive requirements are critical at this time.

Peasant women, like other mammals, lactate easily and naturally. For them it is a normal part of life and a natural experience from which they derive emotional satisfaction. Both urban life and wealth brought distractions, and for many women they made lactation difficult and sometimes distasteful. The decline in breast feeding was marked. In the decade 1960-70, fewer than 15% of infants in the United States and most of Europe, and among the wealthy of Africa and Asia, were breast fed.

Both working away from home and social life are obstacles to lactation. Fashion, advertisements for artificial milks, and greater convenience for staff in obstetrical wards all mitigate against natural feeding. Among many of the elite there is a taboo against suckling except in strictest privacy. Also, many mothers who start breast feeding switch to bottle feeding because they feel that their baby is not getting sufficient milk. Whatever the reasons for not breast feeding, if it is to be successful, the advantages must be sold to mothers early in pregnancy.

During the first 6 months of life, human milk with an initial supplement of vitamin K and daily supplements of vitamin D, iron, and fluoride is regarded by the authors of this book as the best source of nourishment for the young infant. Alternatives to breast feeding include milk- or soy-based commercially prepared formulas, evaporated milk formulas, or cow's milk.

There is considerable controversy regarding the necessity of vitamin and mineral supplements for the exclusively breast fed infant. For the newborn, an injection of vitamin K is generally recommended, because the vitamin K content of human milk is low. To avoid a vitamin D deficiency, particularly when exposure to sunlight is limited, supplementation with 400 I.U. of vitamin D per day is advocated. The question of iron supplementation is debated, but the authors recommend 7 mg of iron daily during the first 6 months. There are also opposing views concerning fluoride supplementation. Because the fluoride content of human milk is low, the authors recommend a supplement of 0.25 mg of fluoride per day, dispensed with a dropper, beginning at 2 weeks of age.

Breast feeding may continue into the second year of life. But after the first 6 months, the energy needs of the infant may exceed those that can be met by breast feeding; so, the addition of solid foods is desirable at this time.

Unfortunately, there are many gaps in our present knowledge regarding infant nutritional needs and the appropriateness of various feeding regimens. As a result, opinions and practices of infant feeding differ. So, in arriving at a choice between dietary regimens—breast feeding versus formula feeding—and in deciding on the mineral and vitamin supplementation, each mother is admonished to seek and follow the counsel of her pediatrician.

THE MAMMARY GLANDS. The mammary glands are highly specialized female secretory organs. They also exist in the male, but only in a rudimentary state unless their growth is excited by peculiar circumstances.

The Breasts. The breasts are large, paired hemispherical eminences on the front of the chest. Their weight and dimensions vary at different periods of life and in different individuals. Before puberty, they are small, but they enlarge as the generative organs become more completely developed. They increase during pregnancy, especially after delivery, and they become atrophied in old age. The left breast is generally a little larger than the right. Their base is nearly circular, and flattened or slightly concave; and the long diameter is directed upward and outward. The outer surface of the breast is convex and is adorned by small conical prominence, the nipple.

Structurally, the breasts are composed of glandular, fat, and connective tissues. The glandular tissue is composed of 15 to 20 lobes, each containing smaller units called *lobules.* In the lobules, rounded secretory cells called *alveoli* form milk from the materials supplied to them by a rich capillary system. The alveoli open into a system of converging branches of ducts which lead to reservoir spaces under the areola, the pigmented area of the skin surrounding the nipple. These reservoir spaces for milk are called *ampullae.* From 15 to 20 excretory lactiferous ducts carry the milk from the ampullae out to the surface of the nipple.

The fibrous tissue surrounds the entire surface of the breast and sends out dividing walls between the lobes, connecting them together.

The fatty tissue surrounds the surface of the gland and occupies the interval between the lobes. It usually exists in considerable abundance, and it determines the form and size of the breasts.

- **The nipple**—This is a cylindrical or conical eminence, capable of undergoing a sort of erection from mechanical excitement, through which milk is released. It is characterized by a pink or brownish hue, a surface that is wrinkled and provided with papillae; and it is perforated by numerous openings, the apertures of the laciferous ducts. The nipple is surrounded by an areola of a colored tint. In the virgin, the areola is a delicate rosy hue; in pregnancy, it acquires a darker tinge, and becomes dark-brown to black as pregnancy advances. The change in color of the areola is an indication of first pregnancy.

- **Blood vessels and nerves**—Milk production places high demands on the circulatory system—the arteries and veins. The mammary glands are dependent upon the blood flowing through them for their source of energy and precursors of milk constituents. The blood enters the breasts by arteries and leaves by veins.

The nervous system plays a role in the suckling stimulus and in the removal of milk from the mammary glands. It also controls the blood flow through the mammary glands, thereby regulating the supply of hormones and the milk precursors to the glands.

Milk Synthesis.

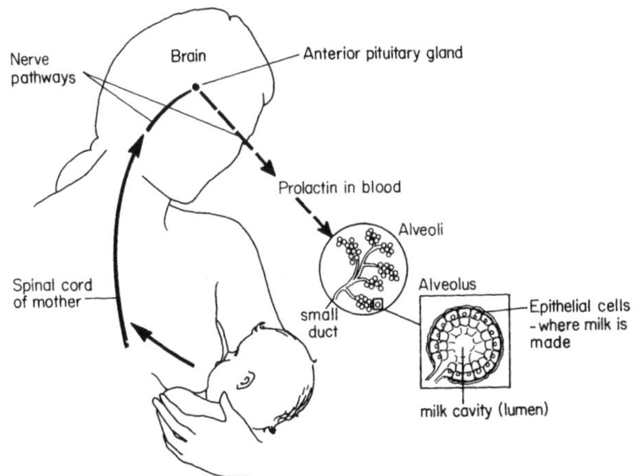

Fig. B-85. The synthesis of milk.

During the interval between nursings, milk is produced in the epitelial cells under the stimulation of the hormone, prolactin, which is produced by the anterior pituitary gland and carried in the blood. The basic milk-producing unit of the breast is the very small bulb-shaped structure with a hollow center called the *alveolus*. Each alveolus is lined with a single layer of epithelial cells which are responsible for secreting milk. Their functions are threefold: (1) remove nutrients from the blood, (2) transform these nutrients into milk, and (3) discharge the milk into the milk cavity (lumen). Groups of alveoli empty into a duct thereby forming a functional unit called a lobule. Several lobules empty into another duct system forming a larger unit called a lobe.

Milk Ejection Or "Let Down."

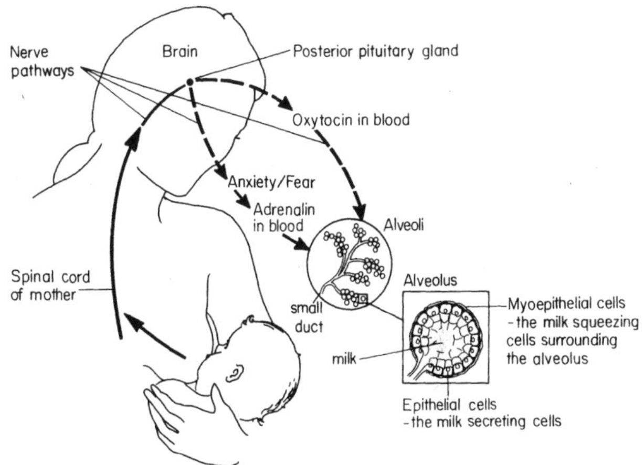

Fig. B-86. Milk ejection or "let-down."

As has already been noted, the milk is stored in alveoli. Before it is available to the baby, it has to be forced from the alveoli into the larger ducts and cisterns. This process is known as ejection or let-down. Here is how it works (see Fig. B-86): When the breast (especially the nipple) is stimulated by the initial sucking of the baby, (1) impulses are conducted along the nerves to the posterior pituitary gland at the base of the mother's brain; (2) the posterior pituitary gland stores and releases the hormone oxytocin into the bloodstream; (3) the blood transports oxytocin to the breast; and (4) the oxytocin causes the smooth, musclelike myoepithelial cells surrounding each alveolus to contract, thereby forcing the milk out of them into the duct system. The myoepithelial cells function like a rubber band; they stretch out as the alveolus balloons out with milk, and they contract, forcing out the milk during let-down. A second hormone, prolactin, is discharged by the anterior pituitary by the same stimulation that releases oxytocin. It is also carried in the blood to the breast where it stimulates the milk secreting cells (the epithelial cells) to begin immediately the manufacture of milk as quickly as the milk already formed is removed.

It is noteworthy that milk ejection or let-down is enhanced by confidence and inhibited by anxiety or anger. Inhibition is due to an overriding hormone action. Upon such occasions, another hormone, adrenalin, is released into the bloodstream. This hormone prevents the let-down hormone, oxytocin, from reaching the myoepithelial cells and causing their contraction—so the milk cannot be removed.

Care Of The Breasts. Plain water is best for cleaning the breasts and nipples. Soap and alcohol are too drying; and boric acid should never be used.

If difficulty with the nipples should occur (cracking or infection), the milk may be expressed with a hand breast pump and fed to the baby in a bottle until the nipples heal. Sometimes a nipple shield is used satisfactorily until healing occurs.

Lactating mothers should wear a properly fitted brassiere day and night to provide adequate support. It should be changed daily to a clean one. A breast shield should be placed in the brassiere to absorb any milk that may leak between feedings.

BREAST FEEDING BY MAMMALS ADAPTED TO NEEDS. During his million years of existence, like other mammals, man has, until the past 2 generations, reared his young almost exclusively on his own milk. The causation of the shift from breast feeding to bottle feeding during the past 50 years is complex. But, in large measure it was related to thinking of infant feeding only in terms of mathematical calculations. The current resurgence of interest in breast feeding takes cognizance that infant feeding involves much more than merely supplying nutrients; that it includes provision for anti-infective bridging (providing substantial immunological protection to the very young infant against certain bacteriologic infections until their own immunological process develops) and the development of mother-baby bonding. In this concept, there are lessons to be learned from man's fellow mammals. Indeed, there is abundant evidence that "adaptive suckling" has developed appropriate to each mammalian species over the 200 million years of evolution, with the milk and the feeding of their young adapted to the needs and way of life of different mammal species, particularly to the maturity of the newborn, the special nutritional requirements of the young, and the lactatory apparatus and number of offspring. Yet, in contrast to all other species (including birds, who also care for their young), each mammalian mother, regardless of species, except for man, remained the sole supplier of food for her young in the form of milk produced by her body.

- **Maturity of the newborn**—Newborn mammals vary in their maturity at birth—from the egg of the spiny anteater, to the bean-sized, blind, pouch-reared kangaroo, to the more mature young of most species. Human newborn are immature; they're incapable of locomotion and completely dependent on the mother.

- **Special nutritional requirements**—Mammals vary in the composition of their milk and in the length of their lactation period. Generally speaking, the protein content of the milk varies with the rate of growth of the offspring. For example, the human with 1.5% protein in the milk takes 150 days to double birth weight, whereas the horse with 6% protein in the milk takes 60 days, and the rabbit with 13 to 15% protein in the milk doubles the birth weight in only 6 days.

Also, protein content of milk may be related to frequency of feeding. For example, the low protein content of human milk (1.5%) calls for frequent feeding, whereas rabbits, whose milk contains about 13-15% protein, usually feed their young only once daily.

High fat concentration of milk is usually related both to animal size and low environmental temperatures whence the young have to develop and maintain a thick layer of insulating subcutaneous fat. For example, elephant milk contains only about 16.6% fat, while polar bear milk and gray seal milk contain 33.1% fat and 53.2% fat, respectively.

- **Lactatory apparatus and number of offspring**—The lactatory apparatus of mammals has also become adapted to the needs of the species. For example, the mammae of seals and walruses are flattened beneath the animal's blubber, so that the efficient streamlining is not affected; and the teats of the whale are recessed under overlying skin folds (mammary slits). Also, the number of mammary glands is related to the number of offspring; in the human, which generally produce singletons or twins, there are two mammary glands, whereas litter-bearing sows and dogs usually have ten or more mammary glands.

THE DECLINE AND RESURGENCE OF BREAST FEEDING. In the opening section, some of the reasons for the shift to bottle feeding in the developed countries are given. During the past 30 years, bottle feeding has increased greatly in the less developed countries, especially in urban areas. The reasons are various and include the following:

1. The changing to an urban lifestyle
2. Mothers having to work away from home
3. Ill-informed health professionals
4. The promotion of the infant formula industry

The present resurgence of interest in breast feeding started about 20 years ago. In the United States (and subsequently in other industrialized countries) it began by mothers themselves questioning the mechanized, unnatural method of infant feeding in vogue. Paralleling this movement, and about the same period of time, pediatricians working in developing, largely tropical countries became increasingly aware of declining breast feeding and the rising problems of diarrhea and marasmus in infancy. Also, recent scientific research is increasingly causing specialists in other fields, such as allergy, emotional development, etc., to join in the crusade.

ADVANTAGES AND DISADVANTAGES OF BREAST FEEDING. There is much evidence that the earliest experiences of the newborn baby are of great importance in its later adjustment to the world in which it must live. This is particularly true of feeding, and it applies whether the baby is breast fed or bottle fed. If the mother is relaxed and confident, the baby will respond to her and, through her, to the world about it with trust and confidence. Conversely, if the mother is tense and radiates anxiety and fear, and if the feeding is hurried, the baby becomes aware of the situation; and there may be fretfulness or crying, which may prevent it from taking the food that it needs.

A mother should be encouraged to breast feed her baby. This is best accomplished in early pregnancy by pointing out the advantages and disadvantages of breast feeding vs bottle feeding. Then, if the mother decides to breast feed her infant, instructions for the preparation of her breasts prior to delivery should be given by the doctor or nurse. Likewise, she must be well informed relative to the importance of proper diet and rest during the rigors of lactation.

Despite what has been said, a mother should not be made to feel guilty if there are circumstances, such as mothers not producing sufficient milk or being ill, which favor bottle feeding. If the baby is cuddled and made comfortable when it is being fed, whether by bottle or breast, its feelings will be those of warmth and comfort.

Also, it is noteworthy that breast feeding is of greater importance in the developing countries than in the developed countries with more affluent societies.

Although not all the advantages or disadvantages herein ascribed to breast feeding apply to a particular circumstance, all of them have been pertinent at one case or another; hence; they are important when deciding between breast feeding and bottle feeding.
The *advantages* of breast feeding are:

• **Presence of colostrum**—Although scientists have improved upon nature's product, milk, they have not yet learned how to formulate a synthetic product that will replace colostrum.

• **Most nutritive requirements are met**—The nutritive composition and certain nutritive interactions of human milk are suited to infant needs.

• **Nutrient availability**—Protein and iron are more readily available from human milk than from prepared formulas.

• **Less obesity**—The low calorie content of human milk, along with the lack of maternal concentration of feedings, are important factors in the uncommonness of infantile obesity in breast-fed infants.

• **Lessened risks**—Breast feeding reduces risk of neonatal tetany, hypertonic dehydration, allergy to cow's milk, constipation, and certain other hazards.

• **Immunization**—Through antibodies received in the mother's milk, it imparts immunization to the baby against certain infectious diseases.

• **Fewer diseases and less mortality**—It lessens gastroentritis and protein-energy malnutrition, which inevitably make for increased infant mortality in the developing countries.

• **Safer**—It is bacteriologically safer than a poor formula unhygienically prepared.

• **Freedom from contamination**—It is free from contamination.

• **Ease of preparation**—It eliminates preparation of the formula with special utensils and sterilized bottles and nipples; and the milk is available at proper temperature and without errors of calculation and formula preparation. (Of course, ready bottled commercial formulas are available today.)

• **Alleviates space and equipment needs**—It does not require space and equipment to prepare each feed in a hygienic manner—an important consideration in developing countries.

• **Economical**—It is more economical; it takes money to buy formula products.

• **Alleviates dependence on availability**—It is not dependent on the availability of high quality formula foods, an important consideration in some of the developing countries.

• **A comforting and satisfying experience**—It provides a natural part of the life cycle for both mother and baby, which may be a comforting and satisfying experience.

• **Psychological effects on the mother**—Many mothers derive satisfaction from knowing that they are the source of their baby's food.

• **Enhances mother-child relationship**—It permits an early mother-child bonding.

• **Birth control**—Continued lactation tends to delay the resumption of ovulation and increase birth spacing, an important consideration in developing countries. It is not, however, a reliable method of contraception.

• **Some speculations**—It has been speculated, without proof, that the use of unnatural foods in the first months of life may sow seeds which appear later as chronic disorders such as allergies, atherosclerosis, hypertension, and diabetes.

The following *disadvantages* have been attributed to breast feeding at one time or another:

• **Milk is not the perfect food**—It is deficient in iron, copper, and vitamin D. In addition to correcting these deficiencies, formulas for bottle feeding may be fortified with other minerals and vitamins.

• **Time out from work for breast feeding**—It necessitates that women who work away from home take time out to breast feed their babies.

• **It curbs social life**—Breast feeding is an obstacle to a full social life.

• **Mother must give adequate milk**—The mother must supply at least half the infant's needs; otherwise, breast feeding isn't practical. It should be added that it is unusual for a healthy woman with no disease of the breast to fail to produce enough food for her baby.

• **Subject to being discontinued**—Breast feeding must be discontinued when any of the following conditions arise:

1. When chronic illnesses—such as heart disease, tuberculosis, severe anemia, nephritis, epilepsy, insanity, or chronic fevers—strike the mother.
2. When another pregnancy ensues.

3. When it is necessary for the mother to return to employment away from home.
4. When the infant is weak and unable to nurse because of cleft palate or harelip.
5. When the mother acquires a long-lasting infection which the infant has not yet had.

• **Breast pumping may be necessary**—If breast feeding must be discontinued temporarily, for such reasons as cracked nipples or the mother having an acute infection, the breasts must be pumped at regular intervals so that the milk supply will not dwindle.

In summary it may be said that (1) if the environmental sanitation and food supply factors are favorable, either method of feeding will meet the physical need; and (2) if the emotional and psychological character of the mother and the mother-child relationship factors are unfavorable, breast feeding may be contraindicated. The mother should come to a thoughtful decision between breast and bottle feeding early in pregnancy so that intelligent ways and means of implementing her choice may be explored with her.

COLOSTRUM. During pregnancy, the breast is prepared for lactation. The alveoli enlarge and multiply, and toward the end of the pregnancy period, secrete a thin yellowish fluid called colostrum. After delivery, colostrum is secreted for 2 to 4 days, following which regular milk is produced. Colostrum contains more protein, minerals, and vitamin A than regular milk, but less carbohydrate and fat. In addition, it also contains helpful antibodies which confer to the newborn an immunity to certain infections during the first few months.

FEEDING TECHNIQUE. Breast feeding should be initiated within 24 to 48 hours after birth.

The rooting reflex of the newborn, its oral need for sucking, and its basic hunger drive make breast feeding easy for the healthy relaxed mother whose diet meets lactation requirements. Nevertheless, the suggestions that follow may be helpful:

1. The mother must be comfortable and relaxed; she not only feeds, but she talks and smiles. At first she will probably hold the baby cradled in her arms in a semireclining position against the breast while reclining on and supported by pillows. Later, she should sit in a comfortable chair with armrests—perhaps a rocker; preferably, with a footstool to support her feet.
2. The warm touch of the breast on the baby's cheek will stimulate the natural rooting reaction, causing it to turn its head in the direction of the touch and to begin sucking motions with its mouth. Stroke the cheek nearest the nipple, and the baby will turn toward it. (Never touch the cheek furthest from the nipple, as it will cause the baby to turn its head in that direction—away from the nipple.) If necessary to get the baby started, press a little milk onto its lips. The baby should grasp most of the nipple in its mouth, not merely the outer tip.
3. In the beginning, the baby may get enough food by emptying one breast, but if it is still hungry it should be offered the other breast. After lactation is established, alternate breasts may be used for each feeding.
4. Usually, a hungry baby will get its fill of milk in about 5 minutes of nursing, but some will continue as long as 20 minutes. When the infant has had sufficient milk and is satisfied, it will stop nursing and become disinterested in more feeding.
5. After each feeding (and sometimes during the feeding), the baby should be "burped," to expel any air that it may have swallowed during nursing. This may be accomplished (a) by holding the baby over the shoulder and patting it on the back, (b) by laying it, stomach down, across the knee, and patting its back.

INTERVALS AND ADEQUACY OF FEEDING. Although there is not full agreement, most authorities subscribe to the following thinking relative to the frequency of feeding and the amount to feed:

1. Feedings may be given according to the hunger needs of the baby, sometimes referred to as *self-demand feeding*. The very young infant may require feeding every 2 to 3 hours, but it soon establishes a rhythm of feeding at 3- to 4-hour intervals. After the second month, the night feeding usually may be discontinued.
2. About 2.5 oz of human milk per pound of body weight results in satisfactory weight gain. The baby is getting enough milk (a) if it is satisfied and falls asleep after feeding; and (b) if it is making satisfactory gains as determined by weighing each week, at the same time and with the same amount of clothing. Also, by weighing the baby before and after feeding, a mother can get a pretty good idea of how much milk it is getting.

Insufficient milk is indicated when the baby is not satisfied at the completion of feeding, or is restless and either fails to fall asleep quickly after nursing or awakens frequently. In such cases, the physician may recommend adding a supplemental food, or replacing one or more of the breast feedings with bottle feeding.

3. After lactation is established, an occasional bottle feeding may replace breast feeding if the mother so desires or has to be away.

MOTHER'S LACTATION DIET. The lactating mother requires more food and more nutritious food than a nonlactating woman, with the amount of such increases dependent upon the quantity of milk secreted. Under normal conditions, a mother produces approximately ⅓ oz (*10 g*) of milk on the 1st day, and 33 oz (*1,000 g*) per day on the 24th week.

Lactating mothers need to increase their daily energy intake by 600 Calories (kcal), and more if they are doing a lot of housework or are employed in a factory. They may also draw upon stores of fat laid down during pregnancy for additional energy. A weekly weighing is the best way to determine if the food intake is sufficient.

The extra protein and calcium for lactation are most conveniently provided by drinking a minimum of 1 pt (*470 ml*) of milk daily.

The mother's diet determines the pattern of fatty acids and the concentration of vitamin A, thiamin, riboflavin, vitamin C, and vitamin B-12.

Coffee, tea, beers, and wines may be taken in moderation and do not alter the quality of the milk. But most drugs are excreted in the milk, although they are usually in such small amounts as to cause no ill-effects to the baby. However, all drugs are potentially harmful if taken in large amounts. Most authorities recommend against the use of the following drugs while breast feeding: anticoagulants, antithyroid drugs, atrophine, diuretics, morphine, oral contraceptives, radioactive preparations, reserpine, and steroids.

Fortunately, women can store up body reserves of certain nutrients before and during the pregnancy period, to be drawn upon following birth. Here is how this phenomenon works: When properly fed before and during pregnancy, certain nutrient deposits are made in the body. Then, during lactation when the demands may be greater than can be obtained from the food, the mother draws from the stored body reserves. Thus, maternal stores of energy in the form of subcutaneous fat are laid down; and both calcium and phosphorus can be stored in the bones, then withdrawn during early lactation when milk production is at its peak. Of course, if there hasn't been proper body storage, something must "give"—and that something will be the mother, for nature ordained that the growth of the fetus, and the lactation that follows, shall take priority over the maternal requirements. Hence, when there is a nutrient deficiency, the mother's body will be deprived, or even stunted if she is young, before the developing fetus or milk production will be materially affected.

Even for a healthy woman on a good diet, providing sufficient milk to meet all the demands of a rapid-growing infant is a physiological strain. So, supplemental feeding, i.e., the substitution of a bottle feed for one or more breast feeds and/or the early introduction of solid foods, may be indicated. Also, it is important that the mother's hemoglobin should be checked during the period between childbirth and the return of the uterus to its normal size. Iron deficiency anemia is rather common at this time, especially if blood loss was excessive at or after delivery.

It is noteworthy, too, that fetal stores of iron and vitamin A are made during pregnancy, provided the mother's nutritional status during pregnancy is adequate.

WEANING THE BABY. Weaning is the process whereby feeding from the breast or bottle is replaced by other foods, usually pasteurized cow's milk and solid foods. Except in an emergency, it should be started during the 5th to 9th month, and it should take place gradually. Weaning is usually accomplished by substituting a cup feeding for the breast or bottle feeding one period daily. When the baby has become accustomed to this—after about 4 to 5 days—the second cup feeding daily may be offered. Subsequent increases of cup feedings should be offered until the baby is entirely weaned. Weaning usually requires a period of 2 to 3 weeks.

(Also see BABY FOODS; INFANT DIET AND NUTRITION; and NUTRIENTS: REQUIREMENTS, ALLOWANCES, FUNCTIONS, SOURCES.)

BREATHING RATE

Normal breathing rates per minute are: men, 18; women, 20; children, 25; and infants, 35. It varies under different conditions of health and disease. Breathing increases on exertion and excitement, and in case of fever, asthma, and heart disease. It slows down during rest and sleep.

BREWERS' YEAST

Brewers' yeast, which is usually dried (the fresh form spoils quickly), is the nonfermentative, nonextracted yeast of the botanical classification, *Saccharomyces*, derived as a by-product from the brewing of beer and ale. It is used primarily as a rich supplemental source of the B vitamins and unidentified factors. It is also an excellent source of protein (it contains a minimum of 35% crude protein) of good quality. When irradiated with ultraviolet light, it also provides vitamin D.

BREWING

- The process of making malt beverages such as beer and ale.

- A process of preparation, or a concoction, such as teas and herbal teas.

(Also see BEERS AND BREWING.)

BRIGHT'S DISEASE

This term is commonly used to designate kidney disease in general. However, the types of kidney disease vary greatly in terms of their causes and their effects on the rest of the body. Depending upon the disease, there are various therapeutic diets.

(Also see MODIFIED DIETS.)

BRISKET

A cut of meat consisting of the breast.

(Also see BEEF AND VEAL, Fig. B-39.)

BRITISH GUMS

These are formed by modifying starch with high temperatures and possibly a little acid. The glucose molecules forming the starch rearrange under these conditions, and become highly branched molecules. British gums are used as carriers for food flavors.

(Also see STARCH.)

BRITISH THERMAL UNIT (BTU)

The amount of energy required to raise 1 pound of water 1°F; equivalent to 252 calories.

BRIX

A term used to express the sugar (sucrose) content of molasses, and of syrups used in canned fruits.

BROAD BEAN (FAVA BEAN) *Vicia faba*

This legume, which is referred to in the Bible as a vetch, was a major bean consumed throughout the Old World until the common bean was brought from the newly discovered Americas in the 16th century. Broad beans are a little larger than large lima beans, which they resemble. Fig. B-87 shows this legume.

Fig. B-87. Broad bean (fava bean). (Courtesy, Minnesota Agricultural Experiment Station, University of Minnesota)

ORIGIN AND HISTORY. Botanical studies have led to the conclusion that the plant was native to northern Africa and the eastern Mediterranean region. Ancient records indicate that the Chinese used these beans for food almost 5,000 years ago. Also, they were cultivated in Biblical times by the Hebrews, and a little later by the Egyptians, Greeks, and Romans. It is reported that the Greek philosopher Pythagoras (6th century B.C.), who is known to students for his theorem in geometry, met his death at the hands of an angry mob when he balked at an opportunity to escape through a field of fava beans. The misfortune of the renowned Greek is depicted in Fig. B-88.

Fig. B-88. Pythagoras allows himself to be caught by his pursuers rather than cross a field of fava beans.

Historians suspect that the philosopher suffered from an inherited tendency to favism, a painful blood condition brought on by (1) eating the broad bean, or (2) inhaling the pollen from the flowering plant.

The bean was brought to the New World shortly after the landing of Columbus, and is now grown in some parts of Latin America.

PRODUCTION. Broad beans are an important crop worldwide, with an annual production of about 4.3 million metric tons. However, the small crop grown in the United States is used almost exclusively for livestock pasturage, hay, and silage. Fig. B-89 shows the leading countries in production of broad beans.

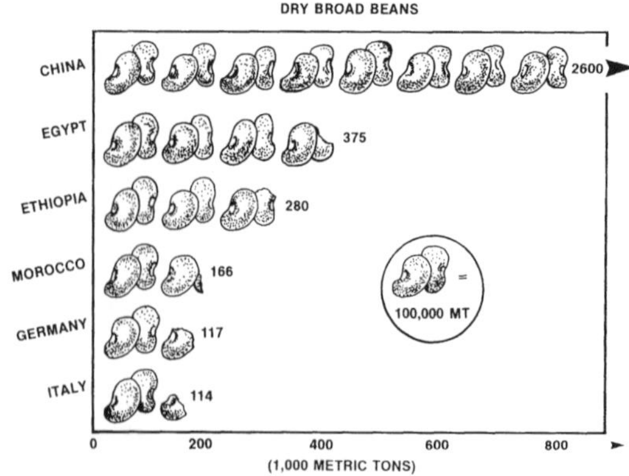

Fig. B-89. Leading broad bean producing countries of the world. (Based on data from *FAO Production Yearbook*, 1990, FAO/UN, Rome, Italy, Vol. 44, p. 102, Table 33)

Propagation and Growing. The crop is grown during the mild seasons in the temperate and subtropical zones, when the temperature ranges from 50)°F (*10°C*) to 86°F (*30°C*). However, the seeds of the broad bean, unlike those of the common bean, may be sown while the soil is still cold, as long as the danger of a heavy frost has passed. The hardiest varieties are sown in the late fall, while the other types are planted in the late winter. Recently, the broad bean has been introduced into the highland areas of Central America, which are much cooler than the lowland areas where the crop does not grow well. (Temperatures over 70°F [*21°C*] may cause the unopened blossoms to drop off.) Nevertheless, the plant is moderately resistant to high temperatures and droughts when exposures to these conditions are of brief duration.

Harvesting. The Chinese often pick the immature pods and seeds for use as green beans. Other producers usually harvest the crop after the seeds are fully mature. The worldwide average yield is about 1,500 lb of beans per acre (*1,347 kg/ha*), but yields as high as 3,100 lb (*3,472 kg*) and 3,200 lb (*3,584 kg*) have been obtained in Argentina and Switzerland, respectively.

PROCESSING. Almost all of the broad bean crop is dried. Small amounts of the production in the Mediterranean region are canned for export.

SELECTION AND PREPARATION. Most Americans purchase broad beans in the form of dried mature seeds, which may be uncooked or cooked and canned. Therefore, it is noteworthy that these beans swell up to 2½ and 3 times their dried weight when cooked in water. Hence, the shopper who purchases 1 lb (*0.45 kg*) of canned beans receives the equivalent of only about ⅓ lb of uncooked dried beans.

The soaking and cooking time for dried broad beans may be shortened by the procedures that follow:

- **One-hour hot soak**—The dried beans plus water (2½ cups per cup [*240 ml*] of beans) is boiled for 2 minutes. Then, the mixture is allowed to soak 1 hour, after which the water is drained off and new water added before cooking. Hard water (which makes the beans tough) may be softened by adding ⅛ tsp to ¼ tsp of baking soda per cup of beans.

- **Pressure cooking**—Soaked beans plus 2 cups of water per cup of beans are added to a pressure cooker with about 1 Tbsp (*15 ml*) of fat to prevent excessive foaming (which may cause a dangerous situation by clogging the vent of the pressure cooker). The pressure is brought up to 15 lb and maintained for 25 minutes, after which it is allowed to drop gradually (without cooling the cooker or removing the counterweight).

Cooked dried broad beans may be prepared and served in much the same ways as mature lima beans. That is, they may be (1) baked in casseroles containing cheese, meat, poultry, and/or other vegetables; (2) added to salads, soups, or stews; or (3) served alone with butter or various sauces.

Immature broad bean seeds and/or pods may be (1) stir-fried to make Chinese style mixed dishes; (2) simmered briefly until tender and served with butter or a sauce; or (3) cooked in casseroles, soups, and stews. (They should be added after cooking of the other ingredients has been underway for some time, so that overcooking is avoided.)

CAUTION: Favism, an inborn error of metabolism, affects up to 35% of some Mediterrean populations and about 10% of American negroes. In sensitive people, ingesting (or even inhaling the pollen) of broad beans, or fava beans (*Vicia faba*), causes the breakdown of red blood cells (an acute hemolytic anemia). The symptoms are dizziness, nausea and vomiting, and sometimes a high fever and collapse, followed by severe anemia. There is no known way of processing broad beans to remove or inactivate the substances that cause favism; hence, people who are susceptible to this type of hemolytic anemia should avoid broad beans.

More recently, it has been discovered that persons with the defect that causes favism develop a similar anemia when treated with oxidant drugs, including commonly used antimalarials, sulphonamides, antipyretics, and analgesics.

Broad beans (fava beans) also contain inhibitors of trypsin and red blood cell clumping agents (hemagglutinins)—potentially harmful substances, which may be inactivated by soaking and thorough cooking (steaming, pressure cooking, extrusion cooking). Thus, bean flour should be made from dried beans that have been pressure cooked prior to grinding.

In common with most beans, broad beans contain certain indigestible carbohydrates which are acted upon by gas-producing bacteria in the lower intestine, resulting in gas and even diarrhea in some people. Much of the indigestible carbohydrate may be broken down by sprouting, fermentation, or incubation of raw dried beans in a mildly acid solution at 113°F (*45°C*) to 149°F (*65°C*) for 1 or 2 days. (The acid treatment is not yet carried out on a commercial scale.) People who are greatly bothered by gas should not eat broad beans unless (1) they have been treated, or (2) they are immature (green).

NUTRITIONAL VALUE. The nutrient composition of broad beans is given in Food Composition Table F-36.

The percentage of solids (100 minus water content) in the mature seeds is over three times that of the immature seeds; hence, the mature seeds are much higher in calories, protein, minerals and certain vitamins (except that the immature seeds are a much better source of vitamins A and C). Broad beans, like the common bean, are a good source of calories, protein, carbohydrates, fiber (they contain about twice as much as the common bean), phosphorus, iron, potassium, and the vitamin B complex. However, the ratio of calcium to phosphorus is low, so the calcium might not be well utilized unless other foods rich in calcium and somewhat lower in phosphorus (dairy products and green leafy vegetables) are consumed.

Protein Quantity and Quality. People living in the Mediterranean region and in China may depend upon broad beans to supply much of their dietary protein.

Therefore, some pertinent facts on this subject follow:

1. A 3½ oz (*100 g*) portion (approximately ½ cup) of cooked broad beans supplies about 8 g of protein, equivalent to the protein in 1 oz of lean meat. It is noteworthy that the beans supply between 50 and 100% more calories per gram of protein than most cheeses, eggs, fish, meats, milk, and poultry.
2. The protein quality is lower than that in most animal foods because it is moderately deficient in the amino acids cystine and methionine. However, this problem may be countered by consuming the beans with various cereal products, since the two types of protein are complementary (each supplies some of what the other lacks). The best mixtures are those containing about ⅓ cooked beans and ⅔ cooked grain product. For example, beans and macaroni are a popular dish in certain parts of Italy.

(Also see BEAN[S], Table B-10 Beans of the World; LEGUMES, Table L-2 Legumes of the World; and VEGETABLE[S], Table V-6 Vegetables of the World.)

BROCCOLI *Brassica oleracea*, variety *italica*

Some children think that the edible parts of this vegetable (flower heads and stalks) resemble miniature trees. Apparently, the ancient Romans thought similarly, because the name broccoli is derived from the Latin word *brachium*, which means arm or branch. Broccoli was derived from wild cabbage. Most of its improvement took place in Italy. Fig. B-90 shows the edible portion of this vegetable.

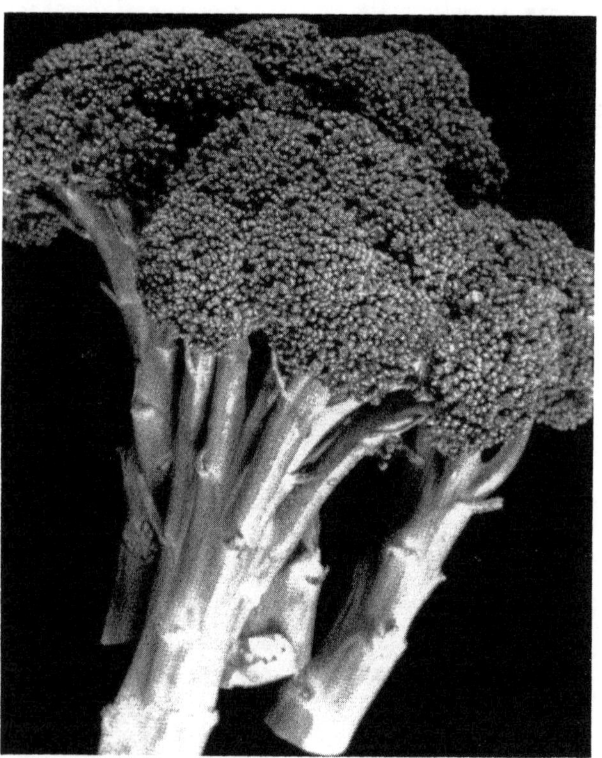

Fig. B-90. Broccoli. (Courtesy, USDA)

ORIGIN AND HISTORY. Broccoli developed from the wild cabbage that was native to coastal Europe, which had spread through the Near East to the Orient at an early date. The ancestral plant resembled modern-day collards in that the leaves did not clump together to form a head.

The Roman botanist Pliny (1st century A.D.) wrote about a vegetable which might have been broccoli or cauliflower, but it is not certain whether either plant had been developed by that time because no other mention of the vegetable is known to have been made until the 17th century, when it was identified as *Italian asparagus.*

It is believed that the early selection of the plant for its edible flower heads may have been done somewhere in Asia Minor, and that sea-going traders brought the plant to the Italian peninsula where further development took place.

Later, Italian immigrants grew the vegetable in the eastern United States and helped to popularize it there.

PRODUCTION. Fig. B-91 shows that California accounts for over 95% of the U.S. production of broccoli.

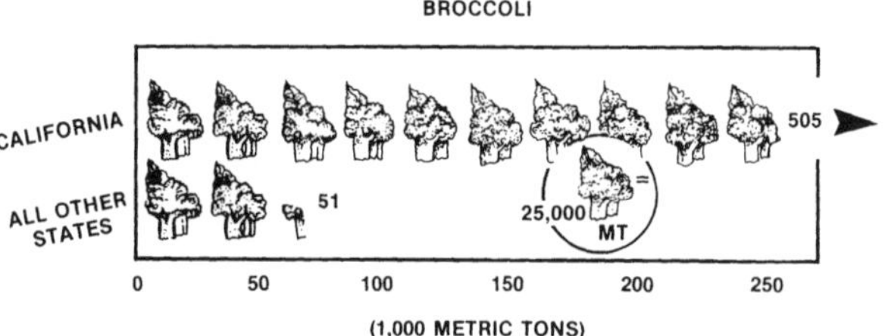

Fig. B-91. U.S. broccoli production. (Data from *Agricultural Statistics*, 1991, USDA, p. 148, Table 206)

The planting of broccoli must be timed so that the vegetable will be ready for harvesting during cool weather because high temperatures are likely to spoil the crop by hastening the ripening and opening of the flower heads. Hence, the crop is planted very early in the spring for harvesting in the late spring, and in the summer for harvesting in the fall. It is also grown throughout the winter in the southern states.

Each broccoli plant develops a large, central flower head which should be harvested before the flower buds open. After the central head and stalk has been removed, several lateral shoots will develop heads, which may be harvested as soon as they are sufficiently mature.

Freshly cut broccoli heads must be cooled immediately, then kept under refrigeration to prevent the flowers from opening or turning yellow.

PROCESSING. A little over one-half of the U.S. broccoli crop is processed by freezing, and the rest is marketed fresh.

SELECTION AND PREPARATION. Good quality broccoli is fresh and clean, with compact bud clusters which have not opened to the extent that the flower color is evident. The general color should be dark-green, deep sage-green, or purplish-green, depending on variety.

Stalks and stem branches should be tender and firm. Toughness and woodiness of stalks and attached branches develop with age, the base portion of the stalk being most susceptible. Yellowed or wilted leaves are indicative of age and lack of freshness. Bud clusters which are sufficiently open to show the distinct yellow color of the blossoms are usually overmature, although an occasional blossom does not indicate overmaturity. Wilted, flabby or noticeably bruised broccoli should be avoided.

Most recipes for broccoli dishes utilize the cooked vegetable. Fresh broccoli requires from 9 to 12 minutes cooking in boiling water, depending on the size of the pieces, whereas frozen broccoli requires 6 to 8 minutes.

NOTE: Frozen broccoli is cooked partially by blanching prior to freezing. Once cooked, the vegetable may, or may not, be sauteed briefly in oil or bacon fat, with or without bits of browned garlic, capers, coriander, crisp bacon, curry powder, pepper, and/or pimento. Broccoli may also be served on toast with a cheese sauce, hollandaise sauce, or a white sauce containing sliced hard-cooked eggs. Finally, pureed broccoli makes an excellent soup.

CAUTION: Broccoli and other vegetables of the mustard family *(Cruciferae)* contain small amounts of goiter-causing *(goitrogenic)* substances which, under certain conditions, may have detrimental effects. These compounds appear to interfere with the utilization of the essential element iodine by the thyroid gland. As a result the tissue of the gland may become enlarged (a condition known as goiter). However, it is believed that the goitrogenic effects of these vegetables may be counteracted by the consumption of ample amounts of dietary iodine. This element is abundantly present in iodized salt, ocean fish, seafood, and seaweeds such as kelp.

NUTRITIONAL VALUE. The nutrient composition of broccoli is given in Food Composition Table F-36.

Some noteworthy observations regarding the nutrient composition of broccoli follow:

1. Broccoli is high in water content (over 90%), and low in calories. Fewer than 50 Calories (kcal) are provided by 1 cup *(240 ml)* of the cooked vegetable.
2. A 1-cup serving of cooked broccoli supplies about the same amount of protein (5 g) as a cup of cooked corn or rice, but less than one-third as many calories (less than 50 kcal, vs 150 to 200 kcal in the cooked cereals).
3. Approximately equal amounts of calcium and phosphorus are furnished by broccoli, which is an ideal dietary ratio between the two essential minerals. The vegetable is also a fair to good source of iron, and it is an excellent source of potassium.
4. Broccoli is an excellent source of vitamins A and C. A cup of the cooked vegetable supplies as much vitamin C as two oranges.

Broccoli is also a good source of bioflavonoids, substances that appear to act along with vitamin C in strengthening the smallest blood vessels (capillaries) against breakage or leakage of fluid into the surrounding tissues.

(Also see VITAMIN[S], section headed "Vitamin Table.")

BROILERS (FRYERS)

Meat-type chickens that are 6 to 13 weeks of age. (Also see POULTRY.)

BROMATOLOGY

The science of foods (from the Greek *broma*—food).

BROMELIN

This is a protein-digesting and milk clotting enzyme isolated from fresh pineapple juice. Bromelin is used in biochemical research, and for tenderizing meat. Because bromelin will digest gelatin, pineapple cannot be used to make gelatin dishes unless it has been canned or heated.

(Also see MEAT[S], section headed "Meat Tenderizing.")

BROMINATED VEGETABLE OILS

These are made by adding bromine to the unsaturated fatty acid component of the vegetable oil. The major use of brominated vegetable oils is the production of stable flavor emulsions for use in citrus flavored soft drinks. They are considered a food additive, and the Food and Drug Administration (FDA) allows citrus and other fruit-flavored beverages to contain only 15 parts per million (ppm).

BRONCHIAL ASTHMA

A serious and often chronic condition in which the bronchial tubes which bring air into the lungs become clogged with mucus so that little air may pass through them. Usually, broncial asthma is due to allergic reactions provoked by one or more of a wide variety of different allergens. Some of the most common allergens are house dust, mold, pollens, ordinary foods, various medicines, or even hair or flakes of skin from house pets. Emotional upset, extreme fatigue, changes in temperature or humidity, or moving to a new climate may also bring on or aggravate bronchial asthma. The best preventive measures are identification and avoidance of the allergens responsible for the condition.

(Also see ALLERGIES.)

BRONCHIECTASIS

A condition in which the bronchial tubes and a portion of the lungs are stretched and dilated. Bronchiectasis is characterized by continuous coughing and the bringing up of a puslike secretion. It may be caused by chronic bronchitis, dust, infections, asthma, and/or malnutrition. If the disorder lasts long enough, respiratory acidosis and a coma may develop because the lungs cannot get rid of the excess carbon dioxide produced in metabolism.

(Also see ACID-BASE BALANCE; and BRONCHITIS.)

BRONCHITIS

An inflammation of the mucous lining of the bronchial tubes which bring air into the lungs. Bronchitis is often accompanied by a slight fever and a dry rasping cough. It is usually caused by a bacterial infection and may occur suddenly after chilling, exposure to irritants in the air, or upon the development of a common cold. If the condition is not treated promptly, it may develop into bronchopneumonia. Therefore, acute bronchitis should be treated with (1) bed rest in a warm room with a vaporizer; (2) small but nutritious meals; and (3) plenty of hot liquids. Bronchitis may be aggravated by very dry air, smoking, excessive use of alcohol, a dusty environment, and obesity.

BRONZE DIABETES

A disorder in which (1) the skin has a bronzelike hue, and (2) there is a loss of function of the liver and/or the pancreas. It results from the abnormal accumulation of iron in these tissues due to some unknown disorder of iron metabolism, or to the absorption of a great excess of iron. One of the conditions under which a great excess of iron may be absorbed is the consumption of iron in alcoholic beverages, such as the beer made by the Bantu tribes of Africa. (The Bantu tribes brew their acid beer in iron pots.) Also, iron overload can be caused by high intakes of iron supplements.

(Also see IRON.)

BROWN RICE

Paddy rice from which the husk has been removed.

BROWN SUGAR

Brown sugar, technically known as "soft" sugar, is a mass of fine crystals covered with a film of highly-refined, colored, molasses-flavored syrup. It is valued primarily for flavor and color. Brown sugar does contain several minerals at levels higher than refined sugar. Still, the energy content is similar to refined sugar, 385 kcal/100 g for granulated refined sugar and 373 kcal/100 g for brown sugar. Lighter types are used in baking and making butterscotch, condiments, and glazes for ham. The dark brown sugar, with its rich flavor, is desirable for gingerbread, mince meat, baked beans, plum pudding, and other full-flavored foods.

(Also see SUGAR.)

BRUCELLA

A family of microorganisms which chiefly affect livestock and cause pregnant females to abort their young. Man is also susceptible to the organism, which may be transmitted in defective meat or milk. However, such transmission may be prevented by pasteurizing milk and cooking meat.

(Also see DISEASES; and BRUCELLOSIS.)

BRUCELLOSIS (UNDULANT FEVER; MALTA FEVER)

A disease caused by bacteria of the genus *Brucella*.

(Also see BACTERIA IN FOOD, Table B-5 Food Infections [Bacterial]; BRUCELLA; and DISEASES.)

BRUNNER'S GLANDS

Duodenal glands that secrete mucus which protects the mucosa from irritation and erosion by the strongly acid gastric juices entering from the stomach. Emotional stress inhibits mucous secretion, which may lead to an ulcer.

BRUSH BORDER

Microscopic bristlelike projections (microvilli) which line the villi of the small intestine and which function in the absorption of nutrients. These cells are also the sources of certain intestinal enzymes which aid in digestion.

(Also see DIGESTION AND ABSORPTION.)

BRUSSELS SPROUTS *Brassica oleracea*, variety *gemmifera*

The edible buds from the stem of this plant resemble miniature cabbages, which is not surprising because Brussels sprouts and the large-headed cabbage are very closely related. Brussels sprouts are derived from the wild cabbage, *B. oleracea*. The variety *gemmifera* refers to the abundant production of the budlike growths on the stem of the plant. Fig. B-92 shows the sprouts on the plant.

Fig. B-92. Brussels sprouts. (Courtesy, USDA)

ORIGIN AND HISTORY. Although this vegetable is a relative of the wild cabbage which has been used as food for thousands of years, it does not appear to have been developed in its present form until a few hundred years ago in the northern part of Europe near Brussels. The exact time that this occurred is uncertain, because the reports of a "many-headed cabbage" in the 13th century might have referred to a different type of cabbage vegetable. Although it most likely attained its present form before the 19th century, it remained a local crop in Belgium until after World War I, when its use spread throughout Europe.

PRODUCTION. Almost all of the U.S. crop of Brussels sprouts comes from California, where 75 million lb *(34 thousand metric tons)* are produced each year.

The planting of the seeds in outdoor seedbeds and the transplanting of the young plants in the field are done in the late spring so that the crop may be harvested in the fall. The young plants tolerate light frosts, but the sprouts are spoiled by hot weather, which makes them soft and causes them to open.

Usually, the lowest mature sprouts are removed from the plant first, and the higher ones removed as they develop.

Fig. B-93. Brussels sprouts ready for harvesting. (Courtesy, Agricultural Chemicals Division of Mobay Chemical Corporation, Kansas City, Mo.)

The fresh vegetable may be stored for as long as 2 months at 32°F *(0°C)* and at 95% relative humidity.

PROCESSING. Over 95% of the crop is frozen, and the rest is marketed fresh.

SELECTION AND PREPARATION. Brussels sprouts of good quality are hard or firm, compact, fresh, bright in appearance, and good green in color. Puffy or soft sprouts are usually poor in quality and flavor. Wilted or yellowing leaves usually indicate aging and probable excessive waste in preparation for use. Brussels sprouts with worm-eaten leaves should be avoided, and also those with a smudgy, dirty appearance. The latter imperfection often indicates that aphids may be found on the inner surfaces of leaves in numbers which will cause undue waste or make the product unfit for food.

Some cooks boil Brussels sprouts only long enough to make them crispy tender (5 to 7 minutes of boiling), while others cook them for as long as 15 minutes. Usually, frozen sprouts require less cooking because they were blanched with boiling water or steam prior to freezing.

Cooked Brussels sprouts are very good when served with butter, cream, meat drippings, white sauce, and/or such seasonings as basil, caraway seed, dill, mustard seed, sage, or thyme.

CAUTION: Brussels sprouts, like other members of the mustard family of vegetables *(Cruciferae)*, contain small amounts of goiter-causing *(goitrogenic)* substances that may cause enlargement of the thyroid gland (a condition called goiter). It appears that the goitrogenic effect may be minimized by the consumption of adequate amounts of the essential mineral iodine, which is abundantly present in iodized salt, ocean fish, seafood, and seaweed.

NUTRITIONAL VALUE. The nutrient composition of Brussels sprouts is given in Food Composition Table F-36.

Some noteworthy observations regarding the nutrient composition of Brussels sprouts follow:

1. Brussels sprouts contain almost 90% water and are low in calories. (They contain from 50 to 60 Calories [kcal] per cup.)
2. A 1-cup *(240 ml)* serving of the cooked vegetable provides 5 g of protein, which is about the same as that provided by a cup of cooked corn or rice. However, the sprouts contain only one-third as many calories.
3. The phosphorus content of the vegetable is more than double the calcium content. Hence, other foods with more favorable calcium to phosphorus ratios (dairy products and green leafy vegetables) should also be consumed.
4. The vegetable is a fair to good source of iron, and it is an excellent source of potassium.
5. Brussels sprouts provide moderate amounts of vitamin A, and they are rich in vitamin C. One cup provides the amount of vitamin C present in two oranges.

(Also see VEGETABLE[S], Table V-6 Vegetables of the World.)

BUCKWHEAT *Fagopyrum*

There are three species of buckwheat: common buckwheat, *F. esculentum*; Tartary buckwheat, *F. Tataricum*; and winged buckwheat, *F. emarginatum*.

Fig. B-94. Buckwheat. (Courtesy, Minnesota Agricultural Experiment Station, University of Minnesota)

ORIGIN AND HISTORY. Buckwheat is a native of Asia. It was cultivated widely in China during the 10th and 13th centuries. From China it was taken to Europe via Turkey and the U.S.S.R. during the 14th and 15th centuries, thence to Great Britain and the United States during the 17th century.

WORLD AND U.S. PRODUCTION. World production of buckwheat is estimated at 3 million tons annually. The U.S.S.R. is the leading producer, followed by Poland. In both countries, buckwheat is a basic food item, consumed mainly as porridge and soup. Currently, U.S. production amounts to less than 30,000 tons annually, most of which is in the North Central and Northeastern states.

THE BUCKWHEAT PLANT. Buckwheat belongs to the *Polygonaceae* family; hence, it is related to rhubarb. Because buckwheat is used for about the same purposes as the cereal grains, it is often incorrectly referred to as a cereal. Unlike the grasses, however, buckwheat develops a strong tap root rather than fibrous roots.

The plant grows about 3 ft (*91 cm*) high. The stem varies from green to reddish, and both the stems and the branches turn brown at maturity.

BUCKWHEAT CULTURE. Buckwheat is a summer annual, adapted to a cool, moist climate. It does well on poor soils, more so than the cereals. It does not require a long growing season, 10 to 12 weeks only. Yields of 25 to 30 bu per acre are considered average. Most of the U.S. buckwheat crop is now harvested with a combine. It is relatively free of ravage by diseases and insects.

KINDS OF BUCKWHEAT. Three kinds or species of buckwheat are grown in different parts of the world; (1) common buckwheat, (2) Tartary buckwheat, and (3) winged buckwheat. Only common buckwheat and Tartary buckwheat are of importance in the United States. Standard varieties of common buckwheat for all purposes are Silverhull and Japanese. Tetratataricum is the leading variety of Tartary buckwheat.

MILLING BUCKWHEAT. Buckwheat is milled into flour by a roller milling process, much like wheat flour is produced. The flour yield ranges from 60 to 80%.

To make buckwheat groats, the grain is passed between two mill stones adjusted to crack the hull without grinding the seed.

NUTRITIONAL VALUE. The nutritive composition of buckwheat grain and flour are given in Food Composition Table F-36.

The gluten content of buckwheat is low; the principal protein is globulin. The profile of essential amino acids of buckwheat shows that it is high in lysine and low in methionine; the amino acid pattern of buckwheat complements the major cereal grains. The carbohydrate is mostly starch.

BUCKWHEAT PRODUCTS AND USES. The chief food use of buckwheat in the United States and Canada is griddle cakes made from the flour, often mixed with wheat flour. Buckwheat cakes are brown, palatable, and nutritious. Groats, which are sometimes sold under the name kasha, are used as breakfast food and in soups.

In Japan, 10 to 50% wheat flour is added to buckwheat flour, and the mixture is used to make Soba (buckwheat noodles). Also, the Japanese process whole buckwheat grain much like parboiled rice; i.e., the grain is soaked, steamed, dried, and then milled for removing the hulls, resulting in a product called Soba-mai. Soba-mai is cooked with cereal, similar to rice cooking.

Buckwheat grain is also fed to livestock. Likewise, middlings, the by-product obtained from milling buckwheat flour, are fed to animals. In England, buckwheat is a popular feed for pheasants.

Buckwheat is sometimes seeded for plowing under as a green manure crop. Because it has a short growing season, it is also used as an emergency food crop where earlier seeded crops have failed.

Buckwheat is also an important honey plant. It blossoms profusely over an extended period of a month or more, and the flowers contain a rich store of nectar which attracts bees. Buckwheat honey has a dark color and a pleasant flavor.

Buckwheat is also used to a limited extent for medicinal purposes. It contains 1 to 6% rutin (formerly called vitamin P) a flavonoid which is used in the treatment of certain kinds of hemorrhage, frostbite, x-ray burns, and exposure to atomic radiation.

(Also see CEREAL GRAINS, Table C-18 Cereal Grains of the World; and FLOURS, Table F-25 Major Flours.)

BUERGER'S DISEASE

This disorder, which is also known as *thromboangiitis obliterans*, is a blockage of the veins and small arteries in the feet, legs, hands, and arms by blood clots. The characteristic symptoms and signs of Buerger's disease are: (1) pain in the feet and legs which is brought on by walking (doctors call this condition intermitten claudication); (2) blanching of the skin on the limbs when they are elevated, but reddening of the skin when they hang down; (3) coldness, numbness, and tingling in the extremities; and (4) inflammation and swelling of the veins (phlebitis). Sometimes, the affected tissues develop gangrene and require amputation.

Most of the victims of Buerger's disease are men who have been heavy smokers, although a few nonsmokers develop the condition after their legs have been chilled. The disease appears to run in certain families.

Measures which slow the worsening of the condition are: (1) immediate cessation of smoking, (2) avoidance of chilling and injury to the limbs by wearing warm and protective clothing; (3) mild exercises to keep the blood vessels open, and (4) administration of vitamin E under a doctor's supervision, because this measure has relieved the pains brought on by walking.

BUFFALO BERRY *Shepherdia argentea*

The small scarlet berry of the shrub (of the family *Eleagnaceae)* which grows wild in the Great Plains and the western United States.

Fig. B-95. Buffalo berry.

Sweetening is sometimes needed to offset the sour taste of the fruit. However, the acidity is reduced somewhat after the berries have been exposed to a frost. The Indians used this fruit in mushes and stews, and the early settlers made the berries into jams and jellies.

BUFFALO HUMP

An abnormal, humplike deposit of fat under the skin of the upper back, which is seen in people who have chronic oversecretion of certain adrenal cortical hormones. The medical name for this disorder is *Cushing's Syndrome*.

(Also see CUSHING'S SYNDROME; and ENDOCRINE GLANDS.)

BUFFER

A substance in a solution that makes the degree of acidity (hydrogen-ion concentration) resistant to change when an acid or base is added. Buffers such as bicarbonate ion help to maintain neutrality in body fluids.

(Also see ACID-BASE BALANCE.)

BULIMAREXIA (BULIMIA NERVOSA; GORGE PURGE)

Bulimarexia (from the Greek words for ox and hunger), an eating disorder, is anorexia's opposite. Bulimarexia is characterized by eating binges, whereas anorexia nervosa is distinguished by self-inflicted voluntary starvation.

One typical case history reads as follows: One night, Nelli F. ate three sandwiches, crackers and dip, a jar of peanut butter, a half jar of jelly, two slices of bread with cheese and mayonnaise, a large pizza, a bowl of cereal, topped off by a candy bar, cookies, and an eclair. Then she made herself throw up.

Some bulimarectics gorge themselves four or five times a week; and they always follow each eating binge by self-induced vomiting or heavy use of laxatives. They have chosen this way to handle stress, just as alcoholics use alcohol.

Like the better known eating disorder anorexia nervosa—the *starvation disease*, almost all bulimarectics are women, many of them from homes where food was a focal point for power struggles and gibes about weight. Anorectics are generally shy, withdrawn females who develop this symptom with the onset of puberty, whereas bulimarectics tend to be extroverts who start the gorging behavior in their late teens. Anorectics are usually thin, while bulimarectics are of normal weights.

Dr. Craig Johnson, director of the Anorexia Nervosa Center at Chicago's Michael Reese Hospital and Medical Center, estimates that up to 20% of women on college campuses are involved in some degree of bulimia and purging; a study at Ohio State University produced an ever higher estimate—30% (*Time*, November 17, 1980, p. 94).

Treatment, which consists of group therapy, individual psychotherapy, or behavior modification, lessens the frequency of the binges, but, to date, cures are unknown.

It is noteworthy that bulimarectics dislike cooking for friends. The reason: They're afraid that they will eat all the food before the guests arrive.

BULIMIA

This refers to an insatiable appetite.
(Also see APPETITE.)

BULK

The undigestible residue from foods which passes unabsorbed through the intestine and acts as a stimulus for bowel movement. Usually, bulk is provided by fruits, vegetables, and whole grains, but not by dairy products, eggs, fish, meats, or poultry. However, tough connective tissue from meat may also provide bulk.
(Also see FIBER.)

BULKING AGENT

This term generally refers to nonfood materials, such as methylcellulose, which may be added to foods so that the diet contains greater amounts of undigestible matter. The purpose of this addition is to (1) provide a greater feeling of fullness so that less food is eaten; or (2) help control constipation.
(Also see CONSTIPATION; and FIBER.)

BURGOO

In times past, this word was applied to different things. Today, it is not used very extensively, but it can mean any one of the following:

• Oatmeal gruel—a thick porridge which was the mainstay of a ship's mess.

• Hardtack (giant soda crackers) and molasses cooked together, also popular ship's fare.

• A savory, highly seasoned stew or thick soup containing several kinds of meats and vegetables. It became associated with Kentucky, and, because it was easy to prepare, it was served at political rallies, group picnics, and other large gatherings, which were sometimes referred to as *burgoos.*

BURN

An injury to the skin most commonly due to the application of excessive heat. However, similar injuries may also be due to chemicals, friction, electricity, or radiation. In severe burns, the skin may be broken so that fluids escape. In such cases, special treatment is required to replace the fluid and salts which have been lost. Furthermore, prompt healing requires a diet rich in protein, calories, and vitamins.
(Also see DISEASES, section headed "Nutritional Therapies or Adjuncts.")

BURNING FEET SYNDROME

A dietary deficiency disease which is usually due to very poor diets that lack protein and the B-complex vitamins. The disease is characterized by a burning in the feet that becomes worse as the condition progresses.
(Also see DEFICIENCY DISEASES.)

BURP

Another name for a belch.
(Also see BELCHING.)

BUSHEL

A unit of capacity equal to 2,150.42 cubic inches (approximately 1.25 cu ft).
(Also see WEIGHTS AND MEASURES.)

BUTT (SIRLOIN)

• The sirloin portion of a full beef loin that has been separated from the short loin.
(Also see BEEF AND VEAL, Fig. B-39.)

• The upper end of a ham, now more correctly termed the rump portion.
(Also see PORK, Fig. P-74.)

BUTTER, BLACK

Butter that has been browned by heating, then vinegar, salt, pepper or other seasoning added; used as a sauce.

BUTTERMILK

• The residue from churning cream to make butter.

• A cultured skim milk.
(Also see MILK AND MILK PRODUCTS.)

BUTTER, MOWRAH (VEGETABLE BUTTER)

This includes a number of vegetable butters, such as cocoa butter (made from the cocoa bean), Borneo tallow or green butter (made from Malayan, an East Indian plant), shea butter (made from an African plant), and Mowrah fat or illipe butter (made from an Indian plant and used for soap and candles).

BUTTER STOOLS

Stools with a high fat content. The condition usually occurs in infants and children, and is due to a malabsorption of dietary fats. The underlying cause may be (1) a lack of pancreatic enzymes which digest fat, (2) a sensitivity to dietary gluten, or (3) a disorder of the liver which reduces the flow of bile.

(Also see DIGESTION AND ABSORPTION.)

BUTYLATED HYDROXYANISOLE (BHA)

An antioxidant (see Fig. B-96) that prevents oxidative rancidity of polyunsaturated fats, used to prevent rancidity in foods.

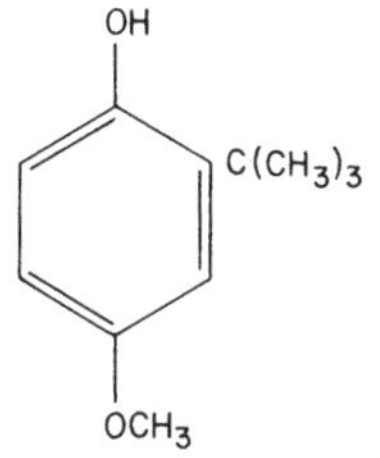

Fig. B-96. Chemical structure of butylated hydroxyanisole.

BHA is considered GRAS (Generally Recognized As Safe) by the Food and Drug Administration (FDA) for addition to edible fats and oils and to foods containing them. The FDA limitation on its use is: 0.02% (200 ppm) of fat or oil content, including essential (volatile) oil content, of the food.

(Also see ADDITIVES, Table A-3.)

BUTYLATED HYDROXYTOLUENE (BHT)

Like BHA, BHT is a phenol (see Fig. B-97), is considered GRAS, and is limited in its use to 0.02% (200 ppm) of fat or oil content, including essential (volatile) oil content, of the food.

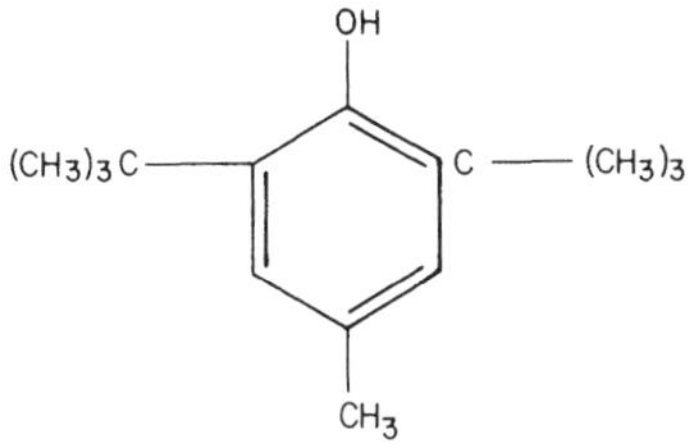

Fig. B-97. Chemical structure of butylated hydroxytoluene.

(Also see ADDITIVES, Table A-3.)

BUTYRIC ACID

One of the volatile fatty acids with the formula $CH_3CH_2CH_2COOH$, commonly found in butter and certain products of microbial action.

(Also see FATS AND OTHER LIPIDS; and MILK AND MILK PRODUCTS.)

BUTYROMEL

A home remedy consisting of a mixture of butter and honey which is given to weakened, underweight people to help them gain weight.

BY-PRODUCT

Secondary products produced in addition to the principal product; often refers to wastes which have a productive use.

Fig. B-98. Corn, leading source of U.S. by-product feeds. For each bushel of corn processed for sweeteners or ethanol, enough corn oil is produced for 2 lb of margarine and enough by-product feedstuffs are produced to feed three broilers to market weight. (Courtesy, USDA)

Which breakfast is better? Four male rats from the same litter. The two on the left had a breakfast of buttered toast and coffee. The others had oatmeal and milk. This advertisement, which was used in the early 1900s, was based on an experiment conducted by Thomas B. Osborne, of The Connecticut Agricultural Experiment Station, and Lafayette B. Mendel, of Yale University, who, in 1911, formed a brilliant partnership and pioneered in protein, mineral, and vitamin studies. (Courtesy, The Connecticut Agricultural Experiment Station, New Haven)

CABBAGE *Brassica oleracea,* family *Cruciferae*

Contents *Page*

Origin and History 285
Production 285
Processing (Sauerkraut) 287
Selection 287
Types of Cabbage 287
Quality 287
Preparation 288
Nutritional Value 288

This vegetable is one of the major crops of the world. It has been used for many centuries in Europe and Asia, although it originally differed considerably from its present form. It is a member of the mustard family *(Cruciferae).* Fig. C-1 shows a head of cabbage.

Fig. C-1. Cabbage, one of the world's leading vegetable crops. (Courtesy, USDA)

ORIGIN AND HISTORY. Our modern day heads of cabbage were developed centuries ago from the wild cabbage, which does not form heads and which bears a much closer resemblance to collards and kale than it does to domesticated cabbage.

It is not certain when the first head-forming cabbages were grown, but it is suspected that this type of plant was brought from Asia Minor into the heartland of Europe by roving bands of Celtic peoples around 600 B.C. Then, it seems likely that it was developed further in the area which later became Germany. Although people grew other cabbages around the Mediterranean in Greek and Roman times, those types did not form solid heads. However, it appears that the Italians developed the soft-headed or Savoy cabbage.

The heading cabbage spread throughout northern Europe because (1) it was adapted to grow in a cool temperate climate, (2) high yields per acre were obtained, and (3) the heads could be stored over the winter in cold cellars or similar facilities. Hence, it became very popular in Austria, Bohemia, Germany, Poland, and Russia.

Sauerkraut (fermented cabbage) helped to prevent scurvy from afflicting Dutch sailors during long voyages of exploration and trading. Later, German settlers brought sauerkraut to Pennsylvania. The long association of this form of cabbage with the Germans eventually led to the designation of German soldiers as "krauts."

At the present time, cabbages are grown in many parts of Europe, North America, and the Orient (mainly China and Japan). Some of the largest individual heads are produced in Alaska during the summer, when very long periods of daylight stimulate unusual growth.

PRODUCTION. Cabbages are among the top 2 dozen vegetable crops of the world, with an estimated annual production of about 39 million metric tons. Fig. C-2 shows where most of the world's cabbage is produced.

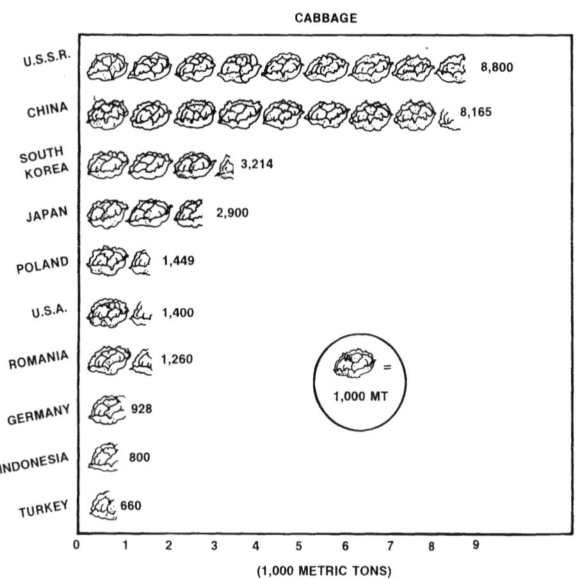

Fig. C-2. The world's leading cabbage-producing countries. (Data from *FAO Production Yearbook,* 1990, FAO/UN, Rome, Italy, Vol. 44, p. 128, Table 50)

The U.S. production of cabbages averages about 1.4 million metric tons per year. Fig. C-3 shows the eight leading cabbage markets of the United States.

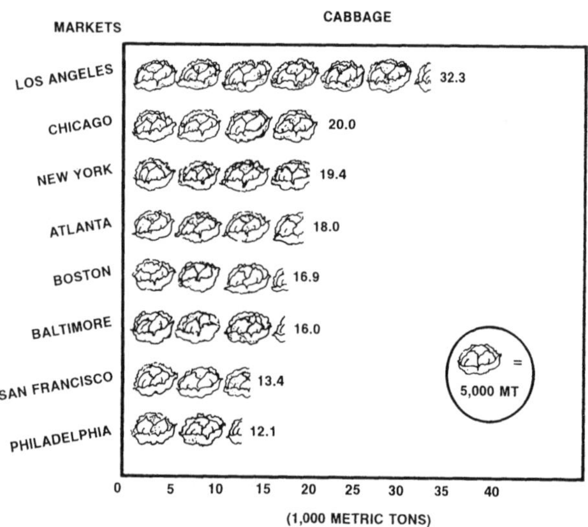

Fig. C-3. Cabbage arrivals at 8 markets in 1990. (Data from *Agricultural Statistics*, 1991, USDA, pp. 167, 168, Tables 242, 243)

Fig. C-4 shows the U.S. monthly cabbage shipments by rail, truck, and air. Note that the vast majority of cabbage is marketed during the cool months.

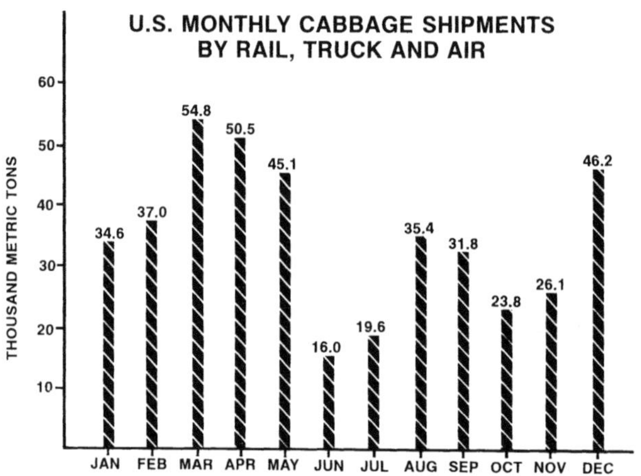

Fig. C-4. Monthly cabbage shipments by rail, truck, and air, 1990. (Data from *Agricultural Statistics*, 1991, USDA, p. 169, Table 244)

Cabbage grows best in a cool, moist climate. However, different varieties have been developed for production under various climatic conditions. Nevertheless, extremes of temperature should be avoided because (1) temperatures under 50° F (*10° C*) may cause the plant to "bolt" (go to seed), and (2) the heads may split if the weather is too warm. Therefore, winter crops are grown mainly in the southern states and California, whereas summer crops are grown in the northern states. Cabbage requires considerable fertilization (nitrogen and potash) throughout its growing period because of the large amount of vegetable matter that is produced. Fig. C-5 shows growing cabbage plants.

Some of the early crop which is grown in the warmer states may be harvested while the heads are still small because good prices are obtained when the supply of the fresh vegetable is limited. On the other hand, the heads

Fig. C-5. Cabbage growing in a field. (Courtesy, Union Pacific Railroad Company, Omaha, Neb.)

of the late crop may be allowed to grow quite large so that they may be used for making sauerkraut. Fig. C-6 shows the harvesting of cabbage.

Fig. C-6. California farm workers harvesting cabbage near Oxnard. (Courtesy, Union Pacific Railroad Company, Omaha, Neb.)

PROCESSING (Sauerkraut). About one-sixth of the cabbage produced in the United States is made into sauerkraut, while most of the rest is marketed fresh. Sauerkraut is made by mixing shredded cabbage with salt, then allowing lactic acid fermentation to occur. A considerable amount of the sauerkraut now sold in supermarkets is canned, although some is also sold in sealed plastic bags kept under refrigeration.

(Also see SAUERKRAUT.)

SELECTION. The use of fresh cabbage in the United States declined from 23.2 lb (*11 kg*) per capita in the 1920s to 8.2 lb (*4 kg*) per capita in 1976.[1] Perhaps, part of the decline in cabbage consumption was due to the selection of poor quality heads by consumers. Therefore, some guidelines for selecting fresh cabbage follow.

Types Of Cabbage. Several types of cabbage are marketed during the year. The general types are pointed, Danish, domestic, Savoy, and red.

- **Pointed type** that develops conical or pointed heads usually includes the crop sold as early or "new" cabbage.

- **Danish type** includes varieties that mature late in the season and develop hard, tight-leaved, compact heads. Danish-type cabbage, even after trimming, appears tight and smooth-leaved around the top portion; when viewed from the stem end, it appears circular and regular in outline. This type of cabbage is most suitable for winter storage. The color is usually white, especially during winter months.

- **Domestic type** includes varieties that develop heads less compactly formed than those of the Danish type, and which are either round or flat in shape. Early, midseason, and medium-late varieties of the domestic type are produced.

- **Savory type** comprises the finely-crumpled-leaved varieties with round or "drumhead" shaped heads.

- **Red type** includes all red varieties.

Early (or new) cabbage shipped from southern areas during the winter months is frequently harvested before the heads are firm and is not trimmed as closely as later production. New cabbage that has not been properly handled soon loses its desirable freshness.

Freshness of appearance in late or main-crop cabbage is as important as in the early crop. Late cabbage out of storage may be trimmed down to completely white heads and still be entirely desirable for most purposes.

Quality. Prime quality heads of cabbage should be reasonably solid, hard, or firm; heavy or fairly heavy in relation to size, and closely trimmed (stems cut close to the head and only three or four outer or wrapper leaves remaining). Early cabbage need not be as solid or firm

---

[1]Brewster, L., and M. F. Jacobson, *The Changing American Diet*, Center for Science in the Public Interest, Washington, D.C., 1978, p. 51.

as late crop cabbage. Fig. C-7 shows a basket full of high-quality heads of cabbage.

Fig. C-7. Cabbage heads of the "Golden Acre" variety. (Photo by J. C. Allen & Son, Inc., West Lafayette, Ind.)

Worm injury, yellowing of leaves, bursted heads, and decay are the most common defects of cabbage and may be easily detected. Cabbage seriously affected by any such defects should be avoided, but if heads are only slightly affected they may be trimmed and utilized. Considerable waste can generally be expected in heads of cabbage with yellow or discolored leaves.

Examination of a head of cabbage may sometimes reveal the base of some outer leaves separated from the stem with the leaves held in place only by the natural folding over the head. Such heads may be undesirably strong in flavor, or course in texture. Soft or puffy heads are usually poorer in quality than firm or hard heads.

PREPARATION. Cooked cabbage goes well with a wide variety of meats, potato dishes, poultry, and other vegetables. The usual ways of cooking cabbage are boiling, frying, sauteing, and stuffing and baking. However, overcooking drives off the pleasant odorous components and intensifies the unpleasant ones. It also reduces the nutritional value greatly. Hence, cabbage should be added near the end of the cooking period when it is to be part of mixed dishes such as casseroles, soups, and stews. Many of these cabbage dishes were invented by the peoples of Europe. Brief descriptions of some typical preparations follow:

• **British "bubble and squeak"**—A large quantity of shredded cabbage is sauteed with thin slices of leftover meat.

• **German red cabbage and apples**—The shredded red cabbage is cooked with apple sections browned in butter, onions, beef stock and various seasonings.

• **Hungarian stuffed cabbage**—Large, blanched cabbage leaves are stuffed with a meat and rice mixture, then baked in a casserole. As with other Hungarian dishes, paprika is used as one of the seasonings.

• **Irish corned beef and cabbage**—Hot corned beef is served with boiled cabbage and boiled potato.

• **Portuguese cabbage and potato soup**—Finely shredded cabbage is cooked briefly in a soupy mixture of mashed potatoes and seasonings.

• **Cole slaw**—Raw cabbage is often shredded and made into cole slaw, which is a mixture of the vegetable with French dressing, mayonnaise, salad dressing, sour cream, and/or vinegar. Usually, sugar and other seasonings are added to the slaw.

• **Sauerkraut (fermented cabbage)** has long been served as an accompaniment to meat dishes such as pig knuckles, roasts, and sausages. The latter combination is best exemplified in America by the frankfurters and sauerkraut that are served from small, umbrella-shaded vending wagons on busy street corners and by the vendors who scurry around ballparks during baseball games.

CAUTION: Normally, cabbage is a wholesome, nutritious food. But, like many other foods, it contains components which may be harmful if certain precautions are not taken. These follow:

• **Goiter-causing (goitrogenic) substances**—Cabbage and other vegetables of the mustard family (*Cruciferae*) contain small amounts of substances called goitrogens that interfere with the utilization of iodine by the thyroid gland. As a result, the thyroid gland becomes enlarged (a goiter is formed) as the body attempts to compensate for the lack of thyroid hormones. It appears that the goitrogenic effects of cabbage may be offset by ample amounts of dietary iodine, which is furnished by iodized salt, ocean fish, seafood, and seaweeds.

• **Laxative effects**—Sauerkraut has a moderate to strong laxative effect because the fermentation produces acetylcholine, a substance that mediates the impulses from certain types of nerves. These nerves stimulate both the secretion of digestive juices and the movements of the digestive tract. Unfermented cabbage has only a mild to moderate laxative effect which is due mainly to its content of fiber.

• **Selenium content of cabbages grown on selenium-rich soils**—One study showed that cabbages grown on high-selenium soils, such as those in Wyoming and the Dakotas, accumulated 150 ppm of selenium.[2] If eaten in sufficient quantity, this level could cause poisoning. Fortunately, cabbage is seldom consumed in such large quantities; besides, most of the U.S. cabbage crop is grown on low-selenium soils.

NUTRITIONAL VALUE. The nutrient composition of cabbage is given in Food Composition Table F-36.

[2]Friedman, L., and S. I. Shibko, "Adventitious Toxic Factors in Processed Foods," *Toxic Constituents of Plant Foodstuffs*, edited by I. E. Liener, Academic Press, New York, N.Y., 1969, p. 351.

Some noteworthy observations regarding the nutrient composition of cabbage follow:

1. Cabbage is high in water content (over 90%) and low in calories (less than 35 Calories [kcal] per cup [*240 ml*]). However, most types of cole slaw contain about five times as many calories as cabbage alone because the slaw usually contains a dressing that is rich in fat. Hence, dieters would be wise to make their own cole slaw with a low-calorie dressing.
2. Sauerkraut is about as nutritious as the various forms of unfermented cabbage, except that the former has a high sodium content.
3. All forms of cabbage are good sources of potassium.

• **Antiulcer factors**—Starting in the 1940s and continuing into the 1950s, the American medical researcher Cheney and his co-workers demonstrated that the juice from raw cabbage speeded up the healing of peptic ulcers.[3] The unidentified therapeutic substance was called *vitamin U*, because of its effect on ulcers. Recently, it was found that vitamin U contains the amino acid methionine, but that methionine alone does not have this effect.

The problem in treating people with cabbage juice is that the antiulcer factor is present in only minute amounts in cabbage, so that it is necessary to drink about l qt (*0.95 liter*) of fresh cabbage juice (the equivalent of about 3 lb [*l.35 kg*] of raw cabbage) per day in order to obtain a noticeable improvement in healing. Therefore, scientists have developed various nondestructive means of concentrating the juice. (Heat destroys "vitamin U.")

CACHEXIA

A term which indicates a severe weight loss and wasting of body tissues—a general lack of nutrition. Frequently, cachexia occurs in the course of a chronic disease; for example, many forms of cancer.

CADMIUM

A widely used industrial metal which may get into air, food and water; and ultimately into the human body where it may have toxic effects. It appears that regions which have high levels of cadmium in the environment are likely to have greater than normal incidences of high blood pressure because cadmium causes damage to the kidneys. Fortunately, certain dietary minerals, such as selenium and zinc, may counteract the effects of cadmium.

(Also see MINERAL[S].)

CAFFEINE ($C_8H_{10}N_4O_2$)

Caffeine is a drug with primarily stimulating effects. It is found in foods, beverages, and medicines and it occurs naturally in plant products such as coffee, tea, cacao beans, kola nuts, mate drink, and guarana paste. More than 63 species of plants growing in all parts of the world contain caffeine in their leaves, seeds, or fruit. Pure caffeine is obtained (1) as a by-product from the manufacture of decaffeinated coffee, (2) from the extraction of coffee bean and tea leaf waste, and (3) from the methylation of theobromine obtained from cocoa waste. Most Americans consume some caffeine. Recently, there has been considerable concern as to the effect of caffeine on health.

CHEMISTRY AND METABOLISM. Caffeine is an organic chemical belonging to a class of chemicals called purines—some of which are constitutents of the nucleic acids, RNA and DNA. The proper chemical name for caffeine is 1, 3, 7,-trimethylxanthine and its chemical structure is shown in Fig. C-8. Pure caffeine is a white odorless powder with a bitter taste. Theophylline and theobromine are chemical cousins of caffeine, differing only in the number and placement of the methyl groups (CH_3), but possessing some of the same drug actions.

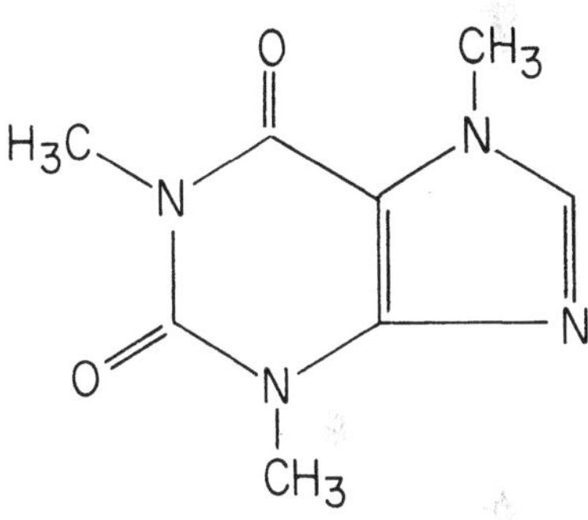

Fig. C-8. The structure of caffeine—1, 3, 7, -trimethylxanthine.

Ingested caffeine is rapidly absorbed from the intestine and within a few minutes caffeine enters all organs and tissues. Within one hour after ingestion, it is distributed in the body tissues in proportion to their water content. The metabolic half-life of caffeine is about 3 hours; hence, there is no day-to-day accumulations as it almost completely disappears from the body overnight. Most of the ingested caffeine is excreted in the urine, and the major excretory product in man is 1-methyl uric acid formed by the demethylation (removal of CH_3 groups) of caffeine.

EFFECTS ON THE BODY. A pharmacologically active dose of caffeine is about 200 mg, depending upon the individual. In the body, caffeine demonstrates the following actions:

1. Stimulates the central nervous system (brain) thereby prolonging wakefulness with alert intellectual facilities.
2. Stimulates the heart action.
3. Relaxes smooth muscle—the type of muscle found in the digestive tract and blood vessels.

[3]Salunkhe, D. K., S. K. Pao, and G. G. Dull, "Assessment of Nutritive Value, Quality, and Stability of Cruciferous Vegetables During Storage and Subsequent to Processing," *Storage, Processing, and Nutritional Quality of Fruit and Vegetables*, edited by D. K. Salunkhe, et al., CRC Press, Inc., Cleveland, Ohio, 1974, p. 23.

4. Increases urine flow—a *diuretic*; and increases the urinary calcium losses, which may contribute to the development of osteoporosis.

5. Stimulates stomach acid secretion.

6. Increases muscle strength and the amount of time a person can perform physically exhausting work.

While the effects of caffeine vary from person to person, a dose of 1,000 mg or more will generally produce adverse effects such as insomnia, restlessness, excitement, trembling, rapid heart beat with extra heart beats, increased breathing, desire to urinate, ringing in the ears, and heartburn. A fatal dose of caffeine appears to be more than 10 g or 170 mg/kg of body weight—about 80 to 100 cups of coffee in one sitting.

SOURCES AND CONSUMPTION. In the United States, about 75% of all the caffeine consumed as food comes from coffee while 15% comes from tea and the remaining 10% comes from soft drinks, cocoa, and chocolate. Additionally, many over-the-counter medications contain caffeine. Table C-1 shows the caffeine content of some sources.

TABLE C-1
CAFFEINE CONTENT IN FOOD AND DRUGS

Item	Measure[3]	Caffeine (mg)
Percolated roasted and ground coffee[1]	cup	76-155
Instant coffee	cup	66
Decaffeinated coffee	cup	2-5
Tea[1]	cup	20-100
Instant tea	cup	24-131
Hot cocoa	cup	5
Soft drinks[2]	12 oz can	26-34
Milk chocolate candy	2 oz	12
Sweet or dark chocolate	2 oz	40
Baking chocolate	2 oz	70
Alertness tablets	tablet	100-200
Pain relievers	tablet	32-65
Cold allergy relief remedies	tablet	15-32

[1]The longer coffee or tea is brewed, the greater its caffeine content.
[2]Includes primarily cola and pepper drinks, but other flavors of soft drinks may also contain caffeine.
[3]One cup = 240 ml; one oz = 30 ml

Over 30 million lbs (*13.5 million kg*) of caffeine are consumed annually, and almost every age group consumes some caffeine. In general, daily consumption increases with age until about 60, and then declines. Men between the ages of 40 and 49 consume more caffeine than any age group, male or female. One survey has shown the average intake of caffeine from coffee to be about 186 mg or 2 cups (*480 ml*) of coffee, but some individuals within the survey drank as much as 8 cups of coffee or about 676 mg of caffeine. Some individuals may even consume more. Users of the over-the-counter drugs may receive substantially more than coffee drinkers.

HEALTH CONCERNS. In the amounts normally consumed, caffeine acts as a drug, indeed much of its popularity is owed to its stimulant effect on the central nervous system. Like many drugs, users develop a dependence on caffeine when consumed in amounts equivalent to more than 4 cups of coffee per day. Furthermore, abstinence from caffeine-containing substances for a day or two may cause the development of withdrawal symptoms—headaches, irritability, restlessness, or fatigue. Everyone responds to any drug differently and some individuals using caffeine may experience mood changes, anxiety, depression, and irritability. Large amounts of caffeine may cause heartburn, upset stomach, and irregular heartbeats.

Over the years, caffeine has been implicated in (1) birth defects, (2) heart disease, (3) high blood pressure (hypertension), (4) cancer, (5) ulcers, and (6) breast disease.

1. **Birth defects.** Numerous animal studies since 1946 have demonstrated in a variety of animals that large doses of caffeine can cause birth defects. However, the amounts of caffeine fed to these animals was equivalent to 20 to 100 cups (*5 to 24 liter*) of coffee per day. Human epidemiologic studies of caffeine consumption and birth defects fail to support the findings of the animal studies. Although more evidence is needed, there does seem to be sufficient cause to caution pregnant women, or soon to be pregnant women, about their caffeine consumption. Indeed, the Food and Drug Administration counsels pregnant women to avoid caffeine-containing foods and drugs.

2. **Heart disease.** Since 1864, there has been a debate as to whether or not caffeine causes heart disease. Although some studies conducted in the early 1970s suggested that heavy coffee drinking was a causative factor in heart disease, it was later found that these studies failed to take into account the fact that heavy coffee drinkers were often heavy smokers. To date, the results of animal and human studies provide no evidence that caffeine consumption is a causal factor in heart attacks.

3. **High blood pressure.** In sensitive individuals, caffeine consumption can temporarily elevate blood pressure. However, in regular users 2 to 3 cups of coffee at a time does not increase blood pressure.

4. **Cancer.** Caffeine *per se* has not been shown to cause cancer. However, coffee drinking has been implicated in the development of bladder cancer and cancer of the pancreas. Epidemiologic studies since 1962, have demonstrated no direct relationship between coffee consumption and bladder cancer. The link between pancreatic cancer and coffee consumption awaits further testing. Oddly enough, caffeine seems to have anticancer properties since in experimental settings it seems to increase the effectiveness of radiation and drug therapy by preventing cancer cells from repairing themselves. Further testing of this indication is needed.

5. **Ulcers.** Two large epidemiologic studies involving more than 60,000 people failed to establish a significant risk of developing ulcers from caffeine consumption. Persons with ulcers can, however, have their situation aggravated since caffeine stimulates acid secretion and relaxes muscles and blood vessels of the digestive system.

6. **Breast disease.** Cystic breast disease—a noncancerous condition—affects about 50% of the American women, most under age 45. The evidence linking cystic breast disease to caffeine consumption comes from one study involving about 50 women. There is no conclusive evidence to link cystic breast disease to caffeine consumption. More research is needed.

Overall, caffeine, as consumed in foods, beverages, and over-the-counter drugs does not seem to represent a threat to the health of most Americans. However, some sensitive individuals who consume large amounts may experience insomnia, chronic headaches, rapid heart beat, anxiety, and upset stomach.

(Also see COFFEE; SOFT DRINKS; and TEA.)

CALAMONDIN *Citrus mitis; C. madurensis*

A small citrus fruit (of the family *Rutaceae*) that looks like a miniature orange and is native to China.

Fig. C-9. The calamondin, a citrus fruit that is grown mainly as an ornamental plant.

The calamondin appears to be a three-species hybrid of the kumquat (*Fortunella* species), sweet orange (*C. sinensis*), and the trifoliate orange (*Poncirus trifoliata*). Although the sour fruit has little commercial value, the tree is grown extensively in the United States and Europe as an ornamental plant in pots and tubs. It is noteworthy that the fruits may be used to make marmalade, or the juice may be sweetened and used to make various drinks.

CALCEMIA (HYPERCALCEMIA)

An abnormally high level of calcium in the blood. This condition may be dangerous because it can lead to calcium deposits in various soft tissues. The condition is also called hypercalcemia.

(Also see CALCIUM, section headed "Calcium Related Diseases.")

CALCIFEROL

Another name for vitamin D_2 or ergocalciferol. Both vitamin D_2 and vitamin D_3 have equal activity for people, but chickens, turkeys, and other birds utilize vitamin D_3 more efficiently than vitamin D_2.

(Also see VITAMIN D.)

CALCIFICATION

The process by which organic tissue becomes hardened by a deposit of calcium salts.

CALCITONIN (OR THYROCALCITONIN)

A hormone secreted by the parafollicular or C cells of the thyroid gland. (The parafollicular or C cells of the mammal are of ultimobranchial origin.) When the calcium level of the blood rises above normal, calcitonin is secreted, which prohibits further release of calcium from the bones. Thus, the effect of this hormone is opposite to that of the parathyroid hormone which removes calcium from bone. However, it appears that the action of calcitonin occurs mainly during the growing years and that it does not have much effect in adulthood.

(Also see CALCIUM; PHOSPHORUS; and ENDOCRINE GLANDS.)

CALCIUM (Ca)

Contents *Page*

History . . . 292
Absorption of Calcium . . . 292
Metabolism of Calcium . . . 294
Excretion of Calcium . . . 295
Functions of Calcium . . . 296
Deficiency Symptoms . . . 296
Interrelationships . . . 297
Recommended Daily Allowance of Calcium . . . 297
Toxicity . . . 300
Calcium Related Diseases . . . 300
Calcium-Phosphorus Ratio and Vitamin D . . . 301
Calcium Losses During Processing, Cooking, and Storage . . . 301
Sources of Calcium . . . 301
Top Calcium Sources . . . 302

Everyone needs calcium, which constitutes about 2% of the body. Calcium gives strength and structure to bones and teeth. Additionally, it controls the heartbeat—the heart could not keep up its continuous alternate contraction and relaxation were it not for calcium; it has a role in the transmission of nerve impulses; it is related to muscle contraction—without calcium, muscles lose their ability to contract; it is necessary for blood clotting—calcium prevents fatal bleeding from any break in the wall of a blood vessel; and it activates a number of the enzymes, including lipase—the fat-splitting enzyme.

Ninety-nine percent of the calcium of the body is present in the bones and teeth, where calcium salts (chiefly calcium phosphate) held in a cellular matrix provide the rigid framework of the body. The bones also furnish the reserves of calcium to the circulation so that the concentration in the plasma can be kept constant at all times. It has been estimated that in an adult man about 700 mg of calcium enter and leave the bones each day. Teeth are similar to bone in chemical composition, but, in comparison with bone, enamel is much harder and is lower in water content—containing only about 5%. The calcium in teeth, unlike that in bone, cannot be replaced; therefore, teeth cannot repair themselves.

CALCIUM, builds bones, teeth, and is needed by all tissues of the body

Rats from same litter 22 weeks old.

This rat did not have enough calcium. Note the short, stuffy body, due to poorly formed bones. It weighed 91 g.

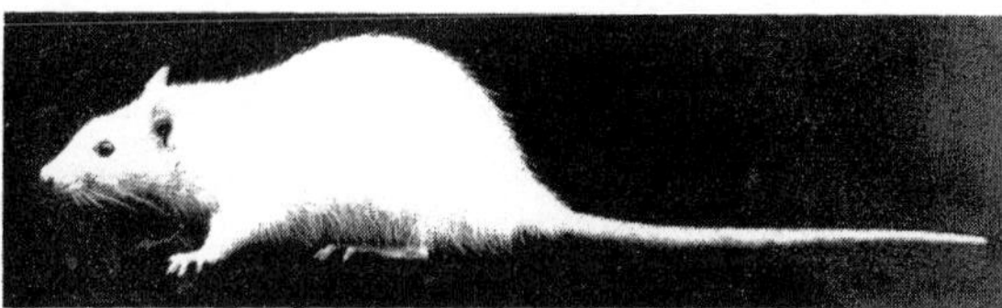

This rat had plenty of calcium. It has reached full size, and its bones are well-formed. It weighed 219 g.

TOP FOOD SOURCES

Fig. C-10. Calcium made the difference! *Left*: Rat on calcium-deficient diet. *Right*: Rat that received plenty of calcium. Note, too, top sources of calcium. (Adapted from USDA sources.)

The remaining 1% of the calcium—about 10 g in an adult—is widely distributed in the soft tissues and the extracellular fluids. The vital role of this small amount of calcium is reflected in the precision with which plasma calcium is regulated, a narrow range of 9 to 11 g that is controlled by the parathyroid hormone and calcitonin.

Obviously, all of the calcium in the body comes from food. Because this element is both a body builder and a body regulator, everyone needs dietary calcium.

HISTORY. Calcium was among the first materials known to be essential in the diet. As early as 1842, Chossat, a Frenchman, showed experimentally that pigeons developed poor bone on a diet low in calcium. When fed wheat alone, the birds died after 10 months; and, on autopsy, the bones were found very much depleted. Calcium carbonate prevented the trouble. In later studies, Chossat also used chickens, rabbits, frogs, eels, lizards, and turtles. But well-controlled experiments with man were not made until recent years.

ABSORPTION OF CALCIUM. Calcium salts are more soluble in acid solution; hence, absorption occurs largely in the upper part (proximal part, or the duodenal area) of the small intestine, where the food contents are still somewhat acidic following digestion in the stomach. An increase in the passage of food through the gastrointestinal tract also decreases the percentage of absorption.

Not all the calcium in food becomes available to the body. Normally, depending on the intake, only 20 to 30% of the calcium in the average diet is absorbed from the intestinal tract and taken into the bloodstream. Calcium absorption is dependent upon the calcium needs of the body, the type of food, and the amount of calcium ingested. Growing children and pregnant-lactating women utilize calcium most efficiently—they absorb 40% or more of the calcium in their diets. Also, the body's need is greatest and the absorption of calcium is relatively more efficient following long periods of low calcium in-

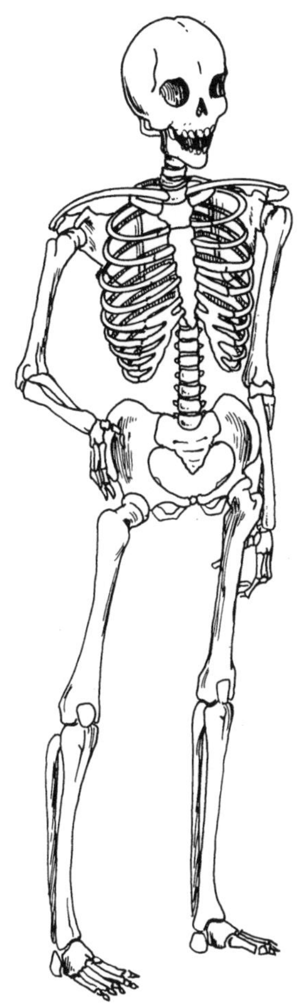

Fig. C-11. Ninety-nine percent of the calcium of the body is in the bones and teeth. A 154-lb (*70 kg*) man's body contains approximately 2.64 lb (*1,200 g*) of calcium—more than any other mineral.

take and body depletion and during healing of bone fractures.

It follows that utilization of calcium is much more efficient in countries where the diet is low in calcium than in the United States where diets are high in calcium.

Thus, the relative amounts of calcium retained in the body vary according to (1) the age and the pregnancy-lactation status of the person, (2) the previous dietary habits, and (3) the level of current supply.

By increasing the absorptive capacity of the gut and by regulating the renal (kidney) excretion, the body can either adapt itself to (1) reduced dietary intakes of calcium, or (2) increased requirements for calcium.

A number of dietary factors in addition to need and the amount of calcium ingested influence calcium absorption; some enhancing it, other interfering with it.

• **Dietary factors enhancing absorption of calcium**—The following dietary factors increase the absorption of calcium:

1. **Vitamin D.** One of the most important factors affecting calcium absorption is an adequate supply of vitamin D, whether from the diet or exposure to ultraviolet radiation of the sun. Vitamin D or its derivative (metabolite), 25-hydroxycholecalciferol (25-HCC), increases calcium absorption by inducing synthesis of a calcium-binding protein that facilitates transport of the calcium through the intestinal walls.
2. **Protein.** Dietary protein increases the rate of calcium absorption from the small intestine. The probable explanation of this phenomenon is that the amino acids, especially lysine and arginine, liberated in the course of protein digestion, form soluble calcium salts which are easily absorbed. But any advantage from increased absorption will likely be more than counterbalanced by the increased urinary loss of calcium on high-protein diets.
3. **Lactose.** Dietary lactose (milk sugar) enhances the rate of calcium absorption from the small intestine. In this connection, it is noteworthy that (1) milk is the only source of lactose, and (2) the improved absorption of calcium is dependent on the activity of intestinal lactase, the enzyme that hydrolyzes lactose.
4. **Acid Medium.** Absorption of calcium is favored in an acid medium (a lower pH) because it keeps the calcium in solution; hence, most of the absorption occurs in the duodenum.

• **Dietary factors interfering with absorption of calcium**—The following dietary factors interfere with the absorption of the calcium:

1. **Vitamin D deficiency.** Insufficient levels of vitamin D depress the amount of the calcium-binding protein that is essential for the absorption of calcium. Thus, in northern latitudes and/or in smoggy cities where ultraviolet radiation is limited or blocked, the dietary source of vitamin D becomes very important.
2. **Calcium-phosphorus imbalance.** A great excess of either calcium or phosphorus interferes with the absorption of both minerals and the increased excretion of the lesser mineral. This is why a certain ratio between them in the diet is desirable. The Ca:P ratio in U.S. diets has been estimated to be 1:1.5 to 1:1.6, while the most desirable ratio is thought to be 1.5:1 in infancy, decreasing to 1:1 at 1 year of age, and remaining at 1:1 throughout the rest of life; although man can tolerate a much wider Ca:p ratio—between 2:1 and 1:2.
3. **Phytic acid.** Phytic acid, found in the outer hulls (bran) of many cereal grains, forms an insoluble salt with calcium—calcium phytate—which prevents the absorption of calcium. But the effect of phytic acid is important only when whole grain cereals comprise a major part of the diet and/or the calcium intake is low.

   Phytin can be split by the enzyme phytase, which has been identified in several cereal grains. The presence of this enzyme may explain why calcium is more available in leavened than in unleavened breads.
4. **Oxalic acid.** This compound can inhibit the absorption of calcium because of the formation of calcium oxalate, a relatively insoluble compound. Oxalic acid is high in only a few foods; among them, spinach, beet tops, swiss chard, cocoa, and rhubarb. But the amount of oxalic acid present in typical American diets is not sufficiently great to interfere seriously with the absorption of calcium.

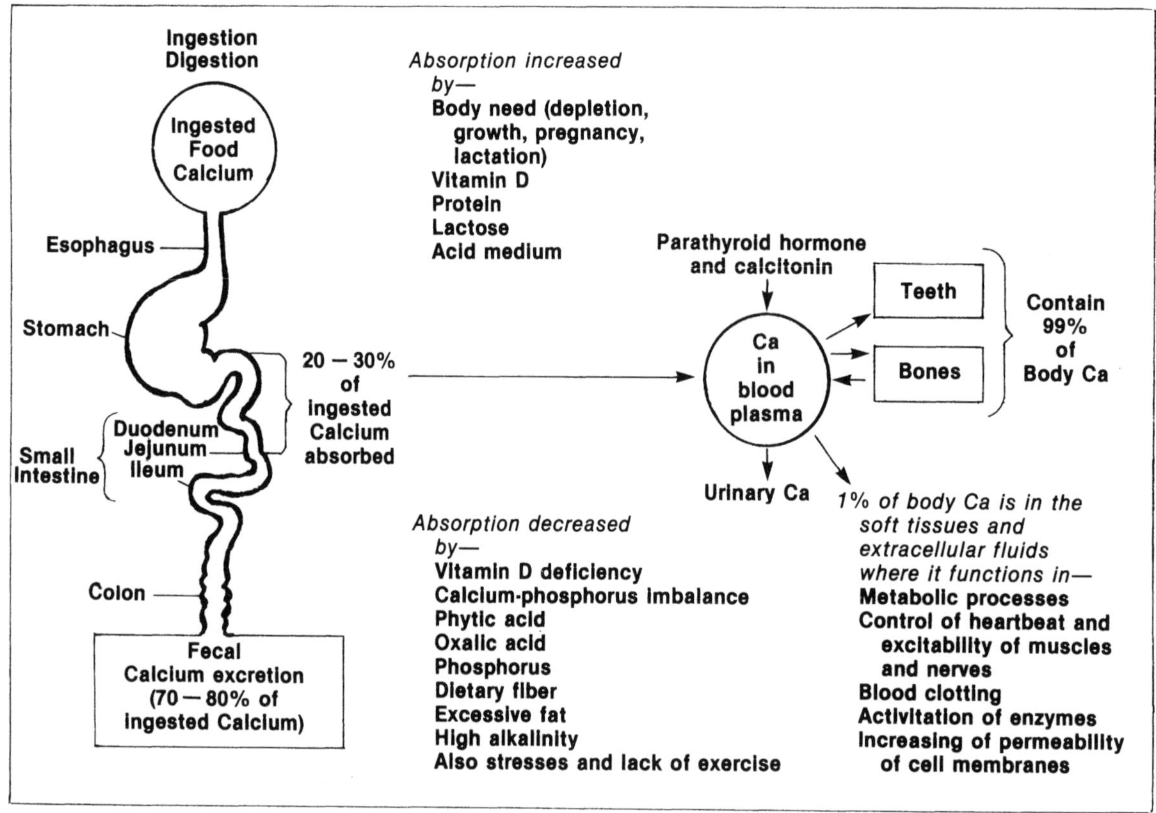

Fig. C-12. Calcium utilization. Note that healthy adults absorb only 20 to 30% of the calcium contained in their food, and that 70 to 80% is excreted in the feces. Note, too, the factors that increase and decrease absorption.

5. **Dietary fiber.** The fiber in plants low in phytate binds calcium in proportion to its uronic acid content. Since uronic acids can be digested by bacteria in the colon, this may be part of the process whereby calcium absorption is increased slowly after a change from a low-fiber to a high-fiber diet.

6. **Excessive fat.** Excessive levels of fat, especially those that are saturated, depress calcium absorption because the fats combine with calcium to form insoluble soaps, a process called *saponification.* These insoluble soaps are excreted in the feces, with consequent loss of the incorporated calcium. (They may also carry with them fat-soluble vitamin D.) This explains why patients with chronic intestinal disorders, such as sprue and celiac disease, leading to increased fat in the feces (steatorrhea) may develop osteomalacia in due time.

7. **High alkalinity.** Calcium is insoluble in an alkaline medium, and is poorly utilized under such conditions.

8. **Other factors.** Stresses and lack of exercise have an important bearing upon the calcium balance; several studies have shown that persons who are under extreme nervous strain or worry or who do not exercise sufficiently have negative calcium balances even when the dietary intake is good. Also, aging appears to decrease the rate of calcium absorption.

METABOLISM OF CALCIUM. Fig. C-13 shows the factors affecting the metabolism of calcium. Normally, the small intestine acts as an effective control and prevents an excess of calcium from being absorbed. Body requirement is the major factor governing the amount of calcium that will be absorbed; growing children and pregnant and lactating women absorb 40% or more of the calcium in their diets.

After absorption through the intestinal wall, most of the calcium is stored in the bones, especially the spongy bone (trabeculae), from which it is withdrawn in time of need.

(Also see BONE, including Figs. B-71 and B-72.)

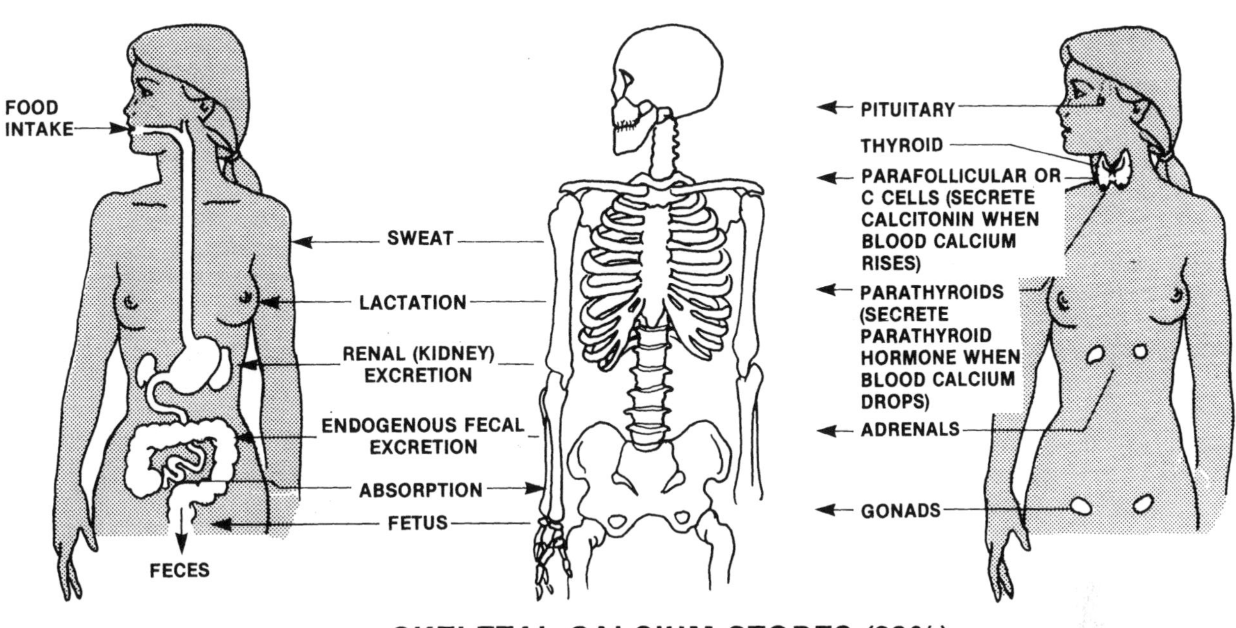

Fig. C-13. Factors involved in calcium metabolism. To maintain a constant extracellular fluid calcium concentration, calcium is excreted through the kidneys and gut when the concentration increases, or is recruited by increased absorption from the gut or resorption from the skeleton when the concentration decreases. These movements of calcium are regulated by the endocrine system. Calcium is also lost through sweating, pregnancy, and lactation.

• **The endocrine glands and hormones**—The calcium concentration in the blood is kept relatively constant primarily by two endocrine glands—the parathyroids which secrete parathyroid hormone, in response to *hypocalcemia* (low blood calcium), and the parafollicular or C cells of the thyroid gland which release calcitonin during *hypercalcemia* (high blood calcium). (The parafollicular or C cells of the mammal are of ultimobranchial origin.) Secretion or release of both hormones (parathyroid hormone and calcitonin) is determined by the level of calcium in the blood. When the calcium level of the blood drops, the parathyroid releases its hormone, which acts in three ways to restore the normal calcium level: (1) it increases calcium absorption from the intestine; (2) it withdraws calcium from the bone; and (3) it causes the kidney to lessen excretion of calcium in the urine. When the calcium level of the blood rises, the secretion of the parathyroid hormone is decreased and calcitonin, which prohibits further release of calcium from the bones, is secreted.

The pituitary growth hormone may influence calcium metabolism indirectly through its action on longitudinal bone growth. The adrenocortical hormones control calcium movements at three sites: (1) at the renal level—calcium excretion is increased; (2) at the gastrointestinal level—calcium absorption may be inhibited; and (3) at the skeletal level—both bone formation and bone resorption are inhibited, with the greater effect on bone formation. The role of the sex hormones in calcium metabolism is less clear, but the major evidence to date suggests that neither androgens nor estrogens have much effect on calcium metabolism.

EXCRETION OF CALCIUM. The quantity of calcium excreted by adults in good nutritional state tends to equal the intestinal absorption. The body disposes of the calcium that it does not need in three ways—in the feces, in the urine, and in the sweat. Also, about 30 g of calcium is passed from the mother to the fetus during gestation; and during lactation the mother excretes as much as 250 mg of calcium daily in her milk.

• **Calcium in the feces**—Calcium is excreted mainly in the feces, most of which is dietary intake that is not absorbed (see Fig. C-13). The remainder, called endogenous fecal calcium (which ranges between 125 and 180 mg per day) comes from shed epithelial cells and the digestive juices (bile and pancreatic juice). As shown in Fig. C-12, approximately 70 to 80% of food calcium is unabsorbed and excreted in the feces.

A recent study suggests that excessive intakes of

phosphorus may lead to increased bone resorption and increased calcium loss in the feces.

• **Calcium in the urine**—Urinary calcium excretion varies widely among individuals, ranging anywhere from 100 to 200 mg per day. But, under most conditions, it appears to be relatively constant for any given individual. Major changes in calcium intake produce little change in the quantity of urine calcium, whereas fecal calcium, by contrast, is highly correlated with calcium intake, indicating that the gut exercises considerable control over calcium absorption.

The amount of calcium excreted in the urine is related to skeletal size, the acid-base regulation of the body, and the dietary protein intake. Urinary excretion of calcium rises when dietary protein is increased and falls when dietary protein is decreased. It appears that calcium losses can be substantial when protein intake is high; hence, if this type of diet is continued for a prolonged period, it could result in a considerable loss of body calcium and even osteoporosis. However, studies show that a high protein intake from a high meat diet has little effect on calcium excretion, possibly because of the high phosphate intake with the meat diet. A recent studys suggests that increased phosphorus intakes reduce urinary excretion of calcium and lower serum calcium levels.

• **Calcium in the sweat**—Normally, the loss of calcium in perspiration amounts to only 15 mg per day and is insignificant. However, persons working in extreme heat may lose over 100 mg of calcium per hour in the sweat—which may approximate 30% of the total calcium output.

• **Calcium losses in pregnancy and lactation**—In the female, there are two other possible losses of calcium. Apporoximately 30 g of calcium is deposited in the fetal skeleton, and during lactation, 250 mg of milk calcium per day may be secreted in the milk.

FUNCTIONS OF CALCIUM. The physiologic function of 99% of the calcium in the body is to build the bones and teeth, and to maintain the bones. The remaining 1% of the calcim controls the following vital physiologic functions:

• **Blood clotting**—Calcium is necessary for blood clotting; it serves as a catalyst in the conversion of prothrombin to thrombin. Thrombin then ploymerizes fibrinogen to fibrin, causing blood clotting.

• **Muscle contraction and relaxation**—Calcium has a vital role in muscle contraction and relaxation. Other elements, such as magnesium and potassium, are also involved in the process. The catalyzing action of calcium ions is particularly important in the contraction-relaxation cycle of the heartbeat.

• **Nerve transmission**—Calcium is required for normal transmission of nerve impulses.

• **Cell wall permeability**—Ionized calcium controls the passage of fluid through the cell walls by affecting cell wall permeability.

• **Enzyme activation**—Calcium ions activate several enzymes, including adenosine triphosphatase (ATPase), lipase, and some of the protein-splitting enzymes.

• **Secretion of hormones**—Calcium has been shown to be necessary for the secretion of a number of hormones and hormone-releasing factors.

DEFICIENCY SYMPTOMS. The most dramatic symptoms of calcium deficiency are manifested in the bones and teeth of the young of all species. Such a deficiency during the growth period is evidenced in one or more of the following ways:

1. Stunting of growth.
2. Poor quality bones and teeth.
3. Malformation of bones—rickets.

When a deficiency of calcium is not too severe, body size may not be affected, but the bones may either be delicate and brittle, or remain soft and pliable because of too little mineral salts being deposited in them.

The teeth are largely formed during the latter part of pregnancy and during infancy. Any lack of calcium during these periods is likely to result in (1) malformed teeth, (2) teeth crowded too closely together in a narrow jaw, and/or (3) poor-quality teeth that are subject to excess decay later in life. Also, it is noteworthy that it is difficult to undo the effects of such deficiencies during the formative periods by consuming a good diet in later life.

In those parts of the world where milk and meat are scarce or unavailable, a large proportion of the diet must come from vegetable sources, especially cereal grains. Such diets are usually low in calcium, as well as deficient in the quantity and quality of the proteins supplied. Under such conditions, calcium deficiencies frequently occur; especially in children, but in adults, also. Populations do adapt to a certain extent to low-calcium diets. However, in comparison with well-nourished children, calcium-deficient children are characterized by (1) retarded growth, and (2) shorter stature at maturity; conditions that strongly suggest that the low level of calcium intake (probably accompanied by other dietary deficiencies) may have prevented them from growing to the full height of which they were genetically capable. When given a more nutritionally adequate diet, children of such parents respond with increased growth rate and greater adult height than their parents.

The clinical manifestations of calcium related diseases are rickets, osteomalacia, osteoporosis, hypercalcemia, tetany, and renal calculi, or kidney stones. Each of these diseases is described under the heading "Calcium Related Diseases," which appears later in this section; hence, the reader is referred thereto.

A clinical test that is sometimes used to assess calcium status and detect deficiency is known as Chvostek's sign (after the Austrian military surgeon who first used it), evidenced by the contraction of the perioral muscles (around the mouth) on tapping the molar (cheekbone) process. Contraction of the muscles around the mouth, as a result of a slight tap over the facial nerve, indicates facial irritability due to a calcium deficiency or a calcium-phosphorus imbalance.

INTERRELATIONSHIPS. Calcium is involved in a number of relationships, the most important of which follow:

• **Calcium-phosphorus ratio and vitamin D**—Without doubt, the best known relationship pertaining to minerals is the calcium-phosphorus ratio, along with vitamin D. Generally speaking, nutritionists recommend a calcium-phosphorus ratio of 1.5:1 in infancy, decreasing to 1:1 at 1 year of age, and remaining at 1:1 throughout the rest of life; although they consider ratios between 2:1 and 1:2 as satisfactory. However, the ratio of calcium to phosphorus is less critical when adequate amounts of vitamin D are present.

• **Fats**—An excessive intake of dietary fats or poor absorption of fats results in an excess of free fatty acids which unite with calcium to form insoluble soaps that are excreted in the feces.

• **Protein intake**—Dietary protein increases the rate of calcium absorption from the small intestine, probably because the amino acids, especially lysine and arginine, liberated in the course of protein digestion, form calcium salts which are easily absorbed. But any advantage from increased absorption is likely more than counterbalanced by the increased urinary loss of calicum on high-protein diets.

• **Excess calcium intake**—Excessive calcium intake (intake much higher than body requirements) will combine with phosphorus to form insoluble tricaclium phosphate, and thus interfere with the absorption of phosphorus. Also, excessive calcium intake may reduce the absorption of magnesium, iron, iodine, manganese, zinc, and copper, particularly when the intake of one of these minerals is borderline in terms of need.

• **Other factors enhancing absorption of calcium**—Absorption of calcium is increased by adequate vitamin D, dietary lactose (milk sugar), and an acid medium (a low pH). (See earlier section headed "Absorption of Calcium.")

• **Other factors interfering with absorption of calcium**—Absorption of calcium is decreased by vitamin D deficiency, phytic acid, oxalic acid, fiber, high alkalinity, stress, and lack of exercise. (See earlier section headed "Absorption of Calcium.")

RECOMMENDED DAILY ALLOWANCE OF CALCIUM. The National Research Council recommended daily dietary allowances, with provision for individual variation, of calcium are given in Table C-2.

TABLE C-2
RECOMMENDED DAILY CALCIUM ALLOWANCES[1]

Group	Age	Weight		Height		Calcium
	(years)	(lb)	*(kg)*	(in)	*(cm)*	(mg)
Infants	0–0.5	13	*6*	24	*60*	400
	0.5–1.0	20	*9*	28	*71*	600
Children	1–3	29	*13*	35	*90*	800
	4–6	44	*20*	44	*112*	800
	7–10	62	*28*	52	*132*	800
Males	11–14	99	*45*	62	*157*	1,200
	15–18	145	*66*	69	*176*	1,200
	19–22	154	*70*	70	*177*	1,200
	23–50	154	*70*	70	*178*	800
	51+	154	*70*	70	*178*	800
Females	11–14	101	*46*	62	*157*	1,200
	15–18	120	*55*	64	*163*	1,200
	19–22	120	*55*	64	*163*	1,200
	23–50	120	*55*	64	*163*	800
	51+	120	*55*	64	*163*	800
Pregnant						1,200
Lactating						1,200

[1]*Recommended Dietary Allowances*, 10th ed., 1989, NRC–National Academy of Sciences, p. 285.

Note that the recommended daily allowances for calcium range from 400 mg to 1,200 mg. Note, too, that the allowances vary according to age, and that provision is made for added allowances for pregnant and lactating females.

The bases used by NRC–National Academy of Sciences for calculating the Table C-2 Recommended Daily Calcium Allowances for Adults are given in Table C-3.

TABLE C-3
BASES FOR CALCULATING THE RECOMMENDED DAILY CALCIUM ALLOWANCES FOR ADULTS

Calcium Bases	Calcium/Day
	(mg)
Calcium in the feces	125
Calcium in the urine	175
Calcium in the sweat	20
Total calcium loss daily	320
Calcium absorbed from food $\frac{320}{X} \times \frac{100}{40} = 800$ mg/day	40% of 800 (or 320 mg)
Recommended daily allowance for adult	800

Note that, in Table C-3, the assumptions are made that (1) an adult will consume 800 mg of calcium per day; and (2) 40% of the dietary calcium will be absorbed, thus fully covering the 320 mg lost in the feces, urine, and sweat each day.

The recommendation of the FAO/WHO Expert Group relative to the calcium requirements of adults is somewhat lower than the levels cited above. They recommend 400 to 500 mg of calcium daily, increased to 1,000 to 1,200 mg of calcium per day during the last trimester of pregnancy and lactation, which is in line with the recommendations of the National Academy of Sciences, Washington, D.C.[4]

[4]*Handbook of Human Nutritional Requirements*, FAO of the UN, Rome, Italy, 1974, p. 51.

In a number of countries, the average daily intake of calcium is higher than the levels recommended by the National Research Council and FAO/WHO, in some cases as high as 1,500 mg per day. It is noteworthy, however, that there is no evidence that such a high intake is undesirable, even if it is not helpful.

• **Infants and children**—For infants, the calcium intake requirements may well be stated in terms of the amount of milk, since this is the chief source of food. Also, it is noteworthy that the proportion of calcium to phosphorus in milk is about the same as it is in the skeleton.

A breast-fed infant receives about 60 mg of calcium per kilogram of body weight (300 mg/liter of milk) and retains about two-thirds of this. By contrast, an infant fed a standard cow's milk formula containing added carbohydrate (600 to 700 mg of calcium per liter) receives about 170 mg of calcium per kilogram but retains 25 to 30%. Although the breast-fed infant has less calcium available, its calcium needs are fully met. Thus, the NRC recommended allowance for infants (to 6 months) is 400 mg/day.

For children 1 to 10 years of age, the recommended NRC calcium allowances are 800 mg/day. Per unit of weight, growing children may need two to four times as much calcium as does an adult.

During the rapid growth that characterizes preadolescence and puberty (10 to 18 years), a higher intake of calcium is recommended, i.e., 1,200 mg/day.

For older children, the requirements are most easily met by including a quart of milk a day or its equivalent in milk products.

During all periods of growth, vitamin D (or sunshine) is essential for the most efficient absorption and utilization of calcium.

• **Recommended calcium level for adults**—In view of the high levels of protein and phosphorus provided by U.S. diets, an allowance of 800 mg of calcium per day is recommended for healthy adults.

• **Recommended calcium intake for the elderly**—Special attention should be given to the calcium intake for the elderly, those at high risk for development of osteoporosis. Because they have a tendency to decrease their intake, and because they absorb calcium less efficiently, some eminent scientists who have done extensive research on the calcium requirements of the aged feel that the Recommended Daily Allowances (RDA) for the elderly made by the National Academy of Sciences are too low. Most of these researchers recommend 1,000 to 1,200 mg of calcium per day as optimum intake for the prevention of osteoporosis.

Even though osteoporosis may not be preventable by increasing calcium intake, there are reports that calcium supplements have induced calcium retention and relieved symptoms. This may reflect the fact that, although the efficiency of absorption decreases with the amount of calcium in the diet, the total amount of calcium actually retained increases.

Collectively, the possible effects of high dietary intakes of protein and phosphate on urinary calcium excretion and enhanced bone resorption, respectively, along with the possibility of reduced calcium absorption with advancing age, argue for recommending an ample intake of calcium.

• **Recommended calcium level during pregnancy and lactation**—Calcium accumulation during pregnancy totals approximately 30 g, nearly all related to calcification of the fetal skeleton in the last third of pregnancy. To meet this need, and to provide for individual variation, an additional allowance of 400 mg of calcium/day (for a total allowance of 1,200 mg/day) is recommended during gestation. An intake of this magnitude is recommended throughout pregnancy, rather than just during the last third of pregnancy when needs are greatest, because of the likelihood that calcium is stored in the skeleton in early gestation. Diets deficient in calcium (as well as in energy and protein) during pregnancy have been associated with decreased bone density in newborn infants.

The calcium content of breast milk averages 300 mg/liter. Assuming a daily production of 850 ml, this makes for a daily yield of approximately 250 mg of milk calcium. To meet this need, a total allowance of 1,200 mg of calcium daily is recommended for the lactating woman. Such an intake should prevent maternal demineralization, which otherwise accompanies lactation. A greater allowance may be necessary for women with a very high production of milk.

Also it is important to remember that, in addition to age and to pregnancy and lactation of women, calcium requirements are affected by the several factors that influence calcium absorption and excretion. Calcium is a good example, therefore, of a nutrient whose requirement cannot be decided on its own; the factors that influence calcium absorption and excretion must also be considered. For this reason, calcium supplementation should be determined by a knowledgeble nutritionist or M.D.

• **Calcium intake in average U.S. diet**—Of the nutrients that must be provided by the food we eat, studies show that calcium is most likely to be lacking.

In 1988, the calcium available for consumption in foods in the United States amounted to 890 mg per day, 75.1% of which was from dairy products, excluding butter.

The U.S. Department of Agriculture made a nationwide food consumption survey for the purpose of determining the nutrient quality of diets of U.S. households.[5] Then they compared, on a percentage basis, the consumption of certain nutrients to the Recommended Daily Allowances (RDA) of the National Academy of Sciences, National Research Council, as a means of determining how well people were eating. These studies revealed the following relative to calcium consumption:

1. That families in about three households in 10 were not consuming sufficient calcium to meet the RDA (see Fig. C-14).

[5]*Nutrient Levels in Food Used for Households in the United States*, USDA, Science and Education Administration.

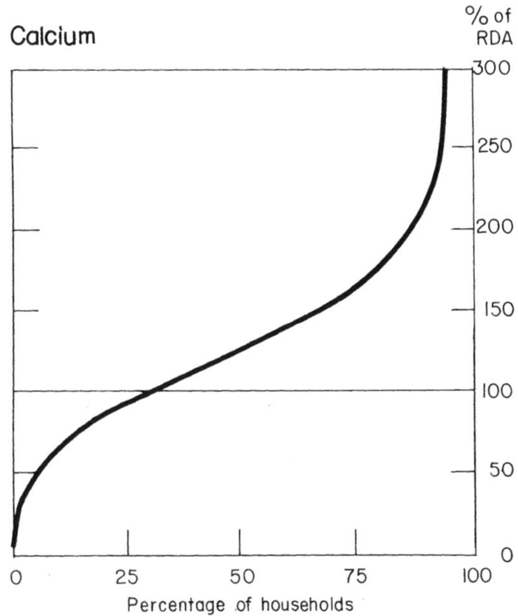

Fig. C-14. Calcium consumption levels of U.S. households. Note that 30% of the households were not consuming sufficient calcium—they were below the 100% of RDA. (From: *Nutrient Levels in Food Used for Households in the United States,* USDA, Science and Education Administration)

2. That high income within itself did not assure that the households met the RDA for calcium. Although high income families consumed slightly more calcium than low income families, more than one-fourth of the households with incomes of $20,000 and over failed to meet the RDA (see Fig. C-15).

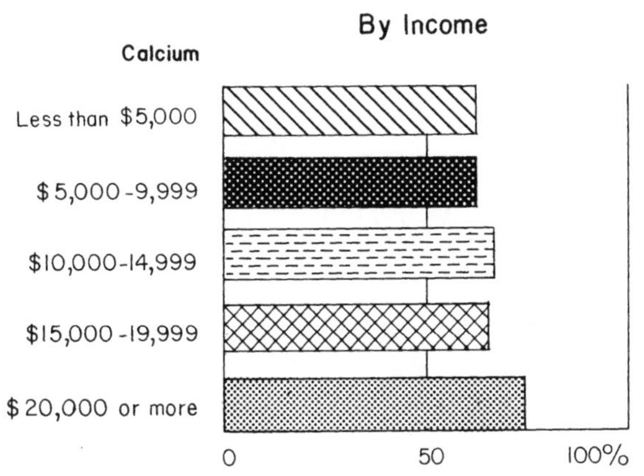

Fig. C-15. Effect of level of income on meeting the RDA for calcium. (From: *Nutrient Levels in Food Used for Households in the United States,* USDA, Science and Education Administration)

3. That suburban families consumed slightly more calcium than central city or nonmetropolitan families, but none of them met the RDA for calcium (see Fig. C-16).

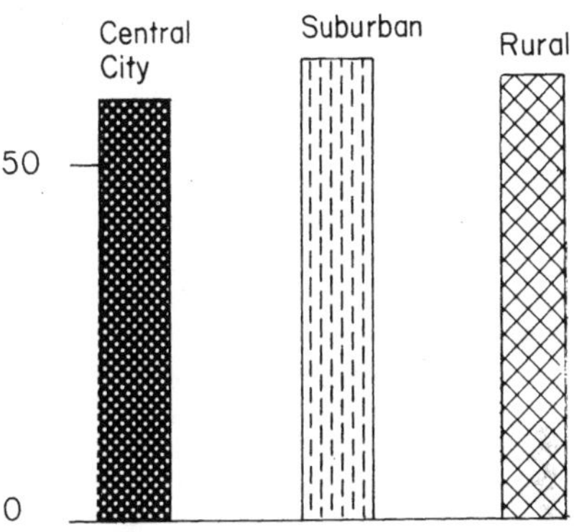

Fig. C-16. Effect of urbanization on meeting the RDA for calcium. (From: *Nutrient Levels in Food Used for Households in the United States,* USDA, Science and Education Administration)

4. That western families consumed slightly more calcium than families in the Northeast, North Central, or the South, but none of them met the RDA for calcium (see Fig. C-17).

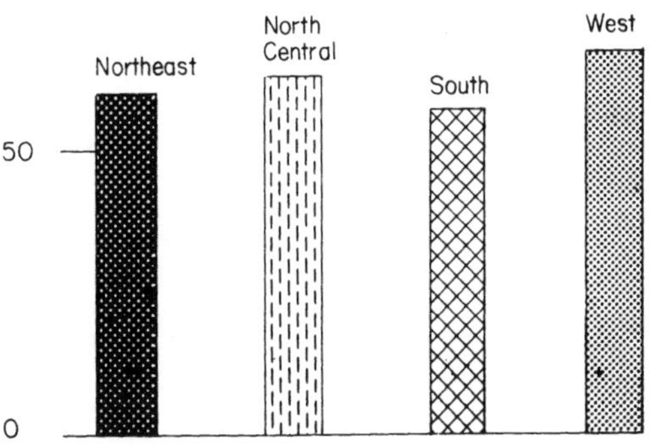

Fig. C-17. Effect of region on meeting the RDA for calcium. (From: *Nutrient Levels in Food Used for Households in the United States,* USDA, Science and Education Administration)

Of course, in evaluating the above data, it is recognized that household food may not have been divided among members of the family according to nutritional needs. Other studies have shown that, within a family, the father and adolescent boys are most likely to eat food that meets the recommended allowances; mothers and adolescent girls are least likely to do so.

Also we must remember that these figures for calcium values of people's diets were averages; hence, some were higher and some were lower. This means that some individuals were eating much less calcium than is indicated by the mean intake of the group. These are the persons that pose the problem.

Undoubtedly some of the people who live on low calcium diets are able to adapt to their low intakes. But many of these may not enjoy radiant good health. In this connection, we need to consider the number of young women who cannot nurse their babies; the prevalence of dental caries in boys and girls; what happens to those on minimum amounts of dietary calcium when stress occurs; and the importance of large amounts of calcium in the diet of today's fast-growing, tall children.

TOXICITY. Normally, the small intestine acts as an effective control and prevents excess calcium from being absorbed. However, a breakdown of this control may raise the level of calcium in the blood and lead to pathological calcification of the kidneys and other internal organs. This may occur in infants who have been fed on artificial foods fortified with excessive amounts of vitamin D and calcium.

Continuous high dietary calcium intake may cause a hypersecretion of calcitonin and bone abnormalities such as *osteopetrosis* (dense bone). High calcium intakes have also been reported to cause the formation of kidney stones.

CALCIUM RELATED DISEASES. Diseases may be caused by (1) inadequate intake of calcium or (2) factors inhibiting its absorption or excretion. The clinical manifestations of calcium related diseases are rickets, osteomalacia, osteoporosis, hypercalcemia, tetany, and renal calculi (kidney stones).

(Also see BONE, section headed "Bone Disorders.")

• **Rickets**—This is a children's disease, which may be caused by a lack of either calcium, phosphorus, or vitamin D; or an incorrect ratio of the two minerals.

Fully developed cases of rickets present the following characteristics: bulging of the forehead, giving it a box-like appearance; pigeon chest, spinal curvature, enlarged wrists and ankles, bowed legs, and knock-knees; enlargement or beading of the ribs at the juncture of the ribs with the breastbone, frequently called "the rachitic rosary"; and stunted growth.

(Also see RICKETS.)

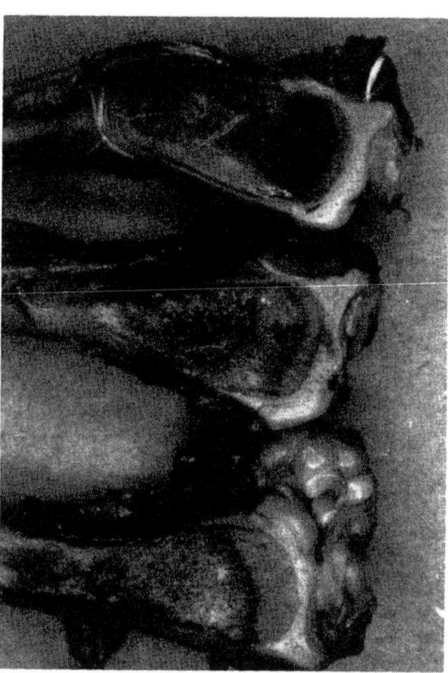

Fig. C-18. Deficiencies of calcium, phosphorus, and/or vitamin D affect the normal calcification and development of the tibia in young chicks. The bone on the top was caused by a deficiency of calcium or vitamin D. The bone on the bottom was caused by a deficiency of phosphorus. The bone in the middle is from a chick that was fed a well-balanced diet. (Courtesy, L. S. Jensen, Ph.D., Department of Poultry Science, The University of Georgia, Athens)

• **Osteomalacia**—This condition is the adult counterpart of rickets. Osteomalacia is caused by a prolonged deficiency of dietary calcium, phosphorus, and/or vitamin D (or of sunlight). It is characterized by poor calcification of the bones with increasing softness and flexibility, leading to deformities of limbs, spine, thorax, and pelvis. Bone changes may be accompanied by rheumatic pains in the lower back and legs, and exhaustion.

• **Osteoporosis**—This condition, which is frequently observed in persons over 50 years of age—especially women after the menopause, occurs when bone resorption exceeds bone formation. So, osteoporosis is a condition of too little bone. Various surveys have indicated that approximately 30% of women over age 55 and men over age 60 have had sufficient mineral loss to have produced at least one fracture. It is believed to be caused by many factors; among them, the age-related decline in secretions of hormones of the sex glands and pituitary glands; a long-continued low intake of calcium in early adult life, or during pregnancy and lactation in women; low fluorine; defective synthesis of the protein matrix; and/or lack of stimulating exercise.

Osteoporosis is characterized by demineralization of the jaw bone; low-back pains; a decreased amount of bone, accompanied by a shrinking in height; and porous bones. Also, the bones are brittle and slow to heal when broken.

(Also see OSTEOPOROSIS.)

• **Hypercalcemia** (*calcemia; high serum calcium*)—This disease occurs in both infants and adults.

In infants, hypercalcemia usually starts between the ages of 5 and 8 months. Affected infants suffer from loss of appetite, vomiting, wasting, constipation, flabby muscles, a characteristic facial appearance, increased calcium in the plasma and urea, increased cholesterol, elevated pressure, and calcification in the heart and kidneys. Also, mental retardation may be marked, and the damage to the brain may be irreparable. Death may follow.

Hypercalcemia in infants results in excessive absorption of calcium, caused by an overdosage of vitamin D, or, perhaps in some cases, by hypersensitivity to the vitamin.

Treatment consists in providing a diet as low in calcium as possible and with no vitamin D.

In adults, hypercalcemia may occur as a result of hyperparathyroidism, excessive doses of vitamin D, or patients with peptic ulcer using excessive alkali therapy along with a large consumption of milk over a period of years. With healthy adults, there is no danger of excessive calcium intake as such (provided that Ca:P ratio is satisfactory), because the body mechanisms ensure that the amount absorbed from the food and retained in the body does not exceed the amount needed to replace the wear and tear of the bones and soft tissues.

- **Tetany**—An abnormal decrease in blood serum calcium causes tetany, characterized by severe, intermittent spastic contractions of the muscles and muscular pain. It is manifested by spasm of the muscles that cause flexion of wrist and thumb, with extension of the fingers.

- **Renal calculi** (*kidney stones*)—The majority of renal stones are composed of calcium. Although the cause of the malady is unknown, it does seem to be nutritional. There appears to be a higher incidence of urinary calculi when there is (1) a high phosphorus-low calcium ratio, (2) a high potassium intake, or (3) a high animal protein (meat, fish, poultry) intake. Also, a deficiency of vitamin A may be a contributing factor.

(Also see KIDNEY STONES.)

CALCIUM-PHOSPHORUS RATIO AND VITAMIN D. When considering the calcium and phosphorus requirements, it is important to realize that the proper utilization of these minerals of the body is dependent upon three factors (1) an adequate supply of calcium and phosphorus in an available form, (2) a suitable ratio between them, and (3) sufficient vitamin D to make possible the assimilation and utilization of the calcium and phosphorus.

Generally speaking, nutritionists recommend a calcium-phosphorus ratio of 1.5:1 in infancy, decreasing to 1:1 at 1 year of age, and remaining at 1:1 throughout the rest of life; although they consider ratios between 2:1 and 1:2 as satisfactory. However, if plenty of vitamin D is present (provided either in the diet or by sunlight), the ratio of calcium to phosphorus becomes less critical. Likewise, less vitamin D is needed when there is a desirable calcium-phosphorus ratio.

It is noteworthy that there is much evidence indicating that calcium-phosphorus ratios of 1:1 to 2:1 for nonruminants (hogs and horses) and 1:1 to 7:1 for ruminants are satisfactory; but that ratios below 1:1 are often disastrous.

The dietary Ca:P ratio is particularly important during the critical periods of life—for children, and for women during the latter half of pregnancy and during lactation.

The most recent estimates indicate that the Ca:P ratio in the United States diet is approximately 1:1.5 to 1.6.

CALCIUM LOSSES DURING PROCESSING, COOKING, AND STORAGE. Homogenization, pasteurization, heating, drying, or acidifying does not reduce availability of calcium. However, when milk is heated, a precipitate of calcium phosphates usually settles on the bottom of the pan. To reduce the loss of both calcium and phosphorus, the milk should be stirred as it heats to incorporate the calcium salts into the liquid.

Calcium losses in the cooking of vegetables can be minimized by using small amounts of water, by keeping the size of the pieces as large as possible to avoid excessive surface exposure, and by cooking with the skins on when practical, since minerals occur in greatest concentration near the skin. Cooking dried fruits and vegetables in the liquid in which they are soaked decreases calcium loss.

SOURCES OF CALCIUM. Among common foods, the milk and milk products group contains many of the richest sources of calcium. Some leafy green vegetables and some fish are also rich sources of calcium. However, in the United States, more than 70% of the calcium intake is derived from milk and milk products. Other foods contribute smaller amounts.

Furthermore, the total quantity of calcium available for absorption is augmented by the calcium in intestinal secretions.

The calcium-phosphorus ratio is determined by the total intake of each of the minerals. For the most part, this means foods, for few people take calcium and/or phosphorus mineral supplements.

Groupings by rank of common sources of calcium follow:

- **Rich sources**—Cheeses, wheat-soy flour, blackstrap molasses.

- **Good sources**—Almonds, Brazil nuts, caviar, cottonseed flour, dried figs, fish with soft edible bones, green leafv vegetables, hazel nuts, ice cream, milk, oysters, sour cream, soybean flour, yogurt.

- **Fair sources**—Beans, bread, broccoli, cabbage, clams, cottage cheese, crab, dehydrated apricots, dehydrated peaches, eggs, legumes, lettuce, lobster, okra, olives, oranges, parsnips, peanut butter, prunes, raisins, rhubarb, spinach, Swiss chard, turnips, wheat germ.

- **Negligible sources**—Asparagus, beef, beets, Brussels sprouts, carrots, cauliflower, corn and cornmeal and other grains and cereals, cucumbers, fats and oils, fish, juices, kohlrabi, most fresh fruits, mushrooms, peas, pickles, popcorn, pork, potatoes, poultry, pumpkin, radishes, roe, squash, sugar, tomatoes, tuna.

- **Supplemental sources**—Calcium carbonate, calcium

gluconate, calcium lactate, dicalcium phosphate, kelp. If calcium is taken as a pill, it should be in the form of some soluble salt, such as calcium lactate.

It is difficult to meet the recommended daily allowances for calcium without including milk and/or milk products in the diet. If the average American diet contained no dairy products, one would be hard pressed to obtain more than 300 mg of calcium daily. Two cups (*480 ml*) of milk are sufficient to meet the daily calcium needs of adults. Children need 3 or more cups daily, whereas adolescents need 4 cups or more.

For additional sources and more precise values of calcium, see Food Composition Table F-36 of this book.

TOP CALCIUM SOURCES. The top calcium sources are listed in Table C-4. *NOTE WELL:* This table lists (1) the top sources without regard to the amount normally eaten (left column), and (2) the top food sources (right column). Note, too, that the phosphorus content, the calcium-phosphorus ratio, and the caloric (energy) content of each food is given.

The following pertinent deductions relative to the calcium-phosphorus ratios of, and contents in, some foods can be made from Table C-4:

1. Milk, ice cream, and yogurt contain desirable calcium-phosphorus ratios. The cheeses are rich in calcium but low in phosphorus.
2. Some seafoods have an acceptable calcium-phosphorus ratio, and are rich sources of both minerals.
3. Green leafy vegetables have an imbalanced calcium-phosphorus ratio. They are a good source of calcium but a poor source of phosphorus. Red meats and poultry are poor sources of calcium, but excellent sources of phosphorus. Thus, a desirable calcium-phosphorus ratio, along with adequate calcium and phosphorus, may be achieved by combining vegetables and meat.
4. Blackstrap molasses has an imbalanced calcium-phosphorus ratio; it is a rich source of calcium, but a poor source of phosphorus.
5. Most fresh fruit is a poor source of calcium.
6. Soy flour has an imbalanced calcium-phosphorus ratio. It is a fair source of calcium, but high in phosphorus. A desirable calcium-phosphorus ratio, along with adequate calcium and phosphorus, may be achieved by combining blackstrap molasses and soy flour.

TABLE C-4
TOP CALCIUM

Top Sources[2]	Calcium	Phosphorus	Calcium-Phosphorus Ratio	Energy
	(mg/100 g)	(mg/100 g)	(Ca:P)	(kcal/100 g)
Baking powder, home use[3]	6,300	—	—	112
Fish flour from filet waste	6,040	4,060	1.5:1.0	305
Baking powder, home use/calcium carbonate	5,778	1,452	4.0:1.0	78
Baking powder, commercial, low sodium	4,816	7,308	.7:1.0	172
Fish flour from whole fish	4,610	3,100	1.5:1.0	307
Savory	2,132	140	4.4:1.0	272
Basil	2,113	490	4.3:1.0	251
Whey, acid dry	2,054	1,348	1.5:1.0	339
Marjoram, dried	1,990	306	6.5:1.0	271
Baking powder, home use/sodium aluminum sulfate	1,932	2,904	.7:1.0	129
Thyme	1,890	200	9.5:1.0	276
Dill weed, dried	1,784	543	3.3:1.0	253
Celery seed	1,767	550	3.2:1.0	392
Sage	1,652	90	18.4:1.0	315
Oregano	1,576	200	7.9:1.0	306
Dill seed	1,516	277	5.5:1.0	305
Rosemary leaves	1,500	70	21.4:1.0	440
Parsley, dried	1,468	351	4.2:1.0	276
Poppy seed	1,448	848	1.7:1.0	530
Cheese, natural Parmesan, grated	1,376	807	1.7:1.0	456

[1]These listings are based on the data in Food Composition Table F-36 of this book. Some top or rich food sources may have been overlooked since some of the foods in Table F-36 lack values for calcium.

Whenever possible, foods are on an "as used" basis, without regard to moisture content; hence, certain high-moisture foods may be disadvantaged when ranked on the basis of calcium content per 100 g (approximately 3½ oz) without regard to moisture content.

7. Egg yolks are a fair source of calcium, but high in phosphorus, resulting in calcium-phosphorus imbalance.

In order that the diet will have a desirable calcium-phosphorus ratio, knowledge of the calcium and phosphorus composition of major foods is necessary. These are given for many foods in Food Composition Table F-36.

(Also see MINERAL[S], Table M-67; and PHOSPHORUS.)

CALCIUM ACETATE

An approved food additive sometimes employed as an antimicrobial or sequestering agent. Calcium acetate will prevent the formation of "ropy" bread.

(Also see ADDITIVES; BREADS AND BAKING; and ROPE.)

CALCIUM ACID PHOSPHATE

An acid-tasting major calcium salt, composed of approximately 16% calcium and 21% phosphorus, which is soluble in the acid environment of the stomach. It is used as the leavening ingredient of baking powder and of self-rising flour as it reacts with bicarbonate of soda to liberate carbon dioxide. Also, it is used in mineral supplements, to control the pH in malt, as a buffer in foods, as a firming agent in foods, and in fertilizers.

(Also see BAKING POWDER AND BAKING SODA; and CALCIUM.)

CALCIUM GLUCONATE ($Ca[C_6H_{11}O_7]_2 \cdot H_2O$).

A tasteless, ordorless, water soluble salt of calcium. It may be administered intravenously for the relief of tetany, caused by an abnormal decrease in blood serum calcium and characterized by cramplike involuntary muscle spasm. Also, calcium gluconate is used as a food additive, buffer, and firming agent, and in some vitamin tablets.

(Also see CALCIUM.)

CALCIUM-PHOSPHORUS RATIO

The proportional relation that exists between calcium and phosphorus is known as the calcium-phosphorus ratio. Generally speaking, nutritionists recommend a

SOURCES[1]

Top Food Sources	Calcium	Phosphorus	Calcium-Phosphorus Ratio	Energy
	(mg/100 g)	(mg/100 g)	(Ca:P)	(kcal/100 g)
Cheese, natural Swiss (domestic)	961	605	1.6:1.0	372
Cheese, natural Cheddar (domestic American)	718	510	1.4:1.0	402
Wheat-soy blend (WSB)/bulgur flour or straight grade wheat flour	685	548	1.3:1.0	365
Molasses, cane, blackstrap[4]	684	84	8.1:1.0	230
Cheese, pasteurized process American	616	745	.8:1.0	375
Sardines, Atlantic, canned in oil, solids/liquids	354	434	.8:1.0	246
Lamb's quarter, raw	309	72	4.3:1.0	43
Milk, cow's, canned, evaporated, skimmed	290	195	1.5:1.0	78
Soybean, flour, defatted	265	655	.4:1.0	326
Turnip greens, raw	246	58	4.2:1.0	28
Almonds	234	504	.5:1.0	598
Trout, cooked	218	272	.8:1.0	196
Salmon, Atlantic, canned, solid/liquid	215	—	—	124
Collard leaves, cooked, small amount water	188	52	3.6:1.0	33
Kale, boiled, drained, leaves	187	58	3.2:1.0	39
Yogurt made partially w/skimmed milk	183	144	1.3:1.0	51
Eggs, chicken, raw yolks, fresh	152	508	.3:1.0	369
Ice cream, chocolate	140	126	1.1:1.0	222
Milk, cow's, 2% fat	122	95	1.3:1.0	50

[1]Listed without regard to the amount normally eaten.
[3]Average value for calcium sulfate and straight phosphate baking powders.
[4]The calcium in blackstrap molasses is due to the addition of lime during the refining process.

calcium-phosphorus ratio of 1:1, although they consider ratios between 2:1 and 1:2 as satisfactory. An excess of either can hinder absorption, however, the normal variation encountered in dietary calcium to phosphorus ratios does not significantly affect absorption from the intestine. If plenty of vitamin D is present (provided either by the diet or by sunlight), the ratio of calcium to phosphorus is less critical.

(Also see CALCIUM, section headed "Calcium-Phosphorus Ratio and Vitamin D.")

CALCULUS

Another name for stones formed in one or more passages or organs of the body. It is believed that many types of stones are formed when various substances come out of solution from the body fluids. Hence, some of the therapeutic measures consist of attempts to redissolve the stones.

(Also see KIDNEY STONES.)

CALIPER

An instrument, which usually has movable jaws, that is used to measure the thickness or diameter of various objects. Special types are used by nutritionists for measuring skin-fold thickness, as part of the assessment of nutritional status, and in judging leanness or fatness.

(Also see OBESITY.)

CALORIC (Energy) EXPENDITURE

Contents *Page*

History of Calorimetry **304**
Utilization of Dietary Energy by the Body **306**
Consequences of Too Few or Too Many Calories **307**
Caloric Deficiencies **307**
Caloric Excesses **307**
Measurement of Energy Expenditures **308**
Direct Calorimetry **308**
Indirect Calorimetry **308**
Basal Metabolic Rate (BMR) **309**
Respiratory Quotient (RQ) **310**
Recommended Caloric (Energy) Intakes **310**
Allowances for Adults **310**
Allowances for Infants, Children, and Adolescents **311**
Resting Energy Expenditure (REE) **311**
Special Circumstances that Affect Energy Requirements **312**
Pregnancy **312**
Lactation **312**
Environmental Stresses **312**
Diseases, Injuries, and Surgery **313**
Summary: Balancing Energy Intakes and Expenditures **313**

There has been a recent growth of worldwide interest in the application of calorimetry in the fields of human medicine and nutrition, exercise and sports physiology, and space exploration. Furthermore, new technological developments have made it possible to transmit physiological measurements via radio or telephone to distant diagnostic centers. Hence, many investigators around the world are currently engaged in modern versions of the calorimetric studies that were first conducted over 200 years ago. Fig. C-19 shows a typical study in progress.

Fig. C-19. Measurement of energy expenditure during physical work. The man pushing the loaded wheelbarrow is exhaling into a Max Planck respirometer which measures and records the volume of air and collects a sample for a subsequent analysis of its composition.

HISTORY OF CALORIMETRY. The French chemist Antoine Lavoisier (1743 to 1794) demonstrated in the 1770s that in the processes of combustion and respiration oxygen gas was consumed and carbon dioxide was produced. Other researchers of his era conducted similar experiments, but they made erroneous interpretations—based upon the cumbersome phlogiston theory that Lavoisier discredited by his work. Then, in 1780 Lavoisier and the mathematician La Place measured the oxygen consumption, carbon dioxide production, and heat emitted from guinea pigs. The exchange or respiratory gases was measured in a bell jar, whereas, the emission of heat was measured in an ice calorimeter. Five years later, Lavoisier studied the respiration of his associate Sequin, whose body was enclosed in a specially constructed bag. This experiment showed that the rate of metabolism, which was measured at normal room temperature after an overnight fast, was increased significantly by (1) exposure to cold, (2) consumption of food, and (3) physical work. Unfortunately, Lavoisier, who is credited with having founded the science of nutrition, went to the guillotine at age 50 because he was a tax collector during the French Revolution. (He was accused of having collected taxes on the water content of tobacco.) It is noteworthy that his work made the Academie des

Sciences in Paris the world center for nutritional science from which the German scientist Liebig drew his inspiration and passed it down through his students to the first American nutritional scientists.

Lavoisier's widow later married the American born Benjamin Thomas (who was made Count Rumford of the Holy Roman Empire) who developed the water calorimeter that was used by the French scientist Dulong in a competition sponsored by the Academie des Sciences in 1823. Dulong found that the ratio of carbon production to oxygen consumption (a quantity that is now called the respiratory quotient) was higher for herbivorous animals than it was for carnivorous animals. This finding was clarified in 1849, by Regnault, who was a professor of chemistry and physics at the University of Paris. Using a closed circuit calorimeter that he and his associate Reiset designed, he showed that differences in the respiratory quotient were due to the types of food that were consumed.

The renowned German chemist Justus von Liebig (1803 to 1873), who is called "the father of organic chemistry," owed his early success to certain members of the Academie des Sciences in Paris, who encouraged him and promoted his interests after his academic training in Germany had left him dissatisfied. In 1840, Liebig published the first edition of his monumental work on the application of organic chemistry to agriculture and physiology. The work of Liebig was largely responsible for the dominant role of chemistry in the research projects of the first agricultural experiment stations in Scotland, Germany, England, and the United States. He was the first to study the individual roles of carbohydrates, fats, and proteins in the body, but he concluded wrongly that muscular work depended upon the metabolism of protein. In 1852, John Lawes and Joseph Gilbert of the Rothamsted Experiment Station in England showed that the nonprotein food supplied most of the energy of muscular work. Five years later, Eduard Smith of England made 1,200 observations of humans engaged in a wide variety of physical activities and found that the expired carbon dioxide (mainly a measure of fat and carbohydrate metabolism) increased in proportion to the intensity of the activity, whereas the urea excretion in the urine (a measure of protein metabolism) remained constant.

In 1862, Liebig's former associates Max von Pettenkofer and Carl von Voit constructed a large roomlike open-circuit calorimeter in which they measured the metabolism of human subjects at rest, while fasting, and during the performance of various types of work. Their findings made Liebig admit in 1870 that he had been wrong about protein furnishing the energy for muscular work. The studies of Pettenkofer and Voit were continued by the German physiologist Max Rubner, who had been taught by the latter. Rubner worked with such precision that he was able to obtain results by direct calorimetry which agreed within a fraction of 1% with those predicted from indirect calorimetry. He discovered that the feeding of protein after a fast resulted in a 30% increase in the metabolic rate. (This effect is now called *specific dynamic action* or *thermogenesis.*) Wilbur Atwater, who was the first American nutritionist, worked with Pettenkofer, Voit, and Rubner in 1887, prior to his appointment as chief of the Office of Experiment Stations of the U.S. Department of Agriculture.

In 1892, Atwater and the physicist Rosa began construction of a calorimeter at the Connecticut Agricultural Experiment Station in New Haven, and in 1899 published tables of food composition which had been prepared with the aid of data obtained in their calorimeter. Atwater was the first to deduct the energy losses which occurred in both the feces and urine from the gross energy obtained by the combustion of the foods.) Francis Benedict, who had worked with Atwater starting in 1896, became the head of the Carnegie Nutrition Laboratory in Boston, where he continued the work started by Atwater. During World War I, Benedict found that fasting male students who had lost 12% of their body weight had a lowering of their basal metabolic rates which averaged about 19%.

In Europe, Nathan Zuntz of Switzerland developed a portable device for measuring respiratory gases that was used in 1895 by his former student, the German physiologist Adolph Magnus-Levy, to measure the metabolic rates of hospital patients who had disorders of the thyroid gland. This was the first application of calorimetry to clinical medicine. Magnus-Levy found that hypothyroid patients had subnormal metabolic rates, whereas those with hyperthyroidism had elevated metabolic rates.

Interest in human calorimetry appeared to have declined somewhat between World Wars I and II since the major use for these techniques during that period was in the measurment of basal metabolic rates in patients suspected of having thyroid disorders. Animal calorimetry was extensively employed at some of the major American agricultural experiment stations during this period. However, the American nutritionist Graham Lusk, who earned his Ph.D. under Voit in 1891, was instrumental in having the calorimeter built at the Cornell Medical College in New York, where his protege Eugene DuBois carried on his metabolic studies for several decades. DuBois measured the surface areas of human subjects and developed formulas for predicting the areas from measurements of height and weights. He also made many studies on the metabolic effects of varying the carbohydrate, fat, and protein contents of diets that were fed to healthy people and those suffering from various diseases.

A renewed interest in human caloric expenditures developed during World War II when food supplies had to be rationed. Studies conducted in Germany showed that the production of both coal and steel dropped significantly when the caloric intakes of the workers in these industries were restricted. In the United States, Ancel Keys and his co-workers at the University of Min nesota studied the effects of semistarvation on a group of conscientious objectors. After consuming approximately 1,600 Calories (kcal) per day for 22 weeks the subjects resembled prisoners of war and had an average reduction in basal metabolism of about 18%. They also expended subnormal amounts of energy while performing various exercises.

After the war, calorimetric studies were made of various groups suffering from protein-energy malnutrition, and of healthy people engaged in various activities. The use of the basal metabolic rate as a diagnostic measure for thyroid disorders was replaced in many hospitals by more specific tests of thyroid hormonal status. By the 1960s, there was renewed interest in the basal metabolism of normal and obese people, since statistics showed that obesity was sometimes associated with various degenerative and fatal diseases. Recently, studies of the specific dynamic action (thermogenesis) due to food consumption have suggested that this effect may play an important role in the regulation of the body weights of some people.

(Also SEE CALORIC VALUES OF FOODS.)

UTILIZATION OF DIETARY ENERGY BY THE BODY. The total heat content of the foods which constitute a diet may be determined by burning representative samples in a bomb calorimeter. However, some of the food energy cannot be utilized by the body because it is present in one or more substances that are not completely digested and absorbed, or if they are absorbed, are not metabolized completely. Unabsorbed materials are excreted in the feces, whereas those that are absorbed, but not metabolized, are excreted in the urine. The utilizable and nonutilizable forms of dietary energy are illustrated in Fig. C-20.

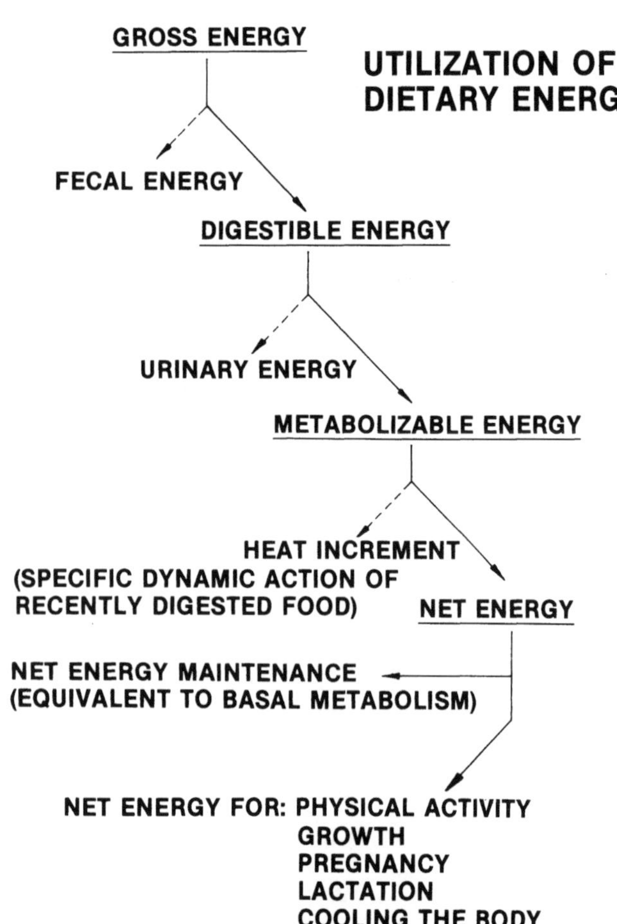

Fig. C-20. Utilization of dietary energy by the body.

Definitions of the parts of the outline shown in Fig. C-20 follow:

- **Gross energy**—This is the total heat released from a food when it is burned in a calorimeter. This value may be considerably larger than the amount of energy that is available for utilization by the body since it does not take into account the unavailable part of the food energy. For example, 1 lb (*0.45 kg*) of corncobs contains about the same amount of gross energy as 1 lb of shelled corn.

- **Fecal energy**—This fraction represents the energy content of the food constituents that are not asorbed in the digestive tract. The most common contributor to fecal energy in the human diet is cellulose and the related carbohydrates that collectively constitute the crude fiber of plant foods. Certain lipids and nitrogenous substances also contribute to the fecal energy.

- **Digestible energy**—This is the energy present in the food constituents that is absorbed by the body. It is determined by subtracting the fecal energy from the gross energy.

- **Urinary energy**—This fraction represents the energy content of the metabolic products that is excreted in the urine. Most of this energy is present in the nitrogenous waste products of metabolism such as urea, uric acid, creatine, and creatinine.

- **Metabolizable energy**—This is the energy that is released from food constituents during metabolism. It is determined by subtracting both the fecal and urinary energy fractions from the gross energy. The Atwater caloric conversion factors of 4, 9, and 4 Calories (*kcal*) represent the average metabolizable energy content of each gram of the dietary carbohydrate, fat, and protein, respectively.

- **Heat increment**—This fraction corresponds to the heat that is lost from the body after the consumption of a meal. It is generally considered to be wasted, except when it contributes to the maintenance of the body temperature in a cold environment. The heat increment varies in magnitude with the composition of the meal that is consumed, since dietary protein generally yields more waste heat than equicaloric amounts of carbohydrate and fat. However, the heat increment also varies among different people who have consumed the same meal. It is noteworthy that the heat increment of the diet has often been referred to as the "specific dynamic action," or more recently as the "thermic effect."

- **Net energy**—This is the energy that is available in a useful form for doing the work of the body. It is determined by subtracting the heat increment (usually assumed to be 10% of the metabolizable energy, since it cannot be measured readily) from the metabolizable energy.

- **Net energy for maintenance**—This fraction of the net energy is that which is required to keep the body in a state of balance; that is, with no net loss or gain of energy in the body tissues. Hence, it corresponds to what is considered to be the basal metabolism in healthy, nonpregnant, nonlactating adults.

• **Net energy for physical activity, growth, pregnancy, lactation, cooling the body, and heating the body**—This is the energy cost of body functions over and above those of maintenance. Values for these incremental energy expenditures range from next to nothing for certain awake, but resting, adults to amounts that greatly exceed the basal metabolism when the various functions other than those of maintenance are operative.

(Also see ENERGY UTILIZATION BY THE BODY.)

CONSEQUENCES OF TOO FEW OR TOO MANY CALORIES.

People may be weakened significantly and made extra vulnerable to disease by either undernourishment or overnourishment. A recent statistical analysis of mortality data from around the world has shown that the minimum mortality occurs in the middle range of body weights and that mortality increases for both lower and higher weights.[6] Discussion of caloric imbalances follow:

Caloric Deficiencies.

Many Americans tend to overlook the possible consequences of underweight because they are exposed to a steady stream of propaganda on the virtues of being slim. However, the older generation seemed to know better since they employed many different dietary measures to ensure that the members of their families maintained adequate body weights. Standards for defining underweight vary, but the condition is said to exist when one is more than 10% below the average weight for his or her height, body build, sex, and age. Therefore, it is worth reviewing some of the consequences from the lack of sufficient food energy.

• **Failure of growth and development in infants, children, and adolescents**—Adequate energy intake is required for efficient utilization of dietary protein, growth, and sexual development during adolescence. Furthermore, even mild energy deficiencies may delay the various stages of growth and sexual development. More severe deficiencies may stunt growth permanently and bring the normal sexual development to a halt. It is well known that stringent dieting by adolescent girls sometimes stops menstruation.

• **Greater susceptibility to chilling and infections**—Heat escapes from the bodies of overly thin people much more rapidly than it does from those who are heavier because layers of fat act as insulation against heat loss. Also, the heat production by the body is diminished in caloric undernutrition. Recent studies of malnourished children in the developing countries have shown that caloric deficiencies may be accompanied by reduced immunity to infection. Furthermore, the effects of infections are likely to be more severe and longlasting due to the lack of energy reserves for the maintenance of the body tissues.

[6] Keys, A., "Overweight, Obesity, Coronary Heart Disease and Mortality," *Nutrition Reviews*, Vol. 38, September, 1980, pp. 297-307.

• **Lack of energy for physical activities**—Many studies have shown that the failure to provide sufficient dietary energy to cover the cost of muscular work usually results in apathy and lack of drive. It is believed that the caloric undernutrition somehow alters the normal patterns of hormonal secretion which promote physical vigor.

(Also see MALNUTRITION; and UNDERWEIGHT.)

Caloric Excesses.

It is well known that the chronic consumption of excessive amounts of food energy leads to obesity. Fig. C-21 shows an extraordinary gain in weight which occurred throughout the life of the heaviest medically weighed human.

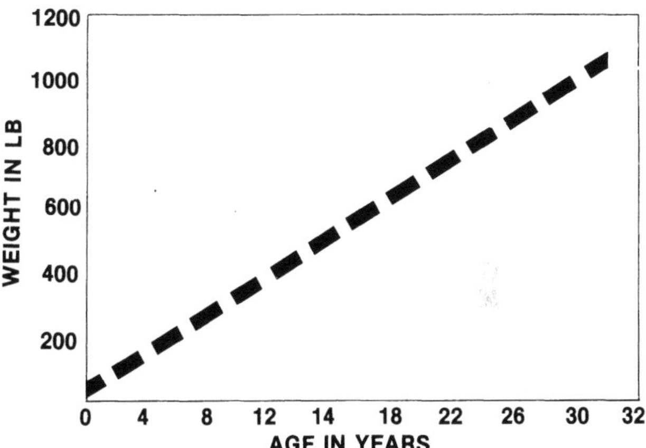

Fig. C-21. The continuous gain in weight of one of the world's heaviest humans. (Plotted from data given in McWhirter, N., *Guinness Book of World Records*, 16th ed., Bantam Books, Inc., New York, N.Y., 1978, pp. 21-22)

The man whose weight record is plotted in Fig. C-21 gained an average of 33.3 lb (*15 kg*) per year over his lifespan of 32 years. This pattern of weight gain can be accomplished by the consumption of as few as 318 excess Calories (*kcal*) per day since approximately 3,500 Calories (*kcal*) are equivalent to a pound of body weight. In terms of extra food, this means an extra hamburger, or a piece of cake or pie each day. Fortunately, there are some poorly understood biological mechanisms by which most overeaters are protected from the extreme results shown in Fig. C-21. Nevertheless, the disorders which may be associated with obesity are noteworthy. (Insurance companies consider a person to be obese when his or her weight is more than 20% greater than that deemed to be appropriate for his or her height, body build, and age.)

(Also see OBESITY, section headed "Height-Weight Tables.")

• **Disorders of the heart and circulation**—The relationship between obesity and these disorders is not simple, since susceptibility also depends upon many other factors such as age, body build, climate, heredity, metabolic disorders, physical activity, sex, and smoking. Even so, obesity often aggravates conditions such as angina pectoris (sharp pain in the chest or shoulder which regularly accompanies physical exertion), coronary heart disease, enlargement of the heart, excessive numbers of red cells and abnormal clotting of the blood, high blood pressure, high levels of fats in the blood, stroke, and sudden death resulting from a heart attack.

• **Increased susceptibility to heat exhaustion**—Layers of body fat slow the escape of heat from the body. Experiments have shown that exercising in a hot environment usually results in a greater elevation of body temperature in the obese than in the nonobese.

• **Kidney diseases**—Insurance company statistics show that obese policyholders have about twice the death rate from kidney diseases as the nonobese, but it is not certain whether obesity is a direct cause of the disease, or whether they result from conditions that may be associated with obesity.

• **Metabolic abnormalities**—Patterns of hormonal secretion and metabolism are sometimes altered in obesity and may be responsible in part for the greater susceptibility of some obese people to diabetes, gallstones, gout and/or hyperuricemia (high blood levels of uric acid), hypoglycemia (low blood sugar), and toxemia of pregnancy. However, obesity may not necessarily be the direct cause of the metabolic problems but rather the result of an acquired or an inhibited abnormality that predisposes the people to both the disorder and the obesity.

• **Physical handicaps**—The workload that is imposed upon the body by a great amount of extra weight may increase the likelihood of breathing difficulties, easy fatigability, hernia, osteoarthritis, and sleepiness in spite of what seems to be adequate sleeping time.

(Also see OBESITY.)

MEASUREMENT OF ENERGY EXPENDITURES. The dangers of underweight and overweight which were discussed in the preceding section show the need for accurate data for estimating caloric requirements of people in various circumstances. These data are obtained by the means which follow:

Direct Calorimetry. In this procedure, the subject is confined to a well-insulated chamber and the heat losses (by radiation, convection, and conduction from the body surface; by evaporation of water from the skin and lungs; and by excretion of urine and feces) are measured either by (1) the increase in temperature of a known volume of water, or (2) electrical current generated as heat passes across thermocouples (gradient layer calorimetry). About ¼ of the heat lost by the body results from moisture vaporiztion. The remaining ¾ of the body heat is lost by radiation, conduction, and convection.

Direct calorimetry is the most accurate method of measuring the heat production of people, but it is costly and arduous. The machine itself is very expensive to build and maintain; and considerable labor is involved in running the machine and analyzing the results. For these reasons, indirect calorimetry is usually the method of choice for the measurement of energy expenditures.

Indirect Calorimetry. In this procedure, heat production is calculated from measurement of the respiratory exchange—the O_2 consumption, and usually the CO_2 production—of the subject. This method is based on the fact that O_2 consumption and CO_2 production are closely related to heat production. Measurement of the respiratory exchange of a person may be made with an open circuit mask system or an automated system for metabolic measurements.

• **Open circuit mask system**—The nose is clamped, and the subject is told to exhale through an expiratory tube for a short period so that all of the exhaled air may be collected in a bag or spirometer for analysis.

• **Automated system for metabolic measurements**—This type of system usually includes an infrared CO_2 analyzer, a high-speed O_2 analyzer, electronic temperature and pressure transducers, a turbine volume meter, and an automated system for sample handling, data collection, computation, and printout.[7]

The operation of this type of system is shown in Fig. C-22.

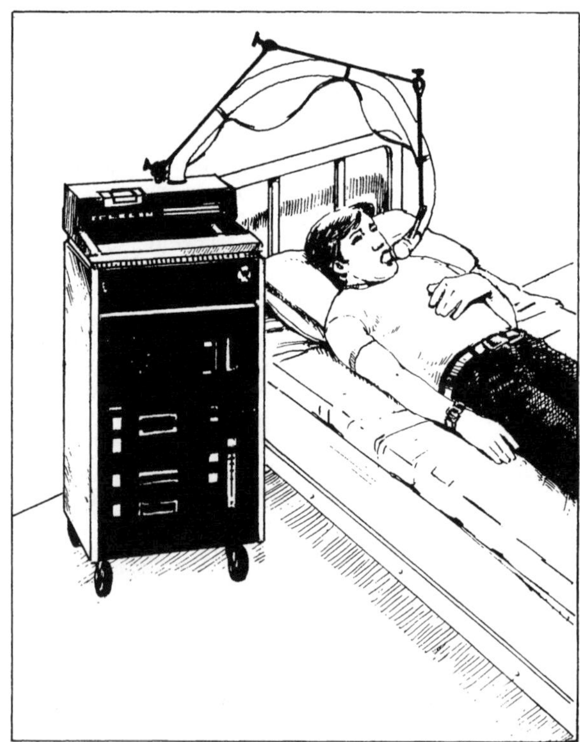

Fig. C-22. An automated system for metabolic measurements. Before the test, the operator enters the program that will control data collection on a magnetic program card which controls the operation of the system.

[7]Norton, A. C., "Portable Equipment for Gas Exchange," *Assessment of Energy Metabolism in Health and Disease,* Ross Laboratories, 1980, pp. 37-41.

The most commonly made calorimetric tests are the measurements of the basal metabolic rate and the respiratory quotient. Details follow:

BASAL METABOLIC RATE (BMR). This test provides a measure of the minimum daily energy expenditure required to maintain the essential functions of the body. Doctors have often ordered BMR tests to determine whether excessive fatness or thinness is due to hypometabolism or hypermetabolism, respectively. The simplest determination consists of measuring the oxygen consumption of the subject for two 6-minute periods, then averaging the results. Sometimes the CO_2 production is also measured. The following conditions are necessary for the accurate measurement of BMR:

1. Measurement of the BMR is usually made in the morning before the subject has had breakfast, since a minimum fasting period of 12 hours guarantees the absence of the specific dynamic action which accompanies food consumption.
2. The subject should be relaxed and awake, but lying down and in a calm state of mind.
3. The temperature of the room should be within the range of 70 to 80°F (*21 to 27°C*) so that no extra work is being done by the body to heat or cool itself.

The result of a BMR determination is usually calculated in terms of Calories (kcal) per square meter of body surface per hour. Body surface values are given in charts and tables as functions of height and weight. Then, the result is compared with standard BMR values for age and sex and expressed as a percentage deviation from the standard value which is appropriate for the subject under consideration. For example, a 47-year-old female who has a basal metabolic rate of 30 Calories (kcal) per square meter per hour would be given a rating of -14. This means that her BMR is 14% below the standard value for her age. The interpretation of the measurement may vary among different physicians, but generally the range from −15 to +15 is considered normal, because it covers about 95% of normal, healthy people.

Some of the factors which may lower or raise the BMR are given in Table C-5.

TABLE C-5
FACTORS WHICH MAY LOWER OR RAISE THE BASAL METABOLIC RATE (BMR)

Type of Factor	Description	Comments
	Factors That May Lower the BMR	
Depression	An emotional and/or physical state in which the activity of the sympathetic nervous system is depressed.	Activity of the sympathetic nervous system is usually associated with the secretion of hormones which tend to raise the BMR.
Glandular deficiencies	Hypothyroidism (subnormal secretions of thyroid hormones).	The BMR has long been used to diagnose this condition. Now, better tests are available.
	Panhypopituitarism (low levels of the pituitary hormones which stimulate the other glands).	Stimulatory hormones from the pituitary are needed for the optimal functioning of the other glands that affect metabolism.
Malnutrition	Chronically inadequate intake of calories, protein, minerals, and/or vitamins.	The most metabolically active tissues are very sensitive to nutrient deficiencies. Also, an insufficient caloric intake depresses the BMR.
Obesity	An abnormally high percentage of body fat.	Fatty tissue has a lower rate of metabolism than lean body tissue.
Sedative drugs	These agents depress the activity of the sympathetic nervous system.	As in depression, the subnormal secretion of stimulatory hormones results in a lowering of the BMR.
	Factors That May Raise the BMR	
Anemia	A deficiency of red blood cells and a thinning (lowered viscosity) of the blood.	The heart has to work harder to pump more than the normal amount of blood to the tissues in order to achieve the normal oxygenation.
Anxiety	The sympathetic nervous system is hyperactive.	Increased activity of the nerves stimulates an increased metabolism.
Cardiovascular disorders	Congestive heart failure (the heart muscle fails to function efficiently).	The heart muscle races in order to compensate for its failure and the effort requires the metabolism of extra nutrients.
	Hypertension (high blood pressure).	The heart has to work harder and burn more food energy to pump at the elevated pressure.
Fever	The body temperature is elevated.	BMR is increased about 7% for each degree (F) of fever (14%/°C).
Glandular oversecretions	Acromegaly (oversecretion of growth hormone).	The metabolism of carbohydrates, fats, and proteins is stepped up in this disorder.
	Cushing's disease (oversecretion of adrenal cortical hormones).	Protein metabolism is stepped up as amino acids are converted into glucose.
	Hyperthyroidism (oversecretion of thyroid hormones).	Energy metabolism is increased significantly and there is a rapid heart rate.
	Pheochromocytoma (oversecretion of epinephrine by an adrenal tumor).	Epinephrine increases oxygen consumption and oxidative metabolism.

(Continued)

TABLE C-5 *(Continued)*

Type of Factor	Description	Comments
Lactation	The production of milk by a mother who nurses her baby.	Milk production requires at least 30 kcal per oz *(30 ml)*, and about 30 oz *(900 ml)* are produced daily.
Lung disorders	Asthma and emphysema.	Clogged or constricted airways make it necessary for the lungs to work harder.
Parkinson's disease	Characterized by trembling and other involuntary movements.	The uncontrollable movements require an increase in energy metabolism.
Pregnancy	The extra tissues of pregnancy add to the metabolic workload.	The BMR is increased by 25% or more.
Stimulant drugs	Aspirin and other salicylates.	Salicylates cause the release of active thyroid hormones from the protein binding which inactivates them.
	Caffeine (usually taken in coffee or in stimulant pills).	Stimulates the sympathetic nervous system, which acts to increase the BMR.
Stress	Hypersecretion of stress hormones.	The metabolism of carbohydrates, fats, and proteins is stepped up.
Thinness	An abnormally high percentage of lean body tissue.	Lean body tissue has a higher rate of metabolism than fatty tissue.
Trauma	An injury, wound, or an event which brings on shock.	Stress hormones raise the metabolic rate as the body attempts to heal the trauma.

RESPIRATORY QUOTIENT (RQ). This measurement may be made as part of the determination of the basal metabolic rate, or it may be measured separately in order to provide other information, such as the amounts and types of the nutrients which are metabolized under various circumstances. The noted American physiologist Eugene DuBois made many determinations of the RQ for the latter purpose.

The ratio of carbon dioxide produced to oxygen consumed, which is known as the respiratory quotient (RQ), is distinctive for each of the three major types of energy-yielding nutrients; hence, it serves to indicate the type(s) of nutrient being metabolized. Respiratory quotient can be determined through the following equation:

$$RQ = \frac{CO_2 \text{ produced}}{O_2 \text{ consumed}}$$

Therefore, if we utilize the equation for the catabolism of carbohydrates, we can calculate a respiratory quotient:

1. $C_6H_{12}O_6 + 6\ O_2 \rightarrow 6\ CO_2 + 6\ H_2O + \text{heat}$
2. $RQ = \frac{6\ CO_2}{6\ O_2} = 1$

The RQ for fat would be less than 1, as seen in the following example:

1. Palmitic acid $(C_{16}H_{32}O_2) + 23\ O_2 \rightarrow 16\ CO_2 + 16\ H_2O + \text{heat}$
2. $RQ = \frac{16\ CO_2}{23\ O_2} = .70$

Most mixed fats have RQ values of about 0.7, while short-chained fatty acids have higher values (about 0.8). Proteins have values intermediate between carbohydrates and fats—about 0.81.

It should be noted that in stress conditions, the RQ value can be greater than 1. This can occur when there is hyperventilation where large quantities of carbon dioxide are exhaled while an oxygen debt exists in the body. Metabolic acidosis creates an excess exhalation of carbon dioxide as a compensatory mechanism.

RECOMMENDED CALORIC (Energy) INTAKES. In lieu of individual calorimetric studies, it is possible to establish daily caloric allowances by utilizing standard data that have been compiled by the leading authorities. A good way to begin this procedure is to consider the recommended energy intakes that have been published by the Food and Nutrition Board, which are shown in Table C-6.

These data are only very rough approximations which may be refined somewhat by a close scrutinizing of the major factors that influence total energy expenditures. Adult needs will be considered first, then those of infants, children, and adolescents, followed by the special circumstances that may also influence the energy expenditures.

Allowance for Adults. In many cases, the energy needs of adult males and females are determined mainly by the average daily expenditure of energy in physical activities. Special allowances have to be made for females who are either pregnant or lactating. Table C-7 shows how the average amounts of time spent engaged in various energy-consuming activities may be utilized in the estimation of daily energy needs.

It may be seen from Table C-7 that the estimated daily expenditures of typical American men and women are based entirely on sedentary, light, and moderate levels of activity. However, increases in the energy allowances should be made from persons who are both larger than the reference persons and more active than shown in the example. For example, people who spend significantly more time in activities requiring moderate, vigorous,

TABLE C–6
MEAN HEIGHTS AND WEIGHTS AND RECOMMENDED CALORIC (ENERGY) INTAKES[1,2]

Category	Age	Weight		Height		Energy Needs	
	(years)	(lb)	*(kg)*	(in.)	*(cm)*	(kcal/kg)	(per day)
Infants	0.0–0.5	13	*6*	24	*60*	108	650
	0.5–1.0	20	*9*	28	*71*	98	850
Children	1–3	29	*13*	35	*90*	102	1300
	4–6	44	*20*	44	*112*	90	1800
	7–10	62	*28*	52	*132*	70	2000
Males	11–14	99	*45*	62	*157*	55	2500
	15–18	145	*66*	69	*176*	45	3000
	19–24	160	*72*	70	*177*	40	2900
	25–50	174	*79*	70	*176*	37	2900
	51+	170	*77*	68	*173*	30	2300
Females	11–14	101	*46*	62	*157*	47	2200
	15–18	120	*55*	64	*163*	40	2200
	19–24	128	*58*	65	*164*	38	2200
	25–50	138	*63*	64	*163*	36	2200
	51+	143	*65*	63	*160*	30	1900
Pregnancy	2nd & 3rd trimester						+300
Lactation							+500

[1]Adapted by the authors from *Recommended Dietary Allowances,* 10th ed., 1989, NRC–National Academy of Sciences, p. 33, Table 3–5.

[2]The data in this table have been assembled from the observed median heights and weights of children, together with desirable weights for adults based on the mean heights of men (70 in., or *178 cm*) and women (64 in., or *163 cm*) between the ages of 18 and 34 years as surveyed in the U.S. population (HEW/NCHS data).

The energy allowances for the young adults are for men and women doing light work.

and/or strenuous energy expenditure may be allowed 300 or more extra Calories (*kcal*) daily.

Allowances for Infants, Children, and Adolescents. It is difficult to determine the caloric needs of individuals within these groups because of the great variations in heights, weights, and participation in physical activities. This is particularly true in the case of older adolescents because differences in the rates of growth are accentuated with the passage of time. For example, 18-year-old boys in the 95th percentile on the growth charts by the National Center for Health Statistics (NCHS) weigh about 210 lb (*96 kg*), whereas those in the 5th percentile weigh about 119 lb (*54 kg*). Furthermore, many children are transported to school and have only minimal exercise, where others engage in strenuous activities for an hour or more daily.

One way of coping with the uncertainty of caloric allowances for our youth has been the derivation of formulas such as the one which follows:

Calories/kg = 100 – (3 × age in years).[8]

In terms of pounds, this formula may be expressed as follows:

Calories/lb = 45 – (1.4 × age in years).

The formula is suitable for predicting the allowances of boys up to 14 years of age, and for girls up to 12. After that, other formulas which may be used are as follows:

[8]Wallace, W.M., "Quantitative Requirements of the Infant and Child for Water and Electrolyte under Varying Conditions," *The American Journal of Clinical Pathology,* Vol. 23.

I. Boys—Calories/lb = 19 (Appropriate for boys from ages 15 to 18)

II. Girls—Calories/lb = 35 – (1.1 × age in years from 13 to 18)

The above formulas are all based upon the concept of feeding growing youth on the basis of their body weights. Unfortunately, the thin person may receive too little and the obese one too much. Nevertheless, the prevailing belief among nutritionists today is that this approach, while faulty, is not as bad as stuffing thin children with food and starving the obese. Severe nutritional deprivation of obese children while they are growing rapidly may result in a reduction in growth and poor development of essential body functions. Furthermore, children who are obviously at either of these extremes should be examined by a pediatrician to make certain that they are not suffering from any metabolic or other disorders. Those who are moderately overweight according to their age, sex, and body build should be encouraged to engage in more vigorous activities, rather than restrict their diets drastically. It is noteworthy that the caloric allowances for children established by the Food and Agriculture Organization (FAO) of the UN are considerably higher than those of the U.S. Food and Nutrition Board. The main reason for this difference is that children in the developing countries expend much more energy in physical activities than American children.

(Also see CHILDHOOD AND ADULT NUTRITION; and INFANT DIET AND NUTRITION.)

Resting Energy Expenditure (REE). Unless levels of physical activity are very high, resting energy expenditure (REE) is the largest component of total energy expenditure. REE represents the energy expended by a

TABLE C-7
ESTIMATION OF DAILY CALORIC (ENERGY) NEEDS OF ADULTS[1]

Type of Activity[2]	Time	Man, 154 lb (70 kg)		Woman, 123 lb (56 kg)	
		Rate	Total	Rate	Total
	(hr)	(kcal/hr)	(kcal)	(kcal/hr)	(kcal)
Sleeping	8	75	600	60	480
Sedentary Reading, writing, watching TV, sewing, typing, or miscellaneous work or games while seated	12	100	1,200	80	960
Light Food preparation, dusting, ironing, washing dishes, walking slowly (2½ to 3 mph), or shopping	3	160	480	110	330
Moderate Walking at 3½ to 4 mph, making beds, sweeping floors, washing clothes in a machine, or table tennis	1	240	240	170	170
Vigorous Gardening, bowling, golfing, scrubbing or waxing floors, or other heavy work	0	350	—	250	—
Strenuous Work with pick and shovel, running, dancing, swimming, tennis, skiing or bicycling (7 mph)	0	350+	—	350+	—
Total	24		2,520		1,940

[1]Based mainly upon data in *Food*, Home and Garden Bull. No. 228, USDA, pp. 14-15.
[2]One mi equals 1.6 km.

person at rest under conditions of thermal neutrality. Basal metabolic rate (BMR) is more precisely defined as the REE measured soon after awakening in the morning, at least 12 hours after the last meal. REE is not usually measured under basal conditions. REE may include the residual thermic effect of a previous meal and may be lower than BMR during quiet sleep. In practice, BMR and REE differ by less than 10%, and the terms are used interchangeably. The values used in Table C-8 were derived from equations published by WHO (1985). These equations take into account age, sex, and weight, but ignore height.

Special Circumstances that Affect Energy Requirements. The major conditions which require adjustment of energy allowances follow.

PREGNANCY. The National Research Council Committee on Maternal Nutrition recommends that women, who start pregnancy with a normal body weight, eat sufficient calories to bring about a weight gain of about 24 lb (*11 kg*). Teenagers who become pregnant should gain this amount of weight *plus* whatever they would normally gain during this time if they were not pregnant. Women who are underweight when they conceive should also gain more than the recommended 24 lb. However, women who are overweight at the time of conception should *not* attempt to gain less than 24 lb. Physiologists have determined that the daily caloric intake of healthy pregnant women should be at least 16 Calories (*kcal*) per lb (*36 kcal per kg*) of pregnant body weight. Normally, meeting this requirement would require the consumption of 300 extra Calories (*kcal*) per day through pregnancy, or about 150 Calories (*kcal*) daily during the first trimester and 350 Calories (*kcal*) daily during the second and third trimesters.

LACTATION. It has been estimated that the production of about 30 oz (*850 to 900 ml*) of breast milk daily expends from 750 to 900 Calories (*kcal*) per day. However, it is expected that most healthy women who consume adequate food energy during pregnancy will have gained from 4 to 9 lb (*2 to 4 kg*) of extra body fat which will furnish some of the extra calories for lactation. Therefore, the Recommended Dietary Allowances specify that about 500 extra Calories (*kcal*) should be consumed daily during lactation.

ENVIRONMENTAL STRESSES. Recently, the traditional concepts regarding the need to make special caloric allowances for climate have been revised significantly. The latest thinking in this regard is that working in a cold environment requires only about 2 to 5% increase in the caloric allowance, which is mainly to

TABLE C–8
CALCULATING REE (RESTING ENERGY EXPENDITURE) USING BODY WEIGHTS[1]

Males		Females	
Age–years	REE–kcal/day	Age–years	REE–kcal/day
0–3	(60.9 × wt)[2] – 54	0–3	(61.0 × wt) – 51
3–10	(22.7 × wt) + 495	3–10	(22.5 × wt) + 499
10–18	(17.5 × wt) + 651	10–18	(12.2 × wt) + 746
18–30	(15.3 × wt) + 679	18–30	(14.7 × wt) + 496
30–60	(11.6 × wt) + 879	30–60	(8.7 × wt) + 829
60+	(13.5 × wt) + 487	60+	(10.5 × wt) + 596

[1]Adapted from *Recommended Dietary Allowances*, 10th ed., 1989, National Research Council, p. 26, Table 3–1.
[2]Weight of person in kilograms.

compensate for the extra work of physical activity while encumbered somewhat by heavy clothing. On the other hand, the energy allowances for people living in warm climates should be about the same as those for the more moderate environments. However, physical work at temperatures over 85°F (*30°C*) requires an extra energy allowance of about 0.5% for each degree centigrade above this temperature because the heart has to work harder to keep the body cool.

DISEASES, INJURIES, AND SURGERY. It is very important to make certain that people in these circumstances receive adequate food energy so that dietary protein is utilized efficiently to promote healing and recovery. Furthermore, the basal metabolic rate may be increased significantly. For example, for each °F of fever, the metabolic rate is increased by about 7%. Multiple fractures may increase the resting energy expenditure from 10 to 30%, and recovery from surgery may involve up to a 10% increase in metabolism.

People who are in any way hobbled by having to use crutches or various prosthetic devices may expend extra energy in normal activities. Victims of cerebal palsy and Parkinson's disease are also likely to require extra food energy when they are ambulatory.

SUMMARY: BALANCING ENERGY INTAKES AND EXPENDITURES. The U.S. Food and Nutrition Board has recommended that, whenever feasible, efforts to reduce excessive body weight should include only a moderate restriction of dietary calories which is accompanied by increased physical activity, because it is difficult to ensure the nutritional adequacy of diets that are too low in energy content (less than 1,800 to 2,000 kcal daily). Therefore, the authors recommend that healthy people step up their activity so that they may continue to eat a well-balanced diet containing a variety of foods. Some guidelines on the exercise equivalents of various foods are given in Fig. C-23.

(Also see OBESITY.)

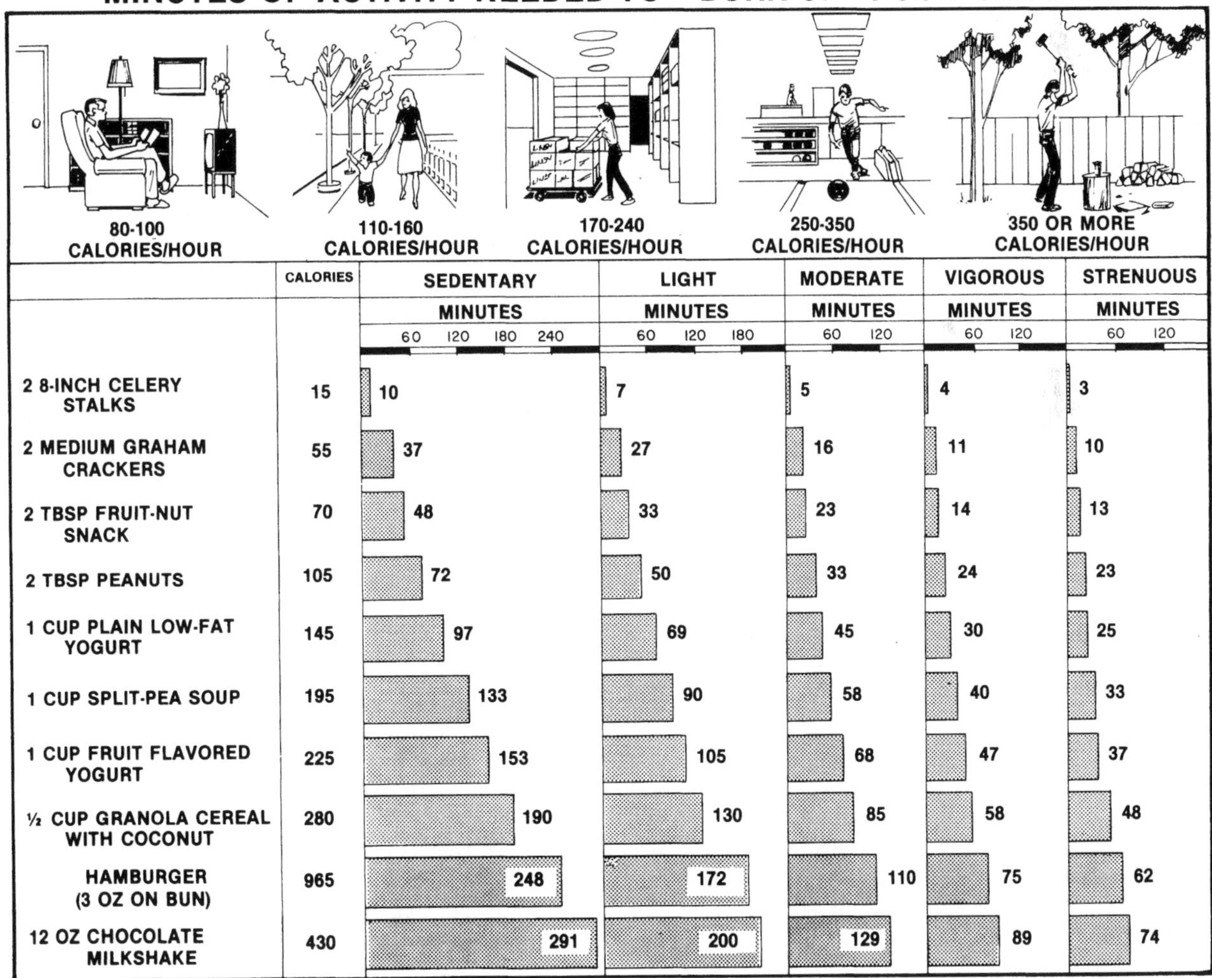

MINUTES OF ACTIVITY NEEDED TO "BURN-UP" FOOD CALORIES

	CALORIES	SEDENTARY (80-100 CALORIES/HOUR) MINUTES	LIGHT (110-160 CALORIES/HOUR) MINUTES	MODERATE (170-240 CALORIES/HOUR) MINUTES	VIGOROUS (250-350 CALORIES/HOUR) MINUTES	STRENUOUS (350 OR MORE CALORIES/HOUR) MINUTES
2 8-INCH CELERY STALKS	15	10	7	5	4	3
2 MEDIUM GRAHAM CRACKERS	55	37	27	16	11	10
2 TBSP FRUIT-NUT SNACK	70	48	33	23	14	13
2 TBSP PEANUTS	105	72	50	33	24	23
1 CUP PLAIN LOW-FAT YOGURT	145	97	69	45	30	25
1 CUP SPLIT-PEA SOUP	195	133	90	58	40	33
1 CUP FRUIT FLAVORED YOGURT	225	153	105	68	47	37
½ CUP GRANOLA CEREAL WITH COCONUT	280	190	130	85	58	48
HAMBURGER (3 OZ ON BUN)	965	248	172	110	75	62
12 OZ CHOCOLATE MILKSHAKE	430	291	200	129	89	74

Fig. C-23. Exercise equivalents of the Calories provided by some common foods. (Based upon data in *Food,* Home and Garden Bull. No. 228, USDA, pp. 14-15)

CALORIC VALUES OF FOODS

It is necessary for nutritionists to obtain accurate estimates of the caloric values of various diets in order to help people avoid the consumption of too little or too much food energy. For example, a chronic dietary deficit of calories may result in the wasting of muscles and other body tissues and a temporary or permanent impairment of certain functions. On the other hand, long-term dietary excesses of food energy may lead to obesity which is believed to increase the risk of diabetes, heart disease, and high blood pressure.

The caloric contents of diets are usually calculated by the use of *Atwater caloric conversion factors* which were derived for the mixed diets consumed by Americans around the turn of the century. These factors are based upon the assumptions that each gram of carbohydrate, fat, and protein in the diet will yield 4, 9, and 4 Calories (kcal), respectively. However, the Atwater factors were not intended to be used for single foods or for mixed diets that differed markedly in composition from those for which they were derived. Today, there are a wide variety of diets that have been drastically modified from the average American dietary pattern. Therefore, it is important that dietary planners understand the basic principles of food calorimetry.

HISTORY. Only a few people are aware that the first measurements of caloric values of food were made in 1866 by the English chemist Frankland, who oxidized samples of 29 commonly consumed items in a crude calorimeter. He also determined the caloric values of products of metabolism such as urea and uric acid. His measurements lacked the precision of those that were made later, but his procedures laid the foundation for the workers who were to follow. It is noteworthy that Frankland worked for a short time under the famous German chemist Liebig, who has been called the "father of organic chemistry." (Liebig trained the German scientist Voit, who in turn trained the German physiologist, Rubner, and the American nutritionist, Atwater.)

Caloric conversion factors—the respective Calorie-yielding values of 1 g of protein, of fat, and of carbohydrate—were first determined by Rubner in 1884. His factors—4.1, 9.3, 4.1—took into account only those losses due to the incomplete oxidation of protein in the body and were therefore, slightly higher than those published by Atwater in 1899, which allowed for all possible losses. (Atwater made these studies in the late 1890s at the Connecticut Agricultural Experiment Station in New Haven.) These first energy standards were based to a large extent upon dietary studies which had already been made by Voit and students in his laboratory before Rubner determined the caloric-value factors.

The Atwater procedure, in brief, was to adjust the heats of combustion (gross calories) of the fat, protein, and carbohydrate in a food to allow for the losses in digestion and metabolism found for human subjects, and to apply the adjusted caloric factors to the amounts of protein, fat, and carbohydrate in the food. The contents of protein and fat were determined by chemical analyis, and the percentage of carbohydrate was obtained by difference; that is, it was taken as the remainder after the sum of the fat, protein, ash, and moisture had been deducted from 100. This so-called total carbohydrate, therefore, included fiber (an all-inclusive term for carbohydrates that are not digested by people) as well as any noncarbohydrate residue present.

CALORIC CONVERSION FACTORS FOR THE MAJOR GROUPS OF FOODS. The energy factors that Atwater derived from his digestibility experiments have been expanded and modified to take into account additional experiments conducted with human subjects since his time. The factors for the major groups of foods are shown in Table C-9.

It may be seen from Table C-9 that there are wide variations in the energy factors that are appropriate for the different types of foods, and that rather large errors may result from the use of the standard Atwater factors without consideration of the types of food which constitute the diet. For example, the protein in cornmeal is only 60% digestible, and the fat is only 90% digestible, whereas the proteins of eggs, fish, meat, and milk are 97% digestible and the fats are 95% digestible. Similarly, wheat bran and various vegetables also have lower digestibility than the more highly digestible foods derived from animals. Hence, the caloric conversion factors for certain plant foods are considerably lower than those for most animals.

USE AND MISUSE OF ATWATER FACTORS. Reasonably accurate estimates of the caloric contents of diets may be obtained when the standard Atwater factors are applied to groups of foods that are similar to those in the average American dietary pattern, which is usually comprised mainly of highly digestible foods. However, these factors will yield an overestimation of the caloric contents of diets based largely on vegetable foods that have significantly lower digestibilities of their carbohydrates, fats, and proteins. Therefore, dietary planners who are evaluating unusual dietary patterns should obtain information regarding the digestibility of the constituent foods before attempting to calculate caloric values.

CALORIE (cal)

This unit, which is always written with a small c, is the amount of energy heat required to raise the temperature of 1 gram of water 1°C (precisely from 14.5 to 15.5°C). It is equivalent to 4.184 joules. *In popular writings, especially those concerned with human caloric requirements, the term "calorie" is frequently used erroneously for the "Kilocalorie" (k-calorie or kcal).*

(Also see CALORIMETRY; and KILOCALORIE.)

CALORIES, EMPTY

A term identifying foods that supply energy (calories) only, while other nutrients such as minerals, vitamins and proteins are missing or present in very low levels. For example, table sugar made from sugarcane or sugar beets is almost pure carbohydrate and provides calories only.

(Also see BREAKFAST CEREALS, section headed "The Sugar-Breakfast-Cereal Binge.")

TABLE C-9
DATA USED FOR CALCULATING ENERGY VALUES OF FOODS OR FOOD GROUPS BY THE ATWATER SYSTEM[1]

Food or food group	Protein			Fat			Carbohydrate		
	Coefficient of digestibility	Heat of combustion less 1.25[2]	Factor to be applied to ingested nutrients	Coefficient of digestibility	Heat of combustion	Factor to be applied to ingested nutrients	Coefficient of digestibility	Heat of combustion	Factor to be applied to ingested nutrients
	%	cal./g	cal./g	%	cal./g	cal./g	%	cal./g	cal./g
Eggs, Meat products, Milk products:									
Eggs	97	4.50	4.36	95	9.50	9.02	98	3.75	3.68
Gelatin	97	4.02	3.90	95	9.50	9.02	—	—	—
Glycogen	—	—	—	—	—	—	98	4.19	4.11
Meat, fish	97	4.40	4.27	95	9.50	9.02	—	—	—[3]
Milk, milk products	97	4.40	4.27	95	9.25	8.79	98	3.95	3.87
Fats, separated:									
Butter	97	4.40	4.27	95	9.25	8.79	98	3.95	3.87
Other animal fats	—	—	—	95	9.50	9.02	—	—	—
Margarine, vegetable	97	4.40	4.27	95	9.30	8.84	98	3.95	3.87
Other vegetable fats and oils	—	—	—	95	9.30	8.84	—	—	—
Fruits:									
All (except lemons, limes)	85	3.95	3.36	90	9.30	8.37	90	4.00	3.60
All fruit juice (except lemon, lime) unsweetened	85	3.95	3.36	90	9.30	8.37	98	4.00	3.92
Lemons, limes	85	3.95	3.36	90	9.30	8.37	90	2.75	2.48
Lemon juice, lime juice, unsweetened	85	3.95	3.36	90	9.30	8.37	98	2.75	2.70
Grain products:									
Barley, pearled	78	4.55	3.55	90	9.30	8.37	94	4.20	3.95
Buckwheat flour, dark	74	4.55	3.37	90	9.30	8.37	90	4.20	3.78
Buckwheat flour, light	78	4.55	3.55	90	9.30	8.37	94	4.20	3.95
Cornmeal, whole-ground	60	4.55	2.73	90	9.30	8.37	96	4.20	4.03
Cornmeal, degermed	76	4.55	3.46	90	9.30	8.37	99	4.20	4.16
Dextrin	—	—	—	—	—	—	98	4.11	4.03
Macaroni, spaghetti	86	4.55	3.91	90	9.30	8.37	98	4.20	4.12
Oatmeal, rolled oats	76	4.55	3.46	90	9.30	8.37	98	4.20	4.12
Rice, brown	75	4.55	3.41	90	9.30	8.37	98	4.20	4.12
Rice, white or polished	84	4.55	3.82	90	9.30	8.37	99	4.20	4.16
Rye flour, dark	65	4.55	2.96	90	9.30	8.37	90	4.20	3.78
Rye flour, whole-grain	67	4.55	3.05	90	9.30	8.37	92	4.20	3.86
Rye flour, medium	71	4.55	3.23	90	9.30	8.37	95	4.20	3.99
Rye flour, light	75	4.55	3.41	90	9.30	8.37	97	4.20	4.07
Sorghum *(kaoliang)*, whole or nearly whole meal	20	4.55	.91	90	9.30	8.37	96	4.20	4.03
Wheat, 97-100 percent extraction	79	4.55	3.59	90	9.30	8.37	90	4.20	3.78
Wheat, 85-93 percent extraction	83	4.55	3.78	90	9.30	8.37	94	4.20	3.95
Wheat, 70-74 percent extraction	89	4.55	4.05	90	9.30	8.37	98	4.20	4.12
Wheat, flaked, puffed, rolled, shredded, whole meal	79	4.55	3.59	90	9.30	8.37	90	4.20	3.78
Wheat bran (100 percent)	40	4.55	1.82	90	9.30	8.37	56	4.20	2.35
Other cereals, refined	85	4.55	3.87	90	9.30	8.37	98	4.20	4.12
Wild rice	78	4.55	3.55	90	9.30	8.37	94	4.20	3.95
Legumes, Nuts	78	4.45	3.47	90	9.30	8.37	97	4.20	4.07
Sugars:									
Cane or beet sugar (sucrose)	—	—	—	—	—	—	98	3.95	3.87
Glucose	—	—	—	—	—	—	98	3.75	3.68
Vegetables:									
Mushrooms	70	3.75	2.62	90	9.30	8.37	85	4.10	3.48
Potatoes and starchy roots	74	3.75	2.78	90	9.30	8.37	96	4.20	4.03
Other underground crops[4]	74	3.75	2.78	90	9.30	8.37	96	4.00	3.84
Other vegetables	65	3.75	2.44	90	9.30	8.37	85	4.20	3.57
Miscellaneous foods:									
Alcohol[5]	—	—	—	—	—	—	—	—	—
Chocolate, cocoa	42	4.35	1.83	90	9.30	8.37	32	4.16	1.33
Vinegar	—	—	—	—	—	—	98	2.45	2.40
Yeast	80	3.75	3.00	90	9.30	8.37	80	4.20	3.35

[1]Watt, B. K., and A. L. Merrill, *Composition of Foods*, Ag. Hdbk. No. 8, USDA, 1963, p. 160.

[2]It is necessary to subtract 1.25 Calories per gram from the heat of combustion of protein because this nutrient is incompletely converted to heat in the body. This correction gives values that are applicable to grams of digested protein (protein intake X % digestibility), and that are identical to the factors derived originally by Atwater.

[3]The small amounts of carbohydrates present in brain, heart, kidney, and liver yield 3.87 Calories per gram; and those in tongue and shellfish yield 4.11 Calories per gram.

[4]Vegetables such as beets, carrots, onions, parsnips, radishes.

[5]Each gram of alcohol yields 6.93 Calories in the body because it is 98% digestible and has a heat of combustion of 7.07 Calories per gram.

CALORIMETER

In nutrition, this term designates any apparatus used for measuring (1) the heat content of food, or (2) the heat given off by living organisms.

(Also see BOMB CALORIMETER; CALORIC [ENERGY] EXPENDITURE; and CALORIC VALUES OF FOODS.)

CALORIMETRY

• **Direct calorimetry**—The following two methods of direct calorimetry are employed; the first one for foods, the second one for people:

1. The heat content of foods is determined by burning them in a bomb calorimeter.

2. The heat given off by a person engaged in various activities is determined by placing the subject in an insulated chamber that contains pipes carrying circulating water. The heat given off by the subject is calculated from the rise in temperature of a fixed quantity of water that enters and leaves the chamber. In a more recent type, the *gradient layer calorimeter*, the quantity of heat is measured electrically as it passes through the wall of the chamber.

• **Indirect calorimetry**—In indirect calorimetry, the heat produced in metabolism is calculated by measuring, in a fixed period of time, the respiratory exchange of the person—the oxygen consumed, and usually the CO_2 exhaled.

(Also see ANALYSIS OF FOODS, section on "Calorimetry and Spectrophotometry"; BOMB CALORIMETER; CALORIC VALUES OF FOODS; CALORIE [ENERGY] EXPENDITURE; and METABOLISM.)

CAMOMILE

This is a member of the daisy family. Its flowers have been touted as a minor complaint cure-all for centuries in Europe. It is a popular folk remedy for mild stomach upset, menstrual cramps, throat sprays, inhalants, ointments, creams, and lotions.

Caution: Although camomile is considered a safe herbal remedy, it may cause an allergic reaction in some individuals.

CAMU-CAMU *Myrciaria paraensis*

The small grapelike fruit of a bush (of the family *Myrtaceae*) that is native to Peru. Camu-camu berries are red-colored, and about 1 in. (*3 cm*) in diameter. They may be eaten fresh, or made into jam, jelly, or wine.

CANADIAN BACON

Bacon made from pork loins, which are given a light cure and smoked.

CANCER

Contents *Page*

Types of Cancer 317
Commonly Afflicted Organs and Mortality Rates 317
Causes of Cancer—Carcinogens 318
Chemical Agents 318
Tobacco 319
Alcohol 319
Naturally Occurring Chemicals 319
Food Additives 319
Air and Water Pollution 320
Physical Agents 320
Age 320
Radiation 320
Continued Irritation 321
Heredity 321
Predisposing Factor 321
Directly Hereditary 321
Viruses 321
Emotions and Stress 321
Nutrition and Diet 322
Detection 322
Signs 323
Tests 323
Therapy—Treatment 323
Surgery 323
Radiation 323
Chemotherapy 324
Nutrition and Diet in Cancer Therapy 324
Status Prior to and During Therapy 324
As a Therapy 326
Prevention 326
Environment 326
Nutrition and Diet in Cancer Prevention 326
The Future 327

Americans have reason to worry about cancer! In 1991, 1,100,000 people in the United States were told that they had a life-threatening malignancy; and that same year 514,000 people died from cancer.

The characteristic of cancer is wild, uncontrolled growth of cells which can arise in any organ or tissue of the body. Cancer cells have a rather primitive appearance when compared to normal cells. They are quite irregular in shape and they have large irregular nuclei. Fig. C-24 illustrates the difference between normal cells and cancer cells under the microscope.

All cells have the capacity to divide. This process is called *mitosis*—one cell becomes two, two cells become four, four cells become eight, and so on. When a cell divides, it goes through a cell cycle and the cell receives some sort of signal that starts and stops this process.

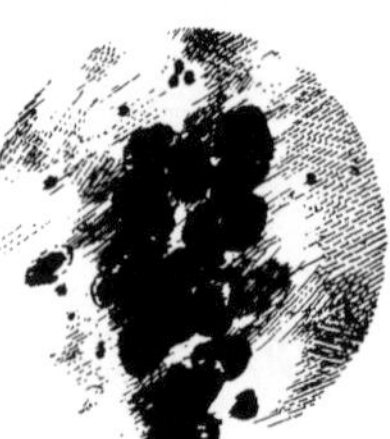

Fig. C-24. Microscopic views of normal cells on the left and cancer cells on the right.

Cancer cells divide more often. Out of 100 cancer cells, 10 or 20 of them will continuously be going through division and many cells pile up. The cancer cell, then, has an apparent autonomy due to some abnormality in the chromosomes. It will continue to grow regardless of the size of the mass it creates. As cancers continue to grow, they interfere with body function by crowding normal organs and outstripping the food supply available to the patient, eventually killing the patient.

TYPES OF CANCER. Tumors formed from rapidly dividing cells are of two types: benign and malignant. Benign tumors are those that cannot invade the surrounding tissues and remain as strictly local growths. The more dreaded malignant tumors or cancers are those that spread from their site of origin and, therefore, can reach the bloodstream and lymphatic system. These cancers are divided into three broad groups:

1. The **carcinomas**, which arise in the epithelial tissues—the sheets of cells covering the surface of the body and lining the various glands.
2. The **sarcomas**, which are rather rare, and which arise in supporting structures such as fibrous tissue or connective tissue and blood vessels.
3. The **leukemias** and the **lymphomas**, which arise in blood-forming cells of the bone marrow and lymph nodes.

Cancers are further classified by the organ in which they originate and by the kind of cell involved. Considered this way, there are 100 or more distinct varieties of cancer. Most of these are rare.

COMMONLY AFFLICTED ORGANS AND MORTALITY RATES. Most new cases of cancer and cancer mortality can be accounted for by considering a fairly short list. Twelve leading cancer sites are shown in Fig. C-25.

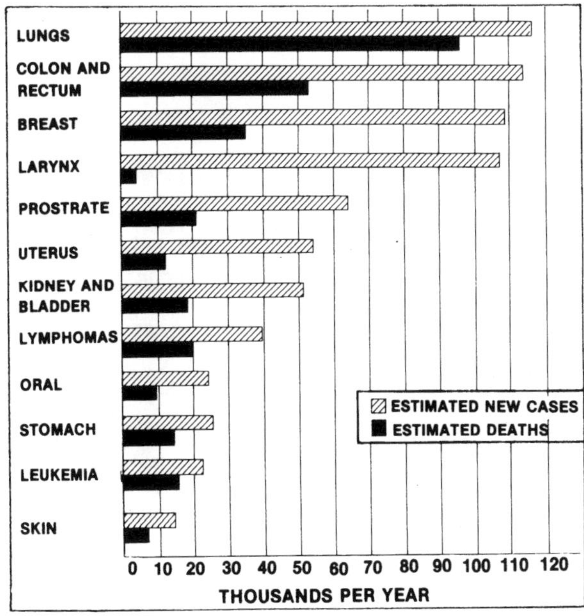

Fig. C-25. Leading cancer sites.

In terms of total deaths, cancers are responsible for about 22% of the deaths in the United States. A majority of the people, 78%, die from some other cause. Of the cancers, roughly half of the deaths are caused by cancer of three organs: the lungs, the colon or rectum, and the breast. Leukemia and lymphomas—cancers of the blood and immune system—are important mainly because they often affect children and young adults. While the estimated new cases is a useful statistic, mortality is the most accessible and reliable indicator of the impact of cancer. Too many variables influence the diagnosis of cancer to make new cases a reliable statistic. Death from cancer is a cold hard fact.

Through the years, several of the cancers have shown site and mortality rate patterns. Fig. C-26 illustrates these patterns.

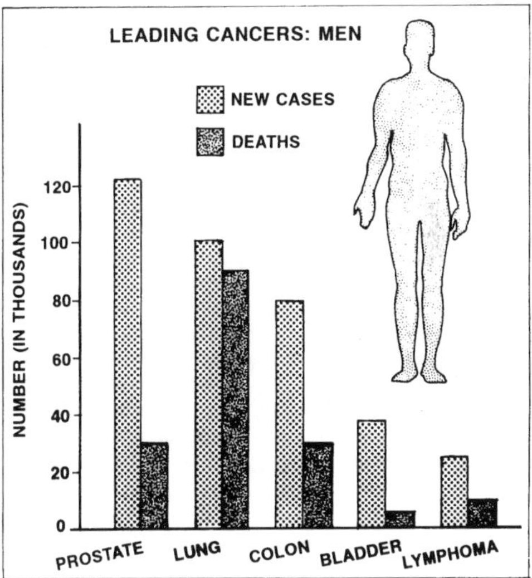

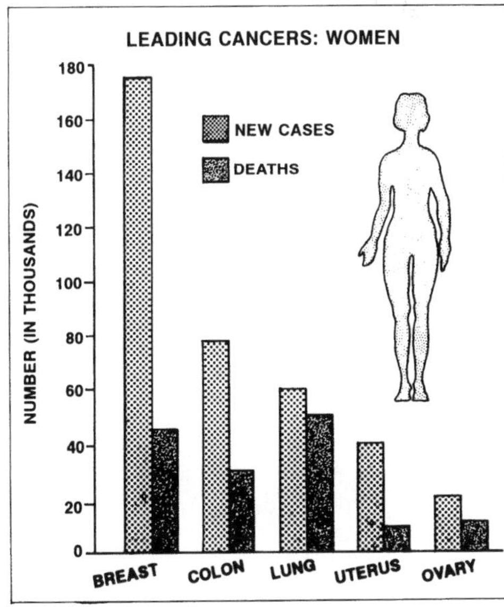

Fig. C-26. Leading cancer sites in (a) men and (b) women, with new cases and deaths of each. (Source: *American Cancer Society Facts and Figures*, 1991)

As shown in Fig. C-26, the most common sites of cancer in men, by rank, are: prostate, lung, and colon, with lung cancer being the leading cause of deaths from cancer.

In women, breast cancer is by far the leading site of cancer, with death from cancer highest from lung cancer, followed in order by breast cancer and colon cancer.

It is noteworthy that lung cancer is the leading cause of deaths from cancer in both men and women.

For unknown reasons, cancer of the pancreas is increasing in incidence. In 1990, an estimated 5% of both males and females in the United States died from cancer of the pancreas.

CAUSES OF CANCER—CARCINOGENS. At times we all may feel that everything we enjoy is either illegal, immoral, fattening, or causes cancer. Indeed, the list of suspected carcinogens continues to grow. Still the cause of cancer is more complex than just the exposure to a carcinogen. Many factors must be considered, including chemical agents, physical agents, heredity, viruses, emotions and stress, and diet and nutrition.

Chemical Agents. Chemicals have long been a fact of life; by no means are they a creation of the 20th century. However, the word "chemical" has been given a negative connotation by some, as if all chemicals or anything containing chemicals causes cancer. Some chemicals do cause cancer, but *not* all chemicals.

In 1775, a prominent English surgeon, Percivall Pott, wrote a book in which he devoted a few pages to the subject of cancer of the scrotum in chimney sweeps. He attributed this cancer to the long exposure and intimate contact with tarry soot. It was not until 1915 that two persistent Japanese scientists demonstrated the formation of cancer by the application of tar to the skin of rabbits. Then, in 1933 a pure chemical, benzopyrene, isolated from tar was shown to produce cancer in mice. Thus, the search for chemical carcinogens was born. Now, numerous chemicals are scrutinized for carcinogenic activity in an effort to help us make decisions about the chemicals we swallow, inhale, or rub on our skin. To do this, mice, rats, hamsters, or sometimes larger animals, are used by researchers to test for carcinogens. Hence, animals are "stand-ins" for humans.

• **Animal testing**—To overcome the prohibitive cost of exposing large numbers of animals to low doses of a test chemical, animals are usually given the highest daily dose of the chemical that they can tolerate without ill effects. The reasoning behind this method is that since only about 500 animals are used to test a chemical, the test could easily miss potential carcinogens unless some method is employed to increase the frequency of cancers in the animals. A high daily dose is currently the method of choice.

• **Animals vs humans**—Once a test demonstrates that a chemical is capable of causing cancer in animals, the next question is whether the chemical will cause cancer in people. We really do not know. The best answer seems to be "possibly." Also, the reverse is often true; that is, human carcinogens will generally cause cancer in animals—but not always. Thus, there are no easy answers, and all testing and interpretations are subject to human error. However, combined observation of human populations and laboratory testing have indicated that there are some chemicals which likely cause human cancers. The chemical structures of some of these carcinogens are shown in Fig. C-27.

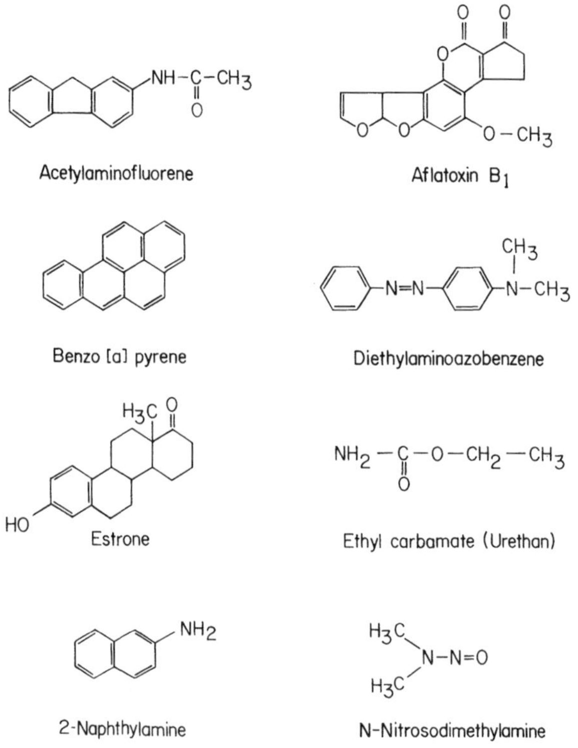

Acetylaminofluorene

Aflatoxin B_1

Benzo [a] pyrene

Diethylaminoazobenzene

Estrone

Ethyl carbamate (Urethan)

2-Naphthylamine

N-Nitrosodimethylamine

Fig. C-27. Chemical structure of some carcinogens.

Following the discovery of some of these carcinogens, there were numerous attempts to relate chemical structure to carcinogenicity. However, the various chemicals causing cancer are not known to share any specific structural similarities as Fig. C-27 illustrates.

While the carcinogenicity of many compounds remains questionable, a list of some carcinogenic hazards to man is presented in Table C-10. The carcinogenic ability of these chemicals has been fairly well documented over the years.

A discussion of the exposure of some of the more important ones follows.

One other concept relating to carcinogens should be defined: Sometimes the presence of other compounds in a mixture, which do not possess carcinogenic activity, will accelerate or stimulate a cancer-producing reaction. These compounds are referred to as cocarcinogens.

TABLE C–10
SOME CARCINOGENIC HAZARDS TO MAN

Chemical	Site of Cancer	Comments
Aflatoxins	Liver	Cancer of the liver is rare in Europe and North America, but common in many parts of Africa. Dietary intakes of aflatoxin in some parts of Africa may be as high as 21 mcg/kg of body weight. The same association exists in Thailand. Aflatoxins are potent carcinogens in laboratory animals.
Alkylating agents	Bone marrow, lung	Included are mustard gas—"war gas"—and derivatives, and melphalan and analogues. *Note well:* These agents are also used for treatment of cancer.
Aromatic amines	Bladder	Aniline dye workers who handle 2-naphthylamine and related chemicals are exposed to this hazard.
Arsenic	Skin, lung	Chemical element used in metallurgy.
Asbestos	Pleura, lung	Asbestos is an industrial chemical with over 3,000 uses, and represents a possible danger to individuals in shipbuilding, mining, construction, and brake lining occupations. The physical nature of asbestos seems to cause the problem. Composed of fibers of calcium-magnesium silicate which are present as dust.
Benzene	Bone marrow	Used in a variety of industrial processes and compounds. Harmful amounts may be absorbed through the skin.
Chrome and nickel ores	Lung, nasal sinus	Miners and workers who inhale demonstrate increased risk of cancer.
Cycasin	Kidney	It is a product of Cycas circinalis palm nut. Natives of the Pacific islands eat it and the bacteria of digestive tract split into methylazoxymethalol, a carcinogen.
Estrogens	Uterus, vagina	Diethylstilbestrol (DES) given to pregnant women to prevent abortion has been linked to cancer in offspring. Estrogen given to women for menopausal symptoms increases risk of uterine cancer.
Immuno-suppressive drugs	Lymphatic system	Immune-deficient individuals develop more cancers. Drugs suppressing the immune system are used in kidney transplants.
Polycyclic hydrocarbons	Lung, skin	Found in soot, tar, oil, and tobacco smoke. Benzopyrene is a specific polycyclic hydrocarbon.
Vinyl chloride	Liver	Used in the plastics industry and used as a refrigerant.

TOBACCO. At the top of the list of substances that contain carcinogens is tobacco. Chemists have isolated and identified at least a dozen carcinogenic chemicals of the hydrocarbon type in the "tars" from tobacco smoke. There is general agreement—and good evidence—that cigarette smoking contributes to a large proportion of cancer in both men and women—especially lung cancer. Although estimates of the number of lung cancer deaths caused by smoking vary, most experts agree that at least 80% and perhaps 90% of lung cancer deaths are the result of cigarette smoking. But smoking acts to increase cancer risks for body sites in addition to the lungs. Tobacco smokers have greater risks than nonsmokers for cancers of the mouth, throat, bladder, pancreas, and kidney. Scientists have estimated that cigarette smoking and other tobacco use is responsible for about 30% of all cancers, and that cigarette smoking causes one out of every five deaths in the United States. Also, it is noteworthy that secondhand smoke is a health hazard to nonsmokers.

ALCOHOL. It is well established that there is an association between excessive alcohol consumption and cancer of the mouth, throat, and esophagus. Furthermore, persons drinking large quantities of alcohol may develop nutritional deficiencies leaving them more susceptible to the action of tobacco, alcohol, and other possible carcinogens.

NATURALLY OCCURRING CHEMICALS. Cancer-producing chemicals also occur in nature in plants and in molds. Exposure to these naturally occurring carcinogenic chemicals may be largely attributed to geographical location and culture. The nut of the *Cycas circinalis* palm used as food in the Pacific islands contains cycasin, a compound that produces cancer of the kidney and other organs in the rat. Alkaloids of *Senecio jacobaes*, a plant used by the African natives in their diet, and chili peppers, lead to the development of liver cancer in rats. While it is generally not a problem for humans in the United States, a common mold, *Aspergillus flavus*, which grows on wet peanuts, corn and other foods, forms potent liver carcinogens—the aflatoxins. Nitrosamines, another group of potent liver carcinogens, can be found in some plants, and some bacteria cause the formation of nitrosamines in foods. Safrole and related compounds, which are components of essential oils such as sassafras, produce liver cancer in rats. Antithyroid compounds found in plants such as turnips, kale, cabbage, and rapeseed produce thyroid tumors in test animals. The above are some of the more important potential carcinogens that occur in nature. There are more, and the list will undoubtedly grow as more information based on human population studies and animal studies is collected.

FOOD ADDITIVES. Food additives and their potential dangers can best be discussed by dividing food additives into two groups—intentional and incidental. Intentional are, of course, those added during the processing of a food to improve, maintain, or stabilize some characteristic of the food. Incidental additives are environmental chemicals of known origin and anticipated use; for example, pesticides, antibiotics, and growth stimulants. Due to the wording in the Delaney Clause of the Federal Food Drug and Cosmetic Act, the occurrence of additives, whether intentional or incidental, has created a dilemma. On the one hand, we hear that anything shown to cause cancer in man or animal shall not exist in the food supply at any level. On the other hand, we need to consider the risks versus the benefits in light of sound scientific judgment.

• **Intentional additives**—As more and more of the food additives are evaluated by animal studies, the list of suspected carcinogens grows. Some are synthetic chemicals while others are naturally occurring chemicals that have been extracted from plants and used in food processing. Thus, a substance can be both a food and a food additive.

• **Incidental additives**—These are governed by the same regulatory requirements as intentional food additives. Incidental additives may be transferred to foods by contact such as ingredients of packaging or of equipment surface coating. Also included are drugs and growth stimulants given to livestock or poultry or pesticides used on a food prior to processing. Undoubtedly, the most famous—or infamous—example is diethylstilbestrol (DES), a synthetic estrogen. It was used in the livestock industry as a growth promotant, and it is still used in human medicine as a "morning-after pill." Panic surrounding DES can be attributed to two things: (1) actual cancer caused by DES in daughters of mothers who received DES to prevent a miscarriage; and (2) technological advances which allow the determination of miniscule amounts—billionths and even trillionths—of chemicals in a food. (One part per trillion is roughly the equivalent of one grain of sugar in an Olympic-sized swimming pool.) Hence, a known carcinogen detected at *any* level in a food must be banned according to the Delaney Clause.

As testing continues and techniques are refined, other chemicals will certainly come under suspicion. Hopefully, we possess the good judgment to evaluate both the benefits and risks in terms of the findings.

(Also see ADDITIVES; DELANEY CLAUSE; and SACCHARIN.)

AIR AND WATER POLLUTION. About 10% of all cases of lung cancer occur in nonsmokers. This may in part be due to air pollution and other environmental factors. Densely populated and industrialized areas have higher mortality rates from cancer of the lung and digestive systems than rural areas. Numerous factors may account for this, but air and water pollution are suspect. Some of the common indices of air pollution are measures of suspended particulates, sulfates, nitrogen oxides, carbon monoxide, and hydrocarbons. Thus far, all attempts to define and measure the effects of air pollution on the incidence of cancer have not resolved a clear cut association.

From a list of organic compounds found in the drinking water in the United States, about two dozen suspected carcinogens have been identified. One group of organic contaminants, the trihalomethanes, may be related to increases in cancer of the urinary bladder and large intestine. There is no evidence to suggest that drinking fluoridated water, whether natural or artificial, increases the risk of cancer.

Physical Agents. Throughout our life we are exposed to numerous physical agents which may elicit some biochemical change in a cell and eventually lead to the development of cancer.

AGE. Although aging involves biochemical processes, it is observed as a physical occurrence. Aging definitely increases the risk of cancer as Fig. C-28 illustrates.

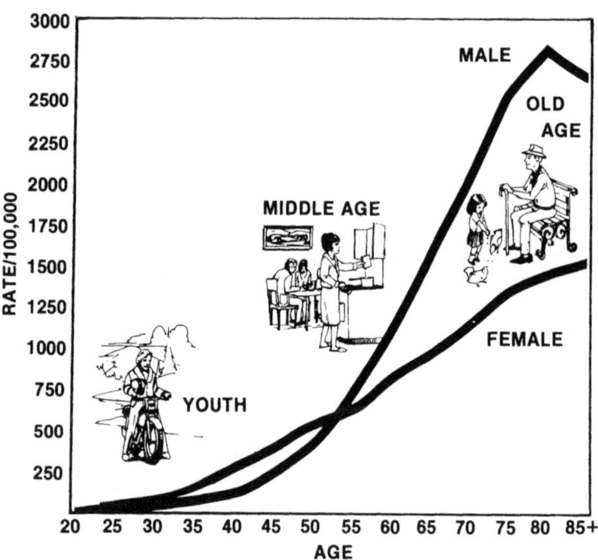

Fig. C-28. Age-specific cancer incidence rates in the United States by sex.

The incidence of almost all forms of cancer increases dramatically with advancing age. With age, cells accumulate damage to their genetic material—the chromosomes. The reason for this accumulated damage seems to be that the repair processes of the cell just slow down. This damage, may eventually lead to the development of cancer. Some other physical agents such as radiation and irritation may contribute to the genetic damage.

RADIATION. There is clear evidence that radiation can cause cancer in humans. Exposure comes from two sources, (1) natural and (2) man-made.

• **Natural**—The sun—the energy source upon which life itself depends—is a cancer-inducing stimuli. In 1894, Paul Unna of Germany related exposure to sunlight with skin cancer. Later, it was proven that the cancer-producing effects of sunlight were due to ultraviolet radiation. The incidence of skin cancer is highest in the white populations of the southern and western part of the United States. Furthermore, skin cancer occurs more frently among people who work outdoors.

• **Man-made**—Exposures, or at least the chance of exposures, to man-made radiation is a hotly debated issue. The people of Hiroshima and Nagasaki who survived the atomic bomb have had a higher incidence of leukemia, and of cancers of the lung, breast, and thyroid. Even closer to home are the people in southern Utah who were exposed to radiation following some of the Nevada tests. These people, too, have demonstrated an increased incidence of cancer. Also, during the 1920s radiation thereapy was considered good medical practice and effective treatment for ringworm of the scalp, enlargement of the thymus, acne, and enlargement of the tonsils and adenoids. Recently, a link has been established between these treatments and tumors of the thyroid. Most people, though, are not concerned with massive doses of radiation, but low-level radiation. It is believed that no level of radiation is entirely without risk; the concern is just how much risk?

CONTINUED IRRITATION. This is a rather vague cause, but there are some known illustrations. The lesions of syphilitic inflammation of the mouth (glossitis) and chronic irritation due to gallstones are two examples known to be associated with cancer. Broken or decayed teeth that continually irritate the mouth and tongue are associated with cancer in that area of the body. Individuals with ulcerative colitis are more apt to develop cancer of the colon. Cancer of the lower lip was common in the days when men smoked clay pipes. Possibly there are others that remain unknown—at this time.

Heredity. The role of heredity seems two-fold. Heredity is a predisposing factor in the development of cancer, and there are some cancers which are directly hereditary.

PREDISPOSING FACTORS. Susceptibility to some cancers seems to run in families. However, it is not easy to determine whether or not genetic factors are truly involved since close relatives also are likely to share the same environment. Of the common cancers, the one with the strongest likelihood to run in families is breast cancer, which is found at about twice the normal frequency in close relatives of patients with cancer of the breast. Recent studies have pinpointed some chromosomal changes which appear related to the development of breast cancer. Cancer of the rectum and lung may also aggregate in families. Furthermore, animal studies demonstrate animals can be selected for their susceptibility to certain cancers.

Some inborn disorders lead to cancer. These include the following: modularity syndromes such as neurofibromatosis and tuberous sclerosis; genetic skin disorders such as albinism; chromosome breakage disorders such as Down's, Bloom's, and Fanconi's syndrome; and immune deficiency syndromes such as agammaglobulinemia.

(Also see INBORN ERRORS OF METABOLISM.)

DIRECTLY HEREDITARY. Some known hereditary cancers of man include: retinoblastoma; multiple polyps of the colon; Gardner's syndrome; chemodectoma; pheochromocytoma; and thyroid cancer. Fortunately, all these are relatively rare and contribute little to the total cancer causes. However, study of these cancers may help determine the actual importance of genetic susceptibility to cancer.

Viruses. Again, from animal studies comes the idea that viruses may be involved in carcinogenesis. Certain leukemialike diseases of chickens, cats, and inbred strains of mice can be induced in young animals by inoculating them with viruses isolated from leukemic animals. If the same is true for some human cancers, then there is the hope of developing a vaccine to prevent the cancer. Unfortunately, no human cancer, or for that matter animal cancer, has proven that simple. As of yet, there is no hard evidence that any class of human cancers is caused by a virus. Still the search continues, and there is some evidence linking perhaps one of the most common of all the viruses—the Epstein-Barr virus—to two types of cancer, a carcinoma and a lymphoma. The future will tell.

Emotions And Stress. Think happy thoughts and avoid or cure cancer. If it were only that simple. There is one popular theory suggesting that destructive emotions—*stress*, or rather *distress*—can weaken the surveillance mechanism of the body—the immune system—which is constantly destroying abnormal cells that may otherwise cause cancer. Furthermore, there is also increased interest in the possibility that healthy emotions and good thoughts can be directed by a cancer patient to fight his disease. While there is no doubt that a holistic approach and the will to live are strongly tied to overcoming any disease, there is no hard evidence to support this position. On the other hand, our mental state, our response to our environment should not be disregarded. Physiological processes are influenced by sociologic and psychologic factors. Therefore, some influence of sociologic and psychologic factors upon the occurrence and growth of cancer should not surprise us. The degree of their importance is, however, a matter of question; nevertheless, it is a promising field of research.

Although there are some very convinçing testimonials, most ideas run something like this: Those people with a depressive response to life and its problems, who react to stress with feelings of despair, and who routinely deny and repress their emotions, appear to have a greater risk of developing cancer. Dr. Lawrence Le Shan, who studied numerous cancer patients, found a significantly higher percentage of the cancer patients experienced (1) the loss of crucial relationship, (2) the inability to express hostility, and (3) tension over the death of a parent.[9] In other studies, scientists have identified some of the following

[9]Le Shan, L. *You Can Fight For Your Life*, Harcourt Brace Jovanovich, New York, N.Y., 1978, pp. 32-36.

personality characteristics in many cancer patients: (1) a tendency to harbor resentment and impairment of his ability to express hostility; (2) tendency toward self-pity; (3) difficulty in developing and maintaining meaningful, long-term relationships; (4) a poor self-image; and (5) a basic sense of rejection by one or both parents. How strong the relationship is between emotions, stress, and cancer may never be resolved. Most physicians will agree that there is some solid evidence that disease in general is affected by emotional distress. Currently, this psychological component of disease—the effects of the brain on the body's ability to resist and fight illness—is receiving more attention.

(Also see DISEASES; ENDOCRINE GLANDS; and STRESS.)

NUTRITION AND DIET. There is substantial scientific evidence that diet may be an important factor in human cancer development. However, the exact roles played by specific aspects or components of the diet are still very uncertain. Also, it is recognized that cancer is *not* caused by any single item, but by an interaction of such things as genetics, life-style, cultural patterns, health status, nutrition and diet. A direct cause-effect relationship between diet and cancer has not been established. The sources of information implicating diet in the formation of cancer include epidemiological surveys, animal experiments, and a few controlled studies in humans. From all of these studies emerge some patterns, which in general seem to be that dietary and nutrient excesses, deficiencies, or imbalances are related to the development of cancers in the stomach, colon, pancreas, liver, and breast. Nutrition and diet appear to modify rather than to initiate the carcinogenic processes. Based on numerous studies there are some hints and guidelines which may be gleaned.

• **Caloric intake**—Many animal studies have shown that caloric restriction generally inhibits cancer formation and increases life expectancy. Conversely, obesity seems to be associated with uterine (endometrial) and kidney cancer in women.

• **Nutrient deficiencies**—Many animal studies have linked nutrient deficiencies, especially of the trace elements, to the development of cancer. Furthermore, deficiencies of vitamins A, C, E, and the B-complex vitamins have been associated with the increased susceptibility of chemically-induced tumors in animals, as have protein, iodine, iron, magnesium, and selenium deficiencies.

• **Foodborne carcinogens**—These have already been mentioned under the section headed, "Chemical Agents." Foodborne carcinogens may occur as naturally occurring, intentionally added, or incidentally added chemicals. The known hazards have been eliminated from our foods, while other unknown possible hazards are continually monitored and tested.

• **Dietary fiber**—Lack of fiber in the American diet has been implicated by some as the sole cause of cancer of the colon. Population studies have shown that in countries where dietary fiber intake is high, such as Africa and Finland, the risk of colon cancer is lower than in countries like the United States and Western Europe. The theory: Fiber binds potential carcinogens in the bowel, making them inactive and promoting rapid removal from the large intestine, thus reducing the time caarcinogens are in contact with the wall of the colon. In the other direction, studies of both Mormons and non-Mormons in Utah show a lower colon cancer incidence than that of the American population, but little difference in fiber consumption.

(Also see FIBER.)

• **Bile acids**—The diet—in particular fat—may have a role in the development of colon cancer. A current hypothesis is that high dietary fat intake (1) increases bile secretion, and (2) influences the composition and the metabolic activity of the intestinal microfloral so that the anaerobic clostridia increase. These microorganisms—clostridia—are capable of converting, by dehydrogenation, bile acids and neutral sterols (primary bile acids) to carcinogens or cocarcinogens—secondary bile acids. There is substantial evidence in experimental animal studies to implicate dietary fat and secondary bile acids as colon cancer related.

• **Cholesterol**—Studies have indicated that low blood levels of cholesterol appear to *correlate* with a modest increase in cancer risk. So, individuals with levels of blood cholesterol less than 180 mg/100 ml should possibly avoid further reductions until the connection between low cholesterol and cancer is resolved.[10]

• **Hormone production**—Nutritional status, nutrient intake and dietary components play a role in determining an individual's hormone profile. A greater risk of breast cancer has been associated with other conditions which influence endocrine status—increased body height and weight, age of menarche and late menopause. Diet may, in turn, affect these conditions. For example, it is known that over the last 100 years or so the age at menarche of girls in developed countries has declined dramatically; presently, in the United States it is about 11 or 12 years of age. The level of nutrition in developed countries is undoubtedly related to this decline in age at menarche. Breast cancer, a hormone-dependent malignancy of the mammary gland, is the leading cause of cancer mortality in developed countries with the exception of Japan. Furthermore, diet may promote or inhibit breast cancer by modifying the levels or actions of hormones like androgens, prolactin, and estrogens.

After two decades of relative neglect, attention is again being focused upon the role of diet and nutrition in the development of cancer. While there seems to be no doubt that diet can and does influence the development of cancer, diet and nutrition should not be considered the sole causative factors.

DETECTION. The most curable cancers are those that have not progressed to the stage of producing symptoms. Hence, the best detection methods are those that are able to determine cancerous growths in their early stages.

[10]*The Harvard Medical School Health Letter*, published by the Department of Continuing Education, Harvard Medical School, October 1981, p. 1.

Signs. Most cancers are diagnosed because an individual becomes aware of certain symptoms or signs, or because a physician suspects that certain symptoms or signs exhibited by a patient suggest cancer.

• **Physician**—Certain physical examinations have stood the test of time and proven worthwhile in the detection of symptomless cancer. Included in an examination is a general physical inspection and palpation of the whole body, with special attention to the cavities of the mouth, the vagina, the bladder, and rectum. Detection of rectal cancer at a curable stage may be accomplished by the use of a proctosigmoidoscope. Furthermore, the employment of fiber optics allows physicians to view inside other areas of the body without major surgery. X ray and computer-assisted tomography scanning (CAT), a recent development in x-ray equipment, both allow visualization of internal masses. Recently, ultrasound has been used to detect cancerous tissues.

• **Self**—Every individual should be aware of his or her body. An excellent health practice is periodic self-examinations, particularly of the mouth, skin, and genitals. In addition, women should examine their breasts once a month and men should check their testes for enlargement, hardness or irregular contour. The American Cancer Society emphasizes Seven Warning Signals:

1. Unusual bleeding or discharge.
2. A lump or thickening in the breast or elsewhere.
3. A sore that does not heal.
4. Change in bowel or bladder habits.
5. Hoarseness or cough.
6. Indigestion or difficulty in swallowing.
7. Change in a wart or mole.

Signals lasting longer than two weeks should be checked by a doctor. The signals do not necessarily mean cancer is present since they can be due to many other causes. Neverthless, they should be checked.

Tests. A wide variety of biochemical tests on urine and blood is available for hormones, enzymes and antigens, some of which are used to detect and follow the development of cancer. Harmless radioactive isotopes may be injected or swallowed to discover abnormalities in the liver, brain, bones, kidneys and other organs. One routine test is the Pap Smear (developed by Dr. George Papanicolaau) which is a vaginal smear test for detecting cancerous cells of the cervix. Furthermore, the cells shed from other locations in the body are helpful for the diagnosis of cancer of the bladder, stomach, nose and mouth. However, the examination of these cells is seldom conclusive. An actual biopsy—a small piece of living tissue—of the suspected organ or tissue is necessary for a definite diagnosis of cancer. Biopsies must be taken expertly from the right spot and then examined under a microscope by a pathologist with the experience and judgment necessary for a correct interpretation.

THERAPY—TREATMENT. Successful treatment of cancer is, or course, entire removal from the body. The indicated thereapy depends upon (1) the type of cancer, (2) the extent of the cancer, (3) location of the cancer, (4) the age of the patient, and (5) the general health of the patient. Generally three types of therapy are employed in the United States. One type of therapy, or a combination, may be used on a patient. The three methods of treatment that produce the most curative results are surgery, radiation, and chemotherapy. Nutrition plays a role in each of these forms of treatment. Furthermore, there are some suggestions that nutritional therapy can provide some relief from cancer.

Often, new miracle cures for cancer are announced. Some may be valid. Others definitely are not valid—the word is *caution*. One should carefully evaluate who is reporting the miracle cure, what is at stake, and what is to be gained. As an aid, the following should be considered:

1. What is the source of the study?
2. Who paid for the study?
3. Have valid testing procedures been employed?
4. What stage of research is reported—test tube, animal, or human?
5. Were the tests extensive or small observations?
6. Does the story report both the good and the bad results or only the successes?

Research on treatment of cancer is directed toward developing and evaluating new and improved methods, minimizing toxicity and avoiding the injury of the more radical procedures required to treat advanced stages. Many advances have been made on the traditional three—surgery, radiation, and chemotherapy. In 1991, the National Cancer Institute reported that 51.1% of cancer patients live 5 years or longer.

Surgery. The hope of surgical therapy is the removal of all cancerous tissue. It is the primary form of treatment for cancer of the lung, and most effective for cancer of the colon and rectum. Cancer of the breast is also treated primarily by surgery. Other cancers may also require surgery depending upon their nature. Control of shock by blood transfusions and the use of antibiotics to control infections has been a great boon to surgical techniques. Furthermore, the recent development of intravenous hyperalimentation, by infusion into the main veins of the body, provides sufficient calories, amino acids and other nutrients to promote weight gain and wound healing following surgery.

A new surgical technique—cryosurgery—destroys tumors by freezing with liquid nitrogen at a temperature of-321°F (*-196°C*). It is still basically in the experimental stages, but it seems to hold hope for special types of cancers, in particular, cancer of the pancreas.

Some cancer surgeries create the need for specialized nutrition and diet therapy.

Radiation. The principle is to encompass total cancerous tumor with a dose of radiation large enough to destroy cancer cells, but not so large as to seriously damage normal tissue and prevent healing. Some cancers are not destroyed by radiation that is within a dose range safe for surrounding normal tissues. Various sources of radiation may be used, and the radiation may be applied to the cancer in a number of ways. It all depends on the type of cancer. The two most familiar sources of radiation are x rays and cobalt-60.

Radiation therapy can create nutritional problems

depending upon the general area of the body exposed. These problems and some approaches to these problems are discussed subsequently in the section headed "Nutrition and Diet."

Chemotherapy. Cancer which has spread—metastasized—or cancer of the blood or blood-forming tissues, such as leukemia, are treated with drugs or chemicals. These drugs or chemicals supposedly hunt out and destroy cancer cells. Unfortunately, most chemotherapeutic agents, except the hormones, act to inhibit one or more of the key intermediary metabolism steps in normal as well as cancerous cells. Chemical agents may be classified into three general groups: alkylating agents, metabolic antagonists, and hormones.

• **Alkylating agents**—In tissues, the effect of alkylating agents is said to mimic radiation. Hence, they are sometimes called radiomimetic agent. Common alkylating agents include: nitrogen mustard, cyclophosphamide, chlorambucid, busulfan and alkeran. These drugs can produce anorexia, nausea, vomiting, and abdominal pain, thus complicating the nutrition of the cancer patient.

• **Metabolic antagonists or antimetabolites**—These chemicals are designed to starve cancer cells by interfering with their metabolic processes. Metabolic antagonists are counterfeit biochemicals which fool the cell into using them. This blocks the activity of the cell. However, metabolic antagonists cannot recognize normal cells, so normal cells are also damaged. The most widely used metabolic antagonist is methotrexate—a counterfeit of the vitamin folic acid. Other antimetabolites used in cancer treatment include: 6-mercaptopurine, a purine counterfeit; fluorinated pyrimidines; and cytosine arabinoside or cytosan. These drugs also have such adverse effects on the digestive processes as malabsorption, anorexia, nausea, and intestinal ulceration. These effects complicate the nutrition of the cancer patient.

• **Hormone**—Hormones are effective for treating some cancers of the breast, prostate, and uterus. For example, male hormones or androgens are sometimes effective in treating breast cancer in young women, or estrogens are employed to treat breast cancer in women who are past menopause. Progesterone is sometimes used to treat cancer of the uterus. Adults and children with acute leukemia or lymphomas have been treated with derivatives of the hormones of the adrenal cortex; for example, prednisone or prednisolone. These hormones induce well-known metabolic effects which include depletion of nitrogen (protein) from muscle and loss of calcium and potassium from the body; thus, requiring dietary adjustments.

• **Other anticancer drugs**—The drugs vincristine and vinblastine are compounds derived from the periwinkle plant, and both are used in cancer treatment. These, too, produce undesirable effects on the digestive processes. Furthermore, antibiotics such as actinomycin D are used in cancer therapy in combination with the other drugs or therapies. Antibiotics also induce a variety of gastrointestinal—digestive—changes.

Still under criticism and study are such compounds as bioflavonoids, pangamic acid, and Laetrile.

Nutrition And Diet In Cancer Therapy. Nutrition may be considered as an adjunct to the current therapies, or as a therapy. Furthermore, good nutrition of the cancer patient (1) improves sense of well-being, (2) preserves tissue function and repair, (3) improves the immune function, and (4) increases the tolerance to the side effects of chemotherapy, radiation and surgery.

STATUS PRIOR TO AND DURING THERAPY. Many cancers adversely affect the nutritional status of the victim before any treatment has been started. Depending on the type of cancer some of the following may be observed:

1. Malnutrition resulting from anorexia—failure to eat.
2. Malnutrition caused by an obstruction which impairs food intake.
3. Malabsorption associated with (a) a deficiency of pancreatic enzymes or bile, (b) infiltration of the small intestine by the cancer, (c) gastric hypersecretion as in the Zollinger-Ellison syndrome, (d) blind loop syndrome caused by obstruction in upper part of the small intestine, or (e) hypoplasia—regression of the villi of the small intestine.
4. Protein-losing malabsorption associated with stomach cancer or lymphatic obstruction.
5. Disturbances in the electrolyte and fluid balance of the body which are associated with (a) persistent vomiting due to an obstructive growth, (b) vomiting due to pressure from a brain tumor, (c) diarrhea induced by hormone secreting cancers and cancer of the colon, (d) cancers which disrupt the release of antidiuretic hormone from the posterior pituitary, and (e) tumors causing excessive secretion of corticotropin or corticosteroid which results in hyperadrenalism.

For the above reasons, and probably others yet unknown, cancer patients often exhibit cachexia—a general lack of nutrition. Hence, steps should be taken to correct the malnutrition and fluid and electrolyte imbalances in order for other therapeutic measures to proceed—and succeed.

• **Nutritional problems following cancer treatment**—Radiation, surgery, and chemotherapy—predispose the cancer patient to nutritional problems. Table C-11 outlines some of these nutritional problems that can be expected from the currently accepted cancer treatments. Obviously, each case is different and dietary adjustments required differ.

Table C-11 demonstrates that during radiation and chemotherapy or following surgical therapy, there needs to be a keen awareness of changes in the nutritional status of the patient. Malnutrition, if allowed to develop, presents the cancer patient an additional problem which will increase morbidity and mortality. Malnutrition is especially important to guard against in children since normal growth and development may be affected.

(Also see DIGESTION AND ABSORPTION; and MALABSORPTION SYNDROME.)

TABLE C-11
NUTRITIONAL PROBLEMS WHICH OCCUR FOLLOWING CANCER TREATMENT

General Treatment	Specific Treatment	Nutritional Problems	Comments
Radiation	Radiation of mouth and throat.	Loss of taste sensation; dry mouth; loss of teeth; chewing and swallowing difficulty.	Use a softer diet; use a blender; use gravies and sauces; make stews; aroma and appearance become of prime importance; nutritional liquid formulas are available to provide supplemental intake. (Also see MODIFIED DIETS.)
	Radiation of lower neck and chest.	Inflammation of the esophagus accompanied by difficult swallowing; possible narrowing of the esophagus.	Inflammation of the esophagus may disappear; narrowing of the esophagus is a delayed effect.
	Radiation of the abdomen and pelvis.	Damage to the intestinal absorptive surface resulting in diarrhea, malabsorption, obstruction, and fistulization.	Problem of both large and small intestine; caused by both internal and external radiation treatments; next to bone marrow the cells of the small intestine are most sensitive to radiation. (Also see MALABSORPTION SYNDROME.)
Surgery	Extensive resection of the mouth and throat area.	Chewing and swallowing difficulties.	Tube feeding may be required; liquid diets can be prepared by blending ordinary foods or simple inexpensive tube formulas can be used.
	Removal or reconstruction of the esophagus.	Malabsorption of fat; steatorrhea; diarrhea; gastric stasis.	Low-fat diet with use of an easily absorbed fat—medium chain triglycerides (MCT)—to reduce steatorrhea; malabsorption of fat seems due to removal of the vagus nerve occurring at time of esophagus removal.
	Stomach removal—gastrectomy—high subtotal or total.	Dumping syndrome; malabsorption; lack of hydrochloric acid; lack of intrinsic factor; hypoglycemia.	Problems increase in proportion to the extent of the surgery; causes of deficient nutrition vary from patient to patient; long term effects result in development of various vitamins and mineral deficiencies; persistent attention should be paid to prevent problems. (Also see DUMPING SYNDROME; and MALABSORPTION SYNDROME.)
	Resection involving the jejunum of the small intestine.	Decreased efficiency of absorption of many nutrients.	The ileum has some reserve absorptive capacity so absorption can be maintained providing adequate nutrients are supplied.
	Resection of the ileum.	Vitamin B-12 deficiency; diarrhea resulting in bile salt losses; hyperoxaluria—high oxalate in the urine.	Intramuscular doses of 100 mcg of vitamin B-12 every month or two; low bile salts can lead to steatorrhea; diet may need to consist of an easily absorbed fat source—medium chain triglycerides (MCT); use of water soluble vitamin K; supplemental source of the other fat soluble vitamins A, D, and E; oxalate kidney stones form due to hyperoxaluria.
	Massive intestinal resection.	Malabsorption may reach life-threatening situation; malnutrition; metabolic acidosis.	Individuals may survive with as little as 8 to 10 in. *(20 to 25 cm)* of small intestine; 3 ft *(90 cm)* or less presents problem of maintaining adequate nutrition; following surgery requires total intravenous nutrition; feedings through gastrostomy as condition improves; reduction in fat intake and use of an easily absorbed energy source—medium chain triglycerides (MCT); acidosis requiring sodium bicarbonate; long-term, meticulous care required.
	Ileostomy and colostomy.	Salt and water balance.	Large sodium and water losses first days following surgery; losses stabilize within 7 to 10 days.
	Pancreatectomy—pancreas removal.	Malabsorption; diabetes mellitus.	Loss of digestive enzymes leading to impaired digestion and absorption of fats, proteins, vitamins and minerals; picture complicated by required partial gastrectomy at the time; total pancreatectomy results in diabetes mellitus; all in all a very complicated situation; easily absorbed medium chain triglycerides (MCT) and carbohydrates used; complex carbohydrates avoided; pancreatic extracts used to aid digestion. (Also see DIABETES MELLITUS.)
Chemotherapy	Corticosteroids or derivatives.	Fluid and electrolyte imbalances; loss of nitrogen and calcium; high blood sugar—hyperglycemia.	Effects depend on type of steroid, dosage, and duration.
	Antimetabolites; alkylating agents; other drugs.	Wide variety of digestive disorders including: anorexia; nausea; vomiting; diarrhea; malabsorption; obstructive constipation.	Effects depend on drugs or combination employed; rapidly dividing cells of the lining of the intestine susceptible to all drugs affecting the rapidly dividing cancer cells. Prolonged treatments result in weight loss and progressive debility.

AS A THERAPY. While clinical studies have shown that a nutritionally balanced patient has a much improved chance of undergoing successful treatment and withstanding the rigors of cancer, no conclusive evidence exists to indicate that reduced or excess amounts of any nutrients have a beneficial role in the treatment of cancer in humans. Still, the role of nutition as a cancer therapy may be viewed as three-fold depending upon the need of the patient: (1) supportive, (2) adjunctive, and (3) definitive. To this end, some current research is aimed at increasing the appetite and food utilization in cancer patients.

- **Supportive**—Cancer victims who, because of malnutrition caused by cancer, are a poor risk for other forms of therapy; for example, surgery. The risk of this type of cancer patient can be clinically lowered by nutritional means before undertaking other forms of therapy.

- **Adjunctive**—In this case nutrition becomes part of the therapy—a complement. Improved nutrition will (1) improve the immune status, (2) allow better adherence to a proposed antitumor therapeutic program, and (3) induce healing.

- **Definitive**—This type of nutrition therapy occurs when adequate nutrition becomes the means of longer term existence for the cancer patient. Definitive nutrition is the employment of special oral, tube, or intravenous programs which allow the survival of the patient in good condition. Intravenous hyperalimentation has recently received attention as a viable means for supplying the much needed nutrition. Definitive nutrition is required by patients who have had a massive bowel resection and/or extensive radiation treatment of the abdomen.

(Also see INTRAVENOUS HYPERALIMENTATION.)

PREVENTION. If the factors causing cancer can be identified, then individuals can avoid, eliminate, or counteract their effects. Environmental factors are now generally thought to play a major role in causing cancer. However, the exact nature of environmental factors needs to be considered.

Environment. The statistic that 60 to 90% of all cancer is related to environmental factors if often quoted. People immediately take this figure to mean that cancer is being caused by synthetic chemicals or air pollution. This is not correct. Environment includes smoking, alcohol, diet, sunlight, background radiation, health, hygiene and pollution. On the basis of facts and sound judgment, we must determine our exposure to carcinogenic substances in the environment and take the necessary steps to counteract, avoid or eliminate. As has been pointed out, levels of exposure and risk following exposure to a carcinogen are difficult to assess. Every individual responds differently to his environment. Factors which should be considered when assessing an individual's exposure to carcinogens are listed in Table C-12.

TABLE C-12
FACTORS WHICH CAUSE CANCER DEATHS

Factor	Percentage of All Cancer Deaths	
	(average)	(range)
Tobacco	30	25 - 40
Diet	35	10 - 70
Infection	10	1 - ?
Reproductive and sexual behavior	7	1 - 13
Occupation	4	2 - 8
Geographical[1]	3	2 - 4
Alcohol	3	2 - 4
Pollution	2	<1 - 5
Medicine and medical procedures	1	.5 - 3
Food additives[2]	<1	−5 - 2
Industrial products	<1	<1 - 2
Unknown	?	?

[1]Actually geographical factors cause more nonfatal cancers due to the importance of ultraviolet light causing the relative nonfatal carcinomas of sunlight exposed skin.

[2]Some food additives may actually exert a protective effect; hence, the range from −5 to 2.

One clear-cut factor that increases the risk of cancer—smoking—shows no sign of being eliminated by people. It is reasonable to assume that if smoking were abolished, lung cancer—the most common form of death from cancer—would largely be eliminated. Unfortunately, new participants—by choice—join each year. Many of these are women, with the result that lung cancer is now reaching epidemic proportions in women. Moreover, secondhand smoke from the smokers is hazardous to nonsmokers. On the other hand, some synthetic chemicals suspected of carcinogenic activity have stirred emotions and have been removed from the market without due consideration. Clearly, known hazards should be removed from the environment. A paradox, however, does seem to exist. We subsidize one industry—the tobacco industry (government payments to growers)—for producing carcinogenic products and penalize another who may have unknowingly produced a suspected carcinogen.

Another environmental agent known to cause cancer is sunlight. Yet, thousands of people—by choice—still sunbathe.

In the final analysis, more prevention can be accomplished on an individual basis. It is up to each individual to control his, or her, own environment. Those known environmental hazards which could expose the population as a whole to carcinogens are either closely monitored, removed, or in the process of being removed from the environment. This includes such things as air and water pollution, low level radiation, diethylstilbestrol, and asbestos.

Still, individuals will be faced with more choices that are matters of economics and morals. How safe is safe? Do the benefits clearly outweight the risks?

Nutrition And Diet In Cancer Prevention. Many epidemiological studies have been conducted which suggest correlation between certain dietary habits or nutritional imbalances. From these studies some individuals have drawn rather exacting dietary recommendations, failing to recognize that epidemiologic correlations do not establish causation. Furthermore, there is no possible way that epidemiological studies can pinpoint carcinogenic agents. Therefore, when an epidemiological study says meat consumption was correlated to the incidence of cancer of the large intestine, this does not mean that meat causes cancer. The number of dresses a woman owns may also be correlated to the incidence of cancer of the large intestine. "Thus," epidemiological studies may provide leads, but not definitive scientific evidence.

Based on all current information from epidemiological studies, animal studies, and case studies, foods and nutrients appear to modify rather than initiate the carcinogenic process. There is no proof that a certain diet will cause cancer. Conversely, no diet will prevent cancer. Furthermore, the role that foodborne carcinogens have in causing human cancers remains to be ascertained. Although, no specific diet can be recommended that will prevent cancer, the following salient points deserve mention:

1. Some studies have shown that persons with a higher vitamin A (carotene) intake or blood level are less susceptible to cancer. Nontoxic vitamin A analogs have been successfully used to prevent cancer in animals.

2. Obesity seems to be associated with cancer of the uterus and of the kidneys in women.

3. Various indoles present in vegetables of the Brassicaeae family—Brussels sprouts, cabbage and broccoli—induce drug metabolizing enzymes in the lining of the digestive tract which inactivate carcinogens.

4. Long-standing deficiencies of vitamins A, C, and E, as well as of most of the B-complex vitamins, have been shown to increase the susceptibility of animals to chemically induced tumors.

5. Minerals appear to have an optimum range, and an intake above or below this range may increase susceptibility to cancer. For example, high levels of selenium may be carcinogenic, while selenium deficiency has been associated with an increased incidence of cancer.

6. Some new studies suggest that while a low cholesterol diet may protect against heart attacks, it may also increase vulnerability to cancer.

7. Antioxidants, such as vitamin C, have been shown to prevent the formation of nitrosamines. It is noteworthy that some individuals attribute the decline in stomach cancer to the increased availability of vitamin C.

8. Restriction of energy intake or underfeeding decreases the development of transplanted or spontaneous tumors in animals.

9. Dietary fats—saturated and polyunsaturated—have been implicated in colon and breast cancer through population studies; but this has not been confirmed by properly designed and well-controlled experiments.

10. In some areas of the world where foods are likely to be stored under warm damp conditions, there is a high incidence of liver cancer. This may be due to the formation of aflatoxins during storage.

11. Population studies indicate that fiber may lessen cancer of the bowel by increasing the rate of food passage, and by altering the types and amounts of microorganisms. But this theory has not been tested in controlled experiments.

12. It has been suggested that coffee may be contributing to cancer of the pancreas. Although this remains unproved, moderation in the use of coffee is recommended—limiting it to 2 cups a day.

13. Foods which contain bioflavonoids may provide protection against cancer, but this is without proof. Currently, studies are in progress to determine the efficacy of bioflavonoids in this regard.

14. Protease inhibitors, compounds found in foods such as soybeans and lima beans, appear to have antitumor activity. They have been suggested as a possible factor in the lowered incidence of cancers among vegetarians. But further studies are needed.

An "anticancer" diet exists for almost every one of the above points. All of the above points contain an element of truth which can contribute—but cannot stand alone—to our overall understanding of carcinogenesis. As for food or diet which prevents cancer, the best recommendations that can be made are: (1) eat less fat, (2) eat more fiber-containing foods, (3) eat fruit and vegetables every day, (4) eat a well balanced diet and maintain your ideal weight, (5) consume adequate minerals and vitamins, and (6) avoid the known carcinogens, particularly tobacco and excessive alcohol.

THE FUTURE. While a pill or injection that will prevent cancer has not been discovered, more is known and victory is a reasonable expectation. In the battle against cancer, research continues in a number of important directions.

• **Detection and screening**—Early detection of cancer improves the chance of cure tremendously. However, the cancer of many patients at the time of diagnosis may have already metastasized—spread. Reliable tests for the early detection of cancer are needed, and they are being sought. Furthermore, some work is being done to develop profiles of risk, so that individuals at a high risk of developing cancer can receive medical surveillance.

• **Immunotherapy**—Immunotherapy is a cancer therapy that harnesses the body's immune response to identify and destroy cancer cells. This type of therapy is promising, but still in the experimental stages. Attempts have been made to stimulate the whole immune system. For example, BCG organisms (*Bacillus Calmette-Guerin*) have been used, with promising results, on a limited basis to stimulate the immune system. One hope of the immune system is interferon, which is naturally produced by the body and which stimulates the immune system. The recombinant DNA technology now allows human interferon to be produced in sufficient quantities for cancer treatment. Also, recombinant DNA technology may allow the development of monoclonal antibodies (proteins) which, when injected into the body could selectively seek out and destroy cancer cells. At present the uses of immunity to combat cancer must be regarded as experimental, but there is hope for the future.

(Also see NUCLEIC ACIDS, section headed "Genetic Engineering.")

• **Novel treatments**—Heat is an old-fashioned treatment of cancer which has found new hope. The initial results of using heat against cancer have been good. Also, microwaves have been used experimentally to destroy tumors. Exercise may turn out to be a form of cancer treatment. Mice that jog experience cancer regression while nonjogging mice do not. Some human testimonials also suggest benefits of exercise. The search continues!

• **Nutrition and diet**—Eventually some dietary manipulation may prevent or treat cancer. Possibly the greatest strides will come as everyone becomes more health conscious and more aware of their bodies. Then each individual will make the correct decisions of what and how much to eat and whether to smoke or drink based on knowledge of the facts. Furthermore, medical authorities are accepting the nutrition approach in the prevention and treatment of cancer. The use of such nutrients as vitamin A (carotene), folic acid, vitamin C, and other vitamins, along with a general supplemental program of minerals, may be significant for cancer treatment. Perhaps, even more important in the long run is the apparent success of nutrients of many kinds in strengthening the body's immune system—as preventives.

While cancer is a terrible disease, a clear majority, 80%, of us will die of other causes. Cancer is not epidemic, though the fear of cancer is epidemic.

The enemy in the cancer battle is ourselves—our own cells in rebellion, and very likely our own habits and lifestyles.

(Also see ADDITIVES; BIOFLAVONOIDS; DEFICIENCY DISEASES; LAETRILE; NITRATES AND NITRITES; OBESITY; and PANGAMIC ACID.)

CANCRUM ORIS

A decaying of the flesh around the mouth, due to a combination of malnutrition and infection. However, the infectious organism responsible for this condition has not been isolated.

CANDY

Confections have been known since ancient times. About 2000 B.C., ancient Egyptian records depict a confection made of honey and flavored with an extract from the mallow plant which grew wild in the marshes of Egypt. The records also showed confectionery processes used to make candy.

The story of candy is the story of sugar. In fact, the word "candy" comes from both an Arabic word *qandi*, meaning "candied," and the Persian word *qand*, meaning "sugar." About the middle of the 1400s, a candymaker in Venice learned to refine sugar for making confections. But it wasn't until the cultivation of sugarcane became worldwide that the candy industry developed into one of major significance. Today, the United States is the leading manufacturer of candies (a dubious honor).

The basic candy-making principle is simply the boiling of sugar; and the temperature to which it is cooked determines whether it will be a fondant (236°F [*113.3°C*]) or a toffee (300°F [*148°C*]). Today, candy manufacturing plants are equipped with every conceivable machine for each step in the manufacture of over 2,000 different kinds of sweets. There are continuous cookers, cooling tunnels, crystallizers, forming machines, chocolate coaters, and taffy-pulling machines.

The annual candy consumption in the United States is over 17 lb (or almost *8 kg*) per person, which ranks us number 11 in world candy consumption. Switzerland and the United Kindom share in the top consumption of about 27.5 lb (*12.5 kg*) per person.

The U.S. Food and Drug Law requires that all candy be made with pure ingredients and nonpoisonous flavorings and colorings. In addition to the sugar(s) and syrups, candies may contain eggs, fats, gelatin, gums, lecithin, milk products, cooked starches, and every kind of fruit, nut, or flavoring. Although these are all good, wholesome food ingredients, the high sugar content makes candy less desirable for consumption in large amounts. Candy should be eaten very sparingly and is not recommended as a source of quick energy. A much more desirable food for quick energy is fresh fruit. In addition to the fact that sugar is being increasingly incriminated into the degenerative diseases, there is also the well-documented fact that sugar produces tooth decay. So a "sweet tooth" may eventually mean "no teeth."

There are some fortified food or breakfast bars that, nutritionally, are superior to unfortified candy bars.

Although there are literally thousands of kinds of candies, they all fall into a few types or classes. These are given in Table C-13 Candy Classifications. For the analysis of some of the candies, see the section on Desserts and Sweets in Table F-36 Food Composition Table.

TABLE C-13
CANDY CLASSIFICATIONS

Class	Popular Examples, Uses	Comments
Candied or crystallized fruits	Cherries Citron Citrus peels Melon Pears Peaches Plums	The fruits are prepared and dropped into a syrup at 234°F *(112°C)*. They are simmered until clear, spread on a screen, and dried until no longer sticky.
Caramels	Vanilla Chocolate Caramel apples	Caramels contain sugar, corn syrup, milk, cream and butter; and are cooked to stiff-ball stage of 246°F *(119°C)*. After pouring into a buttered pan and cooling, they are turned out and cut into squares. Apples are dipped into the warm caramel and allowed to cool on waxed paper.
Chewing gums	Peppermint	Although chewing gum is not eaten, it is chewed; and, because of the sugar content in it, can be classed as a candy. Increasingly, the sugarless gum is growing in popularity because of the pressures brought by the dental industry.

(Continued)

TABLE C-13 (*Continued*)

Class	Popular Examples, Uses	Comments
Chocolates	Chocolate-coated creams and caramels. Chocolate-coated fruits and nuts. Chocolate bars	Chocolate coatings commonly used are bittersweet, sweet milk, skim milk, or buttermilk. They are made by grinding roasted cocoa beans with sugars, cocoa butter, dried powdered milk products, and flavors.
Fondants	Fillings for dipping. Fruits added. Pralines (nuts added).	For fondants, the syrup is cooked to the soft-ball stage (238°F, or *114°C*), poured into large platters, cooled until lukewarm, then stirred and kneaded until smooth.
Fudges	Chocolate Divinity Maple Marshmallow	Fudge is made by gently boiling sugar and corn syrup with milk to 238°F *(114°C)*, adding butter, cooling, then beating until it holds its shape. Spread in buttered pan and when hard, cut into squares.
Glace fruits and nuts	Same as Candied.	Similar to Candied fruits, but the fruit is not cooked. The fruit or nut is dipped into the syrup which has been cooked to 300°F *(149°C)* and dried on a rack.
Hard candies	All shapes, colors, and flavors.	Hard candies are the simplest form of candy—they are made mainly from sugar and syrup and are usually boiled to 300°F *(149°C)*. They come in all shapes, sizes, colors and flavors.
Jellies	Jelly beans Gum drops Turkish delight	The main ingredients for jellies are sugar, corn syrup, and a jellying agent such as gelatin, natural gums, pectin or starch.
Licorice	Usually in long sticks.	Licorice sticks are made with flour, molasses, sugar, and corn syrup, and flavored with licorice extract.
Marshmallows	Large Small Colored Rolled in crumbs. Fillings for Easter eggs.	Marshmallows are made by whipping a combination of sugar, corn syrup, gelatin, and/or egg whites. This makes a light fluffy texture. Marshmallows are used extensively with sweet potatoes, in fruit salads, and in desserts.
Marzipan	Formed into fruit shapes. Chocolate dipped. For stuffing dates and prunes.	Egg whites are beaten and mixed with almond paste plus sugar. It is not cooked. After standing 24 hours, it can be shaped, stuffed, or dipped.
Nougats	Similar to marshmallows.	The syrup is cooked to 246-250° F *(119-121°C)* and then poured over the well beaten egg whites, or gelatin, or both. Fat is added and it is beaten until cool.
Peanut brittle	A hard candy.	The syrup is cooked to 300°F *(149°C)*. It is then poured over the nuts spread out on a pan, and allowed to cool, then broken into pieces.
Popcorn (puffed wheat, puffed rice, nuts)	Popcorn balls	The sugar, corn syrup, and water is cooked to the medium-crack stage (280°F, or *138°C)* flavoring added, then poured over the popcorn, and stirred. Then, with buttered hands, shaped into balls.
Spun sugar	Decorative nests for ice cream, baked Alaska, or Easter eggs.	The syrup is boiled to 310°F *(154°C)*. Then the pan is put into cold water to stop cooking, and back into a warm pan before it crystallizes. Using a wooden spoon, the threads of syrup are wrapped across the greased handles of two wooden spoons which have been anchored into a drawer.
Toffee	Chocolate Vanilla	Toffee is hard caramel. A taffy-pull party is sometimes relished by the younger set; it is good clean, but sticky, fun.

(Continued)

CANE SUGAR

Table sugar—sucrose—recovered from the juice of the sugarcane plant.

(Also see SUGAR, section headed "Sugarcane.")

CANKER

A small ulcer in the mouth around the lips, which may be due to a virus or to digestive disorders.

CANOLA (RAPE)

Canola was created from selected rape by Canadian scientists in the 1970s. In comparison with rape seeds, canola seed is lower in glucosinolates and erucic acid (a long-chain fatty acid).

(Also see RAPE.)

CANTALOUPE (MUSKMELON) *Cucumis melo.*

True cantaloupes are seldom grown in the United States. Correctly speaking, the term "cantaloupe" should be applied only to a rough, scaly fruit with deep vein tracts, grown in Europe and Asia. These true cantaloupe bear the scientific name of *Cucumis melo*, variety *can-*

Fig. C-29. A cantaloupe, also called muskmelon. The edible, juicy pulp is 6 to 8% sugar when ripe and is encased in a hard netted rind. Cantaloupe belong to the same family as cucumbers, squash, pumpkins, gourds, and watermelons. (Courtesy, USDA)

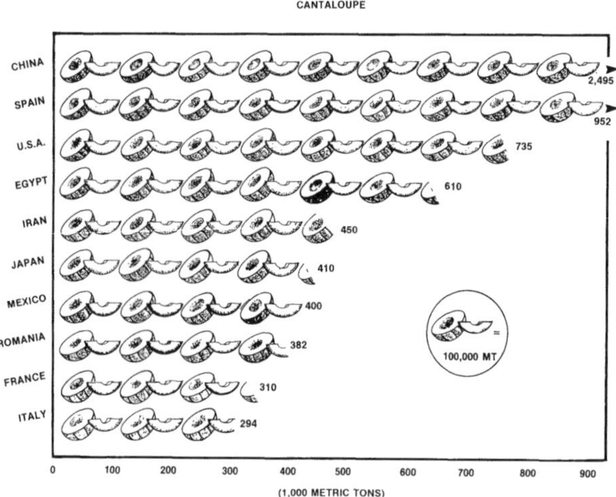

Fig. C-30. Countries which lead in the production of cantaloupe and other melons. (Data from *FAO Production Yearbook*, 1990 FAO/UN, Rome, Italy, Vol. 44, p. 151, Table 64)

talupensis. In the United States, the term cantaloupe is applied to melons with the scientific name *Cucumis melo*, variety *reticulatus*, which are more correctly muskmelons. Herein, the term "cantaloupe" is used to refer to the familiar fruit with a yellowish-brown, netted rind and orange pulp, and with seeds attached to a netlike fiber in a central hollow. Ripe cantaloupes have a distinctive sweet flavor and a musky odor. The plant producing cantaloupe is an annual trailing vine with 3 to 5 runners that may reach 10 to 12 ft (*3 to 3.6 m*) in length, with lobed leaves and small yellow flowers.

ORIGIN AND HISTORY. The exact origin of the cantaloupe is lost in antiquity. Some scientists believe it originated in Africa, whereas others believe that it came from Asia (Persia) or India. The Greeks and Romans are said to have eaten muskmelons. In America, cantaloupes were introduced during colonial days, but they were not grown commercially until about 1890. Now, cantaloupes enjoy worldwide distribution where the growing season is long enough.

PRODUCTION. Each year, the worldwide production of cantaloupes, and other melons like casaba, honeydew, crenshaw, and Persian, totals about 9.5 million metric tons. China is by far the leading producer of melons. Spain, the United States, and Egypt are top producers, as Fig. C-30 shows.

Cantaloupes are grown throughout the United States in many gardens, though they require a warm season. Some varieties can even be grown in northern regions, but the commercial production of cantaloupe, excluding the other melons like casaba, Persian, crenshaw, and honeydew, occurs in Arizona, California, Colorado, Georgia, Indiana, Michigan, South Carolina, and Texas. Each year, U.S. production amounts to over 810 thousand tons (*735 thousand metric tons*). California produces around two-thirds of the crop. Arizona and Texas are leading producers, but they produce only 15 to 25% as many cantaloupes as California.

Honeydew melons are also a muskmelon which is commercially grown, primarily in Arizona, California, and Texas. These states produce around 320 million lb (*145 thousand metric tons*) of honeydew melons each year, worth over $35 million.

Propagation And Growing. Cantaloupes are propagated by seed, which require soil temperatures to be 75 to 85° F (*24 to 29° C*) for successful germination. Cantaloupe may be planted (1) by direct field seeding in rows, (2) by seeding in hills, or (3) by starting plants in hotbeds or greenhouses and then transplanting to the field. After emerging, the plants are thinned so that there are 2 or 3 plants in hills 2 to 4 ft (*0.6 to 1.2 m*) apart, or 12 in. (*0.3 m*) apart in rows 5 to 7 ft (*1.5 to 2.1 m*) apart.

Cantaloupes are sensitive to cold and require 110 to 140 frost-free days. They thrive during long, hot, dry days, when properly watered. Soil must be well-drained and properly fertilized. Pollination of the small yellow flower is aided by bees, but some self-pollination also occurs because flowers are combined male and female. Each vine bears an average of 2 or 3 cantaloupes. In the United States, commercial yields average 13,600 lb/acre (*15,234 kg/ha*).

Harvesting. Mature cantaloupes weigh 2 to 4 lb (*0.9 to 1.8 kg*) and the edible flesh is usually salmon-colored. Furthermore, nature has provided an unmistakable sign which marks a mature cantaloupe. At maturity an abscission layer forms between the stem and fruit. This layer appears as a crack, completely encircling the stem, at the point of attachment of the fruit. If the stem is forcibly separated from the fruit, the fruit is immature. On

Fig. C-31. A field of cantaloupes grown between orchard rows. Each plant will bear 2 to 4 fruits. (Courtesy, Union Pacific Railroad Company, Omaha, Neb.)

mature fruits, the cantaloupe "slips" from the vine when a slight pressure is applied; hence, cantaloupes are harvested at the "full slip" stage as it is known in the trade.

PROCESSING. Most cantaloupes are consumed as fresh fruit. However, they have been canned successfully, but without great acceptance. The only processed product of any significance is frozen cantaloupe balls. These are processed by (1) washing and sorting, (2) cutting into halves, (3) removing seeds, (4) scooping out round balls, and (5) freezing in a sugar syrup. Melon balls may be frozen as only cantaloupe; or as a mixture of cantaloupe, watermelon, and/or honeydew.

SELECTION. Cantaloupes should be mature and ripe. There are three major signs of full maturity: (1) no stem, with a smooth symmetrical, shallow basin ("full slip") at the point of the stem attachment; (2) thick, coarse and corky netting or veining, which stands out in bold relief over some part of the surface; and (3) yellowish-buff, yellowish-gray, or pale yellow skin color (ground color) between the netting. If all or part of the stem base remains or if the stem scar is jagged or torn, the melon is probably not fully matured. A ripe cantaloupe will have a yellowish cast to the rind, have a pleasant cantaloupe odor when held to the nose, and will yield slightly to light thumb pressure on the blossom end of the melon.

Overripeness is indicated by a pronounced yellow rind color, a softening over the entire rind, and soft, watery and insipid flesh.

PREPARATION. For fresh consumption, cantaloupes are halved and the seeds are scooped out. They may be served in the rind, or they may be cut into smaller sections and the rind removed before serving.

Cantaloupes are enjoyed alone as a breakfast fruit or a dessert. Also, cantaloupe can be used in a variety of fruit dishes and salads. It is good with a scoop of ice cream, too.

NUTRITIONAL VALUE. The sweetness of cantaloupe is attributed to the 6 to 8% sugar it contains, but the water content is around 90% so each 100 g (about 3½ oz) of cantaloupe contains only 30 Calories (kcal). The salmon-colored flesh is an excellent source of vitamin A, containing about 3,400 IU per 100 g. Additional details of the nutritive value of cantaloupes are listed in Food Composition Table F-36.

(Also see FRUIT[S], Table F-47 Fruits of the World; and MELON[S].)

CAPE GOOSEBERRY(S) *Physalis peruviansa*

Fruit of a perennial shrub (of the family *Solanaceae*) that is native to the American tropics, and is closely related to the ground cherry and the strawberry tomato, and more distantly related to the common tomato (*Lycopersicon esculentum*). It is called the Cape gooseberry because it had become an important fruit on the Cape of Good Hope by the end the 18th century.

The husk-covered yellow berries, which resemble gooseberries, but are not related to them, are about ½ in. (*1 to 2 cm*) in diameter and have a slightly acid flavor.

The fruits are usually stewed or made into preserves. Raw Cape gooseberries are moderately high in calories (73 kcal per 100 g) and carbohydrates (20%). They are a good source of fiber, iron, vitamin A and vitamin C.

(Also see STRAWBERRY TOMATO; and TOMATO.)

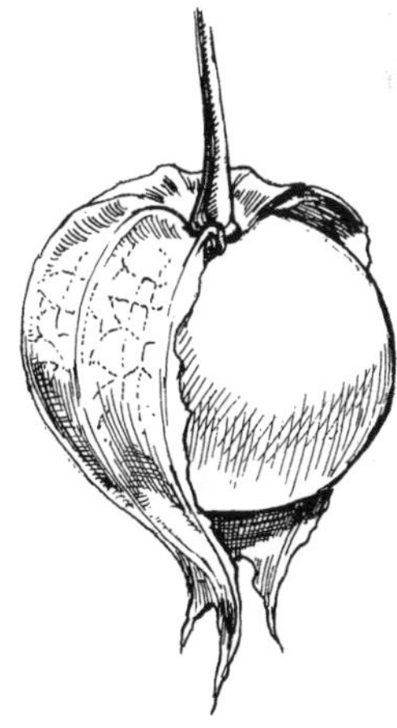

Fig. C-32. The Cape gooseberry, a relative of the tomato.

CAPER (*Capparis spinosa* L.)

The caper bush is a spiny, straggling, vine-like shrub, up to 3 ft high, with round to ovate, deciduous leaves. The buds are pickled in strong vinegar and used as pickles or in sauce.

CAPILLARIES

The tiny blood vessels which (1) deliver fresh, oxygenated blood from the arteries to the tissues; and (2) carry used, deoxygenated blood from the tissues to the veins.

CAPILLARY EXCHANGE MECHANISM

This is the process that controls the movement of water and small molecules in solution (electrolytes, nutrients) between the capillaries and the surrounding interstitial area.

CAPILLARY RESISTANCE TEST

A test which determines the ability of capillaries to resist breaking down and producing small hemorrhages under the skin. Usually, pressure is applied to the arm and the number of minute blood spots in the skin due to ruptured capillaries are noted. A lowering of capillary resistance is suspected as being due to deficiencies of vitamin C and other nutrients such as the bioflavonoids.

(Also see BIOFLAVONOIDS; and VITAMIN C.)

CAPRIC ACID ($CH_3[CH_2]_8COOH$)

A fatty acid, which is so named because in pure form it has a goatlike smell. (The latin name for goat is *caper.*) Capric acid is found in coconut oil, cow's milk, and goat's milk. It is a short chain fatty acid containing only ten carbon atoms, and it is very well absorbed and utilized by the body.

(Also see FATS AND OTHER LIPIDS.)

CAPROIC ACID; HEXANOIC ACID ($CH_3[CH_2]_4COOH$)

A naturally occurring saturated fatty acid normally found as a triglyceride in cow and goat butter, and in coconut and palm oils.

(Also see FATS AND OTHER LIPIDS.)

CAPRYLIC ACID; OCTANOIC ACID ($CH_3[CH_2]_6COOH$)

A naturally occurring rancid-tasting saturated fatty acid found in butter (cow and goat) and coconut oil and used as an intermediary in the manufacture of perfumes and dyes.

(Also see FATS AND OTHER LIPIDS.)

CARAMBOLA *Averrhoa carambola*

The fruit of a tree (of the family *Oxalidaceae*) that grows wild in Indonesia and other tropical areas.

Carambola fruit is yellow when ripe, from 3 to 5 in. (*8 to 12 cm*) long and from 1 to 3 in. (*3 to 6 cm*) in diameter, with a star-shaped cross section. It may be used fresh, or made into jam, jelly, juice, or tarts.

The nutrient composition of carambola is given in Food Composition Table F-36.

Fig. C-33. The carambola.

Some noteworthy observations regarding the nutrient composition of carambola follow:

1. The raw fruit is low in calories (35 kcal per 100 g) and in carbohydrates (8%).
2. Carambola is an excellent source of vitamin A and a good source of potassium, iron, and vitamin C.

CARBOHYDRASES

A general name for all of the enzymes which act on complex carbohydrate molecules and break them down into smaller, simple molecules such as dextrins and sugars.

(Also see CARBOHYDRATE[S].)

CARBOHYDRATE(S)

Contents — *Page*

History . . . 333
Synthesis . . . 334
Classification . . . 334
  Monosaccharides . . . 334
    Pentoses (5-Carbon Sugars) . . . 334
    Hexoses (6-Carbon Sugars) . . . 335
    Derivatives of Monosaccharides . . . 335
      Sugar Alcohols . . . 335
      Amino Sugars . . . 336
      Sugar Acids . . . 336
  Oligosaccharides . . . 336
    Disaccharides . . . 336
    Trisaccharides . . . 337
    Tetrasaccharides . . . 337
  Polysaccharides . . . 337
    Pentosans . . . 337
    Hexosans . . . 337
    Mixed Polysaccharides . . . 338
  Digestion, Absorption, and Metabolism . . . 338
  Functions . . . 339
    Energy . . . 339
    Protein Sparing Action . . . 339
    Regulation of Fat Metabolism . . . 339
    Provide Fiber . . . 339
    Special Functions . . . 340
    Flavor . . . 340
Required Intake . . . 340
Sources . . . 340
  Plant . . . 341
    Cereal Grains . . . 341
    Fruits . . . 342
    Nuts . . . 343
    Vegetables . . . 343
  Animal . . . 343
  Other . . . 343
Fermentation . . . 343

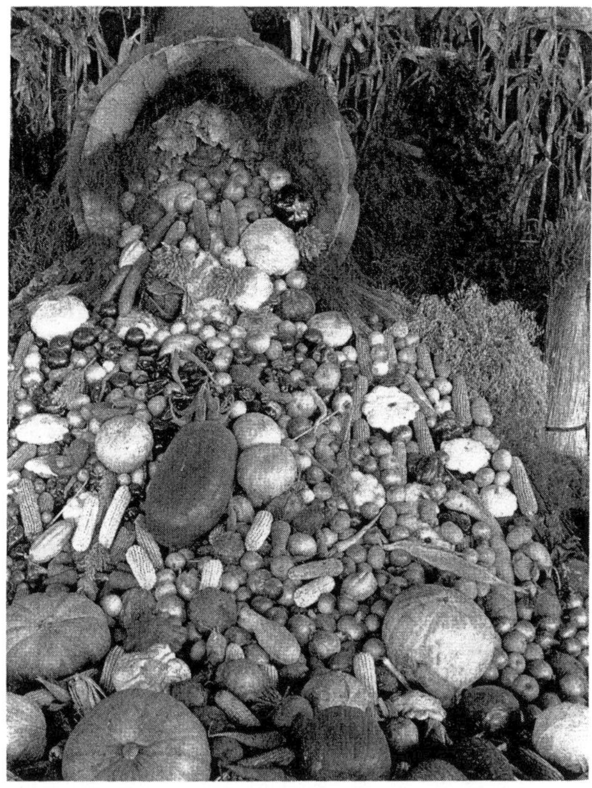

Fig. C-34. Carbohydrates—the harvest from sun, water, and carbon dioxide—the fruits of the earth. (Photo by J. C. Allen & Son, West Lafayette, Ind.)

The carbohydrates are organic compounds composed of carbon (C), hydrogen (H), and oxygen (O). They are one of three main classes of foods essential to the body; the others are proteins and fats.

The name *carbohydrate* was originally assigned to compounds thought to be hydrates of carbon (water of carbon) with the general formula

$$C_n(H_2O)_n$$

With the accumulation of more information, the definition of carbohydrates has been modified and broadened to include numerous compounds. This group includes the sugars, starch, cellulose, gums, and related substances. As far as man is concerned, carbohydrates in food are divided into two categories: (1) available carbohydrates, which man can absorb and utilize; and (2) unavailable carbohydrates like cellulose, which man cannot digest. Available carbohydraytes are used by the body as a source of heat and energy; and any excess is stored in the body as fat.

No appreciable amount of carbohydrate is found in the animal body at any one time, the blood supply being held rather constant at about 0.05 to 0.1% for most animals. However, this small quantity of glucose in the blood, which is constantly replenished by changing the glycogen of the liver back to glucose, serves as the chief source of fuel with which to maintain the body temperature and to furnish the energy needed for all body processes. The storage of glycogen (so-called animal starch) in the liver amounts of 3 to 7% of the weight of that organ.

HISTORY. Long before man knew anything about the chemical nature or the formation of carbohydrates, he made good use of them. He ate them. He fueled his fires with them. He fermented them. He fed them to his animals. He built with them. He wrote about them.

Primitive man lived on carbohydrates—on what fruits, seeds, and roots he could find. In the first chapter of Genesis, the Bible tells us that, in the beginning—nearly 6,000 years ago, the fruit of one tree in the Garden of Eden was forbidden. But Adam yielded to temptation and ate thereof, with the result that man has known but one creed ever since: "In the sweat of thy brow shalt thou eat bread." So, a carbohydrate—a fruit—caused the fall of man.

Early man learned that by adding carbohydrate (wood) fuel to fires, he could keep them burning and provide warmth for his home. Many cultures of the past have had their fermented drinks made from some carbohydrate source. Persian farmers living by the Caspian Sea were some of the first to make wine, and their knowledge spread to Egypt sometime before 3000 B.C. When men tamed animals and became herdsmen of sheep, cattle, and goats, they recognized the need for these animals to graze the land. Somewhere along the course of man's progression, he started making weapons from wood, and then shelters, ships, wagons, and machinery. In his attempt to communicate and preserve the written word, man developed paper from a carbohydrate—cellulose.

Aristotle was one of the first to suggest the origin of carbohydrates when he proposed that plants withdraw food from the soil. Although he was incorrect, his theory persisted until about 1630 A.D., when Jean van Helmont proved Aristotle wrong, though van Helmont's ideas were not entirely correct. Finally, a Dutch physician, Jan Igenhousz, demonstrated that when plants were exposed to light, they made their own food.

The actual chemistry of carbohydrates begins with some Moorish writings of the 12th century which make reference to grape sugar. Then, during the 1700s, Andreas Marggraf, a German pharmacist, isolated pure sugar from sugar beets and glucose from raisins. In the 19th century, chemists determined that carbohydrates contained the elements carbon, hydrogen, and oxygen in a ratio that suggested they were carbon hydrates; hence, the name that persists today is a contraction of this. By about 1900, the chemical structure of some simple carbohydrates was determined by a German chemist named Emil Fischer, who thereby laid the foundations of carbohydrate chemistry.

Considering all of the uses made of carbohydrates of many kinds, they are truly an important and versatile group of chemicals. Furthermore, they will continue to be important in the future since their formation relies on the ultimate source of all energy—the sun.

(Also see PHOTOSYNTHESIS.)

SYNTHESIS. Carbohydrates are formed in the plant by photosynthesis as follows:

$6CO_2$ (carbon dioxide) + $6H_2O$ (water) + energy from sun = $C_6H_{12}O_6$ (glucose) + $6O_2$ (oxygen).

Roughly, more than 100 billion tons of carbohydrates are formed each year from carbon dioxide and water. They are the main products through which the energy of the sun is harnessed.

On the average, the carbohydrates comprise about three-fourths of all the dry matter in plants. They form the woody framework of plants as well as the chief reserve food stored in seeds, roots, and tubers.

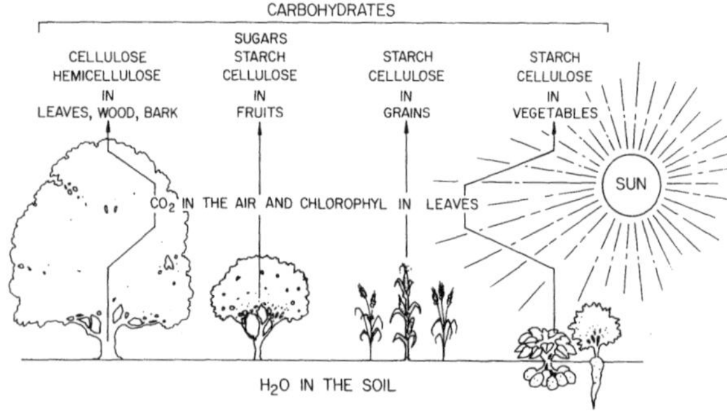

Fig. C-35. Plants trapping the energy of the sun in carbohydrates.

CLASSIFICATION. Carbohydrates are extremely abundant in plants, occurring in a wide variety of forms. Table C-14 lists the types of carbohydrates that are commonly found in nature.

Monosaccharides. The term *saccharide* is derived from the Latin word *Saccharum,* meaning sugar. Monosaccharides, or single sugars, are seldom found free in nature. Rather, they constitute the building blocks of more complex carbohydrate molecules. Simple sugars are classified as to the number of carbon atoms within the molecule. For example, a triose is a sugar containing 3 carbons, and a tetrose is a 4-carbon molecule. While monosaccharides of various lengths are integral to the metabolism of carbohydrates, the pentoses (5-carbon sugars) and hexoses (6-carbon sugars) are of paramount importance. Furthermore, there are some derivatives of these monosaccharides which are also important—the sugar alcohols, amino sugars, and sugar acids.

PENTOSE (5-Carbon Sugars). Very limited amounts of this type of sugar are found in a free form in plants. All pentoses have the same chemical formula, $C_5H_{10}O_5$, but a slightly different structure as Fig. C-36 shows.

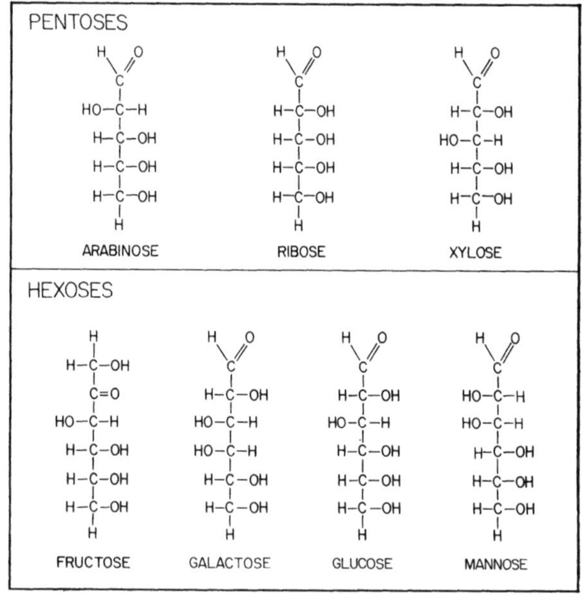

Fig. C-36. Structure of some important monosaccharides.

TABLE C-14
CLASSIFICATION OF CARBOHYDRATES

Monosaccharides		Oligosaccharides[1]			Polysaccharides		
Pentoses $(C_5H_{10}O_5)$	**Hexoses $(C_6H_{12}O_6)$**	**Disaccharides $(C_{12}H_{22}O_{11})$**	**Trisaccharides $(C_{18}H_{32}O_{16})$**	**Tetrasaccharides $(C_{24}H_{42}O_{21})$**	**Pentosans $(C_5H_8O_4)n$[2]**	**Hexosans $(C_6H_{10}O_5)n$[2]**	**Mixed Polysaccharides**
Arabinose Ribose Xylose	Fructose Galactose Glucose Mannose	Lactose Maltose Sucrose Trehalose	Maltotriose Melezitose Raffinose	Stachyose Maltotetrose	Araban Xylan	Cellulose Dextrins Glycogen Inulin Mannan Starch (amylose and amylopectin)	Agar Alginic acid Carrageenan Chitin Hemicelluloses Pectin Vegetable gums
Derivatives of monosaccharides: Sugar alcohols Amino sugars Sugar acids							

[1]Includes compounds that may be 2 to 10 sugar units.
[2]The "n" indicates any number of sugar units greater than 10.

Generally, pentoses are polymerized (combined chemically into very large molecules) to form pentosans. Arabinose is a 5-carbon sugar found in large amounts in gums. When a number of arabinose molecules are linked together, the pentosan, araban, is formed.

Ribose plays a major role in many physiological systems. When it is joined with pyrimidines and purines, nucleosides are formed. When phosphoric acid is esterified with the nucleosides, nucleotides are formed. These compounds are then used in the formation of ribonucleic acid (RNA) and deoxyribonucleic acid (DNA). The nucleotides of adenosine monophosphate (AMP), adenosine disphosphate (ADP), and adenosine triphosphate (ATP) are compounds that are essential to cellular energy metabolism. Ribose is also a constituent of the vitamin riboflavin.

Xylose is a pentose produced upon the hydrolysis—breakdown—a number of roughages and woody material. When polymerized, xylose forms the pentosan xylan.

HEXOSES (6-Carbon Sugars). Four hexoses are found in physiological systems: fructose, galactose, glucose, and mannose. All hexoses have the same chemical formula, $C_6H_{12}O_6$, but a slightly different structure as Fig. C-36 shows. In nature, only two—fructose and glucose—occur in free form. Galactose, together with glucose, forms the disaccharide, lactose, or milk sugar. Mannose is found in the polysaccharide mannan. The structure of glucose, fructose, galactose, and mannose (see Fig. C-36) was established about the year 1900, by the German carbohydrate chemist, Emil Fischer.

• **Fructose**—The hexose occurs in a large number of fruits, as well as in honey and in bull and human semen. Fructose is sometimes called fruit sugar or levulose, since it rotates polarized light to the left. It is the sweetest of the sugars and it does not readily crystallize.

• **Galactose**—The dietary source for galactose is lactose, found in milk. It is seldom found free in nature. Galactose can only be used by the body after being converted to glucose. Some individuals lack the enzyme necessary for this conversion with the result that they suffer from galactosemia, an inborn error of metabolism, with serious consequences if not detected early in life. Nursing mothers, of course, are able to convert glucose to galactose so the mammary gland can manufacture lactose.

• **Glucose**—This hexose is also known as blood sugar, dextrose, corn sugar, and grape sugar. Glucose is probably the best known monosaccharide; and the most thoroughly studied. Moorish writings of the 12th century make reference to grape sugar. Undoubtedly, ancients recognized the need for glucose in the preparation of their fermented drinks. In 1747, the German pharmacist Andreas Marggraf wrote of isolating glucose from raisins, and various researchers after 1811 established that the sugar of grapes, honey, urine of diabetics, and the acid hydrolysis of starch and cellulose were all identical. They all contain glucose—a name given by the French chemist Jean Baptiste Andre Dumas in 1838. Glucose is the main end product of the digestion of disaccharides and starch. It is the carbohydrate used by the cells of the body for the production of energy.

DERIVATIVES OF MONOSACCHARIDES. Some of the monosaccharides possess a slightly altered chemical structure yielding sugar alcohols, amino sugars, and sugar acids. These compounds are important biochemicals of the body or constituents of food.

• **Mannose**—Mannose is not found free in food, but it is derived from ivory nuts, orchid tubers, pine trees, yeasts, molds, and bacteria. It is unimportant to human nutrition, but it is a component of some glycoproteins and mucoproteins of the body.

Sugar Alcohols. Monosaccharides reduced by having a hydroxyl group (OH) replace the aldehyde or ketone group (C = O) are known as sugar alcohols. One of these alcohols, glycerol, is important in the matabolic process, while other sugar alcohols such as galactitol, inositol, mannitol, sorbitol, and xylitol, are present in some foods.

• **Glycerol**—This is a 3-carbon alcohol forming the backbone of triglycerides—a combination of three fatty acids and glycerol. It is manufactured by the body.

• **Galactitol**—Sometimes this alcohol form of galactose is called dulcitol. It is found in Madagascar manna—dried sap of ash trees. Galactitol may be commercially synthesized by the hydrogenation (addition of hydrogen) of galactose. It is one-half as sweet as sucrose, and it may be added to foods.

• **Inositol**—This is a cyclic alcohol closely related to the hexoses. Inositol occurs in many foods, in particular the bran of cereal grains. When combined with six phosphate molecules it forms phytic acid, a compound known to hinder the intestinal absorption of calcium and iron. Inositol is an essential nutrient in the diet of mice and rats. For humans and other animals its essentiality is questionable.

(Also see INOSITOL.)

• **Mannitol**—This sugar alcohol is about 70% as sweet as sucrose; hence, it is used as a sweetener in foods. Furthermore, mannitol is slowly and incompletely absorbed from the intestine with the result that it supplies only about one-half the energy value of glucose. Mannitol is also employed in foods as an anticaking substance and as a stabilizer-thickener. Natural sources of mannitol include pineapples, olives, asparagus, and carrots. Commercially, mannitol is produced by the hydrogenation of mannose.

(Also see MANNITOL.)

• **Sorbitol**—This naturally occurring 6-carbon sugar alcohol tastes about 60% as sweet as table sugar (sucrose). In nature, it occurs in many berries, cherries, plums, pears, apples, and blackstrap molasses. Sorbitol is a GRAS (generally recognized as safe) food additive. For commercial use, it is prepared by the hygrogenation of glucose. Due to a variety of unique properties, sorbitol enjoys widespread use by the food industry. Most often sorbitol is employed as a humectant or sweetener in such foods as confections, chewing gum, dried roasted nuts, meat products, icings, toppings, coconut, dairy products, brown sugar, and beverages.

Because of its sweetness, slow intestinal absorption, and unique matabolism—burned to CO_2 without ever appearing as blood sugar, glucose—sorbitol is used to replace sugar in numerous special dietary foods. However, it does have the same caloric value as glucose. Large intakes of sorbitol cause diarrhea due to slow absorption and osmotic effects in the intestine.

(Also see SORBITOL.)

• **Xylitol**—In nature, xylitol is found in many fruits and vegetables, including berries, yellow plums, and mushrooms. It can be produced commercially from plant products which contain xylans, such as oat hulls, corncobs, and birch wood chips. Xylitol tastes about as sweet as sucrose, and is approved for use in special dietary foods. Xylitol is absorbed slowly and may also produce osmotic diarrhea.

(Also see XYLITOL.)

Amino Sugars. The amino sugars have an amino (NH_2) group substituted for a hydroxyl (OH) group. More than 60 of these have been identified. Possibly these sugars give antibiotics their activity. Two important representatives are galactosamine and glucosamine.

• **Galactosamine**—This amino sugar is formed from galactose and is found in the cartilage, tendons, and aorta (blood vessel) of the body.

• **Glucosamine**—This amino sugar is formed from glucose, and it forms the basis of chitin, the substance which forms the outside covering of insects and crustaceans. Glucosamine occurs throughout the body in connective tissue.

Sugar Acids. These compounds contain an acid group (COOH) in their chemical structure. Familiar sugar acids are gluconic acid, gluconolactone, glucuronic acid, and ascorbic acid (vitamin C).

• **Gluconic acid**—This is a nontoxic compound formed from glucose which is well metabolized and often employed for the introduction of minerals such as calcium into the body.

• **Gluconolactone**—This compound is formed from glucose. It is used as a food additive in the meat-processing industry to speed the color fixing and smoking processes.

• **Glucuronic acid**—Glucuronic acid is formed in the body from glucose, and it is important in the detoxification and excretion of other compounds. Morphine, salicylic acid, and the sulfa drugs are a few of the compounds detoxified by combining with glucuronic acid. The steroid hormones and the bile pigment, bilirubin, are excreted in the urine bound to glucuronic acid. Also, glucuronic acid is a component of the connective tissue in the body.

• **Ascorbic acid**—This sugar acid is not formed in the human body, but it is extremely important to the human body. Most people know it as vitamin C, which, due to its importance, is more completely covered elsewhere.

(Also see VITAMIN C.)

Oligosaccharides. The prefix oligo means "few." Oligosaccharides contain 2 to 10 monosaccharides chemically bonded. Most oligosaccharides result from the partial breakdown of polysaccharides. The three most important oligosaccharides are all disaccharides—lactose, maltose, and sucrose.

DISACCHARIDES. Disaccharides are compound sugars composed of two monosaccharides (see Table C-14). This group includes the first sugar eaten—lactose, and the sugar eaten to excess—sucrose.

• **Lactose**—This disaccharide is composed of one molecule of galactose and one molecule of glucose. It is the only sugar not found in plants. It is found in milk; hence, it is called milk sugar.

An interesting example of metabolic defect of disaccharide digestion is the inability of some individuals to split certain disaccharides into their constituent simple sugars—a carbohydrate intolerance. Lactose, milk sugar, is probably the most troublesome disaccharide as some people lack the enzyme lactase which splits the molecule into glucose and galactose. Thus, if a sensitive individual ingests milk or dairy products, his system cannot properly digest the material. A reaction, generally in the form of cramps and diarrhea, occurs when large amounts of water are drawn into the lumen of the gut due to the osmotic effect exerted by the uncleaved lactose molecule. The metabolic disorder interests the anthropologist as well as the physiologist because certain populations throughout the world show extremely high rates of lactose intolerance. Scandinavian and Western Europeans exhibit the lowest rate of lactose intolerance, about 2 to 8%, while about 9 out of every 10 Japanese, Thais, and Filipinos are intolerant to lactose. In the United States, it has been estimated that about 70% of the Negro population is lactose intolerant. The questions posed by anthropologists are whether the intolerance to lactose evolved due to the inadequate consumption of milk over a period of generations or whether the low consumption of milk was a result of low gut lactase activity originally. That is, did these people once have the ability to digest lactose and eventually lose it, or have they always had this lactase insufficiency?

• **Maltose**—Maltose or malt sugar is chiefly an intermediate product of the breakdown—digestion—of starch. It is comprised of two glucose molecules. In the malting and fermentation of grains, maltose is produced. Maltose is present in beer, breakfast cereals, and some infant formulas. In the body, maltose is easily split into two molecules of glucose.

• **Sucrose**—This sugar is commonly known as table sugar, beet sugar, or cane sugar. Many fruits and some vegetables contain small amounts of sucrose. It is formed from one molecule of glucose and one molecule of fructose. Thus, upon digestion glucose and fructose are released and absorbed into the bloodstream. Sucrose is a major food sugar. In 1990, the per capita consumption of sucrose (refined sugar) in the United States was about 64.2 lb (*29 kg*), considerably lower than the 101.8 lb (*46.2 kg*) for 1970 and 97.6 lb (*44.3 kg*) for 1960.[11] However, that is not the whole story. In 1990, the per capita consumption of caloric sweeteners was 137.5 lb and the consumption of corn sweeteners was 71.9 lb. The world production of sugar rose from 8 million tons in 1900 to nearly 110 million tons in 1990. Humans have a sweet tooth! Few other human foods have shown an increase in production of this magnitude during the same time span. However, the rich countries around the world, like the United States, Britain, Australia, and Sweden, consume about 88 to 100 lb (*40 to 50 kg*) per person, while in the poor countries people consume 11 lb (*5 kg*) or less. Many accusations have been made to the effect that sucrose exerts a detrimental effect on the health of people in these rich (developed) countries. Claims are made that sucrose, at least to some extent, is responsible for such health problems as diabetes, obesity and dental caries (tooth decay). However, as is the case in many attempts to link nutrition to health, there is no clear-cut cause-and-effect relationship between the single food, sucrose, and the disease with which it has been implicated, with the exception of tooth decay.

(Also see SUGAR, section headed "Sugar as a Factor in Human Disease.")

• **Trehalose**—This is the sugar present in many fungi and bacteria. It is sometimes called mushroom sugar. Trehalose is composed of two molecules of glucose, which are joined by a different chemical bond than the two glucose molecules of maltose. Humans possess the enzyme trehalase; hence, they can split trehalose into two glucose units during the digestive processes.

TRISACCHARIDES. Trisaccharides, 3-sugar polymers, are not abundant in nature, but two trisaccharides —melezitose and raffinose—are found in limited amounts in certain plants. Melezitose, a component of sap in some coniferous plants, contains 2 molecules of glucose and 1 of fructose. Raffinose, which is found in sugar beets, molasses, beans, and cottonseed meal, consist of glucose, fructose, and galactose. Enzymes of the digestive tract are not capable of splitting melezitose and raffinose into monosaccharides.

Maltotriose is a trisaccharide which is formed during the digestion of starch; it contains 3 glucose molecules which are eventually split and absorbed during the digestive processes.

TETRASACCHARIDES. Stachylose is a tetrasaccharide composed of 2 molecules of galactose, 1 molecule of glucose, and 1 molecule of fructose. Beans produce gas in the digestive tract—flatulence—due to microbial action on stachylose and raffinose which are not split into monosaccharides by enzymes of the digestive tract. Maltotetrose is formed during the digestion or breakdown of starch. It consists of four molecules of glucose, which the digestive enzymes of the body can break.

[11]*Agricultural Statistics,* 1991, USDA, p. 479, Table 672.

Polysaccharides.

Polysaccharides are large sugar complexes that contain repeating sequences of simple sugars—chains of monosaccharides. In plants and animals they are used either for carbohydrate storage or structural support.

PENTOSANS. These polysaccharides yield pentose sugars upon hydrolysis—breakdown to individual sugars. Some common pentosans include araban, a chain of arabinose molecules; and xylan, a chain of xylose molecules. These pentosans are widely distributed plant polysaccharides.

HEXOSANS. Hexosans are polysaccharide sugars which contain hexoses as their respective repeating units.

Several polymers of glucose are found in plants or animals. The difference among the various compounds results from the linkages between the glucose molecules. Linkages between glucose molecules may be visualized by thinking of a large group of people. These individuals may form a chain by holding hands. However, hands can be held (1) right to left or (2) right to right and/or left to left—two types of linkagae. Furthermore, an individual not in the chain may grasp with one hand the foot of one individual in the chain, and then with the free hand hold to the hand of another individual starting a branch of hand to hand linkage. The chief polymers—long chains—of glucose are cellulose, dextrins, glycogens, inulin, mannan, and starch.

• **Cellulose**—This polymer is, by far, the most abundant polysaccharide in nature—composing close to 50% of the total organic carbon. Cellulose is the principal industrial carbohydrate. Its worldwide use is estimated at 800 million tons per year. It is a straight chain polymer which is extremely resistant to breakdown by acid or alkali. Humans lack the necessary enzymes to cleave the linkages of glucose molecules in cellulose (hands held right to left). Hence, they are unable to derive energy from cellulose. However, the microorganisms in the rumen of ruminants—cattle, sheep and goats—contain the enzyme cellulase; hence, ruminants can effectively utilize cellulose. Cellulose provides fiber in human diets.

• **Dextrins**—Dextrins are metabolic by-products of starch digestion. They are shorter chains of glucose molecules, but no set number of glucose units per molecule. Dextrins are more soluble than starches and they are found in relatively high concentrations in seeds. Baking bread or toasting bread produces some dextrins from flour. Eventually, dextrins are split into glucose units and absorbed in the small intestine.

(Also see STARCH.)

• **Glycogen**—Glyocogen—the primary form of glucose storage in animals—is a highly branched hexosan that is soluble in water. (Linkages are right hand to right hand with some hands grasping a foot forming the branches.) Of course, the linkages of glycogen can be broken down by enzymes within the animal body. Muscle and liver tissue contain extremely high concentrations of glycogen. However, there are few or no dietary sources of glycogen since glyogen in foods rapidly decomposes to lactic acid. Storage of glycogen is controlled by the hormones glucagon, insulin, and epinephrine. When glucose is needed in the body it is quickly released from glycogen.

• **Inulin**—Inulin, a hexosan found in Jerusalem artichokes, consists of repeating units of fructose. Inulin is not important to human nutrition, but it is used in a diagnostic test which determines kidney function.

• **Mannan**—This polysaccharide is a long chain of mannose molecule. It is widely distributed in plants, yeast, molds, and bacteria.

• **Starch**—Most glucose is stored in plants in the form of starch, of which there are two types: (1) amylose, a straight-chained structure of repeating glucose molecules (linked right to right and left to left); and (2) amylopectin, a highly branched compound. The structure of amylose is similar to cellulose—both straight chains of glucose. However, digestive enzymes are capable of splitting the linkage between the glucose molecules of amylose but not cellulose, though the linkage type is only very slightly different. Amylopectin is similar to glycogen but with fewer branched chains and a lower molecular weight. Furthermore, amylopectin is not water soluble as is glycogen.

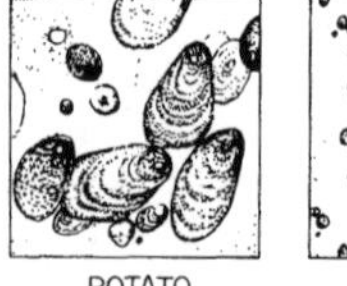

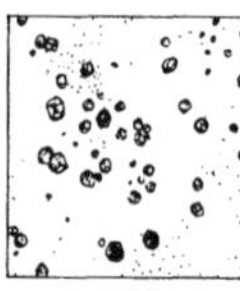

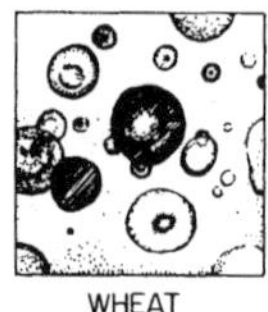

Fig. C-37. Microscopic view of starch granules of different sizes from different sources.

Starch type depends upon the ratio of amylose to amylopectin. Some starches contain virtually no amylose and are called waxy. Fresh starch obtained from plants is in granules and completely insoluble in water (see Fig. C-37). When starch is cooked in moist heat, the granules of starch absorb water, swell, and the walls of the granules are ruptured. This permits more ready access to the digestive enzymes. Ingested starch is split by enzymes into dextrins, maltotetrose, maltotriose, maltose, and eventually glucose. Starch in the diet is commonly referred to as a complex carbohydrate.

(Also see STARCH.)

MIXED POLYSACCHARIDES. A number of complex mixed polysaccharides are found in nature, many of them serving structural or protective functions. These include agar, alginic acid, carrageenan, chitin, hemicelluloses, pectins, and vegetable gums. Man has adapted some of these for use in foods as thickening and stabilizing agents.

(Also see ADDITIVES.)

• **Agar**—This is a polymer of galactose obtained from seaweed. It is used as a thickening agent in foods, especially confectionaires and dairy products. Agar is not digestible.

• **Alginic acid**—This polyaccharide with gelling properties is extracted from seaweed. It is capable of absorbing 200 to 300 times its weight in water. In the manufacture of ice cream, alginic acid serves as a stabilizing colloid ensuring a creamy texture and preventing the growth of ice crystals. Various other food and nonfood uses exist for alginic acid and its sodium salt, sodium alginate.

• **Carrageenan (IRISH MOSS)**—It is a mixture of nondigestible polysccharides, depending upon the seaweed from which it was extracted. Carrageenan is employed as a gelling, emulsifying, stabilizing, and thickening agent in foods and nonfoods. Milk products especially lend themselves to its use.

(Also see CARRAGEENAN.)

• **Chitin**—Chitin forms the thick material of the hard exoskeletons—outside covering—of insects and crustaceans. Chemically, it may be regarded as a derivative of cellulose. It has some industrial uses, and except for cellulose, it is probably the most abundant polysaccharide in nature.

• **Hemicelluloses**—These are a mixed group of polysaccharides which are closely associated with cellulose in plant tissues, though chemically distinct. Hemicelluloses are not digested in the small intestine, but are broken down by microorganisms in the colon.

• **Pectin**—This nondigestible polysaccharide is found in the soft tissue of fruits and vegetables. One of the richest sources is lemon or orange rind. Pectin possesses the ability to gel, and it is often used as a base for fruit jellies.

• **Vegetable gums**—Several gums obtained from plants are employed in a variety of products as water binders, thickeners, and stabilizers. These include arabic, tragacanth, and guar. Food uses consist of ice cream, ice milk, pudding, whipped cream, and yogurt, to name a few.

In the body, there is another group of polysaccharides which have an important role—the mucopolysaccharides. These include hyaluronic acid, chondrotin sulfate, and keratan. Many mucopolysaccharides are synthesized and found in connective tissue where they have structural importance.

DIGESTION, ABSORPTION, AND METABOLISM.

The digestion of starch begins in the mouth and continues in the small intestine. Eventually, most all available carbohydrates are converted, by enzymes, to the monosaccharides galactose, glucose, and fructose, absorbed into the bloodstream and distributed to the tissues of the body. Metabolism of carbohydrates involves primarily glucose, which can be metabolized directly to yield energy or it can be stored as glycogen in the liver and muscle. Liver glycogen is used constantly to replenish blood glucose as it is depleted between meals. Excess glucose may also be stored as fat.

(Also see DIGESTION AND ABSORPTION; and METABOLISM.)

FUNCTIONS. Although carbohydrates comprise only about 1% of the human body, they now account for about 47% of the caloric intake of Americans, as shown in Fig. C-38.

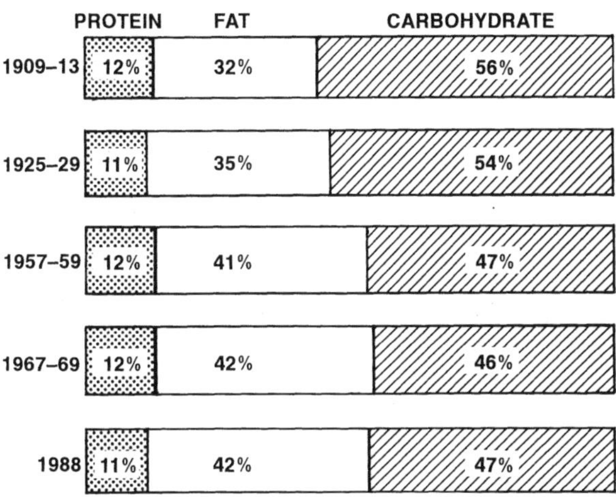

Fig. C-38. The changing pattern of calories derived from protein, fat, and carbohydrates in the American diet. (Courtesy, USDA)

Carbohydrates in the diet make the important contributions shown in Fig. C-39 and detailed in the sections which follow.

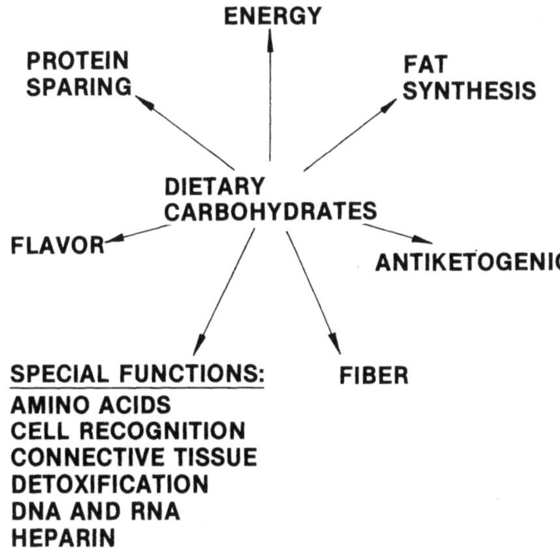

Fig. C-39. Functions of dietary carbohydrates.

Energy. Carbohydrates are the fuel of life. Each gram of carbohydrate provides about 4 Calories (kcal) of energy—the same as protein but less than fat. Following digestion and absorption, available carbohydrates may be (1) used to meet immediate energy needs of tissue cells, (2) converted to glycogen and stored in the liver and muscle for later energy needs, and (3) converted to fat as a larger reserve for energy.

1. **Immediate energy.** Glucose can be used directly by all tissues of the body. However, about one-fifth of the basal metabolic rate is due to the brain. The major source of energy for the brain appears to be carbohydrate (glucose) since it does not possess any store or reserve of energy.

2. **Glycogen.** Although glycogen is a stored form of glucose, there is only enough glycogen stored in the body to meet about one half of a day's energy requirements. Furthermore, exercise such as a walk for 2 to 3 hours without food depletes the glycogen stores of the body. Glucose is released from liver glycogen when needed to maintain the blood glucose levels, while muscle glycogen is available for the energy needs of the muscle but is not available for the regulation of blood glucose. Glycogen storage and release are controlled by the hormones glucagon, insulin, and epinephrine.

3. **Fat.** Once the energy needs of the body have been fulfilled, excess carbohydrate is converted to fat and stored in adipose tissue. As many people can testify, this type of storage seems to be unlimited.

Protein Sparing Action. Meeting the energy needs of the body takes priority over other functions. Thus, any deficiency of energy will be made up by energy yielding reactions involving protein and adipose. When available carbohydrates are supplied in sufficient quantities by the diet, the body uses carbohydrates first as a source of energy. This spares protein for tissue building. Eating 50 to 100 g of digestible carbohydrate daily prevents the excessive breakdown of tissue protein. Most American diets include 200 to 300 g of carbohydrate daily.

Regulation of Fat Metabolism. If carbohydrate intake is very low, fats are not completely utilized. Fat metabolism occurs faster than the body can take care of the intermediate products of fat metabolism—ketones. As ketones accumulate, the individual is said to be in a state of ketosis. If the production of ketones goes unchecked, the serious condition of acidosis occurs. This can be prevented by the ingestion of a minimum of 50 to 100 g of digestible carbohydrate per day. Since the metabolism of adequate amounts of carbohydrate allows the ketone forming substances to be completely metabolized, carbohydrates are said to have an antiketogenic effect.

(Also see METABOLISM.)

Provide Fiber. Not all carbohydrates yield nutrients to the body. Cellulose, gums, hemicelluloses, pectins, and pentosans are such carbohydrates. Collectively, they are referred to as dietary fiber or roughage. These substances absorb water and lend bulk to the intestinal contents, thereby stimulating the peristaltic movements of the digestive tract and reducing the passage time through the bowel. There is no metabolic requirement for fiber. Recently, much has been attributed to the amount of or lack of fiber in the diet; for example, (1) constipation, (2) diverticular disease, (3) irritable bowel syndrome, (4) bowel cancer, (5) inflammatory bowel disease, (6) atherosclerosis, (7) gallstones, (8) dental caries, (9) hemmorrhoids, and (10) ulcers. At times, it seems that every ill can be explained by lack of dietary fiber; hence, many people strive to increase their fiber intake by various means. One popular means is the daily ingestion of bran. Unfortunately, the cause of many of these diseases which are said to be cured or prevented by high fiber diets are complex, poorly understood, and not likely due to any single factor. Therefore, in most cases it is unlikely that high fiber is the magic for all that ails or might ail an individual. Still, the value of fiber should not be completely discounted. Fiber is an aid in weight reduction because it fills without providing extra calories. Among the various disorders mentioned above, there is good evidence that dietary fiber will help constipation and the symptoms of diverticular disease, but the role of fiber in other diseases is largely speculative. For those avid fans of fiber there is no real health hazard in eating more fiber. Currently, it is not possible to recommend a specific intake of fiber; more research is needed. However, dramatic increases in dietary fiber should be avoided since the absorption of minerals may be reduced by high dietary fiber intakes. Moderate increases in the consumption of fiber can be easily achieved by increasing the consumption of nuts, vegetables, fruits, and whole grain cereal products. The fiber content of various foods is given in Food Composition Table F-36.

(Also see FIBER.)

Special Functions. Some carbohydrates and carbohydrate derivatives perform special duties in structure and function. These carbohydrates include (1) glucuronic acid which detoxifies chemicals and bacterial byproducts, and inactivates hormones in the liver prior to their excretion; (2) hyaluronic acid and chondroitin sulfate which are components of connective tissue; (3) the compounds deoxyribonucleic acid (DNA) and ribonucleic acid (RNA) which possess and transfer genetic information; (4) heparin which prevents clotting of blood; (5) carbon skeletons for certain amino acids; (6) those which associate with lipids in nervous tissues; and (7) those which serve as biological markers and provide information for cell recognition and antigen antibody reactions of the immune system. A familiar example of the last special function is that the difference between blood type A and blood type B lies in a single sugar unit that sticks out from the end of a carbohydrate chain of a glycoprotein or glycolipid on the surface of the red blood cell. This small difference is sometimes a matter of life and death since using the wrong type of blood in a transfusion can have fatal results.

Flavor. Foods are often accepted or rejected on the basis of their flavor. Some carbohydrates flavor food. Their flavor is interpreted by our taste buds as sweet, one of the four taste sensations. Sweeteners of any kind, whether nutritive or nonnutritive, are commonly compared to the carbohydrate, sucrose. Table C-15 compares the sweetness of some sweet-tasting substances to sucrose—table sugar.

TABLE C–15
COMPARISON OF SOME SWEET-TASTING SUBSTANCES

Substance	Sweetness[1]	Substance Identification
Saccharin	500	Artificial sweetener
Aspartame	180	Artificial sweetener
Tryptophan	30	Amino acid
Fructose	1.7	Monosaccharide
Invert sugar[2]	1.3	Monosaccharide
Sucrose	1.0	Disaccharide
Glycine	0.8	Amino acid
Glucose	0.7	Monosaccharide
Sorbitol	0.5	Sugar alcohol
Maltose	0.3	Disaccharide
Galactose	0.3	Monosaccharide
Lactose	0.3	Disaccharide
Raffinose	0.2	Trisaccharide
Starch	0	Polysaccharide

[1]The sweetness of sucrose taken as the standard.

[2]Invert sugar is formed by the splitting of sucrose. Thus, it is a mixture of half glucose and half fructose.

Two noteworthy conclusions may be drawn from Table C-15: (1) Carbohydrates vary in sweetness; and (2) carbohydrates are not the sweetest substances. Although sucrose is not the sweetest substance, it is the chief source of sweetening used in most desserts, ice cream, candies, and soft drinks. Moveover, in our attempt to stimulate the sweet-tasting sensation, excessive calories from sucrose consumption may certainly be a contributing factor to the great American problem of obesity, and it does contribute to tooth decay. Meanwhile, the search continues for a safe, nonnutritive sweetener so that we may indulge our "sweet tooth" and not suffer the consequences.

(Also see ADDITIVES; OBESITY; SOFT DRINKS; SUGAR; SWEETENING AGENTS; and TASTE.)

REQUIRED INTAKE. There is no specific dietary requirement for carbohydrate. Nonetheless, it is generally agreed that a reasonable proportion of the caloric intake should be derived from carbohydrate. A diet devoid of carbohydrate is likely to lead to ketosis, excessive breakdown of tissue protein, loss of cations—especially sodium, and involuntary dehydration. These effects, produced by high-fat diets or fasting, can be prevented by the ingestion of 50 to 100 g of digestible carbohydrate per day. However, intakes considerably above this minimal level are desirable. Foods such as fruits, vegetables, and whole-grain cereals provide energy principally from carbohydrate and are generally good sources of other nutrients, such as vitamin and minerals.

SOURCES. The carbohydrate content of various foods is given in Food Composition Table F-36. No specific level of carbohydrate intake is most conducive to health. Furthermore, foods vary in their carbohydrate content, and the relative prominence of carbohydrate-rich foods in the diet varies widely in individuals, and in different parts of the world. A diet consisting of chiefly carbohydrate-rich foods need be no health hazard provided those foods are not lacking in proteins, minerals, and vitamins. In most cases, a country's staple food is its primary source of carbohydrate, and it must serve as the principal source of protein in poor countries. Total protein consumption and total calorie intake increase with income.

Plant. Plants provide the major source of carbohydrates. Among the plants, the cereals are the most important source of available carbohydrates for man. The total carbohydrate contents—fiber, sucrose, fructose, glucose, lactose, and maltose—of a variety of foods are given in Table C-16, Carbohydrates & Sugars in Selected Foods. Additionally, the total carbohydrate content of numerous other foods is given in Food Composition Table F-36.

CEREAL GRAINS. Much of the world is dependent upon calories supplied by a single cereal staple which supplies carbohydrate in the form of starch. For example, in Afghanistan and Pakistan the staple is wheat, and in Mexico and Central America it is corn (maize). In China and southeastern Asia, regions of the world with the largest population, the staple is rice. Annually world production of these three cereals dwarfs the production of all other plant foods. American diets derive carbohydrates from wheat, corn, and rice because they appear in so many forms. Other cereal grains include barley, millet, oats, rye, sorghum, and triticale. The grains and the products made from them have variable water content, ranging from about 10 to 85% carbohydrate.

(Also see BARLEY; CEREAL GRAINS; CORN; MILLET; OATS; RICE; RYE; SORGHUM; TRITICALE; and WHEAT.)

TABLE C-16
CARBOHYDRATES & SUGARS IN SELECTED FOODS[1]

Food Name Portion—100 Gram (3.5 oz)	Water	Food Energy Calories	Total Carbo-hydrate	Crude Fiber	Sugars				
					Sucrose	Fructose	Glucose	Lactose	Maltose
	(g or %)	(kcal)	(g)	(g)	(g)	(g)	(g)	(g)	(g)
BAKERY PRODUCTS									
BREAD, whole wheat	36.4	243	47.7	1.6	—	—	—	1.8	—
ROLL, dinner type (pan), enriched	31.4	298	53.0	0.2	—	—	—	1.4	—
DOUGHNUT, cake type, plain	23.7	391	51.4	0.1	15.9	—	—	2.3	—
CEREALS & FLOURS									
WHEAT GERM, dry	11.5	363	46.7	2.5	8.3	—	—	—	—
SHREDDED WHEAT BISCUIT	6.6	354	79.9	2.3	2.7	—	—	—	—
WHITE FLOUR, all purpose, enriched	12.0	364	76.1	0.3	0.2	—	—	—	0.1
DESSERTS & SWEETS									
CANDY, chocolate, plain	0.9	520	56.9	0.4	43.0	—	—	8.1	—
JELLY, assorted	29.0	273	70.6	0.0	52.0	—	—	—	—
HONEY	17.2	304	82.3	0.1	1.9	40.5	34.2	—	—
MOLASSES, blackstrap	24.0	213	55.0	0.0	36.9	6.8	6.8	—	—
SYRUP, corn, light or dark	24.0	290	75.0	0.0	—	—	21.2	—	26.4
SYRUP, pure maple	33.0	252	65.0	0.0	62.9	—	—	—	—
SUGAR, granulated, white	0.5	385	99.5	0.0	99.5	—	—	—	—
FLAVORINGS & SEASONINGS									
SOY SAUCE	62.8	68	9.5	0.0	—	0.9	—	—	—
FRUITS									
APRICOTS, raw	85.3	51	12.8	0.6	3.5	0.8	1.8	—	—
dried, uncooked	25.0	260	66.5	3.0	—	—	—	—	—
BANANA, yellow, raw	75.7	85	22.2	0.5	8.5	5.9	5.2	—	—
BLACKBERRIES, raw	84.5	58	12.9	4.1	0.9	2.9	3.2	—	—
BLUEBERRIES, raw	83.2	62	15.3	1.5	2.4	—	—	—	—
BOYSENBERRIES, frozen, unsweetened	86.8	48	11.4	2.7	1.1	—	—	—	—
CANTALOUPE, raw	91.2	30	7.5	0.3	4.1	1.3	1.1	—	—
CHERRIES, sour red, raw, w/o pits and stems	83.7	58	14.3	0.2	0.1	7.2	4.7	—	—
sweet, raw, w/o pits and stems	80.4	70	17.4	0.4	1.9	7.2	4.7	—	—
DATES, moisturized or hydrated	22.5	274	72.9	2.3	—	23.9	24.9	—	—
FIGS, dried, uncooked	23.0	274	69.1	5.6	9.9	22.1	32.2	—	—
GRAPEFRUIT, pink, red, or white	88.4	41	10.6	0.2	2.1	1.2	1.9	—	—
pink/red, Texas	87.7	43	11.3	0.2	2.1	1.2	1.9	—	—
white, Texas	87.7	43	11.3	0.2	2.1	1.2	1.9	—	—

(Continued)

TABLE C-16 (*Continued*)

Food Name Portion—100[1] Gram (3.5 oz)	Water	Food Energy Calories	Total Carbo-hydrate	Crude Fiber	Sugars				
					Sucrose	Fructose	Glucose	Lactose	Maltose
	(g or %)	(kcal)	(g)	(g)	(g)	(g)	(g)	(g)	(g)
GRAPES, American type, raw, slip skin (Concord/Delaware)	81.6	69	15.7	0.6	0.2	4.3	4.8	—	—
European type, raw, adherent skin (red/dark blue)	81.4	67	17.3	0.5	1.2	—	—	—	—
green, seedless, raw, (Thompson)	81.4	67	17.3	0.5	1.2	—	—	—	—
HONEYDEW MELON, raw	90.6	33	7.7	0.6	4.1	—	—	—	—
LEMON	90.1	27	8.2	0.4	0.2	—	—	—	—
MANGO, raw	81.7	66	16.8	0.9	7.4	—	—	—	—
NECTARINES, raw	81.8	64	17.1	0.4	5.1	—	—	—	—
ORANGE, raw	86.0	49	12.2	0.5	4.2	1.8	2.5	—	—
PEACH raw	89.1	38	9.7	0.6	5.5	1.0	1.0	—	—
PEAR, raw, w/skin	83.2	61	15.3	1.4	1.2	5.0	2.5	—	—
PINEAPPLE, raw	85.3	52	13.7	0.4	6.9	1.4	2.3	—	—
PLUMS, greengage, canned, water pack	90.6	33	8.6	0.2	2.9	4.0	5.5	—	—
purple, canned, water pack	86.8	46	11.9	0.3	5.4	—	—	—	—
RAISINS, uncooked	18.0	289	77.4	0.9	14.2	—	—	—	—
RASPBERRIES, red, raw	84.2	57	13.6	3.0	2.1	1.3	0.9	—	—
STRAWBERRIES, raw	89.9	37	8.4	1.3	1.0	1.8	1.7	—	—
TANGERINE, raw	87.0	46	11.6	0.5	3.8	—	—	—	—
WATERMELON, raw	92.6	26	6.4	0.3	3.2	—	—	—	—
JUICES									
LEMON JUICE, fresh	91.0	25	8.0	Trace	0.2	0.9	0.5	—	—
LIME JUICE, fresh	90.3	26	9.0	Trace	1.3	—	—	—	—
ORANGE JUICE, fresh, all varieties	88.3	45	10.4	0.1	4.7	2.4	2.4	—	—
NUTS									
CHESTNUTS	52.5	194	42.1	1.1	3.6	—	—	—	—
MACADAMIA NUTS	3.0	691	15.9	2.5	5.5	—	—	—	—
PECANS	3.4	687	14.6	2.3	1.1	—	—	—	—
VEGETABLES									
BEANS, green, raw	90.1	32	7.1	1.0	0.5	—	—	—	—
CABBAGE, raw	92.4	24	5.4	0.8	0.3	—	—	—	—
CARROT, raw	88.2	42	9.7	1.0	1.7	—	—	—	—
CAULIFLOWER, raw	91.0	27	5.2	1.0	0.3	—	2.8	—	—
CELERY, raw	94.1	17	3.9	0.6	0.3	—	—	—	—
CUCUMBER, raw, not pared	95.1	15	3.4	0.6	0.1	—	—	—	—
LETTUCE, Bibb or Boston, raw	95.1	14	2.5	0.5	0.2	—	—	—	—
Iceberg, raw	95.5	13	2.9	0.5	0.2	—	—	—	—
Romaine or Cos, raw	94.0	18	3.5	0.7	0.2	—	—	—	—
ONION, dry yellow, raw	89.1	38	8.7	0.6	7.9	—	—	—	—
PEAS, green, immature, raw	78.0	84	14.4	2.0	5.5	—	—	—	—
RADISHES, raw, common	94.5	17	3.6	0.7	0.3	—	—	—	—
SWEET POTATO, baked in skin	63.7	141	32.5	0.9	7.2	—	—	—	—
TOMATO, raw	93.5	22	4.7	0.5	—	1.2	1.6	—	—
TOMATOES, canned, solids & liquid	93.7	21	4.3	0.4	0.3	—	—	—	—

[1]The authors gratefully acknowledge that these food compositions were obtained from the HVH-CWRU Nutrient Data Base developed by the Division of Nutrition, Highland View Hospital, and the Departments of Biometry and Nutrition, School of Medicine, Case Western Reserve University, Cleveland, Ohio.

FRUITS. Because of their high water content, fruits represent a less concentrated source of carbohydrates than cereals. In fresh fruits, the carbohydrate is mainly in the form of monosaccharides—glucose and fructose—and some disaccharide—sucrose. Canned fruits often contain added sucrose or glucose unless specifically labeled as being canned with added sugar.

In fresh fruits, the sugar—available carbohydrate—content may range from about 6 to 25%. Cantaloupe and watermelon contain 3 to 7% while bananas contain about 20%. The degree of ripeness also influences the sugar content of fruits. Ripe fruits have a higher sugar content.

Dried fruits possess a much higher sugar content. Dried dates, figs, raisins, prunes, and apricots may contain 50 to 90% sugar due to their low water content. Thus, the energy value of fruits, whether fresh or canned, is determined largely by their sugar (monosaccharide and disaccharide) content.

(Also see FRUIT[S]; and SUGAR.)

NUTS. Due to their low moisture content, nuts contain from 10 to 27% carbohydrate. Furthermore, nuts contain from 1 to 3% fiber. However, because of their high protein and high fat content, nuts are not usually thought of as a carbohydrate source.

(Also see NUT[S].)

VEGETABLES. Almost every part of the plant used for food—leaves, stems, seeds, seed pods, flowers, fruits, roots and tubers—has been grouped under the term vegetable. For this reason, the carbohydrate content of vegetables may range from 3 to 35%, and the carbohydrate may be present as starch, sugar, and fiber. Leaf, flower, and stem vegetables contain high levels of water and cellulose and few calories. The root, tuber, and seed vegetables have a high starch and sugar content and low water content, thus providing more calories per unit of weight. Freshly picked vegetables—sweet corn, young peas, and young carrots—contain more sugar and less starch than mature vegetables—just the opposite of fruits.

(Also see VEGETABLE[S].)

Animal. Most animal foods contain little, if any, carbohydrate. When an animal is slaughtered, the glycogen stored in the liver and muscles is rapidly broken down to lactic acid and pyruvic acid. Oysters and scallops may contain some glycogen but the amount is not significant to the diet. Milk is the only animal food, or at least the only animal-produced food, that is an important carbohydrate source. Milk provides the carbohydrate lactose or milk sugar.

(Also see MEAT[S]; and MILK AND MILK PRODUCTS.)

Other. Actually, the other sources of carbohydrate are plant sources. However, they represent an important, but rather specialized, source of carbohydrate. These consist of common table sugar, molasses, maple syrup and sugar, corn syrup, honey, and sorghum syrup. These sources of carbohydrate are extremely concentrated. They provide mainly calories and little else.

- **Sugar**—The refined disaccharide, sucrose, is processed from sugar beets or sugarcane. White and brown sugar are nearly 100% carbohydrate. Sugar consumption occurs mainly in the form of candies, soft drinks, and rich desserts, though it is an important constituent of many foods.

(Also see SUGAR.)

- **Molasses**—Molasses is a by-product of sugar refining. It has a higher mineral content than beet, cane, or refined sugar, but still it is about 55 to 70% carbohydrate.

(Also see MOLASSES.)

- **Maple syrup and sugar**—These are some of the earliest sweeteners. The Indians taught the early American settlers how to use them. They are made by boiling down the sap of the sugar maple. The sugar of maple syrup and maple sugar consist of the disaccharide sucrose. Maple sugar is 90% carbohydrate while maple syrup is 65% carbohydrate.

(Also see MAPLE SYRUP.)

- **Corn syrup**—Corn syrup is formed by breaking down (hydrolyzing) corn starch. It is mostly glucose and maltose. Light and dark corn syrup contains 75% carbohydrate.

(Also see CORN SYRUP.)

- **Honey**—Bees make honey from flower nectars. It is a mixture of the two monosaccharides, fructose and glucose. Fructose gives honey its sweetness. Honey contains about 82% carbohydrate.

(Also see HONEY.)

- **Sorghum syrup**—The sweet juice of the grain sorghum stem is used to make sorghum syrup. Sorghum syrup is not as widely used as the other carbohydrate sources listed above. It contains 68% carbohydrate.

(Also see ADDITIVES; DESSERTS AND SWEETS; DIGESTION AND ABSORPTION; METABOLISM; SOFT DRINKS; SUGAR; and SWEETENING AGENTS.)

FERMENTATION. Many carbohydrate sources may be fermented; for example fruit juices, palm juice, cactus juice, molasses, sugar, honey, milk, potatoes, or cereal grains. Fermentation is the production of ethanol or ethyl alcohol by the action of yeast on carbohydrate as shown in Fig. C-40.

$$\underset{\text{HEXOSE (SUGAR)}}{C_6H_{12}O_6} \xrightarrow{\text{YEAST}} \underset{\text{ETHYL ALCOHOL}}{2C_2H_5OH} + \underset{\text{CARBON DIOXIDE}}{2CO_2}$$

Fig. C-40. The action of yeast on sugar to produce ethyl alcohol—the alcohol of beers, distilled liquors and wine.

Throughout history, most cultures have learned to use and control the process of fermentation. Regardless of the carbohydrate source used, the ethanol produced is the same, while the taste may vary with the carbohydrate source. Hard liquors are those produced by distillation which concentrates the ethanol and separates it from the starting material. Beer and wine contain some of the nutrients present in the malted barley or fruit juice.

Alcohol contains calories. It is metabolized at a fixed rate in the liver and yields about 7 Calories (kcal) per gram. Therefore, the intake of alcoholic beverages must be accounted for when counting calories. When alcohol is consumed faster than can be metabolized, then the level of alcohol in the body tissues builds up and intoxication follows. Upon drinking, alcohol is rapidly absorbed from the stomach and the small intestine. It requires no digestion. Excessive drinking of alcoholic beverages creates many problems, one of which is low intake of minerals and vitamins. In many cases the calories of alcoholic drinks are empty calories. Furthermore, the evidence is mounting that any level of intake during pregnancy may be harmful to the developing child.

(Also see ALCOHOL; ALCOHOLISM; BEERS AND BREWING; and DISTILLED LIQUORS.)

CARBOHYDRATE INTOLERANCE

A digestive disorder due to (1) the lack of one or more enzymes which digest carbohydrates, or (2) other conditions which prevent the absorption of certain sugars. The unabsorbed sugars provide nutrients for bacterial growth and have a laxative effect which often results in diarrhea.

(Also see CARBOHYDRATE[S]; and DISEASES.)

CARBOHYDRATES, AVAILABLE

These are carbohydrates—products of photosynthesis —which provide nourishment to humans and other monogastric animals. They may be digested and absorbed. Included are the sugars, dextrins, and starches.

(Also see CARBOHYDRATE[S]; and DIGESTION AND ABSORPTION.)

CARBOHYDRATES, UNAVAILABLE

These are carbohydrates which do not provide nourishment to humans. Man and other monogastric animals lack the enzymes necessary to breakdown such carbohydrates as cellulose, hemicellulose, and pectin. Often these carbohydrates are referred to as dietary fiber or roughage. Because of the microbial action in the rumen, ruminant animals—cattle, sheep, and goats—can utilize carbohydrates that are unavailable to man and other monogastrics.

(Also see CARBOHYDRATE[S].)

CARBON DIOXIDE (CO_2)

A gas formed in the cells of the body during the production of energy by the metabolism of nutrients. It is carried in the blood to the lungs, where it is exhaled.

(Also see METABOLISM.)

CARBONIC ACID (H_2CO_3)

A weak acid formed by the reaction of carbon dioxide gas with water. Under certain conditions carbonic acid may release the bicarbonate ion which helps to maintain the neutrality of the blood and body fluids.

(Also see ACID-BASE BALANCE.)

CARBONIC ANHYDRASE

An enzyme which contains an atom of zinc in each molecule. It is present in various tissues, and in red blood cells, where, depending upon the circumstances, it promotes either (1) the formation of carbonic acid from carbon dioxide, or (2) the breakdown of carbonic acid to carbon dioxide and water. These reactions facilitate the picking up of carbon dioxide from the tissues by the red blood cells, and the transfer of carbon dioxide from the red blood cells to the lungs, where it is exhaled.

(Also see ACID-BASE BALANCE.)

CARBOXYLASE

A general term for an enzyme that promotes the breakdown of organic acids, such as amino acids, to carbon dioxide and compounds that contain fewer carbon atoms.

(Also see ENZYMES.)

CARBOXYMETHYL CELLULOSE (CMC) (SODIUM CARBOXYMETHYL CELLULOSE; CM CELLULOSE)

A preparation of cellulose, made from the pure cellulose of cotton or wood, which absorbs up to 50 times its weight of water to form a stable mass. Because of this unique characteristic, it is used as a whipping agent in confectioneries, jellies and ice cream, and as an inert food filler in slimming aids. Ethyl methyl cellulose and hydroxyethyl-cellulose are examples of other cellulose derivatives with similar properties.

CARBOXYPEPTIDASE

A digestive enzyme secreted by the pancreas via the pancreatic juice into the small intestine where it digests protein.

(Also see DIGESTION AND ABSORPTION; and ENZYMES.)

CARCASS

The body of an animal without the viscera, and usually without the head, skin, and/or lower legs.

(Also see MEAT[S].)

CARCINOGEN

Any cancer-producing substance or agent.

(Also see CANCER.)

CARCINOGENIC

Cancer producing.

CARCINOMA

A malignant growth of living tissue which invades surrounding tissue.

(Also see CANCER.)

CARDIA

• The heart.

• The region extending from the lower portion of the esophagus to the upper part of the stomach.

Hence, pains coming from the region under the breastbone may originate from (1) the heart or (2) the digestive tract. Therefore, a heart attack may be mistaken for indigestion, or vice versa.

(Also see DIGESTION AND ABSORPTION; and HEART DISEASE.)

CARDIAC

A term which refers to the heart.

CARDIOSPASM

• A fluctuation in the beating of the heart.

• A spasm occurring anywhere between the lower portion of the esophagus and the upper part of the stomach.

CARDIOVASCULAR

Relating to or involving the heart and blood vessels.

(Also see DISEASES; and HEART DISEASE.)

CARDIOVASCULAR DISEASE

A general term which is used to denote a large number of disorders that affect the heart and/or the blood vessels. The disorders which are most common are atherosclerosis, heart disease, high blood pressure, and stroke. Currently, there is a tendency to attribute these disorders to faulty diets. However, nondietary factors may be responsible for many cases of cardiovascular disease.

(Also see HEART DISEASE; and HIGH BLOOD PRESSURE.)

CARDOON (CARDON) *Cynara cardunculus*

A large thistlelike plant related to the artichoke, the blanched leaves and stalks and the thick main roots of which are used as food.

CARIES

Tooth decay—the pathological process of localized destruction of tooth tissue by microorganisms.

(Also see DENTAL HEALTH, NUTRITION, AND DIET.)

CARIOGENIC AGENT

A food, or a food ingredient, which promotes the formation of dental caries.

(Also see DENTAL HEALTH, NUTRITION, AND DIET.)

CARMINATIVE

A substance that helps in the expulsion of gas from the stomach or intestine, or prevents the formation of gas.

(Also see DIGESTION AND ABSORPTION.)

CARNITINE (VITAMIN B_T)

Carnitine, a vital coenzyme in animal tissues and involved in fat metabolism, is a vitaminlike substance that has received much attention recently. It is similar to a vitamin with the exception that under normal conditions higher animals synthesize their total requirement within their bodies; hence,it is unnecessary to supply this substance in food on a daily basis. However, recent studies suggest (1) that synthesis of body carnitine may be inadequate for some individuals, and (2) that a number of diseases alter levels of carnitine in body fluids and tissues. This prompts two questions: (1) Are carnitine needs met adequately by body synthesis to assure buoyant good health; and (2) what role, if any, does carnitine play in certain diseases?

HISTORY. Carnitine was first isolated from meat extract in 1905, but its structure was not established until 1927. Then, another 20 years elapsed before Fraenkel, in 1947, while investigating the role of folic acid in the nutrition of insects, found that the meal worm (*Tenebrio molitor*) required a growth factor present in yeast. Frankel called this factor "Vitamin B_T"; vitamin B because of its water-soluble property, and the T standing for *Tenebrio.* Because of not being recognized as a vitamin, the name was subsequently changed to carnitine.

CHEMISTRY, METABOLISM, PROPERTIES.

• **Chemistry**—Carnitine has the following molecular structure:

$$(CH_3)_3\ N\text{-}CH_2\text{-}\underset{\displaystyle OH}{\underset{|}{C}}H\text{-}CH_2\text{-}COOH$$

• **Metabolism**—Like the water-soluble vitamins, it is believed that carnitine is easily and rather completely absorbed. However, the form of carnitine (free or esterified) which is absorbed, the exact mechanism of absorption across the mucosa, and the site of the absorption are unknown.

It is not known how carnitine is transported in the blood from the gut to tissues.

In addition to being obtained from food, carnitine is synthesized in the liver.

In the rat, the highest concentrations of carnitine are found in the adrenal gland, followed by the heart, skeletal muscle, adipose tissue, and liver. Smaller concentrations are found in the kidney and brain. In humans, skeletal muscle has about 40 times the concentration of carnitine found in the blood.

Free carnitine is excreted in the urine.

• **Properties**—Carnitine is very hygroscopic (it absorbs water readily), and is easily soluble in water and ethanol.

MEASUREMENT/ASSAY. Several different assay procedures for determining the carnitine content of foods and tissues have been developed, but the two most common methods are: (1) bioassy, based on the growth or survival of meal worm *(Tenebrio molitor)* larvae; and (2) enzymatic techniques. Comparisons of carnitine values between studies are made difficult by lack of uniformity of assay procedures.

FUNCTIONS. Carnitine plays an important role in fat metabolism and energy production in mammals. It functions as follows:

1. **Transport and oxidation of fatty acids.** Carnitine plays an important role in the oxidation of fatty acids by facilitating their transport across the mitochondrial membrane.

Carnitine is part of the shuttle mechanism whereby long-chain fatty acids are made into acyl carnitine derivatives and transported across the mitochondrial membrane, which is impermeable to long-chain fatty acids *per se* and to their coenzyme A esters. Once across the membrane, the acyl carnitines are reconverted to their fatty acid CoA form and undergo beta-oxidation to liberate energy.

2. **Fat synthesis.** Although this role is controversial, carnitine appears to be involved in transporting acetyl groups back to the cytoplasm for fatty acid synthesis.

3. **Ketone body utilization.** Carnitine stimulates acetoacetate oxidation; thus, it may play a role in ketone body utilization.

Currently, the main emphasis of research is on the role of carnitine in fatty acid oxidation. Carnitine metabolism is being studied with increasing frequency (1) where there are changes in fat catabolism with changes in physiological state (fasting, exercise, extensive burns, pregnancy, cold adaptation), or (2) where there is disease (hyperthyroidism or hypothyroidism, diabetes, muscle fat storage disease, atherosclerosis, etc.).

DEFICIENCY SYMPTOMS. If carnitine is to be considered an essential dietary nutrient, it must be possible to show that a lack in the diet results in a reproducible deficiency disease in humans. To date, carnitine has not met this criterion.

RECOMMENDED DAILY ALLOWANCE OF CARNITINE. Under normal conditions, there is no dietary requirement for carnitine. However, where a metabolic abnormality exists which inhibits synthesis, interferes with use, or increases catabolism of carnitine, illness may follow, which is sometimes relieved by dietary supplement.

There is need for further research on the dietary role of carnitine in human health and disease.

CARNITINE LOSSES DURING PROCESSING, COOKING, AND STORAGE. Since carnitine is water-soluble, cooking procedures using moist-heat methods will likely result in loss of the free carnitine.

SOURCES OF CARNITINE. Generally speaking, carnitine is high in animal foods and low in plant foods. Few foods have been assayed for carnitine, but, based on available data, the following evaluation of dietary sources of carnitine may be helpful:

- **Rich sources**—Muscle meat, liver, heart, yeast (torula and brewers'), chicken, rabbit, milk, whey.

- **Good sources**—Avocado, casein, wheat germ.

- **Poor sources**—Cabbage, cauliflower, peanuts, wheat.

- **Negligible sources**—Barley, corn, egg, orange juice, spinach.

The lower level of carnitine in plant foods in comparison with animal foods is explainable on the basis that plant materials are most likely deficient in the essential amino acids, lysine and methionine, precursors of carnitine. Thus, a vegetarian diet will likely be low in carnitine, in both preformed carnitine and the amino acid precursors of carnitine.

Man and other higher animals appear to be able to synthesize their total needs within the body. But the mechanism of carnitine synthesis in humans is unknown. Very likely, carnitine is synthesized in humans from lysine and methionine—two essential amino acids which are low in plant foods.

(Also see VITAMIN[S], Table V-9.)

CAROB *Ceratonia siliqua*

This food is the fruit pod of a leguminois evergreen tree (*Ceretonia siliqua*) which grows mainly in the Mediterranean region.

BOTANICAL DESCRIPTION OF THE FOOD PLANT. The carob tree (also called algarroba, locust bean, locust pod, and St. John's bread) grows to heights of 40 to 50 ft (*12 to 15 m*), has dark green leaves, and bears long (3 to 12 in. [*7.5 to 30 cm*] in length), thick, fleshy pods, rich in sugars. It grows readily where water is scarce; hence, it has been planted for shade purposes along some of the avenues in southern California. The carob powder, sold in many health food stores, is produced by grinding the pod after removal of the seeds. Inside the pods are hard, brown seeds which account for about a tenth of the weight of the pod and seed combined. These seeds, which are covered with a tight-fitting brown coat, contain a white, transluscent endosperm, which is the source of carob gum (also known as locust bean gum). Fig. C-41 shows the leaves and pods of the carob plant.

Consumers may see only one product from the carob plant—the powdered pod—in retail stores. But, more than likely, they will consume the gum which is used as an additive by food processors.

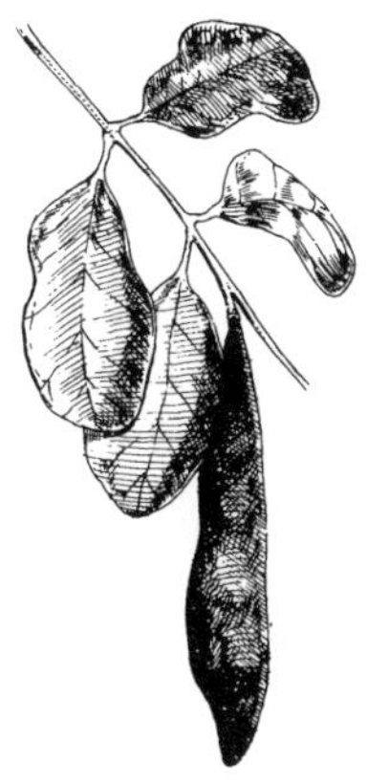

Fig. C-41. Carob, a plant which yields a tasty pod and valuable gum-containing seeds.

HISTORICAL BACKGROUND. The name carob comes from the Arabian word *kharrub,* which means bean pod. Interestingly, the word carat has a similar derivation. The Arabs used carob seeds as standards to weigh gems and precious metals. The Egyptians extracted a glue from the seeds, which they used to bind the cloth stretched around their mummies.

The Bible tells of the early uses of carob pods. Presumably, these pods were the "locusts" that St. John the Baptist ate with honey while traveling in the desert (*Matthew* 3:4). Also, the husks which were fed to the swine and craved by The Prodigal Son, are believed to have been carob pods (*Luke* 15:16).

Eventually, the carob tree was cultivated throughout the entire Mediterranean region, where it was often relied upon to provide food for humans and livestock in times of scarcity. Recently, it was introduced into the Americas and Australia.

PRODUCTION OF CAROB PRODUCTS. The mature pods of the locust tree have a tough, leathery appearance. They are usually dried in the sun prior to being broken open for removal of their seeds.

Extraction of Gum from the Beans. The seed coats may be removed either by abrasion or chemical treatment. Then, bare seeds are split and the endosperm is separated from the germ. After grinding, the powdered endosperm is graded and sold as locust bean gum. (The grade and price of the gum depend mainly upon the completeness of the separation of the endosperm from the other parts of the seed.) An alternative method is to roast and extract the dehulled and pulverized endosperm with boiling water, then dry the extract.

Carob Powder. This product is made by pulverizing the dried pod after the seeds have been removed.

Carob Syrup. This syrup is made by extracting the sugars from the carob powder with water, followed by boiling the resulting solution to the desired thickness.

PROPERTIES OF CAROB PRODUCTS. The carob tree is unique in that it furnishes two distinct and valuable food products: pods whose powder is similar in many ways to cocoa, and beans which yield a gum having properties like those of other vegetable gums. Details of the properties for each product follow.

Carob Bean Gum. This gum has a great similarity to guar gum, which is not surprising because both products consist of long-chain molecules (polymers) made up of alternating units of galactose and mannose (sugar units which contain 6-carbon atoms like glucose and fructose). However, in comparison with guar gum, carob, or locust bean gum, has about half as many branching units of galactose attached to the backbone of the gum molecule. As a result of this small difference in molecular structure, carob gum requires hot water for the preparation of its solutions while guar gum dissolves readily in cold water. The wetting, heating, steaming, drying, and pulverization of a mixture of carob bean gum and 1.5 times its weight of corn sugar yields a product which will dissolve more readily in cold water.

Although highly viscous solutions are formed by mixing the gum with hot water, these mixes do not gel. However, they improve the elasticity of seaweed gels made from agar and carrageenan. Solutions of carob bean gums are stable over a wide range of acidity and alkalinity.

Comparisons of Carob Powder with Cocoa Powder. Although there were great expectations of a large market for carob products in California during the early 1920s, interest in these products soon waned. However there has recently been a resurgence of interest in carob powder, due in part to promotion of carob as a substitute for cocoa products, and in part to the rising costs of cocoa and chocolate. Table C-17 gives a comparison of the nutrient contents of carob powder vs various types of cocoa powder.

The following facts should also be noted in comparing carob and cocoa:

1. Carob may be used in some recipes without additional sweetening, but cocoa always requires the addition of sugar or other sweetener.

2. Carob contains no stimulant, whereas cocoa contains the potent theobromine, which is similar to caffeine in its effects.

3. Both carob and low-fat cocoa powders are low in sodium and high in potassium. This is desirable for diets which restrict sodium and encourage potassium, such as those used in the treatment of congestive heart failure and hypertension. However, there are often much higher levels of sodium in the cocoas having moderate to high-fat contents, since they may have been processed with an alkali in order to increase their mixability in water.

It is noteworthy, too, that carob (pod) powder contains only 3.8% protein, which is much lower than most legumes.

TABLE C-17
NUTRIENT COMPOSITION OF CAROB POWDER AND COCOA PRODUCTS[1]
(Per 100 g Portions)

Food and Description	Water	Food Energy	Protein	Fat	Carbohydrate Total	Carbohydrate Fiber	Calcium	Phosphorus	Sodium	Potassium
	(%)	(kcal)	(g)	(g)	(g)	(g)	(mg)	(mg)	(mg)	(mg)
Carob (pod) powder	11.2	380	3.8	.2	90.6	5.4	290	81	10	800
Cocoa, high-fat or breakfast, alkali-processed	3.0	295	16.8	23.7	45.4	4.3	133	648	717	651
Cocoa, high to medium-fat, alkali-processed	4.1	261	17.3	19.0	48.5	4.3	123	649	717	651
Cocoa, low to medium-fat, alkali-processed	5.2	215	19.2	12.7	50.2	5.2	152	686	717	651
Cocoa, low-fat	4.4	187	20.2	7.9	58.0	5.8	153	752	6	1522

[1]Data from Food Composition Table F-36.

COMMERCIAL USES OF CAROB BEAN GUM. The bean gum is not usually available in retail stores since this product is used almost exclusively as an additive to commercially processed food products. Some of these uses follow:

- **Bakery goods**—Small amounts of this additive (less than 0.5% by the weight of the food) are used to strengthen doughs, improve the textures of baked products and extend their shelf lives, and to stabilize pie fillings and meringues.

- **Dairy products**—The gum is used as a stabilizer to prevent separation of the fat, solids, and water components of processed dairy products such as milk, cheese, and ice cream. It also imparts to these products a smoothness and a sensation of richness, while adding only a trace of calories.

- **Meat products**—These products are stabilized and thickened by the gum. Also, the additive gives a "chewiness" and meatlike texture to vegetable protein analogs of meat products.

- **Miscellaneous uses**—The gum is also used as a stabilizer and thickening agent in confections, frozen desserts, gelatin salads, party dips, salad dressings, and sauces.

USES OF CAROB POWDER IN HOME RECIPES. Carob buffs may take offense at the suggestion that the main uses for this food are as substitutes for chocolate and cocoa products. If one is willing to accept carob as a food in its own right, then various proportions may be used in recipes, depending upon the consistency, color, and taste desired for the final product. However, if one wishes to make a substitution for cocoa or chocolate, then the following suggestions may be helpful:

1. Substitute about 1½ to 2 parts by weight of carob for each part of cocoa or chocolate used in beverages unless spices are used to enhance the taste of the drink (in the latter case, less carob may be used).

2. The taste of the bland-flavored carob may be enhanced with spices such as cinnamon, peppermint, or even coffee.

3. Reduce the amount of sugar or other sweetener when replacing cocoa or chocolate with carob, since the latter has a sweet taste of its own.

New products containing carob pod powder have recently appeared in health food stores and in the dietetic sections of supermarkets. Among these products are carob bits, carob covered peanuts and raisins, carob candies and cookies, and a carob flavored malted milk drink. However, the economy-minded consumer may readily prepare some of these items at home and save money. Health food stores and mail order suppliers of these products sell plain, powdered carob pod, and often provide recipes.

(Also see COCOA AND CHOCOLATE; FLOURS, Table F-26; GUMS; and LEGUME[S], Table L-2 Legumes of the World.)

CAROTENE

This is the counterpart of vitamin A, found in fruits and vegetables. Since the animal body transforms carotene into vitamin A, it is often referred to as "provitamin A."

Carotene derives its name from the carrot, from which it was first isolated 100 years ago. It is the yellow-colored, fat-soluble substance that gives the characteristic color to carrots and butter.

Several of the cartenoids found in plants can be converted with varying efficiencies to vitamin A. Of the group, beta-carotene (see Fig. C-42) has the highest vitamin A activity.

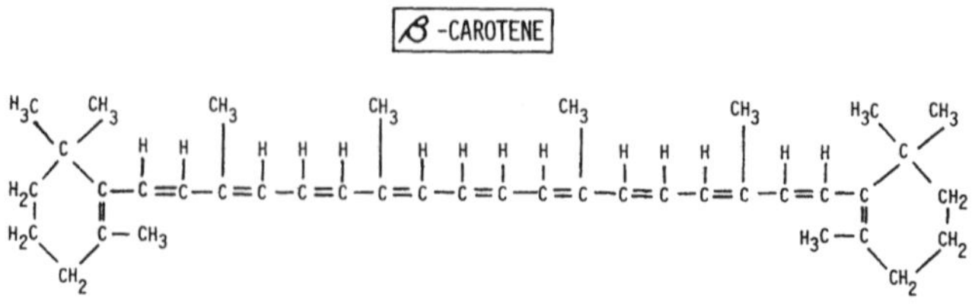

β-CAROTENE

Fig. C-42. The structure of beta-carotene.

Different species of animals convert beta-carotene to vitamin A with varying degrees of efficiency. The conversion rate of the rat is used as the standard value, with 1 mg of beta-carotene equal to 1,667 I.U. of vitamin A. Man is only one-third as efficient as the rat in making the conversion—1 mg of beta-carotene being equal to only 556 I.U. of vitamin A.

(Also see VITAMIN A.)

CAROTENEMIA

A yellowish discoloration of the skin, which appears first on the palms of the hand and the soles of the feet. It is due to a heavy consumption of food rich in carotene. It may sometimes be mistaken for jaundice. However, carotenemia differs from jaundice in that the whites of the eyes and the urine are not discolored.

CAROTENOID PIGMENTS

A group of substances with yellow, orange, and red colors. Many of the pigments have vitamin A activity. They are present in a wide variety of fruits and vegetables.

(Also see VITAMIN A.)

CARRAGEENAN

This naturally occurring food gum is obtained from several related species of red seaweed (algae), of which the best known is *Chondrus crispus,* or Irish moss. These algae are found along the Atlantic coasts of Europe and North America. The seaweed and its gums were first exploited on a commercial basis by the people who lived in the vicinity of Carragheen, Ireland; hence, the names Carrageenan and Irish moss.

Chondrus crispus is mostly reddish-brown, but it also has hues of green, purple, and black. Although a cluster of these plants looks like a bed of moss when viewed from a distance, a closer look reveals a mass of short, flat, branching stems (see Fig. C-43) which are attached to underwater rocks.

Fig. C-43. Irish moss, source of carrageenan, a widely used gum.

HISTORY OF CARRAGEENAN. It is not known when man first used Irish moss as a source of crude carrageenan, although it was likely used as a food centuries before it appeared in the marketplace. Old recipes indicate that the freshly gathered seaweed was used to make meat gels, puddings, and broths like St. Patrick's soup, which helped the Irish to stave off the pangs of hunger during times of famine. Crude carrageenan was extracted from the algae by cooking it with milk or water. Then, the resulting solution was used to prepare the gelled foods.

The early American settlers obtained Irish moss from Ireland until about 1835, when Dr. J. F. C. Smith, an Irish immigrant and the mayor of Boston, discovered that the seaweed could be obtained along the coast of Massachusetts.

Although a patent for the extraction of carrageenan from seaweed was granted to Bourgade of France in 1871, commercial production and marketing of the extract did not begin until 1937. Shortly thereafter, the reduction in overseas trade due to World War II resulted in the establishment of a North American industry which now produces most of the pure carrageenan used in foods in the United States.

PRODUCTION OF CARRAGEENAN. The seaweed is dried, washed in cold water, then broken up and the carrageenen extracted by violent agitation in a hot alkaline solution. If necessary, the extract may be decolorized by passage through charcoal. Then, the gum solution is further clarified by filtration, which is followed by evaporation and treatment with alcohol to precipitate the carrageenan. The precipitated gum is then vacuum dried and ground into a powder.

SPECIAL PROPERTIES OF THE DIFFERENT TYPES OF CARRAGEENAN. Most commercial carrageenan is derived form the algae species of *Chondrus, Gigartina, Eucheuma, Irideae,* and *Hypnea.* The characteristics of a batch of the commercial gum depend upon the species of seaweed used as a source, because there are three types of naturally occurring carrageenan molecules present in varying proportions in the different algae. These giant molecules have molecular weights ranging fom 100,000 to 500,000, and are polysccharides made up of chains of galactose (a 6-carbon sugar like glucose) units. Some of the galactose units have attached sulfate groups, while others are unsulfated. The three types of gum molecule are designated as iota, kappa, or lambda carrageenan. They differ by (1) the types of linkages between the galactose units, and (2) the points of attachment of the sulfate groups to the galactose units. These small differences in chemical constitution and structure make major differences in the properties of each type of molecule. Discussions of each type follow:

• **Iota carrageenan**—This type, which is the least abundant, is not present in *Chondrus crispus*, but is found mainly in *Eucheuma spinosum*. It is the most stable of the carrageenans in acid solution, and it forms strong gels in solutions containing calcium salts. It is insoluble in cold milk, sparingly soluble in sugar solutions, and moderately soluble in hot salt solutions.

• **Kappa carrageenan**—This type, which is the most abundant one (it constitutes 60% of the carrageenans in *Chondrus crispus* and accounts for most of the gum in *Eucheuma cottonii*), is most likely to break down in acid solution. Therefore, products containing kappa carrageenen should not be acidified until the final stages of production. However, after gels have been formed, this carrageenan becomes resistant to degradation. It forms strong gels in solutions which contain potassium salts, is insoluble in milk and salt solutions, but is moderately soluble in hot sugar solutions. This type is prevented from gelling by the addition of sodium salts to the gum solution.

• **Lambda carrageenan**—This type, which ranks second in abundance, is the major gum constituent of *Gigartina acicularis* and *Gigartina pistillata,* and constitutes 40% of the carrageenans in *Chondrus crispus.* Also, it is the second most stable (after iota carrageenan) in acid solutions; it does *not* gel; and it is soluble in cold milk, hot sugar solutions, and cold water. But it is insoluble in salt solutions.

Therefore, a food technologist should know the makeup of his batch of carrageenan so that he may obtain the desired properties in his products.

(Also see CARBOHYDRATE[S], and GUMS.)

USES OF CARRAGEENAN IN FOODS. Most of the food applications of carrageenans are concerned with gelling, thickening, and/or prevention of separation (stabilization) of food mixtures. Some of these applications follow:

• **Batters, doughs, and pastas**—Strengthening of proteins in flour by carrageenan results in (1) better batters for coating chicken and fish prior to frying, (2) elimination of shrinkage in breads due to the addition of nonfat dry milk, and (3) greater resistance of spaghetti to breakdown during cooking.

• **Dairy products**—Carrageenan prevents separation of the fat, protein, and water phases during the processing of milk, and imparts smoothness and a sensation of richness to cheeses, ice creams, and eggless (blanc mange) milk puddings. It even gives evaporated skim milk a consistency like that of cream.

• **Fish, meat, and poultry**—These products may be protected from spoilage during shipment by antibiotic ice containing carrageenan as a stabilizer. Also, the canned products may be cushioned against breakage during transit by carrageenan gels which, unlike gelatin, do not melt at room temperature.

• **Fruit products**—The addition of carrageenan to the packing syrup prior to the freezing of fruits results in better quality of the fruits upon thawing. Carrageenan may be substituted for some or all of the pectin in jams and jellies, particularly in the low-caloric types where pectin is not an effective jelling agent because the sugar content is drastically reduced.

• **Gelled desserts**—Mixtures of carrageenan and carob (locust bean) gum may be used for clear fruit gels which do not melt or soften at room temperature. Also, the viscous solutions of the gums keep fruit ingredients suspended until gelling has occurred.

• **Relishes, salad dressings, and sauces**—Relishes retain water better when carrageenan is used in the recipe. Also, salad dressings and sauces may be stabilized by carrageenan. However, care must be taken to prevent degradation of carrageenan during the processing of acidic sauces. Whenever possible, acidic ingredients should be added after heating.

• **Special dietary products**—Carrageenan is nonnutritive, so it may be used in low-caloric recipes in place of emulsifiers, thickeners, and stabilizers, such as egg, flour, and lecithin.

HOME COOKING WITH CRUDE CARRAGEENAN (Irish Moss). People living on the Atlantic coast of Europe and North America may be able to gather their own seaweed (if they can correctly identify the carrageenan-bearing species, and if the water where it grows is unpolluted.) Others may be able to purchase carrageenan-based gelling powders from health food stores. Recipe suggestions follow:

1. Blanc Mange-type puddings may be prepared by either (a) boiling about ½ cup (*120 ml*) of Irish moss in a quart of milk and then straining out the seaweed, or (b) adding ½ teaspoon (*2.5 ml*) of powdered carrageenan to a quart of milk. Then salt, vanilla, cocoa, molasses, and/or other flavorings may be added and the mixtures allowed to cool and gel.

2. Fruit-flavored gels may be made from a mixture of ½ cup of strongly flavored fruit or fruit juice and 1½ pt (*720 ml*) of water. These ingredients should be boiled together, then the heating stopped, before adding 1 teaspoon of carrageenan powder for gelling. If Irish moss is used, 1 cup of the seaweed should be extracted by boiling in water alone, then the heated fruit juice may be added to the strained gel solution and the mixture allowed to cool. The reason for the separate heating of the fruit juice is that carrageenan should not be boiled in acid solution because this may result in it being degraded.

3. A meat-flavored gel may be prepared by substituting meat broth for the water or milk used in the preceding recipes.

QUESTIONS ABOUT THE SAFETY OF CARRAGEENAN. In the early 1970s there was controversy over the safety of carrageenan as a food additive.[12] The issue now appears to have been resolved by the establishment of the following pertinent facts:

1. There have been many tests in which animals were fed greater amounts of carrageenan than those which might be consumed by man in his food. It was shown by these tests that the naturally occurring gum (with a molecular weight of over 100,000) was not digested by man and nonruminant animals, and that it had no harmful effects for these species. However, it was also shown that ruminants, such as cattle and sheep, were able to digest and absorb carrageenan.

2. A degraded form of carrageenan (with a molecular weight of 10,000, or less than 1/10 that of natural carrageenan), which has never been used in food, but which has been used to treat stomach ulcers, caused irritation and ulceration of the digestive tract of nonruminants, such as rhesus monkeys, and was absorbed and retained in the livers of these animals.

3. Food processors should avoid prolonged cooking of natural carrageenan with other ingredients which contain acids since there may be some degradation under these conditions. (The kappa type is most susceptible and the iota type least susceptible to degradation.) The breakdown of carrageenan results in the loss of some of its desirable properties, so it is in the best interests of food processors to prevent such decomposition of the gum.

4. The U.S. Food and Drug Administration (FDA) considers natural carrageenan to be safe as a food additive, but requires that molecular weight determinations be made on samples of the gums prior to their use in foods. This requirement insures that degraded carrageenan is not used.

(Also see ADDITIVES, Table A-3.)

CARRIER

• A substance that carries one or more nutrients, such as the carrying of fat-soluble vitamins by fat.

• A filler material which is mixed with some of the minerals and/or vitamins required in a diet. The mix, which is sometimes called a premix, makes it easier to measure the correct amounts of the trace nutrients providing that they are homogenously distributed in the carrier material.

• An animal, or a person that carries and transmits disease-germs, but does not show any signs of a disease.

CARROT *Dacus carota*

The edible roots of this plant are one of the world's leading vegetable crops. They are also the richest source of vitamin A among the commonly-used vegetable. Carrots are a member of the parsley family (*Umbelliferae*) of plants. Other well-known members of this family are caraway, celery, dill, fennel, parsley, and parsnips.

Fig. C-44. The carrot, one of the best plant sources of vitamin A. (Courtesy, USDA)

ORIGIN AND HISTORY. Carrots are believed to have originated in the Near East and central Asia, where they were cultivated for thousands of years. However, the ancient ancestors of the modern form of this vegetable were not yellow-orange, but had purplish colors ranging from lavender to almost black. The yellow roots apparently arose from a mutant variety which lacked the purple pigment. Both the purple and yellow varieties spread west to the Mediterranean, where they were used medicinally by the ancient Greeks and Romans.

Europeans probably started using carrots as food sometime during the Middle Ages, since the vegetable became a staple item in their diets by the 13th century. In the 14th century, carrots were taken to China, the country which is now the world's leading producer of the crop. Starting in the early 17th century, European agriculturalists worked on the development of the yellow-orange types of carrots and eventually discontinued the production of the purple ones. However, the latter are still grown in the Near East. About the same time, the vegetable was introduced into Japan and the newly settled colonies in North America.

Carrots were among the first vegetables to be canned following the development of the process by the French food technologist Appert in the early 1800s. During World War II, the British developed some high-carotene carrots so that their aviators might see better at night. Since then, plant breeders have sought to make carrots sweeter and more tender.

[12]"Carrageenan," *Journal of Food Science*, 1973, Vol. 38, p. 367.

PRODUCTION. The world production of carrots is about 13.4 million metric tons. Fig. C-45 shows the amounts contributed by each of the 10 top carrot-producing countries.

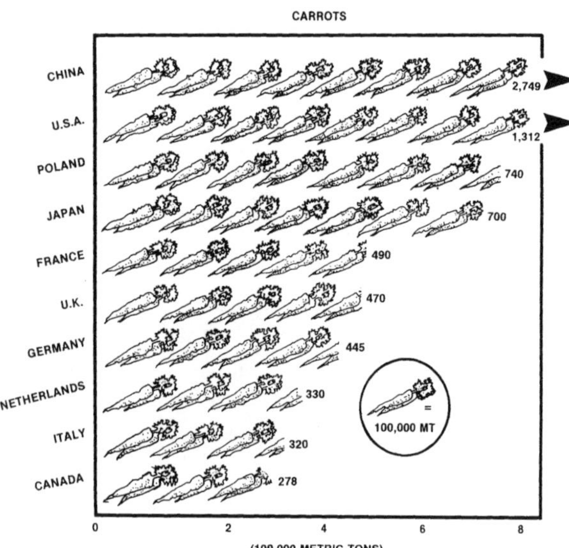

Fig. C-45. The leading carrot-producing countries of the world. (Data from *FAO Production Yearbook*, 1990, FAO/UN, Rome, Italy, Vol. 44, p. 147, Table 62)

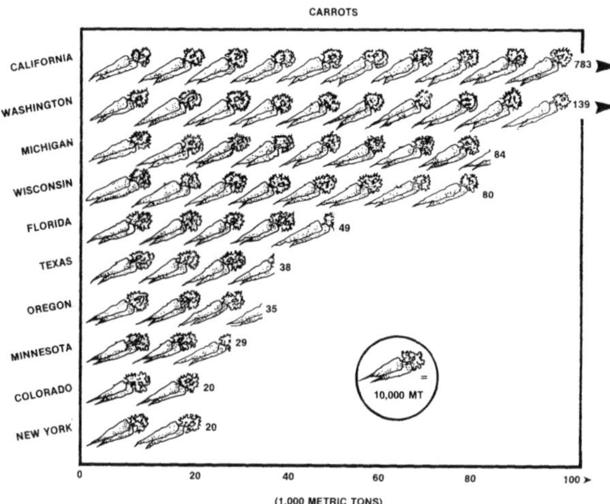

Fig. C-46. The leading carrot-producing states. (Data from *Agricultural Statistics*, 1991, USDA, p. 149, Table 207)

Each year, about 1.3 million metric tons of carrots are produced in the United States. Fig. C-46 gives the amounts produced by each of the 10 top carrot-producing states.

Carrot plants are resistant to frost. Hence, seeds may be planted in the field as soon as the soil can be worked. The best yields of long, tapering roots are obtained when the temperature during the latter part of the growing season ranges from 60° to 70°F (*16° to 21°C*). Therefore, carrots are grown in the north during the period from early spring to fall, and in the south during the fall, winter, and early spring. It is noteworthy that the exposure of young carrot plants to temperatures ranging from 40° to 50°F (*4° to 10°C*) for a period of 2 weeks or longer is likely to induce the premature development of seedstalks, unless the period of chilling is followed by exposure to a temperature of 70°F (*21°C*) or greater.

The soil should be well-cultivated, loose, fertile, moist, and well-drained. Heavy soils are suitable for only short, stubby varieties of carrots, since the longer-rooted varieties are likely to become misshapen in this type of soil. However, heavy clay soils may be loosened somewhat by mixing in sand and humus. (The addition of sand alone to clay soils may result in the formation of a soil as hard as concrete.) Carrots also require a soil that is rich in organic matter, nitrogen, phosphorus, and potassium. Unfortunately, the addition of fresh compost or manure causes the formation of forked roots which are unsuitable for marketing. (The organic materials should be well-rotted or leached to render them safe.) Carrot seeds must be sown very shallowly and the soil kept moist to ensure that the seeds will germinate and that the sprouts will be able to penetrate the upper layer of soil. Sometimes carrots are grown in raised beds of soil so that there is adequate drainage.

Carrot roots are usually ready for harvesting in from 2 to 3 months after planting, when the diameters of the upper ends range from ¾ to 2 in. (*19 to 51 mm*). Much of the U.S. crop is harvested by a machine which lifts the roots, and removes the tops, although some of the smaller fields are still harvested manually.

The worldwide average yield of carrots is 18,793 lb per acre (*21,068 kg/ha*), but yields as high as 53,520 lb per acre (*60,000 kg/ha*) are obtained in The Netherlands. By comparison, the U.S. average yield per acre is 28,964 lb (*32,471 kg/ha*), but yields as high as 48,500 lb (*54,320 kg/ha*) are obtained in Minnesota.

Carrots store well at high humidities (93 to 98%) and at a temperature around 32°F (*0°C*). Some home gardeners leave their carrots in the ground until needed, but they dig them up if a hard frost is expected.

Fig. C-47. Freshly-harvested carrots. (Courtesy, Agricultural Chemicals Division of Mobay Chemical Corporation, Kansas City, Mo.)

PROCESSING. A little less than half of the U.S. carrot crop is processed. The leading type of processing is freezing (over 148,000 metric tons annually), followed by canning. However, there is a steadily growing utilization of other processes, such as those which follow:

• **Dehydration**—Some of the newer types of dried diced carrots, which are sold mainly by wholesalers to food service establishments, may be rapidly rehydrated by soaking in cold water. Most of the older products require boiling for about 15 minutes or more.

• **Juice production**—Carrot juice is made by finely pulverizing the roots to a pulpy suspension, then forcing the liquid through a screen to remove hard pieces. The juice is usually pasteurized at a temperature between 175° and 190°F (*79° to 88°C*) for a few minutes. This treatment also prevents undesirable enzyme actions from occurring during storage. Carrot juice is sometimes mixed with other fruit or vegetable juices and/or seasonings to make juice "cocktails."

• **Miscellaneous processes and products**—There is an ever-growing utilization of carrots in items such as cakes, canned soups and stews, packaged salads, and ready-to-use salad or soup mixtures.

SELECTION AND PREPARATION. Carrots of good quality are firm, fresh, smooth, well-shaped, and generally well-colored. Wilted, flabby, soft, or shriveled carrots are undesirable. Those which are excessively forked, rough, or cracked may cause considerable waste in preparation for use. Decay usually appears as soft or watersoaked areas which may be partially covered with mold.

The mild, sweet flavor of carrots blends well with those of many other foods. Hence, this vegetable enhances a wide variety of mixed dishes.

Grated raw carrots may be (1) added to salads made predominantly from fruits such as chopped apples, pineapple slices and raisins; (2) used in vegetable salads along with raw cabbage, chopped celery, hard-boiled eggs, sliced onions, and chopped green peppers; (3) mixed with peanut butter to make a filling for sandwiches; (4) baked in a carrot cake; or (5) combined with eggs, bread crumbs or flour, milk or water, and grated onions, then formed into patties and fried.

Cooked carrots are often used in casseroles, soups, and stews. However, they are also good when boiled and served with butter, margarine, or a special sauce.

Pureed cooked carrots may be used in cookies, puddings, and souffles. (Baby food purees may be convenient to use when only small amounts of carrot puree are needed.)

Candied or glazed carrots are prepared by steaming whole carrots briefly, cooling, scrubbing off the skin with a stiff brush, and simmering the scrubbed roots in a mixture of melted butter or margarine, brown sugar or orange marmalade, salt, and pepper.

Carrot pickles are prepared by soaking the briefly steamed and deskinned vegetables in a mixture of vinegar, water, sugar, and seasonings.

NUTRITIONAL VALUE. The nutrient compositions of various forms of carrots are given in Food Composition Table F-36.

Some noteworthy observations regarding the nutrient composition of carrots follow:

1. Most forms of carrots are high in water content (88 to 92%) and low in calories (29 to 42 kcal per 100 g). However, they are excellent sources of vitamin A, since a 3½ oz (*100 g*) serving provides more than double the Recommended Dietary Allowance (RDA) for adults.

2. Baby food purees and cooked carrots furnish only ⅔ of the calories supplied by raw carrots and carrot juice. The former items have a higher water content than the latter ones.

3. Dehydrated carrots furnish almost as many calories and about ¾ of the protein supplied by dried forms of corn and rice. It is noteworthy that each ounce (*28 g*) of this product also provides about ¼ of the calcium and phosphorus content of a cup (*240 ml*) of milk and almost 6 times the RDA for vitamin A. Hence, dehydrated carrots may be added to various dishes to increase their nutritional values.

(Also see VEGETABLE[S], Table V-6 Vegetables of the World.)

CARTILAGE

The gristle or connective tissue attached to the ends of bones.

CASAL'S NECKLACE

A sunburnlike rash seen around the collar where the neck is exposed to the sun. These lesions are characteristic of severe niacin deficiency or pellagra. Apparently this deficiency makes the skin extra sensitive to the sun.

(Also see NIACIN; and PELLAGRA.)

CASEIN

Milk protein which is precipitated by acid and/or the gastric enzyme rennin. It is used for making cheese, paint, glue, and plastics.

(Also see MILK AND MILK PRODUCTS; and PROTEIN[S].)

CASSAVA (TAPIOCA; MANIOC; YUCCA) *Manihot esculenta*

This is a small tropical shrub, the roots of which are eaten or used to make tapioca. The starchy tubers of this plant constitute the second leading vegetable crop of the world. (Irish potatoes are the leading one.) Cassava belongs to the same family (*Euphorbiaceae*) as the tung tree (noted for its oil-bearing nuts), rubber tree, and castor bean. There are two main kinds of cassava—bitter cassava (*Manihot esculenta*) which is used to make tapioca, and sweet cassava (*M. dulcis*; *M. aipi*; or *M. utilissima*) which is eaten like potatoes. Fig. C-48 shows a typical cassava plant, whereas Fig. C-49 provides a close-up view of the tubers.

ORIGIN AND HISTORY. Cassava is native to the humid tropics of northeastern Brazil and the low-lying western and southern sections of Mexico. It is conjectured that the modern form of the plant may be the result of accidental or intentional crossbreeding between ancestral wild species that were gathered by the Indians, since this form does not grow wild at the present time.

Fig. C-48. The cassava, a valuable food plant of the tropics. (Courtesy, International Development Research Centre, Ottawa, Canada)

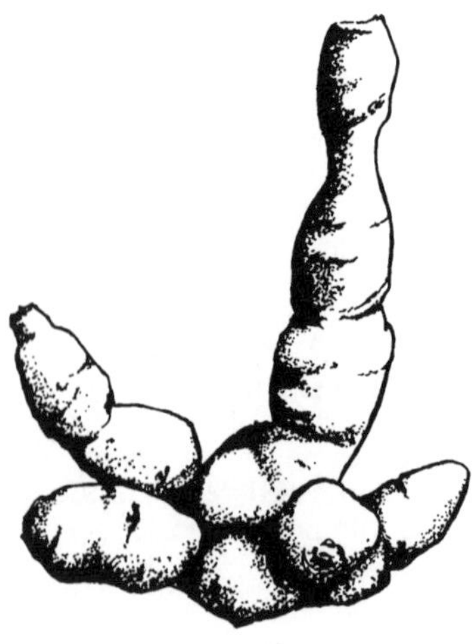

Fig. C-49. Cassava tubers, a rich source of starch.

Flour made from the tubers was traded throughout tropical South America. Perhaps it was the commercial value of the crop that led to the first cultivation of cassava in Peru about 4,000 years ago and in Mexico about 2,000 years ago. Eventually, the tubers were grown throughout a large area of the Americas lying between the Tropics of Cancer and Capricorn.

After the discovery of America, the Portuguese took cassava to Africa, and the Spanish took it to the Philippines. Later, the crop was introduced into India. During the 19th century, various colonial powers in Africa promoted extensive growing of cassava because it was (1) practically immune to attack by locusts and wild animals (most of the commonly grown varieties contained poisonous cyanides which could be removed or rendered ineffective by processing), and (2) drought resistant. Hence, millions of people in the tropical belt extending around the world came to rely on the crop as a staple food.

PRODUCTION. Over 157 million metric tons of cassava are produced worldwide each year. Fig. C-50 shows the amounts contributed by the five top producers. A limited amount of cassava is raised in the southeastern United States.

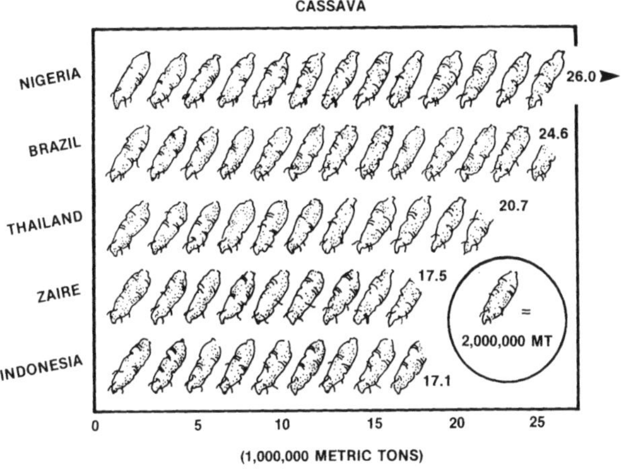

Fig. C-50. The leading cassava-producing countries of the world. (Data from *FAO Production Yearbook*, 1990, FAO/UN, Rome, Italy, Vol. 44, pp. 94-95, Table 28)

The plant is propagated by 6- to 10- in. (*15 to 25 cm*) long pieces of stem that are inserted into the soil horizontally, vertically, or at an angle. Little fertilizer other than phosphorus and potassium is usually required. Too much nitrogenous material, such as manure, will promote excessive growth of the tops at the expense of the tubers. Cassava grow well on poor soil, provided it is adequately drained. In developing countries, it is the last in the rotation of a series of crops before the land is left fallow to regain its fertility. Cassava require little care except (1) some weeding is usually done until the plants have become established, and (2) about 2 to 3 months after planting, the soil is hilled up around the stems so that tubers will develop readily. The tubers are swellings on the adventitious roots that grow out from the base of the stem. Cassava is rarely irrigated because (1) it is grown in humid areas, and (2) drought does not kill the plant, merely slowing its growth until the rains come.

The tubers may be harvested when the plants have reached a height of about 6 ft (*180 cm*), or from 6 months to a year after planting. This is done by (1) cutting the stems off at about 6 in. (*15 cm*) or so above the ground, (2) loosening the soil carefully to avoid damaging the tubers, and (3) pulling up the tubers by the stubs of the stems. About 10 lb (*4.5 kg*) of tubers are obtained from each plant, and yields of over 60,000 lb per acre (*67,2ll kg/ha*) have been achieved under ideal conditions. The tubers may be left in the ground until it is convenient to harvest them, but they may become tough and woody if left there for much more than a year. Fig. C-51 shows some cassava tubers in a market.

Fig. C-51. Cassava tubers (*maniocas*) in a Cuban supermarket. (Photo by A. H. Ensminger)

PROCESSING. Almost all cassava is processed in one or more ways because (1) the moist tubers keep for only a few days after they have been removed from the ground, and (2) the different varieties contain variable amounts of linamarin, a substance that may be converted into poisonous prussic (hydrocyanic) acid by an enzyme present in the plant tissues. The tubers which contain moderate to large amounts of linamarin usually have a bitter taste, whereas those that are low in the toxicant are likely to be slightly sweet. Also, the outer layer (rind) of a tuber of the bitter variety has a notably higher level of linamarin than the inner layers. Nevertheless, all varieties of cassava should be processed to minimize the likelihood of toxicity, which varies according to the local soil and climatic conditions. Some of the most common ways of processing cassava tubers follow:

1. Soaking the peeled or unpeeled roots in water is one of the most commonly used means of detoxifying fresh cassava. However, the removal of the cyanide-releasing substance is more certain if the roots are grated or cut into pieces prior to soaking. Soaked cassava must be dried if it is to be kept for more than a few days without spoiling.

2. Boiling or roasting of whole or cut tubers inactivates the enzyme that converts linamarin into cyanide. Nevertheless, some cyanide may still be produced if the cassava is eaten by a person whose intestine is inhabited by certain bacteria. (*E. coli* and some other species of intestinal bacteria have an enzyme that acts on linamarin.)

3. Drying of whole, cut up, or grated cassava appears to detoxify it. However, whole tubers require a much longer drying period than the pulpy material obtained by grating. It is noteworthy that (1) dried cassava chips may be ground into a flour, and (2) grated cassava may be sieved to remove fibers, then dried on a hot plate to make tapioca, the starchy granules used in the preparation of pie fillings and puddings. Fig. C-52 shows a homemade grater for preparing cassava.

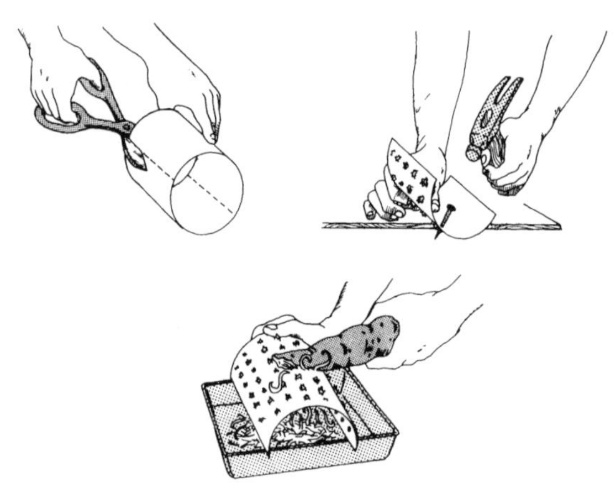

Fig. C-52. A cassava grater made from an empty can. The grating surface is made by punching holes from the inside surface with a heavy nail. (Adapted from *The Health Aspects of Food and Nurition*, World Health Organization; West Pacific Regional Office, Manila, Philippines, 1969, p. 320)

4. Expression of the juice from cassava is a means of detoxification practiced by (1) Africans, who place pieces of tuber in a cloth bag, then press the bag with heavy stones or logs; and (2) Brazilians, who use a vertically suspended basketlike cylinder called a *tepiti*, that is filled with cassava pulp and constricted by pulling on a pole attached to the lower end.

5. Fermentation of a cassava slurry into an African beer involves the cooking of a watery cassava paste with one or more germinated grains (corn, millet, and/or sorghum) in order to convert the cassava starch to fermentable sugar. The cooking drives off the cyanide.

SELECTION AND PREPARATION. It is not safe to buy processed cassava unless it is certain that the processor carried out the appropriate measures for detoxification. Also, bruised or wilted fresh tubers may contain considerable cyanide because damage to the plant tissues brings about the cyanide-releasing reaction. Therefore, the safest course is to select unbruised, fresh tubers that are of a sweet variety. However, all varieties of fresh cassava should be peeled, and washed under running water after peeling. Fresh water should be used for cooking the vegetable in an open pot. A cover may prevent the cyanide fumes from escaping. Also, the water used for cooking should be discarded.

Cassava may be prepared in much the same way as Irish potatoes, providing it has been detoxified properly. Also, it may be used as a thickener for gravies, soups, and stews. Many cassava dishes were given Portuguese names when Brazil and certain parts of Africa were colonies of Portugal. Some typical preparations follow:

- **Cassava flour**—This staple food of northeastern Brazil is made by grinding dried cassava chips, and sieving out the fibers. It is sometimes used to make bread products when wheat flour is scarce, and to thicken various dishes.

- **Farina de mandioca**—The people of Angola prepare this type of cassava meal by grating peeled roots, then drying the pulp on heated stones. Elsewhere in Africa, the cassava pulp may be placed in a cloth bag and allowed to drain and ferment for several days, after which it is pressed with heavy logs or stones, sieved to remove fibers, and dried.

- **Farofa**—Brazilians make a gelled product from cassava flour by (1) mixing it with boiling water and salt, or (2) stirring the flour into a mixture of hot fat, onions, tomatoes, salt, red pepper, olives, diced boiled eggs, fancy meats, and/or small slices of bacon. The first *farofa* is usually eaten with a dried meat called *charque*, while the second is commonly served with various main dishes.

- **Pirao**—This Brazilian porridgelike thick soup is made by sprinkling cassava flour into a boiling mixture of fat and meat stock and allowing it to thicken, while stirring to obtain a uniform distribution of the ingredients. Usually, the dish is served with cooked meat.

- **Virado**—In Brazil, this term refers to a semisolid puree prepared by mixing cassava flour with cooked beans.

- **Wrapped cassava paste**—The Africans of the Congo basin make a variety of dishes from cassava flour by (1) mixing it with water to form a paste, and (2) wrapping the paste in leaves (often banana leaves) and baking or steaming the wrapped paste.

NUTRITIONAL VALUE. The nutrient composition of various cassava products is given in Food Composition Table F-36.

Some noteworthy observations regarding the nutrient composition of cassava follow:

1. Cassava tubers are rich in carbohydrates (mostly starch), but low in proteins and most other nutrients. The

raw roots are lower in water content and higher in calories than the other leading roots and tubers. Also, the various dried forms of cassava tubers are about as rich in food energy as the common cereal flours. Nutritional deficiencies may result if cassava is the predominant food in the diet unless the missing nutrients are obtained by consuming adequate amounts of certain other foods. For example, the Brazilians who can afford to do so mix cassava flour with meats and vegetables.

2. The leaves of the cassava plant are nutritionally superior to the tubers with respect to protein, calcium, phosphorus, iron, vitamin A, the vitamin B complex, and vitamin C. However, the leaves often contain sufficient amounts of cyanide compounds to warrant special processing to make them safe.

CAUTION: It is suspected that some of the commonly used procedures for detoxifying cassava are not fully effective because certain peoples of Africa who consume enormous quantities of cassava show signs of chronic, sublethal cyanide poisoning,[13] typical signs of which are: loss of vision (nutritional amblyopia), nerve deafness, and other neurological disorders. Normally, the body can detoxify small amounts of cyanide by converting it to thiocyanate, which is excreted in the urine. Nevertheless, a building up of thiocyanate in the body may cause the thyroid gland to enlarge. Thus, it is thought that the eating of large quantities of cassava may have been responsible for some of the goiters seen in eastern Nigeria. However, the afflicted cassava eaters had diets deficient in the essential sulfur-containing amino acids and vitamin B-12. These nutrients are required for the detoxification of cyanide. Therefore, a cassava-rich diet should be supplemented with the animal protein foods to guarantee that any residual traces of cyanide are handled safely by the body.

NOTE WELL: The toxic principle in cassava is greatly reduced by proper processing—by soaking, boiling, drying, expression, or fermentation.

FUTURE PROSPECTS FOR CASSAVA. Cassava tubers now represent about two-thirds of the total production of root and tuber crops in the developing countries, where these vegetables help to feed over one-half billion people. Furthermore, the world production of cassava is continually increasing. In 1950, 51 million metric tons were produced; in 1970, 92 million metric tons; and in 1990, 157 million metric tons. Therefore, plant breeders are striving to develop new varieties with desirable characteristics such as a lower linamarin content, higher protein content, greater resistance to diseases and drought, more rapid development of maturity, and an increased yield of tubers. Food scientists and nutritionists are also seeking new ways to utilize cassava as a food while overcoming its nutritional limitations. Two of the most notable developments follow:

- **Bread made from cassava products**—Cereal grains are usually scarce in the humid tropics because (1) many of them do not grow well under such conditions, (2) the yields of those that do grow there are much lower than the yields of cassava, and (3) proper storage facilities for grains are often unavailable. These considerations led to the preparation and testing of breads containing various mixtures of wheat flour, cassava flour, cassava starch, and defatted peanut and soy flours. It was found that the cassava doughs that contained little or no wheat flour did not rise as well as wheaten doughs unless small amounts of glyceryl monostearate (an emulsifying agent) were added. Also, growth studies using laboratory animals showed that the breads made from cassava starch and peanut flour, and from cassava starch and soy flour, contained higher quality protein than a standard bread made from wheat flour alone. Finally, an acceptable bread was made from 70% wheat flour and 30% cassava flour.

- **Protein-fortified cassava meals**—Cooked porridgelike mixtures of cereal grains, legumes, and other ingredients have long been used as staple food preparations in the developing countries. Hence, trials were made of certain cassava-containing mixtures which might be useful in places where cereals are scarce. Some of the ingredients used to fortify the mixtures were soy flour, skim milk powder, sugar, minerals, and vitamins. It is noteworthy that the cassava formulas were found to have a protein quality equal to or greater than that of similar formulas containing cereal flours. A granular mixture of cassava and soy, which was designed for eating in bulk or sprinkling over over foods, was well accepted when test marketed in Brazil without identification or publicity.

(Also see VEGETABLE[S], Table V-6 Vegetables of the World.)

CASTOR OIL

A very strong laxative (also called a cathartic or a purgative) obtained from castor beans (*Ricinus communis*). It is one of the oldest household remedies for constipation. Its use as a purgative was recorded on Egyptian papyri (writing material made from the papyrus plant, used by the ancient Egyptians between the 4th century B.C. and the 4th century A.D.). The Incas of Peru believed that castor oil possessed spiritual powers and used it to drive out demons.

CATABOLISM

The oxidation of nutrients, liberating energy (exergonic reaction) which is used to fulfill the body's immediate demands. Catabolism is the reverse of anabolism.

(Also see METABOLISM.)

CATALASE

An enzyme which splits hydrogen peroxide into water and oxygen. This reaction prevents a buildup of the dangerous peroxide which may harm sensitive cell membranes.

---

[13]Davidson, S., *et al.*, *Human Nutrition and Dietetics*, 7th ed., Churchill Livingstone, Edinburgh, Scotland, 1979, p. 300.

CATALYST

Any substance which speeds up the rate of a chemical reaction without being destroyed or inactivated in the process. Enzymes are catalysts.

(Also see ENZYME.)

CATALYZE

To speed up the rate of a chemical reaction through the action of a catalyst.

CATARACT

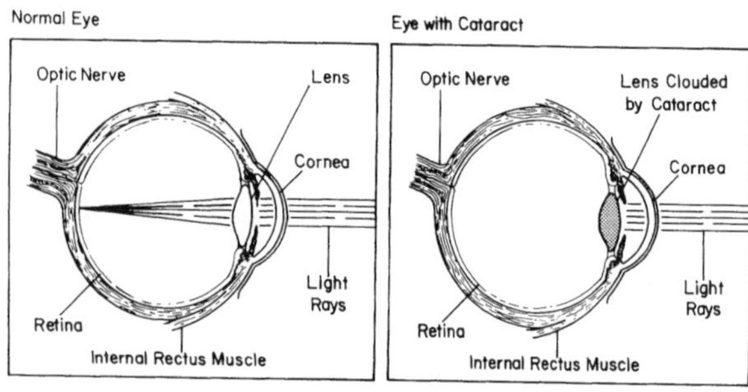

Fig. C-53. Normal eye (left) vs cataract (right). In the normal eye, light rays from objects pass through the clear lens of the eye and transmit the picture to the retina. In the eye with cataracts, light rays from objects are blocked from transmitting a clear picture to the retina by the clouding over of the lens.

Cataract is a clouding over of the lens of the eye due to the formation of certain types of crystals and deposits. When the lens becomes completely opaque, no longer allowing light to pass through to the retina, sight is lost. The disease may start in one eye, but eventually both eyes are affected.

This clouding of the lens is seen most frequently after middle age. But some people are born with cataracts. Various causes have been implicated. Some people appear to inherit the disease or a tendency toward it. Cataracts may also be caused by nutritional deficiencies, diabetes, certain drugs, or radiation.

Surgery, in which the clouded lens of the eye is removed, is the only effective treatment for cataracts once they have developed. After the traditional type of surgery, the patient wears special thick glasses (pop bottle glasses) or contact lenses to replace the clouded lenses that have been removed. A new type of surgery consists in replacing the cataractious lens with a lens made of plastic.

CATECHOLAMINES

These compounds, which are also called biogenic amines, are transmitters of impulses in the nervous system and in the brain. The three principal catecholamines found in the body are dopamine, norepinephrine (also called noradrenaline), and epinephrine (adrenaline). They have strong effects on body movement and balance, the emotional state, heart rate, and blood pressure.

(Also see ENDOCRINE GLANDS.)

CATHARTIC

A strong laxative.

CATION

An ion carrying a positive charge of electricity. Cations include the metals and hydrogen.

(Also see ELECTROLYTE BALANCE.)

CAULIFLOWER *Brassica oleracea*, VARIETY *botrytis*

This vegetable, which is grown for its whitened flower head, is practically a gourmet item, because it sells for several times the price of cabbage. Cauliflower was derived from the wild cabbage, *B. oleracea*. Fig. C-54 shows a cauliflower with a large edible flower head (commonly called a curd).

Fig. C-54. Cauliflower, a high-class relative of the common cabbage.

ORIGIN AND HISTORY. The time and place at which the present form of cauliflower was developed remains a mystery. However, it is known to be very closely related to the wild cabbage which resembles collards and kale.

The Roman Botanist Pliny (1st Century, A.D.) is thought by some to have described cauliflower, but this belief has been questioned because no other writings mentioned anything that even remotely resembled the vegetable.

It is suspected that a primitive form of either broccoli or cauliflower originated somewhere in Asia Minor, and that the plant was taken to Italy where further development occurred. Later, it was taken to northern Europe and the British Isles, where it is now highly favored.

PRODUCTION. The growing of cauliflower requires considerable labor. Hence, it is most likely to be grown where either the labor costs are low, or where harvesting machines are available. Fig. C-55 shows where most of the world's cauliflower is grown.

Fig. C-56 shows the contribution made by California to U.S. cauliflower production.

Because cauliflower is highly susceptible to both frost and hot weather, about 78% of the U.S. crop is grown in California, where the weather is favorable. it is usually planted early in the year, so that it may be harvested before summer; or it is planted in the summer, for harvesting in the fall. The plants often require a nitrogen fertilizer for the development of full size flower heads. Boron may also be needed, if the soil is deficient in this element. Once the flower heads (curds) have developed, they should be protected against sunlight. Some varieties have leaves that naturally curl around the curds, while others require tying of the leaves around the curds.

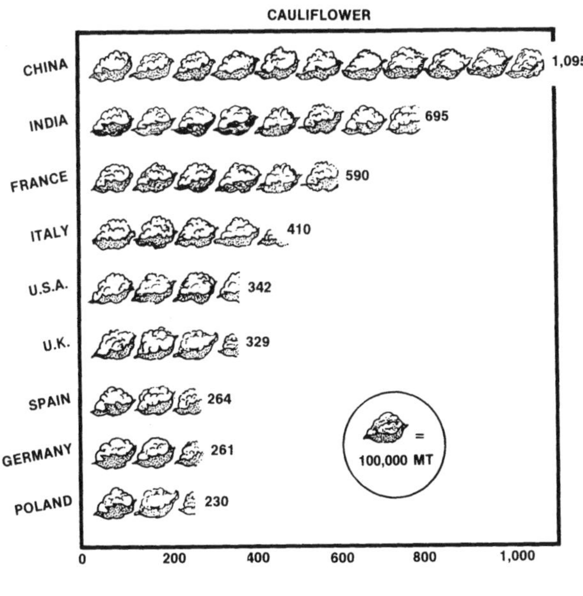

Fig. C-55. The leading cauliflower-producing countries. (Data from *FAO Production Yearbook*, 1990, FAO/UN, Rome, Italy, Vol. 44, p. 133, Table 53)

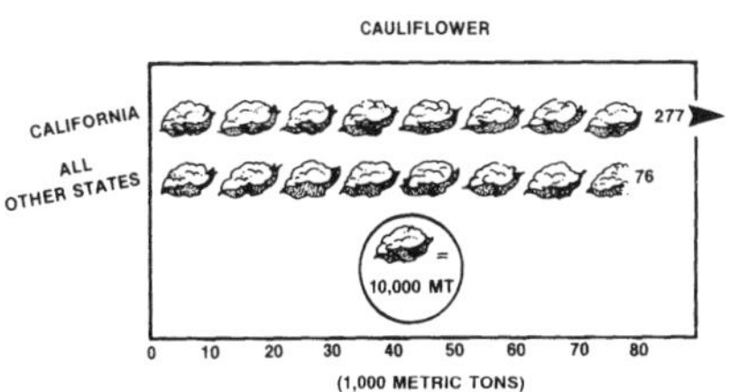

Fig. C-56. U.S. cauliflower production. (Data from *Agricultural Statistics*, 1991, USDA, p. 151, Table 210)

The curds are harvested as soon as they are large enough to be marketed, since those that are allowed to become overmature may develop loose heads, discoloration, or a ricy appearance. After harvesting, cauliflower should be cooled immediately and kept refrigerated.

PROCESSING. About half of the cauliflower grown in the United States is processed. The leading form of processing is freezing, but pickling is also a major means of preservation.

SELECTION AND PREPARATION. Good quality in cauliflower is indicated by white or creamy-white, clean, firm, compact curd, with the "jacket leaves" (outer leaf portions remaining) fresh, turgid, and green. Small leaves extending through the curd do not affect edible quality. Large or small heads, equally mature, are equally desirable. A slightly "ricy" or granular appearance is not objectionable unless the flower clusters are spreading. Spreading occurs when the flower clusters have developed enough to cause a separation of the clusters which makes the curd open or loose.

Spotted, speckled, or bruised curd should be avoided unless it can be trimmed without excessive waste. The appearance of aphids (plant lice) may be indicated by a smudgy or speckled appearance.

Raw cauliflower may be served in a relish tray or a salad, or it may be dipped into well-seasoned sauces before eating.

Cooked cauliflower goes well with a wide variety of cheese or cream sauces, egg dishes, meats, poultry, and other vegetable. Also, pieces of the cooked vegetable may be dipped into an egg batter, then fried in deep fat.

Cauliflower pickles are very good appetizers or accompaniments to cold meat sandwiches.

CAUTION: Cauliflower and other members of the mustard family of vegetables (*Cruciferae*) contain small amounts of goiter-causing (goitrogenic) substances that interfere with the utilization of iodine by the thyroid gland. This effect causes the thyroid to enlarge (form a goiter) as part of the body's attempt to compensate for the lack of thyroid hormones. However, the goitrogenic effect may be counteracted by the consumption of ample amounts of dietary iodine, which is provided by iodized salt, ocean fish, seafood, and seaweeds such as kelp.

NUTRITIONAL VALUE. The nutrient composition of cauliflower is given in Food Composition Table F-36.

Some noteworthy observations regarding the nutrient composition of cauliflower follow:

1. Cauliflower is low in calories (23 to 32 kcal per cup) because it contains over 90% water. Hence, it is filling, but not fattening for people trying to lose weight.
2. The blanching (whitening) of cauliflower makes it low in vitamin A, compared to most of the other cabbage vegetables. However, the purple-headed varieties which turn green when cooked contain much more of the vitamin.
3. A 1-cup (*240 ml*) serving of cooked cauliflower provides about the same amount of vitamin C as a medium size orange. Raw cauliflower provides at least 20% more vitamin C than the cooked vegetable.

(Also see VEGETABLE[S], Table V-6 Vegetables of the World.)

CAVY (GUINEA PIGS)

South American rodents constituting the family *Cavidae*, the best known of which are guinea pigs. Cavy flesh is edible.

CECUM (CAECUM)

A blind sac or pouch located at the junction of the small and large intestines, often considered to be part of the large intestine. A valve at the ileocecal junction allows the products of digestion to flow from the ileum to the cecum, but not in the opposite direction. The nonfunctional appendix is attached to the cecum.

(Also see DIGESTION AND ABSORPTION.)

CELERIAC (TURNIP-ROOTED CELERY) *Apium graveolens*, VARIETY *rapaceum*

This vegetable, which is a type of celery characterized by swollen, edible roots, is little known in the United States. Celeriac, like ordinary celery, is a member of the Parsley family (*Umbelliferae*).

Fig. C-57. Celeriac, a variety of celery grown for its edible root.

ORIGIN AND HISTORY. Little is known about the history of celeriac, other than the fact that it was developed from the wild celery which originated in marshlands around the Mediterranean. Celeriac is thought to have been developed in northern Europe and is sometimes called "German celery."

PRODUCTION. Celeriac is grown mainly in West Germany, Britain, and other European countries, and to a lesser extent in Asia and North America. International statistics for the production of celeriac are not readily available because the data for this vegetable are usually combined with those for celery.

The crop may be grown by (1) planting the seeds in a greenhouse or in hot beds, then transplanting the seeds in a field when the outdoor temperatures are warm enough; or (2) sowing the seeds directly in an outdoor field in areas with mild climates. A fertile soil and plenty of water are required. Growers often add a nitrogenous fertilizer at various times during the growth period. Usually from 3 to 4 months are required from the planting of seed to the stage at which the young plants are suitable for transplanting, and another 3 to 4 months are required for the transplants to grow large enough for harvesting.

Celeriac may be harvested when the roots are at least 2 in. (*5 cm*) in diameter. However, the roots may be left in the ground until after the first frost, which helps to develop the flavor of the vegetable. Some home gardeners in the milder climates leave the roots in the ground after applying a mulch of leaves or straw. Celeriac roots may also be stored in a cool place after the tops have been removed.

PROCESSING. There is little or no processing of celeriac at the present time.

SELECTION. The best quality celeriac roots are those that are heavy in relation to size, of medium size, and with very few fibrous roots. Shriveled or soft roots may be tough when cooked. Large, coarse, overgrown roots, especially if light in weight for their size, are apt to be tough, woody, pithy, hollow, or strong in flavor.

PREPARATION. Celeriac roots should be scrubbed thoroughly, then scraped or pared. It is noteworthy that some chefs prefer to counteract the slightly bitter taste of celeriac by blanching the raw vegetable in salted water and lemon juice prior to preparing it in various dishes. The raw root can be made into juice (in a blender or a juicer), or it can be grated and used in salads. However, this vegetable is usually served cooked in ways such as (1) an ingredient of soups or stews; (2) boiled, pureed, and mixed with mashed potatoes; (3) boiled, diced, and served in salads; (4) cooked, sliced, dipped in batter, and fried; (5) an individual cooked vegetable dish served with butter, cheese sauce, cream sauce, margarine, mustard sauce, or a white sauce.

NUTRITIONAL VALUE. The nutrient composition of celeriac is given in Food Composition Table F-36.

Some noteworthy observations regarding the nutrient composition of celeriac follow:

1. This vegetable is high in water content (88%) and low in calories (40 kcal per 100 g). However, celeriac contains about twice the solids and calories of celery stalks. It is also a good source of phosphorus and potassium, but only a fair source of vitamin C.
2. Herbalists have long believed that celeriac is helpful in eliminating dropsy (the accumulation of excessive water in the tissues) because it is thought to increase the flow of urine.

NOTE: The authors make no claim regarding this belief, but present it for purposes of (1) information, and (2) in order to stimulate further research on the subject.

(Also see VEGETABLE[S], Table V-6 Vegetables of the World.)

CELERY *Apium graveolens,* VARIETY *dulce*

This vegetable, which has edible leafstalks, leaves, and seeds, is the third leading salad crop of the United States. Celery is a member of the Parsley family (*Umbelliferae*), which also includes caraway, carrots, celeriac, dill, fennel, parsley, parsnips and certain lesser known vegetables. Fig. C-58 shows a typical celery plant.

Fig. C-58. Celery, a leading salad crop that has many culinary uses.

ORIGIN AND HISTORY. The modern type of celery was developed from wild celery, which is believed to have originated in marshy areas around the Mediterranean. However, it is not certain whether the place of origin was western Asia, northern Africa, or southern Europe. Furthermore, wild celery has a much higher proportion of leaves to stems than the domesticated type of celery.

It appears that the leaves of wild celery were first used as a medicine because the earliest mention of the plant in ancient literature is in Homer's *Odyssey*, written in the 9th century B.C. Later, just before the Christian Era, celery leaves were used by the Greeks as laurels for their athletes, and in funeral wreaths. The ancient Romans used celery seed as a seasoning. It appears that the Chinese received celery plants as a gift from the rulers of Nepal in the 7th century A.D.

Over the centuries, the wild plant was domesticated for either (1) the thick, fleshy leaf stems that characterize modern celery, or (2) the swollen, edible taproot which is the outstanding feature of celeriac ("turnip-rooted celery"). However, it was not until the Middle Ages that celery came into widespread use as a vegetable, and even then it was utilized mainly for flavoring. Its use as a raw vegetable probably began in Europe in the 18th century. Celery was brought to the United States early in the 19th century. Now, the U.S. production of the vegetable is sufficient to meet both national and foreign needs.

PRODUCTION. The U.S. production of celery in 1990 was about 891 thousand metric tons. Fig. C-59 shows the states which contributed most of the production.

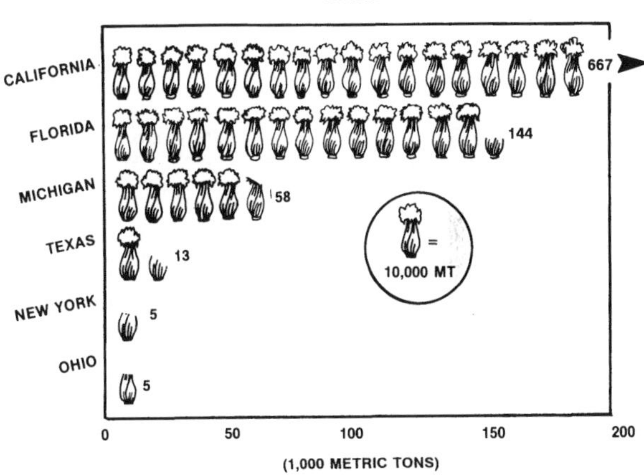

Fig. C-59. The leading celery-producing states. (Data from *Agricultural Statistics*, 1991, USDA, p. 151, Table 211)

The modern type of celery was developed from a wild plant that thrived in the rich soils, mild climate, and long growing season of the marshy areas around the Mediterranean. Hence, similar conditions favor the production of the domesticated vegetable. For example, it is a fall and winter crop in the southern United States, and a summer and fall crop in certain northerly areas where the summer weather is milder than average due to the nearness of large bodies of water such as the Atlantic Ocean, Pacific Ocean, or the Great Lakes.

From 2 to 3 weeks are required for celery seed to germinate, and an additional 8 or 9 weeks are needed for the young plants to grow large enough for transplanting. Transplanted seedlings require from 3 to 4 months to reach maturity. The long growing season and the slowness of seed germination make it desirable to raise young celery plants in a greenhouse or in hot beds, then to transplant them outdoors in a field. However, some seed is planted outdoors in areas which have many months of mild weather. It is noteworthy that exposure of the young plants to temperatures below 50°F (*10°C*) stimulates bolting (the premature development of seedstalks), which is undesirable unless the grower wishes to produce new seed. Seed may also be produced by leaving celery plants in the ground over the winter, providing that the temperatures does not get cold enough to kill the plants.

The soil used to grow celery must be kept fertile and moist because this vegetable has very high requirements for nutrients and water. Therefore, most growers fertilize with nitrogen, phosphorus, and potassium once every 2 to 3 weeks during the early growing season and one or more times per week later in the season. Sometimes the foliage is sprayed with solutions containing boron, calcium, and/or magnesium, since spraying the leaves appears to be more effective than adding these nutrients to the soil. Furthermore, irrigation is used wherever the rainfall is inadequate.

Fig. C-60. Young celery plants growing in California. Irrigation ditches run between the rows of plants to insure that sufficient water is provided. (Courtesy, Union Pacific Railroad Company, Omaha, Neb.)

Today, only a small part of the U.S. celery crop is blanched. Green and golden varieties are now widely accepted.

Blanching is the lightening of the color of the celery stalks brought about by shielding them from light to prevent chlorophyll from forming in the growing plant. It may be accomplished by placing boards, paper, black plastic, or soil against the sides of the plant. Where soil-blanching is practiced, it is important to keep dirt from

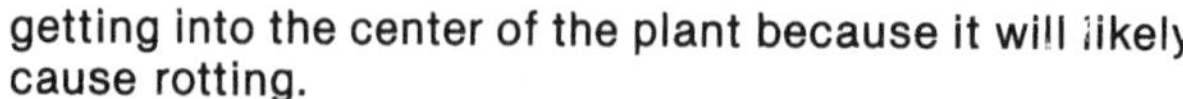

getting into the center of the plant because it will likely cause rotting.

During harvesting, the tops of the celery are cut so that the stalks will fit into boxes 16 in. (*41 cm*) long. Then, the stalks are severed from the roots and packed directly into fiberboard cartons.

Fig. C-61. Workers harvesting and packing celery in the field in Oxnard, California. (Courtesy, Union Pacific Railroad Company, Omaha, Neb.)

Fig. C-62. Loading cartons of freshly harvested celery onto a truck for a quick trip to the cooler. (Courtesy, Union Pacific Railroad Company, Omaha, Neb.)

Cartons of fresh harvested celery are cooled rapidly so that the vegetable will retain its freshness during shipment to distant markets.

Fig. C-63. Celery entering and emerging from the cooler which reduces the temperature of the vegetable to 34°F (*1°C*) in 15 to 40 minutes. (Courtesy, Union Pacific Railroad Company, Omaha, Neb.)

PROCESSING. Most of the celery crop is marketed fresh. However, small but steadily increasing amounts of this vegetable are processed into forms such as (1) canned celery hearts, (2) other canned items that contain celery, and (3) dried celery, which can be rehydrated to 16 times its dried weight.

Celery seed is dried and sold in whole or ground form as a seasoning agent, or the ground seed may be mixed with (1) table salt to make celery salt, a product that may not legally contain more than 75% salt, (2) ground black pepper to make celery pepper, which may not contain more than 70% pepper, or (3) various other products such as bouillon preparations, salad dressings, and vegetable juices.

SELECTION. Best quality celery is fresh, crisp, and clean, of medium length, thickness, and density, with good heart formation, and branches that are brittle enough to snap easily. Pithy, woody, or excessively stringy celery is undesirable. Soft, somewhat pliable branches often indicate pithiness, and small hard branches are frequently very stringy or woody.

Celery is sometimes affected by "blackheart," which can be detected as brown or black discoloration of one or more of the small heart branches or leaves. Blackheart or the presence of insects or insect injury can be seen by separating the branches and examining the heart. "Seedstems" appear as a solid, somewhat round stem replacing the desirable typical heart formation. Celery that has formed a large seedstem may have a somewhat bitter, undesirable flavor.

PREPARATION. Raw celery should be washed well to remove any sand or soil. Then, the leaves should be removed for immediate use. The stalks may be stored for 1 to 2 weeks in a refrigerator.

- **Celery leaves**—Although this part of the plant is often discarded, it is very nutritious and useful for flavoring or garnishing various dishes.
- **Celery seeds**—This item is quite expensive. Hence, it is used sparingly in (1) dishes such as casseroles, chicken loaf, sauces, soups, and stews; and (2) mixtures with other seasoning agents such as pepper and salt.
- **Celery stalks**—Raw celery can be used for (1) dunking in cold or hot dips; (2) relish trays or salads; (3) hors d' oeuvres in which the stalks are stuffed with fillings such as cheese spreads, chopped or pureed cooked beans, eggs, fish, meats, poultry, or seafood, fruit butters or purees, mashed banana, or peanut butter; or (4) making celery juice or a mixture of vegetable juices.

Celery stalks are also good when cooked in soups or stews, with stewed tomatoes, stir-fried in Chinese dishes, or as an individual vegetable dish served with butter or margarine and various seasonings, or with a special sauce.

NUTRITIONAL VALUE. The nutrient compositions of various forms of celery are given in Food Composition Table F-36.

Some noteworthy observations regarding the nutrient composition of celery follow:

1. Celery stalks have a high water content (94%) and a low caloric content (17 kcal per 100 g). They are an excellent source of potassium, but only a fair source of vitamins A and C. It is noteworthy that the white varieties of the vegetable and the green ones which have been blanched have a much lower vitamin A content than the unblanched green varieties. Whenever possible, celery stalks should be eaten with the leaves on because the leaves are much richer than the stalks in calcium, iron, potassium, vitamin A and vitamin C.

2. Celery seeds are among the tiniest vegetable seeds; 72,000 seeds weigh only 1 oz. (*28 g*). Nevertheless, they are very rich in calories (392 kcal per 100 g), protein, fat, fiber, calcium, phosphorus, magnesium, potassium, iron, and zinc. However, they are not likely to make a substantial contribution to the diet because only small amounts are consumed.

3. Herbalists have long used celery to treat a wide variety of ailments, of which the most common is dropsy (an accumulation of excess water in the tissues) because consumption of the stalks or roots is thought to increase urination. (The authors are not aware of any research confirming or contradicting this belief. Hence, it is presented for purposes of (1) information, and (2) in order to stimulate research on the subject.)

(Also see VEGETABLE[S], Table V-6 Vegetables of the World.)

CELIAC

Pertaining to the abdomen.

CELIAC DISEASE

A rare metabolic disorder of children, sometimes called "Malabsorption Syndrome," which appears to run in families and, therefore, may be due to a genetic defect. The disease results from a sensitivity to gluten, a protein found in wheat, rye, and several other—but not all—grains. Affected children are unable to absorb fats, certain starches, and some sugars. Symptoms include recurrent diarrhea, severe cramps, a pale, foul-smelling stool containing large amounts of fat, distended abdomen, stunted growth, anemia, irritability and susceptibility to infection. Therapy evolved during World War II when a Dutch scientist observed improvement in affected children when bread was unavailable. Treatment is now based on a diet low in high gluten foods, but high in protein. Buckwheat, rice, corn, and soybean flours are substituted for wheat, rye, barley, and oat flours, and fruit sugars are used to replace milk sugars. Once the symptoms disappear, a fuller diet is achieved by adding one food at a time to the diet and then monitoring its effects. Celiac disease is similar to an adult disease called sprue.

(Also see ALLERGIES, section headed "Wheat Allergy"; DISEASES; and SPRUE.)

CELL MEMBRANE

The thin layer of tissue that encloses the cell and restricts the entry and exit of various substances.

CELLOBIOSE

A sugar formed by the partial breakdown of cellulose.

(Also see CARBOHYDRATE[S].)

CELLULASE

The general name given to an enzyme which breaks down cellulose. Cellulases are present in certain intestinal bacteria and in various other microorganisms such as those which inhabit the digestive tract of grazing animals (cattle, goats, and sheep).

(Also see ENZYME.)

CELLULOSE

The principal carbohydrate constituent of the cell wall of plants. It is not digested by man.

(Also see ADDITIVES, Table A-3; CARBOHYDRATE[S], UNAVAILABLE; and FIBER.)

CELL WALL

The outer layer of plant and animal cells. Plant cell walls are made up mainly of cellulose plus other complex carbohydrates; hence, they are poorly digested by man.

(Also see FIBER.)

CEMENTUM

The calcified portion covering the root of the tooth.

CENTIGRADE (CELSIUS)

A thermometer scale on which the interval between the two standard points, the freezing point and the boiling point of water, is divided into 100°, with 0° representing the freezing point and 100° the boiling point. To convert Centigrade to Fahrenheit multiply by 9/5 and add 32.

(Also see WEIGHTS AND MEASURES.)

CENTRAL NERVOUS SYSTEM

The brain and spinal cord in vertebrates. It is that part of the nervous system to which sensory impulses are transmitted and from which motor impulses pass out; it supervises and coordinates the activity of the entire nervous system.

CENTRIFUGE

A machine for whirling fluids rapidly, which exerts a pull many times stronger than gravity. It is used to separate two liquids of different densities; e.g., cream from milk—the Babcock Test. Also, specimens of blood, urine, and other substances are centrifuged to separate the sediment or solid constituents from the liquid.

(Also see BABCOCK TEST.)

CEPHALIN

A group of phospholipids; associated with lecithins found in brain tissue, nerve tissue, and egg yolk.

CEREAL GRAINS

Contents *Page*

Botanical Characteristics ... 365
History ... 366
  Origins of Grain Farming ... 366
  The Spread of Grain Crops Throughout the World .. 367
Production of Grains and Cereal Products ... 368
  Growing, Harvesting, and Storage of the Common Grains ... 368
  Production of Beverages, Feeds, and Foods from Grains ... 373
    Malting (Sprouting) ... 373
    Brewing ... 374
    Distilling ... 374
    Milling ... 374
    Baking ... 375
    Miscellaneous Processes ... 375
Nutritive Values of the Various Grains ... 377
  Limitations in the Protein Quality of Grains ... 378
  Cereals That Have Been Enriched or Fortified with Certain Nutrients ... 378
  New and Improved Varieties of Grains ... 379
  Sprouted Grains ... 379
Commercial Uses of Grains and Cereal Products 380
  Beverages and Foods ... 380
  Feeds for Livestock ... 381
Economical and Nutritious Meals Based Upon Cooked Cereals ... 381
Future Prospects for the Cereal Grains ... 383

This term denotes the seeds from cereal plants—members of the grass family *Gramineae*—which are used as food for man and animals. The word cereal is derived from Ceres, the Roman goddess of agriculture.

Grains, and the products made from them, are literally the "staff of life" for many peoples around the world. Fig. C-64 shows that a large part of the world's crop production is represented by grains.

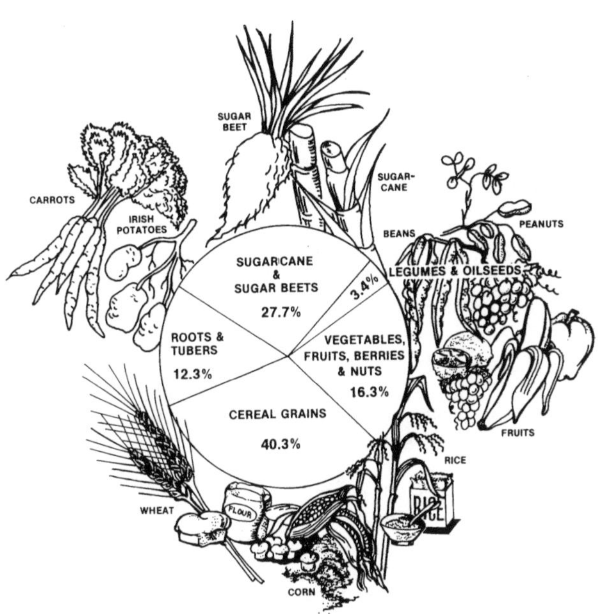

Fig. C-64. Major food crops of the world. (Data from *FAO Production Yearbook*, 1990, FAO/UN, Rome, Italy, Vol. 44)

Fig. C-65. Wheat, a typical member of the grain-bearing family of plants. (Courtesy, Union Pacific Railroad Company, Omaha, Neb.)

Wheat Barley

Fig. C-66. Samples of wheat (left) and barley (right). Individual kernels of grain are called *caryopses.* Grains lacking husks, like wheat, are referred to as naked caryopses, whereas grains which contain husks, like barley, are called covered caryopses. (Courtesy, J. I. Case Company, Racine, Wisc.)

BOTANICAL CHARACTERISTICS. The grain-bearing grasses are tall, slender plants with long, narrow leaves and jointed stems. Fig. C-65 shows wheat, which is a typical member of this family.

It is noteworthy that the wild cereal grasses are characterized by (1) seed heads which shatter and scatter after ripening; and (2) tight-fitting husks which cover and protect the inner seed kernels until they sprout. These characteristics allow the wild plants to spread throughout the areas where conditions are favorable for their growth.

On the other hand, the varieties of grains which have been selected and propagated by man often have (1) seed heads which remain attached to the plant after ripening;

and (2) loose-fitting husks that are readily removed by threshing. (Some of the domesticated grains have tight-fitting husks which can be removed only after roasting and/or grinding.) Hence, the grains which have been domesticated usually require the help of man for their propagation because they have lost the ability to shed their seeds.

Archaeologists are often able to determine whether certain ancient peoples (1) merely gathered wild grains, or (2) were grain farmers by noting the characteristics of the plant remains that have been found at the excavated sites of early settlements.

Details of the characteristics of the various grains are given in the separate articles covering these items.

(Also see BARLEY; CORN; MILLET; OATS; RICE; RYE; SORGHUM; TRITICALE; WHEAT; and WILD RICE.)

HISTORY. The development of the great ancient civilizations was spurred by man's cultivation of grains, which made it possible for many more people to be fed from the food produced on a given amount of land. For example, an average of 10 square miles (*26 square km*) was needed to supply animals and plants to feed each person in the nomadic hunting and gathering groups, whereas 1 square mile (*2.6 square km*) of productive agricultural land fed 25 people in the primitive agricultural settlements.[14] Hence, the growing and storage of grains paved the way for the development of highly populated communities in China, India, Babylon, Egypt, Africa, Europe, and the Americas.

Origins Of Grain Farming. Archaeological evidence shows that wild varieties of barley and wheat were used as early as 10,000 B.C. by the peoples living in the *Levant—the eastern coastal region of the Mediterranean which extends from Greece to Egypt*. These wild grains grow abundantly in this region because there is sufficient rain during the growing season for grasses, but not enough to promote the growth of forests. However, the first cultivation of these crops by man appears to have occurred in an area east of the Levant called the *Fertile Crescent, which was a broad, crescent-shaped area that curved northward and eastward from what is now the eastern border of Egypt, to the Taurus Mountains of southern Turkey, across to the Zagros Mountains of western Iran, and down to the Persian Gulf.*

One of the earliest agricultural villages excavated in this region was Jarmo, a small settlement about 150 miles (*240 km*) north of Baghdad, Iraq. It appears that by around 7000 B.C. the people of Jarmo had domesticated several types of wild barley and wheat. The irrigation of crops was probably first used in this area around 5000 B.C. This practice eventually led to a rise in the salt content of the poorly drained soils, as a result of which wheat could no longer be grown. Barley, however, was not so affected, because it tolerates salt better. Hence, it is not surprising that beers and wines made from barley were important beverages in ancient Babylon—the empire which developed between 3000 and 2000 B.C. in the area around Baghdad. The evidence for the importance of these beverages consists of (1) a recipe for barley wine which was engraved on a Babylonian brick dated about 2800 B.C., and (2) the mention of a public beerhouse on a clay tablet dated about 2225 B.C.[15]

It is certain that grain farming along the Nile River in Egypt was well established by 5000 B.C., because the brewing of beer was practiced at that time. Irrigation was more successful here than in Babylon because the water from the Nile River was almost salt-free.

Most authorities believe that rice production originated in the tropics of Asia, somewhere between southern China and eastern India, because the climate of this region favors the growth of this crop. Also, the oldest archaeological evidence of the cultivation of rice by man dates back to about 4500 B.C. in Thailand.

By 4000 B.C., millet farming was well established along the upper reaches of the Yellow River in the area of northern China where the modern provinces of Honan, Shansi, and Shensi come together. At about the same time, rice was grown in the region around Shanghai.

There is no doubt about corn having originated in the Americas. Sampling of the soil 227 ft (*69 m*) below Mexico City uncovered fossil grains of corn pollen more than 80,000 years old. More recent evidence of man's use of corn consists of (1) miniature corn cobs, dating between 5200 and 3400 B.C., found buried in a cave in the Tehuacan Valley in southern Mexico; and (2) corn cobs unlike those discovered in Mexico found in a cave located near Ayacucho in the Andes Mountains of Peru, and dating between 4300 and 2800 B.C. Neither of these types of primitive corn bore a close resemblance to modern-day types. Therefore, it is suspected that at about 1500 B.C. the early types of corn were crossed by wind pollination with a wild grass called teosinte (*Euchaena mexicana*) which grew near to where the Mexican Indians planted their primitive corn. Hence, corn as we know it is only about 3,000 years old.

The first cultivation of sorghum and of certain millets might have occurred at various times after 4000 B.C. within a wide area of Africa north of the equator. However, agriculture on the African continent has long been characterized by (1) a great diversity of crops, and

[14]Leonard, J. N., *The First Farmers*, Time-Life Books, 1973, p. 25.

[15]Lehner, E. and J. Lehner, *Folklore and Odysseys of Food and Medicinal Plants*, Tudor Publishing Company, New York, N.Y., 1962, p. 32.

(2) its adoption by societies which also retain their ancestral practices of hunting and gathering.

Oats and rye appear to have been domesticated from the wild grasses native to the mountainous area running from the southern border of Turkey to the border of the Soviet Union with Afghanistan. However, there is no mention of these grains in the ancient writings of such Middle Eastern peoples as the Babylonians, Hebrews, or the Egyptians. Furthermore, their first cultivation by man appears to have occurred at sometime between 3000 and 1000 B.C. in northern Europe, because seeds of the domesticated varieties were found in Switzerland in the excavations of sites occupied by the primitive lake dwellers. It seems that the seeds of oats and rye were often mixed in with seeds of barley or wheat because the former grains grew as weeds among the latter. The first cultivation of oats and rye occurred in a cool, damp climate because these grains thrive under such conditions, whereas they do poorly in the warmer and dryer areas of the Middle East.

The Spread Of Grain Crops Throughout The World. Dried grains keep well, so it is not surprising that the cultivation of these crops was spread rapidly by people who traveled to distant places. For example, it seems likely that rice farming, which began in Southeast Asia, spread northward to China and westward to India. Likewise, the cultivation of barley and wheat spread southeastward from Iraq to Egypt, and northeastward through Turkey to Macedonia—the area in the southern Balkan Peninsula which is now divided between Bulgaria, Greece, and Yugoslavia. In the New World, corn (maize) farming spread slowly northward from Mexico to the region that is now the United States.

Among the factors which hastened the spread of grain agriculture were (1) changes in climate; (2) population explosions in the early farming settlements that forced groups of people to move on to less densely populated areas; (3) the domestication of wild animals such as donkeys, camels, cattle, and buffalo—which were used to carry loads of grain on trading trips and/or to pull crude plows for the cultivation of the soil; and (4) commerce in products made from grain, such as beer, breads, and the like. Some of the more noteworthy examples of these factors follow:

• **Changes in climate**—An unusually warm climate between 5500 and 3000 B.C. caused considerable melting of the snow and ice in the mountains which stretched across Europe from the Soviet Union to Switzerland. This condition allowed some of the grain farmers from Macedonia to migrate northward to the Balkan Peninsula, and to travel through the mountain passes into a broad area of Central Europe which is now made up of the countries of Hungary, Austria, Czechoslovakia, Poland, Germany, Belgium, Luxembourg, and the Netherlands. Barley and wheat were grown almost exclusively until the climate cooled somewhat. Then, oats and rye, which had been brought to the area as weed seeds mixed in the barley and wheat, were cultivated by man, who had noticed that the "weeds" thrived in the cool, damp climate of Central Europe.

The period of warm weather was also accompanied by a melting of much of the polar ice caps, which caused a rise in sea level around the world and the submergence of low-lying coastal areas. Rice farmers in lowland areas of Asia were forced to migrate to places with higher elevations when their land became covered with water.

There were also several periods of unusually moist weather in North Africa—rainfall peaked around 7000, 5000, and 3000 B.C., respectively—which fostered the growth of a grassy belt in the area that is now occupied by the Sahara Desert. The archaeological evidence shows that wild grain was gathered in this region around 6000 B.C., until the weather became drier and the vegetation disappeared. Perhaps, the grain collectors moved southward to the area north of the equator in Central Africa, where they cultivated the millets and the sorghums.

• **Population explosions**—It seems likely that various factors might have acted to limit the populations of the prehistoric peoples who lived by hunting and gathering. For example, only the able-bodied could keep on the move from place to place in the search for food. Hence, those handicapped in various ways might have been abandoned to die. However, other practices, such as the restriction of sexual activity, probably existed in certain cultures.

Some of the factors which promoted overpopulation in the early agricultural societies were: (1) the assurance of a steady supply of food; (2) encouragement of childbearing in order to have more people for tilling the soil, and for supporting the aged and infirm; and (3) increased survival of the handicapped, who mated and reproduced.

Overpopulation fostered the spread of agriculture because, from time to time, it forced people to move out of their settlements and to establish new ones. Eventually, the cultivation of larger and larger expanses of land made man increasingly dependent upon agriculture because the natural habitat of the wild animals and plants was altered so that they dwindled to insignificance in his diet. It is noteworthy that each of the densely populated ancient civilizations developed because the people were grain farmers.

• **The domestication of animals and their employment in farming and trading**—It is believed that an ancient Asian religious cult castrated bulls routinely in the performance of their rites, as a result of which they discovered that docile, but strong, animals were produced. This taming of heavily muscled males led to yoking them as oxen for pulling the plows for grain farming, depicted in ancient drawings from Iraq and Egypt, dating about 3000 B.C.

The donkey was domesticated in Egypt around 3400 B.C., and the camel in Arabia at about the same time. Hence, beasts of burden became available for carrying grains and other items over the long trading routes.

• **Commerce in the grain products of ancient times**—Although bread wheat dating between 6000 and 5500 B.C. has been found at archaelogical sites along the northern borders of the Fertile Crescent, the peoples in this region were not the first to develop commercial bakeries because (1) the salt in their irrigation water made the soils salty, and (2) barley—which is better for brewing than for breadmaking—replaced wheat as the major cultivated grain because it tolerated saltiness better than wheat. However, starting about 5000 B.C., the ancient peoples of Mesopotamia utilized their barley for brewing beer; and beginning about 2200 B.C., the Babylonians had public beerhouses.

Later, the Egyptians became the first people to establish commercial bakeries because the salt-free irrigation water they drew from the Nile River enabled wheat to thrive. By 2000 B.C., there were combined bakeries and breweries in Egypt which turned out both bread and beer. Apparently, the grain technology of the Egyptians surpassed that of other regions, because in 1300 B.C. caravans from Egypt carried dried beer-cakes which underwent fermentation when they were soaked in water. The beer-cakes were made by malting (sprouting) barley, then pulverizing it and making it into a dough which was baked in an oven. After baking, the cakes were dried in the sun so that spoilage would not occur while they were en route to their destination.

It is noteworthy, too, that brewing spread much more rapidly through the Middle East and Europe than did the baking of yeast-leavened breads; because each species of grain could be sprouted and fermented, but only the bread-type of wheat made good yeast breads.

PRODUCTION OF GRAINS AND CEREAL PRODUCTS. The world production of grains averages about 2,151 million tons (*1,955 million metric tons*)[16] which should be sufficient to meet the food energy and protein needs of approximately 10.8 billion people, if there is an annual per capita consumption of 400 lb (*180 kg*) of grain. (The estimated population of the world in 1990 was 5.3 billion.) However, Fig. C-67 shows that there are some major discrepancies between the countries which produce the most grain and those which have the most people. For example, the United States and Canada, together, produce about 19% of the world's grain, yet they have only 5.2% of the world's people; whereas Asia produces 44% of the grain, but feeds 59% of all the people in the world. Unfortunately, it is not always economically feasible to distribute grain from areas of high per capita production to those of low per capita production. Hence, if self-help programs are to be instituted in order to increase the supplies of foods in the countries where they are needed most, it is important that those responsible for such programs be familiar with the basic principles of grain farming and cereal technology.

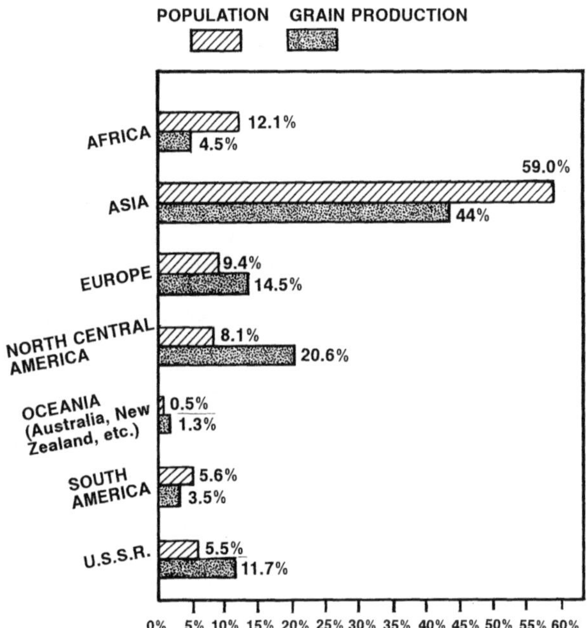

Fig. C-67. The world production of grains vs the distribution of the world's population. (Data from *FAO Production Yearbook*, 1990, FAO/UN, Rome, Italy, Vol. 44)

Growing, Harvesting, and Storage of the Common Grains. It was not by accident that different grains came to be cultivated by the various peoples who were the first farmers. Each of these crops requires certain soil and climatic conditions which may be present in some parts of the world, but not in others. Table C-18 gives the cultural characteristics of the cereal grains.

[16]*FAO Production Yearbook*, 1990, FAO/UN, Rome, Italy, Vol. 44.

TABLE C-18
CEREAL GRAINS OF THE WORLD

Grains	Geographic Areas Where Grown	Soil and Climatic Requirements	Other Comments
Adlay (Job's Tears) *Coix lachryma-jobi*	Southeastern Asia and other tropical areas of the world.	One of the few grains which grows well in tropical rain forests.	Has long been used as an emergency food in Asia. The grains may be boiled whole, or they may be milled into a flour.
Barley *Hordeum vulgare* (Also see BARLEY.)	In the northern areas of the world, or at high elevations farther south. It is grown in the U.S. from the northern Great Plains to California; and in the Old World from the Arctic Circle to the equator (on the high plateaus of Ethiopia).	Highest yields are obtained on well-drained, moderately fertilized soils. Heavy fertilization may result in the plants lodging (falling over). It tolerates salty soils better than the other grains. Adapted to cool areas. Does poorly in hot, humid weather.	Overproduction of protein in the grain due to over-fertilization makes it less suitable for malting by brewers. May be sown in the winter and harvested in the spring in Arizona and California.
Buckwheat *Fagopyrum esculentum* and other species (Also see BUCKWHEAT.)	The U.S.S.R. is the leading producer, followed by Poland. U.S. production amounts to only about 30,000 tons annually while world production is about 3 million tons annually.	Buckwheat is a summer annual, adapted to a cool, moist climate. It does well on poor soils, and requires only a 10-to 12-week growing season.	Buckwheat is *not* a cereal grain, but it is used for the same purposes as cereal grains and most people think of it as a cereal. Actually it is a relative of rhubarb. Buckwheat is a major food item in the U.S.S.R. and Poland. It is also an important honey plant.
Corn *Zea mays* (Also see CORN.)	The U.S. produces 42% of the world production. Most of this is grown in the Corn Belt, an area stretching from Ohio to the eastern Great Plains and from Minnesota to Missouri. Many other areas of the world grow corn, but the yields per acre are usually lower than those in the U.S.	Does best on soils which contain ample organic matter and moisture, but which have good drainage. Requires temperature over 60°F *(16°C)* for growth, but does not do too well in the hot southern states.	New hybrids may be grown in areas which were not previously suitable for corn. Corn removes more nutrients from the soil than any other grain crop. Hence, heavy fertilization is required if it is to be grown on the same land year after year.

(Continued)

TABLE C-18 (*Continued*)

Grains	Geographic Areas Where Grown	Soil and Climatic Requirements	Other Comments
Emmer *Triticum dicoccum*	Although it was grown in ancient times throughout most of Asia Minor, it is now produced mainly in the Dakotas of the U.S. and in the mountainous areas of Europe.	Does best on fertile, well-drained soils in cool, dry climates.	It appears to be closely related to modern day wheat, although it looks more like barley. Valuable for plant breeding, because it has resistance to certain diseases which attack wheat.
Indian ricegrass *Oryzopsis hymenoides*	Grows wild throughout western U.S.	Usually found on dry, sandy soils; but it also tolerates alkali soils and drought.	The seed from this plant was harvested by the American Indians when the corn crop failed.
Millet *Panicum milaceum* and other species (Also see MILLET.)	Most of the world's production is outside the U.S. in areas such as Africa, China, India, Latin America, and the Soviet Union. Some millet is grown in the northern Great Plains of the U.S. (mostly in North Dakota, South Dakota, and Colorado).	Needs a well-drained, fertile soil. A good crop for areas where the warm growing season is short, because it is ready for harvest in from 60 to 70 days after sowing.	The name millet is given to various species of grasses which have numerous small seeds on each ear of grain. Fertilizers are not used for the millet grown in the Great Plains.
Oats *Avena sativa* (Also see OATS.)	In Europe and Asia, most production is between 50° and 52° latitude; in North America, it is between 40° and 45°. There is some production at latitudes as low as 30°, but only at the higher elevations.	Grows well on soils which are less fertile than those needed for other grains, but it requires more moisture than all other cereals except rice. A good crop for cool climates.	Should not be rotated after legumes, because high soil nitrogen causes the plant to grow too tall and fall over (lodge). Oats are not as winter hardy or as salt tolerant as barley.

(Continued)

TABLE C-18 *(Continued)*

Grains	Geographic Areas Where Grown	Soil and Climatic Requirements	Other Comments
Rice *Oryza sativa* (Also see RICE.) 	Although the most important area of rice production is Southeast Asia, the crop is grown as far north as Hokkaido, Japan; and as far south as Argentina and Chile. Production in the U.S. is limited mainly to California, Arkansas, Texas, Louisiana, and Mississippi.	Rice is an aquatic plant which is usually grown on flooded soils. Most of the varieties of this grain require a long, warm weather growing season (about 180 days). Irrigation is usually needed in areas which do not have heavy rainfall.	Short season rices—which mature in 85 to 145 days—may be planted in the temperate climates. These varieties may be grown in cold irrigation water. Rice may be grown on alkaline soils which are not suitable for other grains.
Rye *Secale cereale* (Also see RYE.) 	More than half of the world's rye is produced in the Soviet Union. Significant amounts are also produced in Poland, E. Germany, W. Germany, Canada, and Argentina. Most of the U.S. production is in the North Central states.	Grows well on soils which are not fertile enough for other grain crops. Also, little preparation of the soil is needed for sowing. Rye is extremely cold resistant, so it may be sown in September and harvested the following August or September, depending upon the latitude.	The grain heads on rye tend to shatter, so it should be harvested as soon as its moisture content drops low enough to ensure storage without spoilage. Rye may be grown on the same land for several years in succession, but crop rotation is usually desirable.
Sorghum *Sorghum vulgare* (Also see SORGHUM.) 	This grain is grown on all continents of the world below 45° latitude. The leading areas of production are Africa, Asia, and the U.S.—where it is grown primarily in the central and south plains states.	Does best on fertile soils which contain ample organic matter and moisture. It withstands heat and lack of moisture much better than the other grains. However, irrigation is usually required to produce good yields under such conditions.	Some varieties of sorghum "wait for rain," that is, they become dormant during dry spells. Drought-resistant varieties should be sown where there are likely to be long periods without rain, since these types will produce grain under such conditions.
Spelt *Triticum spelta* 	A limited amount of spelt is grown in eastern U.S. and in northern Spain. It was once the major cereal grain of central Europe.	Grows well in a wider variety of environments than most of the modern varieties of wheat.	Grain is of no value for milling. Hence, it is used as a feed for livestock.
Teff *Eragrostis tef*	Grown mainly in Ethiopia and the eastern African highlands.	Needs a well-drained, fertile soil.	Used mainly in Ethiopia for grinding into a flour, for making a beer, and as an ingredient of a high-protein food for growing children.

(Continued)

TABLE C-18 (*Continued*)

Grains	Geographic Areas Where Grown	Soil and Climatic Requirements	Other Comments
Teosinte *Euchlaena mexicana*	Limited to latitudes where there are 12 hours of daylight during the growing season (such as in southern Mexico).	Requires rich, moist soil and a tropical climate.	It is believed to be the primitive ancestor of corn. The edible kernel of the grain is encased in a hard, tight-fitting husk.
Triticale (a hybrid of wheat and rye) (Also see TRITICALE.)	This grain is being grown experimentaly on grasslands around the world—in such places as the Great Plains in the U.S., Canada, Mexico, Argentina, South Africa, northern Europe, and the Soviet Union.	Requires fertile, well-watered soil for best growth; but is more drought-resistant than wheat. Certain varieties lack winter hardiness, while other varieties approach rye in tolerance to cold weather.	Triticale has created much interest because of having higher quality protein than wheat. However, many varieties yield less grain per acre than wheat, and are more susceptible to ergot—a poisonous fungus which replaces the grain. Hence, research is underway to overcome these shortcomings.
Wheat *Triticum vulgare* *(Also see WHEAT.)*	Most of the world's wheat is grown in Asia, Europe, and North America. The main wheat growing areas of the U.S. are the Great Plains and the North Central states, although some is grown in the northwestern and northeastern states.	Needs fertile, well-drained soils because it cannot tolerate long periods of wet ground. Does best in a cool, dry climate, although varieties have been developed for other climatic conditions.	Most important grain crop in the world. Second only to corn in the U.S. in total acreage and production.
Wild rice *Zizania aquatica* (Also see WILD RICE.)	Although this grain grows wild on mud flats all over the eastern U.S. and southeastern Canada, the major production comes from Wisconsin and Minnesota.	This aquatic plant grows well on flooded soils and in marshes. No soil preparation or sowing is needed if the previous crop was harvested by the Indian method, because under such conditions the crop seeds itself.	Most of this crop is harvested by Indians of the Chippewa tribe. The production is small, compared to that of other grains, since it is used mainly as a delicacy.

The American Indians, like certain other primitive peoples, harvested wild grains by gently knocking them off the plants with paddles or sticks so that they fell into baskets. This method of gathering seeds—which is still used by the Chippewa Indians to harvest wild rice in Minnesota—works because the wild grasses usually have seed heads which shatter easily when the grains are ripe. However, the early grain farmers ended up domesticating nonshattering varieties of the wild grasses because these varieties retain their grains on their stalks at harvest time, whereas the shattering types lost most of their ripened seeds. The primitive farmers also selected varieties of wheat, corn, rye, and sorghum which had seeds with loose-fitting husks that might be removed by threshing. Hence, the grains came to be harvested by (1) cutting the plants, and (2) threshing the grain—an operation which was often performed by having animals trample the cut plants. Fig. C-68 shows grain being threshed by tramping.

Fig. C-68. Ethiopians threshing teff with the help of oxen. Threshing is done by driving the oxen round and round on the teff, to "tread-thresh" the grain. After tramping out the grain, the straw is thrown into the air by means of a wooden-tined fork in order to separate the grain from the straw. The grain of teff is ground into flour and used in making bread. The straw is fed to cattle. (Photo by A. H. Ensminger)

Primitive methods of harvesting are still used in many of the developing countries, although most of the large-scale grain farmers in the developed countries now use machines called combines which both cut and thresh the grain. Sometimes, grain is harvested when it contains too much moisture for safe storage. At such times, it may be cut and bundled into sheaves for drying in the field, provided that the wheather remains dry or protection is provided against rain.

Grains are almost always stored for use over long periods of time. Hence, they must either be dry at harvest time (usually the moisture content should not exceed 13 to 16%) or they must be dried before storage. Most drying systems other than in the field are based upon the circulation of warm or hot air through the grain. Care is taken when the grains are to be used as seeds or for malting, because overheating may reduce their ability to germinate when the conditions are otherwise suitable. Furthermore, moist grains heat up in storage and pose the danger of fire. However, moist grains which are to be fed to livestock may be preserved with an acid or stored in a silo.

Production of Beverages, Feeds, and Foods from Grains. Cereal technology around the world is directed mainly towards the processing of wheat, rice, corn, and barley, because these four grains account for over 85% of the world's grain production. However, products are also made from oats, rye, millets, and sorghums, particularly in those countries where these crops are of greater than average importance. Fig. C-69 shows the proportion of the world grain production contributed by each of the major cereal crops.

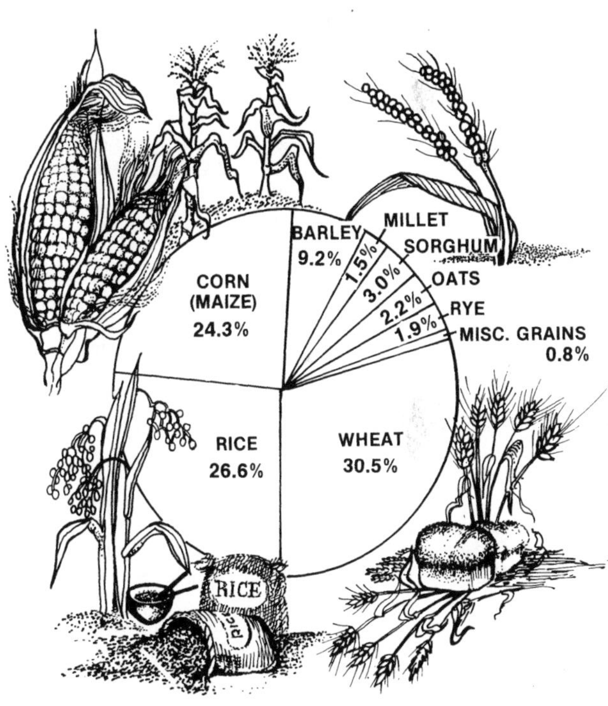

Fig. C-69. Contributions of the major cereal crops to world grain production. (Data from *FAO Production Yearbook*, 1990, FAO/UN, Rome, Italy, Vol. 44)

Details of the major types of cereal technology follow.

MALTING (Sprouting). In this process the germination of grains is started by soaking them in water—for the purpose of activating certain enzymes that are useful in baking, brewing, and the production of distilled alcoholic beverages. (The enzymes activated during sprouting digest the stored carbohydrates and proteins in the grains to sugars and amino acids which support the growth of the embryonic plant.) Usually the malt producer stops the germination right after the emergence of sprouts by careful heating and drying so as to retain the desired enzyme activity. Then, the sprouts are removed from the malted grain by passing them through revolving sets of wire screens. The term malt may be applied to any grain which has been so treated. However, it is generally understood that when the term malt is used alone (without indicating the grain source) it refers to malt from barley.

(Also see BARLEY; and MALT.)

BREWING. This process consists of a series of operations in which (1) starchy gruels are prepared by cooking crushed grains; (2) the gruels are digested or "mashed" with malted grains or other sources of enzymes in order to produce sugary solutions; and (3) the sugary solutions are filtered, flavored with hops or other substances, and fermented with yeast to yield a beverage which contains about 4% alcohol (ale, beer, lager, porter, stout, etc.). Most commercial brewers now use malted barley as the source of the starch-digesting enzymes, and other grains or "adjuncts" such as corn, rice, sorghum, and wheat to prepare the starchy gruels. Nevertheless, it is noteworthy that beers have been brewed from (1) wheat by the Germans, (2) rice by the Chinese and the Japanese, (3) sorghums and millets by African tribes, (4) rye by the Russians, and (5) corn by the American Indians.

The Indians used a unique process to make their beer, or chicha. They (1) chewed the corn to produce a pulpy mass which was well mixed with their saliva; and (2) took the mass from their mouths and placed it in a large vessel to ferment. It appears that the starch-digesting enzyme in their saliva converted some of the corn starch to sugar which was fermented by the wild yeasts which are ever present in the air and on plants. Although many of the early American colonists had a craving for beer, few could bring themselves to produce it by the Indians' method.

(Also see BEERS AND BREWING.)

DISTILLING. Distilled alcoholic beverages are produced by distilling fermented mashes which are prepared like those used for making beer. Most of these beverages are produced from corn, rye, and malted barley.

All of the by-products from malting, brewing, and distilling are suitable for feeding to livestock. Hence, both man and animals reap the benefits of these types of cereal technology.

(Also see DISTILLATION; DISTILLED LIQUORS; and FERMENTATION.)

MILLING. The term milling refers to various processes of grinding and/or rolling which are applied to grains for the purposes of (1) removing fibrous hulls, branny seed coats, and oily embryos (germs); (2) altering their sizes and shapes, as in the pearling of barley; and (3) reducing the size of grain particles to meal or flour. "Meal" denotes particles like grains of sand, whereas "flour" denotes a fine powder.

It appears that the earliest farmers began to employ a crude form of milling soon after they started to cultivate grains because archaeological studies have shown that (1) the teeth of these people had no signs of the wear which would have resulted from the chewing of whole grains, and (2) they had stones which might have been used for the grinding of grain. Grinding the grains made the baking of breads and similar items possible because it produced starchy particles which, when wetted, bound together to form a dough or a paste.

Fig. C-70. Pounding rice. The way it used to be done and still is in some parts of the world.

Today, the various milling procedures provide a variety of foods and livestock feed ingredients. Table C-19 provides a guide to the main products of milling which may be used as foods.

(Also see BARLEY; CORN; MILLETS; OATS; RICE; SORGHUM; TRITICALE; WHEAT; and WILD RICE.)

BAKING. The first breads were made by mixing grain meals with water and baking the resulting dough over hot coals. They were flat, heavy, and unleavened until the ancient Egyptians and the Hebrews learned how to use sourdough cultures for leavening. The Egyptians soon went a step further and cultivated yeast, which they sometimes dried for later use. However, it was not until the end of the 18th century that a crude baking powder made from wood ashes was developed in the United States. Now, breads leavened with baking powder are called "quick breads" because this type of leavening takes place much more rapidly than the production of gas by yeast fermentation. Brief descriptions of the major types of bread leavening follow:

TABLE C-19
CEREAL BY-PRODUCTS

Cereal Product	Description	Grain(s)	Comments
Bran	The coarse outer covering of the grain kernal in the form of tiny flakes.	Corn, oats, rice, wheat.	High in water-absorbing fiber which has a laxative action.
Flour	Soft, powdery product derived mainly from the inner portion of the grain kernel.	Barley, corn, oats, rice, rye, wheat.	Consists mostly of starch and protein. Low in fiber. Used to make bread.
Germ	Embryo of the seed.	Corn, rice, wheat.	Rich in fat and protein.
Gluten	Tough, elastic protein which remains when flour is washed to remove the starch.	Wheat	Imparts elasticity to doughs, thereby enabling them to be leavened with yeast.
Grits (hominy)	Coarsely ground deskinned grain from which the bran and germ have been removed.	Corn	Usually cooked in water and used as a cereal or a side dish.
Groats	Grain from which the hulls have been removed.	Barley, oats.	Usually cooked in water and used as a cereal or in soups.
Meal	Particles of ground grain that are larger than those in flour.	Corn, oats, rice, wheat.	Usually cooked in water and used as a breakfast cereal.
Polishings	Soft, fine residue obtained by polishing brown rice to make white rice.	Rice	Added to infant cereals to increase the content of minerals and vitamins.

• **Unleavened breads**—These items, which in their simplest forms consist of baked doughs made from flour and water, are the most palatable when they are thin and crackerlike, such as the Jewish matzo, made from wheat flour, and the Mexican tortilla. The latter is prepared by (1) treating kernels of corn with limewater to soften them and to facilitate the removal of their skins, (2) rubbing the skins off of the softened kernels, (3) crushing the skinned kernels with additional water to form a dough, and (4) baking thin, flat circles of the dough on a hot plate. (French crepes and Chinese egg roll wrappers are similar, but they are not crisped.) There is usually a slight leavening of the so-called unleavened breads when some of the water in the dough is converted to steam and trapped within the bread during baking. This effect is most likely to occur when the dough is placed in a very hot oven so that there is a rapid formation of a crust which prevents the steam from escaping.

• **Sourdough breads**—Generally, the leavening agent for these products is a species of bacteria such as *Lactobacilli* that produces (1) carbon dioxide which leavens the bread, and (2) lactic acid which gives it a sour taste. The bacteria is provided in the form of either a dried, pure culture of the organism; or as a component of a "starter," which is usually a portion taken from a batch of previously fermented dough. Sometimes, both bacteria and yeast are present in the starter so that two types of fermentation occur, such as in the production of San Francisco sourdough breads. Rye breads are often leavened with sourdough cultures. Typical ingredients of these breads are sourdough, rye flour, white flour, yeast, sugar, shortening, and flavorings.

• **Yeast-leavened breads**—Wheat flours rich in gluten are required for these breads. Because the yeast fermentation is slow and a strong, elastic dough is needed to trap the bubbles of carbon dioxide. However, the leavening process may be speeded up by (1) adding sugar to hasten the growth of the yeast, and (2) using double or triple the amount of yeast. Too much yeast produces a sticky, stringy texture in the bread. Excess sugar slows the growth of yeast. Usually, these breads contain flour, water, yeast, salt, sugar, and a small amount of shortening.

• **Quick breads**—Flours made from any of the grains may be used in these breads when eggs are added to the dough, because the inelasticity due to lack of gluten is compensated for by the protein in egg white. The leavening action results from one or more chemical reactions between the baking powder and the water in the dough. Sometimes, a combination of an acidic food, such as a soured dairy product and baking soda, is used for leavening. These breads usually contain variable amounts of flour, water or milk, egg, sugar, and shortening.

(Also see BREADS AND BAKING; and FLOUR.)

MISCELLANEOUS PROCESSES. Many of these processes are used in the production of breakfast cereals and snack foods. Both types of products appear to be increasing in their popularity, so food technologists are ever seeking new applications. It even seems likely that better nutrition in the developing countries may be promoted through the development of similar items based upon whatever grains might be available locally. Hence, the basic principles underlying these processes are noteworthy.

(Also see BREAKFAST CEREALS; and SNACK FOODS.)

• **Canning**—Cooked sweet corn is the only commonly used canned grain in the United States. However, cooked rice and bulgur wheat are also available in cans.

• **Cracking**—Breakfast cereals which require cooking often contain cracked grains, which cook faster than whole grains. Also, limited amounts of cracked grains are added to certain breads to give them a nutty flavor and texture. (Only limited amounts may be added to breads because their sharp edges cut the strands of gluten which make the doughs elastic.) Bulgur wheat differs from other cracked grains in that it has been parboiled prior to partial milling and cracking.

• **Extruding**—In cereal technology this term denotes the process of forcing doughs through a die to form pellets, ribbons, rods, tubing, etc. After extrusion, the shaped dough is cut to an appropriate length and either dried or cooked, so that (1) a strengthened product is formed which does not readily break upon handling, and (2) the moisture content is reduced to the point where the product keeps well when stored for prolonged periods at room temperature. The various forms of pasta (macaroni, noodles, shells, etc.) are the best-known products of this type of processing.

• **Flaking**—The flake types of ready-to-eat breakfast cereals are usually prepared by (1) cooking whole grains or partially milled pieces of grain in water with added flavorings such as sugar, salt, and malt; (2) drying the cooked grains to the desired moisture content; (3) forming flakes by passing the grains between rollers; and (4) drying or toasting the flakes. Sometimes, flakes are rolled from pellets of cooked doughs prepared from various flour mixtures.

• **Oil extraction**—The oils which are present in the various grains may be extracted in connection with the milling operations. Corn oil is by far the leading grain oil produced in the United States; the amounts of oil extracted from wheat and rice are small by comparison. The germ of the corn is extracted by a combination of heating, pressing, and solvent extraction—after the separation of the germ in either the dry milling or wet milling process.

• **Parboiling**—Both rice and wheat grains may be cooked in water prior to milling. Parboiled rice requires longer cooking than ordinary white rice because parboiling followed by drying makes the grain harder. However, the parboiled rice is higher in vitamins than unenriched white rice and it is less likely to stick when cooked. Most parboiled wheat is used to produce bulgur wheat, a product which is cracked into small pieces after parboiling and light milling. Bulgur wheat, unlike parboiled rice, requires considerably less cooking than whole grains of untreated wheat.

• **Pearling**—This process is applied mainly to barley grains after the hulls have been removed. The dehulled kernels or "groats" are polished into rounded shapes by an abrasive action.

• **Popping**—Although most people are familiar with the application of this process to popcorn (a special variety of corn), few realize that it is also applied to rice, sorghum, and wheat. The success of this procedure requires that grains of appropriate moisture content be heated rapidly so that the moisture changes to steam, which literally causes the starchy, inner portion of the grains to explode. Anthropologists believe that the ancient Indians of Mexico popped the heavy-hulled teosinte, which was the ancestor of our present-day corn.

• **Puffing**—Special puffing guns—consisting of heated, rotating cylinders that shoot out puffed grains—have long been used by the makers of breakfast cereals. A newer development is the puffing of dough mixtures rather than whole grains or pieces. In the latter process, the doughs are expanded by forcing them through dies. Sometimes, grains are cooked in water with sugar, salt, and flavorings; then, they are puffed by tumbling in a hot oven.

• **Rolling**—Rolled oats are the best-known example of grains which have been compressed into flat pieces (flakes) by passing through rollers. The oats are prepared for rolling by (1) removal of their hulls; and (2) steaming them to destroy enzymes which might cause deterioration, and to condition them for rolling. Other grains are also rolled into flakes, but they are usually subjected to more preliminary processing than oats.

• **Shredding**—This process dates back to the end of the 19th century, when it was first patented. Whole grain wheat is converted into shredded wheat biscuits by (1) cooking and drying to the desired moisture content; (2) squeezing into long, slender strands by passage through very closely spaced, grooved rolls; and (3) forming the strands into biscuits and baking them in an oven. Similar type of biscuits may also be made from corn, oats, and rice.

• **Sprouting**—Sometimes, grains may be sprouted for use as a vegetable, rather than for the production of malt. Hence, it is noteworthy that special equipment is now available for sprouting grains by hydroponics—a process in which the grain is sprouted by immersion in a water solution of essential nutrients.

• **Starch extraction**—Corn is the leading grain used as a source of starch in the United States, followed by sorghum, wheat, and rice. Starch is extracted by (1) soaking the grains in water containing small amounts of sulfur dioxide or sodium hydroxide; (2) grinding of the soaked kernels to free the germs and the fibrous components; and (3) flotating and centrifuging to separate the other components from the starch. A newer process for obtaining wheat starch involves the milling of wheat into flour, followed by the washing of the starch from the dough prepared by mixing the flour with water.

NUTRITIVE VALUES OF THE VARIOUS GRAINS. Many people tend to regard cereal products solely as starchy foods that are rich in calories, yet these items may provide substantial amounts of protein, minerals, and vitamins—depending upon such factors as the type of grain and the degree of refining. Food Composition Table F-36 of this book gives the nutrient composition for the commonly used grain products. Fig. C-71 shows the level of certain minerals in some selected cereal products; Fig. C-72 shows the level of certain vitamins and of one vitaminlike substance (inositol) in some selected cereal products; and Fig. C-73 shows the amount of lysine and protein in some of the major grains in comparison with cow's milk. Similar comparisons of breads, ready-to-eat breakfast cereals, flour, pasta, and related products are given in the separate articles dealing with these subjects.

(Also see BISCUITS AND CRACKERS; BREADS AND BAKING; BREAKFAST CEREALS; FLOUR; and MACARONI AND NOODLE PRODUCTS.)

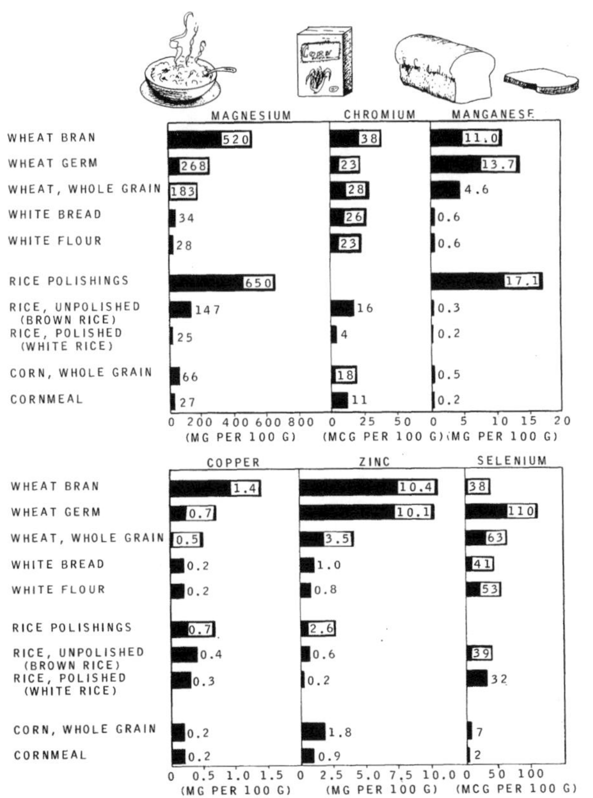

Fig. C-71. The level of certain minerals in some selected cereal products. (Graphs were constructed by the authors primarily from data in Schroeder, H. A., "Losses of Vitamins and Trace Minerals Resulting from Processing and Preservation of Foods," *The American Journal of Clinical Nutrition,* 1971, Vol. 24, p. 569, Tables 5 and 6)

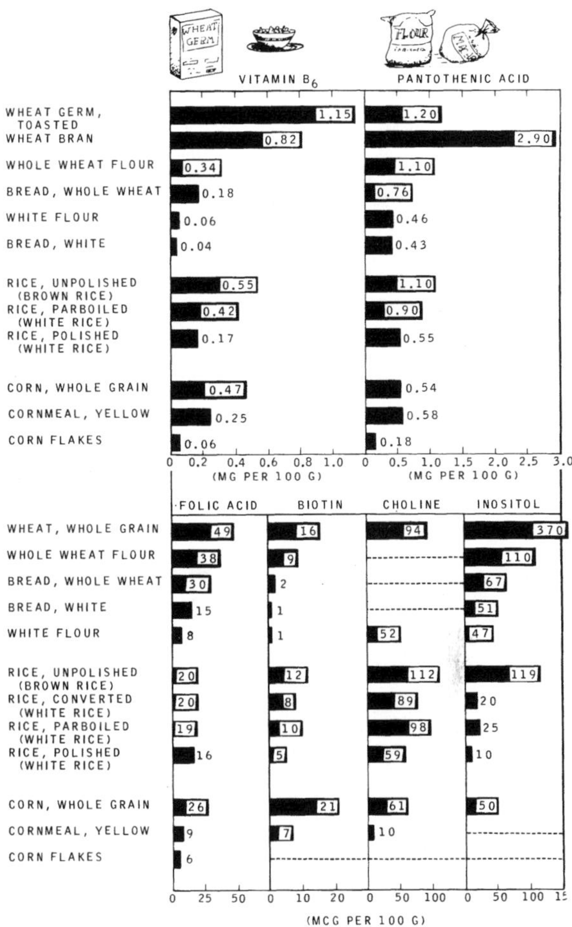

Fig. C-72. The level of certain vitamins and of one vitaminlike substance (inositol) in some selected cereal products. (Graphs were constructed by the authors from data in Schroder, H.A., "Losses of Vitamins and Trace Minerals Resulting from Processing and Preservation of Foods," *The American Journal of Clinical Nutrition,* 1971, Vol. 24, pp. 566, 568, Tables 2 and 4)

Based on the data presented in the Food Composition Table F-36 of this book and Figs. C-71 and C-72, the following conclusions may be drawn:

1. The cereal grains are rich in carbohydrates, which makes them good sources of calories. For example, 1 lb (*0.45 kg)* of raw, dried grain supplies from 1,550 to 1,750 Calories (kcal). Cooked cereals supply much fewer calories because of their higher water content.
2. Although grains may serve as the main source of dietary protein for adults, young growing children also need protein from other sources, because most grains are low in the amino acid lysine.
3. Most grain products are low in fat, except for rice polish and wheat germ, which are good sources of polyunsaturated fats.

4. Milling lowers the fiber content of grains, making them more digestible but less filling.

5. Grains contain only small amounts of calcium compared to much greater quantities of phosphorus, but much of the latter element may be in the form of unavailable phytates—which bind with minerals such as calcium, iron, and zinc so as to hinder their absorption. It is noteworthy that the yeast leavening of breads breaks down some of the phytates. (Whole grains contain more phytates than milled grains.)

6. Removal of the outer layer of grain by milling reduces the levels of essential minerals and vitamins because the inner, starchy portion of the kernel contains less of the nutrients than the outer layers. Hence, by-products of milling such as rice polish, wheat bran, and wheat germ are good sources of the minerals and vitamins lost in milling.

Although not shown in Fig. C-71, or Fig. C-72, it is noteworthy that most cereals are very poor sources of vitamins A and C, unless they have been sprouted. However, yellow corn contains about 490 IU of vitamin A per 100 g.

Limitations in the Protein Quality of Grains. The people of southeastern Asia whose diets consist mainly of rice are usually short of stature, but the children of those who immigrated to the United States are often taller than their parents. It appears that the higher quantity and quality of protein in the American diet is largely responsible for the taller children, because cereal grains alone do not provide sufficient amounts of lysine and other amino acids needed for optimal growth. Fig. C-73 shows the lysine and protein content of the major grains in comparison with cow's milk.

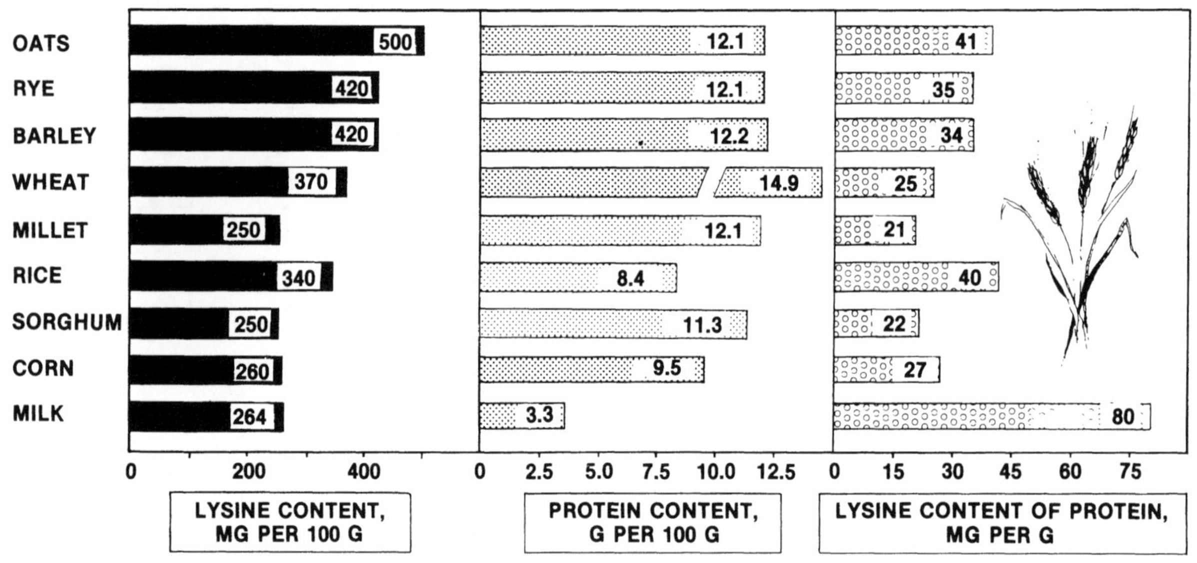

Fig. C-73. The amounts of lysine and protein in some of the major grains in comparison with cow's milk.

It may be seen from Fig. C-73 that the proportion of lysine to protein in the cereal grains is only about half that of cow's milk, which is almost an ideal source of protein for growing children. The amino acid lysine is much more essential for growing children than it is for adults, because it is required for the growth of the tissues such as bone, cartilage, and muscle. For this reason, the protein in most cereal grains is poorly utilized by infants and children.

The nutritional quality of diets based on cereals may be improved by adding higher quality protein in the form of (1) animal products or (2) legumes, both of which contain proportionately more lysine than grains. Other means of overcoming the problems of protein quality in cereal products are described in the sections that follow.

(Also see INFANT DIETS AND NUTRITION; LYSINE; and PROTEIN[S].)

Cereals that have been Enriched or Fortified with Certain Nutrients. Cereal products are often the mainstay of emergency feeding programs designed to serve large numbers of people, because they are less expensive than many of the other foods which provide calories and protein. However, it is well known that the prolonged consumption of a diet-consisting mainly of grains may lead to multiple nutritional deficiencies, because cereals fail to supply sufficient amounts of certain other essential nutrients. Nevertheless, it may cost less to enrich or fortify simple mixtures of grains, legumes, and a few other ingredients—with amino acids, minerals, and vitamins—than it would to provide these nutrients from a wide variety of foods. Details of some typical measures follow:

• **Amino acids**—These nutrients—which are provided in the form of pure, white powders—are usually added to flours or meals prior to their incorporation into various products. In certain cases, increased amounts should be added because heat processing by baking and/or toasting renders some of the lysine and other amino acids unavailable.

• **Minerals**—Iron has frequently been added to grain products, particularly those that have been highly milled. However, the milling of grains into refined flours removes large percentages of many of the other essential minerals, but they are rarely replaced by enrichment. Calcium is deficient in grains, but it is seldom added to cereal products because of certain technical difficulties; although calcium salts are added to flour in Great Britain, Denmark, and Israel, and to certain breakfast cereals in the United States.

• **Vitamins**—Thiamin, riboflavin, and niacin are often added to white flour and breakfast cereals, and sometimes to white rice. Milling also removes much of the other B-complex vitamins present in whole grains. Vitamins A, C, and D are not always added to cereal products; so, other sources of the vitamins should be consumed along with the cereal products.

New and Improved Varieties of Grains. Some of the limitations in the protein quality of the most important cereal grains have been overcome by crossbreeding them with mutant varieties so as to obtain new hybrids which have higher contents of both lysine and protein. Fig. C-74 shows how the new high-lysine and high-protein varieties compare with some of the older, more common varieties of the major grains.

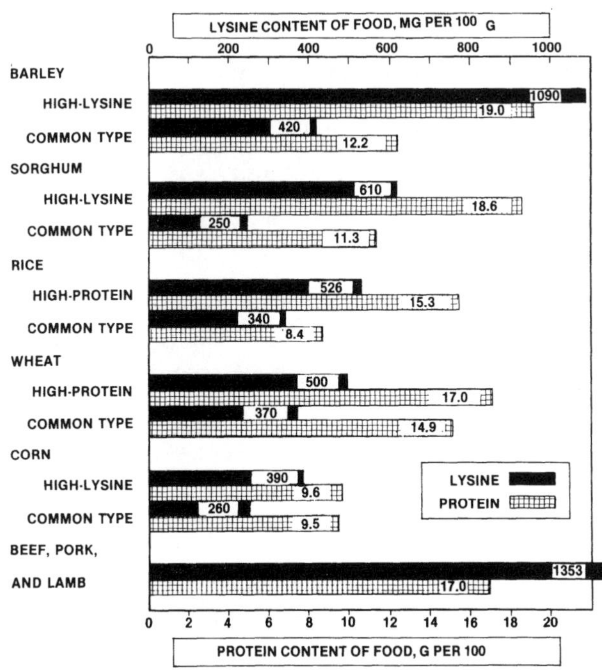

Fig. C-74. The amounts of lysine and protein supplied by hybrid and common types of the major grains in comparison with red meats—beef, pork, and lamb.

It may be seen from Fig. C-74 that the new, improved hybrids are notably higher in lysine and/or protein than the more common types of cereals which are presently grown around the world. For example high-lysine barley contains 81% as much lysine as the red meats (beef, pork, and lamb), and has a slightly higher protein content. However, farmers are reluctant to grow the hybrid varieties because they yield from 10 to 15% less grain per acre than the more common varieties. Hence, their margin of profit would shrink drastically if both types grain brought the same price. Certain agricultural economists and nutritionists believe that higher prices for the high-protein, hybrid cereals are justified because (1) they may be used to feed people who have high protein requirements—such as infants, children, and pregnant and lactating women—without much need for additional protein from eggs, legumes, fish, meats, milk, nuts, or poultry; and (2) providing the other protein foods is much more expensive in terms of energy, land, and money than paying a higher price for the hybrids in order to give farmers an incentive to grow them.

Sprouted Grains. During World War II, people were advised to use sprouted seeds as sources of vitamin C in the event that the supplies of fresh produce became limited. Nevertheless, few people are aware of the other improvements in nutritive value which result from the sprouting of the cereal grains. Table C-20 shows how the sprouting of oats in a special nutrient solution (the process known as hydroponic sprouting (affects the level of nutrients.

TABLE C-20
NUTRIENT COMPOSITION OF OAT GRAIN VS SPROUTED OATS

Nutrient or Other Constituent	Oat Grain	Sprouted Oats
	(2 oz dry) equals	(1 lb sprouted)[1]
Water . . . . (g)	7	404
Dry matter . . . . (g)	50	50
Protein . . . . (g)	7.5	10.5
Fat . . . . (g)	2.1	2.6
Carbohydrate, available. (g)	32.9	21.4
Energy[2] . . . . (kcal)	181	151
Fiber . . . . (g)	5.9	13.1
Ash . . . . (g)	1.6	2.0
Calcium . . . . (mg)	32	119
Phosphorus . . . . (mg)	180	254
Vitamin A[3] . . . . (IU)	0	3,039
Vitamin E . . . . (mg)	0.9	2.4
Niacin . . . . (mg)	0.9	5.2
Riboflavin . . . . (mg)	0.1	1.1
Thiamin . . . . (mg)	0.2	0.6
Vitamin C . . . . (mg)	0	10.9

[1]Approximately 1 lb *(0.45 kg)* of fresh sprouts may be obtained from 2 oz *(56 g)* of grain.
[2]Calculated by allowing 4 Calories (kcal) per gram of protein, 9 Calories (kcal) per gram of fat, and 4 Calories (kcal) per gram of carbohydrate.
[3]The vitamin A equivalent of the carotene content, considering 1 mg of carotene to be equivalent to 1,667 IU of vitamin A.

It may be seen from Table C-20 that the hydroponic sprouting of oats results in (1) great increases in fiber, water, vitamin A, and bulk (1 lb [*0.45 kg*] of sprouts is obtained from 2 oz [*56 g*] of grain); (2) moderate increases in protein, fat, total minerals (ash), calcium, phosphorus, vitamin E, vitamin C, and the vitamin B complex; and (3) decreases in carbohydrate and calories. Therefore, sprouted grains make a satisfactory vegetable supplement to diets when fresh fruits and vegetables may not be available. Unfortunately, the large amounts of water and fiber that they contain make them very filling, so consumption is limited.

COMMERCIAL USES OF GRAINS AND CEREAL PRODUCTS. The cereal products that people eat are derived from only a small portion of the grain plants. Likewise, hardly any of the by-products from the brewing and distilling industries are consumed by people. Therefore, the greater part of the vegetable matter from these crops may go to waste unless arrangements are made to feed the surplus materials to livestock. The efficient utilization of grain by-products for the feeding of farm animals has helped to keep the prices of meats, poultry, and dairy products within the reach of most American families. Uses of some of the major cereal products follow. (Additional details regarding the products made from each of the major grains are given in the separate articles covering each of these items.)

(Also see BARLEY; CORN; MILLET; OATS; RICE; RYE; SORGHUM; TRITICALE; WHEAT; and WILD RICE.)

Beverages and Foods. Most of the people who eat minimally processed grains do so because of either sheer necessity or long-standing custom. Even the most primitive cultures devised ways of converting the rather bland and chewy raw grains into more appealing beverages and foods.

• **Alcoholic beverages**—These drinks are the basis of popular social institutions such as the "pub" (short for public house) in Great Britain and the "bar" in the United States. Beer and grain whiskeys are the leading beverages served in these establishments. Home consumption of beer and liquor also account for much of the usage of these grain-based beverages, which may be produced from a wide variety of different cereals.

(Also see BEERS AND BREWING; and DISTILLED LIQUORS.)

• **Alimentary pastes (pasta)**—Macaroni, noodles, and spaghetti—which probably originated in China, but came to the United States by way of Europe—are often used as the main course of low cost dinners. Most of these products are made from durum wheat, which imparts to the pasta a great resistance to disintegration in boiling water. However, a few American food companies have developed special macaronis that are made from various combinations of corn, soy, and wheat flours. These products have higher protein quality than those made from wheat flour alone.

(Also see MACARONI AND NOODLE PRODUCTS.)

• **Breads**—Although the total consumption of all types of breads in the United States has declined steadily since the beginning of the 20th century, the use of specialty bread products such as hero or submarine sandwiches, hamburgers, and pizza has increased. There are now fast food restaurants which serve a variety of these products. Also, sandwiches of all types appear to be among the most popular lunch items for Americans, whether they carry a "bag lunch" from home or eat out. Furthermore, a wide variety of breads and related products are now available in frozen forms which require only baking.

The predominant grain used for breads in the United States is wheat, but recent innovations in baking technology have made it possible to use almost any type of flour to produce appealing products. Perhaps larger quantities of wheatless baked goods will soon appear on the American scene. An unknown number of people suffer from various types of sensitivity to wheat protein or gluten.

(Also see BREADS AND BAKING; and GLUTEN.)

• **Breakfast cereals**—Archaeological evidence suggests that the preparation of grain porridges goes back thousands of years to the times of the earliest grain farmers, but the use of ready-to-eat breakfast cereal is less than a hundred years old. Most of the impetus for the development of the latter items came from certain members of the newly founded Seventh-Day Adventist Church who sought a meatless substitute for the traditional American breakfast of meat (usually bacon or ham) and eggs. It is noteworthy that the Kellogg brothers, who later founded the cereal company that bears their name, invented a granola-type of ready-to-eat breakfast while they were on the staff of the Battle Creek Sanitarium in Michigan. Mr. Post, who was one of their patients, also developed a cereal and founded another company in Battle Creek. Lately, there has been rising interest in so-called "natural cereals" which bear a close resemblance to their early forerunners.

(Also see BREAKFAST CEREALS.)

• **Flours**—Any of the major grains may be milled into flour, but people still prefer baked goods made from the traditional type of white flour that is derived from wheat. Nevertheless, food technologists have recently developed some new additives which impart good leavening characteristics to a wide variety of flours. Hence, new types of breads may soon be available.

(Also see FLOURS.)

• **Infant cereals**—These items account for only a small fraction of the total consumption of grain products in the United States, yet they represent a nutritious group of fortified cereal products that might be used profitably by people of all ages. However, some mothers may feed their babies excessive amounts of cereals, milk, and other foods in the belief that if a little is good, more is better. Overfed infants may become afflicted with lifelong obesity.

(Also see INFANT DIETS AND NUTRITION.)

• **Snacks**—Many processes have been invented for turning grains into such tempting snacks as biscuits, cakes, cookies, corn chips, crackers, popcorn, pretzels, and similar items. These tidbits usually supply some of the daily nutrient requirements, except that they are often overloaded with fat, salt, and/or sugar. Therefore, some of the unsugared, ready-to-eat breakfast cereals might make more nutritious snacks, particularly when they are taken with milk.

(Also see BREAKFAST CEREALS; and SNACK FOODS.)

• **Vegetable dishes**—Many peoples around the world prepare cooked grains as vegetable side dishes like potatoes, or in mixed dishes like stews and casseroles. For example, cooked corn grits are a regular accompaniment to meals in the southeastern United States, whereas cooked fresh sweet corn often serves a similar function in the Midwest. Likewise, most southeastern Asians have cooked rice with their meals.

Feeds for Livestock. Cereal grains are excellent sources of energy and protein for all types of farm animals and pets, provided that supplementary protein from sources like alfalfa and soybeans are given to young, rapidly growing animals. It is noteworthy that the leading ingredient of a popular brand of dried dog food is ground yellow corn.

Recently, various groups of lay and professional people have questioned the practice of feeding grains to farm animals when cereals are needed for distribution to large numbers of malnourished people around the world. However, the feeding of grains to needy people requires that ways be found to (1) reimburse the grain farmers for their products, and (2) distribute the cereals in the areas where the needs are greatest.

A few of the main arguments for feeding grains to farm animals are as follows:

• **Animal foods contain more and better protein than grains**—Grains, such as corn, are much lower in protein content than meat, milk, or eggs. On a dry basis, the protein contents of selected products are: corn, 10.4%; beef, 30.7%; milk, 26.4%; and eggs, 47.0%. Also, animal proteins generally are of higher quality (biological value) than the proteins in grains.

• **Livestock provide elasticity and stability to grain production**—Livestock feeding provides a large and flexible outlet for the year-to-year changes in grain supplies. When there is a large production of grain, more can be fed to livestock, with the animals fed to heavier weights and greater finish. On the other hand, when grain supplies are low, herds and flocks can be maintained by cutting down on grain feeding and increasing the forages and by-products in the ration.

• **Stockmen usually supplement grain feeds with by-products that are not suited for human consumption**—Animals provide a practical outlet for the by-products of beverage and food production, many of which are obtained from grain agriculture and technology. For example, the parts of the plants which remain after the harvesting and threshing of grain may be fed to cattle and sheep. Likewise, the by-products of the various steps in the production of alcoholic beverages are usually utilized as ingredients of livestock feeds. These residues include malt sprouts and hulls, brewers' grains, brewers' yeast, distillers' grains and distillers' solubles. Finally, the parts of the grain removed in flour milling—such as the hull, seed coat and adjoining layers, and the germs—are also used in animal rations. (Additional information on the by-products from the processing of the cereal grains is given in the separate articles dealing with each of these grains.)

(Also see BARLEY; CORN; MILLET; OATS; RICE; SORGHUM; TRITICALE; WHEAT; and WORLD FOOD.)

ECONOMICAL AND NUTRITIOUS MEALS BASED UPON COOKED CEREALS. The use of grains as major dietary staples usually picks up when times get tight, and decreases with rising affluence. Hence, it is not surprising that some of the low income groups of American people may obtain more than three-quarters of their calories from various cereal products. Even so, most Americans—whether rich or poor—consume most of their grain in the form of breads made from wheat flour. However, it would be less expensive for people to utilize other types of cereal products, which generally cost less than half as much as bread. Unfortunately, many people believe that cooked cereals are fattening, because of their starchiness. Therefore, it is worth noting the nutrient composition of various cooked cereals, which is presented in Table C-21. (The nutritive values of the products not covered in this table are given in other acticles or in Table F-36 Food Compositions.)

(Also see BISCUITS AND CRACKERS; BREADS AND BAKING; BREAKFAST CEREALS; FLOUR; and MACARONI AND NOODLE PRODUCTS.)

Table C-21 shows that the cooking of cereals in water results in great increases in their water content and bulk. For example, 1 lb (*0.45 kg*) of uncooked wheat farina yields 15.1 cups (*3.6 liter*) of cooked cereal which weigh 8.6 oz (*241 g*) each, for a total yield of 8.1 lb (*4 kg*). Similarly, 1 lb of dry cornmeal cooks up to 7.1 lb. Fortunately, few people rely on either cooked wheat or cornmeal alone for their dietary energy, because it would be difficult for most people to eat enough of these items to meet their needs. On the other hand, large numbers of low income people around the world rely heavily on rice and various types of pasta, which cook up to only about three times their original bulk. Therefore, it is noteworthy that cooked cereals alone are not fattening—since it takes from 7 to 15 cups (*1.7 to 3.6 liter*) of the various items to meet the minimum caloric needs of most adults. (Most sedentary adults require between 1,600 and 2,800 Calories [kcal] per day.) Usually, obesity in cereal eaters is due to the fatty and sugary foods consumed along with the grain products.

TABLE C-21
NUTRITIVE VALUES OF COOKED

Product	Amount of Uncooked Cereal Which Yields 1 Cup of Cooked Cereal[2]		Yield of Cooked Cereal from 1 Lb of Uncooked Cereal	Weight of 1 Cup of Cooked Cereal[3]		Composition of 1 Cup of Cooked Cereal Product: Water	Calories	Protein	Fat	Carbohydrate: Total	Fiber
	(oz)	(g)	(cups)	(oz)	(g)	(g)	(kcal)	(g)	(g)	(g)	(g)
Corn, sweet, on-the-cob[4]	10.6	300	1.5	5.8	165	125	137	5.3	1.7	31.0	1.2
Cornmeal, degermed, enriched	1.2	34	13.5	8.4	240	211	120	2.6	0.5	25.7	0.2
Macaroni, elbow, enriched[5]	1.5	42	10.8	4.9	140	102	155	4.8	0.6	32.2	0.1
Noodles, egg, enriched	1.8	52	8.8	5.6	160	113	200	6.6	2.4	37.3	0.2
Oats, rolled[6]	1.2	35	12.8	8.4	240	208	132	4.8	2.4	23.3	0.5
Popcorn, large kernel, without oil[7]	0.2	6.4	70.6	0.2	6	—	23	0.8	0.3	4.6	0.1
Rice, brown[8]	2.3	66	6.9	6.9	195	137	232	4.9	1.2	49.7	0.6
Rice, white, enriched[8]	2.2	64	7.1	7.2	205	149	223	4.1	0.2	49.6	0.2
Spaghetti, enriched[5]	1.5	42	10.8	4.9	140	102	155	4.8	0.6	32.2	0.1
Wheat, farina, quick cooking, enriched	1.1	30	15.1	8.6	245	218	105	3.2	0.2	21.8	trace
Wheat, rolled	1.9	54	8.4	8.4	240	192	180	5.3	1.0	40.6	1.2
Wheat, whole grain	1.2	34	13.5	8.6	245	215	110	4.4	0.7	23.0	0.7

[1]From Food Composition Table F-36, which also contains further composition data on these cereal products. One cup equals 240 ml and 1 lb equals 454 g.
[2]The amount of uncooked cereal is calculated from the weight (wt) and dry matter content (dm) of 1 cup of the cooked cereal, by the following formula:

$$\text{Wt uncooked cereal (g)} = \frac{\text{wt cooked cereal (g)} \times \text{dm of cooked cereal (\%)}}{\text{dm of uncooked cereal (\%)}}$$

[3]Weight is that of the freshly cooked, hot cereal.
[4]The weight of the kernels cut from the cob represents 55% of the weight of the uncut cob.
[5]Cooked to the tender stage.
[6]Data apply to either the regular or the instant-cooking product.
[7]Small, popped kernels pack together more closely. Hence, a cup of the small variety may weigh up to twice as much as one of the large variety, and the nutrients will be increased proportionately.
[8]Long grain variety.

Another important consideration in deciding which cereals are the best buys is the amounts of protein supplied by these items. The most nutritious products are those which supply the maximum protein per pound and per calorie. Fig. C-75 gives these data for some of the most commonly used cereal items on an as-purchased basis.

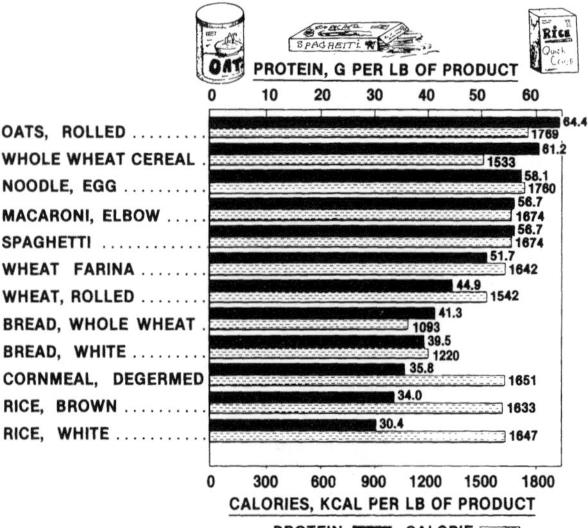

Fig. C-75. Protein and caloric contents of 1-lb (*0.45 kg*) packages of selected cereal products. (Values are from Food Composition Table F-36 of this book)

It may be seen from Fig. C-75 that rolled oats and whole wheat cereal provide the most protein per pound; whereas brown and white rice provide the least. Also, the oat and wheat products provide about twice as much protein per calorie as the rice products.

Finally, it is not desirable for people to attempt to eat the quantities of cereals which might completely meet their protein needs, because they would have little appetite left for the foods that supply the other essential nutrients. Hence, it is necessary to plan for the use of other protein-containing foods. Table C-22 lists the minimal portion sizes and caloric contents of foods which are considered to be good protein supplements for cereal dishes.

Items from Table C-22 may be incorporated in economical daily menus that are based upon the liberal use of cereals. Details follow.

1. Provide ½ cup (*120 ml*) or more of cooked cereal or its equivalent in bread (½ cup cereal = 1 slice bread) per person per meal. (Yields 6 or more grams of protein per day
2. Allow a minimum of one serving of a milk product or of a cheese for each person at each meal. (This adds about 24 g of protein daily.)
3. Serve at least one fruit or vegetable item per meal, for a total of four servings per day.
4. Provide two or more servings of high-quality protein foods such as eggs, fish or seafood, legumes, meats, poultry, or specialty meats for each person each day. (This will add from 12 to 20 g of protein to the day's protein allowance, for a total of at least 42 g per day.)
5. Allow extra servings of cereals and supplementary

CEREAL PRODUCTS[1]

Composition of 1 Cup of Cooked Cereal Product

Minerals			Vitamins		
Calcium	Phosphorus	Iron	Thiamin	Riboflavin	Niacin
(mg)	(mg)	(mg)	(mg)	(mg)	(mg)
5	147	1.0	0.2	0.2	2.1
2	34	1.0	0.1	0.1	1.2
11	70	1.3	0.2	0.1	1.5
16	94	1.4	0.2	0.1	1.9
22	137	1.4	0.2	0.1	0.2
1	17	0.2	trace	trace	0.1
23	142	1.0	0.2	trace	2.7
21	57	1.8	0.2	trace	2.1
11	70	1.3	0.2	0.1	1.5
147	162	12.3	0.1	0.1	1.0
19	182	1.7	0.2	0.1	2.2
17	127	1.2	0.2	0.1	1.5

protein foods for children over 10 years of age and for adults, with the quantity varying according to age, health status, and physical activity. Note that the high-quality animal and vegetable protein foods contain about twice as much protein per calorie as the cereals. Hence, people who are trying to restrict their energy intake should eat the more concentrated sources of protein, while others may eat more of the cereals.

For those who are unaccustomed to serving grain products other than breads or ready-to-eat breakfast cereals, some suggestions for "beefing up" the nutritive values of cooked cereals follow:

1. Add nonfat dry milk and/or rice polish or wheat germ to bowls of watery cooked cereals such as farina in order to increase the quality and quantity of protein for growing children, pregnant women, and others who have above average needs.

2. Cook cereals with milk instead of water, but either take care to keep the cooking temperature low enough or use a double boiler, to avoid scorching the milk.

3. Prepare casserole dishes which utilize cooked pasta or rice, plus supplementary protein foods, fruits or vegetables, and spices. This means items such as beef or cheese with macaroni, lasagna, rice pudding, or tuna-and-noodle casserole.

4. Toss cooked elbow macaroni with cottage cheese, sour cream, or yogurt; then mix in spices plus chopped, raw vegetables like onions, celery, peppers, and carrots.

Additional suggestions for dishes based upon grain products are given in the article dealing with the individual grains and related subjects.

(Also see BARLEY; BISCUITS AND CRACKERS; BREADS AND BAKING; BREAKFAST CEREALS; CORN; FLOUR; MACARONI AND NOODLE PRODUCTS; MILLET; OATS; RICE; RYE; SORGHUM; TRITICALE; WHEAT; and WILD RICE.)

FUTURE PROSPECTS FOR THE CEREAL GRAINS. It seems likely that the worldwide demand for cereal grains will continue to rise along with the steady growth in population. However, not much land is left for expansion of the acreage which is presently allotted to grain crops. Therefore, certain other measures have been either instituted, or are under consideration, so that additional supplies of grain may be made available to prevent malnutrition and starvation from occurring on a massive scale.

TABLE C-22
AMOUNTS OF SELECTED HIGH-QUALITY PROTEIN FOODS WHICH SUPPLY FROM 6 to 10 GRAMS OF PROTEIN PER SERVING[1]

Food	Household Measure[2]	Weight		Calories
		(oz)	*(g)*	(kcal)
Cereal products				
Rice polish	½ cup	1.8	*52*	139
Wheat germ	¼ cup	0.8	*24*	92
Cheeses				
American, Cheddar, or Swiss	1 Slice	1.0	*28*	113
Cottage, creamed	¼ cup	1.8	*52*	56
Cottage, dry curd	¼ cup	1.3	*36*	31
Eggs				
Egg, medium	1	1.8	*50*	72
Fish and seafood				
Clams, raw	4 or 5	2.5	*70*	56
Codfish, flounder, or halibut, cooked	½ filet	1.0	*29*	58
Fish sticks, cooked	2	1.3	*38*	75
Oysters, raw	6	3.0	*85*	51
Sardines	3 medium	1.3	*36*	72
Shrimp	10 medium	1.1	*32*	37
Legumes				
Beans, lima, navy, etc., cooked	½ cup	3.3	*95*	120
Lentils or split peas, cooked	½ cup	3.5	*100*	110
Peanut butter	2 Tbsp	1.1	*32*	188
Meats				
Beef patty, cooked	1 small	1.2	*33*	93
Heart, beef, cooked	1 slice	1.0	*28*	53
Kidneys, sliced, cooked	¼ cup	1.2	*35*	88
Lamb, cooked	1 slice	1.0	*28*	79
Liver, cooked	1 slice	1.0	*28*	65
Pork, cooked	1 slice	1.0	*28*	106
Veal, cooked	1 slice	1.0	*28*	66
Milk products				
Buttermilk	1 cup	8.6	*245*	88
Milk, evaporated	½ cup	4.4	*126*	173
Milk, nonfat, powdered	¼ cup	1.1	*30*	109
Milk, skim	1 cup	8.8	*245*	88
Milk, whole	1 cup	8.6	*245*	159
Yogurt, low fat, plain	1 cup	8.6	*245*	123
Poultry				
Chicken, cooked	1 slice	1.0	*28*	48
Turkey, cooked	1 slice	1.0	*28*	53
Specialty Meats				
Cold cuts	2 slices	2.0	*57*	172
Frankfurter	1 (8 per lb)	1.0	*56*	176

[1]Calculated from Food Composition Table F-36.
[2]One cup equals *240 ml.*

- **Grain reserves**—There is a great need to establish reserve supplies of grains which may be used when adverse conditions in some countries reduce crop production below the levels needed to feed people adequately. The details of such a program still need to be worked out, especially who will pay for it and where it will be stored.

• **The Green Revolution**—This term designates the development and utilization of special, high-yielding varieties of rice and wheat. The effect of these breakthroughs has been the doubling and tripling of the yields of these popular grains so that more people might be fed by the crops grown on fixed acreages.
(Also see GREEN REVOLUTION.)

• **High-lysine and high-protein varieties of grain**—Special, new hybrid varieties of grains have been developed which have markedly higher protein quality and quantity than the older, more common types. These new varieties better meet the amino acid and protein requirements of growing children and pregnant mothers, so that there is less need to supplement their diets with other protein-containing foods. Usually, much more land is required to produce high-quality protein foods such as eggs, legumes, meats, milk, and poultry than is required to produce cereal protein.

• **Less grain for livestock and more for people**—Mounting demands for grains are likely to make it increasingly expensive to feed these crops to livestock. Fortunately, grazing animals—cattle, goats, and sheep—can be fed rations based mainly on forage crops and various by-products, whereas pigs and poultry require more readily digestible feeds such as grains and certain by-products from food processing. Therefore, the diversion of corn and other feed grains from animals to people may not cause any reduction in the supplies of beef, lamb, and milk, but it may lessen pork, chicken, turkey, and eggs.

• **Increased utilization of crops which provide the nutrients that are lacking in grains**—Attempts to cope with world food shortages by utilization of the high-yielding varieties of grains should allow for measures that provide the nutrients lacking in grains. For example, legumes and oilseeds are often used to supplement the protein of cereals. Therefore, the production of the required amounts of grains on fewer acres of land might make it possible to shift some of the acreage now devoted to corn, rice, and wheat to other crops such as peanuts, soybeans, and green leafy vegetables.

Another approach would be to promote the cultivation of lesser known species such as *Amaranth* and *Chenopodium*. The grainlike seeds of these plants have long been used as food by the native peoples of Latin America. It is noteworthy that the amino acid patterns of these seeds, which contain about 13% protein, resemble those of the legumes and oilseeds, and that they complement those of the major cereals. Furthermore, the leaves of *Amaranth* and *Chenopodium* are rich in vitamin A, a nutrient that is lacking in the grains. Finally, it has been reported that cultivated stands of *Amaranth* have yielded more seeds per acre than corn.

Perhaps, the achievement of world order and peace will depend upon how well the production and utilization of cereal grains is improved through the efforts of scientists, agriculturalists, and food technologists—on whether the hungry get their share of the "staff of life."
(Also see WORLD FOOD.)

CEREBRAL

Pertaining to the cerebrum, the largest part of the human brain.
(Also see TISSUES OF THE BODY.)

CEREBRAL HEMORRHAGE

This condition, which is commonly called stroke or apoplexy, results from a breaking of a blood vessel in the brain. A hemorrhage is more likely to occur when the blood pressure is elevated, than when it is normal. A stroke is often followed by a loss of consciousness and/or partial paralysis. The diagnosis of a cerebral hemorrhage may be confirmed by finding of blood in the cerebrospinal fluid, which bathes both the brain and the spinal cord. If blood is absent from the cerebrospinal fluid, the stroke was most likely due to the blockage of blood flow to the brain (cerebral infarct).
(Also see CEREBRAL INFARCT; CEREBROSPINAL FLUID; and HIGH BLOOD PRESSURE.)

CEREBRAL INFARCT

A blockage of the blood flow to the brain, which is usually due to a clot lodged within an artery serving the brain. The consequences of this mishap are essentially the same as those of a cerebral hemorrhage (loss of consciousness and partial paralysis). Hence, both conditions are commonly referred to as "strokes." An infarct is most likely to occur in people afflicted with atherosclerosis and/or congestive heart failure. Also, it may take place when the blood pressure is normal or subnormal. This conditions is distinguished from a cerebral hemorrhage by the absence of blood in the cerebrospinal fluid.
(Also see CEREBRAL HEMORRHAGE; CEREBROSPINAL FLUID; and HEART DISEASE.)

CEREBROSPINAL FLUID

A clear fluid secreted by the ventricles of the brain, which bathes both the brain and the spinal cord. It serves to (1) act as a medium for the exchange of nutrients and metabolic products with the nerve cells; (2) regulate the fluid content and pressure within the brain and spinal cord; and (3) protect the tissues of the brain against injury. Doctors may withdraw some of the cerebrospinal fluid from the spine (spinal tap) in order to diagnose certain diseases.
Also see BRAIN, NUTRITION AND THE; and TISSUES OF THE BODY.)

CERIMAN *Monstera deliciosa*

The conelike fruit of a vine (of the family *Araceae*) that is native to Mexico.

Fig. C-76. The ceriman, a tropical fruit of the Americas.

Cerimans are usually about 8 in. (*20 cm*) long and are suitable for eating only if they are fully ripe, when the flavor is like that of a mixture of bananas and pineapple. The unripe fruit contains acid crystals that burn the mouth.

CEROID

A type of brownish pigment found in clogged arteries (such as in atherosclerosis), the intestinal lining, and in various types of disorders where polyunsaturated fats are abnormally oxidized. The accumulation of ceroid pigment appears to be associated with deficiencies of selenium and vitamin E.

(Also see FATS AND OTHER LIPIDS; SELENIUM; and VITAMIN E.)

CERULOPLASMIN

A protein which transports copper in the blood. When there is a deficiency of ceruloplasmin, copper is not efficiently utilized. Hence, there may be certain types of anemia, or deposits of copper in various tissues.

(Also see COPPER.)

CERVELAT (CERVELET; CHERVELAS)

Sausage of several regional kinds made of varying proportions of pork and beef with added fat and spices, stuffed into casings, and smoked.

CHAFF

Glumes, husks, or other seed coverings, together with other plant parts, separated from seed in threshing or processing.

(Also see CEREAL GRAINS.)

CHARCOAL, ACTIVATED

This is an antidote for many types of poisons. Poison centers commonly use activated charcoal as a follow-up measure after doctors induce vomiting with syrup of ipecac or pump the patient's stomach. Charcoal works by a process called *adsorption*; its surface locks on to the surface of the poison molecule. The charcoal/poison combination then safely works its way through the digestive tract and is eliminated.

CHARD; SWISS CHARD *Beta vulgaris*, var. *circla*

This is a garden vegetable plant the leaves of which are eaten as greens. Chard is related to the common beet plant, which it resembles except it does not have a large fleshy root. Chard is grown more widely in Europe than in the United States. (Also see SWISS CHARD; and VEGETABLE[S], Table V-6 Vegetables of the World.)

CHARQUI (CHARQUE)

Dried meat; jerked meat.

CHAYOTE *Sechium edule*

This tropical plant, which is a member of the gourd and melon family (*Cucurbitaceae*), bears edible fruit which is similar in taste and texture to summer squash. Edible, tuberous roots are also formed in tropical regions where there is an extended dry season. Also, people sometimes use the young shoots for greens. Fig. C-77 shows the flattened pear-shaped ripe fruit of the chayote, which varies in color from whitish to dark green. Each fruit contains only one seed.

ORIGIN AND HISTORY. The chayote originated in southern Mexico and Central America. It was grown as a food plant by the Aztec Indians of Mexico before the Spaniards conquered them. Now, it is grown in West Indies and as far north as the U.S. Sunbelt.

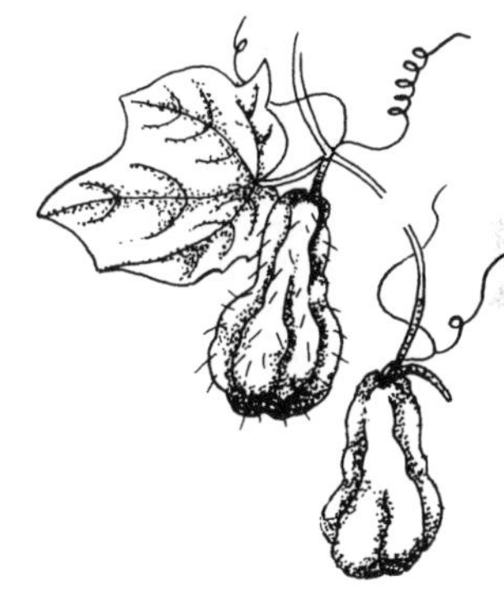

Fig. C-77. The chayote, a tropical food plant that may be grown in the U.S. Sunbelt, where the winters are mild.

PRODUCTION. Although this vegetable is an important crop in the tropical regions of Latin America, few statistics are kept on its production. It is also grown in California, the Gulf states, and the South Atlantic states, where the mild winters allow the cold-sensitive perennial vine to survive when its roots are protected by a heavy mulch.

Chayote is propagated from stem cuttings or from putting mature fruits in the ground. Many growers support the heavy vines on trellises or fences in order to make the most efficient use of the available land.

The plant requires abundant water through its life cycle. It will bear fruit continuously in tropical areas where the rainfall is abundant. Like certain other tropical plants, it will stop bearing fruit and form large tuberous roots during a dry season. Hence, irrigation may be required for maximum fruit production.

The fruits are ready for harvesting about a month after the flowers have been pollinated. They may be stored for 2 to 3 months at 50° to 60° (*10° to 16°C*).

PROCESSING. There is little or no processing of chayote at the present time.

SELECTION AND PREPARATION. Chayote should be fresh, fairly heavy in relation to size, free from cuts and noticeable bruises, and tender. The rind should be soft; a hard rind denotes that the flesh will likely be fibrous and tough.

If tender, the edible rind may be left on the fruit; if hard, the rind should be peeled off. The seed is also edible, as are the shoots which may emerge from the seed when the fruit is stored. Chayotes may be (1) baked with or without a stuffing, or in custards, pastries, and puddings; (2) boiled or steamed, then mashed and served with butter and Mexican seasonings; (3) dipped in batter and fried; or (4) pickled.

Chayote roots may be used in the same ways as Irish potatoes and similar tuberous vegetables.

NUTRITIONAL VALUE. The nutrient composition of chayote fruit is given in Food Composition Table F-36.

Some noteworthy observations regarding the nutrient composition of chayote follow:

1. Chayote fruits are high in water content (91%) and low in calories (28 kcal per 100 g) and other nutrients. However, they are a fair to good source of potassium and vitamin C.

2. The roots of chayote are nutritionally similar to the Irish potato with respect to the content of water, calories, protein, carbohydrates, and vitamin C. Therefore, it is noteworthy that these roots are a staple food in the dry tropics, as much so as the potato once was for the peoples of northern Europe.

(Also see VEGETABLE[S], Table V-6 Vegetables of the World.)

CHEDDARING

After whey is drained and curds have knit together sufficiently, slabs are cut from the curd. These are turned and piled, then repiled, to induce matting of the curd and expel most of the whey. This turning and stacking technique, known as "cheddaring," helps give the cheese its characteristic body.

CHEILITIS

A soreness of the lip, that is sometimes due to a riboflavin deficiency.

(Also see CHEILOSIS.)

CHEILOSIS

Cracks at the corners of the mouth that often signify a riboflavin deficiency. However, this disorder may also be due to badly chapped lips, particularly when the victim has been exposed to a cold or windy environment.

(Also see RIBOFLAVIN.)

CHELATE

The word *chelate* is derived from the Greek word meaning "claw." It refers to a cyclic compound which is formed between an organic molecule and a metallic ion, the latter being held within the organic molecule as if by a claw. Examples of naturally occurring chelates are the cholorophylls, cytochromes (respiratory enzymes), hemoglobin, and vitamin B-12. The addition to foods and mineral supplements of a chelating agent, such as ethylenediaminetetracetic acid (EDTA), may in some cases improve the availability of the mineral elements.

CHELATION

The process of chelating—the formation of a bond between an organic molecule and a metallic ion.

CHEMICAL BONDING

The attachment of various chemical elements to form chemical compounds. The chemical bonds that hold the elements of a compound together consist of stored potential energy, which is released to do body work when the compound is broken into parts.

CHEMICAL REACTION

A chemical reaction that may occur by (1) combination, (2) replacement, (3) decomposition, or (4) some modification of the first three; the process may either require energy or yield it.

CHEMICAL SCORE

A means of evaluating proteins on the basis of their content of amino acids. The score is the percentage of the most limiting amino acid in the test protein compared to the amount in a standard reference protein such as whole egg. For example, in wheat protein, the most limiting amino acid is lysine. A chemical score of 60 would indicate that wheat protein contains only 60% as much lysine as whole egg protein.

(Also see PROTEIN, Section headed "Quantity and Quality.")

CHERRY *Prunus* spp.

Cherries are a small round fruit of certain species of trees belonging to the genus *Prunus*—a group which also includes the plum, peach, nectarine, apricot, and almond. These are all stone fruits. Several hundred varieties and several species of cherries grow in the United States; hence, there is a large variation in the fruits. Cherries of some trees are nearly black, while others are nearly white. The taste of cherries can be extremely bitter or very

Fig. C-78. Cherries ready for harvesting. Cherries are a varied fruit. They may be wild or cultivated, sweet or sour, and nearly black or yellowish-white in color. (Photo by J. C. Allen & Son, West Lafayette, Ind.)

sweet. Both wild and cultivated trees grow in almost all countries in the Temperate Zone.

Besides their fruit, cherry trees are esteemed for their beautiful, springtime clusters of small pink and white blossoms, and for their beautiful wood, particularly the wild black cherry tree. Some varieties are grown solely for ornamental purposes. Commercially, three types of cherry trees are important for the production of edible fruit.

KINDS OF CHERRIES. The three basic kinds of cultivated cherry trees are (1) sweet cherry, (2) sour cherry, and (3) the Duke. However, the sweet cherry and sour cherry are of prime importance in the United States. The Duke is a hybrid resulting from a cross between the sour and the sweet cherry.

- **Sweet cherry**—There are over 500 varieties of the sweet cherry, but only about 15 are important commercially. The scientific name for sweet cherry trees is *Prunus avium.* Some familiar varieties include Bing, Black Tartarian, Black Republican, Lambert, Schmidt, Windsor, and Napolean (Royal Ann). Sweet cherry trees require a mild climate. Both extremes in winter or summer harm trees. Cultivated trees are kept at heights of only about 15 to 20 ft (*4.6 to 6.1 m*). Sweet cherries require cross-pollination for fruit to set; hence, growers plant several different varieties in orchards.
- **Sour cherry**—Sometimes these are also referred to as tart cherries, pie cherries, or red cherries. There are about 270 varieties of sour cherries, but only a few are of commercial importance. The scientific name for the sour cherry is *Prunus cerasus.* They are smaller than sweet cherry trees and they are more tolerant of heat and cold. Sour cherry trees self-pollinate or cross pollinate with trees of the same variety. Montmorency, Richmond, and English Morello are the commercially important varieties.

ORIGIN AND HISTORY. The original home of the cherry is lost in antiquity. Some kinds are native to North America, while others may have originated from the region between the Black and the Caspian seas. Around 300 B.C., the Greek botanist, Theophrastus, described the cherry, though it was probably cultivated several centuries earlier. Then around 70 A.D. Pliny indicated that the cherry tree was in Rome, Germany, England, and France. The cherry was one of the first fruits brought to the New World by the colonists, and it moved west across the United States with the settlers. By 1847, cherries were being cultivated in Oregon. Now, some type of cherry is grown in every state in the United States proper.

PRODUCTION. The type of cherry produced depends upon the climate. Most of the world's cherries are sweet cherries, though a few sour cherries are grown in Canada, Austria, Germany, Greece, Yugoslavia, Turkey, and the United States. In overall yearly cherry production, the United States is among the leading producers, followed very closely by Germany, as shown in Fig. C-79.

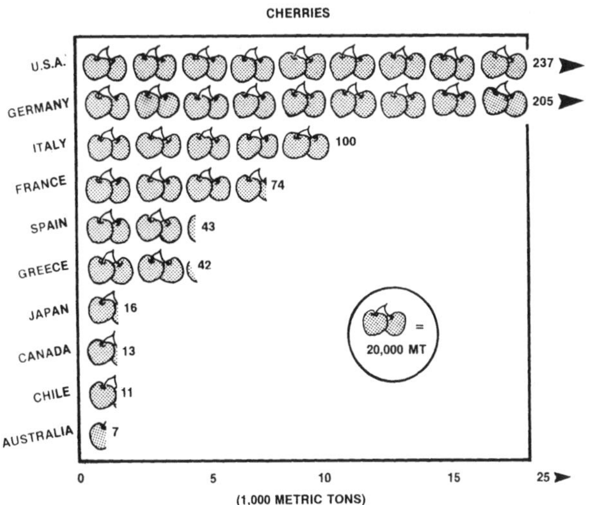

Fig. C-79. Some leading cherry-producing countries of the world. (Based on data from *Agricultural Statistics*, 1991, USDA, p.186, Table 273)

In the United States, most sweet cherries are raised along the eastern shore of Lake Michigan, in the Hudson Valley of New York, and on the Pacific Coast. Yearly production of sweet cherries in the United States averages 156,730 tons (*141,057 metric tons*), with a cash value of $140 million. States contributing to the yearly U.S. crop are shown in Fig. C-80.

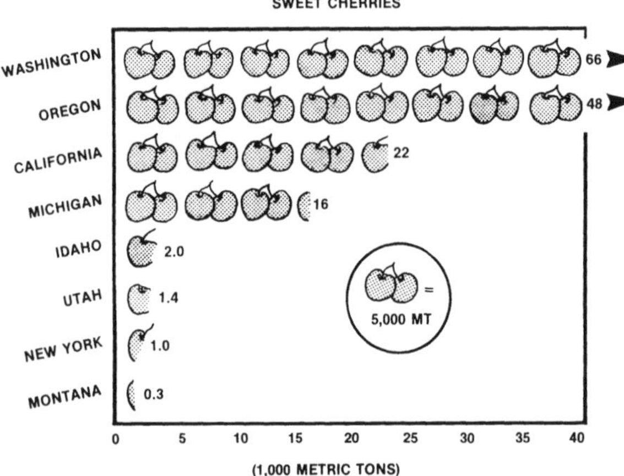

Fig. C-80. Leading cherry-producing states. (Based on data from *Agricultural Statistics*, 1991, USDA, p. 185, Table 272)

Only about 100,000 tons (*90,000 metric tons*) of sour cherries are produced each year in the United States. Michigan produces 77% of this crop, but, by rank, the following states also contribute: New York, Utah, Oregon, Wisconsin, Pennsylvania, and Colorado. The value of the sour cherry crop is estimated at $37 million.

Propagation And Growing. Both sweet and sour cherry trees are propagated by grafting buds to seedling rootstocks of the wild sweet cherry (mazzard) or of the perfumed cherry of southern Europe, *Prunus mahaleb*. When trees are 1 year old they are transplanted from the nursery to the orchard, about 15 to 30 ft (*4.6 to 9.2 m*) apart depending on the type.

Cherry trees will not tolerate wet soils. They do best in well-drained sandy loam soils. With proper pruning, fertilizing, and pest control, sweet cherries begin bearing some fruit after having been in the orchard 5 to 6 years while sour cherries often bear 1 to 2 years earlier. By the time an orchard is 10 years old, the trees are usually at full bearing capacity. Although some cherry trees occasionally live to 100 years, the profitable life-span of cherry trees is 20 to 30 years. Well-managed orchards can produce 3 to 5 tons per acre.

Harvesting. Cherries must be picked ripe, since they do not ripen after picking. The time of picking cherries varies with the area. In New York, the harvest begins around July 1 and continues for 4 to 6 weeks. Historically, cherry picking was a hand operation, which amounted to one of the most expensive operations associated with cherries. Today, much of the sour cherry crop is harvested mechanically—using tree shakers and catching frames. Sweet cherries are more difficult to harvest mechanically. Also, if the fruit is to be shipped and marketed, fresh cherries need to be hand picked with stems. For some varieties, mechanical harvesting may yield fruits with stems.

Fig. C-81. Upon reaching maturity, cherries must be harvested and processed quickly. (Courtesy, Agricultural Chemicals Division of Mobay Chemical Corporation, Kansas City, Mo.)

Harvested cherries were traditionally transferred in lugs, but now many are placed in chilled water in large tanks and then transported to the processing plant.

PROCESSING. In the United States, 54% of the sweet cherries are sold fresh, 10% frozen, 17% canned, and 30% brined (maraschino or confectionery use). Of the sour cherries, 3% are sold fresh, 35% are canned, 59% are frozen, and 3% are brined. Cherries sold fresh include cherries sold for home use.

At processing plants cherries are grated, sorted, and pitted before being canned, frozen, juiced, or brined. Some sour cherries are water-packed; others are canned as ready-to-use pie filling; still others are canned as cherry sauce and jellied cherry sauce containing spices. Sweet cherries are canned in syrup, with or without pits.

The pitted sour cherry is one of the most satisfactory fruits for freezing, with or without sugar. A few sour cherries are pressed fresh or frozen for juice extraction. Brined cherries are used for the production of maraschino cherries, and for candied and glaceed cherries. Brine contains sulfur dioxide to bleach cherries to a uniform white or yellowish-white, and lime or calcium salt to firm or toughen the fruit.

• **Maraschino**—Maraschino cherries are sweetened, dyed, given maraschino flavoring, packed and canned in syrup. The original maraschino flavoring was obtained from the fruits and leaves of Marasca cherry trees grown in Yugoslavia. However, today, most maraschino flavoring is made from bitter almond oil, neroli oil, and vanilla extract.

• **Candied and glaceed cherries**—The candying process consists essentially of slowly impregnating the fruit with syrup until the sugar concentration in the fruit is high enough to prevent growth of microorganisms. Following impregnation, the fruit is washed and dried and sold as candied fruit. For glaceed fruit, the washed and dried candied fruit is coated with a thin transparent layer of heavy syrup that dries to a more or less firm texture.

SELECTION. Bing, Black Tartarian, Schmidt, Chapman, and Republican varieties should range from deep maroon or mahogany red to black, for richest flavor. Lambert cherries should be dark red. Good cherries have bright, glossy, plump-looking surfaces and fresh-looking stems.

Overmature cherries lack flavor, condition indicated by shrivelling, dried stems, and a generally dull appearance. Decay is fairly common at times on sweet cherries, but because of the normal dark color, decayed areas are often inconspicuous. Soft, leaking flesh, brown discoloration and mold growth are all indications of decay.

PREPARATION. Served alone, sweet cherries, fresh or canned, are an excellent dessert fruit, or they may be used in sauces, fresh fruit dishes, gelatin desserts, and ice cream. Sour cherries are used primarily in cooked desserts. Cherry pie is an all-time favorite, but there are other baked goods requiring sour cherries. Maraschino cherries are used as an accent on other desserts or toppings and in cocktails. They are available with or without stems. Candied cherries are used in baked goods.

NUTRITIONAL VALUE. Fresh cherries are 80 to 84% water. Fresh sweet cherries contain 70 Calories (kcal) per 100 g (about 3 ½ oz), while fresh sour cherries contain 58 Calories (kcal) per 100 g. When either type is canned in water they contain fewer calories than the fresh cherries. When cherries are canned in a syrup, the calories per serving increase. Calories in the fresh cherries are derived primarily from the natural sugars (carbohydrate). Cherries contain small amounts of minerals and vitamins. However, sour cherries possess 1,000 IU of vitamin A while sweet cherries possess only 110 IU of this vitamin. Candied cherries contain only 12% water but they are packed with 339 Calories (kcal) per 100 g and little else. They are higher in calories due to the added sugar. More detailed information regarding the nutritional value of fresh or canned sour and sweet cherries, frozen sour cherries, candied cherries and maraschino cherries is provided in Food Composition Table F-36.

(Also see FRUIT[S], Table F-47 Fruits of the World.)

CHERRY, WEST INDIAN (BARBADOS CHERRY) *Malpighia glabra*

These are other names for the acerola berry.

(Also see ACEROLA; and FRUIT[S], Table F-47 Fruits Of The World—"Acerola.")

CHICK ANTIDERMATITIS FACTOR

This name was formerly given to pantothenic acid, one of the B-complex vitamins.

(Also see PANTOTHENIC ACID.)

CHICKPEA (GARBANZO BEAN) *Cicer arietinum*

Although the names chickpea and garbanzo bean are most widely used, the plant also has various local names, such as Bengal gram (in India) and Egyptian pea. Most Americans associate the use of the legume with people of Mediterranean ancestry, although it is now much more utilized on the Indian subcontinent where it serves as a major source of dietary protein. Fig. C-82 shows a portion of the growing plant.

The leaves of the chickpea are divided into very small leaflets, and the small round pods contain one or two seeds each.

Fig. C-82. Chickpea (garbanzo bean). (Courtesy, Minnesota Agricultural Experiment Station, University of Minnesota)

ORIGIN AND HISTORY. The first evidence of the utilization of chickpeas as food dates back to about 5000 B.C. in the region that is now Turkey. It appears to have been first cultivated in the Mediterranean between 5,000 and 6,000 years ago, and in India about 2000 B.C. The people in Ethiopia first grew the crop around 1000 B.C. This vegetable was also grown by the ancient Egyptians, Hebrews, Greeks, and Romans. (The Roman name Cicero was derived from the Latin name of the chickpea.)

Later, chickpeas were spread around the world by 16th century Spanish and Portuguese explorers, and by immigrants from India to other subtropical areas. Hence, they are now grown in the semidry subtropics of the Americas, Africa, and Asia.

PRODUCTION. Chickpeas are among the top dozen vegetable crops of the world; about 6.9 million metric tons are produced annually. Over 74% of the crop is produced in India and Turkey. Fig. C-83 shows the leading producers of chickpeas.

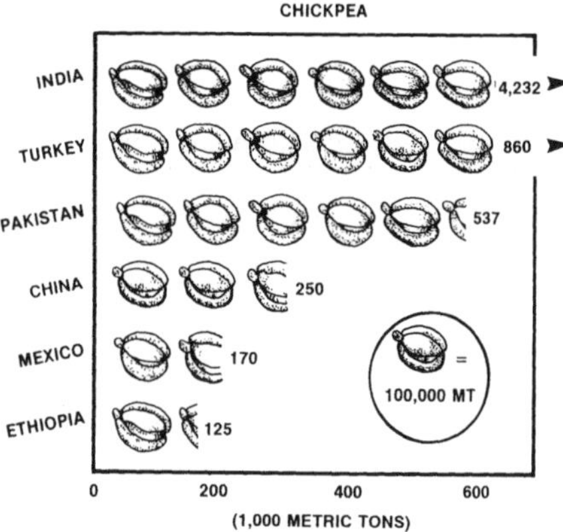

Fig. C-83. Leading chickpea-producing countries. (*FAO Production Yearbook*, 1990, FAO/UN, Rome, Italy, Vol. 44, p. 105, Table 35)

The United States now imports chickpeas, mostly from Mexico. Most of the commercial production in the United States, which is very limited, is in California. It is noteworthy, however, that there is both potential and interest in chickpea production in the Pacific Northwest at the present time.

This vegetable became one of the leading crops of India because (1) it tolerates drought well, and (2) many of the Indian people are vegetarians who eat no flesh foods. It is grown in subtropical areas as a winter crop and in cool temperate areas as a summer crop because cool nights with dew are required to produce an optimal yield. (The temperatures tolerated by the growing crop range from 36°F *[2°C]* to 86°F *[30°C]*.) Also, flowering is impaired by hot, wet weather. However, the crop may require irrigation in some of the dryer areas. Garbanzo beans are usually harvested from 4 to 6 months after sowing by cutting, drying, and threshing. Yields of dried seeds range from 400 to 1,600 lb per acre (*448 to 1,792 kg/ha*).

PROCESSING. The chickpeas produced in India are usually made into "dhals"; that is, seeds with the outer skins removed. They are also ground into a flour that has many uses in baking and cooking. Outside of India the legume is sold as dried seeds or as canned cooked seeds.

PREPARATION. There are many ways in which chickpeas are prepared in different parts of the world. However, most recipes start with the cooked legumes, which have an appealing, nutlike flavor. (Each pound *[0.45 kg]* of dried chickpeas yields a little over 6 cups of cooked legumes, or 1 cup *[240 ml]* of dried chickpeas yields about 2 ¾ cups of cooked legumes.) However, most supermarkets sell cans of cooked chickpeas. Nevetheless, it is more economical to buy dried chickpeas when they are available and to cook them at home. The preparation time may be shortened considerably by the procedure that follows:

- **One-hour hot soak**—A pound of dried legumes is placed in a sauce pan with 8 cups of water, heated to boiling and boiled for 2 minutes, then allowed to stand and soak without further heating for about an hour.
- **Pressure-cooking**—This method saves fuel because only about 30 minutes of heating are required. However, a tablespoon of fat or oil should be added to prevent foaming of the chickpeas and clogging of the vent of the cooker. (Three cups of water are added for each cup of the dried legume.) Furthermore, the pressure should be allowed to drop without either cooling the cooker or removing the counterweight from the vent.

A typical dish prepared from chickpeas is shown in Fig. C-84.

Fig. C-84. Chickpea and vegetable vinaigrette. The vegetables are marinated overnight in salad oil, lemon juice, vinegar, sugar, onion salt, and pepper; then served either in a salad, or used as snack food. (Courtesy, USDA)

Other ways of preparing and serving cooked chickpeas follow:

1. Some people make a dip for bread or crackers by mixing the mashed legume with other ingredients such as sesame seed paste, lemon juice, garlic, chopped peppers, and seasonings.
2. Cooks in the Middle East prepare *falafel*, which are fried balls made from coarsely chopped chickpeas, eggs, bread crumbs or flour, baking soda, chopped parsley, minced garlic, and various other seasonings.

CAUTION: Raw chickpeas, like the other major legumes, contain gas-producing substances that make it necessary to soak and cook them thoroughly. This means that flour should be prepared from cooked chickpeas which have been redried after cooking, rather than from raw dried legumes.

NUTRITIONAL VALUE. The nutrient composition of chickpeas is given in Food Composition Table F-36.

The data for uncooked chickpeas which is given in Food Composition Table F-36 may be converted to values for cooked legumes by dividing by 2.75 (for simplicity, 3 is close enough to be used as a divisor), since each cup (*240 ml*) of dried chickpeas yield 2 ¾ cups of cooked legumes. For example, a cup of the cooked product furnishes 262 Calories (kcal) and 15 g of protein, which is similar to the content of cooked navy beans. Other noteworthy observations follow:

1. A 3 ½ oz (*100 g*) portion (about ½ cup) of cooked chickpeas furnishes 7 g of protein, or about the same amount as 1 oz (*28 g*) of cooked lean beef. However, the legume furnishes about 2 ½ times as many calories as the beef.
2. Chickpeas contain less than half as much calcium as phosphorus, so other foods rich in calcium and lower in phosphorus (dairy products and green leafy vegetables) should be consumed in order to make certain that the calcium is well utilized.

Fig. C-85. Middle Eastern pocket bread stuffed with falafel balls.

Protein Quantity and Quality. Chickpeas are a dietary staple in India and Pakistan, where they are often referred to as Bengal gram. Therefore, certain facts regarding the quantity and quality of protein in this food are noteworthy.

1. Chickpeas supply 5.7 g of protein per 100 Calories (kcal), whereas animal foods such as cheeses, eggs, fish, meats, milk, and poultry contain an average of 9 g of protein per 100 Calories (kcal), and cereal grains about 3 g of protein per 100 Calories (kcal).
2. The protein in chickpeas is deficient in the amino acids methionine and cystine, but contains ample amounts of lysine which is deficient in the cereal grains. However, the latter contains ample methionine and cystine. Therefore, mixtures of chickpeas and grains yield a higher quality of protein than either food alone because the amino acid patterns of the two foods are complementary.

(Also see BEAN[S], Table B-10 Beans of the World; LEGUMES, Table L-2 Legumes of the World; and VEGETABLE[S], Table V-6 Vegetables of the World.)

CHICORY (BELGIUM ENDIVE; FRENCH ENDIVE; SUCCORY; WITLOOF) *Cichorium intybus*

This plant, which provides edible leaves and roots, is much more appreciated in Europe than in the United States. Chicory and endive are closely related and appear to have been developed from the same ancestral species. Both vegetables are members of the Sunflower family (*Compositae*), which also includes the artichoke, celery, dandelions, the Jerusalem artichoke, and lettuce. Just about everyone has seen the bright blue flowers of wild chicory because the plant frequently "escapes" from cultivation and grows as a weed that spreads readily.

The three major products obtained from various preflowering growth stages of chicory are (1) chicons, the blanched, elongated creamy-white heads of tightly folded leaves that are produced by "forcing" defoliated chicory rootstocks to grow new buds in a darkened environment; (2) leafy greens, which are usually grown from seeds; and (3) roots, the item that is usually roasted and ground to make a coffee extender or substitute.

Fig. C-86. Wild chicory, an attractive weedy plant that has been domesticated to yield three different vegetable products.

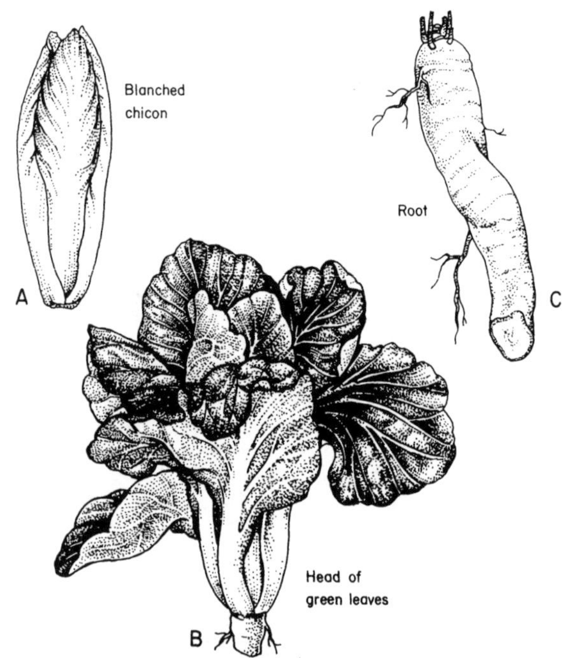

Fig. C-87. The three major food products obtained from chicory (A) chicons, (B) head of green leaves, and (C) roots.

ORIGIN AND HISTORY. Chicory is native to the Mediterranean region. It is believed to have been used since ancient Greek and Roman times because the Roman naturalist Pliny the Elder (23-79 A.D.) described the curative powers of the plant and its value as a food. However, the ancient peoples probably gathered wild chicory which grew in abundance along roadsides and as a weed in cultivated fields.

The time and place where chicory was first cultivated is not known for certain, since there is no mention of growing this vegetable until 1616, when its production in Germany was described. Later, it was grown in other European countries. The immigrants to the United States who brought chicory with them to grow in their gardens were responsible for the wide distribution of the plant as a weed in this country.

Production of chicory was stepped up after an unknown, but enterprising, person discovered that the roasted and ground roots of the plant could be used as an extender for coffee. The French brought the practice of mixing chicory with coffee to Louisiana, where it became a tradition in Creole cookery. This type of coffee mixture is now marketed throughout the United States, but it is most popular in the South.

The technique for producing the tender, mild-flavored chicons appears to have been discovered by Flemish gardeners in the mid-1800s. Hence, this form of chicory is now designated as Belgium endive, French endive, or Witloof chicory. (The French became the leading producers of chicons, and the Witloof variety of chicory is the major one that is used.) Recently, new labor-saving techniques have been developed for the production of chicons, among which is hydroponic culture (the production of plants on soilless media that are irrigated with nutrient solutions).

PRODUCTION. Most of the world's chicory production, which has amounted to approximately one-half million metric tons annually in recent years, is produced in Italy, France, Belgium, and Holland. About 40% of the crop is comprised of the greens that are grown mainly in Italy, whereas the other 60% consists of the yellow-white chicons from the other three countries. Small amounts of chicory are also grown in the United States and elsewhere.

The cultural practices used to produce each of the three forms of the vegetable differ in various aspects; hence, they are discussed separately.

• **Chicons**—The first stage of chicon production consists of the growing of clean, well-shaped roots on light, fertile soils that are free of rocks. Seed is sown in the spring when the danger of extended periods of temperatures below 50°F (*10°C*) has passed, because low temperatures may trigger bolting (the premature production of seedstalks). Generous amounts of phosphorus and potassium are added to the soil, but the application of nitrogen is limited to just what is needed in order to avoid the growth of luxuriant foliage at the expense of the roots. Normally, rainfall is adequate and irrigation is unnecessary in the major areas of chicon production. After several weeks of growth, the plants are thinned to provide sufficient space for the development of large roots. Harvesting of the roots is done in the fall, after about 4 months of growth. Then, the freshly harvested roots are stored outdoors for one or more months so that the cold weather may stimulate the metabolic processes which render the roots more suitable for chicon production. (Chicory roots contain stored carbohydrate in the form of inulin. Exposure to low temperatures activates the enzyme inulase, which converts inulin to the more readily utilized sugar fructose.)

During the second stage of chicon production the roots are trimmed of unwanted foliage and extraneous branches prior to their placement in specially prepared forcing beds. (Forcing is the stimulation of plant growth by artificial heating and other special conditions.) Light is excluded, the temperature is maintained at about 60°F (*16°C*) by heating coils or other means, and the beds are watered when necessary. The chicons are usually large enough for harvesting after 3 to 4 weeks.

• **Greens**—An early crop of chicory greens that is grown in a cool climate may be started from seed in a greenhouse or in hot beds, then transplanted in a field when there is no longer any danger of a hard frost. A later crop or one grown in a warmer climate may be planted directly in a field. The soil should be fertilized with plenty of nitrogen, phosphorus, and potassium. Irrigation should be provided if the rainfall is inadequate.

Normally, the plants are thinned after several weeks of growth. The unblanched leaves may be harvested about 2 months after planting, or they may be allowed to grow to a length of about 10 in. (*25 cm*), when the outer leaves are usually tied together at their tops in order to blanch the inner leaves. Blanching lightens the color and makes the leaves taste milder than unblanched leaves.

• **Roots**—The seeds of special large-rooted varieties of chicory are planted in the field at the usual time of the last frost. The soil is fertilized with ample nitrogen, phosphorus, and potassium, and the rainfall is supplemented by irrigation when necessary. Once the plants are well established, they are thinned to insure that there will be adequate space for the development of large roots. About 4 months are required from the planting of seed to the harvesting of the roots. However, roots should always be dug before the first frost because chilling brings about the conversion of the stored carbohydrate to sugar and alters their physical and chemical characteristics.

PROCESSING. Chicons and chicory greens are marketed as fresh vegetables. However, chicory roots may be marketed either fresh or after conversion into a coffee substitute by cleaning, peeling, dicing, roasting, and grinding.

SELECTION. Selection emphasis is determined primarily by intended use, whether the plant part is to be used for: (1) chicon—the forced top for salads; (2) greens; or (3) roots as a coffee supplement or substitute.

- **Chicons**—The blanched chicons (heads) should be creamy white, elongated, and about 2 in. (*5 cm*) in diameter and 5 to 7 in. (*12 to 18 cm*) long.
- **Greens**—The greens should be crisp, fresh, and tender. Tough, coarse-leaved plants are undesirable because they are apt to be bitter. Toughness or tenderness can be determined by breaking or twisting a leaf. Decay, which is shown by a browning of the leaves or a shiny appearance, should be avoided.
- **Roots**—The roots should be long, fleshy, and milky.

PREPARATION. There are some differences in the ways that each of the forms of chicory are prepared. Hence, they are discussed individually.

- **Chicons**—These creamy-white heads are most commonly cut up and served in salads, with care taken not to use excessive seasonings that might mask their delicate flavor. They may also be cut in half lengthwise and braised briefly in hot fat or oil.
- **Greens**—The mild-flavored blanched greens are often used as the major raw vegetable in salads, whereas the stronger-flavored unblanched greens may be mixed with other more bland-flavored greens such as lettuce. Both types of chicory leaves may also be added to soups and stews, braised in fat or oil, or simmered in a broth. It is noteworthy that the fresh greens deteriorate rapidly. Hence, they should be used within a day or so after they are purchased or harvested from the garden.
- **Roots**—This part of the plant is usually cut into pieces, roasted carefully, ground, and used as a coffee extender or replacement. Pure-food laws in the United States forbid the mixing of ground chicory root with coffee unless the label is plainly marked. Chicory can be detected in coffee by putting a spoonful of the mixture into a glass of cold water; the coffee will float on the surface, whereas the chicory will separate from the coffee and color the water. Some persons prefer coffee that is flavored with chicory.

The roasted root powder may also be used to bring out the flavor of various cocoa preparations.

Sometimes the roots are used as a cooked starchy vegetable. For this purpose, they may be boiled or sauteed, then served with lemon juice and a salad oil.

NUTRITIONAL VALUE. The nutrient compositions of chicons (Witloof variety) and chicory greens are given in Food Composition Table F-36.

Some noteworthy observations regarding the nutrient composition of chicons and chicory greens follow:

1. Both chicons and chicory greens have a high water content (95 and 93%, respectively) and a low calorie content (15 to 20 kcal per 100 g, respectively). However, the greens are much richer in nutrients than the chicons, the former being an excellent source of potassium and vitamin A, and a moderately good source of calcium and vitamin C.
2. Herbalists have long recommended chicory leaves for stimulating the flow of digestive juices and the movements of the bowels.

NOTE: The authors are not aware of any recent research that confirms or contradicts these beliefs. Hence, the ideas are presented for the purpose of encouraging some studies in this area.

(Also see VEGETABLE[S], Table V-6 Vegetables of the World.)

CHIEF CELLS

Cells in the stomach lining which secrete (1) pepsinogen, an enzyme that begins the digestion of protein; and (2) lipase, an enzyme that splits fatty acids from fat. These enzymes only start the digestive process which is finished by the pancreatic enzymes secreted into the small intestine. However, the partially digested products from the stomach trigger the release of hormones from the small intestine, which (1) stimulate the pancreas to secrete digestive juices, (2) slow the rate at which the stomach empties its contents into the intestines, and (3) cause the gallbladder to expel its bile.

(Also see DIGESTION AND ABSORPTION; and ENZYME.)

CHILDHOOD AND ADOLESCENT NUTRITION

Contents *Page*

History ... 394
Childhood Growth and Its Measurements ... 395
Growth and Development in Pubescence and Adolescence ... 397
Nutrient Allowances ... 398
Dietary Guidelines ... 399
Selection of Foods ... 399
Nutritional Supplements ... 400
School Lunch and Breakfast Programs ... 400
Common Nutrition-Related Problems ... 400
Adolescent Pregnancy ... 400
Anemia ... 400
Anorexia Nervosa ... 400
Constipation ... 400
Diarrhea ... 401
Food Preferences and Aversions ... 401
Gastrointestinal Disorders ... 401
Overweight and/or Obesity ... 401
Skin Blemishes ... 402
Tooth Decay ... 402
Tooth Decay Preventative Measures ... 402
Underweight ... 403
Modification of Children's and Adolescent's Diets ... 403

Fig. C-88. Growing children need a highly nutritious diet for proper growth and development. (Courtesy, Gerber Products Company)

The early nutrition of children has major effects on the physical and mental well-being in later years. Furthermore, each young person develops important habits of food selection and consumption as he or she progresses from infancy to adolescence and adulthood. Therefore, nutrition during both childhood and adolescence is covered in this article since these stages of physical, mental, and emotional growth usually occur within the same family setting.

HISTORY. Even today children in some parts of the world live precarious existences because their nutritional needs are high in relation to their size, yet they often receive only a meager share of the available food. In many societies, the male heads of families receive the largest shares. This situation has existed for countless centuries, as evidenced by the disproportionate number of skeletons of children aged 14 years and younger which have been found at various archaeological sites that were occupied by prehistoric peoples. However, not all primitive societies fared so poorly with their children since there is much evidence that well fed, healthy young people have had the innate biological capacities to adapt to a wide range of environments, ranging from the frigid arctic region to the hot, humid tropics or the dusty, dry deserts.

Fig. C-89. It takes a lot of good nutrition for a little girl to grow into a healthy, attractive young woman. (Courtesy, USDA)

The children of the nomadic Eskimos of the Canadian Arctic once had an excellent diet because their mothers breast fed them for 2 to 3 years, and gave them supplements of raw meat and fish.[17] However, the diets of Eskimo children have recently changed rapidly for the worse as a result of the establishment of military bases and trading posts throughout the region. Now, fewer children are breast fed because nursing bottles, evaporated milk, and similar items can be obtained from nearby trading posts. Breast feeding tends to increase birth spacing, and possibly its decline led to an over 50% rise in the birth rate among the Eskimos between the mid-1950s and the mid-1960s. Furthermore, the children now consume large quantities of candy and carbonated soft drinks, which are thought to be mainly responsible for the high incidence of acne and tooth decay. This high consumption of refined carbohydrates by many Eskimos appears to be undesirable for them because (1) their bodies do not utilize the sugars as well as white people; (2) the incidence of calcification of the leg arteries is much higher in those who live in settlements than in those who are nomadic hunters; and (3) of the increased incidence of gallbladder disease. High levels of blood fats and obesity are becoming increasingly prevalent among the sugar eaters.

The diets of children in the United States, in contrast to those of the Eskimo children, have apparently improved over the past century as evidenced by the fact that the average height of 10-year-old children increased.[18] Furthermore, there has been a continuous decline in childhood mortality since 1925. Although there was a slight increase in 1989. This trend is shown in Fig. C-90.

Some authorities on childhood and adolescent growth have questioned whether feeding children nutrient-rich diets to promote rapid growth is as desirable as it seems, since the early attainment of a large body mass is suspected of predisposing people to the premature development of degenerative diseases. Although the total death rate for people aged 55 to 64 years has dropped from 22/100,000 in 1940 to 12.4/100,000 in 1989, heart disease and cancer deaths have both increased.[19] The most notable changes in the American diet in the 20th century have been (1) an 80% increase in the use of sugar and related sweeteners, and (2) a 27% increase

[17]Schaeffer, O., "When the Eskimo Comes to Town," *Nutrition Today*, Vol. 6, November-December 1971, pp. 8-16.

[18]*Health, United States 1975*, U.S. Dept, HEW.

[19]*Statistical Abstract of the United States*, 1949 and 1991, U.S. Department of Commerce.

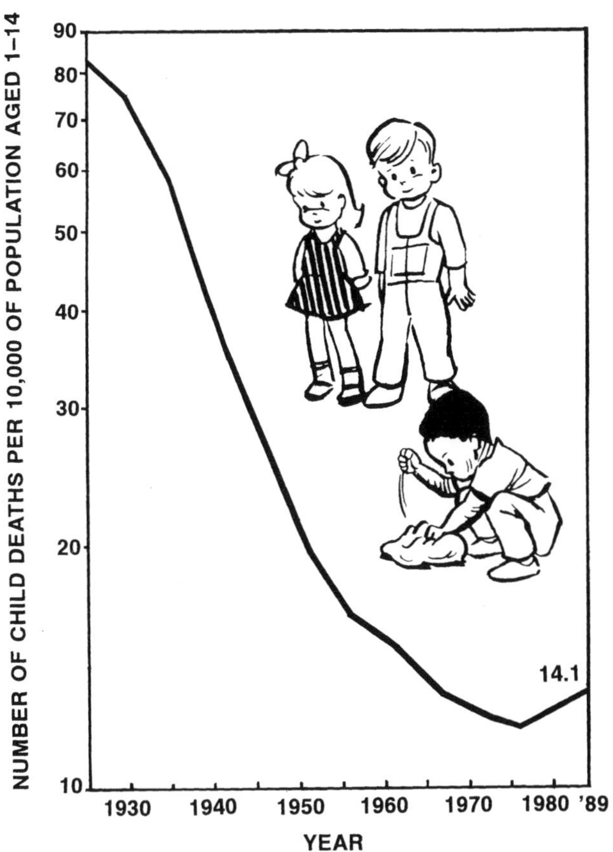

Fig. C-90. Childhood mortality in the United States from 1925 to 1989. (Source: *Health, United States, 1975*, U.S. Dept. HEW, pp. 359, 361, Tables CD. II.8, CD II.9 and *Statistical Abstract of the United States 1991*, p. 75, Table 109)

in fat consumption. These changes did not raise the caloric content of the average diet because the use of starchy foods declined sharply. However, it seems likely that the great reductions in strenuous physical activities since the early 1900s may account in part for the current high mortality from certain degenerative diseases which are more likley to afflict the obese than the nonobese.

CHILDHOOD GROWTH AND ITS MEASUREMENT. Some parents may be dismayed when the appetite of their toddler declines considerably from what it was during infancy. However, it is noteworthy that the average rates of growth for children during their second and third years are only about one-third of the rate for the first year of life, when the birth weight is tripled. For example, the average gain in weight from birth to age 1 is between 14 and 15 lb (*6.5 to 7.0 kg*), whereas only 10 lb (*4.5 kg*) are gained between ages 1 and 3. Standard growth curves for children from birth to 36 months of age are given elsewhere.

(Also see INFANT DIET AND NUTRITION.)

Standard growth curves for girls and boys aged 2 to 18 years are presented in Figs. C-91 through C-96.

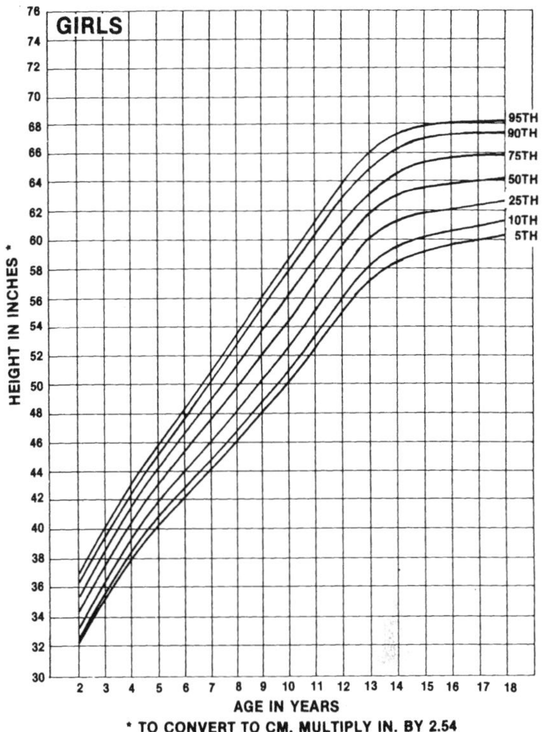

Fig. C-91. Heights of girls by age percentiles from 2 to 18 years. (Source: *NCHS Growth Curves for Children, Birth–18 Years, United States*, U.S. Dept. HEW)

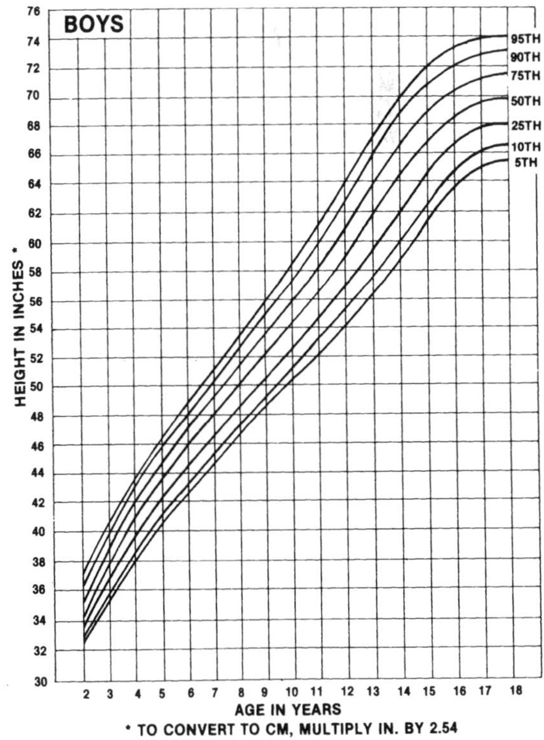

Fig. C-92. Heights of boys by age percentiles from 2 to 18 years. (Source: *NCHS Growth Curves for Children, Birth–18 Years, United States*, U.S. Dept. HEW)

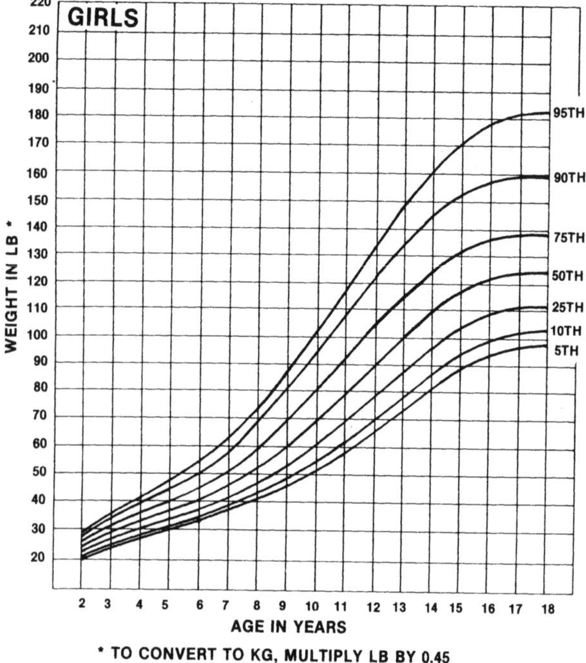

Fig. C-93. Weights of girls by age percentiles from 2 to 18 years. (Source: *NCHS Growth Curves for Children, Birth–18 Years, United States*, U.S. Dept. HEW)

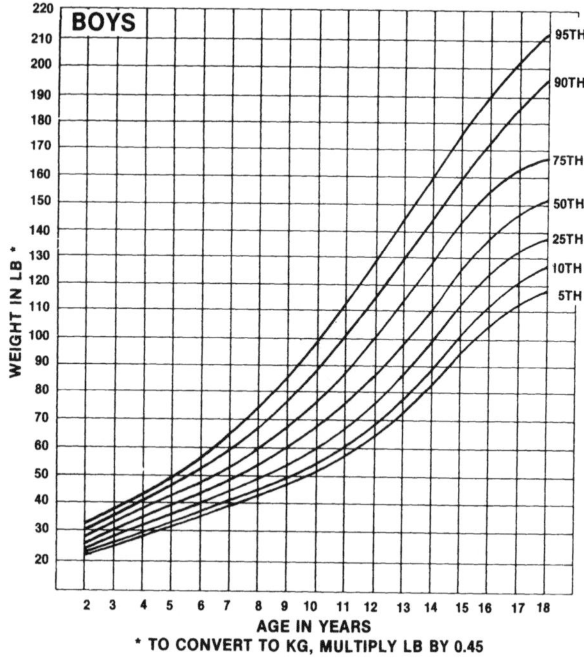

Fig. C-94. Weights of boys by age percentiles from 2 to 18 years. (Source: *NCHS Growth Curves for Children, Birth–18 Years, United States*, U.S. Dept. HEW)

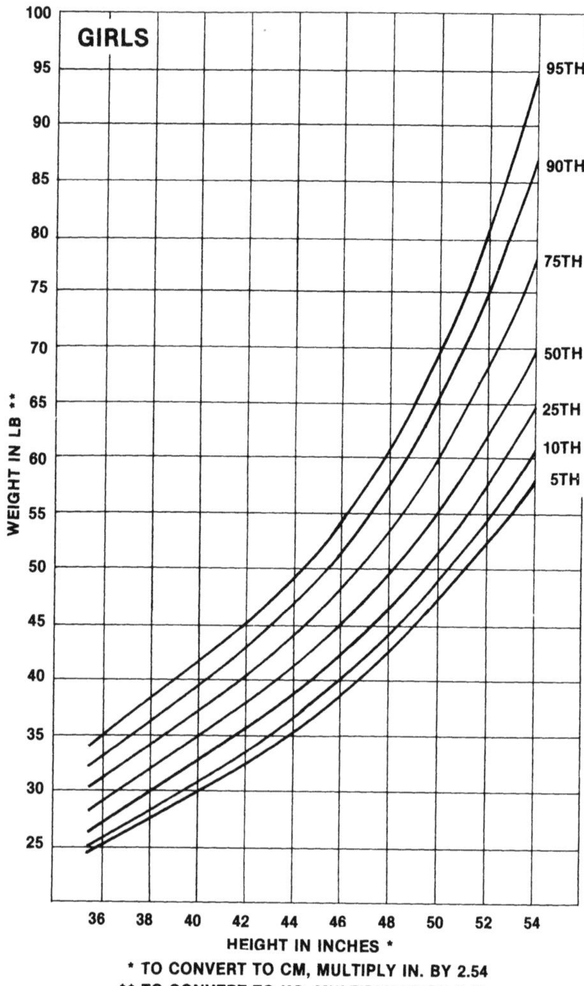

Fig. C-95. Weight by height percentiles for prepubescent girls. (Source: *NCHS Growth Curves for Children, Birth–18 Years, United States*, U.S. Dept. HEW)

To use the growth curves shown in Figs. C-91 through C-96, it is necessary to determine the percentiles which correspond most closely to the data from the girl or boy that was measured. A simple procedure follows:

1. Locate and mark the height or weight of the child on the vertical scale in the left margin of the appropriate chart. If the value of the measurement falls between the values marked on the scale, its placement should be estimated as accurately as possible.
2. With the aid of a ruler draw a light, horizontal line across the chart, starting from the marked value for height or weight.
3. Using a procedure similar to that in Step I, locate and mark the age of the child on the horizontal scale at the bottom of the chart.
4. Draw a vertical line on the chart, starting from the marked value for age.
5. Circle the point at which the horizontal and vertical lines intersect on the chart and note the percentile curve which is closest to the intersection.

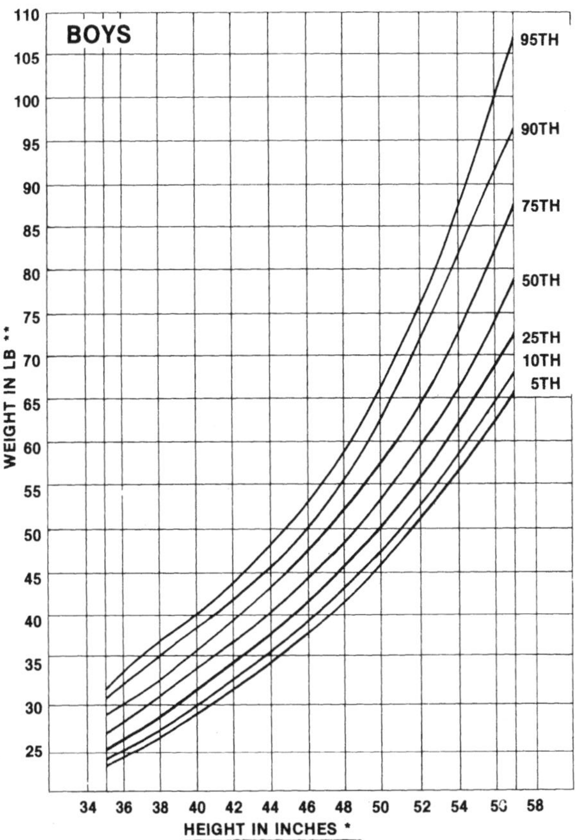

Fig. C-96. Weight by height percentiles for prepubescent boys. (Source: *NCHS Growth Curves for Children, Birth–18 Years, United States*, U.S. Dept. HEW)

Values which fall between the 5th and 95th percentiles are considered to be within the normal range. However, there should not be a wide discrepancy between the percentiles for height and weight. Sometimes, a pediatrician will not diagnose and treat an otherwise healthy girl or boy until a clear-cut trend of growth abnormality is indicated by measurements taken at two or more consecutive monthly or bimonthly visits. Some typical interpretations of deviate growth measurements follow:

• **Short height and/or low weight for age**—When either height or weight falls below the 5th percentile for the age of the child there may be undernutrition or other conditions which impair growth. Impaired growth may be diagnosed by x rays of the bones. However, subnormal values may also indicate a slow rate of development due to (1) heredity (one or both parents may be small), or (2) past conditions that are no longer present. Therefore, it is important to consider the long-term record of growth since the boy or girl may be growing normally at the time of measurement. Furthermore, some smaller-than-average children "catch up" with their peers during the growth spurts of pubescence and adolescence.

• **Low weight for height**—When the weight for height is less than the 5th percentile the child is likely to be excessively thin as a result of (1) an insufficient caloric intake, or (2) a nutritional deficiency, an illness, or other condition that interferes with the utilization of nutrients. Hence, it is desirable for the physician, nutritionist, social worker, and/or other child care professionals to review the medical and social background of the child and her or his family. For example, chronic conditions such as heart murmurs or digestive disorders may require different dietary therapy than uncomplicated undernutrition.

• **Above normal height and/or weight for age**—Height or weight that exceeds the 95th percentile may indicate an excessively rapid growth due to overconsumption of protein and/or calories. It is suspected that certain types of high-carbohydrate, high-protein diets may overstimulate the secretion of growth hormone, which may in some cases be detrimental to a child. There is evidence that the development of atherosclerosis and diabetes may be hastened by the oversecretion of growth hormone in susceptible people. On the other hand, some children grow large at an earlier age than normal, then stop growing at a younger age than is common for those of the same sex.

• **Excess weight for height**—In the case of children whose weight for height is greater than the 95th percentile it is likely that excessive amounts of body fat have accumulated. Nevertheless, some pediatricians will wish to confirm the presence of obesity by a physical exam that may include the measurements of skinfolds on the triceps and under the shoulder blades. It is noteworthy that weight alone is a poor indicator of body composition, since each pound of body protein in muscle is accompanied by 3 to 4 lb (*1.35 to 1.8 kg*) of water, whereas fat is not accompanied by water. Hence, a heavily muscled child may be overweight, but not obese.

GROWTH AND DEVELOPMENT IN PUBESCENCE AND ADOLESCENCE. The period of growth and development between childhood and adulthood is commonly called adolescence, although it is more accurate to divide this interval into two phases that are referred to as pubescence and adolescence.

Pubescence, which occurs first and usually lasts for 2 to 3 years, begins with the initial appearance of secondary sex characteristics such as pubic hair and ends with the attainment of reproductive capability. This phase begins around 10 to 11 years of age in girls, and at about 12 to 13 years of age in boys, although it may occur one or more years earlier or later in early-maturing or late-maturing children. It is noteworthy that the maximum rate of growth of pubescent girls is about 3 in (*8 cm*) in a single year, whereas boys may grow as much as 4 in. (*10 cm*) during the growth spurt.

Adolescence extends from the end of pubescence to the completion of physical growth. Growth in height is usually completed between 16 and 18 years of age in girls, and around 18 to 21 in boys, although some people continue to grow in height up to age 30. During adolescence, females gain a considerable amount of body fat, whereas males gain less fat, but more muscle. Accumulation of fat in adolescent females serves the purpose of supplying extra calories to meet the high demands of pregnancy and lactation. Hence, many adolescent girls may begin to consider themselves too fat, and attempt to lose weight by going on stringent diets. On the other hand, adolescent boys often consider themselves too slender, and may gorge themselves with food to become heavier.

The growth curves shown in Figs. C-91 through C-96 provide some general guidelines for evaluating the progress of girls and boys during pubescence and adolescence. However, the ranges of normal values for height and weight grow wider with age, with the result that it become increasingly difficult to detect deviations from normal. Sometimes, doctors or other health professionals will assess excessive fatness or slenderness by skinfold measurements, since weight gain alone gives little indication of the gain in fat, bone, and muscle. Furthermore, pediatricians can usually judge the growth and development status of adolescents better than most other people because they are experienced in observing the sequence of physical changes during pubescence and adolescence. The growth spurts which are accompanied by significant gains in weight often occur in conjunction with certain stages of sexual development.

NUTRIENT ALLOWANCES. Growing children and adolescents need a highly nutritious diet to meet the requirements for the growth and development of their body tissues. For example, dietary protein should supply from 8 to 10% of the calories, depending upon the rate of growth. Poor diets during these states of life may have consequences that cannot be overcome later. Thus, a poorly nourished female during childhood and adolescence may bear weakened babies and have a gradual deterioration in health during childbearing. A male may be so physically and/or mentally handicapped from a poor diet in childhood and adolescence that he makes a poor breadwinner, husband, and father.

The recommended nutrient allowances of children and adolescents are given in Table C-23.

Meeting the allowances listed in Table C-23 requires the consumption of foods that are rich in protein, minerals, and vitamins in proportion to the calories that are provided. However, it is not expected that many parents will attempt to check the composition of their children's diets in detail; rather, they will teach and guide them to make intelligent food choices.

The nutrient allowances by age groups, including children and adolescents, along with nutrient functions

TABLE C-23
RECOMMENDED DAILY NUTRIENT ALLOWANCES FOR CHILDREN AND ADOLESCENTS[1]

Category	Children			Males		Females	
	1–3 yrs	4–6 yrs	7–10 yrs	11–14 yrs	15–18 yrs	11–14 yrs	15–18 yrs
Weight . . . . . . lb (*kg*)	29 (*13*)	44 (*20*)	62 (*28*)	99 (*45*)	145 (*66*)	101 (*46*)	120 (*55*)
Height . . . . . in. (*cm*)	35 (*90*)	44 (*112*)	52 (*132*)	62 (*157*)	69 (*176*)	62 (*157*)	64 (*163*)
Energy . . . . kcal	1300	1800	2000	2500	3000	2200	2200
Protein . . . . g	16	24	28	45	59	46	44
Minerals							
Calcium . . . . . . mg	800	800	800	1200	1200	1200	1200
Phosphorus . . . . mg	800	800	800	1200	1200	1200	1200
Sodium . . . . . . mg	225	300	400	500	500	500	500
Chloride . . . . . . mg	350	500	600	750	750	750	750
Magnesium . . . . mg	80	120	170	270	400	280	300
Potassium . . . . mg	1000	1400	1600	2000	2000	2000	2000
Chromium[2,3] . . . mcg	20–80	30–120	50–200	50–200	50–200	50–200	50–200
Copper[2,3] . . . . . mg	0.7–1.0	1.0–1.5	1.0–2.0	1.5–2.5	1.5–2.5	1.5–2.5	1.5–2.5
Fluoride[2,3] . . . . . mg	0.5–1.5	1.0–2.5	1.5–2.5	1.5–2.5	1.5–2.5	1.5–2.5	1.5–2.5
Iodine . . . . mcg	70	90	120	150	150	150	150
Iron . . . . mg	10	10	10	12	12	15	15
Manganese[2,3] . . . mg	1.0–1.5	1.5–2.0	2.0–3.0	2.5–5.0	2.5–5.0	2.5–5.0	2.5–5.0
Molybdenum[2,3] . . mcg	25–50	30–75	50–100	75–250	75–250	75–250	75–250
Selenium . . . . . mcg	20	20	30	40	50	45	50
Zinc . . . . mg	10	10	10	15	15	15	15
Fat-Soluble Vitamins							
Vitamin A . . . mcg RE	400	500	700	1000	1000	800	800
Vitamin D . . . . . mcg	10	10	10	10	10	10	10
Vitamin E . . . mg α TE	6	7	7	10	10	8	8
Vitamin K . . . . . mcg	15	20	30	45	65	45	55
Water-Soluble Vitamins							
Biotin . . . . mcg	20	25	30	30–100	30–100	30–100	30–100
Folate . . . . mcg	50	75	100	150	200	150	180
Niacin . . . . . mg NE	9	12	13	17	20	15	15
Pantothenic acid . mg	3	3–4[2]	4–5[2]	4–7[2]	4–7[2]	4–7[2]	4–7[2]
Riboflavin . . . . . mg	0.8	1.1	1.2	1.5	1.8	1.3	1.3
Thiamin . . . . . . mg	0.7	0.9	1.0	1.3	1.5	1.3	1.3
Vitamin B-6 . . . . mg	1	1.1	1.4	1.7	2	2.0	2.0
Vitamin B-12 . . . mcg	0.7	1.4	1.4	2	2	2.0	2.0
Vitamin C . . . . . mg	40	45	45	50	60	50	60

[1]Source: *Recommended Dietary Allowances*, 10th ed. 1989, NRC–National Academy of Sciences.

[2]These figures are given in the form of ranges of recommended intakes because there is less information on which to base allowances.

[3]The upper values of the range should not be habitually exceeded since the toxic levels for many trace elements may be only several times the usual intakes.

and best food sources of each nutrient, are more fully covered in this book in the section on Nutrients: Requirements, Allowances, Functions, Sources, and in its accompanying Table N-6; hence, the reader is referred thereto.

DIETARY GUIDELINES. Many people are reluctant to accept the fact that too much of a good thing may be detrimental. In this case, the "good thing" is the nutrient-rich foods that may produce dietary imbalances when consumed in excessive amounts. For example, eggs and liver are rich in protein, nucleic acids, and cholesterol. Some of the nutrients that are provided abundantly by these two foods serve important functions in the body, but they may also contribute to ill effects in people who are susceptible to certain disorders.

- **Protein**—Overconsumption of this nutrient may tax the ability of the body to utilize and dispose of the flood of amino acids provided. Furthermore, the brain levels of the essential neurotransmitter serotonin may drop significantly when the passage of the amino acid tryptophan into the brain is restricted by large quantities of other amino acids. Serotonin is produced from tryptophan. Research studies have shown that depression and compulsive sexual activity may be due to lack of sufficient serotonin in the brain.

- **Nucleic acids**—Some people develop dangerously high blood levels of uric acid when they consume too much nucleic-acid rich foods, and occasionally gout results from this condition.

- **Cholesterol**—This substance may contribute to atherosclerosis if consumed in excessive amounts by people who tend to have high blood levels of lipids.

Therefore, nutrient-rich foods should be selected cautiously, so that excesses of certain substances are avoided.

Selection Of Foods. Parents should help their children to make wise food choices as soon as possible, because there are many influences that encourage bad eating habits. Furthermore, most children, adolescents, and adults lead sedentary lives and can spare little room in their diets for "empty calories" without risking the development of obesity. (The term "empty calories" refers to the food energy that is obtained from highly refined foods that are rich in carbohydrates and fats and furnish little else in the way of nutrients; for example, sugar.) Table C-24 gives some guidelines for daily food patterns.

(Also see FOOD GROUPS.)

TABLE C-24
DAILY FOOD GUIDE FOR CHILDREN AND ADOLESCENTS[1]

Type of Food		Each Day
Meat and Meat Substitutes		2 or more servings
Include—		
Meat, poultry, fish, shellfish, or eggs.		
Dried beans or peas, peanut butter, and nuts can be used as meat substitutes.		
Milk Group		
Milk (fluid whole), evaporated (diluted 50:50 with water), skim, buttermilk, nonfat dry milk	Children under 9	2 to 3 cups[2]
	Children 10 to 12	3 or more cups[2]
	Teenagers	4 or more cups[2]
Dairy products such as:		
Cheddar cheese, cottage cheese, ice cream, and yogurt		May sometimes be used in place of milk
Vegetable-Fruit Group		5 or more servings
Include—		
A fruit or vegetable that contains a high amount of vitamin C: grapefruit, oranges, cantaloupe, and tomatoes (whole or in juice), raw cabbage, green or sweet red pepper, broccoli, and fresh strawberries.		
A dark green or deep yellow vegetable or a yellow fruit for vitamin A: broccoli, spinach, greens, cantaloupe, apricots, carrots, pumpkin, sweet potatoes, and winter squash.		
Other vegetables and fruits, including potatoes.		
Breads and Cereals		6 or more servings
Whole grain, enriched, or restored bread and cereals or other grain products such as cornmeal, grits, macaroni, noodles, spaghetti, and rice.		
Plus Other Foods		
To round out meals and to satisfy the appetite, many children will eat more of these foods, and other foods not specified will be used, such as butter, margarine, other fats, oils, sugars, and unenriched refined grain products. These "other" foods are frequently combined with the suggested foods in mixed dishes, baked goods, desserts, and other recipe dishes. They are a part of daily meals, even though they are not stressed in the food plan.		

[1]Adapted by the authors from *Your Child from 6 to 12,* Children's Bureau, U.S. Dept. HEW, Pub. No. 324.
[2]Or, equivalent in nonfat dry milk (usually, about 1/4 cup of dry milk powder is equivalent to a cup of fluid milk). One cup equals 240 ml.

Nutritional Supplements. Although it is possible for many children and adolescents to obtain the nutrients they need from the foods they ordinarily consume, there are a number of situations in which nutritional supplements may be needed. Some examples follow:

• **Modified diets**—Diets for weight reduction and/or the correction of certain metabolic disorders may involve the restriction of certain foods that would otherwise provide essential nutrients. For example, reduction in caloric intake often means that the consumption of items rich in energy and essential nutrients is curtailed drastically. Furthermore, the restriction of fat and cholesterol intake to treat high blood lipids may eliminate egg yolks and other dietary sources of minerals and vitamins.

• **Food allergies**—Unfortunately, some of the common allergy-provoking foods are also good sources of nutrients.

• **Nutritional deficiencies**—The correction of these conditions may require greater amounts of certain nutrients than may be provided by ordinary foods. For example, the treatment of iron deficiency anemia often requires an iron supplement.

• **Use of oral contraceptives**—Special supplements are now available for girls and women who use these agents, which may increase the needs for certain nutrients.

Therefore, it may be advisable for some children and adolescents to take nutritional supplements that have been recommended by a doctor.

SCHOOL LUNCH AND BREAKFAST PROGRAMS.

This refers to the noon and morning meals served to children in schools.

The School Lunch program was initiated in the 1930s, followed by the addition of the School Breakfast program in 1966. The program expanded rapidly in response to (1) more children being transported to schools greater distances from their homes, (2) more mothers working away from home, (3) the desire of people and government to provide improved nutrition to the children of the poor, and (4) low-cost, subsidized meals.

A good understanding of the School Lunch Program is important to parents and teachers who influence the child's eating habits.

(Also see SCHOOL LUNCH PROGRAM.)

COMMON NUTRITION-RELATED PROBLEMS.

Most of these problems may be dealt with successfully if the appropriate dietary measures are used without undue delay. If this is done, then only minor changes need be made in the diet of the child or adolescent. It is often wise to play down minor health problems in order to avoid creating excessive concern, or perhaps even hypochondria. Nevertheless, a doctor should be consulted whenever an abnormal condition persists for more than a few days. Therefore, the information which follows is presented mainly to foster better communication between parents, patients, and doctors; rather than to encourage an excessive reliance on home remedies.

Adolescent Pregnancy. The number of adolescent pregnancies has increased in recent years, although the overall maternity rate for females of all ages has declined. Furthermore, adolescent females in the United States have given birth to a disproportionately large share of low birth weight babies. Hence, the young mothers should be given the appropriate prenatal dietary care as soon as pregnancy is diagnosed. The nutrient requirements of pregnant adolescents who are still growing rapidly are quite high, as a result of their own needs for growth combined with those of the developing fetus. Details of diets for pregnancy are given elsewhere.

(Also see PREGNANCY AND LACTATION DIETS.)

Anemia. The most common form of this condition, which is characterized by an insufficient number of red blood cells, is iron deficiency anemia. Anemic children and adolescents may become tired easily and have a lower than normal resistance to infectious diseases. Anemia occurs during the various stages of growth as a result of different factors such as (1) overuse of milk as a food for toddlers, (2) insufficient consumption of iron-containing foods, (3) excessive loss of blood during menstruation of adolescent females, and (4) very rapid rates of growth of adolescent males.

Treatment of this condition consists of the consumption of iron-rich foods and/or the taking of iron supplements. However, it is doubtful that nonanemic children receive any benefits from the latter measure. Hence, the diagnosis and treatment of anemia should be made only by a physician, since the symptoms of anemia are similar to those of other disorders that may require different types of treatments.

Anorexia Nervosa. The beginning stages of this behavioral and nutritional disorder may be unnoticed by parents. It begins when a child starts to restrict drastically her or his food intake so that a considerable loss of weight occurs.

This condition is about 10 times more prevalent in females than in males. One of the most common theories regarding the reason for the abnormal eating behavior is that the patient becomes obsessed with the desirability of being extremely slender. Often, the menstrual flow of adolescent females ceases when semistarvation develops. Hence, some psychologists believe that the female patient may also find it difficult to cope with the rapid development of sexual characteristics. Nevertheless, a few medical researchers suspect that in some cases there may be abnormal loss of appetite due to glandular disturbances.

It is difficult to treat anorexia nervosa once it has become well established. Furthermore, there are other disorders that affect appetite and eating behavior. Hence, parents should promptly seek the assistance of both medical and psychiatric professionals if the condition appears to be present in one of their children.

(Also see ANOREXIA NERVOSA.)

Constipation. It is a wonder that every child of school age is not constipated, considering the usual restrictions about using the toilets in schools and the rather frantic morning schedules of many families. Also, the fiber content of the American diet has declined steadily. Therefore, the occurrence of constipation in a child now and then is of little cause for concern, but the chronic occurrence may lead to more severe conditions such as colitis, diverticulosis, and hemorrhoids.

Some of the most common means of correcting constipation are (1) allowing enough time to eat breakfast in a relaxed manner and to use the toilet at home in the morning, (2) consuming a diet containing sufficient fiber and liquid, and (3) light exercise that stimulates bowel movements.

(Also see FIBER.)

Diarrhea. This condition may be much more serious than constipation because frequent loose, watery stools represent excessive losses of water and other vital substances from the body. Furthermore, the cause of the diarrhea may be difficult to identify. Therefore, a physician should be consulted whenever the condition lasts for more than a day or two, or if there are traces of blood in the stools. Young children are highly vulnerable to dehydration from severe diarrhea.

Mild diarrhea may be treated by measures such as (1) reduction in the fiber content of the diet by using only cooked, peeled fruits and vegetables instead of raw, unpeeled fruits and vegetables, whole grain products, beans, cabbage, corn, nuts, and onions; (2) avoidance of gassy foods, carbonated beverages,and fruit juices; (3) elimination of fatty and spicy items; (4) consumption of frequent, small meals rather than infrequent, large ones; and (5) drinking of fluids between meals instead of with meals.

Sometimes, children have chronic cramps and diarrhea due to lactose (milk sugar) intolerance or various allergies Therefore, parents should try to identify the foods that cause digestive problems in their children.

Food Preferences And Aversions. The food likes and dislikes that are developed in childhood may have far-reaching effects on one's health throughout life. For example, preferences for refined, sweet, or fatty foods combined with aversions to fresh fruits and vegetables may lead to the chronic consumption of a diet excessively high in calories, but low in the other essential nutrients. This problem is not new to our times, since many ancient cultures had deeply ingrained prejudices regarding the foods deemed suitable and unsuitable. Even when a wide variety of foods are available at reasonable costs, many people adhere to nutritionally imbalanced diets based upon their early eating experiences. It seems that everytime the cooks in the schools prepare broccoli for lunch, most of the vegetable ends up in the garbage pails, yet the serving of hamburger, hot dogs, and pizza is greeted enthusiastically. The latter foods are good sources of calories, protein, and some other nutrients, but they alone do not constitute a nutritionally balanced diet.

One of the solutions to this problem appears to be the introduction of a variety of foods to youngsters as soon as they are able to eat them. Sometimes, the parents of a child are the source of the poor eating habits that are acquired at an early age since children are very observant of parental reactions to foods. Recently, there have been many attempts by educators to make youngsters in day care centers and elementary schools more nutrition conscious. However, parental cooperation is needed to ensure that students receive a variety of nutritious, tasty foods at home.

Gastrointestinal Disorders. The incidences of the early stages of colitis and stomach ulcers in children and adolescents are higher than most people suspect. This unfortunate state of affairs may be the result of factors similar to those which cause the disorders in adults. For example, emotional stress coupled with a poor diet may make the gastrointestinal tract more susceptible to injury. Hence, eating on the run should be avoided, as should the gulping down of either ice cold, highly acidic, soft drinks or boiling hot coffee or tea. Sometimes, the conditions in school lunchrooms are overly stressful because too little time may be left for eating after waiting in line to be served. In such cases, it would be better for children to bring their lunches from home.

Most of the traditional treatments for gastrointestinal disorders have utilized bland diets. However, the modern trend is towards advising most patients to eat a varied, well balanced nutritious diet, and to avoid only the highly irritating foods such as black pepper, chili powder, coffee, tea, cocoa, carbonated soft drinks, and beverages with a high content of alcohol. In the cases of children and adolescents who have mild forms of certain digestive problems, it is best for them to eat all of the foods other than those just mentioned, but to avoid (1) overeating, (2) going too long between meals, (3) overly cold or hot foods, and (4) drinking acidic fruit and vegetable juices on an empty stomach.

Overweight And/Or Obesity. Because excessive gain in weight represents an imbalance between food energy intake and energy needs, prevention of obesity cannot be concerned only with energy intake. Energy needs are determined by the basal metabolic rate and by energy requirements for activity and growth. After infancy, energy requirements for growth represent a small percentage of the total energy requirements. Basal metabolic rates vary from child to child but are more or less fixed in a particular child. The balance therefore depends primarily on the relation between energy intake and expenditures of energy in activity.

Imbalance between energy intake and expenditure leading to excessive gain in weight may result from abnormally high intake, unusually low expenditure, or from a combination of the two. Relatively small excesses of energy, accumulating day after day, year after year, are the common factor leading to obesity. A common pattern for children is an excess gain of about 4 to 5 lb (*2 kg*) in a year, representing an excess of from 35 to 50 Calories (kcal) per day.

Consumption of an amount of food that is barely adequate for one child may represent overeating for another. Not only do energy expenditures at rest vary from individual to individual but activity patterns are also widely different. There is reason to believe that highly active children may be at relatively low risk of developing obesity. In a society in which food is readily available, nearly all children are likely to eat at times for reasons other than satisfying energy requirements—because of social pressures, or merely because irresistibly attractive food is at hand. Under these circumstances, the child with high activity and consequent large energy expenditures seems likely to be at less risk from overeating than the inactive child whose energy needs are readily met.

A number of measures may be suggested for prevention of obesity:

1. Parents should be educated about the dangers of overfeeding during infancy and early childhood because of the possibility that habits of overeating may be

established and persist into childhood and adult life.

2. Breast feeding should be encouraged and the introduction of supplementary foods should be delayed at least until 3 months of age.

3. Vigorous physical activity should be encouraged on a regular basis.

4. Efforts to develop appropriate community facilities for year-round physical activity of children (and adults) are important.

5. Several smaller meals are more effective in preventing obesity than 1 or 2 large meals providing the same energy intake.

6. When one or both parents are obese, the child is at greater risk of obesity than are most other children. Therefore, children from families with an obese parent should be identified and given particular guidance with respect to avoidance of excess weight gain.

Skin Blemishes. Although various types of skin blemishes may be signs of nutritional deficiencies, the most common skin imperfection found in American children and adolescents is acne, which occurs on the face, chest, and back. It is believed to result mainly from an inflammation of the sebaceous glands of the skin due to (1) external irritation from dirt, cosmetics, soap, or other agents; (2) changes in hormonal secretion that occur during pubescence; and/or (3) intolerances to foods such as chocolate, fats, iodized salt, peanut butter, soft drinks, or sugars. However, restriction of the latter foods does not always result in improvement.

Prevention or treatment of acne should involve skin care, good nutritional habits, and the use of topical preparations for the inflammation. There is little evidence that large doses of vitamin A are beneficial to those who already receive sufficient amounts of the vitamin in their diets. Furthermore, excesses of vitamin A are accumulated in the body and can cause toxic effects such as increased intracranial pressure which has symptoms like those of a brain tumor.

Tooth Decay. Dental caries consist of localized, progressive decay of the teeth, initiated by demineralization of the outer surface of the tooth. This demineralization is caused by organic acids produced locally by bacteria that ferment deposits of dietary carbohydrates. The conditions that are usually required for production of dental caries are (1) the presence of a susceptible tooth, (2) an oral environment conducive to the abundant growth and implantation of cariogenic microorganisms and a diet providing adequate food for the microorganisms. Dental caries is usually a chronic disease requiring considerable time for the destructive process to become clinically evident.

Rather convincing circumstantial evidence implicates table sugar (sucrose) as a causative agent of dental caries in human subjects. First, epidemiologic studies have demonstrated a significant decrease in the dental caries rate among school children in occupied countries during the second half of World War II and the return of the caries rate to prewar levels after the war. These changes in caries rate coincided with the wide availability of sucrose before and after the war and its relative unavailability during the second half of the war. Second, in a study conducted under controlled conditions in an institution, frequent and long-term consumption of sticky candy was associated with increase in the dental caries rate. Finally, children in Bowral, Australia who had lived from infancy to puberty in Hopewood House, where a lacto-vegetarian diet with almost complete absence of refined carbohydrate was consumed, demonstrated significantly fewer decayed, missing and filled teeth than did children living under similar socioeconomic conditions in state schools but not adhering to the same type of diet.

Because refined carbohydrates exert their effect in promoting dental caries by serving as a substrate for caries-producing streptococci, it is apparent that for older children as well as for infants not only the total quantity but the form of the carbohydrate and the frequency of consumption are important. A single piece of sticky candy may adhere to the teeth for almost an hour. In the case of sugars that are not in sticky form, a specified amount consumed at one time is likely to be less conducive to formation of dental caries than the same amount consumed in small snacks throughout the day.

TOOTH DECAY PREVENTATIVE MEASURES. Some means of preventing tooth decay follow:

• **Avoidance of sugary snacks**—It is evident that forbidding children to eat between meals is likely to meet with quite limited success. A better approach may be development of a list of foods to be avoided between meals and another list of foods to be permitted. Foods to be avoided are the following: sugar, honey, corn syrup, candies, jellies, jams, sugared breakfast cereals, cookies, cakes, chewing gum, sweetened beverages—including flavored milks, carbonated drinks, sweetened fruit juices and fruit, and fruit-flavored drinks.

• **Cleaning the teeth**—Parents should be instructed to begin cleaning the teeth of toddlers as soon as it is practical to do so. Fine gauze wrapped around the parent's finger is satisfactory at first. Brushing by the parent may be possible with some children even before 18 months of age. Toothpaste should be considered optional and its use should depend on acceptance by the child.

• **Fluoridation of drinking water**—When fluoridated water (0.7 to 1.2 mg of fluoride per liter of water) is consumed throughout life, the dental caries attack rate is reduced by 50 to 60% in permanent teeth and slightly less in primary teeth. The most likely explanation for the small reduction in dental caries attack rate in primary teeth is that many infants and small children consume relatively little drinking water—suggesting the desirability of administering fluoride supplements.

• **Single daily dose of fluoride**—Because of the large number of individuals who do not have access to fluoridated drinking water, the effectiveness of single daily dose of fluoride is a matter of considerable interest. There can be no doubt that single daily dosage of fluoride in drop or tablet form reduces the caries attack rate. However, combined fluoride-vitamin preparations appear to be more readily accepted than vitamin-free fluoride preparations. Unfortunately, such combination preparations do not permit the adjustment of fluoride dosage without alteration of the vitamin dosage. A large chewable tablet or lozenge will impart additional topical benefits to the teeth and is preferable to small tablets that are usually quickly swallowed.

When single dose fluoride supplements are administered, it is, of course, desirable to achieve maximal protection against dental caries with minimal risk of fluorosis of the enamel. Considerable uncertainty exists, particularly in the case of infants and toddlers, with respect to the dose most likely to accomplish this end. Newer information from several sources suggests that previous recommendations for fluoride supplementation of infants and small children may lead to somewhat greater intakes of fluoride than are desirable. Therefore, no fluoride supplementation is recommended during the first 6 months of life, a maximum dosage of 0.25 mg daily between 6 and 18 months, 0.5 mg daily between 18 and 36 months, and a maximum of 0.75 mg daily between 3 and 6 years of age. It is anticipated that these recommendations may require modification as additional evidence becomes available.

Underweight. The current concerns about obesity have led some people to pay insufficient attention to children and adolescents who may be underweight. Therefore, it is noteworthy that the very thin child is likely to be chilled easily, have a lowered resistance to infection, and a tendency to tire easily from physical activity. Excessive slenderness in adolescent females should keep teachers and parents on the watch for anorexia nervosa, because the female sex hormones which are secreted during this period normally promote the deposition of fat when the diet is adequate. Fortunately, most boys wish to avoid excessive slenderness, and are likely to seek means of gaining weight.

Underweight is best treated with a diet that contains abundant amounts of calories, protein, minerals, and vitamins. The diet should provide from 500 to 1,000 Calories (kcal) more than is needed for the maintenance of body weight. It may be a problem for the thin person to eat sufficient food, unless it is apportioned among frequent small meals. Finally, mild outdoor exercise may enhance the appetite, providing that it is done more than ½ hour before meals. Also, strenuous activity should be avoided for at lease an hour after a meal because the increased flow of blood to the skeletal muscles during exercise decreases the circulation to the digestive organs.

MODIFICATION OF CHILDREN'S AND ADOLESCENT'S DIETS. It is generally recognized (1) that the early nutrition of children and adolescents has major effects on their physical and mental well-being later in life, and (2) that habits of food selection and consumption firmly established at a young age are likely to remain throughout a lifetime. Thus, the meals of children and adolescents, both at home and at school, should be closely linked to an educational program for better health. This should include a balanced diet, along with reduced sugar and salt and increased whole grains and raw fruits and vegetables.

(Also see NUTRIENTS: REQUIREMENTS, ALLOWANCES, FUNCTIONS, SOURCES.)

CHINESE CABBAGES *Brassica chinensis* AND *Brassica pekinensis*

These vegetables, although called cabbages, are more closely related to mustards and turnips than they are to head-forming cabbages. Chinese cabbages, like the cabbage vegetables, mustards, rutabaga, and turnip, belong to the mustard family (*Cruciferae*). The most commonly known of the Chinese cabbage-type vegetables are (1) pak-choy or Chinese mustard (*Brassica chinensis*), and (2) pe-tsai or celery cabbage (*Brassica pekinensis*). Figs. C-97 and C-98 show these vegetables.

Fig. C-97. Pak-choy, which is also known as Chinese mustard. (Courtesy, USDA)

Fig. C-98. Pe-tsai (celery cabbage).

ORIGIN AND HISTORY. The Chinese cabbages are believed to have been developed from *Brassica campestris*, which is the wild ancestor of the turnip and the rape. Various varieties of rapes were developed for their oil-bearing seeds thousands of years ago in the Mediterranean region, Middle East, and India. Therefore, it is suspected that about 3,000 years ago a rapelike plant was taken to China, where the peasants developed several new varieties with large, succulent leaves. Pak-choy was developed into a nonheading plant which resembles chard and spinach, whereas pe-tsai became a heading variety that resembles cos lettuce.

Many varieties and hybrids of these vegetables have been developed over the centuries, but most of them are used mainly in the Far East.

PRODUCTION. It is difficult, if not impossible, to determine the extent to which Chinese cabbages are grown worldwide or in the United States because (1) the Food and Agriculture Organization of the United Nations combines the data for all types of cabbages under a single statistic, and (2) production of these vegetables is usually on a small scale.

Both pak-choy and pe-tsai are planted so that they will be ready for harvesting in fairly cool weather, because hot weather causes the mature plants to go to seed. Hence, the crops must be planted during the winter in the southern states, and during the summer in the northern states. At the present time, much of the U.S. crop is grown in California and Arizona. However, growers in the northeastern United States can produce Chinese cabbages which withstand frost down to 22°F (*-5°C*) and grow well at 47°F (*8°C*).

PROCESSING. The lack of production statistics for the Chinese cabbages makes it difficult to determine whether any significant portion of the crop is processed. However, it seems likely that producers of Chinese foods may utilize some of these cabbages in various types of canned and pickled products.

SELECTION AND PREPARATION. Best quality Chinese cabbages are fresh, young, tender, and green. Those which show insect injury, coarse stems, dry or yellowing leaves, excessive dirt, or poor development, are usually lacking in quality and may cause excessive waste. Flabby, wilted plants are generally undesirable.

Raw pe-tsai is excellent when shredded and made into a slaw. Also, it may be cut into wedges and cooked like common cabbage, then served with butter, cheese sauce, sour cream, or a white sauce.

Pak-choy may be used like spinach in casseroles, salads, and soups. It is particularly good in *wonton* soup. (*Wonton* is a Chinese type of dumpling made by wrapping thin sheets of wheat flour and egg dough around filling ingredients such as minced fish, meat, poultry, seafood, and/or vegetables.)

CAUTION: Chinese cabbages and other vegetables of the mustard family (*Cruciferae*) contain small amounts of goiter-causing (goitrogenic) substances that interfere with the utilization of the essential mineral iodine by the thyroid gland. Hence, people who eat very large amounts of these vegetables while consuming an iodine-deficient diet may develop an enlargement of the thyroid, commonly called a goiter. The best insurance against this potentially harmful effect is the consumption of ample amounts of dietary iodine. This element is abundantly present in iodized salt, ocean fish, seafood, and edible seaweeds (kelp).

NUTRITIONAL VALUE. The nutrient composition of Chinese cabbages (pak-choy and pe-tsai) is given in Food Composition Table F-36.

Some noteworthy observations regarding the nutrient compositions of Chinese cabbages follow:

1. Chinese cabbages are very low in calories (11 to 24 kcal per cup [*240 ml*]) because they contain about 95% water. Hence, they are excellent foods for "weight watchers."
2. A 1-cup serving of cooked pak-choy provides almost as much calcium as a cup of milk. Since the people in China consume few dairy products, they can obtain much of their calcium requirement by eating sufficient quantities of this vegetable. (Soybeans are also a moderately good source of calcium. However, the ratio of calcium to phosphorus is low in the beans, but high in pak-choy. Hence, the two foods are complementary.)
3. Pak-choy is a fair source of iron, and a good source of potassium. It is an excellent source of vitamin A.

(Also see VEGETABLE[S], Table V-6 Vegetables of the World.)

CHINESE RESTAURANT SYNDROME

The name given to a set of transient symptoms noted after dining on Chinese food or other highly flavored foods. The symptoms include headache, neck and chest pain, palpitations, and numbness. Monosodium glutamate (MSG), a food additive in high favor in most such restaurants, is the cause. An estimated 30% of consumers are affected. Sufferers soon learn to avoid excess consumption of highly flavored foods (especially the soup course). Although MSG has remained on the GRAS list of the Food and Drug Administration, it was deleted voluntarily from baby foods and many Chinese restaurants either reduced their dependence on the chemical, or eliminated it entirely. (Some Chinese restaurants even include a statement on the menu conveying the following message: *Our foods do not contain monosodium glutamate.*)

CHIPOLATA

A small spicy sausage which is used as a garnish or as an hors d'oeuvre. Sometimes the name is used for a dish that contains such sausages.

CHIPPED BEEF

Thinly sliced dried beef.

CHITIN

A complex nitrogen-containing carbohydrate which is a major component of the cuticles or shells of insects, lobsters, shrimp, crayfish, and similar animals. It is not digestible in the human digestive tract. Therefore, it provides mainly bulk to those who eat the chitinous membranes.

CHITTERLING

The term chitterling applies only to the large intestine of swine. If the large intestine of young bovine is used, it must be identified as "veal chitterling" or "calf chitterling."

CHIVES *Allium schoenoprasum*

This plant differs from most of the other onion vegetables (*Allium* genus) in that the edible part is the leaves rather than the bulbs. Also, the bulbs are clumped closely together in bunches or tufts. Chives, like its close relatives, is a member of the *Alliaceae* family. Fig. C-99 shows the plant together with a typical flower head (encircled).

Fig. C-99. Chives, a close relative of the common onion.

ORIGIN AND HISTORY. Chives grow wild around the world in a wide belt extending from the Arctic Circle to the North Temperate Zone. It is not known when the plant was first cultivated, but it was listed in the catalog of plants that was compiled in 812 A.D. at the request of Charlemagne (Charles the Great), King of the Franks from A.D. 768 to 814 and Emperor of the Romans from 800 to 814.

PRODUCTION. Statistics on the total production of this crop are not readily available. However, much of the small commercial crop is sold to processors and hotel and restaurant suppliers. Also, the vegetable is produced in many home gardens.

Chives may be propagated by planting (1) seed, or (2) individual bulbs that have been separated from the bunches. The plants are usually started in the spring, except where there are likely to be long periods of hot summer weather when the temperature exceeds 90°F (*32°C*). In the latter areas, planting is done in the late summer or early fall. Once established, the plant will multiply and spread. However, most growers take up the clumps every 2 or 3 years in order to divide and replant them. The plants thrive on soils rich in organic matter. The edible leaves may be harvested at regular intervals by cutting them off close to the ground. It is noteworthy that cutting the leaves encourages the growth of the plant.

PROCESSING. The most common ways in which chives are processed are (1) chopping and flash freezing; (2) chopping and blending into cottage cheese, cream cheese, or salad dressings; and (3) drying.

SELECTION AND PREPARATION. Fresh chives may be kept only a few days in a refrigerator without spoiling. However, the leaves may be chopped, placed in a plastic container, and stored in a freezer.

The chopped vegetable adds an appetizing flavor to casseroles, cheese spreads, omelets, salads, sandwiches, sauces, soups, and stews.

NUTRITIONAL VALUE. The nutrient composition of raw chives is given in Food Composition Table F-36.

Some noteworthy observations regarding the nutrient composition of chives follow:

1. Chives are high in water and low in calories and protein. They are a fair source of calcium; a good source of iron, potassium, and vitamin C; and an excellent source of vitamin A.
2. Even though chives have a milder flavor than onions, only small amounts of the raw vegetable can be comsumed by most people. However, much of the pungency is lost in cooking, while most of the nutritional value is retained. Therefore, it may be desirable to eat cooked chives rather than raw chives in order to be able to consume greater amounts of the vegetable and the nutrients it provides.

(Also see VEGETABLE[S], Table V-6 Vegetables of the World.)

CHLORELLA

A species of green freshwater algae that has been considered as a potential food source for people. However, chlorella is high in undigestible fiber. Hence, it might be better used to feed grazing animals such as cattle and sheep, because these species can digest many types of fiber.

(Also see ALGAE; and FIBER.)

CHLORIDE

One of the forms in which the element chlorine occurs in chemical compounds. The chloride ion serves such functions as (1) regulation of acid-base balance, (2) a constituent of stomach (hydrochloric) acid, and (3) a regulator of the fluid balance between the blood, tissue fluid, and the cells.

(Also see CHLORINE; and MINERALS.)

CHLORIDE SHIFT (HAMBERGER SHIFT)

The exchange process of bicarbonate and chloride between plasma and red blood cells is referred to as the chloride shift, or Hamberger shift, after the Dutch physiologist.

CHLORINE DIOXIDE (ClO_2)

A chemical used to "age" flour, to impart to it the desirable characteristics of flour stored for weeks. Also chlorine dioxide is used to bleach organic material such as cellulose, flour, fats and oils, and to act as a bacteriocide and antiseptic in water purification.

(Also see ADDITIVES.)

CHLORINE OR CHLORIDE (Cl)[20]

Chlorine, which is a strong-smelling, greenish-yellow gas, is never found in nature as an individual element. But its compounds, such as sodium chloride (common salt), are widespread.

Chlorine is an essential mineral, with the special function of forming the hydrochloric acid (HCl) present in gastric juice. The body contains approximately 100 g of chlorine, which represents about 0.15% of the weight of an average person, most of which is combined with sodium or potassium. It is found mainly in the extracellular fluids of the body, but it is present to some extent in red blood cells and to a lesser extent in the cells of other tissues. The blood contains 0.25% chlorine, 0.22% sodium, and 0.02% to 0.22% potassium; thus, the chlorine content of the blood is higher than that of any other mineral.

HISTORY. Carl Wilhelm Scheele, a Swedish chemist, discovered chlorine in 1774 by treating hydrochloric acid with manganese dioxide. A few years later Antoine Lavoisier, a French chemist, concluded that all acids contained oxygen. It followed that chlorine was called oxymuriatic acid gas because chemists believed that it also contained oxygen. In 1810, Sir Humphrey David, the English chemist, determined that chlorine was an element (further, he showed that muriatic acid itself contains only hydrogen and the new element, chlorine), which he named *chloros,* a Greek word meaning greenish-yellow.

ABSORPTION, METABOLISM, EXCRETION. Chloride from foods and from gastric juice is absorbed chiefly from the small intestine.

During digestion some of the chloride of the blood is used for the formation of hydrochloric acid in the gastric glands and is secreted into the stomach where it functions temporarily with the gastric enzymes and is then reabsorbed into the blood stream with other nutrients.

The highest concentrations of body chlorine are found in the gastric juice and in the cerebro spinal fluid.

Excessive chlorine in the diet is excreted via the urine. Additional losses occur in sweating, vomiting, and diarrhea. Excreted chlorine is usually accompanied by excess sodium or potassium, unless the body has need to conserve base; in the latter case, the ammonium ion accompanies the chloride ion.

FUNCTIONS OF CHLORINE. Chlorine, in the form of the chloride ion which is negatively charged, plays a major role in the regulation of osmotic pressure, water balance, and acid-base balance. The chloride ion is also required for the production of hydrochloric acid by the stomach. This acid is necessary for proper absorption of vitamin B-12 and iron, for the activation of the enzyme that breaks down starch, and for suppressing the growth of microorganisms that enter the stomach with food and drink.

Chloride is a normal constituent of extracellular (between cells) rather than intracellular (within cells) fluid. However, in erythrocytes (red blood cells), chloride crosses the cell membrane rapidly to establish an equilibrium between cell contents and the extracellular fluids to aid in minimizing fluid shifts. The ability of the chloride to pass readily from the erythrocytes into the blood plasma (known as the chloride shift) enhances the ability of the blood to carry large amounts of carbon dioxide to the lungs.

Some cities add chlorine to drinking water to kill harmful bacteria. Opponents of chlorinated drinking water argue (1) that chlorine is a highly reactive chemical and may join with inorganic minerals and other chemicals to form substances that may be harmful, (2) that chlorine in drinking water destroys vitamin E, and (3) that chlorine destroys many of the intestinal flora that help in the digestion of food.

DEFICIENCY SYMPTOMS. Deficiencies of chloride may develop from prolonged or severe vomiting, diarrhea, pumping the stomach, injudicious use of diuretic drugs, or strict vegetarian diets used without salt. Severe deficiencies may result in alkalosis (an excess of alkali in the blood), characterized by slow and shallow breathing, listlessness, muscle cramps, lack of appetite, and occasionally, by convulsions.

INTERRELATIONSHIPS. Loss of chloride generally parallels that of sodium. When sodium choloride intake is restricted, the chlorine level in the urine falls, followed by a drop in tissue chloride levels. Increased losses of sodium that occur with sweating or diarrhea result in concurrent losses of chloride.

Chloride also aids in the conservation of potassium.

RECOMMENDED DAILY ALLOWANCE OF CHLORINE. There are no recommended dietary allowances for chlorine because the average person's intake of 3 to 9 g daily from foods and added table salt easily meets the requirements. Also, diets that provide sufficient sodium and potassium provide adequate chlorine.

The body's requirement for chlorine is approximately half that of sodium.

Estimated safe and adequate intakes of chloride are given in Table C-25.

TABLE C–25
ESTIMATED SAFE AND ADEQUATE DAILY DIETARY INTAKES OF CHLORIDE[1]

Group	Age	Chloride
	(years)	(mg)
Infants	0–0.5	180
	0.5–1.0	300
Children and Adolescents	1	350
	2–5	500
	6–9	600
	10–18	750
Adults	18+	750

[1]*Recommended Dietary Allowances,* 10th ed., 1989, NRC–National Academy of Sciences, p. 253, Table 11-1.

[20]Chlorine exists in the body almost entirely as the choloride ion. The two terms—chlorine and chloride—are used interchangeably.

TOXICITY. An excess of chlorine ions is unlikely when the kidneys are functioning properly.

SOURCES OF CHLORINE. Chlorine is provided by table salt (sodium chloride) and foods that contain salt.

Persons whose sodium intake is severely restricted (owing to diseases of the heart, kidney, or liver) may need an alternative source of chloride; a number of chloride-containing salt substitutes are available for this purpose.

(Also see MINERAL[S], Table M-67.)

CHLOROPHYLL

The green pigment of plants which permits them to manufacture foodstuffs from simple salts and carbon dioxide with energy derived from sunlight; i.e., photosynthesis. This makes for a true distinction between plants and animals; plants can manufacture food, animals must by supplied ready-made foods.

(Also see PHOTOSYNTHESIS.)

CHLOROPHYLLASE

An enzyme present in all green plants which removes a chemical group from the chlorophyll molecule, producing a chlorophyllide and phytol.

(Also see PHOTOSYNTHESIS.)

CHLOROPHYLLIDE

A water soluble, green colored substance resulting from the chemical or enzymatic removal of a portion of the chlorophyll molecule. It is responsible for the green color of water after cooking certain vegetables.

(Also see CHLOROPHYLLASE; and PHOTOSYNTHESIS.)

CHLOROPHYLLIN COPPER

This is a synthetic version of chlorophyll, the substance that makes plants green. It is most commonly used for its deodorant qualities. Chlorophyllin copper tablets are sometimes given to incontinent patients in nursing homes to help neutralize odors. Sometimes, they are used for the same purpose by people who have had a colostomy or ileostomy. Chlorophyllin copper tables are available without prescription.

CHLOROPHYLLINS

These are chemically prepared salts of chlorophyll used commercially in food coloring, dyes, breath and body deodorants, and medicine.

CHLOROSIS ("GREEN SICKNESS")

A type of anemia, formerly found in young women, characterized by a large reduction of hemoglobin in the blood, but only a slight diminution of the number of red blood cells. The symptoms are a greenish color to the skin, due to iron deficiency. Today, chlorosis has almost completely disappeared because of increased knowledge of the role of iron in the diet.

CHLORPROPAMIDE

A drug given by mouth in the treatment of diabetes.

(Also see DIABETES MELLITUS.)

CHLORTETRACYCLINE (AUREOMYCIN)

An antibiotic which is sometimes used (1) in the treatment of certain diseases and (2) as a food additive to prevent spoilage in meat, fish, and poultry.

(Also see ADDITIVES.)

CHOKING

Normally, when food is swallowed, it is guided down the esophagus by a series of automatic muscular contractions. Whenever food touches the surface of the throat (pharynx), the vocal cords in the larynx (the "voice box") close strongly, and the epiglottis, a triangular plate of cartilage, swings backward over the glottis, the opening to the windpipe. These two actions prevent food from entering the trachea (windpipe). Occasionally, food or some other object becomes lodged in the throat or larynx, resulting in blocking the air passage and choking. When the air passage is blocked, the person turns bluish-gray, and unconsciousness and death can result within 4 to 6 minutes.

Each year, more than 3,000 Americans choke to death, most of them while eating. Choking victims are usually (1) the elderly who do not chew their food thoroughly, (2) people who have several alcoholic drinks before eating, or (3) children who run or play with food in their mouths. Since choking causes collapse, and since it occurs during eating, it is sometimes referred to as a "cafe coronary." However, the hallmarks of the choking individual are excitement and attempted speech, but the inability to speak. Often the victim will point to his mouth or his throat. In contrast an individual experiencing a heart attach may be excited, but he is usually able to speak.

When total blockage of the airflow occurs, immediate action is necessary. While emergency medical help is on the way, first aid may include any or all of the following:

1. The Heimlich Maneuver, named for Dr. Henry J. Heimlich of Cincinnati, is a first aid technique for choking victims. It is performed by (a) grabbing the victim from behind with both arms, (b) making a fist with one hand, (c) placing the fist near the navel but just below the rib cage, and (d) using the other hand to pull the fist quickly upward towards the victim's diaphragm. The maneuver may be repeated several times, and it can be applied with the victim sitting in a chair or lying down. The Heimlich Maneuver pushes air out of the lungs with a sudden burst of pressure, thereby popping the obstruction out of the airway.
2. With the victim upside down or with his head lowered, a pound on the back *may* dislodge the object. However, this should never be done with the victim in the upright position, as the obstruction may fall further down the trachea.
3. There are devices for removing material from the

back of the throat, which many restaurants have available. These are effective in trained hands. Furthermore, a finger may be used to dislodge the object. However, blind and frantic attempts to reach into the back of the throat should be avoided, as this may only serve to push the blockage further down the air passage.

Note: In 1985, the U.S. Surgeon General endorsed the Heimlich Maneuver as the only safe method for helping a choking victim.

CHOLAGOGUE

A food, nutrient, or similar substance that brings about expulsion of bile from the gallbladder. Most cholagogues act by stimulating the lining of the duodenum to release the hormone cholecystokinin-pancreozymin which travels via the blood to the gallbladder and triggers its contraction. Egg yolk and fats are considered to be the most potent cholagogues, although it seems likely that dairy products, fish, meat, and poultry have similar effects.

(Also see CHOLECYSTOKININ-PANCREOZYMIN; DIGESTION AND ABSORPTION; and GALLSTONES.)

CHOLANGITIS

An inflammation of the bile ducts.

(Also see GALLBLADDER.)

CHOLECALCIFEROL (VITAMIN D)

Cholecalciferol may be obtained either from food or irradiation of the skin. Absorption of dietary cholecalciferol takes place in the duodenum in much the same way as other fats and fat-soluble compounds.

The body synthesizes 7-dehydrocholesterol, a substance with no vitamin activity, which is found in the skin. When the skin is exposed to the action of ultraviolet radiation from the sun, or from a sunlamp, 7-dehydrocholesterol is converted to vitamin D_3. Therefore, people who are exposed to sunlight for extended periods do not need dietary supplementation of vitamin D. But people whose skin has only limited exposure to ultraviolet light require a dietary source of vitamin D.

Fig. C-100 shows what happens to cholecalciferol in the body.

(Also see VITAMIN D.)

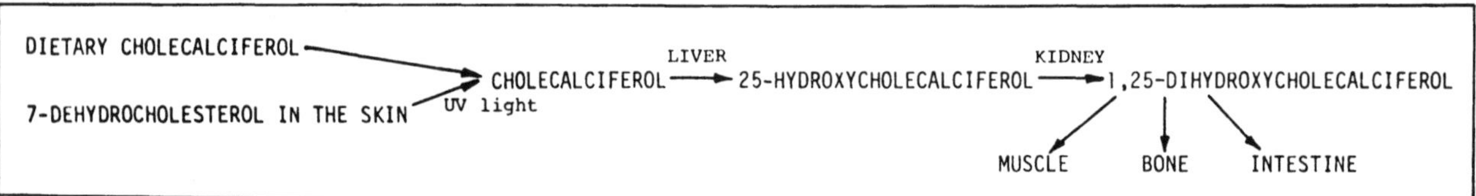

Fig. C-100. Fate of cholecalciferol in the body.

CHOLECYSTITIS

An inflammation of the gallbladder due to such causes as (1) infection, (2) trapped bile which cannot be secreted due to some type of blockage in the bile duct, (3) a tumor, or (4) adhesions of the walls of the gallbladder. It is characterized by a sharp pain under the breastbone which may travel to the shoulder and/or the lower abdomen. It may also be accompanied by vomiting, chills and fever, and jaundice.

(Also see GALLSTONES.)

CHOLECYSTOKININ-PANCREOZYMIN

A hormone secreted by the lining of the duodenum in response to stimulation by partially digested fats and proteins, stomach acid, or calcium ions. Cholecystokinin-pancreozymin was once thought to be two distinct hormones (hence, the hyphenated name), but recent evidence points to only a single substance. The hormone travels in the blood, and has such effects as (1) triggering the contraction of the gallbladder so that bile is expelled into the intestine, and (2) stimulating the pancreatic secretion of a digestive juice rich in enzymes.

(Also see DIGESTION AND ABSORPTION; and GALLSTONES.)

CHOLELITH

Another name for a gallstone.

(Also see GALLSTONES.)

CHOLELINTHIASIS

The presence of stones in the gallbladder and/or one of the bile ducts.

(Also see GALLSTONES.)

CHOLERA

A severe diarrheal disease. It is caused by the *Vibrio comma* organism and is transmitted by flies, mice, cockroaches, polluted water, and contaminated foods. Epidemics of cholera usually result in many fatalities in the countries where large numbers of people are undernourished. However, the loss of considerable fluid and mineral salts in the stool can also be fatal to many otherwise healthy people, if prompt treatment is not received.

(Also see DIARRHEA; BACTERIA IN FOOD, section headed "Food Infections [bacterial].")

CHOLERTIC AGENT

A substance that stimulates the liver to produce and secrete bile. The most potent cholertic agents appear to be protein, corn oil, and olive oil. Also, the entry of partially digested food and stomach acid into the duodenum has a cholertic effect.

(Also see DIGESTION AND ABSORPTION.)

CHOLESTEROL ($C_{27}H_{46}O$)

Contents *Pages*

Chemistry 409
History 409
Blood Levels 409
Metabolism 410
Digestion and Absorption 410
Synthesis 410
Functions 410
Excretion 410
The Risk—Heart Disease 410
Sources 412

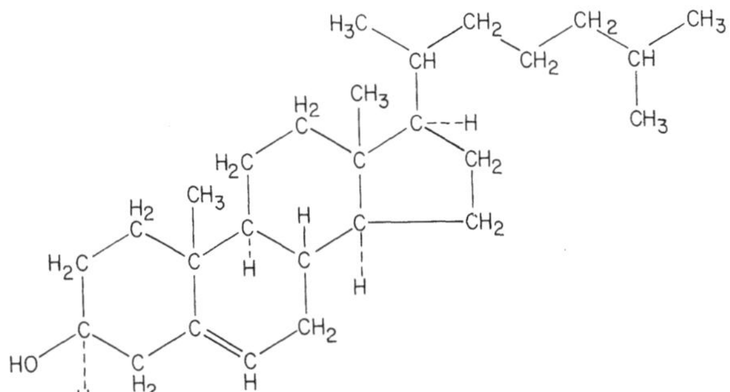

Fig. C-101. The chemical structure of cholesterol.

Currently, cholesterol is a household word, but originally the name was given to a substance isolated from gallstones in the late 1700s. Two Greek words form the word cholesterol: *Chole* is the Greek word for bile and *stereos* is the word for solid. Gallstones are solid granules formed in the bile, and very often are composed of the substance now called cholesterol. Although isolated from gallstones, cholesterol is found in all body tissues, especially the brain and spinal cord. And, as most everyone is now aware, cholesterol is found in many of the animal food products—eggs, dairy products, poultry, fish, lard, and other fats. Further, cholesterol is an essential ingredient for certain biochemical processes, including the production of sex hormones in humans.

CHEMISTRY. Chemically, cholesterol is a fatlike compound—actually an alcohol—which in its pure form appears as pearly flakes. It is composed of 27 carbon atoms which form three fused cyclohexane (6-carbon) rings, a cyclopentane (5-carbon) ring and a side chain of 8 carbon atoms (see Fig. C-101). Cholesterol is practically insoluble in water. Since it contains an alcoholic hydroxyl group (OH), cholesterol can participate in esterification reactions—the combination of an alcohol and an organic acid. Often cholesterol is esterified to unsaturated fatty acids in foods and in the body.

HISTORY. As obvious as it may seem to some, cholesterol appears to have always been with man in his food (except for vegetarian cultures) and in his body. It is not a curse of the 20th century. However, man's familiarity with cholesterol came mostly during this century.

More than 100 years ago, cholesterol was found in plaque formation on the walls of arteries. During the depression of the 1930s, a high incidence of plaque formation was noted in the poorly nourished bodies on which autopsies were performed. Following World War II, scientists directed their attention toward an ever increasing rate of heart disease.

Many investigators reached a conclusion that atherosclerosis, with its generalized thickening of the inner arterial wall, was a consequence of abnormal cholesterol metabolism, whatever the metabolic failure. This view was supported by studies with experimental animals—rabbits, chickens, mice, rats, guinea pigs, dogs, and monkeys—in which elevated blood cholesterol produced by diet, drugs, and other means tended to result in arterial damages resembling in many respects those found in human atherosclerosis. The cholesterol heart disease connection was officially opened when, in 1953, Dr. Ancel Keys at the University of Minnesota reported a positive correlation between the consumption of animal fat and the occurrence of atherosclerosis in humans. Subsequently, other studies correlated high blood levels of cholesterol with increased incidence of atherosclerosis in humans. In 1964, the American Heart Association recommended that the general public reduce cholesterol intake to 300 mg/day. Subsequently, various government and health agencies around the world have followed suit. Thus attention was, and still is, centered on cholesterol.

BLOOD LEVELS. Cholesterol is present in both free and esterified forms in the blood. From birth, blood cholesterol increases throughout life as shown in Table C-26. The moderate risk category in Table C-26 includes large numbers of people with elevated blood cholesterol due, in part, to their diet. The high risk category in Table C-26 includes individuals with hereditary forms of high blood cholesterol which require the most aggressive treatment.

Since cholesterol is insoluble in a water-based medium such as blood, it is transported in blood as a lipoprotein. Primarily, two types of lipoproteins are involved in cholesterol transport. Low-density lipoproteins—LDL—transport cholesterol from the liver to the cells, while high-density lipoproteins—HDL—transport cholesterol from the tissue cells to the liver. Current research suggests that measuring HDL and LDL may be a more accurate means of assessing the blood cholesterol level. Regardless of whether total cholesterol, the HDL, and/or LDL are measured in the blood of an individual, one should be aware that just a single determination on a blood sample is far from adequate for evaluating the cholesterol status. Factors such as (1) age, (2) time of day, (3) physical condition, (4) stress, (5) genetic back-

TABLE C–26
AGE AND CHOLESTEROL CONCENTRATION OF MEN AND WOMEN AT MODERATE AND HIGH RISK[1]

Age	Moderate Risk	High Risk
(years)	(mg/dl)	(mg/dl)
2 to 19	Greater than 170	Greater than 185
20 to 29	Greater than 200	Greater than 220
30 to 39	Greater than 220	Greater than 240
40 and over	Greater than 240	Greater

[1]National Insitute of Health guidelines of moderate and high risk levels of blood cholesterol as measured in milligrams per deciliter (mg/dl).

ground, and/or (6) laboratory expertise may all affect the determination.

METABOLISM. Cholesterol in the body arises from two sources: (1) that from the diet—exogenous cholesterol; and (2) that manufactured in the body—endogenous cholesterol. Cholesterol in the blood reflects the overall cholesterol metabolism—that derived from both sources.

Digestion and Absorption. The average individual ingests between 500 and 800 mg of cholesterol each day. Dietary fat aids the absorption of cholesterol. Moreover, absorption is dependent upon the availability of bile acids from the liver, and pancreatic cholesterol esterase. Also, the absorption of cholesterol depends upon the amount eaten. Increasing intake decreases the percentage absorbed. At high levels, man absorbs less than 10%, and the remainder leaves the body via the feces. Interestingly, many of the animals used in cholesterol-atherosclerosis studies—rabbits, dogs, and cats—are far more efficient than man in absorbing cholesterol from the digestive tract. Initially, dietary cholesterol enters the body as chylomicrons which are eventually converted to the cholesterol-containing lipoproteins—LDL and HDL. About 2 to 4 hours after eating, the cholesterol in the blood which came from the food is indistinguishable from that synthesized in the body.

Synthesis. Most tissues, except possibly the brain, are capable of manufacturing—synthesizing—cholesterol. However, the liver is the major site of synthesis. The entire cholesterol molecule can be synthesized from 2-carbon units called acetate. In the whole metabolic scheme, acetate can be derived from the breakdown of carbohydrates, proteins (amino acids), and of course, fats. Each day the body manufactures 1,000 to 2,000 mg of cholesterol. However, on a day-to-day basis the synthesis and metabolism of cholesterol is controlled by such factors as (1) fasting, (2) caloric intake, (3) cholesterol intake, (4) bile acids, (5) hormones, primarily the thyroid hormones and estrogen, and (6) disorders such as diabetes, gallstones and hereditary high blood cholesterol—hypercholesterolemia. Control of cholesterol synthesis by cholesterol intake is important, since this means that when intake is high then synthesis is low and vice versa.

Functions. Cholesterol is vital to the body. Its primary importance concerns tissues, bile acids, and hormones.

• **Tissues**—Cholesterol is a structural component of (1) the cell membranes, particularly the skin and intestines, (2) the myelin sheath which surrounds nerves, and (3) the brain, though its function in these areas is not completely understood.

• **Bile acids**—This is a major pathway for the degradation of cholesterol. Approximately 80% of cholesterol is used by the liver to form the bile acids. To form bile acids, enzymes modify the cholesterol molecule by chopping off 3 carbon atoms from the side chain (bile salts contain 24 carbon atoms), and adding some hydroxyl (OH) groups to the ring structure. Cholic acid is the most abundant bile acid in human bile. Eventually, bile acids are coupled to glycine and taurine. The salts of these substances are water soluble and are powerful detergents important in the digestion of fat.

(Also see DIGESTION AND ABSORPTION.)

• **Hormones**—By chopping off more carbon atoms and rearranging some oxygen and hydrogen atoms on the cholesterol molecule, some important hormones are formed. Cholesterol is the chemical ancestor of the hormones, progesterone, cortisone, cortisol, corticosterone, aldosterone, testosterone, and estrogen, which contain 21, 19, or 18 carbon atoms. These hormones are formed in the ovaries, testes, and adrenal gland. They are vital. Furthermore, cholesterol is a precursor of vitamin D.

Excretion. Removal from the body occurs primarily via the conversion of cholesterol to bile acids. About 0.8 g of cholesterol are degraded daily by this method. Also, a minor amount is converted to the above mentioned hormones. Additionally, some cholesterol is never digested, and hence, excreted via the feces, particularly when the intake is high.

THE RISK—HEART DISEASE. High blood cholesterol is one of the three major modifiable risk factors for coronary heart disease (CHD); the other two are high blood pressure and cigarette smoking. Approximately 25% of the adult population 20 years of age and older has high blood cholesterol levels—levels that are high enough to need intensive medical attention. More than half of all adult Americans have a blood cholesterol level that is higher than desirable.

The principal nutritional factors identified with high blood cholesterol and the development of CHD are dietary fat, particularly saturated fatty acids and cholesterol, and energy imbalance leading to obesity. Other dietary constituents, such as fiber or alcohol, may interact with these factors in ways that are not clearly understood.

The relationship of dietary fat and cholesterol to CHD is supported by extensive and consistent clinical, epidemiologic, metabolic, and animal evidence. These studies strongly indicate that the formation of atherosclerotic lesions in coronary arteries—contributing to the risk for CHD—is increased in proportion to levels of total and LDL (low-density lipoprotein) cholesterol in blood, which, in turn are increased by diets high in total and saturated fat but decreased by diets containing polyunsaturated and/or monounsaturated fat. International epidemiologic comparisons and migration studies have revealed strong associations of fat, especially saturated fat, intake to development of elevated blood cholesterol levels, atherosclerosis, and CHD. Evidence from studies within a given population has been less consistent but points in a similar direction. Dietary intervention trials in men with elevated blood cholesterol levels have demonstrated small but significant proportionate improvements such that each 1% reduction in total blood cholesterol is accompanied by about a 1.5% reduction in heart disease risk. Intervention to lower elevated blood cholesterol levels has been shown in both human and animal studies to reduce CHD risk and to slow lesion progression. Animal studies have shown lesion regression, and there is suggestive evidence from some clinical studies that this also occurs in humans.

Taken together, these studies provide strong support for recommendations for an overall considerable decrease in dietary fat intake by the general public from the present level of 37% of total caloric intake and decrease in saturated fat from the present level of about 13% of total caloric intake.

Although the effect of dietary cholesterol on blood cholesterol is somewhat weaker and more variable among individuals than that for dietary saturated fatty acids, a reduction in the amount of cholesterol consumed by the general public from present average levels of approximately 305 mg/day for women and 440 mg/day for men seems appropriate.

Obesity is associated with such CHD risk factors as elevated LDL and total blood cholesterol, lower HDL (high-density lipoprotein) cholesterol, high blood pressure, and diabetes mellitus. It is also a significant independent predictor of CHD, especially in women and in persons under age 50. Thus, current evidence suggests that an overall decrease in the prevalence and severity of overweight in the population, through both a decrease in caloric intake and an increase in caloric expenditure, is advisable on the basis of the relationship of obesity to heart disease risk.

Studies of animal protein, coffee, and sugar have shown variable associations with increased blood lipid levels, but present evidence of their relationship to CHD, if any, is too weak and insufficient to draw implications for changes in the consumption of these substances. Likewise, evidence from some studies that certain components of dietary fiber and omega-3 fatty acids from fish oils reduce blood cholesterol levels and heart disease risk is too preliminary to recommend changes in average intake of these substances.

• **Special populations**—There is a need to identify those individuals with high cholesterol levels, who are therefore at greatest risk. For individuals whose high total and LDL cholesterol levels warrant treatment, the first line of intervention is diet therapy. The recently released National Cholesterol Education Program (NCEP) guidelines on the treatment of high blood cholesterol in adults recommend that intensive dietary treatment should generally be carried out for at least 6 months. Only after this period of time, and if the cholesterol level remains significantly high, should the addition of drugs to the dietary regimen be considered. Even then, continuation of diet therapy can reduce the need for drugs and thus their risk of side effects and cost. Furthermore, studies in persons with CHD suggest that diets low in fat, saturated fat, and cholesterol can retard the progression of the disease, including recurrent heart attacks, and perhaps induce regression of atherosclerotic lesions. Persons with such high blood cholesterol levels should receive dietary guidance by qualified health professionals.

Adults with total cholesterol levels of 240 mg/dl or above (whose LDL cholesterol levels are also significantly elevated), and those with total cholesterol levels of 200 to 239 mg/dl with CHD or two or more CHD risk factors should begin a program of supervised dietary treatment. The NCEP guidelines recommend starting dietary therapy with a step-one diet, in which the intake of total fat is less than 30% of calories, saturated fat is less than 10% of calories, and cholesterol is less than 300 mg/day. If after three months on this diet cholesterol lowering is insufficient, the person should progress to a step-two diet, in which saturated fat is further reduced to less than 7% of total calories and cholesterol intake is further reduced to less than 200 mg/day.

Although in epidemiologic studies light to moderate alcohol consumption is associated with reduced heart disease risk, a cause-and-effect relationship has not been proved. Since heavy drinking has numerous adverse health consequences, including several on the cardiovascular system, the use, even in moderate quantities, of alcohol for its possible beneficial effects on CHD is not recommended.[21]

• **Recommended diet to lower blood cholesterol**—The first treatment that the doctor will prescribe for high blood cholesterol is a change in the patient's diet. Generally, blood cholesterol levels begin to drop two to three weeks after a person begins a cholesterol-lowering diet. Over time, cholesterol levels may drop 30 to 55 mg/dl.

The following seven changes should be made in the diet of persons having high blood cholesterol:

1. **Eat less high-fat food.** There are two major types of dietary fat—saturated and unsaturated. Unsaturated fats are further classified as either polyunsaturated or monounsaturated fats. All fat is a mixture of saturated and unsaturated fats.

Note: Eating less total fat is an effective way to eat less saturated fat and fewer calories.

2. **Eat less saturated fat.** Saturated fat raises the blood cholesterol level more than anything else in the diet. So, the best way to reduce the blood cholesterol level is to reduce the amount of saturated fat that is eaten.

Note: Saturated fats are found primarily in animal products. But a few vegetable fats and many commercially processed foods also contain saturated fat. Read labels carefully. Choose foods wisely.

3. **Substitute unsaturated fat for saturated fat.** Unsaturated fats (polyunsaturated and monounsaturated fats) should be substituted for saturated fat to the extent practical.

Note: Unsaturated fats lower blood cholesterol levels when substituted for saturated fats.

4. **Eat less high-cholesterol food.** Dietary cholesterol can raise the blood cholesterol level. Therefore, it is important to eat less food that is high in cholesterol. (See Table C-28, Cholesterol Content of Some Common Foods.)

Note: There is very little cholesterol in low-fat dairy foods like skim milk and no cholesterol in food from plants, like fruits, vegetables, vegetable oils, grains, cereals, nuts, and seeds.

5. **Substitute complex carbohydrates for saturated fats.** Breads, pasta, rice, cereals, dried peas and beans, fruits, and vegetables are good sources of complex carbohydrates (starch and fiber). They are excellent substitutes for foods that are high in saturated fat and cholesterol. The type of fiber found in foods such as cereal brans, some fruits like apples and oranges, and in some dried beans may even help reduce blood cholesterol levels.

---

[21]In this section, the authors drew heavily from the following authoritative publication: *The Surgeon General's Report on Nutrition and Health*, U.S. Department of Health and Human Services, Public Health Service, DHHS (PHS) Publication No. 88-50211, 1988.

Note: Foods that are high in complex carbohydrates, if eaten plain, are low in saturated fat and cholesterol as well as being good sources of minerals, vitamins, and fiber.

6. **Maintain a desirable weight.** People who are overweight frequently have higher blood cholesterol levels than people of desirable weight. Weight reduction can be achieved by consuming fewer calories and increasing exercise on a regular basis.

Note: To achieve or maintain a desirable weight, caloric intake must not exceed the number of calories the body burns.

7. **Eat foods that are high in soluble fiber, such as oat bran.** In the June, 1992 *Journal of the American Medical Association*, a University of Minnesota, Minneapolis, researcher reported that, based on an analysis of 12 previous studies, eating a large bowl of ready-to-eat oat bran cereal or three packets of instant oatmeal daily can, on the average, reduce the total cholesterol in the blood by 2 to 3%. For people with high cholesterol levels, the reduction can be as much as 6 to 7%.

Note: Basically, there are two basic types of fibers: water soluble fiber (like oat bran) and non-soluble fiber (like wheat bran). Only soluble fiber is effective in lowering cholesterol. Soluble fiber dissolves in water.

All of the above indicate that the development of atherosclerosis is not just a simple matter of eating too much cholesterol. Moreover, these recommendations encompass accepted measures of good health—cessation of smoking, normal blood pressure, ideal weight, exercise, control of stress, and awareness of family history. Each of the recommendations complement the others and contribute to decreasing the risks of atherosclerosis and heart disease.

(Also see HEART DISEASE; and HYPERLIPOPROTEINEMIAS.)

SOURCES. The following information is presented for those individuals who wish to reduce their cholesterol intake for personal reasons or who require cholesterol restriction as part of a control measure in cases of hyperlipoproteinemias. Table C-27 indicates the foods in the American diet which supply cholesterol and their relative contribution to the total cholesterol available.

TABLE C–27
AMOUNT OF CHOLESTEROL AVAILABLE PER PERSON PER DAY IN THE UNITED STATES[1]

Food Source	Year	
	1967–1969	1988
	(mg)	(mg)
Meat, poultry, fish	183.3	207.2
Eggs	239.6	144.0
Dairy products	73.9	67.0
Fats and oils	28.8	21.0
Animal sources	525.6	440.0
Vegetable sources	0	0

[1]*Agricultural Statistics 1991*, USDA, p. 478.

Table C-28 which follows provides a ranking of the cholesterol levels of some common foods. This is provided for informational purposes only—not to encourage or discourage consumption of certain foods by the general public. Rather, each individual should consider his or her own case.

Information on the cholesterol levels of more foods is given in Fats and Fatty Acids in Selected Foods, Table F-9 in the article entitled Fats and Other Lipids.

(Also see EGGS; and HEART DISEASE.)

TABLE C-28
CHOLESTEROL CONTENT OF SOME COMMON FOODS[1]

Food	Cholesterol
	(mg/100 g)
Egg yolks, chicken	1602
Kidneys, beef, braised	804
Liver, chicken or turkey, simmered	615
Eggs, fried, poached or hard cooked	540
Sweetbread (thymus), braised	466
Liver, beef, calf, or pork, fried	438
Kidneys, calf, lamb, or pork, raw	375
Roe, salmon, sturgeon, or turbot, raw	360
Heart, beef, braised	274
Butter	219
Shrimp, canned	150
Sardines, canned	140
Whipping cream, 37.6% fat	133
Cream cheese	111
Beef tallow	109
Cheddar cheese	106
Turkey, roasted	105
Veal	101
Beef and lamb, variety of cuts	95
Swiss cheese	92
Chicken, roasted	90
Pork, variety of cuts	89
Frankfurter	62
Ice cream, 12% fat	60
Tuna, canned	55
Milk, evaporated	31
Milk, whole	14
Milk, 2% fat	8
Yogurt	6
Cottage cheese	4
Milk, 1% fat	4

[1]Values are approximate and they are for foods in Food Composition Table F-36.

CHOLESTYRAMINE

A drug which binds bile salts and causes them to be excreted in the stool. By this effect, it tends to reduce blood cholesterol because a more rapid breakdown of cholesterol is needed to produce bile acids that are lost in the stool. Under certain conditions cholestyramine may also help to dissolve gallstones.

(Also see CHOLESTEROL; and GALLSTONES.)

CHOLIC ACID ($C_{24}H_{40}O_5$)

The most abundant bile acid, which is synthesized from cholesterol in the liver, and is then combined chemically with the amino acids glycine and taurine.

(Also see DIGESTION AND ABSORPTION.)

CHOLINE

Contents Page

History 413
Chemistry, Metabolism, Properties 413
Measurement/Assay 413
Functions 414
Deficiency Symptoms 414
Recommended Daily Allowance of Choline 414
Toxicity 414
Choline Losses During Processing, Cooking, and Storage 414
Sources of Choline 414
Top Food Sources of Choline 415

Choline, which is a key part of the constituent of lecithin, is vital for the prevention of fatty livers, the transmitting of nerve impulses, and the metabolism of fat. The classification of choline as a vitamin is debated, however, because it does not meet all the criteria for vitamins, especially those of the B vitamins. Of course, where to draw the line between an essential nutritional factor with vitaminlike activity and a vitamin is no easier than setting any boundary in nature.

The following facts favor classifying choline as an essential nutritional factor with vitaminlike activity, and *not* a vitamin:

1. Specific deficiency symptoms have not been observed in man.
2. The body can synthesize considerable choline, thereby reducing the need for dietary supplementation.
3. Choline is used in much larger quantities than any of the known vitamins.
4. Choline serves as a structural component of fat and nerve tissue, rather than a catalytic role characteristic of the vitamins.
5. It does not share the common ability of B vitamins to support the growth of microorganisms; only a very few require it.

On the other hand, the following biological roles favor classifying choline as a vitamin:

1. Although the need by man has not been established, choline is required for the growth of the young of many animal species; and deficiencies, evidenced primarily by liver and kidney damage, have been produced in rats, mice, dogs, chicken, pigs, hamsters, guinea pigs, rabbits, calves, ducklings, and monkeys.
2. Although choline can be synthesized in the body from serine and methionine with the aid of vitamin B-12 and folacin as coenzyme factors, it is not made fast enough and in sufficient quantity for many animal species—especially the young; besides, adequate building materials—serine, methionine, vitamin B-12, and folacin—must be present.
3. Choline is a key component of the phospholipid lecithin and is present in sphingomyelins, the principal bound forms of choline, which together make up 70 to 80% of the phospholipids in the body. Lecithin is important in the metabolism of fat by the liver, whereas sphingomyelin is involved in brain and nerve tissue. Choline also serves as a precursor of acetylcholine, which is physiologically important in the transmission of nerve impulses.

HISTORY. In 1844 and 1846, N. T. Gobley isolated a substance from egg yolk, which he called lecithin (from the Greek, *lekithos,* egg yolk. Choline is a basic constituent of lecithin).[22]

In 1849, Strecker, a German chemist, isolated a compound from hog bile, to which he subsequently (in 1862) applied the name choline (after *chole*, the Greek word for bile).[23]

In 1866-67, Baeger and Wurtz, working independently, determined the correct structure of choline and carried out the first synthesis of it.[24] But the compound did not attract the attention of nutrition investigators at the time.

In 1932, Best and co-workers at Toronto University reported that choline prevented fatty livers in rats fed high fat diets.

In 1940, Sure, and Gyorgy-Goldblatt, working independently, reported that choline is essential for growth of rats, thereby indicating its vitamin nature.

By 1942, the vitamin nature of choline was fully confirmed by many other workers, who used the rat, chicken and turkey as experimental animals.

CHEMISTRY, METABOLISM, PROPERTIES.

• **Chemistry**—Structurally, choline ($C_5H_{15}NO_2$) is a relatively simple molecule, containing three methyl groups (CH_3-) (see Fig. C-102).

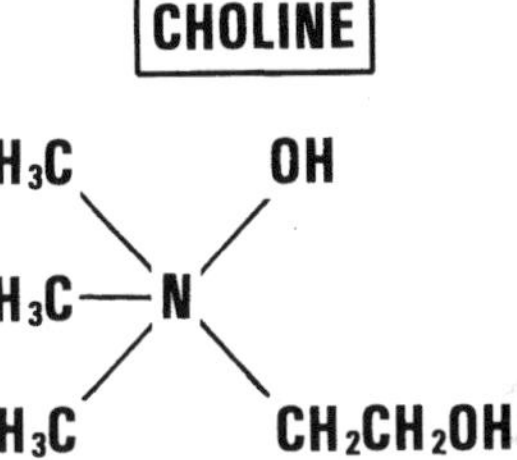

Fig. C-102. Structure of choline.

• **Metabolism**—Choline is absorbed in the small intestine.

• **Properties**—It is a colorless, bitter-tasting, water-soluble white syrup that takes up water rapidly on exposure to air (hygroscopic) and readily forms more stable crystalline salts with acids such as choline chloride or choline bitartrate. It is fairly stable to heat and storage, but unstable to strong alkali. It exists in all foods in which phospholipids occur liberally.

MEASUREMENT/ASSAY. The activity of choline and choline chloride is expressed in grams and milligrams of the chemically pure substances.

[22]McCollum, Elmer V., *A History of Nutrition,* Houghton Mifflin Company, Boston, 1957, p. 33.

[23]Edited by Sebrell, W. H., Jr. and Robert S. Harris, *The Vitamins,* Academic Press, New York and London, 1971, p. 4.

[24]Ibid, p. 12.

Choline content of foods is usually determined by a colorimetric method or by microbiological assay. Recent assay techniques include: fluorometric enzyme assay, enzymatic radioisotopic assay, and gas chromatography.

FUNCTIONS. Choline has several important functions in the body. As a constituent of several phospholipids (primarily lecithin), it prevents fatty livers through the transport and metabolism of fats; as a constituent of acetylcholine, it has a role in nerve transmission; and by a phenomenon known as *transmethylation*, it serves as a source of labile methyl groups, which facilitate metabolism.

1. **Prevention of fatty livers.** Choline is a "lipotropic agent"; lipotropic means "having an affinity for fat." In this role, choline prevents the abnormal accumulation of fat in liver (fatty liver) by promoting its transport as lecithin or by increasing the utilization of fatty acids in the liver itself. Without choline, fatty deposits build up inside the liver, blocking its hundreds of functions and throwing the whole body into a state of ill health.

2. **Nerve transmission.** Choline combines with acetate to form acetylcholine, a substance which is needed to jump the gap between nerve cells so that impulses can be transmitted.

3. **Facilitation of metabolism.** When the methyl group (CH_3-) is present in such form that it can be transferred within the body from one compound to another, it is called a labile methyl group and the process is called transmethylation. The body has a pool of labile methyl groups, which it uses for such purposes as (a) the formation of creatine (important in muscle metabolism), (b) methylating certain substances for excretion in the urine, and (c) the synthesis of several hormones, such as epinephrine. Among the dietary sources of labile methyl are choline (and related substances), the amino acid methionine, and the vitamins folacin and vitamin B-12. Of the related substances that can replace choline, the primary one is betaine (which derives its name from the Latin word *beta*—the beet family, which is a rich source). Each of these sources of a labile methyl group serves to "spare" (or partially make up for the shortage of) one of the others. Thus, choline can be fully replaced by betaine in preventing fatty liver in some species; in other species, betaine can only supplement choline. To some extent, methionine and vitamin B-12 can also spare choline in certain animal species.

Thus, the metabolic needs for choline can be supplied in either of two ways: by dietary choline as such, or by choline synthesis in the body which makes use of labile methyl groups. But the synthesis in the body cannot take place fast enough to meet the choline needs for rapid growth; hence, the symptoms of deficiency may result.

DEFICIENCY SYMPTOMS. Poor growth, fatty livers, and hemorrhagic kidney damage are the common deficiency symptoms in mammals. Chickens and turkeys develop slipped tendons (perosis). In young rats, choline deficiency produces hemorrhagic lesions in the kidneys and other organs.

RECOMMENDED DAILY ALLOWANCE OF CHOLINE. Experimental deficiency of choline in humans has not been produced. Of course, methionine can completely eliminate the mammalian requirements for choline, but not those for young birds.

Fatty infiltration of the liver is common in chronic alcoholics and in persons on very low protein diets (e.g., in children with kwashiorkor). But treatments of these disorders with choline have been inconsistent and disappointing.

The requirement for choline is influenced by the amounts of methionine, folacin, and vitamin B-12 in the diet, plus the growth rate of the individual, its energy intake and expenditure, the amount and type of dietary fat, the type of carbohydrate eaten, the total amount of protein in the diet, and possibly the amount of dietary cholesterol. As a result, the exact human requirement has not been established.

It is noteworthy, however, that a Committee of the American Academy of Pediatrics recommends that choline be added to infant formulas in amounts equivalent to breast milk. Human milk contains about 145 mg of choline per liter, nearly 0.1% of total solids.

It is noteworthy, too, that choline is added to most animal feeds in this country, and that high levels of dietary protein tend to increase the choline requirement of birds.

• **Choline intake in average U.S. diet**—It is estimated that an average mixed diet for adults in the United States contains 400 to 900 mg of choline per day.

TOXICITY. No toxic effects have been observed. However, oral pharmacologic doses of up to 20 g per day of choline chloride used for periods of several weeks in the treatment of fatty liver, alcoholism, and kwashiorkor have caused some patients to experience dizziness, nausea, and diarrhea.

CHOLINE LOSSES DURING PROCESSING, COOKING, AND STORAGE. Choline is heat-stable; hence, little of it is lost in processing and cooking.

Choline remains at nearly a constant level in dried foods when stored over long periods.

SOURCES OF CHOLINE. Choline is widely distributed in foods.

• **Rich sources**—Egg yolk, eggs, liver (beef, pork, lamb).

• **Good sources**—Soybeans, potatoes (dehydrated), cabbage, wheat bran, navy beans, alfalfa leaf meal, dried buttermilk and dried skimmed milk, rice polish, rice bran, whole grain (barley, corn, oats, rice, sorghum, wheat), hominy, turnips, wheat flour, blackstrap molasses.

• **Negligible sources**—Fruit, fruit juices, milk, vegetables.

• **Supplemental sources**—Yeast (brewers', torula), wheat germ, soybean lecithin, egg yolk lecithin, and synthetic choline and choline derivatives.

Lecithin, of which choline is a basic constituent, is a rich source of choline. Soybean lecithin and egg yolk lecithin have been used as natural concentrates of choline for dietary supplementation.

Also, the body manufactures choline from methionine, with the aid of folacin and vitamin B-12 as coenzymes. So, the needs for choline can be supplied in two ways: (1) by dietary choline, and/or (2) by body synthesis through transmethylation.

TOP FOOD SOURCES OF CHOLINE. Choline values are not given in Food Composition Table F-36 of this book, simply because few such values are available. So, the top food sources are given in the accompanying boxed list.

(Also see VITAMIN[S], Table V-9.)

TOP FOOD SOURCES OF CHOLINE along with energy values

Food	Choline	Energy
	(mg/100 g)	(kcal/100 g)
Eggs	527	157
Brewers' yeast	408	283
Liver (beef, pork, lamb)	356	136
Wheat germ	306	391
Soybean seeds	290	405
Torula yeast	289	277
Potatoes, dehydrated	262	334
Cabbage, raw	254	24
Soybean flour	225	386
Wheat bran	188	353
Navy beans, dried	168	340
Buttermilk, dehydrated	167	387
Alfalfa leaf meal, dehydrated	150	215
Milk, skim, dehydrated	139	360
Rice polish	122	265
Rice bran	122	276
Oats, whole grain	101	283
Wheat, whole grain	101	360
Hominy	99	358
Rice, whole grain	93	363
Turnip greens	91	30
Barley, whole grain	90	305
Wheat flour	80	361
Blackstrap molasses	74	230
Sorghum, whole grain	63	339
Corn, whole grain	54	348

CHOLINERGIC

A type of nerve fiber which utilizes acetylcholine as a transmitter. Some of these nerves stimulate the secretion of digestive juices and the movement of the digestive tract. Also, they slow the heart.

(Also see DIGESTION AND ABSORPTION.)

CHOLINESTERASE

An enzyme which breaks down the nerve transmitter acetylcholine at the nerve junction. The breakdown of the transmitter is necessary to restore the nerve to its resting state so that it is made ready to transmit another impulse. Certain foods contain cholinesterase inhibitors, which may produce toxic effects when eaten by man or animals.

(Also see POISONOUS PLANTS.)

CHONDROITIN SULFATE

A mucopolysaccharide which is an important component of cartilage.

CHOP SUEY

This is a favorite Chinese dish containing chicken, bean sprouts, onion, bamboo shoots, and mushrooms; cooked and served with a Chinese gravy.

CHOWCHOW

- Consisting of several kinds mingled together; hodgepodge, assorted, or mixed.
- A Chinese preserve or confection of ginger, fruits, and peels in heavy syrup.
- A spicy relish of chopped mixed pickles in mustard sauce.

CHOWDER

Chowder is a word of French origin, derived from the French word *chaudiere*, the vessel in which the French of the coastal regions cooked fish soups and stews.

Chowder is a thick soup or stew of seafood (like clams or white-fleshed sea fishes), usually made with milk and containing salt pork or bacon, onion, potatoes, and sometimes other vegetables.

CHROMATIN

The part of the cell nucleus that stains most intensely; it contains the chromosomes.

CHROMATOGRAPHY

A method used in analytical chemistry first employed by a Russian botanist in 1903. Initially, the technique was used to separate colored substances; hence, the name chromatography, from the Greek meaning "color writing." Today, many of the compounds separated by chromatography are colorless, so the name no longer relates to the underlying principles.

There are a number of variations of the general chromatographic technique. Partition chromatography was introduced in 1941, paper chromatography in 1944, gas chromatography in 1952, and thin-layer chromatography in 1958.

Fig. C-103. Gas chromatograph being injected with a sample of extracted fruit to analyze for pesticide residues. The toxicity of pesticides and their effect on nutrient content of food must be determined in order that safe standards can be set. (Courtesy, University of Maine)

Numerous materials from foods, such as proteins, amino acids, sugars, fatty acids, minerals, coffee aroma, and many other components, are routinely identified and quantitated through this analytical technique. In addition to nutrient analysis, chromatography can be adapted to the detection of drug residues, hormones, pesticides and other food contaminants, smog, and cigarette smoke.

CHROMIUM (Cr)

Contents *Pages*

History 416
Metabolism of Chromium 417
Functions of Chromium 418
Deficiency Symptoms 419
Diagnosis of Chromium Deficiency 419
Evidence of Chromium Deficiencies 419
Disorders Which May Result From Chromium Deficiency 419
Treatment and Prevention of Chromium Deficiency 420
Interrelationships 421
Recommended Daily Allowance of Chromium 421
Conserving the Body's Supply of Chromium 421
Toxicity 421
Sources of Chromium 422

The mention of this chemical element makes most people think of the "chrome" plating on the bumpers and body trim of their automobiles. However, it was recently discovered that the shiny metal may also exist in forms which function as (1) an essential element, (2) a hormone, (3) a vitamin, and (4) a poison. A very comprehensive treatment of the nutritional aspects of this mineral is presented because (1) there is a lack of detailed literature on this subject for the lay person, and (2) few people are aware of the health consequences which might result from chromium deficiencies.

HISTORY. Chromium was discovered by the French chemist Vauquel in 1797, while he was studying the properties of crocoite, an ore which is rich in lead chromate. Its common name of chrome was derived from the Greek word *chroma,* which means color, because the element is present in many different colored compounds. These compounds have long been used as pigments in dyeing, and in the tanning of leather. In the early 1900s, chromium became an important ingredient of corrosion-resistant metals—a use which has increased to the present. Because of this use, people living in or near industrialized areas are likely to be exposed to air, food, and water contaminated by traces of chromium compounds. Unfortunately, much of the inorganic chromium in the environment is harmful rather than helpful to the body, because certain compounds of the element injure body tissues.

It was not until 1959 that the medical scientists W. Mertz and K. Schwarz—who came to the United States from Germany—discovered that the feeding of chromium salts corrected the abnormal metabolism of sugar in rats which resulted from the feeding of diets based upon torula yeast. Later work by these researchers, and by H. Schroeder of the Dartmouth Medical School, established chromium as a cofactor with insulin, necessary for normal glucose utilization and for growth and longevity in rats and mice. (Dr. Schwarz had long studied the nutritional effects of various types of yeasts; in 1957, he had discovered that selenium was present as a vital factor in American type of brewers' yeast.) Shortly thereafter, it was found that the inorganic salts of chromium were utilized poorly, compared to an organically bound form of chromium present in brewers' yeast which was utilized well by both animal and man. The chromium-containing substance from yeast was named the *glucose tolerance factor* (GTF), because it sometimes restored the metabolism of sugars to normal when diabeticlike tendencies were present. (The term *glucose tolerance* denotes the ability of the body tissues to take the sugar glucose from the blood. Hence, it is measured by the rate at which the blood sugar drops back to normal after a test dose of the sugar has been administered to the patient. The blood sugar of diabetics remains abnormally high during the test of glucose tolerance.) Diabetic animals and people were *not* always helped by either inorganic chromium or the glucose tolerance factor because their disease may have been caused by factors other than a deficiency of chromium.

Recently, evidence was found which suggested that other disorders such as atherosclerosis, cataracts, and high blood fats might be the results of prolonged deficiencies of chromium. Details of these discoveries are given in the sections which follow.

METABOLISM OF CHROMIUM. Further studies are needed to determine chromium's role in metabolism and its nutritional significance. Nevertheless, current thinking is reflected in the sections that follow.

• **Absorption**—Studies have shown that only about 1%, or less, of the dietary intake of inorganic chromium is absorbed, whereas as much as 10 to 25% of the GTF-chromium may be absorbed. Chromic oxide, an inorganic form of the mineral, is so insoluble that it has found wide application as a "marker" for determining the digestibility of components of the diet and of feed intakes by grazing livestock. (All of the marker which is put into the feed appears in the feces—it is not absorbable or degradable. Hence, when used as an indicator, chromic oxide provides a quick, indirect way of determining the digestibility of feed.)

(Also see DIGESTIBILITY.)

• **Age**—It appears that the ability of the body to utilize inorganic salts of chromium decreases with aging. The problem appears to be related to the metabolism of chromium in the tissues as well as with its absorption because tissue levels in older experimental animals were half those of younger ones, after both groups had been injected with chromium.[25] (The injections bypassed the absorptive processes.)

• **Alkalinity of the intestinal contents and the blood**—One of the reasons for the poor absorption and utilization of the common inorganic forms of chromium is that they are converted in the intestine and elsewhere in the body to triply charged chromium ions, which react with alkali ions to form insoluble masses that are biologically inactive. Furthermore, both the intestinal contents and the blood are mildly alkaline. Hence, it is noteworthy that the binding of inorganic chromium by substances such as (1) oxalates (present in rhubarb and spinach), (2) phosphates (occur naturally or are present as additives in many foods), or (3) tartrates (in grapes and other fruits, baking powders, and wines) prevents the formation of the insoluble masses and helps to keep chromium in solution so that it might be more readily absorbed and transported in the blood under alkaline conditions.

Chromium which is found in certain organic molecules, such as the glucose tolerance factor, appears to be much better absorbed and utilized by the body because it is protected against adverse chemical reactions which affect inorganic chromium.

• **Anemia**—Anemic people, or those who consume diets low in iron, may have better utilization of inorganic chromium because both elements are carried in the blood by the same protein, transferrin. These people have less than normal amounts of iron bound to their transferrin, so the protein has a greater capacity for carrying chromium.

Inorganic chromium is also carried in the blood by the protein albumin, whereas organically-bound chromium is carried in the blood as a component of the glucose tolerance factor. These forms of chromium do not appear to be affected by the iron nutritional status.

[25]Underwood, E. J., *Trace Elements in Human and Animal Nutrition,* 4th ed., Academic Press, Inc., New York, N.Y., 1977, p. 261.

• **Chelating agents**—Other molecules that bind (chelate) mineral elements may alter the amounts of chromium which are absorbed and retained within the body. For example, oxalate (occurs in spinach and rhubarb) increases, while phytate (present in whole grains and legumes) decreases the absorption of chromium salts. However, the feeding of oxalate also causes a great increase of chromium in the urine.

• **Diabetes**—When other conditions are similar, diabetics absorb a greater percentage of inorganic chromium than normal people, but they also excrete much more of the element in their urine.

• **Dietary carbohydrate**—It is suspected that overconsumption of low-chromium, high-carbohydrate foods such as white flour and white sugar may rapidly deplete the body's stores of chromium by overstimulating both the release of the mineral from the tissues and its subsequent loss in the urine.

• **Fats**—One way in which dietary fat may reduce the absorption of chromium is by stimulating the flow of bile, which is an alkaline secretion. Alkaline substances tie up inorganic chromium in soluble masses that cannot be absorbed.

• **Infant malnutrition**—A diabeticlike disorder of metabolism is often found in malnourished infants and children around the world. Therefore, it is noteworthy that this disorder was promptly corrected by the administration of chromium salts to young victims of malnutrition in Jordan, Nigeria, and Turkey. However, malnourished Egyptian children apparently did not benefit from chromium supplements, because they had already received adequate amounts of the mineral in their diets.

• **Intestinal microorganisms**—It has been suggested that the unusually good utilization of inorganic chromium by malnourished infants may be due to its conversion to the glucose tolerance factor by microorganisms present in their intestines. This idea is supported by the finding that chromium stimulates the growth of *Aerobacter aerogenes*—a bacteria found in foods such as cereals, soured dairy products, and vegetables; and present in the human intestine. Perhaps, people eat the foods which contain this bacteria and it gets into the intestine where it converts inorganic chromium to a more biologically active form (like the way in which brewers' yeast synthesizes the glucose tolerance factor from chromium).

• **Other dietary minerals**—Zinc-deficient animals absorb five times as much inorganic chromium as those given zinc supplements. Other minerals that appear to interfere with the absorption and/or the utilization of chromium are calcium, iron, and manganese. Thus, the taking of supplements which contain large doses of these minerals may contribute to the development of a chromium deficiency.

• **Pregnancy**—The unborn child draws chromium from its mother, as evidenced by the fact that the hair of newborn infants contains about 2½ times the level of chromium as the hair of their mothers. There is evidence that chromium content of hair may be a good indicator of the chromium present in other tissues such as the liver. Furthermore, the levels of chromium in the hair of women who had borne children were found to be much lower than those in women who were childless. Hence, it appears that pregnant women may not consume enough readily available chromium to meet both their own needs and those of their unborn children. Perhaps, chromium deficiency is responsible for some of the cases of diabetes in women which appear to have been brought on by pregnancy.

Other studies have produced evidence that most of the chromium loss from the mother is likely to occur during the first pregnancy, since the chromium content of the mother's hair does not appear to decrease further during subsequent pregnancies. If such is the case, then it leads one to speculate whether children other than the firstborn begin life with an adequate supply of chromium in their bodies.

• **Storage of chromium in the body**—The storage of chromium by the body may provide insurance against the development of diabeticlike disorders at times when the requirements for the glucose tolerance factor (GTF) exceed the amounts which (1) are supplied preformed in the diet, or (2) are synthesized from dietary inorganic chromium. However, animal studies have shown that a greater proportion of a given dose of chromium is stored in the liver when it is supplied as GTF, than when it is supplied as an inorganic salt. Furthermore, chromium in the liver has been shown to have GTF activity. Next to the liver, the kidneys appear to be one of the best sources of GTF-chromium. Finally, it is noteworthy that diabetics were found to have had only about two-thirds as much chromium in their hair and their liver as normal, healthy people.

• **Stresses**—Various stressful conditions, such as malnutrition and loss of blood, have long been known to impair the body's utilization of sugars, or even to bring on diabetes. Likewise, it seems that the chromium needs of the body become more critical under such conditions. For example, studies on animals have shown that the effects of (1) low-protein diets accompanied by controlled exercise, and (2) the withdrawal of measured amounts of blood were more severe in the chromium-deficient groups than in those given supplements containing the mineral.

• **Excretion of chromium**—The predominant route of excretion of endogenous chromium is the urine. The average daily loss is about 7 to 10 mcg.

FUNCTIONS OF CHROMIUM. Identifying functions is difficult because this essential element does not do its work by itself; rather, it appears to act cooperatively with other substances that contol metabolism, such as (1) a hormone—insulin, (2) various enzymes, and (3) the genetic material of the cell—DNA and RNA. Chromium has a variety of functions, including (1) component of the glucose tolerance factor (GTF), (2) activator of certain enzymes, (3) stabilizer of nucleic acids (DNA and RNA), and (4) formation of fatty acids and cholesterol. Details follow.

• **Component of the glucose tolerance factor (GTF)**—The complete identity of this hormonelike agent is not yet known, although it is certain that it contains chromium and the vitamin niacin, and perhaps amino acids such as glycine, glutamic acid, and cysteine.

Chromium in the form of the GTF is released into the blood—from perhaps the liver, kidneys, or other tissues which store chromium—whenever there is a marked increase in the blood levels of sugar (glucose) and/or of insulin. *Hence, the GTF might be considered to behave like a hormone.* It, along with insulin, acts in making it easier for amino acids, fatty acids, and sugars to pass from the blood into the cells of various tissues. It also promotes the metabolism of the nutrients within the cells. Much more insulin is required to accomplish these tasks when GTF is lacking; but GTF does not have any effect when insulin is absent. (Some types of diabetes are caused by lack of insulin.) Other noteworthy actions which are jointly promoted by GTF and insulin are (1) utilization of amino acids for the synthesis of proteins, (2) improvement in the ability of germ eating (Phagocytic) white blood cells to search out noxious bacteria (this function of the cells is impaired in diabetics), and (3) utilization of glucose by the lens of the eye.

Each time that the GTF is called upon to do its work, there is a corresponding rise in the amount of chromium excreted in the urine. Apparently, the GTF is utilized less efficiently by diabetics who need injections of insulin than it is by normal, healthy people. Hence, diabetics might have above average needs for this factor.

Various experimental studies have produced evidence that the bodies of many people are able to convert inorganic chromium into the GTF, but that the speed of this conversion slows with aging. For example, the impaired sugar metabolism in a group of malnourished Jordanian infants was corrected overnight by a chromium supplement; whereas, from 1 to 3 months were required for noticeable improvement in sugar metabolism to be achieved in elderly people.

Pregnant women and diabetics, who appear to have above average requirements for the GTF, may not be able to synthesize adequate amounts of this factor from the inorganic chromium that is supplied in their diets. Hence, they may need dietary sources of preformed GTF in order to avoid the depletion of the limited supplies of this factor in their bodies. Furthermore, the results of studies in animals suggest that GTF, but *not* inorganic chromium, passes through the placenta from the mother to her unborn child.

• **Activator of certain enzymes**—Chromium is the spark plug which "fires up" certain enzymes into vigorous metabolic activity. Most of these enzymes are involved in the production of energy from carbohydrates, fats, and proteins. However, some of these enzymes may also be activated by other metal elements such as aluminum, iron, manganese, and tin. Similarly, chromium activates the digestive enzyme trypsin; although here, too, other metals may act as substitutes. Hence, the activities of these enzymes may not be noticeably depressed when there is a chromium deficiency.

• **Stabilizer of nucleic acids (DNA and RNA)**—Chromium and certain other essential elements appear to stabilize nucleic acids (mainly RNA) against distortions of their structures. Hence, it is tempting to speculate that these elements may help to prevent mutations of the genetic material within cells, and thereby prevent the development of cancer and similar diseases. However, there is little evidence either to support or refute this speculation.

• **Formation of fatty acids and cholesterol**—Chromium is a stimulator of synthesis of fatty acids and cholesterol in the liver.

DEFICIENCY SYMPTOMS. Chromium deficiency, which is believed to be relatively common in the United States, is manifested by impaired glucose tolerance, which may be accompanied by high blood sugar and the spilling of sugar into the urine. It is seen especially in older persons, in maturity-onset diabetes, and in infants with protein-calorie malnutrition.

There are *no* clear-cut signs or symptoms of the early stages of chromium deficiency because it appears that tissue stores are drawn upon to meet the needs of the body. Even after the body's reserves have been exhausted, it may take time before any signs or symptoms of a deficiency appear, because enough extra insulin may be secreted to compensate for the reduced effectiveness of the hormone which occurs in chromium deficiency. However, an excessive secretion of insulin is thought to be undesirable, because it may lead to low blood sugar, unhealthy fattening, atherosclerosis, or even exhaustion and damaging of the insulin-secreting cells of the pancreas. Thus, diabetes—accompanied by the usual signs or symptoms of impaired growth in children, sugar in the urine, frequent and excessive urination, abnormal thirst, increased hunger, loss of weight, nausea, and tiredness—may develop if the pancreas is overstressed by chromium deficiency and/or other conditions. Chromium deficiency is also characterized by disturbances in lipid and protein metabolism.

Diagnosis of Chromium Deficiency. The morbid consequences of chromium deficiencies make it desirable to detect them before there is irreparable damage to body tissues. However, the laboratory tests which provide the most information regarding chromium nutritional status are expensive and inconvenient for the patient. The screening of patients according to their recent dietary histories is of little value because (1) there is a lack of information regarding the chromium content of the various foods eaten by people in the United States; (2) such information tells little about the body stores of chromium, even if the recent chromium intakes could be determined; and (3) other conditions such as intestinal microorganisms and stresses may have altered the utilization of dietary chromium.

Several laboratory tests may be used to diagnose chromium deficiency. Most likely, a combination of two or more tests will be needed.

• **Analysis of hair samples**—It is hoped that this test may serve at least as a screening method to identify people whose body stores of chromium are low because (1) the levels of the mineral in hair appear to parallel, roughly, the levels in other body tissues, and (2) extracts of hair have been found to act like the glucose tolerance factor (GTF). People whose hair levels of chromium are found to be low might then be given sugar and chromium metabolism tests.

• **Blood sugar test (glucose tolerance test)**—The term "glucose tolerance" refers to the speed at which an elevated level of blood sugar—produced by feeding or injecting glucose—drops back to normal as the cells of the body remove the sugar from the blood. Both insulin and chromium are needed to facilitate the passage of sugar and other nutrients from the blood into the cells, so an abnormally slow return of an elevated blood sugar level to normal may be indicative of a deficiency of insulin and/or of chromium. Hence, blood levels of insulin may also have to be determined if the glucose tolerance test is to be used to diagnose chromium deficiency; unless the glucose tolerance test is given on two separate occasions, both before and after the administration of chromium. An improvement in glucose tolerance after one or more doses of chromium may mean that the patient has a deficiency of the mineral.

(Also see HYPOGLYCEMIA, section headed "Glucose Tolerance Test.")

• **Chromium in the urine**—Normally, only small amounts of chromium are excreted in the urine. However, the administration of a test dose of glucose usually causes a marked increase in the amount of the mineral in the urine, unless the patient happens to be chromium deficient. Hence, the use of this test requires that the urine be collected and analyzed for chromium content both before and after the administration of the sugar. The test is not valid for diabetics who require injections of insulin, because they usually excrete abnormally high amounts of chromium in the urine.

Evidence of Chromium Deficiencies. Evidence suggesting that many Americans might have various degrees of chromium deficiency began to accumulate in the 1960s when it was found that some people who showed diabetic tendencies benefited from chromium supplementation. It was concluded that the people who showed improvement after receiving chromium had suffered from a deficiency which aggravated the disease.

Further evidence of chromium deficiencies in the United States is provided by the low chromium levels found in the tissues obtained from deceased persons of various ages and nationalities, and of both sexes. Also, it was found that people from Europe, Africa, and Asia had, on the average, several times as much chromium in their aortas, hearts, kidneys, livers, and spleens as the average American.

Disorders Which May Result from a Chromium Deficiency. Much of the evidence of the role of chromium deficiency in the disorders that follow is circumstantial. Nevertheless, some current concerns are presented, in order that readers may (1) be alerted to the dangers of low chromium diets, and (2) make special efforts to plan their diets so as to obtain sufficient chromium.

• **Adult-onset diabetes**—People who have this disease—which occurs mainly in obese people, and is characterized by an abnormal metabolism of sugar in spite of an apparently adequate secretion of insulin—may sometimes be helped by supplemental chromium in the form of either inorganic salts or the glucose tolerance factor (usually provided as brewers' yeast, or as a special extract of this substance). Perhaps, some of the people who developed this disease after having had an apparently healthy childhood underwent a gradual depletion of their body stores of chromium, which brought on the abnormal condition.

• **Atherosclerosis**—This disorder—which is the buildup of abnormal patches or "plaques" on the walls of arteries—has been produced by making animals chromium deficient, and it has been prevented by chromium supplementation. Researchers have also found that chromium cannot be detected in the aortas of people who die from atherosclerosis, whereas it is almost always present in the aortas of people who die accidently.

• **Cloudy spots on the lens of the eye (cataracts)**—Diabetics have an above-average likelihood of developing cataracts because their abnormally slow utilization of carbohydrates allows unmetabolized sugars to accumulate in the lens of the eye. Studies of the lenses removed from the eyes of rats showed that, although insulin causes a slight speedup of sugar metabolism in these tissues, a much greater effect was obtained by the administration of chromium along with insulin.

• **Growth retardation in children**—It has long been known that diabetic children may be stunted in growth due to lack of insulin, one of the hormones which promotes the synthesis of protein for the building up of body tissues. However, even children with metabolic abnormalities that are milder than diabetes may also suffer from growth retardation. Therefore, it is noteworthy that chromium supplementation of diets promotes better growth, both in rats fed low-protein diets and in children suffering from protein-energy malnutrition.

• **Heart Disease**—The Framingham study of the risk factors associated with heart disease—which was conducted from 1949 to 1969 on over 5,000 men and women, aged 30 to 62, in Framingham, Massachusetts—showed that people with diabetic tendencies had a much higher risk of heart disease than those without such tendencies. Hence, it appears that chromium might be linked in various ways to heart disease, because (1) lack of this mineral results in an abnormal metabolism of sugars, like that which occurs in diabetes; and (2) hearts obtained from deceased people in countries with low rates of heart disease contained several times as much chromium as the hearts of deceased Americans.

• **High blood fats (cholesterol and triglycerides)**—Diabetes and other disorders of sugar metabolism are often accompanied by high blood levels of fatty substances such as cholesterol and triglycerides. Researchers report that a group of people with these conditions had lower levels of blood fats after their diets were supplemented with either inorganic chromium, or with brewers' yeast containing the glucose tolerance factor. However, the declines in the blood fats due to chromium therapy averaged only about 14%.

• **Impotence and frigidity**—These problems might be the first signs of unsuspected diabetes because they are sometimes due to impairments of the nerves which usually accompany this disease. Occasionally, chromium deficiency may contribute to the development of diabetes.

• **Low resistance to infections**—People who have diabetes may be overly susceptible to infections because their germ-eating white blood cells may have lost some of their power to locate and engulf infectious microorganisms. However, the addition of small amounts of inorganic chromium to samples of these weakened cells in test tubes has been shown to restore their vitality for fighting infections.

• **Neurological disorders**—Although it is common knowledge that disorders of the nerves often develop in diabetics, recently doctors learned that chromium deficiency alone may sometimes be the cause of this problem, judging from the case of a 38-year-old female whose sole source of nourishment for more than 3 years was intravenous feedings.[26] The feedings had supposedly furnished all of her nutrient requirements, but she lost weight and developed various other signs of diabetes, such as abnormal sugar metabolism and disorders of the nervous system. The addition of insulin to her intravenous formula was ineffective. Analysis of her blood and hair for chromium revealed severely deficient levels. Thereupon, an inorganic chromium salt was added to her intravenous feedings, and the insulin was withdrawn. The diabetic condition was corrected within 2 weeks, but it took about 5 months for the normal functioning of her nervous system to be restored.

• **Unexpected loss of weight**—The inability to maintain body weight when sufficient dietary energy is consumed may be a sign of uncontrolled diabetes, or, in some cases, of an unsuspected deficiency of chromium. Furthermore, chromium supplements have been found to increase the weight gain of infants undergoing treatment for protein-energy malnutrition.

TREATMENT AND PREVENTION OF CHROMIUM DEFICIENCY. The treatment of chromium deficiency is strictly a job for the physician because he or she will have to decide on the form of chromium which is most suitable for the patient. Very ill people may even require injections of the mineral.

The prevention of chromium deficiency requires that application of such information as (1) the daily needs of various types of people, (2) food sources of well-utilized forms of the mineral, and (3) the maximum amounts of chromium which may be taken without toxic effects. Unfortunately, there is a paucity of information on this subject.

[26]Jeejeebhoy, K. N., *et al.*, "Chromium Deficiency, Glucose Intolerance, and Neuropathy Reversed by Chromium Supplementation, in a Patient Receiving Long-term Total Parenteral Nutrition," *The American Journal of Clinical Nutrition*, Vol. 30, 1977, p. 531.

INTERRELATIONSHIPS. The following interrelationships are pertinent:

1. Chromium functions best in the body when it is in the form of the GTF.
2. Diets rich in carbohydrates may cause the supply of GTF-chromium to be depleted.
3. Inorganic chromium is utilized by many people, but less efficiently than that in GTF.
4. The absorption of chromium is impeded by oxalates (in rhubarb and spinach, and phytates (present in whole grains).
5. Zinc and vanadium antagonize the effects of chromium.

RECOMMENDED DAILY ALLOWANCE OF CHROMIUM. A normal healthy adult loses about 1 mcg (microgram) of chromium daily in his or her urine. The dietary intake of chromium needed to replace this loss ranges from 4 mcg of GTF-chromium in brewers' yeast (as much as 25% of this form may be absorbed) to 200 mcg of chromium from an inorganic salt (as little as 0.5% of this form may be absorbed).

It is difficult to establish an allowance for this mineral, because of the limited amount of data regarding its availability from various food sources. Therefore, some intakes that have been estimated to be suitable by the National Academy of Sciences are presented in Table C-29.

TABLE C–29
ESTIMATED SAFE AND ADEQUATE
DAILY DIETARY INTAKES OF CHROMIUM[1]

Group	Age	Chromium Intake[2]
	(years)	(mcg)
Infants	0–½	10–40
	½–1	20–60
Children	1–3	20–80
	4–6	30–120
Adolescents	7–10	50–200
	11+	50–200
Adults		50–200

[1]Adapted by the authors from *Recommended Dietary Allowances*, 10th ed., 1989, NRC–National Academy of Sciences, p. 284.
[2]The intakes are expressed as ranges because the available data do not permit the definition of a single intake.

• **Chromium intake in average U.S. diet**—A range of chromium intakes between 50 and 200 mcg per day is tentatively recommended for adults. This range is based on the absence of signs of chromium deficiency in the major part of the U.S. population consuming an average of 50 mcg/day.[27]

CONSERVING THE BODY'S SUPPLY OF CHROMIUM. Although the present evidence is mainly circumstantial, some researchers suspect that cardiovascular diseases and diabetes may be more severe in people who have very low body stores of chromium. Therefore, it is important to prevent excessive urinary excretion of this essential element, while also making every effort to obtain adequate amounts of dietary chromium to replace the daily losses. Some dietary measures which may help to reduce the urinary losses of chromium follow. *NOTE:* It is not known for certain whether these measures are effective in cutting down the loss of chromium in the urine, since very little research has been performed on this subject. Nevertheless, all of the recommendations represent good nutritional practices, so their application may bring other benefits.

• **Avoidance of excessive amounts of highly refined sugars**—The object of this practice is to avoid the sudden burdening of the body with large loads of readily available sugars which may overstimulate the secretion of both insulin and glucose tolerance factors (GTF), since these conditions tend to increase the amount of chromium lost in the urine. This means that it is better to take dietary carbohydrate in the form of starches combined with fiber, such as occurs in legumes and whole grain products, than it is to rely largely on foods rich in simple sugars.

• **Maintenance of an appropriate body weight**—People who are markedly underweight or overweight may be troubled with an abnormal carbohydrate metabolism, so efforts should be made—with the help of a doctor and/or a dietitian when necessary—to attain the weight which is normal for one's age, sex, height, and body build. "Crash" or semistarvation diets are not recommended, because they may promote the tearing down of vital protein tissues in the body (such as the heart, kidneys, muscles, and liver), which are believed to store much of the body's supply of chromium. The sudden release of large amounts of chromium into the blood is likely to result in the copious loss of this element in the urine.

TOXICITY. It is unlikely that people will get too much chromium in their diets because (1) only minute amounts of the element are present in most foods, (2) the body utilizes it poorly, and (3) there is a wide range of safety between the helpful and the harmful doses of chromium (the toxic dose is about 10,000 times the lowest effective medicinal dose). However, the taking of chromium in the form of inorganic salts may cause a reduction in the absorption of other trace elements, as evidenced by the fact that doses of inorganic chromium reduce the absorption of zinc in experimental animals. *NOTE: Under no conditions should a lay person take an inorganic chromium salt without the prior advice of a doctor, because certain types of these compounds—which are readily available for nonmedicinal uses—may be deadly poisons.* Finally, the administration of chromium in the form of the glucose tolerance factor (GTF) to diabetics who require injections of insulin may cause their blood sugar levels to drop to dangerously low levels because the GTF enhances the effects of insulin. Certain researchers have even speculated that perhaps some of the diabetics who find it difficult to manage their disease might have a highly variable dietary intake of both chromium and glucose tolerance factor. Thus, the effects of a regular dose of insulin may vary from day to day.

(Also see DIABETES; MINERAL[S]; and POISONS.)

[27]*Recommended Dietary Allowances*, 10th ed., NRC–National Academy of Sciences, 1989, p. 242.

SOURCES OF CHROMIUM. Groups by rank of some common food sources of chromium follow:

• **Rich sources**—Blackstrap molasses, cheese, eggs, liver.

• **Good sources**—Apple peel, banana, beef, beer, bread, brown sugar, butter or margarine, chicken, cornflakes, cornmeal, flour, oysters, potatoes, vegetable oils, wheat bran, whole wheat.

• **Fair sources**—Carrots, green beans, oranges, spinach, strawberries.

• **Negligible sources**—Milk, most fruits and vegetables, sugar.

• **Supplemental sources**—Brewers' yeast, dried liver.

For additional sources and more precise values of chromium, see Table C-30, The Chromium Content of Some Foods, which follows.

TABLE C-30
THE CHROMIUM CONTENT OF SOME FOODS[1]

Food	Chromium
	(mcg/100 g)
Dried liver	170
Brewers' yeast	118
Blackstrap molasses	115
Eggs	52
Cheese	51
Liver	50
Wheat bran	40
Beef	32
Wheat, whole grain	29
Apple peel	25
Wheat germ	25
Potatoes	24
White flour	23
White bread	20
Oysters	20
Brown sugar	18
Butter or margarine	15
Chicken	14
Cornflakes	14
Vegetable oils	13
Cornmeal	11
Banana	11
Spinach	9
Carrots	8
Orange	5
Green beans	4
Strawberries	3
Mushrooms	3
White sugar	2
Milk	1

[1]These values represent the total amount of chromium present in a food, and not the amount available to the body.

NOTE WELL: The content and/or availability of chromium in foods may be affected by the following:

1. **The chromium content of the soil.** Soil content of chromium affects the chromium content of plants grown therein.

2. **The processing of grains.** As with other essential minerals, there is much more chromium in the outer layers of grains—the branny layers which are usually removed—than there is in the inner layers.

3. **The refining of molasses.** Generally, the darker the color (and the less the refinement), the richer the chromium content of molasses.

4. **The type of cooking utensil.** The heating of acid foods—such as tomato sauce—in stainless steel equipment may result in some of the chromium content of the stainless steel being leached out into the food.

5. **Fermentation.** Brewers' yeast converts inorganic chromium into a highly active organic substance—the glucose tolerance factor. Thus, beer, which is produced through the action of brewers' yeast, may have (a) much of its chromium in the form of the glucose tolerance factor, and (b) all of its chromium content is in an alcohol-extractable form, which has a high level of biological activity.

Other fermented food which also may be good sources of glucose tolerance factor are: wine, apple cider, cider vinegar, wine vinegar, root beers, yeast-leavened whole grain breads, pickles, summer sausages, cheeses and their derivatives, and sauerkraut.

6. **The alcohol-extractable fraction.** A close correlation appears to exist between the amount of chromium in the alcohol-extractable fraction and the level of GTF activity. Therefore, it is tempting to speculate that the alcohol-extractable fraction is an organically bound form that might be more readily absorbed and utilized than the chromium which cannot be so extracted. This might mean that the chromium in such alcoholic beverages as beer and wine is highly available to the body. Of course, Table C-30, The Chromium Content of Foods, like most food composition tables, gives only the total chromium content of foods. At this time, the relationship between total chromium, alcohol-extractable chromium, and the human dietary requirement is uncertain; hence, further research in this area is needed.

(Also see MINERAL[S], Table M-67; and POISONS, Table P-11 Some Potentially Poisonous [Toxic] Substances.)

CHRONIC DISORDER

A long-lasting disorder, in contrast to one that is acute.

CHUTNEY

The word *chutney* is derived from an East Indian word meaning to taste or lick. Chutneys are either like a sweet pickle or a condiment—with the consistency of jam. They are highly seasoned and contain a variety of chopped fruits and vegetables. Chutneys are served with cold meats, sausages, or stews.

CHYLE

The globules of emulsified fat which are formed in the small intestines as a result of digestion, and which are transported as chylomicrons via the lymphatics.

(Also see DIGESTION AND ABSORPTION.)

CHYLOMICRON

Complex molecules of protein and fats which are (1) synthesized in the intestinal mucosa, (2) carried by the lymphatics to the blood, and (3) circulated in the blood until metabolized by the liver or other tissues. After a meal high in fat, the blood plasma has a milky appearance due to the large number of circulating chylomicrons. However, after an overnight fast, very few of the chylomicrons should be present and the plasma should be clear, unless there is a disorder of fat metabolism called hyperchylomicronemia (also called Type I-Hyperlipoproteinemia), in which clearing is impaired due to the lack of substance called "clearing factor."

(Also see CLEARING FACTOR; FATS AND OTHER LIPIDS; and HYPERLIPOPROTEINEMIAS.)

CHYME

The mixture of partially digested food and gastric juice which passes from the stomach into the small intestine.

(Also see DIGESTION AND ABSORPTION.)

CHYMOSIN

An out-of-date name for the milk curdling enzyme rennin.

(Also see RENNIN.)

CHYMOTRYPSIN

A digestive enzyme secreted by the pancreas, which acts on proteins.

(Also see DIGESTION AND ABSORPTION; and ENZYME.)

CIBOPHOBIA

Another name for sitophobia, an abnormal fear of eating. Cibophobia differs from anorexia since appetite may persist but the person fears eating because of some associated or subsequent discomfort.

CIDER

In the United States this term usually refers to unfermented, unpasteurized, fresh apple juice; whereas, in Europe it means apple juice that has been fermented. Europeans also make a fermented cider from pears, which is called "perry." Unpasteurized cider will begin to ferment within a few days unless it is kept from doing so by the addition of a preservative such as sodium benzoate or potassium sorbate.

Americans call fermented cider "hard cider." In the eastern United States, hard cider is sometimes called "Jersey lightning." The alcohol content of the cider may be increased by storing it outside in the winter so that some of the water content freezes to ice. Then, the alcoholic cider that remains liquid can be drawn off and the process repeated until the alcohol content is sufficiently high to prevent freezing.

(Also see APPLE; and PEAR.)

CIEDDU (FERMENTED MILK)

A type of fermented milk in Italy.

CIRCULATORY SYSTEM

The system of blood, blood vessels, lymphatics, and heart concerned with the circulation of the blood and lymph.

CIRRHOSIS

A disorder in which functioning liver tissue is replaced by a type of scar tissue. Cirrhosis may be due to (1) alcoholism, (2) infections, (3) disorders of the biliary tract, or (4) nutritional deficiencies. People may recover from mild cases of cirrhosis with only a small loss of liver function. However, advanced cases usually result in liver failure and death.

(Also see DIGESTION AND ABSORPTION.)

CISSA

A craving for and eating of unnatural substances (such as chalk, hair, dirt, or sand) that sometimes occurs in nutritional deficiencies during pregnancy, or in children (in which it is a recurrence of the natural tendency to put everything in the mouth).

(Also see ALLOPRIOPHAGY; and PICA.)

CITRAL

A constituent of orange peel that is suspected of causing damage to blood vessels when large amounts are consumed. Orange peel is present in such food products as marmalade, fruit juices, and fruit drinks. However, it is not known whether many people are affected by the small amount of citral that is usually consumed.

CITRIC ACID

An acid that is (1) present in citrus fruits, and (2) formed during metabolism in the body. Citric acid is used as a flavoring agent in foods, carbonated beverages, and pharmaceuticals.

(Also see CITRUS FRUITS; and METABOLISM.)

CITRIC ACID CYCLE (KREBS CYCLE)

This oxygen-requiring (aerobic) metabolic process, which is also known as the Krebs cycle, is the means by which metabolic products of fats, carbohydrates, and protein are converted to water and carbon dioxide, plus energy. The citric acid cycle operates in the mitochondria of cells, where the appropriate enzymes are located.

(Also see METABOLISM.)

CITRIN

A substance from citrus fruits which has a vitaminlike activity. It was originally designated as vitamin T. Now, it is classified among substances called bioflavonoids.

(Also see BIOFLAVONOIDS.)

CITRON *Citrus Medica*

Contents *Page*

Origin and History 424
Production 424
Processing 424
Selection and Preparation 424
Nutritional Value 425

A large egg-shaped fruit—the largest in the citrus family. Most Americans are familiar with the candied peel of this fruit. The rind of the citron is much thicker than those of the other common citrus fruits. Citron, like the other citrus fruits, is a member of the rue family (*Rutaceae*).

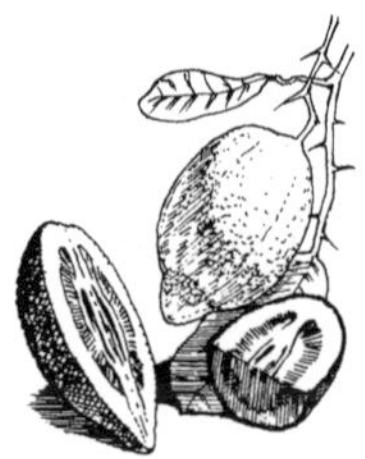

Fig. C-104. The citron, an ancient fruit which has a very thick rind that is used mainly in making candied peel. The fruit is 6 to 10 in. (*15 to 25 cm*) long.

ORIGIN AND HISTORY. The citron originated in Southeast Asia, where it has been grown since ancient times. It appears that traders brought the fruit from the Orient to Persia, but it is not certain when this happened. Historians disagree as to when citrons were first grown in the Middle East. Perhaps, they were brought there by (1) the Persians under Cyrus the Great when they conquered Babylon in 539 B.C., or (2) the Greeks under Alexander the Great when they conquered Persia in 330 B.C. It is noteworthy that in 310 B.C. the citron was called the "Persian apple" in the writings of the Greek naturalist Theophratus (372 to 287 B.C.).

There has been much speculation as to when the Hebrews first planted the citron in Palestine and when it was used as an offering in their Feast of Tabernacles. However, the fruit was apparently well known to the Jews by the time of their revolt against the Romans which began in 66 A.D. because it was depicted on their coins of that era. (The citron was a symbol of Jewish resistance; it replaced the picture of the Roman emperor Nero on the coins.) The dispersal of the Jews after the suppression of their revolt was one of the major factors which led to the widespread cultivation of the fruit in the eastern Mediterranean by the 2nd century A.D.

The Romans first grew citron trees in southern Italy at the time of Augustus (27 B.C. to 14 A.D.); and they soon learned how to protect the trees during the winter in northern Italy by means of enclosures that utilized glasslike panes of mica which kept out the cold, but admitted the sunlight. Over the centuries, the enclosures were improved until they began to resemble modern day greenhouses. However, production of the fruit in the latter area was disrupted by the barbarians who invaded Italy around the end of the 4th century A.D. Nevertheless, the cultivation of the citron continued without interruption in Naples, Sardinia, and Sicily. Hence, Italy continued to supply citrons to northern Europe for many centuries, since the fruit was not grown in France and elsewhere in Europe until the 15th century A.D.

In modern times, citron has been introduced into many tropical areas of the world, but it is of commercial importance only in the Mediterranean region.

PRODUCTION. Statistics on the size of the citron crop are not readily available because the worldwide production of citron is dwarfed by the production of oranges and the other more commonly grown citrus fruits. Furthermore, the fruit is very sensitive to cold weather and can be grown only in subtropical or tropical climates such as those of the areas bordering on the Mediterranean. Hence, most of the crop is grown in France, Corsica, Sicily, Italy, Greece, and Israel, although small amounts of the fruit are now produced in the West Indies, California, and Florida. The general details of citrus fruit growing are given elsewhere.

(Also see CITRUS FRUITS, section headed "Production.")

PROCESSING. Almost all of the citron crop is made into candied citron by a process which consists essentially of (1) cutting the fruit into halves and soaking them in a water solution of salt, sulfur dioxide (a preservative), and calcium chloride (keeps the peel firm), (2) removal of the salt by soaking the fruit in hot water, and (3) candying the fruit by soaking it in a series of sugar solutions.

SELECTION AND PREPARATION. Many people who believe that they have obtained citron may be mistaken, because they may have purchased candied preserving melon (*Citrullus lanatus*, var. *citrioides*), a less expensive fruit that is closely related to the watermelon, and is also called citron. Generally speaking, if candied "citron" is sold for about the same price as candied lemon peel, there is reason to suspicion that it may be preserving melon. The peel of the citrus fruit citron has a distinctive aromatic flavor that differs from the flavors of the other citrus fruits and the preserving melon. Usually, products bearing a list of ingredients will have the correct identification of the source of the peel. Candied citrus citron is most likely to be found in gourmet shops and in stores that sell products imported from Italy and other Mediterranean countries.

Citron may be used to enhance cakes, cookies, holiday breads, marmalades, mincement, and steamed puddings. However, care should be taken that the other ingredients of the preparations do not mask the flavor of the citron.

CAUTION: Some people are allergic to one or more constituents of citrus peel. When such allergies are suspected, it would be wise to consume only very small amounts of candied citron peel unless it is certain that the peel is well tolerated.

NUTRITIONAL VALUE. The nutrient composition of candied citron peel is given in Food Composition Table F-36.

Candied citron peel contains 314 Calories (kcal) per 100 g due to its high sugar content—about 80% carbohydrate. Also, it is a fair source of potassium, but moderately high in sodium (290 mg per 100 g).

Citrus peel contains citral, an aldehyde that antagonizes the actions of vitamin A. Hence, people should make certain that their dietary supply of this vitamin is adequate before consuming large amounts of candied citron peel.

(Also see FRUIT[S], Table F-47 Fruits of the World.)

CITRON MELON (STOCKMELON) *Citrullus vulgaris*

This is a variety (*Citroides*) of watermelon that is not as palatable as the commonly used varieties of the fruit because it has a hard white flesh. However, the rind may be candied or pickled. It is noteworthy that the candied "citron" sold in many American stores is often made from the citron melon rather than from the peel of the citrus fruit citron, which is much more expensive and has a more distinctive flavor.

(Also see CITRON; and FRUIT[S], Table F-47 Fruits of the World—"Watermelon.")

CITROVORUM FACTOR (FOLINIC ACID)

A form of the vitamin folic acid.

(Also see FOLIC ACID.)

CITRULLINE

A nonessential amino acid formed in the urea cycle. It was first isolated from the juice of watermelon.

(Also see METABOLISM.)

CITRUS FRUITS *Citrus* spp

Contents *Page*

Origin and History 425
Production 426
Processing 427
Selection and Preparation 431
Drinks, Ready-to-Serve 431
Juices, Ready-to-Serve 432
Dry Mixes for Drinks or Juices 432
Frozen Drink Concentrates 432
Frozen Juice Concentrates 432
Fruit Pieces or Sections 432
Flavorings 432
Nutritional Value 432

These juicy subtropical fruits, which together constitute a major part of U.S. fruit production, are members of the rue family (*Rutaceae*).

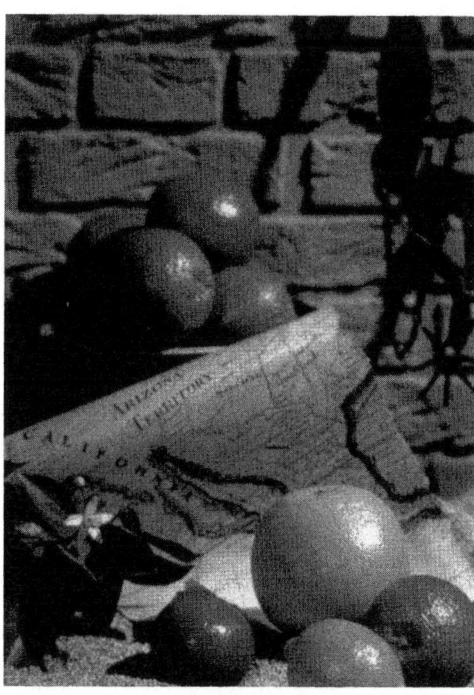

Fig. C-105. The major types of commercially grown citrus fruits. (Courtesy, Sunkist, Van Nuys, Calif.)

Details regarding the individual fruits are given in the separate articles dealing with each fruit.

(Also see CITRON; GRAPEFRUIT; KUMQUAT; LEMON; LIME; MANDARIN ORANGE; ORANGE, SWEET; and SEVILLE ORANGE.)

ORIGIN AND HISTORY. The ancestral species of citrus fruits apparently originated somewhere in Southeast Asia between eastern India and the islands off the Pacific coast. It is believed that the sacs of watery pulp which enclose the seeds of these fruits is an evolutionary adaptation to dry monsoon regions such as India, although the ancient Asian farmers have long grown some of these crops in the more humid areas. The introduction of certain citrus species into new habitats, such as the various regions of China, resulted in the crossbreeding (hybridization) of the new plants with those that were native to the areas. Hence, it is now difficult to classify some of the hybrid varieties because of the amount of crossing which has occurred over the centuries.

Citrus fruits were first cultivated in China about 4,000 years ago. It is uncertain when (1) these crops were first brought from the Orient to India, and (2) Arab traders and others brought the fruits to the Middle East. However, the Chinese remained the unsurpassed masters of citrus breeding and cultivation for many centuries, as evidenced by a book called *Citrus Fruit Records* which was compiled by Han Yen-Chih in 1178 A.D. He covered 27 varieties of citrus in great detail, describing their distribution, cultivation, and orchard management, and explaining methods of picking, storing, processing, and utilization.

Citrons were the first of the citrus fruits to be grown by the peoples of the Middle East and the Mediterranean region, as evidenced by their description around 310 B.C. The ancient Romans first grew citron trees and other citrus trees in the 1st century B.C., but the trees did not bear fruit at that time because the climate was cooler than that of the Middle East. Undaunted by the failure of their initial attempts to produce citrus fruit, the Romans developed special enclosures with glasslike panes of mica which kept out the cold, but admitted the rays of the sun. Within 3 or 4 centuries, the fruit was produced throughout Italy. In the centuries that followed, the Italians improved upon their prototypes of modern day greenhouses so that by the end of the 14th century, the buildings were heated artificially and called "orangeries." The first orangery to be constructed in northern Europe was erected at Amboise, France by the order of King Charles VIII, after he had returned from an expedition to Italy in 1495.

Lemons, sour oranges, and sweet oranges were brought westward to Persia and Arabia by traders from the Orient. The spread of Islam and Arabic culture, which began in the 7th century A.D. and continued for the four centuries that followed, led to the growing of the citrus fruits throughout the countries of North Africa, from which they were eventually introduced into Spain and the rest of Europe.

Columbus planted orange seeds on Hispaniola (the island that is now comprised of Haiti and the Dominican Republic) in 1493. In the years that followed, the trees were planted elsewhere among the islands of the Caribbean. By the end of the 16th century, citrus crops had been introduced by the Spanish to Mexico, Florida, and Peru; and by the Portuguese to Brazil.

On the other side of the world, the Dutch sea captains gave their sailors citrus fruit to prevent scurvy, but another two centuries elapsed before all British sailors received similar rations. British naval authorities and private ship owners were not convinced of the value of citrus fruit or juice until after the Scottish naval surgeon Lind conducted his well-known demonstration in 1753. Furthermore, only lemons or limes kept well enough to be carried on long voyages. The issuance of daily rations of lemon juice to sailors in the British Navy was not instituted until the latter part of the 18th century. Although citrus trees were planted by Franciscan monks in San Diego, California in 1769, scurvy continued to plague the sailors of Spanish ships plying the waters off the California coast for a considerable time thereafter. Sporadic outbreaks of scurvy occurred in California right up to the time of the gold rush of 1849.

The early growth of the U.S. citrus industry was due in large part to (1) the construction of the transcontinental railway system which brought the fruit to the heavily populated cities in the Midwest and the Northeast; (2) the large scale breeding experiments which were initiated by the U.S. Department of Agriculture in Florida in 1893, and by the University of California at Riverside in 1914. Production increased steadily between the First and Second World Wars until grapefruit, lemons, and oranges became common foods instead of treats for special occasions such as Christmas. However, the most phenomenal spurt in citrus production was triggered by the development of frozen citrus juice concentrates soon after the end of World War II. U.S. production after the war increased to more than 1½ times the prewar levels. Since then, an ever increasing share of the U.S. citrus crop has been processed and the per capita consumption of the fresh fruit has dropped sharply. Almost ¾ of the crop is processed.

PRODUCTION. The world production of all types of citrus fruits is about 50 million metric tons, with about one-fourth of the total supplied by the United States.[28] Almost all of the U.S. citrus production comes from Florida, California, Texas, and Arizona.

Oranges account for about two-thirds of the world citrus crop and two-thirds of the U.S. citrus crop. Details regarding the production of the individual citrus fruits are given in the separate articles on the different fruits.

(Also see CITRON; GRAPEFRUIT; KUMQUAT; LEMON; LIME; MANDARIN ORANGE; ORANGE, SWEET; and SEVILLE ORANGE.)

Most of the citrus fruits and their close relatives originated in the subtropics or the tropics. Therefore, the major zone of world citrus production lies between 35°N and 35°S latitudes, except near the Mediterranean, where it extends to 44°N latitude. It is noteworthy that the kumquat (*Fortunella* spp.) and the trifoliate orange (*Poncirus trifoliata*) are (1) closely related to the citrus fruits, (2) more resistant to cold than most of the citrus species, and (3) sometimes crossbred with citrus to obtain hybrid trees and fruits that tolerate low temperatures better than the parent citrus species.

Citrus trees are propagated by grafting buds of the desired species onto the appropriate root stocks, which are grown from seeds in prepared beds. In California, the young citrus trees are transplanted in an orchard within 1 to 2 years after grafting. The soil should be deep, fertile, and well-drained. Growers usually make sure that the trees receive sufficient nutrients by (1) fertilizing the soil with nitrogen, magnesium, potassium, calcium, and sulfur; (2) adding extra phosphorus if the soil is alkaline;

[28]*World Agricultural Production,* USDA, Foreign Agricultural Service, Circular Series, WAP 6-90, June 1990.

(3) increasing the availability of soil iron by adding ammonium sulfate, or inserting iron-containing pellets into the trunks of the trees; and (4) applying zinc, copper, manganese, boron, and molybdenum in foliar sprays. Wind breaks may be needed to protect the trees in windy areas.

The trees are usually pruned on a regular basis to (1) remove all suckers from the stock and the graft, (2) cut out dead wood and parasites, and (3) encourage the formation of sturdy, low branches for easy picking of the fruit.

Citrus fruits are usually ready for harvesting within 8 to 16 months after the blossoming of the trees. However, each state where the fruits are produced has its own maturity standards that specify the contents of sugar, acid, and juice which should be present in the fruit before picking. Hence, the fruits are usually sampled and analyzed for the levels of these components. It is noteworthy that the color of the peel is *not* a good indicator of fruit maturity because the yellow and orange colors which are so desirable to consumers are brought out only if the fruit is exposed to night temperatures between 40° and 50°F (*4° to 10°C*). Otherwise, the color of the peel is likely to be green. Freshly harvested fruits may be "degreened" by exposing them to ethylene gas in a warm room. (Ethylene gas is a natural substance that is produced by many fruits as they ripen.)

PROCESSING. The rapid growth of the citrus industry in the United States was due in large part to the continual development of ever more efficient ways of utilizing the crop so that even the peel and the seeds may be processed profitably.

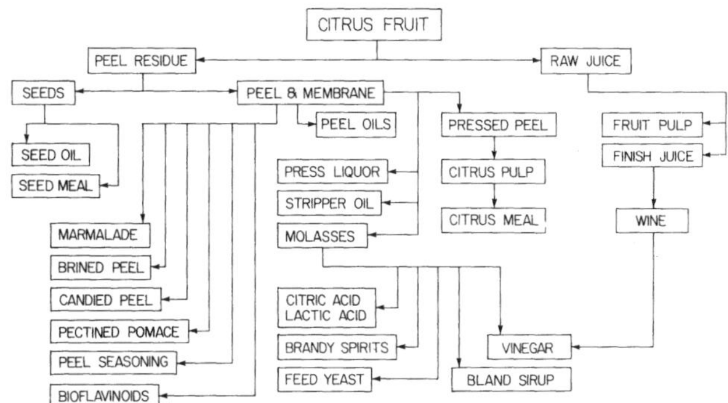

Fig. C-106. Nothing is wasted during the processing of citrus fruits since even the seeds and the peel yield valuable products. (Adapted from R. Hendrickson and J. W. Kesterson, *By-Products of Florida Citrus*, Bull. 698, University of Florida Ag. Exp. Sta., Gainsville)

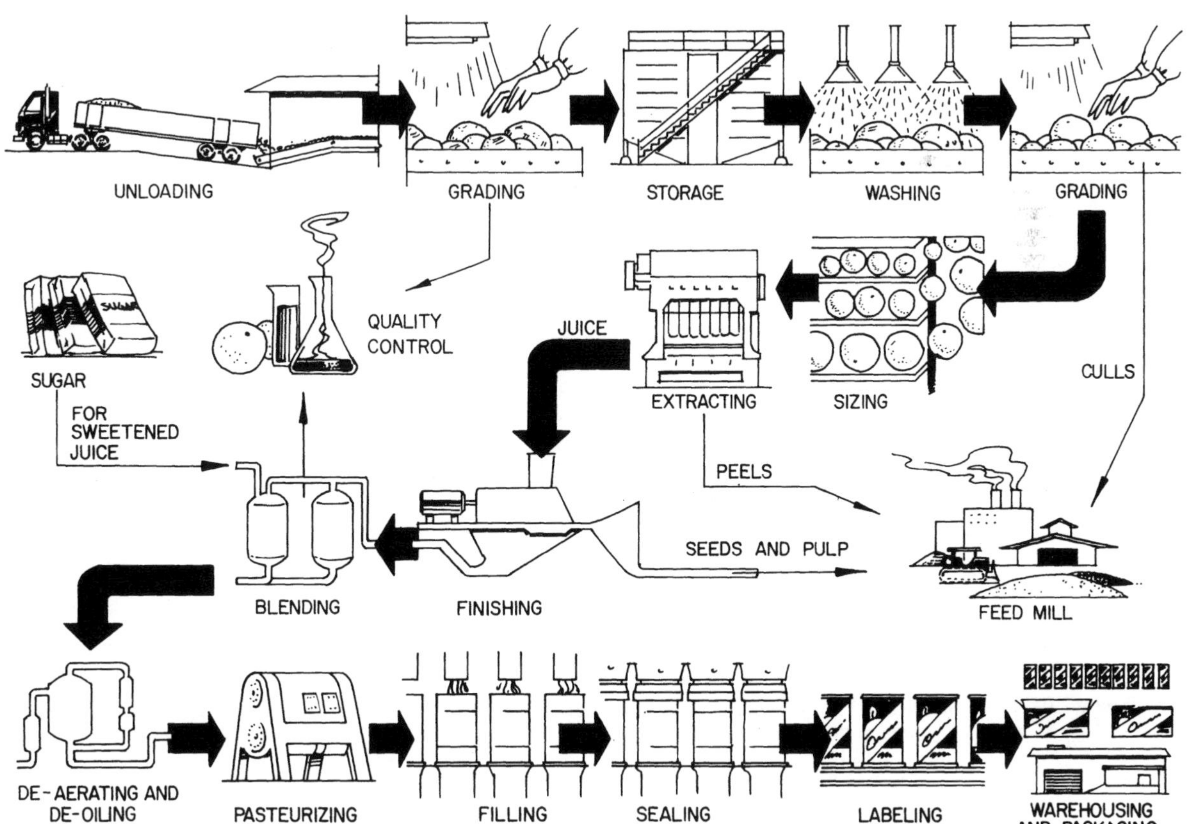

Fig. C-107. Process used to produce juice from citrus fruit. (Courtesy, USDA)

Descriptions of the major products derived from citrus fruits are given in Table C-31.

TABLE C-31
PROCESSED CITRUS FRUIT PRODUCTS

Product	Description	Uses	Comments
	I. BEVERAGES, FOODS, AND OTHER EDIBLE ITEMS		
Alcoholic beverages **Brandies**	Fermented and distilled citrus juices that may have been sweetened to promote fermentation.	As brandies, or in cordials and liqueurs. In mixed drinks or in various dessert items.	The flavor of the original citrus fruit juice is more apparent when the brandy has been aged.
Cordials and Liquers	Prepared as brandy, except that they usually contain additional flavorings such as peel oil and various sweeteners.	After dinner drinks to aid digestion. In mixed drinks, or in dessert dishes such as cakes and ice creams.	Products made from citrus brandies are superior to those made from noncitrus brandies (citrus juice imparts extra flavor).
Wines	Fermented citrus juices that may have been sweetened and aged in plain or in charred oak barrels.	As wines, or in mixed drinks. In the production of brandies, cordials, or liqueurs.	Fruit juice must be extracted with care to avoid an excessive content of peel oil, which inhibits fermentation and imparts bitterness.
Desserts **Dessert gels**	Products vary considerably in their content of natural and/or synthetic citrus flavorings and ingredients, gelling agents, sweeteners, and additives.	Desserts Salads Unbaked chiffon pies	May be purchased in prepared forms or as dry mixes to be prepared at home. The most nutritious products are made from fresh citrus juices, pulp, and/or pieces of the fruit(s).
Frozen desserts	Most items contain juice or frozen concentrated juice with or without ice cream, sweeteners, and additives.	Ice cream Ices Sherbets Mixtures of vanilla ice cream and sweetened citrus juices.	The Florida Citrus Commission provides rebates to producers who utilize unsweetened citrus juice products.
Flavorings **Essences**	Volatile citrus flavor components from fresh juices that are collected during the production of various juice concentrates by vacuum evaporation.	Citrus-flavored drinks and juices.	Available in alcoholic or nonalcoholic solutions which range in strength from dilute to concentrated. Similar flavorings may be produced from peel and other citrus wastes.
Peel oils	Strongly-flavored oils that are usually obtained during the extraction of juice from fresh fruit.	Enhancement of flavors in cordials, drinks, juices, and other citrus products.	More strongly-flavored than essences obtained from fruit juices. May contain bitter-flavored components.
Fruit drink preparations **Dry drink mixes**	Mixtures of dehydrated citrus juices and/or other citrus products (oil, peel, pulp, etc.), plus sweetener(s), natural or synthetic colorings and flavorings, and/or instant tea.	Alcoholic and nonalcoholic drinks. Instant tea drinks (spiced and unspiced).	Purchasers should check labels carefully to determine economic, esthetic, and nutritional values of the various products (some items are comprised mainly of sugar plus synthetic colorings and flavorings).

(Continued)

TABLE C-31 (*Continued*)

Product	Description	Uses	Comments
	I. BEVERAGES, FOODS, AND OTHER EDIBLE ITEMS (*Continued*)		
Frozen drink concentrates	Citrus juice concentrates plus sweeteners and/or colorings and flavorings.	Alcoholic and nonalcoholic drinks.	These items may not be as nutritious as unsweetened frozen citrus juice concentrates.
Fruit drinks, ready-to-serve	One or more citrus juices plus sweetener(s) and/or coloring(s), flavoring(s), and various additives.	Alcoholic and nonalcoholic drinks.	It may be more economical for consumers to dilute pure fruit juices to the desired strengths, because drinks are essentially sweetened, watered down juices that may contain various additives.
Thirst quenchers	Citrus drinks fortified with potassium, calcium and/or sodium salts. May contain sugar or an artificial sweetener.	Fluid and salt replacement after heavy losses of perspiration, such as occurs after strenuous exercise.	Used mainly by athletes. Compositions of different products may vary considerably.
Fruit pieces **Canned fruit sections**	Citrus fruit sections packed in various combinations of water, sweeteners, and/or citrus juices.	Fruit dish served at breakfast, or as an appetizer at other meals.	Caloric content may vary somewhat, depending upon the compostion of the packing medium.
Frozen fruit sections	Frozen sections of citrus fruit(s), coated with dry sugar or a syrup.	Fruit dish for breakfast, or an appetizer for other meals.	Caloric content varies with the type of coating material (used to prevent the sections from sticking together).
Jam, Jellies, and Marmalades **Jams**	Semisolid mixture of sugar or other sweetener(s) and a citrus fruit ingredient with or without pectin and/or other additives as deemed necessary.	Spread for biscuits, bread, crackers, etc. In fillings of layer cakes and jelly rolls. Topping for various desserts.	Low-calorie products may contain artificial sweeteners, or they may be made with special pectins and a reduced amount of natural sweetener(s), and labeled as "imitation jam."
Jellies	Gelled mixture of sugar or other sweetener(s) and a clarified citrus fruit juice with or without pectin and/or other additives.	Spread for biscuits, bread, crackers, etc. In jelly rolls. Topping for various desserts.	Low-calorie products may contain artificial sweeteners, or the content of natural sweetener(s) may be reduced.
Marmalades	Semisolid mixture of sugar or other sweeteners, and a citrus fruit ingredient containing pieces of peel, with or without pectin and/or other additives.	Spread for biscuits, bread, crackers, etc. Topping for various desserts.	The best tasting orange marmalades are made from Seville (sour) oranges *(Citrus aurantium).*
Juice products **Blends**	Various mixtures of 2 or more citrus fruit juices that are available in ready-to-drink form, or as frozen concentrates.	Beverage Mixer for alcoholic drinks.	Flavors of the combinations may be different from those of the separate components. Citrus oil(s) may be added to enhance the flavor.

(Continued)

TABLE C-31 (*Continued*)

Product	Description	Uses	Comments
I. BEVERAGES, FOODS, AND OTHER EDIBLE ITEMS (*Continued*)			
Crystals and powders	Dehydrated citrus juices that contain about 98% solids.	Dry beverage mixes. Production of juice tablets.	A 1-lb. *(0.45 kg)* package of a dry mix usually reconstitutes to about a gallon *(3.8 liter)* of juice.
Frozen juice concentrates	Concentrates of 1 or more citrus juices produced by vacuum evaporation of fresh juice(s).	Alcoholic and nonalcoholic drinks. Ingredient of frozen desserts.	Most products may be reconstituted by adding 3 parts of water to 1 part of concentrate. More highly-concentrated products may be available soon.
Juice tablets	Dehydrated citrus juices that have been pressed into tablets.	Candylike snack food.	Tablets are equivalent to between 8 and 10 times their weight of fresh juice.
Protein-fortified juices	Citrus juices that contain between 3.0 and 3.5% protein in the form of cheese whey protein concentrate.	Nutrient fortified beverages. Frozen desserts.	May soon be available as (1) ready-to-drink juices, (2) frozen juice concentrates, (3) dry juice mixes, and/or (4) frozen dessert bars.
Single strength juices	Ready-to-drink citrus juices that are usually bottled or canned with or without sweetener and/or peel oil.	Beverages (alcoholic and nonalcoholic). Ingredients of gelatin desserts and salads. Production of frozen juice concentrates and various other citrus juice products.	Canning juices were the earliest major means of processing citrus fruits. Canned juices declined in popularity after the introduction of (1) chilled juices in bottles and cartons, and (2) frozen juice concentrates.
Miscellaneous items **Bioflavonoids (formerly called vitamin P)**	Vitaminlike substances present in the juice, membranes, and peel of citrus fruits.	Treatment of capillary fragility, red blood cell sedimentation, and other circulatory disorders.	The medicinal use of bioflavonoids remains controversial in the United States, although it has long been practiced in the U.S.S.R.
Candied peel	Citrus peel cured in a concentrated salt solution or by other means, then desalted and candied by immersion in a syrup for an extended period of time.	Candylike snack food. Ingredient of cakes, cookies, marmalades, mincemeat, and mixed candied fruits.	Citrus peel contains the aldehyde citral, which appears to be an antagonist of vitamin A. However, the effect is overcome by increasing the vitamin A content of the diet.
Natural coloring agents	Colored pigments which are naturally present in the peel of citrus fruits.	Deepening the color of citrus juices and related products.	Deeply-colored juices sell better than faintly-colored ones.
Pectin	Jellying agent present in citrus peel. Extracted commercially from peel that has had the oil removed.	Jams, jellies, marmalades. Emulsifier and thickener in various processed foods. Component of a popular remedy for diarrhea.	A variety of pectin derivatives may be produced by chemical modification to suit various food and nonfood applications.
Peel seasoning	Ground, dehydrated citrus peel that is preserved by treatment with antioxidants.	Production of baked goods, candies, cordials and liqueurs, medicines, and seasonings.	Some of the volatile aromatic constituents of the peel are lost during drying.

(Continued)

TABLE C-31 (*Continued*)

Product	Description	Uses	Comments
I. BEVERAGES, FOODS, AND OTHER EDIBLE ITEMS (*Continued*)			
Syrups	Bottled syrups prepared from pulp-free citrus juice concentrates.	Fruit syrups with natural colors and flavors. Production of jams, jellies, and marmalades.	Produced on a limited commercial scale, but expansion of output is expected in the future.
Vinegars	Produced from citrus juices or citrus peel press liquors by alcoholic fermentation followed by acetic acid fermentation.	Seasoning agent for a wide variety of foods.	Vinegar made from citrus juices contains considerable citric acid, whereas that made from press liquors is darker colored, has a fruity aroma, and contains no citric acid.
II. LIVESTOCK FEEDSTUFFS			
Citrus molasses	The viscous liquid obtained by evaporating the liquor released from limed, cured, and pressed citrus residue (the peel, pulp, rag, and seeds left after juice has been extracted from citrus fruit).	Feedstuff for cattle (occasionally, it has also been fed to swine). Raw material for the production of beverage alcohol by fermentation and distillation. Source of bioflavonoids (vitaminlike substances).	Contains about 45% sugars as fed (it is useful as an energy feed which may replace some of the corn in a ration.) Best utilized when amount fed is limited to between 10 to 15% of the total ration. May be sold in a mixture with dried citrus pulp.
Citrus pulp, dried	The dried meal obtained from limed, cured, and pressed citrus residue (the peel, pulp, rag, and seeds left after juice has been extracted from citrus fruit).	Feedstuff for cattle (the fiber content is too high for nonruminant species).	Contains about 62% carbohydrate as fed (it is useful mainly as an energy feed.) Ammoniated pulp has double the protein value of nonammoniated pulp. High levels in the ration may depress milk production in cows.
Feed yeast	Torula yeast grown on the nutrient-fortified liquor released from limed, cured, and pressed citrus residue (the peel, pulp, rag, and seeds left after juice has been extracted from citrus fruit).	Feedstuff for all species of livestock and poultry. Protein and vitamin supplement for animal feeds.	Between 44 and 48 lb *(20 and 22 kg)* of dry yeast may be obtained from each 100 lb *(45 kg)* of sugar present in the citrus residue press liquor. Contains about 55% protein, which makes it useful as a protein supplement in feeds.

SELECTION AND PREPARATION. Criteria of quality for the selection of fresh citrus fruits are given in the separate articles on the individual fruits, as are some of the special ways for utilizing them and items made from them.

(Also see CITRON; GRAPEFRUIT; KUMQUAT; LEMON; LIME; MANDARIN ORANGE; ORANGE, SWEET; and SEVILLE ORANGE.)

The selection of processed citrus fruit products in order to obtain the greatest economic and nutritional values should be based on considerations such as (1) the quality and cost of the pure fruit ingredient(s), and (2) convenience in serving and storing. Some guidelines on the major types of products follow.

Drinks, Ready-To-Serve. These items vary considerably in citrus fruit juice content. For example, an "orange juice drink" usually contains about 50% orange juice, whereas "orangeade" has 25 to 50% juice, products labeled "orange drink" contain from 10 to 30% juice, and an "orange flavored drink" has less than 10% orange juice. Hence, certain products are mainly sweetened water plus coloring, flavoring, and other additives. Furthermore, the highly diluted citrus drinks do not contribute much to mixtures containing other beverages, except to those with considerable pulp (such as apricot, peach, and pear nectars), in which a thinning of the consistency and strengthening of the flavor of the beverage might be desirable.

Juices, Ready-To-Serve. Only a few people squeeze juice from fresh citrus fruits because it is time consuming, and often it is more expensive (3 to 4 medium-size oranges yield only 1 cup [*240 ml*] of juice) than prepared single strength juices that come in bottles, cans, and cartons. However, the popularity of canned citrus juices has declined recently, because the products in bottles and cartons usually taste better.

Pure citrus juices may often be better buys than citrus drinks, which are often merely watered down versions of the juices. It is easy for consumers to dilute and sweeten pure citrus juices, when it is so desired. Also, the pure juices mix well with club soda, ginger ale, gin, rum, vodka, whiskey, brandies, and liqueurs.

Dry Mixes For Drinks Or Juices. The compositions of these products range from—mixtures of sugar, artificial coloring and flavoring, and a little ascorbic acid (vitamin C)—to pure citrus juice crystals or powders. Hence, consumers should read the list of ingredients carefully before buying an unfamiliar product.

It is noteworthy that people taking diuretic drugs (often for conditions such as congestive heart failure and/or hypertension) are frequently advised to drink orange juice in order to replace the potassium that is lost in the urine as a result of the actions of the drugs. However, some of the drinks which resemble orange juice contain little or no potassium. Furthermore, pure citrus juices are likely to contain other nutrients and beneficial substances which are lacking in some of the drinks.

Frozen Drink Concentrations. These products may be convenient for the homemaker or host who is in a great hurry to prepare lemonade, limeade, or orangeade, but it would be better to use frozen juice concentrates because the latter usually contain more citrus juice and no added high caloric sweetener(s).

Frozen Juice Concentrates. Concentrated juices are more convenient than the single strength juices because they require less storage space, and they can be used as (1) sauces for fish, meats, and poultry; (2) syrups to be poured over ice cream, ices, sherbets, fruit cakes, plum puddings, and pound cakes; (3) flavorings for herb teas, iced tea, or hot tea; (4) mixes for alcoholic drinks or nonalcoholic drinks such as club soda, ginger ale, and quinine water; (5) ingredients in cookies, cup cakes, fruit bars, fruit cakes, fruit puddings, muffins, pastries, and pie fillings; and (6) components of sandwich fillings or spreads, when mixed with cottage cheese, cream cheese, mashed bananas, peanut butter, or yogurt.

Fruit Pieces Or Sections. The preparation of sections from fresh fruit may take too much of the precious little time that is usually available for breakfast. Hence, it may be worth paying more to obtain canned or frozen citrus pieces or sections. The various brands of broken pieces of fruit usually cost less than whole sections. However, calorie counters should take note of the packing medium for the fruit since it may be one of a variety of mixtures of syrup, juice, water, and/or sweetener(s).

Citrus pieces or sections may be utilized by (1) serving them plain or with a little sweetener, (2) adding them to gelatin dessert or salad mixtures, (3) broiling them with a little honey plus a pinch or so of cinnamon, cloves, or allspice, (4) incorporating them in salads, and (5) serving them with a scoop or two of ice cream, ices, or sherbet.

Flavorings. Citrus flavorings and extracts have often been derived from the peel of the fruits. Hence, economy minded homemakers may make their own flavorings by grating the peel of fruits they have purchased. However, some experimentation may be required to determine the right amounts of peel to be used, if recipes calling for peel are not at hand because too much peel may impart a bitter flavor to the preparation.

CAUTION: Some people are allergic to one or more constituents of citrus peel. When such allergies are suspected, neither the peel nor products containing it should be consumed,and juice from citrus fruits should be extracted gently to avoid squeezing the oil and other substances from the peel into the juice. It is noteworthy that some of the commercially prepared citrus juices and citrus drinks may contain peel and/or the substances extracted from it, but that prepared citrus juice products for infants contain little or none of the peel constituents.

NUTRITIONAL VALUE. The nutrient compositions of various citrus fruits and citrus products are given in Food Composition Table F-36.

Noteworthy observations regarding the nutrient composition of the individual citrus fruits and their products are given in the separate articles on these fruits.

(Also see CITRON; GRAPEFRUIT; KUMQUAT; LEMON; LIME; MANDARIN ORANGE; ORANGE, SWEET; and SEVILLE ORANGE.)

Some general remarks on the overall nutritional merits of citrus fruits and related products follow:

1. Fresh citrus fruits are low in calories and sugars, but they are good to excellent sources of fiber, pectin, potassium, vitamin C, inositol, and bioflavonoids (vitaminlike substances). They are also fair sources of folic acid.
2. Citrus juices are not as nutritious as the whole edible portion of the fruits (includes all but the seeds and the peel) because substances such as fiber, pectin, and bioflavonoids are present mainly in the peel and the membranes which surround the segments of fruit.
3. More of the vitamin C is likely to be retained in juices packed in bottles or cartons, or those prepared from concentrates, than in canned juices which are heated strongly during canning, then stored at room temperature for extended periods of time.
4. Citrus peel contains citral, an aldehyde that antagonizes the actions of vitamin A. Hence, people should make certain that their dietary supply of the vitamin is adequate before consuming large amounts of the peel from citrus fruits.

(Also see FRUIT[S], Table F-47 Fruits of the World; and PECTIN.)

CLARIFICATION

Means of making a cloudy liquid clear by such processes as filtration, centrifugation, or allowing the particles which cause cloudiness to settle out upon standing.

CLARIFIXATION

A method of homogenizing milk. First the cream is separated and homogenized, then it is remixed with the milk in a clarifixator.

(Also see HOMOGENIZATION.)

CLAUDICATION

Weakness and pain in the leg which cause limping and difficulty in walking.

CLAUDICATION, INTERMITTENT

Soreness of the calf muscles in the leg due to lack of blood resulting from blockage or constriction of the blood vessels. This condition has been helped by the administration of therapeutic amounts of vitamin E.

(Also see HEART DISEASE; BUERGER'S DISEASE; and VITAMIN E.)

CLEARING FACTOR

An enzyme which helps to clear the fat from the blood after a meal. This is demonstrated by the disappearance of the milky color of blood several hours after eating. Clearing factor appears to act together with a complex carbohydrate called heparin, but it also requires calcium and magnesium as activators.

(Also see CARBOHYDRATE[S]; ENZYME; and FATS AND OTHER LIPIDS.)

CLIMACTERIUM (CLIMACTERIC)

The combined phenomena accompanying cessation of the reproductive function in the female—the menopause; and the corresponding phenomena of reduced sexual activity and potency in the male.

CLOD

This is an English term for the cuts of beef made from what is known as the chuck in the United States; it's the shoulder of beef, or of the neck near the shoulder. This is a cheaper and less tender cut, usually used in roasts or stews.

CLOFIBRATE

A drug used to lower the blood levels of both cholesterol and triglycerides.

(Also see FATS AND OTHER LIPIDS.)

CLOSTRIDIUM BOTULINUM

A species of bacteria which causes the type of food poisoning called botulism. Botulism may be fatal if not immediately treated.

(Also see BACTERIA IN FOOD, section headed "Food Poisonings [bacterial toxin]"); BOTULISM; and FOOD-BORNE DISEASE.)

CLOUDBERRY *Rubus chamaemorus*

A species of raspberry (a fruit of the family *Rosaceae*) that grows in the swampy areas of Alaska, northern Canada, northern Europe, Asia, and as far north as the Arctic Circle.

Fig. C-108. The cloudberry.

The fruit has an orange-yellow color and is much esteemed in the Scandinavian countries where it is used to make a liqueur. It is also good in pies and puddings.

(Also see FRUIT[S], Table F-47 Fruits of the World—"Raspberry"; and RASPBERRY.)

COAGULATED

Curdled, clotted, or congealed, usually brought about by the action of a coagulant.

COARSE-BOLTED

Separated from its parent material by means of a coarsely woven bolting cloth.

(Also see FLOURS.)

COARSE-SIFTED

Passed through coarsely woven wire sieves for the separation of particles of different sizes.

(Also see FLOURS.)

COB

The fibrous inner portion of an ear of corn (maize) from which the kernels have been removed.

COBALAMIN

Another name for vitamin B-12.

(Also see VITAMIN B-12.)

COBALT (CO)

This element is an essential constituent of vitamin B-12 and must be ingested in the form of the vitamin molecule inasmuch as humans synthesize little of the vitamin. No other function of cobalt has been established.

The body of an average adult human contains about 1.1 mg of cobalt.

HISTORY. The word cobalt is derived from the German word *kobold*, meaning goblin or mischievous spirit. The term originated in the 16th century, when arsenic-containing cobalt ores were dug up in the silver mines of the Harz Mountains. Believing that the ores contained copper, miners heated them and were injured by the toxic arsenic trioxide vapors that were released. These evils were attributed to the goblin or kobold. George Brandt, the Swedish chemist, first isolated the element in 1742, although cobalt had been used for centuries for the blue color in decorative glass and pottery.

The discovery, in 1948, that vitamin B-12 (cyanocobalamin) contains 4% cobalt (Co) proved this element to be an essential nutrient for man. It is noteworthy that cobalt's essential role in ruminant animal nutrition was known much earlier. In 1935, Australian scientists discovered that lack of cobalt, resulting from its deficiency in the soil and thus in the herbage grazed, produced a wasting disease. Much earlier, and long before the cause was known, stockmen in different areas of the world learned that this peculiar malady could be prevented and/or cured by transferring animals from "sick" to "healthy" areas.

ABSORPTION, METABOLISM, EXCRETION. Cobalt is readily absorbed in the small intestine. But the retained cobalt serves no physiological function since human tissues cannot synthesize vitamin B-12.

Most of the absorbed cobalt is excreted in the urine; very little of the element is retained—only small amounts are concentrated in the liver and kidneys.

Vitamin B-12, of which cobalt is a component, is synthesized by *E. coli*, a species of bacteria commonly found in the human colon. But the microorganisms of the colon do not make sufficient B-12 to meet human requirements; besides, very little of the vitamin can be absorbed past the small intestine.

FUNCTIONS OF COBALT. The only known function of cobalt is that of an integral part of vitamin B-12, an essential factor in the formation of red blood cells.

DEFICIENCY SYMPTOMS. A cobalt deficiency as such has never been produced in humans. The signs and symptoms that are sometimes attributed to cobalt deficiency are actually due to lack of vitamin B-12, characterized by pernicious anemia, poor growth, and occasionally neurological disorders.

A deficiency of cobalt in ruminants (cattle and sheep) results in loss of appetite, impaired growth, listlessness, and progressive emaciation.

A cobalt deficiency in soils has been reported in Australia, western Canada, and in many of the eastern and midwestern states of the United States.

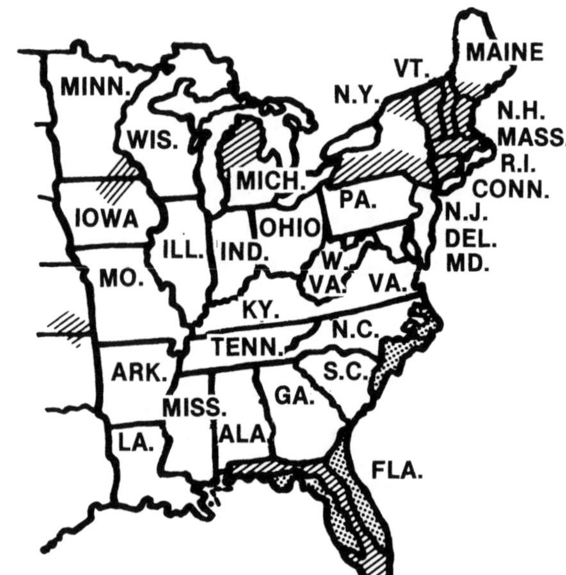

Fig. C-109. Cobalt-deficient areas in eastern United States, resulting from its deficiency in the soil and thus in the herbage produced thereon.

INTERRELATIONSHIPS. Cobalt is a component of vitamin B-12.

RECOMMENDED DAILY ALLOWANCE OF COBALT. There is no known human requirement for cobalt, except for that contained in vitamin B-12.

- **Cobalt intake in average U.S. diet**—Estimates of the average cobalt intake of adults in the United States range from 0.14 to 0.58 mg per day.

TOXICITY. Cobalt toxicity is not likely to result from the consumption of normal foods and beverages, because there is a very wide margin between essential and harmful levels. Excess cobalt intake in man results in an increase in the number of red blood cells, a disorder known as polycythemia.

SOURCES OF COBALT. Cobalt is present in many foods. However, the element must be ingested in the form of vitamin B-12 in order to be of value to man; hence, a table showing the cobalt content of human foods serves no useful purpose. Instead, some rich sources of vitamin B-12 are shown in Table C-32.

(Also see MINERAL[S], Table M-67; and VITAMIN B-12.)

TABLE C-32
SOME RICH SOURCES OF VITAMIN B-12 (COBALAMINS)[1]

Food	Vitamin B-12 (Cobalamins)
	(mcg/100 g)
Beef liver	111
Clams	98
Lamb kidneys	63
Turkey liver	48
Calf kidneys	25
Chicken liver	19
Beef pancreas	16
Roe	15
Oysters	15
Crab	10
Sardines	10
Pork kidneys	10
Mozzarella cheese	8
Herring	8
Salmon	7

[1]Data from Food Composition Table F-36.

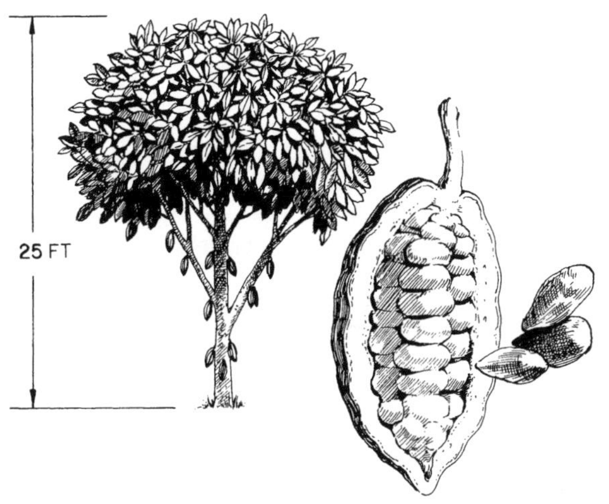

Fig. C-110. The cacao tree—a tropical evergreen. The seeds from the seedpods are the source of cocoa and chocolate.

COCARBOXYLASE

A thiamin-containing enzyme; it is the key substance in decarboxylation (removal of the carboxyl group [COOH]), an energy-producing reaction in the body.

COCCI

A spherical type of bacteria. Various species of cocci are responsible for human illnesses such as strep throat (streptococci), boils and similar infections (staphylococci), gonorrhea (gonococci), and meningitis (meningococci). Unsanitary practices of food handlers may result in the contamination of beverages and foods by these microorganisms.

(Also see DISEASES; and FOODBORNE DISEASE.)

COCKLE

A common edible European bivalve mollusk of the family *Cardiidae* that has a shell with radial ribs. It is boiled and eaten with a variety of condiments.

COCOA AND CHOCOLATE

Through different processing, cocoa and chocolate are derived from the seeds of the cacao tree or *Theobroma cacao*—its scientific name. The Swedish botanist, Carolus Linnaeus, must have been familiar with the pleasant flavor of cocoa or chocolate when he classified it in the 1700s, since the Latin term *theobroma* means "food of the gods." This pleasant chocolate flavor develops during processing.

Chocolate liquor, formed by grinding the roasted, cracked cacao bean, is the base for cocoa and chocolate products. To make cocoa, the chocolate liquor is put under pressure and most of the cocoa butter (fat) is squeezed out leaving a cake that is pulverized and sifted into a powder. Alkalizing cocoa produces Dutch cocoa. While cocoa butter is removed in the making of cocoa, it is added in the making of chocolate. Chocolate is a combination of chocolate liquor, sugar, cocoa butter, and vanilla or some other added flavor. Chocolate may be light, dark, sweet, semisweet, or milk chocolate.

ORIGIN AND HISTORY. The exact origin of the cacao tree is uncertain, but it is believed to have originated in the Amazon-Orinoco river basin in South America. Nonetheless, cacao trees are native to Central and South America. Man introduced cacao trees into Africa. The Maya and Aztec Indians cultivated the cacao tree before Columbus discovered America. In some areas, the beans were even used as currency; in others, they were used as food. When Hernando Cortes and his men conquered the Aztecs in Mexico, they found the Indians using the beans to make a drink called *chocolatl*, an Aztec word meaning acrid or sour water. Cortes returned to Spain with some cacao beans, and introduced the Spaniards to *choclatl*, who added sugar to the beverage. The Spanish tried to keep the cacao a secret, and sold it very profitably as chocolate to the wealthy of Europe. However, in the early 1600s, cacao beans and chocolate drink were introduced into Italy. Soon people in Austria, France, and England began to use cacao beans. During the 17th century, "chocolate houses" became fashionable meeting places in England and throughout Europe. By 1765, cacao beans from the West Indies were being manufactured into chocolate at Dorchester, Massachusetts, in a factory founded by Dr. James Baker.

Gradually, processing changes were made, resulting in the familiar cocoa and chocolate products of today. In 1828, C. J. van Houten of the Netherlands, revolutionized the chocolate industry by inventing a press that squeezed the cocoa butter (fat) out of the cacao beans, producing cocoa powder. Then, in 1847, the English company of Fry and Sons mixed cocoa butter, the by-product of cocoa powder, with chocolate liquor, and sugar, and molded it into bars. People were now able to drink and eat chocolate. In 1876, one more familiar refinement was added to cacao bean processing. M. Daniel Peter of

Switzerland added milk to his chocolate products, producing a new flavor—milk chocolate. Over the years, there has been a proliferation of cocoa and chocolate products and chocolate flavored foods enjoyed by most individuals. Milton Snavely Hershey, who opened a factory in Lancaster, Pennsylvania in 1888, was largely responsible for the peak development of cocoa and chocolate in the United States. In fact, a Hershey bar is synonymous to a chocolate bar. Americans now rank tenth among the world's chocolate eaters. The Swiss hold first place.

Fig. C-111. The first automobile seen in Lancaster, Pennsylvania—a Riker Electric. The vehicle was purchased in 1900 to attract attention and "make people think the chocolate company was modern and up-to-date." After the people in Lancaster became accustomed to its presence, the automobile was sent to other Pennsylvania cities manned by salesmen with orderbooks in hand. (Courtesy, Hershey Food Corporation, Hershey, Pa.)

PRODUCTION. Cacao trees grow best in a warm, moist climate. Therefore, most cacao trees are grown within an area 20° latitude north and south of the equator. Table C-33 indicates the major cacao bean producing countries. Most cacao beans come from small farms in West Africa and Brazil.

Propagation And Growing. Cacao trees may be grown from seeds or cuttings, planted from 5 to 15 ft (*1.5 to 4.5 m*) apart. Often the trees are planted under the shade of taller trees. After about 4 years of growing, the cacao tree bears fruit in the form of seedpods which are about 1 ft (*30 cm*) long and 4 in. (*10 cm*) thick, with a hard leathery shell which may be red, yellow, gold, or pale green. Each seedpod contains up to 40 almond-shaped seeds or beans, and some may be as wide as 1 in. (*2.5 cm*). Mature, cultivated cacao trees grow to a height of about 25 ft (*7.6 m*), and they produce leaves, flowers, and seedpods the year round. Flowers, followed by seedpods, occur singly or in clusters on the main stem of the branches and the trunk. As the seedpods reach maturity, workers harvest them with knives.

TABLE C–33
TEN MAJOR CACAO BEAN GROWING COUNTRIES[1]

Country	1990 Production	
	(thousand tons)	*(thousand metric tons)*
Cote Divoire	770	*700*
Brazil	396	*360*
Malaysia	275	*250*
Ghana	270	*245*
Nigeria	171	*155*
Indonesia	165	*150*
Camaroon	127	*115*
Ecuador	105	*95*
Dominican Republic	65	*59*
Colombia	61	*55*

[1]*FAO Production Yearbook,* 1990, FAO/UN, Rome, Italy, Vol. 44, p. 176, Table 79. Most values are unofficial estimates.

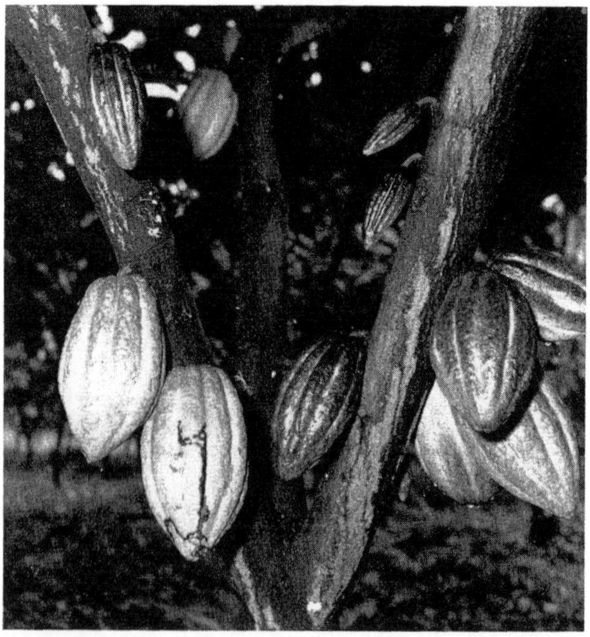

Fig. C-112. Fruiting cacao tree. The cacao tree is unique in that the flowers and the pods (fruit) grow on both the trunk and older branches. (Courtesy, Chocolate Manufacturers Association of the United States of America, McLean, Va.)

Harvesting. The knives used for harvesting are machetes for pods on the trunk, or knives on long poles for high pods. Following the harvest, the pods are opened and the beans are removed, placed (1) in piles, and covered with banana leaves, or (2) in leaf-lined holes in the ground, or (3) in boxes with perforated bottoms, and allowed to ferment—the first step in processing, and necessary for the development of the chocolate flavor.

Fig. C-113. Chocolate cake, cookies, and ice cream. (Courtesy, Hershey Foods Corp., Hershey, Penn.)

PROCESSING. Fermentation continues for 3 to 9 days during which time the temperature may reach 122°F (*50°C*) thus killing the germ and activating enzymes responsible for producing substances which eventually yield the chocolate flavor during roasting. Fermented beans are reddish brown inside, and possess a heavy sharp fragrance. After fermentation, the beans are allowed to sun dry until their moisture content is 6 to 8%. Then they are bagged for shipment.

Chocolate manufacturers sort beans by type and origin, then clean them before roasting. Roasting develops the flavor in the cacao bean, which ultimately determines the flavor and color of the final product. Depending on the ultimate use of the product and the kind of bean, roasting varies from a very light roast to a very high roast. Most modern roasters are continuous, rather than batch roasters.

Following the roasting process cacao beans are broken or cracked, which necessitates the separation of the shell from the pieces of the cotyledon, or nibs as they are called in the industry. A winnower separates the shells from the nibs in an air current. The nibs are then ground, and the heat of grinding causes the fat in the nibs to melt, producing a mixture of melted fat—cocoa butter—and finely ground nibs. This mixture is called chocolate liquor. The typical composition of chocolate liquor is about 55% cocoa butter (fat), 18% carbohydrate, 12% protein, 6% tannin compounds, 3% minerals, 3% organic acids, and 2% water with traces of caffeine and theobromine, an alkaloid related to caffeine. From this point, the chocolate liquor may be processed into cocoa or chocolate. Federal standards define several kinds of cocoa and chocolate products.

• **Cocoa**—To make cocoa the chocolate liquor may be Dutched first, if Dutch chocolate is desired. Dutching modifies the flavor and color of the cocoa by the addition of alkalizing chemicals such as the carbonates or hydroxides of sodium, potassium, ammonium, or magnesium. Next, the liquor is pressed in large hydraulic presses to squeeze out the cocoa butter. Pressing time depends on the desired fat content of the final product. Longer pressing is required for a lower fat content. Cocoas are classified according to their fat content.

After pressing, the press cake is cooled, then ground and sifted—forming cocoa powder, which may contain between 10 to 22% fat. In the United States, breakfast cocoa must contain a minimum of 22% fat, while low fat cocoa contains less than 10% fat.

• **Chocolate**—Depending upon the final product desired, sugar, milk, and flavors are added to the chocolate liquor and mixed. Also, the formation of chocolate requires more fat than there is in the chocolate liquor. Therefore, cocoa butter is added during chocolate processing. Next, the mixture is refined by being crushed between a series of rollers. This reduces the particles in the mixture to microscopic fineness, producing the typical smooth chocolate. Next, this smooth paste is "conched." Conching completely mixes the chocolate at high temperatures (130° to 190°F [*54 to 80°C*]) while exposing it to a blast of air. This develops the chocolate flavor, darkens the color, covers all particles with cocoa butter, and lowers the water content. This process is named for the shell-shaped machine or "conche" that was originally used. From this point the chocolate may be molded into bars, or other shapes, or used as a coating on some type of confection.

The quality of the final products depends upon (1) the blend of beans used, (2) the kind and amount of milk or other ingredients used, and (3) the degree of roasting, refining, and conching.

The whole process of cocoa and chocolate manufacturing is summarized in Fig. C-114.

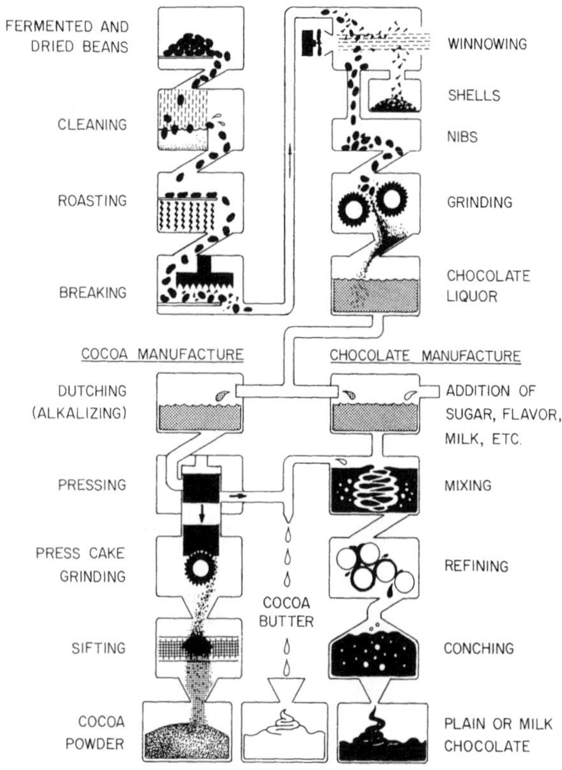

Fig. C-114. Processes in cocoa and chocolate manufacturing.

BY-PRODUCTS. Shells from the cacao bean are the major by-product of cocoa and chocolate manufacturing. These are used for fertilizer, mulch, cattle feed, and fuel. Also, cocoa butter can be considered a by-product. It is widely used in the pharmaceutical and other industries, due to its melting point; it is solid at room temperature, but melts at body temperature. However, the major use for cocoa butter is chocolate.

NUTRITIONAL VALUE. Because of their fat content, cocoa and chocolate are high in calories. One hundred grams, or about 3 ½ oz, of chocolate (sweet or semisweet) provides about 470 to 528 Calories (kcal) of energy, since it contains 40 to 53% fat. Cocoa powders contain about 215 to 300 Calories (kcal) per 100 g (*3 ½ oz*), depending upon their fat content. Cocoa and chocolate also supply some carbohydrate and protein as well as significant amounts of some minerals such as chromium, iron, magnesium, phosphorus, and potassium. More complete information on the composition of cocoa powder (high, medium, and low fat) and chocolate (bitter, candy, and semisweet) may be obtained in Food Composition Table F-36 of this book.

The theobromine and caffeine content of cocoa and chocolate produce a mild stimulating effect.

USES. The major products manufactured from the chocolate liquor include cocoa, baking chocolate, milk chocolate, and sweet and semisweet chocolate. Each products has some specific uses.

- **Cocoa**—Whether regular or Dutch, and whether high, medium, or low fat, cocoa powder is used to flavor milk drinks, candy, baked goods, ice cream, syrups, and pharmaceuticals.

- **Baking chocolate**—This is the commercial form of the chocolate liquor. It is cooled and formed into cakes. Home use of baking chocolate, which is bitter and unsweetened, includes many baked products.

- **Milk chocolate**—Milk chocolate is probably the most popular of the chocolate products. It is sold directly to the consumer in a variety of solid bars or chocolate coated candy bars or other coated candies.

- **Sweet and semisweet chocolate**—Sweet chocolate is usually dark colored and contains sugar, added cocoa butter, and flavorings. Semisweet contains less sugar. Confectioners use the sweet and semisweet chocolate for making chocolate covered candies. Home use includes cookies, candy, cakes, and a variety of other items.

Fig. C-115. Conching semisweet chocolate—a 72-hour process. (Courtesy, Hershey Foods Corp., Hershey, Penn.)

Recently, carob powder, a product obtained from the bean of a leguminous evergreen tree, *Ceratonia siliqua*, has gained attention as a possible substitute for cocoa. Carob is grown in the United States, contains no stimulants, and may be used without additional sweetening.

(Also see CAROB.)

COCOA BUTTER

Fat pressed from the roasted cacao bean (cocoa bean) is known as cocoa butter. It is yellow and possesses a slight chocolate flavor. It consists primarily of triglycerides (three fatty acids bonded to glycerol) of the fatty acids palmitic, stearic, and oleic acids. This fatty acid composition causes cocoa butter to be solid at room temperature but to melt at near body temperature. Since it melts near body temperature, cocoa butter has been used for making suntan products, toilet soaps, suppositories, ointments, creams, and lotions. However, most of the cocoa butter is used in the manufacture of chocolate.

(Also see COCOA AND CHOCOLATE.)

COCONUT (COCONUT PALM) *Cocos nucifera*

Fig. C-116. Coconut palm tree, a tall (up to 100 ft *[30 cm]* high), graceful tree that produces coconuts.

ORIGIN AND HISTORY. The native habitat of the coconut palm is unknown, but it appears to have originated in Southeast Asia and on the islands of Melanesia in the Pacific Ocean. Through the centuries, however, coconuts were distributed by water (coconut fruits float readily) and man from continent to continent and island to island to all the tropical and subtropical parts of the world, where they became the "staff of life" of the natives. Marco Polo was among the first Europeans to describe coconuts.

PRODUCTION. The world production of coconuts averages 42.1 million metric tons each year.

Coconut palms are found in the tropical, sea coastal areas of the Old World. They thrive best on rich, sandy soil. Salty water is tolerated provided the water table fluctuates so as to give the roots aeration. Propagation is accomplished by half burying the coconut in a horizontal position. The tree will begin bearing fruit in 7 to 8 years, and continue to 80 years old. A well-tended tree will produce about 100 coconuts a year. Ripe coconuts fall from the tree. But on plantations, the nuts are harvested every 2 or 3 months. Because of the height of the trees, harvesting is difficult. It is accomplished by (1) letting the ripe fruits fall to the ground; (2) using a skilled climber with a rope to reach the treetop; (3) standing on the ground and cutting fruits with a knife at the end of a long bamboo pole; or (4) using a trained monkey, wearing a collar and a long lead, to send up the tree and throw the fruit down.

The leading coconut-producing countries of the world, by rank, are Indonesia, Philippines, India, Sri Lanka, Thailand, and Mexico.

PROCESSING. The fresh nut of the coconut contains about 50% water and 30 to 40% oil. The first step in processing is to reduce whole coconut to copra, the dried meat of the coconut from which coconut oil, the most important commercial product of the coconut palm, is obtained. In the production of copra, the nuts are husked, opened, and dried, to separate the oil-bearing meat from the shell and to prevent spoilage. The primitive, and still-used, system of drying consists of splitting the nuts open and drying the meat either in the sun or in a kiln heated by burning coconut shells. Copra of better and more uniform quality is produced by hot air drying, in which the copra is conveyed slowly through a heated tunnel or oven. Copra is fairly resistant to mold, rancidity, and putrefaction; and, when protected from insects and rodents by packing, it can be stored for many months or shipped to remote sections of the world.

In general, 1,000 coconuts will yield about 500 lb *(225 kg)* of copra and 25 gal *(95 liter)* of oil.

Before processing, the copra is (1) passed through a cleaner which removes sand and trash, (2) passed through a crusher which breaks it into small pieces, (3) moved through a magnetic separator, which removes pieces of iron (bolts, nails, etc.), and (4) ground finely.

Most of the world production of coconut oil is obtained from copra by use of the continuous mechanical screw presses. Sometimes solvent extraction is employed following the preliminary compressing in screw presses. In some areas, hydraulic cage and box presses are still used.

PREPARING. Except for local use by the natives, the preparation of coconut products is done commercially. It varies according to (1) the part of the nut, and (2) the desired end product. For these reasons, preparation is presented under the section on "Uses" at the end of this article.

NUTRITIONAL VALUE. The nutritional value varies according to the coconut product. The nutritive composition of some coconut products is given in Food

Composition Table F-36. Pertinent information relative to each of the main food products follows:

1. **Coconut oil.** It has the highest percentage of saturated fatty acids of all common food oils. The chief fatty acids of coconut oil are: lauric (45%), myristic (18%), palmitic (9.5%), oleic (8.2%), caprylic (7.8%), capric (7.6%), and stearic (5%).

The high content of lauric, caprylic, and capric acids is a nutritional asset because these medium and short chain fatty acids are very useful in the dietary treatment of certain digestive disorders. However, the high degree of saturation of coconut oil may raise the blood cholesterol in some people, even though the oil itself contains no cholesterol.

The high degree of saturation imparts long stability to coconut oil. Also, when fed to milk cows, it produces rather hard (highly saturated) butter; and when fed to hogs, it produces firm pork (whereas peanuts produce soft pork).

2. **Coconut meal or copra meal.** Mechanically extracted coconut meal averages about 21.2% crude protein, 7% moisture, 6.8% ash, 11.5% crude fiber, 6.5% ether extract (fat), and 47.0% nitrogen free extract. It is low in the amino acids lysine (0.54%) and methionine (0.33%). Hence, it should either be (a) used for ruminants, or (b) supplemented with the amino acids or more complete proteins (such as fish meal) for nonruminants.

USES. The coconut palm is one of the most useful trees. The Indonesians claim that coconuts have as many uses as there are days in the year—365, of which about 200 are foods which have been extolled for centuries. The following six uses are important in world trade:

1. **Whole coconuts.** These may be either green or mature.

Green coconuts are harvested when the meat is soft and rubbery. At this stage, the husk and shell can be cut with a heavy knife or saw, and the meat can be easily scooped out of the shell with a spoon. It is eaten unseasoned, sometimes with a scoop of ice cream added. Also, the meat of green coconuts may be removed and chipped, then used in ice cream, pies, cakes, or cookies, to which it imparts excellent flavor and texture.

Mature coconuts are harvested after the shell is hard and the meat is firm. The "eyes" are punched out and the sugary liquid (called coconut milk) is drained off for subsequent adding to the ground product; the shell is broken by striking; the meat is removed, hand peeled, and ground in a food chopper; then the liquid that was drained off is added to the ground meat. This product is used in salads, desserts, puddings, candies, toppings, cakes, pies, ices, and ice cream.

2. **Copra.** This is the dried meat of the coconut from which the oil is to be extracted, yielding coconut oil and coconut meal or copra meal. The oil is generally extracted by either (a) continuous mechanical screw presses, or (b) hydraulic presses.

3. **Coconut oil.** Because of the high degree of saturation and long stability of coconut oil, it is one of the most desirable oils for confections, bakery goods, deep fat frying, and candies. It is also used in shortening, oleomargarines, soaps, shampoos, and detergents. Minor uses of coconut oil include filled milk, imitation milk, lotions, rubbing creams, prepared flours and cake mixes, and pressurized toppings.

Recently, coconut oil has served as the raw material for producing medium chain triglycerides, which are very useful in the treatment of certain digestive disorders.

4. **Coconut meal or copra meal.** This is the by-product of the production of oil from the dried meats of coconuts. Coconut meal averages about 21% crude protein. If fed to nonruminants, it should be supplemented with the amino acids lysine and methionine. The fat of coconut meal (which averages about 6.9% on an as-fed basis in mechanically extracted meal) is very low in unsaturated fatty acids. Hence, the feeding of copra meal to dairy cows produces rather hard (highly saturated) butterfat, and the feeding of copra meal to swine produces firm body fat.

The high fiber content of regular coconut meal makes it unsuited as the major source of protein for people and nonruminant animals. However, through various extraction processes it is possible to separate the protein from the fiber and to produce a product that contains about 32% protein and negligible amounts of fiber.

5. **Desiccated or shredded coconut.** This is the familiar product on grocers' shelves. It is processed in seven main steps: shelling, paring, shredding, blanching, drying, sieving, and packing. Sometimes, shredded coconut is sweetened or toasted. One ton of fresh coconuts in the shells will produce about 335 lb *(151 kg)* of desiccated coconut. Desiccated or shredded coconut is used primarily for the following purposes: topping agent, bulking agent, nutmeat, or flavoring.

6. **Coir.** The husks yield coir, a fiber which is highly resistant to salt water and is used in the manufacture of ropes, mats, baskets, brushes, and brooms.

Additional uses of coconuts include: a refreshing and palatable drink obtained from green coconuts; in the treatment of diarrhea, to replace body fluid and provide some electrolytes; toddy, a beverage which is consumed either fresh or fermented as an intoxicating palm wine, obtained from the unopened flower stalks; and palm cabbage, a salad made from the delicate young buds cut from the top of the tree.

(Also see FATS AND OTHER LIPIDS; NUT[S] Table N-8 Nuts of the World; and OILS, VEGETABLE; Table O-6.)

COCONUT MILK

This is the clear or whitish, sugar fluid contained inside of coconuts. It is good to drink, and has some nutritional value. Each cup of coconut milk contains about 57 Calories (kcal) of energy, 0.7 g of protein, 0.5 g of fat, and 12 g of carbohydrate, as well as some minerals and vitamins. However, the coconut milk listed in Food Composition Table F-36 of this book is actually liquid squeezed from a mixture of grated coconut meat and water. Therefore, it has a different composition, primarily higher fat and protein content. This type of coconut milk contains about 615 Calories (kcal) of energy, 8 g of protein, 61 g of fat, and 13 g of carbohydrate per cup. The natural sugary fluid contained inside of coconuts is referred to as coconut water in Food Composition Table F-36 of this book.

(Also see COCONUT.)

CODDLE

- To cook very gently, just below boiling point, such as coddled eggs (also referred to as the 3-minute egg).

- To pamper or to treat with excessive care.

COD-LIVER OIL

The oil obtained from the liver of the cod and related fishes, used for its medicinal properties long before vitamins were discovered. Cod-liver oil is an important supplemental source of vitamins A and D. Also, it is a fair source of vitamin E. Each 100 g provides approximately 85,000 IU of vitamin A, 8,500 IU of vitamin D, and 20 mg of vitamin E. (See Food Composition Table F-36.)

Unlike salmon and herring, the cod stores excess energy from its diet in the liver, not the muscles. As a result, the cod's liver is a relatively large organ of high fat content. This source of oil was noticed by the Vikings, whose diet has included cod-liver oil for centuries. In ancient Norway, it was the custom to keep a basin of cod-liver oil and a dipper by the inner door of each house so that the family and visitors could have a swig as they entered and departed. In retrospect, it has been determined that a fair swig for a Norwegian contained about 200,000 IU of vitamin A and 20,000 IU of vitamin D, approximately 50 times as much as the National Research Council recommends today. It was said that cod-liver oil protected the hearty Nordics from the cold blasts of winter and kept them healthy.

Cod-liver oil, along with the vitamins therein, is easily oxidized and destroyed; hence, it should be protected from strong light and stored air-tight.

(Also see FATS AND OTHER LIPIDS.)

COEFFICIENT OF DIGESTIBILITY

The percentage value of a food nutrient that is absorbed. For example, if a food contains 10 g of nitrogen and it is found that 9.5 g are absorbed, the digestibility is 95%.

(Also see DIGESTIBILITY.)

COENZYME

A substance whose presence is required for the activity of an enzyme. Coenzymes usually contain vitamins as part of their structure, and they may contain a mineral element (metal ions) as the activator. Coenzymes resist breakdown by heat. (They're heat-stable.) It is noteworthy that the same coenzyme may be used in different enzyme systems; it is the protein molecule that gives an enzyme its particular specificity.

(Also see ENZYME; and METABOLISM.)

COENZYME I (Co I)

An old name for a coenzyme employed as an electron transporter in biological oxidations. Following the recommendation of the International Commission on Nomenclature, it is more frequently called nicotinamide adenine dinucleotide or NAD.

(Also see NICOTINAMIDE ADENINE DINUCLEOTIDE.)

COENZYME II (Co II)

An old name for a coenzyme employed as an electron transporter in biological oxidations. Following the recommendation of the International Commission on Nomenclature, it is more frequently called nicotinamide adenine dinucleotide phosphate or NADP.

(Also see NICOTINAMIDE ADENINE DINUCLEOTIDE PHOSPHATE.)

COENZYME A (CoA)

A key coenzyme which contains the vitamin pantothenic acid. Its combining with two-carbon fragments from fats, carbohydrates, and certain amino acids to form acetyl coenzyme A is an essential step in their complete metabolism, because the coenzyme enables these fragments to enter the citric acid cycle. The combination of acetyl group and coenzyme A is commonly referred to as acetyl CoA.

(Also see KREBS CYCLE.)

COENZYME Q (UBIQUINONE)

Coenzyme Q, or ubiquinone, is a collective name for a number of ubiquinones—lipidlike compounds that are chemically somewhat similar to vitamin E.

Coenzyme Q is found in most living cells where it seems to be concentrated in the mitochondria. Because it is synthesized in the cells, it cannot be considered a true vitamin.

HISTORY. In 1957-58, coenzyme Q (ubiquinone) was discovered independently at Liverpool (England) and Madison (Wisconsin) via two different routes; in the one case via the study of the fat-soluble vitamins, and in the other via enzymatic processes in mitochondria.

In 1958-59, the structures of the ubiquinones were proved jointly by the Liverpool and Madison groups.

CHEMISTRY, METABOLISM, PROPERTIES. The ubiquinones consist of a basic quinone ring structure to which between 30 and 50 carbon atoms are attached in a side chain. Differences in properties are due to the difference in length of the side chain. The

50-carbon side chain occurs exclusively in mammalian tissues (see Fig. C-117).

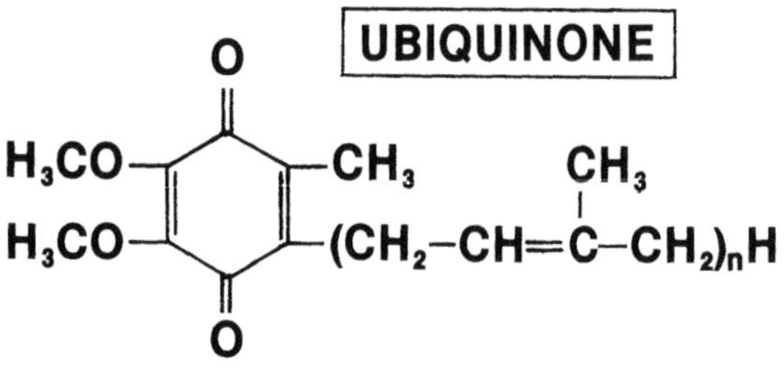

UBIQUINONE

$$H_3CO,\ H_3CO,\ CH_3\ \text{ring with two }=O;\ \text{side chain }(CH_2{-}CH{=}C(CH_3){-}CH_2)_nH$$

Fig. C-117. Structure of ubiquinones (coenzymes Q). The "n" in the formula varies according to the source—it varies from 6 in some yeasts to 10 in mammalian liver.

Coenzyme Q is present in all cell nuclei and microsomes. It is concentrated in the mitochondria.

FUNCTIONS. Coenzyme Q functions in the respiratory chain in which energy is released from the energy-yielding nutrients as ATP.

There is evidence that specific ubiquinones function in the remission (prevention) of some of the symptoms of vitamin E deficiency.

SOURCES OF COENZYME Q. Quinones occur widely in aerobic organisms, from bacteria to high plants and animals. Because they are synthesized in the body, they cannot be considered a true vitamin.

The entire series of ubiquinones has been prepared synthetically.

CONCLUSION. The importance of coenzyme Q as a catalyst for respiration imparts status as an essential metabolite. It may have other significant roles. For man and other higher animals a simple precursor substance with an aromatic ring may have vitaminlike status, but dietary ubiquinone seems, on the whole, to be unimportant unless it provides the aromatic nucleus for endogenous (body) synthesis.

(Also see VITAMIN[S], Table V-9.)

COENZYME R

An out-of-date name for the vitamin biotin.

(Also see BIOTIN; and METABOLISM.)

COFACTOR

A nonprotein substance which is a part of an enzyme and is needed for the full functioning of the enzyme. A cofactor may consist of an organic compound, e.g., a vitamin derivative; or a metallic ion, e.g., potassium, manganese, magnesium, calcium, or zinc. In the latter case, the metals are usually activators of the enzymes to which they become attached.

(Also see ENZYME.)

COFFEE *Coffea arabica* [29]

Contents — Page

Origin and History 442
The Coffee Plant 443
Coffee Culture 443
Preparation for Market 444
World Production and Exports 444
U.S. Production and Imports 444
Processing Coffee 445
Roasting 445
The Cup Test 445
Unblended Coffee 445
Blended Coffee 446
Chicory 446
Instant Coffee 446
Decaffeinated Coffee 446
Grinding Coffee 446
Packing and Storing Coffee 447
How to Make Good Coffee 447
World and U.S. Consumption 447
Coffee, Caffeine, and Health 448
Summary 450

Coffee is the beverage made from the roasted and ground beans of the coffee tree. It is the favorite drink in almost every country in temperate or cold climates. It is noteworthy, too, that the coffee break has become an integral part of the business world.

The coffee bean is the world's most valuable agricultural commodity. Among natural commodities in international trade, coffee usually ranks second only to petroleum in dollar value, accounting for $12 billion annually.

ORIGIN AND HISTORY. According to legend, coffee was discovered in Ethiopia when Kaldi, an Arabian goatherder, who lived in that part of Africa now known as Ethiopia in about the 3rd century A.D., noticed his goats prancing and cavorting on their hind legs after eating coffee berries. Being an adventuresome sort and needing a lift, he decided to try the fruit himself. Soon thereafter, the abbot of a nearby monastery passed that way and was astonished to see the herder and his goats merrily dancing together in a meadow. The whimsical story goes on to say that the abbot decided to test the power of the berries on himself. So, he poured boiling water on them to make a brew. When he drank it, he found that he could stay awake for hours on end. From that time forward, he and the monks drank the berry liquid so that they could stay awake during the long hours of evening prayers.

---

[29]In the preparation of this section, the authors adapted selected material from: Schapira, Joel, David, and Karl, *The Book of Coffee & Tea*, St. Martin's Press, New York, N.Y., 1975.

Fig. C-118. The Coffee Vendor decked out with the paraphernalia of his calling, from a copper engraving of nearly 18th century. (Courtesy, The Bettmann Archive, Inc., New York, N.Y.)

Coffee was first used as a food, rather than as a beverage. African tribes crushed ripe cherries from wild coffee trees with stone mortars, added animal fat, mixed this exotic blend and fashioned it into round balls, which they ate at their war parties. The food filled two needs: (1) the animal fat and the protein of raw coffee beans (the protein is lost when coffee is prepared as a beverage) provided concentrated nourishment; and (2) the caffeine of the coffee served as a stimulant to spur the warriors to greater heights of savagery.

Later, African tribes made wine from coffee, by fermenting the ripe berries and adding water to the juice.

Coffee reached Yemen about 1000 A.D., where the ingenious Arabs stirred up the first "bean broth" from the cherry's agreeable seed. Soon, its popularity perked across all Arabia. For teetotal Muslims, it became an integral part of religious and secular life. Coffee moved from Arabia to Turkey during the 1500s, where it quickly acquired such importance that Turkish law permitted a wife to divorce her husband for failing to keep the family *ibrik* (pot) filled. Merchants of Venice carried their first cargo of coffee from Constantinople to Italy in 1615. Coffeehouses, where people met for serious discussions, sprang up all over Europe in the 1600s. But, some devout Catholics denounced it as the drink of infidels, and therefore sinful. Before committing himself, Pope Clement—so it's said—tried a cup of coffee and became an instant convert. He settled the matter by baptizing the brew in order to give it Christian status. Coffee probably came to America in the 1660s. Coffee-growing was introduced in Brazil in the 1700s.

THE COFFEE PLANT.

Fig. C-119. Coffee tree, loaded with coffee cherries. Underneath is a catching frame for a piece of equipment designed to harvest the cherries. (Courtesy, USDA)

The coffee plant belongs to the genus *Coffea,* species *C. arabica.* It originally grew wild in Ethiopia.

The coffee plant is an evergreen shrub or small tree, which is propagated from seed. It reaches a height of 14 to 20 ft (*4.3 to 6.1 m*) when fully grown. However, coffee growers generally prune it to under 12 ft (*3.7 m*). It bears its first fruit in 5 to 6 years, and annually yields more than 5 lb (*2 kg*) of fruit—the red, seed-bearing coffee "cherries"—from which 1 lb (*0.5 kg*) of green coffee seeds or beans, is obtained. The tree will be productive through the fifteenth year.

Many varieties of the coffee plant are known to exist. But only two species are of commercial importance; namely, *Coffea arabica,* indigenous to Ethopia, and *Coffea robusta,* which was first found growing in the Belgian Congo; together, these two varieties constitute 99% of the total world output.

COFFEE CULTURE. Coffee thrives best in a tropical rainforest habitat. Rainfall in its Ethiopian homeland exceeds 60 in. (*150 cm*) per year, much of it from April to September. Coffee prefers the cool temperature of the highlands, but it cannot stand freezing. The crop grows best where temperature varies little from 70°F (*21°C*). Well-drained volcanic soils are very productive in Africa, as are the red soils rich in iron and potassium in southern Brazil. Mulching is effective for increasing yields; typically, alternate rows are mulched, with the rows reversed each year. In Africa, coffee plants are usually started from seed in nurseries and transplanted to the field when they reach a height of 18 to 24 in. (*45 to 60 cm*). In Brazil, direct field planting is common, with 3 to 4 seeds dropped in each hole. Recommended spacing for mature multiple-stemmed trees is 10 to 12 feet (*3 to 3.7 m*). In Africa, the tendency is to plant shade trees in coffee groves, thereby limiting sun exposure of the coffee trees for part of the day, and providing shade for the rest of the day. In Brazil, plantings are generally in the open. Weeds are generally controlled by hand pulling, and fertilization is limited.

The desirable flavor in coffee seems to result more from the location in which the coffee is grown than from the kind and amount of care. Colombian, Blue Mountain Jamaican, and highland Central American coffees are usually considered superior to African coffees.

The vast majority of cherries are harvested by hand. If the dry method of processing is to be used, all the cherries are stripped from the branches at once—green, partially ripe, ripe, and overripe. If the wet method is to be used, only ripe cherries are picked.

PREPARATION FOR MARKET. After harvesting, the cherries are dried and the covering around the beans is removed by one of the following methods:

1. **The dry method.** In the older dry method, which is used in the Near East and wherever water is scarce, the gathered berries are dried in the open for 2 to 3 weeks, with cover kept ready to protect the hand turned piles in case of rain. During this time, fermentation occurs. When thoroughly dry, they are transferred to milling machines for removal of the dried husk, parchment, and thin, inner silver skin. The result is the green coffee bean.

2. **The wet method.** This method, which is used where moisture is abundant, works like this: Defective fruit and debris are first separated by water flotation. (The good berries settle to the bottom of the flotation tank.) Then, as soon as possible, the berries are put into pulping machines to remove the outside fruit pulp, exposing the sticky substance which surrounds the parchment. Fragments of remaining pulp and mucilage are removed by either (1) controlled fermentation over a 12- to 24-hour period, or (2) chemical or mechanical methods. Then follows a thorough washing in running water and drying in trays, on paved areas, or on mats, much as in the dry method. After thorough drying, the beans are transferred to hulling machines for removal of the tough parchmentlike skin and the silver skin. The final product is the same as in the dry process: the green coffee bean.

Coffees processed by the wet method tend to be more highly flavored than those processed by the dry method, and they generally bring higher prices in world markets. The superior quality of wet processed coffee is attributed primarily to two factors: (1) only ripe cherries are picked for preparation by the wet method, and (2) the fermentation process is controlled.

Coffee processed by either the dry method or the wet method is then graded for size, type and quality, by hand or by machine; sorted by hand to remove impurities and defective beans; packed in burlap bags of about 132 lb (*60 kg*) each; and transported to the nearest port for shipment to processors, usually in the importing countries.

WORLD PRODUCTION AND EXPORTS. The world production of coffee totals 5 to 6 million metric tons annually (5,964,000 metric tons in 1990), of which about ⅓ is grown in Brazil.

Brazil and Colombia dominate the world's coffee production, which they have done for many years. But some of the old coffee-producing countries are again active, and new countries are making inroads on the supremacy of Brazil and Colombia. The ten leading coffee-producing countries in the world, by rank, in 1990, are shown in Fig. C-120.

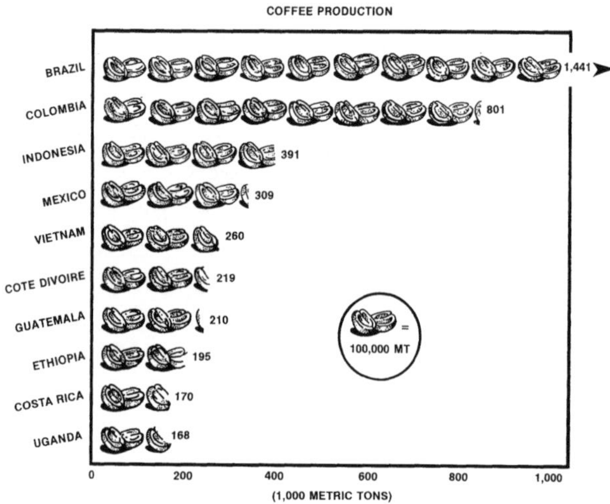

Fig. C-120. Leading coffee-producing countries of the world in 1990. (Source: *FAO Production Yearbook*, 1990, FAO/UN, Rome, Italy, Vol. 44, pp. 174-175, Table 78)

All 50 coffee exporting countries—led by Brazil, Colombia, Indonesia, and Mexico—rely on coffee as a major source of foreign exchange. Unlike Brazil, which drinks a third of what it raises, most coffee producing nations consume coffee sparingly; the bean brings more leaving home than staying there.

World coffee prices are strongly influenced by production shortfalls in Brazil, the leading producing and exporter. Frost may severely affect Brazil's coffee output.

U.S. PRODUCTION AND IMPORTS. Hawaii grows the only U.S. coffee. This premium bean, started from imported Brazilian trees in 1825, is produced in small quantities. Complex production methods are required due to the hard volcanic soil of the island. Its flavor is excellent and it commands high prices wherever coffee is sold. In 1990/91, Hawaii produced 2,700,000 lb (*1,225,800 kg*) of coffee, which sold at $2.60 per pound, and had a total value of $7,020,000.

In 1990, the Latin American countries shipped 15.4 million bags of coffee to the United States, or nearly 79% of the 19.6 million bags imported. Brazil was the leading supplier, followed by Mexico and Colombia (see Fig. C-121).

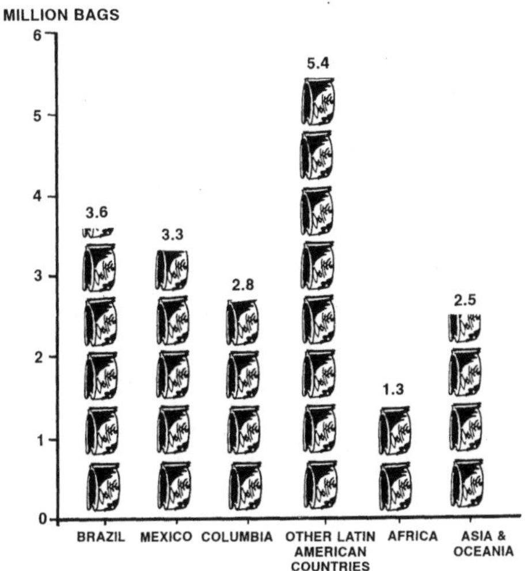

Fig. C-121. Origin of U.S. coffee imports, 1990/91, in bags. Each bag contains 132 lb (*60 kg*). (From: *World Agricultural Production*, June 1990, USDA, FAS)

PROCESSING COFFEE. The flavor of the coffee is determined by the variety, by where and how the coffee was grown, by the method of preparation for market and by the roasting.

At the roasting plant, the beans are emptied into chutes leading from the upper to the lower floor. An air-suction device removes dust and other materials. The coffee may then go to the blending machine, a cylinder that mixes different types of coffee. From the blender, the beans flow by gravity to storage bins, then to roaster ovens.

Roasting. Roasting brings about both physical and chemical changes in coffee beans. When coffee is roasted, it shrinks about 16% in weight (up to 20% for the darkest Italian roast), doubles in volume, turns from pale green to a rich brown in color, and develops characteristic coffee taste and aroma. The resulting chemical composition differs radically from that of the green beans.

In continuous roasting, hot air (400° to 500° F, or *200° to 260°*) is forced through small quantities of beans for a 5-minute period. In batch mixing, much larger quantities of beans are roasted for a longer time.

The longer coffee is roasted, the darker it gets. From light to dark are a variety of roasts all tasting like coffee, each appealing to a particular taste.

Through the years, and from around the globe, has emerged a charming, although somewhat unwieldy, array of names to designate roast color, of which a few select ones follow:

- **City roast (American roast, brown roast)**—This is the most widely used style in the United States.

- **Brazilian**—This coffee is somewhat longer roasted than city roasts. It has a faint dark roast flavor; and a trace of oil shows on the beans.

- **French roast (New Orleans roast)**—With this roast, oiliness appears on the surface; and the color is burnt amber. It approaches Espresso flavor. Louisiana style coffee is French roasted coffee to which chicory is added.

- **Spanish; Cuban; French; Italian roast**—This roast is darker than French roast, but not as dark as Italian. It's like Espresso without the bite.

- **Italian; Espresso roast**—This is darkest of all roasts—it's almost to carbonization. The bean surface is shiny and oily, and the color is black.

The Cup Test. Before purchasing, most coffee buyers like to examine both green and roasted samples. Green beans are examined for imperfections, including beans that are broken, unripe, blighted, or underdeveloped; excesses of which will affect cup quality. Next, the appearance or "style" of the roasted beans is judged, with consideration given to bean development, color uniformity, and swelling.

Although the appearance of the green and roasted beans is useful, the most significant tests are those for flavor, aroma, and body, qualities which are best evaluated by the human senses of smell and taste—the cup test.

For cup testing, the only ingredients are clean white cups, measured amounts of coffee of each sample, and freshly boiled water. A measured amount of drip-ground coffee (equal to the weight of a nickel) of each sample is placed in a 6-oz (*180 ml*) white china cup; and boiling water is poured over it. Expert tasters evaluate the coffee; first for its aroma, then for taste.

Unblended Coffee. Two species provide most of the world's commercial coffee; namely, *Coffea arabica* and *Coffea robusta*. A third species, *Coffea liberica*, provides about 5% of American coffee. As a rule, only *arabica* coffee, which possesses finer aroma, flavor, and body than robusta, is offered unblended. *Robusta* coffee, which is neutral in the cup, is widely used in commercial blends.

Besides the above specie, or botannical, classifications there are commercial classifications for marketing purposes, such as Brazils, Milds, and Robustas. The most telling and useful classifications for coffee are geographical, with different kinds of coffee named according to where they are grown, and further identified by the region where grown or the port from which shipped; e.g., Colombian Armenia, Mexican Coatepec, Jamaican Blue Mountain.

The growing demand for unblended coffees in indicative of the desire of people to experience the distinctive qualities of some of the world's great varieties and types of coffee, along with their disenchantment of "nothing much" blends. There's magic in such names as Kona, Jamaican, Mocha, and Java.

The opereator of a specialty coffee store can provide high-quality unblended coffees.

Blended Coffee. Almost all of the coffee brought into, and marketed in, the United States is blended by the major packers. Although the proportion of unblended coffee is less in other countries than in the United States, blended coffee predominates throughout the world.

Today, brand name coffees are promoted in the United States, with little or no attempt made to identify the species or place where grown. Coffee brands are extolled on the basis of being richer, coffee-er, good to the last drop, mountain grown, tree ripened, or hand-picked.

The art of blending coffees was born because coffee men found that, by blending various coffees, they could create a single coffee with an aggregate of delightful traits. Nature provides beans possessing taste, aroma, and body in varying degrees, but unblended coffees rarely possess these qualities in the most desirable proportions. So, the job of the blender is to combine these qualities in a balanced coffee. Not only that, the blender of commercial coffee must, from month to month, be able to find suitable replacements for coffees that cease to satisfy the requirements of his blend.

The operator of a specialty coffee store can provide a "house blend" of excellent quality. His reputation and livelihood depend on his maintaining the highest standards in his selection of coffees; hence, he will not sacrifice quality.

Chicory. Sometimes, chicory, which is also known as French endive or witloof, is added to coffee to reduce the cost. In New Orleans, where the Creole influence is strong, chicory blended coffee is widely used. Also, to economize, many European immigrants learned to use chicory in the old country; some never switched to pure coffee because they came to prefer the flavor of the chicory additive.

Chicory is the perennial turnip-shaped vegetable the tops of which are commonly used as a salad green and the young roots of which may be boiled and eaten like carrots. The roots are also kiln dried, then roasted and ground; it then looks like ground coffee. Coffee with chicory additive is more bitter, heavier, and darker than pure coffee.

The United States imports chicory from Central Europe. Some is also grown in Michigan.

Instant Coffee. Instant coffee entails the extraction of ground coffee with warm water to form a concentrated liquid brew, followed by removal of the water from the brew by some method of dehydration. The residue which is left is instant coffee.

Instant coffee, which constitutes about one-fifth of all coffee sold, is prepared in either of the following ways:

1. By forcing an atomized spray of very strong coffee extract through a jet of hot air. This evaporates the water in the extract and leaves dried coffee particles, which are packaged as instant, or soluble, coffee. Some of the processors of instant coffee have added aroma to instant coffee in the form of oil removed from the roasted beans prior to extraction. Still other processors have used the process of aglomeration (steaming instant coffee so that the particles are sticky and will lump together, followed by redrying the lumps by reheating) to make instant coffee chunkier and darker, so that it looks more like ground coffee.
2. By freeze-drying, which results in a product that looks somewhat like ground roasted coffee and retains some of coffee's aroma.

The processing of instant coffee has gone through many changes through the years, and the product has been improved, but to coffee connoisseurs instant coffee lacks the cup quality of ground roasted coffee.

Decaffeinated Coffee. Decaffeinated coffee is made by removing the caffeine from the green beans as follows: The green, unroasted beans are softened by steam and water; flushed with a solvent, usually containing chlorine, which soaks through all parts of the beans; agitated in the solvent to cause the caffeine from the beans to be drawn in combination with the chlorine; drained of the solvent, then heated and blown with steam to evaporate all traces of solvent. To obtain a product that is 97% caffeine-free, this process must be repeated up to 24 times. In addition to removing caffeine, the process removes some of the coffee's oils and waxes.

Most coffee-drinkers agree that decaffeinated coffee does not have the wonderful flavor of real coffee, but it isn't bad. Of course, some people have reason to switch to decaffeinated coffee—some are made uncomfortable by coffee's stimulation, still others complain that coffee keeps them awake. On the other hand, most people find that the stimulation produced by the caffeine in 1 or 2 cups *(240 to 480 ml)* of coffee is pleasant to them; and they experience no side effects.

During the decade of the 70s, the consumption of decaffeinated coffee more than doubled. But most of the caffeine removed from coffee was added to soft drinks and pharmaceuticals. Today, soft drinks are the No. 2 source of caffeine in our diet; more than two-thirds of the soft drinks consumed in the United States have the stimulant added to them. One recent study revealed that of the 24 soft drinks tested, including the top 10 in national sales, some of them contained almost as much caffeine as a cup of instant coffee.[30]

So, depending on what kind and how much soda they drink, those who switched to decaffeinated coffee may be consuming what they thought they were avoiding.

Grinding Coffee. It would take hours to brew coffee from whole roasted beans, and the beverage would

[30]*Consumer Reports*, Consumers Union, Box 461, Radio City Station, New York, N.Y. 10019.

be flat. Ground coffee permits rapid absorption of water and extraction of the soluble components (solids and volatiles) that comprise flavor; and the absorption of water and extraction take place at a rate inversely proportional to particle size.

The desirable 19% extraction is obtained with the following grind-time-pot combinations:

Grind	*Brewing Time*	*Type of Pot*
	(min.)	
Fine	*1 to 4*	*Vacuum and pressure*
Drip	*4 to 6*	*Drip pot*
Regular	*6 to 8*	*Percolator*

Actually, all grinds are a mixture of particles of various sizes. However, a mixture containing a high percentage of small particles is designated "fine," whereas a mixture containing a high percentage of larger particles is designated "regular."

The grind should be selected so as to match the pot. Generally, this calls for fine grind for vacuum and pressure pots, drip grind for drip pots, and regular or perc for percolators.

Packing And Storing Coffee. After roasting, the beans are usually ground and vacuum-packed in cans. The vacuum pack will keep coffee fresh for many months without refrigeration. However, once the can is opened, coffee will keep well for only 7 to 10 days at room temperature. So, if it will take longer than this period to use up a can of coffee, it should be transferred to a glass jar with a good seal and put into the refrigerator. Ground coffee may be stored in the freezer for up to a month.

Since the flavor of coffee deteriorates rapidly after it is ground, or after a sealed can is opened, many coffee drinkers buy whole roasted beans and grind them at home. Coffee beans stored in a tightly covered glass jar in the refrigerator can be kept for several months.

HOW TO MAKE GOOD COFFEE.

A good cup of coffee is one in which taste, aroma, body, color, and stimulation are present in pleasing proportions. Making a good cup of coffee is an art, which requires the same careful attention as the preparation of other articles of food. Observance of the following rules will assure good coffee:

1. Select a pot which makes it easy to control the variables of brewing.
2. Keep the coffee maker scrupulously clean.
3. Use high-quality coffee—coffee that has been properly grown, shipped, roasted, blended (if it is blended coffee), and vacuum packed.
4. Use fresh coffee. Buy coffee in the size can or package that can be used in 7 to 10 days. Refrigerate after opening a can to preserve freshness.
5. Start with fresh, cold water.
6. Use the full capacity of the coffee maker; for lesser quantities, use a smaller coffee maker.
7. Measure both coffee and water accurately. The standard recommendation for full-strength beverage is 1 to 2 Tbsp (*15 to 30 ml*) of coffee to each cup (6 oz, or *180 ml*) of water. Vary the strength (percentage of solids in the brew) according to individual preference, by increasing the amount of coffee or decreasing the amount of water.
8. Use the proper grind-time-pot combination, so as to extract about 19% of the weight of the coffee in the brew.
9. Never boil coffee; boiling produces an undesirable flavor change. Most coffee makers on the market are planned on this principle.
10. Serve coffee piping hot as soon as possible after brewing; freshly brewed coffee always tastes best.
11. Remove grounds after brewing; and never repour coffee through the spent grounds.
12. Never reheat brewed coffee that has been allowed to cool.

WORLD AND U.S. CONSUMPTION.

Reports on per capita coffee consumption for any given year are estimates, based on annual imports of green beans, without consideration of changes in inventory levels at the beginning and end of the year, and on population figures, which for some of the coffee-consuming areas are uncertain. However, estimates of per capita coffee consumption are given in Table C-34 and Fig. C-122.

TABLE C-34
LEADING COFFEE-CONSUMING COUNTRIES OF THE WORLD[1]

Country	Per Capita Coffee Consumption[2]			
	Annual Consumption			Daily Consumption
	Green Beans	Roasted Coffee	Cups Coffee	Cups Coffee
	(lb)	(lb)	(No.)	(No.)
Finland	28	24	1,920	5.3
Sweden	27	23	1,840	5
Denmark	23	19	1,520	4
Norway	22	18.5	1,480	4
Iceland	19	16	1,280	3.5
Belgium	18	15	1,200	3.3
Netherlands	18	15	1,200	3.3
West Germany	15	13	1,040	2.8
Costa Rica	14	12	960	2.6
Austria	13	11	880	2.4
France	13	11	880	2.4
United States	12	10	800	2.2
Switzerland	11	9	720	2.0
Canada	10	8	640	1.8

[1]Computations in this table based on the following: green beans lose 16% of weight in roasting; 1 lb *(0.45 kg)* of roasted coffee makes 80 cups *(19 liter)* of coffee.
[2]One lb = 0.45 kg; one cup = 0.24 liter.

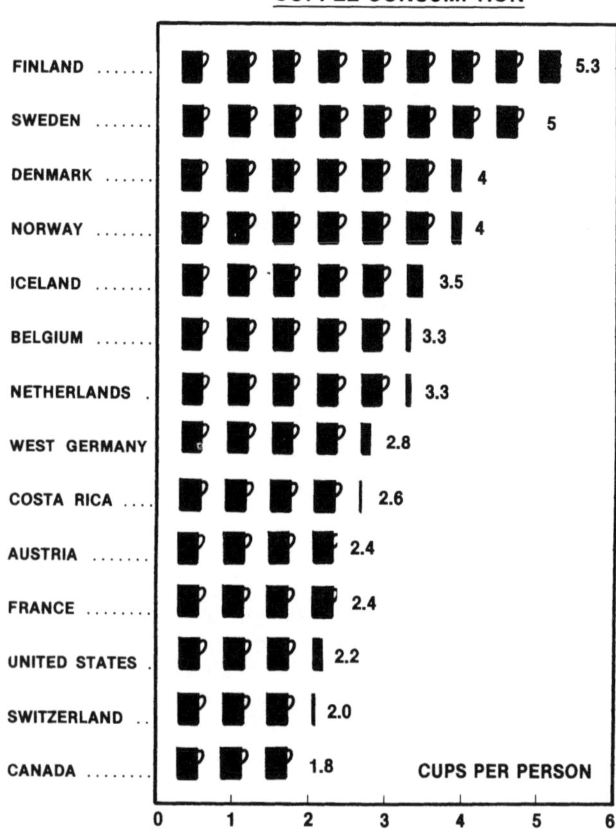

Fig. C-122. World's coffee drinkers. (One cup = *240 ml.*)

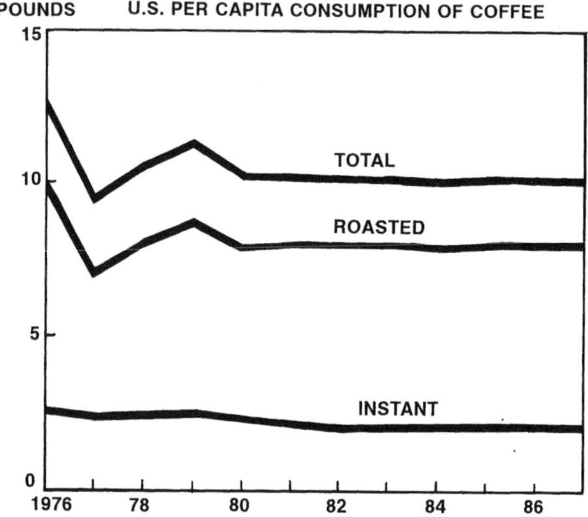

Fig. C-123. Green bean equivalent. Instant is quantity processed for soluble use minus net exports. (From: *1989 Handbook of Agricultural Charts*, Agricultural Handbook No. 684, USDA, p. 104, Chart 253)

Note that Finland, top per capita coffee consumer, downs 5.3 cups (*1.3 liter*) a day for every man, woman, and child. By contrast, the United States averages less than half that much per person.

Coffee consumption (per capita green-bean basis) in the United States reached a peak of 19.9 lb (*9 kg*) in 1946, and has declined since. On a per capita green-bean basis, U.S. consumption in 1960 was 15.8 lb (*7 kg*); in 1970, 13.7 lb (*6 kg*); and in 1990, 10.2 lb (*4.6 kg*). Fig. C-123 shows U.S. coffee consumption since 1976; total, and with a breakdown into roasted and instant.

In a 1976 Health Interview Survey, conducted by the U.S. Department of Health, Education and Welfare, a representative sample of the U.S. population was asked questions about coffee and certain other suspect health related substances. Pertinent findings relative to coffee follow:[31]

1. Coffee was drunk by 101 million people 20 years of age and over in the United States—80% of the civilian population of this age range.
2. Of those who drank coffee, more people drank 2 cups (*480 ml*) per day than any other amount, although 1 to 3 cups of coffee per day were fairly common. The average amount consumed by coffee drinkers was 3.2 cups (*800 ml*) per day, which is an average of 2.6 cups (*600 ml*) per day from all Americans 20 years of age or older.
3. Most coffee drinkers drank regular coffee (74.6%); the rest drank either decaffeinated coffee (18.4%) or both decaffeinated and regular coffee (6.5%).

Based on a 1991 survey, the National Coffee Association of U.S.A. reported (1) that 51% of American over age 10 drink coffee, and (2) that coffee drinkers average about 3½ cups per day.

COFFEE, CAFFEINE, AND HEALTH. Various accusations have been leveled at coffee, including that it causes heart attacks, high blood pressure, pancreatic cancer, peptic ulcers, hypoglycemia, fetal malformations, cystic breast disease, and nervousness; and that it shortens life. But confirmed coffee drinkers know better!

Increasingly, the debate has focused on caffeine, most of which comes from coffee, although headache and cold remedies also contain caffeine. The two major varieties of coffee beans, arabica and robusta, contain 1 and 2% caffeine, respectively. Brewed coffee contains 80 to 120 mg of caffeine per cup (*240 ml*), usually about 85 mg; tea contains about one-half this amount per cup, and cola

[31]Bonham, G. S., Ph.D., and P. E. Leaverton, Ph.D., Coffee Use Habits Among Adults, *Sourcebook of Food and Nutrition*, Marquis Academic Media, Marquis Who's Who, Inc., Chicago, Ill., 1980, pp. 334-338.

beverages about 30 mg per cup. Instant coffee has about 65 to 75 mg of caffeine per cup and decaffeinated coffee about 1 to 6 mg per cup. Caffeine is absorbed readily and reaches peak levels in the body about 1 hour after ingestion.

The question is whether caffeine does any real harm. The accusation and fact of each case against coffee follow:

• Coffee causes heart attacks?—In the early 1970s, heavy coffee drinkers were shaken by certain research reports linking coffee with heart attacks.

Fact. Later, it was found that the studies indicating that heavy coffee drinking was a causative factor in heart disease did not take into account the fact that heavy coffee drinkers were often heavy smokers.[32]

The famed Framingham study, conducted in Framingham, Massachusetts, from 1949 to 1969, on 5,209 males and females, ages 30 to 62 showed that when cigarette smoking was held constant, the association of coffee consumption with coronary heart disease ceased to exist.[33]

Since 1968, the IBM Corporation has conducted a voluntary health testing, or screening, study of its employees above 35 years of age. At the time of the reported results which follow, 92,000 initial and 48,000 follow-up examinations had been done, for a total of 140,000. One of the questions in the questionnaire concerns coffee, and the question is divided as follows: Do you regularly take 0 cups per day, 1 to 3 cups (*240 to 720 ml*), 4 to 8 cups, or 9 or more cups per day? Using the abnormal electrocardiogram as an index of heart disease, IBM also did a univarite analysis of coffee vs the abnormal electrocardiogram. They reported that *there was no relationship whatsoever between coffee consumption and the electrocardiogram.*[34]

In summary, there is no convincing evidence that coffee causes heart attacks. However, protracted abuse of coffee drinking appears responsible for arrhythmia (an alteration in the rhythm of the heartbeat either in time or force) in a small percent of the population.

• Coffee causes high blood pressure?—In the January 26, 1978 issue of *The New England Journal of Medicine*, Robertson *et al.* reported that, after 250 mg of caffeine (roughly the amount of caffeine in 3 cups of coffee) was given, there was a modest rise, of rather short duration, in blood pressure of 14/10. Then the article concluded with this statement: "Further investigation into the effects of chronic caffeine use and caffeine use in hypertensive subjects is needed."[35]

Fact. The short-duration rise in blood pressure of a mean figure 14/10, as reported above, is not significant. Almost any individual may, for example, experience a rise in blood pressure of from 30 to 60 mm of mercury when he becomes nervous and excited such as (1) when watching a sports event, or (2) when having an argument with his wife.

In the IBM study described under " **• Coffee causes heart attacks?**," the blood pressure was also taken on all participants. The IBM researchers concluded that, "it is clear that there was no specific relationship between daily coffee consumption and blood pressure in those studied." It is noteworthy, too, that the IBM study included 72,100 chronic coffee drinkers.[36]

It may be concluded, therefore, that man apparently adapts well to coffee, even in fairly large amounts, and chronic coffee use is not a cause of high blood pressure.

• Coffee causes pancreatic cancer?—It has been suggested that coffee may be contributing to cancer of the pancreas. This disease has been increasing in frequency for about 50 years. Now, at around 20,000 deaths a year, it is the fifth most common cause of cancer.

Fact. U.S. coffee consumption has been decreasing since 1946. Besides, the case remains unproved.

Of two recent studies, one implicated *decaffeinated* coffee and the other made no distinction; neither study was regarded as conclusive. Meanwhile, moderation in the use of coffee is suggested—limiting to 2 cups (*480 ml*) a day in any form. An estimated 5 million Americans take upward of 10 cups a day. But drinkers of coffee (with caffeine) may find that they have trouble cutting back because they have become dependent on the caffeine and suffer from headache, muscle tension, sleepiness, and other symptoms when they experience withdrawal. These individuals should try to taper off on coffee and perhaps switch to tea (1 cup of tea supplies about a third of the caffeine of a cup of coffee). Further studies are indicated.[37]

• Coffee causes peptic ulcers?—The possible relationship between caffeine and peptic ulcers has long been debated. Stimulation of gastric secretion by caffeine has been shown in several experimental animals and with human subjects in single dose experiments.

Fact. A variety of studies have failed to establish a clear-cut cause and effect relationship between caffeine ingestion and peptic ulcer.[38] Nevertheless, coffee is known to cause acidosis (sourness) of the stomach. So, when a patient has an ulcer, most internists recommend that coffee be eliminated.

• Coffee causes hypoglycemia (low blood sugar)?—This condition, characterized by an abnormally low concen-

[32]Kannel, W. B., "Coffee, Cocktails and Coronary Candidates," *The New England Journal of Medicine,* Vol. 297, 1977, p. 443.

[33]Bertrand, Charles A., M.D., F.A.C.C., F.A.C.P., On Coffee and Your Heart, *Sourcebook on Food and Nutrition*, Marquis Academic Media, Marquis Who's Who, Inc., Chicago, Ill., 1980, pp. 301-303.

[34]*Ibid.*

[35]*Ibid.*

[36]*Ibid.*

[37]*The Harvard Medical School Health Letter*, June 1981, Department of Continuing Education of Harvard Medical School, Cambridge, Mass.

[38]Graham, D. M., Ph.D., Caffeine—Its Identity, Dietary Sources, Intake and Biological Effects, *Sourcebook on Food and Nutrition*, Marquis Academic Media, Marquis Who's Who, Inc., Chicago, Ill., 1980, pp. 329-333.

tration of sugar in the blood, may be associated with excess coffee.

Fact. Low blood sugar may be associated with a number of disorders; among them, alcoholism, allergies, behavior problems, brain damage in infants, depression, diabetes, drug addiction, fatigue, high blood pressure, impotency, obesity, ulcers, underachievement, and coffee.

Coffee gives a temporary lift in energy and sharpens mental acuity, but some individuals experience a let down feeling later. The reason for the lift is that caffeine stimulates the central nervous system and promotes the breakdown of glycogen to glucose in the liver, which raises the blood sugar level. However, the elevation of blood glucose may be short lived in some people, because their pancreas reacts by secreting insulin. Hence, they may feel let down due to a drop in their blood sugar.

If excess coffee (caffeine) is the cause of hypoglycemia, recommended treatment is to lessen or eliminate the coffee.

• **Coffee causes fetal malformations?**—There have been reports that pregnant women who drink caffeine—containing substances may be increasing the risk of birth defects. Also, it is known that transfer of caffeine across the placental barrier occurs readily.

Fact. After evaluating thousands of studies, the American Council on Science and Health (ACSH) concluded: "There is no evidence from human studies to support the belief that the use of coffee or cola is harmful to the fetus." Dr. Elizabeth Whelan of the ACSH con tinued: "Use some common sense. People who drink 10 cups of coffee a day should realize that they are overdoing it. In the case of pregnant women, it's been found that they don't metabolize caffeine as well as those who are not pregnant. Another reason why pregnant women shouldn't drink too much coffee, or other caffeine-containing drinks, is that it may keep them awake during a time when they need their rest."[39]

Although more evidence is needed, there does appear to be sufficient reason to caution pregnant women, or soon to be pregnant women, about their coffee consumption.

• **Coffee causes cystic breast disease?**—Cystic breast disease—a noncancerous condition—affects about 50% of American women, most of them under age 45. One study, involving about 50 women, indicated that caffeine consumption may increase the incidence of cystic breast disease.

Fact. There is no conclusive evidence to link cystic breast disease to caffeine consumption. More research is needed.

• **Coffee causes nervousness, tremors, insomnia, and gastric symptoms; withdrawal symptoms?**—With excessive coffee intake, some individuals develop untoward nervous system effects from caffeine, such as nervousness, tremors, and insomia. In rare cases, there may be gastric symptoms such as heartburn, abdominal discomfort, or diarrhea; but generally such gastrointestinal effects occur with excessive coffee consumption. Also, some experience headaches with the withdrawal of caffeine, beginning about 12 to 16 hours after the last dose of caffeine.

Fact. Such adverse effects as the above are rare with moderate intakes of coffee. Rather, the nervous system is stimulated agreeably—there is enhanced alertness and wakefulness, increased energy, and elevated mood. Of course, those who experience gastric distress from minimal amounts of coffee should abstain from it. Two remedies for caffeine withdrawal headaches are suggested: (1) a gradual decrease in coffee consumption—a weaning period of about a week; and (2) the use of a suppository containing 150 mgm of caffeine.[40]

• **Coffee drinking shortens life?**—Creative people have long sworn by, and occasionally at, the heavenly brew. Typical are the experiences of three Frenchmen-of-letters.

Balzac, one of the most important novelists of the 1800s, gave this description of what coffee did for him: "Everything becomes agitated; ideas are set in motion like army battalions on the battlefields, and then the battle begins. Memories charge forward, banners flying; the light cavalry of comparisons progresses at a magnificent gallop; the artillery of logic hastens into the fray with its cannons and cartridges...." Balzac often wrote more than 16 hours a day, keeping himself awake with countless cups of black coffee, kept hot over the flame of a spirit lamp in his studio. But he died in 1850 at age 51. His death was immediately attributed to the gradual accumulation of coffee poisons in his system.

Fact. Caffeine doesn't accumulate in the human body; it's metabolized at the rate of 15% an hour and rapidly excreted. Moreover, in addition to writing and drinking black coffee, Balzac filled his life with wild money-making schemes and affairs with women. So, there could have been other reasons than coffee for his relatively early demise. Besides, what of Voltaire, French author and philosopher, who a century earlier drank 50 cups (*12 liter*) a day and lasted, still vigorous, into his middle eighties? Even more arresting is the case of a third French writer, Bernard de Fontenelle, who on his hundredth year (in 1757) attributed his long life to coffee.

Whatever explanations for their longevity centenarians make, its unlikely that coffee drinking prolongs life. But not many think it shortens life, either.

SUMMARY. While individual tolerances vary, up to 200 or 300 mg of caffeine, which is equivalent to 3 to 4 cups of coffee per day, seems to be a mild stimulant helpful in relieving minor fatigue and boredom, with little risk of any harmful effects. This was substantiated in a nationwide survey, conducted by the U.S. Department of Health and Human Services, from which it was concluded that: "There is no evidence that heavy coffee

[39]*American Council on Science and Health*, Update, Fall 1981, p. 16.

[40]Bertrand, Charles A., M.D., F.A.C.C., F.A.C.P., On Coffee and Your Heart, *Sourcebook on Food and Nutrition*, Marquis Academic Media, Marquis Who's Who, Inc., Chicago, Ill., 1980, pp. 301-303.

drinking—5 or more cups (*1.2 liter*) per day—is related to poor health."[41]

• **The confession of a coauthor**—I have a confession. I am addicted to coffee. Some people say that I drink too much of the stuff. Maybe I do. But if I didn't drink coffee, I would write less. Besides, I wonder how I would feel were I to drink 20 cups of either orange juice or milk daily. So, at age 84, I side with Voltaire, who reportedly drank 50 cups of coffee a day and said of the brew: "It is a poison, certainly—but a slow poison, for I have been drinking it these eighty-four years." Pass me a cup of that poison-laden coffee. M.E.E.

(Also see CAFFEINE.)

COFFEE ESSENCE

Concentrated extract of coffee. Not less than 4 lb of coffee must be used to prepare 1 gal (*3.8 liter*).

COGNAC

A brandy produced in southern France from special varieties of grapes.

COHUNE NUT

This is the nut from a Central American palm. The shells of the nut are made into ornaments, and the meat yields an oil resembling coconut oil.

COLD STERILIZATION

Large doses of radiation are capable of destroying the bacteria in food. However, doses sufficient to destroy bacteria only slightly elevate the temperature of the food, hence, the term cold sterilization.

(Also see RADIATION PRESERVATION OF FOOD.)

"COLD STORE" BACTERIA

Microorganisms that survive and may even thrive at the usual refrigeration temperatures (40°F, or *3 to 4°C*).

(Also see PSYCHROPHILIC BACTERIA; and BACTERIA IN FOOD, section headed "Bacterial Control Methods," point No. 4.)

COLIC

A severe indigestion, which causes abdominal discomfort.

(Also see INFANT DIET AND NUTRITION.)

[41]Bonham, G. S., Ph.D., and P. E. Leaverton, Ph.D., Coffee Use Habits Among Adults, *Sourcebook of Food and Nutrition*, Marquis Academic Media, Marquis Who's Who, Inc., Chicago, Ill., 1980, pp. 334-338.

COLIFORM BACTERIA

A term which refers to a group of intestinal tract bacteria which can survive with or without air. *Escherichia coli* is the most important member of the group. Most coliform bacteria are disease-causing only under special circumstances, but their presence in food, and particularly in water, suggests fecal contamination.

(Also see BACTERIA IN FOOD, Table B-5 Food Infections [Bacterial]—"*Escherichia coli*.")

COLITIS

Contents *Page*

Causes of Colitis 452
Diagnostic, Preventative, and Therapeutic Measures 453
G.I. Series (Fluoroscopic Examination of the Digestive Tract) 453
Modified Diets 453
Acute Stages of Ileocolitis or Ulcerative Colitis 453
Mild and Chronic Forms of Colitis 454
Special Dietary Products 454
Drugs and Other Therapeutic Agents 456
Surgery 457
Psychological Measures 457

This is an inflammation of the large intestine (colon) which may be accompanied by abdominal pain, constipation or diarrhea, passage of bloody stools, weakness, and/or weight loss. Although it may seem that most of those afflicted with colitis are either middle-aged or elderly, the various types of the disorder afflict people at all ages. Hemorrhages, malnutrition, and other consequences sometimes result if treatment is delayed too long. Also, the chances of developing cancer of the colon increase greatly with the duration of ulcerative colitis, so prompt treatment is a good preventative measure. For these reasons, the characteristics of the different forms of colitis are noteworthy. Fig. C-124 shows the parts of the digestive tract where colitis is likely to occur.

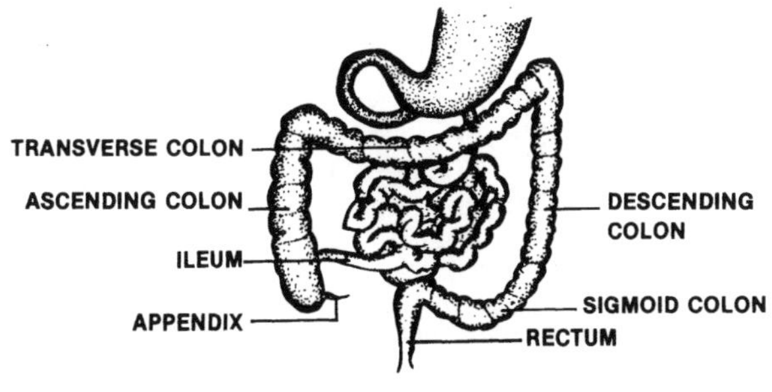

Fig. C-124. Colitis may strike anywhere between the ileum and the rectum.

• **Ileocolitis**—This condition affects the lower part of the small intestine (ileum) and the adjoining part of the colon. It occurs commonly in children, adolescents, and young adults. They may have pain or tenderness on the lower right side of the abdomen which is similar to that of appendicitis. Other characteristics of ileocolitis are (1) partial obstruction of the intestine by scar tissue; (2) anemia, fatty stools, lack of appetite, and weight loss due to malabsorption; and (3) inflammation and fever.

• **Mucous colitis**—Abdominal cramps, along with the passage of large amounts of mucus with dry, hard stools characterize this disorder, which is thought to result from (1) food intolerances or other allergic conditions; and/or (2) emotional disturbances.

• **Spastic colitis**—The main sign of this condition, which is also called irritable colon, is irregular, abnormally rapid movements of the bowels. The most common symptoms are (1) various types of abdominal pain, ranging from a dull ache to cramps; (2) lack of appetite, nausea, and vomiting; and (3) heartburn. Often, mucous colitis and spastic colitis are considered to be essentially the same disorder because their characteristics are similar. Together, they constitute the most prevalent digestive disturbances.

• **Ulcerative colitis**—Young adults who strive hard to conform to others' expectations may be afflicted with this disorder, which is indicated by (1) ulcers of the colon that bleed from time to time, (2) a recurring, crampy diarrhea, (3) abdominal pains of variable severity and duration, (4) formation of scar tissue which sometimes blocks the intestine, (5) lack of appetite and vomiting, and (6) fever. The condition is somewhat unpredictable, in that the acute attacks may occur at irregular intervals over periods of 20 years or more. Often, the attacks coincide with emotional distress. However, an almost identical disorder develops in swine fed diets deficient in the vitamin pantothenic acid.

Various studies have shown that as many as two-thirds of the patients who have had chronic ulcerative colitis for many years die from either its direct effects or its complications. Nevertheless, total recovery without recurrence of the condition is possible, thanks to recent advances in dietary therapy, anti-inflammatory drugs, and certain surgical procedures.

CAUSES OF COLITIS. Any discussion of the causes of colitis is bound to be controversial because of great variations in susceptibility among people who appear to be exposed to the same environmental factors. It might be that some people have an inherited tendency to develop one or more of the forms of colitis when certain provocative conditions are present. Some of the factors which are generally thought to be responsible for attacks of colitis follow:

• **Allergic conditions**—Some doctors suspect that allergies may be responsible for ulcerative colitis, because certain patients feel better when eggs, milk, tomatoes, and wheat are eliminated from their diets. Also, these patients have more than their share of arthritis, asthma, and hay fever. Furthermore, allergic reactions are accompanied by the release of histamine, a substance which provokes inflammation in various tissues.

It is noteworthy that a group of children with chronic, noninfectious diarrhea were found to have antibodies to milk, barley, wheat, rye, oats, and soybeans.[42] Their bowel habits returned to normal when the antibody-stimulating substances (allergens) were eliminated from their diets.

• **Cold or hot environments**—Uncomfortable temperatures may stress the body, depending upon how well an individual adjusts to such conditions. For example, it is well known that chilling may stimulate abnormally rapid movements of the digestive tract, whereas heat and humidity may make the bowels sluggish so that constipation ensues.

• **Constipation**—The habit of straining to pass hard stools may be responsible for irritation of the bowel lining. Also, chronic constipation may itself be an indication of a hidden bowel disorder when the diet normally contains sufficient bulk and water.

(Also see CONSTIPATION.)

• **Diarrhea**—This condition, if severe and prolonged, may cause the intestine to become raw, inflamed, or even ulcerated; depending upon the causative factor(s).

(Also see DIARRHEA.)

• **Drugs**—The anti-inflammatory steroid drugs work well in allaying the symptoms of an acute attack of ulcerative colitis, but they may also cause lesions of the intestinal lining if their administration is prolonged. Other drugs may irritate the bowel or cause diarrhea. Even mineral and vitamin supplements containing high levels of iron, potassium, or vitamin C may be highly irritating to the intestinal lining.

• **Emotional upsets**—These events receive much of the blame for attacks of colitis, although the disturbance may follow the attack. Nevertheless, it is well known that strong emotions such as anxiety, anger, and fear cause both a deep reddening of the bowel, and more rapid movements of the intestines.

• **Immune reactions against one's own body tissues (autoimmunity)**—Some medical scientists suspect that certain cases of ulcerative colitis result from attacks on the intestinal lining by antibodies to the lining cells, because similar attacks on body tissues occur in Addison's disease, rheumatoid arthritis, and Hashimoto's thyroiditis.

• **Infectious microorganisms**—Severe diarrhea may often be due to the consumption of foods contaminated with certain bacteria or viruses. Diarrheal diseases claim many lives in developing countries, where sanitation is poor. Therefore, it is noteworthy that bacillary dysentery (an infection with *Shigella* bacilli) produces lesions in the colon which are similar to those of ulcerative colitis.

• **Irritating drinks, foods, and spices**—Certain colitis-prone people may have acute attacks of the disorder

---

[42]Self, T. W., *et al.*,, "Gastrointestinal Protein Allergy," *The Journal of The American Medical Association*, 1969, Vol. 207, p. 2393.

when they consume excessive amounts of (1) alcoholic beverages; (2) caffeine-containing drinks such as coffee, colas, and strong tea; (3) whole-grain breads and cereals; (4) raw fruits and vegetables; (5) fried onions and cooked vegetables of the cabbage family; (6) gas-producing legumes; or (7) pungent herbs, seasonings, and spices.

• **Lack of rest**—The digestive tract is extra susceptible to disturbances when one is extremely tired. Unfortunately, many people eat heavy meals when they are fatigued.

• **Laxatives and enemas**—These treatments stimulate the movement of the bowels and may irritate them, particularly strong laxatives and soapsuds enemas. On the other hand, certain medication-containing enemas may be used to soothe the intestine during acute attacks of ulcerative colitis.

• **Nutritional deficiencies**—The intestinal lining has high nutritional requirements because (1) the lining cells grow rapidly, and (2) the old cells are continually sloughed off and replaced by new ones. Hence, the lining surface (intestinal mucosa) is susceptible to damage by the malnutrition which may result from (1) chronic consumption of a deficient diet, (2) prolonged illness or infection, (3) loss of appetite (anorexia) that persists for a long time, and (4) deprivation of food during hospitalization for extensive laboratory tests. Furthermore, the initial damage to the intestinal mucosa may lead to further injury if the absorption of essential nutrients is impaired to a significant extent.

However, the bowel may also be irritated or weakened by milder nutritional deficiencies such as iron deficiency anemia, nontropical sprue, pantothenic acid deficiency, pellagra, pernicious anemia, and vitamin E deficiency. Also, various stresses may greatly increase the needs for certain nutrients.

• **Parasitic infectations**—Certain parasites may only irritate the intestine, whereas hookworm causes ulceration and bleeding. Furthermore, the competition of parasites with their human host for certain nutrients may bring about one or more nutritional deficiencies that inflame or weaken the bowel.

• **Smoking**—The nicotine which gets into the body from smoking tobacco may disrupt the normal functioning of the digestive tract. Also, smoking increases nervousness, which is not good for people with colitis.

• **Strenuous exercise right after eating**—A few people habitually work hard or play hard right after meals. Heavy exercise diverts blood from the digestive tract to the muscles, which sometimes provokes cramps and diarrhea. Usually, foods high in fats and proteins are more difficult to digest than those rich in carbohydrates.

• **Surgery**—Almost any type of major surgery is sufficiently stressful to cause an abundant secretion of adrenal cortical hormones which may cause ulceration of the digestive tract. Furthermore, it takes time for the intestines to recover from the effects of anesthesia. Digestive complaints are common after anesthesia. Finally, an operation on the digestive tract may itself alter the secretions and the muscle tone of the various sections.

DIAGNOSTIC, PREVENTATIVE, AND THERAPEUTIC MEASURES. Many people have long been afflicted with mild forms of mucous or spastic colitis without ever developing either ileocolitis or ulcerative colitis. Nevertheless, it is wiser to take certain preventive measures at the first sign of bowel trouble, than to suffer needless anxiety and pain at a later time. Hence, some common diagnostic, preventive and therapeutic measures follow.

G.I. Series (Fluoroscopic Examination Of The Digestive Tract). This diagnostic procedure is used to detect various abnormalities of the gastrointestinal tract by means of fluoroscopy, which consists of the placing of the patient between a source of x rays and a fluorescent screen that displays a picture of the organs outlined by the x rays. Usually, a series of photographs are taken of various parts of the digestive tract. There are two types of G.I. series.

• **Upper G.I. series**—In this series, the patient swallows a chalky suspension of barium sulfate (sometimes called a barium meal) which fills the esophagus, stomach, and small intestine so that they may be seen in x-ray photographs.

• **Lower G.I. series**—This series requires that the patient be given a barium-containing enema to outline his or her colon. X-ray photographs are then taken before and after the enema is expelled.

The diagnosis of colitis may require both upper and lower G.I. series, because ileocolitis often affects both the small intestine and the colon.

Modified Diets. The dietary measures used to relieve the symptoms and signs of colitis should be selected carefully, because the characteristics of the disorder vary widely from patient to patient. Furthermore, there is uncertainty regarding the diets which may prevent even the mildest forms of colitis, because some people can eat highly irritating foods from infancy to old age without developing the disorder, whereas others are afflicted after only the briefest exposures to an irritant or a stress factor. Therefore, modified diets for the most critical cases are presented first, following which dietary guidelines are given for the milder forms of colitis.

CAUTION: The dietary information which follows is given only for the purpose of illustrating certain dietary principles, because the selection of a patient's diet is strictly the responsibility of his or her attending physician and of the consulting dietitian.

ACUTE STAGES OF ILEOCOLITIS OR ULCERATIVE COLITIS. Patients who are critically ill with these disorders may be badly malnourished because (1) they reduced their food consumption drastically in order to avoid aggravating their condition; or (2) chronic diarrhea prevented absorption of nutrients, while also causing the loss of the gastrointestinal secretions that are rich in protein and mineral salts. Therefore, it is sometimes necessary to resort to special medical measures such as hyperalimentation (feeding a liquid formula rich in nutrients through a vein at the base of the neck) or nasogastric tube feeding (a tube is passed up through the nose and down through the throat and the esophagus to the stomach) until the patient is well enough to take food by mouth.

It may be necessary to eliminate the carbohydrate in the diets of patients on oral feedings until diarrhea has been brought under control.[43] In these cases, the minimum requirement for sugar (glucose) may be met by intravenously administered solutions of dextrose. Usually, the ability of the patient to tolerate carbohydrates improves as malnutrition is corrected and the damaged intestinal lining is healed. Sometimes, wheat protein (gluten) is tolerated poorly and has to be eliminated from the diet.

Hence, therapeutic diets should be designed so that they are (1) very well digested and absorbed, and contain only minimal amounts of undigestible residue; (2) rich in calories and protein; (3) free from foods which commonly cause allergic or other digestive disturbances; and (4) soft in texture and bland. These criteria are best met by elemental diets, followed by diets containing low-residue foods.

• **Elemental diets**—These liquids and powders (intended for mixing with water), which have recently been marketed by certain pharmaceutical companies, usually contain (1) predigested protein, or a mixture of individual amino acids; (2) sugar, and/or partially digested starch; (3) small amounts of special fats (medium-chain triglycerides); and (4) supplemental minerals, vitamins, and flavoring. The undigestible residue content of elemental diets is so low that the bowel movements of patients are smaller than normal and occur only about once every 5 days. The results obtained with these products have generally been favorable, since prompt healing and gains in weight have often occurred. Unfortunately, elemental diets may become monotonous to patients as they begin to feel better, because (1) the products come in only a few flavors, and (2) drinking liquids is not as enjoyable as eating solid foods. However, the manufacturers of the dietary preparations now provide recipes for varying the ways in which their products are served.

• **Diets containing selected, low-residue foods**—The doctor may recommend that the first foods to be given be limited to those low in residue, such as clear broth, weak tea, toasted white bread, small amounts of butter (which should be immediately discontinued if fatty, frothy stools are produced), soft cooked eggs, custards, gelatin desserts, and bland cooked cereals like cream of rice and cream of wheat. Milk and items made from it are often barred from these diets because there is a high incidence of milk intolerance in patients with ulcerative colitis. The digestive problems which follow the consumption of milk or milk-containing foods may be due to (1) an intolerance to milk sugar, which may disappear after the intestine has healed, and/or (2) a sensitivity to milk protein that makes it necessary to eliminate these foods from the patient's diet. If milk is not allowed, calcium supplements should be taken to prevent weakening of the bones. Extra calories may be obtained from moderate amounts of sugar, hard candies, honey, and syrups, as long as there is no laxative effect from these items.

If improvement continues, the next step is to add (1) tender cooked fish, meats, and poultry (providing that the fat content is low); and (2) baked or boiled, white potatoes without skins. After that, canned fruits and cooked vegetables may be added, a little at a time; but they should be discontinued immediately if watery stools are passed. It is noteworthy that many people who have recovered from acute attacks of ulcerative colitis may have relapses if they attempt to eat raw fruits and vegetables. Hence, these items may have to be restricted permanently.

CAUTION: The elimination of certain foods, such as raw fruits and vegetables, from low residue diets makes it advisable for people following these recommendations to take mineral and vitamin supplements. Most doctors and dietitians will advise their patients regarding the products best suited to their needs.

MILD AND CHRONIC FORMS OF COLITIS. People who are prone to develop various forms of colitis may have to avoid irritating foods for the rest of their lives, particularly at times when they are likely to have emotional stress. However, they must also avoid dietary extremes which may lead to constipation, diarrhea, or nutritional deficiencies because these conditions make colitis worse. Therefore, it is necessary to eat a wide variety of nutritious, nonirritating foods.

CAUTION: The information which follows is for illustrative purposes only, because individual diets to control colitis must be designed to meet the specific needs of the patient. For example, a mildly alcoholic drink such as a small glass of table wine diluted 50/50 with water may help tense people to relax. However, no forms of alcoholic beverages are allowed when certain medicines are taken. Table C-35 gives some general guidelines for the selection of nonirritating foods.

SPECIAL DIETARY PRODUCTS. The dietary recommendations in Table C-35 mention a few special products which may be substituted for the ordinary foods that may aggravate colitis. Brief descriptions of these items follow.

• **Baby food purees**—Many of these products are useful for people who are recovering from bowel problems because (1) they are designed for infants, who have sensitive intestines; (2) their contents of fat, fiber, and irritating seasonings are low; and (3) little or no preparation is required.

• **Medium-chain triglycerides (MCT)**—This item, which is an ingredient of certain commercial dietary formulations, is made from coconut oil by a special process. MCT is useful in the treatment of various disorders of fat absorption because under such conditions it is much more readily digested, absorbed, and utilized than ordinary dietary fats.

(Continued p. 456)

---

[43]Coale, M.S., and J. R. K. Robson, "Dietary Management of Intractable Diarrhea in Malnourished Patients," *Journal of the American Dietetic Association*, Vol. 76, May 1980, pp. 444-450.

TABLE C-35
DIETARY GUIDELINES FOR PEOPLE WHO ARE LIKELY TO DEVELOP COLITIS

Foods That Are Allowed	Foods that Are *Not* Allowed
Beverages (other than fruit juices or milk)	
Caffeine-free coffee substitutes; carob drinks; decaffeinated coffee; herb teas; mildly carbonated soft drinks (carbonation may be weakened by allowing to stand in a warm place after opening); weak coffee and tea; and table wine diluted to half strength with water.	Beer; boiling hot or ice-cold drinks; cocoa; distilled liquors; highly carbonated soft drinks; and strong coffee. All alcoholic beverages (including wine) when certain medications are taken.
Breads and cereal products	
Breads and cereal products made from refined flours, without added fibrous substances. Yeast-raised doughnuts baked in an oven.	Items made from bran, candied fruits, coconut, nuts, pie crust, raisins, seeds, wheat germ and whole grain flour; doughnuts fried in fat.
Cheeses	
American cheese, cottage cheese, cream cheese; and other mild flavored items; providing that milk products are well tolerated.	Strongly flavored cheeses, such as those aged for long periods, and those ripened with molds.
Desserts	
Bread pudding, cakes and cookies (of low to moderate fat content), cornstarch puddings, custards, gelatin desserts, junket puddings (if milk is well tolerated), rice pudding without raisins, and sherbets (must be eaten slowly).	Desserts containing candied or dried fruits, coconut, nuts, peel, pie crust, seeds, whole grain products, and large amounts of fats or oils. Sugary items when diarrhea occurs.
Eggs	
Boiled or poached (medium to hard cooked); soft scrambled in a double boiler with a minimum of fat.	Fried, hard scrambled, or raw eggs.
Fats	
Butter, lard, margarine, mild flavored mayonnaise, and vegetables. (In the event of diarrhea these items should be eliminated temporarily.) It is noteworthy that medium-chain triglycerides (MCT, a special dietetic fat derived from coconut oil) does not aggravate diarrhea as ordinary dietary fats do.	Spicy mayonnaise, salad dressings, and sandwich spreads—particularly those similar to relishes and tartar sauce.
Fish, meats, poultry, and seafood	
Baked or broiled with visible fat removed. Canned fish or seafood without bones or skin. Certain canned meat products such as hash may contain sufficient fat to cause diarrhea.	Cured, fatty, fried, pickled, or smoked items; luncheon meats; sausages; skins from fish, meats, poultry, and seafood.
Fruits and fruit juices	
Baby food purees; canned or cooked items without seeds or skin; all juices (it might be wise to dilute citrus products and similar acid juices to half strength with water). Citrus fruits should be eaten at the end of a meal (one such item per day is sufficient).	Berries; candied fruit or peel; canned or cooked fruits with seeds or skin; canned grapefruit, orange; or pineapple; dried fruits, unless pureed; and fresh or frozen raw fruits other than avocado or ripe banana.
Herbs, seasonings, and spices	
Cinnamon, lemon, salt, and vanilla. Other items as recommended by the doctor or dietitian.	Coarsely textured or pungent products such as chili, garlic, green and red peppers, horseradish, mustard, pepper, vinegar, and Worchestershire sauce.
Milk	
Small amounts of various products should be tried to see if they are tolerated. Some patients may have to use nondairy substitutes.	In the event that a milk intolerance is diagnosed, all forms of this food may have to be eliminated from the diet.
Nuts, nut products, and seeds	
Smooth style nut butters (amounts should be limited because of the high fat content).	Chunk style nut butters; nuts and seeds.
Potatoes	
Baked or boiled sweet or white (Irish) potatoes without skins.	Any form of fried potatoes; potato chips; and skins of potatoes.

(Continued)

TABLE C-35 *(Continued)*

Foods That Are Allowed	Foods that Are *Not* Allowed
Sugar and other sweeteners	
Honey, molasses, sugar, and syrups; confections such as gumdrops, hard candies, and marshmallows; and strained cranberry sauce. Products which contain milk sugar (lactose) should be used cautiously.	Jams and preserves; sauce containing whole cranberries. All sugars may have to be restricted greatly if diarrhea occurs (particularly, molasses).
Vegetables	
Baby food purees; canned or soft cooked items without seeds or skins; and juices without seasonings other than salt. Whole cooked beans, kernels of corn, lentils, and peas may have to be mashed and strained in order to be tolerated by some extra sensitive people.	Any forms of broccoli, Brussels sprouts, cabbage, cauliflower, kale, and turnips; beans, if they produce excessive gas; fried onions; pickles; raw vegetables other than finely shredded lettuce.

• **Nondairy substitutes for milk**—Soybean derivatives are commonly used in formulas for infants who have an intolerance to cow's milk. The infants who cannot tolerate soybean protein may be given a formula based upon pureed beef heart. Therefore, it might be useful for some colitis patients to try one or more of these milk substitutes.

• **Water-soluble derivatives of vitamins A, D, E, and K**—The fat-soluble vitamins may not be well absorbed when ileocolitis or ulcerative colitis is accompanied by disorders of fat absorption. Therefore, it may be necessary for the patient to take water-soluble derivatives of these vitamins.

Drugs And Other Therapeutic Agents. Each of the medications commonly used in treating colitis has some undesirable effects, so it is essential that they be used only under a doctor's supervision. Hence, the information which follows is presented so that patients may recognize the need for careful selection of medications.

• **Antidiarrheal agents**—One of the long used remedies is a mixture of finely powdered clay (kaolin) and pectin, a gummy substance extracted from apple skins and the white membranes of oranges. However, its effectiveness in stopping infectious diarrhea is uncertain, unless an antibiotic is added to the mixture.

Another old remedy is paregoric (camphorated tincture of opium), which drastically slows the motion of the intestine.

Finally, one of the newer preparations which many people take with them on trips out of the country is a mixture of (1) a neuromuscular blocking agent (diphenoxylate hydrochloride); and (2) atropine, an antispasmodic drug.

The antidiarrheal agents should be used cautiously, because they may have such unwanted side effects as abdominal discomfort and nausea. Occasionally, intestinal obstruction occurs when intestinal movements are slowed excessively.

(Also see ATROPINE; and DIARRHEA.)

• **Bulking agents for constipation**—The low-residue diets which are often used in the treatment of colitis may not provide sufficient bulk to stimulate normal bowel movements. Therefore, the doctor or dietitian may recommend that the patient take certain nonirritating, bulking agents. These undigestible substances increase the bulk of the intestinal contents by binding large amounts of water. Some typical preparations are (1) agar-agar, a seaweed derivative; and (2) refined, psyllium seed powder. It is noteworthy that plenty of water should be taken with these substances because there have been cases in which certain bulking agents formed hard, dry masses in the colon, and did more harm than good.

(Also see CONSTIPATION; and FIBER.)

• **Medicated enemas and suppositories**—Rectal administration of medication is not new, since herbal enemas have long been used as "natural cures" for various ailments. These measures treat the bowel directly, and avoid any interference with the normal functioning of the rest of the body. (Medications given by mouth are likely to be absorbed and carried in the blood to all of the body tissues.) A medicated enema or suppository may contain (1) belladonna drugs which reduce the secretions and slow the movements of the intestine, and (2) hydrocortisone, an anti-inflammatory steroid drug. Unfortunately, the long term use of anti-inflammatory steroids may render the small and large intestines extra susceptible to ulcers, hemorrhages, and perforation.

• **Orally-administered, anti-inflammatory steroids**—These substances are usually identical, or at least similar, to the adrenal cortical hormones secreted by the body in stressful situations. They may be administered orally in the treatment of colitis—even though their absorption exposes the whole body to their effects—because (1) they counteract inflammation elsewhere in the body, which may indirectly cause agitation of the bowels via the nerves; and (2) their protein-degrading actions throughout the body raise the blood level of amino acids so that these nutrients become more available for the healing of the intestines. However, the long term administration of these steroids is risky because of such side effects as (1) weakening of the bones and muscles, and (2) thinning of the digestive tract so that it becomes extra susceptible to ulceration and/or rupture.

• **Sedatives and tranquilizers**—These drugs are sometimes useful in treating people whose bowels are continually agitated as a result of their chronic anxiety and nervous tension. Nevertheless, they are only stopgap measures in critical situations, since they do not remove the underlying cause(s) of the emotional distress. Furthermore, some patients may become so dependent upon the drugs that they cannot function normally without them.

Surgery. Surgical removal of highly inflamed portions of the intestines may be necessary in certain cases of ileocolitis or ulcerative colitis when there is a deterioration due to (1) obstruction of the intestinal opening, (2) bleeding or diarrhea which is difficult to stop, (3) continuous pain, or (4) tumors which may become cancerous. Fortunately, recent advances in surgical techniques and supportive measures have reduced the risks of such operations.

A typical surgical procedure (a colectomy) may involve (1) removal of the lower portion of the colon and attachment of the upper portion to an artificial opening (stoma) on the abdomen (this operation is called a colostomy); or (2) removal of the upper portion of the colon and/or the lower ileum so that the remaining end of the ileum is connected to a stoma (an ileostomy). Sometimes, the upper end of the remaining colon or ileum is connected to the rectum, so that a stoma is unnecessary.

Many people have had these operations and are able to live normal lives. Furthermore, moral support is provided by members of local chapters of the United Ostomy Association to those who have undergone this type of surgery.

Psychological Measures. Attacks of colitis sometimes follow events which produce anger, anxiety, fear, or resentment. Therefore, certain patients may benefit from being taught how to modify their behavior and control their emotions. Such instruction requires an understanding of the underlying causes of unhealthy reactions to ordinary circumstances like delays in traffic and disagreements with other people. Few amateurs are able to provide much help with such matters, so it would seem best for easily aggravated, colitis-prone people to obtain either professional counseling or to participate in group therapy which is conducted by a well-trained leader.

(Also see ELEMENTAL DIETS; LIQUID DIETS; and MODIFIED DIETS.)

COLLAGEN

A protein which forms the major part of the connective tissue. It forms gelatin when heated with water.

(Also see BODY TISSUE.)

COLLARDS *Brassica oleracea,* var. *acephala*

This highly nutritious vegetable, derived from the wild cabbage, is grown mainly in the southeastern part of the United States, where the green leaves have long been cooked with fatback. *Acephala* means that the leaves do not form a head. Collards and kale are essentially the same, except that (1) the leaves of the former have smooth edges, while those of the latter have curly edges; and (2) collards tolerate hot weather better than kale. Fig. C-125 shows a rosette of collard leaves.

Fig. C-125. Collards, a close relative of the wild cabbage.

ORIGIN AND HISTORY. Collards, unlike most of the other cabbage vegetables, has been changed very little by domestication. Hence, this crop was probably grown in Greek and Roman times, when it was referred to as cabbage because of its very close resemblance to the wild cabbage. At that time, many people believed that eating the vegetable before a banquet prevented the mind from being befuddled by wine.

Collards and kale, which were grown extensively throughout Europe for many centuries, were used both as feed for animals and as food for people. Later, immigrants brought collards to the southeastern United States, where they are now grown from Delaware to Florida.

PRODUCTION. Statistics for the total U.S. production of collards are not readily available. In 1990, 20.7 million lb (*9.4 thousand metric tons*) were frozen.[44]

Collards are more tolerant of hot weather than any of the other cabbage vegetables. Hence, they are grown mainly in the southeastern part of the United States. Usually, they are planted in the summer or fall in the coastal areas where only light freezes are likely to occur, and in the spring in the inland areas where frosts are more severe.

[44]*Agricultural Statistics 1991*, USDA, p. 173, Table 249.

The vegetable is harvested by cutting off entire rosettes of leaves, or by cutting off the individual large lower leaves as they become sufficiently mature, but before they are too fibrous. Harvested collards are tied in bunches for marketing.

PROCESSING. At one time, most of the collards crop was marketed fresh along the East Coast. Today, much of the crop is frozen, and some is canned.

SELECTION AND PREPARATION. High quality collards are fresh, young, tender, and green. Greens which show insect injury, coarse stems, dry or yellowing leaves, excessive dirt, or poor development, are usually lacking in quality and may cause excessive waste. Flabby, wilted plants and leaves are generally undesirable.

Southern cooks usually boil collards, because the raw leaves are not as appetizing as the cooked greens. However, too many Southerners have acquired the taste for an overcooked, mushy form of the vegetable. Nutritionally, it would be better merely to wilt the young fresh greens by tossing them with hot bacon fat or a heated salad oil. Also, collards may be added to casseroles, omelets, and soups, or cooked briefly and served with a mild flavored sauce.

CAUTION: Collards and other vegetables of the mustard family (*Cruciferae*) contain small amounts of goiter-causing (goitrogenic) substances that interfere with the utilization of the essential mineral iodine by the thyroid gland. Hence, people who eat very large amounts of these vegetables while consuming an iodine-deficient diet may develop an enlargement of the thyroid, commonly called a goiter. The best insurance against this potentially harmful effect is the consumption of ample amounts of dietary iodine. This element is abundantly present in iodized salt, ocean fish, seafood, and edible seaweeds.

NUTRITIONAL VALUE. The nutrient composition of collards is given in Food Composition Table F-36.

Some noteworthy observations regarding the nutrient composition of collards follow:

1. Collards are low in calories (42 to 63 kcal per cup) because they contain almost 90% water.
2. A 1-cup (*240 ml*) serving of cooked collards supplies about the same amount of protein (4 to 7 g) as a cup of a cooked cereal grain, but only about one-third as many calories.
3. The calcium content of collards is higher than that of any of the other commonly consumed vegetables, since a cup of these greens contains about as much calcium as a cup of milk. Therefore, it is noteworthy that people in the southeastern United States have long obtained much more of their dietary calcium from collards and other greens than from dairy products. Furthermore, collards contain over three times as much calcium as phosphorus, whereas most other foods contain much less calcium than phosphorus. (Excessive phosphorus may reduce calcium utilization in the body.) Therefore, these greens complement the other foods.
4. Collards are a good source of iron and an excellent source of potassium.
5. The vitamin A content of these greens is exceptionally high, and the vitamin C content moderately high.

(Also see VEGETABLE[S], Table V-6 Vegetables of the World.)

COLLOID

A substance (as gelatin, albumin, or starch) dispersed through another medium. Colloids are larger than crystalline molecules, but not large enough to settle out; and they are incapable of passing through an animal membrane.

COLLOIDAL

Having the properties of a colloid.

COLLOIDAL OSMOTIC PRESSURE

The pressure exerted by the protein molecules in the plasma and in the cell. Because proteins are large molecules, they do not pass through the separating membranes of the capillary cell walls; hence, they remain in their respective compartments, producing a constant osmotic pull that protects vital plasma and cell fluid volumes in these compartments.

COLLOP

- A small slice of meat made tender by beating
- A rasher of bacon
- A fold of fat flesh
- An Irish measure of land based on the grazing requirements for one year, of a beef animal, a horse, or the equivalent in sheep.

COLON

A term used to designate a portion or all of the large intestine.

(Also see DIGESTION AND ABSORPTION.)

COLORING OF FOOD

Contents — Page

Why Food Colors 459
Regulation of Food Colors 459
Synthetic Food Colors 459
Natural Food Colors 460
General Uses 460

The eyes are involved in eating, since the appearance is the first judgment passed on a food. Except under dire circumstances, foods which are unappealing to the eye will often be rejected. It's the color of a food that's eye catching first, then association with flavor, texture, nutritive value, and wholesomeness follow. For example, individuals are taught color associations: dark green vegetables, yellow vegetables, white potatoes, yellow egg yolks, and yellow butter and margarine. These associations and many others are then related to the quality of a food. Furthermore, color contributes to the attractiveness of food, which, in turn, adds to its enjoyment.

Of all the food additives, perhaps the addition of colors is the hardest to justify. Many people feel that foods have a natural color, so why tamper with it. Indeed, color additives have met strong opposition; and regulations have been passd banning some and restricting the use of others. Most of the opposition and regulation involves the synthetic colors, while the natural colors receive less attention. However, 90% of the colors used in food come from a small group of synthetic colors, while the diverse group of natural color sources is used in about 10% of foods only.

WHY FOOD COLORS. "Mother Nature" uses a variety of colors to give foods their typical appearance. The so-called anthocyanin pigments lend their brilliant colors to grapes, cherries, plums, strawberries, and other deciduous fruit. Another class of pigments known as carotenoids colors citrus fruits and a number of vegetables, including carrots, sweet potatoes, and rutabagas. Also, the carotenoids account for the color of butter, poultry, and meat fat, and they are present in all dark-green leafy vegetables, though chlorophyll may mask them. Carotenoids are yellow, except lycopene, the characteristic coloring matter of tomatoes and pink grapefruit. The most familiar green pigment is chlorophyll, the universal coloring matter in plants, which is prized in peas, beans, and other green vegetables. Many other classes of pigment occur naturally in foods. The end result is that food consumers associate color, quality and freshness. Color is a means of identity.

Modern food processing, preservation, and manufacturing has created a need for the addition of colors to foods to (1) overcome color changes occurring during processing and preservation, (2) standardize the appearance of a product which may vary depending upon season, and (3) gain consumer acceptance. Unfortunately, because of the associations of color with quality, added color can also deceive by implying richness, flavor, or maturity in foods where these traits do not exist. Hence, laws and regulations have been established governing the use of coloring agents in foods, primarily the synthetic dyes.

REGULATION OF FOOD COLORS. At the turn of the century, about 80 synthetic dyes were used in the United States for coloring food, and regulations regarding the nature and purity of these dyes were nonexistent. Then, the Food and Drug Act of 1906 listed seven dyes which could be used in food: orange I, erythrosine, ponceau 3R, amaranth, indigotine (indigo carmine), napthol yellow, and light green. These colors had been examined in physiological tests with no unfavorable results. Also, a system was established for the certification of these colors which acknowledged conformance to chemical specifications, and was carried out by the Department of Agriculture. Then, to fill the needs of the food industry, eight more colors were tested and approved for foods: tartrazine, yellow AB, yellow OP, Guinea green, fast green, ponceau SX, sunset yellow, and brilliant blue. Therefore, at the time of the Food, Drug, and Cosmetic Act of 1938, 15 synthetic dyes were approved for usage in foods. Again, these 15 dyes were tested and declared safe. Also, under the new Act the common names of the dyes were no longer employed, but the color prefix FD&C, plus a number were used to designate the dyes. For example, orange I became FD&C orange No. 1; erythrosine became FD&C red No. 3; fast green became FD&C green No. 3; and so on. In 1950, an unfortunate incident occurred which cast a shadow on use of dyes in food. The overuse of color in candy and on popcorn resulted in a number of cases of diarrhea in children. This incident along with some additional toxicity tests in animals led to the delisting of three synthetic dyes, and to additional legislation. In 1960, the Color Additives Amendment became law. This gave the FDA the authority to impose limits on the amounts of color used in foods, drugs, and cosmetics. Furthermore, the new Amendment placed the onus of establishing safety of food dyes upon industry. Undoubtedly, some manufacturers will shift to the less functional and more expensive natural food colorings which have not been scrutinized as rigorously.

SYNTHETIC FOOD COLORS. In the past, the synthetic dyes were commonly referred to as coal tar dyes of which there are about 700. Because of the effectiveness of the synthetic dyes, only exceedingly small amounts—parts per million—are required to color foods. Also, the synthetic dyes are generally much more stable than natural food colorings, resisting breakdown in air, light, or heat, or by interacting with the other food components. Table C-36 lists the synthetic dyes and the guidelines for their use.

**TABLE C–19
SYNTHETIC COLOR ADDITIVES
APPROVED FOR FOOD USE**

Coloring Compound		
Color Prefix and Number	**Common Name**	**Restriction in Use[1]**
Citrus Red No. 2		Skins of oranges at 2 ppm.
FD&C Blue No. 1	Brilliant blue FCF	No specific limitations or restrictions on use.
FD&C Blue No. 2	Indigo carmine	No specific limitations or restrictions on use.
FD&C Green No. 3	Fast green FCF	No specific limitations or restrictions on use.
FD&C Red No. 4	Ponceau	Maraschino cherries at levels not exceeding 150 ppm.
FD&C Red No. 40	Allura Red AC	No specific limitations or restrictions on use.
FD&C Yellow No. 5	Tartrazine	No specific limitations or restrictions on use.
FD&C Yellow No. 6	Sunset yellow FCF	No specific limitations or restrictions on use.
Orange B		Sausage casings at 150 ppm.

[1]*Code of Federal Regulations*, 1975, Title 21, Parts 1-9.

NATURAL FOOD COLORS. Colors abound in the plant and animal kingdom which can be used for coloring foods. None of these require batch certification. Some are nutrients in their own right, others are spices, and still others fall under the general classification of carotenoids—a product of which nature makes about 400 million pounds (*180 million kg*) annually. Table C-37 lists most of the more common approved naturally-occurring color compounds, and indicates their source and use.

TABLE C-37
COMMON NATURALLY-OCCURRING COLOR ADDITIVES APPROVED FOR FOOD USE[1]

Substance	Source	Restriction on Use
Algae meal, dried	Algae	Chicken feed to promote yellow skin and egg color.
Alkanet	Roots of *Alkanna tinctoria.*	Sausage casings, oleomargarines, and shortenings.
Annatto	Seed pods of *Bixa orellana.*	No limitations or restrictions on use.
Aztec marigold (Tagetes meal)	Flower petals of Aztec marigold.	Chicken feed to promote yellow skin and egg color.
Beets, dry powder	Beets	No limitations or restrictions on use.
Beta-apo-8′-cartenal	Fruits, vegetables	Not to exceed 15 mg per lb *(0.45 kg)* or pint *(0.47 liter)* of food.
Beta-carotene	Fruits, vegetables	No limitations or restrictions on use.
Canthaxanthin	Fruits, vegetables	Not to exceed 30 mg per lb *(0.45 kg)* or pint *(0.47 liter)* of food.
Caramel	Heated sugars	No limitations or restrictions on use.
Carmine (cochineal)	Female insects, *Coccus cacti.*	No limitations or restrictions on use.
Carrot oil	Carrots	No limitations or restrictions on use.
Chlorophyll	Green leaves	Sausage casings, margarine, and shortening.
Corn endosperm oil	Corn	Chicken feed to promote yellow skin and egg color.
Cottonseed flour partially defatted, cooked, toasted.	Cottonseed	No limitations or restrictions on use.
Ferrous gluconate	Iron salt of gluconic acid.	Ripe olives
Fruit juice	Fruit	No limitations or restrictions on use.
Grape skin extract	Grapes	No limitations or restrictions on use.
Iron oxide (ferric oxide)	Hematite	Dog and cat foods not to exceed 0.25% by weight of finished product.
Paprika	Plant *Capsicum annum*	No limitations or restrictions on use.
Paprika oleoresin	Plant *Capsicum annum*	Not for use on fresh meat or ground fresh meat products except chorizo sausage and Italian brand sausage and other meats as specified by USDA.
Riboflavin (vitamin B-2)	Plant and animal sources.	No limitations or restrictions on use.
Saffron	Stigmas of *Crocus sativus.*	No limitations or restrictions on use.
Titanium dioxide	Found in the minerals—rutile, titanite, anatase, brookite, almenite.	Not to exceed 1.0% by weight of food.
Turmeric and oleoresin	Rhizome of *Curcuma longa*	No limitations or restrictions on use.
Ultramarine blue	Mineral, lapis lazuli	Animal feed, not to exceed 0.5% by weight of salt.
Vegetable juice	Vegetables	No limitations or restrictions on use.

[1]*Code of Federal Regulations,* 1974, Title 21, Parts 1-9.

GENERAL USES. First, not all foods contain added coloring agents. Major staple foods like bread, meat, potatoes, vegetables, and fruits are not colored. Second, all color additives, synthetic or natural (uncertified), are designated on the label as artificial color. The general classes of food making the greatest use of food colors follow:

• **Beverages**—One of the largest users of the certified colors (FD&C) is the beverage industry. Most of the fruit-type beverages contain single or combinations of certified color additives while colas and root beers are colored with caramel as indicated in Table C-38.

TABLE C-38
COLOR ADDITIVES AND THEIR CONCENTRATION IN SOFT DRINKS

Flavor	Color	Concentration
		(ppm)
Orange	FD&C Yellow #6	75
	FD&C Yellow #6 ,[1] FD&C Red #40	50
Cherry	FD&C Red #40 ,[1] FD&C Blue #1	100
Grape	FD&C Red #40 ,[1] FD&C Blue #1	75
Strawberry	FD&C RED #40	60
Lime	FD&C Yellow #5 ,[1] FD&C Blue #1	20
Lemon	FD&C Yellow #5	20
Cola	Caramel color	400
Root beer	Caramel color	400

[1]The two colors are combined to produce the flavor to the left (in first column).

• **Bakery products**—Bakery items which employ color additives include dough products, cookies, sandwich fillings, icings, coatings, and ice cream cones. To obtain the desired depth of color in such baked goods as dark chocolate pieces, various combinations of certified colors and natural colors are used.

• **Candy**—With the wide variety of candies available, the use of color additives is extremely important to their desired appearance.

• **Dairy Products**—Most all ice creams and sherbets contain artificial color. Chocolate ice cream may be the exception. Annatto and beta-carotene are used in cheese, since the certified colors are not sufficiently stable. Margarine and butter are also colored with annatto and beta-carotene.

• **Dry mix products**—Products such as gelatin desserts, puddings, cake mixes, doughnut mixes, and pancake mixes depend on proper color blends for their desired appearance.

• **Pet foods**—Although not for human consumption, pet food coloring is under the jurisdiction of the Food, Drug, and Cosmetic Act. Traditionally, iron oxide has been used to color pet foods because of its stability during the retorting operation. However, the certified colors are used extensively in dry extruded pet foods.

Since 1970, food colors, especially the synthetic colors, have received tremendous publicity—mostly bad. Consumer activists claim food colors are for cosmetic reasons only. Many food processors openly advertise that their products are natural and free of colors. A large share of the bad publicity surrounding food colors is due to the claims made by a few that hyperactivity and learning difficulties are attributed to colors. While there are no hard data supporting the need of colors in food, the elimination of food colors would be the end to such items as soft drinks, gelatin desserts, hard candy, and many others.

(Also see ADDITIVES; and BEHAVIOR AND DIET.)

COLOSTOMY

An operation in which an artificial opening (a stoma) from the colon to the outside of the body is placed on the abdomen. This type of surgery is performed when the lower part of the intestines must be removed or closed off because of a disease.

(Also see DIGESTION AND ABSORPTION.)

COLOSTRUM

The first secretion of a mammal (animal or human) after childbirth. Colostrum is valuable to the newborn because it is high in protein and it contains antibodies which transfer some of the immunity of the mother to the newborn.

(Also see BREAST FEEDING, section headed "Colostrum.")

COMA

A prolonged state of unconsciousness. The two most common types of nutritionally related comas follow:

• **Diabetic coma**—This condition is usually brought on by the combination of acidosis and ketosis which occurs when diabetes is (1) out of control; or (2) aggravated by failure to eat, the stress of infection, or the failure to take insulin when it is needed.

• **Hepatic coma**—In this condition the brain is intoxicated by an excess of ammonia which accumulates in the blood during liver failure. Normally, the liver converts the ammonia formed in protein metabolism to urea which may be excreted safely in the urine.

(Also see DIGESTION AND ABSORPTION; and METABOLISM.)

COMFIT

A confection consisting of a solid center, such as a piece of fruit or a nut, which is preserved by covering with layers of sugar.

COMPETITIVE INHIBITION

This refers to the inhibition of enzyme action, which causes the rate of enzyme-catalyzed reactions to decrease. Competitive inhibition occurs when a foreign molecule becomes tightly bound to the active site of an enzyme, preventing the binding of the normal substrate (the substance acted on by the enzyme). Many of the "nerve gases" manufactured as chemical warfare agents act by inhibiting enzymes of the body in this manner.

(Also see ENZYME.)

COMPOTE

A dish of fresh, canned, or stewed fruit served in a syrupy liquid that may be flavored with grated peel, a liqueur, and/or various spices. The caloric content of compotes varies considerably since the liquid medium may be solely the liquid in which dried fruit has been stewed, or, at the other extreme, it may contain a heavy syrup or jelly. Fruit compotes are best when made from a mixture of mildly and strongly flavored fresh fruits such as sliced bananas and citrus fruits.

COMPOUND

• In chemistry, a substance formed by the intimate chemical union of two or more elements.

• In pharmacy, a preparation containing several ingredients as distinguished from one of the same name containing one or a few.

CONALBUMIN

A protein of the white of an egg which may combine with iron salts to form a red iron-protein complex. This accounts for the pinkish color resulting when eggs are stored in rusty containers.

CONGENITAL

Malformation existing at birth; acquired during development in the uterus and not through heredity. Sometimes, congenital defects in newborns are the result of (1) the administration of drugs during pregnancy, or (2) maternal deficiencies of nutrients such as folic acid.

CONJUNCTIVA

The membrane which lines the eyeball and the eyelid. Certain nutritional deficiencies cause conspicuous abnormalities of the conjunctiva.

(Also see DISEASES; and DEFICIENCY DISEASES.)

CONJUNCTIVAL HEMORRHAGES

Hemorrhages in the membranes that line the eyelids and cover the eyeball.

CONNECTIVE TISSUE

The name given to the types of tissues which hold the body together or give it structural support. Bones, muscles, tendons, ligaments, adipose tissue, and teeth are typical connective tissues.

(Also see TISSUES OF THE BODY.)

CONSTIPATION

An excessively long retention of the stool within the lower bowel. Also, difficulty may be encountered in moving the bowel. Two major types of constipation follow:

• Atonic constipation is due to the lack of tone in the muscle around the lower bowel. This condition may occur because there is insufficient bulk or fluid in the diet. It also may be due to lack of sufficient exercise or activity.

• Spastic constipation is characterized by irregular movements (spasms) of the bowel which may be painful. It may be due to various nervous disorders, irritating foods, excessive smoking, and/or obstruction in the bowel itself. The stools are small, and very slender.

Only the doctor can safely outline the measures to be followed to rectify constipation. In the absence of any illness that might be causative, most cases of constipation can be alleviated by embarking upon a program of good eating habits combined with proper exercise and regular time for elimination. Common treatments follow:

1. **Exercise.** Regular exercise, such as a walk before breakfast or daily exercise of the abdominal muscles, may be necessary, especially for a sedentary person.

2. **Foods.** Wheat bran (preferably unprocessed and without sugar) may be the best natural fiber and bowel regulator. It soaks up water like a sponge, increasing its weight approximately eightfold when thoroughly moistened.

Persons who cannot tolerate bran, or who wish to get their fiber from different sources, may select from the following: stewed prunes or apples, prune juice, fresh fruits, green vegetables and salads, whole-grain bread, blackstrap molasses, olives, or chocolate.

3. **Water.** Persons subject to constipation should drink sufficient water to prevent dehydration of the material in the colon and consequent difficulty in passing a dry mass. It is noteworthy, for example, that bran alone isn't effective unless sufficient water is consumed to let it do its work.

4. **Vitamins.** B-complex vitamins appear to give relief to some constipated patients.

5. **Drugs.** Many drugs are available for treating various kinds of constipation. Vegetable and salt cathartics, organic and mineral medicines, substances that act mechanically, sedatives or tranquilizers, and water in various forms are most common. When taken habitually, drugs may defeat their intended purpose and make elimination more difficult. Some have a useful function, but they had best be prescribed by a doctor.

Constipation in infancy is not infrequent, especially in formula-fed babies. Usually, the condition may be corrected by changing the type of sugar used in the formula, or by giving prune juice or strained prunes once or twice daily.

CONSUMER ADVOCATES

This refers to those who take it upon themselves to plead the causes of consumers; presumably, they protect consumers, usually by protesting.

In the 1960s, many people became concerned about the effect of chemical fertilizers and pesticides on the environment and on our food. Major influences in the thinking were the book *Silent Spring* by Rachel Carson and several environmental organizations. Conditions were ripe for consumer advocates; and they popped up like toadstools after a rain. Once hooked, they usually became lifetime advocates—even militants. Consumer advocates do alert responsible scientists and the food team (producers-processors-marketers) to the need for constant vigilance, controls, and research.

Today, consumer-protection laws exist at the local, state, and federal levels. Chief among the federal agencies that act to protect and serve the consumer are the Food and Drug Administration, the Federal Trade Commission, the U.S. Public Health Service, and the Consumer and Marketing Service of the U.S. Department of Agriculture.

CONVENIENCE AND FAST FOODS

Contents *Page*

Historical Background and Recent Trends **463**
Major Types of Convenience Foods Sold in Retail Stores **466**
Utilization of Convenience Foods by Food Service Operators **469**
Fast Foods **471**
Summary **476**

These two types of products, convenience foods and fast foods, are closely related, because both were developed to save considerable time in food preparation. However, they may also differ in certain aspects. Hence, some commonly accepted definitions follow.

- **Convenience foods**—These are partially or fully prepared items that have been combined, processed, and/or cooked by the manufacturer and/or the distributor so that only minimal amounts of preparation time are required in the home. Some of the old standbys in this category are baby foods in jars, baking mixes, canned fruits and vegetables, and ready-to-eat breakfast cereals.

- **Fast foods**—This term refers to ready-to-eat items that are dispensed by commercial establishments which may or may not have accommodations for eating on the premises. The hallmark of these operations is that there is little or no waiting time between the time the food is ordered and when it is served. This speedy handling of orders may be the result of (1) using commercial types of convenience foods, and/or (2) cooking the unprocessed food item well in advance and keeping it warm (or cold, if necessary to prevent spoilage) until it is purchased.

Both convenience and fast foods usually cost more than similar items prepared from fresh, unprocessed foods at home because the manufacturer and/or the distributor may have invested considerable labor, time, and fuel in producing or preparing the former types of items.

Fig. C-126. A modern fast-food establishment is usually simple but clean, and characterized by fast service.

HISTORICAL BACKGROUND AND RECENT TRENDS. Some of the basic processes for producing convenience and fast foods were developed by primitive peoples who exemplified the principle "necessity is the mother of invention." For example, Stone Age hunters in Siberia stored frozen and dried meat for later use. Similarly, the American Indians of the Great Plains dried strips of buffalo meat in the sun and wind, then mixed the dried meat with fat and berries to make pemmican. The latter product was taken on long journeys into areas where game was likely scarce. (It is noteworthy that the U.S. military services used a modified canned type of pemmican as a survival ration in World War II.) Aborigines in other parts of the world dried and salted fish, fruits, and meats; brewed mildly alcoholic beverages; and made flours and meals from acorns, grains, roots, and tubers. These food products were sometimes traded, but in most cases they served the group that made them.

The commercial production and sale of processed foods began shortly after the first agricultural and trading settlements were founded thousands of years ago in China, India, the Middle East, and Egypt. Dietary staples such as beer and bread were produced for city dwellers and traders who often traveled long distances. In the latter case, the availability of food and lodging at certain places on the caravan routes allowed the traders to dispense with bringing their own food and made it possible for them to carry more articles for trading. It is noteworthy that in Egypt, which was at the western end of certain trade routes, baker-brewers made dried half-baked loaves of bread that could be carried on the caravans and fermented into beer when soaked in water. Furthermore, some of the yeasty mass which grew considerably during the brewing of beer was saved for the subsequent baking of bread. (The commercial production of bread in ancient times required knowledge and skill in milling the grain to obtain the proper flours, cultivating the yeast so as to avoid contamination by less desirable microorganisms, and using the crude ovens carefully to avoid burning or undercooking the dough.) The art of baking later spread from Egypt to all of the large civilizations around the Mediterranean.

Baking helped to stimulate the development of other food service industries in ancient Rome because the affluent people entertained both lavishly and often, and required the preparation of a wide variety of dishes. Hence, the baker of the household became an all around gourmet cook. Some cooks eventually left their employers and established their own restaurants. However, the need for food services diminished when the cities of Rome and the neighboring civilizations began to decline a few centuries after the dawn of the Christian era and the city dwellers migrated to the countryside where they grubbed out a living on small farms or feudal estates. Several more centuries passed before the production of beverages and foods began to regain some of its former status. Culinary arts were preserved in many monasteries, but the recipes and preparation procedures were often kept secret to ensure the livelihood of the monks. Meanwhile, the millers, who had formed trade guilds in Ancient Rome, united again in the Middle Ages to have laws enacted that prohibited individual households from owning their own mills. Soon, they monopolized the milling of grain, and became even more powerful when windmills were introduced in the 11th and 12th centuries.

Fig. C-128. The development of canning in France during the 19th century and the use of canned foods in the Franco-Prussian War (1870-1871) inspired the prediction of an almost total mechanization of food services, that would be characterized by the dispensing of fish, soups, beef steak, and roasts through pipelines from a giant community kitchen. (Courtesy, The Bettmann Archive, Inc., New York, N.Y.)

Fig. C-127. A sausage vendor in France. The young man is purchasing a sausage to be eaten with the piece of bread broken off the large loaf under his arm. (Courtesy, The Bettmann Archive, Inc., New York, N.Y.)

The Crusades which began around the end of the 11th century fostered the reinstitution of communication and trade between Europe and the Middle East and the Orient. This era initiated a use of exotic foods by the more affluent people and expert cooks once again acquired a high status. Spices were brought great distances to enhance the flavors of rather mediocre foods. In time, trading led to the growth of new large cities such as Venice, Florence, Paris, and London. At first, the cities of northern Europe were merely gathering places for marketing and trading of domestic and imported goods, but soon many people resided in them permanently. The living quarters of all but the affluent were very crowded and had little space for food preparation. Therefore, urban laborers often ate foods dispensed from small stands in the marketplaces. Also, vendors moved through the streets dispensing their wares.

The quest for spices and other valuable items of commerce led to extensive exploration and colonization of new lands around the world and the trading of nonperishable items between the peoples. This in turn led to the growth of food preservation and processing for the production of commodities that could be shipped long distances. For example, the North American colonies shipped corn, onions, potatoes, pickled beef and pork, and salted cod fish to British colonies in the Caribbean in exchange for rum and sugar. However, the greatest advances in the production of convenience and fast foods came after the Industrial Revolution brought about the processing of foods in large factories.

The modern age of food production and food processing dawned in the 18th century as a result of such developments as (1) the establishment of guilds of bakers and cooks by the king of France for the purpose of providing high quality cooked meals to those who could afford them, (2) the opening of the first "restaurant" in Paris, which was actually a small soup shop operated for the common people by a man named Boulanger, and (3) the development of a safe canning process by the French confectioner Nicholas Appert. Nevertheless, canning was not utilized widely until the latter part of the 19th century, when the explanation of food spoilage was given by the French microbiologist Louis Pasteur. The Civil War in America provided the major impetus for the utilization of canning in the United States, whereas the Franco-Prussian War did the same for France.

During the 20th century, the migration of people from the country to the cities continued at a rapid pace and created a growing need for foods in the urban areas. At first, horse-drawn trucks conveyed all foods, ice, and other commodities, but within a few years motorized vehicles appeared. The first ones were limited in size by their small engines and served mainly as transportation for street vendors and deliverymen.

Fig. C-129. A popcorn vendor in the early 1900s. (Courtesy, The Bettmann Archive, Inc., New York, N.Y.)

A great demand was created for canned foods by World War I, which broke out shortly after the modern type of "sanitary" can had been developed. Similarly, World War II, spurred the growth of the fledgling frozen foods industry which was started by the experiments of Clarence Birdseye in the 1920s. The war also fostered the use of convenience foods at home and eating out in restaurants as large numbers of homemakers went out to work in factories to help the war efforts. Frozen orange juice concentrate was marketed right after World War II ended and now costs considerably less than fresh squeezed juice, due to the large volume of oranges that are processed. Since then, there has been a steady proliferation of new convenience and fast-food items such as canned gourmet foods, dry mixes, frozen entrees (often known as "TV dinners"), freeze-dried foods, and powdered beverage mixes. It is noteworthy that table restaurants now use many convenience items which are difficult to distinguish from items prepared from fresh foods by professional chefs.

Fast-food establishments date back to 1906, when the Sears and Roebuck mail-order firm in Chicago opened a restaurant for their employees which could feed 8,400 people in 1 hour and 20 minutes. An automatic machine washed the dishes and an artificial ice maker provided 1 ton of ice per day. Franchising (the selling of individual food service units by a food service chain to investors who own and operate their units under the name of the chain) originated with the A & W Corporation which started a small root beer stand in 1919. In 1955, McDonald's one and only fast-food restaurant was located in Des Plaines, Illinois. McDonald's now has over 12,000 restaurants, in more than 59 countries, worldwide. The chain competes with at least 15 other national chains.

The life-style of the American consumer in the 1990s is moving steadily towards the expenditure of more time at work and play, and less at home. Despite inflation, more income is being spent on leisure activities. The daily three-meal schedule has changed because population mobility has increased, eating patterns have become more individualistic, and more time is used in traveling to work and to recreation areas. Because people are eating away from home more often, the traditional family gathering at the dinner table has become less frequent. Social mobility has given rise to hamburger and fried chicken chains, the steak and seafood restaurant, and other fast-food chains.

With proper planning and management of kitchen design and food preparation equipment, convenience items may reduce preparation time and need for storage facilities, and lower food costs. (Some convenience foods require more freezer space but less preparation area.) The market for convenience foods will continue to grow as long as the food processors provide product variety with uniformity, and high quality items.

MAJOR TYPES OF CONVENIENCE FOODS SOLD IN RETAIL STORES. The main types of convenience items that are steadily replacing their homemade counterparts are described in Table C-39.

TABLE C-39
MAJOR TYPES OF CONVENIENCE FOODS SOLD IN RETAIL STORES

Type of Food	Typical Convenience Products	Comparison With Nonconvenience Items (Costs of food, fuel, and preparation times)	Comments
Baby foods (Also see BABY FOODS.)	Pureed meats, fruits, and vegetables in 4½ oz *(127 g)* and 7½ oz *(213 g)* glass jars.	The convenience foods cost considerably less than the home prepared items, especially when the homemade food for the baby is cooked separately from that for the rest of the family.	These types of foods are also convenient for use by adults in beverages, broths, desserts, salads, sandwiches, and soups.
Bakery products, breads	Boxed mixes, brown-and-serve, chilled, frozen, and ready-to-serve forms of baking powder biscuits, bread stuffing, corn muffins, pancakes, waffles, and yeast rolls.	Generally, the overall costs of the convenience items are less than those of similar homemade items. However, based upon purchase price alone, the ready-to-serve, frozen, and chilled products are the most expensive; and the products made from complete baking mixes are less expensive than homemade products or those made from mixes that require the addition of eggs, milk, and other ingredients.	Many of the convenience products are made from enriched flour or contain added B vitmins.
Beverages	Instant coffee, instant tea, lemon-flavored teas in cans, and tea bags.	Instant coffee is less expensive than that brewed from regular ground coffee, but instant tea and tea bags cost more per cup *(240 ml)* than tea brewed from the leaves. Canned lemon-flavored tea is the most costly of these types of products.	Processing raw coffee and tea to instant forms reduces their bulk, weight, and marketing costs.
Cakes, cookies, and pies	Boxed mixes, chilled, frozen, and ready-to-serve forms of angel food, bundt, devil's food, pound, and yellow cakes; brownies, chocolate chip, peanut butter, and sugar cookies; and apple, cherry, coconut cream, lemon, mince, and pumpkin pies.	The overall costs of the convenience items are less than those of homemade versions. However, based upon purchase price alone, the frozen products are the most expensive, homemade ones are less expensive, and those made from complete mixes are the least costly.	Consumers should read labels carefully because some convenience products contain imitations of berries, cream filling, and nuts that are made from much cheaper ingredients than those they replace.
Cheese products	Processed American cheese in loaves, slices, individually wrapped slices, and in an aerosol can.	There is little variation in cost between the loaves and slices of American cheese, but the product in the aerosol can is much more expensive.	Inexpensive cheese loaves may be high in water content and low in butterfat. Low fat cheeses do not cook or melt well, and they usually have a bland taste and rubbery texture.
Egg substitutes	Low cholesterol products that contain egg whites, coloring, and other ingredients and are usually sold in liquid and powdered forms.	Low cholesterol egg substitutes cost from 1½ to 2 times as much per serving as fresh eggs. The dried egg white products are even more expensive.	

(Continued)

TABLE C-39 *(Continued)*

Type of Food	Typical Convenience Products	Comparison With Nonconvenience Items (Costs of food, fuel, and preparation times)	Comments
Entrees	Canned and frozen main dishes that contain fish, meats, pasta products, poultry, and seafood with sauces, and/or vegetables.	The convenience products are usually more expensive than the homemade versions, even when the costs of cooking and preparation time are considered. Frozen entrees are more expensive than canned items.	Some convenience entrees may contain only 1/8 to ½ as much fish, meat, or poultry as homemade products. (They may be extended considerably by the use of excessive amounts of breading, sauces, and/or soy protein.)
Fish and seafood	Canned and frozen fish and seafood items that may be plain, breaded, or packed in a sauce.	Most of these items cost more than fresh fish or seafood, although frozen and canned shrimp may cost less than fresh shrimp, which have considerable waste.	Large pieces of fish and seafood have considerably lower proportions of breading than smaller pieces.
Fruits	Canned, dried, and frozen fruits that may be packaged with or without a packing medium such as fruit juice, low-calorie fluid, natural sugar syrup, or water.	Although canned and frozen fruits are generally 1½ to 2 times as costly as fresh items, large bags of frozen fruits cost only slightly more than fresh fruit. Also, the convenience items may be good buys when the fresh fruits are out of season.	Canned and frozen fruits are often more tender than their fresh counterparts. Drained weights of canned fruits vary considerably from brand to brand. Dried fruits contain little or no vitamin C.
Juices	Bottled or canned ready-to-serve juices, and frozen concentrates that require reconstitution with water.	Orange juice prepared from a frozen concentrate costs only about ⅓ as much as that squeezed from fresh oranges. (Canned orange juice is about ½ as costly as fresh orange juice.)	Frozen fruit juice concentrates may be used thawed, but undiluted, as syrups in baked products, cocktails, and desserts.
Legumes	Canned and frozen beans and peas with or without a packing medium such as salted water or a sauce.	The purchase prices of the convenience products are about the same as those for the fresh legumes. Hence, they are less costly when fuel and preparation time are considered.	A concern in Michigan produces a processed bean that cooks much faster than other forms of dried beans.
Meats	Canned and frozen forms of beef and pork that may be raw, cured (bacon, corned beef, ham, etc.), precooked, and/or extended with a cereal product or soy protein.	Frozen beef generally costs about 1½ to 2 times as much as the fresh meat, but ground beef that contains soy protein costs about the same or slightly less than a similar product prepared at home. Canned hams and frozen, precooked link sausages cost a little less than items prepared from the fresh meat.	Some of the meat patties which contain soy protein may be dry and have a cereal-like flavor. Frozen meats may not cook as well as fresh items unless they are thawed prior to cooking. (The thawing should be done in a refrigerator to minimize the growth of undesirable microorganisms.)

(Continued)

TABLE C-39 (*Continued*)

Type of Food	Typical Convenience Products	Comparison With Nonconvenience Items (Costs of food, fuel, and preparation times)	Comments
Milk	Dried and evaporated forms of buttermilk, skim milk, and whole milk.	Reconstituted evaporated milk and nonfat dry milk cost less than the fluid counterparts, but fresh buttermilk sometimes costs less than reconstituted dried buttermilk.	Dried and evaporated milks are concentrated sources of protein that may be added without reconstitution to various dishes.
Pizza	Boxed mixes, chilled, and frozen pizzas.	The product made from the boxed mix costs about the same as one made from all homemade ingredients. However, the chilled and frozen products that require only baking cost from 1½ to 2 times as much.	The crust of cooked frozen pizza is often heavier and less tasty than that of a freshly prepared dough. Frozen appetizer (miniature) pizzas cost about twice as much per oz or g as chilled or frozen items in the usual sizes.
Poultry	Canned or frozen precooked chicken which may be in a broth, or coated with breadcrumbs or a batter.	On the average, the convenience items cost about twice as much as the homemade ones. However, the large family size cans and packages of frozen chicken are more economical than the smaller units.	Convenience chicken products are less costly than ready-to-eat fried chicken from a fast-foods establishment; the latter cost about 3 times as much as the homemade version.
Puddings	Pudding mixes that are instant or which require cooking, and preprepared puddings in cans or in refrigerated cartons.	Puddings made from mixes are less costly than totally homemade items. Preprepared products cost about twice as much as those made from mixes.	Sometimes, the flavors of convenience puddings are weaker than those of homemade puddings.
Salads	Preprepared items such as bean, carrot and raisins, cole slaw, gelatin, macaroni, and potato salads sold in prepacked refrigerated cartons or dispensed from a refrigerated tray. Salad mixtures in plastic bags that require the addition of a dressing.	The purchase price of the convenience items are usually 2 to 3 times those of the totally homemade items, but the former may be more economical when the values of the preparation times are considered.	Take out salads have long been sold by delicatessens in large American cities, and are now available in certain supermarkets. Salad mixtures sold in plastic bags are often too dry and tough.
Soups	Canned condensed, canned single strength, dried, and frozen soup mixtures.	Based upon the cost of canned condensed soup (the most economical convenience item), dried soup costs about 1½ times as much, canned single-strength soup about 2½ times as much, and frozen soup from 3 to 4 times as much.	Few homemakers prepare their soups totally from fresh ingredients. Certain products may be too watery when prepared as directed.

(Continued)

TABLE C-39 (*Continued*)

Type of Food	Typical Convenience Products	Comparison With Nonconvenience Items (Costs of food, fuel, and preparation times)	Comments
Vegetables	Canned, dehydrated, and frozen vegetables that may be packed with or without a sauce or salted water.	Many canned and frozen single-ingredient items cost about the same or slightly less than those prepared from fresh vegetables. However, frozen vegetable mixtures with special seasonings and sauces cost more than homemade items. Most dehydrated potato products cost more than the homemade versions.	Frozen vegetables may often be of higher quality than fresh vegetables which have been held for awhile or shipped a considerable distance from their place of production.

UTILIZATION OF CONVENIENCE FOODS BY FOOD SERVICE OPERATORS. The ever-increasing use of convenience foods by fast-food outlets, restaurants, and other types of commercial and nonprofit establishments that serve foods is largely a result of the very high costs of labor, fuel, equipment, food ingredients, and the space for food preparation. Furthermore, the wide variety of these foods that is now available makes it possible for the food service operator to offer special diet foods, such as those low in cholesterol, salt, and carbohydrates, or high in protein. Therefore, current trends in the use of convenience foods by various types of establishments are noteworthy.

• **Fast-food operations**—These firms generally have a limited menu using pre-prepared items that require little garnishing, but which will result in an acceptable and palatable meal. Sheer convenience remains a major selling point, but some fast-food outlets now have salad bars and self-service tables. Others feature hostess service and dining by candlelight. There are growing possibilities for adding more variety to the fast-food segment of the foodservice industry.

• **Table restaurants**—To attract families and keep them coming back, food establishments must continuously have new ideas, new ways of presenting old ideas, and good quality. These restaurants generally have big kitchen facilities and a full-course menu. The major problem facing this type of restaurant is the high cost of labor. According to some food processors, labor turnover is estimated at 60 to 75% annually. Therefore, consideration is being given to increased use of convenience foods and automation in table restaurants.

• **Airlines**—Foodservice on the airlines is mainly for passengers who would otherwise not be able to eat because of flight schedulings. Most travelers now take foodservice aboard planes for granted. The airlines are very receptive to new ideas that are different and unique from those of their competitors. Airlines are reluctant to do away with meals; therefore, they must develop appetizing and tasty new dishes. Since most passengers prefer to eat on planes, meal flights have increased over beverage-only flights.

Most airlines cannot afford to maintain their own in-flight kitchens so they depend on caterers. These caterers often supply one or more companies at the same time by offering a line of items on an exclusive basis for a menu cycle and then offering the items to another airline. With higher food and service costs, some airlines are expressing an interest in appetizing, less expensive foods such as casseroles and cold dishes.

• **Industrial food services**—The successful in-plant food service manager is interested in maintaining quality and service while holding down costs. He or she generally uses a 4- to 5-week menu cycle, rotating popular items with new ideas. To hold their customers, in-plant feeders must add flexibility and variety to the menu.

• **Hospitals, schools, and colleges**—These foodservice operators must hold food costs within tight budgets and must vary menus for clients who eat all, or nearly all, meals at the institution. The managers of these operations place greater emphasis on the menu cycle and new trends than other foodservice categories.

Hospitals and institutions are changing from the notion of the "Basic Four" menu planning to the concept of the recommended dietary allowances. They are looking for new ways to improve menu planning and cycling. Furthermore, new ideas along modified diet lines such as low-sodium, low-calorie, and low-fat foods are needed. Many institutions are now using catering services. The caterers provide both tray and cafeteria service, and may often use the cafeteria as a testing ground for new products. If a product is popular in the cafeteria, it may be used in the patients' tray service.

Schools and colleges seem somewhat more willing to experiment with new products. By using unique foods, plating techniques, and efficient service, they are able to attract more students to the dining hall. They need new products and must have great flexibility in menu planning to keep up with food fads.

• **Military**—Now that the Army is directed toward volunteer service, the military must have appealing foods to attract young people. Volunteers want the same food options they had as civilians. The military has taken some measures to improve their foodservices. All types of food, including food fads, have been introduced and incorporated in menu planning. It appears that convenience foods have a promising future in military feeding.

• **Correctional institutions**—Since about 1970, the American Correctional Food Service Association has sought to improve foodservice in correctional institutions with the premise that well-fed inmates are more cooperative. Many institutions are on a 30- to 50-day menu cycle, but most inmates eat all meals at the institution, creating a need for variety in foods and menus.

FAST FOODS. The fast-food industry has enjoyed a phenomenal growth. The U.S. National Restaurant Association estimates that 45.8 million people eat at fast-food restaurants each day; in 1989, the typical American ate 1 out of every 12 meals at a fast-food restaurant.[45] Fast foods are appealing because they are quick, reasonably priced, and readily available.

As time becomes increasingly valuable, more and more Americans will rely on fast foods. So, it is important that nutrient compositions be available to enable people to make more healthful food choices. To the latter end, the nutritional analyses of items served at some of the leading fast-food chains are presented in Table C-40.

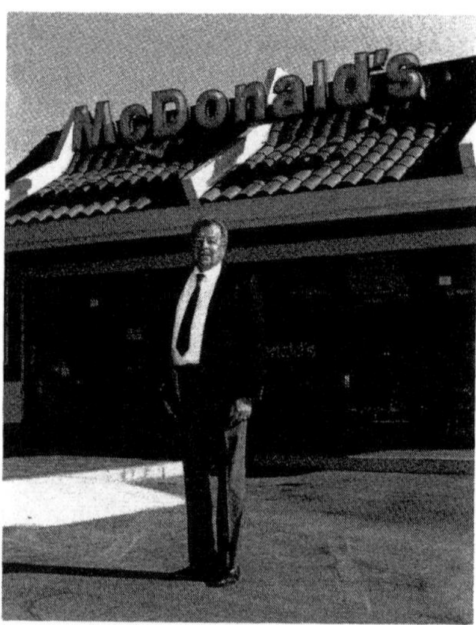

Fig. C-129a. Leonard Vanderhoop, who majored in animal science at the University of Massachusetts-Amherst, is now a successful owner-operator of four McDonald's in California.

[45]*Dietetic Currents*, Vol. 18, No. 4, 1991, Ross Laboratories, Columbus, Ohio 43216.

TABLE C-40. NUTRITIONAL ANALYSIS OF FAST FOODS[1]

ITEM	SERVING SIZE		NUTRIENTS PER SERVING																		
			Calories	Protein	Carbohydrate	Dietary Fiber	Total Fat	Unsaturated Fatty Acids		Saturated Fatty Acids	Cholesterol	Sodium	Potassium	Vitamin A		Vitamin C		Iron		Calcium	
	(oz)	(g)		(g)	(g)	(g)	(g)	Poly (g)	Mono (g)	(g)	(mg)	(mg)	(mg)	(IU)	(% US RDA)	(mg)	(% US RDA)	(mg)	(% US RDA)	(mg)	(% US RDA)
Restaurant: Arby's[+]																					
Regular Roast Beef	5	147	353	22.2	31.6	*1*	14.8	2.4	5.1	7.3	39	588	368	0	0	1.2	2	3.6	20	80	8
Beef 'N Cheddar	7	197	455	25.7	27.7	*1*	26.8	7.1	12.1	7.6	63	955	335	400	8	0.0	0	3.6	20	60	6
Chicken Breast Sandwich	7	184	493	23.0	47.9	*1*	25.0	10.3	9.6	5.1	91	1019	330	0	0	4.8	8	3.6	20	80	8
Roast Chicken Club	8	234	610	31.0	40.0	*1*	33.0	14.0	11.0	8.0	80	1500	430	0	0	0.0	0	3.6	20	150	15
Turkey Deluxe	7	197	375	23.8	32.5	*2*	16.6	7.8	4.7	4.1	39	1047	346	300	6	4.8	8	2.7	15	80	8
Ham 'N Cheese	6	156	292	22.9	19.2	*1*	13.7	2.7	6.3	4.7	45	1350	312	250	5	0.0	0	2.7	15	200	20
Super Roast Beef	8	234	501	25.1	50.4	*1*	22.1	5.4	8.2	8.5	40	798	503	750	15	36.0	60	4.5	25	100	10
French Fries	3	71	246	2.1	29.8	*2*	13.2	4.7	5.5	3.0	0	114	240	0	0	0.0	0	1.1	6	0	0
Potato Cakes	3	85	204	1.8	19.8	*1*	12.0	4.3	5.5	2.2	0	397	289	0	0	21.0	35	1.4	8	0	0
Jamocha Shake	12	326	368	9.3	59.1	*2*	10.5	1.6	6.4	2.5	35	262	525	300	6	2.4	4	2.7	15	250	25
Restaurant: Arthur Treacher's Fish & Chips[+]																					
Chicken Sandwich	6	156	413	16.2	44.0	*1*	19.2	6.7	*7.7*	2.8	32	708	*173*	117	2	19.0	32	1.7	9	59	6
Fish Sandwich	6	170	440	16.4	39.4	*1*	24.0	6.9	*8.1*	4.2	42	836	*223*	117	0	2.0	2	1.5	8	89	8
Chicken	5	147	369	27.1	16.5	*1*	21.6	5.6	*9.1*	3.5	65	495	*221*	102	2	2.0	3	0.8	4	11	0
Coleslaw	4	107	123	1.0	11.1	*3*	8.2	4.4	*1.9*	1.1	7	266	*346*	170	5	59.0	99	0.2	0	24	2
Lemon Luvs™	3	83	276	2.6	35.1	*2*	13.9	1.9	*5.9*	2.2	0	314	*68*	64	0	0.9	2	0.9	5	10	0
Krunch Pups™	3	74	203	5.4	12.0	*1*	14.8	2.2	*6.2*	3.7	25	446	*101*	43	0	4.0	6	0.6	3	8	0
Chips	4	113	276	4.0	34.9	*3*	13.2	2.7	*5.9*	2.3	1	39	*597*	85	0	6.0	10	0.5	3	12	0
Shrimp	4	106	381	13.1	27.2	*1*	24.4	6.2	*11.0*	3.3	93	538	*128*	86	0	1.0	2	0.6	4	57	6
Fish	5	153	355	19.2	25.4	*1*	*20.1*	5.6	*9.0*	2.8	56	450	*255*	111	0	2.0	3	0.6	4	15	0

Restaurant: Burger King[*]																					
Whopper®/Everything	10	270	628	27.0	46.0	2	36.0	13.0	11.0	12.0	90	880	545	678	14	12.0	19	5.0	28	84	8
Whopper/Cheese	10	294	706	32.0	47.0	2	43.0	13.0	13.0	17.0	113	1164	568	1061	21	12.0	19	5.0	28	215	21
Hamburger	4	108	272	15.0	29.0	1	12.0	1.0	5.0	4.0	37	509	235	150	3	3.0	5	3.0	16	37	4
Cheeseburger	4	121	318	17.0	30.0	1	15.0	1.0	6.0	7.0	48	651	247	341	7	3.0	5	3.0	16	102	10
Bacon Double Cheeseburger	6	160	515	33.0	27.0	1	31.0	2.0	13.0	15.0	104	728	363	384	8	0.0	0	4.0	21	168	17
Hamburger Deluxe	5	138	344	15.0	30.0	1	17.0	7.0	6.0	6.0	41	486	275	296	6	6.0	10	3.0	16	40	4
Cheeseburger Deluxe	5	151	390	17.0	31.0	2	20.0	7.0	7.0	8.0	52	628	287	488	10	6.0	10	3.0	16	105	11
Ocean Catch Filet	7	194	488	19.0	45.0	2	25.0	13.0	6.0	4.0	77	592	369	36	1	0.0	0	2.0	9	46	5
Chicken Specialty Sandwich	8	229	685	26.0	56.0	2	40.0	20.0	11.0	8.0	82	1423	375	126	3	0.0	0	3.0	16	79	8
Chicken Tenders™	3	90	236	20.0	10.0	1	10.0	3.0	5.0	2.0	47	636	200	95	2	0.0	0	1.0	7	18	2
French Fries	4	111	341	3.0	24.0	3	13.0	0.0	9.0	7.0	14	160	360	0	0	0.0	0	1.0	7	0	0
Onion Rings	3	86	302	4.0	28.0	2	16.0	4.0	8.0	3.0	0	665	173	0	0	0.0	0	1.0	7	124	12
Breakfast Croissan'wich™/Bacon	4	118	355	14.0	20.0	1	24.0			8.0	249	762	182	426	9	0.0	0	2.0	9	136	14
Breakfast Croissan'wich/Sausage	6	159	538	20.0	20.0	1	40.0	5.0	20.0	13.0	293	1042	285	426	9	0.0	0	3.0	16	145	15
Breakfast Croissan'wich/Ham/Egg/Cheese	5	144	346	19.0	19.0	1	21.0	2.0	11.0	7.0	241	962	256	426	10	0.0	0	2.0	11	136	15
Breakfast Bagel Sandwich/Bacon	6	169	438	20.0	46.0	2	20.0	4.0	7.0	7.0	274	905	202	426	9	0.0	0	3.0	16	124	12
Breakfast Bagel Sandwich/Sausage/Egg/Cheese	7	210	626	27.0	49.0	2	36.0	6.0	15.0	12.0	318	1137	305	426	10	0.0	0	4.0	21	135	14
Breakfast Bagel Sandwich/Ham/Egg/Cheese	7	196	418	23.0	46.0	2	15.0	4.0	6.0	6.0	287	1130	276	426	9	0.0	0	3.0	16	125	12
Scrambled Egg Platter	7	211	549	17.0	44.0	3	30.0	6.0	18.0	9.0	370	808	487	375	7	3.0	5	3.0	16	101	10
Scrambled Egg Platter/Sausage	9	260	768	26.0	47.0	3	53.0	8.0	30.0	15.0	412	1271	623	375	7	5.0	9	4.0	21	120	12
Scrambled Egg Platter/Bacon	8	221	610	21.0	44.0	3	39.0	7.0	20.0	11.0	373	1043	532	375	22	5.0	9	3.0	16	110	11
French Toast Sticks	5	141	538	10.0	53.0	2	32.0	11.0	15.0	5.0	80	537	126	0	0	0.0	0	3.0	16	77	8
Great Danish	3	71	500	5.0	40.0	3	36.0	11.0	15.0	23.0	6	288	116	94	2	0.0	0	2.0	9	91	9
Vanilla Shake	10	284	334	9.0	51.0	0	10.0	0.0	3.0	6.0	39	205	505	0	0	0.0	0	0.0	0	295	29
Chocolate Shake	10	284	326	9.0	49.0	1	10.0	0.0	4.0	6.0	33	202	567	0	0	0.0	0	2.0	9	262	26
Apple Pie	4	125	311	3.0	44.0	2	14.0	1.0	8.0	4.0	4	412	122	0	0	5.0	8	1.0	7	0	0
Chicken Salad	9	258	142	20.0	8.0	2	4.0	1.0	1.0	1.0	50	440	630	4742	90	20.0	35	1.0	7	44	4
Chef Salad	10	273	178	17.0	7.0	2	9.0	1.0	3.0	4.0	120	570	550	4889	100	14.0	23	2.0	9	158	16
Garden Salad	8	223	95	6.0	8.0	2	5.0	0.0	1.0	3.0	15	125	440	5143	100	34.0	60	1.0	7	147	15
Side Salad	5	135	25	1.0	5.0	2	0.0	0.0	0.0	0.0	0	20	230	2495	50	11.0	17	1.0	7	21	2
Thousand Island Dressing	2	63	290	1.0	15.0	0	26.0	14.0	5.0	5.0	36	470	45	157	64	2.0	0	0.0	0	11	0
Bleu Cheese Dressing	2	59	300	3.0	2.0	0	32.0	16.0	7.0	7.0	40	600	23	108	2	0.0	0	0.0	0	58	6
Reduced Calorie Italian	2	59	170	0.0	3.0	0	18.0	10.0	5.0	3.0	3	762	6	0	0	0.0	0	0.0	0	0	0
French Dressing	2	64	290	0.0	23.0	1	22.0	8.0	10.0	3.0	2	400	70	237	31	0.0	0	0.0	0	0	0
Bacon Bits	0.1	3	16	1.0	0.0	0	1.0	0.0	1.0	0.0	5	1	30	0	0	0.0	0	0.0	0	0	0
Croutons	0.3	7	31	1.0	5.0	0	1.0	0.0	0.0	0.0	0	90	10	0	1	0.0	0	0.0	0	0	0
Restaurant: Dairy Queen[*]																					
Cone, Small	3	85	140	3.0	22.0	0	4.0	0.1	0.9	1.7	10	45	146	100	2	0.0	0	0.4	2	100	10
Cone, Regular	5	142	240	6.0	38.0	0	7.0	0.2	1.5	3.1	15	80	252	200	4	0.0	0	0.7	4	150	15
Cone, Large	8	213	340	9.0	57.0	0	10.0	0.3	2.6	5.2	25	115	417	400	8	0.0	0	1.4	8	250	25
Cone, Small, Chocolate-Dipped	3	92	190	3.0	25.0	2	9.0	0.3	2.3	4.2	10	55	179	100	2	0.0	0	0.4	2	100	10
Cone, Regular, Chocolate-Dipped	6	156	340	6.0	42.0	3	16.0	0.5	4.1	7.8	20	100	312	200	4	0.0	0	0.7	4	150	15
Cone, Large, Chocolate-Dipped	8	234	510	9.0	64.0	4	24.0	0.7	6.2	11.8	30	145	473	400	8	0.0	0	1.4	8	250	25
Chocolate Sundae, Small	4	106	190	3.0	33.0	1	4.0	0.1	1.0	2.2	10	75	215	100	2	0.0	0	0.4	2	100	10
Chocolate Sundae, Regular	6	177	310	5.0	56.0	1	8.0	0.2	1.8	3.8	20	120	370	200	4	0.0	0	1.1	6	200	20
Chocolate Sundae, Large	9	248	440	8.0	78.0	2	10.0	0.3	2.6	5.6	30	165	530	400	8	0.0	0	1.4	8	250	25
Chocolate Shake, Small	9	241	409	8.0	69.0	1	11.0	0.5	3.8	7.8	30	150	445	500	10	0.0	0	1.8	10	300	30
Chocolate Shake, Regular	15	418	710	14.0	120.0	2	19.0	0.7	6.0	12.4	50	260	708	750	15	0.0	0	2.7	15	450	45
Chocolate Shake, Large	17	489	831	16.0	140.0	2	22.0	0.8	6.9	14.2	60	304	809	1000	20	0.0	0	3.6	20	550	55
Chocolate Malt, Small	9	241	438	8.0	77.0	2	10.0	0.5	3.1	6.5	30	150	416	500	10	0.0	0	2.7	15	300	30
Chocolate Malt, Regular	15	418	760	14.0	134.0	3	18.0	0.8	5.4	11.2	50	260	711	750	15	0.0	0	4.5	25	450	45
Chocolate Malt, Large	17	489	889	16.0	157.0	3	21.0	0.9	6.4	13.1	60	304	828	1000	20	0.0	0	5.4	30	550	55
Float	14	397	410	5.0	82.0	0	7.0	0.2	1.6	3.5	20	85	270	200	4	0.0	0	1.1	6	200	20
Peanut Buster Parfait[*]	11	305	740	16.0	94.0	6	34.0	4.0	11.6	14.1	30	250	759	300	6	0.0	0	1.8	10	250	25
Parfait	10	283	430	8.0	76.0	1	8.0	0.3	2.2	5.8	30	140	379	400	8	0.0	0	1.4	8	250	25
Freeze	14	397	500	9.0	89.0	0	12.0	0.4	3.3	7.0	30	180	532	400	8	0.0	0	1.8	10	300	30
Mr Misty®, Small	9	248	190	0.0	48.0	0	0.0	0.0	0.0	0.0	0	10	4	0	0	0.0	0	0.0	0	0	0
Mr Misty, Regular	12	330	250	0.0	63.0	0	0.0	0.0	0.0	0.0	0	10	6	0	0	0.0	0	0.0	0	0	0
Mr Misty, Large	16	439	340	0.0	84.0	0	0.0	0.0	0.0	0.0	0	10	8	0	0	0.0	0	0.0	0	0	0
Mr Misty Kiss	3	89	70	0.0	17.0	0	0.0	0.0	0.0	0.0	0	10	2	0	0	0.0	0	0.0	0	0	0
Mr Misty Freeze	15	411	500	9.0	91.0	0	12.0	0.4	3.3	7.0	30	140	532	400	8	0.0	0	1.4	8	300	30
Mr Misty Float	15	411	390	5.0	74.0	0	7.0	0.2	1.6	3.5	20	95	269	200	4	0.0	0	0.7	4	200	20
Buster Bar	5	149	448	10.0	41.0	6	29.0	5.9	12.0	8.9	10	175	485	100	2	0.0	0	1.1	6	100	10
Fudge Nut Bar	5	142	406	8.0	40.0	2	25.0	3.2	6.3	11.3	10	167	226	100	2	0.0	0	1.1	6	100	10
Dilly Bar	3	85	210	3.0	21.0	1	13.0	0.3	1.1	11.1	10	50	179	100	2	0.0	0	0.4	2	100	10
DQ Sandwich	2	60	140	3.0	24.0	0	4.0	0.3	1.5	2.0	5	40	105	0	0	0.0	0	0.0	0	60	6
Chipper Sandwich	4	113	318	5.0	56.0	0	7.0	0.5	3.0	5.7	13	170	388	100	2	1.0	2	2.7	15	100	10
Heath[*] Blizzard, Regular	14	404	800	15.0	125.0	3	24.0	0.8	6.8	13.8	65	325	1009	300	6	5.0	8	2.7	15	500	50

(Continued)

TABLE C-40 *(Continued)*

ITEM	SERVING SIZE (oz)	SERVING SIZE (g)	Calories	Protein (g)	Carbohydrate (g)	Dietary Fiber (g)	Total Fat (g)	Unsaturated Fatty Acids Poly (g)	Unsaturated Fatty Acids Mono (g)	Saturated Fatty Acids (g)	Cholesterol (mg)	Sodium (mg)	Potassium (mg)	Vitamin A (IU)	Vitamin A (% US RDA)	Vitamin C (mg)	Vitamin C (% US RDA)	Iron (mg)	Iron (% US RDA)	Calcium (mg)	Calcium (% US RDA)
Dairy Queen (cont'd)																					
Single Hamburger	5	148	360	21.0	33.0	1	16.0	1.4	6.2	5.7	45	630	328	100	2	0.0	0	3.6	20	100	10
Double Hamburger	7	210	530	36.0	33.0	1	28.0	1.9	11.4	10.6	85	660	546	100	2	0.0	0	6.3	35	100	10
Triple Hamburger	10	272	710	51.0	33.0	1	45.0	2.3	16.5	15.4	135	690	763	200	4	0.0	0	9.0	50	100	10
Single Hamburger/Cheese	6	162	410	24.0	33.0	1	20.0	1.6	7.5	8.5	50	790	351	200	4	0.0	0	3.6	20	200	20
Double Hamburger/Cheese	8	239	650	43.0	34.0	1	37.0	2.2	14.0	16.1	95	980	592	400	8	0.0	0	6.3	35	350	35
Triple Hamburger/Cheese	11	301	820	58.0	34.0	1	50.0	2.6	19.1	21.0	145	1010	809	400	8	0.0	0	9.0	50	350	35
Hot Dog	4	100	280	11.0	21.0	1	16.0	2.2	8.5	6.7	45	830	133	0	0	0.0	0	1.4	8	80	8
Hot Dog/Chili	5	128	320	13.0	23.0	1	20.0	2.2	9.2	7.4	55	985	215	0	0	0.0	0	1.4	8	80	8
Hot Dog/Cheese	4	114	330	15.0	21.0	1	21.0	2.3	9.7	9.4	55	990	156	100	2	0.0	0	1.4	8	150	15
DQ Hounder	5	151	480	16.0	21.0	1	36.0	3.7	16.2	12.8	80	1800	227	0	0	0.0	0	4.5	25	150	15
DQ Hounder/Chili	7	208	575	22.0	25.0	1	41.0	4.0	18.6	14.9	89	1900	489	0	0	0.0	0	4.5	25	200	20
DQ Hounder/Cheese	6	165	533	19.0	22.0	1	40.0	3.8	17.4	15.5	89	1995	250	0	0	0.0	0	4.5	25	250	25
Super Hot Dog	6	175	520	17.0	44.0	1	27.0	3.5	13.0	10.2	80	1365	218	0	0	0.0	0	2.7	15	150	15
Super Hot Dog/Chili	8	218	570	21.0	47.0	2	32.0	3.7	14.7	11.8	100	1595	405	0	0	0.0	0	2.7	15	150	15
Super Hot Dog/Cheese	7	196	580	22.0	45.0	1	34.0	3.7	14.9	14.4	100	1605	253	100	2	0.0	0	1.4	8	250	25
Fish Filet	6	177	430	20.0	45.0	1	18.0	4.9	6.4	5.4	40	674	259	0	0	0.0	0	3.6	20	150	15
Fish Filet/Cheese	7	191	483	23.0	46.0	1	22.0	4.4	7.9	9.4	49	870	291	500	10	0.0	0	3.6	20	250	25
Chicken Breast Filet	7	202	608	27.0	46.0	2	34.0	12.6	9.7	8.4	78	725	284	100	2	2.4	4	5.4	30	150	15
Chicken Breast Filet/Cheese	8	216	661	30.0	47.0	2	38.0	11.6	11.2	12.2	87	921	318	750	15	2.4	4	5.4	30	250	25
All White Chicken Nuggets	4	99	276	16.0	13.0	1	18.0	2.4	6.6	6.1	39	505	166	0	0	0.0	0	1.1	6	0	0
BBQ Nugget Sauce	1	28	41	0.0	9.0	0	0.7	0.1	0.3	0.3	0	130	72	750	15	0.0	0	0.4	.2	20	2
French Fries, Small	3	71	200	2.0	25.0	2	10.0	0.2	4.4	4.4	10	115	416	0	0	9.0	15	0.4	2	0	0
French Fries, Large	4	113	320	3.0	40.0	3	16.0	1.3	7.0	7.1	15	185	675	0	0	15.0	25	1.1	6	0	0
Onion Rings	3	85	280	4.0	31.0	3	16.0	1.3	6.7	6.7	15	140	336	0	0	2.4	4	0.7	4	0	0
Restaurant: Domino's Pizza® (2 slices of each pizza)																					
Cheese Pizza	6	168	376	21.6	56.3	6	10.0	1.2	3.3	5.5	19	483	364	374	7	0.5	0	2.8	13	221	17
Pepperoni Pizza	7	187	460	24.1	55.6	5	17.5	1.9	7.3	8.4	28	825	423	482	7	0.6	0	3.1	17	239	19
Sausage/Mushroom Pizza	7	200	430	24.2	55.3	8	15.8	1.8	6.3	7.7	28	552	418	488	8	0.5	0	3.0	17	227	20
Veggie Pizza	9	261	498	31.0	60.0	8	18.5	1.7	6.6	10.2	36	1035	537	529	10	0.3	0	4.7	26	435	39
Deluxe Pizza	8	234	498	26.7	59.2	7	20.4	2.2	8.9	9.3	40	954	616	468	9	0.8	0	4.7	23	233	23
Double Cheese/Pepperoni Pizza	8	227	545	32.1	55.2	8	25.3	2.2	9.9	13.3	48	1042	529	472	9	1.0	2	4.0	22	459	45
Ham Pizza	7	186	417	23.2	58.0	2	11.0	1.3	3.8	5.9	26	805	485	242	0	0.6	0	4.6	19	226	19
Restaurant: Kentucky Fried Chicken®																					
Nuggets	1	16	46	2.8	2.2	0	2.9	0.3	1.3	0.7	12	140	23	100	2	1.0	0	0.1	0	2	0
Barbeque Sauce	1	28	35	0.3	7.1	0	0.6	0.3	0.4	0.1	0	450	98	370	7	0.0	0	0.2	0	6	0
Sweet & Sour Sauce	1	28	58	0.1	13.0	0	0.6	0.3	0.0	0.1	0	148	12	100	2	0.0	0	0.2	0	5	0
Honey	1	14	49	0.0	12.1	0	0.0	0.0	0.0	0.0	0	15	9	0	0	0.0	0	0.1	0	1	0
Mustard	1	28	36	0.9	6.0	0	0.9	1.1	0.2	0.1	0	346	37	100	0	0.0	0	0.3	0	10	0
Chicken Littles™ Sandwich	2	47	169	5.7	13.8	0	10.1	3.4	3.2	2.0	18	331	61	100	0	0.0	0	1.7	9	23	2
Buttermilk Biscuit	2	65	235	4.5	28.0	1	11.9	2.2	6.0	3.2	1	655	78	100	0	0.0	0	1.6	9	95	10
Mashed Potatoes/Gravy	4	98	71	2.4	11.9	1	1.6	0.2	0.5	1.6	0	339	217	100	0	0.0	0	0.4	2	22	2
French Fries, Regular	3	77	244	3.2	31.1	2	11.9	0.7	6.1	2.6	2	139	519	100	0	16.0	26	0.6	3	13	0
Corn-on-the-cob	5	143	176	5.1	31.9	7	3.1	1.5	1.3	0.5	0	21	307	272	5	2.0	3	0.8	4	7	0
Coleslaw	3	91	119	1.5	13.3	1	6.6	3.4	1.5	1.0	5	197	161	310	6	22.0	36	0.2	0	33	3
Original Recipe Chicken																					
Wing	2	55	178	12.2	6.0	0	11.7	1.8	4.5	3.0	64	372	9	100	0	1.0	0	1.2	6	48	4
Breast	4	115	283	27.5	8.8	0	15.3	2.0	6.8	3.8	93	672	242	100	0	1.0	0	1.0	5	36	3
Drumstick	2	57	146	13.1	4.2	0	8.5	1.3	3.4	2.2	67	275	122	100	0	1.0	0	1.1	5	21	2
Thigh	4	104	294	17.9	11.1	1	19.7	3.1	9.3	5.3	123	619	164	104	2	1.0	0	1.3	7	65	6
Extra Crispy Chicken																					
Wing	2	65	254	12.4	9.3	0	18.6	2.5	9.8	4.0	67	422	116	100	2	0.0	0	0.5	2	21	2
Breast	5	135	342	33.0	11.7	1	19.7	2.0	8.9	4.8	114	790	298	100	0	0.0	0	0.8	4	33	3
Drumstick	2	69	204	13.6	6.1	0	13.9	1.7	6.3	3.4	71	324	127	100	0	0.0	0	0.6	4	13	1
Thigh	4	119	406	20.0	14.4	1	29.8	4.2	14.6	7.7	129	688	193	100	3	0.0	0	1.2	6	46	4

Restaurant: Long John Silver's® Seafood Shoppe																					
Three-piece Fish Light/Paprika (Baked)	5	134	120	28.0	1.0	*0*	1.0	*0.7*	*0.4*	*0.4*	110	120	*395*	*44*	*43*	*0.0*	*0*	*0.4*	*2*	*21*	*2*
Three-piece Fish/Lemon Crumb (Baked)	5	141	150	29.0	4.0	*0*	1.0	*0.7*	*0.4*	*0.4*	110	370	*409*	*45*	*44*	*0.0*	*0*	*0.6*	*3*	*24*	*2*
Three-piece Fish/Scampi Sauce (Baked)	5	148	170	28.0	2.0	*0*	5.0	*0.9*	*1.5*	*2.9*	110	270	*396*	*195*	*19*	*0.0*	*0*	*0.4*	*2*	*22*	*2*
Shrimp/Scampi Sauce (Baked)	5	148	120	15.0	2.0	*0*	5.0	*0.5*	*1.6*	*3.2*	205	610	*143*	*351*	*35*	*1.7*	*3*	*2.4*	*13*	*32*	*3*
Chicken Light/Herbs (Baked)	4	117	140	25.0	1.0	*0*	4.0	*1.2*	*1.4*	*1.2*	70	670	*234*	*12*	*1*	*0.0*	*0*	*1.0*	*5*	*15*	*1*
Rice Pilaf	5	142	210	5.0	43.0	*1*	2.0	*0.6*	*1.5*	*0.3*	0	570	*110*	*5*	*0*	*0.5*	*1*	*1.9*	*10*	*33*	*3*
Green Beans	4	113	30	1.0	6.0	*3*	1.0	*0.1*	*0.0*	*0.0*	5	540	*113*	*534*	*53*	*8.3*	*14*	*0.8*	*4*	*46*	*4*
Garden Vegetables	4	113	120	4.0	16.0	*5*	6.0	*2.0*	*2.5*	*1.2*	5	95	*233*	***6148***	***100***	*4.4*	*7*	*1.1*	*6*	*36*	*3*
Coleslaw	3	98	140	1.0	20.0	*1*	6.0	*3.8*	*1.4*	*1.0*	15	260	*152*	*91*	*9*	*28.2*	*47*	*0.4*	*2*	*30*	*3*
Breadstick	1	34	110	3.0	18.0	*1*	3.0	*1.0*	*1.3*	*0.6*	0	120	*27*	*155*	*15*	*0.0*	*0*	*0.2*	*1*	*7*	*0*
Small Salad	2	54	8	1.0	2.0	*1*	0.0	*0.1*	*0.0*	*0.0*	0	0	*109*	*1056*	*100*	*5.3*	*9*	*0.3*	*1*	*8*	*0*
Restaurant: McDonald's®																					
Egg McMuffin®	5	138	290	18.2	28.1	*1*	11.2	1.3	6.1	3.8	226	740	*469*	499	10	0.0	0	2.8	15	256	25
Hotcakes/Butter/Syrup	6	176	410	8.2	74.4	*2*	9.2	2.5	3.1	3.7	21	640	*168*	173	4	0.0	0	2.1	10	114	10
Scrambled Eggs	4	100	140	12.4	1.2	*0*	9.8	1.4	5.0	3.3	399	290	*121*	518	10	0.0	0	2.1	10	57	6
Pork Sausage	2	48	180	8.4	0.0	*0*	16.3	1.9	8.5	5.9	48	350	*175*	0	0	0.0	0	0.7	4	8	0
English Muffin/Butter	2	59	170	5.4	26.7	*1*	4.6	0.5	1.7	2.4	9	270	*319*	122	2	0.0	0	1.1	8	151	15
Hashbrown Potatoes	2	53	130	1.4	14.9	*2*	7.3	0.4	3.7	3.2	9	330	*328*	0	0	1.6	2	0.3	0	6	0
Biscuit/Biscuit Spread	3	75	260	4.6	31.9	1	12.7	0.6	8.6	3.4	1	730	*82*	0	0	0.0	0	1.3	8	75	8
Biscuit/Sausage	4	123	440	13.0	31.9	*1*	29.0	2.5	17.2	9.3	49	1080	*323*	0	0	0.0	0	2.0	10	83	8
Biscuit/Sausage/Egg	6	180	520	19.9	32.6	*1*	34.5	3.4	20.0	11.2	275	1250	*293*	294	6	0.0	0	3.2	20	116	10
Biscuit/Bacon /Egg/Cheese	6	156	440	17.5	33.3	*1*	26.4	2.0	16.1	8.2	253	1230	*215*	534	10	0.0	0	2.6	15	185	20
Sausage McMuffin®	4	117	370	16.5	27.3	*1*	21.9	2.4	11.7	7.8	64	830	*487*	240	4	0.0	0	2.3	15	235	25
Sausage McMuffin/Cheese	6	167	440	22.6	27.9	*1*	26.8	3.2	14.2	9.5	263	980	*548*	499	10	0.0	0	3.3	20	263	25
Apple Danish	4	115	390	5.8	51.2	*2*	17.9	2.0	10.8	3.5	25	370	*76*	115	0	16.1	25	1.4	8	14	0
Iced Cheese Danish	4	110	390	7.4	42.3	*1*	21.8	1.8	12.1	6.0	47	420	*83*	188	4	1.1	0	1.4	8	33	4
Cinnamon Raisin Danish	4	110	440	6.4	57.5	*2*	21.0	1.6	13.0	4.2	34	430	*130*	110	0	3.2	6	1.8	10	35	4
Raspberry Danish	4	117	410	6.1	61.5	*2*	15.9	1.1	10.2	3.1	26	310	*84*	117	0	3.2	6	1.5	8	14	0
Apple Bran Muffin	3	85	190	5.0	46.0	*2*	0.0	0.0	0.0	0.0	0	230	*251*	*7*	*0*	*0.7*	*1*	*0.6*	*3*	*31*	*3*
Blueberry Muffin	*3*	*85*	170	3.0	40.0	*1*	0.0	0.0	0.0	0.0	0	220	*136*	*157*	*15*	*1.2*	*2*	*0.6*	*3*	*81*	*8*
Chicken McNuggets®	4	113	290	19.0	16.5	*1*	16.3	1.8	10.4	4.1	65	520	*180*	0	0	0.0	0	1.0	6	13	0
Hot Mustard Sauce	1	30	70	0.5	8.2	*0*	3.6	1.9	1.2	0.5	5	250	*26*	16	0	0.5	0	0.2	0	15	2
Barbeque Sauce	1	32	50	0.3	12.1	*0*	0.5	0.2	0.2	0.1	0	340	*126*	153	4	2.3	4	0.3	2	13	0
Sweet & Sour Sauce	1	32	60	0.2	13.8	*0*	0.2	0.1	0.1	0.0	0	190	*13*	324	6	0.6	0	0.2	0	11	0
Honey	1	14	45	0.0	11.5	*0*	0.0	0.0	0.0	0.0	0	0	*7*	0	0	0.1	0	0.1	0	1	0
Hamburger	4	102	260	12.3	30.6	*1*	9.5	0.8	5.1	3.6	37	500	*221*	152	4	2.2	4	2.3	15	122	10
Cheeseburger	4	116	310	15.0	31.2	*1*	13.8	0.9	7.7	5.2	53	750	*244*	392	8	2.2	4	2.3	15	199	20
McLean Deluxe™	7	203	310	20.0	34.0	*2*	10.0	*1.2*	*5.0*	*4.6*	37	650	*404*	493	*49*	*7.2*	*12*	3.8	*21*	145	*14*
Quarter Pounder®	6	166	410	23.1	34.0	*1*	20.7	1.2	11.4	8.1	86	660	*392*	223	4	3.2	6	3.7	20	142	15
Quarter Pounder®/Cheese	7	194	520	28.5	35.1	*1*	29.2	1.5	16.5	11.2	118	1150	*438*	703	15	3.2	6	3.7	20	296	30
Big Mac®	8	215	560	25.2	42.5	*1*	32.4	1.5	20.9	10.1	103	950	*361*	352	8	1.7	2	4.0	20	256	25
Filet-O-Fish®	5	142	440	13.8	37.9	*1*	26.1	10.8	10.2	5.2	50	1030	*217*	146	2	0.1	0	1.8	10	165	15
McD.L.T.®	8	234	580	26.3	36.0	*2*	36.8	8.5	16.7	11.5	109	990	*480*	754	15	7.4	10	3.9	20	225	25
McChicken®	7	190	490	19.2	39.8	*1*	28.6	11.6	11.5	5.4	43	780	*245*	104	2	2.4	4	2.6	15	143	15
Chef Salad	10	283	230	20.5	7.5	*2*	13.3	0.9	6.5	5.9	128	490	*572*	4114	80	13.6	20	1.5	8	256	25
Garden Salad	8	213	110	7.1	6.2	*2*	6.6	0.5	3.2	2.9	83	160	*325*	3915	80	13.5	20	1.3	6	149	15
Chicken Salad Oriental	9	244	140	23.1	5.0	*2*	3.4	0.5	2.0	0.9	78	230	*492*	3625	70	11.0	20	1.0	6	33	4
Side Salad	4	115	60	3.7	3.3	*1*	3.3	0.3	1.6	1.5	41	85	*194*	2173	45	7.4	10	0.7	4	76	8
Bleu Cheese Dressing	1	14	70	0.5	1.2	*0*	6.9	3.8	1.8	1.3	6	150	*6*	18	0	0.0	0	0.0	0	15	0
French Dressing	1	14	58	0.1	2.7	*0*	5.2	3.1	1.3	0.8	0	180	*12*	15	0	0.3	0	0.0	0	2	0
Ranch Dressing	1	14	83	0.2	1.3	*0*	8.6	5.1	2.2	1.4	5	130	*35*	11	0		0	0.0	0	7	0
1000 Island Dressing	1	14	78	0.2	2.4	*0*	7.5	4.4	1.9	1.2	8	100	*22*	49	0	0.5	0	0.1	0	3	0
Lite Vinaigrette Dressing	1	14	15	0.2	2.0	*0*	0.5	0.3	0.1	0.1	0	75	*2*	92	0	0.4	0	0.0	0	3	0
Oriental Dressing	1	14	24	0.2	5.8	*0*	0.1	0.0	0.1	0.0	0	180	*10*	28	0	0.4	0	0.2	0	6	0
Red French Reduced Calorie Dressing	1	14	40	0.1	5.2	*0*	1.9	1.1	0.5	0.3	0	110	*5*	74	0	0.6	0	0.1	0	3	0
Caesar Dressing	1	14	60	0.4	0.6	*0*	6.1	3.5	1.5	1.1	7	170	*5*	13	0	0.1	0	0.1	0	13	0
Peppercorn Dressing	1	14	80	0.2	0.5	*0*	8.7	5.2	2.2	1.4	7	85	*5*	9	0	0.1	0	0.1	0	3	0
French Fries, Small	2	68	220	3.1	25.6	*2*	12.0	0.5	6.5	5.1	9	110	*390*	0	0	8.2	15	0.5	2	10	0
French Fries, Medium	3	97	320	4.4	36.3	*3*	17.1	0.7	9.2	7.2	12	150	*571*	0	0	11.6	20	0.7	4	14	0
French Fries, Large	4	122	400	5.6	45.9	*3*	21.6	0.9	11.6	9.1	16	200	*701*	0	0	14.6	25	0.9	6	18	0
Apple Pie	3	83	260	2.2	30.0	*2*	14.8	0.9	9.1	4.8	6	240	*72*	0	0	11.4	20	0.7	4	110	0
Vanilla Lowfat Milk Shake	11	293	290	10.8	60.0	*0*	1.3	0.1	0.7	0.6	10	170	*510*	306	6	0.0	0	0.1	0	327	35
Chocolate Lowfat Milk Shake	11	293	320	11.0	66.0	*1*	1.7	0.1	0.9	0.8	10	240	*552*	306	6	0.0	0	0.9	0	332	35
Strawberry Lowfat Milk Shake	11	293	320	10.7	67.0	*0*	1.3	0.1	0.6	0.6	10	170	*635*	306	6	0.0	0	0.1	0	327	35
Soft Serve Cone	3	86	140	3.9	21.9	*0*	4.5	0.2	2.3	2.1	16	70	*152*	128	2	0.0	0	0.2	0	11	10
Strawberry Sundae	6	171	210	5.7	49.2	*1*	1.1	0.0	0.4	0.6	5	95	*298*	214	4	1.3	2	0.2	0	191	20

(Continued)

TABLE C-40 *(Continued)*

ITEM	SERVING SIZE (oz)	SERVING SIZE (g)	Calories	Protein (g)	Carbohydrate (g)	Dietary Fiber (g)	Total Fat (g)	Unsaturated Fatty Acids Poly (g)	Unsaturated Fatty Acids Mono (g)	Saturated Fatty Acids (g)	Cholesterol (mg)	Sodium (mg)	Potassium (mg)	Vitamin A (IU)	Vitamin A (% US RDA)	Vitamin C (mg)	Vitamin C (% US RDA)	Iron (mg)	Iron (% US RDA)	Calcium (mg)	Calcium (% US RDA)
McDonald's (cont'd)																					
Hot Fudge Sundae	6	169	240	7.3	50.5	*1*	3.2	0.1	0.8	2.4	6	170	*323*	214	4	0.0	0	0.5	2	235	25
Hot Caramel Sundae	6	174	270	6.6	59.3	*0*	2.8	0.1	1.2	1.5	13	180	*297*	291	6	0.0	0	0.1	0	222	20
McDonaldland Cookies	2	56	290	4.2	47.1	*1*	9.2	0.5	6.8	1.8	0	300	*181*	0	0	0.0	0	2.1	10	9	0
Chocolaty Chip Cookies	2	56	330	4.2	41.9	*0*	15.6	0.4	10.2	5.0	4	280	*43*	0	0	0.0	0	2.2	10	24	2
Restaurant: Pizza Hut®																					
Pan Pizza, 2 slices																					
Cheese	7	205	492	30.0	57.0	5	18.0	*0.7*	*4.2*	*9.3*	34	940	320	450	9	7.2	12	5.4	30	630	63
Pepperoni	8	211	540	29.0	62.0	5	22.0	*1.5*	*6.8*	*10.9*	42	1127	405	500	10	8.4	14	6.3	35	520	52
Supreme	9	255	589	32.0	53.0	7	30.0	*1.6*	*7.3*	*11.6*	48	1363	580	600	12	9.6	16	5.0	28	500	50
Super Supreme	9	257	563	33.0	53.0	6	26.0	*1.5*	*7.0*	*11.1*	55	1447	532	600	12	10.8	18	6.7	37	540	54
Thin 'n Crispy Pizza, 2 slices																					
Cheese	5	148	398	28.0	37.0	4	17.0	*0.7*	*4.6*	*10.2*	33	867	261	350	7	4.8	8	3.2	18	660	66
Pepperoni	5	146	413	26.0	20.0	4	20.0	*1.5*	*7.3*	*8.9*	46	986	287	350	7	6.0	10	3.2	18	450	45
Supreme	7	200	459	28.0	41.0	5	22.0	*1.2*	*5.7*	*9.0*	42	1328	544	500	10	9.6	16	5.9	33	430	43
Super Supreme	7	203	463	29.0	44.0	5	21.0	*1.2*	*5.8*	*9.3*	56	1336	463	500	10	8.4	14	4.9	27	460	46
Hand-Tossed Pizza, 2 slices																					
Cheese	8	220	518	34.0	55.0	7	20.0	*0.8*	*4.5*	*9.9*	55	1276	396	500	10	9.6	16	5.4	30	750	75
Pepperoni	7	197	500	28.0	50.0	6	23.0	*1.4*	*6.4*	*10.2*	50	1267	415	500	10	7.2	12	5.0	28	440	44
Supreme	8	239	540	32.0	50.0	7	26.0	*1.5*	*7.0*	*11.1*	55	1470	578	550	11	12.0	20	8.0	45	480	48
Super Supreme	9	243	556	33.0	54.0	7	25.0	*1.5*	*7.1*	*11.4*	54	1648	516	550	11	12.0	20	6.8	38	440	44
Personal Pan Pizza, 1 pizza																					
Pepperoni	9	256	675	37.0	76.0	8	29.0	*1.8*	*8.6*	*13.7*	53	1335	408	600	12	10.2	17	5.8	32	730	73
Supreme	9	264	647	33.0	76.0	9	28.0	*1.8*	*8.3*	*13.2*	49	1313	487	600	12	10.8	18	6.7	37	520	52
Restaurant: Taco Bell®																					
Bean Burrito/Red Sauce	7	191	356	13.1	54.4	*5*	10.2	2.0	4.0	2.9	9	888	428	353	7	53.0	87	3.5	19	147	14
Beef Burrito/Red Sauce	7	191	403	22.5	39.1	*3*	17.3	2.0	8.6	7.4	57	1051	313	504	10	2.0	2	3.7	20	114	11
Burrito Supreme/Red Sauce	9	241	413	18.0	46.6	*4*	17.6	2.0	7.2	7.7	33	921	434	876	17	26.0	42	3.6	19	153	15
Double Beef Burrito Supreme/Red Sauce	9	255	457	23.7	41.7		21.8	2.1		10.1	57	1053	431	952	19	9.0	14	4.0	21	145	14
Tostada/Red Sauce	6	156	243	9.5	26.6	*6*	11.1	0.8	4.4	4.1	16	596	401	668	13	45.0	75	1.5	8	179	17
Enchirito®/Red Sauce	8	213	382	19.8	30.9	*4*	19.7	1.5	6.9	9.3	54	1243	423	965	19	28.0	46	2.8	15	269	26
Cinnamon Crispas	2	47	259	2.7	27.5	*2*	15.3	0.8	7.4	3.7	1	127	36	3	0	1.0	0	1.3	6	37	3
Pintos & Cheese/Red Sauce	5	128	191	9.0	19.0	*4*	8.7	0.8	3.8	3.6	16	642	384	441	8	52.0	86	1.4	7	156	15
Nachos	4	106	346	7.5	37.5	*4*	18.5	1.5	8.1	5.7	9	399	159	564	11	2.0	3	0.9	5	191	19
Nachos Bellgrande®	10	287	649	21.6	60.6	*7*	35.3	2.6	14.3	12.3	36	997	674	1137	22	58.0	96	3.5	19	297	29
Taco	28	778	183	10.3	11.0	*1*	10.8	0.8	4.0	4.6	32	276	159	327	6	1.0	0	1.1	5	84	8
Taco Bellgrande®	6	163	355	18.3	17.7	*2*	23.1	1.3	8.2	11.0	56	472	334	845	16	5.0	9	1.9	10	182	18
Taco Light	6	170	410	19.0	18.1	*2*	28.8	5.4	11.2	11.6	56	594	316	662	13	5.0	7	2.4	13	155	15
Soft Taco	3	92	228	11.8	17.9	*1*	11.9	5.4	4.6	5.4	32	516	178	213	4	1.0	2	2.3	12	116	11
Soft Taco Supreme	4	124	275	12.6	19.1	*1*	16.3	1.2	7.1	8.1	32	516	225	440	8	3.0	4	2.3	12	142	14
Taco Salad/Salsa	21	595	941	36.0	63.1	*10*	61.3	12.1	26.6	18.7	80	1662	1212	2958	59	77.0	128	7.1	39	398	39
Taco Salad/Salsa Without Shell	19	530	520	30.6	30.0	*6*	31.4	1.7	*9.8*	14.4	80	1431	1151	3024	60	76.0	126	5.1	28	367	36
Taco Salad Without Shell	19	530	520	29.5	26.3	*6*	31.3	1.7	*9.8*	14.4	80	1056	988	1906	38	74.0	123	4.5	25	331	32
Mexican Pizza	8	223	575	21.3	39.7	*6*	36.8	9.7	16.7	11.4	52	1031	408	984	19	31.0	51	3.7	20	257	25
Taco Sauce	< 1	< 1	2	0.1	0.4	*0*	0.0	0.0	0.0	0.0	0	126	13	185	3	0.0	0	0.0	0	2	0
Hot Taco Sauce	< 1	< 1	3	0.1	0.3	*0*	0.1	0.0	*0.0*	0.0	0	82	14	147	2	1.0	0	0.1	0	2	0
Salsa	3	10	18	1.1	3.6	*1*	0.1	0.0	*0.0*	0.0	0	376	376	1120	22	2.0	3	0.6	3	36	3
Ranch Dressing	4	74	236	1.7	1.5	*0*	24.9	13.8	*6.6*	4.6	36	571	44	240	4	0.0	0	0.5	2	29	2
Jalapeno Peppers	5	100	20	1.0	4.0	*1*	0.2	0.0	*0.0*	0.1	0	1370	110	250	5	2.0	3	0.3	0	40	4
Steak Fajita	5	135	234	14.6	19.5	*1*	10.9	1.1	*5.1*	4.8	14	485	207	560	11	3.0	4	3.0	16	118	11
Chicken Fajita	1	135	226	13.6	19.8	*1*	10.2	2.1	*5.2*	3.7	44	619	201	721	14	4.0	6	2.1	11	123	12
Sour Cream	1	21	46	0.6	0.9	*0*	4.4	*0.2*	*1.3*	2.7	*16*	*10*	31	168	3	0.0	0	0.0	0	25	2
Pico De Gallo	1	28	8	0.3	1.1	*0*	0.2	0.1	*0.0*	0.0	1	88	0	546	10	2.0	2	0.2	0	6	0
Guacamole	4	21	34	0.4	3.0	*1*	2.3	0.4	*1.6*	0.4	0	113	110	116	2	3.0	5		0	10	0
Meximelt®	4	106	266	12.9	18.7	*1*	15.4	1.5	*5.6*	7.9	38	689	114	811	16	2.0	2	2.0	10	247	24

Restaurant: Wendy's Old Fashioned Hamburgers[*]																				
Junior Hamburger	3	104	260	15.0	32.0	1	9.0	1.1	3.7	3.3	35	570	205	100	2	1.2	2	2.7	20	100	10
Junior Cheeseburger	3	116	300	18.0	33.0	1	13.0	1.3	5.1	6.4	35	770	205	100	2	1.2	2	2.7	20	100	10
Small Hamburger	4	111	260	15.0	33.0	1	9.0	1.2	3.7	3.4	34	570	215	100	2	2.4	4	2.7	20	100	10
Small Cheeseburger	4	125	310	18.0	33.0	1	13.0	1.4	5.3	7.2	34	770	215	100	2	2.4	4	2.7	20	100	10
Chicken Sandwich	8	219	430	26.0	41.0	2	19.0	7.2	7.3	4.6	60	725	390	500	2	500.0	8	14.4	80	100	10
Big Classic/Cheese	10	295	640	30.0	46.0	4	38.0	7.0	12.6	12.5	100	1370	590	1000	20	15.0	25	5.4	30	150	15
Plain Single	4	126	340	24.0	30.0	1	15.0	1.5	7.0	6.4	65	500	275	0	0	0.0	0	4.5	25	100	10
Single/Everything	8	210	420	25.0	35.0	1	21.0	4.3	7.3	5.6	70	890	430	750	15	15.0	25	4.5	30	100	10
Plain Single/Cheese	5	137	410	25.0	29.0	1	22.0	1.7	8.9	10.6	80	710	265					4.5	25	100	10
Garden Salad (Take-Out)	10	227	102	7.0	9.0	4	5.0	0.4	1.1	2.3	0	110	560	5500	110	66.0	110	1.8	10	200	20
Chef Salad (Take-Out)	12	331	180	15.0	10.0	4	9.0	1.0	3.1	5.2	120	140	590	5500	110	66.0	110	2.7	15	250	25
New Chili	9	256	220	21.0	23.0	7	7.0	0.6	3.2	2.9	45	750	495	1000	15	9.0	15	4.5	35	60	8
Taco Salad	28	791	660	40.0	46.0	9	37.0	2.4	13.9	15.9	35	1110	1330	4000	80	48.0	80	6.3	35	800	80
French Fries, Small	3	91	240	3.0	33.0	2	12.0	0.8	8.0	4.6	15	145	510	0	0	0.0	10	0.7	4	0	0
Salad Bar Items																					
Creamy Peppercorn Dressing	1	15	80	0.0	0.0	0	8.0	4.3	1.9	1.5	100	135	5	1250	25	1.2	2	0.0	0	100	10
Hidden Valley Ranch Dressing	1	15	50	0.0	0.0	0	6.0	2.9	1.6	1.0	5	115	15	0	0	0.0	0	0.0	0	0	0
Thousand Island Dressing	1	15	70	0.0	2.0	0	7.0	3.1	1.3	0.9	5	105	15	0	0	0.0	0	0.0	0	0	0
French Dressing	1	15	60	0.0	4.0	0	6.0	3.4	1.3	1.5	0	190	25	100	2	0.0	0	0.0	0	0	0
Sweet Red French Dressing	1	15	70	0.0	5.0	0	6.0	2.7	1.0	1.2	0	130	20	0	0	0.0	0	0.0	0	0	0
Italian Caesar Dressing	1	15	80	0.0	0.0	0	8.0	4.1	1.7	1.0	0	125	5	0	0	0.0	0	0.0	0	0	0
Blue Cheese Dressing	1	15	80	0.0	0.0	0	9.0	4.3	1.9	1.5	10	90	5	0	0	0.0	0	0.0	0	0	0
Celery Seed Dressing	1	15	70	0.0	3.0	0	6.0	3.1	1.3	0.9	5	65	10	0	0	0.0	0	0.0	0	0	0
Golden Italian Dressing	1	15	45	0.0	3.0	0	4.0	3.4	1.3	1.5	0	260	10	0	0	0.0	0	0.0	0	0	0
Reduced Calorie Italian Dressing	1	15	25	0.0	2.0	0	2.0	0.6	0.2	0.1	0	180	10	0	0	0.0	0	0.0	0	0	0
Reduced Calorie Bacon & Tomato Dressing	1	15	45	0.0	3.0	0	4.0	0.9	1.1	0.4	0	190	15	0	0	0.0	0	0.0	0	0	0
Alfredo Sauce	2	56	50	2.0	7.0	0	2.0	0.1	0.6	1.3	0	300	80	0	0	0.0	0	0.0	0	60	6
Cheese Sauce	2	56	50	1.0	8.0	0	2.0	0.1	0.7	1.4	0	420	260	0	0	0.0	0	0.0	0	60	6
Spaghetti Sauce	2	56	30	1.0	7.0	1	0.0	0.1	0.0	0.0	0	345	95	0	0	0.0	0	0.0	0	0	0
Deluxe Three Bean Salad	2	57	60	1.0	13.0	2	0.1	0.1	0.0	0.0	0	15	55	200	4	0.0	0	0.4	2	0	0
California Coleslaw	2	57	60	0.0	9.0	2	6.0	1.5	1.7	0.6	10	140	95	750	15	15.0	25	0.0	0	20	2
Red Bliss Potato Salad	2	57	110	0.0	6.0	1	9.0	4.8	2.3	1.3	10	265	150	0	0	6.0	10	0.4	2	0	0
Old Fashioned Corn Relish	2	57	35	0.0	9.0	0	0.0	0.1	0.0	0.1	0	205	30	200	4	0.0	0	0.0	0	0	0
Pasta Deli Salad	2	57	35	2.0	6.0	1	0.0	0.9	0.5	0.3	0	120	88	0	0	0.0	0	0.4	2	0	0
Sliced Peppperoni	1	28	140	5.0	2.0	0	12.0	1.2	6.0	4.6	30	450	95	0	0	0.0	0	0.4	2	0	0
Cheddar Chips	1	28	160	3.0	12.0	1	11.0	2.7	4.0	3.3	11	445	34	0	0	0.0	0	0.4	2	60	6
Crushed Red Peppers	1	28	120	5.0	15.0	7	4.0	2.4	0.8	1.0	0	5	571	10,000	200	9.0	15	2.7	15	20	2
Imitation Parmesan Cheese	1	28	80	9.0	4.0	0	3.0	0.1	1.9	4.1	trace	410	95	1000	20	0.0	0	0.0	0	500	50
Shredded Imitation Cheese (Salad Bar)	1	28	90	6.0		0	6.0	2.3	2.7	1.0	2	120	135	300	6	0.0	0	0.0	0	200	20
Potato Chili Cheese	1	28	100	6.0	1.0	0	8.0	0.3	2.5	5.6	27	130	50	300	6	0.0	0	0.0	0	200	20
Picante Sauce	2	56	18	< 1.0	4.0	0	< 1.0	0.0	0.0	0.0	0	5	155	500	10	18.0	30	0.4	2	0	0
Imitation Sour Topping	1	28	45	< 1.0	2.0	0	5.0	0.0	0.2	5.0	0	30	40	0	0	0.0	0	0.0	0	20	2
Country Crock® Spread	1	14	70	0.0	0.0	0	7.0	3.0	3.0	1.3	0	100	trace	0	0	0.0	0	0.0	0	1	0
Taco Chips	1	40	260	4.0	40.0	3	10.0	3.8	4.1	1.5	38	20	79	0	0	0.0	0	0.7	4	80	8
Taco Shells	< 1	11	45	< 1.0	8.0	1	2.0	0.6	0.5	1.0	0	trace	28	0	0	0.0	0	0.0	0	20	2
Flour Tortilla	1	37	110	3.0	19.0	1	3.0	0.3	1.3	1.1	3	220	37	0	0	0.0	0	0.4	2	80	8
Chocolate Pudding	2	57	90	0.0	12.0	1	4.0	1.1	1.3	0.5	trace	70	99	0	0	0.0	0	0.4	2	150	15
Butterscotch Pudding	2	57	90	1.0	11.0	0	4.0	1.1	1.3	0.5	trace	85	54	0	0	0.0	0	0.4	2	60	6
Restaurant: White Castle[*]																					
Hamburger	2	58	161	5.9	15.4	2	7.9	0.7	3.0	2.8	18	266	86	0		0.0		1.1		44	
Cheeseburger	2	65	200	7.8	15.5	3	11.2	0.8	4.0	4.9	28	361	103	130		0.0		1.1		98	
Fish Sandwich Without Tartar Sauce	2	59	155	5.8	20.9	1	5.0	1.2	2.2	1.2	8	201	111	9		0.2		1.0		49	
Sausage/Egg Sandwich	3	96	322	12.6	16.1	3	22.0	4.3	9.4	6.4	151	698	193	543		0.7		1.5		69	
Sausage Sandwich	2	49	196	6.7	13.3	2	12.3	2.2	5.2	3.7	22	488	117	155		0.5		0.8		44	
Chicken Sandwich	2	64	186	8.0	20.5	2	7.5	2.1	2.9	1.6	19	497	87	17		0.0		1.2		47	
French Fries	3	97	301	2.5	37.7	5	14.7	5.6	6.3	2.3	0	193	597	0		13.5		0.6		15	
Onion Rings	2	63	246	2.9	26.6	3	13.4	5.3	5.9	2.1	0	566	112	0		2.6		1.4		16	
Onion Chips	3	93	329	3.7	38.8	4	16.6	5.6	6.3	2.3	0	823	597	0		13.5		0.6		15	

[*]*Dietetic Currents*, Vol. 18, No. 4, 1991, Ross Laboratories, Columbus, Ohio 43216, Division of Abbott Laboratories, USA, with the permission of Ross Laboratories.

Through the years, convenience and fast-food menus have reflected current health concerns as they worked to meet the demand of consumers and public interest groups. In the 1960s, Americans were urged to eat less sugar. Sugar substitutes were soon produced and made available in the food industry. Nearly all restuarants now offer diet beverages and sugar substitutes. In the 1970s, Americans were advised to increase the amount of fiber in their diets to reduce the risk of developing certain types of cancer. Salad bars and take-out salads quickly appeared on the menus.

Today, the dietary "culprit" is fat. An estimated 35 million Americans are considered to be obese. Obesity is connected with a higher incidence of diabetes, hypertension, osteoarthritis, gallbladder disease, and some types of cancer. Additionally, high blood lipid levels are a major risk factor for heart disease. So, currently convenience and fast-food menus reflect consumer concern about the amount and type of dietary fat they are eating. For example, McDonald's "McLean Deluxe" is a low-fat ground beef patty that is 91% fat free, and contains only 310 calories and 10 grams of fat.

SUMMARY. Convenience and fast foods are now used by a substantial number of Americans. These items may make significant nutritional contributions to the diet when the purchaser knows enough about their nutrient composition to make wise selections. Furthermore, many products are economical in terms of their overall cost, which includes the purchase price, and the costs of fuel for heating (or electricity for cold storage) and labor for preparing.

Today, some convenience and fast-food chains publish pamphlets listing the nutrient content of their menu items. In the future, more and more of them will include values for saturated fat, cholesterol, and fiber.

(Also see FAST FOODS.)

CONVULSION

A violent involuntary contraction or series of contractions of the voluntary muscles.

COPPER

Contents *Page*

History 476
Absorption, Metabolism, Excretion 477
Functions of Copper 477
Deficiency Symptoms 477
Interrelationships 477
Recommended Daily Allowance of Copper 477
Toxicity 477
Sources of Copper 478

Fig. C-130. Copper is required for normal pigmentation of hair. *Top:* Rabbit reared on a diet adequate in copper. *Bottom:* Rabbit reared on a copper-deficient diet, showing graying of hair. (Courtesy, Sedgwick E. Smith, Cornell University)

HISTORY. Copper was discovered and first used by neolithic man during the late Stone Age. The exact date of this discovery will probably never be known, but it is believed to have been about 8000 B.C. The late Bronze Age (3000 to 1000 B.C.) takes its name from the use during this period of bronze, an alloy of copper and tin.

In ancient time, the chief source of copper for the people near the Mediterranean Sea was the island of Cyprus. As a result, the metal became known as *Cyprian metal.* Both the word copper and the chemical symbol for the element, *Cu,* come from *cuprum,* the Roman name for Cyprian metal.

As a result of a series of studies beginning in 1925 (and reported in 1928), Hart and associates at the University of Wisconsin discovered that a small amount of copper is necessary, along with iron, for hemoglobin formation.[46] Then, in 1931, Josephs, found that copper was more effective than iron alone in overcoming the anemia of milk-fed infants. Today, copper is considered as an essential nutrient for all vertebrates and some lower animal species.

ABSORPTION, METABOLISM, EXCRETION. The site of copper absorption depends on the specie of animal; it's primarily in the stomach and the upper part of the small intestine in man. Of the intake, about 30% is absorbed.

After absorption in the intestine, copper reaches the bloodstream where most of it (80% or more) becomes bound to ceruloplasmin, a protein (globulin-copper) complex. The rest is loosely bound to albumen and transported to various tissues.

FUNCTIONS OF COPPER. The role of copper in hemoglobin formation is generally recognized as that of facilitating the absorption of iron from the intestinal tract and releasing it from storage in the liver and the reticuloendothelial system. Although it is not a part of the hemoglobin as such, copper is essential for the formation of hemoglobin. Also, copper is a constituent of several enzyme systems required for normal energy metabolism and its many ramifications; is essential for the development and maintenance of skeletal structures (bones, tendons, and connective tissue); is necessary for the development of the aorta and vascular system; and is required for the formation and functioning of the brain cells and the spinal cord. Additionally, copper is required for normal pigmentation of the hair. Also, a number of important copper containing proteins and enzymes have been identified; some which are essential for the proper utilization of iron.

DEFICIENCY SYMPTOMS. Dietary copper deficiency is not known to occur in adults under normal circumstance, but it has been diagnosed in Peru in malnourished children; and in the United States in premature infants fed exclusively on modified cow's milk and in infants breast fed for an extended period of time. Fortunately, the liver of newborn babies contains 5 to 10 times as much copper as the liver of adults—a reserve which is drawn upon during the first year of life.

Copper deficiency leads to a variety of abnormalities, including anemia, skeletal defects, demyelination and degeneration of the nervous system, defects in the pigmentation and structure of hair, reproductive failure, and pronounced cardiovascular lesions.

Recent animal experiments have shown that a mild copper deficiency can result in elevated serum cholesterol levels, particularly in the presence of high zinc intakes, and epidemiological studies have postulated a positive correlation between the zinc-to-copper ratio in the diet and the incidence of cardiovascular disease.[47, 48]

[46]Hart, E. B., *et al.*, Iron in Nutrition, VII, Copper as a supplement to iron for hemoglobin building in the rat, *Journal of Biological Chemistry,* Vol. 77, 1928, p. 792.

INTERRELATIONSHIPS. Copper, along with certain vitamins, is involved in iron metabolism.

Dietary excesses of cadmium, calcium, iron, lead, molybdenum plus sulfur, silver, and zinc reduce the utilization of copper.

RECOMMENDED DAILY ALLOWANCE OF COPPER. The estimated copper requirement of man is based on balance studies. On the basis of such studies, and in order to allow for a margin of safety, the National Academy of Sciences, National Research Council, recommends a daily copper intake of 1.5 to 3 mg for adults. The requirement for infants and children has been estimated at between 0.4 and 2.0 mg per day. It is emphasized, however, that intake of copper at this level may be too low for the premature infant, who is always born with low copper reserves. It is suggested that infant bottle formulas contain sufficient copper to furnish 100 mcg/kg of body weight per day.

Estimated safe and adequate intakes of copper are given in Table C-41.

TABLE C–41
ESTIMATED SAFE AND ADEQUATE DAILY DIETARY INTAKES OF COPPER[1]

Group	Age	Copper
	(years)	(mg)
Infants	0–0.5	0.4–0.6
	0.5–1.0	0.6–0.7
Children and adolescents	1–3	0.7–1.0
	4–6	1.0–1.5
	7–10	1.0–2.0
	11+	1.5–2.5
Adults		1.5–3.0

[1]*Recommended Dietary Allowances*, 10th ed., 1989, NRC–National Academy of Sciences, p. 284.

• **Copper intake in average U.S. diet**—Ordinary varied U.S. diets supply from 2.5 to 5 mg per day.

TOXICITY. Copper is relatively nontoxic to monogastric species, incuding man. A FAO/WHO Expert

[47]Murphy, L., E. O'Flaherty, and H. G. Petering, Effect of Low Levels of Copper and Zinc on Lipid Metabolism, *Report on Center of Environmental Health,* University of Cincinnati, Cincinnati, Ohio, 1973, p. 35.

[48]Klevay, L. M., Hypercholesterolemia in Rats Produced by an Increase of the Ratio of Zinc to Copper Ingested, *American Journal of Clinical Nutrition,* 1973, Vol. 26, pp. 1060-1068.

Committee has stated that no deleterious effects in man may be expected from a copper intake of 0.5 mg/kg of body weight per day. Usual diets in the United States rarely supply more than 5 mg/day. In order to provide for a margin of safety, the National Academy of Sciences, National Research Council, recommended copper intake for adults is in the range of 1½ to 3 mg/day. Daily intakes of more than 20 to 30 over extended periods would be expected to be unsafe.

Most cases of copper poisoning, which are rare, result from drinking water or beverages that have been stored in copper tanks and/or that pass through copper pipes.

Wilson's disease, a genetically determined inborn error of metabolism, is characterized by a marked reduction in blood ceruloplasmin and greatly increased deposits of copper in the liver, brain, and other organs. The excess copper leads to hepatitis, renal malfunction, and neurologic disorders. The condition can be reversed by controlling dietary copper intake and by administering chelating agents to bind the free copper and excrete it in the urine.

(Also see POISONS, Table P-11 Some Potentially Poisonous [Toxic] Substances.)

SOURCES OF COPPER. Copper is widely distributed in foods. The contribution of drinking water to copper intake varies with the type of piping and the hardness of water. Milk is a poor source of copper. Cow's milk ranges from 0.015 to 0.18 mg/liter; and human milk ranges from 1.05 mg/liter at the beginning of lactation to 0.15 mg/liter at the end.

Groupings by rank of common food sources of copper follow:

• **Rich sources**—Black pepper, blackstrap molasses, Brazil nuts, cocoa, liver, oysters (raw).

• **Good sources**—Lobster, nuts and seeds, olives (green), soybean flour, wheat bran, wheat germ (toasted).

• **Fair sources**—Avocado, banana, beans, beef, breads and cereals, butter, Cheddar cheese, coconut, dried fruits, eggs, fish, granola, green vegetables (turnip greens, collards, spinach), lamb, peanut butter, pork, poultry, turnip, yams.

• **Negligible sources**—Fats and oils, milk, and most milk products (ice cream, cottage cheese), other fruits and vegetables, sugar.

• **Supplemental sources**—Alfalfa leaf meal, brewers' yeast, copper carbonate, copper sulfate. (*NOTE WELL:* Copper carbonate or copper sulfate should only be taken on the advice of a physician or nutritionist.)

For additional sources and more precise values of copper, see (1) Table C-42, Some Food Sources of Copper, which follows, and (2) Food Composition Table F-36.

TABLE C-42
SOME FOOD SOURCES OF COPPER[1]

Food	Copper
	(mg/100 g)
Oysters, raw meat only, Pacific and Eastern	13.7
Liver, beef or calf, fried	12.0
Liver, lamb, broiled	10.0
Molasses, blackstrap	6.0
Cocoa powder, low fat	4.3
Pepper, black	4.3
Brewers' yeast	3.3
Liver, hog, fried in margarine	2.5
Brazil nuts, shelled	2.4
Lobster	1.8
Sunflower kernels, dry hulled	1.8
Soybean flour, defatted	1.7
Olives, green	1.6
Walnuts, Persian or English	1.4
Almonds, dried, shelled	1.4
Wheat germ, toasted	1.3
Wheat bran	1.3
Pecans	1.1
Alfalfa leaf meal (powder) dehydrated	1.0
Granola	.9
Wheat germ	.7
Lamb, leg, loin or rib cuts, lean, roasted or broiled	.7
Coconut, dry	.7
Peanut butter	.6
Parsley, dried	.6
Shredded wheat	.6
Apricots, dried, sulfured, uncooked	.5
Beans, common, dry, red, cooked	.5
Butter	.4
Turnip greens, raw	.4
Avocado, raw, all varieties	.4
Pork	.3
Collards	.3
Beef	.2
Turnip	.2
Bread, whole wheat	.2
Spinach, raw	.2
Yam cooked in skin	.2
Bananas, raw	.2
Turkey, roasted	.2
Cheese, natural, Cheddar	.2
Salmon, broiled or baked	.2
Eggs	.1

[1]This listing is based on the data in Food Composition Table F-36. Some food sources may have been overlooked since many of the foods in Table F-36 lack values for copper.

(Also see MINERAL[S], Table M-67; and POISONS, Table P-11 Some Potentially Poisonous [Toxic] Substances.)

COPPER (Cu) TOXICITY

Copper is a required trace mineral, necessary in iron metabolism and the production of red blood cells, in certain enzyme systems, for hair growth and pigmentation, for the development of bones and other tissues, and for reproduction and lactation. Although the incidence of copper toxicity is rare, it may occur in diets rich in the mineral, but low in other minerals which counteract its effects.

(Also see MINERAL[S]; COPPER; and POISONS.)

COPRA

This is the dried meat of the coconut from which the oil is to be extracted, yielding coconut oil and coconut meal or copra meal. The oil is generally extracted by either continuous mechanical screw presses or hydraulic presses. One thousand coconuts will yield about 500 lb *(225 kg)* of copra and 25 gal *(95 liter)* of oil.

(Also see COCONUT.)

COPROPHAGY

The ingestion of fecal material. This constitutes normal behavior among many insects, birds and other animals. But in man it is indicative of some form of behavioral disorder.

(Also see PICA.)

CORI CYCLE

The series of reactions which (1) break down muscle glycogen to blood glucose, or (2) convert blood glucose to muscle glycogen.

(Also see CARBOHYDRATE[S]; and GLYCOGEN.)

CORN (MAIZE) *Zea mays*

Contents *Page*

Origin and History ... 479
World and U.S. Production ... 480
The Corn Plant ... 481
Hybrid Corn ... 481
Corn Culture ... 483
Kinds of Corn ... 483
Federal Grades of Corn ... 484
Processing of Corn ... 484
Corn Milling ... 485
Wet Milling ... 485
Dry Milling ... 485
Distilling and Fermentation ... 486
Nutritional Value of Corn ... 486
Children Cannot Live By Corn Alone ... 487
High-Lysine Corn (Opaque 2, or O_2) ... 487
Supplementing Corn ... 488
Corn Products and Uses ... 489

The word "maize" is preferred in international usage because in many countries the term "corn," the name by which the plant is known in the United States, is synonymous with the leading cereal grain; thus, in England "corn" refers to wheat, and in Scotland and Ireland it refers to oats.

(Also see CEREAL GRAINS.)

Fig. C-131. Corn—as high as an elephant's eye and climbing clear up to the sky—has long been the most important U.S. cereal grain. (Photo by J. C. Allen & Son, Inc., West Lafayette, Ind.)

ORIGIN AND HISTORY. Corn is indigenous to America. Fossilized pollens, estimated to be 80,000 years old, have been found in soil profiles near Mexico City; and ears of corn about the size of strawberries, estimated to be 3,000 years old, have been discovered in Mexico. Archeological discoveries almost as ancient as those in Mexico have been found in South America, causing postulation of another locus of domestication.

During the 13th century in southwestern United States and in Mexico, corn or maize meant life itself; and the cultivation of corn was an act of worship. From birth through death, the economic, social, and religious activities of the Hopi Indians of Arizona were bound to the growing of corn. No child could be born with security and survive the first 20 hazardous days of life without corn. For this reason, a special ear of corn was dedicated to each newborn baby as its "corn mother."

No one in Europe knew about corn until Columbus sailed to America in 1492. On November 5, 1492, two Spaniards whom Christopher Columbus had dispatched to explore the interior of Cuba returned with a report of "a sort of grain called maize, which is very well tasted when boiled, roasted, or made into porridge." Later explorers to the new world found corn being grown by the Indians in all parts of America, from Canada to Chile. The Indians grew all the main types of corn that are raised today. They prized corn with colorful kernels—red, blue, pink, and black, or with bands, spots, or stripes. The kernels ranged in size from as small as a grain of wheat to as large as a quarter. Corn played an important part in the religious life of many of the tribes; they held elaborate ceremonies when planting and harvesting it. They often used fish as a fertilizer; various rituals were carried out to ward off evil spirits; and human sentries were stationed to discourage pilfering by birds and animals. Also, they used corn patterns to decorate pottery, sculpture, and other works of art.

Fig. C-132. Indian observing the growth of corn (maize). (After a painting by Frederick Remington. Courtesy, The Bettmann Archive, Inc., New York, N.Y.)

The importance of corn in the early years of the English settlements is taught to every school child—how a Narraganset Indian named Squanto taught the struggling Plymouth Pilgrims how to cultivate corn in the rocky Massachusetts soil; how he insisted that each hill be fertilized with three herring heads, pointing inward like the spokes of a wheel.

The colonists often used corn as money; with it, they paid their rent, taxes, and debts. They even traded corn for marriage licenses. In many settlements, corn kept people from starving in difficult times.

Fig. C-133. Pilgrims discovering corn. (From a contemporary woodcut. Courtesy, The Bettmann Archive, Inc., New York, N.Y.)

Fig. C-134. Corn husking in Vermont. The finder of the "red ear" claims his reward. (Courtesy, The Bettmann Archive, New York, N.Y.)

WORLD AND U.S. PRODUCTION. North America has always been the center for corn production. In 1990, it accounted for 48% of the total world production of 475,429,000 metric tons; the United States alone produced 42% of the world total. The other prominent corn-growing countries are shown in Fig. C-135, although not approaching the United States in quantity, nor usually in yields.

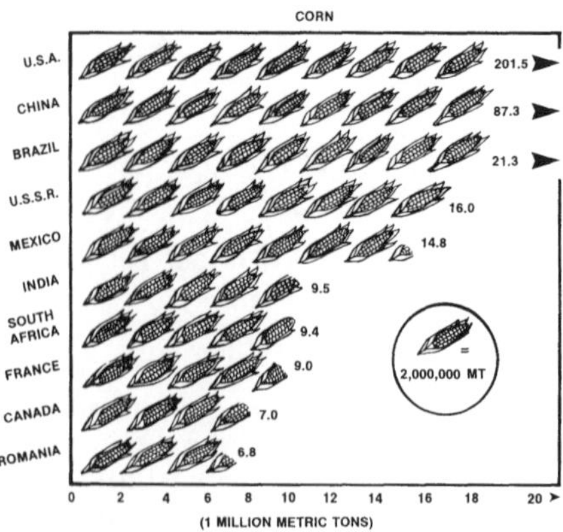

Fig. C-135. Leading corn-growing countries of the world. (*FAO Production Yearbook*, 1990, FAO/UN, Rome, Italy, Vol. 44, p. 79, Table 20)

Although corn is grown throughout the United States, the greatest production is in an area of the Midwest called the Corn Belt consisting of the seven states of Illinois, Indiana, Iowa, Kansas, Missouri, Nebraska, and Ohio. About two-thirds of the corn grown for grain is produced in the Corn Belt, and more than a third of it is produced in Illinois and Iowa. The ten leading corn-growing states of the nation are shown in C-136.

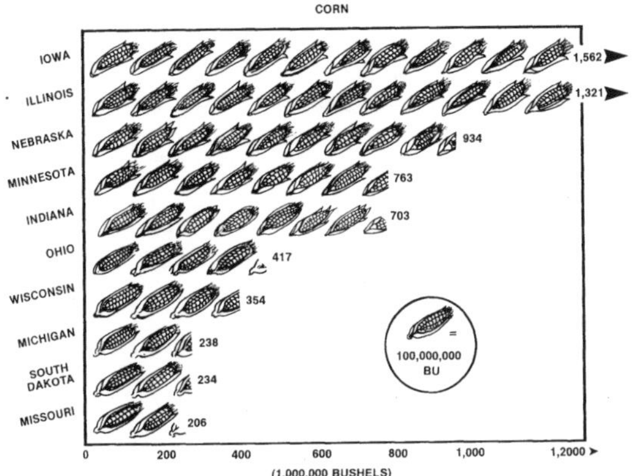

Fig. C-136. Leading corn-growing states of the United States. (One bushel equals 56 lb, or *25 kg*. Source: *Agricultural Statistics*, 1991, USDA, p. 33, Table 41.)

THE CORN PLANT. Corn or maize (Indian corn, *Zea mays*) is a plant of the tribe Maydea of the grass family (*Gramineae*).

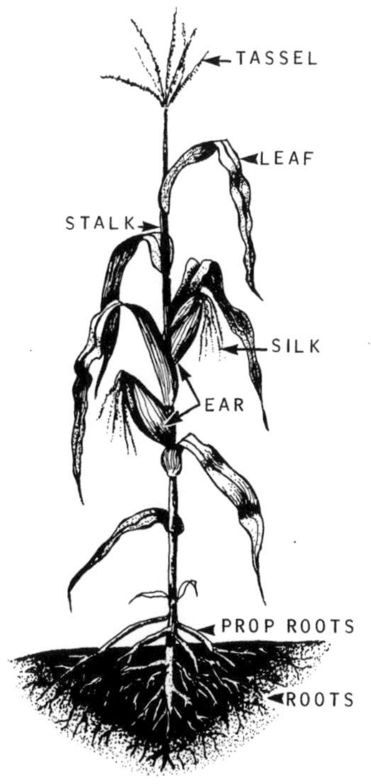

Fig. C-137. Parts of the corn plant.

The corn plant is a tall, annual grass with a fibrous root system; a stout, erect, solid stem; large narrow leaves with waxy margins, spaced alternately on opposite sides of the stem; and male and female flowers borne on separate inflorescenses (see Fig. C-137). Corn belongs to the group of six true grains, or cereals, which includes wheat, barley, oats, rice, and rye.

Unlike other major grain crops, the corn plant has separate male and female flowering parts. The tassel is the male flowering structure of the corn plant. In its anthers, it produces somewhere between 2 and 5 million pollen grains. This means that there are 20,000 to 50,000 pollen grains produced for each silk. The anthers split open and shed their pollen into the wind.

The corn ear, or female flowering structure, is a cylindrical group of female flowers, each capable of producing a kernel if pollinated at the proper time. On a well-developed ear, there are 750 to 1,000 potential kernels—"ovules"—arranged in an even number of rows around the cob. The familiar "silks" which emerge from the end of the ear are the styles attached to the ovaries. There is a single silk for each potential seed; and the number of seeds on the ear may equal but can never exceed the number of silks which have emerged.

The wind shakes pollen out and some of it falls on the sticky threads of corn silk hanging out of the young ears. Upon landing on the silk, the pollen grain sends out a tiny pollen tube that grows down the silk to an egg cell in the young kernel. There, a male cell from the pollen grain fertilizes, or unites with, the egg cell. As the fertilized egg cell grows, the whole kernel becomes the seed. It is noteworthy that, under field conditions, 76 to 94% or more of the kernels produced by each plant are cross-pollinated by other plants in the field.

The corn plant is one of nature's most amazing energy-storing devices. From a seed which weighs little more than one-hundreth of an ounce (*28g*), a plant 7 to 10 ft (*2 to 3 m*) tall develops in about 9 weeks. In another 2 months, this plant produces 600 to 1,000 seeds similar to the one from which it started. For perspective, the corn plant's achievement can be compared with wheat, which produces a 50-fold yield per seed planted.

HYBRID CORN. Beginning about 1920, a new method of corn improvement involving the production of hybrid corn, rather than open-pollinated varieties, was initiated, based on the genetic discoveries of G. H. Shull of the Carnegie Institution and E. M. East and D. F. Jones of the Connecticut Agricultural Experiment Station. But there is a time lag between research findings and application! In 1933, only 0.1% of the U.S. corn acreage was planted to hybrid seed. By 1943, it was 52%; in 1961, it was 96%; and today it is virtually 100%. Hybrid corn has also become important in the other corn-producing countries of the world. Hybrids produce up to one-third more grain than open-pollinated varieties, because they are bred to (1) have greater ability to withstand drought, diseases, and pests; (2) have stronger roots and stalks; and (3) ripen at the same time. Although no hybrid has all of these desirable qualities, each has some of them.

The first step in producing hybrid corn is to develop pure lines of corn. This is accomplished by inbreeding, or fertilizing a plant with pollen from the same plant. To produce inbreds, breeders take pollen from the tassel of a corn plant and place it on the silk of an ear of the same plant. The process of inbreeding is continued for 5 to 10 years, during which time the inbred strains become more and more alike. With each succeeding generation of inbreeding, however, the plants lose vigor. Because of their small yields, inbreds are used only for breeding.

The next step in hybrid-corn production is cross-fertilizing plants from two different inbred lines to produce plants that are better than either parent. This is done by planting the two inbreds in an alternating series of rows, then removing the tassels of one inbred while allowing the tassels on the other inbred to remain. Since each corn plant produces a very large amount of pollen, normally two rows of the inbred that are to be detasseled are alternated with one row of the inbred which is to produce the pollen. Thus, the pollen from one inbred fertilizes the silks of the detasseled inbred and the two inbreds are crossed. The seed produced by the detasseled rows is known as a single-cross hybrid. In addition to single-cross hybrids, three-way hybrids, and double or four-way hybrids are grown for commercial hybrid seed corn production. Three-way hybrids are produced by crossing a single-cross hybrid with an inbred of different breeding than that of the two inbreds used in producing the single-cross hybrid. Double or four-way hybrids are produced by crossing two single hybrids. The procedure used in producing single, three-way, and double or four-way hybrids is illustrated diagrammatically as follows:

Single hybrid:

Inbred A × Inbred B
\ /
Single-cross hybrid AB

Three-way hybrid:

Inbred A × Inbred B
\ /
Single-cross hybrid AB
|
AB × Inbred C
\ /
Three-way hybrid ABC

Double or four-way hybrid:

Inbred A × Inbred B Inbred C × Inbred D
\ / \ /
Single-cross hybrid AB × Single-cross hybrid CD
\ /
Double or four-way hybrid ABCD

The big difference between the three hybrids is that the single cross has two parental inbred lines, the three-way cross has three, and the double cross has four. Also, the hybrid produced when only two lines are crossed is very uniform. This leads to uniformly good performance if growing conditions are favorable for that particular genetic combination. However, if conditions are unfavorable for the single-cross hybrid, the uniformity can be a serious handicap. Attack by disease or insects, drought, high plant populations, and other unfavorable conditions can result in a more uniformly poor growth in a susceptible single-cross than in almost any three-way or double-cross hybrid. As a result of the devasting outbreak of southern corn blight in 1970, corn breeders and producers learned the never-to-be-forgotten hazard of dependence on a single cytoplasm, or on vast acreages of similar hybrids.

Fig. C-138 shows the steps in producing a double or four-way hybrid.

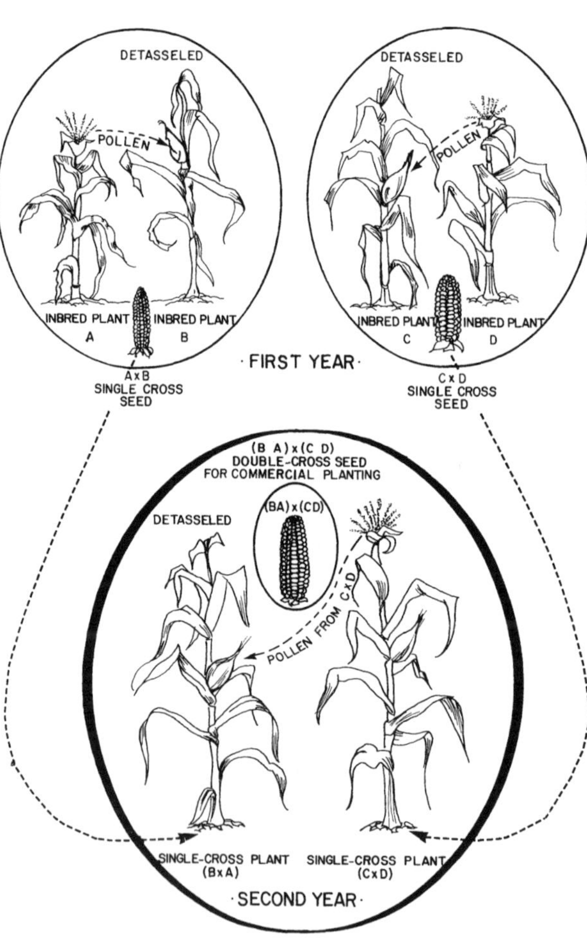

Fig. C-138. Hybrid seed produced by double crossing. Fertilizing plant B with pollen from plant A and plant C with pollen from plant D results in single-cross hybrid seeds, BA and CD. Fertilizing plant BA with pollen from plant CD results in the double-cross seed corn, (BA) X (CD).

In the 1950s, breeders found a way to eliminate detasseling and save considerable labor. They discovered male-sterility factors which cause a plant to be sterile to its own pollen, but which will produce seed with another line that carries the fertility-restoring factor.

The hybrid vigor achieved by crossing inbred lines is

not carried over from year to year. Hence, seed for planting should not be saved from hybrid-corn fields, for both the quality and yield may be reduced 10 to 20% or more. New hybrid corn seed must be produced and used every year. As a result, the production of hybrid seed has become an important commercial industry.

CORN CULTURE. Corn does best in areas with ample moisture, fertile soil, and warm nights. It is grown primarily in the north central states, but it has expanded over a wider area with the development of new hybrids.

The growing of corn requires a high level of technology. It follows that production costs are high. These costs are offset by high yields, but this potential is reached only through adequate inputs of all the key factors, including: selecting hybrids adapted to the area; using adequate amounts of fertilizer of the right formulation as determined by soil tests; planting enough seed to establish a good, uniform stand; applying effective weed control to deal with the species present; and using specialized machinery to carry out the necessary operations from planting to harvesting and storing effectively, on time, and with a minimum of labor.

KINDS OF CORN.

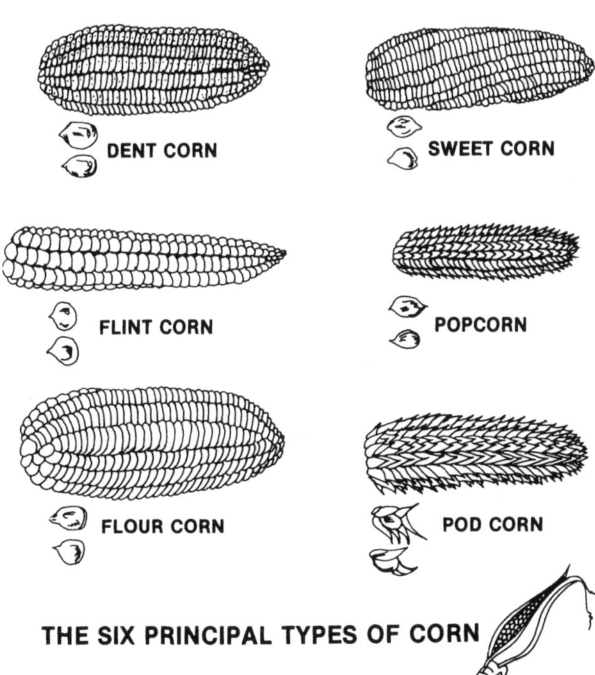

Fig. C-139. The six principal types (races) of corn.

The many kinds or varieties of corn can be grouped into six different categories based on types of kernels: (1) dent corn, (2) sweet corn, (3) flint corn, (4) popcorn, (5) flour corn, and (6) pod corn. Breeders can produce hybrids of any of the different kinds of corn, but most of the breeding of hybrid corn has been done with dent corn.

• **Dent corn**—This is the most common variety. It accounts for about 95% of all the corn grown in the United States. Each kernel of dent corn has a small dent on top, caused by the hard and soft starch in the kernel shrinking unequally in drying. Dent corn may be yellow, white, or red. Most of the corn grown commercially is yellow, although white corn is preferred for the manufacture of cereals. Most of the dent corn is fed to livestock, although some of it is sold to manufacturers who make many food and industrial products from it.

• **Sweet corn**—Sweet corn is grown chiefly for human food and is harvested at an immature stage. The kernels are relatively high in sugar at the time they are canned, frozen, or eaten as corn on the cob. Most of the sweet corn is grown for the canning industry, but some is grown by truck gardeners and almost all home gardeners. Practically all varieties of sweet corn are white or yellow. Upon maturity, the kernels are characterized by their wrinkled, caramelized appearance.

Unless it is kept refrigerated after harvesting, sweet corn quickly loses its flavor. This happens because heat turns the sugar in the kernels to starch. Refrigerated trucks make it possible to carry fresh sweet corn far from the places where it is grown. Canned and quick-frozen sweet corn lasts indefinitely and can be shipped throughout the world. The leading states in producing sweet corn for processing are Minnesota, Wisconsin, Oregon, Illinois, and Washington. Florida supplies most of the fresh sweet corn for the winter, spring, and fall markets. Several states produce a substantial amount of fresh sweet corn for the summer markets, with New York and Ohio leading.

Sweet corn is a good source of energy. It contains twice as much sugar as dent corn. Also, it is a good source of vitamins A (if yellow) and C.

• **Flint corn**—The corn that the Indians taught the early settlers to grow was flint corn. The kernels of flint corn are comprised of small amounts of soft starch completely surrounded by a large quantity of hard starch. Because the kernels are very hard, grinding is usually necessary before feeding this type of corn to livestock. Flint corn makes a good quality corn meal. There is less spoilage than with dent corn when flint corn is shipped overseas, because the hard seed absorbs less moisture. Also, it is more resistant to corn weevil damage. Argentina, which exports a large proportion of its corn crop, produces mostly flint corn.

Compared to dent corn, most flint corn varieties mature earlier, and their seeds germinate better in cold, wet soil in the spring. For the latter reason, they are grown farther north than the dent types. Flint corn comes in many colors—white, yellow, red, blue, and variable.

• **Popcorn**—This is a favorite "fun food" of Americans. It is eaten plain with salt and butter, or covered with caramel or white sugar syrup.

Fig. C-140. A bowl of seasoned buttery popcorn (far left), alone or with other goodies, is a favorite snack or fun food of armchair sports fans. (Courtesy, United Dairy Industry Assn., Rosemont, Ill.)

The kernels of popcorn are smaller than flint corn. When heated rapidly, the moisture inside the kernels turns to steam and builds up a great pressure. This pressure bursts the outer shell, and the entire inside of the kernel puffs out into a mass of flaky starch. Upon popping, the volume is 25 to 30 times greater than the original kernel.

Based on the shape of the kernel, all varieties of popcorn are classed as either pearl or rice. Pearl varieties have short, thick kernels which may be round or flat, and a smooth, rounded crown or cap. Rice varieties are either round or flat, relatively long and slender, and have pointed crowns which are sometimes hooked. The kernels are generally white or yellow, although there are red, blue, calico, and brown varieties. Regardless of the color of the kernel, when popped the flake is either yellow or white.

The leading popcorn-producing states are Nebraska, Indiana, Ohio, Iowa, and Kentucky.

- **Flour corn**—Flour corn, or soft corn as it is sometimes called, has soft, starchy kernels. Because the kernels can be easily ground into flour by hand, flour corn is popular with the Indians of southwestern United States. Also, it is grown in the warm areas of South America. White and blue are the most common colors, although numerous other colors are found. Flour corn ripens late in the season.

- **Pod corn**—Each kernel of pod corn is enclosed within a separate pod or husk; and the entire ear is surrounded by a large husk similar to that found on other types of corn. Some scientists believe that pod corn may be the ancestor of all other types of corn. Pod corn is of no commercial importance in the United States; it is grown only as a curiosity and for use in breeding and related studies.

FEDERAL GRADES OF CORN. Those who buy and sell corn for processing into human food should be familiar with the federal grades of corn. Table C-43 gives the standards by which corn is graded.

TABLE C-43
UNITED STATES STANDARDS FOR CORN[1]

Grade	Minimum Test Weight per Bushel		Maximum Limits of				
			Moisture	Broken Grain	Foreign Material	Heat Damaged Kernels	Total Damaged Kernels
	(lb)	(kg)	(%)				
U.S. No. 1	56.0	25.4	14.0	—[2]	—[2]	.1	3.0
U.S. No. 2	54.0	24.5	15.5	—[2]	—[2]	.2	5.0
U.S. No. 3	52.0	23.6	17.5	—[2]	—[2]	.5	7.0
U.S. No. 4	49.0	22.2	20.0	—[2]	—[2]	1.0	10.0
U.S. No. 5	46.0	20.9	23.0	—[2]	—[2]	3.0	15.0
U.S. Sample grade	U.S. Sample grade shall be corn which does not meet the requirements for any of the grades from U.S. No. 1 to U.S. No. 5, inclusive; or which contains stones; or which is musty, or sour, or heating; or which has any commercially objectionable foreign odor; or which is otherwise of distinctly low quality.						

[1]Adapted by the authors from *The Official United States Standards for Grain*, USDA, December 1975.
[2]In the grading of corn, the criteria of broken kernels and foreign material are combined. Therefore, corn of grades U.S. Nos. 1, 2, 3, 4, and 5 shall have no more than 2, 3, 4, 5, and 7% broken corn and foreign material, respectively.

PROCESSING OF CORN. Most of the corn crop goes directly into animal feed uses, much of it right on the farm where it is grown. But each day of the year corn refiners process over a million bushels of corn, producing ingredients vital to innumerable food and industrial products. Of the 94 supermarket items that go into most grocery carts, over a quarter contain one or more ingredients from the corn refining industry. Corn-derived feed ingredients are important to our animal economy. And industrial products ranging from cast metals to automobile tires depend on the use of specially designed corn starches.

Fig. C-141. Ears of corn. (Courtesy, USDA)

Corn Milling. In order to understand corn milling and corn by-products, it is first necessary to know the composition of the different parts of the corn kernel. Each year, about 6% of the U.S. corn crop is milled. Corn is wet milled for the production of starch, sweeteners, and oil; and corn is dry milled for the production of grits, flakes, meal, oil, and feeds.

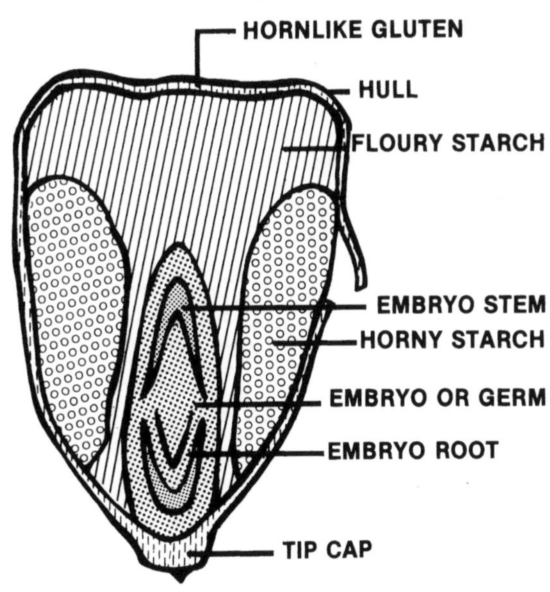

Fig. C-142. Part of kernel of dent corn.

WET MILLING. Before storage, shelled corn (mostly yellow dent) intended for wet milling is cleaned to remove all foreign matter, such as insects, rocks, and trash. When it is taken from storage, it is cleaned again, then soaked in water—swelling the kernels. In the soaking (or steeping) process, nutrients are absorbed by the water (steepwater). When the steeping is complete, this water is drawn off and concentrated.

During the subsequent milling process, the corn germ

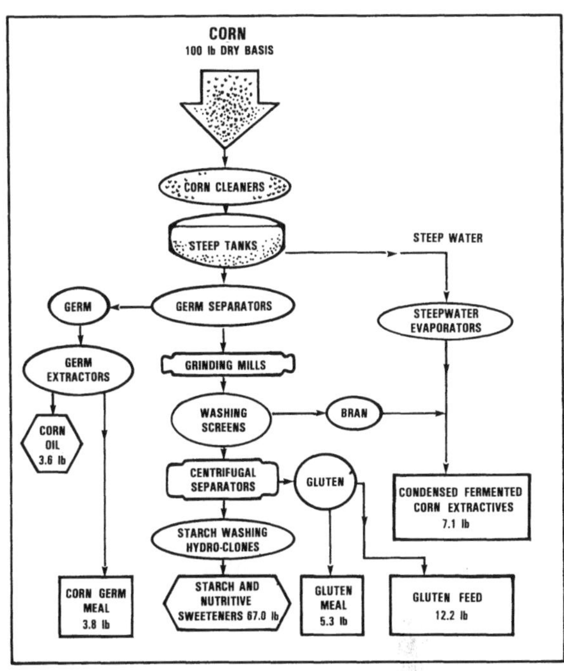

Fig. C-143. Schematic outline of the corn wet milling process.

is separated from the kernel. The germ is further processed to remove oil, and the remaining germ meal is isolated for later use.

After the germ has been removed, the rest of the corn kernel, containing starch, gluten, and bran (the outer hull of the kernel) is screened, and the bran is removed. The remaining mixture of starch and gluten is separated by centrifugal action. The starch portion is either dried, or modified and dried, and sold to the food, paper, and textile industries, or is further converted into various sweeteners.

Thus, the following four major ingredients are isolated during the production of starch, sweeteners, and oil: steepwater, germ, bran, and gluten. From these substances come the refining industry's four major feed ingredients: corn gluten meal, corn gluten feed, corn germ meal, and condensed fermented corn extractives—commonly known as corn steepwater. Thus, while the corn refiners primarily produce corn starch, corn sweeteners (sugar and syrup), and corn oil, 25 to 30% of their output is livestock feed ingredients.

DRY MILLING. The dry milling of corn is generally less involved than wet milling. Two basic processes—degerming and nondegerming—are used extensively.

In the degerming process, the hull, germ, and endosperm are separated before milling. This process is used for the production of grits, flakes, meal, flour, oil, and feeds.

The entire kernel is ground intact in the nondegerming process of dry milling. The resulting product is an oily flour which is subsequently used in baked products. Although some hulls and germs sift out in this process, the quantity of by-products is too insignificant to be considered as a reliable source of animal feed.

Dry millers produce cornmeal by either of two methods; the *old process*, or the *new process*. In the oil process, the germ is left in the meal. This improves the flavor and nutritional value, but it limits the time that the meal stays fresh. Most commercial cornmeal is made by the new process, from hulled corn kernels with the germ removed.

Distilling And Fermentation. The corn distilling and fermentation industries include manufacturers of ethyl and butyl alcohols, acetone, and whiskey.

- **Alcohol**—Malt converts corn starch to sugar, which is fermented by yeast to ethyl alcohol and carbon dioxide. The grain that remains is used as livestock feed.

(Also see BEERS AND BREWING; and DISTILLED LIQUORS.)

The current fuel shortage has sparked much interest in "gasohol," a mixture of 10% anhydrous ethyl alcohol and 90% gasoline; with the ethyl alcohol manufactured from corn (and wheat, sorghum, and other agricultural products). Although the term "gasohol" is new, the idea of using grain alcohol as a motor fuel is as old as the internal combustion engine. In the past, the main deterrent to the use of alcohol motor fuels has been unfavorable economics. But the combined effects of increasing petroleum prices, increasing energy demands, needs for liquid fuels, requirements for pollution abatement, and the need for rural assistance could make corn and other agricultural products a significant factor in the future overall energy situation, including motor fuels.

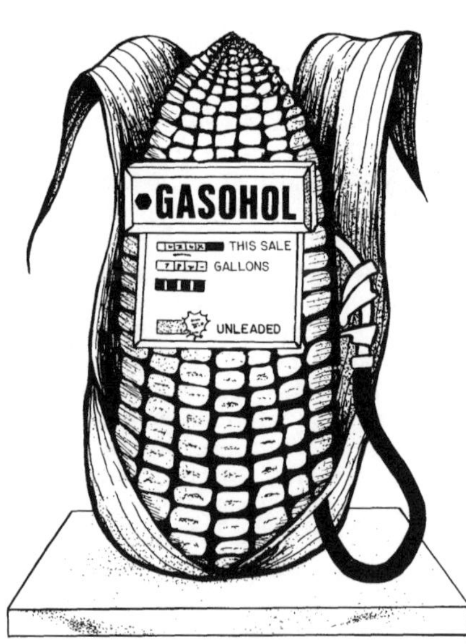

Fig. C-144. Corn power.

A little over 2.5 gal (*9.5 liter*) of anhydrous alcohol can be produced from a bushel of most cereal grains. However, from an energy balance standpoint, more energy is now required at the distillery to process the grain, ferment it, recover the alcohol, and recover the by-products than is contained in the alcohol produced. But coal or agricultural residues, which are not suitable for powering motors, may be used as boiler fuels in such distilleries to produce needed liquid motor fuels.

Alcohol costs depend on many factors, including grain prices, conversion costs, and by-product feed values. Currently, the basic cost of fermentation alcohol is higher than the refinery price of gasoline. However, this difference is narrowing due to the rapid increase in oil prices, and the difference may disappear.

- **Industrial products**—Manufacturers allow bacteria to ferment cornmeal to produce acetone and butyl alcohol, which are used in manufacturing rayon, plastics, and other products.

The by-products of distilling and fermentation of corn include carbon dioxide, corn oil, vitamin and protein concentrates, and grain germ mixtures.

NUTRITIONAL VALUE OF CORN. Corn is palatable, nutritious, and rich in the energy-producing carbohydrates (starch) and fats. It is higher in fat than rice or wheat (4% vs 2%). The caloric values of these three cereal grains are: cornmeal, whole-ground, 1,610 Calories (kcal) per pound (*0.45 kg*); rice, brown, 1,633 Calories (kcal) per pound (*0.45 kg*); and wheat grain, 1,497 Calories (kcal) per pound (*0.45 kg*). But corn has certain limitations! It lacks quantity of protein (8 to 11%) and quality of protein (being especially low in the amino acids, lysine and tryptophan), and it is deficient in minerals, particularly calcium, and in the vitamin niacin. White corn is also low in carotene (the precursor of vitamin A). However, when vitamin A is added, white corn and yellow corn are equal in nutritive value. It is noteworthy that the carotene content of yellow corn decreases with storage; about 25% of its original vitamin A value may be lost after 1 year of storage, and 50% after 2 years.

The nutritive value of different forms of corn and corn products is given in Food Composition Table F-36.

The amino acid composition of corn warrants careful consideration (see Table C-44). In countries where high corn diets prevail, the amino acid profile can be used as a basis to determine what protein supplement to add.

TABLE C-44
PROFILE OF ESSENTIAL AMINO ACIDS IN CORN COMPARED TO MILK—A HIGH-QUALITY PROTEIN[1]

Amino Acid	Whole Corn	Cow's Milk
	← (mg/g protein) →	
Histidine	27	27
Isoleucine	38	47
Leucine	133	95
Lysine	27	78
Methionine plus cystine	41	33
Phenylalanine plus tyrosine	92	102
Threonine	37	44
Tryptophan	9	14
Valine	46	64

[1]*Recommended Dietary Allowances*, 10th ed. 1989, NRC–National Academy of Sciences, p. 67.

Although corn contains about 10% protein, half of the protein consists of zein, which is especially poor in lysine and tryptophan, essential amino acids which people must get from food (see Table C-44).

In 1944, researchers at the Connecticut Agricultural Experiment Station induced starvation in laboratory rats by feeding them generous helpings of corn. Further, it was found that rats could be restored to health by supplementing the high-corn diet with two protein fractions—the amino acids lysine and tryptophan.

Children Cannot Live By Corn Alone. Corn alone will not keep weanling infants alive. They will develop the disease kwashiorkor, primarily due to protein deficiency; and they may develop pellagra, due to a deficiency of the vitamin niacin.

The disease *pellagra*, caused by a deficiency of niacin, is prevalent among peoples who rely on corn for a large proportion of their daily food. Although corn contains niacin, 50 to 80% of it occurs in a bound form as niacytin, which is biologically unavailable. It is noteworthy that pellagra is not suffered by Mexicans who consume cornmeal in the form of tortillas. The explanation: tortillas are made by mixing cornmeal with lime water, creating an alkaline condition which releases the bound niacin and makes it available.

In Colombia, corn grain is pounded with potash and a little water is added, in the preparation of cornmeal. Although the primary purpose of the potash is to loosen the bran, the alkaline reaction is effective in unbinding the bound niacin.

The quantity of protein provided by the diet depends both on the amount of food consumed and on the concentration of protein in the food. Foods with low concentrations of protein, such as corn, may provide useful amounts of protein provided enough of them are eaten. There is, however, a limit to consumption, since the amount of food eaten is determined largely by the requirement for calories, rather than the requirement for protein. Also, the stomach can hold just so much. For example, to meet the energy requirements (1,200 kcal/day) of a 2-year old, 30 lb (*13.6 kg*) child from cornmeal would necessitate the consumption of about 3 ½ lb (*1.6 kg*) per day of a gruel or dough—a physical impossibility. Even if this large amount of corn cereal could be consumed, it would still fall short of meeting the child's protein requirements by about 18%.

Also, corn lacks quality of protein. It is deficient in lysine and typtophan, amino acids that are essential for all single-stomached creatures, such as humans and pigs.

High-Lysine Corn (Opaque-2, or O_2). It has been known for many years that corn, the world's third most important human food after rice and wheat, is nutritionally inadequate.

Although normal corn contains about 10% protein, half of it is locked up in the fraction zein, which is poorly utilized by man. Moreover, normal corn is especially poor in lysine and tryptophan, essential amino acids that the human body cannot manufacture and must get from food.

This deficiency of corn shows up in people wherever corn is a major source—if not the only source—of protein in the diet. Known by the name kwashiorkor, this nutritional deficiency disease is the leading cause of mortality among infants and children in many parts of the world.

For years, plant scientists assayed the world's corn varieties one by one, looking for a strain with more nutritionally balanced protein. Finally, in 1963, a Purdue University team headed by biochemist Edwin T. Mertz analyzed an odd group of corns characterized by soft, floury endosperm inside an opaque, chalk-white kernel. The Purdue scientists found that the opaque characteristic of corn, which had been noted for years without exciting much scientific interest, is associated with a recessive gene that replaces some of the kernel's amino acid deficient zein with other protein higher in the needed lysine and tryptophan. The mutant—routinely labeled opaque-2, or O_2 for short—had a lysine content of 3.4%, compared to 2.0% for normal corn. Additionally, opaque-2 showed higher levels of tryptophan and other amino acids.

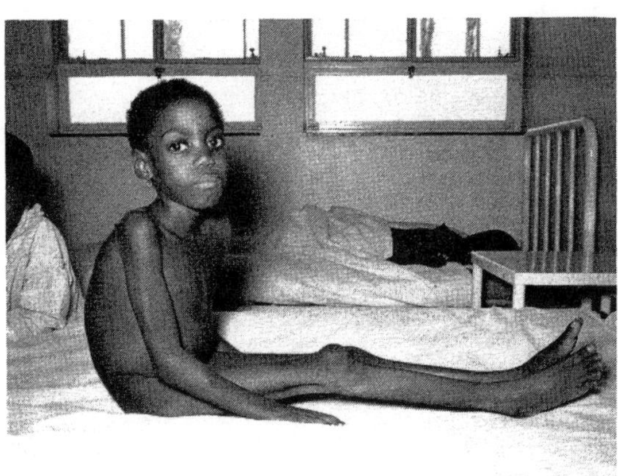

Fig. C-145. Kwashiorkor, due to a protein deficiency. Note enlarged stomach and knees of this 9-year-old girl. (Courtesy, FAO and United Nations Children's Fund)

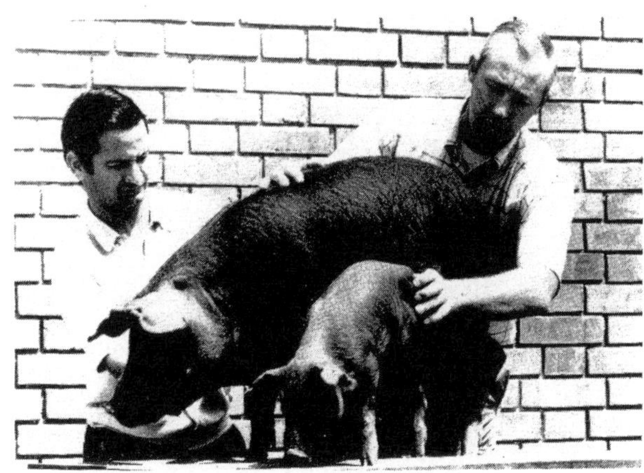

Fig. C-146. Lysine made the difference! These littermate pigs were started on test at weaning, at a weight of 20 lb (*9 kg*). The only difference in their ration was the kind of corn. The big pig (left) received high-lysine corn; the little pig (right) got regular corn. During the 130-day trial, the pig fed opaque-2 gained a respectable 73.2 lb (*33 kg*), whereas the pig eating ordinary corn gained only 6.6 lb (*3 kg*). Lysine is as essential for growing children as it is for growing pigs. (Courtesy, The Rockefeller Foundation, New York, N.Y.)

But the millenium that seemed so near with the discovery of opaque-2 corn has remained frustratingly out of reach. Although the nutritional value of the high-lysine corn is recognized, two major hurdles between discovery and application must yet be overcome: (1) The mutant gene is linked to opaque-2's soft, floury kernel, which is both light in weight and vulnerable to pest attacks, producing lower yields for farmers; and (2) opaque-2 has not been accepted by the majority of consumers, who are accustomed to the harder "flint" or "dent" kernels with a deeper, translucent color. But the need is great—human lives are at stake. So, plant breeders have set about crossing the opaque-2 gene on corn varieties that better meet the demands of farmers and consumers.

Supplementing Corn. Corn, which is the basic human food in many parts of the Americas, is an excellent energy food when properly supplemented. Zein, the principal protein in corn, which forms about half of the total protein in the whole grain, is an imperfect protein; its biological value is about 60 (vs biological values of 94 and 85 for whole eggs and milk, respectively), it is almost devoid of lysine, and it contains very little tryptophan. As a result, kwashiorkor is altogether too common among children subsisting on high corn diets. Also, most of the niacin (nicotinic acid) in corn is in bound form and unavailable; for this and other reasons, pellagra is associated with high corn diets.

In the new mutant corn, opaque-2, the ratio between the protein fractions is changed, with the zein content lowered from about 50% to less than 30% of the protein. The net result of this changed ratio is that opaque-2 has a higher content of lysine and tryptophan than regular corn. For this reason, corn of this type can supply a larger part of the protein requirement in the diets of people than ordinary corn.

Fortunately, man does not normally derive protein from a single source; rather, he gets protein from a variety of sources. Moreover, it is well recognized that proteins of different sources mutually supplement each other; hence, blends of two or more protein sources usually possess a higher nutritive value than a single source protein. In most parts of the world, corn-based diets traditionally include small amounts of pulses—the seeds of leguminous plants. Pulses contain more protein than corn (beans, 20-28%; soybeans, 39-45%; corn, 8-11%), and usually they supply the amino acids not provided by corn, especially lysine. Hence, diets based on corn and a pulse possess a nutritive value that is significantly higher than those based on corn alone.

Also, as people prosper, their diet becomes more varied and the consumption of part of the corn is replaced by animal protein foods—meat, fish, milk, and eggs. Animal proteins have a significant supplemental value in relation to corn; they provide added protein, along with increased lysine, tryptophan, and niacin. Thus, animal protein can be used to supplement corn-based diets effectively.

CORN PRODUCTS AND USES.

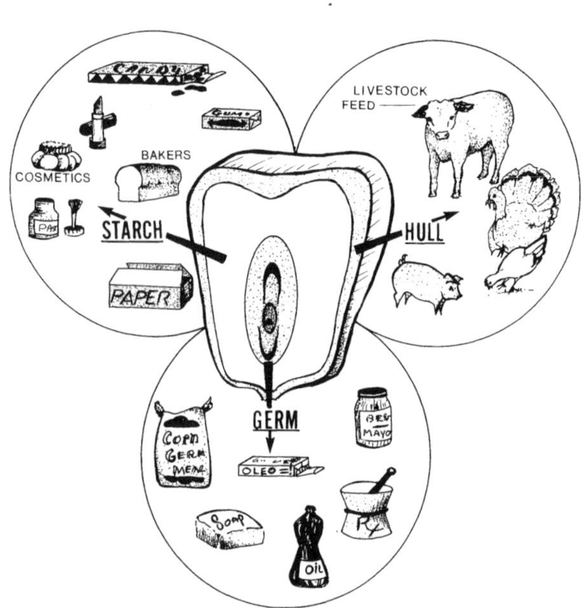

Fig. C-147. Some corn products and uses.

Fig. C-148. Sweet corn, a rich source of carbohydrates and fats. (Photo by J. C. Allen & Son, West Lafayette, Ind.)

Corn is used as food for man, for fermentation, for feed for livestock, and for industrial purposes. The U.S. annual supply of corn is used about as follows: livestock feed, 87%; human food, 11%; alcohol, 1.5%; and seed, 0.5%.

In one form or another, corn makes up more of our diet than any other farm crop. Corn-derived products are used in more than 800 different kinds of processed foods. Additionally, corn provides food indirectly; we eat it in the form of meat, milk, and eggs—products that come from animals raised on corn.

At the present time, few of the developing countries outside Latin America or Africa produce or utilize much corn as human food. However, these circumstances could change dramatically in the future, as the demand for cereals in the densely populated emerging nations continues to spiral and the great potential of corn becomes more fully appreciated. For example, the average corn yield in the United States is 2.5 tons per acre, compared to 0.9 tons per acre for wheat and 2.2 tons per acre for rice. Therefore, some special fortified foods, based on corn products, for use as low cost staple foods in developing countries are noteworthy (see Table C-45 section headed "Fortified Foods, Based on Corn Products, For Developing Countries.")

Corn grain is fed to all classes of animals. It accounts for about 12% of all U.S. livestock feeds. Additionally, about 116 million tons of corn silage and enormous quantities of corn by-products are fed to animals each year.

The corn distilling and fermentation industries manufacture ethyl and butyl alcohols, acetone, and whiskey; and factories manufacture hundreds of nonedible corn products.

Fig. C-149 and Table C-45 present, in summary form, the story of corn products and uses.

(Also see CEREAL GRAINS, Table C-18 Cereal Grains of the World; FLOURS, Table F-25 Major Flours; OILS, VEGETABLE, Table 0-6; and VEGETABLE[S], Table V-6 Vegetables of the World.)

CORN PRODUCTS

Food For Man

Baked goods	Cornmeal	Salad oil
Baking powder	Corn starch	Sausage
Breakfast foods	Grits	Shortenings
Brewing	Ice cream	Soft drinks
Candy	Icings and candy	Syrup
Catsup	Jams and jellies	Tamales
Chewing gum	Margarine	Vinegar
Cooking oil	Mayonnaise	Yeast
Corn flakes	Popcorn	
Corn flour	Pudding	

Feed For Livestock

Condensed fermented corn extractives	Corn gluten feed	Ground ear corn
Corn bran	Corn gluten meal	Husklage
Corn cob	Corn grain	Hominy feed
Corn feed meal	Corn grits	Oil cake and meal
Corn germ meal	Cracked corn	Silage
	Flaked corn	Stalklage

Other Products

Adhesives	Fertilizer	Pipes
Antibiotics	Fuel	Plastics
Antifreeze	Furfural	Rubber substitutes
Antiseptics	Gasohol	Safety glass
Ceramics	Insulating material	Soaps
Cork substitute	Leather dressing	Solvents
Cosmetics	Paints	Synthetic fibers
Dyes	Paper and paperboard	Varnishes
Ether	Pastes	Whiskey
Explosives	Photographic film	
Felt	Pharmaceuticals	

Fig. C-149. Some uses of corn.

TABLE C-45
CORN PRODUCTS AND USES

Product	Description	Uses	Comments
HUMAN FOODS			
Corn grain, whole **Corn on the cob**	This is the whole grain, on or removed from the cob.	Roasted or boiled, freshly picked corn on the cob. Whole grain removed from the cob is consumed as succotash, chowder, pudding, popcorn, fritters, and parched corn.	Americans eat about 50 lb *(22.7 kg)* of corn per person per year. Corn kernels are a rich source of carbohydrates and fats. One lb or 0.5 kg (about 3 ears) of sweet corn on the cob contains about 240 Calories (kcal). Also, the kernels on four 5 in. *(12.5 cm)* ears of corn are about equal to the contents of one 12-oz *(360 ml)* can of vacuum packed corn.
Corn flakes	Manufactured from hominy by flavoring, rolling, and toasting.	Ready-to-eat breakfast cereal.	
Corn flour	The smallest particles of ground corn.	Pancake mixes, as a filler in various meat products, and as a substitute for wheat flour in baking.	Doughs made from corn flour lack sufficient elasticity (because of their low-gluten content) for making yeast-leavened breads.

(Continued)

TABLE C-45 *(Continued)*

Product	Description	Uses	Comments
		HUMAN FOODS	
Corn grits	Hominy ground into large particles.	Cooked cereal or starchy vegetable.	Grits were long served in one or more forms with almost every meal in southeastern United States. They are still served at many breakfasts, with a pat of butter and a dab of jam or jelly.
Corn, hominy	Deskinned, degerminated kernels of soaked dried corn. Hominy may be left whole and canned or frozen. Usually, it is ground into large particles (grits).	Starchy vegetable or cooked cereal.	Hominy is prepared from a hard type of corn (Dent or Flint) that is (1) dried on the cob; (2) shelled; (3) soaked in a solution of baking soda, lime, lye, or wood ashes (each of which is an alkali that can burn the skin); and (4) deskinned and degermed by rubbing. Whole hominy requires long soaking and cooking. Generally, it is served with butter.
Cornmeal	Whole or degerminated corn grain coarsely ground.	Corn bread, cooked cereal, cookies, tamales, tortillas, griddlecakes, and waffles.	Whole grain cornmeal spoils readily because of the high oil content of the germ. Hence, it should be kept in a freezer or refrigerator. Degerminated cornmeal serves well in most recipes. One pound *(0.45 kg)* of uncooked cornmeal yields about 13½ cups *(3 liter)* of cooked cereal.
Cornmeal enriched	Degerminated cornmeal with added thiamin, riboflavin, niacin, and iron.	Corn bread, cooked cereal, cookies, tamales, tortillas, griddlecakes, and waffles.	The federal standards for enriched cornmeals and enriched corn grits are: **Ingredient** — **Required/lb *(0.45 kg)* of Cornmeal or Grits** Thiamin .. 2.0— 3.0 mg Riboflavin . 1.2— 1.8 mg Niacin .... 16.0— 24.0 mg Iron ...... 13.0— 26.0 mg Calcium .. 500.0—750.0 mg* *(optional)

(Continued)

TABLE C-45 *(Continued)*

Product	Description	Uses	Comments
		HUMAN FOODS	
Cornmeal, self-rising	Whole or degerminated cornmeal with added baking soda, salt, and one or more acidic salts such as monocalcium phosphate.	Corn bread, cookies, griddlecakes, and waffles.	Self-rising cornmeal contains its own leavening. Hence, no baking powder is required. One cup *(0.24 liter)* of this product may contain up to 412 mg of calcium and 859 mg of phosphorus, depending upon the types and amounts of acidic salts that are added. The federal standards for enriched self-rising cornmeal are: **Ingredient** — **Required/lb *(0.45 kg)* of Self-rising Cornmeal** Thiamin . . 2.0— 3.0 mg Riboflavin . 1.2— 1.8 mg Niacin . . . . 16.0— 24.0 mg Iron . . . . . . 13.0— 26.0 mg Calcium . . 500.0—1750.0 mg* *(optional)
Corn oil (Also see vegetable oils and fats.)	Corn oil is extracted from the germ of the kernel.	Salad dressings and cooking oil; and in such products as margarine and shortenings.	The oil is rich in essential fatty acids and contains moderate amounts of vitamin E.
Cornstarch	A white, odorless, tasteless, granular or powdery material.	Thicken puddings, gravies, and sauces; and in such products as candy, chewing gum, and baked goods.	The first step in the production of cornstarch involves the soaking of the corn in water and sulfur dioxide for 24 hours. This liquor is called corn steep liquor.
Corn sugar (dextrose) (Also see SUGAR.)	Converted from cornstarch.	Used in ice cream, frozen fruits, jams and jellies, bakery products, candy, carbonated beverages, many canned foods, and some meat products.	Dextrose (glucose) is the sugar present in the human bloodstream. Corn sugar is not as sweet as either honey or table sugar, because the latter items also contain fructose —one of the sweetest of the common sugars.
Corn syrup (Also see SYRUPS.)	Made by heating cornstarch with acid in closed tanks.	To sweeten many foods, and as a spread for bread, pancakes, and waffles.	Recently, a process was developed for converting some of the dextrose in corn syrup into fructose, thereby resulting in a sweeter syrup.

(Continued)

TABLE C-45 *(Continued)*

Product	Description	Uses	Comments
FORTIFIED FOODS, BASED ON CORN PRODUCTS, FOR DEVELOPING COUNTRIES			
CSM (corn-soy-milk)	A finely ground mixture of corn, soybean flour, skim milk powder, soybean oil, minerals, and vitamins. Contains 22% protein.	Substitute for milk.	Developed by the American Corn Millers' Federation in cooperation with the U.S. Department of Agriculture for use in the U.S. Food for Peace Program. On a par with the best of the cereal-based products for feeding young children. It may be supplemented with extra calories in the form(s) of sugar and/or peanut oil.
Golden elbow macaroni	Macaroni made with corn flour, soy flour, wheat flour, minerals, and vitamins. It contains about 20% protein.	Replacement for conventional macaroni products used as staple foods.	Developed by General Foods and test marketed in Brazil, where it was well accepted. Protein quality equals that of milk protein (casein).
Incaparina	Mixture of corn flour, cotton-seed flour, torula yeast, minerals, and vitamins. It has a protein content of about 25%.	Cereal for weanling infants and young children.	Developed by the Institute of Nutrition of Central America and Panama (INCAP) with the aid of funds from Central American governments, Kellogg Foundation, Rockefeller Foundation, and several other foundations. Test marketed in Nicaragua, El Salvador, Guatemala, and Colombia.
Pronutro	A finely ground mixture of corn, skim milk powder, peanuts, soybeans, wheat germ, sugar, salt, minerals, vitamins, and flavoring(s). It has a protein content of about 22%.	Cereals, soups, gravies, beverages, and chocolate bars.	Developed without any governmental or international assistance by Hind Brothers, a private food firm in Natal, South Africa. Effective as skim milk powder in the rehabilitation of infants suffering from severe protein-energy malnutrition (kwashiorkor).
FERMENTED PRODUCTS (Also see BEERS AND BREWING; DISTILLED LIQUORS)			
Corn grain, whole	Cleaned, whole corn	Bourbon whiskey	Each year, about 1.5% of the U.S. corn crop is used by the fermentation processors. Whiskey is obtained from the distillation of fermented mash of cereal grain, and is aged in wooden barrels 2 to 8 years before sale. Bourbon whiskey is made from at least 51% corn. The rise of the whiskey industry in Kentucky dates to a time when it was easier and more profitable for a farmer to ship his corn east over the mountains in the form of whiskey rather than in sacks of grain.

(Continued)

TABLE C-45 *(Continued)*

Product	Description	Uses	Comments
		FERMENTED PRODUCTS	
Corn and other cereal grains	Cleaned, whole corn, wheat, rye, or barley.	Canadian whiskey	
Cornstarch	Malt converts cornstarch to sugar, which is fermented by yeast to ethyl alcohol and carbon dioxide.	1. Ethyl alcohol, used largely for beverages. 2. Denatured alcohol for industrial purposes. (This is ethyl alcohol which has been rendered unfit for human consumption as a beverage by the addition of a denaturant.) 3. Gasohol (10% anhydrous ethanol/90% gasoline) a motor fuel.	
	Cornstarch, mostly in the form of grits, is used as a brewing adjunct, providing up to 40% of the mash.	Beer	The Indians brewed beer from corn before Columbus discovered America. Certain African tribes consume about 40% of their corn supply in the form of beer, thereby receiving substantially greater amounts of the B vitamins than they would receive by consuming their corn in unfermented form.
		LIVESTOCK FEEDS[1]	
Corn; whole, unprocessed	The kernels of corn that have not been broken.	Fed to all classes of animals. Normally, about 1/4 of the U.S. corn crop is fed to hogs. Also, corn is the most common, and the most desirable, grain for finishing cattle and lambs.	Corn is the most important concentrate feed in the U.S. It is palatable and rich in energy-producing carbohydrates and fats, and low in fiber; and it is easily stored.
Corn; whole, processed	The kernels of corn that have been processed by various methods, including: grinding, dry rolling (cracking, crushing), steam rolling (crimping), roasting, flaking, high-moisture (early harvested), or reconstituted (moisture added).	The use to be made of the processed corn is a determining factor in choosing the processing method.	In addition to intended use, available equipment and cost must be considered in arriving at the method for processing corn.
Corn bran	The outer coating of the corn kernel, with little or none of the starchy part or germ.	Most corn bran is incorporated in corn gluten feed and corn gluten meal.	Corn bran does not resemble wheat bran in composition.
Corn cob, ground	The ground entire cob.	Roughage for cattle and litter for poultry.	Ground corn cobs are comparable to poor grass hay. They should be properly supplemented with proteins, minerals, and vitamins. About 10 lb *(4.5 kg)* of cobs are obtained for every bushel *(56 lb, or 25 kg)* of corn shelled.
Corn; cracked, screened	The coarse portion of cracked corn from which most of the fine particles have been removed.	Fed to swine and poultry.	

footnote at end of table

(Continued)

TABLE C-45 *(Continued)*

Product	Description	Uses	Comments
		LIVESTOCK FEEDS[1]	
Corn feed meal	The fine siftings obtained from screened cracked corn.	Fed to swine and poultry.	
Corn flour	The fine-sized hard, flinty portions of ground corn containing little or none of the bran or germ.	Primarily for human food. Used for livestock feed when the quality is off or the price is right.	
Corn germ meal (dry milled)	Ground corn germ with other parts of the corn kernel from which part of the oil has been removed; obtained in the dry milling manufacture of corn meal, corn grits, hominy feed, and other corn products.	Most of the corn germ meal that is marketed separately goes into poultry or swine rations.	Corn germ meal with a high content of fat may cause soft pork if fed in large quantities.
Corn germ meal (wet milled)	Ground corn germ from which most of the solubles have been removed by steaming and most of the oil removed by extraction; obtained in the wet milling manufacture of cornstarch, corn syrup, or other corn products.	Primarily in poultry and swine rations.	Corn germ meal with a high content of fat may cause soft pork if fed in large quantities.
Corn gluten feed (gluten feed)	That part of shelled corn which remains after the extraction of the larger portion of the starch, gluten, and germ by the wet milling manufacture of cornstarch or syrup.	As a medium protein feed for dairy and beef cattle, poultry (layers and turkey breeders), and swine. It is not very palatable.	When corn gluten is fed to swine or poultry, the amino acids must be balanced by including a higher quality protein.
Corn gluten meal (gluten meal)	The dried residue from corn after the removal of the larger part of the starch and germ, and the separation of the bran by the process employed in the wet milling manufacture of cornstarch or syrup, or by enzymatic treatment of the endosperm.	Corn gluten meal is a high-protein (60%), high-energy ingredient. Much of it is fed to dairy cattle. The high xanthophyll makes it particularly valuable as a pigmenting ingredient of poultry feeds.	When corn gluten meal is fed to swine or poultry, the amino acids must be balanced by including a higher quality protein. When made from yellow corn, corn gluten meal is high in carotene and xanthophyll.
Corn grits	The medium-sized, hard, flinty portions of ground corn containing little or none of the bran or germ.	Primarily for human food. Used for livestock feed when the quality is off or the price is right.	
Ground ear corn (corn-and-cob meal)	The ground entire ear of corn, including the cobs, but without husks.	For ruminants and horses.	Ground ear corn has 85 to 90% the feeding value of corn grain for ruminants and horses.
Ground ear corn, with husks (ground snapped corn)	The ground or chopped entire ear of corn, with husks.	For ruminants.	Ground ear corn with husks has 75 to 80% the feeding value of corn grain.

footnote at end of table

(Continued)

TABLE C-45 *(Continued)*

Product	Description	Uses	Comments
		LIVESTOCK FEEDS[1]	
Hominy feed	A mixture of corn bran, corn germ, and part of the starchy portion of either white or yellow corn kernels or mixtures thereof, as produced in the manufacture of pearl hominy, hominy grits, or table meal. It must contain not less than 4% fat.	Primarily for dairy cattle and poultry, for which it is equal to corn.	Hominy feed will produce soft pork if it constitutes more than ½ the grain ration.
Corn cobs	The axis on which kernels of corn are arranged.	Ground and substituted for cork; used in making felts, cleaning compounds, and plywood; made into meal and used for soil conditioner, fertilizer, cleaning furs, and polishing metal; used as an ingredient of furfural, a chemical used in making plastics, nylon, and other industrial products; burned as a fuel; or manufactured into corncob pipes—made famous by the late General Douglas MacArthur.	
Corn oil	Extracted from the germ of the kernel.	In soaps, glycerine, paints, varnishes, and rubber substitutes.	
Cornstalks	The stalk of the corn plant.	Wallboards and certain kinds of paper.	
Cornstarch	A white, odorless, tasteless granular or powdery material.	In coating paper and paperboard, as adhesives, and as sizings in the manufacture of textiles. In cosmetics, explosives, electric batteries, and drugs. Cornstarch is sometimes used in pills and capsules.	
Corn sugar	Converted from cornstarch.	In leather tanning, rayon, and in paper manufacturing. Doctors sometimes give sterilized dextrose (refined corn sugar) intravenously to hospitalized patients as a quick source of energy. Dextrose may also be an ingredient of certain drugs. The microorganism *Penicillium chrysogenum,* which secretes the lifesaving antibiotic Penicillin, lives in a huge fermentation vat and thrives on a diet of which dextrose is the chief carbohydrate source of energy.	
Corn syrup	Made by heating cornstarch with acid in closed tanks.	To sweeten cough syrup and other liquid medicines.	

[1]Many other corn products than those listed are used for livestock feed, including: condensed fermented corn extractives, corn fodder, corn husklage (shucklage), corn silage, sweet corn silage (cannery waste), and corn stover.

CORNEA

The transparent circular membrane which covers the front of the eyeball. Good nutrition is important for corneal function. Hence, nutritional deficiencies may be indicated by abnormalities such as the appearance of tiny blood vessels at the edge of the cornea (corneal vascularization).

(Also see CORNEAL VASCULARIZATION; and DEFICIENCY DISEASES.)

CORNEAL VASCULARIZATION

The formation of tiny capillaries on the surface of the cornea (the projecting part of the eyeball, which is usually transparent). The cause(s) of this disorder may be riboflavin deficiency and/or some type of inflammation or irritation.

(Also see DEFICIENCY DISEASES.)

CORNED BEEF

Boneless cuts of brisket, plate, chuck, and round usually are used in making corned beef.

The preservation of beef by the use of salt is termed *corning*. Lexicographers explain the origination of the word by referring back to the 16th century when the word *corn* was synonymous with the word *grain*. At that time manufacturers of gun powder used the word *corning* to indicate that their product had been spread out and allowed to dry in single grains. The term *corned* was later applied to the process of curing beef by sprinkling it with grains of salt. Of course, the making of corned beef today entails more than the application of salt. A popular formula consists of dissolving 8 lb of salt, 3 lb of sugar, 4 oz of baking soda, and ¼ oz of sodium nitrite in 4 gal of water. This is sufficient to cure 100 lb of beef.

CORN STEEPWATER (CORN STEEPLIQUOR)

When corn is wet milled to make starch, sweeteners, and oil, the first step is to soften or condition the kernel for grinding. This is accomplished by soaking the grain in a sulfurous acid steep for 28 to 48 hours. This liquor is called corn steepwater or corn steepliquor. When concentrated, steepwater is excellent for growing the mold that produces penicillin, because it contains a biochemical precursor of penicillin. It is also used in animal feeds.

(Also see CORN, section headed "Wet Milling," Fig. C-143 schematic outline of the corn wet milling process.)

CORN SUGAR

Starch is composed of numerous molecules of the sugar glucose. Acids or enzymes may be employed to break down—hydrolyze—starch into glucose. Hence, glucose derived in this manner from corn starch is sometimes referred to as corn sugar.

(Also see CORN; CORN SYRUP; DEXTROSE; GLUCOSE; and STARCH.)

CORN SYRUP (GLUCOSE SYRUP)

Corn syrup is a viscous liquid containing dextrose, maltotriose, maltotetrose, maltose, dextrin, and sometimes fructose—depending on the manufacturer. Unlike sugar, it has a distinct flavor other than sweetness. Corn syrup is the product of the partial hydrolysis of corn starch. It is usually obtained by heating cornstarch with a dilute acid or by enzymatic action. The degree of conversion is expressed by the "dextrose equivalent" (DE) which is, in effect, a measure of sweetness in the syrup. High conversion syrups are substantially less sweet than sugar. Since corn syrups can (1) control crystallization in candy making, (2) retain moisture, (3) ferment, and (4) produce a high osmotic pressure solution for preservation, they are employed in many food products including the following: baby foods, bakery products, canned fruits, carbonated beverages, confections, dry bakery mixes, fountain syrups and toppings, frozen fruits, fruit juice drinks, ice cream and frozen desserts, jams, jellies, and preserves, meat products, pickles and condiments, and table syrups (Karo syrup).

(Also see ADDITIVES; CARBOHYDRATE[S]; CORN; DEXTROSE; STARCH; and SUGAR.)

CORN, WAXY (MAIZE, WAXY)

Corn high in amylopectin, a starch which is used in the paper industry and as a coating or sizing in the fabrication of woven fiber glass.

CORONARY

A term referring to the heart and certain blood vessels that serve it.

CORONARY HEART DISEASE

A type of heart disease resulting from an obstruction in a coronary artery that reduces the flow of blood to the heart muscles. However, other types of disorders may be diagnosed as coronary heart disease. For example, the failure of diseased heart muscles to pump sufficient blood may result in a stagnation of blood flow, the formation of a clot, and a blockage of the coronary artery. In the latter case, the heart muscle weakness leads to the problem in the coronary artery rather than vice versa.

(Also see HEART DISEASE.)

CORTEX

The outside layer(s) of a gland or a special tissue, in contrast to the inner layers or medulla. For example, the adrenal cortex secretes steroid hormones; whereas, the adrenal medulla secretes catecholamine hormones (adrenaline and noradrenaline). However, the most important cortex is the cerebral cortex, which is the outer layer or thinking part of the human brain.

CORTICOSTEROIDS

A group of organic compounds secreted by the outside layer (cortex) of the adrenal gland in response to starvation and other stresses.

(Also see ENDOCRINE GLANDS.)

CORTICOSTERONE

A steroid hormone secreted by the adrenal cortex. Corticosterone promotes the formation of carbohydrate from protein and the breakdown of glycogen to glucose.

(Also see ENDOCRINE GLANDS.)

CORTICOTROPIN

Another name for ACTH, a hormone secreted by the pituitary gland.

(Also see ENDOCRINE GLANDS.)

CORTISOL

The major hormone from the adrenal cortex which raises the blood sugar through the conversion of glycogen and protein to glucose. Cortisol also counteracts the effect of insulin so that there is a reduction in the rate at which glucose in the blood is picked up by the muscles and the fatty tissues of the body. However, it does not seem to slow the rate at which glucose is used by the brain. Hence, the net effect of the hormone is to ensure that there is a continuous supply of blood sugar for the brain.

(Also see ENDOCRINE GLANDS.)

CORTISONE

A hormone which is formed from adrenal cortical hormones in the liver. It is now produced synthetically for various medical applications.

CORYZA

An inflammation of the membranes in the nose that is often accompanied by a watery secretion. This disorder may be due to an allergy, an infection, or hay fever.

(Also see ALLERGY.)

COSTIVENESS

A type of constipation due to lack of sufficient water in the bowels. It is usually due to insufficient water-binding bulk in the diet. Hence, it may often be corrected by increasing the fiber content of the diet and drinking plenty of water.

(Also see CONSTIPATION.)

COTTONSEED *Gossypieum*

Contents — *Page*

Origin and History 498
World and U.S. Production 498
The Cotton Plant 499
Cotton Culture 500
Kinds of Cotton 500
Processing Cottonseed 501
Nutritional Value of Cottonseed 502
Cottonseed Products and Uses 503

Fig. C-150. Cottonseed stored at mill, showing screw conveyer (above) which distributes the seed. (Courtesy, Rancher's Cotton Oil, Fresno, Calif.)

Cottonseed is the seed of the cotton plant, and cotton is the white fluff that grows attached to the seed. Cotton is grown primarily for its fiber (lint), which is used in making textiles, while the seed is used in making food for man and feed for livestock. Because this book pertains to food, the rest of this article will be devoted primarily to a discussion of cottonseed.

Cottonseed is a by-product of fiber production. With each 100 lb (*45 kg*) of fiber, the cotton plant yields approximately 170 lb (*77 kg*) of cottonseed.

Cotton, with a production value of $5.8 billion in 1990, ranked fifth in value among U.S. crops, exceeded only by corn, soybeans, wheat, and hay. Also, it thrives in tropical regions, which includes Africa, Central and South America, India and other parts of Asia.

The cottonseed, once largely wasted, is now converted into food, feed, and many other useful products. Oil is the most important product made from cottonseed; meal ranks second in importance.

For many years, the primary deterrent to the use of the seed as a protein source was gossypol, a toxic yellow pigment. Following intensive research, methods of processing were developed that resulted in a protein supplement suitable and safe for monogastric animals. This led to the use of the protein concentrate for humans, particularly for children suffering from protein deficiency in some of the developing countries.

Of course, it would be better were it possible to alleviate or remove gossypol without compromising the amino acid content. Two approaches may eventually solve this problem: (1) breeding, by infusing glandless wild strains of cotton without pigment glands and, therefore, free of gossypol; and (2) extracting in a manner that would remove the gossypol without damaging the amino acid availability or reducing or impairing the yield or quality of the oil. If either or both of these efforts are successful, it is predicted that cottonseed meal will be processed primarily for human food. There are many areas in the world, and there are many types of foods, where a satisfactory cottonseed protein concentrate could assist in increasing the protein content of the human diet. Besides, cotton is indigenous to many of the countries that need protein; hence, the protein concentrate would be low cost. Improvement of the protein, along with the value of the oil, may make cottonseed a major source of human food in the future.

ORIGIN AND HISTORY. Cotton has been grown for fiber for centuries. The Aztec Indians of Mexico grew cotton for textile purposes nearly 8,000 years ago. Record of cotton textiles, dating back about 5,000 years, was found in the Indus River Valley in what is now Pakistan. Excavations in Peru have uncovered cotton cloth identified as 4,500 years old. Cotton fabrics have also been found in the pueblo ruins in Arizona and in the ruins of some of the early civilizations of Egypt. The ancient peoples used cotton for clothing, and for many other necessities—ranging from bindings for sandals to elephant harnesses.

One of the first written references to cotton appears in Hindu literature, written 1,500 years before the birth of Christ. In the 3rd century B.C., the army of Alexander the Great found cotton widely used in India and introduced it into the countries on the eastern and southern borders of the Mediterranean Sea. From there, its use spread to Europe.

Before the time of Christ, Greek and Roman travelers described cotton plants as "the fleece of tiny lambs growing on trees." Herodotus (484 to 424 B.C.), the first Greek historian, wrote about a tree in Asia that bore cotton "exceeding in goodness and beauty the wool of any sheep."

The desire for cotton and cotton fabrics was one of the factors that led to the many explorations of the 15th and 16th centuries. When Columbus landed in the West Indies, he found the natives wearing cotton cloth. A short time later, Cortez found cotton production and manufacture well developed in Mexico.

In the early 1600s, the southern colonists grew a limited amount of cotton, but, for many years, the crop remained of minor importance because: (1) many of the varieties were not well adapted to the soil and climate; (2) the seed and fiber of the available varieties clung so closely together that existing ginning machinery could not effectively separate them; and (3) the British discouraged cotton manufacture in the colonies, since it would compete with their home textile industry.

But conditions changed for the better in the 1700s; and the cotton industry expanded. The American Revolution lifted government restrictions. This was followed by two very important events that gave tremendous impetus to the cotton industry: (1) the building, by Samuel Slater in 1790, of the first successful textile mill in the United States; and (2) the invention of the cotton gin by Eli Whitney in 1793.

Although cotton was grown for its fiber for centuries, extensive use of the seed is of relatively recent origin. The ancient Hindus and Chinese developed crude methods for obtaining oil from cottonseed, using the principal of the mortar and the pestle. They used the oil for their lamps and fed the remainder of the processed seed to their cattle.

During the first part of the 19th century, mills in Europe began to crush Egyptian cottonseed on a limited scale. However, it remained for American machinists and chemists to transform cottonseed into useful products. With more cotton, there was more seed; and with more seed, there was the challenge to find a profitable outlet for it. Crushing mills were built, but all of them failed due to several factors; among them, (1) the seed of the principal varieties was covered by short fibers, known as linters, which were not removed at the gin; (2) whole seed was crushed, with the result that the linters and hulls absorbed much of the oil; (3) the remaining feedstuff was of low quality; and (4) transportation facilities were poor. Despite these difficulties and failures, the stakes were high. So, others tried to find a profitable commercial use for cottonseed. Eventually, they succeeded. For the most part, the oil came to be used for human food, and the meal and hulls were fed to cattle and sheep.

WORLD AND U.S. PRODUCTION. China, the United States, and the U.S.S.R. are the leading cotton-producing countries of the world, with the ranking of the three countries shifting back and forth according to growing conditions during the year.

Production figures are in metric tons. The world total production averages 54 million metric tons. The ten leading cotton-producing countries by rank are: China, United States, U.S.S.R., India, Pakistan, Brazil, Turkey, Egypt, Australia, and Argentina. China, the United States, and the U.S.S.R. account for 57% of the world total. The worldwide average yield per acre is 1,424 lb (*1,596 kg/ha*), while the U.S. average yield is 1,679 lb per acre (*1,882 kg/ha*).

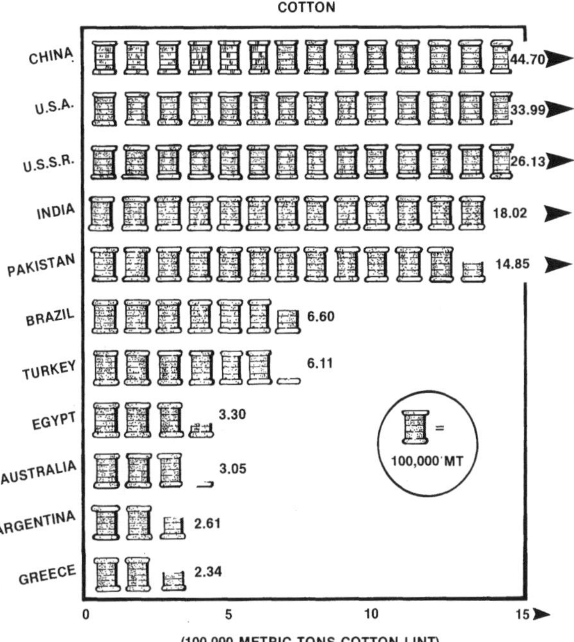

Fig. C-151. Leading cotton-producing countries of the world. (Data from *FAO Production Yearbook*, 1990, FAO/UN, Rome, Italy, Vol. 44, p. 185, Table 87)

The ten leading U.S. cotton-producing states, in total bales produced, by rank, are: Texas, California, Mississippi, Louisiana, Arkansas, Arizona, Tennessee, Georgia, Alabama, and Oklahoma. Texas and California produce 51% of the U.S. crop.

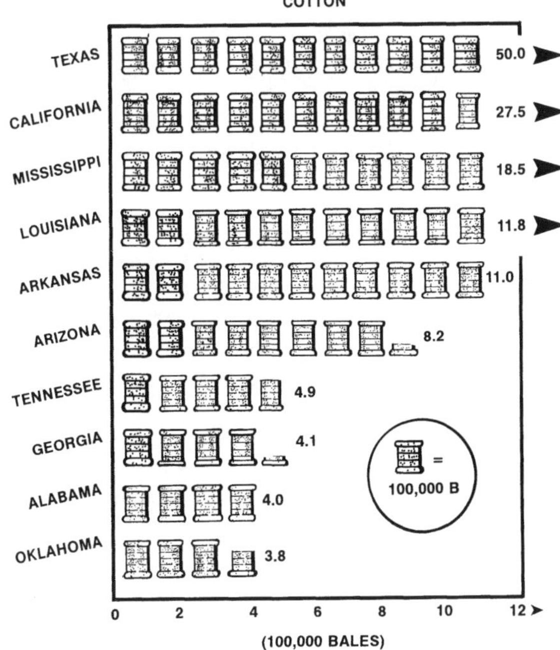

Fig. C-152. Leading cotton-producing states of the United States, in bales. One bale weighs 480 lb, or *218 kg*. (Source: *Agricultural Statistics*, 1991, USDA, p. 62, Table 82)

THE COTTON PLANT.

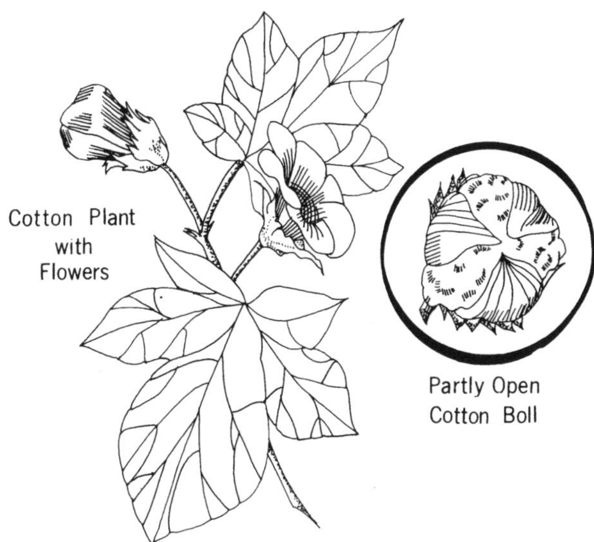

Fig. C-153. The cotton plant.

Cotton belongs to the genus *Gossypieum*, of which there are about 20 species, only four of which are cultivated. The four cultivated species are: *G. **hirsutum***, American upland cotton; *G. **barbadense***, Egyptian and Sea Island cottons; *G. **herbaceum*** and *G. **arboreum***, the Asiatic cottons. Each of the four species embraces several varieties.

Most cotton is grown as an annual, although not infrequently in North Africa and Brazil the plantings are maintained several years as perennials.

The plant described herein is American upland cotton, the most widely grown kind, which resembles other kinds of cotton.

Fig. C-154. Cotton bolls: left, unopened; right, fully opened mature cotton—one of the most beautiful and useful gifts of nature to man. (Courtesy, National Cottonseed Products Association, Inc., Memphis, Tenn.)

The American upland cotton plant grows upright and has branches spreading in all directions. It has broad leaves with 3 to 5 lobes; a taproot (main root) which may grow 4 ft (*1.2 meters*) into the ground; and creamy white or yellow flowers, which turn pink, blue, and finally purple as they wither and fall off. As each flower withers, the boll, which contains the cotton fibers, begins to form. At maturity, the boll splits and reveals many seeds, each with its mass of cellulosic seed hairs.

COTTON CULTURE. United States cotton production is highly mechanized. However, in many parts of the world cotton is still grown and harvested in a relatively unmechanized fashion. In Ethiopia, for example, cotton still supports a cottage (family-type) industry.

Cotton thrives best in fertile, well-drained soil that gets adequate moisture during the growing season. It requires a warm to hot climate with about 180 frost-free days during the growing season.

In the United States, cotton is planted in April or May, using planters to open furrows, drop the cotton seeds into the ground, apply fertilizer, and press down the soil. Many farmers also apply weed-control chemicals and soil fungicides at planting time. Rows are usually about 40 in. (*100 cm*) apart. Preferably, there should be 30,000 to 60,000 plants per acre (*75,000 to 150,000 plants per hectare*).

Cotton is subject to many pests and diseases. It is estimated that, in an average year, a cotton farmer loses one bale of every eight because of insect damage, and one bale of every seven or eight because of disease.

The most harmful insects are: aphid, boll weevil, bollworm, lygus, pink bollworm, and thrips. Insecticides can control all of these pests except the pink bollworm.

Diseases of seedlings are controlled by treating the seeds or spraying the seedbed with fungicides. Diseases of older cotton plants, such as wilt and blight, may be partially controlled by using disease-resistant varieties.

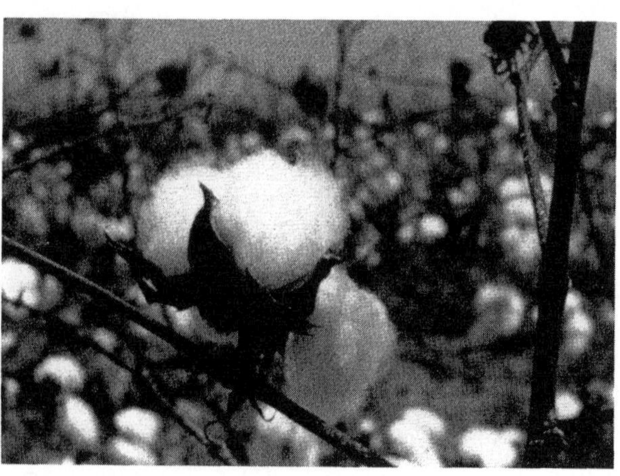

Fig. C-155. Cotton ready for harvest. (Courtesy, USDA)

Most cotton harvest in the United States is done by stripping or picking. Usually a chemical defoliant precedes picking to free the plants of interfering foilage. In areas where the cotton plant is relatively small, such as in Texas and Oklahoma, stripping machines are used. Farmers in other parts of the Cotton Belt use picking machines. Some of the U.S. cotton crop, and much of the crop in other parts of the world, is picked by hand. The pickers remove cotton fibers and seeds from the burrs by hand, then put the cotton in large sacks that they drag along the ground behind them.

Fig. C-156. Cotton picking machine (two-row). (Courtesy, USDA)

After harvest, the cotton is taken to the gin, which removes the fibers (lint) from the seeds. The fibers are then baled under heavy compresssion into 500-lb (*227 kg*) bales, measuring about 56 × 28 × 45 in. (*142 × 71 × 114 cm*). Each bale is bound by six steel ties and covered with jute or other material. In succession, the bales are: (1) normally sold at the gin; (2) transported from the gin to a warehouse for storage; (3) classified on the basis of grade, staple (length of the fiber), and preparation; and (4) sent to a textile mill for making into cloth.

In most cases, the seeds are sold to the ginner, who, in turn, sells to the crushing mill.

KINDS OF COTTON. The various kins of cotton plants resemble each other in most ways, but they differ in such characteristics as time of blooming, color of flowers, and character of fibers. The four main kinds of cotton are:

1. **American Upland cotton.** This kind of cotton, which makes up ⅔ to ¾ of the world's cotton crop, is raised

in almost every cotton-producing country. The plant grows 1 to 7 ft (*30 to 213 cm*) tall; has creamy-white flowers; and has white fibers 7/8 to 1 1/4 in. (*22 to 32 mm*) long. It is made into many kinds of fabrics, from coarse to fine; and it is used both for heavy canvas and fine shirts.

2. **Egyptian cotton.** This kind of cotton was developed from stocks that originated in South and Central America. it has lemon-colored flowers and long, silky, light-tan fibers. It is made into fine fabrics.

3. **Sea Island cotton.** This kind of cotton first grew on the Sea Islands off the coast of South Carolina, Georgia, and northern Florida, from which it takes its name. It is now grown in the West Indies. Sea Island cotton is characterized by brilliant yellow flowers, small bolls, low yields, and silky fibers about 1 3/4 in. (*44 mm*) long. It is one of the most valuable and costly kinds of cotton. Sea Island cotton is made into high-quality textiles.

4. **Asiatic cotton.** This kind of cotton is grown mainly in China, India, and Pakistan. It is characterized by short, coarse, harsh fibers, and low yields. It is used for blankets, padding, filters, and coarse cloth.

• **Glandless cottonseed**—In 1954, in a major scientific breakthrough, a stalk of wild cotton with no glands in the seed was found on an Arizona Indian reservation.

Fig. C-157.Cross section of glanded (left) and glandless (right) cottonseeds. The dark spots in the cottonseed on the left are the glands that hold the gossypol. (Courtesy, USDA)

By infusing glandless wild strains, geneticists have produced commercial varieties of cotton with glandless seed characterized by the absence of the gossypol-containing pigment glands common to glanded cottonseed. Although there is need for further improvement in glandless cotton, this development makes the future prospects for the use of cottonseed in human food very promising. The rate of transition from glanded to glandless varieties in the United States, and in other cotton-producing countries, will be determined by (1) the yield and quality of fiber of the glandless varieties, including resistance to diseases, parasites, birds, etc.; (2) government policies; and (3) incentives in the form of higher prices paid to cotton farmers for oil and meal from glandless seed. Although glandless cotton varieties are now being grown to a limited extent, it will be a number of years before all the problems associated with the development of a new variety of cotton are solved and large-scale production is underway.

Glandless cottonseed kernels have been shown to have properties characteristic of nutmeats. When toasted, roasted, or fried, they have a pleasant flavor and texture and can be used either as a snack or as an ingredient in cookies, candies, etc.

The flour produced from glandless cottonseed has a very light cream color and contains in excesss of 50% protein.

PROCESSING COTTONSEED. Most cottonseeds arrive at the crushing mill by truck, where they are unloaded and stored. As seeds are received at the mill, samples are usually taken and analyzed for moisture and free fatty acid content. On the basis of such analyses, seeds are placed in storage in accordance with their keeping quality. Seeds that are most likely to spoil are processed first. The usual steps in processing are:

1. **Cleaning.** Seeds are cleaned to remove any leaves, twigs, bolls, or dirt.

2. **Delinting.** The cleaned seeds are conveyed to the lint room where the short fibers, known as linters, are removed. The delinting machines use a series of circular saws designed to cut the short fibers. The linters are then collected and pressed into bales. Most mills run the seed through the delinting machine twice.

3. **Hulling.** Hulling is accomplished by a machine that employs a series of knives which cut the leathery hull and loosen it from the meat, following which the seeds are passed through a series of beaters, shakers, and separators which separate the loosened hulls from the meats.

4. **Removing oil.** The oil is removed by either of three methods. The methods, along with the proportion of the U.S. cottonseed production processed by each, follow:

Processing Method	Proportion of U.S. Cottonseed Production Processed by Each Method (%)
Screw pressing.........	40
Prepress solvent extraction...........	28
Direct solvent extraction.	32

Less than 1% of U.S. cottonseed is extracted by the old hydraulic press method; hence, the latter is insignificant.

The type of processing affects the usefulness of the meal. In comparison with other processing methods, screw pressing reduces free gossypol and nitrogen solubility, but results in higher oil content and gross energy. For hens, the higher oil content may contribute to discoloration of eggs during storage; for cattle, the higher oil content makes for higher energy and is desirable.

Direct solvent extraction produces a meal which contrasts most to that produced by screw processing. Usually, it is higher in free gossypol and in nitrogen solubility. From the standpoint of protein quality, it is excellent for monogastrics.

Prepress solvent extraction produces a meal that is intermediate between screw pressed and direct solvent extraction meals. It is low in free gossypol content, medium-high to high in nitrogen solubility, and the lowest in fat content. Because of these characteristics, this is the meal of choice for laying hens.

Cottonseed is processed into 36%, 41%, 44%, and 48 to 50% protein meals. The 36% protein meal is screw press; the high protein meals (44% and higher) are prepress solvent extraction and direct solvent extraction.

• **Liquid Cyclone Process (LCP).** The LCP process, developed by USDA's Southern Regional Reseach Center, is designed to remove gossypol-containing pigment glands from glanded cottonseed and produce a food-grade flour for human consumption. To date, this processing method has not reached the point of being a commercial reality on a large scale.

Fig. C-158. Two cottonseed flours. No. 1, made from glandless cottonseed. No. 2, made by the Liquid Cyclone Process. (Courtesy, USDA)

NUTRITIONAL VALUE OF COTTONSEED.[49]

Typically, cottonseed is comprised of about 10% linters, 35% hull, and 55% kernels (meats). The kernels, from which food and feed are obtained, contain about 7% moisture, 30% oil, 30% crude protein, 24% nitrogen-free extract, 4.8% crude fiber, and 4.4% ash.

Cottonseed oil is classified as a polyunsaturated vegetable oil. Linoleic acid, its principal fatty acid, comprises about 47 to 50% of the total fatty acids. Its unique crystalline properties result from the presence of about 26% palmitic acid. Its good flavor is generally attributed to the absence of linolenic acid.

Table C-46 shows the composition of two defatted cottonseed flours that are suitable for human consumption—glandless (direct solvent) and glanded (liquid cyclone). Presently, neither product is available commercially on a large scale; nevertheless, they are feasible, and it is expected that supplies will increase with demand and price.

TABLE C-46
COMPOSITION OF GLANDLESS AND GLANDED COTTONSEED FLOURS[1]

	Glandless Direct Solvent	Glanded Liquid Cyclone
	(%)[2]	(%)[2]
Moisture	9.5	3.8
Crude fat	2.4	1.2
Crude fiber	3.3	2.3
Total gossypol	.06	.06
Free gossypol	.04	.02
Protein (N X 6.25)	54.7	66.2
Available lysine (g/16gN)	3.9	4.0
Nitrogen solubility	95.7	99.6

[1]Martinez, W.H., L.C. Berardi, and L.A. Goldblatt, Southern Regional Research Laboratory, New Orleans, Louisiana. "The Potential of Cottonseed: Products, Composition, and Use," paper presented at the Third International Congress of Food Science and Technology, Washington, D.C., August 9-14, 1970.

[2]All values in percent except lysine. The latter is in grams/16 grams of nitrogen.

Flour is but one of three major classes of food protein products which can be obtained from cottonseed; protein concentrates and isolates can also be produced. That is, the product development has been researched; with the incentives of demand and price, they will become available commercially. The quality of the flour, therefore, is important not only from the standpoint of its edible use, but also as the raw material for the production of cottonseed concentrates and isolates.

Table C-47 gives the composition and yield of the various concentates.

[49]The authors are grateful to Mr. George C. Cavanagh, Director of Research, Rancher's Cotton Oil, Fresno, California, for providing most of the data used in this section.

TABLE C-47
COMPOSITION AND YIELD OF THREE COTTONSEED PROTEIN CONCENTRATES:
(1) AIR CLASSIFIED, (2) ALCOHOL LEACHED, AND (3) DILUTE CALCIUM EXTRACTED[1]

Method	Air Classification	90% Ethanol	.008 M $CaCl_2$ H_2O
	(%)	(%)	(%)
Composition:[2]			
Protein (N X 6.25) . . .	73.7	71.9	75.9
Lipid	.7	.1	2.6
Ash	9.4	8.7	8.5
Crude fiber	1.7	2.7	3.7
Total sugar (as invert)	4.7	2.1	.3
Yield:[3]			
% of total weight . . .	56.0	84.0	60.0
% of total nitrogen . .	63.0	93.0	77.0

[1]Martinez, W.H., L.C. Berardi, and L.A. Goldblatt, Southern Regional Research Laboratory, New Orleans, Louisiana. "The Potential of Cottonseed: Products, Composition, and Use," paper presented at the Third International Congress of Food Science and Technology, Washington, D.C., August 9-14, 1970.
[2]Dry weight basis.
[3]"As is" basis.

To prepare an isolate, the proteins must be extracted from the defatted flour. The classical procedure for the preparation of cottonseed isolates is similar to the isolate procedure used in the soybean industry. Other methods for the preparation of cottonseed isolates have been devised.

Table C-48 gives the proximate and amino acid composition of glandless cottonseed flour and three different isolates.

Tables C-46, C-47, and C-48 show that the protein quality of cottonseed is inherently high. However, the effects of protein fractionation and processing techniques on amino acid availability must be recognized and balanced in the final food product. Lysine will probably be the major deficiency; and the limiting amino acid in cottonseed-cereal mixtures used for human food in some of the developing countries. Preservation of the existing lysine content by careful processing becomes one of the means of improving the nutritive potential of cottonseed protein.

As is true in any food formulation, the manufacturer will utilize the protein product which provides the most advantageous compromise between end-use functional requirements and cost; and each country must decide which use of its oilseed resources provides the most advantageous route toward an adequate supply of protein for its people. In some instances, the best route may be to use the oilseed proteins as feed for animals; in others the best use will be as food for man. If the need for protein becomes critical, food not feed should be the dominant course; and the potential of cottonseed is available to fulfill this need.

TABLE C-48
PROXIMATE AND AMINO ACID COMPOSITION OF GLANDLESS COTTONSEED FLOUR AND THREE ISOLATES—(1) CLASSICAL METHOD (EP and SP), (2) FUNCTIONAL PROTEIN (FP), AND (3) STORAGE PROTEIN (SP)[1]

	Isolates		
Flour	Classical Method (EP and SP)	Selective (FP)[2]	Extraction (SP)
Composition, %:[3]			
Nitrogen 10.73	15.58	13.08	17.24
Crude fiber . . . 2.2	.5	.5	.2
Lipid 9	1.1	3.0	.2
Ash 7.8	3.4	14.1	1.0
Phosphorus . . 2.19	.69	3.13	.26
Total sugar . . . 7.3	.5	.5	0
Amino Acid, g/16gN:			
Alanine 3.7	3.6	3.2	3.5
Arginine 12.4	10.0	10.4	11.3
Aspartic 9.1	9.0	6.7	8.4
½ Cystine . . . . —	—	2.6	.3
Glutamic 20.4	16.4	21.8	18.9
Glycine 4.1	3.5	3.2	3.7
Histidine 2.9	2.9	2.6	3.0
Isoleucine 3.4	3.4	2.6	3.1
Leucine 5.8	5.7	5.1	5.8
Lysine 4.4	3.4	6.0	3.0
Methionine . . . 1.3	1.4	1.7	1.0
Phenylalanine 5.5	5.7	3.7	6.3
Proline 3.6	3.4	3.1	3.1
Serine 4.1	4.0	3.4	4.5
Threonine 3.0	2.9	2.9	2.7
Tyrosine 3.1	2.8	3.3	2.6
Valine 4.6	4.7	3.3	4.4

[1]Martinez, W.H., L.C. Berardi, and L.A. Goldblatt, Southern Regional Research Laboratory, New Orleans, Louisiana. "The Potential of Cottonseed: Products, Composition, and Use," paper presented at the Third International Congress of Food Science and Technology, Washington, D.C., August 9-14, 1970.
[2]Neutralized.
[3]Dry weight basis.

COTTONSEED PRODUCTS AND USES. Of the four primary products of cottonseed—oil, cake or meal, hulls, and linters—oil is the most valuable. On the average, it accounts for slightly more than half of the total value of the four products. Cottonseed meal, the second most valuable product of cottonseed, accounts for 30 to 35% of the total product value.

In recent years, industry-wide yields of products per ton of seed have averaged 336 lb (*151 kg*) of oil, 936 lb (*421 kg*) of meal, 460 lb (*207 kg*) of hulls, and 168 lb (*76 kg*) of linters, with manufacturing loss of 100 lb (*45 kg*) per ton. These figures vary from area to area and from mill to mill, depending on the character of the seed, the type of process used, and the marketing practices.

Scientists and technologists are convinced that, because of its unique properties, the prospective new markets for cottonseed are food-oriented. Without doubt, this time will be hastened by the perfection and wide production of glandless cottonseed.

Figs. C-159, C-160, and C-161 show cottonseed products and uses.

Fig. C-159. Mayonnaise made from cottonseed oil, with 6% cottonseed protein added. (Courtesy, USDA)

Fig. C-160. Foods fortified with cottonseed protein: top left, bread; top right, cookies; bottom left, meatballs; bottom right, Rio punch with strawberries. (Courtesy, USDA)

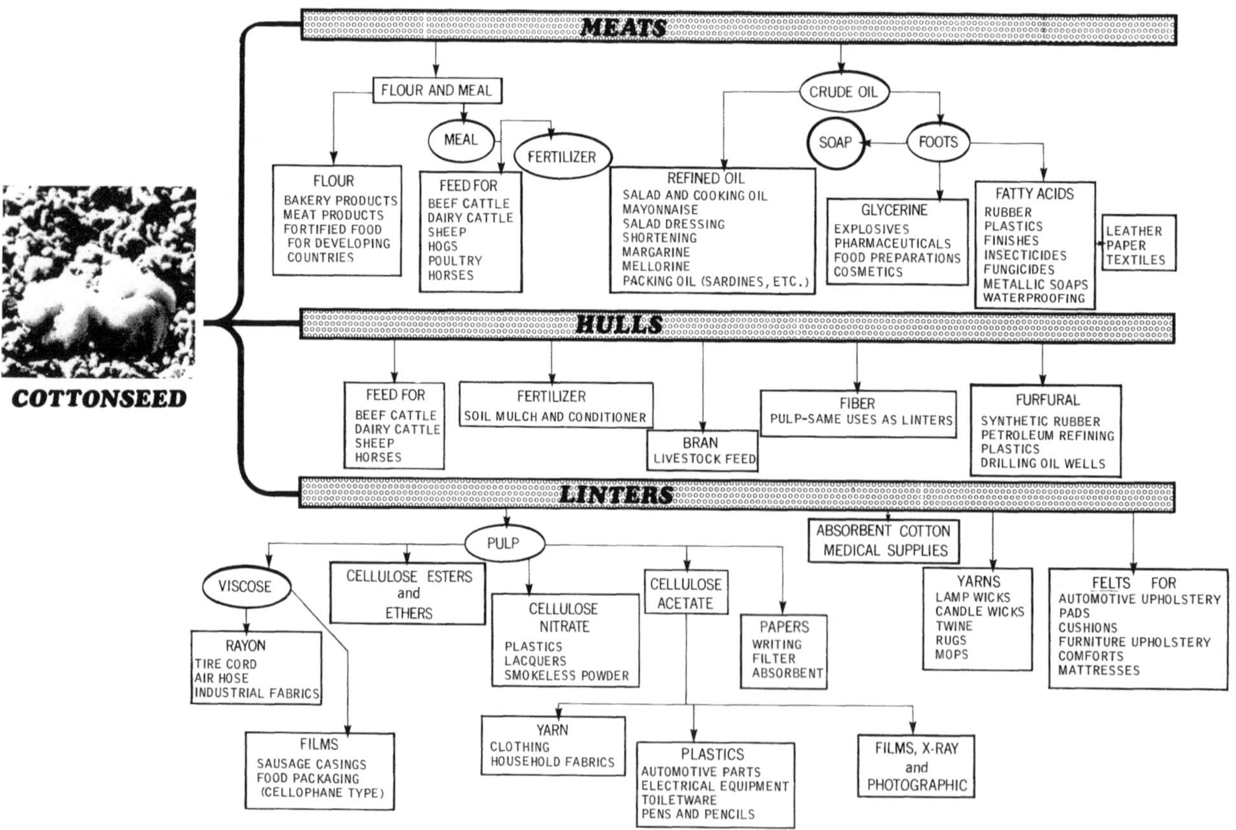

Fig. C-161. Cottonseed products and their many uses. (Adapted from: *Cottonseed and Its Products*, National Cottonseed Products Association, Inc., Memphis, Tenn., pp. 12-13)

Table C-49 presents in summary form the story of cottonseed products and uses.

(Also see OILS, VEGETABLE, Table O-6.)

TABLE C-49
COTTONSEED PRODUCTS AND USES

Product	Description	Uses	Comments
		HUMAN FOODS	
Cottonseed flour (Also see FLOURS, Table F-26 Special Flours.)	Finely ground cottonseed meal, containing 48 to 60% protein. Concentration to 70% protein can be achieved by extraction or particle classification.	In the bakery trade; to increase shelf life (because of its capacity to bind water), reduce fat absorption (especially in doughnuts), improve mixing and machining properties of the dough, and, in some cases, to impart a rich golden color to the bakery product. In meat products, such as hot dogs, luncheon meats, hamburger, sausage, and meat loaf; to retain fat and natural juices in the product, bind and hold the ground meat products together, and minimize shrinkage during cooking. In improving the diets of undernourished populations in the developing countries.	When the hulls are completely removed in the dehulling operation, cottonseed flour will contain substantially in excess of 50% protein. The presence of gossypol has been a major deterrent in the direct food utilization of cottonseed flour. Nevertheless, a food grade cottonseed flour, specialy processed to minimize the toxicological properties of the gossypol, has been marketed for many years. In the future, meat analogs may be created from the protein of cottonseed flour.
Cottonseed oil	A highly quality vegetable oil, processed from cottonseed and used primarily for human food, preferred because of its flavor stability. The crude oil extracted from the seed is usually subjected to the following processes: (1) refined, resulting in a clear yellow oil; (2) bleached to remove the color pigments, resulting in a clear white oil, preferred especially for shortening and margarine; (3) deodorized by exposing to steam under partial vacuum; and (4) winterized by reducing the temperature to 38 to 40°F *(3 to 4°C)*, then removing the portion of the stearine that crystallizes or solidifies.	Salad oil, shortening, margarine, and mellorine. The principal use of cottonseed oil in the U.S. is in salad and cooking oils.	Cottonseed oil is one of the most important food oils, both in the U.S. and elsewhere. It is the second most widely consumed vegetable oil in the U.S., being exceeded only by soybean oil. Mellorine is a frozen dessert that is comparable to ice cream in appearance and nutritive value. The fat content of this product is provided by vegetable oil. In the production of margarine, oil is hydrogenated.
Cottonseed linters	The short fibers removed from the seed after ginning.	Sausage casings.	
		LIVESTOCK FEEDS	
Whole cottonseed	Whole unprocessed cottonseed. On an as-fed basis, it averages about 21.7% protein, 18.2% fiber, and 90% TDN.	For ruminants, especially lactating dairy cows and fattening cattle. For fattening lambs.	Whole cottonseed may have a laxative effect if cattle consume more than 7 to 8 lb *(3 to 4 kg)* per head per day. Because cottonseed is subject to mold, dairymen should use only seed that has been tested and certified aflatoxin-free.

(Continued)

TABLE C-49 *(Continued)*

Product	Description	Uses	Comments
		LIVESTOCK FEEDS	
Cottonseed hulls	The outer covering of the cottonseed.	The major source of roughage for cattle and sheep in cotton production areas.	Hulls are palatable to cattle and sheep. Hulls are easy to handle, require no grinding or final processing, and mix easily. Cottonseed hulls are now being pelleted on a limited scale.
Cottonseed meal	The product which remains after removal of most of the oil from dehulled cottonseed. According to the American Feed Control Officials, it must contain more than 36% protein. Cottonseed meals are designated according to the method of processing as follows: 1. Cottonseed meal, mechanical extracted 2. Cottonseed meal, solvent extracted 3. Cottonseed cake, mechanical extracted 4. Cottonseed flakes, mechanical extracted 5. Cottonseed flakes, solvent extracted	The preferred protein supplement for ruminants over 3 to 4 months of age. In addition to providing protein, it is an excellent source of energy and minerals. To provide a portion of the dietary needs of swine and poultry, provided attention is given (1) to meeting their requirements for all necessary nutrients, and (2) to the tolerance of the species for gossypol.	Cottonseed meal is the second most valuable product from cottonseed; on the average, accounting for 30 to 35% of the total value. It is common practice to remove sufficient hulls so as to produce a cottonseed meal containing better than 41% protein. Cottonseed meal may be marketed in the form of cake, meal, cubes, or pellets. Animals with functioning rumens are not affected by gossypol. Poultry are less sensitive to gossypol than swine, but discoloration of eggs stored longer than 3 months can occur. Addition of iron salts increases gossypol tolerance. Cottonseed meal has a constipating effect on cattle.
Gin trash	A by-product of the ginning of cotton, consisting of burrs, stems, leaf fragments, dirt, and a small amount of lint.	As a roughage for ruminants.	It has about the same feeding value as cottonseed hulls. There is some hazard of pesticide residues from feeding gin trash.
Low gossypol cottonseed meal, mechanical extracted	A mechanical extracted cottonseed meal in which the free gossypol does not exceed 0.04%.	Swine, poultry, and fish.	No gossypol problem exists in fish foods if meal containing no more than 0.04% free gossypol is used to provide up to ½ of the supplemental protein.
Low gossypol cottonseed meal, solvent extracted	A solvent extracted cottonseed meal in which the free gossypol does not exceed 0.04%.	Swine, poultry, and fish.	
Whole-pressed cottonseed, mechanical extracted	A feed product composed of sound, mature, clean, delinted, and *unhulled* cottonseed, from which most of the oil has been removed by mechanical pressure. It must be designated and sold by its crude protein content.	A protein supplement for ruminants.	

(Continued)

TABLE C-49 *(Continued)*

Product	Description	Uses	Comments
		OTHER USES	
Cottonseed hulls	The outer covering of the cottonseed.	In making a plastic, incorporated in the mud used in drilling oil wells; in the production of furfural—a selective solvent used in the production of synthetic rubber, and in petroleum refining.	
Cottonseed linters	The short fibers removed from the seed as the first step in processing.	Absorbent cotton, medical pads, gauze, twine, wicks, and carpet yarn. Felts (or batting) for use in mattresses and other bedding products and in cushioning for furniture and motor cars. Smokeless gun powder, photographic and x-ray films, in many molded products—such as automotive steering wheels and instrument panels, radio cases, flexible pipe, and handles for toothbrushes and tools, manufacture of high grade bond paper, in fibers—viscose fiber is used in industrial fabrics and acetate fiber is used in clothing, draperies, and other household articles. Production of celluose lacquers, laminates, filters, battery separators, and printed electrical circuits.	Markets for linters fall into two broad classes: (1) non-chemical, and (2) chemical. Generally, "first-cut" linters enter nonchemical uses, while "second cut" linters enter chemical products.
Cottonseed meal	The material remaining after hulling and extracting the oil of cottonseed.	As a fertilizer, either alone or in a mixed fertilizer.	Especially for lawns, gardens, and tobacco. Being an organic material, it releases its nutrients more slowly than the inorganic fertilizers.
Cottonseed oil	Usually produced from off-grade oil and soapstock. Both crude oil and soapstock are used to produce fatty acids which, in turn, are used in making a wide range of industrial products.	Emulsifiers, pharmaceuticals, insecticides and fungicides, cosmetics, rubber, plastics, and finishes for leather, paper, and textiles.	

COUMARIN

Coumarin is found in certain plants, including sweet clover. During the process of spoiling, the natural coumarin in sweet clover is converted to dicoumarol (bishydroxycoumarin). Coumarin was formerly used in the synthesis of dicoumarol, but most dicoumarol is made synthetically today.

Coumarin was previously used as an anitclotting agent and as a flavor for tobacco, butter, and medicines. Today, it is used as a deodorizing and odor-enhancing agent, in perfumery and soap. Presently, the use of coumarin in food and feed products is prohibited by the Food and Drug Administration.

The properties and uses of coumarin and dicoumarol are very dissimilar. Today, coumarin is not used as an anticlotting agent, whereas the primary use of dicoumarol is for this purpose.

COUNTY AGRICULTURAL AGENT (FARM ADVISOR)

The county agricultural agent (called Farm Advisor in California) is qualified to give the latest findings of the state agricultural college and the U.S. Department of Agriculture, and to help in applying these findings to the individual farm or ranch.

COW MANURE FACTOR

This unaesthetic name was one applied to what is now known as vitamin B-12.

Before the discovery and synthesis of vitamin B-12, it was recognized that there was an unidentified factor(s) essential for the growth of chicks and pigs fed diets entirely of plant origin, and that the missing nutrient(s) could be supplied by several different animal products, including cow manure, fish meal, meat scrap, etc. This association was responsible for it once being designated "cow manure factor," "animal protein factor," or "chick growth factor." Because it develops rapidly in fecal material, the eating of feces by poultry and pigs was undoubtedly one method of obtaining it prior to the synthesis of vitamin B-12.

COWPEA (BLACK-EYED PEA) *Vigna sinensis; V. unguiculata*

Fig. C-162. The cowpea. (Courtesy, Minnesota Agricultural Experiment Station, University of Minnesota)

Despite its name, this vegetable is more akin to the bean than the pea. Traditionally, black-eyed peas are eaten on New Year's Day in the southeastern United States. The plant has a trailing or bushy vine, but does not climb. The flowers, which are usually in pairs, are yellowish-white to purple.

Yard-long beans (*Dolichos sesquipedalis* or *Vigna sesquipedalis*) are thought by some to be a variety of cowpea, whereas others consider them a different species.

Fig. C-162 shows a cowpea plant and a close up of a pod filled with seeds.

ORIGIN AND HISTORY. Cowpeas are native to the tropical belt north of the equator in Africa, where they have been used as food for thousands of years. Apparently, they were taken from West Africa to India about a millenium or so before Christian times. They also spread northward to the Mediterranean, where they were well known to the ancient Greeks and Romans.

Different varieties of cowpeas were developed and cultivated in India and brought eastward to Southeast Asia, where the variety known as yard-long beans came to be grown.

Much later, the cowpea was brought to the New World by both the Spanish explorers and the African slaves. Hence, they came to be grown in the West Indies and in southeastern United States.

PRODUCTION. Statistics relative to the worldwide production of cowpeas are not available, but it is believed that the total production may be as high as hundreds of thousands of tons per year. The leading areas of production are the tropics of Africa, India, and Southeast Asia, where the temperature ranges between 68°F (*20°C*) and 95°F (*35°C*). Similar climatic conditions have encouraged the growing of cowpeas in the West Indies, southeastern United States, and the Central Valley of California. However, the soil must be well drained if the crop is to do well. In general, cowpeas are more resistant to diseases, drought, and pests than other leading legumes.

Most of the cowpeas grown in Southeast Asia, the West Indies, and the southeastern United States are harvested when the pods and seeds are immature, whereas those grown in Africa, India, and California are harvested after the seeds have matured fully. The yields of dry beans in the developing countries amount to only 400 to 600 lb per acre (*448 to 672 kg/ha*), while yields in California may be as high as 2,500 lb per acre (*2,800 kg/ha*).

PROCESSING. Some of the U.S. crop is canned and frozen, while the rest is marketed as dry beans. In Africa, the beans are used (1) as a staple food (after the appropriate cooking), (2) to make a meal, and (3) as a substitute for coffee beans in the production of a beverage.

SELECTION AND PREPARATION. Canned cooked cowpeas are available in most supermarkets, but it is more economical to cook the dry beans at home since a 1-lb (*0.45 kg*) can of the cooked item yields about 2 cups (*480 ml*), whereas a 1-lb package of the uncooked item yields about 6 cups when cooked.

Soaking time may be reduced greatly by placing the beans plus 3 cups of water per cup of beans in a saucepan, boiling for about 2 minutes, then allowing the beans to soak for an hour, after which they will be ready for cooking.

The expenditures of both fuel and time for cooking cowpeas may be cut considerably by using a pressure cooker instead of a saucepan. About a tablespoon of melted fat or oil should be added to the beans and water (3 cups water per cup of beans) to prevent foaming. The pressure should be maintained for about 30 minutes, then allowed to drop without opening the cooker after cessation of heating.

In the United States, cowpeas, which are also called "black-eyed peas," are well known in the deep south, where some people eat them as part of a dietary pattern called "soul food." (In addition to cowpeas, soul food includes collard greens and turnip greens cooked with fatback pork, cornbread, sweet potatoes, fish, and certain variety meats.) Traditionally, black-eyed peas are cooked with pork. However, there are many other ways to serve them, some of which are shown in Figs. C-163 through C-168.

Fig. C-164. Pickled black-eyed peas. This appetizing bean and pickle combination contains black-eyed peas, cider vinegar or pickle juice, diced sweet or dill pickles, chopped onion, sugar, diced pimiento or sliced raw carrot, salt, and pepper. (Courtesy, The California Dry Bean Advisory Board, Dinuba, Calif.)

Fig. C-163. "Chilada" beans. A hearty dish made from cooked black-eyed peas and a frozen cheese enchilada. The beans and the enchilada are baked together, then sprinkled liberally with grated cheese, red wine, and/or green chili salsa. (Courtesy, The California Dry Bean Advisory Board, Dinuba, Calif.)

Fig. C-165. Barbecue black-eyes. An entree prepared in a skillet from black-eyed peas, chopped onions, sliced celery, diced carrots, bacon, barbecue sauce and mustard. Crushed barbecue potato chips and diced green pepper are mixed in just before serving. (Courtesy, The California Dry Bean Advisory Board, Dinuba, Calif.)

Fig. C-166. Hashed black-eyes. Black-eyed peas cooked with corned beef hash, chopped onion, bacon drippings or butter, soy sauce, salt, and pepper. (Courtesy, The California Dry Bean Advisory Board, Dinuba, Calif.)

Fig. C-168. Black-eyes and shrimp creole. Expensive shrimp goes much farther when it is combined with black-eyed peas, Cheddar cheese, and a special creole sauce on a bed of cooked rice. The sauce is made from tomato sauce, chopped onion, chopped green pepper, minced garlic, butter or margarine, salt, and pepper. (Courtesy, The California Dry Bean Advisory Board, Dinuba, Calif.)

Fig. C-167. Black-eye chicken salad. This dish may be served either hot or cold. It is made from black-eyed peas, cooked pieces of chicken or turkey, sliced celery, sliced green onions, diced green pepper, French dressing, and diced or sliced pimiento. (Courtesy, The California Dry Bean Advisory Board, Dinuba, Calif.)

NUTRITIONAL VALUE. The nutrient composition of cowpeas is given in Food Composition Table F-36.

Some noteworthy observations regarding the nutritive value of cowpeas follow:

1. The cooked immature seeds of the cowpea are lower in calories, but higher in protein than the cooked mature seeds. Both types of seeds are poor sources of calcium, but good sources of iron and phosphorus. However, the immature seeds are a much better source of vitamins A and C. It is noteworthy that the dietary pattern of "soul food" in the southeastern United States usually provides for ample amounts of greens which compensate for nutrients lacking in cowpeas; the greens are (a) rich in calcium and low in phosphorus, and (b) are a good source of vitamins A and C.
2. One cup (*240 ml*) of cooked immature or mature beans contains about the same amount of protein (14 g) as 2 oz (*56 g*) of cooked lean meat, but the beans contain from 50 to 100% more calories.

Protein Quantity And Quality. Cowpeas may provide a major part of the protein in the diets of people in certain developing countries. Therefore, some facts on this subject are noteworthy.

1. The cooked legume provides about twice as much protein per gram of food and per calorie as such cooked cereal grains as rice and corn. Also, the protein quality of legume and grain combinations is significantly higher than that of either food alone.

2. Cowpea protein is superior to that of most other nonoilseed legumes in the content of methionine and cystine, which are the most deficient amino acids in this class of plant foods.

(Also see BEAN[S], Table B-10 Beans of the World; LEGUMES, Table L-2 Legumes of the World; and VEGETABLE[S], Table V-6 Vegetables of the World.)

CRACKLINGS

The term applied to pressed, rendered pork fat remaining after the lard has been extracted.

CRANBERRY *Vaccinium* spp

This berry type of fruit grows on viny plants that belong to the family *Ericaceae*, which also includes bilberries, blueberries, huckleberries, and certain nonedible plants such as azaleas, heathers, mountain laurel, and rhododendrons.

Fig. C-169. American cranberry.

Most Americans associate cranberries with the celebration of Thanksgiving and Christmas, although many people now consume them in one form or another throughout the year.

Brief descriptions of some of the fruits that are commonly called cranberries follow:

• **American cranberry** (*V. macrocarpon*)—The major species that is grown commercially in North America. It is native to the swamps of the northern United States and southern Canada and bears larger berries than the other species.

• **European cranberry; northern cranberry** (*V. oxycoccus*)—A very cold hardy plant that is found in the marshes of northern Canada, northern Europe, and northern Asia. It has smaller berries than the American cranberry.

• **Highbush cranberry; cranberry tree** (*Viburnum tribolum*)—This ornamental plant is not a true cranberry; rather, it is a member of the family *Caprifoliaceae* and is more closely related to the elderberry. It is grown mainly in the northern United States and Canada, where the berries are sometimes used as a substitute for true cranberries.

• **Mountain cranberry; rock cranberry; lingonberry** (*V. vitis-idaea*)—A northern European species that is much esteemed by the people of Scandinavia.

• **Southern cranberry** (*V. erythrocarpum*)—A wild species that is native to the mountains of the eastern United States, from Virginia to Georgia. The fruit has a notably good flavor.

ORIGIN AND HISTORY. Although various species of cranberries grow wild in Europe, North Asia, and North America, it is not certain how long they have been used as a food by people. Most species are quite tart and require liberal amounts of a sweetener to make them palatable. Hence, the inadequate supply of sugary substances may have been a factor in the limited utilization of this fruit until the development of the beet sugar and cane sugar industries in recent times.

Lack of refined sugar did not prevent the American Indians from consuming cranberries since they sometimes cooked them with honey or maple sugar. The Indians also used the fruit as a source of dye, and as a poultice on wounds. (Raw cranberries have an astringent effect which contracts tissues and stops bleeding.) Perhaps, the primitive cranberry sauce made by the Indians was served with turkey at the early Thanksgiving feasts, since the New England colonists exported the berries to England early in the 18th century. At that time the fruit was picked from the wild plants which grew in many of the local bogs.

It is believed that first commercial cultivation of cranberries was initiated in the early 1800s at Dennis, Massachusetts by Henry Hall, who observed that abundant amounts of large fruits were produced when the prevailing winds and tides swept some sand into his bog. A moderate amount of sand in the bog helps to stifle the growth of weeds without interfering with the cranberry plants, because the latter have deeper roots.

The growth of the cranberry industry in the late 1800s was aided by the increased availability of sugar for sweetening the berries. However, cranberry utilization reached a plateau and remained there for many years because sales of the sauce boomed around Thanksgiving and Christmas, but were in a slump for the rest of the year. This situation improved markedly when various cranberry juice drinks were marketed in the 1960s. As a result, there is now a fairly steady utilization of the fruit throughout the year, since the juice drinks have become the best-selling cranberry product.

PRODUCTION. Almost all of the world cranberry crop is produced in the United States, where over 154 thousand metric tons are grown annually. Fig. C-170 shows the amounts contributed by the leading cranberry producing states.

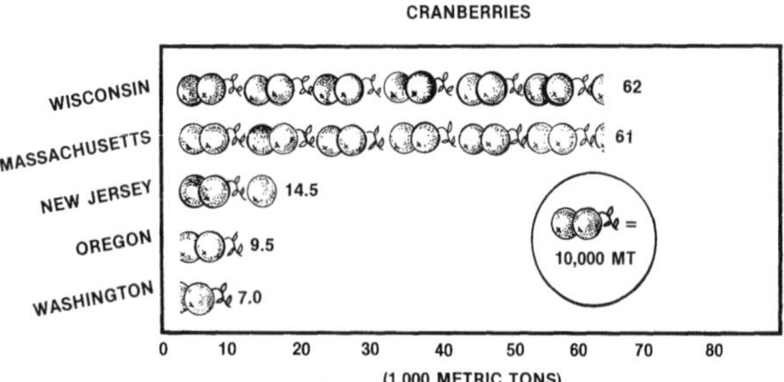

Fig. C-170. Cranberry production in the United States in 1990. (Data from *Agricultural Statistics*, 1991, USDA, p. 195, Table 289)

Most of the Canadian cranberry crop comes from the provinces of British Columbia and Nova Scotia, but small amounts are also grown in New Brunswick, Prince Edward Island, Quebec, and Ontario.

The production of American cranberries is limited to certain areas because (1) best results are achieved when the plants are grown in low-lying acid bogs that may be flooded when necessary, and (2) the climate of the growing region should not be colder than Nova Scotia nor warmer than New Jersey.

Bogs are prepared for planting by removal of unwanted vegetation followed by leveling of the soil, then covering with a layer of sand. Cranberry stem cuttings are planted by inserting them down through the sand into the underlying peat of the bog. About 3 years are required for new cranberry plantings to start bearing fruit.

The bogs are flooded at various times during the year in order to (1) protect the plants against frost during the winter and spring, and (2) provide sufficient water and prevent damage by heat and insects during hot, dry spells in the summer. Growers apply various mixtures of nitrogen, phosphorus and potassium judiciously when needed, but avoid overstimulating growth of the vines, which reduces the yield of the fruit. Cranberries require nitrogen in the form of ammonium salts since the plants cannot utilize nitrates.

The fruit is usually harvested in September and October. Deeply red-colored berries (those with a high content of anthocyanin pigments) bring a premium price because they are most suitable for processing into the various cranberry juice cocktails. Hence, it often pays for some growers to leave the fruit on the vine until it has ripened fully. On the other hand, the color can be intensified in paler berries that were picked prior to full ripening by exposing them to a combination of ethylene gas and light. (Ethylene is produced naturally by certain fruits when they ripen.)

Cranberry harvesting has long been done by hand with the aid of specially designed, toothed, wooden scoops that the pickers use to rake the berries from the vines. Another means of harvesting involves flooding the bogs, then using a machine that knocks the berries loose from the plants so that they float in the water. Then, the floating fruit is pushed to a corner of the bog where a conveyor loads it onto trucks. Other types of mechanical pickers are used on dry (nonflooded) bogs. Cranberries are sometimes referred to as "bounceberries" because one of the methods of sorting out the high-quality fruit from the culls involves bouncing the berries against barriers consisting of slanted boards. The high-quality fruit bounces over the barriers, while the culls roll into the rejected pile. After bouncing, the sound fruit is further sorted to remove green, irregular, and decayed berries.

PROCESSING. About 80% of the cranberry crop is processed. Some of the processed berries are made directly into cranberry juice drinks and the traditional types of cranberry sauces (mainly those made from either pureed or whole berries). The rest of the processed berries are frozen in bulk for later production of these items.

The ever-increasing production of cranberry juice drinks in recent years has sparked research on utilization of the fruit residue that remains after juice has been pressed from the berries. One of the most promising developments has been the production of cranberry cell wall material (dubbed "cranberry sponge" by researchers at Massachusetts Institute of Technology (MIT) in Cambridge, Massachusets. The material has an extraordinary water binding capacity which makes it suitable for imparting juiciness to fabricated foods. For example, it may be used to overcome the dryness that is now a problem in the various meat analogs made from vegetable protein.

SELECTION. Good quality in cranberries is indicated by a fresh, plump appearance, combined with high luster and firmness. Poor quality is indicated by shriveling, dull appearance, and softness. Cranberries will show more or less moisture shortly after removal from cold storage. Such dampness does not indicate poor quality and soon disappears, but cranberries that show moisture caused by injury or damage are usually sticky, leathery, or tough, and the flesh may be discolored.

Fresh cranberries can be stored in a refrigerator for several months, or they may be kept in deep-freeze storage for several years with only minimal loss of moisture. However, the frozen berries become very soft upon thawing and should be utilized immediately in order to avoid spoilage.

PREPARATION. The various cranberry products may be utilized in ways such as those which follow:

1. Fresh cranberries may be made into an uncooked relish by chopping them in a food grinder with quartered oranges (after removal of any seeds), apples, and/or other raw fruit. The chopped fruit mixture should be sweetened with honey or sugar and stored in a refrigerator until it is used.

2. The fruit is good when cooked and sweetened in preparations such as gelatin desserts, jam, jelly, pies (cranberries alone or in combination with apples, mincemeat, or peaches), and sauces.

3. Canned cranberry sauce (pureed or whole berry) is not only good with meats or poultry, but also when substituted for the fillings, spreads, and toppings commonly added to baked products, ice cream, puddings, and other desserts.

4. Cranberry juice cocktails may be used alone or when mixed with unflavored soda water or gingerale.

NUTRITIONAL VALUE. The nutrient compositions of various cranberry products are given in Food Composition Table F-36.

Some noteworthy observations regarding the nutrient composition of some of the more common cranberry products follow:

1. Raw cranberries are fairly low in calories (46 kcal per 100 g) and carbohydrates (11%). They are a good source of fiber and bioflavonoids, and a fair source of potassium and vitamin C. Hence, they may be sweetened with an artificial sweetener and used in various low calorie dishes.

2. Cranberry-orange relish that has been sweetened with a caloric sweetener (corn syrup, honey, sugar, etc.) usually contains about four times the calories and carbohydrates that are present in raw cranberries, but the levels of the other nutrients are approximately the same in the relish and the raw fruit.

3. Canned cranberry sauce is high in calories (145 kcal per 100 g) and carbohydrates (38%), and low in most other nutrients.

4. Most types of cranberry juice drinks are sweetened with caloric sweeteners and fortified with added vitamin C. Hence, they are likely to contain moderate amounts of calories (about 60 kcal per 3½ oz or *100 g*) and carbohydrates (about 15%) and to be good sources of vitamin C. It is noteworthy that artificially sweetened cranberry drinks and unsweetened juices are available which contain only about ¼ of the calories and carbohydrates of the beverages that contain added caloric sweetener(s).

5. Cranberry juice has long been prescribed by some doctors to reduce the likelihood that calcium phosphate stones will form in the urinary tract. (Cranberries contain quinic acid which is not broken down in the body but is excreted unchanged in the urine. Hence, it serves to make the urine mildly acid, a condition that keeps calcium and phosphate ions from forming insoluble stones. Some doctors also believe that in certain cases stones already formed may be dissolved.)

6. New England folk medicine has it that a mixture of boiled cranberries and seal oil will reduce the severity of a gallbladder attack. The truth of this belief has not been confirmed by medical science, but it is noteworthy that acid (amply present in cranberries) and fat (seal oil) stimulate the release of bile from the gallbladder, thereby preventing the stagnation of bile that sometimes leads to the formation of gallstones.

7. The color of ripe cranberries is due mainly to anthocyanin pigment(s) (anthocyanins are a type of bioflavonoid) that certain European researchers have found to (a) accelerate the regeneration of visual purple (the pigment in the eyes that is involved in vision), (b) aid in the adaptation of the eyes to the dark, and (c) inhibit the formation of tumors.

NOTE: These findings have not been confirmed by researchers in the United States. Hence, the authors make no claims regarding the benefits of consuming foods that contain anthocyanins; rather, the findings are reported (1) for informational purposes, and (2) in order to stimulate interest in additional research.

(Also see BIOFLAVONOIDS; and FRUIT[S], Table F-47 Fruits of the World.)

CREAM, CLOTTED (DEVONSHIRE CREAM)

This is a high fat content cream, prepared by allowing cream to rise on milk, setting it by heating and cooling then skimming it from the underlying skim milk.

CREAM, CORNISH

This is a high fat content cream, similar to clotted cream or Devonshire cream. But it is prepared by scalding the cream alone, not floated on a layer of milk.

CREAM LINE

The place where the risen cream meets the milk.

CREAM LINE INDEX

This is the ratio of the percentage cream layer to the percentage of fat in the milk. It is used as a test of the milk. In ordinary bulk milk it is about 1:7.

CREAM OF TARTAR (POTASSIUM BITARTRATE, ACID TARTRATE, ACID TARTRATE OF POTASH [$KHC_4H_4O_6$])

Take your choice of name, but whatever you call it, this acid is a white odorless powder with an acid taste. Besides its use in leavenings, it can be used as a diuretic and to neutralize alkaline urine.

(Also see BAKING POWDER AND BAKING SODA.)

CREAM, PLASTIC

Cream that has been centrifuged at high speed causing it to form an oil-in-water emulsion; used for preparation of cream cheese and whipped cream.

CREATINE

A nitrogenous compound formed in the liver, which is converted to phosphocreatine in muscle, where it serves as a source of high energy phosphate for muscle contraction. Greater than normal amounts of creatine are excreted in the urine when muscle disorders and other abnormal conditions are present. In muscular dystrophy, the muscle is unable to accept creatine, and it is excreted in the urine.

The anhydride of creatine is creatinine, in which form it is found in urine. Changes in the excretory function of the kidneys are reflected in plasma urea and creatinine concentrations. Glomerular filtration rate (GFR) and overall kidney function is conveniently estimated by measuring creatinine clearance in the volume of urine excreted in 1 minute. Its value is about 125 ml/minute in healthy young men. It is usually a little less in young women; and at 70 years of age it is about 75% of the value in youth. Kidney failure rarely produces symptoms until the GFR falls below 30 ml/minute.

CREATININE

A nitrogenous compound arising from protein metabolism and excreted in the urine.

(Also see PROTEINS.)

CREATININE COEFFICIENT

The amount of urinary creatinine excreted in 24 hours divided by the body weight in kilograms. Creatinine is formed from phosphocreatine in muscle and is excreted in fairly constant amounts each day. Hence, the amount of urinary creatinine which is excreted under uniform conditions may be taken as an indication of the total muscle mass.

CREME

Cream or cream sauce as used in cookery.

CRESS (GARDEN CRESS) *Lepidium sativum*

The young leaves and stems of this tasty and nutritious plant are commonly used in salads. Cress, which is also called garden cress to distinguish it from watercress, is an annual and a member of the mustard family of vegetables (*Cruciferae*). Fig. C-171 shows garden cress.

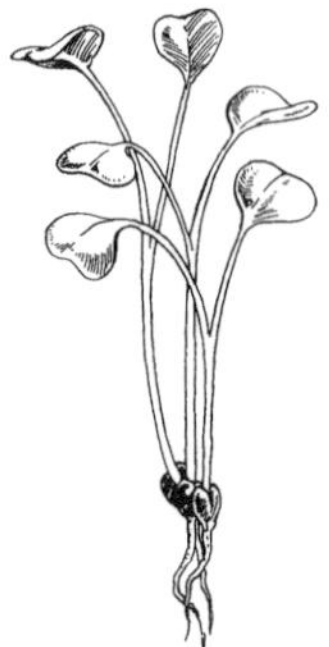

Fig. C-171. Garden cress, a tasty salad green.

ORIGIN AND HISTORY. Cress appears to be native to the broad area that includes western Asia, the Near East, Ethiopia, and perhaps even parts of Europe. There is evidence that the Europeans have cultivated the plant since Greek and Roman times. Now, it is grown on a small scale in many parts of the world.

PRODUCTION. No statistics on the annual production of cress are available because it is grown mainly on a small scale for local markets.

Cress is grown in the spring, because hot weather causes the plant to go to seed rapidly. Usually, the seeds are planted as early in the year as possible. A highly fertile soil is needed to produce the rapid growth that results in a high quality vegetable.

The young shoots may be harvested continuously, starting at about 6 weeks after planting, providing that the crown of the plant is not damaged. If some of the crop goes to seed because of the onset of hot weather, it is likely that the seeds dropped on the soil will sprout the following spring.

PROCESSING. There is no indication that any of the cress crop is processed. However, it would seem to be suitable for freezing, as is done with chives and parsley.

SELECTION AND PREPARATION. High quality cress is fresh, young, tender, and green; and free from insect injury, coarse stems, dry or yellowing leaves, excessive dirt, or poor development.

Raw cress may be used as a garnish for various hot or cold dishes, and as an ingredient of salads. Also, it may be added to casseroles, omelets, soups, and stews, to which it adds a refreshing flavor.

CAUTION: Cress, like other members of the mustard family of vegetables (*Cruciferae*), contains small amounts of substances called goitrogens that interfere with the utilization of iodine by the thyroid gland. This effect may result in an enlargement of the thyroid if very large amounts of cress are consumed while on an iodine-deficient diet. Any possible goitrogenic effects of cress may be offset by ample amounts of dietary iodine, the best sources of which are iodized salt, ocean fish, seafood, and seaweeds.

NUTRITIONAL VALUE. The nutrient composition of cress is given in Food Composition Table F-36.

Some noteworthy observations regarding the nutrient value of cress follow:

1. It is high in water content (about 90%) and low in calories. (One cup [*240 ml*] contains about 30 kcal.)

2. The raw vegetable is an excellent source of vitamin A and a good source of vitamin C. However, cooking destroys about 20% of the vitamin A and about 50% of the vitamin C.

(Also see VEGETABLES[S], Table V-6 Vegetables of the World.)

CRETINISM

A condition originating during pregnancy or early infancy characterized by stunted physical and mental development, caused by severe thyroid deficiency.

(Also see DEFICIENCY DISEASES, Table D-1 Major Dietary Deficiency Diseases; and GOITER.)

CROHN'S DISEASE (REGIONAL ENTERITIS)

This mysterious and frustrating intestinal disorder goes by a variety of names, including Crohn's Disease (after the New York physician who first described it), regional enteritis, ileitis, and granulomatous colitis, depending on which part of the intestinal tract is affected. It is an intermittent and chronic inflammation of the small and large intestines, in which hardening, thickening, and eventual ulceration of parts of the bowel lining occur. Further details about the disease follow.

• **Symptoms**—Sufferers experience abdominal cramps, diarrhea, rectal bleeding, fever, lack of appetite, weight loss, and vomiting.

• **Cause**—The cause is unknown, although researchers suspect that a virus or a flaw in the body's immune system may be involved. As with many gastrointestinal disorders, emotional stress may be a contributing cause, however, when the condition is chronic, it is difficult to tell whether emotional factors are a cause or a result.

• **Incidence and distribution**—In the United States, the exact incidence is unknown. However, it is estimated that Crohn's Disease afflicts one million Americans, including 100,000 children. The disorder is worldwide in distribution and occurs among all races. The disease is more common in certain families than in others.

• **Treatment**—In a disease of unknown cause and unpredictable course, a cure is similarly elusive. In treating the disease, doctors use drugs that supress inflammation and the immune system. In severe cases, they must resort to surgery, cutting away diseased portions of bowel. The case fatality rate for all patients with Crohn's Disease, regardless of treatment, probably ranges from 5 to 10%.

CROQUETTES

Croquettes can be made from chopped beef, chicken, crab, egg, fish, ham, lobster, oyster, salmon, sweetbreads, or veal. The chopped, cooked meat is mixed with seasonings and flour, egg etc. This mixture is formed into the traditional croquette shape, which is like a small cone. They are rolled in crumbs, dipped in beaten egg, rolled in crumbs again, and fried in deep fat. They are served with or without a sauce, depending on the recipe of choice.

CROUTONS

Dried cubes of bread, seasoned or unseasoned, homemade or commercially manufactured, which are added to soups and salads.

CROWDIES

A type of fermented milk in Scotland.

CRUDE FAT

Material that is extracted from moisture-free foods by ether. It consists largely of fats and oils with small amounts of waxes, resins, and coloring matter. In calculating the energy value of a food, the fat is considered to have 2.25 times as much energy as either the nitrogen-free extract or the crude protein.

(Also see ANALYSIS OF FOOD; and FATS AND OTHER LIPIDS.)

CRUDE FIBER (CF)

That portion of foods composed of cellulose, hemicellulose, lignin, and other polysaccharides which serve as structural and protective parts of plants. It is determined by its insolubility in dilute acids and alkalis.

(Also see ANALYSIS OF FOODS; CARBOHYDRATE[S], UNAVAILABLE; and FIBER.)

CRUDE PROTEIN (CP)

Refers to all the nitrogenous compounds in a food. The nitrogen content of protein averages about 16% (100 ÷ 16 = 6.25). Crude protein is determined by finding the nitrogen content and multiplying the result by 6.25.

(Also see ANALYSIS OF FOODS; and PROTEINS.)

CRYPTOXANTHIN

A yellow pigment which may be converted to vitamin A in the body. It is found mainly in such foods as corn, oranges, and paprika.

(Also see VITAMIN A.)

CUCUMBER AND GHERKIN *Cucumis* spp

These crops together rank among the top ten vegetable crops of the world, although in the developed countries they are used more as condiments than as vegetables. The cucumber (*C. sativus*) is long and cylindrical, whereas the closely-related gherkin ((*C. anguria*) is much smaller and is shaped more like a miniature football. Sometimes, small cucumbers are erroneously called gherkins.

However, the growing, processing, preparation, and nutritional values of the two vegetables are sufficiently similar to warrant their joint coverage here without undue concern over the minor differences between them. Both species belong to the gourd or melon family (*Cucurbitaceae*), which also includes pumpkins and squashes. Only the immature fruits of these plants are commonly consumed, but some peoples in Southeast Asia also eat the young leaves and the seeds of the mature fruit.

Fig. C-172. A cucumber and a gherkin.

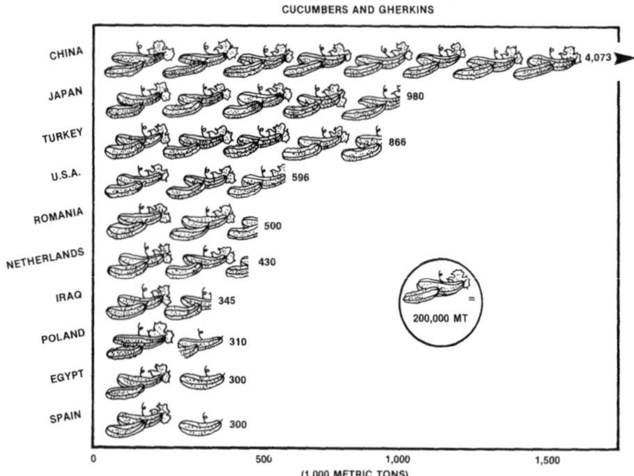

Fig. C-173. The ten top cucumber- and gherkin-producing countries. (Data from *FAO Production Yearbook*, 1990, FAO/UN, Rome, Italy, Vol. 44, p. 136, Table 55)

ORIGIN AND HISTORY. Both the cucumber and the gherkin are tropical plants. However, the modern form of the cucumber is not found growing wild. Hence, it must have been developed from a wild ancestor. Present thinking is that the cucumber did not originate in India or Africa as was first thought, but that it originated in Southeast Asia where seeds judged to be about 12,000 years old have been found.[50] Perhaps the cucumber was brought from the Far East to Central Asia and India by ancient land travelers, or by the seafaring peoples who migrated as far as Madagascar. However, the gherkin appears to have been developed from a wild plant that is native to Africa.

The ancient Egyptians ate cucumbers and/or gherkins, for it is recorded that the Hebrews longed for them while being led out of Egypt by Moses. Cucumbers were also highly esteemed by the ancient Greeks and Romans, as evidenced in their writings. It is not certain when the fruit was first pickled, but it is noteworthy that Roman emperors had pickles imported from Spain. Gherkins were brought to the New World during the early days of African slave trading and became an important crop in the West Indies.

In recent times, the English have developed seedless varieties of cucumbers that set fruit without pollination and are grown exclusively in greenhouses from which bees are excluded. Pollination of these varieties results in seedy, bitter fruit that is unsuitable for marketing.

PRODUCTION. The world production of cucumbers and gherkins is about 13.3 million metric tons. Fig. C-173 shows the amounts contributed by each of the leading producing countries.

About 591,000 metric tons of cucumbers were produced in the United States in 1990. This statistic includes an unspecified quantity of gherkins. Fig. C-174 shows the amounts contributed by each of the leading producing states.

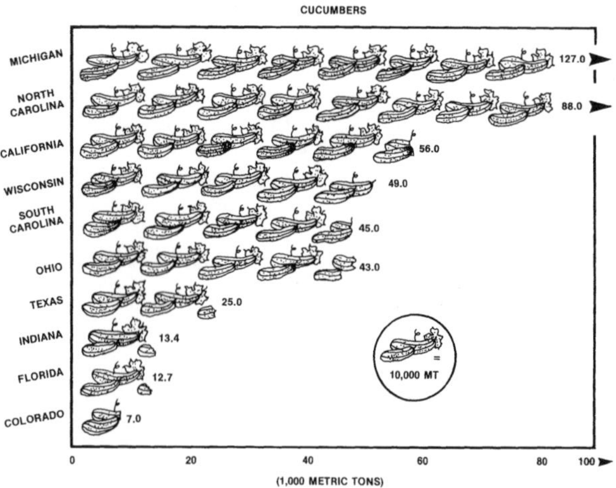

Fig. C-174. The ten top cucumber-producing states. (Data from *Agricultural Statistics*, 1991, USDA, p. 154, Table 216)

[50]Solheim, W. G., "New Light on a Forgotten Past," *National Geographic*, Vol. 139, 1971, pp. 338-339.

Cucumber seeds are planted in the field when there is no longer any danger of frost and the temperature ranges from 60° to 75°F (*16° to 24°C*). Furthermore, the plants grow best when the daytime temperature is around 85°F (*30°C*) and the night temperature ranges from 65° to 70°F (*19° to 22°C*). Hence, production of this crop during the winter is limited to the warmest areas of the southern states. In the northern states, the seeds may be planted early in the spring in greenhouses or hotbeds, then the seedlings may be transplanted in the field when the weather warms up. Elsewhere, cucumbers are grown in the field from spring to fall.

The soil should be deeply cultivated, well drained, neutral to mildly alkaline, and rich in both inorganic and organic nutrients. Therefore, many growers may find it necessary to add lime, nitrogen, phosphorus, potassium, and compost or manure. However, compost should not be made from the residues of plants belonging to the gourd or melon family, because diseases may be spread among the different species of this family.

Irrigation is often needed when the crop is grown commercially. Furrow irrigation is the best way to water the plants because sprinklers wet the leaves and increase the likelihood of disease.

Many home gardeners and greenhouse producers of cucumbers (in Great Britain, the entire crop is grown in greenhouses) train the vines to climb trellises or special wire frameworks so that the maximum production of fruit per unit of land may be obtained. One vine may yield 100 cucumbers.

The fruit is ready for picking in about 2 months after planting. Removal of the cucumbers allows the vines to set more fruit, whereas the vines stop fruiting if the fruit is not picked. Cucumbers are rarely allowed to reach maturity because the development of mature seeds requires large amounts of nutrients and stresses the vines excessively. The worldwide average yield of cucumbers is 12,856 lb per acre (*14,413 kg/ha*), but yields as high as 265,000 lb per acre (*296,800 kg/ha*) are obtained in British greenhouses. In comparison, the U.S. average yield per acre is 11,319 lb (*12,689 kg/ha*), but yields as high as 31,600 lb (*35,392 kg/ha*) are obtained in California fields.

PROCESSING. Over 70% of the U.S. cucumber and gherkin crop is made into pickles. The major operations utilized in the production of cucumber pickles are (1) fermentation of whole or sliced cucumbers in a concentrated salt solution called a brine; (2) soaking of the pickles in hot water to remove some of the salt; and (3) canning the pickles in a mixture of vinegar and various seasonings.

SELECTION. Cucumbers should be firm, fresh, bright, well-shaped, and medium to dark green in color. A small amount of whitish green color at the tip and on the ridged seams is not objectionable in certain varieties. The flesh should be firm and the seeds immature.

Withered or shriveled cucumbers should be avoided as they are apt to be tough and somewhat bitter. Overmature cucumbers are generally dull or yellowed in color, and have an overgrown, puffy appearance. Cucumbers in this condition are excellent for certain kinds of pickles, but are not suited to fresh use. Decay usually appears on the surface first as water-soaked spots, but later as sunken, irregular areas.

PREPARATION. Sometimes, the skins of the fresh cucumbers sold in retail stores are coated with wax. Hence, it might be a good idea to peel them before using them. Furthermore, a wax coating makes them unsuitable for making pickles, because it prevents the absorption of the pickling solution.

Raw cucumbers or pickles may be diced or sliced and added to relish trays, salads, and sandwiches. It is noteworthy that the slightly bitter flavor of the raw vegetable is improved by dressing with acid-flavored items such as lemon juice, mayonnaise, sour cream, vinegar, and yogurt; whereas pickles are acid-flavored to begin with, and have no need for such dressing.

Although cucumbers are rarely cooked, they are an adventure in good eating when (1) boiled and served with a special sauce; (2) grated and added to a soup; (3) stewed or roasted with meat; (4) sliced, dipped in egg batter or flour, and fried; (5) stewed with tomatoes; and (6) hollowed out by removal of the seeds, stuffed with bread crumbs, cheese, chopped eggs, meat, nuts, or poultry, and baked.

NUTRITIONAL VALUE. The nutrient compositions of various forms of cucumbers are given in Food Composition Table F-36.

Some noteworthy observations regarding the nutrient composition of cucumbers follow:

1. Fresh cucumbers have a very high water content (95%), and they are low in calories (15 kcal per 100 g) and most other nutrients. However, they are a fair source of iron, potassium, vitamin A, and vitamin C. Most of the vitamin A is in the rind., Hence, peeled cucumbers contain only traces of this vitamin. Many people find that the raw vegetable is cool and refreshing. It is also filling, but not fattening.

2. The different types of cucumber pickles and relishes vary widely in their compositions, which depend mainly upon the ingredients and procedures used in processing. For example, dill pickles, sour pickles and sour relishes contain from 93 to 95% water and only 10 to 19 Calories (kcal) per 3½ oz (100 g) serving, whereas the sweet pickles and relishes contain on the average only 62% water and a whopping 142 Calories (kcal) per 100 g. The much higher caloric contents of the sweet items are due mainly to the sugar and/or other sweeteners that are used in their production.

It is noteworthy that the nutrient compositions of pickles differ from that of fresh cucumbers in that the former generally have a high sodium content and are poor sources of vitamins A and C. Sour pickles contain much more sodium and iron than sweet pickles. Some types of dietetic pickles are very low in sodium.

3. Practitioners of herbal medicine have long held that the consumption of cucumbers induces copious urination, which may be helpful in ridding the body of excessive water and purging the blood of potentially harmful substances such as uric acid. However, pickles would be more likely to induce retention of water in the body by virtue of their high sodium content.

NOTE: The authors make no claim regarding the veracity of these beliefs. Rather, they are presented here to stimulate further research on the subject.

(Also see VEGETABLE[S], Table V-6 Vegetables of the World.)

CURD

The coagulated or thickened part of milk. Curd from whole milk consists of casein, fat, and whey, whereas curd from skim milk contains casein and whey, but only traces of fat.

(Also see MILK PRODUCTS.)

CURD TENSION

A measure of toughness of the curd formed from milk by the digestive enzymes and used as an index of the digestibility of the milk.

CURRANT *Ribes* spp.

This berry type of fruit grows on bushes that are members of the family *Saxifragaceae.*

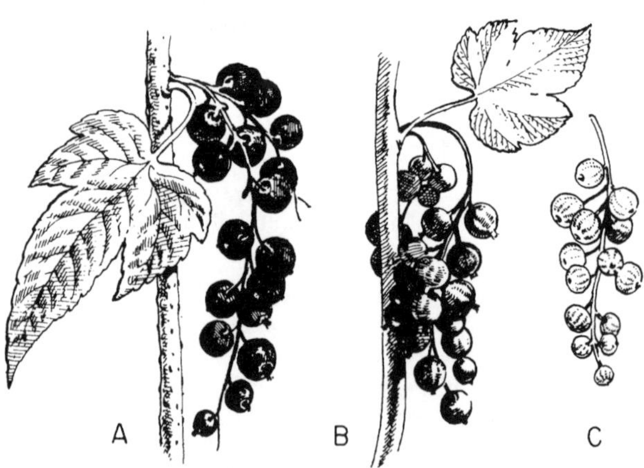

Fig. C-175. The major types of currants: (A) black currant (*R. nigrum*), (B) red currant (*R. sativum*) and (C) white currant (a pigmentless variety of red currant).

Currants are closely related to gooseberries, except that they are smaller and grow in clusters. Furthermore, the two types of plants do not crossbreed (hybridize) readily.

ORIGIN AND HISTORY. Various species of currants grow wild in northern Europe, northern Asia, and northern United States. Most likely the fruit was gathered from wild plants in prehistoric times, since seeds were found at the sites of ancient settlements in Denmark.

It is believed that currants were first cultivated in the 15th century in the countries that border on the Baltic Sea. They were grown in England, France, and Holland in the 1600s and were brought to the New World by the early settlers.

The European species of the fruit were grown throughout the northern United States and Canada until the end of the 1800s when an outbreak of white pine blister rust was traced to black currants, which are an intermediate host of the disease organism. After that, the growing of black currants was prohibited in many of the pine-growing areas of the United States.

Now, only red currants are grown commercially in the United States, but black currants may still be found growing wild in the northern part of the country and as cultivated plants in small gardens.

PRODUCTION. The world production of currants averages about 527,194 metric tons, about 81% of which comes from Europe.

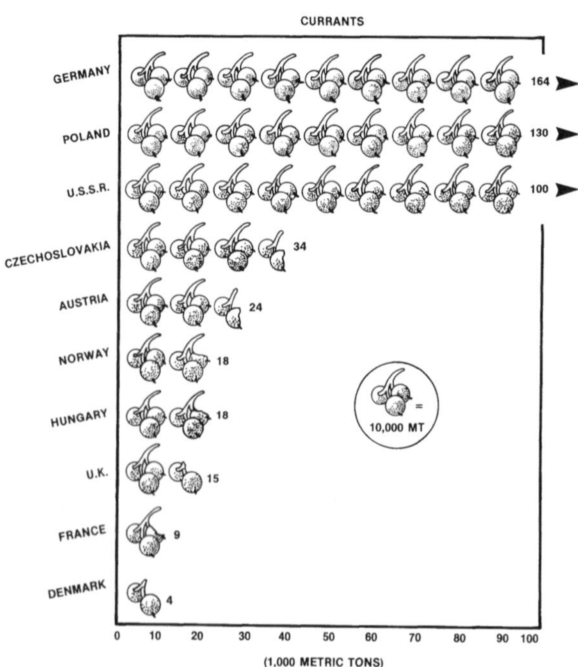

Fig. C-176. The leading currant-producing countries of the world. (Data from *FAO Production Yearbook*, 1990, FAO/UN, Rome, Italy, Vol. 44, p. 171, Table 75)

Currants grow best in areas with cool moist climates like those of northern Europe, the northern United States, and Canada. Furthermore, they are among the most cold hardy of the commercially grown berry crops and can withstand temperatures as low as -40°F (*-40°C*). The plants are propagated by hardwood cuttings taken in the fall and rooted in a nursery. Well-rooted cuttings are planted in the field in the fall or early spring. Sometimes, they are planted in orchards between the rows of fruit trees because moderate amounts of shade protect the currant bushes against sun scalding in hot, dry weather.

The soil should be a fertile, well-drained, clay or silt that has been prepared for planting by removal of the weeds. Some type of nitrogenous fertilizer is usually applied, or a green manure or cover crop is grown and turned under prior to the planting of the currant bushes. Competitive weeds are eliminated by shallow cultivation or mulching with straw. Mulching also helps to keep the soil cool during hot weather. Irrigation is required during hot dry weather.

Currants that are to be used for making jam or jelly are usually picked before they are completely ripe because their pectin content diminishes with ripening. (Pectin is the substance which brings about gelling.) On the other hand, the fruit that is to be eaten raw is picked after it has ripened fully and developed its maximum sweetness. The usual picking procedure is to remove entire clusters of the berries since they keep much better when attached to the cluster stem than when separated from it.

PROCESSING. Much of the currant crop is frozen in order to preserve the fruit for sale to bakeries, and for subsequent processing into jam, jelly, juice, and wine. It is noteworthy that in the past dried currants were used liberally in coffee cakes, holiday breads, and hot cross buns, but are now often replaced by Zante currants (small raisins that resemble dried currants).

SELECTION AND PREPARATION. Most people are not likely to find fresh currants available in their locality unless (1) commercial growers are nearby, or (2) they know someone who grows the bushes in a garden. When the fresh fruit is available, it makes an excellent dessert dish with cream and a little honey or sugar.

Dried currants make tasty additions to baked desserts and puddings. They may be softened a little by soaking them in hot water before using them in these dishes.

NUTRITIONAL VALUE. The nutrient compositions of fresh and dried currants are given in Food Composition Table F-36.

Some noteworthy observations regarding the nutrient composition of currants follow:

1. Fresh black currants contain moderate amounts of calories (54 kcal per 100 g) and carbohydrates (13%). They are rich in fiber, potassium, vitamin C, and bioflavonoids (vitaminlike substances that have beneficial effects when consumed along with vitamin C), and are a fair to good source of calcium, phosphorus, iron, and vitamin A.
2. Fresh red currants and white currants have about the same caloric and carbohydrate contents as black currants, but contain only about one-fifth as much vitamin C. The levels of the other nutrients noted above are about 60% of those in black currants, except that the fiber content of red and white currants is about 50% higher.
3. Dried currants are rich in calories (243 kcal per 100 g) and carbohydrates, but contain almost no vitamin C. However, they are rich in fiber and potassium, and are a good source of calcium and iron.
4. The colors of black and red currants are due mainly to their content of anthocyanin pigments, which are bioflavonoids.

(Also see FRUIT[S], Table F-47 Fruits of the World.

CUSHING'S SYNDROME

A group of symptoms associated with Cushing's disease, first described in 1932 by Dr. Harvey Williams Cushing, famous American brain surgeon. The disease, which seems to affect women primarily, is due to the oversecretion of adrenocortical hormone. Excesses of these hormones promote obesity of the trunk, face, and buttocks. It also produces a weakening of the muscles, because the protein in muscle is broken down and converted to sugar in the blood. Hence, there may also be a diabeticlike condition. Older people who have had this condition for a long time tend to develop a hump on their back (buffalo hump) which results from an abnormal deposit of fat, and they also develop a rounded face (moon face). Women with Cushing's disease develop excessive hair growth, such as mustaches and beards.

If the diagnosis is made relatively early, before the onset of heart and kidney involvement, a patient can be helped. When the adrenal gland becomes enlarged, a cure may be effected by surgery. In a third to a half of the cases where the pituitary gland is involved, x-ray therapy may halt or slow the progress of the disease.

(Also see ENDOCRINE GLANDS.)

CYANOCOBALAMIN

Same as vitamin B-12.

(Also see VITAMIN B-12.)

CYANOSIS

A dark bluish or purplish coloration of the skin and mucous membrane due to deficient oxygenation of the blood (as in some congenital heart defects).

CYCLAMATES

Compounds which were formerly used as artificial sweeteners. Their use for this purpose was banned by

FDA because it was found that toxic metabolites were sometimes formed. However, cyclamates were reclassified as drugs so that they might be used under medical supervision.

CYSTATHIONINURIA

An inborn error of metabolism in which cystathionine is at high levels in the blood and urine. The condition appears to be due to a defect in the metabolism of the sulfur-containing amino acids. Sometimes, cystathioninuria may be corrected by the administration of vitamin B-6 (pyridoxine).

(Also see INBORN ERRORS OF METABOLISM.)

CYSTEINE

A nonessential, sulfur-containing amino acid. However, it may to a limited degree spare the essential amino acid methionine. Furthermore, it is converted in the body to the amino acid *cystine*, an important constituent of hair.

(Also see AMINO ACID[S]; and TISSUES OF THE BODY.)

CYSTIC DUCT

The tube between the gallbladder and the common bile duct, through which the gallbladder discharges its bile to the small intestine. Occasionally, gallstones may lodge in the cystic duct and block the flow of bile from the gallbladder. (Also see GALLSTONES.)

CYSTIC FIBROSIS

A disease characterized by a general malfunctioning of various secretory glands in the body. It usually results in death unless a combined therapy is undertaken immediately. The therapy involves (1) a special diet which provides missing enzymes in purified form, (2) treatment of digestive disorders, and (3) careful control of respiratory infections.

(Also see DISEASES.)

CYSTINE

One of the nonessential amino acids. It is sulfur-containing and may be used in part to meet the need for methionine.

(Also see AMINO ACID[S].)

CYSTINURIA

An inborn error of metabolism in which there is an abnormally high excretion of the amino acid cystine in the urine. Cystinuria results from a failure of the kidneys to prevent the urinary excretion of this amino acid, as it does with many of the other amino acids.

(Also see INBORN ERRORS OF METABOLISM.)

CYSTOLITHIASIS

The presence of stones in the urinary bladder. The condition is usually treated with a special diet, and medication to dissolve the stones.

(Also see KIDNEY STONES.)

-CYTE

Suffix meaning cell.

CYTOCHROME

A class of iron-porphyrin proteins of great importance in cell metabolism.

CYTOLOGY

The anatomy, chemistry, pathology, and physiology of the cell.

CYTOPLASM

Substance within the cell exclusive of the nucleus.

CYTOSINE

One of the nitrogenous bases in nucleic acid.

(Also see NUCLEIC ACIDS.)

DADHI

A type of fermented milk in India.

DAIRY PRODUCTS

This generally refers to all the products made from milk.

DAMSON PLUM *Prunus institia*

An oval-shaped, dark-blue plum (a fruit of the family *Rosaceae)* that grows throughout Europe and is best when cooked or made into jam.

Fig. D-1. Damson plum.

Certain varieties of damsons are used to make dried prunes.

The nutrient composition of damson plums is given in Food Composition Table F-36.

Some noteworthy observations regarding the nutrient composition of damson plums follow:

1. The raw fruit is moderately high in calories (66 kcal per 100 g) and carbohydrates (18%).
2. Damson plums are a good source of potassium, a fair source of iron and vitamin A, and a poor source of vitamin C.

(Also see PLUMS AND PRUNES.)

DANDELION *Taraxacum officinale*

This wild plant, which is often considered to be an undesirable weed, has edible roots, leaves, and flowers. Its name is derived from the French word *dent-de-lion,* which means "lion's tooth," and refers to the toothlike edges of the leaves. The dandelion is a member of the Daisy family (*Compositae*). Fig. D-2 shows the plant with its long tapering root.

Fig. D-2. The dandelion, a hardy wild plant which is a troublesome weed, yet it supplies edible roots, leaves, and flowers.

ORIGIN AND HISTORY. The dandelion appears to be native to Europe, North Africa, north and central Asia, and North America. It may be that there are many different species of dandelion, since the leaves and other characteristics of the different varieties vary considerably.

Little was written about the plant until the Middle Ages, when it was noted that the famous Arab physician Avicenna (980-1037) used it medicinally to regulate menstruation. It was not very popular as a food until the 19th century when several more palatable varieties were developed in Europe. Today, the tender leaves of cultivated plants are sold in markets, but many people of European descent are ever on the watch for an opportunity to gather the dandelions that grow wild.

PRODUCTION. There are no statistics on the commercial production of dandelions, since the crop is grown on a small scale in many widely separated regions.

The seeds are usually planted early in the spring so that the plant may be harvested before the weather becomes too hot. Once established, the plants will propagate themselves, provided that their roots are left intact. Dandelions do best in a light, well-drained fertile soil. Sometimes, the plants are grown during the late fall in winter greenhouses or hotbeds to meet the demand during the winter and early spring. The leaves should be picked before the flowers appear.

PROCESSING. The leaves are not usually processed, because they are at their best when consumed fresh. However, the roots may be roasted in an oven until brown, after which they can be ground and used as a substitute for coffee. The drink made from this preparation tastes like coffee, but contains no caffeine. In addition, the flowers may be made into wine.

SELECTION AND PREPARATION. High quality dandelion leaves are fresh, young, tender, and green. Those which show insect injury, coarse stems, dry or yellowing leaves, excessive dirt, or poor development, are usually lacking in quality and may cause excessive waste. Flabby, wilted leaves are generally undesirable.

The washed leaves may be served raw in salads or sandwiches, where the slightly bitter (alkaline) taste may be offset by a small amount of an acidic ingredient such as lemon juice, pickles, salad dressing, tomato or tomato juice, or vinegar; or by various other seasonings. Cooked leaves may be served with a little butter or margarine. They are also good in a wide variety of soups.

Some natural foods enthusiasts cook the young flowers alone in fat or oil, or in fritters and omelets.

CAUTION: Dandelions growing wild along a busy highway or road may have a dangerously high lead content, as a result of exposure to automobile exhaust fumes.

NUTRITIONAL VALUE. The nutrient compositions of raw and cooked dandelion greens are given in Food Composition Table F-36.

Some noteworthy observations regarding the nutrient composition of dandelions follow:

1. Dandelion greens have a high water content (85 to 90%), and are low in calories (33 to 45 kcal per 100 g). They are an excellent source of iron, potassium, and vitamin A; and a good source of fiber, calcium, and vitamin C. However, cooking reduces the vitamin content and causes the escape of minerals into the cooking water.
2. Ever since the Middle Ages the leaves and roots of dandelions have been used to treat conditions such as constipation and other digestive disorders, irregularity of the menstrual cycle, and water retention in the tissues.

NOTE: The authors make no claims regarding the efficacy of dandelion preparations in the treatment of these conditions. Rather, they are presented as a matter of historic interest.

(Also see VEGETABLE[S], Table V-6 Vegetables of the World.)

DARK ADAPTATION

The rate at which a person's eyes adapt to a change from bright light to darkness. A slower than normal rate may indicate a vitamin A deficiency.

(Also see VITAMIN A.)

DATE *Phoenix dactylifera*

Fig. D-3. Clusters of dates that may have up to 200 dates each hanging from a palm tree.

Next to the coconut palm, the date palm is the most useful tree of the palm family. It produces the fruit that provides one of the chief articles of food in North Africa and the Middle East.

Fruits of the date palm, *Phoenix dactylifera,* are over 1 in (*2.5 cm*) long, sweet, fleshy, oblong, and contain a single seed. They grow in clusters that contain up to 200 dates and weigh up to 25 lb (*11 kg*). While hanging on the tree, dates are a rich red to golden brown.

ORIGIN AND HISTORY. According to Muslim legend, the date palm was made from the dust left over after the creation of Adam; hence, it is considered "the Tree of Life." The date palm was probably the first cultivated tree in history; it has been under cultivation in the Holy Land for at least 8,000 years. Sun-baked bricks, made in Mesopotamia (Iraq) more than 5,000 years ago, record directions for growing the tree. The Bible tells of the date palm, and the poetry and proverbs of the East often mention it. About 1,700 years ago, it was introduced into China from Iran. Then, in the 17th century the Spanish took it to California.

PRODUCTION. Dates rank among the top 20 fruits of the world. Worldwide, about 3.4 million metric tons of dates are produced each year, primarily in countries of North Africa and the Middle East. Egypt, Iran, and Saudi Arabia lead. Other important date-producing countries are indicated in Fig. D-4.

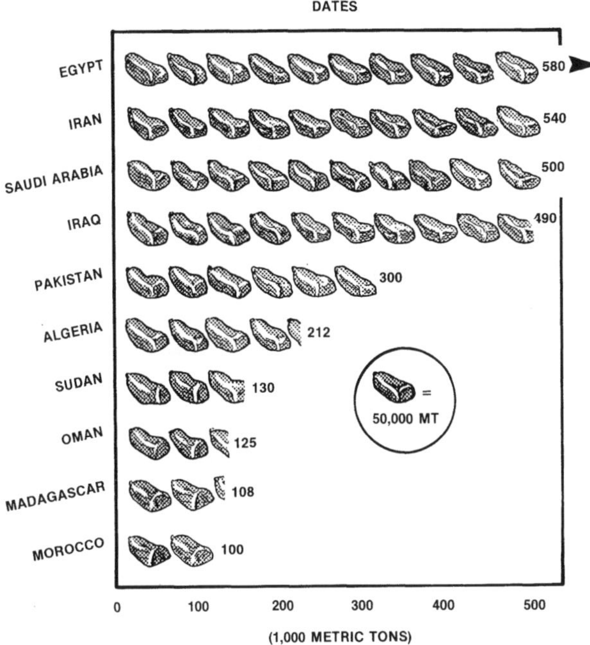

Fig. D-4. Leading date-producing countries of the world. (Based on data from *FAO Production Yearbook*, 1990, FAO/UN, Rome, Italy, Vol. 44, p. 155, Table 66)

Some dates are grown in the United States in hot areas of California, Arizona, and Texas. Recent crops have amounted to about 24 thousand metric tons each year, worth around $20.9 million.

Propagation and Growing. Mature date palms reach heights of 40 to 100 ft (*12 to 30 m*), and their trunk stands straight and about the same thickness clear to the top. A crown of large feather-shaped leaves grows on top, and the flowers grow inconspicuously among the leaves.

Though dates may be propagated from seed, purity of variety in commercial cultivation is ensured by propagating offshoots or suckers from the base of mature trees. Plantings are spaced about 30 × 30 ft (*9 × 9 m*). They require a hot dry climate and grow well in sandy alkaline soil. Irrigation is often used by date palm growers.

The trees begin to bear at 4 to 5 years of age, are in full bearing by age 15 years, and may continue fruiting to about 80 years of age. Date palms exist as male and female trees. Therefore, the pollen from the male blooms is transferred to the female blooms. This used to be accomplished by placing several male blooms among the females—a technique requiring much climbing. Now, machines blow pollen up through a tube to the female blooms. One male date palm provides sufficient pollen to polinate abut 50 females. After pollination, the blossoms are bagged to protect the fruits from damage by birds, insects and rain.

Fig. D-5. Date palms. (Courtesy, Field Museum of Natural History, Chicago, Ill.)

Harvesting. A yield of 100 to 200 lb (*45 to 91 kg*) of dates per tree is considered good. Dates are hand picked by cutting the large bunches and lowering them to the ground.

PROCESSING. Three categories of dates are marketed: (1) dates harvested when still soft and unripe—soft dates; (2) semidry dates from the firmer fleshed cultivars, also picked unripe; and (3) dry dates, sun dried on the tree.

PREPARATION AND USES. Dates are served alone, fresh or dried, or used in various dishes such as fruit dishes, baked goods, candies, ice cream, salads and syrups.

In the Arab world, dates are commonly consumed with milk products which bolster the protein content.

A fermented maceration of dates yields an alcoholic beverage called *arrak*, which Pedro Texeira, a 16th century traveler, described as, "the strongest and most dreadful drink ever invented." The date pits are roasted, ground and used as a coffee substitute in some countries. Additionally, like many other palms, date palms provide a food supply from the sugary sap which can be obtained by tapping the crown of the plant. This can be boiled down to provide sugar, or it can be fermented to make palm wine or toddy. When the tree is cut down, the tender terminal bud is eaten as a vegetable or salad. But the date palm tree is not used for food alone. The trunk furnishes timber for houses; the midribs of the leaves supply material for crates and furniture; the leaves are woven into baskets, mattings, and bags; pagans, Jews, and Christians have used date leaves in religious ceremonies from earliest times; the leaf bases are used for fuel; the fruit stalks are used for rope; the fiber is used for cordage and packing material; and the seeds are sometimes ground for stock feed.

NUTRITIONAL VALUE. The chief nutritional value of dates is their high sugar content, which varies from about 60% in soft dates to as much as 70% in some dry types. Most varieties contain glucose or fructose sugar, but one variety, Deglet Noor, grown primarily in California, contains only sucrose sugar. The dried fruit contains about 2% each of protein, fat, and mineral matter. Dates also contain some vitamin A, B-1, B-2, and nicotinic acid. Complete nutritional information on dates may be found in Food Composition Table F-36.

(Also see FRUIT[S], Table F-47 Fruits of the World.)

DEAERATION

Removal of air, without removing moisture. In the processing of milk, it refers to vacuumization for removal of undesirable volatile substances that impart odors.

DEAMINATION

A process in protein metabolism whereby an amino group is removed from an amino acid through the action of an enzyme called deaminase. This reaction occurs in the liver and in the kidneys.

(Also see METABOLISM, section on "Protein Catabolism.")

DEBILITY

Weakness.

DECALCIFICATION

A loss of calcium from the bones and teeth which may occur when (1) the diet contains too little calcium to replace the daily losses of the mineral in the urine and the stool, or (2) overactivity of the parathyroid glands causes demineralization.

(Also see CALCIUM; and OSTEOPOROSIS.)

DECARBOXYLATION

A metabolic reaction in which a decarboxylace enzyme brings about the removal of carbon and oxygen from a molecule.

(Also see METABOLISM.)

DECIDUOUS TEETH

Another name for the baby teeth which are lost in the early years of life and replaced by the permanent teeth.

(Also see DENTAL HEALTH, NUTRITION, AND DIET.)

DECORTICATION

• Removal of the bark, hull, husk, or shell from a plant, seed, or root.

• Removal of portions of the cortical substance of a structure or organ, as in the brain, kidney, and lung.

DEFIBRILLATION

Any means by which an irregular beating of the heart muscle fibers is brought back to normal.

(Also see HEART DISEASE.)

DEFIBRILLATOR

An electrical device which converts an irregular beating of the heart muscle fibers to a normal rhythmic pattern.

(Also see HEART DISEASE.)

DEFICIENCY DISEASES

Contents *Page*

Eating Patterns Which Lead to Dietary Deficiencies . 524
Stages in the Development of Dietary Deficiency Diseases . . . 526
Types of Dietary Deficiency Diseases . . . 527
Primary Deficiencies . . . 527
Major Dietary Deficiency Diseases—Table D-1 . . 530
Minor Dietary Deficiency Disorders—Table D-2 . . 536
Secondary or Conditioned Deficiencies . . . 527
Diagnosis of Subclinical Deficiencies . . . 527
Treatment and Prevention of Dietary Deficiencies . . . 528

These diseases result from dietary shortages of one or more essential nutrients. They may be prevented or cured by the administration of the missing nutrient(s), except when there is irreparable damage to vital tissues of the body.

EATING PATTERNS WHICH LEAD TO DIETARY DEFICIENCIES. Although hunger due to a lack of sufficient quantity of food often causes a deficiency disease(s), the factors which lead to this problem are discussed elsewhere. (See HUNGER, WORLD; MALNUTRITION; PROTEIN-ENERGY; POPULATION, WORLD; and WORLD FOOD.) Herein, the dietary patterns which may cause deficiency diseases even though sufficient quantity of food is available are identified and discussed. Descriptions of such dietary patterns follow.

• **Lack of a sufficient variety of foods**—Even when a wide variety of foods is available at a cost most people can afford, some people will eat rather limited diets. For example, a middle-aged banker was found to have developed scurvy.[1] He lived alone because his wife had died the year before. His diet consisted mainly of meat, milk, bread, coffee, and alcoholic beverages. Thus, deficiency diseases are not limited to poverty-stricken people.

Occasionally, it is difficult to detect deficiencies in people who subsist on only a few foods. One would expect to find deficiencies of vitamins A and C in the few Eskimos who still eat the primitive diet of meat and fish, and little else. The secret of their survival may lie in the eating of raw meat which contains enough vitamin C to prevent scurvy, and in the occasional nibbling of a piece of liver which is rich in vitamin A. However, most people would find the food of the primitive Eskimos to be unpalatable. Besides, a diet consisting largely of well-cooked muscle meats would likely cause deficiencies, unless fruit and vegetables were included.

• **Use of highly refined foods**—It is well known that beriberi was the scourge of some Asian peoples who ate large quantities of highly milled rice, while other groups who ate rice which had been parboiled prior to milling were not as susceptible to this disease. Likewise, white sugar and white flour are poor sources of minerals and vitamins in comparison with the unrefined products from which they are made. However, well-fed populations who eat these sources of "empty calories" do not develop deficiencies if they are able to obtain the nutrients they need from other foods.

• **Dependence on starchy fruits and vegetables**—Multiple deficiencies are usually found in areas where the staple foods consist of such starchy items as bananas, yams, cassava, and sago. These crops are usually easy to cultivate and are grown extensively in tropical areas, where the rain-leached soils may require special treatments if they are to support a more diversified agriculture. However, such diets may also be preferred for cultural reasons, even when the local environment is favorable for growing other crops.

• **Foods containing substances which interfere with nutrient utilization**—Vegetables of the cabbage family contain goitrogenic substances—so named because of their interference with the utilization of iodine. Likewise, phytates in whole grains bind such mineral elements as calcium, magnesium, iron, and zinc so that the absorption of these nutrients may be significantly reduced. Other substances, such as oxalates from rhubarb and spinach, also reduce the absorption of trace minerals.

• **Destruction or loss of nutrients during cooking**—Vitamins A, C, E, and thiamin may be destroyed by cooking foods at high or prolonged temperatures in the presence of oxygen. Also, some minerals and water-soluble vitamins may be lost when they are extracted from foods by the cooking water, which is often discarded. Finally, the overheating of cooking oils until they begin to smoke results not only in the almost total destruction of their vitamin E content, but also in the formation of oxidation products which are likely to accelerate the destruction of vitamin E in the body.

• **Foods grown in mineral-deficient soils**—There has been much recent debate as to the importance of this factor as a cause of human deficiency diseases other than goiter. The latter condition is more common in areas where the soil and water have abnormally low levels of iodine. However, there is little doubt that mineral-deficient soils may be reponsible for lower levels of minerals in certain food plants, and in the livestock fed these plants.

The controversy on this subject arises because some soils naturally contain an abundant supply of most of the elements needed by plants, animals, and man, whereas other soils may have an abundant supply of most required elements and yet be deficient in one or more essentials. For example, southwestern United States is known as a phosphorus-deficient area; northwestern United States and the Great Lakes area are iodine-deficient areas, and southeastern United States is a cobalt-deficient area.

Also, differences between plant species in their tendency to accumulate different elements are often important in determining the mineral status of livestock that eat these plants. For example, in many places in the United States such forages as alfalfa and clover contain adequate levels of cobalt for cattle and sheep (these animals convert the mineral to vitamin B-12,) whereas grass species in the same fields or pastures do not contain enough cobalt to meet the requirements of these animals.

Furthermore, at every step in the chain from soils to man, the essential mineral elements interact with other elements, and these interactions may profoundly affect (1) the availability of essential elements to plants, animals, or man, or (2) the amount of the essential element required for normal growth or metabolic function. For example, the availability of zinc to animals and man may be depressed if their diets are high in calcium, and high levels of molybdenum may interfere with their copper metabolism.

Fortunately, much of the American diet is made up of products from farm animals that have been fed mineral supplements to compensate for mineral deficiencies in their feeds and forages. In many cases, the minerals present in animal foods are in forms which are more efficiently utilized by man than the mineral compounds present in plants. Two leading examples of such cases are the iron in hemoglobin and the chromium in the glucose tolerance factor. (Liver is an excellent source of the factor.) Strict vegetarians who eat no animal foods are more likely to develop deficiencies from eating nutrient-deficient plants than people who eat liberal amounts of animal foods. (See Fig. D-6.)

• **Protein-deficient diets**—Many of the essential minerals and vitamins are cofactors for vital enzyme-controlled processes. These nutrients perform their roles in combination with protein. Also, protein is the vehicle which carries many of the nutrients in the blood and which helps to prevent their excretion in the urine. Thus, protein deficiencies may greatly reduce the efficiency of utilization of many essential nutrients.

---

[1]Sanstead, H. H., J. P. Carter, and W. J. Darby, "How to Diagnose Nutritional Disorders in Daily Practice," *Nutrition Today,* Vol. 4, Summer 1969, p. 20.

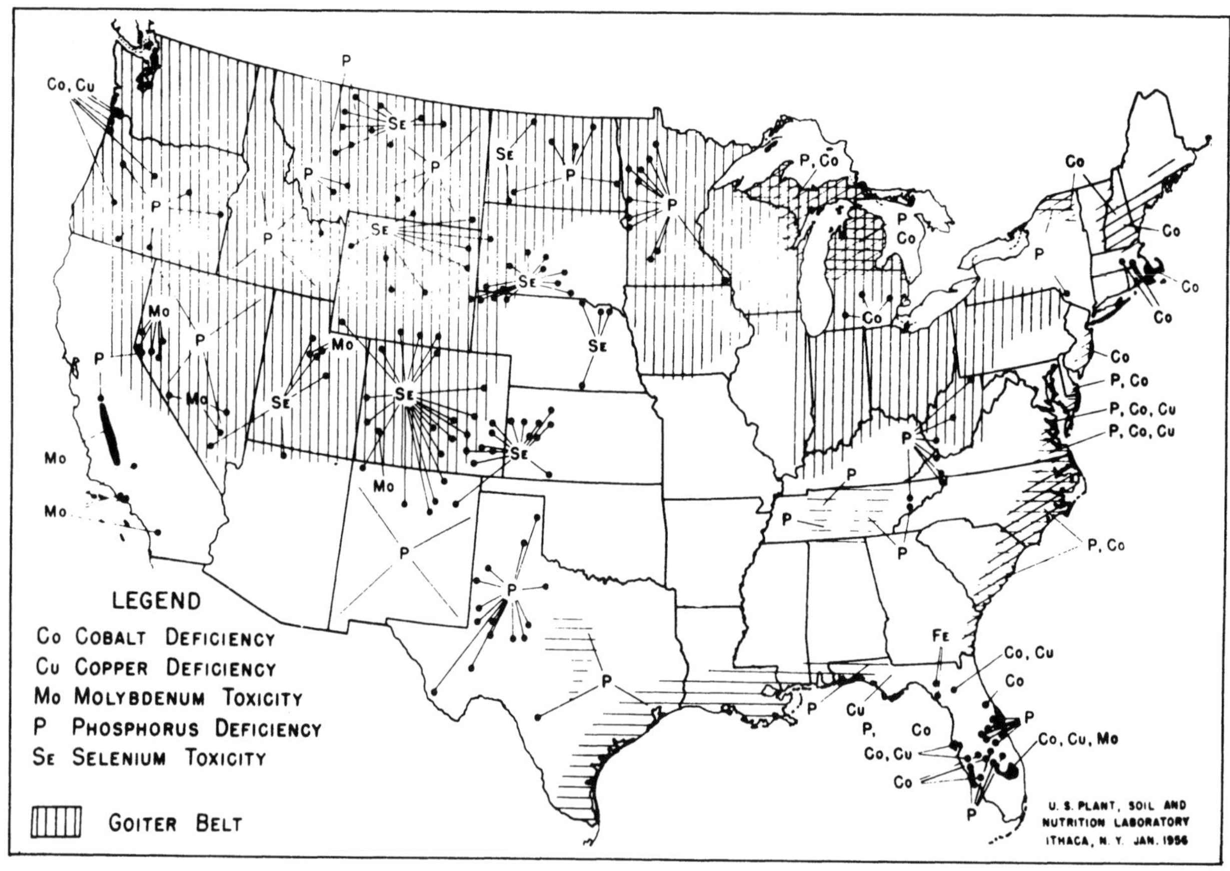

Fig. D-6. Mineral deficiency areas of the U.S. and the excess selenium area of the northern and central Great Plains. (Courtesy, USDA)

STAGES IN THE DEVELOPMENT OF DIETARY DEFICIENCY DISEASES. Depending upon the degree of dietary inadequacy, and/or the physiological importance of the deficient nutrient(s), a variable period of time usually elapses before outward signs of a deficiency are noticeable. It is convenient to view the development of such a disorder as taking place in several stages, as outlined in Fig. D-7.

A discussion of these stages follows.

1. **Inadequate supply of dietary nutrient(s).** The first stage of a dietary deficiency begins when the amount of a certain nutrient(s) present in the daily food pattern is consistently less than the amount of the nutrient which is either broken down in metabolism or excreted in the feces, urine, and sweat. For example, the feeding of a milk formula containing vegetable oils (which are high in polyunsaturated fats) to a newborn, premature baby is likely to cause a deficiency of vitamin E, unless extra amounts of the vitamin are provided.

Sometimes, a dietary shortage of a particular nutrient may be partially compensated for by other nutrients which serve similar functions. Hence, the development of a vitamin E deficiency may be delayed when the diet contains sufficient amounts of the essential mineral, selenium. Other pairs of nutrients which apparently work together in certain functions are ascorbic acid (vitamin C) and the bioflavonoids, calcium and vitamin D, folic acid and vitamin B-12, and vitamins A and C.

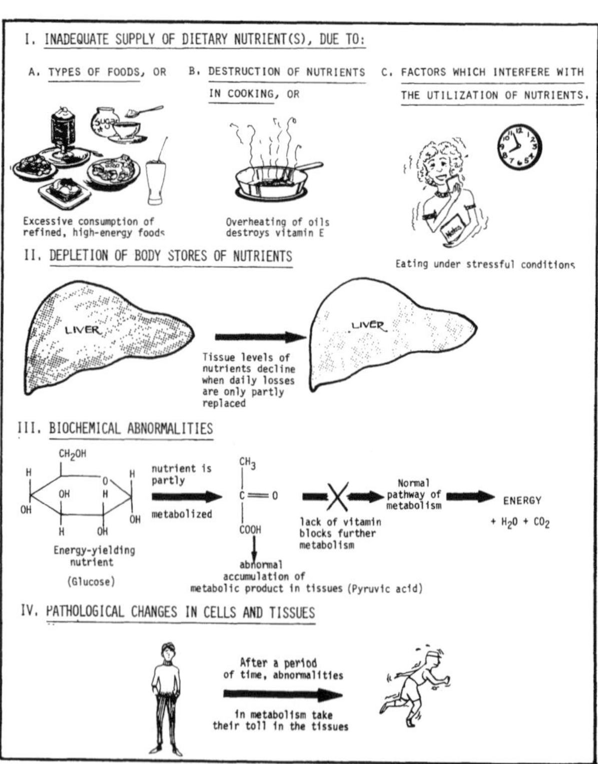

Fig. D-7. Stages in the development of dietary deficiency diseases.

2. **Depletion of body stores of nutrient(s).** Often the body's response to a shortage of one or more nutrients is a reduction in the excretion of its limited supply of the nutrient(s). However, if the dietary deficiency continues, there is a steady depletion of the nutrient(s). In the case of the premature baby given vegetable oils, there may be a drop in blood levels of vitamin E.

3. **Biochemical abnormalities.** A reduced supply of such key nutrients as vitamins may block certain steps in metabolic pathways and give rise to biochemical abnormalities like deficiencies or excesses of various products of metabolism. For example, vitamin E deficiency is often accompanied by a buildup of dangerous products (peroxides) which result from the abnormal metabolism of polyunsaturated fats.

4. **Pathological changes in cells and tissues.** Biochemical abnormalities may cause damage to tissues by either depriving them of essential substances, or by overwhelming their ability to cope with potentially harmful agents. A newborn baby with a vitamin E deficiency induced by vegetable oils may have damaged red blood cells which may be highly susceptible to breaking down (hemolysis). Anemia results when the rate of red cell destruction outstrips the rate of cell production. This type of anemia is made worse, rather than helped, by extra iron because the effects of the vitamin are somehow antagonized by iron.

TYPES OF DIETARY DEFICIENCY DISEASES.

It is common practice to classify deficiency diseases according to the ways in which they are produced. Thus, diseases directly due to deficient levels of dietary nutrients are said to be *primary deficiencies,* while those due to substances which interfere with utilization of marginal to adequate levels of nutrients are called *secondary deficiencies.* Discussion of each class of deficiency follows.

Primary Deficiencies.

These deficiencies are often divided into two groups: The *major dificiency diseases* where there are severe impairments of vital functions such as heart action, vision, energy metabolism, or tissue growth; and the *minor deficiency disorders* which are characterized by irritating and/or unsightly conditions such as skin lesions, loss of ankle and knee reflexes, or mottled teeth. The features of each group of deficiencies are outlined in Tables D-1 and D-2 beginning on page 530.

Secondary or Conditioned Deficiencies.

These disorders result from some type of interference with what otherwise might be adequate intakes of nutrients. The end results of such interferences are essentially the same as those outlined under primary deficiencies. Discussions of interfering factors follow.

• **Conditions which interfere with nutrient absorption**—Examples of these actions are (1) binding of trace minerals by phytates from whole grains, (2) formation of insoluble mineral oxalates by oxalic acid from rhubarb and spinach, and (3) general malabsorption caused by diarrhea (such as in celiac disease or sprue) or overuse of strong laxatives.

• **Factors which interfere with nutrient utilization**—Dietary imbalances between amino acids result in poor overall utilization of the total dietary protein, but some of the details of this effect are still uncertain. There also appears to be antagonisms between various mineral elements such as calcium and phosphorus, calcium and magnesium, iron and copper, copper and zinc, and molybdenum and sulfur (in the form of sulfate). Finally, large doses of vitamin E apparently antagonize the effects of vitamin K.

• **Substances which increase nutrient requirements**—In general, the requirements for vitamins are raised by large increases in the nutrients whose metabolism they control. For example, extra carbohydrate requires more thiamin, extra protein more vitamin B-6 (pyridoxine), and extra polyunsaturated fats more vitamin D.

• **Factors which increase the rate of nutrient destruction**—Smoking and air pollution appear to increase the rates of destruction of such oxidizable nutrients as vitamins A, C, and E. Likewise, rancid fats and oils increase the rate of vitamin E destruction, both in the diet and in the body.

• **Conditions resulting in increased excretion of nutrients**—Any factor that promotes excessive urination (diuresis) will cause increased losses of water-soluble nutrients (mainly mineral salts and vitamins). Likewise, there may be heavy sweat losses of these nutrients in hot climates. Also, starvation and other stresses cause elevations in urinary losses of nitrogen, which means that amino acids and proteins are used less efficiently.

DIAGNOSIS OF SUBCLINICAL DEFICIENCIES.

The sections on major and minor deficiency diseases described the diagnostic features which accompany the advanced stages of nutrient deficiencies. However, it is desirable, whenever possible, to detect the early stages of nutrient deficiencies so that the more advanced stages may be prevented. Unfortunately, it is not feasible to give everyone laboratory tests for such deficiencies, so some type of screening system has to be used to identify the people who should be thoroughly tested. An outline for such a system follows.

1. **Epidemiological or statistical studies.** Statistics on causes of death, and on the incidences of various diseases are published annually by state departments of health. These data are usually broken down by counties or other geographic regions, so that public health workers might identify populations that should be given special attention. For example, an abnormally high rate of children's deaths from infectious diseases, such as measles, might indicate that these and other children in the area have had nutritional deficiencies which rendered them more susceptible to infection.

Other circumstances which suggest that there might be dietary deficiencies are large numbers of low-income or unemployed people, the presence of many who are either foreign born or who have unusual ethnic food patterns, and the lack of public assistance programs. The average level of education may also be a factor affecting dietary patterns, although fairly well-educated people may sometimes give less thought to their diets than some less educated people.

It may be worthwhile for nutritionists and other health professionals to visit different neighborhoods in large cities so that they may become more informed concerning the variations in the cost and availability of nutritious foods. Investigators have found that grocery chains sometimes charge higher prices and offer less variety in their inner city stores than in their suburban stores.

2. **Surveys of food consumption patterns.** After a population likely to have dietary deficiencies has been selected, dietary data may help to rule out the possibility of certain deficiencies, thereby allowing health professionals to concentrate their resources on obtaining detailed data concerning the most critical nutrients. Individuals in the population under consideration may be chosen according to the principles of statistical sampling and asked to list the types and amounts of foods they ate in the previous 24-hour period. Then nutrient consumption may be calculated from the dietary data through the use of food composition tables. An often-used measure of dietary adequacy is the Recommended Dietary Allowances (RDAs).[2] It is not wise to conclude that an individual is nutritionally deficient when his or her diet fails to meet the levels of the RDAs. First of all, a dietary recall does not necessarily reflect the long-term dietary pattern of an individual. Second, most studies of nutrient requirements have shown that there is great variation among different persons in a population. However, such information is useful for selecting individuals who might be given more thorough assessments. Also, dietary histories should always be taken when there are clinical signs which suggest dietary deficiencies.

3. **Laboratory tests: nutrient levels in blood and urine.** Laboratory tests of nutrient levels in the blood are useful for detecting subnormal levels of nutrients; but urinary levels of nutrients are not too meaningful unless they are measured against some standard, such as creatinine. Urinary excretion of creatinine is fairly constant on a day-to-day basis when the diet does not vary. However, it should be noted that blood levels of nutrients may be different from those in the tissues which may become depleted sooner than the blood. One way around this difficulty is to measure nutrient levels in the red and white blood cells, which are more representative of the nutritional status of tissues than the fluid part of the blood.

4. **Laboratory tests: measurements of physiological functions.** These tests are more useful for measuring the severity of a deficiency than the analysis of nutrient levels in body fluids, since they show the extent to which deficiencies have altered biochemical reactions. For example, the ability of red blood cells to resist hemolysis by hydrogen peroxide is correlated with the vitamin E status of the blood cell membranes. Many of these tests are still in the category of research tools, since the statistical limits of normal function need to be established.

(Also see NUTRITIONAL STATUS.)

---

[2]*Recommended Dietary Allowances*, 10th ed., 1989, NRC-National Academy of Sciences.

TREATMENT AND PREVENTION OF DIETARY DEFICIENCIES. Severe deficiencies may require therapeutic amounts of nutrients which are many times the normal physiological requirements. It is best to use purified forms of nutrients so that the dosage is accurately known. When absorption is poor, the nutrients should be injected rather than given orally. See Tables D-1 and D-2 for the treatments used for various deficiency diseases.

There are several basic approaches to the prevention of dietary deficiency diseases. Discussions of the merits and demerits of each approach follow.

• **Promotion of dietary patterns based upon food groups**—The aim of this approach is to encourage people to select a variety of foods from among such designated food groups as milk products, meats (and meat substitutes), breads (and cereals), and fruits-vegetables. These groups are also known as the "Basic Four." Details follow.

1. Two or more servings should be taken from the *milk group* each day. A serving consists of a cup (*240 ml*) of milk or its equivalent in cheese, ice cream, yogurt, or other dairy products. Children should receive at least three servings, while teenagers should receive four. These foods are the best dietary sources of calcium, but they also supply liberal amounts of protein, the vitamin B complex, and vitamin A. However, only small amounts of the latter vitamin are present in low-fat dairy products.

2. Each serving of the *meat group* should consist of two or more ounces of cooked, lean meat, fish, poultry, or an equivalent amount of protein which may be supplied by such items as cheese, dried beans or peas, eggs, and peanut butter. At least two servings of these foods should be taken each day, because they are good sources of iron and other minerals, protein, and the vitamin B complex.

3. The *breads and cereals group* supplies liberal amounts of carbohydrate, and moderate amounts of iron, protein, and the vitamin-B complex. However, the products made from whole grains are richer in a wide variety of minerals and vitamins than the highly milled products. It is recommended that four servings be taken daily of bread (1 slice is a serving), cereals, grains such as rice, or pasta. (One-half cup [*120 ml*] of cooked grains or pasta is equivalent to a slice of bread.)

4. At least four servings of the *fruits-vegetables group* should be taken daily. There should be at least one serving each of a raw item rich in vitamin C (citrus fruit), and of a cooked or raw item rich in vitamin A. (Such items usually are deep green or yellow-orange in color.)

One weakness of this system is that cost or particular food preferences may lead some people to avoid eating certain groups of foods. For example, some low-income families in the southeastern United States obtain much more of their dietary calcium from green, leafy vegetables than they do from dairy products. Furthermore, there are wide variations in the nutrient contents of foods within each group so it is possible that some people may consistently choose the least nutritious foods from certain groups (such as fruits and vegetables low in vitamin A).

• **Enrichment and fortification of preferred foods**—The aim here is either to restore some of the nutrients removed during processing (this action is called *enrichment*), or to add nutrients in amounts which were not present in the foods in the first place (the process of *fortification*). Recently, there was much controversy over whether the USDA-sponsored School Breakfast Program should give children a highly fortified snack cake in lieu of more traditional breakfast foods like cereals, eggs, meat, etc. Opponents of the snack cake program contended that it might encourage children to select so-called "junk foods" when foods which are naturally more nutritious are available. However, the addition of extra nutrients to popular foods which are low in vitamins and minerals may be the most practical approach for areas where the more nutritious foods are either scarce or too expensive for many consumers. For example, sugar is fortified with vitamin A in Guatemala.[3]

• **Increased use of special food products which are high in nutrients**—During the years of the Great Depression (the early 1930s) in the southeastern United States, brewers' yeast was distributed to families in order to prevent them from developing pellagra. Since then, nutritionists have been reluctant to promote the use of such special foods for fear of being called "food faddists" by their fellow professionals. However, it is noteworthy that before purified mineral and vitamin supplements became available, deficiency diseases were treated and prevented in high risk groups by giving them special products like rice bran extract, rice polishings, liver extract, yeast extract, and cod-liver oil.

Some primitive societies insist that couples planning marriage should eat fish roe so as to increase their chances of having healthy children. Finally, it is of interest to note that in the developed countries, livestock are fed the "lion's share" of the highly nutritious by-products of food processing like brewers' yeast, distillers' grains, wheat germ, bran, whey, and blackstrap molasses. While it would be wrong to overemphasize these items (dietary excesses of any concentrated source of nutrients may lead to nutritional problems), perhaps there should be more efforts to promote their incorporation into such food items as breads, buns, cakes, drinks, and ice cream.

• **Production of new, fabricated foods**—There have been suggestions that certain foods, like vegetables, should be processed into forms which resemble popular foods. For example, spinach might be dehydrated, powdered, then pressed into a form like that of a potato chip. Production of such food analogs has been underway for a long time as evidenced by such items as meat substitutes made from soybeans and imitation milks in which the butterfat is replaced with hydrogenated vegetable oils. The fabrication of such foods provides the opportunity to add such nutrients as minerals, vitamins and amino acids. Food technologists have recently begun to use the term "nutrification" to denote the addition of nutrients to fabricated foods. What is often overlooked in this approach is the possibility that processing alters, destroys, or removes nutrients or other beneficial substances which are not replaced by enrichment or fortification. For example, some of the totally synthetic orange drink powders contain such ingredients as sugar, citric acid, ascorbic acid (vitamin C), artificial flavoring, and artificial coloring. These items may replace orange juice which contains such additional constituents as potassium and the bioflavonoids. Although the latter substances are not considered essential nutrients, they have been shown to augment the protective effects of vitamin C. One manufacturer has recently added potassium to their orange drink powder so that it may be used by persons on diets which call for supplemental potassium.

• **Recommendation of mineral and vitamin supplements**—This might be the least expensive way to cover nutrient deficiencies in ordinary diets. Judging from the volume of over-counter (nonprescription) sales of such preparations, it appears that large numbers of people are deciding on their own initiative to take these supplements. Unfortunately, such decisions are more often based upon advertising claims than on any type of nutritional evaluation of one's diet or health status.

Nutritionists are concerned that the promotion of nutrient supplements in the forms of pills or tonics might lead the public away from selecting foods which are good sources of the essential nutrients. Also, these dosage forms are similar to those used for cold and cough remedies, pain killers, sleeping pills, and stimulants—substances which are heavily used by people to obtain relief from unpleasant symptoms. Hence, overdoses of nutrients might be taken by those who feel ill or weak, while others might tend to be neglectful in taking their supplements. Overdoses of such nutrients as selenium, vitamin D, and vitamin A have been shown to be very dangerous. The U.S. Food and Drug Administration has waged an uphill struggle for about 20 years in trying to establish regulations governing the potency of nutrient supplements.

• **Periodic megadoses of certain nutrients**—Some of the outlying areas of Africa, Asia, and Latin America are inhabited by people suffering from deficiency diseases; yet public health personnel can pay only infrequent visits to those areas. These conditions have led to the development of types of nutritional therapy which have long-term effects and which need only periodic administration. For example, goiter may be prevented in regions where there are scarcities of iodine-containing foods by the periodic injection (as infrequently as once every 4 years) of iodized oil.[4] Likewise, injections of megadoses of vitamin A (200,000 IU every 6 months) have been used in India, Asia, and Latin America.[5]

Finally, monthly injections of vitamin B-12 are the only practical means of treating people who lack the intrinsic factor which is required for the intestinal absorption of this vitamin.

(Also see HUNGER, WORLD; and MALNUTRITION, PROTEIN-ENERGY.)

---

[3]**Arroyave, G., "Distribution of Vitamin A to Population Groups,"** in *Proceedings Western Hemisphere Nutrition Congress III*, Edited by Philip L. White, Futura Publishing C., Inc., Mount Kisco, N.Y., 1972, p. 68.

[4]Kevany, J., and J. G. Chopra, "The Use of Iodized Oil in Goiter Prevention," American Journal of Public Health, Vol. 60, 1970, p. 919.

[5]Reddy, V., "Vitamin A Deficiency in Children," *Indian Journal of Medical Research*, Vol. 54, 1969, p. 54.

TABLE D-1
MAJOR DIETARY DEFICIENCY

Disease	Cause	Signs and Symptoms	Distribution[1]	Treatment
Anemia, nutritional (Also see ANEMIA.)	Iron deficiency in most cases; but it may also be caused by deficiencies of copper, folic acid, protein, pyridoxine, vitamin B-12, and vitamin E.	Difficulty in performance of strenuous activity, shortness of breath, fatigue, light headedness, paleness of skin and gums, rapid heartbeat, insomnia, headache, and blurred vision.	Worldwide. The highest incidences are in infants, children, women in their reproductive years, and elderly persons.	Administer orally, or by injection, therapeutic amounts of the deficient nutrient(s). The emergency treatment of a severe iron deficiency may require the administration of from 60 mg to 180 mg of iron per day.
Beriberi (Also see BERIBERI.)	Deficiency of vitamin B-1 (thiamin).	Lack of appetite, loss of weight, fatigue, weakness, depression, mental confusion, irritability, neurological impairment, edema, hypertension, rapid heartbeat, and heart failure.	Rare in developed countries. Common in countries where milled, unenriched rice is the major food.	Furnish therapeutic amounts of thiamin in the diet through vitamin pills or food sources rich in this vatamin. Wet (edematous) beriberi should be promptly treated with intramuscular injections of 25 mg of thiamin twice daily. Also complete rest is necessary until the crisis passes.
Blindness due to vitamin A deficiency (Xerophthalmia) (Also see BLINDNESS due to vitamin A deficiency; and XEROPHTHALMIA in this table.)	Deficiency and/or poor utilization of vitamin A due to low-fat diets, and/or disorders of fat absorption.	Dryness or wrinkling of the conjunctiva, softening of the cornea, loss of the lens, infection and scarring of the whole eye leading to total blindness.	Occurs in most developing countries, but there are higher incidences in South Asia and East Asia.	Immediate administration of therapeutic amounts of vitamin A (100,000 IU daily for the first few days, then 30,000 IU per day). If there are disorders of fat absorption, then a water-soluble form of vitamin A should be injected into the muscles.
Cretinism (Also see GOITER.)	Failure of the fetal thyroid to develop fully. It may be due to a maternal deficiency of thyroid hormones, or perhaps an inherited lack of a thyroid protein factor.	Congenital retardation of thyroid development, stunted growth, mental retardation, deafmutism, flat nose, thick lips, enlarged tongue, underdeveloped gonads, obesity, and dry skin.	Worldwide, particularly in mountainous regions and areas subjected to the action of glaciers. In some areas, it may may affect 4% of the population.	Daily administration of thyroid extract (120 mg) or 0.2 mg of thyroxine. Treatment should be initiated at the first sign of the disorder, and continued for the lifetime of the patient. some or all of the developmental defects may be incurable, particularly when there is a delay in diagnosis and treatment.
Goiter (Also see GOITER.)	Dietary deficiency of iodine (simple goiter) in most cases.	Enlargement of the thyroid gland.	About 200 million persons worldwide. The highest incidence is in Brazil, India, and Mexico.	Administration 3 times a day of thyroid extract (60 mg) or hormones (0.1 mg of thyroxine) and/or surgery, depending upon the severity of the disorder. After reduction of the goiter, dietary iodine supplementation (in table salt, or as 0.1 mg of potassium iodide per day) to prevent further thyroid growth.

footnote at end of table

DISEASES

Prevention	Remarks
Supply dietary sources of iron, copper, folic acid, protein, pyridoxine, vitamin B-12 and vitamin E. Foods such as fish, fruits and vegetables, legumes, meats, nuts, vegetable oils, and whole grains are rich natural sources. Also, mineral and vitamin supplements may be taken.	Anemia is a condition in which the blood is either deficient in quality or quantity. A deficient quality refers to a deficiency in hemoglobin (an oxygen-carrying pigment) and/or red cells. The heart may have to work harder in order to deliver sufficient oxygen to the tissues.
Enrichment of milled grains with thiamin. Dietary supplementation with whole grains, nuts, legumes, yeast, wheat germ, rice polishings, liver, and fresh green vegetables.	Attacks of beriberi may be precipitated by rapid depletion of thiamin due to alcoholism and/or a diet high in carbohydrates and low in vitamin. Also strenuous activity aggravates the deficiency.
Supply dietary sources of vitamin A and/or carotene such as animal livers, fish-liver oil, green and yellow vegetables, butter, or fortified margarine. Make certain to supplement feedings of nonfat dry milk with sources of the vitamin.	There is evidence that the requirements for vitamin A may be higher in tropical areas where intense sunlight rapidly destroys visual pigment in the eyes.
Special efforts should be made to detect thyroid deficiencies in females during their reproductive years. Newborn children should be carefully scrutinized for signs of cretinism, since early treatment may prevent some of the development defects, such as mental retardation.	There seems to be a higher incidence of cretinism in areas of endemic goiter, and where there are close kinship marriages.
Use of iodized salt has been shown to be an effective preventative measure. Injections of iodized oil have been used in some areas. People on low salt diets may have to be given supplemental iodine.	Simple goiter should be distinguished from other disorders (toxic goiter, thyroiditis, thyroid tumors) which also cause enlargement of the thyroid gland.

(Continued)

A little African boy suffering from "hidden hunger"—a lack of protein in this case, the African name for which is Kwashiorkor. (Courtesy, FAO, Rome, Italy)

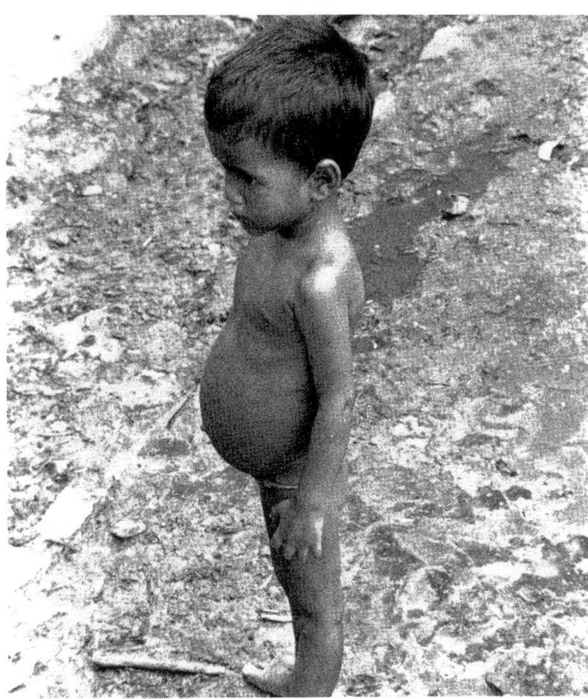

The swollen stomach of this youngster is a symptom of serious malnutrition. For millions of children throughout the world, food is needed to save them from permanent mental and physical damage that can be caused by lack of essential nutrients in their early years. (Courtesy, CARE, 660 First Avenue, New York, NY)

TABLE D–1

Disease	Cause	Signs and Symptoms	Distribution[1]	Treatment
Hemolytic anemia of the newborn (Also see ANEMIA.)	Vitamin E deficiency which may be aggravated by milk formulas which contain vegetable oils.	Increased susceptibility of red blood cells to hemolysis.	Found mainly in premature infants who were fed formulas containing vegetable oils (sources of polyunsaturated fats).	Oral administration of 25 IU water-soluble vitamin E per day (use of the water-soluble form avoids problems due to disorders of fat absorption).
Hemorrhagic disease of the newborn	Impairment of blood clotting due to a deficiency of vitamin K, which is needed by the baby for synthesis of clotting factors.	Bleeding of infant in various tissues such as the brain, skin, nervous system, peritoneal cavity, gastrointestinal tract; or excessive bleeding by boys upon circumcision.	Occurs mainly in newborn infants during the first week of life.	Intramuscular injection of 1 mg of vitamin K.
Iron-deficiency anemia (Also see ANEMIA.)	Dietary deficiency of iron and/or factors which interfere with its absorption or utilization.	Paleness of the gums and mucous membranes, easy tiring, shortness of breath, light headedness, rapid heartbeat, and red blood cells which are subnormal in color, size, and numbers.	Millions of persons worldwide. The highest incidences are in infants, children, women in their reproductive years, and elderly persons.	Administration of therapeutic doses of iron (orally or by injection). Severe cases may require 60-180 mg of iron per day.
Keratomalacia (Also see KERATOMALACIA; and BLINDNESS due to vitamin A deficiency.)	Vitamin A deficiency.	Softening and perforation of the cornea leading to loss of lens, infection and scarring of the eye, and blindness.	Occurs in most developing countries, but there are higher incidences in South Asia and East Asia.	Immediate administration of 100,000 IU of vitamin A per day. Injection of a water-soluble form is the most effective therapy when there is severe malnutrition.
Kwashiorkor (Also see KWASHIORKOR; and MALNUTRITION, PROTEIN-ENERGY.)	Deficiency of protein is a major cause, together with energy deficiency in many cases.	Growth failure, underweight, edema, anemia, diarrhea, malabsorption, apathy, and discoloring of the hair.	Mostly in the tropical areas of the world. Affects mainly children from weaning to age 4.	Provision of 1-2 g of high-quality protein per pound of body weight, along with sufficient energy. Cautious replacement of mineral salts may also be needed. In case of severe diarrhea, the protein intake should be temporarily restricted to 0.5 g per pound of weight.
Marasmus (Also see MARASMUS.)	Calorie-protein malnutrition.	Wasting of muscles, loss of subcutaneous fat, dry baggy skin, general appearance of old age, low body weight, large sunken eyes, diarrhea, subnormal body temperature, and malabsorption.	It is found in infants and young children in many areas of the world, particularly in the developing and over-populated countries.	Treatment consists of feeding an easily digested diet, such as milk and oil, which supplies at least 1 g of protein and 50-60 kcal per lb of body weight.
Osteomalacia (Also see RICKETS.)	Dietary deficiencies, or factors which interfere with the utilization of calcium, phosphorus, or vitamin D.	Softening and deformity of bones, bone tenderness and pain, muscular weakness, and tetany.	More prevalent in areas where either climate or clothing practices limit exposure of skin to sunlight, or where the diet does not supply correct proportions of calcium, phosphorus, and vitamin D.	Administration of vitamin D (1,000 IU daily) and calcium (1-2 g per day). Care should be taken that vitamin D toxicity does not develop during treatment. Massive doses of vitamin D (50,000 or more IU per day) may be required in certain kidney diseases.

Footnote at end of table

(Continued)

Prevention	Remarks
Supplementation of all infant formulas with 8-12 IU vitamin E per quart (*0.95 liter*), particularly those which contain polyunsaturated fats.	Avoid unnecessary iron supplementation (iron interferes with the utilization of vitamin E).
Administration of 5 mg per day vitamin K to the pregnant woman just prior to childbirth.	Synthetic water-soluble analogs of vitamin K should *not* be given to premature newborns because of their susceptibility to hemolytic anemia, jaundice, and brain damage.
Provide foods rich in iron and the other nutrients needed for the production of red blood cells (meats, fish, legumes, nuts, fresh fruits and vegetables, and whole grains).	Care should be exercised in the administration of therapeutic doses of iron since excessive iron intake can lead to the deposition of iron in tissues.
Foods which contain good sources of vitamin A or carotene (animal livers, fish-liver oil, green and yellow vegetables, butter or fortified margarine).	Victims of protein-energy malnutrition may develop the disease during their rehabilitation, particularly when they are given nonfat dry milk.
Feeding children an adequate amount of protein (particularly after weaning) from such sources as fish, legumes, meats, milk, and nuts. (Milk supplies 1 g of protein per oz [*30 ml*]).	Kwashiorkor differs from marasmus in terms of the greater role of protein deficiency as a cause of the former.
Prevention consists in feeding children an adequate amount of calories and protein.	The word *Marasmus* is from the Greek *Marasmos*, meaning *a dying away.* The disease is characterized by a progressive wasting and emaciation.
Exposure of skin to sunlight or sunlamps. Fortification of milk with vitamin D. Provision of extra calcium when diets are high in phytates or oxalates, or low in calcium (dietary phosphorus is usually adequate).	Osteomalacia is an adult form of rickets. Phytates (in whole grains) and oxalates (spinach, rhubarb), bind calcium and reduce its absorption.

(Continued)

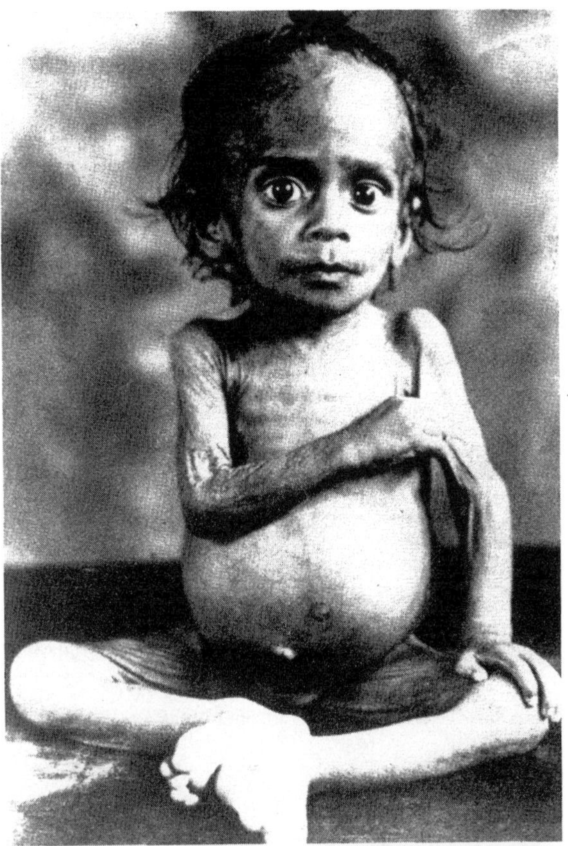

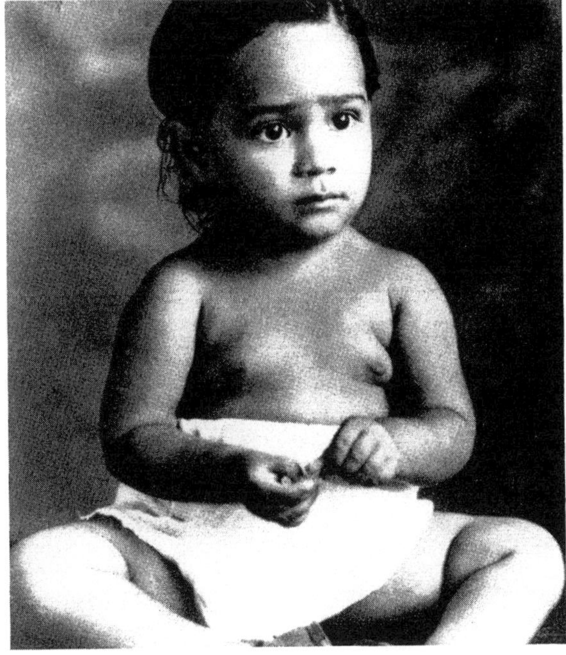

Marasmus, due to lack of energy foods.
Upper: A 2-year-old girl, weighing 11 ½ lb.
Lower: Same little girl after 10 months of treatment.
(Courtesy, FAO, Rome, Italy)

TABLE D-1

Disease	Cause	Signs and Symptoms	Distribution[1]	Treatment
Osteoporosis (Also see OSTEOPOROSIS.)	Loss of calcium from bone during aging. The cause of this loss may be lack of calcium.	Brittle bones which fracture easily, reduced bone density detectable by x rays; loss of alveolar bone from the jaw; and weakening of the spinal column and the long bones.	Found extensively in the aged with women more affected than men.	Administration of calcium, fluoride, and vitamin D may be helpful in some cases. Such treatment requires close medical supervision in order to avoid fluoride toxicity.
Pellagra (Also see PELLAGRA.)	Deficiency of niacin. Excessive content of leucine in grains low in tryptophan may be a contributing factor.	Skin discoloration similar to sunburn, dermatitis, diarrhea, inflamed tongue, disordered nervous and mental functions (such as dementia).	Prevalent in developing countries dependent upon corn (maize) or sorghum as staple foods.	Administration of niacin (up to 100 mg every 4 hours) or of vitamin B complex supplements to severe cases. Otherwise, diets rich in animal protein which are tolerated by the gastrointenstinal tract.
Pernicious anemia (Also see ANEMIA; and ANEMIA, PERNICIOUS.)	Vitamin B-12 deficiency due to inadequate amounts in the diet, or lack of intrinsic factor.	Megaloblastic anemia, inflammation of the tongue and mouth, degeneration of the nervous system, and abnormal brain wave patterns.	Found in strict vegetarians (who do not eat any animal foods) and in the aged who lack intrinsic factor.	Intramuscular injections of vitamin B-12 (1 month's supply of 100 mcg may be administered this way), or massive oral doses of vitamin B-12. Lifelong therapy is usually required.
Protein-energy malnutrition (Also see KWASHIORKOR; MALNUTRITION, PROTEIN-ENERGY; MARASMUS; and STARVATION.)	Deficiency of energy and/or protein due to intake of low-quality protein or insufficient amounts of food.	Growth failure (less than expected height for age, and subnormal weight for height and age), anemia, weakness, apathy, gastrointestinal disturbances, and increased susceptibility to infectious disease.	Occurs mainly in children under 5 years of age.	Provide 50 kcal energy and 1 g protein per pound *(0.45 kg)* of body weight. Vitamin and/or mineral supplementation may also be required.
Rickets (Also see RICKETS).	A lack of calcium, phosphorus, or vitamin D; or an incorrect ratio of the two minerals.	Enlarged epiphyses, beaded ribs, bowed legs, deformed chest cage, distended abdomen, and misshapen skull.	More prevalent in areas where environment, clothing, or lifestyle limits exposure to sunlight. Dark-skinned people who live in northern regions are more susceptible than light-skinned people in these areas.	Administration of vitamin D (1,000 IU daily) and calcium (500-1,000 mg per day). Care should be taken that vitamin D toxicity does not develop during treatment.
Scurvy (Also see SCURVY.)	Deficiency of vitamin C.	Swollen, bleeding gums; loosening and loss of teeth; anemia; capillary hemorrhages into the skin; weakness; and pain in the joints.	Rare in developed countries, except among the aged or chronic alcoholics.	Administration of 250 mg doses of vitamin C, 4 times daily. Citrus fruits, or their juices, (orange, grapefruit, lemon, lime) are excellent sources of the vitamin for mild cases of the disease.

footnote at end of table

(Continued)

Prevention	Remarks
Adequate amounts of dietary calcium (at least 800 mg per day) plus vitamin D from diet or sunshine to promote optimal calcification and minimum loss from bone with aging. Avoid excessive intakes of phosphorus from diets extra high in protein, or from processed foods containing phosphate additives.	Recent research suggests that diets high in protein result in increased urinary losses of calcium, and, therefore, an increased requirement for dietary calcium.
Provision of protein foods containing adequate niacin and tryptophan (eggs, fish, meats, milk). Avoid such meats as ham hocks, which are high in connective tissue.	Anemia and other dietary deficiencies may occur with pellagra. Meats which contain large amounts of connective tissues tend to be deficient in tryptophan.
Diet rich in animal protein. Strict vegetarians should take vitamin B-12 supplements.	Vitamin B-12 deficiency may be masked in its early stages by injudicious use of folic acid supplements which hide early signs).
Promotion of breast-feeding. Food supplementation for vulnerable groups of pregnant women, infants, and children.	The various types of protein-energy malnutrition may require different therapies.
Exposure of skin to sunlight or sunlamps. Fortification of milk with vitamin D. Provision of extra calcium when diets are high in phytates and oxalates, or low in calcium (dietary phosphorus is usually adequate when liberal amounts of eggs, fish, legumes, meat, milk, and nuts are consumed).	Rickets may also be caused by excessive excretion of phosphates in kidney disease, or reduction of calcium absorption by phytates (whole grains) and oxalates (spinach, rhubarb).
Diet containing fresh or frozen fruits and vegetables. Fortification of canned fruit and vegetable products with vitamin C. Emergency sources of the vitamin are sprouted seeds and rose hips.	It has been suggested that megadoses of vitamin C to pregnant women might condition the fetus to high requirements for this nutrient.

(Continued)

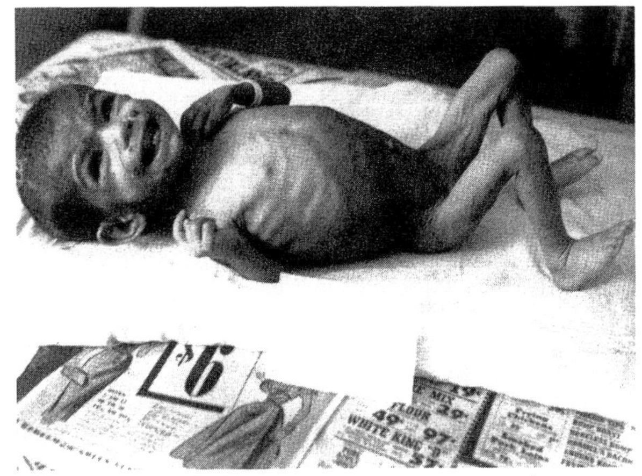

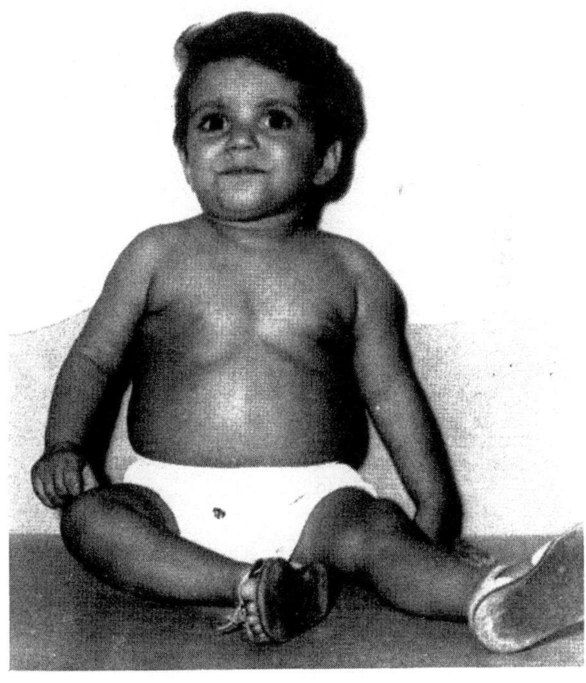

Hunger and malnutrition claim increasing numbers.
Upper: Sheila at 20 months of age and weighing 10.8 lb.
Lower: Same little girl after 10 months' treatment, weighing 22.7 lb. (Courtesy, Public Health Dept., Iran; issued by FAO)

TABLE D-1

Disease	Cause	Signs and Symptoms	Distribution[1]	Treatment
Starvation (Also see STARVATION.)	Insufficient food or impairment in the utilization of nutrients from food.	Weight loss, wasting of tissue, loose skin, ketoacidosis, dehydration, hypoglycemia, weakness, diarrhea, hypotension, irritability, and edema.	Worldwide	Provision of suitable food, using care not to aggravate gastrointestinal tract. In some cases, intravenous feeding and/or hyperalimentation may be required.
Wernicke's Disease	Thiamin deficiency, and, in some cases, deficiencies of other B-complex vitamins.	Lack of appetitie, nausea, vomiting, failing vision, poor balance in walking, and wrist drop, apathy, confusion, and delusions.	Prevalent among prisoners of war and others living on a very poor diet and among chronic alcoholics.	Intramuscular injection of 25 mg does of thiamin twice a day followed by supplements rich in the vitamin B complex (such as yeast) during convalescence.
Xerophthalmia (Also see VITAMIN A.)	Vitamin A deficiency.	Drying and pigmentation of the conjunctival membranes of the eyes; opacity of the cornea.	Occurs in areas where the diet is lacking in whole milk, butter, green or yellow vegetables.	Administer 100,000 IU doses of vitamin A. If there are disorders of fat absorption, then intramuscular injections of water-soluble form of vitamin A should be used.

[1]The estimated incidence of disease figures given under the column headed "Distribution" are from Aykroyd, W.R., *Conquest of Deficiency Diseases,* WHO, Geneva, Switzerland, 1970.

TABLE D-2
MINOR DIETARY DEFICIENCY DISORDERS

Disorders	Cause	Signs and Symptoms	Treatment	Prevention	Remarks
Eyes					
Angular blepharo-conjunctivitis	A combination of pyridoxine (vitamin B-6) deficiency and a bacterial infecttion *(Hemophilus duplex).*	Inflammation of the lining of the eyelid.	Administration of 10 mg of vitamin B-6 per day.	Provision of food sources of vitamin B-6 (eggs, fish, legumes, meats, milk, nuts, and whole grains).	Although the inflammation is caused by the bacterial infection, it is much more likely to occur when there is a vitamin B-6 deficiency.
Bitot's spots (Also see VITAMIN A.)	Deficiency of vitamin A. However, Bitot's spots have been observed when there is adequate vitamin A, indicating that other nutritional deficiencies may be responsible for this condition.	Foamy patches on the white portion of the eyes.	Immediate administration of from 25,000-100,000 IU of vitamin A per day until condition improves (injection of water-soluble forms should be used to treat patients suffering from any type of malabsorption).	Diets containing carotene (provitamin A found in deep-green and dark-yellow fruits and vegetables) or vitamin A (butter, fortified margarine, or fish-liver oils).	Neglect of this early sign of vitamin A deficiency could lead to xerophthalmia and possible blindness.
Corneal vascularization	Riboflavin (vitamin B-2) deficiency.	Invasion of cornea by capillaries (usually seen only under magnification). However, dilated blood vessels may often be readily seen in the neighboring conjunctiva (mucous membrane covering the eyeball).	Administration of 5 mg of riboflavin per day until condition improves.	Provision of food sources of riboflavin (cereal products from whole or enriched grains, fish, meats, and milk).	Patients may also have photophobia (aversion to light), making examination difficult.

(Continued)

(Continued)

Prevention	Remarks
Continual surveillance of the nutritional status of individuals and populations, with provision of basic subsistence requirements where necessary.	Postoperative patients are sometimes placed on restricted diets which may lead to starvation. Limited value as a means of weight reduction.
Nutritional surveillance of persons subject to thiamin deficiency (alcoholics; diabetics; toxemic, pregnant women; patients with chronic gastrointestinal disease).	The severe psychiatric disorders which accompany this disease constitute *Korsakoff's psychosis.*
Supply dietary sources of vitamin and/or carotene (see BLINDNESS). Fortify margarine with vitamin A.	The signs of dryness and pigmentation of the conjunctival membranes may sometimes be due to irritants such as dust and glare.

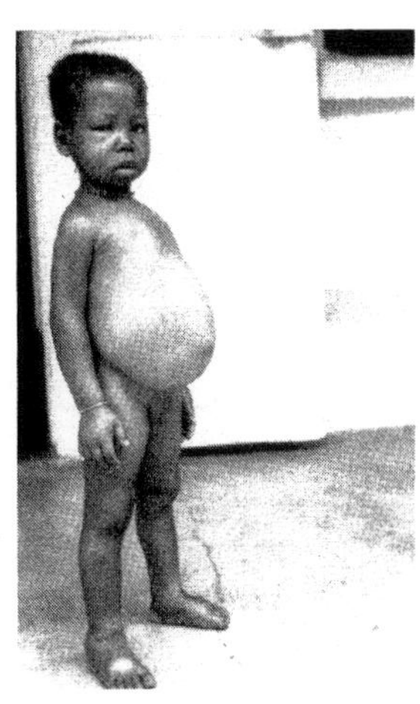

Kwashiorkor, a protein deficiency disease. Note characteristic bloated belly. Other usual symptoms are: stunted growth, diarrhea, brittle hair with an abnormal reddish color, and retarded mentality. (Courtesy, FAO, Rome, Italy)

TABLE D-2 *(Continued)*

Disorders	Cause	Signs and Symptoms	Treatment	Prevention	Remarks
Night blindness (Also see BLINDNESS due to vitamin A deficiency; and VITAMIN A.)	Deficiency of vitamin A	An abnormally long time is required for adaptation from vision in strong light to vision in dim light or darkness.	Administration of from 25,000-100,000 IU of vitamin A per day until condition improves.	Diets containing food sources of vitamin A. (Also see BITOT'S SPOTS.)	This disorder may also result from emotional disturbances and tiredness.
Pigmentation of the conjunctiva	May sometimes be due to vitamin A deficiency.	Abnormal pigmentation (coloring in the whites of the eyes or in the lower eyelid).	Administration of from 25,000-100,000 IU vitamin A per day until condition improves.	Plant and/or animal foods containing vitamin A. (Also see BITOT'S SPOTS.)	
Xerosis of the conjunctiva (Also see VITAMIN A.)	Vitamin A deficiency.	Dryness and thickening of the conjunctiva.	Immediate administration of 100,000 IU of vitamin A orally (or by injection in cases of malabsorption) for 3 days, then 30,000 IU per day until condition improves.	Diets containing food sources of vitamin A. (Also see BITOT'S SPOTS.)	This condition may also be caused by environmental irritants such as dust and bright light, or by infections. (Also see BITOT'S SPOTS.)

(Continued)

TABLE D-2 *(Continued)*

Disorders	Cause	Signs and Symptoms	Treatment	Prevention	Remarks
Glands **Parotid enlargement**	Deficiency of protein and/or the B-complex vitamins.	Swelling of the parotid gland.	Provide sources of high-quality protein and B-complex vitamins (eggs, fish, legumes, meats, milk, and nuts).	Same as Treatment.	There may be temporary parotid enlargement during the rehabilitation of victims of malnutrition. This condition should be distinguished from mumps.
Gums **Bleeding gums**	Vitamin C deficiency.	Gums may be bleeding and swollen.	Administer 4 daily doses of 250 mg of ascorbic acid (vitamin C) per day for a week.	Diets containing fresh fruits and vegetables.	May also be caused by lack of dental hygiene or certain drugs.
Periodontal disease	In some cases, deficiency of calcium.	Teeth fall out of gums, which may be inflamed. In calcium deficiency there may be loss the underlying (alveolar) bone in the jaw.	Administer daily doses of calcium (1,000 mg) and vitamin D (1,000 IU).	Calcium-containing foods, such as dairy products and green, leafy vegetables.	May also be caused by bacterial infections (pyorrhea alveolaris).
Hair **Discoloration and early pluckability** (Also see HAIR ANALYSIS.)	Protein and/or energy deficiency.	Light-colored hair turns dirty-brown, dark hair becomes lighter in color (reddish brown to white or grey). Hair may be plucked without pain.	Provide adequate dietary protein (eggs, fish, legumes, meats, milk, and nuts).	Same as Treatment.	May also be due to deficiency of thyroid hormones such as in cretinism and myxedema).
Loss of hair in patches (alopecia)	Uncertain in man. Lack of riboflavin or inositol in various animal species.	Hair falls out in patches from various parts of the body (different from male pattern baldness).	Uncertain in man. Some have been helped by various mineral and vitamin supplements.	Provide food sources of essential fatty acids, minerals, protein, and vitamins.	This condition may also be due to hormone deficiencies or the side effects of certain drugs.
Internal organs **Liver enlargement (hepatomegaly)**	Deficiencies of protein, methionine, and/or choline.	An enlarged liver which may be felt by touching the abdomen.	Provide adequate dietary protein (eggs, fish, meats and milk).	Same as Treatment.	May also be due to chronic alcoholism and the toxic effects of chemicals and drugs.
Racing heartbeat (tachycardia)	Anemia (lack of iron) or thiamin deficiency. This disorder may also be due to emotional stress or excessive secretion of thyroid hormones.	Heart rate is over 100 beats per minute.	First, confirm the nature of the deficiency; then administer iron or thiamin.	Foods containing iron and thiamin (eggs, fish, legumes, liver, muscle meats, nuts, and whole grains).	Nutritional tachycardia should be distinguished from that due to other causes (anxiety, heart disease, drug effects, etc).
Lips **Angular fissures** (stomatitis)	Often due to riboflavin deficiency, but may also be due to lack of iron, niacin, or pyridoxine (vitamin B-6).	Skin is broken at the corners of the mouth (the condition often occurs along with cheilosis).	Provide a multivitamin and mineral supplement which contains riboflavin, niacin, pyridoxine, and iron.	Foods rich in the B-complex vitamins and minerals, such as eggs, fish, legumes, meats, nuts, green vegetables, and whole grains.	This condition may also be caused by herpes infection, various irritants, chapping, or hot and dry environments.

(Continued)

TABLE D-2 *(Continued)*

Disorders	Cause	Signs and Symptoms	Treatment	Prevention	Remarks
Cheilosis	Often due to deficiency of riboflavin or niacin.	Lips are swollen and puffed out. In some cases, the lining of the inside of the mouth may be exposed (eversion of buccal membrane).	Provide vitamin supplement containing niacin and riboflavin.	Same as for Angular fissures.	Same as for Angular fissures.
Mouth and tongue **Dry mouth** (xerostoma)	Prolonged deficiency of vitamin A.	Lack of saliva which may be accompanied by dried patches on the mucous membranes. Often accompanied by signs of vitamin A deficiency in the eyes.	Administration of 7,500 IU vitamin A per day, until condition improves.	Dietary sources of carotene (provitamin A present in dark-green and deep-yellow fruits and vegetables) or vitamin A (butter, fortified margarine, and fish-liver oils).	Any form of vitamin A deficiency should be immediately treated so as to avoid permanent damage to the eyes. However, dry mouth may also be caused by emotional distress, various medical treatments, or disorders of the salivary glands.
Glossitis	Deficiencies of folic acid, iron, niacin, pyridoxine (vitamin B-6), riboflavin, tryptophan, and/or vitamin B-12.	Tongue is inflamed and shiny, with atrophy of the nipplelike projections (papillae). Fissures may also be present. Often accompanies anemia.	Provide a supplement containing vitamin B-complex and iron, together with high-quality protein.	Foods rich in iron, protein, and the vitamin B-complex (eggs, fish, legumes, meats, nuts, green vegetables, and whole grains).	May also be present in nonnutritional anemias, uremia, and infections.
Muscles **Wasting and flabbiness**	Deficiency of protein and/or calories.	Muscles are subnormal in size (as determined by measurements) and lack firmness when flexed.	Provide sufficient protein and calories.	Same as Treatment.	Other wasting diseases (like cancer) may produce similar effects.
Nails **Spoon nails** (koilonychia) (Also see NAILS.)	Iron deficiency.	Nails are spoonlike (concave).	Administration of iron.	Foods rich in iron (eggs, fish, legumes, meats, nuts, vegetables).	May also be a sign of a cardiac defect.
Nervous system **Peripheral neuropathies**	Deficiencies of the vitamin B-complex, particularly thiamin and vitamin B-12.	Absence of vibratory sense in the feet, aching, burning, or throbbing in the feet, lack of ankle and tendon reflexes, and tender calf muscles.	Provide a supplement containing the vitamin B-complex.	Foods containing the B-complex vitamins (eggs, fish, legumes, meats, milk, nuts, and whole grains).	These conditions may also result from toxic substances in foods, or from nonnutritional causes.
Skeleton **Misshapened bones**	Deficiencies of calcium and/or vitamin D, or interference with their utilization.	Bossing of the skull, bowing of the legs, enlarged wrists, and beaded ribs.	Provide a daily supplement containing 1,000 mg calcium and 1,000 IU vitamin D.	Adequate dietary calcium (milk products and deep-green vegetables) plus vitamin D (dietary and/or by exposure of skin. to sunlight).	These conditions may also be produced by low blood levels of phosphate (a rare disorder characterized by excessive excretion of the mineral).

(Continued)

TABLE D-2 *(Continued)*

Disorders	Cause	Signs and Symptoms	Treatment	Prevention	Remarks
Skin **Atrophy of the skin**	Starvation	Skin is drawn taut over the bones.	Provide sufficient quantity and quality of food.	Same as Treatment.	Care should be taken to avoid injury to atrophied skin, since healing of lesions may be greatly impaired.
Discoloration	Various nutritional deficiencies.	Skin may have darkened areas (as in pellagra), lightened patches (protein-energy malnutrition), or it may be pale (as in anemia).	First establish diagnosis of specific deficiencies; then provide missing nutrients.	Diets containing adequate amounts of of all essential nutrients.	Color changes are most likely to be detected on the parts of the body normally exposed to sunlight.
Dryness of the skin (xeroderma)	Lack of essential fatty acids and/or vitamin A	Dry skin which is lined and flaky.	Administer vitamins A and E and vegetable oils. Although vegetable oils contain vitamin E, the losses in processing are uncertain.	Foods containing vitamin A, such as dark-green and deep-yellow fruits and vegetables, butter, fish-liver oils, and fortified margarine (the latter usually contains the essential fatty acids which should also be provided).	The condition may also be caused by poor hygiene, excessive washing of the skin with soap, hypothyroidism, uremia, and poor circulation.
Easy bruising (ecchymosis)	Deficiency of vitamin C or vitamin K.	Bruises occur where there is only slight pressure or trauma.	After suitable diagnostic tests, give either vitamin C or vitamin K.	Diets containing fresh fruits and vegetables (sources of vitamins C and K).	Other causes are nonnutritional blood disorders and trauma.
Eczema (Also see ECZEMA.)	Sometimes caused by lack of essential fatty acids and/or pyridoxine (vitamin B-6).	Skin eruptions which may be dry or moist, and may be accompanied by burning or itching.	Administer vegetable oils with pyridoxine (large doses of the vitamin [10 mg or more per day] have sometimes been used in such therapy).	At least 1 tsp of vegetable oil per day plus foods containing pyridoxine (eggs, fish, legumes, meats, milk, nuts, and whole grains).	It is generally believed that the majority of cases of this condition are caused by sensitivities to foods and other substances rather than by dietary deficiencies.
Elephant skin (pachyderma)	It is believed to be due to a vitamin A deficiency.	Skin is created like that of an elephant in such areas as the elbow, knee, and instep.	Administration of 7,500 IU of vitamin A per day.	Foods containing vitamin A. (Also see Dryness of the skin.)	The condition may be aggravated by diet and other irritating factors.
Flaky paint dermatosis	Protein-energy malnutrition.	Darkening and flaking of the skin.	Diets containing high-quality protein (eggs, fish, legumes, meats, milk, nuts) and sufficient energy.	Same as Treatment.	Amino acid supplementation may be required for people whose major sources of protein are cereal grains.
Follicular hyperkeratosis, toad-skin or phrynoderma	May be due to deficiencies of vitamin A, essential fatty acids, and/or pyridoxine (vitamin B-6).	The hair follicles are plugged with a hardened material (keratin) which may project from the skin and gives it a bumpy appearance like the skin of a toad.	Provide vitamin A, essential fatty acids, and pyridoxine.	Diets containing vitamin A (also see Dryness of the skin), essential fatty acids (vegetable oils) and pyridoxine (also see Eczema).	A similar condition occurs in secondary syphilis.

(Continued)

TABLE D-2 *(Continued)*

Disorders	Cause	Signs and Symptoms	Treatment	Prevention	Remarks
Nasolabial seborrhea	Deficiencies of niacin, pyridoxine, and/or riboflavin.	Plugging of the follicles around the nose with a yellow, fatty material.	Provide a supplement containing the vitamin B-complex.	Diets containing sources of the B-complex vitamins (eggs, fish, legumes, meats, milk, nuts, and whole grains).	Dried brewers' yeast or non-fat dry milk may be useful in areas where food sources of the vitamin B-complex are scare.
Orogenital syndrome	Deficiency of the B-complex vitamins.	Angular fissures of the lips plus inflammation of the lining inside the mouth. There may also be dermatitis around the eyes, ears, and genital areas.	Provide a supplement containing the vitamin B-complex.	Diets containing vitamin B-complex. (Also see Nasolabial seborrhea.)	May often occur along with corneal vascularization and nasolabial seborrhea.
Tropical ulcer (jungle rot)	Believed to be caused by a poor diet plus environmental conditions in hot, humid areas.	Ulcers of the skin on the legs. These lesions may be infected with such organisims as *Treponema vincenti,* staphylococci, and streptococci.	Provide an adequate diet and proper hygiene (prompt disinfection and treatment of skin lesions).	Same as Treatment.	The condition may be precipitated by debilitating diseases like dysentery and malaria.

DEGENERATION

• Deterioration—a sinking from a higher to a lower level or type.

• A worsening of physical or mental qualities.

• A retrogressive pathological change in cells or tissues as a result of which the functioning power is lost.

DEHYDRATE

To remove most of all moisture from a substance for the purpose of preservation, primarily through artificial drying.

DEHYDRATION

The removal of water from a material. In nutrition, dehydration generally means one or both of the following:

• Drying of a beverage, food, or tissue of the body.

• A state in which the body lacks sufficient fluids to carry out its normal function. If it is not corrected this condition may result in death.

(Also see WATER AND ELECTROLYTES.)

DEHYDROASCORBIC ACID ($C_6H_6O_6$)

The oxidized (hydrogen removed) form of ascorbic acid (vitamin C). In the body it readily reverts to ascorbic acid; thus, it is utilized as such.

(Also see ASCORBIC ACID; and VITAMIN C.)

DEHYDROCHOLESTEROL

A substance produced by the body, and found in or on the skin. It is converted to vitamin D by ultraviolet radiation from the sun or from another source, such as a sun lamp.

(Also see VITAMIN D.)

DEHYDROGENASES

Enzymes that accelerate oxidation by transferring hydrogen to a hydrogen acceptor and thus play an important role in biological oxidation-reduction processes.

DEHYDRORETINOL

One of the forms of vitamin A.

(Also see VITAMIN A.)

DELANEY CLAUSE

In 1958, the food additives amendment, better known for its Delaney Clause (named after the congressman who sponsored it), was passed. This bill has proven to be one of the most controversial pieces of legislation ever to affect American producers and consumers. The Delaney Clause states:

> "*Provided,* That no additive shall be deemed safe if it is found to induce cancer when ingested by man or animal, or if it is found, after tests which are appropriate for the evaluation of the safety of food additives, to induce cancer in man or animals . . ."

This clause gave rise to the policy of "zero tolerance"—that is, no substance can be used as a food additive, even in miniscule amounts, if it has been, in any way, implicated as an inducer of cancer in either man or beast. What, at the time, appeared to be a well-intentioned law aimed at protecting the consumer from potential health hazards proved to be a nightmare for the drug industry and livestock producers. The manufacturers must now prove a negative hypothesis which many feel is impossible. That is, a new drug must be demonstrated to be 100% noncarcinogenic. Unfortunately, the lawmakers are in a tenuous position because, to repeal the Delaney Clause, they run the risk of being accused of supporting the addition of cancer-causing drugs to our food supply.

• **Diethylstilbestrol (DES)**—In 1954, prior to passage of the Delaney Clause, based on Burrough's work at Iowa State University, diethylstilbestrol (DES), a synthetic estrogenlike compound, was approved by the Food and Drug Administration (FDA) for oral use as a growth promotant in cattle finishing rations; and 2 years later, FDA approved the use of stilbestrol implants for steers. Both producers and consumers benefitted, for, on the average, DES increased the rate of gain by 10 to 15%, and improved the feed efficiency by a like amount. Not only that, the carcasses had more lean meat and less fat. By the late 1960s, it was estimated that more than 75% of the nation's feedlot finishing cattle and feedlot finishing lambs were receiving stilbestrol, either in their feed or as an ear implant. All went well until the early 1970s, when it was disclosed that the daughters of women who, in the 1950s, had been given DES to prevent miscarriages, had a high rate of vaginal cancer. By 1973, the assay techniques for DES had become extremely sensitive. (It could measure residues down to 0.5 ppb) and trace amounts of DES (less than 2 ppb) were detected in the livers and kidneys of some beef cattle. So, in keeping with the Delaney Clause, FDA was required by law to prohibit the use of DES as an additive. The ban on DES in feed first became effective on January 1, 1973; and the ban on DES implants followed on April 25, 1973. But FDA did not ban the use of DES as a "morning after pill"; they argued that the contraceptive use of DES by women was voluntary, whereas the meat consumer had no control over what was given to cattle or sheep to promote growth. Emotions waxed hot! Those favoring the continued use of DES estimated the risk of cancer from eating beef liver at the usual consumption rates to be of the order of one case in 1,500 years in the entire U.S. population; meanwhile, other causes of cancer will kill 525 million Americans during the same period of time. FDA countered that DES had proven to be carcinogenic in large doses; then the argument was clinched by stating that the law called for zero tolerance. A 6-year struggle between the livestock industry and the FDA ensued, punctuated by on-again and off-again bans, extensions and court orders. Finally FDA imposed two cutoff dates: (1) the manufacture and shipment of DES became illegal July 13, 1979; and (2) the use of DES (as feed or implants) for cattle and sheep became illegal November 1, 1979; with severe penalties imposed for violations thereafter.

• **Saccharin**—In March 1977, FDA called for a ban on saccharin, when Canadian studies showed that the artificial sweetener caused cancer in laboratory animals. Since a ban on saccharin would affect diabetics, weight-conscious consumers, and the soft drink industry, there was an immediate and indignant outcry. Controversy over the proposed ban was so great that U.S. Congress intervened and kept the prohibition from taking effect, leaving saccharin in use. In the meantime, FDA requires that any saccharin-containing food or beverage carry a warning label, stating that the product contains saccharin which has been determined to cause cancer in laboratory animals.

In 1981, the Secretary of the Department of Health and Human Services, Richard Schweiker, called for a rewrite of the Delaney Clause, which requires the banning of any substance shown to cause cancer in laboratory animals; and which had previously prompted FDA to attempt a ban on saccharin and to threaten a ban on the use of sodium nitrite in bacon and processed meats.

The Delaney Clause will, in all likelihood, be modified, but the debate of how safe is "safe" will probably continue without end.

(Also see FEDERAL FOOD, DRUG, AND COSMETIC ACT; and FOOD AND DRUG ADMINISTRATION.)

DELIRIUM TREMENS

A condition which occurs in alcoholics. It is characterized by mental confusion, trembling, loss of appetite, and nausea.

(Also see ALCOHOLISM.)

DEMENTIA

A mental disorder characterized by serious mental impairment and deterioration.

DEMINERALIZATION

The loss of calcium and phosphorus from the bones.

(Also see OSTEOPOROSIS; and TISSUES OF THE BODY.)

DEMOGRAPHY

The study of the statistical characteristics of human population. Such statistics may be useful in determining the need for certain types of health services.

DEMULCENT

A material which soothes mucous membranes. Certain mucilaginous gums such as psyllium and tragacanth have long been used as herbal remedies to soothe the digestive tract. Some of these gums also have a laxative effect because they absorb water.

DEMYELINATION

A process by which the protective sheath around nerve endings breaks down.

(Also see TISSUES OF THE BODY.)

DENATURATION

A chemical (acid, base, detergent, organic solvent, salt, urea) or physical (heating, shaking, whipping) process which alters the nature of a substance. In biology the substance is often protein. Two examples of denaturation are: (1) the coagulation of egg white (albumen) by heating, and (2) the formation of meringue by beating egg white.

DENDRITIC SALT

A structurally different form of ordinary table salt (sodium chloride; NaCl). Normally table salt crystals are cubes. Dendritic crystals are branched or star-like, hence the term dendritic. The following advantages are listed for dendritic salt: lower bulk density, rapid solution, and an unusual capacity for absorbing moisture before becoming wet.

DENSITY

The weight of a given volume of a substance. For example, the weight of 1 ml of water at 39°F (*4°C*) is 1 g. Hence, the density of water at that temperature is 1 g per milliliter. Objects that are less dense than water (such as the human body) will float in water, whereas those that are more dense will sink in water.

DENTAL CARIES

The conventional name for tooth decay or "cavities," caused by the bacterial breakdown of sugars to lactic acid, which occur mainly at contact points between teeth, at the gum margins, or in pits and fissures of posterior teeth. Eventually the lactic acid erodes the tooth enamel, which cannot regenerate, and a *cavity* is formed. Up to this point the whole process is usually quite painless and may continue to be so until the decay process passes through the dentin and approaches the tender pulp. Several factors determine the amount of dental caries a person will experience: (1) the bacterial flora of the mouth; (2) the level of sugar (carbohydrate) in the diet, or the eating habits; (3) oral hygiene, including brushing, flossing, and dental checkups; (4) resistance of the teeth due to genetic influences and adequate dietary amounts of protein, calcium, phosphorus, and vitamins A, C, and D; and (5) fluoride in the drinking water. Fig. D-8 illustrates the structure of teeth and the stages of tooth decay.

(Also see DENTAL HEALTH, NUTRITION, AND DIET.)

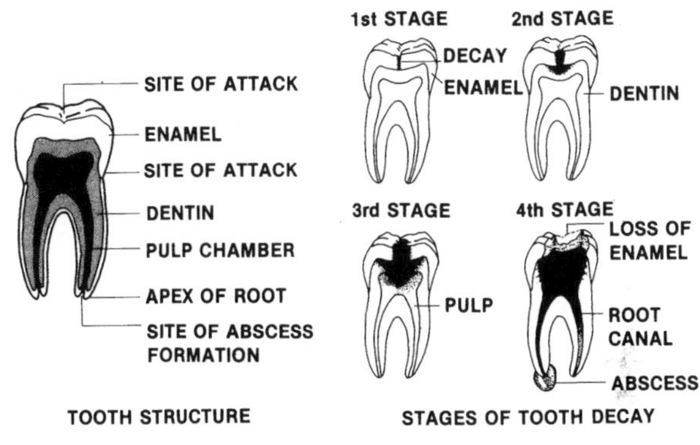

Fig. D-8. Tooth structure and dental caries.

DENTAL FLUOROSIS

Although very low levels (1 ppm) of fluoride in the drinking water help prevent dental caries (tooth decay), higher levels (3 to 5 ppm) can cause mottling or dental fluorosis of the permanent teeth during their formation. This mottling is characterized by rough enamel without luster and brown pigmentation separated by chalk white patches. In areas where it is known that the fluoride content of the water is naturally high, dental fluorosis can be prevented by treating the water supply with phosphate.

(Also see FLUORINE.)

DENTAL HEALTH, NUTRITION, AND DIET

Contents — *Page*

Scope of Dental Health 544
History 544
Development and Anatomy of Teeth 544
Role of Nutrition and Diet 545
Development 545
Dental Caries (Tooth Decay) 546
Causative Factors 546
Susceptible Teeth 546
Microbiology of the Mouth 546
Diet and Nutrition 546
Passage of Time 548
Periodontal Disease 548
Good Dental Health 548
Nutrition and Diet 548
Care of the Teeth 548
Fluoride 548

A complementary relationship exists between nutrition and dental health. Good nutrition is necessary for the development of sound healthy teeth and gum structures, while healthy teeth and gum structures are needed so that a nutritious diet may be eaten.

Except for the common cold, tooth decay is the most prevalent disease in the United States. More than 95% of all Americans have decayed teeth by the time they become adults. Each year, the bill for dental care exceeds $30 billion. But more than money is involved! Tooth decays results in lost time, pain, and if it leads to loss of teeth, there is impaired chewing, speech, appearance, and general well-being. However, since no one dies of tooth decay, people generally do not get overly excited about it, as they do heart disease or cancer. Overall, dental health is viewed indifferently by the general public.

SCOPE OF DENTAL HEALTH. Dental health begins in the womb, since during pregnancy the primary or baby teeth begin to form. Sometime after birth these teeth erupt or push through the gums. Also, important to dental health are all of the oral structures such as the jaw, palate, salivary glands, and supporting tissues. Following birth, a set of 20 teeth gradually appears. Although these are temporary and will be lost, their care is important for good dental health. As the primary or baby teeth fall out, they are replaced by permanent teeth which can serve from 6 to 7 years of age until death, depending upon dental health. Before age 35, tooth decay is the leading cause of tooth loss, but after age 35 the main cause of tooth loss is periodontal disease which attacks the gums, connective tissue, and bone that surround and support the teeth.

Good dental health depends upon (1) nutrition and diet, (2) care of the teeth, and (3) fluoride. Although, closely related, nutrition and diet influence dental health and disease in different ways. The diet is defined as all nutrients that enter the body through the mouth; and nutrition is defined as all of the body processes that nutrients undergo after ingestion.

HISTORY. Tooth decay is one of the oldest diseases of mankind. In the 7th century B.C., the Assyrians believed that tooth decay was caused by a tooth worm—an idea which persisted in some areas until 1850. During Greek and Roman times, many people believed that dental decay resulted from an imbalance of the evil humors over good humors. However, Aristotle noted that when people ate soft sweet figs and some of the fruit sticking to their teeth putrefied, their teeth had a greater tendency to decay. Between the end of the 1700s and the middle of the 1800s, it was widely believed that decay began inside a tooth. In 1819, the suggestion was made that the tooth was attacked from the outside by a chemical agent. During the 1830s, other investigators found that sulfuric and nitric acids could decompose teeth. Around 1845, a Dresden physician, Ficinus, found various bacteria (microorganisms) in material removed from cavities, and he suggested that these bacteria caused the tooth enamel to disintegrate. In 1861, a British dentist, W. K. Bridgeman, concluded that the flow of electricity in the mouth caused the teeth to disintegrate. (It is noteworthy that Bridgeman was fascinated by the electrical experiments of Michael Faraday.) An American dentist, Willoughby D. Miller, published a classic paper in 1890, "The Microorganisms of the Human Mouth," in which he outlined the theory of tooth decay that is still most widely accepted. Miller concluded that decay is produced when acids, such as lactic acid, on the surface of teeth demineralize the enamel by removing calcium, and that these acids are produced when bacteria form acids from sugars and other carbohydrates which cling to the teeth. Then, in 1924, J. K. Clarke at St. Mary's Hospital in London reported that the type of bacteria most frequently found in cavities and most likely to cause tooth decay was *Streptococcus mutans*. Subsequent work with germ-free rats, has demonstrated that *Streptococcus mutans* is one of the most virulent decay-producing organisms.

DEVELOPMENT AND ANATOMY OF TEETH. Teeth, particularly the enamel, begin forming in the womb and continue to develop during the first 20 years of life. Tooth development is a long process of growth and calcification in the jaw, before teeth appear in the mouth. The first indications of tooth development are apparent in embryos only 6 to 7 weeks old. Enamel producing tissue is evident in embryos after 2 ½ months of development. At birth, the first set of teeth are well formed in the jaw and they begin to appear in the mouth after about 6 months.

Humans grow two sets of teeth. The first set—the one formed in the womb—is called the primary teeth. These are also known as the baby teeth, deciduous teeth, or milk teeth. Primary teeth appear—erupt—in the mouth when a baby is from 6 to 30 months old. Twenty teeth—ten on the top and ten on the bottom—make a complete set of primary teeth. Next, the permanent teeth begin appearing in the mouth at about 6 years of age and continue to appear until about age 21, though calcification began in the jaw near the time of birth. There are 28 to 32 permanent teeth, depending upon whether or not the four wisdom teeth appear. Table D-3 lists the primary and permanent teeth and the approximate ages at which they appear or erupt in the mouth.

The anatomy of a tooth and the supporting tissues is shown in Fig. D-9. A tooth is composed of four separate tissues:

1. **Enamel.** The outermost layer is enamel—the hardest substance in the body. It is mainly inorganic material, consisting of only 2% organic matter. The major constituents of enamel are calcium (Ca) and phosphate ($-PO_4$) existing as a derivative known as hydroxyapatite, $Ca_{10}(PO_4)_6(OH)_2$. Hydroxyapatite forms billions of tiny crystallites, which, in turn, form prisms. Small amounts of fluoride, chloride, carbonate, and magnesium are also present in the hydroxyapatite.

2. **Dentin.** Beneath the enamel is the dentin, which contains about 70% inorganic material—calcium and phosphorus—and 30% organic material. Dentin extends almost the entire length of the tooth.

3. **Pulp.** The pulp occupies the center of the tooth and contains nerves, arteries, and veins (blood vessels), and fibrous tissue.

4. **Cementum.** This is also a calcified tissue (bonelike). It acts as a surface for the attachment that holds the tooth to the surrounding tissues.

Other tissues called the periodontal tissues make up

TABLE D-3
APPROXIMATE AGE AT THE TIME OF ERUPTION OF PRIMARY TEETH AND PERMANENT TEETH

Primary Teeth (Baby Teeth)	Approximate Eruption Age	Permanent Teeth	Approximate Eruption Age
	(months)		(years)
Upper Teeth		**Upper Teeth**	
Central incisor . .	7.5	Central incisor . .	7-8
Lateral incisor . .	9	Lateral incisor . .	8-9
Cuspid (canine or eyetooth) . .	18	Cuspid (canine or eyetooth) . .	11-12
First molar	14	First bicuspid (premolar) . . .	10-11
Second molar . .	24	Second bicuspid (premolar) . . .	10-12
Lower Teeth		First molar	6-7
Central incisor . .	6	Second molar . .	12-13
Lateral incisor . .	7	Third molar (wisdom tooth)	17-21
Cuspid (canine or eyetooth) . .	16	**Lower Teeth**	
First molar	12	Central incisor . .	6-7
Second molar . .	20	Lateral incisor . .	7-8
		Cuspid (canine or eyetooth) . .	9-10
		First bicuspid (premolar) . . .	10-12
		Second bicuspid (premolar) . . .	11-12
		First molar	6-7
		Second molar . .	11-13
		Third molar (wisdom tooth)	17-21

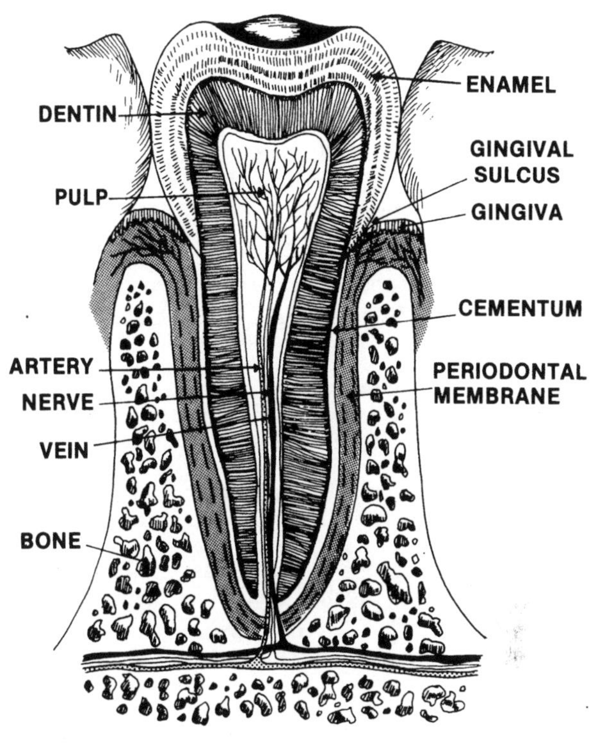

Fig. D-9. The structure of a tooth and adjacent gum (gingiva) area, and the cementum, periodontal membrane, and bone which hold the tooth in place.

the gums and tissue which hold the tooth in place. Fig. D-9 shows the structure of a tooth and gum (gingiva) area.

ROLE OF NUTRITION AND DIET. Nutrition and diet have a role in dental health throughout life—starting in the womb during embryonic and fetal life. Although dental caries (tooth decay) and periodontal disorders are influenced by the interaction of (1) tooth resistance or supporting structures, (2) bacteria in contact with the oral tissues, and (3) the oral environment, the quality, quantity, physical consistency of foods, and the frequency of food intake further interact with these three items.

Development. Since teeth are largely formed during pregnancy and childhood, deficiencies of calcium, phosphate, vitamins A, D, or C, proteins, calories, or fluoride may enhance the susceptibility of teeth to decay-promoting factors. A well-planned diet, providing good nutrition, is necessary for optimal tooth development—chemical composition, size, shape, and possibly time of eruption. While the relationship of the nutritional status of the pregnant or lactating woman to the subsequent dental health of her child is not well understood, the following points indicate the importance of nutrition to dental health during pregnancy and childhood:

1. Any deficiency of calcium during the formation of the teeth and jaws is likely to cause malformed teeth, teeth crowded too closely together in a narrow jaw, and/or poor quality teeth.
2. Calcium for the formation of teeth (and bones) during pregnancy is derived from the mother's diet, not her teeth. The "loss of a tooth for every child" is a popular notion, but false.
3. The calcium-phosphorus ratio is important to the proper utilization of these minerals, as is adequate vitamin D.
4. The growth and proper mineralization of teeth requires vitamin D. Teeth of vitamin D-deficient children have thin, poorly calcified enamel, with pits and fissures, and are especially prone to decay.
5. Excessive vitamin D causes misshapen jaws in the offspring.
6. Vitamin A is essential for optimal development of the enamel. Vitamin A-deficiency results in enamel with fissures and pits, and poorly formed dentin.
7. Prolonged periods of subclinical vitamin A-deficiency may cause inadequate bone growth patterns resulting in misalignment of teeth which further increases the susceptibility of teeth to attack by bacteria.
8. Normal formation of the dentin requires vitamin C.
9. In rats, low protein and high carbohydrate diets have been shown to be associated with reduction in the size of molars, delay in eruption and increased susceptibility to caries.
10. The prevalence of protein-calorie malnutrition in less affluent societies suggests an explanation for the relatively high incidence of dental caries found in the primary teeth of children.

11. Fluoride strengthens the enamel during its formation.

During pregnancy and lactation, extra energy, protein, calcium, phosphorus, and vitamins A, D, and C in the diet are recommended.

(Also see NUTRIENTS: REQUIREMENTS, ALLOWANCES, FUNCTIONS, SOURCES; and PREGNANCY AND LACTATION NUTRITION.)

Dental Caries (Tooth Decay). Before the age of 35, the primary reason for loss of a tooth is dental caries. Moreover, it is estimated that 96% of all high school students—those at the halfway mark between birth and age 35—have tooth decay. Dental caries is the conventional name for tooth decay or cavities, caused by the bacterial breakdown of sugars to an acid, which occur mainly at contact points between teeth, at the gum margins, or in pits and fissures of posterior teeth. Eventually, the acid erodes the tooth enamel, which cannot regenerate, and a cavity is formed. Up to this point, the whole process is usually quite painless, and it may continue to be so until the decay process passes through the dentin and approaches the tender pulp where the nerves are located.

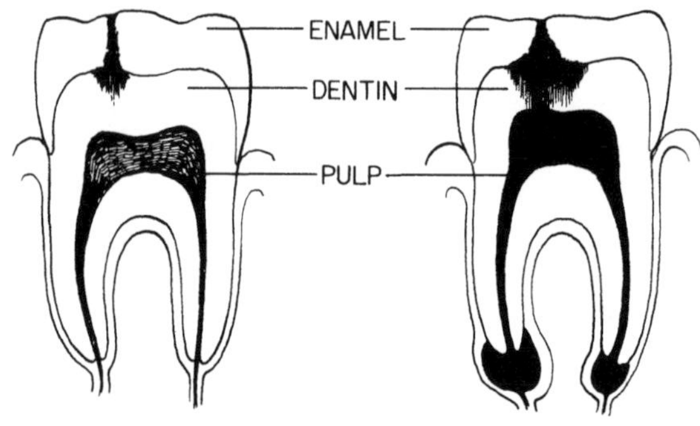

Fig. D-10. The progress of dental caries. On the left, acid from bacteria has eroded the enamel and bacteria have penetrated and attacked the softer dentin. On the right, the bacteria have moved into the dentin, killed the pulp, and infection has spread to the root. Such a tooth may have to be removed.

CAUSATIVE FACTORS. No single factor is involved in causing tooth decay. Rather, tooth decay requires the interaction of (1) susceptible teeth, (2) microbiology of the mouth (bacteria), (3) nutrition and diet, and (4) passage of time.

Susceptible Teeth. A small portion, probably less than 2%, of the American people appear immune to caries. As already pointed out, general nutrition knowledge, animal studies, and some population studies of protein-calorie malnutrition suggest that the availability of nutrients such as protein, energy, vitamins A and D, calcium, phosphorus and fluoride influence the mineralization and formation of teeth; hence, the susceptibility of teeth. The vulnerability of teeth also depends on their spacing and shape, which can encourage food particles to collect between or on them. Furthermore, although not directly related to the teeth, the salivary glands influence the susceptibility of teeth to tooth decay. Anytime the salivary glands quit functioning or markedly reduce their function, the formation of saliva, rampant tooth decay usually develops. This is evidenced when salivary glands are removed from rats or when salivary gland function is markedly reduced in individuals receiving radiation treatment for head or neck cancer. Normally, saliva washes away food particles, bacteria and acids, and neutralizes acids.

Microbiology of the Mouth. Tooth decay does not develop in germ-free animals, but unfortunately this is an unrealistic situation for humans. Decay-promoting bacteria live in the mouth and adhere to the tooth surfaces forming sticky colorless masses called dental plaque. Today, it is generally accepted that the bacteria *Streptococcus mutans*, (*S. mutans*), is one of the most important caries causing bacteria for the following reasons:

1. *S. mutans* contains the enzyme lactic acid dehydrogenase which converts pyruvic acid to lactic acid; then lactic acid demineralizes tooth enamel.
2. By polymerizing glucose to extracellular glucans—polysaccharides—the *S. mutans* and other bacteria are able to cling together and stick to the teeth since glucans are sticky.
3. Colonies of *S. mutans* and glucans are the major components of dental plaque.
4. *S. mutans* can form intracellular polysaccharides or starch from sugars. This starch is stored in the cells and then converted to acid later.

Overall, it is the ability of bacteria in the mouth to produce organic acids, primarily lactic acid, which contributes to the formation of dental caries in susceptible teeth, given the right diet and length of time. The acids produced decompose the hydroxyapatite in the enamel by removing calcium and phosphate, allowing bacteria entry inside the tooth—a cavity is born.

Diet and Nutrition. Diet is highly influential in caries development. Certain diets enhance the activity of cariogenic bacteria, and the formation of plaque.

• **Plaque**—Plaque varies in thickness, and is composed of about 70% bacteria with the remaining 30% being polysaccharides,enzymes, and acids. Unless a tooth has been thoroughly cleaned, the bacteria colonies in plaque continue to grow, particularly near the gum line (gingival sulcus). A thin organic layer, the pellicle, between the tooth enamel and the plaque is formed when proteins in the saliva are adsorbed on the enamel. It seems that the pellicle promotes the attachment of bacteria to the teeth and influences the transport of acids into the enamel and diffusion of calcium and phosphate out of the enamel. Accumulations of plaque are implicated in caries and periodontal disease. (See Fig. D-11.)

• **Sugar and carbohydrates**—Dietary carbohydrates, in particular sugar (sucrose), are prime contributors to the bacterial colonization of the teeth and the production and support of tooth decay. Bacteria in the plaque act on sucrose to produce lactic acid, shown in Fig. D-12.

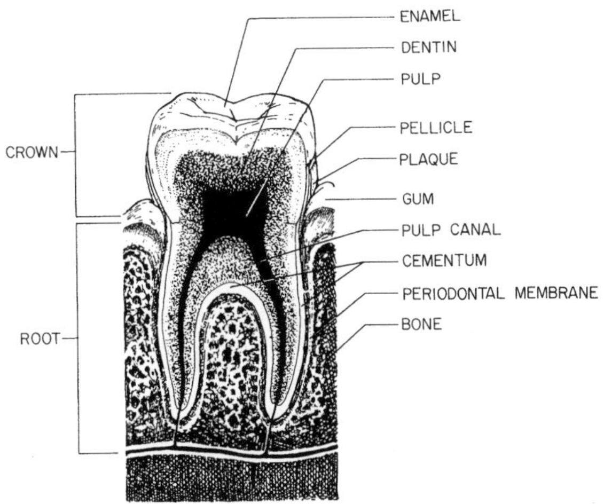

Fig. D-11. Accumulation of plaque at the gum line showing the location of the pellicle.

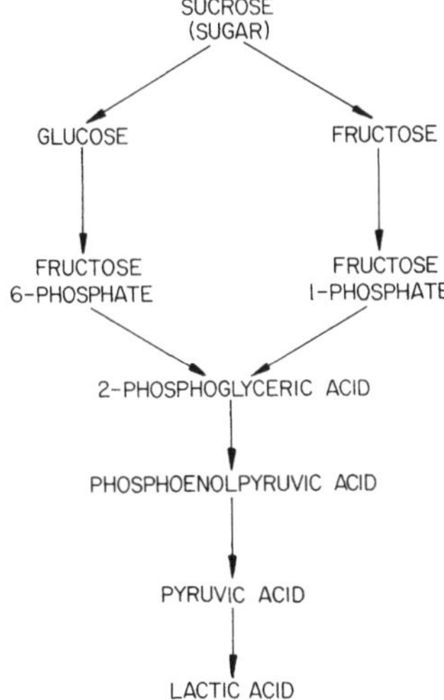

Fig. D-12. The formation of lactic acid from sucrose by bacteria fermentation. Lactic acid is the strongest acid formed by the fermentation of sugar.

The following points demonstrate the role of sucrose in the development of dental caries:

1. Dental caries dropped significantly 2 years after sugar consumption was severely restricted in Europe during World War II. The incidence of caries returned to the prewar levels when sugar intake increased.
2. Nursing caries is observed in infants and young children who are allowed to fall asleep suckling on a bottle of fruit juice or milk with sugar. This continuously bathes the upper front teeth—the four incisors—with sugar. Eventually rampant caries destroy these four teeth.
3. In the classical Vipeholm, Sweden study, various diets were examined for their ability to produce caries. Over a 5-year period, those individuals who ate sticky toffee several times a day demonstrated 12 times more caries than those individuals who were not exposed frequently to sugar.
4. In rats, it has been shown that fructose, glucose, lactose, and maltose are less cariogenic (caries-causing) than sucrose (sugar).
5. In countries where sugar intake is low, such as China and Ethiopia, the incidence of caries is low, while in areas where the intake of sugar is high, such as Australia, Hawaii, and French Polynesia, there is a very high incidence of tooth decay.
6. Studies, which have subsituted nonfermentable sugar alcohols or polyols such as sorbitol or xylitol for some or all of the sweeteners (sugar) used in the diet, have shown a reduction in caries.
7. People with the genetic defect called fructose intolerance, avoid all sucrose-containing foods since sucrose (sugar) is comprised of a molecule of glucose and a molecule of fructose. These people display excellent dental health.
8. Sugar consumption in England has increased from 20 lb (*9 kg*) per person to over 110 lb (*149.5 kg*) per person in the last 100 years. There has been a nearly parallel rise in the incidence of dental caries.
9. Sucrose (sugar) is used in the formation of the extracellular polysaccharides or glucans which contribute to plaque formation. Maltose, lactose, fructose, and glucose cannot be used for this purpose.

Sucrose is not the only culprit carbohydrate. The other sugars—glucose, fructose, lactose, and maltose—and even starch, contribute to acid formation, though starchy foods seem the least offensive.

• **Physical properties of foods**—Foods that stick to tooth surfaces and that are slowly soluble promote caries. Liquid foods, in comparison to sticky or retentive foods, tend to have lower cariogenicity. Therefore, honey and the sugars of dried fruits can contribute to dental caries as well as refined sugar and the products made from it. Sometimes, raw fibrous foods are recommended for between meal snacks. These foods generally require more chewing and stimulate saliva flow, but they do not clean the teeth. Increased saliva flow is beneficial. Also, fats, which produce a protective oily film on the surface of the teeth serve as a barrier of acid penetration to the enamel.

• **Nutritional factors**—Adequate, balanced nutrition is essential for all tissues to grow, develop, and remain healthy. Nutrition is an ally to good oral health which depends upon the health of the salivary glands, the bones of the face, periodontal tissues, and the immune system. Moreover, nutrients in the food contribute to the environment in the mouth by stimulating or inhibiting caries-producing bacteria in the dental plaque. Some trace minerals, other than fluorine, may increase the resistance of the enamel or change the properties of the saliva or plaque, while other minerals may promote caries. Dietary phosphates seem to prevent caries, but copper and lead may be cariogenic. More research is needed to determine the effects of specific nutrients on the growth, development, and maintenance of the teeth and supporting structures.

Passage of Time. One to two years are necessary for a small area of decalcified enamel to progress to clinical caries. But if preventative dietary and oral hygiene measures are taken the development of the decay may be stopped, though the decalcified area will never return to normal.

New teeth are more susceptible to decay than older teeth, since 2 to 4 years after eruption the enamel is in the final stages of maturation. Therefore, individuals with new primary or permanent teeth are more susceptible; hence, young children and teen-agers have a higher incidence of caries.

The more frequent the exposure to sugar, the longer the acid levels are elevated in the mouth, thus increasing the incidence of caries. This also relates to sticky foods which stay in contact with the teeth for long periods and are not washed away by the saliva. Therefore, snacking on sugary foods is not good, but snacking on sticky sugary foods is worse.

Periodontal Disease. An inflammatory disease of the gums and of the bones that support the teeth is called periodontal disease, because it involves structures surrounding the teeth. Periodontal disease is the major cause of tooth loss in adults. The exact cause of periodontal disease is unknown, but several local factors may be involved in its initiation. The primary factor is poor oral hygiene, since bacterial plaque near the gum line (gingival sulcus) (see Fig. D-11) is a prime initiating factor. Rates of periodontal disease are the highest in populations where oral hygiene is the poorest. Plaque causes the inflammation of the gingiva (gingivitis) resulting in red swollen gums (gingiva). As the bacteria continue to invade the sulcus, they secrete toxins that cause pockets of pus to form. Eventually, the disease reaches the bone and results in demineralization. Reabsorption or loss of the bone supporting the teeth results in loose teeth which eventually fall out.

Nutrition and diet have been widely studied as possible causative factors in periodontal disease, but there does not seem to be a direct relationship. Nutritional deficiences do not initiate periodontal disease. The diet is, however, the source of all nutrients necessary for the health of the periodontal tissues that are resistant to disease. Dietary restriction of sugars, especially those in sticky foods, is important, due to the role of sugar in the formation of dental plaque.

GOOD DENTAL HEALTH. Disregarding some genetic defect, most tooth decay and periodontal disease is preventable through sound nutrition and diet, proper care of the teeth, and fluoride treatment.

Nutrition And Diet. Teeth form during a prolonged period, beginning 7 months before birth, continuing through childhood and adolescent years, and not ending until the formation of the wisdom teeth is complete. Like any other developing structure, the teeth require that all nutrients be supplied in adequate amounts. Of prime importance are energy (calories), protein, calcium, magnesium, phosphate, fluoride, and vitamins A, C, and D, which ensure the development of teeth with a rigid structure and increased resistance to decay. Therefore, proper nutrition during pregnancy, lactation, infancy, childhood, and adolescence contributes to good dental health.

High incidence of dental caries is associated with dietary carbohydrates, primarily the monosaccharides (glucose and fructose) and the disaccharides (sucrose and lactose). But more important than their intake is the frequency and their form; therefore, for good dental health the following should be practiced:

1. Sticky sugary snack foods should be avoided; for example, caramel, dried fruits, bread and butter with honey, honey, cookies, chewing gum, jams and jellies, and syrups.

2. Snacking should be confined to foods such as nuts, popcorn, raw fruit and vegetables, and cheeses.

3. Sweets and carbohydrates should be consumed in conjunction with meals. Even starch, if allowed to remain on the teeth long enough, will release glucose molecules which bacteria will convert to acid.

The use of foods containing artificial sweetening agents such as saccharin and aspartame, and the use of polyols like mannitol, sorbitol, and xylitol, may decrease the incidence of caries, since bacteria of the mouth cannot ferment these sweeteners to acids.

(Also see CHILDHOOD AND ADOLESCENT NUTRITION; INFANT DIET AND NUTRITION; PREGNANCY AND LACTATION NUTRITION; and SWEETENING AGENTS.)

Care Of The Teeth. Good dental hygiene is important. Proper tooth brushing along with flossing breaks up the plaque before the underlying enamel begins to decalcify. A daily routine is necessary. It is especially important to brush before going to sleep since less saliva is secreted and, therefore, bacterial acids are less diluted. Brushing following a meal is not always feasible, but a thorough rinsing of the mouth with just plain water is helpful.

Visits to the dentist help maintain good oral health. Dentists may remove plaque from teeth, or fill cavities to prevent further decay. Also, a dentist may apply a plastic sealant to the chewing surfaces of decay-free teeth as a preventative measure.

Fluoride. Fluoride is the most effective agent available for strengthening tooth resistance to acid demineralization. The mechanism by which fluorine increases caries resistance of the teeth is not fully understood. However, it appears that crystals of fluoroapatite can replace some of the calcium phosphate crystals of hydroxyapatite that are normally deposited during tooth formation, and that it may also replace some of the carbonate normally found in the tooth. Apparently these fluoride substances are more resistant to mouth acids. Fluorine may also inactivate oral bacterial enzymes which create acids from carbohydrates.

The fluoride ion, at proper levels of intake, assists in the prevention of dental caries. When children under 9 years of age consume drinking water containing 1 part per million of fluorine, the teeth have fewer dental caries in childhood, adolescence, and throughout life. This has led to the fluoridation of water supplies in many countries. Fluoridation of water supplies to bring the concentration of fluoride to 1 ppm (one part of fluorine to a million parts of water) has proved to be safe, economical, and an efficient way to reduce tooth decay—a highly important public health measure in areas where natural water supplies do not contain this amount. Extensive medical and public health studies have clearly demonstrated the safety and nutritional advantages that result from fluoridization of water supplies. In communities in which fluoridation has been introduced, the incidence of tooth decay in children has been decreased by 50% or more.

The concentration of fluoride in public water supplies should be adjusted slightly to allow for differences in water consumption with seasonal temperature changes.

In areas where water supplies are not fluoridated, fluoride supplements are available as drips, tablets, and mouth rinses. A physician or dentist can prescribe a supplement appropriate for each individual and provide instructions for its use. Generally, drops are prescribed for infants until they are old enough to use the tablets. Older children and teen-agers use tablets or mouth rinses, or both. Dentists sometimes recommend a fluoride mouth rinse for adults with tooth decay problems. Also, about 80% of all toothpaste sold in the United States contains fluoride.

Because fluoride is so effective against tooth decay, many states have established school fluoridation programs in which schools with independent water supplies fluoridate their water through the use of special equipment. In addition, schools in some areas of the country have adopted programs that provide fluoride tablets or a mouth rinse to children in the schools. The procedure takes only a few minutes each day and is supervised by someone trained to administer the supplement, usually the teacher.

(Also see FLUORIDATION; FLUORINE; and MINERALS, Table M-67.)

Although the future may hold a vaccine which will prevent dental caries, dental health, including dental caries prevention, requires life long dedication of an individual who understands and incorporates into daily habits the benefits to be derived from the combined effects of nutrition and diet, oral hygiene (care of the teeth and supporting structures), and fluoride.

(Also see CALCIUM; PHOSPHORUS; VITAMIN A; VITAMIN C; and VITAMIN D.)

DENTIFRICE

A substance used to clean the teeth. Usually, a dentifrice is a paste or a powder.

DENTIN

The major calcified portion of a tooth which lies between the tooth pulp and the enamel.

(Also see DENTAL HEALTH, NUTRITION AND DIET.)

DENTURE

This term generally refers to false teeth.

DEOXYPYRIDOXINE

A compound with similar structure to pyridoxine but which is antagonistic (opposes or neutralizes) to the action of pyridoxine.

DEOXYRIBONUCLEIC ACID (DNA)

The main chemical substance of the genes (chromosomes), which are located in the nucleus of cells. Hence, DNA and similar substances such as RNA are called nucleic acids. It is a complex molecule which varies in composition from species to species. The individual constituents of DNA determine the characteristics which are passed down from generation to generation of living organisms. DNA also governs the metabolic processes which take place during the life of an organism.

(Also see NUCLEIC ACIDS.)

DEPOT FAT

The body fat which is stored in the adipose tissues.
(Also see TISSUES OF THE BODY.)

DERMATITIS

An inflammation or irritation of the skin which may be accompanied by flaking, itching, redness, or swelling. Dermatitis may result from contact with irritants, from drugs or foods which have been ingested, or from various types of nutritional deficiencies.

DERMATOLOGY

The medical specialty which is concerned with the diseases of the skin.

DERMIS

The inner layer of skin which is just inside the outer layer or epidermis. It is much more sensitive than the epidermis, so a wound in the skin which exposes the dermis may be quite painful.

(Also see TISSUES OF THE BODY.)

DESICCATE

To dry.

DESSERTS

Such adages as "sweets for the sweet" and "sugar and spice and everything nice" indicate that many people look on desserts as a delightful and romantic conclusion to each meal. Special occasion meals would not be considered proper unless the traditional desserts were served; for example, pumpkin pie for Thanksgiving, and mincemeat pie or plum pudding for Christmas. And what would a birthday party be without a birthday cake?

Although, as children, all of us were persuaded to eat our vegetables so that we could have some dessert, our love for desserts should be curbed, because we now know that too much sugar can be counterproductive to continued buoyant good health. However, it is not recommended that 100% of the sugar or sweetening be eliminated from the diet. Even when on a reducing diet, it is recommended that some dessert be eaten, but it should be a considerably reduced portion—like ½ or ⅓ of the average serving.

A small amount of sweet food may help digest a meal. Many of our grandmothers used warm sweetened water to ease the infant's colic. In addition, desserts containing milk, eggs, fruit, whole wheat flour and honey or molasses do contribute important nutrients to the diet.

There is an endless parade of wonderful dessert recipes. Table D-4 lists the basic types of desserts. For the composition of the various desserts see the sections on Bakery Products, Desserts & Sweets, and Milk and Products in Table F-36 Food Composition Table.

TABLE D-4
TYPES OF DESSERTS

Type of Dessert	Definition and/or Basic Ingredients
Cakes (Courtesy, American Dairy Association, Rosemont, Ill.)	A cake may be defined as a sweeter, richer quick bread. *Shortening:* It may be made without shortening—angel and sponge cakes. It may be made with shortening—conventional cakes. It may be a combination of the above two—chiffon cakes. *Leavening:* It may be raised by using baking powder. It may be raised by using yeast—Baba au Rhum. It may be raised by stiffly beaten egg whites folded in. *Flavoring:* It may be flavored with any one or a combination of a myriad of flavorings and spices, with or without added fruits and nuts. Chocolate is considered the all-time favorite.
Candies 	Nothing compares to the fancy decorated candy eggs of Easter! And what would Halloween be without candy treats? Candies, of which there are as many varieties and recipes as there are of cakes, should be considered in the dessert category. But, because of their concentrated sweetness, they should be very limited in the diet. (Also see CANDY.)
Charlottes	This is a dessert made by lining a dish with pieces of bread, cake or the most commonly used ingredient, ladyfingers, then filling it with fruit custard, whipped cream or other filling.
Cheesecakes, chocolate  (Courtesy, Hershey Foods Corporation, Hershey, Penn.)	Cheesecakes are incorrectly named. They do contain cheese, but they do not contain cake. They have a crumb crust, with a filling made of cream cheese, or part cottage cheese, cream, eggs and flavoring. Some recipes are baked and some are not. They are usually served with a fruit topping.

(Continued)

TABLE D-4 *(Continued)*

Type of Dessert	Definition and/or Basic Ingredients
Cobblers 	Fruit desserts prepared with a pastry or biscuit dough top. Some cobblers have the flour mixture just sprinkled on top, rather than made into a rolled out pastry. A Brown Betty is a typical example.
Cookies  (Courtesy, Hershey Foods Corporation, Hershey, Penn.)	Americans use cookies for special occasions, similar to cakes. For instance, it wouldn't be Christmas without several batches of cookies in the cookie jar, and the traditional gingerbread house made from gingerbread cookie dough. A cookie may be defined as an unleavened cake, or a sweetened pastry. Some cookies have the same proportion of fat to flour as a pastry, but, because of the addition of other ingredients, they are no longer pastries. There are as many recipes and varieties of cookies as there are of cakes.
Crepes or pancakes  (Courtesy, American Egg Board, Park Ridge, Ill.)	Crepe is the French word for very thin pancakes. Each country seems to have some variation of the pancake; for instance, the tortillas of Mexico, and the egg rolls of Chinese cuisine. The very thin pancake or crepe can be rolled up with any of many sauces, and topped with whipped cream or ice cream, and chopped nuts or more sauce.
Custards  (Courtesy, California Honey Advisory Board)	A custard is perhaps one of the most nutritious desserts; it is used in many convalescent diets. Custard contains only eggs, milk, and a very little sweetening and flavoring. Sometimes it is baked with a little caramelized sugar in the bottom of the custard cup (a flan). Custard sauces are also used extensively in other desserts.
Eclairs and cream puffs	Eclairs are made by cooking water, butter, and flour together; cooling, and adding eggs; then forming the mixture into either rounds for cream puffs or long fingers for chocolate eclairs and baking them. These desserts have always seemed so difficult for the young bride, but, in reality, they are one of the simplest and easiest desserts to make. They just look complicated.

(Continued)

TABLE D-4 *(Continued)*

Type of Dessert	Definition and/or Basic Ingredients
Frozen desserts  (Courtesy, United Dairy Industry Association, Rosemont, Ill.)	There are many ice box desserts that lend themselves to freezing, thus they can be made in advance of a dinner party and removed from the freezer at the desired time. They are a delight for the busy cook.
Fruits  (Courtesy, California Table Grape Commission, Fresno, Calif.)	Fruit, either fresh, stewed, or added to gelatins, should be on the dessert menu each day. Fruit, without added sugar, makes a wonderful snack food.
Ice Cream  (Courtesy, United Dairy Industry Association, Rosemont, Ill.)	Ice cream is a favorite dessert. It usually contains milk, cream, eggs, and flavorings. A family picnic without homemade ice cream is almost unthinkable for many people. Here again, the flavorings and toppings are unlimited in number. Perhaps one of the simplest and most elegant desserts is ice cream topped with a liqueur with or without chopped nuts. Another easy dessert, that elevates the cook to the gourmet level, is the Baked Alaska. A piece of cake is placed on a board, topped with ice cream, and covered completely with a meringue, then baked at 500°F until the peaks turn brown. The trick is in having the oven at 500°F so that the meringue browns quickly, before the ice cream melts. This "king" of desserts is a delicious contrast between hot and cold.
Meringues	Meringues are a delight in texture contrast between the crunchy meringue, and the smooth cream of the filling. Meringues are made by beating egg whites with sugar and baking in whatever shape is desired. Considerable skill is required to produce a perfect meringue topping on a pie.
Milk 	Milk, which is used in many types of desserts, adds a great deal to the nutritive value. Milk desserts not mentioned in other categories include junket, cornstarch puddings (Blanc Manges) and Bavarian creams (gelatin plus cream or milk).

(Continued)

TABLE D-4 (*Continued*)

Type of Dessert	Definition and/or Basic Ingredients
Mousses	A mousse is a frozen whipped cream dessert with fruits and nuts. It does not require beating during freezing and is often put into a mold, which can be unmolded, decorated, and brought to the table for a festive dessert.
Parfaits	A parfait is similar to a mousse, but it is usually served by layering different mixtures in a special tall parfait glass which is then frozen and removed from the freezer a short time ahead of serving. It makes an elegant dessert.
Pastries  (Courtesy, United Dairy Industry Association, Rosemont, Ill.)	As American as Apple Pie! A popular saying such as this indicates that apple pie would win the dessert popularity contest. Here again, the varieties of pies are almost unlimited. A pastry basically must have a crust made with flour, shortening and a small amount of water, which is mixed and rolled out into a thin sheet.
Puddings	This category includes any dessert that does not fall into some other division. The word "pudding" does not conjure up a very elegant dessert, but some puddings are most delicious and nutritious.
Sherbets	A frozen mixture of fruit juice, egg white, milk or water, plus some sugar. It is usually beaten as it freezes.
Shortcakes	Shortcakes can be made with any kind of fruit, but it is a favorite way to serve strawberries in the spring and early summer months. The cake can be either a traditional cake with eggs, or it can be made like a baking powder biscuit without eggs and very little sugar. The cake is layered with the sugared, fresh, sliced fruit and whipped cream between.
Souffles	Souffles are made by beating egg whites and gently folding in the fruit puree. They are served warm from the oven with whipped cream and sauce.
Tortes  (Courtesy, American Dairy Association, Chicago, Ill.)	Torte is an old German name for a rich dessert made with many eggs, plus sugar and either bread crumbs, grated nuts, or flour, in addition to flavoring. It is usually compact and flat rather than high and fluffy. There are many variations on this theme of tortes.
Trifles	Trifles originated in England and were quite possibly developed to use up stale cake. A trifle is a delicious mixture of plain butter or pound cake, fruit sauce or jam, custard, sometimes with added whipped cream and liqueur. The mixture is usually made at least 24 hours before serving.

DETOXICATION

A process by which poisonous substances are rendered less harmful. Most of the detoxication reactions in the body take place in the liver.

DEVELOPMENT

As applied to humans, it generally denotes the series of changes by which the embryo becomes a mature organism; for example, the replacement of skeletal cartilage by calcified bone in the growing child.

DEWBERRY *Rubus baileyanus*

A trailing type of blackberry (a fruit of the family *Rosaceae*) that is grown in the eastern United States. The drupes of the berries are soft and less closely joined together than those of the other blackberries.

Fig. D-13. The dewberry.

Dewberries are good fresh (especially when sweetened a little) and in jams, jellies, pies, and sauces.

(Also see BLACKBERRY[S]; and FRUIT[S], Table F-47 Fruits of the World "Blackberry.")

DEXTRAN

A gummy substance which is produced by the action of certain microorganisms on sugar (sucrose). It may be dissolved in water to make an intravenous solution which is used for the treatement of hemorrhage or severe burns when blood plasma is unavailable.

(Also see GUMS.)

DEXTRIMALTOSE

A mixture of dextrins and maltose prepared from cornstarch by acid treatment or by digestion with enzymes. It is used mainly in milk formulas for infants.

(Also see CARBOHYDRATE[S].)

DEXTRIN

An intermediate polysaccharide, formed during the partial breakdown of starch.

(Also see CARBOHYDRATE[S]; and STARCH.)

DEXTROSE ($C_6H_{12}O_6$)

Since it rotates polarized light to the right (dextrorotatory), the industrial name for the simple sugar, glucose, is dextrose or corn sugar. It is made commercially from starch by the action of heat and acids, or enzymes. Two types of refined dextrose are available commercially: (1) dextrose hydrate, containing 9% by weight of water of crystallization and 426 kcal/100 g; and (2) anhydrous dextrose, containing 366 kcal/100 g and less than 0.5% water. Both are very soluble in water, but only 74% as sweet as sugar. However, dextrose acts synergistically with other sweeteners. Due to the following properties, dextrose is used in many food products: browning, fermentability, flavor enhancement, osmotic pressure, sweetness, humectancy (prevention of drying), hygroscopicity (moisture absorption), viscosity, and reactivity. The major users of dextrose are the confection, wine, and canning industries. In medicine, various concentrations of glucose (dextrose) are utilized for intravenous administration. *Example:* A 5% sterile dextrose solution or "D5W".

(Also see ADDITIVES; CARBOHYDRATE[S]; GLUCOSE; and SUGAR.)

DEXTROSE EQUIVALENT VALUE

Dextrose is the commercial name for glucose. During the conversion of corn starch to corn syrup, dextrins, maltotetrose, maltotriose, maltose, and dextrose are formed. As the process continues, more and more dextrose is formed. If the process is stopped at any given point a corn syrup of certain dextrose equivalent (DE) is produced. The longer the conversion process, the higher the DE and the sweeter the syrup. Pure dextrose has a DE of 100.

(Also see CORN SYRUP [Glucose Syrup].)

DIABETES, ALLOXAN

Diabetes experimentally induced in animals by the administration of the chemical alloxan, which preferentially damages the insulin producing cells of the pancreas. Alloxan diabetes provides a useful animal model for the study of diabetes mellitus.

(Also see DIABETES MELLITUS.)

DIABETES INSIPIDUS

An unusual form of diabetes (the term means the passage of an excessive amount of urine) in which large quantities of water are lost in the urine. This disorder leads to dehydration and excessive thirst. Diabetes Insipidus differs from ordinary diabetes in that it has no connection with the blood sugar, but is due to an abnormality of the pituitary gland.

(Also see ENDOCRINE GLANDS.)

DIABETES MELLITUS

Contents Page

History of Diabetes ... 555
Metabolic Abnormalities in Diabetes ... 556
Factors Which May Precipitate or Aggravate Diabetes ... 558
Complications of Diabetes ... 562
Diagnosis of Diabetes ... 565
Signs and Symptoms of Diabetes ... 565
Tests of Blood and Urine ... 565
Glucose Tolerance Test (GTT) ... 565
Detection of Diabetes in Children ... 566
Treatment and Prevention of Diabetes ... 566
Emergency Treatments ... 566
Diabetic Coma ... 567
Hypoglycemia (Insulin Shock) ... 567
Dietary Measures ... 567
Types of Diets ... 568
Special Foods Which are Low in Energy (Kilocalories) ... 570
Prevention of Nutritional Deficiencies in Diabetics ... 571
Insulin ... 571
Meal Patterns for Different Types of Insulin ... 571
Oral Drugs ... 572
Sulfa-type Drugs (Sulfonylureas) ... 573
Phenformin ... 573
Foot Care ... 573
Exercise ... 573
Promising Research Developments in the Fight Against Diabetes ... 574
Pancreatic Transplants ... 574
Psychosomatic Therapy ... 574
Therapeutic Agents ... 575

This disease is a collection of disorders which result from either lack of insulin (a hormone secreted by the pancreas) or factors which interfere with the actions of this hormone. Most of the tissues in the body depend upon insulin for the promotion of the flow of sugars, fats, and amino acids into cells. The untreated diabetic literally starves to death in the midst of abundance since his or her blood may be loaded with nutrients which cannot get into the cells where they are needed. Diabetes is now the fifth leading cause of death by disease in the United States. There are two major forms in which the disease usually appears: adult-onset diabetes, and juvenile diabetes. Brief descriptions follow.

• **Adult-onset diabetes**—This form, which is by far the most common, may also be called *obesity diabetes* because it is found mainly in obese adults, and often it may be treated successfully by the reduction of body weight. There may be many adults who are free from symptoms, yet who have an early stage of this disease since the complications may develop very slowly. Sometimes, the pathological changes develop over several decades of life. Other characteristics of this form of diabetes are (1) victims are resistant to the development of ketoacidosis, except under conditions of stress; (2) persons over 40 years are the most likely to have the disease, although it sometimes occurs in children; (3) the various disorders develop slowly; (4) injections of insulin may *not* be needed when the early stages are successfully controlled by diet and/or oral drugs; and (5) most deaths from the disease are now due to degenerative disorders of the blood vessels, brain, kidneys, and heart, rather than to diabetic coma.

• **Juvenile diabetes**—In contrast to the adult-onset form, this type accounts for only about 10% of all cases of diabetes, yet it is much harder to control, and severe ketoacidosis is common. Other characteristics of this form are (1) oral drugs are not suitable, so injections of insulin are required; (2) its victims are often underweight rather than obese; (3) progress of the disease may be very rapid; (4) symptoms of excessive urination, unusual thirst, and overeating are common features; and (5) it usually strikes young people (under 25 years). This form is believed to originate from a genetic defect in the pancreas. The tendency to develop diabetes is passed along as a recessive trait, but it appears that as many as 20% of the world's people may carry the gene.

HISTORY OF DIABETES. This disease has apparently plagued man for a very long time, since the writings from the earliest civilizations (Asia Minor, China, Egypt, and India) refer to boils and infections, excessive thirst, loss of weight, and the passing of large quantities of a honeysweet urine which often drew ants and flies. (The term diabetes is derived from the Greek word meaning siphon, or the passing through of water; while mellitus is Latin for honeysweet.) For example, the *Papyrus Ebers*, an Egyptian document dated about 1550 B.C., recommended that those afflicted with the malady go on a diet of beer, fruits, grains, and honey, which was reputed to stifle the excessive urination. Indian writings from the same era attributed the disease to overindulgence in food and drink.

Accounts of the diets of the middle class in northern European countries during the 15th, 16th, and 17th centuries described meals consisting of many courses of roast meats dripping with fat, rich and sugary pastries, and plenty of butter and cream, but little coarse bread or green, leafy vegetables. It is, therefore, not surprising that many cases of diabetes were reported during these times of abundance. It is noteworthy, too, that during this period doctors had to taste the urine of patients for sweetness in order to detect the disease.

Soon, there emerged two schools of thought concerning diets. One school believed in dietary replacement of the sugar lost in the urine, while the other believed in restriction of carbohydrate so as to reduce the effects which were attributed to an excess of sugar. The first school was exemplified by the British physician Willis, who, in 1675, recommended a diet limited to milk, barley water, and bread. This diet was high in carbohydrate, but low in calories. On the other hand, the British military surgeon Rollo, in 1797, started a long-lasting trend towards high-fat, high-protein, and low-carbohydrate diets by prescribing mainly meat and fat. None of the physicians of those times knew much about the nature of the abnormality, since various writers referred to it as a disease of the blood, kidneys, liver, or stomach. Nevertheless, some of the patients appeared to have been helped by the diets which were prescribed, as evidenced by reductions in the amounts of sugar spilled in the urine. The restriction of the caloric intake appears to have been the most effective therapy, since the French physician Bouchardat observed that the limited availability of food in Paris during the Franco-Prussian war of 1870 to 1871 resulted in marked reductions in the sugar spilled by his diabetic patients.

The major breakthroughs in understanding the pathology of diabetes occurred in the latter part of the 19th century. In 1869, while examining a piece of pancreas under a microscope, Langerhans, a German doctor, discovered tiny cells which were different from the rest of the pancreatic tissue. Later, these cells were named *Isles of Langerhans*. Subsequent experimentation with animals showed that if these cells were spared when the rest of the pancreas was destroyed, diabetes could be prevented. Quite by accident in 1889, it was discovered that surgical removal of the pancreas from dogs caused their early death from diabetes. Two German physiologists, Minkowski and von Mering, had been studying the role of the pancreas in the digestion of fats when their animal caretakers noted that the depancreatized dogs passed large quantities of urine which attracted flies. It was then found that the dogs' urine contained sugar, so the relationship between pancreatic function and diabetes was established. Following this discovery, many researchers tried to cure diabetes with extracts of the pancreatic islets, but these attempts were unsuccessful. We now know that the extracts were contaminated with digestive juices from the pancreas that destroyed the activity of insulin, which is a protein.

Finally, in 1921, Banting and Best, physiologists who were working at the University of Toronto, discovered that they could obtain biologically active insulin from the pancreases of dogs provided (1) they first tied off the pancreatic duct so as to cause degeneration of the cells that secreted digestive juices, or (2) they used embryonic pancreases from fetal pups, since the insulin-secreting cells develop before the digestive cells. The insulin they obtained cured the diabetes of depancreatized dogs. Following many other tests on dogs, the hormone was administered to a male diabetic human, who experienced a remarkable recovery. Banting was awarded the Nobel Prize for the discovery, but Best was not even considered for the award because he held only an undergraduate degree at the time. However, Banting shared the money with him.

Ten years before the first use of insulin, Dr. Allen of the Rockefeller Institute in New York started treating diabetics by using the well-known starvation diet which contained very little carbohydrate and only about one-half of the energy requirement. His patients became very emaciated, but some lived to benefit from insulin, which restored their health.

The use of insulin in the treatment of diabetes resulted in a dramatic drop in the number of deaths due to diabetic coma and greatly increased the years of survival after detection of the disease. However, insulin as first used caused sharp drops in the blood sugar (hypoglycemia) which were accompanied by distressing symptoms. Therefore, new forms of insulin were developed by chemically modifying the hormone so as to slow its action. One modification was developed by Hagedorn of Denmark, in 1936, who added protamine, a proteinlike substance. This, and other modifications of insulin, made it possible to use only one injection per day, instead of the three or four originally required.

(Also see HYPOGLYCEMIA.)

Even though several types of insulin are now available, it is still necessary to inject the hormone, since it is a protein which is broken down by digestive juices and, therefore, cannot be given by mouth. (It is noteworthy that it takes pancreatic glands from 150 cattle or 750 hogs to supply 1 year's supply of insulin for the average diabetic who requires injections of this hormone.)

Doctors have known since about 1942 that certain sulfa drugs, which may be given orally, will lower blood sugar. However, the first drugs that were available were too dangerous for the type of use required by diabetes. Their occasional use against infectious disease was safe, but treatment of diabetes often required daily, lifelong therapy. Eventually, safe forms of these drugs were developed, and the first oral antidiabetic medication (tolbutamide) was used in the United States in 1956. Since then, there has been some concern as to the long-term effects of oral drugs, which do not substitute for insulin, but only heighten the secretion of insulin. Therefore, they should be used with caution in cases of adult-onset diabetes, but never for juvenile diabetes.

About the same time as the initiation of the use of oral drugs, medical researchers discovered that adult-onset diabetics secreted approximately normal amounts of insulin, but that the action of the hormone was somehow impaired. It was then discovered that obesity is a major contributing factor to this type of diabetes, and that a great improvement in the utilization of glucose often follows a reduction in body weight. It seems that the distended fat-containing cells found in obese people are resistant to the effects of insulin. Other researchers have recently suggested that it is wise to have strict dietary control in diabetes so as to minimize the need for extra insulin. Excessive insulin may accelerate the development of cardiovascular lesions. Hence, it appears that some of the strict dietary measures which were used in the preinsulin days are still useful for the treatment of adult-onset diabetes.

METABOLIC ABNORMALITIES IN DIABETES. Many of the deadly complications of uncontrolled diabetes can be traced to the metabolic disorders which accompany advanced stages of the disease. (*NOTE:* Many of these disorders may be absent in mild cases of adult-onset diabetes. Nevertheless, it is important for all diabetics to recognize the consequences of poor control of their disease. These disorders are outlined in Fig. D-14 and discussed in the sections that follow.

• **Abnormal appetite**—Many diabetics overeat. It is suspected that they may have an impairment in the utilization of glucose (a simple sugar) by the satiety center in the brain. A high rate of metabolism in the satiety center tends to depress appetite.

• **Secretion of hormones by the small intestine**—The entry of food into the small intestine from the stomach stimulates the release of intestinal hormones which trigger the release of insulin from the pancreas. It is suspected that abnormalities in the release of the intestinal hormones might be partially responsible for some

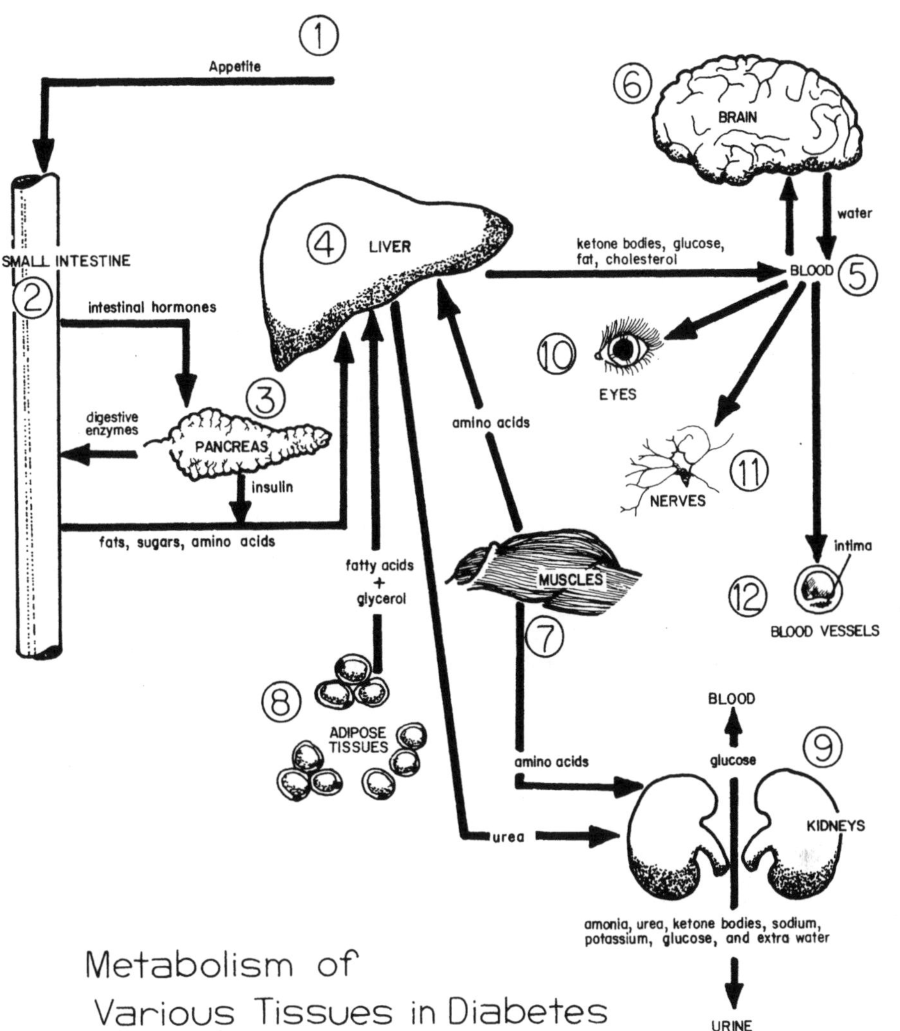

1. Appetite: Diabetic overeats due to excessive appetite.
2. Small Intestine: There may be excessive secretion of intestinal hormones.
3. Pancreas: Secretion of insulin may be deficient (juvenile diabetes) or action of insulin or tissues may be impaired (adult diabetes).
4. Liver: Contributes to elevated blood levels of cholesterol, fats, glucose, glycoproteins, and ketone bodies
5. Blood: May be thickened by elevated levels of glycoproteins and other materials
6. Brain: Structure and functions may be impaired by abnormal carbohydrate metabolism.
7. Muscles: Net breakdown of protein and release of amino acids into the blood
8. Adipose Tissues: Net breakdown of fat and release of fatty acids into the blood
9. Kidneys: Contribute glucose to blood. Excrete large volume of urine containing glucose, ketone bodies and mineral salts. There may be damage to the filtering structures and the blood vessels.
10. Eyes: Damage to retina and blood vessels may occur. Cataracts may also develop.
11. Nerves: There may be damage and loss of functions.
12. Blood Vessels: Thickening of lining and formation of plaques

Fig. D-14. Metabolism of various tissues in diabetes.

of the defects in the secretion and/or action of insulin in diabetes.

• **Secretion of insulin by the pancreas**—There is usually a lack of sufficient insulin in juvenile diabetes, while in adult-onset diabetes there appears to be a delay in the release of insulin in response to dietary carbohydrate. The normal pattern of insulin release consists of two phases: an early discharge of the hormone, and a late discharge. Adult-onset diabetics often lack the early discharge. They may also have "insulin resistance," a condition in which greater than normal amounts of the hormone are required for the entry of bloodborne nutrients into the tissues. Hence, their blood sugar builds up to abnormally high levels until a late and heavy discharge of insulin brings it down to lower levels.

• **Disorders of metabolism in the liver**—This organ contributes to much of the abnormal blood chemistry associated with diabetes because it is the main site for the syntheses of (1) ketones (substances formed during the incomplete metabolism of fats), (2) cholesterol, (3) glycoproteins (complex molecules containing sugars and proteins), and (4) glucose (synthesized from amino acids). Also, there may be an accumulation of fat in the liver due to an accelerated release of fatty acids from the body's fat stores. Finally, there are reductions in the synthesis of carbohydrate reserves (glycogen) and in the utilization of carbohydrate for energy, because these are insulin-dependent processes.

• **Abnormal blood chemistry**—Nutrients and their metabolites tend to accumulate in the blood because they are not fully utilized by the body tissues due to lack of insulin, and/or a reduced effectiveness of the hormone. Hence, there may be elevated blood levels of such substances as cholesterol and other fatty materials, glucose, glycoproteins, and ketones. There may even be a greater than normal tendency for the blood to clot.

• **Altered metabolism in the muscles**—Lack of insulin action on muscle cells results in a reduced uptake of amino acids, fatty acids, and glucose by these cells. Hence, the muscles gradually deteriorate due to lack of energy for their metabolism and lack of amino acids for their maintenance. In many cases, there is a net flow of amino acids out of muscle, resulting from the oversecretion of the hormones which promote the breakdown of tissue proteins.

• **Altered metabolism in adipose (fatty) tissues**—As in the case of the muscles, the adipose tissues cannot take up the nutrients they need from the blood without the action of insulin. Thus, there is a gradual breakdown of these tissues and a net release of fatty acids and glycerol into the blood. The mobilization of large amounts of fat from the body stores is believed to be the reason for the accumulation of excessive fat in both the blood and the liver.

• **Disorders involving the kidneys**—The ketoacidosis of severe diabetes provokes the kidneys to convert amino acids to ammonia (in order to neutralize the excess acid prior to its excretion in the urine), and to glucose, which is released into the blood. Also, some of the excess glucose in the blood spills into the urine. Mineral salts may also be lost in the urine since the various metal ions (calcium, magnesium, potassium, and sodium) are also used by the body to neutralize the ketoacid ions which are excreted via the urine. Occasionally, there is the spilling of protein from the blood into the urine, which may be a sign that there has been damage to the filtering apparatus of the kidneys.

• **Metabolism in the eyes**—Elevated blood levels of fats and glucose alter metabolism of the tissues of the eyes because these substances apparently accumulate in the lens, blood vessels, and nerves. While glucose itself is not so harmful, the other sugars derived from it (fructose and sorbitol) are very harmful to the eyes when they are present in excessive amounts.

• **Metabolism in the nerves**—Like the eyes, the nerves are exposed to abnormally high levels of fats and glucose, against which they have few natural defenses. Normally, this situation would exist for only short periods after meals, but in diabetes the periods of such exposure are greatly lengthened.

• **Metabolism in the blood vessels**—These tissues, like the eyes and the nerves, are ill-equipped to deal with great excesses of bloodborne fats and glucose. There is a formation of abnormal connective tissue in the inner lining of the smaller blood vessels. The abnormality consists of an excessive proportion of glucose to protein in the tissue, which results in a thickened and irregular lining. The larger blood vessels, such as the arteries, develop abnormal patches (plaques) which are made up of both fibrous and fatty materials.

FACTORS WHICH MAY PRECIPITATE OR AGGRAVATE DIABETES. Many doctors have observed that diabetes seems to be more prevalent in certain families than others, so they have concluded that the tendency to develop the disease is an inherited trait. Support for this theory is found in studies of twins who originate from a single, fertilized egg cell (identical or monozygotic twins). Identical twins have identical genes, so they should also have the same susceptibility to hereditary diseases; whereas fraternal twins (which originate from two separate, fertilized eggs) have no greater similarity in their heredity than brothers or sisters born at different times. A study revealed that 70% of the identical twins of known diabetics had the disease, while only 10% of the fraternal twins of known diabetics were diabetic.[6] The fact that fewer than 100% of the identical twins of diabetics had the disease showed that environmental factors also played a role in its development.

Another study showed that about 25% of the nontwin brothers and sisters of diabetics might be expected to develop either the insulin-requiring form, or the milder adult-onset form sometime in their lifetime.[7]

---

[6] Gottlieb, M. S., and H. F. Root, "Diabetes Mellitus in Twins, "*Diabetes*, Vol. 17, 1968, p. 693.

[7] Kobberling, J., "Studies on the Genetic Heterogeneity of Diabetic Mellitus," *Diabetologia*, Vol. 7, 1971, p. 46.

The strong evidence for hereditary susceptibility to diabetes leads many health professionals to believe that for diabetes-prone families, the potential to become diabetic is present in all members at the time of their birth. Whether or not they develop clinical signs of the disease depends upon their exposure to factors which may precipitate or aggravate the disease. If this concept is correct, then heredity is the single "cause" of diabetes. Therefore, it is important to be aware of the more common factors which may bring on the clinical features of the disease. These factors follow.

• **Antibodies against insulin and/or pancreatic tissue**—It is well known that diabetics treated with beef (bovine) insulin may develop antibodies which render this insulin ineffective, making it necessary to switch them to pork (porcine) insulin. However, it is uncertain whether the apparant ineffectiveness of some individuals' own insulin is due to antibodies in their blood. Antibodies against one's own tissue are called *autoantibodies*, and this phenomenon is called autoimmunity. It is noteworthy that diabetics who require injections of insulin (a sign that their pancreas does not secrete enough of the hormone) are more likely to have autoantibodies to their stomach and thyroid tissues than are nondiabetics.[8] Perhaps these diabetics have had autoimmune damage to their pancreatic cells which secrete insulin. Autoantibodies against a particular tissue may attack and destroy that tissue.

• **Chronic hypoglycemia (low blood sugar)**—Many people who start out with hypoglycemia end up with adult-onset diabetes.[9] Some tentative explanations for this observation are that (1) hypoglycemia is an early sign of the metabolic disorders leading to clinical diabetes; (2) people with this condition are likely to overeat in order to rid themselves of annoying symptoms, so they become obese, adult-onset diabetics; (3) the low-carbohydrate diets traditionally used to treat the condition cause deterioration of carbohydrate utilization; and/or (4) the chronic release of hormones which raises the blood sugar when it is low may overtax insulin-secreting cells of the pancreas, or disrupt carbohydrate metabolism in another manner.

(Also see HYPOGLYCEMIA.)

• **Damage to the pancreas by an iron overload**—People suffering from iron overload (hemochromatosis—a disorder characterized by deposits of iron in such tissues as the heart, liver, and pancreas) often have "bronze diabetes," which is impairment of pancreatic function by the iron deposits. The majority of patients in one study of hemochromatosis had either clinical diabetes or impaired utilization of carbohydrate.[10] It is thought that iron overload is caused by either (1) a large excess of dietary iron which somehow gets past the intestinal barrier that usually blocks the absorption of such excesses, or (2) excessive breakdown of red blood cells with the release of iron which had been incorporated in hemoglobin. It is noteworthy that the babies born in certain Bantu tribes of Africa have hemochromatosis due to their mother's consumption of large quantities of beer brewed in iron pots. These findings make one wonder about the advisability of the practice of giving pregnant women "shotgun" doses of iron in order to prevent anemia. Little is known about the ultimate disposition of the iron absorbed by the pregnant mother.

• **Deficiencies of essential minerals**—Attention has recently been focused on the role of essential mineral elements in carbohydrate metabolism, because of the discovery of glucose tolerance factor containing chromium.[11] This factor, which has been isolated from brewers' yeast grown on a nutrient medium rich in chromium, enhances the glucose utilization of elderly, adult-onset diabetics. Other research has shown that juvenile diabetics have lower levels of chromium in their hair.[12]

Studies in animals have shown that there is impairment of glucose utilization in deficiencies of either manganese or zinc, and that supplements containing these minerals apparently correct the problem.

Likewise, deficiencies of potassium are known to cause defects in the release of insulin which was corrected by feeding this mineral. However, another mineral, magnesium, is required for operation of the "potassium-sodium pump," a biochemical mechanism which transfers potassium from the fluid outside to the interior of the cells, and in opposite fashion draws sodium out of the cells. Hence, the vital role of potassium in the secretion of insulin depends to some extent upon an adequate supply of magnesium.

It is ironic that the production of refined grains by milling results in the removal of about three quarters of the chromium, magnesium, manganese, and zinc; since these minerals tend to be concentrated in the bran and the germ.

Calcium is also required for the secretion of insulin, but its level in the blood is usually normal, even when the dietary supply may be inadequate.

• **Excessive secretion of diabetogenic hormones**—The hormones which counter the effects of insulin are said to be diabetogenic; that is, they act so as to maintain or raise the blood sugar, while insulin lowers it. These hormones are adrenocorticotropic hormone (ACTH) and a growth hormone from the pituitary; adrenalin, noradrenalin, and adrenal cortical hormones (mainly cortisol) from the adrenals; glucagon from the pancreas; and

[8]Whittingham, S., *et al.*, "Diabetes Mellitus, Autoimmunity and Aging," *Lancet*, Vol. 1, 1979, p. 763.

[9]Seltzer, H. S., S. S. Fajans, and J. W. Conn, "Spontaneous Hypoglycemia as an Early Manifestation of Adult Onset Diabetes," *Diabetes*, Vol. 5, 1956, p. 437.

[10]Dymock, I. W., *et al.*, "Observations on the Pathogenesis, Complications, and Treatment of Diabetes in 115 Cases of Haemochromatosis," *American Journal of Medicine*, Vol. 52, 1972, p. 203.

[11]Mertz, W., "Effects and Metabolism of Glucose Tolerance Factor," *Nutrition Reviews*, Vol. 33, 1975, p. 129.

[12]Hambidge, K. M., D. O. Rodgerson, and D. O'Brien, "Concentration of Chromium in the Hair of Normal Children and Children with Juvenile Diabetes Mellitus," *Diabetes*, Vol. 19, 1968, p. 517.

thyroid hormones from the thyroid.

(Also see HYPOGLYCEMIA, section headed "Factors Which Raise or Lower the Blood Sugar in Healthy People.")

Excessive secretion of these hormones may be provoked by factors such as emotional stress, endocrine disorders like acromegaly (excessive growth hormone), Cushing's syndrome (over-secretion of cortisol), and pheochromocytoma (a hormone-secreting tumor of the adrenals), and, in one way or the other, by most of the other factors which are discussed in this section.

• **Fasting or starvation**—The body adapts to the lack of food so as to protect the brain and the nerves from lack of glucose. Hence, the normal pattern of endocrine secretion is shifted towards a predominantly diabetogenic pattern, insulin secretion drops, and the utilization of carbohydrate by the fatty tissues and muscles is greatly reduced.

• **Fever**—This condition is the bane of juvenile diabetes, since it may lead to ketoacidosis in normal children; but it may be the cause of coma, and sometimes death, in diabetic children. The reason for these effects is that metabolism is greatly stepped up in the febrile state, carbohydrate stores (glycogen) are soon exhausted, and the body must depend almost entirely upon its fat and protein for energy. Ketosis results from the incomplete metabolism of fat, the complete metabolism of which requires at least a small amount of carbohydrate. There is usually sufficient conversion of protein to carbohydrate in adults to forestall ketoacidosis, but some children appear to lack such protection.

• **High-fat diets**—These diets are usually based upon the substitution of fat for carbohydrate so as to avoid overtaxing the pancreas. However, it has been found that in many cases these diets cause further impairment of carbohydrate metabolism. Also, there is now considerable concern as to the possibility that the long-term use of such diets may lead to an increased incidence of cardiovascular disease, which is at present the leading cause of death of diabetics in the United States.

• **High-protein diets**—It has long been thought that diabetics require extra protein because in uncontrolled diabetes there is excessive breakdown of protein plus elevated urinary excretion of the nitrogenous waste from protein. However, the control of diabetes by diet and/or insulin curtails the pathological degradation of protein. Besides, it is now known that extra dietary protein may stimulate the secretion of excessive amounts of glucagon and growth hormone, two hormones which counter the effects of insulin.

• **High-sugar diets**—These diets, which are rich in fruit sugar (fructose), grape sugar (glucose), milk sugar (lactose), and table sugar (sucrose)—tend to overstimulate the pancreas to secrete insulin. Furthermore, it is believed that they also contribute to elevations in the blood level of triglycerides, a situation associated with a reduction in the utilization of carbohydrate; and possibly, with the development of cardiovascular disease. High-sugar diets should be distinguished from the type of high-carbohydrate diet that is based upon starches and other complex carbohydrates which are utilized more slowly. However, slow utilization of carbohydrate is best achieved with the use of high-fiber foods such as fruits, vegetables, and whole grains.

• **Infections**—The close association between infectious diseases and the first signs of diabetes in children has led some medical scientists to believe that the diseases might cause pancreatic damage, or somehow stimulate the formation of antibodies against insulin. So far, there is no hard evidence to prove this hypothesis. Furthermore, virulent diseases like influenza produce sufficient stresses to unmask a hidden tendency to develop diabetes.

• **Irregular patterns of eating, exercising, and sleeping**—Maximum harmony between the various processes of the body is achieved when there is a regular, daily pattern (circadian rhythm) of the secretion of hormones which is coordinated with the rise and fall in nutrient metabolism. However, this harmony may be disrupted by practices which are inconsistent with a regular biorhythm. For example, skipping meals triggers the release of the diabetogenic hormones which in some persons may cause a delayed, but excessive release of insulin at the next meal; while overeating may lead to excessive secretion of insulin, an increased production of fat, and/or diabetic characteristics like high blood sugar. Although the role of sleep in metabolism is not fully understood, it is well known that erratic patterns of sleep and wakefulness are associated with emotional disorders. These disorders may precipitate "stress diabetes." Finally, diet and exercise need to be balanced so as to prevent large gaps between the intake and expenditure of energy.

• **Ketosis**—This condition, which results from the metabolism of fat in the absence of sufficient carbohydrate, causes impairment of glucose utilization by the fatty tissues and the muscles. Part of this effect is due to the stimulation of the secretion of cortisol, a diabetogenic hormone from the adrenal cortex. Mild forms of this condition, such as those provoked by fasting and special diets, may not be harmful, except in the cases of overly susceptible children and diabetics. However, chronic or severe ketosis may lead to coma or even death. Therefore, ketogenic diets (those high in fat and protein, but very low in carbohydrate) should only be used under the supervision of a doctor.

• **Lack of exercise**—Some obese people eat about the same as other people, but are much less active, so their energy intake is constantly in excess of their energy expenditure. It is well known that obesity may lead to adult-onset diabetes. However, some slender people also have a tendency to have high blood levels of glucose and/or fats. Sometimes these abnormalities are corrected by a regular program of exercise, even when the energy intake is increased so as to match the increased energy expenditure.

• **Lack of Fiber**—Diets low in fiber (indigestible carbohydrates found in plant foods) may allow too rapid digestion and absorption of carbohydrates. Also, foods low in bulk (refined foods which are rich sources of energy) may circumvent one of the mechanisms for the control of food intake—the distention of the small intestine. The distention provokes the release of intestinal hormones which slow the emptying of the stomach into the intestine. There may also be some coordination of this response with the operation of the satiety center in the brain, which depresses appetite. It is noteworthy that animals accustomed to diets high in fiber, such as the sand rat, develop diabetes when fed a low-fiber laboratory chow.[13] It is noteworthy, too, that few cases of diabetes are found among the rural Bantu people of South Africa who eat high-fiber diets, but that the incidence of the disease among these people increases when they move to cities like Cape Town where they adopt the urban lifestyle. However, many other factors change when people move from the countryside to urban areas.

• **Large meals**—First of all, the habitual consumption of large amounts of food at individual meals may condition the body to somehow override the normal mechanisms of appetite control. Second, there may be chronic overstimulation of the secretion of insulin and eventual enlargement of the insulin-secreting cells of the pancreas. Finally, many studies have shown that two or three large meals tend to stimulate a greater amount of fat synthesis by the bodv than five or six smaller meals.[14]

• **Obesity**—This condition often leads to adult-onset diabetes which may sometimes be corrected by reduction of the body weight to the value appropriate for height and body build. It seems that the cells of the fatty tissues become resistant to the effects of insulin when they are engorged with fat, so it takes longer for the blood sugar to return to normal after meals.

• **Oral contraceptives (birth control pills)**—Some women using this means of birth control may develop an abnormal utilization of carbohydrate like that seen in the early stages of adult-onset diabetes. It is believed that this condition occurs because the hormones present in the pills mimic those of pregnancy, when diabetic tendencies are aggravated. Doctors are usually reluctant to prescribe these pills to women from families with histories of diabetes.

• **Pancreatitis**—Inflammation of the pancreas may cause damage to both the digestive cells and the insulin-secreting cells. This inflammation may occur when the common bile duct is blocked by a gallstone, causing the backflow of both bile and pancreatic digestive juices. Chronic alcoholism may also be a cause of pancreatitis.

• **Pregnancy**—The first signs of diabetes may be noticed in some women when they are pregnant. Apparently, the pattern of hormonal secretion in pregnancy is such that resistance of the woman's tissues to insulin increases with the duration of pregnancy. The result of this condition is that nutrient levels in the mother's blood are elevated, a situation which guarantees that the rapidly growing fetus receives adequate nourishment. However, diabetic women tend to have babies that are larger than normal, due to an exaggeration of this effect. An entirely different cause of pregnancy diabetes is described in the item which follows.

• **Pregnancy diabetes due to a deficiency of pyridoxine (vitamin B-6)**—Recently, it was discovered that some pregnant women developed a so-called *gestational diabetes* due to a deficiency of vitamin B-6. Women suspected of having diabetes during pregnancy were checked relative to their utilization of a test dose of sugar (they were given glucose tolerance tests).[15] Thirteen out of fourteen women whose tests showed a diabetic tendency were found to have deficiencies of vitamin B-6. The vitamin deficiency was confirmed by a special laboratory test of urine. After treatment with the vitamin to correct the deficiency, only 2 women out of the 13 who were pyridoxine deficient still showed diabetic tendencies. Perhaps, the test for vitamin B-6 deficiency should be given to other groups, including men and nonpregnant women, who show diabetic tendencies. For example, it is well known that additional amounts of the vitamin may be required when protein intakes are high, or when women take birth control pills for long periods of time.

• **Protein-energy malnutrition**—This condition is often accompanied by potassium depletion, which may often be responsible for an impairment of insulin secretion.[16] This type of malnutrition has been found to be much more common in hospital patients than hitherto suspected. Impairment of insulin secretion may retard recovery, so there may be cases where potassium supplementation is advisable.

• **Stress diabetes**—Hidden (latent) diabetes has often been brought out in susceptible people by such emotional stress situations as family or personal illness; other problems at home, school, or work; or squabbles with friends or relatives. This type of diabetes is called "stress diabetes," because it may disappear when the stress factor is eliminated, but it may also reappear at a later date. It is noteworthy that the stresses which may bring out latent diabetes are also useful indicators for predicting the probability that one may have a heart attack in the near future. Each stress factor is given a numerical value which is then used to compute a total stress score for a person. Scores higher than specified values indicate a strong likelihood for a heart attack. The two stress-induced diseases appear to be closely related because a significant proportion of people suffering from chronic

[13]Trowell, H. C., "Dietary-Fiber Hypothesis of the Etiology of Diabetes Mellitus," *Diabetes,* Vol. 24, 1975, p. 762.

[14]Leveille, G. A., "Effect of Periodicity of Eating and Diet Composition on Enzymatic Adaptations," in *Proceedings of the Ninth International Congress of Nutrition,* edited by A. Chavez, published by S. Krager, Basel, Switzerland, 1975, p. 13.

[15]Coelingh Bennick, H. J. T., and W. H. P. Schreurs, "Improvement of Oral Glucose Tolerance in Gestational Diabetes by Pyridoxine," *British Medical Journal,* Vol. 3, 1975, p. 13.

[16]"Potassium Depletion and Impaired Insulin Release in Protein-Energy Malnutrition," *Nutrition Reviews,* Vol. 34, 1976, p. 16.

pains in the chest and shoulder (angina pectoris) due to cardiovascular disease may have hidden diabetes.[17]

The generally accepted explanation for stress diabetes is that the abnormalities in carbohydrate utilization are due to the stress-provoked oversecretion of hormones from both the pituitary and the adrenal glands. Doctors employ this principle as the basis of a test for hidden diabetes.[18]

• **Thiazide diuretics (water pills)**—These drugs are used to flush excessive water from the body in the treatment of congestive heart failure, high blood pressure, and water accumulation which occurs prior to menstruation. However, they may cause potassium depletion, because this mineral is lost with the water. Hence, there may be impairment of insulin secretion due to lack of sufficient potassium. Also, some of the drugs in this category may cause damage to the pancreatic cells which secrete insulin.

COMPLICATIONS OF DIABETES. The purpose of this section is not to dwell on morbid details, nor to frighten the reader; rather, it is for the purpose of demonstrating the need for careful control of diabetes so as to prevent deadly complications. Good control requires much more than the prevention of the spilling of sugar into the urine. The other factors which need to be controlled are the blood levels of cholesterol, free fatty acids (these come from the body's fat stores), glucose, insulin, ketones, and triglycerides (this is the main form of fat carried by lipoproteins in the blood). Failure to control these factors may result in some of the complications which follow.

• **Alcoholism**—Overconsumption of alcoholic beverages may result in a drastic lowering of the blood sugar which is difficult to correct.[19] Hence, it may be dangerous to treat alcoholic diabetics with insulin, because the combination of excessive alcohol and insulin may produce effects similar to an overdose of insulin.

• **Atherosclerosis**—This degenerative disease is characterized by the buildup of abnormal patches or "plaques" on the walls of arteries. The plaques are made up of fibrous material which is mainly protein, and a mushy material composed of cholesterol and cell debris. The currently accepted belief that this condition is due mainly to long-term effects of high levels of blood fats is probably an oversimplification since it leaves many questions unanswered (such as the reason for the buildup of fibrous material, and the increased susceptibility of the diabetics compared to nondiabetics with similar characteristics). This disease endangers life when it obstructs the flow of blood to such organs as the brain and the heart.

(Also see HYPERLIPOPROTEINEMIAS.)

• **Birth defects in children of diabetic mothers**—Women who suffer from such complications of diabetes as deterioration of blood vessels, impairment of kidney function, high blood pressure, or ketoacidosis run a high risk of having babies with congenital defects or fatal disorders.[20] Hence, the following facts are noteworthy for diabetic or prediabetic women (those believed to have an inherited tendency to develop the disease) who are planning families:

1. Careful medical control of diabetes results in less risk in pregnancy.
2. Older women are more likely to have complications of diabetes than younger women, unless the latter are juvenile diabetics. There is considerable concern as to whether two parents who have the juvenile type of diabetes should have children, since there is a strong likelihood that such children would have a tendency to develop the disease.
3. The tendency for the development of severe diabetes in a woman increases with the number of pregnancies.
4. Many of the problems of the fetus occur during the last stages of pregnancy in a diabetic woman, so close medical supervision is needed at this time.

• **Blindness**—Diabetes is the leading cause of blindness in the United States. The main cause of the tissue changes leading to damage to the retina (retinopathy) and blindness in diabetics is the prolonged elevation of the blood sugar which leads to the thickening of membranes in the lining of small blood vessels. The thickening eventually blocks the flow of blood in the capillaries and stimulates the abnormal growth of new, more fragile blood vessels which readily break and produce hemorrhages in the retina and in the fluid of the eyeball. The clouding of the lens of the eye due to cataracts formation may also result in loss of vision. However, this complication will be discussed separately.

• **Blood clots**—It is well known that "thickened" blood has a greater tendency to clot than normal blood. Thickening of blood is usually due to an increased proportion of solids to liquids. The scientific term for such thickening is "increased viscosity," which usually means that the blood flow through the smaller blood vessels may be abnormally slow. The blood of some diabetics contains extra amounts of protein-bound carbohydrates (glycoproteins) which tend to thicken the blood. There appear to be higher rates of synthesis of these compounds in diabetics who have had chronic, high blood sugar.[21] Hence it follows that these diabetics may have an increased susceptibility to blood clots.

[17]Fabrykant, M., and M. L. Gelfand, "Symptom-Free Diabetes in Angina Pectoris," *American journal of Medical Science,* Vol. 247, 1964, p. 665.

[18]Alexander, R. W., P. H. Forsham, and G. M. Grodsky, "Early Insulin Response to Intravenous Glucose in Steroid-Stress Diabetes," *Metabolism,* Vol. 18, 1969, p. 248.

[19]Bondy, P. K. and P. Felig, "Disorders of Carbohydrate Metabolism," Duncan's Diseases of Metabolism, 7th ed., edited by P. K., Bondy and L. E. Rosenberg, W. B. Saunders Company, Philadelphia, Pa., 1974, p. 312.

[20]Oakley, W., "Prognosis in Diabetic Pregnancy," *British Medical Journal,* Vol. 1, 1953, p. 1413.

[21]Anderson, J. W., "Metabolic Abnormalities Contributing to Diabetic Complications. 1. Glucose Metabolism in Insulin-Insensitive Pathways," *The American Journal of Clinical Nutrition,* Vol. 28, 1975, p. 273.

• **Bone disorders**—These problems are likely to occur in diabetics only when they have had frequent and severe complications, like ketoacidosis accompanied by excessive loss of calcium in the urine, or excessive urinary excretion of phosphate due to kidney disease.

• **Brain damage**—Severe low blood sugar may cause brain damage. This condition is likely to occur from an overdose of either insulin or an oral drug which lowers blood sugar, although it sometimes occurs between meals in adult-onset diabetics who are attempting to control their disease by diet alone.

(Also see HYPOGLYCEMIA.)

• **Cloudy spots on the lens of the eye (cataracts)**—These occluded areas on the lens block the rays of light from passing through the eye. Most medical people believe that the major cause of this disorder in diabetics is the entry of excessive glucose into the lens, where it is converted to fructose and sorbitol, which may then accumulate in amounts sufficient to disrupt the normal physiological processes.

• **Death of tissue (gangrene)**—The major factors contributing to this condition in diabetics are (1) blockage of blood flow to the affected area(s) due to atherosclerosis in arteries and/or the thickening of the walls of the smaller blood vessels, and (2) infections which are not checked by the normal defense mechanisms of the body. Gangrene usually starts in the toes and moves through the foot and up the leg. Surgical removal of the affected part (amputation) is often necessary.

• **Dehydration (excessive loss of body water)**—This problem is likely to occur only in diabetics whose disease is out of control. Excessive loss of water usually accompanies the urinary excretion of large quantities of sugar and ketones.

• **Diabetic coma**—Juvenile diabetics are much more likely to have this condition than adult-onset diabetics, who appear to be more resistant to its development. However, adults whose diabetes is poorly controlled have an increased susceptibility. The coma is usually brought on by a stress factor such as an infection and/or a fever which may provoke the oversecretion of the stress hormones, which, in turn, aggravates diabetes. This condition may also result from severe deprivation of food. Occasionally, children slip into this condition without having had prior signs of diabetes. Usually, the onset of a diabetic coma is preceded by the spilling of ketones and sugar into the urine together with excessive urination and extreme thirst.

• **Digestive disorders**—Gastrointestinal abnormalities, such as acute abdominal pain, difficulty in swallowing, easily provoked vomiting, malabsorption, and diarrhea, may be the result of damage to visceral nerves by diabetes, or they may indicate an impending diabetic crisis like ketoacidosis and/or diabetic coma.

• **Dizziness when standing up (postural hypotension)**—Diabetes-damaged blood vessels and nerves may not be capable of promptly adjusting the flow of blood to counter the effects of a sudden change to an upright position from sitting or lying down. The dizziness which accompanies postural hypotension may be mistaken for a symptom of low blood sugar.

• **Excessive loss of essential minerals in the urine**—An elevated urinary loss of calcium, magnesium, and potassium is likely to occur whenever there is a need for neutralization of excessive acid production by the body. These conditions occur when diabetics have ketoacidosis of lacticacidosis. These losses may be greatly aggravated by kidney disease, which often accompanies a long-standing diabetic condition.

• **Fatty liver**—many unregulated diabetics have fatty livers because a reduction in the utilization of carbohydrate triggers the release of fatty acids from the fat stores. Much of this fat accumulates in the liver when the rate of release greatly exceeds the rate at which the liver is able to turn out the lipoproteins that carry fat in the blood.

• **Heart failure due to ventricular fibrillation**—The ventricles are the large, muscular chambers of the heart which contract and pump blood out to the body. Efficient pumping depends upon the coordinated beating of each of the tiny fibers which constitute the muscle mass. Uncoordinated beating of the individual fibers (fibrillation) causes the heart to fail in its pumping action. Fibrillation is the leading heart-related cause of sudden death in middle-aged men. It is suspected that the main cause of this disorder is an abnormality in the conduction of nerve impulses through the heart muscle due to imbalances between the levels of such ions (charged particles) as calcium, hydrogen (acid ion), hydroxyl (alkaline ion), magnesium, potassium, and sodium. Impaired electrical conductivity in heart muscle may also result from tissue damage to such factors as chronic overwork, nutrient deficiencies, poor utilization of carbohydrate, elevated blood levels of free fatty acids, obstruction of the arteries which supply blood to the heart muscle (coronary arteries), and severe stress. Uncontrolled diabetics are likely to have many of the conditions listed above, so it is not surprising that they are highly susceptible to heart failure due to ventricular fibrillation.

(Also see ACID-BASE BALANCE; ELECTROLYTES; and HEART DISEASE.)

• **Heavy babies**—Diabetic mothers tend to have heavy babies. Sometimes, these babies have low blood sugar (hypoglycemia) at birth because the insulin-secreting cells of their pancreases were overstimulated by the mother's high blood sugar while they were in the womb. It is noteworthy that oversecretion of insulin may promote excessive fat synthesis, which may then lead to obesity.

• **High blood fats**—The high rate of release of fatty acids from the body's fat stores in uncontrolled diabetes provokes the liver to produce a flood of lipoproteins which transport fats in the blood. This situation is aggravated by a diminished rate of fat utilization by the fatty tissues and muscle when there is not sufficient insulin to promote such utilization. Hence, there are likely to be elevations in the blood levels of cholesterol, free fatty acids, and triglycerides.

(Also see HYPERLIPOPROTEINEMIAS.)

• **High blood pressure (hypertension)**—The main factors responsible for the development of hypertension in diabetes are (1) thickening of the capillary walls, which increases the resistance to the flow of blood; (2) blocking the flow of blood by plaques in atherosclerotic arteries; (3) kidney disease due to factors (1) and (2); (4) swelling of the blood volume due to the water-retaining effects of the elevated levels of solutes (glucose, glycoproteins, ketones, etc.); and (5) thickening of the blood due to high levels of glycoproteins.

• **Impotence and other sexual problems**—Impotence may be the first indication of hidden diabetes in a male because nerve disorders may develop from chronic high blood sugar. Similar nerve problems may make it difficult for a diabetic woman to have an orgasm.

• **Infections**—The old idea that the high blood sugar of diabetics promotes the growth of microorganisms is no longer accepted, although some believe that the sugar from the urine which sometimes collects on underwear may promote skin infections. However, infections within the body are believed to spread with little control because the thickening of the capillary walls may prevent the germ-fighting, white blood cells from reaching the sites of infection in the tissues.

• **Ketoacidosis**—This condition is characterized by the overacidity of the blood due to the accumulation of ketone acids. The respiration rate is increased as the body attempts to reduce the acidity of the blood by getting rid of carbon dioxide. Prolonged ketoacidosis may lead to diabetic coma, or even death.

• **Ketosis**—In contrast to ketoacidosis, ketosis is characterized mainly by the excretion of ketones in the urine, since the acidity of the blood in this condition is normal. However, ketones in the urine may be a sign that control of a diabetic condition is failing.

• **Kidney disease**—The kidneys, like the eyes, are vulnerable to diabetic changes in the blood vessels. Thickening and other abnormalities of the filtering apparatus may allow the escape of protein from the blood into the urine (albuminuria or proteinuria). Often, considerable kidney damage occurs before the first signs of this damage become evident. Research has shown that there may be up to 95% loss of kidney function before the first symptoms appear.

• **Lacticacidosis**—In contrast to ketoacidosis, the acidity of the blood in lacticacidosis is due to the overproduction and accumulation of lactic acid. This condition is very dangerous and often leads to death. The reason for its occurrence is not certain, but it is more likely to occur in diabetics than in nondiabetics.

• **Loss of sensation in the feet**—Diabetics are prone to burns on their feet from stepping into hot bath water, because they often fail to feel pain. This loss of sensation is caused by nerve disorders (this problem will be separately discussed). However, there is also an enhanced susceptibility to injury caused by a reduced flow of blood to the tissues in the feet. This abnormality may be caused in part by atherosclerosis, and in part by thickening of capillary walls due to chronic elevations in the blood sugar.

• **Loss of weight**—Even though diabetics may sometimes be obese when the disease is first discovered, loss of weight is a hallmark of severe, untreated diabetes, particularly in juvenile diabetics. The loss of weight occurs in spite of the consumption of what might normally be more than enough food. What happens is that the major nutrients (carbohydrates, fats, and proteins) are poorly utilized, due to a lack of effective insulin, which is needed for the entry of these nutrients into the cells of the body. The urine is an indicator of the extent to which diabetes is out of control since it may contain glucose from unused carbohydrate or degraded protein, ketones from partly metabolized fats, and elevated levels of nitrogenous waste products from degraded protein.

• **Muscle wasting**—This condition is more a feature of severe diabetes than it is of mild adult-onset diabetes. The loss of muscle tissue is due both to the lack of entry of amino acids into muscle cells because of lack of insulin, and to the secretion of the diabetogenic hormones which promote the breakdown of muscle protein and the release of amino acids into the blood.

• **Nerve disorders**—Chronic high blood sugar leads to abnormalities in nerve cells, just as it does to the eyes, kidneys, and small blood vessels. Excess glucose is metabolized to fructose and sorbitol which accumulates in nerve cells and disrupts the conduction of the nerve impulses. Also, it is often observed that these nerve disorders may be accompanied by damage to the myelin sheaths (which are somewhat like the insulation around an electric wire) of nerves. The cause of this damage is unknown.

• **Pains in the chest and shoulder (angina pectoris)**—These pains are due to an inadequate flow of blood to these parts. Hence, the taking of nitroglycerin pills at the onset of pain relieves the symptoms because the drug causes the blood vessels to dilate (relax and open up). However, the relief is only temporary because the major causes of this disorder in diabetics are the obstruction of arteries by atherosclerotic plaques, and the thickening of capillary walls due to the effects of chronic high blood sugar.

• **Pancreatic deterioration and the worsening of diabetes**—The usual explanation of the deterioration of pancreatic function in diabetes is that chronic high blood sugar overtaxes the ability of the insulin-secreting cells to produce and secrete insulin; hence, they become exhausted and disintegrate. While this concept seems valid for juvenile diabetes, it may not be as applicable to adult-onset diabetes. In the latter disorder, the insulin response to a surge in blood glucose is usually delayed, but the amount of hormone which is eventually secreted may be greater than normal. Also, there appears to be inappropriate pancreatic secretion of glucagon while the blood sugar is high. High blood sugar usually causes the secretion of this hormone to stop.

• **Skin disorders**—Some of these disorders are (1) black-and-blue marks due to capillary fragility, (2) a bluish color of the skin on the feet when the supply of blood is insufficient, (3) wrinkling of the skin as the underlying fat is lost, and (4) irritation around the urinary orifice when poor control results in the spilling of sugar into the urine. Ulcers of the skin also plague diabetics, particularly if the sense of pain is diminished, since the beginning of an injury may hardly be noticed.

• **Spontaneous abortion**—Women who have been diagnosed as diabetics prior to becoming pregnant are more likely than nondiabetics to have either spontaneous abortions (loss of the baby early in pregnancy) or miscarriages (loss of the baby late in pregnancy). The reason for this fetal loss is not well understood.

• **Stroke**—The cause of this occurrence in diabetics, like that in nondiabetics, is probably advanced atherosclerosis. However, the capillary abnormalities which occur in diabetes may accelerate the development of high blood pressure, which is a major factor in the precipitation of a stroke.

DIAGNOSIS OF DIABETES.

The ways in which diabetes is diagnosed have changed in recent years, from those with a major emphasis on examination for overt signs of the disease to those used to detect the early stages which are present prior to the appearance of clinical signs. Therefore, it is of interest to note the stages of development of the disease, since the earlier the disease is found, the better the prognosis.[22]

1. **Prediabetes.** The widespread belief that there is a hereditary basis for diabetes leads to the classification of close kin of diabetics (usually the nondiabetic identical twins of diabetics and the children of two diabetic parents) as prediabetics, even though these people may show no signs of the disease. Although it is not inevitable that those so classified will eventually develop the disease, there is a high probability that this may be the case. Hence, these people should be watched for the early signs of the disease.

2. **Subclinical or "stress" diabetes.** This stage, where clinical signs are absent, is characterized by a diabeticlike utilization of carbohydrate under stress, but a normal utilization when the stress is removed. The utilization of carbohydrate is assessed by measuring the fasting blood sugar, or by giving a glucose tolerance test. Typical stresses which provoke abnormal carbohydrate metabolism are emotional disturbances, fever, infections, and pregnancy. The diagnosis of subclinical diabetes is based upon giving the patient a dose of a stress hormone, such as cortisone, prior to the glucose tolerance test. This test determines whether the pancreas has sufficient reserve capacity to meet the challenge imposed by stress situations where there may be greater than normal secretion of diabetogenic hormones.

3. **Chemical or latent diabetes.** Unlike the preceding stages, the utilization of carbohydrate in this stage is clearly abnormal, even when no stress factors are present. However, clinical signs, such as sugar in the urine, are not usually present unless there is some type of stress. Nevertheless, the abnormal glucose tolerance is a sign that there will most likely be a further development of the disease if treatment is not promptly started.

4. **Clinical, or overt, diabetes.** In this stage, there are both chemical and clinical signs of diabetes such as excessive urination, sugar in the urine, and occasional bouts of ketosis or ketoacidosis.

The progression of diabetes to the overt stage may be rapid, particularly in the cases of juvenile diabetics. Regression of diabetes to a less severe stage also occurs in significant numbers of people, so there is almost always hope for victims of this disease. Therefore, it is worth noting the diagnostic information which follows.

Signs and Symptoms of Diabetes.

These signs are usually present only in the overt diabetic, but they may sometimes be seen in latent diabetics. Hence, their absence does *not* eliminate the possibility of diabetes. The signs are sugar in the urine, frequent and excessive urination, abnormal thirst, increased hunger, loss of weight, nausea, and tiredness. People who repeatedly have these signs should go to a doctor. Symptoms of hypoglycemia may also be early signs of diabetes.

(Also see HYPOGLYCEMIA.)

Tests of Blood and Urine.

A markedly elevated fasting blood sugar in the absence of unusual stress is a strong indicator of diabetes. (Usually, this test is conducted on a sample of blood drawn after an overnight fast.) However, the glucose tolerance test provides more specific information as to the severity of the disease.

The repeated spilling of sugar into the urine is another strong indicator of diabetes. Often young people spill sugar without being diabetic. For example, it was not uncommon for young men summoned for physical examination prior to induction in the U.S. Army to have sugar in their urine. The examiners in New York City would sometimes detain the potential inductees for repeated testing under controlled conditions since a high intake of carbohydrate in the evening may lead to a spilling of sugar in the urine voided in the morning. On the other hand, older diabetics with high blood sugar levels may fail to spill sugar in their urine, particularly if their kidney function is partially impaired. Therefore, the test for sugar in the urine has its limitations for the detection of diabetes, but it may be useful to diabetics as a rough indicator of the control of their disease.

Glucose Tolerance Test (GTT).

This test is considered to be the most reliable of those available for the evaluation of carbohydrate utilization by the body. The term "glucose tolerance" literally means the rate at which the cells of the body take up glucose from the blood.

---

[22]Goodkin, G., "How Long Can a Diabetic Expect to Live?" *Nutrition Today*, Vol. 6, May/June 1971, p. 21.

Details of the administration of the test are given elsewhere. An abnormal and/or prolonged rise in the blood sugar after a test dose of glucose is interpreted as a sign that the utilization of carbohydrate by the cells is somehow impaired. Diabetes is the most likely cause of this abnormality because insulin is required for the passage of glucose and other nutrients from the blood into the cells. Fig. D-15 shows curves for normal glucose tolerance, early adult-onset diabetes, and advanced diabetes.

(Also see HYPOGLYCEMIA, section headed "Glucose Tolerance Test.")

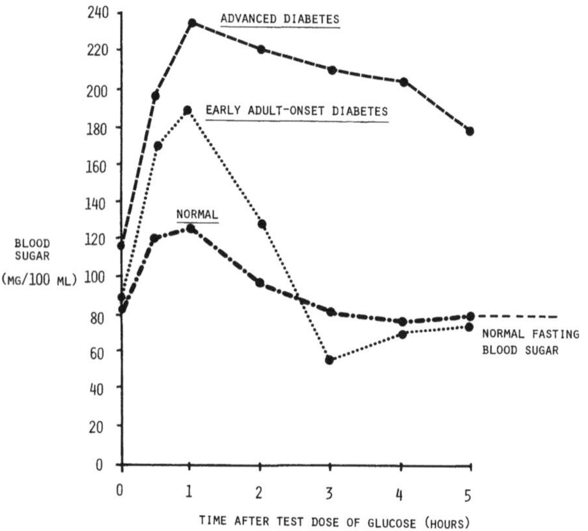

Fig. D-15. Glucose tolerance curves.

Explanations of these curves follow.

The fasting blood sugar of a normal person is usually around 80 mg per 100 ml of blood. A test load of glucose produces a moderate rise in the blood sugar which reaches a peak after about an hour. Then, the values decline, and reach the normal fasting level after about 3 hours.

In early adult-onset diabetes, the fasting blood sugar may be at least slightly elevated, and there is an abnormally high peak value of the blood sugar after a test load of glucose. However, in some people the blood sugar may decline to normal, or even hypoglycemic levels after about 3 hours, while in others the return to fasting levels may be delayed by an hour or more.

All values of the blood sugar are likely to be greatly elevated in advanced diabetes, which shows that there is very poor utilization of carbohydrate without injections of insulin.

The criteria used by various clinicians for differentiating between normal utilization of carbohydrate and borderline diabetes vary. For example, the doctors in the Department of Medicine of the University of Washington derive their criteria from the standards of the U.S. Public Health Service.[23]. These standards call for a normal fasting level of blood glucose of less than 110 mg/100 ml, and for a blood sugar level 2 hours after an oral test load of glucose of less than 120 mg/100 ml.

Detection of Diabetes in Children. The development of diabetes in children may be very rapid; so, the earlier that it is detected, the better the outcome. Sometimes, children in certain families develop the juvenile form of the disease well in advance of the time that the disease is detected in its adult-onset form in older relatives such as grandparents. At present, no one knows why juvenile diabetes may suddenly appear in a family with no apparent history of the disease.

The most common signs of juvenile diabetes are excessive urination (it may be almost impossible to stop some children from wetting their beds), unusual thirst, unexplainable weight loss, tiredness, and increased appetite. A negative test for sugar in the urine does not rule out the disease, nor does a positive test confirm it. Blood tests (fasting blood sugar and/or glucose tolerance) are the only certain means of diagnosis. However, it might be wise to apply the tests for sugar and ketones to the urine of a child with a fever, particularly if his or her breath smells like acetone (a rotten, fruity odor, like that noticed after the drinking of alcoholic beverages). The fever may be a sufficient stress to bring out a hidden diabetic tendency. (Specially treated paper strips and tablets for testing urine are sold in most drug stores.)

TREATMENT AND PREVENTION OF DIABETES. One of the great medical achievements of our time is the successful treatment of diabetics so as to give those so afflicted many trouble-free years of life, an almost impossible task prior to the discovery of insulin. Many highly successful people and even some celebrities have had lifelong diabetes, yet they have not let the disease prevent them from enjoying life. An entire book has recently been written on this subject.[24] However, what is unique in diabetes, compared to most other diseases, is that the diabetic is literally his or her own doctor, since most of the therapeutic measures have to be self-administered on a day-to-day basis. Therefore, careful attention should be paid to the treatments which follow.

Emergency Treatments. There are two life-threatening crises which may occur in diabetes: diabetic coma, and hypoglycemia. The measures to be taken depend upon the correct identification of the critical condition, since the wrong treatment may do more harm than good. The characteristics of each condition are presented in Table D-5.

[23]Brunzell, J. D., *et al.*, "Improved Glucose Tolerance with High Carbohydrate Feeding in Mild Diabetes," *The New England Journal of Medicine*, Vol. 284, 1971, p. 521.

[24]Biermann, J., and B. Toohey, *The Diabetic's Sports and Exercise Book*. J. B. Lippincott Co., Philadelphia, Pa., 1977.

TABLE D-5
DIAGNOSTIC FEATURES OF DIABETIC COMA AND HYPOGLYCEMIA

Feature[1]	Diabetic Coma	Hypoglycemia
Signs and symptoms:		
Abdomen	Pain is often present.	Pain is **not** usually present.
Blood chemistry	Acidosis (bicarbonate is low) sugar is elevated.	Normal bicarbonate, low sugar level.
Blood pressure	Low	Normal to high.
Breathing	Deep and labored.	Normal or shallow.
Digestive system	Vomiting often occurs.	Occasional vomiting.
Emotional status	Apathetic	Anxious and irritable.
Mouth	Dry	Moist
Odor of breath	Fruity (acetone)	Normal
Pulse	Slow and weak.	Rapid
Reflexes	Slow	Normal
Skin	Dry, warm	Moist, cool
Urine, characteristics[2]	Contains ketones and sugar; volume may be greater than normal.	Normal composition and volume.
History of the patient:		
Control of the disease	Usually not very good.	May be either good or poor.
Insulin status	Too little (dose may have been skipped).	Too much
Previous infection	Often brings on condition.	Not usually present.
Skipping of meals	Sometimes	Often brings on condition.
Strenuous exercise	Not usually	Often brings on condition.
Time required for development of condition	There may be a gradual deterioration over several days.	Onset is usually rapid, and often occurs between meals.

[1]It is not always easy to distinguish between diabetic coma and hypoglycemia because (a) vigorous treatment of the coma may lead to hypoglycemia, (b) prolonged hypoglycemia causes brain abnormalities which have many features like those of diabetic coma, and (c) the metabolic reaction to hypoglycemia (increased secretion of diabetogenic hormones) in severe diabetes may lead to diabetic coma.

[2]Has diagnostic value only when it represents metabolism over the past few hours. Sometimes, urine has been retained in the bladder so long that it represents an earlier metabolic state.

Treatments for diabetic coma and hypoglycemia follow.

DIABETIC COMA. The acetone breath which accompanies this condition has often led to the mistaken identification of a comatose diabetic as a drunk. Subsequent confinement in a prison has sometimes resulted in severe complications, and in a few cases, death. Therefore, a diabetic should always *wear* an identification tag. (Purses or wallets are often lost or stolen during emergency situations.) A doctor should be immediately summoned, since attempts at first aid by lay persons may be ineffective, or even dangerous. The family of a diabetic should *not* attempt to administer insuin unless the doctor so advises, since the injection of insulin may be fatal to a person suffering from hypoglycemia.

HYPOGLYCEMIA (Insulin Shock). All diabetics should have available at all times a source of readily available .sugar for use in case of hypoglycemia (low blood sugar). Convenient sources of sugar are items such as soft candy (hard candy may dissolve too slowly), cubes or paper packets of table sugar, dextrose tablets (a form of pure glucose which is sold in certain health foods stores and pharmacies), and malted milk tablets. If a diabetic is at home, he or she may drink a glass of fruit juice. However, nothing should be placed in the mouth of an unconscious person. In the latter case, a doctor should be called immediately.

Some doctors recommend that diabetics keep home supplies for the injection of glucagon, and that family members be instructed in its administration in the event that emergency care is not available. Each of these emergency measures should be used with great discretion, for the rapid raising of the blood sugar may worsen diabetes. Hence, each diabetic should be very familiar with the specific symptoms of hypoglycemia, and should refrain from using these measures for getting a lift from feelings of depression or tiredness.

(Also see HYPOGLYCEMIA, section headed "Signs and Symptoms.")

Dietary Measures. In 1986, the American Diabetes Association (ADA) revised its dietary guidelines for only the second time in more than 25 years. It is noteworthy that these guidelines are similar to those of the American Heart Association and the National Cancer Institute. The revised ADA nutritional recommendations follow:

1. **Calories.** Calories should be prescribed according to energy needs and to achieve and maintain a desirable body weight.

2. **Carbohydrates.** The amount of carbohydrate intake should be liberalized to 55–60% of daily total calories. Whenever possible, substitute unrefined carbohydrates high in fiber (such as whole grain bread and cereal) for highly refined foods.

3. **Fat.** Lower the fat intake to 30% or less of the total daily calories; and limit saturated fat to 10% of the total calories.

4. **Cholesterol.** Limit cholesterol in the diet to 300 mg/day or less.

5. **Sweeteners.** The use of nutritive and nonnutritive sweeteners is not encouraged, but is acceptable in diabetes management.

6. **Salt.** The limitation of salt intake is highly recommended in most circumstances.

7. **Protein intake.** The recommended daily allowance for adults is 0.8 g/kg body weight.

8. **Alcohol.** Preferably, drinking of alcohol should be avoided; otherwise, drink in moderation.[25]

The control of blood sugar also involves the maintenance of the body weight which is appropriate for age, body build, height, and sex. This means that obese diabetics should make every effort to lose weight in a medically approved manner, but not by crash diets which may aggravate their disease.

Finally, the diet should help to prevent rather than promote atherosclerosis. (There is growing concern that the traditional practice of replacement of dietary carbohydrate with fat in diets has been at least partly responsible for the high rate of cardiovascular disease.)

The newer thinking in diabetic dietetics was summarized in the *Mayo Clinic Nutrition Letter*.[26] It is noteworthy that the authors gathered evidence, from their research and that of others, which shows that people with mild adult-onset diabetes, who do not require injections of insulin, may show improvement in their glucose tolerance when placed on high-carbohydrate, low-fat diets. However, this may not be the case for all such diabetics, since some show a tendency to have elevations in their blood levels of triglycerides when their dietary carbohydrate is increased. Therefore, the choice of a diet which is most suitable for a particular person still requires a careful evaluation of his or her individual characteristics by a physician and dietitian team who are experienced in the management of diabetes.

TYPES OF DIETS. The three basic types of diets which are currently being prescribed for diabetics are (1) a moderately low-carbohydrate diet which restricts mainly simple sugars and alcohol, for those who have high blood levels of triglycerides (Type IV Hyperlipoproteinemia), (2) the most commonly used diabetic diet which has from 45 to 55% of its calories supplied by carbohydrate, and (3) a recently tested high-carbohydrate, low-fat diet for people who have high blood levels of cholesterol and/or other fats (Type II and other types of hyperlipoproteinemias). However, the selection of diets to combat the various types of hyperlipoproteinemia is still somewhat controversial, so doctors tend to rely on special laboratory tests for guidance in selecting a diet. Some typical diets and their compositions are given in Table D-6.

(Also see HYPERLIPOPROTEINEMIAS.)

Brief descriptions of the exchanges used by most dieticians follow.

• **Bread**—The standard unit for these exchanges is one slice of bread (25 g), which contains 15 g of carbohydrate, 2 g of protein, and 70 Calories (kcal). Other food in this group are biscuits, cereals, crackers, flour, macaroni, noodles, and starchy vegetables. A few items, like biscuits, muffins, potato chips, and waffles, contain extra fat; so it is necessary to subtract the appropriate number of exchanges from the fat allowance when these items are selected.

• **Fat**—Each exchange in this group contains 5 g of fat and 45 Calories (kcal). Typical items are bacon, butter, cream, lard, margarine, mayonnaise, nuts, and oils.

• **Fruit**—The various fruits which comprise these exchanges may be canned (unsweetened), dried, fresh, frozen, or in the form of juices. Each exchange, such as a small apple, furnishes 10 g of carbohydrate and 40 Calories (kcal).

• **Meat**—A lean meat exchange is equivalent to 1 oz (*28 g*) of cooked meat, which contains 3 g of fat, 7 g of protein, and 55 Calories (kcal). Similar, high-protein foods included in this group are low-fat cheeses, fish and other seafood, and poultry (without the skin). However, medium-fat meat exchanges consist of eggs, cheeses, and meats which contain sufficiently more fat to yield an energy value of about 78 Calories (kcal) per ounce. Hence, the dieter should deduct one-half fat exchange for each of these meat exchanges. Similarly, the high-fat meat exchanges are equivalent to a lean meat exchange plus one fat exchange for a total of about 100 Calories (kcal) per ounce.

• **Milk**—The standard unit in these exchanges is an 8-oz (*240 ml*) cup or glass of nonfat (skim) milk, which contains 12 g of carbohydrates, negligible fat, 8 g of protein, and 80 Calories (kcal). One fat exchange should be deducted from the amounts allowed each time that 2% milk or plain yogurt is used instead of nonfat milk. Likewise, two fat exchanges should be deducted for each exchange of whole milk or of a product (buttermilk, evaporated milk, or yogurt) made from whole milk.

• **Optional beverages**—These fluids contain few calories, so for all practical purposes they may be taken as desired.

• **Alcoholic beverages**—Many doctors and dietitians may discourage the use of alcoholic beverages by diabetics, because alcohol tends to lower the blood sugar. Hence, the control of diabetes may be made more difficult by the consumption of these beverages because the combined effects of alcohol and insulin may be similar to those of an overdose of insulin.

It is noteworthy that 7 oz of beer, 3½ oz of table wine, 2 oz of dessert wine, or 1 oz of gin, rum, vodka, or whiskey are each equivalent in caloric value to about 4½ tsp of sugar (80 kcal). Hence, dietitians generally do not allow more than one serving of *one* of these alcoholic beverages per day for diabetics.

• **Vegetable, raw, low-calorie**—The items in this group are chicory, Chinese cabbage, endive, escarole, lettuce, parsley, radishes, and watercress, all of which yield few calories when they are eaten raw. Therefore, they are sometimes called "free vegetables" since one may eat as much of them as desired. However, more of their carbohydrate content is made available by cooking, so they

[25]These recommendations were developed by a task force of the American Diabetes Association, chaired by Aaron I. Vinik, M.D.; published in *Nutrition Today*, January/February, 1987, pp. 29-30.

[26]*Mayo Clinic Nutrition Letter*, Vol. 1, No. 8, November 1988.

TABLE D-6
TYPICAL MENUS FOR VARIOUS TYPES OF DIETS

Types of Food Exchanges (used for each meal or snack)[1]	Number of Exchanges per Meal (based upon daily energy requirements)					
	Moderately Low-Carbohydrate, High-Protein Diet		Standard Diet (50% carbo-hydrate)		High Carbohydrate, Low-Fat Diet	
	1,800 kcal/ day	2,400 kcal/ day	1,800 kcal/ day	2,400 kcal/ day	1,800 kcal/ day	2,400 kcal/ day
Breakfast:						
Bread (or biscuits, cereals, crackers, muffins, pancakes, waffles, etc.)	1	2	2	2	2	3
Fat (bacon, butter, coffee lightener, margarine, mayonnaise, etc.)[2]	1	3	2	2	1	2
Fruit (or its equivalent in juice)	1	1	1	2	1	1
Lean meat (or low-fat cheese, fish, fowl, etc.)	2	2	1	1	1	1
Milk, nonfat (if whole milk products are used, deduct 2 servings of fat from the day's allowance)	0	0	0	0	1	1
Optional beverage: bouillon (fat free), clear broth, club soda, coffee, herb tea, tea, water	Any amount, but limit sweetener.					
Mid-morning snack:						
Bread	1	1	1	1	2	2
Fat	2	2	0	2	0	1
Fruit (or juice)	0	0	0	0	0	0
Meat	0	1	0	0	0	0
Milk or yogurt, plain, nonfat	1	1	1	1	0	0
Optional beverage (see Breakfast)	Any amount, but limit sweetener.					
Lunch:						
Bread	1	2	2	3	3	3
Fat	1	3	2	3	1	2
Fruit	0	0	1	1	1	1
Meat, lean (see Breakfast)	3	4	3	3	2	3
Optional beverage (see Breakfast)	Any amount, but limit sweetener.					
Raw vegetable, or salad w/o dressing	1	1	1	1	1	1
Mid-afternoon snack:						
Bread	1	1	1	1	2	2
Fat	2	2	0	2	0	1
Fruit (or juice)	0	0	0	0	0	0
Meat	0	1	0	0	0	0
Milk, nonfat	1	1	1	1	1	1
Optional beverage (see Breakfast)	Any amount, but limit sweetener.					
Supper:						
Bread	1	1	2	4	2	4
Fat	2	4	3	3	2	2
Fruit	1	1	1	2	1	2
Meat, lean (see Breakfast)	4	4	3	3	3	3
Optional beverage	Any amount, but limit sweetener.					
Raw vegetable, or salad w/o dressing	1	1	1	1	1	1
Nonstarchy vegetable, cooked	1	1	1	1	1	1
Late evening snack:						
Bread	1	1	1	2	1	2
Fat	2	2	1	1	0	0
Fruit	0	1	1	1	1	2
Meat	1	1	0	0	0	0
Milk, nonfat	1	1	1	1	1	1
Optional beverage	Any amount, but limit sweetener.					

[1]Exchanges are groups of foods which may be substituted for each other because they contain similar proportions of carbohydrate, fat, protein, and calories.

[2]The total day's allowance for fat plus the distributing of fat at meals and snacks may have to be adjusted so as to compensate for foods which contain considerably less or more fat than the amounts which are typical for their exchange groups. Also, doctors generally recommend that as many as possible of the fat exchanges be chosen from polyunsaturated fats (salad oils and soft margarines) because ample amounts of saturated fats are provided by the meat exchanges.

should be counted as an exchange of a cooked, nonstarchy vegetable when more than a cup of cooked item(s) is eaten at a single meal.

• **Vegetable, cooked, nonstarchy**—The standard measure for these exchanges is ½ cup (*120 ml*) of asparagus, beets, broccoli, cabbage, carrots, greens, kale, mushrooms, okra, onions, pumpkin, rutabagas, squash, string beans, or tomatoes which furnishes 5 g of carbohydrate, 2 g of protein, and 25 Calories (kcal). Other more starchy vegetables are grouped under the bread exchanges.

NOTE: A more complete description of the items which comprise the various exchanges is given elsewhere in this book.

(Also see MODIFIED DIETS, section headed" Food Exchanges.")

Table D-6 is intended for the guidance of adult-onset diabetics who do *not* require injections of insulin, since each type of insulin requires a specific meal pattern for its maximum effectiveness. Also, most juvenile diabetics require special meal patterns based upon their age, sex, body build, rate of growth, type of insulin, etc.

The procedure for using the table follows:

1. Determine your desirable body weight in pounds by referring to a standard height-weight table, or by estimating it from the following formula for (a) adult males, by allowing 106 lb for the first 5 ft of height and 6 lb for each additional inch of height over 5 ft, or (b) adult females, by allowing 100 lb for the first 5 ft of height and 5 lb for each additional inch.
2. Convert your ideal weight from pounds to kilograms by dividing the former by 2.2.
3. Calculate your energy needs in calories by using one of the following allowances:
   a. 20 Calories (kcal)/kg ideal body weight if age 65 or more, overweight, and activity level is sedentary
   b. 25 Calories (kcal)/kg ideal body weight if under age 65, overweight and sedentary
   c. 30 Calories (kcal)/kg ideal body weight if normal weight and sedentary; or if overweight and moderately active
   d. 35 Calories (kcal)/kg ideal body weight if underweight and sedentary; if normal weight and moderately active, or if overweight and very active
   e. 40 Calories (kcal)/kg ideal body weight if underweight and moderately active; or if normal weight and very active
   f. 45 Calories (kcal)/kg ideal body weight if underweight and very active (Most adult-onset diabetics fit the criteria for one of the first three allowances.)
4. Select the *standard* dietary pattern which comes closest to meeting your energy needs. Additional standard patterns for various energy levels are given elsewhere.

(Also see HYPOGLYCEMIA, section headed "Diet Therapy," Table H-7.)

NOTE: Low- or high-carbohydrate diets should be used *only* when they are prescribed by a doctor or a dietitian. Additional plans for these diets are given elsewhere.

(Also see HYPERLIPOPROTEINEMIAS.)

EXAMPLE: An overweight 50-year-old male with adult-onset diabetes is 5 ft, 9 in. tall (*1.7 m*), has a medium body build, and does not exercise very much. His ideal body weight is 106 + 54 = 160 lb or *73 kg*. Considering that he is sedentary, his daily energy needs are 73 × 25 = 1,825 Calories (kcal). Some typical meals and snacks for an 1,800 Calorie (kcal) daily plan follow:

			Energy (kcal)
Breakfast	Soft-boiled egg	1	78
	Whole wheat toast	2 slices	110
	Butter or margarine	1 tsp	36
	Orange sections	½ cup	40
	Coffee	1 c with 1 tsp of light cream	22
Mid-morning snack	Rye wafers	three 2" × 3½"	70
	Skim milk	1 c or 8 oz glass	84
Lunch	Salad made with lettuce	no limit	0
	Peach halves	2 medium	40
	French dressing	2 tbsp	90
	Sandwich made with cooked, lean roast beef	3 slices	188
	dill pickle slices	no limit	0
	horseradish sauce	1 tbsp	0
	pumpernickel bread	2 thin slices	157
	Hot, nonfat bouillon	1 c	0
Mid-afternoon snack	Whole wheat crackers	three 2-in. square	48
	Skim milk	1 c or 8 oz glass	84
Supper	Assorted raw vegetables—chicory, parsley, radishes	as much as desired	0
	Whole wheat bread	1 slice	55
	Butter or margarine	3 tsp	108
	Roast chicken	3 slices (3 oz)	140
	Mashed potatoes	½ c	70
	Cooked green peas	½ c	53
	Figs, water pack	2	30
	Oatmeal cookies	2 med.	109
	Herb tea	no limit	0
Late evening snack	Cottage cheese (2% fat)	¼ c	50
	Graham crackers	two 2½ in. squares	54
	Apple, fresh	1 medium	84
Total food energy			**1,800 Calories (kcal)**

SPECIAL FOODS WHICH ARE LOW IN ENERGY (kilocalories). One of the major problems of obese diabetics is that of sticking to a diet low in calories. Whether they are taking insulin or not, they must lose weight because neither their own nor the various insulins which may be injected have maximum effectiveness in obesity. Thus, they are faced with the problem of selecting foods that are not only low in energy, but also palatable and satisfying. There are several ways to accomplish these objectives. Suggestions follow.

The most common foods which are filling but not fattening are high-fiber, low-energy fruits and vegetables such as blackberries, bean sprouts, cabbage, greens, and string beans.

(Also see FIBER.)

Less common, but also useful, are special low-energy foods like Jerusalem artichoke tubers which contain inulin, a starchlike carbohydrate which cannot be utilized by man. The floury material obtained by drying and pulverizing these tubers is used to replace all or part of the wheat flour in dietetic noodles and spaghetti. Although these products are sold in many health food stores, they are seldom labeled as to their content of available carbohydrate or energy.

Some other special food items are agar-agar (a non-nutritive gelling agent), fruits canned in water instead of sugar syrup, gluten flour (low in starch, high in protein), and low-fat salad dressings. While these items were once available only in health food stores or pharmacies, they are now sold in many supermarkets. Also, materials like agar-agar may be purchased for incorporation in recipes prepared at home. Greater detail as to the use of these and other similar products is presented elsewhere in the book.

(Also see FOODS, LOW-ENERGY.)

PREVENTION OF NUTRITIONAL DEFICIENCIES IN DIABETICS. It was mentioned earlier that diabetes may be brought on and/or aggravated by deficiencies of the essential nutrients chromium, manganese, potassium, pyridoxine (vitamin B-6), and zinc. This is *not* to say that the basic nutrient requirements of diabetics are significantly different from nondiabetics; it is only to note that (1) a few people may show diabetic tendencies due to deficiencies of various nutrients, (2) diabetes may impair the utilization and/or increase the urinary excretion of certain nutrients and so create a need for higher dietary levels of such substances, and (3) diets made up mainly of highly refined foods may be lacking in nutrients.

However, information as to the incidence of nutritional deficiencies is lacking, since the appropriate diagnostic tests are rarely performed, except in a few research studies. Although the widespread use of such tests might be objected to on the grounds of increased costs of health care, the ultimate cost of disabilities due to severe deficiencies might be much greater.

A good start towards the prevention of the various deficiencies might be made if doctors would take more dietary histories from their patients, since many medical centers now have computers programmed to translate food patterns into the nutrients which are consumed. Once patients on deficient diets have been identified, the next step is to use such tests as those for pyridoxine (vitamin B-6) deficiency, and/or the analysis of hair for such trace minerals as chromium, manganese, and zinc.

While mineral and vitamin pills may be purchased without a prescription, it would be better for most people to obtain all of the nutrients they need from minimally processed foods such as dairy products, fish, fresh or frozen vegetables, legumes (beans and peas), meats, and whole grain products. Each of the plant foods is also a good source of fiber, which helps to control overeating by providing a feeling of fullness.

(Also see FIBER; MINERALS; and VITAMINS.)

Insulin. This hormone has proved to be a lifesaver for many diabetics. The current thinking is that injections of insulin may not be necessary for some adults with mild diabetes, but that this therapy is almost always necessary for juvenile diabetics and adult diabetics whose insulin secretion is inadequate. Sometimes, insulin-dependent diabetics may appear to have a remission of their disease so that it may seem that injections of the hormone are no longer necessary. It is believed that injections of insulin reduce the stress on the few insulin-secreting cells which remain functional. Thus, the "rested" pancreatic cells may be able to cope for a while on their own. However, many doctors believe it is wise to continue the administration of insulin throughout temporary periods of remission so as to keep the insulin-dependent diabetic in a consistent pattern of regulation.

Like many other therapies, injections of insulin have their drawbacks. First of all, it is not always easy for the doctor to determine the dose that is needed. Too little gives poor control, which is still better than no control, while too much may produce hypoglycemia (low blood sugar). Second, it was mentioned earlier that some doctors suspect that excessive insulin might accelerate the buildup of fatty materials in atherosclerotic plaques, since insulin promotes the uptake of fat by cells and the synthesis of storage fat. Finally, some diabetics develop various types of physiological blocks against the actions of the hormone. It is now suspected that occurrence of these problems might be greatly curtailed if the requirements for insulin were to be lowered in many diabetics by strict dietary control and regular programs of moderate to vigorous physical activity.

Determination of the amounts and types of insulin to be administered is the responsibility of the attending physician. Likewise, the choice of a diet to be used by an insulin-dependent diabetic must also be determined by the doctor who usually asks a dietitian to work out the details. Although this situation differs from that of the mild diabetic treated by diet alone (where the dietary prescription may merely be to consume less energy so as to lose weight), extra precautions are necessary in order to coordinate the effects of both diet and insulin injections (and, in some cases, the added efffects of strenuous exercise). It is with this in mind that Fig. D-16, Types of Insulin and Their Characteristics, is presented.

MEAL PATTERNS FOR DIFFERENT TYPES OF INSULIN. Fig. D-16 shows that each type of insulin has specific characteristics relating to its speed and duration of action. Hence, meal patterns must be coordinated so that carbohydrate feedings throughout the day correspond with the times of maximum action of the injected insulin(s). Typical patterns follow.

• **Mixtures of regular and slow-acting insulins**—The injection is usually given before breakfast, so about 2/5 of the day's allowance of carbohydrate is assigned to that meal because of the early effect of regular insulin; only 1/5 is assigned to lunch when there is a lull in insulin ac-

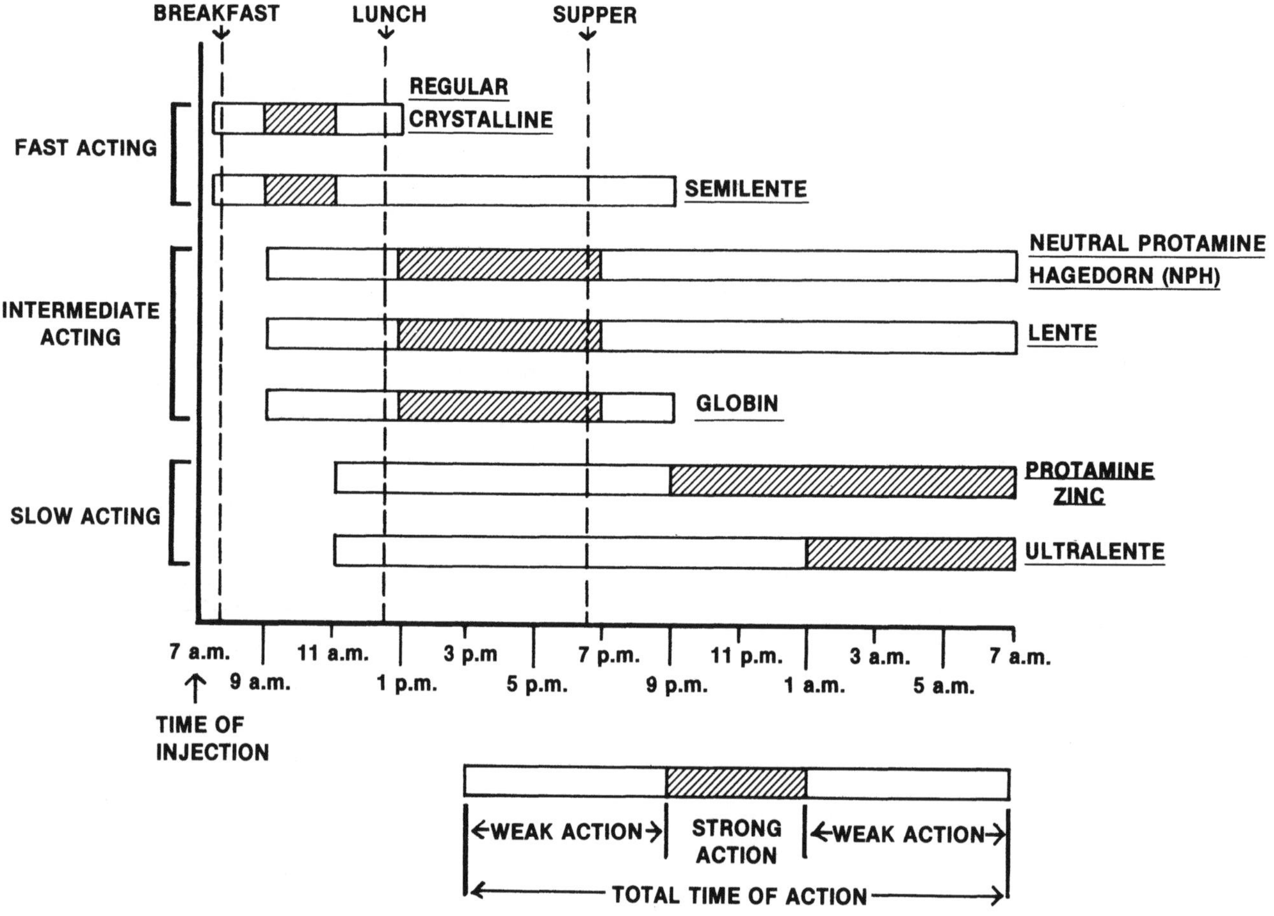

Fig. D-16. Types of insulin and their characteristics.

tivity, and the remaining 2/5 is assigned to dinner when the slow-acting insulin begins to take effect.

• **Slow-acting insulins**—In this case there is little insulin action in the morning, so only 1/6 of the carbohydrate is allotted to breakfast, 2/6 to lunch, 2/6 to dinner, and 1/6 to a late evening snack. This pattern has most of the carbohydrate assigned to the afternoon and evening because most of the insulin action occurs late in the day.

• **Intermediate-acting insulins**—The effects of these insulins peak in the afternoon; hence, only 1/6 of the carbohydrate allowance for the day is given at breakfast, 2/6 at lunch, 1/6 at a mid-afternoon snack to counteract a tendency for low blood sugar to develop at that time, and 2/6 at dinner.

Oral Drugs. These drugs are effective in lowering the blood sugar *only* when the diabetic is able to secrete sufficient insulin. They are *not* usually effective for either juvenile or adult-onset diabetics who are dependent upon injections of insulin, nor do they help much where there has been a history of ketosis or acidosis. Hence, they are mainly used to treat adult-onset diabetics who are unable to control their disease by diet alone, but who may be able to use these drugs in lieu of injections of insulin. Some doctors believe that when patients secrete adequate insulin it is best to avoid giving injections of extra insulin provided oral drugs reduce the high blood sugar levels and stop the spilling of sugar in the urine.[27] The reason for avoiding extra insulin is that this hormone stimulates the production of fat from carbohydrate, and most adult-onset diabetics already have too much body fat. There are two main types of oral antidiabetic drugs, descriptions of which follow.

[27]Fineberg, S. K., "Obesity-Diabetes," *Nutrition Today*, Vol. 1, September 1966, p. 16.

SULFA-TYPE DRUGS (SULFONYLUREAS). The main action of these compounds is stimulation of the pancreatic cells to release insulin. Some, like tolbutamide are short acting (6 to 12 hours), while others like acetohexamide and tolazamide are intermediate acting (12 to 24 hours), and chlorpropamide is long acting (24 to 72 hours). These drugs sometimes cause hypoglycemia, which may be prolonged in the case of chlorpropamide, particularly when given to elderly people who eat poorly or who have kidney disorders. In the latter case, the drug may not be excreted from the body, so the patient may require dialysis in order to correct the low blood sugar.

NOTE: People who are receiving tolbutamide should avoid alcoholic beverages because the drug may interfere with alcohol metabolism to the extent that they become nauseous and feel poorly.[28] The symptoms are very similar to those which occur when a person drinks an alcoholic beverage after taking Antabuse.[29]

PHENFORMIN. This drug slows the rate of glucose absorption from the intestine (which results in a slowing in the rise of blood sugar after meals) and increases the utilization of carbohydrate. Its disadvantages are that its effects are short lived (4 to 6 hours), it may irritate the gastrointestinal tract, and occasionally it may be the cause of lacticacidosis. Hence, the dosage level of this drug has to be limited. However, it works well in combination with the sulfa-type drugs, since the antidiabetic effects of the two agents are additive.

Foot Care. Too many diabetics have sustained permanent, but often preventable, damage to their feet and legs from what may have started out as such minor problems as cuts and infections. The reasons why the lower limbs of diabetics are so vulnerable is that (1) nerve disorders result in the dulling of the sense of pain; (2) the reduction in the circulation through the small blood vessels is greatest in the feet because the force of gravity on the circulating blood produces the greatest pressure there and, hence, the greatest thickening of capillary membranes;[30] (3) the feet and legs bear the weight of the rest of the body, which in most adult diabetics is greater than normal; and (4) infections spread more rapidly in diabetics. Therefore, it is essential that diabetics give their feet special care. Some suggestions follow.

1. The toenails should be trimmed very carefully so as to prevent ingrowth and/or infection. Diabetics whose hands are unsteady or whose vision is poor should have someone else do this job for them.
2. Many burns have occurred on the feet of diabetics who did not feel pain from hot water. Therefore, bath water should always be drawn and the temperature tested with a thermometer *before* a diabetic steps into the tub or foot pan.
3. It is wise for diabetics to wear safety shoes to protect their feet from accidental injury.
4. One should *never* abrade or cut calluses or corns on the feet of diabetics. It is well worth the money for a diabetic to have such problems attended to by a chiropodist, or by a podiatrist.
5. Shoes and stockings should always fit well. In hot, humid weather, a diabetic should always have on hand a clean pair of stockings since wet stockings may rub the skin and cause blisters. Shoe linings should be regularly inspected for creases, looseness, and other irregularities.
6. A diabetic should *never* walk anywhere in bare feet, not even in his or her own home.

Exercise. Physical activity may be a form of therapy for the diabetic (since it may make it easier to lose weight, and it may also lower the blood levels of both fats and glucose), or, it may cause such problems as low blood sugar. Therefore, it is worthwhile for diabetics to try and anticipate in advance any change in their customary level of activity. Although it is well known that increases in energy expenditures by diabetics usually lower their requirements for insulin, most doctors prefer that patients who are dependent upon such injections adjust their food intake to correspond with changes in activity, rather than change their doses of insulin. If such increased energy expenditure becomes a regular part of a diabetic's life, then the doctor will explain how to change the dose of insulin.

NOTE: Any diabetic who is thinking of trying to increase greatly his or her level of physical activity should first have a thorough examination by a doctor since hidden cardiovascular weaknesses may be present. However, many diabetics benefit from well-planned exercise programs.

It is simple enough to add food to the diet so as to replace the extra energy expended in additional physical activity. The best foods for this purpose are those high in carbohydrate—like bread, fruits, and vegetables. People who are trying to lose weight may not wish to replace all of the energy spent in extra exercise, but it might be wise for them to eat at least a little more food (about half of the equivalent of exercise) so that they do not develop low blood sugar. The extra food should be taken between meals, and prior to the activity. The food equivalents of various types of exercise are given in Table D-7.

Table D-7 shows that an extra slice or two of bread or a few pieces of fruit cover the energy cost of most additional activities which are likely to be performed by sedentary adults. However, children and teenagers may have wide variations in their day-to-day activities, so they should always bring with them a source of sugar in the event that their blood sugar might drop excessively. Another use for Table D-7 is in determining the amount of additional exercise which must be performed in order to expend the energy furnished by extra items of food added to the daily allotment. More detailed information on energy expenditure for various activities is given elsewhere in this book.

---

[28]Podgainy, H., and R. Bressler, "Biochemical Basis of the Sulfonylurea-Induced Antabuse Syndrome," *Diabetes*, Vol. 17, 1968, p. 679.

[29]Antabuse® is the brand of disulfiram produced by Ayerst Laboratories, New York, N.Y.

[30]Williamson, J.R., and C. Kilo, "Basement Membrane Thickening and the Mystery of Diabetes," *Hospital Practice*, Vol. 6, January 1971, p. 108.

TABLE D-7
FOOD EQUIVALENTS OF VARIOUS ACTIVITIES

Description of Activity	Time Spent in Activity	Total Energy Expenditure[1]	Bread Exchanges	Food Equivalents[2] Fruit Exchanges	Vegetables, B Group
	(min.)	(kcal)	(70 kcal)	(40 kcal)	(35 kcal)
Dishwashing	30	30[3]	½	1	1
Jogging (running alternated with walking)	10	100[4]	1½	2½	3
Riding a bicycle (7 miles *[11.2 km]* per hour)	20	130[5]	2	3	4
Sweeping floors	60	110[3]	1½	3	3
Swimming (30 yards *[21 m]* per minute)	20	170[4]	2½	4	5
Walking (3½ to 4 miles *[5.6 to 6.4 km]* per hour)	30	156[5]	2	4	4

[1]Based upon the performance of the activity by a 154-lb *(69.3 kg)* man.
[2]Approximate amount of each food = total energy expenditure ÷ energy per exchange.
[3]Burton, B.T., *The Heinz Handbook of Nutrition,* 2nd ed., McGraw-Hill Book Company, New York, N.Y., 1965, p. 27, Table 3.
[4]Konishi, F., *Exercise Equivalents of Foods,* Southern Illinois University Press, Carbondale, Ill., 1973, p. 28.
[5]*Ibid.,* p. 27.

Promising Research Developments In The Fight Against Diabetes. Although injections of insulin have been used for more than 50 years in the treatment of diabetes, the various complications of the disease still cause many disabilities and deaths. One problem with the present tests of blood sugar is that it is time-consuming and expensive to conduct analyses of blood for glucose, so the tests are infrequent events in the treatment of most diabetics. Even if these tests could be conducted weekly, they would still represent only a small sample of a diabetic's daily metabolism. Urine tests are simpler and may be easily done by the patient, but the absence of spillage of sugar does not necessarily indicate good control because the blood sugar may be chronically elevated to a value just below the level where passage into the urine occurs. Hence, the disease may be out of control more often than is suspected. What is needed is some means of imitating the close control of blood sugar which is achieved by a healthy pancreas. A promising approach to this program has been the development of devices which act like a pancreas, by Boston's Joslin Diabetes Foundation, and by Miles Laboratories of Elkhart, Indiana.[31] Descriptions of these devices follow.

- **Implantable glucose sensor**—Dr. Soeldner, associate director of the Joslin Foundation Research Laboratory, has developed a dime-shaped disk which, when implanted in the stomach area, will generate small electric currents in response to changes in the level of blood glucose. These signals are picked up by a radio receiver located in a special belt. A dial on the belt indicates the blood sugar, and an alarm sounds whenever the level is either too low or too high. Then, the diabetic knows when to eat extra food or take extra insulin. Some diabetics might engage in extra exercise in order to bring down their blood sugar.

- **Implantable insulin-releasing device**—Scientists at the Joslin Diabetes Foundation are also perfecting an implantable, artificial pancreas which may be connected to the glucose sensor. The device contains a reservoir of insulin plus a pump for discharging the hormone into the blood.

- **Pancreas machine for hospitals**—Miles Laboratories has developed and tested a pancreas machine for hospitalized diabetics. The machine monitors the blood sugar and releases either insulin or glucose as necessary.

PANCREATIC TRANSPLANTS. Various types of pancreatic transplants have been tried in diabetic patients receiving kidney transplants.[32] Details follow.

- **Whole pancreas transplants**—Pancreases obtained from donors have been transplanted at the same time that kidney transplants were made. Although the pancreases have functioned well for up to 16 months, it has been difficult to determine the long-range value of such transplants because the patients have usually died from kidney complications.

- **Transplantation of pancreatic islets**—There would be much less foreign tissue to be rejected by the body if only the pancreatic islet cells which secrete insulin were to be transplanted. Dr. Ballinger, chairman of the Surgery Department at the Washington University School of Medicine in St. Louis, and his co-workers have isolated both animal and human islets. The isolated islets from animals have been transplanted into diabetic monkeys and rats where they have functioned well in such tissues as the liver, peritoneum, and thigh muscles.

PSYCHOSOMATIC THERAPY. Doctors are often reluctant to insist upon strict dietary control for diabetics because they fear that the difficulties associated with such regimens may create emotional stresses which might aggravate the disease. Therefore, it is noteworthy that psychologists have recently been employing modified versions of ancient oriental ways of teaching people how to gain control of their nerves when they are

[31]Lublin, J. S., "New Hope: Cures for Diabetes Appear on the Way, Researchers Report," *The Wall Street Journal*, Vol. XCI, November 4, 1974, p. 1.

[32]Bondy, P. K., and P. Felig, "Disorders of Carbohydrate Metabolism: Recent Developments," *Duncan's Diseases of Metabolism*, 7th ed., edited by P. K. Bondy and L. E. Rosenberg, W. B. Saunders Co., Philadelphia, Pa., 1974, p. 340A.

in trying situations. A typical technique is called "biofeedback" where such measurements as the blood pressure and heart rate for an individual are displayed on a screen where he or she can see them. The person then tries to change these functions by mental concentration while watching the results of these efforts. It seems likely that there will soon be attempts to control blood sugar levels by such techniques.

THERAPEUTIC AGENTS. At the present time, drug therapy for diabetes is limited to various types of insulin, along with a few oral drugs, most of which stimulate insulin secretion. However, each increment of insulin, whether produced by the body or injected, tends to stimulate fat synthesis in the body. Therefore, any therapeutic agent which avoids this problem would be very useful for diabetics. Such agents follow.

- **Glucose tolerance factor**—This chromium-containing substance, which was mentioned earlier in this article, has been shown to be helpful for some elderly, adult-onset diabetics. It appears that in some manner this factor increases the effectiveness of insulin. Additional information should soon be available, since tests of the factor are now underway.

- **Hypoglycin**—Although this substance from unripe akee fruit (the akee tree grows in Jamaica, West Indies, and in West Africa) is highly toxic, it lowers the blood sugar by blocking production of glucose in the liver.[33] Perhaps some means will be found to reduce the toxic effects of this material so that it might be used to treat diabetes.

- **Somatostatin**—This proteinlike material, which is isolated from extracts of sheep hypothalamus glands, lowers the blood sugar by blocking the secretion of some of the hormones which raise the blood sugar (pancreatic glucagon and pituitary growth hormone). However, there is a delay in the testing of this substance in the United States while questions of its safety are being studied by the Food and Drug Administration.

DIABETES, RENAL

An older term for renal glucosuria.
(Also see RENAL GLUCOSURIA.)

DIABETES TESTS

These are diagnostic procedures which determine the amount of sugar (glucose) in the blood and urine. Abnormally high levels of blood sugar and of sugar in the urine (glucosuria) are usually characteristic of diabetes mellitus. They may appear when no other symptoms are present. Specific tests include fasting and after-meal blood sugar determinations and the glucose tolerance test (GTT).

(Also see DIABETES MELLITUS, section headed "Diagnosis of Diabetes.")

[33]Bressler, R., "The unripe Akee—Forbidden Fruit," *Nutrition Reviews*, Vol. 34, 1976, p. 349.

DIABETIC COMA

A complication of diabetes mellitus due to lack of insulin. Blood sugar becomes elevated, ketone bodies build up in the blood causing "acetone breath," the symptoms of acidosis appear—nausea, vomiting, dizziness, deep labored breathing—and finally coma. A diabetic in this condition should receive a doctor's care immediately as death could occur shortly. A diabetic should always carry some medical identification so that he will not be mistaken for a "drunk" if found in a coma. Another coma-producing complication of diabetes is insulin shock (hypoglycemia), but its causes are just the opposite—excess insulin and low blood sugar.

(Also see DIABETES MELLITUS, section headed "Treatment and Prevention of Diabetes"; and HYPOGLYCEMIA.)

DIABINASE

A brand name for a form of sulfonylurea called chlorpropamide, a long-acting oral drug used to stimulate the pancreatic cells of a diabetic to release insulin. The drug is only effective when the diabetic has some insulin secreting ability remaining in the pancreas. Other brand names include Adiaben, Asucrol, Catanil, Chloronase, Diabechlor, Diabenal, Diabetoral, Melitase, Millinese, Oradian, and Stabinol.

(Also see DIABETES MELLITUS.)

DIACETYL

A substance produced by the action of microorganisms on butter fat. It is responsible for the distinct taste of butter which has been ripened. However, it may be produced artificially and used as an additive to margarine.

(Also see ADDITIVES.)

DIALYSIS

A process in which certain substances in a solution are removed by the passage of the solution through a membrane. Dialysis is used to purify the blood of patients who suffer from kidney disease, when the kidneys are unable to perform this function.

DIAPHYSIS

A shaft of a long bone, such as is found in the arms and legs. During the growth of the long bones the diaphyses are separated from the ends of the bones (epiphyses) by a plate of cartilage. When growth ceases during maturity, the diaphyses fuse with the epiphyses.

(Also see BONE.)

DIARRHEA

Rapid movement of the fecal matter through the digestive system producing frequent, watery stools. It may be caused by a wide variety of disorders, including bacterial contamination of food, virus infection, allergy, nervous reaction, as well as various serious and chronic ailments.

The discomfort of mild cases of diarrhea can be eased by various home remedies. Stomach cramps may be relieved by the application of heat. Dehydration can be prevented by consuming fluids such as tea, soup, and ginger ale. Bland and nourishing solid foods, such as boiled or poached eggs, rice pudding, cottage cheese, and/or toast, should be eaten as soon as possible.

Prompt medical attention is advised (1) when a diabetic has diarrhea, or (2) when diarrhea is chronic.

DIASTASE

The general name given to any of the enzymes which digest starch to smaller molecules of carbohydrate.

(Also see ENZYMES.)

DIASTOLE

A period in which the heart fills up with blood. During this time it is in a relaxed state so that the blood pressure is at its lowest point.

(Also see BLOOD PRESSURE.)

DICOUMAROL

A chemical compound found in spoiled sweet clover hay or made synthetically. It is an antagonist of vitamin K and thus an anticoagulant of blood. Medically it is used to prevent blood clots. Other names include Dicoumarin, Dicumol, Dufalone, and Melitoxin.

(Also see VITAMIN[S], section headed, "Vitamin K"; and VITAMIN K.)

DIET

The beverages and foods normally consumed by a person. However, most people take the term diet to mean specially designated foods and beverages to be taken in measured amounts, while other items are restricted.

(Also see MODIFIED DIETS.)

DIETARY GUIDELINES FOR AMERICANS

Americans need to make changes in their eating habits. Dietary guidelines follow:

1. Reduce total fat consumption to 30% or less of total daily calorie intake.
2. Reduce cholesterol intake to less than 300 mg a day.
3. Increase the amount of starches and other complex carbohydrates in the daily diet.
4. Increase the amount of fiber in the diet.
5. Increase the amount of fruits and vegetables eaten daily.
6. Avoid drinking alcohol, especially during pregnancy. If abstaining is impossible, drink in moderation.
7. Reduce the daily intake of sodium to 6 g or less.
8. Maintain desirable body weight through prudent diet and regular exercise.

DIETETIC FOODS

Foods which have been produced so that they conform to the requirements of certain modified diets. For example, canned fruits may be made with little or no sugar, or with artificial sweeteners. In many cases dietetic foods are lower in fat or sugar than the food they are designed to replace. Hence, the cooks in homes may prepare dietetic foods by simply reducing the amounts of high calorie ingredients which are used.

DIETITIAN; DIETETICS

A dietitian is a person trained in nutrition and dietetics (the science that deals with the relationship of food to health) who plans menus and supervises the preparation of food. Dietitians work in hospitals, industrial food services, restaurants, schools, universities, and in many other areas.

Over half the professionally trained dietitians work in hospitals, where they plan diets and supervise the preparation of food, and help patients plan and understand the diets prescribed for them. Others work in educational and research programs.

To become a dietitian, a student must complete a prescribed course of study leading to a bachelor's, or higher, degree, followed by serving an internship in dietetics work.

(Also see AMERICAN DIETETIC ASSOCIATION [ADA]; and NUTRITIONIST.)

DIETS, MODIFIED

The term *diet* as such refers to food and drink regularly provided or consumed. When the consumption of the regular diet is inadvisable, the diet should be changed. For guidance at such times, an extensive section on Modified Diets is presented.

(Also see MODIFIED DIETS.)

DIETS OF THE WORLD

Contents *Page*

Eating Habits and Patterns **577**
Mealtime Around the World **578**

The diet is the usual kind of food and drink of an individual. Worldwide, the diets of people are similar in one respect—everyone eats foods from plants and/or animals. Until very recently in the history of mankind, the type and amount of plant food and animal food available for eating depended upon the area. Then, based on the foods available, customs, economics, social patterns and religious beliefs, the diet of a country or area

Fig. D-17. Diets vary according to where an individual lives. Many diets of the Old World have influenced those of the New World.

developed. As time passed, the diet changed as foods from other countries or areas were introduced. Despite outside influence, however, many diets have remained traditional over the years, particularly in Europe and Asia where there are still well defined nationalities, or where the food supply is largely controlled by that produced within the country. Countries like the United States and Canada have been influenced by the diets of people from many nations. Moreover, these countries have a plentiful and varied food supply.

Overall, the diet of an individual depends upon where he or she lives; and even when the foods available are similar, they are often prepared differently. The following are some examples of worldwide diets from the four food groups:

1. **Meat, poultry, fish and beans.** Most people in the United States would include beef, lamb, pork, chicken, turkey, goose, and duck, along with some common sea foods like fish, clams, lobsters, oysters, and shrimp in this group. In some countries, people may eat meat from other domestic animals. For example, some Belgian and French people eat horsemeat, and some Greeks and Japanese enjoy goat meat. Hunters in almost all countries seek wild animals for food, as well as for sport. Favorite game meats in the United States include deer, opossum, rabbit, raccoon, squirrel, bear, and such wild fowl as duck, grouse, partridge, pheasant, and quail. In other countries, people may hunt baboons, caribou, elk, elephants, gazelles, monkeys, and snakes for meat. They may also eat ants, grasshoppers, locusts, and other insects.

Eggs from chickens and ducks are an important food almost everywhere. Some people eat the eggs of such birds as emus, gulls, ostriches, plovers, and penguins. In many countries, people enjoy the eggs of alligators, crocodiles, iguanas, turtles, and other reptiles.

In Mexico, beans are a popular food; in China, wide use is made of soybeans.

2. **Milk and milk products.** In the United States, Canada, and some other countries, cows provide most of the milk. But in southern Europe and the Middle East, the goat furnishes much milk. In Asia, sheep are often milked. People in India drink buffalo milk, and Laplanders enjoy reindeer milk. The yak furnishes milk in Tibet, and Arab herdsmen milk camels. From milk, countless forms of cheese have been developed in many countries.

3. **Vegetables and fruits.** Many of the vegetables and fruits eaten today had their origin in specific areas or countries where they became an important component of the diet. The cucumber, eggplant, soybean and squash came from China. The cowpea and yam came from Africa, while the chickpea, lentil and mung bean came from Asia. Corn, beans, potatoes and tomatoes are New World vegetable crops.

Numerous fruits, such as the citrus fruits, apple, peach, apricot, banana, pear, and plum, originated in Asia. The watermelon, akee, and tamarind originated in Africa. Berries were common fruits of northwestern Europe and North America. Avocados, acerola and papaya come from Central America, while cherimoya, custard apples, guava, and passion fruits are native to South America. Today, many regions still grow vegetables and fruits which are unique to them.

4. **Breads and cereals.** Grains—wheat, rice, barley, rye and oats—are the largest single food item used throughout the world. The methods of preparing the grains for consumption vary widely, but almost everyone in the world eats some kind of bread, either English *graham bread*, Scottish *bannocks*, Finnish *tunnbrod*, German *pumpernickel*, Latin-American *arepas*, or Swedish *hardtack*. Worldwide, breads may be sweet or sour, brown or white, heavy or light, and raised or flat. In addition, grains are cooked in a variety of dishes or processed to make such products as macaroni, spaghetti, or breakfast cereals.

Condiments also vary from country to country—olives from Greece, malt vinegar from England, spices from the Far East, and maple sugar and syrup from North America. Every country has its favorite flavorings and seasonings.

EATING HABITS AND PATTERNS. Eating habits and patterns often vary between individuals and families—let alone countries or regions. Many factors, such as economics, religion, customs, and fads, influence eating habits and patterns, as well as individual preferences.

Economics inside a country, and those directly relating to an individual, determine the types of foods available. In general, the poorer nations and individuals buy less food and less variety, which limits their diet. People in highly developed industrial nations generally have the most varied diets, because they have money to buy different kinds of food. Also, they eat the greatest amount of animal foods, which are the most costly to produce and the most expensive to buy.

(Also see INCOME, PROPORTION SPENT FOR FOOD; and WORLD FOOD.)

Religions and diets have been closely connected for a long time. The ancient Greeks and Romans worshiped gods and goddesses who ruled over agriculture and hunting. The word "cereal" comes from *Ceres*, the Roman goddess of fruits and grains. The Bible tells of the early use of food as religious offerings.

Many religions do not permit their followers to eat certain foods. Buddhists are not allowed to eat any meat. Moslems can eat no pork. Hindus consider the cow sacred, and eat no beef. Some Hindus consume no animal foods except milk and dairy products, because their belief forbids them to kill animals. Orthodox Jews avoid certain kinds of foods, including pork, nonkosher meats, and shellfish, such as shrimp and lobsters. The Seventh-Day Adventists promote ovolactovegetarianism.

(Also see RELIGIONS AND DIETS.)

Customs influence the ways people eat, and to some extent what they eat. Most Americans and Europeans eat from individual plates, using knives, forks, and spoons. Arabs use only their right hands to spoon foods from a central bowl. Chinese and Japanese use chopsticks to pick up food from a small bowl held close to the mouth. Many Orientals sit on the floor and eat from low tables.

Most persons eat at least three meals a day—breakfast, lunch, and dinner. The English and some other Europeans eat a fourth meal, supper, late at night. Meals vary in different countries. For example, breakfast in the United States, may include fruit or fruit juice, coffee, toast, and a choice of cereal or bacon and eggs. Some people like pancakes or meat and potatoes for breakfast. In rural areas, families usually have their big meal at noon and a lighter meal in the evening. In many areas, families and groups enjoy outdoor meals, such as picnics, corn roasts, clambakes, and backyard barbecues.

Continental Europeans sometimes have an early breakfast of sweet rolls and coffee or hot chocolate, and eat a second breakfast later in the morning. English breakfasts often include *kippers* (salted, smoked herring), or such meat as kidneys or sausage; cooked porridge; toast and marmalade; fresh or stewed fruit; and tea. Another English tradition, tea or "tiffin," provides an extra meal served in the late afternoon. Its simplest menu usually includes tea and special tea cakes such as crumpets and scones, or biscuits with jam.

Some people choose their diet on the basis of fads and myths. Eating habits have long been partly controlled by beliefs about what is fit to eat. For example, some people once thought that tomatoes were poisonous, and refused to eat them. Today, some persons avoid white bread or processed foods, because they fear that substances added during flour milling or processing make these foods impure. Others, eat "natural" or organically grown food to ensure a healthy diet. Still others think that certain foods should not be eaten together, such as orange juice and milk, or starches and proteins. Some fad groups have encouraged strict vegetarianism for the purpose of achieving a certain "spiritual" level. Many of these faddish diets are based on misinformation—they are reminiscent of some ancient food practices such as eating a bull's tail for bravery, or eating brains for wisdom.

In areas where a wide variety of foods are abundantly available, people may not eat certain foods simply because they do not like them. Many areas of the world, however, cannot afford this luxury.

(Also see FOOD MYTHS AND MISINFORMATION.)

MEALTIME AROUND THE WORLD. Eating is a very complicated process. Whole books are written concerning the foods of a country or the lack of food in a country, as are whole books written concerning the preparation and consumption of foods in various countries and even various regions. The following discussion of countries serves only to acquaint the reader with some basic points of the diets in various countries. Diets of some countries have had a profound affect on the diets of Americans, since people immigrating to America brought their diets with them.

Africa. Most rural African families grow their own crops and raise their own animals. A few peoples of Africa still get their food by hunting and gathering. But most Africans are either farmers or cattle raisers.

Desert dwellers of the north and east get food and drink from the camels and goats they raise. Their plant food comes from the oases.

The grassy plains of the northern savannas are well suited to raising cattle. The people who live there have meat to eat and milk to drink. They also grow millet (a kind of grain) and groundnuts (peanuts), as well as yams, beans, rice, and peas. Fruits are not readily available to these people.

Fruit is very common in the southern rain forest. Here the people eat oranges, guavas, grapefruit, pineapples, limes, mangoes, and bananas. They drink the milk of the coconut. They also eat yams, rice, cassava, potatoes, and maize (corn).

Corn, sorghum, and millet are important foods in many parts of Africa. The women grind the grain into flour. Then they boil the flour in a big kettle. When it is cooked they have a corn porridge, which is often served with beans.

Cassava is another important plant because it can be stored for a long time and prepared in several different ways. One West African cassava dish is called *gari*.

Australia. Not surprisingly, because of their heritage, Australian food is much like that of England. Tea is a favorite, and considerable beef and mutton are consumed. Australia is an industrialized nation with a varied agriculture, therefore it has a wide food selection. Kangaroo tail soup is definitely Australian.

Canada. Canada is a rich country, characterized by industrialization and a well-developed agriculture. A wide variety and an abundant supply of foods exist. In many areas, the diet and cooking vary according to which country a family originated. Primarily, European cookery forms the basis of Canadian cuisine. In Labrador, and on the shores of Newfoundland, specialties of Mecklenburg, Swabia, and Switzerland may be found. The French Canadians cling to the traditional regional dishes of France, especially those of Normandy.

China. Chinese cookery is an art, passed down from generation to generation (mother to daughter) and chef to chef (professional to apprentice). Until recently, recipes were secret and by word of mouth. Properly mastered, the art should appeal to the senses, or as the Chinese say: "A properly prepared dish should appeal to the eye by its coloring and form; to the nose by its aroma; to the ears by its sound; and to the mouth by its flavor."

Every part of China has a distinctive style of cooking, varying according to the farm products, natural conditions, and way of life of the different regions. The people in the south of China prefer their dishes sweet; in the north, salty; in the east, sour; in the west, hot.

The Chinese diet is short on meat and milk. Soybean milk is substituted for cow's milk. In fact, wide use is made of soybean products. Among the staple foods, rice is the main grain eaten by the people south of the Yangtze River. Further north, wheat and ground corn, which are more common than rice, are made into steamed rolls, noodles, dumplings, and flat steamed or baked cakes. Chinese vegetables are chopped, and never over-cooked.

Baked buns of wheat flour are the staple food of the people of Sinkiang. On special occasions, rice steamed with a little oil replaces the buns. The herdsmen of Inner Mongolia eat large quantities of mutton, beef, and dairy products, including *koumiss* (fermented mare's milk) and dried milk cakes.

Some well-known foods of China include the following:

- **Chow mein**—This dish may consist of veal, chicken, or shrimp, with celery, mushrooms, water chestnuts, and bamboo shoots added; served with soy sauce.

- **Egg rolls**—Shrimp or meat and vegetable filling are rolled in thin dough, and fried in deep fat to make egg rolls.

- **Egg foo yung**—This is a combination of eggs, chopped chicken, mushrooms, scallions, celery, and bean sprouts; cooked similar to an omelet.

- **Sweet and sour pork**—This is formed from pork cubes coated with batter and fried in oil; then simmered in a sauce of green pepper, cubed pineapple, molasses, brown sugar, vinegar, and seasonings.

- **Tofu**—Tofu is soybean curd, like cheese, and it is used in many dishes.

England. Many English dishes are based on such products as beef, mutton, lamb, pork, pheasant, grouse, partridge, duck, salmon, sole, plaice, turbot, halibut, butter and cheese, which have always been available.

Puddings, pies and tarts are another important part of English cooking. Most recipes for baked, boiled and steamed puddings make use of flour, fat, eggs, and fruit. The main meal of the day usually includes one of these puddings or a pudding made with milk as the main ingredient, or it may be a pie or tart having pastry as one of its main ingredients.

Cakes, scones and biscuits are also featured in the English diet, and are usually eaten at the tea meal in the late afternoon. There are many varieties, a number of which are fairly plain, having a smaller proportion of fat, sugar and eggs than those of other countries.

Popular dishes for which England is justly famous include pork pies, Lancashire hotpot, and Cornish pasties. Other delicious specialties are Christmas pudding, mince pies, stew with dumplings, Welsh rarebit, Yorkshire pudding, steak and kidney pudding, and kippers (salted smoked herring).

France. No country has added as much to the art of cookery as France. Leading restaurants throughout the world feature French style of cooking, known as *haute cuisine*.

French cooks are famed for mixing herbs, seasonings, and wines—with a touch of imagination. Fine sauces for meat, poultry, and fish result. France is also famous for its pastries and breads. Moreover, the French are great cheese eaters; and their cheeses are mong the world's finest.

Thomas Jefferson learned to make ice cream while he was living in Paris as America's minister to France. He brought the recipe, written in his own hand, back to the United States. He was among the first to serve ice cream as a dessert at a state dinner.

Well-known French specialties are: truffles (mushroomlike) in the Guyenne region; snails in Burgundy; sausages in Arles and Lyon; omelets in Mont St. Michel; and pressed duck in Paris.

Germany. From soups to desserts, German food is filling. Famed German soups include: *Hamburger Aalsuppe* (eel soup), named after the great German port; *Haferschleimsuppe*, a consomme made of oats; and *Arfensuppe mit Snuten und Poren*, which is a pea soup flavored with pig's snout and trotters. Another Hamburg speciality is brisket of beef garnished with veal, semolina quenelles and diced seasonal vegetables.

In Wurttemberg and Baden, there is *Spatzle* soup. (Spatzle consists of squares of fresh noodle paste, plain, or filled with liver, ham or mushrooms.) In Swabia, there are many vegetable soups: cauliflower soup, bean soup, lentil soup, pea soup, and potato soup.

Main dishes may include *Gefullte Schweinerippchen*, which consists of pork cutlets stuffed with currants, apples, biscuits and rum. In Berlin, *Sauerkraut* if served with *Eisbein*, which is boiled salted knuckle of pork accompanied by pea puree. The staple diet of the people of Wurttemberg is boiled flour quenelles, of which there are many varieties, ranging from the Spatzle with added liver, ham, etc., to the *Maultasche*, a kind of highly seasoned canelloni stuffed with brains, veal and pork. Upper Bavaria is the country of *Knodel* (potato balls) which are varied by the addition of white breadcrumbs, liver stuffing, or bacon.

There are many pastries throughout Germany. For example, in Berlin there is *Streuselkuchen*, a yeast cake made with butter, flour and sugar; and *Pfannkuchen*, jam fritters dusted with sugar. In Dresden, there are custard tarts speckled with raisins.

Hawaii. Although Hawaii is a state, it represents a unique mixture of East and West and native foods. The foods on supermarket shelves reflect the tastes of Hawaiians, Chinese, Japanese, Americans, and others who live there.

Hawaiians have their own cuisine. *Poi*, a paste made of taro root, is often eaten by Hawaiians. It has a slightly sour taste which people new to the Islands often find unpleasant. Fine fruits and vegetables are grown. The pineapples are world famous. *Laulau* is generally served at the *Luau*, or outdoor feast. The *luau* is the Hawaiian version of a New England clambake or a picnic. *Laulau* is a steamed dish made of pork and fish wrapped in ti plant leaves. A suckling pig roasted in a pit is a feature of the *luau*.

Hungary. Hungarians love good food in large quantities, especially on holidays and other festive occasions. Hungary has added its *goulash* to the world's finest dishes. It is a thick stew of chicken or beef, cooked in a sauce seasoned with Hungary's favorite spice, *paprika*. Hungarians also use paprika in many of their other dishes. Noodles, potatoes and small dumplings are popular side dishes.

Hungarian pastries are famous, particularly *retes* or *strudel*, which consists of a thin flaky crust filled with fruit or cheese.

India. Much of the Indian diet and cooking has not changed since ancient times. Furthermore, religion has a strong influence on diet. The chief foods of Indians include such grains as barley, rice, and wheat, and such legumes as beans and peas. Indians eat grain flour cooked into flat bread cakes. In northern India, the people make bread by grinding the grain into flour, and then adding water to make dough. They roll the dough flat and bake it. The people of southern India make bread from rice or legumes, which they first soak in water. Then they make a paste from the rice or seeds and cook it like dumplings. A well-known Indian dish is *curry*—made of eggs, fish, meat or vegetables cooked in spicy sauces. Wealthy Indians cook with *ghee*—clarified butter. Hindus do not eat beef, and Muslims do not eat pork.

Indonesia. The primary food in Indonesia is rice, though corn is a major food in some areas. Rice is boiled or fried in various ways. Indonesia is known for its *rijsttafel*, or rice-table which serves for midday or evening meals. It consists of at least five or six dishes made of vegetables, meat, poultry, fish, and boiled rice. A helping from each of these dishes is placed on a mound of rice and eaten separately. Foods are so highly spiced they are often mouth-burning.

People of the cities eat a more varied diet than those in farm areas.

Ireland. Irish cooking is simple. Favorite dishes include chicken, lamb, potatoes, sausage, and steak. For hundreds of years, potatoes have been an important food in Ireland. The Irish especially enjoy potato pancakes and boiled potatoes. One of Ireland's famous dishes, *Irish Stew*, is made by boiling layers of potatoes, onions and pieces of mutton in a covered pot.

Israel. Most people living in Israel are Jewish; hence, kosher food predominates, and many foods are forbidden. Some typical Jewish foods include: *gefilte fish*, which is chopped, highly seasoned fish; *knishes*, pastries filled with ground meat; *kreplach*, noodle dough filled with ground meat or cheese filling; *bagel*, a doughnut-shaped hard yeast roll; *farfel*, noodle dough grated for soup; and *kuchen*, coffee cake of many varieties. Meats are kosher and some are forbidden.

(Also see RELIGIONS AND DIETS.)

Italy. *Pasta* is the basic food throughout Italy. Pasta includes spaghetti, macaroni, ravioli, lasagne and vermicelli served with various sauces. *Pizza* is a popular Italian dish which has been popularized in the United States. Some favorite Italian cheeses include mozzarella, ricotta, parmesan, romano, and gorgonzola. Almost every region of Italy has its own specialty. This includes such foods as the *sausages* of Bologna, the *prosciutto* (salted ham) or Parma, and the *minestrone* (thick vegetable soup) of Milan. Italians often cook with butter, olive oil and tomatoes.

Japan. The abundance of seafood explains why fish form the basis of Japanese cuisine. The people are skillful cooks, and have devised many ways of serving food of marine origin. They use not only all the types of fish and shellfish found in their seas, but also edible seaweed and many other marine products unknown in the West, which they eat both in their fresh and preserved states. Tinned sea urchins, crabs, mussels, and fish quenelles are widely available in Japan.

Dried fish, which the Japanese call *sashimi*, is very popular. They also eat solid-fleshed fish such as sea bream, tuna, carp, turbot and, less frequently, mackerel; usually seasoned with soy sauce.

The following are some popular seafood dishes in Japan:

• **Kabayaki**—This is grilled eel filets (*unagi*). It is prepared in a very special way in Japan. The eel is gorged with beer or spirits, bled, then immediately split in half lengthwise with one blow from a chopper and completely deboned. This requires much skill. It is first steamed, then grilled over a wood fire and served sprinkled with sauce.

• **Katsus-buski**—This is dried bonito (a species of tuna). When the drying process is completed, the fish is like a piece of wood. It is then grated as it is required. Bonito powder is added to certain soups for the dual purpose of seasoning them and making them more nutritious. It is widely used in Japan.

• **Musi-yaki**—This is fish cooked in the oven in a sauce made of potato flour or cooked dry and sprinkled afterwards with the same sauce.

• **Maguro-no-suhki**—This is a bowl of cooked rice containing powdered egg wrapped in a thin slice of raw tuna, or other substance.

• **Nosi-maki**—This is a bowl of rice garnished with seaweed.

• **Nuta**—This dish is a mixture of raw fish and vegetables. It is strongly seasoned with horseradish, mustard and other condiments.

• **Sukiyaki**—This is another national dish. Sukiyaki means "roasted on the spit." When served, each guest prepares the dish according to his taste, using the foods that are available: thin slices of meat, raw diced vegetables, mushrooms, bamboo shoots, raw egg, etc.

During the past 20 years, there have been spectacular changes in Japanese food habits, influenced by Western culture. Today, instant and frozen food items are available in Japan.

Mexico. Mexico's food is highly seasoned with spices, as is the food in all Latin American countries. Corn and beans are important foods in Mexico. Corn is generally softened in hot limewater, boiled, and then ground into a meal. Beans in the Mexican diet are eaten as *frijoles*—boiled, mashed and then fried and refried in lard. The following are the names and descriptions of some common Mexican dishes:

• **Atole**—This is a cornmeal gruel.

• **Chile con carne**—This dish consists of beef with garlic seasonings, beans, and chili peppers.

• **Enchiladas**—These are tortillas filled with cheese, onion, shredded lettuce, and rolled.

• **Sopaipillas**—These are puffs of deep-fried dough, which are used as bread or dessert, usually with honey.

• **Taco**—A taco is a tortilla filled with seasoned ground meat, lettuce, and served with chili sauce.

• **Tamales**—Tamales are seasoned ground meat placed on masa (dried ground corn) wrapped in corn husks, steamed, and served with chili sauce.

• **Tortilla**—This is the main cornmeal food in Mexico. The tortilla is a thin pancake shaped by hand or machine and cooked on an ungreased griddle. It is the bread of the poor people. Tortillas can be eaten plain or as part of the taco, the enchilada, or the tostada.

Middle East. The inhabitants of the Middle East are outdoor people. Many of them are farmers. Therefore, they raise their own sheep, goats, cattle, chickens, ducks, and geese; and, wherever water is plentiful, they produce their own grains and grow fruits and vegetables in abundance. Grains, usually rice or wheat, furnish the major source of energy. Bread is baked on griddles in round flat loaves; and such pastries as *baklava, kadaif,* and *boreg* are favorites. Whole wheat is parboiled and cracked for use as a staple starchy food at the main meal. Eggs, butter, and cheese are also produced on the farm. Lamb is the favorite meat. *Moussaka* consists of ground lamb, eggplant, tomato sauce, and cheese; while *shashlik* (*shish kebab*) consists of mutton or lamb marinated in garlic, oil, vinegar and roasted on skewers with tomato and onion slices. The food is not highly spiced, but it is rich in fat. The fat is cooked with the food and this takes the place of butter. Matzoon, leban, or yogurt—which are sour-milk preparations—are used almost universally by these people. Hard and soft cheeses, such as kacher and feta (white goat's milk cheese), are enjoyed.

Black coffee, heavily sweetened, in which the pulverized bean is retained—often called Turkish coffee—is the preferred beverage in many countries of the Middle East.

Netherlands. The land of wooden shoes and tulips is also a land where people like good, hearty food. All kinds of fish and seafood are enjoyed, particularly herring. Holland's rich dairy produce goes into its superb cheeses, which represent one of the country's largest exports. Edam, Purmerend and Alkmar cheeses are spherical and painted red. Another famous cheese is Gouda, a cream cheese manufactured mostly in the south of Holland. It is shaped like a flattened cylinder with rounded ends. Dutch chocolates are shipped everywhere. The Dutch are very fond of spices and *rijsttafel*, the Indonesian rice-table dishes. They developed a taste for these highly seasoned dishes when their early explorers found their way to the islands of Indonesia.

Poland. Polish meals feature meat and dairy products. Fish, ham, pork, sour cream, dark breads, mushrooms, sauerkraut, noodles, dumplings, kasha or groats, and rich cakes are the principal foods. A yeast-dough cake called *babka*, which has bits of candied fruits, nuts and raisins scattered through it, is a great favorite.

Puerto Rico. Rice, legumes and *viandas*—starchy vegetables—are basic to all diets. *Viandas* include primarily green bananas and green plantains, and also *batata amarillo* (yellow sweet potato), *batata blanca* (white sweet potato), ripe plantain, some breadfruit and cassava. Dried codfish and milk are used by some, but most milk is used in strong coffee (*cafe con leche*). *Viandas* are boiled and served hot with oil, vinegar and some codfish. Fruits on the island are abundant and include acerola, grapefruit, guava, mango, orange, papaya and pineapple. *Sofrito*, a sauce made of tomatoes, onions, garlic, thyme, and other herbs, salt pork, green pepper and fat, is basic for much of the cookery.

Russia. A distinguishing feature of Russian cuisine is the use of *smetana* (whey) and beetroot, which forms the base of *bortsch*, without which no Russian meal would be complete. Russian soups, like those of Poland, are very nourishing. Besides vegetables, they contain a variety of meats. The following soups are most commonly served in Russia: bortsch, stschi a la russe, cucumber rossolnick, oukha de sterlet, meschanski soup, nettle soup, Livonian soup, botwinia (served iced), kliotskis soup, cream of ham soup, game okrochka, morel soup, kalstchale (a macedoine of glace fruits), calves kidney soup, and iced ochrochka fish soup.

The Russians make excellent pies, the best-known of which is the *coulibiac*, which dates to the 16th or 17th century. Formerly, it was filled with cabbage, but nowadays it usually contains salmon or other fish.

Vodka accompanies a meal and should traditionally be drunk in one gulp. It is odorless and colorless, and made from fermented wheat or occasionally corn or potatoes.

Scandinavia. The Scandinavian neighbors—Finland, Sweden, Norway, and Denmark—speak different languages, but their foods are much alike. They eat a good deal of fish and seafood. Wheat, rye and barley bread are served with the meals. The Finnish people are very fond of herring prepared in many different ways.

Sweden is the home of the *smorgasbord.* This is a long buffet table loaded with meat and fish dishes, cheeses, salads, hot dishes, and, of course, bread and butter. *Smorgasbord* means "bread and butter table." Some evening meals in Swedish homes consist of *artsoppa*, a thick soup of yellow dried peas seasoned with spices and ham.

Reindeer roasts and steaks, rabbit, bear, and other game are popular in Norway. The Norwegian *flattbrod*, a thin, crisp cracker, is shipped to many countries.

One of Denmark's greatest food inventions is the open sandwich on one slice of bread, known as *smorrebrod.* There are more than one hundred different kinds of *Smorrebrod.* Danish *saucisson* resembles Italian salami, but it is fattier and more savory. In Denmark, Danish pastry is known as *wienerbrot*, which means Vienna pastry.

Scotland. Most Scottish cooking is simple. Favorite dishes of the people include herring, mutton stew, roast beef, and roast lamb. Other popular foods include haggis, kippers, oatmeal and salmon. *Haggis* is a famous national dish made from the heart, liver and lungs of a sheep or calf. These ingredients are chopped with animal fat (suet), onions, oatmeal, and seasonings, and then boiled in a bag made from a sheep's stomach. *Kippers* are smoked herring. Oatmeal is used in many Scottish dishes including porridge, and oatcakes (craker)—popular breakfast foods.

South And Central America. Each country in South and Central America has important dishes, but all the foods are highly seasoned, and in South America in particular, they are generally Spanish or Portuguese. Seafoods are important in many countries and corn, beans and rice are often served. Nearly all South American dishes are served with rice. Vegetables include a variety of ways of preparing potatoes, and plantains, yams, cassava, and gombos. Many fruits are available, including the avocado, banana, custard apple, anonas, sapota, guavas, mangoes, citron, and kumquats.

Spain. Food in Spain is colorful and spicy. Seafood is enjoyed by many in a variety of dishes. Often fish and shellfish are fried in oil. A popular dish is *paella*. It consists of such foods as shrimp, lobster, chicken, ham, and vegetables, all combined with rice that has been cooked with a flavoring called *saffron*. Simple paellas are made in the home. A popular dish during warm weather is *gazpacho*, a cold soup made from tomatoes, green pepper, cucumber, and other raw vegetables.

In the streets of Spain, vendors sell *churros* (hot fritters), watermelon, and *touron* (a kind of almond paste). Watermelon is offered with a glass of Manzanilla wine. Spaniards drink wine with all meals except breakfast. They also enjoy a drink called *sangria*, which consists of wine, soda water, fruit juice and fruit.

Switzerland. Cheese is a basic ingredient of Swiss cuisine. There are only two types: (1) *Gruyere*, a pleasantly piquant, fatty cheese (minimum 45% fat), and (2) *Emmenthal*, fatty and sticky with large vacuoles, a rarer cheese with a less pronounced flavor. However, there are infinite variations of quality in these two cheeses, depending on their provenance (origin) and age.

One of the best-known cheese dishes is *fondue*, for which there are many recipes. The fondue begins in the kitchen in a glazed earthenware casserole or cast-iron enamelled pan which is rubbed with a clove of garlic. The cheese is finely grated. Then, while the fondue is being prepared, the guests cut their bread into small pieces. The fondue is brought in on a spirit heater, and at the last moment, a small glass of pure fruit kirsch (a clear sweet brandy) is added. Everyone eats from the same dish, dipping their forkful of bread right to the bottom to ensure that the *fondue* does not stick to the pan while cooking.

Gastronomic customs vary from region to region. In French-speaking Switzerland, customs are more or less the same as in France, though with a tendency to make the main meal of the day the midday meal, as they do in German-speaking Switzerland. For most families, the evening meal consists of a cup of white coffee and potatoes browned in butter (*rosti*).

Fribourg is renowned for its salted and smoked beef. Smoked cows' udders are considered a delicacy. The fondue of this canton is made with Vacherin cheese.

Berne produces the renowned *cochonnailles* (cooked meats) such as the saveloy, ham, smoked bacon, and smoked pork shoulder, which, with the addition of cabbages and sauerkraut, form a rich stew called *bernerplatte.*

Basle is principally a town of pastry-makers, and the *leckerlis* of Basle, with their hard, brittle, full-flavored base are well-known. There is a wide variety of homemade pastries, and these increase in number at carnival time.

Finally, Swiss chocolates enjoy a world-wide reputation for quality. The raw materials are all imported, since Switzerland does not produce cocoa or a sufficient quantity of sugar.

United States. It is fitting that the United States should be the last country discussed, since the diets of the United States have grown out of the customs of many peoples from many different lands. Moreover, the climatic and geographical conditions of the United States are so varied that from the fertile valleys, vast forests and plains, lakes, rivers and streams, and extensive coastal regions, almost every food known to man can be produced within its borders. With modern technology, the produce of any one section of the country can be enjoyed in any other. But there are still special dishes that certain localities claim as their own. To mention a few: baked beans and brown bread belong to Boston; lobster stew to Maine; fried chicken, hominy grits and pecan pie are at home south of the Mason-Dixon line; roast pork tastes best in Iowa; and New Orleans is where one finds the best Creole cooking. Furthermore, in most large cities in the United States one can find restaurants representing almost every country in the world. In New York City or Los Angeles, for example, one can order a dinner in any language—Rumanian mushk steak, Japanese sukiyaki, Italian antipasto, French hors d'oeuvres, or Swedish smorgasbord. The menus in the restaurants that serve foreign foods can be the magic carpet that whisks one to faraway places.

In the United States, food technology, which concerns itself in part with the processing and packaging of food, takes much of the hard work of food preparation out of the home or restaurant kitchen. This, together with country-wide distribution, tends to standardize the American diet. Presently, the whole country is highly nutrition-conscious. Great emphasis is placed on the importance of eating foods with high vitamin and mineral content. Food is advertised and sold as much for its food value as for its enjoyment. Food fads extolling the health-giving properties of one food or another come and go. Peculiar to America are the eating-places dedicated to eating as quickly as possible for as little money as possible. These include lunch counters where quick-order meals are served, cafeterias where customers serve themselves from a wide variety of foods, the ubiquitous hot-dog and hamburger stands, and drive-ins where meals are served to patrons in their cars. This fast-food industry is now catching on worldwide. So, as Old World diets influenced the New World, the New World now influences the Old World.

DIFFUSE

Not localized.

DIFFUSION

The process by which particles (as molecules and ions) in soulution spread throughout the solution and across separating membranes from the places of highest concentration to lowest concentration.

DIGESTIBLE NUTRIENT

The part of each food nutrient that is digested or absorbed by man.

Digestible nutrients are computed by multiplying the percentage of each nutrient in the food (protein, fiber, nitrogen-free extract, and fat) by its digestion coefficient. The result is expressed as digestible protein, digestible fiber, digestible NFE, and digestible fat. For example, if corn contains 8.9 percent protein of which 77 percent is digestible, the percent of digestible protein is 6.9.

DIGESTIBILITY

The proportion of a nutrient absorbed from the digestive tract into the bloodstream. It is determined by the difference between the nutrients consumed and the nutrients excreted, expressed as a percentage of the nutrients consumed. The digestibility for most foods is 90-95%.

(Also see DIGESTION AND ABSORPTION.)

DIGESTIBILITY, APPARENT

An approximate measure of digestibility, determined by the difference between intake and fecal output without considering the part of the feces not derived from undigested food. For example, the feces also contains shed lining cells of the digestive tract, bacteria, and digestive juice residues. These items are considered when the true digestibility is determined.

(Also see DIGESTION AND ABSORPTION.)

DIGESTION AND ABSORPTION

Contents *Page*

Hunger and Appetite ... 584
Eating Behavior ... 584
Anatomy of the Digestive System ... 585
- **Mouth** ... 585
- **Esophagus** ... 585
- **Stomach** ... 585
- **Small Intestine** ... 586
- **Large Intestine** ... 586
- **Rectum** ... 586
- **Salivary Glands** ... 586
- **Pancreas** ... 586
- **Liver** ... 586

Physiology of Digestion ... 587
- **Mouth** ... 587
- **Esophagus** ... 588
- **Stomach** ... 588
- **Small Intestine** ... 588
  - **Pancreas** ... 589
  - **Liver** ... 589
- **Large Intestine** ... 590

Process of Absorption ... 590
- **Mechanism of Absorption** ... 590
- **Nutrient Carriers** ... 591
  - **Lymph** ... 591
  - **Blood** ... 592

Chemical Digestion and Absorption of Nutrients ... 592
- **Carbohydrates** ... 592
- **Lipids** ... 594
- **Proteins** ... 594
- **Minerals and Vitamins** ... 595
- **Water** ... 595

Control of the Digestive Tract ... 595
- **Neurological Control** ... 595
- **Hormonal Control** ... 596

Food Factors Affecting Digestion and Absorption ... 596
Dysfunctions of the Digestive Tract ... 597
- **Vomiting** ... 597
- **Diarrhea** ... 597
- **Gastritis** ... 597
- **Malabsorption** ... 597
- **Ulcers** ... 597

Car engines will not run on crude oil. However, chemical changes and refining produce a product which cars can burn. Likewise, the protein, fat, and starch in food is of little use to the body until foods undergo some rather drastic changes so that they may be absorbed, assimilated, and burned by the body.

Digestion is the process by which food is broken down into smaller particles, or molecules, for use in the human body.

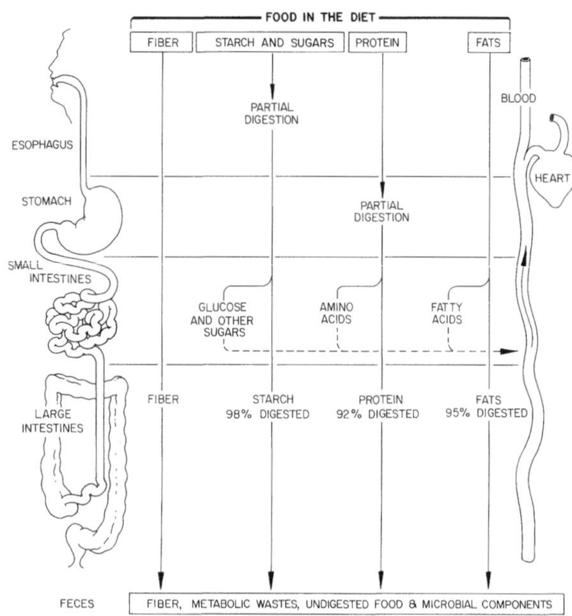

Fig. D-18. An overview of digestion and absorption from food to nourishment of the body.

Absorption is the process of moving the nutrients from the digestive system into the circulation where they can be distributed to the cells of the body.

Humans eat because of the sensation of hunger, and they eat certain foods, partially due to learned responses and partially due to the anatomy and physiology of their digestive system. Moreover, the digestive tract has certain controls which affect digestion and absorption. Also, some foods influence digestion and absorption. Like any system, however, there are functional problems which can occur.

HUNGER AND APPETITE. *Hunger is the physiological desire for food following a period of fasting. Appetite, on the other hand, is a learned or habitual response to the presence of food.* An individual that is extremely hungry may not have an appetite for a type of food that it deems undesirable. Conversely, if the food is of a desirable nature, an individual may have an appetite for it in spite of the fact that he is not hungry. If a food is extremely high in nutrients but is refused by an individual because he does not have an appetite, its value is nil. On the other hand, many individuals eat only the foods they like. This may lead to a poor selection of foods, particularly in children.

The *hypothalamus* (derived from the terms *hypo* meaning below, and *thalamus*, meaning a region of the brain)—a structure in the ventral region of the diencephalon—has been implicated as one of the major control centers of appetite regulation. Within the hypothalamus, certain areas can be differentiated. Two of these are of particular importance in the regulation of appetite. The first area is that of the lateral hypothalamus. It is commonly called the *feeding center*, because, upon stimulation of this region, the person commences to eat whether or not he is hungry. If this area is damaged, the person loses all desire to eat and will eventually starve. The ventro-medial area of the hypothalamus functions as the ***satiety center***. Stimulation of this region will depress appetite. If the ventro-medial nuclei are destroyed, there is no inhibition of food intake, and the person will have an uncontrollable appetite. These concepts have been verified on experimental animals. It is believed that there is a chronic activity in the lateral hypothalamus which is kept in check by the inhibitory influence of the ventro-medial area.

Several theories have been advanced as to the exact physiological mechanism which triggers the hypothalamus to tell the animal when to eat. While each theory has its merits, there is no conclusive proof in favor of any one of them. In the long run, it seems probable that a combination of a number of factors will provide the answer.

The two theories concerning the hypothalamic control of appetite that have received the most attention are: (1) the chemostatic hypothesis, and (2) the thermostatic hypothesis. The chemostatic hypothesis reasons that the hypothalamus is sensitive to circulating blood nutrient levels, such as sugar or lipid. When these levels become too low, the hypothalamus sends signals to begin eating. Once the blood nutrient level is elevated, stimuli from the feeding center are inhibited and the person feels full. The second theory of appetite control, the thermostatic hypothesis, theorizes that the hypothalamus plays an important role in heat regulation within the body, and that a decrease in hypothalamic temperature will induce eating.

Appetite is also influenced by (1) distention of the digestive tract; (2) various sociological factors such as habits, customs, education, cultural background, and environment; (3) psychological factors such as stress and anxiety; (4) physiological factors such as illness, pregnancy, and age; and (5) drugs, alchohol, and tobacco.

(Also see APPETITE.)

EATING BEHAVIOR. Throughout the animal kingdom, the feeding behavior of animals is correlated with the various anatomical adaptions of the gastrointestinal tract. It is logical to assume that animals which consume food that can be easily digested and absorbed would have gastrointestinal tracts that are smaller and simpler in structure than those animals which utilize foods that are complex in chemical composition.

Based on the kind of food eaten, animals are classified as follows:

1. **Carnivores.** These are the flesh eaters. They feed mostly on the flesh of other animals. The organization of the digestive tract is relatively simple. When a wild carnivore kills an animal, it will first consume the contents of the gastrointestinal tract which contains digested and partially digested plant material. However, the great bulk of the diet consists of meat, and it is this type of food that the gastrointestinal tract is best equipped to handle. The food that is consumed is, for the most part, composed of fat and protein—both of which are readily digested and absorbed. For this reason, the length of the digestive tract is short. Examples of this type of animal are bears, cats, dogs, and mink.
2. **Herbivores.** These are vegetarians. They depend

entirely upon plants for their food supply. Since plant material is difficult to digest, anatomical adaptations enabling the animal to digest the material more efficiently have evolved. The gastrointestinal tract in these animals tends to be long, and changes in the structure of the stomach and/or cecal-colon areas enable these animals to digest and utilize plant material efficiently. Examples of herbivores are cattle, goats, sheep, and horses.

3. **Omnivores.** These consume both flesh and plants. Anatomically, these animals represent an intermediate stage in the types of the gastrointestinal tract. The digestive tracts of omnivores are generally longer than those of carnivores but shorter than those of herbivores. Plant material can be utilized by omnivores, but the efficiency of utilization does not approach that of herbivorous animals. Humans and swine are examples of omnivores.

ANATOMY OF THE DIGESTIVE SYSTEM.

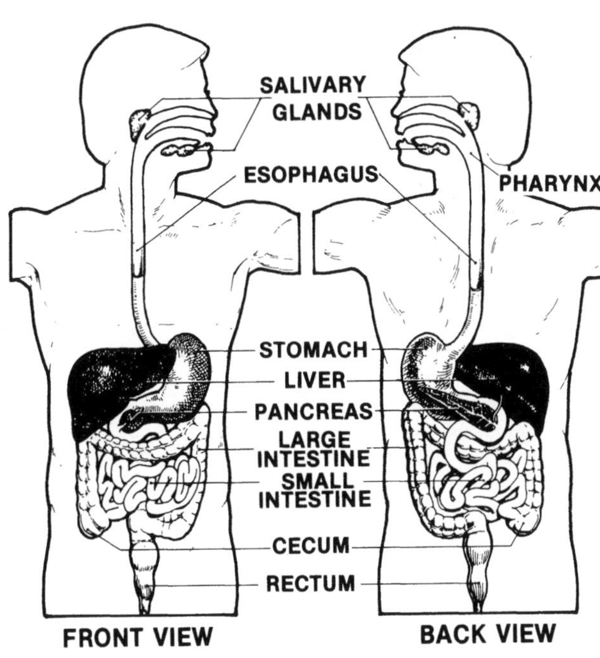

Fig. D-19. Anatomy of the human digestive tract.

Basically, the digestive system consists of a tube which courses internally from the lips to the anus. In the adult human, it is a 25 to 30 ft (*7.5 to 9 m*) long tube, through which food passes following consumption. At prescribed intervals, it becomes specialized in regions called the mouth, esophagus, stomach, small intestine, large intestine, rectum, and anus. Protruding along the way are the salivary glands, gallbladder, liver, and the pancreas, which provide essential secretory products for digestion. Based on the anatomy of the digestive system, man belongs to the group of animals called nonruminants. Furthermore, man's digestive system does not possess a functional cecum; hence, he has a limited capacity and limited microbial action and fiber digestion. Table D-8 compares the capacity of man's digestive tract to that of other species.

TABLE D-8
AVERAGE CAPACITIES OF DIGESTIVE TRACTS OF SELECTED ANIMALS

Animal Species	Volume	
	(gal)	*(liter)*
Man	1.6	6.0
Pig	7.2	27.5
Sheep	11.7	44.2
Horse	56.0	211.3
Cattle	93.9	356.0

Mouth. Two very moveable folds known as the lips surround the opening to the mouth, while the side walls of the mouth are called the cheeks. Together, the lips and cheeks play an important role in moving and holding food. Inside the mouth the teeth are embedded in the gums. Powered by the muscles of the jaws, each type of tooth is adapted to its function—incisors for cutting, canines for grasping, and molars for grinding. The floor of the mouth contains the tongue—a mass of muscle fibers winding in three different directions. Thus, the tongue is a very moveable structure. On the tongue are the numerous taste buds which sense sour, sweet, salt, and bitter. Emptying into the mouth are the ducts from three pairs of salivary glands. The roof of the mouth is called the palate which is hard in the front portion of the mouth and soft in the back. The soft portion ends in a flap of tissue known as the uvula. This is certain to have been seen by anyone looking at their throat in a mirror. At the back of the mouth just beyond the uvula, lies the pharynx—a common passageway for both air (nasal) and food.

Esophagus. This is a long, straight muscular tube which conducts food from the pharynx to the stomach—a distance of about 10 in. (*25 cm*). Passage of food down the esophagus is facilitated by the force of gravity and muscular contractions of the esophagus.

About 2 in. (*5 cm*) above the entry point of the esophagus into the stomach, the muscular wall of the esophagus is thickened and stronger. This area of the esophagus, which is called the gastroesophageal constrictor, functions to prevent the contents of the stomach from reflexing into the esophagus.

Stomach. This is the most dilated portion of the digestive tract. Often the shape of the stomach is described as a J-shaped organ. Actually, it varies in shape depending upon its contents and the condition of the neighboring intestines. The stomach is a very muscular organ. Its wall is composed of three layers of muscles, while the inside surface is thrown into folds called rugae.

There are three parts of the stomach: (1) the fundus, an upper portion ballooning toward the left, (2) the body, the central portion, and (3) the pylorus, a constricted portion just before entry into the small intestine. Gastric glands are found in all three parts of the stomach. Muscles in the pyloric end of the stomach form a valve called the pyloric sphincter which controls the entry of food into the small intestine. Fig. D-20 illustrates the anatomy of the stomach.

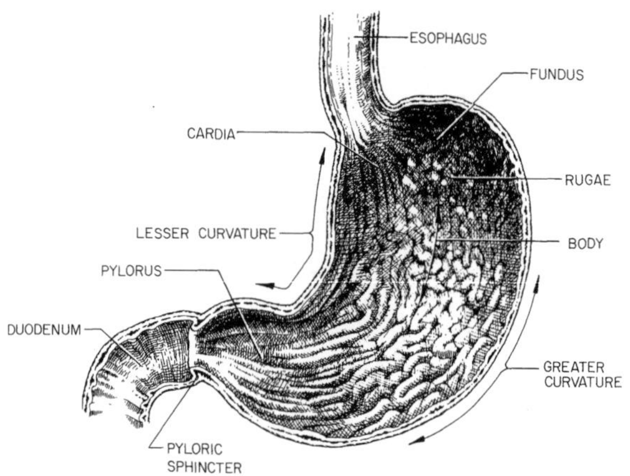

Fig. D-20. Anatomy of the stomach.

Small Intestine. This tube leaving the pyloric end of the stomach is about 18 ft (*5.5 m*) long and it is divided into three sections: (1) the duodenum, (2) jejunum, and (3) the ileum. The whole length of the small intestine is surrounded by two muscular layers which run in opposite directions. One layer runs longitudinally along the small intestine while the other runs circular around the small intestine. Contractions of these muscular layers propel food. Also, along the course of the small intestine are numerous secretory glands and cells which dump their secretions into the lumen—the cavity of the small intestine, and the ducts of the gallbladder and pancreas empty into the small intestine. The inside surface of the small intestine is thrown into folds and studded with millions of hairlike projections called villi. Thus, the surface area of the small intestine is greatly increased.

At the junction of the small and large intestine is a short, blind, pouch called the cecum. In some animals the cecum is large and plays an important role in digestion. A small fingerlike tubular projection extends from the lower part of the cecum. This small organ has contributed its share to human discomfort. It is the appendix.

Large Intestine. It is about 5 ft (*1.5 m*) long. Actually, the large intestine is divided into (1) the cecum, (2) the colon, and (3) the rectum. However, the terms large intestine and colon are used interchangeably. A valve—the ileocecal valve—controls the flow of the intestinal contents between the small and large intestine. The large intestine is also surrounded by circular and longitudinal layers of muscle which move digested food products toward the rectum. There are no villi on the lining of the large intestine. From the junction of the large and small intestine, the large intestine travels upward toward the liver (ascending colon), thence crosses the abdominal cavity from right to left (transverse colon), and thence drops down the left side (descending colon), angling off toward the rectum and anus.

Rectum. The rectum is the last 6 in. (*14 cm*) of the large intestine (colon) before ending at the anal opening. It stores the products of digestion prior to defecation. Crescent-shaped valves inside the rectum support the weight of fecal material until distension initiates the desire to defecate. The rectum is surrounded and supported by a group of veins which may dilate and produce hemorrhoids (piles).

Salivary Glands. Three pairs of salivary glands secrete a clear, somewhat viscid solution into the mouth. According to their location they are named: (1) the parotid, (2) the sublingual, and (3) the submaxillary. The functions of saliva are manyfold, including the following:

1. **Lubricant.** These secretions act as aids in mastication and swallowing. Without this moisture, swallowing would be extremely difficult. This lubricating ability is attributed to the presence of a glycoprotein.
2. **Enzymatic activity.** The enzyme alpha-amylase (ptyalin) is found in the saliva. It acts to break down starch.
3. **Buffering capacity.** A large quantity of bicarbonate is secreted in saliva, thus serving as a buffer in the ingesta, and in the mouth.
4. **Taste.** Saliva solubilizes a number of the chemicals in the food which, once in solution, can be detected by the taste buds.
5. **Protection.** The membranes within the mouth must be kept moist in order to remain viable. Saliva provides one means by which this is accomplished. In addition, the saliva may also possess some antibacterial action, which protects the teeth against dental caries.

Pancreas. This organ is a large lobulated gland which lies in close association with the duodenum of the small intestine. Moreover, it is a dual purpose gland. It secretes substances important to the body through ducts and without ducts (ductless). In other words, the pancreas has both exocrine (duct) and endocrine (ductless) functions. The exocrine function of the pancreas is directly related to digestion. This secretion, which is dumped into the duodenum, contains sodium bicarbonate and a variety of digestive enzymes. The endocrine function of the pancreas is the release of the two hormones, insulin and glucagon. Interestingly, the secretion of the pancreas is controlled by other hormones.

Liver. The liver is the largest organ in the body. It is divided into a large right lobe and smaller left lobe. The whole organ weighs about 3.5 lb (*1.6 kg*). Microscopically, the structure of the liver is a series of blood filled sinusoids and little canals (canaliculi) which transport bile. Nutrient-laden blood from the digestive tract enters these sinusoids. Although it is considered as a structural and functional part of the digestive system, the liver functions in many activities not directly concerned with digestion. Among these are blood formation, coagulation, phagocytosis, and detoxification. A major digestive function of the liver includes the production of bile. Bile is a yellowish-green fluid which is stored in the gallbladder—a saclike structure on the under side of the liver. Bile contains waste products being excreted from the body and bile salts. Bile salts are produced by the liver from the cholesterol, and they act as powerful detergents in the intestine aiding the digestion and absorption of fats. Bile empties into the duodenum of the small intestine via a duct.

Other influences of the liver on digestion and absorption include: Storage and release of glucose; storage of vitamins A, D, E, K, and B-12, and other chemical transformations of the products of digestion following their absorption.

PHYSIOLOGY OF DIGESTION. On a physiologic basis, the principles of digestion are quite similar among the animal species. Man is no exception. In discussing the digestive process in the sections that follow, events occurring in the various structures are considered in the order that food passes through them—mouth, esophagus, stomach, small intestine, and large intestine.

Mouth. Three physical processes occur in the mouth region: (1) prehension, (2) mastication, and (3) the initiation of deglutition.

Prehension can be defined as the act of bringing food into the mouth. Numerous modes of prehension can be found. Animals such as the raccoon and man, use their forelimbs to bring food to the mouth, while other types of animals rely on the structures of the mouth, such as the tongue, lips, and teeth.

Mastication is the act of chewing food. Most species chew their food immediately following prehension. One notable exception is the fowl which swallows its food whole. Mastication involves the physical grinding and tearing of the food in addition to the admixture of saliva which lubricates the food as well as initiates a limited amount of enzymatic digestion. Food that has been masticated and formed into a small compact ball for passage down the digestive tract is called a bolus.

Deglutition is the act of swallowing. This process involves both voluntary and involuntary reflexes. Upon completion of mastication, the bolus is lifted by the tongue and moved to the back of the mouth. The bolus passes through the pharynx. The pharynx is the structure which controls the passage of air and food. In this organ, the openings of the mouth, esophagus, posterior nares, Eustachian tubes, and larynx come together. During the act of swallowing, the opening into the larynx is reflexly closed by the arytenoid cartilages; and the epiglottis is passively folded over the opening of the larynx. This forces food into the esophagus, thus preventing it from passing into the respiratory tract.

In the mouth, the teeth, tongue, and salivary glands aid these physical processes.

• **Teeth**—The teeth serve primarily as a mechanical aid for mastication. By tearing and grinding the food, they provide a means whereby a large surface area is created which can be exposed effectively to the digestive fluids of the tract. There are four types of teeth, each serving a specialized function: incisors, canines, premolars, and molars. The teeth on the front of the jaw are called incisors and are used for the tearing and slicing of food. Moving progressively to the back of the jaw, the next teeth are called canines. These teeth—sometimes called the eye teeth—are also used for tearing. Following the canines are the premolars and molars, both types of which are used for grinding. Generally grinding can occur on only one side of the jaw at a time.

• **Tongue**—In many species of domestic animals, the tongue is the primary structure for prehension. For example, the tongue of the cow is elongated and covered with rough papillae, making it adapted to wrapping around grass and other forages, and then bringing the forage into the mouth.

Throughout the process of mastication, the tongue serves a threefold purpose. First, movement of the tongue transports the food to the various areas of the mouth to be torn and ground. While doing this, the tongue is also mixing the food with the various secretions of the mouth, ultimately forming a bolus. Secondly, the presence of taste buds on the tongue provides a neurological control for food selection and intake. If the food is bitter or unpalatable, as determined by the taste buds, the food may be rejected. Finally, the tongue initiates the process of deglutition. When the bolus has been adequately prepared, the tongue moves it to the back of the mouth where nerves are stimulated, and swallowing commences.

• **Salivary glands**—The salivary glands represent a network of accessory structures which are essential to digestion. Three pairs of salivary glands are of primary importance—parotid, submaxillary, and sublingual. Fig. D-21 illustrates the location of these glands.

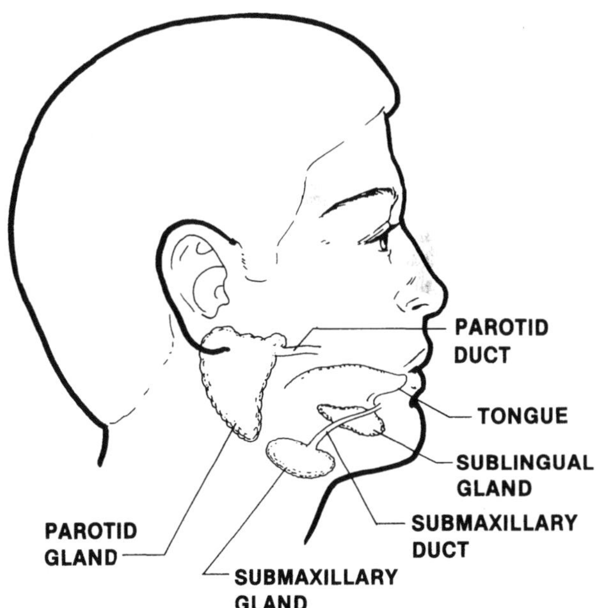

Fig. D-21. Location of the salivary glands.

Saliva, the secretion from these glands, is highly variable in chemical composition. Two basic types of saliva are produced. The first type is extremely thick, being rich in the glycoprotein mucin. In this type of saliva, there is very little enzymatic activity. The second type of saliva is serous in composition; that is, it is watery and thin, containing various proteins and enzymes but little mucin. Saliva from the parotid glands tends to be serous in composition, while the submaxillary glands normally secrete a mixture of the two. The sublingual glands secrete saliva that is primarily mucous in composition.

The uses of saliva in digestion are manyfold. Saliva acts as a lubricant and aids in mastication, the formation of the bolus, and swallowing. Without this moisture, swallowing would be extremely difficult. The enzyme amylase or ptyalin in the serous saliva begins the digestion of starch. A large quantity of bicarbonate in the saliva serves as a buffer. The saliva solubilizes the chemicals in food so that bitter, sweet, sour, and salt taste can be detected by the taste buds on the tongue. Finally, the saliva keeps the membranes within the mouth moist, thus keeping them viable.

Esophagus. Upon swallowing, gravity and unidirectional peristaltic waves move the bolus through the esophagus and down to the stomach. Travel time is about 7 seconds. Peristaltic waves are formed in the esophagus and throughout the digestive tract by the coordinated contraction and relaxation of two layers of smooth muscle which help form the tubes of the digestive tract. These layers run in opposite directions. One layer is longitudinal, the other is circular. In humans, peristalsis normally moves from the mouth to the stomach. Belching and vomiting, reverse peristalsis, are usually dysfunctions of normal digestion. It is noteworthy that reverse peristalsis is a normal process in cattle, sheep, and goats—the ruminants. It enables them to chew their "cud" and rid themselves of a tremendous amount of gas generated in their digestive tract.

Stomach. Secretions of the stomach amount to about 2 liters each day. Two specialized types of cells in the stomach provide the gastric secretions needed for the initial stages of digestion. The parietal cells, located in the fundic region, secrete hydrochloric acid. Hydrochloric acid hydrolyzes a limited amount of protein, but its main function is to establish an acid environment conducive to the activity of certain hormones and enzymes. The second cell type is called the chief cell or peptic cell. These cells secrete the enzyme pepsinogen. When pepsinogen is secreted in an acid environment (pH of 1.6 to 3.2), the proenzyme is activated forming pepsin—an enzyme that hydrolyzes certain peptide bonds. In addition, rennin is secreted by the stomach of young mammals, including babies; it coagulates milk and is important to their nutrition. Apparently, rennin is absent from the gastric secretions of adults.

Mucus is secreted by cells in the stomach lining. This secretion provides protection for the lining of the stomach. If there is a malfunction in the secretion of mucus, the stomach can digest itself—resulting in an ulcer.

Two types of motility have been observed in the stomach. The first type is that of peristalsis whereby food is moved toward the duodenum of the small intestine. Tonic contractions, the second type of motility, churn and knead the ingesta to ensure thorough mixing, but do not propel the food from one end of the stomach to the other.

Food ingested by carnivores is passed through the stomach at an extremely fast rate (on the order of a couple of hours), while ingesta of herbivores and omnivores tend to remain in the stomach for relatively long periods, sometimes in excess of 24 hours. The rate of passage depends largely on the nutrient composition of the diet. Carbohydrates pass through the stomach faster than either proteins or fats, with proteins being intermediate in rate of passage. Water can pass directly through to the small intestine, spending very little time in the stomach.

Intrinsic factor, a protein necessary for proper absorption of vitamin B-12, is produced in the parietal cells of the stomach. If there is a malfunction in the production of this protein, a condition called pernicious anemia results. Quite often, people suffering from this condition eat amounts of dietary vitamin B-12 that are normally considered adequate, but the vitamin has no means of being absorbed—thus, pernicious anemia develops.

Small Intestine. The small intestine is the primary site of both digestion and absorption. In normal individuals, 90 to 95% of all absorption occurs in the first half of the small intestine. The structure, both visible and microscopic, contributes to the absorptive capacity of the small intestine, and to the mechanical processes of digestive physiology.

• **Mechanical digestion**—Throughout the luminal surface—the inside surface of the small intestine lies an extensive network of fingerlike projections called villi. Moreover, the surface is thrown into folds called mucosal folds. In the human, there are about 20 to 40 of these projections per square millimeter of intestine, each one being from 0.5 to 1.0 mm long. Each villus contains a lymph vessel called a lacteal and a series of capillary vessels. On the surface of the villi are cells which possess a great number of microvilli and provide further surface area for absorption. These microvilli form the brush border, which contains high concentrations of digestive enzymes.

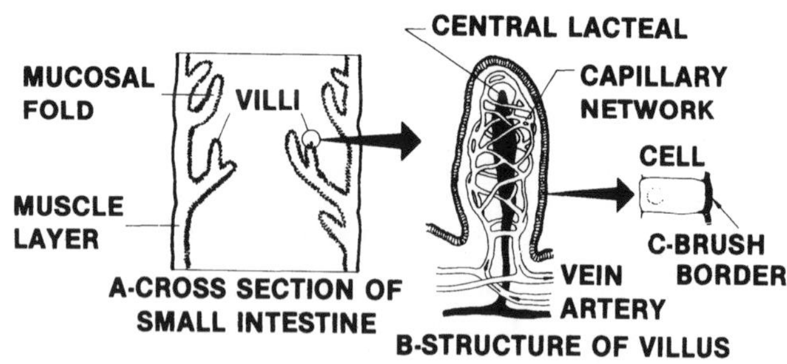

Fig. D-22. Three views of the small intestine structure. (A) Mucosal folds studded with numerous villi. (B) A villus containing capillaries and a central lacteal which transport absorbed nutrients. (C) The brush border on a cell like those covering each villus.

Constant motion of the small intestine (1) mixes food with secretions, and (2) exposes new villi surfaces for absorption.

Three types of motility can be observed in the small intestine. The first type is called pendular motion. These waves do not advance down the intestine. Rather, they are merely a localized shortening and lengthening of the intestine which produces a mixing action. Segmentation contractions are the second type of intestinal motility. These intestinal movements are ringlike contractions at regular intervals which periodically relax, whereupon the area that has been previously relaxed contracts. This type of motility provides a means of mixing in addition to the pendular contractions. Peristalsis, a form of motility that has been previously discussed, is the third type of intestinal motility, providing a means for movement of chyme (intestinal contents) down the tract—much like squeezing the tube of toothpaste in the center forces toothpaste out of the open end.

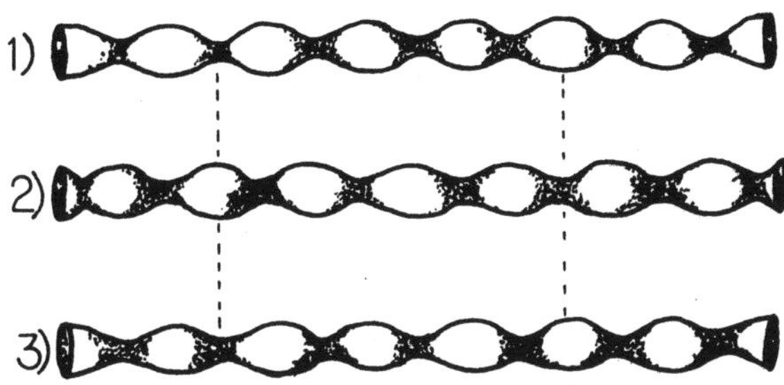

Fig. D-23. Segmentation waves provide a mixing and churning action by alternately relaxing and contracting in localized areas.

PANCREAS. The pancreas secretes a digestive fluid directly into the small intestine via the pancreatic duct. This digestive fluid produced by the pancreas is clear and alkaline and consists of two phases—an aqueous phase and an organic phase. Being rich in bicarbonate, the aqueous phase serves primarily to neutralize the highly acid chyme produced in the stomach and passed on to the small intestine. In the organic phase, enzymes produced in the acinar cells of the pancreas are transported to the duodenum. These enzymes are stored as granules in the pancreas and are secreted from the cells through the process of emelocytosis (cell vomiting). This process is sometimes called reverse pinocytosis because the granule fuses with the cell membrane, followed by a breakdown of the membrane and the evacuation of the granule. A listing of the composition and function of pancreatic fluids is given in Table D-9.

Many of the pancreatic enzymes are stored and secreted in an inactive form to be activated at the site of digestion. Trypsinogen is a proteolytic enzyme that is activated in the small intestine by the enterokinase, an enzyme secreted from the intestinal mucosa. When activated, trypsinogen becomes trypsin. Trypsin, in turn, can then activate chymotrypsinogen to chymotrypsin.

The nucleases, lipases, and pancreatic amylase are secreted in their active form. Many of the enzymes require a specific environment before they will function. For example, amylase requires a pH of about 6.9 and the presence of inorganic ions before it will digest complex carbohydrates.

LIVER. The hepatic region includes the liver, gallbladder, and bile duct.

In addition to the panreas and salivary glands, the liver is an indispensable accessory organ of the gastrointestinal tract. From the stomach and small intestine, most of the absorbed nutrients travel through the portal vein to the liver—the largest gland in the body. The liver not only plays an important part in nutrient metabolism and storage, but also forms bile, a fluid essential for lipid absorption in the small intestine. The numerous physiological functions of the liver follow:

1. Secretion of bile.
2. Detoxification of harmful compounds.
3. Metabolism of proteins, carbohydrates, and lipids.
4. Storage of vitamins.
5. Storage of carbohydrates.
6. Destruction of red blood cells.
7. Formation of plasma proteins.
8. Inactivation of polypeptide hormones.
9. Urea formation.

The primary role of the liver in digestion and absorption is the production of bile. Bile facilitates the solubilization and absorption of dietary fats and also aids in the excretion of certain waste products such as cholesterol and by-products of hemoglobin degradation. The greenish color of bile is due to the end products of

TABLE D-9
COMPOSITION AND FUNCTION OF PANCREATIC SECRETIONS

Composition	Function
Proteolytic enzymes:	Splits proteins into peptides and amino acids.
Trypsin Chymotrypsin A Chymotrypsin B Carboxypeptidase A Carboxypeptidase B	
Lipolytic enzymes:	Breakdown of lipids.
Phosphorolipase A Pancreatic lipase Cholesterol esterase	Esterification of cholesterol to fatty acids.
Nucleolytic enzymes:	Breakdown of nucleic acids.
Ribonuclease Deoxyribonuclease	
Amylolytic enzymes:	Breakdown of starch.
Pancreatic amylase	
Cations:	Buffers; cofactors; osmotic regulators.
Sodium (Na^+) Potassium (K^+) Calcium (Ca^{++}) Magnesium (Mg^{++})	
Anions:	Buffers; osmotic regulators.
Bicarbonate (HCO_3^-) Choride ($C1^-$) Sulfate ($SO_4^=$) Phosphate ($HPO_4^=$)	
Proteins:	Buffers
Albumin Globulin	

red blood cell destruction—biliverdin and bilirubin. Bile contains a number of salts resulting from the combination of sodium and potassium with bile acids. There are four types of bile acids: cholic acid, deoxycholic acid, chenodeoxycholic acid, and lithocholic acid. These salts combine with lipids in the small intestine to form micelles. *Micelles are colloidal complexes of monoglycerides and insoluble fatty acids that have been emulsified and solubilized for absorption.* When the micelle has been formed, the lipid can be digested and the resulting products (fatty acids and glycerol) can cross the mucosal barrier of the small intestine and enter the lymphatic system. Bile salts, however, do not travel with the lipid; rather, they are reabsorbed in the intestine and then excreted again by the liver. This recycling process is termed enterohepatic circulation.

The volume of bile production is highly variable. An animal that has been starved produces little bile. Conversely, an animal that is fed a high-fat diet will produce substantial quantities in order to keep up with absorptive requirements. Generally, the volume of bile is dependent on (1) blood flow, (2) nutritive state of the individual, (3) type of diet, and (4) the enterohepatic bile salt circulation.

In many animals—including humans—the gallbladder is the storage site for bile. Several species of livestock and animals, however, do not have gallbladders; among them, horses, rats, gophers, deer, elk, moose, giraffes, camels, elephants, pigeons, and doves.

Large Intestine. The large intestine is composed of several layers of muscle. There is a circular layer of muscle that forms the basic tube of the colon and facilitates movement. In addition to this layer of muscle, there are three strips of longitudinal muscle which form the taenia coli. These strips form a series of pouches or sacculations throughout the colon which are called haustrae. Ingesta are held in these saclike structures to facilitate the removal of water. Subsequently, the feces generally take on the shape of the haustrae. Numerous mucous-secreting goblet cells can be found in the colon, but villi, such as the type that are found in the small intestine, are absent. The large intestine performs no digestive functions. Movement through the large intestions is much slower than in all other parts of the digestive system. Movement from the stomach to the end of the small intestine may require 30 to 90 minutes while passage through the large intestine may require 1 to 7 days. Three types of motility can be observed in the colon: (1) haustral contractions, (2) massive peristalsis, and (3) defecation.

1. Haustral contractions. This type of motility creates a mixing action of the ingesta; hence, the absorption of water from the material is facilitated. These contractions are localized in the various portions of the colon with no coordinated wave movement traveling along the organ.

2. Massive peristalsis. These waves are slow, strong movements that propel the digesta down the colon.

3. Defecation. When massive peristalsis moves fecal material into the rectum a reflex reaction is created called the defecation reflex. If an individual relaxes the muscles of the anal sphincter then defecation occurs. However, if an individual tightens the anal sphincter then the reflex usually dies out after a few minutes.

The large intestine contains a dense population of bacteria of predominantly *Escherichia coli.* These bacteria affect the color and odor of the stool. Those food components not acted upon by digestive processes may be altered or digested by the bacteria. Thus, some complex polysaccharides or a few simple carbohydrates such as stachyose (four sugar) or raffinose (three sugar) will be converted to hydrogen, carbon dioxide and short-chain fatty acids. Nondigestible protein residues are converted to odorous compounds by the bacteria. Additionally, the bacteria synthesize vitamin K, biotin, and folacin, though it is not known for sure how much or how important these vitamin sources are to the individual.

PROCESS OF ABSORPTION. Once the various nutrients have been adequately digested, several modes of absorption can occur. These modes are dependent on the chemical nature of the nutrient and the site of absorption. Virtually no absorption takes place before the food enters the stomach; and very little absorption occurs in the stomach. The primary site for the absorption of most nutrients is in the small intestine. For the most part, absorption in the large intestine is restricted to water and electrolytes.

Mechanisms of Absorption. Several mechanisms of nutrient absorption are found in animals. The particular mechanism used is dependent on the physical size of the particle, the chemical properties of the nutrient, and the site of absorption. The four basic mechanisms of absorption are: (1) diffusion, (2) osmosis, (3) active transport, and (4) pinocytosis.

1. Diffusion. *Diffusion is the mechanism whereby the molecules of a solvent expand to fill all of the available volume of a given area.* If two solutions are separated by a permeable membrane, and one solution has either a higher chemical or electrical concentration than the other, diffusion will occur until there is a uniform concentration throughout both solutions.

The rate of diffusion depends on the size, shape, charge, and polarity of the particle that is to be absorbed.

2. Osmosis. *Osmosis is a form of diffusion whereby there is a migration of water molecules (the solvent) through a membrane—a semipermeable membrane—that will permit the passage of water but not the passage of dissolved substances.* There is a tendency for solutions separated by such a membrane to become equal in concentration. Thus, water will flow from the weaker to a stronger solution to equalize concentrations. The driving force created by this situation is called osmotic pressure. The effect of the osmotic pressure of a solution relative to that of plasma is called tonicity. If the solution is of a higher concentration than that of plasma, it is referred to as a hypertonic solution. Conversely, if the solution is of a lower concentration relative to plasma, it is hypotonic.

Fig. D-24 illustrates diffusion and osmosis.

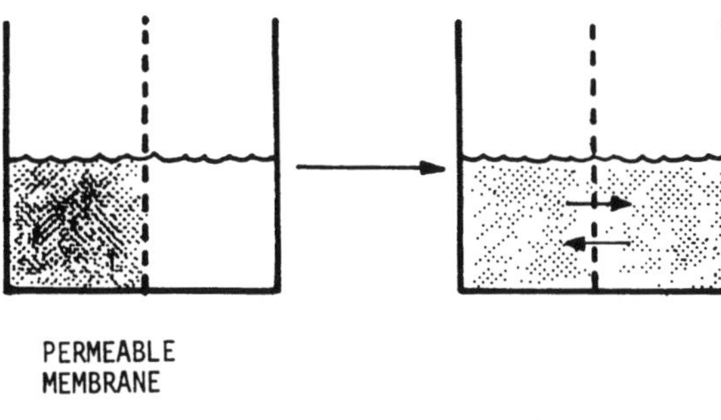

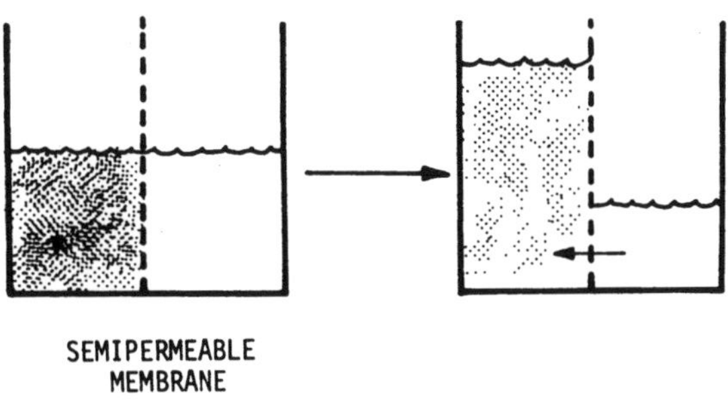

Fig. D-24. Simple diffusion occurs when a permeable membrane separates two solutions of differing concentrations; there is movement across both sides of the membrane until an equilibrium of concentration is reached. Osmosis occurs when two solutions of unequal concentrations are separated by a semipermeable membrane; movement of water, the solvent, is unilateral.

3. **Active transport.** In the process of diffusion, no energy is expended. In many cases, though, a nutrient must cross a membrane against a concentration gradient (chemical or electrical). This process—called active transport—requires the expenditure of a certain amount of energy. Diffusion is a balancing of the solution concentrations on both sides of a membrane, while active transport works against a concentration gradient.

The exact mechanism as to how active transport is achieved has not been elucidated, but there is strong evidence indicating the involvement of a carrier system. The nutrient to be absorbed is believed to combine with a specific membrane-bound carrier with the resulting complex moving the nutrient across the membrane. Once the nutrient passes through the membrane, the carrier and the nutrient dissociate. Fig. D-25 illustrates the basic mechanism.

4. **Pinocytosis.** *Pinocytosis (cell drinking) is the process by which dissolved materials are taken into the cell through an invagination and subsequent dissolving of a part of the cell membrane.* This process enables the cell to absorb certain lipids and proteins intact—a critical

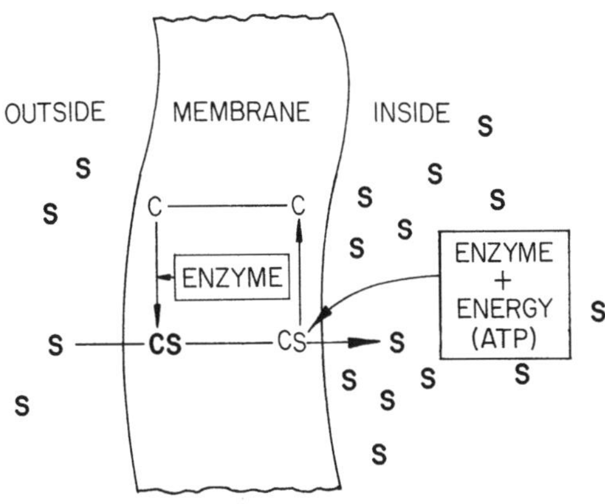

Fig. D-25. Active transport showing substance S entering the outside surface of the membrane where it combines with the carrier C with the help of an enzyme. At the inside surface of the membrane, S separates from C by using energy and another enzyme. C then moves back to the outside surface of the membrane to pick up more S.

factor in the newborn animal which must absorb antibodies from colostrum. The material to be absorbed comes into contact with the cell membrane. The membrane then invaginates to surround the material. Once the material has been completely surrounded, the membrane fuses, and the invaginated section of the membrane is dissolved by lysosomal enzymes releasing the intact substance in the cell. Absorptive cells in the small intestine are capable of using this mechanism.

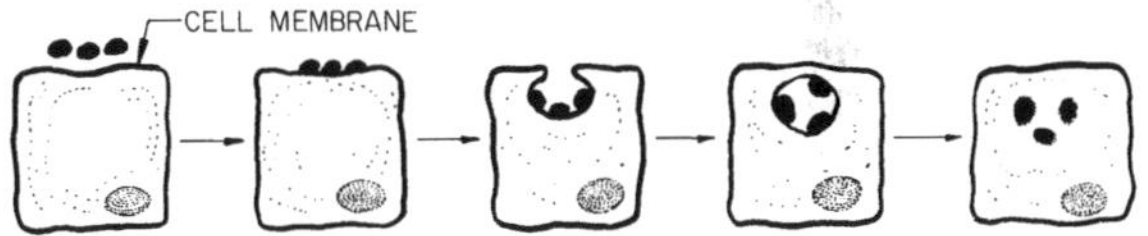

Fig. D-26. The mechanism of pinocytosis—cell drinking.

Nutrient Carriers. When the nutrients have been digested and absorbed, they must be transported to tissues that either have an immediate demand for them or can store them for later use. Lymph and blood are the primary transport media for the nutrients which have been absorbed.

LYMPH. Within the membranes of the intestinal tract there is a capillary network of lymph vessels. Cholesterol, water, long-chain fatty acids, and some proteins are picked up by this system and transported through a series of larger vessels which ultimately empty into the venous system just before the heart.

Since the immune system of the newborn baby is not well developed, it is essential that it receives colostrum from its mother as soon as possible. Through this intake of colostrum, antibodies are passed from the mother to the newborn, imparting a certain degree of immunity to stress and disease which will last for the first critical days of the life of the young. Many of these antibodies are absorbed intact in the newborn and transported via the lymphatic system.

BLOOD. Most of the low molecular weight (small) products of digestion are absorbed and transported by the blood. These nutrients include water, salts, glycerol, amino acids, short-chain fatty acids, monosaccharides, and certain vitamins. These materials are absorbed into the capillary system of the intestine. The capillary network drains into the venous system, eventually entering the portal vein of the liver. From the liver, the nutrients then travel through the hepatic veins which, in turn, enter the main systemic vein—the vena cava.

CHEMICAL DIGESTION AND ABSORPTION OF NUTRIENTS. *Chemical digestion is the process whereby proteins, fats, and complex carbohydrates are broken down into units that are of small enough size to be absorbed from the digestive tract into the circulation and distributed to the cells of the body.*

Early in the 19th century many theories of digestion were entertained. Then in the 1820s an accident occurred which significantly contributed to the understanding of the digestive processes. Results of this accident were published in 1833 by William Beaumont, a frontier surgeon, in an article titled "Experiments and Observations on the Gastric Juice and the Physiology of Digestion." The subject of this study was Alexis St. Martin, a Canadian voyageur who had been accidently wounded by the discharge of a musket. Beaumont described the shot which "entered posteriorly, and in an oblique direction, forward and inward, literally blowing off integuments and muscles the size of a man's hand, fracturing and carrying away the anterior half of the sixth rib, fracturing the fifth, lacerating the lower portion of the left lobe of the lungs, the diaphragm, and perforating the stomach." After surgical repair and healing, there remained an opening 2½ in. (*6.25 cm*) in circumference in both the wall of the stomach and the side of the patient. Beaumont attempted to close this wound, but failed. Subsequently, the natural protrusion of the layers of the stomach in a sort of fistula (an abnormal passage leading from the abdominal wall to the inside of the stomach) produced a permanent valve, which prevented the escape of stomach contents even when the stomach was full, but which could easily be depressed to permit the entrance of a tube or other instrument and the introduction of food substances. The interior of the stomach could be seen with the naked eye. Realizing the unique opportunity presented, Beaumont conducted a series of experiments of digestive physiology between 1825 and 1833, during which time St. Martin enjoyed normal, robust health, and food, drink, and lodging at Beaumont's expense.

Now it is realized that the process of digestion is accomplished primarily through the action of digestive enzymes, though mechanical events such as chewing and mixing are also important. Enzymes are organic catalysts produced by certain cells within the body. They speed biochemical reactions at ordinary body temperatures without being used up in the process. Enzymatic activity is responsible for most of the chemical changes occurring in foods as they move through the digestive tract. Many of the digestive enzymes are stored in an inactive form. In this state, they are called zymogens or proenzymes. Once secreted into a favorable environment for digestion, generally governed by activators such as pH or other enzymes, these inactive enzymes "turn on" and perform their specific digestive function. A summary of the enzymatic digestion of carbohydrates, fats, and proteins is presented in Table D-10.

From this point, the final digestion and absorption events are best discussed as they relate to the general classes of nutrients—carbohydrates, proteins, lipids (fats), and minerals and vitamins.

Carbohydrates. The digestion and absorption of most carbohydrates occurs in the small intestine, though it starts in the mouth. Such intestinal enzymes as sucrase, maltase, and lactase split carbohydrates—maltose, sucrose, lactose—into monosaccharides, whereupon absorption takes place. These enzymes are located on the surface of the cells lining the villi in the brush border. Of course, sugars ingested as monosaccharides, primarily glucose and fructose, do not require digestion. Sugar absorption takes place in the duodenum and jejunum of the small intestine. Glucose and galactose are absorbed through an active transport mechanism somehow tied to the active transport of sodium. Sodium ion concentration within the intestinal contents has been shown to be critical in this mechanism. A high sodium ion (Na^+) concentration will facilitate rapid absorption of these sugars while a low Na^+ concentration will reduce the rate of absorption. Some pentoses (5-carbon sugars) and other hexoses are absorbed through diffusion—a process considerably slower than that of active transport. Fig. D-27 summarizes the digestion and absorption of starch, lactose, and sucrose.

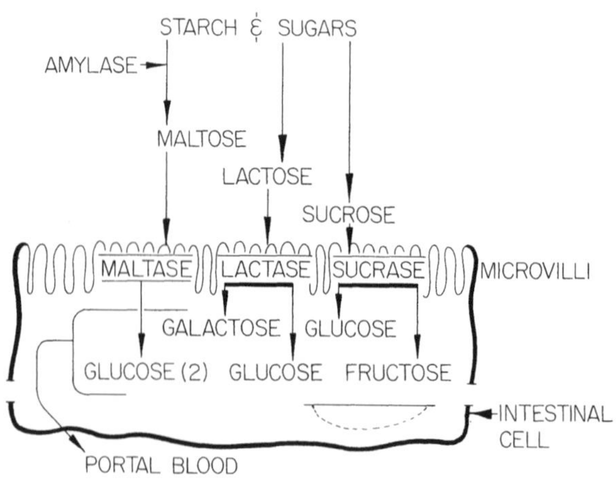

Fig. D-27. Summary of carbohydrates digestion and absorption.

TABLE D-10
ENZYMATIC DIGESTION OF CARBOHYDRATES, FATS, AND PROTEINS

Source (secretion)	Enzyme	Activator	Substrate (substance acted upon)	Catalytic Function or Products
Salivary glands (saliva)	Salivary amylase (ptyalin)	—	Starch	Produces dextrins, maltotriose, and maltose.
Stomach (gastric juice)	Pepsins (pepsinogens)[1] Rennin[2]	HC1 (hydrochloric acid) —	Proteins and polypeptides Casein	Cleave peptide bonds adjacent to aromatic amino acids. Clots the milk protein casein.
Pancreas (pancreatic juice)	Trypsin (trypsinogen)[1]	Enterokinase	Proteins and polypeptides	Cleaves peptide bonds adjacent to the amino acids arginine or lysine.
	Chymotrypsins (chymotrypsinogens)[1]	Trypsin	Proteins and polypeptides	Cleave peptide bonds adjacent to aromatic, large, hydrophobic amino acids.
	Elastase (proelastase)[1]	Trypsin	Elastin, some other proteins	Cleave peptide bonds adjacent to the amino acids alanine, glycine, or serine.
	Carboxypeptidases (procarboxypeptidase)[1]	Trypsin	Proteins and polytides	Cleaves terminal amino acids.
	Pancreatic lipase Cholesterol esterase	Emulsifying agents (bile) —	Triglycerides Cholesterol and fatty acids	Di- and monoglycerides and fatty acids. Joins cholesterol and fatty acids before absorption.
	Pancreatic amylase	$C1^-$ (chloride ion)	Starch	Same as salivary amylase.
	Ribonuclease	—	RNA	Nucleotides
	Deoxyribonuclease	—	DNA	Nucleotides
	Phospholipase A (prophospholipase A)[1]	Trypsin	Lecithin	Lysolecithin (one fatty acid removed).
Small intestine (intestinal juice)	Enterokinase	—	Trypsinogen	Trypsin
	Aminopeptidases	—	Polypeptides	Cleave terminal amino acid from peptide.
	Dipeptidases	—	Dipeptides	Two amino acids.
	Maltase	—	Maltose, maltotriose	Glucose
	Lactase	—	Lactose	Galactose and glucose.
	Sucrase	—	Sucrose	Fructose and glucose.
	Isomaltase	—	Limit dextrins	Glucose
	Nucleases and related enzymes.	—	Nucleic Acids	Pentoses and purine and pyrimidine bases.
	Intestinal lipase	—	Monoglycerides	Clycerol, fatty acids.

[1]The corresponding proenzyme.
[2]Present in infants and other young mammals.

Carbohydrates are 98% digested. However, this figure refers only to the digestion of starch, sucrose, and lactose, and not to those carbohydrate components of the diet called fiber. While fiber includes a number of polysaccharides, cellulose, the most abundant carbohydrate in nature, is a component of the diet. It is composed of glucose molecules, but man lacks the enzymes necessary to break cellulose down into glucose. Cellulose and other nondigestible polysaccharides contribute bulk to the diet and intestinal contents. Collectively these carbohydrates are termed unavailable carbohydrates, while those carbohydrates which are digested and absorbed are termed available carbohydrates.

The whole process of digestion is dependent upon enzymes which are specific for the reaction they catalyze. Interestingly, many individuals lack the enzyme lactase which is necessary for the conversion of lactose—milk sugar— to glucose and galactose. Hence, lactose is not digested or absorbed. This leads to diarrhea, bloating, and flatulence (gas) after ingesting lactose (milk). In most mammals and many races of humans, intestinal lactase is high at birth and declines to low levels during childhood, remaining low in adulthood. Lactose intolerance is uncommon among individuals with a European background. However, it occurs among 72 to 77% of the blacks in North America.

(Also see CARBOHYDRATE[S]; and FIBER.)

Lipids. Lipids (fats) are digested and absorbed primarily in the upper part of the small intestine, but considerable absorption can take place as far down as the ileum. Fats are not water soluble. Since enzymatic reactions of the body occur in water-base solution, fats must be emulsified for digestion to occur. Bile salts from the gallbladder play two roles in fat digestion: (1) they act as a detergent decreasing the surface tension, and (2) they transport the end-products of digestion away from fat globules in water soluble micelles which are absorbed. It is the detergent action which allows the mixing movements of the intestines to break the fat globules into very finely emulsified (water oil mixture) particles with a greatly increased surface area on which the water soluble enzymes—lipases— can act. When lipids, emulsified by bile salts, come into contact with the various lipases that are found in the duodenum, they are broken down into diglycerides, monoglycerides, fatty acids, and glycerol. Short-chain fatty acids, less than 10 to 12 carbon atoms, are absorbed directly into the mucosa—lining— of the small intestine and are transported to the portal circulation of the liver. Monoglycerides and insoluble fatty acids are emulsified by bile salts, forming micelles. By attaching to the surface of the epithelial cells, the micelles enable these components to be absorbed into the intestinal cells. Once inside these cells, the long-chain fatty acids are reesterified (joined to the alcohol, glycerol) to form triglycerides. Triglycerides then combine with cholesterol, lipoproteins, and phospholipids to form chylomicrons—minute fat droplets. The chylomicrons are then passed into the lymphatic circulatory system via the central lacteal of the villi. Eventually these chylomicrons dump into the blood. After a fatty meal, the level of chylomicrons circulating in the blood reaches a maximum about 2 to 4 hours after a meal and may give blood a cloudy appearance. Within 2 to 3 hours, they disappear—having been deposited in the fat tissue or liver.

Fats are about 95% digestible. When fats are not digested due to lack of lipase or bile, absorption does not take place and a fatty stool, or steatorrhea, results. Fig. D-28 summarizes fat digestion and absorption.

(Also see CHOLESTEROL; FATS AND OTHER LIPIDS; and MALABSORPTION SYNDROME.)

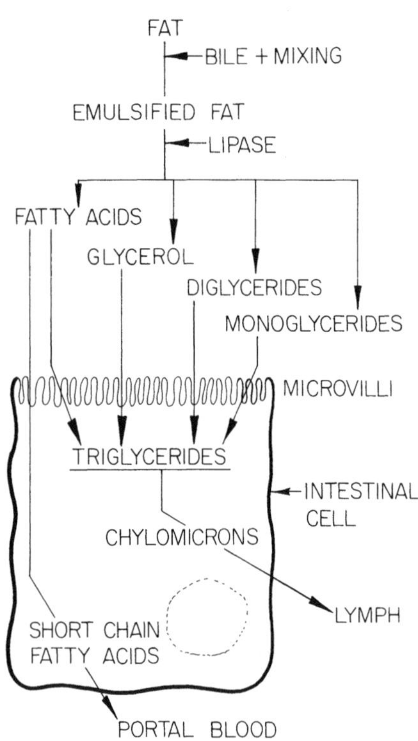

Fig. D-28. Summary of fat digestion and absorption.

Proteins. While protein digestion is initiated in the stomach, most digestion and absorption occurs in the small intestine. Numerous pancreatic and intestinal enzymes split proteins into proteoses, peptones, polypeptides, and finally their constituent amino acids, which are subsequently absorbed. In humans, it has been estimated that 50% of the digested protein comes from the diet, 25% from the proteins in the digestive fluids, and the remaining 25% from sloughed cells of the gastrointestinal tract. The rate of turnover of mucosal intestinal cells is extremely rapid—1 to 3 days—thereby giving an excellent source of recyclable protein. Overall, about 92% of the dietary protein is digested. The digestibility of vegetable protein is 80 to 85% while that of animal protein is about 97%.

Amino acid absorption is not clearly understood; but an active transport mechanism involving sodium ions (Na^+), similar to that of glucose absorption, has been implicated. Amino acids are rapidly absorbed in the duodenal and jejunal segments, but are poorly absorbed in the ileum. The digestion and absorption of protein is summarized in Fig. D-29.

A limited amount of the absorption of protein can occur, especially in the newborn. This mechanism of absorption, pinocytotic in nature, facilitates the passage of antibodies from the colostrum of the mother to her young. Additionally, this may contribute to allergic reactions which infants, and some adults, develop after eating certain food. Presumably, these individuals are capable of absorbing whole proteins, which then provoke an antigen-antibody reaction—an allergic reaction.

(Also see PROTEIN[S].)

The intestine also contains enzymes capable of digesting the nucleic acids—ribonucleic acid (RNA) and deoxyribonucleic acid (DNA). Pancreatic nucleases split nucleic acids into nucleotides (purine or pyrimidine base, sugar, and phosphoric acid), and the nucleotides are split into nucleosides (purine or pyrimidine base and a sugar), and phosphoric acid. These nucleosides are then split into their constituent sugar (pentoses), purine (adenine or guanine) and pyrimidine (cytosine, uracil, or thymine)

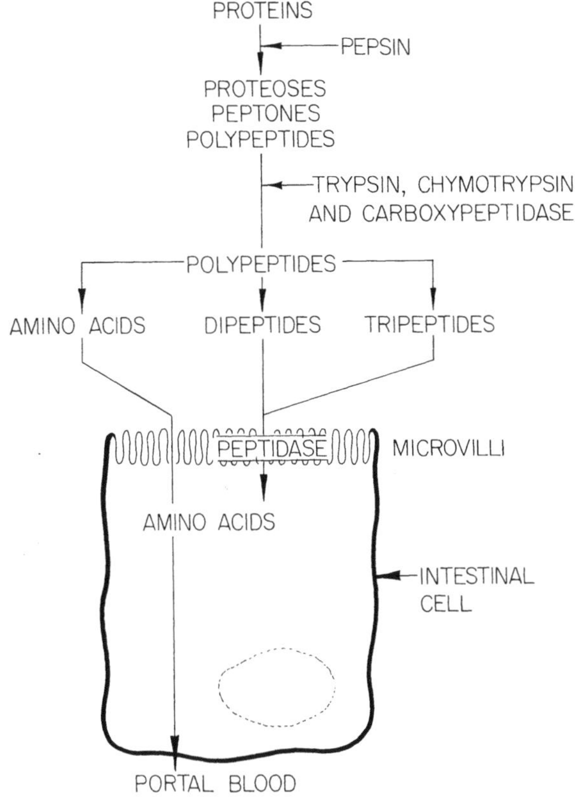

Fig. D-29. Summary of protein digestion and absorption.

bases. These bases are then absorbed by active transport.

Minerals and Vitamins. Mineral absorption occurs throughout the small and large intestines, with the rate of absorption depending on a number of factors—pH, carriers, diet composition, etc. Numerous mechanisms of mineral absorption have been elucidated. Many minerals, for example, iron and sodium, require active transport systems. Others, such as calcium, utilize both carrier proteins and diffusion mechanisms. Moreover, vitamin D is required for calcium absorption, and vitamins C and E favor the absorption of iron.

Most of the vitamins are absorbed in the upper portion of the intestine; vitamin B-12, which is an exception, is absorbed in the ileum. Water-soluble vitamins are rapidly absorbed, but the absorption of fat-soluble vitamins relies heavily on the fat absorption mechanisms which are generally slow.

(Also see MINERAL[S]; and VITAMIN[S].)

Water. Water moves freely across the membranes of the digestive tract—from inside the digestive tract to inside the cells lining the digestive tract. It's movement is via diffusion and osmosis. As the products of digestion—sugars, amino acids, and minerals—are actively transported out of the intestine, they create an osmotic gradient, also causing water to move out of the intestine and into the cells. In the large intestine sodium ions (Na^+) are actively pumped out of the large intestine and again water moves out following the concentration gradient. The amount of water moved is considerable when one considers that the water in foods and in the digestive secretions entering the digestive tract amount to about 10,000 to 12,000 ml per day, and that the water lost in the feces only amounts to 150 to 200 ml per day.

(Also see WATER; and WATER AND ELECTROLYTES.)

CONTROL OF THE DIGESTIVE TRACT. Like any system in the body, control is exerted over the functioning of the digestive tract in an effort to maintain the status quo of the body. The digestive tract is under neurological and hormonal control.

Neurological Control. Fear, anger, irritation, worry, all may exert unfavorable influences on the digestive system, while the thoughts, smell and presence of food cause secretion and motility necessary for digestion. Both types of changes are controlled via the nerves or nervous system of the body. The nervous system can be divided into two anatomical systems—the somatic nervous system, and the autonomic nervous system. The somatic nervous system enables the body to adapt to stimuli from the external environment. Various stimuli, such as touch, are perceived by specialized receptors within this system, and the body responds accordingly. The autonomic system involves the maintenance of homeostasis—the internal environment of the body. This is the system that controls the gastrointestinal tract.

The autonomic system can be further divided into (1) the sympathetic nervous system, and (2) the parasympathetic nervous system. The sympathetic system is generally associated with the traditional "fight or flight" response, and the parasympathetic system is usually associated with routine integration of normal activity.

When the sympathetic system is stimulated, there is a need for large amounts of blood in peripheral tissues, such as skeletal muscle. In order to accommodate this need, blood is shunted from the gastrointestinal tract, resulting in reduced digestive activity. For the most part, salivation ceases, and the mouth becomes dry. Secretions from the digestive glands are inhibited as well as peristalsis throughout the tract. The various sphincters of the gastrointestinal tract contract in response to sympathetic stimulation.

Stimulation of the parasympathetic system induces increased gastrointestinal activity. Generally, the parasympathetic system is stimulatory to the gastrointestinal system during rest and normal activity.

With the knowledge of the action of the sympathetic and parasympathetic autonomic systems, one can understand the action of certain drugs. When acute diarrhea is encountered, sympathetic-type drugs or parasympathetic depressant drugs are often used. Drugs that act as parasympathetic stimulators are frequently used as laxatives.

Hormonal Control. Chemical substances, known as hormones,are secreted by a number of ductless—endocrine—glands throughout the body. Hormones control a variety of body functions. Several areas of the digestive tract secrete hormones which act as chemical messengers on other areas of the digestive tract to control the process of digestion. Their secretion is a well-planned concert directed by the passage of foods through the gastrointestinal tract. A number of hormones have been isolated and characterized from the gastrointestinal tract. Gastrointestinal endocrinology is a very recent area of study; and new hormones are being found and chemically identified. Table D-11 lists the gastrointestinal hormones, and gives their site of origin, signal for release, and action.

Some other hormones, not of gastrointestinal orgin, also influence digestion. Glucocorticoids from the cortex of the adrenal gland may increase stomach secretions, while epinephrine from the medulla of the adrenal gland inhibits stomach secretions. Thyroid hormones stimulate the motility of the intestines.

(Also see ENDOCRINE GLANDS.)

TABLE D-11
GASTROINTESTINAL HORMONES[1]

Hormone	Origin	Mechanism of Release	Physiological Function
Gastrin	Antral portion of gastric mucosa; pancreatic islets.	Distension of stomach; presence of proteins and polypeptides; alcohol; caffeine; stimulation of vagus nerve.	Stimulates gastric acid (HC1) and pepsin secretion; stimulates gastric motility.
Enterogastrone	Duodenum	Presence of fats.	Inhibits gastric acid (HC1) secretion and motility.
Cholecystokinin-Pancreozymin (cholecystokinin)	Duodenum	Presence of fats and products of protein digestion.	Contraction of gallbladder and secretion of pancreatic enzymes.
Secretin	Duodenum	Presence of acid and protein.	Stimulates secretion of aqueous pancreatic fluid (high in bicarbonate).
Enterocrinin	Duodenum	Presence of chyme.	Increases secretion of enzyme containing intestinal fluids.
Villikinin	Duodenum	Presence of chyme.	Increases contractions of villi.
Glucagonlike immunoreactive factor (GLI)	Wall of small intestine.	—	Stimulates insulin secretion.

[1]Other gastrointestinal hormones have been proposed to exist, but their existence has not yet been proven.

FOOD FACTORS AFFECTING DIGESTION AND ABSORPTION. Most people believe that some foods are hard to digest, that others are easy to digest, or that still others are irritating. In many cases the foods cannot be blamed, but the functions of the digestive tract can be blamed. However, there are some food factors which are known to alter digestion and absorption; among them, the following:

• **Alcohol and caffeine**—Both alcohol and caffeine act directly on the lining of the stomach and stimulate gastric secretion. The use of alcohol for this purpose has been known since ancient times.

• **Fiber**—Diets consisting of unrefined cereals and containing large amounts of fiber decrease the digestibility of proteins and increase the loss of protein in the feces.

• **Liquids and fine foods**—Since chewing increases the surface area for enzyme action and is necessary for the digestive processes to proceed, it stands to reason that finely divided foods would be easily digested. Some food processing can divide foods into much finer particles than chewing; for example, a puree. Liquids are also rapidly handled by the digestive tract. Fat foods, especially fats mixed with protein and introduced into the digestive tract in large chunks, are hard to digest and require time.

• **Chelates**—Chelates are formed when certain organic molecules combine with metallic ions to form cyclic compounds. These chelated complexes possess different solubility characteristics than unbound metallic ions. By this binding, certain minerals may be more readily or less readily absorbed in the gastrointestinal tract. Several naturally-occurring chelating agents are chlorophylls, cytochromes, hemoglobin, ascorbic acid, vitamin B-12, and some amino acids. The most commonly used synthetic chelating agent is EDTA (ethylenediamine tetraacetic acid).

• **Phytic acid**—Phytic acid is a hexaphosphoric acid ester of inositol. When the acid form is combined with a cation (a positively charged ion) to form a salt, the compound is referred to a phytin. More than 50% of the phosphorus in mature seeds is in the form of phytin. Numerous studies have shown that animals vary greatly in their ability to absorb phytins. Sheep have little trouble breaking down phytins and absorbing the phosphorus. In the dog and man, phytic acid may combine with calcium, thus making the calcium less available for absorption.

• **Oxalic acid**—Oxalic acid, a compound present in certain leafy plants, may interfere with calcium absorption. The acid precipitates calcium and renders it less available for absorption. Spinach has a high oxalic acid content, which may subsequently tie up substantial portions of its calcium.

(Also see MINERAL[S].)

DYSFUNCTIONS OF THE DIGESTIVE TRACT. The digestive tract, though relatively simple in structure, is complex in function. In order for proper digestion and absorption to take place, various humoral and neural mechanisms coordinate the movement of ingesta throughout the tract. If anything goes wrong either physically or chemically in the digestive organs, the entire integrated process can break down, causing a variety of disorders. However, some rather common and general disorders include vomiting, diarrhea, gastritis, malabsorption, and ulcers.

Vomiting. When an irritating substance enters the gastrointestinal tract, the first effect is an increased rate of mucous secretion in the localized area. At the same time contractions running backwards toward the mouth, or antiperistalsis, occur. These can move food all the way from the end of the small intestine into the stomach. Upon reaching the stomach the irritation created causes nervous impulses to the brain which are interpreted as nausea. If this sensation is strong enough the vomiting reflex is initiated. This reflex consists of (1) closure of the airway into the trachea, (2) relaxation of gastroesophageal constrictor at the bottom of the esophagus, (3) violent contraction of the diaphragm and abdominal muscles, and (4) squeezing action of the stomach. The final results being expulsion of the stomach contents up the esophagus and out the mouth.

Causes of vomiting are diverse, and may include acute fevers, drugs, motion sickness, brain disorders, pregnancy, intestinal obstruction, emotional upsets, and viral, bacterial, and parasitic infections of the digestive tract.

Prolonged vomiting is dangerous, especially in children, due to the loss of water and electrolytes. Moreover, vomit may be inhaled with serious consequences.

Diarrhea. Frequent passage of loose or watery, unformed stool is diarrhea. It is a symptom of either an acute or a chronic disorder—not a disease itself. Severe diarrhea can be dangerous due to the loss of large amounts of sodium, potassium, and water, resulting in dehydration.

The most common cause of acute diarrhea is Salmonella food poisoning. Diarrhea is a symptom of dysentery caused by Shigella bacteria, and this diarrhea is severe in cholera.

Chronic diarrhea is a symptom of many diseases, including tumors, parasitic infections, celiac disease, ulcerative colitis, Graves' disease, antibiotic treatment, and emotional upsets.

(Also see BACTERIA IN FOOD; and WATER AND ELECTROLYTES.)

Gastritis. The acute form of gastritis is probably the most common stomach ailment. It is an inflammation of the stomach lining. It arises from a variety of causes, many of which can be summarized in one word—overindulgence. Some specific causes include alcohol, rich foods, spicy foods, coffee, strong tea, hot foods, and mechanically irritating foods (like popcorn) in excess—all can result in acute gastritis. Some drugs (aspirin, sulfonamides, antibiotics, and quinine), and some viral and bacterial infections, cause acute gastritis. By far the most common drug causing acute gastritis is aspirin. Loss of appetite, heartburn, pain in the upper abdomen, nausea, vomiting, cramps, and discomfort due to distension are all common symptoms. Vomiting and/or diarrhea removes the irritant and often corrects acute gastritis. Other measures may be purely supportive.

(Also see GASTRITIS.)

Malabsorption. At times the absorption of one or more nutrients is prevented. This is malabsorption. Individuals suffering from malabsorption display in varying degrees the following symptoms: (1) diarrhea; (2) steatorrhea (fatty diarrhea) due to impaired fat absorption; (3) progressive weight loss and muscle wasting; (4) abdominal distension; and (5) evidence of vitamin and mineral difficiencies. The causes of malabsorption are diverse, but they fall into seven general categories: (1) heart and blood vessel disorders, (2) endocrine disorders, (3) inadequate absorptive surface, (4) inadequate digestion, (5) lymphatic obstruction, (6) defects in the absorptive surface, and (7) reduced bile salts. Treatment is directed toward eliminating, insofar as possible, the abnormality causing the malabsorption, vitamin and mineral supplements, and dietary modifications.

(Also see MALABSORPTION SYNDROME.)

Ulcers. Ulcers—open sores—develop due to the lessened ability of the digestive tract to withstand the digestive action of pepsin and hydrochloric acid. Ulcers can occur in any area of the digestive tract exposed to the action of these two substances. However, most ulcers occur in the first portion of the duodenum of the small intestine or in the stomach. The cause of ulcers is obscure. It is known that, physiologically, there is a disturbance in the acid-pepsin secretion and the tissue resistance of the digestive tract lining. Other factors such as chemicals, heredity, and emotions seem to be involved.

The hallmark symptom of ulcers is the upper abdominal pain when the stomach is empty, which is relieved by eating.

Treatment varies with each individual and the severity of the ulcer, but dietary changes, drugs, and rest are often prescribed.

(Also see ULCERS, PEPTIC.)

DIGESTION COEFFICIENT (COEFFICIENT OF DIGESTIBILITY)

The difference between the nutrients consumed and the nutrients excreted expressed as a percentage.

DIGESTIVE JUICES

A broad term which includes the secretions of the salivary glands, the stomach, the intestine, the pancreas, and the gallbladder, all of which aid the process of digestion.

(Also see DIGESTION AND ABSORPTION.)

DIGESTIVE SYSTEM (ALIMENTARY CANAL)

Consists of a tube which courses internally from the lips to the anus. In the adult, it's a 25 to 30 ft (*7.5 to 9 m*) long tube, through which food passes following consumption. At prescribed intervals, it becomes specialized in regions called the mouth, esophagus, stomach, small intestine, large intestine, rectum, and anus. Protruding along the way are the salivary glands, gallbladder, liver, and the pancreas, which provide essential secretory products for digestion.

(Also see DIGESTION AND ABSORPTION.)

DIGITALIS

A drug which converts a feeble, irregular heartbeat to a more rhythmic, stronger beat. However, it has toxic effects, particularly when the body supply of potassium is low.

(Also see HEART DISEASE.)

DIGLYCERIDE

A fat containing two fatty acid molecules.

DIGUANIDE

A drug given by mouth to lower the blood sugar of a diabetic.

(Also see DIABETES MELLITUS.)

DIPEPTIDE

The name given to two amino acids chemically linked together. It is a final stage in the digestion and absorption of proteins or an initial step in protein synthesis.

(Also see DIGESTION AND ABSORPTION; and PROTEIN[S].)

DIPHENYL

A highly toxic chemical used in minute amounts to prevent mold growth on fruit following harvesting.

(Also see ADDITIVES.)

DIPSOGEN

An agent which causes the sensation of thirst.

(Also see THIRST.)

DIPSOMANIA

An uncontrollable urge to overindulge in alcoholic drinks.

(Also see ALCOHOLISM.)

DIPSOSIS

Excessive thirst, or a longing for certain unusual forms of drink.

DISACCHARIDASE

An enzyme which splits double sugars (disaccharides) such as lactose, maltose, and sucrose into single sugar units (monosaccharides) such as fructose, galactose, and glucose.

(Also see CARBOHYDRATE[S]; and ENZYME.)

DISACCHARIDE

Any of a class of double sugars which yield two single sugars (monosaccharides) upon hydrolysis. Sucrose, lactose, and maltose are the most common.

(Also see CARBOHYDRATE[S]; and SUGAR.)

DISEASES

Contents *Page*

Classification of Diseases 599
Causes of Diseases 599
Abnormal Growths 600
Aging and Degeneration of Tissues 601
Allergies and Other Disorders of Immunity 602
Congenital and Genetic Disorders 602
Endocrine and Metabolic Disorders 602
Infectious Diseases and Parasitic Infestations 603
Infectious Types of Organisms 603
Parasites 603
Ways in Which Diseases Are Spread 604
Nutritional Deficiencies and Infection 604
Injuries from Physical Agents 611
Malnutrition 611
Stresses 612
Toxic Substances 612
Chemicals and Drugs 612
Constituents of Foods 612
Environmental Pollutants 613
Essential Nutrients 613
Poisonous Animals and Plants Which Are Sometimes Used as Foods 614
Toxicants Which Arise from Metabolism 614
Signs and Symptoms of Diseases 615
Obvious Abnormalities in Vital Functions 615
Subtle Disorders Which May Require Special Diagnostic Procedures for Their Detection 615
Treatment and Prevention of Diseases 615
Lessening the Chances of Abnormal Growth (Tumors) 616

(Continued)

DISEASES *(Continued)*

Contents *Page*

Prevention of the Degenerative Diseases Which Accompany Aging 616
Minimizing Allergic Reactions 616
Prevention of Congenital Disorders 616
Treatment of Inborn Errors of Metabolism (Genetic Disorders) 616
Maintenance of Optimal Endocrine Function and Metabolism 617
Building of Resistance to Infectious Diseases 617
Colostrum and Breast Milk—Protectors of the Newborn 617
Immunization Procedures 618
Vaccination 618
Special Dietary Means of Promoting Resistance to Infectious Diseases 618
Sanitary Measures Which Help to Prevent the Spread of Diseases 619
Food Sanitation 619
Other Sanitary Measures 620
Healing of Injuries from Physical Agents 620
Prevention of Malnutrition 620
Reduction of the Effects of Stress 621
Detoxification of Poisonous Substances 621
Health and Nutrition Functions of Government Agencies 621
U.S. Department of Health, and Human Services . 621
U.S. Department of Agriculture (USDA) 621
State and Local Government Agencies 622
International and Voluntary Organizations Engaged in Health and/or Nutrition Activities 622

This term denotes harmful disorders in the normal structures and functions of one or more parts of the body. The consequences of disease depend upon (1) the parts of the body which are affected; (2) the amount of impairment of normal functions; and (3) the characteristics of the diseased person, such as previous state of health, age, sex, and temperament. For example, small, benign tumors of the skin may have no more than a cosmetic effect. Conversely, a heart attack may cause rapid deterioration and death in an overweight, quick-tempered, elderly person.

The purpose of this article is to put foods and nutrition in their proper perspectives with regard to the various types of diseases. Keeping healthy requires an understanding of the relationship between diets and nutrition-related disorders. Resistance to disease, and the ability to make a fast recovery from a disease or an injury, depend to a large extent upon one's nutritional status. Hence, each of the major causes of disease will be briefly discussed, along with what is known concerning the influence of food and nutrition thereon. The common diet-related diseases will be covered in greater detail in the articles noted in the cross references.

Data compiled by the U.S. government and published in the Surgeon General's report on *Nutrition and Health* in 1988 warned the nation that many of its top causes of death and chronic diseases—including heart disease, cancer, stroke, diabetes, intestinal disease, and osteoporosis—are linked to diet. The National Academy of Sciences echoed this warning and admonished Americans to make changes in their eating habits. (See section on "Dietary Guidelines for Americans" for the recommended dietary guidelines.)

Diseases are responsible for many of the deaths of people up to age 45, and for most of the deaths of those over 45. Table D-12 gives the leading causes of death and the death rates for people of various age groups in the United States.

CLASSIFICATION OF DISEASES. Diseases are classified according to the following bases:

1. **Infectiousness** as either (a) infectious—diseases caused by the presence in or on the human body of a foreign living organism that creates disturbances and leads to symptoms; or (b) noninfectious.
2. **Communicability or contagiousness** as (a) communicable, or (b) noncommunicable. A communicable disease is one which may be transmitted from an infected person or animal to a noninfected person or animal.
3. **Manner of occurrence** as (a) sporadic—one which occurs in isolated cases or outbreaks, like Rocky Mountain spotted fever; (b) epidemic—one which appears suddenly and affects many people over a large area at the same time, like influenza; or (c) endemic—one which affects certain people of a given area year after year, like goiter.
4. **Anatomic** as (a) respiratory, (b) nervous, (c) urogenital, etc.
5. **Course and duration** as (a) acute—one that runs a rapid course of a few days; (b) subacute—one which runs a slower course and lasts 2 or 3 weeks; or (c) chronic—one which lasts from 4 weeks to an indefinite period.
6. **Prognosis** as (a) curable, (b) incurable, (c) malignant, or (d) benign.
7. **Origin** as (a) inherited, (b) acquired, (c) prenatal, or (d) postnatal.
8. **Causes** as (a) infectious organisms, (b) malnutrition, (c) trauma, etc. See the section that follows.

CAUSES OF DISEASES. Any agent which adversely alters the harmonious balance between the body's processes may be the cause of disease. This is not to say that any exposure to a harmful agent will inevitably result in disease. Whether or not disease develops depends upon (1) the potency or severity of the injurious agent, and (2) the ability of the body's defenses to counter the effects of the injurious agent.

Of the many factors which affect resistance to disease, one of the best known is chilling, and its lowering of man's resistance to infections of the respiratory tract. (The respiratory tract is lined with tiny, undulating hairlike projections called cilia which help to sweep out infectious microorganisms. Chilling slows the movements of the cilia.) On the other hand, chilling may help to save the life of a person with a high fever (104°F [40°C] or higher), which accompanies an infectious disease. Thus, it is not always easy to predict the outcome when two or more potentially harmful agents act on the body at the same time. However, it is worthwhile to consider each

TABLE D–12
CAUSES OF DEATH BY LIFE STAGES[1]

Problem	Age Groups						
	Infants (Under 1)	**Children (1–14)**	**Adolescents/ Young Adults (15–24)**	**Adults (25–44)**	**Adults (45–64)**	**Adults (Over 65)**	**Total Deaths (All Ages)**
Developmental:							
Gestation & birth	7,241						7,241
Respiratory	3,822						3,822
Congenital	8,141						8,141
Sudden infant death	5,476						5,476
All other causes	13,294						13,294
Trauma:							
Accidents	936	8,000	18,500	28,300	15,200	27,000	97,100
Suicide		200	4,900	11,900	6,900	6,400	30,400
Homicide		1,200	5,800	10,800	2,800	1,300	22,000
Infectious diseases:		(1-12 yrs)	(13-29 yrs)	(30-39 yrs)	(40-59 yrs)	(over 60)	
AIDS[2]		287	4,039	9,865	6,731	753	21,675
Influenza & pneumonia	(not available by age groups)						77,700
Septicemia	(not available by age groups)						7,700
Chronic diseases:							
Heart		1,500	1,100	15,700	119,300	627,400	765,200
Cancer		1,700	1,900	20,800	136,500	324,100	485,000
Stroke		300	300	3,400	15,800	130,700	150,200
Bronchitis, emphysema, and asthma		200	200	900	13,000	67,600	82,900
Diabetes mellitus		<50>	100	2,100	8,600	28,500	40,400
Cirrhosis of the liver		<50>	100	4,700	11,800	10,000	26,400

[1]*Statistical Abstract of the United States, 1991*, U.S. Dept. of Commerce, pp. 79, 80, and 83.

[2]The age groups for AIDS differ from the other causes of deaths. *Note:* The age groups for AIDS are given in the parentheses immediately above each listing of deaths.

of the basic factors which produce disease so that a wide variety of preventative measures may be identified. The major causes of disease are depicted in Fig. D-30 and discussed in the sections that follow.

Abnormal Growths. Various cells within the body may start to grow abnormally and spread throughout the body, because they have escaped the controls which regulate their growth. When such abnormal growth (neoplasia) occurs, disease may result from pressure on neighboring tissues, competition for vital nutrients by cancer cells, and other pathological effects. Tissues in aging persons appear to be more susceptible to tumorous growths than those in growing children, although certain types of cancer are leading causes of death in teenagers.

Medical scientists have long known that trace contaminants such as asbestos, coal tar, dyes, and hydrocarbons in air, food, and water may cause cancer in man. However, recent studies of the diets consumed by various peoples around the world provide evidence which suggests that excesses of ordinary, uncontaminated foods might somehow promote abnormal growths.[34] It appears that obesity and diets high in fats are associated with higher than average incidences of bowel and breast cancers. This is not to say that fat or other foods are carcinogens, but rather that excesses of certain nutrients might act indirectly by (1) providing nourishment for microorganisms whch convert harmless substances and their metabolites into potentially carcinogenic substances; and (2) stimulating abnormal patterns of hormone secretion. Additional information on this subject

[34]Rose, D. P., "Update: Diet, Nutrition and Cancer," *The Professional Nutritionist*, Vol. 8, Fall 1976, pp. 1-2.

is given under the section on the treatment and prevention of various diseases.
(Also see CANCER.)

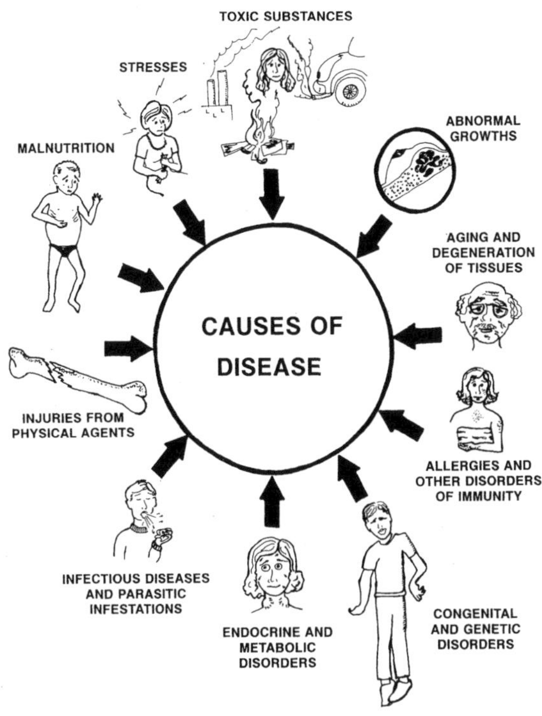

Fig. D-30. The major causes of diseases.

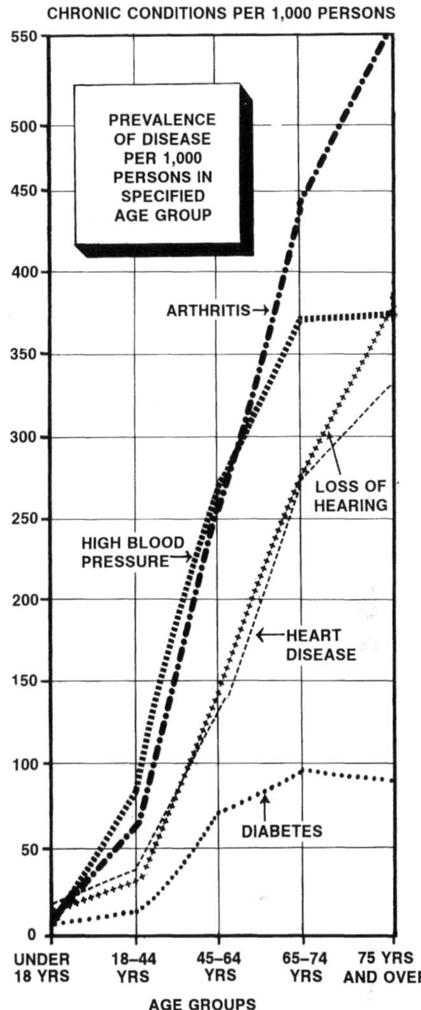

Fig. D-31. The prevalence of selected chronic disorders in people of various ages in the U.S. for each of the diseases or conditions. (*Statistical Abstract of the United States, 1991*, U.S. Dept. of Commerce, p. 120, Table 195)

Aging and Degeneration of Tissues. For reasons not well understood, many of the tissues of the body undergo degenerative changes with aging. Thus, certain disorders, such as arthritis, bowel lesions, bone loss (osteoporosis), cancer, diabetes, hardening of the arteries (arteriosclerosis), and high blood pressure (hypertension), are found more frequently in persons over 50 years of age than in younger persons. However, some older people may show fewer signs of deterioration than some younger people since the rate of aging may vary considerably from person to person.

Fig. D-31 shows how the prevalence of various chronic diseases and conditions increases with aging.

The role of nutrition in aging is heatedly debated by various groups of health scientists whose views differ widely.

A few biochemists—notably Drs. Pauling and Williams—promote the concept of "supernutrition," which means the taking of special mineral and vitamin supplements for the purpose of forestalling the degeneration of tissues that is usually associated with the aging process.

Another school of thought holds that longevity may be promoted by the lifelong consumption of less food than that required by the body. This view is derived from the following observations:

1. The most long-lived people in the world eat sparingly, and lead vigorous lives which tend to keep them slender.[35]

2. Pioneering animal studies conducted by Dr. McCay of Cornell in the 1930s showed that sharp restriction of food during the period of growth markedly slowed the rate of growth, but greatly extended the life span and reduced the susceptibility of the animals to arteriosclerosis, cancer, diabetes, and kidney disease.[36]

[35] Leaf, A., "Observations of a Peripatetic Gerontologist," *Nutrition Today*, Vol. 8, 1973, pp. 4-12.

[36] Ross, M. H., "Dietary Behavior and Longevity," *Nutrition Reviews*, Vol. 35, 1977, pp. 257-264.

However, it is noteworthy that these studies were conducted in the protected confines of the laboratory, whereas it is well documented that nutritional deprivation of growing animals and of children in their natural environments frequently renders them very susceptible to infectious diseases. More information on the latter point will be given in the section dealing with infectious diseases.

Also see GERONTOLOGY AND GERIATRIC NUTRITION.)

Occasionally, children may have a rare disease which causes rapid aging (progeria). Their teenage years may be a period of growing old rather than of adolescence. Less dramatic, but more common, is the physical and mental deterioration of both young and middle-aged adults in the impoverished groups of people who have long lacked adequate food, health care, sanitation, and other amenities of life that are taken for granted by the more affluent groups.

Allergies and Other Disorders of Immunity. The body attempts to protect itself from foreign substances and organisms (antigens) by various responses designed to reduce its vulnerability to these invaders. For example, a person who has been exposed to an infectious disease may have developed bloodborne antibodies against the disease organism. *The state of protection afforded the body by antibodies against disease is called immunity.*

An allergy is considered to be a disorder of the process(es) by which immunity is produced in that it is an overreaction (hypersensitivity) to one or more mildly irritating substances which may occur in the air, on one's clothing, or in certain foods. Some people have such a hypersensitivity to certain antigens or other irritants that they may develop hives, asthma, dermatitis, diarrhea, or even severe shock which sometimes leads to death. Occasionally, these disorders may also be provoked by chilling of the body, or by various emotional stresses. Many of the unpleasant signs of hypersensitivity result from the body's overproduction of histamine, a substance which is synthesized from the amino acid histidine.

Recent research has shown that even diets which are free of antigens may lead the body to overproduce histamine by setting up conditions such as those which follow.

(Also see ALLERGIES.)

• **Hypoglycemia**—This condition, which is also known as "low blood sugar," appears to provoke a greater than normal production of histamine in response to an antigen.[37] Low blood sugar may result from lack of sufficient food, or from the eating of excessive amounts of simple sugars or refined starches.

(Also see HYPOGLYCEMIA.)

• **Production of histamine by bacteria in the bowel**—*E. coli,* a type of bacteria which may inhabit the bowel, converts the amino acid histidine to histamine. The growth of *E. coli* is promoted by diets rich in meat, milk protein, and refined carbohydrates.[38]

Congenital and Genetic Disorders. *Abnormalities which are present at birth are said to be congenital.* They may result from inherited (genetic) traits, or they may have been induced by environmental factors during either fetal development or the birth process itself. It is noteworthy that congenital disorders are second to accidents as a leading cause of death for children ages 1 to 4 in the United States.[39]

Not all genetic disorders are evident at birth—some may be manifested later in life. For example, sickle cell anemia may not be evident until a victim of this disorder attempts to exercise strenuously, at which time there is a crisis due to lack of oxygen in the blood.

(Also see ANEMIA.)

Inborn errors of metabolism are inherited, physiological defects which interfere with the normal utilization of nutrients by the body. For example, phenylketonuria (PKU) may cause mental retardation due to the accumulation of phenylpyruvic acid, which is derived from the incomplete metabolism of the amino acid phenylalanine. Sometimes, permanent damage from these disorders may be prevented by restricting the dietary content of the nutrients which give rise to the harmful products of metabolism. In other cases, it may be necessary to provide extra amounts of certain nutrients to people who have a genetic defect which leads to poor utilization of these nutrients.

(Also see GENETIC DISEASES; and INBORN ERRORS OF METABOLISM.)

Endocrine and Metabolic Disorders. The secretions of the endocrine glands (hormones) are in many ways the switches that turn metabolic processes on or off. Therefore, disorders involving one or more of these glands are likely to produce corresponding abnormalities of metabolism. Conversely, impairment of almost any aspect of metabolism usually triggers some type of endocrine response.

For example, excessive secretion of hormones by the thyroid gland may overstimulate energy metabolism and other functions to the extent that there is loss of weight, a rapid heartbeat, and extreme nervousness. On the other hand, victims of protein-energy malnutriton may have markedly subnormal thyroid functions, so that they always feel cold, even in the tropics.

In other situations, stress factors such as emotional distubances, fasting, starvation, or diabetes, are signals to the pituitary gland to release adrenocorticotropic hormone (ACTH), which acts both alone on tissues and stimulates secretion of other stress hormones by the adrenal glands. The net effects of these stress-responsive

[37]Adamkiewicz, V. W., and P. J. Sacra, "Histamine and Sugars," *Federation of American Societies for Experimental Biology Proceedings,* Vol. 26, 1967, p. 209.

[38]Mount, J. L., *The Food and Health of Western Man,* John Wiley & Sons, Inc., New York, N.Y., 1975, pp. 184, 257.

[39]*Health: United States 1975,* Pub. No. (HRA) 76-1232, Dept. of HEW, p. 359.

hormones are the tearing down (catabolism) of fat, muscle, and/or bone tissues, so as to provide essential nutrients for metabolism.

Finally, malnutrition, or even the administration of sex hormones, may inhibit the stimulation of the sex glands by the pituitary, so that there is a weakening of these glands. For example, atrophy of the testes has occurred in men who have been severely deprived of food, or who have regularly taken testosterone for the purpose of accelerating the development of their muscles.

(Also see ENDOCRINE GLANDS.)

Infectious Diseases and Parasitic Infestations. The invasion of the body by bacteria, protozoa, or viruses, or by larger organisms such as worms, can lead to disease when the invading species secretes toxins, carries other disease organisms, feeds on body tissues, or competes with its host for essential nutrients. Some people tolerate parasitism better than others. Also, the severity of the disease depends upon the type and the numbers of the infectious or parasitic organisms which are present in the host. Fig. D-32 shows the ranking of the leading communicable diseases which were reported in the United States.

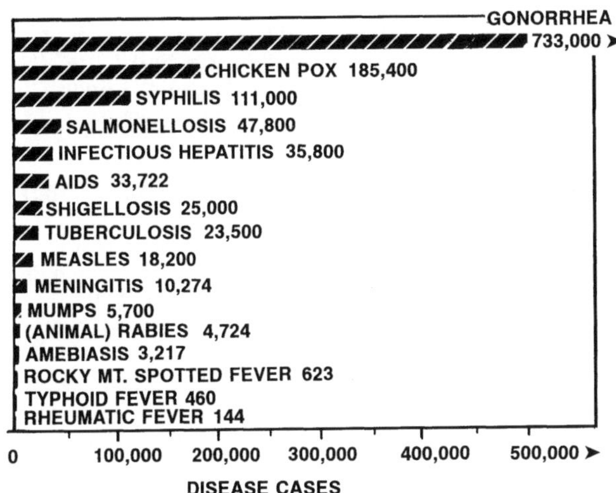

Fig. D-32. The leading communicable diseases in the United States, according to the number of cases reported to the Center for Disease Control. It is suspected that there was a failure to diagnose and report many cases of gonorrhea and syphilis, which may have been several times as prevalent as the number of reported cases. (The graph was plotted from data given in *Statistical Abstract of the United States 1991* [U.S. Bureau of the Census], Washington, D.C., p. 118, Table 190)

Descriptions of various types of disease organisms follow.

INFECTIOUS TYPES OF ORGANISMS. The principal types of infectious organisms that cause disease may be grouped as follows:

1. **Bacteria** are miscroscopic forms of plant life which possess just one cell, vary in shape, multiply by splitting into two new cells, and possess no chlorophyll.

2. **Chalamydia** (*C. psittachi and C. trachomatis)* are bacteria which lack some important mechanisms for production of metabolic energy; hence, they must live within the cells of other forms of life.

3. **Flukes (trematodes)** are soft, flat, leaf-shaped parasitic worms.

4. **Insect larvae** are the immature, wingless form of insects. For example, fly larvae (maggots) may infest open wounds.

5. **Moldlike bacteria** are somewhat higher in the evolutionary scale than ordinary bacteria.

6. **Molds (fungi)** are fungi distinguished by the formation of mycelium (a network of filaments or threads), or by spore masses.

7. **Mycoplasmas (PPLO)** are microscopic organisms intermediate between viruses and bacteria.

8. **Protozoa** are the simplest and most primitive form of animal life; they consist of only a single cell.

9. **Rickettsiae** appear to be intermediate between the bacteria and the viruses.

10. **Roundworms (nematodes)** are unsegmented worms, usually cylindrical and elongated in shape and with tapered ends.

11. **Tapeworms (cestodes)** have bodies made up of flattened segments joined together to make a chain. Each segment contains a set of male and female reproductive organs.

12. **Viruses** may be defined as disease-producing agents that (a) are so small that they cannot be seen through an ordinary microscope (they can be seen by using an electron microscope), (b) are capable of passing through the pores of special filters which retain ordinary bacteria, and (c) propagate only in living tissue. They are generally classified according to the tissues they invade, although this is a very arbitrary method, as some viruses invade many tissues.

13. **Yeastlike fungi** are characterized by budding, yeastlike cells.

PARASITES. *Broadly speaking, parasites are organisms living in, on, or at the expense of another living organism.*

People may harbor a wide variety of internal and external parasites. They include fungi, protozoa (unicellular animals), arthropods (insects, lice, ticks, and related forms), and helminths (worms).

Any animal or person that serves as a residence for a parasite is referred to as a host. In order to complete their life span (cycle), some parasites require only one host while others need more.

While in residence, parasites usually seriously affect the host, but there are notable exceptions. Among the ways in which parasites may do harm are (1) absorbing food, (2) sucking blood or lymph, (3) feeding on the tissue of the host, (4) obstructing passages, (5) causing nodules or growths, (6) causing irritation, and (7) transmitting diseases. Fig. D-33 depicts two common infestations of man by parasites from undercooked meats.

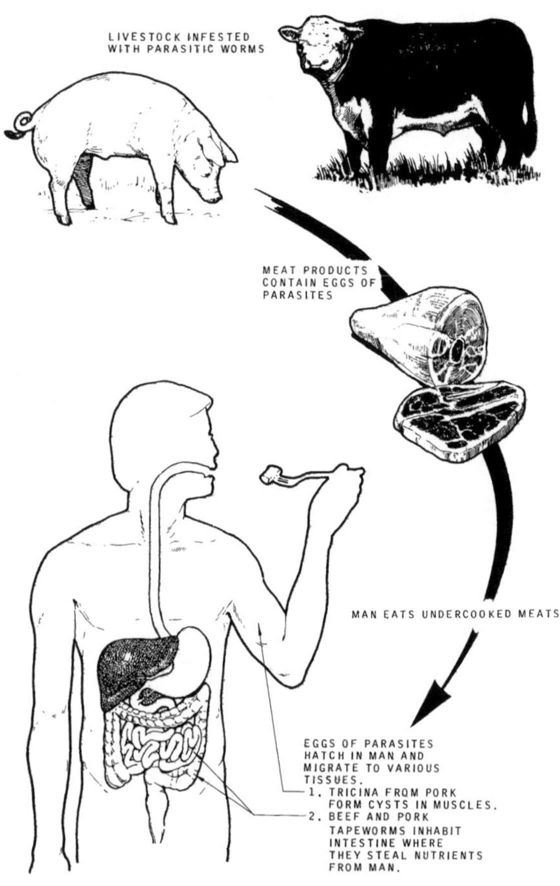

LIVESTOCK INFESTED WITH PARASITIC WORMS
MEAT PRODUCTS CONTAIN EGGS OF PARASITES
MAN EATS UNDERCOOKED MEATS
EGGS OF PARASITES HATCH IN MAN AND MIGRATE TO VARIOUS TISSUES.
1. TRICINA FROM PORK FORM CYSTS IN MUSCLES.
2. BEEF AND PORK TAPEWORMS INHABIT INTESTINE WHERE THEY STEAL NUTRIENTS FROM MAN.

Fig. D-33. Trichinosis and beef tapeworm, two parasitic diseases that may be transmitted from animals to man.

(Also see BEEF TAPEWORM (Measles); PARASITE INFECTIONS; and TRICHINOSIS.)

WAYS IN WHICH DISEASES ARE SPREAD. Infectious diseases may be spread from one person to another in a variety of ways; among them, the following:

1. **Direct contact** with diseased animals or people in which the infected host actually touches the susceptible person and transmits the disease. Venereal diseases are spread in this manner.

2. **Indirect contact,** such as (a) by susceptible people touching infected animals or people's excretions or secretions, like feces, urine, blood, pus, or saliva; or (b) by susceptible people breathing airborne infected droplets exhaled from the nose and mouth of infected people.

3. **Contaminated items,** including beverages, foods, water, kitchen utensils, dishes, silverware, counters and sinks, dish cloths and towels, bedding, clothing, etc. Gastrointestinal diseases, such as parasitic worms and salmonellosis (a bacterial infection which causes intestinal inflammation) are often transmitted in contaminated foods. Table D-13 summarizes various ways in which certain common diseases and parasites are spread by means of contaminated foods and water.

4. **Carriers** (also called disease vectors) which include insects, mites, ticks, and snails. In some cases, transmission of the agent by carrier is purely mechanical; for example, a biting fly serves as a "flying needle." In other cases, a stage of development, or part of the life cycle, of the infectious agent in the carrier is actually necessary before it may be passed on to the new host.

The term "carrier"may also be applied to a person who is infected with a disease organism, but who shows no signs of the disease. "Typhoid Mary" was a cook in New York City who appeared to be healthy although she carried the typhoid bacillus.[40] An investigation showed that she transmitted typhoid fever to at least 100 other people before it was discovered that she was a carrier.

5. **Carrion feeders** (flesh eaters, such as rats, cats, dogs, foxes, or birds) may serve as reservoirs of infectious organisms or of parasites such as trichina, which they may spread to people by means of direct contact or by infecting swine (which may eat dead rats and birds) or other farm animals.

The rapidity of spread of different infections, their geographical and seasonal distributions, and the relative ease of their prevention and control may all depend, at least in part, upon their means of transmission.

NUTRITIONAL DEFICIENCIES AND INFECTION. The infection of a tissue and the production of a disease by a living agent is not always easily accomplished. The agent must first gain entrance into the body through one of the openings or through the skin. Usually, it then multiplies and attacks the tissues. To accomplish this, it must be sufficiently powerful (virulent) to overcome the defenses of the human body. The defenses of the human body vary, and they may be weak or entirely lacking, especially under conditions of malnutrition and poor hygienic practices. Nutritional deficiencies are widespread in many of the developing countries, so it is not surprising that many of the children under age 5 in such countries die from infectious and parasitic diseases. For example, a survey of childhood mortality in 13 selected communities in Latin American countries revealed that infective and parasitic diseases were responsible for over half of the deaths of children under 5 years of age.[41] Furthermore, nutritional deficiencies were found to be contributory factors to about 60% of the deaths from these diseases. The most prevalent deficiencies were those of protein and energy. It is noteworthy, too, that the survey

[40]Guthrie, R. K., *Food Sanitation*, The Avi Publishing Company, Inc., Westport, Conn., 1972, p. 44.

[41]Puffer, R. R., and C. V. Serrano, "The Role of Nutritional Deficiency in Mortality: Findings of the Inter-American Investigation of Mortality in Childhood," *Boletin de la Oficina Sanitaria Panamericana* (English Edition), Vol. VII, 1973, p. 21, Table 10.

TABLE D-13
INFECTIOUS AND PARASITIC DISEASES WHICH MAY BE TRANSMITTED BY CONTAMINATED FOODS AND WATER[1]

Disease	Disease Organism	Signs and Symptoms	How Disease is Transmitted	Preventative Measures	Remarks
Infectious diseases:					
Amoebic dysentery	*Entamoeba histolytica* (An amoeba is a microscopic one-celled animal organism.)	Diarrhea may occur anytime from several days to four weeks after consuming contaminated food or water. Symptoms are: unexplained fatigue, low grade fever, and recurrent diarrhea.	Cysts of the amoeba are ingested in contaminated food or water, particularly those which are uncooked when eaten. Flies may carry the organism to foods.	Prevention of sewage contamination of water supply. Avoidance of drinking cold water, use of ice, or eating cold foods in places with poor sanitation.	Infection is most likely to occur in warm climates. An infected person with no apparent signs of the disease may spread the organism to other people.
Bacillary dysentery (Shigellosis)	Various types of *Shigella* bacteria.	Diarrhea and/or fever from 1-7 days after ingesting infected food or water.	Bacteria shed from infected people get into food or water because of poor sanitary practices. Some types of *Shigella* may infect chickens.	Hand washing by food handlers. Protection of water supply, and chlorination of water when feasible. Protection of foods against contamination by flies.	Young children and sickly adults may become dehydrated, so care must be taken to maintain the balances of mineral salts and water.
Brucellosis (Bang's disease, undulant fever, malta fever)	*Brucella abortus* *Brucella suis* *Brucella melitensis* (bacteria)	Chills, headache, fever, severe night sweats, and extreme weakness (6-30 days after infection).	The organisms may pass from infected animals to man through meat and milk.	Cooking meat and pasteurizing milk make these foods safe for human consumption.	Apparently this disease is *not* spread from person to person, but only from animals to man.
Cholera	*Vibrio cholerae* (bacterium)	Severe diarrhea, abdominal cramps, vomiting, and copious sweating may occur anytime from a few hours to a day or so after infection. Death is caused by severe dehydration.	Food and drinking water may be contaminated by flies, fecal material, or vomitus carrying the organism. It is also spread by direct contact with infected persons.	Protection of water against contamination by sewage. Sanitary handling of foods and their protection against flies. Vaccination of persons going into cholera areas.	Epidemics of cholera regularly occur in the Philippines, southeast Asia, India, Iraq, Egypt, and southern China. Vaccination is good for less than 6 months.
Diphtheria	*Corynebacterium diphtheriae*	Fever accompanied by running nose and sore throat which occur from 1-6 days after exposure. Death by suffocation has resulted from a membrane which sometimes forms in the throat.	Usually by contact with nose or throat discharges from an infected person. Epidemics have been spread through the use of raw milk.	Immunization of all young children, plus booster shots when necessary. Disinfection of all articles which were in contact with infected persons.	The disease organisms produce a toxin which may travel through the body and cause damage to the heart, nerves, and kidneys. Human diphtheria has no relationship to calf diphtheria. The two diseases are caused by entirely different organisms.
Epidemic diarrhea of the newborn	Unknown virus	A marked change in the infant's stool pattern towards more frequent, watery, and unformed stools.	It is not certain whether the infant picks up the virus from its food or from direct contact with others. However, people handling the infant may infect the food of others.	Scrupulous sanitation in the preparation of milk formulas for bottle-fed babies. Hand washing by people in contact with the infant. Disinfection of infant's clothing and bedding.	Many infants around the world—including a few in the U.S.—die from the dehydration which may result from severe diarrhea.

footnote at end of table

(Continued)

TABLE D-13 *(Continued)*

Disease	Disease Organism	Signs and Symptoms	How Disease is Transmitted	Preventative Measures	Remarks
Epidemic gastro-enteritis (intestinal flu)	Unidentified virus	Diarrhea, nausea, and and vomiting which last for a few hours.	Possibly in food handled by infected persons. More likely by personal contact, contaminated water, and discharge from the nose and throat of the infected persons.	Persons with this disease should *not* handle food until all signs have disappeared. Protection of food against airborne droplets by "sneeze shields" at serving counters. Protection and disinfection of water supply.	Measures should be taken to stop diarrhea and/or vomiting which last for more than a few hours so that severe dehydration may be prevented.
Infectious hepatitis	Type-A hepatitis virus	Yellowing of the skin and whites of the eyes (jaundice), nausea, lack of appetite, and weakness. Signs may not appear until more than a month after infection.	The virus is carried in secretions and excretions from the digestive tract. Hence, it may be spread by personal contact, sewage-contaminated water or shellfish, or poorly handled food.	Hand washing by food handlers. Protection and/or disinfection of the drinking water supply, and avoidance of shellfish (mainly clams and oysters taken from polluted waters).	The recovery period may be as long as several months, so prolonged use of sanitary precautions may be needed for those in contact with the patient.
Poliomyelitis (infantile paralysis)	Polio viruses—types I, II and III.	Sore throat and/or diarrhea, fever, nausea, and vomiting.	Mainly by personal contact, contaminated water, and secretions and excretions from the digestive and respiratory tracts. It is suspected that the viruses may be borne in foods.	Immunization of all young children and food handlers. Protection of foods and drinking water against contamination.	Most infected persons do *not* become paralyzed. However, such persons, with barely noticeable symptoms, may be carriers of the disease.
Salmonellosis	Various *Salmonella* bacteria (other than *Salmonella typhi* or *paratyphi*)	Diarrhea, nausea, and vomiting beginning about one-half day after infection.	Bacteria shed from man grow rapidly in such cooked foods as eggs, meat, milk, custards, salads—which have been left unrefrigerated for several hours. It may also be transmitted by sewage-polluted water.	Foods likely to be contaminated held at temperatures below 40°F *(4.4°C)*. Hand washing by food handlers. Scrupulous cleaning of food processing equipment. Avoidance of the use of cracked eggs unless thoroughly cooked.	This is one of the most common foodborne infections in the U.S. Sometimes, single outbreaks involve thousands of people.
Scarlet fever	*Streptococcal* bacteria	Fever, vomiting, sore throat, and characteristic rash (which appears only in certain susceptible people). Occurs mainly in children.	Usually by direct contact or airborne droplets from infected people. However, food may occasionally be contaminated by infected people.	Disinfection of personal articles from infected people. Use of the Dick test to confirm disease, since it may be necessary to treat the patient so as to prevent the spread of the disease.	Scarlet fever is now uncommon in the U.S. Treatment with various antibiotics eliminates the ability of the patient to transmit the disease.

footnote at end of table

(Continued)

TABLE D-13 *(Continued)*

Disease	Disease Organism	Signs and Symptoms	How Disease is Transmitted	Preventative Measures	Remarks
Strep throat	*Streptococcus pyogenes*	Fever, vomiting, and sore throat. Sometimes, related bacteria are the cause of communicable skin infections.	Organisms may get into foods from infected handlers since they are carried on airborne droplets from the respiratory tract of infected people who may sneeze or cough on food. Likewise, the causative organism may be spread directly from person to person through the air.	Protection of foods against coughs and sneezes, and against contamination by handlers wth skin infections.	Occasionally, there may be such complications as rheumatic fever, or infections of the bones and the kidneys. Penicillin and other antibiotics are effective against this disease.
Tuberculosis (TB)	*Mycobacterium tuberculosis*	Chronic coughing, extreme fatigue, loss of appetite, and eventually the coughing up of blood.	Mainly by airborne droplets discharged during the coughing and sneezing of infected persons. The organism may live in various foods from whence it may infect people.	Pasteurization and safe handling of milk. Protection of foods against coughs and sneezes. Testing and disposal of tubercular animals.	There is a high incidence of TB in skid-row alcoholics, so areas where they gather are likely to be contaminated.
Tularemia	*Francisella tularensis*	Headache, chills, fever, and vomiting, accompanied by irregular fever, which lasts for several weeks.	Most cases of the disease come from the eating of undercooked meat from wild rabbits. However, the hunter or the cook may get the disease from handling an infected rabbit. Occasionally, water supplies may be contaminated by dead animals.	Thoroughly cooking the rabbit meat kills the organism. Wearing of rubber gloves by those handling wild game.	The disease is named after Tulare County, California, where it was discovered in 1911.
Typhoid fever	*Salmonella typhi*, or *Salmonella paratyphi*.	Fever, nausea, headache, and loss of appetite.	Contamination of food or water by sewage, flies, or infected persons. Direct contact with patients or symptomless carriers of the disease. Shellfish (mainly oysters and clams) from polluted waters may carry the disease.	Protection and disinfection (chlorination) of drinking water. Finding and curing carriers of the disease. Sanitary food handling practices.	The bacteria which cause the disease may live in the gallbladder for long periods of time and be gradually shed in the feces. Antibiotics have greatly reduced the number of deaths from this disease.

footnote at end of table

(Continued)

TABLE D-13 (Continued)

Disease	Disease Organism	Signs and Symptoms	How Disease is Transmitted	Preventative Measures	Remarks
Parasitic diseases: **Beef tapeworm** ("beef measles")	*Taenia saginata*	Often, the infested person is free of symptoms. Occasionally there may be anemia, nausea, vomiting, or diarrhea, alternating with constipation. There may be some discomfort when the worm segments are passed in the stool.	Deposition of feces from infested humans in water, feeds, or pastures causes infestation of cattle with eggs. The eggs hatch in cattle and the larvae invade tissues used for meat. Man picks up live larvae from undercooked beef. (Also see Fig. D-33.)	Cooking of beef until well done. Enforcement of the use proper sanitary facilities by help on cattle farms. Protection of pastures, feeds, and cattle from contamination by human feces or sewage. Where possible, protection of feedlots against flooding. Barring of hobos and hunters from farms and feedlots.	This problem is confined largely to the four states which border Mexico—Texas, New Mexico, Arizona, and California. However, it may be spread to other states by animals, feeds, meats, and people that are infested with the parasite.
Dwarf tapeworm	*Hymenolepis nana*	Usually there are no symptoms unless the infestation is massive. Then, there may be abdominal pains, nausea, or vomiting.	Contamination of foods with feces from infested mice, rats, and people—all of whom may harbor the parasite for its entire life cycle. Sometimes, people pick up the worm eggs from the foodborne insects such as grain beetles.	Control of rats, mice, and insects, and protection of foods from their contamination.	Infestation is very common in areas of the world which abound with rats. However, few cases appear in the U.S.
Fish tapeworm	*Diphyllobothrium latum*	Often, an infested person has no symptoms; but there may eventually be pernicious anemia due to the worm(s) consuming most of the host's dietary vitamin B-12.	Newly hatched tapeworm larvae are first eaten by waterfleas, which are then ingested by freshwater fish. Man becomes infested by eating the living worm in raw fish. The tapeworm also infests dogs, cats, polar bears, and sea lions.	Cooking of all freshwater fish. Freezing of fish for several days at 0°F *(−17.8°C)* will also kill the parasite.	Many of the freshwater fish in northern European lakes and in the U.S. Great Lakes contain fish tapeworms. However, regular microscopic examination of these fish is not feasible.
Liver flukes	*Clonorchiasis*	Digestive disturbances and eventually liver damage.	Eggs of flukes from infested people are shed via feces into fresh water where they are ingested by snails. Fish become infested by eating the snails, while man picks up living parasites from undercooked fish.	Thorough cooking of all fish, whether fresh, dried, salted, or pickled. Prevention of contamination of lakes and ponds by human feces.	Infestations with liver flukes affect many people in the crowded areas of eastern Asia. However, there are few deaths, and some people carry flukes for more than 20 years.

footnote at end of table

(Continued)

TABLE D-13 *(Continued)*

Disease	Disease Organism	Signs and Symptoms	How Disease is Transmitted	Preventative Measures	Remarks
Pinworms (seat worms, thread worms)	*Enterobius vermicularis*	Itching of the anal region. Sometimes there is also lack of appetite, irritability, and loss of weight.	Usually the eggs are spread from the human anal region to other people through the air, on clothing, or on furniture. However, food is sometimes contaminated by fecal material, or by unhygienic food handlers.	Sanitary disposal of human wastes. Daily bathing by infested people. Elimination of worms for the patient(s) by medical means. Hand washing by food handlers.	This is the most common parasite found in U.S. children. White people appear to be more susceptible than other races.
Pork tapeworm	*Taenia solium*	Symptoms may range from mild digestive disturbances to serious disorders when larvae invade the heart, brain, eyes, or nerves.	Swine pick up worm eggs from eating matter contaminated with human feces. The larvae from the hatched eggs invade the muscles. Man becomes infested by (1) eating the eggs on raw vegetables grown in soils fertilized with human feces, and (2) undercooked pork which contains the larvae.	Avoidance of the use of untreated human wastes as fertilizer for vegetables eaten raw. Thorough cooking of all forms of pork (fresh pork, hams, sausages, luncheon meats, frankfurters, etc.). Cooking of all vegetables grown in soils fertilized by human wastes.	Many deaths are caused by this parasite in countries where there is poor sanitation.
Round worms	*Ascaris lumbricoides*	Often, there are no symptoms of this infestation. Occasionally, there is nausea, fever, loss of weight, and allergiclike irritation of the nose and throat due to the presence of the worms in these places.	The eggs are spread by unhygienic practices of food handlers, use of human wastes to fertilize vegetables which are eaten raw, or when there is cesspool or sewage contamination of drinking water.	Sanitary disposal of human wastes. Careful washing of vegetables to be eaten raw, although in areas where infestation is common, it might be best to cook all foods thoroughly and to boil drinking water.	It is believed that as many as 2/3 of the world's people may be infested with these worms. The worms may migrate to the liver, heart, lungs, nose, and Eustacian tubes (which connect the throat and the ears).
Trichinosis	*Trichinella spiralis*	The first symptoms of intestinal infestation may appear as early as a day after eating parasitized meat. They are: abdominal pain, diarrhea, nausea, and vomiting. About a week later, when the larvae have migrated to various tissues, there may be severe muscle pain, difficulty in breathing, swollen eyelids, and sometimes death (when the heart muscle is infected).	Swine become infested by eating contaminated garbage or infested rats. (Other meat-eating animals, such as bears, may become infested by similar means.) The larvae of the worms migrate from the pig's intestine to its muscles where they form cysts. Man may become infected by eating raw or undercooked pork (or bear meat) which contains trichina larvae. (Also see Fig. D-33.)	Thorough cooking of pork (until the last trace of pink in the meat disappears), or freezing the meat continuously for not less than 20 days at a temperature not higher than 5°F *(−15°C)*. (The meat of wild carnivores, such as bears, should be similarly treated.) Destruction of all rats on the farm. Cooking of all garbage and slaughterhouse by-products fed to swine.	This disease is not often found in the U.S. or in the developed countries of Europe, where the laws governing the sanitary feeding and handling of swine are strictly enforced. However, the disease is common in the developing countries where pork is is eaten.

[1]Also see BACTERIA IN FOOD; FOODBORNE DISEASE; FOOD POISONING; PRESERVATION OF FOOD; and SPOILAGE OF FOOD.

showed that diarrheal disease was responsible for about 40% of childhood deaths from all causes in Latin America.

Table D-14 gives some of the ways in which nutritional deficiencies may weaken the body's defenses against infectious and parasitic diseases.

(Also see DEFICIENCY DISEASES; MALNUTRITION; and MALNUTRITION, PROTEIN-ENERGY.)

TABLE D-14
THE EFFECTS OF NUTRITIONAL DEFICIENCIES ON THE BODY'S DEFENSES AGAINST INFECTION[1]

Location and Nature of the Defenses	How Infection is Prevented	Ways in Which Nutritional Deficiencies May Weaken the Defense(s)	Remarks
Outer layer of the skin: Horny layer, oil, and sweat.	The horny layer acts as a barrier to prevent invasion of underlying tissues. Secretion of oil and sweat tend to protect the skin against dust, heat, dryness, and infectious organisms.	Drying, cracking, peeling, and slow healing of wounds may result from lack of protein, essential fatty acids, and such vitamins as A, C, E, and those in the Vitamin B complex.	The skin is affected by a wide range of nutritional deficiencies because it receives the nutrients which are left in the blood after the innermost tissues of the body have withdrawn their shares
Respiratory tract: Mucus, cilia, and antibodies.	More than a quart *(liter)* of mucus is secreted daily. This viscous fluid traps foreign bodies and is swept along by the motion of cilia (tiny, hairlike projections from the tract). Also, the mucus contains antibodies against infectious organisms.	Mucus-secreting cells are very sensitive to deficiencies of vitamin A. The production of antibodies is markedly curtailed by protein-energy malnutrition. Even a mild deficiency of vitamin C may aggravate the symptoms of a common cold.	The motions of the cilia and the actions of the antibodies are greatly impaired by chilling. Also, the hormones secreted during stress block the production of antibodies.
Eyes: Protective film and tears.	The protective film on the eyeball and the copious secretion of tears help to prevent infection.	Most injury to the eye's defenses results from lack of vitamin A, the requirements for which may be increased by prolonged exposure to bright sunlight. However, lack of the vitamin B complex may also be harmful.	Nonnutritional factors which may weaken the eye's defenses are bacteria, diet, dry air, and excessively bright sunlight.
Digestive tract: Saliva, stomach acid, antibodies, and protective lining cells.	Saliva and the antibodies it contains kill some bacteria, whereas stomach acid has a stronger effect and may even kill some parasites. The protective lining (epithelial cells) prevents many types of infectious organisms and the toxins they secrete from passing through the intestinal wall into the blood.	Vitamin A deficiency may cause a lack of saliva and a weakening of the lining cells. Irritation of the entire tract occurs in deficiencies of iron, the amino acid tryptophan and the vitamin B complex. Iron deficiency may also be responsible for lack of stomach acid.	There is some evidence that helpful microorganisms such as lactobacilli may protect against the infection of the body by harmful species.
Genital tract (female): Lining cells, population of helpful microorganisms.	The lining cells are a barrier to infection. Estrogens (female hormones) encourage the vaginal production of glycogen, a nutrient for lactic acid-producing bacteria. Lactic acid retards the growth of many harmful microorganisms.	The female genital tract is particularly vulnerable to vitamin A deficiency, which causes abnormalities in the vaginal lining cells. Deficiency of this vitamin may also reduce the secretion of estrogens, thus indirectly it may reduce the production of lactic acid in the vagina.	Douching may do more harm than good because it washes out the lactic acid-producing bacteria and renders the tract more susceptible to invasion by harmful microorganisms.
Blood: Antibodies and phagocytic white blood cells.	Bloodborne antibodies attack and render noninfective the organisms which have breached the other defenses. The white blood cells phagocytize (ingest) many harmful organisms.	Antibody production is greatly reduced in protein-energy malnutrition and in deficiencies of the amino acid tryptophan; the vitamins A, C, D, and B complex. Also, the phagocytic ability of white cells may be impaired by such deficiencies.	Certain artificial means of producing immunity such as vaccination, are less effective for people with severe nutritional deficiencies. Various stresses may also reduce the production of antibodies.

[1]Also see DEFICIENCY DISEASES; MALNUTRITION; and MALNUTRITION, PROTEIN-ENERGY.

Injuries from Physical Agents. Both acute and chronic injuries may result from cold, electric shock, heat, mechanical forces, radiation, or sunlight. Such injuries to cells and tissues may be difficult to heal or may lead to other problems such as infections. Sometimes, these injuries cause disfiguration or deformity.

The human body has a limited ability to resist injury from physical agents, by virtue of the innate strength of such tissues as skin, muscle, and bone. However, people whose tissues have been previously weakened by poor nutrition may be badly hurt by even the mildest of injurious agents. For example, older people may have bones which are easily broken, due to long subsistence on a diet low in calcium, vitamin D, and/or protein. Some of the other ways in which nutritional factors affect susceptibility to injury follow.

• **Heat exhaustion**—This disorder results from the overtaxing of the body's mechanisms for keeping cool, and for maintaining the balance between body water and various mineral salts. It has been found that lack of sufficient potassium in the body renders it more susceptible to heat exhaustion and heat stroke.[42] Of course, it is well known that lack of sufficient salt is also a major contributor to heat exhaustion, as evidenced by the practice of giving extra salt to people working in hot environments.

(Also see HEAT EXHAUSTION.)

• **Radiation injury**—People who are deficient in iodine are more susceptible to damage to their thyroid glands by radioactive iodine (I-131), an atmospheric contaminant which results from the testing of nuclear weapons. The reason for the increased susceptibility is that iodine deficiency causes the thyroid gland to increase the rate of its uptake of both normal and radioactive iodine.[43]

Radiation may also cause cellular damage by triggering the body to produce free radicals—highly reactive chemical fragments which may deform the membranes and nuclei of cells. Injury to tissues by free radicals is enhanced by deficiencies of selenium and vitamin E.

Diet may also play a role in the susceptibility of the digestive tract to radiation injury. Animal studies have shown that a given dose of radiation in the form of x rays is more likely to be fatal if there is a large population of coliform bacteria in the intestine.[44] The growth and multiplication of *E. coli* appears to be encouraged by diets rich in protein. On the other hand, animals fed laboratory chow—which is made up mainly of unrefined cereal grains—have much fewer coliform bacteria in their intestines and are more likely to survive radiation. Many other studies have shown that milk sugar (lactose) promotes the growth of lactobacilli that make the intestinal contents acid so that the growth of *E. coli* is inhibited.

[42]Snively, W. D., Jr., and D. R. Beshear, *Textbook of Pathophysiology*, J. B. Lippincott Co., Philadelphia, Pa., 1972, p. 83.

[43]Robbins, J., J. E. Rall, P. Gorden, "The Thyroid and Iodine Metabolism," *Duncan's Diseases of Metabolism*, 7th ed., edited by P. K. Bondy and L. E. Rosenberg, W. B. Saunders Co., Philadelphia, Pa., 1974, p. 1020.

[44]Klainer, A. S., S. Gorback, and L. Weinstein, "Studies of Intestinal Microflora VII. Effect of Diet and Fecal Microflora on Survival of Animals Exposed to X Irradiation," *Journal of Bacteriology*, Vol. 94, 1967, p. 383.

Malnutrition. In the discussion that follows, malnutrition is typed as subclinical, dietary deficiency, or dietary excess.

Subclinical malnutrition is the stage of malnutrition which precedes the appearance of specific signs of disease. However, there may be subnormal levels of minerals or vitamins in the blood or urine. Other indicators of such malnutrition might be elevations in blood levels of such substances as cholesterol, triglycerides, glucose, uric acid, urea, or bilirubin.

(Also see MALNUTRITION.)

Dietary deficiency diseases are those whose major features are due to the lack of one or more nutrients. They are often characterized by specific signs of abnormalities, such as discoloration of the hair or skin, bleeding gums, swollen glands, etc. One of the major causes of such diseases is the consumption of a diet which does not contain sufficient amounts and varieties of foods. For example, young children fed mainly cereals in the form of watery gruels may develop either protein-energy malnutrition, and/or blindness, due to vitamin A deficiency.

(Also see DEFICIENCY DISEASES.)

Dietary excess of certain nutrients may also cause disease when (1) such excesses are toxic, or (2) they create imbalances.

Vitamin D poisoning, for example, results from excessive dietary intake and/or production in the body of excessive amounts of this nutrient. The latter situation may be caused by overexposure of the skin to light from the sun or a sunlamp.

(Also see VITAMIN D.)

Another example of dietary excess is the storage of surplus dietary energy as fat when there is chronic overeating. Normally, fattening is a natural activity of the body. However, if the deposition of fat results in marked obesity, there may be some pathological deviations from normal metabolism. For example, some obese persons have a diabeticlike underutilization of carbohydrates, fats, and proteins. They seem to secrete normal amounts of insulin, but the effects of the hormone appear to be hampered. Eventually, the pathology of diabetes may develop.

(Also see OBESITY.)

The three leading causes of death in the United States—heart disease, cancer, and strokes—are believed to be at least partly associated with obesity and related dietary factors. However, it also seems that there are other factors—such as stress, heredity, smoking, and environmental pollutants—which may affect one's susceptibility to these diseases. Therefore, poor nutrition may be only one of the multiple factors which cause certain diseases.

What constitutes good nutrition under one set of circumstances may mean malnutrition in others since nutrient needs may vary according to age, sex, environmental conditions, amount of physical activity, and inherited traits which influence the metabolism of nutrients. Furthermore, it may be difficult to detect some types of malnutrition since their features may not be evident until there has been noticeable damage to the body.

Stresses. Chilling, overheating, exhaustion, starvation, dehydration, emotional stress, or even overeating, may trigger the oversecretion of the endocrine glands, particularly the adrenals, thyroid, pancreas, and pituitary, as the body attempts to counter the threats to its harmonious function. Stresses which are mild and short lasting can be beneficial when they tone up the body and its functions, but prolonged stress may lead to upsetting of the balance between the processes of the body.

Some health scientists believe that hardening of the arteries and high blood pressure are the end result of long-term exposure to various stress factors. However, it seems that the stresses have different effects on different people since some people experience many hardships over a lifetime, yet live to ripe old ages, while others succumb to disease by middle age.

(Also see STRESS.)

Toxic Substances. This designation applies to a large number of substances, including some of the otherwise beneficial nutrients, such as several of the trace minerals and vitamins A and D, which are potentially harmful or poisonous at high levels. Thus, toxicity is relative rather than absolute, since the harmful level of a poisonous substance depends upon a person's size, weight, genetic traits, metabolic activity, and overall health status.

Toxic substances may be inhaled while breathing, ingested with food or drink, taken as medications, absorbed through the skin or mucous membranes, or produced in the body as products of the normal metabolic processes. These substances produce their effects by attacking tissues and cells, by disrupting physiological processes, or by interfering with the utilization of vital nutrients. A discussion of several different toxicants follows.

(Also see POISONS.)

CHEMICALS AND DRUGS. The chemical industry has developed many products which contribute to both our exceptionally high rate of agricultural productivity and the attractiveness of our homes, meals, and personal appearances. Hand in hand with this development, there has been increased public concern over the use of these products, for fear of human poisoning.

For example, farmers may use insecticides, rodent killers, weed killers, fertilizers, disinfectants, solvents,and petroleum products, some of which may be toxic under certain circumstances. When properly used, agricultural chemicals are an important adjunct to providing food for people and feed for animals. However, improper use may result in the chemical contamination of food plants, eggs, meats, and dairy products. This type of contamination occurred a few years ago, when some samples of milk from a dairy farm in New York were found to contain excessive amounts of the pesticide dieldrin.[45] An investigation revealed that the source of the pesticide was alfalfa hay secured from a dealer in Pennsylvania. From time to time, other pesticides have been involved in similar incidents involving the contamination of feeds and/or various food products.

Other regular users of various chemical products are the home handymen and the homemakers, whose work is made easier by cleaners, polishes, paints, paint thinners, bleaching agents, dyes, etc. Poisoning from these substances often results from their storage in food and drink containers, particularly by young children who are prone to sample just about everything they can lay their hands on.

Finally, there are many poisonings from the accidental or intentional misuse of such popular drugs as antibiotics, birth control pills, headache remedies, laxatives, antidepressants and other stimulants, mineral and vitamin pills, narcotic-containing cough medicines, sleeping pills, tranquilizers, arthritis remedies, heart drugs, water pills (diuretics), and worm killers.

The hazards associated with the commonly used chemicals and drugs underscore the need for the accurate labeling and use of these products.

(Also see POISONS.)

CONSTITUENTS OF FOODS. There have been many scare stories concerning the potentially toxic effects of a wide variety of food constituents which have been administered to experimental animals in grossly abnormal ways. For example, even a naturally occurring substance such as cholesterol may cause tumors when it is *injected* under the skin. However, findings which demonstrate potential toxicity are most likely to be relevant when the test substance has been *fed* in quantities which might be consumed under normal circumstances. Foodborne toxicants may fall into one or more of the categories which follow.

• **Naturally occurring toxicants**—Through trial and error, man has usually learned of the animals and plants which are either acutely poisonous or are safe to eat. However, mildly toxic effects are more difficult to detect. For example, such vegetables of the *Brassicae* (cabbage family) as rutabaga, turnip, cabbage, and kale contain substances (goitrogens) which interfere with the utilization of iodine by the thyroid gland. Hence, eating large amounts of these items may cause goiter when the dietary level of iodine is low.

(Also see GOITER.)

• **Products of deterioration or of spoilage**—Foods which go bad may often be identified by discoloration, disagreeable odors, and peculiar tastes, unless such characteristics are masked by added colorings or flavorings. However, certain food-contaminating organisms produce deadly toxins which may cause such diseases as aflatoxin poisoning, botulism, or staphyllococcal food poisoning, even though the contaminated foods appear to be normal in all respects. It is noteworthy that the botulism toxin, produced by the bacteria *Clostridium botulinum*, is one of the most potent poisons known to man. Outbreaks of this type of food poisoning usually result from foods that have been improperly canned.

Even readily recognizable types of contamination, such

[45]Zaki, M. H., *et al.*, "Dieldrin in Milk: The Experience of Suffolk County, New York," *American Journal of Public Health*, Vol. 68, 1978, pp. 260-261.

as ergotism on grains of rye (see Fig. D-34), may be concealed when the contaminated grain is ground into flour and baked into bread.

(Also see BACTERIA IN FOOD, Table B-5 Food Infections; BOTULISM: FOOD POISONING; POISONS, Table P-11 Mycotoxins; and PRESERVATION OF FOOD.)

Fig. D-34. A seed head of rye which is infested with ergot, a parasitic fungus which replaces the seed in the heads of certain grains. The ergot appears as a purplish-black, hard banana-shaped dense mass from ¼ to ¾ in. (*6-19 mm*) long.

• **Toxicants which result from the cooking or processing of foods**—Recently, it has been found that certain members of a group of compounds called *nitrosamines* are carcinogens. These substances may be formed in the chemical reactions between the amines (breakdown products of amino acids and proteins), which occur naturally in foods, and the nitrites that are added to certain meats to prevent discoloration and botulism. The foods most likely to contain such carcinogens are various types of smoked fish (which are believed to be responsible for the high rates of stomach cancer in Iceland and Japan), sausages, and bacon (the high cooking temperature of bacon may be a factor in the formation of nitrosamines).

(Also see NITRATES AND NITRITES; and NITROSAMINES.)

• **Substances added to improve the characteristics of foods**—Foods may often have slight deviations from optimal color, odor, and taste which render them less desirable to consumers, even though they are safe and wholesome. The wholesale rejection of these items by shoppers could result in considerable financial losses by farmers, processors, shippers, and merchants. Therefore, it has become common practice to add a wide variety of substances to foods for the purpose of enhancing their appeal to consumers; among them, preservatives, spices and other flavorings, artificial and natural coloring agents, sweeteners, emulsifying agents to prevent separation of various components, nutritional supplements, air, and water.

The United States and the other developed countries have strict governmental regulation of "intentional food additives" so that cases of acute poisoning from these substances are rare, if not nonexistent. Nevertheless, a few additives are suspected of having subtle toxic effects over long periods of use. Then, too, small bakeries, restaurants, and other establishments, which mainly serve local clientele, occasionally deviate from acceptable practices of food preparation, with the result that there may be sporadic outbreaks of poisonings.

(Also see ADDITIVES.)

• **Trace contaminants of food**—Toxic or otherwise undesirable materials may get into foods when contaminants such as lead, cadmium, and mercury are present in the environment, equipment, or containers with which the foods come in contact during their production, processing, packaging, or shipping. Likewise, minute traces of pesticides are occasionally found in various farm commodities. Occasionally, traces of plastics and antioxidants used to treat packaging materials seep into food products that are stored for long periods. But, like the use of the intentional additives, there are also strict governmental regulations concerning substances which may get into foods accidentally.

(Also see POISONS; and POISONS, CHEMICAL.)

ENVIRONMENTAL POLLUTANTS. For centuries, most of man's activities were conducted on such a small scale that their impact on local environments was limited. However, the Industrial Revolution, which began in the late 18th century, led to steady increases in the populations of cities, and in the discharge of waste products into the environment. At first, these products were promptly dissipated by the natural processes of rain, winds, and percolation through the soil. Now, it seems that in certain heavily industrialized areas there are likely to be persistent accumulations of toxicants in the air, water, and some types of foods. Hence, it is believed that the higher incidence of such diseases as cancer, lung disorders, and kidney disease found in urban areas than in rural areas may be partly due to various pollutants.

(Also see POISONS; and POISONS, CHEMICAL.)

ESSENTIAL NUTRIENTS. Most well-informed people know that eating too much carbohydrate and/or fat may bring on obesity, and perhaps a host of other ills. However, many of those who take various types of nutritional supplements might be surprised to learn that even excesses of proteins, minerals, and vitamins may be the causes of disorders such as those which follow.

• **Protein overload effect(s)**—This term designates the combination of (1) excessive urinary loss of water, and (2) accumulation of the products of protein metabolism in the blood which may result from diets overly rich in protein.[46] These diets might be hazardous for such vulnerable groups as infants, people with liver or kidney disorders, and those prone to dehydration and gouty arthritis (where there is a buildup of uric acid in the blood).

(Also see PROTEIN.)

• **Mineral imbalances and toxicities**—There is still considerable controversy regarding the optimal amounts of various essential mineral elements because dietary excesses of certain elements may interfere with the utilization of others. Furthermore, almost every one of the nutritionally essential minerals has potential toxicity at high levels. However, these toxicities are not as likely to result from eating unfortified foods, as they are from special circumstances such as (1) the taking of mineral supplements; (2) contamination of food and water by environmental factors such as containers, piping, and airborne dusts; and/or (3) fortification of foods with minerals.

It is noteworthy that there are only small margins of safety between beneficial and hazardous doses for (1) fluoride, which may cause severe defects of teeth and bones; and (2) selenium, which is suspected of being a cause of liver damage and tooth decay.

(Also see MINERAL[S].)

• **Vitamin overdoses (hypervitaminoses)**—The best known vitamin toxicities are those which result from vitamin A and vitamin D, because (1) these vitamins are fat soluble, (2) small amounts of them have strong effects, and (3) they tend to accumulate in the liver. Toxic effects do not occur so readily with vitamins E and K, which are also fat-soluble, unless high potency supplements are taken.

Excesses of water-soluble vitamins (vitamin C and the vitamin B complex) are not stored in the body to any great extent, so toxicities from food sources of these nutrients are rare. However, people who take very large doses (megadoses) in the form of supplements run the risk of dangerous, druglike effects.

Additional information on vitamin toxicities is provided in the general article on vitamins and in the separate articles dealing with the individual vitamins.

POISONOUS ANIMALS AND PLANTS WHICH ARE SOMETIMES USED AS FOODS. Many people around the world know that certain animals and plants might cause severe poisoning, yet they often take their chances on eating these items if they know of someone else who has escaped drastic consequence. Therefore, it is noteworthy that, although certain animals and plants are poisonous under all circumstances, others are variable in their toxicities because (1) the animal or plant is poisonous only when it acquires the toxic principle from its food or its environment; (2) certain types of processing either develop or eliminate the toxicity in the food; (3) the development of poisoning in man depends upon the actions of microorganisms, special enzymes, or other processes which occur in the food or in the human body; or (4) only certain stages of maturity in the animal or plant are toxic. Some examples of such variable toxicities follow.

• **Poisonous quail**—Throughout history there have been sporadic outbreaks of poisoning when people have eaten quail that have consumed the seeds of poison hemlock. Normally, quail eat these seeds only when other more favored foods are not available.

• **Seleniferous wheat**—Wheat may contain dangerous levels of selenium when it has been grown in high-selenium soils, such as those in the Great Plains of the United States, where "selenium converter plants" such as *Astragalus racemosus* have grown, died, and decayed for several years. Utilization of selenium from the soil by the converter plants, followed by the return of this mineral to the soil when the plants decay, results in its conversion from a form not readily picked up by wheat to one that is much more absorbable. Hence, high selenium soils alone will *not* yield wheat containing toxic levels of selenium, unless the mineral is in a form which is available to the growing wheat.

(Also see SELENIUM.)

• **Raw fish toxicity**—Various species of freshwater fish (such as carp) and crustaceans contain an enzyme (thiaminase) which destroys thiamin (vitamin B-1). Hence, persons eating these items raw may develop thiamin deficiencies, particularly if the rest of their diets are rich in carbohydrates and/or alcohol which raise the requirement for thiamin. However, thorough cooking destroys the enzyme, but leaves the thiamin intact.

(Also see THIAMIN, section headed "Antithiamin Factors in Food".)

• **Cassava poisoning**—Cassave (*Manihot utilissima*) is a staple food for many people who live in the hot, humid areas of Africa, Asia, and Latin America. However, some varieties of this tuberous plant contain a toxic principle which is greatly reduced by proper processing—soaking, boiling, drying, expression, or fermentation.

(Also see CASSAVA, sections headed "Processing" and "Nutritional Value".)

• **Cyanide-containing lima beans**—Some varieties of lima beans, such as those that are native to the Caribbean, and which are usually colored, contain harmful levels of cyanide-releasing glucosides. Fortunately, soaking and cooking the beans releases much of the toxic principle. Anyhow, the varieties grown in the United States contain only negligible amounts of cyanide. Besides, U.S. law prohibits the marketing of lima beans that contain harmful amounts of this toxic factor.

(Also see LIMA BEANS, section headed "Selection and Preparation".)

• **Irish potato poison**—Green (unripened) or sprouted potatoes may contain toxic levels of solanine, an alkaloidal substance which may bring on headache, vomiting, abdominal pain, diarrhea, slowing of the heart and breathing, and occasionally, death.

(Also see POTATO, section headed "Processing.")

TOXICANTS WHICH ARISE FROM METABOLISM. The human body, like a factory, requires a suitable means of getting rid of the waste products which are produced during its operations. Even excess sugar in the blood, which often occurs in diabetes, has harmful effects, par-

---

[46]Albanese, A. A., and Orto, L. A., "The Proteins and Amino Acids," *Modern Nutrition in Health and Disease*, 5th ed., edited by R. S. Goodhart and M. E. Shils, Lea & Febiger, Philadelphia, Pa., 1973, p. 58.

ticulary when this condition is prolonged or chronic in nature. The failure of the kidneys to excrete urea and uric acid may also result in illness due to the accumulation of these toxins in the blood and tissues. Furthermore, the presence of excessive amounts of fat in the blood (hyperlipoproteinemia) has come to be regarded as dangerous to health. Perhaps there are other little known ways in which the body may poison itself.

SIGNS AND SYMPTOMS OF DISEASE. Often, the first signs of disease are abnormalities like fever, rashes, swelling, pallor, diarrhea, vomiting, or discoloration of parts of the body such as the hair, face, eyes, nose, lips, tongue, teeth, gums, throat, glands, skin, or nails. The patient may also have lack of appetite, dizziness, fatigue, headache, pains in muscles or joints, or tingling in the hands and feet. However, these signs and symptoms may sometimes be present when there is no disease. Hence, the doctor or nurse will most likely check some of the vital functions. Details follow.

(Also see HEALTH, sections headed "Signs of Ill Health" and "Signs of Good Health.")

Obvious Abnormalities In Vital Functions. Any marked and persistent deviations from the normal ranges of body temperature, pulse, and breathing rate may indicate ill health. Normal values for these functions follow.

- **Body temperature**—The normal oral temperature of healthy adults ranges from 97.3° to 99.1°F (*36.3° to 37.3°C*), whereas normal rectal temperature ranges from 97.3° to 99.8°F (*36.3° to 37.7°C*). Active children may have normal oral or rectal temperatures as high as 101°F (*38.3°C*). Some adults may have equally high oral values following strenuous activity, and their rectal values sometimes run as high as 104°F (*40°C*). Although the normal body temperature marked on oral thermometers is 98.6° F (*37°C*), it is noteworthy that readings which are 1° or 2° above this point constitute only slight fevers. Generally, a moderate fever is indicated by a temperature of 101°F (*38.3°C*) or greater in adults or children who have been at rest. A few adults may have temperatures as low as 96°F (*35.6°C*) when they arise on cold mornings, but values below that indicate circulatory troubles or other disorders.

- **Pulse rate**—Resting pulse rates of adults normally range from 68 to 85 beats per minute, although a few highly trained athletes have rates as low as 40. (Physical training increases the efficiency of the cardiovascular system so that fewer heartbeats are needed to supply the tissues with the blood they require.) It is noteworthy that the pulse increases at a rate of from 8 to 10 beats per minute for each degree of temperature above normal. Hence, the body temperature of sick people might be judged if their normal pulse rates are known.

- **Rate of breathing**—The normal range of breathing rates for people at rest is from 15 to 20 breaths per minute. Strenuous exercise may cause rapid breathing for a while after the activity is stopped because the extra carbon dioxide generated by the increased metabolism stimulates the part of the brain which controls breathing. Also, nervous people may breathe more rapidly than normal. Finally, the deep, rapid breathing which accompanies severe acidosis (Kussmaul breathing) is a characteristic sign of diabetic coma.

(Also see HEALTH, sections headed "Signs of Ill Health" and "Signs of Good Health.")

Subtle Disorders Which May Require Special Diagnostic Procedures For Their Detection. Some people may have inherited tendencies to develop cardiovascular disease, diabetes, and low blood sugar (hypoglycemia). However, the early stages of these disorders produce few signs or symptoms, so doctors often suggest that special diagnostic tests be given to those whose close relatives have had these problems.

Others may be afflicted with (1) undiagnosed allergies which cause emotional, mental, or physical troubles; or (2) chronic, low-grade infections which sap their vitality.

Finally, physical and/or emotional stresses may take their toll of otherwise healthy people unless these conditions are identified so that the patient may be taught how to cope with them.

Therefore, people who complain of feeling poorly, yet who have no obvious signs of disease may need to consult a physician.

First, the doctor might inquire about such items as (1) the medical histories of the patient and of his or her relatives; (2) the amounts and types of foods which were recently eaten by the patient; (3) the possibility that the patient might have been exposed to a communicable disease or a toxic agent at home or at work; and (4) the daily physical and emotional stresses of the patient.

Then, the physician might order (1) chemical tests on the blood and urine; (2) a microscopic examination of the blood and the counting of red cells and white cells; (3) x rays; (4) electrocardiograms; or (5) preparation of cultures from the blood, saliva, or stool so as to identify infectious microorganisms or parasites.

Sometimes, the patient may be hospitalized for a brief period so that it will be more convenient to perform certain diagnostic tests.

TREATMENT AND PREVENTION OF DISEASE. Many diseases may cause irritation, inflammation, or damage to one or more tissues. Hence, the treatment of disease, which is basically a counteraction of the effects of the disease-causing agent, may involve one or more approaches designed to augment the body's own healing processes. For example, the healing of burns appears to be speeded by the administration of extra vitamin C (ascorbic acid).[47]

The prevention of disease not only involves the minimization of exposure to injurious agents, but also the building up of maximum resistance by suitable health practices. It is next to impossible to avoid all of the disease-causing factors. One should keep medical records of close relatives, since susceptibility or resistance to certain diseases may be inherited. Often, preventive measures may be instituted early if a disease is suspected; for example, if there is a family tendency

[47]Klasson, D. H., "Ascorbic Acid in the Treatment of Burns," *New York State Journal of Medicine*, Vol. 51, 1951, p. 2388.

to diabetes. However, each type of preventive action should be designed so as to be in harmony with the body's natural defenses. Some principles for countering specific types of diseases follow.

Lessening The Chances Of Abnormal Growths (Tumors). All food should be (1) of high quality, and (2) cooked properly so as to avoid the formation of tumor-causing substances or the destruction of nutrients such as vitamin E which may partially protect cells against the formation of tumors. For example, nuts, oilseeds, and grains which are contaminated with the common mold *Aspergillus flavus* may contain aflatoxin, which causes some types of cancer. Also, the heating of fats until they smoke may result in the formation of cancer-causing agents, and in the destruction of the vitamin E which may be present.

Smoking and air pollution are believed to cause lung cancer; hence, they should be minimized. It is noteworthy, too, that vitamins A and E may exert at least a partially protective effect on lung tissue, particularly against automobile exhaust fumes. Also, statistical (epidemiological) studies have shown that as trace levels of selenium increase in the soil and water of an area the incidence of cancer decreases. However, poisoning may result when the amounts of selenium in foods are in excess of those which have beneficial effects.

(Also see CANCER; FATS AND OTHER LIPIDS; NITRATES AND NITRITES; NITROSAMINES; POISONS; and SELENIUM.)

Prevention Of The Degenerative Diseases Which Accompany Aging. Needless irritation or other stress to body tissues should be avoided, and special efforts should be made to practice good health habits over a lifetime. For example, the poorly nourished person who is subjected to above average stresses—like the childbearing woman, the soldier, or the athlete—may sustain damages to flesh and bone which worsen with aging. One type of disability may lead to another; for example, the loss of teeth often leads to digestive disorders or malnutrition. Also, research has shown that lifelong diets which meet, but do not exceed, the body's requirements for the major nutrients (carbohydrates, fats, and proteins) confer longer life than diets which provide excesses of these nutrients. Finally, there may be increased needs for minerals and vitamins by older persons because their efficiency of absorption diminishes with aging.

(Also see GERONTOLOGY AND GERIATRIC NUTRITION)

Minimizing Allergic Reactions. Many of the unpleasant symptoms of allergic reactions, such as runny nose and weeping eyes, are due to the release of histamine which heightens the sensitivity of various tissues to the allergy-provoking agents. Hence, antihistaminic drugs are widely used to reduce the severity of allergic reactions. These drugs produce drowsiness and are, therefore, limited in their usefulness. Large doses of ascorbic acid also have an antihistaminic action without producing drowsiness, but they may irritate the digestive system.

It is noteworthy that the production of histamine by intestinal bacteria is less likely to occur when the predominant organisms are lactobacilli rather than *E. coli*. Recent research suggests that it might be possible to bring about modifications in the numbers and types of intestinal bacteria by the regular drinking of unfermented milk to which lactobacilli have been added.[48] Milk of this type tastes sweet rather than sour, so some people may find it to be more acceptable than buttermilk, sour milk, or yogurt, which are also good sources of lactobacilli.

Steroid drugs may be a last resort when severe inflammation of the tissues accompanies allergies, although there is the hazard that they may cause a gradual disintegration of the bones. In certain types of allergies, the physician may administer a series of "desensitizing" injections.

(Also see ALLERGIES.)

Prevention Of Congenital Disorders. This goal is best accomplished by adequate nutrition and prenatal care of pregnant women. Special attention needs to be given to women from underprivileged areas, teenage mothers, and women who have conceived right after long term use of birth control pills (oral contraceptives). Oral contraceptive agents may deplete a woman of folic acid and pyridoxine—two vitamins which are needed in pregnancy.

Also, pregnant women should be blood tested. For example, it is important to know if an Rh-negative mother is carrying an Rh-positive fetus, since she may have developed antibodies against her child's blood. In such cases, exchange transfusions may save the child's life and prevent disability.

Finally, the birth process itself must be carefully handled to avoid injury to the newborn. There is a trend towards performing a Caesarian section whenever the attending physician believes that the normal vaginal delivery may injure the child.

(Also see GENETIC DISEASES.)

Treatment Of Inborn Errors Of Metabolism (Genetic Disorders). It may soon be possible to detect, and to begin the treatment of, inherited defects in metabolism while the fetus is in the mother's womb. A start in this direction was made when methylmalonicaciduria was treated in a fetus by the administration of vitamin B-12.[49] When born, the child was free of the nervous system damage which usually accompanies this disorder. (Methylmalonicaciduria is an inherited disorder in which abnormal amounts of the metabolite methylmalonic acid accumulates in tissues and are excreted in the urine. Some types of the disorder respond to the administration of vitamin B-12.)

Prompt dietary treatment of newborns who have phenylketonuria (PKU) has successfully prevented mental retardation. (Treatment of PKU involves the provision

---

[48]Gilliland, S. E., *et al.*, "Influence of Consuming Nonfermented Milk Containing Lactobacillus acidophilus on Fecal Flora of Healthy Males," *Journal of Dairy Science*, Vol. 61, 1978, p. 1.

[49]Ampola, M. G., *et al.*, "Prenatal Therapy of a Patient with Vitamin B-12 Responsive Methylmalonic Acidemia," *The New England Journal of Medicine*, Vol. 293, 1975, p. 313.

of special foods which are low in the amino acid phenylalanine because the victim lacks the enzyme needed to metabolize this compound.) Soon, it may be possible to alter the genes which produce many of these abnormalities.

(Also see GENETIC DISEASES; and INBORN ERRORS OF METABOLISM.)

Maintenance Of Optimal Endocrine Function And Metabolism. The ability of the body to maintain harmony among its diverse functions is best promoted by regular and balanced patterns of eating, sleeping, and exercising. Then, the intensities of the endocrine secretions and of the metabolic processes rise and fall in a predictable daily cycle (sometimes called a *circadian rhythm*). Disruption of one's biorhythm impairs physical and mental performance, and may be a cause of disease. For example, the regulation of the blood sugar (glucose) level is very important for physical and mental health. The eating of large meals may unduly stress the pancreas to secrete excessive insulin, setting into motion a pernicious cycle of low blood sugar, followed by the secretion of other hormones which elevate blood sugar and block the effects of insulin. Smaller, more frequent meals may help to alleviate such a condition.

The body's natural rhythms may also be disrupted by the abuse of alcohol, barbiturates, and/or caffeine. Each of these substances may augment or antagonize the effects of various hormones.

Endocrine functions, particularly the secretion and effectiveness of insulin in regulating blood sugar, appear to diminish with aging. It is possible that chronic overtaxing of these glands results in their deterioration.

(Also see ENDOCRINE GLANDS.)

Building Of Resistance To Infectious Diseases. The human body is remarkably equipped to fight disease. Chief among this equipment are large white blood cells, called phagocytes, which are able to overcome many invading organisms.

The body also has the ability, when properly stimulated by a given organism or toxin, to produce antibodies and/or antitoxins. When a person has enough antibodies for overcoming particular (disease-producing) organisms, he or she is said to be immune to that disease.

Another type of defense against infection, previously mentioned, is colonization of the intestine and of the female genital tract (mainly the vagina) by lactobacilli. These beneficial bacteria prevent the growth of harmful microorganisms. One species, lactobacillus acidophilus, even secretes an antibiotic.[50]

Although the natural means of resistance were for many millenia all that stood between the human race and its extinction, these defenses often failed when they were either weakened by certain other factors, or when the infectious organisms were very virulent. Hence, man has delved into the mysteries of nature so as to learn how best to utilize or even enhance the body's defenses under various circumstances. Details follow.

[50]Hamdan, I. Y., and E. M. Mikolajcik, "Acidolin: An Antibiotic Produced by Lactobacillus acidophilus," *Journal of Antibiotics*, Vol. 27, 1974, p. 631.

COLOSTRUM AND BREAST MILK—PROTECTORS OF THE NEWBORN. Large numbers of infants and young children in the developing countries of the world die from diarrheal diseases. However, breast fed babies have a lower incidence of these diseases than those that are bottle fed. Although it seems likely that the better health of the breast fed infants is at least partly due to the avoidance of cow's milk which has sometimes been handled under poor sanitary conditions, many research studies have shown that human milk contains certain beneficial substances that are not present in cow's milk. Hence, expectant mothers should be informed regarding the merits of colostrum and breast milk so they may have a sound basis for deciding whether or not to breast feed their infants.

• **Colostrum**—This term refers to the fluid which is secreted from the mammary glands during the first few days after childbirth. Human colostrum contains maternal antibodies against various infectious organisms, plus factors which encourage the growth of beneficial microorganisms in the intestine. Hence, it might be desirable for all child-bearing women to try and nurse their babies for at least the first few days after birth, even if they do not wish to continue with breast feeding. Another reason for starting infants on colostrum is that it is believed that most allergies to cow's milk likely develop right after birth, when undigested milk proteins may be absorbed into the blood from the intestine.

(Also see BREAST FEEDING, section headed "Colostrum".)

• **Breast milk**—One of the reasons why human milk offers babies more protection against diarrhea than cow's milk is that the former contains the "bifidus factor" which promotes the growth of lactobacillus bifidus. Somehow, this beneficial species of bacteria inhibits the colonization of the intestine by harmful types of microorganisms. Other substances in breast milk which protect the infant against infection via the digestive system are (1) antibodies produced by the mother; (2) white blood cells; (3) lysozyme, an enzyme capable of splitting the molecules of carbohydrate in the cell walls of bacteria; and (4) lactoferrin, a protein which binds iron so that it is not available to support the growth of harmful microorganisms.[51] A disadvantage of breast feeding is that a wide variety of chemicals and drugs may be transmitted from the mother to the baby through breast milk. Hence, a nursing mother should ask her physician whether the drugs she customarily uses might pose any hazard to her infant.

(Also see BREAST FEEDING.)

• **Alteration of cow's milk so as to obtain colonization of the intestine by lactobacilli**—The most direct approach to this problem would be to feed the infant a fermented (soured) milk product which contains liberal amounts of lactobacilli. In fact, it has been shown that the feeding of yogurt to infants is an effective treatment for infec-

[51]Bezkorovainy, A., "Human Milk and Colostrum Proteins: A Review," *Journal of Dairy Science*, Vol. 60, 1977, p. 1023.

tious types of diarrhea.[52] Another approach is the feeding of unfermented (sweet) milk which has been fortified with lactobacilli. Finally, a breast milk substitute has been prepared from lactose, whey powder, predigested protein, and cream.[53] This preparation, when fed to babies, promotes the growth of an intestinal population of microorganisms which is similar to that in breast fed infants.

(Also see INFANT DIET AND NUTRITION.)

IMMUNIZATION PROCEDURES. While it is not within the scope of this book to cover immunization procedures fully, a brief presentation is made so that the reader may be informed about some of the nondietary protective measures against infectious diseases.

The artificial or natural process by which a person acquires sufficient antibodies to be immune to one or more diseases is called immunization. For example, an unborn child receives antibodies from its mother through the placenta. Additionally, antibodies may be received if the child is fed colostrum and/or breast milk. In these cases, the infant is said to have *passive immunity* since the antibodies were preformed outside of the infant's body. This type of immunity lasts for only a short time. Longer lasting *active immunity* is that type which results from antibodies being produced within the body in response to the introduction of a weakened or dead disease organism. Fig. D-35 depicts the artificial immunization of an infant.

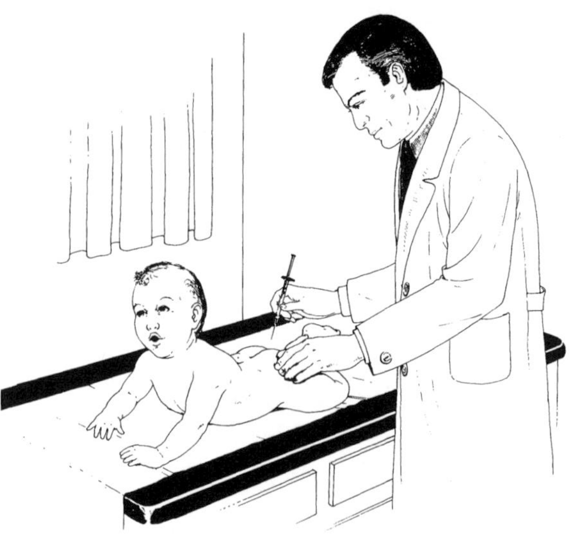

Fig. D-35. Immunization of an infant against infectious disease.

[52]Mount, J. L., *The Food and Health of Western Man*, John Wiley & Sons, Inc., New York, N.Y., 1975, p. 256.

[53]"The Effect of a Breast Milk Substitute on Stool Flora," *Nutrition Reviews*, Vol. 32, 1974, p. 135.

Brief descriptions of some common types of artificial immunizations follow.

Vaccination. This procedure may be defined as the injection or oral administration of some agent (such as a serum or vaccine) for the purpose of preventing disease.

In regions where a disease, such as measles, appears season after season, it is advised that healthy susceptible people be vaccinated before they are exposed, and before there is a disease outbreak, because (1) it takes time to produce an active immunity, and (2) some people may be about to be infected with the disease. Descriptions of the most common types of vaccinations or "shots" follow.

• **Serums**—These agents, which are obtained from the blood of animals (often horses), are used for the protective nature of the antibodies that they contain, which stop the action of an infectious agent or neutralize a product of that agent. They give an immediate, but passive, immunity. Among the serums that have proved successful are those for Rocky Mountain spotted fever, tetanus, and typhus.

• **Toxoids (or antitoxins)**—A toxoid is a "tamed" toxin which is treated chemically so that it loses the poisonous or toxic properties but still retains the power to stimulate the body cells to form the appropriate antibody. Typical toxoids are diphtheria toxoid and tetanus toxoid.

• **Vaccines**—Usually these agents consist of suspensions of live microorganisms (bacteria or viruses) or microorganisms that have had their disease-causing properties removed but their antibody-stimulating properties retained. Examples are smallpox vaccine, Salk (injected) and Sabin (oral) types of polio vaccines, and measles vaccine.

It is noteworthy that the ability of the body to form antibodies against disease appears to be related to nutritional status since it has been shown that various forms of malnutrition, particularly protein deficiency, are accompanied by both subnormal levels of antibodies and increased susceptibility to infection.

Hence, various types of immunizations may not be fully effective when they are given to malnourished people.

SPECIAL DIETARY MEANS OF PROMOTING RESISTANCE TO INFECTIOUS DISEASES. Information regarding the effects of nutritional deficiencies on the body's defenses against infectious organisms has been presented in Table D-14. Hence, it stands to reason that an adequate diet helps to reduce one's susceptibility to communicable diseases. However, it is still uncertain whether special dietary measures—that go beyond merely meeting nutrient requirements—confer any additional protection against these diseases. Nevertheless, it is worthwhile to consider the interesting, but controversial, ideas which follow.

• **High-protein diets**—There is direct evidence from animal studies and indirect evidence from research on humans that extra dietary protein—over and above that needed to meet requirements—confers increased resistance to infectious diseases. For example, growing chicks which were given a "super-normal" amount of protein had greater immunity to Newcastle virus than those fed a normal amount.[54]

Similarly, adult male college students on high-protein diets had a notable absence of respiratory ailments during a winter when there was a high incidence of influenza among their fellow students.[55] It is well known that high levels of dietary protein speed up protein metabolism in the body. Hence, it might be that these diets also speed up the production of antibodies—which are made up of protein—so that there is an increased resistance to infectious diseases.

• **Lactobacillus acidophilus**—Although the colonization of the intestine by these bacteria has been shown to protect infants and young children against infection by intestinal bacteria which cause diarrhea, it is not certain whether older children and adults might be afforded the same protection. Apparently, the consumption of large amounts of milk encourages the growth of lactobacilli in young children, whereas the mixed diets consumed by older people appear to favor the growth of other types of organisms. It seems that the numbers of intestinal lactobacilli decline when diets rich in meat are consumed, and rise when meat is eliminated.[56] However, other components of the diet, besides meat and milk, may affect the growth of lactobacilli.

• **Megadoses of vitamin C (ascorbic acid)**—There have been conflicting reports regarding the possibility that large doses (megadoses) of the vitaimin —from 10 to 100 or more times the amounts obtained in normal diets—might somehow augment the body's defenses against the common cold and influenza. Two of the strongest arguments for these benefits are (1) vitamin C levels in the blood and in the white blood cells drop sharply during colds and other infections, and (2) the findings from a carefully controlled study which was conducted during a winter in Canada.[57] The latter study showed that taking 1,000 mg of vitamin C daily, plus an additional 4,000 mg when the first signs of a cold appeared, resulted in 30% fewer days of disability for those who received the vitamin. (A control group received placebo pills.)

A tentative explanation for the action of vitamin C against colds and other viral diseases is that the vitamin acts indirectly by promoting the production of *interferon*, an antiviral agent which is released from cells when they are attacked by a virus.[58]

NOTE: Megadoses of vitamin C may be hazardous for some people because there have been reports of (1) false positive tests for sugar in the urine of diabetics; (2) an elevated excretion of oxalate in the urine, a condition which may lead to stones in the bladder or kidneys; and (3) over-acidification of the urine.[59] Therefore, one should first consult with a doctor *before* taking megadoses of vitamin C.

(Also see VITAMIN C.)

Resistance to infection may also be lowered by various stresses. This effect is attributed in part to the hypersecretion of the adrenal glands, since the hormones released during stress reduce inflammatory responses and cause shrinkage of the tissues which produce antibodies. It is, therefore, important to minimize the possibility of stresses by eating balanced meals at regular intervals, getting sufficient rest, avoiding extremes of temperature, and learning to minimize emotional upsets.

Sanitary Measures Which May Help To Prevent The Spread Of Diseases. Disease epidemics are most likely to occur in densely populated areas, particularly when certain sanitary measures fall short of what is needed. Hence, the sanitation measures which follow are pertinent.

FOOD SANITATION. Outbreaks of food poisoning from microorganisms or parasites still occur on a regular basis in the United States and in other parts of the world. For example, in September 1974, approximately 3,400 persons at a Navajo Indian fair in Arizona were stricken with foodborne salmonellosis.[60] An investigation of the outbreak by the Public Health Service showed that its probable cause was the improper handling of the ingredients of a potato salad. Similar outbreaks take place each year at large banquets, picnics, and other affairs where the problems of food preparation and storage are complicated by the large numbers of people to be fed. (The cases of food poisoning which come to the attention of health authorities may only be a fraction of those which occur, since many of those stemming from the feeding of small groups of people may not be reported.)

Opportunities for the contamination of foods are present at the various points between their production on farms or factories, and the serving of them at meals. Hence, the following sanitary measures should be observed:

1. Raw fruits and vegetables should be washed before using.

---

[54]Fisher, H., *et al.*, "Protein Reserves: Evidence for Their Utilization under Nutritional and Disease Stress Conditions," *The Journal of Nutrition*, Vol. 83, 1964, p. 165.

[55]Fisher, H., *et al.*, "Nitrogen Retention in Adult Man: A Possible Factor in Protein Requirements," *The American Journal of Clinical Nutrition*, Vol. 20, 1967, p. 932.

[56]Maier, B. R., *et al.*, "Effects of a High-Beef Diet on Bowel Flora: A Preliminary Report," *The American Journal of Clinical Nutrition*, Vol. 27, 1974, p. 1471.

[57]Anderson, T. W., D. B. Reid, and G. H. Beaton, "Vitamin C and the Common Cold. A Double-Blind Trial," *Canadian Medical Association Journal*, Vol. 107, 1972, p. 503.

[58]Lewin, S., "Recent Advances in the Molecular Biology of Vitamin C," *Vitamin C*, edited by G. G. Birch and K. J. Parker, Applied Science Publishers, Ltd., Essex, England, 1974, p. 234.

[59]Lamden, M. P., "Dangers of Massive Vitamin C Intake," *The New England Journal of Medicine*, Vol. 284, 1971, p. 337.

[60]Horwitz, M.A., *et al.*, "A Large Outbreak of Foodborne Salmonellosis on the Navajo Nation Indian Reservation, Epidemiology and Secondary Transmission," *American Journal of Public Health*, Vol. 67, 1977, p. 1071.

2. All meats should be thoroughly cooked, particularly hamburger and pork, which are the most susceptible to contamination.

3. Dairy products should be made from pasteurized milk unless the dairy has been certified to produce raw milk.

4. Food handlers should be free of contagious disease; and they should be required to conform to recommended food sanitation practices.

5. Perishable foods should not be kept at room temperatures between 40° and 140°F (*4.4° and 60°C*) (the range where most microorganisms thrive) for more than a few hours.

6. Frozen foods which have been allowed to thaw completely so that they no longer contain ice crystals should not be refrozen for continued storage.

7. Cracked or broken eggs should *not* be used for the dishes which require short cooking times at low temperatures because they may contain Salmonella organisms. These eggs are safe only when they have been heated to at least 160°F (*71.1°C*) for 20 minutes or longer.

(Also see BACTERIA IN FOOD; FOODBORNE DISEASE; FOOD POISONING; POISONS; and PRESERVATION OF FOOD.)

OTHER SANITARY MEASURES. In addition to the sanitary handling of food, other measures are necessary to prevent the spread of communicable diseases. Hence, the practices which follow are noteworthy.

• **Adequate Disposal of Wastes**—In the distant past, when people lived in small and widely scattered villages, there was little spread of disease. Today, much of the world's population is crowded into areas where accumulations of garbage, sewage, and other wastes provide fertile breeding grounds for disease. Hence, in these places there may be a need for more efficient sewage treatment, better waste disposal, control of pests, and perhaps tougher laws to prevent air, land, and water pollution.

• **Disinfection**—Dishes and eating utensils which have been in contact with diseased persons, or which may have somehow been contaminated with germs or parasites, should either be disposed of or disinfected. Unfortunately, there is no one best germ killer, nor is there anything like a general disinfectant that is effective against all types of microorganisms and parasites under all conditions. However, the application of heat by steam, by hot water, by burning, or by boiling is an effective method of disinfection.

• **Quarantine**—This term refers to the regulation of movement of people, food, animals, or plants by either agricultural or health authorities so as to prevent the spread of infectious diseases, parasites, insects, and other disease agents.

Today, strict quarantines are rarely applied to people in their homes, except in the cases of a few diseases like diphtheria. it is more common for school authorities to bar children who may lack evidence of having had certain immunizations.

• **Ventilation**—This term refers to the circulation of air through buildings with the objective of supplanting foul air with fresh air containing needed oxygen. Generally, moist air is a more favorable medium for the existence of microorganisms than dry air, thus lending itself well to the transmission of contagious diseases. However, excessively dry air, which is present in certain heated buildings during the winter months, may irritate the nose, throat, and lungs and increase susceptibility to respiratory infections. This problem might be overcome by introducing small amounts of moisture into the air with humidifiers.

Healing Of Injuries From Physical Agents. It is well known that some people heal more rapidly than others, which suggests that good nutrition plus other health-promoting factors—like fresh air, sunshine, and adequate rest—may be responsible. Additional vitamin C, vitamin A, protein, and energy may be required for the formation of new connective tissue. Also, the building up and maintenance of abundant muscle through diet and exercise is important. The protein in muscle is often an emergency source of amino acids needed for the synthesis of protein during healing. Most injuries from physical agents involve a stress response where there is an elevated secretion of adrenal cortical hormones. These stress hormones promote the breakdown of muscle protein and the release of amino acids into the blood where they are available for the repair of tissue. However, unnecessary stresses should be avoided, as they may retard healing.

Recently, the mineral element zinc has been shown to promote healing. Also, it has been claimed that vitamin E has similar effects.

It is noteworthy that diabetics heal poorly, so it is important to be ever watchful for signs of this disease.

(Also see DIABETES MELLITUS.)

Prevention Of Malnutrition. Reaching this goal requires more than the consumption of sufficient food from each of the four food groups. It also requires a recognition of individual needs which may vary according to heredity and the special circumstances of individual life-styles. When sufficient food and the incentive to eat are present, there must be considerable self-discipline to avoid overeating of the high-energy foods. People who have great appetites should eat foods which fill, but do not fatten, such as low-calorie fruits and vegetables which contain liberal amounts of fiber.

Some scientists suggest that there may be widespread subclinical malnutrition due to deficiencies of the trace elements, like chromium and selenium, which are removed during the refining of grains and other foods. When the refined foods comprise much of the diet, it may be necessary either to enrich them with the missing elements or to provide some type of mineral supplement.

(Also see MALNUTRITION.)

Persons whose families are prone to certain diet-related diseases should have physicians look for early indicators of these disorders. Early dietary treatment of these diseases may help to forestall their more severe stages. For example, many persons have literally had years added to their lives by the prompt adoption of modified diets for anemia, diabetes, gluten sensitivity, and ulcers of the stomach or duodenum.

Reduction Of Stress. Some stress is inevitable, and may even be beneficial. However, situations which provoke the secretion of stress hormones also cause marked reductions in the vitamin C content of the adrenals. Therefore, it has been suggested that supplemental ascorbic acid might be helpful in meeting the demands of stress. Also, multiple mineral and vitamin capsules which are supposedly beneficial in counteracting stresses are available, but their effectiveness is uncertain.

The best policy seems to be to avoid, or to minimize, stresses whenever they may be anticipated. For example, much of the stress from cold weather may be reduced by wearing protective clothing. Emotional stresses may often be alleviated when one develops a new set of responses to problem situations. Although the evidence is vague, it has been assumed that regular exercise helps hard-pressed executives to overcome their emotional stresses.

Finally, a simple, but helpful principle is to try, whenever possible, to help the body to cope with unavoidable stresses. This means that the treatment of a taxing, infectious disease should include good nutrition and rest.

Detoxification Of Poisonous Substances. A healthy liver has at its disposal many chemical reactions for the detoxification of poisonous substances. These reactions rely on the presence of enzymes whose activity depends in part on the level of dietary protein, and in part on the need to detoxify substances such as alcohol and various drugs. Therefore, protein-deficient persons may have heightened susceptibility to the undesirable side effects of certain medications. Chronic abuse of alcohol may also impair liver function since it may promote fatty degeneration. In addition to protein, other nutrients which promote liver function are the B-complex vitamins, methionine, choline, and lecithin.

Health And Nutrition Functions Of Government Agencies. The welfare of a nation is dependent upon the health of its citizens. Hence, it is often necessary that the various levels of government within a country provide both assistance and regulation in health- and nutrition-related areas such as (1) the prevention and treatment of disease, (2) the production and distribution of foods and drinking water, and (3) other activities and conditions which may affect the health of the people. Some noteworthy functions of government agencies in disease prevention and nutrition follow.

U.S. DEPARTMENT OF HEALTH AND HUMAN SERVICES. This department is made up of agencies that have responsibility for public health, education, and economic security. The health agencies come under the jurisdiction of the U.S. Public Health Service (USPHS). Brief descriptions of these agencies follow.

- **Alcohol, Drug Abuse, and Mental Health Administration**—Programs concerned with the prevention and treatment of alcoholism, drug addiction, and mental illness are conducted by this agency.

- **Center for Disease Control (CDC)**—This agency, whose initials originally stood for "Communicable Disease Center" was established to conduct programs dealing with the causes and the control of contagious disease. Its name was changed to the present one when its mission was expanded to include noninfectious diseases resulting from environmental, nutritional, and occupational factors.

- **Food and Drug Administration (FDA)**—The FDA is charged with the responsibility of safeguarding American consumers against injury, unsanitary food, and fraud. It also protects industry against unscrupulous competition. It inspects and analyzes samples and conducts independent research in fields such as contamination of food by insects, microorganisms, and other unsanitary items; toxicity (using laboratory animals); disappearance curves for pesticides; and long-range effects of drugs.

- **Health Resources Administration**—The programs conducted by this agency are concerned with (1) the training of health professionals, (2) collection of data on health problems in the United States, and (3) research designed to evaluate and improve health services.

- **Health Services Administration**—This agency either provides or oversees (1) health care for American Indians and merchant seamen, (2) family planning, (3) improvement of maternal and child health, and (4) community health programs.

- **National Institutes of Health (NIH)**—The NIH is composed of the following nine sister institutes: the National Cancer Institute, the National Heart Institute, the National Institute of Allergy and Infectious Diseases, the National Institute of Arthritis and Metabolic Diseases, the National Institute of Dental Research, the National Institute of Mental Health, the National Institute of Neurological Diseases and Blindness (including multiple sclerosis, epilepsy, cerebral palsy, and blindness), the National Institute of Child Health and Human Development, and the National Institute of General Medical Science. In addition to its own research program, the NIH provides grants for health-related research at many universities and research institutes in the United States.

(Also see U.S. DEPARTMENT OF HEALTH AND HUMAN SERVICES.)

U.S. DEPARTMENT OF AGRICULTURE (USDA). The following eight divisions of this department have primary responsiblility in the areas of human health and nutrition.

- **The Agricultural Research Service**—This agency is the main research division of the USDA. A considerable amount of its resources and staff are committed to the development of new food products, human nutrition, the nutrient composition of foods, and protection of foods against spoilage.

• **The Animal and Plant Health Inspection Service**—This division is charged with maintaining the wholesomeness and safety of meats processed in packing plants that ship meat and meat products, including poultry and poultry products, interstate.

• **The Consumer and Marketing Service**—The major responsibility for the federal grading of foods is assigned by law to this division. Butter, cheese, eggs, fresh produce, processed fruits and vegetables, grains, legumes, meats, nonfat dry milk, and poultry are inspected and graded according to their desirability for consumers. Also, the safety of certain items is checked by means of bacterial and mold counts, and chemical analyses.

This agency has established standards of identity for meat and poultry products such as breaded items, frankfurters, hash, pot pies, and soups. The standards specify (1) the minimum amounts of meat and poultry which must be present; and (2) the maximum allowable amounts of breading, fat, fillers, sweeteners, and water.

• **The Cooperative State Research Service**—This agency is charged with administering the funds for the research projects which are conducted by the state agricultural experiment stations. Periodically, staff from the experiment stations in various parts of the United States work together in regional projects which study the status of health and nutrition of selected groups within their respective states.

• **The Federal Extension Service**—The function of this division is to teach people how to make better use of agricultural and community resources. Much of their current educational work involves the promotion of better dietary practices. Also, some of the staff are professional dietitians who advise people on the dietary modifications which are necessary in the treatment and prevention of various diseases. The federal extension workers participate in cooperative programs with their counterparts in state and local extension services.

• **The Food and Nutrition Service**—This division provides the means for better nutrition through programs such as (1) distribution of surplus agricultrual commodities, (2) food stamps, and (3) reimbursement to schools for the operation of breakfast and lunch programs.

• **The Labeling and Registration Section**—This section has responsibility for the proper labeling and safe use of pesticides. Manufacturers of pesticides must present new products with their proposed labels for approval before they are authorized to sell them.

It is the responsibility of the FDA, however, to set legal tolerances for pesticides on or in raw agricultural products. Also, it sets the "safe" interval between last application of the insecticide and the time of harvest of the crop or the slaughter of the animal.

Thus, through the cooperative supervision of the USDA and the FDA, both the pesticide user and the consumer of the product are safeguarded.

• **The Veterinary Services Division (VSD)**—This division is responsible for programs to control and eradicate (if possible) certain diseases of livestock; e.g., brucellosis, tuberculosis, scabies, and hog cholera.

(Also see U.S. DEPARTMENT OF AGRICULTURE [USDA].)

STATE AND LOCAL GOVERNMENT AGENCIES. State departments of health and agriculture often perform functions similar to the federal counterparts of these agencies. Some overlapping of activities is unavoidable because the federal government does not have the authority to intervene in matters within the jurisdictions of individual states. However, it is often necessary to delegate further authority to county units such as extension offices and boards of health. Most local units maintain good relations with the state and federal authorities in the event that it might be necessary to call on them for assistance with such problems as epidemics of communicable diseases, and the occurrence of harmful substances in foods or drinking water.

International And Voluntary Organizations Engaged In Health And/Or Nutrition Activities.

The governments of many of the developing countries lack the resources needed to cope with their widespread health and nutritional problems. Hence, it may often be necessary for them to seek help from the World Health Organization (WHO) and the Food and Agriculture Organization (FAO). Also, sometimes assistance may be obtained from voluntary relief organizations like CARE, or from private nonprofit foundations such as the Ford Foundation and the Rockefeller Foundation.

Even the developed countries—whose peoples and governments contribute generously to the support of the United Nations and other international agencies—may benefit from the activites of these organizations. Typically, the various types of nongovernmental programs serve to (1) call attention to the urgency and scope of worldwide health problems; (2) demonstrate that there may be simpler and less expensive ways of meeting basic health and nutritional needs; (3) provide help with problems that are not adequately met by the activities of the governmental agencies; and (4) mobilize people and resources at the grass-roots level.

The many types of voluntary organizatons engaged in health and nutrition activities are too numerous for inclusion in this article; thus, only a few well-known and noteworthy examples follow.

• **Groups sponsored by private industry**—The best-known organizations in this category are the National Dairy Council and the National Live Stock and Meat Board. Generally, these groups engage in educational and public relations activities, which are intended for a wide audience, ranging from lay persons to highly educated professionals.

• **Professional organizations**—Usually, dentists, dietitians, doctors, health educators, nurses, and public health workers belong to professional groups which (1) may help to set and maintain the standards of practice, (2) hold yearly national meetings to discuss current issues, (3) publish scholarly journals, and (4) otherwise promote the interests of the profession. These activities all help to improve the level of health care received by the patient.

• **Private social or charitable agencies and church-sponsored groups**—These organizations often obtain their operating funds from private contributions and/or endowments. They may engage in activities such as health education, providing food to the needy, operating lunch programs for senior citizens, sending out visiting nurses and/or homemakers to the homebound, and referring their clients to other private and public services. Some typical organizations are the Red Cross, March of Dimes, Salvation Army, visiting nurse associations, settlement houses, and the agencies operated by the major religious denominations. It is noteworthy that the establishment of hospices for the sick poor was one of the major missions of the Franciscan Order which was established by Francis of Assisi around the end of the 12th century.[61]

• **Civic and fraternal organizations**—Long before the governments of nations got into the business of looking after the welfare of their citizens, people from local communities joined together to promote their interests. For example, the forerunners of our present-day chambers of commerce were groups of businessmen in the days of the Roman Empire.[62] Today, civic clubs work with fraternal and charitable organizations such as Elks, Lions, Rotarians, and Shriners to raise funds to provide hospitals, care for the blind, homes for the aged, and other services for the needy. These groups provide settings within which physicians, business people, and others from various walks of life may pool their talents for the improvement of the health and welfare of their community.

(Also see NUTRITION EDUCATION.)

DISINFECTION

The process of destroying harmful microorganisms in or on items which come in contact with foods by means such as (1) exposure to sunlight or other ultraviolet radiation, (2) treatment with chemicals, or, (3) application of dry or moist heat.

(Also see DISEASES.)

DISORDER

A derangement of function; an abnormal physical or mental condition; sickness, ailment, malady.

DISTAL

That part of a structure farthest from the point of attachment.

[61]Snively, W. D., and D. R. Beshear, *Textbook of Pathophysiology*, J. B. Lippincott Co., Philadelphia, Pa., 1972, pp. 47-48.

[62]Griffin, A., "Chamber of Commerce," *The World Book Encyclopedia*, Vol. 3, 1976, p. 283.

DISTILLATION

A procedure in which desired substances such as alcohol or water are separated from undesired materials by boiling, followed by condensation and collection of the vaporized liquid in a separate vessel.

(Also see DISTILLED LIQUORS.)

DISTILLED LIQUORS

Contents *Page*

Origin and History of Spirits 624
Production of the Major Types of Distilled Alcoholic Beverages 624
Raw Materials 624
Fermentation 625
Distillation 625
Aging and Mellowing 625
Blending 625
Less Common, but Highly Regarded Liquors 627
Nutritional and Medicinal Effects of Spirits 627
Selection and Utilization of Distilled Liquors 627
Serving Spirits 627
Age, Proof, and Size of Bottle or Glass 627
Age 628
Proof 628
Size of Bottle or Glass 628
Miscellaneous Uses of Spirits and the By-Products of Distillation 628

These alcoholic beverages are made by some of the same processes used to make beers and wines, except that liquors contain more alcohol because they have been distilled. Alcoholic distillation is a process in which (1) a dilute solution of alcohol in a boiler (still) is heated to a temperature above the boiling point of alcohol, but below the boiling point of water, (2) the alcohol vapor produced in the still is led through a condenser where it is converted back into a liquid, and (3) the liquid alcohol (distillate) is caught in a receiving vessel. Fig. D-36 depicts a simple distillation apparatus.

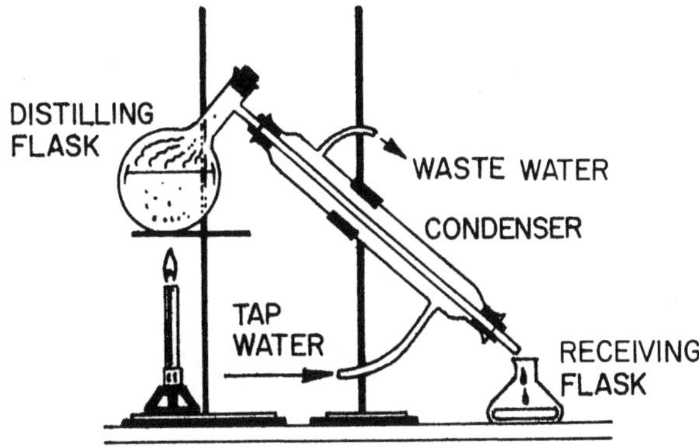

Fig. D-36. Distillation of alcohol on a small scale. This type of apparatus is used mainly in chemistry laboratories. However, commercial distillation equipment embodies the same principles, although it is larger and more complex.

It is noteworthy that distillation of fermented mixtures does not yield pure alcohol, because small amounts of water and other substances are carried over with the alcohol vapor.

Distilled liquors are called *spirits* because, poetically speaking, they represent the spirit or soul of the fermentation mixture. The different types of spirits vary considerably in their characteristics because, in addition to alcohol and water, they contain the volatile substances called congeners, which contribute aroma, color, and taste. The professional distiller is an artist of sorts, who controls the process conditions in order to obtain the desired characteristics in the final product.

ORIGIN AND HISTORY OF SPIRITS. It is not known where or when the first spirits were prepared by distillation. Perhaps, the process was developed independently by different peoples in different parts of the world. It is known that, by 800 B.C., the peoples of both China and the East Indies were making liquors from rice. The Arabs were also ancient distillers, who spread their art among the peoples with whom they traded. However, certain Celtic tribes in Ireland and Scotland appear to have developed the art of distilling fermented grain mashes on their own. In fact, the word whiskey is derived from the gaelic word meaning "water of life."

The distillation practices of the Celts were improved greatly after the conquest of Britain by the Romans, who introduced metalworking techniques. Prior to the use of metal for stills, they were made from pottery, which was difficult to seal so that the alcoholic vapors did not escape. The improvement of distillation equipment led to an ever increasing production of spirits, which, in turn, led various governments to obtain revenues by taxing the alcoholic beverages.

A thirst for liquors helped to bring about early American sea trading because the difficulty in growing barley in New England led the colonists to use imported molasses to make rum. Some historians have even speculated that the Molasses Act passed by England in 1733 was a major cause of the colonists' becoming alienated from their mother country. This law was designed to force the Americans to buy their molasses from the British colonies in the Caribbean, since it levied a high tax on molasses purchased elsewhere. The need for molasses also stimulated slave trading because the American shipowners found it profitable to deliver New England rum to Africa, where they picked up slaves to exchange for molasses in the West Indies. Finally, the attempt of George Washington to obtain needed revenue by taxing whiskey led to the Whiskey Rebellion of 1791. It was necessary to send a force of militia to Pennsylvania after some of the tax agents were tarred and feathered. Fortunately, the rebellion was quelled without bloodshed.

The taxation of whiskey spurred a movement of distillers westward to the states beyond the jurisdiction of the tax collectors. Hence, new sites of production were established in Indiana and Kentucky. It is noteworthy that the product known as *bourbon* was first brewed from corn mash in Bourbon County, Kentucky.

Per capita consumption of spirits in the United States reached a peak around 1830, when it was approximately 10 gal (*38 liter*) per year. (At the present time, it is about 2 gal [*7.6 liter*] per year.) The subsequent decline in consumption has been attributed to the Temperance Movement, which spread rapidly throughout the newly settled Midwestern farming areas in the first half of the 19th century. However, the attempt to prohibit the use of alcoholic beverages from 1920 to 1933 resulted in considerable amounts of illegal beverages being produced and sold by "bootleggers." Also, an important source of tax revenue was lost during this period.

Shortly after the repeal of Prohibition in 1933, the formula for the leading brand of Russian vodka was brought to the United States. However, this drink was not well accepted by Americans until after World War II, when it became fashionable to use it in mixed drinks. One of the early inducements to drinking vodka instead of other liquors was that the former was so highly refined that it left no telltale aroma on the breath.

Today, the distilling industry is more efficient than ever, since the spent grains and other by-products now find a ready market as animal feeds.

PRODUCTION OF THE MAJOR TYPES OF DISTILLED ALCOHOLIC BEVERAGES. Many liquor manufacturers utilize procedures which are similar in many aspects. Fig. D-37 outlines the basic processes used in Great Britain and in the United States for the production of distilled beverages, along with by-products that are used as animal feeds.

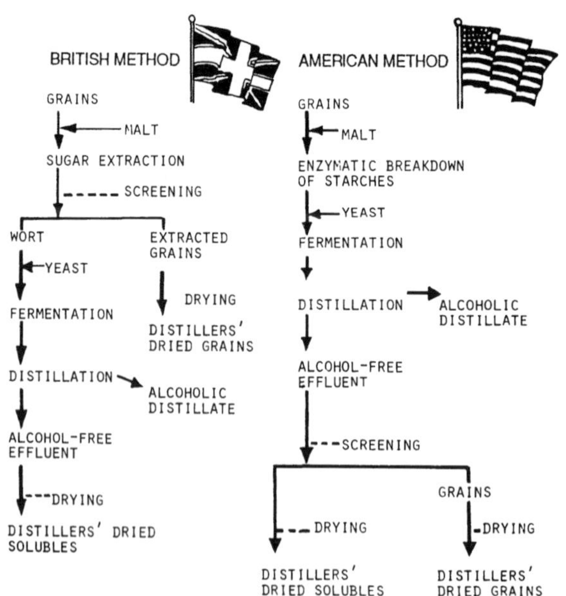

Fig. D-37. An outline of the British and American processes of distillation.

Raw Materials. Alcohol is usually produced by the yeast fermentation of sugars. Hence, the major raw materials may be (1) sugary items such as fruit juices or fruit by-products, honey, sugar or molasses, or the sap from certain trees; or (2) starchy foods such as grains, roots, and tubers. However, starches require conversion to sugar by the enzymes from germinated grains (malts), or from the appropriate microorganisms.

Fermentation. This process is usually carried out at a temperature that promotes the maximum activity of the yeast cells. In most cases, this means from 75° to 85°F (*24° to 29°C*). The product of fermentation is called a mash.

Distillation. The type of still used for this process is selected on the bases of (1) the amount of mash to be distilled, and (2) the substances to be collected in the distillate. For example, certain flavored liquors are produced in stills that are constructed so that the alcoholic vapors extract flavored components from materials such as aniseed, juniper berries, and lemon or orange peel.

Aging And Mellowing. Some liquors which contain harshly flavored components (congeners) when freshly distilled are improved by aging. Aging for several years or more in a wooden cask produces a more mellow flavor because certain congeners are either altered chemically or are absorbed by the wood. Another means of dealing with unwanted congeners consists of the passage of the liquor through absorbent charcoal.

Blending. Strongly flavored liquors are sometimes blended with highly refined alcohol that is almost flavorless (neutral spirits) in order to (1) obtain the desired characteristics in the product offered for sale, and (2) stretch the supply of the flavored product, which is usually more expensive to produce. For example, bourbons and Scotch malt whiskys are usually blended with neutral spirits.

Table D-15 gives some of the more notable details regarding the production and characteristics of the major types of distilled alcoholic beverages.

TABLE D-15
PRODUCTION AND CHARACTERISTICS OF THE MAJOR TYPES OF DISTILLED LIQUORS

Type of Liquor	Source(s) of Sugars for Fermentation	Important Details of Production	Flavor and Other Characteristics
Brandy	Apples, apricots, blackberries, cherries, elderberries, grapes, pears, pineapple, plums, raspberries, and strawberries.	Brandies are distilled from either a wine or a fermented mash of fruit. They are often aged for 2 or more years to mellow the harshly-flavored constituents common to distilled liquors.	A fruit flavor blended with other flavors which result from fermentation, distillation, and aging.
Gin	Mixtures of grains and malt that are similar to those used to make beers and whiskies.	The fermented grain mixture is distilled to yield a strongly alcoholic mixture, which is then either redistilled or mixed with flavoring derived from juniper berries, lemon or orange peel, or other botanical substances.	Flavored like the botanical flavoring material and the liquor obtained by fermentation and distillation.
Liqueur or Cordial	Mixtures of grains, malt, and/or sugar that are similar to those used in the production of the other distilled liquors.	A distilled liquor is either mixed or redistilled with one or more flavoring materials such as fruits, fruit peels, herbs, spices, flowers, cocoa, coffee, or roots. Liqueurs may contain from 2½% (minimum) to 35% added sugar.	Sweet and mellow. They are believed to aid the digestive processes.
Rum	Juice, syrup, molasses, or other products from sugarcane. Molasses is the most commonly used raw material, since it is a by-product of sugar production.	Rum is distilled from a fermented sugarcane product. The stronger flavored rums are usually aged for 3 or more years. Sometimes, the distillation residue is added to subsequent rums to produce a strongly flavored product.	Flavor ranges from mild to very strong, depending upon the mode of production.

(Continued)

TABLE D-15 *(Continued)*

Type of Liquor	Sources(s) of Sugars for Fermentation	Important Details of Production	Flavor and Other Characteristics
Vodka	Starchy materials such as potatoes and/or grains (the starch is converted to sugar by enzymes).	The distillation process used for vodka yields a product high in alcohol and low in congeners. Then, the distillate is usually run through charcoal to remove any unwanted components.	Flavorless, unless special flavorings (usually herbs, honey, or spices) have been added.
Whiskey: 1. All types	Malted and unmalted grains. (The enzymes in the former convert the starch in the latter to sugar.)	Distillation of the fermented grain mash is conducted so as to yield a product rich in congeners. The newly distilled liquor is usually aged in wooden barrels. Various whiskeys are often blended.	The flavor depends upon the raw materials (grains and type of malt), distillation, aging, and blending.
2. Bourbon	Malted grain and unmalted grain which is mainly corn.	The distillation mixture is a sour mash containing about ¼ old mash (previously fermented) and ¾ new mash. Aging is done is charred barrels.	A distinctive flavor due to the corn mash plus the aging process.
3. Canadian	Malted barley and unmalted corn, rye, and wheat.	Produced in Canada. Distilled as other whiskeys, then aged for 3 or more years. Usually, Canadian whiskeys of different characteristics are blended.	Delicately flavored with a light body.
4. Irish	Malted barley and unmalted barley, corn, oats, rye, and wheat.	Product of either the Republic of Ireland or Northern Ireland. The fermented mash is distilled in 3 stages. (Hence, it is "triple distilled.") Irish whiskeys are aged for at least 4 years. Many are blended.	A unique, clean flavor of its own, with considerable smoothness.
5. Rye	The predominant grain in the mash is rye.	Distilled to produce a high content of congeners, then aged in charred barrels.	Distinctive, sweet flavor with a full body.
6. Scotch	Mainly malted barley. The malt is dried on screen over a smoky peat fire.	Produced in Scotland. The fermented mash is distilled in 2 stages. (Hence, it is "double distilled.") Newly distilled Scotch whisky is aged from 3 to 4 years in barrels previously used for whisky or wine, then it is blended with other whiskeys.	The special flavor of Scotch whisky is due in large part to the smoky peat fire used to dry the malted barley.

LESS COMMON, BUT HIGHLY REGARDED LIQUORS. Many peoples around the world have shown great ingenuity in developing liquors from various types of starchy or sugary plants. Although many of these items are produced mainly for local consumption, a few of them have gained international acceptance. Some of the most popular items follow.

• **Aquavit (Akvavit in Denmark)**—This liquor is produced in Denmark, Iceland, Norway, and Sweden by distillation of a fermented mash of barley malt plus grain or potatoes. The resulting distillate, which contains 95% alcohol and is essentially free of congeners, is then mixed with water and redistilled with caraway seeds and one or more of the optional flavorings: anise, cardamon, or lemon or orange peel. However, much of the flavor of Aquavit is derived from the caraway seeds.

• **Bitters**—The formulas and means of preparation for most of these liquors are closely guarded secrets. Generally, they consist of highly refined alcohol that is flavored with an extract of one or more aromatic and bitter plant materials such as bark, flowers, fruit, leaves, roots, seeds, or stems. Bitters are thought to aid the digestive processes.

• **Okolehao**—It is claimed that this liquor is distilled solely from a fermented mash of the roots from the sacred ti plant (*Cordyline australis*) of Hawaii. Aging is not required because the harshly flavored components of the distilllate are removed by passage through charcoal. Okolehao has a unique flavor that is difficult to describe.

• **Tequila**—This liquor is double distilled from the fermented sap of the mezcal plant (a species of *agave*) which is native to Mexico. It is noteworthy that fermented mezcal sap, which is called pulque, was the first alcoholic beverage produced in North America. Many brands of tequila are strongly flavored, but others are more mellow because they have been aged in oak vats for periods ranging from 2 to 4 years.

NUTRITIONAL AND MEDICINAL EFFECTS OF SPIRITS. Food Composition Table F-36 lists a variety of alcoholic beverages—beers, wines, and liquors. In contrast to beers and wines, which contain certain minerals and vitamins, distilled liquors are so highly refined that they supply mainly "empty calories." For example, the caloric content of 1 oz (*29.6 ml*) of gin, rum, vodka, or whiskey ranges from 65 Calories (kcal) for 80-proof spirits to 83 Calories (kcal) for 100-proof spirits. Furthermore, other beverage spirits such as brandies, cordials, liqueurs, and mixed drinks may contain sufficient added sugar to make their caloric values much higher than the unsweetened spirits. (Most of the liqueurs sold in the United States contain from 100 to 120 Calories [kcal] per ounce.) Hence, heavy drinkers may obtain too much of their caloric requirement from liquor, and too little from the foods which furnish essential nutrients. It is well documented that chronic alcoholics often suffer from various types of malnutrition.[63]

On the other hand, the light to moderate use of spirits may indirectly confer nutritional benefits. A drink of liquor taken before a meal may help to insure that optimal amounts of foods are consumed because its calming effect relaxes those who (1) cannot eat enough because of jitters, or (2) are nervous, compulsive eaters. Also, small amounts of brandies, cordials, or liqueurs are thought to aid digestion.

Some of the other medicinal benefits of alcohol are: alleviation of minor aches and pains, inducement of sleep, dilation of blood vessels so that more blood reaches the tissues, warming of the skin, and relaxation of sexual and social inhibitions. However, there is always the danger that one may become dependent on these effects, and that excessive drinking may result.

(Also see ALCOHOLISM.)

SELECTION AND UTILIZATION OF DISTILLED LIQUORS. Three of the most important considerations for the discriminating consumer are (1) choosing the most appropriate liquor for the use intended, (2) serving it so that its qualities may be appreciated fully; and (3) getting the best buy for the desired product in terms of age, level of alcohol, and cost per ounce. Some general guidelines on these matters follow.

Serving Spirits. Some of the customary ways of serving each of the more popular types of liquor are given in Table D-16.

Age, Proof, And Size Of Bottle Or Glass. The larger retail stores may offer two or more variations of a given brand of spirits. Then, the customer has to decide which is the best buy in terms of his or her particular needs. Even those who buy their liquor by the drink in bars may wish to make comparisons between the different establishments. Therefore, the items that follow are noteworthy.

[63]Stone, O. J., "Alcoholic Malnutrition and Skin Infections," *Nutrition Today*, Vol. 13, November/December 1978, pp. 6-10, 27-30.

TABLE D-16
COMMON WAYS OF SERVING SPIRITS

Name of Spirit	How Served[1]
Aquavit	Ice cold in a 1-oz *(30 ml)* glass, with a beer chaser, at meals.
Bitters	Mixed with other spirits, or alone as an aid to digestion after meals.
Brandy	Straight with water as a chaser; mixed with coffee, soda, or water; or added to cakes, fruit compote, ice cream, mincemeat pie, puddings, or sauces.
Gin	Straight; with bitters; or in mixed drinks.
Liqueurs (cordials)	After dinner to aid digestion; as a frappe (poured over finely shaved ice); in mixed drinks; mixed with other spirits; or added to baked items, candies, ice cream, or sauces.
Okolehao	Straight; on the rocks; or in mixed drinks.
Rum	Straight; on the rocks; mixed with a cola soft drink, a citrus juice, gingerale, soda, or water; or added to candies, ice cream, mincemeat pie, or sauces.
Tequila	Straight; mixed with a citrus juice, or in other mixed drinks.
Vodka	Ice cold in a 1-oz *(30 ml)* glass; on the rocks; or in mixed drinks.
Whiskey	Straight; on the rocks; or in mixed drinks.

[1]Many of the retail stores in which spirits are sold provide free booklets of recipes for preparing various types of mixed drinks.

AGE. Liquors with a high congener content should be aged so that the harshly flavored components have a chance to mellow. Hence, straight Scotch whiskys and similar unblended, strongly flavored products require at least 4 years of aging, whereas liqueurs and vodka require no aging because they are distilled so as to contain only negligible amounts of congeners. The American consumer does not have to be overly concerned about aging because the U.S. Standards of Identity for Distilled Spirits specify minimal aging requirements which must be met if the product is to bear a certain name on the label.

PROOF. The numerical value of this term is twice the percent alcohol (by volume) in the liquor. For example, 80 proof means that the product contains 40 percent alcohol. Both the government tax and the selling price are pegged to the alcoholic content of the product. Hence, the lower proof variety of a given brand costs less than one with a higher proof.

SIZE OF BOTTLE OR GLASS. The most common retail sizes are given in Table D-17.

TABLE D-17
COMMON GLASS OR BOTTLE SIZES FOR RETAIL SALES OF SPIRITS

Designation of Size	Contents (oz)	Contents (ml)[1]	Comments
Shot glass	.75 to 2.0	*22 to 59*	Watch out for "cheaters," which are shot glasses marked at 5/8 oz instead of ¾ oz.
Pony	1.0	*30*	A standard size for single drinks in the U.S.
Jigger	1.5	*44*	A standard size for single drinks in the U.S.
Miniature bottle	1.7	*50*	A standard metric size.
Fifth of a pint	3.2	*95*	May be served on airlines.
Fifth of a quart	6.4	*189*	
Quarter-bottle	6.4	*189*	
Split	6.4	*189*	
Pint	16.0	*474*	
Imperial pint	20.0	*592*	Unit of measurement in Canada and Great Britain.
Bottle	25.6	*758*	About the same as the regular size metric bottle.
Fifth	25.6	*758*	About the same as the regular size metric bottle.
Quart	32.0	*948*	
Liter	33.8	*1,000*	Metric unit of measurement.
Imperial quart	40.0	*1,180*	Unit of measurement in Canada and Great Britain.
Half gallon	64.0	*1,890*	
Gallon	128.0	*3,790*	About 3 ¾ liter.
Imperial gallon	160.0	*4,740*	Unit of measurement in Canada and Great Britain.

[1]Imported liquors may be sold in metric size bottles such as 50 ml (miniature), 200 ml (small), 500 ml (medium), 750 ml (regular), 1 liter (large), and 1.75 liter (extra large). For price comparisons, the American consumer should check the nearest size bottle that is used in the United States.

Miscellaneous Uses Of Spirits And The By-Products Of Distillation. Many people are unaware of the other uses of the products of alcoholic distillation. These uses (1) help to lower the cost of spirits, since they promote the most efficient use of the raw materials, and (2) provide a boost to the production of nutritious animal foods such as dairy products, eggs, meats, and poultry. A summary of some of the most important of these uses is given in Table D-18.

TABLE D-18
MISCELLANEOUS USES OF SPIRITS AND THE BY-PRODUCTS OF DISTILLATION

Product	Description	Uses	Comments
		HUMAN FOODS	
Neutral spirits	Pure solution of 95% alcohol and 5% water that is free of congeners.	Solvent for dissolving extracts of flavoring agents.	Used to make extracts of almond, cinnamon, lemon, lime, orange, peppermint, vanilla, and wintergreen. The flavoring extracts are used in baked good, ice cream, icings, soft drinks, and other foods.
Distillers' protein concentrate (DPC)	Protein concentrate (containing more than 80% protein) obtained by extraction of the residue from distillation.	Additive used to increase the protein quantity and quality of snack foods, and as a meat extender.	Corn DPC has been used experimentally to fortify cookies and as an emulsifier of oil in meat products. Wheat DPC has been used experimentally to fortify bread. It is also being tested in pasta products. Both types of DPC require further development and testing before they can be used commercially.
		DRUGS AND TONICS	
Neutral spirits	Pure solution of 95% alcohol and 5% water that is free of congeners.	Alcoholic base for certain fluid forms of medications and tonics.	Long used for pharmaceutical preparations since it is a good mixer with the flavored syrups and the medicinal ingredients.
		FEEDS FOR LIVESTOCK AND POULTRY	
Distillers' grains	The grain residue that remains on the screens after distillation and screening. Contains about twice as much protein as the original grain.	Energy and protein supplement added to feeds for livestock and poultry.	Most suitable for cattle and sheep, because these ruminants utilize it about as well as soybean meal. The high fiber content limits the amounts that may be added to rations for swine and poultry.
Distillers' solubles	The filtrate from the distillation residue after the spent grains have been removed by screening.	Energy, protein, and vitamin supplement for all species of livestock and poultry.	The fluid form known as thin stillage has long been used on farms near distilleries. However, it spoils readily, so it is usually dehydrated to condensed distillers' solubles, which are a good source of B vitamins and unidentified growth factors for swine and poultry. May contain a rumen-stimulating factor that facilitates cellulose digestion by cattle and sheep.
Grain distillers' dried yeast	Yeast separated from the mash either before or after fermentation, then dried.	Protein and vitamin supplement.	Contains at least 40% protein. Extremely rich in all of the B-complex vitamins except vitamin B-12.

DIURESIS

The passage of more than the normal amount of urine. It may be caused by drinking large quantities of water, coffee, or alcoholic beverages, or by nervous tension.

(Also see WATER BALANCE.)

DIURETIC

An agent that increases the flow of urine.

DIVERTICULA

Small pouches which are formed in the wall of the large intestine because of abnormally high pressures that develop during the passage of the intestinal contents. It is suspected that diverticulitis may result from diets low in fiber.

(Also see DIVERTICULITIS; and FIBER.)

DIVERTICULITIS

Inflammation of the diverticular sacs (outpouchings) often occurring in the colon of middle or advanced age persons. Its cause is the mechanical failure of these outpouchings to empty undigested food and bacteria. Attacks of diverticulitis have symptoms similar to acute appendicitis—generalized abdominal pain, nausea, sometimes vomiting, fever, and elevation of the white cell count. Most cases respond to bed rest, liquid diet, and a broad spectrum antibiotic. Occasionally, diverticulitis progresses to the point of perforation of the colon wall or hemorrhage, and surgery may be required.

DNA

The abbreviation for deoxyribonucleic acid, a constituent of the nuclear material of cells.

(Also see NUCLEIC ACIDS.)

DOCK (SORREL; SOUR GRASSES) *Rumex, spp.*

This is the name given to several plants of the genus *Rumex* of the buckwheat family, *Polygonaceae.* All of them have juicy leaves and stems that contain oxalic acid. This gives them a sour taste.

In the Northern Hemisphere, most docks are common weeds that thrive in acid soils; hence, their presence in a meadow indicates that the land needs lime. The leaves of curly docks (*R. crispus*) are used in soups and salads, like spinach.

DOCKAGE

This refers to deductions made in the liveweight of market animals because of excessive dressing losses, or because part of the product is low quality.

DOLOMITE

This mineral, which is also known as dolomitic limestone, is a major ingredient in several antacids, and has long been used as a source of calcium and magnesium for animal feeding. It is a mixture of approximately equal parts of calcium carbonate and magnesium carbonate. Dolomitic limestone contains about five times as much magnesium, and about five-eighths as much calcium as ordinary limestone. A typical analysis of dolomite follows:[64] calcium, 22.30%; chlorine, 0.12%; iron, 0.08%; magnesium, 9.99%; phosphorus, 0.04%; and potassium, 0.36%. Dolomite is sometimes used as a calcium supplement for humans.[65]

VALUE OF DOLOMITE AS A FOOD SUPPLEMENT. Recently, dolomite-containing tablets or wafers (some of which have mint or cherry flavor) have been promoted as supplementary sources of calcium and magnesium. Such dietary supplementation may be desirable under circumstances such as those which follow:

• **When there are dietary deficiencies of calcium and magnesium**—Dietary patterns which do not include sufficient amounts of calcium-rich foods (dairy products and green, leafy vegetables) or good sources of magnesium (fish; green, leafy vegetables; legumes; nuts; and whole grains) may fail to provide at least the RDAs for these minerals. It is often more convenient to give a mineral supplement than to attempt to provide the full allowances for these nutrients in the form of foods, particularly when these allowances are high, such as for pregnant or lactating women.

• **When there is poor absorption of calcium and magnesium**—The absorption of these minerals is impaired by such factors as (1) diets high in phosphorus, phytates (found mainly in whole grains), or oxalates (present in beet greens, chard, rhubarb, and spinach); (2) lack of sufficient stomach acid or bile; (3) loss of gastrointestinal secretions as a result of diarrhea or vomiting; or (4) chronic use of strong laxatives. Also, a deficiency of vitamin D is likely to result in a decreased absorption of calcium, but its effect on magnesium absorption is uncertain. The presence of any of these conditions may warrant the use of a calcium-magnesium supplement such as dolomite.

• **When there is increased urinary excretion of calcium and magnesium**—There may be excessive excretion of these minerals in the urine as a result of alcoholism, diabetic acidosis, diuresis caused by drugs or a heavy consumption of caffeine, kidney disease, or stress. In addition to these factors, diets high in protein (2 to 3 times

---

[64]*Atlas of Nutritional Data on United States and Canadian Feed,* NRC-National Academy of Sciences, and Dept of Ag., Canada, 1971, p. 409.

[65]Because dolomite may contain lead and other impurities, it should be tested and found to be free of such impurities before use.

the RDA) cause an elevated urinary excretion of calcium. The prolonged existence of any of these factors may make it advisable to provide supplemental calcium and magnesium in order to avoid the disorders which may result from depletion of these minerals. Excessive loss of calcium from bone often results in osteoporosis, while depletion of magnesium may result in hyperirritability of muscles and nerves.

Even though dietary supplementation with dolomite may be indicated under the circumstances noted above, no one should take therapeutic doses of dolomite without first consulting a physician. A therapeutic dose is one which supplies more than 1,500 mg of calcium per day, or the upper limit of the U.S. RDA for this mineral for adults other than pregnant or lactating women.[66]

(Also see ANTACIDS; CALCIUM; MAGNESIUM; and OSTEOPOROSIS.)

DOPAMINE

A nerve transmitter in the brain, the lack of which is believed to be responsible for the symptoms of Parkinson's Disease. It is produced from the amino acid tyrosine, and it in turn is converted in the brain to noradrenaline (norepinephrine) and adrenaline (epinephrine).

DOUGLAS BAG

A rubber bag used to collect exhaled air in metabolism studies. It is light enough to be strapped onto a person and permits freedom of movement. After the study is completed the carbon dioxide content of the bag is analyzed, because this measurement may be used to calculate the metabolic rate.

(Also see CALORIE [ENERGY] EXPENDITURE.)

DOVE

This is the common name for various species of birds in the family *Columbidae*—the family that also includes pigeons. Generally, the term dove refers to the smaller members of the family, and pigeon to the larger ones. Doves live throughout the world. In most countries, they are considered a game bird and are hunted.

Doves are drawn (eviscerated), prepared for cooking, and cooked like pigeons of similar ages. The flesh is much esteemed by gastronomes.

DPN

The abbreviation for diphosphopyridine nucleotide, an out-of-date name for nicotinamide adenine dinucleotide (NAD).

(Also see METABOLISM; and NICOTINAMIDE ADENINE DINUCLEOTIDE.)

[66]Lachance, P. A., "A Commentary on the New F.D.A. Nutrition Labeling Regulations," *Nutrition Today*, Vol. 8, January/February 1973, p. 20, Table 11.

DRESSING

Dressing of animals, poultry, game, and fish is a "disassembly" operation—usually on-the-rail, whereas automobile manufacturing is an assembly line procedure. Dressing refers to the skinning of animals and the plucking (removing of the feathers) of birds, along with removing parts other than the carcass—the viscera, feet, head, etc.

DRIED MILK

See MILK.

DRIPPINGS

Juices and melted fat that remain in the pan after meat has been cooked by dry heat.

DROOLING

The dribbling of saliva from the mouth. It is common in most babies and also occurs in some older people.

DROPSY

An abnormal accumulation of fluid in one or more parts of the body.

(Also see WATER AND ELECTROLYTES.)

DRUPES

Fruits which have a single seed that is enclosed in a stony shell, which, in turn, is surrounded by the flesh and skin of the fruits. Almonds, apricots, cherries, peaches, and plums are the best known drupes. These fruits are sometimes referred to as "stone fruits." It is noteworthy that blackberries and raspberries are aggregate fruits made up of many tiny drupes or drupelets.

(Also see FRUIT[S], Table F-47 Fruits of the World.)

DRY FRYING

The frying of foods without the use of fat, either in a nonstick pan or by using an antisticking agent.

DRY ICE

A solid carbon dioxide (CO_2) used as a refrigerant. The name comes from the fact that solid carbon dioxide does not return to liquid form when it melts; it changes directly into a gas. Dry ice is much colder than ordinary ice; it sometimes reaches a temperature as low as -112°F (*-80°C*). Dry ice can be used to ship perishable foods by parcel post, because it cannot melt.

Dry ice is made by compressing carbon-dioxide gas to a liquid, then cooling. Some of the cold liquid is evaporated to make carbon-dioxide snow, then machines compress the snow into blocks of solid dry ice.

DRY MATTER

That part of a food which is not water. Dry matter is found by determining the percentage of water and subtracting the water content from 100%.

(Also see ANALYSIS OF FOODS.)

DUCT

A tube which carries glandular secretions away from the gland to another part of the body. For example, the bile duct carries bile from the liver and gallbladder to the small intestine.

(Also see DIGESTION AND ABSORPTION.)

DUCTLESS GLAND

A gland which secretes directly into the blood stream. Hence, all parts of the body may be affected by the secreted substances.

(Also see ENDOCRINE GLAND.)

DUMPING SYNDROME

This disorder is a physiological response occurring 10 to 15 minutes after ingestion of a meal by a person who has a gastrectomy (partial or complete surgical removal of the stomach). The symptoms are rapid pulse, rapid breathing, weakness, trembling, pallor, cold sweating, and abdominal fullness and distention. In some cases, there may also be nausea, vomiting, and fainting. Another set of reactions may occur about 2 hours later, due to the onset of hypoglycemia (low blood sugar accompanied by mental confusion, double vision, weakness, hunger, faintness, dizziness, headache, sweating, sleepiness, trembling, and rapid pulse). Another name for this disorder is *jejunal hyperosmolic syndrome.*

CAUSES OF DUMPING SYNDROME. The term "dumping" refers to the rapid deposition of food into the small intestine, which usually occurs in persons part or all of whose stomachs are missing. By contrast, a normal, intact stomach usually releases food, diluted by gastric juice, rather slowly into the intestine. One cause of the symptoms is a high concentration in the jejunum of soluble materials in proportion to the amount of water (also known as *hyperosmolality*). This condition results in a flow of water into the intestine from the plasma and extracellular fluid (contributed by the intestinal tissue and blood flowing through it). Loss of water from the circulatory system results in decreases in blood volume and blood pressure leading to the symptoms associated with *cardiac insufficiency* (the heart is an inefficient pump when a reduction in the blood volume prevents its filling to capacity). The circulatory changes lead to a response by the sympathetic nervous system (resulting in rapid heart rate, sweating, and weakness). *Distention of the jejunum* (by the volume of food, by fluids taken with the meal, and/or by the fluid originating from the blood and tissues) probably helps to stimulate the set of troublesome physiological responses.

Hypoglycemia is more likely to occur when the meal contains a moderate to large proportion of easily digested carbohydrate. Rapid absorption of sugars, the digestion products of carbohydrate, results in an elevation of the blood sugar level, which in turn stimulates the pancreatic release of a large amount of insulin (temporary *hyperinsulinism*). The net effect is a rapid drop in the blood sugar level due to the action of insulin (which enhances the rate of sugar passage from the blood into cells).

Persons with normal stomachs may also experience symptoms similar to dumping syndrome when they have rapid emptying of gastric contents into the small intestine, such as may occur with (1) a large amount of food and fluid that is rich in carbohydrates, or (2) hyperactivity of the stomach due to other factors such as nervousness. It appears likely, although not confirmed, that at least some of the symptoms attributed to *Chinese Restaurant Syndrome* (a condition which occurs after eating a heavy meal of Chinese food) are due to factors like those that are responsible for dumping syndrome.

PREVENTION OF DUMPING SYNDROME. The best approach is to minimize the factors responsible for the conditions. This means that meals should be planned according to the rules which follow:

1. Select meals containing moderate to liberal amounts of protein and fat, and use only small portions of foods high in carbohydrate.
2. Avoid foods high in sugars such as soft drinks, candy, cake, cookies, syrups, honey, and molasses.
3. Take frequent small meals or snacks (5 to 6 per day) instead of eating 2 or 3 large meals.
4. Do not take beverages with meals, but have them about an hour before or after meals.
5. Allow foods that are boiling hot or ice cold to approach room temperature before eating.
6. Avoid emotionally exciting situations at mealtimes.
7. Restrict salt and such seasonings as soy sauce and monosodium glutamate.

(Also see HYPOGLYCEMIA; and ULCERS, PEPTIC.)

DUODENUM

The upper portion of the small intestine which extends from the stomach to the jejunum.

(Also see DIGESTION AND ABSORPTION.)

DURRA

The name given to sorghum in the Sudan.

DUTCH COCOA

The addition of an alkali (the carbonates or hydroxides of sodium potassium, ammonium, or magnesium) during the processing of cocoa from the cacao bean causes the cocoa flavor to become milder and the color darker, yielding a product called Dutch cocoa. Dutching of cocoa originated in Holland; hence, the name.

(Also see COCOA AND CHOCOLATE.)

DUTCH OVEN

• A metal utensil with shelves and one open side which can be placed close to the fire.

• A cast iron pot with three legs and a lid on which coals can be placed. It is used for baking in an open fire.

• A heavy pot with a tight, domed cover which is used for braising, steaming, or baking, either in the oven or on top of the stove.

DWARFISM

An abnormal condition in which the affected person is much shorter in height than other individuals of the same age, sex and ethnic group. In most cases dwarfism is due to an undersecretion of growth hormone(s) by the pituitary gland, although other metabolic disorders may also be responsible.

(Also see ENDOCRINE GLANDS.)

DYMELOR

A brand name for a form of sulfonylurea called *acetohexamide*, an intermediate-acting oral drug used to stimulate the pancreatic cells to release insulin. The drug is only effective for the treatment of diabetes when some insulin secreting ability remains in the pancreas. Other brand names are Dimelor, Dimelin, and Ordimel.

(Also see DIABETES, section headed "Oral Drugs"; DIABINASE; and ORINASE.)

DYS-

A prefix conveying the idea of bad or difficult.

DYSENTERY

An inflammation of the mucous membrane of the large intestine resulting from infection of bacteria, protozoa, or virus, and characterized by the appearance of blood and mucus in the stools, severe diarrhea, cramps, and fever.

DYSENTERY, AMOEBIC

A disease of the intestines, primarily the colon, caused by the one-celled microorganism (amoeba), *Entamoeba histolytica.* The amoeba is often acquired from food and water contaminated by sewage. Amoebic dysentery is difficult to diagnose and may require repeated visits to a physician. Symptoms, of which the individual should be aware, are: unexplained fatigue, low grade fever, and recurrent diarrhea. The disease is serious. But the seriousness is not from the dysentery itself, but the amoeba. If not controlled, the amoeba may move into the blood and travel to other organs (to the liver, and even the brain) and produce abcesses. Effective preventative measures are personal hygiene, cleanliness, and sanitary sewage disposal.

(Also see DISEASES, Table D-13 Infectious and Parasitic Diseases Which May Be Transmitted By Contaminated Foods and Water—"Amoebic Dysentery.")

DYSGEUSIA

Perverted sense of taste; a "bad" taste.

DYSINSULINISM

An abnormality in which the blood sugar after a meal or after a test dose of sugar stays abnormally high for an hour or two, then drops to an abnormally low level. Hence, the blood sugar pattern appears to be a combination of those which are found in both diabetes (the high blood sugar level) and hypoglycemia (the low blood sugar level). It is thought to be due to an abnormally slow release of insulin by the pancreas in response to an elevated blood sugar.

(Also see DIABETES; and HYPOGLYCEMIA.)

DYSOSMIA

Impaired sense of smell; an obnoxious odor.

DYSPEPSIA

The general term used to describe any of a variety of disorders which occur in the upper part of the digestive tract after food has been taken. Dyspepsia is sometimes referred to as indigestion, which is a rather vague term for these complaints. For example, some typical symptoms of dyspepsia are abdominal pain, chest pain, belching, excessive gas, heartburn, nausea, and an uncomfortable feeling of fullness.

(Also see DIGESTION AND ABSORPTION.)

DYSPHAGIA

A disorder of the esophagus which makes it difficult to swallow.

DYSPNEA

Abnormal breathing characterized by shortness of breath, difficulty in breathing, or an undue awareness of the act of breathing.

DYSSEBACIA

An oily flaking of the skin which may often be due to a lack of riboflavin or other essential nutrients.

(Also see RIBOFLAVIN.)

DYSTROPHY

One or more defects in body tissues due to such causes as poor nutrition, heredity, or a degenerative disease.

The "backbone" of every meal should be a good entree—containing a complete protein with some of each and every one of the essential amino acids. (Courtesy, National Live Stock & Meat Board, Chicago, Ill.)

EARTH-EATING

The abnormal practice of eating dirt and/or clay, which is also called *geophagia*. It is suspected that the reasons for this bizarre practice are: (1) a need for a nutrient which is lacking in the diet, and (2) an irritation of the digestive tract which is soothed by the earth which is ingested.

(Also see PICA.)

EBERS' PAPYRUS

An ancient Egyptian medical paper, dated about 1500 B.C., which was found by the German Egyptologist George Ebers in 1872. Ebers' Papyrus is of great interest to those studying the history of dietetics and medicine because it describes the ancient treatment of such diseases as (1) diabetes, for which a diet of beer, fruits, grains, and honey was recommended; and (2) goiter, in which the thyroid gland was removed surgically.

ECLAMPSIA

A toxemia of late pregnancy characterized by excessive retention of water, high blood pressure, passage of protein in the urine, convulsions, and coma. Fortunately, the incidence of this disorder has declined greatly in recent years. A milder toxemia without the convulsions and coma is called preeclampsia. The causes of eclampsia and preeclampsia are uncertain although it is suspected that nutrition and/or the amount of weight gain in pregnancy might be important factors.

ECTOMORPH

A person with a tall, slender physique and long arms and legs. Ectomorphs live the longest in countries such as the United States where most occupations are sedentary and little physical exertion is required.

(Also see BODY TYPES; and OBESITY.)

ECZEMA

A condition involving the inflammation of the skin with lesions of either a dry or weeping nature. It is most commonly caused by allergies or exposure to chemicals or other irritants, but it may be due to deficiencies of such nutrients as essential fatty acids and vitamins.

(Also see DEFICIENCY DISEASES, Table D-2 Minor Dietary Deficiency Disorders.)

EDEMA

Swelling of a part or all of the body due to the accumulation of excess water.

EDIBLE PORTION

The part of an animal or plant food which is commonly eaten, in contrast to refuse portions such as bones, gristle, peels, and seeds which are normally discarded.

EDTA

A chelating agent that is added to certain foods to bind with trace minerals which are responsible for the development of off colors and flavors in the food. It may also be given orally or by injection to tie up excesses of certain trace minerals and draw them out of the body via the urine and/or the stool.

(Also see ADDITIVES; and MINERAL[S].)

EEG

Abbreviation for electroencephalogram.

(Also see ELECTROENCEPHALOGRAM.)

EFA

Abbreviation for essential fatty acid, a type of nutrient formerly designated as vitamin F.

(Also see ESSENTIAL FATTY ACID.)

EFFERVESCENT

A term used to describe a liquid that is bubbling because it is giving off bubbles of a gas. A common effervescent liquid is a solution of baking soda, which releases carbon dioxide. (The latter solution is used mainly as an antacid or a leavening agent in baking.)

(Also see ANTACIDS.)

EGG(S)

Contents — *Page*

History ... 636
Production ... 637
Processing ... 638
Physical Characteristics of the Egg and Grading ... 638
Structure ... 638
Size ... 640
Shape ... 640
Color ... 641
Abnormalities ... 641
The Versatile Egg ... 642
Consumption of Eggs ... 643
Health Problems Attributed to Eggs ... 643
Allergies ... 643
Cholesterol ... 643
Nutritional Value ... 644
Nutritional Misconceptions ... 645

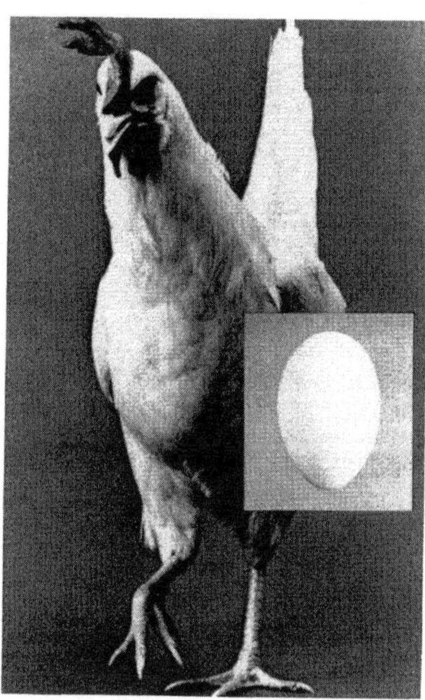

Fig. E-1. What it's all about! A layer and an egg. (Courtesy, Hy-Line International, Division of Hy-Line Indian River Company, Johnston, Iowa)

The bird egg is a marvel of nature. It's one of the most complete foods known to man, as evidenced by the excellent balance of proteins, fats, carbohydrates, minerals, and vitamins which it provides during the 20-day in-the-shell period when it serves as the developing chick's only source of food. Also, the egg is one of the few foods that is produced in prepackaged form. Not only that, it is the reproductive cell (ovum) of the hen. Upon fertilization by the male's reproductive cell (sperm), the egg will develop into a chick when incubated properly.

HISTORY. Ancient man persuaded chickens to live and produce near his abode. It is not known exactly when this happened, but it's obvious that chickens were domesticated at a remote period. The keeping of poultry was probably contemporary with the keeping of sheep by Abel and the tilling of the soil by Cain. Chickens were known in ancient Egypt, and they had already achieved considerable status at the time of the Pharaohs, because artificial incubation was then practiced in crude ovens resembling some that are still in use in that country.

The use of poultry and eggs as food goes back to very early times in the history of man. Methods of slaughter and preparation for consumption have varied with succeeding civilizations and cultures. Not until fairly recent times did these operations become a matter of great commercial importance, or of serious concern to consumers, public health officials, and governments alike.

The American poultry industry had its humble beginning when chickens were first brought to this continent by the early settlers. Small home flocks were started at the time of the establishment of the first permanent homes at Jamestown in 1607. For many years thereafter, chickens were tenderly cared for by the farmer's wife, who fed them on table scraps and the unaccounted-for grain from the crib.

As villages and towns were established, and increased in size, the nearby farm flocks were also in-

Fig. E-2. Cage system of housing layers, with three decks of cages. (Courtesy, DEKALB AgResearch, Inc., DeKalb, Ill.)

creased. Surplus eggs and meat were sold or bartered for groceries and other supplies in the nearby towns. Eventually, grain production to the West, the development of transportation facilities, the use of refrigeration, and artificial incubation further stimulated poultry production in the latter part of the 1800s.

Since World War II, changes in poultry and egg production and processing have paced the whole field of agriculture. Practices in all phases of poultry production—breeding, feeding, management, housing, marketing and processing—have become very highly specialized. The net result is that more products have been made available to consumers at favorable prices, comparatively speaking, and per capita consumption has increased.

PRODUCTION. The production of eggs relates to the fecundity of the chicken. The term *fecundity* is used to describe the inherent capacity of an organism to reproduce rapidly. In higher animals, reproduction is possible only after the ovum (female gamete) is fertilized or united with the spermatozoon (male gamete). In chickens, fertilization is not a necessary preliminary to egg laying. Thus, the hen can lay eggs continuously without being mated or without being stimulated by the presence of a male. This biological phenomenon has been advantageously utilized by man in producing infertile eggs for food. Infertile eggs are of more economic value for food than fertile eggs, because there is no danger of loss through development of the embryo. Table E-1 shows the annual egg production of various species of poultry.

TABLE E–1
EGG PRODUCTION OF POULTRY

Species	Age of Sexual Maturity	Eggs/Year
	(mo)	(no.)
Chicken:		
Light-type	5–6	252
Broiler-type		170
Turkey	7	105
Goose	24	15–60
Duck (Pekin)	7–8	110–175
Pheasant	8–10	40–60
Quail (bobwhite)	8–10	150–200
Pigeon	6	12–15
Guinea fowl	10–12	40–60

Feeding, breeding, management and disease control practices of light-type chickens have improved since 1940. Now, fewer chickens lay more eggs; the number of eggs per chicken (light-type) per year has increased from 134 to 252 eggs as shown in Table E-2. Furthermore, commercial egg production is accomplished on a large scale. Some production units—flocks of layers—consist of 1 million or more chickens. The overall result of scientific egg production is reflected by the more than 67.8 billion eggs produced in the United States in 1990.

TABLE E–2
EGG PRODUCTION HISTORY IN THE UNITED STATES[1]

Year	Layers Producing	Eggs per Layer	Eggs Produced
	(thousands)		(millions)
1940	296,595	134	39,695
1945	369,356	151	55,858
1950	339,540	174	58,954
1955	309,297	192	59,526
1960	294,662	209	61,491
1965	301,058	218	66,560
1970	312,759	218	68,212
1975	278,101	232	64,626
1980	287,705	242	69,686
1985	277,592	247	68,645
1990	271,631	252	67,832

[1]*Agricultural Statistics,* USDA, 1949, p. 473, Table 590; 1965, p. 426, Table 619; 1980, p. 412, Table 592; and 1991, p. 351, Table 523.

In the United States, all 50 states report some egg production, but the ten leading states indicated in Table E-3 are responsible for 61% of the eggs produced. California is the leading egg-producing state by a considerable margin. However, the heaviest concentration of layers is in the southeastern United States.

TABLE E–3
TEN LEADING EGG-PRODUCING STATES, 1990[1]

States	Eggs Produced
	(millions)
California	7,472
Indiana	5,445
Pennsylvania	4,976
Ohio	4,667
Georgia	4,302
Arkansas	3,620
Texas	3,317
North Carolina	2,986
Florida	2,586
Alabama	2,206

[1]*Agricultural Statistics,* 1991, USDA, p. 351, Table 524.

Worldwide, eggs are an important food. Table E-4 shows the leading egg-producing countries for which reliable data exists.

TABLE E–4
AVERAGE YEARLY EGG PRODUCTION IN SPECIFIED COUNTRIES[1]

Country	Eggs
	(millions)
China	158,920
U.S.S.R.	82,000
United States	67,919
Japan	39,850
Mexico	18,040
Germany	16,800
France	14,629
Brazil	13,454
United Kingdom	12,352
Italy	11,454

[1]*Agricultural Statistics,* 1991, p. 350, Table 521.

PROCESSING. The processing of quality eggs begins at the farm level. The poultryman must carefully select for egg production and egg quality in his breeding program. Once a good flock of layers has been established, good feeding and management of birds is required to maximize the genetic potential of the birds.

Frequent gathering of eggs maintains high quality because if eggs are exposed to ambient temperatures for extended periods, quality declines rapidly. By collecting eggs at frequent intervals, the producer is also able to reduce breakage and the incidence of dirty eggs. Once collected, the eggs are sprayed with edible mineral oil to preserve quality and then cooled rapidly to prevent spoilage. After being transported to a central location, eggs are washed. Then, on the basis of a number of physical characteristics of the egg, the eggs are graded, sized and marketed.

Fig. E-3. Modern egg handling is highly automated. The operation pictured above shows automatic egg grading, sizing, and cartoning. (Courtesy, DEKALB AgResearch, Inc., DeKalb, Ill.)

PHYSICAL CHARACTERISTICS OF THE EGG AND GRADING. Consumers demand a superior product even from "Mother Nature"; hence, eggs are graded.

The grading of eggs involves their sorting according to quality, size, weight, and other factors that determine their relative value. U.S. standards for quality of individual shell eggs have been developed on the basis of such interior quality factors as condition of the white and yolk, the size of the air cell, and the exterior quality factors of cleanliness and soundness of the shell, These standards cover the entire range of edible eggs.

Eggs are also classified according to weight (or size), expressed in ounces per dozen.

Egg grading, then, is the grouping of eggs into lots according to similar characteristics as to quality and weight. Although color is not a factor in the standards of grades, eggs are usually sorted for color and sold as either "whites" or "browns."

Four sets of grades, based on the quality standards for individual shell eggs, are used in this country: (1) consumer grades—used in the sale of eggs to individual consumers; (2) wholesale grades—used in the wholesale channels of trade; (3) U.S. Procurement Grades—used for institutional buying and Armed Forces purchases; and (4) U.S. Nest Run Grade—which is also used in wholesale channels of trade.

The U.S. standards for quality of individual shell eggs are applicable only to eggs of the domesticated chicken that are in the shell. These are given in Table E-5. The basis for the egg grades given in Table E-5 is resemblance to normal new-laid eggs.

Consumer grades are those lots of eggs that have been carefully candled and graded for retail trade. These are given in Table E-6.

Size, shape, color, and various abnormalities all affect egg quality and consumer acceptance. A brief discussion of each of these follows, but first a description of egg structure.

Structure. A schematic side view of an egg is shown in Fig. E-4 with the various parts labeled in their normal position.

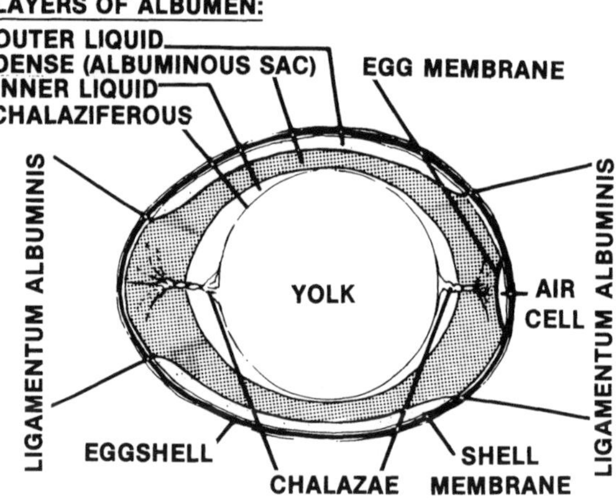

Fig. E-4. Structure of the egg.

The protective covering, known as the shell, is composed primarily of calcium carbonate, with 6,000 to 8,000 microscopic pores permitting transfer of volatile components. The air cell, located in the large end of the egg, is formed when the cooling egg contracts and pulls the inner and outer shell membranes apart. The cordlike chalazae holds the yolk in position in the center of the egg. As shown, the egg is surrounded by a membrane, known as the vitelline membrane. The germinal disc, a normal part of every egg, is located on the surface of the yolk. Embryo formation begins here only in fertilized eggs. For most individuals the two most familiar parts of the egg are (1) the white, and (2) the yolk.

1. **Egg white.** The egg white or albumen surrounds the yellow yolk. It is comprised of four layers: (a) the outer

TABLE E-5
SUMMARY OF U.S. STANDARDS FOR QUALITY OF INDIVIDUAL SHELL EGGS[1]

Quality Factor	Specifications for Each Quality Factor			
	AA Quality	A Quality	B Quality	C Quality
Shell	Clean Unbroken Practically normal	Clean Unbroken Practically normal	Clean to slightly stained Unbroken May be slightly abnormal	Clean to moderately stained Unbroken May be abnormal
Air cell	⅛ in. (*3.2 mm*) or less in depth. May show unlimited movement and may be free or bubbly.	3/16 in. (*4.8 mm*) or less in depth. May show unlimited movement and may be free or bubbly.	⅜ in. (*9.5 mm*) or less in depth. May show unlimited movement and may be free or bubbly.	May be over ⅜ in. (*9.5 mm*) in depth. May show unlimited movement and may be free or bubbly.
White[2]	Clear Firm (72 Haugh units or higher).	Clear May be reasonably firm (60 to 72 Haugh units).	Clear May be slightly weak (31 to 60 Haugh units).	May be weak and watery Small blood clots or spots may be present (less than 31 Haugh units).*
Yolk	Outline slightly defined. Practically free from defects.	Outline may be fairly well defined. Practically free from defects.	Outline may be well defined. May be slightly enlarged and flattened. May show definite but not serious defects.	Outline may be plainly visible. May be enlarged and flattened. May show clearly visible germ development but no blood. May show other serious defects.

*If they are small (aggregating not more than ⅛ in. (*3.2 mm*) in diameter).

For eggs with dirty or broken shells, the standards of quality provide three additional qualities. These are:

Dirty	Check	Leaker
Unbroken May be dirty.	Checked or cracked but not leaking.	Broken so contents are leaking.

[1]*United States Standards, Grades, and Weight Classes for Shell Eggs,* Agricultural Marketing Service, Poultry Division, USDA, June 30, 1975, p. 8.
[2]The Haugh unit is an expression relating egg weight and height of thick albumen. The higher the Haugh value the better the albumen quality of the egg.

TABLE E-6
SUMMARY OF U.S. CONSUMER GRADES FOR SHELL EGGS[1]

U.S. Consumer Grade (origin)	Quality Required[1]	Tolerance Permitted[2]	
		Percent	Quality
Grade AA or Fresh Fancy quality	85% AA	Up to 15 Not over 5	A or B C or Check
Grade A	85% A or better	Up to 15 Not over 5	B C or Check
Grade B	85% B or better	Up to 15 Not over 10	C Checks

U.S. Consumer Grade (destination)	Quality Required[1]	Tolerance Permitted[3]	
		Percent	Quality
Grade AA or Fresh Fancy quality	80% AA	Up to 20 Not over 5	A or B C or Check
Grade A	80% A or better	Up to 20 Not over 5	B C or Check
Grade B	80% B or better	Up to 20 Not over 10	C Checks

[1]*United States Standards, Grades, and Weight Classes for Shell Eggs,* Agricultural Marketing Service, Poultry Division, USDA, July 1, 1974, p. 3.
[2]For the U.S. Consumer grades (at origin), a tolerance of 0.30% Leakers or loss (due to meat or blood spots) in any combination is permitted. No Dirties or other type loss are permitted.
[3]For the U.S. Consumer grades (destination), a tolerance of 0.50% Leakers, Dirties, or loss (due to meat or blood spots) in any combination is permitted, except that such loss may not exceed 0.30%. Other types of loss are not permitted.

thin or liquid white; (b) the dense or thick white; (c) the inner thin or liquid white; and (d) the inner thick white or chalaziferous layer. Water is the major constituent of egg white, amounting to about 88%. The proteins of egg white account for about 11% of the composition. Egg white proteins include ovalbumin, conalbumin, ovomucoid, globulins, ovomucin, flavoprotein, ovoglycoprotein, and avidin. Egg white is often used as an indicator of egg quality. It must be firm with a rather clear demarcation between the thin and thick albumen.

2. **Egg yolk.** The yolk contains about 48% water, 18% protein, and 33% fat. Proteins of the yolk include phosvitin, a nonlipid phosphoprotein, and lipovitellins and low-density lipoproteins (LDL)—both lipid-protein complexes. Lipids of the yolk are triglycerides, cholesterol, and phospholipids. Yolk color is primarily due to xanthophylls.

Size. There is an enormous range in size among the eggs of different species of birds. Ostrich eggs average about 3.1 lb (*1,400 g*), whereas hummingbird eggs weigh only 0.001 lb (*0.5 g*). The average size of eggs laid by a number of domestic species of birds is shown in Table E-7.

TABLE E-7
EGG SIZES OF SOME DOMESTIC BIRDS

Species	Egg Size	
	(oz)	(*g*)
Goose	7.5	*215*
Turkey	3.0	*85*
Duck (Pekin)	2.9	*80*
Chicken (light-type)	2.1	*58*
Guinea fowl	1.4	*40*
Pheasant	1.1	*32*
Pigeon	.6	*17*
Quail (Bobwhite)	.3	*9*

Also, within a species egg size varies considerably. Thus, in chickens, the eggs of Dark Brahmas are more than twice as heavy as those of Japanese Bantams. The eggs of most of the heavy-laying breeds weigh from 0.1 to 0.14 lb (*45 to 64 g*), with the eggs of the large meat-producing breeds at the upper end of size and those of the smaller game birds at the lower end.

Also, the size of the eggs layed by one individual may differ widely from those laid by another of the same species and breed. Even laying chickens produce eggs of varying sizes, and commercially produced eggs are classified by size according to their weight per dozen, following the guidelines of Table E-8. It is not unusual for the price of medium size eggs to be 5 to 10¢ lower per dozen than the price of large eggs. Egg size (weight) is separate and distinct from egg quality.

The most important factors affecting egg size are:

1. **Breeding.** Egg size is an inherited trait. Although environmental factors may result in smaller eggs, the upper limit in egg size is determined when the chicken is hatched.

TABLE E-8
U.S. WEIGHT CLASSES FOR CONSUMER GRADES FOR SHELL EGGS[1]

Size or Weight Class	Minimum Net Weight per Dozen	Minimum Net Weight per 30 Dozen	Minimum Weight for Individual Eggs at Rate per Dozen
	(oz)[2]	(lb)[3]	(oz)
Jumbo	30	56	29
Extra large	27	50½	26
Large	24	45	23
Medium	21	39½	20
Small	18	34	17
Peewee	15	28	—

[1]*United States Standards, Grades, and Weight Classes for Shell Eggs*, Agricultural Marketing Service, Poultry Division, USDA, July 1, 1974.
[2]One oz = 28.35 g.
[3]One lb = 0.45 kg.

2. **Age of bird.** When chickens start to lay, 80% or more of their eggs will be under 21 oz per dozen. Egg size should increase gradually until the birds are around 12 to 14 months of age, following which some decrease in egg size may be expected.

3. **Clutch order.** Chickens lay in clutches—a series of eggs and then a brief rest. The order of an egg in the clutch affects its weight. The first egg of a clutch is usually the heaviest, and there is usually a progressive decrease in the weight of the rest of the eggs of the clutch.

4. **Total eggs laid in a year.** There is a tendency toward a decline in egg size as the total number of eggs laid in a year increases.

5. **Age at maturity.** Delay in maturity will usually result in larger eggs at the start of production.

6. **Temperature.** The size of eggs declines during the hot summer months.

7. **Type of housing.** Caged birds lay eggs about ½ oz (*14 g*) per dozen larger than birds on the floor, and birds on slats lay about 10% more large eggs than birds on floors with part litter.

8. **Feed and water.** Normally, feed is a minor factor in egg size so long as a well-balanced ration is fed.

9. **Disease.** Some diseases affect egg size dramatically, and in some cases the effect persists for months after the layers appear to be healthy.

10. **Grain fumigants.** Ethylene dibromide, an ingredient of a fumigant commonly used on seed oats, will decrease egg size. A drop in egg size to as small as 16 oz (*454 g*) per dozen from an initial 23.8 oz (*566 g*) average has been reported.

Shape. Eggs differ considerably in shape. Although many are truly ovate, some are nearly spherical, whereas others are elongated. Some eggs are almost equally pointed or rounded at both ends; others taper sharply from the large end to the small end. The eggs laid by birds of the same species resemble each other in shape, but they are not identical; nor are all eggs of a particular bird alike.

Aristotle (384-322 B.C.) believed that the cock hatched from the more pointed chicken egg and the hen from the rounder type. Early 19th century naturalists argued that egg contours indicated the general body form of the bird that would develop within. Somewhat later, natural selection advocates theorized that, through adaptation, the eggs of different birds had assumed the shape most likely to ensure the survival of each species in the particular environment.

It is now generally agreed that physiological factors largely determine the diversities in the form of the egg, but that the shape may be modified by certain conditions. Among the causations of egg shape variations within species, or of a particular bird, are the following: (1) muscle tone of the reproductive tract of the chicken; (2) heredity; (3) first eggs laid by a young chicken; and (4) first egg laid after a pause in laying.

Color. The shell color of eggs of the domestic hen may be white, many shades of brown, or yellow. One breed lays blue green eggs. Sometimes very small, dark flecks are present on the shell, especially if it is brown.

Among domestic fowl, the color of the egg is peculiar to the breed, though tinted eggs occasionally appear in breeds that ordinarily lay white eggs. Of the four races recognized in the United States, only the Mediterranean (comprised of Leghorns, Minorcas, Anconas, Black Spanish, and Blue Andalusian) lays white eggs. The other three races—Asiatic, English, and American—lay tinted eggs, with the exception of two or three breeds. Cochin China hens lay eggs that range from bright yellow to dark yellow—speckled with small red dots. Langshans lay dark yellow eggs. Brahmas lay reddish-yellow eggs. Most of the continental European breeds lay white egs. The Araucana of South America lays light bluish-green eggs. Hens normally lay eggs of the same color, but there may be considerable variation in color among hens of the same breed.

Colored eggs occur because pigment is deposited in the shell as it is formed in the reproductive tract. The hereditary nature of such pigmentation is evidenced by the manner in which egg color is identified with species and breed. However, the mode of inheritance of shell color is difficult to determine, because most varieties of fowl are the product of many years of crossbreeding.

Egg color often assumes economic importance, as there are numerous local prejudices in favor of certain shell tints. However, shell color does not alter the nutritional value of the egg.

Abnormalities. Periodically, a deviation in the mechanics of egg laying will create abnormal eggs. Some of these abnormalities are as follows:

1. **Double-yolked eggs.** It is the result of two yolks ripening on the ovary at the same time.

2. **Blood spots.** When a small blood vessel breaks in the reproductive tract, blood spots are formed in the egg.

3. **Meat spots.** Meat spots are degenerated blood clots from the ovary or oviduct that are deposited in the egg.

4. **Yolkless eggs.** Occasionally, foreign material may get into the oviduct and stimulate the secretion of albumen in much the same manner as the yolk.

5. **Dented eggshells.** When one egg is kept too long in the reproductive tract, a second egg may pass down the tract and actually touch the first egg, thereupon creating an indentation in the shell of the second egg.

6. **Soft-shelled eggs.** These eggs result when no shell is secreted.

From the above discussion it is evident that numerous factors are involved in the production of top quality eggs for the consumer. Many of these factors are judged in order to separate those eggs which can be marketed. Eggs are judged for quality through three sets of criteria; (1) external appearance, (2) candling, and (3) samples of eggs broken out to judge internal characteristics.

1. **External appearance.** The initial evaluation of eggs begins with the examination of the external appearance. The size and shape of the egg, as well as the color and texture, are extremely important. A dozen grade A eggs must be of uniform size and shape. Additionally, the shells must be clean and free from cracks.

In some specific localities, consumers show a preference for a particular color of eggshell. People in New York City customarily choose eggs with white shells, while those in Boston prefer eggshells that are brown.

2. **Candling.** When eggs are graded for quality, they are routinely held against a light source—candling—so that the grader can determine certain quality characteristics of the internal parts of the egg without breaking the shell. When an egg is candled the following criteria are evaluated: (1) texture of the shell; (2) size of the air cell; (3) firmness of the white; and (4) detection of meat and blood spots.

3. **Breakouts.** Periodically, samples from a large batch of eggs are broken out in order that the contents can be closely evaluated. Color of the white and yolk, odor, and general appearance can be readily evaluated. The yolk of a high-quality egg must be round and firm and the white must be firm with a rather clear demarcation between the thin and thick albumen. Egg albumen quality is routinely measured in Haugh units—units determined by a micrometer that measures albumen height. This height is then compared to a chart listing the weights of eggs and Haugh units can then be derived. A higher Haugh value indicates a better egg.

The pH (hydrogen ion concentration) of the egg white is another means whereby egg quality is determined. The pH of a freshly layed egg is about 7.6 to 8.2. As the egg ages, carbon dioxide is lost, and the pH increases to as high as 9.5.

Shell strength is an additional criterion for evaluation. Breaking strength, thickness, and specific gravity are generally measured to determine the quality of the eggshell.

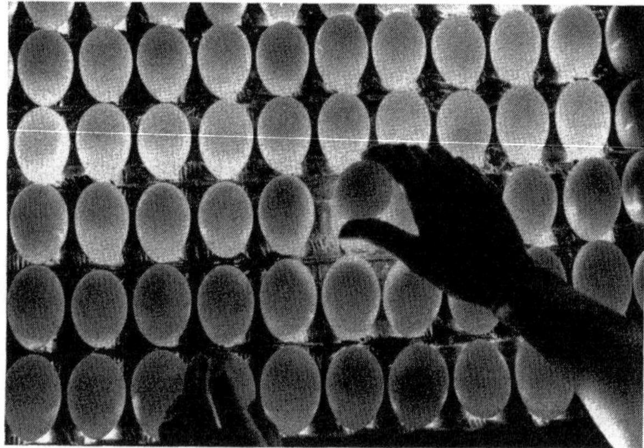

Fig. E-5. Candling. Eggs with defective shells, blood spots, or meat spots are removed in this egg-candling darkened curtain booth as eggs move from the washer to the grading and sizing machine. The eggs are rotated over a bank of high-wattage lights making any defect in them more easily seen by the person doing the candling work. (Courtesy, DEKALB AgResearch, Inc., DeKalb, Ill.)

THE VERSATILE EGG. Eggs serve as the basis for an extraordinarily wide range of uses—a fact emphasized by the large number of food products that contain eggs. Eggs can be eaten scrambled, fried, poached, shirred, baked, soft-boiled, hard-boiled, or coddled as a food. Furthermore, they can be curried, deviled, masked, whipped and variously molded. Eggs are necessary for a good dough—whether for bread, pasta, pie crust, cakes, pancakes, crepes or waffles. Mayonnaise, hollandaise and bernaise sauces depend on the egg. Cosmopolitan dishes such as Frittata, Scotch Woodcock, Egg Foo Young, and Broccoli Timbale all feature the egg—a truly versatile product of nature.

The popularity of the egg is due to several unique features—functional properties.

- **Coagulation**—Coagulation is a change in the structure of egg protein that results in the loss of solubility and thickening, or in other words, a change from fluid to a solid or semisolid. It may be brought about by heat, mechanical means, salts, acids or alkalis. The success of many cooked foods is dependent upon the coagulation of egg protein. Both the egg white and the yolk are utilized because of their ability to bind foods together. The ability to coagulate on heating is used in custards and pie fillings.

- **Foaming**—When egg white is beaten, air bubbles are trapped in the albumen and a foam is formed—a gas phase dispersed in a liquid phase. This foaming power of the egg white—albumen—makes it useful as a carrier of air for leavening; hence, it contributes to the lightness of certain foods. For example, meringues, angel cakes, souffles, omelets, some candies, and sponge cakes all depend upon eggs for their lightness.

- **Emulsification**—The egg yolk is a dispersion of oil droplets in water—an emulsion. Hence, the yolk is an efficient emulsifying agent. The emulsifying components of egg yolks are lecithin, cholesterol, lipoproteins and proteins. As an emulsifier, egg yolk or whole egg is an essential ingredient in mayonnaise (65-75% oil) and important in foods such as cake batter containing shortening, cream puffs, and hollandaise sauce, or wherever eggs are used with fats and oils.

- **Control of crystallization**—In candies, egg white controls the growth of sugar crystals, thus serving as a "sugar doctor."

- **Color**—The naturally occurring pigment of egg yolk—xanthophylls, lutein, and zeaxanthin—contribute color to foods. In some foods, their acceptance is determined by the visual impression of egg yolk. For example, egg yolk pigments contribute a pleasing yellow color to baked products, noodles, ice cream, custard sauces and omelets. In eggs alone, preference has been shown for gold or lemon-colored yolks. It is, however, rare for eggs to be used as an ingredient in food products for only their color contribution.

- **Flavor**—No one component of eggs can be determined as responsible for the characteristic flavor of eggs. The flavor of eggs has been described as fresh, mild, sweet, earthy and musty. Some of the flavor is imparted to baked goods, and eggs are enjoyed by themselves for their distinct flavor.

- **Nutrition**—Last but not least, eggs contain many nutrients and their addition to any food contributes to the quality of that food. A complete discussion of the nutritional value of eggs is presented elsewhere in this article.

Eggs usually perform more than one function in a food. Table E-9 presents a summary of the functions eggs serve in foods.

TABLE E-9
SUMMARY OF THE USE AND FUNCTION OF EGGS IN FOOD

Food Item	Functional Property					
	Coagulation	Foaming	Emulsification	Control of Crystallization	Color	Flavor
Angel cake	•	•				
Custards	•				•	•
Divinity		•		•		
Eggs[1]	•				•	•
Fondant				•		
Mayonnaise			•			
Meringues	•	•				
Salad dressing			•			
Sponge cake	•	•			•	

[1]Cooked in shell, fried, scrambled or poached.

Due to the wide variety of uses and the high demand by food institutions and food industy, a large number of the eggs produced become processed eggs—frozen, liquid, or dried eggs. This segment of the egg industry was originally developed as an outlet for soiled, cracked, and abnormal eggs unsuitable for marketing as shell eggs. Today, many quality eggs are being broken for processed eggs—about 13%.

The advantages of processed eggs are numerous when compared to shell eggs and include the following:

1. Less storage space is involved.
2. Quality of frozen and dried eggs is preserved longer.
3. Packaging is facilitated.
4. Processed eggs are cheaper than shell eggs.
5. Less labor is involved in the egg-breaking industry.
6. One can choose specific parts of the egg for a particular need.

Almost all egg-breaking operations are automated. After the eggs are mechanically broken, they are checked for abnormalities and odor. The broken eggs can then be processed whole or separated into yolks and whites. Whole eggs are mixed and strained before packaging. Yolks are also mixed and strained. Additionally, in frozen yolks, salt or sugar is often added to improve the rubbery consistency of the yolk after freezing and defrosting. Glycerin, molasses, or honey may also be used. Whites are strained to remove the chalazae, meat spots, blood spots, and broken shells, and may be passed through a chopper to homogenize the product. Of the three processed forms—frozen, liquid, and dried eggs—dried eggs have the longest shelf life and require the lowest expense for storage, since refrigeration is not needed. However, the vacuum spray processing of drying incurs additional costs.

Bakeries use about 50% of the processed eggs; food manufacturers about 40% and institutions about 10%.

Nutritional information for many of the various processed egg products is listed in Food Composition Table F-36.

(Also see EGG, DEHYDRATED [Dried].)

CONSUMPTION OF EGGS. Eggs are used chiefly for human food. In 1990, the U.S. per capita consumption of eggs was 234. Table E-10 illustrates the downward trend of per capita consumption of eggs since 1960. This is an alarming trend for the poultry producer. It can be largely attributed to the cholesterol controversy and the development of egg substitutes that are low in fat and have no cholesterol. From Table E-10 one could make a reliable guess as to the year the cholesterol controversy started. In 1964 the American Heart Association recommended that the general public reduce cholesterol intake to 300 mg/day. One egg contains about 213 mg of cholesterol. Thus, the elimination of eggs from the diet was singled out as a means of reducing cholesterol intake.

TABLE E-10
PER CAPITA CONSUMPTION OF EGGS[1]

Year	Per Capita Number of Eggs Consumed
1960	335
1965	314
1970	311
1975	277
1980	272
1985	256
1990	234

[1]Source: USDA.

HEALTH PROBLEMS ATTRIBUTED TO EGGS. Americans are becoming more and more health and diet conscious. Unfortunately, much of the nutritional information that they receive is not authoritative, and the facts are not presented objectively. Worse yet, passions and prejudices sometimes trigger such charges as "poultry products cause allergies and heart disease." Such accusations must be answered by more than simple denial. To this end, sections on "Allergies" and "Cholesterol" follow.

Allergies. Occasionally, a child exhibits an allergic sensitivity to eggs. In most cases, the white of the egg is the portion that creates the reaction, and the yolk is generally readily tolerated. In highly sensitive individuals, the diet must be totally devoid of eggs. However, these cases are rare, and most infants and children allergic to eggs can follow a diet that provides for some heat-treated or cooked eggs.

(Also see ALLERGIES.)

Cholesterol. In recent years, the attention of the American public has been focused on heart disease—the leading killer in the United States.

The cholesterol-heart disease relationship first began when, in 1953, Dr. Ancel Keys, at the University of Minnesota, reported a positive correlation between the consumption of animal fat and the occurrence of atherosclerosis.[1]

Subsequent research indicated that individuals with high serum cholesterol levels have a higher rate of atherosclerosis than people with normal levels. Increased serum cholesterol levels can be induced in susceptible individuals when animal fats (saturated fats) and foods high in cholesterol (notably eggs) are consumed.

That cholesterol is implicated in atherosclerosis is well documented by unbiased, controlled experiments; but research indicates that it is not the sole cause. Rather, a number of factors enters into the cause of heart disease; among them, stress, heredity, hypertension, diabetes mellitus, smoking, lack of exercise, and obesity.

It is noteworthy that eggs contain 22% less cholesterol than previously believed—213 mg instead of 274 mg per large egg. Also, they are low in saturated fat—only 1.5 g per large egg.

(Also see CHOLESTEROL; HEART DISEASE; and HYPERLIPOPROTEINEMIAS [Hyperlipidemias].)

[1]Keys, A., *Journal of Chronic Diseases*, Vol. 4, 1956, p. 364.

Salmonellosis. Salmonellosis may be a concern in eggs and egg products unless they are handled properly, as follows:

1. Store eggs at 45°F *(7.2°C)*, or lower, until they are used; and do not keep them longer than 3 to 4 weeks.

2. Never use eggs that are cracked, contain blood spots, or have pinpoint holes in them.

3. Use pasteurized eggs, rather than raw eggs, as ingredients of uncooked, ready-to-eat menu items such as Caesar salad, eggnog, ice cream, milk shakes and other egg-fortified beverages, and freshly prepared mayonnaise, hollandaise, and bearnaise.

4. Do not pool or comingle eggs that are out of their shells unless they are cooked immediately.

5. Cook individual eggs so that all parts reach temperatures of at least 140°F *(60°C)*.

6. Maintain cooked eggs requiring holding (such as those in breakfast buffets) at an internal temperature of at least 140°F *(60°C)*.

7. Cook home-prepared eggs until the white is thoroughly coagulated and the yolk starts to thicken.

(Also see SALMONELLA; and SALMONELLOSIS.)

NUTRITIONAL VALUE. Of all the foods available to people, the egg most nearly approaches a perfect balance of all the nutrients. This is evidenced by the fact that the egg is the total source of nutrition for the developing chick embryo. Unlike mammals, the chick embryo cannot secure needed nutrients from the reserves of the mother; rather, it lives in a closed system which must contain all the food needed for development. If a single nutrient were to be lacking, the developing embryo would die.

Eggs contain an abundance of proteins, vitamins, and minerals. The protein fraction of eggs is highly digestible and of high quality, having a biological value of 94 on a scale of 100, the highest rating of any food. The reason for this high quality of protein is that egg proteins are complete proteins; that is, they contain all the essential amino acids required to maintain life and promote growth and health. Additionally, eggs are a good source of iron; phosphorus; trace minerals; vitamins A and E, and most of the B vitamins, including vitamin B-12. Eggs are second only to fish-liver oils as a natural source of vitamin D. Eggs are moderate from the standpoint of calorie content, a medium-size egg containing about 75 Calories (kcal).

Overall, eggs provide a well balanced source of nutrients for persons of all ages. During the rapid growth of infants, children and teenagers, eggs can contribute significantly to the body's nutrient needs. They make good therapeutic diets. Eggs are acceptable and valuable for older people whose caloric needs are lower and who sometimes have problems chewing certain foods. Table E-11 shows the nutritive value of eggs.

TABLE E–11
NUTRIENT VALUE OF TWO MEDIUM SIZED EGGS

Nutrient	Amount per Egg[1]	Daily Dietary Recommendations[2]
		(%)
Calories	75.4 kcal	4.2
Protein	6.1 g	9.7
Amino Acids[3]		
Isoleucine	364.0 mg	46.1
Leucine	512.0 mg	46.3
Lysine	394.0 mg	41.6
Methionine and cystine	327.0 mg	31.8
Phenylalanine and tyrosine	572.0 mg	51.7
Threonine	286.0 mg	51.7
Tryptophan	93.0 mg	33.6
Valine	420.0 mg	53.2
Vitamin A	75 mcg RE	7.5
Vitamin D	0.6 mcg	8.0
Thiamin	0.42 mg	28.0
Riboflavin	0.144 mg	8.5
Niacin	0.030 mg	1.6
Vitamin B-6	0.058 mg	2.9
Folate	31.2 mcg	15.6
Vitamin B-12	0.743 mcg	37.2
Calcium	25.4 mg	3.2
Phosphorus	86.4 mg	10.8
Magnesium	5.8 mg	1.7
Iron	1.104 mg	11.0
Zinc	0.69 mg	4.6

[1]Values obtained from the Proteins and Amino Acids in Selected Foods, Table P-37 and the Food Composition Table F-36.

[2]Daily dietary recommendations for a male age 25 to 50, weighing 174 lb *(79 kg)*, and 70 in. *(178 cm)* tall, based on the *Recommended Dietary Allowances*, 10th ed., NRC–National Academy of Sciences, 1989.

[3]The essential amino acids for an adult based on the *Recommended Dietary Allowances*, 10th ed., NRC–National Academy of Sciences, 1989, p. 57, Table 6-1.

Most nutrition programs can be improved by the inclusion of eggs. Eggs are of the highest nutritional quality and are marketed at prices that even the poor can afford. The relative economy at any given time of different protein sources (foods) compared to a serving (3 oz) of beef without bone, which supplies about 20 g of protein (about ⅓ the recommended protein allowance for a 20-year-old male), can be obtained by filling in the three blank columns (market price, cost of amount to give 20 g of protein, and rank) in Table E-12.

TABLE E-12
COST OF *20 GRAMS* OF PROTEIN
FROM SELECTED FOOD SOURCES

Food	Market Price	Amount to Give 20 g Protein	Size of Serving to Supply 20 g Protein	Cost of Amount to Give 20 g Protein	Rank
Beef (boneless)		3 oz			
Fresh whole milk		18 fl oz	2¼ cup		
Cheddar cheese		approx. 3 oz	1 × 1 × 3 in.		
Cottage cheese		4 oz	½ cup		
Eggs		3 large eggs	—		
Salmon (canned)		approx. 3 oz	aprox. ½ cup		
Lima beans		3.3 oz (dried)	1¼ cup (cooked)		
Peanut butter		2½ oz	⅓ cup		
Tuna (canned)		2½ oz	⅓ cup		
Enriched white bread		½ lb	10 slices (Not complete protein)		

Fig. E-6. Commercial brown-egg layer. (Courtesy, Babcock Industries, Ithaca, N.Y.)

Notwithstanding the high nutritive value of eggs, millions of Americans routinely eat eggs for breakfast each morning simply because they like them.

NUTRITIONAL MISCONCEPTIONS. Rumors still persist concerning the factors affecting the nutritional value of eggs.

• **Shell color**—Some individuals are willing to pay more for eggs of a certain shell color believing it to be an indication of quality or nutritive value. As pointed out, the color of the shell—whether brown, white, or blue—is directly related to the breed or strain of chicken; and it has no effect on the nutrient composition of the eggs.

• **Yolk color**—Xanthophyll is the major substance causing yolks to have a deeper yellow color. It has no nutritive value. Years ago when chickens ran freely on the farm they ate grass which contains xanthophyll. Modern production units put enough xanthophyll in the ration of chickens to produce a medium-yellow yolk.

• **Fertile eggs**—Some people feel that fertile eggs—eggs from hens mated to roosters—are more nutritious than nonfertile eggs. There is no scientific proof of this fact. Furthermore, infertile eggs are of more economic value, since there is no danger of loss through development of the embryo.

• **"Organic" eggs**—Organic eggs are of no higher nutritional value than commercial eggs. In fact, if the diet of an "organic" hen is not well-balanced the nutritive value of her eggs tends to be lower then that of the commercial laying hen fed a balanced diet. Moreover, fewer "organic" eggs are produced; hence, they are more expensive than commercially produced eggs.

• **Digestibility**—Some individuals consider raw eggs more digestible. Cooked eggs are more readily digested, but both are very completely digested and absorbed.

NOTE WELL: Raw egg white contains an antivitamin, a protein called avidin, which binds to biotin and renders it unavailable. However, avidin is denatured when eggs are cooked, thus making biotin available.

(Also see AMINO ACID[S]; DIGESTION AND ABSORPTION; METABOLISM; POULTRY; PROTEIN[S]; and REFERENCE PROTEIN.)

EGG, DEHYDRATED (Dried)

Dehydration—removing the water from foods to a level low enough to slow chemical reactions and stop the growth of microorganisms—is a successful way of preserving eggs. Dried or dehydrated eggs offer the following advantages:

1. Less storage space is involved, and storage space is less expensive.
2. Transportation cost is less.
3. Dried eggs have a long shelf life.
4. One can choose specific parts of the egg for a particular need.

Egg drying began in the United States in about 1880, when W. O. Stoddard started a commercial operation in

St. Louis, Missouri. However, shortly after the turn of the century, China had a large supply of eggs available at a low cost. Both Americans and Germans stepped in and set up drying plants in China. Therefore, due to the low cost of egg products from China, there was little incentive for drying eggs in the United States. Then, about 1920 tariffs on eggs from China were increased and egg drying started again in the United States. Next, the need for dried eggs created by World War II and by the Lend Lease Program spurred the expansion of the egg drying industry.

Eggs can be divided into three categories on the basis of the drying methods employed: (1) egg whites, (2) egg yolks, and (3) whole eggs.

1. **Egg whites.** Before drying, egg whites are pasteurized, and a small amount of free glucose must be removed by enzymatic action (glucose oxidase) to prevent the Maillard reaction—a chemical reaction responsible for brown color and off-odors. Often a whipping aid such as sodium lauryl sulfate is added to help whites retain their high whipping ability following drying. Then, after the adjustment of the pH the whites are spray-dried and packaged in 150, 50, or 25 lb (*68, 23, or 11 kg*) containers. Some egg whites are air-dried where the whites are poured in thin layers on trays. This produces flakes of albumen.
2. **Egg yolks.** Egg yolks are pasteurized and then spray-dried, or stabilized by removing the free glucose with the enzyme glucose oxidase and then spray-dried. Often dried egg yolk is converted to a free-flowing powder by the addition of anticaking agents such as silicon dioxide or sodium silicoaluminate.
3. **Whole egg.** Whole eggs are handled in a manner similar to yolks and whites. However, carbohydrates such as sucrose or corn-syrup solids are added before drying to improve or maintain some of the functional qualities like flavor, foaming and emulsifying. Dried whole eggs may also be converted into a free-flowing product by adding anticaking agents.

Overall, the important functional properties of eggs—whipping, emulsifying, coagulation, flavor, and color—are maintained, or in some cases improved, through drying, providing there has been proper processing and storage. Moreover, the nutritional value of dehydrated eggs has been found to be essentially the same as fresh eggs.

In 1979, more than 77 million pounds (*35 million kg*) of dried egg products were produced in the United States. However, the highest levels ever of dried egg production occurred during World War II, the years 1942 to 1946. The average yearly production for these 5 years was more than 209 million pounds (*94 million kg*).

Dehydrated egg products are employed in cake mixes, mayonnaise, salad dressings, noodles, candies and all types of baking products.

(Also see EGG[S].)

EGG FRUIT (CANISTEL) *Lucuma bifera; L. mervosa*

The yellow egg-shaped fruit of an evergreen tree (of the family *Sapotaceae*) that is native to northern South America. It is 2 to 4 in. (*5 to 10 cm*) long and has a yellow, mealy flesh. Egg fruit appears to have been a staple food of the ancient Peruvians since it has often been found at archaeological sites.

The fruit is moderately high in calories (140 kcal per 100 g) and carbohydrates (36%). It is a good source of iron and vitamin C.

EGGPLANT (AUBERGINE) *Solanum melongena*

This vegetable is known as eggplant because the fruit of many varieties resemble a large egg.

Although eggplant has long been underappreciated in the United States, it has slowly, but steadily, gained in popularity as a result of the proliferation of Greek, Italian, and Middle Eastern restaurants which serve it in various tasty dishes. The eggplant, like chili peppers, the Irish potato, sweet peppers and the tomato, is a member of the nightshade family (*Solanaceae*).

Fig. E-7. The eggplant, a vegetable that is much appreciated by the peoples of the Mediterranean region and the Middle East.

ORIGIN AND HISTORY. It is believed that the modern eggplant is an improved version of one or more closely related wild species which are native to India. However, it may also have been domesticated independently in China, where it has been grown since the 5th century B.C. The plant was introduced into Africa by the Arabs and the Persians before the Middle Ages. A little later (in the 14th century A.D.), it was introduced into Italy from Africa. Cultivation of the eggplant spread throughout the Mediterranean region and the Middle East.

The European explorers took the eggplant with them on their voyages to new lands. For example, it was grown in Brazil by the end of the 17th century. This vegetable is now cultivated in the warm areas throughout both the Old World and the New World.

PRODUCTION. Fig. E-8 shows the countries which contribute most of the world's eggplant production.

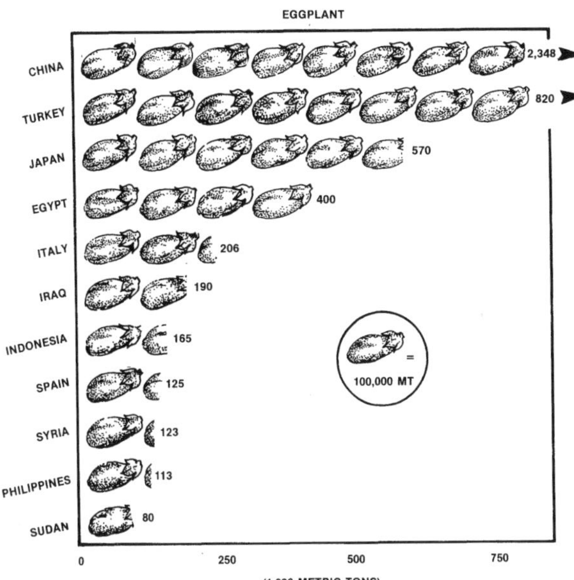

Fig. E-8. The leading eggplant-producing countries of the world. (Based upon data from *FAO Production Yearbook*, 1990, FAO/UN, Rome, Italy, Vol. 44, p. 138, Table 56)

Most of the U.S. eggplant production is in Florida and New Jersey. The total tonnage of fresh eggplants shipped is shown in Fig. E-9.

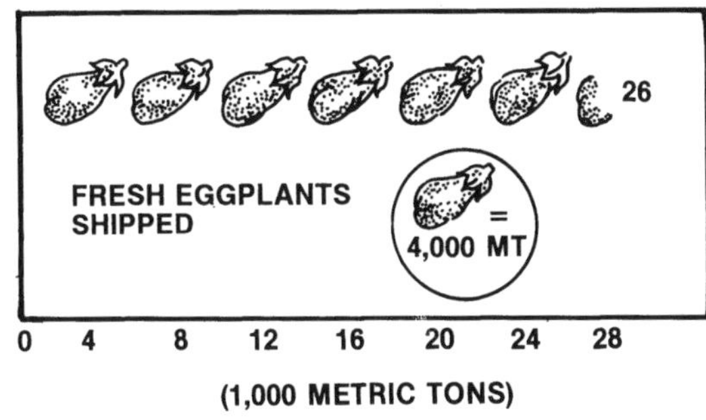

Fig. E-9. Fresh eggplants shipped by rail, truck, and air. (Data from *Agricultural Statistics*, 1991, USDA, p. 169, Table 244)

The eggplant requires a frostfree growing season of at least 5 months and does best when the night temperatures are about 68°F (*20°C*). Hence, much of the U.S. crop is grown in Florida, where these conditions prevail. Also, the plants that are to be grown in the cooler climates are usually started in greenhouses or hotbeds, then transplanted in the field when the warm weather has arrived. Commercial growers usually fertilize the soil with nitrogen, phosphorus, and potassium fertilizer. Eggplant needs an abundant supply of water; hence, it may require irrigation in areas where the rain fall is not sufficient to keep the plants moist.

Usually, the fruits are cut from the vines when the skin is glossy, but before they have reached their full size. By these criteria, the eggplants will be ready for picking about 145 days after planting seeds, or about 70 days after young plants have been transplanted in the field. The average yield of eggplant per acre in the United States is about 20,300 lb (*22,736 kg/ha*) but yields as high as 25,500 lb (*28,560 kg/ha*) have been obtained in Florida.

PROCESSING. Most of the eggplant crop is sold on the fresh market. However, a small portion of the production is made into such items as canned eggplant cubes stewed into a tomato sauce, eggplant appetizer (cooked pieces of eggplant mixed with onions and other vegetables) which is sold in small glass jars, frozen eggplant parmigiana, and frozen fried eggplant slices.

SELECTION. A good quality eggplant is firm, heavy in relation to size, with a uniform dark rich purple color, and free from scars or cuts. A wilted, shriveled, soft, or flabby eggplant will usually be bitter or otherwise poor in flavor. Worm injury can be seen on the surface and, if severe, will cause excessive waste. Decay appears as dark brown spots on the surface, and progresses very rapidly.

PREPARATION. The greatest variety of eggplant dishes were developed in the countries of the Mediterranean region and western Asia where the climatic conditions favor the production of this vegetable. Hence, some typical preparations are listed by country of origin.

- **Egypt**—A firm omelet is made by baking a mixture of eggs, cubed eggplant, chopped onion, crushed garlic, butter, salt, and black pepper.

- **France**—The well known *Ratatouille* is a stew containing sliced eggplant, sliced zucchini, sweet peppers cut into strips, sliced onions, chopped garlic, chopped peeled tomatoes, olive oil, and various seasonings.

- **Greece**—*Moussaka* is a baked dish consisting of layers of ground lamb or veal interspersed with fried slices of eggplant, plus eggs, cheese, wine, onions, garlic, parsley, a medium cream sauce, and seasonings.

- **India**—The country in which the eggplant was first cultivated is known for the various curries and purees that contain the vegetable. Usually these dishes contain other ingredients such as onions, garlic, chili peppers, butter or ghee, and certain spices.

- **Israel**—One of this country's most tasty dishes is fried slices of eggplant baked with fried onions and garlic, grated Cheddar cheese, and tomato sauce.

- **Italy**—*Parmigiana di melanzane* is known in the United States as "Eggplant Parmigiana," and is prepared by baking sliced eggplant in layers with mozzarella and Parmesan cheeses, plus tomato sauce. Americans of Italian descent also relish *melanzane sott'olio,* which is eggplant preserved in oil.

- **Syria**—A salad is made from chopped baked eggplant, sliced tomatoes, sliced green peppers, chopped onion, crushed garlic, chopped parsley, olive oil, and seasonings.

• **Tunisia**—Cubed eggplant is cooked with sliced zucchini, chopped peeled tomatoes, sliced sweet peppers, chopped chili peppers, crushed garlic, vegetable oil and seasonings. Then, the dish is allowed to cool so that it may be served like a salad.

• **Turkey**—The baked stuffed eggplant usually contains fried onions, peppers, and garlic, mixed with tomatoes, tomato puree, parsley, and various spices.

CAUTION: Unripe eggplant, like green or sprouted potatoes, may contain potentially harmful amounts of the alkaloid solanine, which (1) prolongs the action of cholinergic nerves, and (2) hastens the breakdown of red blood cells. Also, a recent research report suggests that the consumption of vegetables of the Nightshade family (capsicum peppers, eggplant, the Irish potato, and the tomato) may aggravate arthritis in susceptible people. Therefore, it is suggested that arthritis-prone people and others who are unaccustomed to eating these vegetables try eating only small amounts of each vegetable alone, until it is certain that there is no sensitivity to one or more of them.

NUTRITIONAL VALUE. The nutrient composition of cooked eggplant is given in Food Composition Table F-36.

Some noteworthy observations regarding the nutrient composition of eggplant follow:

1. Cooked eggplant is high in water content (94%), and low in calories and most other essential nutrients. However, it is a fair to good source of iron and potassium.
2. Eggplant has been used as a medicinal plant since ancient times because of the belief that it is effective in (a) stimulating the flow of bile, and (b) increasing urination to rid the body of excess water. (*NOTE:* It does not appear that medical scientists have investigated these beliefs. Hence, they are herein presented for two purposes: (1) information, and (2) stimulation of research.)
3. It has been shown that a substance in eggplant prevents the breakdown of the nerve transmitter acetylcholine. Therefore, the consumption of eggplant may result in an increase in (a) the secretion of digestive juices, and (b) movements of the digestive tract. (Both actions are brought about by the nerves which release acetylcholine.)

SUMMARY: Much has to be learned regarding the potentially beneficial or harmful substances present in eggplant. It might well be that the ancient peoples, who often compensated for their lack of modern medical instrumentation and knowledge by making painstaking observations, knew something about this vegetable that has yet to be discovered by modern man.

(Also see VEGETABLE[S], Table V-6 Vegetables of the World.)

EGG PROTEINS

Proteins comprise 11.8% of the chemical composition of eggs. This protein fraction of eggs is highly digestible and of high quality, having a biological value of 94 on a scale of 100, the highest rating of any food. The reason for this high quality of protein is that egg proteins are complete proteins. They contain all the essential amino acids required to maintain life and promote growth and health. In fact, the amino acid content of eggs is employed as a reference for evaluating the completeness of other proteins. Proteins of the egg white include ovalbumin, conalbumin, ovomucoid, globulins, ovomucin, flavoprotein, ovoglycoprotein, and avidin. The yolk contains the proteins phosvitin, a nonlipid phosphoprotein, and lipovitellins and low-density lipoproteins (LDL)—both lipid-protein complexes.

(Also see EGG[S]; PROTEIN[S]; and REFERENCE PROTEIN.)

EGG SIZES

Eggs are classified by size according to their weight per dozen. The size classification is independent of quality. Market size of eggs and the minimum weight per dozen is indicated in Table E-13.

(Also see EGG[S], section headed "Size.")

TABLE E-13
U.S. WEIGHT CLASSES FOR CONSUMER GRADES FOR SHELL EGGS[1]

Size or Weight Class	Minimum Net Weight per Dozen	
	(oz)	*(g)*
Jumbo	30	*850*
Extra large	27	*765*
Large	24	*680*
Medium	21	*595*
Small	18	*510*
Peewee	15	*425*

[1]*United States Standards, Grades, and Weight Classes for Shell Eggs*, Agricultural Marketing Service, Poultry Division, USDA.

EGG WHITE

In the shell, the egg white or albumen surrounds the yellow yolk. It is comprised of four layers: (1) the outer thin or liquid white; (2) the dense or thick white; (3) the inner thin or liquid white; and (4) the inner thick white or chalaziferous layer. Water is the major constituent of egg white, amounting to about 88%. The proteins of egg white account for about 11% of the composition. Egg white proteins include ovalbumin, conalbumin, ovomucoid, globulins, ovomucin, flavoprotein, ovoglycoprotein, and avidin.

The unique foaming property of egg white is due to its proteins. When egg white is beaten, air bubbles are trapped in the albumen and a foam is formed—a gas phase dispersed in a liquid phase. This foaming power of the egg white—albumen—makes it useful as a carrier of air for leavening; hence, it contributes to the lightness of certain foods. For example, meringues, angel cakes, souffles, omelets, some candies, and sponge cakes all depend upon eggs for their lightness. Furthermore, these proteins of the white contribute to binding and thickening properties of eggs employed in a number of foods. As foods containing eggs cook, these proteins coagulate—a change in the stucture of egg protein that results in the loss of solubility and increased thickening. The coagulating property of eggs is important to custards and pie fillings.

Last but not least, the proteins of egg white contribute to the nutritional value of eggs. The protein fraction of eggs is highly digestible and of high quality, having a biological value of 94 on a scale of 100, the highest rating of any food. The reason for this high quality of protein is that egg proteins are complete proteins. They contain all the essential amino acids required to maintain life and promote growth and health.

Egg white is often used as an indicator of egg quality. It must be firm with a rather clear demarcation between the thin and thick albumen. Egg albumen quality is routinely measured in Haugh units—units determined by a micrometer that measures albumen height. This height is then compared to a chart listing the weights of eggs and Haugh units can then be derived. As eggs age, the firmness of the white decreases, and when broken, the contents spread and flatten. Also, the process of candling eggs determines the firmness of egg white—hence the quality of the egg. The pH (hydrogen concentration) of the egg white is another means whereby egg quality is determined. The pH of a freshly layed egg is about 7.62 to 8.2. As the egg ages, carbon dioxide is lost, and the pH increases to as high as 9.5.

(Also see EGG[S]; and EGG PROTEINS.)

EGG WHITE INJURY

A disorder which results from the binding of the vitamin biotin by a substance called avidin which is present in raw egg white. Egg white injury is characterized by shedding of flakes of skin, lack of appetite, loss of weight, and pains in the muscles. However, this disorder is rare in humans because (1) they do not normally eat sufficient amounts of raw egg white to bring about this condition; and (2) most people cook eggs before eating them.

(Also see BIOTIN.)

EKG

Abbreviation for electrocardiogram, because in German, cardiogram is spelled with a K.

(Also see ELECTROCARDIOGRAM.)

ELASTASE

An enzyme which digests the connective tissue protein—elastin.

(Also see ELASTIN.)

ELASTIN

A connective tissue protein which is present in ligaments, lung tissue, and the walls of blood vessels. This protein is synthesized mainly during the years of rapid growth. It is rich in lysine, so the lysine needs of growing children are high compared to those of adults.

(Also see PROTEIN[S].)

ELECTROCARDIOGRAM

The tracing made by the heartbeat on an electrocardiograph used to determine abnormality in the heart muscle.

(Also see HEART DISEASE, section headed "Special Diagnostic Procedures For Cardiovascular Problems.")

ELECTROENCEPHALOGRAM (EEG)

This printed record, which is sometimes called a brain wave pattern, shows the fluctuations of electrical charges in the brain. It is commonly used in the diagnosis of such disorders as brain tumors and epilepsy.

ELECTROLYTE

The name given to a substance which when dissolved in water enables the resulting solution to conduct an electric current. The most common electrolytes in the human body are salts of such minerals as sodium, potassium, magnesium, calcium, phosphate, sulfate, and chloride.

(Also see ACID-BASE BALANCE.)

ELECTRON

A negatively charged particle, which is responsible for the various chemical reactions that occur in matter.

ELECTRON MICROSCOPE

A powerful microscope which magnifies the object from 150,000 to 300,000 times, thereby permitting visualization of cellular structure.

ELECTRON TRANSPORT CHAIN

A connected series of chemical reactions which occur within the energy generating part (mitochondria) of cells. The electron transport chain provides the means by which the intermediate products of carbohydrate, fat, and protein metabolism are converted to carbon dioxide, water, and energy.

(Also see METABOLISM.)

ELECTROPURE PROCESS

A technique of pasteurizing milk utilizing a low frequency alternating current.

ELEMENTAL DIETS (SEMISYNTHETIC FIBER-FREE LIQUID DIETS)

These dietary preparations are free of fiber and other undigestible substances, but contain most of the essential nutrients in forms that require little or no digestion and which may be absorbed completely. Hence, little or no residue from the diet reaches the large intestine. The term "elemental" means that the nutrients are present in their most "elemental" or simple chemical forms, although the diets sometimes contain complex forms of certain nutrients. Similarly, the term "semisynthetic" means that the products contain mainly pure chemical compounds rather than the more complex natural food materials. It is noteworthy that the recent development of elemental diets has made it possible to provide much more nutritious diets for people who cannot tolerate normal diets, but can utilize these preparations.

TYPES OF PRODUCTS. The products that are available vary considerably in their major ingredients although the nutrient compositions are similar. However, the types of the ingredients used as sources of carbohydrates, fats, and amino acids or proteins are very important because they determine a characteristic of the elemental diet called the osmolality. *Osmolality is a measure of the number of solute molecules (dissolved materials) in a given amount of water.* Solutions that contain many small molecules such as amino acids have higher osmolalities than those containing a few large molecules such as proteins, even though the total weights of solutes per weight of water may be the same for both types of solution. The higher the osmolality the more likely the dietary preparation will absorb and retain water from the digestive tract and cause distention and diarrhea.

Examples of representative elemental diets follow.

• **Low osmolality dietary preparation**—A typical product contains 60% carbohydrate in the forms of glucose oligosaccharides (low molecular weight polymers) and sucrose (table sugar), 12% protein as egg albumen, 28% fat as vegetable oil, and added minerals and vitamins. It has an osmolality of 300 units.

• **High osmolality dietary preparation**—The high osmolalities (810 units is typical) of these types of products are due mainly to (1) a high carbohydrate content (over 80%) in the forms of glucose oligosaccharides and glucose, and (2) provision of more than 15% protein as pure amino acids. (It is necessary to provide pure amino acids rather than protein when the condition of the patient makes it likely that the latter may not be digested completely.) These products may also contain small amounts of vegetable oil and added minerals and vitamins.

Many of these preparations contain added flavorings such as banana, beef broth, cherry, citrus fruit, fruit punch, grape, lemon, lime, orange, strawberry, tomato, and vanilla. The flavored products were developed to provide variety for patients who must consume only these solutions for extended periods of time.

INDICATIONS FOR THE USE OF ELEMENTAL DIETS. These products are therapeutic preparations which are employed for the treatment of patients with conditions that make feeding with ordinary types of foods unadvisable. Furthermore, they are much more expensive than other dietary formulas. Hence, the decision to give a patient an elemental diet should be made only by a doctor, although an experienced dietitian can usually provide considerable guidance in the selection of the appropriate product.

Generally, elemental diets are indicated when oral feeding is permitted, but it is necessary to have maximal ease of digestion and absorption and only minimal passage of undigestible matter into the large intestine. These requirements are most commonly associated with the following circumstances:

1. Preparation of the bowel for surgery or a barium enema.
2. After surgery on the digestive tract, as an adjunct to or substitute for, intravenous feedings until normal foods or formulas can be consumed.
3. Rehabilitation of severely malnourished patients who cannot tolerate a normal diet.
4. Treatment of certain types of diarrheal diseases. (A product with a low osmolality would be most suitable for these cases.)
5. Nourishment of burn victims who have high metabolic needs. (In these cases, high-nitrogen [or high protein] products should be used.)
6. Tube feedings. It is noteworthy that elemental dietary preparations are less viscous and require smaller bore tubing than ordinary tube feeding preparations because the danger of clogging the tubing is reduced greatly by the low viscosity and the absence of substances that may come out of solution.
7. Inflammation and/or bacterial infection of the bowel resulting from chemotherapy, radiation treatments, or other factors.
8. Fistulae of the digestive tract, which heal more rapidly when the dietary bulk and stimulation of peristalsis are minimal.
9. Pancreatitis, because pancreatic secretions are reduced greatly on these diets.

However, the consumption of an elemental diet may result in certain noteworthy changes in normal physiological functions.

Beneficial Effects. These consequences of elemental diet therapy are considered to be beneficial short-term effects, but some of them may not be desirable over the long term.

• **Reduced secretion of digestive juices**—It appears the secretion of both gastric juice and pancreatic juice is reduced somewhat, which gives the stomach and pancreas some rest, when healing and/or reduction of inflammation is needed.

• **Slower movements of the digestive tract**—Elemental diets are emptied slowly from the stomach because they have a higher solute content (osmolality) than the body fluids. The delayed gastric emptying is accompanied by the slowing of peristalsis in the rest of the digestive tract.

Normally, beverages and foods with high osmolalities are diluted by the gastric juices, but the amounts of these secretions are reduced when elemental diets alone are consumed. The slowing of the digestive tract speeds up healing after inflammation, injury, or surgery.

• **Decreased volume of colonic residue**—Much less than normal amounts of undigested residue reach the colon when elemental diets are consumed. (For example, a small stool may be passed only 1 in 5 days.) Hence, the lower bowel is rested before or after surgery. It is noteworthy that studies showed there were fewer postoperative bowel movements, less pain, and fewer complications when elemental diets were consumed by patients prior to hemorrhoidectomies.

• **Sparing of body protein**—Sometimes, the lean body tissues are catabolized during the recovery from burns, severe injuries, and surgery because large amounts of protein are needed for healing. However, it appears that elemental diets reduce the tissue wastage as evidenced by reduced levels of nitrogen excretion in the urine.

Undesireable and Potentially Harmful Effects. The alteration of normal digestive functions by the hyperosmolality of elemental diets may sometimes be harmful, as noted in the items which follow.

• **Digestive disturbances**—The delayed emptying of the stomach may make a patient nauseous if an elemental diet is consumed too rapidly. Also, diarrhea may result if the dietary fluid is delivered to the intestine at a rate faster than that at which the nutrients and water may be absorbed.

• **Dehydration**—Diarrhea or even moderate looseness produced by elemental diets may cause dehydration that may result in a coma in infants, young children, grossly malnourished adults, and older people who are extra sensitive to changes in the fluid content of the body.

• **Aspiration of the diet into the lungs**—This misfortune is most likely to occur if the patient lies down shortly after consuming the diet and the stomach contents are regurgitated. Foreign materials in the lungs cause pneumonitis, which is an inflammatory reaction of the bronchial tissues that can lead to a pneumonialike infection.

• **Deficiencies of microminerals and vitamins**—Elemental diets may not contain chromium, selenium, and/or vitamin K. The lack of these nutrients may not be serious if the diets are used for only short periods of time (not more than 1 to 2 weeks) because there may be some tissue stores of these nutrients. However, the missing nutrients must be supplied by other means if the artificial feeding is continued for several weeks or longer.

• **Fluctuations of the blood sugar level**—All of the elemental dietary preparations contain large amounts of sugars and other rapidly absorbed carbohydrates which cause elevations of the blood sugar promptly after ingestion of the products. Hence, they may present some problems for (1) diabetics who require insulin to control their blood sugar, and (2) people with reactive hypoglycemia (the development of low blood sugar due to oversecretion of insulin in response to eating rapidly absorbed carbohydrates).

CONTRAINDICATIONS. Elemental diets should *not* be given to infants under 3 months of age, and older infants should be given *only* diluted diets. The diets are unsuitable also when patients have diseases of the kidneys and the liver in which the metabolism and excretion of nutrients is impaired. Furthermore, they should not be given to people with intestinal stasis of various types, and they should be used with great caution in cases of insulin-dependent diabetes. Finally, it may be preferable to use intravenous hyperalimentation when a jejunal fistula is present.

(Also see HYPERALIMENTATION, INTRAVENOUS.)

PREPARATION AND ADMINISTRATION. A key to successful use of elemental diets is their gradual introduction to permit patients to become accustomed to their flavors and any feelings of fullness or discomfort that may occur. It is noteworthy that elemental diets may be fed via the normal oral route or through a tube introduced into the digestive tract. (*NOTE:* Elemental diets are *never* used for intravenous feedings.)

Normal Feeding. Most manufacturers recommend that their formulas be fed only at one-half strength the first day and that the full day's allotment not be fed until several days have passed. Also, many preparations are more palatable if they are served chilled. (Pure amino acids have peculiar and somewhat nauseating tastes that are accentuated in warm solutions.) Usually, individual feedings are packaged in separate packets that contain from 2 to 3 oz (*56 to 85 g*) of powder to be mixed with 8 to 10 fluid ounces (*240 to 300 ml*) of water. However, an entire day's supply may be mixed at one time and kept refrigerated until used. These nutrient solutions are ideal culture mediums for microorganisms. Hence, the prepared diets should not be kept at room temperature or stored in a refrigerator for longer than a day. For variety, the diets may be served as broths, frozen slushes, or puddings. Each manufacturer provides instructions for using his particular products.

Tube Feeding. Although problems of palatability are overcome by tube feeding, the effects of the high solute concentration (hyperosmolality) will be similar to those encountered in normal oral feeding. Hence, the solution should be only one-half strength when fed initially, and the rate of feeding by gravity drip or pumping should be limited to 1 to 2 oz (*40 to 60 ml*) per hour. Furthermore, not more than 1 to 2 pt (*500 to 1,000 ml*) should be kept in the reservoir at one time and the solution should be kept chilled. Finally, the external tubing and fluid reservoir should be changed daily to prevent microbial contamination.

(Also see LIQUID DIETS.)

SUMMARY. Elemental diets are a recent development in dietetics and medicine that can be lifesaving when used properly. As in the case of other therapies which may alter the normal physiology of the body, patients given these diets should be monitored carefully for signs and symptoms of abnormalities. It is noteworthy that patients have done well when given elemental diets as the sole source of nutrients for as long as 1 year.

ELEMENTS, TRACE

Minerals which are needed in small amounts, ranging from millionths of a gram (micrograms) to thousandths of a gram (milligrams) per day.

(Also see MINERAL[S]; and TRACE ELEMENTS.)

EMACIATED

An excessively thin condition of the body.

(Also see MARASMUS; STARVATION; and UNDERWEIGHT.)

EMBDEN-MEYERHOF PATHWAY

The initial stage of metabolism by which carbohydrates are partially converted into energy by living cells. The Embden-Meyerhof Pathway is the part of energy metabolism which takes place without the presence of oxygen. However, only a limited amount of energy may be produced in this pathway. Therefore, it is necessary for the products of this pathway to enter into oxygen-requiring metabolism (respiration) which takes place in the electron transport chain.

(Also see METABOLISM.)

EMBLIC (INDIAN GOOSEBERRY; MALACCA TREE) *Emblica officinalis*

The fruit of a large tree (of the family *Euphorbiaceae*) that is native to southeastern Asia. Its yellowish fruits are 1 to 2 in. (*2.5 to 5 cm*) in diameter and are very sour. They are consumed raw or preserved, and are sometimes used in baked goods. The emblic tree was recently introduced into Florida as a potential source of tannin.

Emblic fruit is moderately high in calories (81 kcal per 100 g) and carbohydrates (22%). It is very rich in vitamin C and is a good source of fiber.

EMBOLISM

A clogging of a blood vessel by materials such as air, blood clots, cellular debris, fat, or microorganisms.

EMBRYO

The earliest stage of development in which a newly formed animal or plant is recognizable. The term embryo usually designates an organism which arises from sexual reproduction.

EMETIC

Any substance which is given by mouth to induce vomiting.

-EMIA

Suffix that denotes a condition of the blood; for example, leukemia.

EMPHYSEMA

A lung disease which is characterized by a stretching and/or rupture of some of the air cells (alveoli) of the lungs. Emphysema greatly limits the amount of exercise the afflicted person may take without having difficulty in breathing. Also severe, chronic emphysema can lead to heart failure because of the extra work required to pump blood through diseased lungs. Some of the factors which may bring on or aggravate emphysema are (1) air pollution, (2) allergies such as hay fever, (3) cigarette smoking, and (4) infectious repiratory diseases. It is not certain regarding the role that nutrition might play in the prevention of emphysema, although recent research has shown that vitamins A and E apparently protect lung tissue against damage from airborne pollutants.

EMPTY CALORIES

The designation given to foods which supply mainly calories. These foods are usually rich in carbohydrates and/or fats, but contain only low levels of proteins, minerals, and vitamins. Some examples of such foods are the various candies and carbonated beverages.

(Also see BREAKFAST CEREALS, Section headed "The Sugar-Breakfast-Cereal Binge.")

EMULSIFY

To disperse small drops of one liquid into another liquid, e.g., mayonnaise and hollandaise sauce.

(Also see EMULSIFYING AGENTS.)

EMULSIFYING AGENTS.

Food additives, some natural and some synthetic, used in a wide variety of food products to create emulsions. Glycerol monostearate, mono- and diglycerides, gums, egg yolk, albumin, alginate, lecithin, and carrageenan (Irish moss) are all examples of common emulsifying agents. Advantages of these agents are: (1) maintenance of homogeneity; (2) more economical use of fats; (3) improved volume uniformity; (4) improved whipping properties; and (5) improved keeping qualities. Emulsifiers are used in such foods as baked goods, cake mixes, confectionery products, frozen desserts, ice cream, margarine, salad dressings, etc. In many foods the emulsifier and the stabilizer are inseparable. By law, these agents must be listed on the package or label.

(Also see CARRAGEENAN; EMULSION; ADDITIVES; and STABILIZER.)

EMULSIFYING SALTS

Food additives, commonly the sodium salts of citrate, phosphate, and tartrate, used as emulsifiers in the manufacture of cheese and evaporated and powdered milk. These salts may also serve as sequestering agents (substances that combine with a metal ion or acid radicle and render it inactive) in some foods.

(Also see EMULSION; ETHYLENEDIAMINE TETRA-ACETIC ACID; and ADDITIVES.)

EMULSION

An intimate mixture of two liquids; for example, oil and water, which do not normally stay mixed unless constantly stirred. With the addition of an emulsifying agent, one liquid becomes suspended in the other as fine droplets. Foods such as shortenings, margarines and mayonnaise are emulsions.

(Also see EMULSIFYING AGENTS; EMULSIFYING SALTS; and STABILIZER.)

ENAMEL

The hard, mineralized material that covers the tooth. (Also see DENTAL HEALTH, NUTRITION AND DIET.)

ENCEPHALOMALACIA

A term which means softening of the brain. Generally, this condition is seen more in livestock than it is in man. However, there have been a few reports of encephalomalacia in infants fed vitamin E-deficient diets and large quantities of vegetable oils.

(Also see VITAMIN E.)

ENCEPHALOPATHY

Any degenerative disease of the brain.

ENDEMIC

A term referring to a disease that is continually present in a given population. It is in contrast to the word epidemic which refers to the type of disease that suddenly appears in a given area. Certain types of nutritional deficiencies, such as goiter, are endemic in the regions which lack particular nutrients in the soil or water.

ENDERGONIC

Chemical reactions occurring in living cells that require an input of energy; for example, the synthesis of protein.

(Also see EXERGONIC; and METABOLISM.)

ENDIVE (ESCAROLE) *Cichorium endivia*

This lettucelike vegetable species is comprised of (1) endive, a variety with very slender, deeply-toothed (dentate) curly leaves; and (2) escarole, a variety with much broader and only slightly-toothed leaves. The two varieties of the vegetable are shown in Figs. E-10 and E-11.

Fig. E-10. Endive.

Fig. E-11. Escarole.

ORIGIN AND HISTORY. Most botanists believe that endive originated in the eastern Mediterranean region. It is very closely related to chicory; hence, it is speculated that it may be the result of crossbreeding (hybridization) between wild chicory and a related species. The Roman naturalist Pliny the Elder (23-79 A.D.) wrote that it was long used by the Egyptians in salads and cooked dishes. Little else was written about the use of this vegetable in ancient times. However, it is well established that endive was introduced some time later into France, where it was used medicinally until it was adopted as a food in the 14th century. Over the centuries, the two distinct varieties were developed into important salad crops in Europe. They are not as well known in the United States, but their use has increased somewhat in recent years as a result of growing popularity of tossed salads.

PRODUCTION. Most of the world production of endive and escarole, which has amounted to approximately one-half million metric tons annually in recent years, is produced in the European countries. Italy produces about half of the world's crop, while France contributes one quarter. Most of the remainder is produced in Holland, Germany, Belgium, and the United States. Most of the U.S. winter crop is grown in Florida, and most of the summer crop is grown in New Jersey and Ohio. Fig. E-12 shows endive and escarole (1) total shipments by rail, truck, and air in 1990, and (2) arrivals at Chicago, Los Angeles, and San Francisco in 1990.

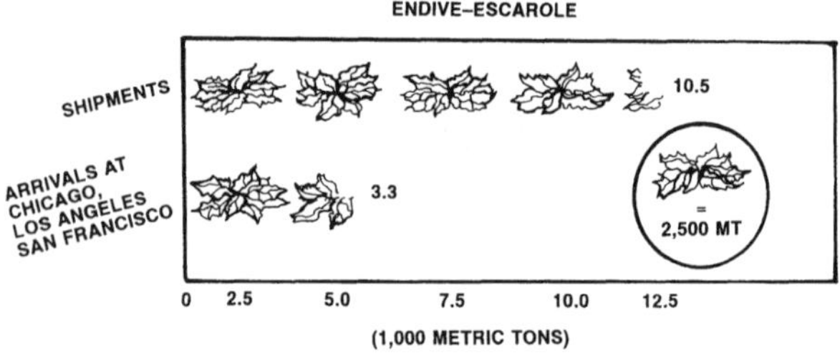

Fig. E-12. Shipments and marketing of endive and escarole. (Data from *Agricultural Statistics*, 1991, USDA, pp. 168-169, Tables 243 and 244)

Endive and escarole are like lettuce in that they are cool weather plants that grow best when the temperature ranges from 60 to 70°F (*16 to 21°C*). Furthermore, the greens harvested in cool weather have a milder flavor than those picked in hot weather. Hence, the crop is planted in winter in warm climates, and in spring, summer, and fall in cool climates. Although young plants tolerate temperatures as low as 28°F (*-2°C*), prolonged exposure to cold causes bolting (the premature development of seedstalks). Therefore, early plantings in the northern United States are usually started in greenhouses or in hot beds, then transplanted in the field after the last hard frost.

Commercial growers usually add ample nitrogen, phosphorus, and potassium to the soil, and use irrigation when necessary to insure that tender, succulent leaves will be produced. When the outer leaves of the plants are about 10 in. (*25 cm*) long, they are tied together at their tops in order to blanch the inner leaves. Blanching lightens the color and imparts a milder taste. About 2 to 3 weeks are required for blanching, after which the heads are harvested and the outer leaves discarded.

PROCESSING. Currently, neither endive nor escarole is processed—they are marketed exclusively as fresh vegetables. However, this situation could change any time in the future because the growing demand for convenience packs of vegetable ingredients for making tossed salads may make it profitable to develop a processing industry similar to the one which produces prepacked shredded lettuce.

SELECTION. Crispness, freshness, and tenderness are the essential factors of quality. Wilted plants can often be freshened in cold water, but this may entail excessive waste. Tough, coarse-leaved plants are undesirable since the usually delicate bitter flavor is apt to be so intensified as to be objectionable. Toughness or tenderness can be determined by breaking or twisting a leaf. When unblanched, the leaves should be green, but if blanched the center leaves should be creamy-white. Decay appears as a browning of the leaves or as a slimy condition.

PREPARATION. Heads of endive or escarole should be washed well under cold running water, then drained well before storing in a refrigerator for not more than a week. These greens are usually served raw in salads because most people prefer them crisp rather than wilted. The stronger-flavored unblanched greens may be flavored with garlic, onions, and other pungent spices, but the blanched leaves should be seasoned more cautiously in order to avoid masking their delicate flavor.

Other ways of preparing these vegetables are: (1) adding the chopped greens to stir-fried dishes near the end of the cooking period; (2) simmering them in meat or poultry broths; (3) wrapping escarole leaves around fillings of bread crumbs, cheese, chopped eggs, legumes, meat, nuts, poultry, and/or rice and baking in an appropriate sauce; (4) braising chunks of the heads briefly in hot fat or oil; and (5) adding the greens to soups and stews.

NUTRITIONAL VALUE. The nutrient compositions of endive and escarole are given in Food Composition Table F-36. Some noteworthy observations regarding their compositions are:

1. These vegetables are high in water content (93%) and low in calories (20 kcal per 100 g). They are also excellent sources of vitamin A (the unblanched leaves contain much more than the blanched leaves); moderately good sources of calcium, phosphorus, iron, and potassium; and only fair sources of vitamin C.
2. The old-time herbalists believed that endive and escarole stimulated digestive secretions and the flow of urine. However, these beliefs have neither been confirmed nor contradicted in recent times because little or no research has been done on the subject. Hence, the authors make no claims regarding the accuracy of these views, but present them here as a matter of historic interest.

(Also see VEGETABLE[S], Table V-6 Vegetables of the World.)

ENDO-

Prefix meaning inner or within.

ENDOCRINE

Pertaining to glands that produce secretions that pass directly into the blood or lymph instead of into a duct (secreting internally). Hormones are secreted by endocrine glands.

(Also see ENDOCRINE GLANDS.)

ENDOCRINE GLANDS

Ductless glands that secrete hormones (chemical messengers) directly into the bloodstream. These hormones then travel via the blood to various organs and tissues where they exert necessary, and sometimes profound, control over such bodily functions as skeletal and sexual development, growth, metabolism, mineral and water balance, and reproduction. Fig. E-13 indicates the location and names of the major endocrine glands, and Table E-14 outlines the origin, names, actions, and associated diseases of the various major hormones.

Secretion of the hormones should never be viewed as a single event, but as a concert. As one hormone comes into play, another may fade out; or one hormone may cause the secretion of another, or the action of one may complement the action of another. Furthermore, the brain or nervous system in many cases acts as the conductor by signaling the proper time for increased or decreased secretion of a hormone.

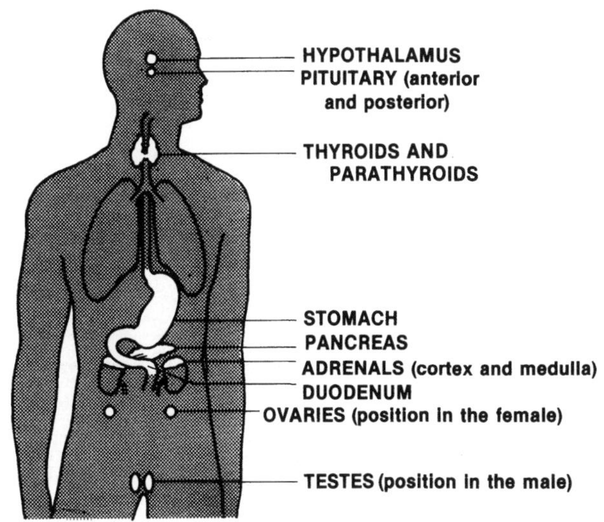

Fig. E-13. The approximate location of the major endocrine glands.

NOTE WELL: Hormone therapy for whatever reason, and whether oral or by injection, should always be under the direction of a physician.

TABLE E-14
HORMONES OF THE ENDOCRINE GLANDS[1]

Hormone	Origin	Mechanism of Release	Physiological Functions	Associated Diseases	Comments
Corticotropin releasing hormone (CRH)	Hypothalamus	The hypothalamic hormones, sometimes called the neurohormones, provide the link between the nervous system and the endocrine system. Thus, such things as stress, nutritional status, emotions, nursing, time of day, season of the year, etc. may manifest themselves as disruptions in bodily function. The hypothalamus acts like a "switchboard" by "plugging in" the proper neurohormone in response to a variety of nervous stimuli received by the brain. Also, other hormones act on the hypothalamus to alter the release of the neurohormone—feedback control.	Releases adrenocorticotropin (ACTH) from the pituitary.	Tumors in the hypothalamus may alter the secretion of any or all of the hypothalamic hormones, then ultimately affect the pituitary, thyroid, adrenals, ovaries, and testes.	So far only GHRIH, TRH, and LRH have been isolated and their chemical nature determined. The presence of other hypothalamic hormones is only suggested by experimental evidence. Much research remains to be done on the hormones of the hypothalamus. Sometimes FRH and LRH are considered one releasing hormone called gonadotropin releasing hormone (GnRH) which controls both LH and FSH. Presently, LRH (GnRH), TRH, and GHRIH are commercially available. LRH (GnRH) offers new hope as a means of fertility control in humans.
Follicle stimulating hormone, releasing hormone (FSHRH or FRH)	Hypothalamus		Releases follicle stimulating hormone from the anterior pituitary.		
Growth hormone, releasing hormone (GHRH; somatotropin releasing hormone, SRH)	Hypothalamus		Releases growth hormone (GH) from the anterior pituitary.		
Growth hormone, release-inhibiting hormone (GHRIH; somatostatin; somatotrophic release-inhibiting hormone, SRIH)	Hypothalamus		Prevents the release of growth hormone (GH) from the anterior pituitary.		
Luteinizing hormone, releasing hormone (LRH)	Hypothalamus		Releases luteinizing hormone (LH) from the anterior pituitary.		
Prolactin releasing hormone (PRH)	Hypothalamus		Releases prolactin from the anterior pituitary.		
Prolactin release-inhibiting hormone (PRIH or PIH)	Hypothalamus		Prevents prolactin release from the anterior pituitary.		
Thyrotropin releasing hormone (TRH)	Hypothalamus		Releases thyroid stimulating hormone (TSH) from the anterior pituitary.		

Footnotes at end of table

(Continued)

TABLE E-14 *(Continued)*

Hormone	Origin	Mechanism of Release	Physiological Functions	Associated Diseases	Comments
Adrenocorticotropic hormone (ACTH)	Anterior pituitary[2]	Corticotropin releasing hormone from the hypothalamus.	Synthesis and release of hormones from the adrenal cortex, mainly the glucocorticoids. ACTH acts directly on fat tissue to liberate free fatty acids into the blood.	Cushing's syndrome; Addison's disease.	ACTH exercises little, if any, control over aldosterone secretion. The effects of stress mediated through the hypothalamus cause increased release. (Also see STRESS.)
Follicle stimulating hormone (FSH)	Anterior pituitary	Folicle stimulating releasing hormone from the hypothalamus.	Stimulation of follicle growth and estrogen secretion in the female; sperm production in the male.	Impaired sexual development or function.	In the female, levels of FSH in the blood rise and fall with each menstrual cycle.
Growth hormone (GH; somatotropin, STH)	Anterior pituitary	Growth hormone releasing hormone and inhibition of growth hormone release-inhibiting hormone from the hypothalamus.	Growth of all tissues; protein synthesis; mobilization of fats for energy while conserving glucose.	Dwarfism; gigantism; acromegaly.	Over secretion of GH in an adult causes acromegaly—excessive growth of the lower jaw, feet, and hands.
Luteinizing hormone (LH)	Anterior pituitary	Luteinizing hormone releasing hormone from the hypothalamus.	Ovarian follicle rupture and egg release, and progesterone secretion in the female; testosterone secretion and sperm production in the male.	Impaired sexual development or function.	In the male this hormone is called interstitial cell stimulating hormone (ICSH). Together with FSH in the female it acts to control events of the menstual cycle.
Oxytocin	Posterior pituitary[2]	Stimulation of the breasts, genitals, and uterus.	Milk "let down," uterine contractions aiding birth process and possibly sperm transport at the time of mating.		There are no known functions in the male. See comment for vasopressin. (Also see BREAST FEEDING.)
Prolactin	Anterior pituitary	Prolactin releasing hormone and inhibition of prolactin release-inhibiting hormone from the hypothalamus.	Formation of milk by the mammary glands by stimulating breast growth and secretory activity.		A variety of nervous stimuli can cause the release of prolactin, but the reason for this is not clear. (Also see BREAST FEEDING.)
Thyroid stimulating hormone (TSH; thyrotropin)	Anterior pituitary	Thyrotropin releasing hormone from the hypothalamus.	Manufacture and release of thyroxine and triiodothyronine from the thyroid gland.	Myxedema; hypothyroidism; goiter.	When the pituitary fails to secrete TSH, the thyroid gland becomes completely nonfunctional.
Vasopressin (antidiuretic hormone; ADH)	Posterior pituitary	Stimulation of special nerves in the hypothalamus which detect the concentration of the body fluids.	Acts on the kidneys to reduce urine volume and conserve body water thus preventing body fluids from becoming too concentrated.	Diabetes insipidus.	Actually, oxytocin and vasopressin originate in the hypothalamus, then they are stored in the posterior pituitary. (Also see WATER BALANCE.)
Calcitonin (thyrocalcitonin)	Thyroid	High calcium level in the blood.	Decreases the level of calcium in the blood by reducing release of calcium from the bones; increases the excretion of calcium in the urine and deposition of calcium in the bones.	Hypocalcemia (low blood calcium).	Some calcitonin may come from the thymus gland. In the thyroid it actually comes from the C-cells. (Also see CALCIUM.)

Footnotes at end of table

(Continued)

TABLE E-14 *(Continued)*

Hormone	Origin	Mechanism of Release	Physiological Functions	Associated Diseases	Comments
Parathyroid hormone (PTH; parathormone)	Parathyroid glands	Low calcium level in the blood.	Elevates blood calcium by increasing the rate of calcium absorption from the kidney and intestine and activating vitamin D; stimulating calcium release from the bones.	Parathyroid tetany (hypoparathyroidism); hyperparathyroidism (osteitis fibrosa cystica).	Hyperfunction of the parathyroid glands results in brittle bones. (Also see CALCIUM.)
Triiodothyronine (T_3)	Thyroid	Release of thyroid stimulating hormone (thyrotropin) from the anterior pituitary.	Both hormones have similar actions, however, triiodothyronine is more potent, and its actions are noted faster. Step up metabolic rate; increases heart performance; increases nervous system activity; stimulates protein synthesis; increases motility and secretion of gastrointestinal tract.	Hypothyroidism (myxedema or cretinism). Hyperthyroidism (exophthalmos or Graves' disease); goiter.	Thyroxin has 1 more iodine atom than triiodothyronine, otherwise they are the same. Many tissues possess an enzyme capable of removing an iodine from thyroxin thus providing a source of T_3 other than the thyroid. Increased blood levels during cold-adaptation. Decreased blood levels during starvation. (Also see METABOLISM.)
Thyroxin (T_4)	Thyroid				
Gastrin	Antral portion of gastric mucosa; pancreatic islets.	Distension of stomach; presence of proteins and polypeptides; alcohol; stimulation of vegas nerve.	Stimulates gastric acid (HCl) and pepsin secretion; stimulates gastric motility.	The Zollinger-Ellison syndrome is caused by a gastrin secreting tumor. However, for the most part it is not known what role the hormones of the digestion system play in digestive diseases.	The isolation and chemical nature of enterocrinin, glucagonlike immunoreactive factor and villikinin have not been determined. Cholecystokinin-pancreozynim is sometimes referred to as just cholecystokinin or pancreozymin. Injections of cholecystokinin may provide a means of treating obesity. (Also see DIGESTION AND ABSORPTION; and ZOLLINGER-ELLISON SYNDROME.)
Cholecystokinin-pancreozymin	Duodenum	Presence of fats and products of protein digestion.	Contraction of gallbladder and secretion of pancreatic enzymes.		
Enterocrinin	Duodenum	Presence of chyme.	Increases secretion of intestinal fluids.		
Enterogastrone	Duodenum	Presence of fats.	Inhibits gastric acid secretion and motility.		
Glucagonlike immunoreactive factor (GLI)	Wall of small intestine.		Stimulates insulin secretion.		
Secretin	Duodenum	Presence of acid and protein.	Stimulates secretion of aqueous pancreatic fluid (high in bicarbonate).		
Villikinin	Duodenum	Presence of chyme.	Increases contraction of villi.		
Glucagon	Pancreas (alpha cells of islets of Langerhans)	Low blood glucose (sugar) level.	Mobilizes glucose from liver glycogen; increases the formation of glucose from proteins—gluconeogenesis.		The important role of glucagon is to keep glucose high enough to prevent hypoglycemic convul-, sions or coma. (Also see DIABETES MELLITUS; and HYPOGLYCEMIA.)

Footnotes at end of table

(Continued)

TABLE E-14 *(Continued)*

Hormone	Origin	Mechanism of Release	Physiological Functions	Associated Diseases	Comments
Insulin	Pancreas (beta cells of islets of Langerhans)	High blood glucose (sugar) level.	Causes transport of glucose from the blood into the cells, and increases protein synthesis in cells.	Diabetes mellitus; hypoglycemia.	In the absence of insulin, blood glucose reaches dangerously high levels. (Also see DIABETES MELLITUS; and HYPOGLYCEMIA.)
Aldosterone (mineralocorticoid)	Adrenal gland (cortex)	Increased potassium concentration in the blood and blood flow through the kidneys; decreased sodium intake; decreased extracellular fluid volume.	Stimulates kidneys to excrete potassium into the urine, and to conserve sodium.	Conn's syndrome, or primary aldosteronism.	Aldosterone accounts for 95% of the mineral–corticoid activity. Renin release from the kidney controls its secretion. (Also see WATER AND ELECTROLYTES.)
Cortisone, corticosterone, and cortisol (glucocorticoids)	Adrenal gland (cortex)	Adrenocorticotropic hormone (ACTH) from the anterior pituitary.	Causes glucose production from protein and fats; makes amino acids available for use wherever needed; stimulates protein synthesis in the liver; mobilizes fats for energy; prevents self-destruction of cells.	Androgenital syndrome; Cushing's syndrome; Addison's Disease.	Necessary for individuals to combat stress; exposure to stress results in increased release of ACTH from the pituitary. The effects of stress are mediated through the hypothalamus. (Also see INBORN ERRORS OF METABOLISM; and STRESS.)
Epinephrine (adrenaline), Norepinephrine (noradrenaline), or Catecholamines	Adrenal gland (medulla)	A variety of nervous stimuli such as surprise, fright, shock, etc.	Both hormones alter heart output, dilate or constrict blood vessels, elevate blood pressure, release free fatty acids into the blood, stimulate the brain, increase heat production, and release glucose from the liver.	Hypertension (high blood pressure).	Epinephrine prevents the action of oxytocin. In the human, 80% of the secretion from the adrenal medulla is epinephrine. Thyroid hormones are required for action of catecholamines. (Also see HIGH BLOOD PRESSURE.)
Estrogens	Ovary	Follicle stimulating hormone from the anterior pituitary.	Deposition of protein and fat for the development and maintenance of feminine characteristics and sex organs; involved in monthly uterine changes.	Female hypogonadism; menopause.	There are actually 3 estrogens: estradiol, estriol, and estrone, and levels rise and fall with each menstrual cycle.
Progesterone	Ovary (corpus luteum)	Luteinizing hormone from the anterior pituitary.	Increases secretory activity in the glands of the breasts and uterus; maintains pregnancy by preventing the uterus from contracting.		During pregnancy, progesterone also comes from the placenta. Progesterone levels also reach high and low levels during the menstrual cycle.

Footnotes at end of table

(Continued)

TABLE E-14 (*Continued*)

Hormone	Origin	Mechanism of Release	Physiological Functions	Associated Diseases	Comments
Relaxin	Ovary (corpus luteum)	Associated with the advanced stages of pregnancy.	Believed to be involved in softening of the cervix and relaxation of the pubic ligament.		Sometimes found in the placenta, but its exact role is still quite unclear.
Testosterone (androgen)	Testes (interstitial cells)	Luteinizing hormone from the anterior pituitary, but called interstitial cell stimulating hormone (ICSH) in the male.	Necessary for sperm production; develops and maintains the male sex organs and masculine characteristics; increases the deposition of protein in the muscles.	Male hypogonadism.	Cholesterol is the chemical ancestor of testosterone, progesterone, and estrogens. Removal of the testes, castration, results in an individual with rather feminine appearance. For centuries, adolescent boys were castrated to work in harems, and choir boys were castrated to preserve their soprano voices. Anabolic steroids used by athletes are derivates of testosterone.

[1]Kidneys, pineal gland, and thymus may also be considered by some as endocrine glands, and during pregnancy, the placenta has some endocrine functions.
[2]The anterior pituitary is also called the adenohypophysis, while the posterior pituitary may be called the neurohypophysis.

ENDOCRINOLOGY

The study of the ductless glands and the effects of their secretion on the various tissues of the body.

(Also see ENDOCRINE GLANDS.)

ENDOGENOUS

Originating within the body, e.g., hormones and enzymes.

ENDOMORPH

A person with a body build characterized by a large round abdomen, and short arms and legs.

(Also see BODY TYPES; and OBESITY.)

ENDOPEPTIDASE

An enzyme which splits the central portions of protein molecules.

(Also see ENZYMES.)

ENDOPLASMIC RETICULUM

The more or less continuous network of small channels present in the cytoplasm of nearly all cells which transport nutrients and their breakdown products (metabolites) throughout the cytoplasm.

ENDOSPERM

The starchy part of the seed which surrounds the embryo or germ. When a seed sprouts, enzymes that develop in the seed break the starch down to sugar which may be used by the embryonic plant. It is noteworthy that man has domesticated plants with seeds that have large, starchy endosperms.

(Also see CEREAL GRAINS.)

ENDOSPERM OIL

Oil obtained from the endosperm of the seed.

(Also see CEREAL GRAINS.)

ENDOTOXIN

A poison (toxin) that remains inside the microorganism which produced it. The toxicity is noted only when the bacterial cells are ruptured by mechanical or chemical means. *Escherichia coli*, *Salmonella typhi* and *Shigella dysenteriae* bacteria are known to form endotoxins.

(Also see BACTERIA IN FOOD; and DISEASES.)

ENEMA

The injection of a liquid into the intestines by way of the anus.

(Also see NUTRIENT ENEMA.)

ENERGY BALANCE (EB)

Energy balance is the difference between the gross energy intake and the energy output. Fattening results when the intake exceeds the output, while a loss of body fat, and usually a loss of body weight, follows a dietary deficit of energy.

(Also see CALORIC EXPENDITURE.)

ENERGY REQUIRED FOR FOOD PRODUCTION

Everyone knows that cars and airplanes are powered by fuel. By now most people know that the developed nations have used fossil fuels to supplement the natural energy that comes directly from the sun on a day-to-day basis, and that they have grown dependent upon them. But few people realize that modern, mechanized food production in the developed countries requires an extra input of fuel, which is mostly of fossil origin.

Of course, the direct input of fuel into food production is of rather recent origin. It all began in a very small way about 1840, when fuel-powered ships transported fertilizer (guano, and later bone meal) from South America to Europe. Then, after 1910, transportation vehicles relied almost exclusively on fossil fuels. But the direct use of fossil fuels in agriculture started with the manufacturing of chemical fertilizer beginning about 1922. Following closely in period of time, the tractor was substituted for horses, mules, and oxen—eventually almost completely replacing them.

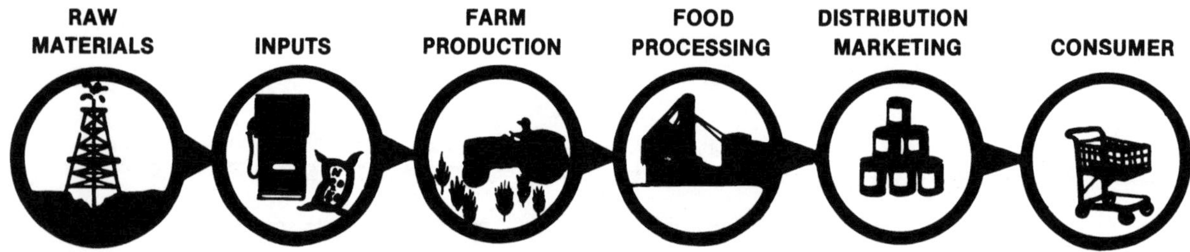

Fig. E-14. Energy use in the food system.

When considering the energy required to produce food products, it is insufficient to focus entirely on energy use on the farm. Energy is consumed during the extraction of raw materials, in the manufacture of farm inputs (fertilizer, farm machinery, feed, pesticides, petroleum products), in processing, in the distribution of food products, and in cooking. Table E-15 demonstrates the changing picture of energy usage for food production.

TABLE E-15
MODERN FOOD PRODUCTION IS INEFFICIENT IN ENERGY UTILIZATION—THE STORY FROM PRODUCER TO CONSUMER[1]

Year	On the Farm	Food Processing	Marketing and Home Cooking	Total/ Person/ Year
1940[2]				
Million kcal ...	0.9	2.2	2.1	5.2
Percent	18.0	42.0	40.0	100.0
1990[3]				
Million kcal ...	2.8	5.7	4.6	13.1[4]
Percent	21.4	43.5	35.1	100.0
Increase, times, 1940–1990 ...	3.1	2.6	2.2	2.5

[1]Energy in million kcal used per capita to produce one million kcal of food in the United States.

[2]Values from Borgstrom, G., "The Price of a Tractor," Ceres, FAO of the U.N., Rome, Italy, Nov.-Dec., 1974, p. 18, Table 3.

[3]Authors' estimate based on several reports detailing trends in energy usage.

[4]This means that in 1990, it required 13.1 million kcal to produce 1 million kcal of food for each person, a daily consumption of 2,740 kcal (1,000,000 ÷ 365 = 2,740).

Table E-15 points up the increasing drain that modern food production is putting on the energy supply. In 1990, U.S. farms put in 2.8 calories of fuel per calorie of food grown, 3.1 times more than the on-farm energy input in 1940.

Table E-15 also shows that, in the United States in 1990, a total of 13.1 calories were used in the production, food processing, and marketing-cooking for every calorie of food consumed, with a percentage distribution of the total cost of energy at each step from producer to consumer as follows: on the farm, 21.4%; food processing, 43.5%; and marketing and home cooking, 35.1%. In 1940, it took only 5.2 calories—about 40% of the 1990 figure—to get 1 calorie of food on the table. It is noteworthy, too, that more energy is required for food processing and marketing-home cooking than for growing the product; and that, from 1940 to 1990, the on-the-farm energy requirement increased by 3.1 times, in comparison with an increase of 2.6 and 2.2 times for each of the other steps—processing and marketing-home cooking.

Prior to the advent of machines and fuel in crop production, 1 calorie of energy input on the farm produced about 16 calories of food energy. Today, on the average, U.S. farms put in about 2.8 calories of fuel per calorie of food grown; hence, to produce a daily intake of 3,000 calories of edible food from cultivated crops may require 8,400 calories of energy from fossil fuels—an exhaustible source. It is more surprising yet—and thought-provoking—to know that, even today in the poorer or developing countries, it takes only 1 calorie to produce each 10 calories of food consumed. The Oriental wet rice peasant uses only 1 unit of energy to produce 50 units of food energy. This gives the Orientals a favorable position among the major powers as the energy crisis worsens.

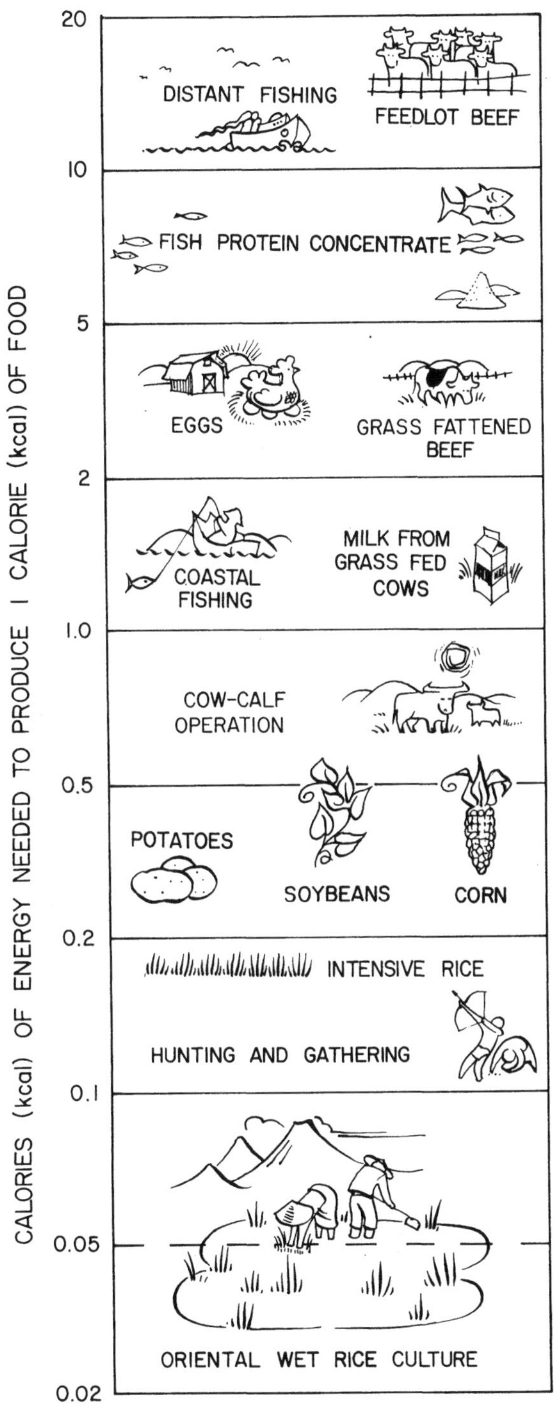

Fig. E-15. The energy cost of producing food.

Fig. E-15 shows that the Oriental wet rice peasant uses only 1 Calorie (kcal) of energy to produce 50 Calories (kcal) of food, or in other words, 0.02 kcal per 1 kcal of food. This gives the Oriental a favorable position among the major powers as the energy crisis worsens. Even hunting and gathering cultures are able to return 10 Calories (kcal) of food for every Calorie (kcal) of input. In the developed countries, production of rice, potatoes, soybeans, corn, and beef from grass and milk (cow-calf operation) requires less than 1 Calorie (kcal) of energy to produce 1 Calorie (kcal) of food. Milk from grass-fed cows and coastal fishing are about at the break-even point—1 Calorie (kcal) of input for 1 Calorie (kcal) of food. Eggs and grass-fattened beef require an overall average of 2.5 Calories (kcal) of energy per Calorie (kcal) of food, while the production of fish protein concentrate, distant fishing, and feedlot beef production consume a considerable amount of energy.

It is noteworthy that the U.S. food system required 6,200 trillion BTUs of energy in 1970, or about 10% of the nation's total energy consumption. About ⅔ of this energy was needed off the farm—in processing, marketing, and distribution. The United States now uses about ⅓ of the world's energy.

The type of energy is very important, especially should substitutions become necessary. Presently, the cultural energy inputs in agriculture include petroleum, natural gas, and electricity. About half of the energy consumed in the food system is petroleum, and another 30% is natural gas. Almost 90%, of the energy used in the manufacture of fertilizer and other inputs is natural gas. Gasoline and diesel fuels for tractors, combines, and other farm machinery comprise ⅘ of the direct energy requirements for crop and livestock production. Electricity, which is a secondary form of energy, has many applications throughout the food industry.

Fossil fuels—the stored photosynthates of previous millenia—are like a bank account. There is nothing wrong with drawing upon either of them, but neither is inexhaustible. It is highly imprudent not to be aware of big withdrawals and not to cover them. Within a short span of a few years, the world made the transition from a positive energy balance based upon the capture of the energy of the sun via green plants, crops, and forests to an imbalance, or even a negative balance, by resorting primarily to the bank of trapped sun energy of fossil fuels that accumulated over millions of years.

Currently, the global use of resources is increasing at an alarming rate—far faster than population. But it took the energy crisis of the early 1970s to cause the world to face up to this dilemma.

The following additional points are pertinent to the future of the energy required for food production.

1. **Photosynthesis fixes energy.** Photosynthesis is by far the most important energy-producing process. But currently only about 1% of the solar energy falling on an area is fixed by photosynthesis; and only 5% of this captured energy is fixed in a form suitable as food for man. Thus, (a) man's manipulation of plants for increased efficiency of solar energy conversion, and (b) converting a greater percentage of total energy fixed as chemical energy in plants (the other 95%) into a form available to man would appear to hold great promise in solving the future food problems of the world.

Crops vary in their return of captured solar energy per unit of cultural energy input. Fig. E-16 shows that grazing land is highly efficient in the capture of solar energy—requiring little input of energy for a high return. Hay and silage rank second in energy return, followed by feed grains and oil crops. For the most part, these efficient capturers of solar energy are not captured in a form available to man.

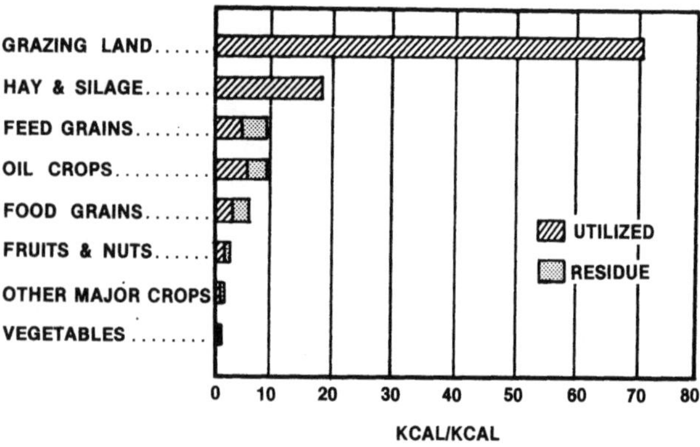

Fig. E-16. Energy output per unit of cultural energy input (kcal/kcal) for production of food, feed, and fiber crops. (Adapted by the authors from *American Society of Agricultural Engineers*, St. Joseph, Mich., paper No. 75-7505, p. 10, Fig. 5, prepared by Nelson, L. F., W. C. Burrows, and F. C. Stickler, Deere & Company, Moline, Ill.)

2. **Animals step up energy.** Grazing land and hay-silage far outrank the other crops in efficiency of capturing solar energy (see Fig. E-16). It follows that ruminants—cattle, sheep, and goats—which utilize grazing land, hay, and silage (feeds not suitable for human consumption), offer the best means of stepping up and storing energy for man. Petroleum is not required to produce beef, mutton, and wool. Also, dairy cattle are extremely efficient converters of energy to food (milk). Thus, ruminants represent a renewable resource, whereas it takes thousands of years to create coal, oil, and natural gas—longer than any of us can wait. Also, animals perpetuate themselves through their offspring. The 887 million acres (*359 million ha*) of pasture and grazing land of continental United States, as well as the vast acreages of grass and browse throughout the world, are converters of solar energy par excellence. Despite energy shortages, there will always be grazing land and ruminants.

3. **Crop residues contain energy.** Crop residues left in the field, above or below the ground surface, may well constitute four to five times more energy than is harvested. Increasingly, this potential source of added feed and organic fertilizer will be utilized in the future.

4. **Other factors.** Other factors that should be considered to meet the future energy needs are the conservation of energy, elimination of food waste, and the development of other energy sources.

Increasing consideration will be given to the conservation of our fossil energy bank account. Farmers will conserve energy by using minimum tillage techniques, and by switching to fuel-conserving diesel tractors which use approximately 73% as much fuel as gasoline tractors in performing the same work. Also, up to a point, big farms utilize energy more efficiently than little farms. Hence, energy shortages favor the trend to bigness.

Food waste caused by a variety of pests such as plant disease, insects, weeds, rats, and birds represent an estimated 30% annual loss in worldwide food production. By eliminating these wastes, the world food supply can be increased by nearly one-third.

Among the energy sources which could be developed, the most abundant and basic source is the sun. Photosynthesis is by far the *most* important energy-producing process, yet only a small fraction of solar energy is fixed in a form available to man. It should, therefore, be possible to increase the effectiveness of this process. To do this, three approaches are suggested: (a) increasing the amount of photosynthesis on earth, (b) manipulating plants for increased efficiency of solar energy conversion, and (c) converting a greater percentage of the total energy fixed as chemical energy in plants to a form available to man. As already indicated, ruminants provide a solution to the latter approach. In addition to the above, the oceans, which cover more than two-thirds of the earth's surface, receive a proportionate amount of all the solar energy, and their potential to provide food could be dramatically increased by learning to farm them. Also, the wind is being harnessed to supply electricity.

(Also see ENERGY UTILIZATION BY THE BODY; PHOTOSYNTHESIS; and WORLD FOOD.)

ENERGY UTILIZATION BY THE BODY

Contents *Page*

Major Forms of Energy 662
The Biological Energy Chain 664
Utilization of Food Energy for the Work of the Body 665
Maintenance of the Body Temperature 666
Synthesis of Essential Substances 666
Active Transport of Substances Across Cell Membranes 666
Muscle Contraction 666
Summary 667

One of the major reasons why people around the world strive to obtain food, clothing, shelter, and fuel is that each of these items plays an important role in the body's utilization of energy, which sustains life and human activities.

Energy is defined as the ability to do work. Work is generally thought of as the application of a force in order to produce the motion of one or more masses or energy waves against an opposing force or a natural tendency. For example, the throwing of a ball into the air requires work to overcome the effect of gravity. The maintenance of a body at a temperature higher than that of its surroundings also requires work (an input of heat energy) to replace the heat that flows out of the object into the surroundings.

MAJOR FORMS OF ENERGY. The various forms of energy are interconvertible. Therefore, it is noteworthy that the Law of Conservation of Energy is that energy can neither be created nor destroyed. This statement must be modified to include "energy and its equivalent in matter," since nuclear reactions involve the interconversion of energy and matter. Furthermore, heat energy can never be totally converted into other forms of energy, but the other forms are totally convertible into heat. The reason for this limitation is that every type of work that is done is accompanied by the release of some heat.

Food energy, which is the chemical energy present in nutrients, is by far the major source of the energy that is utilized by the human body. Nevertheless, few people base their lives on this form of energy alone. Furthermore, the utilization of heat, light, mechanical energy, nuclear energy, and solar energy reduces the need for food energy, which is fortunate because there are certain circumstances that would otherwise require the consumption of enormous amounts of food. Therefore, it is appropriate to consider the major forms of energy which are described in Table E-16.

TABLE E-16
MAJOR FORMS OF ENERGY

Form of Energy	Description	Actions in the Body	Relationships to Other Forms of Energy	Comments
Chemical	The energy that is released or taken up by substances when they undergo chemical changes. Its release in the body is usually under tight control which prevents harmful effects and minimizes wastage.	Only the chemical energy which is present in the nutrients is useful to the body. Chemical energy from other sources may be harmful.	The chemical energy from food is converted to heat, mechanical energy, and electrical energy. The last two forms are ultimately converted to heat which escapes from the body.	The major type of energy used by the body. It is required to sustain life. Body fat is a reservoir of chemical energy which may be drawn upon in times of need.
Electrical	The flow of charged particles or waves from a region of high electrical potential to one of low electrical potential. Electricity may flow through vacuums and various types of matter. The materials which readily permit the flow of electricity are called conductors, whereas, those that offer great resistance to its flow are called insulators.	Low voltage electrical currents are required for the regulation of the heartbeat, transmission of the impulses in the brain and nerves, and various other physiological functions. High voltage and/or high intensity (high amperage) electrical currents may stop the heart and do other types of damage to the body.	Electrical potentials are produced within the body by the active transport of charged particles (ions) into and out of cells so that a net imbalance of electric charges results. (The inside of nerve membranes is usually more negatively charged than the outside.) Man-made devices convert other forms of energy into electricity and vice versa.	The body derives little benefit from the external application of electricity. There is still some uncertainty regarding the functions of electricity within the body and on the surface of the skin.
Heat (Thermal)	Radiant (electromagnetic) waves of energy that are similar to light, but are not visible. (They are also called infrared waves.) Heat flows from warmer regions to cooler ones, through vacuums (such as outer space), and all types of matter. However, some materials (such as metals) are much better conductors of heat than others.	Usually, the production of heat within the body is controlled so that the body temperature is maintained within narrow limits. (The vital functions of the body are likely to be impaired by either too little or too much heat.)	Heat is produced whenever chemical, mechanical, or electrical energy is utilized by the body. However, the body cannot convert heat into these other forms of energy.	Food energy that is released during metabolism is the source of most of the heat in the body.
Light	Visible radiant energy waves that are similar to heat waves. (Both types of waves are part of the spectrum of electromagnetic radiation, which also includes radio waves, microwaves, ultraviolet waves, x rays, gamma rays, and cosmic rays.)	Required for vision. However, excessive exposure of the eyes to strong light destroys the visual pigment more rapidly than it can be resynthesized. The daily cycle of light and darkness helps to maintain biological rhythms such as the menstrual cycle, sleep, and the daily rise and fall of hormonal secretions.	Converted into chemical energy by photosynthetic processes in green plants. May be derived from chemical energy, electrical energy or heat by means of man-made devices.	Ultraviolet radiation, which is similar to light waves but is not visible, converts provitamin D on the skin to vitamin D. (Both visible and ultraviolet "light" are present in sunlight and certain types of artificial light.)

(Continued)

TABLE E-16 — MAJOR FORMS OF ENERGY — *(Continued)*

Form of Energy	Description	Actions in the Body	Relationships to Other Forms of Energy	Comments
Mechanical	Energy of motion that is capable of moving masses of matter in opposition to gravity and other impeding forces.	Required for the actions of the muscles involved in locomotion, breathing, pumping of the heart, movements of the digestive tract, etc.	In the body it is derived from food energy, and its production by the muscles is controlled by the nerves, which operate on electrical energy.	Except in special circumstances (such as during the use of mechanical heart pumps and respiration devices), the body cannot utilize mechanical energy from an external source.
Nuclear	The energy that is released or taken up by the nuclei of atoms when they are radioactive and/or involved in nuclear reactions. (It is noteworthy that some of the earth's geothermal energy results from the radioactive decay of underground rocks.)	Very small doses may be used to destroy cancer cells, and for making certain medical diagnoses. Exposure to larger doses may result in short term effects such as radiation sickness, burns, nausea, and various lesions; and long term effects such as cancer, mutations of reproductive cells, and a general shortening of life.	It is very difficult to convert other forms of energy into nuclear energy, but nuclear energy may be converted into the other forms by means of various devices.	Outside of the sun, nuclear fuels are the most concentrated sources of energy known to man. One pound of enriched uranium fuel contains 3 million times the energy of l lb (*0.45 kg*) of coal.
Solar	An entire spectrum of radiant (electromagnetic) energy which includes radiowaves, microwaves, heat (infrared waves), light, ultraviolet radiation, x rays, gamma rays, and cosmic rays. However, the radiation which reaches the earth is mainly heat, light, and ultraviolet rays. Solar energy is generated by a chain of nuclear reactions that convert hydrogen atoms into helium atoms.	Same as those of heat, light, and ultraviolet radiation. The sun's rays may cause severe sunburn in people with lightly pigmented skins. It is dangerous to look directly at the sun because the eyes may be damaged.	Solar energy heats the air to produce wind power (a source of mechanical energy), and is converted to chemical energy by photosynthesis. Man-made devices convert sunlight to heat and electricity. The energy in the fossil fuel (coal, gas, and oil) originated as the solar energy that was incorporated by photosynthesis into the plants (and into the animals which ate the plants) which later decayed and were compressed under layers of earth for countless centuries.	The sun is the ultimate source of almost all of the energy that is available on the earth; except for geothermal energy, which comes from radioactive decay and other geological processes.

THE BIOLOGICAL ENERGY CHAIN. The rapid rise in energy costs which occurred in the 1970s and early 1980s led to studies of the energy used for food production. Some analysts of this situation have made oversimplified generalizations regarding the overall merits of plants vs animals as foods, without sufficient consideration of (1) the types of foods which best meet the energy needs of people at various stages of life and in a variety of circumstances, and (2) some of the potentially useful products and by-products of food production. Hence, an overview of biological energy utilization is presented in Fig. E-17.

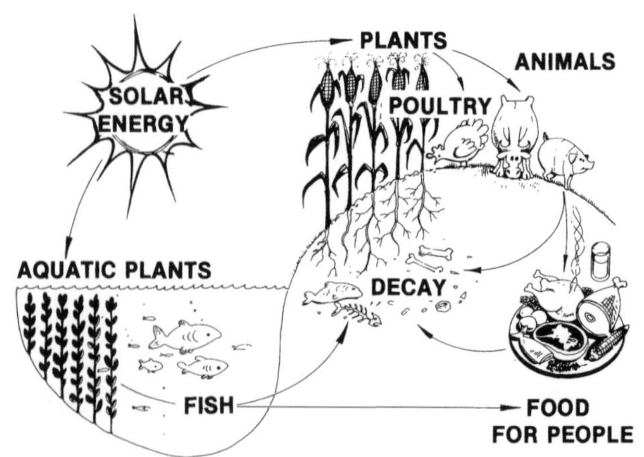

Fig. E-17. The biological energy chain by which energy from the sun is converted into food energy for people.

Details of the various links in the biological energy chain follow:

• **Solar energy**—Although less than 1% of the sun's energy which reaches the earth is converted to chemical energy by photosynthesis in plants, a much larger amount of this energy heats up the atmosphere so that plants, animals, and people can live on this planet. For example, people who live near the equator where the sun's rays are the strongest need little in the way of clothes or shelter to slow the loss of heat from their bodies. As a result, their basal metabolic rates are about 10% lower than those of people living in the North or South Temperate Zones.

• **Photosynthesis by plants**—The chemical energy that is produced in photosynthesis is equivalent to the caloric needs of about 1.2 trillion people, which is almost 280 times the present population of the world. However, the percentages of "plant energy" which are utilizable by the human body as food energy vary considerably for the various types of plant tissue. For example, many types of grasses are high in undigestible fiber that is very poorly utilized by people, but is better utilized by grazing animals such as buffaloes, cattle, goats, and sheep. It is noteworthy that the grazing crops yield an average of 70 Calories (kcal) of total plant energy for each kcal invested by man in their production, whereas the major food crops yield from 1 to 10 Calories (kcal) for each kcal of cultural energy input. Furthermore, even the cereal grains that are favored by man contain large amounts of roots, stems, and leaves that are unsuitable for human food. Hence, there is a need to convert the energy that is present in nonedible plants and plant residues into forms that people can utilize more readily.

• **Conversion of plant products into animal products**—Man has long depended upon various members of the animal kingdom to convert plants into more appetizing types of food. Dairy cows, which are at present the most efficient means of converting forages to food, convert about one-sixth of the energy of grasses into the energy of milk. According to most archaeological evidence, primitive peoples consumed many types of animals, birds, fishes, insects, and reptiles when these foods could be obtained readily. These items were supplemented by the edible parts of wild plants.[2] It might have been the pursuit of wild game which led Asian hunters of the Stone Age to cross what was then a land bridge (now, the Bering Strait) between the Asian and American continents.[3] Therefore, it is noteworthy that the prodigious labors of the primitive peoples of the northern lands were fueled mainly by the energy from fishes and meats which contained ample amounts of fats. Even today, most investigations show that diets rich in animal products may lead to the storage of more food energy in the form of body fat than predominantly vegetarian diets because, on the average, animal products are higher in calories than most vegetables. Unfortunately, the stored energy which once was almost always a great asset is now often a liability because most people lead sedentary lives. Nevertheless, the conversion of vegetable foods into animal foods continues to be essential to man, whether it takes place in the lactating mother or in livestock and poultry. The problem is to select the means of conversion that result in the most efficient uses of energy inputs, agricultural land, and critical raw materials.

• **A diet containing both plant and animal foods ensures the best nutrition for people**—The caloric needs of most people can be met by diets consisting mainly of grains and/or legumes. However, those consuming these diets may develop various disorders due to deficiencies of amino acids, minerals, and vitamins because a wide variety of plant foods must be consumed to ensure that vegetarian diets are nutritionally adequate. The nutrient compositions of most plants differ from the pattern of nutrients that is required by the human body. On the other hand, less variety is required when ample amounts of dairy products, eggs, fish, meats (both muscles and organs), and poultry are consumed. The animal foods are generally good sources of the nutrients required by people. However, only the milk and bones of animals contain sufficient calcium to meet human needs.

• **Plant and animal wastes yield energy**—The energy that is present in inedible plant and animal wastes need not be wasted, since some of these materials release combustible gases and heat when they undergo fermentation, or are burned. Only limited use was made of these materials until the recent energy crisis made it profitable to do so. Now, there are a growing number of locations at which plant and/or animal materials are fermented in closed chambers so that the gases produced may be collected. After fermentation, the residues may be dried and used as feeds or fertilizers. Another development in this field is the production of hot water, steam, and/or electricity by burning agricultural wastes and municipal garbage.

UTILIZATION OF FOOD ENERGY FOR THE WORK OF THE BODY. Although the end products and the amounts of energy released are the same, the oxygen-consuming metabolism of carbohydrates, fats, and proteins in the body, which occurs in the process called respiration, differs from the combustion of these substances in a calorimeter in that the energy is released under much more strictly controlled conditions in respiration. The utilization of food energy in the body involves a chain of closely linked chemical reactions in which the energy of the nutrients is passed from one substance to the next with small leakages of heat occurring at various points in the chain. Consequently, the major end products of energy metabolism are carbon dioxide, water, chemical energy in the form of certain organic phosphates, and heat. Some urea is also formed from the metabolism of protein.

Over one-half of the energy produced in nutrient metabolism is the heat which helps to maintain the body temperature, and most of the rest is contained in the organic phosphates, the most important of which is adenosine triphosphate (ATP). Hence, respiration in the mitochondria of cells throughout the body provides much of the energy to operate the functions shown in Fig. E-18.

[2]Robson, J. R. K., "Foods from the Past—Lessons for the Future," *The Professional Nutritionist*, Spring, 1976, pp. 13-15.

[3]Claiborne, R., *The First Americans*, Time, Inc., 1973, pp. 8-33.

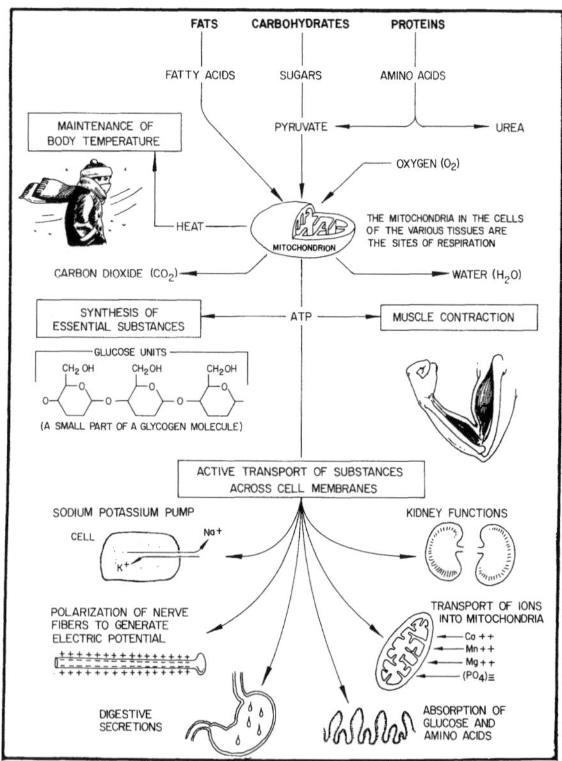

Fig. E-18. The utilization of food energy for the work of the body.

Details of the major energy utilizing processes follow.

Maintenance Of Body Temperature. In the absence of chilling, the heat that is released during the normal functioning of the body is sufficient to maintain the body temperature. Heat is produced whenever chemical, mechanical, or electrical energy is utilized because the processes of the body operate against various types of opposing forces, which, like friction, cause some of the useful energy to be converted into heat. Also, heat escapes from various linkages in the metabolic chain which produces ATP. When chilling does occur, the body temperature is maintained by (1) shivering (involuntary contractions of the skeletal muscles that bring about the conversion of ATP into mechanical energy and heat), and (2) increased rates of secretion of thyroid hormones, epinephrine (adrenalin), and norepinephrine (nor-adrenaline) which bring about the production of a greater proportion of heat energy and a lesser proportion of ATP energy for each unit of food energy that is metabolized.

Certain conditions may overwhelm the processes that help to maintain the body temperature so that dangerously low or high internal temperatures result. For example, an unclothed body in a cold windy environment is likely to lose heat more rapidly than it can be generated by metabolism. Hence, people living in such environments need protective clothing, shelter, and fuel to keep warm. On the other hand, the dissipation of body heat can occur only to a limited extent in hot humid environments. Therefore, these circumstances limit the amount of physical work which can be done in the tropics because increased muscle activity results in increased heat production.

Synthesis Of Essential Substances. The maintenance, repair, and growth of body tissues require energy to synthesize large molecules and related structures. For example, the synthesis of a molecule of glycogen (the storage form of carbohydrate) which contains about 10,000 units of glucose requires the energy that is derived from 20,000 molecules of ATP. This energy is later released when glycogen breaks down to meet the body's need for sugar. Similarly, an average protein molecule requires about 1,500 molecules of ATP energy for its synthesis. When the food energy intake of the diet yields more energy than is needed to fuel the body's essential functions, the surplus ATP energy is utilized to synthesize fats (lipids). The synthesis of each molecule of a typical phospholipid such as lecithin requires the energy from 8 molecules of ATP.

Active Transport Of Substances Across Cell Membranes. *The use of metabolic energy to move substances across membranes against concentration gradients is called active transport.* Normally, the substances which can pass through the various membranes in the body can be expected to flow in the directions that will equalize their concentrations on both sides of the membranes. However, the essential functions of the body often require that gradients in the concentrations of many substances be maintained within and without cells. For example, the concentration of potassium ions must be kept higher within cells than in the surrounding medium, whereas the concentration of sodium ions must be kept higher outside of cells. Therefore, it is necessary for energy to be used to pump out the sodium ions which leak into cells, and to pump back in the potassium ions which leak out.

It is noteworthy that the active transport of sodium and potassium across the membranes of nerve cells establishes the conditions required for generating the electric potential that conducts the nerve impulses. After the nerves have "fired" their impulses, the electric potential must be regenerated with the aid of additional active transport.

Some of the other functions which utilize active transport are (1) the secretion of saliva, gastric juice, and pancreatic juice during digestion; (2) absorption of glucose and amino acids from the intestine; (3) accumulation of calcium, manganese,magnesium, and phosphate ions within the mitochondria of cells; and (4) reabsorption of glucose and amino acids from the kidney tubules.

(Also see DIGESTION AND ABSORPTION, section headed "Mechanism of Absorption, 3. Active Transport.")

Muscle Contraction. The utilization of metabolic energy in muscle differs from that in other cells and tissues in that some of the ATP energy is used to synthesize phosphocreatine, which is a reserve source of the energy used in muscle contraction. Then, when the supply of ATP runs out, it may be replenished rapidly from the phosphocreatine. Respiration is normally required to generate ATP, but the phosphocreatine provides a means of generating it rapidly without respiration. The ATP brings about muscular movement by causing the thick and thin filaments of the muscle fibers to slide past each other so that shortening of the fibers occurs.

Heavy work may require the consumption of large amounts of food to provide sufficient energy for muscle contraction. The lumberjacks who worked in Wisconsin and Minnesota in the 1920s consumed about 9,000 Calories (kcal) per day.[4] Much of this food energy was provided by large servings of bacon, butter, cheese, cream, eggs, fish, meat, and milk. However, it now may be more practical to utilize machine-powered tools than to rely upon the consumption of large quantities of moderately expensive foods to provide energy for muscular work that is done with only hand-powered tools.

SUMMARY. The human body performs its vital functions with the help of energy that originates in the nuclear reactions in the sun, and is trapped within the chemical substances synthesized by plants during photosynthesis. Animals, which eat the plants, convert the vegetable constituents into animal foods that are often more useful and more palatable sources of food energy for people. Although the energy costs of producing animal foods are sometimes higher than those for vegetable foods, the utilization of both plant and animal wastes as sources of energy helps to keep energy wastage to a minimum. Finally, the utilization of other forms of energy by man for heating, transportation, and the performance of work spares the body from having to use very large quantities of food energy.

(Also see CALORIC EXPENDITURE; CALORIC VALUES OF FOODS; ENERGY REQUIRED FOR FOOD PRODUCTION; and METABOLISM.)

ENRICHMENT (FORTIFICATION, NUTRIFICATION, RESTORATION)

Contents — *Page*

History 667
Problems in the Philippines 667
The Bataan Experiment 668
Other Enrichment Programs 671
Enriched Foods 671
Reasons for Enrichment 671
Principles of Enrichment 671
Guidelines for Enrichment 672
Labels and U.S. RDA 672
Problems 674

In foods, enrichment and fortification refer to the addition of vitamins, minerals, and/or protein to raise the nutritive value.

NOTE WELL: Originally, the FDA differentiated between enrichment and fortification, but now the two terms are used interchangeably. Nevertheless, the following differentiations between the terms persist:

- **Enriched**—This refers to the addition of specific nutrients to a food *as established in a federal standard of identity and quality*. The amounts added generally are moderate and include those commonly present at even lower levels.

- **Fortified**—This refers to the addition to foods of specific nutrients. The amounts added are usually in excess of those normally found in the food because of the importance of providing additional amounts of the nutrients to the diet. Some foods are selected for fortification because they are an appropriate carrier for the nutrient.

In the following discussion enrichment and fortification will be used interchangeably unless noted otherwise. Two more terms may be employed as synonyms to enrichment or fortification:

- **Restoration**—This is the replacement of nutrients lost in processing foods.

- **Nutrification**—This is the practice of adding a proportion of vitamins and minerals to a formulated or fabricated, or grouping of foods marketed as a meal or meal replacer; for example, infant formulas and instant breakfast foods.

HISTORY. A classical, pioneering rice enrichment experiment in the Philippines, conducted in the early 1900s (see the section that follows), demonstrated how an action program can combat a widespread deficiency disease.

Compulsory enrichment of bread and flour introduced in 1941 as a wartime measure was replaced after World War II by a program in which the FDA sets standards specifying the nutrients and their amounts to be added to certain foods, but state agencies, not the FDA, are responsible for requiring that specific foods be enriched or fortified. Currently, 34 states require the enrichment of bread and flour, with thiamin, riboflavin, niacin, and iron mandatory and calcium optional within prescribed limits. Ninety percent of the bread and cereal sold in the United States is enriched. Some states require the addition of these same nutrients to cornmeal, corn grits, farina, macaroni, and noodles, and the addition of thiamin, niacin, and iron to rice. Standards are also set for the addition of vitamins A and D to milk products and of vitamin A to margarine. In an effort to enhance effectiveness in preventing iron-deficiency anemia, the standards for iron enrichment were raised up to 25% in 1983, to 20 mg/lb of flour.

Problems In The Philippines.[5] The experimental program of rice enrichment in the Philippines was singularly appropriate for the reason that it was in the Philippines, in 1910, that Dr. Robert R. Williams' attention was first focused on the beriberi problem, and it was there that he found the clue which started him on the 26-year search that culminated in the first successful synthesis of thiamin (vitamin B-1).

---

[4]Cooley, A. C. J., "Paul Bunyan's Cook," *Nutrition Today*, Vol. 5, Spring 1970, pp. 24-25.

[5]The narrative in the sections headed, "Problems in the Philippines," and "The Bataan Experiment," was adapted by the authors from *Better Health Through Better Rice*, published by Williams-Waterman Fund, Research Corporation, New York, N.Y., 1950.

In 1910, Dr. E. B. Vedder of the U. S. Army Board of Tropical Medicine, in Manila, pointed out to Dr. Williams that the bran and other "polishings" of rice which were discarded during milling apparently contained a life-giving substance for chickens which otherwise became paralyzed by feeding on white rice. Neither scientist suspected at that time the real nature or potency of this substance, but, working together, they found that rather crude extracts of rice polishings could frequently cure even advanced stages of the human disease. They treated several cases and inaugurated the preparation of rice polish extract for dispensary use before Dr. Vedder was called home. A chemist by training, Dr. Williams was subsequently called in emergencies, time after time, to the native houses of Manila when no medical doctor was available. He often saw hope return to a Filipino mother's eyes when her gasping baby, blue and near death from beriberi, started nursing again a few hours after he forcibly fed it a few drops of the extract.

In the years of research it took to isolate and identify the life-giving factor, Dr. Williams kept his own hopes high that one day the substance could be made available in effective form to beriberi sufferers all over the world. In time, the factor was established as the first of the B vitamins, and finally, in 1936 he and his colleagues were able to produce it by practical synthetic methods. This paved the way for its commercial manufacture on a scale that would within a few years lower the price to the point where enrichment of cereal grains would be economically feasible. By 1946, when peace had come and the Philippine experiment could be considered, a process had been developed by the pharmaceutical house of Hoffmann-La Roche, Inc., for adding thiamin, niacin, and iron to rice. By this time, the effectiveness of the technique of premix fortification had also been established.

The Bataan Experiment. The scene of the experiment was to be an area better known to the world for a different reason—the Bataan Peninsula, where early in 1942 the joint Filipino-American forces made their gallant stand.

It was the combination of the factors of isolation, uniformity of conditions, and rice self-sufficiency, in addition, of course, to the considerable beriberi death rates, that led to the selection of Bataan as the site of the field trials in which, for the first time, a whole population's entire rice supply would be enriched.

In the summer of 1946, Dr. Williams went to the Philippines with a plan for the combat of beriberi, along with (1) the promise of financial support for the necessary clinical surveys from the Williams-Waterman Fund, and (2) assurance of the premix for the experiment from Hoffmann-LaRoche. At the outset, Dr. Williams very wisely obtained the enthusiastic support of various agencies and leaders, both government and private, and conducted preliminary trials on people.

Pilot feeding trials were conducted on government employees, members of Armed Forces units, and children in welfare institutions—a total of almost 3,500 individuals. These trials showed that the enriched rice was completely satisfactory with regard to its color, taste, odor, palatability, and digestibility. There were also indications that enriched rice had brought about a beneficial effect on beriberi in the Armed Forces units, and general health benefits for the children in the institutions.

The Philippine health authorities were now ready to take the next step—full-scale experimental rice enrichment for tens of thousands of people in a region of the Philippines where the beriberi mortality rate was reported to be one of the highest in the world.

The specific objectives of the mass nutritional feeding experiment were fourfold: (1) to determine to what extent beriberi in the Philippines could be alleviated by substitution of fortified rice for ordinary white rice; (2) to test the feasibility of the practice in the channels of the rice trade; (3) to inaugurate and test practical inspection systems to insure that only fortified rice is sold; (4) to popularize enriched rice with the people and to provide a basis for consideration of the use of fortification throughout the Philippine Islands under adequate government controls.

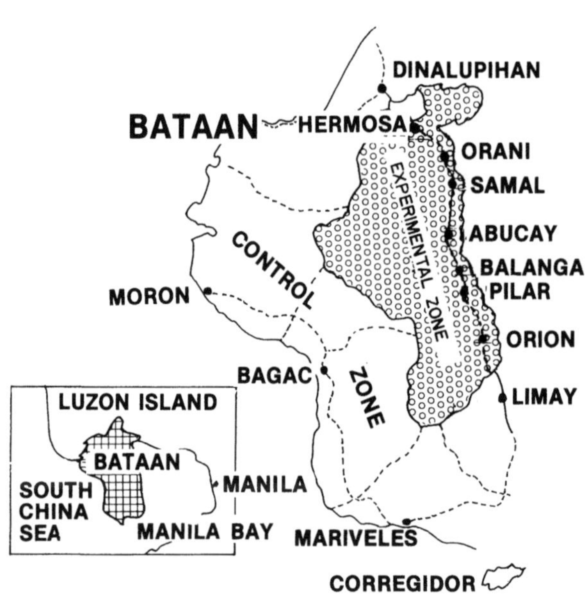

Fig. E-19. Rice enrichment and control areas. All Bataan was divided into two areas for the enrichment project: the experimental zone of smaller but more populous municipalities on the Manila Bay side, which received the enriched rice; and the control zone of the remaining municipalities, which continued to receive ordinary white rice.

For purposes of accurately measuring the effects of rice enrichment, the Province of Bataan was divided into two areas, an experimental zone and a control zone. The seven more populous municipalities on the east coast were in the experimental zone, and the remainder of the province served as a control area. Enriched rice would be made available only to the 63,000 people in the experimental zone; the other 29,000 in the control zone would continue to get their customary supply of rice.

To form a scientific baseline for the experiment, a

clinical beriberi survey was started in July 1947. The survey revealed that, among the 12,384 people examined, there were 1,580 cases of beriberi—a total of nearly 13%, or an astounding one-eighth of the people. There was no significant difference in these figures between urban and rural populations.

The number of infants found with beriberi symptoms was surprisingly small when compared with extremely high infant deaths, which in 1947 and 1948 made up some 90% of all beriberi deaths in Bataan. Because of difficulty in recognizing premonitory symptoms in babies, a relatively small proportion of infants was included in the clinical survey, but the great number of infant deaths recorded while it was actually in progress gave dramatic evidence of the rapidity with which death comes when the disease strikes the nursing child.

Fifteen months before the field tests, the collection of mortality figures was reestablished. Statistics for the year immediately previous to rice enrichment, October 1, 1948, showed mortality rates of 254 per 100,000 people in the experimental zone, and 152 for the control zone. Compared to existing national figures, these indicated that beriberi was considerably more prevalent in the experimental zone than in the Philippines as a whole, while the rate in the control zone was about average for the nation.

Also, in preliminary surveys, studies were made of the dietary intakes of presumably normal subjects for a 9-day period (see Table E-17). Earlier studies had established the thiamin content of the rice, vegetables and other foods generally eaten in the province. Thus, it was possible to arrive at estimated average figures for total thiamin intake in the daily diet of the people being studied.

TABLE E-17
PER CAPITA DAILY INTAKE OF FOODS AND THIAMIN

Food Class	Daily Food Intake		Daily Thiamin Intake
	(g)		(mg)
	Philippines[1]	Bataan[2]	Bataan[2]
Rice	408	400	.248
Leafy vegetables . .	25	20	.010
Nonleafy vegetables	140	100	.088
Dry legumes and nuts	4	4	.014
Sugar	18	16	.000
Lean meat, fish . . .	166	150	.330
Tomatoes, fruits . .	19	20	.008
Total	780	710	.698

[1]Unpublished survey by Dr. Manuel L. Roxas.
[2]Estimated from survey results.

The average figure of approximately 0.7 mg of thiamin daily was considerably below the standards set by the Section on Nutrition, National Research Council of the Philippines. The recommendations of this body called for daily intakes of 1.3 to 2.0 mg for adult Filipino males and 1.1 to 1.8 for females, except during pregnancy and nursing when 2.0 mg of thiamin per day is desirable.

• **Thiamin requirements met by enrichment**—It seemed reasonable to fortify the enriched rice with enough thiamin to bring the average daily intake of the vitamin to nearly 2 mg from the rice alone, based on an estimated daily adult consumption of 400 g of rice, and allowing for losses of thiamin in washing and cooking. Other foods were expected to contribute another 0.45 mg of thiamin daily, bringing the total for the experimental zone well above the minimum levels recommended.

For this purpose, rice fortified to the same levels as enriched white wheat flour in the United States seemed appropriate. As distributed, each pound of the product contained 2 mg of thiamin, 16 mg of niacin, and 13 mg of iron.

Actual enrichment of the rice was planned to take place at the 23 cone rice mills and three hullers—the latter used as small production mills—in the experimental zone. There the highly fortified premix, shipped from the United States, was to be mixed with the native rice customarily processed for consumers in the area. The mixing in each mill was to be done by two inexpensive feeders, such as are widely used for blending in Western grain mills. One would furnish a constant stream of rice and the other would provide the premix, in the weight ratio of 200 to 1.

Inspection was provided for the experimental zone of seven municipalities to insure that the rice would be enriched according to specifications, that only the fortified product reached the native markets, and that consumers would get none but enriched rice for their own use.

Long before the mass feeding experiment began, the medical teams making the clinical surveys started the groundwork of education to prepare the people for enrichment.

The groundwork was well laid by October 1, 1948, when enrichment began. Thereafter, only fortified rice was sold in the markets or used in the homes of the experimental area.

• **Beriberi deaths decrease**—The mortality figures soon began to show the influence of the enrichment program.

The figures for the first full year of enrichment were startling (see Table E-18). There was a decline of 67.3% in beriberi deaths in the experimental zone, and an increase of 2.4% in the control area, as compared to the year immediately prior to enrichment. In the experimental area, enriched rice had apparently saved the lives of 111 people in one year.

The results that were shown as the enrichment project went into its second year were even more impressive. In the experimental zone, beriberi deaths decreased again in the fifth and sixth quarters; in the seventh quarter they reached zero. For the 3 months from April 1 to June 30, 1950, not one death from beriberi was recorded in any of the seven municipalities. It appeared that enrichment saved the lives of some 220 people, mostly babies, in Bataan.

TABLE E-18
EFFECT OF RICE ENRICHMENT ON BATAAN MORTALITY RATE

MUNICIPALITY	Oct. 1, 1947 to Sept. 30, 1948[1]	Oct. 1, 1948 to Sept. 30, 1949[2]	DECREASE
	(per 100,000)	(per 100,000)	(%)
Experimental zone			
Abucay	289.7	23.7	91.9
Balanga	325.0	137.3	57.7
Hermosa	314.2	47.4	84.9
Orani	233.6	55.0	76.4
Orion	234.9	114.7	51.2
Pilar	69.5	.0	100.0
Samal	156.3	157.3	.6 (Increase)
Entire experimental zone	246.2	80.4	67.3
Control zone			
Bagac	180.8	182.1	.7 (Increase)
Dinalupihan	166.8	118.5	28.9
Limay	112.3	199.6	77.9 (Increase)
Mariveles	.0	44.8	— (Increase)
Moron	330.1	389.7	18.0 (Increase)
Entire control zone	152.7	156.5	2.4 (Increase)

[1]Ordinary white rice consumed by all people in both experimental and control zones.
[2]Enriched rice consumed by people in experimental zone; ordinary rice by people in control zone.

Enrichment would have shown its full effect sooner had it not been for the extra time it took to reach infants too young to eat the enriched rice. During the first year when the whole mortality rate had dropped 67%, infant deaths decreased only 50%. Because the mothers' tissues had to be built up before their milk could carry all the needed vitamins to their babies, there was naturally a considerable lag before infant mortality rates showed the full effect.

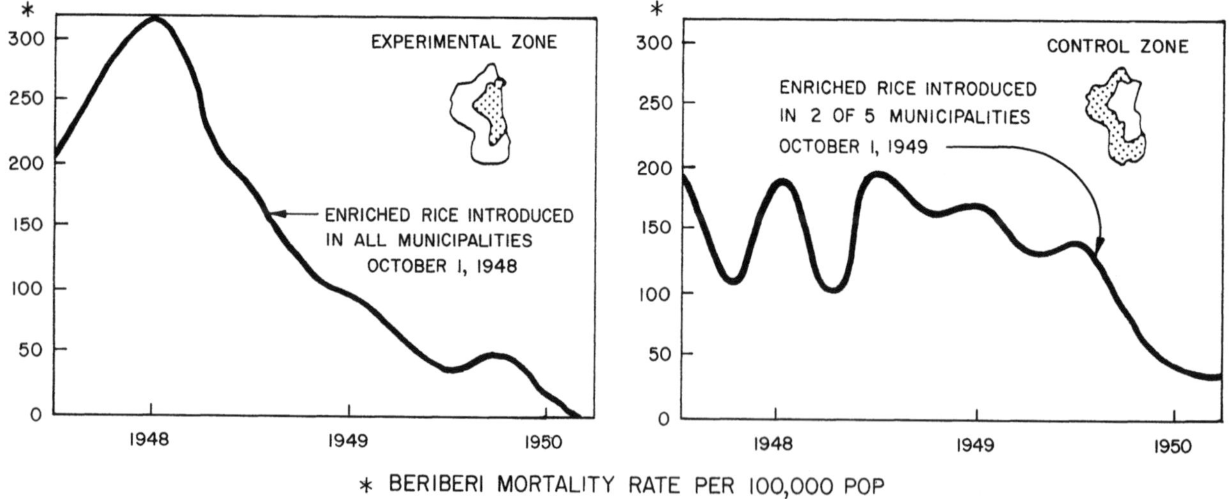

Fig. E-20. Reduction in beriberi death rate. After rice enrichment began, beriberi mortality continued to drop in the Bataan experimental area until it reached zero. In the control zone the rate remained high until enriched rice began to creep in. Full benefits of enrichment reached adults first, infants last; only after the mothers' tissues had been built up could their milk carry normal levels of vitamins to their babies.

• **Better health in Bataan**—Decrease in mortality, however, was not the most important consequence of enrichment. When the clinical resurvey was begun in July 1949, some 12,000 people, mostly adults, were examined—nearly 11,000 of them were the same individuals that were inspected in 1947. Nearly 90% of those who revealed symptoms 2 years previous were much better, and most of them were symptom free. Since these people were a fair sample of the entire population, this means that at least 3,500 people in the experimental area were measurably improved in health. Thousands more must have been benefited to some degree.

These people were school children, adolescents, fathers, mothers, the breadwinners, homemakers, and citizens of the province. What their reinvigoration, if sustained, may have meant to the subsequent productivity and prosperity of Bataan is quite beyond calculation.

(Also see RICE, section headed "Better Rice For Millions.")

Other Enrichment Programs. Enrichment of salt with iodine in 1924 was the first essential nutrient to be added to a consumer product. This enrichment program successfully decreased the incidence of simple goiter in the United States. For example, a survey of the school children in four Michigan counties indicated that the incidence of goiter dropped from 38.6% in 1924 to 1.4% in 1951.

Hazel K. Stiebeling was one of the scientists who considered the problem of the hungry 1930s and set up a working pattern of amounts of nutrients and food required daily for individuals of different ages, sex, and activities. Dr. Stiebeling also guided a nationwide study of food consumption of a representative sampling of the population of the United States in the mid-1930s. The diets were evaluated in terms of adequacy as compared with the standard of requirements. The publication, "Are We Well Fed? A Report on the Diet of Families in the United States," aroused widespread concern. By 1941, considerable evidence had accumulated that many American families were consuming diets that were inadequate in thiamin, riboflavin, niacin, and iron. Also, because of the concern with the problems of the war in Europe, President Franklin D. Roosevelt called the National Nutrition Conference for Defense in May 1941. One year earlier, the National Academy of Sciences-National Research Council appointed a Committee on Food and Nutrition (later called the Food and Nutrition Board) to develop a table of "Recommended Daily Allowances for Specific Nutrients." This committee proposed the use of the term "enriched" and set up minimum and maximum limits for the enrichment of bread and flour with thiamin, riboflavin, niacin, and iron. With the support of the millers, enriched flour became available to the public and was used by the Army and Navy. Enrichment was made possible in the 1940s by the chemists' ability to prepare nutrients in inexpensive forms.

The Danes first realized the need for adding vitamin A to a food. During World War I, practically all the butter of Denmark was exported. Subsequently, an eye ailment was observed in young children and was recognized as a vitamin A deficiency. As a preventive measure, vitamin A concentrates were added to margarine. Other countries adopted the practice. The Council on Foods and Nutrition of the American Medical Association approved it in 1939, and it has since been advocated by the Food and Nutrition Board.

The discovery of vitamin D as the antirachitic vitamin and the recognition of fish-liver oils as a potent source led to the advice that babies should receive cod-liver oil or some concentrate of vitamin D. Prevention of rickets and development of strong bones in young children depend on an adequate intake of calcium and phosphorus, as well as vitamin D, and therefore fortification of milk with vitamin D was begun in the early 1930s. The Council on Foods and Nutrition of the American Medical Association approved the fortification of milk.

(Also see VITAMIN[S].)

The enrichment of salt with iodine, the fortification of milk with vitamin D, and the start of the thiamin, riboflavin, niacin, and iron, grain enrichment program in 1941, have played a significant role in the practical elimination of the following deficiency diseases: simple goiter, rickets, beriberi, ariboflavinosis, pellagra and simple iron-deficiency anemia. The average American receives approximately 40% of his thiamin, 25% of his iron, 20% of his niacin, and 15% of his riboflavin from enriched foods.

ENRICHED FOODS. Consumers will find a variety of enriched foods, and at times it is difficult to differentiate between the use of a nutrient for enrichment or as an additive. Indeed, nutrients—minerals, vitamins, and amino acids—are a class of food additives. The commonly enriched foods are: salt, milk, margarine, cereals, and cereal products. Additionally, a variety of other foods may be enriched or fortified with vitamins and minerals to maintain or improve the nutritional value. There are reasons for enriching foods, and principles of enrichment to follow when considering the value of enrichment. Moreover, once it has been determined to enrich food, there are guidelines, standards, and labeling practices to follow for enrichment practices.

Reasons For Enrichment. The prime reason for enrichment is public health so that diets can be nutritionally improved without trying to change food habits. People are usually adverse to changing food habits, and in some populations, the foods creating the nutritional deficiency may be the only foods available. Other reasons for the ingestion of an inadequate diet, thus necessitating enrichment, include (1) lack of interest in nutrition, (2) meals eaten away from home, (3) snacking, (4) poverty, (5) reduced energy expenditure, hence, reduced caloric requirements, (6) lack of knowledge concerning nutrients present in foods, (7) weight control and fad diets, and (8) selection of foods with diminished nutrient content as compared to the raw product.

Principles Of Enrichment. Many foods have a diminished nutritive content as a result of loss in refining or processing and can benefit from the enrichment. Also, some foods provide effective vehicles for distribution of a nutrient. Other principles which may be followed when considering enrichment include the following which are guidelines issued in 1968 by the Council on Foods and Nutrition of the American Medical Association and the Food and Nutrition Board of the National Research Council:

1. The intake of the nutrient(s) is below the desirable level in the diets of a significant number of people.
2. The food(s) used to supply the nutrient(s) is likely to be consumed in quantities that will make a significant contribution to the diet of the population in need.
3. The addition of the nutrient(s) is not likely to create

an imbalance of essential nutrients.

4. The nutrient(s) added is stable under proper conditions of storage and use.

5. The nutrient(s) is physiologically available from the food.

6. There is reasonable assurance against excessive intake to a level of toxicity.

The function of fresh fruits, vegetables, fresh meat, poultry and fish in a balanced diet is well established and understood by the public. There is no reason to add nutrients to these foods. Also, it is inappropriate to fortify snack foods such as candies and carbonated beverages.

Guidelines For Enrichment. In January 1980, the Food and Drug Administration (FDA), set forth guidelines to follow when nutrients are added to foods. The new guidelines were designed to provide a sensible set of principles for adding minerals, vitamins, and protein to foods in order to achieve a balance of nutrients—not to encourage widespread enrichment. Enrichment is permissible in the following conditions:

1. When suitable to overcome a nutritional deficiency in a particular population group, such as the addition of iodine to salt to prevent goiter and vitamin D to milk to prevent rickets in children.

2. To restore nutrients which have been lost in the storage, handling, and processing of foods. All nutrients in all amounts lost, including protein must be considered.

3. In proportion to the total calories—a basis for adding nutrients to fabricated foods.

4. To foods that substitute for and resemble traditional foods so that they will not be nutritionally inferior.

5. To meet nutritional standards which are required or permitted by existing FDA regulations—the standards of enrichment.

• **Standards of enrichment**—While enrichment is not required by law, certain enriched foods must meet established standards. The Food and Drug Administration (FDA) standards for enrichment include iodine, vitamin D, vitamin A, thiamin, riboflavin, niacin, iron, and calcium in some foods.

• **Foods**—Currently, the FDA has standards for the enrichment of the following:

1. Iodine is added to salt at the level of 0.5 to 1 part in 10,000 or 7.6 mg/100 g.

2. Vitamin D is added to milk at the level of 400 IU per qt (*0.95 liter*) of fluid milk or large can of evaporated milk (1 ⅔ cups or *400 ml*).

3. Vitamin A is added to margarine at the level of 15,000 IU per pound (*0.45 kg*). This is the year-round average for the vitamin A content of butter.

4. Enrichment levels prescribed by federal standards of identity for flour and other cereal products are listed in Table E-19.

TABLE E-19
ENRICHMENT LEVELS OF CEREAL PRODUCTS

Product	Thiamin		Riboflavin		Niacin		Iron		Calcium[1]	
	(mg/lb)	*(mg/100 g)*	(mg/lb)	*(mg/100 g)*	(mg/lb)	*(mg/100 g)*	(mg/lb)	*(mg/100 g)*	(mg/lb)	*(mg/100 g)*
Flour	2.9	*.64*	1.8	*.40*	24	*5.3*	13–16.5	*2.87–3.64*	960	*212*
Flour, self-rising	2.9	*.64*	1.8	*.40*	24	*5.3*	13–16.5	*2.87–3.64*	960	*212*
Bread, rolls, and buns	1.8	*.40*	1.1	*.24*	15	*3.3*	8–12.5	*1.76–2.76*	600	*136*
Cornmeal, corn grits	2.0–3.0	*.44– .66*	1.2–1.8	*.26– .40*	16–24	*3.53–5.29*	13–26.0	*2.87–5.73*	500– 750	*100–165*
Cornmeal, self-rising	2.0–3.0	*.44– .66*	1.2–1.8	*.26– .40*	16–24	*3.53–5.29*	13–26.0	*2.87–5.73*	500–1,750	*110–385*
Macaroni and noodles	4.0–5.0	*.88–1.10*	1.7–2.2	*.37– .48*	27–34	*5.95–7.50*	13–16.5	*2.87–3.64*	500– 625	*110–138*
Farina	2.0–2.5	*.44– .55*	1.2–1.5	*.26– .33*	16–20	*3.53–4.41*	13[2]	*2.87*	500[2]	*110*
Rice, milled	2.0–4.0	*.44– .88*	1.2–2.4[3]	*.26– .53*	16–32	*3.53–7.05*	13–26.0	*2.87–5.73*	500–1,000	*110–220*

[1]The addition of calcium is optional.
[2]No maximum level established.
[3]The addition of riboflavin is feasible but the requirement was stayed many years ago since its addition colors the rice yellow; hence, rice may or may not be enriched.

Today, enrichment is required in about two-thirds of the states and Puerto Rico. In practice, however, all family flour is enriched, and about 90% of all commercially baked standard white bread is enriched.

It is noteworthy that England, Canada, and a few other countries have enrichment programs somewhat similar to the United States, but that France forbids enrichment.

Labels And U.S. RDA. Whenever products are labeled "enriched," or a food product has added nutrients, or a nutritional claim is made for a product, the FDA requires that the nutritional content be listed on the label. In addition, many manufacturers put nutrition information on products when not required to do so in order to make their product competitive.

Nutrition labels list how many calories and how much protein, carbohydrate, and fat are in a serving of the product. They also list the percentage of the U.S. Recommended Daily Allowances (U.S. RDAs) of protein and seven important minerals and vitamins that each serving of the product contains.

• **Nutrition labels**—At the top, the nutrition labeling panel is identified as "Nutrition Information."

Nutrition information is given on a per serving basis. The label tells of a serving; for example, 1 cup (*0.45 kg*), 2 oz (*60 ml*), 1 Tbsp (*15 ml*); the number of servings in the container, the number of calories per serving, and the amounts in grams of protein, carbohydrate, and fat per serving.

The lower part of the nutrition label must give the percentages of the U.S. Recommended Daily Allowances (U.S.RDA) of protein and seven vitamins and minerals in a serving of the product, in the following order: protein, vitamin A, vitamin C, thiamin, riboflavin, niacin, calcium, and iron. The listing of 12 other vitamins and minerals, and of cholesterol, fatty acid, and sodium content is optional—for now. Nutrients present at levels less than 2% of the U.S. RDA may be indicated by a zero or an asterisk which refers to the statement, "contains less than 2% of the U.S. RDA of these nutrients."

• **U.S. RDAs**—These allowances are guides to the amounts of vitamins and minerals an individual needs each day to stay healthy. They were set by the FDA as nutritional standards for labeling purposes. The U.S. RDAs are *based* on the Recommended Dietary Allowances established by the Food and Nutrition Board of the National Academy of Sciences-National Research Council. For practical purposes, the many categories of dietary allowances for males and females of different ages were condensed to as few as nutritionally possible for labeling. Generally, the highest values for the ages combined in a U.S. RDA were used. For example, the U.S. RDAs for adults and children over 4 years are representative, generally, of the dietary allowances recommended for a teen-age boy.

There are four groupings of the U.S. RDAs. (See NUTRIENTS: REQUIREMENTS, ALLOWANCES, FUNCTIONS, SOURCES.) The best known, and the one that will be used on most nutrition information panels and most mineral and vitamin supplements, is for adults and children over 4 years of age, shown in Table E-20. The second is for infants up to 1 year, and the third is for children under 4 years. These two will be used on infant formulas, baby foods, and other foods appropriate for these ages as well as vitamin-mineral supplements intended for their use. The fourth is for pregnant women or women who are nursing their babies.

TABLE E–20
UNITED STATES RECOMMENDED DAILY ALLOWANCES (U.S. RDA), THE MOST COMMONLY USED GROUPING

Nutrient	Adults and Children 4 Years or Older[1]
Required on labels:	
Protein . . . . g	59
Vitamin A . . . . mcg RE	1000
Vitamin C . . . . mg	60
Thiamin . . . . mg	1.5
Riboflavin . . . . mg	1.8
Niacin . . . . mg	20
Calcium . . . . mg	1200
Iron . . . . mg	12
Optional on labels:	
Vitamin D . . . . IU	400
Vitamin E . . . . IU	10
Vitamin B-6 . . . . mg	2
Folate . . . . mcg	200
Vitamin B-12 . . . . mcg	2
Phosphorus . . . . mg	1200
Iodine . . . . mcg	150
Magnesium . . . . mg	400
Zinc . . . . mg	15
Copper . . . . mg	2
Biotin . . . . mcg	60
Pantothenic acid . . . . mg	6

[1] *Federal Register*, Vol. 45, No. 18, January 1980, p. 6323. U.S. RDA taken from *Recommended Dietary Allowances*, 10th ed., 1989, National Research Council.

Many foods today are manufactured into products that are different from traditional foods. Some classes of these foods include frozen dinners; breakfast cereals; meal replacements; noncarbonated breakfast beverages; and main dishes such as macaroni and cheese, pizzas, stews, and casseroles. Nutrients may be added to these foods.

Nutritional labeling allows consumers to select foods that are a particularly good source of a specific nutrient, and to determine whether newly introduced products are as nutritious as their familiar counterparts.

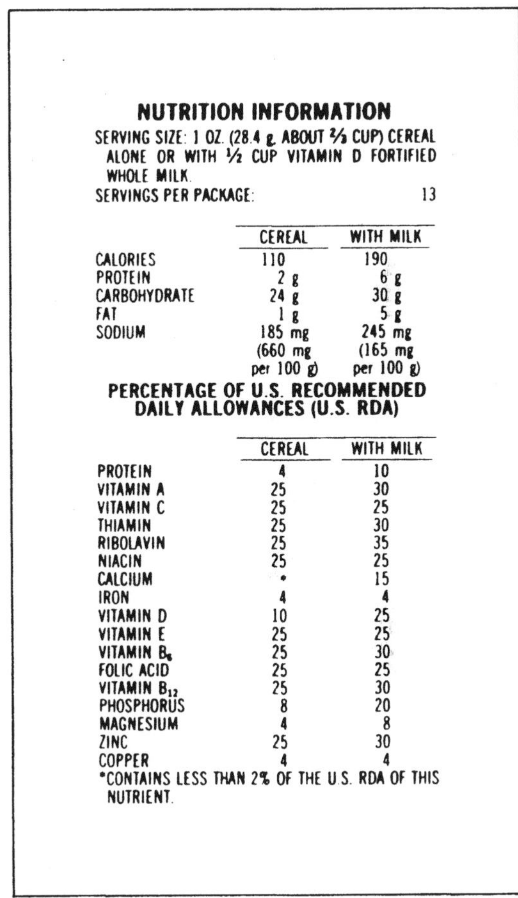

NUTRITION INFORMATION

SERVING SIZE: 1 OZ. (28.4 g, ABOUT ⅔ CUP) CEREAL ALONE OR WITH ½ CUP VITAMIN D FORTIFIED WHOLE MILK.

SERVINGS PER PACKAGE: 13

	CEREAL	WITH MILK
CALORIES	110	190
PROTEIN	2 g	6 g
CARBOHYDRATE	24 g	30 g
FAT	1 g	5 g
SODIUM	185 mg (660 mg per 100 g)	245 mg (165 mg per 100 g)

PERCENTAGE OF U.S. RECOMMENDED DAILY ALLOWANCES (U.S. RDA)

	CEREAL	WITH MILK
PROTEIN	4	10
VITAMIN A	25	30
VITAMIN C	25	25
THIAMIN	25	30
RIBOLAVIN	25	35
NIACIN	25	25
CALCIUM	*	15
IRON	4	4
VITAMIN D	10	25
VITAMIN E	25	25
VITAMIN B_6	25	30
FOLIC ACID	25	25
VITAMIN B_{12}	25	30
PHOSPHORUS	8	20
MAGNESIUM	4	8
ZINC	25	30
COPPER	4	4

*CONTAINS LESS THAN 2% OF THE U.S. RDA OF THIS NUTRIENT.

Fig. E-21. An example of a nutritional label from an enriched ready-to-eat breakfast cereal.

On the basis of the nutritional label and the U.S. RDA in Fig. E-21, one serving of the cereal contains 2 g of protein, 1,250 IU of vitamin A, 15 mg of vitamin C, 0.38 mg of thiamin, 0.43 mg of riboflavin, 5 mg of niacin, less than 20 mg of calcium, 0.72 g of iron, 40 IU of vitamin D, 0.5 mg of vitamin B-6, 0.1 mg of folic acid, 1.5 mcg of vitamin B-12, 80 mg of phosphorus, 16 mg of magnesium, 3.8 mg of zinc, and 0.08 mg of copper.

Aside from the labels of many products which are purchased, nutritional information in Food Composition Table F-36 identifies the enriched products. Furthermore, it should be realized that the nutritional information on any fabricated food represents enrichment, since these foods are literally pieced together; for example, breakfast bars and powders, and meat analogs.

PROBLEMS. Enrichment is not a panacea—only a tool. All nutrient requirements cannot be ensured with any enrichment program. Rather, to be sure people receive adequate amounts of the 45 to 50 nutrients required for good nutrition, the best advice is sound nutrition education so individuals will choose from a variety of nutritious foods. Moreover, people's eating habits change as do their nutrient intakes; hence, relying on enrichment to cope with malnutrition would require constant re-evaluation of enrichment levels. Another problem with enrichment is manipulation for economic benefit. Unscrupulous manufacturers may boast of a product with a much higher content of some nutrient due to enrichment, or may boast of the "perfect" food and charge consumers a disproportionate price. Currently, some ready-to-eat breakfast cereals are enriched to 100% of the U.S. RDAs for many of the vitamins and minerals; and, from certain brand products, one can get B vitamins from soda pop, vitamin C from canned fruits and drinks, and calcium from orange juice. But consumption of a fortified food will not ensure a complete or nutritionally sound diet.

Many people are poorly nourished because of (1) lack of interest in nutrition, (2) inadequate information about the role of foods, and (3) economic deprivation. In cultures where dietary patterns are simple and based on a limited number of commodities, one is able to identify which food could be a carrier for an enrichment program. Traditionally, staples such as wheat, rice, and corn have served as vehicles for niacin, riboflavin, thiamin, iron, and calcium. Milk has been a reliable carrier of vitamin D, as table salt has been a carrier of iodine. Also, water is employed as a carrier of fluorine.

(Also see ADDITIVES; CEREAL GRAINS, section headed "Cereals That Have Been Enriched or Fortified with Certain Nutrients"; CORN, Table C-45 Corn Products and Uses; FLOURS, section headed "Enrichment and Fortification of Flours"; HIGHLY PROCESSED FOODS; IRON, section headed "Sources of Iron"; NIACIN, section headed "Sources of Niacin"; NUTRIENTS: REQUIREMENTS, ALLOWANCES, FUNCTIONS, SOURCES; RIBOFLAVIN, section headed "Sources of Riboflavin"; RICE, section headed "Nutritional Value"; THIAMIN, section headed "Sources of Thiamin"; and WHEAT, section headed "Enriched Flour.")

ENTERIC

Relating to the small intestine.

ENTERITIS

Inflammation of the intestines.
(Also see DISEASES.)

ENTERO-

A prefix denoting intestinal.

ENTEROCRININ

A hormone secreted by the intestinal lining that stimulates the intestinal glands to release their digestive juices.

(Also see DIGESTION AND ABSORPTION.)

ENTEROGASTRONE

A hormone, the release of which is triggered by the presence of fat in the small intestine. The hormone inhibits further gastric secretion and slows the motions of the stomach.

(Also see DIGESTION AND ABSORPTION.)

ENTEROHEPATIC CIRCULATION

This refers to the continuous circulation of the bile from the liver to the gallbladder, then into the intestine, from which it is absorbed and carried by the blood back to the liver to be returned to the circulation. This recycling conserves the bile; of the 20 to 30 g of bile used daily, only about 0.8 g is actually eliminated in the feces and must be replenished by the liver.

ENTEROKINASE

An enzyme which converts the inactive form of the enzyme trypsin to the active form.

(Also see DIGESTION AND ABSORPTION.)

ENTEROSTOMY

The general name given to an operation which connects a piece of the intestine to an artificial outlet on the abdominal wall called a stoma.

(Also see COLITIS.)

ENVIRONMENT

The forces and conditions, both physical and biological, that (1) surround an individual and (2) interact with heredity to determine behavior, growth, and development. Air quality, food supply, lighting, noise, other people, and weather are some of the many factors that make up an individual's environment. Extremes or alterations in the environment may subject an individual to stress.

(Also see STRESS.)

ENZYMATIC

Related to an enzyme.

ENZYME

Complex protein compounds produced in living cells

which speed biochemical reactions without being used up in the process. They are organic catalysts. Many names of enzymes end with *ase* and begin with the name of the substance on which they act (substrate) or the type of action they perform. For example, sucrase splits sucrose (table sugar), and transferase transfers a chemical group. Thousands of different enzymes are manufactured in the human body. Fig. E-22 illustrates how enzymes perform their function as organic catalysts, and the role of a coenzyme.

(Also see METABOLISM.)

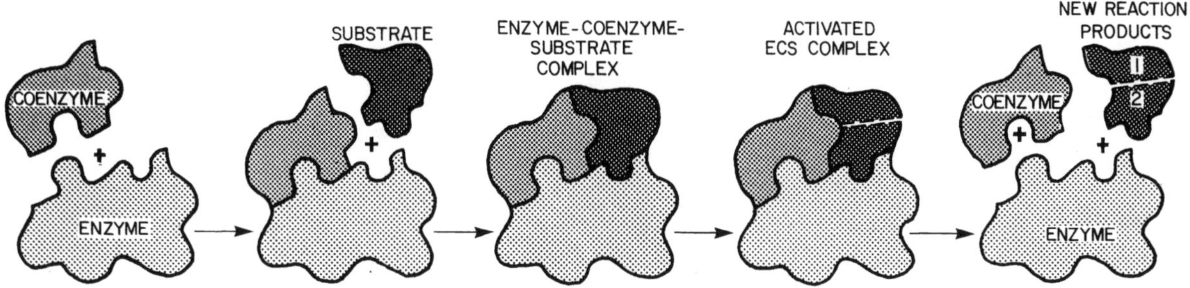

Fig. E-22. The formation of an enzyme-coenzyme-substrate (ECS) complex, followed by the biochemical reaction.

EPIDEMIC

An outbreak of a disease in an area where it does not normally occur, in contrast to the term endemic which refers to a disease regularly occurring in a certain locality.
(Also see ENDEMIC.)

EPIDEMIOLOGY

The study of the various factors responsible for the presence or absence of diseases in populations.

EPIDERMIS

The outer horny layer of skin which serves as a protection for the underlying layer called the dermis.

EPIGASTRIC

The upper central region of the stomach.

EPIGLOTTIS

A flap of tissue which covers the windpipe and prevents food from getting into the bronchial tubes.

EPILEPSY

A disorder of the nervous system characterized by convulsions and/or loss of consciousness. Special ketogenic diets are sometimes used in the treatment of epilepsy.
(Also see DISEASES.)

EPINEPHRINE

The hormone commonly called *adrenalin*, which is secreted by the adrenal medulla. The hormone epinephrine is secreted in larger than normal amounts when certain emotional states such as anger, fear, and pain cause emotional upsets. It is also secreted in response to low blood sugar, a large loss of blood, and vigorous physical exercise. The effects of the hormone are to prepare the body for increased activity because it accelerates the heart beat, increases the flow of blood to the skeletal muscles, and increases the breakdown of stored glycogen to glucose, a sugar which may rapidly be utilized by the brain, muscles, and other tissues of the body.
(Also see ENDOCRINE GLANDS.)

EPIPHYSIS

The end of a long bone which during the growing years is separated from the shaft of the bone by a plate of cartilage. When the growth of the long bone stops, the epiphyses are fused to the shafts of the bone (diaphyses).
(Also see BONE.)

EPITHELIAL

Refers to those cells that form the outer layer of the skin and other membranes.

EPSOM SALT

The common name for magnesium sulfate, a mineral salt which is used as a strong laxative. It is also injected into pregnant women to prevent toxemia of pregnancy.

However, care must be used in this use of the salt because an excessive amount may cause sedation in the unborn child.

ERGOCALCIFEROL

A substance which is formed from ergosterol, a plant substance that is converted into vitamin D by the action of sunlight or other forms of ultraviolet radiation.

(Also see VITAMIN D.)

ERGOMETER

An instrument used to measure the amount of work done under controlled conditions of time, rate, and resistance.

ERGOSTEROL

A plant sterol which, when activated by ultraviolet rays, becomes vitamin D_2. It is also called provitamin D_2 and ergosterin.

(Also see VITAMIN D.)

ERGOT

A poisonous species of fungi that grows on small grains (rye, wheat, barley, oats, and triticale), with rye being particularly susceptible. It has been responsible for many outbreaks of poisoning (ergotism) throughout northern Europe, where the poor growing conditions favor the development of ergot. Ergotism is characterized by an extreme constriction of the blood vessels and muscles which greatly reduces the flow of blood to the arms and legs so that they may become gangrenous. However, small doses are sometimes given to women after childbirth or miscarriage to bring about a shrinking of the uterus (involution).

(Also see DISEASES, section headed "Constituents of Food"; POISONS; SPOILAGE OF FOOD; and PRESERVATION OF FOOD.)

ERUCIC ACID

A very long chain fatty acid with one double bond, found in rapeseed oil and mustardseed oil. It has been found that when large amounts of rapeseed oil (50% of the total energy) are fed to experimental animals fatty changes occur in the heart muscle. This is because erucic acid enters the myocardial cells, but is oxidized more slowly than other fatty acids; so, it accumulates intracellularly in triglycerides. Geneticists have now produced strains of rape that produce seed oil with less than 1% erucic acid.

(Also see RAPE; and CANOLA.)

ERYTHORBIC ACID

A derivative of vitamin C, which is used as an antioxidant in food products to prevent rancidity, to eliminate browning of fruit, and to preserve the red color of meats. It is poorly absorbed and has little antiscorbutic activity.

ERYTHROCYTES

The red blood cells, which carry oxygen to the tissues. An insufficient quantity or quality of these cells is called anemia, a condition characterized by (1) paleness of the mucous membranes in the eyes and in the mouth, and (2) tiredness and weakness. On the other hand, an excess of erythrocytes is called *polycythemia*, which is dangerous because it makes one more susceptible to a blockage of blood vessels by the cells, similar to that caused by blood clots.

(Also see ANEMIA.)

ERYTHROPOIESIS

The process by which red blood cells are formed.

(Also see ANEMIA.)

ERYTHROPOIETIN

The hormone which stimulates the formation of red blood cells.

(Also see ANEMIA.)

ERYTHROTIN

An out-of-date name for vitamin B-12.

(Also see VITAMIN[S].)

ESCHERICHIA COLI *E. coli*

A species of bacteria that inhabits the intestines. It normally is harmless, but under certain conditions it may change to a harmful form.

(Also see BACTERIA IN FOOD, Table B-5 Food Infections [Bacterial]; DIARRHEA; and FOODBORNE DISEASE.)

ESOPHAGITIS

An inflamation of the esophagus, the tube that leads from the mouth to the stomach. Esophagitis may be a dangerous condition if it is caused by the reflux of acid from the stomach because the esophagus is not very well protected against irritating substances.

(Also see DIGESTION AND ABSORPTION.)

ESOPHAGUS

The passageway leading from the mouth to the stomach. Sometimes called the gullet.

(Also see DIGESTION AND ABSORPTION.)

ESSENTIAL AMINO ACIDS

Those amino acids which cannot be made in the body from other substances or which cannot be made in sufficient quantity to supply the individual's needs.

(Also see AMINO ACIDS; and PROTEIN.)

ESSENTIAL FATTY ACID

A fatty acid that cannot be synthesized in the body or that cannot be made in sufficient quantities for the body's needs.

(Also see FATS AND OTHER LIPIDS.)

ESSENTIAL OILS

A large class of volatile, odoriferous oils secured from various parts of certain plants, such as the flowers, the seeds, the leaves, the bark, or the roots. These oils are usually obtained either by (1) steam distillation, (2) expression, or (3) extraction (using a solution that will dissolve out the oil). Essential oils are used in flavoring materials, perfumes, and pharmaceutical preparations. They are called "essential oils" to distinguish them from "fatty oils."

ESTER

The chemical term applied to any combination of an organic acid and an alcohol. An ester holds the position in organic chemistry that a salt holds in inorganic chemistry. *Example*: ethyl alcohol and acetic acid yield ethyl acetate—an ester.

(Also see ETHYL FORMATE.)

ESTROGEN

The name for certain natural or synthetic female sex hormones. The natural hormones are secreted mainly by the ovaries after puberty occurs. Some major effects of the estrogens are the development of the female sex organs, enlargement of the mammary glands, and development of other female characteristics. Furthermore, estrogen promotes the deposition of body fat under the skin in females. Hence, females tend to have more subcutaneous fat than males. There is also an effect of estrogen on bone growth. It is noteworthy that after the menopause, women tend to lose calcium from their bones. Therefore, some doctors have given these hormones to postmenopausal women in order to prevent the weakening of their bones, and other problems associated with the lack of estrogens.

(Also see ENDOCRINE GLANDS.)

ETHANOL

The type of alcohol which is present in alcoholic drinks. It is also called ethyl alcohol.

(Also see ALCOHOLISM; and DISTILLED LIQUORS.)

ETHER EXTRACT (EE)

Fatty substances of foods that are soluble in ether.

(Also see FATS AND OTHER LIPIDS.)

ETHNIC

This term refers to the characteristics which distinguish certain national and racial groups of people from others. Hence, ethnic characteristics may include social customs, food patterns, character and physical traits of the various peoples.

(Also see ETHNIC.)

ETHYLENE

A sweet-smelling gas found in ripening fruit and used commercially to accelerate the ripening of fruit. Green lemons stored where the concentration of ethylene is 0.05% will become yellow in one week.

ETHYLENEDIAMINE TETRA-ACETIC ACID (EDTA; VERSENE)

An organic chemical capable of "tying up" metalic ions such as calcium, copper, iron, and zinc, and thus preventing them from reacting with other compounds. Chemicals that possess this property are called chelating or sequestering agents. A variety of EDTA salts exist; and all are sequestering agents, many of which are used as food additives. EDTA and its salts stabilize and maintain the color, freshness, and flavor of oils, fats, fruits, vegetables, fish, shell fish, dairy, and meat products, and vitamin preparations. In pharmacy and medicine, EDTA is used in the treatment of lead poisoning.

(Also see ADDITIVES.)

ETHYL FORMATE

An ester with a pleasant odor, made from the chemical combination of ethyl alcohol and formic acid. In the food industry, it is valuable as (1) a fungicide and larvicide for cereals and dried fruits; (2) an ingredient of synthetic flavors such as lemon and strawberry; and (3) a chemical intermediate in the synthesis of thiamin (vitamin B-1).

ETIOLOGY

The causes of disease or disorder.

EVAPORATION, FLASH

A rapid application of superheated steam which quickly distills off a small volume, 1% of the liquid being condensed. This flash distillate carries the volatile flavor constituents. Later, these flavors are added back to the concentrate. This process is employed in the production of concentrated fruit juices.

EXACERBATION

The intensification or aggravation of a disease or painful condition, usually because of something other than the original cause; for example, peptic ulcer is exacerbated by alcoholic beverages.

EXCHANGE LIST (FOOD EXCHANGES)

A grouping of similar foods by serving sizes which provide essentially equivalent nutritive value in terms of calories, carbohydrates, fats, and proteins. This type of system is designed so that people on modified diets might select foods which fit their dietary prescriptions. The most commonly used exchange lists are those for diabetics, because they are also useful for weight control.

(Also see MODIFIED DIETS.)

EXCIPIENT

An inert ingredient added to a medication for the purpose of making the dosage form more convenient to take. For example, excipients such as calcium carbonate are added to tablets to give them bulk.

EXCRETION

The process of eliminating the waste products of metabolism from the body, chiefly in the urine and sweat.

EXERGONIC

Chemical reactions occurring in living cells that produce energy; for example, the breakdown of carbohydrates or fats.

(Also see ENDERGONIC; and METABOLISM.)

EXOGENOUS

Originating or produced from the outside.

EXOPEPTIDASE

An enzyme which digests protein.

(Also see ENZYME.)

EXOPHTHALMIC GOITER

An enlargement of the thyroid gland which is accompanied by protruding eyeballs. This type of goiter is also called a toxic goiter because it is usually accompanied by a dangerous oversecretion of thyroid hormones, whereas an iodine deficiency goiter is associated with a normal or a subnormal secretion of thyroid hormones.

(Also see ENDOCRINE GLANDS; and GOITER.)

EXOTOXIN

A poison (toxin) produced inside a bacterial cell and then liberated into the outside environment. *Clostridium botulinum,* the bacteria that causes botulism, produces a very potent exotoxin.

(Also see BACTERIA IN FOOD, section headed "Food Poisonings [bacterial toxins.]")

EXPELLER CAKE (CAKE OR PRESSCAKE).

Expellers or presses are sometimes used to squeeze the oil from oilseeds such as cottonseeds, coconuts, peanuts, soybeans, and sunflowers. The mass remaining following the oil extraction is called expeller cake, cake, or presscake. It is a protein-rich food or feed.

(Also see OILS, VEGETABLE.)

EXPERIMENT

The word "experiment" is derived from the Latin *experimentum,* meaning proof of experience. It is a procedure used to discover or to demonstrate a fact or general truth.

EXTENSOMETER

In general, extensometers are instruments used for measuring the degree of expansion, contraction, or deformation produced in a substance under an applied stress. An extensometer can be designed to measure the stretching strength of dough, an index of baking quality.
(Also see BREADS AND BAKING.)

EXTRACELLULAR FLUID

The fluid outside the cells; it comprises about one-third of the total body fluid, and includes tissue fluid, blood plasma, cerebrospinal fluid, fluid in the eye, and the fluid of the gastrointestinal tract.
(Also see BODY FLUIDS.)

EXTRACT

In nutrition, this term generally has one of the meanings which follow:

- A component which has been removed from a mixture by means of treating the mixture with a solvent in which the component is soluble. For example, the beverage tea is a water extract of tea leaves. Alcohol, oil, and water are the solvents most commonly used to prepare edible extracts.

- The process by which an extract is made.

EXTRACTION RATIO (EXTRACTION)

The percentage of a grain that is converted to flour. For wheat, high extraction (95 to 100%) flours such as whole wheat contain more bran and germ than low extraction (80% or less) or white flours.
(Also see FLOURS; and WHEAT.)

EXTREMITIES

The medical name given to the limbs. The upper extremities are the hand, arm and shoulder. The lower extremities are the foot, leg, thigh and hip. This term is used mainly to designate blood flow because in chilling there is often less blood flow to the extremities than there is in the central portion of the body.

EXTRINSIC FACTOR

A dietary substance, now known to be vitamin B-12, which was formerly thought to interact with the intrinsic factor of the gastric secretion to produce the antianemic factor.
(Also see INTRINSIC FACTOR; and VITAMIN B-12.)

EXUDATE

A fluid discharge into the tissues or any cavity.

EXUDATIVE DIATHESIS

The accumulation of fluid in subcutaneous tissues, muscles, or connective tissues, caused by the escape of plasma from the capillaries.

Every meal should include an item from each of the food groups. Such meals are what nutritionists call *well balanced*. (Courtesy, The California Avocado Advisory Board, Los Angeles, Calif.)

FAGOT (FAGGOT)

Besides meaning a bundle of sticks or pieces of wrought iron, this word is used in two different culinary ways as follows:

• Another term for *Bouquet Garni*—a herbal mixture added to foods during the cooking process.

• A pork product composed of hog livers, hearts, fresh pork, onions, salt, pepper, sweet marjoram, and hog caul fat (fat surrounding the stomach and intestines). The meat is thoroughly ground up, seasoned, then molded in 6-oz (*170 g*) balls. The caul fat is cut into approximately 7-in. squares (*45 cm²*) into which the meat balls are placed and encased, then baked in an oven for 45 minutes.

FAHRENHEIT

A thermometer scale on which the interval between the two standard points, the freezing point and the boiling point of water, is divided into 180°, with 32° representing the freezing point and 212° the boiling point. To convert Fahrenheit to Centigrade subtract 32 and multiply by 5/9.

(Also see WEIGHTS AND MEASURES.)

FAMILIAL

Common to a family

FAO

Abbreviation for the Food and Agriculture Organization of the United Nations.

(Also see FOOD AND AGRICULTURE ORGANIZATION OF THE UNITED NATIONS.)

FARINA

• A general term for starch.

• In the United States, it refers to a breakfast cereal, popularly known as "Cream of Wheat," which consists of the granulated endosperm of wheat other than Durham wheat. (A similar product produced from Durham wheat is called semolina.) The term may also be applied to any flour made from cereal grains, nuts, or sea moss.

• In Italy, a flour made from dried chestnuts is called *farina dolce.*

FAST FOODS

This refers to eating establishments that serve ready-to-eat foods, with little or no waiting time from ordering to serving. This speedy handling of orders is accomplished by (1) using commercial types of convenience foods, and/or (2) by cooking the foods well in advance and keeping them warm (or cold, if necessary to prevent spoilage) until they are sold. Fast food places may or may not have accommodations for eating on the premises.

The most common fast foods are: hamburgers, french fries, and shakes; pizza and cola; fried chicken and slaw; fish and chips; roast beef sandwiches; tacos; hot dogs; and other mass-produced and mass-served quickie meals.

Fast foods appeal to the younger set because (1) they are an important feature of their life-style, and (2) they fit their limited spending money. In 1989, Americans spent 43% of their food dollar on meals away from home, which cost a total of $156.4 billion. That same year, the typical American ate 1 out of every 12 meals at a fast-food restaurant.

The American Council of Science and Health (ACSH), a non-profit, independent educational association of scientists, made a study of fast foods and the American diet (*ACSH Media Update*, Spring 1982). *Note well:* Since that time, in response to consumers searching for fitness and long life, most fast-food restaurants have made their menus more healthful. But, in 1982, the ACSH reported as follows about fast foods:

• **Fast foods contribute to good nutrition**—Fast-food meals can make a significant contribution to good nutrition if they are properly incorporated into a varied diet.

• **Known nutritional shortcomings of fast foods and how to compensate**—By knowing the nutritional shortcomings of fast-food menus and how to compensate for them, you can eat at a fast-food establishment several times a week without compromising your nutritional well-being.

• **High in calories and fat**—For the amount of nutrients provided, fast-food meals are usually too high in calories and saturated fat. A simple fast-food meal may total more than half your daily calories. For example, the calories in a fast-food meal may add up as follows: two pieces of Kentucky Fried Extra Crispy Chicken, 765 calories; a Burger King Whopper, 630 calories; and a McDonald's chocolate shake, 383 calories. Also fat is a major component of fast foods; and when they are deep-fried, their calorie value can more than double.

Pizza is reasonably well-balanced when it comes to fat and protein. A slice of thick pepperoni pizza contains 18 grams of fat and 31 grams of protein.

• **High protein of good quality**—Fast food establishments get high marks for protein. Such animal products as beef, fish, chicken, pork, cheese, and milk are excellent protein sources. Even a small hamburger supplies 25% of the Recommended Daily Allowance (RDA) of protein for an adult man and 33% of a woman's or child's protein; while a large hamburger topped with cheese, or a shake, will fulfill 60 to 100% of the protein RDA.

• **Mixed marks on vitamins**—Most fast food meals are notoriously low in vitamin A, the best sources of which are dark green, leafy vegetables, yellow or orange vegetables and fruits, whole milk, fortified nonfat dairy products and liver. To assure an adequate amount of vitamin A in your diet on the days you eat fast food, you should balance eating at home with fruit, salad, enriched grains, or milk.

The B-complex vitamins are found in most fast foods in varying amounts. Most fast food meals provide adequate amounts of thiamin (B-1), riboflavin (B-2), niacin, vitamin B-6, and vitamin B-12.

Vitamin C can be gotten from orange juice, sold at some fast food places for breakfast, and, in limited amounts, from cole slaw and french fries.

The addition of salads to some fast food menus has increased the available source of vitamins A and C.

• **High salt**—Fast food is high in sodium. An average fast food meal may contain three-fourths of your recommended daily salt intake. People on a salt-restricted diet will have a problem at a fast food establishment.

• **Low fiber**—Fast food menu items are generally low in fiber, needed in most diets to aid digestion. Fiber is found primarily in fresh fruits and vegetables and in whole grain breads and cereals, most of which are not included on fast food menus.

Some fast food establishments have added salad and cole slaw to their menus, which should provide a source of fiber.

• **Cost**—One big drawback to fast food meals is their cost. A comparable meal prepared at home costs about half as much as at a fast food establishment.

• **Limited selection of items**—A nutritional drawback of fast foods, particularly for frequent consumers, is the limited number of items on many fast food establishment menus, because variety is important for good nutrition. For this reason, meals eaten at fast-service restaurants should be incorporated into a varied diet that includes many other food choices.

From the above, it may be concluded that the good news about fast foods is that it's not all bad news.

(Also see CONVENIENCE AND FAST FOODS.)

FASTING

This is the act of abstaining from food, or of limiting food intake, often by choice. Starvation is to perish from the absence or restriction of food intake, but not always by choice. People who fast do so for a variety of reasons, but whatever the reason, fasting elicits certain metabolic changes, most of which are controlled by hormones.

REASONS FOR FASTING. Actually, everyone fasts day-to-day between meals. Other reasons for fasting include weight loss, religious, medical, and protest.

• **Day-to-day**—For most people, breakfast is just that, the meal that breaks the overnight fast of 10 to 12 hours.

• **Weight loss**—Total fasting may be used to treat obesity. Protein and fat are lost from the bodies of those who fast. While indiscriminate fasting programs for weight loss are generally unwise, programs closely supervised by a physician may prove more effective in extreme cases of obesity than weight reduction clinics or gradual weight loss through reducing diets.

(Also see OBESITY, section headed, "Treatment of Obesity.")

• **Religious**—Fasting is common to many religions of the world. In some religions, one food may be substituted for another; in other religions, foods may be abstained from altogether. For example, the Mormons abstain from food and water for a day each month, and individuals of the Jewish faith fast for 24 hours on Yom Kippur and possibly before the Feast of Purim.

• **Medical**—At times, to make medical diagnosis or for medical procedures fasting is required. For example, glucose (sugar) level is determined on blood samples from fasting individuals; before surgery there is generally a brief fast; and certain tests such as the Schilling test (vitamin B-12 absorption) and glucose tolerance test require fasting before performing them.

• **Protest**—Fasting has been, and continues to be, a means of protest. Mohandas K. Gandhi (1869-1948) often made use of this form of protest. Gandhi was thrown into jail many times by the British for some peaceful disobedience or boycott. While in jail, he fasted, whereupon, the British, fearing he would starve to death, usually released him quickly. Gandhi also fasted for better treatment of the untouchables in India and to keep peace between warring Hindus and Muslims. During 1981, the British were again subject to fasting protests. This time members of the Irish Republican Army fasted—to death—to protest the British presence in Northern Ireland.

METABOLIC CHANGES. As always, the prime purpose of the metabolic machinery of the body is to maintain an adequate energy supply to carry on life processes. The major energy source for cells of the body is the carbohydrate, glucose. Without a dietary source of carbohydrate, the liver reserves of glycogen are adequate for maintaining blood glucose levels for only a few hours. The brain and contracting skeletal muscle require a continuous supply of glucose from the blood, which is normally maintained within very narrow limits—even during fasting. Hence, the body does primarily two things during fasting: (1) changes energy sources, and (2) reduces energy expenditures. These changes are hormonally controlled and summarized in Table F-1.

TABLE F-1
METABOLIC CHANGES DURING FASTING AND HORMONES RESPONSIBLE FOR CHANGES

Energy Source	Tissue			Hormones Responsible
	Liver	Fat	Muscle	
Glucose	Increased breakdown of glycogen and release of glucose; increased formation of glucose from amino acids—gluconeogenesis; decreased breakdown of glucose.	Decreased uptake and use of glucose.	Increased breakdown of glycogen; decreased breakdown of glucose and decreased uptake of glucose.	Glucagon from the pancreas. Epinephrine from the medulla of the adrenal gland. Glucocorticoids from the cortex of the adrenal gland. (Also see ENDOCRINE GLANDS.)
Fat	Decreased formation of fat; increased uptake of fatty acids; increased beta-oxidation of fatty acids and ketone formation.	Decreased synthesis of triglycerides; increased breakdown of fat and release of fatty acids.	Increased uptake of fatty acids with increased utilization for energy.	
Protein	Increased amino acid uptake and amino acid breakdown.		Increased breakdown of protein and increased release of amino acids.	

During the initial stages of fasting, blood glucose is maintained by the breakdown of liver glycogen, thus releasing glucose which is critical for the brain and nervous system. At the same time the hormone sensitive lipase (an enzyme) causes the release of free fatty acids, from body fat, into the circulation. They provide fuel for much of the body's tissues, and glycerol released at the same time may be converted to glucose. As liver glycogen becomes depleted, additional glucose is produced from primarily amino acids of the muscle proteins. Alanine is the preferred amino acid for conversion to glucose. Furthermore, the breakdown of muscle glycogen contributes to blood glucose formation via the lactic acid or Cori cycle wherein lactic acid released from the muscles is converted to glucose by the liver. As fasting continues, increased blood levels of free fatty acids are maintained resulting in the formation of ketone bodies from fatty acids—ketosis. Most tissues can use ketone bodies—acetoacetic acid, beta-hydroxybutyric acid, and acetone—as fuel, including the brain. Hence, the need for glucose is further reduced. Shifting to the production and utilization of ketone bodies requires about 3 days of fasting. Individuals may have a sweetish acetone odor to their breath due to the production of ketone bodies.

Should fasting continue, there is an actual reduction in the metabolic rate which reduces the energy requirements of the body. This change is produced by a reduction in the secretion of the thyroid hormones. All of the metabolic changes induced by fasting represent beneficial changes which ensure survival, by providing sufficient energy to maintain life processes. Fasting individuals vary in their susceptibility to low blood sugar (hypoglycemia) and ketosis. Some individuals may become weak and feel dizzy after fasting one day, while others feel few effects after several days. For healthy individuals no harm results from short term fasting. However, when fasting is prolonged it enters the realm of starvation.

(Also see ENDOCRINE GLANDS; HYPOGLYCEMIA; METABOLISM; and STARVATION.)

FAT

The term fat is frequently used in a general sense to include both fats and oils, or a mixture of the two. Both fats and oils have the same general structure and chemical properties, but they have different physical characteristics. The melting points of most fats are such that they are solid at ordinary room temperatures, while oils have lower melting points and are liquids at these temperatures.

(Also see FATS AND OTHER LIPIDS.)

FAT BACK

Solid, relatively uniformly shaped, rectangular fat slabs removed from the surface of the pork loin.

(See PORK, Fig. P-75.)

FAT, BLOOD

Refers to the total lipid (fat) content of the blood plasma or serum, which is usually 500 to 600 mg/ml.

(Also see BLOOD FATS; and FATS AND OTHER LIPIDS.)

FATS AND OTHER LIPIDS

Contents Page

History 684
Synthesis 684
Classification 684
Characteristics of Fats 685
Fatty Acids 685
Other Lipids 688
Digestion, Absorption, and Metabolism 688
Functions 688
Required Intake 690
Recommended Fat Intake 690
Essential Fatty Acid 691
Polyunsaturated vs Saturated 692
Sources 692
Visible and Invisible 693
Animal and Vegetable 693
Medium-Chain Triglycerides (MCT) 694
Nonnutritive Oils 694
Fat and Health Problems 694
Fats in the Blood 694
The Lipid Hypothesis 695
Fat and Health Summary 695
Fats and Fatty Acids in Selected Foods 695

Fat is easily recognized when it accumulates on the body, but the chemical or technical definitions of fats and lipids are more difficult. Not all lipids are fats but all fats are lipids, so the two words are often used interchangeably. Like carbohydrates, lipids—fats and fatlike substances, contain the three elements—carbon, hydrogen, and oxygen, but in different proportions—a larger proportion of carbon and hydrogen. Lipids are soluble in such organic solvents as ether, chloroform, and benzene, but they are not soluble in water. For this reason, grease spots on clothing are usually removed with a solvent. As a food, fats function much like carbohydrates in that they serve as a source of heat and energy and contribute to the formation of body fat.

Because of the larger proportion of carbon and hydrogen, fats liberate more heat—or energy—than carbohydrates or proteins. Upon oxidation, fats liberate approximately 2.25 times as much heat or energy per pound as do carbohydrates or proteins. Hence, they are concentrated energy sources—high in calories, about 900 kcal/100 g.

The general term lipid includes fats, oils, and waxes. Fats are esters (alcohol and acid combinations) of glycerol and fatty acids that are solid at room temperature, while oils are glycerol esters that are liquid at room temperature. However, the term fats is often employed for both, to avoid confusion with essential oils and petroleum oils. Waxes are esters of fatty acids with alcohols other than glycerol.

HISTORY. Fats and oils were recognized and used by mankind even before recorded history. Man used fats and oils for food, for soapmaking, for lubrication, for fuel (light and heat), and for cosmetics and medicines. Numerous Biblical passages refer to fats and oils. Indeed, the use of animal fat and oils in cooking was common in Old Testament times (Leviticus 1:8-12 and 2:4-7). Also, the records of the early Egyptians attest to the use of fats and oils. Egyptians used oil preparations from almonds, castor, lettuce seed, linseed, olive, radish seed, safflower seed, and sesame.

The terms lipid, fat, and oil are very old, showing that man was familiar with these substances from very early times. The word *Lipid* is derived from the Greek word *lipos* meaning fat, while the word *fat* comes from the Old English word *faett*. The word *oil* comes through several derivations from the Greek word *elaion,* referring to olive oil.

One of the earliest chemical examinations of fats was made by a man named Dioscorides who, in the first century A.D., described boiling of fat with lead oxide or zinc oxide to make plasters for medicinal purposes. It was M.E. Chevreul, a French chemist, who made great pioneering studies on the separation and description of fatty acids. He published his investigations into the chemical nature of fats in 1823. But it remained until the late 1800s before physiologists proved that excess dietary carbohydrates could be converted to fat in the body and stored—though farmers and laymen had suspected this for years.

SYNTHESIS. Basically, fat synthesis is a process of combining 2-carbon units called *acetyl CoA* into long chains and then adding hydrogen. Acetyl CoA is a step in carbohydrate, protein, and fat metabolism. While a limited amount of fat synthesis occurs in the mitochondria of the cell, most synthesis takes place in the microsomes of adipose (fat) tissue, liver, and many other tissues. Following synthesis, fats may combine with other compounds such as alcohols, proteins, carbohydrates, or phosphorus.

(Also see METABOLISM.)

CLASSIFICATION. Lipids are often classified into three major groups: (1) simple lipids; (2) compound lipids; and (3) derived lipids. When fatty acids are esterified with alcohols, simple lipids result. If compounds such as choline or serine are esterified to alcohols in addition to fatty acids, compound lipids result. The third type of lipid, derived lipids, result from the hydrolysis or enzymatic breakdown of simple and compound lipids. Table F-2 classifies lipids and provides some examples and characteristics of each of them.

TABLE F-2
CLASSIFICATION OF LIPIDS

Type of Lipid	Example	Chemistry	General Comments
Simple lipids:			
Neutral fats	Triglycerides (triacylglycerols)	Esters of fatty acids with glycerol; ratio of 3 fatty acids to 1 glycerol.	Most abundant lipids in nature. Mixed triglycerides (those in which at least two fatty acids are different) account for 98% of the fats in foods and over 90% of fat in the body.
Waxes	Beeswax	Esters of fatty acids with high-molecular-weight alcohols other than glycerol. This group includes the esters of cholesterol, vitamin A, and vitamin D.	More important in commerce than in human nutrition; occur widely in cuticle of leaves and fruit.
Compound lipids:			
Phospholipids	Lecithins Cephalins Lipositols	Compounds of neutral fat, a phosphoric acid, and a nitrogenous base (choline, ethanolamine, or serine); water-soluble.	Lecithins are largest group of phospholipids. Lecithin may be obtained from egg yolks or soybeans.
Glycolipids	Cerebrosides Gangliosides	Sugar-(carbohydrate) containing fatty acids plus nitrogen.	Sugar can be glucose or galactose; found in nervous tissue; component of cell membrane.
Lipoproteins	Chylomicrons Very low density lipoproteins (VLDL). Low density lipoproteins (LDL). High density lipoproteins (HDL).	They all contain protein, triglycerides, phospholipids, and cholesterol; but in varying amounts.	The lipoproteins, synthesized in the liver, are composed of about ¼ to ⅓ protein, with the remainder lipids. Means of transporting lipids in the blood.
Derived lipids:			
Fatty acids	Palmitic acid Oleic acid Stearic acid Linoleic acid	Generally have one acid group (COOH); may be saturated, or unsaturated—contain 1 or more double bonds.	In most cases, there is an even number of carbon atoms in the naturally-occurring fatty acids. There are few odd-numbered carbon atom fatty acids in nature. Release of fatty acids from triglyceride releases glycerol.
Steroids	Cholesterol Ergosterol Cortisol Bile acids Vitamin D Androgens, estrogens, and progesterone.	Derivatives of the perhydrocyclopentano-penanthrene nucleus (chemical structure is a series of rings).	One of the most studied classes of lipids. Collectively many of these are referred to as steroid hormones—hormones of the adrenal gland, testes, and ovaries.
Hydrocarbons	Terpenes	Compounds of hydrogen and carbon only.	Includes a series of oils (such as camphor), resin acids, and plant pigments. Beta-carotene is an example of an important terpene.

CHARACTERISTICS OF FAT. Triglycerides—the combination of three fatty acid molecules and a glycerol molecule—account for about 98% of the fats in our foods and over 90% of the fat in the body. The remainder of the diet and body fats is comprised primarily of phospholipids and cholesterol. Since triglycerides and their component fatty acids are so abundant in the diet and body, the discussion which follows centers around fatty acids and triglycerides.

Fatty Acids. Fatty acids are key components of lipids. They are called acids because of the organic acid group (COOH) which they contain. Their degree of saturation and the length of their carbon chain determine many of the physical characteristics of lipids. Numerous triglycerides exist due to the variety of fatty acids which may bind with glycerol.

- **Saturation**—This refers to the ratio of hydrogen atoms to carbon atoms. The backbone of the fatty acid consists of a chain of carbon atoms joined by chemical bonds. When a single bond joins each carbon atom, carbon atoms within the chain have two hydrogen atoms joined to them and carbons at the end of the chain have three hydrogens. When carbon atoms are joined by double bonds, these carbon atoms within the chain are only able to have one hydrogen bound to them. Therefore, saturated fatty acids contain all possible hydrogen and no double bond between carbon atoms. Unsaturated fatty acids contain at least one double bond within the carbon chain (monounsaturated) or two or more double bonds within the carbon chain (polyunsaturated). Therefore, unsaturated fatty acids contain the same number of carbon atoms, but fewer hydrogen atoms than their saturated counterparts. Fig. F-1 illustrates the concept of saturated and unsaturated.

SATURATED

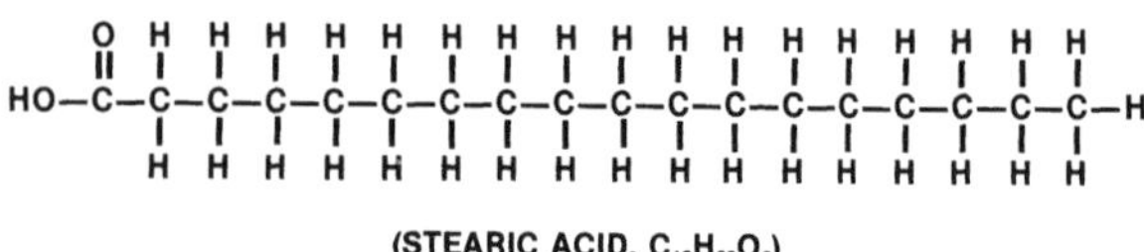

(STEARIC ACID, $C_{18}H_{36}O_2$)

MONOUNSATURATED

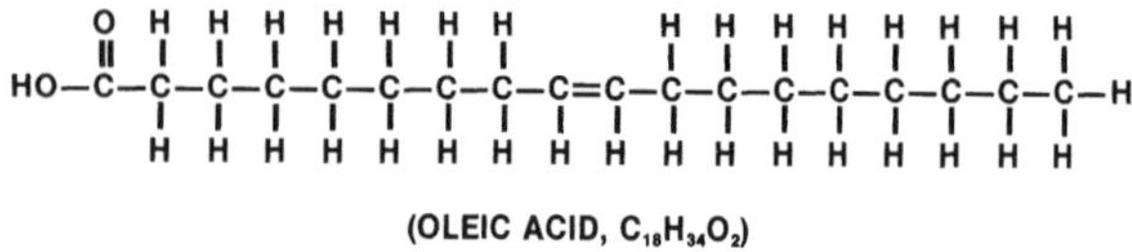

(OLEIC ACID, $C_{18}H_{34}O_2$)

POLYUNSATURATED

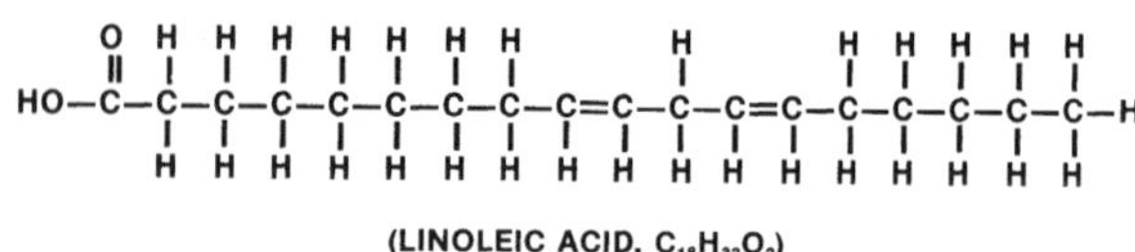

(LINOLEIC ACID, $C_{18}H_{32}O_2$)

Fig. F-1. Three fatty acids all composed of 18 carbons but different degrees of saturation or unsaturation. The = indicates a double bond and C stands for carbon, H for hydrogen, and O for oxygen.

- **Iodine number**—Unsaturated fat readily unites with iodine; two atoms of this element are added to each double bond. Thus, in experimental work, the number of grams of iodine absorbed by a hundred grams of fat—the iodine number—is an excellent criterion of the degree of unsaturation. In the past, the iodine test was commonly used when studying the soft pork problem—a problem caused when pigs are fattened on feeds rich in unsaturated fats, such as peanuts or soybeans. At the present time, the chief measure used in such determinations is the refractive index, as determined by a refractometer.

Fatty acids that are unsaturated have the ability to take up oxygen or certain other chemicals. This presents both advantages and disadvantages. The value of linseed oil and varnish is due to their high content of unsaturated fatty acids, by virtue of which oxygen is absorbed when they are exposed to air, resulting in a tough, resistant coating. On the other hand, because of their unsaturation fats often become rancid through oxidation, resulting in disagreeable flavors and odors which lessen their desirability as foods.

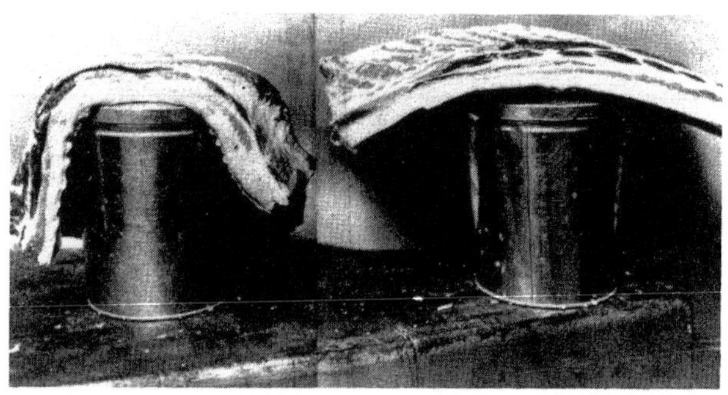

Fig. F-2. Soft pork. The bacon belly on the left came from a hog liberally fed on soybeans. Feed fats affect body fats. (Courtesy, University of Illinois)

- **Rancidity**—This is the oxidation (decomposition) primarily of unsaturated fatty acids resulting in disagreeable flavors and odors in fats and oils. This process occurs slowly and spontaneously, and it is accelerated by light, heat, and certain minerals. Rancidity may be prevented through proper storage and/or the addition of antioxidants such as sodium benzoate. Some fats are naturally protected from oxidation due to the presence of vitamin E. Hydrogenation of fats (adding hydrogen to unsaturated fatty acids) also lessens the threat of rancidity. This process has been used to improve the keeping qualities of vegetable shortenings and lard.

  (Also see ADDITIVES.)

- **Effects of heat**—*Acrolein* is a pungent compound produced from glycerol. It is especially irritating to the lining of the gastrointestinal tract and it may be produced by excessive heating of fats. However, normal home or commercial frying usually does not represent a hazard.

- **Hydrogenization**—This process adds hydrogen to the double bonds of unsaturated fatty acids. It is accomplished with hydrogen gas in the presence of a nickel catalyst. The result of hydrogenization is a harder fat since adding hydrogen increases the melting temperature. It may be used on animal or vegetable fats to produce fats with a desired hardness. Many vegetable oils are converted into a solid or semisolid form for use in shortenings and margarines. Not surprisingly, hydrogenization is also known as hardening.

Hydrogenization may have one drawback. It converts the naturally-occurring *cis* fatty acids to *trans* fatty acids.

The prefixes *cis* and *trans* refer to the orientation of the atoms around the double bond. The *trans* form of essential fatty acids does *not* function as an essential fatty acid in the body. Also, some researchers have found that (1) *trans* fatty acids are not as effective as their *cis* analogs in lowering blood cholesterol, and (2) fats rich in *trans* fatty acids appear to promote atherosclerosis.[1]

The content of *trans* fatty acids generally increases with the extent to which a vegetable oil has been hydrogenated. For example, hard sticks of vegetable oil margarines may contain from 25 to 35% of *trans* fats, whereas lightly hydrogenated liquid oils usually contain 5% or less of these fats.

(Also see TRANS FATTY ACIDS.)

• **Carbon chain length**—Another variable factor in the makeup of fatty acid molecules is the number of carbon atoms. Fatty acids are designated as having (1) short chains when the number of carbon atoms is 6 or less, (2) medium chains when there are 8 or 10 carbon atoms, and (3) long chains when there are 12 or more carbon atoms. In most cases, naturally-occurring fatty acids contain an even number of carbon atoms.

Together, the degree of saturation and the length of the carbon chain, influence the melting point of fats. The melting points of fats are very important in nutrition, since fats that remain solid in the digestive tract are poorly utilized. Furthermore, the melting points are important to food processors because (1) consumers expect certain items to be solid at ordinary room temperatures, and others to be liquid; and (2) most shortenings should be at least semisolid at room temperature because flakiness of pastry depends upon the production of layers of solid fat.

As can be seen in Table F-3, the melting points of fatty acids are highly dependent on the length of the carbon chain and the degree of unsaturation of the molecule.

[1]"Newer Concepts of Coronary Heart Disease," *Dairy Council Digest*, Vol. 45, 1974, p. 33.

TABLE F-3
CHEMISTRY OF SOME FATTY ACIDS

Name	Structural Formula	Chain Length	Melting Point		Example of Source
		(no. C atoms)	(°F)	(°C)	
Saturated:					
Acetic	CH_3 COOH	2	—	—	Vinegar
Propionic	CH_3 CH_2 COOH	3	—	—	Dairy products
Butyric	CH_3 $(CH_2)_2$ COOH	4	19	−7	Coconut butter
Caproic	$CH_3(CH_2)_4COOH$	6	18	−8	Coconut butter
Caprylic	$CH_3(CH_2)_6COOH$	8	62	16	Coconut butter
Capric	$CH_3(CH_2)_8COOH$	10	88	31	Coconut, butter
Lauric	$CH_3(CH_2)_{10}COOH$	12	112	44	Coconut, palm
Myristic	$CH_3(CH_2)_{12}COOH$	14	129	54	Coconut, butter, whale blubber.
Palmitic[1]	$CH_3(CH_2)_{14}COOH$	16	146	63	Palm, beef tallow butter, lard, cottonseed oil.
Stearic[1]	$CH_3(CH_2)_{16}COOH$	18	157	70	Beef tallow, butter, lard.
Arachidic	$CH_3(CH_2)_{18}COOH$	20	170	76	Peanut oil
Lignoceric	$CH_3(CH_2)_{22}COOH$	24	187	86	Beechwood tar
Monounsaturated:					
Palmitoleic[1]	$CH_3(CH_2)_5CH = CH(CH_2)_7COOH$	16	31	-	Menhaden (fish), chicken, beef tallow.
Oleic[1]	$CH_3(CH_2)_7CH = CH(CH_2)_7COOH$	18	56	13	Olive oil, peanut oil, egg.
Polyunsaturated:					
Linoleic[1]	$CH_3(CH_2)_4CH = CHCH_2CH = CH(CH_2)_7COOH$	18	23	−5	Safflower, soybean oil, corn oil.
Linolenic	$CH_3CH_2CH = CHCH_2CH = CHCH_2CH = CH(CH_2)_7COOH$	18	12	−11	Linseed oil, soybean oil.
Arachidonic[1]	$CH_3(CH_2)_4CH = CHCH_2CH = CHCH_2CH = CHCH_2CH = CH(CH_2)_3COOH$	20	−57	−50	Liver, egg.

[1]Most abundant in animal lipids.

Short fatty acids tend to be more volatile. In fact, acetic, propionic, and butyric acid are collectively called volatile fatty acids (VFA). There is a steady rise in melting point as the chain lengths increase. However, as the number of double bonds increases, the melting point decreases.

• **Saponification**—The combination of a fatty acid with a cation such as potassium or sodium forms soap. This reaction is called saponification. Besides forming soap, it is a method of evaluating the average length of carbon chain in the fatty acids which constitute a fat. The test is performed by reacting fats with potassium hydroxide (an alkali). The saponification number, or value, is the number of milligrams of potassium hydroxide required for the complete saponification of 1 g of the fat. A high saponification value signifies a short chain length and vice versa.

Saponification may also occur in the alkaline medium of the intestine. For example, calcium may combine with free fatty acids.

• **Emulsification**—Fats (oils) and water do not stay mixed, but often it is desirable for them to do so. Therefore, fats are often emulsified. Minute droplets of fats or oils are evenly distributed throughout a water-based solution. Emulsions are essential for the digestion, absorption, and transport of fats in the body. Furthermore, emulsification is employed in homogenized milk and other products containing fats or oils such as mayonnaise. Emulsifying agents used to create emulsions include some fatlike and fat-derived substances such as monoglycerides (glycerol with one fatty acid), diglycerides (glycerol with two fatty acids), lecithin and the bile salts.

Other Lipids. Other important lipids include primarily the phospholipids, lipoproteins, and cholesterol. All cells contain phospholipids. They are structural compounds found in cell membranes and in the blood. The brain, nerves, and liver contain particularly high levels. Lecithin is one of the most abundant phospholipids in the diet and the body. Phospholipids are powerful emulsifying agents. Lipoproteins are the primary vehicle for lipid transport in the blood. There are four main types: chylomicrons; very low density lipoproteins (VLDL); low density lipoproteins (LDL); and high density lipoproteins (HDL). Cholesterol is derived from the diet or synthesized in the body. It is necessary for the formation of hormones and bile salts. These will all be discussed in sections which follow.

(Also see CHOLESTEROL.)

DIGESTION, ABSORPTION, AND METABOLISM.

Compared to the digestion of carbohydrates and proteins, the digestion of fats is unique. Before the enzymes of the pancreas and intestine can act on the fats, they are mixed with bile. Fats are not water-soluble. Since enzymatic reactions of the body occur in water-based solution, fats must be emulsified for digestion to occur. Bile salts from the gallbladder play two roles in fat digestion: (1) they act as a detergent decreasing the surface tension, and (2) they form micelles which are colloidal complexes of monoglycerides and insoluble fatty acids that have been emulsified and solubilized for absorption. It is the detergent action of bile which allows the mixing movements of the intestines to break the fat globules into very finely emulsified (water-oil mixture) particles with a greatly increased surface area on which the water-soluble enzymes—lipases—can act. When lipids, emulsified by bile salts, come into contact with the various lipases that are found in the small intestine, they are broken down into diglycerides, monoglycerides, fatty acids, and glycerol. Short-chain fatty acids, less than 10 to 12 carbon atoms, are absorbed directly into the mucosa—lining—of the small intestine and are transported to the portal circulation of the liver. Monoglycerides and insoluble fatty acids are emulsified by bile salts, forming micelles. By attaching to the surface of epithelial cells, the micelles enable these components to be absorbed into the intestinal cells. Once inside these cells, the long-chain fatty acids are reesterified (joined to the alcohol, glycerol) to form triglycerides. Triglycerides then combine with cholesterol, lipoproteins, and phospholipids to form chylomicrons—minute fat droplets. The chylomicrons are then passed into the lymphatic system. Eventually these chylomicrons enter the blood where they are transported to the tissues of the body. Adipose tissue (fat) is the major site for the removal of chylomicron triglycerides. Here the fatty acids are cleaved by the enzyme lipoprotein lipase and taken into the cells. Also, chylomicrons, phospholipids, free fatty acids cholesterol and proteins are transported to the liver and formed into other lipoproteins, which provide the major vehicle for the transport of fat in the blood. These lipoproteins include the following: (1) prebeta lipoproteins or very low-density lipoproteins (VLDL); (2) beta lipoproteins or low-density lipoproteins (LDL); and (3) alpha lipoproteins or high-density lipoproteins (HDL).

Fat provides energy to the body. The free fatty acids (FFA) transported in the blood can be oxidized by most cells to release energy. The final products of oxidation are water and carbon dioxide. Dietary fats, carbohydrates, and protein may all contribute to the synthesis of fats within the body.

(Also see CHOLESTEROL; DIGESTION AND ABSORPTION; and METABOLISM.)

FUNCTIONS.

In general, fat performs four functions in the body. It provides (1) energy, (2) essential fatty acids, (3) structural components, and (4) regulatory functions.

1. **Energy.** The cells of the body, except for red blood cells and those of the central nervous system, can utilize fatty acids directly as a source of energy. Although carbohydrate (glucose) is the energy source normally used by the nervous system, the brain can utilize ketone bodies that are formed from fatty acids during a period of fasting.

Most food fats are triglycerides, but some also contain phospholipids and cholesterol. Regardless of the fat type, fats are readily digested by healthy individuals. Any excess of energy, whether derived from dietary carbohydrate, protein or fat, is stored as triglycerides within the adipose cells of the body. Adipose cells are found beneath the skin, between muscle fibers, around abdominal organs and their supporting structures called mesenteries, and around joints. As an energy storage form, it stores in the same weight 2.25 times as much energy as protein or carbohydrate. Fat or adipose is in a continuous turnover state—a dynamic state—which is controlled by the hormones: insulin, growth hormone, epinephrine, adrenocorticotropic hormone (ACTH), and glucagon.

The immediate sources of energy for the body are the free fatty acids in the circulation liberated from adipose triglycerides by the enzyme lipase. These free fatty acids are oxidized or burned by a systematic process called beta-oxidation, whereby 2-carbon fragments are successively cleaved from the fatty acid molecule to form acetyl CoA which releases energy upon completing the Krebs cycle (TCA cycle).

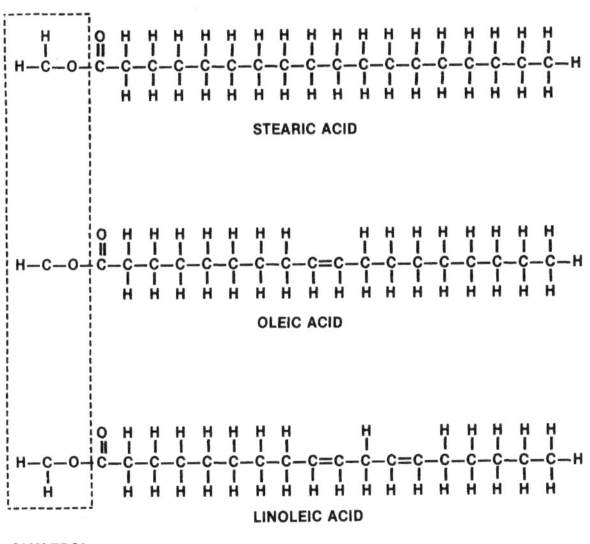

Fig. F-3. A triglyceride containing a saturated, a monosaturated, and a polyunsaturated fatty acid.

2. **Essential fatty acid.** The body has a remarkable ability to synthesize many compounds, and, as has already been pointed out, excess protein and carbohydrates can be converted to fat. However, in 1929, G. M. Burr and M. M. Burr at the University of Minnesota, reported that rats fed a fat-free diet (the diet was adequate in all other nutrients) failed to grow, lost weight, developed scaly skin and kidney damage, and eventually died. These conditions could be prevented or reversed if linoleic acid was added to their diet. Similarly, Hansen found that the inclusion of essential fatty acids in the diet of infants who had eczema led to improvement in the skin and to greater gains in weight (see Fig. F-4). Thus, the concept of essential—not synthesized by the body—fatty acids was introduced.

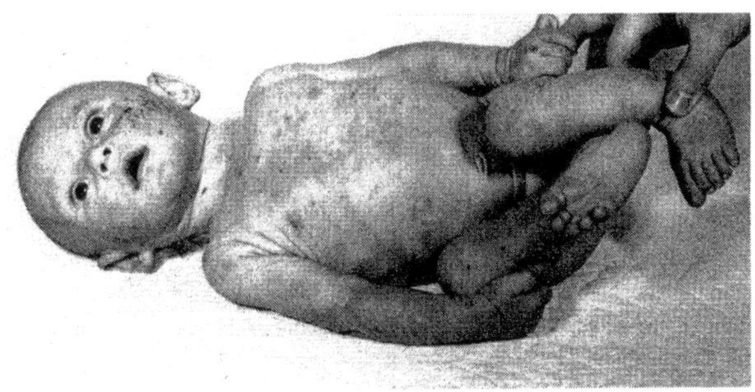

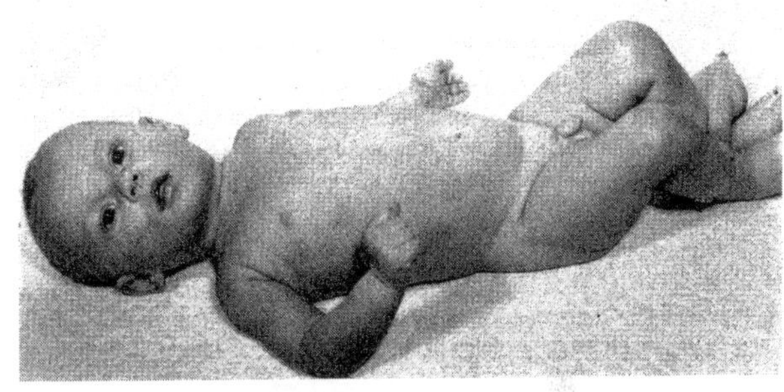

Fig. F-4. Fatty acid made the difference! Infant with very resistant eczema (eruption of the skin) since 2 ½ months of age. The same child 1 month later, after adding linoleic acid (fresh lard) to the diet. (Courtesy, Dr. A. E. Hansen, University of Texas, Medical Branch, Galveston, Tex.)

For many years, three polyunsaturated fatty acids—linoleic acid, linolenic acid, and arachidonic acid—were considered essential fatty acids because they are essential for certain body functioning, and because it seemed that the body could not synthesize them. However, recent research has shown (1) that linolenic acid has relatively little effect in relieving the skin lesions originally associated with a deficiency of essential acids, and (2) that arachidonic acid can be synthesized by the body from linoleic acid and, therefore, does not have to be supplied as such in the diet. (It is noteworthy, however, that recent work indicates that members of the cat family may require arachidonic acid.) So, presently, linoleic acid, the 18-carbon polyunsaturated acid with two double bonds, found widely in foods of both plant and animal origin, may be considered as the only essential fatty acid; it cannot be synthesized in the body and must be present in the diet. Nevertheless, the three fatty acids—linoleic, linolenic, and arachidonic—serve important functions in the body. Deficiency symptoms of these fatty acids have been observed in human infants, and in dogs, guinea pigs, mice, poultry, and swine. Depending on the type of animal, numerous manifestations of these deficiencies are seen; among them, dermatitis, reduced growth, increased water consumption and retention, impaired reproduction and increased metabolic rate.

In the body, linoleic acid is converted to longer-chain fatty acids with three, four, and five double bonds, which are essential components of membranes. In infants fed formulas, the primary symptom noted in essential fatty acid deficiency is drying and flaking of the skin. A fatty acid deficiency in adult humans was unknown until recently. In the past several years, there have been numerous reports of such deficiency being produced inadvertently in hospitalized patients, both infants and adults, fed exclusively by intravenous fluids not containing fat.

Fatty acids in phospholipids are important for maintaining the function and integrity of cellular and subcellular membranes. These acids also play a role in the regulation of cholesterol metabolism—especially its transport, breakdown, and ultimate excretion. In addition, fatty acids have been shown to be the precursors for a group of hormonelike compounds called prostaglandins, which are important in the regulation of widely diverse physiological processes.

(Also see the subsequent section headed, "Essential Fatty Acid.")

3. **Structural components.** Although excessive storage of fat becomes unsightly and burdensome, some body fat is necessary. Indeed, the feminine features owe much to the proper amounts and locations of fat deposits. Body fat (1) holds organs in place, (2) absorbs shock, and (3) insulates the body against rapid temperature changes or excessive heat loss. Furthermore, lipids, in particular phospholipids, have an important role in the structural integrity of the cell membrane. Phospholipids are water soluble and may aid in the transport of other fats in and out of cells.

4. **Regulatory functions.** Often the acceptance of food and its palatability depends upon flavor and aroma. Although triglycerides in the pure state are relatively tasteless, they absorb and retain flavors of foods. Furthermore, in combination with other nutrients, fats provide a texture that enhances palatability, and they delay emptying of the stomach and contribute to a feeling of satiety. Cholesterol is a fatlike compound with a rather sordid history, but it is the chemical ancestor of the hormones of the adrenal glands, ovaries and testes—the steroid hormones—and the bile salts. The prostaglandins are fatlike, hormonelike substances derived from the essential fatty acids and involved in the control of reproduction and circulation. Finally, dietary fat serves as a carrier for the fat-soluble vitamins A, D, E, and K, and as an aid to their absorption in the intestine.

Fig. F-5 summarizes the various functions of dietary fat.

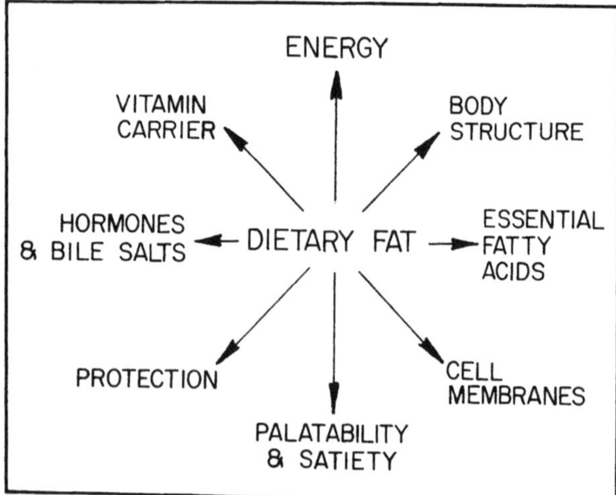

Fig. F-5. Functions of dietary fat.

(Also see CHOLESTEROL; DIGESTION AND ABSORPTION; ENDOCRINE GLANDS; METABOLISM; and VITAMINS.)

REQUIRED INTAKE. There is no required intake of fats and other lipids, and it seems that the body is capable of adapting to a wide range of intakes. During World War II, the fat intake of Japanese soldiers amounted to only 3% of their total energy intake. Worldwide, fat intake varies dramatically. In many developing countries, fat intake is, and has been for many generations, 10 to 20% of the energy intake, while in developed countries dietary fat intake ranges from 35 to 45% of the total energy intake. In general, the consumption of fat increases with increasing per capita income. This observation probably contributed to such expressions as "Living off the fat of the land," "fat cat," and "fat city." In the United States, fat consumption has increased since the turn of the century, and now amounts to about 42% of the total calories, as Table F-4 shows.

TABLE F-4
CALORIES FROM PROTEIN, CARBOHYDRATE, AND FAT[1]

Year	Protein	Carbo-hydrate	Fat		
			Total	Saturated	Unsaturated
	(%)	(%)	(%)	(%)	(%)
1909-1913	12	56	32	13	17
1925-1929	11	54	35	14	18
1957-1959	12	47	41	16	22
1967-1969	12	46	42	15	24
1988	12	46	42	16	26

[1] *1980 Handbook of Agriculture Charts*, USDA, p. 47, and *Agricultural Statistics 1991*, USDA, p. 474, Table 669.

Dietary fats are mainly triglycerides composed of fatty acid of varying chain length and degree of saturation or unsaturation. The proportion of saturated fat and unsaturated fat in the diet has shifted. Vegetable fats are a major source of the increase in unsaturated fatty acids.

Recommended Fat Intake. Many nutritionists and physicians believe that the health of a significant proportion of the United States population could be improved by changes in life style, including dietary modifications. Although some of the proposed changes in diet are currently controversial, there is sufficient evidence to support some recommendations for dietary changes that would be consonant with better health. It should be emphasized that most chronic or degenerative diseases have a number of contributing factors, only one of which may be diet. Changes in diet only, without consideration of measures to alter other risk factors—heredity, smoking, stress, high blood pressure and obesity—will probably have minimal desirable effects.

Associated with diet and nutritional practices is the problem of obesity and general overweight. In the Ten State Nutrition Survey conducted by the U.S. Department of Health, Education, and Welfare, it was found that up to 25% of adult males and 42% of females are classified as obese. Reduction in body weight to the desirable level is considered to be one of the most beneficial measures related to diets that the U.S. population could implement. For much of the U.S. population, maintenance of desirable body weight could be achieved most readily by controlling caloric intake and increasing physical activity. Since fat has the highest caloric density (kcal/g) of the nutrients, a decrease in fat consumption can produce the greatest change in dietary energy.

The Food and Nutrition Board's Committee on Diet and Health recommends that the fat content of the U.S. diet not exceed 30% of caloric intake, that less than 10% of the calories should be provided from saturated fatty acids, and that dietary cholesterol should be less than 300 mg/day.

• **Ketogenic diets**—At times an increased fat intake is recommended. Diets high in fat and low in carbohydrate are termed ketogenic since they cause the accumulation of ketone bodies—acetoacetic acid, beta hydroxybutyric acid and acetone—in the tissues. These diets eliminate carbohydrate sources such as breads, cereals, fruits, desserts, sweets, and sugar-containing beverages, while foods high in fat such as butter, cream, bacon, mayonnaise, and salad dressing are eaten in generous amounts. Ketones are produced from the breakdown (oxidation) of fats. When the ratio of fatty acids to available glucose in the diet exceeds 2:1, ketosis occurs. It is the metabolism of glucose that normally allows fatty acids to be completely oxidized to water and carbon dioxide rather than to ketone bodies. A ketogenic diet is considered monotonous, unpalatable, and most times undesirable. However, a ketogenic diet may be prescribed to control epilepsy, if drugs prove ineffective. Recently, medium chain triglycerides (MCT) have been shown to induce ketosis more readily than regular fats.

Essential Fatty Acid. The amount of dietary linoleic acid found to prevent both biochemical and clinical evidence of deficiency in several animal species and also in man is 1 to 2% of the dietary calories. As vegetable oils consumed in the United States are particularly rich sources of linoleic acid, this requirement level is more than met by normal mixed diets in this country. Several recent reports have shown a range of 5 to 10% of calories as linoleic acid in diets providing 25 to 50% of energy as fat. The U.S. Department of Agriculture estimates that about 23 g of linoleic acid, about 6% of total dietary energy, are available per person per day in the current United States food supply.

Past editions of the *Recommended Dietary Allowances* by the Food and Nutrition Board of the NRC have not proposed a recommended intake of essential fatty acid other than that needed to meet the requirement of 1 to 2% of calories. The American Academy of Pediatrics has recommended that infant formulas provide at least 2.7% of energy as linoleic acid.[2] For the average adult, a minimally adequate intake of linoleic acid is 3 to 6 g/day. This level is more than met by diets in the United States, since most vegetable oils are particularly rich sources of linoleic acid. In several studies, linoleic acid has been found to range from 5 to 10% of calories in diets providing 25 to 50% of energy as fat. For those consuming diets high in fat, there is evidence that a higher intake of linoleic acid may have beneficial health effects for a significant fraction of the population. These health effects relate to the possible prevention of atherosclerosis, coronary heart disease, and elevated blood lipids. Some rich sources of linoleic acid are listed in Table F-5.

TABLE F-5
RICH SOURCES OF THE ESSENTIAL FATTY ACID LINOLEIC ACID[1]

Source	Linoleic Acid
	(g/100 g)
Safflower oil	73
Corn oil	57
Cottonseed oil	50
Soybean oil	50
Sesame oil	40
Black walnuts	37
English walnuts	35
Sunflower seeds	30
Brazil nuts	25
Margarine[2]	22
Pumpkin and squash seeds	20
Spanish peanuts	16
Peanut butter	15
Almonds	10

[1]Values from Fat & Fatty Acids in Selected Foods, Table F-9 of this book.
[2]May vary depending upon the source of oil used.

The role of linolenic acid in human nutrition is not yet clear. Most vegetables and meats contain small amounts of linolenic acid and its derivatives, and the vegetable oils that are the primary sources of linoleic acid in the United States also contain small amounts of linolenic acid. If there is a dietary need for linolenic acid, it would appear to be met by diets that provide adequate linoleic acid.

Arachidonic acid can be synthesized by the body from linoleic acid; hence, it does not have to be supplied in the diet.

(Also see the earlier section headed "Functions, 2. Essential fatty acid.")

[2]*Recommended Dietary Allowances,* National Research Council, 10th ed., National Academy Press, 1989, p. 48.

Polyunsaturated vs Saturated. Some studies have shown that increasing the amount of polyunsaturated (PUFA) fat—linoleic acid, linolenic acid, and arachidonic acid—in the diet while reducing the saturated fat sometimes promotes a modest drop in blood cholesterol and a reduction in the tendency of blood to clot. Of greater importance than increased PUFA intake, however, is the ratio of polyunsaturated fats to saturated fatty acids—the P/S ratio—seems important. Currently, the recommendation is for a P/S ratio ranging from 1:1 to 2:1. The basis for this recommendation is that some studies have demonstrated increased blood cholesterol levels when the intake of saturated fatty acids was high, and that increasing the linoleic acid counteracted this effect. In all types of hyperlipoproteinemias—conditions of elevated blood lipids—an increase in PUFA is recommended. However, increasing the PUFA in the diet is not without possible problems; among them, the following should be considered:

1. Arachidonic acid, a polyunsaturated fatty acid, found mainly in peanut oil, promotes clotting.

2. The effects of polyunsaturated fatty acids on the heart muscle are uncertain, since it has been found that the lifetime feeding of corn, cottonseed, or soybean oils produced more heart lesions in rats than beef fat, butter, chicken fat, or lard.

3. The drop in blood cholesterol may be due mainly to a shifting of the cholesterol from the blood to the tissues.

4. Polyunsaturated fats cause an increase in the amount of cholesterol secreted in the bile, a condition that sometimes leads to the formation of cholesterol gallstones.

5. Extra vitamin E is required to prevent the formation of toxic peroxides in the body when the dietary level of polyunsaturated fats is raised. However, diets containing ample amounts of the essential trace mineral selenium, which is a part of the enzyme that breaks down the peroxides, may also help to offset this danger. The formation of peroxides may contribute to the cocarcinogenic activity observed in some animal studies.

To date, the National Research Council has not established Recommended Daily Allowances for PUFA. However, an upper limit of 10% of the dietary energy as PUFA is considered safe. For a man consuming 2,700 Calories (kcal) or a woman consuming 2,000 Calories (kcal), this translates to 30 g and 22 g per day, respectively.

For individuals interested in determining or altering their saturated and polyunsaturated intake, Fats & Fatty Acids of Selected Foods Table F-9, at the end of this article gives values for a variety of common foods. Table F-6 lists common food sources for saturated and polyunsaturated fats.

NOTE WELL: As a rule of thumb, animal fats are traditionally considered as containing high levels of saturated fat. However, Table F-6 shows some exceptions. Palm oil and coconut oil—both vegetable fats—contain high levels of saturated fats while cod liver oil, an animal fat, contains a high level of polyunsaturated fat.

(Also see HYPERLIPOPROTEINEMIAS.)

TABLE F-6
SOME COMMON FOOD SOURCES OF SATURATED AND POLYUNSATURATED FATS[1]

Food Source	Saturated Fats	Food Source	Polyunsaturated Fats
	(g/100 g)		(g/100 g)
Coconut oil	87	Safflower oil	75
Butter	51	Sunflower-seed oil	66
Palm oil	49	Wheat germ oil	62
Beef tallow	48	Corn oil	59
Mutton tallow	47	Soybean oil	58
Lard	39	Cottonseed oil	52
Baking chocolate	30	Cod liver oil[3]	50
Chicken fat	30	Walnuts	42
Shortening[2]	25	Peanut oil	32
Cream cheese	22	Shortening[2]	26
Cheddar cheese	21	Brazil nuts	25
Pasteurized process cheese	20	Chicken fat	20
Beef, rib cuts	19	Margarine	18
Margarine	19	Peanuts	14
Brazil nuts	17	Peanut butter	12
Peanut oil	17	Lard	11
Lamb chops	16	Almonds	10
Soybean oil	14	Palm oil	9
Olive oil	14	Olive oil	8
Sour cream	13	Mutton tallow	8
Pork sausage	12	Wheat germ	7
Pork chops	11	Beef tallow	4
Hamburger	10	Pork chops	4

[1]Values are for foods contained in Food Composition Table F-36 of this book.
[2]Household shortening made from hydrogenated soybean and cottonseed oil.
[3]Primarily contains polyunsaturated fats with carbon chain lengths of 20 and 22 carbons.

SOURCES. Worldwide, soybean oil is the world's largest volume oil product, palm oil is the second largest, and rapeseed oil ranks third.

In 1990, the average U.S. per capita consumption of fats from all sources—visible sources—was about 63 lb (*29 kg*), or in terms of energy about 261,000 Calories (kcal). This figure includes butter, margarine, lard, shortening, and salad and cooking oils.

Visible and Invisible. Visible fats are those which have been separated from their source and can be readily identified and measured; for example, lard, butter, margarine, shortening, and salad and cooking oils. Invisible fats are those which are not separated from their source. Sometimes these are called hidden fats. Invisible fats includes the fats of meats, eggs, cheese, milk, nuts, and cereals. Thus, the total fat consumed by an individual is at best a good estimate. The fat content of those foods with invisible fat is given in Food Composition Table F-36 for numerous foods. The Fats & Fatty Acids In Selected Foods Table F-9 at the end of this article also lists a variety of foods and gives their saturated, polyunsaturated, oleic, linoleic, and cholesterol content. Table F-7 provides examples of the hidden fat in some foods.

TABLE F-7
HIDDEN FAT OF SOME FOODS[1]

Food	Fat	Food	Fat
	(%)		(%)
Brazil nuts	67	Lamb roast	19
Walnuts	61	Avocado	16
Almonds	54	Ice cream	13
Peanuts	50	Herring	12
Sunflower seeds	47	Poached eggs	11
Pork sausage	44	Tuna, canned	8
Pork roast	30	Poultry, dark meat	7
Cheese	30	Oatmeal, dry	7
Bologna	28	Salmon	6
Beef roast	25	Whole milk	4
Ham, cured	22	Poultry, light meat	4
Hamburger	20	Shredded wheat cereal	2

[1]Amounts are calculated values from Fats & Fatty Acids in Selected Foods Table F-9 of this book.

Thus, depending upon the diet, the total fat can vary tremendously from individual to individual.

Animal and Vegetable. Fats in the diet are derived from either animal sources or plant—vegetable—sources. Both are comprised mainly of mixtures of triglycerides. A breakdown of animal fat sources and the amounts of animal and vegetable fat available for consumption is given in Table F-8. Furthermore, Table F-8 shows how the availability has changed the past 20 years. Vegetable sources have been gaining on animal sources.

TABLE F–8
AMOUNT OF FAT AVAILABLE
PER PERSON PER DAY IN THE UNITED STATES[1]

Food Source	Year	
	1967–1969	1988
	(g)	(g)
Vegetable oils	54.7	—
Butter	19.3	—
Fats & oils (includes butter)	—	78.1
Meat, poultry, fish	55.6	53.3
Dairy products	20.2	19.7
Eggs	5.1	3.4
Legumes	—	6.6
Vegetables & fruits	—	3.0
Grains & misc.	—	3.9
Total	154.9	168.0

[1]*1980 Handbook of Agricultural Charts*, USDA, p. 47, and *Agricultural Statistics 1991*, USDA, p. 478, Table 671.

The comparative food value of animal fats and vegetable oils has long been a stormy issue, particularly in regard to the relative merits of butter vs oleomargarine and of lard vs vegetable shortenings. In general, except for the vitamins that they carry, there is no experimental work to indicate that, as a source of fatty acids, animal fats are superior to vegetable oils. There is reason to believe that margarine, when fortified with vitamins A and D, is— from the nutritional standpoint—just as effective as butter in promoting growth, good health, reproduction, and lactation.

Fats or oils, regardless of their origin from animals or vegetables, furnish about 9 Calories (kcal) per gram. They are all concentrated energy sources containing 2.25 times more energy than carbohydrates or protein. The major differences between plant and animal fats are: (1) plant fats or vegetable oils are higher in polyunsaturated fatty acids, and (2) plant fats do not contain cholesterol. However, there are several exceptions. Cocoa butter, coconut oil, and the palm oils, contain high levels of saturated fats. Marine animals and fish may be rich in unsaturated fats.

Over the years, consumption patterns for vegetable and animal fats have shifted as illustrated in Fig. F-6. While total fat consumption has risen a little, there has been a marked drop in the consumption of animal fat with a corresponding rise in vegetable oil consumption. These shifts may, in part, be explained by (1) educational and media programs encouraging consumers to cut back on saturated fats and cholesterol, (2) changing tastes, and (3) increased availability of vegetable oils such as corn, soybean, peanut, safflower, sunflower, and coconut. Additionally, vegetable oils are the best food source of the natural antioxidant, vitamin E. For example, the amount of vitamin E, in mg per 100 g of oil, supplied by some typical oils are as follows: wheat germ, 150; safflower, 34; cottonseed, 35; peanut, 12; corn, 14; and soybean, 11.

(Also see MEAT[S]; OILS, VEGETABLE; and VITAMIN E.)

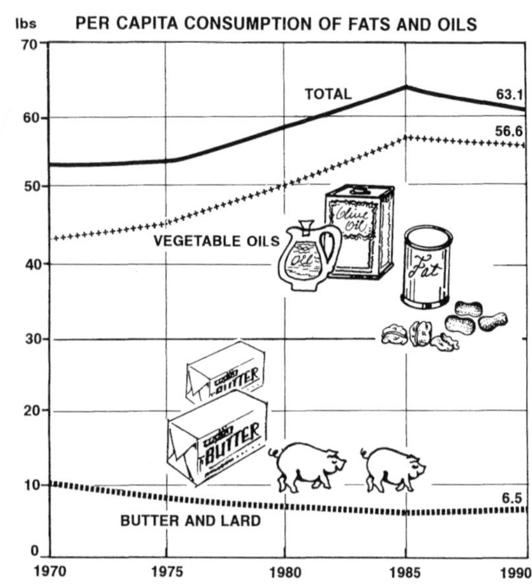

Fig. F-6. Per capita consumption of fats and oils. Animal fats includes butter and lard. (Courtesy, USDA)

Medium-chain Triglycerides (MCT). This is a special dietary product. It is made from coconut oil by (1) steam and pressure hydrolysis of the oil into free fatty acids and glycerol, (2) fractionation of the resulting hydrolysate into medium-chain and long-chain fatty acids, and (3) recombination of the medium-chain fatty acids with glycerol to form MCT oil, which is made up of about ¾ caprylic acid (a saturated fatty acid containing 8 carbon atoms) and about ¼ capric acid (a saturated fatty acid containing 10 carbon atoms). The special nutritive values of this product are as follows:

1. It is much more readily digested, absorbed, and metabolized than either animal fats or vegetable oils which contain mainly long-chain triglycerides. Hence, MCT oil is valuable in the dietary treatment of fatty diarrhea and other digestive disorders in which the absorption of fat is impaired—malabsorption syndrome.

2. Medium-chain triglycerides, unlike other saturated fats, do *not* contribute to a rise in blood cholesterol. Hence, they are used in the treatment of some hyperlipoproteinemias.

At the present time, MCT oil is used almost exclusively for special dietary formulations in the United States, but it is also available in various nonmedical consumer products in Europe.

Nonnutritive Oils. Mineral oils are lubricating oils derived from petroleum and hydrocarbons—hydrogen and carbon-containing—but completely indigestible. In the past, mineral oil has been used to replace other digestible fats in low-calorie diets, and as a laxative. Its use is discouraged because it interferes with the absorption of the fat-soluble vitamins from the intestine.

FAT AND HEALTH PROBLEMS.

Fats either cause heart disease or cancer—or so it seems.[3] As for coronary heart disease, the blame on fats arises primarily from two factors: (1) atherosclerotic deposits in blood vessels are composed of cholesterol and other fatty substances; and (2) increases in certain fat components of the blood contribute to atherosclerosis. Fats are transported in the blood in the form of lipid-protein complexes—lipoproteins. It is the blood level of cholesterol, triglycerides, and certain lipoproteins which are considered risk factors in the development of heart disease.

Fats in the Blood. Fats are not very soluble in the blood. Therefore, they are transported in the blood as water-soluble lipoproteins, all of which contain phospholipids, triglycerides and cholesterol. These lipoproteins are classified by weight using the technique of ultracentrifugation, or by their migration on a paper subjected to an electric field— process called electrophoresis. Fig. F-7 illustrates the composition of these lipoproteins.

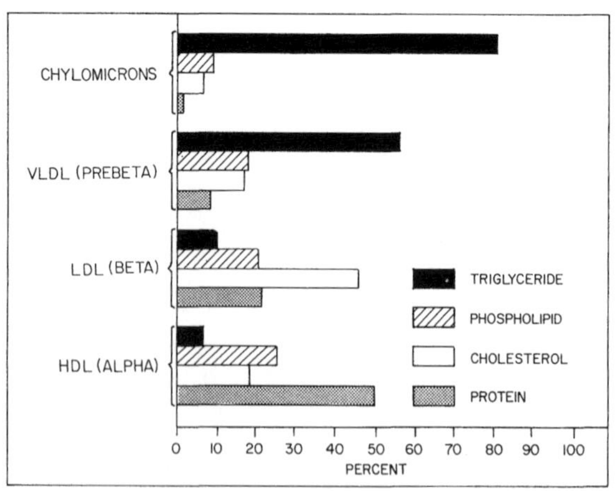

Fig. F-7. The composition of the blood lipoproteins chylomicrons, very low density (VLDL or prebeta), low density (LDL or beta) and high density (HDL or alpha).

---

[3]Kolata, G. B., "Data Sought on Low Cholesterol and Cancer," *Science*, Vol. 211, 1981, pp. 1410-1411.

Levels of lipoproteins in the blood are controlled by (1) diet, (2) a number of hormones, (3) age, (4) weight change, (5) emotions and stress, (6) exercise, (7) drugs, (8) illness, and (9) heredity. Excessive concentration of one or more of the lipoproteins is termed hyperlipoproteinemia. There are six of these: (1) hyperchylomicronemia; (2) hyperbetalipoproteinemia or hypercholesterolemia; (3) hyperlipidemia or hypercholesterolemia with hyperglyceridemia; (4) broad beta; (5) hyperbetalipoproteinemia; and (6) mixed hyperlipidemia. These hyperlipoproteinemias cause numerous health problems; among them, heart disease and accelerated atherosclerosis.

(Also see HYPERLIPOPROTEINEMIAS.)

The Lipid Hypothesis. The lipid hypothesis is based on the following observations:

1. Plaques are rich in cholesterol.
2. Populations with a high average serum cholesterol level tend to have a higher rate of coronary heart disease (CHD).
3. When fed a high cholesterol diet, some laboratory animals develop a type of arterial disease.

The basic assumption underlying this hypothesis is that high serum cholesterol levels cause infiltration of cholesterol into the inner lining of the artery (*intima*) where it causes plaque development. The cholesterol is thought to initiate migration of smooth muscle cells and accumulation of other substances, such as lipids and calcium, which eventually lead to protrusion of a plaque into the artery.

Fat and Health Summary. That the form of dietary fat is implicated in atherosclerosis is well documented by unbiased, controlled experiments stemming from the research initiated by Dr. Ancel Keys.[4] The chart that follows shows a comparison of common dietary fats on the following health-related bases: (1) cholesterol, (2) saturated fat, (3) monounsaturated fat, (4) polyunsaturated fat, and (5) other fats; with the fats ranked from least to most saturated fat. Pertinent definitions follow:

Cholesterol: A sterol found in animal tissue, synthesized in the body, and consumed in the diet. High cholesterol increases coronary heart disease (CHD) risk.

Saturated fat: Fatty acids with no double bonds, such as stearic and palmitic acids, which increase serum cholesterol along with increased CHD risk.

Monounsaturated fat: Fatty acids with one double bond, such as oleic acid, which has a favorable effect on CHD risk, because they lower the harmful low density lipoproteins (LDL) without affecting the protective high density lipoproteins (HDL).

Polyunsaturated fat: Fatty acids with two or more double bonds, such as the essential fatty acids linoleic and alpha-linolenic acids, which should be consumed in moderation, because they lower both the harmful LDL and the beneficial HDL cholesterol.

Other fats: This includes the saturated and polyunsaturated fatty acids that are outside the above definitions.

(Also see CANCER; CHOLESTEROL; and HEART DISEASE.)

FATS AND FATTY ACIDS IN SELECTED FOODS. Table F-9, Fats & Fatty Acids in Selected Foods, provides information on the total fat, animal fat, plant fat, saturated fat, polyunsaturated fat, oleic acid, linoleic acid, and cholesterol content of a variety of foods. As with any food composition data, the values contained in this table are not absolutes, but guidelines. They are meant to aid individuals who must or wish to alter or control their dietary fat intake.

---

[4]"HDL's: Possible Role," *Science News*, Vol. 118, 1980, p. 246.

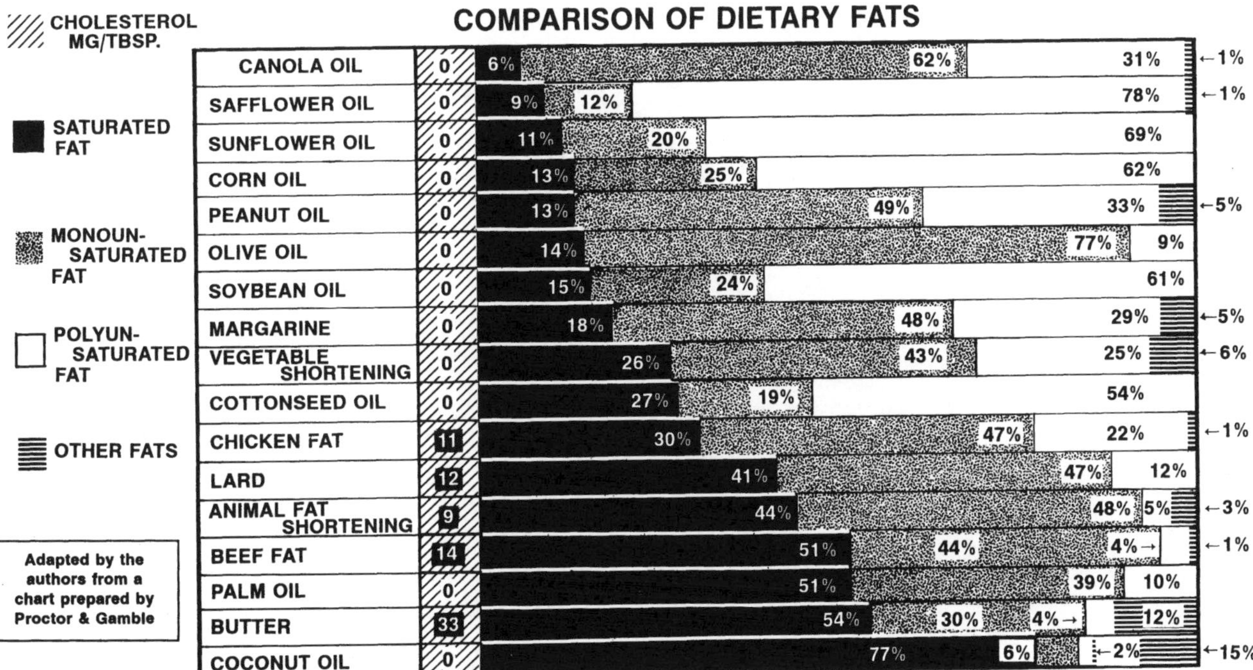

TABLE F-9 FATS & FATTY ACIDS IN SELECTED FOODS[1]

Food Name—100 Gram (3.5 oz) Portion	Water	Food Energy Calories	Total Fat	Animal Fat	Plant Fat	Saturated Fat	Polyun-saturated Fat	Fatty Acids		Chol-esterol	Choline
								Oleic Acid	Linoleic Acid		
	(g)	(kcal)	(g)	(g)	(g)	(g)	(g)	(g)	(g)	(mg)	(mg)
BAKERY PRODUCTS											
BREADS											
BREADCRUMBS, dry,grated	6.5	392	4.6	0	4.6	1.0	—	3.0	1.0	0	—
BREAD STUFFING, dry, crumbly, made w/ water	33.2	358	21.8	0	21.8	11.0	—	8.0	1.0	—	—
CRACKED WHEAT BREAD	34.9	263	2.2	0	2.2	0.5	—	0.9	0.7	—	—
CRACKED WHEAT TOAST	22.5	313	2.6	0	2.6	0.5	—	0.9	0.7	—	—
FRENCH BREAD	30.6	290	3.0	0	3.0	0.7	—	1.3	0.8	0	—
ITALIAN BREAD	31.8	276	0.8	0	0.8	Trace	—	0.2	0.4	0	—
RAISIN BREAD	35.3	262	2.8	0	2.8	0.6	—	1.3	0.7	0	—
RAISIN TOAST	22.0	316	3.4	0	3.4	0.6	—	1.3	0.7	0	—
VIENNA BREAD	30.6	290	3.0	0	3.0	0.7	—	1.3	0.8	0	—
WHITE BREAD, enriched	35.6	270	3.2	0	3.2	1.0	—	2.0	Trace	0	—
TOAST, enriched	25.1	314	3.7	0	3.7	0.8	—	1.7	1.0	0	—
WHOLE WHEAT BREAD	36.4	243	3.0	0	3.0	0.6	—	1.2	1.0	—	—
WHOLE WHEAT TOAST	24.3	289	3.6	0	3.6	0.6	—	1.2	1.0	—	—
CAKES											
BOSTON CREAM PIE	34.5	302	9.4	4.7	4.7	2.9	—	4.3	1.6	—	—
CHOCOLATE OR DEVIL'S FOOD CAKE OR CUPCAKE, w/chocolate icing	23.6	339	12.3	—	—	4.8	—	5.2	1.4	48	—
w/o icing	24.6	366	17.2	—	—	6.2	—	7.7	2.6	—	—
FRUITCAKE, dark, made w/enriched flour	18.1	379	15.3	0.2	15.1	3.2	—	8.1	3.0	45	—
GINGERBREAD, made w/water	37.0	276	6.8	0.5	6.3	1.7	—	3.3	1.6	1	—
LADYFINGER	19.2	360	7.8	—	—	2.5	—	3.0	0.6	356	—
POUNDCAKE	19.4	411	18.7	18.5	0.2	5.0	—	9.0	3.4	—	—
SPICE & HONEY CAKE, w/caramel icing	22.7	352	10.8	1.1	9.7	3.2	—	5.0	1.9	—	—
SPONGE CAKE	31.8	297	5.7	5.6	0.1	1.8	—	2.4	0.3	246	—
WHITE CAKE OR CUPCAKE, w/uncooked white icing	20.0	375	12.9	—	—	4.0	—	6.0	2.4	—	—
w/o icing	24.2	375	16.0	—	—	4.2	—	7.8	3.4	—	—
YELLOW CAKE OR CUPCAKE, w/chocolate icing	21.2	365	13.0	—	—	4.5	—	5.7	2.0	48	—
w/o icing	23.5	363	12.7	—	—	3.5	—	6.1	2.5	—	—
COOKIES											
ASSORTED packaged	2.6	480	20.2	1.0	19.2	—	—	—	—	—	—
BROWNIES, w/nuts, home recipe, w/enriched flour	9.8	485	31.3	16.7	14.6	7.0	—	16.5	5.5	83	—
BUTTER, Christmas, rolled	4.5	496	25.0	25.0	0	6.2	—	12.5	5.4	—	—
CHOCOLATE CHIP	2.7	471	21.0	0.6	20.4	6.4	—	7.9	5.2	—	—
COCONUT BAR	3.8	494	24.5	2.2	22.3	9.4	—	9.8	4.0	—	—
FIG BARS	13.6	358	5.6	0.3	5.3	1.5	—	2.7	1.1	—	—
GINGERSNAPS	3.1	420	8.9	—	—	2.3	—	4.3	1.9	—	—
MACAROONS	4.4	475	23.2	1.2	22.0	16.0	—	5.0	1.0	—	—
MARSHMALLOW, coconut-coated	9.8	409	13.2	—	—	8.0	—	5.0	1.0	—	—
MOLASSES	4.0	422	10.6	1.0	9.6	2.7	—	5.2	2.3	—	—
OATMEAL, w/raisin	2.8	451	15.4	1.1	14.3	4.0	—	7.3	3.5	—	—
PEANUT, sandwich type, w/peanut filling	2.3	473	19.1	—	—	4.0	—	10.0	1.0	—	—
SANDWICH TYPE, chocolate or vanilla	2.2	495	22.5	1.1	21.4	6.1	—	11.0	4.6	—	—
SHORTBREAD	3.0	498	23.1	1.2	21.9	5.8	—	11.4	5.1	—	—
SUGAR COOKIES, soft, home recipe w/ enriched flour	7.9	444	16.8	0.8	16.0	4.5	—	8.1	3.5	—	—
SUGAR WAFERS	1.4	485	19.4	1.0	18.4	4.9	—	9.6	4.1	—	—
VANILLA WAFERS	2.8	462	16.1	0.8	15.3	4.0	—	10.0	1.0	—	—

Footnote at end of table

(Continued)

TABLE F-9 *(Continued)*

Food Name—100 Gram (3.5 oz) Portion	Water	Food Energy Calories	Total Fat	Animal Fat	Plant Fat	Saturated Fat	Polyunsaturated Fat	Fatty Acids: Oleic Acid	Fatty Acids: Linoleic Acid	Cholesterol	Choline
	(g)	(kcal)	(g)	(g)	(g)	(g)	(g)	(g)	(g)	(mg)	(mg)
CRACKERS											
ANIMAL CRACKERS	3.0	429	9.4	—	—	2.4	—	4.6	1.9	—	—
BREAD STICKS	5.0	384	2.9	0	2.9	0.6	—	1.2	0.9	—	—
GRAHAM CRACKERS	6.4	384	9.4	0	9.4	2.3	—	4.5	2.3	—	—
GRAHAM CRACKERS, chocolate covered	1.9	475	23.5	0	23.5	7.0	—	14.4	1.5	—	—
RUSKS	4.8	419	8.7	0	8.7	2.6	—	3.9	1.6	—	—
SALTINE CRACKERS	4.3	433	12.0	0	12.0	2.9	—	5.8	2.9	—	—
DOUGHNUTS											
DOUGHNUT, cake type, plain	23.7	391	18.6	1.5	17.1	4.0	—	12.0	1.0	—	—
DOUGHNUT, raised type, plain	28.3	414	26.7	2.1	24.6	6.0	—	17.0	2.0	—	—
MUFFINS											
BLUEBERRY	39.0	281	9.3	2.8	6.5	3.0	—	5.0	1.0	53	—
BRAN	35.1	261	9.8	2.9	6.9	5.0	—	3.0	1.0	53	—
CORNMEAL	32.7	314	10.1	3.0	7.1	4.0	—	4.0	1.0	53	—
PLAIN, enriched	38.0	294	10.1	3.0	7.1	2.0	—	6.0	1.0	53	—
PANCAKES & WAFFLES											
BUCKWHEAT PANCAKE, made w/egg & milk	57.9	200	9.1	2.0	7.1	3.0	—	4.0	1.5	74	—
PLAIN OR BUTTERMILK PANCAKE, made w/egg & milk	50.6	225	7.3	2.0	5.3	3.0	—	4.0	1.0	74	—
WAFFLE, made w/egg & milk	41.7	275	10.6	6.4	4.2	4.0	—	5.0	1.0	60	—
PIES											
APPLE	47.6	256	11.1	0	11.1	2.9	—	5.4	2.4	0	—
BANANA CUSTARD	54.4	221	9.3	—	—	3.0	—	4.2	1.6	—	—
BLACKBERRY	51.0	243	11.0	0	11.0	2.7	—	5.5	2.5	—	—
BLUEBERRY	51.0	242	10.8	0	10.8	2.7	—	5.3	2.5	—	—
BUTTERSCOTCH	45.1	267	11.0	—	—	3.9	—	4.7	1.6	—	—
CHERRY	46.6	261	11.3	0	11.3	3.0	—	5.5	2.5	—	—
CHOCOLATE CHIFFON	33.0	328	15.3	—	—	5.2	—	7.0	2.2	—	—
CHOCOLATE MERINGUE	48.4	252	12.0	4.1	7.9	4.5	—	5.3	1.6	—	—
COCONUT CUSTARD	55.4	235	12.5	—	—	5.0	—	5.0	1.6	—	—
CUSTARD	58.1	218	11.1	5.6	5.6	3.8	—	4.9	1.6	105	—
LEMON CHIFFON	35.6	313	12.6	—	—	3.3	—	6.0	2.2	169	—
LEMON MERINGUE	47.4	255	10.2	—	—	3.0	—	4.8	1.7	93	—
MINCE	43.0	271	11.5	4.0	7.5	3.0	—	5.7	2.5	—	—
PEACH	47.5	255	10.7	0	10.7	2.6	—	5.3	2.4	0	—
PECAN	19.5	418	22.9	8.9	14.0	3.2	—	13.2	4.6	—	—
PINEAPPLE CHIFFON	41.1	288	12.1	—	—	3.2	—	5.8	2.2	—	—
PUMPKIN	59.2	211	11.2	4.5	6.7	4.0	—	4.9	1.6	61	—
RAISIN	42.5	270	10.7	0	10.7	2.6	—	5.4	2.4	—	—
RHUBARB	47.4	253	10.7	0	10.7	2.6	—	5.3	2.4	—	—
STRAWBERRY	58.4	198	7.9	0	7.9	1.9	—	4.0	1.8	—	—
SWEET POTATO	59.3	213	11.3	—	—	4.0	—	4.9	1.7	—	—
PIE CRUST OR PLAIN PASTRY PIE SHELL, made w/enriched flour	14.9	500	33.4	0	33.4	8.3	—	16.6	7.4	—	—
QUICK BREADS											
BAKING POWDER BISCUIT	27.4	369	17.0	1.4	15.6	4.0	—	11.0	1.0	0	—
CORNBREAD OR JOHNNYCAKE	37.9	267	5.2	—	—	2.0	—	3.0	1.0	70	—
CORN FRITTER	29.1	377	21.5	0.9	20.6	5.0	—	13.0	2.0	—	—
CORN STICK	—	314	10.1	—	—	3.0	—	4.6	1.9	53	—
POPOVER, baked from home recipe	54.9	224	9.2	4.4	4.8	3.0	—	4.0	1.0	147	—
ROLLS											
BROWN & SERVE ROLL, enriched	26.9	328	7.8	0	7.8	1.0	—	3.0	1.0	0	—
CINNAMON BUN, w/raisins	32.0	275	2.9	0.4	2.5	1.0	—	3.0	1.0	—	—
DANISH PASTRY, plain, w/o fruit or nuts	22.0	422	23.5	—	—	7.0	—	14.0	2.0	—	—
DINNER TYPE (PAN) ROLL, enriched	31.4	298	5.6	0	5.6	1.0	—	3.0	1.0	0	—
FRANKFURTER OR HAMBURGER BUN	31.4	298	5.6	—	—	1.0	—	3.0	1.0	0	—
HARD ROLL, enriched	25.4	312	3.2	0	3.2	1.0	—	3.0	1.0	0	—
HOAGIE OR SUBMARINE ROLL	30.6	290	3.0	0	3.0	0.7	—	1.3	0.8	0	—
SWEET ROLL	31.5	316	9.1	7.7	1.4	2.0	—	5.0	1.0	—	—
WHOLE WHEAT ROLL	32.0	257	2.8	0	2.8	1.0	—	3.0	1.0	0	—

Footnote at end of table

(Continued)

TABLE F-9 *(Continued)*

Food Name—100 Gram (3.5 oz) Portion	Water	Food Energy Calories	Total Fat	Animal Fat	Plant Fat	Saturated Fat	Polyun-saturated Fat	Fatty Acids		Chol-esterol	Choline
								Oleic Acid	Linoleic Acid		
	(g)	(kcal)	(g)	(g)	(g)	(g)	(g)	(g)	(g)	(mg)	(mg)
CEREALS & FLOURS											
CEREALS											
BARLEY, pearled, light, uncooked	11.1	349	1.0	0	1.0	Trace	—	1.0	1.0	0	—
BULGUR (parboiled wheat), canned, unseasoned	56.0	168	0.7	0	0.7	0.1	0.3	0.1	0.3	0	—
CORN GRITS, degermed, enriched, cooked	87.1	51	0.1	0	0.1	Trace	0.1	Trace	0.1	0	—
CORN MEAL, white or yellow, degermed, enriched, cooked	87.7	50	0.2	0	0.2	Trace	0.1	Trace	0.1	0	10.0
MACARONI, w/cheese, enriched, home recipe	58.2	215	11.1	10.8	0.3	5.0	—	5.0	1.0	21	—
NOODLES, egg, cooked, any shape	70.4	125	1.5	1.3	0.2	0.6	—	0.6	Trace	31	—
OATMEAL OR ROLLED OATS, regular & instant, cooked	86.5	55	1.0	0	1.0	0.2	0.4	0.4	0.4	0	—
OATMEAL, dry	8.3	390	7.4	0	7.4	1.4	3.1	2.6	2.9	0	—
RICE, BROWN, long grain, cooked	70.3	119	0.6	0	0.6	0.2	0.3	0.2	0.3	0	112.0
RICE, WHITE, enriched, long grain, cooked	72.6	109	0.1	0	0.1	0.1	0.1	0.1	0.1	0	59.0
SPAGHETTI, enriched, w/meatballs & tomato sauce, canned	78.0	103	4.1	—	—	0.9	—	1.3	1.6	9	—
SPAGHETTI, enriched, w/meatballs & tomato sauce, home recipe	70.0	134	4.7	—	—	1.0	—	3.0	Trace	30	—
SPAGHETTI, enriched, in tomato sauce w/ cheese, canned	80.1	76	0.6	0.2	0.4	0.4	—	0.4	0.4	—	—
SPAGHETTI, enriched, in tomato sauce w/ cheese, home recipe	77.0	104	3.5	—	—	1.0	—	2.0	Trace	—	—
WHEAT GERM, dry	11.5	363	10.9	0	10.9	1.9	6.6	1.5	5.9	0	—
BREAKFAST CEREALS											
SHREDDED WHEAT, biscuit or spoon size	6.6	354	2.0	0	2.0	0.4	1.3	0.4	1.2	0	—
WHOLE MEAL WHEAT CEREAL, cooked	87.7	45	0.3	0	0.3	—	—	—	—	0	94.0
FLOURS											
CORN FLOUR	12.0	368	2.6	0	2.6	0.3	1.4	0.6	1.3	0	—
WHITE FLOUR, all purpose, enriched	12.0	364	1.0	0	1.0	0.2	0.6	0.1	0.6	0	52.0
DESSERTS & SWEETS											
APPLE BROWN BETTY, made w/enriched bread	64.5	151	3.5	—	—	1.5	—	1.2	0.3	—	—
BREAD PUDDING, w/raisins, made w/enriched bread	58.6	187	6.1	—	—	2.9	—	2.2	0.4	64	—
CANDY											
ALMONDS, CHOCOLATE COVERED	2.0	569	43.7	Trace	43.7	7.4	—	29.0	5.5	—	—
BUTTERSCOTCH CANDY	1.5	397	3.4	—	—	2.0	—	1.0	Trace	—	—
CANDY BAR, CHOCOLATE, plain	0.9	520	32.3	Trace	32.3	18.2	—	11.8	0.7	—	—
CHOCOLATE, w/almonds	1.5	532	35.6	Trace	35.6	16.1	—	15.7	2.1	—	—
CHOCOLATE, w/peanuts	1.0	543	38.1	Trace	38.1	16.1	—	15.0	5.4	—	—
CANDY COATED CHOCOLATE DISKS	1.2	466	19.7	Trace	19.7	11.0	—	7.1	0.4	—	—
CANDY CORN	7.6	364	2.0	0	2.0	0.5	—	1.0	0.5	—	—
CARAMEL, plain or chocolate	7.6	399	10.2	Trace	10.2	7.0	—	4.0	1.0	—	—
CHOCOLATE BITS	1.1	507	35.7	Trace	35.7	20.0	—	13.0	1.0	—	—
CHOCOLATE CLUSTERS, w/peanuts & caramel	7.0	459	23.1	Trace	23.1	6.0	—	12.5	3.6	—	—
CHOCOLATE-COATED COCONUT CENTER CANDY	6.6	438	17.6	Trace	17.6	10.0	—	6.0	Trace	—	—
CHOCOLATE-COATED CREAMS	7.5	435	17.1	Trace	17.1	5.0	—	10.0	1.8	—	—
CHOCOLATE-COATED HONEYCOMB, w/ peanut butter	1.7	463	19.5	Trace	19.5	6.0	—	11.0	2.0	—	—
CHOCOLATE-COATED NOUGAT & CARAMEL CANDY	7.7	416	13.9	Trace	13.9	4.3	—	8.6	1.1	—	—
CHOCOLATE-COATED PEANUTS	1.0	561	41.3	Trace	41.3	11.0	—	22.0	7.0	—	—
CHOCOLATE-COATED RAISINS	4.8	425	17.1	Trace	17.1	10.0	—	6.0	0.4	—	—
CHOCOLATE KISS	0.9	520	32.3	Trace	32.3	18.2	—	11.8	0.7	—	—
FONDANT PATTIE, mint	7.6	364	2.0	0	2.0	0.5	—	1.0	0.5	—	—
FUDGE CHOCOLATE, plain	8.2	400	12.2	Trace	12.2	4.3	—	5.7	1.8	—	—

Footnote at end of table

(Continued)

TABLE F-9 *(Continued)*

Food Name—100 Gram (3.5 oz) Portion	Water	Food Energy Calories	Total Fat	Animal Fat	Plant Fat	Saturated Fat	Polyunsaturated Fat	Fatty Acids		Cholesterol	Choline
								Oleic Acid	Linoleic Acid		
	(g)	(kcal)	(g)	(g)	(g)	(g)	(g)	(g)	(g)	(mg)	(mg)
CANDY *(Continued)*											
FUDGE CHOCOLATE, w/nuts	7.8	426	17.4	Trace	17.4	4.3	—	6.1	5.7	—	—
FUDGE VANILLA, plain	10.0	398	11.1	Trace	11.1	2.9	—	5.7	2.5	—	—
FUDGE, w/nuts	9.4	424	16.4	Trace	16.4	2.9	—	6.1	6.1	—	—
PEANUT BAR	1.5	515	32.2	0	32.2	7.0	—	14.0	9.0	—	—
PEANUT BRITTLE	2.0	421	10.4	0	10.4	2.1	—	4.6	3.2	—	—
PEPPERMINT PATTIE, chocolate-covered	5.8	410	10.5	Trace	10.5	3.1	—	6.9	0.6	—	—
TOOTSIE ROLL	5.6	396	8.2	0	8.2	2.5	—	3.9	1.8	—	—
CHOCOLATE PUDDING MIX, regular, dry form	1.7	361	2.1	—	—	1.2	—	0.8	0.6	—	—
CHOCOLATE PUDDING, instant, dry form	0.7	357	1.6	—	—	0.9	—	0.6	0.1	—	—
CHOCOLATE PUDDING, instant	68.7	125	2.5	—	—	1.4	—	0.8	0.1	—	—
CREAM PUFF, w/custard filling	58.3	233	13.9	9.3	4.6	4.3	—	6.4	2.2	144	—
CUSTARD, baked	77.2	115	5.5	5.5	0	2.5	0.5	1.5	0.3	105	—
ECLAIR, w/custard filling & chocolate icing	56.2	239	13.6	—	—	4.4	—	6.2	2.1	—	—
ICINGS											
CARAMEL	14.1	360	6.7	—	—	3.7	—	2.2	0.2	—	—
CHOCOLATE	14.3	376	13.9	—	—	7.7	—	5.0	0.3	—	—
COCONUT	15.0	364	7.7	—	—	6.6	—	0.5	Trace	—	—
WHITE, uncooked	11.1	376	6.6	—	—	3.6	—	2.2	Trace	—	—
RENNIN DESSERT, made w/milk, caramel, fruit, or vanilla flavor	79.7	95	3.6	—	—	2.0	—	1.2	Trace	—	—
made w/milk, chocolate flavor	77.9	102	3.8	—	—	2.1	—	1.3	0.1	—	—
RICE PUDDING, w/raisins	65.8	146	3.1	—	—	1.7	—	1.0	0.1	11	—
SYRUPS											
CHOCOLATE, fudge type	25.4	330	13.7	0	13.7	7.6	—	4.9	0.3	—	—
CHOCOLATE, thin type	31.6	245	2.0	0	2.0	1.1	—	0.7	Trace	—	—
TAPIOCA CREAM PUDDING	71.8	134	5.1	—	—	2.3	0.5	1.4	0.3	97	—
TOPPINGS, DESSERT											
non-dairy, powdered	1.5	577	39.9	0	39.9	36.7	0.5	0.6	0.5	0	—
powdered, made w/whole milk	66.7	189	12.4	—	—	10.7	0.2	0.8	0.2	10	—
pressurized	60.4	264	22.3	—	—	18.9	0.2	1.9	0.2	0	—
frozen, semisolid	50.2	318	25.3	0	25.3	21.8	0.5	1.4	0.3	0	—
EGGS & SUBSTITUTES											
EGG, WHOLE, fresh & frozen, raw	74.6	158	11.1	11.1	0	3.4	1.5	4.1	1.2	548	—
WHITE, fresh & frozen, raw	88.1	49	Trace	Trace	0	0	0	0	0	0	—
YOLK, fresh, raw	48.8	369	32.9	32.9	0	9.9	4.3	12.1	3.7	1602	—
WHOLE, FRIED in butter	71.9	180	13.9	13.9	0	5.2	1.5	4.7	1.3	534	—
WHOLE, HARD-COOKED	74.6	158	11.1	11.1	0	3.4	1.5	4.1	1.2	548	—
OMELET, made w/one large egg, butter & milk	76.3	148	11.1	11.1	0	4.4	1.1	3.7	0.9	388	—
WHOLE, POACHED	74.3	157	11.1	11.1	0	3.3	1.4	4.1	1.2	545	—
WHOLE, SCRAMBLED w/butter & milk	76.3	148	11.1	11.1	0	4.4	1.1	3.7	0.9	388	—
WHOLE, DRIED	4.1	594	41.8	41.8	0	12.6	5.4	15.3	4.6	1918	—
WHOLE, DRIED, stabilized	1.9	615	44.0	44.0	0	13.2	5.7	16.1	4.9	2017	—
EGGNOG	74.4	135	7.5	7.5	0	4.4	0.3	2.0	0.2	59	—
DUCK EGG, whole, fresh, raw	70.8	185	13.8	13.8	0	3.7	1.2	6.1	0.6	884	—
GOOSE EGG, whole, fresh, raw	70.4	185	13.3	13.3	0	3.6	1.7	5.4	0.7	—	—
QUAIL EGG, whole, fresh, raw	74.3	158	11.1	11.1	0	3.6	1.3	3.9	0.9	844	—
TURKEY EGG, whole, fresh, raw	72.5	171	11.9	11.9	0	3.6	1.7	3.9	1.2	933	—
EGG SUBSTITUTE, frozen, made w/egg white, corn oil, nonfat dried milk	73.1	160	11.1	0	11.1	1.9	6.2	2.4	6.2	2	—
liquid, made w/egg white, soy oil, protein	82.8	84	3.3	0	3.3	0.7	1.6	0.9	1.4	1	—
powder	3.9	444	13.0	—	—	3.8	1.7	4.8	1.4	572	—

Footnote at end of table

(Continued)

TABLE F-9 *(Continued)*

Food Name—100 Gram (3.5 oz) Portion	Water	Food Energy Calories	Total Fat	Animal Fat	Plant Fat	Saturated Fat	Polyunsaturated Fat	Fatty Acids: Oleic Acid	Fatty Acids: Linoleic Acid	Cholesterol	Choline
	(g)	(kcal)	(g)	(g)	(g)	(g)	(g)	(g)	(g)	(mg)	(mg)
ENTREES											
BEEF											
CHILI CON CARNE W/BEANS, canned	72.4	133	6.1	2.5	3.6	3.0	—	3.0	Trace	—	—
CHOP SUEY W/MEAT, home recipe	75.4	120	6.8	—	—	3.0	—	2.0	Trace	—	—
CHOP SUEY, canned	85.5	62	3.2	—	—	1.0	—	1.0	Trace	—	—
DRIED, CHIPPED, creamed	72.0	154	10.3	9.1	1.2	6.0	—	3.0	Trace	27	—
GREEN PEPPER STUFFED W/BEEF & crumbs	63.1	170	5.5	—	—	—	—	—	—	—	—
MEAT LOAF, cooked	64.1	200	13.2	12.8	0.4	—	—	—	—	—	—
POTPIE, home prepared, baked	55.1	246	14.5	3.2	11.3	4.0	—	9.0	1.0	21	—
STEW W/VEGETABLES, home recipe, w/o salt	82.4	89	4.3	2.4	1.9	2.0	—	2.0	0.1	26	—
STEW W/VEGETABLES, canned	82.5	79	3.1	—	—	—	—	—	—	14	—
CHICKEN											
A LA KING, creamed	68.2	191	14.0	4.9	9.1	5.0	—	7.0	1.0	76	—
CHOW MEIN, w/o noodles, home recipe	78.0	102	4.0	3.0	1.0	1.0	—	1.0	1.0	31	—
CHOW MEIN, w/o noodles, canned	88.8	38	0.1	0.1	0	—	—	—	—	3	—
FRICASSEE	71.3	161	9.3	6.5	2.8	3.0	—	4.0	2.0	40	—
POTPIE, home recipe, baked	56.6	235	13.5	4.5	9.0	5.0	—	7.0	1.0	31	—
HAM CROQUETTE, pan fried	54.0	251	15.1	—	—	6.0	—	7.0	1.0	—	—
LOBSTER NEWBURG	64.0	194	10.6	5.3	5.3	—	—	—	—	182	—
PIZZA W/CHEESE & SAUSAGE, home recipe, w/enriched flour	43.0	282	13.3	—	—	3.4	—	4.7	1.2	—	—
PIZZA W/CHEESE, frozen-type baked, made w/ enriched flour	45.3	245	7.1	—	—	2.0	—	3.0	Trace	—	—
TURKEY POTPIE, cooked	56.2	237	13.5	4.5	9.0	4.0	—	7.0	1.0	31	—
FATS & OILS											
LARD	0	902	100.0	100.0	0	39.6	15.2	40.9	10.0	95	5.0
MARGARINE, regular, soft, whipped, or unsalted	15.5	720	81.0	0	81.0	14.8	—	41.4	22.2	0	5.0
OIL											
CORN	0	884	100.0	0	100.0	12.7	58.4	24.6	57.4	0	5.0
COTTONSEED	0	884	100.0	0	100.0	26.1	51.5	18.1	50.3	0	5.0
OLIVE	0	884	100.0	0	100.0	14.2	9.9	71.5	8.2	0	5.0
PEANUT	0	884	100.0	0	100.0	17.4	31.0	45.6	31.0	0	5.0
SAFFLOWER	0	884	100.0	0	100.0	9.4	74.3	11.9	73.3	0	5.0
SESAME	0	884	100.0	0	100.0	15.2	41.3	39.1	40.0	0	5.0
SOYBEAN	0	884	100.0	0	100.0	14.3	50.3	23.5	50.3	0	5.0
SHORTENING, vegetable, hydrogenated	0	884	100.0	0	100.0	23.0	—	65.0	7.0	0	—
FISH & SEAFOODS											
COD, cooked	64.6	129	0.7	0.7	0	0.1	0.3	0.1	—	50	—
CRAB, all varieties, steamed whole	78.5	93	1.9	1.9	0	0.3	0.6	0.2	Trace	100	—
CRABMEAT, canned	77.2	101	2.5	2.5	0	—	—	—	—	101	—
CRAB LEG, steamed in shell	78.5	93	1.9	1.9	0	0.3	0.6	0.2	Trace	100	—
CRAB, IMPERIAL	71.9	147	7.6	—	—	—	—	—	—	140	—
FLOUNDER, baked	58.1	140	1.3	1.3	0	0.3	0.6	0.1	Trace	50	—
HADDOCK, cooked	66.3	90	0.7	0.7	0	0.1	0.2	0.1	Trace	60	—
HALIBUT, cooked	66.6	130	2.4	2.4	0	0.3	0.6	0.2	Trace	50	—
HERRING											
plain, canned	62.9	208	13.6	13.6	0	2.6	—	—	2.6	97	—
pickled	59.4	223	15.1	15.1	0	2.9	—	—	2.9	85	—
smoked, kippered, canned	61.0	211	12.9	12.9	0	2.4	—	—	2.4	85	—
canned w/tomato sauce	66.7	176	10.5	9.5	1.0	2.0	—	—	2.0	97	—
LOBSTER, meat, cooked	76.8	95	1.5	1.5	0	—	—	—	—	85	—
OCEAN PERCH, Atlantic, cooked, piece	59.0	92	1.3	1.3	0	0.4	0.9	0.4	Trace	—	—
OYSTER, eastern, raw	84.6	66	1.8	1.8	0	0.5	0.7	0.1	Trace	50	—
OYSTER, canned	82.2	76	2.2	2.2	0	—	—	—	—	45	—

Footnote at end of table

(Continued)

TABLE F-9 *(Continued)*

Food Name—100 Gram (3.5 oz) Portion	Water	Food Energy Calories	Total Fat	Animal Fat	Plant Fat	Saturated Fat	Polyun-saturated Fat	Fatty Acids Oleic Acid	Fatty Acids Linoleic Acid	Chol-esterol	Choline
	(g)	(kcal)	(g)	(g)	(g)	(g)	(g)	(g)	(g)	(mg)	(mg)
SALMON											
broiled or baked w/butter or margarine	63.4	182	7.4	7.4	0	1.3	2.9	1.3	0.1	47	—
COHO, SILVER, canned	69.3	153	7.1	7.1	0	2.2	2.8	2.3	0.1	—	—
HUMPBACK PINK, canned	70.8	141	5.9	5.9	0	1.5	—	1.4	0.1	—	—
SOCKEYE, RED, canned	67.2	171	9.3	9.3	0	0.8	4.3	0.8	0.2	35	—
CHUM, canned, low sodium	70.8	139	5.2	5.2	0	—	—	—	—	35	—
SARDINE											
ATLANTIC, canned, w/oil	61.8	203	11.1	—	—	—	—	—	—	140	—
PACIFIC, canned, w/mustard sauce	64.1	196	12.0	—	—	—	—	—	—	120	—
PACIFIC, canned, w/tomato sauce	64.3	197	12.2	—	—	—	—	—	—	120	—
SCALLOP, bay & sea, steamed	73.1	112	1.4	1.4	0	—	—	—	—	53	—
SHRIMP, canned	70.4	116	1.1	1.1	0	—	—	—	—	150	—
TUNA, canned, w/oil	60.6	197	8.2	8.2	0	3.0	2.0	2.0	2.0	65	—
WHITEFISH, lake, cooked	63.2	125	7.3	7.3	0	0.9	2.0	1.3	0.3	—	—
FLAVORINGS & SEASONINGS											
ALLSPICE, ground	8.5	263	8.7	0	8.7	2.6	2.4	0.7	2.3	0	—
ANISE SEED	9.5	337	15.9	0	15.9	—	3.2	9.8	3.2	0	—
BARBECUE SAUCE	80.9	91	6.9	0	6.9	0.7	—	1.9	3.7	0	—
BAY LEAF, crumbled	5.4	313	8.4	0	8.4	2.3	2.3	1.5	1.2	0	—
CARAWAY SEED	9.9	333	14.6	0	14.6	0.6	3.3	7.0	3.1	0	—
CARDAMOM, ground	8.3	311	6.7	0	6.7	0.7	0.4	0.9	0.3	0	—
CELERY SEED	6.0	392	25.3	0	25.3	2.2	3.7	15.5	3.5	0	—
CHOCOLATE, bitter or baking	2.3	505	53.0	0	53.0	29.7	—	19.6	1.1	0	—
CINNAMON, ground	9.5	261	3.2	0	3.2	0.7	0.5	0.5	0.5	0	—
CLOVES, ground	6.9	323	20.1	0	20.1	4.4	—	—	—	0	—
CORIANDER SEED	8.9	298	17.8	0	17.8	1.0	1.8	13.5	1.8	0	—
CORNSTARCH	12.0	362	Trace	0	Trace	0.1	0.3	0.1	0.3	0	—
DILL SEED	7.7	305	14.5	0	14.5	0.7	1.0	9.4	1.0	0	—
FENNEL SEED	8.8	345	14.9	0	14.9	0.5	1.7	9.9	1.7	0	—
GINGER, ground	9.4	347	5.9	0	5.9	1.9	1.3	1.0	1.0	0	—
MACE, ground	8.2	475	32.4	0	32.4	9.5	4.4	10.6	4.3	0	—
MUSTARD SEED, yellow	6.9	469	28.8	0	28.8	1.5	5.4	5.9	2.6	0	—
NUTMEG, ground	6.2	525	36.3	0	36.3	25.9	0.4	159.0	0.4	0	—
OREGANO, ground	7.2	306	10.3	0	10.3	2.7	5.2	0.5	1.1	0	—
PAPRIKA	9.5	289	12.9	0	12.9	2.1	8.3	1.1	7.4	0	—
PEPPER, BLACK	10.5	255	3.3	0	3.3	1.3	1.5	1.4	1.3	0	—
PEPPER, RED OR CAYENNE	8.1	318	17.3	0	17.3	3.3	8.4	2.5	7.7	0	—
POPPY SEED	6.8	533	44.7	0	44.7	4.9	30.8	6.2	30.5	0	—
SAGE, ground	8.0	315	12.7	0	12.7	7.0	1.8	1.8	0.5	0	—
TARTAR SAUCE, regular	34.4	531	57.8	0	57.8	7.0	—	7.0	29.0	51	—
THYME, ground	7.8	276	7.4	0	7.4	2.7	1.2	0.5	0.5	0	—
VINEGAR, CIDER	93.8	14	0	0	0	0	0	0	0	0	—
VINEGAR, DISTILLED	95.0	12	0	0	0	0	0	0	0	0	—
FRUITS											
AVOCADO, raw	74.0	167	16.4	0	16.4	3.0	—	7.0	2.0	—	—
GRAVIES & SAUCES											
WHITE SAUCE, medium	73.3	162	12.5	3.6	8.9	7.0	—	4.0	Trace	13	—
JUICES											
ORANGE JUICE, fresh, all varieties	88.3	45	0.2	0	0.2	—	—	—	—	—	12.0
MEATS											
BEEF											
BRISKET, cooked	36.0	474	42.8	42.8	0	14.6	4.2	14.1	0.8	94	—
CHUCK BLADE STEAK, braised	40.3	427	36.7	36.7	0	15.3	4.5	14.8	0.9	—	—
GROUND, cooked	49.4	327	23.9	23.9	0	8.0	2.4	7.7	0.5	94	—
POT ROAST, braised	53.0	289	19.2	19.2	0	8.0	2.4	7.7	0.5	94	—
STEW MEAT, braised	49.4	327	23.9	23.9	0	8.0	2.4	7.7	0.5	94	—

Footnote at end of table

(Continued)

TABLE F-9 *(Continued)*

Food Name—100 Gram (3.5 oz) Portion	Water	Food Energy Calories	Total Fat	Animal Fat	Plant Fat	Saturated Fat	Polyun-saturated Fat	Fatty Acids		Chol-esterol	Choline
								Oleic Acid	Linoleic Acid		
	(g)	(kcal)	(g)	(g)	(g)	(g)	(g)	(g)	(g)	(mg)	(mg)
BEEF *(Continued)*											
CUBED STEAK, cooked	54.7	261	15.4	15.4	0	6.3	1.8	5.9	0.4	94	—
DRIED, CHIPPED, uncooked	47.7	203	6.3	6.3	0	3.0	—	3.0	Trace	—	—
FLANK STEAK, braised	61.4	196	7.3	7.3	0	3.1	0.9	2.7	0.3	91	—
HAMBURGER (GROUND BEEF), cooked	54.2	286	20.3	20.3	0	9.5	2.8	9.1	0.6	94	—
HEART, cooked	61.3	188	5.7	5.7	0	1.8	1.3	1.3	0.7	274	—
KIDNEY, braised	53.0	252	12.0	12.0	0	—	—	—	—	804	—
LIVER, fried in margarine	—	229	10.6	4.8	5.8	2.9	2.0	4.1	1.1	438	—
MINUTE STEAK, cooked	54.7	261	15.4	15.4	0	6.3	1.8	5.9	0.4	94	—
PLATE SHORT RIBS, cooked	36.0	474	42.8	42.8	0	17.8	5.3	17.5	0.9	94	—
PORTERHOUSE STEAK, broiled	37.2	465	42.2	42.2	0	17.6	5.1	17.3	0.9	94	—
RIB ROAST, roasted	40.0	440	39.4	39.4	0	16.5	4.7	16.0	0.9	94	—
RIB STEAK, broiled	40.0	440	39.4	39.4	0	16.5	4.7	16.0	0.9	94	—
ROAST BEEF, canned	60.0	224	13.0	13.0	0	6.3	1.8	5.9	0.4	91	—
ROUND, GROUND, cooked	54.7	261	15.4	15.4	0	6.3	1.8	5.9	0.4	94	—
RUMP ROAST, roasted	48.1	347	27.3	27.3	0	11.4	3.4	11.0	0.7	94	—
STEAK, boneless, cooked	54.7	261	15.4	15.4	0	6.3	1.8	5.9	0.4	94	—
TIP, ROAST (SIRLOIN), roasted	35.1	487	44.9	44.9	0	13.3	3.9	13.1	0.7	94	—
SIRLOIN, GROUND, cooked	43.9	387	32.0	32.0	0	13.3	3.9	13.1	0.7	94	—
STEAK, broiled	43.9	387	32.0	32.0	0	13.3	3.9	13.1	0.7	94	—
SWEETBREADS, braised	49.6	320	23.2	23.2	0	—	—	—	—	466	—
T-BONE STEAK, broiled	36.4	473	43.2	43.2	0	18.0	5.3	17.7	0.9	94	—
TENDERLOIN STEAK, broiled	57.9	223	10.3	10.3	0	4.4	1.2	3.8	0.4	91	—
TOP LOIN STEAK (CLUB), broiled	37.9	454	40.6	40.6	0	17.6	5.1	17.3	0.9	94	—
TOP LOIN (STRIP) STEAK, broiled	36.4	473	43.2	43.2	0	18.0	5.3	17.7	0.9	94	—
CHICKEN											
ONE-HALF WHOLE, w/skin, roasted	57.0	248	14.7	14.7	0	4.2	4.5	4.3	3.1	87	—
DARK MEAT, w/o skin, roasted	64.4	176	6.3	6.3	0	2.7	2.9	2.7	1.9	91	—
LIGHT MEAT, w/o skin, roasted	63.8	166	3.4	3.4	0	1.0	1.0	0.8	0.6	79	—
FRYER, BACK, fried in vegetable fat	40.5	347	21.2	13.9	7.3	6.5	—	8.8	4.5	91.	—
BREAST, half, fried in vegetable fat	58.4	203	6.4	3.8	2.6	1.9	—	2.8	1.3	80	—
DRUMSTICK, fried in vegetable fat	55.0	235	10.2	5.7	4.5	3.0	—	4.4	2.1	91	—
SKIN, fried in vegetable fat	32.5	419	28.9	—	—	9.0	—	12.0	5.0	—	—
THIGH, fried in vegetable fat	55.8	237	11.4	8.6	2.8	4.0	—	4.0	2.0	91	—
WING, fried in vegetable fat	52.6	268	14.8	5.4	9.4	4.6	—	6.2	3.0	91	—
GIZZARD, simmered	68.0	148	3.3	3.3	0	—	—	—	—	195	—
HEART, cooked	66.7	173	7.2	7.2	0	—	—	—	—	231	—
LIVER, simmered	65.0	165	4.4	4.4	0	—	—	—	—	746	—
CANNED	65.2	198	11.7	11.7	0	4.0	—	4.0	2.0	85	—
CORNISH HEN, roasted, whole	71.0	136	3.8	3.8	0	1.0	1.0	0.8	0.6	—	—
LAMB											
GROUND, cooked	54.0	279	18.9	18.9	0	9.5	1.4	7.9	0.8	98	—
LEG ROAST, roasted	62.2	186	7.0	7.0	0	4.0	0.7	3.4	0.4	100	—
LOIN CHOP, broiled	47.0	359	29.4	29.4	0	15.1	2.3	12.4	1.2	98	—
RIB CHOP, broiled	42.9	407	35.6	35.6	0	16.8	2.5	13.8	1.4	98	—
SHOULDER CHOP OR ROAST, broiled	49.6	338	27.2	27.2	0	12.6	1.9	10.3	1.0	98	—
LUNCHEON MEATS											
BOLOGNA	56.2	304	27.5	27.5	0	11.9	15.3	12.4	0.7	62	60.0
BRAUNSCHWEIGER (smoked liver sausage)	52.6	319	27.4	27.4	0	10.0	—	12.0	2.0	62	—
CORNED BEEF, cooked	43.9	372	30.4	30.4	0	15.0	—	13.0	1.0	94	—
canned	59.3	216	12.0	12.0	0	6.0	—	5.0	Trace	94	—
hash w/potato, canned	67.4	181	11.3	4.0	7.3	5.0	—	5.0	Trace	—	—
HAM, boiled	59.1	234	17.0	17.0	0	5.9	2.6	7.3	1.6	89	—
deviled, canned	50.5	351	32.3	32.3	0	12.0	—	14.0	3.0	—	—
minced	61.7	228	16.9	16.9	0	6.0	—	7.0	2.0	89	—

Footnote at end of table

(Continued)

TABLE F-9 *(Continued)*

Food Name—100 Gram (3.5 oz) Portion	Water	Food Energy Calories	Total Fat	Animal Fat	Plant Fat	Saturated Fat	Polyun-saturated Fat	Fatty Acids Oleic Acid	Fatty Acids Linoleic Acid	Chol-esterol	Choline
	(g)	(kcal)	(g)	(g)	(g)	(g)	(g)	(g)	(g)	(mg)	(mg)
LUNCHEON MEATS *(Continued)*											
HEAD CHEESE	58.8	268	22.0	22.0	0	4.5	2.4	6.7	1.4	—	—
LIVERWURST	53.9	307	25.6	25.6	0	—	—	—	—	105	—
PATE DE FOIE GRAS, canned	37.0	462	43.8	43.8	0	—	—	—	—	—	—
PORK/HAM LOAF TYPE LUNCHEON MEAT	54.9	294	24.9	24.9	0	10.7	4.7	13.2	3.1	89	—
SALAMI, BEEF, cooked	51.0	311	25.6	25.6	0	8.7	10.6	8.4	0.6	62	—
SALAMI, BEEF, hard	29.8	450	38.1	38.1	0	—	—	—	—	62	—
SPICED LUNCHEON MEATS, pork/ham type	54.9	294	24.9	24.9	0	10.7	4.7	13.2	3.1	89	—
VIENNA SAUSAGE	63.0	240	19.8	19.8	0	9.4	3.3	11.1	1.6	—	—
PORK											
BACON, cooked	8.1	611	52.0	52.0	0	17.0	—	25.0	5.0	—	—
CANADIAN BACON, broiled or fried, drained	49.9	277	17.5	17.5	0	5.9	2.7	7.3	1.4	88	—
CURED HAM, roasted	53.6	289	22.1	22.1	0	7.8	3.4	9.6	2.1	89	—
CURED PORK SHOULDER (PICNIC), roasted	48.8	323	25.2	25.2	0	8.8	3.9	10.9	2.4	89	—
GROUND PORK, cooked	45.2	373	30.6	30.6	0	10.8	—	13.2	2.4	89	—
LEG (FRESH HAM) ROAST, roasted	45.5	374	30.6	30.6	0	10.8	—	12.0	2.4	89	—
LOIN CHOP, broiled	42.3	391	31.7	31.7	0	11.4	4.9	13.3	2.8	89	—
LOIN ROAST, roasted	45.8	362	28.5	28.5	0	9.8	4.4	12.1	2.6	89	—
LOIN TENDERLOIN, cooked	55.0	254	14.2	14.2	0	4.7	2.2	5.8	1.2	88	—
PICKLED PIGS FEET	66.9	199	14.8	14.8	0	5.0	—	6.0	1.0	—	—
RIB CHOP, broiled	42.3	391	31.7	31.7	0	11.4	4.9	13.3	2.8	89	—
SHOULDER, simmered	60.3	212	9.8	9.8	0	3.2	—	4.8	1.6	88	—
SHOULDER, BOSTON BUTT ROAST, cooked	48.1	353	28.5	28.5	0	11.7	5.1	14.0	3.0	89	—
SHOULDER, PICNIC ROAST, cooked	45.7	374	30.5	30.5	0	8.7	3.9	10.7	2.3	89	—
SPARERIB, braised	39.7	440	38.9	38.9	0	13.5	6.0	16.6	3.5	89	—
SAUSAGES											
COUNTRY-STYLE SAUSAGE, link	49.9	345	31.1	31.1	0	11.0	—	13.0	3.0	89	—
FRANKFURTER OR WIENER, cooked	57.3	304	27.2	27.2	0	11.2	15.8	12.3	0.9	62	—
ITALIAN SAUSAGE, cooked	49.9	345	31.1	31.1	0	11.0	—	13.0	3.0	89	—
KNOCKWURST, pork	57.6	278	23.2	23.2	0	—	—	—	—	62	—
POLISH SAUSAGE/KOLBASSI, cooked	53.7	304	25.8	25.8	0	—	—	—	—	89	—
PORK SAUSAGE, cooked	34.8	476	44.2	44.2	0	11.7	5.0	14.1	3.4	89	—
SUMMER SAUSAGE/THURINGER/ CERVELAT, beef	48.5	307	24.5	24.5	0	11.0	14.4	11.2	0.8	62	—
TURKEY											
DARK MEAT, w/o skin, roasted	60.5	203	8.3	8.3	0	1.6	1.8	1.1	1.1	101	—
LIGHT MEAT, w/o skin, roasted	62.1	176	3.9	3.9	0	0.7	0.8	0.5	0.5	77	—
DRUMSTICK, w/o skin, roasted	60.5	203	8.3	8.3	0	1.6	1.8	1.1	1.1	101	—
THIGH, w/o skin, roasted	60.5	203	8.3	8.3	0	1.6	1.8	1.1	1.1	101	—
WING, w/o skin, roasted	60.5	203	8.3	8.3	0	—	—	—	—	101	—
CANNED	64.9	202	12.5	12.5	0	4.0	—	5.0	3.0	89	—
VEAL											
CUTLET, braised or broiled	60.4	216	11.1	11.1	0	4.7	1.1	4.0	0.5	101	—
LOIN, CHOP OR ROAST, cooked	58.9	234	13.4	13.4	0	5.7	1.3	4.9	0.6	101	—
RIB CHOP, cooked	54.6	269	16.9	16.9	0	7.1	1.1	4.0	0.5	101	—
ROUND, GROUND, cooked	60.4	216	11.1	11.1	0	4.7	1.2	4.6	0.5	101	—
SHOULDER ARM ROAST, cooked	58.5	235	12.8	12.8	0	5.4	1.6	6.0	0.7	99	—
MILK & PRODUCTS											
BUTTER, salted or unsalted	15.9	717	81.1	81.1	0	50.5	3.0	20.4	1.8	219	—
BUTTERMILK, cultured	90.1	40	0.9	0.9	0	0.6	Trace	0.2	Trace	4	—

Footnote at end of table

(Continued)

TABLE F-9 *(Continued)*

Food Name—100 Gram (3.5 oz) Portion	Water	Food Energy Calories	Total Fat	Animal Fat	Plant Fat	Saturated Fat	Polyun-saturated Fat	Fatty Acids		Chol-esterol	Choline
								Oleic Acid	Linoleic Acid		
	(g)	(kcal)	(g)	(g)	(g)	(g)	(g)	(g)	(g)	(mg)	(mg)
CHEESE											
AMERICAN, pasteurized process	39.2	375	31.3	31.3	0	$9.7	1.0	7.5	0.6	94	—
AMERICAN CHEESE FOOD, cold pack	43.1	331	24.5	24.5	0	15.4	0.7	6.0	0.5	64	—
pasteurized process	43.2	328	24.6	24.6	0	15.4	0.7	6.1	0.5	64	—
AMERICAN CHEESE SPREAD, pasteurized process	47.7	290	21.2	21.2	0	13.3	0.6	5.2	0.4	55	—
BLUE	42.4	353	28.7	28.7	0	18.7	0.8	6.6	0.5	75	—
BRICK	41.1	371	29.7	29.7	0	18.8	0.8	7.4	0.5	94	—
BRIE	48.4	334	27.7	27.7	0	—	—	—	—	100	—
CAMEMBERT	51.8	300	24.3	24.3	0	15.3	0.7	5.8	0.5	72	—
CHEDDAR	36.8	403	33.1	33.1	0	21.1	0.9	7.9	0.6	105	—
CHESHIRE	37.7	387	30.6	30.6	0	—	—	—	—	103	—
COLBY	38.2	394	32.1	32.1	0	20.2	1.0	7.8	0.7	95	—
COTTAGE CHEESE, creamed, small, or large curd	79.0	103	4.5	4.5	0	2.9	0.1	1.1	0.1	15	—
creamed, w/fruit added	72.1	124	3.4	3.4	0	2.2	0.1	0.8	0.1	11	—
dry curd	79.8	85	0.4	0.4	0	0.3	Trace	0.1	Trace	7	—
lowfat, 2% fat	79.3	90	1.9	1.9	0	1.2	0.1	0.5	Trace	8	—
lowfat, 1% fat	82.5	72	1.0	1.0	0	0.6	Trace	0.2	Trace	4	—
CREAM	53.8	349	34.9	34.9	0	22.0	1.3	8.4	0.8	110	—
EDAM	41.6	357	27.8	27.8	0	17.6	0.7	6.9	0.4	89	—
FETA, made from sheep's milk	55.2	264	21.3	21.3	0	15.0	0.6	4.0	0.3	89	—
FONTINA	37.9	389	31.1	31.1	0	19.2	1.7	7.1	0.9	116	—
GJETOST, made from goat's & cow's milk	13.4	466	29.5	29.5	0	19.2	0.9	7.0	0.5	—	—
GOUDA	41.5	356	27.4	27.4	0	17.6	0.7	6.4	0.3	114	—
GRUYERE	33.2	413	32.3	32.3	0	18.9	1.7	8.6	1.3	110	—
LIMBURGER	48.4	327	27.3	27.3	0	16.8	0.5	7.2	0.3	90	—
MOZZARELLA	54.1	281	21.6	21.6	0	13.2	0.8	5.7	0.4	78	—
low moisture	48.4	318	24.6	24.6	0	15.6	0.8	5.9	0.6	89	—
part skim	53.8	254	15.9	15.9	0	10.1	0.5	3.9	0.3	58	—
low moisture, part skim	48.6	280	17.1	17.1	0	10.9	0.5	4.2	0.4	54	—
MUENSTER	41.8	368	30.0	30.0	0	19.1	0.7	7.3	0.4	96	—
NEUFCHATEL	62.2	260	23.4	23.4	0	14.8	0.7	5.7	0.5	76	—
PARMESAN, grated	17.7	456	30.0	30.0	0	19.1	0.7	7.7	0.3	79	—
hard	29.2	392	25.8	25.8	0	16.4	0.6	6.7	0.3	68	—
PIMIENTO, pasteurized process	39.1	375	31.2	—	—	19.7	1.0	7.5	0.6	94	—
PORT DU SALUT	45.5	352	28.2	28.2	0	16.7	0.7	8.1	0.4	123	—
PROVOLONE	41.0	351	26.6	26.6	0	17.1	0.8	6.2	0.5	69	—
RICOTTA, made w/whole milk	71.7	174	13.0	13.0	0	8.3	0.4	2.9	0.3	51	—
made w/part skim milk	74.4	138	7.9	7.9	0	4.9	0.3	1.9	0.2	31	—
ROMANO	30.9	387	26.9	26.9	0	—	—	—	—	104	—
ROQUEFORT, made from sheep's milk	39.4	369	30.6	30.6	0	19.3	1.3	7.5	0.6	90	—
SWISS	37.2	376	27.5	27.5	0	17.8	1.0	6.0	0.6	92	—
pasteurized process	42.3	334	25.0	25.0	0	16.0	0.6	5.9	0.3	85	—
SWISS CHEESE FOOD, pasteurized process	43.7	323	24.1	24.1	0	—	—	—	—	82	—
TILSIT, made w/whole milk	42.9	340	26.0	26.0	0	16.8	0.7	6.1	0.4	102	—
CREAM											
heavy whipping	57.7	345	37.0	37.0	0	23.0	1.4	9.3	0.8	137	—
light whipping	63.5	292	30.9	30.9	0	19.3	0.9	7.7	0.6	111	—
medium, 25% fat	68.5	244	25.0	25.0	0	15.6	0.9	6.3	0.6	88	—
light, coffee or table	73.8	195	19.3	19.3	0	12.0	0.7	4.9	0.4	66	—
half & half, milk & cream	80.6	130	11.5	11.5	0	7.2	0.4	2.9	0.3	37	—
imitation sour cream, non-dairy cultured	71.2	208	19.5	0	19.5	17.8	0.1	0.6	0.1	0	—
sour cream, cultured	70.9	214	21.0	21.0	0	13.1	0.8	5.3	0.5	44	—
sour half & half, cultured	80.1	135	12.0	12.0	0	7.5	0.5	3.0	0.3	38	—
whipped cream topping, pressurized	61.3	257	22.2	22.2	0	13.8	0.8	5.6	0.5	76	—

Footnote at end of table

(Continued)

TABLE F-9 *(Continued)*

Food Name—100 Gram (3.5 oz) Portion	Water	Food Energy Calories	Total Fat	Animal Fat	Plant Fat	Saturated Fat	Polyunsaturated Fat	Fatty Acids: Oleic Acid	Fatty Acids: Linoleic Acid	Cholesterol	Choline
	(g)	(kcal)	(g)	(g)	(g)	(g)	(g)	(g)	(g)	(mg)	(mg)
CREAM SUBSTITUTES											
COFFEE WHITENER, non-dairy liquid w/ hydrogenated vegetable oil & soy protein	77.3	136	10.0	0	10.0	1.9	Trace	7.6	Trace	0	—
non-dairy liquid w/lauric acid oil & casein	77.3	136	10.0	0	10.0	9.3	Trace	—	Trace	0	—
non-dairy powdered	2.2	546	35.5	0	35.5	32.5	Trace	1.0	Trace	0	—
ICE CREAM											
FRENCH VANILLA ice cream, soft serve	59.8	218	13.0	—	—	7.8	0.6	3.4	0.4	89	—
VANILLA ice cream, rich, 16% fat, hardened	58.9	236	16.0	16.0	0	10.0	0.6	4.0	0.4	59	—
regular, 10% fat, hardened	60.8	202	10.8	10.8	0	6.7	0.4	2.7	0.2	45	—
VANILLA ICE MILK, hardened	68.6	140	4.3	4.3	0	2.7	0.2	1.1	0.1	14	—
soft serve	69.6	128	2.6	—	—	1.6	0.1	0.7	0.1	8	—
ORANGE SHERBET	66.1	140	2.0	2.0	0	1.2	0.1	0.5	Trace	7	—
MILK											
WHOLE MILK, 3.7% fat	87.7	64	3.7	3.7	0	2.3	0.1	0.9	0.1	14	—
3.3% fat	88.0	61	3.3	3.3	0	2.1	0.1	0.8	0.1	14	—
evaporated, canned	74.0	134	7.6	7.6	0	4.6	0.2	2.1	0.2	29	—
low sodium	88.2	61	3.5	3.5	0	2.2	0.1	0.9	0.1	14	—
dry	2.5	496	26.7	26.7	0	16.7	0.7	6.2	0.5	97	—
LOWFAT MILK, 2% fat	89.2	50	1.9	1.9	0	1.2	0.1	0.5	Trace	8	—
2% fat w/nonfat milk solids added	88.9	51	1.9	1.9	0	1.2	0.1	0.5	Trace	8	—
2% fat, protein fortified	87.7	56	2.0	2.0	0	1.2	0.1	0.5	Trace	8	—
1% fat	90.1	42	1.1	1.1	0	0.7	Trace	0.3	Trace	4	—
1% fat w/nonfat milk solids added	89.8	43	1.0	1.0	0	0.6	Trace	0.2	Trace	4	—
1% fat, protein fortified	88.7	48	1.2	1.2	0	0.7	Trace	0.3	Trace	4	—
SKIM MILK	90.8	35	0.2	0.2	0	0.1	Trace	Trace	Trace	2	—
evaporated, canned	79.4	78	0.2	0.2	0	0.1	Trace	0.1	Trace	4	—
protein fortified	89.4	41	0.3	0.3	0	0.2	Trace	0.1	Trace	2	—
w/nonfat milk solids added	90.4	37	0.3	0.3	0	0.2	Trace	0.1	Trace	2	—
NONFAT DRY MILK POWDER, non-instantized	3.2	362	0.8	0.8	0	0.5	Trace	0.2	Trace	20	—
instantized	4.0	358	0.7	0.7	0	0.5	Trace	0.2	Trace	18	—
SWEETENED CONDENSED MILK, canned	27.2	321	8.7	8.7	0	5.5	0.3	2.2	0.2	34	—
CHOCOLATE MILK, whole, 3.3% fat	82.3	83	3.4	3.4	0	2.1	0.1	0.9	0.1	12	—
2% fat	83.6	72	2.0	2.0	0	1.2	0.1	0.5	0.1	7	—
1% fat	84.5	63	1.0	1.0	0	0.6	Trace	0.3	Trace	3	—
HOT CHOCOLATE SWEETENED MIX (powdered)	3.1	392	10.6	0	10.6	6.0	—	4.0	Trace	0	—
COCOA POWDER, high fat/breakfast, plain	3.0	299	23.7	0	23.7	13.0	—	9.0	Trace	0	—
HOT COCOA, homemade w/whole milk	81.6	87	3.6	—	—	2.2	0.1	0.9	0.1	13	—
MALTED MILK powder, natural flavor	2.6	411	8.5	—	—	4.2	1.2	—	—	20	—
beverage, natural flavor	81.2	89	3.8	—	—	2.3	0.2	—	—	14	—
powder, chocolate flavor	2.0	396	4.5	—	—	2.2	0.6	—	—	5	—
beverage, chocolate flavor	81.2	88	3.4	—	—	2.1	0.2	—	—	13	—
MILKSHAKE, vanilla flavor, thick type	74.4	112	3.0	—	—	1.9	0.1	0.8	0.1	12	—
chocolate flavor, thick type	72.2	119	2.7	—	—	1.7	0.1	0.7	0.1	10	—
GOAT MILK, whole	87.0	69	4.1	4.1	0	2.7	0.2	1.0	0.1	11	—
HUMAN MILK, whole	87.5	70	4.4	4.4	0	2.0	0.5	1.5	0.4	14	—
YOGURT											
plain	87.9	61	3.3	3.3	0	2.1	0.1	0.7	0.1	13	—
plain, lowfat	85.1	63	1.5	1.5	0	1.0	Trace	0.4	Trace	6	—
plain, skim milk	85.2	56	0.2	0.2	0	0.1	Trace	Trace	Trace	2	—
coffee & vanilla varieties, lowfat	79.0	85	1.3	1.3	0	0.8	Trace	0.3	Trace	5	—
fruit varieties, lowfat (9g protein/8oz)	75.3	99	1.1	1.1	0	0.7	Trace	0.3	Trace	4	—
fruit varieties, lowfat (10g protein/8oz)	74.5	102	1.1	1.1	0	0.7	Trace	0.3	Trace	4	—
fruit varieties, lowfat (11g protein/8oz)	74.1	105	1.4	1.4	0	0.9	Trace	0.3	Trace	6	—

Footnote at end of table

(Continued)

TABLE F-9 *(Continued)*

Food Name—100 Gram (3.5 oz) Portion	Water	Food Energy Calories	Total Fat	Animal Fat	Plant Fat	Saturated Fat	Polyunsaturated Fat	Fatty Acids: Oleic Acid	Fatty Acids: Linoleic Acid	Cholesterol	Choline
	(g)	(kcal)	(g)	(g)	(g)	(g)	(g)	(g)	(g)	(mg)	(mg)
NUTS & SEEDS											
ALMONDS	4.7	598	54.2	0	54.2	4.3	10.5	36.5	9.9	0	—
BRAZIL NUTS	4.6	654	66.9	0	66.9	17.4	25.7	22.2	25.4	0	—
CASHEW NUTS	5.2	561	45.7	0	45.7	9.2	7.7	26.2	7.3	0	—
CHESTNUTS	52.5	194	1.5	0	1.5	0.4	1.1	1.0	0.9	0	—
COCONUT, dried, shredded, sweetened	3.3	548	39.1	0	39.1	34.0	0.3	3.0	Trace	0	—
COCONUT, fresh	50.9	346	35.3	0	35.3	31.2	0.8	2.0	0.7	0	—
FILBERTS (HAZELNUTS)	5.8	634	62.4	0	62.4	4.6	6.9	49.8	6.6	0	—
HICKORY NUTS	3.3	673	68.7	0	68.7	6.0	12.7	47.0	12.0	0	—
MACADAMIA NUTS	3.0	691	71.6	0	71.6	11.0	16.3	43.2	1.2	0	—
PEANUTS, roasted & salted, Spanish	1.6	585	49.8	0	49.8	9.6	16.1	20.5	16.1	0	162.0
roasted & salted, Virginia	1.6	585	49.8	0	49.8	8.3	12.2	24.1	12.2	0	162.0
roasted in shell, w/skins	1.8	582	48.7	0	48.7	8.6	13.3	23.5	13.3	0	162.0
PEANUT BUTTER	1.8	581	49.4	0	49.4	10.5	15.1	23.1	15.1	0	145.0
PECANS	3.4	687	71.2	0	71.2	6.1	18.3	42.9	17.0	0	50.0
PINENUTS (PIGNOLIAS)	5.6	552	47.4	0	47.4	6.2	22.8	19.0	22.2	0	—
PISTACHIO NUTS	5.3	594	53.7	0	53.7	7.4	7.3	36.0	6.8	0	—
PUMPKIN OR SQUASH, SEEDS	4.4	553	46.7	0	46.7	8.0	—	17.0	20.0	0	—
SUNFLOWER SEEDS, hulled	4.8	560	47.3	0	47.3	6.0	—	9.0	30.0	0	—
WALNUTS, BLACK	3.1	628	59.3	0	59.3	5.1	40.9	10.7	36.8	0	—
ENGLISH	3.5	651	64.0	0	64.0	6.9	42.0	9.7	34.9	0	—
PICKLES & RELISHES											
OLIVE, green	78.2	116	12.7	0	12.7	1.4	—	9.6	0.9	0	—
ripe	73.0	184	20.1	0	20.1	2.2	—	15.3	1.4	0	—
ripe, salt-cured (Greek)	43.8	338	35.8	0	35.8	3.9	—	27.2	2.5	0	—
SALADS											
COLESLAW, w/salad dressing	82.9	99	7.9	0	7.9	1.0	1.6	2.0	4.0	—	—
POTATO, w/mayonnaise or french dressing & hard-cooked egg	72.4	145	9.2	1.1	8.1	2.0	—	2.0	4.0	65	—
TUNA, w/celery, mayonnaise, pickle, onion, egg	69.8	170	10.5	6.0	4.5	3.0	—	3.0	3.0	—	—
SALAD DRESSINGS											
BLUE OR ROQUEFORT CHEESE	32.3	504	52.3	6.3	46.0	11.0	—	11.0	25.0	—	—
low cal, (5 cal/tsp, or *5 ml*)	83.7	76	5.9	—	—	3.0	—	2.0	Trace	—	—
FRENCH DRESSING	38.8	410	38.9	0	38.9	7.0	—	8.0	20.0	—	—
low cal, (10 cal/tsp, *or 5 ml*)	78.8	156	16.9	0	16.9	3.0	—	4.0	9.0	—	—
low cal, (5 cal/tsp, or *5 ml*)	77.3	96	4.3	0	4.3	1.0	—	1.0	2.0	—	—
ITALIAN DRESSING	27.5	552	60.0	0	60.0	10.0	—	13.0	31.0	—	—
ITALIAN DRESSING, low cal, (2 cal/tsp, or *5 ml*)	90.1	50	4.7	0	4.7	1.0	—	1.0	2.0	—	—
MAYONNAISE	15.1	718	79.9	14.4	65.5	14.0	—	17.0	40.0	70	—
RUSSIAN DRESSING	34.5	494	50.8	0	50.8	9.0	—	11.0	26.0	—	—
SALAD DRESSING, mayonnaise type	40.6	435	42.3	7.6	34.7	8.0	—	9.0	21.0	50	—
SALAD DRESSING, mayonnaise type, low cal, (8 cal/tsp, or *5 ml*)	80.7	136	12.7	—	—	2.0	—	3.0	6.0	—	—
THOUSAND ISLAND	32.0	502	50.2	5.0	45.2	9.0	—	11.0	25.0	—	—
low cal, (10 cal/tsp, or *5 ml*)	68.2	180	13.7	—	—	2.0	—	3.0	7.0	—	—
SNACK FOODS											
PEANUT BUTTER & CHEESE CRACKER SANDWICH	2.4	491	23.9	—	—	6.3	—	11.0	5.6	—	—
POPCORN, plain	4.0	386	5.0	0	5.0	1.0	—	1.0	3.0	0	—
POPCORN, popped in coconut oil, w/salt	3.1	456	21.8	0	21.8	15.0	—	2.0	2.0	0	—
POTATO CHIPS	1.8	568	39.8	0	39.8	9.9	—	8.3	19.9	—	—
POTATO CHIPS, salt free	1.8	568	39.8	0	39.8	9.9	—	8.3	19.9	—	—
POTATO STICKS	1.5	544	36.4	0	36.4	9.0	—	8.0	18.0	—	—
PRETZELS	4.5	390	4.5	0	4.5	1.0	—	—	—	—	—
SOUPS & CHOWDERS											
BEAN W/PORK, canned, diluted w/equal volume water	84.4	67	2.3	1.7	0.6	0.6	0.8	0.8	0.6	—	—
BEEF NOODLE, canned, diluted w/equal volume water	92.8	29	0.9	—	—	0.5	0.3	0.5	0.2	—	—
CHICKEN GUMBO, canned, diluted w/equal volume water	93.5	24	0.6	—	—	0.1	0.2	0.2	0.1	—	—

Footnote at end of table

(Continued)

TABLE F-9 *(Continued)*

Food Name—100 Gram (3.5 oz) Portion	Water	Food Energy Calories	Total Fat	Animal Fat	Plant Fat	Saturated Fat	Polyunsaturated Fat	Fatty Acids: Oleic Acid	Fatty Acids: Linoleic Acid	Cholesterol	Choline
	(g)	(kcal)	(g)	(g)	(g)	(g)	(g)	(g)	(g)	(mg)	(mg)
CHICKEN NOODLE, canned, diluted w/equal volume water	93.1	27	0.8	—	—	0.3	0.2	0.4	0.2	—	—
CHICKEN W/RICE, canned, diluted w/equal volume water	94.7	22	0.5	—	—	0.2	0.2	0.4	0.2	—	—
CLAM CHOWDER, Manhattan, w/tomato, frozen made w/equal volume water	92.1	32	1.1	—	—	0.2	0.5	0.2	0.5	—	—
CREAM OF CELERY, canned, diluted w/equal volume milk	85.8	69	3.8	—	—	1.6	1.2	0.9	1.0	—	—
canned, diluted w/equal volume water	92.3	36	2.1	—	—	0.6	1.1	0.5	1.0	—	—
condensed, canned	84.6	72	4.2	—	—	1.1	2.1	0.9	1.9	—	—
CREAM OF CHICKEN, canned, diluted w/equal volume milk	85.4	73	4.2	—	—	1.9	0.9	1.6	0.6	—	—
canned, diluted w/equal volume water	92.2	38	2.2	—	—	0.9	0.7	1.2	0.6	—	—
condensed, canned	83.8	79	4.8	—	—	1.7	1.5	2.4	1.1	—	—
CREAM OF MUSHROOM, canned diluted w/ equal volume milk	83.4	92	6.4	—	—	2.1	2.0	1.1	1.8	—	—
canned diluted w/equal volume water	89.9	58	4.4	—	—	1.1	1.9	0.7	1.8	—	—
condensed, canned	79.3	111	8.0	—	—	2.1	3.7	1.4	3.6	—	—
CREAM OF POTATO, frozen made w/equal volume milk	83.4	79	4.1	—	—	2.0	—	1.0	Trace	—	—
MINESTRONE, canned, diluted w/equal volume water	91.2	36	1.2	0	1.2	0.2	0.5	0.3	0.4	—	—
OYSTER STEW, frozen, made w/equal volume milk	83.4	84	4.9	—	—	2.1	0.2	0.8	0.1	—	—
SPLIT PEA, canned, diluted w/equal volume water	85.4	59	1.3	1.0	0.3	1.0	—	2.0	Trace	—	—
TURKEY NOODLE, canned, diluted w/equal volume water	92.7	32	1.3	—	—	0.3	0.3	0.4	0.3	—	—
TOMATO, canned, diluted w/equal volume milk	84.5	69	2.8	—	—	1.2	0.5	0.6	0.4	—	—
canned, diluted w/equal volume water	90.5	36	1.0	0	1.0	0.2	—	0.3	0.4	—	—
condensed, canned	81.0	72	2.1	0	2.1	0.3	0.7	0.3	0.6	—	—
VEGETABLE-BEEF, canned, diluted w/equal volume water	91.9	33	0.9	—	—	0.3	0.7	0.4	0.6	—	—
VEGETARIAN VEGETABLE, canned, diluted w/equal volume water	91.8	32	0.8	0	0.8	0.2	0.4	0.2	0.3	—	—
VEGETABLES											
BEANS, COMMON WHITE, canned, w/pork & sweet sauce	66.4	150	4.7	3.8	0.9	2.0	—	2.0	0.4	—	—
COMMON WHITE, canned, w/pork & tomato sauce	70.7	122	2.6	2.1	0.5	1.0	—	1.0	0.2	—	—
GREEN, raw	90.1	32	0.2	0	0.2	—	—	—	—	—	42.0
CABBAGE, raw	92.4	24	0.2	0	0.2	—	—	—	—	—	23.0
CARROT, raw	88.2	42	0.2	0	0.2	—	—	—	—	—	13.4
PEAS, GREEN, immature, raw	78.0	84	0.4	0	0.4	—	—	—	—	—	75.0
POTATOES, WHITE											
french fried	44.7	274	13.2	0	13.2	3.0	—	3.0	7.0	—	—
french fried, frozen, ovenheated	52.9	220	8.4	0	8.4	1.8	—	1.8	3.6	—	—
fried from raw, w/vegetable shortening	46.9	268	14.2	0	14.2	3.0	—	9.0	1.0	—	—
hashed brown in vegetable shortening	54.2	229	11.7	0	11.7	3.0	—	8.0	1.0	—	—
mashed, w/milk & margarine	79.8	94	4.3	0.6	3.7	2.0	—	1.0	Trace	—	—
mashed, dehydrated flakes, made w/water, milk, margarine	79.3	93	3.2	—	—	0.9	—	1.5	0.7	—	—
mashed, dehydrated granules, made w/water, milk, margarine	78.6	96	3.6	—	—	0.9	—	1.7	0.8	—	—
scalloped & au gratin, w/cheese	71.1	145	7.9	—	—	4.0	—	3.0	Trace	15	—
scalloped & au gratin, w/o cheese	76.7	104	3.9	1.2	2.7	—	—	—	—	6	—
SOYBEANS, mature seeds, cooked	71.0	130	5.7	0	5.7	1.0	—	1.0	3.0	—	—
SPINACH, raw	90.7	26	0.3	0	0.3	—	—	—	—	6	22.0
SWEET POTATO, candied	60.0	168	3.3	0	3.3	2.0	—	1.0	Trace	—	—

1 The authors gratefully acknowledge that these food compositions were obtained from the HVH-CWRI Nutrient Data Base developed by the Division of Nutrition, Highland View Hospital, and the Departments of Biometry and Nutrition, School of Medicine, Case Western Reserve University, Cleveland, Ohio.

FATS, HYDROGENATED (HYDROGENATED OILS)

Fats and oils containing polyunsaturated fatty acids can be hardened and turned into solid fats by the addition of hydrogen in the presence of a catalyst, which is usually nickel. The process is called hydrogenation since it adds hydrogen to the double bond linkages of unsaturated fatty acids and converts them to saturated fatty acids. Hydrogenated fats are of commercial importance, since vegetable oils can be converted into a solid form and used as shortenings and margarine.

(Also see FATS AND OTHER LIPIDS; and MARGARINE.)

FAT-SOLUBLE VITAMINS

These vitamins are stored in appreciable quantities in the body, whereas the water-soluble vitamins are not. Thus, vitamin A and/or carotene may be stored by a person in his (her) liver and fatty tissue in sufficient quantities to meet the body requirements for a period of 6 months or longer. The fat-soluble vitamins are: A (carotene), D, E, and K.

(Also see VITAMIN[S]; VITAMIN A; VITAMIN D; VITAMIN E; and VITAMIN K.)

FATTENING

The deposition of unused energy in the form of fat within the body tissues.

FATTY ACIDS

Fatty acids, which are the key components of lipids, are composed entirely of carbon, hydrogen, and oxygen. They are called acids because they contain an organic acid (COOH) in their structure. The backbone of the chain consists of carbon atoms connected by either single (saturated) or double (unsaturated) linkages (valence bonds). Saturated linkages are more stable than unsaturated linkages. Fatty acids which contain two or more unsaturated bonds per molecule (known as polyunsaturated fatty acids or PUFA) are prone to rancidity—oxidative deterioration.

The properties of the various fatty acids depend mainly upon the combination of (1) the number of unsaturated bonds present, and (2) the length of the carbon chain (6 or less carbon atoms to 12 or more). These chemical characteristics affect melting points which are very important in nutrition because fats that remain solid in the digestive tract are utilized poorly. There is a steady rise in melting points as the chain lengths of fatty acids increase. However, for a given chain length, the melting points decrease as the degree of unsaturation increases.

(Also see FATS AND OTHER LIPIDS; and OILS, VEGETABLE.)

FDA

Food and Drug Administration, a Federal regulatory agency.

(Also see FOOD AND DRUG ADMINISTRATION.)

FEBRILE

Feverish.

FEDERAL FOOD, DRUG AND COSMETIC ACT

The first law enacted to protect the food we eat dates back to 1784, in Massachusetts. In the period from 1891 to 1895, laws were passed by the U.S. Congress requiring the inspection of animals for disease before slaughter. The furor over the quality of meat came to a head in 1906 with the passage of the Meat Inspection Act, which came as a direct result of shocking disclosures of conditions under which meat was processed. On the same day that the Meat Inspection Act was passed, the Food and Drug Act of 1906 was signed. This act was designed to prevent the use of poisonous preservatives and dyes in foods, and to regulate the sales of patent medicines.

In 1927, the Food, Drug, and Insecticide Administration, later to be named the Food and Drug Administration (FDA), was created. In 1938, the Federal Food, Drug, and Cosmetic Act of 1938 was signed, broadening the scope of the Food and Drug Act of 1906 to include (1) coverage of cosmetics and devices; (2) requirements for predistribution clearance for the safety of new drugs; (3) provisions for tolerances for unavoidable poisonous substances; (4) standards of identity, quality, and fill of food containers; (5) authorization for factory inspections; and (6) provision for court injunctions in matters of seizures and possession. With its numerous amendments, the Federal, Food, Drug and Cosmetic Act is probably the most extensive law of its kind in the world.

(Also see DELANEY CLAUSE; and FOOD AND DRUG ADMINISTRATION.)

FEED ADJUVANTS

An adjuvant is a substance which facilitates or enhances the effectiveness of any process. Feed adjuvants such as antibiotics, hormones, and hormonelike substances alter digestion, absorption, or metabolism of meat-producing animals, thus increasing feed efficiency and stimulating growth. Usually these substances are called feed additives, implants, or growth stimulators in the livestock industry. Some consumers are concerned lest these substances show up in animal products and have harmful effects to humans. The controversy relative to one such substance—diethylstilbestrol or DES—caused the FDA to ban its use in 1973.

(Also see ADDITIVES; CANCER; and DELANEY CLAUSE.)

FEEDBACK MECHANISM

The mechanism that regulates the production and secretion by an endocrine gland of its hormone, which, in turn, stimulates another endocrine gland to produce its hormone. Example: The anterior pituitary secretes thyroid-stimulating hormone (TSH), which, in turn, stimulates the thyroid to secrete thyroxin. When the blood level of thyroxin reaches optimum, the anterior pituitary ceases to secrete TSH; and, thereupon , the thyroid stops secreting thyroxin, and the blood level of thyroxin falls.

FEINGOLD DIET

An elimination diet, which seeks to remove all sources of synthetic food coloring, synthetic flavors, salicylates (chemicals related to aspirin), as well as the antioxidants butylated hydroxytoluene (BHT) and butylated hydroxyanisole (BHA). Reasoning behind this diet is based on the observations of Dr. Benjamin F. Feingold, who relates the intake of these chemicals to hyperactive (hyperkinetic) behavior in children. As of yet, there are many convincing observations and testimonials but no conclusive scientific evidence of "cause and effect" linking these chemicals to hyperactive behavior in children. Dr. Feingold's thesis has, however, stimulated awareness and research of the problem.

(Also see ADDITIVES; and MODIFIED DIETS.)

FENNEL *Foeniculum vulgare,* var. *dulce*

This is a perennial plant, which is cultivated as an annual. It is closely related to parsley. The bulb-like bases and feathery leaves of *finoccho,* or Florence fennel, are eaten as a vegetable; the bases are generally cooked, and the leaves are used as a raw salad, either alone or mixed with the bases. The plant is native to southern Europe and widely cultivated in temperate and subtropical regions.

(Also see FLORENCE FENNEL; and VEGETABLE[S], Table V-6 Vegetables of the World.)

FERMENTATION

A gradual chemical change brought about by the enzymes of some bacteria, molds, and yeasts. These chemical changes are usually the acidulation of milk, the decomposition of starches and sugars to alcohol and carbon dioxide, or the oxidation of nitrogenous organic compounds. The three main types of industrial fermentation are (1) bacterial fermentation of carbohydrates, (2) bacterial fermentation of alcohol to acetic acid, and (3) yeast fermentation. In general, the requirements of greatest importance for fermentation are the media, temperature, salt, acidity, the culture container for fermentation, and the time involved. Production of breads, beers, wines, cheeses, yogurts, and antibiotics, to name a few, depends upon the process of fermentation.

(Also see BEERS AND BREWING; BREADS AND BAKING; DISTILLED LIQUORS; MILK AND MILK PRODUCTS; and VINEGAR.)

FERMENTATION PRODUCT

Product formed as a result of an enzymatic transformation of organic substrates.

(Also see FERMENTATION.)

FERMENTATION SOLUBLES (DISTILLERS' SOLUBLES)

That portion of the stillage which passes through screens, consisting mostly of water, water-soluble substances, and the fine particles from the fermentation process, used primarily as feeds for livestock.

(Also see DISTILLED LIQUORS, Table D-18 Miscellaneous Uses of Spirits and the By-Products of Distillation.)

FERMENTED

Acted upon by yeasts, molds, or bacteria in a controlled aerobic or anaerobic process in the manufacture of such products as alcohols, acids, vitamins of the B-complex group, or antibiotics.

(Also see FERMENTATION.)

FERRITIN

The iron-protein complex in which iron is stored, particularly in the cells of the liver, spleen, and bone marrow.

(Also see IRON.)

FERTILIZER

Any material, natural or manufactured, added to the soil to improve the supply of one or more plant nutrients is known as a fertilizer. These applied materials are carriers of compounds or chemical substances usually not found in sufficient quantities in a soil to maintain satisfactory plant growth. All of the chemical elements required by a plant are technically called plant nutrients. In the fertilizer trade these nutrients are called plant foods. The three principal nutrients, or plant foods, are nitrogen, phosphorus, and potassium. Complete fertilizers contain all three nutrients. But, on some soils and under some conditions, a two-nutrient fertilizer may be best; and, under still other conditions, a fertilizer containing only one plant nutrient may be best.

Animal manure also contains nitrogen, phosphorus, and potassium. With the increasing cost of commercial fertilizers, a growing number of American farmers are returning to organic farming; they are using more manure, and they are discovering that they are just as good reapers of the land and far better stewards of the soil.

(Also see ORGANICALLY GROWN FOOD.)

FIBER

Contents Page

History 710
Kinds of Fiber 711
Fibrous Substances in Common Foods 711
Food Additives 711
Determination of the Fiber Content of Foods 712
Proximate Analysis 712
Digestibility Studies 712
Analysis of Cell Wall Constituents 713
Physiological Actions of Dietary Fiber 713
Beneficial Effects 713
Potentially Harmful Effects 714
Dietary Fiber Recommendations 714
Selection and Preparation of Foods High in Fiber 715
Special Products for Dietetic and Therapeutic Uses 716
Future Prospects for Dietary Fiber 716

This term designates the complex of carbohydrates and other substances that are present mainly in the cell walls of the plants used as foods. Fiber includes cellulose, hemicelluloses, gums, mucilages, pectic substances, and lignin. These substances are poorly digested by humans and have little nutritional value. However, they appear to have important physiological effects in the digestive tract; hence, the term "Nonnutritive fiber." Other terms used to designate fiber are "crude fiber," "Unavailable carbohydrates" (although lignin is not a carbohydrate), "bulk," "roughage," and "undigestible residue." The last term may also be applied to certain animal products, such as feathers and hair. Nevertheless, almost all of the undigestible matter in the human diet now comes from plant foods.

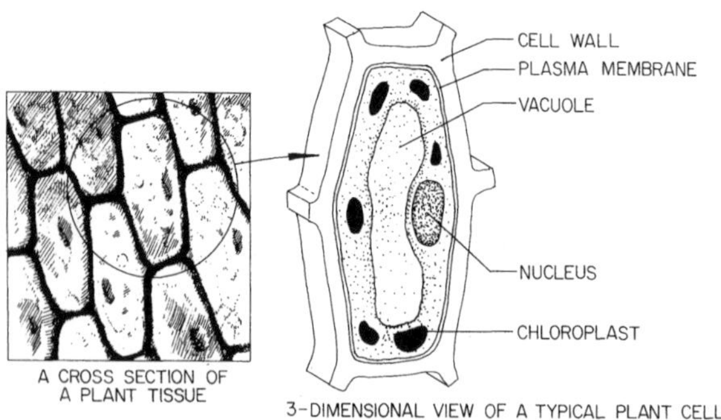

Fig. F-8. A typical plant cell. Most of the fiber in the human diet originates in the rigid cell walls of the various food plants.

The food industry now uses many types of natural, semisynthetic (natural materials altered by man), and totally synthetic carbohydrates having effects similar to naturally occurring nonnutritive components of food. These materials are used (1) as dispersing agents—to keep dissimilar food ingredients well mixed; (2) as thickening agents—to increase viscosity or stiffness of mixtures; and (3) as gelling agents. Furthermore, some of these substances are now added to foods solely to increase the fiber content. It is noteworthy that food technologists have also developed processing techniques for removal of the fiber present in certain foods so that they may be more suitable for people who cannot tolerate much of this substance. Infants and young growing children require highly digestible sources of nutrients. Likewise, people who have impaired digestion and/or absorption should be given diets that are low in fiber. Therefore, many people may benefit from more knowledge of this subject.

HISTORY. Although it was long believed that prehistoric peoples ate mainly animal foods that they obtained by hunting and scavenging, recent archaeological studies have shown that the human diet also contained large amounts of high fiber plant foods. New techniques for the isolation of plant fragments from dried fecal remains, called coprolites, have made it possible to identify the wild plants that were consumed along with the animal foods. For example, the diets of the Indians who lived in the Great Basin area of the American West contained about 25% fiber.

After the development of the first agricultural societies about 10,000 years ago, man started to select plants for higher fat and starch content, and for lower fiber content. Even then, hungry people often consumed grains and legumes without removing the fibrous outer husks, pods, and seed coats. In time, techniques for the crude processing of grains were invented and the long trend towards consuming ever more refined foods began. The most notable example of this kind was the early preference for breads made from white flour, which apparently was first produced in Egypt by sifting the stone ground flour through sieves made from woven strands of papyrus.

Later, the health conscious ancient Greeks noted that large stools were produced when brown (whole grain) breads were consumed, but that white bread was tolerated better by people afflicted with diarrhea. Nevertheless, it was the lot of most people to eat coarse, dark bread because the sieving procedure was too time-consuming and wasteful of the grains, which were usually in short supply. Often, the bread of the poor was heavily loaded with the husks that remained after the production of white flour for the wealthy people.

During the Middle Ages, there were many years in which poor yields were obtained from the grain crops of Europe and it was necessary to extend cereal flours with flourlike materials obtained by grinding acorns, bark, beans, nuts, peas, and twigs. The diets of the poor in Europe were also rich in vegetables such as cabbage, onions, and starchy tubers.

The early American colonists consumed plenty of beans, unrefined corn and cornmeal, and various fruits and vegetables, since large supplies of dairy products, eggs, meats, and poultry did not become available until the 19th century, when (1) the westward movement began, (2) agricultural techniques were improved greatly by the application of the research done at home and abroad, and (3) transportation systems for bringing the farm products to heavily populated areas were established. The opening of the level, fertile soils of the Midwest and the Great Plains altered the American diet by providing land for the production of wheat for flour, and corn for feeding livestock and poultry. Also, the increased availability of animal foods at reasonable prices eventually led to a decline in the use of beans and peas in main dishes.

It is noteworthy that throughout the early 1800s health and religious concerns fired up certain advocates of diets based mainly on whole grains, vegetables, and fruits. These concerns were motivated in part by the high consumption of alcoholic beverages during the 1820s and 1830s. Sailors and laborers often received part of their wages in the form of liquor, and children were encouraged to drink hard cider, which was safer than the drinking water. One of the earliest and most notable of the health reformers was a former Presbyterian minister named Sylvester Graham (1794-1851), after whom graham flour and graham crackers are named, who promoted vegetarianism, abstinence from alcoholic beverages, and the use of coarse, unbolted flour for the making of bread with a high-fiber content. Mother Ellen Harmon White, who was a leader of the Adventists, was influenced considerably by the dietary ideas of Reverend Graham. Hence, whole grains were important items in the diets of the patients at the famous Battle Creek Sanitarium, which was established by Mrs. White and managed by Dr. John Harvey Kellogg, who, with his brother, Will, developed the first ready-to-eat breakfast cereals. However, the work of the 19th century health reformers was offset by the development of large scale roller milling of wheat in the 1870s. This was the first time in history that large supplies of white flour were made available at prices that even the poor could afford.

Interest in whole grain breads and vegetarianism declined steadily in 20th-century America until World War I. The governments of Great Britain and Denmark made the use of high extraction wheat flours mandatory during war in order to make the fullest possible use of the limited supplies of grain. However, the use of highly refined flours resumed in these countries after the war. Similarly, increased use was made of whole grain flours during World War II, but this salutory practice went out of style in the postwar period.

It was not until the 1960s that the current interest in dietary fiber was stimulated by the reports of several British physicians (mainly Drs. Denis Burkitt, Peter Cleave, Neil Painter, and Hugh Trowell) who noted that certain African peoples who consumed high fiber diets had much lower rates of certain diseases than the people of Great Britain and the United States. The average daily intake of crude fiber in developed nations has been estimated to be about 4 grams, compared with as much as 30 grams in some developing countries. Many other investigations and reports followed, with the result that some new high fiber products appeared in supermarkets.

KINDS OF FIBER.

Fiber may be obtained from diverse sources, such as grains, vegetables, fruits, nuts, seaweed, gum trees,and the laboratory of the organic chemist. Therefore, it is worthwhile to consider the plant constituents that comprise most of the fiber in the human diet.

Fibrous Substances in Common Foods.

Most unrefined vegetable foods contain variable amounts of the materials which follow:

- **Cellulose**—This compound is the most abundant polysaccharide found in the cell walls of plants. Its molecule is made up of glucose units joined together in a chainlike structure. Cellulose differs from starch (also a polysaccharide consisting of glucose units) in that it is not digested by the enzymes secreted by the human digestive system. Ruminants (such as cattle, sheep, and goats) digest cellulose fairly effectively, but it is poorly digested by humans, pigs, and chickens. However, the three latter species may harbor intestinal microorganisms that digest small amounts of the dietary cellulose.

- **Hemicelluloses**—These polysaccharides, which are closely associated with cellulose in cell walls, are composed mainly of sugar units containing 5- or 6-carbon atoms (pentoses or hexoses). They are insoluble in water, but soluble in alkali. Thus, the cooking of vegetables in water containing baking soda (an alkali) may result in a mushy texture because the alkaline solution extracts the hemicelluloses from the cell walls, resulting in a loss of rigidity.

- **Gums**—Originally, the term "gum" designated the highly viscous and sticky substances that were exuded by various types of plants. Now, gums are considered to be a large class of substances that may be dissolved or dispersed in water to give a gelling or a thickening effect. However, plants having a high content of gummy material are not often used as food sources; rather, they're used as sources of gums for food additives. Some of the most common natural sources of gums are (1) plant exudates (which yield gum arabic, gum karaya, and gum tragacanth); (2) seeds (the sources of locust bean gum and guar gum); (3) seaweeds (such as those which furnish agar, algin derivatives, and carrageenan); and (4) various other plant extracts such as larch gum.

- **Mucilages**—Presently, these substances are usually classified under gums, since they have many of the same properties. Formerly, mucilages were considered to be the plant polysaccharides which readily formed sticky, slimy (mucilagenous)solutions in water. Two of the most familiar examples of these substances are those that may be extracted from flax seeds and psyllium seeds. Various products made from these seeds have long been used as laxatives. It is noteworthy that ladies used to set their hair in curls or waves with a solution obtained by steeping flax seeds in hot water.

- **Pectic substances**—These polysaccharides are made up mainly of chains of galacturonic acid (a derivative of galactose) units. The individual units may be modified by attachments of methyl and other chemical groups. Some of these compounds are water-soluble. Pectins, which are the best known of these materials, form gels with sugar and acid. The pectin present in apples is partly responsible for the crispy texture (a decrease in pectin content after picking results in mealiness). The leading commercial sources of pectin are citrus peel and apple pomace (the residue that remains after juice extraction).

- **Lignin**—Although included under the categories of crude fiber and unavailable carbohydrate, lignin is not a true carbohydrate. Like carbohydrates, however, it contains carbon, hydrogen, and oxygen; but the proportion of carbon is much higher than in carbohydrates, and nitrogen may also be present. Lignin is found in plant cell walls in intimate association with cellulose, to which it imparts rigidity. It increases in amounts with plant maturity; hence, vegetables, such as cabbage and peas, that are eaten before maturity are low in lignin content.

- **Miscellaneous polysaccharides**—This includes compounds that are undigestible, but which are not likely to be detected by the procedures usually employed in the analyses of foods for fiber content. Examples of such compounds are *inulin,* a water-soluble polysaccharide made up of fructose units and found in various roots and tubers such as the Jerusalem artichoke, and *galactan,* a galactose polysaccharide found in agar, a material obtained from seaweed.

Food Additives.

The use of various types of nonnutritive carbohydrates in foods is increasing as food manufacturers find new ways to produce the characteristics sought by the consumer. Thanks to these materials, mixtures can be kept from separating, thickening can be achieved without increasing the energy content of foods, and jams gel easily. Some examples follow:

• **Natural materials**—These are generally the gums and pectic substances that are derived from various types of land and sea plants. However, cellulose produced from wood chips is sometimes used to increase the fiber content of foods.

• **Semisynthetic materials**—This includes modified celluloses (microcrystalline, carboxymethyl, and methyl), modified gums (derivitives of pectin, alginates, locust bean gum, and guar gum), and microbial fermentation gums (dextran and xanthan).

• **Synthetic materials**—The principle members of this class are polymers of polyvinylpyrrolidone (PVP), ethylene oxide, carboxyvinyl, and methyl vinylether maleic acid.

DETERMINATION OF THE FIBER CONTENT OF FOODS. The human or animal nutritionist has long been interested in fiber mainly from a negative point of view; that is, it has been necessary to know to what extent the carbohydrate fractions of foods or feeds are unavailable for energy production in the body (burning these materials in a calorimeter gives only the total energy content, rather than that available for metabolism). Also, food inspectors have long used fiber as a measure of adulterants and contaminants, such as husks and shells in ground spices and cocoa. Finally, much of the current interest in fiber content of foods stems from reports of its beneficial effects. However, it is surprising that many recent studies used procedures over 100 years old for obtaining estimates of unavailable carbohydrate or fiber, and for the percent digestibility of plant materials and their nutrient fractions. the most widely used procedures are described in the sections that follow.

Proximate Analysis. This term is used to designate a series of chemical analyses which yield an estimate of food or feed values in terms of water, ash, crude protein, crude fat, available carbohydrate, and crude fiber. The food or feed to be analyzed for crude fiber is finely chopped or ground, dried to a constant weight in a low temperature oven, and the fat removed by extraction of the dried sample with solvents. The fat-free residue is first boiled in dilute sulfuric acid, then rinsed with hot water and boiled in dilute sodium hydroxide. Crude fiber is the organic portion of the dry, acid and alkali-extracted residue (organic material is represented by the loss of weight occurring when the residue is ashed). The value for fiber given in food composition tables is that for crude fiber measured by the method just described. It is noteworthy that the percentage of available carbohydrate is determined by the following formula: Available Carbohydrate = 100 − (% water + % ash + % crude protein + % crude fat + % crude fiber).

The residue of crude fiber contains much of the cellulose and lignin, but little of the hemicelluloses, gums, mucilages, and pectic substances present in the original sample of food. The latter components are extracted by cooking in acid solution. Thus, the values obtained in this determination represent only part of the total fiber content and should be used mainly for making comparisons between foods that have similar chemical compositions. For example, the types of nuts with the highest contents of crude fiber are likely the richest in total fiber content, also. However, crude fiber values for nuts are not strictly comparable to crude fiber values for green leafy vegetables because the chemical compositions of the two types of foods are quite different.

(Also see ANALYSIS OF FOODS.)

Digestibility Studies. Another basis for apportioning the total carbohydrate content of a food into digestible and undigestible fractions is the percent digestibility of the carbohydrate. (100 − % digestibility of carbohydrates = % of total carbohydrates which are undigestible.) The percent digestibility of the carbohydrates in a food may be determined by feeding weighed amounts of the material to be tested. During the test period the feces are collected, weighed, and dried. Percent digestibility is the percent disappearance of carbohydrates (on a moisture-free basis) in the digestive tract (food carbohydrates − fecal carbohydrates = disappearance of carbohydrates). Plant foods usually contain at least small amounts of indigestible crude fat and crude protein, so most studies include proximate analyses of both the food and the feces in order to partition the digestible and undigestible components among the major nutrients (carbohydrates, fats, and proteins).

There is a moderate error in this determination due to methods of calculating the total carbohydrate contents of the food and feces. They are assumed to be the quantities which remain after subtraction of the amounts of water, fat, crude protein, and ash from the weights of the food and fecal samples. Thus, the errors in the values for carbohydrates are the sums of the errors in the measurements of the other nutrients. Nevertheless, the greater accuracy of digestibility studies in estimating percent of unavailable carbohydrates (compared to the proximate analysis procedure for crude fiber) has recently been demonstrated by a chemical and enzymatic fractionation of unavailable carbohydrates in plant foods.[5]

For example, the data obtained by Southgate show that the fiber present in wheat contains 10% cellulose, 21% lignin, and 65% hemicelluloses. Hence, the determination of cellulose and lignin contents by the crude fiber procedure measures only about one-third of the total fiber contents of wheat. The U.S. Department of Agriculture food composition tables show that crude fiber comprises only about 15% of the total carbohydrate content of bran, whereas the coefficient of carbohydrate digestibility is listed as 56%.[6] A digestibility coefficient of 56% indicates that bran contains 44% undigestible carbohydrate, which is about three times the crude fiber content. Therefore, there is a close correlation between the data of Southgate and the digestibility data used by the USDA.

The data of Southgate also showed that an average of 71% of the total fiber from grains (wheat and rye) would not be detected by the crude fiber analysis, while an average of only 30% of the total fiber from vegetables and fruits would be undetected. These data help to ex-

---

[5]Southgate, D. A. T., "Determination of Carbohydrates in Foods. II. Unavailable Carbohydrates," *Journal of the Science of Food and Agriculture*, Vol. 20, 1969, p. 331.

[6]*Composition of Foods*, Hdbk. No. 8, USDA, 1963, p. 66, Table 1, and p. 160, Table 6.

plain the many observations of significantly greater physiological effects per gram of crude fiber from grains compared to that from vegetables and fruits. The grains contain greater amounts of hemicelluloses and related substances.

Unfortunately, the procedures used for both digestibility studies and chemical fractionation are expensive compared to analyses for crude fiber by the proximate analysis scheme.

(Also see ANALYSIS OF FOODS.)

Analysis Of Cell Wall Constituents. Another approach to the problem of measuring indigestible components of plant materials is the determination of percent cell wall in the dried food.[7] This procedure is outlined in Fig. F-9. The Van Soest procedure was developed mainly for estimating the nutritive values of forage plants for cattle, goats, and sheep, but it is also applicable to the determination of fiber in feeds and foods eaten by humans, chickens, and pigs; since for the latter three species most of the cell wall components of plants are undigestible.

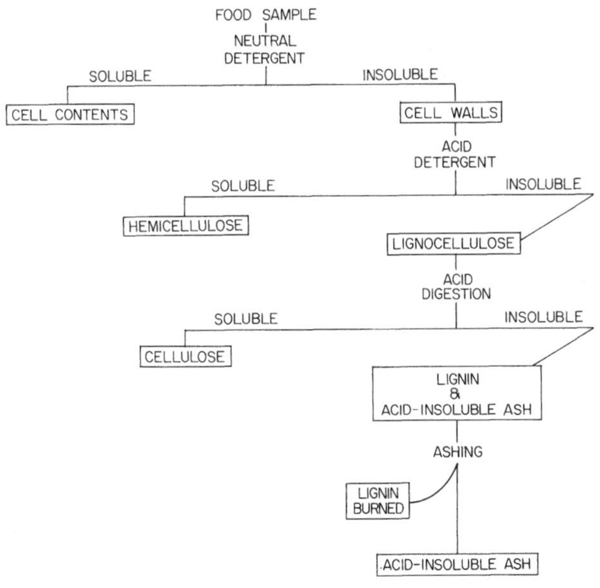

Fig. F-9. Van Soest procedure for determination of the cell wall contents of foods.

The procedure involves the digestion of a dried, pulverized sample in a buffered solution of a neutral detergent such as sodium laurel sulfate. Then, the digest is filtered and the dried residue is taken to represent most of the cell wall material (small amounts of pectic substances are extracted during the digestion), or neutral detergent fiber (NDF). Although this analysis might be carried further by the fractionation of NDF into hemicelluloses, cellulose, and lignin, such an extensive analysis of foods for fiber would not always be necessary, since neutral detergent fiber is by itself a better measure of total undigestible carbohydrate for man than is crude fiber.

(Also see ANALYSIS OF FOODS, section on "Van Soest Analysis for Fiber.")

[7]Van Soest, P. J., "Development of a Comprehensive System of Feed Analysis and Its Application to Forages," *Journal of Animal Science*, Vol. 26, 1967, p. 119.

PHYSIOLOGICAL ACTIONS OF DIETARY FIBER. Although man has long known about some of the beneficial and harmful effects of roughage in the digestive tract, evidence which suggests that there may be other important physiological actions of cell wall constituents has been found recently. It has even been suggested that fiber may soon be classified as an essential nutrient. Discussions of some of these effects follow.

Beneficial Effects. There is still considerable uncertainty regarding certain potentially beneficial effects of dietary fiber. The most controversial and hypothetical of the benefits are followed by a question mark to indicate that hard evidence is lacking.

- **Antidiarrheal actions**—The gums, which are generally soothing to the digestive tract, may act against diarrhea by (1) coating inflamed surfaces, and (2) undergoing microbial fermentation to organic acids which inhibit the growth of pathogenic microorganisms. One of the popular remedies for diarrhea is a mixture of pectin and finely powdered clay (kaolin) in a water suspension.

- **Cancer prevention?**—Population studies in Africa and Finland indicate that fiber may help to prevent cancer of the bowel by increasing the rate of food passage (allowing less time for contact of carcinogens with the tissue in the bowel), and by altering the types and amounts of microorganisms (some of which may degrade bile salts to potential carcinogens). This theory, however, has not yet been tested in controlled experiments with humans or animals.

(Also see CANCER, section on "Nutrition and Diet.")

- **Constipation remedy**—The best known effect of fiber in food is the prevention, or correction, of constipation by the stimulating effect of bulk (undigested material) on the colon. Many of the carbohydrates classified under "fiber" absorb and hold water in the bowel, thereby adding to the volume of the stool and helping to prevent the contraction of the bowel in small segments. The latter function is important because it is believed that such contraction leads to the development of diverticuli, or small pouches, in the wall of the large intestine.

Another means by which fiber may help to correct constipation is by undergoing microbial fermentation to organic acids which stimulate the peristaltic movements of the colon.

NOTE WELL: It is very important to consume plenty of fluids with the water-absorbing bulking agents so that soft, smooth masses of fecal material pass through the colon. Otherwise, hard dry masses which can damage the intestinal tissue may be formed.

- **Detoxifying actions?**—It is suspected that various types of dietary fiber may bind potentially harmful substances that originate (1) in foods, or (2) from the action of intestinal microorganisms on the constituents of foods. Hence, a high-fiber intake may help to prevent toxicants from being absorbed, or from acting upon the tissues of the digestive tract.

- **Diabetes therapy?**—A high carbohydrate diet that is rich in fiber and other complex carbohydrates lowers the fasting blood sugar levels of adult male diabetics to significantly lower values.[8] Also, such a diet lowers the blood cholesterol and triglycerides, and reduces the requirements for oral diabetic drugs and insulin.

• **Diverticular disease prevention and therapy**—Diverticula are small pouches that form in the wall of the colon as a result of overly forceful peristaltic contractions in small segments as shown in Fig. F-10.

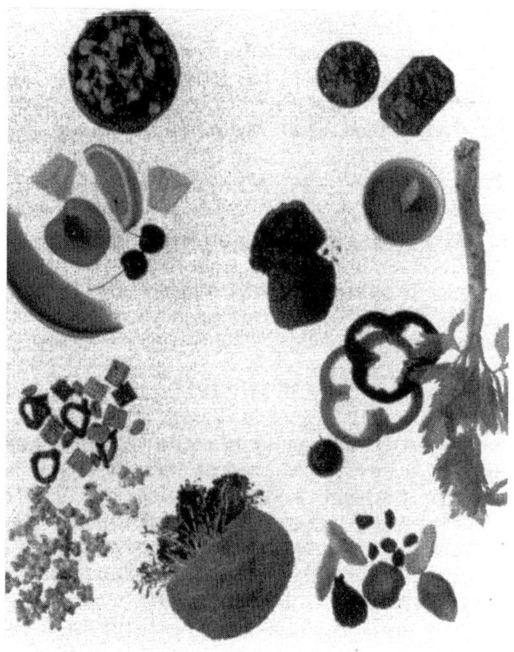

Fig. F-10. High fiber foods lessen diverticulitis. (Courtesy, USDA)

It is believed that chronic constipation is a major cause of this disorder, which may become very painful if the pouches become inflamed. The inflammation is called diverticulitis. Although it was once thought that a low-fiber diet was an appropriate treatment, it has recently been found that a high-fiber diet is the best means of both prevention and treatment.

• **Gallstone therapy?**—Certain types of dietary fiber alter the composition of the bile by binding one or more of the bile salts that would otherwise be absorbed and recirculated. The altered bile is a better solvent for cholesterol (a major constituent of gallstones) than the unaltered bile. However, further tests of this dietary therapy are needed to confirm that it is both effective and safe.

• **Lipid-lowering agents?**—Recently, it was discovered that some types of dietary fiber lower the blood levels of cholesterol and triglycerides. This effect is believed to take place by the binding of bile salts and dietary lipids (mainly cholesterol and triglycerides) to the fiber. Normally, from 95 to 99% of these compounds are absorbed and transported to the liver. When, however, the absorption of bile salts and lipids is reduced by their binding to fiber and their losses in the stool, the turnover of those substances in the body is accelerated to compensate for the losses. Thus, the overall effect of fiber is to reduce the blood levels of cholesterol and triglycerides. However, food sources of fiber are not as effective in treating elevated levels of blood lipids as some of the drugs designed for this purpose.

• **Soothing of gastrointestinal irritation**—Mucilagenous types of fiber, such as the water-dispersible gums present in psyllium seed, okra, carob, and quince seed, coat and soothe irritated areas, particularly mucous surfaces.

• **Weight control adjunct**—Eating a diet high in fiber is likely to result in the consumption of less food energy because the digestive tract becomes full sooner. Normally, appetite is diminished by the distention of the gastrointestinal tract. Food plants consumed by prehistoric Indians in Mexico have been found to have been very high in fiber (as determined by analysis of present day plants which matched the dried specimens found in the caves inhabited by the Indians). Anthropologists believe that the prehistoric Mexicans may have had enlarged digestive tracts as a result of regular consumption of such fibrous food plants.

Potentially Harmful Effects. High-fiber diets which contain seed coats (rich in phytates) and other undigestible substances, may interfere with the absorption of minerals and other nutrients since the fibrous materials accelerate the flow of food through the digestive tract and bind certain nutrients and carry them into the stool. For example, digestibility studies have shown that high levels of dietary fiber reduce the digestibility of fat and protein. Also, people with severe inflammation of the lower bowel, and those recovering from gastrointestinal surgery, often need to eat low-fiber diets until they have sufficient healing. Finally, the consumption of excessive amounts of fiber may lead to enlargement and twisting of the sigmoid colon (the condition called volvulus).

DIETARY FIBER RECOMMENDATIONS. People who are unaccustomed to high-fiber diets should eat just enough fiber to ensure the regularity of bowel movements. Studies with human subjects have shown that there is a relationship between the degree of laxative effect and the proportion of dietary fiber per unit of body weight.[9] It was found that the passage of one stool per day may be ensured for most persons by the consumption of 0.02 to 0.03 g of crude fiber per pound of body weight. Persons with a tendency towards constipation may require 0.04 to 0.05 g of fiber per pound (*0.45 kg*) of body weight. Thus, a 160-lb (*72 kg*) man, who has a tendency to be constipated, may require as much as 8.0 g of crude fiber per day to ensure his regularity. It is not unusual for populations consuming mainly plant foods to each as much as 25 g of fiber per day.

[8]Kiehm, T. G., *et al.*, "Beneficial Effects of a High Carbohydrate, High Fiber Diet on Hyperglycemic Diabetic Men," *The American Journal of Clinical Nutrition*, Vol. 29, August 1976, pp. 895-899.

[9]Cowgill, G. R., and W. E. Anderson, "Laxative Effects of Wheat Bran and 'Washed Bran' in Healthy Men. A Comparative Study," *Journal of the American Medical Association*, Vol. 98, 1932, p. 1866.

SELECTION AND PREPARATION OF FOODS HIGH IN FIBER. Although there has been selective cultivation by man of plant varieties low in fiber (more succulent and tender), there is still available a wide range of foods with moderate to high fiber content. The crude fiber contents of a large number of foods are given in Food Composition Table F-36. In order to help the reader identify the leading fiber rich foods, the crude fiber and energy contents of selected items are presented in Table F-10, which follows.

TABLE F-10
TOP SOURCES OF CRUDE FIBER AND THEIR CALORIC CONTENT[1]

Top Sources[2]	Crude Fiber	Energy
	(mg/100 g)	(kcal/100 g)
Coriander seed	29.1	298
Bay leaves	26.3	313
Pepper, red	24.9	318
Cinnamon	24.4	261
Chili powder	22.2	314
Allspice	21.6	263
Dill seed	21.1	253
Paprika, domestic	20.9	289
Rosemary leaves	19.0	331
Thyme	18.6	276
Marjoram, dried	18.1	271
Sage	18.1	315
Basil	17.8	251
Rosemary, ground	17.7	331
Curry powder	16.3	325
Fennel seed	15.7	345
Savory	15.3	272
Oregano	15.0	306
Pumpkin pie spice	14.9	342
Anise seed	14.6	337
Pepper, black, ground	13.1	255
Caraway seeds	12.7	333

Top Food Sources	Crude Fiber	Energy
	(mg/100 g)	(kcal/100 g)
Parsley, dried	10.3	276
Wheat bran	9.1	350
Pepper, hot chili, mature, red, raw, pods and seeds	9.0	93
Carob flour	7.7	180
Bran breakfast cereals, average of several types	7.6	239
Pears, dried, uncooked	6.3	268
Peppers, hot chili, mature, red, dried pods	6.2	321
Cocoa powder, low fat	5.8	187
Raspberries, black, raw	5.1	73
Figs, dried, uncooked	4.7	274
Onions, mature, dehydrated flakes	4.4	350
Cocoa powder, high and medium fat	4.3	280
Blackberries, raw	4.1	58
Coconut meat, average of several products	4.0	519
Apples, dehydrated, uncooked	3.8	353
Olives, ripe, cured, Greek style	3.0	338
Bran flakes with raisins	3.0	287
Raspberries, red, raw	3.0	57
Almonds, unsalted	3.0	598
Apricots, dried, uncooked	3.0	260

[1]These listings are based on the data in Food Composition Table F-36. Some top or rich sources may have been overlooked since some of the foods in Table F-36 lack values for crude fiber.

Whenever possible, foods are on an "as used" basis, without regard to moisture content; hence, the fiber content of certain high-moisture foods may be misleading when ranked on the basis of crude fiber content per 100 g (about 3½ oz) without regard to moisture content.

[2]Listed without regard to the amount normally eaten.

To use the data on fiber content, select the basic types of foods desired for a meal or a diet (breakfast cereal, fruit, bread, etc.). Then identify the items with the desired fiber contents, note the amounts provided by 100 g (about 3 ½ oz) of the foods and calculate the fiber supplied by the amounts of foods to be consumed. For example, a 1 oz (*28 g*) serving of a bran breakfast cereal furnishes about 2.2 g of fiber (7.6 g per 3½ oz ÷ 3½ = 2.2 g per oz). (The food energy values may be used similarly to make choices between the various items within each of the basic food groups.)

For those who do not wish to bother with calculations of fiber and energy contents, some suggestions for the selection and preparation of foods high in fiber follow:

1. All plant foods are likely to contain at least small amounts of fiber, while animal foods (meat, eggs, fish, dairy products) are almost totally digestible (unless they have been made less digestible by improper cooking).
2. Fresh, unrefined food usually contains more fiber than processed food, except when the processor concentrates a product by drying, or adds less digestible materials, such as the addition of bran to breads or cereals. For example, on an equal weight basis (pound for pound), dried fruits contain from 3 to 7 times the calories, fiber, and other nutrients of fresh fruits.
3. Green leafy vegetables and stems of plants are higher in fiber than starchy roots and tubers. An exception to this rule is the inulin-containing roots and tubers, such as those of dandelions, chicory, globe artichokes, Jerusalem artichokes, and salsify. Although inulin is not classified as crude fiber, it is undigestible. However, long storage of inulin-containing roots and tubers may result in conversion, by enzymes in the plant, of inulin to fructose, which is completely utilized by man.
4. Seed coats of grains and legumes are high in fiber, as are the skins and peels of fruits and vegetables.
5. The sprouting of seeds results in more fiber and less food energy.
6. Whole fruits and vegetables usually contain more nonnutritive carbohydrate than the juices extracted from them, since juice extraction generally involves removal of the fiber by straining or other means.
7. The pulpy residue from the juicing of oranges and other fruits may be added to dishes such as biscuits, muffins, puddings, and sauces.
8. Pureed cooked vegetables such as beans, beets, carrots, peas, spinach, squash, sweet potatoes, and turnips may be used to thicken sauces and soups.

CAUTION: Although it is possible to select a diet low in calories and very high in fiber, a sudden change from a low-fiber diet to a high-fiber diet may result in digestive problems such as bowel pains, excessive gas (flatulence), and diarrhea. Also, liberal amounts of water should be ingested, particularly when dried, fibrous foods (bran, dried fruits, dehydrated vegetables) are eaten. There have been reports of distention and pain of the esophagus, and blockage, irritation, and perforation of the colon and rectum resulting from the consumption of dried cellulosic materials without sufficient fluid.

It is also noteworthy that, throughout their lives, many people consume diets which are low in fiber, yet they appear to be healthy and live to ripe old ages. There is a possibility that some of these people have considerable amounts of friendly bacteria in their bowels. Bacterial cell walls are high in fiber and thereby contribute to the bulk of the stool. Diets high in milk and dairy products, for example, may encourage an intestinal growth of lactic acid producing bacteria which apparently stimulate peristalsis by various means. There are also people who cannot tolerate high-fiber diets because of diarrhea or various bowel troubles.

SPECIAL PRODUCTS FOR DIETETIC AND THERAPEUTIC USES. Although a diet high in fiber may be readily achieved by people who shop in supermarkets and prepare most of their meals at home, it is more difficult for those who eat much of their food in restaurants to consume such a diet regularly. Furthermore, some people do not care to eat the natural foods which are high in fiber. Finally, dietitians and physicians may wish to recommend diets of known fiber content or special low-energy recipes. Therefore, it is desirable in many of these cases to be able to prepare meals where forms of semipurified fiber could be added in known amounts of refined foods. Examples of some special high-fiber products follow:

• **Agar (agar-agar)**—This nonnutritive seaweed gum, called *kanten* by the Japanese, is used to prepare low-calorie recipes, such as jellies, candies, and gelled salads (agar is used by strict vegetarians in place of gelatin). Although the purified flakes are costly (a dollar or so per ounce, or *28 g*), a little goes a long way (1 tsp or 5 g may be used with a pint or 500 ml of hot water).

• **Alfalfa flour**—This is a highly nutritious item (it is rich in protein, fiber, minerals, and vitamins). Usually, alfalfa flour is added in small amounts to (1) the batters or doughs used to prepare various types of baked goods, and (2) sauces or soups.

• **Bran**—This product is well known for its laxative effect and is one of the cheapest sources of food fiber since it is a by-product (seed coat) of the milling of wheat and other cereals. Although not very appetizing by itself, it mixes well in cereals and baked goods. However, it is noteworthy that bran has caused gastrointestinal irritation in some people.

• **Dehydrated onion flakes**—A little of this product goes a long way in that small amounts exert a moderately strong laxative action. On a dry weight basis, onions produce from 2 to 3 times as much intestinal gas as beans.

• **Flax seed**—These nutlike seeds are tasty additions to grain products. Like bran, flax seed is an effective laxative. In addition to being available as seeds, flax seeds are an ingredient of certain ready-to-eat breakfast cereals.

• **Pectin**—This material is extracted from citrus peel and apple pomace (the residue after extraction of juice). It is available in pure form, or in mixtures that contain added sugar and acid and are used to prepare jams and jellies. Although pure citrus pectin has been given orally in the powdered form (1 or 2 tsp [*5 or 10 ml*] per day) to counteract the effects of dietary cholesterol, a more desirable use for it is in low-calorie recipes such as imitation or eggless mayonnaise, tomato aspic, fruit desserts, and pie filling. It is also used in antidiarrheal preparations.

• **Psyllium seed**—Although there are several important varieties of psyllium, the blond variety from India (*Plantago ovata*) is used the most because of its high content of colorless gum. The powdered form (pulverized material from the seeds) swells in water and yields a mucilaginous mixture which is useful as a demulcent (an agent that soothes and protects the lining of the digestive system), a stopping-up agent against diarrhea (the powder is taken with only a small amount of fluid so that a gummy substance is formed in the intestine), or a laxative (the powder is taken with a large amount of water to produce a soft bulk which also lubricates the bowel). A gel is formed when a 1% solution (a teaspoon of psyllium in a pint of water) is heated to 195°F (*90°C*) and allowed to cool. Like agar and pectin, there are potential uses for psyllium in various recipes. This material also helps to counteract the effects of dietary cholesterol.

FUTURE PROSPECTS FOR DIETARY FIBER. Laxative powders made from dried psyllium seeds now cost over $5.00 per pound in retail drug stores. Most of these products require dispersal in water or fruit drinks to make them suitable for oral dosage. It is more convenient and less expensive for many people to obtain extra dietary fiber in the form of appealing food products such as cereals, crispy snacks, frozen desserts, and milk shakes. Food technologists have already developed many new high-fiber products and it is only a matter of time until they become available to most consumers. However, many of the current items are based upon the outmoded concept of merely supplying added amounts of fiber without regard for the unique physiological effects of the different constituents of the cell wall.

The next generation of high-fiber foods will likely be tailored to yield specific beneficial effects. For example, one product could be designed to lower blood lipids, whereas another might be constituted so as to encourage the growth of certain helpful intestinal microorganisms. Greater control over the effects of dietary fiber would be useful to the practitioners of dietary therapy who must achieve a delicate balance between undertreatment and overtreatment of various disorders.

Another approach to obtaining selective utilization of certain fiber fractions is to render them more digestible rather than attempting to remove the undesirable components from the foods. Thus, canned or frozen foods might be treated with cellulose-digesting enzymes (cellulases) or pectin-digesting enzymes (pectinases) to render the foods more digestible for groups of people who cannot tolerate the fiber well. For example, infants, young children, patients with malabsorptive disorders, and older people with impaired digestion are often required to eat low-fiber diets in order to avoid excessive losses of essential nutrients in their stools.

Although the nutrition counselor should be aware of the possible role of fiber in the maintenance of buoyant good health, the importance of fiber in the human diet will finally be determined by additional clinical and experimental observations.

FIBRIN

The insoluble fibrous strands of protein essential to the clotting of blood, derived from fibrinogen by the action of the enzyme thrombin.

FIBRINOGEN

A protein in blood, which, through the action of enzyme thrombin, is converted to fibrin.

FIBROUS

High in cellulose and/or lignin content.

(Also see CARBOHYDRATES, UNAVAILABLE; and FIBER.)

FIBROUS PROTEINS (ALBUMINOIDS, SCLEROPROTEINS)

Proteins may be classified according to their structure, which is important to their function in the body. Those proteins which form long chains bound together in a parallel fashion are called fibrous proteins. They are the proteins of connective tissue, elastic tissue, and hair, or specifically collagen, elastin, and keratin. Fibrous proteins are not very digestible.

(Also see PROTEINS, section headed "Classification.")

FIG *Ficus carica*

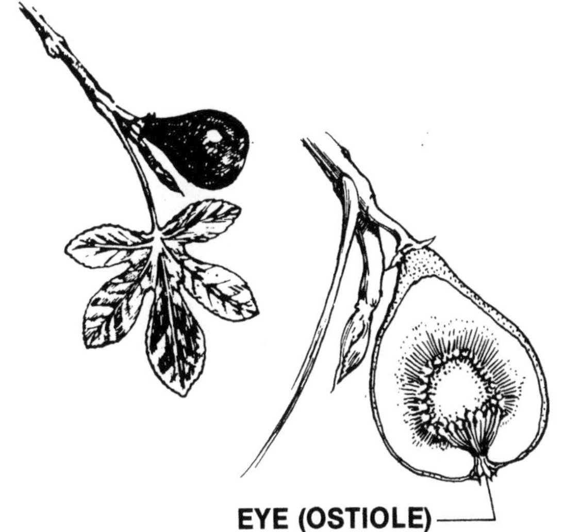

Fig. F-11. The common orchard fig showing the deeply lobed leaves, and the fleshy fruit with its seedy interior.

The fig tree is a deciduous bush or tree which grows in the subtropical regions of Asia, Africa, Europe, and the United States. The fleshy fruit, the fig, is sometimes called the fruit without a flower. But figs do have flowers, though they are hidden in the hollow center of the fig. Figs are unique among fruits in having an opening in the end of the fruit not connected to the tree. This is called the apical orifice, ostiole, or "eye," and it serves for the communication of the flowers with the exterior (see Fig. F-11). Figs are among the most ancient cultivated fruit trees, and man has enjoyed eating figs since early times.

ORIGIN AND HISTORY. Figs are native to southwestern Asia and the eastern Mediterranean. They were an important fruit to the people of western Asia in early civilization, being mentioned in the Bible and other ancient writings. They were probably first cultivated in Arabia and Egypt and were known in Crete by 1600 to 1500 B.C. From there, figs were introduced to Greece around the 9th century B.C., and soon afterward to Italy. Later, through conquests, figs were carried to Spain and Portugal. They were introduced into America, at Hispaniola, by the Spaniards in 1520, and before the end of the 16th century they were reported abundant in Peru.

In California, where most of the U.S. crop is grown, the fig, like many other fruits, dates from the establishment of the mission at San Diego in 1769. Commercial culture started in 1885. Dried Adriatic figs were shipped east in 1889, but these were inferior to imported Smyrna-type figs. Smyrna figs were introduced into California in 1880, but it was not until about 1900 that the fig wasp, *Blastophaga psenes*, was established and used successfully to transfer caprifig pollen to Smyrna-type figs to obtain fruit-set—a process called caprification. This success stimulated interest in commercial production of Calimyrna (California Smyrna) figs in California, and the acreage expanded in the early 1900s.

PRODUCTION. Figs are grown in numerous areas of the world with moderate climates. These areas include Italy, Algeria, Morocco, Tunis, Libya, Egypt, Palestine, Syria, France, Cyprus, Russia, Asia, Africa, Australia, Great Britain, and Central and South America. The five leading countries, in order of production, are Turkey, Greece, United States, Portugal, and Spain. California produces 99% of the U.S. fig crop.

During the last decade, U.S. fig production has not changed much. In 1980, 45,450 short tons (*40,905 metric tons*) were produced, and in 1990, 46,000 short tons (*41,400 metric tons*) were produced.

Almost all California production is consumed in the United States and Canada. A few countries export figs into the United States, mostly as paste figs, but these imports only total 5,437 metric tons. The United States exports about 2,200 metric tons.

Propagation And Growing. Fig trees are propagated by root cuttings from 2- or 3-year-old branches of other fig trees. Following planting, these new trees will produce a few fruits within the second or third year.

Figs trees have soft, pithy wood, and the bark along the tree trunk often has tubers or swellings that are pronounced, especially Calimyrna. Some common type cultivars (varieties) of figs, like Mission and Brown Turkey, bear two crops during one growing season. The first crop (breba) is produced from overwintering fruit buds and matures in June. The second crop is produced in leaf axils of spring and early summer shoot growth and matures in August and September. Other common type varieties, such as Adriatic, bear few first-crop fruits and many more second-crop fruits. Smyrna-type figs, such as the Calimyrna variety, also produce mainly second-crop figs.

In contrast to common-type figs, which set fruit parthenocarpically (without fertilization) the Smyrna-type requires pollination (caprification) for fruit-set. A third horticultural group of figs, known as caprifigs, is grown in California to provide pollen for Smyrna-type figs. This group produces three series of fruit buds each year. Caprifigs are inedible, but they act as continual hosts for the fig wasp, which transmits pollen to Smyrna-type figs. A fourth group of figs, White San Pedro, produces first-crop figs wihtout pollination, but requires pollination for the second crop. The King variety, which is produced for fresh shipment, is an example of this type.

• **Caprification**—Caprification is a horticultural word used to identify the pollination process in figs. The fig wasp, *Blastophaga psenes*, acts as an agent of pollen transfer from caprifigs to Smyrna-type figs that require pollination and fertilization for their fruits to set and mature. Fruits of the Calimyrna variety, without caprification, grow to ½ to ¾ in. (*1.3 to 1.9 cm*) in diameter and then turn yellow, shrivel, and drop before maturing. Common-type figs such as Adriatic, Kadota, and Mission do not require caprification.

The fig fruit, as described, is a hollow peduncle bearing numerous pistillate (female) flowers in the inner wall. For Calimyrna fruits to mature, these flowers need to be pollinated. The pollen must be provided from an external source, and nature has provided the very specialized fig wasp to transfer pollen from profichi caprifigs (which contain male flowers with pollen) to Calimyrna fruits.

The fig wasp relies on caprifigs to reproduce and complete its life cycle. It completes three life cycles per year, coinciding with the three caprifig crops (profichi, mammoni, and mamme). Female adults emerge from each maturing caprifig crop and enter fruit of the succeeding, overlapping crop. Here they deposit eggs on modified female flowers (gall flowers) suited to fig wasp egg laying. Larvae hatch from the eggs and develop in the ovaries of these flowers. Male wasps hatch first and fertilize the females before they leave the gall. After mating, adult female wasps migrate to the succeeding caprifig crop.

The caprification process, where wasps go from pollen-bearing profichi caprifigs to Calimyrna figs, occurs from the end of May into June. Adult female wasps emerge from caprifigs and enter Calimyrna figs, seeking egg-laying sites. In the process, pollen carried on the wasp bodies is spread to female flowers and results in fertilization and the production of viable seeds. Eggs are not laid in female flowers of Calimyrna figs because the flower structure is not suited for use by the wasp.

The female flowers are receptive in young Calimyrna figs that are 3/8 to 1/2 in. (*0.9 to 1.3 cm*) in diameter. Caprifigs are ready to be used for pollination as soon as pollen begins to shed and wasps emerge. Calimyrna figs are receptive in most years before caprifigs from the same distict are shedding pollen. Thus, it is common practice to produce or obtain caprifigs from an earlier district. Caprifigs are transported directly to Calimyrna orchards where they are placed in wire baskets or perforated paper bags hanging in the trees. Mature Calimyrna trees will require 4 to 5 caprifigs per tree every 3 to 5 days for a 2- to 3-week period. The number of figs dispensed, frequency of application, and length of caprification period are dependent on weather conditions and past history of the specific orchard. Wasp activity can be monitored by inspecting the inside of Calimyrna fruits for their presence. Excessive numbers of wasp per fig results in overcaprification which produces too many seeds and increases the potential fruit splitting and spoilage at maturity.

Fig. F-12. Caprification in a California fig orchard. Note the arrow pointing to the bag containing caprifigs. (Photo by A. H. Ensminger)

Fig trees are usually planted about 30 ft (*9.2 m*) apart. They thrive under conditions of low humidity, intense sunshine and hot temperatures during the summer months. Care of fig orchards includes proper pruning, irrigation, cultivation, and fertilization for a good crop.

Harvesting. Figs grown for dried utilization go through normal senescence in the orchard. Thus, figs fully ripen and partially dry on the tree. Then, they fall to the ground, where further drying occurs. Most figs are mechanically harvested from the ground, although some hand labor harvest is still used. A typical operation involves mechanical sweeping of figs from under the trees into a windrow in the center of each row. Further drying may take place here before mechanical harvesters remove the fruit from the ground and place it in bins which hold approximately 1,000 lb (*454 kg*) of fruit. Leaves, small branches, small stones, and other debris are separated from the figs during harvest. Harvest continues for 4 to 6 weeks at weekly intervals. After harvest, the fruit is transported to a collection site, dehydrator, or storage area, where it is fumigated to eliminate insects.

Fresh market or canning figs are picked from trees as they become fully colored and while they are still firm. Harvest is by hand, often from ladders, and the figs are placed in buckets or shallow flats. Pickers wear gloves because latex exuded from freshly harvested fig stems causes skin irritations. Actual fruit removal from the branches is usually accomplished by grasping the fig in the hand while twisting and pulling in one motion. Fresh figs are highly perishable, so it is exceedingly important to transport the harvested product without delay to a cold storage facility.

Figs for drying often require additional moisture removal following harvest. This can be done by sun drying or dehydration. Usually the fruit is subjected to a rapid water rinse before it is dried. With sun drying, the fruit is placed in single layers on wooden trays and spread in direct sun. Fruit to be dehydrated is also placed on trays, which are stacked and moved into dehydrating tunnels. The objective of drying is to reduce fruit moisture to about 17%. This may take a few days in the sun or 6 to 12 hours at 140°F (*60°C*) in dehydrators. After drying, the fruit is again fumigated to control insects.

PROCESSING. About 90% of the fig crop is utilized as dried figs, while the rest is sold fresh or canned.

Upon receipt of fruit, processors hold it under fumigation until processing. Varieties tending to darken, such as Calimyrna and Adriatic, are held in cold storage if facilities are available. Highest quality figs are used whole as "package stock," while other figs are sliced and ground into "fig paste" and a limited quantity is used for fig juice or "concentrate." Processing begins by sizing the fruit, passing the figs through a mechanical washer, and then through a retort filled with hot water and steam. This cleans the fruit and increases the moisture content. In the preparation of paste the figs are sliced and then refrigerated to prevent darkening and spoilage. They are maintained under refrigeration until ground into paste.

Marketing dried figs produced in California varies with variety.

• **Calimyrna**—Whole figs of suitable size and quality are prepared in consumer packages (polyethylene bags or plastic cartons), 30 lb (*13.6 kg*) bulk packages, and "finger-and-pulled" packs. All other marketable fruit is utilized as fig paste for the bakery industry.

• **Mission**—Whole figs are also packaged in consumer packs and bulk containers. Some figs are used in fig paste, but they are usually blended with white varieties. This variety is the main one used for juice or concentrate.

• **Adriatic**—These figs are utilized almost entirely as fig paste.

• **Kadota**—This variety is probably best known in California as a canning fig. There are, however, some orchards that are maintained exclusively for dried fruit production. When dried, it is used as fig paste and sulfured fruit for specialty export markets.

PREPARATION. Most figs are eaten dried but they may also be eaten fresh, canned, preserved, or pickled. Figs are eaten alone or in fruit dishes, baked goods, jams, relishes, or salads. Bakers use the fig paste in fig bars, fig filled cookies, and other bakers products.

NUTRITIONAL VALUE. Raw and canned figs contain 65 to 80 Calories (kcal) of energy per 100 g (about 3 ½ oz), while dried figs contain 266 Calories (kcal) per 100 g. Dried figs are about 60% sugar—glucose, fructose, and sucrose. Furthermore, dried figs are a good source of calcium, potassium, and iron, containing 150 mg, 773 mg and 2.1 mg/100 g, respectively. More complete nutritional data regarding raw, canned, or dried figs and fig cookies is presented in Food Composition Table F-36.

(Also see FRUIT[S], Table F-47 Fruits of the World.)

FIGS, DRIED

The major types of dried figs available in the United States are: (1) light tan Adriatic or Greek string; (2) black Mission; and (3) golden brown Turkish Smyrna and Calimyrna (a California variety developed from the Smyrna fig). Usually, a Smyrna type sells for a much higher price than either the Greek or Mission types because it has a distinctive honey flavor and requires a special type of pollination in its production. Recently, there has been a trend away from the production of the dry, tough figs of the past to moister, more tender products that are protected from spoilage by treatment with potassium sorbate.

Dried figs are high in calories, sugar, fiber, and potassium. They are also good sources of calcium and iron. Many people find that the liberal consumption of dried or stewed figs helps to promote regularity of bowel movements.

(Also see FIG.)

FINELY GROUND

Reduced to very small particle size by impact, shearing, or attrition.

FINES

Any material which will pass through a screen whose openings are immediately smaller than the specified minimum size.

FINISH

To fatten a slaughter animal. Also, the degree of fatness of such an animal.
(Also see MEAT.)

FIRELESS COOKER OR HAYBOX

A tight box packed with hay as insulation, in which food that has been cooked for a short time is placed, and where it continues to remain hot for several hours.

FISH AND SEAFOOD(S)

Contents *Page*

History of Fishing 720
Fishing Industry 722
Fisheries 722
Catching Fish 722
Fish Catch 724
Fish Farming 725
Processing Fish and Seafood 726
Products and Preservation 726
Industrial Products 728
Buying Fish and Seafood 728
Handling and Cooking Fish and Seafood 729
Nutritive Qualities of Fish and Seafood 730
Types of Fish and Seafood 732

Fig. F-13. Fishes of the sea. This is a selection of the wide variety of fish harvested by Canadian fishermen. (Courtesy, Department of Fisheries and Oceans, Canada)

Millions of tons of fish and seafood are taken from the seas, rivers, lakes, and ponds every year. Fish and seafood serve as food and they are also used in manufacturing numerous industrial products.

Fish are classified as freshwater or saltwater fish. Freshwater fish live in the lakes, streams and ponds on land, while the saltwater fish live in the oceans which cover more than 70% of the earth's surface. Saltwater fish are often further classified in the fishing industry as anadromous, groundfish, or pelagic fish.

- **Anadromous fish**—These fish are born in fresh water, swim to the ocean as adults, and return to their birthplace to spawn. The alewife, American shad, and salmon are examples of anadromous fish.

- **Groundfish or demersal fish**—These fish live near the bottom of the ocean. Fish of the cod family, flatfish family, the mullet, the pompano, the porgy, and sea bass are groundfish.

- **Pelagic fish**—These fish stay in the upper layers of the ocean. Fish of the herring family and the mackerel family are classified as pelagic fish.

Seafood also includes shellfish—which is commercial, rather than a scientific classification. Shellfish refers to hard-covered, edible, mostly marine animals from two different groups: the mollusks (oysters, clams, mussels) and the crustaceans (crabs, lobsters, and sometimes shrimp). Among the important mollusk species, harvested from both wild and cultivated stocks, are oysters, hard clams, surf clams, quahogs, soft-shell clams, razor clams, and horse clams.

Fig. F-14. Some examples of shellfish harvested by fishermen. (Courtesy, Department of Fisheries and Oceans, Canada)

In addition, octopus, squid, sea turtles, and possibly other foods are taken from the seas, depending upon the area and culture.

HISTORY OF FISHING. Fishing is one of the oldest and most important activities of man. Ancient remains of spears, hooks, and fishnets have been found in ruins of the Stone Age. The people of early civilizations drew pictures of nets and fishermen lines in their art. Through the ages, men wrote about fishing, used fish in exchange for services, and even learned to fish farm.

Early man's hooks were made from the upper bills of eagles and from bones, shells, horns, and thorn plants. His spears were tipped with the same materials, or sometimes with flints. Lines and nets were made from leaves, plant stalks, and cocoon silk. Ancient fishing nets were rough in design and material, but they were amazingly like some now in use.

There are many examples of fishing shown in art or mentioned in writing. An Egyptian tomb more than 4,000 years old contains a picture of fishermen. An old Chinese proverb recognized the value of fishing: "Give a man a fish and he will live for a day, teach him to fish and he will have food for life."

Fish were often used as a medium of exchange or as payment for services rendered. In the twenty-ninth year of Ramses III, the Union of Grave Diggers in Egypt filed a petition with the royal authorities for higher wages. As part of their wages, these workers received large amounts of fish four times each month. The petition requested a pay increase, pointing out that the petitioners came to the authorities without clothes and ointments—and even without fish.

As early as the 5th century B.C., it was recognized that fish could be farmed, just as hunters had previously come to realize that certain land animals could be domesticated. Ponds were constructed and stocked with fish by both the rich and the poor. Legal documents of the time indicated the importance of these food reservoirs.

In many areas that were too rocky or hilly for farming, fishing became an important industry. Bigger fish in deeper waters tempted early sea fishermen and encouraged them to build larger boats—venturing farther from home. Much of the world was first explored by lost fishing ships that came across strange lands by accident. In fact, some authorities believe that European fishing ships reached North America in the 12th century.

The herring industry grew up around the Baltic Sea in the 12th century and was controlled by the Hanseatic League, a group of German cities whose merchants traded all over northern Europe in fish, timber, cloth, salt, and many other goods.

In the 15th century the herring mysteriously disappeared from the Baltic Sea. Fishermen had to seek their herring in the North Sea and the Atlantic Ocean. The Dutch took over the herring fisheries and led commercial fishing of all kinds until the end of the 17th century.

A 14th century discovery by a Dutchman named Beukelszoon helped the Dutch fishing industry. Buekelszoon pickled herring in brine instead of preserving them in dry salt.

Commercial fishing on the North American continent started over 300 years ago—with the first colonists. So many fish were close to shore that the colonists did not have to build large sailing vessels as the Europeans did. Instead, the colonists followed the Indians' example and fished from small boats. Some fish were caught in traps and weirs (brush fences) set in the mouths of rivers and harbors. Shore fishermen used nets or, when the tide had gone out, searched the rocks and sand for shellfish.

Fig. F-15. To the first colonists of Maine, fishing was the most common occupation since it was easy and profitable. The products of fishing could be bartered for food from other colonies or from England. (Courtesy, The Bettmann Archive Inc., New York, N.Y.)

As colonization progressed, fishermen began sailing farther out to sea to find enough fish for a good catch. They sailed for months as they worked the fishing banks off Canada, and northeastern United States. Many of the early houses along the coast in colonial America had a walk around the roof so that the family could watch for returning ships. Many ships did not return, and this walk became known as the widow's walk.

As ships grew larger and fishing methods were developed and refined, the success of fishing voyages

Fig. F-16. Commercial fishing in the early 1800s was hard work—fighting rough seas and handling large catches. (Courtesy, The Bettmann Archive, Inc., New York, N.Y., from an 1875 engraving.)

and the types of fish and seafood increased. Like other commercial operations, the fishing industry became mechanized. Today, such technologies as airplane spotters, radar, radio-telephones, sonar, and long-range radio navigation (LORAN), are employed in the fishing industry. Some fishing fleets have a factory ship with equipment on board for fileting and freezing or canning. With all of these new developments, ships were able to find new fishing grounds by sailing farther from port and returning safely—loaded with fish. During the 60 years or so between 1900 and the late 1960s, the world fish catch increased by 27 times.

Fishing rights of a country have been, for some time, a source of concern and agitation. As early as 1377, there are records of lawsuits against fishermen who used a net called the wondrychoun—a large net that fishermen dragged through the water. The net caught little fish as well as big, and some people were afraid that soon there would be no fish left. Fishermen, however, continued over the years to catch as many fish as they could.

As early as the 1860s, it was acknowledged that fishery resources are limited, and that they must be managed through international agreements. In 1902, the International Council for the Exploration of the Sea (ICES) was formed by the major European fishing countries. Other nations joined, including the United States in the mid-1960s. ICES led to several conventions for the regulation of fisheries by mesh size of nets, and by quotas, in order to obtain the highest yields consistent with the maintenance of fish stocks. Although such conventions have been effective in the Northeast Atlantic, they have not operated as well in other regions.

FISHING INDUSTRY. As a fishing nation, the United States has lost ground. Historically, the United States was one of the major producers of fishery products in the world. Before World War II (1939–1945) and until 1956, the United States was second only to Japan in size of catch. United States fisheries were in third place in 1957, behind the expanding fisheries of China. At that time, the fisheries of Peru and the U.S.S.R. surged ahead. In 1970, the United States dropped to fifth among the fishing nations of the world, catching a few less fish than Norway. As shown in Table F-11, the latest statistics place the United States in fifth place among the fishing nations. Japan and the U.S.S.R. are still in the lead. The fish catch for the People's Republic of China places it in third place.

TABLE F–11
CATCH OF FISH
OF FIVE MAJOR FISHING COUNTRIES[1]

Country	Year		Percent of 1988 World Catch
	1984	1988	
	(billion lb)		(%)
Japan	26.5	26.2	12.1
U.S.S.R.	23.4	25.0	11.5
China	13.1	22.8	10.5
Peru	7.4	14.6	6.7
U.S.A.	10.6	13.3	6.1
Total World Catch	183.2	216.9	

[1]Based on data from *Statistical Abstract of the United States*, 1991, U.S. Department of Commerce, p. 857, Table 1478.

Today, the U.S. fishing fleet is overshadowed by the new large ocean fishing craft and factory ships operated by foreign nations. Some of these foreign vessels are 275 ft (*83 m*), or more, in length and register up to 4,000 gross tons. Foreign factory ships and other mother ships are as long as 500 ft (*153 m*) and register displacements of up to 20,000 tons. By way of contrast, nearly all of America's fishing vessels are less than 70 ft (*21 m*) in length; only a small number exceed 200 gross tons.

Still, U.S. fishermen land 13.3 billion pounds of fishery products—a catch made by 274,000 fishermen, who operate about 23,000 vessels, each of 5 net tons or more, and 70,000 boats of less than 5 net tons. Furthermore, to meet its needs, the United States imports fishery products in an amount nearly equal to its catch.

Distant fishing ranks as one of the most energy-consuming methods of producing food. Each calorie of food energy gained by distant fishing requires an input of about 11 Calories (kcal) of energy.

(Also see ENERGY REQUIRED FOR FOOD PRODUCTION.)

Fisheries. The term *fishery*, when applied commercially, refers to a place for catching fish—a fishing ground. The most important world fisheries are located close to land in waters less than 650 ft (*200 m*) in depth. The major fishing grounds are in (1) the North Atlantic, including the Grand Banks and the Georges Banks off the New England coast; (2) the North Sea, the waters over the continental shelves of Iceland and Norway, and the Barents Sea; (3) the North Pacific, particularly the Bering Sea, the gulf of Alaska, and the coastal areas around Japan; and (4) the Pacific waters off the coasts of China and Malaysia. Other important fishing grounds are found off the Peruvian coast and off the coast of southeastern United States.

The Fishery Conservation and Management Act of 1976 established a Fisheries Conservation Zone (FCZ) which gives the Federal Government exclusive authority over domestic and foreign fisheries within 200 nautical miles of U.S. shores and over certain living marine resources beyond the FCZ. Within the FCZ, the total allowable level of foreign fishing, if any, is that portion of the "optimum yield" not harvested by U.S. vessels.

Catching Fish.

Fig. F-17. Hupa Indian fishing. (Courtesy, Field Museum of Natural History, Chicago, Ill.)

While local names for the gear are different throughout the world, there are several chief methods of fishing. These are broadly classified as entrapment, luring, attacking, and dredging.

• **Entrapment**—This is the most important method used commercially. There are several types of entrapment but most involve some sort of net. Some fish traps are a series of brush fences, called *weirs*, or huge interlocking nets, called *pound nets*. As the fish move along, looking for food or migrating to a different part of the sea, they are stopped by a long fence, called the *leader*. The fish follow the leader, which guides them into a wing-shaped trap, called the *heart*, and finally into the last trap, the *pocket*, or *crib*. There the fishermen lift the fish out of the sea with large nets, called *seines*.

The *otter trawl* is one of the most widely used nets today because it holds so many fish. The otter trawl is made out of twine webbing and is shaped like a bag. When the trawl is towed along the bottom of the ocean, its mouth opens up like a huge funnel. After enough fish have been scooped into the trawl, the open end is drawn together. The nets are dragged by special boats called *otter trawlers*.

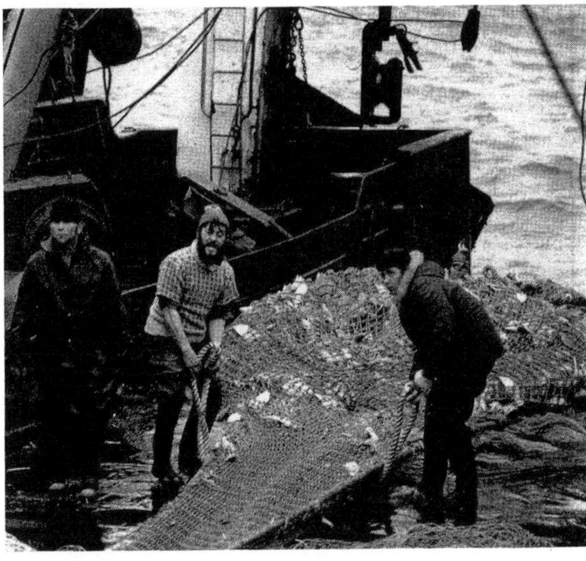

Fig. F-18. A catch of groundfish being hauled aboard a Canadian stern trawler operating off the coast of Newfoundland. (Courtesy, Department of Fisheries and Oceans, Canada)

The *purse net* is often used by commercial fishermen for encircling fish. It is shaped like a pouch purse with its drawstring at the bottom instead of the top. When the net is full, the drawstring is pulled tight by mechanical power and the fish are caught. The *haul seine*, a similar kind of net, has long been used by shore fishermen. The haul seine is a long sheet of netting that is pulled around schools of fish.

Fig. F-19. Shore fishermen using a seine to net carp in Malaysia. (Courtesy, International Development Research Centre, Ottawa, Canada)

Entanglement nets can be fixed like weirs or can float in the water like a curtain. With either method the fish become tangled in the nets and cannot swim away. The entanglement nets used most often are *gill nets* and *trammel*. As the fish start to swim through the net their gills become snarled in the nets.

Scooping nets lift the fish out of the water after they have been trapped. There is a variety of these small nets. *Hoop nets*, *dip nets*, and *cast nets* are among the most popular nets that take fish out of the water.

Baited traps are another kind of entrapment. The lobster pots, which are set out in shallow waters, are probably the best-known traps of this sort. Lobster pots look like cages with nets covering two ends. A piece of rotten meat or fish is put into the pot to attract the lobsters. In one end of the lobster pot is a kind of funnel made of netting. The lobster can get into the trap through the funnel, but it cannot get out again.

Fig. F-20. Heading out to set lobster pots off the coast of Nova Scotia, Canada. (Courtesy, Department of Fisheries and Oceans, Ottawa, Canada)

• **Luring**—Baited lines thrown out by hand or dragged through the water are used by sportsmen and commercial fishermen alike. Various species of tuna, salmon, and many others may be caught by this method from the shore or from a boat. Another method of luring is the longline. Longlines have many hooks on them. As they float or are dragged through the sea the fish are attracted by the bait or flashing lures. After a certain time the lines are retrieved. Japanese tuna longlines have been known to be 50 miles (*80 km*) long.

• **Attacking**—*Spears and harpoons* are also used on large fish, such as swordfish, tuna, and whale. The harpoons are usually shot out of a gun, but the spears are still thrown by hand. Ropes attached to the spears and harpoons keep the plunging fish close to the boat.

• **Dredging**—In this method, equipment is dragged along the bottom of the ocean to harvest oysters, clams, and scallops.

A new fish-harvesting method utilizes powerful lights that are lowered into the ocean to attract schools of fish, which are then sucked up into the ship by powerful vacuum pumps.

Fish Catch. Reporting of the fish catch in the United States is generally broken down into the following regions: (1) New England, which includes Maine, New Hampshire, Massachusetts, Rhode Island, and Connecticut; (2) Middle Atlantic, which includes New York, New Jersey, and Delaware; (3) Chesapeake Bay, which includes Maryland and Virginia; (4) South Atlantic, which includes North Carolina, South Carolina, Georgia, and the east coast of Florida; (5) Gulf States, which includes the west coast of Florida, Alabama, Mississippi, Louisiana, and Texas; (6) Great Lakes, which includes each lake—Superior, Michigan, Huron, and Erie; (7) Pacific Coast, which includes Washington, Oregon, California, and Alaska; (8) Mississippi River and tributaries; and (9) Hawaii.

Table F-12 shows the fish catch in the United States by regions.

TABLE F–12
FISH AND SHELLFISH CATCH IN THE UNITED STATES BY REGIONS[1]

Region	Catch
	(million pounds)
Pacific Coast	4,840
Gulf States	1,789
Chesapeake Bay	778
New England	565
South Atlantic	256
Middle Atlantic	172
Great Lakes	38
Hawaii	24

[1]Based on data from *Statistical Abstract of the United States*, 1991, U.S. Department of Commerce, p. 685, Table 1200. To convert to kg divide by 2.2

A few species of fish and shellfish comprise the majority of the catch in these regions. These species and their catch for the past 24 years are indicated in Table F-13.

TABLE F–13
CATCH OF PRINCIPAL SPECIES FROM U.S. FISHERIES BY SELECTED YEARS[1]

Species	Yearly Catch					
	1965	1970	1975	1980	1985	1989
	(million pounds)					
Fish:						
Cod, Atlantic	36	53	56	118	83	78
Flounder	180	169	162	217	196	203
Haddock	134	27	16	55	14	4
Halibut	40	35	21	19	61	75
Herring, sea	110	79	131	291	142	200
Jack mackerel	67	47	37	44	21	28
Menhaden	1,726	1,837	1,803	2,497	2,739	1,989
Ocean perch, Atlantic	84	55	32	24	10	2
Pollock, Atlantic	12	9	21	43	43	23
Salmon, Pacific	327	410	202	614	727	786
Tuna	319	393	393	399	83	89
Whiting	83	45	42	36	45	39
Shellfish:						
Clams	71	99	113	95	151	138
Crabs	335	277	306	523	338	458
Lobsters, American	30	34	30	37	46	53
Oysters	55	54	53	49	44	32
Scallops	23	11	14	30	38	42
Shrimp	244	367	347	340	333	331

[1]Data from *Statistical Abstract of the United States*, 1980, U.S. Department of Commerce, p. 744, Table 1315; 1984, p. 704, Table 1252; and 1991, p. 686, Table 1202. To convert to kg divide by 2.2.

With the exception of tuna, and a few lobsters and shrimp, those fish and shellfish shown in Table F-13 are caught within 200 miles (*320 km*) of the U.S. shores. Most tuna, 94%, are caught at distances over 200 miles (*320 km*) from U.S. shores.

In terms of dollars, the U.S. fishing industry's catch is worth more than $3.2 billion. The average price per pound (*0.45 kg*) for the overall catch increased from 2.4¢/lb in 1940 to 38.3¢/lb in 1989. Pacific Salmon is the most valuable catch, worth $591 million, and shrimp is next, being worth $468 million. The crab catch is valued at about $414 million, and tuna at about $104 million. Alaska pollock has become a valuable catch, worth $187 million. Table F-14 indicates the values of the principal species for the most recent year available.

TABLE F–14
VALUE OF PRINCIPAL SPECIES FROM U.S. FISHERIES ON A LIVEWEIGHT BASIS EXCEPT WHERE INDICATED[1]

Species	Catch	Value	Price per Pound[3]
	(million lb)[2]	($ million)	($)
Fish:			
Cod, Atlantic	78	48	.61
Flounder	202	120	.59
Haddock	4	5	1.19
Halibut	75	85	1.13
Herring, sea	209	29	.14
Jack mackerel	28	1	.07
Menhaden	1,989	84	.04
Ocean perch, Atlantic	1	.9	.66
Pollock	23	10	.43
Salmon, Pacific	786	591	.75
Tuna	89	104	1.16
Whiting	39	9	.24
Shellfish:			
Clams	138	135	.98
Crabs	458	414	.90
Lobsters, American	53	149	2.82
Oysters	30	84	2.79
Scallops	40	138	3.43
Shrimp	352	468	1.33

[1]Based on data from *Statistical Abstract of the United States*, 1991, U.S. Department of Commerce, p. 686, Table 1202.

[2]To convert to kg divide by 2.2.

[3]Average price paid to the fisherman.

Fish Farming. This is the raising of food fish and other aquatic life in protected enclosures or controlled natural environments.

Although recent research has shown the advantages of fish farming as an alternative type of livestock production, aquaculture—the science of fish farming—is by no means a newly discovered science. Evidence of fish farming dates back 4,000 years to China, Japan, and Egypt. The practice of fish culture can be found in the societies of India and Java 3,000 years ago, and of Europe 2,500 years ago. Through the years, most of the progress in fish farming has evolved from trial and error. In the United States, it is still a small industry, but new methods and technologies will likely spur its development.

Aquaculture can be divided into two types of production—freshwater and marine water. Within these two types of production the following four systems of management are practiced:

1. **Hatchery operation.** Young fish are hatched and raised to a size where they can be released into natural populations to grow and reproduce.

2. **Capture of young fish.** In this system, young fish are captured and grown to market weight on supplemental feeding or on natural feeds in the environment aided by active fertilization of the water.

3. **Growing of young fish to market weight.** In this system, the farmer is only concerned with growing fish and does not attempt to select breeding stock. Shrimp farming is an example of this system.

4. **Management of entire life cycle.** Catfish and trout production in the United States are examples of this system. Young fish are hatched and grown either to be marketed or kept as replacement breeder stock.

In the United States, about 10.5% of the fish harvest is the product of fish farming. Catfish is the largest crop, grown primarily in Mississippi, Alabama, Texas, and Missouri. Catfish cultivation is relatively simple because the fish are omnivorous feeders and reproduce easily. Carp farming is also a possibility in the United States; it was an important business in some areas in the past.

Trout culture requires large quantities of pure, cool water. Of the cultured-trout crop, 90% is produced in Idaho where there is an abundant supply of fast-running river water. Fertilized eggs are obtained from brood stock and the fish are raised in long, shallow tanks. They can be held there until it is time to harvest them, or they may be used as game fish for stocking local streams.

The production of Pacific salmon in hatcheries, and their release into the ocean to augment wild stocks, has been an industry in the United States for more than a century. Hatcheries raise the spawn to young smolts that are released into the ocean and caught when they return to spawn. They may also be raised in pens and harvested after 1 year, when they weigh about 1 lb (*0.45 kg*).

Culture systems for other desirable aquatic species are also being developed. Freshwater shrimp are raised successfully in Hawaii, and the Japanese have established a small marine shrimp industry. In addition, attempts have been made to raise lobsters, abalone, crabs, and oysters, but commercial success has been limited. Some species of marine fish, especially yellowtail and sea bream, are farmed in Japan, but the high cost of raising these fish limits their sale to the luxury market.

PROCESSING FISH AND SEAFOOD.

Fig. F-21. Channel catfish ready to be processed. (Courtesy, USDA)

Once caught, the fish must be processed for preservation and for consumer consumption. This includes fish packaged fresh and frozen, canned, or cured (salted, smoked, or dried). In addition to the edible products for humans, some fish are caught solely for industrial purposes. Also, there are a number of by-products from the edible fish industry.

Products And Preservation. Of the U.S. catch, 26% is marketed fresh and frozen; 43% is canned; and the remaining 31% becomes by-products and bait. A fully equipped factory ship will have machinery on board for fish filleting and freezing or canning. Fish fillets are often frozen at sea into large blocks weighing up to 100 lb (*45 kg*), which are later reprocessed on shore into individual portions. Some ships may also have facilities for drying and grinding fish into fish meal. The U.S.S.R. pioneered in the development of factory ships, which are often huge vessels operated by crews of 500 to 650, and accompanied by their own fleets of smaller ships called catcher boats.

Large fishing vessels on long voyages will often be equipped to keep their catch edible either by storing it in refrigerated facilities or by quick freezing it. Fish may be chilled in refrigerated sea water and held in it. This permits more rapid chilling; and by careful control, the fish may be held at their freezing point or about 30°F (*−1°C*), only a few degrees colder than iced fish. Holding at this termperature increases satisfactory storage time by several days. A recent variant is to chill and hold the fish at about 27°F (*−3°C*), a temperature at which a small part of the water is frozen. In several fisheries, notably the U.S. tuna industry, where fish must be kept for several months before landing, brine freezing followed by dry frozen storage is employed. Tuna handled in this way is thawed or partially thawed at sea just before reaching port to facilitate unloading and to make possible immediate canning.

Some of the details of canning vary for the different species of fish but the fundamental process is the same. Fish are prepared by grading, cleaning, and cutting for the filling machines or for hand packing. After the fish are packed and seasoned, the air is exhausted from the cans, which are then sealed and cooked. Some oily fish, such as sardines and tuna, are precooked before packing. Specialty products, such as fish soups, require more preparation.

Curing fish includes drying, smoking, salting, and pickling. Methods of drying fish by blowing warm, dry air on them have replaced sun-drying in some areas. Herring, salmon, smelt, and mackerel are commonly smoked. The cleaned fish is salted, washed, and drained in preparation for smoking. It is then partially dried and hung in the smokehouse for curing. Producers of salted or pickled fish develop their own characteristic flavor. The process consists of salting, draining, and drying.

Besides being sold whole for fileted, fresh, or frozen, many processors also make breaded fish sticks and meal-sized portions of breaded fish that are sold frozen cooked or ready to cook.

(Also see CONVENIENCE AND FAST FOODS; and PRESERVATION OF FOOD.)

- **Shellfish**—Seafood items falling in this category are handled somewhat differently than fin fish. Because of great susceptibility to spoilage, such species are often kept alive during a part of their commercial handling. Oysters and clams should be alive until shucked and crab and lobster until cooked. A considerable portion of the U.S. shrimp is beheaded (usually at sea), washed and graded ashore, and packaged and frozen in the shell usually in institution-sized packages. For the ultimate consumer's use, shrimp is peeled and deveined, often by machine, and then packaged and frozen. A considerable portion is breaded after peeling and then packaged and frozen. Crab are cooked in the shell and the meat picked by hand. Sometimes, especially in the Chesapeake Bay area, the picked meat is pasteurized by raising the temperature of the meat to 170°F (*77°C*) for 1 minute, thereby greatly extending the storage life of the refrigerated product. Much of the East Coast blue crab and the Pacific Coast dungeness crab is marketed unfrozen, but a considerable portion of the more recently developed Alaskan king crab is sold as a frozen product. The crab is cooked in the shell and the meat removed from the shell by passing through rubber rollers. The meat is then usually frozen in large blocks which are ice glazed and later sawed into consumer sized units. Some king crab is cut in sections and frozen in the shell.

Table F-15 shows the types and amounts of fishery products processed in the United States. Table F-16 indicates the types and amounts of processed fishery products imported into the United States.

TABLE F–15
TYPES AND AMOUNTS OF FISHERY PRODUCTS PROCESSED IN THE UNITED STATES[1]

Product	Year				
	1970	1975	1980	1985	1989
	(million pounds)				
Packaged fresh and frozen:[2]	1,158	1,113	1,403	1,662	879
Fish not breaded:					
Fillets and steaks, raw	134	132	203	246	389
Dressed	32	55	40	132	—
Other[3]	10	26	48	N/A	—
Fish breaded, raw or cooked:					
Sticks	116	91	88	96	89
Fillets, portions, and steaks	296	286	326	330	280
Shellfish: not breaded	328	315	497	N/A	—
Breaded, raw and cooked	119	128	113	N/A	121
Specialties, fish, and shellfish	123	80	88	N/A	—
Canned products:[4]	1,346	1,386	1,516	1,161	1,455
Fish and shellfish	805	802	1,010	913	1,110
Animal food	540	583	506	248	345
Cured fish:[5]	71	55	66	58	—
Salted	37	29	43	32	—
Smoked	31	24	22	25	—
Dried (cod) lutefisk	2	2	1	1	—

[1]Data from *Statistical Abstract of the United States,* 1980, U.S. Department of Commerce, p. 748, Table 1324; 1987, p. 649, Table 1146; and 1991, p. 689, Table 1209. To convert to kg divide by 2.2.
[2]Includes products not shown separately.
[3]Includes bait and unbreaded portions.
[4]Includes salmon eggs for bait.
[5]Includes sun-dried and freeze-dried.

TABLE F–16
TYPES AND AMOUNTS OF FISHERY PRODUCTS IMPORTED INTO THE UNITED STATES[1]

Product	Year				
	1970	1975	1980	1985	1989
	(million pounds)				
Fresh or Frozen:					
Whitefish[2]	8.4	8.4	8.2	—	—
Halibut[2]	18.2	8.0	6.3	12.8	—
Salmon[2]	7.4	9.2	5.5	27.0	98.8
Sea herring	103.9	99.2	22.0	—	—
Smelt	8.5	8.8	11.3	—	—
Tuna	442.8	478.6	722.7	478.8	649.7
Groundfish, fillets, blocks[3]	458.8	513.8	557.1	639.8	484.4
Other fillets and steaks	136.1	167.6	148.2	231.0	252.6
Scallops (meats)	16.8	19.7	20.9	42.0	40.9
Lobster, American and spiny	54.9	58.1	50.5	77.4	72.1
Shrimp and prawn	214.8	201.5	215.1	342.8	491.6
Canned:					
Anchovies, in oil	5.5	3.6	4.5	5.0	—
Salmon	2.4	3.3	0.2	1.9	—
Sardines, in oil	34.1	18.5	18.2	23.0	—
Sardines, herring, not in oil	25.3	24.9	38.5	39.8	40.1
Tuna	72.3	51.7	63.6	213.9	348.2
Bonito and yellowtail	1.2	0.1	0.8	—	—
Crab meat	2.8	1.4	5.0	7.6	—
Oysters	15.0	12.4	17.0	28.9	20.4
Lobster, American and spiny	2.5	2.1	2.2	—	—
Pickled or salted:					
Cod, haddock, hake, pollock, cusk	41.2	33.6	33.0	34.7	16.6
Herring	30.6	20.6	16.7	—	—

[1]Data from *Statistical Abstract of the United States,* 1980, U.S. Department of Commerce, p. 749, Table 1326; 1984, p. 709, Table 1263; 1987, p. 650, Table 1149; and 1991, p. 689, Table 1211.
[2]Excludes fillets.
[3]Includes cod, cusk, haddock, hake, pollock, Atlantic perch and whiting.

Industrial Products. The commerical value of fish is not limited to its use as food. Various industrial products contain fish parts or fish by-products. For example, manufacturers make shoes from the skins of sharks. Abalone shells are made into buttons. Other manufacturers use by-products to make glue. A kind of gelatin called *isinglass* is made from the air bladders of certain fish. Menhaden, sardines, herrings, and sharks produce valuable oils. These oils are used in making paints and varnishes, in tanning leather, and in the manufacture of linoleum and synthetic materials.

• **Fish flour or fish protein concentrate**—Scraps and waste from fish contain much protein. Much of this waste is processed into dried meal and concentrates that are used for animal feeds. During the early 1970s, many nations sought ways to convert fish waste into food for human beings. They worked to develop an odorless, tasteless fish flour or fish protein concentrate that could be added to foods in parts of the world where people lack sufficient protein in their diets. Fish protein concentrate is prepared through the solvent extraction processing of clean, undecomposed whole fish or fish cuttings.

(Also see FLOURS, Table F-26 Special Flours.)

BUYING FISH AND SEAFOOD. Fish and seafood is marketed in several forms, and buyers should be aware of these forms. Fresh and frozen fish are marketed in various forms or cuts. The following are the best known market forms:

• **Whole**—This is fish as they come from the water. Before cooking, the fish must be scaled and eviscerated—usually the head, tail, and fins are removed. The fish may then be cooked, fileted, or cut into steaks or chunks.

• **Dressed**—These are fish with scales and entrails removed, and usually the head, tail, and fins are removed. The fish may then be cooked, fileted, or cut into steaks or chunks. The smaller size fish are called pan dressed and are ready to cook as purchased.

• **Filets**—Filets are the sides of the fish cut lengthwise away from the backbone. They are ready to cook as purchased. A filet cut from one side of a fish is called a single filet. This is the type most generally available on the market. The filets may or may not be skinless.

The two sides of the fish cut lengthwise away from the backbone and held together by the uncut flesh and skin of the belly are referred to as butterfly filets.

• **Steaks**—Steaks are cross section slices from large dressed fish cut 5/8 to 1 in. (*1.6 to 2.5 cm*) thick. A cross section of the backbone is the only bone in a steak. They are ready to cook as purchased.

• **Chunks**—Chunks are cross sections of large dressed fish. A cross section of the backbone is the only bone in a chunk. They are ready to cook as purchased.

• **Raw breaded fish portions**—Portions are cut from frozen fish blocks, coated with a batter, breaded, packaged, and frozen. Raw breaded fish portions weigh more than 1½ oz (*42 g*), are at least 3/8 in. (*0.9 cm*) thick, and must contain not less than 75% fish. They are ready to cook as purchased.

• **Fried fish portions**—Portions are cut from frozen fish blocks, coated with a batter, breaded, partially cooked, packaged, and frozen. Fried fish portions weigh more than 1½ oz (*42 g*), are at least 3/8 in. (*0.9 cm*) thick, and must contain not less than 65% fish. They are ready to heat and serve as purchased.

• **Fish sticks**—Sticks are cut from frozen fish blocks, coated with a batter, breaded, partially cooked, packaged, and frozen. Fried fish sticks weigh up to 1½ oz (*42 g*), are at least 3/8 in. (*0.9 cm*) thick, and must contain not less than 60% fish. They are ready to heat and serve as purchased.

Some shellfish are marketed alive. Other market forms, depending on the variety, include cooked whole in the shell, headless, shucked or fresh meat, cooked meat, breaded, and canned. The following descriptions apply to shellfish:

• **Live in the shell**—Crabs, lobsters, clams, and oysters are available live in their shells. Shellfish purchased live in the shell must be kept alive until it is served or cooked.

• **Cooked in the shell**—Crabs and lobsters are available cooked in the shell either chilled or frozen.

• **Headless**—The shrimp tail is the only part that is sold. Spiny lobster tails are also a common market form. These are ready to cook.

• **Shucked or fresh meat**—Clam, oyster, and scallop meats are available without the shells, either fresh or frozen.

• **Cooked meat**—The cooked meat is picked from the shell of cooked crab, lobster, and shrimp and is available either fresh, frozen, or canned.

• **Breaded**—Frozen raw, breaded or frozen fried clams, oysters, scallops, and shrimp are available. These shellfish are ready to cook or heat as purchased.

• **Canned**—Clams, crabs, lobsters, and shrimp are available canned. They are ready to serve or use as purchased.

Once a person is familiar with the various forms in which fish and seafood can be purchased, there are a few guidelines to follow when buying fresh fish, frozen fish, and shellfish:

• **Fresh fish**—The flesh is firm and elastic and not separating from the bones. In buying filets and steaks, look for a fresh-cut appearance and color that resembles freshly dressed fish. Odor is fresh and mild. A fish just taken from the water has practically no "fish" odor. Eyes are bright, clear, transparent, full, and often protruding. Gills are red in color and free from slime. Skin is shiny and with color that has not faded.

• **Frozen fish**—The flesh is solidly frozen. The flesh should have no discoloration, brownish tinge, or white cottony appearance. Odor is not evident or is slight. Wrapping in which the fish is packaged is moisture-vapor-proof; there is little or no air space between the fish and the wrapping and there has been no damage to the package. Glazing of ice (used to protect shrimp, salmon, and halibut steaks or whole fish frozen in the round or dressed form against drying out) is present on these forms of frozen fish.

• **Shellfish**—The clams and oysters in the shell are alive and the shells close tightly when tapped. Gaping shells indicate that the clam or oyster is dead and not edible. Shucked oysters are plump and have a mild odor with usually a creamy color and a clear liquor or nectar. Cooked crabs and lobsters are bright red with no disagreeable odor. Fresh shrimp have a mild odor and their meat is firm. Cooked shrimp have red in their shells and their meat has a reddish tint. Scallops have a sweetish odor and they are free of excess liquid when purchased in packages.

Fishery products are voluntarily inspected. By contrast, beef and poultry, as well as many other perishable food items, are federally inspected and graded at various stages of processing to ensure buyers of a safe, wholesome, acceptable quality product. There is no similar mandatory federal inspection program for fishery products. The U.S. Department of Commerce (USDC)—provides a voluntary inspection program for fishery products. The program is voluntarily subscribed to by processors, packers, brokers, and seafood users interested in having USDC inspect their products. Inspection service users pay fees for USDC inspectors to evaluate their raw materials, to ensure the hygienic preparation of products—and to certify final product quality and condition. The USDC inspector functions as an objective observer in evaluating processing techniques and product quality and condition. Products packed in plants under USDC inspection can carry marks for easy consumer identification. When displayed on product labels, these marks signify that federal inspectors of the Department of Commerce inspected, graded, and certified the products as having met all the requirements of the inspection regulations, and have been produced in accordance with official U.S. grade standards or approved specifications.

Fig. F-22. The United States Department of Commerce (USDC) makes available a voluntary inspection service which permits processors of inspected seafoods to display official USDC grade or inspection shields on their labels. Only those firms that process fishery products under federal inspection are permitted to use these emblems.

HANDLING AND COOKING FISH AND SEAFOOD.

Because seafood, like many other food products will spoil easily if not handled with care, certain procedures must be followed:

• **Fresh fish and shellfish**—Fresh fish should be placed in the refrigerator in their original leakproof wrapper immediately after they are received, and stored at 35° to 40°F (*1.7° to 4.4°C*) for no longer than a day or two before cooking. Shellfish should be stored at approximately 32°F (*0°C*), but it is a good practice to eat fresh shellfish the day they are purchased.

• **Frozen fish and shellfish**—To maintain quality, commercially packaged frozen fishery products should be placed in the freezer, in their original moisture-vapor-proof wrapper, immediately after purchase. Store in the freezer to 0°F (*-18°C*), or lower. At a temperature above 0°F (*-18°C*), chemical changes cause the fish to lose color, flavor, texture, and nutritive value.

When fish is to be frozen, it should be wrapped and sealed in moisture-vapor-proof materials. Do not freeze fish which has been wrapped only in wax paper, parchment paper, or polyethylene materials which are not moisture-vapor-proof.

Storage time should be limited in order to enjoy the optimum flavor of the frozen fish (see Table F-17). It is a good practice to date the packages as they are put in the freezer.

When fish thaws, it should be cooked immediately. Any premature thawing lowers quality because it reduces the natural moisture and flavor. Never refreeze fish.

TABLE F-17
APPROXIMATE STORAGE LIFE OF FROZEN FISH AND SHELLFISH HELD AT 0°F *(−18°C)*

Type	Species	Months
Fat fish . . .	Mackerel, salmon, tuna, etc.	3
Lean fish .	Haddock, cod, swordfish, etc.	6
	Lobsters and crabs (meat)	2
Shellfish .	Shrimp	6
	Oysters, scallops, clams (shucked)	3 to 4

• **Cooked seafood**—This can be stored in the refrigerator or freezer. When stored in the refrigerator, seafood should be placed in covered containers and not held longer than 2 or 3 days. When stored in the freezer, seafood should be packaged in a moisture-vapor-proof material, and not held longer than 3 months.

• **Canned fishery products**—Canned products should be stored in a cool dry place for not longer than a year.

There are endless recipes and preparation techniques for fish and seafoods. However, they are cooked by using the basic cookery techniques—baking, pan frying, deep-fat frying, steaming, broiling, oven frying, poaching, or planking. Fish are cooked when the flesh loses its translucent appearance and becomes opaque. The flesh flakes easily when pierced with a fork. Cooking fish at too high a temperature or for too long a time toughens, dries, and destroys the flavor. There is just one basic rule of seafood cookery to remember: *Don't overcook!*

(Also see FOOD—BUYING, PREPARING, COOKING, AND SERVING; and MEAT[S].)

Fig. F-23. Poached fish. (Courtesy, USDA)

NUTRITIVE QUALITIES OF FISH AND SEAFOOD. A prime source of protein, fish and fishery products are among the most nutritious foods available to the consumer. Among those foods that provide the best in nutrition, fish and seafood are at the top right along with meat, milk, and eggs. Seafoods contain *complete protein*. This means the protein in seafood supplies the essential amino acids, those the body cannot manufacture and must get from the foods. Most fish contain 18 to 20% protein, which is 85 to 95% digestible.

Fish oil has been shown to slow the formation of arterial deposits which cause heart attacks and strokes. Most fish are low in total fat, low in saturated fat, and low in cholesterol. Fish oils, which are rich in unsaturated fats called omega-3 fatty acids, change the chemistry of the blood; they increase the high density lipoproteins (HDL) and decrease the low density lipoproteins (LDL), they lower the level of triglycerides, they make platelets (the cells involved in clotting) less sticky, and they make the red blood cells less rigid.

Fish and shellfish are rich in vitamins. Varieties of "fat" fish are rich in vitamins A, D, and K. An average serving of these fish gives 10% of the daily adult vitamin A requirement and 50% of the vitamin D requirement. The important B complex vitamins are present in seafoods, too. A serving of lean or fat fish or shellfish, yields about 10% of the thiamin, 15% of the riboflavin, and 50% of the niacin required each day.

The most common minerals found in fish are iodine, magnesium, calcium, phosphorus, iron, potassium, copper, and fluoride. Shellfish are particularly rich in minerals.

Specific details regarding the nutritional value for a wide variety of fish and seafood may be obtained from Food Composition Table F-36.

Despite all of the good nutrition packed in seafood, the major concern is that most people do not eat enough to contribute significantly to their diet. Table F-18 presents an historical account of the per capita consumption of fishery products. Total consumption has increased slightly since 1950, and it now stands at about 15.7 lb. The consumption of meats and poultry is considerably greater.

Table F-19 shows the contributions of meat, poultry, and fish, combined, to the dietary intake in the United States.

TABLE F–18
THE PER CAPITA CONSUMPTION OF FISHERY PRODUCTS FOR SELECTED YEARS, SINCE 1950[1]

Product	Year								
	1950	1955	1960	1965	1970	1975	1980	1985	1989
	(lb)								
Fresh and frozen	6.3	5.9	5.7	6.0	6.9	7.6	8.0	9.0	10.4
Fish[2]	4.7	4.2	3.8	3.8	4.5	5.0	5.5	5.2	N/A
Shellfish	1.6	1.7	1.9	2.2	2.4	2.6	2.5	3.8	N/A
Canned	4.9	3.9	4.0	4.3	4.5	4.3	4.5	5.2	5.0
Tuna	1.1	1.4	2.0	2.3	2.5	2.9	2.9	3.3	3.9
Salmon	1.4	1.0	0.7	0.9	0.7	0.4	N/A	N/A	N/A
Shellfish	0.4	0.4	0.4	0.5	0.5	0.4	N/A	N/A	N/A
Sardines	1.4	0.6	0.4	0.3	0.4	0.2	N/A	N/A	N/A
Other	0.6	0.5	0.5	0.3	0.4	0.4	N/A	N/A	N/A
Cured	0.6	0.7	0.6	0.5	0.4	0.4	0.3	0.3	0.3
Total	11.8	10.5	10.3	10.8	11.8	12.3	12.8	14.4	15.7

[1]Data from *Statistical Abstract of the United States*, 1980, U.S. Department of Commerce, p. 744, Table 1317; 1988, p. 114, Table 184; and 1991, p. 126, Table 207. To convert to kg divide by 2.2.
[2]Beginning 1973, includes catfish from fish farming.

TABLE F–19
CONTRIBUTIONS OF MEAT, POULTRY, AND FISH TO THE DIETARY INTAKE IN THE UNITED STATES[1]

Nutrition	Average per Capita Quantities Available per Day	Total Consumption Contributed by Meat-Poultry-Fish, Combined[2]
		(%)
Energy	3,600 kcal	18.8
Carbohydrate	425 g	0.1
Fat	168 g	31.7
Protein	105 g	43.3
Calcium	890 mg	3.9
Phosphorus	1,540 mg	29.2
Magnesium	330 mg	15.4
Copper	1.7 mg	17.5
Iron	17.1 mg	22.1
Potassium	3,480 mg	19.3
Zinc	12.7 mg	47.2
Vitamin A (R.E.)/carotene	2,400 mcg	20.6
Folate	284 mg	9.8
Niacin	26 mg	44.8
Riboflavin	2.4 mg	22.2
Thiamin	2.2 mg	25.0
Vitamin B-6	2.2 mg	40.5
Vitamin B-12	9.1 mcg	76.1
Vitamin C	118 mg	2.2
Vitamin E (mg alpha-te)	16.7	5.9

[1]Data from *Agricultural Statistics, 1991*, pp. 474–479. Data is for 1988.
[2]The yearly per capita consumption amounted to 119.5 lb (*54.3 kg*) red meat, 57.4 lb (*26.1 kg*) of poultry, and 15.2 lb (*6.9 kg*) of fish that was consumed in 1988.

Table F-20 shows the contributions of meat, poultry, and fish, combined, to the recommended Dietary Allowances (RDA) based on the per capita consumption per day.

(Also see FATS AND OTHER LIPIDS; MEAT[S]; and PROTEIN[S].)

TABLE F–20
CONTRIBUTIONS OF MEAT, POULTRY, AND FISH TO THE RECOMMENDED NUTRIENT ALLOWANCES IN THE UNITED STATES[1]

Nutrient	United States Recommended Daily Allowance (RDA)[2]	Quantities Available for Consumption per Capita per Day	Contribution from Meat-Poultry-Fish, Combined	
			Total Amt/Day[3]	RDA[4]
				(%)
Energy	2,300–3,100 (2,700) kcal	3,600 kcal	677 kcal	25.1
Carbohydrate	Not established	425 g	0.4 g	—
Fat	Not established	168 g	53.3 g	—
Protein	63 g	105 g	45.5 g	72
Calcium	800 mg	890 mg	35 mg	4.4
Phosphorus	800 mg	1,540 mg	450 mg	56
Magnesium	350 mg	330 mg	51 mg	14.6
Copper	2–3 mg	1.7 mg	0.3 mg	15
Iron[5]	10 mg	17.1 mg	3.8 mg	38
Zinc	15 mg	12.7 mg	6.0 mg	40
Vitamin A (R.E.)[6]	1,000 mcg	1,630 mcg	336 mcg	34
Folate	200 mcg	284 mcg	28 mcg	14
Niacin	19 mg	26 mg	12 mg	63
Riboflavin	1.7 mg	2.4 mg	0.6 mg	35
Thiamin	1.5 mg	2.2 mg	0.6 mg	40
Vitamin B-6	2.0 mg	2.2 mg	0.9 mcg	45
Vitamin B-12	2.0 mcg	9.1 mcg	7.0 mcg	35
Vitamin C	60 mg	118 mg	3.0 mg	5

[1]Data from *Agricultural Statistics 1991*, USDA, pp. 474–478.
[2]National Research Council (1989).
[3]USDA.
[4]Calculated as follows: % RDA = $\frac{\text{Av. daily per capita intake}}{\text{Minimum RDA}}$
[5]RDA value used for iron based on adult male, whereas, adult premenopausal female has an RDA of 15 mg/day.
[6]Vitamin A is given in terms of retinol equivalents (R.E.), one R.E. = 3.3 IU of vitamin activity, which was assumed to be the predominant form in meat, poultry, and fish.

TYPES OF FISH AND SEAFOOD. With over 240 species of fish and shellfish on the U.S. market from which to choose, and all the ways they may be prepared, menus may be planned that are nutritious, and very tasty. For a better understanding of the types of fish and seafood available, Table F-21 lists the common types, their description, habitat, fishing, processing, and uses.

TABLE F-21
FISH AND SEAFOOD(S) OF THE WORLD

Common and Scientific Name(s)	Description	Habitat; Fishing	Processing; Uses	Remarks
American Pollock (Boston Bluefish) *Pollachius virens*	A greenish hue, usually deep olive or brownish-green above, paling to yellowish or smoky gray on the sides and silvery gray on the belly; deep and plump body tapering to a pointed nose and tail section; average size of 2 to 3 ft, *(0.6 to 0.9 m)* weighing 4 to 12 lb *(1.8 to 5.4 kg).*	**Habitat:** Cold Atlantic waters from Nova Scotia to Virginia, at any level depending on their food supply. **Fishing:** Gulf of Maine, Georges Bank and off Chatham, Mass. with otter trawl at midwater mostly during early winter months.	**Processing:** Generally filleted and available fresh or frozen. **Uses:** Baked, broiled, pan fried, steamed or poached.	Although they are similar in taste, flavor and texture to haddock and cod, pollock are meatier and firmer fleshed.
American Shad (Buck Roe; White Shad) *Alosa sapidissima*	Silvery colored with a bluish-green metallic luster on the back; a deep body; large scales; prominent dark spot behind gills; rarely over 9 lb *(4 kg).*	**Habitat:** Native to the Atlantic from the Gulf of St. Lawrence and along the coast over to the Gulf of Mexico; introduced to the Pacific, and now range from Mexican border to Alaska. **Fishing:** Caught with seines, gill nets, dip nets, and fishing pole.	**Processing:** Marketed as whole, dressed or filleted fish, fresh or canned; shad roe. **Uses:** Baked, broiled, planked, stuffed, or sauteed.	Shad are anadromous, like salmon, spending most of their lives in the ocean but returning to their natal streams to spawn.
Atlantic Croaker (Croaker; Hardhead) *Micropogon undulatus*	Small dark spots on upper body; lower third of body white; row of little barbels on each side of lower jaw; range from 1 to 4 lb *(0.5 to 1.8 kg).*	**Habitat:** Atlantic coast from Mass. to Fla. and into the Gulf of Mexico around to Tex.; bottom-feeding fish. **Fishing:** Most taken in March to May with pound nets and trawls; some caught by sport-fishermen.	**Processing:** Dressed, filleted, raw minced, frozen breaded or cooked portions. **Uses:** Baked, broiled, poached, pan fried, deep fried, or oven fried.	It is called "croaker" and "drum" because of the distinctive drumming croak it produces at various times.
Blue Crab *Callinectes sapidus*	Five pair of legs with first pair as pincers; 5 to 7 in. *(12.5 to 17.5 cm)* when fully grown; brownish green or dark green shell; white on underside; tops of claws blue.	**Habitat:** Primarily shallow water along Atlantic and Gulf coasts from Mass. to Tex. **Fishing:** Caught with trawls, dredges, baited lines such as trotline, and crab pot.	**Processing:** Marketed live in shell; fresh or frozen cooked; frozen breaded, raw or cooked; canned. **Uses:** Steamed (boiled), cakes, patties, deviled, stuffed, casseroles, salads, appetizers.	Blue crab are caught and marketed in both the hard-shelled and soft-shelled stages. Soft-shelled crabs are considered a delicacy.
Carp *Cyprinus carpio*	Scale carp, mirror carp, and leather carp; few scales on leather carp; scales golden-colored; average 3 to 5 lb *(1.4 to 2.3 kg),* but may reach 80 lb *(36 kg).*	**Habitat:** Inland lakes and streams in most parts of the Northern Hemisphere; introduced to U.S. from Asia in late 1800s, spread rapidly; prolific breeders. **Fishing:** Not a sport fish in U.S.; high-ranking sport fish in England; some farmed in Europe and China.	**Processing:** Fresh, smoked, or canned. **Uses:** Not a favorite of Americans; eaten by Europeans.	The goldfish is a kind of carp. Carp flourish in muddy stagnant or polluted water. They prefer large rivers, slow-moving streams, and backwaters.

(Continued)

TABLE F-21 (Continued)

Common and Scientific Name(s)	Description	Habitat; Fishing	Processing; Uses	Remarks
Channel Catfish *Ictalurus punctatus*	Long barbels (whiskers) about the mouth for locating food; scaleless with heavy, sharp pectoral and dorsal spines; deeply forked tail; small irregular spots on sides.	**Habitat:** Warm quiet, slow-moving waters; prefer large rivers and lowland lakes; originally found mainly in Miss. basin waters, now in many waters throughout the U.S. **Fishing:** From catfish farming in states west from Fla. into Tex. and extending north to Kansas and Missouri; some caught as sport fish.	**Processing:** Purchased as whole dressed, skinned dressed, steaks or fillets; fresh or frozen. **Uses:** Baked, broiled, grilled, barbecued, smoked, sauteed or stuffed.	For many years, the catfish market was adequately supplied by fishermen who harvested wild catfish. Catfish farmers can get fish to reach 1 ½ lb *(0.7 kg)* in a 210 day season.
Cod (Codfish; Scrod) *Gadus morhua*	Vary in color from shades of gray to green, brown, or reddish tints depending on background; back and sides covered with roundish brown or reddish spots; 3 dorsal and 2 anal fins; 30 to 40 in. *(75 to 100 cm)* long and 10 to 25 lb *(4.5 to 11.3 kg)* in weight.	**Habitat:** Atlantic waters from Virginia to the Arctic; abundant in Gulf of St. Lawrence and off Newfoundland. **Fishing:** Originally line trawl; otter trawl; gill and pound nets; varies from season to season and ground to ground.	**Processing:** Filleted or steaked, sold fresh or frozen; breaded fish portions and sticks. **Uses:** Baked, broiled, poached, fried, or steamed; oven finish or deep fry breaded portions and sticks.	Salt cod was one of the first export items from the American colonies. Cod exploitation began in the 16th century when French and Portuguese vessels fished the Grand Banks off Newfoundland.
Crayfish (Crawfish) *Cambarus virilis* and *Cambarus bartoni*	Freshwater crustaceans which look like small lobsters; usually brownish green but may be white, pink, or blue; seldom over 6 in. *(15 cm)* long.	**Habitat:** Streams and ponds throughout the world. **Fishing:** Baited poles and traps.	**Processing:** None. **Uses:** Like lobster; thick soup, crayfish bisque in La.	Crayfish are eaten as food and used as live fish bait in many parts of the world.
Dungeness Crab *Cancer magister*	Light reddish-brown on back with a pattern of lighter streaks and spots; up to 10 in. *(25 cm)* across back and may weigh 1 ¾ to 3 ½ lb *(0.8 to 1.6 kg).*	**Habitat:** Along the Pacific coast from Alaska to southern Calif. in sandy and grassy bottoms below the tidal range. **Fishing:** In offshore waters from 12 to 120 ft *(3.7 to 36.6 m)* with pots and ring nets.	**Processing:** Live in shell; fresh or frozen, cooked meat, sections, claws; breaded raw or cooked; canned. **Uses:** Steamed, baked, broiled, simmered, casseroles, salads, appetizers, cocktails, and sauces.	The Dungeness crab fishery is the oldest known shellfishing of the North Pacific coast. It is named for a small fishing village on the Strait of Juan de Fuca in Wash.
Geoduck Clam (King Clam; Gweduc; Gwec Duk; Gooey Duck) *Panope generosa*	Shell too small to contain entire clam; average 3 lb *(1.4 kg)* with 1 ½ lb *(0.7 kg)* of meat.	**Habitat:** All along U.S. West Coast but most abundant in Washington's Puget Sound; 18 in. to 6 ft *(.5 to 1.8 m)* below surface of beach or beneath surface when underwater. **Fishing:** Divers and hand operated equipment working 10 to 60 ft *(3 to 18 m)* deep.	**Processing:** Frozen and canned minced meats or chunks; canned smoked chunks. **Uses:** Steaks fried; minced as dip or chowder; party snacks.	Geoducks are believed to be second in size only to the giant clam found in the East Indies. Geoducks are available on a regional basis.

(Continued)

TABLE F-21 *(Continued)*

Common and Scientific Name(s)	Description	Habitat; Fishing	Processing; Uses	Remarks
Flounder (Blackback; Fluke; Summer Flounder; Winter Flounder) *Pseudopleuronectes americanus* or *Paralichthys dentatus*	Variety of sizes; all bizarre in appearance; after birth, skulls twist and one eye moves to other side; mouth gives crooked pained look; white underside and dark top; actually swim on their sides.	**Habitat:** Along every coastline in the U.S., Canada, Denmark, England, Iceland, Japan. **Fishing:** Most caught with otter trawls; sport fishermen using pole or rod-type spear called *a gig.*	**Processing:** Dressed; filleted; frozen breaded raw or cooked fillets; sticks and portions. **Uses:** Baked, broiled, poached, fried, steamed; oven finish or deep fry breaded fillets sticks or portions.	Flounder resemble flying saucers as they ripple and glide through the water. Flounder are part of a large group known as flatfish.
Haddock (Scrod) *Melanogrammus aeglefinus*	Black lateral line and black shoulder blotch called the "Devil's thumbprint" or "St. Peter's mark"; weighs 3 to 17 lb *(1.4 to 8 kg).*	**Habitat:** Along the northeastern Atlantic coast off Newfoundland, Nova Scotia, the Gulf of Maine and on Georges Bank. **Fishing:** Otter trawling; some caught with gill nets, fish traps and longlines.	**Processing:** Same as Flounder. **Uses:** Same as Flounder.	Haddock belongs to a large, important group known as ground fish. Scrod is a market category which weighs 1 ½ to 2 ½ lb *(0.7 to 1.1 kg).*
Halibut (North Pacific Halibut) *Hippoglossus stenolepsis*	Both eyes on same side of head due to migration shortly after birth; dark top and white belly side; mouth wears crooked pained look; range from 5 to over 80 lb *(2.3 to over 36.3 kg);* swim with eyeless side parallel to the bottom.	**Habitat:** Along the continental shelf and slope of the North Pacific adjacent to Alaska, British Columbia and Wash. **Fishing:** Longlines from trawlers.	**Processing:** Dressed, or steaked; fresh or frozen. **Uses:** Baked, broiled, poached, fried or steamed.	Largest member of the flatfish family, and sometimes called king of the flatfishes.
Herring (Pacific Sea Herring) *Clupea harengus* and *Clupea pallasii*	Bony streamlined blue-green body; loosely attached iridescent scales; reach maximum length of 13 in. *(33 cm).*	**Habitat:** Off coast of southern Calif. to the Bering Sea. **Fishing:** Purse seines, gill nets and dip nets; powerful lights used to attract herring.	**Processing:** Fresh, salted whole, split (dressed); pickled or smoked; canned; fish meal and oil. **Uses:** Bait, oil, fertilizer; cooked or eaten as kippered herring.	Herring which have been smoked but not salted are known in England as Yarmouth bloaters.
Jonah Crab *Cancer borealis*	Brick red to purplish on top and yellowish underneath; about 6 in. *(15 cm)* wide weighing 1 lb *(0.45 kg);* biggest pieces of meat in claw.	**Habitat:** Low tide depths to depths of several hundred feet; New England coast from Maine to Cape Hatteras, N.C. **Fishing:** Commonly caught along with lobsters in lobster pots; modified lobster pot for only catching Jonah.	**Processing:** Live in shell; fresh or frozen, cooked meat, sections, claws; breaded raw or cooked; canned. **Uses:** Steamed, simmered or boiled; casseroles, salads, appetizers, cocktails, and sauces.	It is a walking crab so the Jonah crab does not have the lump meat of swimming crabs.

(Continued)

TABLE F-21 *(Continued)*

Common and Scientific Name(s)	Description	Habitat; Fishing	Processing; Uses	Remarks
King Crab *Paralithodes camtschatica*	Second, third, and fourth pairs of legs larger than first (pincers); fifth pair small and often inserted under shell; rough, heavy, stony-appearing shell; average 11 lb *(5 kg);* some span tip-to-tip 6 ft *(1.8 m)* and weigh 24 lb *(11 kg).*	**Habitat:** Cold waters of North Pacific ocean, Bering Sea and Okhotsk Sea; move in shore in spring to water depths of 60 to 200 ft *(18 to 61 m)* and offshore to depths of 1,000 ft *(305 m)* in summer and fall. **Fishing:** Baited crab pots.	**Processing:** Fresh or frozen in the shell; cooked meat, sections; canned. **Uses:** Used interchangeably with other crab meat recipes; casseroles, salads, appetizers, cocktails, and sauces.	The peak of the Alaskan king crab fishery was reached in 1966 with a total catch of 159.2 million pounds *(71.6 million kg).* The catch declined since 1966.
Lake Trout *Salvenlinus namaycush*	Vary widely in color with shades of gray and olive predominating; body mottled; flesh varies from ivory to deep pink; for commercial purposes range from 4 to 5 lb *(1.8 to 2.3 kg)* ; size varies from one body of water to the next.	**Habitat:** Principally the large, cool, freshwater lakes of northern North America. **Fishing:** Commercial fish in the Great Lakes with gill nets.	**Processing:** Whole, dressed, filleted; steaks; fresh or frozen. **Uses:** Baked, broiled, poached, fried, steamed, or sauteed.	One of the largest trout on record was 49.5 in. *(1.3 m)* long and weighed 102 lb *(46.3 kg).* It was caught in Lake Athabaska, Saskatchewan in 1961.
Lingcod *Ophiodon elongatus*	Coloration variable and associated with habitat, ranging from mottled brown to bluish-green with cream colored underside; 2 large, fleshy flaps above eyes; long, deeply notched dark dorsal fin, range from 5 to 20 lb *(2.3 to 9.1 kg).*	**Habitat:** Bottom-dwelling along the entire Pacific coast from the Baja peninsula of Calif. to northwest Alaska; most abundant in colder waters. **Fishing:** Most with trawling using otter trawl; some with set lines, handlines and troll lines.	**Processing:** Fresh or frozen as dressed, filleted, or steaked. **Uses:** Broiled, sauteed, baked, poached, or deep fried.	The lingcod is not a true cod. Flesh of generally the immature fish is green or bluish-green, which disappears upon cooking.
Lobster *Homarus* sp or *Panulirus* sp (Spiny lobster)	Most shells, dark green or dark blue with spots; bright red when cooked; 12 to 24 in. *(30 to 61 cm)* long weighing 1 to 20 lb *(0.5 to 9 kg);* 4 pair thin walking legs; 1 pair of large claws.	**Habitat:** Atlantic and Pacific oceans at depths of 6 to 120 ft *(1.8 to 37 m),* under rocks and in holes. **Fishing:** Lobster pots.	**Processing:** Live in shell; fresh or frozen; cooked meat; tails raw; canned; usually sold in shell. **Uses:** Baked, broiled or simmered; variety of recipes for use.	There are three main types of lobster: (1) Northern, (2) Rock, and (3) Spiny. Best lobster comes from the North Atlantic.
Mackerel *Scomber scombrus*	Iridescent greenish-blue on most of upper body, turning to blue-black on head and silvery white on belly; satiny skin with small scales; smooth tapering heads and streamlined bodies; 16 in. *(40 cm)* long and 2 lb *(0.9 kg).*	**Habitat:** The Atlantic coast from Labrador to Cape Hatteras, N.C. **Fishing:** Off the coast of Nova Scotia, the Bay of Fundy, Gulf of St. Lawrence to Middle Atlantic region; pound nets, traps, purse seines and gill nets.	**Processing:** Fresh or frozen, whole, dressed, filleted, or steaked. **Uses:** Baked, broiled, fried, poached, or steamed.	Mackerel belong to the same family as tunas.
Menhaden (Pogy; Fatback) *Brevoortia tyrannus,* and *Brevoortia patronus*	Vary from dark blue to green, blue gray or blue brown with silvery sides, belly and fins and yellow or brassy luster; body about three times deep as long, scaleless large head; average around 1 lb *(0.45 kg)* and 14 in. *(35 cm)* long.	**Habitat:** Along the coast of North America from Nova Scotia to central Florida and from southern Florida over to Veracruz, Mexico. **Fishing:** Purse seines; incidental catch in pound nets.	**Processing:** Fish meal, oils, and solubles. **Uses:** Seldom for human consumption.	The oils and solubles are used to manufacture paints, putties, resins, lubricants, soaps, and cosmetics.

(Continued)

TABLE F-21 *(Continued)*

Common and Scientific Name(s)	Description	Habitat; Fishing	Processing; Uses	Remarks
Mullet (Black or Striped Mullet) *Mugil cephalus*	Dark bluish on top, with silvery sides; bodies elongated and rather stout; large scales give appearance of stripes; range up to 2.5 ft *(0.8 m)* long and weigh 2 to 3 lb *(0.9 to 1.4 kg).*	**Habitat:** Worldwide, in coastal tropical and subtropical waters; in the Americas from Maine to Brazil and from Monterey, California to Chile. **Fishing:** Gill nets, trammel nets, haul seines, and cast nets; often fished at night.	**Processing:** Fresh or frozen whole or filleted; smoked; salted Mullet dip. **Uses:** Deep fried, oven fried, baked, or broiled.	Mullets were cultivated by Romans thousands of years ago in overflow deltas of the Nile River.
Ocean Perch (Redfish; Rockfish; Rosefish) *Sebastes marinus* (Atlantic) or *Sebastodes clutus* (Pacific)	Range in color from orange to flame red from Atlantic; various colors from Pacific; spiny projections; continuous back fin.	**Habitat:** From southern Labrador to the Gulf of Maine in Atlantic from the Bering Sea to lower California in the Pacific; prefer cold waters; generally deep offshore waters. **Fishing:** Net trawling; otter trawl.	**Processing:** Filleted and frozen, breaded, raw, or cooked as portions. **Uses:** Baked, broiled, poached, or steamed fillets; oven fried or deep fried, breaded, raw, or cooked portions.	Until the middle of the 1930s the ocean perch market was unexploited; then a fish cutter discovered the quality of fillets they yielded. Approximately ¼ of the world's known fresh and saltwater fish are perchlike or belong to the perch family.
Ocean Quahog Clam (Mahogany Quahog; Black Quahog) *Arctica islandica*	Clam with black or chestnut colored shell; 3 ½ to 4 in. *(9 to 10 cm)* long and weighing about ½ lb *(227 g).*	**Habitat:** From the Arctic Ocean to Cape Hatteras, N.C. at depths of 36 to 540 ft *(11 to 165 m);* generally occur in beds. **Fishing:** Hydraulic sea clam dredge.	**Processing:** Marketed live in shell; fresh or frozen shucked; frozen breaded, raw, and cooked; canned. **Uses:** Deep fried; pan fried patties; deviled clams, Manhattan clam chowder; clam cakes and rolls.	Early in 1943, a small ocean quahog fishery was established in the waters off Rhode Island, which peaked in 1946 and then declined because of insufficient demand.
Oysters (Eastern or Atlantic Oyster; Pacific Oyster; Western Oyster) *Crassostrea virginica* (Eastern); *Crassostrea gigas* (Pacific);and *Ostrea lurida* (Western)	Dissimilar lower and upper shells; upper shell flat and lower concave; adductor muscle near center of oyster's body controls shell.	**Habitat:** In beds along the temperate and tropical coastlines of all continents; between tidal levels or in shallow waters of bays and estuaries. **Fishing:** Depends on area; handpicking at low tide or use of tongs; hand operated dredges; suction dredges; machine-hoisted dredge; some oyster farming.	**Processing:** Live in shell market; shucked fresh and frozen; frozen breaded, raw, or fried; canned; smoked. **Uses:** Steamed, baked, sauteed, or used in variety of dishes.	Oysters should be cooked just enough to heat through and remain plump and tender. Many oyster beds have been depleted by overfishing.

(Continued)

TABLE F-21 *(Continued)*

Common and Scientific Name(s)	Description	Habitat; Fishing	Processing; Uses	Remarks
Pacific Cod *Gadus macrocephalus*	Ranges in color from brown to gray fading to white or grayish-white on the outer margins of all fins; may reach 3 ft *(1 m)* and weigh 30 to 40 lb *(14 to 18 kg).*	**Habitat:** Colder waters of the Pacific Ocean from California to northern Alaska. **Fishing:** Primarily trawl nets.	**Processing:** Dressed, steaked, or filleted; frozen, breaded raw, or cooked, fillet portions or sticks. **Uses:** Baked, broiled, poached, fried, or steamed; oven finish breaded, cooked portions or sticks, deep fried frozen breaded cuts.	First taken in British Columbia waters in 1881, thus it is a late comer compared to the Atlantic Cod.
Pompano (Cobblerfish; Butterfish; Palmenta) *Trachinotus carolinus*	Silvery body shading to metallic blue above and golden yellow below; deep-bodied with deeply forked caudal and dorsal fins; ranging from 1 ½ to 3 lb *(0.7 to 1.4 kg).*	**Habitat:** Mediterranean; Gulf of Mexico; Atlantic coast from Massachusetts to Brazil. **Fishing:** Netted with trammel or gill nets; haul seines and otter trawls may be used.	**Processing:** Whole or dressed. **Uses:** Baked, broiled, planked, pan fried, or deep fried.	Florida pompano offer potential for farming because of their limited supply and great demand.
Porgy (Scup) *Stenotomus chrysops*	Dull silvery color and iridescent; white on belly and side; sides and back marked with 12 to 15 indistinct stripes; body about half as deep as long; average of 1 to 2 lb *(0.45 to 0.9 kg)* and 12 to 14 in. *(30 to 35 cm)* long.	**Habitat:** Coastal waters of the Atlantic between Cape Cod, Mass., and Cape Hatteras, N.C. **Fishing:** Floating traps; otter trawls; purse and haul seines; sport fishermen.	**Processing:** Whole, dressed. **Uses:** Baked, pan fried, or sauteed.	Scup or porgies are a member of the vast perch family.
Rainbow Trout *Salmon gairdneri*	Broad reddish band or "rainbow" running along their side; olive green on their back and pure white or silvery on their belly.	**Habitat:** Native to the Pacific slope of the Sierras from California to Alaska; now transplanted to almost every state; prefer cool, clear, freshwater. **Fishing:** Popular fish for sport fishermen; commercial production from trout farms.	**Processing:** From trout farms, cleaned and packaged for fresh or frozen market; boned, and boned and breaded. **Uses:** Baked, broiled, pan fried, poached, or steamed.	Some rainbow trout migrate to the ocean and then return to their stream to spawn. These are called steelheads.
Red Crab *Geryon quinquedens*	Shell color ranges between red and deep orange; square body; long slender walking legs; weigh between 1 ¼ and 3 lb *(0.6 and 1.4 kg)* depending on sex.	**Habitat:** New England and Middle Atlantic coast where water temperature ranges between 38 and 41°F *(3 and 4°C)* at depths of 1,200 to 6,000 ft *(366 to 1,830 m).* **Fishing:** Traps or pots.	**Processing:** Cooked upon removal from ships and cold water tanks; whole cooked fresh or frozen; some specialty products; frozen picked meat. **Uses:** Broiled, baked, steamed, sauteed, or served cold; suitable in any recipe for crab.	Prior to the development of proper processing methods, red crabs were generally tossed back into the sea when brought in nets with fish. Refrigerated sea water tanks keep them alive until they are processed.

(Continued)

TABLE F-21 *(Continued)*

Common and Scientific Name(s)	Description	Habitat; Fishing	Processing; Uses	Remarks
Red Snapper *Lutianus blackford*	Brilliant red color; rather flat from side to side; large mouth with strong teeth; slightly forked tail; size ranges up to 30 lb *(13.6 kg)* and 2 ½ ft *(0.8 m)* in length.	**Habitat:** Around Fla., the West Indies and the shores of the Gulf of Mexico, and the Caribbean region. **Fishing:** Taken in water several hundred feet deep; most of the catch with hook and line; some use of a modified otter trawl.	**Processing:** Fresh dressed, fresh fillets, frozen fillets, and frozen portions. **Uses:** Broiled, baked steamed or boiled.	Red snapper industry centered in Fla. Some snappers live in the Pacific around the tropical coral atolls of the East Indies and Philippines.
Rockfish *Sebastodes* sp	Range from black or drab green to bright orange or crimson; stout heavily constructed bodies; large and broad heads; usually have prominent ridges and spines; fins heavily spined; 8 to 10 years to reach commercial size.	**Habitat:** North temperate seas from Calif. to Alaska, along rocky shores, tide water to depths of 1,200 ft *(366 m).* **Fishing:** Otter trawls and set line boats; some by trolling or hand lines from the shore.	**Processing:** Sold whole dressed, filleted, fresh or frozen. **Uses:** Baked, broiled, fried, and in chowders.	Four rockfish are among the best-eating varieties and are commercially important: (1) orange rockfish, (2) yellow-tail rockfish, (3) bocaccio, and (4) red rockfish.
Sablefish (Black Cod) *Anoplopoma fimbria*	Slaty black to greenish on top, shading to lighter gray on the belly; streamlined fish having compressed body with wide separation of dorsal fins; reach more than 3 ft *(1 m)* in length and weigh more than 40 lb *(18 kg);* average of 30 in. *(75 cm)* and 8 lb *(3.6 kg).*	**Habitat:** Range from Bering Sea to Calif., in depths of intertidal waters to 1,200 ft *(366 m).* **Fishing:** Otter trawlers and longline; possibly crab pots.	**Processing:** Dressed, frozen, and cured for smoking. **Uses:** Ready-to-eat; steamed; used in casseroles or salads.	The name black cod is misleading because it is a member of the skilfish family.
Salmon *Oncorhynchus* sp	Five important species: (1) Chinook, (2) Sockeye, (3) Pink, (4) Coho, and (5) Chum; coloration varies from red and green to bluish-green with silvery sides; weighing 3 to 20 lb *(1.4 to 9.1 kg)* and reaching 3 ft *(1 m)* in length; deep salmon to almost white flesh.	**Habitat:** Range from Monterey Bay, Calif. to Alaska; migrate from freshwater streams to ocean after hatch; spend most of life in ocean; some travel 3,000 miles to spawn in freshwater streams. **Fishing:** Purse seining, trolling, and gill netting; some with poles.	**Processing:** Dressed, steaked, filleted, fresh or frozen, smoked; canned. **Uses:** Baked, broiled, barbecued, fried, steamed, or poached; variety of recipes and dishes.	Salmon has nourished the human race since ancient times. The Roman, Pliny, wrote in 77 A.D. that "the river salmon is preferred to all fish that swim the sea." Salmon do not eat once they begin their spawning run. Salmon return to the streams in which they hatched, spawn, and die.
Sardines (Atlantic Herring) *Clupea harengus*	Actually, the immature Atlantic herring, which are greenish blue with a silvery cast on the sides and belly; reach about 4 in. *(10 cm)* by end of a year.	**Habitat:** From Virginia north to Labrador and Greenland. **Fishing:** Easily caught in dark of moon during their feeding time; purse seine; stop seine; some still may use brush weirs; otter trawl.	**Processing:** Packed in oil as well as mustard and tomato sauces; canned. **Uses:** Ready-to-eat snack; convenience food.	The word "sardine" is not the name of just one species of fish, but rather a collective name that represents a variety of tiny soft-boned fish of the herring family.

(Continued)

TABLE F-21 *(Continued)*

Common and Scientific Name(s)	Description	Habitat; Fishing	Processing; Uses	Remarks
Scallop *Placepecten magellanicus*	Young shells pink and white with some darker color variations; fan-shaped shell with fluted (scalloped) edges; active swimmers by snapping shell together with oversized muscle called the "eye"; shell grows to 8 in. *(20 cm)* in diameter with "eye" of 2 in. *(5 cm)* in diameter.	**Habitat:** Middle and South Atlantic, New England, Gulf of Mexico; deep waters for sea scallops; bays and estuaries for bay scallops. **Fishing:** Dredging; dip nets, rakes or by hand in shallow waters.	**Processing:** Shucked soon after harvesting; fresh or frozen shucked, frozen breaded, raw or cooked; specialty products. **Uses:** Boiled or sauteed; cocktails.	Buildings in ancient Pompeii were ornamented with scallop shell designs. During the Crusades, scallop shells were the symbol of the holy pilgrimages.
Sea Bass (Striped Bass) *Roccus saxatilis*	Usually dark olive-green vary to bluish or black above, paling to silver on the sides and white on the belly; 7 or 8 horizontal stripes; elongated body rather stout; 2 spiny dorsal fins; a slightly projecting lower jaw; large scales; a 5-yr-old may weigh 6 to 7 lb *(2.7 to 3.2 kg).*	**Habitat:** Native to the Atlantic coast of North America; range from St. Lawrence River to northern Fla. and in streams along Gulf of Mexico from western Fla. to La.; successfully introduced on the Pacific coast and range from San Diego, Calif. to the Columbia River, Ore. **Fishing:** Shoal areas of open bays and sounds or in large tributaries along Atlantic coast; 50% caught with gill net remainder with pound nets, and haul seines; some incidental catches on Pacific coast.	**Processing:** Filleted, steaked, chunked or whole dressed. **Uses:** Baking, broiling, pan frying, oven frying, planking or poaching.	Sea bass were a vital fishery resource for the early colonists. Sport fishermen catch striped bass by surf casting. Largest striped bass recorded, weighed 125 lb *(56.7 kg)* and was 6 ft *(1.8 m)* long. Most fishing, however, is for pan-sized fish weighing about 1 lb *(0.45 kg).*
Shrimp (Northern Shrimp; North Pacific Shrimp; Southern Shrimp) *Penaeus* sp and *Pandalus* sp	Body shape resembles a small lobster or crayfish without pincers; color from gray, brown, white, pink, yellow, red, to blue; move rapidly backward by flipping tail.	**Habitat:** Offshore waters of Maine and Mass. for Northern shrimp; along the coastlines of Calif., Ore., Wash., and Alaska for North Pacific shrimp; waters of Gulf and South Atlantic States for Southern shrimp. **Fishing:** Otter trawl along bottom.	**Processing:** Raw in shell; peeled, fresh or frozen; deveined, fresh or frozen, frozen raw or cooked breaded; cooked, peeled and deveined, fresh or frozen; canned. **Uses:** Simmered, baked broiled, fried or oven finish; cocktail; hundreds of uses such as casseroles, salads, and sauces.	Three species of Southern shrimp are commercially important: (1) white shrimp, (2) brown shrimp, and (3) pink or brown spotted shrimp. Southern shrimp are usually the largest and North Pacific shrimp are the smallest.
Smelt (Whitebait; Eulachon) *Osmerus mordax*	Resemble a midget salmon; small, slender, silvery fish with olive green coloring along the back; 7 to 8 in. *(17.5 to 20 cm)* long; 10 to 11 in. per pound *(58 cm/kg);* large scales.	**Habitat:** North Atlantic; Pacific coast; Columbia river; bays from Mexico to Canada; Great Lakes. **Fishing:** Pound nets, gill nets, and modified shrimp trawl for commercial fishing.	**Processing:** Fresh or frozen, whole, dressed, breaded, precooked. **Uses:** Broiled, fried, baked or prepared in casserole.	Originally anadromous, but smelt have adapted to their freshwater habitats. They were introduced near the Great Lakes in 1906 as food for salmon.

(Continued)

TABLE F-21 *(Continued)*

Common and Scientific Name(s)	Description	Habitat; Fishing	Processing; Uses	Remarks
Snow Crab (Tanner; Queen) *Chionoecetes* sp	Long slender legs and rounded body; average shell diameter of 6 in. *(15 cm)* and weighing 2½ lb *(1.1 kg).*	**Habitat:** North Pacific coast to Alaska in the Bering Sea in waters no deeper than 1,560 ft (476 m). **Fishing:** Crab pots.	**Processing:** Live in shell; fresh or frozen, cooked meat, sections, claws; specialities, frozen breaded, raw or cooked (cakes, patties, deviled, stuffed); canned. **Uses:** Interchangeably with other crabmeat recipes.	Snow crab fisheries were developed after those of the Dungeness and king crab.
Sole (Gray Sole; Witch Flounder) *Glyptocephalus cynoglossus*	One eye migrates to the opposite side shortly after hatching; swims with eyeless side down; underside white; top color of sea floor where they live; small mouth; flying saucer appearance.	**Habitat:** Along the continental shelf and slope of Atlantic coast. **Fishing:** Otter trawl; also sport fishing with poles, spears, and nets.	**Processing:** Whole, filleted, breaded portions, stuffed or stuffed breaded. **Uses:** Baked, broiled, fried, steamed or deep fried.	The term sole includes several types of fish: (1) lemon sole, *Pseudopleuronectes americanus,* (2) petrale sole (brill sole), *Eopsetta jordani,* and (3) English sole, *Parophrys vetulus.* There are others and they live in the Atlantic and Pacific oceans.
Spanish Mackerel *Scomberomorus maculatus*	Beautifully colored, dark blue on upper part paling until silvery on the belly; many small yellowish or olive oval spots on sides, average fish 2 to 3 lb *(0.9 to 1.4 kg).*	**Habitat:** Along both coasts of North America from San Diego, Calif. south on the Pacific coast and from Monhegan Island, Maine south on Atlantic coast. **Fishing:** Gill nets for most; some troll lines and pound nets; favorite of sport fishermen.	**Processing:** Filleted, fresh or frozen. **Uses:** Baked, broiled or smoked.	Spanish mackerel are known for their spectacular leaps out of the water.
Squid (Inkfish; Bone Squid; Taw Taw; Calamari; Sea Arrow) *Loligo opalescens*	Small edible variety; 10 to 12 in. *(25 to 30 cm)* long; elongated body; 8 muscular arms and 2 tentacles; milky translucent color; 2 fins opposite end of body from arms.	**Habitat:** Both the Atlantic and Pacific coasts and in the Gulf of Mexico; commercial type range from Puget Sound to San Diego. **Fishing:** Lampara nets off the coast of southern Calif.	**Processing:** Whole squid fresh or frozen. **Uses:** Fried or baked with a stuffing; salads; sauces; combination dishes.	Squid is considered a gourmet or specialty item and it has long been popular with Mediterranean, Oriental, and Mexican cooks.
Sturgeon *Acipenser* sp	Slender bodies covered with rows of bony plates; long snout; barbels around small tubelike mouth; white sturgeon reach 286 lb *(130 kg).*	**Habitat:** Marine species anadromous; some entirely freshwater; both Atlantic and Pacific types. **Fishing:** Nets and hook and line.	**Processing:** Flesh usually smoked; eggs (roe) for caviar; swim bladder for isinglass. **Uses:** Specialty item.	There are four genera of sturgeon but most North American sturgeons belong to *Acipenser.* Sturgeon are the only surviving member of their family.
Sunray Venus Clam *Macrocallista nimbosa*	Elongated shell 5 to 7 in. *(12.5 to 17.5 cm)* long; dull pink to bluish purple shell.	**Habitat:** South Carolina to Fla. and the Gulf States. **Fishing:** Dredging which uses water pressure to dig clams.	**Processing:** Shucked and frozen. **Uses:** Chowder, fritters, patties, dips, and clam loaf.	Commercial harvesting began in 1967.

(Continued)

TABLE F-21 *(Continued)*

Common and Scientific Name(s)	Description	Habitat; Fishing	Processing; Uses	Remarks
Surf Clam *Spisula solidissima*	Concentric rings laid down on shell as clams grow; retractable neck or siphon; foot for digging in soft sand.	**Habitat:** Atlantic shores. **Fishing:** Dredges with high-pressure water spray; some by tongs, rakes, forks, and shovels.	**Processing:** Shucked and canned; alive in the shell; shucked and fresh. **Uses:** Steamed, fried, or broiled; in chowders, fritters, sauces, dips, or salads.	Clam farming may be a possibility in the future.
Swordfish *Xiphias gladius*	Vary in color from a dark metallic purplish cast on the upper surfaces to white on sides and belly; weigh average of 200 to 400 lb *(91 to 181 kg)*; absence of scales.	**Habitat:** Tropical waters around the Americas; around Hawaiian Islands. **Fishing:** Longlines, harpooning, or tuna-mackerel traps; highly prized game fish.	**Processing:** Steaked, fresh or frozen. **Uses:** Baked, broiled, fried, poached, or steamed.	In 1953, a swordfish which weighed 1,182 lb *(536 kg)* was caught off the coast of Chile.
Tuna *Thunnus* sp	Metallic steely-blue, yellowish, or deep blue or green depending on type of tuna; streamlined bodies; fins set in grooves on the body; range from 4 to 150 lb *(2 to 68 kg)* depending on type; skipjack smallest; yellowfin largest.	**Habitat:** Atlantic and Pacific coasts; worldwide. **Fishing:** Purse seine; formerly by pole and line from famed tuna clippers.	**Processing:** Catch frozen until processed; most canned. **Uses:** As it comes from the can; variety of recipes.	Four kinds of tuna are commercially important: (1) albacore, (2) yellowfin, (3) bluefin, and (4) skipjack, *Katsuwonus pelamis.*
Weakfish (Gray Sea Trout; Squeteagues) *Cynoscion regalis*	Dark olive green above, with back and sides lustrous with purple, lavender, green, blue, golden or coppery tints; lower surface white to silvery; average size from 1 to 3 lb *(0.45 to 1.4 kg).*	**Habitat:** East coast of Fla. to Mass. **Fishing:** Otter trawls; some gill nets, haul seine and floating traps; popular sport fishing with hook and line.	**Processing:** Whole or dressed; some filleted. **Uses:** Baked, broiled, sauteed, or pan fried.	Rapid contraction of their abdominal muscles against a resonating air bladder produces a drumming which boaters can hear.
Whiting (Frost fish; Hake; Silver Hake) *Merluccius bilinearis*	Silvery iridescent color with brown or dark gray tints on upper part of the body; long body with small scales; large eyes and mouth; reach length of 30 in. *(75 cm)* and weight of 5 lb *(2.3 kg).*	**Habitat:** On the continental shelf of eastern North America from the Newfoundland banks to Cape Hatteras, North Carolina in shallow waters to depths of over 3,000 ft *(915 m).* **Fishing:** Otter trawl; pound and trap nets in the past.	**Processing:** Whole, drawn, dressed or filleted; frozen breaded raw or cooked sticks and portions; some smoked. **Uses:** Baked, broiled, pan fried, poached or steamed; portions and sticks oven finished.	Whiting are members of the cod family. A century ago there was no market for whiting and they were discarded. Real interest in whiting began in the 1920s when they were used in the fried fish shops around St. Louis.
Yellow Perch *Perca flavescens*	Golden yellow color accentuated by greenish-black; pelvic fins yellow to red tinged; distinctly deeper than wide; seldom exceeds 14 in. *(35 cm)* or 1 lb *(0.45 kg).*	**Habitat:** Native to freshwaters east of the Continental Divide in large lakes and rivers. **Fishing:** Commercially from the Great Lakes with trap, gill, fyke, hoop, and pound nets in deep waters.	**Processing:** Whole, dressed, or filleted; frozen, fresh or smoked. **Uses:** Baked, broiled, pan or deep fried, and planked.	In many areas, the prolific yellow perch are in constant danger of overpopulating their environment. Sport fishing regulations are liberal in most bodies of water.

FISH PROTEIN CONCENTRATE (FPC)

This term refers to fish protein processed for human consumption. It is produced from types of fish that are not popular in the usual channels of fresh fish trade, such as Menhaden. The fish are extracted to remove oil, dried, and ground to make a bland meal containing about 80% protein, 0.2% fat, and 13% mineral.

(Also see FISH AND SEAFOOD[S].)

FISTULA

A tubelike passage from some part of the body to another part or to the exterior—sometimes surgically made.

FLASH-PASTEURIZATION (HIGH TEMPERATURE-SHORT TIME METHOD; HTST)

In flash-pasteurization, the material to be pasteurized is held briefly at a temperature well above that normally required for batch (vat) pasteurization. For milk, the flash process at 161°F (*71.7°C*) requires at least 15 seconds of holding at this temperature for pasteurization. Conventional batch (vat) pasteurization of milk needs 145°F (*63°C*) for 30 minutes. Flash-pasteurization is widely used in the food industry. However, time and temperature employed vary according to the needs of the different foods.

(Also see MILK AND MILK PRODUCTS, section headed "Processing Milk"; PASTEURIZATION; and ULTRA-HIGH TEMPERATURE STERILIZATION.)

FLAT SOURS

A type of canned food spoilage characterized by the production of acid without gas formation. The food becomes sour but the ends of the can remain flat since no gas is formed. Low-acid foods such as peas, corn, beans, and greens are more likely to have flat sour spoilage than other vegetables. Thermophilic bacteria are responsible for flat sours. Hence, this type of spoilage occurs when foods are not cooled quickly after canning, or are held at too high storage temperatures.

(Also see PRESERVATION OF FOOD; and THERMOPHILES.)

FLATULENCE

Excessive gases in the stomach and intestines causing discomfort, bloating, rumblings, and gas removal via the mouth (belching) or anus (passing wind). Foods such as apples, beans, cabbage, turnips and carbonated drinks may produce flatulence. The habit of swallowing air during eating is also a common cause. Sensible eating habits and closing the mouth while chewing provide the best preventative measures, unless it is caused by some diseased condition.

FLAVEDO

The colored, oily outer layer of the peel of citrus fruits, which is also called the *zest*. Although the color of the peel often changes from green to yellow or orange as the fruit matures, the color is *not* always a good indicator of the degree of ripeness, because for some citrus fruits the "degreening" of the rind requires cool evenings, which may not occur in locations such as Florida. Furthermore, a yellow or orange color may revert back to green when late maturing fruits are left on the tree until the following spring. Some citrus fruits are degreened artificially by exposing them to ethylene gas in a warm room.

It is noteworthy that there is now a process for recovery of the colored pigment from citrus peel after the juice has been extracted from the fruit. The extracted pigment may be added to pale-colored juices to make them more attractive to consumers.

The oil which is present in the oil glands of the flavedo may be extracted by pressing the peel. It is used for imparting a strong flavor to diluted and sweetened citrus drinks. However, the oil may be oxidized readily and become bitter flavored. Hence, it is not usually added to products that are heated during processing.

(Also see CITRUS FRUITS, section headed "Processing.")

FLAVIN ADENINE DINUCLEOTIDE (FAD)

A highly active coenzyme consisting of riboflavin and adenosine diphosphate required in many reactions that affect amino acids, glucose, and fatty acids.

FLAVIN MONONUCLEOTIDE (FMN)

A riboflavin-containing coenzyme involved in the deamination of certain amino acids.

FLAVOPROTEIN

A conjugated protein that contains riboflavin and is involved in tissue respiration.

FLAVORINGS AND SEASONINGS

Contents *Page*

History of Flavorings and Seasonings **743**
Flavorings and Seasonings Table **743**
Why, When, and How Much Flavoring to Use **760**
Wines and Liqueurs **763**

Although the distinctions are not always clear-cut, the following terms and definitions are pertinent to an understanding of flavorings and seasonings:

• **Flavorings**—*These are the substances that stimulate the senses of taste and/or smell.* With the exception of the four primary sensations—sweet, bitter, salty, and sour—flavor characteristics are the result of our perception of odor; the difference between flavor and fragrance is in large part only a semantic distinction. Thus, a substance that provides an odor in perfumes may also be used to add flavoring to a food.

Most natural flavorings, with the exception of common salt, are derived from plant substances—either from the aromatic, volatile vegetable oils known as essential oils, or from the nonvolatile plant oils called resins. But some are derived from synthetics which resemble the natural products.

• **Seasonings**—*These are ingredients, such as flavorings, condiments, herbs, spices, or extracts, that are added to foods primarily for the savor that they impart.* The imaginative use of these ingredients can lead to highly palatable and interesting flavors.

Salt, or sodium chloride, is the most commonly used seasoning, as well as an essential body mineral. Pepper ranks next to salt as a common seasoning. Herbs, spices, and extracts (including vanilla, almond, and fruit extracts—such as lemon and orange) are also used as seasonings.

• **Condiments**—*This refers to something pungent, acid, salty, or spicy added to or served with food to enhance its flavor or to give added flavor.* They are useful in helping to make certain foods palatable and easier to digest. All condiments should be used sparingly and in such proportions that they do not spoil the natural flavors of other ingredients.

• **Herbs**—*This refers to the fragrant leaves and flowers of certain plants grown in temperate climates.*

(Also see HERB.)

• **Spices**—*Generally, these are products of tropical and subtropical trees, shrubs, or vines and are characterized by highly pungent odors or flavors.*

(Also see SPICES.)

So, take your choice: flavorings, seasonings, condiments, herbs, or spices. The presence of one or more of them makes the difference between the enjoyment of eating and the necessity of eating.

(Also see PALATABILITY.)

HISTORY OF FLAVORINGS AND SEASONINGS. The quest in far off lands for new and different flavorings and seasonings is ages old. Explorations, wars, and conquests were a part of their history.

Herbs and spices were valued for flavoring and seasoning foods and drinks, for medicinals, and for making cosmetic oils and perfumes. Many romantic legends, beliefs, and superstitions revolved around spices and herbs. Many were reputed to have magical powers: thyme was considered a source of courage, and tansy and sesame were associated with immortality.

As early as 3500 B.C., the Egyptian queen, Hatshepsut, used cinnamon as an aromatic. And in 3000 B.C., history records that the gods of the Mesopotamian peoples drank sesame wine, one of the products for which this popular seed was used.

The Bible also makes many references to spices beginning with the Book of Genesis, Chapter 37. Some of the spices mentioned throughout the good book are not familiar today, but many are still used. In the Revelation of St. John the Divine, Chapter 18, we see how distressed the merchants of the world became at the loss of the rich spice market when Babylon fell.

Both the Bible and historical records reveal that spices and aromatics were held in high esteem, and a gift of spices was greatly prized.

In days of yore, many of the noble, as well as the not so noble, had their own herb garden, sometimes formal and sometimes informal. Many were works of art. But whether the cook had a few pots of herbs on the kitchen windowsill, or a formal classic 16th century herbal knot garden, fresh herbs were considered far superior to dried.

Fig. F-24. An interesting garden of herbs. (Courtesy, Brooklyn Botanic Garden, Brooklyn, N.Y.)

FLAVORINGS AND SEASONINGS TABLE. Table F-22 is a summary of Flavorings and Seasonings. It describes them, tells about their production, and gives their uses.

Where available, the nutrient content of flavorings and seasonings is given in Food Composition Table F-36, under the section, Flavorings and Seasonings. In general, herbs and spices are high in fiber and minerals, but, because of the minute quantities consumed, they do not make any significant contribution to the daily requirement.

TABLE F-22
FLAVORINGS AND SEASONINGS

Name	Description; Production	Uses
Ajowan *(Carum ajowan)*	A seed closely related to caraway and cumin. Cultivated in India, Egypt, Persia, and Afghanistan.	Ajowan is used extensively in India to impart a thymelike flavor.
Allspice *(Pimenta officinalis)* 	The purple, pea-size fruit of a tree native to the West Indies, Central and South America. It grows in most tropical areas, but Jamaica produces most of the world's supply. When dried, the fruit turns brown and resembles large peppercorns.	Allspice combines the flavor of cloves, cinnamon, and nutmeg. Whole allspice is used for pickling, gravies, broiled fish, and meats. Ground allspice is used for baked goods, fruit preserves, puddings, and relishes.
Almond, sweet *(Prunus amygdalus dulcis)* **Almond, bitter** *(Prunus amygdalus amara)*	The seeds of a tree related to the peach. Trees reach a height of 9 to 22 ft *(3 to 7 m)* and have pink or white flowers that bloom early in the spring. There are two major types: sweet and bitter. The almond is native to western Asia and North Africa. But today it is grown in most temperate regions. In California, the largest U.S.-producing state, 100 varieties are grown. Almond trees require more than one variety for pollination.	Almonds can be used in every dish from soup to dessert. Bitter almonds, from which almond oil is extracted, should be used only in dishes that so specify, as the flavor is very powerful. Almond extract is used in cookies, confections, and Chinese cuisine.
Aloe vera *(Aloe barbadensis)* 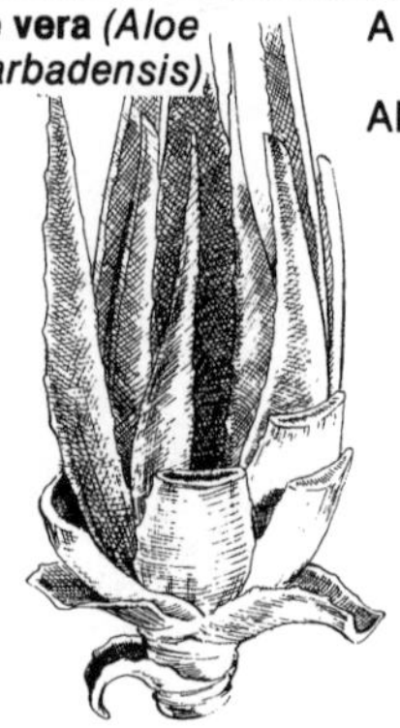	A succulent plant with leathery sword-shaped leaves, 6 to 24 in. *(15 to 60 cm)* long. Aloe vera is a semidesert plant which must be moved inside if the temperature drops below 50°F *(10°C)*. It is grown in Mexico and Hawaii.	The bitter, honeylike sap turns black when boiled down and is used in small amounts in bitters.
Anchovy	A small fish, about 5½ in. *(14 cm)* long, found mostly in the Mediterranean. It runs in extensive schools.	The fresh fish, when cooked, has a delicious white meat with good flavor, not at all like the canned anchovies. The dark brown meat and flavor of canned anchovies is the result of several months of pickling in salt, a process which goes back thousands of years in the Mediterranean area. Anchovies may be served in any number of ways, as an hors d'oeuvre, but they are also added in small quanties to many fish, beef, and mutton dishes, which they complement and to which they impart a richer flavor.
Angelica *(Angelica archangelica)* 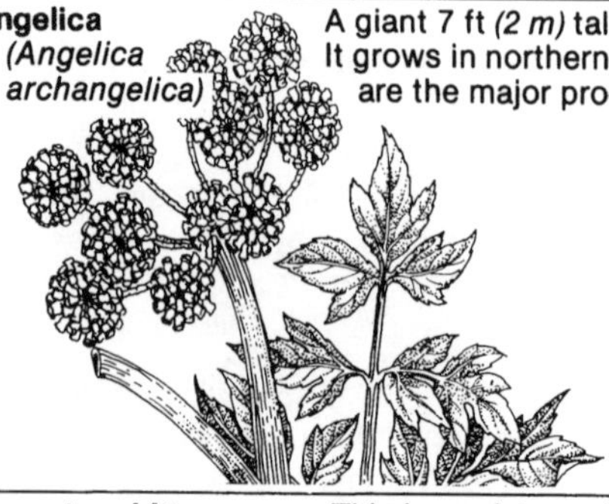	A giant 7 ft *(2 m)* tall member of the parsley family. It grows in northern climates. Germany and France are the major producers today.	All parts of the plant are used. It can be cooked and eaten as a vegetable. The young stems are candied for cake and dessert decoration. The raw blanched young shoots are used in salads, and the leaves are used in chopped herb mixtures. It is used as a flavoring for rhubarb and orange marmalade. It is added to many liqueurs and aperitifs.
Angostura bitters	This is a mixture of cinnamon, cloves, mace, nutmeg, orange and lemon peel, and prunes crushed with their stones, plus quinine and rum.	It is used to flavor iced fruit dishes and ice cream, and is added to some alcoholic beverages.

(Continued)

TABLE F-22 *(Continued)*

Name	Description; Production	Uses
Anise **Aniseseed** *(Pimpinella anisum)* 	The fruit of a small (1 to 2 ft, or *30 to 60 cm*) annual plant grown all over the world. The plant belongs to the carrot family *(Umbelliferae).* Anise grows wild in the countries around the Mediterranean. The U.S. imports about 500 tons annually.	Anise is a popular favorite for a few gourmet dishes such as Oysters Rockefeller. Also, it is used in bakery products, candies (especially licorice candy), certain kinds of cheese, pickles, and many liqueurs, and cordials, including anisette and absinthe.
Anise-pepper *(Xanthoxylum pipesitum)*	The dried red berries of a small feathery-leaved spiny tree, which give a hot and aromatic flavor. Anise-pepper is not used in western cuisine, but it is important in China.	It is one of the ingredients in Chinese Five Spices and is commonly used for fish and strongly flavored foods.
Asafoetida *(Ferula asafoetida)* 	This plant grows from 6 to 12 ft *(1.8 to 36 m)* high, mostly in southern Europe, western Asia, and North Africa. The milky juice from the stem, when dried, forms a gummy mass, which can be bought as a powder or as an extract.	Locally, the whole plant is used as a vegetable. Minute quantities will give fish a delicious flavor, and it may be added to beans, pickles, and vegetables.
Balm **Balm-mint** **Lemon Balm** *(Melissa officinalis)* 	Balm is a fragrant herb of the mint family, which grows over 2 ft *(60 cm)* high, and, although a perennial, dies down in winter. Lemon balm is native to southern Europe, where it has been cultivated for 2,000 years.	Balm has a pleasant lemon scent and can be chopped and combined with other herbs for use in omelets and salads, and in the production of several liqueurs. Also, balm leaves are used to flavor soups and dressings.

(Continued)

TABLE F-22 *(Continued)*

Name	Description; Production	Uses
Basil, sweet	(See Sweet Basil.)	
Bay leaves *(Laurus nobilis)*	The bay or sweet laurel is the laurel of history and poetry. It is a large evergreen shrub, sometimes reaching a height of 60 ft *(18 m)*, but rarely assuming a treelike character. The leathery, 2 to 4 in. *(5 to 10 cm)* long leaves possess an aromatic and slightly bitter flavor. The bay laurel is native to Italy, Greece, and North Africa, but it has spread throughout the world.	Bay leaves can be used either fresh or dried. They are one of the ingredients in bouquet garni, and are used in bouillon, marinades, olives, and pickles. They combine well with fish, potatoes, or tomatoes. The ancient Greeks used bay leaves to crown victorious athletes.
Beer	A fermented somewhat bitter beverage which is generally made from malted barley, although other cereal grains are used. Hops are added to beer during the brewing process. Beer contains 2 to 6% alcohol. Historians believe that the first grain beers were brewed before 5000 B.C. by the ancient Chaldeans and Egyptians. The most common types of beers and related drinks are: *Lager beer,* the light, foaming beer that is popular in the U.S.; *ale,* a strong beer that is popular in England and Ireland; *Porter,* a dark English beer; *stout,* another dark Irish and English beverage; and *Bock beer,* a dark beer that was first brewed in Germany.	In northern Europe, beer is used in cooking in the same way that wine is used in the grape-growing countries. It is important to use the correct beer for the recipe, depending on the country where the recipe originated. In Germany, there is a beer soup; in Flanders, beer is used in a delicious beef and onion stew; and in the U.S., beer is sometimes used to make pancakes.
Bouquet garni	This is a French term meaning a "bundle of sweet herbs." The herbs are tied together, or put in a piece of cheesecloth, along with spices. Then the "bundle" is removed when sufficient flavor has been imparted to the dish or at the end of cooking and before serving.	The bouquet garni is used in soups and stews, or any dish in which there is sufficient liquid to absorb the flavors. There are many recipes for bouquets garni. Possibly, the classic one is three stalks of parsley, a sprig of thyme, and a bay leaf. But many cooks experiment and develop their own secret formula.
Caper *(Caparris rupestris)*	The caper is a small bush with tough oval leaves and white or pink flowers the size of a wild rose. The caper is native to the Mediterranean region. It is grown as a greenhouse plant in northern U.S. and outdoors in warmer areas. Capers are difficult to cultivate and require much hand labor, which is the reason for their relatively high cost. The buds must be picked at the right stage each morning; then after wilting for a day, they are put in casks of strong, salted, white vinegar. After curing, they are graded and packed in bottles.	Capers have been used as a condiment for thousands of years. Capers are much used in European cuisine. They are commonly used in making caper sauce, which is usually eaten with boiled lamb. Also, they are combined with pickles (chopped), with anchovies, and with herbs; and grated lemon rind is often added. In addition to lamb, capers go well with fish dishes and with casseroles of chicken and rabbit.
Caraway *(Carum carvi)*	Caraway is a herb of the parsley family, famous for its spicy seeds. It is a biennial which grows over 2 ft *(60 cm)* high and is found wild in Europe and and temperate Asia. However, it is easy to culivate. Although it is cultivated in Europe, England, and the U.S., most of the seeds found in the marketplace are imported from Holland.	The seeds (actually the dried whole fruits) are used in cakes, cheese, confections, fresh cabbage, meat dishes, rye bread, salads, and sauerkraut. The chopped green leaves can be used in soups and salads. The roots can be cooked and eaten as a vegetable. Commercially, a caraway flavoring is made in the form of the essential oil obtained by steam distillation of the dry ripe seed. This oil is used to flavor chewing gum, candy, and liqueurs such as kummel.

(Continued)

TABLE F-22 *(Continued)*

Name	Description; Production	Uses
Cardamom seed *(Elettaria cardamomum)*	This perennial grows 6 to 16 ft *(1.8 to 4.8 m)* high, has long frondlike leaves, and has pods ¼ in. to 1 in. long, in which are found the seeds. Cardamom is native to India, but, today, it is cultivated in many tropical areas around the world. It is usually grown in a partially cleared forest because the plants require some shade.	Freshly ground cardamom has many uses including: breads, cakes, cookies, cheese, curries, custard, liver sausage, meat dishes, pilaus, pork sausage, and punches.
Cassia *(Cinnamomum cassia)*	Cassia is the aromatic bark derived from *Cinnamomum cassia,* sometimes termed Chinese cinnamon. The bark is much thicker than that of true cinnamon; and the taste is more pungent and the flavor less delicate than, though similar to, cinnamon. When ground as a spice, it is difficult to distinguish cassia from cinnamon; so, it is common practice to substitute the cheap cassia spice for the more valuable article. The tree is native to China, where its presence was recorded in 2500 B.C. It was taken to Europe over the spice routes from the East. Even today, most of our cassia still comes from the East. The bark is peeled off and cleaned, at which time it curls into what is usually called "stick cinnamon."	The stick cinnamon can be used in dishes to impart flavor, and then removed before serving; for example, some punches are flavored in this manner. Powdered cassia is used in combination with allspice, nutmeg, and cloves for spicing mincemeat, curries, pilaus, meat dishes, desserts and cakes. It is one of the ingredients of the famous Chinese Five Spices.
Cayenne pepper *(Capsicum frutescens)*	Cayenne pepper is supposed to have come originally from Cayenne in French Guiana; hence, the orgin of the name. This hot spice comes from red peppers ground fine. They are grown in Africa, Japan, Mexico, South America, and the U.S. Cayenne is more finely ground than chili powder.	Considerable care and caution should be exercised in using cayenne. A little goes a long way, but it is a spice that adds considerable interest to egg dishes, fish, and meat recipes.
Celery salt	Celery salt is made by grinding celery seed and combining it with salt. It is a common commercial condiment.	This spice is slightly bitter, but it combines well with bouillon, eggs, fish, potato salad, and salad dressing.
Celery seed *(Apium graveolens)*	Minute seedlike fruits of the celery plant. Celery is normally a biennial which produces the fleshy stalks the first year, and a flower stalk the second year. Sometimes the seeds are dried for use as a condiment.	Celery seeds have a slightly bitter taste, but they contribute a useful flavoring. They add special interest to many salads and salad dressings.
Chervil, garden *(Anthriscus cerefolium)*	An aromatic annual with lacy leaves, like parsley (there are also curled varieties for garnishing), which grows about 18 in. *(45 cm)* high and has tiny white flowers. Chervil is native to the Caucasus area of Russia. It is easily grown either in the garden or in a window-box inside. However, it does not like hot, dry conditions. The roots should not be eaten as they are poisonous.	Chervil, which has a mild anise-caraway flavor, is one of the ingredients of Fines herbes, a mixture of chopped fresh herbs extensively used in French recipes. Chervil is used in omelets, soups, salads, sauces, and white wine vinegar. It should not be cooked, instead, it must be added at the last minute; otherwise, it loses its flavor.

(Continued)

TABLE F-22 *(Continued)*

Name	Description; Production	Uses
Chinese Five Spices	This is a blend of anise pepper, star anise, cassia, cloves, and fennel seed.	Chinese Five Spices are an integral part of some of the recipes from the Far East. Also, it can be used to good advantage in flavoring pork dishes.
Chives *(Allium schoenoprasum)* 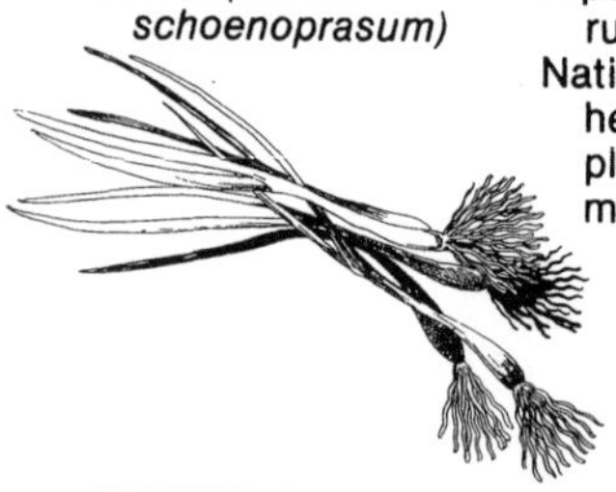	A perennial herb related to the onion, with slender rushlike leaves of bright green color. Native to Europe and Asia, chives are cultured in herb and vegetable gardens, as an indoor pot plant, and occasionally as a commercial crop, mostly for dehydration.	Chives are ideal as a garnish because of their delicate onion flavor and bright green color. Chives add interest and flavor to buttered beets, eggs, cottage or cream cheese, potato and other salads, sliced tomatoes, and soups. Chives contain a fairly large amount of vitamin C.
Chocolate *(Theobroma cacao)* 	Chocolate is the food made by combining the roasted ground kernel of the cacao bean with sugar and cocoa butter. (Cocoa butter is the fat released when the bean is ground.) Chocolate may also contain natural or artificial flavors, emulsifiers, and—in the case of milk chocolate—milk solid. The cocao tree grows 15 to 25 ft *(4.5 to 7.5 m)* high, and has dark green, shiny, leathery leaves. The seed or "bean" is found inside the fruit, which must go through a fermentation process to develop the flavor. The word chocolate comes from the Aztec Indian words chocolatl and theobroma which means "food of the gods." Chocolate was part of the Aztec diet long before Columbus came to America. Although originally wild and native to Central and South America, cacao trees are now cultivated. They flourish in a warm, moist climate. Most of the world's cacao beans come from West Africa, where Ghana and Nigeria are the largest producers.	Chocolate is used as a favorite flavoring for cakes, cookies, candies, drinks, ice cream, and sauces. In Spain and Italy, it is used with onion, garlic, tomato, and spices in meat dishes. A truscan sauce, like the one above, makes a gourmet dish out of sliced cooked tongue.
Cinnamon *(Cinnamomum zeylanicum)* 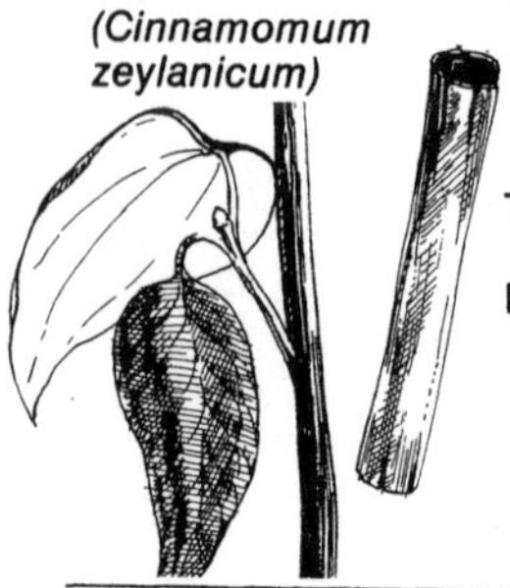	True cinnamon, which is an evergreen bush or tree, is native to Ceylon (now Sri Lanka), India, and Malaysia. The cinnamon tree grows as high as 30 ft *(9 m)*, but the cultivated tree is kept pruned to shrub height so as to facilitate harvesting. The bark of the lower branches is peeled and dried in the sun. Both stick (the dried bark) and ground cinnamon are used extensively as a spice.	Cinnamon has a more delicate flavor than cassia and is more suitable for sweet dishes, cakes, and cookies. The essential oil, distilled from the shoots and bark, is used to flavor candy and cola drinks. In Arab countries, a stick of cinnamon is often added to curries, mutton stew, and pilaus, which makes for a distinctive and pleasant flavor combination.
Citron *(Citrus medica)* 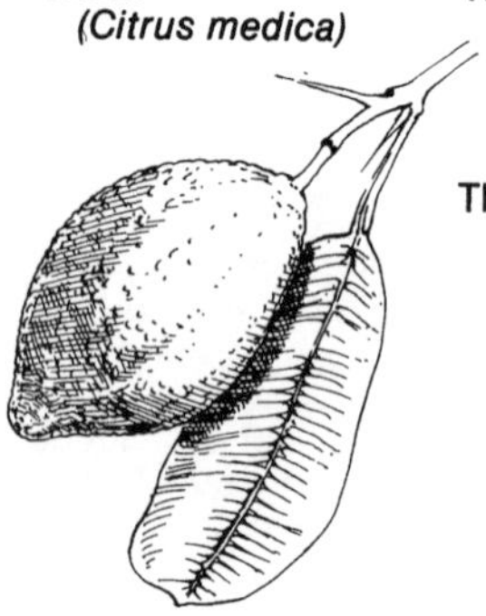	The citron tree is small and thorny, and the fruit looks like a rough-skinned lemon. The skin, which is the most important part, is very thick. The pulp is small, very acid and not of much importance. The citron tree is cultivated in the Mediterranean region and in the West Indies, California, and Florida.	Citron peel has a peculiar taste, quite different from the other citrus. About half of the world production is used in the U.S. as candied peel to be added to cakes, cookies, candies, and desserts. The *etrog*, a variety of citron with small fruits, is grown for use in the Jewish ceremony called the Feast of the Tabernacles.

(Continued)

TABLE F-22 *(Continued)*

Name	Description; Production	Uses
Cloves *(Syzygium aromaticum)*	Cloves, one of the best known spices, are the dried flower bud of a tropical tree; hence, they are picked before the flower opens. The name *clove* comes from the French word for *nail* because of the shape of the flower bud. Cloves are produced by 30-ft *(9 m)* high, red-flowered tropical evergreen trees, native to the Moluccas (the Spice Islands of Indonesia) but now widely cultivated in the tropics.	Whole cloves are used in many meat dishes, but a little goes a long way. Cloves are stuck into lemon slices for tea, into onions, and into hams for baking; they are also popular for apple cookery and pickle making. In the East, they go into many of the curry dishes. Whole cloves are also included in recipes for spiced wine and some liqueurs. Ground cloves are used in baked goods, borscht (beet soup), chocolate puddings, potato soup, and stews.
Cola *(Cola acuminata)*	Cola is made from powdering and boiling the kola nut, the fleshy seeds of tropical trees of the *Cola* genus. The cola tree is native to the rain forests of tropical West Africa. Cultivated trees reach 19 to 32 ft *(5.7 to 9.6 m)* in height and bear fruit for up to 50 years. Each fruit contains several nuts. Nigeria produces most of the world crop. Kola nuts contain two stimulants—caffeine and theobromine.	Cola is used in many soft drinks, and for coloring and flavoring some wines. The essence of cola nuts can be obtained for flavoring drinks or creams. In Africa and South America, kola nuts are chewed to dispel hunger and to alleviate fatigue.
Coriander *(Coriandrum sativum)*	The dried fruit of a small (3 ft, or *1 m*) annual plant native to the Mediterranean area, which is now grown all over the world. Most of the imported coriander comes from Morocco.	Fresh or dried coriander leaves (known as Cilantro) can be used for seasoning, and are popular in Near, Middle, or Far East recipes, as well as Mexico and South America. The seeds are a principal ingredient of curry. Whole coriander seeds can be used in cakes, cookies, biscuits, gingerbread, green salads, pickles, and poultry stuffing. Ground seeds are added to many meat and sweet dishes.
Cress **Watercress** *(Nasturtium officinale)* **Garden cress** *(Lepidium sativum)*	The cresses include several plants of the mustard, or *Cruciferae,* family that have a hot, pepperlike taste. Watercress is a native of Europe, but it is now grown all over the U.S. It is called watercress because it grows best in running water. Garden cress, or peppergrass, is an annual which grows 12 to 24 in. *(30 to 60 cm)* high. It has a sharper taste than watercress. Garden cress, which is of Asiatic origin, is popular in Europe and is becoming so in the U.S.	The cresses are used primarily in salads and sandwiches, but they can be used to flavor soup, cooked greens, or sauces for fish dishes, and to garnish meats.
Cumin *(Cuminum cyminum)*	This small plant is an annual with weak stems and orchid or pink and white flowers, which is grown mostly around the Mediterranean. The seeds, which resemble caraway, are dried to bring out the flavor. The resemblance in shape to caraway has led to confusion between the two in the European languages.	Cumin has been used as a spice since Biblical times. Cumin's principal use is in curry powder. It is also used to flavor bread, stuffed eggs, meats, rice dishes, and soups. Commercially, it may be found in cheese, chutney, pickles, meats, and sausage.
Curry (powder or sauce)	Curry originated in India, where it is a favorite seasoning. It is becoming increasingly popular in the Americas. Each manufacturer has his secret recipe, but curry powder may contain any or all of the following: allspice, cardamom, cayenne, cloves, cinnamon, coriander, cumin, curry leaves, fennel, fenugreek, ginger, red, black or white pepper, and turmeric.	Curry powder may be added to eggs, chicken, fish, meats, rice, soups, or to a salad made of sweet potatoes and pineapple. Traditional curry recipes are served with an array of toppings, such as raisins, slivered or chopped nuts, grated coconut, chutney, chopped green onions, chopped coriander. In India, the chutneys are made fresh from fruits and vegetables at hand, plus an array of seasonings.

(Continued)

TABLE F-22 *(Continued)*

Name	Description; Production	Uses
Dill *(Anethum graveolens)* 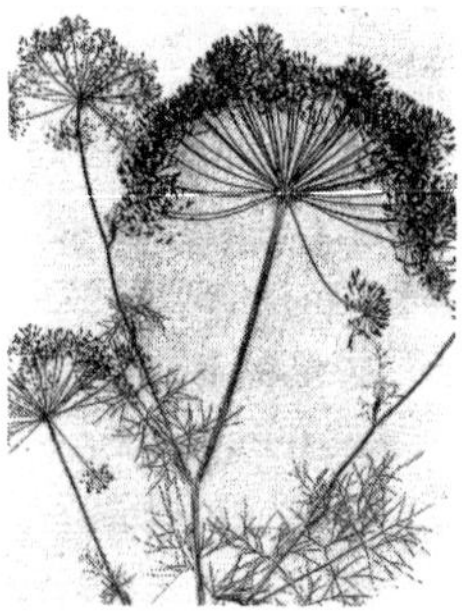	Dill is important for both its leaves and its seed. This annual plant of the parsley family, grows to 3 ft *(1 m)* high, has feathery leaves, and is relatively easy to grow in all moderate climates. Dill is native to Southwest Asia, and is cultivated in India, Europe, and the U.S.	Dill loses its flavor when cooked, so it should be added at the last minute. Fresh dill leaves can be used for dishes containing chicken, mushrooms, or spinach. The seeds are used in dill pickles and dill vinegar, but they can be added to meat dishes, meat and fish sauces, sauerkraut, salads, and borscht (beet soup). Dill is not used much in Mediterranean cuisines. The essential oil, obtained from both the fruits and leaves, it used in pickles, chewing gums, and candies.
Fennel *(Foeniculum vulgare)* 	Fennel is the dried fruit (fennel seed) of a 3 ft *(1 m)* high, perennial plant belonging to the Carrot family. The plant is native to southern Europe and widely cultivated in temperate and subtropical regions. The plant is harvested when the fruits are mature; air dried; then threshed.	Fennel has been used as a food flavoring since ancient times. Both the leaves and the seeds are used. The fennel plant has often been referred to as a symbol of strength and valor. Fennel has an aniselike flavor and is good with several foods: apple pie, candies, fish, liqueurs, pastries, pork, soups, and sweet pickles.
Fenugreek seed *(Trigonella foenum-graceum)* 	This annual plant is native to the Mediterrean region and northern India. It is widely cultivated there and in China. It is similar to clover in size, but it belongs to the Pea family. Fenugreek is not difficult to grow, but it needs lots of sun.	Fenugreek seeds are usually used in Indian curries and chutneys. Fenugreek has a celerylike or burnt sugar flavor. Fenugreek extract is used in imitation maple syrups and in butterscotch and rum flavorings.
Fines herbes	This is a combination of several herbs such as basil, chervil, chives, marjoram, oregano, parsley, rosemary, sage, tarragon, and thyme.	Fines herbes can be used in many dishes such as fish sauces, meat stuffings, omelets, salads, salad dressing, and soups.
Garlic *(Allium sativum)* **Garlic powder** **Garlic salt** 	Garlic is a perennial herb related to the onion grown for its pungently flavored bulb, which is used to season foods. The bulb consists of wedge-shaped pieces commonly called cloves. Garlic is native to central Asia, but it is now widely grown everywhere. Garlic is easy to grow. In northern climates, it can be moved indoors during the winter months.	Garlic may be used fresh, as a powder (ground dehydrated garlic), or as a salt (a mixture of ground, dehydrated garlic and salt). Garlic blends well with a wide range of dishes such as fish, game, meats, and vegetables. It is sometimes combined with parsley and mushrooms as it accentuates the latter. Garlic, along with onion, is a universal favorite for flavoring and is featured in all the countries noted for having the best foods. It gives a vitality to foods that cannot be surpassed by any other seasoning currently being used.

(Continued)

TABLE F-22 *(Continued)*

Name	Description; Production	Uses
Ginger *(Zingiber officinale)*	This is the root, or rhizome, of a perennial plant—the ginger plant—native to southeastern Asia, which has been used since the earliest recorded history. It is one of the best known of all spices, and is now grown all over the tropics. Both fresh and dried ginger are readily available. The dried is more spicy and intense in flavoring, and the fresh more subtle.	Ginger is used in numerous foods including beverages, biscuits, cakes, cookies, fish, gingerbread, ginger beer, ginger wine and cordials, puddings, sauces, and spice mixtures. It is used mostly in sweet preparations in European and North American cooking, but the Orient uses it extensively for chutney, fish, meat, and pickles.
Horseradish *(Armoracia rusticana)*	Horseradish, a member of the mustard family, is native to eastern Europe and is now cultivated in the northern parts of the globe. It will grow anywhere except in hard clay; it soon becomes difficult to eradicate once it takes hold. The pungent, fleshy root of the plant is usually the only part used. However, young leaves are sometimes used in salads.	Many cooks limit the use of horseradish to a sauce used on meats, but it merits a much wider sphere. It should not be added before cooking as the volatile oils (the flavor) are soon lost. Horseradish can be added to chicken salads, egg dishes, mayonnaise for use on fish dishes, or tomato combinations. It is said to have been one of the bitter herbs eaten by the Jews during the Feast of the Passover.
Leek *(Allium porrum)*	The leek is an ancient vegetable, resembling the onion, whose land of origin is obscured in antiquity. It is the root of a biennial and is considered a vegetable in England and America, but the continental countries consider it more as a flavoring. Today, the leek is an important crop in many countries.	The leek is rather like a very mild onion. It is used mostly in soups and chowders. However, the leek may also be used as a bouquet for pork or lamb. The leek is the national flower of Wales. Many Welshmen wear a sprig of leek on St. David's Day, March 1.
Lemon *(Citrus Limon)*	The tangy lemon is a fruit, but it is used as a flavoring in a wide range of dishes. The lemon, a native of southeastern Asia, is grown commercially chiefly in the countries around the Mediterranean Sea and in southern California.	Lemon juice can be used on salads instead of vinegar, and it is the predominant favorite for serving with most fishes. Grated lemon rind is added to many dishes such as cakes, cookies, desserts, and sauces, to give an added taste dimension.
Licorice *(Glycyrrhiza glabra)*	This is an ancient perennial herb of the pea family, growing to a height of 3 to 4 ft *(1 to 1.2 m)*, whose root is sweet. The word *licorice* refers both to the plant and to the flavoring produced from its roots. Today, Spain is the largest producer of licorice, but the plant is grown throughout the Mediterranean area, in parts of Asia, and in California and Louisiana.	Licorice is used to flavor candy, chewing gum, and soft drinks. ***CAUTION:*** Licorice raises the blood pressure of some people dangerously high, due to retention of sodium.

(Continued)

TABLE F-22 *(Continued)*

Name	Description; Production	Uses
Lime *(Citrus aurantifolia)* 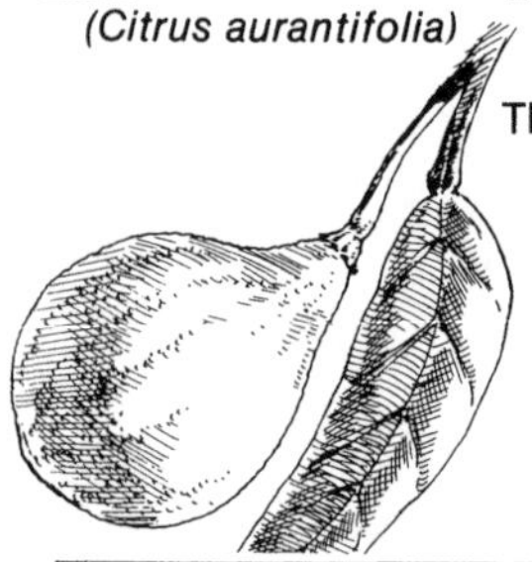	Lime is a rounded fruit that is pointed at both ends. It is greener than the lemon to which it is related. It grows on a small citrus tree. The lime tree came from India. Because it is frost-tender, most limes are grown in warmer countries, especially in Mexico, Egypt, Florida, and the West Indies.	Limes impart a unique taste to dishes, which cannot be replaced by lemons. For instance, scallops, slices of bacon and lime with parmesan cheese sprinkled over and broiled on skewers is a delicious and sensational gourmet dish, but when done with lemon it is of no consequence. Fish is often marinated in lime juice before cooking, to great advantage. For centuries, limes have been used as a preventive against scurvy; hence, "limey," the slang name for a British sailor, whose diet aboard ship included lime juice.
Lovage *(Levisticum officinale)* 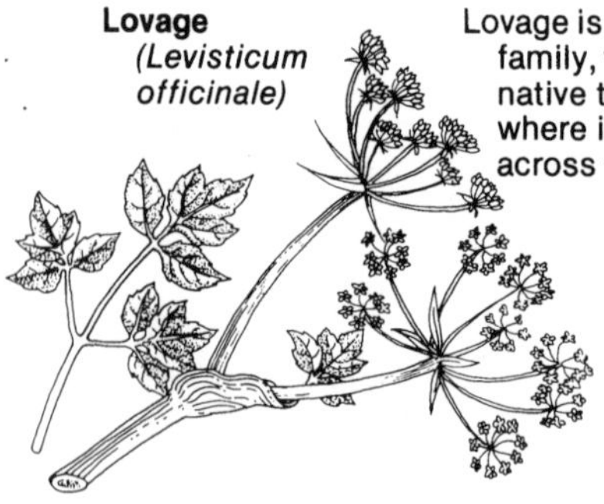	Lovage is a branching perennial herb of the parsley family, which grows to a height of 6 ft *(1.8 m)*. It is native to the mountains of southern Europe, where it is now cultivated. It also grows wild across the U.S. from New Jersey to New Mexico.	The stalks and leaves are used as an alternative to celery. They are also used for a tea. The seeds and flowering tops are used for flavoring and for confections. The roots of the *Ligustrum* herbs are large and aromatic and the essential oils obtained from them are used in flavoring various foods.
Mace *(Myristica fragrans)*	Mace is the fleshy part which grows between the nutmeg shell and the outer husk, which is used as a spice. The flavor resembles nutmeg, but is more subtle.	Mace can be added to apple dishes, beets, cakes, hot chocolate, coffee cakes, cookies, custards, eggnog, gingerbread, and muffins. The alcohol extract of mace is used in pickles and sauces.
Marjoram or **Sweet marjoram** *(Majorana hortensis)* 	Native to western Asia and the Mediterranean, this perennial herb belongs to the mint family and grows to 1 to 2 ft *(30 to 60 cm)* high. The leaves are used fresh as well as dried. There are several closely related plants in the oregano specie grouping, including both the sweet marjoram, and the wild marjoram, commonly known as oregano. Marjoram is commercially cultivated in France, Chile, Peru, and California.	The flavor of marjoram is related to thyme; hence, they are often used together or to replace each other. Marjoram is one of the most important herbs, as it can be added to almost every dish to advantage. It should be added immediately before serving as the flavor is easily lost in cooking. Marjoram is used with egg dishes, lamb, poultry, sausage, soups, stews, and vegetables.
Mint **Peppermint** *(Mentha piperita)* **Spearmint** *(Mentha spicata)* 	Mint refers to any of the various aromatic perennial plants with square stems constituting the family *Labiatae*, especially a member of the genus *Mentha*—peppermint and spearmint. There are about 3,200 different kinds of mint. Mint originated in the temperate regions of the Old World, where most of it is still grown. The leaves can be picked at any time.	Peppermint flavoring is used mostly for candies, cordials, desserts, icings, and liqueurs. Spearmint is the preferred mint for lamb, as well as for iced tea, and mint juleps. It can also be used in soups, stews, fish and meat sauces. Lamb without mint sauce in England would be unthinkable. In this country, it is also to be found as mint jelly.

(Continued)

TABLE F-22 *(Continued)*

Name	Description; Production	Uses
Monosodium glutamate (MSG)	Monosodium glutamate (MSG) is the sodium salt of glutamic acid just as common salt is the sodium salt of hydrochloric acid. Glutamic acid is one of the amino acids and is a component of gluten (a wheat protein). MSG occurs naturally in seaweed, soybeans, and sugar beets.	MSG does not have any flavor of its own, but it intensifies and enhances the flavor in other foods. It is widely used in commercially prepared foods containing meat and fish, as well as in home kitchens. Optimal concentrations in food are 0.2 to 0.5%. At levels of over 1% a sweetish taste may occur.
Mixed whole spice or **Pickling spice**	This is a mixture of various spices such as allspice, bay leaves, cardamom, chilis, cinnamon, cloves, coriander seed, dill seed, ginger, mustard seed, and pepper.	This blend of spices is used in pickling, in meats, in relishes, and in vegetables. It can be added to boiled whole beets, cabbage, stews, and boiled tongue.
Mustard **Black mustard** *(Brassica nigra)*  **Brown mustard** *(Brassica juncea)* **White** or **Yellow mustard** *(Brassica hirta; Brassica alba)* 	Mustard is the name given to certain annual plants of the cabbage family that are grown for their pungent seeds and for their leaves, which are eaten as greens and in salads. It is a native of southern Europe and southwestern Asia and has been cultivated since ancient times. (The mustard referred to in the Bible, Matthew 13:31,32 was probably the black mustard.) Today, mustards are grown in temperate regions throughout the world. The following varieties of mustard are commonly cultivated: • *Black mustard (B. nigra)*—Black mustard, whose seeds produce the strongest flavor of all mustards, grows to a height of 12 ft *(3.6 m)* in Israel. The seeds shatter easily, which makes it unsuitable for mechanical harvesting; hence, it is now grown only in those countries that have plenty of field labor. • *Brown, or Indian, mustard (B. juncea)*—This plant is approximately 4 to 5 ft *(1.2 to 1.5 m)* high and can be harvested mechanically. However, the seed is not as pungent as the black mustard. • *White or yellow mustard (B. hirta; B. alba)*—These plants grow only 2 to 3 ft *(60 to 90 cm)* in height and have yellow seeds. White or yellow mustard is cultivated commercially, but it has gone wild in both Europe and America. Because of the large number of plants required to produce mustard seed, along with the difficulty of processing, it not practical to produce mustard for spice purposes in home gardens. However, it is often grown for the leaves, for use in salads and greens.	Mustard is the most popular U.S. spice. When dry, mustard seeds have no odor or flavor. The pungency is due to an essential oil which develops when the crushed seed is mixed with water. An enzyme then causes a glucoside (a bitter substance chemically related to sugar) to react with the water, and the hot taste of mustard emerges. *Note well:* In general, enzymes are "killed" by boiling water and inhibited by salt and vinegar. Also, the volatile oils are easily lost during cooking, so mustard should be added at the last. Whole mustard seeds are used as a pickling flavoring and to add pungency to many foods, including pickles, meats, and salads. Powdered dry mustard is a common kitchen spice. Its sharp, hot flavor develops when the powder is moistened; so, the resulting paste must be used immediately. Powdered dry mustard is used for roast beef, mustard pickles, sauces, and gravies. Prepared mustard is a mixture of powdered mustard with salt, spices, and lemon juice, with wine or vinegar to preserve the mustard's pungency. It may be used with ham, hamburgers, hot dogs, and sandwich spreads. The leaves of mustard may be harvested while tender and eaten as greens. Mustard greens are an excellent source of vitamins A, B, and C.
Noyeau	The cordial or liqueur called Creme de Noyeau is made by crushing the kernels of apricots, cherries, bitter almond or peach in strong alcohol and then redistilling to eliminate the cyanide. The result of the complete process is a sweet liqueur with a strong almond flavor.	Noyeau can be used instead of almond extract to flavor cakes and desserts or as a topping for ice creams whose flavors blend well with almond. It is an occasional welcome flavoring alternative, useful to the gourmet cook.
Nutmeg *(Myristica fragrans)* 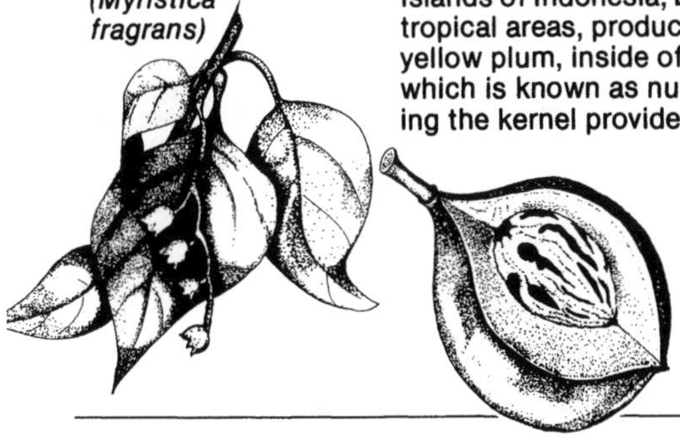	This large evergreen tree, native to the Spice Islands of Indonesia, but now cultivated in other tropical areas, produces a fruit resembling a yellow plum, inside of which is the kernel (seed) which is known as nutmeg. A membrane covering the kernel provides mace, another spice.	Nutmeg is traditionally used in sweet foods such as cakes, custards, doughnuts, eggnog, pies, and puddings, but it goes very well with meat, sausage, spinach, sweet potatoes, and vegetables.

(Continued)

TABLE F-22 *(Continued)*

Name	Description; Production	Uses
Onion *(Allium cepa)* 	The onion is a biennial (a plant that lives for 2 years); it stores food in the bulb during the first year, and it flowers in the second year. The upper part of the plant is a set of leaves growing inside each other. The lower parts of the leaves become very thick. The flowers are small and white, and grow rounded clusters. The bulbs are enclosed in a thin papery covering made up of dried outer leaves. The plant has a few shallow roots. Thought to have originated in Asia, the onion has been cultivated since ancient times. Today, onions are grown all over the world. California is the eading onion-growing state of the U.S. Onions vary in sizes, colors, and shapes.	Onions are used both in the immature stage, as green onion, and in the mature stage, as a bulb. Onions are used either as a separate vegetable or as a flavoring for other foods. The leaves of the onion, along with the bulb, are used in salad. Onion powder is produced by drying and grinding the bulb. However, dehydrated products do not compare to fresh onions for use in flavoring foods. Onion salt is also available for flavoring. Dried onion flakes or minced onion have now been added to the spice shelf. The onion is never used for sweet dishes, but its use in all other recipes is unlimited.
Oregano *(Origanum vulgare)* 	Oregano is the name given to several species of perennial plants whose dried leaves impart a particular flavor to food. *Origanum vulgare* is "officially" the source of oregano, but commercially it appears that the spice is derived from other plants, too. Whatever specie oregano may be, it seems to have a southern European origin, and was known as early as the Greek civilization. The oregano that is grown in cooler climates is not as strong as that grown in such climates as southern Italy.	Oregano is used extensively in Italian cooking and can be added to cheese dishes, chili beans, fish, gravies, meats, sauces, sausage, salads, and soups.
Paprika (Hungarian Pepper, Spanish pepper) *(Capsicum tetragonum)*	Paprika is prepared by grinding the dried pods of the sweet red pepper, from which the core and seeds are usually removed in order to reduce the pungency. It is the national spice of Hungary and is used extensively in Spanish cooking.	Paprika is essential for Hungarian Goulash. It is used in many dishes both for its flavor and as a garnish. It can be added to chicken, sweet corn, fish, meats, sausages, tomato catsup, and tomato juice.
Parsley *(Petroselinum crispum)* 	Although parsley originated in the Mediterranean area, it is now grown worldwide. Parsley is a biennial but it is usually grown as an annual which will bear several crops of leaves during the year. Both fresh and dried leaves are used in cookery.	Parsley can be added to fish and fish sauces, meats, sauces, soups, and vegetables. It is commonly used as decoration for buffet dishes and in restaurant meal service, where, unfortunately, it is usually ignored. Parsley is an excellent source of vitamins A and C, and is rich in minerals, especially iron. (Also see PARSLEY.)
Pepper (spice) *(Piper nigrum)* 	The vine peppers are from a different family than the red or capsicum peppers. Black pepper is the dried, unripe berry, or peppercorn, of a tropical perennial vine indigenous to the East Indies; now cultivated in Indonesia, Malaysia, India, Brazil, Madagascar, and Sri Lanka. White pepper, which has a milder flavor than black pepper, produced by either (1) allowing the peppercorns to ripen, stacking to ferment off the pulpy outer layers, and then rinsing off under water; or (2) using the black peppercorns and removing the outer covering by mechanical abrasion or by soaking, and rubbing them off by hand.	Pepper loses much in aroma when ground, so freshly ground pepper should be used whenever possible. Likewise, pepper loses much in flavor by cooking. Both black and white pepper are universally popular. Whole peppercorns can be purchased, as well as cracked, and coarsely or finely ground. Except for sweet dishes, pepper can be added to all other dishes according to the taste preferences of the family.

(Continued)

TABLE F-22 *(Continued)*

Name	Description; Production	Uses
Poppy seed *(Papaver somniferum)* 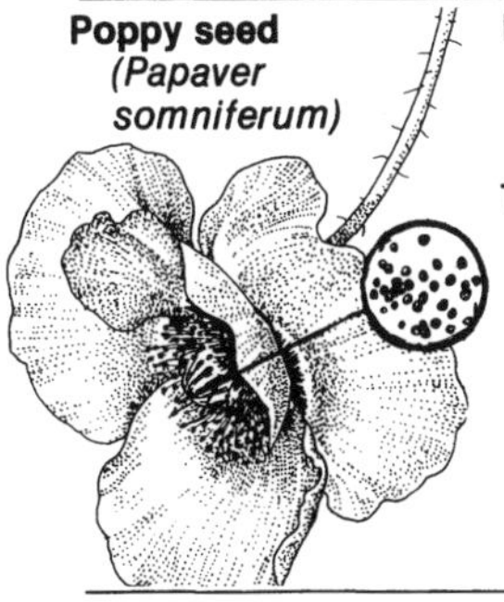	Poppy seeds come from the Opium Poppy, a native of Asia, now grown throughout much of the world except in the U.S., where it is illegal. The plant is an annual with large white, pink, or lilac flowers. The opium is found in the gummy latex which is in the seed pod. It contains the alkaloids of interest to the medical profession, but the seeds when ripe contain no alkaloids; hence, the seeds have no drug properties.	Poppy seeds have a pleasant nutlike flavor and aroma. The main use for poppy seeds is in baked goods, the tops of rolls and bread, in cakes and pastries. However, they are used in confections, fruit salad dressings, and curries.
Poultry seasonings	This is a mixture of herbs and spices which may include black pepper, marjoram, nutmeg, rosemary, sage, and thyme.	As the name suggests, it is usually used for seasoning poultry and poultry stuffings. But it may also be added to fish, veal, or pork, and, in addition, it can add a new interest to biscuits or dumplings that are baked on top of stews containing poultry, pork, or veal.
Rosemary *(Rosmarinus officinalis)* 	Rosemary is a low-growing, perennial, evergreen shrub belonging to the mint family. It is native to the Mediterranean region, but it is now cultivated in most of Europe and the Americas. The leaves may be gathered anytime and used fresh or dried. If dried, they are best gathered just before blooming. Rosemary is an emblem of fidelity and remembrance. In *Hamlet,* Ophelia remarks, "There's rosemary, that's for remembrance."	Rosemary is good with soups, on broiled steaks, or with other meat dishes, sauces, and vegetables. The taste is aromatic, pungent, and slightly bitter.
Saffron *(Crocus sativus)* 	Saffron consists of the dried stigmas of the flower of this bulb, which is a low-growing perennial, possibly native to Greece, Asia Minor, and Iran. The stigmas are hand-gathered, then dried. They are very light in weight after drying; hence, it takes 225,000 stigmas to make a pound of saffron, which accounts for its high cost.	Saffron is used as a flavoring and coloring (yellow) spice in biscuits, confections, boiled fish, fish soup, fancy rolls, and rice, and in some European dishes such as the Spanish "Arroz Con Pollo"— their famous chicken-rice dish.
Sage *(Salvia officinalis)* 	Sage comes from the dried leaves of a perennial plant belonging to the mint family, native to the the Mediterranean area, where it grows wild as well as being cultivated. There are many different sages. So, if cultivated in the home garden, it is important to get one that you like; otherwise, get rid of it and try another.	Sage is available whole, rubbed, or ground. It is used for baked fish, meats and meat stuffings, sausages, cheeses, and sauces, and can be found in some of the mixed spices.

(Continued)

TABLE F-22 *(Continued)*

Name	Description; Production	Uses
Savory **Summer savory** *(Satureja hortensis)* 	Savory is another member of the mint family which originated in southern Europe. The plant is a hardy annual, which grows to about a foot high. The dried leaves and flowering tops are used, and are cut just before flowering time.	Savory is available whole or ground, and is often combined with other herbs to flavor meats. Also, it can be used in beans, scrambled eggs, peas, salads, sauces, and sausages.
Winter savory *(Satureja montana)*	Winter savory is a low (1 ft, or *30 cm*) perennial shrub with purple flowers. The leaves are used.	The flavor of winter savory is similar to summer savory; hence, it can be used in the same manner.
Sesame seed *(Sesamum indicum)* 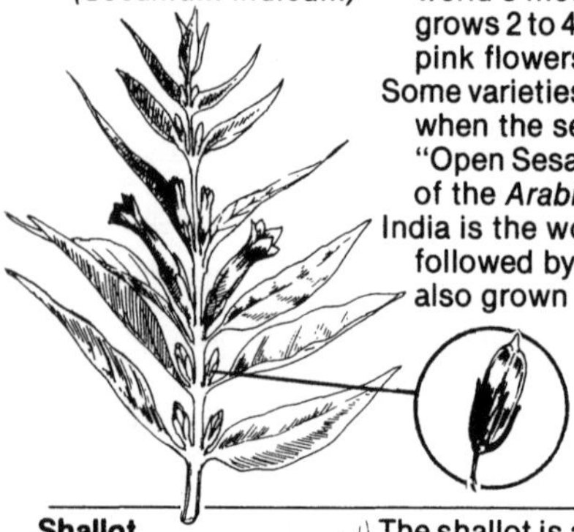	Sesame, and annual native to India, is one of the world's most important oil seeds. The plant grows 2 to 4 ft *(60 to 120 cm)* high and has white or pink flowers within which the seed pods form. Some varieties bear seed capsules that spring open when the seeds are ripe—hence, the phrase "Open Sesame," the magic password in the tales of the *Arabian Nights.* India is the world's largest sesame producer, followed by China and the Sudan. The plant is also grown in southwestern U.S.	Seasame seeds develop a beautiful nutty taste when sprinkled on buns,rolls, or cakes, and then baked. Sesame seeds are used in confections. Also they are ground and used in various products.
Shallot *(Allium ascalonicum)* 	The shallot is another member of the lily family that is similar to the onion in flavor, but which forms a cluster of bulbs instead of a single one. It is extremely popular in northern France, but is not often found in American markets.	Shallots should never be browned as they turn bitter. They can be used in the same way as the onion, although the flavor is much more subtle.
Soy sauce *(Glycine max)*	Soy sauce, a salty brown sauce made by fermenting soybeans, is an essential ingredient for Asian cusine. The "starter" for making soy sauce is prepared from rice inoculated with the spores of special molds. In Japan, about equal quantities of soybeans and wheat are used in making soy sauce. In China, more soybeans and less wheat are used. The flavor is like a salty meat extract.	Soy sauce can be used in a wide array of dishes, especially with beef, chicken, fish, soups, turkey, and vegetable dishes.
Star anise *(Illicium verum)* 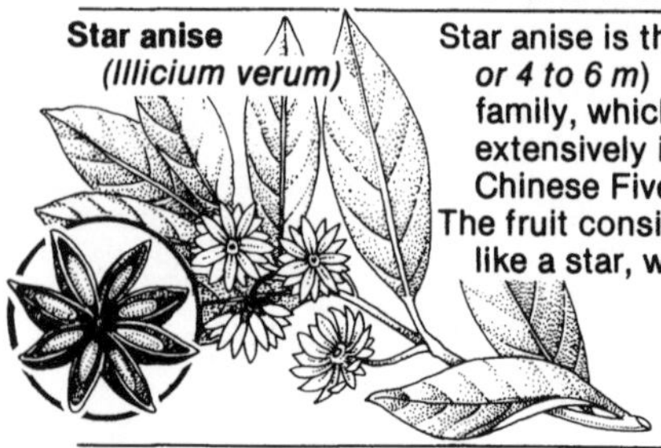	Star anise is the dried fruit of a small (13 to 20 ft, *or 4 to 6 m*) evergreen tree of the Magnolia family, which is native to China, where it is used extensively in their cooking. It is one of the Chinese Five Spices. The fruit consists of about eight carpels arranged like a star, with a single seed in each one.	Star anise has a strong flavor similar to anise, but slightly more bitter and pungent. In Chinese cooking, it is used for duck and pork recipes.

(Continued)

TABLE F-22 (Continued)

Name	Description; Production	Uses
Sweet basil *(Ocimum basilicum)*	Sweet basil is an annual plant about 2 ft *(60 cm)* high. Originally native to India and Iran, it now grows everywhere. The plant will not tolerate frost. Dried basil cannot ever replace the fresh herb.	Basil and tomatoes go together like bread and butter. And what would turtle soup be without basil? Basil can be used for green beans, fish, soups, squash, stews, and vinegar.
Sweet cicely (Sweet chervil) *(Myrrhis odorata)*	It is a perennial plant which grows to 3 ft *(1 m)* high, has white flowers, and large, lacy leaves. Cicely is a member of the parsley family. This old-fashioned herb grows wild in northern Europe, England, and Scotland. Sweet cicely is grown for its pungent, aromatic roots, seeds, and leaves.	The plant smells and tastes somewhere between anise and licorice. The tap root can be boiled and used for salads, and the green fruit can be served with salad dressing. Europeans use the leaves in soups and salads. The plant is also used for flavoring desserts and liqueurs.
Tabasco sauce	This popular Spanish condiment is made from thin red peppers known as chili peppers.	Tabasco sauce has many uses at the table, in the kitchen and at the bar.
Tarragon *(Artemesia dracunculus)*	Tarragon is thought ot have originated in Siberia. There are two varieties, Russian and French. The tarragon produced in the U.S. is French. It is a bushy, 3-ft-high *(1 m)*, perennial plant with narrow leaves which are picked and dried under controlled conditions to ensure retention of flavor. The plant has to be mulched in the winter; and divided every 3 to 4 years.	Tarragon is best known for flavoring vinegar, but it is also used for beef, chicken, eggs, fish, pickles, cookies, salads, and tartar sauce. It has a slightly anise flavor.
Thyme *(Thymus vulgaris)*	Garden thyme has been used from the beginning of time. It originated in southern Europe, where it grows wild in the mountains along the Riviera. Thyme is a small, bushy perennial, with tiny gray-green leaves and purple flowers. It belongs to the mint family. Thyme prefers sun and fairly dry conditions and, therefore, it is easy to grow. It is cultivated in France, Spain, and the U.S. The leaves and tender stems are picked, cleaned, and dried.	Thyme is an important herb in European cooking. Although thyme is a pungent herb, it blends with many foods and flavors without overpowering everything. It can be used in combination with other herbs. Thyme is used with fish dishes, meats, poultry, sauces, tomato dishes, and vegetables. Thyme is an ingredient of bouquet garni *("herb bundle")*.

(Continued)

TABLE F-22 *(Continued)*

Name	Description; Production	Uses
Turmeric *(Curcuma longa)*	Turmeric is native to southeast Asia and is of the same family as ginger. It is a perennial with large lilylike leaves and pale yellow flowers, but the spice comes from the root or rhizome. The rhizomes are boiled, dried in the sun for 2 weeks, then the outer skin is polished off and the root is then ground. Today, turmeric is cultivated in China, Indonesia, India, Jamaica, Haiti, and Peru.	Turmeric and mustard are inseparable partners (it is used to color mustard); and turmeric is an indispensable ingredient in curry powder. Turmeric is superb for almost every meat and egg dish, for pickles, and for curries. Although it adds a good yellow color to everything, it should not be used as a flavorless coloring agent as it has a distinctive flavor.
Vanilla *(Vanilla planifolia)*	Vanilla comes from the pod of a climbing orchid native to the tropical forests of the Americas. It was taken to Europe by the Spanish explorers, who had found it used by the Aztecs. The best vanilla comes from crushing the vanilla pods and then adding alcohol.	Vanilla is almost always used in sweet dishes such as bakery products and desserts. The real vanilla is superior to synthetic, and vanilla pods are superior to the extract. The pods can be reused several times, provided they are washed and dried in between times.
Verbena *(Verbena hybrida)*	Verbena is easily grown in any warm, sunny part of the garden. It is native to South America and was taken to Europe by the Spaniards.	It contributes a lemon flavor to sweet dishes, fruit salads, herbal drinks, and soft drinks.

(Continued)

Spices and flavorings add zest even to the simplest meals, such as this salad plate and garlic French bread. (Courtesy, United Dairy Industry Assoc., Rosemont, Ill.)

TABLE F-22 *(Continued)*

Name	Description; Production	Uses
Wine	Wine has long been used in the daily cooking of the Mediterranean area. Adding wine gives a distinctive flavor that many parts of the world have not yet discovered. When wine is used in cooking, the alcohol is driven off very quickly, but the flavor is still there when the dish is served. Oddly enough the flavor is not the same as it is in the uncooked wine.	Usually wines have been used in the specific area where they are produced, and special recipes evolved for their use. Port and sherry are used in many English recipes. Madeira is used on the island of Madeira, where it originated. Marsala, produce in Sicily, is used in Italian dishes. Rice wines are much used in Oriental cooking. In the U.S., the use of wine in cooking is in its infancy, but those who discover that it can add a new dimension to the flavors of even the plainest of foods, will soon find that it plays an important role in their food preparation.
Wintergreen *(Gaultheria procumbens)*	Wintergreen is a plant of northern latitudes, with small round evergreen leaves. It has white bell-shaped flowers followed by spicy red berries, but it is the green leaves that produce the oil with the wintergreen flavor. The plant received its name because its leaves remain green all winter.	Wintergreen oil is used mainly for candies and lozenges.
Worcestershire sauce	Worcestershire sauce originated in Worcester, England, and, according to one story, quite by accident. A retired Governor General of Bengal took the recipe to his local drugstore in Worcester, owned by Lea and Perrins, ana asked that it be custom made for him. But the product did not pass muster with the ex-Governor General, who rejected it. It remained in storage for several years, until Lea and Perrins were cleaning out their cellar and came across the barrels. On sampling the product, they discovered it to be very tasty. They still had the recipe, so they started making it. It was almost and instant success; and it was soon spread around the world by the captains and crews of the sailing ships.	The sauce is a combination of over 20 ingredients including: anchovies, corn syrup, fruits, garlic, salt, spices, soy sauce, sugar, tamarind and vinegar. It is used for seasoning meats, especially when broiling or grilling chicken, hamburgers, spareribs, and steaks.

Barbecued spare ribs without barbecue sauce would be like bread without butter, or mashed potatoes without gravy. Seasonings add the gourmet touch. (Courtesy, California Beef Council)

WHY, WHEN, AND HOW MUCH FLAVORING TO USE. "Variety is the spice of life," as the old saying goes. This is certainly true in the diet. Not only is life more interesting when meals are varied with different foods, delicately flavored with herbs or spices, but there is an added bonus to your health for it carries out the *Golden Rule of Eating*, which is: "the greater the variety of foods you eat, the greater your chances of getting everything you need."

Fig. F-25. Seasonings are a matter of personal choice, but don't forget that a little seasoning goes a long way. (Courtesy, National Live Stock and Meat Board, Chicago, Ill.)

It is important to keep spices and herbs relatively fresh. If spices have been used only a few times each year, chances are they have become stale. Some of the spices deteriorate with age, with heat, and with exposure to the air. So get rid of the old and stock up on the new. The results of your culinary prowess will be more dramatic with fresh spices.

Any flavoring can be used singly, or two or more may be combined. This is where the cook's daring experimentation can produce just the right effect, and result in delicious food. Sometimes the taste buds have to be educated to new flavorings. The package label usually recommends and suggests how much and when to use. The flavorings and seasonings should always enhance the flavor of the main ingredient, rather than completely mask it. But tastes differ, so, in the final analysis, it is up to the cook to decide.

Flavorings are usually better added the last hour of cooking. But, as the delicate perfume of marjoram or the "pep" of pepper are easily lost in cooking, add them very shortly before serving.

For easy and quick reference, Table F-23 Food and Flavoring Combinations is presented. It is not intended that this be an absolutely complete and final listing. Rather, the sky's the limit; you can add to, mark your favorites, and come up with your own winners, but the format of this table simplifies life for the busy cook. When she flies around the kitchen preparing supper in a hurry, it is not always possible to have several recipes out and to follow them. Thus, this table is an easy reference; as you boil the potatoes and fry the hamburgers, just refer to the table for compatible seasonings. The flavorings are listed alphabetically.

TABLE F-23
FOOD AND FLAVORING COMBINATIONS

Foods	Suggested Flavorings[1]
Appetizers	Anise (½ tsp in 8 oz cream cheese spread), basil, cayenne, mustard, thyme
Apple	Allspice, caraway seed, cinnamon, cloves, coriander, mace, pumpkin pie spice
Baked	Fennel seed
Green apple pie	Dill
Apricots	Mace
Artichokes	Thyme
Asparagus	Basil, bay leaf, sesame seeds
Avocado	Chili powder
Barbecue sauce	Allspice, basil, cardamon, cumin, thyme (¼ tsp. in 6 c)
Beans	Basil, chili powder, cumin, curry, dill seed, nutmeg, oregano, sesame seeds, savory
Bean soup	Cloves, dill weed, ginger
Beef	Basil, curry, horseradish, paprika, peppercorns, marjoram, mustard, rosemary, sage, savory, thyme, turmeric
Meatloaf	Nutmeg, savory
Meat sauce	Dill
Pot roast	Ginger, cloves, cumin (½ tsp to 3 lb), dill weed, mustard, soy sauce tarragon
Roasts	Bay leaves, cloves, ginger, marjoram
Cold roast beef	Mustard, sage
Steak	Ginger, rosemary
Stew	Caraway seeds, chili powder, cinnamon, cumin (½ tsp to 3 lb), dill weed, oregano, saffron
Swedish meat balls	Nutmeg
Beets	Anise, caraway seeds, celery seeds, cloves, dill weed, ginger, thyme
Beverages	Anise, cardamom, cinnamon, coriander seeds
Milk, chocolate, spiced	Nutmeg
Biscuits	Caraway seeds (1 Tbsp in 2 c mix), marjoram (½ tsp in 2 c), sage, thyme (½ tsp in 2 c)
Borsch	Cloves, thyme
Bouillon	Oregano
Bread or rolls	Caraway seed, garlic, oregano (1 tsp in 3 c flour), poppy seeds, sesame seeds
Rye bread	Caraway seed
Toast	Cinnamon
Broccoli	Mace, oregano
Brussels sprouts	Mace, nutmeg, sage

Footnote at end of table

(Continued)

TABLE F-23 (*Continued*)

Foods	Suggested Flavorings[1]
Cabbage	Allspice, caraway, celery seeds, cumin, dill seed, mace, oregano, savory, sesame seeds, thyme
Cakes	Allspice, cardamon, caraway seeds, cloves, mace, nutmeg, poppy seeds, pumpkin pie spice, sesame seeds, vanilla, etc.
Candy	Anise seed, cardamon, mace, sesame seeds
Carrots	Basil, caraway seed, celery seeds, cinnamon, cloves (1/8 tsp in 2 c), curry, dill, ginger (¼ tsp to 2 c), mace, marjoram, nutmeg, sesame seeds, thyme
Casserole topping	Sesame seeds
Catsup	Celery seeds, cinnamon
Cauliflower	Caraway seed, chili powder, curry, dill seed, mace, nutmeg, rosemary
Celery	Marjoram
Cheese	Basil, cayenne, chili powder, cumin, marjoram, mustard, oregano, sage
Fondue	Chili powder, sage
Sauces	Chili powder, mustard, sage, turmeric (¼ tsp to 2 c)
Spreads	Celery seeds
Cheesecake	Poppy seeds
Cherries	Mace
Chicken	Cayenne, curry, dill, ginger, marjoram, nutmeg, oregano, saffron, sage, sesame seeds, savory, tarragon
Fricassee	Allspice, mace, savory, tarragon
Fried	Basil, chili powder, nutmeg, thyme
Roast	Cloves, ginger
Salad	Majoram, sage, thyme
Stewed	Cinnamon
Chili	Cloves, chili powder
Chili con carne	Chili powder, oregano
Chocolate cake and desserts	Cloves, mace
Chutneys	Cinnamon
Clam	Oregano
Chowder	Basil, caraway seed, celery salt or seed, sage, thyme (½ tsp in 3 c)
Coffee	Cardamon, cinnamon, mustard
Coffeecakes	Anise, cardamon, cinnamon, mace, nutmeg
Cole slaw	Celery salt or seed, dill seed, dill weed
Consomme	Sage
Cookies	Allspice, anise seed, caraway seed, cardamon, cloves, ginger, nutmeg, poppy seeds, sesame seeds
Corn	Chili powder (½ tsp in 2 c creamstyle), nutmeg, oregano
Cornbread	Caraway (1 Tbsp in 2 c), curry (1 tsp in 2 c), marjoram, thyme
Corn soup	Caraway seed
Cottage cheese	Allspice, caraway seed (2 Tbsp crushed to 1 c cheese), cayenne, chili powder, dill weed, paprika, poppy seeds
Crabmeat	Basil
Cranberries	Allspice, cloves
Cucumbers	Basil, dill
Custard	Cardamon, cinnamon, mace, nutmeg, vanilla, poppy seeds, thyme
Doughnuts	Cinnamon, mace, nutmeg
Duck	Caraway, curry, rosemary, sesame seeds, tarragon
Roast	Celery salt or seed, ginger, thyme
Dumplings	Poppy seeds, saffron, parsley
Eggplant	Allspice, basil, marjoram, oregano, sage
Eggnog	Nutmeg
Eggs	Allspice, basil (½ tsp to 6 eggs), cayenne, celery salt or seed, chervil, chili powder, cloves, cumin, curry, dill, fennel seed, marjoram, mustard, oregano, paprika, savory, tarragon, turmeric
Boiled	Celery salt, paprika
Creamed	Basil
Deviled eggs	Dill, mustard
Omelets	Sage, thyme
Salad	Horseradish, marjoram, thyme
Scrambled	Chili powder, marjoram, thyme, turmeric (1/8 tsp to 6 eggs)
Fish	Curry, fennel seed, oregano, paprika, rosemary, sage, tarragon
Baked	Basil, chervil, cloves, curry, dill weed, fennel seeds, ginger, mace, nutmeg, paprika, saffron, sage, sesame seeds, thyme
Boiled	Bay leaf, dill weed, marjoram, rosemary
Poached	Allspice, bay leaf, rosemary
Stuffing	Rosemary
French dressing	Allspice, basil, celery seed, chili powder (¼ tsp in 1 ½ c), curry powder, dill, ginger, marjoram, paprika, sage
Frostings	Allspice, mace, pumpkin pie spice
Fruits	Allspice, cinnamon, curry, ginger, rosemary
Fruit cake	Allspice, cardamon, cinnamon, cloves
Fruit cup	Marjoram, rosemary
Fruit salad	Cardamon, mace, poppy seeds
Game	Orange sauce
Stuffing	Basil, sage, thyme
Gingerbread	Cinnamon, cloves, ginger, mace, pumpkin pie spice
Goose	Caraway seed
Stuffing	Cinnamon and sage
Graham cracker pie crust	Allspice (1 tsp)
Grapefruit, broiled	Cinnamon, ginger
Gravies	Bay leaves, celery salt or seeds, chili powder
Green beans	Basil, thyme
Green salad	Marjoram
Guacamole dip	Cayenne, chili powder, oregano
Ham	Allspice, cinnamon, cloves, rosemary, mustard
Hamburger	Basil, cardamon, chili powder, cloves, cumin (¼ tsp to 1 lb), dill (½ tsp dill seed + ¼ c chopped olives or pickles to 1 lb), mustard (½ tsp mustard + ¼ tsp nutmeg to 1 lb), savory

Footnote at end of table

(*Continued*)

TABLE F-23 (*Continued*)

Foods	Suggested Flavorings[1]
Ice cream	Nutmeg, vanilla
Jellies	Allspice, mace, mint
Grape and plum	Cardamon
Lamb	Allspice, basil, capers, cinnamon, cloves, curry, dill seed, dill weed, ginger, mace, marjoram, mint, oregano, rosemary, sage, tarragon.
Lasagne	Oregano
Lentils	Cumin, oregano
Lima beans	Sage
Liver	Celery salt, horseradish, onion, oregano, rosemary
Lobster	Curry, oregano
Marinades	Cloves, cumin, fennel seed, garlic, oregano, rosemary, sage
Meat	Basil, bay leaf, cumin, garlic, mustard
Meat balls	Allspice, curry, mace, nutmeg
Meat loaf	Allspice, chili powder, cumin, mace, nutmeg
Meat stews	Savory, thyme
Mincemeat	Cloves
Muffins	Mace, nutmeg, pumpkin pie spice
Mushrooms	Fennel seed, marjoram, oregano, rosemary, thyme
Mushroom sauce	Nutmeg
Noodles	Caraway seed, paprika, poppy seeds, sesame seeds
Nut bread	Cinnamon, cloves, nutmeg
Onions	Basil, chili powder, cinnamon, cloves, curry, ginger, nutmeg, oregano, sage, thyme
Oranges	Cardamon
Oysters	Allspice, fennel seed, mace, nutmeg
Pancakes	Cinnamon, mace, nutmeg
Parsnips	Allspice, cinnamon
Pastries	Anise, cardamon, coriander seeds, cumin, poppy seeds
Peas	Basil, chili powder, curry, dill, marjoram, mint, savory, thyme
Pea soup	Caraway seed, chili powder, cloves
Split pea soup	Dill, savory
Pears	Cloves, ginger
Pickles	Allspice, anise seed, bay leaf, celery seeds, cinnamon, coriander seeds, dill, mace, savory, tarragon, thyme
Pie, apple, peach, cream, and custard	Cinnamon or nutmeg
Pumpkin pie	Allspice, cinnamon, cloves, ginger, nutmeg, pumpkin pie spice
Pizza	Basil, oregano (1 tsp for 12″ pizza
Plums	Cloves, cardamon
Pork	Caraway seed, cloves, curry, marjoram, oregano, rosemary, sage, savory, thyme
Chops	Cinnamon, dill weed, garlic, paprika
Roast	Rub with mustard and sage
Salad	Sage
Potatoes	Cumin, rosemary
Boiled	Bay leaves, dill weed, mace, marjoram, paprika, poppy seed, sesame seeds, thyme
New potatoes	Basil, mint
Salad	Caraway (2 tsp to 3 c salad), chili powder, dill, turmeric
Scalloped	Mix caraway seeds with breadcrumb topping
Soup	Allspice, ginger, marjoram
Pot-roasts	Allspice, nutmeg
Poultry	Basil, thyme, mustard
Dressing	Sage, savory
Gravy	Sage, savory
Prunes	Cinnamon, mace
Puddings	Allspice, caraway seeds, cloves, ginger, mace, nutmeg, pumpkin pie spice
Chocolate, cottage, or custard	Mace, nutmeg, vanilla
Steamed, bread, tapicoa or rice	Allspice, cardamon, cinnamon, coriander, ginger
Rabbit	Mace, rosemary
Rhubarb	Cloves (1/4 tsp in 4 c), cinnamon
Rice	Basil, bay leaf, caraway seed, cinnamon, cumin, curry, garlic, saffron (1/4 tsp to 1 c uncooked rice), sesame seeds, turmeric (1/4 tsp to 1 c uncooked rice)
Rolls	Caraway seed, dill, garlic, poppy seeds, sesame seeds
Rum drinks	Nutmeg
Russian dressing	Basil
Salads	Anise, basil, celery seeds, chervil, chives, cumin, dill weed, garlic, oregano, poppy seeds, sesame seeds, savory, tarragon, thyme
Salad dressings	Anise, cayenne, cumin (1/8 tsp in 1 c), curry, garlic, ginger, horseradish, mustard, oregano, paprika, poppy seeds, sesame seeds, tarragon
Salmon souffle	Rosemary
Sauces	Seasoning will depend on what the sauce will accompany. Allspice, basil, bay leaf, celery salt or seeds, curry, dill weed, mace, marjoram, mint sauce, mustard, nutmeg, oregano, paprika, saffron, savory, tarragon, thyme, vanilla
Sauerkraut	Caraway seed, cumin, dill, fennel seed
Sausage	Coriander seeds, cumin, mace, sage
Seafood	Basil, saffron, turmeric
Cocktail	Basil, dill
Salad	Curry (1 tsp to 2 c), dill, fennel seed, marjoram, oregano
Shrimp	Curry, oregano
Souffles	Basil

Footnote at end of table

(*Continued*)

TABLE F-23 (*Continued*)

Foods	Suggested Flavorings[1]
Soups	Allspice, anise, basil, bay leaf, caraway seeds, cayenne, celery salt or seeds, chervil, cumin (1/8 tsp in 4 c), dill, marjoram, oregano, rosemary, saffron, sage (¼ tsp in 3 c), savory, thyme
Chowder	Tarragon (¼ tsp to 2 c)
Creamed soups	Rosemary
Spaghetti	Cloves
Sauce	Basil, cayenne, fennel seed
Spinach	Allspice, basil, cinnamon, curry, mace (dash in 1 c cooked), marjoram, nutmeg, oregano, rosemary, sesame seeds
Squash	Allspice, basil, cinnamon, cloves (1/8 tsp to 2 c), dill weed, ginger, pumpkin pie spice, sage
Summer squash	Marjoram, oregano
Winter squash	Ginger, mace, cinnamon
Stews	Marjoram, rosemary
Stuffings	Celery salt or seed, nutmeg, oregano, sesame seeds, savory, thyme
Succotash	Mace, nutmeg
Sweetbreads	Tarragon
Sweet potatoes	Allspice, cinnamon, cloves (1/8 tsp in 2 c), ginger, pumpkin pie spice
Sweet rolls	Allspice, cinnamon, cloves, nutmeg
Swiss chard	Mace, oregano
Tea	Cinnamon, sage
Tomatoes	Basil, bay leaf, celery salt, chili powder, dill weed, garlic, oregano, sage (¼ to 3 c)
Salad	Thyme
Sauce	Allspice
Tongue	Allspice, bay leaves, celery leaves, cloves
Tuna	Celery salt or seed, tarragon, thyme
Turkey	Ginger, marjoram, sage, savory
Gravy	Sage
Salad	Sage
Turnips	Allspice, caraway seed, dill weed
Veal	Allspice, cloves, curry, ginger, marjoram, oregano, paprika, rosemary, saffron, sage, sesame seeds, savory, tarragon
Vegetables	Basil (½ tsp to 2 c), celery salt or seed, cloves (1/8 tsp in 2 c green veg.), mustard, nutmeg (¼ tsp in 2 c), oregano, sage (½ tsp to 2 c), thyme
Soup	Basil, mace
Vinegar	Dill, tarragon
Waffles	Allspice (2 tsp to 2 c flour), caraway (1 Tbsp in 2 c mix), cinnamon, majoram, nutmeg, poppy seeds, pumpkin pie spice, sage (2 tsp in 2 c mix)
White sauce	Mace (1/8 tsp in 2 c), tarragon (½ tsp in 2 c), turmeric (¼ tsp to 2 c)
Wine	Cloves
Yams	Sesame seeds
Zucchini	Caraway seed, marjoram, oregano

[1]To convert to metric, see WEIGHTS AND MEASURES.

WINES AND LIQUEURS. Wines, sauterne, vermouth, and sherry, etc., all give a distinctive flavor to foods. When added during cooking, they lose their alcoholic content, but still impart an interesting flavor. They can be added to cream soups, and to almost any meat dish, casseroles, stews, etc.

Fish is especially good when cooked in a white wine sauce with butter and almonds. Sole is excellent cooked in vermouth.

Red wines are the best for marinades, and for cooking the red meats. A very good marinade consists of red wine, garlic salt, and rosemary or a herb mix of thyme, oregano, sage, rosemary, marjoram and basil.

Liqueurs add zest to many desserts, for instance, ice cream, with a tablespoon of a liqueur, and nuts, macaroons, grated chocolate, or other topping, makes a simple, but elegant, party dessert. Cointreau can be used with almost all fruit desserts, including fresh fruits.

This is not the end, but just the beginning, of the fun you will have creating new and interesting foods for yourself and your family. Any therapeutic value that these condiments may have will be an added dividend.

FLAVOR POTENTIATOR

A substance which enhances the flavor of a food without contributing a flavor of its own. Monosodium glutamate (MSG) is a familiar example.

(Also see ADDITIVES.)

FLAVORS, SYNTHETIC

With the exception of the four primary taste sensations—sweet, bitter, salty, and sour—food flavors are the result of our sense of smell. Today, chemists can make chemicals in the laboratory which alone or in various combinations can imitate many of the natural food flavors. These are synthetic flavors. In many cases the synthetic flavors are superior to natural flavors in terms of (1) withstanding processing, (2) cost, (3) availability, and (4) consistent quality. Synthetic flavors may be substances that are prepared in the laboratory but chemically identical to those found in nature, or substances prepared in the laboratory which as yet have not been found to occur in nature but which produce familiar aromas.

Flavor, regardless of whether it is naturally occurring or created from synthetic chemicals, often results from a complex mixture of chemicals in the proper proportion. Hence, the vast number of synthetic compounds can be employed to create endless flavor formulations. Creating flavors is a science.

Since there are hundreds of synthetic flavor compounds, a complete listing is not within the scope of this book. Instead, a few examples of approved synthetic flavors are listed in Table F-24.

In the labeling of food products, the term "artificial flavor" is a requirement for all food products to which flavor has been added regardless of its origin.

(Also see FLAVORINGS AND SEASONINGS.)

TABLE F-24
SYNTHETIC FLAVORS[1]

Name	Flavor Characteristics	Flavor Use
Aconitic acid	Winey-sour taste.	Fruit flavors, rum, brandy.
Allyl anthranilate	Distinct Concord grape odor with long-lasting grape taste.	Grape, pineapple, citrus.
Allyl butyrate	Fruity apple, pineapple odor with apple taste.	Apple, pineapple, peach, apricot.
Allyl cinnamate	Sweet spicy odor with fruity-spicy taste.	Berry flavors, grape, peach.
Allyl mercaptan	Pungent onion, leek odor and taste.	Spice, onion, garlic.
Allyl 10-undecenoate	Sweet anise, fennel odor with sassafras, root beer taste.	Root beer, pineapple, coconut.
Anisole	Sweet aniselike odor and sweet taste.	Vanilla, anise, root beer.
Benzenethiol	Unpleasant burnt odor which diluted tastes like meat.	Meat, coffee.
Benzyl phenylacetate	Weak floral odor with a honey taste.	Honey, caramel, butter.
Bornyl acetate	Pinelike minty odor and taste.	Mint flavors, some in spice flavors, pineapple.
1-Butanethiol	Cabbage skunklike odor with bitter taste.	Vegetable flavors.
Butan-3-one-zyl butanoate	Yogurt odor with woody, buttery taste.	Yogurt, butterscotch, butter.
***d*-Camphor**	Minty odor with slight minty cool taste.	Essential oil imitations, mint flavors.
Caruacrol	Medicinal smoke odor excellent smoke taste.	Smoke flavors, spice meat flavors.
Caryophyllene alcohol	Earthy, spicy odor, with minty, woody taste.	Mushroom flavor.
Cinnamaldehyde	Spicy cinnamon cassin odor with spicy sweet taste.	Cassia, cinnamon, cola spice blends, cream soda.
Citronellyl vallerate	Floral fruity, dried leaves odor, fruity, honey taste.	Honey, apricot.
Cyclohexyl propionate	Banana odor and banana-rum taste.	Banana, rum.
Diethyl malate	Pleasant cherry, apple odor and taste.	Apple, cherry, banana.
***m*-Dimethoxybenzene**	Earthy, nutlike odor, and taste.	Nut flavors, walnut, vanilla.
2,6 Dimethyl-5-heptenal	Powerful green odor with melon rind taste.	Melon, cucumber, tropical fruit.
2 Ethyl butyraldehyde	Strong chocolate odor with chocolate-cocoa taste.	Chocolate, cocoa.
Geranyl isovalerate	Floral, fruity, apple odor and taste.	Apple, peach, pear, tobacco.
Guaiacyl acetate	Subdued smoke aroma, sweet, smoky, maple taste.	Smoke, maple, currant, coffee.
2-Heptanone	Fruity, cheese odor with blue-roquefort taste.	Cheese, banana, butter, coconut.
Hexyl hexanoate	Green bean odor with green fruity taste.	Strawberry, vegetable flavors.
5-Hydroxy-4-octanone	Sweet, sharp, buttery odor with oily buttery taste.	Butter, cheese, nut.
Isoamyl alcohol	Pleasant, whiskey, applelike odor, mild apple-banana taste.	Apple, banana.
Isobutyl salicylate	Harsh, sweet, floral odor and taste.	Traces in fruit flavors, root beer.
Isoeugenyl formate	Woody, perfumy odor with clovelike taste.	Spice flavors.
Isopulegol	Minty odor with bitter minty taste.	Mint flavors.
Lactic acid	Mild creamy odor with pleasant sour taste.	Dairy flavors, fruit flavors.
***d*-Limonene**	Fresh, citrus odor and taste.	Citrus flavor, artificial essential oils.
Linalyl benzoate	Fruity, citrus odor and taste.	Berry flavors, citrus, mixed fruit.
Maltol	Sweet cotton candy odor and sweet jammy taste.	Strawberry, all fruits.
Menthol	Clean, minty odor with cool minty taste.	Mint flavors, toothpaste, mouthwash.
Methyl anisate	Sweet, floral anise odor, spicy, green, anise taste.	Licorice, root beer, melon.
***p*-Methyl anisole**	Sharp, fruity sweet odor, with nutty taste.	Black walnut, maple, berry.
Methyl butyrate	Strong cheese odor with a fatty cheese taste.	Milk, cheese, apple.
Methyl disulfide	Powerful, onion, cabbage, canned corn odor and taste.	Meat, butter, potato, cocoa, garlic, onion.
Methyl delta-ionone	Sweet, woody, floral odor with berry taste.	Raspberry, strawberry, blackberry.
Methyl propionate	Sweet, fruity rumlike odor and taste.	Rum, black currant, fruit blends.
Myrcene	Resinous, terpeney odor with sweet citrusy taste.	Citrus imitation, fruit blends.
Neryl formate	Carrot odor with vegetable earthy taste.	Vegetable flavors, fruit flavors.

Footnote at end of table

(Continued)

TABLE F-24 (*Continued*)

Name	Flavor Characteristics	Flavor Use
Nonanoic acid	Fatty coconut aroma and excellent coconut taste.	Coconut, berry.
Nonyl isovalerate	Fatty, fried food aroma with nutty taste.	Nut, fatty flavors.
Ocimene	Sweet, terpeney odor and taste.	Essential oil imitations, citrus flavors.
3-Octanol	Mushroom aroma with cheesy taste.	Cheese, mushroom.
3-Octyl acetate	Flowery, citrus aroma and taste.	Citrus flavors.
Perillaldehyde	Spicy, cinnamonlike aroma with oily peanut taste.	Spice flavors, peanut.
Phenethyl isovalerate	Fruity, resinous odor, sweet green fruity taste.	Apple, pineapple, pears, peach, fruit mixtures.
Phenethyl tiglate	Sweet, floral, leafy green odor with nutty taste.	Nut flavors.
4-Phenyl 2-butanol	Fruity, melon spice odor with spicy melon taste.	Melon flavors.
Piperine	Odorless with burning taste.	Black pepper imitations, spice blends.
Pyridine	Powerful, fishlike odor with amine taste.	Seafood flavors, smoke flavors.
Rhodinyl formate	Leafy, green floral aroma with fruity taste.	Apricot, strawberry, raspberry, peach, cherry.
Skatole	Animal-note odor with overripe fruit taste.	Cheese, grape, berry, nut, egg.
Succinic acid	Odorless but sour, acid taste.	Fruit, wine flavors.
Terpinyl acetate	Mild flowery citrus peel odor with spicy, citruslike taste.	Citrus, spice, cherry, raspberry, peach.
2-Thienyl mercaptan	Burnt-rubber, roasted coffee odor with burnt or roasted taste.	Coffee roasted flavors.
***p*-Tolylacetaldehyde**	Watermelon, cherrylike aroma and taste.	Watermelon, cherry, nut, honey.
9-Undecenal	Sweet, citrus odor and taste.	Citrus flavors, nut.
Valeraldehyde	Pleasant, chocolate odor and taste.	Chocolate, coffee, nut flavors.
Vanillin	Characteristic vanilla flavor.	Vanilla, chocolate, root beer, butter.
Zingerone	Sweet, spicy aroma and taste.	Spice, root beer, raspberry.

[1]Adapted by the authors from Fischetti, Frank, Jr., "Natural and Artificial Flavors," in *Handbook of Food Additives,* 2nd ed., Vol. 2, T.E. Furia, editor, CRC Press, Inc., 1980, pp. 256-306.

FLAX (LINSEED) *Linum usitatissimum*

This is a herbaceous annual which originated in the Mediterranean region in prehistoric times. The Abyssinians first used flax for food—they roasted and ate the oval, shiny seeds. Although the seed contains a toxic glucoside, it is detoxicated by heating.

Fig. F-26. Flax plant.

Flax is grown primarily (1) for its fibers, which lie inside the bark of the plant next to the woody core and which are processed and spun into linen cloth; and (2) for its seed, which is processed for oil and meal. The oil is used in the manufacture of varnishes, linoleum, oilcloth, soap, and leather; and the meal is used as livestock feed.

Today, whole flax seed is sold in many health food stores; and certain high-fiber, ready-to-eat breakfast cereals contain whole flax seed. The whole seed contains 24% crude protein, 38% oil, and 6% fiber. The availability of carbohydrates and/or energy is uncertain, as no studies have been made on this subject. More nutritional information is given in Food Composition Table F-36.

An excellent tea, said to be effective in the treatment of colds, may be made with 8 oz (*224 g*) each of seed and rock candy; 3 lemons paired and sliced; added to 2 qt (*1.9 liter*) of boiling water; then strained after cooling. The usual infusion is made with 1 tsp (*5 ml*) of seed steeped in 1 cup (*240 ml*) of boiling water. Also, the mucilage obtained by infusing the whole seed in boiling water, in the proportion of ½ oz (*14 g*) to 1 pt (*480 ml*), is said to be helpful in alleviating constipation, dysentery, catarrh, and other inflammatory affections of the respiratory tract, intestines, and urinarypassages. Where any of these problems are serious or persistent, those so afflicted are admonished to see a medical doctor.

FLORENCE FENNEL (FINOCCHIO) *Foeniculum vulgare* variety *dulce*

This variety of fennel has been developed for its fleshy leaf stalks and feathery leaves, which are used individually or together as raw or cooked vegetables. (The variety of fennel used for seasoning is grown for its seeds, leaves, and roots.) Fennel belongs to the parsley family (*Umbelliferae*), which also includes caraway, carrots, celeriac, celery, dill, and parsnips. Fig. F-27 shows the swollen leaf bases that characterize Florence fennel.

Fig. F-27. Florence fennel, a mildly anise-flavored vegetable which has leaf stalks that resemble celery.

ORIGIN AND HISTORY. Fennel is native to southern Europe. The sweet, anise-flavored varieties that are now in common use were developed from a wild, bitter-flavored variety that has no anise flavor. Originally, the seeds, leaves, and roots of the plant were used medicinally, as evidenced by (1) the descriptions in an ancient Egyptian papyrus dating back to about 1500 B.C., and (2) the writings of ancient Greeks and Romans. Writers often referred to the fennel plant as a symbol of strength and valor.

The Florence variety was developed by the Italians who carefully selected many generations of the plants for enlarged leaf bases. For a long time this vegetable was used only in Italy, but it is now highly esteemed throughout Europe.

PRODUCTION. Florence fennel is produced on only a small scale, and few statistics are available on the amounts that are grown.

This crop should be planted so as to reach maturity in cool weather because high temperatures promote bolting (the premature growth of seedstalks). Hence, planting is done during the fall and winter in the south, and in the early spring and summer in the north. The soil should be fertile and supplied with adequate water. It is noteworthy that the plants tolerate light frosts. Usually, the young plants have to be thinned so that there is enough room for the remaining ones to grow to an adequate size. Sometimes, the stalks are blanched by hilling up the soil around them. The vegetable is ready for harvesting within 2½ to 3 months after planting.

PROCESSING. The thickened leaf-stem bases are processed and used as either a cooked vegetable or a raw salad. The leaves are used as a raw salad, either alone or combined with the thickened stems.

SELECTION AND PREPARATION. High quality Florence fennel has white to light green leafstalks that are joined at their bulblike bases, and bright green feathery leaves.

The bases of the stalks may be cut into slices and served raw in relish trays and salads, or they may be cooked in casseroles, soups, and stews.

Fennel leaves can be served raw in salads or used in a wide variety of cooked dishes. The flavor of fennel has long been used to complement fish, meats, and poultry.

NUTRITIONAL VALUE. Data on the nutrient composition of Florence fennel are not readily available. However, the edible stalks of this plant are likely to be similar in composition to the stalks of its close relative, celery, whereas the leaves of this variety of fennel are probably very similar to the leaves of the common type of fennel used for seasoning. Data for celery stalks and common fennel leaves are given in Food Composition Table F-36.

Some noteworthy observations regarding the probable nutrient composition of Florence fennel follow:

1. The stalks are likely to be high in water content (more than 90%) and low in calories (probably less than 20 kcal per 100 g) and other nutrients. Although the green stalks may contain fair amounts of carotene, the blanched stalks probably lack carotene.

2. Common fennel leaves, and most likely those of Florence fennel, too, are an excellent source of iron, potassium, and carotene, and a good source of calcium and vitamin C.

(Also see VEGETABLE[S], Table V-6 Vegetables of the World.)

FLORIDIAN STARCH

A granular carbohydrate obtained from red algae (*Florideae*), which in several respects resembles glycogen rather than starch.

FLOURS

Contents *Page*

The History of Flours 767
Production and Processing 769
Major Flours 769
Special Flours 770
Nutritive Value of Flours 772
Enrichment and Fortification of Flours 772
Uses of Flours and the By-Products of Milling 773
Major Cereal Flours 773
Special Flours 774
By-Products 775
Recent Developments in Flour Technology 776
Summary 777

Generally, *a flour is considered to be a dry powder produced by the grinding of one or more cereal grains.* A more coarsely ground product is called a *meal.* However, the term may also be applied to similar powders obtained by the grinding of other types of seeds, such as legumes and nuts, or even the dried roots, tubers, or leaves of food plants. Although flours were first used by some of the most primitive peoples, food technologists are ever busy developing new ways of altering the characteristics of the various flours, so that the items made from them may be more appealing and nutritious. Fig. F-28 shows some of the many foods from which flours are made.

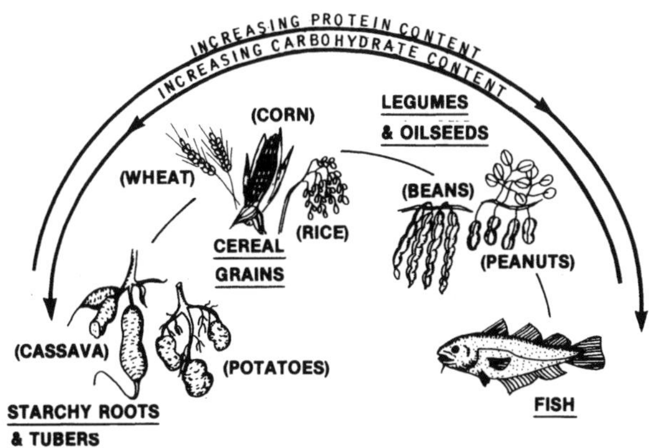

Fig. F-28. Fish, legumes, grains, tubers, and other natural foods comprise the raw materials for man's many flours.

THE HISTORY OF FLOURS. It appears that crude flours and meals were prepared as early as 75,000 years ago, when people obtained their foods solely by fishing, hunting wild game, and gathering wild plants. At that time, edible seeds were roasted, then crushed between two rocks. Sometime between then and the beginnings of agriculture 10,000 years ago, man first made a paste from a grain flour and cooked it on a heated rock to make a crude bread. Although only a few of the North American Indian tribes were agriculturalists, almost all of them prepared flours from various seeds and dried roots and tubers. However, the major grain of the Americas was corn (maize), which produced quite a heavy bread. The development of lighter breads occurred elsewhere in the world, where other grains were available.

The first cultivation of wheat apparently occurred at several places in the Middle East, around 8000 B.C., although it had likely been gathered as a wild grain much earlier. Some of the first grain farmers used primitive mortars and pestles to make flour. The grain was placed in a large saucer-shaped stone, and a smaller rounded stone was used to crush it. Sometime between 6000 and 5000 B.C., breads were made from wheat or barley flours in both Egypt and Mesopotamia (an ancient civilization which developed around the Tigris and Euphrates Rivers in what is now eastern Iraq). However, the Mesopotamians eventually had to abandon the growing of wheat when their irrigation practices made the soil too salty for this grain. Then, only barley—which makes a heavy bread—could be grown. There was no such difficulty in growing wheat in Egypt where the salt-free waters of the Nile River enhanced the fertility of the soil. Hence, the Egyptians became the undisputed leaders in the use of wheat flour for baking. They even made a refined flour by passing it through papyrus sieves to remove the branny particles.

The Hebrews, who had migrated back and forth across the fertile crescent of land between Babylon (the civilization which replaced that of Mesopotamia) and Egypt, adopted whatever flours might be available in the regions they settled. Hence, their breads were sometimes made from mixtures of grain and legume flours. The Bible (Ezekiel 4:9) tells that the Lord instructed His peoples to make bread from wheat, barley, beans, lentils, millet, and fitches (which is interpreted as being one of the types of vetches). It is noteworthy, too, that similar mixtures of flours are presently being promoted for use by impoverished peoples in the Middle East.

After their conquest of Egypt, the Romans adopted many of the milling and baking practices of the Egyptians. However, the Romans used sieves made of animal hair rather than of papyrus. It appears that the Romans were the first people to develop a large mill that could be driven by a horse or a donkey. They also developed water-driven mills. However, it is fortunate that many Romans kept small hand mills at home, because the supply of flour was cut off when (1) the emperor Caligula commandeered the mill animals for his own use, and (2) the flow of water to the large mills was cut off by the invading Goths.

Fig. F-29. Grinding grain and making bread in Pompeii. (Courtesy, The Bettmann Archive, Inc., New York, N.Y.)

In the Middle Ages, the milling of flour became the monopoly of a few professional millers who were granted this concession by the wealthy landholders who built the large mills. Furthermore, laws were passed that made it a crime to own a small mill and grind one's own grain. Also, only affluent persons in northern Europe could enjoy bread from wheat flour, because the cool, damp climate favored the cultivation of rye. Even the best wheaten breads were often a mixture of wheat and rye, called maslin. The poor considered themselves fortunate to have bread made from rye flour that was mixed with oats, peas, and beans. In some ways they were fortunate, because the latter breads were likely more nutritious than those made solely from wheat and rye. However, dependence upon rye brought the risk of a terrifying condition called ergotism, caused by consuming grain (mainly rye) in which the kernels were replaced by the poisonous fungus, ergot. People who consumed bread made from the diseased grain suffered from a burning sensation in the hands, feet, and chest; some even had convulsions and died.

The first wind-powered mills were built in England in the 11th century, and in Belgium, Luxembourg and the Netherlands in the 12th century. By the middle of the 14th century, a tall, tower type windmill had been invented. The top of the latter type could be turned to face the wind.

Both water-driven mills and windmills were used by the early settlers of America to grind corn (maize) for the making of cornbread. The colonies along the Atlantic Coast of America considered wheaten flours to be a luxury, because wheat did not grow well in that climate. Hence, the limited amounts of wheat flour that were available were often mixed with other grains such as oats. It was not until the settling of the Midwest and the Great Plains that sufficient wheat became available to make it the predominant grain for flour in the United States.

Meanwhile, important developments in the production of flour occurred on both sides of the Atlantic. The Scottish inventor James Watt developed the first practical steam engine. A crude prototype had been invented by the ancient Greeks, but little attention was paid to it at the time. At a demonstration given in London in 1784, Watt showed how his engine could drive a flour mill. About the same time in America, Oliver Evans developed a highly mechanized mill which eliminated much of the need for human labor. Evans' mill allowed wheat to be ground into flour in a continuous operation that was made possible by a series of gears, conveyor belts, and elevators. Watt's earlier demonstration led Evans to build a steam-driven flour mill in Pittsburgh in 1810.

Even though the 19th century was ushered in by the use of steam engines to drive flour mills and other machines, the way in which flour was ground between two stones had not changed in principle since the Stone Age. Stone grinding produced a flour that contained bits of bran and oily germ because these parts were crushed during grinding. However, the Europeans had already modified their milling practices by placing the grindstones farther apart, so that the grain was broken into pieces rather than being crushed and smeared between the stones. Also, the grain was broken and sieved in a series of steps designed to separate the particles of white, starchy endosperm from the bran and the germ. The flour obtained by the modified procedure was much whiter and kept better than that produced by the older methods of milling.

Perhaps the most revolutionary change in the production of flour was the development of a practical roller mill over a period of almost 50 years in the 19th century. It is noteworthy that the first publication of an idea for such a device was in some drawings made by the Italian engineer Agastino Ramelli, in 1588. The idea was not taken seriously by the milling profession until the early 1800s in Switzerland, where the first steam-driven roller mill was tested in 1834. Yet, the mill was impractical because the iron rollers wore out too soon. About a half-century later, a mill with porcelain rollers was developed in Hungary. In 1878, Governor Washburn of Minnesota promptly arranged for the Hungarian mill to be installed at the site in Minneapolis where his milling company had recently been destroyed by a fire and an accompanying explosion. The rebuilt "Washburn's mill" prospered, and, in 1928, became the nucleus for General Mills, Inc.

The age old practice of adulterating flour with cheap and hard to detect materials persisted through the 19th century. Even the lowly potato was dried, ground, and mixed with wheat flour, after it had been introduced into Europe by the Spaniards who brought it from South America. Flour was also whitened and extended with alum, bone ashes, and chalk. Finally, the British Parliament passed laws requiring that bread be made from flours or meals derived from wheat, barley, rye, oats, buckwheat, Indian corn (maize), peas, beans, rice, or potatoes, plus salt, pure water, eggs, milk, and the appropriate leavening agent(s). The British Food and Drug Act was passed in 1860, but it was not until 46 years later that a similar law was passed in the United States. The U.S. Pure Food and Drug Act of 1906 established the Food and Drug Administration (FDA) as the agency with the responsibility for preventing the adulteration of foods. Hence, products labeled as "flour" or "bread" which enter interstate commerce must conform to the standards of identity established by the FDA.

In 1929, the per capita consumption of wheat flour was 171 lb. By 1970, with a greater variety of foods on the grocery shelf, the per capita consumption of wheat flour had dropped to a low of 110 lb. But food technologists developed a variety of new products and all baked goods became increasingly popular, so by 1990 the per capita consumption had risen to 137.8 lb.

PRODUCTION AND PROCESSING. Basically, flour production is the grinding of a dried food material, but the details of processing vary somewhat for each of the different flours. Figs. F-30, F-31, F-32, and F-33 show some of the modern equipment used in the production of flours.

Fig. F-30. Flour milling machines. The whole grain enters the tops of the machines through the cylindrical glass hoppers. (Courtesy, Union Pacific Railroad Company, Omaha, Nebr.)

Fig. F-31. A flour milling machine with the cover opened to give a closeup view of the rollers which grind the grain. (Courtesy, Union Pacific Railroad Company, Omaha, Nebr.)

Fig. F-32. Flour sifters. (Courtesy, Union Pacific Railroad Company, Omaha, Nebr.)

Fig. F-33. Bagging flour. (Courtesy, Union Pacific Railroad Company, Omaha, Nebr.)

Major Flours. The American homemaker is most familiar with the flours made from wheat. But there are other important flours.

Pertinent information relative to the major flours is summarized in Table F-25. (Additional details on the flour sources; their origin and history, world and U.S. production, and growing and harvesting, are given in the articles dealing with the individual grains. For example, major treatment on the production of wheat flours is given under WHEAT.)

TABLE F-25
MAJOR FLOURS

Type of Flour	Production	Characteristics
Barley (Also see BARLEY.)	Dehulled grain (groats) may be (1) milled directly into a course flour, or (2) pearled, then milled into a fine flour.	Bland flavored, white colored. Produces tender baked goods. Not strong enough for yeast leavening.
Buckwheat (Also see BUCKWHEAT.)	Groats may be milled into a high extraction[1] (dark colored) flour, or a low extraction (light colored) flour.	Strongly flavored, with a trace of bitterness; light to dark brown in color. Not suitable for yeast-leavened doughs.
Corn (Also see CORN.)	Most flour is made from degerminated white or yellow dent corn.	Mild, sweet flavor, white or yellow colored. May be gritty if not ground finely. Produces tender baked goods that readily dry out and become crumbly. Flour is too weak for making yeast-leavened products.
Rice (Also see RICE.)	Flour is made from broken pieces of polished, dehulled grain.	Bland flavored, white colored. May be gritty if not ground finely. Baked goods may dry out and crumble. Too weak for yeast leavening.
Rye (Also see RYE.)	Milling is similar to that for wheat. Flour may be of high extraction (dark colored), medium extraction (medium colored) or low extraction (light colored).	Distinctive flavor; color varies. May be used with yeast leavening, but dough is denser and moister than that madė from wheat flour.
Wheat		
Cake	Low extraction flour, sometimes bleached, milled from soft wheat.	Bland, white, and soft in texture. Not suitable for yeast leavening.
Gluten	A mixture of 45% guten (produced by washing the starch from white flour), and 55% white flour.	Bland, white, and high in protein. A very strong, high-protein flour which may be mixed with weak flours to impart sufficient strength for yeast leavening.
High ratio flour	Flour of very fine and uniform particle size, treated with chlorine to reduce the gluten strength.	It is possible to add up to 140 parts of sugar to 100 parts of this flour. High ratio flour makes excellent cakes.
Self-rising	White flour to which a dry leavening acid and baking soda have been added.	Needs no additional leavening. Products may have a slightly bitter taste.
White	A flour of about 80% extraction, which may be milled from soft, medium, or hard wheats.	Most white flours, with the exception of cake flour, may be leavened with yeast.
Whole wheat (Also see WHEAT.)	A flour of from 95 to 100% extraction, which contains all or most of the bran and germ from the orginal grain.	Light brown color, may be speckled with bits of bran and germ. May be used for yeast-leavened doughs, but baked products may be heavier and moister than those made from white flours.

[1]Extraction refers to the percentage of the grain that is converted to a flour. High extraction flours contain more bran and germ than low extraction flours.

Special Flours. Many of the plant materials that ancient peoples used to make flour and meals are once again serving this purpose as food tehnologists seek new ways to feed exploding populations. Also, these special flours may be used in dietetic foods. It is noteworthy that each of the special flours lacks gluten, the protein in wheat flours which imparts elasticity to doughs and enables them to be leavened with yeast. Therefore, these nonwheat flours are best suited for making baked goods leavened with baking powder or baking soda, the details of which are given in the articles—BAKING POWDER AND BAKING SODA; and QUICK BREADS. The more common special flours are described briefly in Table F-26. (Additional details on the flour sources; their origin and history, world and U.S. production, and growing and harvesting, are given in the articles dealing with the individual food materials.)

TABLE F-26
SPECIAL FLOURS

Type of Flour	Production	Characteristics
Arrowroot (Also see ARROWROOT.)	Ground tubers are mixed with water, then screened to remove fibrous matter. The starchy particles are then allowed to settle out, so they may be removed and dried.	Tasteless white powder that is almost all starch. When heated with water, forms a clear, thick paste which gels upon cooling.
Carob (Also see CAROB.)	After removal of the seeds, the pods are finely ground and sieved.	Chocolate colored powder with a mildly sweet taste. Blends well with various cereal flours.
Cottonseed (Also see COTTONSEED.)	The seeds are dehulled and the oil is extracted. Then, the meal is ground into a flour. Cottonseed which contains gossypol requires a special treatment.	Color varies from very light cream color to dark brown (when degossypolized). Toasted cottonseed has a pleasant flavor. The flour may contain 50% or more protein.
Fish	Whole fish are dried and extracted with a solvent to remove the oil. The meal is then ground.	Tasteless, odorless, tan colored powder. Fish flour may contain about 75% protein.
Lima bean (Also see LEGUMES.)	Dried, mature lima beans are ground into a flour.	Bland, white colored powder.
Malt (Also see MALT.)	Malted barley or wheat is dehulled, dried, and ground into flour.	Light to dark colored powder. Contains starch-digesting enzymes.
Millet (Also see MILLET.)	Usually, the flour is made from whole grains of millet. Sometimes, the grain is debranned before milling.	Light yellow powder with a bland taste.
Oat (Also see OATS.)	Cleaned oats are dried, dehulled, cut into granules, steamed, and rolled into flakes. Then, the flakes are ground into flour.	A white powder with a bland taste.
Pea (Also see LEGUMES.)	The flour may be made from either cooked or uncooked dried peas.	Yellow or green powder which has a bland, pleasant taste. The color of green pea flour usually bleaches out during baking.
Peanut (Also see PEANUTS.)	Defatted peanut meal is ground into a flour.	Pleasant tasting, light cream colored powder.
Potato (Also see POTATOES.)	Pieces of raw or cooked skinned potato tubers are pulverized to a pulpy mass, dried and ground into flour.	A white powder with a bland taste. Forms thick pastes when heated with water. Flour blends well with other flours.
Soybeans (Also see SOYBEANS.)	Heat treated, dehulled soybeans are usually defatted to varying degrees before grinding to a flour.	Light tan powder with a beany flavor. The flour may contain up to 50% protein.
Sunflower seed (Also see SUNFLOWERS.)	Dehulled seed is ground, and some or all of the fat extracted. The remaining pulp is ground into a flour.	Grey-brown powder with slightly bitter taste.
Tapioca (cassava) (Also see CASSAVA.)	Cassava roots are washed, peeled, and ground. The pulp is mixed with water, passed through a screen and the starch allowed to settle out. The starch is then dried.	Tasteless white powder that is almost all starch. Forms thick gels when cooked with water.

NUTRITIVE VALUE OF FLOURS. In general, flours are more concentrated sources of nutrients than the plants (or animals, in the case of fish flour) from which they have been made because much water has been removed. Sometimes, large amounts of fat have also been removed. Hence, most flours keep well without refrigeration. Food Composition Table F-36 gives the nutrient composition of the major cereal flours, and of special flours.

Some noteworthy facts relative to the nutritive value of flours follow:

1. The darker colored flours are richer in essential nutrients than the lighter colored ones, as exemplified by the different flours made from buckwheat, rye, and wheat.

2. The reduction in the water content of a food during the production of a flour may make it more nearly possible to use the food as a major dietary source of calories and protein. For example, it would take almost 7 lb (*3.14 kg*) of boiled potatoes to furnish the recommended daily per capita allowance of 2,385 Calories (kcal) established by WHO and FAO of the U.N., whereas only 1½ lb (*0.68 kg*) of potato flour would be required. Hence, there is much current interest in the utilization of various vegetable flours for feeding people in the developing countries.

3. Flours made from defatted products are rich in nutrients and comparatively low in calories. For example, defatted soybean flour contains ¼ more protein and ⅓ more calcium, but only ¾ as many calories as undefatted soybean flour.

4. One cup (*125 g*) of self-rising white wheat flour, which contains monocalcium phosphate as a leavening acid, supplies more than ⅓ of the Recommended Daily Allowances (RDA) for calcium, and more than ½ of that for phosphorus. Hence, items baked with this flour may be an important source of these minerals for people who fail to eat sufficient amounts of dairy products.

5. A mere 0.6 oz (*17.3 g*) of fish flour provides the entire RDA for calcium (*800 mg*). The same amount of this flour provides more than half of the daily iron allowance for an adult male.

6. The flours highest in protein are those made from defatted fish and from the defatted oilseed meals (cottonseed, peanut, soybean, and sunflower). It is noteworthy that these flours are currently being used around the world as protein supplements.

7. Two other important observations regarding flours, which are not apparent from Food Composition Table F-36, are that (a) the cereal flours are deficient in the amino acid lysine, which is supplied abundantly by the noncereal flours (with the exceptions of arrowroot and tapioca flours, which are almost all starch); and (b) many of the noncereal flours are deficient in the sulfur-containing amino acids (cysteine, cystine, and methionine), which are supplied in greater proportions by the cereal flours. Therefore, it makes good sense to use combinations of cereal and noncereal flours, so that there might be optimal utilization of the proteins in these foods.

Enrichment and Fortification of Flours. People around the world have strived for thousands of years to produce white, soft flours. Hence, the present dominance of flour production by white flours is not the outcome of a recent "conspiracy" by the flour industry; rather, it is the response to an age old public demand. Nevertheless, it is important that appropriate measures be taken to offset the nutritional shortcomings of highly refined flours, particularly in circumstances where these items are relied upon to furnish much of the caloric and protein requirements. The most widely employed measures have been enrichment and fortification.

NOTE WELL: Originally, the FDA differentiated between enrichment and fortification, but now the two terms are used interchangeably. Nevertheless, the following differentiations between the terms persist:

"Enrichment is a term which refers to the addition of specific nutrients to a food *as established in a federal standard of identity and quality* (for example: enriched bread). The amounts added generally are moderate and include those commonly present at even lower levels."[10]

"Fortified is a term which refers to the addition to foods of specific nutrients. The amounts added are usually in excess of those normally found in the food because of the importance of providing additional amounts of the nutrients to the diet. Some foods are selected for fortification because they are appropriate carriers for the nutrient. Example: Milk is frequently fortified with vitamin D."[11]

It is noteworthy that the FDA regulations for the enrichment and fortification of flours[12] apply to products with labeling that utilizes these terms. Examples of how enrichment and fortification affect the nutrient composition of some typical wheat flours are given in Table F-27. (For additional details, see WHEAT.)

TABLE F-27
MINERAL AND VITAMIN CONTENTS OF SELECTED FLOURS[1]

Type of Flour	Calcium	Iron	Thiamin	Riboflavin	Niacin
(1 lb, or *0.45 kg*)	(mg)	(mg)	(mg)	(mg)	(mg)
Whole wheat	186	15.0	2.49	.54	19.7
White, 80% extraction, unenriched	109	5.9	1.16	.33	9.3
White, enriched[2]	109	13.0-16.5	2.90	1.80	24.0
White, self-rising, enriched[3]	960[4]	13.0-16.5	2.90	1.80	24.0

[1]Values are from Food Composition Table F-36 of this book.
[2]Enriched according to *Code of Federal Regulations*, Title 21, Revised 4/1/78, Section 137.165.
[3]According to *Code of Federal Regulations*, Title 21, Revised 4/1/78, Section 137.185.
[4]This level of calcium is usually equaled or exceeded when calcium monophosphate is used as the leavening acid.

[10]"Nutrition Labeling—Terms you Should Know," *FDA Consumer Memo*, DHEW Publication No. 74-2010, 1974, p. 2.

[11]Ibid., p. 3.

[12]*Code of Federal Regulations*, Title 21, Revised 4/1/78, Sections 137.160, 137.165, and 137.185.

It may be seen from Table F-27 that the milling procedures used to make white flour result in a product containing much less calcium, iron, thiamin, riboflavin, and niacin than whole wheat flour. However, enrichment of white flour restores the iron, thiamin, riboflavin, and niacin to levels equaling or slightly exceeding those in whole wheat flour. The use of calcium monophosphate in self-rising flour constitutes calcium fortification because it raises the level of this mineral so that it greatly exceeds the calcium content of whole wheat flour.

The calcium content of breads made from whole grain flours may not always be well utilized because the whole grains contain phosphorus compounds called phytates, which interfere with the absorption of calcium and other essential minerals such as iron and zinc. Therefore, it is noteworthy that British nutritionists learned about this effect in the early 1940s and that they also found that removal of phytates restored the availability of calcium in whole grain breads.[13]

More recent research has shown that much of the phytates present in whole grain flours may be destroyed if there is a considerable amount of yeast fermentation during the leavening of bread doughs.[14]

There are no FDA standards of identity for the enrichment or fortification of nonwheat flours. Instead, there are some standards for macaroni and noodle products made mainly from wheat flour, but nutritionally enhanced by the addition of nonwheat flours such as those derived from other grains or from oilseeds (usually cottonseeds, peanuts, soybeans, or sunflowers).

(Also see ENRICHMENT.)

USES OF FLOURS AND THE BY-PRODUCTS OF MILLING.

Modern technology has made it possible to produce a wide variety of flours with great efficiency because only negligible amounts of the original raw materials are wasted. The soft, inner tissues of seeds, roots, and tubers yield the flours used in human foods, whereas the highly fibrous outer tissues are fed to animals. By-products intermediate in fiber are utilized in both human foods and feeds for all classes of livestock. Therefore, the uses of various products are briefed in the sections which follow. (Baking procedures are described in the article BREADS AND BAKING.)

Major Cereal Flours. Some of the more important uses of these items are given in Table F-28.

[13]Davidson, S., *et al., Human Nutrition and Dietetics,* 7th ed., Churchill Livingstone, Edinburgh, Scotland, 1979, pp. 94-95, Table 11.2.

[14]"Zinc Availability in Leavened and Unleavened Bread," *Nutrition Reviews,* Vol. 33, January 1975, p. 18.

TABLE F-28
USES OF MAJOR CEREAL FLOURS

Flour	Uses
Barley	Infant cereals; substitute for wheat flour when there is an allergy to wheat; biscuits, muffins, and pancakes; and for making a starchy broth that is well tolerated by people who have digestive disturbances.
Buckwheat[1]	In the U.S., most buckwheat flour is used in pancake mixes. In Japan, it is mixed with wheat flour and made into noodles (Soba).
Corn	Ingredient of ready-to-eat breakfast cereals, snack foods, baking mixes (biscuits, muffins, and pancakes); and a meat extender. Replacement for wheat flour in hypoallergenic and low-gluten products.
Rice	Thickening agent in gravies and sauces; and for imparting crispiness and tenderness to biscuits, cookies, and crackers. Replacement for wheat flour in hypoallergenic and low-gluten products.
Rye[2]	Alone or mixed with wheat flour in breads, biscuits, crackers, and pancakes. (Commercial baked goods made with rye flour keep fresh longer than those made with wheat flour.) Thickener for sauces, soups, and custard powders.
Wheat:	
Cake	Soft-textured and tender baked goods leavened with baking powders.
Gluten	High-protein and other types of yeast-leavened breads, macaroni and noodle products, and strengthening agent for mixing with weaker flours.
Self-rising	Mainly for biscuits, cookies, muffins, pancakes, and waffles. Greatest use is in the southeastern U.S. Elsewhere, this product has been replaced by special mixes.
White, 80% extraction	Most baked goods. Well suited for either baking powder or yeast leavening. Thickener for gravies, puddings, and sauces.
Whole wheat	Well suited for making all types of baked goods (products tend to be heavier and moister than those made with white flour) and cereal products.

[1]Dark and light flours are available. The former contains more nutrients per calorie.

[2]Dark, medium, and light flours are available. The darkest contains the most nutrients per calories.

Special Flours. Major uses of these products are given in Table F-29.

TABLE F-29
USES OF SELECTED SPECIAL FLOURS

Flour	Uses
Arrowroot	Special low-protein baked goods and other items for people with failure of the liver and/or the kidneys, or with certain allergies. An ingredient of biscuits that are easily digested by infants and invalids.
Carob	Substitute for chocolate or cocoa in baked goods, confections, drinks, frozen desserts, and puddings.
Cottonseed	Source of complementary protein for mixtures containing cereal flours. As such, it is being test marketed in substitute "milks" and other cereal mixtures for weanling infants in the developing countries.
Fish	A high-protein ingredient (also known as fish protein concentrate or FPC) for raising the nutritive values of baked goods, cereals, flour, and macaroni and noodle products.
Lima bean	Dry soup mixes; and an ingredient in products made from cereal flours, where it raises the quantity and quality of protein.
Malt	Production of malted milk powder; and in baked goods (the enzymes present in malt flours convert the starch of the dough into sugars), breakfast cereals, coffee substitutes, confections, and infant foods.
Millet	Brewing of beers (mainly in Africa); and an ingredient of breads, porridges, and soups. Mixed with legume flours in special protein foods for infants and children in the developing countries.
Oat	Ingredient of ready-to-eat breakfast cereals and infant cereals. Popular in Scotland for making oatmeal scones.
Pea	Dry soup mixes; and an ingredient in products made from cereal flours, where it raises the quantity and quality of protein.
Peanut	Production of peanut milk, peanut cheese, and peanut protein concentrate. Mixed with cereal flours in special protein foods for infants and children in the developing countries.
Potato	Ingredient of baked goods, snacks and soups. Thickening agent. (Potato starch thickens better than any of the other common food starches. Hence, only small amounts are required.)
Soybean	Added to baked goods, breakfast cereals, confections, dietetic foods, infant foods, macaroni and noodles, and meat products (as an extender and moisturizer). Replacement for wheat flour in hypoallergenic and low-gluten products. Mixed with cereal flours in special protein foods for infants and children in the developing countries.
Sunflower	Source of complementary protein for mixtures containing cereal flours. As such, it is presently being tested for use in baked goods, cereals, desserts, imitation milks, and macaroni and noddle products.
Tapioca	Thickening agent for fruit pies, puddings, and soups.

By-Products. These items are usually rich in fiber, minerals, and vitamins. Their major uses are given in Table F-30.

TABLE F-30
BY-PRODUCTS OF FLOUR PRODUCTION

Product	Description	Uses	Comments
HUMAN FOODS			
Bran (corn, rice, wheat)	Outer coarse coat of grain (beneath the hull) removed during milling into flour.	High-fiber ingredient of baked goods and breakfast cereals.	Good source of protein (9 to 18%), fiber (10 to 14%), minerals, and vitamins. Contains phytates which interfere with mineral absorption. Products containing large amounts of bran may have a laxative effect.
Germ (corn, rice, wheat)	Embryo or sprouting part of the seed.	Enrichment or fortification of baked goods and breakfast cereals.	Excellent source of protein (20 to 27%), essential fatty acids, fiber (3 to 12%), minerals, and vitamins. May turn rancid unless kept refrigerated.
Germ oil (corn, rice, wheat)	Oil extracted from germ.	Nutritional supplement.	Rich in essential fatty acids and vitamin E. Maximum nutritional value is obtained when kept refrigerated and used cold (heating destroys vitamin E).
Middlings (rye, wheat)	A mixture of bran, germ and adhering particles of endosperm.	High fiber ingredient of "natural" breakfast cereals.	Good source of protein (18 to 20%) and fiber (4 to 8.5%). Rye middlings and wheat middlings cannot contain more than 8.5% crude fiber and 7% crude fiber, respectively.
Polishings (rice)	Finely powdered material removed in the conversion of brown rice into white rice (prior to the milling of the latter into flour).	In many baby foods. To prevent and cure beriberi. In certain baked goods and breakfast cereals.	Contains 10 to 15% protein and 3 to 4% fiber. Rich in essential fatty acids, minerals, and the vitamin B complex (should be stored in a refrigerator to prevent rancidity). Gives baked goods a soft texture and makes them more moist (too much may produce heavy, soggy items).
FEEDS FOR LIVESTOCK			
Bran (Corn, rice, wheat)	Outer coarse coat of grain (beneath the hull) removed during milling into flour.	Fed to all classes of livestock, especially horses and cattle.	Good sources of protein (9 to 18%), fiber (10 to 14%), minerals, and vitamins. Large amounts have a laxative action.
Germ (Corn, rice, wheat)	Embryo or sprouting part of the seed removed during milling.	Germ meal is usually mixed with bran and middlings or shorts. Dog foods, mink feeds, and feeds for laboratory animals.	Excellent source of protein (20 to 27%), essential fatty acids, fiber (3 to 12%), minerals, and vitamins. May turn rancid unless kept refrigerated. Sometimes defatted.
Germ meal, defatted (Corn, rice, wheat)	The product that remains after removal of part of the oil or fat.	Fed to all classes of livestock.	Rich in protein (30% or more), but much lower in essential fatty acids and vitamin E than undefatted germ meal.

(Continued)

TABLE F-30 (*Continued*)

Product	Description	Uses	Comments
FEEDS FOR LIVESTOCK (*Continued*)			
Germ oil (corn, rice, wheat)	Oil extracted from germ meal.	A nutritional supplement added to feeds.	Rich source of essential fatty acids and vitamin E. Exposure to air, light, and heat may result in rancidity and the destruction of vitamin E. Store in a refrigerator.
Hulls (grains, malted grains, and oilseeds)	Outer covering of certain seeds.	Fed mainly to cattle.	Low quality roughages.
Malt sprouts (barley, rye, wheat)	Sprouts removed from malted and steamed whole grain before it is milled into malt flour.	Added to feeds for dairy cattle. Not usually fed alone.	Excellent source of protein (24% or more) and fiber (about 15%). Large amounts of malt sprouts in dairy feeds may impart off-flavors to milk.
Middlings (rye, wheat)	The most common by-product of flour mills. It contains a mixture of endosperm, bran, and germ.	Chiefly for calves, swine, and poultry.	Good source of protein (18 to 20%) and fiber (4 to 8.5%). Rye middlings and wheat middlings cannot contain more than 8.5% crude fiber and 7% crude fiber, respectively.
Polishings (rice)	Finely powdered material removed in the conversion of brown rice into white rice (prior to the milling of the latter into flour).	In swine and poultry rations.	Contains 10 to 15% protein and 3 to 4% fiber. Rich in essential fatty acids, minerals, and the vitamin B complex. The high fat content may result in the development of rancity, unless stored in a refrigerator.
Red dog (wheat)	Mixture of fine particles of bran, germ, and flour.	Fed chiefly to poultry, young calves, and swine (especially to young pigs).	Good source of protein (17 to 20%). Cannot contain more than 4% fiber.
Shorts (wheat)	Mixture of fine particles of bran, germ, and flour. Generally from winter wheat.	Used chiefly for swine, calves, and poultry.	Good source of protein (17 to 20%). Cannot contain more than 7% fiber.

RECENT DEVELOPMENTS IN FLOUR TECHNOLOGY. The high rates of grain and legume production in the United States have spurred food technologists to develop new ways of utilizing these foods in processed forms such as flours because then the financial return is much greater than that obtained from selling the unprocessed items. Also, wastage due to pests and spoilage is reduced by conversion of the foods into flours. Hence, the American consumer reaps the benefits of (1) lower costs due to more efficient utilization of foods, and (2) a saving of food preparation time because modern flour products are designed for maximum convenience. Most of the developments which follow are still undergoing testing. Nevertheless, they are noteworthy as indicators of the recent trends in flour technology.

- **Air classification of flour fractions**—This process uses swirling streams of air to separate very finely ground flour into fractions which are (1) high in carbohydrate and low in protein, and (2) low in carbohydrate and high in protein. The high-carbohydrate flour is suitable for cakes, cookies, and pastries, whereas the high-protein flour may be used alone or to fortify low-protein flours made from soft wheats. Blending permits a miller to tailor a flour of exact protein value to a buyer's specifications.

- **Corn germ**—Until recently, almost all of this product was utilized in livestock feeds, because its high oil content rendered it susceptible to the development of rancidity while standing on supermarket shelves. Now this problem has been eliminated by a special process that includes toasting as a final step.

• **Dustless (agglomerated) flour**—The older types of flour were very dry and dusty, and tended to ball up in large masses when mixed with liquids. Furthermore, it is believed that many professional bakers developed allergies to wheat as a result of the chronic inhalation of airborne flour dust. The newer, dustless, easier-to-mix flours are made by a process which involves (1) wetting the flour to make the individual particles stick together (the process of agglomeration), and (2) drying the agglomerated particles to the desired moisture content. Agglomerated flours are free-flowing and may be mixed into batters or doughs without prior sifting.

• **Gluten-free, low-calorie flour**—Gluten-free baked goods are usually made from corn, rice, or soybean flours. The corn and rice product tend to dry out and become crumbly upon standing; whereas, the soybean items are heavy, moist, and have an objectionable beany flavor. These problems have not been encountered in a new type of gluten-free baked product made from a low-calorie flour (about half as high in calories as wheat flour) containing starch, cellulose, and small amounts of special gums derived from cellulose. The very low protein content of the special flour also permits its use in protein restricted diets, such as those employed in the treatment of liver failure or kidney failure.

• **Imitation cheeses made from flour**—Flours (wheat, malted barley, and soy) are the major ingredients of a line of newly developed, dried, imitation cheese products designed for institutional food service. The products contain from 15 to 19% of high quality protein because the flour proteins are complemented by those in the dried buttermilk, dried cheese, and dried whey ingredients. They may be stored at room temperature for up to a year without losing their quality.

• **Pea flour**—Legume flours contain from 2 to 4 times as much protein as the common grain flours. Hence, the addition of legume flours to grain flours raises both the quantity and quality of protein. (The amino acid patterns of legumes are complementary to those of grains. Hence, mixtures have a higher protein quality than either type of flour alone.) However, many legumes impart undesirable characteristics, such as a beany flavor. Therefore, it is noteworthy that USDA scientists have recently found that flours made from green or yellow peas may replace up to 15% of the wheat flour in a loaf of bread without reducing its acceptability. Furthermore, the color of green pea flour is apparently bleached out during baking.

• **Replacement of enzymes lacking in certain flours**—Flours normally contain enzymes which digest starch. These enzyme actions provide sugar for the growth of yeast. When certain flours lack the enzymes, they must be supplied by other ingredients or added in purified form. European bakers have long depended upon malts made from germinated barley or wheat to supply them, but American bakers are now beginning to use pure enzyme preparations derived from various microorganisms.

• **Rice bran derivative**—Protex (trademark of Food Engineering International, Inc.) is a specially processed defatted mixture of rice bran, rice germ, and rice polishings which may be used in baked goods, breakfast cereals, macaroni and noodle products, and milklike beverages. It contains between 17 and 21% protein, plus minerals and vitamins present in the outer layers of the rice grain.

• **Seed coat flours**—The seed coats of grains and legumes are high in fiber and low in calories, so there have been attempts to replace part of the cereal flours used in baking with various seed coat flours. However, yeast-leavened breads made with large amounts of these materials are too heavy and have poor flavors, odors, and textures. Researchers have recently discovered that the seed coat flours may be improved by treating the seed coats with an acid before grinding them into flour.

SUMMARY. Flours have served man for many millennia. Although the major flours now used in the developed countries are made from cereal grains, attempts to improve the nutrition of peoples in these countries have resulted in the utilization of legume flours. It seems likely that in the future a greater variety of flours will be utilized in attempts to feed the burgeoning world population.

FLUID BALANCE

In the body, it is a closely regulated equilibrium between water intake and water output.

(Also see WATER BALANCE.)

FLUID, BODY

This refers to the water and associated dissolved substances within the body. Well known body fluids are: digestive juices, blood, lymph, urine, and perspiration.

(Also see BODY FLUIDS; BODY WATER; and WATER BALANCE.)

FLUMMERY

• A dessert made with stewed fruit, toasted bread, and honey—baked in the oven.

• A soft jelly or porridge made with flour or meal.

FLUORIDATION

This refers to the addition of trace amounts (0.5 to 1.0 parts per million) of fluoride, usually as the sodium salt, to drinking water, for the purpose of providing protection against dental caries. This has proved to be a safe, economical, and efficient way in which to reduce tooth decay—a highly important public health measure in areas where natural water supplies do not contain sufficient fluoride. Extensive medical and public health studies have clearly demonstrated the safety and nutritional advantages that result from fluoridization of water supplies. In communities in which fluoridation has been introduced, the incidence of tooth decay in children has been decreased by 50% or more.

(Also see DENTAL CARIES; FLUORINE OR FLUORIDE; MINERAL[S]; and POISONS.)

FLUORINE OR FLUORIDE[15] (F)

Contents *Page*

History 778
Absorption, Metabolism, Excretion 778
Functions of Fluorine 778
Deficiency Symptoms 778
Interrelationships 778
Recommended Daily Allowance of Fluorine 779
Toxicity 779
Sources of Fluorine 779

Flourine is present in small but widely varying concentrations in practically all soils, water supplies, plants, and animals. It is therefore a constituent of all normal diets. Also, fluorine is one of the atmospheric contaminants of industries which use coal, ore, or earthy phosphates.

The adult human contains less than 1.4 mg of fluorine, most of which is in the bones and teeth. In small amounts, fluorides help develop strong bones and teeth, but in excessive amounts bones become porous and soft and teeth become mottled and easily worn down.

A proper intake of fluorine is essential for maximum resistance to dental caries (tooth decay), a beneficial effect that is particularly evident during infancy and early childhood, and which persists through adult life.

HISTORY. Fluorine was first isolated in 1886 by Henri Moissan, a Frenchman, who obtained it by the electrolysis of anhydrous hydrogen fluoride containing dissolved potassium fluoride. The name fluorine is derived from the Latin *fluo,* meaning to flow, because until 1500 A.D. it was used as a flux in metallurgy.

ABSORPTION, METABOLISM, EXCRETION. A large proportion (about 90%) of ingested fluorine is normally absorbed, primarily from the small intestine.

That portion of absorbed fluorine which is not taken up by the bones and teeth (50% or more of the fluorine absorbed) is excreted mainly in the urine (although small amounts are excreted in the sweat and the feces), with the result that the level of fluoride in the blood plasma is quite constant.

[15]Fluoride is the term for the ionized form of the element fluorine, as it occurs in drinking water. The two terms—fluorine or fluoride—are used interchangeably.

FUNCTIONS OF FLUORINE. Although fluorine is found in various parts of the body, it is particularly abundant in bones and teeth; it normally constitutes 0.02 to 0.05% of these tissues. Fluorine is necessary for sound bones and teeth.

The fluoride ion, at proper levels of intake, assists in the prevention of dental caries. When children under 9 years of age consume drinking water containing 1 part per million of fluorine, the teeth have fewer dental caries in childhood, adolescence, and throughout life. This has led to the fluoridation of water supplies in many countries.

The mechanism by which fluorine increases caries resistance of the teeth is not fully understood. However, it appears that crystals of fluoroapatite can replace some of the calcium phosphate crystals of hydroxyapatite that are normally deposited during tooth formation, and that it may also replace some of the carbonate normally found in the tooth. Apparently these fluoride substances are more resistant to mouth acids. Fluorine may also inactivate oral bacterial enzymes which create acids from carbohydrates.

Early attempts to demonstrate that fluorine is essential for growth were unsuccessful, but recent studies with rats raised in isolation units and fed diets containing less than 0.04 ppm fluorine responded with improved growth when fluorine was added to the diet.[16] Mice fed diets containing low levels of fluorine developed anemia and suffered impaired reproduction.[17]

Increased retention of calcium accompanied by a reduction in bone demineralization has been observed in patients receiving fluoride salts. This indicates the possibility that dietary fluorine is essential for optimal bone structure and prevention of osteoporosis in man. This has prompted some medical doctors to use fluorine therapeutically in the treatment of osteoporosis in the aged.

DEFICIENCY SYMPTOMS. A deficiency of fluorine, results in excess dental caries. Also, there is indication that a deficiency of fluorine results in more osteoporosis in the aged. However, excesses of fluorine are of more concern than deficiencies.

INTERRELATIONSHIPS. Large amounts of dietary calcium, aluminum, and fat will lower the absorption of fluorine.

Fluorine is a cumulative poison; hence, chronic fluoride toxicity, known as fluorosis, may not be noticed for sometime. The enamel of the teeth will likely lose luster

[16]Schwarz, K., and D. B. Milne, "Fluorine Requirement for Growth in the Rat," *Bioinorganic Chemistry,* Vol. 1, 1972, pp. 355-362.

[17]Messer, H. H., W. D. Armstrong, and L. Singer, "Influence of Fluoride Intake on Reproduction in Mice," *Journal of Nutrition,* Vol. 103, 1973, pp. 1319-1326.

and become chalky and mottled when one of the following conditions prevails: (1) when the fluoride content of the drinking water exceeds 2.5 ppm; (2) when the amount of fluorine ingested exceeds 30 to 40 ppm of the dry matter of the diet; or (3) when a person consumes (in food and water) fluorine in excess of 20 mg/day over an extended period of time. The degree of mottling depends upon the level of fluorine intake and individual susceptibility. Mottling of teeth in children has been observed at fluoride concentrations in the diet and drinking water of 2 to 8 ppm.

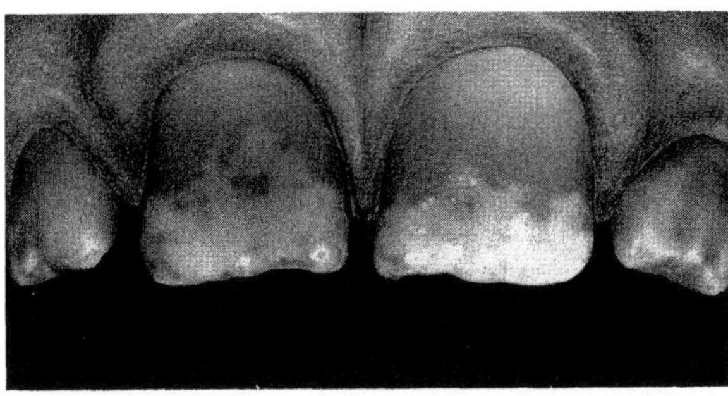

Fig. F-34. Fluorosis of human teeth, showing characteristic change in the enamel—loss of luster, and a chalky mottled appearance. This occurred in a lifetime resident of a community in which the drinking water naturally contained fluoride at concentrations 2 to 3 times greater than the level considered optimal. (Courtesy, Department of Health and Human Services, Public Health Service, National Institute of Health, Bethesda, Md.)

Fluorosis also affects the bones; they lose their normal color and luster, become thickened and softened, and break more easily.

There is no scientific evidence that fluoridation of water has been harmful to anybody at any age, but the effects of fluoride intakes from several sources may be cause for concern. The largest high fluorine area in the United States is the West Texas Panhandle. Also, the soils in some volcanic areas of the world contain large amounts of fluoride, with the result that foods grown in such areas may contain 2 to 3 times more fluoride than foods grown elsewhere.

(Also see POISONS, Table P-11 Some Potentially Poisonous [Toxic] Substances—"Fluorine [F].")

RECOMMENDED DAILY ALLOWANCES OF FLUORINE. Estimated safe and adequate intakes of fluoride are given in Table F-31.

The ranges suggested in Table F-31 are obtained without difficulty in areas with a water supply containing at least 1 mg/liter of fluoride, either naturally or through fluoridation.

TABLE F-31
ESTIMATED SAFE AND ADEQUATE DAILY DIETARY INTAKES OF FLUORIDE[1]

Group	Age	Fluoride
	(years)	(mg)
Infants	0–0.5	0.1–1.0
	0.5–1.0	0.2–1.0
Children and adolescents	1–3	0.5–1.5
	4–6	1.0–2.5
	7–10	1.5–2.5
	11+	1.5–2.5
Adults		1.5–4.0

[1]*Recommended Dietary Allowances,* 10th ed., 1989, NRC-National Academy of Sciences, p. 238–239.

The daily fluoride intake in many areas of the United States is not sufficient to afford optimal protection against dental caries. Fluoridation of water supplies is the simplest and most effective method of providing such added protection, although it is possible to add fluorine to milk, cereals, and salt, or to take it in tablet form as sodium fluoride. The range in safety in fluoride intake is wide enough to accommodate normal fluctuations in the fluoride content of foods without risk of inducing mottling.

TOXICITY. Fluorine has a small safety range. Yet, the range is wide enough so that the normal fluctuations in the fluoride content of foods poses no risk of excessive intake.

When people or animals consume fluoride concentrations in the diet and drinking water of 2 to 8 ppm (or more) over long periods of time, or when there is environmental contamination, it can result in toxicity (fluorosis), manifested by deformed teeth and bones, and softening, mottling, and irregular wear of the teeth.

SOURCES OF FLUORINE. Fluorine is widely, but unevenly, distributed in nature. It is found in many foods, but seafoods and tea are the richest dietary sources. Table F-32 shows the fluorine content in ppm of some common fluorine sources. (To convert ppm to mcg/g multiply by 1.)

TABLE F-32
SOME FLUORINE SOURCES

Source	Fluorine	Source	Fluorine
	(ppm)		(ppm)
Dried seaweed	326.0	Chicken	1.5
Tea	32.0	Butter	1.5
Mackerel	19.0	Soybeans	1.4
Sardines	11.0	Eggs	1.3
Salmon	6.0	Beef	1.2
Shrimp	4.5	Lamb	1.2
Smoked herring	3.8	Spinach	1.0
Wheat germ	2.4	Parsley	.9
Crab	2.2	Whole wheat	.8
Cheese	1.7	Pork	.7

NOTE WELL: The fluorine content of foods varies widely and is affected by the fluorine content of the environment (the air, soil and/or water) in the areas in which they were produced. Nevertheless, some foods do cumulate and contain more fluorine than others. Remember that it's the total fluorine ingested over an extended period of time that's important. There is hazard of fluorisis (evidenced by mottling of the teeth) (1) when the fluorine content of drinking water exceeds 2.5 ppm; (2) when the amount of fluorine ingested exceeds 30 to 40 ppm of dry matter of the diet; or (3) when a person consumes (in food and water) fluorine in excess of 20 mg/day over an extended period of time.

An average daily diet provides 0.25 to 0.35 mg of fluorine. In addition, the average adult may ingest 1.0 to 1.5 mg daily from drinking and cooking water that contains 1 ppm of fluorine. For children 1 to 12 years old, water may contribute anywhere from 0.4 to 1.1 mg of fluorine per day.

Fluorine tablets and fluorine toothpastes can serve as reliable, though expensive, sources of this element.

• **Fluoridation of water supplies**—Fluoridation of water supplies to bring the concentration of fluoride to 1 ppm (one part of fluorine to a million parts of water) has proved to be a safe, economical, and efficient way to reduce tooth decay—a highly important public health measure in areas where natural water supplies do not contain this amount. Extensive medical and public health studies have clearly demonstrated the safety and nutritional advantages that result from fluoridization of water supplies. In communities in which fluoridation has been introduced, the incidence of tooth decay in children has been decreased by 50% or more.

The concentration of fluoride in public water supplies should be adjusted slightly to allow for differences in water consumption with seasonal temperature changes.

(Also see MINERAL[S], Table M-67; POISONS, Table P-11 "Some Potentially Poisonous [Toxic] Substances.")

FLUOROSIS

A chronic disease resulting from the accumulation of toxic levels of the mineral fluorine in the teeth and bones. It is a crippling disease charactertized by bone overgrowth, brittle bones, stiff joints, weakness, weight loss and anemia. Mottling of the teeth may occur if exposure was during the formation of enamel. Contaminated water and food are the principal sources of excessive fluoride.

(Also see DENTAL FLUOROSIS; FLUORINE OR FLUORIDE; MINERAL[S]; and POISONS.)

FOIE GRAS

Livers of fattened geese and ducks. In cookery, the name *foie gras* is applied only to goose or duck liver from birds fattened by force feeding. The livers of Toulouse and Strasbourg geese sometimes weigh 4.4 lb *(2 kg)*. The finest foie gras comes from geese raised in Alsace and in southwestern France. Duck foie gras is also very delicate, but it has a tendency to disintegrate in cooking.

Foie gras is regarded as one of the greatest delicacies available. The quality of foie gras can be judged primarily by its color and texture; it should be creamy white, tinged with pink, and very firm.

FOLACIN (FOLATE/FOLIC ACID)[18]

Contents *Page*

History **781**
Chemistry, Metabolism, Properties **782**
Measurement/Assay **782**
Functions **782**
Deficiency Symptoms **783**
Recommended Daily Allowance **783**
Toxicity **784**
Folacin Losses During Processing, Cooking, and Storage **784**
Sources of Folacin **784**
Top Food Sources of Folacin **785**

There is no single compound vitamin with the name *folacin;* rather, the term *folacin* is used to designate folic acid (pteroylmonoglutamic acid, or PGA) and a group of closely related substances which are essential for all vertebrates, including man, for normal growth and reproduction, for the prevention of blood disorders, for important biochemical mechanisms within each cell, and for the prevention of a variety of deficiency symptoms in different species.

The research work related to folacin is considered to be the most complicated chapter in the story of the B-complex vitamins.

---

[18]Although the word *folate* is used in biochemical literature for salts or folic acid, it lacks specificity.

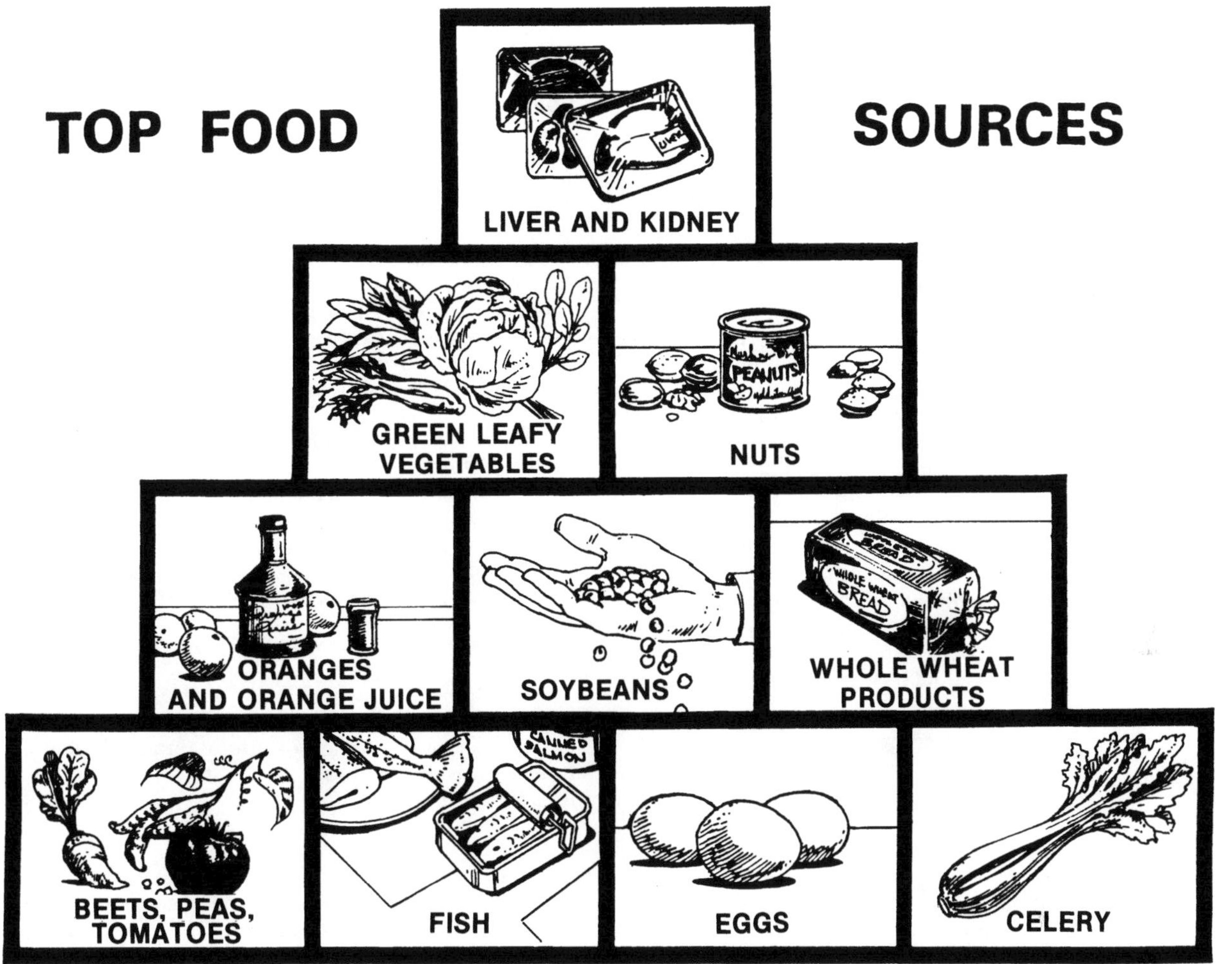

Fig. F-35. Top food sources of folacin.

HISTORY. Long before folacin was isolated or synthesized, its deficiency symptoms had been described in humans, animals, and microorganisms. In 1931, while working in a maternity hospital in Bombay, India, Dr. Lucy Wills described a macrocytic anemia in pregnant women which improved when they were given extracts of yeast. Subsequent work showed that the factor in yeast that was effective in curing Dr. Wills' patients was a group of chemically related vitamins with folic acid activity, given the generic name *folacin.*

In 1941, the name *folic acid,* the forerunner of the term *folacin,* was suggested by Mitchell, Snell, and Williams, of Texas, for the growth factor for bacteria that was found in spinach and known to be widely distributed in green leafy plants. The word *folic* is derived from the Latin word *folium,* meaning foliage or leaf.

In 1945, Angier and co-workers isolated and synthesized folic acid. That same year, Dr. Tom Spies showed that folic acid was effective in the treatment of megaloblastic anemia of pregnancy and of tropical sprue.

Through the years, several names other than folacin and folic acid have been given to this vitamin, including: Wills' factor, pteroylmonoglutamic acid, antianemia factor PGA, vitamin M, vitamin Bc, SLR factor, factor R, factor U, vitamin U, *Lactobacillus casei* factor; citrovoram factor (CF); yeast Norit eluate factor, vitamin B-9, vitamin B-10, and vitamin B-11.

CHEMISTRY, METABOLISM, PROPERTIES.

• **Chemistry**—This vitamin exists in several different forms in nature, making up the folacin group of compounds. All forms have similar activity when fed to man and other higher animals, but they have widely different activities as growth factors for microorganisms. The parent compound, folic acid (known chemically as pteroylglutamic acid [PGA]), which probably does not exist free in nature, consists of three linked components: *pteridine* (a yellow phosphorescent pigment related to the yellow pigment in butterfly wings [the Greek word for wing is *pteron*]); *para-aminobenzoic acid* (a growth factor for bacteria)[19]; and *glutamic acid* (an amino acid commonly found in proteins of foods and body tissues). Folic acid may consist of one, three, or seven glutamate groupings; designated as mono-, tri-, or hepta-pteroylglutamate. These conjugated forms serve as the major precursors of the vitamin in the diet. The coenzyme form, tetrahydrofolic acid, the most common form in the body, is also widely distributed in foods. The structural formula of pure folic acid (pteroylmonoglutamic acid) is given in Fig. F-36. If the parent molecule (consisting of pteridine, para-aminobenzoic acid, and glutamic acid) is broken, nutritional activity is lost. The biological active form of folacin is a reduction product called tetrahydrofolic acid.

FOLIC ACID

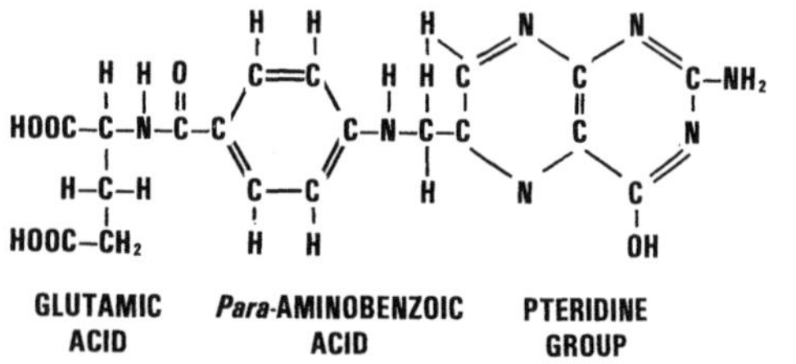

GLUTAMIC ACID | *Para*-AMINOBENZOIC ACID | PTERIDINE GROUP

Fig. F-36. Structure of folic acid (pteroylmonoglutamic acid).

• **Metabolism**—Wide variation exists in the absorption of folates from different foods, with a low of about 10% for yeast to a high of about 80% for eggs and liver. One study showed that only 31% of the folate in orange juice was absorbed, whereas 82% of the folate in bananas was absorbed. It is thought that the number of glutamic acid molecules in the folates affect the rate of absorption.

Folate is absorbed by active transport and by diffusion, mainly in the upper part of the small intestine, although some is absorbed along the entire length of the small intestine; glucose, ascorbic acid, and some antibiotics facilitate its absorption.

Folacin occurs in food in two forms—free folates and bound folates (polyglutamates). The free folates (which account for about 25% of the food intake) are readily absorbed in the intestinal tract. Before polyglutamates can be absorbed, excess glutamates (all but one) must be removed form the side chain of the molecule by conjugase, an enzyme. It is not known whether the conjugase acts in the lumen of the small intestine or in the intestinal wall. The absorption of folic acid is thus controlled by the deconjugating mechanism which, in turn, may be affected by conjugase inhibitors in the food; i.e., yeast. The rate of absorption of conjugated folates appears to be related to chain length.

Folate, bound to protein, is transported in the blood to the liver. There it is methylated and carried to the bone marrow cells, the maturing red blood cells, and perhaps to other cells. Methyl-folate seems to be the chief form of the vitamin in body tissues. Serum levels of folacin range from 7 to 16 nanograms per milliliter of serum. The total body stores of folate normally range between 5 and 12 mg, about half of which is in the liver. The measurement of folacin levels in both blood serum and red blood cells is the procedure used to evaluate folacin nutriture in human beings.

Some folate is excreted in the bile as well as the urine.

• **Properties**—Folic acid is a bright yellow crystaline powder, slightly soluble in water, unstable in acid solution, relatively unstable to heat, and rather easily destroyed upon exposure to light.

MEASUREMENT/ASSAY.

Folic acid is measured in micrograms or nanograms (ng, millimicrograms).

Folic acid in food is usually assayed either biologically by chick or rat growth tests, or microbiologically with *L. casei* or another suitable microorganism. Larger quantities of folacin activity may be measured chemically, fluorometrically, or by paper and thin-layer chromatography.

FUNCTIONS.

Following absorption, folic acid is changed, by a number of reduction reactions that require niacin, to at least five active coenzyme forms, the parent form being tetrahydrofolic acid. The principal function of these coenzymes is the transfer of single carbon units from one compound to another; the one carbon unit can be formyl, forminino, methylene, or methyl groups.

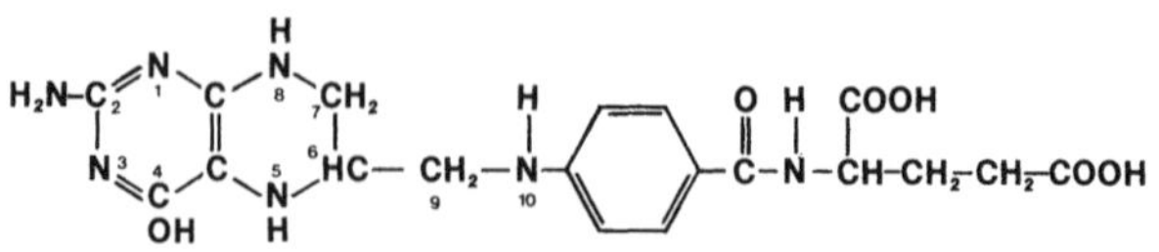

TETRAHYDROFOLIC ACID

Fig. F-37. Structure of tetrahydrofolic acid.

10

[19]In addition to having activity as a growth factor for bacteria, para-aminobenzoic acid has considerable folacin activity when fed to deficient animals in which intestinal synthesis of folacin takes place. For example, for rats and mice, para-aminobenzoic acid can completely replace the need for a dietary source of folacin. This explains why para-aminobenzoic acid was once considered to be a vitamin in its own right.

Folacin coenzymes are responsible for the following important functions:

1. The formation of purines and pyrimidines which, in turn, are needed for the synthesis of the nucleic acids DNA (deoxyribonucleic acid), and RNA (ribonucleic acid), vital to all cell nuclei. This explains the important role of folacin in cell division and reproduction.
2. The formation of heme, the iron-containing protein in hemoglobin.
3. The interconversion of the three-carbon amino acid serine from the two-carbon amino acid glycine.
4. The formation of the amino acids tyrosine from phenylalanine and glutamic acid from histidine.
5. The formation of the amino acid methionine from homocysteine.
6. The synthesis of choline from ethanolamine.
7. The conversion of nicotinamide to N^1-methylnicotinamide, one of the metabolites of niacin that is excreted in the urine.

Ascorbic acid, vitamin B-12, and vitamin B-6 are essential for the activity of the folacin coenzymes in many of these metabolic processes; again and again pointing up the interdependence of various vitamins.

• **Clinical application of folic acid**—The following clinical applications of folic acid are noteworthy:

1. **Nutritional megaloblastic and macrocytic anemias.** These types of anemias, which occur in infancy (megaloblastic anemia) and pregnancy (macrocytic anemia) and are usually due to simple folic acid deficiency, respond rapidly to treatment with folic acid, without vitamin B-12. It is possible that in some cases these anemias are due to an unknown metabolic defect in the production of the folacin enzymes.

2. **Leukemia.** *Aminopterin*, a folic acid antagonist or antivitamin, has been used in the treatment of leukemia.

Folic acid is involved in the normal synthesis of nucleic acid within the cell nucleus, which is responsible for cell growth. As an antagonist, aminopterin can fill in for folic acid, without activity. As a result, it is able to block the rapid production of leucocytes (white blood cells) characteristic of leukemia. Unfortunately, with continued use, the leukemic cells seem to develop a resistance to the antagonist and its effectiveness is lost.

3. **Cancer.** Methotrexate (amethopterin), a drug closely related to aminopterin, is currently being used in cancer chemotherapy. Its action is to bind the enzyme dihydrofolate reductase and thus inhibit the C_1 fixing function of folic acid, thereby preventing synthesis of DNA and purine in the cell.

4. **Sprue.** Folic acid is effective in the treatment of sprue, a gastrointestinal disease characterized by intestinal lesions, malabsorption of food, diarrhea, stools containing large amounts of fat, macrocytic anemia, and general malnutrition.

DEFICIENCY SYMPTOMS. Folacin deficiency in man may result in megaloblastic anemia (of infancy), also called macrocytic anemia (of pregnancy), in which the red blood cells are larger and fewer than normal, and also immature. The anemia is due to inadequate formation of nucleoproteins, causing failure of the megaloblast (young red blood cells) in the bone marrow to mature. The hemoglobin level is low because of the reduced number of red blood cells. Also, the white blood cell, blood platelet, and serum folate levels are low.

Other symptoms include a sore, red, smooth tongue (glossitis), disturbances of the intestinal tract (diarrhea), and poor growth. Also, recent observations suggest that there may be mental deterioration.

A deficiency of folacin may be caused by inadequate dietary intake, impaired absorption or utilization, or an unusual need (caused by increased losses or requirements) by the body's tissues.

The administration of folic acid to patients with megaloblastic anemia brings about dramatic recovery. But folacin is not a substitue for vitamin B-12 in the treatment of pernicious anemia; although it will alleviate the anemia, only vitamin B-12 will cure the neurologic symptoms.

Man, monkeys, chicks, turkeys, fox, and mink must have folacin supplied in the food in order to avoid deficiency symptoms. Rats, dogs, rabbits, and pigs can meet their need for this vitamin through bacterial synthesis in the intestine.

RECOMMENDED DAILY ALLOWANCE. The Food and Nutrition Board (FNB) of the National Research Council (NRC) recommended daily allowances of folacin (folate) are given in Table F-33. It should be noted, however, that stress, such as disease, increases the requirement.

TABLE F-33
RECOMMENDED DAILY
FOLACIN (FOLATE) ALLOWANCES[1]

Group	Age	Weight		Height		Folate
	(yr)	(lb)	*(kg)*	(in)	*(cm)*	(mcg)[2]
Infants	0.0–0.5	13	*6*	24	*60*	25
	0.5–1.0	20	*9*	28	*71*	35
Children	1–3	29	*13*	35	*90*	50
	4–6	44	*20*	44	*112*	75
	7–10	62	*28*	52	*132*	100
Males	11–14	99	*45*	62	*157*	150
	15–18	145	*66*	69	*176*	200
	19–24	160	*72*	70	*177*	200
	25–50	174	*79*	70	*176*	200
	51+	170	*77*	68	*173*	200
Females	11–14	101	*46*	62	*157*	150
	15–18	120	*55*	64	*163*	180
	19–24	128	*58*	65	*164*	180
	25–50	138	*63*	64	*163*	180
	51+	143	*65*	63	*160*	180
Pregnant						400
Lactating						280–260

[1]*Recommended Dietary Allowances*, 10th ed., 1989, NRC-National Academy of Sciences, p. 285.

2The RDA are expressed in terms of "total" folacin; that is, the amount of folic acid activity available from all food folates.

• **Infants and children**—The daily folacin requirements for (1) infants and (2) children have been estimated at 25 to 35 mcg and 50 to 100 mcg, respectively (see Table F-33). Since human and cow's milk contain approximately 2 to 3 mcg/100 ml and most of the folic acid is in an absorbable form, these needs can be met by a milk diet.

CAUTIONS: (1) Boiling destroys folacin in cow's milk, so infants receiving boiled formulas prepared from pasteurized, sterilized, or powdered cow's milk should receive additional folacin to assure an adequate intake; and (2) if the diet consists of goat's milk, folic acid supplementation should be given because of the low content and poor availability of folacin in goat's milk.

• **Adults**—The bulk of the evidence suggests that 25 to 50% of dietary folacin is nutritionally available. Thus, the RDA for adult males is set at 150 to 200 mcg, and for adult females it is 150 to 180 mcg (see Table F-33).

• **Pregnancy and lactation**—The added burden of pregnancy is known to increase the risk and incidence of folacin deficiency among populations with low or marginal intakes of the vitamin. The RDA for folacin is set at 400 mcg/day during pregnancy (see Table F-33).

The RDA allowance of folate during lactation is 280 mcg/day during the first 6 months, and 260 mcg/day during the second 6 months (see Table F-33).

• **Oral contraceptives**—Low tissue folate levels and macrocytic anemia have been found in women taking oral contraceptives. So, higher levels of folacin may be indicated under such circumstances.

• **Folacin intake in average U.S./Canadian diet**—The total folacin of mixed diets varies over a rather wide range. The average folate intake of both U.S. and Canadian populations is about 3 mcg/kg body weight.[20]

Average American intakes of folacin are well over the recommended daily allowances, although low intakes or low tissue levels are often seen in sprue, in infants of low birth weight, in infants on unsupplemented milk diets, in various disease states, in longtime hospitalized patients, in alcoholism, and in pregnancy. Also, it must be remembered that there are large losses of folacin from storage and cooking. So, deficiency of folacin is being found with increasing frequency.

TOXICITY. Normally, folic acid has no adverse effects. However, when it is used to treat megaloblastic anemia secondary to the use of antiepileptic drugs, the epilepsy may be aggravated.

[20]*Recommended Dietary Allowances* 10th ed., 1989, NRC-National Academy of Sciences, p. 153.

FOLACIN LOSSES DURING PROCESSING, COOKING, AND STORAGE. Although folic acid is present in most common foods, it is subject to storage and cooking losses.

Raw vegetables stored at room temperature for 2 to 3 days lose as much as 50 to 70% of their folate content. But storage in a refrigerator for as long as 2 weeks results in little or no loss of folacin.

Between 50 and 95% of food folate is destroyed by cooking or canning. The losses are greatest when high temperatures, long cooking periods, and large volumes of water are used. Most of the folic acid activity of milk is destroyed in processing dried milk.

Exposure to light also reduces the amount of folacin available in foods.

Foods high in vitamin C tend to lose less folacin than those low in vitamin C; vitamin C protects folates from oxidative destruction.

SOURCES OF FOLACIN. Information relative to the folacin activity of foods for humans is incomplete because of the difficulty of assaying the many different forms of folacin in terms of activity. Nevertheless, the general guides that follow will help ensure ample folacin in the daily diet.

• **Rich sources**—Liver and kidney.

• **Good sources**—Avocados, beans, beets, celery, chickpea, eggs, fish, green leafy vegetables (such as asparagus, broccoli, Brussels sprouts, cabbage, cauliflower, endive, lettuce, parsley, spinach, turnip greens), nuts, oranges, orange juice, soybeans, and whole wheat products.

• **Fair sources**—Bananas, brown rice, carrots, Cheddar cheese, cod, halibut, rice, sweet potatoes.

• **Poor sources**—Chicken, dried milk, milk, most fruits, muscle meats (beef, pork, lamb), products made from highly refined cereals (including white flour), and most root vegetables (including Irish potatoes).

• **No folacin**—Fats and oils, and sugar supply no folacin.

• **Supplemental sources**—Yeast, wheat germ, and commercially synthesized folic acid (pteroylglutamic acid, or PGA).

Intestinal bacteria synthesize folacin. Although this source may be important in man, the amount of folacin produced and absorbed has not been determined.

For additional sources and more precise values of folacin (folic acid), see Food Composition Table F-36 of this book.

NOTE WELL: Tables of folacin composition in foods, especially in the raw state, should be used with discretion because folacin compounds differ in their ability to be absorbed in the intestine depending on (1) the number of glutamic acid molecules in the folates, and (2) the nature of other types of binding—generally with protein.

Folacin deficiency may result from: (1) a shortage of the folacin itself; (2) very considerable storage, processing, and cooking losses; (3) the folacin not being absorbed in the intestine; and/or (4) substances that interfere with the normal function of the folacin enzymes. As a result, folacin deficiencies are thought to be a health problem in this country and throughout the world. Infants (especially those born prematurely, of low birth weight, or on unsupplemented milk diets), adolescents, and pregnant women seem particularly vulnerable. Tropical sprue, certain genetic disturbances, cancer, and parasitic infection increase the folate requirement. Chronic alcoholism is an important cause of folate deficiency; alcohol interferes with folate absorption and transportation from storage sites in the body. Oral contraceptives may also interfere with folate absorption.

It is noteworthy, too, that folacin fortification of corn, rice, and bread among population groups in South Africa has been shown to be an effective measure to prevent the development of macrocytic anemia in pregnancy.

So, dietary supplementation with folacin may be indicated more generally than has been thought necessary in the past.

CAUTION: Vitamin preparations without prescription containing more than 0.1 mg of folacin in a daily dose are prohibited by law. Excessive folacin can mask certain symptoms of pernicious anemia.

TOP FOOD SOURCES OF FOLACIN. The top food sources of folacin are listed in Table F-34.

NOTE WELL: This table lists (1) the top sources without regard to the amount normally eaten (left column), and (2) the top food sources (right column); and the caloric (energy) content of each food.

(Also see VITAMIN[S], Table V-9.)

TABLE F-34
TOP SOURCES OF FOLACIN (Folic Acid) AND THEIR CALORIC CONTENT[1]

Top Sources[2]	Folacin (Folic Acid)	Energy
	(mcg/100 g)	(kcal/100 g)
Yeast, baker's, dry (active)	4,090	282
Yeast, brewers', debittered	3,909	283
Yeast, torula	3,000	277
Liver, chicken, broiler/fryer, simmered	770	165
Wheat germ, toasted	420	391
Chicken, broiler/fryer, giblets, fried, flour coated	379	252
Wheat-soy blend (WSB)/straight grade wheat flour	329	365
Wheat germ, crude	328	391
Soybean flour[3]	318	356
Wheat bran	258	353
Eggs, chicken, dried, yolk	214	687
Eggs, chicken, dried, whole, stabilized	193	609
Spinach, raw	193	26
Rice, polish	192	265
Eggs, chicken, dried, whole	184	593
Spinach, frozen, leaf or chopped	153	24
Eggs, chicken, raw, yolks, fresh	152	369
Liver, beef, calf, hog, and lamb, cooked[4]	145	254
Endive (curly and escarole), raw	142	20
Soybeans, fermented natto	126	167

Top Food Sources	Folacin (Folic Acid)	Energy
	(mcg/100 g)	(kcal/100 g)
Liver, chicken, broiler/fryer, simmered	770	165
Chicken, broiler/fryer, giblets, fried, flour coated	379	252
Spinach, raw	193	26
Liver, beef, calf, hog, and lamb, cooked[4]	145	254
Endive (curly and escarole), raw	142	20
Parsley, raw	116	44
Spinach, frozen, chopped or leaf, boiled, drained	108	23
Peanuts, roasted/skins, whole and salted	106	582
Asparagus, green, canned, drained, liquid	102	11
Chickpeas, dry cooked, or canned	102	179
Almonds, unsalted and shelled	96	598
Turnip greens, raw	95	28
Brussels sprouts, frozen, cooked	88	33
Beets, red, fresh, boiled, drained	78	32
Cabbage, common, raw	66	24
Eggs, chicken, raw, whole, fresh	65	157
Beans, lima, baby, frozen, boiled, drained	65	118
Cauliflower, frozen	63	22
Broccoli, frozen, spears, cooked	60	26
Wheat flour, whole (from hard wheats)	54	361

[1]These listings are based on the data in Food Composition Table F-36 of this book. Some top or rich food sources may have been overlooked since some of the foods in Table F-36 lack values for folacin (folic acid).

Whenever possible, foods are on an "as used" basis, without regard to moisture content; hence, certain high-moisture foods may be disadvantaged when ranked on the basis of folacin content per 100 g (approximately 3½ oz) without regard to moisture content.

[2]Listed without regard to the amount normally eaten.

[3]Includes low fat, high fat, full fat and defatted.

[4]Beef and calf liver fried, lamb liver broiled, hog liver fried in margarine.

FOLIC ACID (FOLACIN)

The name folic acid was the forerunner of the term folacin. It was first suggested for the growth factor found in spinach and known to be widely distributed in green plants. The word *folic* is derived from the Latin word *folium*, meaning foliage or leaf.

(Also see FOLACIN.)

FOLINIC ACID ($C_{20}H_{23}N_7O_7$)

Folinic acid is an important metabolite of folic acid and may be the active form in cellular metabolism. Vitamin C and vitamin B-12 are essential for the conversion of folic acid to folinic acid.

FOOD AND AGRICULTURE ORGANIZATION OF THE UNITED NATIONS (FAO)

This is a specialized agency of the United Nations established by 44 countries in 1945 in Quebec, Canada, partially in response to the awareness of world nutrition created by the War. It now has more than 120 member countries. Headquarters are in Rome, Italy, with a North American office in Washington, D.C. Major objective of the FAO are (1) elimination of hunger, and (2) betterment of nutrition by improving the production, distribution, and use of food and other products of farms, forest, and fisheries throughout the world. Activities of the FAO include conducting research, providing technical assistance, conducting educational programs, maintaining statistics on world food, publishing reports, and working closely with the World Health Organization (WHO) all over the world. The FAO/WHO Expert Committee on nutrition has produced nine reports which recommend and encourage work on the assessment of nutritional status, standards of requirement for nutrients, protein-energy malnutrition, nutritional anemia, endemic goiter, xerophthalmia, food technology, and toxicology.

(Also see DISEASES, section headed "International and Voluntary Organizations Engaged in Health and/or Nutrition Activities"; NUTRITION EDUCATION, section headed "Nutrition Education Sources"; and WORLD HEALTH ORGANIZATION.)

FOOD AND DRUG ADMINISTRATION (FDA)

This is the federal agency in the Department of Health and Human Services that is charged with the responsibility of safeguarding American consumers against injury, unsanitary food and fraud. Specifically, the FDA enforces (1) the Federal Food, Drug, and Cosmetic Act with its numerous amendments, (2) the Fair Packaging and Labeling Act, (3) those sections of the Public Health Service Act relating to biological products, and (4) the Radiation Control for Health and Safety Act. Also, the FDA protects industry against unscrupulous competition; and it inspects and analyzes samples, and conducts independent research on such things as toxicity (using laboratory animals), disappearance curves for pesticides, and the long-range effects of drugs.

The FDA began in 1927 as the Food, Drug, and Insecticide Administration. Through the years, it has established rigorous testing policies requiring proof relative to product safety and efficacy. The requirements for new product approval by the FDA have become more rigid, and enforcement has become more unrelenting. In 1962, a new drug could be developed, tested, and approved for a cost of about $1 million—all in a period of 1 to 2 years. Currently (in the 1990s), development to approval of a new drug takes 5 to 6 years and costs $10 to $12 million. It is understandable, therefore, why manufacturers resist the threat of a ban on a product in which they have such an investment.

(Also see DELANEY CLAUSE; DISEASES, section headed "U.S. Department of Health, and Human Services"; and FEDERAL FOOD, DRUG, AND COSMETIC ACT.)

FOODBORNE DISEASE

Stated simply, any disease that is transmitted through food is known as a foodborne disease. Because eating is an activity in which all people engage, the risk of obtaining a disease-causing agent from contaminated food is extremely high. Applying a rather broad definition to disease, a foodborne disease may be caused by (1) contamination of food by bacteria, parasites, viruses, fungi and molds, (2) some naturally occurring toxicant, (3) contamination of food by toxic industrial chemicals or radioactive waste, or (4) food allergies. More specifically, foodborne diseases include the familiar examples of thyphoid, tapeworm, botulism, brucellosis, tuberculosis, salmonella, hepatitis, and "tourist diarrhea"—bacterial, parasitic and viral infections—to name only a few. The greatest deterrent to foodborne diseases is scrupulous attention to cleanliness from the production to the consumption of food, since foodborne diseases usually enter the food chain via (1) infected animals or plants, (2) organisms transmitted by flies, roaches or rodents, (3) contact with sewage-polluted water, or (4) food handlers who lack good personal hygiene and/or fail to follow acceptable food handling practices.

(Also see ALLERGIES; BACTERIA IN FOOD; DISEASES; POISONS; POISONOUS PLANTS; and RADIOACTIVE FALLOUT.)

FOOD—BUYING, PREPARING, COOKING, AND SERVING

Contents *Page*

Meal Planning 787
The Kitchen and Equipment 788
Food Buying 789
Food Storing 790
Food Properties 790
Water & Solutions 791
Carbohydrates 791
Fats & Emulsions 792
Proteins 793
Minerals 796
Vitamins 796
Enzymes 797
Color 797
Flavor 799
Texture 799
Bacteria 799
Cooking 799
Reasons for Cooking 800
Methods and Media of Cooking 800
Mealtime 801

Legally (according to the Federal Food, Drug, and Cosmetic Act, as Amended, Sec. 201 [f]), *the term food means (1) articles used for food or drink for man or other animals, (2) chewing gum, and (3) articles used for components by any such article.*

Because each step in food preparation is important to the nutritive value of the finished product, consideration should be given to what happens to foods from the time the consumer purchases them in the marketplace to the time they reach the dinner table.

With the development of processed, ready-to-serve products, the consumer is now questioning the nutritive value of these foods. Originally, the food manufacturer was concerned only with making the tastiest, most appealing product possible. The profit potential was his main objective, but with a general awakening of the public to the nutritive value of such products, the emphasis has been shifting to making the product more nutritional. Manufacturers are looking twice at some of the food additives, because the news media have publicized the fact that there are many chemicals in our foods. It matters little whether some of these chemicals exist in natural foods; they are suspect unless the public is educated and reassured that they are not harmful.

The discussion in this section will deal primarily with the retention of nutrients as they go from raw products to finished in the average home kitchen. In most instances, the correct preparation and cooking not only produces a better product nutritionally, but the psychological advantages to a nicely prepared meal are immeasureable.

(Also see ADDITIVES.)

MEAL PLANNING. Planning the menu is perhaps the most important step in putting food on the table. Once the meal is planned, the purchasing, storing, and preparing of food follow in a more or less automatic sequence. Unfortunately, many shoppers go to the grocery store without a "blueprint" and end up purchasing either more than they need, or not the right things. A menu does not have to be rigid, but it can be used as a guide.

If the marketing is done daily, meals can be planned one day at a time. But, if the marketing is done on a weekly basis, about 5 days' meals should be planned, allowing a little leeway for meals eaten outside the home and for leftovers.

The following goals should be considered in meal planning:

1. **Balancing the meal nutritionally for all age groups by using the Six Food Groups.** The simplest way to plan any meal is to:

   a. Select the entree—the meat, fish, poultry, eggs, legumes, or nuts. This becomes the center of the plan, and the other foods are grouped around it.
   b. Add some milk and milk products. There are an endless number of ways to do this—milk to drink, a white sauce, cheese in a casserole, yogurt as a salad dressing, ice cream or a milk dessert.
   c. Add vegetables, either raw or cooked, alone or in a salad. Fruits can be included in salads or desserts.
   d. Include cereals as cooked cereal, breads, or desserts.

Meals planned for a growing family will have more than one dish from each of these food groups, but for those trying to lose weight, or for older adults, one selection from each group is adequate for a meal.

(Also see FOOD GROUPS.)

2. **Keeping within the budget.** If money is no problem, then this point can be skipped over, but for the vast majority this is important. Remember that the bottom line in saving money on food purchases is that—*the more preparation that goes into a product, the more costly that product is likely to be.* It is best to purchase only raw products and prepare them yourself. If you have any doubts about this point, check the price of fresh mushrooms as compared to canned. The canned mushrooms are usually 2 to 3 times more expensive. Above all else, never, never throw out any food, for that is the same as throwing away money. There will never be enough money if you throw away food.

If you are really serious about saving food pennies, you can conduct little studies like this one: Based on \$1.20/dozen for jumbo eggs and \$2.50/lb for hamburger, 20 g of protein in eggs (2.5 eggs) costs 25¢ but 20 g of protein in the hamburger (¼ lb) costs 62.5¢. Combining a cereal with a legume (i.e., peas, beans, lentils), to give a complete protein is perhaps the cheapest source of protein on the market.

Another way to save money is to use the fruits and vegetables that are in season. There is a wide range in the price between the first product of the season, and later—when it is at the height of the season; so, watch for bargains.

Buy in quantity when a food is on sale, but remember a bargain is not a bargain if any of it has to be discarded. Make certain that either the family can eat all of it, or the surplus can be stored for later use.

3. **Balancing the flavors, colors, and textures.** This is a tall order, but not impossible. Plan each meal individually, as you would for a company meal. Make certain that in any one meal you neither have too many highly seasoned foods nor all bland dishes. Also, bear in mind that some people enjoy sweet and sour together, whereas others like hot and cold.

Color is important to the color conscious. Most cooks do not have difficulty with this point. But avoid such a meal as creamed chicken, white potatoes, and white turnips, with vanilla ice cream for dessert. After the menus are planned, check them and picture the plate for each meal. Vegetables and fruits are the most colorful foods. Perhaps the only caution would be to avoid some of the reds and oranges together on one plate; e.g., red tomatoes and beets, or grated carrot and red cabbage in a salad together.

Texture combinations should be considered. When the entree is smooth with little chewing required, there should be a crunchy salad, or a vegetable with nuts, or crispy Chinese noodles. Many people enjoy the contrast of smooth dishes with ones that contain something firm or crisp.

Fig. F-38. Party food at any time lifts the spirits and adds to the enjoyment of life. (Courtesy, United Dairy Industry Assn., Rosemont, Ill.)

4. **Managing the available materials, time, and energy efficiently.** This can be tied into point no. 2—the budget, because some of these pointers can also save money.

a. When planning a dish which has to be baked in the oven, plan to have more than one dish that can be baked at the same time; for instance, a casserole, baked potatoes, and rolls can all be baked together.

b. When it is necessary to be gone at meal preparation time, plan a meal that can be prepared ahead of time, or one that can cook slowly while you are gone, or one that can be prepared in short order.

5. **Considering the family preferences.** This point is listed last because if every like and dislike of each family member is catered to, then point no. 1, a balanced meal, will be virtually impossible to achieve. As much as possible, the family should be encouraged to enjoy all foods, because of the golden rule of eating, which is: *The greater the variety of foods that you eat, the greater your chances of getting everything you need.*

THE KITCHEN AND EQUIPMENT. Before the food is purchased at the market, the kitchen and equipment should be considered. It is not necessary to have fancy equipment, but the following are rather basic: stove, refrigerator, sink, mixing bowls and spoons, measuring cups and spoons, pans, and baking dishes.

Today, the available kitchen gadgets literally bogle the mind. There are electrical mixers, which stir, mix, or whip; blenders, which blend, puree, liquefy, or frappe; and food processors which grate, slice, chop, or shred. Everything is "coming up electrical." There are electrical broilers, carving knives, can openers, deep fryers, frying pans, juicers, popcorn poppers, roasters, toasters, waffle irons, and woks.

No matter how much or how little equipment, it should be conveniently arranged in the kitchen, for ease of preparation. There are as many different kitchen arrangements as there are houses in the world. Sometimes, an inconvenient arrangement can be made more workable simply by having a movable (chopping block) table; at other times, it is a case of rearranging the storage areas. For example, when making coffee: Is it necessary to open two cupboards or just one? It is easy to plan a center for baking; a center for sandwich or salad making; and a center for the dishes—which is convenient either for setting the table and/or for putting the dishes away after washing them. If possible, the work centers should follow a logical sequence from bringing the groceries into the kitchen, to storing in refrigerator or pantry; to sink and preparation center; to stove and/or oven; to serving onto the plates; to dinner table; and, finally, from clearing the table back to the sink, thence to the storage cupboard. These activities should follow in sequence around the room and not be crisscrossing the room.

By analyzing the work procedures, it may be possible to come up with some minor changes that will make the work area more convenient, thereby facilitating preparing, cooking, and serving. Even when the opportunity is presented to remodel or build a new kitchen, it takes considerable planning to evolve with the most convenient kitchen because different cooks have different cooking needs. Every step in the process of preparing the meal must be studied.

FOOD BUYING. Food markets have come a long way since the advent of railways, motor freight, and air freight. Years ago, most of the food available was produced locally. The farmers brought their produce to town by horse and wagon, and the customers collected at the wagons to purchase their daily food. These were the first peddlers. Village fairs and town markets evolved when several peddlers banded together for the convenience of the buyer. In those days, there was no refrigeration, so shopping for food was a daily process for the city dweller, as it still is in many parts of the world.

Fig. F-40. Alpha Beta's first self-service *Grocerteria*, in 1915. (Courtesy, Alpha Beta Company, La Habra, Calif.)

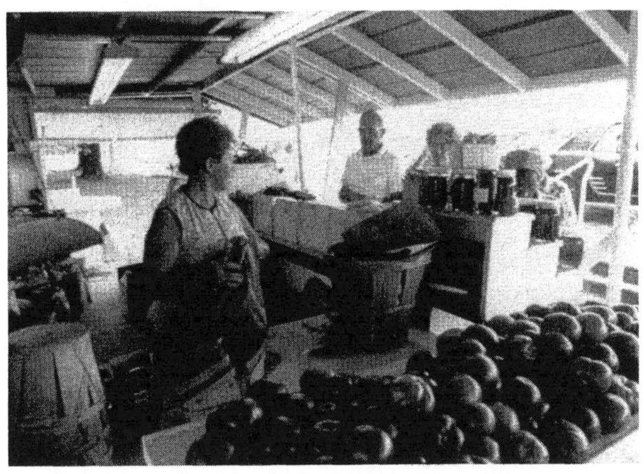

Fig. F-39. Roadside market. (Courtesy, USDA)

By the beginning of the 20th century, the local grocery store was displaying items from many parts of the country. The proprietor was a well-known figure in the community, and a personal friend to all of his customers. However, as cities grew in size, it became increasingly difficult to service all of the grocery store customers with such personalized service. A new concept in marketing came about with the first Alpha Beta store which opened in Pomona, California in 1915. The foods were arranged in alphabetical order (hence its name), so that the customer could find each item easily; and, by the customer serving herself, the proprietor could sell to many more customers at one time.

Although the grocery store was developing in this direction, the custom of the local growers coming to town with their fresh produce persisted. Even today, many cities still have a farmers' market.

Fig. F-41. Farmers' market. (Courtesy, University of Minnesota, St. Paul, Minn.)

Today, the food customer has a wide selection, almost year round. In order to secure maximum nutritive qualities, it is important that perishable foods be as fresh as possible. The produce manager can usually tell from whence most of his fresh foods come; and from this, he can judge how much time has elapsed between picking and marketing.

Fig. F-42. A modern supermarket, with its attractive display of everything from soup to nuts, plus tools, detergents, film, and many other items of both software and hardware. With its electronic reading of the labels, it is a high-speed marketing "machine". (Courtesy, Alpha Beta Company, La Habra, Calif.)

Fig. F-43. Refrigeration revolutionized the marketing of foods and the consumer's purchasing pattern. No longer was it necessary that food be supplied to the grocer, sold to the consumer, and eaten by the family—all in 1 day. The shelf life of perishable foods was extended dramatically. (Courtesy, The Bettman Archive, Inc., New York, N.Y.)

FOOD STORING. The goal in food storage is to prevent food spoilage from microbial decompositions, enzymatic or nonenzymatic chemical changes, or other losses such as may be caused by insects and/or rodents.

Because most spoilage occurs in the presence of moisture and at temperatures between 68° to 131°F (*20° to 55°C*), foods are preserved, both commercially and in the kitchen, by alleviating these conditions. The most popular methods of preservation are canning, freezing, freeze drying, drying, or pickling.

After purchasing the freshest foods available, it is important to get them into refrigeration as soon as possible. As a general rule of thumb, fresh vegetables and fruits can be stored under refrigeration for 1 week. Before storing, to wash or not to wash fruits and vegetables is the question? Possibly the best rule of thumb is to wash those fruits and vegetables that are to be used in salads. Until they are to be used, it is not necessary to wash those fruits and vegetables that eventually will be peeled or prepared in some other way for cooking.

Fresh meats can be stored for only 4 days on the average. If a freezer is not available to freeze part of the week's supply, canned products, dehydrated meats, eggs, cheese, beans, or nuts, can be used for the last meals of the week before the weekly shopping trip for groceries. For those who can get to the grocery store easily, this presents no problem. But for those who live some distance from the store, it requires some planned strategy when a freezer is not available.

Eggs and milk can be kept under refrigeration for about 10 days, without any undue spoilage. The spoilage time depends a great deal on the condition of the item when the consumer purchases it.

Frozen foods that are allowed to thaw and then are refrozen usually lose considerable in quality, and possibly some in nutritive value. In addition, it means that the food cannot be held as long in the freezer. When freezing fresh, unfrozen foods, divide into portions that will be needed, so that it is not necessary to thaw the whole piece in order to use a portion of it. The common rule of thumb is to freeze foods only twice. This means that the food can be thawed, prepared and refrozen until it is needed. Or it can be thawed, prepared and served, and any large amount left over can be refrozen to be used at a later date. Foods cannot be held as long if frozen twice as they can if frozen only once.

Cereals and baked goods should be stored in airtight containers. If they contain whole grains or wheat germ, they should be refrigerated or frozen to prevent deterioration and rancidity.

When keeping leftovers in the refrigerator, there are two very important points to remember:

1. Refrigerate the leftover as quickly as possible. Do not let it cool to room temperature first.
2. Store it in an airtight container so that it does not dry out.

(Also see MEAT[S], section headed "Meat Preservation, Table M-20 Storage Time for Refrigerated and Frozen Meats"; and MILK AND MILK PRODUCTS, section headed "Care of Milk in the Home," Table M-64 Approximate Storage Life of Milk Products at Specific Temperatures.)

FOOD PROPERTIES. Before we plunge into the mixing bowl, some of the basic properties of foods and their ingredients should be understood.

Water And Solutions. There is no substitute for water. It is as indispensable to cooking foods as a match is to lighting a fire. Water fills a multifaceted role. Here are some of its properties and uses:

1. **Water as a method of cooking.** Water is used for simmering foods such as soup bones, less tender cuts of meats, dried and fresh fruits and vegetables, starches and cereals. However, the more water and the greater the cooking time, the more nutritive value that is lost. This fact makes it important to utilize the liquid left from this method of cookery, with the exception of cured meats. Fortunately, cured meats will lose some of their high salt content through this method of cooking, which is a definite advantage.

2. **Water as an absorbable agent.** Foods that have been dried, such as beans, peas, lentils, fruits, soups, yeast, etc. can be restored to their original condition by soaking and/or cooking in water. Even stale dried bread or rolls can be revived by the addition of water (which they have lost by dehydration), followed by heating. Broadly speaking, these foods are *hydrates* (they take up water) and this process is called *hydration*.

Starches also have the ability, when mixed with water and cooked, to swell and thicken a liquid—gravies, sauces, and soups.

3. **Water as a part of certain chemical changes.** When water is added to dry baking powder, carbon dioxide—the gas that leavens quick breads—is released. The change is accelerated when heat is applied.

4. **Water in foods may cause food spoilage.** Foods containing water will develop molds and other bacterial growths if held at temperatures which favor such growths. That is why, in the prevention of food spoilage, either the moisture should be removed or the food should be stored at a temperature which discourages the growth of molds and bacteria.

5. **Water as a solvent.** Water dissolves flavorings, sugar, and salt. This occurs when making teas, coffee, and sugar drinks. Salt is also dissolved in water to produce a brine, which is used to preserve meats and vegetables. Unfortunately, water also dissolves some minerals and vitamins, either when food is soaked in water or during cooking. This means that fresh vegetables should not be soaked in water for any great length of time. Carrots, celery, and radishes are sometimes soaked in water in order to produce fancy curls, sticks, and fan shapes for relish dishes, but this procedure is not recommended.

Solutions are constantly found in cookery. A solution involves the *solvent* (usually water) in which the *solute* (sugar, salt, etc.) is dissolved. Solutions are completely homogeneous. They cannot be separated unless the solvent evaporates or the temperature of the solution changes. The most common solutions are liquids in which gases, other liquids, or solids are dissolved. Solutions can be dilute or concentrated. Most solutes will dissolve in greater quantities at higher temperatures. The solubility of gases in liquids is the opposite; for instance, carbonated beverages contain more carbon dioxide when cold than when warm. Also, the greater the pressure, the more gas will be dissolved in the liquid. Once the cap is removed and the pressure released, the gas will escape. In addition, shaking decreases the solubility of the gas.

The freezing point of a solution is lower than that of the pure solvent in proportion to the molecular weight concentration of the solute. A solution containing a solid dissolved in a liquid has a reduced rate of evaporation—in other words, a syrup, when cooked, will decrease in volume more slowly than an equal volume of water.

A solution can be separated into its two components with a nonvolatile solvent by a process known as *distillation*. If the solution is allowed to evaporate slowly, the solute will form crystals, with each substance having its own distinctive crystal shape.

Carbohydrates. Plants are composed of 60 to 90% carbohydrates. For the detailed classification and nutritional aspects of these organic foodstuffs, the reader is referred to the section on CARBOHYDRATES.

Monosaccharides (glucose, fructose, galactose) and disaccharides (sucrose, lactose, and maltose) are crystalline, water-soluble, sweet compounds, which are used for sweetening foods. They are found in fruits, honey, milk, maple syrup, sugarcane, and sugar beets, and are digestible without cooking.

The polysaccharides (starches, dextrins, and cellulose) are noncrystalline, and have little taste. The polysaccharide, starch, is used extensively as a thickening agent. Cereal grains are composed mainly of starch, but unlike the sugars, starch is more digestible when cooked. During the cooking process, the starch granules swell and become a soft paste. This is called gelatinization. A starch mixture starts to thicken at 165° to 190°F (*73.9° to 87.8°C*), but complete gelatinization does not occur until the temperature approaches the boiling point. When using starch for thickening, it is important to remember a few basic rules. Starch added to a hot liquid will form lumps, which can be removed by putting through a sieve or by whirling in the blender. It is better, however, to mix the starch with a cold liquid or fat before adding to the hot liquid. Also, it is important that there be sufficient liquid and heat to obtain the maximum thickening from the starch, otherwise only partial gelatinization occurs.

Starches also change when exposed to dry heat. This process is known as dextrinization. When these dextrins are dissolved in water they have a sweet taste. Slow toasting of cereals will change a portion of the starch to dextrin, and when vegetables are browned slightly they will have the same distinctive flavor. This effect can also be observed in the crusts of baked goods, or when flour is browned when making gravy.

The most important carbohydrates found in the diet are, of course, the cereal grains. Vegetables contain starch, but the vast majority of them are not considered as primary carbohydrate sources. (For an expanded discussion of cereal grains, the reader is referred to the section on CEREAL GRAINS.)

In this article, we will cover a few helpful hints on cereal cookery.

• **Yeast breads and rolls**—In making yeast breads and rolls, the following procedures make better products:

1. Mix the yeast with a lukewarm liquid. If the liquid is too hot, the yeast will be killed and the dough will never rise.

2. When adding the flour, add only half of it, then stir thoroughly. This allows the starch granules to soak up moisture, which results in less flour being required to make the dough workable, and, in turn, this makes for a lighter final product.

3. When kneading the dough, it is important to develop the gluten thoroughly, for this gives the bread its form as the gas produced by the yeast slowly raises the dough.

4. When placing the pans of bread to rise, make sure they are in a warm spot and covered, to prevent the top from forming a dry crust, which inhibits expansion of the dough. A good place is in the oven. Warm the oven, ever so slightly, put in the rolls or pans of bread and close the door.

• **Quick breads**—These are quite different from yeast breads. In making quick breads, the secret to success is to mix all ingredients and keep the stirring and beating to a minimum. The more stirring, the less tender will be the muffins, biscuits, etc.

• **Cream puffs**—The method of preparing *cream puffs* is quite different from that of yeast breads or quick breads. The water and butter are heated to boiling and the flour is stirred in until it forms a ball. The ball stage is the only possible point of failure; at this particular point, the ball of dough must *cool* before beating in the eggs. Otherwise, eclairs and cream puffs are very easy to prepare.

Fats And Emulsions. A fat consists of one molecule of glycerol and three molecules of fatty acids. Technically, they can be called *triglycerides*. Fats and oils are the same substance, but a fat is solid at room temperature whereas an oil is a liquid.

Fats and oils have many uses in food preparation. They add flavor and nutritive value to the foods; they prevent foods from sticking to pans; they tenderize batters and doughs; they hold in air which is incorporated when beating a mixture; and they form emulsions in such foods as mayonnaise and Hollandaise sauce.

The choice of fats and/or oils is a personal one, for each cook and each family has its favorites. Whatever fats are used, care should be taken to keep them fresh and to prevent rancidity. The best method is to keep them cold, and exclude light and air. For the same reason, foods that have a high fat content should also be refrigerated; this includes whole cereal products, nuts, fat-rich biscuits and crackers.

Fig. F-44. There is a wide variety of fats and oils on the market. (Courtesy, American Soybean Association, St. Louis, Mo.)

There are two methods of frying foods in fat. One is by using a shallow pan with a small amount of fat; the other is by using a large kettle filled with the heated fat, into which the food to be cooked is dropped. It is easy to control the temperature in the deep pan by using a thermometer, but the shallow pan must be temperature controlled to prevent smoking of the fat. Allowing the fat or oil to smoke will cause deterioration of the oil.

If the fat is not allowed to overheat, it can be used successfully several times. When cooking in deep fat, the temperature of the fat is important. At too low a temperature, the food will absorb too much fat; at too high a temperature, not only will the oil smoke, but the food will be browned on the outside before the inside has had a chance to cook. Thus, it is very important to follow the recommended temperature when frying foods.

An emulsion can be defined as a liquid dispersed in another liquid, without either liquid being dissolved in the other. Most emulsions contain an oil and a liquid, such as water or vinegar. Some emulsions are more permanent than others. For instance, a French dressing is a mixture of oil and vinegar, plus seasonings. When shaken together, an emulsion is formed, but on standing the oil particles reunite; thus, this is a temporary emulsion. Mayonnaise, on the other hand, is also a mixture of oil and vinegar, but it contains egg, which is an emulsifying agent. An emulsifying agent prevents the oil from reuniting. Other common emulsifying agents are: casein, gelatin, mustard, paprika, pectin, and starch.

The most important single factor in producing a stable permanent emulsion is to mix the ingredients together very slowly.

Emulsions can be broken by freezing, by storing at too high temperatures, by agitation, or by storing in an open container.

(Also see FATS AND OTHER LIPIDS.)

Proteins. Protein, *per se*, is covered under the section entitled PROTEIN; hence, the reader is referred thereto for information on proteins other than cooking. In this section, the basic rules for cooking protein foods—meat, fish, poultry, eggs, milk, and cheese—will be covered.

• **Preparing and cooking meat**—The lean part of meat consists of bundles of muscle fibers held together with connective tissue. When the fibers and the bundles are small with minimum connective tissue, the meat is tender. This is usually found in the muscles that are not used to any extent by the animal, which also explains why younger animals produce more tender meat than older animals.

Tender cuts can be broiled, but the less tender cuts must be stewed or braised. Regardless of the method of cookery, it is a well-known fact that either cooking at too high temperature or overcooking will make the meat tougher, even the more tender cuts.

Fig. F-45. One of the best ways to avoid under- or overcooking meat is by using a thermometer. (Courtesy, National Live Stock and Meat Board, Chicago, Ill.)

Because of the lower cost of the less tender cuts, attempts have been made to change the less tender cuts into more tender meat. One method is the use of one or all three enzymes (papin, bromelin, and ficin) that tenderize meat. Because they work during the cooking process, it is important to get as much of these into the meat as possible. The meat tenderizer powder can be pounded into the meat, and/or the meat can be marinated for about 48 hours in a mixture of pineapple juice and soy sauce. These and other methods of tenderizing meats are given in the article on MEAT, section headed "Meat Tenderizing."

(Also see MEAT[S], section headed "Meat Cooking.")

• **Preparing and cooking fish (finfish or shellfish)**—Fish is a highly perishable and very delicate product. It cannot be stored long; thus, the fresher the better. It is best when caught and cooked the same day, which illustrates the urgent need to cook it as soon as possible (within 2 days). If it is necessary to hold fish, it should be frozen or canned.

Fish is a tender product which can be cooked by either dry or moist heat methods. Because fish is bland in flavor, the dry-heat method is preferred as it tends to concentrate the flavor. Large fish are baked whole, but filets and steaks can be breaded and either fried or baked. Fish cooked by moist heat is usually served with a sauce to give it added flavor.

It is important to cook fish completely, which means it should flake with a fork. Raw fish is not only unpalatable, but it may be unsafe (fish from some localities may be a source of tapeworm). Also, it is important (1) not to cook it at too high a temperature, (2) not to overcook it, and (3) to handle it gently in order to retain its form.

(Also see FISH AND SEAFOOD[S].)

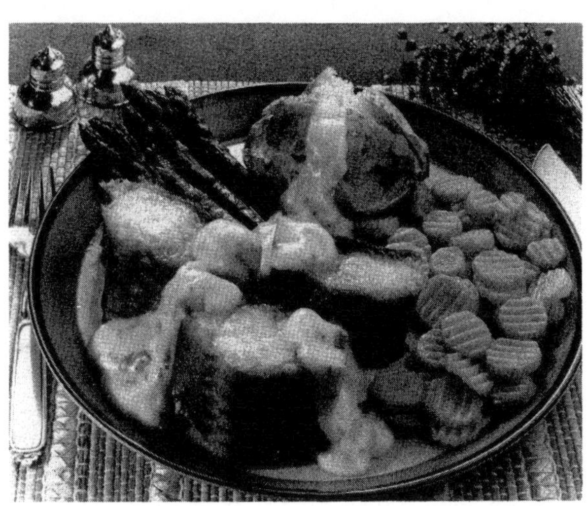

Fig. F-46. Baked cod steaks—a meal fit for a king. (Courtesy, Department of Fisheries and Oceans, Ottawa, Canada)

• **Preparing and cooking poultry**—Poultry is probably the most widely used protein food in the world, with few, if any, social or religious restrictions attached to it.

Chicken isn't just for Sunday anymore. Except for people who live on farms, poultry was formerly considered a delicacy, eaten only on festive occasions. Today, it is usually the cheapest form of muscle meat available—thanks to improved technology and production methods.

The same rules that applied to meat cookery, apply also to poultry. These are:

1. Intense heat will toughen the protein and cause considerable shrinkage and loss of juice.

2. The more tender, younger birds can be cooked by dry heat methods.

3. The less tender birds should be cooked by moist heat.

4. When poultry is to be creamed, used in soups, or used in casseroles, there is no advantage to roasting the bird over moist-heat cooking—a point which casserole lovers should keep in mind. In fact, stewed older chickens are superior in flavor to younger birds.

(Also see POULTRY.)

Fig. F-47. Stuffed roast turkey is a dish for all seasons. (Courtesy, National Turkey Federation, Mt. Morris, Ill.)

• **Preparing and cooking eggs**—Eggs are the most versatile protein product in the diet, and certainly the cheapest source of high-quality protein. Not only are eggs cooked and eaten as eggs, but they adapt to a multitude of other uses in various dishes; among them, the following:

1. As an emulsifying agent in mayonnaise and ice cream.
2. As a thickening agent in custard sauces.
3. As a gel in baked custards.
4. As a coating agent for fish, pork and veal chops, vegetables, and croquettes, etc.
5. As a leavening in many baked products such as pancakes, muffins, cakes (especially sponge and angel food cakes), cream puffs, and popovers.
6. As a binding agent in meat loaves and croquettes.
7. As a foam for omelets, souffles, meringues, beaten cake frostings, marshmallows, divinity and nougat candy.
8. As a clarifying agent in broths or coffee.
9. As a decorative agent—hard-boiled eggs are used to decorate many buffet dishes, either as slices, or as deviled eggs.

When cooking eggs, the same basic principle for all protein dishes applies. They should be cooked at low to medium temperatures; high temperatures produce a hard, leathery product. The whites coagulate at a lower temperature than the yolks—a point to remember when making custard sauces.

Egg white cookery requires a little skill; hence, it is important to know some of the reasons behind the production of perfect, and other than perfect, results. Meringues can "weep"; souffles will flop; popovers won't pop; and cream puffs won't puff unless certain basic rules are followed.

Hard boiled eggs are perhaps one of the most difficult procedures that a novice cook has to learn in order to avoid "green" eggs. A layer of green around the yolk is the result of the hydrogen sulfide in the white combining with the iron in the yolk. What should you do? Should you use cold or hot water; put the pan lid on or take it off; leave the pan on the element or take it off? Whichever method is used, the single most important factor is not to overcook or overheat the eggs. They can be put in either cold or hot water. They can be put on a high heat or a medium heat, as long as they are cooked at a temperature just below the boiling point. The lid can be on or off. If the pan is left on the hot element, it is best not to put the lid on. If the pan is removed from the element, the lid should be put on. The eggs should not stand in the hot water longer than 15 minutes, then they should be chilled under running cold water. The temperature and size of the eggs are both involved, but with a little experimentation, beautiful hard boiled eggs can be produced.

Fried eggs should be fried at a low temperature so that they will not become tough and overly greasy. Here again, a little experience is all that is needed.

Foamy omelets are trickier to produce than souffles. The whites and yolks are beaten separately. The liquid and other ingredients are added to the yolk. Then the yolk mixture is gently folded into the whites. This is poured into a greased pan, which should be hot enough to start coagulation of the egg, but not hot enough to make them tough or to brown them. The top of the omelet can be cooked by one of several methods:

1. Cover the pan, but do not let the top of the omelet touch the lid.
2. When the bottom is partially cooked, the pan can be popped into a 325°F (*180°C*) oven.
3. When the bottom is partially cooked, the pan can be put under the broiler, but with extreme caution, as it

is easy to make the top tough, and the whole product collapse.

4. The foamy omelet can be cooked completely in a 325°F (*180°C*) oven for the whole time.

Souffles are easier to produce than foamy omelets, because of the base of white sauce. They are baked for the entire time in the oven, and, like baked custards, the baking dish should be set in a pan of hot water to prevent overheating and a poor product.

Custards consist only of eggs, milk, sugar and flavoring. There is no starch. A baked custard should be firm enough to stand if it is unmolded. When baking custard, it is important to put the custard cups into a pan and add water to prevent overcooking. Too hot an oven produces bubbles of air and an undesirable texture.

Fig. F-48. Baked custard—a dessert for all ages. (Courtesy, American Egg Board, Park Ridge, Ill.)

Custard sauce should be of the consistency of thick cream, and should be cooked in a double boiler to prevent overcooking and scorching on the bottom of the pan.

Meringues are impressive and look difficult, but they only require some basic knowledge and a little experimentation. Here are some cogent facts:

1. Thin egg whites (usually from older eggs) produce more volume, but the meringue is less stable.
2. Egg whites at room temperature produce more volume, but here again the meringue is less stable.
3. Bowls with small rounded bottoms and straight sides give better meringues when using an electric mixer, because the beater can pick up more of the mass, but bowls with sloping sides are better when using a wire egg beater.
4. Underbeating and overbeating both produce meringues that will break down and "weep." Meringues should be beaten until just the tips of the peaks fall over and the top shines like satin.
5. Adding any oil (egg yolk, milk, oil, cream, etc.) will prevent the formation of a perfect meringue.
6. Adding salt will inhibit a good meringue.
7. Adding ¼ tsp (*1.25 ml*) cream of tarter aids in the stability of the meringue.
8. Adding sugar to egg whites prolongs the time it requires to whip up a good meringue, but once the sugar is added, it is almost impossible to overbeat it.

Soft meringues for pie toppings require three things to avoid either a tough meringue, or one that "weeps" (little beads of moisture on top).

1. The meringue should be properly beaten.
2. The meringue must be placed on a *hot* pie and spread to the edges.
3. The meringue requires the correct baking temperature—a hot oven of 400°F(*222°C*) requires only 4 to 5 minutes; a slower oven of 375°F (*209°C*) requires 10 minutes, and an oven of 350°F (*195°C*) requires 15 minutes. Above all, the meringue should not be overcooked because it will result in a tough meringue, and it will "weep."

Hard meringues which are used for desserts, along with fruits, ice cream, sauces, etc., are a different story. This type of meringue is baked in a very slow oven at about 250°F (*139°C*); and, after the heat is turned off, it is left in the oven to dry until it is cool. If this type of meringue is gummy, it is the result of one or more of the following: too much sugar, underbaking, too rapid baking, or baking too long.

(Also see EGG[S].)

- **Using milk in cookery**—Milk, like water, has no adequate substitute. In some cases, the two can be used interchangeably in a recipe. Nutritionally, milk can substitute for water, but water could never substitute for milk. Most milk is consumed in the fluid state, but it adapts to a myriad of uses in food preparation. Because of its excellent nutritive value, it should be included in the diet in some form. Many adults drift away from their childhood habit of drinking a glass of milk with each meal, but, fortunately, milk can be added to many dishes and can substitute for water in some recipes. Here are a few of the many uses of milk and milk products (e.g., canned evaporated and condensed milk, and dry milk solids)—

1. **For beverages.** Milk may be used in malted milk shakes, eggnogs, and hot or cold drinks.

2. **For sauces.** Milk may be included in soups, gravies, white sauce, cream sauces, or dessert sauces.

3. **For entrees.** Milk can be added to meat loaf, casseroles, croquettes, pasta dishes, curry dishes, etc.

4. **For many bakery products.** Milk may be used as a moistening agent instead of water in many bakery products.

5. **For desserts.** Milk may be used in making many desserts, such as custard, junket, bread pudding, rice pudding, parfaits, cornstarch puddings, and ice creams.

Because milk is a high-protein food, the same principles apply when cooking with milk as with other high-protein foods. Both flavor and texture are adversely affected by too high or too prolonged cooking temperatures.

Milk tends to form a coating on the bottom of the pan and scorches easily, so it is best to heat it either in a double boiler or over a very low heat while stirring constantly. Scalding milk produces a scum, which is a combination of coagulated albumin, some salts, and fat globules. The scum can be prevented by keeping a lid on, or by stirring rapidly.

Curdling is the result of the formation of casein salts (such as casein chloride and casein lactate). Some milk and egg mixtures will curdle if cooked too long. To regain a smooth consistency, cool the mixture and beat or whirl in the blender. When an acid is added to milk, it must be done slowly, so as not to cause curdling. In addition, thickening the mixture helps to prevent curdling.

The following products can be whipped and used for toppings on desserts: thick cream with a minimum of 30% fat, chilled canned evaporated milk, and chilled nonfat dry milk powder. In whipping any of these, the first requisites are a chilled bowl and whipper and icy cold whipping cream or milk. It is important not to overwhip cream as this is the first step in producing butter. Adding sugar before whipping the cream decreases the volume and stiffness and increases the time required to whip it. Add the sugar after the cream is whipped. Whipped cream will be most stable if (1) the cream is whipped as stiff as possible without forming butter, (2) the cream is of optimum (36%) fat, and (3) the cream is held at a cold temperature.

Chilled evaporated milk can be whipped for a topping, but it is not very stable. Stability can be increased by adding a small amount (1 Tbsp/cup of milk) of lemon juice. If the topping is not completely used and it breaks down before the next serving, it can be rewhipped. Nonfat dry milk can also be whipped, but here again it is quite unstable. The stability can be increased by adding a little lemon juice. The evaporated-and dry milk-whipped toppings are best if used immediately.

(Also see MILK AND MILK PRODUCTS.)

• **Using cheese in cookery**—Cheese and nuts are the only two foods that can be found in every course of the meal. The popularity of cheese has been growing in the United States as indicated by the consumption figures. For a full discussion of cheese see MILK AND MILK PRODUCTS.

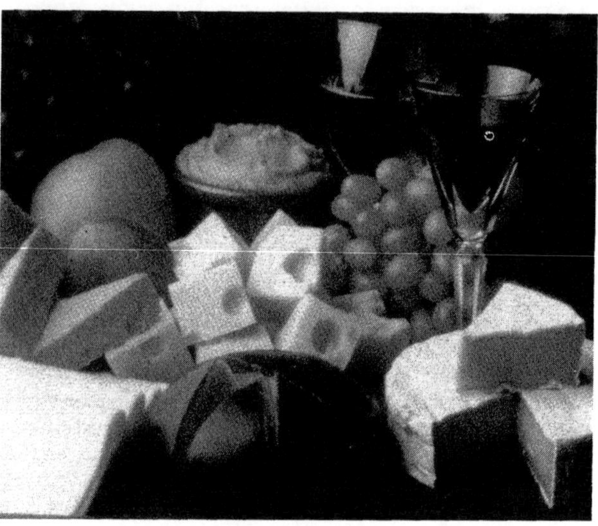

Fig. F-49. Cheese, one of the most versatile foods in the world. (Courtesy, American Dairy Assn., Rosemont, Ill.)

Cheese is a concentrated, high fat, high protein food and, therefore, should be cooked at a lower temperature. A high temperature will make the cheese tough and rubbery. When it is necessary to melt cheese, grating will reduce the time required to melt it and, consequently, will reduce the likelihood of overcooking it. Because cheese is so concentrated, it is best when mixed into other foods; indeed, many dishes are improved with a little added cheese.

Minerals. Some food processing results in loss of minerals, which is covered in the article entitled MINERALS. But mineral losses in the kitchen occur mostly by allowing foods to soak in water and then discarding the water. Otherwise, minerals are not lost to any appreciable extent during ordinary cooking methods.

Vitamins. The loss of vitamins in food preparation is of prime concern to the nutritionist. If the vitamins present in the raw food are lost during preparation and cooking, it makes it that much more difficult to secure the daily recommended vitamins and minerals.

The fat-soluble vitamins, A, D, E, and K, are not easily lost by ordinary cooking methods. On the other hand, the water-soluble vitamins (B complex and C) are dissolved easily in cooking water, and a portion may be destroyed by heating—thus cooking in as little water as possible, and cooking as short a time as possible is, in general, the best procedure.

(Also see VITAMIN[S], section headed "Vitamin Content of Foods.")

Enzymes. *An enzyme may be defined as a complex protein, produced by living cells, which acts as an organic catalyst to accelerate the rate of a chemical reaction in a wide range of processes, without being used in the action.*

Enzymes are present in all living plant and animal material. They are usually classified into two major categories: (1) those which have a basic reaction which involves the addition of water and the breakdown of larger compounds to smaller, called hydrolytic enzymes; and (2) those which cause a breakdown of a molecule without the additon of water or oxygen. Enzymes are activated by normal physiological processes, such as the destruction of the cell by such processes as grinding or crushing. Enzymes can be inactivated by heat.

The following summary gives some of the enzymes and their action which are pertinent to foods and their preparation.

Enzyme	Action
Amylase or diastase	A starch splitting enzyme which is present in flour.
Bromelin	A proteolytic enzyme found in fresh pineapple. Because it will split the protein, gelatin, it prevents the formation of a gel. Pineapple must be cooked or canned for it to be used in gelatin recipes. Bromelin will dissolve or degrade collagen and elastin, action which occurs in tenderizing meat.
Ficin	Obtained from figs. This enzyme is a proteolytic one which is used for tenderizing meat.
Fish enzymes	These enzymes cause rapid spoilage of fish, unless fish are kept icy cold. Fish should not be held longer than 2 days.
Fruit enzymes	Certain fruits turn brown when peeled and cut, and thus exposed to the air. This reaction is halted when the fruit is heated or when acid is present to inactivate the enzyme.
Invertase	This enzyme is used in the production of creamy fondants, or the soft centers in chocolates. It is added to the fondant when it is melted for encasing in a chocolate coating. During storage as some of the sucrose is changed to a liquid, this liquid will dissolve more sucrose and thus produce a soft center. The pH, cooking time, sugar, syrup, water content, and casting temperature all have to be carefully controlled.
Lipase	An enzyme found in fats and oils which can break a fat down into free fatty acids and glycerol. Lipases are found also in the fat of meat, fish, eggs, milk, and cereals.
Meat enzymes	Meat contains over 50 different enzymes. This is to be expected because of its complex structure and activity.

Enzyme	Action
Milk enzymes	Milk contains a number of enzymes including lipases, phosphatases, xanthine oxidase, amylase, and protease. The enzymes are killed during the pasteurization process.
Papain	A proteolytic enzyme found in the papaya leaf that is used for meat tenderizing.
Pectin-esterase	This enzyme is widely distributed in plant tissue. Many studies have been done without any solution to the question of why or how fruit changes in texture as it ripens.
Phosphatase	Phosphatase, one of the enzymes found in milk, is destroyed by pasteurization, hence a test for this enzyme has been common practice for 30 years as a means of quality control.
Rennin	Used to coagulate the casein of milk in the production of rennet desserts and cheeses.
Vegetable enzymes	Vegetables contain enzymes which will bring about undesirable chemical changes in canned or frozen products, unless destroyed by heat. Also, some vegetables, such as eggplant, will turn brown when cut and exposed to the air—the same reaction as in fruits.
Wheat enzymes	The enzymes present in wheat (amylases, lipases, oxidases, proteinases) are only a small portion of the total flour protein, but they may influence flour properties. For instance, the lipase accounts for the rancidity of whole grain flours, and the reason that these flours should be refrigerated.
Yeast enzymes	The production of carbon dioxide in yeast breads is catalyzed by many different enzymes found in the yeast.

Color. Food color means to a cook what paint means to an artist. Without the bright colors of vegetables and fruits, our meals would be a series of brown and white pictures. More about food color combinations is covered in an earlier subsection on "Meal Planning." Also, synthetic colors are covered in the section on "Coloring of Food"; this section pertains to natural colors.

Sometimes the natural colors of foods are modified with vegetable coloring materials which are obtained from tree bark, fruits, leaves, blossoms, roots, and mosses. Beets contain betanin, a food coloring which is used to intensify the color of tomato soup and tomato sauce. A red food color is also used to intensify the dark brown of chocolate products. Some of the other coloring materials that are used include: alkanet, annatto, caramel, carotene, chlorophyll, saffron, and turmeric. Saffron is popular for coloring bakery products and commercially prepared poultry pies. It imparts a good rich pale yellow color, which looks as though the product contains eggs and butter. Caramel is used to give a richer color to meat gravies, cola, and root beer drinks. The added colors are listed on the container, along with the other ingredients.

Some of the common food colors and the changes that may occur during storage, preparation, and cooking are given in Table F-35.

(Also see COLORING OF FOOD.)

TABLE F-35
NATURAL COLORS IN FOODS

Food	Color Problems and Pointers
Chocolate	The gloss on chocolate is rapidly lost and changed to a dull, mottled gray called *bloom,* by heat and moisture. This bloom is likely caused by the melting of some of the chocolate and recrystallization on the surface. Although the proper temperature and timing, and the use of stabilizers during production will help retard this development, chocolates are best when stored at a temperature between 60°-70°F *(16°-21°C)* and a relative humidity between 50-65%.
Eggs	The vitamin A content of yolks cannot be predicted by the color of the yolk. Deep-colored yolks are high in vitamin A, however, if the chicken is given a vitamin A supplement because it does not have access to green or yellow feeds, the yolk may be pale in color, yet be high in vitamin A. The predominant yellow pigment in the yolk is xanthophyll, which is not converted to vitamin A in the body.
Fruit	Fruits and vegetables contain the same color pigments. The common ones found in fruits are listed as follows:

Color(s)	Pigment(s)	Solubility	Acid, Alkaline, or Other Reactions
Yellow and orange	*Carotenoid*	Insoluble	Not affected by acid or alkali.
Red, purple blue	*Flavonoids:* Anthocyanin	Very soluble	Acid intensifies the colors. They turn blue or purple in alkaline solutions.
Light or colorless	Flavonols		

Not all colors play the rules of the game. Pineapple juice changes red or purple fruit to blue regardless of acidity. Orange juice is best omitted with red or blue fruits, as it often changes them to brown.

Food	Color Problems and Pointers
Meat	When first cut, meat appears purplish red due to the reduced form of myoglobin. As it combines with oxygen the myoglobin becomes oxymyoglobin, a bright red color. But on further standing, it oxidizes to metomyoglobin, a brownish color. When cooking, the color of the interior of meat changes from bright red, or pink to grayish pink, and finally to grayish brown. During curing, the nitrite reacts with the myoglobin, the red pigment of meat, and the pigment changes to the characteristic pink color of cured meats.
Milk	The white color of milk is the result of the reflection of light by the colloid casein and calcium phosphate dispersed throughout. Also, the two yellow pigments, carotene and riboflavin, impart the creamy color to milk. The carotene is fat soluble and is found in cream and butter. Riboflavin is a greenish-yellow fluorescent color quite easily detected in the whey of milk.
Vegetables	The bright colors of vegetables are a great asset when planning a meal. Without them it would be difficult to make attractive dishes. The vegetable pigments are given in the following summary:

Color(s)	Pigment(s)	Solubility	Acid, Alkaline or Other Reactions
Green	*Chlorophyll*	Mostly insoluble	During cooking, acid changes chlorophyll to pheophytin, which is a dull olive-green color. If baking soda is added to make it alkaline, the chlorophyll changes to chlorophyllin which is bright green in color.
Yellow and Orange	*Carotenoids:* Carotene Xanthophyll	Insoluble	Acid, alkali, and heat have little effect on the carotenoids.
Red	Lycopene		Red of tomato.
Red, or Purple, or Blue	*Flavonoids:* Anthocyanins	Water soluble	Flavonoids are red in acid and change to purple, then to blue, as the acid changes to alkaline.
White	Flavonols		

Flavor. If the flavor is not desirable, the food will be left on the platter; and, no matter how important nutritionally, it cannot contribute to the diet unless consumed.

It is noteworthy that it took 40 scientists working since 1920 to segregate out the 80 chemical components of apple flavor. Cocoa has been found to have over 125 basic chemical elements, and coffee has nearly 200.

The food tastes experienced by the eating of foods, is a complex combination of taste, smell, texture, and temperature. Basically, there are just four taste sensations: sweet, salt, sour, and bitter.

Flavor research technology has been rapidly changing with the development of very sensitive instruments capable of identifying ingredients, found in very small amounts, in very complex mixtures. Some of these instruments include: gas-liquid chromatography, infrared spectroscopy, mass spectrometry, and nuclear magnetic resonance.

Texture. Texture is not much easier to evaluate than flavor. Although everyone has a preconceived idea, from past experience, just what the texture of each food should be, it is hard to say just how much deviation from this will be acceptable.

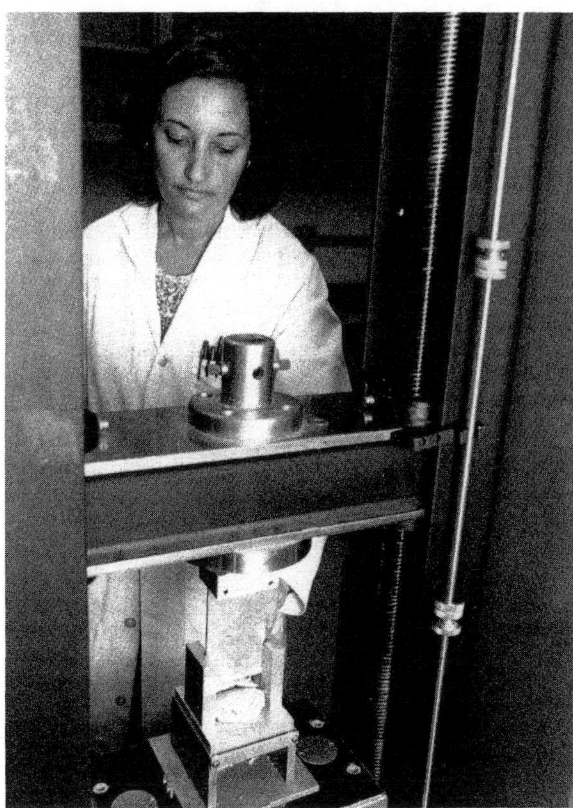

Fig. F-50. This machine simulates teeth biting a bread sample, and thus measures the tenderness of the sample. (Courtesy, Michigan State University)

There are some mechanical devices for measuring volume, crumb firmness, tenderness of meat, and viscosity of syrups and sauces, but usually, when and if necessary, textures are judged by a panel of experts.

Bacteria. In addition to the many chemical and physical aspects of foods that have been covered in this section, there are the bacteriological reactions. Some are desirable, such as the action of yeast in leavening, and in beers and wines; mold in making cheese; acetic acid bacteria in vinegar; and *acidophilus* in yogurt. But others are undesirable and can cause great discomfort and even death; included in the latter group are *Staphylococcus, Streptococcus, Salmonella,* and the dreaded, *Clostridium botulinum.* Undesirable bacteria can be controlled by—

1. Sanitary food handling
2. Holding food at refrigerator temperatures until ready to cook
3. Cooking thoroughly to the center of the product
4. Refrigerating immediately after the meal is finished
5. Canning meats and vegetables under pressure, and cooking for 20 minutes any home-canned foods where the method of canning is unknown.

(Also see BACTERIA IN FOOD.)

COOKING. The history of cooking, as it evolved through the ages, like many other things, is not well documented. Archeological finds indicate that food was likely first cooked by hanging it over the open fire; later, food was cooked by rolling it in wet leaves and steaming, or by using hot rocks, or shells. No doubt, the introduction of pottery in the Neolithic Period revolutionized cooking techniques. During this period, the nomads began to settle in little communities, domesticate livestock, and farm the land—civilization was settling down.

Fig. F-51. A prehistoric family preparing a meal on the caveman's outdoor grill. (Painting by Paul Jamin. Courtesy, The Bettman Archive, Inc., New York, N.Y.)

The development of different cuisines was dependent on local conditions. For instance, in northern Europe where they had many trees and it was necessary to have open fires to keep warm, they used a spit for roasting large pieces of meat and cooked cauldrons of soups, stews, and sauces. Because there were abundant pastures, they raised large herds of cows for milk, and, therefore, used many dairy products, including butter as their basic fat.

In southern Europe, on the other hand, trees, and therefore firewood, were scarce, and pastures were limited. But the hills grew olive trees; thus, a cuisine developed using oils, small cuts of meats, and the saucepan or frying pan.

The birth of French cuisine, which is perhaps the most complex and refined in the western world, is attributed to Catherine de Medicis who arrived in France in 1533, with a retinue of master Italian chefs, along with the staples of Italian cuisine: milk-fed veal, baby peas, broccoli, artichokes, and pasta. Antonin Careme (1784-1833), however, is credited with revolutionizing the cuisines, not only of France, but of all Europe. He developed a logical sequence for the dishes to be served; he insisted on cleanliness; he organized the cooking procedures; he introduced a new and exciting awareness for gourmet foods; he published many cookbooks; and he became famous for his fantastic massive pastry reproductions of classical architecture.

A discussion of cuisines would not be complete without considering the influence of the Chinese. Like the Italians, the Chinese faced a chronic shortage of firewood. This led to the development of the wok in which they could fry, saute, and braise, and to the procedure of cutting everything into bite-size pieces which cooked quickly, and which was also necessary because of their use of chopsticks. The Chinese cuisine is a symphony of color, texture, and flavor, with plenty of variety, subtlety, and sophistication.

As yet, cooking in North America has not evolved into a national cuisine, partly due to the fact that the Americas have been a melting pot for every country of the world. Many towns can boast of their Chinese or Japanese dining rooms, French restaurants, Scandinavian smorgasbords, Italian pizzerias, and/or Mexican taco stands, as well as the restaurants which serve the national dishes of the European countries. With the advent of instant communication, and rapid travel, cuisines can no longer remain local. But even though cuisines may blend together in the future, the basic cooking methods have not changed since the Neolithic Age.

Reasons for Cooking. Cooking is the art and the science of applying heat or microwaves to food in order to make it more nutritious, palatable, and flavorful. Today, some people recommend that all foods be consumed raw. Although most raw foods are not harmful, there are several reasons to cook them. There is no question that raw fruits and vegetables are good—and good for you, but some of them are easier to eat and more digestible when cooked. Therefore, foods are cooked for the following reason:

1. To make available the maximum nutritive value
2. To improve the digestibility
3. To increase the palatability or eating quality
4. To develop the flavor, or to add to the flavor
5. To destroy any harmful ingredients.

Methods and Media of Cooking. *Cooking may be defined as the application of a force (heat or microwaves) to a substance which results in a higher atomic and molecular activity.* Thus, the temperature of a substance is parallel to the agitation and motility of its atoms and molecules. The higher the temperature, the faster the atoms and molecules bombard each other. It follows that the freezing and boiling point of a substance are related to its molecular structure.

Freezing, melting, evaporation, and condensation are physical changes rather than temperature changs, even through heat brings about the change. For instance, the temperature of melting ice and the water it forms are both 32°F (*0°C*).

Until the advent of the magnetron tube, which changes electricity into microwaves, the only method of cooking was heat; and, until about 100 years ago, when electricity was developed, this meant by a fire.

Heat is transmitted by means of conduction, convection and radiation. Most cooking methods are a combination of these methods. Heat always travels from the hotter substance to the cooler substance.

Conduction involves the heating of a body (the pan) by the transfer of heat from the element to the bottom of the pan, and within the pan from molecule to molecule.

Convection occurs only in such materials as gases and liquids, where the molecules can move freely from one area to another. The material closest to the heat will warm first, and become less dense.

Radiation involves waves of energy that travel rapidly through space, and heat the surface they strike; they do not penetrate the substance. The rest of the inside of the object is heated by conduction.

Microwave cooking has been available since the 1950s, but, like the electric typewriter, the microwave oven took 20 years to become a common item on the market.

The magnetron tube is a vacuum tube which can convert electricity into electromagnetic energy radiation. The microwaves travel in straight lines.

The media and the various cooking methods are listed below:

Medium	Methods of Cooking
Air	Baking, broiling, roasting.
Water	Boiling, simmering, stewing.
Steam	Steamer, pressure cooker, wrapping in foil.
Fat	Sauteeing, pan and deep fat frying.
Combinations	Braising, fricasseeing, pot roasting.

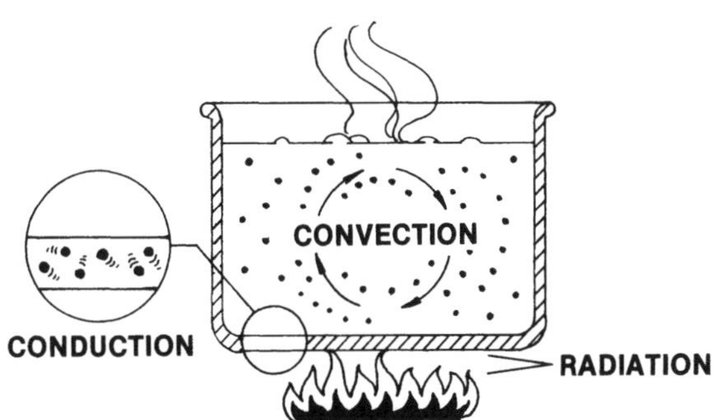

Fig. F-52. A schematic drawing of the three methods of cooking—conduction, convection, and radiation.

MEALTIME. At least one meal of the day should be a time when all members of the family can gather together around the dinner table and have a good visit.

Fig. F-53. Make attractive table settings an every-meal occurrence. (Courtesy, American Dairy Assn., Rosemont, Ill.)

FOOD COMPOSITIONS

Food analyses, such as those shown in Table F-36 Food Compositions, have long provided the bases used by human nutritionists and food processors in the evaluation of foods. Additionally, most of the procedures used by dietitians and home economists in planning meals and modified diets are based on food compositions. Also, and most importantly, consumers may refer directly to Table F-36 Food Compositions to obtain information for selecting foods most suitable for their preferences and health needs.

The authors wish to express their gratitude to all those who provided data for Table F-36 Food Compositions, especially Robin Spencer Palmisano, who at the time was Systems Dietitian, University Hospitals, Ohio State University Hospital Nutrient Data Base Catalogue, The Ohio State University, Columbus, Ohio (presently a lawyer and a member of the firm of McGlinchey, Stafford, Cellini & Lang, New Orleans, Louisiana); and Dr. Harold B. Houser, M.D. and Ms. Grace J. Petot, Division of Nutrition, Highland View Hospital, and the Department of Biometry and Nutrition, School of Medicine, Case Western Reserve University, Cleveland, Ohio. We are especially indebted to Robin Spencer Palmisano for her untiring and dedicated effort in the preparation of this table; no one could have done more.

To facilitate quick and easy use, Table F-36 Food Compositions is divided into the following categories:

- Baby Foods
- Bakery Products
- Beverages
- Cereals & Flours
- Desserts & Sweets
- Eggs & Substitutes
- Entrees
- Fats & Oils
- Fish & Seafoods
- Flavorings & Seasonings
- Food Supplements
- Fruits
- Gravies & Sauces
- Juices
- Legumes & Products
- Meats
- Milk & Products
- Nuts & Seeds
- Pickles and Relishes
- Salads
- Salad Dressings
- Snack Foods
- Soups & Chowders
- Vegetables

Where foods fit into more than one category, they are either cross-referenced or listed in more than one category.

Values for each food are given for both (1) approximate serving size, and (2) 100 grams.

With the exception of the footnoted values, the data in this table were supplied by the Ohio State University Hospital Nutrient Data Base Catalogue, Columbus, Ohio. The source of each of the footnoted foods is indicated by the appropriate prefix number from among the following:

[1]Data from the HVH-CWRU Nutrient Data Base developed by the Division of Nutrition, Highland View Hospital, and the Department of Biometry and Nutrition, School of Medicine, Case Western Reserve University, Cleveland, Ohio.

[2]Data from *Composition of Foods*, Ag. Hdbk No. 8–5, USDA.

[3]Data from *Food Composition Table For Use In Africa*, FAO, United Nations, 1968.

[4]Data from *Food Composition Table For Use In East Asia*, FAO, United Nations, 1972.

[5]Data from *Food Composition Table For Use In Latin America*, Institute of Nutrition of Central America & Panama, and National Institutes of Health, USA, 1961.

[6]Footnote 6 and added values in parentheses are from reliable literature sources.

NOTE WELL:

1. The Food Data Bank from which Ohio State University initially provided most of the Table F-36 Food Compositions in this book is now the property of Unisoft Systems Associates, Ron Bruce, President, 1340 Dublin Road, Columbus, Ohio 43215; hence, Table F-36 Food Compositions cannot be reproduced without the permission of Unisoft Systems Associates.
2. Trade and brand names are used for information and identification purposes only.

TABLE F-36 FOOD COMPOSITIONS* (See notes at end of table)

Item No.	Food Item Name	Approximate Measure	Weight	Moisture	Food Energy Calories	Protein	Fats	Carbo-hydrates	Fiber	Minerals (Macro) Calcium	Phos-phorus	Sodium	Mag-nesium	Potas-sium
			(g)	(%)	(kcal)	(g)	(g)	(g)	(g)	(mg)	(mg)	(mg)	(mg)	(mg)
	BAKED PRODUCTS													
1	**ARROWROOT COOKIES,** w/enriched flour	1 AVG	**6**	**5.6**	**26.5**	**.5**	**.9**	**4.3**	—	**1.9**	**8.0**	**20.7**	**1.3**	**9.4**
			100	5.6	442.0	7.6	14.3	71.2	—	32.0	133.0	345.0	22.0	156.0
2	**COOKIES,** w/enriched flour, w/nutrients	1 AVG	**7**	**5.9**	**30.3**	**.8**	**.9**	**4.7**	**.1**	**7.1**	**12.5**	**13.4**	**3.4**	**35.1**
			100	5.9	433.0	11.8	13.2	67.1	.4	101.0	179.0	192.0	49.0	501.0
3	**PRETZELS,** w/enriched flour	1 AVG	**6**	**4.0**	**23.8**	**.6**	**.1**	**4.9**	**Trace**	**1.5**	**6.6**	**16.1**	**1.7**	**8.2**
			100	4.0	397.0	10.8	2.0	82.2	.4	25.0	110.0	269.0	28.0	137.0
4	**TEETHING BISCUIT** w/enriched flour	1 AVG	**11**	**6.4**	**43.1**	**1.2**	**.5**	**8.4**	**.1**	**28.9**	**18.0**	**39.8**	**3.9**	**35.5**
			100	6.4	392.0	10.7	4.2	76.4	.5	263.0	164.0	362.0	35.0	323.0
5	**ZWIEBACK**	1 PIECE	**7**	**4.5**	**29.8**	**.7**	**.7**	**5.2**	**Trace**	**1.4**	**3.9**	**16.2**	**1.0**	**21.4**
			100	4.5	426.0	10.1	9.7	74.2	.2	20.0	55.0	232.0	14.0	305.0
	CEREALS													
6	**BARLEY** prepared w/whole milk	1 TBLS	**14**	**74.7**	**15.5**	**.6**	**.5**	**2.3**	**Trace**	**32.2**	**21.0**	**6.9**	**4.2**	**26.9**
			100	74.7	111.0	4.6	3.3	16.3	.2	230.0	150.0	49.0	30.0	192.0
7	**CEREAL & EGG YOLKS**	1 TBLS	**14**	**88.7**	**7.3**	**.3**	**.3**	**1.0**	**Trace**	**3.4**	**5.6**	**4.6**	**.4**	**4.9**
			100	88.7	52.0	1.9	1.8	7.1	.1	24.0	40.0	33.0	3.0	35.0
8	**CEREAL, EGG YOLKS, & BACON**	1 TBLS	**14**	**85.9**	**11.1**	**.4**	**.7**	**.9**	**Trace**	**3.9**	**8.4**	**6.7**	**.7**	**4.9**
			100	85.9	79.0	2.5	5.0	6.2	.1	28.0	60.0	48.0	5.0	35.0
9	**GRITS & EGG YOLKS**	1 TBLS	**14**	**88.0**	—	**.3**	**.3**	—	—	**3.9**	**5.0**	—	**.7**	**7.8**
			100	88.0	—	1.8	2.3	—	—	28.0	36.0	—	5.0	56.0
10	**HIGH PROTEIN,** prepared w/whole milk	1 TBLS	**14**	**74.5**	**15.5**	**1.2**	**.5**	**1.6**	**.1**	**30.5**	**24.8**	**6.9**	**6.7**	**48.9**
			100	74.5	111.0	8.7	3.8	11.6	.4	218.0	177.0	49.0	48.0	349.0
11	**HIGH PROTEIN,** w/apple & orange, prepared w/whole milk	1 TBLS	**14**	**74.4**	**15.7**	**1.0**	**.5**	**1.9**	**Trace**	**31.2**	**23.2**	**8.1**	**5.2**	**48.4**
			100	74.4	112.0	6.9	3.9	13.4	.2	223.0	166.0	58.0	37.0	346.0
12	**MIXED CEREAL,** prepared w/whole milk	1 TBLS	**14**	**74.6**	**15.8**	**.7**	**.5**	**2.2**	**Trace**	**30.8**	**19.9**	**6.6**	**3.8**	**27.9**
			100	74.6	113.0	4.8	3.5	15.9	.2	220.0	142.0	47.0	27.0	199.0
13	**MIXED CEREAL,** w/honey, prepared w/whole milk	1 TBLS	**14**	**74.2**	**16.1**	**.7**	**.5**	**2.2**	—	**41.2**	**25.8**	**6.7**	—	**23.9**
			100	74.2	115.0	5.0	3.6	15.9	—	294.0	184.0	48.0	—	171.0
14	**MIXED CEREAL,** w/applesauce & bananas, w/nutrients	1 TBLS	**14**	**79.6**	**11.6**	**.2**	**.1**	**2.6**	**Trace**	**.6**	**4.1**	**5.0**	**1.0**	**4.5**
			100	79.6	83.0	1.2	.4	18.4	.3	4.0	29.0	36.0	7.0	32.0
15	**MIXED CEREAL,** w/bananas, prepared w/whole milk	1 TBLS	**14**	**74.3**	**16.1**	**.6**	**.5**	**2.3**	**Trace**	**30.0**	**19.3**	**8.4**	**3.6**	**33.2**
			100	74.3	115.0	4.5	3.6	16.6	.1	214.0	138.0	60.0	26.0	237.0
16	**OATMEAL,** prepared w/whole milk	1 TBLS	**14**	**74.5**	**16.2**	**.7**	**.6**	**2.1**	**Trace**	**30.8**	**22.4**	**6.4**	**4.9**	**28.6**
			100	74.5	116.0	5.0	4.1	15.3	.2	220.0	160.0	46.0	35.0	204.0
17	**OATMEAL,** w/applesauce & bananas, w/nutrients	1 TBLS	**14**	**81.8**	**10.5**	**.2**	**.1**	**2.2**	**.1**	**.8**	**5.7**	**4.3**	**1.5**	**6.7**
			100	81.8	75.0	1.3	.7	15.7	.4	6.0	41.0	31.0	11.0	48.0
18	**OATMEAL,** w/bananas, prepared w/whole milk	1 TBLS	**14**	**74.3**	**16.2**	**.7**	**.5**	**2.2**	**Trace**	**28.8**	**21.1**	**8.5**	**4.2**	**34.6**
			100	74.3	116.0	4.7	3.8	16.0	.2	206.0	151.0	61.0	30.0	247.0
19	**OATMEAL,** w/honey, prepared w/whole milk	1 TBLS	**14**	**74.5**	**16.1**	**.7**	**.5**	**2.1**	—	**40.5**	**27.7**	**6.9**	—	**23.8**
			100	74.5	115.0	5.0	3.9	15.3	—	289.0	198.0	49.0	—	170.0
20	**RICE,** prepared w/whole milk	1 TBLS	**14**	**74.6**	**16.1**	**.5**	**.5**	**2.3**	**Trace**	**33.5**	**24.5**	**6.4**	**6.3**	**26.6**
			100	74.6	115.0	3.9	3.6	16.7	.1	239.0	175.0	46.0	45.0	190.0
21	[2]**RICE,** w/applesauce & bananas	1 TBLS	**14**	—	**11.1**	**.2**	**0**	**2.4**	**Trace**	**2.4**	**1.7**	**3.9**	**.4**	**3.9**
			100	—	79.0	1.2	.4	17.1	.2	17.0	12.0	28.0	3.0	28.0
22	**RICE,** w/bananas, prepared w/whole milk	1 TBLS	**14**	**74.3**	**16.4**	**.6**	**.5**	**2.4**	**Trace**	**29.8**	**20.3**	**8.0**	**4.8**	**35.4**
			100	74.3	117.0	4.2	3.5	17.0	.1	213.0	145.0	57.0	34.0	253.0
23	**RICE,** w/honey, prepared w/whole milk	1 TBLS	**14**	**74.4**	**16.1**	**.5**	**.5**	**2.4**	—	**40.7**	**25.3**	**6.9**	—	**19.7**
			100	74.4	115.0	3.9	3.3	17.1	—	291.0	181.0	49.0	—	141.0
24	**RICE,** w/mixed fruit, w/nutrients	1 TBLS	**14**	**79.6**	**11.8**	**.1**	**0**	**2.6**	**Trace**	**2.8**	**3.2**	**1.5**	**.7**	**4.6**
			100	79.6	84.0	1.0	.2	18.7	.2	20.0	23.0	11.0	5.0	33.0
	DESSERTS													
25	**APPLE BETTY,** w/vitamin C	1 TBLS	**14**	**80.5**	**9.8**	**.1**	**0**	**2.7**	—	**2.2**	—	**1.3**	—	**7.4**
			100	80.5	70.0	.4	0	19.0	—	16.0	—	9.0	—	53.0
26	**CARAMEL PUDDING**	1 TBLS	**14**	**80.4**	**11.1**	**1.0**	**.1**	**2.4**	—	**7.7**	—	**3.9**	—	**8.1**
			100	80.4	79.0	1.4	.9	17.0	—	55.0	—	28.0	—	58.0
27	**CHERRY VANILLA PUDDING**	1 TBLS	**14**	**81.0**	**9.7**	**0**	**0**	**2.6**	**Trace**	**.7**	**1.0**	**2.1**	**.3**	**4.6**
			100	81.0	69.0	.2	.2	18.4	.1	5.0	7.0	15.0	2.0	33.0
28	**CHOCOLATE CUSTARD PUDDING**	1 TBLS	**14**	**79.9**	**11.8**	**.3**	**.2**	**2.3**	**Trace**	**8.5**	**6.9**	**3.2**	**1.4**	**12.0**
			100	79.9	84.0	1.9	1.7	16.1	.2	61.0	49.0	23.0	10.0	86.0
29	**COTTAGE CHEESE & PINEAPPLE,** w/vitamin C	1 TBLS	**14**	**80.2**	**10.9**	**.4**	**.1**	**2.2**	**.1**	**4.3**	**5.5**	**7.1**	**.6**	**5.9**
			100	80.2	78.0	3.0	.7	15.9	1.0	31.0	39.0	51.0	4.0	42.0
30	**DUTCH APPLE,** w/vitamin C	1 TBLS	**14**	**82.1**	**9.7**	**0**	**.1**	**2.4**	**Trace**	**.6**	**.6**	**2.2**	**.3**	**5.2**
			100	82.1	69.0	0	1.0	16.8	.3	4.0	4.0	16.0	2.0	37.0

Item No.	Minerals (Micro)			Fat-Soluble Vitamins			Water-Soluble Vitamins								
	Iron	Zinc	Copper	Vitamin A	Vitamin D	Vitamin E (Alpha Tocopherol)	Vitamin C	Thiamin	Ribo-flavin	Niacin	Panto-thenic Acid	Vit. B-6 (Pyri-doxine)	Folacin (Folic Acid)	Biotin	Vitamin B-12
	(mg)	(mg)	(mg)	(IU)	(IU)	(mg)	(mg)	(mg)	(mg)	(mg)	(mg)	(mg)	(mcg)	(mcg)	(mcg)
1	**.18**	**.03**	–	–	–	–	**.33**	**.03**	**.03**	**.34**	–	**Trace**	–	–	**Trace**
	3.00	.53	–	–	–	–	5.50	.50	.43	5.74	–	.04	–	–	.07
2	**.29**	**.08**	–	**1.6**	–	–	**.49**	**.10**	**.23**	**1.12**	–	**.41**	–	–	**.32**
	4.18	1.10	–	23.0	–	–	7.00	1.46	3.23	15.97	–	.90	–	–	4.59
3	**.23**	**.05**	–	**.5**	–	–	**.22**	**.03**	**.02**	**.21**	–	**.01**	–	–	–
	3.77	.78	–	8.0	–	–	3.80	.46	.36	3.56	–	.08	–	–	–
4	**.39**	**.10**	–	**12.8**	–	–	**1.00**	**.03**	**.06**	**.48**	–	**.01**	–	–	**.01**
	3.55	.93	–	116.0	–	–	9.10	.23	.54	4.33	–	.11	–	–	.07
5	**.04**	**.04**	–	**4.1**	–	–	**.37**	**.02**	**.02**	**.09**	–	**.01**	–	–	–
	.60	.54	–	58.0	–	–	5.30	.21	.24	1.32	–	.08	–	–	–
6	**1.73**	**.12**	–	**14.7**	–	–	**.15**	**.07**	**.08**	**.84**	–	–	**1.25**	–	**13.44**
	12.34	.83	–	105.0	–	–	1.10	.48	.58	5.98	–	–	8.90	–	96.00
7	**.07**	**.04**	**Trace**	**20.2**	–	–	**.10**	**Trace**	**.01**	**.01**	–	**Trace**	**.46**	–	–
	.51	.29	.02	144.0	–	–	.70	.01	.05	.05	–	.02	3.30	–	–
8	**.07**	**.04**	**Trace**	**13.2**	–	–	**.13**	**Trace**	**.01**	**.04**	–	**.02**	**.57**	–	**Trace**
	.47	.27	.02	94.0	–	–	.90	.05	.08	.27	–	.11	4.10	–	.01
9	**.07**	**.03**	–	**17.1**	–	–	**.07**	**Trace**	**.01**	**.04**	–	**Trace**	**.41**	–	**.01**
	.50	.23	–	122.0	–	–	.50	.03	.07	.29	–	.03	2.90	–	.04
10	**1.70**	**.15**	–	**14.7**	–	–	**.15**	**.07**	**.08**	**.79**	–	**.02**	**4.96**	–	–
	12.14	1.04	–	105.0	–	–	1.10	.47	.58	5.67	–	.11	35.40	–	–
11	**2.02**	**.10**	–	**13.6**	–	–	**.18**	**.09**	**.12**	**.56**	–	**.01**	–	–	**.05**
	14.42	.76	–	97.0	–	–	1.30	.66	.85	3.99	–	.09	–	–	.35
12	**.46**	**.10**	–	**14.7**	–	–	**.17**	**.06**	**.08**	**.81**	–	**.01**	**1.57**	–	–
	10.43	.71	–	105.0	–	–	1.20	.43	.58	5.78	–	.07	11.20	–	–
13	**1.58**	–	–	**12.9**	–	–	**.11**	**.06**	**.08**	**.88**	–	–	–	–	–
	11.27	–	–	92.0	–	–	.80	.45	.58	6.26	–	–	–	–	–
14	**.79**	**.03**	–	–	–	–	**1.27**	**.04**	**.05**	**.57**	–	**.02**	**.52**	–	–
	5.61	.22	–	–	–	–	9.10	.29	.36	4.04	–	.14	3.70	–	–
15	**1.56**	**.08**	–	**15.1**	–	–	**.20**	**.09**	**.10**	**.48**	–	**.01**	–	–	**0.05**
	11.15	.55	–	108.0	–	–	1.40	.65	.72	3.45	–	.10	–	–	0.34
16	**1.70**	**.13**	–	**14.7**	–	–	**.18**	**.07**	**.08**	**.84**	–	**.01**	**1.40**	–	–
	12.14	.92	–	105.0	–	–	1.30	.51	.56	5.93	–	.06	10.00	–	–
17	**.77**	**.05**	**.01**	**3.9**	–	–	**2.67**	**.03**	**.07**	**.47**	–	**.03**	**.49**	–	–
	5.51	.33	.08	28.0	–	–	19.10	.24	.48	3.35	–	.24	3.50	–	–
18	**1.57**	**.09**	–	**13.9**	–	–	**.22**	**.09**	**.10**	**.47**	–	**.02**	–	–	**.05**
	11.23	.63	–	99.0	–	–	1.60	.62	.75	3.38	–	.10	–	–	.33
19	**1.55**	–	–	**12.9**	–	–	**.11**	**.07**	**.08**	**.86**	–	–	–	–	–
	11.09	–	–	92.0	–	–	.80	.49	.60	6.03	–	–	–	–	–
20	**1.71**	**.09**	–	**14.7**	–	–	**.17**	**.07**	**.07**	**.73**	–	**.02**	**1.15**	–	–
	12.19	.64	–	105.0	–	–	1.20	.47	.50	5.21	–	.11	8.20	–	–
21	**.94**	**.01**	–	**2.9**	–	–	**4.42**	**.04**	**.06**	**.56**	–	**.03**	**.35**	–	–
	6.73	.08	–	21.0	–	–	31.60	.26	42	4.02	–	.23	2.50	–	–
22	**1.55**	**.08**	–	**12.9**	–	–	**.15**	**.10**	**.11**	**.55**	–	**.02**	–	–	**.05**
	11.04	.56	–	92.0	–	–	1.10	.68	.76	3.90	–	.15	–	–	.33
23	**1.51**	–	–	**12.9**	–	–	**.11**	**.07**	**.09**	**.85**	–	–	–	–	–
	10.81	–	–	92.0	–	–	.80	.48	.61	6.08	–	–	–	–	–
24	**.66**	**.03**	–	**2.1**	–	–	**2.83**	**.04**	**.08**	**.38**	–	**.04**	–	–	–
	4.71	.18	–	15.0	–	–	20.20	.25	.59	2.73	–	.25	–	–	–
25	**.03**	–	–	**2.2**	–	–	**3.79**	**Trace**	**.01**	**.01**	–	–	**.06**	–	–
	.20	–	–	16.0	–	–	27.10	.01	.05	.05	–	–	.40	–	–
26	**.02**	–	–	**4.6**	–	–	**.31**	**Trace**	**.01**	**.01**	–	–	**.13**	–	–
	.16	–	–	33.0	–	–	2.20	.01	.07	.04	–	–	.90	–	–
27	**.02**	**Trace**	–	**28.0**	–	–	**.15**	**Trace**	**Trace**	**.01**	–	**Trace**	**.04**	–	–
	.17	.03	–	200.0	–	–	1.10	Trace	.01	.04	–	.01	.30	–	–
28	**.05**	**.06**	–	**6.4**	–	–	**.21**	**Trace**	**.02**	**.01**	–	**Trace**	**.63**	–	**Trace**
	.38	.32	–	46.0	–	–	1.50	.01	.10	.10	–	.01	4.50	–	.01
29	**.02**	**.03**	–	**2.2**	–	–	**3.33**	**Trace**	**.01**	**.01**	–	**Trace**	**.71**	–	**.01**
	.13	.16	–	16.0	–	–	23.80	.02	.05	.05	–	.01	5.10	–	.07
30	**.03**	**Trace**	–	**7.0**	–	–	**3.00**	**Trace**	**Trace**	**.01**	–	**Trace**	**.10**	–	–
	.20	.03	–	50.0	–	–	21.40	.01	.01	.05	–	.01	.70	–	–

(Continued)

TABLE F-36 *(Continued)*

Item No.	Food Item Name	Approximate Measure	Weight	Moisture	Food Energy Calories	Protein	Fats	Carbo-hydrates	Fiber	Minerals (Macro) Calcium	Phos-phorus	Sodium	Mag-nesium	Potas-sium
			(g)	(%)	(kcal)	(g)	(g)	(g)	(g)	(mg)	(mg)	(mg)	(mg)	(mg)
31	**FRUIT DESSERT,** w/o vitamin C	1 TBLS	14	82.2	8.8	0	0	2.4	0	1.3	1.1	1.8	.7	13.3
			100	82.2	63.0	.3	0	17.2	.3	9.0	8.0	13.0	5.0	95.0
32	**ORANGE PUDDING**	1 TBLS	14	79.8	11.2	.2	.1	2.5	.1	4.5	3.9	17.9	.7	12.0
			100	79.8	80.0	1.1	.9	17.7	.4	32.0	28.0	128.0	5.0	86.0
33	**PEACH COBBLER,** w/vitamin C	1 TBLS	14	81.2	9.4	0	0	2.6	—	.6	.8	1.3	—	7.8
			100	81.2	67.0	.3	0	18.3	—	4.0	6.0	9.0	—	56.0
34	**PEACH MELBA,** w/vitamin C	1 TBLS	14	83.0	8.4	0	0	2.3	—	1.5	—	1.3	—	13.0
			100	83.0	60.0	.3	0	16.4	—	11.0	—	9.0	—	93.0
35	**PINEAPPLE-ORANGE,** w/vitamin C	1 TBLS	14	80.5	9.8	0	0	2.7	—	1.5	.6	1.4	.6	6.6
			100	80.5	70.0	.2	0	19.1	—	11.0	4.0	10.0	4.0	47.0
36	**PINEAPPLE PUDDING,** w/vitamin C	1 TBLS	14	76.1	12.2	1.0	.1	3.0	.1	4.8	4.2	3.1	1.3	12.6
			100	76.1	87.0	1.4	.4	21.6	.8	34.0	30.0	22.0	9.0	90.0
37	**TROPICAL FRUIT,** w/vitamin C	1 TBLS	14	83.2	8.4	0	0	2.3	—	1.4	—	1.0	—	8.1
			100	83.2	60.0	.2	0	16.4	—	10.0	—	7.0	—	58.0
38	**VANILLA CUSTARD PUDDING**	1 TBLS	14	79.4	12.5	.2	.3	2.3	—	7.8	6.3	4.1	.7	8.7
			100	79.4	89.0	1.6	2.3	16.2	—	56.0	45.0	29.0	5.0	62.0
	DINNERS													
39	**BEEF & EGG NOODLES**	1 TBLS	14	88.6	7.4	.3	.2	1.0	Trace	1.3	4.1	4.1	1.0	6.6
			100	88.6	53.0	2.3	1.7	7.0	.3	9.0	29.0	29.0	7.0	47.0
40	**BEEF & EGG NOODLES,** low sodium	1 TBLS	14	87.8	8.0	.4	.3	1.0	Trace	1.1	4.2	2.4	1.0	6.4
			100	87.8	57.0	2.5	1.9	7.4	.2	8.0	30.0	17.0	7.0	46.0
41	**BEEF & RICE**	1 TBLS	14	81.9	11.5	.7	.4	1.2	Trace	1.5	4.9	50.0	11.2	16.8
			100	81.9	82.0	5.0	2.9	8.8	.3	11.0	35.0	357.0	80.0	120.0
42	**BEEF LASAGNA**	1 TBLS	14	82.3	10.8	.6	.3	1.4	Trace	2.5	5.6	63.6	1.5	17.1
			100	82.3	77.0	4.2	2.1	10.0	.2	18.0	40.0	454.0	11.0	122.0
43	**BEEF STEW**	1 TBLS	14	86.9	7.1	.7	.2	.8	Trace	1.3	6.2	48.3	1.5	19.9
			100	86.9	51.0	5.1	1.2	5.5	.3	9.0	44.0	345.0	11.0	142.0
44	**BEEF,** w/vegetables	1 TBLS	14	83.2	11.9	.9	.6	.7	Trace	1.7	7.3	4.6	1.1	20.9
			100	83.2	85.0	6.3	4.6	5.3	.3	12.0	52.0	33.0	8.0	149.0
45	**CHICKEN & NOODLES**	1 TBLS	14	88.7	7.1	.3	.2	1.1	.1	2.4	3.4	2.4	1.3	4.9
			100	88.7	51.0	1.9	1.4	7.5	.6	17.0	24.0	17.0	9.0	35.0
46	**CHICKEN CREAM SOUP**	1 TBLS	14	87.1	8.1	.4	.2	1.2	Trace	4.9	4.1	2.7	.8	10.9
			100	87.1	58.0	2.5	1.6	8.4	.3	35.0	29.0	19.0	6.0	78.0
47	**CHICKEN SOUP**	1 TBLS	14	89.1	7.0	.2	.2	1.0	—	5.2	—	2.2	—	—
			100	89.1	50.0	1.6	1.7	7.2	—	37.0	—	16.0	—	—
48	**CHICKEN,** w/vegetables	1 TBLS	14	82.7	12.9	1.0	.8	.6	Trace	6.0	7.6	3.6	1.0	8.7
			100	82.7	92.0	7.0	5.5	4.2	.2	43.0	54.0	26.0	7.0	62.0
49	**COTTAGE CHEESE,** w/pineapple	1 TBLS	14	72.0	16.2	.9	.3	2.6	.1	9.1	10.2	20.9	1.0	13.3
			100	72.0	116.0	6.3	2.2	18.9	.9	65.0	73.0	149.0	7.0	95.0
50	**HAM,** w/vegetables	1 TBLS	14	83.6	10.8	.9	.5	.9	Trace	1.4	7.8	3.1	1.3	22.7
			100	83.6	77.0	6.4	3.3	6.1	.2	10.0	56.0	22.0	9.0	162.0
51	**LAMB & NOODLES**	1 TBLS	14	86.5	9.1	.3	.3	1.2	—	2.5	—	2.5	—	10.8
			100	86.5	65.0	2.3	2.2	8.7	—	18.0	—	18.0	—	77.0
52	**MACARONI & BACON**	1 TBLS	14	85.1	10.5	.4	.5	1.2	—	9.9	—	10.9	—	11.8
			100	85.1	75.0	2.5	3.3	8.6	—	71.0	—	78.0	—	84.0
53	**MACARONI & CHEESE**	1 TBLS	14	86.5	8.5	.4	.3	1.1	Trace	7.1	8.3	10.6	1.0	6.2
			100	86.5	61.0	2.6	2.0	8.2	.1	51.0	59.0	76.0	7.0	44.0
54	**MACARONI & HAM**	1 TBLS	14	86.5	8.4	.4	.2	1.2	—	10.5	—	6.6	—	14.8
			100	86.5	60.0	3.2	1.4	8.5	—	75.0	—	47.0	—	106.0
55	**MACARONI, TOMATO & BEEF**	1 TBLS	14	86.7	8.3	.4	.2	1.3	Trace	2.0	6.2	2.4	1.0	10.1
			100	86.7	59.0	2.5	1.1	9.4	.3	14.0	44.0	17.0	7.0	72.0
56	**MACARONI, TOMATO & BEEF,** made w/soy & yeast	1 TBLS	14	87.3	7.7	.3	.2	1.2	.1	2.2	5.9	2.4	1.3	13.4
			100	87.3	55.0	2.2	1.1	8.8	.4	16.0	42.0	17.0	9.0	96.0
57	**MIXED VEGETABLES**	1 TBLS	14	90.6	4.6	.1	0	1.1	—	2.4	—	1.3	—	15.7
			100	90.6	33.0	1.0	0	7.9	—	17.0	—	9.0	—	112.0
58	**SPAGHETTI, TOMATO & MEAT**	1 TBLS	14	85.5	8.8	.4	.2	1.4	.1	2.5	5.2	2.8	—	15.1
			100	85.5	63.0	2.5	1.3	10.1	.4	18.0	37.0	20.0	—	108.0
59	**SPLIT PEAS & HAM**	1 TBLS	14	83.5	9.9	.5	.2	1.6	Trace	3.2	6.9	2.0	—	19.0
			100	83.5	71.0	3.3	1.3	11.[illegible]	.3	23.0	49.0	14.0	—	136.0
60	**TOMATO SOUP**	1 TBLS	14	83.4	7.6	.3	0	1.9	Trace	3.4	7.3	41.2	—	42.0
			100	83.4	54.0	1.9	.1	13.5	.2	24.0	52.0	294.0	—	300.0
61	**TURKEY & RICE**	1 TBLS	14	89.3	6.9	.3	.2	1.0	Trace	3.2	2.4	2.1	—	4.8
			100	89.3	49.0	1.8	1.4	7.2	.2	23.0	17.0	15.0	—	34.0

BABY FOODS

Item No.	Minerals (Micro) Iron	Zinc	Copper	Fat-Soluble Vitamins Vitamin A	Vitamin D	Vitamin E (Alpha Tocopherol)	Water-Soluble Vitamins Vitamin C	Thiamin	Ribo-flavin	Niacin	Panto-thenic Acid	Vit. B-6 (Pyri-doxine)	Folacin (Folic Acid)	Biotin	Vitamin B-12
	(mg)	(mg)	(mg)	(IU)	(IU)	(mg)	(mg)	(mg)	(mg)	(mg)	(mg)	(mg)	(mcg)	(mcg)	(mcg)
31	.03	.01	Trace	33.5	—	—	.42	Trace	Trace	.02	—	.01	.49	—	—
	.21	.05	.03	239.0	—	—	3.00	.02	.01	.14	—	.03	3.50	—	—
32	.01	.02	—	16.1	—	—	1.27	.01	.01	.02	—	Trace	1.09	—	—
	.10	.17	—	115.0	—	—	9.10	.04	.06	.12	—	.03	7.80	—	—
33	.01	Trace	—	19.9	—	—	2.87	Trace	Trace	.04	—	Trace	.15	—	—
	.10	.03	—	142.0	—	—	20.50	.01	.02	.26	—	.01	1.10	—	—
34	.04	—	—	27.4	—	—	3.64	Trace	Trace	.04	—	—	.27	—	—
	.30	—	—	196.0	—	—	26.00	.01	.03	.27	—	—	1.90	—	—
35	.03	—	Trace	8.0	—	—	2.00	Trace	Trace	.01	—	—	.36	—	—
	.18	—	.02	57.0	—	—	14.30	.02	.02	.05	—	—	2.60	—	—
36	.03	.03	—	5.2	—	—	3.74	.01	.01	.02	—	Trace	.78	—	0.1
	.19	.19	—	37.0	—	—	26.70	.04	.05	.12	—	.04	5.60	—	0.6
37	.04	—	—	2.8	—	—	2.63	Trace	Trace	.01	—	—	—	—	—
	.26	—	—	20.0	—	—	18.80	.01	.03	.08	—	—	—	—	—
38	.04	.04	.01	5.0	—	—	.11	Trace	.01	.01	.04	Trace	.87	—	—
	.26	.28	.05	36.0	—	—	.60	.01	.08	.04	.26	.02	6.20	—	—
39	.06	.05	Trace	115.2	—	—	.17	.01	.01	.10	.03	.01	.71	—	.01
	.41	.38	.03	823.0	—	—	1.20	.04	.04	.72	.21	.06	5.10	—	.09
40	.06	.06	Trace	91.8	—	—	.20	Trace	.01	.08	.03	Trace	.77	—	.01
	.43	.40	.03	656.0	—	—	1.40	.03	.07	.58	.23	.03	5.50	—	.10
41	.10	.13	.01	70.3	—	—	.55	Trace	.01	.19	—	.03	—	—	—
	.69	.92	.05	502.0	—	—	3.90	.02	.07	1.34	—	.14	—	—	—
42	.12	—	—	162.8	—	—	.27	.01	.13	.19	—	.01	—	—	—
	.87	—	—	1163.0	—	—	1.90	.07	.89	1.35	—	.07	—	—	—
43	.10	.12	—	230.9	—	—	.42	Trace	.01	.18	—	.01	—	—	—
	.72	.87	—	1649.0	—	—	3.00	.01	.07	1.31	—	.07	—	—	—
44	.11	.20	.01	110.9	—	—	.27	.01	.01	.20	.04	.01	.91	—	.08
	.79	1.40	.09	792.0	—	—	1.90	.04	.08	1.42	.27	.09	6.50	—	.58
45	.06	.04	.01	125.3	—	—	.17	Trace	Trace	.07	—	Trace	.74	—	—
	.39	.29	.04	895.0	—	—	1.20	.03	.03	.52	—	.03	5.30	—	—
46	.04	.04	—	101.6	—	—	.18	Trace	.01	.05	—	.01	—	—	.01
	.30	.26	—	726.0	—	—	1.30	.01	.04	.38	—	.04	—	—	.06
47	.04	—	—	193.8	—	—	.14	Trace	Trace	.04	—	—	.73	—	—
	.27	—	—	1384.0	—	—	1.00	.02	.03	.29	—	—	5.20	—	—
48	.10	.14	—	117.2	—	—	.15	Trace	.01	.14	.05	.01	.15	—	.02
	.74	1.00	—	837.0	—	—	1.10	.03	.07	.99	.34	.04	1.10	—	.16
49	.01	.04	—	10.8	—	—	.20	.01	.02	.02	—	.01	—	—	.03
	.10	.29	—	77.0	—	—	1.40	.04	.14	.11	—	.05	—	—	.23
50	.08	.15	—	36.4	—	—	.27	.02	.01	.16	.06	.01	.91	—	.04
	.59	1.08	—	260.0	—	—	1.90	.11	.09	.16	.40	.10	6.50	—	.28
51	.05	—	—	109.6	—	—	.27	.01	.01	.09	—	—	—	—	—
	.36	—	—	783.0	—	—	1.90	.04	.07	.67	—	—	—	—	—
52	.05	—	—	162.4	—	—	.29	.01	.01	.09	—	—	—	—	—
	.38	—	—	1160.0	—	—	2.10	.05	.08	.63	—	—	—	—	—
53	.04	.05	Trace	1.8	—	—	.18	.01	.01	.08	—	Trace	.21	—	—
	.30	.32	.02	13.0	—	—	1.30	.06	.06	.55	—	.02	1.50	—	.03
54	.05	—	—	73.6	—	—	.31	.01	.01	.11	—	—	—	—	—
	.38	—	—	526.0	—	—	2.20	.06	.10	.81	—	—	—	—	—
55	.05	.05	.01	96.7	—	—	.21	.01	.01	.11	—	.01	—	—	.03
	.36	.36	.04	691.0	—	—	1.50	.05	.06	.75	—	.05	—	—	.24
56	.07	.04	.01	74.3	—	—	.21	.01	.01	.12	—	.01	2.81	—	.03
	.49	.31	.04	531.0	—	—	1.50	.07	.06	.82	—	.04	20.10	—	.23
57	.04	—	—	342.2	—	—	.46	Trace	Trace	.06	—	—	.94	—	—
	.31	—	—	2444.0	—	—	3.30	.01	.02	.41	—	—	6.70	—	—
58	.08	.06	—	97.0	—	—	.31	.01	.01	.15	.03	.01	—	—	—
	.55	.42	—	693.0	—	—	2.20	.07	.07	1.10	.18	.06	—	—	—
59	.07	—	—	84.4	—	—	.27	.01	.01	.07	—	.01	—	—	.01
	.50	—	—	603.0	—	—	1.90	.05	.05	.48	—	.04	—	—	.05
60	.06	—	—	140.0	—	—	.42	.01	.02	.10	—	—	—	—	—
	.40	—	—	1000.0	—	—	3.00	.05	.12	.70	—	—	—	—	—
61	.04	—	—	148.3	—	—	.17	Trace	Trace	.04	—	Trace	.43	—	—
	.29	—	—	1059.0	—	—	1.20	.01	.03	.28	—	.03	3.10	—	—

(Continued)

BABY FOODS

TABLE F-36 *(Continued)*

Item No.	Food Item Name	Approximate Measure	Weight	Moisture	Food Energy Calories	Protein	Fats	Carbo-hydrates	Fiber	Minerals (Macro) Calcium	Phos-phorus	Sodium	Mag-nesium	Potas-sium
			(g)	(%)	(kcal)	(g)	(g)	(g)	(g)	(mg)	(mg)	(mg)	(mg)	(mg)
62	**TURKEY,** w/vegetables	1 TBLS	14	**82.5**	**12.6**	**.8**	**.7**	**.8**	**Trace**	**9.9**	**8.8**	**6.0**	**1.1**	**15.0**
			100	82.5	90.0	5.9	5.0	5.9	.2	71.0	63.0	43.0	8.0	107.0
63	**VEAL,** w/vegetables	1 TBLS	14	**84.3**	**10.2**	**.9**	**.4**	**.8**	**Trace**	**1.5**	**7.6**	**3.5**	**1.3**	**22.0**
			100	84.3	73.0	6.1	3.1	5.8	.2	11.0	54.0	25.0	9.0	157.0
64	**VEGETABLES & BACON**	1 TBLS	14	**86.2**	**9.9**	**.3**	**.5**	**1.1**	**Trace**	**1.5**	**5.3**	**6.3**	—	**12.0**
			100	86.2	71.0	1.8	3.9	7.6	.2	11.0	38.0	45.0	—	86.0
65	**VEGETABLES & BEEF**	1 TBLS	14	**87.9**	**7.4**	**.3**	**.2**	**1.0**	**Trace**	**1.4**	**6.0**	**3.4**	**.8**	**14.7**
			100	87.9	53.0	2.4	1.7	7.4	.2	10.0	43.0	24.0	6.0	105.0
66	**VEGETABLES & CHICKEN**	1 TBLS	14	**88.2**	**7.0**	**.3**	**.2**	**1.2**	**Trace**	**2.0**	**3.6**	**1.3**	—	**3.6**
			100	88.2	50.0	1.9	1.1	8.5	.2	14.0	26.0	9.0	—	26.0
67	**VEGETABLES, DUMPLINGS & BEEF**	1 TBLS	14	**88.6**	**6.7**	**.3**	**.1**	**1.1**	—	**2.0**	—	**7.3**	**1.0**	—
			100	88.6	48.0	2.1	.8	8.0	—	14.0	—	52.0	7.0	—
68	**VEGETABLES & HAM**	1 TBLS	14	**88.4**	**7.3**	**.3**	**.2**	**1.0**	**Trace**	**1.1**	**3.6**	**2.5**	**1.0**	**12.9**
			100	88.4	52.0	2.4	1.7	7.0	.2	8.0	26.0	18.0	7.0	92.0
69	**VEGETABLES & LAMB**	1 TBLS	14	**88.6**	**7.1**	**.3**	**.2**	**1.0**	**Trace**	**1.8**	**6.9**	**1.8**	**1.0**	**13.3**
			100	88.6	51.0	2.1	1.7	7.1	.2	13.0	49.0	13.0	7.0	95.0
70	**VEGETABLES & LIVER**	1 TBLS	14	**88.9**	**6.2**	**.3**	**.1**	**1.1**	**Trace**	**1.4**	**5.3**	**1.8**	—	**12.5**
			100	88.9	44.0	1.8	.6	8.2	.3	10.0	38.0	13.0	—	89.0
71	**VEGETABLES, LIVER, BACON & CEREAL**	1 TBLS	14	**87.2**	**8.0**	**.3**	**.3**	**1.1**	**Trace**	**1.5**	**5.9**	**39.8**	—	**18.3**
			100	87.2	57.0	2.4	1.9	7.5	.3	11.0	42.0	284.0	—	131.0
72	**VEGETABLES, NOODLES & CHICKEN**	1 TBLS	14	**86.2**	**9.0**	**.2**	**.3**	**1.3**	**Trace**	**3.6**	**4.6**	**3.6**	—	**8.3**
			100	86.2	64.0	1.7	2.2	9.1	.2	26.0	33.0	26.0	—	59.0
73	**VEGETABLES, NOODLES & TURKEY**	1 TBLS	14	**88.7**	**7.3**	**.3**	**.2**	**1.1**	**Trace**	**4.5**	**4.1**	**2.4**	**1.3**	**10.2**
			100	88.7	52.0	1.8	1.5	7.6	.2	32.0	29.0	17.0	9.0	73.0
74	**VEGETABLES & TURKEY**	1 TBLS	14	**89.0**	**6.6**	**.2**	**.2**	**1.1**	**Trace**	**1.8**	**2.7**	**2.4**	—	**3.5**
			100	89.0	47.0	1.7	1.2	7.7	.2	13.0	19.0	17.0	—	25.0
	FRUITS													
75	**APPLESAUCE,** w/vitamin C	1 TBLS	14	**89.5**	**5.2**	**0**	**0**	**1.4**	**.1**	**.7**	**.8**	**.3**	**.4**	**10.8**
			100	89.5	37.0	0	0	10.3	.6	5.0	6.0	2.0	3.0	77.0
76	**APPLESAUCE & APRICOTS,** w/vitamin C	1 TBLS	14	**86.9**	**6.6**	**0**	**0**	**1.7**	**.1**	**.8**	**1.4**	**.4**	**.6**	**15.3**
			100	86.9	47.0	.2	.2	12.4	.7	6.0	10.0	3.0	4.0	109.0
77	**APPLESAUCE & BLUEBERRY,** w/vitamin C	1 TBLS	14	**82.8**	**8.7**	**0**	**0**	**2.3**	**Trace**	**.7**	**1.0**	**1.8**	—	**9.1**
			100	82.8	62.0	.2	.2	16.6	.2	5.0	7.0	13.0	—	65.0
78	**APPLESAUCE & CHERRIES,** w/vitamin C	1 TBLS	14	**86.3**	**6.7**	**0**	**0**	**1.8**	—	**1.3**	—	**.4**	—	**13.6**
			100	86.3	48.0	.3	0	13.2	—	9.0	—	3.0	—	97.0
79	**APPLESAUCE & PINEAPPLE,** w/vitamin C	1 TBLS	14	**89.1**	**5.5**	**0**	**0**	**1.5**	—	**.6**	**.8**	**.3**	**.6**	**10.6**
			100	89.1	39.0	.1	.1	10.5	—	4.0	6.0	2.0	4.0	76.0
80	**APPLESAUCE & RASPBERRY,** w/vitamin C & sugar	1 TBLS	14	**84.0**	**8.1**	**0**	**0**	**2.2**	—	**.7**	**1.1**	**.3**	—	**10.1**
			100	84.0	58.0	.2	.2	15.5	—	5.0	8.0	2.0	—	72.0
81	**APRICOTS,** w/tapioca, & vitamin C	1 TBLS	14	**82.1**	**8.8**	**0**	**0**	**2.4**	**.1**	**1.1**	**1.4**	**.8**	**.6**	**17.5**
			100	82.1	63.0	.3	0	17.3	.5	8.0	10.0	6.0	4.0	125.0
82	**BANANAS & PINEAPPLE,** w/tapioca & vitamin C	1 TBLS	14	**81.7**	**9.1**	**0**	**0**	**2.5**	—	**1.0**	**.7**	**.8**	**.8**	**9.5**
			100	81.7	65.0	.2	0	17.8	—	7.0	5.0	6.0	6.0	68.0
83	**BANANAS,** w/tapioca & vitamin C	1 TBLS	14	**81.5**	**9.4**	**.1**	**0**	**2.5**	**Trace**	**1.1**	**1.3**	**1.3**	**1.7**	**15.1**
			100	81.5	67.0	.4	.2	17.8	.2	8.0	9.0	9.0	12.0	108.0
84	**GUAVA & PAPAYA,** w/tapioca & vitamin C	1 TBLS	14	**82.5**	**8.8**	**0**	**0**	**2.4**	**.1**	**1.0**	**.8**	**.6**	**.7**	**10.4**
			100	82.5	63.0	.2	.1	17.0	.5	7.0	6.0	4.0	5.0	74.0
85	**GUAVA,** w/tapioca & vitamin C	1 TBLS	14	**81.2**	**9.4**	**0**	**0**	**2.6**	**.1**	**1.0**	**.7**	**.3**	**.3**	**10.2**
			100	81.2	67.0	.3	0	18.3	1.0	7.0	5.0	2.0	2.0	73.0
86	**MANGO,** w/tapioca & vitamin C	1 TBLS	14	**77.7**	**11.2**	**0**	**0**	**3.0**	**Trace**	**.6**	**.8**	**.6**	**.6**	**8.3**
			100	77.7	80.0	.3	.2	21.6	.2	4.0	6.0	4.0	4.0	59.0
87	**PAPAYA & APPLESAUCE,** w/tapioca & vitamin C	1 TBLS	14	**80.6**	**9.8**	**0**	**0**	**2.6**	**.1**	**1.0**	**.7**	**.7**	**.7**	**11.1**
			100	80.6	70.0	.2	.1	18.9	.4	7.0	5.0	5.0	5.0	79.0
88	**PEACHES,** w/vitamin C & sugar	1 TBLS	14	**80.1**	**9.9**	**.1**	**0**	**2.6**	**.1**	**.7**	**1.5**	**.7**	**.7**	**21.7**
			100	80.1	71.0	.5	.2	18.9	.7	5.0	11.0	5.0	5.0	155.0
89	**PEARS,** w/vitamin C	1 TBLS	14	**87.8**	**6.0**	**0**	**0**	**1.6**	—	**1.1**	**1.7**	**.3**	**1.3**	**16.1**
			100	87.8	43.0	.3	.1	11.6	—	8.0	12.0	2.0	9.0	115.0
90	**PEARS & PINEAPPLE,** w/vitamin C	1 TBLS	14	**88.5**	**6.2**	**0**	**0**	**1.6**	**Trace**	**1.4**	**1.4**	**.1**	**1.0**	**16.5**
			100	88.5	44.0	.3	.2	11.4	.3	10.0	10.0	1.0	7.0	118.0
91	**PLUMS,** w/tapioca, w/o vitamin C	1 TBLS	14	**79.2**	**10.4**	**0**	**0**	**2.9**	—	**.8**	**.8**	**1.1**	**.6**	**11.6**
			100	79.2	74.0	.1	0	20.4	—	6.0	6.0	8.0	4.0	83.0
92	**PRUNES,** w/tapioca, w/o vitamin C	1 TBLS	14	**80.1**	**9.8**	**.1**	**0**	**2.6**	**Trace**	**2.1**	**2.1**	**.3**	**1.4**	**22.7**
			100	80.1	70.0	.6	.1	18.7	.3	15.0	15.0	2.0	10.0	162.0

Item No.	Minerals (Micro) Iron	Zinc	Copper	Fat-Soluble Vitamins Vitamin A	Vitamin D	Vitamin E (Alpha Tocopherol)	Water-Soluble Vitamins Vitamin C	Thiamin	Ribo-flavin	Niacin	Panto-thenic Acid	Vit. B-6 (Pyri-doxine)	Folacin (Folic Acid)	Biotin	Vitamin B-12
	(mg)	(mg)	(mg)	(IU)	(IU)	(mg)	(mg)	(mg)	(mg)	(mg)	(mg)	(mg)	(mcg)	(mcg)	(mcg)
62	.11	.13	.01	88.6	–	–	.18	Trace	.01	.11	–	.01	1.39	–	.06
	.78	.91	.04	633.0	–	–	1.30	.01	.07	.81	–	.04	9.90	–	.45
63	.13	.15	.01	59.5	–	–	.24	Trace	.01	.22	.04	.01	–	–	.06
	.89	1.10	.10	425.0	–	–	1.70	.02	.08	1.54	.25	.07	–	–	.46
64	.06	–	–	220.6	–	–	.15	.01	Trace	.08	–	.01	1.26	–	–
	.41	–	–	1576.0	–	–	1.10	.05	.03	.55	–	.06	9.00	–	–
65	.07	.06	.01	198.0	–	–	.21	Trace	Trace	.09	.02	.01	.69	–	.04
	.47	.41	.04	1414.0	–	–	1.50	.03	.03	.66	.13	.06	4.90	–	.26
66	.04	–	–	167.3	–	–	.20	Trace	Trace	.05	.03	.01	.53	–	–
	.30	–	–	1195.0	–	–	1.40	.01	.02	.33	.24	.04	3.80	–	–
67	.07	.05	–	92.4	–	–	.11	.01	.01	.07	–	–	1.04	–	–
	.47	.33	–	660.0	–	–	.80	.04	.04	.49	–	–	7.40	–	–
68	.03	.03	0	86.1	–	–	.18	.01	Trace	.05	–	Trace	.74	–	–
	.22	.22	.03	615.0	–	–	1.30	.04	.02	.35	–	.03	5.30	–	–
69	.05	.03	Trace	207.6	–	–	.24	Trace	Trace	.08	.02	.01	.50	–	.02
	.34	.22	.03	1483.0	–	–	1.70	.02	.03	.55	.16	.04	3.60	–	.16
70	.25	–	–	550.1	–	–	.25	Trace	.03	.16	–	.01	4.48	–	–
	1.81	–	–	3929.0	–	–	1.80	.02	.23	1.16	–	.10	32.00	–	–
71	.36	–	–	644.0	–	–	.28	Trace	.05	.18	–	.01	–	–	.13
	2.60	–	–	4600.0	–	–	2.00	.03	.33	1.30	–	.10	–	–	.89
72	.07	.05	.01	147.1	–	–	.11	.01	.01	.10	.03	Trace	.48	–	.01
	.49	.32	.06	1051.0	–	–	.80	.04	.04	.68	.20	.02	3.40	–	.09
73	.04	.04	–	139.2	–	–	.11	Trace	.01	.04	–	Trace	.39	–	.02
	.26	.30	–	994.0	–	–	.80	.02	.04	.30	–	.02	2.80	–	.12
74	.05	.04	Trace	125.4	–	–	.15	–	–	–	.03	Trace	.41	–	–
	.32	.25	.03	896.0	–	–	1.10	–	–	–	.21	.03	2.90	–	–
75	.03	.01	.01	1.3	–	–	5.29	Trace	Trace	.01	.01	Trace	.24	–	–
	.22	.04	.04	9.0	–	–	37.80	.01	.03	.06	.10	.03	1.70	–	–
76	.04	Trace	–	47.5	–	–	2.51	Trace	Trace	.02	–	Trace	.20	–	–
	.26	.03	–	339.0	–	–	17.90	.01	.03	.15	–	.03	1.40	–	–
77	.06	–	–	5.9	–	–	1.95	Trace	.01	.01	–	.01	.49	–	–
	.40	–	–	42.0	–	–	13.90	.02	.04	.10	–	.04	3.50	–	–
78	.06	–	–	5.7	–	–	3.25	Trace	.01	.01	–	–	.06	–	–
	.41	–	–	41.0	–	–	23.20	.02	.05	.10	–	–	.40	–	–
79	.01	Trace	–	2.9	–	–	3.75	Trace	Trace	.01	–	.01	.28	–	–
	.10	.03	–	21.0	–	–	26.80	.02	.03	.08	–	.04	2.00	–	–
80	.03	–	–	4.2	–	–	4.05	Trace	Trace	.02	.01	.01	.46	–	–
	.22	–	–	30.0	–	–	28.90	.01	.03	.10	.09	.03	3.30	–	–
81	.04	.01	–	101.2	–	–	2.51	Trace	Trace	.03	.02	Trace	.22	–	–
	.27	.04	–	723.0	–	–	17.90	.01	.01	.28	.14	.03	1.60	–	–
82	.03	.01	.01	5.7	–	–	2.69	Trace	Trace	.02	–	.01	.76	–	–
	.23	.04	.04	41.0	–	–	19.20	.02	.02	.17	–	.08	5.40	–	–
83	.04	.01	.01	6.2	–	–	3.60	Trace	Trace	.03	.02	.02	.90	–	–
	.30	.07	.05	44.0	–	–	25.70	.02	.02	.22	.17	.14	6.40	–	–
84	.03	.01	–	25.8	–	–	11.33	Trace	Trace	.04	–	Trace	–	–	–
	.20	.06	–	184.0	–	–	80.90	.01	.02	.26	–	.01	–	–	–
85	.03	.01	–	41.9	–	–	10.56	Trace	.01	.05	–	.01	–	–	–
	.20	.08	–	299.0	–	–	75.40	.01	.07	.39	–	.04	–	–	–
86	.01	.01	–	93.1	–	–	17.42	Trace	Trace	.04	–	.02	–	–	–
	.10	.06	–	665.0	–	–	124.40	.02	.03	.25	–	.12	–	–	–
87	–	Trace	–	10.6	–	–	15.83	Trace	Trace	.02	–	Trace	.76	–	–
	–	.03	–	76.0	–	–	113.10	.01	.03	.11	–	.02	5.40	–	–
88	.04	.01	.01	24.9	–	–	2.65	Trace	Trace	.09	.02	Trace	.55	–	–
	.27	.06	.05	178.0	–	–	18.90	.01	.03	.65	.13	.02	3.90	–	–
89	.04	.01	.01	4.8	–	–	3.08	Trace	Trace	.03	.01	Trace	.53	–	–
	.25	.08	.08	34.0	–	–	22.00	.01	.03	.19	.10	.01	3.80	–	–
90	.03	.02	.02	4.5	–	–	2.35	Trace	Trace	.03	–	Trace	.41	–	–
	.21	.13	.11	32.0	–	–	16.80	.02	.02	.18	–	.01	2.90	–	–
91	.03	.01	.01	13.2	–	–	.11	Trace	Trace	.03	.02	Trace	.13	–	–
	.22	.08	.04	94.0	–	–	.80	.01	.03	.21	.11	.03	.90	–	–
92	.05	.01	–	57.0	–	–	.11	Trace	.01	.07	.02	.01	.03	–	–
	.33	.10	–	407.0	–	–	.80	.02	.08	.53	.14	.09	.20	–	–

(Continued)

BABY FOODS

TABLE F-36 *(Continued)*

										Minerals (Macro)				
Item No.	Food Item Name	Approximate Measure	Weight	Moisture	Food Energy Calories	Protein	Fats	Carbo-hydrates	Fiber	Calcium	Phos-phorus	Sodium	Mag-nesium	Potas-sium
			(g)	(%)	(kcal)	(g)	(g)	(g)	(g)	(mg)	(mg)	(mg)	(mg)	(mg)
	FRUIT JUICES													
93	**APPLE,** w vitamin C	1 OZ	**30**	**88.0**	**14.1**	**0**	**0**	**3.5**	—	**1.2**	**1.5**	**.9**	**.9**	**27.3**
			100	88.0	47.0	0	0.1	11.7	—	4.0	5.0	3.0	3.0	91.0
94	**APPLE-CHERRY,** w vitamin C	1 OZ	**30**	**89.5**	**12.3**	**0**	**.1**	**3.0**	—	**1.5**	**1.8**	**.9**	**.9**	**29.4**
			100	89.5	41.0	.1	.2	9.9	—	5.0	6.0	3.0	3.0	98.0
95	**APPLE-GRAPE,** w/vitamin C	1 OZ	**30**	**88.1**	**13.8**	**0**	**.1**	**3.4**	—	**1.8**	**1.5**	**.9**	**.9**	**27.0**
			100	88.1	46.0	.1	.2	11.4	—	6.0	5.0	3.0	3.0	90.0
96	**APPLE-PEACH,** w/vitamin C	1 OZ	**30**	**89.0**	**12.6**	**.1**	**0**	**3.2**	—	**.9**	**1.2**	—	**.9**	**29.1**
			100	89.0	42.0	.2	.1	10.5	—	3.0	4.0	—	3.0	97.0
97	**APPLE-PLUM,** w/vitamin C	1 OZ	**30**	**87.3**	**14.7**	**0**	**0**	**3.7**	—	**1.5**	**.9**	—	**.9**	**30.3**
			100	87.3	49.0	.1	0	12.3	—	5.0	3.0	—	3.0	101.0
98	**APPLE-PRUNE,** w vitamin C	1 OZ	**30**	**81.3**	**21.9**	**.1**	**0**	**5.4**	—	**2.7**	**4.5**	**1.5**	—	**44.4**
			100	81.3	73.0	.2	.1	18.0	—	9.0	15.0	5.0	—	148.0
99	**MIXED FRUIT,** w/vitamin C	1 OZ	**30**	**87.9**	**14.1**	**0**	**0**	**3.5**	—	**2.4**	**1.5**	**1.2**	**1.5**	**30.3**
			100	87.9	47.0	.1	.1	11.6	—	8.0	5.0	4.0	5.0	101.0
100	**ORANGE,** w vitamin C	1 OZ	**30**	**88.5**	**13.2**	**.2**	**.1**	**3.1**	—	**3.6**	**3.3**	**.3**	**2.7**	**55.2**
			100	88.5	44.0	.6	.3	10.2	—	12.0	11.0	1.0	9.0	184.0
101	**ORANGE-APPLE,** w vitamin C	1 OZ	**30**	**88.9**	**12.9**	**.1**	**.1**	**3.0**	—	**3.0**	**2.1**	**.9**	**1.5**	**41.4**
			100	88.9	43.0	.4	.2	10.1	—	10.0	7.0	3.0	5.0	138.0
102	**ORANGE-APPLE-BANANA,** w vitamin C	1 OZ	**30**	**87.6**	**14.1**	**.1**	**0**	**3.5**	—	**1.5**	**2.4**	**1.2**	**1.8**	**40.2**
			100	87.6	47.0	.4	.1	11.5	—	5.0	8.0	4.0	6.0	134.0
103	**ORANGE-APRICOT,** w vitamin C	1 OZ	**30**	**87.8**	**13.8**	**.2**	**0**	**3.3**	—	**1.8**	**3.6**	**1.8**	**2.1**	**59.7**
			100	87.8	46.0	.8	.1	10.9	—	6.0	12.0	6.0	7.0	199.0
104	**ORANGE-BANANA,** w vitamin C	1 OZ	**30**	**86.9**	**15.0**	**.2**	**0**	**3.6**	—	**5.1**	—	**.9**	—	**60.0**
			100	86.9	50.0	.7	.1	11.9	—	17.0	—	3.0	—	200.0
105	**ORANGE-PINEAPPLE,** w/vitamin C	1 OZ	**30**	**87.3**	**14.4**	**.2**	**0**	**3.5**	—	**2.4**	**2.7**	**.6**	**2.7**	**42.3**
			100	87.3	48.0	.5	.1	11.7	—	8.0	9.0	2.0	9.0	141.0
106	**PRUNE-ORANGE,** w/vitamin C	1 OZ	**30**	**81.9**	**21.0**	**.2**	**.1**	**5.0**	—	**3.6**	**3.0**	**.6**	**2.4**	**54.3**
			100	81.9	70.0	.6	.3	16.8	—	12.0	10.0	2.0	8.0	181.0
	MEATS & EGGS													
107	**BEEF**	1 TBLS	**14**	**79.9**	**14.8**	**2.0**	**.7**	**0**	**0**	**1.1**	**10.1**	**9.2**	**1.3**	**26.6**
			100	79.9	106.0	14.5	4.9	0	0	8.0	72.0	66.0	9.0	190.0
108	**BEEF,** w beef heart	1 TBLS	**14**	**82.5**	**13.2**	**1.8**	**.6**	**0**	**Trace**	**.6**	**13.2**	**8.8**	**1.7**	**28.0**
			100	82.5	94.0	12.7	4.4	0	.1	4.0	94.0	63.0	12.0	200.0
109	**CHICKEN**	1 TBLS	**14**	**76.0**	**20.9**	**2.1**	**1.3**	**0**	—	**7.7**	**12.6**	**7.1**	**1.5**	**17.1**
			100	76.0	149.0	14.7	9.6	0	—	55.0	90.0	51.0	11.0	122.0
110	**CHICKEN STICKS**	1 AVG	**10**	**68.3**	**18.8**	**1.5**	**1.4**	**.1**	**Trace**	**7.3**	**12.1**	**47.9**	—	**10.6**
			100	68.3	188.0	14.6	14.4	1.4	.2	73.0	121.0	479.0	—	106.0
111	**EGG YOLKS**	1 TBLS	**14**	**70.6**	**28.4**	**1.4**	**2.4**	**.1**	**0**	**10.6**	**40.2**	**5.5**	**1.0**	**10.8**
			100	70.6	203.0	10.0	17.3	1.0	0	76.0	287.0	39.0	7.0	77.0
112	**EGG YOLKS** w/ham or bacon	1 TBLS	**14**	**70.3**	**29.1**	**1.4**	**2.5**	**0**	**0**	**9.9**	**25.9**	**43.8**	—	**11.5**
			100	70.3	208.0	10.0	18.1	.3	0	71.0	185.0	313.0	—	82.0
113	**FORMULA MEAT BASE,** normal dilution	1 OZ	**30**	—	**19.8**	**.8**	**.9**	**1.2**	—	**30.9**	**20.7**	**8.4**	—	**14.1**
			100	—	66.0	2.7	3.1	4.0	—	103.0	69.0	28.0	—	47.0
114	**HAM**	1 TBLS	**14**	**78.5**	**17.5**	**2.1**	**.9**	**0**	—	**.7**	**12.5**	**9.4**	**1.5**	**29.4**
			100	78.5	125.0	15.1	6.7	0	—	5.0	89.0	67.0	11.0	210.0
115	**LAMB**	1 TBLS	**14**	**79.6**	**15.7**	**2.1**	**.7**	**0**	**0**	**1.0**	**12.7**	**10.2**	**1.4**	**29.5**
			100	79.6	112.0	15.2	5.2	0	0	7.0	91.0	73.0	10.0	211.0
116	**LIVER**	1 TBLS	**14**	**79.3**	**14.1**	**2.0**	**.5**	**.2**	**0**	**.6**	**28.4**	**10.4**	**1.8**	**31.8**
			100	79.3	101.0	14.3	3.8	1.4	0	4.0	203.0	74.0	13.0	227.0
117	**LIVER & BACON**	1 TBLS	**14**	**77.0**	**17.2**	**1.9**	**.9**	**.2**	**0**	**.8**	**22.0**	**42.3**	—	**26.9**
			100	77.0	123.0	13.7	6.6	1.3	0	6.0	157.0	302.0	—	192.0
118	**MEAT STICKS**	1 AVG	**10**	**69.5**	**18.4**	**1.3**	**1.5**	**.1**	**Trace**	**3.4**	**10.3**	**54.7**	—	**11.4**
			100	69.5	184.0	13.4	14.6	1.1	.2	34.0	103.0	547.0	—	114.0
119	**PORK**	1 TBLS	**14**	**74.3**	**18.8**	**2.6**	**.8**	**0**	**0**	**1.1**	**20.2**	**33.2**	—	**29.4**
			100	74.3	134.0	18.6	6.0	0	0	8.0	144.0	237.0	—	210.0
120	**PORK,** low sodium	1 TBLS	**14**	**78.4**	**17.4**	**2.0**	**1.0**	**0**	**0**	**.7**	**13.2**	**5.9**	**1.4**	**31.2**
			100	78.4	124.0	14.0	7.1	0	0	5.0	94.0	42.0	10.0	223.0
121	**TURKEY**	1 TBLS	**14**	**77.5**	**18.1**	**2.2**	**1.0**	**0**	**0**	**3.9**	**13.3**	**10.1**	**1.7**	**25.2**
			100	77.5	129.0	15.4	7.1	0	0	28.0	95.0	72.0	12.0	180.0
122	**TURKEY STICKS**	1 AVG	**10**	**69.8**	**18.2**	**1.4**	**1.4**	**.1**	**.1**	**7.2**	**10.3**	**48.3**	—	**9.1**
			100	69.8	182.0	13.7	14.2	1.4	.5	72.0	103.0	483.0	—	91.0
123	**VEAL**	1 TBLS	**14**	**79.8**	**15.4**	**2.1**	**.7**	**0**	**Trace**	**.8**	**13.7**	**9.7**	**1.5**	**33.0**
			100	79.8	110.0	15.3	5.0	0	.2	6.0	98.0	69.0	11.0	236.0

Item No.	Minerals (Micro)			Fat-Soluble Vitamins			Water-Soluble Vitamins								
	Iron	Zinc	Copper	Vitamin A	Vitamin D	Vitamin E (Alpha Tocopherol)	Vitamin C	Thiamin	Ribo-flavin	Niacin	Panto-thenic Acid	Vit. B-6 (Pyri-doxine)	Folacin (Folic Acid)	Biotin	Vitamin B-12
	(mg)	(mg)	(mg)	(IU)	(IU)	(mg)	(mg)	(mg)	(mg)	(mg)	(mg)	(mg)	(mcg)	(mcg)	(mcg)
93	**.17**	**.01**	—	**5.4**	—	—	**17.37**	**Trace**	**.01**	**.03**	—	**.01**	**.03**	—	—
	.57	.03	—	18.0	—	—	57.90	.01	.02	.08	—	.03	.10	—	—
94	**.20**	**.01**	—	**1.5**	—	—	**17.49**	**Trace**	**.01**	**.03**	—	**.01**	**.09**	—	—
	.66	.03	—	5.0	—	—	58.30	.01	.02	.09	—	.03	.30	—	—
95	**.12**	**.01**	—	**1.8**	—	—	**16.08**	**Trace**	**.01**	**.03**	—	**.01**	**.09**	—	—
	.39	.04	—	6.0	—	—	53.60	.01	.02	.11	—	.03	.30	—	—
96	**.17**	**.01**	—	**18.9**	—	—	**17.55**	**Trace**	**Trace**	**.06**	—	**.01**	**.39**	—	—
	.56	.03	—	63.0	—	—	58.50	.01	.01	.21	—	.02	1.30	—	—
97	**.19**	**.01**	—	**12.9**	—	—	**17.46**	**.01**	**.01**	**.06**	—	**.01**	**.06**	—	—
	.62	.03	—	43.0	—	—	58.20	.02	.02	.20	—	.03	.20	—	—
98	**.29**	—	—	—	—	—	**20.25**	**Trace**	**Trace**	**.09**	—	**.01**	**.03**	—	—
	.95	—	—	—	—	—	67.50	.01	Trace	.30	—	.04	.10	—	—
99	**.10**	**.01**	—	**12.6**	—	—	**19.08**	**.01**	**Trace**	**.04**	—	**.01**	**2.01**	—	—
	.34	.03	—	42.0	—	—	63.60	.02	.01	.12	—	.04	6.70	—	—
100	**.05**	**.02**	—	**16.5**	—	—	**18.75**	**.01**	**.01**	**.07**	—	**.02**	**7.92**	—	—
	.17	.06	—	55.0	—	—	62.50	.05	.03	.24	—	.05	26.40	—	—
101	**.06**	**.01**	—	**21.9**	—	—	**23.07**	**.01**	**.01**	**.06**	—	**.01**	**3.66**	—	—
	.20	.03	—	73.0	—	—	76.90	.04	.03	.19	—	.04	12.20	—	—
102	**.11**	**.01**	—	**8.1**	—	—	**9.63**	**.01**	**.01**	**.08**	—	**.02**	**2.91**	—	—
	.35	.03	—	27.0	—	—	32.10	.04	.03	.26	—	.06	9.70	—	—
103	**.11**	**.01**	—	**64.8**	—	—	**25.77**	**.02**	**.01**	**.08**	—	**.02**	**5.97**	—	—
	.38	.04	—	216.0	—	—	85.90	.06	.03	.27	—	.06	19.90	—	—
104	**.03**	—	—	**13.8**	—	—	**10.20**	**.02**	**.01**	**.05**	—	—	**7.32**	—	—
	.11	—	—	46.0	—	—	34.00	.05	.04	.18	—	—	24.40	—	—
105	**.13**	**.01**	—	**9.3**	—	—	**16.02**	**.02**	**.01**	**.06**	—	**.02**	**5.58**	—	—
	.42	.04	—	31.0	—	—	53.40	.05	.02	.19	—	.06	18.60	—	—
106	**.26**	**.01**	—	**39.3**	—	—	**19.14**	**.01**	**.04**	**.12**	—	**.02**	**3.93**	—	—
	.87	.04	—	131.0	—	—	63.80	.04	.12	.40	—	.06	13.10	—	—
107	**.23**	**.28**	**.01**	**14.4**	—	—	**.27**	**Trace**	**.02**	**.46**	**.05**	**.02**	**.80**	—	**.21**
	1.65	2.00	.09	103.0	—	—	1.90	.01	.16	3.28	.35	.12	5.70	—	1.47
108	**.28**	**.26**	**.02**	**17.6**	—	—	**.29**	**Trace**	**.05**	**.54**	—	**.02**	**.70**	—	**.96**
	1.99	1.84	.13	126.0	—	—	2.10	.02	.36	3.89	—	.12	5.00	—	6.82
109	**.14**	**.14**	**.01**	**26.0**	—	—	**.21**	**Trace**	**.02**	**.48**	**.10**	**.03**	**1.55**	—	—
	.99	1.01	.05	186.0	—	—	1.50	.01	.16	3.42	.73	.19	11.10	—	—
110	**.16**	—	—	**317.8**	—	—	**.17**	**Trace**	**.02**	**.20**	—	**.01**	—	—	—
	1.56	—	—	3178.0	—	—	1.70	.02	.20	2.01	—	.10	—	—	—
111	**.39**	**.27**	**.01**	**175.1**	—	—	**.20**	**.01**	**.04**	**Trace**	**.30**	**.02**	**12.89**	—	**.22**
	2.76	1.92	.07	1251.0	—	—	1.40	.07	.27	.03	2.14	.16	92.10	—	1.54
112	**.39**	—	—	**266.0**	—	—	—	**.01**	**.03**	**.07**	—	**.07**	—	—	**.18**
	2.80	—	—	1900.0	—	—	—	.10	.23	.50	—	.53	—	—	1.26
113	**.30**	—	—	**47.7**	**12.60**	**.31**	**1.26**	**.01**	**.04**	**.32**	—	**.02**	—	—	—
	1.00	—	—	159.0	42.00	1.04	4.20	.04	.14	1.06	—	.05	—	—	—
114	**.14**	**.24**	—	**4.5**	—	—	**.29**	**.02**	**.03**	**.40**	**.07**	**.03**	**.29**	—	—
	1.01	1.70	—	32.0	—	—	2.10	.14	.19	2.84	.53	.20	2.10	—	—
115	**.23**	**.36**	**.01**	**3.8**	—	—	**.24**	**Trace**	**.03**	**.45**	**.06**	**.03**	**.28**	—	**.32**
	1.66	2.60	.06	27.0	—	—	1.70	.02	.19	3.19	.42	.18	2.00	—	2.27
116	**.74**	**.42**	**.28**	**5338.9**	—	—	**2.70**	**.01**	**.25**	**1.17**	—	**.05**	**47.24**	—	**.30**
	5.29	2.98	1.99	38135.0	—	—	19.30	.05	1.81	8.33	—	.34	337.40	—	2.16
117	**.59**	—	—	**3080.0**	—	—	**.98**	**.01**	**.28**	**1.09**	—	—	—	—	—
	4.20	—	—	22000.0	—	—	7.00	.05	1.99	7.80	—	—	—	—	—
118	**.14**	—	—	**6.9**	—	—	**.24**	**.01**	**.02**	**.15**	—	**.01**	—	—	—
	1.38	—	—	69.0	—	—	2.40	.06	.17	1.48	—	.08	—	—	—
119	**.17**	—	—	—	—	—	**.22**	**.03**	**.03**	**.39**	—	**.07**	—	—	**.15**
	1.20	—	—	—	—	—	1.60	.23	.23	2.80	—	.49	—	—	1.06
120	**.14**	**.32**	**.01**	**5.3**	—	—	**.25**	**.02**	**.03**	**.32**	—	**.03**	**.27**	—	**.14**
	1.00	2.27	.07	38.0	—	—	1.80	.15	.20	2.27	—	.21	1.90	—	.99
121	**.19**	**.25**	—	**79.5**	—	—	**.34**	**Trace**	**.04**	**.49**	**.09**	**.02**	**1.69**	—	**.15**
	1.35	1.80	—	568.0	—	—	2.40	.02	.25	3.48	.61	.17	12.10	—	1.07
122	**.12**	—	—	**22.8**	—	—	**.15**	**Trace**	**.02**	**.18**	—	**.01**	—	—	—
	1.24	—	—	228.0	—	—	1.50	.01	.16	1.75	—	.08	—	—	—
123	**.16**	**.35**	**.01**	**7.0**	—	—	**.29**	**Trace**	**.03**	**.53**	**.06**	**.02**	**.94**	—	**.18**
	1.25	2.52	.08	50.0	—	—	2.10	.02	.18	3.81	.45	.12	6.70	—	1.30

(Continued)

TABLE F-36 *(Continued)*

Item No.	Food Item Name	Approximate Measure	Weight	Moisture	Food Energy Calories	Protein	Fats	Carbo-hydrates	Fiber	Minerals (Macro) Calcium	Phos-phorus	Sodium	Mag-nesium	Potas-sium
			(g)	(%)	(kcal)	(g)	(g)	(g)	(g)	(mg)	(mg)	(mg)	(mg)	(mg)
	VEGETABLES													
124	**BEANS, GREEN**	1 TBLS	**14**	**92.5**	**3.5**	**.2**	**0**	**.8**	**.1**	**9.1**	**2.7**	**.3**	**3.1**	**17.9**
			100	92.5	25.0	1.2	.1	5.7	.9	65.0	19.0	2.0	22.0	128.0
125	**BEANS, GREEN,** buttered	1 TBLS	**14**	**91.2**	**4.5**	**.2**	**.1**	**.9**	—	**9.7**	—	**.3**	—	**23.9**
			100	91.2	32.0	1.3	.9	6.1	—	69.0	—	2.0	—	171.0
126	**BEANS, GREEN,** creamed	1 TBLS	**14**	**90.9**	**4.5**	**.1**	**.1**	**1.0**	—	**4.5**	**2.7**	**1.7**	**1.0**	**9.1**
			100	90.9	32.0	1.0	.4	7.2	—	32.0	19.0	12.0	7.0	65.0
127	**BEETS**	1 TBLS	**14**	**90.1**	**4.8**	**.2**	**0**	**1.1**	**.1**	**2.0**	**2.0**	**11.6**	**2.0**	**25.5**
			100	90.1	34.0	1.3	.1	7.7	.8	14.0	14.0	83.0	14.0	182.0
128	**CARROTS**	1 TBLS	**14**	**91.0**	**4.5**	**.1**	**0**	**1.0**	**.1**	**3.2**	**2.8**	**6.9**	**1.5**	**28.3**
			100	91.0	32.0	.8	.2	7.2	.8	23.0	20.0	49.0	11.0	202.0
129	**CARROTS,** buttered	1 TBLS	**14**	**91.4**	**4.6**	**.1**	**.1**	**.9**	—	**4.9**	—	**2.2**	—	**20.3**
			100	91.4	33.0	.8	.6	6.7	—	35.0	—	16.0	—	145.0
130	**CORN,** creamed	1 TBLS	**14**	**81.4**	**9.1**	**.2**	**.1**	**2.3**	**Trace**	**2.5**	**4.6**	**7.3**	**1.1**	**11.3**
			100	81.4	65.0	1.4	.4	16.3	.1	18.0	33.0	52.0	8.0	81.0
131	**GARDEN VEGETABLES**	1 TBLS	**14**	**90.0**	**5.2**	**.3**	**0**	**1.9**	**.1**	**3.9**	**3.9**	**4.9**	**2.9**	**23.5**
			100	90.0	37.0	2.3	.2	6.8	.9	28.0	28.0	35.0	21.0	168.0
132	**MIXED VEGETABLES**	1 TBLS	**14**	**89.4**	**5.7**	**.2**	**.1**	**1.1**	**.1**	**1.5**	**3.5**	**5.0**	—	**23.8**
			100	89.4	41.0	1.4	.4	8.2	.5	11.0	25.0	36.0	—	170.0
133	**PEAS**	1 TBLS	**14**	**87.5**	**5.6**	**.5**	**0**	**1.1**	**.2**	**2.8**	**6.0**	**.6**	**2.1**	**15.7**
			100	87.5	40.0	3.5	.3	8.1	1.2	20.0	43.0	4.0	15.0	112.0
134	**PEAS,** buttered	1 TBLS	**14**	**83.5**	**8.4**	**.5**	**.2**	**1.6**	—	**6.3**	—	**.7**	—	**16.4**
			100	83.5	60.0	3.5	1.3	11.3	—	45.0	—	5.0	—	117.0
135	**PEAS,** creamed	1 TBLS	**14**	**86.5**	**7.4**	**.3**	**.3**	**1.2**	**.1**	**1.8**	**4.3**	**2.0**	—	**12.3**
			100	86.5	53.0	2.2	1.9	8.9	.4	13.0	31.0	14.0	—	88.0
136	**SPINACH,** creamed	1 TBLS	**14**	**88.2**	**5.9**	**.4**	**.2**	**.9**	**.1**	**15.8**	**6.9**	**7.7**	**8.8**	**30.9**
			100	88.2	42.0	3.0	1.4	6.4	.5	113.0	49.0	55.0	63.0	221.0
137	**SQUASH**	1 TBLS	**14**	**92.8**	**3.4**	**.1**	**0**	**.8**	**.1**	**3.4**	**2.2**	**.1**	**1.7**	**25.9**
			100	92.8	24.0	.8	.2	5.6	.7	24.0	16.0	1.0	12.0	185.0
138	**SQUASH,** buttered	1 TBLS	**14**	**91.8**	**4.2**	**.1**	**.1**	**.9**	—	**4.3**	—	**.3**	—	**18.9**
			100	91.8	30.0	.7	.6	6.4	—	31.0	—	2.0	—	135.0
139	**SWEET POTATOES**	1 TBLS	**14**	**84.1**	**8.4**	**.2**	**0**	**1.9**	**.1**	**2.2**	**3.4**	**3.1**	**1.7**	**34.0**
			100	84.1	60.0	1.1	.1	13.9	.6	16.0	24.0	22.0	12.0	243.0
140	**SWEET POTATOES,** buttered	1 TBLS	**14**	**85.6**	**8.0**	**.1**	**.1**	**1.7**	—	**3.9**	—	**1.1**	—	**30.2**
			100	85.6	57.0	.8	.7	12.2	—	28.0	—	8.0	—	216.0
	BREADS													
141	**BOSTON BROWN**	1 SLICE	**35**	**45.0**	**73.9**	**1.9**	**.5**	**16.0**	**.3**	**31.5**	**56.0**	**87.9**	—	**102.2**
			100	45.0	211.0	5.5	1.3	45.6	.7	90.0	160.0	251.0	—	292.0
142	**BREADCRUMBS,** dry, grated	1 CUP	**100**	**6.5**	**392.0**	**12.6**	**4.6**	**73.4**	**.3**	**122.0**	**141.0**	**736.0**	**34.0**	**152.0**
			100	6.5	392.0	12.6	4.6	73.4	.3	122.0	141.0	736.0	34.0	152.0
143	**CINNAMON**	1 SLICE	**25**	**35.6**	**67.5**	**2.2**	**.8**	**12.6**	**.1**	**21.0**	**24.3**	**126.8**	—	**26.3**
			100	35.6	270.0	8.7	3.2	50.5	.2	84.0	97.0	507.0	—	105.0
144	**CRACKED WHEAT**	1 SLICE	**25**	**34.9**	**65.8**	**2.2**	**.6**	**13.0**	**.1**	**22.0**	**32.0**	**132.3**	**8.8**	**33.5**
			100	34.9	263.0	8.7	2.2	52.1	.5	88.0	128.0	529.0	35.0	134.0
145	**CRACKED WHEAT,** toasted	1 SLICE	**21**	**22.5**	**65.7**	**2.2**	**.5**	**13.0**	**.1**	**22.1**	**31.9**	**132.3**	**7.4**	**33.6**
			100	22.5	313.0	10.4	2.6	62.0	.6	105.0	152.0	630.0	35.0	160.0
146	**FLAT (GREEK, SYRIAN, & PITA)**	1 LG	**113**	—	**267.8**	**8.2**	**.8**	**55.8**	—	**14.7**	**89.3**	—	**22.6**	—
			100	—	237.0	7.3	.7	49.4	—	13.0	79.0	—	20.0	—
147	**FRENCH OR VIENNA,** enriched	1 SLICE	**20**	**30.6**	**58.0**	**1.8**	**.6**	**11.1**	**Trace**	**8.6**	**17.0**	**116.0**	**4.4**	**18.0**
			100	30.6	290.0	9.1	3.0	55.4	.2	43.0	85.0	580.0	22.0	90.0
148	**FRENCH OR VIENNA,** enriched, toasted	1 SLICE	**17**	**19.3**	**57.5**	**1.8**	**.6**	**10.9**	**Trace**	**8.5**	**16.8**	**114.6**	**3.4**	**17.9**
			100	19.3	338.0	10.6	3.5	64.4	.2	50.0	99.0	674.0	20.0	105.0
149	**FRENCH TOAST**	1 OZ	**28**	**41.1**	**51.3**	**2.3**	**1.4**	**7.1**	**Trace**	**16.6**	**29.3**	**32.2**	**4.5**	**26.1**
			100	41.1	183.3	8.3	5.1	25.5	.1	59.2	104.7	115.1	16.2	93.1
150	**HAMBURGER OR WIENER BUN,** enriched	1 WHOLE	**48**	**31.0**	**144.0**	**3.6**	**2.4**	**25.2**	—	**36.0**	—	—	**11.0**	—
			100	31.0	300.0	7.5	5.0	52.5	—	75.0	—	—	23.0	—
151	**ITALIAN,** enriched	1 SLICE	**20**	**31.8**	**55.2**	**1.8**	**.2**	**11.3**	**Trace**	**3.4**	**15.4**	**117.0**	—	**14.8**
			100	31.8	276.0	9.1	.8	56.4	.2	17.0	77.0	585.0	—	74.0
152	**LIGHT HOLLYWOOD**	1 SLICE	**57**	**38.0**	**139.7**	**6.3**	**2.3**	**26.2**	**.2**	**80.4**	—	**330.6**	—	—
			100	38.0	245.0	11.0	4.0	46.0	.3	141.0	—	580.0	—	—
153	**LIGHT PROFILE**	1 SLICE	**57**	**38.0**	**151.1**	**5.1**	**2.3**	**26.2**	**.2**	**40.5**	—	**342.0**	—	—
			100	38.0	265.0	9.0	4.0	46.0	.3	71.0	—	600.0	—	—

BABY FOODS

Item No.	Minerals (Micro) Iron (mg)	Zinc (mg)	Copper (mg)	Fat-Soluble Vitamins Vitamin A (IU)	Vitamin D (IU)	Vitamin E (Alpha Tocopherol) (mg)	Water-Soluble Vitamins Vitamin C (mg)	Thiamin (mg)	Riboflavin (mg)	Niacin (mg)	Pantothenic Acid (mg)	Vit. B-6 (Pyridoxine) (mg)	Folacin (Folic Acid) (mcg)	Biotin (mcg)	Vitamin B-12 (mcg)
124	**.15** 1.08	**.03** .19	**.01** .05	**60.6** 433.0	— —	— —	**1.18** 8.40	**Trace** .02	**.01** .10	**.05** .32	**.02** .15	**.01** .04	**4.58** 32.70	— —	— —
125	**.16** 1.15	— —	— —	**53.5** 382.0	— —	— —	**1.20** 8.60	**Trace** .01	**.02** .11	**.04** .32	— —	— —	**3.82** 27.30	— —	— —
126	**.04** .26	**.02** .16	— —	**21.0** 150.0	— —	— —	**.38** 2.70	**Trace** .02	**.01** .05	**.03** .23	— —	**Trace** .01	— —	— —	— —
127	**.05** .32	**.02** .12	**.01** .07	**4.6** 33.0	— —	— —	**.34** 2.40	**Trace** .01	**.01** .04	**.02** .13	— —	**Trace** .02	**4.31** 30.80	— —	**0** 0
128	**.06** .39	**.03** .18	**.01** .05	**1653.4** 11810.0	— —	— —	**.77** 5.50	**Trace** .02	**.01** .04	**.07** .50	**.04** .28	**.01** .08	**2.42** 17.30	— —	— —
129	**.04** .31	— —	— —	**1377.6** 9840.0	— —	— —	**1.06** 7.60	**Trace** .02	**.01** .06	**.07** .52	— —	— —	**1.20** 8.60	— —	— —
130	**.04** .27	**.03** .23	**.01** .04	**10.8** 77.0	— —	— —	**.31** 2.20	**Trace** .01	**.01** .05	**.07** .50	**.05** .33	**.01** .04	**1.78** 12.70	— —	**Trace** .02
131	**0.12** .83	**.04** .26	**.01** .07	**849.4** 6067.0	— —	— —	**.80** 5.70	**.01** .06	**.01** .07	**.11** .78	— —	**.01** .10	**5.63** 40.20	— —	— —
132	**0.06** 0.41	— —	— —	**587.3** 4195.0	— —	— —	**.35** 2.50	**Trace** .03	**Trace** .03	**.09** .67	**.04** .26	**.01** .08	**.57** 4.10	— —	— —
133	**.13** .96	**.05** .35	**.01** .06	**79.1** 565.0	— —	— —	**.97** 6.90	**.01** .08	**.01** .06	**.14** 1.02	**.04** .28	**.01** .07	**3.63** 25.90	— —	**0** 0
134	**.15** 1.04	— —	— —	**57.4** 410.0	— —	— —	**1.78** 12.70	**.01** 0.7	**.01** .07	**.19** 1.38	— —	— —	**5.07** 36.20	— —	— —
135	**.08** .56	**.05** .39	**.01** .05	**12.0** 86.0	— —	— —	**.22** 1.60	**.01** .09	**.01** .06	**.11** .81	— —	**.01** .05	**3.18** 22.10	— —	**.01** .08
136	**.20** 1.40	**.05** .35	**.01** .07	**514.6** 3676.0	— —	— —	**.50** 3.60	**Trace** .02	**.01** .09	**.04** .26	— —	**.01** .06	**9.63** 68.80	— —	— —
137	**.05** .35	**.01** .08	**.01** .05	**282.0** 2014.0	— —	— —	**1.09** 7.80	**Trace** .01	**.01** .07	**.05** .38	**.03** .22	**.01** .07	**2.16** 15.40	— —	— —
138	**.06** .42	— —	— —	**214.2** 1530.0	— —	— —	**1.06** 7.60	**Trace** .01	**.01** .07	**.04** .32	— —	— —	**1.64** 11.70	— —	— —
139	**.06** .39	**.02** .11	**.01** .10	**929.0** 6636.0	— —	— —	**1.34** 9.60	**Trace** .03	**.01** .03	**.05** .38	**.06** .41	**.02** .11	**1.44** 10.30	— —	— —
140	**.06** .40	— —	— —	**846.0** 6043.0	— —	— —	**1.30** 9.30	**Trace** .02	**0.1** .05	**.04** .30	— —	— —	**1.88** 13.40	— —	— —

BAKERY PRODUCTS

Item No.	Minerals (Micro) Iron (mg)	Zinc (mg)	Copper (mg)	Fat-Soluble Vitamins Vitamin A (IU)	Vitamin D (IU)	Vitamin E (Alpha Tocopherol) (mg)	Water-Soluble Vitamins Vitamin C (mg)	Thiamin (mg)	Riboflavin (mg)	Niacin (mg)	Pantothenic Acid (mg)	Vit. B-6 (Pyridoxine) (mg)	Folacin (Folic Acid) (mcg)	Biotin (mcg)	Vitamin B-12 (mcg)
141	**.67** 1.90	— —	— —	**0** 0	— —	— —	**0** 0	**.04** .11	**.02** .06	**.42** 1.20	— —	— —	— —	— —	— —
142	**3.60** 3.60	— —	— —	— —	— —	— —	— —	**.22** .22	**.30** .30	**3.50** 3.50	— —	— —	— —	— —	— —
143	**.63** 2.50	— —	— —	— —	— —	— —	— —	**.06** .25	**.05** .21	**.60** 2.40	— —	— —	— —	— —	— —
144	**.28** 1.10	— —	— —	— —	— —	— —	— —	**.03** .12	**.02** .09	**.33** 1.30	**.15** .61	**.02** .09	— —	— —	**0** 0
145	**.27** 1.30	— —	— —	— —	— —	— —	— —	**.02** .11	**.02** .11	**.32** —	— 1.50	— —	— —	— —	— —
146	**2.37** 2.10	— —	— —	**0** 0	— —	— —	**0** 0	**.29** .26	**.28** .25	**2.94** 2.60	— —	**.08** .07	— —	— —	**0** 0
147	**.44** 2.20	**(.28)** (1.40)	**(.05)** (.23)	— —	— —	**(.02)** (.10)	— —	**.06** .28	**.04** .22	**.50** 2.50	**.08** .38	**.01** .05	— —	— —	**0** 0
148	**.44** 2.60	— —	— —	— —	— —	— —	— —	**.10** .58	**.06** .36	**.49** 2.90	— —	**.01** .03	— —	— —	— —
149	**.55** 1.96	**.21** .73	— —	**4.6** 16.4	**4.42** 15.77	— —	**Trace** 0.01	**.10** .34	**.13** .45	**.34** 1.21	**.20** .71	**.01** .04	**10.54** 37.63	— —	**.14** .48
150	**.96** 2.00	**.29** .60	**.12** .25	— —	— —	— —	— —	**.13** .27	**.08** .17	**1.08** 2.25	— —	— —	— —	— —	— —
151	**.44** 2.20	**.01** .05	— —	**0** 0	— —	— —	**0** 0	**.06** .29	**.04** .20	**.52** 2.60	— —	— —	— —	— —	— —
152	**1.82** 3.20	— —	— —	**0** 0	— —	— —	**0** 0	**.23** .40	**.17** .30	**2.00** 3.50	— —	— —	— —	— —	— —
153	**1.82** 3.20	— —	— —	**0** 0	— —	— —	**0** 0	**.23** .40	**.14** .24	**2.00** 3.50	— —	— —	— —	— —	— —

(Continued)

TABLE F-36 *(Continued)*

Item No.	Food Item Name	Approximate Measure	Weight	Moisture	Food Energy Calories	Protein	Fats	Carbo-hydrates	Fiber	Minerals (Macro) Calcium	Phos-phorus	Sodium	Mag-nesium	Potas-sium
			(g)	(%)	(kcal)	(g)	(g)	(g)	(g)	(mg)	(mg)	(mg)	(mg)	(mg)
154	PORTUGUESE	1 SLICE	36	—	—	—	—	—	—	—	—	96.8	—	31.7
			100	—	—	—	—	—	—	—	—	269.0	—	88.0
155	RAISIN	1 SLICE	25	35.3	65.5	1.7	.7	13.4	.2	17.8	21.8	91.3	6.0	58.3
			100	35.3	262.0	6.6	2.8	53.6	.9	71.0	87.0	365.0	24.0	233.0
156	RAISIN, toasted	1 SLICE	21	22.0	66.4	1.7	.7	13.6	.2	18.1	22.1	92.4	6.5	59.0
			100	22.0	316.0	8.0	3.4	64.6	1.1	86.0	105.0	440.0	31.0	281.0
157	ROMAN MEAL	1 SLICE	57	38.0	139.7	6.3	2.3	27.4	.7	60.4	—	342.0	—	—
			100	38.0	245.0	11.0	4.0	48.0	1.3	106.0	—	600.0	—	—
158	RYE, AMERICAN	1 SLICE	23	35.5	55.9	2.1	.3	12.0	.1	17.3	33.8	128.1	9.7	33.4
			100	35.5	243.0	9.1	1.1	52.1	.4	75.0	147.0	557.0	42.0	145.0
159	RYE, AMERICAN, toasted	1 SLICE	20	25.0	56.4	2.1	.3	12.1	.1	17.4	34.2	129.6	10.0	33.8
			100	25.0	282.0	10.6	1.3	60.5	.5	87.0	171.0	648.0	50.0	169.0
160	RYE, PUMPERNICKEL	1 SLICE	32	34.0	78.7	2.9	.4	17.0	.4	26.9	73.3	182.1	22.7	145.3
			100	34.0	246.0	9.1	1.2	53.1	1.1	84.0	229.0	569.0	71.0	454.0
161	SALT FREE	1 SLICE	25	35.6	67.5	2.2	.8	12.6	.1	21.0	24.3	30.0	6.3	26.3
			100	35.6	270.0	8.7	3.2	50.5	.2	84.0	97.0	120.0	25.0	105.0
162	SALT RISING	1 SLICE	25	36.5	66.8	2.0	.6	13.1	.1	5.8	17.3	66.3	—	16.8
			100	36.5	267.0	7.9	2.4	52.2	.2	23.0	69.0	265.0	—	67.0
163	SALT RISING, toasted	1 SLICE	22	29.4	65.3	1.9	.6	12.8	Trace	5.7	16.9	64.7	—	16.3
			100	29.4	297.0	8.8	2.7	58.0	.2	26.0	77.0	294.0	—	74.0
164	STUFFING MIX, stove top rice	1 OZ	28	6.7	106.1	3.1	1.0	20.9	.1	27.2	28.3	491.1	—	40.9
			100	6.7	379.0	11.0	3.4	74.8	.3	97.0	101.0	1754.0	—	146.0
165	STUFFING BREAD (made from mix) dry crumbly w butter, w/o egg	1 CUP	141	33.2	504.8	9.2	30.7	50.2	.6	93.1	136.8	1263.4	—	126.9
			100	33.2	358.0	6.5	21.8	35.6	.4	66.0	97.0	896.0	—	90.0
166	STUFFING BREAD (made from mix) moist, w butter & eggs	1 CUP	190	61.4	395.2	8.4	24.3	37.4	.4	76.0	125.4	957.6	—	110.2
			100	61.4	208.0	4.4	12.8	19.7	.2	40.0	66.0	504.0	—	58.0
167	WIENER BUN	1 WHOLE	36	—	108.0	2.7	1.8	18.9	—	27.0	—	—	—	—
			100	—	300.0	7.5	5.0	52.5	—	75.0	—	—	—	—
168	WHITE, enriched w 1-2% nonfat dried milk	1 SLICE	23	35.8	61.9	2.0	.7	11.6	Trace	16.1	20.0	116.6	3.9	19.6
			100	35.8	269.0	8.7	3.2	50.4	.2	70.0	87.0	507.0	17.0	85.0
169	WHITE, enriched w 1-2% nonfat dried milk, toasted	1 SLICE	20	25.3	62.8	2.0	.7	11.7	Trace	16.2	20.2	118.0	4.0	19.8
			100	25.3	314.0	10.1	3.7	58.7	.2	81.0	101.0	590.0	20.0	99.0
170	WHITE, enriched w 3-4% nonfat dried milk	1 SLICE	23	35.6	62.1	2.0	.7	11.6	Trace	19.3	22.3	116.6	5.8	24.2
			100	35.6	270.0	8.7	3.2	50.5	.2	84.0	97.0	507.0	25.0	105.0
171	WHITE, enriched w 3-4% nonfat dried milk, toasted*	1 SLICE	20	25.1	62.8	2.0	.7	11.8	Trace	19.6	22.6	118.0	4.0	24.4
			100	25.1	314.0	10.1	3.7	58.8	.2	98.0	113.0	590.0	20.0	122.0
172	WHITE, enriched w 5-6% nonfat dried milk	1 SLICE	23	35.0	63.3	2.1	.9	11.5	Trace	22.1	23.5	113.9	6.0	27.8
			100	35.0	275.0	9.0	3.8	50.2	.2	96.0	102.0	495.0	26.0	121.0
173	WHITE, enriched w 5-6% nonfat dried milk, toasted*	1 SLICE	20	24.4	64.0	2.1	.9	11.7	Trace	22.4	23.8	115.2	6.0	28.2
			100	24.4	320.0	10.5	4.4	58.4	.2	112.0	119.0	576.0	30.0	141.0
174	WHOLE WHEAT, made w water	1 SLICE	23	36.4	55.4	2.1	.6	11.3	.3	19.3	58.4	121.9	13.6	58.9
			100	36.4	241.0	9.1	2.6	49.3	1.5	84.0	254.0	530.0	59.0	256.0
175	WHOLE WHEAT, made w water, toasted	1 SLICE	19	24.3	54.5	2.1	.6	11.2	.3	19.0	57.4	119.9	—	58.0
			100	24.3	287.0	10.8	3.1	58.7	1.8	100.0	302.0	631.0	—	305.0
176	WHOLE WHEAT, made w 2% nonfat dried milk	1 SLICE	23	36.4	55.9	2.4	.7	11.0	.4	22.8	52.4	121.2	44.9	62.8
			100	36.4	243.0	10.5	3.0	47.7	1.6	99.0	228.0	527.0	195.0	273.0
177	WHOLE WHEAT, made w 2% nonfat dried milk, toasted	1 SLICE	19	24.3	54.9	2.4	.7	10.8	.3	22.4	51.5	119.1	17.9	61.8
			100	24.3	289.0	12.5	3.6	56.7	1.9	118.0	271.0	627.0	94.0	325.0
178	ZWIEBACK	1 PIECE	7	5.0	29.6	.7	.6	5.2	Trace	.9	4.8	17.5	—	10.5
			100	5.0	423.0	10.7	8.8	74.3	.3	13.0	69.0	250.0	—	150.0
	CAKES													
179	ANGELFOOD	1 PIECE	45	31.5	121.1	3.2	.1	27.1	0	4.1	9.9	127.4	(6.8)	39.6
			100	31.5	269.0	7.1	.2	60.2	0	9.0	22.0	283.0	(15.0)	88.0
180	APPLE COFFEE CAKE	1 WHOLE	383	39.1	1129.6	13.1	47.4	166.6	—	211.2	287.8	1242.9	—	—
			100	39.1	294.9	3.4	12.4	43.5	—	55.1	75.1	324.5	—	—
181	APPLE-RAISIN	1 OZ	28	3.6	72.3	.9	1.7	13.3	—	11.8	19.0	90.5	2.3	25.9
			100	3.6	258.0	3.4	6.1	47.4	.1	42.2	67.9	323.2	8.3	92.6
182	BANANA OR APPLESAUCE, w/o icing	1 OZ	28	—	98.0	.8	3.7	16.2	—	6.7	12.6	—	3.9	—
			100	—	350.0	2.8	13.1	57.9	—	24.0	45.0	—	14.0	—
183	BLUEBERRY CRUMB	INDV	40	23.2	148.0	2.0	5.4	22.8	—	24.8	—	141.2	—	—
			100	23.2	370.0	5.0	13.6	56.9	—	62.0	—	353.0	—	—
184	CARAMEL, w caramel icing	1 PIECE	79	20.9	299.4	2.9	11.7	46.7	.0	66.4	75.1	199.1	—	50.6
			100	20.9	379.0	3.7	14.8	59.1	0	84.0	95.0	252.0	—	64.0

Item No.	Minerals (Micro)			Fat-Soluble Vitamins			Water-Soluble Vitamins								
	Iron	Zinc	Copper	Vitamin A	Vitamin D	Vitamin E (Alpha Tocopherol)	Vitamin C	Thiamin	Ribo-flavin	Niacin	Panto-thenic Acid	Vit. B-6 (Pyri-doxine)	Folacin (Folic Acid)	Biotin	Vitamin B-12
	(mg)	(mg)	(mg)	(IU)	(IU)	(mg)	(mg)	(mg)	(mg)	(mg)	(mg)	(mg)	(mcg)	(mcg)	(mcg)
154	—	—	—	—	—	—	—	—	—	—	—	—	—	—	—
	—	—	—	—	—	—	—	—	—	—	—	—	—	—	—
155	.33	(.30)	(.05)	—	—	—	—	.01	.02	.18	—	—	—	—	—
	1.30	(1.20)	(.23)	—	—	—	—	.05	.09	.70	—	—	—	—	—
156	.34	—	—	—	—	—	—	.01	.02	.17	—	—	—	—	—
	1.60	—	—	—	—	—	—	.05	.11	.80	—	—	—	—	—
157	1.82	—	—	0	—	—	0	.23	.17	2.00	—	—	—	—	—
	3.20	—	—	0	—	—	0	.40	.30	3.50	—	—	—	—	—
158	.37	.37	.04	0	—	(.07)	0	.04	.02	.32	.10	.02	5.29	—	0
	1.60	1.60	.17	0	—	(.30)	0	.18	.07	1.40	.45	.10	23.00	—	0
159	.38	.32	—	0	—	—	0	.03	.02	.32	—	—	—	—	—
	1.90	1.60	—	0	—	—	0	.17	.08	1.60	—	—	—	—	—
160	.77	.38	.05	0	—	—	0	.07	.05	.38	.16	.05	7.36	—	0
	2.40	1.20	.17	0	—	—	0	.23	.14	1.20	.50	.16	23.00	—	0
161	.63	.24	.03	—	—	.03	—	.06	.05	.60	.11	.01	9.75	.28	—
	2.50	.97	.11	—	—	.10	—	.25	.21	2.40	.43	.04	39.00	1.10	—
162	.15	—	—	2.5	—	—	—	.01	.01	.15	—	—	9.75	—	—
	.60	—	—	10.0	—	—	—	.04	.05	.60	—	—	39.00	—	—
163	.15	—	—	2.2	—	—	—	.01	.01	.13	—	—	—	—	—
	.70	—	—	10.0	—	—	—	.04	.05	.60	—	—	—	—	—
164	1.18	—	—	0	—	—	—	.16	.06	1.32	—	—	—	—	—
	4.20	—	—	0	—	—	—	.56	.22	4.70	—	—	—	—	—
165	2.26	—	—	916.5	—	—	—	.13	.17	2.12	—	—	—	—	—
	1.60	—	—	650.0	—	—	—	.09	.12	1.50	—	—	—	—	—
166	1.90	—	—	798.0	—	—	—	.10	.17	1.52	—	—	—	—	—
	1.00	—	—	420.0	—	—	—	.05	.09	.80	—	—	—	—	—
167	.72	.22	—	—	—	—	—	.1	Trace	.81	—	—	—	—	—
	2.00	.60	—	—	—	—	—	.2	.17	2.25	—	—	—	—	—
168	.55	.14	.05	—	—	.02	—	.20	.13	.53	.10	.01	8.05	.25	—
	2.40	.60	.23	—	—	.10	—	.88	.58	2.30	.43	.04	35.00	1.10	—
169	.56	.12	—	—	—	—	—	.05	.04	.54	.09	.01	4.60	—	—
	2.80	.60	—	—	—	—	—	.23	.20	2.70	.43	.04	23.00	—	—
170	.58	.14	.03	—	—	.02	—	.20	.13	.55	.10	.01	8.05	.25	—
	2.50	.60	.11	—	—	.10	—	.88	.58	2.40	.43	.04	35.00	1.10	—
171	.58	.12	—	—	—	—	—	.05	.05	.56	.09	.01	4.60	—	—
	2.90	.60	—	—	—	—	—	.23	.24	2.80	.43	.04	23.00	—	—
172	.58	.14	Trace	—	—	.02	—	.20	.13	.55	.10	.01	8.05	.25	—
	2.50	.60	.02	—	—	.10	—	.88	.58	2.40	.43	.04	35.00	1.10	—
173	.58	.12	Trace	—	—	—	—	.05	.05	.56	.09	.01	4.60	—	—
	2.90	.60	.02	—	—	—	—	.25	.23	2.80	.43	.04	23.00	—	—
174	.53	.41	.04	—	—	.10	0	.07	.02	.64	.18	.04	13.34	.44	0
	2.30	1.80	.17	—	—	.45	0	.30	.10	2.80	.76	.18	58.00	1.90	0
175	.51	.34	—	—	—	—	—	.06	.02	.63	—	—	13.49	—	—
	2.70	1.80	—	—	—	—	—	.29	.12	3.30	—	—	71.00	—	—
176	.53	.41	.12	—	—	.10	—	.06	.03	.64	.18	.04	14.26	.25	0
	2.30	1.80	.51	—	—	.45	—	.26	.12	2.80	.76	.18	62.00	1.10	0
177	.51	.34	—	—	—	—	—	.05	.03	.65	—	—	13.49	—	—
	2.70	1.80	—	—	—	—	—	.25	.15	3.40	—	—	71.00	—	—
178	.04	—	—	2.8	—	—	0	Trace	.01	.06	—	—	—	—	—
	.60	—	—	40.0	—	—	0	.05	.07	.90	—	—	—	—	—
179	.09	.05	(.02)	0	—	(.05)	0	.01	.06	.09	—	—	—	—	—
	.20	.11	(.04)	0	—	(.10)	0	.01	.14	.20	—	—	—	—	—
180	2.84	—	—	1907.7	—	—	7.95	.14	.30	1.52	1.27	.14	—	—	.86
	.74	—	—	498.1	—	—	2.08	.04	.08	.40	.33	.04	—	—	.23
181	.26	.03	.01	17.1	—	—	—	Trace	.04	.40	—	—	—	—	—
	.93	.10	.05	61.1	—	—	—	.02	.14	1.42	—	—	—	—	—
182	.62	—	—	14.0	—	—	0	.04	.03	.31	—	.02	—	—	.02
	2.20	—	—	50.0	—	—	0	.13	.11	1.10	—	.06	—	—	.07
183	.44	—	—	32.4	—	—	—	.03	.07	.52	—	—	—	—	—
	1.10	—	—	81.0	—	—	—	.08	.17	1.30	—	—	—	—	—
184	1.19	—	—	158.0	—	—	—	.02	.06	.08	—	—	—	—	—
	1.50	—	—	200.0	—	—	—	.02	.07	.10	—	—	—	—	—

(Continued)

TABLE F-36 *(Continued)*

Item No.	Food Item Name	Approximate Measure	Weight	Moisture	Food Energy Calories	Protein	Fats	Carbo-hydrates	Fiber	Minerals (Macro) Calcium	Phos-phorus	Sodium	Mag-nesium	Potas-sium
			(g)	(%)	(kcal)	(g)	(g)	(g)	(g)	(mg)	(mg)	(mg)	(mg)	(mg)
185	CARROT	1 OZ	28	—	100.0	1.1	5.3	12.6	.1	10.1	28.0	121.2	—	35.3
			100	—	357.0	4.1	18.8	45.1	.5	36.0	100.0	433.0	—	126.0
186	CHEESE CRUMB	INDV	40	30.0	142.8	3.0	6.6	18.0	—	19.2	—	134.8	—	—
			100	30.0	357.0	7.6	16.4	45.0	—	48.0	—	337.0	—	—
187	CHOCOLATE CUPCAKE, w/icing	1 AVG	36	22.2	128.9	1.6	4.5	21.3	.1	46.8	70.9	120.6	—	42.1
			100	22.2	358.0	4.5	12.6	59.2	.3	130.0	197.0	335.0	—	117.0
188	CHOCOLATE (DEVIL'S FOOD), w/uncooked icing	1 PIECE	74	21.3	273.1	2.8	10.8	43.8	.1	43.7	78.4	173.2	17.8	81.4
			100	21.3	369.0	3.8	14.6	59.2	.2	59.0	106.0	234.0	24.0	110.0
189	CHOCOLATE FUDGE	1 PIECE	86	—	313.9	3.6	15.0	43.3	.3	35.3	124.7	298.4	31.0	89.4
			100	—	365.0	4.2	17.4	50.3	.4	41.0	145.0	347.0	36.0	104.0
190	CHOCOLATE (DEVIL'S FOOD), no icing	1 PIECE	40	24.6	146.4	1.9	6.9	20.8	.1	29.6	54.8	117.6	9.6	56.0
			100	24.6	366.0	4.8	17.2	52.0	.3	74.0	137.0	294.0	24.0	140.0
191	CHOCOLATE (DEVIL'S FOOD), w/chocolate icing	1 PIECE	75	22.0	276.8	3.4	12.3	41.9	.2	52.5	98.3	176.3	18.0	115.5
			100	22.0	369.0	4.5	16.4	55.8	.3	70.0	131.0	235.0	24.0	154.0
192	CHOCOLATE, double layer	1 PIECE	85	37.4	281.4	3.3	15.2	33.3	.3	57.8	120.7	170.0	13.6	90.1
			100	37.4	331.0	3.9	17.9	39.2	.3	68.0	142.0	200.0	16.0	106.0
193	CHOCOLATE ROLL	1 OZ	28	25.3	100.5	1.2	4.9	14.2	.1	19.9	26.0	61.3	2.2	34.2
			100	25.3	359.0	4.4	17.5	50.6	.2	71.0	93.0	219.0	8.0	122.0
194	CINNAMON COFFEE CAKE	1 PIECE	28	—	105.0	1.5	4.0	15.5	—	—	—	230.0	(4.2)	—
			100	—	375.0	5.4	14.3	55.4	—	—	—	821.4	(15.0)	—
195	COCONUT	1 PIECE	87	—	324.5	2.8	15.2	46.3	.1	25.2	92.2	253.2	8.7	57.4
			100	—	373.0	3.2	17.5	53.2	.2	29.0	106.0	291.0	10.0	66.0
196	CREAM CHEESE	1 PIECE	89	48.8	232.3	3.0	11.5	30.3	.1	36.5	33.8	102.4	17.8	75.7
			100	48.8	261.0	3.4	12.9	34.1	.1	41.0	38.0	115.0	20.0	85.0
197	CREAM CHEESE, no bake	1 SERV	71	—	133.5	5.3	4.0	19.3	—	—	—	102.2	—	142.7
			100	—	188.0	7.4	5.6	27.2	—	—	—	144.0	—	201.0
198	DEVIL'S FOOD, frozen, w/chocolate icing	1 PIECE	85	21.0	323.0	3.7	15.0	47.3	.3	45.9	78.2	357.0	(20.4)	101.2
			100	21.0	380.0	4.3	17.6	55.6	.3	54.0	92.0	420.0	(24.0)	119.0
199	DEVIL'S FOOD CUPCAKE, w/chocolate icing	INDV	40	17.2	162.8	1.6	6.5	24.2	—	14.4	—	170.4	—	—
			100	17.2	407.0	4.0	16.3	60.5	—	36.0	—	426.0	—	—
200	FRUITCAKE, dark	1 SLICE	15	18.1	56.9	.7	2.3	9.0	.1	10.8	17.0	23.7	2.4	74.4
			100	18.1	379.0	4.8	15.3	59.7	.6	72.0	113.0	158.0	16.0	496.0
201	FRUITCAKE, light	1 SLICE	15	18.7	58.4	.9	2.5	8.6	.1	10.2	17.3	29.0	2.4	35.0
			100	18.7	389.0	6.0	16.5	57.4	.7	68.0	115.0	193.0	16.0	233.0
202	GERMAN CHOCOLATE	1 PIECE	49	25.1	191.1	2.6	10.0	23.0	.2	28.4	67.1	179.3	7.4	70.1
			100	25.1	390.0	5.4	20.5	47.0	.4	58.0	137.0	366.0	15.0	143.0
203	GOLDEN	1 PIECE	85	—	320.5	3.1	15.6	43.6	.2	27.2	105.4	251.6	18.7	66.3
			100	—	377.0	3.6	18.4	51.3	.2	32.0	124.0	296.0	22.0	78.0
204	HONEY-NUT COFFEE CAKE	1 PIECE	55	—	218.4	3.2	9.7	30.9	.2	45.7	67.1	110.0	—	130.9
			100	—	397.0	5.9	17.6	56.1	.3	83.0	122.0	200.0	—	238.0
205	HONEY-SPICE, w/caramel icing	1 PIECE	77	22.7	271.0	3.2	8.3	46.9	.2	54.7	148.6	188.7	—	63.1
			100	22.7	352.0	4.1	10.8	60.9	.2	71.0	193.0	245.0	—	82.0
206	LOW SODIUM CAKE	1 PIECE	50	—	239.5	2.7	14.5	24.5	—	—	—	5.0	—	28.0
			100	—	479.0	5.4	29.0	49.0	—	—	—	10.0	—	56.0
207	MARBLE LAYER	1 PIECE	81	19.8	349.1	1.8	18.7	43.4	—	67.7	—	—	—	—
			100	19.8	431.0	2.2	23.1	53.6	—	83.6	—	—	—	—
208	MARBLE, w/boiled white icing	1 PIECE	65	23.6	215.2	2.9	5.7	40.3	.1	50.7	111.2	168.4	—	79.3
			100	23.6	331.0	4.4	8.7	62.0	.1	78.0	171.0	259.0	—	122.0
209	ORANGE, w/icing	1 SERV	100	—	367.2	3.0	13.4	59.9	Trace	74.8	112.3	293.4	—	158.5
			100	—	367.2	3.0	13.4	59.9	Trace	74.8	112.3	293.4	—	158.5
210	ORANGE CUPCAKES	1 WHOLE	42	18.0	163.8	1.3	5.9	26.5	—	19.7	51.9	168.0	—	—
			100	18.0	390.0	3.0	14.0	63.0	—	47.0	123.5	400.0	—	—
211	PINEAPPLE UPSIDE DOWN CAKE	1 SERV	100	41.5	300.0	3.0	13.0	42.0	.2	31.0	67.0	170.0	—	101.0
			100	41.5	300.0	3.0	13.0	42.0	.2	31.0	67.0	170.0	—	101.0
212	PLAIN OR CUPCAKE, w/o icing	1 PIECE	86	24.5	313.0	3.9	12.0	48.1	.1	55.0	87.7	258.0	—	67.9
			100	24.5	364.0	4.5	13.9	55.9	.1	64.0	102.0	300.0	—	79.0
213	PLAIN OR CUPCAKE, w/boiled white icing	1 PIECE	114	22.9	401.3	4.3	12.0	70.5	—	55.9	87.8	298.7	—	73.0
			100	22.9	352.0	3.8	10.5	61.8	—	49.0	77.0	262.0	—	64.0
214	PLAIN OR CUPCAKE, w/chocolate icing	1 PIECE	123	21.4	452.6	5.2	17.1	73.1	.2	77.5	127.9	281.7	—	140.2
			100	21.4	368.0	4.2	13.9	59.4	.2	63.0	104.0	229.0	—	114.0
215	PLAIN OR CUPCAKE, w/uncooked white icing	1 PIECE	121	20.6	444.1	4.1	14.3	76.6	—	60.5	90.8	274.7	—	73.8
			100	20.6	367.0	3.4	11.8	63.3	—	50.0	75.0	227.0	—	61.0
216	COTTAGE PUDDING, w/strawberry sauce	1 PIECE	70	36.6	204.4	3.6	6.2	33.9	.2	51.1	65.1	163.1	—	65.1
			100	36.6	292.0	5.1	8.8	48.4	3	73.0	93.0	233.0	—	93.0

Item No.	Minerals (Micro)			Fat-Soluble Vitamins			Water-Soluble Vitamins								
	Iron	Zinc	Copper	Vitamin A	Vitamin D	Vitamin E (Alpha Tocopherol)	Vitamin C	Thiamin	Ribo-flavin	Niacin	Panto-thenic Acid	Vit. B-6 (Pyri-doxine)	Folacin (Folic Acid)	Biotin	Vitamin B-12
	(mg)	(mg)	(mg)	(IU)	(IU)	(mg)	(mg)	(mg)	(mg)	(mg)	(mg)	(mg)	(mcg)	(mcg)	(mcg)
185	**.28**	**.13**	**.05**	**779.5**	–	–	**.56**	**.03**	**.03**	**.17**	–	–	–	–	–
	1.00	.46	.17	2784.0	–	–	2.00	.11	.09	.60	–	–	–	–	–
186	**.40**	–	–	**87.2**	–	–	–	**.01**	**.06**	**.12**	–	–	–	–	–
	1.00	–	–	218.0	–	–	–	.03	.14	.30	–	–	–	–	–
187	**.29**	–	–	**61.2**	–	**.05**	–	**.01**	**.04**	**.07**	–	–	–	–	–
	.80	–	–	170.0	–	.14	–	.04	.11	.20	–	–	–	–	–
188	**.52**	–	–	**133.2**	–	–	–	**.02**	**.06**	**.15**	**.15**	–	**4.44**	–	–
	.70	–	–	180.0	–	–	–	.02	.08	.20	.20	–	6.00	–	–
189	**.20**	–	–	**86.0**	–	–	–	**.11**	**.07**	**1.12**	–	–	**5.16**	–	–
	.40	–	–	100.0	–	–	–	.13	.08	1.30	–	–	6.00	–	–
190	**.36**	**.26**	**.16**	**60.0**	–	–	–	**.01**	**.04**	**.08**	**.08**	–	–	–	–
	.90	.65	.41	150.0	–	–	–	.02	.10	.20	.20	–	–	–	–
191	**.75**	**.95**	–	**120.0**	–	–	–	**.02**	**.08**	**.15**	**.15**	–	**4.50**	–	–
	1.00	1.26	–	160.0	–	–	–	.02	.10	.20	.20	–	6.00	–	–
192	**.77**	–	–	**136.0**	–	–	**.26**	**.05**	**.12**	**.51**	–	–	**5.10**	–	–
	.90	–	–	160.0	–	–	.30	.06	.14	.60	–	–	6.00	–	–
193	**.28**	–	–	**51.2**	**2.24**	–	–	**.01**	**.02**	**.06**	–	–	–	–	–
	1.00	–	–	183.0	8.00	–	–	.03	.07	.22	–	–	–	–	–
194	–	**(.17)**	**(.02)**	–	–	**(.06)**	–	–	–	–	**(.06)**	**(.01)**	–	–	–
	–	(.60)	(.08)	–	–	(.20)	–	–	–	–	(.20)	(.04)	–	–	–
195	**1.22**	–	–	**87.0**	–	–	–	**.11**	**.07**	**1.13**	–	–	–	–	–
	1.40	–	–	100.0	–	–	–	.13	.08	1.30	–	–	–	–	–
196	**.80**	–	–	**281.2**	–	–	**8.10**	**.07**	**.08**	**.27**	–	–	–	–	–
	.90	–	–	316.0	–	–	9.10	.04	.09	.30	–	–	–	–	–
197	–	–	–	–	–	–	–	–	–	–	–	–	–	–	–
	–	–	–	–	–	–	–	–	–	–	–	–	–	–	–
198	**.68**	**(.43)**	**(.26)**	**365.5**	–	**(.09)**	–	**.02**	**.07**	**.17**	**(.17)**	**(.04)**	**5.10**	–	–
	.80	(.50)	(.31)	430.0	–	(.10)	–	.02	.08	.20	(.20)	(.05)	6.00	–	–
199	**.84**	–	–	**29.6**	–	–	–	**.01**	**.07**	**.12**	–	–	**2.40**	–	–
	2.10	–	–	74.0	–	–	–	.02	.18	.30	–	–	6.00	–	–
200	**.39**	[6]**.08**	–	**18.0**	–	–	–	**.02**	**.02**	**.12**	–	–	–	–	–
	2.60	.50	–	120.0	–	–	–	.13	.14	.80	–	–	–	–	–
201	**.24**	[6]**.08**	–	**10.5**	–	–	–	**.02**	**.02**	**.11**	–	–	–	–	–
	1.60	.50	–	70.0	–	–	–	.10	.11	.70	–	–	–	–	–
202	**.69**	–	–	**75.5**	–	–	**.10**	**.04**	**.06**	**.34**	–	–	**2.94**	–	–
	1.40	–	–	154.0	–	–	.20	.09	.12	.70	–	–	6.00	–	–
203	**1.19**	–	–	**85.0**	–	–	–	**.11**	**.07**	**1.11**	–	–	–	–	–
	1.40	–	–	100.0	–	–	–	.13	.08	1.30	–	–	–	–	–
204	**.99**	–	–	**46.2**	–	–	**.11**	**.06**	**.06**	**.39**	–	–	–	–	–
	1.80	–	–	84.0	–	–	.20	.11	.11	.70	–	–	–	–	–
205	**.62**	–	–	**123.2**	–	–	–	**.02**	**.07**	**.15**	–	–	–	–	–
	.80	–	–	160.0	–	–	–	.02	.09	.20	–	–	–	–	–
206	–	–	–	–	–	–	–	–	–	–	–	–	–	–	–
	–	–	–	–	–	–	–	–	–	–	–	–	–	–	–
207	**1.62**	–	–	–	–	–	–	–	–	–	–	–	–	–	–
	2.00	–	–	–	–	–	–	–	–	–	–	–	–	–	–
208	**.52**	–	–	**58.5**	–	–	–	**.01**	**.05**	**.13**	–	–	–	–	–
	.80	–	–	90.0	–	–	–	.02	.08	.20	–	–	–	–	–
209	**.28**	–	–	**573.7**	–	–	**.50**	**.02**	**.05**	**.84**	–	–	–	–	–
	.28	–	–	573.7	–	–	.50	.02	.05	.84	–	–	–	–	–
210	**1.05**	–	–	**0**	–	–	**0**	**.09**	**.10**	**.80**	**.10**	**.01**	–	–	**.06**
	2.50	–	–	0	–	–	0	.21	.24	1.90	.24	.02	–	–	.14
211	**1.50**	**.20**	**.12**	**245.0**	–	–	**2.00**	**.07**	**.06**	**1.30**	–	–	–	–	–
	1.50	.20	.12	245.0	–	–	2.00	.07	.06	1.30	–	–	–	–	–
212	**.34**	–	–	**146.2**	–	–	–	**.02**	**.08**	**.17**	–	**.03**	–	–	–
	.40	–	–	170.0	–	–	–	.02	.09	.20	–	.04	–	–	–
213	**.34**	–	–	**148.2**	–	–	–	**.02**	**.08**	**.23**	–	**.05**	–	–	–
	.30	–	–	130.0	–	–	–	.02	.07	.20	–	.04	–	–	–
214	**.74**	–	–	**221.4**	–	–	–	**.03**	**.11**	**.25**	–	**.05**	–	–	–
	.60	–	–	180.0	–	–	–	.02	.09	.20	–	.04	–	–	–
215	**.36**	–	–	**242.0**	–	–	–	**.02**	**.09**	**.12**	–	**.05**	–	–	–
	.30	–	–	200.0	–	–	–	.02	.07	.10	–	.04	–	–	–
216	**.84**	–	–	**84.0**	–	–	**8.40**	**.08**	**.11**	**.77**	–	–	–	–	–
	1.20	–	–	120.0	–	–	12.00	.12	.15	1.10	–	–	–	–	–

(Continued)

TABLE F-36 *(Continued)*

Item No.	Food Item Name	Approximate Measure	Weight	Moisture	Food Energy Calories	Protein	Fats	Carbo-hydrates	Fiber	Minerals (Macro) Calcium	Phos-phorus	Sodium	Mag-nesium	Potas-sium
			(g)	(%)	(kcal)	(g)	(g)	(g)	(g)	(mg)	(mg)	(mg)	(mg)	(mg)
217	**POUND CAKE** (old fashioned), w/butter & eggs	1 PIECE	**30**	**17.2**	**141.9**	**1.7**	**8.9**	**14.1**	**Trace**	**6.3**	**23.7**	**33.0**	**(3.9)**	**18.0**
			100	17.2	473.0	5.7	29.5	47.0	.1	21.0	79.0	110.0	(13.0)	60.0
218	**SHORTCAKE**	1 PIECE	**25**	—	**85.8**	**1.0**	**2.0**	**12.2**	—	**10.3**	**13.0**	—	—	—
			100	—	343.0	3.8	8.1	48.6	—	41.0	52.0	—	—	—
219	**SPICE,** w/icing	1 OZ	**28**	—	**98.6**	**1.1**	**3.0**	**17.1**	—	**19.9**	**54.0**	—	**5.0**	—
			100	—	352.0	4.1	10.8	60.9	—	71.0	193.0	—	18.0	—
220	**SPONGE**	1 PIECE	**33**	**31.8**	**98.0**	**2.5**	**1.9**	**17.9**	**0**	**9.9**	**37.0**	**55.1**	**4.3**	**28.7**
			100	31.8	297.0	7.6	5.7	54.1	0	30.0	112.0	167.0	13.0	87.0
221	**WALNUT,** layer	1 PIECE	**85**	**39.5**	**279.7**	**3.2**	**16.7**	**30.3**	**.1**	**40.8**	**133.5**	**191.3**	**14.5**	**51.9**
			100	39.5	329.0	3.8	19.6	35.7	.1	48.0	157.0	225.0	17.0	61.0
222	**WHITE,** w/o icing	1 PIECE	**50**	**24.2**	**187.5**	**2.3**	**8.0**	**27.0**	**.1**	**31.5**	**45.5**	**161.5**	**4.0**	**38.0**
			100	24.2	375.0	4.6	16.0	54.0	.1	63.0	91.0	323.0	8.0	76.0
223	**WHITE,** w/chocolate icing	1 PIECE	**71**	**21.1**	**249.2**	**2.8**	**7.6**	**44.6**	**.1**	**70.3**	**127.1**	**161.2**	—	**82.4**
			100	21.1	351.0	3.9	10.7	62.8	.2	99.0	179.0	227.0	—	116.0
224	**YELLOW,** w/o icing	1 PIECE	**50**	**23.5**	**181.5**	**2.3**	**6.4**	**29.1**	**.1**	**35.5**	**56.0**	**129.0**	**4.0**	**39.0**
			100	23.5	363.0	4.5	12.7	58.2	.1	71.0	112.0	258.0	8.0	78.0
225	**YELLOW,** w/caramel icing	1 PIECE	**81**	**21.8**	**293.2**	**3.2**	**9.5**	**49.7**	**.1**	**62.4**	**83.4**	**183.1**	**6.5**	**59.1**
			100	21.8	362.0	4.0	11.7	61.3	.1	77.0	103.0	226.0	8.0	73.0
226	**YELLOW,** w/chocolate icing	1 PIECE	**69**	**25.6**	**232.5**	**2.8**	**7.8**	**39.7**	**.1**	**62.8**	**125.6**	**156.6**	—	**75.2**
			100	25.6	337.0	4.1	11.3	57.6	.2	91.0	182.0	227.0	—	109.0
227	**YELLOW CUPCAKES,** w/chocolate icing	INDV	**40**	**18.4**	**154.8**	**1.6**	**5.3**	**25.2**	—	**17.6**	—	**133.2**	—	—
			100	18.4	387.0	4.1	13.2	62.9	—	44.0	—	333.0	—	—
	COOKIES													
228	**ARROWROOT**	INDV	**4**	—	**17.3**	**.2**	**.4**	**3.2**	—	**1.5**	**4.0**	**12.0**	—	—
			100	—	432.4	5.9	9.9	80.4	—	37.8	100.0	300.0	—	—
229	**ASSORTED,** commercially packaged	1 AVG	**20**	**2.6**	**96.0**	**1.0**	**4.0**	**14.2**	**Trace**	**7.4**	**32.6**	**73.0**	—	**13.4**
			100	2.6	480.0	5.1	20.2	71.0	.1	37.0	163.0	365.0	—	67.0
230	**BROWNIE,** w/chocolate & nuts, frozen	1 SERV	**100**	**12.0**	**510.0**	**6.0**	**33.0**	**48.0**	**.5**	**30.0**	**142.0**	**145.0**	—	**162.0**
			100	12.0	510.0	6.0	33.0	48.0	.5	30.0	142.0	145.0	—	162.0
231	**BROWNIE MIX,** enriched, w/eggs & nuts	1 AVG	**20**	**10.7**	**85.6**	**1.0**	**4.0**	**12.6**	**.1**	**9.0**	**27.4**	**33.2**	—	**33.6**
			100	10.7	428.0	5.0	20.1	63.1	.6	45.0	137.0	166.0	—	168.0
232	**BUTTER,** thin & rich	1 AVG	**11**	**4.5**	**50.3**	**.7**	**1.9**	**7.8**	**Trace**	**13.9**	**10.3**	**46.0**	—	**6.6**
			100	4.5	457.0	6.1	16.9	70.9	.1	126.0	94.0	418.0	—	60.0
233	**CHOCOLATE CHIP,** home recipe	1 AVG	**11**	**3.0**	**56.8**	**.6**	**3.3**	**6.6**	**Trace**	**3.7**	**10.9**	**38.3**	**1.7**	**12.9**
			100	3.0	516.0	5.4	30.1	60.1	.4	34.0	99.0	348.0	15.0	117.0
234	**CHOCOLATE CHIP,** w/coconut	INDV	**16**	—	**81.5**	**.9**	**4.4**	**10.0**	—	**7.6**	**21.7**	—	—	—
			100	—	509.6	5.4	27.4	62.2	—	47.8	135.7	—	—	—
235	**COCONUT BARS**	1 AVG	**22**	**3.8**	**108.7**	**1.4**	**5.4**	**14.1**	**.1**	**15.8**	**26.4**	**32.6**	**3.3**	**50.2**
			100	3.8	494.0	6.2	24.5	63.9	.6	72.0	120.0	148.0	15.0	228.0
236	**FIG BARS**	1 AVG	**14**	**13.6**	**50.1**	**.5**	**.8**	**10.6**	**.2**	**10.9**	**8.4**	**35.3**	**2.1**	**27.7**
			100	13.6	358.0	3.9	5.6	75.4	1.7	78.0	60.0	252.0	15.0	198.0
237	**FORTUNE**	1 AVG	**8**	**3.5**	**31.8**	**.5**	**.2**	**6.9**	**Trace**	**1.5**	**8.3**	**15.7**	**1.5**	**2.6**
			100	3.5	397.3	6.2	2.9	86.6	.1	18.5	104.1	196.5	18.7	32.8
238	**GINGERSNAPS**	1 SM	**4**	**3.1**	**16.8**	**.2**	**.4**	**3.2**	**Trace**	**2.9**	**1.9**	**22.8**	**.6**	**18.5**
			100	3.1	420.0	5.5	8.9	79.8	.1	73.0	47.0	571.0	15.0	462.0
239	**GRANOLA BAR,** honey'n oats	1 BAR	**24**	—	**111.7**	**2.0**	**4.1**	**16.2**	—	—	—	—	—	—
			100	—	465.3	8.5	16.9	67.7	—	—	—	—	—	—
240	**GRANOLA BAR,** peanut	1 BAR	**24**	—	**121.8**	**3.0**	**5.1**	**15.2**	—	—	—	—	—	—
			100	—	507.6	12.7	21.2	63.5	—	—	—	—	—	—
241	**LADYFINGERS**	1 LG	**14**	**19.2**	**50.4**	**1.1**	**1.1**	**9.0**	**Trace**	**5.7**	**23.0**	**9.9**	**2.1**	**9.9**
			100	19.2	360.0	7.8	7.8	64.5	.1	41.0	164.0	71.0	15.0	71.0
242	**MACAROONS**	1 AVG	**14**	**4.4**	**66.5**	**.7**	**3.2**	**9.3**	**.3**	**3.8**	**11.6**	**4.8**	—	**64.8**
			100	4.4	475.0	5.3	23.2	66.1	2.1	27.0	83.0	34.0	—	463.0
243	**MARSHMALLOW**	1 AVG	**28**	**9.8**	**114.5**	**1.1**	**3.7**	**20.2**	**.1**	**5.9**	**16.0**	**58.5**	**4.2**	**25.5**
			100	9.8	409.0	4.0	13.2	72.3	.3	21.0	57.0	209.0	15.0	91.0
244	**MOLASSES**	1 AVG	**33**	**4.0**	**139.3**	**2.1**	**3.5**	**25.1**	**Trace**	**16.8**	**27.4**	**127.4**	—	**45.5**
			100	4.0	422.0	6.4	10.6	76.0	.1	51.0	83.0	386.0	—	138.0
245	**OATMEAL**	INDV	**13**	—	**59.8**	**.7**	**2.3**	**9.2**	—	**2.9**	**16.1**	—	—	—
			100	—	460.3	5.7	17.9	70.7	—	22.2	123.8	—	—	—
246	**OATMEAL,** w/raisins	1 LG	**14**	**2.8**	**63.1**	**.9**	**2.2**	**10.3**	**.1**	**2.9**	**14.3**	**22.7**	**2.1**	**51.8**
			100	2.8	451.0	6.2	15.4	73.5	.4	21.0	102.0	162.0	15.0	370.0
247	**PEANUT**	1 AVG	**12**	**2.3**	**56.8**	**1.2**	**2.3**	**8.0**	**.1**	**5.0**	**13.9**	**20.8**	**1.8**	**21.0**
			100	2.3	473.0	10.0	19.1	67.0	.8	42.0	116.0	173.0	15.0	175.0

Item No.	Minerals (Micro)			Fat-Soluble Vitamins			Water-Soluble Vitamins								
	Iron	Zinc	Copper	Vitamin A	Vitamin D	Vitamin E (Alpha Tocopherol)	Vitamin C	Thiamin	Ribo-flavin	Niacin	Panto-thenic Acid	Vit. B-6 (Pyri-doxine)	Folacin (Folic Acid)	Biotin	Vitamin B-12
	(mg)	(mg)	(mg)	(IU)	(IU)	(mg)	(mg)	(mg)	(mg)	(mg)	(mg)	(mg)	(mcg)	(mcg)	(mcg)
217	**.24** .80	**(.18)** (.60)	**(.02)** (.06)	**84.0** 280.0	— —	**.33** 1.10	**0** 0	**.01** .03	**.03** .09	**.06** .20	**(.09)** (.30)	**(.01)** (.04)	— —	— —	— —
218	**.13** .50	— —	— —	**84.8** 339.0	— —	— —	**.25** 1.00	**.03** .13	**.03** .11	**.10** .40	— —	— —	— —	— —	— —
219	**.36** 1.30	— —	— —	**44.8** 160.0	— —	— —	**0** 0	**.03** .12	**.05** .16	**.28** 1.00	— —	**.01** .04	— —	— —	**.01** .02
220	**.40** 1.20	— —	— —	**148.5** 450.0	— —	— —	— —	**.02** .05	**.05** .14	**.07** .20	— —	— —	**2.31** 7.00	— —	— —
221	**.94** .10	— —	— —	**174.3** 205.0	— —	— —	**.17** .20	**.06** .07	**.09** .11	**.51** .60	— —	— —	— —	— —	— —
222	**.10** .20	**.10** .20	— —	**15.0** 30.0	— —	— —	— —	**.01** .01	**.04** .08	**.10** .20	**.15** .30	— —	— —	— —	— —
223	**.36** .50	— —	— —	**42.6** 60.0	— —	— —	— —	**.01** .02	**.06** .08	**.14** .20	— —	— —	— —	— —	— —
224	**.20** .40	**(.25)** (.50)	**(.05)** (.10)	**75.0** 150.0	— —	**(.25)** (.50)	— —	**.01** .02	**.04** .08	**.10** .20	**(.15)** (.30)	**(.02)** (.04)	— —	— —	— —
225	**.57** .70	— —	— —	**137.7** 170.0	— —	— —	— —	**.02** .02	**.07** .08	**.16** .20	— —	— —	— —	— —	— —
226	**.41** .60	— —	— —	**96.6** 140.0	— —	— —	— —	**.01** .02	**.06** .08	**.14** .20	— —	— —	— —	— —	— —
227	**.60** 1.50	— —	— —	**30.8** 77.0	— —	— —	— —	**.04** .09	**.07** .17	**.20** .50	— —	— —	— —	— —	— —
228	**.10** 2.43	— —	— —	— —	— —	— —	— —	— —	— —	— —	— —	— —	— —	— —	— —
229	**.14** .70	— —	— —	**16.0** 80.0	— —	**(.10)** (.50)	— —	**.01** .03	**.01** .05	**.08** .40	**(.08)** (.40)	**(.01)** (.05)	— —	— —	— —
230	**2.20** 2.20	— —	— —	**10.0** 10.0	— —	— —	**2.00** 2.00	**.12** .12	**.10** .10	**.50** .50	— —	— —	— —	— —	— —
231	**.38** 1.90	— —	— —	**20.0** 100.0	— —	— —	— —	**.03** .13	**.02** .10	**.14** .70	— —	— —	— —	— —	— —
232	**.06** .50	— —	— —	**71.5** 650.0	— —	— —	**0** 0	**Trace** .03	**.01** .06	**.04** .40	— —	— —	— —	— —	— —
233	**.23** 2.10	**.11** .96	— —	**12.1** 110.0	— —	— —	— —	**.01** .11	**.01** .11	**.10** .90	— —	— —	**.99** 9.00	— —	— —
234	**.35** 2.17	— —	— —	— —	— —	— —	— —	— —	— —	— —	— —	— —	— —	— —	— —
235	**.31** 1.40	— —	— —	**35.2** 160.0	— —	— —	**0** 0	**.01** .04	**.01** .06	**.09** .40	— —	— —	— —	— —	— —
236	**.14** 1.00	**.09** .64	**.04** .29	**15.4** 110.0	— —	— —	— —	**.01** .04	**.01** .07	**.04** .30	— —	— —	— —	— —	— —
237	**.12** 1.45	**.05** .58	**.01** .10	— —	— —	— —	— —	— —	— —	— —	— —	— —	— —	— —	— —
238	**.09** 2.30	— —	— —	**2.8** 70.0	— —	— —	— —	**Trace** .04	**Trace** .06	**.02** .40	— —	— —	— —	— —	— —
239	— —	— —	— —	— —	— —	— —	— —	— —	— —	— —	— —	— —	— —	— —	— —
240	— —	— —	— —	— —	— —	— —	— —	— —	— —	— —	— —	— —	— —	— —	— —
241	**.21** 1.50	— —	— —	**91.0** 650.0	— —	— —	**0** 0	**.01** .06	**.02** .14	**.03** .20	— —	— —	— —	— —	— —
242	**.13** .90	— —	— —	**0** 0	— —	— —	**0** 0	**.01** .04	**.02** .15	**.08** .60	— —	— —	— —	— —	— —
243	**.14** .50	— —	— —	**72.8** 260.0	— —	— —	— —	**.01** .02	**.02** .06	**.06** .20	— —	— —	— —	— —	— —
244	**.69** 2.10	— —	— —	**26.4** 80.0	— —	— —	**0** 0	**.01** .04	**.02** .06	**.23** .70	— —	— —	— —	— —	— —
245	**.30** 2.30	— —	— —	— —	— —	— —	— —	— —	— —	— —	— —	— —	— —	— —	— —
246	**.41** 2.90	— —	— —	**7.0** 50.0	— —	— —	— —	**.02** .11	**.01** .08	**.07** .50	— —	— —	— —	— —	— —
247	**.11** .90	— —	— —	**24.0** 200.0	— —	— —	— —	**.01** .07	**.01** .08	**.34** 2.80	— —	— —	— —	— —	— —

(Continued)

TABLE F-36 *(Continued)*

Item No.	Food Item Name	Approximate Measure	Weight	Moisture	Food Energy Calories	Protein	Fats	Carbo-hydrates	Fiber	Minerals (Macro) Calcium	Phos-phorus	Sodium	Mag-nesium	Potas-sium
			(g)	(%)	(kcal)	(g)	(g)	(g)	(g)	(mg)	(mg)	(mg)	(mg)	(mg)
248	**PEANUT BUTTER,** prepared	INDV	12	—	54.9	.7	3.1	6.5	—	—	—	75.4	—	—
			100	—	457.1	5.7	25.7	54.3	—	—	—	628.6	—	—
249	**RAISIN**	1 AVG	15	8.2	56.9	.7	.8	12.1	.1	10.7	23.6	7.8	2.3	40.8
			100	8.2	379.0	4.4	5.3	80.8	.9	71.0	157.0	52.0	15.0	272.0
250	**SANDWICH TYPE**	1 AVG	14	2.2	69.3	.7	3.2	9.7	Trace	3.6	33.7	67.6	2.1	5.3
			100	2.2	495.0	4.8	22.5	69.3	.1	26.0	241.0	483.0	15.0	38.0
251	**SHORTBREAD**	1 AVG	7	3.0	34.9	.5	1.6	4.6	Trace	4.9	10.9	4.2	1.1	4.6
			100	3.0	498.0	7.2	23.1	65.1	.2	70.0	156.0	60.0	15.0	66.0
252	**SUGAR,** soft, thick, home recipe, w/enriched flour	1 AVG	8	7.9	35.5	.5	1.3	5.4	Trace	6.2	8.2	25.4	—	6.1
			100	7.9	444.0	6.0	16.8	68.0	.1	78.0	103.0	318.0	—	76.0
253	**SUGAR WAFERS**	1 AVG	6	1.4	29.1	.3	1.2	4.4	Trace	2.2	4.8	11.3	.9	3.6
			100	1.4	485.0	4.9	19.4	73.4	.1	36.0	80.0	189.0	15.0	60.0
254	**VANILLA WAFERS**	1 SM	4	2.8	18.5	.2	.6	3.0	Trace	1.6	2.5	10.1	.6	2.9
			100	2.8	462.0	5.4	16.1	74.4	.1	41.0	63.0	252.0	15.0	72.0
	CRACKERS													
255	**ANIMAL**	1 AVG	2	3.0	8.6	.1	.2	1.6	Trace	1.0	2.3	6.1	—	1.9
			100	3.0	429.0	6.6	9.4	79.9	.1	52.0	114.0	303.0	—	95.0
256	**BISCUIT, ARROWROOT**	1 AVG	5	—	21.5	0	1.1	4.3	—	2.9	6.4	1.2	—	5.7
			100	—	429.2	0	21.5	85.8	—	57.0	128.0	23.9	—	113.0
257	**BUTTER**	1 AVG	4	4.6	18.3	.3	.7	2.7	Trace	5.9	10.4	43.7	1.0	4.5
			100	4.6	458.0	7.0	17.8	67.3	.3	148.0	260.0	1092.0	25.0	113.0
258	**CHEESE**	1 AVG	1	3.9	4.8	.1	.2	.6	Trace	3.4	3.1	10.4	.3	1.1
			100	3.9	479.0	11.2	21.3	60.4	.2	336.0	309.0	1039.0	29.0	109.0
259	**CHEESE & PEANUT BUTTER SANDWICH**	1 AVG	7	2.4	34.4	1.1	1.7	3.9	Trace	3.9	12.5	69.4	—	15.8
			100	2.4	491.0	15.2	23.9	56.1	.5	56.0	179.0	992.0	—	226.0
260	**CHEESE STRAWS**	1 PIECE	6	21.7	27.2	.7	1.8	2.1	Trace	15.5	12.4	43.3	—	3.8
			100	21.7	453.0	11.2	29.9	34.5	.1	259.0	206.0	721.0	—	63.0
261	**GRAHAM,** plain	1 AVG	7	6.4	26.9	.6	.7	5.1	.1	2.8	10.4	46.9	1.3	26.9
			100	6.4	384.0	8.0	9.4	73.3	1.1	40.0	149.0	670.0	18.0	384.0
262	**GRAHAM,** sugar & honey-coated	1 AVG	7	3.3	28.8	.5	.8	5.3	.1	6.2	23.0	35.3	—	18.9
			100	3.3	411.0	6.7	11.4	76.4	.8	88.0	329.0	504.0	—	270.0
263	**GRAHAM,** chocolate-coated	1 AVG	13	1.9	61.8	.7	3.1	8.8	.1	14.7	26.5	52.9	—	41.6
			100	1.9	475.0	5.1	23.5	67.9	.8	113.0	204.0	407.0	—	320.0
264	**MATZO**	1 PIECE	20	—	78.0	2.1	.2	17.3	—	—	—	—	—	—
			100	—	390.0	10.5	1.0	86.5	—	—	—	—	—	—
265	**MEAL**	1 CUP	128	—	537.6	12.3	12.3	93.1	—	25.6	122.9	—	—	—
			100	—	420.0	9.6	9.6	72.7	—	20.0	96.0	—	—	—
266	**RITZ**	1 AVG	3	2.1	16.1	.2	.9	1.9	—	4.4	7.1	29.1	—	2.3
			100	2.1	535.7	7.1	28.6	64.3	—	147.0	236.0	970.0	—	75.0
267	[2]**RUSK OR RUSKETT**	1 OZ	28	4.8	117.3	3.9	2.4	19.9	.1	5.6	33.3	68.9	—	45.1
			100	4.8	419.0	13.8	8.7	71.0	.2	20.0	119.0	246.0	—	161.0
268	**RYE WAFERS,** whole grain	1 AVG	7	6.0	24.1	.9	.1	5.3	.2	3.7	27.2	61.7	—	42.0
			100	6.0	344.0	13.0	1.2	76.3	2.2	53.0	388.0	832.0	—	600.0
269	**RYE WAFERS**	INDV	6	—	21.5	.8	Trace	4.5	—	—	—	.8	—	31.2
			100	—	358.0	13.5	.8	74.2	—	—	—	13.6	—	520.0
270	**SODA**	1 AVG	7	4.0	30.7	.6	.9	4.9	Trace	1.5	6.2	77.0	2.0	8.4
			100	4.0	439.0	9.2	13.1	70.6	.2	22.0	89.0	1100.0	29.0	120.0
271	**WHEAT & RYE THINS**	1 AVG	2	—	9.5	.2	.4	1.3	—	.8	3.1	—	.2	—
			100	—	476.0	7.7	20.1	66.1	—	40.0	157.0	—	12.0	—
272	**WHOLE WHEAT**	1 AVG	4	6.9	16.1	.3	.6	2.7	.1	.9	7.6	21.9	—	—
			100	6.9	403.0	8.4	13.8	68.2	2.4	23.0	190.0	547.0	—	—
	DOUGHNUTS													
273	**CAKE-TYPE,** enriched	1 AVG	32	23.7	125.1	1.5	6.0	16.4	Trace	12.8	60.8	160.3	6.7	28.8
			100	23.7	391.0	4.6	18.6	51.4	.1	40.0	190.0	501.0	21.0	90.0
274	**CREAM-FILLED**	1 AVG	35	—	121.8	1.8	4.7	16.2	0	15.1	22.8	—	8.6	—
			100	—	348.0	5.2	13.5	46.2	0	43.0	65.0	—	24.6	—
275	**YEAST-LEAVENED,** enriched	1 AVG	30	28.3	124.2	1.9	8.0	11.3	.1	11.4	22.8	70.2	6.0	24.0
			100	28.3	414.0	6.3	26.7	37.7	.2	38.0	76.0	234.0	20.0	80.0
	LEAVENINGS													
	BAKING POWDER													
276	home use, calcium carbonate	1 TSP	3	1.0	2.3	.0	Trace	.6	—	173.3	43.6	348.5	(.3)	(1.5)
			100	1.0	78.0	.1	.2	18.9	—	5778.0	1452.0	11618.0	(9.0)	(49.0)

Item No.	Minerals (Micro)			Fat-Soluble Vitamins			Water-Soluble Vitamins								
	Iron	Zinc	Copper	Vitamin A	Vitamin D	Vitamin E (Alpha Tocopherol)	Vitamin C	Thiamin	Ribo-flavin	Niacin	Panto-thenic Acid	Vit. B-6 (Pyri-doxine)	Folacin (Folic Acid)	Biotin	Vitamin B-12
	(mg)	(mg)	(mg)	(IU)	(IU)	(mg)	(mg)	(mg)	(mg)	(mg)	(mg)	(mg)	(mcg)	(mcg)	(mcg)
248	—	—	—	—	—	—	—	—	—	—	—	—	—	—	—
	—	—	—	—	—	—	—	—	—	—	—	—	—	—	—
249	.32	—	—	31.5	—	—	—	.01	.01	.09	—	—	—	—	—
	2.10	—	—	210.0	—	—	—	.04	.08	.60	—	—	—	—	—
250	.10	—	—	0	—	—	0	.01	.01	.07	—	—	—	—	—
	.70	—	—	0	—	—	0	.04	.04	.50	—	—	—	—	—
251	.04	.03	—	5.6	—	.03	0	Trace	Trace	.04	—	—	.63	—	—
	.50	.42	—	80.0	—	.46	0	.04	.05	.50	—	—	9.00	—	—
252	.11	.05	.01	8.8	—	—	—	.01	.01	.10	—	—	—	—	—
	1.40	.57	.17	110.0	—	—	—	.16	.16	1.30	—	—	—	—	—
253	.02	—	—	8.4	—	.03	0	Trace	Trace	.03	—	—	—	—	—
	.30	—	—	140.0	—	.53	0	.01	.04	.50	—	—	—	—	—
254	.02	.01	—	5.2	—	—	0	Trace	Trace	.01	—	—	—	—	—
	.40	.30	—	130.0	—	—	0	.02	.07	.30	—	—	—	—	—
255	.01	—	—	2.6	—	—	—	Trace	Trace	.01	—	—	—	—	—
	.50	—	—	130.0	—	—	—	.04	.10	.30	—	—	—	—	—
256	.13	—	—	—	—	—	—	.02	.02	.11	—	—	—	—	—
	2.50	—	—	—	—	—	—	.38	.30	2.25	—	—	—	—	—
257	.02	—	—	8.8	—	—	0	0	Trace	.04	—	—	—	—	—
	.60	—	—	220.0	—	—	0	.01	.04	1.00	—	—	—	—	—
258	.01	—	—	3.6	—	—	0	0	Trace	.01	—	—	—	—	—
	.90	—	—	360.0	—	—	0	.01	.10	.80	—	—	—	—	—
259	.04	—	—	2.8	—	—	0	Trace	.01	.25	—	—	—	—	—
	.60	—	—	40.0	—	—	0	.03	.07	3.50	—	—	—	—	—
260	.04	—	—	23.4	—	—	0	Trace	.01	.02	—	—	—	—	—
	.60	—	—	390.0	—	—	0	.02	.16	.30	—	—	—	—	—
261	.11	.08	Trace	0	—	(.01)	0	Trace	.02	.11	(.04)	(.01)	—	—	—
	1.50	1.10	.04	0	—	(.10)	0	.04	.21	1.50	(.50)	(.07)	—	—	—
262	.11	—	—	—	—	—	0	Trace	Trace	.07	—	—	—	—	—
	1.60	—	—	—	—	—	0	.03	.02	1.00	—	—	—	—	—
263	.34	—	—	7.8	—	—	0	.01	.04	.16	—	—	—	—	—
	2.60	—	—	60.0	—	—	0	.07	.28	1.20	—	—	—	—	—
264	—	—	—	—	—	—	—	—	—	—	—	—	—	—	—
	—	—	—	—	—	—	—	—	—	—	—	—	—	—	—
265	1.41	—	—	0	0	—	0	.08	.06	1.40	—	—	—	—	—
	1.10	—	—	0	0	—	0	.06	.05	1.09	—	—	—	—	—
266	.08	.02	.01	—	—	—	—	.01	.01	.09	—	—	—	—	—
	2.75	.51	.18	—	—	—	—	.39	.35	2.90	—	—	—	—	—
267	.36	—	—	64.4	—	—	Trace	.02	.06	.31	—	.03	—	—	—
	1.30	—	—	230.0	—	—	Trace	.08	.22	1.10	—	.09	—	—	—
268	.27	—	—	0	—	—	0	.02	.02	.08	—	—	—	—	—
	3.90	—	—	0	—	—	0	.32	.25	1.20	—	—	—	—	—
269	—	—	—	—	—	—	—	—	—	—	—	—	—	—	—
	—	—	—	—	—	—	—	—	—	—	—	—	—	—	—
270	.11	—	—	0	—	—	0	Trace	Trace	.07	—	—	—	—	—
	1.50	(.70)	(.17)	0	—	—	0	.01	.05	1.00	—	—	—	—	—
271	.08	—	—	.2	—	—	0	.01	.01	.07	—	Trace	—	—	0
	3.90	—	—	10.0	—	—	0	.58	.37	3.60	—	.08	—	—	0
272	.01	.12	.04	0	—	—	0	Trace	Trace	.04	—	—	—	—	—
	.30	3.04	.87	0	—	—	0	.06	.04	.90	—	—	—	—	—
273	.45	.16	.05	25.6	—	(.12)	—	.05	.05	.38	.12	(.01)	2.56	—	—
	1.40	.50	.17	80.0	—	(.40)	—	.16	.16	1.20	.39	(.04)	8.00	—	—
274	.42	—	—	65.1	—	—	0	.06	.05	.48	—	—	—	—	—
	1.20	—	—	186.0	—	—	0	.18	.15	1.38	—	—	—	—	—
275	.45	(.21)	(.03)	18.0	—	(.21)	0	.05	.05	.39	.12	—	6.60	—	—
	1.50	(.70)	(.11)	60.0	—	(.70)	0	.16	.17	1.30	.39	—	22.00	—	—
276	Trace	—	Trace	0	0	Trace	0	0	0	0	Trace	Trace	Trace	Trace	0
	Trace	—	Trace	0	0	Trace	0	0	0	0	Trace	Trace	Trace	Trace	0

(Continued)

TABLE F-36 *(Continued)*

Item No.	Food Item Name	Approximate Measure	Weight	Moisture	Food Energy Calories	Protein	Fats	Carbo-hydrates	Fiber	Minerals (Macro) Calcium	Phos-phorus	Sodium	Mag-nesium	Potas-sium
			(g)	(%)	(kcal)	(g)	(g)	(g)	(g)	(mg)	(mg)	(mg)	(mg)	(mg)
	BAKING POWDER (Continued)													
277	home use, calcium sulfate	1 TSP	**3**	**1.3**	**3.1**	**.0**	—	**.8**	—	**189.6**	**46.8**	**300.0**	—	—
			100	1.3	04.0	.1	—	25.1	—	6320.0	1560.0	10000.0	—	—
278	home use, sodium aluminum sulfate	1 TSP	**3**	**1.6**	**3.9**	**.0**	—	**.9**	—	**58.0**	**87.1**	**328.6**	—	**4.5**
			100	1.6	29.0	.1	—	31.2	—	1932.0	2904.0	10953.0	—	150.0
279	home use, straight phosphate	1 TSP	**3**	**1.6**	**3.6**	**.0**	—	**.9**	—	**188.4**	**283.1**	**246.6**	—	**5.1**
			100	1.6	21.0	.1	—	29.3	—	6279.0	9438.0	8220.0	—	170.0
280	home use, tartrate	1 TSP	**3**	**1.0**	**2.3**	**.0**	—	**.6**	—	**0**	**0**	**219.0**	—	**114.0**
			100	1.0	78.0	.1	—	18.9	—	0	0	7300.0	—	3800.0
281	commercial, low sodium	1 TSP	**3**	**2.2**	**5.2**	**.0**	**Trace**	**1.2**	—	**144.5**	**219.2**	**.2**	—	**328.4**
			100	2.2	72.0	.1	.6	41.6	—	4816.0	7308.0	6.0	—	10948.0
282	commercial, w/o additional leavening acid	1 TSP	**3**	**1.4**	**3.3**	**.0**	—	**.8**	—	**0**	**358.6**	**504.1**	—	—
			100	1.4	09.0	.1	—	26.5	—	0	11954.0	16804.0	—	—
283	commercial, pyrophosphate	1 TSP	**3**	**1.4**	**3.2**	**.0**	—	**.8**	—	**27.0**	**367.4**	**486.3**	—	—
			100	1.4	05.0	.1	—	25.5	—	900.0	12245.0	16210.0	—	—
284	commercial, pyrophosphate, w/calcium lactate	1 TSP	**3**	**1.3**	**3.1**	**.0**	—	**.8**	—	**29.8**	**347.4**	**478.4**	—	—
			100	1.3	03.0	.1	—	25.0	—	993.0	11580.0	15947.0	—	—
285	**BAKING SODA**	1 TSP	**3**	—	**0**	**0**	**0**	**0**	—	—	—	**821.4**	—	—
			100	—	0	0	0	0	—	—	—	27380.0	—	—
	YEAST													
286	**BAKER'S,** compressed, unfortified	1 CAKE	**12**	**71.0**	**10.3**	**1.5**	**Trace**	**1.3**	—	**1.6**	**47.3**	**1.9**	**7.1**	**73.2**
			100	71.0	86.0	12.1	.4	11.0	—	13.0	394.0	16.0	59.0	610.0
287	**BAKER'S,** compressed, fortified	1 CAKE	**12**	**71.0**	**10.3**	**1.5**	**Trace**	**1.3**	—	**1.6**	**47.3**	**1.9**	**7.1**	**73.2**
			100	71.0	86.0	12.1	.4	11.0	—	13.0	394.0	16.0	59.0	610.0
288	**BAKER'S,** dry (active)	1 TBLS	**8**	**5.0**	**22.6**	**3.0**	**.1**	**3.1**	—	**3.5**	**103.3**	**4.2**	**4.7**	**159.8**
			100	5.0	82.0	36.9	1.6	38.9	—	44.0	1291.0	52.0	59.0	1998.0
	MUFFINS													
289	**BLUEBERRY,** made w/enriched flour	1 AVG	**40**	**39.0**	**112.4**	**2.9**	**3.7**	**16.8**	**.1**	**33.6**	**52.8**	**252.8**	**10.0**	**46.0**
			100	39.0	281.0	7.3	9.3	41.9	.3	84.0	132.0	632.0	25.0	115.0
290	**BRAN,** made w/enriched flour	1 AVG	**40**	**35.1**	**104.4**	**3.1**	**3.9**	**17.2**	**.7**	**56.8**	**162.0**	**179.2**	—	**172.4**
			100	35.1	261.0	7.7	9.8	43.1	1.8	142.0	405.0	448.0	—	431.0
291	**CORN**	1 OZ	**28**	**43.8**	**77.4**	**1.7**	**3.2**	**10.7**	**.1**	**15.0**	**24.1**	**59.8**	**4.9**	**36.3**
			100	43.8	276.3	6.0	11.3	38.1	.2	53.5	85.9	213.7	17.6	129.5
292	**CORN,** made w/whole ground cornmeal	1 AVG	**45**	**37.8**	**129.6**	**3.2**	**4.6**	**19.1**	**.2**	**50.4**	**97.2**	**222.8**	**22.0**	**59.4**
			100	37.8	288.0	7.2	10.3	42.5	.5	112.0	216.0	495.0	48.9	132.0
293	**CORN,** made w/enriched, degermed, cornmeal	1 AVG	**45**	**32.7**	**141.3**	**3.2**	**4.5**	**21.6**	**.1**	**47.3**	**76.1**	**216.5**	**22.0**	**60.8**
			100	32.7	314.0	7.1	10.1	48.1	.2	105.0	169.0	481.0	48.9	135.0
294	**CORN MIX,** made w/cake flour, nonfat dried milk & eggs	1 AVG	**40**	**33.1**	**118.8**	**1.8**	**3.1**	**20.8**	**Trace**	**59.2**	**91.2**	**138.4**	—	**41.6**
			100	33.1	297.0	4.5	7.8	51.9	.1	148.0	228.0	346.0	—	104.0
295	**ENGLISH**	1 WHOLE	**56**	—	**137.8**	**5.3**	**1.4**	**28.3**	—	—	—	**203.3**	—	—
			100	—	246.0	9.5	2.5	50.6	—	—	—	363.0	—	—
296	**PLAIN,** made w/enriched flour	1 AVG	**40**	**38.0**	**117.6**	**3.1**	**4.0**	**16.9**	**Trace**	**41.6**	**60.4**	**176.4**	**11.0**	**50.0**
			100	38.0	294.0	7.8	10.1	42.3	.1	104.0	151.0	441.0	27.5	125.0
	PANCAKES & WAFFLES													
297	**PANCAKES,** home recipe, enriched	1 AVG	**45**	**50.1**	**104.0**	**3.2**	**3.2**	**15.3**	**Trace**	**45.5**	**62.6**	**191.3**	**11.0**	**55.4**
			100	50.1	231.0	7.1	7.0	34.1	.1	101.0	139.0	425.0	24.4	123.0
298	**PANCAKE & WAFFLE MIX, PLAIN & BUTTERMILK,** enriched, made w/milk	1 AVG	**45**	**53.9**	**90.9**	**2.7**	**2.5**	**14.4**	**Trace**	**99.5**	**108.9**	**203.0**	—	**70.2**
			100	53.9	202.0	6.1	5.6	31.9	.1	221.0	242.0	451.0	—	156.0
299	**PANCAKE & WAFFLE MIX, PLAIN & BUTTERMILK,** enriched, made w/milk, & eggs	1 AVG	**45**	**50.6**	**101.3**	**3.2**	**3.3**	**14.6**	**Trace**	**96.8**	**117.0**	**253.8**	—	**69.3**
			100	50.6	225.0	7.2	7.3	32.4	.1	215.0	260.0	564.0	—	154.0
300	**PANCAKE & WAFFLE MIX, BUCKWHEAT,** prepared w/egg & milk	1 AVG	**45**	**57.9**	**90.0**	**3.1**	**4.1**	**10.7**	**.2**	**99.0**	**151.7**	**208.8**	—	**110.3**
			100	57.9	200.0	6.8	9.1	23.8	.4	220.0	337.0	464.0	—	245.0
301	**WAFFLES,** home recipe, enriched	1 MED	**75**	**41.4**	**209.3**	**7.0**	**7.4**	**28.1**	**.1**	**84.8**	**129.8**	**356.3**	—	**108.8**
			100	41.4	279.0	9.3	9.8	37.5	.1	113.0	173.0	475.0	—	145.0
302	**WAFFLE MIX** (pancake), enriched, made w/egg & milk	1 MED	**75**	**41.7**	**206.3**	**6.6**	**8.0**	**27.2**	**.2**	**179.3**	**257.3**	**514.5**	—	**146.3**
			100	41.7	275.0	8.8	10.6	36.2	.2	239.0	343.0	686.0	—	195.0
303	**WAFFLE MIX,** enriched, made w/water	1 MED	**75**	**38.6**	**228.8**	**3.6**	**10.5**	**30.2**	**.1**	**57.0**	**95.3**	**420.0**	—	**41.3**
			100	38.6	305.0	4.8	14.0	40.2	.1	76.0	127.0	560.0	—	55.0
304	**WAFFLES,** enriched, frozen	1 MED	**34**	**42.1**	**86.0**	**2.4**	**2.1**	**14.3**	**.1**	**41.5**	**70.7**	**219.0**	—	**53.7**
			100	42.1	253.0	7.1	6.2	42.0	.2	122.0	208.0	644.0	—	158.0
	PIES (MADE W/VEGETABLE SHORTENING)													
305	**APPLE**	1 PIECE	**118**	**47.6**	**302.1**	**2.6**	**13.1**	**45.0**	**.5**	**9.4**	**26.0**	**355.2**	**7.1**	**94.4**
			100	47.6	256.0	2.2	11.1	38.1	.4	8.0	22.0	301.0	6.0	80.0

Item No.	Minerals (Micro)			Fat-Soluble Vitamins			Water-Soluble Vitamins								
	Iron	Zinc	Copper	Vitamin A	Vitamin D	Vitamin E (Alpha Tocopherol)	Vitamin C	Thiamin	Ribo-flavin	Niacin	Panto-thenic Acid	Vit. B-6 (Pyri-doxine)	Folacin (Folic Acid)	Biotin	Vitamin B-12
	(mg)	(mg)	(mg)	(IU)	(IU)	(mg)	(mg)	(mg)	(mg)	(mg)	(mg)	(mg)	(mcg)	(mcg)	(mcg)
277	—	—	—	**0**	—	—	**0**	**0**	**0**	**0**	—	—	—	—	—
	—	—	—	0	—	—	0	0	0	0	—	—	—	—	—
278	—	—	—	**0**	—	—	**0**	**0**	**0**	**0**	—	—	—	—	—
	—	—	—	0	—	—	0	0	0	0	—	—	—	—	—
279	—	—	—	**0**	—	—	**0**	**0**	**0**	**0**	—	—	—	—	—
	—	—	—	0	—	—	0	0	0	0	—	—	—	—	—
280	**0**	—	—	**0**	—	—	**0**	**0**	**0**	**0**	—	—	—	—	—
	0	—	—	0	—	—	0	0	0	0	—	—	—	—	—
281	—	—	—	**0**	—	—	**0**	**0**	**0**	**0**	—	—	—	—	—
	—	—	—	0	—	—	0	0	0	0	—	—	—	—	—
282	—	—	—	**0**	—	—	**0**	**0**	**0**	**0**	—	—	—	—	—
	—	—	—	0	—	—	0	0	0	0	—	—	—	—	—
283	—	—	—	**0**	—	—	**0**	**0**	**0**	**0**	—	—	—	—	—
	—	—	—	0	—	—	0	0	0	0	—	—	—	—	—
284	—	—	—	**0**	—	—	**0**	**0**	**0**	**0**	—	—	—	—	—
	—	—	—	0	—	—	0	0	0	0	—	—	—	—	—
285	—	—	—	—	—	—	—	—	—	—	—	—	—	—	—
	—	—	—	—	—	—	—	—	—	—	—	—	—	—	—
286	**.59**	—	—	—	—	—	—	**.09**	**.20**	**1.34**	**.42**	**.07**	—	—	**0**
	4.90	—	—	—	—	—	—	.71	1.65	11.20	3.50	.60	—	—	0
287	**.59**	—	—	—	—	—	—	**1.66**	**.20**	**17.28**	**.42**	**.07**	—	—	**0**
	4.90	—	—	—	—	—	—	13.80	1.65	144.00	3.50	.60	—	—	0
288	**1.29**	—	**.14**	—	—	—	—	**.19**	**.43**	**2.94**	**.88**	**.16**	**327.20**	—	**0**
	16.10	—	1.78	—	—	—	—	2.33	5.41	36.70	11.00	2.00	4090.00	—	0
289	**.64**	—	—	**88.0**	—	—	**.40**	**.06**	**.08**	**.48**	—	—	—	—	—
	1.60	—	—	220.0	—	—	1.00	.16	.20	1.20	—	—	—	—	—
290	**1.48**	—	—	**92.0**	—	—	—	**.06**	**.10**	**1.60**	—	—	—	—	—
	3.70	—	—	230.0	—	—	—	.14	.24	4.00	—	—	—	—	—
291	**.39**	**.17**	—	**51.3**	**5.34**	—	**.10**	**.06**	**.06**	**.42**	**.13**	**.03**	**3.81**	—	**.07**
	1.41	.62	—	183.2	19.08	—	.35	.20	.20	1.50	.46	.09	13.59	—	.26
292	**.63**	—	—	**139.5**	—	—	—	**.08**	**.08**	**.45**	—	—	—	—	—
	1.40	—	—	310.0	—	—	—	.17	.17	1.00	—	—	—	—	—
293	**.77**	—	—	**135.0**	—	—	—	**.09**	**.10**	**.72**	—	—	—	—	—
	1.70	—	—	300.0	—	—	—	.20	.23	1.60	—	—	—	—	—
294	**.44**	—	—	**60.0**	—	—	—	**.05**	**.05**	**.52**	—	—	—	—	—
	1.10	—	—	150.0	—	—	—	.12	.12	1.30	—	—	—	—	—
295	—	—	—	—	—	—	—	—	—	—	—	—	—	—	—
	—	—	—	—	—	—	—	—	—	—	—	—	—	—	—
296	**.64**	**(.48)**	**(.09)**	**40.0**	—	**(.08)**	—	**.07**	**.09**	**.56**	**(.20)**	**(.02)**	—	—	—
	1.60	(1.20)	(.22)	100.0	—	(.20)	—	.17	.23	1.40	(.50)	(.05)	—	—	—
297	**.59**	**.37**	**(.02)**	**54.0**	—	—	—	**.15**	**.10**	**.59**	—	**.01**	—	—	—
	1.30	.82	(.05)	120.0	—	—	—	.32	.22	1.30	—	.02	—	—	—
298	**.41**	—	—	**54.0**	—	—	—	**.06**	**.10**	**.36**	—	—	—	—	—
	.90	—	—	120.0	—	—	—	.14	.23	.80	—	—	—	—	—
299	**.54**	—	—	**112.5**	—	—	—	**.07**	**.11**	**.36**	—	—	—	—	—
	1.20	—	—	250.0	—	—	—	.15	.24	.80	—	—	—	—	—
300	**.59**	—	—	**103.5**	—	—	—	**.05**	**.07**	**.32**	—	—	—	—	—
	1.30	—	—	230.0	—	—	—	.12	.16	.70	—	—	—	—	—
301	**1.28**	—	—	**247.5**	—	—	—	**.36**	**.50**	**.98**	—	**.03**	—	—	—
	1.70	—	—	330.0	—	—	—	.48	.66	1.30	—	.04	—	—	—
302	**.98**	—	—	**172.5**	—	—	—	**.11**	**.17**	**.68**	**.49**	—	—	—	—
	1.30	—	—	230.0	—	—	—	.14	.23	.90	.65	—	—	—	—
303	**.75**	—	—	**60.0**	—	—	—	**.09**	**.08**	**.75**	**.49**	—	—	—	—
	1.00	—	—	80.0	—	—	—	.12	.11	1.00	.65	—	—	—	—
304	**.61**	—	—	**44.2**	—	—	—	**.06**	**.05**	**.41**	**.22**	—	—	—	—
	1.80	—	—	130.0	—	—	—	.17	.16	1.20	.65	—	—	—	—
305	**.35**	**.11**	—	**35.4**	—	**2.95**	**1.18**	**.02**	**.02**	**.47**	**.13**	—	**4.72**	—	**0**
	.30	.09	—	30.0	—	2.50	1.00	.02	.02	.40	.11	—	4.00	—	0

(Continued)

TABLE F-36 *(Continued)*

Item No.	Food Item Name	Approximate Measure	Weight	Moisture	Food Energy Calories	Protein	Fats	Carbo-hydrates	Fiber	Minerals (Macro) Calcium	Phos-phorus	Sodium	Mag-nesium	Potas-sium
			(g)	(%)	(kcal)	(g)	(g)	(g)	(g)	(mg)	(mg)	(mg)	(mg)	(mg)
306	**APPLE,** frozen, baked	1 PIECE	**92**	**47.3**	**233.7**	**1.7**	**9.3**	**36.8**	**.3**	**7.4**	**19.3**	**196.0**	—	**66.2**
			100	47.3	254.0	1.9	10.1	40.0	.3	8.0	21.0	213.0	—	72.0
307	**APPLE, DUTCH**	1 SERV	**200**	—	**519.5**	**2.6**	**21.0**	**82.0**	**.6**	**37.1**	**31.1**	**153.1**	—	**176.4**
			100	—	259.7	1.3	10.4	41.0	.3	18.6	15.6	76.6	—	88.2
308	**APPLE TURNOVER,** frozen	1 PIECE	**79**	**34.6**	**296.3**	**3.2**	**19.0**	**29.2**	**.2**	**2.4**	**18.2**	**217.3**	—	**55.3**
			100	34.6	375.0	4.0	24.0	37.0	.2	3.0	23.0	275.0	—	70.0
309	**BANANA CUSTARD**	1 PIECE	**114**	**54.4**	**251.9**	**5.1**	**10.6**	**35.0**	**.2**	**75.2**	**93.5**	**221.2**	**14.8**	**231.4**
			100	54.4	221.0	4.5	9.3	30.7	.2	66.0	82.0	194.0	13.0	203.0
310	**BLACKBERRY**	1 PIECE	**118**	**51.0**	**286.7**	**3.1**	**13.0**	**40.6**	**2.2**	**22.4**	**30.7**	**316.2**	**7.1**	**118.0**
			100	51.0	243.0	2.6	11.0	34.4	1.9	19.0	26.0	268.0	6.0	100.0
311	**'BLUEBERRY**	1 PIECE	**114**	**51.0**	**275.9**	**2.7**	**12.3**	**39.8**	**.8**	**12.5**	**26.2**	**305.5**	**6.8**	**74.1**
			100	51.0	242.0	2.4	10.8	34.9	.7	11.0	23.0	268.0	6.0	65.0
312	**BOSTON CREAM,** w/powdered sugar topping	1 PIECE	**69**	**34.5**	**208.4**	**3.5**	**6.5**	**34.4**	**0**	**46.2**	**69.7**	**128.3**	—	**61.4**
			100	34.5	302.0	5.0	9.4	49.9	0	67.0	101.0	186.0	—	89.0
313	**BUTTERSCOTCH**	1 PIECE	**114**	**45.1**	**304.4**	**5.0**	**12.5**	**43.7**	—	**85.5**	**92.3**	**244.0**	**14.8**	**108.3**
			100	45.1	267.0	4.4	11.0	38.3	—	75.0	81.0	214.0	13.0	95.0
314	**CHERRY**	1 PIECE	**118**	**46.6**	**308.0**	**3.1**	**13.3**	**45.3**	**.1**	**16.5**	**29.5**	**358.7**	**7.1**	**123.9**
			100	46.6	261.0	2.6	11.3	38.4	.1	14.0	25.0	304.0	6.0	105.0
315	**CHERRY,** frozen, baked	1 PIECE	**97**	**40.6**	**282.3**	**2.1**	**11.6**	**43.1**	**.1**	**11.6**	**22.3**	**222.1**	—	**79.5**
			100	40.6	291.0	2.2	12.0	44.4	.1	12.0	23.0	229.0	—	82.0
316	**CHOCOLATE CREAM**	1 SERV	**160**	—	**362.2**	**5.6**	**19.7**	**43.0**	**.3**	**96.6**	**119.2**	**169.0**	—	**192.7**
			100	—	226.4	3.5	12.3	26.9	.2	60.4	74.5	105.6	—	120.4
317	**CHOCOLATE BAVARIAN CREAM,** frozen	1 PIECE	**85**	**41.0**	**299.2**	**2.5**	**21.3**	**25.5**	**.2**	**52.7**	**67.2**	**68.0**	**16.2**	**119.0**
			100	41.0	352.0	2.9	25.0	30.1	.2	62.0	79.0	80.0	19.0	140.0
318	**CHOCOLATE CHIFFON**	1 PIECE	**81**	**33.0**	**265.7**	**5.5**	**12.4**	**35.4**	**.2**	**19.4**	**78.6**	**204.1**	—	**89.1**
			100	33.0	328.0	6.8	15.3	43.7	.2	24.0	97.0	252.0	—	110.0
319	**CHOCOLATE MERINGUE**	1 PIECE	**114**	**48.4**	**287.3**	**5.5**	**13.7**	**38.2**	**.2**	**78.7**	**111.7**	**291.8**	—	**158.5**
			100	48.4	252.0	4.8	12.0	33.5	.2	69.0	98.0	256.0	—	139.0
320	**COCONUT CUSTARD**	1 PIECE	**114**	**55.4**	**267.9**	**6.8**	**14.3**	**28.4**	**.2**	**107.2**	**132.2**	**281.6**	**14.8**	**185.8**
			100	55.4	235.0	6.0	12.5	24.9	.2	94.0	116.0	247.0	13.0	163.0
321	**COCONUT CUSTARD,** mix, w/egg yolk & milk, baked	1 PIECE	**100**	**57.6**	**203.0**	**4.3**	**7.9**	**29.1**	**.3**	**93.0**	**103.0**	**235.0**	—	**154.0**
			100	57.6	203.0	4.3	7.9	29.1	.3	93.0	103.0	235.0	—	154.0
322	**COCONUT CUSTARD,** frozen & baked	1 PIECE	**100**	**51.2**	**249.0**	**6.0**	**12.0**	**29.5**	**.2**	**95.0**	**115.0**	**252.0**	—	**172.0**
			100	51.2	249.0	6.0	12.0	29.5	.2	95.0	115.0	252.0	—	172.0
323	**CUSTARD**	1 PIECE	**114**	**58.1**	**248.5**	**7.0**	**12.7**	**26.7**	—	**109.4**	**128.8**	**327.2**	—	**156.2**
			100	58.1	218.0	6.1	11.1	23.4	—	96.0	113.0	287.0	—	137.0
324	**LEMON CHIFFON,** homemade filling	1 PIECE	**81**	**35.2**	**264.1**	**4.5**	**10.2**	**38.7**	—	**18.6**	**67.2**	**211.4**	—	**65.6**
			100	35.2	326.0	5.5	12.6	47.8	—	23.0	83.0	261.0	—	81.0
325	**LEMON MERINGUE,** homemade filling	1 PIECE	**105**	**50.9**	**251.0**	**2.8**	**10.7**	**37.2**	—	**14.7**	**51.5**	**296.1**	—	**52.5**
			100	50.9	239.0	2.7	10.2	35.4	—	14.0	49.0	282.0	—	50.0
326	**MINCE**	1 PIECE	**118**	**43.0**	**319.8**	**3.0**	**13.6**	**48.6**	**.5**	**33.0**	**44.8**	**528.6**	**21.2**	**210.0**
			100	43.0	271.0	2.5	11.5	41.2	.4	28.0	38.0	448.0	18.0	178.0
327	**PEACH**	1 PIECE	**118**	**47.5**	**305.6**	**3.0**	**12.6**	**45.1**	**.5**	**11.8**	**34.2**	**316.2**	**7.1**	**175.8**
			100	47.5	259.0	2.5	10.7	38.2	.4	10.0	29.0	268.0	6.0	149.0
328	**PECAN**	1 PIECE	**103**	**19.5**	**430.5**	**5.3**	**23.6**	**52.8**	**.5**	**48.4**	**106.1**	**227.6**	—	**126.7**
			100	19.5	418.0	5.1	22.9	51.3	.5	47.0	103.0	221.0	—	123.0
329	**PINEAPPLE**	1 PIECE	**118**	**48.0**	**298.5**	**2.6**	**12.6**	**45.0**	**.2**	**15.3**	**24.8**	**319.8**	—	**85.0**
			100	48.0	253.0	2.2	10.7	38.1	.2	13.0	21.0	271.0	—	72.0
330	**PINEAPPLE CHIFFON**	1 PIECE	**81**	**41.1**	**233.3**	**5.3**	**9.8**	**31.7**	**.1**	**19.4**	**61.6**	**207.4**	—	**79.4**
			100	41.1	288.0	6.6	12.1	39.1	.1	24.0	76.0	256.0	—	98.0
331	**PINEAPPLE CUSTARD**	1 PIECE	**114**	**54.3**	**250.8**	**4.6**	**9.9**	**36.6**	**.1**	**57.0**	**74.1**	**212.0**	**14.8**	**110.6**
			100	54.3	220.0	4.0	8.7	32.1	.1	50.0	65.0	186.0	13.0	97.0
332	**PUMPKIN**	1 PIECE	**114**	**59.2**	**245.1**	**4.6**	**12.8**	**27.9**	**.6**	**58.1**	**78.7**	**244.0**	**14.8**	**182.4**
			100	59.2	215.0	4.0	11.2	24.5	.5	51.0	69.0	214.0	13.0	160.0
333	**RAISIN**	1 PIECE	**118**	**42.5**	**318.6**	**3.1**	**12.6**	**50.7**	**.4**	**21.2**	**47.2**	**336.3**	—	**226.6**
			100	42.5	270.0	2.6	10.7	43.0	.3	18.0	40.0	285.0	—	192.0
334	**RHUBARB**	1 PIECE	**118**	**47.4**	**298.5**	**3.0**	**12.6**	**45.1**	**.7**	**75.5**	**30.7**	**318.6**	—	**187.6**
			100	47.4	253.0	2.5	10.7	38.2	.6	64.0	26.0	270.0	—	159.0
335	**STRAWBERRY**	1 PIECE	**93**	**58.4**	**184.1**	**1.8**	**7.3**	**28.7**	**.7**	**14.9**	**23.3**	**180.4**	—	**111.6**
			100	58.4	198.0	1.9	7.9	30.9	.8	16.0	25.0	194.0	—	120.0
336	**SWEET POTATO**	1 PIECE	**160**	**59.3**	**340.8**	**7.2**	**18.1**	**37.9**	**.3**	**110.4**	**134.4**	**348.8**	—	**260.8**
			100	59.3	213.0	4.5	11.3	23.7	.2	69.0	84.0	218.0	—	163.0
337	**PIECRUST,** enriched, baked	1 AVG	**180**	**14.9**	**900.0**	**11.0**	**60.1**	**78.8**	**.4**	**25.2**	**90.0**	**1099.8**	**25.2**	**90.0**
			100	14.9	500.0	6.1	33.4	43.8	.2	14.0	50.0	611.0	14.0	50.0

Item No.	Minerals (Micro)			Fat-Soluble Vitamins			Water-Soluble Vitamins								
	Iron	Zinc	Copper	Vitamin A	Vitamin D	Vitamin E (Alpha Tocopherol)	Vitamin C	Thiamin	Ribo-flavin	Niacin	Panto-thenic Acid	Vit. B-6 (Pyri-doxine)	Folacin (Folic Acid)	Biotin	Vitamin B-12
	(mg)	(mg)	(mg)	(IU)	(IU)	(mg)	(mg)	(mg)	(mg)	(mg)	(mg)	(mg)	(mcg)	(mcg)	(mcg)
306	.28	—	—	9.2	—	—	—	.02	.02	.18	—	—	—	—	—
	.30	—	—	10.0	—	—	—	.02	.02	.20	—	—	—	—	—
307	2.04	—	—	118.2	—	—	2.40	.13	.08	1.38	—	—	—	—	—
	1.02	—	—	59.1	—	—	1.20	.07	.04	.69	—	—	—	—	—
308	1.11	—	—	7.9	—	—	2.37	.16	.06	.63	—	—	—	—	—
	1.40	—	—	10.0	—	—	3.00	.20	.08	.80	—	—	—	—	—
309	.57	—	—	285.0	—	—	1.14	.05	.15	.34	—	—	—	—	—
	.50	—	—	250.0	—	—	1.00	.04	.13	.30	—	—	—	—	—
310	.59	—	—	106.2	—	—	4.72	.02	.02	.35	.22	—	—	—	0
	.50	—	—	90.0	—	—	4.00	.02	.02	.30	.18	—	—	—	0
311	.14	.01	—	34.2	—	20.18	3.42	.09	.07	.91	.21	—	—	—	—
	1.00	.01	—	30.0	—	17.70	3.00	.08	.06	.80	.18	—	—	—	—
312	.35	—	—	144.9	—	—	—	.02	.08	.14	—	—	—	—	—
	.50	—	—	210.0	—	—	—	.03	.11	.20	—	—	—	—	—
313	.03	—	—	296.4	—	—	—	.03	.11	.23	—	—	—	—	—
	.90	—	—	260.0	—	—	—	.03	.10	.20	—	—	—	—	—
314	.35	.05	—	519.2	—	—	—	.02	.02	.59	—	—	—	—	—
	.30	.04	—	440.0	—	—	—	.02	.02	.50	—	—	—	—	—
315	.29	—	—	281.3	—	—	1.94	.02	.02	.19	—	—	—	—	—
	.30	—	—	290.0	—	—	2.00	.02	.02	.20	—	—	—	—	—
316	1.02	—	—	179.2	—	—	.70	.09	.18	1.56	—	—	—	—	—
	.64	—	—	112.0	—	—	.44	.06	.11	.98	—	—	—	—	—
317	.60	—	—	701.3	—	—	.60	.03	.09	.26	—	—	—	—	—
	.70	—	—	825.0	—	—	.70	.04	.11	.30	—	—	—	—	—
318	.97	—	—	251.1	—	—	0	.02	.08	.16	—	—	—	—	—
	1.20	—	—	310.0	—	—	0	.03	.10	.20	—	—	—	—	—
319	.80	—	—	216.6	—	—	—	.03	.14	.23	—	—	—	—	—
	.70	—	—	190.0	—	—	—	.03	.12	.20	—	—	—	—	—
320	.80	—	—	262.2	—	—	0	.07	.22	.34	—	—	—	—	—
	.70	—	—	230.0	—	—	0	.06	.19	.30	—	—	—	—	—
321	.40	—	—	210.0	—	—	—	.03	.14	.20	—	—	—	—	—
	.40	—	—	210.0	—	—	—	.03	.14	.20	—	—	—	—	—
322	.60	—	—	160.0	—	—	—	.04	.16	.20	—	—	—	—	—
	.60	—	—	160.0	—	—	—	.04	.16	.20	—	—	—	—	—
323	.68	—	—	262.2	—	—	0	.06	.18	.34	1.08	—	—	—	0
	.60	—	—	230.0	—	—	0	.05	.16	.30	.95	—	—	—	0
324	.73	—	—	137.7	—	—	2.43	.02	.07	.16	—	—	—	—	—
	.90	—	—	170.0	—	—	3.00	.03	.08	.20	—	—	—	—	—
325	.53	—	—	178.5	—	—	3.15	.03	.08	.21	—	—	—	—	—
	.50	—	—	170.0	—	—	3.00	.03	.08	.20	—	—	—	—	—
326	1.18	—	—	—	(.94)	(.94)	1.18	.08	.05	.47	(.12)	(.09)	(5.90)	(1.18)	0
	1.00	—	—	—	(.80)	(.80)	1.00	.07	.04	.40	(.10)	(.08)	(5.00)	(1.00)	0
327	.59	.11	—	861.4	—	—	3.54	.02	.05	.83	.14	—	—	—	0
	.50	.09	—	730.0	—	—	3.00	.02	.04	.70	.12	—	—	—	0
328	2.88	—	—	164.8	—	—	—	.17	.07	.31	—	—	—	—	—
	2.80	—	—	160.0	—	—	—	.16	.07	.30	—	—	—	—	—
329	.59	—	—	23.6	—	—	1.18	.05	.02	.47	—	—	—	—	—
	.50	—	—	20.0	—	—	1.00	.04	.02	.40	—	—	—	—	—
330	.73	—	—	283.5	—	—	.81	.03	.07	.32	—	—	—	—	—
	.90	—	—	350.0	—	—	1.00	.04	.09	.40	—	—	—	—	—
331	.46	—	—	205.2	—	—	1.14	.05	.10	.46	—	—	—	—	—
	.40	—	—	180.0	—	—	1.00	.04	.09	.40	—	—	—	—	—
332	.57	—	—	2815.8	—	—	—	.03	.11	.57	.59	—	—	—	0
	.50	—	—	2470.0	—	—	—	.03	.10	.50	.52	—	—	—	0
333	1.06	—	—	—	—	—	1.18	.04	.04	.35	—	—	—	—	—
	.90	—	—	—	—	—	1.00	.03	.03	.30	—	—	—	—	—
334	.83	—	—	59.0	—	—	3.54	.02	.05	.35	.15	—	—	—	0
	.70	—	—	50.0	—	—	3.00	.02	.04	.30	.13	—	—	—	0
335	.65	—	—	37.2	—	—	23.25	.02	.04	.37	—	—	—	—	—
	.70	—	—	40.0	—	—	25.00	.02	.04	.40	—	—	—	—	—
336	.80	—	—	3840.0	—	—	6.40	.08	.19	.48	—	—	—	—	—
	.50	—	—	2400.0	—	—	4.00	.05	.12	.30	—	—	—	—	—
337	306	—	—	0	—	.88	0	.36	.25	3.24	—	—	—	—	—
	1.70	—	—	0	—	.49	0	.20	.14	1.80	—	—	—	—	—

(Continued)

TABLE F-36 *(Continued)*

Item No.	Food Item Name	Approximate Measure	Weight	Moisture	Food Energy Calories	Protein	Fats	Carbo-hydrates	Fiber	Minerals (Macro) Calcium	Phos-phorus	Sodium	Mag-nesium	Potas-sium
			(g)	(%)	(kcal)	(g)	(g)	(g)	(g)	(mg)	(mg)	(mg)	(mg)	(mg)
338	**PIECRUST,** graham cracker, homemade	1 SERV	32	8.2	158.6	1.2	10.0	17.5	.2	8.3	23.7	183.6	2.8	57.3
			100	8.2	495.5	3.8	31.2	54.7	.5	25.8	73.9	573.7	8.6	179.1
	QUICK BREADS (MISCELLANEOUS)													
339	**BISCUIT MIX,** w/enriched flour & milk	1 AVG	28	28.5	91.0	2.0	2.6	14.6	.1	19.0	65.0	272.4	(6.7)	32.5
			100	28.5	325.0	7.1	9.3	52.3	.2	68.0	232.0	973.0	(24.0)	116.0
340	**BISCUITS, BAKING POWDER,** w/enriched flour	1 AVG	28	27.4	103.3	2.1	4.8	12.8	.1	33.9	49.0	175.3	6.2	32.8
			100	27.4	369.0	7.4	17.0	45.8	.2	121.0	175.0	626.0	22.0	117.0
341	**CORN FRITTERS**	1 MED	35	29.1	132.0	2.7	7.5	13.9	.2	22.4	54.3	167.0	–	46.6
			100	29.1	377.0	7.8	21.5	39.7	.5	64.0	155.0	477.0	–	133.0
342	**CORN PONE,** made w/white whole ground cornmeal	1 AVG	60	51.8	122.4	2.7	3.2	21.7	.5	37.2	97.8	237.6	–	36.6
			100	51.8	204.0	4.5	5.3	36.2	.8	62.0	163.0	396.0	–	61.0
343	**CORNBREAD**	1 OZ	28	43.8	77.4	1.7	3.2	10.7	.1	15.0	24.1	59.8	4.9	36.3
			100	43.8	276.3	6.0	11.3	38.1	.2	53.5	85.9	213.7	17.6	129.5
344	**CORNBREAD DRESSING**	1 CUP	190	–	704.9	24.4	10.8	137.4	2.1	267.9	461.7	2606.8	–	372.4
			100	–	371.0	12.8	5.7	72.3	1.1	141.0	243.0	1372.0	–	196.0
345	**CORNBREAD,** southern style, w/enriched, white, degermed cornmeal	1 PIECE	83	50.2	185.9	5.9	5.0	28.8	.2	90.5	129.5	490.5	–	130.3
			100	50.2	224.0	7.1	6.0	34.7	.2	109.0	156.0	591.0	–	157.0
346	**CORNBREAD,** southern style, w/white whole ground cornmeal	1 AVG	78	53.9	161.5	5.8	5.6	22.7	.4	93.6	164.6	489.8	–	122.5
			100	53.9	207.0	7.4	7.2	29.1	.5	120.0	211.0	628.0	–	157.0
347	**CORNBREAD,** southern style, w/yellow whole ground cornmeal	1 PIECE	45	53.9	93.2	3.3	3.2	13.1	.2	54.0	95.0	282.6	–	70.7
			100	53.9	207.0	7.4	7.2	29.1	.5	120.0	211.0	628.0	–	157.0
348	**CORNBREAD,** southern style, w/unenriched, yellow, degermed cornmeal	1 PIECE	45	50.2	100.8	3.2	2.7	15.6	.1	49.1	70.2	266.0	–	70.7
			100	50.2	224.0	7.1	6.0	34.7	.2	109.0	156.0	591.0	–	157.0
349	**DUMPLINGS**	1 OZ	28	–	37.8	1.0	1.0	6.2	–	29.1	17.4	–	2.8	–
			100	–	135.0	3.6	3.4	22.2	–	104.0	62.0	–	10.0	–
350	**EGGROLL**	1 AVG	20	–	58.8	–	4.7	–	–	–	–	–	–	–
			100	–	294.0	–	23.6	–	–	–	–	–	–	–
351	**GINGERBREAD,** homemade	1 PIECE	117	30.8	370.9	4.4	12.5	60.8	.1	79.6	76.1	277.3	–	531.2
			100	30.8	317.0	3.8	10.7	52.0	.1	68.0	65.0	237.0	–	454.0
352	**GINGERBREAD,** made from mix	1 PIECE	63	37.0	173.9	2.0	4.3	32.2	–	56.7	63.0	191.5	–	172.6
			100	37.0	276.0	3.1	6.8	51.1	–	90.0	100.0	304.0	–	274.0
353	**JOHNNYCAKE,** northern style, made w/yellow, degermed, enriched cornmeal	1 AVG	40	37.9	106.8	3.5	2.1	18.2	.1	44.4	62.0	276.0	–	75.2
			100	37.9	267.0	8.7	5.2	45.5	.3	111.0	155.0	690.0	–	188.0
354	**POPOVERS,** home recipe, enriched, baked	1 AVG	40	54.9	89.6	3.5	3.7	10.3	Trace	38.4	56.0	88.0	–	60.0
			100	54.9	224.0	8.8	9.2	25.8	.1	96.0	140.0	220.0	–	150.0
355	**SPOONBREAD,** w/white, whole-ground cornmeal	1 AVG	96	–	187.2	6.4	10.9	16.2	.3	92.2	157.4	462.7	–	126.7
			100	–	195.0	6.7	11.4	16.9	.3	96.0	164.0	482.0	–	132.0
356	**TACO SHELL**	1 AVG	11	–	48.0	.9	2.2	7.6	–	38.2	–	–	–	–
			100	–	436.0	8.3	20.3	68.8	–	347.0	–	–	–	–
357	**TORTILLA,** lime-treated, yellow corn, 6″ in diameter	1 AVG	30	47.5	63.0	1.4	.5	13.6	.2	58.8	41.4	–	(32.1)	–
			100	47.5	210.0	4.6	1.8	45.3	.8	196.0	138.0	–	(107.0)	–
358	**TORTILLA CHIPS**	1 SERV	100	–	456.0	8.0	26.6	59.9	–	445.0	–	–	–	–
			100	–	456.0	8.0	26.6	59.9	–	445.0	–	–	–	–
	ROLLS													
359	**BAGEL,** made w/egg	1 AVG	55	32.0	165.0	6.0	2.0	28.0	–	8.8	–	–	–	–
			100	32.0	300.0	10.9	3.6	50.9	–	16.0	–	–	–	–
360	**BAGEL,** made w/water	1 AVG	55	29.0	165.0	6.0	2.0	30.0	–	8.3	–	–	–	–
			100	29.0	300.0	10.9	3.6	54.5	–	15.0	–	–	–	–
361	**BROWN & SERVE,** enriched, unbrowned	1 AVG	28	33.0	83.7	2.2	1.9	14.2	.1	13.2	23.0	143.6	–	25.5
			100	33.0	299.0	7.9	6.8	50.6	.2	47.0	82.0	513.0	–	91.0
362	enriched, browned	1 AVG	26	26.9	85.3	2.3	2.0	14.2	.1	13.3	23.1	146.1	6.0	26.0
			100	26.9	328.0	8.7	7.8	54.8	.2	51.0	89.0	562.0	23.0	100.0
363	**CINNAMON**	1 AVG	55	31.5	173.8	4.7	5.0	27.1	.1	46.8	58.9	214.0	18.0	68.2
			100	31.5	316.0	8.5	9.1	49.3	.2	85.0	107.0	389.0	32.7	124.0
364	**DANISH CINNAMON,** w/raisins	1 WHOLE	39	–	135.0	2.0	5.0	21.0	–	–	–	315.0	–	–
			100	–	346.2	5.1	12.8	53.8	–	–	–	807.7	–	–
365	**DANISH PASTRY,** ready to serve	1 SM	35	22.0	147.7	2.6	8.2	16.0	Trace	17.5	38.2	128.1	8.0	39.2
			100	22.0	422.0	7.4	23.5	45.6	.1	50.0	109.0	366.0	22.9	112.0
366	**HARD,** enriched, ready to serve	1 AVG	25	25.4	78.0	2.5	.8	14.9	.1	11.8	23.0	156.3	5.7	24.3
			100	25.4	312.0	9.8	3.2	59.5	.2	47.0	92.0	625.0	22.9	97.0
367	**PLAIN,** enriched, ready to serve	1 AVG	28	31.4	83.4	2.3	1.6	14.8	.1	20.7	23.8	141.7	10.3	26.6
			100	31.4	298.0	8.2	5.6	53.0	.2	74.0	85.0	506.0	36.8	95.0

Item No.	Minerals (Micro)			Fat-Soluble Vitamins			Water-Soluble Vitamins								
	Iron	Zinc	Copper	Vitamin A	Vitamin D	Vitamin E (Alpha Tocopherol)	Vitamin C	Thiamin	Ribo-flavin	Niacin	Panto-thenic Acid	Vit. B-6 (Pyri-doxine)	Folacin (Folic Acid)	Biotin	Vitamin B-12
	(mg)	(mg)	(mg)	(IU)	(IU)	(mg)	(mg)	(mg)	(mg)	(mg)	(mg)	(mg)	(mcg)	(mcg)	(mcg)
338	**.24**	**.12**	**.05**	**326.4**	**0**	**.19**	**0**	**.01**	**.03**	**.22**	**0**	**0**	**.32**	–	**.02**
	.76	.36	.15	1019.9	0	.60	0	.02	.10	.68	0	Trace	1.00	–	.05
339	**.64**	**(.34)**	**(.09)**	–	–	**(.06)**	–	**.08**	**.07**	**.56**	**(.11)**	**(.01)**	–	–	–
	2.30	(1.20)	(.31)	–	–	(.20)	–	.27	.25	2.00	(.40)	(.04)	–	–	–
340	**.45**	–	–	–	–	–	–	**.06**	**.06**	**.50**	–	–	–	–	–
	1.60	–	–	–	–	–	–	.21	.21	1.80	–	–	–	–	–
341	**.60**	–	–	**140.0**	–	–	**.70**	**.06**	**.07**	**.56**	–	–	–	–	–
	1.70	–	–	400.0	–	–	2.00	.16	.20	1.60	–	–	–	–	–
342	**.72**	–	–	–	–	–	**0**	**.09**	**.03**	**.54**	–	–	–	–	–
	1.20	–	–	–	–	–	0	.15	.05	.90	–	–	–	–	–
343	**.39**	**.17**	–	**51.3**	**5.34**	–	**.10**	**.06**	**.06**	**.42**	**.13**	**.03**	**3.81**	–	**.07**
	1.41	.62	–	183.2	19.08	–	.35	.20	.20	1.50	.46	.09	13.59	–	.26
344	**6.08**	–	–	**136.8**	–	–	–	**.63**	**.49**	**5.99**	–	–	–	–	–
	3.20	–	–	72.0	–	–	–	.33	.26	3.15	–	–	–	–	–
345	**1.16**	–	–	**124.5**	–	–	**.83**	**.39**	**.22**	**.91**	–	**.07**	**8.30**	–	–
	1.40	–	–	150.0	–	–	1.00	.47	.26	1.10	–	.08	10.00	–	–
346	**.86**	–	–	**117.0**	–	–	**.78**	**.10**	**.15**	**.47**	–	–	–	–	–
	.10	–	–	150.0	–	–	1.00	.13	.19	.60	–	–	–	–	–
347	**.50**	–	–	**139.5**	–	–	**.45**	**.21**	**.12**	**.27**	–	**.04**	–	–	–
	.10	–	–	310.0	–	–	1.00	.47	.26	.60	–	.08	–	–	–
348	**.32**	–	–	**139.5**	–	–	**.45**	**.03**	**.08**	**.18**	–	–	–	–	–
	.70	–	–	310.0	–	–	1.00	.07	.17	.40	–	–	–	–	–
349	**.22**	–	–	**8.4**	–	–	**0**	**.04**	**.04**	**.36**	–	**.01**	–	–	**.02**
	.80	–	–	30.0	–	–	0	.14	.14	1.30	–	.02	–	–	.07
350	–	–	–	–	–	–	–	–	–	–	–	–	–	–	–
	–	–	–	–	–	–	–	–	–	–	–	–	–	–	–
351	**2.69**	–	–	**105.3**	–	–	**0**	**.14**	**.13**	**1.05**	–	–	–	–	–
	2.30	–	–	90.0	–	–	0	.12	.11	.90	–	–	–	–	–
352	**1.01**	–	–	–	–	–	–	**.02**	**.06**	**.50**	–	–	–	–	–
	1.60	–	–	–	–	–	–	.03	.09	.80	–	–	–	–	–
353	**.72**	–	–	**136.0**	–	–	**.40**	**.08**	**.12**	**.60**	–	–	–	–	–
	1.80	–	–	340.0	–	–	1.00	.20	.30	1.50	–	–	–	–	–
354	**.60**	–	–	**132.0**	–	–	–	**.06**	**.10**	**.40**	–	–	–	–	–
	1.50	–	–	330.0	–	–	–	.14	.25	1.00	–	–	–	–	–
355	**.96**	–	–	**278.4**	–	–	–	**.09**	**.17**	**.38**	–	–	–	–	–
	1.00	–	–	290.0	–	–	–	.09	.18	.40	–	–	–	–	–
356	**.34**	**.14**	**.04**	**5.0**	–	–	**.01**	**Trace**	**.01**	**.21**	–	–	–	–	–
	3.06	1.29	.32	45.0	–	–	.10	.01	.06	1.86	–	–	–	–	–
357	**.78**	**(.03)**	**(.06)**	**6.0**	–	**(.03)**	**0**	**.05**	**.02**	**.30**	**(.03)**	**(.02)**	–	–	–
	2.60	(.10)	(.19)	20.0	–	(.10)	0	.15	.05	1.00	(.10)	(.07)	–	–	–
358	**3.40**	**2.15**	**.29**	**25.0**	–	–	**.50**	**.02**	**.07**	**.62**	–	–	–	–	–
	3.40	2.15	.29	25.0	–	–	.50	.02	.07	.62	–	–	–	–	–
359	**1.21**	**.53**	–	**30.3**	–	–	**0**	**.14**	**.10**	**1.20**	–	–	–	–	–
	2.20	.96	–	55.0	–	–	0	.25	.18	2.18	–	–	–	–	–
360	**1.21**	–	–	**0**	–	–	**0**	**.15**	**.11**	**1.40**	–	–	–	–	–
	2.20	–	–	0	–	–	0	.27	.20	2.54	–	–	–	–	–
361	**.50**	–	–	–	–	–	–	**.07**	**.06**	**.59**	**.09**	**.01**	–	–	–
	1.80	–	–	–	–	–	–	.24	.20	2.10	.31	.04	–	–	–
362	**.52**	–	–	–	–	–	–	**.07**	**.06**	**.60**	**.08**	**.01**	–	–	–
	2.00	–	–	–	–	–	–	.26	.22	2.30	.31	.04	–	–	–
363	**.44**	–	–	**38.5**	–	–	–	**.04**	**.08**	**.44**	**.17**	**.02**	–	–	–
	.80	–	–	70.0	–	–	–	.07	.15	.80	.31	.04	–	–	–
364	–	–	–	–	–	–	–	–	–	–	–	–	–	–	–
	–	–	–	–	–	–	–	–	–	–	–	–	–	–	–
365	**.32**	–	–	**108.5**	–	–	–	**.03**	**.05**	**.23**	**.11**	**.01**	–	–	–
	.90	–	–	310.0	–	–	–	.07	.15	.80	.31	.04	–	–	–
366	**.58**	**.28**	–	–	–	–	–	**.07**	**.06**	**.68**	**.08**	**.01**	–	–	–
	2.30	1.13	–	–	–	–	–	.26	.23	2.70	.31	.04	–	–	–
367	**.53**	**.34**	**.07**	–	–	–	–	**.08**	**.05**	**.62**	**.09**	**.01**	–	–	–
	1.90	1.22	.25	–	–	–	–	.28	.18	2.20	.31	.04	–	–	–

(Continued)

BAKERY PRODUCTS BEVERAGES

TABLE F-36 *(Continued)*

Item No.	Food Item Name	Approximate Measure	Weight	Moisture	Food Energy Calories	Protein	Fats	Carbo-hydrates	Fiber	Minerals (Macro) Calcium	Phos-phorus	Sodium	Mag-nesium	Potas-sium
			(g)	(%)	(kcal)	(g)	(g)	(g)	(g)	(mg)	(mg)	(mg)	(mg)	(mg)
368	**ROLL DOUGH,** enriched, frozen & baked	INDV	24	28.3	74.6	2.0	1.3	13.4	.1	9.4	21.1	134.4	—	23.0
			100	28.3	311.0	8.5	5.4	56.0	.3	39.0	88.0	560.0	—	96.0
369	**ROLL MIX,** enriched, made w/water	1 AVG	38	30.6	113.6	3.4	1.7	20.7	.1	21.3	36.9	118.9	—	46.7
			100	30.6	299.0	9.0	4.5	54.5	.2	56.0	97.0	313.0	—	123.0
370	**RAISIN,** ready to serve	1 AVG	60	32.0	165.0	4.1	1.7	33.8	.5	45.0	54.6	230.4	12.0	147.0
			100	32.0	275.0	6.9	2.9	56.4	.9	75.0	91.0	384.0	20.0	245.0
371	**SWEET,** ready to serve	1 AVG	55	31.5	173.8	4.7	5.0	27.1	.1	46.8	58.9	214.0	18.0	68.2
			100	31.5	316.0	8.5	9.1	49.3	.2	85.0	107.0	389.0	32.7	124.0
372	**WHOLE WHEAT,** ready to serve	1 AVG	35	32.0	90.0	3.5	1.0	18.3	.6	37.1	98.4	197.4	40.0	102.2
			100	32.0	257.0	10.0	2.8	52.3	1.6	106.0	281.0	564.0	114.3	292.0
373	**RUSK OR RUSKETT**	1 PIECE	9	4.8	37.7	1.2	.8	6.4	Trace	1.8	10.7	22.1	—	14.5
			100	4.8	419.0	13.8	8.7	71.0	.2	20.0	119.0	246.0	—	161.0
374	**SALT STICKS,** regular type	INDV	2	5.0	7.7	.2	.1	1.5	Trace	.6	2.0	33.5	—	1.8
			100	5.0	384.0	12.0	2.9	75.3	.3	28.0	99.0	1674.0	—	92.0
375	**SALT STICKS,** Vienna bread type	INDV	35	25.0	106.4	3.3	1.1	20.3	.1	15.8	31.2	547.8	—	32.9
			100	25.0	304.0	9.5	3.1	58.0	.2	45.0	89.0	1565.0	—	94.0
376	**SHAKE'N BAKE**	1 TBLS	5	4.4	20.3	.5	.8	3.1	Trace	2.8	2.4	169.8	—	3.1
			100	4.4	406.0	9.4	15.0	61.7	.7	55.0	47.0	3395.0	—	62.0
	ALCOHOLIC													
377	**ALE,** mild	1 CUP	230	—	98.9	1.1	—	8.0	—	29.9	41.4	—	—	—
			100	—	43.0	.5	—	3.5	—	13.0	18.0	—	—	—
378	**BEER,** near beer	1 CUP	230	—	41.4	.4	—	—	—	—	—	—	—	—
			100	—	18.0	.2	—	—	—	—	—	—	—	—
379	**BEER,** natural light	1 CUP	240	—	66.7	.2	0	4.0	—	—	—	—	—	—
			100	—	27.8	.1	0	1.7	—	—	—	—	—	—
380	**BEER,** 4.5% alcohol by volume	1 CUP	240	92.1	100.8	.7	0	9.1	0	12.0	72.0	16.8	24.0	60.0
			100	92.1	42.0	.3	0	3.8	0	5.0	30.0	7.0	10.0	25.0
381	**BRANDY**	1 OZ	30	—	72.9	—	—	—	—	—	—	—	—	—
			100	—	243.0	—	—	—	—	—	—	—	—	—
	CORDIAL													
382	**ANISETTE**	1 CORDIAL	20	—	74.0	—	—	7.0	—	—	—	—	—	—
			100	—	370.0	—	—	35.0	—	—	—	—	—	—
383	**APRICOT BRANDY**	1 CORDIAL	20	—	64.0	—	—	6.0	—	—	—	—	—	—
			100	—	320.0	—	—	30.0	—	—	—	—	—	—
384	**BENEDICTINE**	1 CORDIAL	20	—	69.0	—	—	6.6	—	—	—	—	—	—
			100	—	345.0	—	—	33.0	—	—	—	—	—	—
385	**CREME DE MENTHE**	1 CORDIAL	20	—	67.0	—	—	6.0	—	—	—	—	—	—
			100	—	335.0	—	—	30.0	—	—	—	—	—	—
386	**CURACAO**	1 CORDIAL	20	—	54.0	—	—	6.0	—	—	—	—	—	—
			100	—	270.0	—	—	30.0	—	—	—	—	—	—
387	**DAIQUIRI**	1 CUP	240	—	292.8	.2	26.9	12.5	—	9.6	7.2	—	—	—
			100	—	122.0	.1	11.2	5.2	—	4.0	3.0	—	—	—
388	**GIN/RUM/VODKA/WHISKEY,** 80-proof	1 OZ	30	66.6	69.3	0	—	—	0	—	—	.3	0	.6
			100	66.6	231.0	0	—	—	0	—	—	1.0	0	2.0
389	**GIN/RUM/VODKA/WHISKEY,** 86-proof	1 OZ	30	64.0	74.7	0	—	—	0	—	—	.3	0	.6
			100	64.0	249.0	0	—	—	0	—	—	1.0	0	2.0
390	**GIN/RUM/VODKA/WHISKEY,** 90-proof	1 OZ	30	62.1	78.9	0	—	—	0	—	—	.3	0	.6
			100	62.1	263.0	0	—	—	0	—	—	1.0	0	2.0
391	**GIN/RUM/VODKA/WHISKEY,** 94-proof	1 OZ	30	60.3	82.5	0	—	—	0	—	—	.3	0	.6
			100	60.3	275.0	0	—	—	0	—	—	1.0	0	2.0
392	**GIN/RUM/VODKA/WHISKEY,** 100-proof	1 OZ	30	57.5	88.5	0	—	—	0	—	—	.3	0	.6
			100	57.5	295.0	0	—	—	0	—	—	1.0	0	2.0
393	**HIGHBALL**	1 CUP	240	—	165.6	—	—	—	0	—	—	—	—	—
			100	—	69.0	—	—	—	0	—	—	—	—	—
394	**MANHATTAN**	1 COCKTAIL	100	—	164.0	—	—	7.9	0	1.0	1.0	—	—	—
			100	—	164.0	—	—	7.9	0	1.0	1.0	—	—	—
395	**MARTINI**	1 COCKTAIL	100	—	140.0	.1	15.4	.3	0	5.0	1.0	—	—	—
			100	—	140.0	.1	15.4	.3	0	5.0	1.0	—	—	—
396	**MINT JULEP**	1 OZ	30	—	21.0	—	—	.3	0	—	—	—	—	—
			100	—	70.0	—	—	.9	0	—	—	—	—	—
397	**OLD FASHIONED**	1 OZ	30	—	53.7	—	—	1.1	0	—	—	—	—	—
			100	—	179.0	—	—	3.5	0	—	—	—	—	—

Item No.	Minerals (Micro) Iron	Zinc	Copper	Fat-Soluble Vitamins Vitamin A	Vitamin D	Vitamin E (Alpha Tocopherol)	Water-Soluble Vitamins Vitamin C	Thiamin	Ribo-flavin	Niacin	Panto-thenic Acid	Vit. B-6 (Pyri-doxine)	Folacin (Folic Acid)	Biotin	Vitamin B-12
	(mg)	(mg)	(mg)	(IU)	(IU)	(mg)	(mg)	(mg)	(mg)	(mg)	(mg)	(mg)	(mcg)	(mcg)	(mcg)
368	**.48**	—	—	—	—	—	—	**.07**	**.05**	**.55**	**.07**	**.01**	—	—	—
	2.00	—	—	—	—	—	—	.27	.22	2.30	.31	.04	—	—	—
369	**.76**	—	—	—	—	—	—	**.10**	**.10**	**.84**	**.12**	**.01**	—	—	—
	2.00	—	—	—	—	—	—	.25	.25	2.20	.31	.04	—	—	—
370	**.84**	—	—	—	—	—	—	**.04**	**.06**	**.42**	—	—	—	—	—
	1.40	—	—	—	—	—	—	.06	.10	.70	—	—	—	—	—
371	**.44**	—	—	**38.5**	—	—	—	**.04**	**.08**	**.44**	**.17**	**.02**	—	—	—
	.80	—	—	70.0	—	—	—	.07	.15	.80	.31	.04	—	—	—
372	**.84**	—	—	—	—	—	—	**.12**	**.05**	**1.05**	—	—	—	—	—
	2.40	—	—	—	—	—	—	.34	.13	3.00	—	—	—	—	—
373	**.12**	—	—	**20.7**	—	—	—	**.01**	**.02**	**.10**	—	**.01**	—	—	—
	1.30	—	—	230.0	—	—	—	.08	.22	1.10	—	.09	—	—	—
374	**.02**	—	—	—	—	—	—	**Trace**	**Trace**	**.02**	—	—	—	—	—
	.90	—	—	—	—	—	—	.06	.07	1.00	—	—	—	—	—
375	**.28**	—	—	—	—	—	—	**.02**	**.03**	**.28**	—	—	—	—	—
	.80	—	—	—	—	—	—	.05	.08	.80	—	—	—	—	—
376	**.06**	—	—	**1.0**	—	—	**0**	**Trace**	**Trace**	**.05**	—	—	—	—	—
	1.10	—	—	19.0	—	—	0	.06	.07	.90	—	—	—	—	—
377	**.23**	—	—	**0**	**0**	—	**0**	—	**.07**	**.48**	—	—	—	—	—
	.10	—	—	0	0	—	0	—	.03	.21	—	—	—	—	—
378	—	—	—	—	—	—	—	—	—	—	—	—	**13.80**	—	—
	—	—	—	—	—	—	—	—	—	—	—	—	6.00	—	—
379	—	—	—	—	—	—	—	—	—	—	—	—	**14.40**	—	—
	—	—	—	—	—	—	—	—	—	—	—	—	6.00	—	—
380	—	**.07**	**.09**	**0**	—	—	**0**	—	**.07**	**1.44**	**.19**	**.14**	**14.40**	—	**0**
	—	.03	.04	0	—	—	0	—	.03	.60	.08	.06	6.00	—	0
381	—	**.02**	**.01**	—	—	—	—	—	—	—	—	—	—	—	—
	—	.07	.05	—	—	—	—	—	—	—	—	—	—	—	—
382	—	—	—	—	—	—	—	—	—	—	—	—	—	—	—
	—	—	—	—	—	—	—	—	—	—	—	—	—	—	—
383	—	—	—	—	—	—	—	—	—	—	—	—	—	—	—
	—	—	—	—	—	—	—	—	—	—	—	—	—	—	—
384	—	—	—	—	—	—	—	—	—	—	—	—	—	—	—
	—	—	—	—	—	—	—	—	—	—	—	—	—	—	—
385	—	—	—	—	—	—	—	—	—	—	—	—	—	—	—
	—	—	—	—	—	—	—	—	—	—	—	—	—	—	—
386	—	—	—	—	—	—	—	—	—	—	—	—	—	—	—
	—	—	—	—	—	—	—	—	—	—	—	—	—	—	—
387	**.24**	—	—	**0**	**0**	—	**19.20**	**0**	**0**	—	—	—	—	—	—
	.10	—	—	0	0	—	8.00	0	0	—	—	—	—	—	—
388	—	**.01**	**.01**	**0**	—	—	**0**	**0**	**0**	**0**	—	—	—	—	—
	—	.03	.02	0	—	—	0	0	0	0	—	—	—	—	—
389	—	**.01**	**.01**	**0**	—	—	**0**	**0**	**0**	**0**	—	—	—	—	—
	—	.03	.02	0	—	—	0	0	0	0	—	—	—	—	—
390	—	**.01**	**.01**	**0**	—	—	**0**	**0**	**0**	**0**	—	—	—	—	—
	—	.03	.02	0	—	—	0	0	0	0	—	—	—	—	—
391	—	**.01**	**.01**	**0**	—	—	**0**	**0**	**0**	**0**	—	—	—	—	—
	—	.03	.02	0	—	—	0	0	0	0	—	—	—	—	—
392	—	**.01**	**.01**	**0**	—	—	**0**	**0**	**0**	**0**	—	—	—	—	—
	—	.03	.02	0	—	—	0	0	0	0	—	—	—	—	—
393	—	—	—	—	—	—	—	—	—	—	—	—	—	—	—
	—	—	—	—	—	—	—	—	—	—	—	—	—	—	—
394	—	—	—	**35.0**	—	—	**0**	**0**	**0**	**0**	—	—	—	—	—
	—	—	—	35.0	—	—	0	0	0	0	—	—	—	—	—
395	**.10**	—	—	**4.0**	—	—	**0**	**0**	**0**	**0**	—	—	—	—	—
	.10	—	—	4.0	—	—	0	0	0	0	—	—	—	—	—
396	—	—	—	—	—	—	—	—	—	—	—	—	—	—	—
	—	—	—	—	—	—	—	—	—	—	—	—	—	—	—
397	—	—	—	—	—	—	—	—	—	—	—	—	—	—	—
	—	—	—	—	—	—	—	—	—	—	—	—	—	—	—

(Continued)

TABLE F-36 *(Continued)*

Item No.	Food Item Name	Approximate Measure	Weight	Moisture	Food Energy Calories	Protein	Fats	Carbo-hydrates	Fiber	Minerals (Macro) Calcium	Phos-phorus	Sodium	Mag-nesium	Potas-sium
			(g)	(%)	(kcal)	(g)	(g)	(g)	(g)	(mg)	(mg)	(mg)	(mg)	(mg)
398	**SCOTCH**	**1 OZ**	**30**	—	**67.8**	—	—	—	**0**	—	—	—	—	—
			100	—	226.0	—	—	—	0	—	—	—	—	—
399	**TOM COLLINS**	**1 OZ**	**30**	—	**18.0**	**Trace**	**1.6**	**9**	**0**	**.6**	**.6**	—	—	—
			100	—	60.0	.1	5.3	3.0	0	2.0	2.0	—	—	—
400	**WHISKEY SOUR**	**1 COCKTAIL**	**75**	**65.3**	**138.0**	**.2**	**0**	**7.7**	**0**	**1.5**	**3.0**	**.8**	—	**93.8**
			100	65.3	184.0	.2	0	10.3	0	2.0	4.0	1.0	—	125.0
	WINES													
401	**CHAMPAGNE,** domestic	**WINE GLASS**	**120**	—	**84.0**	**.2**	**7.9**	**3.0**	**0**	—	—	—	—	—
			100	—	70.0	.2	6.6	2.5	0	—	—	—	—	—
402	**DESSERT,** 18.8% alcohol by volume	**1 CUP**	**240**	**76.7**	**328.8**	**.2**	**0**	**18.5**	**0**	**19.2**	—	**9.6**	**21.6**	**180.0**
			100	76.7	137.0	.1	0	7.7	0	8.0	—	4.0	9.0	75.0
403	**MUSCATELLE OR PORT**	**WINE GLASS**	**100**	—	**158.0**	**.2**	**11.2**	**14.0**	**0**	**8.0**	—	**4.0**	—	**75.0**
			100	—	158.0	.2	11.2	14.0	0	8.0	—	4.0	—	75.0
404	[a]**ROSE**	**WINE GLASS**	**100**	—	**71.5**	**.1**	**0**	**2.8**	**0**	**12.3**	**6.2**	**4.0**	**7.3**	**75.0**
			100	—	71.5	.1	0	2.8	0	12.3	6.2	4.0	7.3	75.0
405	**SAUTERNE,** California	**WINE GLASS**	**100**	—	**84.0**	**.2**	—	**4.0**	**0**	—	—	—	—	—
			100	—	84.0	.2	—	4.0	0	—	—	—	—	—
406	**SHERRY,** domestic	**1 OZ**	**30**	—	**42.0**	**.1**	**3.6**	**2.4**	**0**	**2.4**	**(3.0)**	**1.2**	**(3.3)**	**22.5**
			100	—	140.0	.3	11.9	8.0	0	8.0	(10.0)	4.0	(10.9)	75.0
407	**TABLE,** 12.2% alcohol by volume	**1 CUP**	**240**	**85.6**	**204.0**	**.2**	**0**	**10.1**	**0**	**21.6**	**24.0**	**12.0**	**21.6**	**220.8**
			100	85.6	85.0	.1	0	4.2	0	9.0	10.0	5.0	9.0	92.0
408	**VERMOUTH,** French	**WINE GLASS**	**100**	—	**105.0**	**.1**	**0**	**1.0**	**0**	**8.0**	**(7.6)**	**4.0**	**(5.8)**	**75.0**
			100	—	105.0	.1	0	1.0	0	8.0	(7.6)	4.0	(5.8)	75.0
409	**VERMOUTH,** Italian	**WINE GLASS**	**100**	—	**167.0**	**Trace**	**0**	**12.0**	**0**	**(6.2)**	**(6.1)**	**26.5**	**(4.5)**	**32.1**
			100	—	167.0	Trace	0	12.0	0	(6.2)	(6.1)	26.5	(4.5)	32.1
	NONALCOHOLIC													
410	**ATOLE**	**1 CUP**	**240**	**94.7**	**50.4**	**1.0**	**.2**	**11.3**	**.5**	**33.6**	—	—	—	—
			100	94.7	21.0	.4	.1	4.7	.2	14.0	—	—	—	—
411	**CLUB SODA,** carbonated, unsweetened	**1 CUP**	**240**	**100.0**	**0**	**0**	**0**	**0**	**0**	**7.2**	—	—	—	—
			100	100.0	0	0	0	0	0	3.0	—	—	—	—
	COFFEE													
412	**GROUND,** prepared	**1 CUP**	**240**	—	**7.2**	**.4**	**.2**	**1.1**	**0**	**7.2**	**7.2**	**Trace**	**(15.1)**	**159.1**
			100	—	3.0	.2	.1	.4	0	3.0	3.0	Trace	(6.3)	66.3
413	**INSTANT,** dry	**1 TSP**	**2**	—	**2.0**	**.3**	**.1**	**(.2)**	**0**	**(3.2)**	**(70.0)**	**(.8)**	**7.6**	**80.0**
			100	—	100.1	15.0	5.0	(10.9)	0	(160.1)	(350.3)	(41.2)	380.0	3998.9
414	**INSTANT,** prepared	**1 CUP**	**240**	**98.1**	**2.4**	—	—	—	—	**4.8**	**9.6**	**2.4**	**16.1**	**86.4**
			100	98.1	1.0	—	—	—	—	2.0	4.0	1.0	6.7	36.0
415	**DECAFFEINATED,** instant, powder	**1 TSP**	**2**	**2.6**	**2.6**	**Trace**	**Trace**	**.7**	**Trace**	**3.6**	**7.7**	**1.7**	—	**79.6**
			100	2.6	129.0	Trace	Trace	35.0	Trace	179.0	383.0	85.0	—	3980.0
416	**DECAFFEINATED,** freeze-dried powder	**1 TSP**	**2**	**2.6**	**2.6**	**Trace**	**Trace**	**.7**	**Trace**	**3.6**	**7.7**	**1.7**	—	**79.6**
			100	2.6	129.0	Trace	Trace	35.0	Trace	179.0	383.0	85.0	—	3980.0
417	**DECAFFEINATED,** instant, prepared	**1 CUP**	**240**	**98.1**	**2.4**	**Trace**	**Trace**	**Trace**	**Trace**	**4.8**	**9.6**	**2.4**	—	**86.4**
			100	98.1	1.0	Trace	Trace	Trace	Trace	2.0	4.0	1.0	—	36.0
418	**COLA**	**1 CUP**	**240**	**90.0**	**85.7**	**0**	**0**	**24.0**	**0**	**7.2**	**(36.7)**	**(19.7)**	**(2.9)**	**(2.6)**
			100	90.0	35.7	0	0	10.0	0	3.0	(15.3)	(8.2)	(1.2)	(1.1)
419	**COLA,** diet	**1 SERV**	**360**	**99.6**	**0**	**0**	**0**	**0**	—	**35.0**	—	**50.4**	**4.0**	**4.0**
			100	99.6	0	0	0	0	—	9.7	—	14.0	1.1	1.1
420	**CREAM SODA**	**1 CUP**	**240**	**89.0**	**103.2**	**0**	**0**	**26.4**	**0**	**7.2**	—	—	—	—
			100	89.0	43.0	0	0	11.0	0	3.0	—	—	—	—
421	**DIET DRINKS,** w/artificial sweetener, 1 cal/oz	**1 CUP**	**240**	**100.0**	—	**0**	**0**	—	**0**	**24.0**	—	—	—	—
			100	100.0	—	0	0	—	0	10.0	—	—	—	—
422	**FRUIT-FLAVORED SODAS,** 10-13% sugar	**1 CUP**	**240**	**88.0**	**110.4**	**0**	**0**	**28.8**	**0**	**7.2**	—	—	**1.4**	—
			100	88.0	46.0	0	0	12.0	0	3.0	—	—	.6	—
423	**GINGER ALE**	**1 CUP**	**240**	**92.0**	**74.4**	**0**	**0**	**19.2**	**0**	**7.2**	—	**16.8**	—	—
			100	92.0	31.0	0	0	8.0	0	3.0	—	7.0	—	—
424	**ICE CREAM SODA**	**1 CUP**	**345**	—	**262.2**	**2.3**	**7.2**	**48.7**	—	**69.0**	**55.2**	—	—	—
			100	—	76.0	.7	2.1	14.1	—	20.0	16.0	—	—	—
425	**MIX,** orange flavor, prepared	**1 CUP**	**240**	—	**135.1**	**0**	**.1**	**33.8**	**0**	**165.4**	**149.6**	**16.9**	**0**	**99.8**
			100	—	56.3	0	0	14.1	0	68.9	62.3	7.0	0	41.6
426	**QUININE SODA,** carbonated, sweetened	**1 CUP**	**240**	**92.0**	**74.4**	**0**	**0**	**19.2**	**0**	**24.0**	—	—	—	—
			100	92.0	31.0	0	0	8.0	0	10.0	—	—	—	—
427	**ROOT BEER**	**1 CUP**	**240**	**89.5**	**98.4**	**0**	**0**	**25.2**	**0**	**7.2**	—	—	—	—
			100	89.5	41.0	0	0	10.5	0	3.0	—	—	—	—

Item No.	Minerals (Micro) Iron	Zinc	Copper	Fat-Soluble Vitamins Vitamin A	Vitamin D	Vitamin E (Alpha Tocopherol)	Water-Soluble Vitamins Vitamin C	Thiamin	Ribo-flavin	Niacin	Panto-thenic Acid	Vit. B-6 (Pyri-doxine)	Folacin (Folic Acid)	Biotin	Vitamin B-12
	(mg)	(mg)	(mg)	(IU)	(IU)	(mg)	(mg)	(mg)	(mg)	(mg)	(mg)	(mg)	(mcg)	(mcg)	(mcg)
398	—	**.01**	—	—	—	—	—	—	—	—	—	—	—	—	—
	—	.02	—	—	—	—	—	—	—	—	—	—	—	—	—
399	**0**	—	—	**0**	**0**	—	**2.10**	**0**	**0**	**0**	—	—	—	—	—
	0	—	—	0	0	—	7.00	0	0	0	—	—	—	—	—
400	—	—	—	**3.8**	**0**	—	**7.50**	**.02**	**0**	—	—	—	—	—	—
	—	—	—	5.0	0	—	10.00	.02	0	—	—	—	—	—	—
401	—	—	—	—	—	—	—	—	—	—	—	—	**.60**	—	—
	—	—	—	—	—	—	—	—	—	—	—	—	.50	—	—
402	—	—	—	—	—	—	—	**.02**	**.05**	**.48**	**.07**	**.10**	**.72**	—	**0**
	—	—	—	—	—	—	—	.01	.02	.20	.03	.04	.30	—	0
403	—	—	—	—	—	—	—	**.01**	**.01**	**.17**	—	—	**2.00**	—	—
	—	—	—	—	—	—	—	.01	.01	.17	—	—	2.00	—	—
404	**.96**	**.05**	**.02**	**Trace**	**0**	—	**0**	**Trace**	**.01**	**.08**	**.03**	**.02**	**.21**	**0**	**Trace**
	.96	.05	.02	Trace	0	—	0	Trace	.01	.08	.03	.02	.21	0	Trace
405	—	—	—	—	—	—	—	—	—	—	—	—	**.50**	—	—
	—	—	—	—	—	—	—	—	—	—	—	—	.50	—	—
406	**(.12)**	—	**(.04)**	**Trace**	**0**	**0**	**0**	**Trace**	**Trace**	**.05**	—	**Trace**	**.60**	—	**Trace**
	(.38)	—	(.12)	Trace	0	0	0	.01	.01	.17	—	.01	2.00	—	Trace
407	**.96**	**.24**	—	—	—	—	—	—	**.02**	**.24**	**.07**	**.10**	**4.80**	—	**0**
	.40	.10	—	—	—	—	—	—	.01	.10	.03	.04	2.00	—	0
408	**(.35)**	**(.05)**	**(.06)**	**0**	**0**	**0**	**0**	**.01**	**.01**	**.17**	—	**Trace**	**.50**	—	**Trace**
	(.35)	(.05)	(.06)	0	0	0	0	.01	.01	.17	—	Trace	.50	—	Trace
409	**(.36)**	**(.04)**	**(.05)**	**0**	**0**	**0**	**0**	**Trace**	**Trace**	**(.04)**	—	**Trace**	**.50**	—	**Trace**
	(.36)	(.04)	(.05)	0	0	0	0	Trace	Trace	(.04)	—	Trace	.50	—	Trace
410	—	—	—	**0**	—	—	**0**	**.05**	**0**	**.24**	—	—	—	—	—
	—	—	—	0	—	—	0	.02	0	.10	—	—	—	—	—
411	—	—	—	**0**	—	—	**0**	**0**	**0**	**.00**	—	—	—	—	—
	—	—	—	0	—	—	0	0	0	.00	—	—	—	—	—
412	**.24**	**.07**	**Trace**	**0**	**0**	—	**0**	**0**	**0**	**1.20**	—	—	—	—	—
	.10	.03	Trace	0	0	—	0	0	0	.50	—	—	—	—	—
413	**Trace**	**.01**	**.01**	**0**	**0**	—	**0**	**0**	**Trace**	**.61**	**.01**	**Trace**	—	—	**0**
	4.3	.60	.28	0	0	—	0	0	.21	30.60	.40	.03	—	—	0
414	**.24**	**.07**	**.01**	**0**	—	**0**	**0**	**0**	—	**.72**	**.01**	—	—	—	**0**
	.10	.03	Trace	0	—	0	0	0	—	.30	Trace	—	—	—	0
415	**.11**	**.01**	—	**0**	—	—	**0**	**0**	**Trace**	**.61**	—	—	—	—	—
	5.60	.60	—	0	—	—	0	0	.21	30.60	—	—	—	—	—
416	**.11**	**.01**	—	**0**	—	—	**0**	**0**	**Trace**	**.61**	—	—	—	—	**0**
	5.60	.60	—	0	—	—	0	0	.21	30.60	—	—	—	—	0
417	**.24**	**.02**	—	**0**	—	—	**0**	**0**	**Trace**	**.72**	—	—	—	—	—
	.10	.01	—	0	—	—	0	0	Trace	.30	—	—	—	—	—
418	**Trace**	**Trace**	**(.07)**	**0**	**0**	**0**	**0**	**0**	**0**	**0**	**0**	**0**	**0**	**0**	**0**
	Trace	Trace	(.03)	0	0	0	0	0	0	0	0	0	0	0	0
419	**.20**	—	—	**0**	—	—	**75.00**	**0**	**0**	**0**	—	—	—	—	—
	.06	—	—	0	—	—	20.83	0	0	0	—	—	—	—	—
420	—	—	—	**0**	—	—	**0**	**0**	**0**	**0**	—	—	—	—	—
	—	—	—	0	—	—	0	0	0	0	—	—	—	—	—
421	—	—	—	**0**	—	—	**0**	**0**	**0**	**0**	—	—	—	—	—
	—	—	—	0	—	—	0	0	0	0	—	—	—	—	—
422	—	—	**.05**	**0**	—	—	**0**	**0**	**0**	**0**	—	—	—	—	—
	—	—	.02	0	—	—	0	0	0	0	—	—	—	—	—
423	—	—	—	**0**	—	—	**0**	**0**	**0**	**0**	—	—	—	—	—
	—	—	—	0	—	—	0	0	0	0	—	—	—	—	—
424	**0**	—	—	**296.7**	**0**	—	**1.04**	**.04**	**.10**	**.07**	—	—	—	—	—
	0	—	—	86.0	0	—	.30	.01	.03	.02	—	—	—	—	—
425	**0**	—	—	**2621.3**	—	—	**125.95**	**0**	**0**	**0**	**0**	**0**	—	—	**0**
	0	—	—	1092.2	—	—	52.48	0	0	0	0	0	—	—	0
426	—	—	—	**0**	—	—	**0**	**0**	**0**	**0**	—	—	—	—	—
	—	—	—	0	—	—	0	0	0	0	—	—	—	—	—
427	—	**.02**	—	**0**	—	—	**0**	**0**	**0**	**0**	—	—	—	—	—
	—	.01	—	0	—	—	0	0	0	0	—	—	—	—	—

(Continued)

TABLE F-36 *(Continued)*

Item No.	Food Item Name	Approximate Measure	Weight	Moisture	Food Energy Calories	Protein	Fats	Carbo-hydrates	Fiber	Minerals (Macro) Calcium	Phos-phorus	Sodium	Mag-nesium	Potas-sium
			(g)	(%)	(kcal)	(g)	(g)	(g)	(g)	(mg)	(mg)	(mg)	(mg)	(mg)
428	**SEVEN-UP**	**1 CUP**	**230**	—	**92.0**	**0**	**0**	**23.0**	**0**	**0**	**0**	**2.3**	—	**0**
			100	—	40.0	0	0	10.0	0	0	0	1.0	—	0
429	**TEA BAG, ORANGE PEKOE**	**INDV**	**3**	—	**0**	**0**	**0**	**0**	—	**14.0**	**8.3**	**1.1**	**5.8**	**53.9**
			100	—	1.0	.1	0	.2	—	465.0	277.0	37.0	192.0	1795.0
430	**TEA,** instant, water-soluble solids	**1 CUP**	**240**	**99.4**	**4.8**	—	—	**1.0**	—	**16.8**	**0**	**2.4**	**10.1**	**60.0**
			100	99.4	2.0	—	—	.4	—	7.0	0	1.0	4.2	25.0
431	**TONIC WATER**	**1 CUP**	**240**	—	**88.0**	**0**	**0**	**22.0**	—	—	—	**0**	—	—
			100	—	36.7	0	0	9.2	—	—	—	0	—	—
	CEREALS													
	ACHITA, AMARANTH													
432	**WHOLE GRAIN**	**1 OZ**	**28**	**12.3**	**100.2**	**3.6**	**2.0**	**18.2**	**1.9**	**69.2**	**140.0**	—	—	—
			100	12.3	358.0	12.9	7.2	65.1	6.7	247.0	500.0	—	—	—
433	**TOASTED**	**1 OZ**	**28**	**4.3**	**108.1**	**3.8**	**2.3**	**19.9**	**2.8**	**81.8**	**144.8**	—	—	—
			100	4.3	386.0	13.5	8.2	71.1	10.0	292.0	517.0	—	—	—
434	**ARROWROOT**	**1 TBLS**	**8**	—	**29.0**	**0**	**0**	**7.0**	**0**	**0**	**0**	**4.0**	—	**1.4**
			100	—	362.0	0	0	87.5	0	0	0	50.0	—	18.0
	BARLEY, PEARLED													
435	light, raw	**1 TBLS**	**14**	**11.1**	**48.9**	**1.1**	**.1**	**11.0**	**.1**	**2.2**	**26.5**	**.4**	**5.2**	**22.4**
			100	11.1	349.0	8.2	1.0	78.8	.5	16.0	189.0	3.0	37.0	160.0
436	[6]light, boiled	**1 OZ**	**28**	—	**33.6**	**.8**	**.2**	**7.7**	**Trace**	**.9**	**19.0**	**.3**	**2.0**	**1.2**
			100	—	120.0	2.8	.6	27.5	.2	3.2	71.0	1.0	7.3	40.0
437	**BUCKWHEAT, WHOLE GRAIN**	**1 CUP**	**100**	**11.0**	**360.0**	**11.7**	**2.4**	**72.9**	**9.9**	**114.0**	**282.0**	—	**252.6**	**448.0**
			100	11.0	360.0	11.7	2.4	72.9	9.9	114.0	282.0	—	252.6	448.0
	BULGUR													
438	**DRY** (from Club Wheat)	**1 TBLS**	**14**	**9.0**	**51.1**	**1.2**	**.2**	**11.1**	**.2**	**4.2**	**44.7**	—	—	**36.7**
			100	9.0	365.0	8.7	1.4	79.5	1.7	30.0	319.0	—	—	262.0
439	**DRY** (from Hard Red Winter Wheat)	**1 TBLS**	**14**	**10.0**	**50.5**	**1.6**	**.2**	**10.6**	**.2**	**4.1**	**47.3**	**.6**	**22.4**	**32.1**
			100	10.0	361.0	11.2	1.5	75.7	1.7	29.0	338.0	4.0	160.0	229.0
440	**DRY** (from White Wheat)	**1 TBLS**	**14**	**9.0**	**51.0**	**1.4**	**.2**	**10.9**	**.2**	**5.0**	**42.0**	—	—	**43.4**
			100	9.0	364.0	10.3	1.2	78.1	1.3	36.0	300.0	—	—	310.0
441	**CANNED,** unseasoned	**1 CUP**	**135**	**56.0**	**230.9**	**8.4**	**.9**	**47.3**	**1.1**	**27.0**	**270.0**	**808.7**	—	**117.5**
			100	56.0	171.0	6.2	.7	35.0	.8	20.0	200.0	599.0	—	87.0
442	**CANNED,** seasoned	**1 CUP**	**135**	**56.0**	**245.7**	**8.4**	**4.5**	**44.3**	**1.1**	**27.0**	**263.3**	**621.0**	—	**151.2**
			100	56.0	182.0	6.2	3.3	32.8	.8	20.0	195.0	460.0	—	112.0
	CANIHUA (CHENOPODIUM)													
443	[5]**WHOLE GRAIN,** flakes	**1 OZ**	**28**	**8.1**	**102.5**	**4.9**	**2.3**	**17.3**	**3.1**	**47.9**	**138.9**	—	—	—
			100	8.1	366.0	17.6	8.3	61.7	11.0	171.0	496.0	—	—	—
444	[5]**WHOLE GRAIN,** yellow	**1 OZ**	**28**	**12.5**	**91.6**	**4.0**	**.8**	**18.5**	**3.1**	**37.5**	**108.1**	—	—	—
			100	12.5	327.0	14.2	2.7	66.0	11.0	134.0	386.0	—	—	—
	CORNMEAL													
445	**DEGERMED,** unenriched, cooked	**1 CUP**	**238**	**87.7**	**119.0**	**2.6**	**.5**	**25.5**	**.2**	**2.4**	**33.3**	—	**16.7**	**38.1**
			100	87.7	50.0	1.1	.2	10.7	.1	1.0	14.0	—	7.0	16.0
446	**DEGERMED,** enriched, cooked	**1 CUP**	**238**	**87.7**	**119.0**	**2.6**	**.5**	**25.5**	**.2**	**2.4**	**33.3**	—	**16.7**	**38.1**
			100	87.7	50.0	1.1	.2	10.7	.1	1.0	14.0	—	7.0	16.0
447	**SELF-RISING,** degermed, enriched, w/o wheat flour	**1 CUP**	**145**	**11.3**	**504.6**	**10.9**	**1.6**	**109.2**	**.9**	**1631.3**	**759.8**	**2001.0**	—	**163.9**
			100	11.3	348.0	7.5	1.1	75.3	.6	1125.0	524.0	1380.0	—	113.0
448	**SELF-RISING,** degermed, enriched, w/soft wheat flour	**1 CUP**	**145**	**11.3**	**504.6**	**11.2**	**1.6**	**108.9**	**.7**	**1631.3**	**759.8**	**2001.0**	**66.7**	**158.1**
			100	11.3	348.0	7.7	1.1	75.1	.5	1125.0	524.0	1380.0	46.0	109.0
449	**SELF-RISING,** whole ground, enriched, w/o wheat flour	**1 CUP**	**118**	**11.3**	**409.5**	**10.0**	**3.8**	**84.5**	**1.1**	**1327.5**	**756.4**	**1628.4**	—	**276.1**
			100	11.3	347.0	8.5	3.2	71.6	.9	1125.0	641.0	1380.0	—	234.0
450	**SELF-RISING,** whole ground, enriched, w/soft wheat flour	**1 CUP**	**118**	**11.3**	**409.5**	**10.1**	**3.4**	**84.8**	**1.1**	**1327.5**	**736.3**	**1628.4**	**123.9**	**250.2**
			100	11.3	347.0	8.6	2.9	71.9	.9	1125.0	624.0	1380.0	105.0	212.0
	GRITS													
451	**CORN,** degermed, unenriched, cooked	**1 CUP**	**242**	**87.1**	**121.0**	**2.9**	**.2**	**26.6**	**.2**	**2.4**	**24.2**	—	**7.3**	**26.6**
			100	87.1	50.0	1.2	.1	11.0	.1	1.0	10.0	—	3.0	11.0
452	**CORN,** degermed, enriched, cooked	**1 CUP**	**242**	**87.1**	**121.0**	**2.9**	**.2**	**26.6**	**.2**	**2.4**	**24.2**	—	**7.3**	**26.6**
			100	87.1	50.0	1.2	.1	11.0	.1	1.0	10.0	—	3.0	11.0
453	**HOMINY,** enriched, dry	**1 CUP**	**156**	**10.7**	**558.5**	**13.3**	**1.4**	**123.4**	**.8**	**3.1**	**121.7**	**1.6**	**31.2**	**166.9**
			100	10.7	358.0	8.5	.9	79.1	.5	2.0	78.0	1.0	20.0	107.0
454	[4]**JOB'S TEARS,** whole seed, hulled	**1 OZ**	**28**	**15.0**	**85.7**	**3.4**	**1.9**	**18.2**	**.2**	**12.9**	**41.4**	—	—	**61.0**
			100	15.0	306.0	12.0	6.7	64.9	.8	46.0	148.0	—	—	218.0
455	**MALT,** dry	**1 TBLS**	**15**	**5.2**	**55.2**	**2.0**	**.3**	**11.6**	**.9**	—	—	—	—	—
			100	5.2	368.0	13.1	1.9	77.4	5.7	—	—	—	—	—

Item No.	Minerals (Micro) Iron	Zinc	Copper	Fat-Soluble Vitamins Vitamin A	Vitamin D	Vitamin E (Alpha Tocopherol)
	(mg)	(mg)	(mg)	(IU)	(IU)	(mg)
428	**0** 0	— —	— —	**0** 0	**0** 0	**0** 0
429	**.99** 33.00	**.16** 5.40	**.14** 4.80	**0** 0	— —	— —
430	**0** 0	**.10** .04	**.02** .01	— —	— —	— —
431	— —	— —	— —	— —	— —	— —
432	**.95** 3.40	— —	— —	**0** 0	— —	— —
433	**.45** 1.60	— —	— —	— —	— —	— —
434	**0** 0	**.01** .07	**Trace** .01	**0** 0	**0** 0	**Trace** Trace
435	**.28** 2.00	— —	**.11** .75	**0** 0	**0** 0	— —
436	**.08** .30	**.02** .07	**.01** .05	**0** 0	**0** 0	**Trace** Trace
437	**3.10** 3.10	— —	**.82** .82	**0** 0	— —	— —
438	**.66** 4.70	— —	— —	**0** 0	— —	— —
439	**.52** 3.70	— —	— —	**0** 0	— —	— —
440	**.66** 4.70	— —	— —	**0** 0	— —	— —
441	**1.76** 1.30	— —	— —	**0** 0	— —	— —
442	**1.89** 1.40	— —	— —	**0** 0	— —	— —
443	**4.20** 15.00	— —	— —	— —	— —	— —
444	**2.94** 10.50	— —	— —	— —	— —	— —
445	**.48** .20	**.24** .10	— —	**142.8** 60.0	— —	**1.52** .64
446	**.95** .40	**.24** .10	— —	**142.8** 60.0	— —	**1.52** .64
447	**4.21** 2.90	— —	— —	**609.0** 420.0	— —	— —
448	**4.21** 2.90	— —	— —	**507.5** 350.0	— —	— —
449	**3.42** 2.90	— —	— —	**531.0** 450.0	— —	— —
450	**3.42** 2.90	— —	— —	**448.4** 380.0	— —	— —
451	**.24** .10	— —	— —	**145.2** 60.0	— —	**.75** .31
452	**.73** .30	— —	— —	**145.2** 60.0	— —	**.75** .31
453	**4.46** 2.86	**0** 0	**.06** .04	**0** 0	**0** 0	— —
454	**.20** .70	— —	— —	— —	— —	— —
455	**.60** 4.00	— —	— —	— —	— —	— —

Item No.	Water-Soluble Vitamins Vitamin C	Thiamin	Riboflavin	Niacin	Pantothenic Acid	Vit. B-6 (Pyridoxine)	Folacin (Folic Acid)	Biotin	Vitamin B-12
	(mg)	(mg)	(mg)	(mg)	(mg)	(mg)	(mcg)	(mcg)	(mcg)
428	**0** 0	**0** 0	**0** 0	**0** 0	**0** 0	**0** 0	**0** 0	**0** 0	**0** 0
429	**.02** .50	**0** 0	**Trace** .02	**Trace** .05	— —	— —	— —	— —	— —
430	— —	— —	**.02** .01	— —	— —	— —	— —	— —	— —
431	— —	— —	— —	— —	— —	— —	— —	— —	— —
432	**.84** 3.00	**.04** .14	**.09** .32	**.28** 1.00	— —	— —	— —	— —	— —
433	— —	**0** 0	**.09** .32	**.31** 1.10	— —	— —	— —	— —	— —
434	**0** 0	**0** 0	**0** 0	**0** 0	— —	**Trace** Trace	**Trace** Trace	**Trace** Trace	**0** 0
435	**0** 0	**.02** .12	**.01** .05	**.43** 3.10	**.07** .50	**.03** .22	— —	— —	**0** 0
436	— —	— —	— —	— —	— —	— —	— —	— —	— —
437	**0** 0	**.60** .60	— —	**4.40** 4.40	— —	— —	— —	— —	— —
438	**0** 0	**.04** .30	**.01** .10	**.59** 4.20	— —	— —	— —	— —	— —
439	**0** 0	**.04** .28	**.02** .14	**.63** 4.50	— —	**.03** .23	**5.74** 41.00	— —	**0** 0
440	**0** 0	**.04** .30	**.01** .10	**.59** 4.20	— —	— —	— —	— —	— —
441	**0** 0	**.07** .05	**.04** .03	**3.24** 2.40	— —	— —	— —	— —	— —
442	**0** 0	**.08** .06	**.05** .04	**4.05** 3.00	— —	— —	— —	— —	— —
443	**.28** 1.00	**.16** .57	**.27** .95	**.45** 1.60	— —	— —	— —	— —	— —
444	**.56** 2.00	**.24** .86	**.23** .83	**.31** 1.10	— —	— —	— —	— —	— —
445	**0** 0	**.05** .02	**.02** .01	**.24** .10	**1.64** .69	— —	**21.42** 9.00	— —	— —
446	**0** 0	**.14** .06	**.10** .04	**1.19** .50	**1.64** .69	— —	**21.42** 9.00	— —	— —
447	**0** 0	**.64** .44	**.38** .26	**5.08** 3.50	**.84** .58	**.36** .25	**34.80** 24.00	— —	**0** 0
448	**0** 0	**.64** .44	**.38** .26	**5.08** 3.50	— —	— —	— —	— —	— —
449	**0** 0	**.52** .44	**.31** .26	**4.13** 3.50	**.68** .58	**.30** .25	— —	— —	**0** 0
450	**0** 0	**.52** .44	**.31** .26	**4.13** 3.50	— —	— —	— —	— —	— —
451	**0** 0	**.05** .02	**.02** .01	**.48** .20	— —	— —	— —	— —	— —
452	**0** 0	**.04** .02	**.03** .01	**.40** .20	— —	— —	— —	— —	— —
453	**0** 0	**.69** .44	**.41** .26	**5.49** 3.52	**0** 0	**.19** .12	**18.72** 12.00	**6.24** 4.00	**0** 0
454	**0** 0	**.12** .41	**.03** .10	**.64** 2.30	— —	— —	— —	— —	— —
455	— —	**.07** .49	**.05** .31	**1.35** 9.00	— —	— —	— —	— —	— —

(Continued)

TABLE F-36 *(Continued)*

Item No.	Food Item Name	Approximate Measure	Weight	Moisture	Food Energy Calories	Protein	Fats	Carbo-hydrates	Fiber	Minerals (Macro) Calcium	Phos-phorus	Sodium	Mag-nesium	Potas-sium
			(g)	(%)	(kcal)	(g)	(g)	(g)	(g)	(mg)	(mg)	(mg)	(mg)	(mg)
456	**MALT EXTRACT,** dried	1 TBLS	8	3.2	29.4	.5	—	7.1	—	3.8	23.5	6.4	11.2	18.4
			100	3.2	367.0	6.0	—	89.2	—	48.0	294.0	80.0	140.0	230.0
	MILLET													
457	[3]**BULRUSH, WHOLE GRAIN,** dried	1 OZ	28	12.0	95.5	2.9	1.1	20.0	.5	6.2	80.1	—	—	—
			100	12.0	341.0	0.4	4.0	71.6	1.9	22.0	286.0	—	—	—
458	[3]**BULRUSH, FLOUR,** roasted, 75% extraction	1 OZ	28	16.0	92.4	1.7	1.0	20.0	.2	4.8	47.6	—	—	—
			100	16.0	333.0	5.9	3.5	71.3	.6	17.0	170.0	—	—	—
459	[3]**BULRUSH, WHOLE GRAIN,** germinated	1 OZ	28	—	—	—	—	—	—	—	—	—	—	—
			100	—	—	—	—	—	—	—	—	—	—	—
460	**PROSO (BROOMCORN, HOGMILLET), WHOLE GRAIN**	1 CUP	230	11.8	846.4	22.8	9.4	167.7	7.4	46.0	715.3	—	384.1	989.0
			100	11.8	368.0	9.9	4.1	72.9	3.2	20.0	311.0	—	167.0	430.0
461	[3]**WHOLE GRAIN,** unclassified	1 CUP	230	10.8	795.8	20.5	8.5	172.3	3.9	48.3	641.7	—	—	—
			100	10.8	346.0	8.9	3.7	74.9	1.7	21.0	279.0	—	—	—
462	[3]**MEAL,** unclassified	1 OZ	28	8.6	108.4	2.6	1.2	21.4	.4	4.8	80.9	—	—	—
			100	8.6	387.0	9.3	4.2	76.3	1.6	17.0	289.0	—	—	—
463	[3]**BRAN**	1 OZ	28	18.0	91.0	3.1	2.1	15.0	—	22.4	112.0	—	—	—
			100	18.0	325.0	11.0	7.6	53.4	—	80.0	400.0	—	—	—
464	**OATS, WHOLE GRAIN**	1 CUP	80	—	—	—	—	—	—	—	—	—	135.2	—
			100	—	—	—	—	—	—	—	—	—	169.0	—
	QUINOA (CHENOPODIUM)													
465	[5]**WHOLE GRAIN**	1 OZ	28	11.0	98.3	3.4	1.7	19.0	1.3	31.4	80.1	—	—	—
			100	11.0	351.0	12.3	6.1	67.7	4.6	112.0	286.0	—	—	—
466	[5]**FLAKES**	1 OZ	28	7.0	100.5	2.4	1.0	22.0	1.1	31.9	44.8	—	—	—
			100	7.0	359.0	8.5	3.7	78.7	3.8	114.0	160.0	—	—	—
	RICE													
467	**BRAN**	1 OZ	28	9.7	77.3	3.7	4.4	14.2	.4	21.3	388.1	—	—	418.6
			100	9.7	276.0	13.3	15.8	50.8	1.5	76.0	1386.0	—	—	1495.0
468	**BROWN,** cooked	1 CUP	150	70.3	178.5	3.8	1.2	38.3	.5	18.0	109.5	423.0	43.5	105.0
			100	70.3	119.0	2.5	.8	25.5	.3	12.0	73.0	282.0	29.0	70.0
469	**FRIED**	1 CUP	155	—	353.4	2.6	23.4	31.9	—	12.4	37.2	—	10.9	—
			100	—	228.0	1.7	15.1	20.6	—	8.0	24.0	—	7.0	—
470	**GRANULATED,** w/added nutrients, cooked	1 CUP	245	87.5	122.5	2.0	—	27.4	—	4.9	31.9	431.2	—	—
			100	87.5	50.0	.8	—	11.2	—	2.0	13.0	176.0	—	—
471	**POLISHED**	1 CUP	105	9.8	278.3	12.7	13.4	60.6	2.5	72.5	1161.3	—	—	749.7
			100	9.8	265.0	12.1	12.8	57.7	2.4	69.0	1106.0	—	—	714.0
472	**WHITE,** unenriched, glutinous, cooked	1 OZ	28	13.2	101.1	1.6	.3	22.3	.1	10.1	28.0	2.8	—	36.4
			100	13.2	361.0	5.6	.9	79.8	.3	38.0	100.0	10.0	—	130.0
473	**WHITE,** enriched (all varieties), cooked	1 CUP	150	72.6	160.5	3.0	.3	36.3	.2	15.0	42.0	561.0	12.0	42.0
			100	72.6	107.0	2.0	.2	24.2	.1	10.0	28.0	374.0	8.0	28.0
474	**WHITE,** enriched (long grain), parboiled, cooked	1 CUP	150	73.4	159.0	3.2	.2	35.0	.2	28.5	85.5	537.0	—	64.5
			100	73.4	106.0	2.1	.1	23.3	.1	19.0	57.0	358.0	—	43.0
475	**WHITE,** enriched (long grain), precooked (instant), ready to serve	1 CUP	148	72.9	161.3	3.3	—	35.8	.1	4.4	28.1	404.0	—	—
			100	72.9	109.0	2.2	—	24.2	.1	3.0	19.0	273.0	—	—
476	**RYE, WHOLE GRAIN**	1 CUP	185	12.1	669.7	22.4	4.1	135.8	3.7	70.3	695.6	1.9	246.6	864.0
			100	12.1	362.0	12.1	2.2	73.4	2.0	38.0	376.0	1.0	133.3	467.0
	SAFFLOWER SEED													
477	**KERNELS,** dry	1 OZ	28	5.0	172.2	5.3	16.7	3.5	—	—	—	—	—	—
			100	5.0	615.0	19.1	59.5	12.4	—	—	—	—	—	—
478	**MEAL,** partially defatted	1 CUP	150	9.1	582.0	63.2	12.3	54.8	11.1	112.5	930.0	—	—	—
			100	9.1	388.0	42.1	8.2	36.5	7.4	75.0	620.0	—	—	—
479	**SEMOLINA,** low protein, cooked	1 OZ	28	10.0	95.2	.1	0	23.9	1.1	—	—	2.2	—	.6
			100	10.0	340.0	.5	.1	85.3	4.0	—	—	8.0	—	2.0
480	**SORGHUM GRAIN,** all types	1 OZ	28	11.0	102.5	3.1	.9	20.4	.5	7.8	80.4	—	—	98.0
			100	11.0	366.0	11.0	3.3	73.0	1.7	28.0	287.0	—	—	350.0
481	**SOYBEAN PROTEINATE**	1 OZ	28	5.5	87.4	22.6	0	2.2	.2	—	—	336.0	—	—
			100	5.5	312.0	80.6	.1	7.7	.6	—	—	1200.0	—	—
	TEFF													
482	[3]**WHOLE GRAIN,** red	1 OZ	28	12.5	91.8	2.7	.7	20.2	.8	48.2	87.6	—	—	—
			100	12.5	328.0	9.6	2.6	72.0	2.7	172.0	313.0	—	—	—
483	[3]**MEAL** from red variety	1 OZ	28	13.2	97.4	2.8	.5	19.8	.7	41.7	75.6	—	—	—
			100	13.2	348.0	10.0	1.9	70.8	2.6	149.0	270.0	—	—	—
484	[3]**WHOLE GRAIN,** white	1 OZ	28	11.3	93.0	2.4	.5	21.1	.6	40.9	106.4	—	—	—
			100	11.3	332.0	8.7	1.9	75.5	2.3	146.0	380.0	—	—	—

Item No.	Minerals (Micro)			Fat-Soluble Vitamins			Water-Soluble Vitamins								
	Iron	Zinc	Copper	Vitamin A	Vitamin D	Vitamin E (Alpha Tocopherol)	Vitamin C	Thiamin	Ribo-flavin	Niacin	Panto-thenic Acid	Vit. B-6 (Pyri-doxine)	Folacin (Folic Acid)	Biotin	Vitamin B-12
	(mg)	(mg)	(mg)	(IU)	(IU)	(mg)	(mg)	(mg)	(mg)	(mg)	(mg)	(mg)	(mcg)	(mcg)	(mcg)
456	**.70**	—	—	—	—	—	—	**.03**	**.04**	**.78**	—	—	—	—	—
	8.70	—	—	—	—	—	—	.36	.45	9.80	—	—	—	—	—
457	**5.80**	—	—	**0**	—	—	**.84**	**.08**	**.06**	**.48**	—	—	—	—	—
	20.70	—	—	0	—	—	3.00	.30	.22	1.70	—	—	—	—	—
458	**10.92**	—	—	—	—	—	—	**.05**	**.06**	—	—	—	—	—	—
	39.00	—	—	—	—	—	—	.18	.22	—	—	—	—	—	—
459	—	—	—	—	—	—	—	**.06**	**.09**	**.84**	—	—	—	—	—
	—	—	—	—	—	—	—	.21	.32	3.00	—	—	—	—	—
460	**15.64**	—	**.54**	**0**	—	**.12**	**0**	**1.68**	**.87**	**5.29**	—	—	—	—	—
	6.80	—	.23	0	—	.05	0	.73	.38	2.30	—	—	—	—	—
461	**8.51**	—	—	**38.4**	—	—	—	**.76**	**.32**	**4.14**	—	—	—	—	—
	3.70	—	—	16.7	—	—	—	.33	.14	1.80	—	—	—	—	—
462	**2.80**	—	—	—	—	—	—	**.08**	**.03**	**.36**	—	—	—	—	—
	10.00	—	—	—	—	—	—	.27	.12	1.30	—	—	—	—	—
463	**11.48**	—	—	—	—	—	—	**.28**	—	—	—	—	—	—	—
	41.00	—	—	—	—	—	—	1.01	—	—	—	—	—	—	—
464	—	—	**.03**	—	—	**1.12**	—	—	—	—	—	—	—	—	—
	—	—	.04	—	—	1.40	—	—	—	—	—	—	—	—	—
465	**2.10**	—	—	**0**	—	—	**.84**	**.10**	**.12**	**.39**	—	—	—	—	—
	7.50	—	—	0	—	—	3.00	.36	.42	1.40	—	—	—	—	—
466	**1.32**	—	—	—	—	—	—	**.04**	**.11**	**.31**	—	—	—	—	—
	4.70	—	—	—	—	—	—	.13	.38	1.10	—	—	—	—	—
467	**5.43**	—	—	**0**	—	—	**0**	**.63**	**.07**	**8.34**	**.78**	**.70**	**10.92**	**16.80**	—
	19.40	—	—	0	—	—	0	2.26	.25	29.80	2.77	2.50	39.00	60.00	—
468	**.75**	**.90**	—	**0**	—	**.23**	**0**	**.14**	**.03**	**2.10**	**2.28**	**.93**	—	—	—
	.50	.60	—	0	—	.15	0	.09	.02	1.40	1.52	.62	—	—	—
469	**1.24**	—	—	**0**	—	—	**0**	**.14**	**.02**	**1.24**	—	**.05**	—	—	**0**
	.80	—	—	0	—	—	0	.09	.01	.80	—	.03	—	—	0
470	**1.72**	—	—	**0**	—	—	**0**	**1.47**	**.02**	**1.96**	—	—	—	—	—
	.70	—	—	0	—	—	0	.06	.01	.80	—	—	—	—	—
471	**16.91**	**.37**	—	**0**	—	—	**0**	**1.93**	**.19**	**29.61**	**3.50**	**2.10**	**201.60**	**59.85**	—
	16.10	.35	—	0	—	—	0	1.84	.18	28.20	3.33	2.00	192.00	57.00	—
472	**.56**	—	—	**0**	—	—	**0**	**.02**	**.01**	**.56**	—	—	**2.80**	—	—
	2.00	—	—	0	—	—	0	.07	.04	2.00	—	—	10.00	—	—
473	**1.35**	**.60**	—	**0**	—	**.27**	**0**	**.17**	**.11**	**1.50**	**1.13**	**.06**	**24.00**	**7.50**	—
	.90	.40	—	0	—	.18	0	.11	.07	1.00	.75	.04	16.00	5.00	—
474	**1.20**	**.53**	—	**0**	—	—	**0**	**.17**	**.11**	**1.80**	**2.06**	**.15**	**28.50**	**15.00**	—
	.80	.35	—	0	—	—	0	.11	.07	1.20	1.37	.10	19.00	10.00	—
475	**1.18**	**.30**	—	**0**	—	—	**0**	**.19**	**.10**	**1.48**	—	—	—	—	—
	.80	.20	—	0	—	—	0	.13	.07	1.00	—	—	—	—	—
476	**6.85**	**6.01**	**0.47**	**0**	**0**	**3.05**	**0**	**.80**	**.41**	**2.96**	—	—	—	—	—
	3.70	3.25	0.25	0	0	1.65	0	.43	.22	1.60	—	—	—	—	—
477	—	—	—	—	—	—	—	—	—	—	—	—	—	—	—
	—	—	—	—	—	—	—	—	—	—	—	—	—	—	—
478	—	—	—	—	—	—	**0**	**1.68**	**.60**	**3.30**	—	—	—	—	—
	—	—	—	—	—	—	0	1.12	.40	2.20	—	—	—	—	—
479	—	—	—	—	—	—	—	—	—	—	—	—	—	—	—
	—	—	—	—	—	—	—	—	—	—	—	—	—	—	—
480	**1.23**	—	—	**0**	—	—	**0**	**.11**	**.04**	**1.09**	—	—	**7.56**	—	—
	4.40	—	—	0	—	—	0	.38	.15	3.90	—	—	27.00	—	—
481	—	—	—	—	—	—	**0**	—	—	—	—	—	—	—	—
	—	—	—	—	—	—	0	—	—	—	—	—	—	—	—
482	**21.14**	—	—	**2.3**	—	—	—	**.04**	**.03**	**.39**	—	—	—	—	—
	75.50	—	—	8.3	—	—	—	.13	.12	1.40	—	—	—	—	—
483	**13.66**	—	—	**0**	—	—	—	**.02**	**.03**	**.34**	—	—	—	—	—
	48.80	—	—	0	—	—	—	.06	.09	1.20	—	—	—	—	—
484	**5.85**	—	—	**0**	—	—	—	**.11**	**.04**	**.53**	—	—	—	—	—
	20.90	—	—	0	—	—	—	.38	.14	1.90	—	—	—	—	—

(Continued)

TABLE F-36 *(Continued)*

Item No.	Food Item Name	Approximate Measure	Weight	Moisture	Food Energy Calories	Protein	Fats	Carbo-hydrates	Fiber	Minerals (Macro) Calcium	Phos-phorus	Sodium	Mag-nesium	Potas-sium
			(g)	(%)	(kcal)	(g)	(g)	(g)	(g)	(mg)	(mg)	(mg)	(mg)	(mg)
	TEFF (Continued)													
485	[3]**MEAL** from white variety	1 OZ	**28**	**10.3**	**103.9**	**2.7**	**.7**	**21.0**	**.5**	**36.1**	**83.2**	—	—	—
			100	10.3	371.0	9.8	2.6	75.1	1.7	129.0	297.0	—	—	—
486	[5]**TEOSINTE,** milled	1 OZ	**28**	**12.0**	**93.5**	**6.0**	**.7**	**17.7**	**.1**	**2.5**	**54.3**	—	—	—
			100	12.0	334.0	21.6	2.5	63.1	.4	9.0	194.0	—	—	—
	WHEAT													
487	**WHOLE GRAIN,** Durum	1 CUP	**198**	**14.0**	**714.8**	**25.1**	**6.5**	**138.8**	**3.6**	**73.3**	**764.3**	**5.9**	**316.8**	**861.3**
			100	14.0	361.0	12.7	3.3	70.1	1.8	37.0	386.0	3.0	160.0	435.0
488	**WHOLE GRAIN,** Hard Red Spring	1 CUP	**198**	**14.0**	**706.9**	**27.7**	**5.3**	**136.8**	**4.6**	**71.3**	**758.3**	**5.9**	**316.8**	**732.6**
			100	14.0	357.0	14.0	2.7	69.1	2.3	36.0	383.0	3.0	160.0	370.0
489	**WHOLE GRAIN,** Hard Red Winter	1 CUP	**198**	**14.0**	**710.8**	**24.4**	**5.0**	**142.0**	**4.6**	**91.1**	**700.9**	**5.9**	**316.8**	**732.6**
			100	14.0	359.0	12.3	2.5	71.7	2.3	46.0	354.0	3.0	160.0	370.0
490	**WHOLE GRAIN,** Soft Red Winter	1 CUP	**198**	**14.0**	**714.8**	**20.2**	**4.0**	**142.8**	**4.6**	**83.2**	**792.0**	**5.9**	**316.8**	**744.5**
			100	14.0	361.0	10.2	2.0	72.1	2.3	42.0	400.0	3.0	160.0	376.0
491	**WHOLE GRAIN,** white	1 CUP	**198**	**14.0**	**706.9**	**18.6**	**4.0**	**149.3**	**3.8**	**71.3**	**780.1**	**5.9**	**316.8**	**772.2**
			100	14.0	357.0	9.4	2.0	75.4	1.9	36.0	394.0	3.0	160.0	390.0
492	**WHOLE GRAIN,** cooked	1 CUP	**42**	**87.7**	**18.9**	**.8**	**.1**	**3.9**	**.12**	**2.9**	**21.8**	**89.0**	—	**20.2**
			100	87.7	45.0	1.8	.3	9.4	.30	7.0	52.0	212.0	—	48.0
493	**BRAN,** unprocessed	1 TBLS	**9**	**14.0**	**31.8**	**1.4**	**.4**	**5.6**	**.8**	**10.7**	**114.8**	**.8**	**53.8**	**100.9**
			100	14.0	353.0	16.0	4.6	61.9	9.1	119.0	1276.0	9.0	597.9	1121.0
494	**GERM,** unprocessed	1 TBLS	**9**	**14.0**	**35.2**	**2.4**	**1.0**	**4.2**	**.2**	**6.5**	**100.6**	**.3**	**30.6**	**74.4**
			100	14.0	391.0	26.6	10.9	46.7	2.5	72.0	1118.0	3.0	340.0	827.0
495	**GERM,** toasted	1 TBLS	**6**	**4.2**	**23.5**	**1.5**	**.7**	**3.0**	**.1**	**2.8**	**65.0**	**.1**	**21.9**	**56.8**
			100	4.2	391.0	25.2	11.5	49.5	1.7	47.0	1084.0	2.0	364.9	947.0
496	**WILD RICE,** raw	1 CUP	**112**	**8.5**	**395.4**	**15.8**	**.8**	**84.3**	**1.1**	**21.3**	**379.7**	**7.8**	**144.5**	**246.4**
			100	8.5	353.0	14.1	.7	75.3	1.0	19.0	339.0	7.0	129.0	220.0
497	**WILD-WHITE RICE MIX**	1 CUP	**112**	**4.2**	**228.5**	**6.6**	**1.1**	**47.3**	—	—	—	—	—	—
			100	4.2	204.0	5.9	1.0	42.3	—	—	—	—	—	—
	BREAKFAST CEREALS													
	BRAN													
498	**ALL-BRAN,** Kellogg's	1 CUP	**56**	**.03**	**128.8**	**6.2**	**1.1**	**39.8**	**4.2**	**39.2**	**493.9**	**566.7**	**89.6**	**517.4**
			100	.03	230.0	11.0	2.0	71.0	7.5	70.0	882.0	1012.0	160.0	924.0
499	**BRAN BUDS,** Kellogg's	1 CUP	**85**	—	**209.9**	**9.0**	**0**	**63.0**	**6.0**	**60.0**	**599.6**	**773.5**	**239.9**	**1079.4**
			100	—	246.9	10.6	0	74.1	7.1	70.5	705.5	910.1	282.2	1269.8
500	**BRAN FLAKES** (40% bran), w/added thiamin	1 CUP	**37**	**3.0**	**112.1**	**3.8**	**.7**	**29.8**	**1.3**	**26.3**	**183.2**	**342.3**	**64.8**	**270.1**
			100	3.0	303.0	10.2	1.8	80.6	3.6	71.0	495.0	925.0	175.0	730.0
501	**FLAKES,** w/raisins, added thiamin	1 CUP	**42**	**7.3**	**120.5**	**3.5**	**.6**	**33.3**	**1.3**	**23.5**	**166.3**	**336.0**	—	—
			100	7.3	287.0	8.3	1.4	79.3	3.0	56.0	396.0	800.0	—	—
502	w/sugar & defatted wheat germ	1 CUP	**56**	**3.0**	**133.3**	**6.0**	**1.0**	**44.1**	**3.6**	**40.9**	**547.1**	**274.4**	—	—
			100	3.0	238.0	10.8	1.8	76.8	6.5	73.0	977.0	490.0	—	—
503	w/sugar & malt extract	1 CUP	**56**	**3.6**	**134.4**	**7.1**	**1.7**	**41.6**	**4.4**	**39.2**	**658.6**	**593.6**	—	**599.2**
			100	3.6	240.0	12.6	3.0	74.3	7.8	70.0	1176.0	1060.0	—	1070.0
	CORN													
504	**FLAKES,** w/added nutrients	1 CUP	**25**	**3.8**	**96.5**	**2.0**	**.1**	**21.3**	**.2**	**4.3**	**11.3**	**251.3**	**2.2**	**30.0**
			100	3.8	386.0	7.9	.4	85.3	.7	17.0	45.0	1005.0	8.7	120.0
505	**FLAKES,** sugar-covered, w/added nutrients	1 CUP	**25**	**2.2**	**96.5**	**1.1**	**.1**	**22.8**	**.1**	**3.0**	**6.0**	**193.8**	**4.0**	—
			100	2.2	386.0	4.4	.2	91.3	.4	12.0	24.0	775.0	16.0	—
506	**PUFFED,** w/added nutrients, presweetened	1 CUP	**13**	**5.0**	**49.3**	**.5**	**0**	**11.7**	**Trace**	**1.4**	**3.6**	**39.0**	—	—
			100	5.0	379.0	4.0	.2	89.8	.3	11.0	28.0	300.0	—	—
507	**PUFFED,** w/added nutrients, presweet cocoa flavor	1 CUP	**13**	**2.1**	**50.7**	**.8**	**.3**	**11.3**	**.1**	**2.6**	**11.7**	**110.5**	—	—
			100	2.1	390.0	6.2	2.2	86.7	.5	20.0	90.0	850.0	—	—
508	**PUFFED,** w/added nutrients, presweet fruit flavor	1 CUP	**13**	**2.1**	**51.4**	**.7**	**.4**	**11.4**	**Trace**	**3.9**	**9.1**	**78.0**	—	—
			100	2.1	395.0	5.6	2.7	87.4	.3	30.0	70.0	600.0	—	—
509	**SHREDDED,** w/added nutrients	1 CUP	**38**	**3.9**	**149.7**	**2.7**	**.8**	**33.0**	**.2**	**1.9**	**14.8**	**375.4**	—	—
			100	3.9	394.0	7.0	2.1	86.9	.6	5.0	39.0	988.0	—	—
	FARINA													
510	enriched, instant, dry	1 CUP	**100**	**10.3**	**362.0**	**11.4**	**.9**	**74.9**	**.4**	**500.0**	**396.0**	**7.0**	**25.0**	**83.0**
			100	10.3	362.0	11.4	.9	74.9	.4	500.0	396.0	7.0	25.0	83.0
511	enriched, instant, cooked, w/salt added	1 CUP	**245**	**85.9**	**134.8**	**4.2**	**.2**	**27.9**	**.2**	**188.7**	**147.0**	**460.6**	**9.8**	**31.9**
			100	85.9	55.0	1.7	.1	11.4	.1	77.0	60.0	188.0	4.0	13.0
512	enriched, quick, dry	1 CUP	**100**	**10.3**	**362.0**	**11.4**	**.9**	**74.9**	**.4**	**500.0**	**561.0**	**250.0**	**41.0**	**83.0**
			100	10.3	362.0	11.4	.9	74.9	.4	500.0	561.0	250.0	41.0	83.0

Item No.	Minerals (Micro) Iron	Zinc	Copper	Fat-Soluble Vitamins Vitamin A	Vitamin D	Vitamin E (Alpha Tocopherol)	Water-Soluble Vitamins Vitamin C	Thiamin	Ribo-flavin	Niacin	Panto-thenic Acid	Vit. B-6 (Pyri-doxine)	Folacin (Folic Acid)	Biotin	Vitamin B-12
	(mg)	(mg)	(mg)	(IU)	(IU)	(mg)	(mg)	(mg)	(mg)	(mg)	(mg)	(mg)	(mcg)	(mcg)	(mcg)
485	**6.05**	–	–	**0**	–	–	–	**.10**	**.03**	**.50**	–	–	–	–	–
	21.60	–	–	0	–	–	–	.34	.12	1.80	–	–	–	–	–
486	**.92**	–	–	**0**	–	–	**0**	**.04**	**.02**	**.20**	–	–	–	–	–
	3.30	–	–	0	–	–	0	.14	.07	.70	–	–	–	–	–
487	**8.51**	**5.35**	**1.05**	**0**	–	**2.77**	**0**	**1.31**	**.24**	**8.71**	–	–	**102.96**	**21.78**	–
	4.30	2.70	.53	0	–	1.40	0	.66	.12	4.40	–	–	52.00	11.00	–
488	**6.14**	**6.73**	**1.01**	**0**	–	**2.77**	**0**	**1.13**	**.24**	**8.51**	–	–	**102.96**	**21.78**	–
	3.10	3.40	.51	0	–	1.40	0	.57	.12	4.30	–	–	52.00	11.00	–
489	**6.73**	**6.73**	**1.01**	**0**	–	**2.77**	**0**	**1.03**	**.24**	**8.51**	–	–	**102.96**	**21.78**	–
	3.40	3.40	.51	0	–	1.40	0	.52	.12	4.30	–	–	52.00	11.00	–
490	**6.93**	**5.35**	**.89**	**0**	–	**2.77**	**0**	**.85**	**.22**	**7.13**	–	–	**102.96**	**21.78**	–
	3.50	2.70	.45	0	–	1.40	0	.43	.11	3.60	–	–	52.00	11.00	–
491	**5.94**	**4.36**	**1.05**	**0**	–	**2.77**	**0**	**1.05**	**.24**	**10.49**	–	–	**102.96**	**21.78**	–
	3.00	2.20	.53	0	–	1.40	0	.53	.12	5.30	–	–	52.00	11.00	–
492	**.21**	**.21**	–	**0**	–	–	**0**	**.03**	**.01**	**.25**	**.58**	**.22**	**2.72**	**.58**	–
	.50	.50	–	0	–	–	0	.06	.02	.60	1.37	.53	6.48	1.37	–
493	**1.34**	**.88**	**(.11)**	**0**	–	**.15**	**0**	**.07**	**.03**	**1.89**	**.27**	**.12**	**23.22**	**5.13**	–
	14.90	9.80	(1.27)	0	–	1.63	0	.72	.35	21.00	3.00	1.38	258.00	57.00	–
494	**.85**	**1.29**	**.07**	**0**	–	**(1.22)**	**0**	**.18**	**.06**	**.38**	**.11**	–	**29.52**	**2.08**	–
	9.40	14.30	.74	0	–	(13.50)	0	2.01	.68	4.20	1.20	–	328.00	22.00	–
495	**.53**	**.92**	**.08**	**6.6**	–	–	**.60**	**.10**	**.06**	**.32**	**.07**	**.07**	**25.20**	–	**0**
	8.90	15.40	1.30	110.0	–	–	10.00	1.65	.98	5.30	1.20	1.15	420.00	–	0
496	**4.70**	–	–	**0**	–	–	**0**	**.50**	**.71**	**6.94**	**1.14**	–	–	–	**0**
	4.20	–	–	0	–	–	0	.45	.63	6.20	1.02	–	–	–	0
497	–	–	–	–	–	–	–	–	–	–	–	–	–	–	–
	–	–	–	–	–	–	–	–	–	–	–	–	–	–	–
498	**8.85**	**7.39**	**.51**	**2469.0**	**78.96**	–	**29.68**	**.74**	**.84**	**9.86**	–	**.99**	**196.00**	–	–
	5.80	13.20	.91	4409.0	141.00	–	53.00	1.32	1.50	17.60	–	1.76	350.00	–	–
499	**3.49**	**4.50**	**.90**	**3747.8**	**119.93**	–	–	**1.13**	**1.27**	**14.99**	–	**1.50**	**299.82**	–	–
	5.87	5.29	1.06	4409.2	141.09	–	–	1.32	1.50	17.64	–	1.76	352.73	–	–
500	**1.63**	**1.33**	**.24**	**0**	–	–	**0**	**.15**	**.06**	**2.29**	**.32**	**.14**	–	–	**0**
	4.40	3.60	.64	0	–	–	0	.40	.17	6.20	.88	.38	–	–	0
501	**1.68**	–	–	–	**0**	–	**0**	**.13**	**.06**	**2.23**	–	–	–	–	–
	4.00	–	–	–	0	–	0	.32	.13	5.30	–	–	–	–	–
502	**4.93**	–	–	**0**	–	–	**0**	**.16**	**.12**	**7.84**	–	–	–	–	–
	8.80	–	–	0	–	–	0	.28	.21	14.00	–	–	–	–	–
503	**4.48**	–	–	**0**	–	–	–	**.06**	**.16**	**9.97**	–	–	–	–	–
	8.00	–	–	0	–	–	–	.10	.29	17.80	–	–	–	–	–
504	**.35**	**.08**	**.01**	**0**	**0**	**.02**	**0**	**.11**	**.02**	**.53**	**.05**	**.02**	**1.38**	–	**0**
	1.40	.30	.02	0	0	.09	0	.43	.08	2.10	.19	.07	5.50	–	0
505	**.25**	–	–	**0**	–	**.02**	**0**	**.10**	**.01**	**.48**	**.05**	**.02**	–	–	**0**
	1.00	–	–	0	–	.09	0	.41	.04	1.90	.18	.07	–	–	0
506	**.23**	–	–	**0**	–	**.01**	**0**	**.06**	**.02**	**.27**	**.04**	–	–	–	**0**
	1.80	–	–	0	–	.09	0	.42	.17	2.10	.29	–	–	–	0
507	**.78**	–	–	**0**	–	**.01**	**0**	**.10**	**.02**	**.33**	–	–	–	–	–
	6.00	–	–	0	–	.03	0	.79	.18	2.50	–	–	–	–	–
508	**.65**	–	–	**0**	–	**.01**	**13.78**	**.13**	**.02**	**.33**	–	–	–	–	–
	5.00	–	–	0	–	.09	106.00	.99	.17	2.50	–	–	–	–	–
509	**.91**	–	–	**0**	–	**.03**	**0**	**.16**	**.07**	**.80**	–	–	–	–	–
	2.40	–	–	0	–	.08	0	.42	.18	2.10	–	–	–	–	–
510	**42.40**	–	–	**0**	–	–	**0**	**.44**	**.26**	**3.50**	**.52**	**.07**	–	–	**0**
	42.40	–	–	0	–	–	0	.44	.26	3.50	.52	.07	–	–	0
511	**15.68**	–	–	**0**	–	–	**0**	**.17**	**.10**	**1.23**	–	–	–	–	–
	6.40	–	–	0	–	–	0	.07	.04	.50	–	–	–	–	–
512	**42.40**	**2.20**	**.12**	**0**	–	–	**0**	**.44**	**.26**	**3.50**	**.52**	**.07**	**12.90**	–	**0**
	42.40	2.20	.12	0	–	–	0	.44	.26	3.50	.52	.07	12.90	–	0

(Continued)

TABLE F-36 *(Continued)*

Item No.	Food Item Name	Approximate Measure	Weight	Moisture	Food Energy Calories	Protein	Fats	Carbo-hydrates	Fiber	Minerals (Macro) Calcium	Phos-phorus	Sodium	Mag-nesium	Potas-sium
			(g)	(%)	(kcal)	(g)	(g)	(g)	(g)	(mg)	(mg)	(mg)	(mg)	(mg)
	FARINA (Continued)													
513	enriched, quick, cooked, w/salt added	1 CUP	**245**	**89.0**	**105.4**	**3.2**	**.2**	**21.8**	–	**147.0**	**161.7**	**404.3**	**7.4**	**24.5**
			100	89.0	43.0	1.3	.1	8.9	–	60.0	66.0	165.0	3.0	10.0
514	enriched, regular, dry	1 CUP	**100**	**10.3**	**367.0**	**11.4**	**1.5**	**77.0**	**.4**	**25.0**	**107.0**	**2.0**	**17.0**	**83.0**
			100	10.3	367.0	11.4	1.5	77.0	.4	25.0	107.0	2.0	17.0	83.0
515	enriched, regular, cooked, w/salt added	1 CUP	**245**	**89.5**	**102.9**	**3.2**	**.5**	**21.3**	–	**9.8**	**29.4**	**352.8**	**7.4**	**22.1**
			100	89.5	42.0	1.3	.2	8.7	–	4.0	12.0	144.0	3.0	9.0
	GRANOLA													
516	granola	1 CUP	**112**	–	**480.0**	**12.0**	**16.0**	**76.0**	–	–	–	**160.0**	–	–
			100	–	428.6	10.7	14.3	67.8	–	–	–	142.9	–	–
517	w/cinnamon & raisins	1 SERV	**28**	–	**128.4**	**3.0**	**4.9**	**18.8**	–	–	–	–	–	–
			100	–	458.6	10.6	17.6	67.0	–	–	–	–	–	–
518	mixed-corn, rice, wheat flakes, w/ added nutrients	1 CUP	**36**	**2.8**	**140.0**	**2.7**	**.3**	**31.0**	**.4**	**14.0**	**43.2**	**342.0**	–	–
			100	2.8	389.0	7.4	.7	86.1	1.2	39.0	120.0	950.0	–	–
519	[6]**GRAPENUTS**	1 SERV	**25**	–	**68.3**	**3.8**	**1.4**	**10.8**	–	**18.5**	**60.0**	**417.5**	**92.5**	**267.5**
			100	–	273.0	15.1	5.7	43.0	–	74.0	240.0	1670.0	370.0	1070.0
	OATS													
520	**FLAKES,** maple flavored, instant, cooked	1 CUP	**236**	**83.0**	**162.8**	**6.1**	**1.9**	**30.7**	**.2**	**23.6**	**153.4**	**252.5**	–	–
			100	83.0	69.0	2.6	.8	13.0	.1	10.0	65.0	107.0	–	–
521	**FLAKED,** w/soy flour & rice, & added nutrients, dry	1 CUP	**105**	**3.5**	**416.9**	**15.6**	**6.0**	**74.2**	**.9**	**157.5**	**367.5**	**1260.0**	–	–
			100	3.5	397.0	14.9	5.7	70.7	.9	150.0	350.0	1200.0	–	–
522	**GRANULES,** maple flavored, quick, cooked	1 CUP	**245**	**85.2**	**147.0**	**5.6**	**1.5**	**27.9**	**.5**	**24.5**	**154.4**	**176.4**	–	–
			100	85.2	60.0	2.3	.6	11.4	.2	10.0	63.0	72.0	–	–
523	**PUFFED,** sugar-covered, w/or w/o corn, wheat, but w/added nutrients, dry	1 CUP	**25**	**1.9**	**100.0**	**1.7**	**.9**	**21.4**	**.2**	**18.0**	**50.5**	**147.0**	**28.0**	–
			100	1.9	400.0	6.7	3.4	85.6	.7	72.0	202.0	588.0	112.0	–
524	**PUFFED,** w/ or w/o corn, but w/added nutrients, dry	1 CUP	**25**	**3.4**	**99.3**	**3.0**	**1.4**	**18.8**	**.3**	**44.3**	**102.0**	**316.8**	**28.0**	–
			100	3.4	397.0	11.9	5.5	75.2	1.1	177.0	408.0	1267.0	112.0	–
525	**ROLLED** (oatmeal), cooked	1 CUP	**236**	**86.5**	**112.7**	**4.7**	**2.4**	**22.9**	**.5**	**21.2**	**134.5**	**514.5**	**49.6**	**144.0**
			100	86.5	52.0	2.0	1.0	9.7	.2	9.0	57.0	218.0	21.0	61.0
526	**SHREDDED,** w/protein & other added nutrients	1 CUP	**45**	**3.9**	**170.6**	**8.5**	**.9**	**32.4**	**.8**	**119.3**	**142.7**	**274.5**	–	–
			100	3.9	379.0	18.8	2.1	72.0	1.8	265.0	317.0	610.0	–	–
527	w/wheat, cooked	1 CUP	**245**	**83.6**	**159.3**	**6.4**	**2.2**	**29.6**	**.7**	**27.0**	**183.8**	**411.6**	–	–
			100	83.6	65.0	2.6	.9	12.1	.3	11.0	75.0	168.0	–	–
528	w/toasted wheat germ & soy grits, cooked	1 CUP	**236**	**84.4**	**146.3**	**7.8**	**3.5**	**22.4**	**1.4**	**30.7**	**226.6**	**689.1**	–	–
			100	84.4	62.0	3.3	1.5	9.5	.6	13.0	96.0	292.0	–	–
	RICE													
529	[1]**CREAM OF,** cooked	1 OZ	**28**	**87.5**	**14.0**	**.2**	**Trace**	**3.1**	**Trace**	**.6**	**3.6**	**49.3**	**2.2**	**Trace**
			100	87.5	50.0	.8	Trace	11.2	Trace	2.0	13.0	176.0	8.0	Trace
530	**FLAKES,** w/added nutrients	1 CUP	**32**	**3.2**	**124.8**	**1.9**	**.1**	**28.1**	**.2**	**9.3**	**42.2**	**315.8**	–	**57.6**
			100	3.2	390.0	5.9	.3	87.7	.6	29.0	132.0	987.0	–	180.0
531	w/casein & added nutrients	1 CUP	**85**	**3.0**	**324.7**	**34.0**	**.2**	**46.6**	**.4**	**135.2**	**270.3**	**510.0**	–	–
			100	3.0	382.0	40.0	.2	54.8	.5	159.0	318.0	600.0	–	–
532	w/protein concentrate (mainly wheat gluten), & added nutrients	1 CUP	**20**	**2.5**	**77.2**	**4.0**	**.1**	**14.9**	**.1**	**10.6**	**37.4**	**160.0**	–	–
			100	2.5	386.0	20.0	.3	74.4	.5	53.0	187.0	800.0	–	–
533	**PUFFED,** w/added nutrients, no salt	1 CUP	**13**	**3.7**	**51.9**	**.8**	**.1**	**11.6**	**.1**	**2.6**	**12.0**	**.3**	**5.2**	**13.0**
			100	3.7	399.0	6.0	.4	89.5	.6	20.0	92.0	2.0	40.0	100.0
534	**PUFFED,** presweetened w/honey, w/added nutrients	1 CUP	**13**	**1.8**	**50.4**	**.5**	**.1**	**11.8**	**Trace**	**6.0**	**9.6**	**91.8**	–	–
			100	1.8	388.0	4.2	.7	90.6	.2	46.0	74.0	706.0	–	–
535	**PUFFED,** presweetened w/honey or cocoa, w/ added nutrients	1 CUP	**13**	**3.4**	**52.1**	**.6**	**.5**	**11.3**	**.1**	**6.6**	**10.7**	**46.5**	–	**7.9**
			100	3.4	401.0	4.5	4.0	86.7	.4	51.0	82.0	358.0	–	61.0
536	**SHREDDED,** w/added nutrients	1 CUP	**25**	**3.0**	**98.0**	**1.3**	**.1**	**22.2**	**.1**	**3.5**	**23.8**	**211.5**	–	–
			100	3.0	392.0	5.2	.3	88.8	.3	14.0	95.0	846.0	–	–
	WHEAT													
537	**CREAM OF,** cooked	1 CUP	**245**	**89.5**	**102.9**	**3.2**	**.2**	**21.3**	–	**9.8**	**29.4**	**352.8**	**7.4**	**22.1**
			100	89.5	42.0	1.3	.1	8.7	–	4.0	12.0	144.0	3.0	9.0
538	**FLAKES,** w/added nutrients	1 CUP	**30**	**3.5**	**114.0**	**3.2**	**.5**	**24.2**	**.5**	**12.3**	**92.7**	**309.6**	**42.6**	**96.0**
			100	3.5	380.0	10.8	1.6	80.5	1.6	41.0	309.0	1032.0	142.0	320.0
539	& malted barley flakes, w/added nutrients	1 CUP	**36**	**14.0**	**141.8**	**3.2**	**.9**	**30.3**	**.6**	**17.6**	**90.0**	**280.8**	–	–
			100	14.0	394.0	8.8	2.4	84.3	1.8	49.0	250.0	780.0	–	–
540	& malted barley granules, w/added nutrients	1 CUP	**110**	**2.9**	**430.1**	**11.0**	**.7**	**92.8**	**1.7**	**58.3**	**193.6**	**781.0**	–	**253.0**
			100	2.9	391.0	10.0	.6	84.4	1.5	53.0	176.0	710.0	–	230.0

Item No.	Minerals (Micro)			Fat-Soluble Vitamins			Water-Soluble Vitamins								
	Iron	Zinc	Copper	Vitamin A	Vitamin D	Vitamin E (Alpha Tocopherol)	Vitamin C	Thiamin	Ribo-flavin	Niacin	Panto-thenic Acid	Vit. B-6 (Pyri-doxine)	Folacin (Folic Acid)	Biotin	Vitamin B-12
	(mg)	(mg)	(mg)	(IU)	(IU)	(mg)	(mg)	(mg)	(mg)	(mg)	(mg)	(mg)	(mcg)	(mcg)	(mcg)
513	**12.25**	—	—	**0**	**0**	—	**0**	**.12**	**.07**	**.98**	—	—	—	—	—
	5.00	—	—	0	0	—	0	.05	.03	.40	—	—	—	—	—
514	**2.90**	**.50**	**.11**	**0**	—	—	**0**	**.44**	**.26**	**3.50**	**.52**	**.07**	**24.00**	—	**0**
	2.90	.50	.11	0	—	—	0	.44	.26	3.50	.52	.07	24.00	—	0
515	**.74**	**.15**	—	**0**	**0**	—	**0**	**.10**	**.07**	**.98**	—	—	—	—	—
	.30	.06	—	0	0	—	0	.04	.03	.40	—	—	—	—	—
516	—	**2.35**	**.95**	—	—	—	—	—	—	—	—	—	**51.52**	—	—
	—	2.10	.85	—	—	—	—	—	—	—	—	—	46.00	—	—
517	—	—	—	—	—	—	—	—	—	—	—	—	**12.88**	—	—
	—	—	—	—	—	—	—	—	—	—	—	—	46.00	—	—
518	**.65**	—	—	**0**	—	—	**0**	**.14**	—	**1.15**	—	—	—	—	—
	.80	—	—	0	—	—	0	.39	—	3.20	—	—	—	—	—
519	**1.30**	**.53**	—	—	**.88**	**.40**	**0**	**.30**	**.40**	**4.25**	—	**.70**	**10.25**	—	**1.25**
	5.20	2.10	—	—	3.50	1.60	0	1.20	1.60	17.00	—	2.81	41.00	—	5.00
520	**1.42**	—	—	**0**	—	—	**0**	**.14**	—	—	—	—	—	—	—
	.60	—	—	0	—	—	0	.06	—	—	—	—	—	—	—
521	**8.93**	—	—	**0**	—	—	**0**	**.75**	**.35**	**8.93**	**.97**	**.70**	**23.52**	—	**0**
	8.50	—	—	0	—	—	0	.71	.33	8.50	.92	.67	22.40	—	0
522	**1.47**	—	—	**0**	—	—	**0**	**.15**	—	—	—	—	—	—	—
	.60	—	—	0	—	—	0	.06	—	—	—	—	—	—	—
523	**1.10**	—	—	**0**	—	—	**0**	**.26**	**.03**	**.43**	**.23**	—	**5.60**	—	—
	4.40	—	—	0	—	—	0	1.03	.12	1.70	.92	—	22.40	—	—
524	**1.18**	—	—	**0**	—	—	**0**	**.25**	**.05**	**.48**	**.23**	—	**5.60**	—	—
	4.70	—	—	0	—	—	0	.98	.18	1.90	.92	—	22.40	—	—
525	**1.42**	**1.18**	—	**0**	**0**	**5.36**	**0**	**.20**	**.05**	**.24**	—	—	—	—	—
	.60	.50	—	0	0	2.27	0	.08	.02	.10	—	—	—	—	—
526	**2.39**	—	—	**0**	—	**.27**	**0**	**1.59**	**1.90**	**15.89**	**.41**	—	**10.08**	—	—
	5.30	—	—	0	—	.60	0	3.53	4.23	35.30	.92	—	22.40	—	—
527	**1.72**	—	—	**0**	—	—	**0**	**.22**	**.07**	**1.23**	—	—	—	—	—
	.70	—	—	0	—	—	0	.09	.03	.50	—	—	—	—	—
528	**2.60**	—	—	**0**	—	—	**0**	**.38**	**.07**	**.47**	—	—	—	—	—
	1.10	—	—	0	—	—	0	.16	.03	.20	—	—	—	—	—
529	**.20**	—	—	**0**	—	—	**0**	**.02**	**Trace**	**.22**	—	—	—	—	—
	.70	—	—	0	—	—	0	.06	.01	.80	—	—	—	—	—
530	**.51**	**.45**	—	**0**	—	**.01**	**0**	**.11**	**.02**	**1.73**	**.11**	**.04**	**2.43**	**.42**	**0**
	1.60	1.40	—	0	—	.04	0	.35	.05	5.40	.34	.13	7.60	1.30	0
531	**14.96**	—	—	**0**	—	—	**45.05**	**1.45**	**1.79**	**14.96**	—	—	**19.55**	—	—
	17.60	—	—	0	—	—	53.00	1.70	2.10	17.60	—	—	23.00	—	—
532	**2.48**	—	—	**0**	—	—	**7.00**	**.28**	**.34**	**3.40**	—	—	—	—	—
	12.40	—	—	0	—	—	35.00	1.40	1.70	17.00	—	—	—	—	—
533	**.23**	**.18**	**.02**	**0**	**0**	**.01**	**0**	**.06**	**.01**	**.57**	**.05**	**.01**	**.99**	**.17**	**0**
	1.80	1.40	.17	0	0	.04	0	.44	.04	4.40	.38	.07	7.60	1.30	0
534	**.12**	—	—	**0**	—	**.01**	**0**	**.04**	—	**.60**	**.05**	**.01**	—	—	**0**
	.90	—	—	0	—	.08	0	.33	—	4.60	.38	.07	—	—	0
535	**.43**	—	—	**0**	—	**.01**	**0**	**.06**	**.01**	**.82**	**.05**	**.01**	—	—	**0**
	3.30	—	—	0	—	.08	0	.42	.06	6.30	.38	.08	—	—	0
536	**.45**	—	—	**0**	—	**.01**	**0**	**.10**	—	**1.75**	**.09**	**.03**	**1.90**	**.33**	**0**
	1.80	—	—	0	—	.02	0	.39	—	7.00	.34	.13	7.60	1.30	0
537	**.74**	—	—	**0**	**0**	—	**0**	**.10**	**.07**	**.98**	—	—	—	—	—
	.30	—	—	0	0	—	0	.04	.03	.40	—	—	—	—	—
538	**1.32**	**.69**	**.14**	**0**	—	**.12**	**0**	**.19**	**.04**	**1.47**	**.14**	**.09**	**9.00**	—	**0**
	4.40	2.30	.47	0	—	.40	0	.64	.14	4.90	.47	.29	30.00	—	0
539	**.94**	—	—	**0**	—	**.22**	**0**	**.17**	**.04**	**1.40**	**.26**	—	—	—	**0**
	2.60	—	—	0	—	.61	0	.46	.11	3.90	.71	—	—	—	0
540	**3.08**	—	—	**0**	—	**.67**	**0**	**.51**	**.08**	**5.83**	**.79**	—	**59.40**	—	**0**
	2.80	—	—	0	—	.61	0	.46	.07	5.30	.71	—	54.00	—	0

(Continued)

TABLE F-36 *(Continued)*

Item No.	Food Item Name	Approximate Measure	Weight	Moisture	Food Energy Calories	Protein	Fats	Carbo-hydrates	Fiber	Minerals (Macro) Calcium	Phos-phorus	Sodium	Mag-nesium	Potas-sium
			(g)	(%)	(kcal)	(g)	(g)	(g)	(g)	(mg)	(mg)	(mg)	(mg)	(mg)
	WHEAT (Continued)													
541	& malted barley, toasted, instant, cooked	1 CUP	**245**	**80.0**	**196.0**	**7.4**	**.7**	**39.4**	**.7**	**22.1**	**200.9**	**249.9**	**76.0**	–
			100	80.0	80.0	3.0	.3	16.1	.3	9.0	82.0	102.0	31.0	–
542	& malted barley, toasted, quick, cooked	1 CUP	**245**	**84.1**	**159.3**	**4.9**	**.7**	**32.3**	**.5**	**22.1**	**144.6**	**176.4**	–	–
			100	84.1	65.0	2.0	.3	13.2	.2	9.0	59.0	72.0	–	–
543	**PUFFED,** no added salt, w/added nutrients	1 CUP	**12**	**3.4**	**43.6**	**1.8**	**.2**	**9.4**	**.2**	**3.4**	**38.6**	**.5**	**15.5**	**40.8**
			100	3.4	363.0	15.0	1.5	78.5	2.0	28.0	322.0	4.0	129.0	340.0
544	**PUFFED,** w/sugar & honey, no added nutrients	1 CUP	**12**	**2.8**	**45.1**	**.7**	**.3**	**10.6**	**.1**	**3.1**	**18.0**	**1.2**	–	**11.9**
			100	2.8	376.0	6.0	2.1	88.3	.9	26.0	150.0	10.0	–	99.0
545	**PUFFED,** w/sugar, honey, & added nutrients	1 CUP	**15**	**2.8**	**56.4**	**.9**	**.3**	**13.2**	**.1**	**3.9**	**22 5**	**24.2**	–	**14.9**
			100	2.8	376.0	6.0	2.1	88.3	.9	26.0	150.0	161.0	–	99.0
546	**ROLLED,** cooked	1 CUP	**245**	**79.7**	**183.8**	**5.4**	**1.0**	**41.4**	**1.2**	**19.6**	**186.2**	–	–	**205.8**
			100	79.7	75.0	2.2	.4	16.9	.5	8.0	76.0	–	–	84.0
547	**SHREDDED,** no salt or other added nutrients	1 MED	**22**	**6.6**	**83.2**	**2.2**	**.4**	**17.6**	**.5**	**9.5**	**85.4**	**.7**	**24.0**	**76.6**
			100	6.6	378.0	10.1	2.0	79.9	2.3	43.0	388.0	3.0	109.0	348.0
548	**SHREDDED,** spoon size	1 CUP	**45**	–	**174.2**	**4.8**	**1.6**	**36.4**	–	**0**	**158.4**	**4.5**	**50.7**	–
			100	–	387.2	10.6	3.5	81.0	–	0	352.0	10.0	112.6	–
549	**SHREDDED,** w/malt, salt, & sugar	1 MED	**22**	**14.0**	**84.9**	**2.0**	**.6**	**18.0**	**.5**	**8.6**	**81.4**	**153.3**	–	–
			100	14.0	386.0	9.1	2.5	81.7	2.2	39.0	370.0	697.0	–	–
	FLOURS													
	BUCKWHEAT FLOUR													
550	dark	1 CUP	**100**	**12.0**	**357.0**	**11.7**	**2.5**	**72.0**	**1.6**	**33.0**	**347.0**	–	–	–
			100	12.0	357.0	11.7	2.5	72.0	1.6	33.0	347.0	–	–	–
551	light	1 CUP	**100**	**12.0**	**354.0**	**6.4**	**1.2**	**79.5**	**.5**	**11.0**	**88.0**	**1.0**	**48.0**	**320.0**
			100	12.0	354.0	6.4	1.2	79.5	.5	11.0	88.0	1.0	48.0	320.0
	CORN													
552	**DOUGH,** lime treated, yellow corn	1 OZ	**28**	**62.2**	**43.1**	**1.0**	**.5**	**8.9**	**.2**	**19.6**	**25.5**	–	–	–
			100	62.2	154.0	3.5	1.9	31.8	.6	70.0	91.0	–	–	–
553	**FLOUR**	1 CUP	**110**	**12.0**	**398.2**	**8.6**	**2.9**	**84.5**	**.8**	**6.6**	**180.4**	**1.1**	–	–
			100	12.0	362.0	7.8	2.6	76.8	.7	6.0	164.0	1.0	–	–
554	**CORNSTARCH**	1 TBLS	**8**	**12.0**	**28.6**	**0**	**0**	**7.0**	**Trace**	**0**	**0**	**.3**	**.2**	**.3**
			100	12.0	357.0	.3	.6	87.6	.1	0	0	4.0	2.2	4.0
555	**COTTONSEED FLOUR**	1 CUP	**140**	**6.1**	**498.4**	**67.3**	**9.2**	**46.2**	**2.8**	**396.2**	**1556.8**	–	**910.0**	–
			100	6.1	356.0	48.1	6.6	33.0	2.0	283.0	1112.0	–	650.0	–
556	[4]**JOB'S TEARS FLOUR**	1 OZ	**28**	**10.4**	**101.1**	**3.8**	**1.9**	**19.0**	**.2**	**11.8**	**116.2**	**33.9**	–	**89.3**
			100	10.4	361.0	13.5	6.7	67.8	.6	42.0	415.0	121.0	–	319.0
557	**LIMA BEAN FLOUR**	1 CUP	**110**	**10.5**	**377.3**	**23.7**	**1.5**	**69.3**	**2.2**	–	–	–	–	–
			100	10.5	343.0	21.5	1.4	63.0	2.0	–	–	–	–	–
558	**OAT FLOUR**	1 CUP	**133**	–	–	–	–	–	–	–	–	–	**146.3**	–
			100	–	–	–	–	–	–	–	–	–	110.0	–
559	**PEANUT FLOUR,** defatted	1 CUP	**60**	**7.3**	**222.6**	**28.7**	**5.5**	**18.9**	**1.6**	**62.4**	**432.0**	**5.4**	**216.0**	**711.6**
			100	7.3	371.0	47.9	9.2	31.5	2.7	104.0	720.0	9.0	360.0	1186.0
560	**POTATO FLOUR**	1 OZ	**28**	**7.6**	**98.3**	**2.2**	**.2**	**22.4**	**.4**	**9.2**	**49.8**	**9.5**	–	**444.6**
			100	7.6	351.0	8.0	.8	79.9	1.6	33.0	178.0	34.0	–	1588.0
561	[8]**QUINOA** flour	1 OZ	**28**	**12.0**	**99.1**	**2.9**	**1.1**	**19.9**	**1.1**	**26.3**	**36.1**	–	–	–
			100	12.0	354.0	10.4	4.0	71.1	3.8	94.0	129.0	–	–	–
	RYE													
562	**FLOUR,** dark	1 OZ	**28**	**11.0**	**98.0**	**4.6**	**.4**	**19.1**	**.7**	**15.1**	**150.1**	**.3**	**32.2**	**240.8**
			100	11.0	350.0	16.3	1.4	68.1	2.4	54.0	536.0	1.0	115.0	860.0
563	**FLOUR,** medium	1 OZ	**28**	**11.0**	**100.0**	**3.2**	**.4**	**20.9**	**.3**	**7.6**	**73.4**	**.3**	**20.4**	**56.8**
			100	11.0	357.0	11.4	1.4	74.8	1.0	27.0	262.0	1.0	73.0	203.0
564	**FLOUR,** light	1 CUP	**80**	**11.0**	**289.6**	**7.5**	**1.1**	**62.3**	**.3**	**17.6**	**148.0**	**.8**	**58.4**	**124.8**
			100	11.0	362.0	9.4	1.4	77.9	.4	22.0	185.0	1.0	73.0	156.0
	SOYBEAN													
565	**FLOUR,** full fat	1 CUP	**72**	**8.0**	**303.1**	**25.8**	**14.6**	**21.9**	**1.7**	**143.3**	**401.8**	**.7**	**177.8**	**1195.2**
			100	8.0	421.0	35.9	20.3	30.4	2.4	199.0	558.0	1.0	247.0	1660.0
566	**FLOUR,** high fat	1 CUP	**88**	**8.0**	**334.4**	**37.4**	**10.6**	**29.3**	**1.9**	**211.2**	**572.0**	**.9**	**239.4**	**1562.0**
			100	8.0	380.0	42.5	12.1	33.3	2.2	240.0	650.0	1.0	272.0	1775.0
567	**FLOUR,** low fat	1 CUP	**88**	**8.0**	**313.3**	**39.3**	**5.9**	**32.2**	**2.2**	**231.4**	**557.9**	**.9**	**254.3**	**1635.9**
			100	8.0	356.0	44.7	6.7	36.6	2.5	263.0	634.0	1.0	289.0	1859.0
568	**FLOUR,** defatted	1 CUP	**138**	**8.0**	**449.9**	**64.9**	**1.2**	**52.6**	**3.2**	**365.7**	**903.9**	**1.4**	**427.8**	**2511.6**
			100	8.0	326.0	47.0	.9	38.1	2.3	265.0	655.0	1.0	310.0	1820.0
569	**SUNFLOWER SEED FLOUR,** partially defatted	1 CUP	**130**	**7.3**	**440.7**	**58.8**	**4.4**	**49.0**	**6.0**	**452.4**	**1167.4**	**72.8**	–	**1404.0**
			100	7.3	339.0	45.2	3.4	37.7	4.6	348.0	898.0	56.0	–	1080.0
570	**TORTILLA FLOUR,** refined yellow corn, untreated	1 OZ	**28**	**10.5**	**101.1**	**2.7**	**1.1**	**20.9**	**.4**	**3.6**	**49.0**	–	–	–
			100	10.5	361.0	9.7	4.0	74.8	1.3	13.0	175.0	–	–	–

Item No.	Minerals (Micro)			Fat-Soluble Vitamins			Water-Soluble Vitamins								
	Iron	Zinc	Copper	Vitamin A	Vitamin D	Vitamin E (Alpha Tocopherol)	Vitamin C	Thiamin	Ribo-flavin	Niacin	Panto-thenic Acid	Vit. B-6 (Pyri-doxine)	Folacin (Folic Acid)	Biotin	Vitamin B-12
	(mg)	(mg)	(mg)	(IU)	(IU)	(mg)	(mg)	(mg)	(mg)	(mg)	(mg)	(mg)	(mcg)	(mcg)	(mcg)
541	**2.21**	–	–	**0**	–	–	**0**	**.17**	**.05**	–	**1.75**	–	–	–	**0**
	.90	–	–	0	–	–	0	.07	.02	–	.71	–	–	–	0
542	**.98**	–	–	**0**	–	–	**0**	**.12**	**.03**	–	**1.75**	–	–	–	**0**
	.40	–	–	0	–	–	0	.05	.01	–	.71	–	–	–	0
543	**.50**	**.31**	–	**0**	**0**	**.07**	**0**	**.07**	**.03**	**.94**	–	**.02**	–	–	**0**
	4.20	2.60	–	0	0	.58	0	.55	.23	7.80	–	.17	–	–	0
544	**.26**	–	–	**0**	**0**	–	**0**	**Trace**	**.01**	**.42**	–	**.02**	–	–	**0**
	2.20	–	–	0	0	–	0	.03	.04	3.50	–	.17	–	–	0
545	**.50**	–	–	**0**	**0**	–	**0**	**.07**	**.03**	**.96**	–	**.03**	–	–	**0**
	3.30	–	–	0	0	–	0	.48	.18	6.40	–	.17	–	–	0
546	**1.72**	–	–	**0**	–	–	**0**	**.17**	**.07**	**2.21**	–	–	–	–	–
	.70	–	–	0	–	–	0	.07	.03	.90	–	–	–	–	–
547	**.77**	**.62**	**.13**	**0**	**0**	**.04**	**0**	**.05**	**.02**	**.97**	**.16**	**.05**	**11.00**	–	**0**
	3.50	2.80	.61	0	0	.19	0	.22	.11	4.40	.71	.24	50.00	–	0
548	**1.71**	**.95**	**.19**	**0**	–	–	**0**	**.10**	**0**	**2.53**	–	–	**22.50**	–	–
	3.80	2.11	.42	0	–	–	0	.21	0	5.63	–	–	50.00	–	–
549	**.75**	–	–	**0**	**0**	–	**0**	**.02**	**.03**	**1.06**	–	–	–	–	–
	3.40	–	–	0	0	–	0	.09	.15	4.80	–	–	–	–	–
550	**2.80**	–	–	**0**	–	–	**0**	**.58**	**.15**	**2.90**	**1.45**	**.58**	**44.00**	–	**0**
	2.80	–	–	0	–	–	0	.58	.15	2.90	1.45	.58	44.00	–	0
551	**1.00**	–	**.70**	**0**	–	–	**0**	**.08**	**.04**	**.40**	–	–	**44.00**	–	–
	1.00	–	.70	0	–	–	0	.08	.04	.40	–	–	44.00	–	–
552	**.45**	–	–	**5.6**	–	–	**0**	**.04**	**.01**	**.17**	–	–	–	–	–
	1.60	–	–	20.0	–	–	0	.13	.04	.60	–	–	–	–	–
553	**1.98**	–	–	**374.0**	–	–	**0**	**.22**	**.07**	**1.54**	–	–	–	–	–
	1.80	–	–	340.0	–	–	0	.20	.06	1.40	–	–	–	–	–
554	**0**	**Trace**	**.01**	**0**	**0**	–	**0**	**Trace**	**0**	**0**	–	–	–	–	–
	0	.03	.13	0	0	–	0	Trace	0	0	–	–	–	–	–
555	**17.64**	–	–	**84.0**	–	–	–	**1.69**	**1.18**	**9.10**	**6.05**	**1.37**	–	–	**0**
	12.60	–	–	60.0	–	–	–	1.21	.84	6.50	4.32	.98	–	–	0
556	**1.51**	–	–	**0**	–	–	–	**.01**	**.01**	**.39**	–	–	–	–	–
	5.40	–	–	0	–	–	–	.02	.03	1.40	–	–	–	–	–
557	–	–	–	**0**	–	–	**0**	–	–	–	–	–	–	–	–
	–	–	–	0	–	–	0	–	–	–	–	–	–	–	–
558	–	–	–	–	–	–	–	–	–	–	–	–	–	–	–
	–	–	–	–	–	–	–	–	–	–	–	–	–	–	–
559	**2.10**	–	–	–	–	–	**0**	**.45**	**.13**	**16.68**	–	–	–	–	–
	3.50	–	–	–	–	–	0	.75	.22	27.80	–	–	–	–	–
560	**4.82**	–	–	–	–	–	**5.32**	**.12**	**.04**	**.95**	–	**Trace**	–	–	**0**
	7.20	–	–	–	–	–	19.00	.42	.14	3.40	–	.01	–	–	0
561	**1.57**	–	–	**0**	–	–	**0**	**.05**	**.07**	**.20**	–	–	–	–	–
	5.60	–	–	0	–	–	0	.19	.24	.70	–	–	–	–	–
562	**1.26**	–	**.19**	**0**	–	–	**0**	**.17**	**.06**	**.76**	**.38**	**.08**	**21.84**	–	**0**
	4.50	–	.66	0	–	–	0	.61	.22	2.70	1.34	.30	78.00	–	0
563	**.73**	–	**.12**	**0**	**(1.48)**	–	**0**	**.08**	**.03**	**.70**	**(1.85)**	**(.65)**	**21.84**	**(11.10)**	**0**
	2.60	–	.42	0	(.80)	–	0	.30	.12	2.50	(1.00)	(.35)	78.00	(6.00)	0
564	**.88**	–	**.34**	**0**	–	–	**0**	**.12**	**.06**	**.48**	**.58**	**.07**	**62.40**	–	**0**
	1.10	–	.42	0	–	–	0	.15	.07	.60	.72	.09	78.00	–	0
565	**6.05**	–	–	**79.2**	**0**	–	**0**	**.61**	**.22**	**1.51**	**1.26**	**.41**	**228.96**	**50.40**	**0**
	8.40	–	–	110.0	0	–	0	.85	.31	2.10	1.75	.57	318.00	70.00	0
566	**7.92**	–	–	–	–	–	**0**	**.78**	**.32**	**2.02**	**1.72**	**.56**	**279.84**	**61.60**	**0**
	9.00	–	–	–	–	–	0	.89	.36	2.30	1.95	.64	318.00	70.00	0
567	**8.01**	–	–	**70.4**	–	–	**0**	**.73**	**.32**	**2.29**	**1.83**	**.60**	**279.84**	**61.60**	**0**
	9.10	–	–	80.0	–	–	0	.83	.36	2.60	2.08	.68	318.00	70.00	0
568	**15.32**	**6.72**	**2.39**	**55.2**	–	–	**0**	**1.50**	**.47**	**3.59**	**3.06**	**1.00**	**438.84**	**96.60**	**0**
	11.10	4.87	1.73	40.0	–	–	0	1.09	.34	2.60	2.22	.72	318.00	70.00	0
569	**17.16**	–	–	–	–	–	–	**4.68**	**.60**	**35.49**	–	–	–	–	–
	13.20	–	–	–	–	–	–	3.60	.46	27.30	–	–	–	–	–
570	**.95**	–	–	**47.6**	–	–	**0**	**.07**	**.03**	**.48**	–	–	–	–	–
	3.40	–	–	170.0	–	–	0	.25	.12	1.70	–	–	–	–	–

(Continued)

TABLE F-36 *(Continued)*

Item No.	Food Item Name	Approximate Measure	Weight	Moisture	Food Energy Calories	Protein	Fats	Carbo-hydrates	Fiber	Minerals (Macro) Calcium	Phos-phorus	Sodium	Mag-nesium	Potas-sium
			(g)	(%)	(kcal)	(g)	(g)	(g)	(g)	(mg)	(mg)	(mg)	(mg)	(mg)
571	**TORTILLA FLOUR,** white corn, lime treated	1 OZ	**28**	**10.3**	**103.3**	**2.3**	**1.6**	**20.7**	**.9**	**24.9**	**107.0**	—	—	—
			100	10.3	369.0	8.2	5.8	73.9	3.2	89.0	382.0	—	—	—
572	**TRITICALE FLOUR**	1 CUP	**130**	**12.5**	**426.4**	**17.2**	**2.0**	**92.3**	**3.3**	**57.2**	**900.9**	**13.0**	—	**562.9**
			100	12.5	328.0	13.2	1.5	71.0	2.5	44.0	693.0	10.0	—	433.0
	WHEAT													
573	**FLOUR,** all purpose, enriched	1 CUP	**110**	**12.0**	**400.4**	**11.6**	**1.1**	**83.7**	**.3**	**17.6**	**95.7**	**2.2**	—	**104.5**
			100	12.0	364.0	10.5	1.0	76.1	.3	16.0	87.0	2.0	—	95.0
574	**FLOUR,** bread, enriched	1 TBLS	**8**	**12.0**	**29.2**	**.9**	**.1**	**6.0**	**Trace**	**1.3**	**7.6**	**.2**	**2.0**	**7.6**
			100	12.0	365.0	11.8	1.1	74.7	.3	16.0	95.0	2.0	25.0	95.0
575	**FLOUR,** cake or pastry	1 TBLS	**7**	**12.0**	**25.5**	**.5**	**.1**	**5.6**	**Trace**	**1.2**	**5.1**	**.1**	**1.8**	**6.7**
			100	12.0	364.0	7.5	.8	79.4	.2	17.0	73.0	2.0	26.0	95.0
576	**FLOUR,** cake, self-rising	1 CUP	**112**	**12.2**	**397.6**	**9.1**	**.9**	**85.8**	**.2**	**173.6**	**644.0**	**1344.0**	—	—
			100	12.2	355.0	8.1	.8	76.6	.2	155.0	575.0	1200.0	—	—
577	**FLOUR,** gluten w/45% gluten/55% patent flour	1 CUP	**140**	**8.5**	**529.2**	**58.0**	**2.7**	**66.1**	**.6**	**56.0**	**196.0**	**2.8**	**41.9**	**84.0**
			100	8.5	378.0	41.4	1.9	47.2	.4	40.0	140.0	2.0	29.9	60.0
578	**FLOUR,** self-rising, enriched (anhydrous, monocalcium phosphate)	1 CUP	**114**	**11.5**	**401.3**	**10.6**	**1.1**	**84.6**	**.5**	**302.1**	**531.2**	**1230.1**	**34.1**	**102.6**
			100	11.5	352.0	9.3	1.0	74.2	.4	265.0	466.0	1079.0	29.9	90.0
579	**FLOUR,** straight hard	1 CUP	**133**	**14.0**	**477.5**	**15.7**	**2.0**	**99.1**	**.5**	**26.6**	**129.0**	**2.7**	**35.9**	**126.4**
			100	14.0	359.0	11.8	1.5	74.5	.4	20.0	97.0	2.0	27.0	95.0
580	**FLOUR,** straight soft	1 CUP	**133**	**14.0**	**477.5**	**12.9**	**1.9**	**102.3**	**.5**	**26.6**	**129.0**	**2.7**	**35.9**	**126.4**
			100	14.0	359.0	9.7	1.4	76.9	.4	20.0	97.0	2.0	27.0	95.0
581	**FLOUR,** whole, from hard wheats	1 CUP	**133**	**14.0**	**480.1**	**17.7**	**3.5**	**94.4**	**3.1**	**54.5**	**494.8**	**4.0**	**199.8**	**492.1**
			100	14.0	361.0	13.3	2.6	71.0	2.3	41.0	372.0	3.0	150.2	370.0
582	**FLOUR,** 80% extraction, from hard wheats	1 CUP	**110**	**12.0**	**401.5**	**13.2**	**1.4**	**81.5**	**.6**	**26.4**	**210.1**	**2.2**	**27.5**	**104.5**
			100	12.0	365.0	12.0	1.3	74.1	.5	24.0	191.0	2.0	25.0	95.0
583	**WHEAT-SOY BLEND,** w/bulgur flour	1 CUP	**135**	**8.7**	**486.0**	**26.6**	**8.4**	**81.1**	**2.4**	**924.8**	**758.7**	**399.6**	**272.7**	**842.4**
			100	8.7	360.0	19.7	6.2	60.1	1.8	685.0	562.0	296.0	202.0	624.0
584	**WHEAT-SOY BLEND,** w/straight grade wheat flour	1 CUP	**125**	**9.4**	**456.3**	**26.8**	**8.5**	**71.6**	**1.6**	**855.0**	**666.3**	**371.3**	**211.3**	**802.5**
			100	9.4	365.0	21.4	6.8	57.3	1.3	684.0	533.0	297.0	169.0	642.0
	PASTAS													
	MACARONI													
585	enriched, cooked firm (8–10 minutes)	1 CUP	**140**	**63.6**	**207.2**	**7.0**	**.7**	**42.1**	**.1**	**15.4**	**91.0**	**1.4**	**28.0**	**110.6**
			100	63.6	148.0	5.0	.5	30.1	.1	11.0	65.0	1.0	20.0	79.0
586	enriched, cooked tender (14–20 minutes)	1 CUP	**140**	**72.0**	**155.4**	**4.8**	**.6**	**32.2**	**.1**	**11.2**	**70.0**	**1.4**	**25.2**	**85.4**
			100	72.0	111.0	3.4	.4	23.0	.1	8.0	50.0	1.0	18.0	61.0
	NOODLES													
587	chow mein, canned	1 CUP	**50**	**1.1**	**248.0**	**6.6**	**11.8**	**29.0**	—	—	—	—	—	—
			100	1.1	496.0	13.2	23.5	58.0	—	—	—	—	—	—
588	cooked	1 OZ	**28**	**3.0**	**35.0**	**1.1**	**.6**	**6.2**	**Trace**	**4.8**	**15.9**	**24.2**	**5.1**	**11.7**
			100	3.0	124.9	3.9	2.0	22.1	.1	17.3	56.7	86.3	18.4	41.8
589	egg, enriched, cooked	1 CUP	**160**	**70.4**	**200.0**	**6.6**	**2.4**	**37.3**	**.2**	**16.0**	**94.4**	**3.2**	—	**70.4**
			100	70.4	125.0	4.1	1.5	23.3	.1	10.0	59.0	2.0	—	44.0
590	egg, unenriched, cooked	1 CUP	**160**	**70.4**	**196.8**	**6.6**	**2.4**	**37.3**	**.2**	**16.0**	**94.4**	**3.2**	—	**70.4**
			100	70.4	123.0	4.1	1.5	23.3	.1	10.0	59.0	2.0	—	44.0
	PASTINAS													
591	enriched, dry, egg	1 CUP	**170**	**10.4**	**651.1**	**21.9**	**7.0**	**122.1**	**.5**	**59.5**	**329.8**	**8.5**	—	—
			100	10.4	383.0	12.9	4.1	71.8	.3	35.0	194.0	5.0	—	—
592	enriched, dry, carrot	1 CUP	**170**	**10.0**	**630.7**	**20.2**	**2.7**	**128.7**	**1.0**	**64.6**	**272.0**	—	—	—
			100	10.0	371.0	11.9	1.6	75.7	.6	38.0	160.0	—	—	—
593	enriched, dry, spinach	1 CUP	**170**	**10.1**	**625.6**	**21.1**	**2.7**	**127.2**	**.9**	**107.1**	**294.1**	—	—	—
			100	10.1	368.0	12.4	1.6	74.8	.5	63.0	173.0	—	—	—
	SPAGHETTI													
594	enriched, cooked firm (8–10 minutes)	1 CUP	**146**	**63.6**	**216.1**	**7.3**	**.7**	**43.9**	**.1**	**16.1**	**94.9**	**1.5**	**29.0**	**115.3**
			100	63.6	148.0	5.0	.5	30.1	.1	11.0	65.0	1.0	19.9	79.0
595	enriched, cooked tender (14–20 minutes)	1 CUP	**150**	**72.0**	**166.5**	**5.1**	**.6**	**34.5**	**.2**	**12.0**	**75.0**	**1.5**	**27.0**	**91.5**
			100	72.0	111.0	3.4	.4	23.0	.1	8.0	50.0	1.0	18.0	61.0
	DESSERTS AND SWEETS													
596	**APPLE BROWN BETTY**	1 CUP	**215**	**64.5**	**324.7**	**3.4**	**7.5**	**63.9**	**1.1**	**38.7**	**47.3**	**329.0**	**10.8**	**215.0**
			100	64.5	151.0	1.6	3.5	29.7	.5	18.0	22.0	153.0	5.0	100.0
597	**APPLE BUTTER**	1 TBLS	**20**	**51.6**	**37.2**	**.1**	**.2**	**9.4**	**.2**	**2.8**	**7.2**	**.4**	—	**50.4**
			100	51.6	186.0	.4	.8	46.8	1.1	14.0	36.0	2.0	—	252.0
598	**APPLE STRUDEL**	1 SERV	**100**	**43.8**	**290.0**	**4.0**	**14.3**	**39.0**	**.4**	**3.0**	**19.0**	**130.0**	**10.0**	**74.0**
			100	43.8	290.0	4.0	14.3	39.0	.4	3.0	19.0	130.0	10.0	74.0

Item No.	Minerals (Micro)			Fat-Soluble Vitamins			Water-Soluble Vitamins								
	Iron	Zinc	Copper	Vitamin A	Vitamin D	Vitamin E (Alpha Tocopherol)	Vitamin C	Thiamin	Ribo-flavin	Niacin	Panto-thenic Acid	Vit. B-6 (Pyri-doxine)	Folacin (Folic Acid)	Biotin	Vitamin B-12
	(mg)	(mg)	(mg)	(IU)	(IU)	(mg)	(mg)	(mg)	(mg)	(mg)	(mg)	(mg)	(mcg)	(mcg)	(mcg)
571	**.73**	—	—	**1.4**	—	—	**.28**	**.10**	**.03**	**.53**	—	—	—	—	—
	2.60	—	—	5.0	—	—	1.00	.37	.10	1.90	—	—	—	—	—
572	**7.02**	**4.29**	—	**0**	—	—	**0**	**1.05**	**.35**	**1.82**	—	—	**68.90**	—	—
	5.40	3.30	—	0	—	—	0	.81	.27	1.40	—	—	53.00	—	—
573	**3.19**	**.77**	—	**0**	—	—	**0**	**.48**	**.29**	**3.85**	—	—	**23.10**	—	—
	2.90	.70	—	0	—	—	0	.44	.26	3.50	—	—	21.00	—	—
574	**.23**	**.06**	—	**0**	—	—	**0**	**.04**	**.02**	**.28**	**.04**	**.01**	**2.00**	**.08**	**0**
	2.90	.80	—	0	—	—	0	.44	.26	3.50	.50	.06	25.00	1.00	0
575	**.04**	**.02**	**.01**	**0**	—	—	**0**	**Trace**	**Trace**	**.05**	**.02**	**Trace**	**1.75**	**.07**	**0**
	.50	.30	.08	0	—	—	0	.03	.03	.70	.32	.05	25.00	1.00	0
576	**1.12**	**.34**	—	—	—	—	—	—	—	—	—	—	—	—	—
	1.00	.30	—	—	—	—	—	—	—	—	—	—	—	—	—
577	—	—	—	**0**	—	—	**0**	—	—	—	—	—	—	—	—
	—	—	—	0	—	—	0	—	—	—	—	—	—	—	—
578	**3.31**	—	—	**0**	—	—	**0**	**.50**	**.30**	**3.99**	—	—	—	—	—
	2.90	—	—	0	—	—	0	.44	.26	3.50	—	—	—	—	—
579	**1.86**	**.84**	**.25**	**0**	—	—	**0**	**.16**	**.09**	**1.86**	—	—	**33.25**	—	—
	1.40	.63	.19	0	—	—	0	.12	.07	1.40	—	—	25.00	—	—
580	**1.46**	**.63**	**.21**	**0**	—	—	**0**	**.11**	**.07**	**1.60**	—	—	**33.25**	—	—
	1.10	.47	.16	0	—	—	0	.08	.05	1.20	—	—	25.00	—	—
581	**4.39**	**3.19**	—	**0**	—	—	**0**	**.73**	**.16**	**5.72**	**1.46**	**.45**	**71.82**	**11.97**	**0**
	3.30	2.40	—	0	—	—	0	.55	.12	4.30	1.10	.34	54.00	9.00	0
582	**1.43**	**1.65**	—	**0**	—	—	**0**	**.29**	**.08**	**2.20**	—	—	—	—	—
	1.30	1.50	—	0	—	—	0	.26	.07	2.00	—	—	—	—	—
583	**28.08**	**.37**	—	**487.4**	—	**8.69**	**.92**	**2.01**	**.80**	**.08**	**.31**	**.04**	**5.40**	—	**.35**
	20.80	.27	—	361.0	—	6.44	.68	1.49	.59	.06	.23	.03	4.00	—	.26
584	**25.63**	—	—	**2075.0**	—	**8.80**	**50.00**	**2.53**	**.74**	**10.25**	—	**.83**	**411.50**	—	**5.00**
	20.50	—	—	1660.0	—	7.04	40.00	2.02	.59	8.20	—	.67	329.20	—	4.00
585	**1.54**	—	—	**0**	—	—	**0**	**.25**	**.14**	**1.96**	—	—	—	—	—
	1.10	—	—	0	—	—	0	.18	.10	1.40	—	—	—	—	—
586	**1.26**	**.70**	**(.03)**	**0**	**0**	**(Trace)**	**0**	**.20**	**.11**	**1.54**	**(Trace)**	**(.01)**	**(2.80)**	**(Trace)**	**0**
	.90	.50	(.02)	0	0	(Trace)	0	.14	.08	1.10	(Trace)	(.01)	(2.00)	(Trace)	0
587	—	—	—	—	—	—	—	**.47**	**.17**	—	—	**.04**	—	—	—
	—	—	—	—	—	—	—	.93	.34	—	—	.07	—	—	—
588	**.25**	**.18**	—	**18.9**	**0**	—	**0**	**.08**	**.03**	**.52**	**0**	**.07**	**0**	—	**0**
	.89	.63	—	67.6	0	—	0	.27	.12	1.84	0	.23	0	—	0
589	**1.44**	—	—	**112.0**	—	—	**0**	**.29**	**.25**	**1.92**	—	**.01**	—	—	—
	.90	—	—	70.0	—	—	0	.18	.16	1.20	—	.01	—	—	—
590	**.96**	—	—	**112.0**	—	—	**0**	**.05**	**.03**	**.64**	—	—	—	—	—
	.60	—	—	70.0	—	—	0	.03	.02	.40	—	—	—	—	—
591	**4.93**	—	—	**374.0**	—	—	**0**	**1.50**	**.65**	**10.20**	—	—	—	—	—
	2.90	—	—	220.0	—	—	0	.88	.38	6.00	—	—	—	—	—
592	**4.93**	—	—	**1241.0**	—	—	**0**	**1.50**	**.65**	**10.20**	—	—	—	—	—
	2.90	—	—	730.0	—	—	0	.88	.38	6.00	—	—	—	—	—
593	**4.93**	—	—	**1088.0**	—	—	**0**	**1.50**	**.65**	**10.20**	—	—	—	—	—
	2.90	—	—	640.0	—	—	0	.88	.38	6.00	—	—	—	—	—
594	**1.61**	—	—	**0**	—	—	**0**	**.26**	**.15**	**2.04**	—	—	—	—	—
	1.10	—	—	0	—	—	0	.18	.10	1.40	—	—	—	—	—
595	**1.35**	**.72**	—	**0**	—	—	**0**	**.21**	**.12**	**1.65**	**Trace**	**(.02)**	**3.00**	**Trace**	**0**
	.90	.48	—	0	—	—	0	.14	.08	1.10	Trace	(.01)	2.00	Trace	0
596	**1.29**	—	—	**215.0**	—	—	**2.15**	**.13**	**.09**	**.86**	—	—	—	—	—
	.60	—	—	100.0	—	—	1.00	.06	.04	.40	—	—	—	—	—
597	**.14**	**.01**	**.02**	**0**	**0**	—	**.40**	**Trace**	**Trace**	**.04**	—	—	—	—	—
	.70	.06	.08	0	0	—	2.00	.01	.02	.20	—	—	—	—	—
598	**1.20**	—	—	**10.0**	—	—	**3.00**	**.15**	**.09**	**.60**	—	—	—	—	—
	1.20	—	—	10.0	—	—	3.00	.15	.09	.60	—	—	—	—	—

(Continued)

TABLE F-36 *(Continued)*

Item No.	Food Item Name	Approximate Measure	Weight	Moisture	Food Energy Calories	Protein	Fats	Carbo-hydrates	Fiber	Minerals (Macro) Calcium	Phos-phorus	Sodium	Mag-nesium	Potas-sium
			(g)	(%)	(kcal)	(g)	(g)	(g)	(g)	(mg)	(mg)	(mg)	(mg)	(mg)
599	**BANANA SPLIT**	**INDV**	**411**	—	**580.0**	**11.0**	**16.0**	**97.0**	—	—	—	—	—	—
			100	—	141.1	2.6	3.9	23.6	—	—	—	—	—	—
600	**BLINTZES,** cheese	**1 SERV**	**28**	—	**54.9**	—	**3.9**	—	—	—	—	—	—	—
			100	—	196.0	—	14.1	—	—	—	—	—	—	—
601	**BREAD PUDDING,** w/raisins	**1 CUP**	**220**	**58.6**	**411.4**	**12.3**	**13.4**	**62.5**	**.2**	**239.8**	**250.8**	**442.2**	—	**473.0**
			100	58.6	187.0	5.6	6.1	28.4	.1	109.0	114.0	201.0	—	215.0
602	[1]**CANDIED GINGER ROOT,** piece	**1 OZ**	**28**	**12.0**	**95.2**	**.1**	**.1**	**24.4**	**.2**	—	—	—	—	—
			100	12.0	340.0	.3	.2	87.1	.7	—	—	—	—	—
	CANDY													
603	butterscotch	**1 PIECE**	**5**	**1.5**	**19.9**	—	**.2**	**4.7**	**0**	**.9**	**.3**	**3.3**	—	**.1**
			100	1.5	397.0	—	3.4	94.8	0	17.0	6.0	66.0	—	2.0
604	caramels, chocolate flavored, roll	**1 AVG**	**7**	**5.6**	**27.7**	**.2**	**.6**	**5.8**	**Trace**	**4.8**	**8.3**	**13.8**	—	**8.6**
			100	5.6	396.0	2.2	8.2	82.7	.2	68.0	119.0	197.0	—	123.0
605	caramels, plain or chocolate	**1 MED**	**10**	**7.6**	**39.9**	**.4**	**1.0**	**7.7**	**Trace**	**14.8**	**12.2**	**22.6**	**.2**	**19.2**
			100	7.6	399.0	4.0	10.2	76.6	.2	148.0	122.0	226.0	2.0	192.0
606	caramels, plain or chocolate, w/nuts	**1 MED**	**12**	**7.1**	**51.4**	**.5**	**2.0**	**8.5**	**Trace**	**16.8**	**16.7**	**24.4**	—	**28.0**
			100	7.1	428.0	4.5	16.3	70.5	.4	140.0	139.0	203.0	—	233.0
607	chocolate, bittersweet	**1 OZ**	**28**	**1.8**	**133.6**	**2.2**	**11.1**	**13.1**	**.5**	**16.2**	**79.5**	**.8**	—	**172.2**
			100	1.8	477.0	7.9	39.7	46.8	1.8	58.0	284.0	3.0	—	615.0
608	chocolate coated almonds	**1 OZ**	**28**	**2.0**	**159.3**	**3.4**	**12.2**	**11.1**	**.4**	**56.8**	**96.0**	**16.5**	**11.2**	**152.9**
			100	2.0	569.0	12.3	43.7	39.6	1.5	203.0	343.0	59.0	40.0	546.0
609	chocolate coated chocolate fudge	**1 PIECE**	**30**	**6.2**	**129.0**	**1.1**	**4.8**	**21.9**	**.1**	**30.3**	**33.0**	**68.4**	—	**57.9**
			100	6.2	430.0	3.8	16.0	73.1	.2	101.0	110.0	228.0	—	193.0
610	chocolate coated chocolate fudge, w/nuts	**1 OZ**	**28**	**6.0**	**126.6**	**1.4**	**5.8**	**18.8**	**.1**	**28.3**	**38.4**	**57.4**	—	**61.3**
			100	6.0	452.0	4.9	20.8	67.3	.4	101.0	137.0	205.0	—	219.0
611	chocolate coated coconut center	**1 OZ**	**28**	**6.6**	**122.6**	**.8**	**4.9**	**20.2**	**.2**	**13.4**	**21.6**	**55.2**	—	**46.2**
			100	6.6	438.0	2.8	17.6	72.0	.6	48.0	77.0	197.0	—	165.0
612	chocolate coated fondant	**1 OZ**	**28**	**5.8**	**114.8**	**.5**	**2.9**	**22.7**	**Trace**	**16.0**	**15.1**	**51.8**	—	**25.5**
			100	5.8	410.0	1.7	10.5	81.0	.1	57.0	54.0	185.0	—	91.0
613	chocolate coated fudge, w/caramel & peanuts	**1 OZ**	**28**	**8.3**	**121.2**	**2.2**	**5.1**	**17.9**	**.1**	**50.1**	**52.1**	**57.1**	—	**84.3**
			100	8.3	433.0	7.7	18.1	64.1	.4	179.0	186.0	204.0	—	301.0
614	chocolate coated fudge, w/peanuts & caramel	**1 OZ**	**28**	**7.0**	**128.5**	**2.6**	**6.5**	**16.4**	**.2**	**35.6**	**53.8**	**35.8**	—	**62.2**
			100	7.0	459.0	9.4	23.1	58.7	.7	127.0	192.0	128.0	—	222.0
615	chocolate coated, honeycombed, hard candy, w/peanut butter	**1 OZ**	**28**	**1.7**	**129.6**	**1.8**	**5.5**	**19.8**	**.1**	**22.4**	**37.8**	**45.6**	—	**63.0**
			100	1.7	463.0	6.6	19.5	70.6	.4	80.0	135.0	163.0	—	225.0
616	chocolate coated nougat & caramel	**1 OZ**	**28**	**7.7**	**116.5**	**1.1**	**3.9**	**20.4**	**.1**	**35.6**	**34.4**	**48.4**	—	**59.1**
			100	7.7	416.0	4.0	13.9	72.8	.2	127.0	123.0	173.0	—	211.0
617	chocolate coated peanuts	**1 PIECE**	**2**	**1.0**	**11.2**	**.3**	**.8**	**.8**	**Trace**	**2.3**	**6.0**	**1.2**	—	**10.1**
			100	1.0	561.0	16.4	41.3	39.1	1.2	116.0	298.0	60.0	—	504.0
618	chocolate coated raisins	**1 SM**	**1**	**4.8**	**4.3**	**.1**	**.2**	**.7**	**Trace**	**1.5**	**1.7**	**.6**	—	**6.0**
			100	4.8	425.0	5.4	17.1	70.5	.6	152.0	174.0	64.0	—	603.0
619	chocolate coated vanilla cream	**1 PIECE**	**13**	**7.5**	**56.6**	**.5**	**2.2**	**9.1**	**Trace**	**16.6**	**14.3**	**23.7**	—	**23.1**
			100	7.5	435.0	3.8	17.1	70.3	.1	128.0	110.0	182.0	—	178.0
620	chocolate fudge	**1 PIECE**	**25**	**8.2**	**100.0**	**.7**	**3.1**	**18.8**	**.1**	**19.3**	**21.0**	**47.5**	**12.5**	**36.8**
			100	8.2	400.0	2.7	12.2	75.0	.2	77.0	84.0	190.0	50.0	147.0
621	chocolate fudge, w/nuts	**1 PIECE**	**25**	**7.8**	**106.5**	**1.0**	**4.4**	**17.3**	**.1**	**19.8**	**28.5**	**42.8**	—	**44.3**
			100	7.8	426.0	3.9	17.4	69.0	.4	79.0	114.0	171.0	—	177.0
622	[1]chocolate kiss	**1 OZ**	**28**	**0.9**	**145.6**	**2.2**	**9.0**	**15.9**	**.1**	**63.8**	**64.7**	**26.3**	**16.2**	**107.5**
			100	0.9	520.0	7.7	32.3	56.9	.4	228.0	231.0	94.0	58.0	384.0
623	chocolate, semi-sweet	**1 OZ**	**28**	**1.1**	**142.0**	**1.2**	**10.0**	**16.0**	**.3**	**8.4**	**42.0**	**.6**	**27.2**	**91.0**
			100	1.1	507.0	4.2	35.7	57.0	1.0	30.0	150.0	2.0	97.0	325.0
624	chocolate, sweet	**1 OZ**	**28**	**0.9**	**147.8**	**1.2**	**9.8**	**16.2**	**.1**	**26.3**	**39.8**	**9.2**	**30.0**	**75.3**
			100	0.9	528.0	4.4	35.1	57.9	.5	94.0	142.0	33.0	107.0	269.0
625	cinnamon drops	**1 OZ**	**28**	—	—	—	—	—	—	—	—	—	—	—
			100	—	—	—	—	—	—	—	—	—	—	—
626	fondant	**1 AVG**	**11**	**7.6**	**40.0**	**0**	**.2**	**9.9**	—	**1.5**	**.7**	**23.3**	—	**.6**
			100	7.6	364.0	.1	2.0	89.6	—	14.0	6.0	212.0	—	5.0
627	gum drops, starch jelly pieces	**1 SM**	**1**	**11.7**	**3.5**	**0**	**.0**	**.9**	**0**	**.1**	—	**.4**	—	**.1**
			100	11.7	347.0	.1	.7	87.4	0	6.0	—	35.0	—	5.0
628	hard	**1 PIECE**	**5**	**1.4**	**19.3**	**0**	**.1**	**4.9**	**0**	**1.1**	**.4**	**1.6**	—	**.2**
			100	1.4	386.0	0	1.1	97.2	0	21.0	7.0	32.0	—	4.0
629	jelly beans	**1 PIECE**	**3**	**6.3**	**11.0**	—	**.0**	**2.8**	—	**.4**	**.1**	**.4**	—	**0**
			100	6.3	367.0	—	.5	93.1	—	12.0	4.0	12.0	—	1.0

Item No.	Minerals (Micro)			Fat-Soluble Vitamins			Water-Soluble Vitamins								
	Iron	Zinc	Copper	Vitamin A	Vitamin D	Vitamin E (Alpha Tocopherol)	Vitamin C	Thiamin	Ribo-flavin	Niacin	Panto-thenic Acid	Vit. B-6 (Pyri-doxine)	Folacin (Folic Acid)	Biotin	Vitamin B-12
	(mg)	(mg)	(mg)	(IU)	(IU)	(mg)	(mg)	(mg)	(mg)	(mg)	(mg)	(mg)	(mcg)	(mcg)	(mcg)
599	—	—	—	—	—	—	—	—	—	—	—	—	—	—	—
	—	—	—	—	—	—	—	—	—	—	—	—	—	—	—
600	—	—	—	—	—	—	—	—	—	—	—	—	—	—	—
	—	—	—	—	—	—	—	—	—	—	—	—	—	—	—
601	**2.42**	—	—	**660.0**	—	—	**2.20**	**.13**	**.42**	**.22**	—	—	—	—	—
	1.10	—	—	300.0	—	—	1.00	.06	.19	.10	—	—	—	—	—
602	—	—	—	—	—	—	—	—	—	—	—	—	—	—	—
	—	—	—	—	—	—	—	—	—	—	—	—	—	—	—
603	**.07**	—	—	**7.0**	**0**	—	**0**	**0**	—	—	—	—	—	—	—
	1.40	—	—	140.0	0	—	0	0	—	—	—	—	—	—	—
604	**.13**	—	—	—	—	—	—	**Trace**	**.01**	**.01**	—	—	—	—	—
	1.80	—	—	—	—	—	—	.02	.07	.10	—	—	—	—	—
605	**.14**	—	—	**1.0**	—	—	—	**Trace**	**.02**	**.02**	—	—	—	—	—
	1.40	—	—	10.0	—	—	—	.03	.17	.20	—	—	—	—	—
606	**.18**	—	—	**2.4**	—	—	—	**.01**	**.02**	**.02**	—	—	—	—	—
	1.50	—	—	20.0	—	—	—	.11	.17	.20	—	—	—	—	—
607	**1.40**	—	—	**11.2**	—	**1.48**	**0**	**.01**	**.05**	**.28**	—	—	—	—	—
	5.00	—	—	40.0	—	5.30	0	.03	.17	1.00	—	—	—	—	—
608	**.78**	—	—	**14.8**	**9.80**	—	**.20**	**.03**	**.15**	**.48**	—	—	—	—	—
	2.80	—	—	53.0	35.00	—	.70	.12	.53	1.70	—	—	—	—	—
609	**.39**	—	—	—	—	—	**0**	**.01**	**.04**	**.06**	—	—	—	—	—
	1.30	—	—	—	—	—	0	.04	.13	.20	—	—	—	—	—
610	**.42**	—	—	—	—	—	—	**.02**	**.04**	**.06**	—	—	—	—	—
	1.50	—	—	—	—	—	—	.06	.13	.20	—	—	—	—	—
611	**.31**	—	—	**0**	—	—	**0**	**.01**	**.02**	**.06**	—	—	—	—	—
	1.10	—	—	0	—	—	0	.02	.07	.20	—	—	—	—	—
612	**.31**	—	—	—	—	—	—	**.01**	**.02**	**.03**	—	—	—	—	—
	1.10	—	—	—	—	—	—	.03	.06	.10	—	—	—	—	—
613	**.39**	—	—	—	—	—	—	**.05**	**.06**	**.53**	—	—	—	—	—
	1.40	—	—	—	—	—	—	.16	.22	1.90	—	—	—	—	—
614	**.31**	—	—	—	—	—	—	**.07**	**.04**	**1.04**	—	—	—	—	—
	1.10	—	—	—	—	—	—	.26	.15	3.70	—	—	—	—	—
615	**.50**	—	—	—	—	—	—	**.01**	**.03**	**.81**	—	—	—	—	—
	1.80	—	—	—	—	—	—	.05	.09	2.90	—	—	—	—	—
616	**.45**	—	—	**11.2**	—	—	—	**.02**	**.05**	**.06**	—	—	—	—	—
	1.60	—	—	40.0	—	—	—	.06	.17	.20	—	—	—	—	—
617	**.03**	—	—	—	—	—	—	**.01**	**Trace**	**.15**	—	—	—	—	—
	1.50	—	—	—	—	—	—	.37	.18	7.40	—	—	—	—	—
618	**.03**	—	—	**1.5**	—	—	—	**Trace**	**Trace**	**Trace**	—	—	—	—	—
	2.50	—	—	150.0	—	—	—	.08	.21	.40	—	—	—	—	—
619	**.08**	—	—	—	—	—	—	**.01**	**.01**	**.01**	—	—	—	—	—
	.60	—	—	—	—	—	—	.05	.07	.10	—	—	—	—	—
620	**.25**	—	—	—	—	—	—	**.01**	**.02**	**.05**	—	—	—	—	—
	1.00	—	—	—	—	—	—	.02	.09	.20	—	—	—	—	—
621	**.30**	—	—	—	—	—	—	**.01**	**.02**	**.08**	—	—	—	—	—
	1.20	—	—	—	—	—	—	.04	.09	.30	—	—	—	—	—
622	**.31**	**.13**	**.02**	**75.6**	**46.76**	**1.80**	**.39**	**.02**	**.10**	**.08**	—	—	—	**17.10**	—
	1.10	.46	.07	270.0	167.00	4.20	1.40	.06	.34	.30	—	—	—	30.00	—
623	**.73**	—	—	**5.6**	—	—	**0**	**Trace**	**.02**	**.14**	—	—	—	—	—
	2.60	—	—	20.0	—	—	0	.01	.08	.50	—	—	—	—	—
624	**.39**	—	—	**2.8**	—	—	—	**.01**	**.04**	**.08**	—	—	—	—	—
	1.40	—	—	10.0	—	—	—	.02	.14	.30	—	—	—	—	—
625	—	—	—	—	—	—	—	—	—	—	—	—	—	—	—
	—	—	—	—	—	—	—	—	—	—	—	—	—	—	—
626	**.12**	—	—	**0**	—	—	**0**	—	—	—	—	—	—	—	—
	1.10	—	—	0	—	—	0	—	—	—	—	—	—	—	—
627	**.01**	—	—	**0**	—	—	**0**	**0**	—	—	—	—	—	—	—
	.50	—	—	0	—	—	0	0	—	—	—	—	—	—	—
628	**.10**	—	—	**0**	**0**	—	**0**	**0**	**0**	**0**	—	—	—	—	—
	1.90	—	—	0	0	—	0	0	0	0	—	—	—	—	—
629	**.03**	—	—	**0**	—	—	**0**	**0**	—	—	—	—	—	—	—
	1.10	—	—	0	—	—	0	0	—	—	—	—	—	—	—

(Continued)

TABLE F-36 *(Continued)*

Item No.	Food Item Name	Approximate Measure	Weight	Moisture	Food Energy Calories	Protein	Fats	Carbo-hydrates	Fiber	Minerals (Macro) Calcium	Phos-phorus	Sodium	Mag-nesium	Potas-sium
			(g)	(%)	(kcal)	(g)	(g)	(g)	(g)	(mg)	(mg)	(mg)	(mg)	(mg)
	CANDY (Continued)													
630	[1]licorice	1 OZ	28	11.7	9[illegible].2	0	.2	24.5	0	1.7	Trace	9.1	—	1.4
			100	11.7	347.0	.1	.7	87.4	0	6.0	Trace	35.0	—	5.0
631	Life savers	1 PIECE	2	1.4	7.7	0	Trace	1.9	0	.4	.1	.6	—	.1
			100	1.4	386.0	0	1.1	97.2	0	21.0	7.0	32.0	—	4.0
632	lollypops	1 MED	28	—	108.1	0	0	28.0	0	0	0	—	—	—
			100	—	386.0	0	0	100.0	0	0	0	—	—	—
633	marshmallows	1 AVG	8	17.3	25.5	.2	—	6.4	0	1.4	.5	3.1	—	.5
			100	17.3	319.0	2.0	—	80.4	0	18.0	6.0	39.0	—	6.0
634	milk chocolate, plain	1 BAR	57	0.9	296.4	4.4	18.4	32.4	.2	130.0	131.7	53.6	33.1	218.9
			100	0.9	520.0	7.7	32.3	56.9	.4	228.0	231.0	94.0	58.0	384.0
635	milk chocolate, w/almonds	1 BAR	56	1.5	297.9	5.2	19.9	28.7	.4	128.2	152.3	44.8	46.5	247.5
			100	1.5	532.0	9.3	35.6	51.3	.7	229.0	272.0	80.0	83.0	442.0
636	peanut bars	1 OZ	28	1.5	144.2	4.9	9.0	13.2	.3	12.3	76.4	2.8	—	125.4
			100	1.5	515.0	17.5	32.2	47.2	1.2	44.0	273.0	10.0	—	448.0
637	peanut brittle, no added salt or soda	1 PIECE	25	2.0	105.3	1.4	2.6	20.3	.1	8.8	23.8	7.8	—	37.8
			100	2.0	421.0	5.7	10.4	81.0	.5	35.0	95.0	31.0	—	151.0
638	[1]peppermint pattie, chocolate-covered	1 OZ	28	5.8	114.8	.5	2.9	22.7	Trace	16.0	15.1	51.8	—	25.5
			100	5.8	410.0	1.7	10.5	81.0	.1	57.0	54.0	185.0	—	91.0
639	sugar-coated almonds	1 AVG	4	2.3	18.2	.3	.7	2.8	Trace	4.0	6.6	.8	—	10.2
			100	2.3	456.0	7.8	18.6	70.2	.9	100.0	166.0	20.0	—	255.0
640	sugar-coated chocolate discs	1 OZ	28	1.2	130.5	1.5	5.5	20.4	Trace	37.8	39.2	20.2	—	70.0
			100	1.2	466.0	5.2	19.7	72.7	.3	135.0	140.0	72.0	—	250.0
641	vanilla fudge	1 PIECE	25	10.0	99.5	.8	2.8	18.7	0	28.0	20.8	52.0	12.5	31.8
			100	10.0	398.0	3.0	11.1	74.8	0	112.0	83.0	208.0	50.0	127.0
642	vanilla fudge, w/nuts	1 OZ	28	9.4	118.7	1.2	4.6	19.3	.1	31.1	31.6	52.4	—	31.9
			100	9.4	424.0	4.2	16.4	68.8	.2	111.0	113.0	187.0	—	114.0
643	**CHARLOTTE RUSSE,** w/lady fingers, & whipped cream filling	1 SERV	114	45.5	326.0	6.7	16.6	38.2	—	52.4	103.7	49.0	—	73.0
			100	45.5	286.0	5.9	14.6	33.5	—	46.0	91.0	43.0	—	64.0
644	**CHEWING GUM**	1 PIECE	3	3.5	9.5	—	—	2.9	—	—	—	—	—	—
			100	3.5	317.0	—	—	95.2	—	—	—	—	—	—
645	**CITRON,** candied	1 OZ	28	18.0	87.9	.1	.1	22.5	.4	23.2	6.7	81.2	—	33.6
			100	18.0	314.0	.2	.3	80.2	1.4	83.0	24.0	290.0	—	120.0
646	**COBBLER,** apple cake	1 SERV	100	56.0	195.0	2.0	5.0	37.0	.3	12.0	85.0	135.0	—	50.0
			100	56.0	195.0	2.0	5.0	37.0	.3	12.0	85.0	135.0	—	50.0
647	**CREAM PUFFS,** w/custard filling	1 AVG	105	58.3	244.7	6.8	14.6	21.5	0	85.1	119.7	87.2	13.7	127.1
			100	58.3	233.0	6.5	13.9	20.5	0	81.0	114.0	83.0	13.0	121.0
648	**CUSTARD,** baked	1 CUP	265	77.2	294.2	14.3	13.3	29.4	0	296.8	310.1	209.4	—	386.9
			100	77.2	111.0	5.4	5.0	11.1	0	112.0	117.0	79.0	—	146.0
649	**CUSTARD,** w/whole milk	1 CUP	224	—	286.0	11.2	9.2	39.4	0	284.0	(439.0)	256.0	(31.4)	348.0
			100	—	127.7	5.0	4.1	17.6	0	126.8	(140.0)	114.3	(14.0)	155.4
650	**ECLAIRS,** w/custard filling & chocolate icing	1 AVG	110	56.2	262.9	6.8	15.0	25.5	0	88.0	123.2	90.2	(17.6)	134.2
			100	56.2	239.0	6.2	13.6	23.2	0	80.0	112.0	82.0	(16.0)	122.0
651	**FRUIT FILLING FOR APPLE PIE**	1 SERV	103	—	—	2.0	30.0	—	—	8.9	—	—	—	—
			100	—	—	1.9	29.1	—	—	8.7	—	—	—	—
652	**FRUIT ICE**	1 CUP	184	63.0	261.3	.2	—	65.1	—	1.8	3.7	—	—	27.6
			100	63.0	142.0	.1	—	35.4	—	1.0	2.0	—	—	15.0
	GELATIN													
653	dry, unsweetened	1 TBLS	10	13.0	33.5	8.6	0	0	0	0	0	11.2	2.8	.2
			100	13.0	335.0	85.6	.1	0	0	0	0	112.0	28.3	2.0
654	dessert, powder	1 TBLS	10	1.6	37.1	.9	0	8.8	0	—	—	31.8	—	—
			100	1.6	371.0	9.4	0	88.0	0	—	—	318.0	—	—
655	dessert, prepared	1 CUP	240	—	146.9	2.8	0	34.0	0	0	0	97.2	—	1.1
			100	—	61.2	1.2	0	14.2	0	0	0	40.5	—	.5
656	w/water & fruit	1 CUP	250	81.8	167.5	3.3	.3	41.0	.5	—	—	85.0	—	—
			100	81.8	67.0	1.3	.1	16.4	.2	—	—	34.0	—	—
657	all flavors, dietetic	1 SERV	120	—	10.0	2.0	0	0	—	—	—	5.0	—	—
			100	—	8.3	1.7	0	0	—	—	—	4.2	—	—
658	**GRAPEFRUIT PEEL,** candied	1 OZ	28	17.4	88.5	.1	.1	22.6	.6	—	—	—	—	—
			100	17.4	316.0	.4	.3	80.6	2.3	—	—	—	—	—
	HONEY													
659	[6]comb	1 OZ	28	—	78.4	.1	1.2	20.8	—	2.2	8.9	2.0	.6	9.7
			100	—	280.0	.5	4.3	74.3	—	7.9	31.6	7.3	2.2	34.8
660	strained or extracted	1 TBLS	20	17.2	60.8	.1	0	16.5	Trace	1.0	1.2	1.0	.6	10.2
			100	17.2	304.0	.3	0	82.3	.1	5.0	6.0	5.0	3.0	51.0

DESSERTS & SWEETS

Item No.	Minerals (Micro)			Fat-Soluble Vitamins			Water-Soluble Vitamins								
	Iron	Zinc	Copper	Vitamin A	Vitamin D	Vitamin E (Alpha Tocopherol)	Vitamin C	Thiamin	Ribo-flavin	Niacin	Panto-thenic Acid	Vit. B-6 (Pyri-doxine)	Folacin (Folic Acid)	Biotin	Vitamin B-12
	(mg)	(mg)	(mg)	(IU)	(IU)	(mg)	(mg)	(mg)	(mg)	(mg)	(mg)	(mg)	(mcg)	(mcg)	(mcg)
630	**.14** .50	— —	— —	**0** 0	— —	— —	**0** 0	**0** 0	**Trace** Trace	**Trace** Trace	— —	— —	— —	— —	— —
631	**.04** 1.90	— —	— —	**0** 0	**0** 0	— —	**0** 0	**0** 0	**0** 0	**0** 0	— —	— —	— —	— —	— —
632	**0** 0	— —	— —	**0** 0	**0** 0	— —	**0** 0	**0** 0	**0** 0	**0** 0	— —	— —	— —	— —	— —
633	**.13** 1.60	— —	— —	**0** 0	— —	— —	**0** 0	**0** 0	— —	— —	— —	— —	— —	— —	— —
634	**.63** 1.10	**.06** .11	**.04** .07	**153.9** 270.0	**95.19** 167.00	**.63** 1.10	**.80** 1.40	**.03** .06	**.19** .34	**.17** .30	— —	— —	**3.99** 7.00	**17.10** 30.00	— —
635	**.90** 1.60	— —	— —	**128.8** 230.0	**39.76** 71.00	— —	**.67** 1.20	**.05** .08	**.23** .41	**.45** .80	— —	— —	— —	— —	— —
636	**.50** 1.80	— —	— —	**0** **0**	— —	— —	**0** **0**	**.12** **.43**	**.02** **.08**	**2.63** **9.40**	— —	— —	— —	— —	— —
637	**.58** 2.30	— —	— —	**0** 0	— —	— —	**0** 0	**.04** .16	**.01** .03	**.85** 3.40	— —	— —	— —	— —	— —
638	**.31** 1.10	**.11** .38	— —	**Trace** Trace	— —	— —	**Trace** Trace	**.01** .03	**.02** .06	**.03** .10	— —	— —	— —	— —	— —
639	**.08** 1.90	— —	— —	**0** 0	— —	— —	**0** 0	**Trace** .05	**.01** .27	**.04** 1.00	— —	— —	— —	— —	— —
640	**.36** 1.30	— —	— —	**28.0** 100.0	— —	— —	— —	**.02** .06	**.06** .20	**.08** .30	— —	— —	— —	— —	— —
641	**.13** .50	— —	— —	— —	— —	— —	— —	**.01** .02	**.03** .13	**.03** .10	— —	— —	— —	— —	— —
642	**.22** .80	— —	— —	— —	— —	— —	— —	**.01** .05	**.04** .13	**.03** .10	— —	— —	— —	— —	— —
643	**.80** .70	— —	— —	**843.6** 740.0	— —	— —	— —	**.03** .03	**.11** .10	**.11** .10	— —	— —	— —	— —	— —
644	— —	— —	— —	**0** 0	— —	— —	**0** 0	**0** 0	**0** 0	**0** 0	— —	— —	— —	— —	— —
645	**.22** .80	— —	— —	— —	— —	— —	— —	— —	— —	— —	— —	— —	— —	— —	— —
646	**.70** .70	— —	— —	**10.0** 10.0	— —	— —	**0** 0	**.10** .10	**.04** .04	**.60** .60	— —	— —	— —	— —	— —
647	**.74** .70	— —	— —	**367.5** 350.0	— —	— —	**0** 0	**.04** .04	**.18** .17	**.11** .10	— —	— —	— —	— —	— —
648	**.06** .04	— —	— —	**927.5** 350.0	— —	— —	**1.06** .40	**.11** .04	**.50** .19	**.27** .10	— —	— —	— —	— —	— —
649	**(.22)** (.10)	**(.90)** (.40)	**(.07)** (.03)	— —	**(.83)** (.37)	**(.70)** (.30)	**Trace** Trace	**(.11)** (.05)	**(.56)** (.25)	**(.2)** (.1)	**1.3** .6	**(.11)** (.05)	**(16.00)** (8.00)	**(15.68)** (7.00)	**Trace** Trace
650	**.77** .70	**(.40)** (.40)	**(.17)** (.15)	**374.0** 340.0	**(1.00)** (.91)	**(1.30)** (1.20)	**0** 0	**.04** .04	**.18** .16	**.11** .10	**(.33)** (.30)	**(.04)** (.04)	**(4.40)** (4.00)	**(1.10)** (1.00)	**Trace** Trace
651	**3.64** 3.53	— —	— —	— —	— —	— —	**.75** .73	— —	— —	— —	— —	— —	— —	— —	— —
652	— —	— —	— —	**25.8** 14.0	**0** 0	— —	**0** 0	**.02** .01	— —	**.04** .02	— —	— —	— —	— —	— —
653	**0** 0	**.03** .32	**.04** .39	**0** 0	**0** 0	— —	**0** 0	**0** 0	**0** 0	**0** 0	— —	**Trace** .01	— —	— —	— —
654	— —	— —	— —	— —	— —	— —	— —	— —	— —	— —	— —	— —	— —	— —	— —
655	**3.50** 1.46	**.05** .02	— —	**209.7** 87.4	— —	— —	**34.23** 14.26	**0** 0	**.37** .15	**4.32** 1.80	— —	— —	— —	— —	— —
656	— —	— —	— —	— —	— —	— —	**7.50** 3.00	— —	— —	— —	— —	— —	— —	— —	— —
657	— —	— —	— —	— —	— —	— —	— —	— —	— —	— —	— —	— —	— —	— —	— —
658	— —	— —	— —	— —	— —	— —	— —	— —	— —	— —	— —	— —	— —	— —	— —
659	**.06** .22	— —	**.01** .05	**0** 0	**0** 0	— —	**Trace** Trace	**Trace** Trace	**.01** .04	**.07** .26	— —	— —	— —	— —	**0** 0
660	**.10** .50	**.02** .10	**.04** .20	**0** 0	**0** 0	— —	**.20** 1.00	— —	**.01** .04	**.06** .30	**.04** .20	**Trace** .02	**.60** 3.00	— —	**0** 0

(Continued)

DESSERTS & SWEETS

TABLE F-36 *(Continued)*

Item No.	Food Item Name	Approximate Measure	Weight	Moisture	Food Energy Calories	Protein	Fats	Carbo-hydrates	Fiber	Minerals (Macro) Calcium	Phos-phorus	Sodium	Mag-nesium	Potas-sium
			(g)	(%)	(kcal)	(g)	(g)	(g)	(g)	(mg)	(mg)	(mg)	(mg)	(mg)
661	**ICES,** water, lime	1 CUP	185	66.9	144.3	.7	0	60.3	—	—	—	—	—	5.6
			100	66.9	78.0	.4	0	32.6	—	—	—	—	—	3.0
	ICINGS													
662	caramel	1 SERV	10	14.1	36.0	.1	.7	7.7	0	10.2	6.3	8.3	—	5.2
			100	14.1	360.0	1.3	6.7	76.5	0	102.0	63.0	83.0	—	52.0
663	chocolate	1 SERV	10	14.3	37.6	.3	1.4	6.7	Trace	6.0	11.1	6.1	—	19.5
			100	14.3	376.0	3.2	13.9	67.4	.4	60.0	111.0	61.0	—	195.0
664	coconut	1 SERV	10	15.0	36.4	.2	.8	7.5	.1	.6	3.0	11.8	—	16.7
			100	15.0	364.0	1.9	7.7	74.9	.8	6.0	30.0	118.0	—	167.0
665	mix, chocolate fudge, w/water & margarine	1 OZ	28	15.3	105.8	.6	4.0	18.8	.1	4.5	18.5	43.7	—	17.6
			100	15.3	378.0	2.2	14.4	67.0	.5	16.0	66.0	156.0	—	63.0
666	mix, creamy fudge, w/nonfat dry milk & water	1 OZ	28	15.1	94.9	.8	1.8	20.9	.1	10.9	24.9	65.0	—	27.2
			100	15.1	339.0	2.8	6.5	74.6	.5	39.0	89.0	232.0	—	97.0
667	mix, creamy fudge, w/nonfat dry milk, water & margarine	1 OZ	28	15.1	107.2	.7	4.3	18.5	.1	10.4	22.7	89.9	—	24.9
			100	15.1	383.0	2.6	15.2	65.9	.5	37.0	81.0	321.0	—	89.0
668	white, boiled	1 SERV	10	17.9	31.6	.1	0	8.0	0	.2	.2	14.3	—	1.8
			100	17.9	316.0	1.4	0	80.3	0	2.0	2.0	143.0	—	18.0
669	white, uncooked	1 SERV	10	11.1	37.6	.1	.7	8.2	0	1.5	1.2	4.9	—	1.8
			100	11.1	376.0	.5	6.6	81.6	0	15.0	12.0	49.0	—	18.0
670	**JAMS & PRESERVES,** red cherry or strawberry	1 TBLS	20	29.0	54.4	Trace	Trace	14.0	.2	4.0	1.8	2.4	—	17.6
			100	29.0	272.0	.6	.1	70.0	1.0	20.0	9.0	12.0	—	88.0
671	**JELLY**	1 TBLS	20	29.0	54.6	Trace	Trace	14.1	0	4.2	1.4	3.4	—	15.0
			100	29.0	273.0	.1	.1	70.6	0	21.0	7.0	17.0	—	75.0
672	grape	1 TBLS	20	29.0	54.6	Trace	Trace	14.1	0	4.2	1.4	3.4	.8	15.0
			100	29.0	273.0	.1	.1	70.6	0	21.0	7.0	17.0	4.0	75.0
673	red cherry or strawberry	1 TBLS	20	29.0	54.6	Trace	Trace	14.1	0	4.2	1.4	3.4	—	15.0
			100	29.0	273.0	.1	.1	70.6	0	21.0	7.0	17.0	—	75.0
674	**MARMALADE, CITRUS**	1 TBLS	20	29.0	51.4	.1	Trace	14.0	0	7.0	1.8	2.8	.8	6.6
			100	29.0	257.0	.5	.1	70.1	.4	35.0	9.0	14.0	4.0	33.0
675	***MINCEMEAT**	1 OZ	28	—	66.1	.2	1.2	17.4	1.0	8.5	4.8	39.2	2.9	53.2
			100	—	236.0	.6	4.3	62.0	3.4	30.2	17.3	139.8	10.2	189.8
	MOLASSES													
676	cane, blackstrap	1 TBLS	20	24.0	46.0	.5	0	11.0	0	136.8	16.8	19.2	41.9	585.4
			100	24.0	230.0	2.4	0	55.0	0	684.0	84.0	96.0	209.3	2927.0
677	cane, medium	1 TBLS	20	24.0	50.0	.5	—	12.0	0	58.0	13.8	7.4	16.2	212.6
			100	24.0	250.0	2.4	—	60.0	0	290.0	69.0	37.0	81.0	1083.0
678	cane, light	1 TBLS	20	24.0	54.0	.5	0	13.0	0	33.0	9.0	3.0	41.9	183.4
			100	24.0	270.0	2.4	0	65.0	0	165.0	45.0	15.0	209.3	917.0
679	cane, Barbados	1 TBLS	20	24.0	54.2	—	—	14.0	0	49.0	10.0	—	—	—
			100	24.0	271.0	—	—	70.0	0	245.0	50.0	—	—	—
680	unsulfured	1 TBLS	20	21.0	50.0	0	0	14.4	—	20.0	—	11.0	—	132.0
			100	21.0	250.0	0	0	72.0	—	100.0	—	55.0	—	660.0
681	**PEACH CRISP**	1 OZ	28	—	48.2	1.1	1.8	7.1	.1	2.0	4.3	45.3	.3	24.7
			100	—	172.1	3.9	6.4	25.3	.2	7.0	15.3	161.9	1.0	88.3
682	**POPSICLE**	1 AVG	88	80.0	65.1	0	0	16.7	—	0	—	—	—	—
			100	80.0	74.0	0	0	18.9	—	0	—	—	—	—
683	**PRUNE WHIP,** baked	1 CUP	90	57.3	140.4	4.0	.2	33.2	.5	19.8	29.7	147.6	—	261.0
			100	57.3	156.0	4.4	.2	36.9	.6	22.0	33.0	164.0	—	290.0
	PUDDINGS													
684	**BANANA,** canned	1 CUP	260	71.8	325.0	6.2	9.6	55.1	—	218.4	182.0	507.0	20.8	286.0
			100	71.8	125.0	2.4	3.7	21.2	—	84.0	70.0	195.0	8.0	110.0
685	**BANANA CREAM,** instant, prepared	1 CUP	290	—	360.0	8.8	12.2	55.6	—	—	—	560.0	—	480.0
			100	—	124.1	3.0	4.2	19.2	—	—	—	193.1	—	165.5
686	**BUTTERSCOTCH,** canned	1 CUP	260	70.5	322.4	6.5	9.4	59.8	0	231.4	176.8	444.6	23.4	306.8
			100	70.5	124.0	2.5	3.6	23.0	0	89.0	68.0	171.0	9.0	118.0
687	**BUTTERSCOTCH,** w/skim milk	1 SERV	130	—	116.5	3.6	.2	25.0	Trace	153.3	115.8	642.9	24.7	238.7
			100	—	89.6	2.8	.2	19.2	Trace	117.9	89.1	494.6	19.0	183.6
688	**CHOCOLATE,** canned	1 CUP	227	68.5	297.4	6.8	9.1	52.9	0	215.7	168.0	363.2	52.2	295.1
			100	68.5	131.0	3.0	4.0	23.3	0	95.0	74.0	160.0	23.0	130.0
689	**CHOCOLATE,** diet	1 OZ	28	89.2	10.7	.9	.1	1.6	0	33.4	27.7	17.6	3.0	45.2
			100	89.2	38.3	3.2	.2	5.6	0	119.3	98.9	63.0	10.7	161.3
690	**CHOCOLATE,** w/skim milk	1 SERV	158	—	235.6	4.8	6.1	44.4	.3	147.9	155.1	636.7	47.3	276.5
			100	—	149.1	3.0	3.9	28.1	.2	93.6	98.2	403.0	29.9	175.0

Item No.	Minerals (Micro)			Fat-Soluble Vitamins			Water-Soluble Vitamins								
	Iron	Zinc	Copper	Vitamin A	Vitamin D	Vitamin E (Alpha Tocophorol)	Vitamin C	Thiamin	Ribo-flavin	Niacin	Panto-thenic Acid	Vit. B-6 (Pyri-doxine)	Folacin (Folic Acid)	Biotin	Vitamin B-12
	(mg)	(mg)	(mg)	(IU)	(IU)	(mg)	(mg)	(mg)	(mg)	(mg)	(mg)	(mg)	(mcg)	(mcg)	(mcg)
661	—	—	—	**0**	—	—	**1.85**	—	—	—	—	—	—	—	—
	—	—	—	0	—	—	1.00	—	—	—	—	—	—	—	—
662	**.20**	—	—	**28.0**	—	—	—	**Trace**	**.01**	—	—	—	—	—	—
	2.00	—	—	280.0	—	—	—	.01	.06	—	—	—	—	—	—
663	**.12**	—	—	**21.0**	—	—	—	**Trace**	**.01**	**.02**	—	—	—	—	—
	1.20	—	—	210.0	—	—	—	.02	.10	.20	—	—	—	—	—
664	**.05**	—	—	**0**	—	—	**0**	**Trace**	**Trace**	**.02**	—	—	—	—	—
	.50	—	—	0	—	—	0	.01	.04	.20	—	—	—	—	—
665	**.28**	—	—	**75.6**	—	—	**0**	**Trace**	**.01**	**.06**	—	—	—	—	—
	1.00	—	—	270.0	—	—	0	.01	.04	.20	—	—	—	—	—
666	**.31**	—	—	—	—	—	—	**.01**	**.02**	**.08**	—	—	—	—	—
	1.10	—	—	—	—	—	—	.02	.08	.30	—	—	—	—	—
667	**.28**	—	—	**109.2**	—	—	—	**.01**	**.02**	**.08**	—	—	—	—	—
	1.00	—	—	390.0	—	—	—	.02	.07	.30	—	—	—	—	—
668	—	—	—	**0**	—	—	**0**	—	**Trace**	—	—	—	—	—	—
	—	—	—	0	—	—	0	—	.03	—	—	—	—	—	—
669	—	—	—	**27.0**	—	—	—	—	**Trace**	—	—	—	—	—	—
	—	—	—	270.0	—	—	—	—	.02	—	—	—	—	—	—
670	**.20**	—	—	**2.0**	**0**	—	**3.00**	**Trace**	**.01**	**.04**	—	—	**1.60**	—	—
	1.00	—	—	10.0	0	—	15.00	.01	.03	.20	—	—	8.00	—	—
671	**.30**	—	—	**2.0**	**0**	—	**.80**	**Trace**	**.01**	**.04**	—	—	—	—	—
	1.50	—	—	10.0	0	—	4.00	.01	.03	.20	—	—	—	—	—
672	**.30**	**.04**	**.02**	**2.0**	—	—	**.80**	**Trace**	**.01**	**.04**	—	—	—	—	—
	1.50	.22	.09	10.0	—	—	4.00	.01	.03	.20	—	—	—	—	—
673	**.30**	—	—	**2.0**	**0**	—	**3.00**	**Trace**	**.01**	**.04**	—	—	—	—	—
	1.50	—	—	10.0	0	—	15.00	.01	.03	.20	—	—	—	—	—
674	**.12**	—	**(.03)**	**0**	**0**	**Trace**	**1.20**	**Trace**	**Trace**	**.02**	**Trace**	**0**	**(1.08)**	**Trace**	**0**
	.60	—	(.13)	0	0	Trace	6.00	.02	.02	.10	Trace	.02	(5.40)	Trace	0
675	**.43**	**.06**	**.06**	**.5**	**0**	—	**Trace**	**Trace**	**.01**	**.06**	**Trace**	**.03**	**Trace**	**Trace**	**0**
	1.53	.22	.21	1.7	0	—	Trace	.02	.03	.21	.02	.09	Trace	Trace	0
676	**3.22**	**.44**	**1.20**	**0**	**0**	—	**0**	**.02**	**.04**	**.40**	**.07**	**.04**	**1.90**	**1.80**	**0**
	16.10	2.20	6.00	0	0	—	0	.11	.19	2.00	.35	.20	9.50	9.00	0
677	**1.20**	**.07**	**.11**	**0**	**0**	—	**0**	**.02**	**.02**	**.24**	**.07**	**.04**	**1.90**	**1.80**	**0**
	6.00	.35	.55	0	0	—	0	.08	.12	1.20	.35	.20	9.50	9.00	0
678	**.86**	**.07**	**.11**	**0**	**0**	—	**0**	**.01**	**.01**	**.04**	**.07**	**.04**	**2.60**	**1.80**	**0**
	4.30	.35	.55	0	0	—	0	.07	.06	.20	.35	.20	13.00	9.00	0
679	—	**.07**	**.11**	—	**0**	—	—	**.01**	**.04**	—	**.07**	**.04**	**1.90**	**1.80**	**0**
	—	.35	.55	—	0	—	—	.06	.20	—	.35	.20	9.50	9.00	0
680	**.40**	**.07**	**.11**	**0**	—	—	**0**	**0**	**0**	**0**	—	—	—	—	—
	2.00	.35	.55	0	—	—	0	0	0	0	—	—	—	—	—
681	**.15**	**Trace**	—	**134.1**	—	—	**7.22**	**.01**	**.01**	**.24**	—	**0**	—	—	—
	.53	.01	—	479.0	—	—	25.78	.03	.04	.85	—	0	—	—	—
682	—	—	—	**0**	—	—	**0**	**0**	**0**	**0**	—	—	—	—	—
	—	—	—	0	—	—	0	0	0	0	—	—	—	—	—
683	**1.17**	—	—	**414.0**	—	—	**1.80**	**.02**	**.13**	**.45**	—	—	—	—	—
	1.30	—	—	460.0	—	—	2.00	.02	.14	.50	—	—	—	—	—
684	**.26**	**.75**	**.08**	—	—	—	—	**.05**	**.44**	**4.16**	—	—	—	—	—
	.10	.29	.03	—	—	—	—	.02	.17	1.60	—	—	—	—	—
685	—	—	—	—	—	—	—	—	—	—	—	—	—	—	—
	—	—	—	—	—	—	—	—	—	—	—	—	—	—	—
686	**.26**	**.73**	**.08**	**5.2**	—	—	**2.60**	**.05**	**.31**	**.26**	—	—	—	—	—
	.10	.28	.03	2.0	—	—	1.00	.02	.12	.10	—	—	—	—	—
687	**.59**	**.45**	**.04**	**225.5**	**45.11**	—	**1.11**	**.05**	**.16**	**.13**	**.36**	**.05**	**5.53**	—	**.44**
	.45	.35	.03	173.4	34.70	—	.85	.04	.12	.10	.28	.03	4.25	—	.34
688	**.23**	**1.18**	**.27**	**4.5**	—	—	**2.27**	**.05**	**.30**	**.23**	—	—	—	—	—
	.10	.52	.12	2.0	—	—	1.00	.02	.13	.10	—	—	—	—	—
689	**.01**	**.11**	**8.42**	**55.4**	**11.08**	—	**.27**	**.01**	**.04**	**.02**	**.09**	**.01**	**1.36**	—	**.11**
	.04	.39	30.07	197.9	39.59	—	.97	.04	.14	.09	.32	.04	4.85	—	.38
690	**.82**	**.45**	**.04**	**231.4**	**44.96**	—	**1.10**	**.05**	**.18**	**.26**	**.38**	**.05**	—	—	**.51**
	.52	.28	.03	146.4	28.45	—	.70	.03	.11	.17	.24	.03	—	—	.33

(Continued)

TABLE F-36 *(Continued)*

Item No.	Food Item Name	Approximate Measure	Weight	Moisture	Food Energy Calories	Protein	Fats	Carbo-hydrates	Fiber	Minerals (Macro) Calcium	Phos-phorus	Sodium	Mag-nesium	Potas-sium
			(g)	(%)	(kcal)	(g)	(g)	(g)	(g)	(mg)	(mg)	(mg)	(mg)	(mg)
	PUDDINGS (Continued)													
691	**CHOCOLATE**, starch base	1 CUP	**260**	**65.8**	**384.8**	**8.1**	**12.2**	**66.8**	**.5**	**249.6**	**254.8**	**145.6**	—	**444.6**
			100	65.8	148.0	3.1	4.7	25.7	.2	96.0	98.0	56.0	—	171.0
692	**CUSTARD**, instant, prepared w/milk	1 CUP	**260**	**70.0**	**340.6**	**8.1**	**9.1**	**58.8**	—	**275.6**	**213.2**	**257.4**	**36.4**	**335.4**
			100	70.0	131.0	3.1	3.5	22.6	—	106.0	82.0	99.0	14.0	129.0
693	**LEMON**, canned	1 CUP	**226**	—	**305.1**	**0**	**5.9**	**63.3**	**0**	**13.6**	**42.9**	**282.5**	—	**20.3**
			100	—	135.0	0	2.6	28.0	0	6.0	19.0	125.0	—	9.0
694	**RICE**	1 CUP	**260**	—	**369.2**	**4.7**	**6.2**	**73.6**	**0**	**117.0**	**101.4**	**611.0**	—	**158.6**
			100	—	142.0	1.8	2.4	28.3	0	45.0	39.0	235.0	—	61.0
695	**RICE**, w/raisins	1 CUP	**265**	**65.8**	**386.9**	**9.5**	**8.2**	**70.8**	**.3**	**259.7**	**249.1**	**188.2**	—	**469.1**
			100	65.8	146.0	3.6	3.1	26.7	.1	98.0	94.0	71.0	—	177.0
696	**STRAWBERRY**, canned	1 CUP	**260**	**73.6**	**299.0**	**1.8**	**7.5**	**57.7**	—	**78.0**	**135.2**	**457.6**	**7.8**	**140.4**
			100	73.6	115.0	.7	2.9	22.2	—	30.0	52.0	176.0	3.0	54.0
697	**TAPIOCA**, canned	1 CUP	**260**	**72.6**	**312.0**	**6.2**	**7.8**	**55.1**	—	**306.8**	**205.4**	**462.8**	**20.8**	**408.2**
			100	72.6	120.0	2.4	3.0	21.2	—	118.0	79.0	178.0	8.0	157.0
698	**TAPIOCA**, apple	1 CUP	**244**	**70.1**	**285.5**	**.5**	**.2**	**71.7**	**.2**	**7.3**	**9.8**	**124.4**	—	**63.4**
			100	70.1	117.0	.2	.1	29.4	.1	3.0	4.0	51.0	—	26.0
699	**TAPIOCA CREAM**	1 OZ	**28**	**71.8**	**37.5**	**1.4**	**1.4**	**4.8**	**0**	**29.4**	**30.5**	**43.7**	—	**37.8**
			100	71.8	134.0	5.0	5.1	17.1	0	105.0	109.0	156.0	—	135.0
700	**TAPIOCA**, quick, made w/whole milk	1 CUP	**224**	—	**274.0**	**7.8**	**7.8**	**43.2**	—	—	—	**.2**	—	—
			100	—	122.3	3.5	3.5	19.3	—	—	—	.1	—	—
701	**VANILLA**, canned	1 CUP	**227**	**70.5**	**281.5**	**5.7**	**8.6**	**51.8**	**0**	**202.0**	**154.4**	**388.2**	**18.2**	**267.9**
			100	70.5	124.0	2.5	3.8	22.8	0	89.0	68.0	171.0	8.0	118.0
702	**VANILLA**, made w/skim milk	1 SERV	**132**	—	**123.1**	**3.6**	**.2**	**26.0**	**Trace**	**140.4**	**113.3**	**641.2**	**14.9**	**185.1**
			100	—	93.3	2.8	.2	19.7	Trace	106.3	85.8	485.8	11.3	140.2
703	**VANILLA**, made w/whole milk	1 CUP	**297**	—	**351.3**	**8.2**	**8.4**	**61.1**	**0**	**298.9**	**217.2**	**504.1**	**0**	**372.5**
			100	—	118.3	2.8	2.8	20.6	0	100.6	73.1	169.7	0	125.4
704	**VANILLA**, starch base (Blanc Mange)	1 CUP	**250**	**76.0**	**277.5**	**8.8**	**9.8**	**39.8**	—	**292.5**	**227.5**	**162.5**	—	**345.0**
			100	76.0	111.0	3.5	3.9	15.9	—	117.0	91.0	65.0	—	138.0
	SAUCES													
705	**CUSTARD**	1 TBLS	**18**	—	**21.2**	**.9**	**1.0**	**2.3**	—	**19.4**	**20.5**	—	—	—
			100	—	118.0	5.1	5.3	12.9	—	108.0	114.0	—	—	—
706	**LEMON**	1 TBLS	**14**	—	**34.4**	**0**	**.7**	**7.2**	—	**.7**	**.6**	—	—	—
			100	—	246.0	2	5.2	51.5	—	5.0	4.0	—	—	—
707	**RAISIN**	1 TBLS	**12**	—	**31.6**	**.2**	**.8**	**6.5**	—	**3.7**	**5.9**	—	—	—
			100	—	263.0	1.3	6.3	54.0	—	31.0	49.0	—	—	—
708	**SHERBET**, orange	1 CUP	**193**	**67.0**	**258.6**	**1.7**	**3.8**	**59.4**	**0**	**104.2**	**73.3**	**88.8**	**12.9**	**198.8**
			100	67.0	134.0	.9	2.0	30.8	0	54.0	38.0	46.0	6.7	103.0
	SUGAR													
709	**BEET OR CANE**, brown	1 TBLS	**14**	**2.1**	**52.2**	**0**	**0**	**13.5**	**0**	**11.9**	**2.7**	**4.2**	**8.7**	**48.2**
			100	2.1	373.0	0	0	96.4	0	85.0	19.0	30.0	62.0	344.0
710	**BEET OR CANE**, granulated	1 TSP	**8**	**0.5**	**30.8**	**0**	**0**	**8.0**	**0**	**0**	**0**	**.1**	**Trace**	**.2**
			100	0.5	385.0	0	0	99.5	0	0	0	1.0	.2	3.0
711	**BEET OR CANE**, powdered	1 TBLS	**11**	**0.5**	**42.4**	**0**	**0**	**10.9**	**0**	**0**	**0**	**.1**	**0**	**.3**
			100	0.5	385.0	0	0	99.5	0	0	0	1.0	0	3.0
712	**DEXTROSE**, anhydrous	1 TBLS	**12**	**0.5**	**43.9**	**0**	**0**	**11.9**	**0**	—	—	—	—	—
			100	0.5	366.0	0	0	99.5	0	—	—	—	—	—
713	**DEXTROSE**, crystallized	1 OZ	**28**	**9.0**	**119.3**	**0**	**0**	**25.5**	**0**	—	—	—	—	—
			100	9.0	426.0	0	0	91.0	0	—	—	—	—	—
714	**MAPLE**	1 PIECE	**15**	**8.0**	**52.2**	—	—	**13.5**	—	**21.5**	**1.7**	**2.1**	—	**36.3**
			100	8.0	348.0	—	—	90.0	—	143.0	11.0	14.0	—	242.0
715	**SUGARCANE JUICE**	1 CUP	**240**	**78.8**	**196.8**	**.7**	**.2**	**49.2**	**1.0**	**31.2**	**28.8**	—	—	—
			100	78.8	82.0	.3	.1	20.5	.4	13.0	12.0	—	—	—
	SYRUPS													
716	**CANE**	1 TBLS	**15**	**26.0**	**39.5**	**0**	**0**	**10.2**	**0**	**9.0**	**4.4**	—	—	**63.8**
			100	26.0	263.0	0	0	68.0	0	60.0	29.0	—	—	425.0
717	**CHOCOLATE**, fudge-type	1 TBLS	**20**	**25.4**	**66.0**	**1.0**	**2.7**	**10.8**	**.1**	**25.4**	**31.8**	**17.8**	—	**56.8**
			100	25.4	330.0	5.1	13.7	54.0	.4	127.0	159.0	89.0	—	284.0
718	**CHOCOLATE**, thin-type	1 TBLS	**20**	**31.6**	**49.0**	**.5**	**.4**	**12.5**	**.1**	**3.4**	**18.4**	**10.4**	**12.6**	**56.4**
			100	31.6	245.0	2.3	2.0	62.7	.6	17.0	92.0	52.0	63.0	282.0
719	**MAPLE**	1 TBLS	**20**	**33.0**	**50.4**	**0**	**0**	**13.0**	**0**	**20.8**	**1.6**	**2.0**	**2.0**	**35.2**
			100	33.0	252.0	0	0	65.0	0	104.0	8.0	10.0	10.0	176.0
720	**SORGHUM**	1 TBLS	**20**	**23.0**	**51.4**	—	—	**13.6**	**0**	**34.4**	**5.0**	**4.0**	—	**120.0**
			100	23.0	257.0	—	—	68.0	0	172.0	25.0	20.0	—	600.0

Item No.	Minerals (Micro) Iron	Zinc	Copper	Fat-Soluble Vitamins Vitamin A	Vitamin D	Vitamin E (Alpha Tocopherol)	Water-Soluble Vitamins Vitamin C	Thiamin	Ribo-flavin	Niacin	Panto-thenic Acid	Vit. B-6 (Pyri-doxine)	Folacin (Folic Acid)	Biotin	Vitamin B-12
	(mg)	(mg)	(mg)	(IU)	(IU)	(mg)	(mg)	(mg)	(mg)	(mg)	(mg)	(mg)	(mcg)	(mcg)	(mcg)
691	1.30	—	—	390.0	—	—	—	.05	.36	.26	—	—	—	—	—
	.50	—	—	150.0	—	—	—	.02	.14	.10	—	—	—	—	—
692	—	—	—	364.0	—	—	—	.05	.36	.26	—	—	—	—	—
	—	—	—	140.0	—	—	—	.02	.14	.10	—	—	—	—	—
693	1.13	—	—	40.7	—	—	2.26	.05	.14	.23	—	—	—	—	—
	.50	—	—	18.0	—	—	1.00	.02	.06	.10	—	—	—	—	—
694	1.04	—	—	41.6	—	—	2.60	.08	.26	.26	—	—	—	—	—
	.40	—	—	16.0	—	—	1.00	.03	.10	.10	—	—	—	—	—
695	1.06	—	—	291.5	—	—	—	.08	.37	.53	—	—	—	—	—
	.40	—	—	110.0	—	—	—	.03	.14	.20	—	—	—	—	—
696	.52	.29	.08	—	—	—	59.80	—	.10	.78	—	—	—	—	—
	.20	.11	.03	—	—	—	23.00	—	.04	.30	—	—	—	—	—
697	.52	.73	.08	—	—	—	—	.08	.31	.78	—	—	—	—	—
	.20	.28	.03	—	—	—	—	.03	.12	.30	—	—	—	—	—
698	.49	—	—	24.4	—	—	—	—	—	—	—	—	—	—	—
	.20	—	—	10.0	—	—	—	—	—	—	—	—	—	—	—
699	.11	—	—	81.2	—	—	.28	.01	.05	.03	—	—	—	—	—
	.40	—	—	290.0	—	—	1.00	.04	.18	.10	—	—	—	—	—
700	—	—	—	—	—	—	—	—	—	—	—	—	—	—	—
	—	—	—	—	—	—	—	—	—	—	—	—	—	—	—
701	.23	.61	.07	4.5	—	—	2.27	.05	.27	.23	—	—	—	—	—
	.10	.27	.03	2.0	—	—	1.00	.02	.12	.10	—	—	—	—	—
702	.06	.45	.04	226.5	45.32	—	1.11	.04	.16	.10	.36	.05	5.55	—	.48
	.05	.34	.03	171.6	34.33	—	.84	.03	.12	.07	.28	.04	4.21	—	.36
703	.34	.83	.08	307.4	100.40	.15	3.36	.10	.44	1.16	.77	.10	12.24	7.57	.89
	.11	.27	.03	103.5	33.80	.05	1.13	.03	.15	.39	.26	.03	4.12	2.55	.30
704	—	—	—	400.0	—	—	2.50	.08	.40	.25	—	—	—	—	—
	—	—	—	160.0	—	—	1.00	.03	.16	.10	—	—	—	—	—
705	.09	—	—	58.7	—	—	—	.01	.06	.02	—	—	—	—	—
	.50	—	—	326.0	—	—	—	.05	.33	.13	—	—	—	—	—
706	—	—	—	32.2	—	—	.77	0	—	—	—	—	—	—	—
	—	—	—	230.0	—	—	5.50	0	—	—	—	—	—	—	—
707	—	—	—	40.2	—	—	1.25	.01	.01	.15	—	—	—	—	—
	—	—	—	335.0	—	—	10.40	.06	.04	1.25	—	—	—	—	—
708	.19	1.33	.04	185.3	—	—	3.86	.02	.09	.13	.06	.03	13.51	—	.16
	.10	.69	.02	96.0	—	—	2.00	.01	.05	.07	.03	.01	7.00	—	.08
709	.48	Trace	Trace	0	0	—	0	Trace	Trace	.03	—	—	—	—	—
	3.40	.03	.02	0	0	—	0	.01	.03	.20	—	—	—	—	—
710	.01	.01	Trace	0	0	—	0	0	0	0	—	—	—	—	.02
	.10	.05	.02	0	0	—	0	0	0	0	—	—	—	—	.23
711	.01	Trace	Trace	0	0	—	0	0	0	0	—	—	—	—	.03
	.10	.02	.02	0	0	—	0	0	0	0	—	—	—	—	.23
712	—	—	—	0	0	—	0	0	0	0	—	—	—	—	—
	—	—	—	0	0	—	0	0	0	0	—	—	—	—	—
713	—	—	—	0	0	—	0	0	0	0	—	—	—	—	—
	—	—	—	0	0	—	0	0	0	0	—	—	—	—	—
714	.21	—	—	—	0	—	0	—	—	—	—	—	—	—	—
	1.40	—	—	—	0	—	0	—	—	—	—	—	—	—	—
715	1.68	—	—	0	—	—	4.80	.05	.02	.24	—	—	—	—	—
	.70	—	—	0	—	—	2.00	.02	.01	.10	—	—	—	—	—
716	.54	—	—	0	0	—	0	.02	.01	.02	—	—	—	—	—
	3.60	—	—	0	0	—	0	.13	.06	.10	—	—	—	—	—
717	.26	.18	—	30.0	—	—	—	.01	.04	.08	—	—	—	—	—
	1.30	.90	—	150.0	—	—	—	.04	.22	.40	—	—	—	—	—
718	.32	.18	.09	—	0	—	0	Trace	.01	.08	—	—	—	—	—
	1.60	.90	.43	—	0	—	0	.02	.07	.40	—	—	—	—	—
719	.24	—	—	0	0	—	0	—	—	—	—	—	—	—	—
	1.20	—	—	0	0	—	0	—	—	—	—	—	—	—	—
720	2.50	—	—	—	0	—	—	—	.02	.02	—	—	—	—	—
	12.50	—	—	—	0	—	—	—	.10	.10	—	—	—	—	—

(Continued)

TABLE F-36 *(Continued)*

Item No.	Food Item Name	Approximate Measure	Weight	Moisture	Food Energy Calories	Protein	Fats	Carbo-hydrates	Fiber	Minerals (Macro) Calcium	Phos-phorus	Sodium	Mag-nesium	Potas-sium
			(g)	(%)	(kcal)	(g)	(g)	(g)	(g)	(mg)	(mg)	(mg)	(mg)	(mg)
	SYRUPS (Continued)													
721	**TABLE BLEND OF CANE & MAPLE**	1 TBLS	20	33.0	50.4	0	0	13.0	0	3.2	.2	.4	—	5.2
			100	33.0	252.0	0	0	65.0	0	16.0	1.0	2.0	—	26.0
722	**TABLE BLEND OF CHIEFLY CORN LIGHT & DARK**	1 TBLS	20	24.0	58.0	0	0	15.0	0	9.2	3.2	13.6	.4	.8
			100	24.0	290.0	0	0	75.0	0	46.0	16.0	68.0	2.0	4.0
	TOPPINGS													
723	marshmallow	1 TBLS	8	17.6	25.3	.1	0	6.4	0	2.6	1.0	4.7	—	2.6
			100	17.6	316.0	1.0	0	80.6	0	32.0	12.0	59.0	—	32.0
724	[1]non-dairy, powdered	1 OZ	28	1.5	161.6	1.4	11.2	14.7	0	4.5	20.7	34.2	—	46.5
			100	1.5	577.0	4.9	39.9	52.5	0	17.0	74.0	122.0	—	166.0
725	[1]non-dairy, powdered, made w/whole milk	1 OZ	28	66.7	52.9	1.0	3.5	4.6	0	25.2	24.1	18.5	2.8	42.3
			100	66.7	189.0	3.6	12.4	16.5	0	90.0	86.0	66.0	10.0	151.0
726	whipped, imitation, frozen	1 TBLS	15	52.0	47.7	.2	3.8	3.5	0	1.1	1.2	3.8	.3	2.7
			100	52.0	318.0	1.3	25.3	23.1	0	7.0	8.0	25.0	2.0	18.0
727	whipped, imitation, pressurized	1 TBLS	9	61.0	23.1	.1	2.0	1.4	0	.5	1.6	5.6	.1	1.7
			100	61.0	257.0	1.0	22.2	16.1	0	5.0	18.0	62.0	1.0	19.0
728	whipped, non-dairy, cool whip	1 TBLS	5	47.3	14.0	0	1.0	1.0	—	0	0	1.0	.1	1.0
			100	47.3	280.0	0	20.0	20.0	—	0	0	20.0	1.0	20.0
729	[6]**TRIFLE**	1 SERV (4 OZ)	112	64.5	179.2	3.9	6.8	27.2	0	91.8	97.4	56.0	15.7	91.8
			100	64.5	160.0	3.5	6.1	24.3	0	82.0	87.0	50.0	14.0	82.0
	CHICKEN EGGS													
730	raw, whole, fresh	1 MED	48	73.7	75.4	6.1	5.4	.6	0	25.9	86.4	66.2	5.8	61.9
			100	73.7	157.0	12.8	11.3	1.2	0	54.0	180.0	138.0	12.0	129.0
731	raw whites, fresh & frozen	1 MED	31	87.6	15.8	3.3	0	.4	0	3.4	3.4	45.3	3.1	43.1
			100	87.6	51.0	10.8	0	1.2	0	11.0	11.0	146.0	9.9	139.0
732	raw, yolks, fresh	1 MED	17	48.8	62.7	2.8	5.6	0	0	25.8	86.4	8.8	2.6	15.3
			100	48.8	369.0	16.3	32.9	.2	0	152.0	508.0	52.0	15.0	90.0
733	fried	1 MED	48	68.4	100.9	5.6	8.6	.5	0	24.6	79.4	97.6	5.3	57.2
			100	68.4	210.2	11.7	17.9	1.1	0	51.3	165.5	203.3	11.1	119.2
734	hard-cooked	1 MED	48	73.7	75.4	6.1	5.4	.6	0	25.9	86.4	58.6	5.8	61.9
			100	73.7	157.0	12.8	11.3	1.2	0	54.0	180.0	122.0	12.0	129.0
735	omelet	1 MED	62	76.3	91.8	5.8	6.9	1.5	0	49.6	93.6	159.3	6.8	82.5
			100	76.3	148.0	9.3	11.1	2.4	0	80.0	151.0	257.0	11.0	133.0
736	poached	1 MED	48	73.3	74.4	6.1	5.4	.6	0	26.4	85.9	140.6	5.8	61.4
			100	73.3	155.0	12.8	11.2	1.2	0	55.0	179.0	293.0	12.0	128.0
737	scrambled	1 MED	65	76.3	96.2	6.1	7.2	1.6	0	52.0	98.2	167.1	7.8	86.5
			100	76.3	148.0	9.3	11.1	2.4	0	80.0	151.0	257.0	12.0	133.0
738	dried, whole	1 TBLS	5	4.1	29.7	2.3	2.1	.2	0	10.6	34.0	26.1	2.3	24.5
			100	4.1	593.0	46.8	41.8	4.8	0	212.0	679.0	521.0	46.0	490.0
739	dried, whole, stabilized	1 TBLS	5	2.0	30.5	2.4	2.1	.1	0	11.1	35.8	27.4	2.5	25.8
			100	2.0	609.0	48.2	42.9	2.5	0	222.0	715.0	548.0	49.0	515.0
740	dried, white flakes	1 CUP	45	14.6	158.0	33.8	0	1.9	0	37.4	37.4	520.2	30.2	468.9
			100	14.6	351.0	75.1	0	4.2	0	83.0	83.0	1156.0	67.0	1042.0
741	dried, white powder	1 CUP	107	6.8	398.0	88.2	0	4.8	0	95.2	95.2	1324.7	77.0	1194.1
			100	6.8	372.0	82.4	0	4.5	0	89.0	89.0	1238.0	72.0	1116.0
742	dried, yolks	1 TBLS	4	4.5	27.5	1.3	2.5	0	0	11.0	37.8	4.0	1.1	7.4
			100	4.5	687.0	33.2	61.3	.4	0	275.0	946.0	100.0	28.0	186.0
743	frozen raw, whole	1 MED	48	73.7	75.4	6.1	5.4	.4	0	25.9	98.4	58.6	4.7	61.9
			100	73.7	157.0	12.8	11.3	.9	0	54.0	205.0	122.0	9.8	129.0
744	frozen raw, yolks	1 MED	17	55.5	55.6	2.5	4.9	.1	0	21.3	68.2	10.7	2.2	15.8
			100	55.5	327.0	14.5	28.7	.4	0	125.0	401.0	63.0	13.0	93.0
745	frozen raw, yolks, sugared	1 MED	17	50.7	56.8	2.2	4.3	1.6	0	19.2	60.7	9.7	2.0	14.1
			100	50.7	334.0	12.9	25.5	9.5	0	113.0	357.0	57.0	12.0	83.0
746	**DUCK EGGS,** raw, whole, fresh	1 MED	74	70.4	138.4	9.8	10.2	1.1	0	47.4	162.8	90.3	11.8	164.3
			100	70.4	187.0	13.3	13.8	1.5	0	64.0	220.0	122.0	16.0	222.0
747	**GOOSE EGGS,** raw, whole, fresh	1 MED	144	70.4	260.6	20.0	19.2	1.9	0	—	—	—	—	—
			100	70.4	181.0	13.9	13.3	1.3	0	—	—	—	—	—
748	**QUAIL EGGS,** raw, whole, fresh	1 AVG	9	74.4	14.2	1.2	1.0	0	0	5.8	20.3	—	—	—
			100	74.4	158.0	13.1	11.0	.4	0	64.0	226.0	—	—	—
749	**TURKEY EGGS,** raw, whole, fresh	1 MED	72	72.6	118.8	9.4	8.5	.8	0	71.3	122.4	—	—	—
			100	72.6	165.0	13.1	11.8	1.2	0	99.0	170.0	—	—	—

Item No.	Minerals (Micro)			Fat-Soluble Vitamins			Water-Soluble Vitamins								
	Iron	Zinc	Copper	Vitamin A	Vitamin D	Vitamin E (Alpha Tocopherol)	Vitamin C	Thiamin	Ribo-flavin	Niacin	Panto-thenic Acid	Vit. B-6 (Pyri-doxine)	Folacin (Folic Acid)	Biotin	Vitamin B-12
	(mg)	(mg)	(mg)	(IU)	(IU)	(mg)	(mg)	(mg)	(mg)	(mg)	(mg)	(mg)	(mcg)	(mcg)	(mcg)
721	—	—	—	0	—	—	0	0	0	0	—	—	—	—	—
	—	—	—	0	—	—	0	0	0	0	—	—	—	—	—
722	.82	—	—	0	0	—	0	0	0	0	—	0	0	—	0
	4.10	—	—	0	0	—	0	0	0	0	—	0	0	—	0
723	—	—	—	—	—	—	.80	—	—	—	—	—	—	—	—
	—	—	—	—	—	—	10.00	—	—	—	—	—	—	—	—
724	—	—	—	301.6	—	—	0	0	0	0	0	0	—	—	0
	—	—	—	1077.0	—	—	0	0	0	0	0	0	—	—	0
725	.01	.08	—	101.1	—	—	.191	.01	.03	.02	.06	.01	1.12	—	.07
	.04	.27	—	361.0	—	—	.68	.03	.12	.06	.23	.03	4.00	—	.26
726	.02	.01	.01	129.2	—	—	0	0	0	0	0	0	0	—	0
	.12	.03	.06	861.0	—	—	0	0	0	0	0	0	0	—	0
727	Trace	Trace	—	82.2	—	—	0	0	.06	0	0	0	0	—	0
	.02	.01	—	913.0	—	—	0	0	.65	0	0	0	0	—	0
728	0	Trace	Trace	0	—	—	—	—	—	—	—	—	—	—	—
	0	.05	.06	0	—	—	—	—	—	—	—	—	—	—	—
729	.78	.45	.10	28.0	.19	.34	1.12	.06	.16	.22	.45	.07	6.72	3.36	Trace
	.70	.40	.09	25.0	.17	.30	1.00	.05	.14	.20	.40	.06	6.00	3.00	Trace
730	1.10	.69	.03	249.6	24.00	(.80)	0	.04	.14	.03	.77	.06	31.20	10.80	.74
	2.30	1.44	.05	520.0	50.00	(1.60)	0	.09	.30	.06	1.60	.12	65.00	22.50	1.55
731	.01	.01	Trace	0	0	0	0	Trace	.08	.03	.06	0	4.96	2.17	.02
	.03	.02	.01	0	0	0	0	.01	.27	.09	.20	0	16.00	7.00	.07
732	.94	.58	0	312.6	26.86	(.78)	0	.04	.08	.01	.75	.05	25.84	8.84	.64
	5.50	3.38	.01	1839.0	158.00	(4.60)	0	.22	.44	.07	4.40	.30	152.00	52.00	3.80
733	1.01	.63	.04	365.1	21.77	.08	0	.05	.26	.03	.70	.04	28.44	9.80	.67
	2.11	1.31	.08	760.7	45.36	.17	0	.10	.54	.06	1.45	.08	59.25	20.41	1.40
734	1.10	.69	—	249.6	22.08	—	0	.04	.13	.03	.83	.06	23.52	—	.63
	2.30	1.44	—	520.0	46.00	—	0	.09	.28	.06	1.73	.11	49.00	—	1.32
735	1.05	.68	—	301.3	—	.29	.12	.05	.17	.04	.79	.06	21.70	—	.62
	1.70	1.10	—	486.0	—	.46	.20	.08	.28	.07	1.28	.09	35.00	—	1.00
736	1.06	.69	—	248.6	25.92	(.80)	0	.04	.12	.03	.83	.05	23.52	(12.02)	.59
	2.20	1.43	—	518.0	54.00	(1.60)	0	.08	.25	.05	1.72	.10	49.00	(25.00)	1.23
737	.95	.72	—	315.9	—	.30	.13	.05	.18	.04	.83	.06	22.75	—	.65
	.46	1.10	—	486.0	—	.46	.20	.08	.28	.07	1.28	.09	35.00	—	1.00
738	.39	.27	—	97.5	—	(.28)	0	.02	.06	.01	.31	.02	9.20	—	.50
	7.88	5.43	—	1950.0	—	(5.60)	0	.33	1.20	.20	6.20	.40	184.00	—	10.00
739	.45	.29	—	102.5	—	—	0	.02	.06	.01	.34	.02	9.65	—	.53
	9.00	5.71	—	2050.0	—	—	0	.34	1.25	.20	6.71	.42	193.00	—	10.51
740	.10	.07	—	0	—	—	0	.02	.97	.32	.82	.01	40.05	—	.22
	.23	.15	—	0	—	—	0	.04	2.16	.70	1.83	.02	89.00	—	.49
741	.26	.17	—	0	—	—	0	.04	2.48	.75	2.10	.03	102.72	—	.57
	.24	.16	—	0	—	—	0	.04	2.32	.70	1.96	.02	96.00	—	.53
742	.43	.25	—	136.9	—	—	0	.02	.03	0	.33	.02	8.52	—	.28
	.80	6.15	—	3422.0	—	—	0	.41	.86	.10	8.24	.58	213.00	—	7.08
743	1.10	.86	.03	566.4	24.00	—	0	.05	.14	.05	.72	.06	2.45	10.80	.96
	2.30	1.80	.05	1180.0	50.00	—	0	.11	.30	.10	1.50	.11	5.10	22.50	2.00
744	.75	.45	—	272.0	—	—	0	.03	.07	.01	.60	.04	20.74	—	.51
	4.39	2.66	—	1600.0	—	—	0	.20	.39	.07	3.52	.24	122.00	—	3.00
745	.67	.40	—	242.1	—	—	0	.03	.07	.01	.53	.04	18.53	—	.45
	3.91	2.37	—	1424.0	—	—	0	.18	.38	.06	3.13	.22	109.00	—	2.67
746	2.85	1.04	—	982.7	—	—	0	.13	.22	.07	—	.19	59.20	—	3.99
	3.85	1.41	—	1328.0	—	—	0	.18	.30	.10	—	.25	80.00	—	5.40
747	—	—	—	—	—	—	0	—	—	—	—	—	—	—	—
	—	—	—	—	—	—	0	—	—	—	—	—	—	—	—
748	.33	—	—	27.0	—	—	0	.01	.07	.01	—	.00	—	—	—
	3.65	—	—	300.0	—	—	0	.13	.79	.15	—	.02	—	—	—
749	2.95	—	—	—	—	—	0	.08	.34	.02	—	—	—	—	—
	4.10	—	—	—	—	—	0	.11	.47	.02	—	—	—	—	—

(Continued)

TABLE F-36 *(Continued)*

Item No.	Food Item Name	Approximate Measure	Weight	Moisture	Food Energy Calories	Protein	Fats	Carbo-hydrates	Fiber	Minerals (Macro) Calcium	Phos-phorus	Sodium	Mag-nesium	Potas-sium
			(g)	(%)	(kcal)	(g)	(g)	(g)	(g)	(mg)	(mg)	(mg)	(mg)	(mg)
	EGG SUBSTITUTES													
750	egg beaters, Fleischmanns	1 CUP	240	73.0	400.8	26.4	30.0	7.2	—	192.0	168.0	432.0	—	511.2
			100	73.0	167.0	11.0	12.5	3.0	—	80.0	70.0	180.0	—	213.0
751	[1]frozen, made w/egg white, corn oil, nonfat dry milk	1 OZ	28	73.1	44.8	3.2	3.1	.9	0	20.4	20.2	15.6	—	59.6
			100	73.1	160.0	11.3	11.1	3.2	0	73.0	72.0	199.0	—	213.0
752	[1]liquid, made w/egg white, soy oil, & protein	1 OZ	28	82.8	23.5	3.4	.9	.2	0	14.8	33.9	49.6	—	92.4
			100	82.8	84.0	12.0	3.3	.6	0	53.0	121.0	177.0	—	330.0
753	[1]powder	1 OZ	28	3.9	124.3	15.4	3.6	6.1	0	91.3	133.8	224.0	—	208.3
			100	3.9	444.0	55.5	13.0	21.8	0	326.0	478.0	800.0	—	744.0
754	Second Nature, Avoset Foods	1 SERV	47	—	36.8	3.8	1.6	1.8	—	25.0	57.0	65.8	—	135.0
			100	—	78.2	8.0	3.4	3.8	—	53.2	121.3	139.9	—	287.2
755	**BEANS & FRANKFURTERS,** canned	1 CUP	255	70.7	367.2	19.4	18.1	32.1	2.6	94.4	303.5	1374.5	—	668.1
			100	70.7	144.0	7.6	7.1	12.6	1.0	37.0	119.0	539.0	—	262.0
756	**BEEF DRIED, CHIPPED,** creamed	1 CUP	240	72.0	369.6	19.7	24.7	17.0	—	252.0	336.0	1718.4	—	367.2
			100	72.0	154.0	8.2	10.3	7.1	—	105.0	140.0	716.0	—	153.0
757	**BEEF W/NOODLES,** cooked	1 CUP	240	—	275.3	35.3	9.4	12.5	—	67.2	213.6	—	28.8	—
			100	—	114.7	14.7	3.9	5.2	—	28.0	89.0	—	12.0	—
758	**BEEF STEW**	1 OZ	28	—	26.7	3.3	.7	1.5	.1	4.1	20.1	58.8	3.1	63.5
			100	—	95.4	11.8	2.6	5.5	.3	14.6	71.7	210.1	11.2	226.8
759	**BEEF & VEGETABLE STEW,** w/lean beef chuck, cooked	1 CUP	245	82.4	218.1	15.7	10.5	15.2	1.0	29.4	183.8	90.7	—	612.5
			100	82.4	89.0	6.4	4.3	6.2	.4	12.0	75.0	37.0	—	250.0
760	**BEEF & VEGETABLE STEW,** canned	1 CUP	235	82.5	185.7	13.6	7.3	16.7	.7	28.2	105.8	965.9	—	408.9
			100	82.5	79.0	5.8	3.1	7.1	.3	12.0	45.0	411.0	—	174.0
761	**BOILED DINNER**	1 OZ	28	7.4	32.1	2.6	1.7	1.8	.1	7.0	18.3	8.7	3.1	84.8
			100	7.4	114.8	9.3	6.0	6.6	.5	25.1	65.4	31.1	11.0	303.0
762	**CHEESEBURGER,** w/bun	INDV	105	48.6	258.5	15.5	10.7	25.0	.2	115.5	221.6	664.7	19.7	244.7
			100	48.6	246.2	14.8	10.2	23.8	.2	110.0	211.0	633.0	18.8	233.0
763	**CHEESE FONDUE,** homemade	1 TBLS	16	54.2	42.4	2.4	2.9	1.6	—	50.7	47.0	86.7	—	26.4
			100	54.2	265.0	14.8	18.3	10.0	—	317.0	294.0	542.0	—	165.0
764	**CHEESE SOUFFLE,** homemade	1 CUP	150	65.0	327.0	14.9	25.7	9.3	—	301.5	292.5	546.0	—	181.5
			100	65.0	218.0	9.9	17.1	6.2	—	201.0	195.0	364.0	—	121.0
765	**CHICKEN A LA KING,** homemade, cooked	1 CUP	245	68.2	468.0	27.4	34.3	12.3	—	127.4	357.7	759.5	—	404.3
			100	68.2	191.0	11.2	14.0	5.0	—	52.0	146.0	310.0	—	165.0
766	**CHICKEN FRICASSEE,** homemade, cooked	1 CUP	240	71.3	378.5	36.7	22.3	7.7	—	14.4	271.2	369.6	—	336.0
			100	71.3	157.7	15.3	9.3	3.2	—	6.0	113.0	154.0	—	140.0
767	**CHICKEN W/NOODLES,** homemade, cooked	1 CUP	240	71.1	367.2	22.3	18.5	25.7	—	26.4	247.2	600.0	—	148.8
			100	71.1	153.0	9.3	7.7	10.7	—	11.0	103.0	250.0	—	62.0
768	**CHICKEN W/RICE**	1 CUP	250	—	292.5	18.8	15.8	18.3	—	152.5	55.0	—	—	—
			100	—	117.0	7.5	6.3	7.3	—	61.0	22.0	—	—	—
	CHILI CON CARNE													
769	w/o beans, canned	1 CUP	255	66.9	510.0	26.3	37.7	14.8	.5	96.9	387.5	—	—	—
			100	66.9	200.0	10.3	14.8	5.8	.2	38.0	152.0	—	—	—
770	w/beans, canned	1 CUP	230	72.4	305.9	17.3	14.0	28.1	1.4	73.6	289.8	1221.3	—	535.9
			100	72.4	133.0	7.5	6.1	12.2	.6	32.0	126.0	531.0	—	233.0
771	**CHITTERLINGS**	1 SERV	100	—	335.0	8.6	25.7	—	—	—	—	—	—	—
			100	—	335.0	8.6	25.7	—	—	—	—	—	—	—
	CHOP SUEY													
772	w/meat, homemade, cooked	1 CUP	250	75.4	300.0	26.0	17.0	12.8	1.3	60.0	247.5	1052.5	—	425.0
			100	75.4	120.0	10.4	6.8	5.1	.5	24.0	99.0	421.0	—	170.0
773	w/meat, canned	1 CUP	250	85.5	155.0	11.0	8.0	10.5	2.0	87.5	290.0	1377.5	—	345.0
			100	85.5	62.0	4.4	3.2	4.2	.8	35.0	116.0	551.0	—	138.0
	CHOW MEIN													
774	**CHICKEN,** w/o noodles, homemade, cooked	1 CUP	250	78.0	255.0	31.0	10.0	10.0	.8	57.5	292.5	717.5	—	472.5
			100	78.0	102.0	12.4	4.0	4.0	.3	23.0	117.0	287.0	—	189.0
775	**CHICKEN,** w/o noodles, canned	1 CUP	360	88.8	136.8	9.4	.4	25.6	1.1	64.8	122.4	1044.0	—	601.2
			100	88.8	38.0	2.6	.1	7.1	.3	18.0	34.0	290.0	—	167.0
776	**CORNED BEEF HASH,** canned	1 CUP	230	—	457.7	20.4	33.6	18.6	—	59.8	158.7	1994.1	—	—
			100	—	199.0	8.9	14.6	8.1	—	26.0	69.0	867.0	—	—
777	**CORNED BEEF HASH** w/potato, canned	1 CUP	220	67.4	398.2	19.4	24.9	23.5	1.1	28.6	147.4	1188.0	—	440.0
			100	67.4	181.0	8.8	11.3	10.7	.5	13.0	67.0	540.0	—	200.0

Item No.	Minerals (Micro)			Fat-Soluble Vitamins			Water-Soluble Vitamins								
	Iron	Zinc	Copper	Vitamin A	Vitamin D	Vitamin E (Alpha Tocopherol)	Vitamin C	Thiamin	Ribo-flavin	Niacin	Panto-thenic Acid	Vit. B-6 (Pyri-doxine)	Folacin (Folic Acid)	Biotin	Vitamin B-12
	(mg)	(mg)	(mg)	(IU)	(IU)	(mg)	(mg)	(mg)	(mg)	(mg)	(mg)	(mg)	(mcg)	(mcg)	(mcg)
750	4.32	.77	.46	3240.0	103.20	—	—	.31	1.03	—	—	—	—	—	—
	1.80	.32	.19	1350.0	43.00	—	—	.13	.43	—	—	—	—	—	—
751	.55	.27	—	378.0	—	—	—	.03	.11	—	.46	.04	—	—	—
	1.98	.98	—	1350.0	—	—	—	.12	.39	—	1.66	.13	—	—	—
752	.59	.36	—	604.8	—	—	0	.03	.08	.03	.76	—	—	—	.08
	2.10	1.30	—	2160.0	—	—	0	.11	.30	.11	2.70	—	—	—	.30
753	.89	—	—	343.8	—	—	.21	.06	.49	.16	—	—	—	—	—
	3.16	—	—	1228.0	—	—	.75	.23	1.76	.58	—	—	—	—	—
754	1.00	.61	.09	1015.0	24.00	—	—	.05	.14	.05	1.27	—	—	—	.14
	2.12	1.30	.19	2159.6	51.06	—	—	.11	.30	.11	2.70	—	—	—	.30
755	4.86	—	—	331.5	—	—	—	.18	.15	3.32	—	—	—	—	—
	1.90	—	—	130.0	—	—	—	.07	.06	1.30	—	—	—	—	—
756	1.92	—	—	864.0	4.80	—	—	.14	.46	1.44	—	—	—	—	—
	.80	—	—	360.0	2.00	—	—	.06	.19	.60	—	—	—	—	—
757	3.84	—	—	0	—	—	0	0	.22	5.28	—	.41	—	—	2.18
	1.60	—	—	0	—	—	0	0	.09	2.20	—	.17	—	—	.91
758	.47	.04	—	393.8	0	—	1.20	.02	.03	1.34	.01	Trace	.23	—	0
	1.67	.13	—	1406.6	0	—	4.29	.07	.10	4.79	.04	.01	.81	—	0
759	2.94	—	—	2401.0	—	—	17.15	.15	.17	4.66	—	—	—	—	—
	1.20	—	—	980.0	—	—	7.00	.06	.07	1.90	—	—	—	—	—
760	2.12	—	—	2279.5	0	—	7.05	.07	.12	2.35	—	—	—	—	1.53
	.90	—	—	970.0	0	—	3.00	.03	.05	1.00	—	—	—	—	.65
761	.39	.02	—	372.0	0	—	4.87	.02	.03	.49	.03	.02	7.94	—	0
	1.38	.06	—	1328.6	0	—	17.38	.06	.09	1.74	.11	.07	28.36	—	0
762	1.79	2.42	.26	—	—	—	—	—	—	—	—	—	—	—	—
	1.70	2.30	.25	—	—	—	—	—	—	—	—	—	—	—	—
763	.19	—	—	140.8	—	—	—	.01	.05	.03	—	—	—	—	—
	1.20	—	—	880.0	—	—	—	.06	.34	.20	—	—	—	—	—
764	1.50	—	—	1200.0	—	—	—	.08	.36	.30	—	—	—	—	—
	1.00	—	—	800.0	—	—	—	.05	.24	.20	—	—	—	—	—
765	2.45	—	—	1127.0	—	—	12.25	.10	.42	5.39	—	—	—	—	—
	1.00	—	—	460.0	—	—	5.00	.04	.17	2.20	—	—	—	—	—
766	.16	—	—	168.0	—	—	0	.05	.17	5.76	—	—	—	—	—
	.90	—	—	70.0	—	—	0	.02	.07	2.40	—	—	—	—	—
767	2.16	—	—	432.0	—	—	—	.05	.17	4.32	—	—	—	—	—
	.90	—	—	180.0	—	—	—	,02	.07	1.80	—	—	—	—	—
768	1.75	—	—	42.5	—	—	0	.08	.15	4.25	—	.13	—	—	.05
	.70	—	—	17.0	—	—	0	.03	.06	1.70	—	.05	—	—	.02
769	3.57	—	—	382.5	—	—	—	.05	.31	5.61	—	—	—	—	—
	1.40	—	—	150.0	—	—	—	.02	.12	2.20	—	—	—	—	—
770	3.91	3.75	—	138.0	—	—	—	.07	.16	2.99	.32	.24	—	—	—
	1.70	1.63	—	60.0	—	—	—	.03	.07	1.30	.14	.10	—	—	—
771	—	—	—	—	—	—	—	—	—	—	—	—	—	—	—
	—	—	—	—	—	—	—	—	—	—	—	—	—	—	—
772	4.75	—	—	600.0	—	—	32.50	.28	.38	5.00	—	—	—	—	—
	1.90	—	—	240.0	—	—	13.00	.11	.15	2.00	—	—	—	—	—
773	4.75	—	—	75.0	—	—	5.00	.13	.13	1.75	—	—	—	—	—
	1.90	—	—	30.0	—	—	2.00	.05	.05	.70	—	—	—	—	—
774	2.50	—	—	275.0	—	—	10.00	.08	.23	4.25	—	—	—	—	—
	1.00	—	—	110.0	—	—	4.00	.03	.09	1.70	—	—	—	—	—
775	1.80	—	—	216.0	—	—	18.00	.07	.14	1.44	—	—	—	—	—
	.50	—	—	60.0	—	—	5.00	.02	.04	.40	—	—	—	—	—
776	2.76	—	—	0.0	0	—	0	.07	.32	6.58	—	—	—	—	—
	1.20	—	—	0.0	0	—	0	.03	.14	2.86	—	—	—	—	—
777	4.40	—	—	—	—	—	—	.02	.20	4.62	—	.17	—	—	—
	2.00	—	—	—	—	—	—	.01	.09	2.10	—	.08	—	—	—

(Continued)

TABLE F-36 *(Continued)*

Item No.	Food Item Name	Approximate Measure	Weight	Moisture	Food Energy Calories	Protein	Fats	Carbo-hydrates	Fiber	Minerals (Macro) Calcium	Phos-phorus	Sodium	Mag-nesium	Potas-sium
			(g)	(%)	(kcal)	(g)	(g)	(g)	(g)	(mg)	(mg)	(mg)	(mg)	(mg)
	FROZEN DINNERS													
778	**BEEF POT ROAST,** w/whole potato, peas & corn,	**1 WHOLE**	**205**	**76.3**	**217.3**	**26.9**	**6.6**	**12.5**	**.6**	**20.5**	**155.8**	**531.0**	—	**500.2**
	unheated		100	76.3	106.0	13.1	3.2	6.1	.3	10.0	76.0	259.0	—	244.0
779	**FRIED CHICKEN,** w/mashed potato & mixed	**1 WHOLE**	**312**	**66.1**	**539.8**	**39.9**	**26.5**	**35.3**	**1.2**	**127.9**	**452.4**	**1073.3**	—	**349.4**
	vegetable, unheated		100	66.1	173.0	12.8	8.5	11.3	.4	41.0	145.0	344.0	—	112.0
780	**MEAT LOAF,** w/tomato sauce, mashed potato &	**1 WHOLE**	**312**	**73.7**	**408.7**	**25.0**	**20.9**	**30.6**	**.9**	**59.3**	**365.0**	**1226.2**	—	**358.8**
	peas, unheated		100	73.7	131.0	8.0	6.7	9.8	.3	19.0	117.0	393.0	—	115.0
781	**TURKEY,** w/mashed potato & peas, unheated	**1 WHOLE**	**312**	**74.7**	**349.4**	**26.2**	**9.4**	**39.6**	**.9**	**81.1**	**271.4**	**1248.0**	—	**549.1**
			100	74.7	112.0	8.4	3.0	12.7	.3	26.0	87.0	400.0	—	176.0
782	[1]**GREEN PEPPER STUFFED W/BEEF** & crumbs	**1 OZ**	**28**	**63.1**	**47.6**	**3.6**	**1.5**	**4.7**	**.2**	**11.8**	**33.9**	**87.9**	—	**72.2**
			100	63.1	170.0	13.0	5.5	16.8	.7	42.0	121.0	314.0	—	258.0
783	**HAM CROQUETTE,** panfried	**1 PIECE**	**65**	**54.0**	**163.2**	**10.6**	**9.8**	**7.6**	**.1**	**44.9**	**104.0**	**222.3**	—	**54.0**
			100	54.0	251.0	16.3	15.1	11.7	.1	69.0	160.0	342.0	—	83.0
784	**MACARONI & CHEESE,** w/margarine, enriched,	**1 CUP**	**200**	**58.2**	**430.0**	**16.8**	**22.2**	**40.2**	**.2**	**362.0**	**322.0**	**1086.0**	—	**240.0**
	baked		100	58.2	215.0	8.4	11.1	20.1	.1	181.0	161.0	543.0	—	120.0
785	canned	**1 CUP**	**240**	**80.2**	**228.0**	**9.4**	**9.6**	**25.7**	**.2**	**199.2**	**182.4**	**729.6**	—	**139.2**
			100	80.2	95.0	3.9	4.0	10.7	.1	83.0	76.0	304.0	—	58.0
786	**MEATBALLS,** cooked	**1 OZ**	**28**	—	**78.4**	**5.0**	**5.5**	**1.9**	—	**11.5**	**42.0**	—	**5.6**	—
			100	—	280.0	17.9	19.5	6.9	—	41.0	150.0	—	20.0	—
787	**MEATLOAF,** cooked	**1 OZ**	**28**	—	**78.4**	**5.0**	**5.5**	**1.9**	—	**11.5**	**42.0**	—	**5.6**	—
			100	—	280.0	17.9	19.5	6.9	—	41.0	150.0	—	20.0	—
788	[4]**MOUSSAKA**	**4 OZ**	**112**	—	**218.5**	**10.4**	**15.0**	**11.0**	—	**99.5**	**147.1**	**359.5**	**23.9**	**398.0**
			100	—	195.1	9.3	13.4	9.8	—	88.8	131.3	321.0	21.3	355.4
789	**PEPPERS, SWEET, GREEN, STUFFED W/BEEF** &	**1 AVG**	**185**	**63.1**	**314.5**	**24.1**	**10.2**	**31.1**	**1.3**	**77.7**	**223.9**	**580.9**	—	**477.3**
	crumbs, cooked		100	63.1	170.0	13.0	5.5	16.8	.7	42.0	121.0	314.0	—	258.0
	PIZZA													
790	**BEEF,** ground, frozen, baked	**1 PIECE**	**100**	**47.0**	**248.0**	**6.8**	**6.4**	**41.0**	—	—	—	—	—	—
			100	47.0	248.0	6.8	6.4	41.0	—	—	—	—	—	—
791	**BURGER**	**1 OZ**	**28**	—	**66.0**	**4.8**	**3.6**	**3.3**	**Trace**	**45.3**	**57.9**	**107.1**	—	**73.5**
			100	—	235.8	17.2	12.9	11.6	.1	161.9	206.7	382.5	—	262.6
792	**CHEESE,** frozen, baked	**1 PIECE**	**57**	**47.0**	**134.5**	**3.2**	**3.2**	**23.1**	**.2**	**88.9**	**88.9**	**368.8**	—	**65.0**
			100	47.0	236.0	5.6	5.7	40.6	.3	156.0	156.0	647.0	—	114.0
793	**CHEESE,** homemade, w/cheese topping, enriched,	**1 PIECE**	**65**	**50.0**	**145.0**	**2.7**	**4.4**	**23.7**	**.2**	**143.7**	**126.8**	**456.3**	**(17.6)**	**84.5**
	baked		100	50.0	223.0	4.1	6.7	36.5	.3	221.0	195.0	702.0	(27.0)	130.0
794	**CHEESE,** homemade, w/sausage, enriched, baked	**1 PIECE**	**67**	**50.6**	**156.8**	**5.2**	**6.2**	**19.8**	**.2**	**11.4**	**61.6**	**488.4**	—	**112.6**
			100	50.6	234.0	7.8	9.3	29.6	.3	17.0	92.0	729.0	—	168.0
795	**CHEESE,** mix, baked	**1 PIECE**	**100**	**47.0**	**236.0**	**4.8**	**7.3**	**37.9**	—	—	—	—	—	—
			100	47.0	236.0	4.8	7.3	37.9	—	—	—	—	—	—
796	**CHEESE & PEPPERONI,** baked	**1 SLICE**	**100**	**50.6**	**234.0**	**7.8**	**9.3**	**29.6**	**.3**	**17.0**	**92.0**	**729.0**	—	**168.0**
			100	50.6	234.0	7.8	9.3	29.6	.3	17.0	92.0	729.0	—	168.0
797	**PEPPERONI,** frozen, baked	**1 PIECE**	**100**	**44.0**	**278.0**	**6.2**	**10.0**	**40.7**	—	—	—	—	—	—
			100	44.0	278.0	6.2	10.0	40.7	—	—	—	—	—	—
798	**SAUSAGE,** frozen, baked	**1 PIECE**	**100**	**47.0**	**250.0**	**6.5**	**7.4**	**39.2**	—	—	—	—	—	—
			100	47.0	250.0	6.5	7.4	39.2	—	—	—	—	—	—
	POTPIES													
799	**BEEF,** homemade, baked	**1 INDV**	**227**	**55.1**	**558.4**	**22.9**	**32.9**	**42.7**	**.9**	**31.8**	**161.2**	**644.7**	—	**360.9**
			100	55.1	246.0	10.1	14.5	18.8	.4	14.0	71.0	284.0	—	159.0
800	**BEEF,** commercial, frozen, unheated	**1 INDV**	**227**	**63.3**	**431.3**	**16.6**	**22.5**	**40.9**	**.2**	**22.7**	**109.0**	**830.8**	—	**211.1**
			100	63.3	190.0	7.3	9.9	18.0	.1	10.0	48.0	366.0	—	93.0
801	**CHICKEN,** homemade, baked	**1 HALF**	**116**	**56.6**	**272.6**	**11.7**	**15.7**	**21.2**	**.5**	**34.8**	**116.0**	**297.0**	—	**171.7**
			100	56.6	235.0	10.1	13.5	18.3	.4	30.0	100.0	256.0	—	148.0
802	**CHICKEN,** commercial, frozen, unheated	**1 WHOLE**	**227**	**57.8**	**497.1**	**15.2**	**26.1**	**50.4**	**.9**	**25.0**	**113.5**	**933.0**	—	**347.3**
			100	57.8	219.0	6.7	11.5	22.2	.4	11.0	50.0	411.0	—	153.0
803	**TURKEY,** homemade, baked	**1 INDV**	**227**	**56.2**	**538.0**	**23.6**	**30.6**	**42.0**	**.9**	**61.3**	**229.3**	**619.7**	—	**449.5**
			100	56.2	237.0	10.4	13.5	18.5	.4	27.0	101.0	273.0	—	198.0
804	**TURKEY,** commercial, frozen, unheated	**1 WHOLE**	**227**	**62.3**	**447.2**	**13.2**	**23.6**	**45.6**	**.7**	**27.2**	**127.1**	**837.6**	—	**258.8**
			100	62.3	197.0	5.8	10.4	20.1	.3	12.0	56.0	369.0	—	114.0
805	**QUICHE LORRAINE**	**1 SERV**	**162**	—	**494.7**	**14.6**	**40.5**	**18.0**	—	**256.0**	**309.4**	—	**14.6**	—
			100	—	305.4	9.0	25.0	11.1	—	158.0	191.0	—	9.0	—
	SANDWICHES													
806	**CREAM CHEESE & OLIVE**	**1 OZ**	**28**	—	**172.0**	**5.7**	**8.9**	**18.7**	**.6**	**56.6**	**110.8**	**136.9**	**75.5**	**131.9**
			100	—	614.1	20.2	31.8	66.7	2.2	202.1	395.8	488.8	269.8	471.0
807	**PEANUT BUTTER & JELLY**	**1 INDV**	**75**	**28.0**	**280.6**	**9.2**	**11.3**	**37.2**	**.6**	**54.3**	**126.8**	**356.3**	**46.0**	**193.1**
			100	28.0	374.2	12.3	15.1	49.6	.6	72.3	169.0	475.0	61.3	257.4

Item No.	Minerals (Micro) Iron	Zinc	Copper	Fat-Soluble Vitamins Vitamin A	Vitamin D	Vitamin E (Alpha Tocopherol)	Water-Soluble Vitamins Vitamin C	Thiamin	Ribo-flavin	Niacin	Panto-thenic Acid	Vit. B-6 (Pyri-doxine)	Folacin (Folic Acid)	Biotin	Vitamin B-12
	(mg)	(mg)	(mg)	(IU)	(IU)	(mg)	(mg)	(mg)	(mg)	(mg)	(mg)	(mg)	(mcg)	(mcg)	(mcg)
778	**3.28**	–	–	**225.5**	–	–	**10.25**	**.12**	**.21**	**4.31**	–	–	**49.20**	–	–
	1.60	–	–	110.0	–	–	5.00	.06	.10	2.10	–	–	24.00	–	–
779	**3.74**	–	–	**1840.8**	–	–	**12.48**	**.22**	**.56**	**16.22**	–	–	–	–	–
	1.20	–	–	590.0	–	–	4.00	.07	.18	5.20	–	–	–	–	–
780	**4.06**	–	–	**1341.6**	–	–	**12.48**	**.31**	**.44**	**5.30**	–	–	–	–	–
	1.30	–	–	430.0	–	–	4.00	.10	.14	1.70	–	–	–	–	–
781	**3.43**	–	–	**405.6**	–	–	**12.48**	**.22**	**.28**	**7.18**	–	–	–	–	–
	1.10	–	–	130.0	–	–	4.00	.07	.09	2.30	–	–	–	–	–
782	**.59**	–	–	**78.4**	–	–	**11.20**	**.03**	**.05**	**.70**	–	–	–	–	–
	2.10	–	–	280.0	–	–	40.00	.09	.17	2.50	–	–	–	–	–
783	**1.37**	–	–	**169.0**	–	–	–	**.18**	**.14**	**1.63**	–	–	–	–	–
	2.10	–	–	260.0	–	–	–	.28	.22	2.50	–	–	–	–	–
784	**1.80**	–	–	**860.0**	–	–	–	**.20**	**.40**	**1.80**	–	–	–	–	–
	.90	–	–	430.0	–	–	–	.10	.20	.90	–	–	–	–	–
785	**.96**	–	–	**264.0**	–	–	–	**.12**	**.24**	**.96**	–	–	–	–	–
	.40	–	–	110.0	–	–	–	.05	.10	.40	–	–	–	–	–
786	**.62**	–	–	**14.0**	–	–	**0**	**.02**	**.04**	**.76**	–	**.12**	–	–	**.39**
	2.20	–	–	50.0	–	–	0	.06	.15	2.70	–	.41	–	–	1.38
787	**.62**	**(.80)**	–	**14.0**	–	–	**.28**	**.02**	**.04**	**.76**	–	**.12**	–	–	**.39**
	2.20	(2.86)	–	50.0	–	–	1.00	.06	.15	2.70	–	.41	–	–	1.38
788	**1.52**	**2.07**	**.16**	**40.3**	**.07**	**.37**	**4.60**	**.07**	**.17**	**1.68**	**.60**	**.20**	**9.02**	**2.39**	**1.46**
	1.36	1.85	.14	36.0	.06	.33	4.11	.06	.15	1.50	.54	.18	8.05	2.13	1.30
789	**3.89**	–	–	**518.0**	–	–	**74.00**	**.17**	**.32**	**4.63**	–	–	–	–	–
	2.10	–	–	280.0	–	–	40.00	.09	.17	2.50	–	–	–	–	–
790	–	–	–	–	–	–	–	–	–	–	–	–	–	–	–
	–	–	–	–	–	–	–	–	–	–	–	–	–	–	–
791	**.57**	–	–	**101.6**	–	–	**1.18**	**.03**	**.06**	**1.83**	–	–	–	–	–
	2.02	–	–	362.9	–	–	4.21	.11	.23	6.52	–	–	–	–	–
792	**.51**	–	–	**250.8**	–	–	**3.42**	**.03**	**.10**	**.57**	–	–	–	–	–
	.90	–	–	440.0	–	–	6.00	.06	.17	1.00	–	–	–	–	–
793	**.52**	**(.72)**	**(.22)**	**409.5**	–	**(.20)**	**5.20**	**.08**	**.05**	**.59**	**(.20)**	**(.03)**	–	–	–
	.80	(1.10)	(.34)	630.0	–	(.30)	8.00	.12	.08	.90	(.30)	(.05)	–	–	–
794	**.54**	–	–	**375.2**	–	–	**6.03**	**.08**	**.05**	**.60**	–	–	–	–	–
	.80	–	–	560.0	–	–	9.00	.12	.08	.90	–	–	–	–	–
795	–	–	–	–	–	–	–	–	–	–	–	–	–	–	–
	–	–	–	–	–	–	–	–	–	–	–	–	–	–	–
796	**1.20**	–	–	**560.0**	–	–	**9.00**	**.09**	**.12**	**1.50**	–	–	–	–	–
	1.20	–	–	560.0	–	–	9.00	.09	.12	1.50	–	–	–	–	–
797	–	–	–	–	–	–	–	–	–	–	–	–	–	–	–
	–	–	–	–	–	–	–	–	–	–	–	–	–	–	–
798	–	–	–	–	–	–	–	–	–	–	–	–	–	–	–
	–	–	–	–	–	–	–	–	–	–	–	–	–	–	–
799	**4.09**	–	–	**1861.4**	–	–	**6.81**	**.25**	**.27**	**4.54**	–	–	–	–	–
	1.80	–	–	820.0	–	–	3.00	.11	.12	2.00	–	–	–	–	–
800	**2.27**	–	–	**930.7**	–	–	–	**.07**	**.14**	**2.72**	–	–	–	–	–
	1.00	–	–	410.0	–	–	–	.03	.06	1.20	–	–	–	–	–
801	**1.51**	–	–	**1542.8**	–	–	**2.32**	**.13**	**.13**	**2.09**	–	–	–	–	–
	1.30	–	–	1330.0	–	–	2.00	.11	.11	1.80	–	–	–	–	–
802	**2.27**	–	–	**2065.7**	–	–	**9.08**	**.23**	**.32**	**3.16**	–	–	–	–	–
	1.00	–	–	910.0	–	–	4.00	.10	.14	1.40	–	–	–	–	–
803	**3.18**	–	–	**3019.1**	–	–	**4.54**	**.25**	**.30**	**5.68**	–	–	–	–	–
	1.40	–	–	1330.0	–	–	2.00	.11	.13	2.50	–	–	–	–	–
804	**2.04**	–	–	**2020.3**	–	–	**4.54**	**.20**	**.18**	**3.63**	–	–	–	–	–
	.90	–	–	890.0	–	–	2.00	.09	.08	1.60	–	–	–	–	–
805	**1.78**	–	–	**1365.7**	–	–	**0**	**.13**	**.34**	**.81**	–	**.28**	–	–	**1.33**
	1.10	–	–	843.0	–	–	0	.08	.21	.50	–	.17	–	–	.82
806	**1.16**	**.51**	–	**159.7**	**0**	–	**0**	**.10**	**.09**	**1.08**	**.35**	**.08**	**26.53**	–	**.10**
	4.16	1.83	–	570.4	0	–	0	.37	.32	3.87	1.25	.28	94.76	–	.34
807	**1.77**	**.84**	–	**1.5**	**0**	–	**.60**	**.14**	**.13**	**4.26**	**.69**	**.08**	**26.66**	–	**0**
	2.36	1.12	–	2.0	0	–	.80	.19	.17	5.68	.93	.11	35.54	–	0

(Continued)

TABLE F-36 *(Continued)*

Item No.	Food Item Name	Approximate Measure	Weight	Moisture	Food Energy Calories	Protein	Fats	Carbo-hydrates	Fiber	Minerals (Macro) Calcium	Phos-phorus	Sodium	Mag-nesium	Potas-sium
		.	(g)	(%)	(kcal)	(g)	(g)	(g)	(g)	(mg)	(mg)	(mg)	(mg)	(mg)
	SANDWICHES (Continued)													
808	**TUNA,** grilled	**1 OZ**	**28**	—	**77.5**	**7.0**	**8.5**	**10.6**	**.1**	**25.2**	**71.5**	**239.9**	**11.0**	**59.8**
			100	—	276.6	25.1	30.5	38.0	.4	89.9	255.3	856.8	39.2	213.4
	SPAGHETTI													
809	w/meat sauce	**1 SERV**	**292**	—	**397.1**	**12.7**	**20.7**	**39.4**	**.6**	**26.3**	**148.9**	—	—	—
			100	—	136.0	4.3	7.1	13.5	.2	9.0	51.0	—	—	—
810	w/meatballs in tomato sauce	**1 CUP**	**250**	**78.1**	**275.0**	**10.0**	**12.5**	**27.5**	**.5**	**47.5**	**140.0**	**1150.0**	—	**440.0**
			100	78.1	110.0	4.0	5.0	11.0	.2	19.0	56.0	460.0	—	176.0
811	w/tomato sauce & cheese, homemade, cooked	**1 CUP**	**250**	**77.0**	**260.0**	**8.8**	**8.8**	**37.0**	**.5**	**80.0**	**135.0**	**955.0**	—	**407.5**
			100	77.0	104.0	3.5	3.5	14.8	.2	32.0	54.0	382.0	—	163.0
812	w/tomato sauce & cheese, canned	**1 CUP**	**250**	**80.1**	**190.0**	**5.5**	**1.5**	**38.5**	**.5**	**40.0**	**87.5**	**955.0**	**27.5**	**302.5**
			100	80.1	76.0	2.2	.6	15.4	.2	16.0	35.0	382.0	11.0	121.0
813	**SPANISH RICE**	**1 OZ**	**28**	—	**29.1**	**1.0**	**.8**	**4.6**	**.1**	**3.2**	**11.5**	**31.1**	**3.0**	**28.4**
			100	—	103.8	3.7	2.7	16.5	.2	11.4	41.0	111.1	10.6	101.3
814	homemade, cooked	**1 CUP**	**150**	**78.5**	**130.5**	**2.7**	**2.6**	**24.9**	**.8**	**21.0**	**59.5**	**474.0**	—	**346.5**
			100	78.5	87.0	1.8	1.7	16.6	.5	14.0	39.0	316.0	—	231.0
815	**SWEDISH MEAT BALLS,** frozen	**1 PIECE**	**28**	—	**58.0**	**2.5**	**4.5**	**1.8**	—	**20.1**	**35.0**	**235.0**	—	**3.0**
			100	—	207.1	9.0	16.0	6.5	—	71.8	125.0	839.3	—	10.7
816	**TORTELLINI,** w/tomato sauce	**1 CUP**	**200**	—	**311.6**	**17.0**	**14.8**	**27.6**	—	**4.6**	**164.0**	—	**34.0**	—
			100	—	155.8	8.5	7.4	13.8	—	2.3	82.0	—	17.0	—
817	**TUNA NOODLE CASSEROLE**	**1 CUP**	**200**	—	**280.0**	**17.8**	**11.8**	**25.0**	—	—	—	—	—	—
			100	—	140.0	8.9	5.9	12.5	—	—	—	—	—	—
818	**TURKEY & HAM LOAF**	**1 OZ**	**28**	—	**69.2**	**3.7**	**4.1**	**4.0**	**Trace**	**13.8**	**41.4**	**91.4**	—	**73.6**
			100	—	247.1	13.3	14.7	14.3	Trace	49.2	147.9	326.4	—	262.9
	VEGETARIAN MAIN DISHES													
819	w/peanuts & soybean, canned	**1 OZ**	**28**	**55.3**	**66.4**	**3.3**	**4.7**	**3.8**	**.3**	—	—	—	—	—
			100	55.3	237.0	11.7	16.9	13.4	.9	—	—	—	—	—
820	w/wheat & soy protein, canned	**1 OZ**	**28**	**73.4**	**29.1**	**4.5**	**.3**	**2.1**	**.1**	—	—	—	—	—
			100	73.4	104.0	16.1	1.2	7.6	.3	—	—	—	—	—
821	w/wheat & soy protein & soybean or vegetable oil, canned	**1 OZ**	**28**	**66.6**	**42.0**	**4.5**	**1.6**	**2.7**	**.2**	—	—	—	—	—
			100	66.6	150.0	16.1	5.6	9.5	.6	—	—	—	—	—
822	w/wheat protein, canned	**1 OZ**	**28**	**72.9**	**30.5**	**4.6**	**.2**	**2.5**	**Trace**	—	—	—	—	—
			100	72.9	109.0	16.3	.8	8.8	.1	—	—	—	—	—
823	w/wheat protein & vegetable oil, canned	**1 OZ**	**28**	**63.5**	**52.9**	**5.3**	**2.9**	**1.5**	—	—	—	—	—	—
			100	63.5	189.0	19.1	10.4	5.2	—	—	—	—	—	—
824	w/wheat protein & nuts, canned	**1 OZ**	**28**	**52.4**	**59.4**	**5.7**	**2.0**	**5.0**	**.1**	—	—	—	—	—
			100	52.4	212.0	20.3	7.1	17.7	.4	—	—	—	—	—
825	w/wheat protein & peanuts, canned	**1 OZ**	**28**	**52.4**	**59.4**	**5.7**	**2.0**	**5.0**	**.1**	—	—	—	—	—
			100	52.4	212.0	20.3	7.1	17.7	.4	—	—	—	—	—
826	**BACON**	**1 TBLS**	**14**	—	**126.0**	**0**	**14.0**	—	—	—	—	—	—	—
			100	—	900.0	0	100.0	—	—	—	—	—	—	—
827	**BEEF TALLOW**	**1 TBLS**	**13**	**0.0**	**117.3**	**0**	**13.0**	**0**	**0**	—	—	**0**	—	**0**
			100	0.0	902.0	0	100.0	0	0	—	—	0	—	0
828	**CHICKEN**	**1 TBLS**	**14**	**0.2**	**126.0**	**0**	**14.0**	**0**	—	—	—	—	—	—
			100	0.2	900.0	0	100.0	0	—	—	—	—	—	—
829	**DUCK**	**1 TBLS**	**13**	**0.2**	**117.1**	**0**	**13.0**	**0**	**0**	—	—	—	—	—
			100	0.2	900.4	0	99.8	0	0	—	—	—	—	—
830	**GOOSE**	**1 TBLS**	**13**	**0.2**	**117.1**	**0**	**13.0**	**0**	**0**	—	—	—	—	—
			100	0.2	900.4	0	99.8	0	0	—	—	—	—	—
831	**MUTTON TALLOW**	**1 TBLS**	**13**	**0.0**	**117.3**	**0**	**13.0**	**0**	**0**	—	—	—	—	—
			100	0.0	902.0	0	100.0	0	0	—	—	—	—	—
832	**TURKEY**	**1 TBLS**	**13**	**0.2**	**117.1**	**0**	**13.0**	**0**	**0**	—	—	—	—	—
			100	0.2	900.4	0	99.8	0	0	—	—	—	—	—
833	**LARD, PORK**	**1 TBLS**	**13**	**0.0**	**117.3**	**0**	**13.0**	**0**	**0**	**0**	**0**	**0**	**0**	**0**
			100	0.0	902.0	0	100.0	0	0	0.1	0	0	0	0
	PORK BACKFAT													
834	fresh, raw, fat class (100% fat)	**1 OZ**	**28**	**6.4**	**235.5**	**.5**	**25.9**	**0**	**0**	**.3**	**0**	**19.6**	—	**79.8**
			100	6.4	841.0	1.7	92.4	0	0	1.0	0	70.0	—	285.0
835	fresh, raw, med-fat class (100% fat)	**1 OZ**	**28**	**7.5**	**231.6**	**.6**	**25.4**	**0**	**0**	**.3**	**0**	**19.6**	—	**79.8**
			100	7.5	827.0	2.1	90.7	0	0	1.0	0	70.0	—	285.0

Item No.	Minerals (Micro) Iron	Zinc	Copper	Fat-Soluble Vitamins Vitamin A	Vitamin D	Vitamin E (Alpha Tocopherol)	Water-Soluble Vitamins Vitamin C	Thiamin	Ribo-flavin	Niacin	Panto-thenic Acid	Vit. B-6 (Pyri-doxine)	Folacin (Folic Acid)	Biotin	Vitamin B-12
	(mg)	(mg)	(mg)	(IU)	(IU)	(mg)	(mg)	(mg)	(mg)	(mg)	(mg)	(mg)	(mcg)	(mcg)	(mcg)
808	**.95**	**.23**	—	**65.6**	**3.74**	—	**.88**	**.06**	**.08**	**2.23**	**.30**	**.02**	**10.65**	—	**.44**
	3.38	.81	—	234.1	13.35	—	3.16	.21	.29	7.96	1.06	.06	38.04	—	1.57
809	**2.04**	—	—	**902.3**	—	—	**23.94**	**.12**	**.12**	**2.98**	—	—	—	—	—
	.70	—	—	309.0	—	—	8.20	.04	.04	1.02	—	—	—	—	—
810	**3.50**	—	—	**655.0**	—	—	**2.75**	**.23**	**.20**	**4.50**	—	—	—	—	—
	1.40	—	—	262.0	—	—	1.10	.09	.08	1.80	—	—	—	—	—
811	**2.25**	—	—	**1075.0**	—	—	**12.50**	**.25**	**.18**	**2.25**	—	—	—	—	—
	.90	—	—	430.0	—	—	5.00	.10	.07	.90	—	—	—	—	—
812	**2.75**	—	—	**925.0**	—	—	**10.00**	**.35**	**.28**	**4.50**	—	—	—	—	—
	1.10	—	—	370.0	—	—	4.00	.14	.11	1.80	—	—	—	—	—
813	**.30**	**.01**	—	**82.9**	**0**	—	**3.33**	**.03**	**.02**	**.35**	**.04**	**.02**	**.92**	—	**Trace**
	.06	.05	—	296.0	0	—	11.89	.12	.08	1.26	.13	.07	3.29	—	Trace
814	**.90**	—	—	**990.0**	—	—	**22.50**	**.06**	**.05**	**1.05**	—	—	—	—	—
	.60	—	—	660.0	—	—	15.00	.04	.03	.70	—	—	—	—	—
815	**.24**	—	—	**80.0**	—	—	**0**	**.02**	**.05**	**.33**	—	—	—	—	—
	.86	—	—	285.7	—	—	0	.07	.18	1.18	—	—	—	—	—
816	**2.60**	—	—	**744.0**	—	—	**8.00**	**.26**	**.20**	**3.00**	—	**.18**	—	—	**.28**
	1.30	—	—	372.0	—	—	4.00	.13	.10	1.50	—	.09	—	—	.14
817	—	—	—	—	—	—	—	—	—	—	—	—	—	—	—
	—	—	—	—	—	—	—	—	—	—	—	—	—	—	—
818	**.58**	—	—	**43.0**	—	—	**.30**	**.08**	**.05**	**1.13**	—	—	—	—	—
	2.07	—	—	153.5	—	—	1.05	.27	.18	4.04	—	—	—	—	—
819	—	—	—	—	—	—	—	—	—	—	**.05**	**.01**	—	—	**0**
	—	—	—	—	—	—	—	—	—	—	.19	.04	—	—	0
820	—	—	—	—	—	—	—	—	—	—	**.02**	**.02**	—	—	**0**
	—	—	—	—	—	—	—	—	—	—	.06	.07	—	—	0
821	—	—	—	—	—	—	—	—	—	—	**.01**	**.02**	—	—	**0**
	—	—	—	—	—	—	—	—	—	—	.02	.06	—	—	0
822	—	—	—	—	—	—	—	—	—	—	**.02**	**.02**	—	—	**0**
	—	—	—	—	—	—	—	—	—	—	.06	.06	—	—	0
823	—	—	—	—	—	—	—	—	—	—	**.04**	**.02**	—	—	**0**
	—	—	—	—	—	—	—	—	—	—	.15	.08	—	—	0
824	—	—	—	—	—	—	—	—	—	—	**.01**	**.02**	—	—	**0**
	—	—	—	—	—	—	—	—	—	—	.03	.05	—	—	0
825	—	—	—	—	—	—	—	—	—	—	**.05**	**.02**	—	—	**0**
	—	—	—	—	—	—	—	—	—	—	.19	.07	—	—	0
826	—	—	—	—	—	—	—	—	—	—	—	—	—	—	—
	—	—	—	—	—	—	—	—	—	—	—	—	—	—	—
827	—	—	—	—	—	**.35**	—	—	—	—	—	—	—	—	—
	—	—	—	—	—	2.70	—	—	—	—	—	—	—	—	—
828	—	—	—	**0**	—	—	—	—	—	—	—	—	—	—	—
	—	—	—	0	—	—	—	—	—	—	—	—	—	—	—
829	—	—	—	—	—	—	—	—	—	—	—	—	—	—	—
	—	—	—	—	—	—	—	—	—	—	—	—	—	—	—
830	—	—	—	—	—	—	—	—	—	—	—	—	—	—	—
	—	—	—	—	—	—	—	—	—	—	—	—	—	—	—
831	—	—	—	—	—	—	—	—	—	—	—	—	—	—	—
	—	—	—	—	—	—	—	—	—	—	—	—	—	—	—
832	—	—	—	—	—	—	—	—	—	—	—	—	—	—	—
	—	—	—	—	—	—	—	—	—	—	—	—	—	—	—
833	**0**	**.01**	**.03**	**0**	**364.00**	**.16**	**0**	**0**	**0**	**0**	—	**Trace**	**0**	—	**0**
	0	.11	.26	0	2800.00	1.20	0	0	0	0	—	.02	0	—	0
834	**.08**	—	—	**0**	**0**	—	**0**	**.02**	**Trace**	**.11**	—	—	—	—	—
	.30	—	—	0	0	—	0	.08	.02	.40	—	—	—	—	—
835	**.08**	—	—	**0**	**0**	—	**0**	**.03**	**Trace**	**.14**	—	—	—	—	—
	.30	—	—	0	0	—	0	.10	.02	.50	—	—	—	—	—

(Continued)

ENTREES

FATS & OILS

TABLE F-36 *(Continued)*

Item No.	Food Item Name	Approximate Measure	Weight	Moisture	Food Energy Calories	Protein	Fats	Carbo-hydrates	Fiber	Minerals (Macro) Calcium	Phos-phorus	Sodium	Mag-nesium	Potas-sium
			(g)	(%)	(kcal)	(g)	(g)	(g)	(g)	(mg)	(mg)	(mg)	(mg)	(mg)
	PORK BACKFAT (Continued)													
836	fresh, raw, thin class (100% fat)	1 OZ	28	8.6	227.9	.7	24.9	0	0	.3	0	19.6	—	79.8
			100	8.6	814.0	2.4	89.1	0	0	1.0	0	70.0	—	285.0
837	**SUET (BEEF KIDNEY FAT),** raw	1 TBLS	14	4.0	119.6	.2	13.2	0	0	—	—	—	—	—
			100	4.0	854.0	1.5	94.0	0	0	—	—	—	—	—
	MARGARINES													
838	**HARD,** unsalted, hydrogenated vegetable oils	1 TSP	5	18.5	35.7	0	4.0	0	0	.9	.7	.1	.1	1.2
			100	18.5	714.0	.5	80.3	.5	0	17.4	13.3	2.2	1.5	24.7
839	**HARD,** salted, hydrogenated vegetable oils	1 TSP	5	15.7	35.9	0	4.0	0	0	1.5	1.1	47.2	.1	2.1
			100	15.7	718.7	.9	80.5	.9	0	29.9	22.9	943.4	2.6	42.4
840	**SOFT,** unsalted, unspecified oils	1 TSP	5	17.9	35.8	0	4.0	0	—	1.3	1.0	1.4	.1	1.9
			100	17.9	716.4	.8	80.3	.9	—	26.5	20.3	27.5	2.3	37.7
841	**SOFT** tub, salted, unspecified oils	1 TBLS	14	16.2	100.3	.1	11.3	.1	—	3.7	2.8	151.0	.3	5.3
			100	16.2	716.4	.8	80.4	.5	—	26.5	20.3	1078.7	2.3	37.7
842	**SPREAD** tub, salted, 60% fat, unspecified oils	1 TSP	5	37.0	27.0	0	3.0	0	—	1.0	.8	49.7	.1	1.5
			100	37.0	539.6	.6	60.8	0	—	20.9	16.1	993.9	1.8	29.8
843	**LIQUID,** unsalted, hydrogenated vegetable oils	1 TSP	5	15.5	36.0	0	4.1	0	0	1.0	.8	.5	—	.5
			100	15.5	720.0	.6	81.0	.4	0	20.0	16.0	10.0	—	10.0
844	**LIQUID,** salted, hydrogenated vegetable oils	1 TSP	5	15.8	36.1	.1	4.0	0	0	3.3	2.5	39.0	.3	4.7
			100	15.8	721.0	1.9	80.6	0	0	66.1	50.9	780.9	5.8	94.3
845	**WHIPPED**	1 TSP	4	16.0	28.7	.1	3.2	0	—	.8	—	—	—	—
			100	16.0	717.0	1.3	80.3	0	—	20.0	—	—	—	—
846	**IMITATION,** salted, 40% fat, hydrogenated corn	1 TSP	5	58.1	17.3	0	1.9	0	0	.9	.7	48.0	.1	1.3
			100	58.1	345.2	.5	38.8	.4	0	17.8	13.7	959.6	1.6	25.3
	OILS													
847	**ALMOND**	1 TBLS	14	0.0	123.8	0	14.0	0	0	—	—	—	—	—
			100	0.0	884.0	0	100.0	0	0	—	—	—	—	—
848	**APRICOT KERNEL**	1 TBLS	14	0.0	123.8	0	14.0	0	0	—	—	—	—	—
			100	0.0	884.0	0	100.0	0	0	—	—	—	—	—
849	**BABASSU**	1 TBLS	14	0.0	123.8	0	14.0	0	0	—	—	—	—	—
			100	0.0	884.0	0	100.0	0	0	—	—	—	—	—
850	**COCOA (CACAO) BUTTER**	1 TBLS	14	0.0	123.8	0	14.0	0	0	—	—	—	—	—
			100	0.0	884.0	0	100.0	0	0	—	—	—	—	—
851	**COCONUT**	1 TBLS	14	0.0	123.8	0	14.0	0	0	—	—	—	—	—
			100	0.0	884.0	0	100.0	0	0	—	.1	—	—	—
852	**COCONUT CREAM**	1 TBLS	14	54.1	47.7	.6	4.5	1.2	—	2.1	17.6	.6	—	45.4
			100	54.1	341.0	4.4	32.2	8.6	—	15.0	126.0	4.0	—	324.0
853	**CORN**	1 TBLS	14	0.0	123.8	0	14.0	0	0	0	0	0	.1	0
			100	0.0	884.0	0	100.0	0	0	0	0	0	0.5	0
854	**COTTONSEED**	1 TBLS	14	0.0	123.8	0	14.0	0	0	0	0	0	0	0
			100	0.0	884.0	0	100.0	0	0	0	0	0	.1	0
855	**CUPU ASSU**	1 TBLS	14	0.0	123.8	0	14.0	0	0	—	—	—	—	—
			100	0.0	884.0	0	100.0	0	0	—	—	—	—	—
856	**GRAPESEED**	1 TBLS	14	0.0	123.8	0	14.0	0	0	—	—	—	—	—
			100	0.0	884.0	0	100.0	0	0	—	—	—	—	—
857	**HAZELNUT**	1 TBLS	14	0.0	123.8	0	14.0	0	0	—	—	—	—	—
			100	0.0	884.0	0	100.0	0	0	—	—	—	—	—
858	**LINSEED**	1 TBLS	14	0.0	123.8	0	14.0	0	0	0	0	—	0	—
			100	0.0	884.0	0	100.0	0	0	.1	.2	—	0	—
859	**NUTMEG BUTTER**	1 TBLS	14	0.0	123.8	0	14.0	0	0	—	—	—	—	—
			100	0.0	884.0	0	100.0	0	0	—	—	—	—	—
860	**OLIVE**	1 TBLS	14	0.0	123.8	0	14.0	0	0	0	.2	0	0	0
			100	0.0	884.0	0	100.0	0	0	.2	1.2	0	0	0
861	**PALM**	1 TBLS	14	0.0	123.8	0	14.0	0	0	—	0	—	—	—
			100	0.0	884.0	0	100.0	0	0	—	.2	—	—	—
862	**PALM KERNEL**	1 TBLS	14	0.0	123.8	0	14.0	0	0	—	—	—	—	—
			100	0.0	884.0	0	100.0	0	0	—	—	—	—	—
863	**PEANUT**	1 TBLS	14	0.0	123.8	0	14.0	0	0	0	0	0	0	0
			100	0.0	884.0	0	100.0	0	0	.1	0	.1	0	0
864	**POPPYSEED**	1 TBLS	14	0.0	123.8	0	14.0	0	0	—	—	—	—	—
			100	0.0	884.0	0	100.0	0	0	—	—	—	—	—
865	**RAPESEED,** high (45% & over) erucic acid content	1 TBLS	14	0.0	123.8	0	14.0	0	0	—	—	—	—	—
			100	0.0	884.0	0	100.0	0	0	—	—	—	—	—

Item No.	Minerals (Micro) Iron	Zinc	Copper	Fat-Soluble Vitamins Vitamin A	Vitamin D	Vitamin E (Alpha Tocopherol)	Water-Soluble Vitamins Vitamin C	Thiamin	Ribo-flavin	Niacin	Panto-thenic Acid	Vit. B-6 (Pyri-doxine)	Folacin (Folic Acid)	Biotin	Vitamin B-12
	(mg)	(mg)	(mg)	(IU)	(IU)	(mg)	(mg)	(mg)	(mg)	(mg)	(mg)	(mg)	(mcg)	(mcg)	(mcg)
836	**.11**	—	—	**0**	**0**	—	—	**.03**	**Trace**	**.17**	—	—	—	—	—
	.40	—	—	0	0	—	—	.12	.03	.60	—	—	—	—	—
837	—	—	—	—	—	—	—	—	—	—	—	—	—	—	—
	—	—	—	—	—	—	—	—	—	—	—	—	—	—	—
838	**Trace**	**0**	**Trace**	**165.4**	**(.40)**	**(.40)**	**.01**	**0**	**Trace**	**Trace**	**Trace**	**0**	**.04**	**Trace**	**Trace**
	(.30)	0	(.04)	3307.0	(7.94)	(8.00)	.09	.01	.02	.01	.05	.01	.69	Trace	.066
839	**Trace**	—	—	**165.4**	—	—	**.01**	**Trace**	**Trace**	**Trace**	**Trace**	**0**	**.06**	—	**.01**
	(.30)	—	—	3307.0	—	—	.16	.01	.04	.02	.08	.01	1.18	—	.10
840	—	—	—	**165.4**	—	—	**.01**	**0**	**Trace**	**Trace**	**Trace**	**0**	**.05**	—	**Trace**
	—	—	—	3307.0	—	—	.14	.01	.03	.02	.08	.01	1.05	—	.08
841	**0**	—	—	**463.0**	—	—	**.02**	**Trace**	**Trace**	**Trace**	**.01**	**Trace**	**.15**	—	**.01**
	0	—	—	3307.0	—	—	.14	.01	.03	.02	.08	.01	1.05	—	.08
842	—	—	—	**165.4**	—	—	**.01**	**0**	**Trace**	**Trace**	**Trace**	**0**	**.04**	—	**Trace**
	—	—	—	3307.0	—	—	.11	.01	.03	.02	.06	.01	.83	—	.07
843	**0**	—	—	**165.0**	—	**1.42**	**0**	—	—	—	—	—	**.10**	—	—
	0	—	—	3300.0	—	28.40	0	—	—	—	—	—	2.00	—	—
844	—	—	—	**165.4**	—	**.23**	**.02**	**Trace**	**Trace**	**Trace**	**.01**	**Trace**	**.13**	—	**.01**
	—	—	—	3307.0	—	4.60	.36	.02	.08	.05	.19	.02	2.63	—	.21
845	**0**	—	—	**131.6**	—	—	**0**	—	—	—	—	—	**.08**	—	—
	0	—	—	3289.0	—	—	0	—	—	—	—	—	2.00	—	—
846	**Trace**	**Trace**	**Trace**	**165.4**	**(.40)**	**(.20)**	**.01**	**0**	**Trace**	**Trace**	**Trace**	**0**	**.04**	**Trace**	**Trace**
	Trace	Trace	Trace	3307.0	(7.94)	(4.00)	.09	.01	.02	.01	.05	.01	.71	Trace	.06
847	—	—	—	—	—	**5.49**	—	—	—	—	—	—	—	—	—
	—	—	—	—	—	39.20	—	—	—	—	—	—	—	—	—
848	—	—	—	—	—	**.56**	—	—	—	—	—	—	—	—	—
	—	—	—	—	—	4.00	—	—	—	—	—	—	—	—	—
849	—	—	—	—	—	—	—	—	—	—	—	—	—	—	—
	—	—	—	—	—	—	—	—	—	—	—	—	—	—	—
850	—	—	—	—	—	**.25**	—	—	—	—	—	—	—	—	—
	—	—	—	—	—	1.80	—	—	—	—	—	—	—	—	—
851	**.01**	—	—	—	—	**.06**	—	—	—	—	—	—	—	—	—
	.04	—	—	—	—	.40	—	—	—	—	—	—	—	—	—
852	**.25**	—	—	**0**	—	—	**.14**	**Trace**	**Trace**	**.07**	—	—	—	—	—
	1.80	—	—	0	—	—	1.00	.02	.01	.50	—	—	—	—	—
853	**0**	**.02**	**.03**	—	—	**2.00**	**0**	**0**	**0**	**0**	—	—	—	—	—
	0	.16	.22	—	—	14.30	0	0	0	0	—	—	—	—	—
854	**0**	—	**.02**	**0**	—	**4.94**	**0**	**0**	**0**	**0**	—	—	—	—	—
	0	—	.13	0	—	35.30	0	0	0	0	—	—	—	—	—
855	—	—	—	—	—	—	—	—	—	—	—	—	—	—	—
	—	—	—	—	—	—	—	—	—	—	—	—	—	—	—
856	—	—	—	—	—	—	—	—	—	—	—	—	—	—	—
	—	—	—	—	—	—	—	—	—	—	—	—	—	—	—
857	—	—	—	—	—	—	—	—	—	—	—	—	—	—	—
	—	—	—	—	—	—	—	—	—	—	—	—	—	—	—
858	—	—	—	—	—	—	—	—	—	—	—	—	—	—	—
	—	—	—	—	—	—	—	—	—	—	—	—	—	—	—
859	—	—	—	—	—	—	—	—	—	—	—	—	—	—	—
	—	—	—	—	—	—	—	—	—	—	—	—	—	—	—
860	**.05**	**.01**	**.05**	**0**	—	**1.67**	**0**	**0**	**0**	**0**	—	—	—	—	—
	.38	.06	.32	0	—	11.90	0	0	0	0	—	—	—	—	—
861	**Trace**	—	—	—	—	**2.67**	—	—	—	—	—	—	—	—	—
	.01	—	—	—	—	19.10	—	—	—	—	—	—	—	—	—
862	—	—	—	—	—	—	—	—	—	—	—	—	—	—	—
	—	—	—	—	—	—	—	—	—	—	—	—	—	—	—
863	**Trace**	**Trace**	**.01**	**0**	—	**1.62**	**0**	**0**	**0**	**0**	—	—	—	—	—
	.03	.01	.08	0	—	11.60	0	0	0	0	—	—	—	—	—
864	—	—	—	—	—	—	—	—	—	—	—	—	—	—	—
	—	—	—	—	—	—	—	—	—	—	—	—	—	—	—
865	—	—	—	—	—	—	—	—	—	—	—	—	—	—	—
	—	—	—	—	—	—	—	—	—	—	—	—	—	—	—

(Continued)

TABLE F-36 *(Continued)*

Item No.	Food Item Name	Approximate Measure	Weight	Moisture	Food Energy Calories	Protein	Fats	Carbo-hydrates	Fiber	Minerals (Macro) Calcium	Phos-phorus	Sodium	Mag-nesium	Potas-sium
			(g)	(%)	(kcal)	(g)	(g)	(g)	(g)	(mg)	(mg)	(mg)	(mg)	(mg)
866	**RAPESEED,** medium (30-45%) erucic acid content	1 TBLS	14	0.0	123.8	0	14.0	0	0	—	—	—	—	—
			100	0.0	884.0	0	100.0	0	0	—	—	—	—	—
867	**RAPESEED,** low to 30% erucic acid content	1 TBLS	14	0.0	123.8	0	14.0	0	0	—	—	—	—	—
			100	0.0	884.0	0	100.0	0	0	—	—	—	—	—
868	**RAPESEED,** 0% erucic acid content	1 TBLS	14	0.0	123.8	0	14.0	0	0	—	—	—	—	—
			100	0.0	884.0	0	100.0	0	0	—	—	—	—	—
869	**RICE BRAN**	1 TBLS	14	0.0	123.8	0	14.0	0	0	—	—	—	—	—
			100	0.0	884.0	0	100.0	0	0	—	—	—	—	—
870	**SAFFLOWER,** over 70% linoleic acid	1 TBLS	14	0.0	123.8	0	14.0	0	0	0	0	0	—	0
			100	0.0	884.0	0	100.0	0	0	0	0	0	—	0
871	**SAFFLOWER,** over 70% oleic acid content	1 TBLS	14	0.0	123.8	0	14.0	0	0	—	—	—	—	—
			100	0.0	884.0	0	100.0	0	0	—	—	—	—	—
872	**SESAME**	1 TBLS	14	0.0	123.8	0	14.0	0	0	0	0	0	—	0
			100	0.0	884.0	0	100.0	0	0	0	0	0	—	0
873	**SHEANUT**	1 TBLS	14	0.0	123.8	0	14.0	0	0	—	—	—	—	—
			100	0.0	884.0	0	100.0	0	0	—	—	—	—	—
874	**SOYBEAN**	1 TBLS	14	0.0	123.8	0	14.0	0	0	0	0	0	0	0
			100	0.0	884.0	0	100.0	0	0	0	.3	0	0	0
875	**SOYBEAN,** hydrogenated	1 TBLS	14	0.0	123.8	0	14.0	0	0	—	—	—	—	—
			100	0.0	884.0	0	100.0	0	0	—	—	—	—	—
876	**SOYBEAN & COTTONSEED,** hydrogenated	1 TBLS	14	0.0	123.8	0	14.0	0	0	—	—	—	—	—
			100	0.0	884.0	0	100.0	0	0	—	—	—	—	—
877	**SOYBEAN,** lecithin	1 TBLS	14	0.0	123.8	0	14.0	0	0	—	—	—	—	—
			100	0.0	884.0	0	100.0	0	0	—	—	—	—	—
878	**SUNFLOWER,** hydrogenated	1 TBLS	14	0.0	123.8	0	14.0	0	0	—	—	—	—	—
			100	0.0	884.0	0	100.0	0	0	—	—	—	—	—
879	**SUNFLOWER,** less than 60% linoleic acid	1 TBLS	14	0.0	123.8	0	14.0	0	0	0	—	0	0	—
			100	0.0	884.0	0	100.0	0	0	.2	—	.1	.2	—
880	**SUNFLOWER,** 60% & over linoleic acid	1 TBLS	14	0.0	123.8	0	14.0	0	0	—	—	—	—	—
			100	0.0	884.0	0	100.0	0	0	—	—	—	—	—
881	**TEASEED**	1 TBLS	14	0.0	123.8	0	14.0	0	0	—	—	—	—	—
			100	0.0	884.0	0	100.0	0	0	—	—	—	—	—
882	**TOMATOSEED**	1 TBLS	14	0.0	123.8	0	14.0	0	0	—	—	—	—	—
			100	0.0	884.0	0	100.0	0	0	—	—	—	—	—
883	**UCUHUBA BUTTER**	1 TBLS	14	0.0	123.8	0	14.0	0	0	—	—	—	—	—
			100	0.0	884.0	0	100.0	0	0	—	—	—	—	—
884	**WALNUT**	1 TBLS	14	0.0	123.8	0	14.0	0	0	—	—	—	—	—
			100	0.0	884.0	0	100.0	0	0	—	—	—	—	—
885	**WHEAT GERM**	1 TBLS	14	0.0	123.8	0	14.0	0	0	—	—	—	—	—
			100	0.0	884.0	0	100.0	0	0	—	—	—	—	—
	SHORTENINGS													
886	**SHORTENING,** vegetable	1 TBLS	14	0.0	123.8	0	14.0	0	0	0	0	0	0	0
			100	0.0	884.0	0	100.0	0	0	0	0	0	0	0
887	**SHORTENING,** hydrogenated cottonseed & soybean	1 TBLS	14	0.0	123.8	0	14.0	0	0	0	0	0	—	0
			100	0.0	884.0	0	100.0	0	0	0	0	0	—	0
888	**SHORTENING,** lard & vegetable oil	1 TBLS	13	0.0	117.0	0	13.0	0	0	—	—	—	—	—
			100	0.0	900.0	0	100.0	0	0	—	—	—	—	—
889	**ABALONE,** canned	1 CUP	160	80.2	128.0	25.6	.5	3.7	0	22.4	204.8	—	—	—
			100	80.2	80.0	16.0	.3	2.3	0	14.0	128.0	—	—	—
890	**ALEWIFE,** canned, solids & liquids	1 CUP	160	73.0	225.6	25.9	12.8	0	0	—	—	—	—	—
			100	73.0	141.0	16.2	8.0	0	0	—	—	—	—	—
	ANCHOVY													
891	canned	INDV	4	—	7.0	.8	.4	—	0	6.6	8.3	—	—	—
			100	—	175.0	19.0	10.0	—	0	166.0	208.0	—	—	—
892	pickled	1 OZ	28	58.6	49.3	5.4	2.9	.1	0	47.0	58.8	—	—	—
			100	58.6	176.0	19.2	10.3	.3	0	168.0	210.0	—	—	—
893	paste	1 TSP	7	—	9.0	1.4	.3	.3	0	1.0	15.0	770.0	—	14.0
			100	—	128.0	20.0	3.6	4.0	0	14.0	214.0	11000.0	—	200.0
894	**BARRACUDA,** Pacific, raw	1 OZ	28	75.4	31.6	5.9	.7	0	0	—	—	—	—	—
			100	75.4	113.0	21.0	2.6	0	0	—	—	—	—	—
	BASS													
895	**BLACK SEA,** stuffed, baked	1 OZ	28	52.9	72.5	4.5	4.4	3.2	0	—	—	—	—	—
			100	52.9	259.0	16.2	15.8	11.4	0	—	—	—	—	—

Item No.	Minerals (Micro)			Fat-Soluble Vitamins			Water-Soluble Vitamins								
	Iron	Zinc	Copper	Vitamin A	Vitamin D	Vitamin E (Alpha Tocopherol)	Vitamin C	Thiamin	Ribo-flavin	Niacin	Panto-thenic Acid	Vit. B-6 (Pyri-doxine)	Folacin (Folic Acid)	Biotin	Vitamin B-12
	(mg)	(mg)	(mg)	(IU)	(IU)	(mg)	(mg)	(mg)	(mg)	(mg)	(mg)	(mg)	(mcg)	(mcg)	(mcg)
866	–	–	–	–	–	–	–	–	–	–	–	–	–	–	–
	–	–	–	–	–	–	–	–	–	–	–	–	–	–	–
867	–	–	–	–	–	–	–	–	–	–	–	–	3.99	–	–
	–	–	–	–	–	–	–	–	–	–	–	–	28.50	–	–
868	–	–	–	–	–	–	–	–	–	–	–	–	–	–	–
	–	–	–	–	–	–	–	–	–	–	–	–	–	–	–
869	.01	–	–	–	–	4.52	–	–	–	–	–	–	–	–	–
	.07	–	–	–	–	32.30	–	–	–	–	–	–	–	–	–
870	0	.03	.02	0	–	4.77	0	0	0	0	–	–	–	–	–
	0	.19	.13	0	–	34.10	0	0	0	0	–	–	–	–	–
871	–	–	–	–	–	–	–	–	–	–	–	–	–	–	–
	–	–	–	–	–	–	–	–	–	–	–	–	–	–	–
872	0	–	–	–	–	.20	0	0	0	0	–	–	–	–	–
	0	–	–	–	–	1.40	0	0	0	0	–	–	–	–	–
873	–	–	–	–	–	–	–	–	–	–	–	–	–	–	–
	–	–	–	–	–	–	–	–	–	–	–	–	–	–	–
874	Trace	0	–	0	–	1.54	0	0	0	0	–	–	–	–	–
	.02	0	–	0	–	11.00	0	0	0	0	–	–	–	–	–
875	–	–	–	–	–	1.13	–	–	–	–	–	–	–	–	–
	–	–	–	–	–	8.10	–	–	–	–	–	–	–	–	–
876	–	–	–	–	–	–	–	–	–	–	–	–	–	–	–
	–	–	–	–	–	–	–	–	–	–	–	–	–	–	–
877	–	–	–	–	–	–	–	–	–	–	–	–	–	–	–
	–	–	–	–	–	–	–	–	–	–	–	–	–	–	–
878	–	–	–	–	–	–	–	–	–	–	–	–	–	–	–
	–	–	–	–	–	–	–	–	–	–	–	–	–	–	–
879	Trace	–	–	–	–	–	–	–	–	–	–	–	–	–	–
	.03	–	–	–	–	–	–	–	–	–	–	–	–	–	–
880	–	–	–	–	–	6.29	–	–	–	–	–	–	–	–	–
	–	–	–	–	–	44.90	–	–	–	–	–	–	–	–	–
881	–	–	–	–	–	–	–	–	–	–	–	–	–	–	–
	–	–	–	–	–	–	–	–	–	–	–	–	–	–	–
882	–	–	–	–	–	.53	–	–	–	–	–	–	–	–	–
	–	–	–	–	–	3.80	–	–	–	–	–	–	–	–	–
883	–	–	–	–	–	–	–	–	–	–	–	–	–	–	–
	–	–	–	–	–	–	–	–	–	–	–	–	–	–	–
884	–	–	–	–	–	.01	–	–	–	–	–	–	–	–	–
	–	–	–	–	–	.04	–	–	–	–	–	–	–	–	–
885	–	.53	.08	–	–	20.92	–	–	–	–	–	–	–	–	–
	–	3.80	.60	–	–	149.40	–	–	–	–	–	–	–	–	–
886	0	–	–	0	–	1.39	0	0	0	0	–	0	0	–	0
	0	–	–	0	–	9.93	0	0	0	0	–	0	0	–	0
887	0	–	–	0	–	–	0	0	0	0	–	–	–	–	–
	0	–	–	0	–	–	0	0	0	0	–	–	–	–	–
888	–	–	–	–	–	–	–	–	–	–	–	–	–	–	–
	–	–	–	–	–	–	–	–	–	–	–	–	–	–	–
889	–	–	–	–	–	–	–	.19	–	–	–	–	–	–	–
	–	–	–	–	–	–	–	.12	–	–	–	–	–	–	–
890	–	–	–	–	–	–	–	–	–	–	–	–	–	–	–
	–	–	–	–	–	–	–	–	–	–	–	–	–	–	–
891	–	[6] .13	–	–	–	–	–	–	–	–	–	–	–	–	–
	–	3.20	–	–	–	–	–	–	–	–	–	–	–	–	–
892	–	.50	.02	–	–	–	–	–	–	–	–	.04	2.80	–	–
	–	1.77	.08	–	–	–	–	–	–	–	–	.14	10.00	–	–
893	–	–	–	–	–	–	–	–	–	–	–	–	–	–	–
	–	–	–	–	–	–	–	–	–	–	–	–	–	–	–
894	–	–	–	–	–	–	–	–	–	–	–	–	–	–	–
	–	–	–	–	–	–	–	–	–	–	–	–	–	–	–
895	–	–	–	–	–	–	–	–	–	–	–	–	–	–	–
	–	–	–	–	–	–	–	–	–	–	–	–	–	–	–

(Continued)

TABLE F-36 *(Continued)*

Item No.	Food Item Name	Approximate Measure	Weight	Moisture	Food Energy Calories	Protein	Fats	Carbo-hydrates	Fiber	Minerals (Macro) Calcium	Phos-phorus	Sodium	Mag-nesium	Potas-sium
			(g)	(%)	(kcal)	(g)	(g)	(g)	(g)	(mg)	(mg)	(mg)	(mg)	(mg)
	BASS (Continued)													
896	**SMALLMOUTH & LARGEMOUTH,** raw	1 OZ	28	77.3	29.1	5.3	.7	0	0	13.2	53.8	–	7.6	–
			100	77.3	104.0	18.9	2.6	0	0	47.0	192.0	–	27.0	–
897	**STRIPED,** broiled	1 OZ	28	67.6	63.8	5.7	3.6	2.2	0	13.2	64.4	–	12.0	–
			100	67.6	228.0	20.2	12.8	7.9	0	47.0	230.0	–	43.0	–
898	**STRIPED,** oven-fried	1 OZ	28	60.8	54.9	6.0	2.4	1.9	–	–	–	–	–	–
			100	60.8	196.0	21.5	8.5	6.7	–	–	–	–	–	–
899	**WHITE,** raw	1 OZ	28	78.8	27.4	5.0	.6	0	0	–	–	–	–	–
			100	78.8	98.0	18.0	2.3	0	0	–	–	–	–	–
	BLUEFISH													
900	baked or broiled	1 OZ	28	–	51.0	6.6	2.7	.1	–	7.8	78.1	–	8.1	–
			100	–	182.0	23.5	9.6	.3	–	28.0	279.0	–	29.0	–
901	fried, w/egg, milk or water & bread crumbs	1 OZ	28	60.8	57.4	6.4	2.7	1.3	–	9.8	72.0	40.9	–	–
			100	60.8	205.0	22.7	9.8	4.7	–	35.0	257.0	146.0	–	–
902	**BUFFALOFISH,** raw	1 OZ	28	77.4	31.6	4.9	1.2	0	0	–	–	14.6	–	82.0
			100	77.4	113.0	17.5	4.2	0	0	–	–	52.0	–	293.0
903	**BULLHEAD, BLACK,** raw	1 OZ	28	81.3	23.5	4.6	.4	0	0	–	–	–	–	–
			100	81.3	84.0	16.3	1.6	0	0	–	–	–	–	–
904	**BURBOT,** fried	1 OZ	28	60.5	43.7	10.4	.3	0	0	–	–	–	–	–
			100	60.5	156.0	37.0	1.2	0	0	–	–	–	–	–
	BUTTERFISH													
905	raw, from Gulf waters	1 OZ	28	78.2	26.6	4.5	.8	0	0	–	–	–	–	–
			100	78.2	95.0	16.2	2.9	0	0	–	–	–	–	–
906	raw, from northern waters	1 OZ	28	71.4	47.3	5.1	2.9	0	0	–	–	–	–	–
			100	71.4	169.0	18.1	10.2	0	0	–	–	–	–	–
907	**CARP,** raw	1 OZ	28	77.8	32.2	5.0	1.2	0	0	14.0	70.8	14.0	–	80.1
			100	77.8	115.0	18.0	4.2	0	0	50.0	253.0	50.0	–	286.0
908	**CATFISH,** freshwater, raw	1 OZ	28	78.0	28.8	4.9	.9	0	0	3.9	–	16.8	7.6	92.4
			100	78.0	103.0	17.6	3.1	0	0	14.0	–	60.0	27.0	330.0
	CAVIAR													
909	**STURGEON,** granular	1 TSP	10	46.0	26.2	2.7	1.5	.3	–	27.6	35.5	220.0	30.0	18.0
			100	46.0	262.0	26.9	15.0	3.3	–	276.0	355.0	2200.0	300.0	180.0
910	**STURGEON,** pressed	1 TSP	10	36.0	31.6	3.4	1.7	.5	0	14.0	18.0	220.0	30.0	64.0
			100	36.0	316.0	34.4	16.7	4.9	0	140.0	180.0	2200.0	300.0	640.0
911	**CHUB,** raw	1 OZ	28	74.9	40.6	4.3	2.5	0	0	–	–	–	–	–
			100	74.9	145.0	15.3	8.8	0	0	–	–	–	–	–
	CLAM(S)													
912	raw, hard, soft, meat & liquid	1 CUP	200	85.9	106.0	16.2	1.8	5.0	–	–	394.0	–	–	–
			100	85.9	53.0	8.1	.9	2.5	–	–	197.0	–	–	–
913	raw, hard & soft, meat only	1 CUP	200	81.7	152.0	25.2	3.2	4.0	–	138.0	324.0	240.0	–	362.0
			100	81.7	76.0	12.6	1.6	2.0	–	69.0	162.0	120.0	–	181.0
914	raw, hard or round, meat & liquid	1 CUP	200	86.2	98.0	13.0	.8	8.4	–	–	350.0	–	96.0	–
			100	86.2	49.0	6.5	.4	4.2	–	–	175.0	–	48.0	–
915	raw hard or round, meat only	1 LG	20	79.8	16.0	2.2	.4	1.2	–	13.8	30.2	41.0	17.8	62.2
			100	79.8	80.0	11.1	2.0	5.9	–	69.0	151.0	205.0	89.0	311.0
916	raw, soft, meat & liquid	1 CUP	200	85.8	108.0	17.2	2.0	4.0	–	–	416.0	–	–	–
			100	85.8	54.0	8.6	1.0	2.0	–	–	208.0	–	–	–
917	raw, soft meat only	1 LG	25	80.8	20.5	3.5	.2	.3	–	–	45.8	9.0	–	58.8
			100	80.8	82.0	14.0	.7	1.3	–	–	183.0	36.0	–	235.0
918	fritters, w/flour, baking powder, butter, & egg	1 SERV	90	40.3	279.9	10.3	13.5	27.8	–	68.4	175.5	–	–	132.3
			100	40.3	311.0	11.4	15.0	30.9	–	76.0	195.0	–	–	147.0
919	canned, solids & liquid	1 CUP	200	86.3	104.0	15.8	1.4	5.6	–	110.0	274.0	–	–	280.0
			100	86.3	52.0	7.9	.7	2.8	–	55.0	137.0	–	–	140.0
920	canned, drained solids	1 LG	20	77.0	19.6	3.2	.5	.4	–	–	–	–	–	–
			100	77.0	98.0	15.8	2.5	1.9	–	–	–	–	–	–
921	canned, liquor, bouillon or nectar	1 CUP	200	93.6	38.0	4.6	.2	4.2	0	–	–	–	–	–
			100	93.6	19.0	2.3	.1	2.1	0	–	–	–	–	–
922	***COCKLES,** sand, raw	1 OZ	28	79.2	22.7	4.7	.3	0	0	60.8	21.8	–	–	–
			100	79.2	81.0	16.8	1.0	0	0	217.0	78.0	–	–	–
	COD													
923	broiled	1 OZ	30	64.6	51.0	8.6	1.6	0	0	9.3	82.2	33.0	–	122.1
			100	64.6	170.0	28.5	5.3	0	0	31.0	274.0	110.0	–	407.0

Item No.	Minerals (Micro)			Fat-Soluble Vitamins			Water-Soluble Vitamins								
	Iron	Zinc	Copper	Vitamin A	Vitamin D	Vitamin E (Alpha Tocopherol)	Vitamin C	Thiamin	Ribo-flavin	Niacin	Panto-thenic Acid	Vit. B-6 (Pyri-doxine)	Folacin (Folic Acid)	Biotin	Vitamin B-12
	(mg)	(mg)	(mg)	(IU)	(IU)	(mg)	(mg)	(mg)	(mg)	(mg)	(mg)	(mg)	(mcg)	(mcg)	(mcg)
896	—	—	—	—	—	—	—	.03	.01	.59	.14	—	—	—	—
	—	—	—	—	—	—	—	.10	.03	2.10	.51	—	—	—	—
897	.53	—	—	32.5	—	—	0	.04	.04	.81	—	.07	—	—	.42
	1.90	—	—	116.0	—	—	0	.15	.14	2.90	—	.26	—	—	1.50
898	—	—	—	—	—	—	—	—	—	—	—	—	—	—	—
	—	—	—	—	—	—	—	—	—	—	—	—	—	—	—
899	—	—	—	—	—	—	—	—	—	—	—	—	—	—	—
	—	—	—	—	—	—	—	—	—	—	—	—	—	—	—
900	.20	—	—	90.2	—	—	.28	.03	.03	.50	—	.08	—	—	.48
	.70	—	—	322.0	—	—	1.00	.10	.10	1.80	—	.29	—	—	1.71
901	.25	—	—	—	—	—	—	.03	.03	.50	—	—	—	—	—
	.90	—	—	—	—	—	—	.11	.11	1.80	—	—	—	—	—
902	—	—	—	—	—	—	—	—	—	—	.13	—	—	—	—
	—	—	—	—	—	—	—	—	—	—	.46	—	—	—	—
903	—	—	—	—	—	—	—	—	—	—	—	—	—	—	—
	—	—	—	—	—	—	—	—	—	—	—	—	—	—	—
904	—	—	—	—	—	—	—	.15	.06	1.04	—	—	—	—	—
	—	—	—	—	—	—	—	.54	.23	3.70	—	—	—	—	—
905	—	—	—	—	—	—	—	—	—	—	—	—	—	—	—
	—	—	—	—	—	—	—	—	—	—	—	—	—	—	—
906	—	—	—	—	—	—	—	—	—	—	—	—	—	—	—
	—	—	—	—	—	—	—	—	—	—	—	—	—	—	—
907	.25	—	—	47.6	—	—	.28	Trace	.01	.42	.04	—	—	—	.06
	.90	—	—	170.0	—	—	1.00	.01	.04	1.50	.15	—	—	—	.20
908	.11	—	—	—	—	—	—	.01	.01	.48	.13	—	—	—	.67
	.40	—	—	—	—	—	—	.04	.03	1.70	.47	—	—	—	2.20
909	1.18	—	—	—	—	—	—	—	—	—	—	—	—	—	—
	11.80	—	—	—	—	—	—	—	—	—	—	—	—	—	—
910	—	—	—	—	—	—	—	—	—	—	—	—	—	—	—
	—	—	—	—	—	—	—	—	—	—	—	—	—	—	—
911	—	—	—	—	—	—	—	—	—	—	—	—	—	—	—
	—	—	—	—	—	—	—	—	—	—	—	—	—	—	—
912	—	2.18	.67	—	—	—	—	—	—	—	—	—	30.00	—	—
	—	1.09	.33	—	—	—	—	—	—	—	—	—	15.00	—	—
913	12.20	3.00	—	200.0	—	—	20.00	.20	.36	2.60	1.16	—	—	—	38.20
	6.10	1.50	—	100.0	—	—	10.00	.10	.18	1.30	.58	—	—	—	19.10
914	—	3.00	.67	—	—	—	—	—	.26	—	—	—	30.00	—	—
	—	1.50	.33	—	—	—	—	—	.13	—	—	—	15.00	—	—
915	1.50	—	—	22.0	—	—	—	.02	—	.32	.12	—	—	—	—
	7.50	—	—	110.0	—	—	—	.10	—	1.60	.60	—	—	—	—
916	—	3.00	.67	—	—	—	—	—	—	—	.60	.16	30.00	—	196.00
	—	1.50	.33	—	—	—	—	—	—	—	.30	.08	15.00	—	98.00
917	.85	—	—	—	—	—	—	—	—	—	.08	.02	—	—	24.50
	3.40	—	—	—	—	—	—	—	—	—	.30	.08	—	—	98.00
918	3.15	—	—	—	—	—	—	.03	.11	.81	—	—	—	—	—
	3.50	(1.70)	—	—	—	—	—	.03	.12	.90	—	—	—	—	—
919	8.20	2.40	—	—	—	—	—	.02	.22	2.00	1.18	.17	4.00	—	—
	4.10	1.20	—	—	—	—	—	.01	.11	1.00	.59	.08	2.00	—	—
920	—	—	—	—	—	—	—	—	—	—	—	.02	—	—	—
	—	—	—	—	—	—	—	—	—	—	—	.08	—	—	—
921	—	—	—	—	—	—	—	—	—	—	—	—	—	—	—
	—	—	—	—	—	—	—	—	—	—	—	—	—	—	—
922	.78	—	—	—	—	—	—	—	.04	1.74	—	—	—	—	—
	2.80	—	—	—	—	—	—	—	.15	6.20	—	—	—	—	—
923	.30	—	—	54.0	—	—	—	.02	.03	.90	—	—	—	—	—
	1.00	—	—	180.0	—	—	—	.08	.11	3.00	—	—	—	—	—

(Continued)

FISH & SEAFOODS

TABLE F-36 *(Continued)*

Item No.	Food Item Name	Approximate Measure	Weight	Moisture	Food Energy Calories	Protein	Fats	Carbohydrates	Fiber	Minerals (Macro) Calcium	Phosphorus	Sodium	Magnesium	Potassium
			(g)	(%)	(kcal)	(g)	(g)	(g)	(g)	(mg)	(mg)	(mg)	(mg)	(mg)
	COD (Continued)													
924	canned	1 OZ	**28**	**78.6**	**23.8**	**5.4**	**.1**	**0**	**0**	—	—	—	—	—
			100	78.6	85.0	19.2	.3	0	0	—	—	—	—	—
925	dehydrated, lightly salted	1 OZ	**28**	**12.3**	**105.0**	**22.9**	**.8**	**0**	**0**	—	**249.5**	**2268.0**	—	**44.8**
			100	12.3	375.0	81.8	2.8	0	0	—	891.0	8100.0	—	160.0
926	[6]dried, salted, boiled	1 OZ	**28**	—	**36.4**	**8.1**	**.3**	**0**	**0**	**6.2**	**44.8**	**112.0**	**10.0**	**8.7**
			100	—	130.0	29.0	.9	0	0	22.0	160.0	400.0	35.0	31.0
	CRAB													
927	**(BLUE, DUNGENESS, ROCK, KING),** steamed	1 OZ	**28**	**78.5**	**26.0**	**4.8**	**.5**	**.1**	**0**	**12.0**	**49.0**	**65.8**	**9.5**	**85.4**
			100	78.5	93.0	17.3	1.9	.5	0	43.0	175.0	235.0	34.0	305.0
928	[1]leg, steamed in shell	1 OZ	**28**	**78.5**	**26.0**	**4.8**	**.5**	**.1**	—	**12.0**	**49.0**	**58.8**	—	**50.4**
			100	78.5	93.0	17.3	1.9	.5	—	43.0	175.0	210.0	—	180.0
929	deviled	1 OZ	**28**	**63.3**	**51.8**	**3.2**	**2.6**	**3.7**	—	**13.2**	**37.8**	—	**7.3**	—
			100	63.3	185.0	11.4	9.4	13.3	—	47.0	135.0	—	26.0	—
930	deviled, w/bread, butter, parsley, egg, lemon, & catsup	1 CUP	**240**	**63.3**	**451.2**	**27.4**	**22.6**	**31.9**	—	**112.8**	**328.8**	**2080.8**	—	**398.4**
			100	63.3	188.0	11.4	9.4	13.3	—	47.0	137.0	867.0	—	166.0
931	imperial, w/butter, flour, milk, onion, green pepper, egg, & lemon	1 CUP	**220**	**71.9**	**323.4**	**32.1**	**16.7**	**8.6**	—	**132.0**	**365.2**	**1601.6**	—	**288.2**
			100	71.9	147.0	14.6	7.6	3.9	—	60.0	166.0	728.0	—	131.0
932	soft shell, fried	1 AVG	**107**	—	**348.8**	**19.4**	**21.6**	**19.2**	—	**67.4**	**203.3**	—	**38.5**	—
			100	—	326.0	18.1	20.2	17.9	—	63.0	190.0	—	36.0	—
933	cake	1 OZ	**28**	**64.6**	**46.2**	**5.4**	**2.5**	**.5**	**0**	**15.4**	**58.8**	—	**10.4**	—
			100	64.6	165.0	19.3	9.0	1.7	0	55.0	210.0	—	37.0	—
934	canned	1 OZ	**28**	**77.2**	**28.3**	**4.9**	**.7**	**.3**	**0**	**12.6**	**51.0**	**251.4**	**11.2**	**54.9**
			100	77.2	101.0	17.4	2.5	1.1	0	45.0	182.0	898.0	40.0	196.0
935	**CRAPPIE, WHITE,** raw	1 OZ	**28**	**81.8**	**22.1**	**4.7**	**.2**	**0**	**0**	—	—	—	—	—
			100	81.8	79.0	16.8	.8	0	0	—	—	—	—	—
936	**CRAYFISH,** freshwater & spiny lobster, raw	1 OZ	**28**	**82.5**	**20.2**	**4.1**	**.1**	**.3**	—	**21.6**	**56.3**	—	—	—
			100	82.5	72.0	14.6	.5	1.2	—	77.0	201.0	—	—	—
	CROAKER													
937	**ATLANTIC,** baked	1 OZ	**28**	**71.3**	**37.2**	**6.8**	**.9**	**0**	**0**	—	—	**33.6**	—	**90.4**
			100	71.3	133.0	24.3	3.2	0	0	—	—	120.0	—	323.0
938	**WHITE,** raw	1 OZ	**28**	**79.7**	**23.5**	**5.0**	**.2**	**0**	**0**	—	—	—	—	—
			100	79.7	84.0	18.0	.8	0	0	—	—	—	—	—
939	**YELLOWFIN,** raw	1 OZ	**28**	**79.0**	**24.9**	**5.4**	**.2**	**0**	**0**	—	—	—	—	—
			100	79.0	89.0	19.2	.8	0	0	—	—	—	—	—
	CUTTLEFISH													
940	[4]raw	1 OZ	**28**	**81.0**	**22.7**	**4.5**	**.3**	**.3**	**0**	**7.6**	**40.0**	—	—	**76.4**
			100	81.0	81.0	16.1	.9	1.0	0	27.0	143.0	—	—	273.0
941	[4]canned, salted & seasoned	1 OZ	**28**	**63.6**	**38.4**	**5.8**	**.8**	**1.6**	**0**	**2.8**	**50.4**	**210.0**	—	**26.3**
			100	63.6	137.0	20.8	2.8	5.6	0	10.0	180.0	750.0	—	94.0
942	[4]dried	1 OZ	**28**	**18.1**	**94.6**	**18.9**	**1.2**	**.7**	**0**	**35.0**	**224.3**	—	—	—
			100	18.1	338.0	67.5	4.4	2.5	0	125.0	801.0	—	—	—
943	**DOGFISH,** spiny **(grayfish)**, raw	1 OZ	**28**	**72.3**	**43.7**	**4.9**	**2.5**	**0**	**0**	—	—	—	—	—
			100	72.3	156.0	17.6	9.0	0	0	—	—	—	—	—
944	**DOLLY VARDEN,** raw	1 OZ	**28**	**73.1**	**40.3**	**5.6**	**1.8**	**0**	**0**	—	—	—	—	—
			100	73.1	144.0	19.9	6.5	0	0	—	—	—	—	—
945	**DRUM,** freshwater, raw	1 OZ	**28**	**77.0**	**33.9**	**4.8**	**1.5**	**0**	**0**	—	—	**19.6**	—	**80.1**
			100	77.0	121.0	17.3	5.2	0	0	—	—	70.0	—	286.0
946	**DRUM,** red **(redfish)**, raw	1 OZ	**28**	**80.2**	**22.4**	**5.0**	**.1**	**0**	**0**	—	—	**15.4**	—	**76.4**
			100	80.2	80.0	18.0	.4	0	0	—	—	55.0	—	273.0
947	**EEL,** smoked	1 SERV	**100**	**50.2**	**330.0**	**18.6**	**27.8**	**0**	**0**	—	**210.0**	—	—	—
			100	50.2	330.0	18.6	27.8	0	0	—	210.0	—	—	—
948	**EULACHON (SMELT),** raw	1 OZ	**28**	**79.6**	**33.0**	**4.1**	**1.7**	**0**	**0**	—	—	—	—	—
			100	79.6	118.0	14.6	6.2	0	0	—	—	—	—	—
949	**FINNAN HADDIE (SMOKED HADDOCK)**	1 OZ	**28**	**72.6**	**28.8**	**6.5**	**.1**	**0**	**0**	—	—	—	—	—
			100	72.6	103.0	23.2	.4	0	0	—	—	—	—	—
	FISH													
950	unspecified, fried	1 OZ	**28**	—	**56.8**	**5.1**	**3.1**	**2.2**	—	**8.1**	**59.9**	**77.8**	**9.2**	**75.0**
			100	—	203.0	18.1	11.0	7.9	—	29.0	214.0	278.0	33.0	268.0
951	cakes, fried w/canned fish, potato, & egg	1 CAKE	**100**	**66.0**	**172.0**	**14.7**	**8.0**	**9.3**	—	—	—	—	—	—
			100	66.0	172.0	14.7	8.0	9.3	—	—	—	—	—	—

Item No.	Minerals (Micro)			Fat-Soluble Vitamins			Water-Soluble Vitamins								
	Iron	Zinc	Copper	Vitamin A	Vitamin D	Vitamin E (Alpha Tocopherol)	Vitamin C	Thiamin	Ribo-flavin	Niacin	Panto-thenic Acid	Vit. B-6 (Pyri-doxine)	Folacin (Folic Acid)	Biotin	Vitamin B-12
	(mg)	(mg)	(mg)	(IU)	(IU)	(mg)	(mg)	(mg)	(mg)	(mg)	(mg)	(mg)	(mcg)	(mcg)	(mcg)
924	—	—	—	—	—	—	—	—	.02	—	—	—	—	—	—
	—	—	—	—	—	—	—	—	.08	—	—	—	—	—	—
925	1.01	—	—	0	—	—	—	.02	.13	3.05	.36	.08	—	—	2.80
	3.60	—	—	0	—	—	—	.08	.45	10.90	1.29	.30	—	—	10.00
926	.52	.08	—	Trace	Trace	—	Trace	Trace	Trace	—	.10	.06	Trace	Trace	1.01
	1.84	.27	—	Trace	Trace	—	Trace	Trace	Trace	—	.35	.20	Trace	Trace	3.61
927	.22	1.20	—	607.6	—	—	.56	.05	.02	.78	.17	.08	—	—	2.80
	.80	4.30	—	2170.0	—	—	2.00	.16	.08	2.80	.60	.30	—	—	10.00
928	.22	1.20	—	607.6	—	—	.56	.05	.02	.78	—	—	—	—	—
	.80	4.30	—	2170.0	—	—	2.00	.16	.08	2.80	—	—	—	—	—
929	.34	—	—	151.2	—	—	1.68	.02	.02	.05	—	.04	—	—	1.02
	1.20	—	—	540.0	—	—	6.00	.08	.08	.19	—	.15	—	—	3.64
930	2.88	—	—	—	—	—	14.40	.19	.26	3.60	—	—	—	—	—
	1.20	—	—	—	—	—	6.00	.08	.11	1.50	—	—	—	—	—
931	1.98	—	—	—	—	—	11.00	.13	.26	2.42	—	—	—	—	—
	.90	—	—	—	—	—	5.00	.06	.12	1.10	—	—	—	—	—
932	1.71	—	—	92.0	—	—	2.14	.25	.18	3.42	—	.30	—	—	8.89
	1.60	—	—	86.0	—	—	2.00	.23	.17	3.20	—	.28	—	—	8.31
933	.31	—	—	81.8	—	—	0	.02	.03	.48	—	.08	10.36	—	2.19
	1.10	—	—	292.0	—	—	0	.07	.11	1.70	—	.28	37.00	—	7.83
934	.22	1.01	.08	Trace	Trace	—	Trace	.02	.02	.53	.17	.08	.25	Trace	2.80
	.80	3.60	.27	Trace	Trace	—	Trace	.08	.08	1.90	.60	.30	.90	Trace	10.00
935	—	—	—	—	—	—	—	—	.01	.39	—	—	—	—	—
	—	—	—	—	—	—	—	—	.03	1.40	—	—	—	—	—
936	.42	—	—	—	—	—	—	Trace	.01	.53	.12	.06	—	—	.76
	1.50	—	—	—	—	—	—	.01	.04	1.90	.41	.21	—	—	2.70
937	—	—	—	19.6	—	—	—	.04	.03	1.82	—	—	—	—	—
	—	—	—	70.0	—	—	—	.13	.10	6.50	—	—	—	—	—
938	—	—	—	—	—	—	—	—	—	—	—	—	—	—	—
	—	—	—	—	—	—	—	—	—	—	—	—	—	—	—
939	—	—	—	—	—	—	—	—	—	—	—	—	—	—	—
	—	—	—	—	—	—	—	—	—	—	—	—	—	—	—
940	.22	—	—	0	—	—	0	.02	.01	.73	—	—	—	—	—
	.80	—	—	0	—	—	0	.07	.05	2.60	—	—	—	—	—
941	.45	—	—	98.0	—	—	—	.01	.02	.11	—	—	—	—	—
	.60	—	—	350.0	—	—	—	.02	.07	.40	—	—	—	—	—
942	.95	—	—	0	—	—	0	.02	.10	1.18	—	—	—	—	—
	3.40	—	—	0	—	—	0	.08	.36	4.20	—	—	—	—	—
943	—	—	—	—	—	—	—	.01	—	—	.19	—	—	—	.50
	—	—	—	—	—	—	—	.05	—	—	.69	—	—	—	1.80
944	—	—	—	—	—	—	—	.02	.02	—	—	—	—	—	—
	—	—	—	—	—	—	—	.06	.06	—	—	—	—	—	—
945	—	—	—	—	—	—	—	—	—	—	—	—	—	—	—
	—	—	—	—	—	—	—	—	—	—	—	—	—	—	—
946	—	—	—	—	—	—	—	.04	.01	.98	—	—	—	—	—
	—	—	—	—	—	—	—	.15	.05	3.50	—	—	—	—	—
947	1.00	—	—	2500.0	—	—	—	.14	.07	—	—	.12	—	—	5.60
	1.00	—	—	2500.0	—	—	—	.14	.07	—	—	.12	—	—	5.60
948	—	—	—	—	—	—	—	.01	.01	—	—	—	—	—	—
	—	—	—	—	—	—	—	.04	.04	—	—	—	—	—	—
949	—	—	—	—	—	—	—	.02	.01	.59	—	—	—	—	—
	—	—	—	—	—	—	—	.06	.05	2.10	—	—	—	—	—
950	.36	—	—	21.8	—	—	0	.02	.03	.50	—	.04	—	—	.26
	1.30	—	—	78.0	—	—	0	.07	.09	1.80	—	.15	—	—	.93
951	—	—	—	—	—	—	—	—	—	—	.25	.05	—	—	1.00
	—	—	—	—	—	—	—	—	—	—	.25	.05	—	—	1.00

(Continued)

TABLE F-36 *(Continued)*

Item No.	Food Item Name	Approximate Measure	Weight	Moisture	Food Energy Calories	Protein	Fats	Carbo-hydrates	Fiber	Minerals (Macro) Calcium	Phos-phorus	Sodium	Mag-nesium	Potas-sium
			(g)	(%)	(kcal)	(g)	(g)	(g)	(g)	(mg)	(mg)	(mg)	(mg)	(mg)
	FISH (Continued)													
952	cakes, canned	1 CUP	165	72.1	183.2	40.8	1.0	0	0	80.9	382.8	—	—	—
			100	72.1	111.0	24.7	.6	0	0	49.0	232.0	—	—	—
953	cakes, fried, frozen, reheated	1 CAKE	100	52.9	270.0	9.2	17.9	17.2	—	—	—	—	—	—
			100	52.9	270.0	9.2	17.9	17.2	—	—	—	—	—	—
954	creole	1 OZ	28	81.3	24.1	3.3	.7	1.0	—	3.5	36.6	23.5	6.6	17.3
			100	81.3	85.9	11.6	2.7	3.4	—	12.3	130.9	84.0	23.6	61.9
955	[4]ham	1 OZ	28	63.5	52.1	4.8	2.8	1.7	0	12.6	14.0	308.0	—	—
			100	63.5	186.0	17.0	10.0	6.0	0	45.0	50.0	1100.0	—	—
956	loaf, w/canned fish, bread, egg, tomato & onion, cooked	1 SLICE	150	72.2	186.0	21.2	5.6	11.0	—	—	—	—	—	—
			100	72.2	124.0	14.1	3.7	7.3	—	—	—	—	—	—
957	[4]sausage	1 OZ	28	68.5	41.2	4.2	1.7	2.1	0	28.0	16.8	280.0	—	—
			100	68.5	147.0	15.0	6.0	7.5	0	100.0	60.0	1000.0	—	—
958	sticks, frozen, cooked	1 PIECE	23	65.8	40.5	3.8	2.0	1.5	—	2.5	38.4	—	—	—
			100	65.8	176.0	16.6	8.9	6.5	—	11.0	167.0	—	—	—
959	flour, from whole fish	1 CUP	160	2.0	491.2	121.6	.5	0	—	7376.0	4960.0	272.0	—	688.0
			100	2.0	307.0	76.0	.3	0	—	4610.0	3100.0	170.0	—	430.0
960	flour, from filets	1 CUP	160	3.0	636.8	148.8	.2	0	0	1472.0	976.0	64.0	—	128.0
			100	3.0	398.0	93.0	.1	0	0	920.0	610.0	40.0	—	80.0
961	flour, from filet waste	1 CUP	160	3.0	488.0	113.6	.3	0	0	9664.0	6496.0	352.0	—	864.0
			100	3.0	305.0	71.0	.2	0	0	6040.0	4060.0	220.0	—	540.0
962	[4]sauce, Hong Kong	1 TBLS	15	65.0	7.7	1.6	0	.2	—	1.5	6.9	1020.0	—	43.5
			100	65.0	51.0	10.7	.1	1.1	—	10.0	46.0	6800.0	—	290.0
963	[4]sauce, Philippines	1 TBLS	15	66.3	7.8	1.6	0	.1	0	6.3	4.8	—	—	—
			100	66.3	52.0	10.6	.3	.9	0	42.0	32.0	—	—	—
964	[4]sauce, Thailand	1 TBLS	15	71.3	2.6	.1	.1	.4	0	—	—	—	—	—
			100	71.3	17.0	.6	.4	2.6	0	—	—	—	—	—
965	[4]sauce, Vietnam	1 TBLS	15	75.7	17.6	1.0	.8	1.4	0	—	—	—	—	—
			100	75.7	117.0	6.8	5.4	9.5	0	—	—	—	—	—
966	**FLATFISHES (FLOUNDER/SOLE/SAND DAB),** broiled	1 OZ	28	70.7	37.8	5.7	1.9	.1	0	4.1	63.9	26.6	10.2	112.0
			100	70.7	135.1	20.4	6.7	.3	0	14.6	228.2	95.2	36.6	400.1
967	**FLOUNDER,** w/salt, baked	1 OZ	28	58.1	56.6	8.4	2.3	0	0	6.4	96.3	66.4	8.4	164.4
			100	58.1	202.0	30.0	8.2	0	0	23.0	344.0	237.0	30.0	587.0
968	**GOOSEFISH,** raw	1 OZ	28	79.3	24.4	5.2	.2	0	0	2.0	50.4	—	—	—
			100	79.3	87.0	18.7	.8	0	0	7.0	180.0	—	—	—
969	**GROUPER,** raw	1 OZ	28	79.2	24.4	5.4	.1	0	0	—	—	—	—	—
			100	79.2	87.0	19.3	.5	0	0	—	—	—	—	—
	HADDOCK													
970	broiled	1 OZ	28	70.0	39.5	5.6	1.8	.1	0	3.6	64.5	20.0	9.0	99.6
			100	70.0	141.0	20.1	6.6	.3	0	13.0	230.5	71.4	32.0	355.7
971	coated w/egg, milk, & bread crumbs, fried	1 OZ	28	66.3	46.2	5.5	1.8	1.6	—	11.2	69.2	49.6	—	97.4
			100	66.3	165.0	19.6	6.4	5.8	—	40.0	247.0	177.0	—	348.0
972	frozen	1 OZ	28	—	—	—	—	—	—	2.8	45.9	56.6	6.7	103.6
			100	—	—	—	—	—	—	10.0	164.0	202.0	24.0	370.0
973	smoked, steamed	1 OZ	28	72.6	28.6	6.3	.4	0	0	7.1	70.2	341.8	7.1	81.2
			100	72.6	102.0	22.2	1.4	0	0	25.2	250.5	1220.2	25.5	290.0
974	**HAKE,** broiled	1 OZ	28	71.2	39.5	5.6	1.8	.1	0	3.6	62.4	25.3	9.0	118.9
			100	71.2	141.0	20.1	6.6	.3	0	13.0	223.0	90.3	32.0	424.7
	HALIBUT													
975	**CALIFORNIA,** raw	1 OZ	28	77.8	27.2	5.5	.3	0	0	3.6	(53.5)	23.7	6.4	(74.6)
			100	77.8	97.0	19.8	1.2	0	0	13.0	(191.1)	84.6	23.0	(266.3)
976	**GREENLAND,** raw	1 OZ	28	74.5	48.7	4.6	3.4	0	0	3.6	58.8	—	6.4	—
			100	74.5	174.0	16.4	12.0	0	0	13.0	210.0	—	23.0	—
977	**ATLANTIC & PACIFIC,** broiled	1 OZ	28	66.6	45.6	7.1	2.0	0	0	4.5	69.4	37.5	—	147.0
			100	66.6	163.0	25.2	7.0	0	0	16.0	248.0	134.0	—	525.0
978	**ATLANTIC & PACIFIC,** smoked	1 OZ	28	49.4	62.7	5.8	4.2	0	0	—	—	—	—	—
			100	49.4	224.0	20.8	15.0	0	0	—	—	—	—	—
	HERRING													
979	**PACIFIC,** raw	1 OZ	28	79.4	27.4	4.9	.7	0	0	12.0	63.0	20.7	8.7	117.6
			100	79.4	98.0	17.5	2.6	0	0	43.0	225.0	74.0	31.0	420.0
980	**ATLANTIC,** cooked	1 AVG	85	—	216.8	20.8	14.2	0	0	—	289.9	—	—	—
			100	—	255.0	24.5	16.7	0	0	—	341.0	—	—	—
981	canned, in tomato sauce	1 OZ	28	66.7	49.3	4.4	2.9	1.0	0	—	68.0	—	—	—
			100	66.7	176.0	15.8	10.5	3.7	0	—	243.0	—	—	—

Item No.	Minerals (Micro) Iron (mg)	Zinc (mg)	Copper (mg)	Fat-Soluble Vitamins Vitamin A (IU)	Vitamin D (IU)	Vitamin E (Alpha Tocopherol) (mg)	Water-Soluble Vitamins Vitamin C (mg)	Thiamin (mg)	Ribo-flavin (mg)	Niacin (mg)	Panto-thenic Acid (mg)	Vit. B-6 (Pyri-doxine) (mg)	Folacin (Folic Acid) (mcg)	Biotin (mcg)	Vitamin B-12 (mcg)
952	**1.32**	—	—	—	—	—	—	—	—	—	—	—	—	—	—
	.80	—	—	—	—	—	—	—	—	—	—	—	—	—	—
953	—	—	—	—	—	—	—	—	—	—	—	—	—	—	—
	—	—	—	—	—	—	—	—	—	—	—	—	—	—	—
954	**.13**	**.02**	—	**91.5**	**0**	—	**2.93**	**.02**	**.02**	**.46**	**.05**	**.02**	**.44**	—	**.14**
	.47	.08	—	326.7	0	—	10.50	.06	.06	1.63	.16	.06	1.59	—	.51
955	**.28**	—	—	**4.7**	—	—	**0**	**.06**	**.17**	**1.40**	—	—	—	—	—
	1.00	—	—	16.7	—	—	0	.20	.60	5.00	—	—	—	—	—
956	—	—	—	—	—	—	—	—	—	—	—	—	—	—	—
	—	—	—	—	—	—	—	—	—	—	—	—	—	—	—
957	**.28**	—	—	**4.7**	—	—	—	**Trace**	**Trace**	**.39**	—	—	—	—	—
	1.00	—	—	16.7	—	—	—	.01	.01	1.40	—	—	—	—	—
958	**.09**	—	—	**0**	—	—	—	**.01**	**.02**	**.37**	—	—	—	—	—
	.40	—	—	0	—	—	—	.04	.07	1.60	—	—	—	—	—
959	**65.60**	—	—	—	—	—	—	**.11**	**.99**	**3.52**	—	—	—	—	—
	41.00	—	—	—	—	—	—	.07	.62	2.20	—	—	—	—	—
960	**12.80**	—	—	—	—	—	—	—	—	—	—	—	—	—	—
	8.00	—	—	—	—	—	—	—	—	—	—	—	—	—	—
961	**86.40**	—	—	—	—	—	—	—	—	—	—	—	—	—	—
	54.00	—	—	—	—	—	—	—	—	—	—	—	—	—	—
962	**.18**	—	—	**52.5**	—	—	—	**Trace**	**.02**	**.32**	—	—	—	—	—
	1.20	—	—	350.0	—	—	—	.01	.10	2.100	—	—	—	—	—
963	**1.40**	—	—	—	—	—	—	—	**.01**	**.62**	—	—	—	—	—
	1.30	—	—	—	—	—	—	—	.08	4.10	—	—	—	—	—
964	**.03**	—	—	—	—	—	—	—	—	**.06**	—	—	—	—	—
	.20	—	—	—	—	—	—	—	—	.40	—	—	—	—	—
965	—	—	—	—	—	—	—	**.01**	**.04**	—	—	—	—	—	—
	—	—	—	—	—	—	—	.03	.27	—	—	—	—	—	—
966	**.26**	—	**.05**	**77.3**	—	—	**.28**	**.01**	**.01**	**.48**	**.29**	**.05**	—	—	**.40**
	.91	—	.18	276.0	—	—	1.00	.04	.05	1.70	1.04	.17	—	—	1.44
967	**.39**	—	—	—	—	—	**.50**	**.02**	**.02**	**.70**	—	—	—	—	—
	1.40	—	—	—	—	—	2.00	.07	.08	2.50	—	—	—	—	—
968	**.28**	—	—	**28.0**	—	—	**0**	**.01**	**.02**	**1.40**	—	—	—	—	—
	1.00	—	—	100.0	—	—	0	.03	.06	5.00	—	—	—	—	—
969	—	—	—	—	—	—	—	**.05**	—	—	—	—	—	—	—
	—	—	—	—	—	—	—	.17	—	—	—	—	—	—	—
970	**.14**	—	—	**77.3**	—	—	**.84**	**.01**	**.02**	**.59**	**.05**	**.06**	—	—	**.22**
	.50	—	—	276.0	—	—	3.00	.03	.07	2.10	.16	.22	—	—	.78
971	**.34**	—	—	—	—	**.17**	**.56**	**.01**	**.02**	**.90**	**.04**	—	—	—	—
	1.20	—	—	—	—	.60	2.00	.04	.07	3.20	.13	—	—	—	—
972	**.39**	**.09**	**Trace**	—	—	—	—	—	—	—	—	—	**2.80**	—	—
	1.40	.32	.01	—	—	—	—	—	—	—	—	—	10.00	—	—
973	**.31**	[6]**.08**	—	**Trace**	**Trace**	—	**Trace**	**.02**	**.01**	**.56**	**.05**	**.11**	**1.40**	**.88**	**.59**
	1.10	.30	—	Trace	Trace	—	Trace	.06	.05	2.00	.18	.36	5.00	3.12	2.10
974	**.14**	—	—	**77.3**	—	—	**.84**	**.02**	**.02**	**.59**	—	**.06**	—	—	**.22**
	.50	—	—	276.0	—	—	3.00	.09	.07	2.10	—	.22	—	—	.78
975	**(.15)**	—	**(.02)**	**Trace**	**Trace**	**(.26)**	**Trace**	**(.02)**	**(.03)**	**(1.40)**	**(.10)**	**(.06)**	**(3.36)**	**(2.00)**	**(.31)**
	(.52)	—	(.06)	Trace	Trace	(.93)	Trace	(.08)	(.10)	(5.00)	(.35)	(.22)	(12.00)	(5.05)	(1.11)
976	—	—	—	—	—	—	—	**Trace**	—	—	**.07**	—	—	—	**.42**
	—	—	—	—	—	—	—	.01	—	—	.25	—	—	—	1.50
977	**.22**	—	—	**190.4**	—	—	—	**.01**	**.02**	**2.32**	—	—	—	—	—
	.80	—	—	680.0	—	—	—	.05	.07	8.30	—	—	—	—	—
978	—	—	—	—	—	—	—	—	—	—	—	—	—	—	—
	—	—	—	—	—	—	—	—	—	—	—	—	—	—	—
979	**.36**	—	—	**28.0**	—	—	**.84**	**.01**	**.05**	**.98**	—	—	—	—	**.56**
	1.30	—	—	100.0	—	—	3.00	.02	.16	3.50	—	—	—	—	2.00
980	**1.19**	—	—	**130.1**	—	—	**0**	**.02**	**.15**	**3.30**	—	—	—	—	—
	1.40	—	—	153.0	—	—	0	.02	.18	3.88	—	—	—	—	—
981	—	—	—	—	—	—	—	—	**.03**	**.98**	—	—	—	—	—
	—	—	—	—	—	—	—	—	.11	3.50	—	—	—	—	—

(Continued)

TABLE F-36 *(Continued)*

Item No.	Food Item Name	Approximate Measure	Weight	Moisture	Food Energy Calories	Protein	Fats	Carbo-hydrates	Fiber	Minerals (Macro) Calcium	Phos-phorus	Sodium	Mag-nesium	Potas-sium
			(g)	(%)	(kcal)	(g)	(g)	(g)	(g)	(mg)	(mg)	(mg)	(mg)	(mg)
	HERRING (Continued)													
982	canned, plain, solids & liquid	1 OZ	**28**	**62.9**	**56.6**	**5.6**	**3.8**	**0**	**0**	**41.2**	**83.2**	–	–	–
			100	62.9	202.0	19.9	13.6	0	0	147.0	297.0	–	–	–
983	pickled, Bismarck type	1 OZ	**28**	**59.4**	**62.4**	**5.7**	**4.2**	**0**	**0**	–	–	–	–	–
			100	59.4	223.0	20.4	15.1	0	0	–	–	–	–	–
984	salted or brined	1 OZ	**28**	**53.8**	**61.0**	**5.3**	**4.3**	**0**	**0**	–	–	–	–	–
			100	53.8	218.0	19.0	15.2	0	0	–	–	–	–	–
985	smoked bloaters	1 OZ	**28**	**64.0**	**54.9**	**5.5**	**3.5**	**0**	**0**	–	–	–	–	–
			100	64.0	196.0	19.6	12.4	0	0	–	–	–	–	–
986	smoked hard	1 OZ	**28**	**34.6**	**84.0**	**10.3**	**4.4**	**0**	**0**	**18.5**	**71.1**	**1744.7**	–	**4.0**
			100	34.6	300.0	36.9	15.8	0	0	66.0	254.0	6231.0	–	157.0
987	smoked kippered	1 OZ	**28**	**61.0**	**59.1**	**6.2**	**3.6**	**0**	**0**	**18.5**	**71.1**	–	–	–
			100	61.0	211.0	22.2	12.9	0	0	66.0	254.0	–	–	–
	HORSE MACKEREL													
988	[3]raw	1 OZ	**28**	**66.8**	**40.0**	**7.0**	**1.1**	**0**	**0**	–	–	–	–	–
			100	66.8	143.0	25.0	4.0	0	0	–	–	–	–	–
989	[3]dried, salted	1 OZ	**28**	**8.3**	**86.5**	**17.1**	**1.5**	**0**	**0**	–	–	–	–	–
			100	8.3	309.0	60.9	5.4	0	0	–	–	–	–	–
990	**INCONNU (SHEEPFISH),** raw	1 OZ	**28**	**72.0**	**40.9**	**5.6**	**1.9**	**0**	**0**	–	–	–	–	–
			100	72.0	146.0	19.9	6.8	0	0	–	–	–	–	–
991	**JACK MACKEREL,** raw	1 OZ	**28**	**71.4**	**40.0**	**6.0**	**1.6**	**0**	**0**	–	–	–	–	–
			100	71.4	143.0	21.6	5.6	0	0	–	–	–	–	–
	JELLY FISH													
992	[4]Medusa, raw	1 OZ	**28**	**85.2**	**8.4**	**1.1**	**0**	**.8**	**0**	**7.3**	**7.6**	–	–	**23.8**
			100	85.2	30.0	4.0	.2	2.9	0	26.0	27.0	–	–	85.0
993	[4]Medusa, salted	1 OZ	**28**	**72.4**	**9.2**	**1.5**	**0**	**.6**	**0**	**16.2**	**75.6**	**899.4**	–	**51.8**
			100	72.4	33.0	5.5	.1	2.2	0	58.0	270.0	3212.0	–	185.0
994	**KINGFISH (SOUTHERN, GULF, NORTHERN WHITING),** cooked	1 OZ	**28**	**67.3**	**71.4**	**6.3**	**3.8**	**3.3**	**0**	**22.4**	**80.4**	**28.4**	**15.7**	**81.9**
			100	67.3	255.0	22.3	13.4	11.7	0	80.0	287.0	101.3	56.0	292.5
995	**LAKE HERRING** (cisco), raw	1 OZ	**28**	**79.7**	**26.9**	**5.0**	**.6**	**0**	**0**	**3.4**	**57.7**	**13.2**	**4.8**	**89.3**
			100	79.7	96.0	17.7	2.3	0	0	12.0	206.0	47.0	17.0	319.0
	LAKE TROUT													
996	raw	1 OZ	**28**	**70.6**	**47.0**	**5.1**	**2.8**	**0**	**0**	**5.6**	**66.6**	**22.4**	**7.3**	**93.5**
			100	70.6	168.0	18.3	10.0	0	0	20.0	238.0	80.0	26.0	334.0
997	**SISCOWET,** raw, less than 6.5 lbs	1 OZ	**28**	**64.9**	**67.5**	**4.0**	**5.6**	**0**	**0**	–	–	–	–	–
			100	64.9	241.0	14.3	19.9	0	0	–	–	–	–	–
998	**SISCOWET,** raw, 6.5 lbs and over	1 OZ	**28**	**36.8**	**146.7**	**2.2**	**15.2**	**0**	**0**	–	–	–	–	–
			100	36.8	524.0	7.9	54.4	0	0	–	–	–	–	–
	LAMPREY													
999	[4]raw	1 OZ	**28**	**60.3**	**70.6**	**5.9**	**5.0**	**0**	**0**	**2.8**	**50.4**	–	–	–
			100	60.3	252.0	21.0	18.0	0	0	10.0	180.0	–	–	–
1000	[4]canned, drained solids w/corn starch, egg white, & salt	1 OZ	**28**	**81.6**	**19.3**	**2.6**	**0**	**2.2**	**0**	**.8**	**16.0**	**105.6**	–	**12.6**
			100	81.6	69.0	9.4	.1	7.7	.1	3.0	57.0	377.0	–	45.0
1001	[4]dried	1 OZ	**28**	**36.5**	**113.1**	**9.3**	**8.1**	**0**	**0**	**4.5**	**67.2**	–	–	–
			100	36.5	404.0	33.3	29.0	0	0	16.0	240.0	–	–	–
1002	**LINGCOD,** raw	1 OZ	**28**	**80.0**	**23.5**	**5.0**	**.2**	**0**	**0**	–	–	**16.5**	–	**121.2**
			100	80.0	84.0	17.9	.8	0	0	–	–	59.0	–	433.0
	LOBSTER													
1003	[1]tail, cooked	1 OZ	**28**	**76.8**	**26.6**	**5.2**	**.4**	**.1**	**0**	**18.2**	**53.8**	**58.8**	–	**50.4**
			100	76.8	95.0	18.7	1.5	.3	0	65.0	192.0	210.0	–	180.0
1004	**NORTHERN,** canned or cooked	1 OZ	**28**	**76.8**	**26.6**	**5.2**	**.4**	**.1**	**0**	**18.2**	**53.8**	**58.8**	**(9.8)**	**50.4**
			100	76.8	95.0	18.7	1.5	.3	0	65.0	192.0	210.0	(34.9)	180.0
1005	newburg	1 OZ	**28**	**64.0**	**54.3**	**5.2**	**3.0**	**1.4**	–	**24.4**	**53.8**	**64.1**	–	**47.9**
			100	64.0	194.0	18.5	10.6	5.1	–	87.0	192.0	229.0	–	171.0
	LUNG-FISH													
1006	[3]raw	1 OZ	**28**	**82.2**	**20.2**	**4.6**	**0**	**0**	**0**	–	–	–	–	–
			100	82.2	72.0	16.4	.2	0	0	–	–	–	–	–
1007	[3]smoked, dried	1 OZ	**28**	**10.7**	**105.3**	**23.0**	**.8**	**0**	**0**	–	–	–	–	–
			100	10.7	376.0	82.1	2.8	0	0	–	–	–	–	–
	MACKEREL													
1008	cooked	1 OZ	**28**	–	**64.4**	**6.4**	**4.3**	**.1**	–	**1.7**	**80.6**	–	**9.5**	–
			100	–	230.0	22.9	15.2	.3	–	6.0	288.0	–	34.0	–

Item No.	Minerals (Micro) Iron	Zinc	Copper	Fat-Soluble Vitamins Vitamin A	Vitamin D	Vitamin E (Alpha Tocopherol)	Water-Soluble Vitamins Vitamin C	Thiamin	Ribo-flavin	Niacin	Panto-thenic Acid	Vit. B-6 (Pyri-doxine)	Folacin (Folic Acid)	Biotin	Vitamin B-12
	(mg)	(mg)	(mg)	(IU)	(IU)	(mg)	(mg)	(mg)	(mg)	(mg)	(mg)	(mg)	(mcg)	(mcg)	(mcg)
982	**.50**	—	—	—	—	—	—	—	**.05**	—	**.20**	**.05**	**1.43**	—	**2.24**
	1.80	—	—	—	—	—	—	—	.18	—	.70	.16	5.10	—	8.00
983	—	—	—	—	—	—	—	—	—	—	—	**.03**	—	—	—
	—	—	—	—	—	—	—	—	—	—	—	.12	—	—	—
984	—	—	—	—	—	—	—	—	**.05**	—	**.14**	**.06**	—	—	**1.96**
	—	—	—	—	—	—	—	—	.19	—	.50	.20	—	—	7.00
985	—	—	—	—	—	—	—	—	—	—	**.26**	**.09**	—	—	**.36**
	—	—	—	—	—	—	—	—	—	—	.93	.32	—	—	1.30
986	**.39**	—	—	**0**	—	—	—	—	**.08**	**.81**	**.26**	**.09**	—	—	**.36**
	1.40	—	—	0	—	—	—	—	.28	2.90	.93	.32	—	—	1.30
987	**.39**	—	—	**8.4**	—	—	—	—	**.08**	**.92**	**.29**	**.07**	—	—	**.42**
	1.40	—	—	30.0	—	—	—	—	.28	3.30	1.04	.25	—	—	1.50
988	—	—	—	—	—	—	—	—	—	—	—	—	—	—	—
	—	—	—	—	—	—	—	—	—	—	—	—	—	—	—
989	—	—	—	—	—	—	—	**.04**	**.05**	**3.02**	—	—	—	—	—
	—	—	—	—	—	—	—	.13	.16	10.80	—	—	—	—	—
990	—	—	—	—	—	—	—	—	—	—	—	—	—	—	—
	—	—	—	—	—	—	—	—	—	—	—	—	—	—	—
991	—	—	—	—	—	—	—	—	—	—	—	—	—	—	—
	—	—	—	—	—	—	—	—	—	—	—	—	—	—	—
992	**.22**	—	—	**0**	—	—	**0**	**.02**	**Trace**	—	—	—	—	—	—
	.80	—	—	0	—	—	0	.06	.01	—	—	—	—	—	—
993	**1.79**	—	—	**0**	—	—	**0**	—	—	**0**	—	—	—	—	—
	6.40	—	—	0	—	—	0	—	—	0	—	—	—	—	—
994	**.53**	—	—	**26.0**	—	—	**0**	**.03**	**.04**	**.81**	—	**.08**	—	—	**.45**
	1.90	—	—	93.0	—	—	0	.11	.13	2.90	—	.27	—	—	1.59
995	**.14**	**1.06**	—	—	—	—	—	**.03**	**.03**	**.92**	—	—	—	—	—
	.50	3.80	—	—	—	—	—	.09	.10	3.30	—	—	—	—	—
996	**.22**	**1.71**	—	**3.4**	—	—	—	**.03**	**.03**	**.76**	—	—	—	—	—
	.80	6.10	—	12.0	—	—	—	.09	.12	2.70	—	—	—	—	—
997	—	—	—	—	—	—	—	—	—	—	—	—	—	—	—
	—	—	—	—	—	—	—	—	—	—	—	—	—	—	—
998	—	**.95**	—	—	—	—	—	—	—	—	—	—	—	—	—
	—	3.40	—	—	—	—	—	—	—	—	—	—	—	—	—
999	**2.52**	—	—	**7000.0**	—	—	**0**	**.24**	**1.68**	**1.32**	—	—	—	—	—
	9.00	—	—	25000.0	—	—	0	.85	6.00	4.70	—	—	—	—	—
1000	**1.14**	—	—	—	—	—	**0**	**Trace**	**.01**	**.25**	—	—	—	—	—
	4.10	—	—	—	—	—	0	.01	.04	.90	—	—	—	—	—
1001	**3.92**	—	—	—	—	—	**0**	**.28**	**1.68**	**1.96**	—	—	—	—	—
	14.00	—	—	—	—	—	0	1.00	6.00	7.00	—	—	—	—	—
1002	—	—	—	**0**	—	—	—	**.01**	**.01**	—	—	—	—	—	**1.01**
	—	—	—	0	—	—	—	.05	.04	—	—	—	—	—	3.60
1003	**.22**	**.62**	—	—	—	—	**0**	**.03**	**.02**	**.62**	—	—	—	—	—
	.80	2.20	—	—	—	—	0	.10	.07	2.23	—	—	—	—	—
1004	**.22**	**.62**	**(.49)**	**77.0**	**Trace**	**(.42)**	**0**	**.03**	**.02**	**.62**	**(.46)**	—	**4.76**	**(1.42)**	**(.30)**
	.80	2.20	(1.76)	275.0	Trace	(1.50)	0	.10	.07	2.23	(1.65)	—	17.00	(5.06)	(1.08)
1005	**.25**	—	—	—	—	—	—	**.02**	**.03**	—	—	—	—	—	—
	.90	—	—	—	—	—	—	.07	.11	—	—	—	—	—	—
1006	—	—	—	—	—	—	—	—	—	—	—	—	—	—	—
	—	—	—	—	—	—	—	—	—	—	—	—	—	—	—
1007	—	—	—	—	—	—	—	—	—	—	—	—	—	—	—
	—	—	—	—	—	—	—	—	—	—	—	—	—	—	—
1008	**.34**	—	—	**157.4**	—	—	**.28**	**.04**	**.10**	**2.35**	—	**.19**	—	—	**2.27**
	1.20	—	—	562.0	—	—	1.00	.14	.36	8.40	—	.68	—	—	8.12

(Continued)

TABLE F-36 *(Continued)*

Item No.	Food Item Name	Approximate Measure	Weight	Moisture	Food Energy Calories	Protein	Fats	Carbo-hydrates	Fiber	Minerals (Macro) Calcium	Phos-phorus	Sodium	Mag-nesium	Potas-sium
			(g)	(%)	(kcal)	(g)	(g)	(g)	(g)	(mg)	(mg)	(mg)	(mg)	(mg)
	MACKEREL (Continued)													
1009	**ATLANTIC,** broiled w/butter	1 OZ	**28**	**61.6**	**66.1**	**6.1**	**4.4**	**0**	**0**	**1.7**	**78.4**	–	–	–
			100	61.6	236.0	21.8	15.8	0	0	6.0	280.0	–	–	–
1010	**ATLANTIC,** broiled w/margarine	1 OZ	**28**	**61.6**	**66.1**	**6.1**	**4.4**	**0**	**0**	**1.7**	**78.4**	–	–	–
			100	61.6	236.0	21.8	15.8	0	0	6.0	280.0	–	–	–
1011	**ATLANTIC,** canned, solids & liquid, vitamins based on dried solids	1 CUP	**210**	**66.0**	**384.3**	**40.5**	**23.3**	**0**	**0**	**388.5**	**575.4**	–	–	–
			100	66.0	183.0	19.3	11.1	0	0	185.0	274.0	–	–	–
1012	**PACIFIC,** canned, solids & liquid, vitamins based on dried solids	1 CUP	**210**	**66.4**	**378.0**	**44.3**	**21.0**	**0**	**0**	**546.0**	**604.8**	–	–	–
			100	66.4	180.0	21.1	10.0	0	0	260.0	288.0	–	–	–
1013	salted	1 OZ	**28**	**43.0**	**85.4**	**5.2**	**7.0**	**0**	**0**	–	–	–	–	–
			100	43.0	305.0	18.5	25.1	0	0	–	–	–	–	–
1014	smoked	1 OZ	**28**	**59.4**	**61.3**	**6.7**	**3.6**	**0**	**0**	–	–	–	–	–
			100	59.4	219.0	23.8	13.0	0	0	–	–	–	–	–
	MAHIMAHI													
1015	plain, broiled	1 SERV	**100**	–	–	–	–	–	–	–	–	**120.0**	–	**267.0**
			100	–	–	–	–	–	–	–	–	120.0	–	267.0
1016	breaded, fried	1 SERV	**124**	–	–	–	–	–	–	–	–	**466.2**	–	**285.2**
			100	–	–	–	–	–	–	–	–	376.0	–	230.0
1017	**MENHADEN, ATLANTIC,** canned, solids & liquid	1 OZ	**28**	**67.9**	**48.2**	**5.2**	**2.9**	**0**	**0**	–	–	–	–	–
			100	67.9	172.0	18.7	10.2	0	0	–	–	–	–	–
1018	**MULLET, STRIPED,** breaded, fried	1 OZ	**28**	**53.7**	**82.9**	**6.3**	**4.9**	**3.4**	–	**14.8**	**72.2**	**27.7**	**10.6**	**95.7**
			100	53.7	296.0	22.6	17.4	12.2	–	53.0	258.0	98.8	38.0	341.6
1019	**MUSKELLUNGE,** raw	1 OZ	**28**	**76.3**	**30.5**	**5.7**	**.7**	**0**	**0**	–	**63.6**	–	–	–
			100	76.3	109.0	20.2	2.5	0	0	–	227.0	–	–	–
1020	**MUSSELS, PACIFIC,** canned, drained solids	1 OZ	**28**	**74.6**	**31.9**	**5.1**	**.9**	**.4**	–	–	–	–	**7.0**	–
			100	74.6	114.0	18.2	3.3	1.5	–	–	–	–	25.0	–
	OCTOPUS													
1021	raw	1 OZ	**28**	**82.2**	**20.4**	**4.3**	**.2**	**0**	**0**	**8.1**	**48.4**	**(142.8)**	**(11.9)**	**(73.1)**
			100	82.2	73.0	15.3	.8	0	0	29.0	173.0	(509.8)	(42.6)	(261.0)
1022	dried	1 OZ	**28**	**12.0**	**111.7**	**14.1**	**4.0**	**4.0**	**0**	**39.8**	**42.0**	–	–	–
			100	12.0	399.0	50.3	14.3	14.3	0	142.0	150.0	–	–	–
	OYSTER													
1023	raw meat only, Pacific and Eastern (Olympia)	1 MED	**35**	**79.1**	**30.1**	**3.7**	**.7**	**2.2**	–	**29.8**	**53.6**	–	**8.4**	–
			100	79.1	86.0	10.6	2.0	6.4	–	85.0	153.0	–	24.0	–
1024	fried	1 MED	**15**	**54.7**	**35.9**	**1.3**	**2.1**	**2.8**	–	**22.8**	**36.2**	**30.9**	–	**30.5**
			100	54.7	239.0	8.6	13.9	18.6	–	152.0	241.0	206.0	–	203.0
1025	canned, solids & liquid	1 MED	**15**	**82.2**	**11.0**	**1.3**	**.3**	**.7**	**Trace**	**4.2**	**18.6**	–	**5.0**	**10.5**
			100	82.2	73.0	8.5	2.2	4.9	.1	28.0	124.0	–	33.0	70.0
1026	frozen, solids & liquid	1 MED	**15**	**87.4**	–	**.9**	–	–	–	–	–	**57.0**	**2.3**	**31.5**
			100	87.4	–	6.1	–	–	–	–	–	380.0	15.4	210.0
	PERCH													
1027	**WHITE,** raw	1 MED	**100**	**75.7**	**118.0**	**19.3**	**4.0**	**0**	**0**	–	**192.0**	**79.0**	–	–
			100	75.7	118.0	19.3	4.0	0	0	–	192.0	79.0	–	–
1028	**YELLOW,** raw	1 MED	**100**	**79.2**	**91.0**	**19.5**	**.9**	**0**	**0**	–	**180.0**	**68.0**	–	**230.0**
			100	79.2	91.0	19.5	.9	0	0	–	180.0	68.0	–	230.0
1029	filet, fried	1 OZ	**28**	–	**46.5**	**5.4**	**2.3**	**0**	**0**	**3.9**	**48.7**	–	–	–
			100	–	166.0	19.2	8.2	0	0	14.0	174.0	–	–	–
1030	breaded, fried	1 OZ	**28**	**59.0**	**64.1**	**5.3**	**3.6**	**2.0**	–	**9.2**	–	–	–	–
			100	59.0	229.0	18.8	12.9	7.0	–	33.0	–	–	–	–
1031	**ATLANTIC (REDFISH),** raw	1 OZ	**28**	**79.7**	**24.6**	**5.0**	**.3**	**0**	**0**	**5.6**	**58.0**	**22.1**	**9.0**	**75.3**
			100	79.7	88.0	18.0	1.2	0	0	20.0	207.0	79.0	32.0	269.0
1032	**ATLANTIC (REDFISH),** fried	1 OZ	**28**	**59.0**	**63.6**	**5.3**	**3.7**	**1.9**	**0**	**9.2**	**63.3**	**42.8**	–	**79.5**
			100	59.0	227.0	19.0	13.3	6.8	0	33.0	226.0	153.0	–	284.0
1033	**ATLANTIC (REDFISH),** breaded, fried, frozen, reheated	1 OZ	**28**	**43.2**	**89.3**	**5.3**	**5.3**	**4.6**	–	–	–	–	–	–
			100	43.2	319.0	18.9	18.9	16.5	–	–	–	–	–	–
1034	**PICKEREL,** chain, raw	1 OZ	**28**	**79.7**	**23.5**	**5.2**	**.1**	**0**	**0**	–	–	–	–	–
			100	79.7	84.0	18.7	.5	0	0	–	–	–	–	–
	PIKE													
1035	**WALLEYE,** raw	1 OZ	**30**	**78.3**	**27.9**	**5.8**	**.4**	**0**	**0**	–	**64.2**	**15.3**	**9.3**	**95.7**
			100	78.3	93.0	19.3	1.2	0	0	–	214.0	51.0	31.0	319.0
1036	**NORTHERN,** breaded, fried	1 OZ	**28**	**59.2**	**59.4**	**5.4**	**3.3**	**2.2**	–	**9.2**	**65.0**	**25.6**	**6.2**	**114.7**
			100	59.2	212.0	19.1	11.8	7,9	–	33.0	232.0	91.5	22.0	409.5

Item No.	Minerals (Micro)			Fat-Soluble Vitamins			Water-Soluble Vitamins								
	Iron	Zinc	Copper	Vitamin A	Vitamin D	Vitamin E (Alpha Tocopherol)	Vitamin C	Thiamin	Ribo-flavin	Niacin	Panto-thenic Acid	Vit. B-6 (Pyri-doxine)	Folacin (Folic Acid)	Biotin	Vitamin B-12
	(mg)	(mg)	(mg)	(IU)	(IU)	(mg)	(mg)	(mg)	(mg)	(mg)	(mg)	(mg)	(mcg)	(mcg)	(mcg)
1009	.34	—	—	148.4	—	—	—	.04	.08	2.13	—	—	—	—	—
	1.20	—	—	530.0	—	—	—	.15	.27	7.60	—	—	—	—	—
1010	.34	—	—	148.4	—	—	—	.04	.08	2.13	—	—	—	—	—
	1.20	—	—	530.0	—	—	—	.15	.27	7.60	—	—	—	—	—
1011	4.41	—	—	903.0	—	—	—	.13	.44	12.18	1.05	.59	—	—	16.17
	2.10	—	—	430.0	—	—	—	.06	.21	5.80	.50	.28	—	—	7.70
1012	4.62	—	—	63.0	—	—	—	.06	.69	18.48	.99	.57	1.26	37.80	—
	2.20	—	—	30.0	—	—	—	.03	.33	8.80	.47	.27	.60	18.00	—
1013	—	—	—	—	—	—	—	—	—	—	.15	.12	—	—	3.36
	—	—	—	—	—	—	—	—	—	—	.52	.41	—	—	12.00
1014	—	[6] .31	—	—	—	—	—	—	—	—	—	—	—	—	—
	—	1.10	—	—	—	—	—	—	—	—	—	—	—	—	—
1015	—	—	—	—	—	—	—	—	—	—	—	—	—	—	—
	—	—	—	—	—	—	—	—	—	—	—	—	—	—	—
1016	—	—	—	—	—	—	—	—	—	—	—	—	—	—	—
	—	—	—	—	—	—	—	—	—	—	—	—	—	—	—
1017	.36	—	—	—	—	—	—	—	—	—	—	—	—	—	—
	1.30	—	—	—	—	—	—	—	—	—	—	—	—	—	—
1018	.66	—	—	26.6	—	—	0	.03	.03	1.43	—	.10	—	—	.06
	2.34	—	—	95.0	—	—	0	.10	.10	5.10	—	.37	—	—	.21
1019	.17	—	—	—	—	—	—	—	—	—	—	—	—	—	—
	.60	—	—	—	—	—	—	—	—	—	—	—	—	—	—
1020	—	—	—	—	—	—	—	—	.04	—	—	—	—	—	—
	—	—	—	—	—	—	—	—	.13	—	—	—	—	—	—
1021	(1.74)	(12.44)	(2.10)	(2.1)	Trace	(.24)	Trace	.01	.02	.50	(.15)	.10	—	(2.83)	(4.23)
	(6.20)	(44.40)	(7.50)	(7.5)	Trace	(.86)	Trace	.02	.06	1.80	(.54)	.36	—	(10.10)	(15.11)
1022	.87	—	—	0	—	—	0	.05	.11	—	—	—	—	—	—
	3.10	—	—	0	—	—	0	.18	.38	—	—	—	—	—	—
1023	2.52	52.05	4.80	—	—	—	10.50	.04	—	.46	.09	.02	3.50	—	—
	7.20	148.70	13.71	—	—	—	30.00	.12	—	1.30	.25	.05	10.00	—	—
1024	1.22	—	—	66.0	—	—	—	.03	.04	.48	—	—	—	—	—
	8.10	—	—	440.0	—	—	—	.17	.29	3.20	—	—	—	—	—
1025	.84	—	—	—	—	—	—	Trace	.03	.12	.07	.01	1.70	1.31	—
	5.60	—	—	—	—	—	—	.02	.20	.80	.49	.04	11.30	8.70	—
1026	—	11.21	—	46.5	—	—	—	.02	.03	.38	—	—	—	—	—
	—	74.70	—	310.0	—	—	—	.14	.18	2.50	—	—	—	—	—
1027	—	—	—	—	—	—	—	—	—	—	—	—	—	—	—
	—	—	—	—	—	—	—	—	—	—	—	—	—	—	—
1028	.60	—	—	—	—	—	—	.06	.17	1.70	—	—	—	—	—
	.60	—	—	—	—	—	—	.06	.17	1.70	—	—	—	—	—
1029	.31	.70	—	0	—	—	0	.02	.02	1.16	—	—	—	—	—
	1.10	2.50	—	0	—	—	0	.06	.07	4.15	—	—	—	—	—
1030	.36	—	—	—	—	—	—	.03	.03	.49	—	—	—	—	—
	1.30	—	—	—	—	—	—	.09	.10	1.76	—	—	—	—	—
1031	.28	—	—	—	—	—	—	.03	.02	.53	.10	.06	—	—	.28
	1.00	—	—	—	—	—	—	.10	.08	1.90	.36	.23	—	—	1.00
1032	.36	—	—	—	—	—	—	.03	.03	.50	—	—	—	—	—
	.30	—	—	—	—	—	—	.10	.11	1.80	—	—	—	—	—
1033	—	—	—	—	—	—	—	—	—	—	—	—	—	—	—
	—	—	—	—	—	—	—	—	—	—	—	—	—	—	—
1034	.20	—	—	—	—	—	—	—	—	—	—	—	—	—	—
	.70	—	—	—	—	—	—	—	—	—	—	—	—	—	—
1035	.12	—	—	—	—	—	—	.08	.05	.69	—	—	—	—	—
	.40	—	—	—	—	—	—	.25	.16	2.30	—	—	—	—	—
1036	.25	—	—	13.7	—	—	0	.06	.05	.64	—	.03	—	—	.42
	.90	—	—	49.0	—	—	0	.21	.19	2.30	—	.11	—	—	1.51

(Continued)

TABLE F-36 *(Continued)*

Item No.	Food Item Name	Approximate Measure	Weight	Moisture	Food Energy Calories	Protein	Fats	Carbo-hydrates	Fiber	Minerals (Macro) Calcium	Phos-phorus	Sodium	Mag-nesium	Potas-sium
			(g)	(%)	(kcal)	(g)	(g)	(g)	(g)	(mg)	(mg)	(mg)	(mg)	(mg)
	PIKE (Continued)													
1037	**POLLOCK,** creamed, cooked	1 SERV	**100** 100	**74.7** 74.7	**128.0** 128.0	**13.9** 13.9	**5.9** 5.9	**4.0** 4.0	**0** 0	— —	— —	**111.0** 111.0	— —	**238.0** 238.0
1038	**POMPANO,** broiled	1 OZ	**28** 100	**61.7** 61.7	**79.5** 283.9	**6.4** 22.9	**15.5** 55.4	**0** 0	**0** 0	— —	— —	**16.1** 57.3	— —	**62.5** 223.4
1039	**PORGY & SCUP,** raw	1 OZ	**28** 100	**76.2** 76.2	**31.4** 112.0	**5.3** 19.0	**1.0** 3.4	**0** 0	**0** 0	**15.1** 54.0	**70.0** 250.0	**17.6** 63.0	— —	**80.4** 287.0
1040	[6]**PRAWNS,** boiled	1 OZ	**28** 100	— —	**30.0** 107.0	**6.3** 22.5	**.5** 1.9	**0** 0	**0** 0	**43.5** 155.2	**96.2** 343.3	**442.7** 1580.4	**12.1** 43.3	**72.4** 258.5
1041	**REDHORSE, SILVER,** raw	1 OZ	**28** 100	**78.6** 78.6	**27.4** 98.0	**5.0** 18.0	**.6** 2.3	**0** 0	**0** 0	— —	— —	— —	— —	— —
1042	**ROCKFISH (BLACK, CANARY, YELLOWTAIL, RASPHEAD, BOCACCIO),** cooked	1 OZ	**28** 100	**75.4** 75.4	**30.0** 107.0	**5.1** 18.1	**.7** 2.5	**.5** 1.9	**0** 0	— —	— —	**19.0** 68.0	— —	**124.9** 446.0
	ROE													
1043	**(SALMON, STURGEON, TURBOT)** raw	1 OZ	**28** 100	**61.3** 61.3	**58.0** 207.0	**7.1** 25.2	**2.9** 10.4	**.4** 1.4	— —	— —	— —	— —	— —	— —
1044	**(COD, SHAD),** baked or broiled	1 OZ	**28** 100	**71.3** 71.3	**35.3** 126.0	**6.2** 22.0	**.8** 2.8	**.5** 1.9	**0** 0	**3.6** 13.0	**112.6** 402.0	**20.4** 73.0	**(3.0)** (10.8)	**37.0** 132.0
1045	**(COD, HADDOCK, HERRING),** canned, solids & liquid	1 OZ	**28** 100	**72.4** 72.4	**33.0** 118.0	**6.0** 21.5	**.8** 2.8	**.1** .3	**0** 0	**4.2** 15.0	**96.9** 346.0	— —	— —	— —
1046	**SABLEFISH,** raw	1 OZ	**28** 100	**71.6** 71.6	**53.2** 190.0	**3.6** 13.0	**4.2** 14.9	**0** 0	**0** 0	— —	— —	**15.7** 56.0	— —	**100.2** 358.0
	SALMON													
1047	broiled or baked	1 SMALL	**100** 100	**63.4** 63.4	**182.0** 182.0	**27.0** 27.0	**6.5** 6.5	**0** 0	**0** 0	— —	**414.0** 414.0	**116.0** 116.0	— —	**443.0** 443.0
1048	patty	1 OZ	**28** 100	— —	**32.0** 114.4	**4.4** 15.8	**3.5** 12.4	**4.5** 16.1	**.2** .8	**21.9** 78.1	**29.0** 103.6	**27.0** 96.5	**9.6** 34.1	**24.9** 88.9
1049	patty, cooked	1 AVG	**87** 100	— —	**187.7** 215.8	**11.6** 13.3	**12.0** 13.8	**8.4** 9.6	— —	**121.8** 140.0	**185.3** 213.0	— —	**20.9** 24.0	— —
1050	rice loaf, 3 ¼″ X 2 ½″ X 1 ½″ slice	1 SLICE	**174** 100	**74.4** 74.4	**212.3** 122.0	**20.9** 12.0	**7.8** 4.5	**12.7** 7.3	— —	— —	— —	— —	— —	— —
1051	**ATLANTIC,** canned, solids & liquid	1 OZ	**28** 100	**64.2** 64.2	**34.7** 124.0	**6.1** 21.7	**2.6** 9.3	**0** 0	**0** 0	**60.2** 215.0	— —	— —	**8.1** 29.0	— —
1052	**CHINOOK (KING),** canned w/o salt, solids, liquid, & bones	1 CUP	**250** 100	**64.4** 64.4	**525.0** 210.0	**45.8** 15.3	**39.5** 15.8	**0** 0	**0** 0	**260.0** 104.0	**420.0** 168.0	**112.5** 45.0	**67.5** 27.0	**915.0** 366.0
1053	**CHUM,** canned w/o salt, solids, liquid, & bones	1 OZ	**28** 100	**70.8** 70.8	**34.7** 124.0	**6.0** 21.5	**1.2** 4.2	**0** 0	**0** 0	**55.4** 198.0	**59.9** 214.0	**14.8** 53.0	**8.4** 30.0	**94.1** 336.0
1054	**COHO, (SILVER),** canned, solids, liquid, & bones	1 OZ	**28** 100	**69.3** 69.3	**42.8** 153.0	**5.3** 18.8	**2.3** 8.2	**0** 0	**0** 0	**51.8** 185.0	**69.7** 249.0	**98.3** 351.0	**8.4** 30.0	**94.9** 339.0
1055	**PINK, (HUMPBACK),** canned, solids, liquid, & bones	1 CUP	**250** 100	**70.8** 70.8	**352.5** 141.0	**47.0** 18.8	**14.8** 5.9	**0** 0	**0** 0	**445.0** 178.0	**612.5** 245.0	**967.5** 387.0	**75.0** 30.0	**902.5** 361.0
1056	**SOCKEYE, (RED),** canned, solids, liquid, & bones	1 OZ	**28** 100	**67.2** 67.2	**46.8** 167.0	**5.7** 20.2	**2.4** 8.4	**0** 0	**0** 0	**49.8** 178.0	**60.5** 216.0	**146.2** 522.0	**8.1** 29.0	**96.3** 344.0
1057	smoked	1 OZ	**28** 100	**58.9** 58.9	**49.3** 176.0	**6.0** 21.6	**2.6** 9.3	**0** 0	**0** 0	**3.9** 14.0	**68.6** 245.0	**(560.1)** (1880.2)	**(8.8)** (31.5)	**(118.2)** (422.0)
	SARDINES													
1058	**ATLANTIC,** canned in oil, solids & liquid	1 MED	**12** 100	**50.6** 50.6	**29.5** 246.0	**2.5** 21.1	**2.9** 24.4	**.1** .6	**0** 0	**42.5** 354.0	**52.1** 434.0	**61.2** 510.0	— —	**67.2** 560.0
1059	**ATLANTIC,** canned in oil, drained, solids, skin & bones	1 MED	**13** 100	**61.8** 61.8	**26.4** 203.0	**3.1** 24.0	**1.4** 11.1	**.2** 1.2	**0** 0	**56.8** 437.0	**64.9** 499.0	**107.0** 823.0	— —	**76.7** 590.0
1060	**ATLANTIC,** canned in oil, drained, solids, no skin or bone	1 MED	**13** 100	**61.8** 61.8	**25.5** 196.0	**3.1** 24.0	**1.4** 11.1	**0** 0	**0** 0	**7.0** 54.0	**41.5** 319.0	**107.0** 823.0	**(14.7)** (52.3)	**76.7** 590.0
1061	**PACIFIC,** canned in brine, or mustard, solids & liquid	1 OZ	**28** 100	**64.1** 64.1	**52.1** 186.0	**5.0** 17.7	**3.4** 12.0	**.5** 1.7	**0** 0	**84.8** 303.0	**99.1** 354.0	**212.8** 760.0	**11.5** 41.0	**72.8** 260.0
1062	**PACIFIC,** canned in oil, drained, solids	1 OZ	**28** 100	— —	— —	— —	**3.1** 11.1	— —	**0** 0	— —	— —	— —	— —	— —
1063	**PACIFIC,** canned in tomato sauce, solids & liquid	1 LG	**66** 100	**64.3** 64.3	**123.4** 187.0	**11.7** 17.7	**8.1** 12.2	**1.1** 1.7	**0** 0	**296.3** 449.0	**315.5** 478.0	**264.0** 400.0	**(14.3)** (51.2)	**211.2** 320.0
1064	**SAUGER,** raw	1 OZ	**28** 100	**80.8** 80.8	**23.5** 84.0	**5.0** 17.9	**.2** .8	**0** 0	**0** 0	— —	— —	— —	— —	— —
	SCAD													
1065	[4]raw	1 OZ	**28** 100	**80.6** 80.6	**21.3** 76.0	**4.9** 17.5	**0** .1	**0** 0	**0** 0	**14.0** 50.0	**32.2** 115.0	**17.1** 61.0	— —	**171.9** 614.0
1066	[4]dried, salted	1 OZ	**28** 100	**43.3** 43.3	**49.0** 175.0	**10.6** 37.8	**.4** 1.5	**0** 0	**0** 0	**93.0** 332.0	**228.8** 817.0	— —	— —	— —

Item No.	Minerals (Micro)			Fat-Soluble Vitamins			Water-Soluble Vitamins								
	Iron	Zinc	Copper	Vitamin A	Vitamin D	Vitamin E (Alpha Tocopherol)	Vitamin C	Thiamin	Ribo-flavin	Niacin	Panto-thenic Acid	Vit. B-6 (Pyri-doxine)	Folacin (Folic Acid)	Biotin	Vitamin B-12
	(mg)	(mg)	(mg)	(IU)	(IU)	(mg)	(mg)	(mg)	(mg)	(mg)	(mg)	(mg)	(mcg)	(mcg)	(mcg)
1037	–	–	–	–	–	–	–	.03	.13	.70	–	–	–	–	–
	–	–	–	–	–	–	–	.03	.13	.70	–	–	–	–	–
1038	–	–	–	–	–	–	–	.14	.08	–	–	–	–	–	–
	–	–	–	–	–	–	–	.50	.27	–	–	–	–	–	–
1039	–	–	–	–	–	–	–	–	–	–	–	–	–	–	–
	–	–	–	–	–	–	–	–	–	–	–	–	–	–	–
1040	.31	.40	.20	Trace	Trace	–	Trace	–	–	–	–	–	–	–	–
	1.10	1.60	.70	Trace	Trace	–	Trace	–	–	–	–	–	–	–	–
1041	–	–	–	–	–	–	–	–	–	–	–	–	–	–	–
	–	–	–	–	–	–	–	–	–	–	–	–	–	–	–
1042	–	–	–	–	–	–	.28	.01	.03	–	–	–	–	–	–
	–	–	–	–	–	–	1.00	.05	.12	–	–	–	–	–	–
1043	–	–	–	–	–	–	5.04	.11	.20	.64	–	–	–	–	–
	–	–	–	–	–	–	18.00	.38	.72	2.30	–	–	–	–	–
1044	.64	–	–	Trace	.64	(1.90)	7.0	(.36)	(.25)	(.36)	(.73)	(.08)	–	(4.23)	(3.08)
	2.30	–	–	Trace	2.30	(6.80)	25.0	(1.30)	(.89)	(1.29)	(2.61)	(.27)	–	(15.10)	(10.98)
1045	.34	–	–	–	–	–	.56	–	–	–	.55	.04	–	–	4.20
	1.20	–	–	–	–	–	2.00	–	–	–	1.97	.14	–	–	15.00
1046	–	–	–	–	–	–	–	.03	.03	–	–	–	–	–	–
	–	–	–	–	–	–	–	.11	.09	–	–	–	–	–	–
1047	1.20	–	.20	160.0	–	1.35	–	.16	.06	9.80	–	–	–	–	–
	1.20	–	.20	160.0	–	1.35	–	.16	.06	9.80	–	–	–	–	–
1048	.35	.23	–	18.5	11.82	–	1.04	.03	.06	1.13	.18	.02	3.70	–	.96
	1.24	.84	–	65.9	42.23	–	3.72	.12	.22	4.04	.66	.07	13.22	–	3.42
1049	.78	–	–	90.5	–	–	2.61	.05	.11	3.48	–	.16	–	–	2.48
	.90	–	–	104.0	–	–	3.00	.06	.13	4.00	–	.18	–	–	2.85
1050	–	–	–	–	–	–	–	–	–	–	–	–	–	–	–
	–	–	–	–	–	–	–	–	–	–	–	–	–	–	–
1051	–	.17	–	–	–	–	–	–	–	–	.15	.08	5.60	4.20	1.93
	–	.60	–	–	–	–	–	–	–	–	.55	.30	20.00	15.00	6.89
1052	1.50	1.23	–	625.0	–	–	0	.08	.06	18.25	1.38	.75	50.00	37.50	17.23
	.60	.49	–	250.0	–	–	0	.03	.03	7.30	.55	.30	20.00	15.00	6.89
1053	.20	.16	–	16.8	–	–	0	.03	.01	1.99	.15	.08	5.60	4.20	1.93
	.70	.58	–	60.0	–	–	0	.12	.04	7.10	.55	.30	20.00	15.00	6.89
1054	.17	.17	(.03)	53.2	(3.51)	(.43)	0	.01	.02	2.07	.15	.08	5.60	4.20	1.93
	.60	.61	(.09)	190.0	(12.52)	(1.55)	0	.03	.06	7.40	.55	.30	20.00	15.00	6.89
1055	2.00	1.90	–	175.0	–	–	0	.23	.10	15.85	1.38	.75	50.00	37.50	17.23
	.80	.76	–	70.0	–	–	0	.09	.04	6.34	.55	.30	20.00	15.00	6.89
1056	.17	.19	.02	46.2	–	–	0	.03	.01	2.04	.15	.08	5.60	4.20	1.93
	.60	.66	.08	165.0	–	–	0	.09	.05	7.30	.55	.30	20.00	15.00	6.89
1057	(.17)	(.12)	(.02)	Trace	Trace	–	Trace	(.04)	(.05)	(2.47)	.20	.20	–	–	1.96
	(.60)	(.43)	(.08)	Trace	Trace	–	Trace	(.15)	(.18)	(8.82)	.71	.70	–	–	7.00
1058	.42	–	.01	21.6	–	–	0	Trace	.02	.53	.10	.02	1.92	–	1.20
	3.50	–	.11	180.0	–	–	0	.020	.16	4.40	.85	.18	16.00	–	10.00
1059	.38	[6] .39	.01	28.6	–	–	0	Trace	.03	.70	.11	.02	2.08	–	1.30
	2.90	3.00	.11	220.0	–	–	0	.03	.20	5.40	.85	.18	16.00	–	10.00
1060	.38	(.93)	(.05)	28.6	(2.12)	(.10)	0	Trace	.03	.70	.11	.02	2.08	(1.42)	1.30
	2.90	(3.33)	(.18)	220.0	(7.58)	(.34)	0	.03	.20	5.40	.85	.18	16.00	(5.06)	10.00
1061	1.46	–	–	8.4	–	–	–	–	–	–	.17	.08	4.48	6.72	–
	5.20	–	–	30.0	–	–	–	–	–	–	.60	.28	16.00	24.00	–
1062	–	–	–	–	–	–	–	Trace	.08	2.07	.17	.08	4.48	6.72	–
	–	–	–	–	–	–	–	.01	.30	7.40	.60	.28	16.00	24.00	–
1063	2.71	–	–	19.8	(2.10)	(.15)	Trace	.01	.18	3.50	.26	.15	(3.65)	(1.40)	(3.94)
	4.10	–	–	30.0	(7.52)	(.52)	Trace	.01	.27	5.30	.40	.22	(13.04)	(5.00)	(14.06)
1064	–	–	–	–	–	–	–	–	–	–	–	–	–	–	–
	–	–	–	–	–	–	–	–	–	–	–	–	–	–	–
1065	.11	–	–	–	–	–	–	.01	.02	.89	–	–	–	–	–
	.40	–	–	–	–	–	–	.03	.08	3.20	–	–	–	–	–
1066	1.43	–	–	–	–	–	–	.01	.01	3.78	–	–	–	–	–
	5.10	–	–	–	–	–	–	.02	.05	13.50	–	–	–	–	–

(Continued)

TABLE F-36 *(Continued)*

Item No.	Food Item Name	Approximate Measure	Weight	Moisture	Food Energy Calories	Protein	Fats	Carbo-hydrates	Fiber	Minerals (Macro) Calcium	Phos-phorus	Sodium	Mag-nesium	Potas-sium
			(g)	(%)	(kcal)	(g)	(g)	(g)	(g)	(mg)	(mg)	(mg)	(mg)	(mg)
	SCALLOPS													
1067	bay & sea, steamed	1 OZ	28	73.1	29.4	6.5	.4	Trace	0	32.2	94.6	74.2	(10.9)	133.3
			100	73.1	105.0	23.2	1.4	Trace	0	115.0	338.0	265.0	(38.8)	476.0
1068	bay & sea, breaded, fried, frozen, reheated	1 MED	18	60.2	34.9	3.2	1.5	1.9	0	—	—	—	6.0	—
			100	60.2	194.0	18.0	8.4	10.5	0	—	—	—	33.5	—
1069	**SCAMPI,** fried	1 OZ	28	—	88.7	3.5	4.7	7.8	0	28.0	88.2	106.4	8.8	109.9
			100	—	316.6	12.5	16.9	27.9	0	99.8	315.0	379.9	31.3	392.2
1070	**SEABASS,** white, raw	1 OZ	28	76.3	26.9	6.0	.1	0	0	—	—	—	—	—
			100	76.3	96.0	21.4	.5	0	0	—	—	—	—	—
	SHAD													
1071	baked, w/butter or margarine & bacon slices	1 OZ	25	64.0	50.3	5.8	2.8	0	0	6.0	78.3	19.8	—	94.3
			100	64.0	201.0	23.2	11.3	0	0	24.0	313.0	79.0	—	377.0
1072	creole, w/tomatoes, onion, green pepper, butter, & flour	1 OZ	25	73.3	38.0	3.8	2.2	.4	—	4.8	47.5	18.3	—	70.0
			100	73.3	152.0	15.0	8.7	1.6	—	19.0	190.0	73.0	—	280.0
1073	canned, solids & liquid	1 OZ	28	71.1	42.6	4.7	2.5	0	0	—	—	—	—	—
			100	71.1	152.0	16.9	8.8	0	0	—	—	—	—	—
	SHARK													
1074	[4]raw	1 OZ	28	77.0	28.0	5.8	.4	0	0	9.0	53.8	22.1	—	153.7
			100	77.0	100.0	20.6	1.3	0	0	32.0	192.0	79.0	—	549.0
1075	[4]fins, dried, soaked, and drained	1 OZ	28	73.3	31.6	7.2	.1	0	0	47.6	44.8	5.0	—	15.4
			100	73.3	113.0	25.8	.3	0	0	170.0	160.0	18.0	—	55.0
1076	**SHEEPSHEAD,** Atlantic, raw	1 OZ	28	75.9	31.6	5.8	.8	0	0	—	55.2	28.3	—	65.5
			100	75.9	113.0	20.6	2.8	0	0	—	197.0	101.0	—	234.0
	SHRIMP													
1077	boiled	1 MED	11	—	—	—	—	—	—	—	—	—	5.6	—
			100	—	—	—	—	—	—	—	—	—	51.0	—
1078	french fried, dipped in egg, breadcrumbs, & flour	1 OZ	28	56.9	63.0	5.7	3.0	2.8	—	20.2	53.5	52.1	14.3	64.1
			100	56.9	225.0	20.3	10.8	10.0	—	72.0	191.0	186.0	51.0	229.0
1079	canned, wet pack, solids & liquid	1 CUP	200	78.2	160.0	32.4	1.6	1.6	—	118.0	304.0	4600.0	46.0	326.0
			100	78.2	80.0	16.2	.8	.8	—	59.0	152.0	2300.0	23.0	163.0
1080	canned, dry pack or drained solids of wet pack	1 MED	11	70.4	12.8	2.7	.1	.1	0	12.7	28.9	15.4	8.1	13.4
			100	70.4	116.0	24.2	1.1	.7	.2	115.0	263.0	140.0	74.0	122.0
1081	or lobster paste, canned	1 OZ	28	61.3	50.4	5.8	2.6	.4	—	—	—	—	—	—
			100	61.3	180.0	20.8	9.4	1.5	—	—	—	—	—	—
1082	**SKATE (RAJA FISH),** raw	1 OZ	28	77.8	27.4	6.0	.2	0	0	—	—	—	—	—
			100	77.8	98.0	21.5	.7	0	0	—	—	—	—	—
1083	**SMELT, ATLANTIC, JACK & BAY,** canned, solids & liquid	1 MED	23	62.7	46.0	4.2	3.1	0	—	82.3	85.1	—	—	—
			100	62.7	200.0	18.4	13.5	0	—	358.0	370.0	—	—	—
1084	**SNAPPER,** red & grey, raw	1 OZ	28	78.5	26.0	5.5	.3	0	0	4.5	59.9	18.8	7.8	90.4
			100	78.5	93.0	19.8	.9	0	0	16.0	214.0	67.0	28.0	323.0
	SOLE													
1085	raw	1 OZ	28	81.3	19.0	4.2	.1	0	0	17.1	54.6	15.7	8.4	102.5
			100	81.3	68.0	14.9	.5	0	0	61.0	195.0	56.0	30.0	366.0
1086	[6]fried	1 OZ	28	—	60.5	4.5	3.9	2.5	0	26.8	67.3	39.3	6.2	70.6
			100	—	216.0	16.0	13.8	9.0	0	95.5	240.2	140.2	22.2	252.0
1087	**SPOT,** w salt, baked	1 OZ	28	53.8	82.6	6.4	6.1	0	0	—	—	87.4	—	—
			100	53.8	295.0	22.8	21.9	0	0	—	—	312.0	—	—
	SQUID													
1088	raw	1 OZ	28	80.2	23.5	4.6	.3	.4	—	3.4	33.3	—	—	—
			100	80.2	84.0	16.4	.9	1.5	—	12.0	119.0	—	—	—
1089	dried	1 SERV	4	—	—	—	—	—	—	—	—	183.5	—	32.2
			100	—	—	—	—	—	—	—	—	4587.0	—	805.0
	STURGEON													
1090	steamed	1 OZ	28	67.5	44.8	7.1	1.6	0	0	11.2	73.6	30.2	—	65.8
			100	67.5	160.0	25.4	5.7	0	0	40.0	263.0	108.0	—	235.0
1091	smoked	1 OZ	28	63.7	41.7	8.7	.5	0	0	—	—	—	—	—
			100	63.7	149.0	31.2	1.8	0	0	—	—	—	—	—
1092	**SUCKER CARP,** raw	1 OZ	28	76.2	31.1	5.4	.9	0	0	—	—	—	—	—
			100	76.2	111.0	19.2	3.2	0	0	—	—	—	—	—
1093	**SUCKERS (WHITE & MULLET),** raw	1 OZ	28	76.4	29.1	5.8	.5	0	0	—	61.6	15.7	—	94.1
			100	76.4	104.0	20.6	1.8	0	0	—	220.0	56.0	—	336.0

Item No.	Minerals (Micro)			Fat-Soluble Vitamins			Water-Soluble Vitamins								
	Iron	Zinc	Copper	Vitamin A	Vitamin D	Vitamin E (Alpha Tocopherol)	Vitamin C	Thiamin	Ribo-flavin	Niacin	Panto-thenic Acid	Vit. B-6 (Pyri-doxine)	Folacin (Folic Acid)	Biotin	Vitamin B-12
	(mg)	(mg)	(mg)	(IU)	(IU)	(mg)	(mg)	(mg)	(mg)	(mg)	(mg)	(mg)	(mcg)	(mcg)	(mcg)
1067	.84	—	—	—	Trace	—	Trace	—	—	—	.04	—	4.90	Trace	—
	3.00	—	—	—	Trace	—	Trace	—	—	—	.15	—	17.50	Trace	—
1068	—	—	—	—	—	.11	—	—	—	—	—	—	—	—	—
	—	—	—	—	—	.60	—	—	—	—	—	—	—	—	—
1069	.31	.18	.06	Trace	Trace	—	Trace	.03	.02	.37	—	—	—	—	—
	1.11	.63	.23	Trace	Trace	—	Trace	.09	.06	1.32	—	—	—	—	—
1070	—	—	—	—	—	—	—	—	—	—	—	—	—	—	—
	—	—	—	—	—	—	—	—	—	—	—	—	—	—	—
1071	.15	—	—	7.5	—	—	—	.03	.07	2.16	—	—	—	—	—
	.60	—	—	30.0	—	—	—	.13	.26	8.60	—	—	—	—	—
1072	.15	—	—	112.5	—	—	2.00	.02	.04	1.28	—	—	—	—	—
	.60	—	—	450.0	—	—	8.00	.09	.16	5.10	—	—	—	—	—
1073	.20	—	—	—	—	—	—	—	.05	—	—	—	—	—	—
	.70	—	—	—	—	—	—	—	.16	—	—	—	—	—	—
1074	.39	—	—	0	—	—	—	.01	.01	1.23	—	—	—	—	—
	1.40	—	—	0	—	—	—	.02	.03	4.40	—	—	—	—	—
1075	1.34	—	—	—	—	—	—	Trace	.03	—	—	—	—	—	—
	4.80	—	—	—	—	—	—	.01	.12	—	—	—	—	—	—
1076	—	—	—	—	—	—	—	—	—	—	—	—	—	—	—
	—	—	—	—	—	—	—	—	—	—	—	—	—	—	—
1077	—	.23	—	—	—	—	—	—	—	—	—	—	—	—	—
	—	2.10	—	—	—	—	—	—	—	—	—	—	—	—	—
1078	.56	—	—	—	—	.53	—	.01	.02	.76	—	—	2.80	—	—
	2.00	—	—	—	—	1.90	—	.04	.08	2.70	—	—	10.00	—	—
1079	3.60	3.80	.34	100.0	—	—	—	.02	.06	3.00	.42	.12	32.00	—	—
	1.80	1.90	.17	50.0	—	—	—	.01	.03	1.50	.21	.06	16.00	—	—
1080	.34	.23	.04	6.6	—	—	0	Trace	Trace	.20	.02	.01	1.76	—	—
	3.10	2.10	.40	60.0	—	—	0	.01	.03	1.80	.21	.06	16.00	—	—
1081	—	—	—	—	—	—	—	—	.07	—	—	—	—	—	—
	—	—	—	—	—	—	—	—	.26	—	—	—	—	—	—
1082	—	—	—	—	—	—	—	.01	—	—	—	—	—	—	—
	—	—	—	—	—	—	—	.02	—	—	—	—	—	—	—
1083	.39	—	—	—	—	—	—	—	—	—	—	—	—	—	—
	1.70	—	—	—	—	—	—	—	—	—	—	—	—	—	—
1084	.22	—	—	—	—	—	—	.05	.01	—	—	—	—	—	—
	.80	—	—	—	—	—	—	.17	.02	—	—	—	—	—	—
1085	.22	—	(.03)	Trace	Trace	—	Trace	.02	.01	.48	.08	—	(3.22)	(1.40)	.22
	.80	—	(.10)	Trace	Trace	—	Trace	.06	.05	1.70	.30	—	(11.50)	(5.00)	.80
1086	.34	—	.04	Trace	Trace	—	Trace	—	—	—	—	—	—	—	—
	.20	—	.16	Trace	Trace	—	Trace	—	—	—	—	—	—	—	—
1087	—	.13	—	—	—	—	—	—	—	—	—	—	—	—	—
	—	.48	—	—	—	—	—	—	—	—	—	—	—	—	—
1088	.14	—	—	—	—	—	—	.01	.03	—	—	—	—	—	—
	.50	—	—	—	—	—	—	.02	.12	—	—	—	—	—	—
1089	—	—	—	—	—	—	—	—	—	—	—	—	—	—	—
	—	—	—	—	—	—	—	—	—	—	—	—	—	—	—
1090	.56	—	—	—	—	—	—	—	—	—	—	—	—	—	—
	0	—	—	—	—	—	—	—	—	—	—	—	—	—	—
1091	—	—	—	—	—	—	—	—	—	—	—	—	—	—	—
	—	—	—	—	—	—	—	—	—	—	—	—	—	—	—
1092	—	—	—	—	—	—	—	—	—	—	—	—	—	—	—
	—	—	—	—	—	—	—	—	—	—	—	—	—	—	—
1093	—	—	—	—	—	—	—	—	—	.34	—	—	—	—	—
	—	—	—	—	—	—	—	—	—	1.20	—	—	—	—	—

(Continued)

TABLE F-36 *(Continued)*

Item No.	Food Item Name	Approximate Measure	Weight	Moisture	Food Energy Calories	Protein	Fats	Carbo-hydrates	Fiber	Minerals (Macro) Calcium	Phos-phorus	Sodium	Mag-nesium	Potas-sium
			(g)	(%)	(kcal)	(g)	(g)	(g)	(g)	(mg)	(mg)	(mg)	(mg)	(mg)
	SWORDFISH													
1094	broiled, w/butter or margarine	1 OZ	28	64.6	48.7	7.8	1.7	0	0	7.6	77.0	–	–	–
			100	64.6	174.0	28.0	6.0	0	0	27.0	275.0	–	–	–
1095	canned, solids & liquid	1 OZ	28	78.0	28.6	4.9	.8	0	0	–	–	–	–	–
			100	78.0	102.0	17.5	3.0	0	0	–	–	–	–	–
1096	**TAUTOG (BLACKFISH),** raw	1 OZ	28	79.3	24.9	5.2	.3	0	0	–	63.6	–	–	–
			100	79.3	89.0	18.6	1.1	0	0	–	227.0	–	–	–
1097	**TERRAPIN (DIAMOND BACK),** raw	1 OZ	28	77.0	31.1	5.2	1.0	0	0	–	–	–	–	–
			100	77.0	111.0	18.6	3.5	0	0	–	–	–	–	–
1098	**TILEFISH,** baked	1 OZ	28	71.6	38.6	6.9	1.0	0	0	–	–	–	–	–
			100	71.6	138.0	24.5	3.7	0	0	–	–	–	–	–
1099	**TOMCOD, ATLANTIC,** raw	1 OZ	28	81.5	21.6	4.8	.1	0	0	–	–	–	–	–
			100	81.5	77.0	17.2	.4	0	0	–	–	–	–	–
	TROUT													
1100	cooked	1 OZ	28	–	54.9	6.6	3.1	.1	0	61.0	76.2	–	9.8	–
			100	–	196.0	23.5	11.2	.4	0	218.0	272.0	–	35.0	–
1101	**RAINBOW** OR **STEELHEAD,** canned	1 OZ	28	63.2	58.5	5.8	3.8	0	0	–	–	–	–	–
			100	63.2	209.0	20.6	13.4	0	0	–	–	–	–	–
	TUNA													
1102	**BLUEFIN,** raw	1 OZ	28	70.5	40.6	7.1	1.1	0	0	–	–	–	–	–
			100	70.5	145.0	25.2	4.1	0	0	–	–	–	–	–
1103	**YELLOWFIN,** raw	1 OZ	28	71.5	37.2	6.9	.8	0	0	–	–	10.4	–	–
			100	71.5	133.0	24.7	3.0	0	0	–	–	37.0	–	–
1104	canned in water, solids & liquid, w/o salt	1 OZ	28	70.0	35.6	7.8	.2	0	0	4.5	53.2	11.5	9.4	78.1
			100	70.0	127.0	28.0	8	0	0	16.0	190.0	41.0	33.4	279.0
1105	canned in water, solids & liquid, w/salt	1 CUP	200	70.0	238.0	56.0	1.6	0	0	32.0	380.0	1750.0	–	550.0
			100	70.0	119.0	28.0	.8	0	0	16.0	190.0	875.0	–	275.0
1106	canned in oil, solids & liquid	1 CUP	130	52.6	374.4	31.5	26.7	0	0	7.8	382.2	1040.0	43.4	391.3
			100	52.6	288.0	24.2	20.5	0	0	6.0	294.0	800.0	33.4	301.0
1107	canned in oil, drained, solids	1 OZ	28	60.6	55.2	8.1	2.3	0	0	2.2	65.5	–	9.4	–
			100	60.6	197.0	28.8	8.2	0	0	8.0	234.0	–	33.4	–
	TURTLE													
1108	green, raw	1 OZ	28	78.5	24.9	5.5	.1	0	0	–	–	–	–	–
			100	78.5	89.0	19.8	.5	0	0	–	–	–	–	–
1109	green, canned	1 OZ	28	75.0	29.7	6.6	.2	0	0	–	–	–	–	–
			100	75.0	106.0	23.4	.7	0	0	–	–	–	–	–
	WEAKFISH													
1110	raw	1 OZ	28	76.7	33.9	4.6	1.6	0	0	–	–	21.0	–	88.8
			100	76.7	121.0	16.5	5.6	0	0	–	–	75.0	–	317.0
1111	w/salt, broiled	1 OZ	28	61.4	58.2	6.9	3.2	0	0	–	–	156.8	–	130.2
			100	61.4	208.0	24.6	11.4	0	0	–	–	560.0	–	465.0
1112	**WHALE MEAT,** raw	1 OZ	28	70.9	43.7	5.8	2.1	0	0	3.4	40.3	21.8	–	6.2
			100	70.9	156.0	20.6	7.5	0	0	12.0	144.0	78.0	–	22.0
	WHITEFISH													
1113	lake, stuffed w/bacon, butter, onion, celery, & bread, baked	1 OZ	28	63.2	60.2	4.3	3.9	1.6	–	–	68.9	54.6	–	81.5
			100	63.2	215.0	15.2	14.0	5.8	–	–	246.0	195.0	–	291.0
1114	lake, smoked	1 OZ	28	68.2	43.4	5.9	2.0	0	0	6.2	76.7	–	–	–
			100	68.2	155.0	20.9	7.3	0	0	22.0	274.0	–	–	–
1115	**WRECKFISH,** raw	1 OZ	28	76.5	31.9	5.2	1.1	0	0	13.2	47.9	–	–	79.0
			100	76.5	114.0	18.4	3.9	0	0	47.0	171.0	–	–	282.0
1116	**YELLOWTAIL (PACIFIC COAST),** raw	1 OZ	28	72.7	38.6	5.9	1.5	0	0	–	–	–	–	–
			100	72.7	138.0	21.0	5.4	0	0	–	–	–	–	–
1117	**ALLSPICE**	1 TSP	2	8.5	5.3	.1	.2	1.4	.4	13.2	2.3	1.6	2.7	20.9
			100	8.5	263.0	6.1	8.7	72.1	21.6	661.0	113.0	80.0	135.0	1044.0
1118	**AMARANTH,** raw	1 OZ	28	86.9	10.1	1.0	.1	1.8	.4	74.8	18.8	–	–	115.1
			100	86.9	36.0	3.5	.5	6.5	1.3	267.0	67.0	–	–	411.0
1119	**ANISE SEED**	1 TSP	2	9.5	6.7	.4	.3	1.0	.3	12.9	8.8	.3	–	28.8
			100	9.5	337.0	17.6	15.9	50.0	14.6	646.0	440.0	16.0	–	1441.0
1120	**BASIL**	1 TSP	1	6.4	2.5	.1	0	.6	.2	21.1	4.9	.3	4.2	34.3
			100	6.4	251.0	14.4	4.0	61.0	17.8	2113.0	490.0	34.0	422.0	3433.0

Item No.	Minerals (Micro)			Fat-Soluble Vitamins			Water-Soluble Vitamins								
	Iron	Zinc	Copper	Vitamin A	Vitamin D	Vitamin E (Alpha Tocopherol)	Vitamin C	Thiamin	Riboflavin	Niacin	Pantothenic Acid	Vit. B-6 (Pyridoxine)	Folacin (Folic Acid)	Biotin	Vitamin B-12
	(mg)	(mg)	(mg)	(IU)	(IU)	(mg)	(mg)	(mg)	(mg)	(mg)	(mg)	(mg)	(mcg)	(mcg)	(mcg)
1094	.36 1.30	– –	– –	574.0 2050.0	– –	– –	– –	.01 .04	.01 .05	3.05 10.90	– –	– –	– –	– –	– –
1095	– –	– –	– –	442.4 1580.0	– –	– –	– –	Trace .01	.01 .05	3.19 11.40	– –	– –	– –	– –	– –
1096	– –	– –	– –	– –	– –	– –	– –	– –	– –	– –	– –	– –	– –	– –	– –
1097	.90 3.20	– –	– –	– –	– –	– –	– –	– –	– –	– –	– –	– –	– –	– –	– –
1098	– –	– –	– –	– –	– –	– –	– –	– –	– –	– –	– –	– –	– –	– –	– –
1099	– –	– –	– –	– –	– –	– –	– –	– –	.05 .17	– –	– –	– –	– –	– –	– –
1100	.31 .10	[6].14 .50	– –	89.3 319.0	– –	– –	.28 1.00	.03 .12	.02 .06	.70 2.50	– –	.06 .20	– –	– –	.74 2.63
1101	– –	– –	– –	– –	– –	– –	– –	– –	– –	– –	– –	– –	– –	– –	– –
1102	.36 1.30	– –	– –	– –	– –	– –	– –	– –	– –	– –	.14 .50	.25 .90	– –	– –	.84 3.00
1103	– –	– –	– –	– –	– –	– –	– –	– –	– –	– –	.14 .50	.25 .90	– –	– –	.84 3.00
1104	.45 1.60	– –	– –	– –	– –	– –	– –	– –	.03 .10	3.72 13.30	.09 .32	.12 .43	4.20 15.00	.84 3.00	.62 2.20
1105	3.20 1.60	[6]1.40 .70	– –	– –	– –	– –	– –	– –	.20 .10	26.60 13.30	.64 .32	.85 .43	30.00 15.00	6.00 3.00	4.40 2.20
1106	1.43 1.10	1.30 1.00	(.13) (.10)	117.0 90.0	(1.65) (5.80)	(1.77) (6.33)	Trace Trace	.05 .04	.12 .09	13.13 10.10	.41 .32	.55 .43	19.50 15.00	3.90 3.00	2.86 2.20
1107	.53 1.90	.31 1.10	– –	22.4 80.0	65.00 232.00	(1.76) (6.30)	Trace Trace	.01 .05	.03 .12	3.33 11.90	.09 .32	.12 .43	4.20 15.00	.84 3.00	.62 2.20
1108	– –	– –	– –	– –	– –	– –	– –	– –	– –	– –	– –	– –	– –	– –	– –
1109	– –	– –	– –	– –	– –	– –	– –	– –	– –	– –	– –	– –	– –	– –	– –
1110	– –	– –	– –	– –	– –	– –	– –	.03 .09	.02 .06	.76 2.70	– –	– –	– –	– –	– –
1111	– –	– –	– –	– –	– –	– –	– –	.03 .10	.02 .08	.98 3.50	– –	– –	– –	– –	– –
1112	– –	– –	– –	520.8 1860.0	– –	– –	1.68 6.00	.03 .09	.02 .08	– –	.08 .30	.04 .15	– –	– –	.56 2.00
1113	.14 .50	– –	– –	560.0 2000.0	– –	– –	– –	.03 .11	.03 .11	.64 2.30	– –	– –	– –	– –	– –
1114	– –	– –	– –	– –	– –	– –	– –	– –	– –	– –	– –	– –	– –	– –	– –
1115	.90 3.20	– –	– –	– –	– –	– –	– –	– –	– –	– –	– –	– –	– –	– –	– –
1116	– –	– –	– –	– –	– –	– –	– –	– –	– –	– –	– –	– –	– –	– –	– –
1117	.14 7.06	.02 1.03	.01 .45	10.8 540.0	– –	– –	.78 39.20	0 .10	0 .06	.06 2.90	– –	– –	– –	– –	0 0
1118	1.09 3.90	– –	– –	1708.0 6100.0	– –	– –	22.40 80.00	.02 .08	.05 .16	.39 1.40	– –	– –	– –	– –	– –
1119	.74 36.96	.11 5.30	– –	– –	– –	– –	– –	– –	– –	– –	– –	– –	– –	– –	0 0
1120	.43 42.80	.06 5.83	.01 1.30	93.8 9375.0	– –	– –	.61 61.30	.00 .15	Trace .32	.07 6.90	– –	– –	– –	– –	0 0

(Continued)

TABLE F-36 *(Continued)*

Item No.	Food Item Name	Approximate Measure	Weight	Moisture	Food Energy Calories	Protein	Fats	Carbo-hydrates	Fiber	Minerals (Macro) Calcium	Phos-phorus	Sodium	Mag-nesium	Potas-sium
			(g)	(%)	(kcal)	(g)	(g)	(g)	(g)	(mg)	(mg)	(mg)	(mg)	(mg)
	FLAVORINGS & SEASONINGS (Continued)													
1121	**BAY LEAVES**	1 TSP	1	5.4	3.1	.1	.1	.7	.3	8.3	1.1	.2	1.2	5.3
			100	5.4	313.0	7.6	8.4	74.7	26.3	834.0	113.0	23.0	120.0	529.0
1122	**BRANDY FLAVORING**	1 TSP	5	—	0	—	—	—	—	—	—	—	—	—
			100	—	0	—	—	—	—	—	—	—	—	—
1123	**CARAWAY SEEDS**	1 TSP	2	9.9	6.7	.4	.3	1.0	.3	13.8	11.4	.3	5.2	27.0
			100	9.9	333.0	19.8	14.6	49.9	12.7	689.0	568.0	17.0	258.0	1351.0
1124	**CARDAMOM SEED**	1 TSP	2	8.4	6.2	.2	.1	1.4	.2	7.7	3.6	.4	4.6	22.4
			100	8.4	311.0	10.7	6.7	68.5	11.3	383.0	178.0	18.0	229.0	1119.0
1125	***CAROB POWDER**	2 TBLS	14	—	54.3	.5	Trace	12.9	.8	41.4	—	1.4	—	114.3
			100	—	380.0	3.8	.2	90.6	5.4	290.0	—	10.0	—	800.0
1126	**CATSUP,** bottled	1 TBLS	15	68.6	15.9	.3	.1	3.8	.1	3.3	7.5	156.3	3.2	54.5
			100	68.6	106.0	2.0	.4	25.4	.5	22.0	50.0	1042.0	21.0	363.0
1127	**CATSUP, LOW SODIUM**	1 TBLS	17	68.7	18.0	.3	.1	4.3	.1	3.7	8.5	3.4	—	61.7
			100	68.6	106.0	2.0	.4	25.4	.5	22.0	50.0	20.0	—	363.0
1128	**CELERY SEED**	1 TSP	2	6.0	7.8	.4	.5	.8	.2	35.3	11.0	3.2	8.8	28.0
			100	6.0	392.0	18.1	25.3	41.4	11.9	1767.0	550.0	160.0	440.0	1400.0
1129	**'CHERVIL,** dried	1 OZ	28	7.2	66.4	6.5	1.1	13.7	3.2	376.9	126.0	23.2	36.4	1327.2
			100	7.2	237.0	23.2	3.9	49.1	11.3	1346.0	450.0	83.0	130.0	4740.0
1130	**CHILI POWDER**	1 TBLS	15	7.8	47.1	1.8	2.5	8.2	3.3	41.7	45.5	151.5	25.5	287.4
			100	7.8	314.0	12.3	16.8	54.7	22.2	278.0	303.0	1010.0	170.0	1916.0
	CHILI SAUCE													
1131	regular	1 TBLS	17	68.0	17.7	.4	.1	4.2	.1	3.4	8.8	227.5	—	62.9
			100	68.0	104.0	2.5	.3	24.8	.7	20.0	52.0	1338.0	—	370.0
1132	low sodium	1 TBLS	17	68.0	17.7	.4	.1	4.2	.1	3.4	8.8	3.4	—	62.9
			100	68.0	104.0	2.5	.3	24.8	.7	20.0	52.0	20.0	—	370.0
1133	hot, green	1 TBLS	17	93.9	3.4	.1	0	.9	.2	.9	2.4	—	—	—
			100	93.9	20.0	.7	.1	5.0	1.0	5.0	14.0	—	—	—
1134	hot, red	1 TBLS	17	94.1	3.6	.2	.1	.7	.3	1.5	2.7	—	—	—
			100	94.1	21.0	.9	.6	3.9	1.7	9.0	16.0	—	—	—
1135	**CHOCOLATE,** bitter or baking	1 OZ	28	2.3	141.4	3.0	14.8	8.1	.7	21.8	107.5	1.1	81.8	232.4
			100	2.3	505.0	10.7	53.0	28.9	2.5	78.0	384.0	4.0	292.0	830.0
1136	**CINNAMON**	1 TSP	2	9.5	5.2	.1	.1	1.6	.5	24.6	1.2	.5	1.1	10.0
			100	9.5	261.0	3.9	3.2	79.8	24.4	1228.0	61.0	26.0	56.0	500.0
1137	**CLOVES,** ground	1 TSP	2	6.9	6.5	.1	.4	1.2	.2	.1	2.1	5.0	5.3	22.0
			100	6.9	323.0	6.0	20.7	61.2	9.6	6.5	105.0	250.0	264.0	1102.0
	COCOA													
1138	powder, w/o milk	1 TBLS	7	1.3	24.3	.3	.1	6.3	.1	2.1	12.0	18.8	29.4	35.0
			100	1.3	347.0	4.0	2.0	89.4	1.0	30.0	171.0	268.0	420.0	500.0
1139	powder, low fat	1 TBLS	7	4.4	13.1	1.4	.6	4.1	.4	10.7	52.6	.4	29.4	106.5
			100	4.4	187.0	20.2	7.9	58.0	5.8	153.0	752.0	6.0	420.0	1522.0
1140	powder, low-medium fat, plain	1 TBLS	7	5.2	15.4	1.3	.9	3.8	.4	10.6	48.0	.4	29.4	106.5
			100	5.2	220.0	19.2	12.7	53.8	5.2	152.0	686.0	6.0	420.0	1522.0
1141	powder, high-med fat, plain	1 TBLS	7	4.1	18.6	1.2	1.3	3.6	.3	8.6	45.4	.4	30.0	106.5
			100	4.1	265.0	17.3	19.0	51.5	4.3	123.0	649.0	6.0	428.9	1522.0
1142	powder, high fat, plain	1 TBLS	7	3.0	20.9	1.2	1.7	3.4	.3	9.3	45.4	.4	30.9	106.5
			100	3.0	299.0	16.8	23.7	48.3	4.3	133.0	648.0	6.0	442.0	1522.0
1143	powder, low-medium fat, processed w/alkali	1 TBLS	7	5.2	15.1	1.3	.9	3.5	.4	10.6	48.0	50.2	—	45.6
			100	5.2	215.0	19.2	12.7	50.2	5.2	152.0	686.0	717.0	—	651.0
1144	powder, high-medium fat, processed w/alkali	1 TBLS	7	4.1	18.3	1.2	1.3	3.4	.3	8.6	45.4	50.2	—	45.6
			100	4.1	261.0	17.3	19.0	48.5	4.3	123.0	649.0	717.0	—	651.0
1145	powder, high fat, processed w/alkali	1 TBLS	7	3.0	20.7	1.2	1.7	3.2	.3	9.3	45.4	50.2	29.4	45.6
			100	3.0	295.0	16.8	23.7	45.4	4.3	133.0	648.0	717.0	420.0	651.0
1146	mix for hot chocolate	1 TBLS	7	3.1	27.4	.7	.7	5.2	.1	19.3	20.3	26.7	7.7	42.4
			100	3.1	392.0	9.4	10.6	73.9	.8	275.0	290.0	382.0	110.0	605.0
1147	**CORIANDER LEAF,** dried	1 TSP	1	7.3	2.8	.2	0	.5	.1	12.5	4.8	2.1	6.9	44.7
			100	7.3	279.0	21.8	4.7	52.1	10.4	1246.0	481.0	211.0	694.0	4466.0
1148	**CORIANDER SEED**	1 TSP	3	8.9	8.9	.4	.5	1.7	.9	21.3	12.3	1.1	9.0	38.0
			100	8.9	298.0	12.3	17.7	55.0	29.1	709.0	409.0	35.0	300.0	1267.0
1149	**CUMIN SEED**	1 TSP	2	8.1	7.5	.4	.4	.9	.2	18.6	10.0	3.4	7.3	35.8
			100	8.1	375.0	17.7	22.3	44.6	10.5	931.0	499.0	168.0	366.0	1788.0
1150	**CURRY POWDER**	1 TSP	2	9.5	6.5	.3	.3	1.2	.3	9.6	7.0	1.0	5.1	30.9
			100	9.5	325.0	12.7	13.8	58.2	16.3	478.0	349.0	52.0	254.0	1543.0

Item No.	Minerals (Micro)			Fat-Soluble Vitamins			Water-Soluble Vitamins								
	Iron	Zinc	Copper	Vitamin A	Vitamin D	Vitamin E (Alpha Tocopherol)	Vitamin C	Thiamin	Ribo-flavin	Niacin	Panto-thenic Acid	Vit. B-6 (Pyri-doxine)	Folacin (Folic Acid)	Biotin	Vitamin B-12
	(mg)	(mg)	(mg)	(IU)	(IU)	(mg)	(mg)	(mg)	(mg)	(mg)	(mg)	(mg)	(mcg)	(mcg)	(mcg)
1121	**.43** 43.00	**.04** 3.70	**0** .42	**61.9** 6185.0	— —	— —	**.47** 46.60	**.00** .10	**.00** .42	**.02** 2.00	— —	— —	— —	— —	**0** 0
1122	— —	— —	— —	— —	— —	— —	— —	— —	— —	— —	— —	— —	— —	— —	— —
1123	**.31** 16.23	**.11** 5.50	**.01** .43	**7.3** 363.0	— —	— —	**.24** 12.00	**.01** .38	**.01** .38	**.07** 3.61	— —	— —	— —	— —	— —
1124	**.28** 13.97	**.15** 7.47	— —	**0** 0	— —	— —	**.24** 12.00	**.00** .18	**Trace** .18	**.02** 1.10	— —	— —	— —	— —	**0** 0
1125	**.3** 2.0	— —	— —	— —	— —	— —	— —	— —	— —	— —	— —	— —	— —	— —	— —
1126	**.12** .80	**.04** .26	— —	**210.0** 1400.0	— —	— —	**2.25** 15.00	**.01** .09	**.01** .07	**.24** 1.60	— —	**.02** .11	**.75** 5.00	— —	**0** 0
1127	**.14** .80	— —	— —	**238.0** 1400.0	— —	— —	**2.55** 15.00	**.02** .09	**.01** .07	**.27** 1.60	— —	**.02** .11	**.85** 5.00	— —	**0** 0
1128	**.90** 44.90	**.14** 6.93	— —	**1.0** 52.0	— —	— —	**.34** 17.20	**.01** .41	**.01** .49	— —	— —	— —	— —	— —	**0** 0
1129	**8.95** 31.95	**2.46** 8.80	— —	— —	— —	— —	— —	— —	— —	— —	— —	**.34** 1.23	— —	— —	**0** 0
1130	**2.14** 14.25	**.41** 2.70	**.19** .60	**5239.1** 34927.0	— —	— —	**9.62** 64.14	**.05** .35	**.11** .75	**1.18** 7.89	— —	— —	— —	— —	**0** 0
1131	**.14** .80	— —	— —	**238.0** 1400.0	— —	— —	**2.72** 16.00	**.02** .09	**.01** .07	**.27** 1.60	— —	— —	— —	— —	— —
1132	**.14** .80	— —	— —	**238.0** 1400.0	— —	— —	**2.72** 16.00	**.02** .09	**.01** .07	**.27** 1.60	— —	— —	— —	— —	— —
1133	**.07** .40	— —	— —	**103.7** 610.0	— —	— —	**11.56** 68.00	**.01** .03	**.01** .03	**.12** .70	— —	— —	— —	— —	— —
1134	**.09** .50	— —	— —	**1630.3** 9590.0	— —	— —	**5.10** 30.00	**Trace** .01	**.02** .09	**.10** .60	— —	— —	— —	— —	— —
1135	**1.88** 6.70	— —	— —	**16.8** 60.0	**0** 0	— —	**0** 0	**.01** .05	**.07** .24	**.42** 1.50	**.05** .19	**.01** .04	**2.52** 9.00	**8.96** 32.00	**0** 0
1136	**.76** 38.07	**.04** 1.97	**.01** .23	**5.2** 260.0	— —	— —	**.57** 28.46	**Trace** .07	**Trace** .14	**.03** 1.30	— —	— —	— —	— —	**0** 0
1137	**.17** 8.68	**.09** 4.72	**.01** .35	**10.6** 530.0	— —	— —	**1.62** 80.90	**Trace** .11	**Trace** .27	**.03** 1.46	— —	— —	— —	— —	**0** 0
1138	**.15** 2.10	**.39** 5.60	— —	— —	— —	— —	**0** 0	**Trace** .02	**Trace** .09	**.04** .50	— —	— —	— —	— —	— —
1139	**.75** 10.70	— —	**.30** 4.30	**.7** 10.0	— —	— —	**0** 0	**.01** .11	**.03** .46	**.17** 2.40	— —	— —	— —	— —	— —
1140	**.75** 10.70	— —	**.30** 4.30	**1.4** 20.0	— —	— —	**0** 0	**.01** .11	**.03** .46	**.17** 2.40	— —	— —	— —	— —	— —
1141	**.75** 10.70	— —	**.30** 4.30	**1.4** 20.0	— —	— —	**0** 0	**.01** .11	**.03** .46	**.17** 2.40	— —	— —	— —	— —	— —
1142	**.75** 10.70	**.38** 5.40	**.35** 5.00	**2.1** 30.0	— —	— —	**0** 0	**.01** .11	**.03** .46	**.17** 2.40	— —	— —	— —	— —	— —
1143	**.75** 10.70	**.34** 4.86	— —	**1.4** 20.0	— —	— —	**0** 0	**.01** .11	**.03** .46	**.17** 2.40	— —	— —	— —	— —	— —
1144	**.75** 10.70	**.34** 4.86	— —	**1.4** 20.0	— —	— —	**0** 0	**.01** .11	**.03** .46	**.17** 2.40	— —	— —	— —	— —	— —
1145	**.75** 10.70	**.34** 4.86	— —	**2.1** 30.0	— —	— —	**0** 0	**.01** .11	**.03** .46	**.17** 2.40	— —	— —	— —	— —	— —
1146	**.10** 1.40	— —	— —	**.7** 10.0	— —	— —	**.07** 1.00	**.01** .08	**.03** .41	**.04** .50	— —	— —	— —	— —	— —
1147	**.43** 42.46	— —	— —	— —	— —	— —	**5.67** 566.71	**.01** 1.25	**.01** 1.50	**.11** 10.71	— —	— —	— —	— —	**0** 0
1148	**.49** 16.32	**.14** 4.70	— —	**0** 0	— —	— —	**.36** 12.00	**.01** .24	**.01** .29	**.07** 2.13	— —	— —	— —	— —	**0** 0
1149	**1.33** 66.35	**.10** 4.80	— —	**25.4** 1270.0	— —	— —	**.15** 7.71	**.01** .63	**.01** .33	**.09** 4.58	— —	— —	— —	— —	**0** 0
1150	**.59** 29.59	**.08** 4.05	**(.21)** (1.03)	**19.7** 986.0	**0** 0	— —	**.23** 11.41	**.01** .25	**.01** .28	**.07** 3.47	— —	— —	— —	— —	**0** 0

(Continued)

TABLE F-36 *(Continued)*

Item No.	Food Item Name	Approximate Measure	Weight	Moisture	Food Energy Calories	Protein	Fats	Carbo-hydrates	Fiber	Minerals (Macro) Calcium	Phos-phorus	Sodium	Mag-nesium	Potas-sium
			(g)	(%)	(kcal)	(g)	(g)	(g)	(g)	(mg)	(mg)	(mg)	(mg)	(mg)
	FLAVORINGS & SEASONINGS (Continued)													
1151	**DILL SEED**	1 TSP	**2**	**7.7**	**6.1**	**.3**	**.3**	**1.1**	**.4**	**30.3**	**5.5**	**.4**	**5.1**	**23.7**
			100	7.7	305.0	16.0	14.5	55.2	21.1	1516.0	277.0	20.0	256.0	1186.0
1152	**DILL WEED,** dried	1 TSP	**1**	**7.3**	**2.5**	**.2**	**0**	**.6**	**.1**	**17.8**	**5.4**	**2.1**	**4.5**	**33.1**
			100	7.3	253.0	20.0	4.4	55.8	11.9	1784.0	543.0	208.0	451.0	3308.0
1153	**FENNEL SEED**	1 TSP	**2**	**8.8**	**6.9**	**.3**	**.3**	**1.0**	**.3**	**23.9**	**9.6**	**1.8**	**7.7**	**34.0**
			100	8.8	345.0	15.8	14.9	52.3	15.7	1196.0	480.0	90.0	385.0	1700.0
1154	**FENUGREEK SEED**	1 TSP	**4**	**8.8**	**12.9**	**.9**	**.3**	**2.3**	**.4**	**7.0**	**11.8**	**2.7**	**7.6**	**30.8**
			100	8.8	323.0	23.0	6.4	58.4	10.1	176.0	296.0	67.0	191.0	770.0
1155	**GARLIC POWDER**	1 TSP	**3**	**6.5**	**10.0**	**.5**	**0**	**2.2**	**.1**	**2.4**	**12.6**	**.8**	**1.7**	**33.0**
			100	6.5	332.0	16.8	.8	72.7	1.9	80.0	420.0	26.0	58.0	1100.0
1156	**GARLIC SALT**	1 TSP	**5**	—	—	—	—	—	—	—	—	—	**1.0**	—
			100	—	—	—	—	—	—	—	—	—	20.0	—
	GINGER													
1157	**ROOT,** fresh	1 OZ	**28**	**87.0**	**13.7**	**.4**	**.3**	**2.7**	**.3**	**6.4**	**10.1**	**1.7**	—	**73.9**
			100	87.0	49.0	1.4	1.0	9.5	1.1	23.0	36.0	6.0	—	264.0
1158	**ROOT,** crystallized (candied)	1 OZ	**28**	**12.0**	**95.2**	**.1**	**.1**	**24.4**	**.2**	—	—	—	—	—
			100	12.0	340.0	.3	.2	87.1	.7	—	—	—	—	—
1159	**ROOT,** dried, ground	1 TSP	**2**	**9.4**	**6.9**	**.2**	**.1**	**1.4**	**.1**	**2.3**	**3.0**	**.6**	**3.3**	**26.8**
			100	9.4	347.0	9.1	6.0	70.8	5.9	116.0	150.0	30.0	164.0	1342.0
1160	**HORSERADISH,** prepared	1 TSP	**6**	**87.1**	**2.3**	**.1**	**0**	**.6**	**.1**	**3.7**	**1.9**	**5.8**	—	**17.4**
			100	87.1	38.0	1.3	.2	9.6	.9	61.0	32.0	96.0	—	290.0
1161	**KETCHUP**—See **Catsup**													
1162	**MACE,** ground	1 TSP	**2**	**8.2**	**9.5**	**.1**	**.6**	**1.0**	**.1**	**5.0**	**2.2**	**1.6**	**3.2**	**9.3**
			100	8.2	475.0	6.7	32.4	50.5	4.8	252.0	110.0	80.0	160.0	463.0
1163	**MARJORAM,** dried	1 TSP	**1**	**7.6**	**2.7**	**.1**	**.1**	**.6**	**.2**	**19.9**	**3.1**	**.8**	**3.5**	**15.2**
			100	7.6	271.0	12.7	7.0	60.6	18.1	1990.0	306.0	77.0	346.0	1522.0
1164	**MONOSODIUM GLUTAMATE,** Accent	1 TSP	**5**	**0.1**	**14.4**	**2.3**	**0**	**0**	—	—	—	**645.0**	—	—
			100	0.1	288.0	46.8	0	0	—	—	—	12900.0	—	—
1165	**MUSTARD,** dried	1 TSP	**2**	**3.2**	**11.6**	**.6**	**.9**	**.4**	**Trace**	**6.0**	**15.8**	**.2**	**8.5**	**14.0**
			100	3.2	580.0	31.7	42.6	18.5	1.9	300.0	790.0	10.0	422.5	700.0
1166	prepared, brown	1 TSP	**5**	**78.1**	**4.6**	**.3**	**.3**	**.3**	**.1**	**6.2**	**6.7**	**65.4**	**2.4**	**6.5**
			100	78.1	91.0	5.9	6.3	5.3	1.3	124.0	134.0	1307.0	48.0	130.0
1167	prepared, yellow	1 TSP	**5**	**80.2**	**3.8**	**.2**	**.2**	**.3**	**.1**	**4.2**	**3.7**	**62.6**	**2.4**	**6.5**
			100	80.2	75.0	4.7	4.4	6.4	1.0	84.0	73.0	1252.0	48.0	130.0
1168	**NUTMEG**	1 TSP	**2**	**6.2**	**10.5**	**.1**	**.7**	**1.0**	**.1**	**3.7**	**4.3**	**.3**	**3.6**	**7.0**
			100	6.2	525.0	5.8	36.3	49.3	4.0	184.0	213.0	16.0	180.0	350.0
1169	**ONION POWDER**	1 TSP	**2**	**5.0**	**6.9**	**.2**	**0**	**1.6**	**.1**	**7.3**	**6.8**	**1.1**	**2.4**	**18.9**
			100	5.0	347.0	10.1	1.1	80.5	5.7	363.0	340.0	54.0	122.0	943.0
1170	**OREGANO**	1 TSP	**2**	**7.2**	**6.1**	**.2**	**.2**	**1.3**	**.3**	**31.5**	**4.0**	**.3**	**5.4**	**33.4**
			100	7.2	306.0	11.0	10.3	64.9	15.0	1576.0	200.0	15.0	270.0	1669.0
1171	**PAPRIKA,** domestic	1 TSP	**2**	**9.5**	**5.8**	**.3**	**.3**	**1.1**	**.4**	**3.5**	**6.9**	**.7**	**3.7**	**46.9**
			100	9.5	289.0	14.8	13.0	55.7	20.9	177.0	345.0	34.0	185.0	2344.0
	PEPPER													
1172	**BLACK,** ground	1 TSP	**2**	**10.5**	**5.1**	**.2**	**.1**	**1.3**	**.3**	**8.0**	**3.5**	**.9**	**3.9**	**25.2**
			100	10.5	255.0	10.4	3.3	66.5	13.1	400.0	173.0	44.0	194.0	1259.0
1173	**RED**	1 TSP	**2**	**8.1**	**6.4**	**.2**	**.3**	**1.1**	**.6**	**3.0**	**5.9**	**.6**	**3.0**	**40.3**
			100	8.1	318.0	12.0	17.3	56.6	24.9	148.0	293.0	30.0	152.0	2014.0
1174	**WHITE**	1 TSP	**2**	**11.4**	**5.9**	**.2**	**0**	**1.4**	**.1**	**5.3**	**3.5**	**.1**	**1.8**	**1.5**
			100	11.4	296.0	10.4	2.1	68.6	4.3	265.0	176.0	5.0	90.0	73.0
1175	**POPPY SEED**	1 TSP	**3**	**6.8**	**15.9**	**.5**	**1.3**	**.7**	**.2**	**43.4**	**25.4**	**.6**	**9.6**	**21.0**
			100	6.8	530.0	18.0	44.7	23.7	6.3	1448.0	848.0	21.0	320.0	700.0
1176	**POULTRY SEASONING**	1 TSP	**2**	**9.3**	**6.1**	**.2**	**.2**	**1.3**	**.2**	**19.9**	**3.4**	**.5**	**4.5**	**13.7**
			100	9.3	307.0	9.6	7.5	65.6	11.3	996.0	171.0	27.0	224.0	684.0
1177	**PUMPKIN PIE SPICE**	1 TSP	**2**	**8.5**	**6.8**	**.1**	**.3**	**1.4**	**.3**	**13.6**	**2.4**	**1.0**	**2.7**	**13.3**
			100	8.5	342.0	5.8	12.6	69.3	14.8	682.0	118.0	52.0	136.0	663.0
	PURSLANE													
1178	leaves & stems, raw	1 CUP	**60**	**92.5**	**12.6**	**1.0**	**.2**	**2.3**	**.5**	**61.8**	**23.4**	—	—	—
			100	92.5	21.0	1.7	.4	3.8	.9	103.0	39.0	—	—	—
1179	leaves & stems, boiled & drained	1 CUP	**85**	**94.7**	**12.8**	**1.0**	**.3**	**2.4**	**.7**	**73.1**	**20.4**	—	—	—
			100	94.7	15.0	1.2	.3	2.8	.8	86.0	24.0	—	—	—

Item No.	Minerals (Micro)			Fat-Soluble Vitamins			Water-Soluble Vitamins								
	Iron	Zinc	Copper	Vitamin A	Vitamin D	Vitamin E (Alpha Tocopherol)	Vitamin C	Thiamin	Ribo-flavin	Niacin	Panto-thenic Acid	Vit. B-6 (Pyri-doxine)	Folacin (Folic Acid)	Biotin	Vitamin B-12
	(mg)	(mg)	(mg)	(IU)	(IU)	(mg)	(mg)	(mg)	(mg)	(mg)	(mg)	(mg)	(mcg)	(mcg)	(mcg)
1151	**.33**	**.10**	—	**1.1**	—	—	**.24**	**.01**	**.01**	**.06**	—	—	—	—	**0**
	16.32	5.20	—	53.0	—	—	12.00	.42	.28	2.80	—	—	—	—	0
1152	**.45**	**.03**	**.03**	—	—	—	—	**Trace**	**Trace**	**.03**	—	**0**	—	—	**0**
	44.77	3.30	3.42	—	—	—	—	.41	.28	2.81	—	Trace	—	—	0
1153	**.37**	**.07**	—	**2.7**	—	—	**.24**	**.01**	**.01**	**.12**	—	—	—	—	**0**
	18.54	3.70	—	135.0	—	—	12.00	.41	.36	6.00	—	—	—	—	0
1154	**1.34**	**.10**	—	—	—	—	**.12**	**.01**	**.02**	**.07**	—	—	**2.28**	—	**0**
	33.53	2.50	—	—	—	—	3.00	.32	.36	1.64	—	—	57.00	—	0
1155	**.08**	**.08**	—	**0**	—	—	**.36**	**.01**	**.01**	**.02**	—	—	—	—	**0**
	2.75	2.63	—	0	—	—	12.00	.47	.15	.70	—	—	—	—	0
1156	—	—	—	—	—	—	—	—	—	—	—	—	—	—	—
	—	—	—	—	—	—	—	—	—	—	—	—	—	—	—
1157	**.59**	—	—	**2.8**	—	—	**1.12**	**.01**	**.01**	**.20**	**.06**	—	—	—	**0**
	2.10	—	—	10.0	—	—	4.00	.02	.04	.70	.20	—	—	—	0
1158	—	—	—	—	—	—	—	—	—	—	—	—	—	—	—
	—	—	—	—	—	—	—	—	—	—	—	—	—	—	—
1159	**.23**	**.09**	**.01**	**2.9**	**0**	—	**.24**	**Trace**	**Trace**	**.10**	—	—	—	—	**0**
	11.30	4.72	.32	147.0	0	—	12.00	.05	.19	5.16	—	—	—	—	0
1160	**.05**	**.06**	—	—	—	—	—	—	—	—	—	—	—	—	—
	.90	1.07	—	—	—	—	—	—	—	—	—	—	—	—	—
1161															
1162	**.28**	**.05**	**.05**	**16.0**	—	—	—	**.01**	**.01**	**.03**	—	—	—	—	**0**
	13.90	2.30	2.47	800.0	—	—	—	.31	.44	1.35	—	—	—	—	0
1163	**.83**	**.04**	**.01**	**80.7**	—	—	**.51**	**.00**	**Trace**	**.04**	—	—	—	—	**0**
	82.71	3.60	1.10	8068.0	—	—	51.43	.29	.32	4.12	—	—	—	—	0
1164	—	—	—	—	—	—	—	—	—	—	—	—	—	—	—
	—	—	—	—	—	—	—	—	—	—	—	—	—	—	—
1165	**.17**	**.09**	**.01**	**3.9**	**0**	—	**.04**	**.01**	**.01**	**.17**	—	—	—	—	—
	8.30	4.47	.30	195.0	0	—	2.21	.65	.45	8.50	—	—	—	—	—
1166	**.09**	**.01**	—	—	—	**.09**	—	—	—	—	—	—	—	—	—
	1.80	.21	—	—	—	1.75	—	—	—	—	—	—	—	—	—
1167	**.10**	**.03**	—	—	—	**.09**	—	—	—	—	—	—	**.20**	—	—
	2.00	.63	—	—	—	1.75	—	—	—	—	—	—	3.90	—	—
1168	**.06**	**.04**	**.02**	**2.0**	—	—	**.24**	**.01**	**Trace**	**.03**	—	—	—	—	**0**
	3.04	2.15	1.03	102.0	—	—	12.00	.36	.06	1.30	—	—	—	—	0
1169	**.05**	**.05**	—	**0**	—	—	**.29**	**.01**	**Trace**	**.01**	—	—	—	—	**0**
	2.56	2.32	—	0	—	—	14.70	.42	.06	.65	—	—	—	—	0
1170	**.88**	**.09**	—	**138.1**	—	—	**.24**	**.01**	**Trace**	**.12**	—	—	—	—	**0**
	44.00	4.43	—	6903.0	—	—	12.00	.34	.04	6.20	—	—	—	—	0
1171	**.46**	**.08**	**.01**	**1212.1**	—	—	**1.42**	**.01**	**.04**	**.31**	—	—	—	—	**0**
	23.10	4.06	.61	60604.0	—	—	71.12	.65	.74	15.30	—	—	—	—	0
1172	**.58**	**.03**	**.09**	**3.8**	**0**	—	**.24**	**Trace**	**Trace**	**.02**	—	—	—	—	**0**
	28.86	1.42	4.30	190.0	0	—	12.00	.11	.21	1.14	—	—	—	—	0
1173	**.14**	**.05**	—	**832.2**	—	—	**1.53**	**.01**	**.02**	**.17**	—	—	—	—	**0**
	7.08	2.48	—	41610.0	—	—	76.44	.33	.92	8.70	—	—	—	—	0
1174	**.29**	**.02**	**.02**	**0**	—	—	—	**0**	**Trace**	**Trace**	—	—	—	—	**0**
	14.31	1.13	.80	0	—	—	—	.02	.13	.21	—	—	—	—	0
1175	**.28**	**.31**	**.05**	**0**	—	**.05**	**.36**	**.03**	**.01**	**.03**	—	**.01**	—	—	**0**
	9.40	10.23	1.63	0	—	1.80	12.00	.85	.18	.98	—	.45	—	—	0
1176	**.71**	**.06**	—	**52.6**	—	—	**.24**	**.01**	**Trace**	**.06**	—	—	—	—	**0**
	35.30	3.14	—	2632.0	—	—	11.96	.26	.19	2.97	—	—	—	—	0
1177	**.39**	**.05**	—	**5.2**	—	—	**.47**	**Trace**	**Trace**	**.05**	—	—	—	—	**0**
	19.71	2.37	—	261.0	—	—	23.38	.13	.14	2.24	—	—	—	—	0
1178	**2.10**	—	—	**1500.0**	—	—	**15.00**	**.02**	**.06**	**.30**	—	—	—	—	—
	3.50	—	—	2500.0	—	—	25.00	.03	.10	.50	—	—	—	—	—
1179	**1.02**	—	—	**1785.0**	—	—	**10.20**	**.02**	**.05**	**.34**	—	—	—	—	—
	1.20	—	—	2100.0	—	—	12.00	.02	.06	.40	—	—	—	—	—

(Continued)

TABLE F-36 *(Continued)*

Item No.	Food Item Name	Approximate Measure	Weight	Moisture	Food Energy Calories	Protein	Fats	Carbo-hydrates	Fiber	Minerals (Macro) Calcium	Phos-phorus	Sodium	Mag-nesium	Potas-sium
			(g)	(%)	(kcal)	(g)	(g)	(g)	(g)	(mg)	(mg)	(mg)	(mg)	(mg)
	ROSEMARY													
1180	leaves	1 TSP	1	5.7	4.4	0	.2	.7	.2	15.0	.7	.4	—	10.0
			100	5.7	440.0	4.5	17.4	66.4	19.0	1500.0	70.0	40.0	—	1000.0
1181	ground	1 TSP	1	9.3	3.3	0	.2	.6	.2	12.8	.7	.5	2.2	9.6
			100	9.3	331.0	4.5	15.2	64.1	17.7	1280.0	70.0	50.0	220.0	955.0
1182	**SAFFRON**	1 TSP	1	11.9	3.1	.1	.1	.7	Trace	1.1	2.5	1.5	—	17.2
			100	11.9	310.0	11.4	5.9	65.4	3.9	111.0	252.0	148.0	—	1724.0
1183	**SAGE**	1 TSP	1	8.0	3.2	.1	.1	.6	.2	16.5	.9	.1	4.3	10.7
			100	8.0	315.0	10.6	12.7	60.7	18.1	1652.0	90.0	10.0	428.0	1070.0
	SALT													
1184	table	1 TSP	5	—	0	0	0	0	—	12.7	—	1940.5	—	.2
			100	—	0	0	0	0	—	253.0	—	38809.0	—	4.0
1185	table, iodized	1 PKG	1	.05	0	0	0	0	—	1.0	.4	390.9	0	0
			100	.05	.2	0	0	0	—	100.0	40.0	39090.0	1.4	2.4
1186	**SAVORY**	1 TSP	1	9.0	2.7	.1	.1	.7	.2	21.3	1.4	.2	3.8	10.5
			100	9.0	272.0	6.7	5.9	69.9	15.3	2132.0	140.0	20.0	377.0	1051.0
	SESAME SEEDS													
1187	dry, whole	1 OZ	28	5.4	157.6	5.2	13.7	6.0	1.8	324.8	172.5	16.8	50.7	203.0
			100	5.4	563.0	18.6	49.1	21.6	6.3	1160.0	616.0	60.0	181.0	725.0
1188	dry, decorticated	1 OZ	28	4.8	163.0	7.4	15.3	2.6	.8	36.7	217.3	11.2	97.2	114.0
			100	4.8	582.0	26.4	54.8	9.4	3.0	131.0	776.0	40.0	347.0	407.0
1189	**SOY SAUCE**	1 TBLS	15	62.8	10.2	.8	.2	1.4	0	12.3	15.6	1098.8	—	54.9
			100	62.8	68.0	5.6	1.3	9.5	0	82.0	104.0	7325.0	—	366.0
1190	**STRAWBERRY FLAVORED MILK FLAVORING**	1 TBLS	15	0.5	59.2	0	0	14.8	—	—	—	—	—	—
			100	0.5	394.8	0	.3	98.6	—	—	—	—	—	—
1191	**TARRAGON**	1 TSP	1	7.7	3.0	.2	.1	.5	.1	11.4	3.1	.6	3.5	30.2
			100	7.7	295.0	22.8	7.3	50.2	7.4	1139.0	310.0	62.0	347.0	3020.0
1192	**THYME**	1 TSP	1	7.8	2.8	.1	.1	.6	.2	18.9	2.0	.6	2.2	8.1
			100	7.8	276.0	9.1	7.4	63.9	18.6	1890.0	200.0	55.0	220.0	814.0
1193	**TABASCO SAUCE**	1 TSP	5	87.0	.4	0	0	.1	0	—	—	22.3	—	3.2
			100	87.0	8.1	.9	0	1.1	0	—	—	445.5	—	63.0
1194	**TOMATO CATSUP OR KETCHUP** —See Catsup													
1195	**TURMERIC**	1 TSP	2	11.4	7.1	.2	.2	1.3	Trace	3.6	5.2	.8	3.9	50.0
			100	11.4	354.0	7.8	9.9	64.9	6.9	182.0	260.0	38.0	193.0	2500.0
	VINEGAR													
1196	cider	1 TBLS	15	93.8	2.1	0	0	.9	0	.9	1.4	.2	—	15.0
			100	93.8	14.0	0	0	5.9	0	6.0	9.0	1.0	—	100.0
1197	distilled	1 TBLS	15	95.0	1.8	—	—	.8	—	—	—	.2	.2	2.3
			100	95.0	12.0	—	—	5.0	—	—	—	1.0	1.0	15.0
1198	**WATER CHESTNUTS,** Chinese, raw	1 PIECE	6	78.3	4.7	.1	0	1.1	Trace	.2	3.9	1.2	.7	30.0
			100	78.3	79.0	1.4	.2	19.0	.8	4.0	65.0	20.0	12.0	500.0
1199	**WORCESTERSHIRE SAUCE**	1 TSP	5	70.0	5.9	0	0	.8	Trace	5.3	2.9	48.9	—	40.1
			100	70.0	117.3	.1	.2	16.6	.3	106.0	57.0	977.0	—	802.0
1200	**ALFALFA LEAF MEAL** (powder), dehydrated		—	—	—	—	—	—	—	—	—	—	—	—
			100	—	215.0	20.0	3.1	57.8	18.1	1640.0	230.0	60.0	350.0	2070.0
1201	**COD LIVER OIL**		—	—	—	—	—	—	—	—	—	—	—	—
			100	—	899.0	0	99.9	0	0	Trace	Trace	Trace	Trace	Trace
	SEAWEED													
1202	agar, raw	1 OZ	28	16.3	—	—	.1	—	.2	158.8	6.2	—	—	—
			100	16.3	—	—	.3	—	.7	567.0	22.0	—	—	—
1203	dulse, raw	1 OZ	28	16.6	—	—	.9	—	.3	82.9	74.8	583.8	—	2256.8
			100	16.6	—	—	3.2	—	1.2	296.0	267.0	2085.0	—	8060.0
1204	Irish Moss, (carrageenan), raw	1 OZ	28	19.2	—	—	.5	—	.6	247.8	44.0	809.8	—	796.3
			100	19.2	—	—	1.8	—	2.1	885.0	157.0	2892.0	—	2844.0
1205	kelp, raw	1 OZ	28	21.7	—	—	.3	—	1.9	306.0	67.2	842.0	—	1476.4
			100	21.7	—	—	1.1	—	6.8	1093.0	240.0	3007.0	—	5273.0
1206	laver, raw	1 OZ	28	17.0	—	—	.2	—	1.0	—	—	—	—	—
			100	17.0	—	—	.6	—	3.5	—	—	—	—	—
	YEAST													
1207	brewers', debittered	1 TBLS	10	5.0	28.3	3.9	.1	3.84	.2	21.0	175.3	12.1	23.1	189.4
			100	5.0	283.0	38.8	1.0	38.40	1.7	210.0	1753.0	121.0	231.0	1894.0
1208	brewers', tablet form	6 TABLETS	5	—	—	4.0	—	—	—	—	—	—	—	—
			100	—	—	80.0	—	—	—	—	—	—	—	—

Item No.	Minerals (Micro)			Fat-Soluble Vitamins			Water-Soluble Vitamins								
	Iron	Zinc	Copper	Vitamin A	Vitamin D	Vitamin E (Alpha Tocopherol)	Vitamin C	Thiamin	Ribo-flavin	Niacin	Panto-thenic Acid	Vit. B-6 (Pyri-doxine)	Folacin (Folic Acid)	Biotin	Vitamin B-12
	(mg)	(mg)	(mg)	(IU)	(IU)	(mg)	(mg)	(mg)	(mg)	(mg)	(mg)	(mg)	(mcg)	(mcg)	(mcg)
1180	Trace	—	—	1.8	—	—	.61	.01	0	.01	—	—	—	—	—
	.33	—	—	175.0	—	—	61.30	.51	.04	1.00	—	—	—	—	—
1181	.29	.03	.01	31.3	—	—	.61	.01	—	.01	—	—	—	—	0
	29.25	3.23	.68	3128.0	—	—	61.22	.51	—	1.00	—	—	—	—	0
1182	.11	—	—	—	—	—	—	—	—	—	—	—	—	—	0
	11.10	—	—	—	—	—	—	—	—	—	—	—	—	—	0
1183	.28	.05	—	59.0	—	—	.32	.01	Trace	.06	—	—	—	—	0
	28.12	4.70	—	5900.0	—	—	32.38	.75	.34	5.70	—	—	—	—	0
1184	.01	—	—	0	—	—	0	0	0	0	—	—	—	—	—
	.10	—	—	0	—	—	0	0	0	0	—	—	—	—	—
1185	Trace	0	Trace	0	0	—	0	0	0	0	0	0	0	0	0
	.10	0	.10	0	0	—	0	0	0	0	0	0	0	0	0
1186	.38	.04	—	51.3	—	—	.12	Trace	0	.04	—	—	—	—	0
	37.80	4.30	—	5130.0	—	—	12.00	.37	.04	4.10	—	—	—	—	0
1187	2.94	—	—	8.4	—	—	0	.27	.07	1.51	—	—	26.88	—	—
	10.50	—	—	30.0	—	—	0	.98	.24	5.40	—	—	96.00	—	—
1188	2.18	2.87	—	18.5	—	—	0	.20	.02	1.31	.19	.04	—	—	0
	7.80	10.25	—	66.0	—	—	0	.72	.09	4.68	.68	.15	—	—	0
1189	.72	.08	.02	0	—	—	0	Trace	.04	.06	—	—	4.20	—	—
	4.80	.53	.13	0	—	—	0	.02	.25	.40	—	—	28.00	—	—
1190	—	—	—	—	—	—	—	—	—	—	—	—	—	—	—
	—	—	—	—	—	—	—	—	—	—	—	—	—	—	—
1191	.32	.04	—	42.0	—	—	.12	Trace	.01	.09	—	—	—	—	0
	32.30	3.90	—	4200.0	—	—	12.00	.25	.34	8.90	—	—	—	—	0
1192	1.24	.06	.01	38.0	—	—	.12	.01	Trace	.05	—	—	—	—	0
	123.60	6.40	.86	3800.0	—	—	12.00	.51	.40	4.90	—	—	—	—	0
1193	—	—	—	—	—	—	—	Trace	.01	.02	—	—	—	—	—
	—	—	—	—	—	—	—	.03	.10	.32	—	—	—	—	—
1194															
1195	.83	.09	—	0	—	—	.52	Trace	.01	.10	—	—	—	—	0
	41.42	4.35	—	0	—	—	25.85	.15	.23	5.14	—	—	—	—	0
1196	.09	.02	.01	—	0	—	—	—	—	—	—	0	—	—	0
	.60	.10	.08	—	0	—	—	—	—	—	—	Trace	—	—	0
1197	—	.02	—	—	—	—	—	—	—	—	—	0	.02	—	0
	—	.10	—	—	—	—	—	—	—	—	—	—	.10	—	0
1198	.04	—	—	0	—	—	.24	.01	.01	.06	—	—	—	—	—
	.60	—	—	0	—	—	4.00	.14	.20	1.00	—	—	—	—	—
1199	.28	—	—	5.5	—	—	9.15	Trace	.01	.02	—	—	—	—	—
	5.60	—	—	110.0	—	—	183.00	.06	.16	.44	—	—	—	—	—
1200	—	—	—	—	—	—	—	—	—	—	—	—	—	—	—
	36.0	1.58	1.0	22940.0	35.6	—	—	.55	1.51	3.90	3.35	—	—	33.00	—
1201	—	—	—	—	—	—	—	—	—	—	—	—	—	—	—
	Trace	Trace	Trace	85,000.0	8500.0	20.0	0	0	0	0	0	0	0	0	0
1202	1.76	—	—	—	—	—	—	—	—	—	—	—	—	—	—
	6.30	—	—	—	—	—	—	—	—	—	—	—	—	—	—
1203	—	—	—	—	—	—	—	—	—	—	—	—	—	—	—
	—	—	—	—	—	—	—	—	—	—	—	—	—	—	—
1204	2.49	—	—	—	—	—	—	—	—	—	—	—	—	—	—
	8.90	—	—	—	—	—	—	—	—	—	—	—	—	—	—
1205	—	—	—	—	—	—	—	—	—	—	—	—	—	—	—
	—	—	—	—	—	—	—	—	—	—	—	—	—	—	—
1206	—	—	—	—	—	—	—	—	—	—	—	—	—	—	—
	—	—	—	—	—	—	—	—	—	—	—	—	—	—	—
1207	1.73	—	—	—	—	—	—	1.56	.43	3.79	1.20	.25	390.90	20.00	0
	17.30	—	—	—	—	—	—	15.61	4.28	37.90	12.00	2.50	3909.00	200.00	0
1208	0	—	—	—	—	—	—	.48	.16	1.60	.20	.16	13.00	4.00	—
	12.00	—	—	—	—	—	—	9.60	3.20	32.00	4.00	3.20	260.00	80.00	—

(Continued)

TABLE F-36 *(Continued)*

Item No.	Food Item Name	Approximate Measure	Weight	Moisture	Food Energy Calories	Protein	Fats	Carbo-hydrates	Fiber	Minerals (Macro) Calcium	Phos-phorus	Sodium	Mag-nesium	Potas-sium
			(g)	(%)	(kcal)	(g)	(g)	(g)	(g)	(mg)	(mg)	(mg)	(mg)	(mg)
	YEAST (Continued)													
1209	torula	1 TBLS	10	6.0	27.7	3.9	.1	3.70	.3	42.4	171.3	1.5	16.5	204.6
			100	6.0	277.0	38.6	1.0	37.00	3.3	424.0	1713.0	15.0	165.0	2046.0
	ACEROLA													
1210	raw, pulp & skin	1 AVG	8	92.3	2.2	0	0	.5	Trace	1.0	.9	.6	—	6.6
			100	92.3	28.0	.4	.3	6.8	.4	12.0	11.0	8.0	—	83.0
1211	pitted	1 MED	6	91.1	1.9	0	0	.5	Trace	.5	1.0	—	—	—
			100	91.1	31.0	.7	.2	7.6	.6	8.7	16.2	—	—	—
	APPLES													
1212	raw, fresh, unpared	1 MED	150	84.8	84.0	.3	.9	21.2	1.5	4.5	9.0	1.5	7.2	208.5
			100	84.8	56.0	.2	.6	14.1	1.0	3.0	6.0	1.0	4.8	139.0
1213	raw, fresh, pared	1 MED	145	85.3	76.9	.3	.4	20.2	.9	2.9	8.7	1.5	4.6	176.9
			100	85.3	53.0	.2	.3	13.9	.6	2.0	6.0	1.0	3.2	122.0
1214	raw, freshly harvested & stored, unpared	1 MED	150	84.4	87.0	.3	.9	21.8	1.5	4.5	9.0	1.5	4.8	208.5
			100	84.4	58.0	.2	.6	14.5	1.0	3.0	6.0	1.0	3.2	139.0
1215	raw, freshly harvested & stored, pared	1 MED	145	85.1	78.3	.3	.4	20.4	.9	2.9	8.7	1.5	4.6	176.9
			100	85.1	54.0	.2	.3	14.1	.6	2.0	6.0	1.0	3.2	122.0
1216	raw, stored, unpared	1 MED	150	83.9	90.0	.3	1.1	22.2	1.5	10.5	15.0	1.5	12.0	165.0
			100	83.9	60.0	.2	.7	14.8	1.0	7.0	10.0	1.0	8.0	110.0
1217	raw, stored, pared	1 MED	145	84.8	79.8	.3	.4	20.9	.9	8.7	14.5	1.5	7.3	159.5
			100	84.8	55.0	.2	.3	14.4	.6	6.0	10.0	1.0	5.0	110.0
1218	baked, unpared	1 MED	150	—	187.5	.3	.9	—	—	9.0	15.0	1.5	—	165.0
			100	—	125.0	.2	.6	—	—	6.0	10.0	1.0	—	110.0
1219	canned, sliced	1 CUP	200	82.3	134.0	.4	1.0	33.4	1.0	8.0	10.0	6.0	4.0	136.0
			100	82.3	67.0	.2	.5	16.7	.5	4.0	5.0	3.0	2.0	68.0
1220	canned, cubed, artifical sweetener	1 CUP	200	—	50.0	.4	.2	12.0	—	—	—	6.0	—	—
			100	—	25.0	.2	.1	6.0	—	—	—	3.0	—	—
1221	dried, sulfured, uncooked	1 OZ	28	32.9	66.9	.3	.1	18.1	.8	4.2	11.5	25.8	4.5	113.4
			100	32.9	239.0	.9	.3	64.8	2.8	15.0	41.0	92.0	16.0	405.0
1222	dried, sulfured, cooked w/o sugar	1 OZ	28	78.4	21.8	.1	.1	5.7	.3	2.5	4.2	.3	1.7	45.4
			100	78.4	78.0	.3	.5	20.3	.9	9.0	15.0	1.0	6.0	162.0
1223	dried, sulfured, cooked w/sugar	1 OZ	28	69.7	31.4	.1	.1	8.2	.2	2.2	3.6	.3	—	40.3
			100	69.7	112.0	.3	.4	29.2	.8	8.0	13.0	1.0	—	144.0
1224	dehydrated,sulfured, uncooked	1 OZ	28	2.5	98.8	.4	.6	25.8	1.1	11.2	18.5	2.0	—	204.4
			100	2.5	353.0	1.4	2.0	92.1	3.8	40.0	66.0	7.0	—	730.0
1225	dehydrated, sulfured, cooked w/sugar	1 OZ	28	79.6	21.3	.1	.1	5.5	.1	1.7	2.8	.3	—	29.7
			100	79.6	76.0	.2	.3	19.6	.5	6.0	10.0	1.0	—	106.0
1226	frozen, sliced, sweetened	1 CUP	200	75.1	186.0	.4	.2	48.6	1.4	10.0	12.0	28.0	8.0	136.0
			100	75.1	93.0	.2	.1	24.3	.7	5.0	6.0	14.0	4.0	68.0
1227	pie filling, canned	1 OZ	28	73.2	29.7	0	—	7.3	—	.8	—	—	—	—
			100	73.2	106.0	.1	—	26.2	—	3.0	—	—	—	—
	APPLESAUCE													
1228	canned, unsweetened or artificial sweetener	1 CUP	244	88.3	102.5	.4	.1	27.6	1.3	9.8	24.4	4.9	7.3	190.3
			100	88.3	42.0	.2	.1	11.3	.5	4.0	10.0	2.0	3.0	78.0
1229	canned, sweetened	1 CUP	255	80.0	191.3	.5	.4	49.9	1.1	7.7	17.9	7.7	7.7	153.0
			100	80.0	75.0	.2	.2	19.6	.5	3.0	7.0	3.0	3.0	60.0
	APRICOTS													
1230	raw	1 MED	38	85.3	19.4	.4	.1	4.9	.2	3.0	5.7	.4	3.5	121.2
			100	85.3	51.0	1.0	.2	12.8	.6	8.0	15.0	1.0	9.3	319.0
1231	canned, water pack, w/o artificial sweetener	1 HALF	33	92.5	8.6	.2	0	2.1	.1	2.6	4.3	1.0	2.3	62.4
			100	92.5	26.0	.7	.1	6.3	.4	8.0	13.0	3.0	7.0	189.0
1232	canned, artificial sweetener	1 MED	28	92.5	7.3	.2	0	1.8	.1	2.2	3.6	.8	2.0	52.9
			100	92.5	26.0	.7	.1	6.3	.4	8.0	13.0	3.0	7.0	189.0
1233	canned, juice pack	1 MED	30	86.6	14.4	.2	0	3.7	.1	3.9	6.0	.9	3.0	49.2
			100	86.6	48.0	.7	0	12.4	.4	13.0	20.0	3.0	10.0	164.0
1234	canned,light syrup	1 SERV	120	81.9	79.2	.8	.1	20.2	.5	13.2	18.0	1.2	8.4	286.8
			100	81.9	66.0	.7	.1	16.8	.4	11.0	15.0	1.0	7.0	239.0
1235	canned, extra heavy syrup	1 SERV	120	72.9	121.2	.7	.1	31.2	.5	13.2	18.0	1.2	8.4	276.0
			100	72.9	101.0	.6	.1	26.0	.4	11.0	15.0	1.0	7.0	230.0
1236	canned,unpeeled halves	1 MED	33	77.5	26.4	.2	0	7.1	—	3.0	3.6	2.0	2.3	45.9
			100	77.5	80.0	.5	.1	21.5	—	9.0	11.0	6.0	7.0	139.0

	Minerals (Micro)			Fat-Soluble Vitamins			Water-Soluble Vitamins								
Item No.	Iron	Zinc	Copper	Vitamin A	Vitamin D	Vitamin E (Alpha Tocopherol)	Vitamin C	Thiamin	Riboflavin	Niacin	Pantothenic Acid	Vit. B-6 (Pyridoxine)	Folacin (Folic Acid)	Biotin	Vitamin B-12
	(mg)	(mg)	(mg)	(IU)	(IU)	(mg)	(mg)	(mg)	(mg)	(mg)	(mg)	(mg)	(mcg)	(mcg)	(mcg)
1209	**1.93** 19.30	**.19** 1.85	— —	— —	— —	— —	— —	**1.40** 14.01	**.51** 5.06	**4.44** 44.40	**1.10** 11.00	**.30** 3.00	**300.00** 3000.00	**10.00** 100.00	**0** 0
1210	**.02** .20	— —	— —	**2.6** 33.0	— —	— —	**104.00** 1300.00	**Trace** .02	**.01** .06	**.03** .40	— —	**Trace** .01	— —	— —	**0** 0
1211	**.01** .17	— —	— —	**24.5** 408.0	— —	— —	**139.80** 2330.00	**Trace** .03	**.01** .08	**.02** .34	— —	— —	— —	— —	— —
1212	**.15** .10	**.08** .05	**.06** .04	**135.0** 90.0	— —	**.47** .31	**10.50** 7.00	**.05** .03	**.03** .02	**.15** .10	**.16** .11	**.05** .03	**12.00** 8.00	**1.35** .90	**0** 0
1213	**.15** .10	**.07** .05	**.05** .03	**58.0** 40.0	— —	— —	**5.80** 4.00	**.04** .03	**.03** .02	**.15** .10	**.15** .10	**.04** .03	**11.60** 8.00	— —	**0** 0
1214	**.15** .10	**.08** .05	**.06** .04	**135.0** 90.0	— —	**1.08** .72	**6.00** 4.00	**.05** 0.3	**.03** .02	**.15** .10	**.16** .11	**.05** .03	**12.00** 8.00	**1.35** .90	**0** 0
1215	**.15** .10	**.07** .05	**.05** .03	**58.0** 40.0	— —	— —	**2.90** 2.00	**.04** .03	**.03** .02	**.15** .10	**.15** .10	**.04** .03	**11.60** 8.00	— —	**0** 0
1216	**.45** .30	**.08** .05	**.06** .04	**135.0** 90.0	— —	**1.08** .72	**4.50** 3.00	**.05** .03	**.03** .02	**.15** .10	**.16** .11	**.05** .03	**12.00** 8.00	**1.35** .90	**0** 0
1217	**.44** .30	**.07** .05	**.04** .03	**58.0** 40.0	— —	— —	**2.90** 2.00	**.04** .03	**.03** .02	**.15** .10	**.15** .10	**.04** .03	**11.60** 8.00	— —	**0** 0
1218	**.45** .30	— —	— —	**139.5** 93.0	— —	— —	**4.50** 3.00	**.05** .03	**.03** .02	**.20** .13	— —	— —	— —	— —	— —
1219	**.44** .22	**.05** .03	**.10** .05	**102.0** 51.0	— —	— —	**2.00** 1.00	**.02** .01	**.02** .01	**.14** .07	**.06** .03	**.09** .04	**.60** .30	— —	— —
1220	— —	— —	— —	— —	— —	— —	— —	— —	— —	— —	— —	— —	— —	— —	— —
1221	**.48** 1.72	**.05** .19	**.05** .17	**0** 0	— —	— —	**1.06** 3.80	**0** 0	**.05** .16	**.26** .91	— —	**.04** .14	— —	— —	**0** 0
1222	**.14** .50	— —	— —	**0** 0	— —	— —	— —	**Trace** .01	**.01** .03	**.03** .10	— —	**.04** .14	— —	— —	**0** 0
1223	**.11** .40	— —	— —	— —	— —	— —	— —	**Trace** .01	**.01** .03	**.03** .10	— —	**.04** .14	— —	— —	**0** 0
1224	**.56** 2.00	— —	— —	— —	— —	— —	**2.80** 10.00	— —	**.02** .06	**.17** .60	— —	— —	— —	— —	— —
1225	**.08** .30	— —	— —	— —	— —	— —	**.28** 1.00	— —	**Trace** .01	**.03** .10	— —	— —	— —	— —	— —
1226	**1.00** .50	— —	— —	**40.0** 20.0	— —	— —	**14.00** 7.00	**Trace** .01	**.06** .03	**.40** .20	**.17** .09	**.05** .03	— —	**1.80** .90	**0** 0
1227	**.03** .10	— —	— —	**0** 0	— —	— —	**0** 0	**Trace** .01	**Trace** .01	**.03** .10	— —	— —	— —	— —	— —
1228	**.32** .13	**.06** .03	**.06** .03	**70.8** 29.0	— —	— —	**40.99** 16.80	**.02** .01	**.05** .02	**.10** .04	— —	— —	— —	— —	— —
1229	**.87** .34	**.09** .04	**.11** .04	**28.1** 11.0	— —	— —	**4.08** 1.60	**.03** .01	**.08** .03	**.10** .04	**.12** .05	**.07** .03	**2.55** 1.00	— —	**0** 0
1230	**.19** .50	**.05** .12	**.03** .08	**1026.0** 2700.0	**0** 0	— —	**3.80** 10.00	**.01** .03	**.01** .04	**.23** .60	**.09** .24	**.03** .07	**1.25** 3.30	— —	**0** 0
1231	**.18** .54	**.03** .10	**.01** .04	**597.3** 1810.0	— —	— —	**1.06** 3.20	**.01** .02	**.01** .02	**.13** .40	**.03** .10	**.02** .05	**.66** 2.00	— —	— —
1232	**.15** .54	**.03** .10	**.01** .04	**506.8** 1810.0	— —	— —	**.90** 3.20	**.01** .02	**.01** .02	**.11** .40	**.03** .10	**.02** .05	**.56** 2.00	— —	— —
1233	**.08** .25	**.03** .11	**.02** .05	**444.9** 1483.0	— —	— —	**1.53** 5.10	**.01** .02	**.01** .02	**.09** .31	**.03** .10	**.02** .05	**.15** .50	— —	— —
1234	**.36** .30	**.14** .12	**.05** .04	**2136.0** 1780.0	— —	— —	**4.80** 4.00	**.02** .02	**.02** .02	**.48** .40	**.11** .09	**.07** .05	**2.40** 2.00	— —	**0** 0
1235	**.36** .30	**.14** .12	**.05** .04	**2064.0** 1720.0	— —	— —	**4.80** 4.00	**.02** .02	**.02** .02	**.36** .30	**.11** .09	**.07** .05	**2.40** 2.00	— —	**0** 0
1236	**.10** .30	**.03** .10	**.03** .10	**308.6** 935.0	— —	— —	**1.32** 4.00	**.01** .02	**.01** .03	**.13** .40	— —	— —	**.66** 2.00	— —	— —

(Continued)

TABLE F-36 *(Continued)*

Item No.	Food Item Name	Approximate Measure	Weight	Moisture	Food Energy Calories	Protein	Fats	Carbo-hydrates	Fiber	Minerals (Macro) Calcium	Phos-phorus	Sodium	Mag-nesium	Potas-sium
			(g)	(%)	(kcal)	(g)	(g)	(g)	(g)	(mg)	(mg)	(mg)	(mg)	(mg)
	APRICOTS (Continued)													
1237	candied	1 HALF	**15**	**12.0**	**50.7**	**.1**	**0**	**13.0**	**.1**	**2.0**	**3.0**	—	—	—
			100	12.0	338.0	.6	.2	86.5	.6	13.0	20.0	—	—	—
1238	dried,sulfured, uncooked	1 HALF	**4**	**31.5**	**9.4**	**.2**	**0**	**2.4**	**.1**	**2.0**	**4.8**	**.4**	**2.0**	**56.9**
			100	31.5	236.0	3.9	.5	60.8	2.9	51.0	119.0	9.0	50.0	1422.0
1239	dried, sulfured, cooked w/o sugar	1 CUP	**285**	**75.6**	**242.3**	**4.6**	**.6**	**61.6**	**2.9**	**62.7**	**99.8**	**22.8**	**57.0**	**906.3**
			100	75.6	85.0	1.6	.2	21.6	1.0	22.0	35.0	8.0	20.0	318.0
1240	dried,sulfured, cooked w/sugar	1 CUP	**285**	**66.2**	**347.7**	**4.0**	**.3**	**89.5**	**2.6**	**54.2**	**88.4**	**20.0**	**66.1**	**792.3**
			100	66.2	122.0	1.4	.1	31.4	.9	19.0	31.0	7.0	23.2	278.0
1241	dehydrated, sulfured, uncooked	1 MED	**12**	**3.5**	**39.8**	**.7**	**.1**	**10.2**	**.5**	**10.3**	**16.7**	**4.0**	—	**151.2**
			100	3.5	332.0	5.6	1.0	84.6	3.8	86.0	139.0	33.0	—	1260.0
1242	dehydrated,sulfured, cooked w/sugar	1 MED	**50**	**66.7**	**59.5**	**.7**	**.1**	**15.3**	**.5**	**10.0**	**16.5**	**4.0**	—	**149.5**
			100	66.7	119.0	1.3	.2	30.5	.9	20.0	33.0	8.0	—	299.0
1243	frozen, sweetened	1 CUP	**300**	**73.3**	**294.0**	**2.1**	**.3**	**75.3**	**1.8**	**30.0**	**57.0**	**12.0**	**27.0**	**687.0**
			100	73.3	98.0	.7	.1	25.1	.6	10.0	19.0	4.0	9.0	229.0
	AVOCADOS													
1244	all varieties, raw, halved, fruit served w/skin	1 AVG	**250**	**74.0**	**417.5**	**5.3**	**41.0**	**15.8**	**4.0**	**25.0**	**105.0**	**10.0**	**112.5**	**1510.0**
			100	74.0	167.0	2.1	16.4	6.3	1.6	10.0	42.0	4.0	45.0	604.0
1245	California, raw, halved, fruit served w/skin	1 WHOLE	**240**	**73.6**	**403.2**	**5.5**	**37.3**	**11.3**	**5.8**	**26.4**	**96.0**	**28.8**	**126.0**	**1704.0**
			100	73.6	168.0	2.3	15.6	4.7	2.4	11.0	40.0	12.0	52.5	710.0
1246	Florida, raw, halved, fruit served w/skin	1 AVG	**360**	**78.0**	**460.8**	**4.7**	**39.6**	**31.7**	**5.4**	**46.8**	**140.4**	—	**122.8**	**1756.8**
			100	78.0	128.0	1.3	11.0	8.8	1.5	13.0	39.0	—	34.1	488.0
1247	Beards Lee, pitted	1 HALF	**202**	**67.5**	**470.7**	**.5**	**50.9**	**11.7**	**2.4**	**9.5**	**160.8**	—	—	—
			100	67.5	233.0	.3	25.2	5.8	1.1	4.7	79.6	—	—	—
1248	Kahaluu, pitted	1 HALF	**152**	**69.0**	**335.9**	**1.9**	**35.7**	**8.5**	**2.7**	**11.9**	**32.5**	—	—	—
			100	69.0	221.0	1.3	23.5	5.6	1.8	7.8	21.4	—	—	—
	BANANAS													
1249	Amarilla, long & slender	1 AVG	**115**	**72.2**	**111.6**	**1.4**	**.1**	**29.3**	**.5**	**11.5**	**31.1**	—	—	—
			100	72.2	97.0	1.2	.1	25.5	.4	10.0	27.0	—	—	—
1250	common, yellow, short & thick	1 AVG	**100**	**68.8**	**110.0**	**1.2**	**.2**	**29.0**	**.4**	**7.0**	**28.0**	—	—	—
			100	68.8	110.0	1.2	.2	29.0	.4	7.0	28.0	—	—	—
1251	dwarf	1 AVG	**70**	**72.3**	**67.9**	**1.0**	**.1**	**17.6**	**.4**	**7.0**	**22.4**	—	—	—
			100	72.3	97.0	1.4	.2	25.2	.6	10.0	32.0	—	—	—
1252	red, raw	1 SMALL	**100**	**74.4**	**90.0**	**1.2**	**.2**	**23.4**	**.4**	**10.0**	**18.0**	**1.0**	—	**370.0**
			100	74.4	90.0	1.2	.2	23.4	.4	10.0	18.0	1.0	—	370.0
1253	yellow, raw	1 SMALL	**100**	**75.7**	**85.0**	**1.1**	**.2**	**22.2**	**.5**	**3.0**	**18.0**	**1.0**	**21.8**	**398.0**
			100	75.7	85.0	1.1	.2	22.2	.5	3.0	18.0	1.0	21.8	398.0
1254	flakes	1 CUP	**100**	**3.0**	**340.0**	**4.0**	**1.0**	**89.0**	—	**32.0**	—	—	—	—
			100	3.0	340.0	4.0	1.0	89.0	—	32.0	—	—	—	—
1255	dehydrated, or banana powder	1 OZ	**28**	**3.0**	**95.2**	**1.2**	**.2**	**24.8**	**.6**	**9.0**	**29.1**	**1.1**	**37.0**	**413.6**
			100	3.0	340.0	4.4	.8	88.6	2.0	32.0	104.0	4.0	132.0	1477.0
1256	[5]**BARBADOS CHERRY** (acerola) ripe	1 OZ	**28**	**90.3**	**10.1**	**.1**	**.1**	**2.4**	**.1**	**3.4**	**3.1**	—	—	—
			100	90.3	36.0	.4	.4	8.7	.4	12.0	11.0	—	—	—
	BLACKBERRIES													
1257	raw	1 CUP	**144**	**84.5**	**83.5**	**1.7**	**1.3**	**18.6**	**5.9**	**20.2**	**27.4**	**1.4**	**28.5**	**305.3**
			100	84.5	58.0	1.2	.9	12.9	4.1	14.0	19.0	1.0	19.8	212.0
1258	canned,water pack, solids & liquid	1 CUP	**250**	**89.3**	**100.0**	**2.0**	**1.5**	**22.5**	**7.0**	**10.0**	**32.5**	**2.5**	**9.0**	**287.5**
			100	89.3	40.0	.8	.6	9.0	2.8	4.0	13.0	1.0	3.6	115.0
1259	canned, juice pack, solids & liquid	1 CUP	**260**	**85.8**	**140.4**	**2.1**	**2.1**	**31.5**	**7.0**	**65.0**	**44.2**	**2.6**	—	**442.0**
			100	85.8	54.0	.8	.8	12.1	2.7	25.0	17.0	1.0	—	170.0
1260	canned,light syrup, solids & liquid	1 CUP	**260**	**81.0**	**187.2**	**2.1**	**1.6**	**45.0**	**7.0**	**54.6**	**31.2**	**2.6**	—	**288.6**
			100	81.0	72.0	.8	.6	17.3	2.7	21.0	12.0	1.0	—	111.0
1261	canned, heavy syrup, solids & liquid	1 CUP	**250**	**76.1**	**227.5**	**2.0**	**1.5**	**55.5**	**6.5**	**52.5**	**30.0**	**2.5**	—	**272.5**
			100	76.1	91.0	.8	.6	22.2	2.6	21.0	12.0	1.0	—	109.0
1262	canned, extra heavy syrup, solids & liquid	1 CUP	**260**	**71.2**	**286.0**	**2.1**	**1.6**	**70.5**	**6.8**	**52.0**	**31.2**	**2.6**	—	**278.2**
			100	71.2	110.0	.8	.6	27.1	2.6	20.0	12.0	1.0	—	107.0
	BLUEBERRIES													
1263	raw	1 CUP	**140**	**83.2**	**86.8**	**1.0**	**.7**	**21.4**	**2.1**	**21.0**	**18.2**	**1.4**	**8.4**	**113.4**
			100	83.2	62.0	.7	.5	15.3	1.5	15.0	13.0	1.0	6.0	81.0
1264	canned, water pack, solids & liquid	1 CUP	**242**	**89.3**	**94.4**	**1.2**	**.5**	**23.7**	**2.4**	**24.2**	**21.8**	**2.4**	**9.7**	**145.2**
			100	89.3	39.0	.5	.2	9.8	1.0	10.0	9.0	1.0	4.0	60.0

Item No.	Minerals (Micro) Iron	Zinc	Copper	Fat-Soluble Vitamins Vitamin A	Vitamin D	Vitamin E (Alpha Tocopherol)	Water-Soluble Vitamins Vitamin C	Thiamin	Ribo-flavin	Niacin	Panto-thenic Acid	Vit. B-6 (Pyri-doxine)	Folacin (Folic Acid)	Biotin	Vitamin B-12
	(mg)	(mg)	(mg)	(IU)	(IU)	(mg)	(mg)	(mg)	(mg)	(mg)	(mg)	(mg)	(mcg)	(mcg)	(mcg)
1237	.15	—	—	185.0	—	—	—	—	.01	.10	—	—	—	—	—
	1.00	—	—	1233.0	—	—	—	—	.03	.66	—	—	—	—	—
1238	.19	.03	.02	285.9	0	—	.12	0	.01	.12	.03	.01	.56	—	0
	4.82	.79	.49	7147.0	0	—	3.00	.01	.15	2.99	.75	.17	14.00	—	0
1239	5.13	.31	.34	8550.0	0	—	8.55	Trace	.14	2.85	2.15	.48	Trace	—	0
	1.80	.11	.12	3000.0	0	—	3.00	Trace	.05	1.00	.75	.17	Trace	—	0
1240	4.56	.26	.31	7410.0	0	—	5.70	Trace	.11	2.57	2.15	.48	Trace	—	0
	1.60	.09	.11	2600.0	0	—	2.00	Trace	.04	.90	.75	.17	Trace	—	0
1241	.64	.01	—	1692.0	—	—	1.80	—	.01	.43	—	—	—	—	—
	5.30	.10	—	14100.0	—	—	15.00	—	.08	3.60	—	—	—	—	—
1242	.65	—	—	1400.0	—	—	1.00	—	.01	.40	—	—	—	—	—
	1.30	—	—	2800.0	—	—	2.00	—	.02	.80	—	—	—	—	—
1243	2.70	.30	—	5040.0	—	—	84.00	.06	.12	2.40	.60	.18	—	—	0
	.90	.10	—	1680.0	—	—	28.00	.02	.04	.80	.20	.06	—	—	0
1244	1.50	0	1.00	725.0	—	—	35.00	.28	.50	4.00	2.68	1.05	127.50	13.75	0
	.60	0	.40	290.0	—	—	14.00	.11	.20	1.60	1.07	.42	51.00	5.50	0
1245	4.08	0	.72	1048.0	—	4.03	20.40	.25	.29	4.16	2.38	.54	122.40	7.80	—
	1.70	0	.30	436.7	—	1.68	8.50	.10	.12	1.74	.99	.23	51.00	3.25	—
1246	1.80	0	.90	1044.0	—	—	50.40	.40	.72	5.76	—	—	183.60	—	—
	.50	0	.25	290.0	—	—	14.00	.11	.20	1.60	—	—	51.00	—	—
1247	1.25	—	—	4201.6	—	—	4.65	.08	.44	1.60	—	—	103.02	—	—
	.62	—	—	2080.0	—	—	2.30	.04	.22	.79	—	—	51.00	—	—
1248	.61	—	—	180.9	—	—	11.40	—	—	—	—	—	77.52	—	—
	.40	—	—	119.0	—	—	7.50	—	—	—	—	—	51.00	—	—
1249	.46	—	—	—	—	—	16.10	.05	.06	.69	—	—	—	—	—
	.40	—	—	—	—	—	14.00	.04	.05	.60	—	—	—	—	—
1250	.50	—	—	65.0	—	—	15.00	.04	.04	.70	—	—	—	—	—
	.50	—	—	65.0	—	—	15.00	.04	.04	.70	—	—	—	—	—
1251	.63	—	—	56.0	—	—	5.60	.04	.06	.42	—	—	—	—	—
	.90	—	—	80.0	—	—	8.00	.05	.09	.60	—	—	—	—	—
1252	.80	.20	—	400.0	0	—	10.00	.05	.04	.60	—	—	30.00	—	—
	.80	.20	—	400.0	0	—	10.00	.05	.04	.60	—	—	30.00	—	—
1253	.10	.20	.08	190.0	0	.22	10.00	.05	.06	.70	.26	.51	28.00	4.40	0
	.10	.20	.08	190.0	0	.22	10.00	.05	.06	.70	.26	.51	28.00	4.40	0
1254	2.80	—	—	760.0	0	—	7.00	.18	.24	2.80	—	—	—	—	—
	2.80	—	—	760.0	0	—	7.00	.18	.24	2.80	—	—	—	—	—
1255	.78	—	—	212.8	—	—	1.96	.05	.07	.78	—	—	—	—	—
	2.80	—	—	760.0	—	—	7.00	.18	.24	2.80	—	—	—	—	—
1256	.06	—	—	—	—	—	501.20	.01	.01	.17	—	—	—	—	—
	.20	—	—	—	—	—	1790.00	.03	.05	.60	—	—	—	—	—
1257	.58	0	.16	288.0	0	5.00	30.24	.04	.06	.58	.35	.07	19.73	.59	0
	.40	0	.11	200.0	0	3.50	21.00	.03	.04	.40	.24	.05	13.70	.41	0
1258	1.50	0	.10	350.0	—	—	17.50	.05	.05	.50	.20	.06	35.00	—	—
	.60	0	.04	140.0	—	—	7.00	.02	.02	.20	.08	.02	14.00	—	—
1259	2.34	—	—	390.0	—	—	26.00	.05	.08	.78	.21	.06	36.40	—	—
	.90	—	—	150.0	—	—	10.00	.02	.03	.30	.08	.02	14.00	—	—
1260	1.56	—	—	338.0	—	—	18.20	.03	.05	.52	.20	.06	36.40	—	0
	.60	—	—	130.0	—	—	7.00	.01	.02	.20	.08	.02	14.00	—	0
1261	1.50	—	—	325.0	—	—	17.50	.03	.05	.50	.20	.06	35.00	—	0
	.60	—	—	130.0	—	—	7.00	.01	.02	.20	.08	.02	14.00	—	0
1262	1.56	—	—	338.0	—	—	18.20	.03	.05	.52	.20	.06	36.40	—	0
	.60	—	—	130.0	—	—	7.00	.01	.02	.20	.08	.02	14.00	—	0
1263	1.40	—	.15	140.0	—	—	19.60	.04	.08	.70	.22	.09	8.40	—	0
	1.00	—	.11	100.0	—	—	14.00	.03	.06	.50	.16	.07	6.00	—	0
1264	1.69	.22	.07	96.8	—	—	16.94	.02	.02	.48	.17	.09	10.16	—	0
	.70	.09	.03	40.0	—	—	7.00	.01	.01	.20	.07	.04	4.20	—	0

(Continued)

TABLE F-36 *(Continued)*

Item No.	Food Item Name	Approximate Measure	Weight	Moisture	Food Energy Calories	Protein	Fats	Carbo-hydrates	Fiber	Minerals (Macro) Calcium	Phos-phorus	Sodium	Mag-nesium	Potas-sium
			(g)	(%)	(kcal)	(g)	(g)	(g)	(g)	(mg)	(mg)	(mg)	(mg)	(mg)
	BLUEBERRIES (Continued)													
1265	canned, extra heavy syrup, solids & liquid	1 CUP	240	73.2	242.4	1.0	.5	62.4	2.2	21.6	19.2	2.4	9.6	132.0
			100	73.2	101.0	.4	.2	26.0	.9	9.0	8.0	1.0	4.0	55.0
1266	frozen, unsweetened	1 CUP	160	85.0	88.0	1.1	.8	21.8	2.4	16.0	20.8	1.6	9.6	129.6
			100	85.0	55.0	.7	.5	13.6	1.5	10.0	13.0	1.0	6.0	81.0
1267	frozen,sweetened	1 CUP	150	77.4	127.5	.6	.2	32.9	1.4	9.0	10.5	1.5	6.0	90.0
			100	77.4	85.0	.4	.1	22.0	.9	6.0	7.0	1.0	4.0	60.0
1268	pie filling, canned	1 OZ	28	68.9	35.0	.1	—	8.5	—	1.7	—	—	—	—
			100	68.9	125.0	.2	—	30.4	—	6.0	—	—	—	—
	BOYSENBERRIES													
1269	canned, water pack, solids & liquid	1 CUP	140	89.8	50.4	1.0	.1	12.7	2.7	26.6	26.6	1.4	—	119.0
			100	89.8	36.0	.7	.1	9.1	1.9	19.0	19.0	1.0	—	85.0
1270	frozen,unsweetened	1 CUP	125	86.8	60.0	1.5	.4	14.3	3.4	31.3	30.0	1.3	22.5	191.3
			100	86.8	48.0	1.2	.3	11.4	2.7	25.0	24.0	1.0	18.0	153.0
1271	frozen, sweetened	1 CUP	150	74.3	144.0	1.2	.5	36.6	2.7	25.5	25.5	1.5	18.0	157.5
			100	74.3	96.0	.8	.3	24.4	1.8	17.0	17.0	1.0	12.0	105.0
1272	**BREADFRUIT,** raw	1 MED	350	70.8	360.5	6.0	1.1	91.7	4.2	115.5	112.0	52.5	—	1536.5
			100	70.8	103.0	1.7	.3	26.2	1.2	33.0	32.0	15.0	—	439.0
1273	**CANTALOUPE,** raw	1 WHOLE	770	91.2	231.0	4.6	.8	57.8	2.3	107.8	123.2	92.4	64.7	1932.7
			100	91.2	30.0	.6	.1	7.5	.3	14.0	16.0	12.0	8.4	251.0
1274	**CARAMBOLA,** raw	1 AVG	57	90.4	20.0	.4	.3	4.6	.5	2.3	9.7	1.1	—	109.4
			100	90.4	35.0	.7	.5	8.0	.9	4.0	17.0	2.0	—	192.0
1275	**CASABA** (Golden Beauty), vine ripened	1 WHOLE	850	91.5	272.0	10.2	.9	55.3	4.3	119.0	136.0	102.0	68.0	2133.5
			100	91.5	32.0	1.2	.1	6.5	.5	14.0	16.0	12.0	8.0	251.0
1276	**CHERIMOYA,** raw	1 AVG	488	73.5	458.7	6.3	2.0	117.1	10.7	112.2	195.2	—	—	—
			100	73.5	94.0	1.3	.4	24.0	2.2	23.0	40.0	—	—	—
	CHERRIES													
1277	sweet, raw	1 CUP	200	80.4	140.0	2.6	.6	34.8	.8	20.0	26.0	4.0	32.4	500.0
			100	80.4	70.0	1.3	.3	17.4	.4	10.0	13.0	2.0	16.2	250.0
1278	sweet, canned, water pack	1 CUP	200	87.1	92.0	1.6	.3	23.5	.4	22.0	30.0	2.0	18.0	.3
			100	87.1	46.0	.8	.1	11.8	.2	11.0	15.0	1.0	9.0	.1
1279	sweet, canned, light syrup	1 CUP	200	82.0	130.0	1.8	.4	33.0	.6	30.0	26.0	2.0	18.0	256.0
			100	82.0	65.0	.9	.2	16.5	.3	15.0	13.0	1.0	9.0	128.0
1280	sweet,canned, heavy syrup	1 CUP	200	77.5	166.0	1.2	.3	42.7	.6	18.0	36.0	6.0	18.0	290.0
			100	77.5	83.0	.6	.2	21.3	.3	9.0	18.0	3.0	9.0	145.0
1281	sweet, canned, extra heavy syrup	1 CUP	200	73.0	200.0	1.6	.4	51.2	.6	28.0	24.0	2.0	18.0	246.0
			100	73.0	100.0	.8	.2	25.6	.3	14.0	12.0	1.0	9.0	123.0
1282	Royal Anne, canned, water pack or artificial sweetener	1 CUP	200	87.1	92.0	1.6	.3	23.5	.4	22.0	30.0	2.0	18.0	262.0
			100	87.1	46.0	.8	.1	11.8	.2	11.0	15.0	1.0	9.0	131.0
1283	sour red, raw	1 CUP	200	83.7	116.0	2.4	.6	28.6	.4	12.0	22.0	4.0	23.0	382.0
			100	83.7	58.0	1.2	.3	14.3	.2	6.0	11.0	2.0	11.5	191.0
1284	sour, canned, water pack	1 CUP	200	89.9	72.0	1.5	.2	17.9	1.8	22.0	26.0	4.0	18.0	196.0
			100	89.9	36.0	.8	.1	8.9	.9	11.0	13.0	2.0	9.0	98.0
1285	sour, canned, light syrup	1 CUP	200	80.0	148.0	1.6	.4	37.4	.2	28.0	26.0	2.0	—	252.0
			100	80.2	74.0	.8	.2	18.7	.1	14.0	13.0	1.0	—	126.0
1286	sour, canned, heavy syrup	1 CUP	200	76.0	178.0	1.6	.4	45.4	.2	28.0	24.0	2.0	—	248.0
			100	76.0	89.0	.8	.2	22.7	.1	14.0	12.0	1.0	—	124.0
1287	sour, canned, extra heavy syrup	1 CUP	200	70.1	224.0	1.6	.4	57.2	.2	28.0	24.0	2.0	—	242.0
			100	70.1	112.0	.8	.2	28.6	.1	14.0	12.0	1.0	—	121.0
1288	sour red, frozen, unsweetened	1 CUP	200	84.9	110.0	2.0	.8	26.8	.6	26.0	44.0	4.0	20.0	376.0
			100	84.9	55.0	1.0	.4	13.4	.3	13.0	22.0	2.0	10.0	188.0
1289	sour red, frozen, sweetened	1 CUP	200	70.6	224.0	2.0	.8	55.6	.4	24.0	30.0	4.0	16.0	260.0
			100	70.6	112.0	1.0	.4	27.8	.2	12.0	15.0	2.0	8.0	130.0
1290	candied	1 AVG	4	12.0	13.6	0	0	3.5	Trace	—	—	—	—	—
			100	12.0	339.0	.5	.2	86.7	.5	—	—	—	—	—
1291	maraschino, solids & liquid	1 PIECE	8	70.0	9.3	0	0	2.4	Trace	—	—	—	—	—
			100	70.0	116.0	.2	.2	29.4	.3	—	—	—	—	—
1292	pie filling, canned	1 OZ	28	71.8	31.4	0	—	7.7	—	2.5	—	—	—	—
			100	71.8	112.0	.1	—	27.4	—	9.0	—	—	—	—
1293	**COCONUT** —See Nuts & Seeds													
1294	**CRABAPPLES,** raw 1½" diameter	1 AVG	60	81.1	40.8	.2	.2	10.7	.4	3.6	7.8	.6	—	66.0
			100	81.1	68.0	.4	.3	17.8	.6	6.0	13.0	1.0	—	110.0
	CRANBERRIES													
1295	raw	1 CUP	100	87.9	46.0	.4	.7	10.8	1.4	7.0	6.0	2.0	4.5	67.0
			100	87.9	46.0	.4	.7	10.8	1.4	7.0	6.0	2.0	4.5	67.0
1296	raw,dehydrated	1 CUP	95	4.9	349.6	2.7	6.3	80.1	8.3	77.9	20.9	15.2	—	611.8
			100	4.9	368.0	2.8	6.6	84.3	8.7	82.0	22.0	16.0	—	644.0

Item No.	Minerals (Micro) Iron	Zinc	Copper	Fat-Soluble Vitamins Vitamin A	Vitamin D	Vitamin E (Alpha Tocopherol)	Water-Soluble Vitamins Vitamin C	Thiamin	Ribo-flavin	Niacin	Panto-thenic Acid	Vit. B-6 (Pyri-doxine)	Folacin (Folic Acid)	Biotin	Vitamin B-12
	(mg)	(mg)	(mg)	(IU)	(IU)	(mg)	(mg)	(mg)	(mg)	(mg)	(mg)	(mg)	(mcg)	(mcg)	(mcg)
1265	**1.44**	–	–	**96.0**	–	–	**14.40**	**.02**	**.02**	**.48**	**.16**	**.09**	**10.08**	–	**0**
	.60	–	–	40.0	–	–	6.00	.01	.01	.20	.07	.04	4.20	–	0
1266	**1.28**	–	–	**112.0**	–	–	**11.20**	**.05**	**.10**	**.80**	**.24**	**.10**	**6.72**	–	**0**
	.80	–	–	70.0	–	–	7.00	.03	.06	.50	.15	.06	4.20	–	0
1267	**.59**	**.10**	**.06**	**66.0**	–	–	**1.50**	**.03**	**.08**	**.38**	**.18**	**.12**	**6.30**	–	**0**
	.39	.06	.04	44.0	–	–	1.00	.02	.05	.25	.12	.08	4.20	–	0
1268	**.11**	–	–	**0**	–	–	**1.12**	**Trace**	**.01**	**.03**	–	–	–	–	–
	.40	–	–	0	–	–	4.00	.01	.03	.10	–	–	–	–	–
1269	**1.68**	–	–	**182.0**	–	–	**9.80**	**.01**	**.14**	**.98**	–	–	–	–	–
	1.20	–	–	130.0	–	–	7.00	.01	.10	.70	–	–	–	–	–
1270	**2.00**	–	–	**212.5**	–	–	**16.25**	**.03**	**.16**	**1.25**	**.28**	**.07**	–	–	**0**
	1.60	–	–	170.0	–	–	13.00	.02	.13	1.00	.22	.06	–	–	0
1271	**.90**	–	–	**210.0**	–	–	**12.00**	**.03**	**.15**	**.90**	**.30**	**.06**	–	–	**0**
	.60	–	–	140.0	–	–	8.00	.02	.10	.60	.20	.04	–	–	0
1272	**4.20**	–	–	**140.0**	–	–	**101.50**	**.39**	**.11**	**3.15**	**1.60**	–	–	–	**0**
	1.20	–	–	40.0	–	–	29.00	.11	.03	.90	.46	–	–	–	0
1273	**3.08**	**1.08**	**.11**	**26180.0**	**0**	**1.08**	**254.10**	**.40**	**.26**	**4.62**	**1.93**	**.42**	**231.00**	**23.87**	**0**
	.40	.14	.01	3400.0	0	.14	33.00	.05	.03	.60	.25	.06	30.00	3.10	0
1274	**.86**	–	–	**684.0**	–	–	**19.95**	**.02**	**.01**	**.17**	–	–	–	–	–
	1.50	–	–	1200.0	–	–	35.00	.04	.02	.30	–	–	–	–	–
1275	**3.40**	–	–	**255.0**	–	–	**110.50**	**.34**	**.26**	**5.10**	–	–	–	–	–
	.40	–	–	30.0	–	–	13.00	.04	.03	.60	–	–	–	–	–
1276	**2.44**	–	–	**48.8**	–	–	**43.92**	**.48**	**.54**	**6.34**	–	–	–	–	–
	.50	–	–	10.0	–	–	9.00	.10	.11	1.30	–	–	–	–	–
1277	**.40**	**0**	**.27**	**220.0**	–	–	**20.00**	**.10**	**.12**	**.80**	**.52**	**.06**	**16.00**	–	**0**
	.20	0	.13	110.0	–	–	10.00	.05	.06	.40	.26	.03	8.00	–	0
1278	**.72**	**.15**	**.15**	**250.0**	–	–	**4.40**	**.04**	**.04**	**.40**	**.24**	**.06**	**6.00**	–	**0**
	.36	.08	.08	125.0	–	–	2.20	.02	.02	.20	.12	.03	3.00	–	0
1279	**.60**	**.22**	**.12**	**120.0**	–	–	**6.00**	**.04**	**.04**	**.40**	–	**.06**	–	–	**0**
	.30	.11	.06	60.0	–	–	3.00	.02	.02	.20	–	.03	–	–	0
1280	**.70**	**.20**	**.27**	**304.0**	–	–	**6.80**	**.04**	**.08**	**.80**	–	**.06**	–	–	**0**
	.35	.10	.14	152.0	–	–	3.40	.02	.04	.40	–	.03	–	–	0
1281	**.60**	**.22**	**.12**	**100.0**	–	–	**6.00**	**.04**	**.04**	**.40**	–	**.06**	–	–	**0**
	.30	.11	.06	50.0	–	–	3.00	.02	.02	.20	–	.03	–	–	0
1282	**.72**	**.15**	**.15**	**250.0**	–	–	**4.40**	**.04**	**.04**	**.40**	–	–	–	–	–
	.36	.08	.08	125.0	–	–	2.20	.02	.02	.20	–	–	–	–	–
1283	**.80**	**0**	**.18**	**2000.0**	–	–	**20.00**	**.10**	**.12**	**.80**	**.29**	**.13**	**16.00**	–	**0**
	.40	0	.09	1000.0	–	–	10.00	.05	.06	.40	.14	.07	8.00	–	0
1284	**.60**	**.15**	**.15**	**1508.0**	–	–	**4.20**	**.04**	**.08**	**.36**	**.21**	**.09**	**16.00**	–	**0**
	.30	.08	.08	754.0	–	–	2.10	.02	.04	.18	.11	.04	8.00	–	0
1285	**.60**	–	**.10**	**1320.0**	–	–	**10.00**	**.06**	**.04**	**.40**	**.21**	**.09**	**16.00**	–	**0**
	.30	–	.05	660.0	–	–	5.00	.03	.02	.20	.11	.04	8.00	–	0
1286	**.60**	–	**.10**	**1300.0**	–	–	**10.00**	**.06**	**.04**	**.40**	**.21**	**.09**	**16.00**	–	**0**
	.30	–	.05	650.0	–	–	5.00	.03	.02	.20	.11	.04	8.00	–	0
1287	**.40**	–	**.10**	**1260.0**	–	–	**10.00**	**.06**	**.04**	**.40**	**.21**	**.09**	**16.00**	–	**0**
	.20	–	.05	630.0	–	–	5.00	.03	.02	.20	.11	.04	8.00	–	0
1288	**1.40**	–	–	**2000.0**	–	–	**10.00**	**.08**	**.14**	**.60**	**.20**	**.12**	–	–	**0**
	.70	–	–	1000.0	–	–	5.00	.04	.07	.30	.10	.06	–	–	0
1289	**1.00**	–	–	**960.0**	–	–	**12.00**	**.06**	**.12**	**.60**	**.17**	**.12**	–	–	**0**
	.50	–	–	480.0	–	–	6.00	.03	.06	.30	.08	.06	–	–	0
1290	–	–	–	**8.0**	–	–	**0**	–	–	–	–	–	–	–	–
	–	–	–	200.0	–	–	0	–	–	–	–	–	–	–	–
1291	–	–	–	–	–	–	–	–	–	–	–	–	–	–	–
	–	–	–	–	–	–	–	–	–	–	–	–	–	–	–
1292	**.17**	–	–	**0**	–	–	**0**	**Trace**	**.01**	**.03**	–	–	–	–	–
	.60	–	–	0	–	–	0	.01	.02	.10	–	–	–	–	–
1293															
1294	**.13**	–	–	**24.0**	–	–	**4.80**	**.02**	**.01**	**.06**	–	–	–	–	–
	.30	–	–	40.0	–	–	8.00	.03	.02	.10	–	–	–	–	–
1295	**.20**	**0**	**.06**	**40.0**	**0**	–	**11.00**	**.03**	**.02**	**.10**	**.22**	**.04**	**2.00**	–	**0**
	.20	0	.06	40.0	0	–	11.00	.03	.02	.10	.22	.04	2.00	–	0
1296	**3.23**	–	–	**285.0**	–	–	**30.40**	**.16**	**.11**	**.76**	–	–	–	–	–
	3.40	–	–	300.0	–	–	32.00	.17	.12	.80	–	–	–	–	–

(Continued)

FRUITS

TABLE F-36 *(Continued)*

Item No.	Food Item Name	Approximate Measure	Weight	Moisture	Food Energy Calories	Protein	Fats	Carbo-hydrates	Fiber	Minerals (Macro) Calcium	Phos-phorus	Sodium	Mag-nesium	Potas-sium
			(g)	(%)	(kcal)	(g)	(g)	(g)	(g)	(mg)	(mg)	(mg)	(mg)	(mg)
	CRANBERRIES (Continued)													
1297	sauce, homemade, sweetened, whole, unstrained	1 CUP	**200**	**53.9**	**356.0**	**.4**	**.6**	**91.0**	**1.4**	**14.0**	**10.0**	**2.0**	–	**76.0**
			100	53.9	178.0	.2	.3	45.5	.7	7.0	5.0	1.0	–	38.0
1298	sauce, canned, sweetened, whole	1 CUP	**280**	**61.6**	**439.6**	**.6**	**3.4**	**101.6**	**.3**	**40.0**	**26.6**	**79.0**	**12.6**	**53.8**
			100	61.6	157.0	.2	1.2	36.3	.1	14.3	9.5	28.2	4.5	19.2
1299	sauce, canned, sweetened, strained	1 CUP	**277**	**60.9**	**418.3**	**.6**	**.6**	**106.9**	**.6**	**11.1**	**13.9**	**69.3**	**8.3**	**66.5**
			100	60.9	151.0	.2	.2	38.6	.2	4.0	5.0	25.0	3.0	24.0
	CURRANTS													
1300	Black European, raw	1 CUP	**132**	**84.2**	**71.3**	**2.2**	**.1**	**17.3**	**3.2**	**79.2**	**52.8**	**4.0**	**22.6**	**491.0**
			100	84.2	54.0	1.7	.1	13.1	2.4	60.0	40.0	3.0	17.1	372.0
1301	red & white, raw	1 CUP	**133**	**85.7**	**66.5**	**1.9**	**.3**	**16.1**	**4.5**	**42.6**	**30.6**	**2.7**	**20.0**	**341.8**
			100	85.7	50.0	1.4	.2	12.1	3.4	32.0	23.0	2.0	15.0	257.0
1302	[6]seedless, dried	1 CUP	**140**	–	**394.0**	**5.8**	**.4**	**103.9**	**2.2**	**24.4**	**176.0**	**9.8**	**47.6**	**1262.0**
			100	–	283.0	4.1	.3	74.1	1.6	89.0	126.0	7.0	34.0	902.0
1303	**CUSTARDAPPLE,** bullocksheart, raw, 3" diameter	1 AVG	**180**	**71.5**	**181.8**	**3.1**	**1.1**	**45.4**	**6.1**	**48.6**	**36.0**	–	–	–
			100	71.5	101.0	1.7	.6	25.2	3.4	27.0	20.0	–	–	–
1304	**DATES,** domestic, natural, dry	1 CUP	**178**	**23.8**	**482.4**	**3.5**	**.9**	**128.5**	**4.1**	**58.7**	**64.1**	**1.8**	**60.5**	**1185.5**
			100	23.8	271.0	1.9	.5	72.2	2.3	33.0	36.0	1.0	34.0	666.0
1305	[6]**DURIAN,** civet	1 OZ	**28**	**81.1**	**18.8**	**.6**	**.2**	**4.1**	**.4**	**2.2**	**10.6**	–	–	–
			100	81.1	67.0	2.2	.8	14.8	1.6	8.0	38.0	–	–	–
1306	**ELDERBERRIES,** raw	1 CUP	**458**	**79.8**	**329.8**	**11.9**	**2.3**	**75.1**	**32.1**	**174.0**	**128.2**	–	–	**1374.0**
			100	79.8	72.0	2.6	.5	16.4	7.0	38.0	28.0	–	–	300.0
	FIGS													
1307	raw	1 MED	**41**	**77.5**	**32.8**	**.5**	**.1**	**8.3**	**.5**	**14.4**	**5.7**	**.4**	**8.2**	**95.1**
			100	77.5	80.0	1.2	.3	20.3	1.2	35.0	14.0	1.0	20.0	232.0
1308	canned, water pack	1 MED	**33**	**86.6**	**15.8**	**.2**	**.1**	**4.1**	**.2**	**4.6**	**4.6**	**.7**	–	**51.2**
			100	86.6	48.0	.5	.2	12.4	.7	14.0	14.0	2.0	–	155.0
1309	canned, light syrup	1 MED	**33**	**82.2**	**21.5**	**.2**	**.1**	**5.5**	**.2**	**4.3**	**4.3**	**.7**	–	**50.2**
			100	82.2	65.0	.5	.2	16.8	.7	13.0	13.0	2.0	–	152.0
1310	canned, heavy syrup	1 MED	**33**	**76.3**	**29.0**	**.1**	**0**	**7.6**	**.2**	**8.9**	**3.3**	**.3**	**3.3**	**32.7**
			100	76.3	88.0	.4	.1	22.9	.6	27.0	10.0	1.0	10.0	99.0
1311	canned, extra heavy syrup	1 MED	**33**	**72.3**	**34.0**	**.2**	**.1**	**8.8**	**.2**	**4.3**	**4.3**	**.7**	–	**48.2**
			100	72.3	103.0	.5	.2	26.7	.6	13.0	13.0	2.0	–	146.0
1312	candied	1 OZ	**28**	**21.0**	**83.7**	**1.0**	**.1**	**20.6**	–	–	–	–	–	–
			100	21.0	299.0	3.5	.2	73.7	–	–	–	–	–	–
1313	dried, uncooked	1 MED	**20**	**25.8**	**53.2**	**.6**	**.3**	**13.5**	**.9**	**30.4**	**14.2**	**2.0**	**12.2**	**154.6**
			100	25.8	266.0	3.1	1.5	67.5	4.7	152.0	71.0	10.0	61.0	773.0
	FRUIT													
1314	cocktail, canned, water pack or artificial sweetener, solids & liquid	1 CUP	**256**	**90.8**	**81.9**	**1.1**	**.1**	**21.8**	**1.0**	**12.8**	**28.2**	**10.2**	**17.9**	**240.6**
			100	90.8	32.0	.4	.1	8.5	.4	5.0	11.0	4.0	7.0	94.0
1315	cocktail, canned, light syrup, solids & liquid	1 CUP	**256**	**83.6**	**153.6**	**1.0**	**.3**	**40.2**	**1.0**	**23.0**	**30.7**	**12.8**	**17.9**	**419.8**
			100	83.6	60.0	.4	.1	15.7	.4	9.0	12.0	5.0	7.0	164.0
1316	cocktail, canned, heavy syrup, solids & liquid	1 CUP	**256**	**80.4**	**186.9**	**1.0**	**.2**	**48.4**	**1.2**	**15.4**	**28.2**	**15.4**	**12.8**	**225.3**
			100	80.4	73.0	.4	.1	18.9	.5	6.0	11.0	6.0	5.0	88.0
1317	cocktail, canned, extra heavy syrup, solids & liquid	1 CUP	**256**	**75.6**	**235.5**	**1.0**	**.3**	**60.7**	**1.0**	**23.0**	**30.7**	**12.8**	**17.9**	**407.0**
			100	75.6	92.0	.4	.1	23.7	.4	9.0	12.0	5.0	7.0	159.0
1318	salad, canned, water pack, solids & liquid	1 CUP	**256**	**90.1**	**89.6**	**1.0**	**.3**	**23.3**	**1.3**	**20.5**	**28.2**	**2.6**	–	**355.8**
			100	90.1	35.0	.4	.1	9.1	.5	8.0	11.0	1.0	–	139.0
1319	salad, canned, light syrup, solids & liquid	1 CUP	**256**	**83.9**	**151.0**	**.8**	**.3**	**39.7**	**1.0**	**20.5**	**28.2**	**2.6**	–	**348.2**
			100	83.9	59.0	.3	.1	15.5	.4	8.0	11.0	1.0	–	136.0
1320	salad, canned, heavy syrup, solids & liquid	1 CUP	**256**	**80.3**	**186.9**	**.9**	**.2**	**48.9**	**1.5**	**15.4**	**23.0**	**15.4**	**12.8**	**204.8**
			100	80.3	73.0	.3	.1	19.1	.6	6.0	9.0	6.0	5.0	80.0
1321	salad, canned, extra heavy syrup, solids & liquid	1 CUP	**256**	**76.0**	**230.4**	**.8**	**.3**	**59.9**	**1.0**	**20.5**	**28.2**	**2.6**	–	**335.4**
			100	76.0	90.0	.3	.1	23.4	.4	8.0	11.0	1.0	–	131.0
	GOOSEBERRIES													
1322	raw	1 CUP	**150**	**88.9**	**58.5**	**1.2**	**.3**	**14.6**	**2.8**	**27.0**	**22.5**	**1.5**	**13.5**	**232.5**
			100	88.9	39.0	.8	.2	9.7	1.9	18.0	15.0	1.0	9.0	155.0
1323	canned, water pack, solids & liquid	1 CUP	**200**	**92.5**	**52.0**	**1.0**	**.2**	**13.2**	**2.6**	**24.0**	**20.0**	**2.0**	–	**210.0**
			100	92.5	26.0	.5	.1	6.6	1.3	12.0	10.0	1.0	–	105.0
1324	canned, heavy syrup	1 CUP	**200**	**76.1**	**180.0**	**1.0**	**.2**	**46.0**	**2.4**	**22.0**	**18.0**	**2.0**	–	**196.0**
			100	76.1	90.0	.5	.1	23.0	1.2	11.0	9.0	1.0	–	98.0
1325	canned, extra heavy syrup	1 CUP	**200**	**69.2**	**234.0**	**1.0**	**.2**	**60.0**	**2.4**	**22.0**	**18.0**	**2.0**	–	**190.0**
			100	69.2	117.0	.5	.1	30.0	1.2	11.0	9.0	1.0	–	95.0

FRUITS

Item No.	Minerals (Micro) Iron	Zinc	Copper	Fat-Soluble Vitamins Vitamin A	Vitamin D	Vitamin E (Alpha Tocopherol)	Water-Soluble Vitamins Vitamin C	Thiamin	Ribo-flavin	Niacin	Panto-thenic Acid	Vit. B-6 (Pyri-doxine)	Folacin (Folic Acid)	Biotin	Vitamin B-12
	(mg)	(mg)	(mg)	(IU)	(IU)	(mg)	(mg)	(mg)	(mg)	(mg)	(mg)	(mg)	(mcg)	(mcg)	(mcg)
1297	**.40**	—	—	**40.0**	—	—	**4.00**	**.03**	**.04**	**.20**	—	**.03**	—	—	—
	.20	—	—	20.0	—	—	2.00	.02	.02	.10	—	.01	—	—	—
1298	—	—	—	—	—	—	—	—	—	—	—	—	—	—	—
	—	—	—	—	—	—	—	—	—	—	—	—	—	—	—
1299	**.55**	**.03**	**(.06)**	**55.4**	—	—	**5.54**	**.06**	**.03**	**0**	—	**.04**	—	—	**0**
	.20	.01	(.02)	20.0	—	—	2.00	.02	.01	0	—	.01	—	—	0
1300	**1.45**	[6]**.40**	**(.17)**	**303.6**	**0**	**1.32**	**264.00**	**.07**	**.07**	**.40**	**.53**	**.09**	—	**(3.00)**	**0**
	1.10	.30	(.13)	230.0	0	1.00	200.00	.05	.05	.30	.40	.07	—	(2.30)	0
1301	**1.33**	—	**.15**	**159.6**	**0**	**(.17)**	**54.53**	**.05**	**.07**	**.13**	**.09**	**.05**	—	**(3.42)**	**0**
	1.00	—	.11	120.0	0	(.13)	41.00	.04	.05	.10	.06	.04	—	(2.57)	0
1302	**4.54**	**.93**	**.66**	**102.1**	—	—	**6.44**	**.22**	**.20**	**2.24**	—	—	**15.40**	—	—
	3.24	.66	.47	73.0	—	—	4.60	.16	.14	1.60	—	—	11.00	—	—
1303	**1.44**	**.09**	—	—	—	—	**39.60**	**.14**	**.18**	**.90**	**.24**	—	—	—	**0**
	.80	.05	—	—	—	—	22.00	.08	.10	.50	.14	—	—	—	0
1304	**1.85**	**.52**	**.50**	**89.0**	**0**	—	**0**	**.16**	**.18**	**3.92**	**1.39**	**.35**	**25.45**	—	**0**
	1.04	.29	.28	50.0	0	—	0	.09	.10	2.20	.78	.20	14.30	—	0
1305	**.20**	—	—	—	—	—	**6.72**	**.10**	**.06**	**.20**	—	—	—	—	—
	.70	—	—	—	—	—	24.00	.35	.20	.70	—	—	—	—	—
1306	**7.33**	—	—	**2748.0**	—	—	**164.88**	**.32**	**.28**	**.29**	**.64**	**1.05**	—	—	**0**
	1.60	—	—	600.0	—	—	36.00	.07	.06	.50	.14	.23	—	—	0
1307	**.16**	**0**	**.03**	**32.8**	—	—	**.82**	**.03**	**.02**	**.16**	**.12**	**.05**	**5.74**	—	**0**
	.40	0	.07	80.0	—	—	2.00	.06	.05	.40	.30	.11	14.00	—	0
1308	**.13**	—	—	**9.9**	—	—	**.33**	**.01**	**.01**	**.07**	**.02**	—	—	—	**0**
	.40	—	—	30.0	—	—	1.00	.03	.03	.20	.07	—	—	—	0
1309	**.13**	—	—	**9.9**	—	—	**.33**	**.01**	**.01**	**.07**	**.02**	—	—	—	**0**
	.40	—	—	30.0	—	—	1.00	.03	.03	.20	.07	—	—	—	0
1310	**.09**	**.04**	**.04**	**12.2**	—	—	**.33**	**.01**	**.01**	**.14**	**.02**	—	—	—	**0**
	.28	.11	.11	37.0	—	—	1.00	.02	.04	.43	.07	—	—	—	0
1311	**.13**	—	—	**9.9**	—	—	**.33**	**.01**	**.01**	**.07**	**.02**	—	—	—	**0**
	.40	—	—	30.0	—	—	1.00	.03	.03	.20	.07	—	—	—	0
1312	—	—	—	—	—	—	—	—	—	—	—	—	—	—	—
	—	—	—	—	—	—	—	—	—	—	—	—	—	—	—
1313	**.41**	**.10**	**.06**	**26.6**	**0**	—	**.16**	**.01**	**.02**	**.14**	**.09**	**.05**	**1.80**	—	**0**
	2.06	.50	.28	133.0	0	—	.80	.07	.09	.70	.44	.24	9.00	—	0
1314	**.64**	**.22**	**.16**	**640.0**	—	—	**5.38**	**.05**	**.03**	**.92**	—	**.08**	—	—	**0**
	.25	.09	.06	250.0	—	—	2.10	.02	.01	.36	—	.03	—	—	0
1315	**1.02**	**.38**	**.13**	**358.4**	—	—	**5.12**	**.05**	**.03**	**1.28**	—	**.08**	—	—	**0**
	.40	.15	.05	140.0	—	—	2.00	.02	.01	.50	—	.03	—	—	0
1316	**.72**	**.22**	**.17**	**524.8**	—	—	**4.86**	**.05**	**.05**	**.95**	—	**.13**	—	—	**0**
	.28	.09	.07	205.0	—	—	1.90	.02	.02	.37	—	.05	—	—	0
1317	**1.02**	**.38**	**.13**	**358.4**	—	—	**5.12**	**.03**	**.03**	**1.02**	—	**.08**	—	—	**0**
	.40	.15	.05	140.0	—	—	2.00	.01	.01	.40	—	.03	—	—	0
1318	**.77**	—	—	**1203.2**	—	—	**7.68**	**.03**	**.08**	**1.54**	—	**.08**	—	—	**0**
	.30	—	—	470.0	—	—	3.00	.01	.03	.60	—	.03	—	—	0
1319	**.77**	—	—	**1177.6**	—	—	**5.12**	**.03**	**.08**	**1.54**	—	**.08**	—	—	**0**
	.30	—	—	460.0	—	—	2.00	.01	.03	.60	—	.03	—	—	0
1320	**.72**	**.18**	**.16**	**1290.2**	—	—	**6.14**	**.05**	**.05**	**.90**	—	**.08**	—	—	**0**
	.28	.07	.06	504.0	—	—	2.40	.02	.02	.35	—	.03	—	—	0
1321	**.77**	—	—	**1152.0**	—	—	**5.12**	**.03**	**.08**	**1.54**	—	**.08**	—	—	**0**
	.30	—	—	450.0	—	—	2.00	.01	.03	.60	—	.03	—	—	0
1322	**.75**	**(.17)**	**.12**	**435.0**	**0**	**(.59)**	**49.50**	**(.06)**	**(.05)**	**(.44)**	**.43**	**.02**	—	**(.72)**	**0**
	.50	(.11)	.08	290.0	0	(.39)	33.00	(.04)	(.03)	(.29)	.29	.01	—	(.48)	0
1323	**.60**	—	—	**400.0**	—	—	**22.00**	—	—	—	—	—	—	—	—
	.30	—	—	200.0	—	—	11.00	—	—	—	—	—	—	—	—
1324	**.60**	[6]**.20**	—	**380.0**	—	—	**20.00**	—	—	—	—	—	—	—	—
	.30	.10	—	190.0	—	—	10.00	—	—	—	—	—	—	—	—
1325	**.60**	[6]**.20**	—	**360.0**	—	—	**20.00**	—	—	—	—	—	—	—	—
	.30	.10	—	180.0	—	—	10.00	—	—	—	—	—	—	—	—

(Continued)

TABLE F-36 *(Continued)*

Item No.	Food Item Name	Approximate Measure	Weight	Moisture	Food Energy Calories	Protein	Fats	Carbo-hydrates	Fiber	Minerals (Macro) Calcium	Phos-phorus	Sodium	Mag-nesium	Potas-sium
			(g)	(%)	(kcal)	(g)	(g)	(g)	(g)	(mg)	(mg)	(mg)	(mg)	(mg)
1326	**GRANADILLA, PURPLE (PASSIONFRUIT)**, pulp & seeds, raw	1 AVG	**18** 100	**75.1** 75.1	**16.2** 90.0	**.4** 2.2	**.1** .7	**3.8** 21.2	— —	**2.3** 13.0	**11.5** 64.0	**5.0** 28.0	**5.2** 29.0	**62.6** 348.0
	GRAPEFRUIT													
1327	pink-red-white, all varieties	1 WHOLE	**482** 100	**88.4** 88.4	**197.6** 41.0	**2.4** .5	**.5** .1	**51.1** 10.6	**1.0** .2	**77.1** 16.0	**77.1** 16.0	**4.8** 1.0	**57.8** 12.0	**650.7** 135.0
1328	pink-red-white, Calif. and Ariz.	1 WHOLE	**482** 100	**87.5** 87.5	**212.1** 44.0	**2.4** .5	**.5** .1	**55.4** 11.5	**1.0** .2	**154.2** 32.0	**96.4** 20.0	**4.8** 1.0	**57.8** 12.0	**650.7** 135.0
1329	pink-red-white, Florida	1 WHOLE	**482** 100	**89.1** 89.1	**183.2** 38.0	**2.4** .5	**.5** .1	**47.7** 9.9	**1.0** .2	**72.3** 15.0	**72.3** 15.0	**4.8** 1.0	**57.8** 12.0	**650.7** 135.0
1330	pink-red, seeded, (Foster Pink)	1 WHOLE	**482** 100	**88.6** 88.6	**192.8** 40.0	**2.4** .5	**.5** .1	**50.1** 10.4	**1.0** .2	**77.1** 16.0	**77.1** 16.0	**4.8** 1.0	**57.8** 12.0	**650.7** 135.0
1331	pink-red, seedless, (Pink Marsh Redblush)	1 WHOLE	**482** 100	**88.6** 88.6	**192.8** 40.0	**2.4** .5	**.5** .1	**50.1** 10.4	**1.0** .2	**77.1** 16.0	**77.1** 16.0	**4.8** 1.0	**57.8** 12.0	**650.7** 135.0
1332	pink-red Texas, (medium)	1 MED	**200** 100	**87.7** 87.7	**86.0** 43.0	**1.0** .5	**.2** .1	**22.6** 11.3	**.4** .2	**30.0** 15.0	**30.0** 15.0	**2.0** 1.0	**24.0** 12.0	**270.0** 135.0
1333	pink-red Texas, (large)	1 WHOLE	**482** 100	**87.7** 87.7	**207.3** 43.0	**2.4** .5	**.5** .1	**54.5** 11.3	**1.0** .2	**72.3** 15.0	**72.3** 15.0	**4.8** 1.0	**57.8** 12.0	**650.7** 135.0
1334	white seeded, (Duncan)	1 WHOLE	**482** 100	**88.2** 88.2	**197.6** 41.0	**2.4** .5	**.5** .1	**52.1** 10.8	**1.0** .2	**77.1** 16.0	**77.1** 16.0	**4.8** 1.0	**57.8** 12.0	**650.7** 135.0
1335	white seedless, (Marsh Seedless)	1 WHOLE	**482** 100	**88.9** 88.9	**188.0** 39.0	**2.4** .5	**.5** .1	**48.7** 10.1	**1.0** .2	**77.1** 16.0	**77.1** 16.0	**4.8** 1.0	**57.8** 12.0	**650.7** 135.0
1336	white Texas	1 MED	**200** 100	**87.7** 87.7	**86.0** 43.0	**1.0** .5	**.2** .1	**22.6** 11.3	**.4** .2	**30.0** 15.0	**30.0** 15.0	**2.0** 1.0	**24.0** 12.0	**270.0** 135.0
1337	segments, canned, water pack	1 CUP	**254** 100	**91.3** 91.3	**76.2** 30.0	**1.5** .6	**.3** .1	**19.3** 7.6	**.5** .2	**33.0** 13.0	**35.6** 14.0	**10.2** 4.0	**27.9** 11.0	**365.8** 144.0
1338	segments, canned, syrup	1 CUP	**254** 100	**81.1** 81.1	**177.8** 70.0	**1.5** .6	**.3** .1	**45.2** 17.8	**.5** .2	**33.0** 13.0	**35.6** 14.0	**2.5** 1.0	**27.9** 11.0	**342.9** 135.0
	GRAPES													
1339	American type, slip skin, raw	1 CUP	**153** 100	**81.6** 81.6	**105.6** 69.0	**2.0** 1.3	**1.5** 1.0	**24.0** 15.7	**.9** .6	**24.5** 16.0	**18.4** 12.0	**4.6** 3.0	**19.9** 13.0	**241.7** 158.0
1340	European type, adherent skin, raw	1 CUP	**153** 100	**81.4** 81.4	**102.5** 67.0	**.9** .6	**.5** .3	**26.5** 17.3	**.8** .5	**18.4** 12.0	**30.6** 20.0	**4.6** 3.0	**9.2** 6.0	**264.7** 173.0
1341	Thompson seedless, canned, water pack	1 CUP	**200** 100	**85.5** 85.5	**102.0** 51.0	**1.0** .5	**.2** .1	**27.2** 13.6	**.4** .2	**16.0** 8.0	**26.0** 13.0	**8.0** 4.0	— —	**220.0** 110.0
1342	Thompson seedless, canned, heavy syrup	1 CUP	**200** 100	**79.5** 79.5	**152.0** 76.0	**1.0** .5	**.2** .1	**39.3** 19.7	**.4** .2	**20.0** 10.0	**34.0** 17.0	**10.0** 5.0	**12.0** 6.0	**206.0** 103.0
1343	[5]**GREENSAPOTE**	1 OZ	**28** 100	**68.4** 68.4	**30.8** 110.0	**.4** 1.6	**.1** .2	**8.0** 28.6	**.5** 1.8	**6.4** 23.0	**7.8** 28.0	— —	— —	— —
1344	**GROUNDCHERRIES (POHA** OR **CAPE-GOOSEBERRIES)**, raw	1 CUP	**150** 100	**85.4** 85.4	**79.5** 53.0	**2.9** 1.9	**1.1** .7	**16.8** 11.2	**4.2** 2.8	**13.5** 9.0	**60.0** 40.0	— —	— —	— —
	GUAVAS													
1345	common, whole, raw	1 MED	**100** 100	**83.0** 83.0	**62.0** 62.0	**.8** .8	**.6** .6	**15.0** 15.0	**5.6** 5.6	**23.0** 23.0	**42.0** 42.0	**4.0** 4.0	**13.0** 13.0	**289.0** 289.0
1346	strawberry, whole, raw	1 MED	**100** 100	**81.8** 81.8	**65.0** 65.0	**1.0** 1.0	**.6** .6	**15.8** 15.8	**6.4** 6.4	**23.0** 23.0	**42.0** 42.0	**4.0** 4.0	— —	**289.0** 289.0
1347	[5]**HACKBERRY, SPINY**	1 OZ	**28** 100	**79.5** 79.5	**20.2** 72.0	**.3** 1.0	**.1** .3	**5.1** 18.3	**.3** .9	**17.6** 63.0	**6.2** 22.0	— —	— —	— —
1348	**HAWS,** scarlet, flesh & skin, raw 1″ diameter	1 AVG	**35** 100	**75.8** 75.8	**30.5** 87.0	**.7** 2.0	**.2** .7	**7.3** 20.8	**.7** 2.1	— —	— —	— —	— —	— —
1349	**HONEYDEW MELON,** vine ripened	1 WHOLE	**900** 100	**90.6** 90.6	**297.0** 33.0	**7.2** .8	**2.7** .3	**69.3** 7.7	**5.4** .6	**27.0** 3.0	**90.0** 10.0	**108.0** 12.0	**60.3** 6.7	**2259.0** 251.0
1350	**INDIANFIG,** 1½″ diameter, 2½″ long	1 AVG	**65** 100	**81.4** 81.4	**43.6** 67.0	**.7** 1.1	**.3** .4	**10.8** 16.6	**.7** 1.1	**37.1** 57.0	**20.8** 32.0	— —	— —	— —
1351	**JACKFRUIT**, raw, cubed	1 CUP	**85** 100	**72.0** 72.0	**83.3** 98.0	**1.1** 1.3	**.3** .3	**21.6** 25.4	**.9** 1.0	**18.7** 22.0	**32.3** 38.0	**1.7** 2.0	— —	**346.0** 407.0
	JUJUBE													
1352	**COMMON (CHINESE DATE)**, raw	1 MED	**20** 100	**70.2** 70.2	**21.0** 105.0	**.2** 1.2	**Trace** .2	**5.5** 27.6	**.3** 1.4	**5.8** 29.0	**7.4** 37.0	**.6** 3.0	— —	**53.8** 269.0
1353	**COMMON (CHINESE DATE)**, dried	1 AVG	**8** 100	**19.7** 19.7	**23.0** 287.0	**.3** 3.7	**.1** 1.1	**5.9** 73.6	**.2** 3.0	**6.3** 79.0	**8.0** 100.0	**.2** 3.0	— —	**42.5** 531.0
1354	**KUMQUATS,** raw	1 MED	**20** 100	**81.3** 81.3	**13.0** 65.0	**.2** .9	**Trace** .1	**3.4** 17.1	**.7** 3.7	**12.6** 63.0	**4.6** 23.0	**1.4** 7.0	— —	**47.2** 236.0

Item No.	Minerals (Micro) Iron	Zinc	Copper	Fat-Soluble Vitamins Vitamin A	Vitamin D	Vitamin E (Alpha Tocopherol)	Water-Soluble Vitamins Vitamin C	Thiamin	Riboflavin	Niacin	Pantothenic Acid	Vit. B-6 (Pyridoxine)	Folacin (Folic Acid)	Biotin	Vitamin B-12
	(mg)	(mg)	(mg)	(IU)	(IU)	(mg)	(mg)	(mg)	(mg)	(mg)	(mg)	(mg)	(mcg)	(mcg)	(mcg)
1326	.29	—	—	126.0	—	—	5.40	—	.02	.27	—	—	—	—	—
	1.60	—	—	700.0	—	—	30.00	—	.13	1.50	—	—	—	—	—
1327	1.93	.48	.20	385.6	0	1.21	183.16	.19	.10	.96	1.36	.16	53.02	14.46	0
	.40	.10	.04	80.0	0	.25	38.00	.04	.02	.20	.28	.03	11.00	3.00	0
1328	1.93	.48	.20	48.2	—	1.21	192.80	.19	.10	.96	1.36	.16	53.02	14.46	0
	.40	.10	.04	10.0	—	.25	40.00	.04	.02	.20	.28	.03	11.00	3.00	0
1329	1.93	.48	.19	385.6	—	1.21	178.34	.19	.10	.96	1.36	.16	53.02	14.46	0
	.40	.10	.04	80.0	—	.25	37.00	.04	.02	.20	.28	.03	11.00	3.00	0
1330	1.93	.48	.20	2120.8	—	1.21	187.98	.19	.10	.96	1.36	.16	53.02	14.46	0
	.40	.10	.04	440.0	—	.25	39.00	.04	.02	.20	.28	.03	11.00	3.00	0
1331	1.93	.48	.20	2120.8	—	1.21	173.52	.19	.10	.96	1.36	.16	53.02	14.46	0
	.40	.10	.04	440.0	—	.25	36.00	.04	.02	.20	.28	.03	11.00	3.00	0
1332	.80	.20	.08	880.0	—	—	76.00	.08	.04	.40	.57	.06	22.00	6.00	0
	.40	.10	.04	440.0	—	—	38.00	.04	.02	.20	.28	.03	11.00	3.00	0
1333	1.93	.48	.20	2120.8	—	1.21	183.16	.19	.10	.96	1.36	.16	53.02	14.46	0
	.40	.10	.04	440.0	—	.25	38.00	.04	.02	.20	.28	.03	11.00	3.00	0
1334	1.93	.48	.20	48.2	—	1.21	183.16	.19	.10	.96	1.36	.16	53.02	14.46	0
	.40	.10	.04	10.0	—	.25	38.00	.04	.02	.20	.28	.03	11.00	3.00	0
1335	1.93	.48	.20	48.2	—	1.21	178.34	.19	.10	.96	1.36	.16	53.02	14.46	0
	.40	.10	.04	10.0	—	.25	37.00	.04	.02	.20	.28	.03	11.00	3.00	0
1336	.80	.20	.06	20.0	—	.50	76.00	.08	.04	.40	.57	.07	22.00	6.00	0
	.40	.10	.03	10.0	—	.25	38.00	.04	.02	.20	.28	.03	11.00	3.00	0
1337	.76	.10	.04	25.4	—	—	76.20	.08	.05	.51	.31	.05	22.86	—	0
	.30	.04	.01	10.0	—	—	30.00	.03	.02	.20	.12	.02	9.00	—	0
1338	.76	.10	.04	25.4	—	—	76.20	.08	.05	.51	.31	.05	22.86	—	0
	.30	.04	.01	10.0	—	—	30.00	.03	.02	.20	.12	.02	9.00	—	0
1339	.61	.06	.05	153.0	—	—	6.12	.14	.09	.46	.12	.19	10.71	2.45	0
	.40	.04	.03	100.0	—	—	4.00	.09	.06	.30	.08	.13	7.00	1.60	0
1340	.61	.06	1.45	153.0	—	1.07	6.12	.14	.09	.46	.12	.19	10.71	2.45	0
	.40	.04	.95	100.0	—	.70	4.00	.09	.06	.30	.08	.13	7.00	1.60	0
1341	.60	—	—	140.0	—	—	4.00	.08	.02	.40	—	—	—	—	—
	.30	—	—	70.0	—	—	2.00	.04	.01	.20	—	—	—	—	—
1342	1.88	.10	.11	128.0	—	—	2.00	.06	.04	.24	—	—	—	—	—
	.94	.05	.05	64.0	—	—	1.00	.03	.02	.12	—	—	—	—	—
1343	.20	—	—	7.0	—	—	12.04	Trace	.01	.53	—	—	—	—	—
	.70	—	—	25.0	—	—	43.00	.01	.03	1.90	—	—	—	—	—
1344	1.50	—	—	1080.0	—	—	16.50	.17	.06	4.20	—	—	—	—	—
	1.00	—	—	720.0	—	—	11.00	.11	.04	2.80	—	—	—	—	—
1345	.90	—	—	280.0	—	—	242.00	.05	.05	1.20	.15	—	—	—	0
	.90	—	—	280.0	—	—	242.00	.05	.05	1.20	.15	—	—	—	0
1346	.90	—	—	90.0	—	—	37.00	.03	.03	.60	—	—	—	—	—
	.90	—	—	90.0	—	—	37.00	.03	.03	.60	—	—	—	—	—
1347	.78	—	—	4.2	—	—	14.28	0	0	.08	—	—	—	—	—
	2.80	—	—	15.0	—	—	51.00	0	0	.30	—	—	—	—	—
1348	—	—	—	—	—	—	—	—	—	—	—	—	—	—	—
	—	—	—	—	—	—	—	—	—	—	—	—	—	—	—
1349	.90	0	.36	360.0	—	—	207.00	.36	.27	5.40	1.86	.50	—	—	0
	.10	0	.04	40.0	—	—	23.00	.04	.03	.60	.21	.06	—	—	0
1350	.78	—	—	—	—	—	11.70	.01	.01	.20	—	—	—	—	—
	1.20	—	—	—	—	—	18.00	.01	.02	.30	—	—	—	—	—
1351	—	—	—	—	—	—	6.80	.03	—	.34	—	—	—	—	—
	—	—	—	—	—	—	8.00	.03	—	.40	—	—	—	—	—
1352	.14	—	—	8.0	—	—	13.80	.00	.01	.18	—	—	—	—	—
	.70	—	—	40.0	—	—	69.00	.02	.04	.90	—	—	—	—	—
1353	.14	—	—	3.2	—	—	1.04	.02	.03	.07	—	—	—	—	—
	1.80	—	—	40.0	—	—	13.00	.23	.38	.87	—	—	—	—	—
1354	.08	—	—	120.0	—	—	7.20	.02	.02	—	—	—	—	—	—
	.40	—	—	600.0	—	—	36.00	.08	.10	—	—	—	—	—	—

(Continued)

TABLE F-36 *(Continued)*

Item No.	Food Item Name	Approximate Measure	Weight	Moisture	Food Energy Calories	Protein	Fats	Carbo-hydrates	Fiber	Minerals (Macro) Calcium	Phos-phorus	Sodium	Mag-nesium	Potas-sium
			(g)	(%)	(kcal)	(g)	(g)	(g)	(g)	(mg)	(mg)	(mg)	(mg)	(mg)
	LEMONS													
1355	pulp w/peel, raw	1 SM	100	87.4	20.0	1.2	.3	10.7	(5.1)	61.0	15.0	3.0	(11.9)	145.0
			100	87.4	20.0	1.2	.3	10.7	(5.1)	61.0	15.0	3.0	(11.9)	145.0
1356	pulp w/o peel, raw	1 MED	100	90.1	27.0	1.1	.3	8.2	.4	26.0	16.0	2.0	9.0	138.0
			100	90.1	27.0	1.1	.3	8.2	.4	26.0	16.0	2.0	9.0	138.0
1357	peel, raw	1 TBLS	6	81.6	—	.1	Trace	1.0	—	8.0	.7	.4	—	9.6
			100	81.6	—	1.5	.3	16.0	—	134.0	12.0	6.0	—	160.0
1358	peel, candied	1 OZ	28	17.4	88.5	.1	.1	22.6	.6	—	—	—	—	—
			100	17.4	316.0	.4	.3	80.6	2.3	—	—	—	—	—
1359	**LIMES,** acid type, raw	1 MED	100	89.3	28.0	.8	.2	9.5	.5	33.0	18.0	2.0	—	102.0
			100	89.3	28.0	.8	.2	9.5	.5	33.0	18.0	2.0	—	102.0
	LOGANBERRIES													
1360	raw	1 CUP	150	83.0	93.0	1.5	.9	22.4	4.5	52.5	25.5	1.5	37.5	255.0
			100	83.0	62.0	1.0	.6	14.9	3.0	35.0	17.0	1.0	25.0	170.0
1361	canned, water pack	1 CUP	200	89.2	80.0	1.4	.8	18.8	4.0	48.0	22.0	2.0	22.0	230.0
			100	89.2	40.0	.7	.4	9.4	2.0	24.0	11.0	1.0	11.0	115.0
1362	canned, juice pack	1 CUP	200	85.7	108.0	1.4	1.0	25.4	4.2	54.0	30.0	2.0	22.0	340.0
			100	85.7	54.0	.7	.5	12.7	2.1	27.0	15.0	1.0	11.0	170.0
1363	canned, light syrup	1 CUP	200	81.4	140.0	1.4	.8	34.4	4.0	46.0	22.0	2.0	22.0	222.0
			100	81.4	70.0	.7	.4	17.2	2.0	23.0	11.0	1.0	11.0	111.0
1364	canned, heavy syrup	1 CUP	200	76.5	178.0	1.2	.8	44.4	3.8	44.0	22.0	2.0	22.0	218.0
			100	76.5	89.0	.6	.4	22.2	1.9	22.0	11.0	1.0	11.0	109.0
1365	canned, extra heavy syrup	1 CUP	200	71.5	216.0	1.2	.8	54.4	3.8	44.0	22.0	2.0	22.0	214.0
			100	71.5	108.0	.6	.4	27.2	1.9	22.0	11.0	1.0	11.0	107.0
	LONGANS													
1366	raw 1" diameter	1 AVG	15	82.4	9.2	.2	0	2.4	.1	1.5	6.3	—	—	—
			100	82.4	61.0	1.0	.1	15.8	.4	10.0	42.0	—	—	—
1367	dried 1" diameter	1 AVG	3	17.6	8.6	.1	0	2.2	.1	1.4	5.9	—	—	—
			100	17.6	286.0	4.9	.4	74.0	2.0	45.0	196.0	—	—	—
	LOQUATS													
1368	raw	1 AVG	12	86.5	5.8	0	0	1.5	.1	2.4	4.3	—	—	41.8
			100	86.5	48.0	.4	.2	12.4	.5	20.0	30.0	—	—	348.0
1369	[4]**CANNED, SYRUP PACK**	1 OZ	28	76.7	23.5	.1	0	6.4	.1	6.2	.8	—	—	—
			100	76.7	84.0	.3	.1	22.7	.5	22.0	3.0	—	—	—
	LYCHEES													
1370	raw	1 AVG	9	81.9	5.8	.1	0	1.5	Trace	.7	3.8	.3	(1.0)	15.3
			100	81.9	64.0	.9	.3	16.4	.3	8.0	42.0	3.0	(9.9)	170.0
1371	[4]canned, drained, solids	1 OZ	28	79.7	20.7	.1	.1	5.4	.1	.8	2.8	9.8	—	19.0
			100	79.7	74.0	.3	.4	19.4	.2	3.0	10.0	35.0	—	68.0
1372	dried	1 AVG	3	22.3	8.3	.1	0	2.1	.1	1.0	5.4	.1	—	33.0
			100	22.3	277.0	3.8	1.2	70.7	1.4	33.0	181.0	3.0	—	1100.0
1373	**MANGOS,** raw	1 WHOLE	200	81.7	132.0	1.4	.8	33.6	1.8	20.0	26.0	14.0	17.6	378.0
			100	81.7	66.0	.7	.4	16.8	.9	10.0	13.0	7.0	8.8	189.0
1374	[4]**MANGOSTEEN,** raw	1 OZ	28	84.3	16.0	.1	.1	4.1	1.4	2.8	2.8	.3	—	37.8
			100	84.3	57.0	.5	.3	14.7	5.0	10.0	10.0	1.0	—	135.0
	MELON													
1375	**CASSABA**	1 AVG	874	86.2	445.7	4.4	4.4	109.3	8.7	96.1	96.1	131.1	—	410.8
			100	86.2	51.0	.5	.5	12.5	1.0	11.0	11.0	15.0	—	47.0
1376	**BALL (CANTALOUPE, HONEYDEW),** frozen in syrup	1 CUP	175	83.2	108.5	1.1	.2	27.5	.5	17.5	21.0	15.8	(22.9)	329.0
			100	83.2	62.0	.6	.1	15.7	.3	10.0	12.0	9.0	(13.1)	188.0
	MULBERRIES													
1377	[4]black, raw, pulp and seeds	1 OZ	28	87.9	11.8	.4	.1	2.7	.2	6.7	7.3	8.4	(4.2)	34.4
			100	87.9	42.0	1.4	.3	9.8	.7	24.0	26.0	30.0	(14.9)	123.0
1378	[4]white, raw, pulp and seeds	1 OZ	28	85.0	14.8	.5	.1	3.4	.3	8.4	9.0	10.4	—	42.6
			100	85.0	53.0	1.7	.4	12.2	.9	30.0	32.0	37.0	—	152.0
1379	**NECTAR, GUAVA**	1 CUP	243	—	—	—	—	—	—	—	—	4.9	—	9.7
			100	—	—	—	—	—	—	—	—	2.0	—	4.0
1380	**NECTAR, PASSION-ORANGE**	1 CUP	223	—	—	—	—	—	—	—	—	4.5	—	2.2
			100	—	—	—	—	—	—	—	—	2.0	—	1.0
1381	**NECTARINE,** raw	1 MED	50	81.8	32.0	.3	0	8.6	.2	2.0	12.0	3.0	3.1	147.0
			100	81.8	64.0	.6	0	17.1	.4	4.0	24.0	6.0	6.3	294.0

Item No.	Minerals (Micro)			Fat-Soluble Vitamins			Water-Soluble Vitamins								
	Iron	Zinc	Copper	Vitamin A	Vitamin D	Vitamin E (Alpha Tocopherol)	Vitamin C	Thiamin	Ribo-flavin	Niacin	Panto-thenic Acid	Vit. B-6 (Pyri-doxine)	Folacin (Folic Acid)	Biotin	Vitamin B-12
	(mg)	(mg)	(mg)	(IU)	(IU)	(mg)	(mg)	(mg)	(mg)	(mg)	(mg)	(mg)	(mcg)	(mcg)	(mcg)
1355	**.70**	**(.11)**	**(.26)**	**30.0**	**0**	—	**77.00**	**.05**	**.04**	**.20**	**(.22)**	**(.12)**	**12.00**	**(.49)**	**0**
	.70	(.11)	(.26)	30.0	0	—	77.00	.05	.04	.20	(.22)	(.12)	12.00	(.49)	0
1356	**.60**	—	**.26**	**20.0**	—	—	**53.00**	**.04**	**.02**	**.10**	**.19**	**.08**	—	—	**0**
	.60	—	.26	20.0	—	—	53.00	.04	.02	.10	.19	.08	—	—	0
1357	**.05**	—	—	**3.0**	—	—	**7.74**	**Trace**	**.01**	**.02**	—	—	—	—	—
	.80	—	—	50.0	—	—	129.00	.06	.08	.40	—	—	—	—	—
1358	—	—	—	—	—	—	—	—	—	—	—	—	—	—	—
	—	—	—	—	—	—	—	—	—	—	—	—	—	—	—
1359	**.60**	—	—	**10.0**	—	—	**37.00**	**.03**	**.02**	**.20**	**.22**	—	**4.00**	—	**0**
	.60	—	—	10.0	—	—	37.00	.03	.02	.20	.22	—	4.00	—	0
1360	**1.80**	—	**(.21)**	**300.0**	**0**	**(.44)**	**36.00**	**.04**	**.06**	**.60**	**(.35)**	**(.09)**	—	—	**0**
	1.20	—	(.14)	200.0	0	(.29)	24.00	.03	.04	.40	(.23)	(.06)	—	—	0
1361	**1.60**	—	—	**280.0**	—	—	**16.00**	**.02**	**.04**	**.40**	—	—	—	—	—
	.80	—	—	140.0	—	—	8.00	.01	.02	.20	—	—	—	—	—
1362	**2.40**	—	—	**300.0**	—	—	**24.00**	**.04**	**.06**	**.60**	—	—	—	—	—
	1.20	—	—	150.0	—	—	12.00	.02	.03	.30	—	—	—	—	—
1363	**1.60**	—	—	**260.0**	—	—	**16.00**	**.02**	**.04**	**.40**	—	—	—	—	—
	.80	—	—	130.0	—	—	8.00	.01	.02	.20	—	—	—	—	—
1364	**1.60**	—	—	**260.0**	—	—	**16.00**	**.02**	**.04**	**.40**	—	—	—	—	—
	.80	—	—	130.0	—	—	8.00	.01	.02	.20	—	—	—	—	—
1365	**1.60**	—	**(.08)**	**260.0**	**0**	—	**14.00**	**.02**	**.04**	**.40**	**(.36)**	**(.06)**	—	—	**0**
	.80	—	(.04)	130.0	0	—	7.00	.01	.02	.20	(.18)	(.03)	—	—	0
1366	**.18**	—	—	—	—	—	**.90**	—	—	—	—	—	—	—	—
	1.20	—	—	—	—	—	6.00	—	—	—	—	—	—	—	—
1367	**.16**	—	—	—	—	—	**.84**	**Trace**	—	—	—	—	—	—	—
	5.40	—	—	—	—	—	28.00	.04	—	—	—	—	—	—	—
1368	**.05**	—	—	**80.4**	—	—	**.12**	—	—	—	—	—	—	—	—
	.40	—	—	670.0	—	—	1.00	—	—	—	—	—	—	—	—
1369	**.01**	—	—	**56.0**	—	—	**0**	**Trace**	**0**	**.11**	—	—	—	—	—
	.10	—	—	200.0	—	—	0	.01	0	.40	—	—	—	—	—
1370	**.04**	[6]**.03**	—	—	**0**	—	**3.78**	—	**.01**	—	—	—	—	—	**0**
	.40	.30	—	—	0	—	42.00	—	.05	—	—	—	—	—	0
1371	**.22**	[6]**.06**	—	**0**	—	—	**17.36**	**0**	**.01**	**.03**	—	—	—	—	—
	.80	.20	—	0	—	—	62.00	0	.05	.10	—	—	—	—	—
1372	**.05**	—	—	—	—	—	—	—	—	—	—	—	—	—	—
	1.70	—	—	—	—	—	—	—	—	—	—	—	—	—	—
1373	**.80**	**0**	**.23**	**9600.0**	**0**	—	**70.00**	**.10**	**.10**	**2.20**	**.32**	—	—	—	**0**
	.40	0	.12	4800.0	0	—	35.00	.05	.05	1.10	.16	—	—	—	0
1374	**.14**	—	—	**0**	—	—	**1.12**	**.01**	**.01**	**.17**	—	—	—	—	—
	.50	—	—	0	—	—	4.00	.03	.02	.60	—	—	—	—	—
1375	**6.12**	—	—	**2010.2**	—	—	**122.36**	**.18**	**.35**	**3.50**	**.90**	—	—	—	**0**
	.70	—	—	230.0	—	—	14.00	.02	.04	.40	.10	—	—	—	0
1376	**.53**	**(.23)**	**(.05)**	**2695.0**	**0**	**(.23)**	**28.00**	**.05**	**.04**	**.88**	**(.39)**	**(.09)**	**(51.80)**	—	**0**
	.30	(.13)	(.03)	1540.0	0	(.13)	16.00	.03	.02	.50	(.22)	(.05)	(29.60)	—	0
1377	**.84**	—	**(.01)**	**7.0**	**0**	—	**10.92**	**.01**	**.02**	**.20**	**(.07)**	**(.02)**	—	**(.11)**	**0**
	3.00	—	(.05)	25.0	0	—	39.00	.04	.08	.70	(.24)	(.06)	—	(.39)	0
1378	**1.04**	—	—	**4.7**	—	—	**1.40**	**.01**	**.02**	**.20**	—	—	—	—	—
	3.70	—	—	16.7	—	—	5.00	.03	.06	.70	—	—	—	—	—
1379	—	—	—	—	—	—	—	—	—	—	—	—	—	—	—
	—	—	—	—	—	—	—	—	—	—	—	—	—	—	—
1380	—	—	—	—	—	—	—	—	—	—	—	—	—	—	—
	—	—	—	—	—	—	—	—	—	—	—	—	—	—	—
1381	**.25**	**.04**	**.03**	**825.0**	**0**	—	**6.50**	**(.01)**	**(.03)**	**(.55)**	**(.08)**	**.01**	**2.50**	—	**0**
	.50	.08	.06	1650.0	0	—	13.00	(.02)	(.06)	(1.10)	(.16)	.01	5.00	—	0

(Continued)

TABLE F-36 *(Continued)*

Item No.	Food Item Name	Approximate Measure	Weight	Moisture	Food Energy Calories	Protein	Fats	Carbo-hydrates	Fiber	Minerals (Macro) Calcium	Phos-phorus	Sodium	Mag-nesium	Potas-sium
			(g)	(%)	(kcal)	(g)	(g)	(g)	(g)	(mg)	(mg)	(mg)	(mg)	(mg)
	ORANGES													
1382	**ALL VARIETIES,** w/o peel, raw	1 SM	100	86.0	49.0	.9	.2	12.2	.5	41.0	20.0	1.0	6.7	200.0
			100	86.0	49.0	.9	.2	12.2	.5	41.0	20.0	1.0	6.7	200.0
1383	**CALIFORNIA NAVEL (WINTER),** w/o peel, raw	1 LG	235	85.4	119.9	3.1	.2	29.8	1.2	56.4	37.6	2.4	15.7	455.9
			100	85.4	51.0	1.3	.1	12.7	.5	24.0	16.0	1.0	6.7	194.0
1384	**CALIFORNIA VALENCIA (SUMMER),** w/o peel, raw	1 MED	150	85.6	76.5	1.8	.5	18.6	.8	28.5	19.5	1.5	10.1	246.0
			100	85.6	51.0	1.2	.3	12.4	.5	19.0	13.0	1.0	6.7	164.0
1385	**FLORIDA, ALL VARIETIES,** w/o peel, raw	1 MED	150	86.4	70.5	1.1	.3	18.0	.8	64.5	25.5	1.5	10.1	309.0
			100	86.4	47.0	.7	.2	12.0	.5	43.0	17.0	1.0	6.7	206.0
1386	**MANDARIN,** canned	1 CUP	100	87.0	46.0	.8	.2	11.6	.5	8.0	18.0	2.0	7.0	126.0
			100	87.0	46.0	.8	.2	11.6	.5	8.0	18.0	2.0	7.0	126.0
1387	peel, raw	1 TBLS	6	72.5	—	.1	.0	1.5	—	9.7	1.3	.2	—	12.7
			100	72.5	—	1.5	.2	25.0	—	161.0	21.0	3.0	—	212.0
1388	peel, candied	1 OZ	28	17.4	88.5	.1	.1	22.6	—	—	—	—	—	—
			100	17.4	316.0	.4	.3	80.6	—	—	—	—	—	—
1389	**PAPAWS,** common North American, raw	1 AVG	98	76.6	83.3	5.1	.9	16.5	—	—	—	—	—	—
			100	76.6	85.0	5.2	.9	16.8	—	—	—	—	—	—
1390	**PAPAYAS,** raw	1 MED	300	88.7	117.0	1.8	.3	30.0	2.7	60.0	48.0	9.0	22.8	702.0
			100	88.7	39.0	.6	.1	10.0	.9	20.0	16.0	3.0	7.6	234.0
1391	[6]**PASSION FRUIT,** raw	1 OZ	28	—	9.5	.8	Trace	1.7	4.4	4.5	15.1	7.8	10.0	98.3
			100	—	34.0	2.7	Trace	6.1	15.8	16.2	53.8	27.8	38.8	351.1
	PEACHES													
1392	raw	1 MED	100	89.1	38.0	.6	.1	9.7	.6	9.0	19.0	1.0	6.6	202.0
			100	89.1	38.0	.6	.1	9.7	.6	9.0	19.0	1.0	6.6	202.0
1393	canned, water pack	1 HALF	50	93.1	12.0	.2	0	3.1	0	1.0	5.0	1.5	2.5	49.5
			100	93.1	24.0	.4	.1	6.1	0	2.0	10.0	3.0	5.0	99.0
1394	canned, juice pack	1 HALF	50	87.5	22.5	.3	0	5.8	.1	3.0	8.5	2.0	3.5	64.0
			100	87.5	45.0	.7	0	11.6	.3	6.0	17.0	4.0	7.0	128.0
1395	canned, light syrup	1 HALF	50	84.7	28.0	.2	0	7.3	.2	1.5	5.5	2.5	2.5	48.5
			100	84.7	56.0	.5	0	14.5	.3	3.0	11.0	5.0	5.0	97.0
1396	canned, heavy syrup	1 HALF	50	79.3	38.5	.2	.1	10.0	.1	1.5	5.5	3.0	2.5	45.5
			100	79.3	77.0	.5	.1	19.9	.3	3.0	11.0	6.0	5.0	91.0
1397	canned, extra heavy syrup, solids & liquid	1 HALF	50	74.1	48.5	.2	.1	12.6	.2	2.0	6.0	1.0	3.0	64.0
			100	74.1	97.0	.4	.1	25.1	.4	4.0	12.0	2.0	6.0	128.0
1398	dried, sulfured, uncooked	1 CUP	160	32.5	379.2	5.9	1.2	96.8	4.8	44.8	187.2	9.6	68.8	1572.8
			100	32.5	237.0	3.7	.8	60.5	3.0	28.0	117.0	6.0	43.0	983.0
1399	dried, sulfured, cooked w/o sugar	1 HALF	33	76.5	27.1	.3	.1	7.1	.3	5.0	12.2	1.7	5.0	98.0
			100	76.5	82.0	1.0	.2	21.4	1.0	15.0	37.0	5.0	15.0	297.0
1400	dried, sulfured, cooked w/sugar	1 HALF	33	67.3	39.3	.3	.1	10.2	.3	4.3	10.6	1.3	—	86.1
			100	67.3	119.0	.9	.2	30.8	.9	13.0	32.0	4.0	—	261.0
1401	dehydrated, sulfured, uncooked	1 CUP	200	3.0	680.0	9.6	1.8	176.0	8.0	124.0	302.0	42.0	—	2458.0
			100	3.0	340.0	4.8	.9	88.0	4.0	62.0	151.0	21.0	—	1229.0
1402	dehydrated, sulfured, cooked w/sugar	1 CUP	242	66.6	292.8	2.7	.5	75.7	2.2	36.3	87.1	12.1	—	706.6
			100	66.6	121.0	1.1	.2	31.3	.9	15.0	36.0	5.0	—	292.0
1403	frozen, sliced, sweetened	1 CUP	250	74.7	235.0	1.6	.3	60.0	.8	5.0	27.5	15.0	12.5	325.0
			100	74.7	94.0	.6	.1	24.0	.3	2.0	11.0	6.0	5.0	130.0
	PEARS													
1404	raw, w/skin	1 AVG	200	83.2	122.0	1.4	.8	30.6	2.8	16.0	22.0	4.0	14.0	260.0
			100	83.2	61.0	.7	.4	15.3	1.4	8.0	11.0	2.0	7.0	130.0
1405	[6]raw, peeled	1 MED	182	—	66.4	.5	Trace	17.1	5.1	13.1	27.1	5.3	8.7	184.5
			100	—	36.5	.3	Trace	9.4	2.8	7.2	14.9	2.9	4.8	101.4
1406	canned, water pack	1 HALF	50	91.9	14.5	.1	0	3.9	.3	2.0	3.5	1.5	2.0	25.0
			100	91.9	29.0	.2	0	7.8	.5	4.0	7.0	3.0	4.0	50.0
1407	canned, juice pack	1 HALF	50	86.5	25.0	.2	0	6.5	.2	4.5	6.0	2.0	3.5	48.0
			100	86.5	50.0	.3	.1	12.9	.5	9.0	12.0	4.0	7.0	96.0
1408	canned, light syrup	1 HALF	50	84.5	28.5	.1	0	7.6	.3	2.5	3.5	2.5	2.0	32.5
			100	84.5	57.0	.2	0	15.2	.5	5.0	7.0	5.0	4.0	65.0
1409	canned, heavy syrup	1 HALF	50	80.4	37.0	.1	.1	9.6	.3	2.5	3.5	2.5	2.0	32.5
			100	80.4	74.0	.2	.1	19.2	.6	5.0	7.0	5.0	4.0	65.0
1410	canned, extra heavy syrup	1 HALF	50	75.8	46.0	.1	.1	11.8	.3	2.5	3.5	.5	2.5	41.5
			100	75.8	92.0	.2	.2	23.6	.6	5.0	7.0	1.0	5.0	83.0
1411	candied	1 OZ	28	21.0	84.8	.4	.2	21.3	—	—	—	—	—	—
			100	21.0	303.0	1.3	.6	75.9	—	—	—	—	—	—

Item No.	Minerals (Micro)			Fat-Soluble Vitamins			Water-Soluble Vitamins								
	Iron	Zinc	Copper	Vitamin A	Vitamin D	Vitamin E (Alpha Tocopherol)	Vitamin C	Thiamin	Ribo-flavin	Niacin	Panto-thenic Acid	Vit. B-6 (Pyri-doxine)	Folacin (Folic Acid)	Biotin	Vitamin B-12
	(mg)	(mg)	(mg)	(IU)	(IU)	(mg)	(mg)	(mg)	(mg)	(mg)	(mg)	(mg)	(mcg)	(mcg)	(mcg)
1382	**.40**	**.20**	**Trace**	**200.0**	**0**	**.23**	**50.00**	**.10**	**.04**	**.40**	**.25**	**.06**	**46.00**	**1.90**	**0**
	.40	.20	Trace	200.0	0	.23	50.00	.10	.04	.40	.25	.06	46.00	1.90	0
1383	**.24**	**.47**	**.01**	**470.0**	—	**.54**	**143.35**	**.24**	**.09**	**.94**	**.59**	**.14**	**108.10**	**4.47**	**0**
	.10	.20	0	200.0	—	.23	61.00	.10	.04	.40	.25	.06	46.00	1.90	0
1384	**.15**	**.30**	**.01**	**300.0**	—	**.35**	**73.50**	**.15**	**.06**	**.60**	**.38**	**.09**	**69.00**	**2.85**	**0**
	.10	.20	0	200.0	—	.23	49.00	.10	.04	.40	.25	.06	46.00	1.90	0
1385	**.30**	**.30**	**.01**	**300.0**	—	**.35**	**67.50**	**.15**	**.06**	**.60**	**.38**	**.09**	**69.00**	**2.85**	**0**
	.20	.20	0	200.0	—	.23	45.00	.10	.04	.40	.25	.06	46.00	1.90	0
1386	**.40**	**(.24)**	**.04**	**420.0**	**0**	**Trace**	**31.00**	**.06**	**.02**	**.10**	**(.14)**	**(.03)**	**(8.03)**	**(.79)**	**0**
	.40	(.24)	.04	420.0	0	Trace	31.00	.06	.02	.10	(.14)	(.03)	(8.03)	(.79)	0
1387	**.05**	—	—	**25.2**	—	—	**8.16**	**.01**	**.01**	**.05**	—	—	—	—	—
	.80	—	—	420.0	—	—	136.00	.12	.09	.90	—	—	—	—	—
1388	—	—	—	—	—	—	—	—	—	—	—	—	—	—	—
	—	—	—	—	—	—	—	—	—	—	—	—	—	—	—
1389	—	—	—	—	—	—	—	—	—	—	—	—	—	—	—
	—	—	—	—	—	—	—	—	—	—	—	—	—	—	—
1390	**.90**	**.06**	**.04**	**5250.0**	—	—	**168.00**	**.12**	**.12**	**.90**	**.65**	—	—	—	**0**
	.30	.02	.01	1750.0	—	—	56.00	.04	.04	.30	.22	—	—	—	0
1391	**.31**	—	**.04**	**4.7**	**0**	—	**5.60**	**Trace**	**.03**	**.41**	—	—	—	—	**0**
	1.09	—	.13	16.9	0	—	20.00	Trace	.11	1.48	—	—	—	—	0
1392	**.50**	**.20**	**.07**	**1330.0**	—	—	**7.00**	**.02**	**.05**	**1.00**	**.17**	**.02**	**8.00**	**1.70**	**0**
	.50	.20	.07	1330.0	—	—	7.00	.02	.05	1.00	.17	.02	8.00	1.70	0
1393	**.16**	**.04**	**.03**	**266.0**	—	—	**1.45**	**.01**	**.01**	**.26**	**.03**	**.01**	**1.50**	**.10**	**0**
	.32	.09	.05	532.0	—	—	2.90	.01	.02	.52	.05	.02	3.00	.20	0
1394	**.13**	**.05**	**.03**	**293.5**	—	—	**1.70**	**.01**	**.01**	**.28**	**.03**	**.01**	**.25**	**.10**	**0**
	.25	.11	.05	587.0	—	—	3.40	.01	.02	.55	.05	.02	.50	.20	0
1395	**.18**	**.04**	**.03**	**177.0**	—	—	**1.20**	**.01**	**.01**	**.30**	**.03**	**.01**	**1.50**	**.10**	**0**
	.36	.09	.05	354.0	—	—	2.40	.01	.02	.59	.05	.02	3.00	.20	0
1396	**.14**	**.04**	**.03**	**194.0**	—	—	**2.35**	**.01**	**.01**	**.31**	**.03**	**.01**	**1.50**	**.10**	**0**
	.27	.09	.05	388.0	—	—	4.70	.01	.02	.61	.05	.02	3.00	.20	0
1397	**.15**	**.04**	**.02**	**210.0**	—	—	**1.50**	**.01**	**.01**	**.25**	**.03**	**.01**	**1.50**	**.10**	**0**
	.30	.07	.04	420.0	—	—	3.00	.01	.02	.50	.05	.02	3.00	.20	0
1398	**6.30**	**.93**	**.59**	**3427.2**	**0**	—	**8.32**	**0**	**.34**	**6.93**	**(.05)**	**.12**	**(22.37)**	—	**0**
	3.94	.58	.37	2142.0	0	—	5.20	0	.21	4.33	(.03)	.08	(13.98)	—	0
1399	**.63**	—	—	**402.6**	—	—	**.66**	—	**.02**	**.50**	—	—	—	—	—
	1.90	—	—	1220.0	—	—	2.00	—	.06	1.50	—	—	—	—	—
1400	**.53**	—	—	**353.1**	—	—	**.66**	—	**.02**	**.46**	—	—	—	—	—
	1.60	—	—	1070.0	—	—	2.00	—	.05	1.40	—	—	—	—	—
1401	**7.00**	—	—	**10000.0**	—	—	**28.00**	—	**.20**	**15.60**	—	—	—	—	—
	3.50	—	—	5000.0	—	—	14.00	—	.10	7.80	—	—	—	—	—
1402	**1.94**	—	—	**2153.8**	—	—	**4.84**	—	**.05**	**4.11**	—	—	—	—	—
	.80	—	—	890.0	—	—	2.00	—	.02	1.70	—	—	—	—	—
1403	**.93**	**.13**	**.06**	**710.0**	—	—	**235.50**	**.03**	**.10**	**1.63**	**.33**	**.05**	—	—	**0**
	.37	.05	.02	284.0	—	—	94.20	.01	.04	.65	.13	.02	—	—	0
1404	**.60**	**.16**	**.23**	**40.0**	—	—	**8.00**	**.04**	**.08**	**.20**	**.14**	**.03**	**28.00**	—	**0**
	.30	.08	.12	20.0	—	—	4.00	.02	.04	.10	.07	.02	14.00	—	0
1405	**.40**	**.15**	**.18**	**30.8**	**0**	**Trace**	**5.86**	**.05**	**.05**	**.35**	**.13**	**.04**	**25.48**	**.20**	**0**
	.22	.08	.10	16.9	0	Trace	3.22	.03	.03	.19	.07	.02	14.00	.11	0
1406	**.11**	**.04**	**.03**	**0**	—	—	**.40**	**.01**	**.01**	**.03**	**.01**	**.01**	**.50**	—	**0**
	.21	.08	.05	0	—	—	.80	.01	.01	.05	.02	.01	1.00	—	0
1407	**.12**	**.04**	**.03**	**0**	—	—	**.80**	**.01**	**.01**	**.05**	**.01**	**.01**	—	—	**0**
	.24	.09	.05	0	—	—	1.60	.01	.01	.10	.02	.01	—	—	0
1408	**.14**	**.04**	**.03**	**0**	—	—	**.35**	**.01**	**.01**	**.08**	**.01**	**.01**	**.50**	—	**0**
	.28	.08	.05	0	—	—	.70	.01	.02	.15	.02	.01	1.00	—	0
1409	**.11**	**.04**	**.03**	**0**	**0**	**Trace**	**.55**	**.01**	**.01**	**.12**	**.01**	**.01**	**.50**	**Trace**	**0**
	.22	.08	.05	0	0	Trace	1.10	.01	.02	.24	.02	.01	1.00	Trace	0
1410	**.10**	**.04**	**.02**	—	—	—	**.50**	**.01**	**.01**	**.05**	**.01**	**.01**	**.50**	—	**0**
	.20	.08	.04	—	—	—	1.00	.01	.02	.10	.02	.01	1.00	—	0
1411	—	—	—	—	—	—	—	—	—	—	—	—	—	—	—
	—	—	—	—	—	—	—	—	—	—	—	—	—	—	—

(Continued)

FRUITS

TABLE F-36 *(Continued)*

Item No.	Food Item Name	Approximate Measure	Weight	Moisture	Food Energy Calories	Protein	Fats	Carbo-hydrates	Fiber	Minerals (Macro) Calcium	Phos-phorus	Sodium	Mag-nesium	Potas-sium
			(g)	(%)	(kcal)	(g)	(g)	(g)	(g)	(mg)	(mg)	(mg)	(mg)	(mg)
	PEARS (Continued)													
1412	dried, sulfured, uncooked	1 CUP	160	26.3	422.4	2.9	1.0	112.4	10.2	51.2	91.2	9.6	52.8	852.8
			100	26.3	264.0	1.8	.6	70.2	6.4	32.0	57.0	6.0	33.0	533.0
1413	dried, sulfured, cooked w/o sugar	1 HALF	33	65.2	41.6	.5	.3	10.5	1.0	6.3	7.6	1.0	5.0	88.8
			100	65.2	126.0	1.5	.8	31.7	2.9	16.0	23.0	3.0	15.0	269.0
1414	dried, sulfured, cooked w/sugar	1 HALF	33	59.1	49.8	.4	.3	12.5	.9	5.0	6.6	1.0	—	80.5
			100	59.1	151.0	1.3	.8	38.0	2.6	15.0	20.0	3.0	—	244.0
	PERSIMMONS													
1415	native, raw	1 MED	100	64.4	127.0	.8	.4	33.5	1.5	27.0	26.0	1.0	—	310.0
			100	64.4	127.0	.8	.4	33.5	1.5	27.0	26.0	1.0	—	310.0
1416	Japanese or Kaki, raw	1 MED	100	78.6	77.0	.7	.4	19.7	1.6	6.0	26.0	6.0	8.0	174.0
			100	78.6	77.0	.7	.4	19.7	1.6	6.0	26.0	6.0	8.0	174.0
1417	Hachiya, peeled	1 MED	136	78.7	103.4	.7	.2	27.6	.5	8.0	25.8	—	—	—
			100	78.7	76.0	.5	.2	20.3	.3	5.9	19.0	—	—	—
	PINEAPPLE													
1418	raw	1 CUP	132	85.3	76.6	.4	.3	18.1	.5	22.4	10.6	1.3	17.2	192.7
			100	85.3	58.0	.3	.2	13.7	.4	17.0	8.0	1.0	13.0	146.0
1419	canned, water pack	1 SLICE	100	89.1	39.0	.3	.1	10.2	.3	12.0	5.0	1.0	8.0	99.0
			100	89.1	39.0	.3	.1	10.2	.3	12.0	5.0	1.0	8.0	99.0
1420	canned, juice pack	1 LG	100	83.4	61.0	.4	.1	15.8	.4	20.0	6.0	1.0	14.0	113.0
			100	83.4	61.0	.4	.1	15.8	.4	20.0	6.0	1.0	14.0	113.0
1421	canned, light syrup	1 LG	100	83.9	59.0	.3	.1	15.4	.3	11.0	5.0	1.0	8.0	97.0
			100	83.9	59.0	.3	.1	15.4	.3	11.0	5.0	1.0	8.0	97.0
1422	canned, heavy syrup	1 SLICE	100	79.0	78.0	.4	.1	20.2	.4	14.0	7.0	2.0	16.0	131.0
			100	79.0	78.0	.4	.1	20.2	.4	14.0	7.0	2.0	16.0	131.0
1423	canned, extra heavy syrup	1 LG	100	75.9	90.0	.3	.1	23.4	.3	11.0	5.0	1.0	8.0	94.0
			100	75.9	90.0	.3	.1	23.4	.3	11.0	5.0	1.0	8.0	94.0
1424	candied	1 SLICE	38	18.0	120.1	.3	.2	30.4	.3	—	—	—	—	—
			100	18.0	316.0	.8	.4	80.0	.8	—	—	—	—	—
1425	frozen, chunks, sweetened	1 CUP	264	77.1	224.4	1.1	.3	58.6	.8	23.8	10.6	5.3	26.4	264.0
			100	77.1	85.0	.4	.1	22.2	.3	9.0	4.0	2.0	10.0	100.0
1426	**PITANGA (SURINAM-CHERRY),** raw	1 OZ	28	85.8	14.3	.2	.1	3.5	.2	2.5	3.1	—	—	—
			100	85.8	51.0	.8	.4	12.5	.6	9.0	11.0	—	—	—
1427	**PLANTAIN (BAKING BANANA),** green, raw	1 SM	100	66.4	119.0	1.1	.4	31.2	.4	7.0	30.0	5.0	—	385.0
			100	66.4	119.0	1.1	.4	31.2	.4	7.0	30.0	5.0	—	385.0
1428	**PLANTAIN,** Maiamaoli, peeled	1 MED	236	67.2	271.4	2.2	.1	72.9	.7	8.7	62.1	—	—	—
			100	67.2	115.0	.9	0	30.9	.3	3.7	26.3	—	—	—
	PLUMS													
1429	Caissa, raw	1 AVG	20	80.8	14.0	.1	.3	3.2	.2	—	—	—	—	—
			100	80.8	70.0	.5	1.3	16.0	.9	—	—	—	—	—
1430	Damson, raw	1 MED	50	81.1	33.0	.3	—	8.9	.2	9.0	8.5	1.0	4.5	149.5
			100	81.1	66.0	.5	—	17.8	.4	18.0	17.0	2.0	9.0	299.0
1431	Japanese hybrid, 2⅛" diameter, raw	1 AVG	70	86.6	33.6	.4	.1	8.6	.4	8.4	12.6	.7	6.3	119.0
			100	86.6	48.0	.5	.2	12.3	.6	12.0	18.0	1.0	9.0	170.0
1432	prune type, raw	1 MED	50	78.7	37.5	4	.1	9.9	.2	6.0	9.0	.5	4.5	85.0
			100	78.7	75.0	.8	.2	19.7	.4	12.0	18.0	1.0	9.0	170.0
1433	Greengage, canned, water pack	1 OZ	28	90.6	9.2	.1	0	2.4	.1	2.5	3.6	.3	—	23.0
			100	90.6	33.0	.4	.1	8.6	.2	9.0	13.0	1.0	—	82.0
1434	purple, canned, water pack	1 MED	50	86.8	23.0	.2	.1	6.0	.2	4.5	5.0	1.0	—	74.0
			100	86.8	46.0	.4	.2	11.9	.3	9.0	10.0	2.0	—	148.0
1435	purple, canned, light syrup	1 MED	50	82.4	31.5	.2	.1	8.3	.2	4.5	5.0	.5	2.5	72.5
			100	82.4	63.0	.4	.1	16.6	.3	9.0	10.0	1.0	5.0	145.0
1436	purple, canned, heavy syrup	1 MED	50	76.0	44.5	.2	.1	11.6	.2	4.5	5.0	.5	2.5	71.0
			100	76.0	89.0	.4	.1	23.2	.3	9.0	10.0	1.0	5.0	142.0
1437	purple, canned, extra heavy syrup	1 MED	33	72.4	33.7	.1	0	8.8	.1	2.6	3.0	.3	1.7	45.9
			100	72.4	102.0	.4	.1	26.7	.3	8.0	9.0	1.0	5.0	139.0
1438	**POMEGRANATE,** pulp, raw	1 MED	100	82.3	63.0	.5	.3	16.4	.2	3.0	8.0	3.0	—	259.0
			100	82.3	63.0	.5	.3	16.4	.2	3.0	8.0	3.0	—	259.0
	PRUNES													
1439	dried, softenized, uncooked	1 LG	10	32.4	23.9	.3	.1	6.3	.2	5.1	7.9	.4	4.5	75.4
			100	32.4	239.0	2.6	.5	62.7	2.1	51.0	79.0	4.0	45.0	754.0

Item No.	Minerals (Micro)			Fat-Soluble Vitamins			Water-Soluble Vitamins								
	Iron	Zinc	Copper	Vitamin A	Vitamin D	Vitamin E (Alpha Tocopherol)	Vitamin C	Thiamin	Ribo-flavin	Niacin	Panto-thenic Acid	Vit. B-6 (Pyri-doxine)	Folacin (Folic Acid)	Biotin	Vitamin B-12
	(mg)	(mg)	(mg)	(IU)	(IU)	(mg)	(mg)	(mg)	(mg)	(mg)	(mg)	(mg)	(mcg)	(mcg)	(mcg)
1412	**3.23**	**.62**	**.61**	**4.8**	–	–	**11.20**	**.02**	**.24**	**2.21**	–	–	–	–	–
	2.02	.39	.38	3.0	–	–	7.00	.01	.15	1.38	–	–	–	–	–
1413	**.20**	–	–	**9.9**	–	–	**.66**	–	**.03**	**.10**	–	–	–	–	–
	.60	–	–	30.0	–	–	2.00	–	.08	.30	–	–	–	–	–
1414	**.20**	–	–	**9.9**	–	–	**.66**	–	**.02**	**.07**	–	–	–	–	–
	.60	–	–	30.0	–	–	2.00	–	.07	.20	–	–	–	–	–
1415	**2.50**	–	–	–	–	–	**66.00**	–	–	–	–	–	–	–	–
	2.50	–	–	–	–	–	66.00	–	–	–	–	–	–	–	–
1416	**.30**	–	–	**2710.0**	–	–	**11.00**	**.03**	**.02**	**.10**	–	–	–	–	–
	.30	–	–	2710.0	–	–	11.00	.03	.02	.10	–	–	–	–	–
1417	**.23**	–	–	**1339.6**	–	–	**9.93**	–	**.08**	**.25**	–	–	–	–	–
	.17	–	–	985.0	–	–	7.30	–	.06	.18	–	–	–	–	–
1418	**.66**	**.28**	**.19**	**92.4**	**0**	–	**22.44**	**.12**	**.04**	**.26**	**.21**	**.12**	**14.52**	**Trace**	**0**
	.50	.21	.14	70.0	0	–	17.00	.09	.03	.20	.16	.09	11.00	Trace	0
1419	**.30**	**.08**	**.15**	**50.0**	–	–	**7.00**	**.08**	**.02**	**.20**	**.10**	**.07**	**5.00**	–	**0**
	.30	.08	.15	50.0	–	–	7.00	.08	.02	.20	.10	.07	5.00	–	0
1420	**.30**	**.11**	**.11**	**27.0**	–	–	**8.80**	**.10**	**.02**	**.34**	**.10**	**.07**	**.80**	–	**0**
	.30	.11	.11	27.0	–	–	8.80	.10	.02	.34	.10	.07	.80	–	0
1421	**.30**	**.08**	**.15**	**50.0**	–	–	**7.00**	**.08**	**.02**	**.20**	**.10**	**.07**	**5.00**	–	**0**
	.30	.08	.15	50.0	–	–	7.00	.08	.02	.20	.10	.07	5.00	–	0
1422	**.41**	**.12**	**.10**	**14.0**	–	–	**7.50**	**.09**	**.03**	**.31**	**.10**	**.07**	**5.00**	–	**0**
	.41	.12	.10	14.0	–	–	7.50	.09	.03	.31	.10	.07	5.00	–	0
1423	**.30**	**.08**	**.15**	**40.0**	–	–	**6.00**	**.08**	**.02**	**.20**	**.10**	**.07**	**5.00**	–	**0**
	.30	.08	.15	40.0	–	–	6.00	.08	.02	.20	.10	.07	5.00	–	0
1424	–	–	–	–	–	–	–	–	–	–	–	–	–	–	–
	–	–	–	–	–	–	–	–	–	–	–	–	–	–	–
1425	**1.06**	–	–	**79.2**	–	–	**21.12**	**.26**	**.08**	**.79**	**.28**	**.20**	**15.84**	–	**0**
	.40	–	–	30.0	–	–	8.00	.10	.03	.30	.11	.08	6.00	–	0
1426	**.06**	–	–	**420.0**	–	–	**8.40**	**.01**	**.01**	**.08**	–	–	–	–	–
	.20	–	–	1500.0	–	–	30.00	.03	.04	.30	–	–	–	–	–
1427	**.70**	–	–	**605.0**	–	–	**14.00**	**.06**	**.04**	**.60**	**.37**	–	**16.00**	–	**0**
	.70	–	–	605.0	–	–	14.00	.06	.04	.60	.37	–	16.00	–	0
1428	**1.06**	–	–	**915.7**	–	–	**35.87**	**.13**	**.28**	**1.53**	–	–	–	–	–
	.45	–	–	388.0	–	–	15.20	.05	.12	.65	–	–	–	–	–
1429	–	[6]**.02**	–	**8.0**	–	–	**7.60**	**.01**	**.01**	**.04**	–	–	**1.20**	–	–
	–	.10	–	40.0	–	–	38.00	.04	.06	.20	–	–	6.00	–	–
1430	**.25**	[6]**.05**	–	**150.0**	–	–	–	**.04**	**.02**	**.25**	**.09**	**.03**	**3.00**	–	**0**
	.50	.10	–	300.0	–	–	–	.08	.03	.50	.19	.05	6.00	–	0
1431	**.35**	–	–	**175.0**	–	–	**4.20**	**.02**	**.02**	**.35**	**.13**	**.04**	**4.20**	–	**0**
	.50	–	–	250.0	–	–	6.00	.03	.03	.50	.19	.05	6.00	–	0
1432	**.25**	–	–	**150.0**	–	–	**2.00**	**.02**	**.02**	**.25**	**.09**	**.04**	**3.00**	–	**0**
	.50	–	–	300.0	–	–	4.00	.03	.03	.50	.19	.05	6.00	–	0
1433	**.06**	–	–	**44.8**	–	–	**.56**	**Trace**	**.01**	**.08**	**.02**	**.01**	–	–	**0**
	.20	–	–	160.0	–	–	2.00	.01	.02	.30	.07	.03	–	–	0
1434	**.50**	–	–	**625.0**	–	–	**1.00**	**.01**	**.01**	**.20**	**.04**	**.01**	**1.50**	–	**0**
	1.00	–	–	1250.0	–	–	2.00	.02	.02	.40	.07	.03	3.00	–	0
1435	**.45**	–	–	**615.0**	–	–	**1.00**	**.01**	**.01**	**.20**	**.04**	**.01**	**1.50**	–	**0**
	.90	–	–	1230.0	–	–	2.00	.02	.02	.40	.07	.02	3.00	–	0
1436	**.42**	–	–	**605.0**	–	–	**.20**	**.01**	**.02**	**.15**	**.04**	**.01**	**1.50**	–	**0**
	.84	–	–	1210.0	–	–	.40	.02	.04	.29	.07	.03	3.00	–	0
1437	**.30**	–	–	**389.4**	–	–	**.66**	**.01**	**.01**	**.13**	**.02**	**.01**	**.99**	–	**0**
	.90	–	–	1180.0	–	–	2.00	.02	.02	.40	.07	.03	3.00	–	0
1438	**.30**	–	–	–	–	–	**4.00**	**.03**	**.03**	**.30**	**.60**	–	–	–	**0**
	.30	–	–	–	–	–	4.00	.03	.03	.30	.60	–	–	–	0
1439	**.25**	**.05**	**.04**	**199.4**	**0**	–	**.33**	**.01**	**.02**	**.19**	**.05**	**.03**	**.40**	**Trace**	**0**
	2.47	.53	.44	1994.0	0	–	3.30	.08	.22	1.94	.46	.29	4.00	Trace	0

(Continued)

TABLE F-36 *(Continued)*

Item No.	Food Item Name	Approximate Measure	Weight	Moisture	Food Energy Calories	Protein	Fats	Carbo-hydrates	Fiber	Minerals (Macro) Calcium	Phos-phorus	Sodium	Mag-nesium	Potas-sium
			(g)	(%)	(kcal)	(g)	(g)	(g)	(g)	(mg)	(mg)	(mg)	(mg)	(mg)
	PRUNES (Continued)													
1440	dried, softenized, cooked w/o sugar	1 CUP	270	66.4	321.3	2.7	.8	84.8	2.2	64.8	99.9	10.8	54.0	882.9
			100	66.4	119.0	1.0	.3	31.4	.8	24.0	37.0	4.0	20.0	327.0
1441	dried, softenized, cooked w/sugar	1 CUP	270	53.2	464.4	2.2	.5	121.8	1.6	51.3	81.0	8.1	64.8	707.4
			100	53.2	172.0	.8	.2	45.1	.6	19.0	30.0	3.0	24.0	262.0
1442	dehydrated, uncooked	1 LG	10	2.5	34.4	.3	.1	9.1	.2	9.0	10.7	1.1	—	94.0
			100	2.5	344.0	3.3	.5	91.3	2.2	90.0	107.0	11.0	—	940.0
1443	dehydrated, cooked w/sugar	1 MED	20	50.7	36.0	.2	0	9.4	.2	6.2	7.4	.8	—	65.8
			100	50.7	180.0	1.2	.2	47.1	.8	31.0	37.0	4.0	—	329.0
	PUMPKIN													
1444	raw, diced	1 CUP	185	91.6	48.1	1.9	.2	12.0	2.0	38.9	81.4	1.9	22.2	629.0
			100	91.6	26.0	1.0	.1	6.5	1.1	21.0	44.0	1.0	12.0	340.0
1445	canned	1 CUP	250	90.2	82.5	2.5	.8	19.8	3.3	62.5	65.0	5.0	45.0	600.0
			100	90.2	33.0	1.0	.3	7.9	1.3	25.0	26.0	2.0	18.0	240.0
1446	**QUINCES,** raw, 2½″ diameter, 3½″ long	1 AVG	155	83.8	88.4	.6	.2	23.7	2.6	17.1	26.4	6.2	(9.8)	305.4
			100	83.8	57.0	.4	.1	15.3	1.7	11.0	17.0	4.0	(6.3)	197.0
	RAISINS													
1447	California, Thompson seedless	1 TBLS	10	17.0	28.9	.3	0	7.7	.1	8.6	10.2	1.7	3.5	67.8
			100	17.0	289.0	3.0	.2	77.2	.8	86.0	102.0	17.0	35.0	678.0
1448	natural, uncooked	1 TBLS	10	16.0	28.9	.3	0	7.7	.1	3.5	6.0	2.0	2.7	78.3
			100	16.0	289.0	2.5	.2	77.4	.9	35.2	60.4	19.9	26.8	783.0
1449	natural, cooked w/sugar	1 CUP	170	41.4	362.1	2.0	.2	95.9	.7	49.3	79.9	22.1	27.2	603.5
			100	41.4	213.0	1.2	.1	56.4	.4	29.0	47.0	13.0	16.0	355.0
1450	[4]**RAMBUTAN,** raw	1 OZ	28	82.0	17.9	.3	0	4.6	.3	5.6	4.2	.3	—	17.9
			100	82.0	64.0	1.0	.1	16.5	1.1	20.0	15.0	1.0	—	64.0
	RASPBERRIES													
1451	black, raw	1 CUP	123	80.8	89.8	1.8	1.7	19.3	6.3	36.9	27.1	1.2	36.9	244.8
			100	80.8	73.0	1.5	1.4	15.7	5.1	30.0	22.0	1.0	30.0	199.0
1452	black, canned, water pack	1 CUP	200	86.7	102.0	2.2	2.2	21.4	6.6	40.0	30.0	2.0	—	270.0
			100	86.7	51.0	1.1	1.1	10.7	3.3	20.0	15.0	1.0	—	135.0
1453	red, raw	1 CUP	132	84.2	75.2	1.6	.7	18.0	4.0	29.0	29.0	1.3	26.4	221.8
			100	84.2	57.0	1.2	.5	13.6	3.0	22.0	22.0	1.0	20.0	168.0
1454	red, canned, water pack	1 CUP	200	90.1	70.0	1.4	.2	17.6	5.2	30.0	30.0	2.0	26.0	228.0
			100	90.1	35.0	.7	.1	8.8	2.6	15.0	15.0	1.0	13.0	114.0
1455	red, frozen, sweetened, not thawed	1 CUP	246	72.6	250.9	1.8	.4	64.3	5.4	39.4	41.8	2.5	32.0	280.4
			100	72.6	102.0	.7	.2	26.1	2.2	16.0	17.0	1.0	13.0	114.0
	RHUBARB													
1456	raw	1 CUP	117	94.8	18.7	.7	.1	4.3	.8	112.3	21.1	2.3	13.6	293.7
			100	94.8	16.0	.6	.1	3.7	.7	96.0	18.0	2.0	11.6	251.0
1457	cooked w/sugar	1 CUP	266	62.8	375.1	1.3	.3	95.8	1.6	207.5	39.9	5.3	34.6	540.0
			100	62.8	141.0	.5	.1	36.0	.6	78.0	15.0	2.0	13.0	203.0
1458	frozen, sweetened	1 CUP	120	80.1	90.0	.7	.2	22.2	1.1	111.6	16.8	4.8	14.4	253.2
			100	80.1	75.0	.6	.2	18.5	.9	93.0	14.0	4.0	12.0	211.0
1459	frozen, cooked w/sugar	1 CUP	266	62.6	380.4	1.3	.5	96.3	2.1	207.5	31.9	8.0	—	468.2
			100	62.6	143.0	.5	.2	36.2	.8	78.0	12.0	3.0	—	176.0
1460	**ROSEAPPLES,** raw, ¾″ diameter	1 AVG	20	84.5	11.2	.1	.1	2.8	.2	5.8	3.2	—	—	—
			100	84.5	56.0	.6	.3	14.2	1.1	29.0	16.0	—	—	—
1461	**SAPODILLA,** raw	1 CUP	200	76.1	178.0	1.0	2.2	43.6	2.8	42.0	24.0	24.0	—	386.0
			100	76.1	89.0	.5	1.1	21.8	1.4	21.0	12.0	12.0	—	193.0
1462	**SAPOTES (MARMALADE PLUMS),** raw	1 CUP	200	64.9	250.0	3.6	1.2	63.2	3.8	78.0	56.0	—	—	—
			100	64.9	125.0	1.8	.6	31.6	1.9	39.0	28.0	—	—	—
1463	green, peeled, seeded	1 MED	225	69.8	240.8	3.9	1.1	60.7	.2	78.3	45.7	—	—	—
			100	69.8	107.0	1.7	.5	27.0	.1	34.8	20.3	—	—	—
1464	**SOURSOP,** raw, pureed	1 CUP	225	81.7	146.3	2.3	.7	36.7	2.5	31.5	60.8	31.5	—	596.3
			100	81.7	65.0	1.0	.3	16.3	1.1	14.0	27.0	14.0	—	265.0
	STRAWBERRIES													
1465	raw	1 CUP	150	89.9	55.5	1.1	.8	12.6	2.0	31.5	31.5	1.5	18.0	246.0
			100	89.9	37.0	.7	.5	8.4	1.3	21.0	21.0	1.0	12.0	164.0
1466	canned, water pack, solids & liquid	1 CUP	240	93.7	52.8	1.0	.2	13.4	1.4	33.6	33.6	2.4	—	266.4
			100	93.7	22.0	.4	.1	5.6	.6	14.0	14.0	1.0	—	111.0
1467	frozen, unsweetened	INDV	10	90.0	3.5	0	0	.9	.1	1.6	1.3	.2	1.1	14.8
			100	90.0	35.0	.4	.1	9.1	.8	16.0	13.0	2.0	11.0	148.0
1468	frozen, sweetened, whole	1 CUP	244	75.7	231.8	1.3	.3	59.8	1.5	26.8	29.3	4.9	17.1	239.1
			100	75.7	95.0	.5	.1	24.5	.6	11.0	12.0	2.0	7.0	98.0
1469	frozen, sweetened, sliced	1 CUP	256	74.6	243.2	1.4	.3	62.8	1.6	28.2	30.7	5.1	17.9	250.9
			100	74.6	95.0	.5	.1	24.5	.6	11.0	12.0	2.0	7.0	98.0

Item No.	Minerals (Micro)			Fat-Soluble Vitamins			Water-Soluble Vitamins								
	Iron	Zinc	Copper	Vitamin A	Vitamin D	Vitamin E (Alpha Tocopherol)	Vitamin C	Thiamin	Ribo-flavin	Niacin	Panto-thenic Acid	Vit. B-6 (Pyri-doxine)	Folacin (Folic Acid)	Biotin	Vitamin B-12
	(mg)	(mg)	(mg)	(IU)	(IU)	(mg)	(mg)	(mg)	(mg)	(mg)	(mg)	(mg)	(mcg)	(mcg)	(mcg)
1440	**4.86**	**.89**	**.68**	**2025.0**	**0**	—	**2.70**	**.08**	**.19**	**1.89**	**(.62)**	**(.41)**	**Trace**	**Trace**	**0**
	1.80	.33	.25	750.0	0	—	1.00	.03	.07	.70	(.23)	(.15)	Trace	Trace	0
1441	**4.05**	**.89**	**.68**	**1620.0**	**0**	—	**2.70**	**.08**	**.16**	**1.62**	**(.59)**	**(.38)**	**Trace**	**Trace**	**0**
	1.50	.33	.25	600.0	0	—	1.00	.03	.06	.60	(.22)	(.14)	Trace	Trace	0
1442	**.44**	—	—	**217.0**	—	—	**.40**	**.01**	**.02**	**.21**	—	—	—	—	—
	4.40	—	—	2170.0	—	—	4.00	.12	.22	2.10	—	—	—	—	—
1443	**.30**	—	—	**152.0**	—	—	**.20**	**.01**	**.01**	**.14**	—	—	—	—	—
	1.50	—	—	760.0	—	—	1.00	.03	.07	.70	—	—	—	—	—
1444	**1.48**	—	**.19**	**2960.0**	—	—	**16.65**	**.09**	**.20**	**1.11**	—	—	**66.60**	—	—
	.80	—	.10	1600.0	—	—	9.00	.05	.11	.60	—	—	36.00	—	—
1445	**1.00**	**.48**	**.14**	**16000.0**	—	—	**12.50**	**.08**	**.13**	**1.50**	**1.00**	**.14**	**37.50**	—	**0**
	.40	.19	.05	6400.0	—	—	5.00	.03	.05	.60	.40	.06	15.00	—	0
1446	**1.09**	—	**(.20)**	**62.0**	**0**	—	**23.25**	**.03**	**.05**	**.31**	—	—	—	—	**0**
	.70	—	(.13)	40.0	0	—	15.00	.02	.03	.20	—	—	—	—	0
1447	**.21**	**.02**	**.04**	**2.0**	**0**	—	**.09**	**.01**	**Trace**	**.06**	**.01**	**.02**	**.40**	**.45**	**0**
	2.10	.22	.38	20.0	0	—	.88	.10	.03	.60	.06	.23	4.00	4.50	0
1448	**.29**	**.02**	**.03**	**2.0**	**0**	—	**.10**	**.01**	**.01**	**.05**	**.01**	**.02**	**.40**	**.45**	**0**
	2.85	.18	.28	20.0	0	—	1.00	.11	.08	.50	.10	.24	4.00	4.50	0
1449	**2.72**	—	—	**17.0**	—	—	—	**.07**	**.05**	**.34**	—	—	—	—	—
	1.60	—	—	10.0	—	—	—	.04	.03	.20	—	—	—	—	—
1450	**.53**	—	—	**0**	—	—	**14.84**	**Trace**	**.02**	**.11**	—	—	—	—	—
	1.90	—	—	0	—	—	53.00	.01	.06	.40	—	—	—	—	—
1451	**1.11**	—	**.16**	—	—	—	**22.14**	**.04**	**.11**	**1.11**	**.30**	**.07**	**6.15**	—	**0**
	.90	—	.13	—	—	—	18.00	.03	.09	.90	.24	.06	5.00	—	0
1452	**1.20**	—	—	—	—	—	**12.00**	**.02**	**.08**	**1.00**	**.34**	**.08**	—	—	**0**
	.60	—	—	—	—	—	6.00	.01	.04	.50	.17	.04	—	—	0
1453	**1.19**	**0**	**.08**	**171.6**	**0**	**(.41)**	**33.00**	**.04**	**.12**	**1.19**	**.32**	**.08**	**6.60**	**(2.52)**	**0**
	.90	0	.06	130.0	0	(.31)	25.00	.03	.09	.90	.24	.06	5.00	(1.91)	0
1454	**1.20**	—	**(.42)**	**180.0**	**0**	**(.41)**	**18.00**	**.02**	**.08**	**1.00**	**.34**	**.08**	**Trace**	**Trace**	**0**
	.60	—	(.21)	90.0	0	(.31)	9.00	.01	.04	.50	.17	.04	Trace	Trace	0
1455	**1.72**	**.44**	**.26**	**130.4**	—	—	**37.15**	**.05**	**.10**	**.54**	**.37**	**.08**	**63.96**	—	**0**
	.70	.18	.11	53.0	—	—	15.10	.02	.04	.22	.15	.03	26.00	—	0
1456	**.94**	**0**	**.01**	**117.0**	**0**	**(.27)**	**10.53**	**.04**	**.08**	**.35**	**.10**	**.04**	**8.19**	—	**0**
	.80	0	.01	100.0	0	(.23)	9.00	.03	.07	.30	.09	.03	7.00	—	0
1457	**1.60**	**.29**	—	**212.8**	—	—	—	—	—	—	—	—	—	—	—
	.60	.11	—	80.0	—	—	—	—	—	—	—	—	—	—	—
1458	**.96**	—	—	**96.0**	—	—	**9.60**	**.02**	**.06**	**.24**	**.08**	**.03**	**4.80**	—	**0**
	.80	—	—	80.0	—	—	8.00	.02	.05	.20	.07	.03	4.00	—	0
1459	**1.86**	—	—	**186.2**	—	—	**15.96**	**.05**	**.11**	**.53**	—	—	—	—	—
	.70	—	—	70.0	—	—	6.00	.02	.04	.20	—	—	—	—	—
1460	**.24**	—	—	**26.0**	—	—	**4.40**	**Trace**	**.01**	**.16**	—	—	—	—	—
	1.20	—	—	130.0	—	—	22.00	.02	.03	.80	—	—	—	—	—
1461	**1.60**	—	—	**120.0**	—	—	**28.00**	—	**.04**	**.40**	**.50**	—	—	—	**0**
	.80	—	—	60.0	—	—	14.00	—	.02	.20	.25	—	—	—	0
1462	**2.00**	—	—	**820.0**	—	—	—	—	—	—	—	—	—	—	—
	1.00	—	—	410.0	—	—	—	—	—	—	—	—	—	—	—
1463	**.95**	—	—	**1642.5**	—	—	**65.70**	—	**.12**	**3.53**	—	—	—	—	—
	.42	—	—	730.0	—	—	29.20	—	.05	1.57	—	—	—	—	—
1464	**1.35**	—	—	**22.5**	—	—	**45.00**	**.16**	**.11**	**2.03**	**.57**	—	—	—	**0**
	.60	—	—	10.0	—	—	20.00	.07	.05	.90	.25	—	—	—	0
1465	**1.50**	**0**	**.04**	**90.0**	**0**	**.20**	**88.50**	**.05**	**.11**	**.90**	**.51**	**.08**	**24.00**	**6.00**	**0**
	1.00	0	.02	60.0	0	.13	59.00	.03	.07	.60	.34	.06	16.00	4.00	0
1466	**1.68**	—	—	**96.0**	—	—	**48.00**	**.02**	**.07**	**.96**	—	—	—	—	—
	.70	—	—	40.0	—	—	20.00	.01	.03	.40	—	—	—	—	—
1467	**.08**	—	—	**4.5**	—	—	**4.12**	**.00**	**Trace**	**.05**	—	—	**1.70**	—	—
	.75	—	—	45.0	—	—	41.20	.02	.04	.46	—	—	17.00	—	—
1468	**1.34**	**.13**	**.05**	**63.4**	—	—	**93.70**	**.05**	**.15**	**.93**	**.26**	**.08**	**41.48**	**9.76**	**0**
	.55	.06	.02	26.0	—	—	38.40	.02	.06	.38	.11	.03	17.00	4.00	0
1469	**1.41**	**.14**	**.05**	**66.6**	—	**.54**	**98.30**	**.05**	**.15**	**.97**	**.28**	**.07**	**43.52**	**10.24**	**0**
	.55	.06	.02	26.0	—	.21	38.40	.02	.06	.38	.11	.03	17.00	4.00	0

(Continued)

FRUITS

TABLE F-36 *(Continued)*

Item No.	Food Item Name	Approximate Measure	Weight	Moisture	Food Energy Calories	Protein	Fats	Carbo-hydrates	Fiber	Minerals (Macro) Calcium	Phos-phorus	Sodium	Mag-nesium	Potas-sium
			(g)	(%)	(kcal)	(g)	(g)	(g)	(g)	(mg)	(mg)	(mg)	(mg)	(mg)
1470	**SUGARAPPLES (SWEETSOP),** raw	1 CUP	**250**	**73.3**	**235.0**	**4.5**	**.8**	**59.3**	**4.3**	**55.0**	**102.5**	**27.5**	–	**687.5**
			100	73.3	94.0	1.8	.3	23.7	1.7	22.0	41.0	11.0	–	275.0
1471	**TAMARINDS,** raw	1 OZ	**28**	**31.4**	**92.7**	**5.3**	**.2**	**17.5**	**1.4**	**20.7**	**31.6**	**14.3**	–	**218.7**
			100	31.4	331.0	18.8	.6	62.5	5.1	74.0	113.0	51.0	–	781.0
1472	**TANGELOS,** raw	1 MED	**170**	**89.4**	**39.0**	**.5**	**.1**	**9.2**	–	**27.2**	**20.4**	**1.7**	**19.0**	**295.8**
			100	89.4	22.9	.3	.1	5.4	–	16.0	12.0	1.0	11.2	174.0
1473	**TANGERINES,** Dancy, raw	1 MED	**116**	**87.0**	**53.4**	**.9**	**.2**	**13.5**	**.6**	**19.7**	**11.6**	**2.3**	**14.3**	**187.9**
			100	87.0	46.0	.8	.2	11.6	.5	17.0	10.0	2.0	12.3	162.0
1474	**TOMATO**—See **VEGETABLES**													
1475	**TOWELGOURD,** raw, cubed	1 CUP	**130**	**94.5**	**23.4**	**1.0**	**.3**	**5.3**	**.7**	**24.7**	**42.9**	–	–	–
			100	94.5	18.0	.8	.2	4.1	.5	19.0	33.0	–	–	–
1476	**WATERMELON,** raw	1 CUP	**200**	**92.6**	**52.0**	**1.0**	**.4**	**12.8**	**.6**	**14.0**	**20.0**	**2.0**	**20.4**	**200.0**
			100	92.6	26.0	.5	.2	6.4	.3	7.0	10.0	1.0	10.2	100.0
1477	**WAXGOURD (CHINESE PRESERVING MELON),** raw, cubed	1 CUP	**160**	**96.1**	**20.8**	**.6**	**.3**	**4.8**	**.8**	**30.4**	**30.4**	**9.6**	–	**177.6**
			100	96.1	13.0	.4	.2	3.0	.5	19.0	19.0	6.0	–	111.0
1478	'**WAXGOURD,** sugared	1 OZ	**28**	**20.1**	**79.8**	**.1**	**.1**	**22.2**	**.1**	**26.0**	**4.8**	–	–	–
			100	20.1	285.0	.2	.2	79.2	.2	93.0	17.0	–	–	–
	GRAVIES AND SAUCES													
1479	**GRAVY, BEEF**	1 TBLS	**18**	–	**41.0**	**.3**	**3.5**	**2.0**	**0**	**0**	**2.0**	–	–	–
			100	–	228.0	1.7	19.4	11.1	0	0	11.0	–	–	–
1480	**GRAVY, CHICKEN**	1 OZ	**28**	–	**63.8**	**.5**	**5.4**	**3.1**	**0**	**0**	**3.1**	**55.5**	**.6**	**2.0**
			100	–	228.0	1.7	19.4	11.1	0	0	11.0	198.3	2.1	7.1
	SAUCE													
1481	**BARBECUE**	1 TBLS	**16**	**80.9**	**14.6**	**.2**	**1.1**	**1.3**	**.1**	**3.4**	**3.2**	**130.4**	–	**27.8**
			100	80.9	91.0	1.5	6.9	8.0	.6	21.0	20.0	815.0	–	174.0
1482	**CHEESE**	1 TBLS	**19**	–	**34.0**	**1.5**	**2.6**	**1.2**	–	**44.1**	**32.5**	–	–	–
			100	–	179.0	7.9	13.6	6.3	–	232.0	171.0	–	–	–
1483	**HARD,** medium	1 TBLS	**10**	–	**46.2**	**0**	**2.7**	**5.7**	–	**.9**	**.5**	–	–	–
			100	–	462.0	.5	27.1	57.1	–	9.0	5.0	–	–	–
1484	**HOLLANDAISE,** mock	1 TBLS	**21**	–	**46.8**	**1.0**	**4.1**	**1.6**	–	**21.8**	**27.7**	–	–	–
			100	–	222.6	4.8	19.3	7.4	–	104.0	132.0	–	–	–
1485	**HOLLANDAISE,** true	1 TBLS	**13**	–	**46.8**	**.6**	**4.8**	**.1**	–	**6.0**	**20.3**	–	–	–
			100	–	360.0	4.4	37.0	.8	–	46.0	156.0	–	–	–
1486	**WHITE,** thin, made w/butter	1 TBLS	**15**	**78.7**	**18.2**	**.6**	**1.3**	**1.1**	–	**18.3**	**14.6**	**52.7**	–	**21.9**
			100	78.7	121.0	3.9	8.7	7.2	–	122.0	97.0	351.0	–	146.0
1487	**WHITE,** medium, made w/butter	1 TBLS	**15**	**73.3**	**24.3**	**.6**	**1.9**	**1.3**	–	**17.3**	**14.0**	**56.9**	–	**20.9**
			100	73.3	162.0	3.9	12.5	8.8	–	115.0	93.0	379.0	–	139.0
1488	**WHITE,** thick, made w/butter	1 TBLS	**15**	**67.9**	**29.7**	**.6**	**2.3**	**1.7**	–	**16.1**	**13.5**	**59.9**	–	**20.0**
			100	67.9	198.0	4.0	15.6	11.0	–	107.0	90.0	399.0	–	133.0
1489	**TARTAR**	1 TBLS	**20**	**34.4**	**106.2**	**.3**	**11.6**	**.8**	**.1**	**3.6**	**6.4**	**141.4**	–	**15.6**
			100	34.4	531.0	1.4	57.8	4.2	.3	18.0	32.0	707.0	–	78.0
1490	**TARTAR,** low calorie (10 cal/tsp)	1 TBLS	**20**	**68.1**	**44.8**	**.1**	**4.5**	**1.3**	**.1**	**3.6**	**6.4**	**141.4**	–	**15.6**
			100	68.1	224.0	.6	22.4	6.7	.3	18.0	32.0	707.0	–	78.0
	JUICES													
1491	**ACEROLA JUICE,** raw	100 CC	**100**	**94.3**	**23.0**	**.4**	**.3**	**4.8**	**.3**	**10.0**	**9.0**	**3.0**	–	–
			100	94.3	23.0	.4	.3	4.8	.3	10.0	9.0	3.0	–	–
	APPLE JUICE													
1492	cider	1 CUP	**249**	–	**124.5**	**.2**	**0**	**34.3**	**0**	**14.9**	**27.4**	**10.0**	–	**249.0**
			100	–	50.0	.1	0	13.8	0	6.0	11.0	4.0	–	100.0
1493	canned or bottled	1 CUP	**247**	**87.8**	**116.1**	**.2**	**0**	**29.4**	**.4**	**14.8**	**22.2**	**2.5**	**9.9**	**249.5**
			100	87.8	47.0	.1	0	11.9	.1	6.0	9.0	1.0	4.0	101.0
1494	canned, special dietary	1 CUP	**250**	–	**135.0**	**.3**	**.5**	**32.5**	–	–	–	**5.0**	–	–
			100	–	54.0	.1	.2	13.0	–	–	–	2.0	–	–
1495	w/cherry	1 CUP	**250**	–	**132.5**	**.5**	–	**32.5**	–	**15.0**	**20.0**	**10.0**	–	**225.0**
			100	–	53.0	.2	–	13.0	–	6.0	8.0	4.0	–	90.0
1496	**APRICOT JUICE,** unsweetened	1 CUP	**250**	–	**122.5**	**1.0**	**.3**	**29.3**	–	–	–	**7.5**	–	–
			100	–	49.0	.4	.1	11.7	–	–	–	3.0	–	–
1497	**APRICOT NECTAR,** w/vitamin C added	1 CUP	**247**	**84.6**	**140.8**	**.7**	**.2**	**36.1**	**.5**	**22.2**	**29.6**	–	**12.4**	**373.0**
			100	84.6	57.0	.3	.1	14.6	.2	9.0	12.0	–	5.0	151.0
1498	**BLACKBERRY JUICE,** canned, unsweetened	1 CUP	**250**	**90.9**	**92.5**	**.8**	**1.5**	**19.5**	–	**30.0**	**30.0**	**2.5**	**52.5**	**425.0**
			100	90.9	37.0	.3	.6	7.8	–	12.0	12.0	1.0	21.0	170.0

	Minerals (Micro)			Fat-Soluble Vitamins			Water-Soluble Vitamins								
Item No.	Iron	Zinc	Copper	Vitamin A	Vitamin D	Vitamin E (Alpha Tocopherol)	Vitamin C	Thiamin	Ribo-flavin	Niacin	Panto-thenic Acid	Vit. B-6 (Pyri-doxine)	Folacin (Folic Acid)	Biotin	Vitamin B-12
	(mg)	(mg)	(mg)	(IU)	(IU)	(mg)	(mg)	(mg)	(mg)	(mg)	(mg)	(mg)	(mcg)	(mcg)	(mcg)
1470	1.50	–	–	25.0	–	–	85.00	.25	.35	2.50	.57	–	–	–	0
	.60	–	–	10.0	–	–	34.00	.10	.14	1.00	.23	–	–	–	0
1471	.78	–	–	8.4	–	–	.56	.10	.04	.34	.04	–	–	–	0
	2.80	–	–	30.0	–	–	2.00	.34	.14	1.20	.14	–	–	–	0
1472	.17	0	.06	–	–	–	26.00	–	–	–	–	.09	23.80	–	0
	.10	0	.03	–	–	–	15.29	–	–	–	–	.05	14.00	–	0
1473	.46	.17	.03	487.2	0	–	35.96	.07	.02	.12	.23	.08	24.36	–	0
	.40	.15	.03	420.0	0	–	31.00	.06	.02	.10	.20	.07	21.00	–	0
1474															
1475	1.17	–	–	494.0	–	–	10.40	.04	.05	.52	–	–	–	–	–
	.90	–	–	380.0	–	–	8.00	.03	.04	.40	–	–	–	–	–
1476	1.00	.18	.03	1180.0	–	–	14.00	.06	.06	.40	.60	.14	16.00	7.20	0
	.50	.09	.02	590.0	–	–	7.00	.03	.03	.20	.30	.07	8.00	3.60	0
1477	.64	–	–	0	–	–	20.80	.06	.18	.64	–	–	–	–	–
	.40	–	–	0	–	–	13.00	.04	.11	.40	–	–	–	–	–
1478	.95	–	–	0	–	–	0	0	0	0	–	–	–	–	–
	3.40	–	–	0	–	–	0	0	0	0	–	–	–	–	–
1479	.09	–	–	0	0	–	–	.01	.01	0	–	–	–	–	–
	.50	–	–	0	0	–	–	.05	.04	0	–	–	–	–	–
1480	.14	.02	1.50	43.0	0	–	.01	.01	.18	.05	.01	Trace	.35	–	0
	.50	.06	5.36	153.7	0	–	.02	.02	.64	.18	.02	Trace	1.23	–	0
1481	.13	.02	–	57.6	–	–	.80	Trace	Trace	.05	–	–	.64	–	–
	.80	.13	–	360.0	–	–	5.00	.01	.01	.30	–	–	4.00	–	–
1482	.06	–	–	103.0	–	–	–	.01	.04	.05	–	–	–	–	–
	.30	–	–	542.0	–	–	–	.03	.21	.26	–	–	–	–	–
1483	0	–	–	110.0	–	–	0	0	0	0	–	–	–	–	–
	0	–	–	1100.0	–	–	0	0	0	0	–	–	–	–	–
1484	.17	–	–	206.0	–	–	.25	.01	.03	.05	–	–	–	–	–
	.80	–	–	981.0	–	–	1.20	.05	.15	.23	–	–	–	–	–
1485	.23	–	–	267.0	–	–	–	.01	.01	–	–	–	–	–	–
	1.80	–	–	2054.0	–	–	–	.05	.08	–	–	–	–	–	–
1486	.02	–	–	48.0	–	–	.06	.01	.03	.03	–	–	–	–	–
	.10	–	–	320.0	–	–	.40	.04	.17	.20	–	–	–	–	–
1487	.03	–	–	69.0	–	–	.06	.01	.03	.03	–	–	–	–	–
	.20	–	–	460.0	–	–	.40	.04	.17	.20	–	–	–	–	–
1488	.05	–	–	85.5	–	–	–	.01	.03	.05	–	–	–	–	–
	.30	–	–	570.0	–	–	–	.05	.16	.30	–	–	–	–	–
1489	.18	.20	.02	44.0	–	–	.20	Trace	.01	–	–	–	–	–	–
	.90	.98	.12	220.0	–	–	1.00	.01	.03	–	–	–	–	–	–
1490	.18	–	–	44.0	–	–	.20	Trace	.01	–	–	–	–	–	–
	.90	–	–	220.0	–	–	1.00	.01	.03	–	–	–	–	–	–
1491	.50	–	–	–	–	–	1600.00	.02	.06	.40	.21	.01	–	–	0
	.50	–	–	–	–	–	1600.00	.02	.06	.40	.21	.01	–	–	0
1492	1.25	.08	–	99.6	0	–	2.74	.05	.08	0	–	–	–	–	–
	.50	.03	–	40.0	0	–	1.10	.02	.03	0	–	–	–	–	–
1493	1.48	.17	.06	0	–	–	2.47	.03	.05	.25	–	.08	1.24	–	0
	.60	.07	.02	0	–	–	1.00	.01	.02	.10	–	.03	.50	–	0
1494	–	–	–	–	–	–	–	–	–	–	–	.08	1.25	–	0
	–	–	–	–	–	–	–	–	–	–	–	.03	.50	–	0
1495	1.75	–	–	7.5	–	–	100.00	–	.03	.25	–	–	–	–	–
	.70	–	–	3.0	–	–	40.00	–	.01	.10	–	–	–	–	–
1496	–	–	–	–	–	–	–	–	–	–	–	–	5.00	–	–
	–	–	–	–	–	–	–	–	–	–	–	–	2.00	–	–
1497	.50	.20	.37	2346.5	–	–	103.74	.03	.03	.50	–	.07	4.94	–	0
	.20	.08	.15	950.0	–	–	42.00	.01	.01	.20	–	.03	2.00	–	0
1498	2.25	–	–	–	–	–	25.00	.05	.08	.75	.20	–	–	–	0
	.90	–	–	–	–	–	10.00	.02	.03	.30	.08	–	–	–	0

(Continued)

TABLE F-36 *(Continued)*

Item No.	Food Item Name	Approximate Measure	Weight	Moisture	Food Energy Calories	Protein	Fats	Carbo-hydrates	Fiber	Minerals (Macro) Calcium	Phos-phorus	Sodium	Mag-nesium	Potas-sium
			(g)	(%)	(kcal)	(g)	(g)	(g)	(g)	(mg)	(mg)	(mg)	(mg)	(mg)
1499	**CRANBERRY JUICE COCKTAIL** (33% cranberry juice)	1 CUP	**250**	**83.2**	**162.5**	**.3**	**.3**	**41.3**	—	**12.5**	**7.5**	**2.5**	—	**25.0**
			100	83.2	65.0	.1	.1	16.5	—	5.0	3.0	1.0	—	10.0
	GRAPE JUICE													
1500	canned, or bottled	1 CUP	**250**	**82.9**	**165.0**	**.5**	**0**	**41.5**	—	**27.5**	**30.0**	**5.0**	**30.0**	**290.0**
			100	82.9	66.0	.2	0	16.6	—	11.0	12.0	2.0	12.0	116.0
1501	drink, canned (30%) 30 mg vitamin C/6 fl oz	1 CUP	**250**	**86.0**	**135.0**	**.3**	—	**34.5**	—	**7.5**	**10.0**	**2.5**	—	**87.5**
			100	86.0	54.0	.1	—	13.8	—	3.0	4.0	1.0	—	35.0
1502	drink, canned (30%) 30 mg vitamin C/8 fl oz	1 CUP	**250**	**86.0**	**135.0**	**.3**	—	**34.5**	—	**7.5**	**10.0**	**2.5**	—	**87.5**
			100	86.0	54.0	.1	—	13.8	—	3.0	4.0	1.0	—	35.0
1503	frozen concentrate, sweetened, undiluted	1 CUP	**250**	**52.8**	**457.5**	**1.5**	—	**115.8**	**.3**	**25.0**	**37.5**	**7.5**	**30.0**	**295.0**
			100	52.8	183.0	.6	—	46.3	.1	10.0	15.0	3.0	12.0	118.0
1504	frozen concentrate, sweetened, diluted	1 CUP	**250**	**86.4**	**132.5**	**.5**	**0**	**33.3**	—	**7.5**	**10.0**	**2.5**	**10.0**	**85.0**
			100	86.4	53.0	.2	0	13.3	—	3.0	4.0	1.0	4.0	34.0
	GRAPEFRUIT JUICE													
1505	fresh, pink-red-white, all varieties	1 CUP	**250**	**90.0**	**97.5**	**1.3**	**.3**	**23.0**	—	**22.5**	**37.5**	**2.5**	**30.0**	**405.0**
			100	90.0	39.0	.5	.1	9.2	—	9.0	15.0	1.0	12.0	162.0
1506	fresh, pink-red-white, Calif. and Ariz. (Marsh)	1 CUP	**250**	**89.0**	**105.0**	**1.0**	**.3**	**25.5**	—	**22.5**	**37.5**	**2.5**	**30.0**	**405.0**
			100	89.0	42.0	.4	.1	10.2	—	9.0	15.0	1.0	12.0	162.0
1507	fresh, pink-red-white, Florida	1 CUP	**250**	**90.4**	**92.5**	**1.3**	**.3**	**22.0**	—	**22.5**	**37.5**	**2.5**	**30.0**	**405.0**
			100	90.4	37.0	.5	.1	8.8	—	9.0	15.0	1.0	12.0	162.0
1508	fresh, pink-red, seeded, (Foster Pink)	1 CUP	**250**	**90.0**	**95.0**	**1.3**	**.3**	**22.8**	—	**22.5**	**37.5**	**2.5**	**30.0**	**405.0**
			100	90.0	38.0	.5	.1	9.1	—	9.0	15.0	1.0	12.0	162.0
1509	fresh, pink-red, seedless, (Pink Marsh Redblush)	1 CUP	**250**	**90.0**	**97.5**	**1.0**	**.3**	**23.3**	—	**22.5**	**37.5**	**2.5**	**30.0**	**405.0**
			100	90.0	39.0	.4	.1	9.3	—	9.0	15.0	1.0	12.0	162.0
1510	fresh, pink-red Texas	1 CUP	**250**	**89.2**	**105.0**	**1.3**	**.3**	**25.0**	—	**22.5**	**37.5**	**2.5**	**30.0**	**405.0**
			100	89.2	42.0	.5	.1	10.0	—	9.0	15.0	1.0	12.0	162.0
1511	fresh, white, seeded, (Duncan)	1 CUP	**250**	**89.6**	**100.0**	**1.3**	**.3**	**23.8**	—	**22.5**	**37.5**	**2.5**	**30.0**	**405.0**
			100	89.6	40.0	.5	.1	9.5	—	9.0	15.0	1.0	12.0	162.0
1512	fresh, white, seedless, (Marsh)	1 CUP	**250**	**90.2**	**95.0**	**1.3**	**.3**	**22.5**	—	**22.5**	**37.5**	**2.5**	**30.0**	**405.0**
			100	90.2	38.0	.5	.1	9.0	—	9.0	15.0	1.0	12.0	162.0
1513	fresh, white, Texas	1 CUP	**250**	**89.2**	**105.0**	**1.3**	**.3**	**25.0**	**.3**	**22.5**	**37.5**	**2.5**	**20.0**	**405.0**
			100	89.2	42.0	.5	.1	10.0	.1	9.0	15.0	1.0	8.0	162.0
1514	canned, unsweetened	1 CUP	**247**	**89.2**	**101.3**	**1.2**	**.2**	**24.2**	—	**19.8**	**34.6**	**2.5**	**29.6**	**400.1**
			100	89.2	41.0	.5	.1	9.8	—	8.0	14.0	1.0	12.0	162.0
1515	canned, sweetened	1 CUP	**250**	**86.2**	**132.5**	**1.3**	**.3**	**32.0**	—	**20.0**	**35.0**	**2.5**	**30.0**	**405.0**
			100	86.2	53.0	.5	.1	12.8	—	8.0	14.0	1.0	12.0	162.0
1516	frozen, unsweetened, undiluted	1 CUP	**250**	**62.0**	**362.5**	**4.8**	**1.0**	**86.5**	**.3**	**85.0**	**150.0**	**10.0**	**82.5**	**1510.0**
			100	62.0	145.0	1.9	.4	34.6	.1	34.0	60.0	4.0	33.0	604.0
1517	frozen, unsweetened, diluted w/3 parts water	1 CUP	**250**	**89.3**	**102.5**	**1.3**	**.3**	**24.5**	—	**25.0**	**42.5**	**2.5**	**22.5**	**425.0**
			100	89.3	41.0	.5	.1	9.8	—	10.0	17.0	1.0	9.0	170.0
1518	frozen, unsweetened, reconstituted vitality	1 CUP	**250**	—	**100.0**	—	—	**25.2**	—	**17.5**	**27.5**	**0**	**21.0**	**385.0**
			100	—	40.0	—	—	10.1	—	7.0	11.0	0	8.4	154.0
1519	frozen, sweetened, undiluted	1 CUP	**250**	**57.0**	**412.5**	**4.0**	**.8**	**100.5**	**.3**	**70.0**	**125.0**	**7.5**	—	**1270.0**
			100	57.0	165.0	1.6	.3	40.2	.1	28.0	50.0	3.0	—	508.0
1520	frozen, sweetened, diluted w/3 parts water	1 CUP	**250**	**87.8**	**117.5**	**1.0**	**.3**	**28.5**	—	**20.0**	**35.0**	**2.5**	—	**360.0**
			100	87.8	47.0	.4	.1	11.4	—	8.0	14.0	1.0	—	144.0
1521	dehydrated (crystals), dry	1 OZ	**28**	**1.0**	**105.8**	**1.3**	**.3**	**25.3**	**.1**	**24.4**	**43.4**	**2.8**	—	**440.2**
			100	1.0	378.0	4.8	1.0	90.3	.4	87.0	155.0	10.0	—	1572.0
1522	dehydrated, made w/water, 1 lb = 1 gal	1 CUP	**247**	**89.5**	**98.8**	**1.2**	**.2**	**23.7**	—	**22.2**	**39.5**	**2.5**	—	**412.5**
			100	89.5	40.0	.5	.1	9.6	—	9.0	16.0	1.0	—	167.0
1523	w/orange, canned, unsweetened	1 CUP	**250**	**88.7**	**107.5**	**1.5**	**.5**	**25.3**	**.3**	**25.0**	**37.5**	**2.5**	**20.0**	**460.0**
			100	88.7	43.0	.6	.2	10.1	.1	10.0	15.0	1.0	8.0	184.0
1524	w/orange, canned, sweetened	1 CUP	**250**	**86.9**	**125.0**	**1.3**	**.3**	**30.5**	**.3**	**25.0**	**37.5**	**2.5**	**20.0**	**460.0**
			100	86.9	50.0	.5	.1	12.2	.1	10.0	15.0	1.0	8.0	184.0
1525	w/orange, frozen, unsweetened, undiluted	1 CUP	**250**	**59.1**	**392.5**	**5.3**	**1.3**	**92.8**	**.3**	**72.5**	**117.5**	**5.0**	**80.0**	**1557.5**
			100	59.1	157.0	2.1	.5	37.1	.1	29.0	47.0	2.0	32.0	623.0
1526	w/orange, frozen, unsweetened, diluted w/3 parts water	1 CUP	**250**	**88.4**	**110.0**	**1.5**	**.3**	**26.3**	—	**20.0**	**32.5**	—	**22.5**	**442.5**
			100	88.4	44.0	.6	.1	10.5	—	8.0	13.0	—	9.0	177.0
	LEMON JUICE													
1527	raw	1 TBLS	**15**	**91.0**	**3.8**	**.1**	**0**	**1.2**	—	**1.1**	**1.5**	**.2**	**1.2**	**21.2**
			100	91.0	25.0	.5	.2	8.0	—	7.0	10.0	1.0	8.0	141.0
1528	canned or bottled, unsweetened	1 TBLS	**15**	**91.6**	**3.5**	**.1**	**0**	**1.1**	—	**1.1**	**1.5**	**.2**	**1.5**	**21.2**
			100	91.6	23.0	.4	.1	7.6	—	7.0	10.0	1.0	10.0	141.0
1529	frozen, unsweetened, single-strength	1 TBLS	**15**	**92.0**	**3.3**	**.1**	**0**	**1.1**	—	**1.1**	**1.4**	**.2**	**1.1**	**21.2**
			100	92.0	22.0	.4	.2	7.2	—	7.0	9.0	1.0	7.0	141.0

(Continued)

Item No.	Minerals (Micro)			Fat-Soluble Vitamins			Water-Soluble Vitamins								
	Iron	Zinc	Copper	Vitamin A	Vitamin D	Vitamin E (Alpha Tocopherol)	Vitamin C	Thiamin	Ribo-flavin	Niacin	Panto-thenic Acid	Vit. B-6 (Pyri-doxine)	Folacin (Folic Acid)	Biotin	Vitamin B-12
	(mg)	(mg)	(mg)	(IU)	(IU)	(mg)	(mg)	(mg)	(mg)	(mg)	(mg)	(mg)	(mcg)	(mcg)	(mcg)
1499	.75	—	—	—	—	—	—	.03	.03	—	—	—	—	—	—
	.30	—	—	—	—	—	—	.01	.01	—	—	—	—	—	—
1500	.75	.28	.43	0	—	—	0	.10	.05	.50	.20	.20	5.00	—	0
	.30	.11	.17	0	—	—	0	.04	.02	.20	.08	.08	2.00	—	0
1501	.25	—	—	—	—	—	40.00	.03	.03	.25	—	—	—	—	—
	.10	—	—	—	—	—	16.00	.01	.01	.10	—	—	—	—	—
1502	.25	—	—	—	—	—	40.00	.03	.03	.25	—	—	—	—	—
	.10	—	—	—	—	—	16.00	.01	.01	.10	—	—	—	—	—
1503	1.00	—	—	50.0	—	—	37.50	.15	.25	1.75	.35	.18	—	—	0
	.40	—	—	20.0	—	—	15.00	.06	.10	.70	.14	.07	—	—	0
1504	.25	.28	.43	—	—	—	10.00	.05	.08	.50	.10	.05	5.00	.75	0
	.10	.11	.17	—	—	—	4.00	.02	.03	.20	.04	.02	2.00	.30	0
1505	.50	.25	—	200.0	—	—	95.00	.10	.05	.50	.71	.09	52.50	1.75	0
	.20	.10	—	80.0	—	—	38.00	.04	.02	.20	.28	.04	21.00	.70	0
1506	.50	—	—	25.0	—	—	100.00	.10	.05	.50	.71	.09	52.50	1.75	0
	.20	—	—	10.0	—	—	40.00	.04	.02	.20	.28	.04	21.00	.70	0
1507	.50	.25	—	200.0	—	—	92.50	.10	.05	.50	.71	.09	52.50	1.75	0
	.20	.10	—	80.0	—	—	37.00	.04	.02	.20	.28	.04	21.00	.70	0
1508	.50	.25	—	1100.0	—	—	97.50	.10	.05	.50	.40	.04	52.50	1.75	—
	.20	.10	—	440.0	—	—	39.00	.04	.02	.20	.16	.02	21.00	.70	—
1509	.50	—	—	1100.0	—	—	90.00	.10	.05	.50	.40	.04	52.50	1.75	—
	.20	—	—	440.0	—	—	36.00	.04	.02	.20	.16	.02	21.00	.70	—
1510	.50	.25	—	1100.0	—	—	95.00	.10	.05	.50	.40	.04	52.50	1.75	—
	.20	.10	—	440.0	—	—	38.00	.04	.02	.20	.16	.02	21.00	.70	—
1511	.50	.25	—	25.0	—	—	95.00	.10	.05	.50	.40	.04	52.50	1.75	—
	.20	.10	—	10.0	—	—	38.00	.04	.02	.20	.16	.02	21.00	.70	—
1512	.50	.25	—	25.0	—	—	92.50	.10	.05	.50	.40	.04	52.50	1.75	—
	.20	.10	—	10.0	—	—	37.00	.04	.02	.20	.16	.02	21.00	.70	—
1513	.50	.25	—	25.0	—	—	95.00	.10	.05	.50	.40	.04	52.50	1.75	—
	.20	.10	—	10.0	—	—	38.00	.04	.02	.20	.16	.02	21.00	.70	—
1514	.99	.25	.02	24.7	—	.10	83.98	.07	.04	.50	.32	.03	2.97	1.98	0
	.40	.10	.01	10.0	—	.04	34.00	.03	.02	.20	.13	.01	1.20	.80	0
1515	1.00	.25	.02	25.0	—	.10	77.50	.08	.05	.50	.33	.03	3.00	2.00	0
	.40	.10	.01	10.0	—	.04	31.00	.03	.02	.20	.13	.01	1.20	.80	0
1516	1.00	—	—	75.0	—	—	345.00	.35	.15	1.75	1.51	.13	—	—	0
	.40	—	—	30.0	—	—	138.00	.14	.06	.70	.61	.05	—	—	0
1517	.25	.25	—	25.0	—	—	97.50	.10	.05	.50	.41	.04	52.50	1.75	0
	.10	.10	—	10.0	—	—	39.00	.04	.02	.20	.16	.02	21.00	.70	0
1518	.25	—	—	0	—	—	87.50	.10	0	.60	.38	.03	2.50	—	—
	.10	—	—	0	—	—	35.00	.04	0	.24	.15	.01	1.00	—	—
1519	.75	—	—	50.0	—	—	290.00	.30	.13	1.50	1.51	.13	—	—	0
	.30	—	—	20.0	—	—	116.00	.12	.05	.60	.61	.05	—	—	0
1520	.25	.25	—	25.0	—	—	82.50	.08	.03	.50	.41	.04	52.50	1.75	0
	.10	.10	—	10.0	—	—	33.00	.03	.01	.20	.16	.02	21.00	.70	0
1521	.28	—	—	22.4	—	—	98.00	.10	.05	.48	—	—	—	—	—
	1.00	—	—	80.0	—	—	350.00	.36	.16	1.70	—	—	—	—	—
1522	.25	—	—	24.7	—	—	91.39	.10	.05	.50	—	—	—	—	—
	.10	—	—	10.0	—	—	37.00	.04	.02	.20	—	—	—	—	—
1523	.75	—	—	250.0	—	—	85.00	.13	.05	.50	.35	.05	—	—	0
	.30	—	—	100.0	—	—	34.00	.05	.02	.20	.14	.02	—	—	0
1524	.75	—	—	250.0	—	—	85.00	.13	.05	.50	.35	.05	—	—	0
	.30	—	—	100.0	—	—	34.00	.05	.02	.20	.14	.02	—	—	0
1525	1.00	—	—	950.0	—	—	360.00	.58	.08	2.75	—	—	—	—	—
	.40	—	—	380.0	—	—	144.00	.23	.03	1.10	—	—	—	—	—
1526	.25	—	—	275.0	—	—	102.50	.15	.03	.75	—	—	—	—	—
	.10	—	—	110.0	—	—	41.00	.06	.01	.30	—	—	—	—	—
1527	.03	0	.01	3.0	—	—	6.90	.01	Trace	.02	.02	.01	.15	—	0
	.20	0	.03	20.0	—	—	46.00	.03	.01	.10	.10	.05	1.00	—	0
1528	.03	—	—	3.0	—	—	6.30	.01	Trace	.02	—	.01	1.95	—	0
	.20	—	—	20.0	—	—	42.00	.03	.01	.10	—	.05	13.00	—	0
1529	.05	—	—	3.0	—	—	6.60	.01	Trace	.02	.01	.01	.15	—	0
	.30	—	—	20.0	—	—	44.00	.03	.01	.10	.07	.04	1.00	—	0

(Continued)

TABLE F-36 *(Continued)*

Item No.	Food Item Name	Approximate Measure	Weight	Moisture	Food Energy Calories	Protein	Fats	Carbo-hydrates	Fiber	Minerals (Macro) Calcium	Phos-phorus	Sodium	Mag-nesium	Potas-sium
			(g)	(%)	(kcal)	(g)	(g)	(g)	(g)	(mg)	(mg)	(mg)	(mg)	(mg)
	LEMON JUICE (Continued)													
1530	frozen, unsweetened, concentrate	1 TBLS	**15**	**58.0**	**17.4**	**.3**	**.1**	**5.6**	—	**5.0**	**7.1**	**.8**	—	**98.7**
			100	58.0	116.0	2.3	.9	37.4	—	33.0	47.0	5.0	—	658.0
	LEMONADE													
1531	concentrate, frozen, undiluted	1 CUP	**250**	**48.5**	**487.5**	**.5**	**.3**	**127.8**	**.3**	**10.0**	**15.0**	**5.0**	**12.5**	**175.0**
			100	48.5	195.0	.2	.1	51.1	.1	4.0	6.0	2.0	5.0	70.0
1532	concentrate, frozen, diluted w/4⅓ parts water	1 CUP	**250**	**88.5**	**110.0**	**.3**	**0**	**28.5**	**0**	**2.5**	**2.5**	—	**2.5**	**40.0**
			100	88.5	44.0	.1	0	11.4	0	1.0	1.0	—	1.0	16.0
1533	concentrate, frozen, reconstituted vitality	1 CUP	**250**	—	**107.5**	—	—	**26.9**	—	**2.5**	**5.0**	**0**	**3.1**	**45.0**
			100	—	43.0	—	—	10.8	—	1.0	2.0	0	1.3	18.0
	LIME JUICE													
1534	fresh	1 TBLS	**15**	**90.3**	**3.9**	**0**	**0**	**1.4**	—	**1.4**	**1.7**	**.2**	**.8**	**15.6**
			100	90.3	26.0	.3	.1	9.0	—	9.0	11.0	1.0	5.4	104.0
1535	canned or bottled, unsweetened	1 TBLS	**15**	**90.3**	**3.9**	**0**	**0**	**1.4**	—	**1.4**	**1.7**	**.2**	—	**15.6**
			100	90.3	26.0	.3	.1	9.0	—	9.0	11.0	1.0	—	104.0
	LIMEADE													
1536	concentrate, frozen, undiluted	1 CUP	**250**	**50.0**	**467.5**	**.5**	**.3**	**123.8**	—	**12.5**	**15.0**	—	—	**147.5**
			100	50.0	187.0	.2	.1	49.5	—	5.0	6.0	—	—	59.0
1537	concentrate, frozen, diluted w/4⅓ parts water	1 CUP	**250**	**88.9**	**102.5**	**0**	**0**	**27.5**	—	**2.5**	**2.5**	—	—	**32.5**
			100	88.9	41.0	0	0	11.0	—	1.0	1.0	—	—	13.0
	ORANGE JUICE													
1538	fresh, all varieties	1 CUP	**250**	**88.3**	**112.5**	**1.8**	**.5**	**26.0**	**.3**	**27.5**	**42.5**	**2.5**	**27.0**	**500.0**
			100	88.3	45.0	.7	.2	10.4	.1	11.0	17.0	1.0	10.8	200.0
1539	fresh, California Navel (winter)	1 CUP	**250**	**87.2**	**120.0**	**2.5**	**.3**	**28.3**	**.3**	**27.5**	**45.0**	**2.5**	**27.5**	**485.0**
			100	87.2	48.0	1.0	.1	11.3	.1	11.0	18.0	1.0	11.0	194.0
1540	fresh, California Valencia (summer)	1 CUP	**250**	**87.8**	**117.5**	**2.5**	**.8**	**26.3**	**.3**	**27.5**	**47.5**	**2.5**	**27.5**	**475.0**
			100	87.8	47.0	1.0	.3	10.5	.1	11.0	19.0	1.0	11.0	190.0
1541	fresh, Florida, all varieties	1 CUP	**250**	**88.8**	**107.5**	**1.5**	**.5**	**25.0**	**.3**	**25.0**	**40.0**	**2.5**	**27.5**	**515.0**
			100	88.8	43.0	.6	.2	10.0	.1	10.0	16.0	1.0	11.0	206.0
1542	fresh, Florida, early & midseason	1 CUP	**250**	**89.6**	**100.0**	**1.3**	**.5**	**23.3**	**.3**	**25.0**	**37.5**	**2.5**	**27.5**	**520.0**
			100	89.6	40.0	.5	.2	9.3	.1	10.0	15.0	1.0	11.0	208.0
1543	fresh, Florida, late season (Valencia)	1 CUP	**250**	**88.3**	**112.5**	**1.5**	**.5**	**26.3**	**.3**	**25.0**	**45.0**	**2.5**	**27.5**	**507.5**
			100	88.3	45.0	.6	.2	10.5	.1	10.0	18.0	1.0	11.0	203.0
1544	fresh, Florida Temple	1 CUP	**250**	**88.0**	**135.0**	**1.3**	**.5**	**32.3**	**.3**	**25.0**	**42.5**	**2.5**	**27.5**	—
			100	88.0	54.0	.5	.2	12.9	.1	10.0	17.0	1.0	11.0	—
1545	canned, unsweetened	1 CUP	**250**	**87.4**	**120.0**	**2.0**	**.5**	**28.0**	**.3**	**25.0**	**45.0**	**2.5**	**30.0**	**497.5**
			100	87.4	48.0	.8	.2	11.2	.1	10.0	18.0	1.0	12.0	199.0
1546	canned, sweetened	1 CUP	**245**	**86.5**	**127.4**	**1.7**	**.5**	**29.9**	**.3**	**24.5**	**44.1**	**2.5**	**29.4**	**487.6**
			100	86.5	52.0	.7	.2	12.2	.1	10.0	18.0	1.0	12.0	199.0
1547	canned concentrate, unsweetened, undiluted	1 CUP	**250**	**42.0**	**557.5**	**10.3**	**3.3**	**126.8**	**1.3**	**127.5**	**215.0**	**12.5**	**142.5**	**2355.0**
			100	42.0	223.0	4.1	1.3	50.7	.5	51.0	86.0	5.0	57.0	942.0
1548	canned concentrate, unsweetened, diluted w/5 parts water	1 CUP	**250**	**88.2**	**115.0**	**2.0**	**.8**	**25.8**	**.3**	**25.0**	**45.0**	**2.5**	**30.0**	**480.0**
			100	88.2	46.0	.8	.3	10.3	.1	10.0	18.0	1.0	12.0	192.0
1549	frozen concentrate, unsweetened, diluted w/3 parts water	1 CUP	**250**	**88.1**	**112.5**	**1.8**	**.3**	**26.8**	—	**22.5**	**40.0**	**2.5**	**20.5**	**465.0**
			100	88.1	45.0	.7	.1	10.7	—	9.0	16.0	1.0	8.2	186.0
1550	frozen concentrate, unsweetened, undiluted	1 CUP	**250**	**58.2**	**395.0**	**5.8**	**.5**	**95.0**	**.5**	**82.5**	**137.5**	**5.0**	**92.5**	**1642.5**
			100	58.2	158.0	2.3	.2	38.0	.2	33.0	55.0	2.0	37.0	657.0
1551	dehydrated (crystals), dry	1 OZ	**28**	**1.0**	**106.4**	**1.4**	**.5**	**24.9**	**.2**	**23.5**	**37.5**	**2.2**	—	**483.8**
			100	1.0	380.0	5.0	1.7	88.9	.8	84.0	134.0	8.0	—	1728.0
1552	dehydrated (crystals), made w/water, 1 lb = 1 gal	1 CUP	**240**	**88.0**	**110.4**	**1.4**	**.5**	**25.9**	**.3**	**24.0**	**38.4**	**2.4**	—	**501.6**
			100	88.0	46.0	.6	.2	10.8	.1	10.0	16.0	1.0	—	209.0
1553	w/apricot, canned, 40% juice	1 CUP	**249**	**86.7**	**124.5**	**.7**	**.2**	**31.6**	**.5**	**12.5**	**19.9**	—	—	**234.1**
			100	86.7	50.0	.3	.1	12.7	.2	5.0	8.0	—	—	94.0
1554	w/grapefruit juice, frozen, undiluted	1 CUP	**250**	**59.0**	**392.5**	**4.8**	**1.3**	**92.9**	—	**72.5**	—	—	—	—
			100	59.0	157.0	1.9	.5	37.1	—	29.0	—	—	—	—
1555	w/grapefruit juice, frozen, diluted w/3 parts water	1 CUP	**248**	**88.0**	**109.1**	**1.0**	—	**26.0**	—	**19.8**	—	—	—	—
			100	88.0	44.0	.4	—	10.5	—	8.0	—	—	—	—
1556	**PAPAYA JUICE,** canned	1 CUP	**250**	—	**120.0**	**1.0**	**0**	**30.2**	—	**45.0**	**25.0**	—	—	—
			100	—	48.0	.4	0	12.1	—	18.0	10.0	—	—	—
1557	**PEACH NECTAR,** canned, 40% fruit	1 CUP	**250**	**87.2**	**120.0**	**.5**	**0**	**31.0**	**.3**	**10.0**	**27.5**	**2.5**	—	**195.0**
			100	87.2	48.0	.2	0	12.4	.1	4.0	11.0	1.0	—	78.0
1558	**PEAR NECTAR,** canned 40% fruit	1 CUP	**250**	**86.2**	**130.0**	**.8**	**.5**	**33.0**	**.8**	**7.5**	**12.5**	**2.5**	—	**97.5**
			100	86.2	52.0	.3	.2	13.2	.3	3.0	5.0	1.0	—	39.0

Item No.	Minerals (Micro) Iron	Zinc	Copper	Fat-Soluble Vitamins Vitamin A	Vitamin D	Vitamin E (Alpha Tocopherol)	Water-Soluble Vitamins Vitamin C	Thiamin	Ribo-flavin	Niacin	Panto-thenic Acid	Vit. B-6 (Pyri-doxine)	Folacin (Folic Acid)	Biotin	Vitamin B-12
	(mg)	(mg)	(mg)	(IU)	(IU)	(mg)	(mg)	(mg)	(mg)	(mg)	(mg)	(mg)	(mcg)	(mcg)	(mcg)
1530	**.14**	–	–	**12.0**	–	–	**34.50**	**.02**	**.01**	**.05**	–	–	–	–	–
	.90	–	–	80.0	–	–	230.00	.14	.06	.30	–	–	–	–	–
1531	**.50**	–	–	**50.0**	**0**	–	**75.00**	**.05**	**.08**	**.75**	**.12**	**.05**	–	–	**0**
	.20	–	–	20.0	0	–	30.00	.02	.03	.30	.05	.02	–	–	0
1532	–	–	–	**0**	**0**	–	**17.50**	**.03**	**.03**	**.25**	**.03**	**.01**	**12.50**	–	**0**
	–	–	–	0	0	–	7.00	.01	.01	.10	.01	.01	5.00	–	0
1533	**0**	–	–	–	–	–	**18.25**	**0**	**0**	**.05**	**.03**	**.02**	**12.50**	–	–
	0	–	–	–	–	–	7.30	0	0	.02	.01	.01	5.00	–	–
1534	**.03**	**0**	**.01**	**1.5**	–	–	**4.80**	**Trace**	**Trace**	**.02**	**.02**	**.01**	–	–	**0**
	.20	0	.03	10.0	–	–	32.00	.02	.01	.10	.14	.04	–	–	0
1535	**.03**	–	–	**1.5**	–	–	**3.15**	**Trace**	**Trace**	**.02**	–	–	**1.95**	–	–
	.20	–	–	10.0	–	–	21.00	.02	.01	.10	–	–	13.00	–	–
1536	**.25**	–	–	–	–	–	**30.00**	**.03**	**.03**	**.25**	–	–	–	–	–
	.10	–	–	–	–	–	12.00	.01	.01	.10	–	–	–	–	–
1537	–	–	–	–	–	–	**5.00**	–	–	–	–	–	–	–	–
	–	–	–	–	–	–	2.00	–	–	–	–	–	–	–	–
1538	**.50**	**.05**	**.20**	**500.0**	–	**.10**	**125.00**	**.23**	**.08**	**1.00**	**.48**	**.10**	**137.50**	**.75**	**0**
	.20	.02	.08	200.0	–	.04	50.00	.09	.03	.40	.19	.04	55.00	.30	0
1539	**.50**	**.05**	–	**500.0**	–	**.10**	**152.50**	**.23**	**.08**	**1.00**	**.48**	**.10**	**137.50**	**.75**	**0**
	.20	.02	–	200.0	–	.04	61.00	.09	.03	.40	.19	.04	55.00	.30	0
1540	**.75**	**.05**	–	**500.0**	–	**.10**	**122.50**	**.23**	**.08**	**1.00**	**.48**	**.10**	**137.50**	**.75**	**0**
	.30	.02	–	200.0	–	.04	49.00	.09	.03	.40	.19	.04	55.00	.30	0
1541	**.50**	**.05**	–	**500.0**	–	**.10**	**112.50**	**.23**	**.08**	**1.00**	**.48**	**.10**	**137.50**	**.75**	**0**
	.20	.02	–	200.0	–	.04	45.00	.09	.03	.40	.19	.04	55.00	.30	0
1542	**.50**	**.05**	–	**500.0**	–	**.10**	**127.50**	**.23**	**.08**	**1.00**	**.48**	**.10**	**137.50**	**.75**	**0**
	.20	.02	–	200.0	–	.04	51.00	.09	.03	.40	.19	.04	55.00	.30	0
1543	**.50**	**.05**	–	**500.0**	–	**.10**	**92.50**	**.23**	**.08**	**1.00**	**.48**	**.10**	**137.50**	**.75**	**0**
	.20	.02	–	200.0	–	.04	37.00	.09	.03	.40	.19	.04	55.00	.30	0
1544	**.50**	**.05**	–	**500.0**	–	**.10**	**125.00**	**.23**	**.08**	**1.00**	**.48**	**.10**	**137.50**	**.75**	**0**
	.20	.02	–	200.0	–	.04	50.00	.09	.03	.40	.19	.04	55.00	.30	0
1545	**1.00**	**.18**	–	**500.0**	–	–	**100.00**	**.18**	**.05**	**.75**	**.38**	**.09**	**62.50**	**2.00**	**0**
	.40	.07	–	200.0	–	–	40.00	.07	.02	.30	.15	.04	25.00	.80	0
1546	**.98**	**.27**	–	**490.0**	–	–	**98.00**	**.17**	**.05**	**.74**	**.37**	**.09**	**61.25**	**1.96**	**0**
	.40	.11	–	200.0	–	–	40.00	.07	.02	.30	.15	.04	25.00	.80	0
1547	**3.25**	–	–	**2400.0**	–	–	**572.50**	**.98**	**.30**	**4.25**	–	–	–	–	–
	1.30	–	–	960.0	–	–	229.00	.39	.12	1.70	–	–	–	–	–
1548	**.75**	**.28**	–	**500.0**	–	–	**117.50**	**.20**	**.05**	**.75**	**.33**	**.06**	**87.50**	**2.00**	–
	.30	.11	–	200.0	–	–	47.00	.08	.02	.30	.13	.03	35.00	.80	–
1549	**.25**	**.28**	**.02**	**500.0**	–	–	**112.50**	**.23**	**.03**	**.75**	**.41**	**.07**	**137.50**	**.75**	**0**
	.10	.11	.01	200.0	–	–	45.00	.09	.01	.30	.17	.03	55.00	.30	0
1550	**1.00**	–	**.22**	**1775.0**	–	–	**395.00**	**.75**	**.13**	**3.00**	**1.44**	**.25**	–	–	**0**
	.40	–	.09	710.0	–	–	158.00	.30	.05	1.20	.58	.10	–	–	0
1551	**.48**	–	–	**470.4**	–	–	**100.52**	**.19**	**.06**	**.81**	–	–	–	–	–
	1.70	–	–	1680.0	–	–	359.00	.67	.21	2.90	–	–	–	–	–
1552	**.48**	–	–	**480.0**	–	–	**105.60**	**.19**	**.07**	**.96**	–	–	–	–	–
	.20	–	–	200.0	–	–	44.00	.08	.03	.40	–	–	–	–	–
1553	**.25**	–	–	**1444.2**	–	–	**39.84**	**.05**	**.03**	**.50**	–	–	–	–	–
	.10	–	–	580.0	–	–	16.00	.02	.01	.20	–	–	–	–	–
1554	**1.00**	–	–	**952.5**	–	–	**360.00**	**.55**	**.05**	**2.73**	–	–	–	–	–
	.40	–	–	381.0	–	–	144.00	.22	.02	1.09	–	–	–	–	–
1555	**.25**	–	–	**270.3**	–	–	**101.93**	**.15**	–	**.80**	–	–	**.25**	–	–
	.10	–	–	109.0	–	–	41.10	.06	–	.32	–	–	.10	–	–
1556	**.75**	**.13**	**.20**	**5000.0**	–	–	**102.00**	**.05**	**.03**	**.20**	–	–	–	–	–
	.30	.05	.08	2000.0	–	–	40.80	.02	.01	.08	–	–	–	–	–
1557	**.50**	–	–	**1075.0**	–	–	–	**.03**	**.05**	**1.00**	–	–	–	–	–
	.20	–	–	430.0	–	–	–	.01	.02	.40	–	–	–	–	–
1558	**.25**	–	–	–	–	–	–	–	**.05**	–	–	–	–	–	–
	.10	–	–	–	–	–	–	–	.02	–	–	–	–	–	–

(Continued)

TABLE F-36 *(Continued)*

Item No.	Food Item Name	Approximate Measure	Weight	Moisture	Food Energy Calories	Protein	Fats	Carbo-hydrates	Fiber	Minerals (Macro) Calcium	Phos-phorus	Sodium	Mag-nesium	Potas-sium
			(g)	(%)	(kcal)	(g)	(g)	(g)	(g)	(mg)	(mg)	(mg)	(mg)	(mg)
	PINEAPPLE JUICE													
1559	canned, unsweetened	1 CUP	**250**	**85.6**	**137.5**	**1.0**	**.3**	**33.8**	**.3**	**37.5**	**22.5**	**2.5**	**30.0**	**372.5**
			100	85.6	55.0	.4	.1	13.5	.1	15.0	9.0	1.0	12.0	149.0
1560	frozen concentrate, unsweetened, undiluted	1 OZ	**28**	**53.1**	**50.1**	**.4**	**0**	**12.4**	**.1**	**10.9**	**7.8**	**.8**	**9.8**	**132.2**
			100	53.1	179.0	1.3	.1	44.3	.3	39.0	28.0	3.0	35.0	472.0
1561	frozen concentrate, unsweetened, diluted w/3 parts water	1 CUP	**250**	**86.5**	**130.0**	**1.0**	—	**32.0**	**.3**	**27.5**	**20.0**	**2.5**	**22.5**	**340.0**
			100	86.5	52.0	.4	—	12.8	.1	11.0	8.0	1.0	9.0	136.0
1562	w/grapefruit, canned, 40% fruit	1 CUP	**250**	**86.0**	**135.0**	**.5**	**0**	**34.0**	—	**12.5**	**12.5**	—	**25.0**	**155.0**
			100	86.0	54.0	.2	0	13.6	—	5.0	5.0	—	10.0	62.0
1563	w/orange, canned, 40% fruit	1 CUP	**250**	**86.0**	**135.0**	**.5**	**.3**	**33.8**	—	**12.5**	**15.0**	—	**25.0**	**175.0**
			100	86.0	54.0	.2	.1	13.5	—	5.0	6.0	—	10.0	70.0
1564	**PRUNE JUICE,** canned or bottled	1 CUP	**250**	**80.0**	**192.5**	**1.0**	**.3**	**47.5**	—	**35.0**	**50.0**	**5.0**	**14.5**	**587.5**
			100	80.0	77.0	.4	.1	19.0	—	14.0	20.0	2.0	5.8	235.0
1565	**SAUERKRAUT JUICE,** canned	1 CUP	**240**	**94.6**	**24.0**	**1.7**	—	**5.5**	—	**88.8**	**33.6**	**1888.8**	—	—
			100	94.6	10.0	.7	—	2.3	—	37.0	14.0	787.0	—	—
1566	**TANGELO JUICE,** raw	1 CUP	**250**	**89.4**	**102.5**	**1.3**	**.3**	**24.3**	—	—	—	—	—	—
			100	89.4	41.0	.5	.1	9.7	—	—	—	—	—	—
	TANGERINE JUICE													
1567	raw	1 CUP	**250**	**88.9**	**107.5**	**1.3**	**.5**	**25.3**	**.3**	**45.0**	**35.0**	**2.5**	—	**445.0**
			100	88.9	43.0	.5	.2	10.1	.1	18.0	14.0	1.0	—	178.0
1568	canned, unsweetened	1 CUP	**250**	**88.8**	**107.5**	**1.3**	**.5**	**25.5**	**.3**	**45.0**	**35.0**	**2.5**	—	**445.0**
			100	88.8	43.0	.5	.2	10.2	.1	18.0	14.0	1.0	—	178.0
1569	canned, sweetened	1 CUP	**250**	**87.0**	**125.0**	**1.3**	**.5**	**30.0**	**.3**	**45.0**	**35.0**	**2.5**	—	**445.0**
			100	87.0	50.0	.5	.2	12.0	.1	18.0	14.0	1.0	—	178.0
1570	frozen concentrate, unsweetened, undiluted	1 OZ	**28**	**58.0**	**45.4**	**.5**	**.2**	**10.7**	**.1**	**17.4**	**13.4**	**.6**	—	**171.6**
			100	58.0	162.0	1.7	.7	38.3	.3	62.0	48.0	2.0	—	613.0
1571	frozen concentrate, unsweetened, diluted w/3 parts water	1 CUP	**250**	**88.1**	**115.0**	**1.3**	**.5**	**27.0**	**.3**	**45.0**	**35.0**	**2.5**	—	**435.0**
			100	88.1	46.0	.5	.2	10.8	.1	18.0	14.0	1.0	—	174.0
	TOMATO JUICE													
1572	canned or bottled	1 CUP	**200**	**93.5**	**38.0**	**1.5**	**.1**	**8.7**	**.4**	**18.4**	**34.6**	**400.0**	**20.2**	**454.0**
			100	93.5	19.0	.8	.1	4.4	.2	9.2	17.3	200.0	10.1	227.0
1573	canned or bottled, low sodium	1 CUP	**200**	**94.2**	**38.0**	**1.6**	**.2**	**8.6**	**.4**	**14.0**	**36.0**	**6.0**	**8.2**	**454.0**
			100	94.2	19.0	.8	.1	4.3	.2	7.0	18.0	3.0	4.1	227.0
1574	canned concentrate, undiluted	1 CUP	**200**	**75.0**	**152.0**	**6.8**	**.8**	**34.2**	**1.8**	**54.0**	**140.0**	**1580.0**	—	**1776.0**
			100	75.0	76.0	3.4	.4	17.1	.9	27.0	70.0	790.0	—	888.0
1575	canned concentrate, diluted w/3 parts water	1 CUP	**200**	**93.4**	**40.0**	**1.8**	**.2**	**9.0**	**.4**	**14.0**	**38.0**	**418.0**	**20.0**	**470.0**
			100	93.4	20.0	.9	.1	4.5	.2	7.0	19.0	209.0	10.0	235.0
1576	dehydrated crystals, dry	1 CUP	**200**	**1.0**	**42.0**	**1.4**	**.2**	**10.0**	**.4**	**20.0**	**36.0**	**400.0**	—	**442.0**
			100	1.0	21.0	.7	.1	5.0	.2	10.0	18.0	200.0	—	221.0
1577	dehydrated crystals, made w/water, 1 lb = 1¾ gal	1 CUP	**243**	**93.5**	**48.6**	**1.9**	**.2**	**10.9**	**.5**	**14.6**	**43.7**	**626.9**	—	**561.3**
			100	93.5	20.0	.8	.1	4.5	.2	6.0	18.0	258.0	—	231.0
1578	**VEGETABLE JUICE COCKTAIL,** canned	1 CUP	**200**	**94.1**	**34.0**	**1.8**	**.2**	**7.2**	**.6**	**24.0**	**44.0**	**400.0**	—	**442.0**
			100	94.1	17.0	.9	.1	3.6	.3	12.0	22.0	200.0	—	221.0
	ADZUKI BEAN													
1579	[4]boiled, sweetened	1 OZ	**28**	**45.3**	**61.0**	**.8**	**0**	**14.2**	**.2**	**7.3**	**18.5**	—	—	—
			100	45.3	218.0	3.0	.1	50.7	.6	26.0	66.0	—	—	—
1580	[4]dried	1 OZ	**28**	**15.0**	**90.7**	**5.9**	**.3**	**16.7**	**1.1**	**23.0**	**87.6**	**3.9**	—	**420.0**
			100	15.0	324.0	21.1	1.0	59.5	3.9	82.0	313.0	14.0	—	1500.0
	BEANS, COMMON, DRY													
1581	baked, w/o pork	1 CUP	**300**	**68.5**	**360.0**	**18.9**	**1.5**	**69.0**	**4.2**	**204.0**	**363.0**	**1014.0**	**111.0**	**804.0**
			100	68.5	120.0	6.3	.5	23.0	1.4	68.0	121.0	338.0	37.0	268.0
1582	black, brown, & sayo, cooked	1 CUP	**185**	**11.2**	**224.4**	**14.4**	**.9**	**39.6**	—	**87.0**	**272.0**	—	**103.6**	—
			100	11.2	121.3	7.8	.5	21.4	—	47.0	147.0	—	56.0	—
1583	w/frankfurters, canned	1 CUP	**300**	**71.0**	**429.0**	**22.4**	**21.3**	**37.6**	—	**111.0**	—	—	—	—
			100	71.0	143.0	7.5	7.1	12.5	—	37.0	—	—	—	—
1584	w/pork & sweet sauce, canned	1 CUP	**255**	**66.4**	**382.5**	**15.8**	**12.0**	**53.8**	**4.3**	**160.7**	**290.7**	**969.0**	**118.3**	—
			100	66.4	150.0	6.2	4.7	21.1	1.7	63.0	114.0	380.0	46.4	—
1585	w/pork & tomato sauce, canned	1 CUP	**255**	**70.7**	**311.1**	**15.6**	**6.6**	**48.5**	**3.5**	**137.7**	**234.6**	**1180.7**	**127.5**	**535.5**
			100	70.7	122.0	6.1	2.6	19.0	1.4	54.0	92.0	463.0	50.0	210.0
1586	w/tomato sauce, canned	1 CUP	**250**	**68.5**	**300.0**	**15.8**	**1.3**	**57.5**	**3.5**	**170.0**	**302.5**	**845.0**	**182.5**	**670.0**
			100	68.5	120.0	6.3	.5	23.0	1.4	68.0	121.0	338.0	73.0	268.0

Item No.	Minerals (Micro) Iron	Zinc	Copper	Fat-Soluble Vitamins Vitamin A	Vitamin D	Vitamin E (Alpha Tocopherol)	Water-Soluble Vitamins Vitamin C	Thiamin	Ribo-flavin	Niacin	Panto-thenic Acid	Vit. B-6 (Pyri-doxine)	Folacin (Folic Acid)	Biotin	Vitamin B-12
	(mg)	(mg)	(mg)	(IU)	(IU)	(mg)	(mg)	(mg)	(mg)	(mg)	(mg)	(mg)	(mcg)	(mcg)	(mcg)
1559	**.75**	**.18**	**.01**	**125.0**	–	–	**22.50**	**.13**	**.05**	**.50**	**.25**	**.24**	**57.50**	–	**0**
	.30	.07	.01	50.0	–	–	9.00	.05	.02	.20	.10	.10	23.00	–	0
1560	**.25**	–	–	**14.0**	–	–	**11.76**	**.07**	**.02**	**.25**	**.12**	**.07**	–	–	**0**
	.90	–	–	50.0	–	–	42.00	.23	.06	.90	.43	.26	–	–	0
1561	**.75**	–	–	**25.0**	–	–	**30.00**	**.18**	**.05**	**.50**	**.31**	**.19**	**2.50**	–	**0**
	.30	–	–	10.0	–	–	12.00	.07	.02	.20	.13	.08	1.00	–	0
1562	**.50**	–	–	**25.0**	–	–	**40.00**	**.05**	**.03**	**.25**	–	**.08**	–	–	**0**
	.20	–	–	10.0	–	–	16.00	.02	.01	.10	–	.03	–	–	0
1563	**.50**	–	–	**125.0**	–	–	**40.00**	**.05**	**.03**	**.25**	–	–	–	–	–
	.20	–	–	50.0	–	–	16.00	.02	.01	.10	–	–	–	–	–
1564	**10.25**	**.30**	**.05**	**275.0**	–	–	**5.00**	**.03**	**.03**	**1.00**	–	**.17**	**1.25**	–	**0**
	4.10	.12	.02	110.0	–	–	2.00	.01	.01	.40	–	.06	.50	–	0
1565	**2.64**	–	–	–	–	–	**43.20**	**.07**	**.10**	**.48**	**.29**	**.60**	**7.20**	–	**0**
	1.10	–	–	–	–	–	18.00	.03	.04	.20	.12	.25	3.00	–	0
1566	–	–	–	–	–	–	**67.50**	–	–	–	–	–	–	–	–
	–	–	–	–	–	–	27.00	–	–	–	–	–	–	–	–
1567	**.50**	–	–	**1050.0**	–	–	**77.50**	**.15**	**.05**	**.25**	–	–	–	–	–
	.20	–	–	420.0	–	–	31.00	.06	.02	.10	–	–	–	–	–
1568	**.50**	–	–	**1050.0**	–	–	**55.00**	**.15**	**.05**	**.25**	–	**.08**	**3.75**	–	**0**
	.20	–	–	420.0	–	–	22.00	.06	.02	.10	–	.03	1.50	–	0
1569	**.50**	–	–	**1050.0**	–	–	**55.00**	**.15**	**.05**	**.25**	–	**.08**	–	–	**0**
	.20	–	–	420.0	–	–	22.00	.06	.02	.10	–	.03	–	–	0
1570	**.20**	–	–	**408.8**	–	–	**26.88**	**.06**	**.02**	**.11**	–	–	–	–	–
	.70	–	–	1460.0	–	–	96.00	.20	.06	.40	–	–	–	–	–
1571	**.50**	–	–	**1025.0**	–	–	**67.50**	**.15**	**.05**	**.25**	–	–	–	–	–
	.20	–	–	410.0	–	–	27.00	.06	.02	.10	–	–	–	–	–
1572	**1.10**	**.13**	**.21**	**1380.0**	–	**.44**	**26.80**	**.14**	**.07**	**1.60**	**.50**	**.27**	**52.00**	–	**0**
	.55	.07	.11	690.0	–	.22	13.40	.07	.04	.80	.25	.13	26.00	–	0
1573	**1.80**	**.14**	**.01**	**1600.0**	–	**.44**	**32.00**	**.14**	**.07**	**1.40**	**.50**	**.27**	**52.00**	–	**0**
	.90	.07	.01	800.0	–	.22	16.00	.07	.04	.70	.25	.13	26.00	–	0
1574	**7.00**	–	–	**6600.0**	–	–	**98.00**	**.40**	**.24**	**6.20**	–	–	–	–	–
	3.50	–	–	3300.0	–	–	49.00	.20	.12	3.10	–	–	–	–	–
1575	**1.80**	–	–	**1800.0**	–	–	**26.00**	**.10**	**.06**	**1.60**	**.50**	**.39**	**19.80**	–	**0**
	.90	–	–	900.0	–	–	13.00	.05	.03	.80	.25	.19	9.90	–	0
1576	**1.80**	–	–	**1600.0**	–	–	**32.00**	**.10**	**.04**	**1.20**	–	–	–	–	–
	.90	–	–	800.0	–	–	16.00	.05	.02	.60	–	–	–	–	–
1577	**1.22**	–	–	**2089.8**	–	–	**38.88**	**.07**	**.07**	**2.19**	–	–	–	–	–
	.50	–	–	860.0	–	–	16.00	.03	.03	.90	–	–	–	–	–
1578	**1.00**	–	–	**1400.0**	–	–	**18.00**	**.10**	**.06**	**1.60**	–	–	**32.00**	–	–
	.50	–	–	700.0	–	–	9.00	.05	.03	.80	–	–	16.00	–	–
1579	**1.09**	–	–	–	–	–	**0**	**.01**	**.00**	**.02**	–	–	–	–	–
	3.90	–	–	–	–	–	0	.05	.01	.10	–	–	–	–	–
1580	**1.79**	–	–	–	–	–	**0**	**.13**	**.04**	**.62**	–	–	–	–	–
	6.40	–	–	–	–	–	0	.45	.15	2.20	–	–	–	–	–
1581	**6.00**	–	–	**180.0**	–	–	**6.00**	**.21**	**.12**	**1.80**	–	–	–	–	–
	2.00	–	–	60.0	–	–	2.00	.07	.04	.60	–	–	–	–	–
1582	**5.18**	–	–	**18.5**	–	–	**0**	**.35**	**.13**	**1.48**	–	**.32**	–	–	**0**
	2.80	–	–	10.0	–	–	0	.19	.07	.80	–	.17	–	–	0
1583	**5.70**	–	–	**387.0**	–	–	–	**.21**	**.15**	**3.87**	–	–	–	–	–
	1.90	–	–	129.0	–	–	–	.07	.05	1.29	–	–	–	–	–
1584	**5.87**	–	–	–	–	–	–	**.15**	**.10**	**1.28**	**.16**	–	–	–	–
	2.30	–	–	–	–	–	–	.06	.04	.50	.06	–	–	–	–
1585	**4.59**	**4.34**	**.43**	**331.5**	–	**.36**	**5.10**	**.20**	**.08**	**1.53**	**.24**	–	–	–	–
	1.80	1.70	.17	130.0	–	.14	2.00	.08	.03	.60	.09	–	–	–	–
1586	**5.00**	**(.22)**	**(.07)**	**150.0**	**0**	**(.18)**	**5.00**	**.18**	**.10**	**1.50**	–	**(.04)**	**60.00**	–	**0**
	2.00	(.78)	(.24)	60.0	0	(.66)	2.00	.07	.04	.60	–	(.13)	24.00	–	0

(Continued)

TABLE F-36 *(Continued)*

Item No.	Food Item Name	Approximate Measure	Weight	Moisture	Food Energy Calories	Protein	Fats	Carbo-hydrates	Fiber	Minerals (Macro) Calcium	Phos-phorus	Sodium	Mag-nesium	Potas-sium
			(g)	(%)	(kcal)	(g)	(g)	(g)	(g)	(mg)	(mg)	(mg)	(mg)	(mg)
	BEANS, COMMON, DRY (Continued)													
1587	red, cooked	1 CUP	**250**	**69.0**	**295.0**	**19.5**	**1.3**	**53.5**	**3.8**	**95.0**	**350.0**	**7.5**	–	**850.0**
			100	69.0	118.0	7.8	.5	21.4	1.5	38.0	140.0	3.0	–	340.0
1588	red, canned, solids & liquid	1 CUP	**256**	**76.0**	**230.4**	**14.6**	**1.0**	**42.0**	**2.3**	**74.2**	**279.0**	**7.7**	**94.7**	**675.8**
			100	76.0	90.0	5.7	.4	16.4	.9	29.0	109.0	3.0	37.0	264.0
1589	white, unsalted, cooked	1 CUP	**200**	**69.0**	**236.0**	**15.6**	**1.2**	**42.4**	**3.0**	**100.0**	**296.0**	**14.0**	–	**832.0**
			100	69.0	118.0	7.8	.6	21.2	1.5	50.0	148.0	7.0	–	416.0
1590	**BEANS, LIMA,** mature seeds, dry, cooked	1 CUP	**169**	**64.1**	**233.2**	**13.9**	**1.0**	**43.3**	**2.9**	**49.0**	**260.3**	**3.4**	–	**1034.3**
			100	64.1	133.0	8.2	.6	25.6	1.7	29.0	154.0	2.0	–	612.0
	BROAD BEAN													
1591	[4]whole seeds, dried	1 OZ	**28**	**13.8**	**91.8**	**7.0**	**.3**	**15.9**	**1.4**	**29.1**	**111.2**	**2.2**	–	**314.4**
			100	13.8	328.0	25.0	1.2	56.9	5.1	104.0	397.0	8.0	–	1123.0
1592	[4]flour	1 OZ	**28**	**11.1**	**96.0**	**6.8**	**.5**	**16.7**	**1.0**	**18.5**	**99.1**	–	–	–
			100	11.1	343.0	24.4	1.8	59.7	3.4	66.0	354.0	–	–	–
1593	**GARBANZOS or CHICKPEAS,** dry, cooked or canned	1 CUP	**140**	–	**250.6**	**14.3**	**3.4**	**42.4**	–	**105.0**	**231.0**	–	**75.6**	–
			100	–	179.0	10.2	2.4	30.3	–	75.0	165.0	–	54.0	–
1594	[4]flour	1 OZ	**28**	**2.5**	**103.1**	**5.6**	**1.8**	**16.7**	**.7**	**28.0**	**96.6**	**2.8**	–	**281.8**
			100	9.1	368.0	20.1	6.6	59.6	2.5	100.0	345.0	10.0	–	1006.0
	COWPEAS													
1595	immature seeds, cooked	1 CUP	**160**	**71.8**	**172.8**	**13.0**	**1.3**	**29.0**	**2.9**	**38.4**	**233.6**	**1.6**	–	**606.4**
			100	71.8	108.0	8.1	.8	18.1	1.8	24.0	146.0	1.0	–	379.0
1596	immature seeds, canned, solids & liquid	1 CUP	**160**	**81.0**	**112.0**	**8.0**	**.5**	**19.8**	**1.1**	**28.8**	**179.2**	**377.6**	–	**563.2**
			100	81.0	70.0	5.0	.3	12.4	.7	18.0	112.0	236.0	–	352.0
1597	mature seeds, dry, cooked, unsalted	1 CUP	**250**	**80.0**	**190.0**	**12.8**	**.8**	**34.5**	**2.5**	**42.5**	**237.5**	**20.0**	–	**572.5**
			100	80.0	76.0	5.1	.3	13.8	1.0	17.0	95.0	8.0	–	229.0
	HYACINTH BEANS													
1598	raw, mature seeds, dry	1 CUP	**184**	**11.8**	**621.9**	**40.8**	**2.8**	**112.2**	**12.7**	**134.3**	**769.1**	–	–	–
			100	11.8	338.0	22.2	1.5	61.0	6.9	73.0	418.0	–	–	–
1599	[4]**JACKBEAN,** whole seed, dried	1 OZ	**28**	**11.2**	**97.4**	**5.9**	**.9**	**17.1**	**2.1**	**37.5**	**84.0**	**9.8**	–	**210.0**
			100	11.2	348.0	21.0	3.2	61.0	7.6	134.0	300.0	35.0	–	750.0
1600	**LENTILS,** dry, whole, cooked	1 CUP	**150**	**72.0**	**159.0**	**11.7**	**(.6)**	**29.0**	**1.8**	**37.5**	**178.5**	**(18.9)**	**(38.1)**	**373.5**
			100	72.0	106.0	7.8	(.4)	19.3	1.2	25.0	119.0	(12.6)	(25.4)	249.0
	PEAS													
1601	[6]mature seeds, dry, whole boiled	1 CUP	**248**	–	**256.4**	**16.9**	**.7**	**47.6**	**11.7**	**60.0**	**669.8**	**33.0**	**75.4**	**671.1**
			100	–	103.4	6.8	.3	19.2	4.7	24.2	270.1	13.3	30.4	270.6
1602	mature seeds, dry, split, w/o seed coat, cooked	1 CUP	**200**	**70.0**	**230.0**	**16.0**	**.6**	**41.6**	**.8**	**22.0**	**178.0**	**26.0**	–	**592.0**
			100	70.0	115.0	8.0	.3	20.8	.4	11.0	89.0	13.0	–	296.0
	PIGEON PEAS													
1603	immature seeds, raw	1 CUP	**200**	**69.5**	**234.0**	**14.4**	**1.2**	**42.6**	**6.6**	**84.0**	**254.0**	**10.0**	–	**1104.0**
			100	69.5	117.0	7.2	.6	21.3	3.3	42.0	127.0	5.0	–	552.0
1604	mature seeds, dry, raw	1 OZ	**28**	**10.8**	**95.8**	**5.7**	**.4**	**17.8**	**2.0**	**30.0**	**88.5**	**7.3**	**33.9**	**274.7**
			100	10.8	342.0	20.4	1.4	63.7	7.0	107.0	316.0	26.0	121.0	981.0
	SOYBEANS													
1605	immature seeds, boiled & drained	1 CUP	**150**	**73.8**	**177.0**	**14.7**	**7.7**	**15.2**	**2.1**	**90.0**	**286.5**	–	–	–
			100	73.8	118.0	9.8	5.1	10.1	1.4	60.0	191.0	–	–	–
1606	Immature seeds, canned solids & liquid	1 OZ	**28**	**81.8**	**21.0**	**1.8**	**.9**	**1.8**	**.2**	**15.4**	**28.0**	**66.1**	–	–
			100	81.8	75.0	6.5	3.2	6.3	.7	55.0	100.0	236.0	–	–
1607	immature seeds, canned & drained	1 CUP	**150**	**76.7**	**154.5**	**13.5**	**7.5**	**11.1**	**2.1**	**100.5**	**171.0**	**354.0**	–	–
			100	76.7	103.0	9.0	5.0	7.4	1.4	67.0	114.0	236.0	–	–
1608	**MILK** (see milk & products)													
1609	mature seeds, dry, cooked	1 CUP	**180**	**71.0**	**234.0**	**19.8**	**10.3**	**19.4**	**2.9**	**131.4**	**322.2**	**3.6**	–	**972.0**
			100	71.0	130.0	11.0	5.7	10.8	1.6	73.0	179.0	2.0	–	540.0
1610	**MISO, FERMENTED,** cereal & soybeans	1 OZ	**28**	**53.0**	**47.9**	**2.9**	**1.3**	**6.6**	**.6**	**19.0**	**86.5**	**826.0**	–	**93.5**
			100	53.0	171.0	10.5	4.6	23.5	2.3	68.0	309.0	2950.0	–	334.0
1611	**NATTO, FERMENTED,** soybeans	1 OZ	**28**	**62.7**	**46.8**	**4.7**	**2.1**	**3.2**	**.9**	**28.8**	**51.0**	–	–	**69.7**
			100	62.7	167.0	16.9	7.4	11.5	3.2	103.0	182.0	–	–	249.0
1612	**PROTEIN**	1 OZ	**28**	**8.2**	**90.2**	**21.0**	**Trace**	**4.2**	**.1**	**33.6**	**188.7**	**58.8**	–	**50.4**
			100	8.2	322.0	74.9	.1	15.1	.4	120.0	674.0	210.0	–	180.0
1613	**TOFU** (bean curd)	1 OZ	**28**	**84.8**	**20.2**	**2.0**	**1.2**	**.7**	**Trace**	**35.8**	**35.3**	**2.0**	**31.1**	**11.8**
			100	84.8	72.0	7.0	4.2	2.4	.1	128.0	126.0	7.0	111.0	42.0
1614	[3]**ANTELOPE,** salted	1 OZ	**28**	**60.2**	**42.0**	**8.5**	**0.6**	–	–	**18.2**	**84.6**	–	–	–
			100	60.2	150.0	30.4	2.2	–	–	65.0	302.0	–	–	–

Item No.	Minerals (Micro)			Fat-Soluble Vitamins			Water-Soluble Vitamins								
	Iron	Zinc	Copper	Vitamin A	Vitamin D	Vitamin E (Alpha Tocopherol)	Vitamin C	Thiamin	Ribo-flavin	Niacin	Panto-thenic Acid	Vit. B-6 (Pyri-doxine)	Folacin (Folic Acid)	Biotin	Vitamin B-12
	(mg)	(mg)	(mg)	(IU)	(IU)	(mg)	(mg)	(mg)	(mg)	(mg)	(mg)	(mg)	(mcg)	(mcg)	(mcg)
1587	**6.00** 2.40	**2.50** 1.00	**.90** .36	– –	– –	– –	– –	**.28** .11	**.15** .06	**1.75** .70	**1.63** .65	– –	**92.50** 37.00	– –	– –
1588	**4.61** 1.80	**3.22** 1.26	– –	– –	– –	– –	– –	**.13** .05	**.10** .04	**1.54** .60	– –	– –	**94.72** 37.00	– –	– –
1589	**5.40** 2.70	**2.00** 1.00	– –	**0** 0	– –	**.94** .47	**0** 0	**.28** .14	**.14** .07	**1.40** .70	– –	– –	**60.00** 30.00	– –	– –
1590	**5.24** 3.10	**1.52** .90	– –	– –	– –	– –	– –	**.22** .13	**.10** .06	**1.18** .70	– –	– –	**72.67** 43.00	– –	– –
1591	**1.18** 4.20	– –	– –	**60.7** 216.7	– –	– –	**0** 0	**.13** .45	**.05** .19	**.67** 2.40	– –	– –	– –	– –	– –
1592	**1.76** 6.30	– –	– –	**9.3** 33.3	– –	– –	– –	**.12** .42	**.08** .28	**.28** 2.70	– –	– –	– –	– –	– –
1593	**4.76** 3.40	[6]**.39** 28	– –	**35.0** 25.0	– –	– –	**0** 0	**.21** .15	**.10** .07	**1.40** 1.00	– –	**.38** .27	**142.80** 102.00	– –	**0** 0
1594	**2.0** 7.0	– –	– –	**9.4** 33.4	– –	– –	**0** 0	**0.03** .12	**.09** .33	**.2** .7	– –	– –	– –	– –	– –
1595	**3.36** 2.10	– –	– –	**560.0** 350.0	– –	– –	**27.20** 17.00	**.48** .30	**.18** .11	**2.24** 1.40	– –	– –	**164.80** 103.00	– –	**0** 0
1596	**2.40** 1.50	– –	– –	**96.0** 60.0	– –	– –	**4.80** 3.00	**.14** .09	**.08** .05	**.80** .50	**.26** .16	**.09** .05	**41.60** 26.00	– –	**0** 0
1597	**3.25** 1.30	**3.00** 1.20	– –	**25.0** 10.0	– –	– –	– –	**.40** .16	**.10** .04	**1.00** .40	– –	– –	**200.00** 80.00	– –	**0** 0
1598	**9.38** 5.10	– –	– –	– –	– –	– –	– –	**1.14** .62	**.33** .18	**3.86** 2.10	**2.27** 1.23	– –	– –	– –	**0** 0
1599	**2.41** 8.60	– –	– –	– –	– –	– –	**.56** 2.00	**.18** .65	**.04** .13	**.87** 3.100	– –	– –	– –	– –	– –
1600	**3.15** 2.10	**1.50** 1.00	**(.27)** (.18)	**30.0** 20.0	**0** 0	– –	**0** 0	**.11** .07	**.09** .06	**.90** .60	**(.50)** (.33)	**(.24)** (.16)	– –	– –	**0** 0
1601	**6.71** 1.45	**2.73** 1.10	**.45** .18	**336.0** 135.5	**0** 0	**Trace** Trace	**Trace** Trace	– .12	– .08	**3.22** 1.30	– –	– –	– –	– –	**0** 0
1602	**3.40** 1.70	– –	– –	**80.0** 40.0	– –	– –	– –	**.30** .15	**.18** .09	**1.80** .90	– –	– –	**14.00** 7.00	– –	– –
1603	**3.20** 1.60	– –	– –	**280.0** 140.0	– –	– –	**78.00** 39.00	**.800** .400	**.34** .17	**4.40** 2.20	**1.36** .68	– –	– –	– –	**0** 0
1604	**2.24** 8.00	– –	**(.35)** (1.25)	**22.4** 80.0	**0** 0	– –	**Trace** Trace	**.090** .320	**.05** .16	**.84** 3.00	**.42** 1.50	**.08** .30	**30.80** 110.00	– –	**0** 0
1605	**3.75** 2.50	[6]**1.35** .90	– –	**990.0** 660.0	– –	– –	**25.50** 17.00	**.47** .31	**.20** .13	**1.80** 1.20	– –	– –	– –	– –	– –
1606	**.81** 2.90	– –	– –	– –	– –	– –	**2.24** 8.00	**.03** .09	**.03** .09	– –	– –	– –	– –	– –	– –
1607	**4.20** 2.80	– –	– –	**510.0** 340.0	– –	– –	**3.00** 2.00	**.09** .06	– –	– –	– –	– –	– –	– –	– –
1608															
1609	**4.86** 2.70	– –	– –	**54.0** 30.0	– –	– –	**0** 0	**.38** .21	**.16** .09	**1.08** .60	– –	– –	– –	– –	– –
1610	**.48** 1.70	– –	– –	**11.2** 40.0	– –	– –	**0** 0	**.02** .06	**.03** .10	**.08** .30	– –	– –	– –	– –	– –
1611	**1.04** 3.70	– –	– –	**0** 0	– –	– –	**0** 0	**.02** .07	**.14** .50	**.31** 1.10	– –	– –	**35.28** 126.00	– –	– –
1612	– –	– –	– –	– –	– –	– –	**0** 0	– –	– –	– –	– –	– –	– –	– –	– –
1613	**.53** 1.90	[6]**.20** .70	– –	**0** 0	– –	– –	**0** 0	**.02** .06	**.01** .03	**.03** .10	– –	– –	– –	– –	– –
1614	**.58** 2.10	– –	– –	**0** 0	– –	– –	**0** 0	**.02** .07	**.08** .28	**2.18** 7.80	– –	– –	– –	– –	– –

(Continued)

TABLE F-36 *(Continued)*

Item No.	Food Item Name	Approximate Measure	Weight	Moisture	Food Energy Calories	Protein	Fats	Carbo-hydrates	Fiber	Minerals (Macro) Calcium	Phos-phorus	Sodium	Mag-nesium	Potas-sium
			(g)	(%)	(kcal)	(g)	(g)	(g)	(g)	(mg)	(mg)	(mg)	(mg)	(mg)
	BACON													
1615	Canadian, broiled or fried, drained	1 SLICE	**21**	**49.9**	**58.2**	**5.8**	**3.7**	**.1**	**0**	**4.0**	**45.8**	**536.6**	**5.0**	**90.7**
			100	49.9	277.0	27.6	17.5	.3	0	19.0	218.0	2555.0	24.0	432.0
1616	cured, fried, drained, sliced medium	1 SLICE	**7**	**8.1**	**40.3**	**2.1**	**3.4**	**.2**	**0**	**1.0**	**15.7**	**71.5**	**1.8**	**16.5**
			100	8.1	575.0	30.4	49.0	3.2	0	14.0	224.0	1021.0	25.0	236.0
1617	**BACON-FLAVORED VEGETABLE PROTEIN,** frozen, dried	1 SERV	**85**	—	**299.7**	**37.3**	**6.0**	**23.9**	—	—	—	—	—	—
			100	—	352.6	43.9	7.0	28.1	—	—	—	—	—	—
1618	**BEAVER,** roasted	1 OZ	**28**	**56.2**	**69.4**	**8.2**	**3.8**	**0**	**0**	—	—	—	—	—
			100	56.2	248.0	29.2	13.7	0	0	—	—	—	—	—
	BEEF													
1619	**ARM,** Choice, separable lean, cooked w/o bone	1 OZ	**28**	**61.7**	**54.0**	**8.5**	**2.0**	**0**	**0**	**3.9**	**42.0**	**16.8**	**5.0**	**103.6**
			100	61.7	193.0	30.5	7.0	0	0	14.0	150.0	60.0	18.0	370.0
1620	**BOTTOM SIRLOIN BUTT,** Choice, boneless, trimmed, cooked	1 OZ	**28**	**42.1**	**114.2**	**6.2**	**9.7**	**0**	**0**	**2.8**	**52.1**	**16.8**	**5.9**	**103.6**
			100	42.1	408.0	22.2	34.7	0	0	10.0	186.0	60.0	21.0	370.0
1621	**BOTTOM SIRLOIN BUTT,** Good, boneless, trimmed, cooked	1 OZ	**28**	**45.7**	**102.2**	**6.7**	**8.1**	**0**	**0**	**3.1**	**55.4**	**16.8**	**5.9**	**103.6**
			100	45.7	365.0	24.1	29.1	0	0	11.0	198.0	60.0	21.0	370.0
1622	[1]**BRISKET,** boneless, lean meat, cooked	1 OZ	**28**	**59.1**	**62.2**	**8.3**	**2.9**	**0**	**0**	**3.6**	**40.9**	**16.8**	**5.0**	**103.6**
			100	59.1	222.0	29.7	10.5	0	0	13.0	146.0	60.0	18.0	370.0
1623	**CHUCK,** arm, Choice, total edible, braised, 85% lean, 15% fat, w/o bone	1 OZ	**28**	**53.0**	**78.7**	**7.6**	**5.4**	**0**	**0**	**3.4**	**37.5**	**16.8**	**4.2**	**103.6**
			100	53.0	281.0	27.1	19.2	0	0	12.0	134.0	60.0	15.0	370.0
1624	**CHUCK,** arm, Good, total edible, braised, 88% lean, 12% fat, w/o bone	1 OZ	**28**	**56.3**	**70.8**	**8.0**	**4.1**	**0**	**0**	**3.6**	**39.2**	**16.8**	**4.2**	**103.6**
			100	56.3	253.0	28.4	14.6	0	0	13.0	140.0	60.0	15.0	370.0
1625	**CHUCK,** arm, Good, separable lean, cooked, w/o bone	1 OZ	**28**	**63.1**	**52.4**	**8.7**	**2.0**	**0**	**0**	**3.9**	**42.3**	**16.8**	**5.0**	**103.6**
			100	63.1	187.0	30.9	7.0	0	0	14.0	151.0	60.0	18.0	370.0
1626	**CHUCK,** entire Choice, total edible, braised, 81% lean, 19% fat	1 OZ	**28**	**49.4**	**91.6**	**7.3**	**6.7**	**0**	**0**	**3.1**	**39.2**	**16.8**	**4.2**	**102.6**
			100	49.4	327.0	26.0	23.9	0	0	11.0	140.0	60.0	15.0	370.0
1627	**CHUCK,** entire, Choice, separable lean, cooked	1 OZ	**28**	**59.7**	**59.9**	**8.4**	**2.7**	**0**	**0**	**3.6**	**44.8**	**16.8**	**5.0**	**103.6**
			100	59.7	214.0	30.0	9.5	0	0	13.0	160.0	60.0	18.0	370.0
1628	**CHUCK,** 5th rib, Choice, total edible, cooked, 69% lean, 31% fat	1 OZ	**28**	**40.3**	**117.6**	**6.3**	**10.3**	**0**	**0**	**2.8**	**30.8**	**16.8**	**4.2**	**103.6**
			100	40.3	420.0	22.4	36.7	0	0	10.0	110.0	60.0	15.0	370.0
1629	**CHUCK,** 5th rib, Choice, separable lean, cooked, w/o bone	1 OZ	**28**	**56.5**	**67.5**	**8.1**	**3.9**	**0**	**0**	**3.6**	**40.0**	**16.8**	**5.0**	**103.6**
			100	56.5	241.0	28.9	13.9	0	0	13.0	143.0	60.0	18.0	370.0
1630	**CHUCK,** 5th rib, Good, total edible, braised, 73% lean, 27% fat, w/bone	1 OZ	**28**	**44.8**	**105.6**	**6.8**	**8.5**	**0**	**0**	**2.8**	**33.9**	**16.8**	**4.2**	**103.6**
			100	44.8	377.0	24.2	30.3	0	0	10.0	121.0	60.0	15.0	370.0
1631	**CHUCK,** 5th rib, Good, separable lean, cooked, w/o bone	1 OZ	**28**	**59.2**	**68.3**	**8.3**	**3.9**	**0**	**0**	**3.6**	**41.2**	**16.8**	**5.0**	**103.6**
			100	59.2	244.0	29.8	13.9	0	0	13.0	147.0	60.0	18.0	370.0
1632	**CLUB STEAK,** Choice, total edible, boiled, 58% lean, 42% fat	1 OZ	**28**	**37.9**	**127.1**	**5.8**	**11.4**	**0**	**0**	**2.5**	**49.0**	**16.8**	**5.9**	**103.6**
			100	37.9	454.0	20.6	40.6	0	0	9.0	175.0	60.0	21.0	370.0
1633	**CLUB STEAK,** Choice, separable lean, cooked, w/o bone	1 OZ	**28**	**56.0**	**68.3**	**8.3**	**3.6**	**0**	**0**	**3.4**	**66.6**	**16.8**	**8.1**	**103.6**
			100	56.0	244.0	29.6	13.0	0	0	12.0	238.0	60.0	29.0	370.0
1634	**CLUB STEAK,** Good, total edible, cooked, 64% lean, 36% fat	1 OZ	**28**	**42.8**	**111.4**	**6.4**	**9.3**	**0**	**0**	**2.8**	**53.8**	**16.8**	**5.9**	**103.6**
			100	42.8	398.0	22.9	33.3	0	0	10.0	192.0	60.0	21.0	370.0
1635	**CLUB STEAK,** Good, separable lean, cooked	1 OZ	**28**	**58.5**	**60.8**	**8.5**	**2.7**	**0**	**0**	**3.4**	**68.0**	**16.8**	**8.1**	**103.6**
			100	58.5	217.0	30.5	9.6	0	0	12.0	243.0	60.0	29.0	370.0
1636	**CORNED,** boneless, medium fat, cooked	1 OZ	**28**	**43.9**	**104.2**	**6.4**	**8.5**	**0**	**0**	**2.5**	**26.0**	**487.2**	**8.1**	**42.0**
			100	43.9	372.0	22.9	30.4	0	0	9.0	93.0	1740.0	29.0	150.0
1637	**CORNED,** boneless, fat, canned	1 OZ	**28**	**55.3**	**73.6**	**6.6**	**5.0**	**0**	**0**	**5.3**	**27.4**	—	—	—
			100	55.3	263.0	23.5	18.0	0	0	19.0	98.0	—	—	—
1638	**CORNED,** boneless, medium fat, canned	1 OZ	**28**	**59.3**	**60.5**	**7.1**	**3.4**	**0**	**0**	**5.6**	**29.7**	**268.0**	**(4.2)**	**16.8**
			100	59.3	216.0	25.3	12.0	0	0	20.0	106.0	957.0	(15.0)	60.0
1639	**CORNED,** boneless, lean, canned	1 OZ	**28**	**62.0**	**51.8**	**7.4**	**2.2**	**0**	**0**	**5.9**	**30.8**	—	—	—
			100	62.0	185.0	26.4	8.0	0	0	21.0	110.0	—	—	—
1640	**CUBED STEAK,** Choice, cooked	1 OZ	**28**	**54.7**	**73.1**	**8.0**	**4.3**	**0**	**0**	**3.4**	**70.0**	**16.8**	**5.9**	**103.6**
			100	54.7	261.0	28.6	15.4	0	0	12.0	250.0	60.0	21.0	370.0
1641	**DOUBLE-BONE SIRLOIN STEAK,** Choice, total edible, broiled, 66% lean, 34% fat, w/o bone	1 OZ	**28**	**42.1**	**114.2**	**6.2**	**9.7**	**0**	**0**	**2.8**	**52.1**	**16.8**	**5.9**	**103.6**
			100	42.1	408.0	22.2	34.7	0	0	10.0	186.0	60.0	21.0	370.0
1642	**DOUBLE-BONE SIRLOIN STEAK,** Choice, separable lean, cooked, w/o bone	1 OZ	**28**	**58.5**	**60.5**	**8.6**	**2.7**	**0**	**0**	**3.4**	**68.3**	**16.8**	**8.1**	**103.6**
			100	58.5	216.0	30.6	9.5	0	0	12.0	244.0	60.0	29.0	370.0
1643	**DOUBLE-BONE SIRLOIN STEAK,** Good, total edible, cooked, 67% lean, 33% fat	1 OZ	**28**	**45.7**	**102.2**	**6.7**	**8.1**	**0**	**0**	**3.1**	**55.4**	**16.8**	**5.9**	**103.6**
			100	45.7	365.0	24.1	29.1	0	0	11.0	198.0	60.0	21.0	370.0
1644	**DOUBLE-BONE SIRLOIN STEAK,** Good, separable lean, cooked	1 OZ	**28**	**61.0**	**53.2**	**8.8**	**1.7**	**0**	**0**	**3.6**	**69.7**	**16.8**	**8.1**	**103.6**
			100	61.0	190.0	31.5	6.1	0	0	13.0	249.0	60.0	29.0	370.0

Item No.	Minerals (Micro)			Fat-Soluble Vitamins			Water-Soluble Vitamins								
	Iron	Zinc	Copper	Vitamin A	Vitamin D	Vitamin E (Alpha Tocopherol)	Vitamin C	Thiamin	Ribo-flavin	Niacin	Panto-thenic Acid	Vit. B-6 (Pyri-doxine)	Folacin (Folic Acid)	Biotin	Vitamin B-12
	(mg)	(mg)	(mg)	(IU)	(IU)	(mg)	(mg)	(mg)	(mg)	(mg)	(mg)	(mg)	(mcg)	(mcg)	(mcg)
1615	.86	—	—	0	0	—	0	.19	.04	1.05	—	—	—	—	—
	4.10	—	—	0	0	—	0	.92	.17	5.00	—	—	—	—	—
1616	.23	—	—	0	—	.04	0	.04	.02	.36	—	—	.14	—	—
	3.30	—	—	0	—	.53	0	.51	.34	5.20	—	—	2.00	—	—
1617	—	—	—	—	—	—	—	—	—	—	—	—	—	—	—
	—	—	—	—	—	—	—	—	—	—	—	—	—	—	—
1618	—	—	—	—	—	—	—	.02	.11	—	—	—	—	—	—
	—	—	—	—	—	—	—	.08	.38	—	—	—	—	—	—
1619	1.06	—	—	2.8	—	.26	—	.02	.06	1.29	—	—	1.12	—	—
	3.80	—	—	10.0	—	.94	—	.06	.23	4.60	—	—	4.00	—	—
1620	.81	—	.05	16.8	—	—	0	.02	.05	1.29	—	—	—	—	—
	2.90	—	.18	60.0	—	—	0	.06	.18	4.60	—	—	—	—	—
1621	.87	—	.05	14.0	—	—	0	.02	.05	1.37	—	—	—	—	—
	3.10	—	.18	50.0	—	—	0	.07	.19	4.90	—	—	—	—	—
1622	1.06	1.74	—	5.6	—	—	—	.01	.06	1.26	—	—	1.12	—	—
	3.80	6.20	—	20.0	—	—	—	.05	.22	4.50	—	—	4.00	—	—
1623	.95	—	—	8.4	—	.26	—	.01	.06	1.18	—	—	—	—	(.50)
	3.40	—	—	30.0	—	.94	—	.05	.21	4.20	—	—	—	—	(1.80)
1624	1.04	—	—	8.4	—	—	—	.01	.06	1.20	—	—	—	—	—
	3.70	—	—	30.0	—	—	—	.05	.21	4.30	—	—	—	—	—
1625	1.09	1.74	—	2.8	—	—	—	.02	.06	1.31	—	—	1.12	—	—
	3.90	6.20	—	10.0	—	—	—	.06	.23	4.70	—	—	4.00	—	—
1626	.92	—	—	11.2	—	—	—	.01	.06	1.12	—	—	—	—	—
	3.30	—	—	40.0	—	—	—	.05	.20	4.00	—	—	—	—	—
1627	1.06	1.74	—	5.6	—	—	—	.01	.06	1.29	—	—	1.12	—	—
	3.80	6.20	—	20.0	—	—	—	.05	.23	4.60	—	—	4.00	—	—
1628	.81	—	—	19.6	—	.26	—	.01	.05	.98	—	—	—	—	—
	2.90	—	—	70.0	—	.94	—	.04	.17	3.50	—	—	—	—	—
1629	1.04	1.62	—	5.6	—	—	—	.01	.06	1.26	—	—	1.12	—	—
	3.70	5.80	—	20.0	—	—	—	.05	.22	4.50	—	—	4.00	—	—
1630	.67	—	—	16.8	—	—	—	.01	.05	1.06	—	—	—	—	—
	3.10	—	—	60.0	—	—	—	.04	.19	3.80	—	—	—	—	—
1631	1.06	1.62	—	5.6	—	.26	—	.01	.06	1.29	—	—	1.12	—	.01
	3.80	5.80	—	20.0	—	.94	—	.05	.23	4.60	—	—	4.00	—	.04
1632	.76	—	.05	19.6	—	—	—	.02	.05	1.20	—	—	—	—	—
	2.70	—	.18	70.0	—	—	—	.06	.17	4.30	—	—	—	—	—
1633	1.01	1.62	.05	5.6	—	—	—	.02	.06	1.62	—	—	1.12	—	—
	3.60	5.80	.19	20.0	—	—	—	.08	.23	5.80	—	—	4.00	—	—
1634	.84	—	.05	16.8	—	—	—	.02	.05	1.32	—	—	—	—	—
	3.00	—	.19	60.0	—	—	—	.06	.18	4.70	—	—	—	—	—
1635	1.04	1.62	.05	5.6	—	—	—	.02	.06	1.65	—	—	1.12	—	—
	3.70	5.80	.19	20.0	—	—	—	.08	.23	5.90	—	—	4.00	—	—
1636	.81	—	—	—	—	—	0	.01	.04	0.42	—	.04	—	—	—
	2.90	—	—	—	—	—	0	.04	.14	1.50	—	.15	—	—	—
1637	1.12	—	—	—	—	—	0	0	.06	0.90	—	—	—	—	.52
	4.00	—	—	—	—	—	0	.01	.22	3.20	—	—	—	—	1.84
1638	1.20	(1.57)	(.07)	0	Trace	(.22)	0	.01	.07	0.95	(.12)	(.02)	(.59)	(.58)	.52
	4.30	(5.62)	(.24)	0	Trace	(.77)	0	.02	.24	3.40	(.42)	(.07)	(2.11)	(2.08)	1.84
1639	1.26	—	—	—	—	—	0	.01	.07	0.98	—	—	—	—	—
	4.50	—	—	—	—	—	0	.02	.25	3.50	—	—	—	—	—
1640	.98	—	.05	8.4	—	—	0	.02	.06	1.57	.15	.14	—	—	—
	3.50	—	.19	30.0	—	—	0	.08	.22	5.60	.52	.50	—	—	—
1641	.81	—	.05	16.8	—	—	0	.02	.05	1.29	—	—	—	—	—
	2.90	—	.19	60.0	—	—	0	.06	.18	4.60	—	—	—	—	—
1642	1.04	1.62	.05	5.6	—	—	0	.02	.06	1.68	—	—	1.12	—	—
	3.70	5.80	.19	20.0	—	—	0	.08	.23	6.00	—	—	4.00	—	—
1643	.87	—	.05	14.0	—	—	0	.02	.05	1.37	—	—	—	—	—
	3.10	—	.19	50.0	—	—	0	.07	.19	4.90	—	—	—	—	—
1644	1.06	1.62	.05	2.8	—	—	0	.02	.07	1.71	—	—	1.12	—	—
	3.80	5.80	.19	10.0	—	—	0	.08	.24	6.10	—	—	4.00	—	—

(Continued)

MEATS

TABLE F-36 *(Continued)*

Item No.	Food Item Name	Approximate Measure	Weight	Moisture	Food Energy Calories	Protein	Fats	Carbo-hydrates	Fiber	Minerals (Macro) Calcium	Phos-phorus	Sodium	Mag-nesium	Potas-sium
			(g)	(%)	(kcal)	(g)	(g)	(g)	(g)	(mg)	(mg)	(mg)	(mg)	(mg)
	BEEF (Continued)													
1645	**FAT,** cooked	1 OZ	**28**	**15.7**	**204.1**	**1.6**	**21.9**	**0**	**0**	—	—	**16.8**	—	**103.6**
			100	15.7	729.0	5.7	78.1	0	0	—	—	60.0	—	370.0
1646	**FLANK STEAK,** Choice, cooked, 100% lean	1 OZ	**28**	**61.4**	**52.6**	**8.5**	**2.0**	**0**	**0**	**3.9**	**42.0**	**16.8**	**5.0**	**103.6**
			100	61.4	188.0	30.5	7.3	0	0	14.0	150.0	60.0	18.0	370.0
1647	**FLANK STEAK,** Good, cooked, 100% lean	1 OZ	**28**	**61.8**	**53.5**	**8.6**	**1.8**	**0**	**0**	**3.9**	**42.3**	**16.8**	**5.0**	**103.6**
			100	61.8	191.0	30.8	6.6	0	0	14.0	151.0	60.0	18.0	370.0
1648	**HAMBURGER (GROUND BEEF),** cooked, w/21% fat	1 OZ	**28**	**54.2**	**84.3**	**6.8**	**6.4**	**0**	**0**	**3.1**	**54.3**	**13.2**	**5.9**	**126.0**
			100	54.2	301.0	24.2	22.7	0	0	11.0	194.0	47.0	21.0	450.0
1649	**HAMBURGER (GROUND BEEF),** cooked, w/10% fat	1 OZ	**28**	**60.0**	**61.3**	**7.7**	**3.2**	**0**	**0**	**3.4**	**64.4**	**13.4**	**7.0**	**156.2**
			100	60.0	219.0	27.4	11.3	0	0	12.0	230.0	48.0	25.0	558.0
1650	**HINDSHANK,** Choice, total edible, cooked, 66% lean, 34% fat	1 OZ	**28**	**46.1**	**101.1**	**7.0**	**7.9**	**0**	**0**	**3.1**	**35.0**	**16.8**	**4.2**	**103.6**
			100	46.1	361.0	25.1	28.1	0	0	11.0	125.0	60.0	15.0	370.0
1651	**HINDSHANK,** Choice, separable lean, cooked	1 OZ	**28**	**62.5**	**51.5**	**8.6**	**1.7**	**0**	**0**	**3.9**	**42.3**	**16.8**	**5.0**	**103.6**
			100	62.5	184.0	30.7	5.9	0	0	14.0	151.0	60.0	18.0	370.0
1652	**HINDSHANK,** Good, total edible, cooked, 70% lean, 30% fat	1 OZ	**28**	**51.0**	**86.0**	**7.6**	**5.9**	**0**	**0**	**3.4**	**38.1**	**16.8**	**4.2**	**103.6**
			100	51.0	307.0	27.2	21.1	0	0	12.0	136.0	60.0	15.0	370.0
1653	**HINDSHANK,** Good, separable lean, cooked	1 OZ	**28**	**63.4**	**49.3**	**8.7**	**1.3**	**0**	**0**	**3.9**	**42.3**	**16.8**	**5.0**	**103.6**
			100	63.4	176.0	31.0	4.8	0	0	14.0	151.0	60.0	18.0	370.0
1654	**KNUCKLE,** boneless, Choice, cooked	1 OZ	**28**	**46.1**	**101.1**	**7.0**	**7.9**	**0**	**0**	**3.1**	**35.0**	**16.8**	**4.2**	**103.6**
			100	46.1	361.0	25.1	28.1	0	0	11.0	125.0	60.0	15.0	370.0
1655	[1]**MINUTE STEAK,** lean meat & fat, cooked	1 OZ	**28**	**54.7**	**73.1**	**8.0**	**4.3**	**0**	**0**	**3.4**	**70.0**	**16.8**	**7.8**	**103.6**
			100	54.7	261.0	28.6	15.4	0	0	12.0	250.0	60.0	28.0	370.0
1656	**PORTERHOUSE STEAK,** Choice, total edible, broiled, 57% lean, 43% fat, w/bone	1 OZ	**28**	**37.2**	**128.5**	**5.5**	**11.8**	**0**	**0**	**2.5**	**47.0**	**16.8**	**5.9**	**103.6**
			100	37.2	459.0	19.7	42.2	0	0	9.0	168.0	60.0	21.0	370.0
1657	**PORTERHOUSE STEAK,** Good, total edible, cooked, 58% lean, 42% fat	1 OZ	**28**	**38.9**	**124.9**	**5.7**	**11.1**	**0**	**0**	**2.5**	**48.4**	**16.8**	**5.9**	**103.6**
			100	38.9	446.0	20.5	39.7	0	0	9.0	173.0	60.0	21.0	370.0
1658	**PORTERHOUSE STEAK,** Good, separable lean, cooked	1 OZ	**28**	**60.3**	**61.3**	**8.7**	**2.9**	**0**	**0**	**3.4**	**69.2**	**16.8**	**8.1**	**103.6**
			100	60.3	219.0	31.1	10.5	0	0	12.0	247.0	60.0	29.0	370.0
1659	**RIB,** Choice, cooked	1 OZ	**30**	**40.0**	**132.0**	**6.0**	**11.8**	**0**	**0**	**2.7**	**55.8**	**18.0**	**6.0**	**111.0**
			100	40.0	440.0	19.9	39.4	0	0	9.0	186.0	60.0	20.0	370.0
1660	**RIB,** entire, Choice, total edible, roasted, 64% lean, 36% fat, w/o bone	1 OZ	**28**	**40.0**	**121.5**	**5.6**	**11.0**	**0**	**0**	**2.5**	**52.1**	**16.8**	**5.6**	**103.6**
			100	40.0	434.0	19.9	39.4	0	0	9.0	186.0	60.0	20.0	370.0
1661	**RIB,** entire, Choice, separable lean, cooked, w/o bone	1 OZ	**28**	**57.2**	**65.2**	**7.9**	**3.8**	**0**	**0**	**3.4**	**71.7**	**16.8**	**7.8**	**103.6**
			100	57.2	233.0	28.2	13.4	0	0	12.0	256.0	60.0	28.0	370.0
1662	**RIBEYE,** roll, Choice, cooked	1 OZ	**28**	**40.0**	**123.2**	**5.6**	**11.0**	**0**	**0**	**2.5**	**52.1**	**16.8**	**5.6**	**103.6**
			100	40.0	440.0	19.9	39.4	0	0	9.0	186.0	60.0	20.0	370.0
1663	**RIBEYE** steak	1 OZ	**30**	**40.0**	**132.0**	**6.0**	**11.8**	**0**	**0**	**2.7**	**55.8**	**18.0**	**6.0**	**111.0**
			100	40.0	440.0	19.9	39.4	0	0	9.0	186.0	60.0	20.0	370.0
1664	**ROUND,** Choice, total edible, cooked, 81% lean, 19% fat, w/o bone	1 OZ	**28**	**54.7**	**73.1**	**8.0**	**4.3**	**0**	**0**	**3.4**	**70.0**	**16.8**	**7.8**	**103.6**
			100	54.7	261.0	28.6	15.4	0	0	12.0	250.0	60.0	28.0	370.0
1665	**ROUND,** Choice, separable lean, cooked, w/o bone	1 OZ	**28**	**61.2**	**51.2**	**8.8**	**1.8**	**0**	**0**	**3.6**	**75.0**	**16.8**	**8.1**	**103.6**
			100	61.2	183.0	31.3	6.4	0	0	13.0	268.0	60.0	29.0	370.0
1666	**ROUND CUBES,** 1-inch square, Choice, cooked	1 OZ	**28**	**54.7**	**73.1**	**8.0**	**4.3**	**0**	**0**	**3.4**	**70.0**	**16.8**	**5.9**	**103.6**
			100	54.7	261.0	28.6	15.4	0	0	12.0	250.0	60.0	21.0	370.0
1667	**ROUND STEAK,** Choice, cooked	1 OZ	**28**	**54.7**	**69.7**	**8.0**	**4.2**	**0**	**0**	**3.4**	**70.0**	**16.8**	**5.9**	**103.6**
			100	54.7	249.0	28.6	14.9	0	0	12.0	250.0	60.0	21.0	370.0
1668	**ROUND STRIPS,** Choice, cooked	1 OZ	**28**	**54.7**	**73.1**	**8.0**	**1.6**	**0**	**0**	**3.4**	**70.0**	**16.8**	**5.9**	**103.6**
			100	54.7	261.0	28.6	5.7	0	0	12.0	250.0	60.0	21.0	370.0
1669	**RUMP,** Choice, total edible, roasted, 75% lean, 25% fat, w/o bone	1 OZ	**28**	**48.1**	**95.2**	**6.6**	**7.6**	**0**	**0**	**2.8**	**55.2**	**16.8**	**7.8**	**103.6**
			100	48.1	340.0	23.6	27.3	0	0	10.0	197.0	60.0	28.0	370.0
1670	**RUMP,** Choice, separable lean, cooked, w/o bone	1 OZ	**28**	**60.4**	**56.0**	**8.1**	**2.6**	**0**	**0**	**3.4**	**68.0**	**16.8**	**8.1**	**103.6**
			100	60.4	200.0	29.1	9.3	0	0	12.0	243.0	60.0	29.0	370.0
1671	**RUMP,** Good, total edible, roasted, 76% lean, 24% fat, w/o bone	1 OZ	**28**	**50.7**	**88.8**	**7.0**	**6.6**	**0**	**0**	**3.1**	**58.0**	**16.8**	**7.8**	**103.6**
			100	50.7	317.0	24.9	23.4	0	0	11.0	207.0	60.0	28.0	370.0
1672	**RUMP,** Good, separable lean, cooked, w/o bone	1 OZ	**28**	**62.0**	**53.2**	**8.3**	**2.0**	**0**	**0**	**3.6**	**69.4**	**16.8**	**8.1**	**103.6**
			100	62.0	190.0	29.6	7.1	0	0	13.0	248.0	60.0	29.0	370.0
1673	**SHORT PLATE,** Choice, total edible, cooked, 58% lean, 42% fat	1 OZ	**28**	**36.0**	**131.0**	**5.8**	**12.0**	**0**	**0**	**2.5**	**28.3**	**16.8**	**4.2**	**103.6**
			100	36.0	468.0	20.6	42.8	0	0	9.0	101.0	60.0	15.0	370.0
1674	**SHORT PLATE,** Choice, separable lean, cooked	1 OZ	**28**	**59.1**	**59.6**	**8.3**	**2.9**	**0**	**0**	**3.6**	**40.9**	**16.8**	**5.0**	**103.6**
			100	59.1	213.0	29.7	10.5	0	0	13.0	146.0	60.0	18.0	370.0
1675	**SHORT PLATE,** Good, total edible, simmered, 61% lean, 39% fat, w/o bone	1 OZ	**28**	**39.9**	**132.7**	**6.2**	**12.0**	**0**	**0**	**2.5**	**30.8**	**16.8**	**4.2**	**103.6**
			100	39.9	474.0	22.3	42.7	0	0	9.0	110.0	60.0	15.0	370.0

Item No.	Minerals (Micro)			Fat-Soluble Vitamins			Water-Soluble Vitamins								
	Iron	Zinc	Copper	Vitamin A	Vitamin D	Vitamin E (Alpha Tocopherol)	Vitamin C	Thiamin	Ribo-flavin	Niacin	Panto-thenic Acid	Vit. B-6 (Pyri-doxine)	Folacin (Folic Acid)	Biotin	Vitamin B-12
	(mg)	(mg)	(mg)	(IU)	(IU)	(mg)	(mg)	(mg)	(mg)	(mg)	(mg)	(mg)	(mcg)	(mcg)	(mcg)
1645	—	—	**.02**	—	—	—	—	—	—	—	—	—	—	—	—
	—	—	.08	—	—	—	—	—	—	—	—	—	—	—	—
1646	**1.06**	—	—	**2.8**	—	**.26**	—	**.02**	**.06**	**1.29**	—	—	—	—	—
	3.80	—	—	10.0	—	.94	—	.06	.23	4.60	—	—	—	—	—
1647	**1.09**	—	—	**2.8**	—	—	—	**.02**	**.06**	**1.32**	—	—	**1.12**	—	—
	3.90	—	—	10.0	—	—	—	.06	.23	4.70	—	—	4.00	—	—
1648	**.90**	—	—	**11.2**	—	**.10**	**0**	**.03**	**.06**	**1.51**	**.12**	**.13**	**1.12**	—	**(.50)**
	3.20	—	—	40.0	—	.37	0	.09	.21	5.40	.44	.46	4.00	—	(1.80)
1649	**.98**	**1.23**	—	**5.6**	—	—	**0**	**.03**	**.06**	**1.68**	—	—	**1.12**	—	**(.50)**
	3.50	4.40	—	20.0	—	—	0	.09	.23	6.00	—	—	4.00	—	(1.80)
1650	**.92**	—	—	**14.0**	—	—	—	**.01**	**.05**	**1.09**	—	—	—	—	—
	3.30	—	—	50.0	—	—	—	.05	.19	3.90	—	—	—	—	—
1651	**1.09**	**1.74**	—	**2.8**	—	**.26**	—	**.02**	**.06**	**1.32**	—	—	**1.12**	—	—
	3.90	6.20	—	10.0	—	.94	—	.06	.23	4.70	—	—	4.00	—	—
1652	**1.01**	—	**.02**	**11.2**	—	—	—	**.01**	**.06**	**1.18**	—	—	—	—	—
	3.60	—	.08	40.0	—	—	—	.05	.21	4.20	—	—	—	—	—
1653	**1.09**	**1.74**	—	**2.8**	—	—	—	**.02**	**.06**	**1.32**	—	—	**1.12**	—	—
	3.90	6.20	—	10.0	—	—	—	.06	.23	4.70	—	—	4.00	—	—
1654	**.92**	—	—	**14.0**	—	—	—	**.01**	**.05**	**1.09**	—	—	—	—	—
	3.30	—	—	50.0	—	—	—	.05	.19	3.90	—	—	—	—	—
1655	**.98**	**1.34**	—	**8.4**	—	—	**0**	**.02**	**.06**	**1.57**	**.16**	**.14**	**1.12**	—	—
	3.50	4.79	—	30.0	—	—	0	.08	.22	5.60	.52	.50	4.00	—	—
1656	**.73**	—	—	**19.6**	—	—	—	**.02**	**.05**	**1.18**	—	—	—	—	—
	2.60	—	—	70.0	—	—	—	.06	.16	4.20	—	—	—	—	—
1657	**.73**	—	—	**19.6**	—	—	—	**.02**	**.05**	**1.20**	—	—	—	—	—
	2.60	—	—	70.0	—	—	—	.06	.17	4.30	—	—	—	—	—
1658	**1.04**	**1.62**	—	**2.8**	—	—	—	**.02**	**.07**	**1.68**	—	—	**1.12**	—	—
	3.70	5.80	—	10.0	—	—	—	.08	.24	6.00	—	—	4.00	—	—
1659	**.78**	—	—	**24.0**	—	—	**0**	**.02**	**.05**	**1.08**	—	—	—	—	—
	2.60	—	—	80.0	—	—	0	.05	.15	3.60	—	—	—	—	—
1660	**.73**	—	—	**22.4**	—	—	**0**	**.01**	**.04**	**1.01**	—	—	—	—	**.01**
	2.60	—	—	80.0	—	—	0	.05	.15	3.60	—	—	—	—	.03
1661	**1.01**	**1.62**	—	**5.6**	—	—	**0**	**.02**	**.06**	**1.43**	**.17**	**.13**	**1.12**	—	—
	3.60	5.80	—	20.0	—	—	0	.07	.21	5.10	.60	.48	4.00	—	—
1662	**.73**	—	—	**22.4**	—	—	**0**	**.01**	**.04**	**1.01**	—	—	—	—	**(.50)**
	2.60	—	—	80.0	—	—	0	.05	.15	3.60	—	—	—	—	(1.80)
1663	**.78**	—	—	**24.0**	—	—	**0**	**.02**	**.05**	**1.08**	—	—	—	—	—
	2.60	—	—	80.0	—	—	0	.05	.15	3.60	—	—	—	—	—
1664	**.98**	—	—	**8.4**	—	—	**0**	**.02**	**.06**	**1.57**	**.15**	**.14**	—	—	**(.73)**
	3.50	—	—	30.0	—	—	0	.08	.22	5.60	.52	.49	—	—	(2.60)
1665	**1.04**	**1.74**	—	**2.8**	—	—	**0**	**.02**	**.07**	**1.6**	—	—	**1.12**	—	—
	3.70	6.20	—	10.0	—	—	0	.08	.24	6.0	—	—	4.00	—	—
1666	**.98**	—	—	**8.4**	—	—	**0**	**.02**	**.06**	**1.57**	**.15**	**.14**	—	—	—
	3.50	—	—	30.0	—	—	0	.08	.22	5.60	.52	.50	—	—	—
1667	**.98**	—	—	**8.4**	—	—	**0**	**.02**	**.06**	**1.57**	**.15**	**.14**	—	—	—
	3.50	—	—	30.0	—	—	0	.08	.22	5.60	.52	.50	—	—	—
1668	**.98**	—	—	**8.4**	—	—	**0**	**.02**	**.06**	**1.57**	**.15**	**.14**	—	—	—
	3.50	—	—	30.0	—	—	0	.08	.22	5.60	.52	.50	—	—	—
1669	**.87**	—	—	**14.0**	—	—	**0**	**.02**	**.05**	**1.20**	—	—	—	—	—
	3.10	—	—	50.0	—	—	0	.06	.18	4.30	—	—	—	—	—
1670	**1.04**	**1.74**	—	**5.6**	—	—	**0**	**.02**	**.06**	**1.46**	—	—	**1.12**	—	—
	3.70	6.20	—	20.0	—	—	0	.07	.22	5.20	—	—	4.00	—	—
1671	**.87**	—	—	**11.2**	—	—	**0**	**.02**	**.05**	**1.26**	—	—	—	—	—
	3.10	—	—	40.0	—	—	0	.06	.19	4.50	—	—	—	—	—
1672	**1.04**	**1.74**	—	**2.8**	—	—	**0**	**.02**	**.06**	**1.48**	—	—	**1.12**	—	—
	3.70	6.20	—	10.0	—	—	0	.08	.22	5.30	—	—	4.00	—	—
1673	**.76**	—	—	**22.4**	—	—	—	**.01**	**.05**	**.90**	—	—	—	—	**.01**
	2.70	—	—	80.0	—	—	—	.04	.16	3.20	—	—	—	—	.02
1674	**1.06**	**1.74**	—	**5.6**	—	—	—	**.01**	**.06**	**1.26**	—	—	**1.12**	—	—
	3.80	6.20	—	20.0	—	—	—	.05	.22	4.50	—	—	4.00	—	—
1675	**.81**	—	—	**19.6**	—	—	—	**.01**	**.05**	**.95**	—	—	—	—	—
	2.90	—	—	70.0	—	—	—	.04	.17	3.40	—	—	—	—	—

(Continued)

TABLE F-36 *(Continued)*

Item No.	Food Item Name	Approximate Measure	Weight	Moisture	Food Energy Calories	Protein	Fats	Carbo-hydrates	Fiber	Minerals (Macro) Calcium	Phos-phorus	Sodium	Mag-nesium	Potas-sium
			(g)	(%)	(kcal)	(g)	(g)	(g)	(g)	(mg)	(mg)	(mg)	(mg)	(mg)
	BEEF (Continued)													
1676	**SHORT PLATE,** Good, separable lean, simmered, w/o bone	1 OZ	**28**	**61.1**	**55.7**	**8.5**	**2.2**	**0**	**0**	**3.6**	**41.7**	**16.8**	**5.0**	**103.6**
			100	61.1	199.0	30.3	7.7	0	0	13.0	149.0	60.0	18.0	370.0
1677	**SHORT RIBS,** Good, cooked	1 OZ	**28**	**39.9**	**121.0**	**6.2**	**10.4**	**0**	**0**	**2.5**	**30.8**	**16.8**	**4.2**	**103.6**
			100	39.9	432.0	22.3	37.3	0	0	9.0	110.0	60.0	15.0	370.0
1678	**SIRLOIN, GROUND,** Choice, 10% fat, cooked	1 OZ	**28**	**42.1**	**114.2**	**6.2**	**9.7**	**0**	**0**	**2.8**	**52.1**	**16.8**	**5.9**	**103.6**
			100	42.1	408.0	22.2	34.7	0	0	10.0	186.0	60.0	21.0	370.0
1679	**SIRLOIN STEAK,** Choice, total edible, cooked, 66% lean, 34% fat, w/o bone	1 OZ	**28**	**43.9**	**106.4**	**6.4**	**9.0**	**0**	**0**	**2.8**	**53.5**	**16.8**	**5.9**	**103.6**
			100	43.9	380.0	23.0	32.0	0	0	10.0	191.0	60.0	21.0	370.0
1680	**SIRLOIN STEAK,** Choice, separable lean, cooked, w/o bone	1 OZ	**28**	**58.7**	**44.8**	**9.0**	**2.2**	**0**	**0**	**3.6**	**73.1**	**16.8**	**8.1**	**103.6**
			100	58.7	160.0	32.2	7.7	0	0	13.0	261.0	60.0	29.0	370.0
1681	**SIRLOIN STEAK,** Good, total edible, cooked, 68% lean, 32% fat	1 OZ	**28**	**46.9**	**98.8**	**6.9**	**7.7**	**0**	**0**	**3.1**	**56.6**	**16.8**	**5.9**	**103.6**
			100	46.9	353.0	24.5	27.5	0	0	11.0	202.0	60.0	21.0	370.0
1682	**SIRLOIN STEAK,** Good, separable lean, cooked	1 OZ	**28**	**61.6**	**51.2**	**8.9**	**1.5**	**0**	**0**	**3.6**	**70.0**	**16.8**	**8.1**	**103.6**
			100	61.6	183.0	31.7	5.3	0	0	13.0	250.0	60.0	29.0	370.0
1683	**STEW MEAT,** Choice, 90% lean, 10% fat, cooked	1 OZ	**28**	**54.7**	**73.1**	**8.0**	**4.3**	**0**	**0**	**3.4**	**70.0**	**16.8**	**5.9**	**103.6**
			100	54.7	261.0	28.6	15.4	0	0	12.0	250.0	60.0	21.0	370.0
1684	**T-BONE STEAK,** Choice, total edible, broiled, 56% lean, 44% fat w/bone	1 OZ	**28**	**36.4**	**130.8**	**5.5**	**12.1**	**0**	**0**	**2.2**	**46.5**	**16.8**	**5.9**	**103.6**
			100	36.4	467.0	19.5	43.2	0	0	8.0	166.0	60.0	21.0	370.0
1685	**T-BONE STEAK,** Good, total edible, cooked, 50% lean, 42% fat	1 OZ	**28**	**39.2**	**123.8**	**5.8**	**11.0**	**0**	**0**	**2.5**	**49.0**	**16.8**	**5.9**	**103.6**
			100	39.2	442.0	20.6	39.2	0	0	9.0	175.0	60.0	21.0	370.0
1686	**T-BONE STEAK,** Good, separable lean, cooked	1 OZ	**28**	**60.2**	**55.7**	**8.7**	**2.0**	**0**	**0**	**3.4**	**69.2**	**16.8**	**8.1**	**103.6**
			100	60.2	199.0	31.1	7.3	0	0	12.0	247.0	60.0	29.0	370.0
1687	**TENDERLOIN STEAK,** broiled	1 OZ	**28**	—	**62.7**	**7.2**	**3.5**	**0**	**0**	**4.2**	**58.0**	**12.6**	**6.4**	**122.1**
			100	—	224.0	26.1	12.6	0	0	15.0	207.0	45.0	22.7	436.0
1688	**TOP SIRLOIN BUTT,** Choice, boneless, trimmed, cooked	1 OZ	**28**	**42.1**	**114.2**	**6.2**	**9.7**	**0**	**0**	**2.8**	**52.1**	**16.8**	**5.9**	**103.6**
			100	42.1	408.0	22.2	34.7	0	0	10.0	186.0	60.0	21.0	370.0
1689	**TOP SIRLOIN BUTT,** Good, boneless, trimmed, cooked	1 OZ	**28**	**45.7**	**102.2**	**6.7**	**8.1**	**0**	**0**	**3.1**	**55.4**	**16.8**	**5.9**	**103.6**
			100	45.7	365.0	24.1	29.1	0	0	11.0	198.0	60.0	21.0	370.0
1690	**TOP SIRLOIN BUTT STEAK,** Choice, cooked	1 OZ	**28**	**42.1**	**114.2**	**6.2**	**9.7**	**0**	**0**	**2.8**	**52.1**	**16.8**	**5.9**	**103.6**
			100	42.1	408.0	22.2	34.7	0	0	10.0	186.0	60.0	21.0	370.0
1691	**X-CUT SHANK,** Good, cooked	1 OZ	**28**	**51.0**	**86.0**	**7.6**	**5.9**	**0**	**0**	**3.4**	**38.1**	**16.8**	**4.2**	**103.6**
			100	51.0	307.0	27.2	21.1	0	0	12.0	136.0	60.0	15.0	370.0
1692	**6TH RIB,** Choice, total edible, cooked, 70% lean, 30% fat	1 OZ	**28**	**39.3**	**122.4**	**6.2**	**10.6**	**0**	**0**	**2.8**	**30.5**	**16.8**	**5.6**	**103.0**
			100	39.3	437.0	22.1	38.0	0	0	10.0	109.0	60.0	20.0	370.0
1693	**6TH RIB,** Choice, separable lean, cooked	1 OZ	**28**	**55.1**	**73.6**	**8.0**	**4.4**	**0**	**0**	**3.6**	**39.8**	**16.8**	**7.8**	**103.6**
			100	55.1	263.0	28.5	15.7	0	0	13.0	142.0	60.0	28.0	370.0
1694	**6TH RIB,** Good, total edible, cooked, 76% lean, 24% fat	1 OZ	**28**	**45.1**	**104.4**	**6.8**	**8.4**	**0**	**0**	**2.8**	**33.9**	**16.8**	**5.6**	**103.6**
			100	45.1	373.0	24.3	29.9	0	0	10.0	121.0	60.0	20.0	370.0
1695	**6TH RIB,** Good, separable lean, cooked	1 OZ	**28**	**58.6**	**63.0**	**8.3**	**3.1**	**0**	**0**	**3.6**	**40.9**	**16.8**	**7.8**	**103.6**
			100	58.6	225.0	29.6	10.9	0	0	13.0	146.0	60.0	28.0	370.0
1696	**11TH-12TH RIB,** Choice, total edible, cooked, 55% lean, 45% fat	1 OZ	**28**	**36.3**	**133.3**	**5.1**	**12.5**	**0**	**0**	**2.2**	**42.8**	**16.8**	**5.6**	**103.6**
			100	36.3	476.0	18.3	44.7	0	0	8.0	153.0	60.0	20.0	370.0
1697	**11TH-12TH RIB,** Choice, separable lean, cooked	1 OZ	**28**	**57.3**	**65.2**	**7.9**	**3.7**	**0**	**0**	**3.4**	**66.4**	**16.8**	**7.8**	**103.6**
			100	57.3	233.0	28.2	13.3	0	0	12.0	237.0	60.0	28.0	370.0
1698	**11TH-12TH RIB,** Good, total edible, cooked, 63% lean, 37% fat	1 OZ	**28**	**41.9**	**116.8**	**5.9**	**10.2**	**0**	**0**	**2.5**	**49.0**	**16.8**	**5.6**	**103.6**
			100	41.9	417.0	20.9	36.3	0	0	9.0	175.0	60.0	20.0	370.0
1699	**11TH-12TH RIB,** Good, separable lean, cooked	1 OZ	**28**	**59.7**	**60.2**	**8.1**	**2.9**	**0**	**0**	**3.1**	**67.8**	**16.8**	**7.8**	**103.6**
			100	59.7	215.0	28.9	10.2	0	0	12.0	242.0	60.0	28.0	370.0
1700	**ROAST BEEF,** canned	1 OZ	**28**	**60.0**	**62.7**	**7.0**	**3.6**	**0**	**0**	**4.5**	**32.5**	—	—	**72.5**
			100	60.0	224.0	25.0	13.0	0	0	16.0	116.0	—	—	259.0
1701	[4]**BOAR, WILD,** meat, raw	1 OZ	**28**	**74.1**	**41.2**	**4.7**	**2.3**	**0**	**0**	**3.4**	**33.6**	—	—	—
			100	74.1	147.0	16.8	8.3	0	0	12.0	120.0	—	—	—
1702	**BRAINS, BEEF,** cooked	1 CUP	**140**	—	**483.7**	**34.3**	**38.5**	**0**	**0**	**15.4**	**254.8**	**294.0**	**14.0**	**266.0**
			100	—	345.5	24.5	27.5	0	0	11.0	182.0	210.0	10.0	190.0
1703	[4]**BUFFALO, WATER(CARABAO)** meat, raw	1 OZ	**28**	**76.5**	**33.6**	**5.0**	**1.4**	**0**	**Trace**	**3.9**	**61.9**	**25.5**	—	**76.4**
			100	76.5	120.0	17.7	4.9	0	Trace	14.0	221.0	91.0	—	273.0
1704	**CHEVON**—See **GOAT MEAT**													
	CHICKEN													
	BROILERS—FRYERS													
1705	flesh, skin, giblets, & neck, fried, flour coated	1 OZ	**28**	**51.9**	**76.2**	**8.0**	**4.3**	**.9**	**Trace**	**4.8**	**54.3**	**24.1**	**7.0**	**66.4**
			100	51.9	272.0	28.6	15.3	3.3	.01	17.0	194.0	86.0	25.0	237.0

Item No.	Minerals (Micro)			Fat-Soluble Vitamins			Water-Soluble Vitamins								
	Iron	Zinc	Copper	Vitamin A	Vitamin D	Vitamin E (Alpha Tocopherol)	Vitamin C	Thiamin	Ribo-flavin	Niacin	Panto-thenic Acid	Vit. B-6 (Pyri-doxine)	Folacin (Folic Acid)	Biotin	Vitamin B-12
	(mg)	(mg)	(mg)	(IU)	(IU)	(mg)	(mg)	(mg)	(mg)	(mg)	(mg)	(mg)	(mcg)	(mcg)	(mcg)
1676	**1.06**	**1.74**	—	**2.8**	—	—	—	**.01**	**.06**	**1.29**	—	—	**1.12**	—	—
	3.80	6.20	—	10.0	—	—	—	.05	.23	4.60	—	—	4.00	—	—
1677	**.81**	—	—	**19.6**	—	—	—	**.01**	**.05**	**.95**	—	—	—	—	—
	2.90	—	—	70.0	—	—	—	.04	.17	3.40	—	—	—	—	—
1678	**.81**	**1.23**	—	**16.8**	—	—	**0**	**.02**	**.05**	**1.29**	—	—	—	—	—
	2.90	4.40	—	60.0	—	—	0	.06	.18	4.60	—	—	—	—	—
1679	**.81**	—	**.05**	**14.0**	—	—	**0**	**.02**	**.05**	**1.32**	—	—	—	—	**(.50)**
	2.90	—	.18	50.0	—	—	0	.06	.18	4.70	—	—	—	—	(1.80)
1680	**1.09**	**1.62**	**.05**	**2.8**	—	—	**0**	**.03**	**.07**	**1.79**	—	—	**1.12**	—	—
	3.90	5.80	.18	10.0	—	—	0	.09	.25	6.40	—	—	4.00	—	—
1681	**.87**	—	**.05**	**14.0**	—	—	**0**	**.02**	**.05**	**1.40**	—	—	—	—	—
	3.10	—	.18	50.0	—	—	0	.07	.19	5.00	—	—	—	—	—
1682	**1.06**	**1.62**	**.05**	**2.8**	—	—	**0**	**.02**	**.07**	**1.71**	—	—	**1.12**	—	—
	3.80	5.80	.18	10.0	—	—	0	.08	.24	6.10	—	—	4.00	—	—
1683	**.98**	**(2.43)**	**(.07)**	**8.4**	**Trace**	**(.09)**	**0**	**.02**	**.06**	**1.57**	**.15**	**.14**	**(4.58)**	**Trace**	**(.56)**
	3.50	(8.70)	(.25)	30.0	Trace	(.31)	0	.08	.22	5.60	.52	.50	(16.00)	Trace	(2.00)
1684	**.73**	—	—	**22.4**	—	**.04**	—	**.02**	**.05**	**1.15**	—	—	—	—	—
	2.60	—	—	80.0	—	.13	—	.06	.16	4.10	—	—	—	—	—
1685	**.76**	—	—	**19.6**	—	**.04**	—	**.02**	**.05**	**1.20**	—	—	—	—	—
	2.70	—	—	70.0	—	.13	—	.06	.17	4.30	—	—	—	—	—
1686	**1.04**	**1.62**	—	**2.8**	—	—	—	**.02**	**.07**	**1.68**	—	—	**1.12**	—	—
	3.70	5.80	—	10.0	—	—	—	.08	.24	6.00	—	—	4.00	—	—
1687	—	—	—	**0**	**0**	—	**0**	**.03**	**.13**	**.93**	—	—	—	—	—
	—	—	—	0	0	—	0	.10	.46	3.33	—	—	—	—	—
1688	**.81**	—	**.05**	**16.8**	—	—	**0**	**.02**	**.05**	**1.29**	—	—	—	—	—
	2.90	—	.18	60.0	—	—	0	.06	.18	4.60	—	—	—	—	—
1689	**.87**	—	**.05**	**14.0**	—	—	**0**	**.02**	**.05**	**1.37**	—	—	—	—	—
	3.10	—	.18	50.0	—	—	0	.07	.19	4.90	—	—	—	—	—
1690	**.81**	—	**.05**	**16.8**	—	—	**0**	**.02**	**.05**	**1.29**	—	—	—	—	**(108.68)**
	2.90	—	.18	60.0	—	—	0	.06	.18	4.60	—	—	—	—	(388.00)
1691	**1.01**	—	—	**11.2**	—	—	—	**.01**	**.06**	**1.18**	—	—	—	—	—
	3.60	—	—	40.0	—	—	—	.05	.21	4.20	—	—	—	—	—
1692	**.81**	—	—	**19.6**	—	—	**0**	**.01**	**.05**	**.95**	—	—	—	—	—
	2.90	—	—	70.0	—	—	0	.04	.17	3.40	—	—	—	—	—
1693	**1.04**	**1.62**	—	**8.4**	—	—	**0**	**.01**	**.06**	**1.23**	—	—	**1.12**	—	—
	3.70	5.80	—	30.0	—	—	0	.05	.22	4.40	—	—	4.00	—	—
1694	**.90**	—	—	**16.8**	—	—	**0**	**.01**	**.05**	**1.06**	—	—	—	—	—
	3.20	—	—	60.0	—	—	0	.05	.19	3.80	—	—	—	—	—
1695	**1.06**	**1.62**	—	**5.6**	—	—	**0**	**.01**	**.06**	**1.26**	—	—	**1.12**	—	—
	3.80	5.80	—	20.0	—	—	0	.05	.22	4.50	—	—	4.00	—	—
1696	**.67**	—	—	**25.2**	—	—	**0**	**.01**	**.04**	**.95**	—	—	—	—	—
	2.40	—	—	90.0	—	—	0	.05	.14	3.40	—	—	—	—	—
1697	**1.01**	**1.62**	—	**5.6**	—	—	**0**	**.02**	**.06**	**1.43**	—	—	**1.12**	—	—
	3.60	5.80	—	20.0	—	—	0	.07	.21	5.10	—	—	4.00	—	—
1698	**.76**	—	—	**19.6**	—	—	**0**	**.02**	**.05**	**1.06**	—	—	—	—	—
	2.70	—	—	70.0	—	—	0	.06	.16	3.80	—	—	—	—	—
1699	**1.04**	**1.62**	—	**5.6**	—	—	**0**	**.02**	**.06**	**1.46**	—	—	**1.12**	—	—
	3.70	5.80	—	20.0	—	—	0	.07	.22	5.20	—	—	4.00	—	—
1700	**.67**	—	—	—	—	—	**0**	**.01**	**.06**	**1.18**	—	—	—	—	—
	2.40	—	—	—	—	—	0	.02	.23	4.20	—	—	—	—	—
1701	—	—	—	**0**	—	—	**0**	**.11**	**.03**	**1.12**	—	—	—	—	—
	—	—	—	0	—	—	0	.39	.11	4.00	—	—	—	—	—
1702	**4.34**	**2.10**	**.59**	**70.0**	**Trace**	**3.22**	**0**	**.10**	**.27**	**7.00**	**1.96**	**.14**	**4.20**	**4.20**	**2.52**
	3.10	1.50	.42	50.0	Trace	2.30	0	.07	.19	5.00	1.40	.10	3.00	3.00	1.80
1703	**.92**	—	—	**5.1**	—	—	**0**	**.02**	**43.40**	**.98**	—	—	—	—	—
	3.30	—	—	18.3	—	—	0	.06	155.00	3.50	—	—	—	—	—
1704															
1705	**.56**	**.66**	**.03**	**231.3**	—	—	**.14**	**.03**	**.08**	**2.50**	**.36**	**.12**	**8.12**	—	**.31**
	1.99	2.36	.10	826.0	—	—	.50	.09	.28	8.93	1.28	.42	29.00	—	1.11

(Continued)

TABLE F-36 *(Continued)*

Item No.	Food Item Name	Approximate Measure	Weight	Moisture	Food Energy Calories	Protein	Fats	Carbo-hydrates	Fiber	Minerals (Macro) Calcium	Phos-phorus	Sodium	Mag-nesium	Potas-sium
			(g)	(%)	(kcal)	(g)	(g)	(g)	(g)	(mg)	(mg)	(mg)	(mg)	(mg)
	BROILERS—FRYERS (Continued)													
1706	flesh, w/skin, fried, flour coated	1 OZ	**28**	**52.4**	**75.3**	**8.0**	**4.2**	**.9**	**Trace**	**4.8**	**53.5**	**23.5**	**7.0**	**65.5**
			100	52.4	269.0	28.6	14.9	3.2	.01	17.0	191.0	84.0	25.0	234.0
1707	flesh, fried	1 OZ	**28**	**57.5**	**61.3**	**8.6**	**2.5**	**.5**	**Trace**	**4.8**	**57.4**	**25.5**	**7.6**	**72.0**
			100	57.5	219.0	30.6	9.1	1.7	.01	17.0	205.0	91.0	27.0	257.0
1708	dark meat, w/skin, fried, flour coated	1 OZ	**28**	**50.8**	**79.8**	**7.62**	**4.7**	**1.1**	**Trace**	**4.8**	**49.3**	**24.9**	**6.7**	**64.4**
			100	50.8	285.0	27.2	16.9	4.1	.02	17.0	176.0	89.0	24.0	230.0
1709	dark meat, w/o skin, fried	1 OZ	**28**	**55.7**	**66.9**	**8.1**	**3.2**	**.7**	**Trace**	**5.0**	**52.4**	**27.2**	**7.0**	**70.8**
			100	55.7	239.0	29.0	11.6	2.6	.01	18.0	187.0	97.0	25.0	253.0
1710	light meat, w/skin, fried, flour coated	1 OZ	**28**	**54.7**	**68.9**	**8.5**	**3.4**	**.5**	**Trace**	**4.5**	**59.6**	**21.6**	**7.6**	**66.9**
			100	54.7	246.0	30.5	12.1	1.8	.01	16.0	213.0	77.0	27.0	239.0
1711	light meat, w/o skin, fried	1 OZ	**28**	**60.1**	**53.4**	**9.2**	**1.5**	**.1**	**0**	**4.5**	**64.7**	**23.7**	**8.1**	**73.6**
			100	60.1	192.0	32.8	5.5	.4	0	16.0	231.0	81.0	29.0	263.0
1712	back, w/skin, fried, flour coated	1 OZ	**28**	**44.0**	**92.7**	**7.8**	**5.8**	**1.8**	**Trace**	**6.7**	**46.5**	**25.2**	**6.4**	**63.3**
			100	44.0	331.0	27.8	20.7	6.5	.03	24.0	166.0	90.0	23.0	226.0
1713	breast, w/skin, fried, flour coated	1 PIECE	**90**	**56.6**	**199.8**	**28.6**	**8.0**	**1.4**	**Trace**	**14.4**	**209.7**	**68.4**	**27.0**	**233.1**
			100	56.6	222.0	31.8	8.9	1.6	.01	16.0	233.0	76.0	30.0	259.0
1714	drumstick, w/skin, fried, flour coated	1 PIECE	**35**	**56.7**	**85.8**	**9.5**	**4.8**	**.57**	**Trace**	**4.2**	**61.6**	**31.2**	**8.1**	**80.2**
			100	56.7	245.0	27.0	13.7	1.63	.01	12.0	176.0	89.0	23.0	229.0
1715	giblets, fried, flour coated	1 OZ	**28**	**47.9**	**77.6**	**9.1**	**3.8**	**.9**	**Trace**	**5.0**	**80.1**	**31.6**	**7.0**	**92.4**
			100	47.9	277.0	32.5	13.5	4.4	.02	18.0	286.0	113.0	25.0	330.0
1716	neck, fried, flour coated	1 OZ	**28**	**47.5**	**93.0**	**6.7**	**6.6**	**1.2**	**Trace**	**8.7**	**37.0**	**23.0**	**5.3**	**33.0**
			100	47.5	332.0	24.0	23.6	4.2	.02	31.0	132.0	82.0	19.0	180.0
1717	skin, fried, flour coated	1 OZ	**28**	**28.5**	**140.6**	**5.3**	**11.9**	**2.6**	**.01**	**3.9**	**35.3**	**14.8**	**4.8**	**35.0**
			100	28.5	502.0	19.1	42.6	9.3	.04	14.0	126.0	53.0	17.0	125.0
1718	thigh, fried, flour coated	1 PIECE	**55**	**54.1**	**144.1**	**14.7**	**8.3**	**1.8**	**Trace**	**7.7**	**102.9**	**48.4**	**13.8**	**130.4**
			100	54.1	262.0	26.8	15.0	3.2	.01	14.0	187.0	88.0	25.0	237.0
1719	wing, fried, flour coated	1 PIECE	**20**	**48.6**	**64.2**	**5.2**	**4.4**	**.5**	**Trace**	**3.0**	**30.0**	**15.4**	**3.8**	**35.4**
			100	48.6	321.0	26.1	22.2	2.4	.01	15.0	150.0	77.0	19.0	177.0
1720	**CANNED,** meat only w/o bone	1 OZ	**28**	**52.6**	**55.4**	**6.1**	**3.3**	**0**	**0**	**5.9**	**69.2**	**152.0**	**5.3**	**38.6**
			100	52.6	198.0	21.7	11.7	0	0	21.0	247.0	543.0	19.0	138.0
1721	[6]**CANNED,** chicken liver pate	1 OZ	**28**	—	**72.0**	**3.6**	**5.5**	**1.9**	—	**26.6**	—	**383.6**	—	—
			100	—	257.0	12.9	19.5	6.8	—	95.0	—	1370.0	—	—
1722	[2]**FRANKFURTER**	1 OZ	**28**	—	**56.3**	**3.8**	**3.7**	**1.8**	—	**2.8**	—	—	—	—
			100	—	201.0	13.5	13.1	6.6	—	10.0	—	—	—	—
1723	[2]**MECHANICALLY DEBONED,** meat w/skin, raw	1 OZ	**28**	**62.7**	**76.2**	**3.2**	**6.9**	**0**	**0**	**38.6**	**37.0**	**11.2**	**3.4**	**29.1**
			100	62.7	272.0	11.4	24.7	0	0	138.0	132.0	40.0	12.0	104.0
1724	[2]**MECHANICALLY DEBONED,** meat w/o skin, raw	1 OZ	**28**	**69.3**	**55.7**	**3.9**	**4.3**	**0**	**0**	**34.4**	**43.1**	**14.3**	**3.6**	**35.8**
			100	69.3	199.0	13.8	15.5	0	0	123.0	154.0	51.0	13.0	128.0
	ROASTER													
1725	flesh, skin, giblets & neck, roasted	1 OZ	**28**	**62.3**	**61.6**	**6.7**	**3.7**	**0.1**	**0**	**3.6**	**50.1**	**19.9**	**5.6**	**57.1**
			100	62.3	220.0	24.0	13.1	0.5	0	13.0	179.0	71.0	20.0	204.0
1726	flesh & skin, roasted	1 OZ	**28**	**62.1**	**62.4**	**6.7**	**3.8**	**0**	**0**	**3.4**	**50.1**	**20.4**	**5.6**	**59.1**
			100	62.1	223.0	24.0	13.4	0	0	12.0	179.0	73.0	20.0	211.0
1727	flesh, roasted	1 OZ	**28**	**67.4**	**46.8**	**7.0**	**1.8**	**0**	**0**	**3.4**	**54.0**	**21.0**	**5.9**	**64.1**
			100	67.4	167.0	25.0	6.6	0	0	12.0	192.0	75.0	21.0	229.0
1728	dark meat, w/o skin, roasted	1 OZ	**28**	**67.1**	**49.8**	**6.5**	**2.5**	**0**	**0**	**3.1**	**47.9**	**26.6**	**5.6**	**62.7**
			100	67.1	178.0	23.3	8.8	0	0	11.0	171.0	95.0	20.0	224.0
1729	light meat. w/o skin, roasted	1 OZ	**28**	**67.9**	**42.8**	**7.6**	**1.1**	**0**	**0**	**3.6**	**60.8**	**14.3**	**6.4**	**66.1**
			100	67.9	153.0	27.1	4.1	0	0	13.0	217.0	51.0	23.0	236.0
1730	[2]**ROLL,** light meat	1 OZ	**28**	**68.6**	**44.5**	**5.5**	**2.1**	**.7**	—	**12.0**	**44.0**	**163.5**	**5.3**	**63.8**
			100	68.6	159.0	19.5	7.4	2.5	—	43.0	157.0	584.0	19.0	228.0
	STEWING													
1731	flesh, skin, giblets & neck, stewed	1 OZ	**28**	**54.3**	**77.3**	**7.4**	**5.0**	**Trace**	**0**	**3.6**	**51.0**	**19.9**	**5.6**	**49.6**
			100	54.3	276.0	26.5	18.0	.01	0	13.0	182.0	71.0	20.0	177.0
1732	flesh & skin, stewed	1 OZ	**28**	**53.1**	**79.8**	**7.5**	**5.3**	**0**	**0**	**3.6**	**50.4**	**20.4**	**5.6**	**51.0**
			100	53.1	285.0	26.9	18.9	0	0	13.0	180.0	73.0	20.0	182.0
1733	flesh, stewed	1 OZ	**28**	**56.3**	**66.4**	**8.5**	**3.3**	**0**	**0**	**3.6**	**57.1**	**21.8**	**6.2**	**56.6**
			100	56.3	237.0	30.4	11.9	0	0	13.0	204.0	78.0	22.0	202.0
1734	dark meat, w/o skin, stewed	1 OZ	**28**	**55.1**	**72.2**	**7.9**	**4.3**	**0**	**0**	**3.4**	**52.4**	**26.6**	**6.2**	**57.1**
			100	55.1	258.0	28.1	15.3	0	0	12.0	187.0	95.0	22.0	204.0

Item No.	Minerals (Micro)			Fat-Soluble Vitamins			Water-Soluble Vitamins								
	Iron	Zinc	Copper	Vitamin A	Vitamin D	Vitamin E (Alpha Tocopherol)	Vitamin C	Thiamin	Ribo-flavin	Niacin	Panto-thenic Acid	Vit. B-6 (Pyri-doxine)	Folacin (Folic Acid)	Biotin	Vitamin B-12
	(mg)	(mg)	(mg)	(IU)	(IU)	(mg)	(mg)	(mg)	(mg)	(mg)	(mg)	(mg)	(mcg)	(mcg)	(mcg)
1706	.39	.57	.02	24.9	—	—	0	.03	.05	2.52	.30	.11	1.68	—	.09
	1.38	2.04	.08	89.0	—	—	0	.09	.19	8.99	1.08	.41	6.00	—	.31
1707	.48	.63	.02	16.5	—	—	0	.03	.06	2.71	.33	.13	1.96	—	.10
	1.35	2.24	.08	59.0	—	—	0	.09	.20	9.66	1.17	.48	7.00	—	.34
1708	.42	.73	.03	29.1	—	—	0	.03	.07	1.92	.32	.09	2.24	—	.08
	1.50	2.60	.09	104.0	—	—	0	.10	.24	6.84	1.16	.32	8.00	—	.30
1709	.42	.81	.03	22.1	—	—	0	.03	.07	1.98	.35	.10	2.52	—	.09
	1.49	2.91	.09	79.0	—	—	0	.09	.25	7.07	1.26	.37	9.00	—	.33
1710	.34	.35	.02	19.0	—	—	0	.02	.04	3.37	.27	.15	1.12	—	.09
	1.21	1.26	.06	68.0	—	—	0	.08	.13	12.04	.97	.54	4.00	—	.33
1711	.32	.36	.01	8.4	—	—	0	.02	.04	3.74	.29	.18	1.12	—	.10
	1.14	1.27	.05	30.0	—	—	0	.07	.13	13.37	1.03	.63	4.00	—	.36
1712	.45	.69	.03	34.4	—	—	0	.03	.07	2.04	.31	.08	2.24	—	.08
	1.62	2.47	.09	123.0	—	—	0	.11	.24	7.30	1.09	.30	8.00	—	.28
1713	1.07	.99	.05	45.0	—	—	0	.07	.11	12.30	.09	.52	3.60	—	.31
	1.19	1.10	.06	50.0	—	—	0	.08	.13	13.70	1.00	.58	4.00	—	.34
1714	.47	1.01	.03	29.4	—	—	0	.03	.08	2.10	.43	.12	2.80	—	.11
	1.34	2.89	.08	84.0	—	—	0	.08	.23	6.00	1.22	.35	8.00	—	.32
1715	2.90	1.76	.12	3,340.1	—	—	2.4	.03	.43	3.10	1.25	.17	106.12	—	3.73
	10.30	6.27	.42	11,929.0	—	—	8.7	.10	1.52	11.00	4.45	.61	379.00	—	13.31
1716	.70	.86	.04	53.2	—	—	0	.02	.05	1.50	.27	.07	1.68	—	.07
	2.40	3.07	.13	190.0	—	—	0	.08	.26	5.35	.98	.25	6.00	—	.26
1717	.43	.32	.02	65.0	—	—	0	.03	.05	1.6	.19	.03	1.12	—	.05
	1.52	1.15	.08	232.0	—	—	0	.10	.17	5.8	.67	.10	4.00	—	.18
1718	.82	1.38	.05	—	—	—	0	.05	.13	3.8	.64	.18	4.40	—	.17
	1.49	2.52	.09	98.0	—	—	0	.09	.24	6.9	1.19	.33	8.00	—	.30
1719	.25	.35	.01	25.0	—	—	0	.01	.03	1.3	.17	.08	.60	—	.05
	1.25	1.76	.06	125.0	—	—	0	.06	.14	6.7	.87	.41	3.00	—	.28
1720	.42	—	—	64.4	—	—	1.12	.01	.03	1.23	.24	.08	1.18	—	.22
	1.50	—	—	230.0	—	—	4.00	.04	.12	4.40	.85	.30	4.20	—	.79
1721	.56	—	—	—	—	—	—	.02	.03	.87	—	—	—	—	—
	2.00	—	—	—	—	—	—	.07	.12	3.09	—	—	—	—	—
1722	2.57	—	—	202.7	—	—	2.72	.02	.39	2.10	—	—	—	—	—
	9.19	—	—	724.0	—	—	9.70	.05	1.40	7.52	—	—	—	—	—
1723	.44	.36	.02	—	—	—	—	—	—	—	—	—	—	—	—
	1.57	1.29	.07	—	—	—	—	—	—	—	—	—	—	—	—
1724	.48	.51	.02	—	—	—	—	—	—	—	—	—	—	—	—
	1.73	1.82	.07	—	—	—	—	—	—	—	—	—	—	—	—
1725	.45	.48	.02	167.2	—	—	.11	.02	.05	1.97	.28	.10	6.44	—	.22
	1.61	1.71	.07	597.0	—	—	.40	.06	.19	7.03	1.00	.34	23.00	—	.78
1726	.35	.41	.02	23.2	—	—	0	.02	.04	2.08	.26	.10	1.40	—	.08
	1.26	1.45	.06	83.0	—	—	0	.06	.14	7.42	.92	.35	5.00	—	.27
1727	.34	.43	.02	11.5	—	—	0	.02	.04	2.21	.27	.11	1.40	—	.08
	1.21	1.52	.06	41.0	—	—	0	.06	.15	7.88	.97	.41	5.00	—	.29
1728	.37	.60	.02	15.1	—	—	0	.02	.05	1.61	.29	.09	1.96	—	.08
	1.33	2.13	.07	54.0	—	—	0	.06	.19	5.74	1.03	.31	7.00	—	.27
1729	.30	.22	.01	7.0	—	—	0	.02	.03	2.93	.25	.15	.84	—	.09
	1.08	.78	.04	25.0	—	—	0	.06	.09	10.47	.91	.54	3.00	—	.31
1730	.27	.20	.01	—	—	—	—	.02	.04	1.48	—	—	—	—	—
	.97	.72	.04	—	—	—	—	.07	.13	5.29	—	—	—	—	—
1731	.51	.57	.03	259.0	—	—	.14	.03	.08	1.58	.26	.07	10.08	—	.28
	1.83	2.02	.12	925.0	—	—	.50	.09	.30	5.64	.93	.26	36.00	—	1.00
1732	.38	.50	.03	36.7	—	—	0	.03	.07	1.62	.21	.07	1.40	—	.06
	1.37	1.77	.10	131.0	—	—	0	.09	.24	5.80	.75	.25	5.00	—	.23
1733	.40	.58	.03	31.4	—	—	0	.03	.08	1.79	.24	.09	1.68	—	.07
	1.43	2.06	.12	112.0	—	—	0	.11	.28	6.41	.86	.31	6.00	—	.26
1734	.46	.87	.04	40.6	—	—	0	.04	—	.10	.28	.07	2.24	—	.07
	1.64	3.12	.14	145.0	—	—	0	.13	.35	4.56	1.01	.24	8.00	—	.25

(Continued)

TABLE F-36 *(Continued)*

Item No.	Food Item Name	Approximate Measure	Weight	Moisture	Food Energy Calories	Protein	Fats	Carbo-hydrates	Fiber	Minerals (Macro) Calcium	Phos-phorus	Sodium	Mag-nesium	Potas-sium
			(g)	(%)	(kcal)	(g)	(g)	(g)	(g)	(mg)	(mg)	(mg)	(mg)	(mg)
	STEWING (Continued)													
1735	light meat, w/o skin, stewed	1 OZ	28	57.8	59.6	9.2	2.2	0	0	3.9	63.0	16.2	6.4	55.7
			100	57.8	213.0	33.0	8.0	0	0	14.0	225.0	58.0	23.0	199.0
1736	[1]**CORNISH HEN** roasted, whole	1 OZ	28	71.0	38.1	6.7	1.1	0	0	2.5	56.3	18.5	—	76.7
			100	71.0	136.0	23.8	3.8	0	0	9.0	201.0	66.0	—	274.0
	DUCK													
1737	**DOMESTIC,** total edible, raw	1 OZ	28	54.3	91.3	4.5	8.0	0	0	2.8	49.3	23.0	—	79.8
			100	54.3	326.0	16.0	28.6	0	0	10.0	176.0	82.0	—	285.0
1738	**DOMESTIC,** flesh only, raw	1 OZ	28	68.8	46.2	6.0	2.3	0	0	3.4	56.8	20.7	—	79.8
			100	68.8	165.0	21.4	8.2	0	0	12.0	203.0	74.0	—	285.0
1739	[2]**DOMESTIC,** flesh only, roasted	1 OZ	28	64.2	56.3	6.6	3.1	0	0	3.4	56.8	18.2	5.6	70.6
			100	64.2	201.0	23.5	11.2	0	0	12.0	203.0	65.0	20.0	252.0
1740	**WILD,** total edible, raw	1 OZ	28	61.1	65.2	5.9	4.4	0	0	3.4	56.0	—	—	—
			100	61.1	233.0	21.1	15.8	0	0	12.0	200.0	—	—	—
1741	**WILD,** flesh only, raw	1 OZ	28	70.8	38.6	6.0	1.5	0	0	—	—	—	—	—
			100	70.8	138.0	21.4	5.2	0	0	—	—	—	—	—
	FRANKFURTERS (Wieners)													
1742	all meat, cooked	1 AVG	50	—	124.0	7.0	10.0	1.0	—	3.0	25.0	542.0	—	108.0
			100	—	248.0	14.0	20.0	2.0	—	6.0	50.0	1084.0	—	216.0
1743	w/cereal, raw	1 AVG	50	61.7	124.0	7.2	10.3	.1	—	4.0	—	—	8.0	—
			100	61.7	248.0	14.4	20.6	.2	—	8.0	—	—	16.0	—
1744	w/non fat dehydrated milk, raw	1 AVG	50	54.2	150.0	6.6	12.8	1.7	—	4.0	—	—	8.0	—
			100	54.2	300.0	13.1	25.6	3.4	—	8.0	—	—	16.0	—
1745	w/non fat dehydrated milk & cereal, raw	1 AVG	50	50.5	128.5	7.1	10.3	1.4	—	4.0	50.0	550.0	8.0	110.0
			100	50.5	257.0	14.2	20.5	2.7	—	8.0	100.0	1100.0	16.0	220.0
1746	canned	1 AVG	50	66.0	110.5	6.7	9.1	.1	0	4.5	72.5	—	—	—
			100	66.0	221.0	13.4	18.1	.2	0	9.0	145.0	—	—	—
1747	cooked	1 AVG	50	57.3	153.5	7.0	13.6	.8	—	2.5	51.0	542.0	—	108.5
			100	57.3	307.0	14.0	27.2	1.6	—	5.0	102.0	1084.0	—	217.0
1748	**FROG LEGS,** fried	1 LG	24	—	69.6	4.3	4.8	2.0	0	4.6	38.4	—	—	—
			100	—	290.0	17.9	19.8	8.5	0	19.0	160.0	—	—	—
	GIZZARD													
1749	**CHICKEN,** simmered	1 OZ	28	68.0	41.4	7.6	.9	.2	0	2.5	19.9	16.0	—	59.1
			100	68.0	148.0	27.0	3.3	.7	0	9.0	71.0	57.0	—	211.0
1750	**GOOSE,** raw	1 OZ	28	73.0	38.9	6.0	1.5	0	0	—	—	—	—	—
			100	73.0	139.0	21.4	5.3	0	0	—	—	—	—	—
1751	**TURKEY,** simmered	1 OZ	28	62.7	54.9	7.5	2.4	.3	0	—	—	14.3	—	41.7
			100	62.7	196.0	26.8	8.6	1.1	0	—	—	51.0	—	149.0
1752	[3]**GOAT MEAT, (CHEVON),** carcass	1 OZ	28	71.0	46.2	5.2	2.6	0	—	3.1	—	—	—	—
			100	71.0	165.0	18.7	9.4	0	—	11.0	—	—	—	—
	GOOSE													
1753	total edible, roasted	1 OZ	28	39.1	119.3	6.6	10.1	0	0	3.1	67.2	—	—	—
			100	39.1	426.0	23.7	36.0	0	0	11.0	240.0	—	—	—
1754	flesh w/skin, roasted	1 OZ	28	37.9	123.5	6.4	10.7	0	0	3.6	72.8	—	—	—
			100	37.9	441.0	22.9	38.1	0	0	13.0	260.0	—	—	—
1755	flesh, roasted	1 OZ	28	54.8	65.2	9.5	2.7	0	0	3.9	77.6	34.7	(8.7)	169.4
			100	54.8	233.0	33.9	9.8	0	0	14.0	277.0	124.0	(31.0)	605.0
1756	giblets, raw	1 OZ	28	69.9	43.7	5.9	2.0	.2	0	—	—	—	—	—
			100	69.9	156.0	21.1	7.0	.6	0	—	—	—	—	—
	GUINEA HEN													
1757	total edible, raw	1 OZ	28	69.0	43.7	6.5	1.8	0	0	—	—	—	—	—
			100	69.0	156.0	23.1	6.4	0	0	—	—	—	—	—
1758	flesh & skin, raw	1 OZ	28	68.9	44.2	6.6	1.8	0	0	—	—	—	—	—
			100	68.9	158.0	23.4	6.4	0	0	—	—	—	—	—
1759	giblets, raw	1 OZ	28	69.8	44.0	5.8	2.0	.3	0	—	—	—	—	—
			100	69.8	157.0	20.8	7.0	1.2	0	—	—	—	—	—
1760	[4]**GUINEA PIG,** meat	1 OZ	28	78.2	26.9	5.3	.4	(0)	—	8.1	70.8	—	—	—
			100	78.2	96.0	19.0	1.6	(0)	—	29.0	253.0	—	—	—
1761	**HAMBURGER GRAVY**	1 OZ	28	25.0	46.2	3.2	3.6	.4	0	2.4	25.0	28.2	2.9	57.4
			100	25.0	165.0	11.3	12.7	1.5	Trace	8.4	89.1	100.7	10.3	204.8
1762	**HAMBURGER HELPER,** cooked	1 CUP	226	—	720.9	45.9	37.5	48.4	—	56.5	488.2	—	85.9	—
			100	—	319.0	20.3	16.6	21.4	—	25.0	216.0	—	38.0	—
	HEART													
1763	**BEEF,** lean, w/fat, braised	1 OZ	28	44.4	104.2	7.2	8.1	0	0	—	47.3	—	—	—
			100	44.4	372.0	25.8	29.0	.1	0	—	169.0	—	—	—

(Continued)

Item No.	Minerals (Micro) Iron	Zinc	Copper	Fat-Soluble Vitamins Vitamin A	Vitamin D	Vitamin E (Alpha Tocopherol)	Water-Soluble Vitamins Vitamin C	Thiamin	Ribo-flavin	Niacin	Panto-thenic Acid	Vit. B-6 (Pyri-doxine)	Folacin (Folic Acid)	Biotin	Vitamin B-12
	(mg)	(mg)	(mg)	(IU)	(IU)	(mg)	(mg)	(mg)	(mg)	(mg)	(mg)	(mg)	(mcg)	(mcg)	(mcg)
1735	**.33**	**.23**	**.03**	**20.4**	—	—	**0**	**.03**	**.06**	**2.39**	**.19**	**.11**	**1.12**	—	**.08**
	1.19	.83	.09	73.0	—	—	0	.09	.20	8.54	.69	.39	4.00	—	.27
1736	**.48**	—	—	**25.2**	—	—	—	**.01**	**.05**	**2.46**	—	—	—	—	—
	1.70	—	—	90.0	—	—	—	.05	.19	8.80	—	—	—	—	—
1737	**.45**	—	**.11**	**0**	—	—	**2.24**	**.02**	**.05**	**1.88**	—	—	—	—	—
	1.60	—	.40	0	—	—	8.00	.08	.19	6.70	—	—	—	—	—
1738	**.36**	—	—	**0**	—	—	—	**.03**	**.03**	**2.16**	—	—	—	—	—
	1.30	—	—	0	—	—	—	.10	.12	7.70	—	—	—	—	—
1739	**.76**	**.73**	**.07**	**21.6**	—	—	**0**	**.07**	**.13**	**1.43**	**.42**	**.07**	**2.80**	—	**.11**
	2.70	2.60	.23	77.0	—	—	0	.26	.47	5.10	1.50	.25	10.00	—	.40
1740	**.84**	—	—	**0**	—	—	—	—	—	—	—	—	—	—	—
	3.00	—	—	0	—	—	—	—	—	—	—	—	—	—	—
1741	—	—	—	—	—	—	—	—	—	—	—	—	—	—	—
	—	—	—	—	—	—	—	—	—	—	—	—	—	—	—
1742	**.60**	**.80**	—	**0**	**0**	—	**0**	**.08**	**.09**	**.60**	—	—	—	—	—
	1.20	1.60	—	0	0	—	0	.16	.18	1.20	—	—	—	—	—
1743	—	—	—	—	—	—	**0**	—	—	—	**.21**	**.07**	—	—	—
	—	—	—	—	—	—	0	—	—	—	.43	.13	—	—	—
1744	—	—	—	—	—	—	**0**	—	—	—	**.21**	**.07**	—	—	**.84**
	—	—	—	—	—	—	0	—	—	—	.43	.13	—	—	1.68
1745	**.75**	—	—	**0**	**0**	—	**0**	**.09**	**.10**	**1.40**	**.21**	**.07**	—	—	—
	1.50	—	—	0	0	—	0	.18	.19	2.80	.43	.13	—	—	—
1746	**1.10**	—	—	—	—	—	—	**.02**	**.06**	**1.20**	**.10**	**.02**	—	—	**.25**
	2.20	—	—	—	—	—	—	.03	.12	2.40	.20	.03	—	—	.50
1747	**.75**	—	—	**0**	**0**	—	**0**	**.08**	**.10**	**1.25**	—	—	—	—	—
	1.50	—	—	0	0	—	0	.15	.20	2.50	—	—	—	—	—
1748	**.34**	—	—	**0**	**0**	—	—	**.03**	**.06**	**.30**	—	—	—	—	—
	1.40	—	—	0	0	—	—	.12	.24	1.25	—	—	—	—	—
1749	**.87**	**1.20**	—	—	—	—	—	**.01**	**.06**	**1.43**	—	—	—	—	—
	3.10	4.30	—	—	—	—	—	.02	.21	5.10	—	—	—	—	—
1750	—	—	—	—	—	—	—	—	—	—	—	—	—	—	—
	—	—	—	—	—	—	—	—	—	—	—	—	—	—	—
1751	—	**1.15**	—	—	—	—	—	**.01**	**.04**	**1.62**	—	—	—	—	—
	—	4.10	—	—	—	—	—	.03	.14	5.80	—	—	—	—	—
1752	**.62**	—	—	**0**	—	—	**0**	**(.05)**	**.09**	**(1.57)**	—	—	—	—	—
	2.20	—	—	0	—	—	0	(.17)	.32	(5.60)	—	—	—	—	—
1753	**.59**	—	—	**0**	—	—	—	**.02**	**.07**	**2.27**	—	—	—	—	—
	2.10	—	—	0	—	—	—	.08	.24	8.10	—	—	—	—	—
1754	**.53**	—	—	—	—	—	—	**.03**	**.05**	**2.49**	—	—	—	—	—
	1.90	—	—	—	—	—	—	.09	.16	8.90	—	—	—	—	—
1755	**.48**	—	**(.14)**	—	—	—	—	**.03**	**.05**	**2.60**	—	**(.12)**	—	—	—
	1.70	—	(.49)	—	—	—	—	.11	.16	9.30	—	(.43)	—	—	—
1756	—	—	—	—	—	—	—	—	—	—	—	—	—	—	—
	—	—	—	—	—	—	—	—	—	—	—	—	—	—	—
1757	—	—	—	—	—	—	—	—	—	—	—	—	—	—	—
	—	—	—	—	—	—	—	—	—	—	—	—	—	—	—
1758	—	—	—	—	—	—	—	—	—	—	—	—	—	—	—
	—	—	—	—	—	—	—	—	—	—	—	—	—	—	—
1759	—	—	—	—	—	—	—	—	—	—	—	—	—	—	—
	—	—	—	—	—	—	—	—	—	—	—	—	—	—	—
1760	**.53**	—	—	—	—	—	—	**.02**	**.04**	**1.82**	—	—	—	—	—
	1.90	—	—	—	—	—	—	.06	.14	6.50	—	—	—	—	—
1761	**.42**	**.01**	—	**32.7**	**0**	—	**0**	**.01**	**.03**	**.69**	**0**	**0**	**.11**	—	**0**
	1.49	.02	—	116.6	0	—	.01	.05	.10	2.48	.01	0	.40	—	0
1762	**6.78**	—	—	**176.3**	—	—	**2.26**	**.36**	**.52**	**11.30**	—	**.81**	—	—	**2.53**
	3.00	—	—	78.0	—	—	1.00	.16	.23	5.00	—	.36	—	—	1.12
1763	—	—	—	—	—	—	—	—	—	—	—	—	—	—	—
	—	—	—	—	—	—	—	—	—	—	—	—	—	—	—

(Continued)

TABLE F-36 *(Continued)*

Item No.	Food Item Name	Approximate Measure	Weight	Moisture	Food Energy Calories	Protein	Fats	Carbo-hydrates	Fiber	Minerals (Macro) Calcium	Phos-phorus	Sodium	Mag-nesium	Potas-sium
			(g)	(%)	(kcal)	(g)	(g)	(g)	(g)	(mg)	(mg)	(mg)	(mg)	(mg)
	HEART (Continued)													
1764	**BEEF** lean, braised	1 OZ	**28**	**61.3**	**50.1**	**8.8**	**1.6**	**.2**	**0**	**1.7**	**50.7**	**29.1**	**(8.1)**	**65.0**
			100	61.3	179.0	31.3	5.7	.7	0	6.0	181.0	104.0	(29.0)	232.0
1765	**CALF,** braised	1 OZ	**28**	**60.3**	**58.2**	**7.8**	**2.5**	**.5**	**0**	**1.1**	**41.4**	**31.6**	–	**70.0**
			100	60.3	208.0	27.8	9.1	1.8	0	4.0	148.0	113.0	–	250.0
1766	**CHICKEN,** simmered	1 OZ	**28**	**64.9**	**46.5**	**7.1**	**2.0**	**0**	**0**	**1.1**	**30.0**	**19.3**	–	**39.2**
			100	64.9	165.0	25.3	7.2	.1	0	4.0	107.0	69.0	–	140.0
1767	**HOG,** braised	1 OZ	**28**	**61.0**	**54.6**	**8.6**	**1.9**	**.1**	**0**	**1.1**	**33.9**	**18.2**	–	**35.8**
			100	61.0	195.0	30.8	6.9	.3	0	4.0	121.0	65.0	–	128.0
1768	**LAMB,** braised	1 OZ	**28**	**54.1**	**72.8**	**8.3**	**4.0**	**.3**	**0**	**4.0**	**64.7**	**(42.0)**	**(1.0)**	**(103.6)**
			100	54.1	260.0	29.5	14.4	1.0	0	14.0	231.0	(150.0)	(35.0)	(370.0)
1769	**TURKEY,** simmered	1 OZ	**28**	**64.2**	**58.8**	**6.3**	**3.7**	**.1**	**0**	–	–	**17.1**	–	**59.1**
			100	64.2	210.0	22.6	13.2	.2	0	–	–	61.0	–	211.0
1770	[5]**HORSE MEAT,** carcass	1 OZ	**28**	**76.0**	**33.0**	**5.1**	**1.1**	**.3**	**0**	**2.8**	**42.0**	–	–	–
			100	76.0	118.0	18.1	4.1	.9	0	10.8	150.0	–	–	–
1771	[5]**INTESTINES,** beef	1 OZ	**28**	**69.2**	**61.6**	**3.1**	**5.3**	**0**	**0**	**3.4**	**30.8**	–	–	–
			100	69.2	220.0	11.0	19.1	0	0	12.0	110.0	–	–	–
	KIDNEYS													
1772	**BEEF,** braised	1 OZ	**28**	**53.0**	**70.6**	**9.2**	**3.4**	**.2**	**0**	**5.0**	**68.3**	**70.8**	**(5.3)**	**90.7**
			100	53.0	252.0	33.0	12.0	.8	0	18.0	244.0	253.0	(19.0)	324.0
1773	**CALF,** raw	1 OZ	**28**	**77.4**	**31.6**	**4.6**	**1.3**	**0**	**0**	**2.5**	**47.9**	**66.6**	**2.8**	**67.2**
			100	77.4	113.0	16.6	4.6	.1	0	9.0	171.0	238.0	10.0	240.0
1774	**HOG,** raw	1 OZ	**28**	**77.8**	**29.7**	**4.6**	**1.0**	**.3**	**0**	**3.1**	**61.0**	**32.2**	**4.8**	**49.8**
			100	77.8	106.0	16.3	3.6	1.1	0	11.0	218.0	115.0	17.0	178.0
1775	**LAMB,** raw	1 OZ	**28**	**77.7**	**29.4**	**4.7**	**.9**	**.3**	**0**	**3.6**	**61.0**	**56.0**	**4.5**	**64.4**
			100	77.7	105.0	16.8	3.3	.9	0	13.0	218.0	200.0	16.0	230.0
	LAMB													
1776	**COMPOSITE OF CUTS,** trimmed, Prime, 72% lean, 28% fat	1 OZ	**28**	**56.3**	**86.8**	**4.3**	**7.6**	**0**	**0**	**2.5**	**37.8**	**21.0**	**4.2**	**82.6**
			100	56.3	310.0	15.4	27.1	0	0	9.0	135.0	75.0	15.0	295.0
1777	**COMPOSITE OF CUTS,** trimmed, Choice, 77% lean, 23% fat	1 OZ	**28**	**61.0**	**73.6**	**4.6**	**6.0**	**0**	**0**	**2.8**	**41.2**	**21.0**	**4.2**	**82.6**
			100	61.0	263.0	16.5	21.3	0	0	10.0	147.0	75.0	15.0	295.0
1778	**COMPOSITE OF CUTS,** trimmed, Good, 79% lean, 21% fat	1 OZ	**28**	**62.5**	**69.2**	**4.7**	**5.4**	**0**	**0**	**2.8**	**42.3**	**21.0**	**4.2**	**82.6**
			100	62.5	247.0	16.8	19.4	0	0	10.0	151.0	75.0	15.0	295.0
1779	**FAT,** cooked	1 OZ	**28**	**17.7**	**198.5**	**1.8**	**21.2**	**0**	**0**	–	–	**19.6**	–	**81.2**
			100	17.7	709.0	6.3	75.6	0	0	–	–	70.0	–	290.0
1780	**LEG,** Prime, roasted, 79% lean, 21% fat	1 OZ	**28**	**50.4**	**89.3**	**6.7**	**6.7**	**0**	**0**	**2.8**	**54.6**	**19.6**	**5.9**	**81.2**
			100	50.4	319.0	23.9	24.0	0	0	10.0	195.0	70.0	21.0	290.0
1781	**LEG,** Prime, separable lean, roasted	1 OZ	**28**	**61.6**	**53.8**	**8.0**	**2.2**	**0**	**0**	**3.4**	**66.4**	**21.0**	**5.0**	**82.6**
			100	61.6	192.0	28.6	7.7	0	0	12.0	237.0	75.0	18.0	295.0
1782	**LEG,** Choice, total edible, roasted, 83% lean, 17% fat, w/o bone	1 OZ	**28**	**54.0**	**78.1**	**7.1**	**5.3**	**0**	**0**	**3.1**	**58.2**	**19.6**	**5.9**	**81.2**
			100	54.0	279.0	25.3	18.9	0	0	11.0	208.0	70.0	21.0	290.0
1783	**LEG,** Choice, separable lean, roasted, w/o bone	1 OZ	**28**	**62.2**	**52.1**	**8.0**	**2.0**	**0**	**0**	**3.6**	**66.6**	**19.6**	**5.0**	**81.2**
			100	62.2	186.0	28.7	7.0	0	0	13.0	238.0	70.0	18.0	290.0
1784	**LEG,** Good, total edible, roasted, 85% lean, 15% fat	1 OZ	**28**	**55.1**	**74.5**	**7.2**	**4.8**	**0**	**0**	**3.1**	**59.4**	**19.6**	**5.9**	**81.2**
			100	55.1	266.0	25.8	17.3	0	0	11.0	212.0	70.0	21.0	290.0
1785	**LEG,** Good, separable lean, roasted	1 OZ	**28**	**62.4**	**51.2**	**8.0**	**1.9**	**0**	**0**	**3.4**	**66.6**	**19.6**	**5.0**	**81.2**
			100	62.4	183.0	28.7	6.7	0	0	12.0	238.0	70.0	18.0	290.0
1786	**LOIN,** Prime, chops, total edible, broiled, 61% lean, 39% fat	1 MED	**46**	**41.7**	**193.2**	**9.0**	**17.2**	**0**	**0**	**3.7**	**69.0**	**32.2**	**7.8**	**133.4**
			100	41.7	420.0	19.5	37.3	0	0	8.0	150.0	70.0	17.0	290.0
1787	**LOIN,** Prime, separable lean, broiled	1 OZ	**28**	**61.3**	**55.2**	**7.8**	**2.4**	**0**	**0**	**3.1**	**61.0**	**19.6**	**6.2**	**81.2**
			100	61.3	197.0	28.0	8.6	0	0	11.0	218.0	70.0	22.0	290.0
1788	**LOIN,** Choice, chops, total edible, broiled, 66% lean, 34% fat, w/bone	1 MED	**46**	**47.0**	**165.1**	**10.1**	**13.5**	**0**	**0**	**4.1**	**79.1**	**32.2**	**7.8**	**133.4**
			100	47.0	359.0	22.0	29.4	0	0	9.0	172.0	70.0	17.0	290.0
1789	**LOIN,** Choice, separable lean, broiled, w/bone	1 OZ	**28**	**62.1**	**52.6**	**7.9**	**2.1**	**0**	**0**	**3.4**	**61.3**	**19.6**	**6.2**	**81.2**
			100	62.1	188.0	28.2	7.5	0	0	12.0	219.0	70.0	22.0	290.0
1790	**LOIN,** Good, chops, total edible, broiled, 67% lean, 33% fat	1 MED	**46**	**48.6**	**156.9**	**10.5**	**12.4**	**0**	**0**	**4.6**	**82.3**	**32.2**	**7.8**	**133.4**
			100	48.6	341.0	22.8	27.0	0	0	10.0	179.0	70.0	17.0	290.0
1791	**LOIN,** Good, separable lean, broiled	1 OZ	**28**	**62.5**	**51.5**	**7.9**	**2.0**	**0**	**0**	**3.1**	**61.3**	**19.6**	**6.2**	**81.2**
			100	62.5	184.0	28.2	7.1	0	0	11.0	219.0	70.0	22.0	290.0
1792	**RIB,** Prime, chops, total edible, broiled, 53% lean, 47% fat	1 MED	**41**	**35.5**	**201.7**	**6.9**	**19.1**	**0**	**0**	**2.9**	**52.5**	**28.7**	**10.2**	**118.9**
			100	35.5	492.0	16.9	46.5	0	0	7.0	128.0	70.0	24.9	290.0

Item No.	Minerals (Micro)			Fat-Soluble Vitamins			Water-Soluble Vitamins								
	Iron	Zinc	Copper	Vitamin A	Vitamin D	Vitamin E (Alpha Tocopherol)	Vitamin C	Thiamin	Ribo-flavin	Niacin	Panto-thenic Acid	Vit. B-6 (Pyri-doxine)	Folacin (Folic Acid)	Biotin	Vitamin B-12
	(mg)	(mg)	(mg)	(IU)	(IU)	(mg)	(mg)	(mg)	(mg)	(mg)	(mg)	(mg)	(mcg)	(mcg)	(mcg)
1764	**1.65** 5.90	**(.98)** (3.50)	**(0.20)** (0.73)	**8.4** 30.0	**0** 0	**(0.20)** (0.72)	**.28** 1.00	**.07** .25	**.34** 1.22	**2.13** 7.60	**(.45)** (1.60)	**.03** (.11)	**(.56)** (2.00)	**(1.12)** (4.00)	**(4.20)** (15.00)
1765	**1.23** 4.40	– –	– –	**11.2** 40.0	– –	– –	– –	**.08** .29	**.40** 1.44	**2.27** 8.10	– –	– –	– –	– –	– –
1766	**1.01** 3.60	**1.34** 4.80	– –	**8.4** 30.0	– –	– –	**1.12** 4.00	**.02** .06	**.26** .92	**1.48** 5.30	– –	– –	– –	– –	– –
1767	**1.37** 4.90	– –	– –	**11.2** 40.0	– –	– –	**.28** 1.00	**.18** .64	**.53** 1.89	**1.88** 6.70	– –	– –	– –	– –	– –
1768	**(2.27)** (8.10)	– –	– –	**28.0** 100.0	– –	**(.20)** (.70)	**(3.08)** (11.00)	**.26** .93	**.45** 1.62	**1.79** 6.40	**(1.06)** (3.80)	**(.11)** (.38)	**(1.12)** (4.00)	**(2.24)** (8.00)	**(3.92)** (14.00)
1769	– –	**1.34** 4.80	– –	**8.4** 30.0	– –	– –	**1.12** 4.00	**.07** .25	**.27** .98	**1.60** 5.70	– –	– –	– –	– –	– –
1770	**(.76)** (2.70)	– –	– –	– –	– –	– –	– –	**.02** .07	**.03** .12	**(1.20)** (4.30)	– –	– –	– –	– –	– –
1771	**.50** 1.80	– –	– –	**0** 0	– –	– –	**0** 0	**.01** .04	**.05** .16	**.56** 2.00	– –	– –	– –	– –	– –
1772	**3.67** 13.10	**(.84)** (3.00)	**(.18)** (.66)	**322.0** 1150.0	– –	**(.12)** (.42)	**0** 0	**.19** .67	**1.28** 4.58	**3.00** 10.70	**(.84)** (3.00)	**(.08)** (.30)	**(21.01)** (75.00)	**(6.72)** (24.0)	**(8.68)** (31.00)
1773	**1.12** 4.00	– –	– –	**336.0** 1200.0	– –	– –	**1.66** 6.00	**.07** .26	**.67** 2.40	**2.07** 7.40	**1.12** 4.00	**.12** .41	**22.40** 80.00	– –	**7.00** 25.00
1774	**1.88** 6.70	**.87** 3.12	**.14** .49	**36.4** 130.0	**0** 0	**(.11)** (.38)	**3.36** 12.00	**.16** .58	**.48** 1.73	**2.74** 9.80	**.90** 3.20	**.13** .45	**(11.76)** (42.00)	**(8.96)** (32.0)	**2.66** 9.50
1775	**2.13** 7.60	**(.67)** (2.40)	**.03** .10	**193.2** 690.0	**0** 0	**(.13)** (.45)	**4.20** 15.00	**.14** .51	**.68** 2.42	**2.07** 7.40	**1.26** 4.50	**(.08)** (.30)	**11.76** 42.00	**(10.36)** (37.0)	**17.64** 63.00
1776	**.31** 1.10	– –	– –	**0** 0	**0** 0	– –	**0** 0	**.04** .14	**.05** .19	**1.26** 4.50	**.15** .55	**.08** .28	**1.12** 4.00	– –	**.60** 2.15
1777	**.34** 1.20	– –	– –	**0** 0	**0** 0	– –	**0** 0	**.04** .15	**.06** .20	**1.34** 4.80	**.15** .55	**.08** .28	**1.12** 4.00	– –	**.60** 2.15
1778	**.36** 1.30	[6]**1.12** 4.0	– –	**0** 0	**0** 0	– –	**0** 0	**.04** .15	**.06** .21	**1.37** 4.90	**.15** .55	**.08** .28	**1.12** 4.00	– –	**.60** 2.15
1779	– –	– –	– –	**0** 0	**0** 0	– –	**0** 0	– –	– –	– –	– –	– –	– –	– –	– –
1780	**.45** 1.60	**(1.29)** (4.60)	**(.08)** (.28)	**0** 0	**0** 0	**(.03)** (.11)	**0** 0	**.04** .14	**.07** .25	**1.46** 5.20	**.17** .62	**.09** .32	**.84** 3.00	**(.28)** (1.00)	**(.56)** (2.00)
1781	**.62** 2.20	**1.20** 4.30	– –	**0** 0	**0** 0	– –	**0** 0	**.05** .16	**.08** .30	**1.71** 6.10	– –	– –	– –	– –	– –
1782	**.48** 1.70	– –	– –	**0** 0	**0** 0	– –	**0** 0	**.04** .15	**.08** .27	**1.54** 5.50	– –	– –	**.84** 3.00	– –	– –
1783	**.62** 2.20	**1.20** 4.30	– –	**0** 0	**0** 0	– –	**0** 0	**.05** .16	**.08** .30	**1.74** 6.20	– –	– –	– –	– –	– –
1784	**.50** 1.80	– –	– –	**0** 0	**0** 0	– –	**0** 0	**.04** .15	**.08** .27	**1.57** 5.60	– –	– –	**.84** 3.00	– –	– –
1785	**.62** 2.20	**1.20** 4.30	– –	**0** 0	**0** 0	– –	**0** 0	**.05** .16	**.08** .30	**1.74** 6.20	– –	– –	– –	– –	– –
1786	**.51** 1.10	– –	**.33** .71	**0** 0	**0** 0	**.07** .16	**0** 0	**.05** .11	**.10** .21	**2.07** 4.50	**.27** .59	**.15** .33	**1.38** 3.00	– –	– –
1787	**.56** 2.00	**1.20** 4.30	– –	**0** 0	**0** 0	– –	**0** 0	**.04** .15	**.08** .28	**1.68** 6.00	– –	– –	– –	– –	– –
1788	**.60** 1.30	[6]**1.56** 3.40	**.33** .71	**0** 0	**0** 0	**.07** .16	**0** 0	**.06** .12	**.11** .23	**2.30** 5.00	**.27** .59	**.15** .33	**1.38** 3.00	– –	– –
1789	**.56** 2.00	**1.20** 4.30	– –	**0** 0	**0** 0	– –	**0** 0	**.04** .15	**.08** .28	**1.71** 6.10	– –	– –	– –	– –	– –
1790	**.69** 1.50	**(.95)** (3.40)	**.33** .71	**0** 0	**0** 0	**.07** .16	**0** 0	**.06** .13	**.11** .23	**2.35** 5.10	**(.14)** (.50)	**(.04)** (.15)	**1.38** 3.00	**(.28)** (1.00)	**(.56)** (2.00)
1791	**.56** 2.00	**1.20** 4.30	– –	**0** 0	**0** 0	– –	**0** 0	**.04** .15	**.08** .28	**1.71** 6.10	– –	– –	– –	– –	– –
1792	**.33** .80	– –	**.29** .71	**0** 0	**0** 0	**.07** .16	**0** 0	**.04** .10	**.07** .18	**1.64** 4.00	– –	– –	**1.23** 3.00	– –	– –

(Continued)

TABLE F-36 *(Continued)*

Item No.	Food Item Name	Approximate Measure	Weight	Moisture	Food Energy Calories	Protein	Fats	Carbo-hydrates	Fiber	Minerals (Macro) Calcium	Phos-phorus	Sodium	Mag-nesium	Potas-sium
			(g)	(%)	(kcal)	(g)	(g)	(g)	(g)	(mg)	(mg)	(mg)	(mg)	(mg)
	LAMB (Continued)													
1793	**RIB,** Prime, separable lean, broiled	1 OZ	28	59.1	62.7	7.5	3.4	0	0	3.1	59.1	19.6	—	81.2
			100	59.1	224.0	26.9	12.1	0	0	11.0	211.0	70.0	—	290.0
1794	**RIB,** Choice, chops, total edible, broiled, 62% lean, 38% fat, w/bone	1 MED	41	42.9	166.9	8.2	14.6	0	0	3.7	64.0	28.7	7.0	118.9
			100	42.9	407.0	20.1	35.6	0	0	9.0	156.0	70.0	17.0	290.0
1795	**RIB,** Choice, separable lean, broiled, w/bone	1 OZ	28	60.3	59.1	7.6	2.9	0	0	3.1	59.4	19.6	—	81.2
			100	60.3	211.0	27.2	10.5	0	0	11.0	212.0	70.0	—	290.0
1796	**RIB,** Good, chops, total edible, 64% lean, 36% fat	1 MED	41	45.4	155.0	8.7	13.1	0	0	3.7	67.7	28.7	10.2	118.9
			100	45.4	378.0	21.2	31.9	0	0	9.0	165.0	70.0	24.9	290.0
1797	**RIB,** Good, separable lean, broiled	1 OZ	28	60.7	57.7	7.7	2.8	0	0	3.1	59.6	19.6	—	81.2
			100	60.7	206.0	27.4	9.9	0	0	11.0	213.0	70.0	—	290.0
1798	**SHOULDER,** Prime, total edible, roasted, 71% lean, 29% fat	1 OZ	28	46.2	104.7	5.8	8.9	0	0	2.5	45.6	19.6	4.8	81.2
			100	46.2	374.0	20.7	31.7	0	0	9.0	163.0	70.0	17.0	290.0
1799	**SHOULDER,** Prime, separable lean, roasted	1 OZ	28	60.4	60.2	7.4	3.1	0	0	3.1	60.8	19.6	6.2	81.2
			100	60.4	215.0	26.6	11.2	0	0	11.0	217.0	70.0	22.0	290.0
1800	**SHOULDER,** Choice, total edible roasted, 74% lean, 26% fat, w/o bone	1 OZ	28	49.6	94.6	6.1	7.7	0	0	2.8	48.2	19.6	4.8	81.2
			100	49.6	338.0	21.7	27.2	0	0	10.0	172.0	70.0	17.0	290.0
1801	**SHOULDER,** Choice, separable lean, roasted, w/o bone	1 OZ	28	61.4	57.4	7.5	2.8	0	0	3.4	61.3	19.6	6.2	81.2
			100	61.4	205.0	26.8	10.0	0	0	12.0	219.0	70.0	22.0	290.0
1802	**SHOULDER,** Good, total edible, roasted, 75% lean, 25% fat	1 OZ	28	51.1	90.2	6.2	7.1	0	0	2.8	49.0	19.6	.5	81.2
			100	51.1	322.0	22.1	25.2	0	0	10.0	175.0	70.0	1.7	290.0
1803	**SHOULDER,** Good, separable lean, roasted	1 OZ	28	61.8	56.3	7.5	2.7	0	0	3.1	61.3	19.6	6.2	81.2
			100	61.8	201.0	26.8	9.6	0	0	11.0	219.0	70.0	22.0	290.0
	LIVER													
1804	**BEEF,** fried	1 OZ	28	56.0	62.2	7.4	3.0	1.5	0	3.1	133.3	51.5	5.0	106.4
			100	56.0	222.0	26.4	10.6	5.3	0	11.0	476.0	184.0	18.0	380.0
1805	**CALF,** fried	1 OZ	28	51.4	73.1	8.3	3.7	1.1	0	3.6	150.4	33.0	7.3	126.8
			100	51.4	261.0	29.5	13.2	4.0	0	13.0	537.0	118.0	26.0	453.0
1806	**CHICKEN,** simmered	1 OZ	28	65.0	46.2	7.4	1.2	.9	0	3.1	44.5	17.1	(6.2)	42.3
			100	65.0	165.0	26.5	4.4	3.1	0	11.0	159.0	61.0	(22.0)	151.0
1807	**GOOSE,** raw	1 OZ	28	66.9	51.0	4.6	2.8	1.5	0	—	—	39.2	5.9	64.4
			100	66.9	182.0	16.5	10.0	5.4	0	—	—	140.0	21.0	230.0
1808	**HOG,** fried in margarine	1 OZ	28	54.0	68.5	8.4	3.2	.7	0	4.2	150.9	31.1	6.7	110.6
			100	54.0	241.0	29.9	11.5	2.5	0	15.0	539.0	111.0	24.0	395.0
1809	**LAMB,** broiled	1 OZ	28	50.4	73.1	9.0	3.5	.8	0	4.5	160.2	23.8	6.4	92.7
			100	50.4	261.0	32.3	12.4	2.8	0	16.0	572.0	85.0	23.0	331.0
1810	**TURKEY,** simmered	1 OZ	28	63.3	48.7	7.8	1.3	.9	0	—	—	15.4	—	39.5
			100	63.3	174.0	27.9	4.8	3.1	0	—	—	55.0	—	141.0
	LUNCHEON MEATS													
1811	**BOILED HAM**	1 OZ	28	59.1	65.5	5.3	4.8	0	0	3.1	46.5	—	2.5	—
			100	59.1	234.0	19.0	17.0	0	0	11.0	166.0	—	9.0	—
1812	**BOLOGNA,** all meat	1 SLICE	28	57.4	77.6	3.7	6.4	1.0	0	2.8	—	—	4.5	—
			100	57.4	277.0	13.3	22.8	3.7	0	10.0	—	—	16.0	—
1813	**BOLOGNA,** all samples	1 SLICE	28	56.2	88.5	3.4	8.2	.3	0	2.0	35.8	364.0	4.5	64.4
			100	56.2	316.0	12.1	29.2	1.1	0	7.0	128.0	1300.0	16.0	230.0
1814	**BOLOGNA,** w/cereal	1 SLICE	30	57.9	78.6	4.4	6.2	1.2	0	3.0	—	—	4.8	—
			100	57.9	262.0	14.8	20.6	3.9	0	10.0	—	—	16.0	—
1815	**BOLOGNA,** w/non fat, dehydrated milk	1 SLICE	28	57.1	—	3.8	—	—	0	2.8	—	—	4.5	—
			100	57.1	—	13.4	—	—	0	10.0	—	—	16.0	—
1816	**BRAUNSCHWEIGER**	1 OZ	28	52.6	101.6	4.3	9.1	.6	0	2.8	68.6	—	4.5	—
			100	52.6	363.0	15.4	32.5	2.3	0	10.0	245.0	—	16.0	—
1817	**CAPICCOLA OR CAPACOLA**	1 SLICE	28	26.2	139.7	5.7	12.8	0	0	—	—	—	—	—
			100	26.2	499.0	20.2	45.8	0	0	—	—	—	—	—
1818	**DEVILED HAM,** canned	1 TBLS	13	50.5	45.6	1.8	4.2	0	0	1.0	12.0	—	—	—
			100	50.5	351.0	13.9	32.3	0	0	8.0	92.0	—	—	—
1819	**HEADCHEESE**	1 SLICE	28	58.8	54.6	4.2	4.1	.3	0	2.5	48.4	—	—	—
			100	58.8	195.0	15.0	14.6	1.0	0	9.0	173.0	—	—	—
1820	**LIVERWURST,** fresh	1 OZ	28	53.9	138.9	4.7	9.1	.5	0	2.5	66.6	—	—	—
			100	53.9	496.0	16.7	32.5	1.8	0	9.0	238.0	—	—	—
1821	**LIVERWURST,** smoked	1 OZ	28	52.6	89.3	4.1	7.7	.6	0	2.8	68.6	—	—	—
			100	52.6	319.0	14.8	27.4	2.3	0	10.0	245.0	—	—	—
1822	**MEAT LOAF**	1 SLICE	70	64.1	140.0	11.1	9.2	2.3	0	6.3	124.6	—	—	—
			100	64.1	200.0	15.9	13.2	3.3	0	9.0	178.0	—	—	—
1823	**MINCED HAM**	1 OZ	28	61.7	63.8	3.8	4.7	1.2	0	2.2	24.9	—	—	—
			100	61.7	228.0	13.7	16.9	4.4	0	8.0	89.0	—	—	—

Item No.	Minerals (Micro) Iron	Zinc	Copper	Fat-Soluble Vitamins Vitamin A	Vitamin D	Vitamin E (Alpha Tocopherol)	Water-Soluble Vitamins Vitamin C	Thiamin	Ribo-flavin	Niacin	Panto-thenic Acid	Vit. B-6 (Pyri-doxine)	Folacin (Folic Acid)	Biotin	Vitamin B-12
	(mg)	(mg)	(mg)	(IU)	(IU)	(mg)	(mg)	(mg)	(mg)	(mg)	(mg)	(mg)	(mcg)	(mcg)	(mcg)
1793	.53	1.20	—	0	0	—	0	.04	.08	1.62	—	—	—	—	—
	1.90	4.30	—	0	0	—	0	.15	.27	5.80	—	—	—	—	—
1794	.45	—	.29	0	0	.07	0	.05	.09	1.89	—	—	1.23	—	—
	1.10	—	.71	0	0	.16	0	.12	.21	4.60	—	—	3.00	—	—
1795	.53	1.20	—	0	0	—	0	.04	.08	1.65	—	—	—	—	—
	1.90	4.30	—	0	0	—	0	.15	.27	5.90	—	—	—	—	—
1796	.49	2.19	.29	0	0	.07	0	.05	.09	1.97	—	—	1.23	—	—
	1.20	5.33	.71	0	0	.16	0	.12	.22	4.80	—	—	3.00	—	—
1797	.53	1.20	—	0	0	—	0	.04	.08	1.65	—	—	—	—	—
	1.90	4.30	—	0	0	—	0	.15	.27	5.90	—	—	—	—	—
1798	.34	—	—	0	0	—	0	.03	.06	1.29	—	—	.84	—	—
	1.20	—	—	0	0	—	0	.12	.22	4.60	—	—	3.00	—	—
1799	.50	1.20	—	0	0	—	0	.04	.08	1.60	—	—	—	—	—
	1.80	4.30	—	0	0	—	0	.15	.28	5.70	—	—	—	—	—
1800	.34	(1.20)	(.04)	0	0	(.03)	0	.04	.06	1.32	(.14)	(.04)	.84	(.28)	(.56)
	1.20	(4.30)	(.15)	0	0	(.12)	0	.13	.23	4.70	(.50)	(.16)	3.00	(1.00)	(2.00)
1801	.53	1.20	—	0	0	—	0	.04	.08	1.60	—	—	—	—	—
	1.90	4.30	—	0	0	—	0	.15	.28	5.70	—	—	—	—	—
1802	.36	—	—	0	0	—	0	.04	.06	1.34	—	—	.84	—	—
	1.30	—	—	0	0	—	0	.13	.23	4.80	—	—	3.00	—	—
1803	.53	1.20	—	0	0	—	0	.04	.08	1.60	—	—	—	—	—
	1.90	4.30	—	0	0	—	0	.15	.26	5.70	—	—	—	—	—
1804	2.46	1.12	(3.36)	14952.0	5.32	.18	7.56	.07	1.17	4.62	—	—	40.60	(26.70)	31.18
	8.80	4.00	(12.00)	53400.0	19.00	.63	27.00	.26	4.19	16.50	—	—	145.00	(96.00)	111.34
1805	3.98	1.71	(3.36)	9156.0	3.92	(.14)	10.36	.07	1.17	4.62	(2.47)	(.20)	40.60	(14.85)	(24.37)
	14.20	6.10	(12.00)	32700.0	14.00	(.50)	37.00	.24	4.17	16.50	(8.80)	(.73)	145.00	(53.00)	(87.00)
1806	2.38	.95	(.15)	3444.0	18.76	(.09)	4.48	.05	.75	3.28	(1.54)	(.12)	67.20	(47.62)	(13.73)
	8.50	3.40	(.54)	12300.0	67.00	(.33)	16.00	.17	2.69	11.70	(5.50)	(.44)	240.00	(170.0)	(49.00)
1807	—	—	1.36	—	—	—	—	—	—	—	—	—	—	—	—
	—	—	4.87	—	—	—	—	—	—	—	—	—	—	—	—
1808	8.15	(2.32)	(.70)	4172.0	14.28	(.04)	6.16	.10	1.22	6.24	(1.30)	(.18)	40.60	(7.59)	(7.28)
	29.10	(8.29)	(2.50)	14900.0	51.00	(.16)	22.00	.34	4.36	22.30	(4.60)	(.64)	145.00	(27.08)	(26.00)
1809	5.01	(1.23)	(2.80)	20860.0	6.44	(.09)	10.08	.14	1.43	6.97	(2.13)	(.14)	40.60	(11.49)	(22.71)
	17.90	(4.40)	(9.99)	74500.0	23.00	(.33)	36.00	.49	5.11	24.90	(7.60)	(.49)	145.00	(41.00)	(81.09)
1810	—	.95	—	4900.0	—	—	—	.05	.59	4.00	—	—	—	—	—
	—	3.40	—	17500.0	—	—	—	.16	2.09	14.30	—	—	—	—	—
1811	.78	—	—	0	—	—	—	.12	.04	.73	—	—	1.12	—	—
	2.80	—	—	0	—	—	—	.44	.15	2.60	—	—	4.00	—	—
1812	—	—	—	—	0	.02	0	—	—	—	—	.03	1.40	—	—
	—	—	—	—	0	.06	0	—	—	—	—	.10	5.00	—	—
1813	.50	.42	.01	Trace	0	.02	0	.05	.06	.73	(.14)	.03	1.40	Trace	(.29)
	1.80	1.50	.02	Trace	0	.06	0	.16	.22	2.60	(.51)	.10	5.00	Trace	(1.03)
1814	—	—	—	—	0	.02	0	—	—	—	—	.03	—	—	—
	—	—	—	—	0	.06	0	—	—	—	—	.10	—	—	—
1815	—	—	—	—	0	.02	0	—	—	—	—	.03	—	—	—
	—	—	—	—	0	.06	0	—	—	—	—	.10	—	—	—
1816	1.65	.78	—	1828.4	4.20	—	0	.05	.40	2.30	—	—	—	—	—
	5.90	2.80	—	6530.0	15.00	—	0	.17	1.44	8.20	—	—	—	—	—
1817	—	—	—	—	—	—	—	—	—	—	—	—	—	—	—
	—	—	—	—	—	—	—	—	—	—	—	—	—	—	—
1818	.27	—	—	0	—	—	—	.02	.01	.21	—	—	—	—	—
	2.10	—	—	0	—	—	—	.14	.10	1.60	—	—	—	—	—
1819	.64	—	—	0	0	—	0	.01	.03	.25	—	—	.56	—	—
	2.30	—	—	0	0	—	0	.04	.10	.90	—	—	2.00	—	—
1820	1.51	—	—	1778.0	4.20	.10	0	.06	.36	1.60	.78	.05	8.40	—	3.89
	5.40	—	—	6350.0	15.00	.35	0	.20	1.30	5.70	2.78	.19	30.00	—	13.90
1821	1.65	—	—	1828.4	—	—	—	.05	.40	2.30	—	—	8.40	—	—
	5.90	—	—	6530.0	—	—	—	.17	1.44	8.20	—	—	30.00	—	—
1822	1.26	—	—	—	—	—	—	.09	.15	1.75	—	—	—	—	—
	1.80	—	—	—	—	—	—	.13	.22	2.50	—	—	—	—	—
1823	.59	—	—	0	—	—	—	.10	.06	.95	—	—	—	—	—
	2.10	—	—	0	—	—	—	.37	.22	3.40	—	—	—	—	—

(Continued)

MEATS

TABLE F-36 *(Continued)*

Item No.	Food Item Name	Approximate Measure	Weight	Moisture	Food Energy Calories	Protein	Fats	Carbo-hydrates	Fiber	Minerals (Macro) Calcium	Phos-phorus	Sodium	Mag-nesium	Potas-sium
			(g)	(%)	(kcal)	(g)	(g)	(g)	(g)	(mg)	(mg)	(mg)	(mg)	(mg)
	LUNCHEON MEATS (Continued)													
1824	**PORK,** cured ham or shoulder, chopped, canned	1 SLICE	28	54.9	94.1	4.2	8.4	.4	0	2.5	30.2	345.5	—	62.2
			100	54.9	336.0	15.0	30.1	1.3	0	9.0	108.0	1234.0	—	222.0
1825	**POTTED MEAT** (beef, chicken, turkey)	1 OZ	28	60.7	78.4	4.5	5.4	0	0	—	—	—	—	—
			100	60.7	280.0	16.1	19.2	0	0	—	—	—	—	—
1826	**SALAMI,** cooked	1 OZ	28	51.0	73.1	4.9	5.8	.4	0	2.8	56.0	—	4.5	—
			100	51.0	261.0	17.5	20.6	1.4	0	10.0	200.0	—	16.0	—
1827	**SALAMI,** dry	1 SLICE	28	29.8	126.0	6.7	10.7	.3	0	3.9	79.2	(518.3)	4.5	(45.0)
			100	29.8	450.0	23.9	38.1	1.2	0	14.0	283.0	(1850.3)	16.0	(160.8)
1828	**SPAM,** Hormel	1 OZ	28	52.6	86.8	3.9	7.4	1.1	.1	3.1	35.3	336.0	—	58.5
			100	52.6	310.0	14.0	26.5	3.8	.2	11.0	126.0	1200.0	—	209.0
	LUNGS													
1829	**BEEF,** raw	1 OZ	28	78.8	26.9	4.9	.6	0	0	—	60.5	—	—	—
			100	78.8	96.0	17.6	2.3	0	0	—	216.0	—	—	—
1830	**CALF,** raw	1 OZ	28	77.4	29.7	4.7	1.1	0	0	—	—	—	—	—
			100	77.4	106.0	16.8	3.8	0	0	—	—	—	—	—
1831	**LAMB,** raw	1 OZ	28	76.7	28.8	5.4	.6	0	0	—	50.4	—	—	—
			100	76.7	103.0	19.3	2.3	0	0	—	180.0	—	—	—
1832	**MANAPUA PORK**	1 PIECE	94	—	—	—	28.2	—	—	—	—	308.3	—	43.2
			100	—	—	—	30.0	—	—	—	—	328.0	—	46.0
1833	**MUSKRAT,** roasted	1 OZ	28	67.3	42.8	7.6	1.1	0	0	—	—	—	—	—
			100	67.3	153.0	27.2	4.1	0	0	—	—	—	—	—
1834	**OPOSSUM,** roasted	1 OZ	28	57.3	61.9	8.5	2.9	0	0	—	—	—	—	—
			100	57.3	221.0	30.2	10.2	0	0	—	—	—	—	—
1835	**°OXTAIL,** stewed	4 OZ	112	—	273.4	34.2	15.1	0	0	15.7	156.8	213.7	20.2	191.4
			110	—	244.1	30.5	13.5	0	0	14.0	140.0	190.8	18.0	170.9
	PANCREAS													
1836	**BEEF,** very fat, raw	1 OZ	28	53.0	100.0	3.3	9.5	0	0	—	62.2	—	—	—
			100	53.0	357.0	11.8	34.0	0	0	—	222.0	—	—	—
1837	**BEEF,** fat, raw	1 OZ	28	57.0	68.5	3.8	8.1	0	0	—	74.8	—	—	—
			100	57.0	316.0	13.5	29.0	0	0	—	267.0	—	—	—
1838	**BEEF,** medium fat, raw	1 OZ	28	60.0	79.2	3.8	7.0	0	0	—	75.6	—	—	—
			100	60.0	283.0	13.5	25.0	0	0	—	270.0	—	—	—
1839	**BEEF,** lean only, raw	1 OZ	28	73.0	39.5	4.9	2.0	0	0	2.2	92.4	18.8	—	77.3
			100	73.0	141.0	17.6	7.3	0	0	8.0	330.0	67.0	—	276.0
1840	**CALF,** raw	1 OZ	28	69.7	45.1	5.4	2.5	0	0	—	91.3	—	—	—
			100	69.7	161.0	19.2	8.8	0	0	—	326.0	—	—	—
1841	**HOG,** sweetbread, raw	1 OZ	28	63.4	80.4	4.1	5.6	0	0	3.1	79.0	12.3	4.8	60.8
			100	63.4	287.0	14.5	19.9	0	0	11.0	282.0	44.0	17.0	217.0
1842	**°PARTRIDGE,** roasted	1 OZ	28	54.5	59.4	1.0	2.0	0	0	12.9	86.8	28.0	10.1	114.8
			100	54.5	212.0	36.7	7.2	0	0	46.0	310.0	100.0	36.0	410.0
1843	**PATE de FOIS GRAS,** canned	1 TBLS	15	37.0	69.3	1.7	6.6	.7	0	—	—	—	—	—
			100	37.0	462.0	11.4	43.8	4.8	0	—	—	—	—	—
	PHEASANT													
1844	total edible, raw	1 OZ	28	69.2	42.3	6.8	1.5	0	0	3.9	73.4	—	—	—
			100	69.2	151.0	24.3	5.2	0	0	14.0	262.0	—	—	—
1845	flesh & skin, raw	1 OZ	28	67.8	50.7	6.4	2.6	0	0	3.4	59.9	11.2	5.6	68.0
			100	67.8	181.0	22.7	9.3	0	0	12.0	214.0	40.0	20.0	243.0
1846	flesh only, raw	1 OZ	28	72.8	37.2	6.6	1.0	0	0	3.6	64.4	10.4	5.6	73.4
			100	72.8	133.0	23.6	3.6	0	0	13.0	230.0	37.0	20.0	262.0
1847	giblets, raw	1 OZ	28	71.4	38.9	5.8	1.4	.4	0	—	—	—	—	—
			100	71.4	139.0	20.8	4.9	1.6	0	—	—	—	—	—
1848	**PIGS FEET,** pickled	1 OZ	28	66.9	55.7	4.7	4.1	0	0	—	—	—	—	—
			100	66.9	199.0	16.7	14.8	0	0	—	—	—	—	—
	PORK, fresh													
1849	**ALL CUTS,** fat class, total edible, roasted, 72% lean, 28% fat	1 OZ	28	42.1	114.8	5.9	10.0	0	0	2.5	59.6	18.2	6.4	109.2
			100	42.1	410.0	20.9	35.6	0	0	9.0	213.0	65.0	23.0	390.0
1850	**ALL CUTS,** fat class, separable lean, roasted	1 OZ	28	56.4	68.6	7.7	3.9	0	0	3.4	80.4	18.2	8.1	109.2
			100	56.4	245.0	27.6	14.1	0	0	12.0	287.0	65.0	29.0	390.0
1851	**ALL CUTS,** medium fat, total edible, cooked, 77% lean, 23% fat	1 OZ	28	45.2	104.4	6.3	8.6	0	0	2.8	65.0	18.2	6.4	109.2
			100	45.2	373.0	22.6	30.6	0	0	10.0	232.0	65.0	23.0	390.0
1852	**ALL CUTS,** medium fat, separable lean, roasted	1 OZ	28	57.2	66.1	7.8	3.6	0	0	3.4	81.8	18.2	8.1	109.2
			100	57.2	236.0	28.0	12.9	0	0	12.0	292.0	65.0	29.0	390.0

(Continued)

Item No.	Minerals (Micro)			Fat-Soluble Vitamins			Water-Soluble Vitamins								
	Iron	Zinc	Copper	Vitamin A	Vitamin D	Vitamin E (Alpha Tocopherol)	Vitamin C	Thiamin	Ribo-flavin	Niacin	Panto-thenic Acid	Vit. B-6 (Pyri-doxine)	Folacin (Folic Acid)	Biotin	Vitamin B-12
	(mg)	(mg)	(mg)	(IU)	(IU)	(mg)	(mg)	(mg)	(mg)	(mg)	(mg)	(mg)	(mcg)	(mcg)	(mcg)
1824	**.62**	—	—	**0**	—	—	—	**.09**	**.06**	**.84**	**.15**	—	—	—	—
	2.20	—	—	0	—	—	—	.31	.21	3.00	.55	—	—	—	—
1825	—	—	—	—	—	—	—	**.01**	**.06**	**.34**	—	—	—	—	—
	—	—	—	—	—	—	—	.03	.22	1.20	—	—	—	—	—
1826	**.73**	—	—	—	—	**.03**	—	**.07**	**.07**	**1.15**	—	—	**.56**	—	—
	2.60	—	—	—	—	.11	—	.25	.24	4.10	—	—	2.00	—	—
1827	**1.01**	**(.50)**	**(.07)**	**0**	**0**	**(.08)**	**0**	**.10**	**.07**	**1.48**	**(.22)**	**.03**	**.56**	**(.87).**	**.39**
	3.60	(1.79)	(.25)	0	0	(.27)	0	.37	.25	5.30	(.80)	.12	2.00	(3.09)	1.40
1828	**.48**	**.52**	**.02**	—	—	—	—	**.11**	**.04**	**.59**	—	**.06**	**.90**	—	**.21**
	1.70	1.86	.06	—	—	—	—	.38	.13	2.09	—	.23	3.20	—	.74
1829	—	—	—	—	—	—	—	—	—	**1.74**	**.28**	—	—	—	**.98**
	—	—	—	—	—	—	—	—	—	6.20	1.00	—	—	—	3.50
1830	—	—	—	—	—	—	—	—	—	—	**.28**	—	—	—	**.98**
	—	—	—	—	—	—	—	—	—	—	1.00	—	—	—	3.50
1831	—	—	—	—	—	—	—	—	—	—	**.28**	—	—	—	**.98**
	—	—	—	—	—	—	—	—	—	—	1.00	—	—	—	3.50
1832	—	—	—	—	—	—	—	—	—	—	—	—	—	—	—
	—	—	—	—	—	—	—	—	—	—	—	—	—	—	—
1833	—	—	—	—	—	—	—	**.05**	**.06**	—	—	—	—	—	—
	—	—	—	—	—	—	—	.16	.21	—	—	—	—	—	—
1834	—	—	—	—	—	—	—	**.03**	**.11**	—	—	—	—	—	—
	—	—	—	—	—	—	—	.12	.38	—	—	—	—	—	—
1835	**4.26**	**9.86**	**.32**	**Trace**	**Trace**	**.50**	**0**	**.02**	**.31**	**3.70**	**1.03**	**.16**	**10.14**	**2.26**	**2.24**
	3.80	8.80	.29	Trace	Trace	.45	0	.02	.28	3.30	.92	.14	9.05	2.02	2.00
1836	—	—	—	—	—	—	—	—	—	—	**1.04**	**.06**	—	—	**4.56**
	—	—	—	—	—	—	—	—	—	—	3.70	.20	—	—	16.30
1837	—	—	—	—	—	—	—	—	—	—	**1.04**	**.06**	—	—	**4.56**
	—	—	—	—	—	—	—	—	—	—	3.70	.20	—	—	16.30
1838	—	—	—	—	—	—	—	—	—	—	**1.04**	**.06**	—	—	**4.56**
	—	—	—	—	—	—	—	—	—	—	3.70	.20	—	—	16.30
1839	**.78**	—	—	—	—	—	—	—	—	—	**1.04**	**.06**	—	—	**4.56**
	2.80	—	—	—	—	—	—	—	—	—	3.70	.20	—	—	16.30
1840	—	—	—	—	—	—	—	—	—	—	**1.04**	**.06**	—	—	**4.56**
	—	—	—	—	—	—	—	—	—	—	3.70	.20	—	—	16.30
1841	**.28**	—	**.03**	—	—	—	—	—	—	—	**1.04**	**.06**	—	—	**4.56**
	1.00	—	.09	—	—	—	—	—	—	—	3.70	.20	—	—	16.30
1842	**2.16**	—	—	—	—	—	—	—	—	—	—	—	—	—	—
	7.70	—	—	—	—	—	—	—	—	—	—	—	—	—	—
1843	—	—	—	—	—	—	**0**	**.01**	**.05**	**.38**	—	—	—	—	—
	—	—	—	—	—	—	0	.09	.30	2.50	—	—	—	—	—
1844	**1.04**	—	—	—	—	—	—	—	—	—	—	—	—	—	—
	3.70	—	—	—	—	—	—	—	—	—	—	—	—	—	—
1845	**.32**	**.27**	**.02**	**49.6**	—	—	**1.5**	**.02**	**.04**	**1.80**	**.26**	**.18**	—	—	**.22**
	1.15	.96	.07	177.0	—	—	5.3	.07	.14	6.43	.93	.66	—	—	.77
1846	**.32**	**.27**	**.02**	**46.2**	—	—	**1.7**	**.02**	**.04**	—	**1.89**	**.21**	—	—	**.24**
	1.15	.97	.07	165.0	—	—	6.0	.08	.15	6.76	.96	.74	—	—	.84
1847	—	—	—	—	—	—	—	—	—	—	—	—	—	—	—
	—	—	—	—	—	—	—	—	—	—	—	—	—	—	—
1848	—	—	—	—	—	—	—	—	—	—	—	—	—	—	—
	—	—	—	—	—	—	—	—	—	—	—	—	—	—	—
1849	**.76**	—	—	**0**	**0**	—	**0**	**.13**	**.06**	**1.18**	—	—	—	—	—
	2.70	—	—	0	0	—	0	.47	.21	4.20	—	—	—	—	—
1850	**.98**	**1.06**	—	**0**	**0**	—	**0**	**.17**	**.08**	**1.51**	—	—	**1.40**	—	—
	3.50	3.80	—	0	0	—	0	.60	.28	5.40	—	—	5.00	—	—
1851	**.81**	—	—	**0**	**0**	—	**0**	**.14**	**.06**	**1.37**	—	—	—	—	—
	2.90	—	—	0	0	—	0	.50	.23	4.90	—	—	—	—	—
1852	**1.01**	**1.06**	—	**0**	**0**	—	**0**	**.17**	**.08**	**1.54**	—	—	**1.40**	—	—
	3.60	3.80	—	0	0	—	0	.61	.28	5.50	—	—	5.00	—	—

(Continued)

TABLE F-36 *(Continued)*

Item No.	Food Item Name	Approximate Measure	Weight	Moisture	Food Energy Calories	Protein	Fats	Carbo-hydrates	Fiber	Minerals (Macro) Calcium	Phos-phorus	Sodium	Mag-nesium	Potas-sium
			(g)	(%)	(kcal)	(g)	(g)	(g)	(g)	(mg)	(mg)	(mg)	(mg)	(mg)
	PORK, FRESH (Continued)													
1853	**ALL CUTS,** thin class, total edible, roasted, 81% lean, 19% fat	1 OZ	**28**	**48.0**	**95.5**	**6.7**	**7.4**	**0**	**0**	**3.1**	**69.7**	**18.2**	**6.4**	**109.2**
			100	48.0	341.0	24.0	26.4	0	0	11.0	249.0	65.0	23.0	390.0
1854	**ALL CUTS,** thin class, separable lean, roasted	1 OZ	**28**	**57.9**	**63.8**	**8.0**	**3.3**	**0**	**0**	**3.4**	**83.7**	**18.2**	**8.1**	**109.2**
			100	57.9	228.0	28.6	11.7	0	0	12.0	299.0	65.0	29.0	390.0
1855	**BOSTON BUTT,** fat class, total edible, roasted, 76% lean, 24% fat	1 OZ	**28**	**45.0**	**108.9**	**5.9**	**9.2**	**0**	**0**	**2.5**	**59.4**	**18.2**	—	**109.2**
			100	45.0	389.0	20.9	32.7	0	0	9.0	212.0	65.0	—	390.0
1856	**BOSTON BUTT,** fat class, separable lean, roasted	1 OZ	**28**	**56.2**	**73.1**	**7.3**	**4.6**	**0**	**0**	**3.1**	**74.8**	**18.2**	—	**109.2**
			100	56.2	261.0	26.2	16.5	0	0	11.0	267.0	65.0	—	390.0
1857	**BOSTON BUTT,** medium fat, total edible, roasted, 79% lean, 21% fat	1 OZ	**28**	**48.1**	**98.8**	**6.3**	**8.0**	**0**	**0**	**2.8**	**64.1**	**10.4**	**5.9**	**72.8**
			100	48.1	353.0	22.5	28.5	0	0	10.0	229.0	37.0	21.0	260.0
1858	**BOSTON BUTT,** medium fat, separable lean, roasted	1 OZ	**28**	**57.5**	**68.3**	**7.6**	**4.0**	**0**	**0**	**3.4**	**77.6**	**18.2**	—	**109.2**
			100	57.5	244.0	27.0	14.3	0	0	12.0	277.0	65.0	—	390.0
1859	**BOSTON BUTT,** thin class, total edible, roasted, 83% lean, 17% fat	1 OZ	**28**	**51.2**	**88.8**	**6.8**	**6.6**	**0**	**0**	**3.1**	**69.4**	**18.2**	—	**109.2**
			100	51.2	317.0	24.2	23.7	0	0	11.0	248.0	65.0	—	390.0
1860	**BOSTON BUTT,** thin class, separable lean, roasted	1 OZ	**28**	**58.7**	**64.4**	**7.8**	**3.4**	**0**	**0**	**3.4**	**79.8**	**18.2**	—	**109.2**
			100	58.7	230.0	27.8	12.3	0	0	12.0	285.0	65.0	—	390.0
1861	**FAT,** separated from lean cuts, cooked	1 OZ	**28**	**11.1**	**216.4**	**1.3**	**23.4**	**0**	**0**	—	—	**18.2**	—	**109.2**
			100	11.1	773.0	4.8	83.4	0	0	—	—	65.0	—	390.0
1862	[1]**GROUND,** lean meat & fat, cooked	1 OZ	**28**	**45.2**	**104.4**	**6.3**	**8.6**	**0**	**0**	**2.8**	**65.0**	**18.2**	**6.4**	**109.2**
			100	45.2	373.0	22.6	30.6	0	0	10.0	232.0	65.0	23.0	390.0
1863	**HAM,** fat class, total edible, roasted, 72% lean, 28% fat	1 OZ	**28**	**43.7**	**110.3**	**6.1**	**9.3**	**0**	**0**	**2.8**	**63.0**	**18.2**	—	**109.2**
			100	43.7	394.0	21.9	33.3	0	0	10.0	225.0	65.0	—	390.0
1864	**HAM,** fat class, separable lean, roasted	1 OZ	**28**	**58.2**	**63.0**	**8.2**	**3.1**	**0**	**0**	**3.6**	**84.8**	**18.2**	—	**109.2**
			100	58.2	225.0	29.3	11.1	0	0	13.0	303.0	65.0	—	390.0
1865	**HAM,** medium fat, total edible, roasted, 82% lean, 18% fat	1 OZ	**28**	**45.5**	**76.7**	**6.4**	**5.7**	**0**	**0**	**2.8**	**66.1**	**18.2**	—	**109.2**
			100	45.5	274.0	23.0	20.2	0	0	10.0	236.0	65.0	—	390.0
1866	**HAM,** medium fat, separable lean, roasted	1 OZ	**28**	**58.9**	**56.0**	**8.3**	**2.5**	**0**	**0**	**3.6**	**86.2**	**18.2**	—	**109.2**
			100	58.9	200.0	29.7	9.0	0	0	13.0	308.0	65.0	—	390.0
1867	**HAM,** thin class, separable lean, roasted	1 OZ	**28**	**59.3**	**58.8**	**8.5**	**2.5**	**0**	**0**	**3.6**	**88.2**	**18.2**	—	**109.2**
			100	59.3	210.0	30.2	9.0	0	0	13.0	315.0	65.0	—	390.0
1868	**HAM,** thin class, total edible, roasted, 77% lean, 23% fat	1 OZ	**28**	**47.8**	**96.9**	**6.8**	**7.5**	**0**	**0**	**3.1**	**70.6**	**18.2**	—	**109.2**
			100	47.8	346.0	24.2	26.9	0	0	11.0	252.0	65.0	—	390.0
1869	**LOIN,** fat class, total edible, roasted, 79% lean, 21% fat	1 OZ	**28**	**43.7**	**108.4**	**6.6**	**7.9**	**0**	**0**	**2.8**	**68.6**	**18.2**	**5.6**	**109.2**
			100	43.7	387.0	23.5	28.1	0	0	10.0	245.0	65.0	20.0	390.0
1870	**LOIN,** fat class, separable lean, roasted	1 OZ	**28**	**55.0**	**71.1**	**8.2**	**4.0**	**0**	**0**	**3.6**	**86.8**	**18.2**	**6.2**	**109.2**
			100	55.0	254.0	29.4	14.2	0	0	13.0	310.0	65.0	22.0	390.0
1871	**LOIN,** medium fat, total edible, roasted, 80% lean, 20% fat	1 OZ	**28**	**45.8**	**101.4**	**6.9**	**8.0**	**0**	**0**	**3.1**	**71.7**	**14.0**	**5.6**	**98.6**
			100	45.8	362.0	24.5	28.5	0	0	11.0	256.0	50.0	20.0	352.0
1872	**LOIN,** medium fat, separable lean, roasted	1 OZ	**28**	**55.0**	**71.1**	**8.2**	**3.9**	**0**	**0**	**3.6**	**86.8**	**18.2**	**5.6**	**109.2**
			100	55.0	254.0	29.4	13.9	0	0	13.0	310.0	65.0	20.0	390.0
1873	**LOIN,** thin class, total edible, roasted, 85% lean, 15% fat	1 OZ	**28**	**48.3**	**93.2**	**7.2**	**6.9**	**0**	**0**	**3.1**	**75.6**	**18.2**	**5.6**	**109.2**
			100	48.3	333.0	25.8	24.7	0	0	11.0	270.0	65.0	20.0	390.0
1874	**LOIN,** thin class, separable lean, roasted	1 OZ	**28**	**55.0**	**71.1**	**8.2**	**4.0**	**0**	**0**	**3.6**	**86.8**	**18.2**	**5.6**	**109.2**
			100	55.0	254.0	29.4	14.2	0	0	13.0	310.0	65.0	20.0	390.0
1875	[1]**LOIN TENDERLOIN,** lean meat, cooked	1 OZ	**28**	**55.0**	**71.1**	**8.2**	**4.0**	**0**	**0**	**3.6**	**86.8**	**18.2**	**5.6**	**109.2**
			100	55.0	254.0	29.4	14.2	0	0	13.0	310.0	65.0	20.0	390.0
1876	**PICNIC,** fat class, total edible, roasted, 78% lean, 22% fat	1 OZ	**28**	**41.5**	**117.6**	**6.1**	**7.0**	**0**	**0**	**2.5**	**36.1**	**18.2**	**3.9**	**109.2**
			100	41.5	420.0	21.8	24.9	0	0	9.0	129.0	65.0	14.0	390.0
1877	**PICNIC,** fat class, separable lean, simmered	1 OZ	**28**	**58.8**	**64.7**	**8.0**	**3.4**	**0**	**0**	**3.4**	**48.4**	**18.2**	**5.0**	**109.2**
			100	58.8	231.0	28.5	12.1	0	0	12.0	173.0	65.0	18.0	390.0
1878	**PICNIC,** medium fat, total edible, simmered, 74% lean, 26% fat	1 OZ	**28**	**45.7**	**104.7**	**6.5**	**8.5**	**0**	**0**	**2.8**	**38.9**	—	**3.9**	—
			100	45.7	374.0	23.2	30.5	0	0	10.0	139.0	—	14.0	—
1879	**PICNIC,** medium fat, separable lean, simmered	1 OZ	**28**	**60.3**	**59.4**	**8.1**	**2.7**	**0**	**0**	**3.4**	**49.3**	**18.2**	**5.0**	**109.2**
			100	60.3	212.0	29.0	9.8	0	0	12.0	176.0	65.0	18.0	390.0
1880	**PICNIC,** thin class, total edible, simmered, 78% lean, 22% fat	1 OZ	**28**	**49.7**	**92.1**	**7.0**	**6.9**	**0**	**0**	**3.1**	**41.7**	**18.2**	**3.9**	**109.2**
			100	49.7	329.0	24.9	24.7	0	0	11.0	149.0	65.0	14.0	390.0
1881	**PICNIC,** thin class, separable lean, simmered	1 OZ	**28**	**62.0**	**54.3**	**8.3**	**2.1**	**0**	**0**	**3.6**	**50.4**	**18.2**	**5.0**	**109.2**
			100	62.0	194.0	29.7	7.5	0	0	13.0	180.0	65.0	18.0	390.0
1882	[1]**SHOULDER BLADE STEAK,** bone-in, lean meat, cooked	1 OZ	**28**	**57.5**	**68.3**	**7.6**	**4.0**	**0**	**0**	**3.4**	**77.6**	**18.2**	—	**109.2**
			100	57.5	244.0	27.0	14.3	0	0	12.0	277.0	65.0	—	390.0
1883	[1]**SHOULDER BLADE STEAK,** boneless, lean meat, cooked	1 OZ	**28**	**57.5**	**68.3**	**7.6**	**4.0**	**0**	**0**	**3.4**	**77.6**	**18.2**	—	**109.2**
			100	57.5	244.0	27.0	14.3	0	0	12.0	277.0	65.0	—	390.0

Item No.	Minerals (Micro)			Fat-Soluble Vitamins			Water-Soluble Vitamins								
	Iron	Zinc	Copper	Vitamin A	Vitamin D	Vitamin E (Alpha Tocopherol)	Vitamin C	Thiamin	Ribo-flavin	Niacin	Panto-thenic Acid	Vit. B-6 (Pyri-doxine)	Folacin (Folic Acid)	Biotin	Vitamin B-12
	(mg)	(mg)	(mg)	(IU)	(IU)	(mg)	(mg)	(mg)	(mg)	(mg)	(mg)	(mg)	(mcg)	(mcg)	(mcg)
1853	.88	–	–	0	0	–	0	.15	.07	1.34	–	–	–	–	–
	3.10	–	–	0	0	–	0	.53	.24	4.80	–	–	–	–	–
1854	1.01	1.06	–	0	0	–	0	.18	.08	1.57	–	–	1.40	–	–
	3.60	3.80	–	0	0	–	0	.63	.28	5.60	–	–	5.00	–	–
1855	.76	–	–	0	0	–	0	.13	.06	1.18	–	–	–	–	–
	2.70	–	–	0	0	–	0	.47	.21	4.20	–	–	–	–	–
1856	.92	1.26	–	0	0	–	0	.16	.07	1.43	–	–	1.40	–	–
	3.30	4.50	–	0	0	–	0	.57	.26	5.10	–	–	5.00	–	–
1857	.81	–	–	0	0	–	0	.14	.06	1.23	–	–	–	–	–
	2.90	–	–	0	0	–	0	.50	.23	4.40	–	–	–	–	–
1858	.95	–	–	0	0	–	0	.17	.08	1.46	–	–	–	–	–
	3.40	–	–	0	0	–	0	.59	.27	5.20	–	–	–	–	–
1859	.87	–	–	0	0	–	0	.15	.07	1.31	–	–	–	–	–
	3.10	–	–	0	0	–	0	.53	.24	4.70	–	–	–	–	–
1860	.98	1.26	–	0	0	–	0	.17	.08	1.51	–	–	1.40	–	–
	3.50	4.50	–	0	0	–	0	.60	.27	5.40	–	–	5.00	–	–
1861	–	–	–	–	0	–	0	–	–	–	–	–	–	–	–
	–	–	–	–	0	–	0	–	–	–	–	–	–	–	–
1862	.81	.85	–	0	–	–	0	.14	.06	1.37	–	–	1.40	–	–
	2.90	3.04	–	0	–	–	0	.50	.23	4.90	–	–	5.00	–	–
1863	.81	–	–	0	0	–	0	.21	.06	1.23	0.18	0.07	–	–	–
	2.90	–	–	0	0	–	0	.76	.21	4.40	0.64	0.26	–	–	–
1864	1.04	1.12	–	0	0	–	0	.18	.08	1.57	–	–	1.40	–	–
	3.70	4.00	–	0	0	–	0	.63	.29	5.60	–	–	5.00	–	–
1865	.84	–	–	0	0	–	0	.14	.06	1.29	0.18	0.12	–	–	–
	3.00	–	–	0	0	–	0	.51	.23	4.60	0.64	0.44	–	–	–
1866	1.06	1.12	–	0	0	–	0	.18	.08	1.60	–	–	1.40	–	–
	3.80	4.00	–	0	0	–	0	.64	.29	5.70	–	–	5.00	–	–
1867	1.06	1.12	–	0	0	–	0	.19	.08	1.62	–	–	1.40	–	–
	3.80	4.00	–	0	0	–	0	.66	.30	5.80	–	–	5.00	–	–
1868	.90	–	–	0	0	–	0	.15	.07	1.34	0.18	0.12	–	–	–
	3.20	–	–	0	0	–	0	.54	.25	4.80	0.64	0.44	–	–	–
1869	.87	–	.08	0	0	–	0	.25	.07	1.48	.11	.13	22.40	–	.84
	3.10	–	.30	0	0	–	0	.88	.25	5.30	.40	.48	80.00	–	3.00
1870	1.06	.87	.08	0	0	–	0	.30	.09	1.82	.13	.03	1.40	1.40	.84
	3.80	3.10	.30	0	0	–	0	1.08	.31	6.50	.47	.10	5.00	5.00	3.00
1871	.90	–	.08	0	0	–	0	.26	.07	1.57	.11	.13	–	–	.84
	3.20	–	.30	0	0	–	0	.92	.26	5.60	.40	.48	–	–	3.00
1872	1.05	.67	.08	0	0	–	0	.30	.09	1.82	.13	.03	1.40	1.40	.84
	3.80	3.10	.30	0	0	–	0	1.08	.31	6.50	.47	.10	5.00	5.00	3.00
1873	.95	–	.08	0	0	–	0	.27	.08	1.62	.11	.13	–	–	.84
	3.40	–	.30	0	0	–	0	.96	.28	5.80	.40	.48	–	–	3.00
1874	1.06	.87	.08	0	0	–	0	.30	.09	1.82	.13	.03	1.40	1.40	.84
	3.80	3.10	.30	0	0	–	0	1.08	.31	6.50	.47	.10	5.00	5.00	3.00
1875	1.06	1.12	.08	0	–	–	0	.30	.09	1.82	.13	.03	1.40	1.40	.84
	3.80	4.00	.30	0	–	–	0	1.08	.31	6.50	.47	.10	5.00	5.00	3.00
1876	.78	–	–	–	0	–	0	.14	.06	1.26	–	–	–	–	–
	2.80	–	–	–	0	–	0	.51	.23	4.50	–	–	–	–	–
1877	1.01	–	–	0	0	–	0	.18	.08	1.62	–	–	1.40	–	–
	3.60	–	–	0	0	–	0	.65	.30	5.80	–	–	5.00	–	–
1878	.84	–	–	0	0	–	0	.15	.07	1.34	–	–	–	–	–
	3.00	–	–	0	0	–	0	.54	.25	4.80	–	–	–	–	–
1879	1.01	–	–	0	–	–	0	.19	.08	1.65	–	–	1.40	–	–
	3.60	–	–	0	–	–	0	.66	.30	5.90	–	–	5.00	–	–
1880	.90	–	–	0	–	–	0	.16	.07	1.46	–	–	–	–	–
	3.20	–	–	0	–	–	0	.58	.26	5.20	–	–	–	–	–
1881	1.04	–	–	0	–	–	0	.19	.09	1.68	–	–	1.40	–	–
	3.70	–	–	0	–	–	0	.68	.31	6.00	–	–	5.00	–	–
1882	.95	1.26	–	0	0	–	0	.17	.08	1.46	–	–	1.40	–	–
	3.40	4.50	–	0	0	–	0	.59	.27	5.20	–	–	5.00	–	–
1883	.95	1.26	–	0	0	–	0	.17	.08	1.46	–	–	1.40	–	–
	3.40	4.50	–	0	0	–	0	.59	.27	5.20	–	–	5.00	–	–

(Continued)

TABLE F-36 *(Continued)*

Item No.	Food Item Name	Approximate Measure	Weight	Moisture	Food Energy Calories	Protein	Fats	Carbo-hydrates	Fiber	Minerals (Macro) Calcium	Phos-phorus	Sodium	Mag-nesium	Potas-sium
			(g)	(%)	(kcal)	(g)	(g)	(g)	(g)	(mg)	(mg)	(mg)	(mg)	(mg)
	PORK, FRESH (Continued)													
1884	[1]**SHOULDER**, boneless, lean meat, cooked	1 OZ	28	60.3	59.4	8.1	2.7	0	0	3.4	49.3	18.2	5.0	109.2
			100	60.3	212.0	29.0	9.8	0	0	12.0	176.0	65.0	18.0	390.0
1885	**SPARERIBS**, fat class, total edible, braised	1 MED	15	37.2	70.1	3.0	6.4	0	0	1.2	17.0	9.8	–	58.5
			100	37.2	467.0	19.7	42.5	0	0	8.0	113.0	65.0	–	390.0
1886	**SPARERIBS**, medium fat, total edible, braised	1 MED	15	39.7	66.0	3.1	5.8	0	0	1.4	18.2	9.8	3.8	58.5
			100	39.7	440.0	20.8	38.9	0	0	9.0	121.0	65.0	25.0	390.0
1887	**SPARERIBS**, thin class, total edible, braised	1 MED	15	42.4	61.5	3.3	5.3	0	0	1.4	19.4	9.8	–	58.5
			100	42.4	410.0	21.9	35.1	0	0	9.0	129.0	65.0	–	390.0
1888	**STOMACH**—See **STOMACH, PORK**													
	PORK, CANNED													
1889	w/gravy, 90% pork, 10% gravy, canned	1 OZ	28	56.9	71.7	4.6	5.0	1.8	0	3.6	51.2	–	–	–
			100	56.9	256.0	16.4	17.8	6.3	0	13.0	183.0	–	–	–
	PORK, CURED													
1890	**BOSTON BUTT**, medium fat, total edible, 83% lean, 17% fat, w/o bone & skin	1 OZ	28	47.7	92.4	6.4	7.2	0	0	2.8	51.8	–	–	–
			100	47.7	330.0	22.9	25.7	0	0	10.0	185.0	–	–	–
1891	**BOSTON BUTT**, medium fat, separable lean, cooked	1 OZ	28	53.9	68.0	7.8	3.9	0	0	3.4	61.0	260.4	–	91.3
			100	53.9	243.0	27.8	13.8	0	0	12.0	218.0	930.0	–	326.0
1892	**HAM**, canned	1 OZ	28	65.0	54.0	5.1	3.2	.3	0	3.1	43.7	308.0	–	95.2
			100	65.0	193.0	18.3	11.3	.9	0	11.0	156.0	1100.0	–	340.0
1893	**HAM**, dry, long-cure, country-style, fat	1 OZ	28	36.0	127.4	4.1	12.3	.1	0	–	–	–	–	–
			100	36.0	455.0	14.5	44.0	.3	0	–	–	–	–	–
1894	**HAM**, dry, long-cure, country-style, medium fat	1 OZ	28	42.0	108.9	4.7	9.8	.1	0	–	–	–	–	–
			100	42.0	389.0	16.9	35.0	.3	0	–	–	–	–	–
1895	**HAM**, dry, long-cure, country-style, lean	1 OZ	28	49.0	86.8	5.5	7.0	.1	0	–	–	–	–	–
			100	49.0	310.0	19.5	25.0	.3	0	–	–	–	–	–
1896	**HAM**, light-cure, medium fat, total edible, roasted, 84% lean, 16% fat	1 OZ	28	53.6	80.9	5.9	6.2	0	0	2.5	48.2	–	4.8	–
			100	53.6	289.0	20.9	22.1	0	0	9.0	172.0	–	17.0	–
1897	**HAM**, light-cure, medium fat, separable lean, roasted	1 OZ	28	61.9	52.4	7.1	2.5	0	0	3.1	56.0	260.4	5.6	91.3
			100	61.9	187.0	25.3	8.8	0	0	11.0	200.0	930.0	20.0	326.0
1898	**HAM, PICNIC**, medium fat, total edible, roasted, 82% lean, 18% fat	1 OZ	28	48.8	88.5	6.3	7.1	0	0	2.8	51.0	–	–	–
			100	48.8	316.0	22.4	25.2	0	0	10.0	182.0	–	–	–
1899	**HAM, PICNIC**, medium fat, separable lean, roasted	1 OZ	28	57.2	59.1	8.0	2.8	0	0	3.6	61.6	260.4	–	91.3
			100	57.2	211.0	28.4	9.9	0	0	13.0	220.0	930.0	–	326.0
1900	**SALT PORK**, fried	1 SLICE	25	–	170.5	3.0	17.5	0	0	2.0	30.0	–	–	–
			100	–	682.0	12.0	70.0	0	0	8.0	120.0	–	–	–
	QUAIL													
1901	total edible, raw	1 OZ	28	65.9	47.0	7.0	1.9	0	0	4.2	75.6	11.2	–	49.0
			100	65.9	168.0	25.0	6.8	0	0	15.0	270.0	40.0	–	175.0
1902	flesh & skin, raw	1 OZ	28	66.3	48.2	7.1	2.0	0	0	–	–	11.2	–	49.0
			100	66.3	172.0	25.4	7.0	0	0	–	–	40.0	–	175.0
1903	giblets, raw	1 OZ	28	63.0	49.3	6.1	1.7	1.9	0	–	–	–	–	–
			100	63.0	176.0	21.8	6.2	6.7	0	–	–	–	–	–
1904	**RABBIT, DOMESTICATED**, flesh only, stewed	1 OZ	28	59.8	60.5	8.2	2.8	0	0	5.9	72.5	11.5	–	103.0
			100	59.8	216.0	29.3	10.1	0	0	21.0	259.0	41.0	–	368.0
1905	**RABBIT, WILD**, flesh only, raw	1 OZ	28	73.0	37.8	5.9	1.4	0	0	–	–	13.2	8.1	116.2
			100	73.0	135.0	21.0	5.0	0	0	–	–	47.0	29.0	415.0
1906	**RACCOON**, roasted	1 OZ	28	54.8	71.4	8.2	4.1	0	0	–	–	–	–	–
			100	54.8	255.0	29.2	14.5	0	0	–	–	–	–	–
	REINDEER													
1907	total edible, 84% lean, 16% fat, raw	1 OZ	28	63.3	60.8	5.7	4.0	0	0	–	–	–	–	–
			100	63.3	217.0	20.5	14.4	0	0	–	–	–	–	–
1908	forequarter, 91% lean, 9% fat, raw	1 OZ	28	67.4	49.8	6.1	2.6	0	0	–	–	–	–	–
			100	67.4	178.0	21.8	9.4	0	0	–	–	–	–	–
1909	hindquarter, 78% lean, 22% fat, raw	1 OZ	28	59.6	71.7	5.4	5.4	0	0	–	–	–	–	–
			100	59.6	256.0	19.4	19.2	0	0	–	–	–	–	–
1910	lean only, raw	1 OZ	28	73.3	35.6	6.1	1.1	0	0	–	–	–	–	–
			100	73.3	127.0	21.8	3.8	0	0	–	–	–	–	–
	SAUSAGES													
1911	**BLOOD SAUSAGE** or blood pudding	1 SLICE	60	46.4	236.4	8.5	22.1	.2	0	4.8	96.0	–	–	–
			100	46.4	394.0	14.1	36.9	.3	0	8.0	160.0	–	–	–
1912	**BOCKWURST**	1 OZ	28	61.9	73.9	3.2	6.6	.2	0	–	–	–	–	–
			100	61.9	264.0	11.3	23.7	.6	0	–	–	–	–	–

Item No.	Minerals (Micro)			Fat-Soluble Vitamins			Water-Soluble Vitamins								
	Iron	Zinc	Copper	Vitamin A	Vitamin D	Vitamin E (Alpha Tocopherol)	Vitamin C	Thiamin	Ribo-flavin	Niacin	Panto-thenic Acid	Vit. B-6 (Pyri-doxine)	Folacin (Folic Acid)	Biotin	Vitamin B-12
	(mg)	(mg)	(mg)	(IU)	(IU)	(mg)	(mg)	(mg)	(mg)	(mg)	(mg)	(mg)	(mcg)	(mcg)	(mcg)
1884	**1.01** 3.60	**1.12** 4.00	— —	**0** 0	**0** 0	— —	**0** 0	**.19** .66	**.08** .30	**1.65** 5.90	— —	— —	**1.40** 5.00	— —	— —
1885	**.38** 2.50	— —	— —	**0** 0	— —	— —	**0** 0	**.06** .40	**.03** .19	**.48** 3.20	— —	— —	— —	— —	— —
1886	**.39** 2.60	— —	— —	**0** 0	— —	— —	**0** 0	**.07** .43	**.03** .21	**.51** 3.40	— —	— —	— —	— —	— —
1887	**.42** 2.80	— —	— —	**0** 0	— —	— —	**0** 0	**.07** .45	**.03** .22	**.54** 3.60	— —	— —	— —	— —	— —
1888															
1889	**.67** 2.40	— —	— —	**0** 0	— —	— —	— —	**.14** .49	**.05** .17	**.98** 3.50	— —	— —	— —	— —	— —
1890	**.84** 3.00	— —	— —	**0** 0	— —	— —	**0** 0	**.15** .53	**.06** .21	**1.15** 4.10	— —	— —	— —	— —	— —
1891	**1.01** 3.60	— —	— —	**0** 0	— —	— —	— —	**.18** .64	**.07** .25	**1.40** 5.00	— —	— —	— —	— —	— —
1892	**.76** 2.70	[6] **.64** 2.3	— —	**0** 0	— —	— —	— —	**.15** .53	**.05** .19	**1.06** 3.80	— —	**.10** .36	— —	— —	— —
1893	— —	— —	— —	**0** 0	— —	— —	**0** 0	— —	— —	— —	**.15** .53	**.09** .32	— —	— —	**.14** .50
1894	— —	— —	— —	**0** 0	— —	— —	**0** 0	— —	— —	— —	— —	— —	— —	— —	— —
1895	— —	— —	— —	**0** 0	— —	— —	**0** 0	— —	— —	— —	**.15** .53	**.09** .32	— —	— —	**.14** .50
1896	**.73** 2.60	— —	— —	**0** 0	— —	— —	— —	**.13** .47	**.05** .18	**1.01** 3.60	— —	— —	— —	— —	— —
1897	**.90** 3.20	— —	— —	**0** 0	— —	— —	— —	**.16** .58	**.06** .23	**1.26** 4.50	— —	— —	— —	— —	— —
1898	**.81** 2.90	— —	— —	**0** 0	— —	— —	— —	**.15** .52	**.06** .20	**1.12** 4.00	— —	— —	— —	— —	— —
1899	**1.04** 3.70	— —	— —	**0** 0	— —	— —	— —	**.18** .65	**.07** .26	**1.40** 5.00	— —	— —	— —	— —	— —
1900	**.40** 1.60	— —	— —	**0** 0	**0** 0	— —	**0** 0	**.07** .28	**.03** .10	**.50** 2.00	— —	— —	— —	— —	— —
1901	**1.06** 3.80	— —	— —	— —	— —	— —	— —	— —	— —	— —	— —	— —	— —	— —	— —
1902	— —	— —	— —	— —	— —	— —	— —	— —	— —	— —	— —	— —	— —	— —	— —
1903	— —	— —	— —	— —	— —	— —	— —	— —	— —	— —	— —	— —	— —	— —	— —
1904	**.42** 1.50	— —	— —	— —	— —	— —	**0** 0	**.01** .05	**.02** .07	**3.16** 11.30	**(.22)** (.80)	**(.14)** (.50)	**(1.12)** (4.00)	**(.28)** (1.00)	**(3.36)** (12.00)
1905	— —	— —	— —	— —	— —	— —	**0** 0	— —	— —	— —	**.22** .78	**.12** .44	— —	— —	— —
1906	— —	— —	— —	— —	— —	— —	**0** 0	**.17** .59	**.15** .52	— —	— —	— —	— —	— —	— —
1907	— —	— —	— —	— —	— —	— —	— —	— —	— —	— —	— —	— —	— —	— —	— —
1908	— —	— —	— —	— —	— —	— —	— —	— —	— —	— —	— —	— —	— —	— —	— —
1909	— —	— —	— —	— —	— —	— —	— —	— —	— —	— —	— —	— —	— —	— —	— —
1910	**1.48** 5.30	— —	— —	— —	— —	— —	— —	**.09** .33	**.19** .68	**1.54** 5.50	— —	— —	— —	— —	— —
1911	**1.32** 2.20	— —	— —	— —	**0** 0	— —	**0** 0	— —	— —	— —	— —	**.02** .04	— —	— —	— —
1912	— —	— —	— —	— —	— —	— —	— —	— —	— —	— —	— —	— —	— —	— —	— —

(Continued)

MEATS

TABLE F-36 *(Continued)*

Item No.	Food Item Name	Approximate Measure	Weight	Moisture	Food Energy Calories	Protein	Fats	Carbo-hydrates	Fiber	Minerals (Macro) Calcium	Phos-phorus	Sodium	Mag-nesium	Potas-sium
			(g)	(%)	(kcal)	(g)	(g)	(g)	(g)	(mg)	(mg)	(mg)	(mg)	(mg)
	SAUSAGES (Continued)													
1913	**BROWN & SERVE,** browned	1 OZ	**28**	**39.9**	**118.2**	**4.6**	**10.6**	**.8**	**0**	–	–	–	–	–
			100	39.9	422.0	16.5	37.8	2.8	0	–	–	–	–	–
1914	**CERVELAT,** soft	1 SLICE	**28**	**48.5**	**86.0**	**5.2**	**6.9**	**.4**	**0**	**3.1**	**59.9**	–	–	–
			100	48.5	307.0	18.6	24.5	1.6	0	11.0	214.0	–	–	–
1915	**CERVELAT,** dry	1 SLICE	**28**	**29.4**	**126.3**	**6.9**	**10.5**	**.5**	**0**	**3.9**	**82.3**	–	–	–
			100	29.4	451.0	24.6	37.6	1.7	0	14.0	294.0	–	–	–
1916	**COUNTRY STYLE**	1 OZ	**28**	**49.9**	**96.6**	**4.2**	**8.7**	**0**	**0**	**2.5**	**47.0**	–	**4.5**	–
			100	49.9	345.0	15.1	31.1	0	0	9.0	168.0	–	16.0	–
1917	**KNOCKWURST**	1 OZ	**28**	**57.6**	**77.8**	**3.9**	**6.5**	**.6**	**0**	**2.2**	**43.1**	–	–	–
			100	57.6	278.0	14.1	23.2	2.2	0	8.0	154.0	–	–	–
1918	**LINKS OR BULK,** cooked	1 OZ	**28**	**44.6**	**103.3**	**5.1**	**9.1**	**0**	**0**	**2.0**	**45.4**	**268.2**	**4.5**	**75.3**
			100	44.6	369.0	18.1	32.5	0	0	7.0	162.0	958.0	16.0	269.0
1919	**MORTADELLA**	1 OZ	**28**	**48.9**	**88.2**	**5.7**	**7.0**	**.2**	**0**	**3.4**	**66.6**	–	–	–
			100	48.9	315.0	20.4	25.0	.6	0	12.0	238.0	–	–	–
1920	**POLISH STYLE**	1 OZ	**28**	**53.7**	**85.1**	**4.4**	**7.2**	**.3**	**0**	**2.5**	**49.3**	–	–	–
			100	53.7	304.0	15.7	25.8	1.2	0	9.0	176.0	–	–	–
1921	**PORK & BEEF** (chopped together)	1 OZ	**28**	**53.5**	**94.1**	**4.4**	**8.4**	**0**	**0**	**2.5**	**48.7**	–	–	–
			100	53.5	336.0	15.6	29.9	0	0	9.0	174.0	–	–	–
1922	**PORK,** solids & liquid, canned	1 CAKE	**50**	**42.1**	**207.5**	**6.9**	**19.2**	**1.2**	**0**	**4.0**	**75.0**	**(525.1)**	**8.0**	**(100.2)**
			100	42.1	415.0	13.8	38.4	2.4	0	8.0	150.0	(1050.2)	16.0	(200.4)
1923	**PORK,** drained solids, canned	1 CAKE	**50**	**43.2**	**149.5**	**7.7**	**13.0**	**0**	**0**	**4.5**	**83.0**	**370.0**	–	**70.0**
			100	43.2	299.0	15.4	25.9	0	0	9.0	166.0	740.0	–	140.0
1924	**SCRAPPLE**	1 SLICE	**57**	**61.3**	**122.6**	**5.0**	**7.8**	**8.3**	**.1**	**2.9**	**36.5**	–	–	–
			100	61.3	215.0	8.8	13.6	14.6	.1	5.0	64.0	–	–	–
1925	**SOUSE**	1 SLICE	**60**	**70.3**	**108.6**	**7.8**	**8.0**	**.7**	**0**	**4.2**	**85.2**	–	–	–
			100	70.3	181.0	13.0	13.4	1.2	0	7.0	142.0	–	–	–
1926	**THURINGER**	1 OZ	**28**	**48.5**	**86.0**	**5.2**	**6.9**	**.4**	**0**	**3.1**	**59.9**	–	–	–
			100	48.5	307.0	18.6	24.5	1.6	0	11.0	214.0	–	–	–
1927	**VIENNA,** canned	1 OZ	**28**	**63.0**	**67.2**	**4.4**	**5.5**	**.1**	**0**	**2.2**	**42.8**	–	–	–
			100	63.0	240.0	15.8	19.8	.3	0	8.0	153.0	–	–	–
1928	**SLOPPY JOE**	1 OZ	**28**	**59.1**	**66.2**	**5.1**	**4.7**	**1.0**	**Trace**	**4.1**	**42.1**	**75.7**	**5.3**	**106.0**
			100	59.1	236.6	18.2	16.8	3.4	.1	14.7	150.4	270.2	18.8	378.4
1929	**SNAIL,** giant African, raw	1 OZ	**28**	**82.2**	**20.4**	**2.8**	**.4**	**1.2**	–	–	–	–	–	–
			100	82.2	73.0	9.9	1.4	4.4	–	–	–	–	–	–
1930	**SNAIL,** raw	1 OZ	**28**	**79.2**	**25.2**	**4.5**	**.4**	**.6**	–	–	–	–	–	–
			100	79.2	90.0	16.1	1.4	2.0	–	–	–	–	–	–
	SPLEEN													
1931	**BEEF & CALF,** raw	1 OZ	**28**	**76.9**	**29.1**	**5.1**	**.8**	**0**	**0**	–	**76.2**	–	–	–
			100	76.9	104.0	18.1	3.0	0	0	–	272.0	–	–	–
1932	**HOG,** raw	1 OZ	**28**	**77.4**	**30.0**	**4.8**	**1.1**	**0**	**0**	–	**83.4**	–	–	–
			100	77.4	107.0	17.1	3.8	0	0	–	298.0	–	–	–
1933	**LAMB,** raw	1 OZ	**28**	**74.4**	**32.2**	**5.3**	**1.1**	**0**	**0**	–	–	–	–	–
			100	74.4	115.0	18.8	3.9	0	0	–	–	–	–	–
	SQUAB (PIGEON)													
1934	total edible, raw	1 OZ	**28**	**58.0**	**78.1**	**5.2**	**6.2**	**0**	**0**	**4.8**	**115.1**	–	–	–
			100	58.0	279.0	18.6	22.1	0	0	17.0	411.0	–	–	–
1935	flesh & skin, raw	1 OZ	**28**	**56.6**	**82.3**	**5.2**	**6.4**	**0**	**0**	–	–	–	–	–
			100	56.6	294.0	18.5	23.0	0	0	–	–	–	–	–
1936	flesh only, raw	1 OZ	**28**	**72.8**	**39.8**	**4.9**	**2.1**	**0**	**0**	–	–	–	–	–
			100	72.8	142.0	17.5	7.5	0	0	–	–	–	–	–
1937	light meat, no skin, raw	1 OZ	**28**	**74.0**	**35.0**	**5.7**	**1.2**	**0**	**0**	–	–	–	–	–
			100	74.0	125.0	20.4	4.2	0	0	–	–	–	–	–
1938	giblets, raw	1 OZ	**28**	**69.8**	**43.1**	**5.5**	**2.0**	**.3**	**0**	–	–	–	–	–
			100	69.8	154.0	19.8	7.2	1.2	0	–	–	–	–	–
1939	**SQUIRREL,** cooked	1 OZ	**28**	–	**60.5**	**8.2**	**2.8**	**0**	–	**5.9**	**72.5**	–	**6.4**	–
			100	–	216.0	29.3	10.1	0	–	21.0	259.0	–	23.0	–
1940	**STOMACH,** pork, scalded	1 OZ	**28**	**74.0**	**42.6**	**4.6**	**2.5**	**0**	**0**	–	**33.0**	–	–	–
			100	74.0	152.0	16.5	9.0	0	0	–	118.0	–	–	–
	SWEETBREAD (THYMUS)													
1941	**BEEF,** yearling, braised	1 OZ	**28**	**49.6**	**89.6**	**7.3**	**6.5**	**0**	**0**	–	**101.9**	**32.5**	–	**121.2**
			100	49.6	320.0	25.9	23.2	0	0	–	364.0	116.0	–	433.0

(Continued)

Item No.	Minerals (Micro) Iron	Zinc	Copper	Fat-Soluble Vitamins Vitamin A	Vitamin D	Vitamin E (Alpha Tocopherol)	Water-Soluble Vitamins Vitamin C	Thiamin	Ribo-flavin	Niacin	Panto-thenic Acid	Vit. B-6 (Pyri-doxine)	Folacin (Folic Acid)	Biotin	Vitamin B-12
	(mg)	(mg)	(mg)	(IU)	(IU)	(mg)	(mg)	(mg)	(mg)	(mg)	(mg)	(mg)	(mcg)	(mcg)	(mcg)
1913	–	–	–	–	–	–	–	–	–	–	–	–	–	–	–
	–	–	–	–	–	–	–	–	–	–	–	–	–	–	–
1914	**.78**	–	–	–	**0**	–	**0**	**.03**	**.07**	**1.18**	–	**.04**	–	–	–
	2.80	–	–	–	0	–	0	.11	.26	4.20	–	.14	–	–	–
1915	**.76**	–	–	**0**	**0**	–	**0**	**.08**	**.06**	**1.54**	–	**.04**	–	–	–
	2.70	–	–	0	0	–	0	.27	.23	5.50	–	.14	–	–	–
1916	**.64**	–	–	–	–	–	–	**.06**	**.05**	**.87**	–	–	**3.92**	–	–
	2.30	–	–	–	–	–	–	.22	.19	3.10	–	–	14.00	–	–
1917	**.59**	–	–	–	–	–	–	**.05**	**.06**	**.73**	–	–	–	–	–
	2.10	–	–	–	–	–	–	.17	.21	2.60	–	–	–	–	–
1918	**.67**	–	–	**0**	**0**	**.05**	**0**	**.22**	**.10**	**1.04**	–	–	–	–	–
	2.40	–	–	0	0	.16	0	.79	.34	3.70	–	–	–	–	–
1919	**.87**	–	–	–	–	–	–	–	–	–	–	–	–	–	–
	3.10	–	–	–	–	–	–	–	–	–	–	–	–	–	–
1920	**.67**	–	–	**0**	**0**	–	**0**	**.10**	**.05**	**.87**	–	–	–	–	–
	2.40	–	–	0	0	–	0	.34	.19	3.10	–	–	–	–	–
1921	**.64**	–	–	**0**	**0**	–	**0**	–	–	–	–	–	–	–	–
	2.30	–	–	0	0	–	0	–	–	–	–	–	–	–	–
1922	**1.05**	**(.85)**	**(.19)**	**0**	**Trace**	**(.14)**	**0**	**.10**	**.10**	**1.65**	**(.31)**	**(.04)**	**(1.03)**	**(1.55)**	**(.54)**
	2.10	(1.70)	(.38)	0	Trace	(.28)	0	.19	.19	3.30	(.62)	(.07)	(2.05)	(3.10)	(1.08)
1923	**1.15**	–	–	**0**	**0**	–	**0**	**.10**	**.12**	**1.50**	–	–	–	–	–
	2.30	–	–	0	0	–	0	.20	.24	3.00	–	–	–	–	–
1924	**.68**	–	–	–	**0**	–	**0**	**.11**	**.05**	**1.03**	–	–	–	–	–
	1.20	–	–	–	0	–	0	.19	.09	1.80	–	–	–	–	–
1925	**1.20**	–	–	**0**	**0**	–	**0**	–	–	–	–	–	–	–	–
	2.00	–	–	0	0	–	0	–	–	–	–	–	–	–	–
1926	**.78**	–	–	–	–	–	–	**.03**	**.07**	**1.18**	–	–	–	–	–
	2.80	–	–	–	–	–	–	.11	.26	4.20	–	–	–	–	–
1927	**.59**	–	–	–	**0**	–	**0**	**.02**	**.04**	**.73**	–	**.02**	–	–	–
	2.10	–	–	–	0	–	0	.08	.13	2.60	–	.08	–	–	–
1928	**.71**	**.01**	–	**52.1**	**0**	–	**.54**	**.11**	**.05**	**1.18**	**0**	**Trace**	**.53**	–	**0**
	2.52	.05	–	186.0	0	–	1.94	.38	.16	4.21	0	.01	1.89	–	0
1929	–	–	–	–	–	–	–	–	–	–	–	–	–	–	–
	–	–	–	–	–	–	–	–	–	–	–	–	–	–	–
1930	**.98**	–	–	–	–	–	–	–	–	–	–	–	–	–	–
	3.50	–	–	–	–	–	–	–	–	–	–	–	–	–	–
1931	**2.97**	–	–	–	–	–	–	–	**.10**	**2.30**	**.35**	**.03**	–	–	**1.48**
	10.60	–	–	–	–	–	–	–	.37	8.20	1.25	.12	–	–	5.30
1932	**8.23**	–	–	–	–	–	–	–	**.11**	–	**.35**	**.03**	–	–	**1.48**
	29.40	–	–	–	–	–	–	–	.40	–	1.25	.12	–	–	5.30
1933	–	–	–	–	–	–	–	–	–	–	**.35**	**.03**	–	–	**1.48**
	–	–	–	–	–	–	–	–	–	–	1.25	.12	–	–	5.30
1934	–	–	–	–	–	–	**0**	**.03**	**.07**	**1.57**	–	–	–	–	–
	–	–	–	–	–	–	0	.10	.24	5.60	–	–	–	–	–
1935	–	–	–	–	–	–	–	–	–	–	–	–	–	–	–
	–	–	–	–	–	–	–	–	–	–	–	–	–	–	–
1936	–	–	–	–	–	–	–	–	–	**1.85**	–	–	–	–	–
	–	–	–	–	–	–	–	–	–	6.60	–	–	–	–	–
1937	–	–	–	–	–	–	–	–	–	**2.13**	–	–	–	–	–
	–	–	–	–	–	–	–	–	–	7.60	–	–	–	–	–
1938	–	–	–	–	–	–	–	–	–	–	–	–	–	–	–
	–	–	–	–	–	–	–	–	–	–	–	–	–	–	–
1939	**.42**	–	–	**0**	–	–	**0**	**.01**	**.02**	**3.16**	–	**.07**	–	–	**.20**
	1.50	–	–	0	–	–	0	.05	.07	11.30	–	.26	–	–	.72
1940	–	–	–	–	–	–	–	–	–	–	–	–	–	–	–
	–	–	–	–	–	–	–	–	–	–	–	–	–	–	–
1941	–	–	–	–	–	–	–	–	–	–	–	–	–	–	–
	–	–	–	–	–	–	–	–	–	–	–	–	–	–	–

(Continued)

MEATS

TABLE F-36 *(Continued)*

Item No.	Food Item Name	Approximate Measure	Weight	Moisture	Food Energy Calories	Protein	Fats	Carbo-hydrates	Fiber	Minerals (Macro) Calcium	Phos-phorus	Sodium	Mag-nesium	Potas-sium
			(g)	(%)	(kcal)	(g)	(g)	(g)	(g)	(mg)	(mg)	(mg)	(mg)	(mg)
	SWEETBREAD (THYMUS) (Continued)													
1942	**CALF,** braised	1 OZ	**28**	**62.7**	**47.0**	**9.1**	**.9**	**0**	**0**	—	—	—	—	—
			100	62.7	168.0	32.6	3.2	0	0	—	—	—	—	—
1943	**LAMB,** braised	1 OZ	**28**	**64.6**	**49.0**	**7.9**	**1.7**	**0**	**0**	—	**57.1**	—	—	—
			100	64.6	175.0	28.1	6.1	0	0	—	204.0	—	—	—
	[4]**SWIFTLET NEST**													
1944	(used in bird's nest soup), dried	1 OZ	**28**	**12.6**	**96.6**	**15.0**	**.1**	**8.2**	—	**131.6**	**5.0**	**58.8**	—	**15.4**
			100	12.6	345.0	53.4	.4	29.3	—	470.0	18.0	210.0	—	55.0
1945	dried, soaked, drained	1 OZ	**28**	**83.1**	**19.6**	**3.3**	**.1**	**1.2**	—	**30.8**	**1.4**	**12.0**	—	**5.6**
			100	83.1	70.0	11.9	.3	4.2	—	110.0	5.0	43.0	—	20.0
	TONGUE													
1946	**BEEF,** medium fat, braised	1 OZ	**28**	**60.8**	**68.3**	**6.0**	**4.7**	**.1**	**0**	**2.0**	**32.8**	**17.1**	**4.5**	**45.9**
			100	60.8	244.0	21.5	16.7	.4	0	7.0	117.0	61.0	16.0	164.0
1947	**BEEF,** smoked	1 SLICE	**20**	**48.9**	—	**3.4**	**5.8**	—	**0**	—	—	—	—	—
			100	48.9	—	17.2	28.8	—	0	—	—	—	—	—
1948	**CALF,** braised	1 SLICE	**20**	**68.5**	**32.0**	**4.8**	**1.2**	**.2**	**0**	—	—	—	—	—
			100	68.5	160.0	23.9	6.0	1.0	0	—	—	—	—	—
1949	**HOG,** braised	1 SLICE	**20**	**59.4**	**49.4**	**4.4**	**3.5**	**.1**	**0**	**5.2**	**23.8**	—	—	—
			100	59.4	247.0	22.0	17.4	.5	0	26.0	119.0	—	—	—
1950	**LAMB,** braised	1 SLICE	**20**	**60.2**	**50.8**	**4.1**	**3.6**	**.1**	**0**	—	**20.4**	—	—	—
			100	60.2	254.0	20.5	18.2	.5	0	—	102.0	—	—	—
1951	**SHEEP,** braised	1 SLICE	**20**	**51.6**	**64.6**	**4.0**	**5.1**	**.5**	**0**	—	—	—	—	—
			100	51.6	323.0	19.8	25.3	2.4	0	—	—	—	—	—
1952	canned or cured, beef/lamb/etc, whole canned or pickled	1 SLICE	**20**	**56.6**	**53.4**	**3.9**	**4.1**	**.1**	**0**	—	—	—	—	—
			100	56.6	267.0	19.3	20.3	.3	0	—	—	—	—	—
1953	canned or cured, beef/lamb/etc, potted or deviled	1 OZ	**28**	**52.8**	**81.2**	**5.2**	**6.4**	**.2**	**0**	—	—	—	—	—
			100	52.8	290.0	18.6	23.0	.7	0	—	—	—	—	—
	TRIPE, BEEF													
1954	commercial	1 OZ	**28**	**79.1**	**28.0**	**5.3**	**.6**	**0**	**0**	**35.6**	**24.1**	**20.2**	**(4.3)**	**2.5**
			100	79.1	100.0	19.1	2.0	0	0	127.0	86.0	72.0	(15.4)	9.0
1955	pickled	1 OZ	**28**	**86.5**	**17.4**	**3.3**	**.4**	**0**	**0**	—	—	**12.9**	—	**5.3**
			100	86.5	62.0	11.8	1.3	0	0	—	—	46.0	—	19.0
	TURKEY													
1956	total edible, roasted	1 SLICE	**40**	**55.4**	**105.2**	**10.8**	**6.6**	**0**	**0**	—	—	—	**11.2**	—
			100	55.4	263.0	27.0	16.4	0	0	—	—	—	28.0	—
1957	meat & skin, roasted	1 OZ	**28**	**61.7**	**58.2**	**7.9**	**2.7**	**0**	**0**	**7.3**	**56.8**	**19.0**	**7.0**	**78.4**
			100	61.7	208.0	28.1	9.7	0	0	26.0	203.0	68.0	25.0	280.0
1958	meat only, roasted	1 OZ	**28**	**64.9**	**47.6**	**8.2**	**1.4**	**0**	**0**	**7.0**	**60.0**	**20.0**	**7.3**	**83.4**
			100	64.9	170.0	29.3	5.0	0	0	25.0	213.0	70.0	26.0	298.0
1959	skin only, roasted	1 OZ	**28**	**39.7**	**124**	**5.5**	**11.1**	**0**	**0**	**9.8**	**38.4**	**14.8**	**4.5**	**44.8**
			100	39.7	442	19.7	39.7	0	0	35.0	137.0	53.0	16.0	160.0
1960	[2]light & dark meat, diced, seasoned	1 OZ	**28**	**71.7**	**38.6**	**5.2**	**1.7**	**.3**	—	**.3**	**67.2**	**238.0**	—	**86.8**
			100	71.7	138.0	18.7	6.0	1.0	—	1.0	240.0	850.0	—	310.0
1961	light meat, roasted	1 OZ	**28**	**62.1**	**49.3**	**9.2**	**1.1**	**0**	**0**	—	—	**23.0**	**7.8**	**115.1**
			100	62.1	176.0	32.9	3.9	0	0	—	—	82.0	28.0	411.0
1962	dark meat, roasted	1 OZ	**28**	**60.5**	**56.8**	**8.4**	**2.3**	**0**	**0**	**8.4**	**112.0**	**27.7**	**7.8**	**111.4**
			100	60.5	203.0	30.0	8.3	0	0	30.0	400.0	99.0	28.0	398.0
1963	[2]breast, meat & skin, roasted	1 OZ	**28**	**63.2**	**52.3**	**8.0**	**2.1**	**0**	**0**	**5.9**	**58.8**	**17.6**	**7.6**	**80.1**
			100	63.2	189.0	28.7	7.4	0	0	21.0	210.0	63.0	27.0	288.0
1964	[2]thigh, meat & skin, prebasted, roasted	1 OZ	**28**	**70.6**	**44.0**	**5.3**	**2.4**	**0**	**0**	**2.2**	**47.9**	**122.4**	**4.8**	**67.5**
			100	70.6	157.0	18.8	8.5	0	0	8.0	171.0	437.0	17.0	241.0
1965	[2]roll, light & dark meat	1 OZ	**28**	**70.1**	**41.7**	**5.1**	**2.0**	**.6**	—	**9.0**	**47.0**	**164.1**	**5.0**	**75.6**
			100	70.1	149.0	18.1	7.0	2.1	—	32.0	168.0	586.0	18.0	270.0
1966	[2]roll, light meat	1 OZ	**28**	**71.6**	**41.2**	**5.2**	**2.0**	**.1**	—	**11.2**	**51.2**	**136.9**	**4.5**	**70.3**
			100	71.6	147.0	18.7	7.2	.5	—	40.0	183.0	489.0	16.0	251.0
1967	giblets, some gizzard fat, simmered	1 OZ	**28**	**61.0**	**65.2**	**5.8**	**4.3**	**.4**	**0**	—	—	—	—	—
			100	61.0	233.0	20.6	15.4	1.6	0	—	—	—	—	—
1968	meat only, canned	1 OZ	**28**	**64.5**	**56.6**	**5.9**	**3.5**	**0**	**0**	**2.8**	—	—	—	—
			100	64.5	202.0	20.9	12.5	0	0	10.0	—	—	—	—
1969	[2]roast, boneless, light & dark meat, frozen	1 OZ	**28**	**67.9**	**43.4**	**6.0**	**1.6**	**.9**	—	**1.4**	**68.3**	**190.4**	**6.2**	**83.4**
			100	67.9	155.0	21.3	5.8	3.1	—	5.0	244.0	680.0	22.0	298.0

Item No.	Minerals (Micro)			Fat-Soluble Vitamins			Water-Soluble Vitamins								
	Iron	Zinc	Copper	Vitamin A	Vitamin D	Vitamin E (Alpha Tocopherol)	Vitamin C	Thiamin	Ribo-flavin	Niacin	Panto-thenic Acid	Vit. B-6 (Pyri-doxine)	Folacin (Folic Acid)	Biotin	Vitamin B-12
	(mg)	(mg)	(mg)	(IU)	(IU)	(mg)	(mg)	(mg)	(mg)	(mg)	(mg)	(mg)	(mcg)	(mcg)	(mcg)
1942	—	—	—	—	—	—	—	**.03**	**.08**	**.81**	—	—	—	—	—
	—	—	—	—	—	—	—	.09	.27	2.90	—	—	—	—	—
1943	—	—	—	—	—	—	—	—	—	—	—	—	—	—	—
	—	—	—	—	—	—	—	—	—	—	—	—	—	—	—
1944	**1.20**	—	—	**0**	—	—	**0**	**.014**	—	—	—	—	—	—	—
	4.30	—	—	0	—	—	0	.05	—	—	—	—	—	—	—
1945	**.28**	—	—	**0**	—	—	—	**.00**	**.15**	**Trace**	—	—	—	—	—
	1.00	—	—	0	—	—	—	.01	.55	Trace	—	—	—	—	—
1946	**.62**	—	—	**0**	**0**	**.10**	**0**	**.01**	**.08**	**.98**	**(.15)**	**(.03)**	**(1.41)**	**(.86)**	**(1.12)**
	2.20	—	—	0	0	.35	0	.05	.29	3.50	(.52)	(.09)	(5.05)	(3.06)	(4.00)
1947	—	—	—	—	—	—	—	**.01**	**.04**	**.60**	**.12**	—	—	—	—
	—	—	—	—	—	—	—	.04	.21	3.00	.60	—	—	—	—
1948	—	—	—	—	—	—	—	—	—	—	—	—	—	—	—
	—	—	—	—	—	—	—	—	—	—	—	—	—	—	—
1949	**.28**	—	—	—	—	—	—	**.01**	**.06**	**.70**	—	—	—	—	—
	1.40	—	—	—	—	—	—	.07	.29	3.50	—	—	—	—	—
1950	—	—	—	—	—	—	—	—	—	—	—	—	—	—	—
	—	—	—	—	—	—	—	—	—	—	—	—	—	—	—
1951	**.68**	—	—	—	—	—	—	—	—	—	—	—	—	—	—
	3.40	—	—	—	—	—	—	—	—	—	—	—	—	—	—
1952	—	—	—	—	—	—	—	—	—	—	—	**.05**	—	—	—
	—	—	—	—	—	—	—	—	—	—	—	.24	—	—	—
1953	—	—	—	—	—	—	—	**.01**	**.03**	**.36**	—	**.07**	—	—	—
	—	—	—	—	—	—	—	.04	.11	1.30	—	.24	—	—	—
1954	**.45**	**(.65)**	**(.04)**	**0**	**0**	**(.03)**	**0**	**Trace**	**.04**	**.45**	**(.07)**	**Trace**	**(.28)**	**(.57)**	**Trace**
	1.60	(2.31)	(.16)	0	0	(.09)	0	Trace	.15	1.60	(.24)	(.02)	(1.00)	(2.02)	Trace
1955	—	—	—	—	—	—	—	—	—	—	—	—	—	—	—
	—	—	—	—	—	—	—	—	—	—	—	—	—	—	
1956	—	—	—	—	—	—	—	—	—	—	—	—	—	—	—
	—	—	—	—	—	—	—	—	—	—	—	—	—	—	—
1957	**.5**	**.83**	**.03**	**0**	—	—	—	**.02**	**.05**	**1.43**	**.24**	**.12**	**2.0**	—	**.10**
	1.8	2.96	.09	0	—	—	—	.06	.18	5.09	.86	.41	7.0	—	.35
1958	**.5**	**.87**	**.03**	**0**	—	—	**0**	**.02**	**.05**	**1.52**	**.26**	**.13**	**2.0**	—	**.01**
	1.8	3.10	.09	0	—	—	0	.06	.18	5.44	.94	.46	7.0	—	.04
1959	**.5**	**.58**	**.03**	**0**	—	—	**0**	**Trace**	**.04**	**.74**	**.09**	**.02**	**1.1**	—	**.07**
	1.8	2.07	.09	0	—	—	0	.02	.15	2.66	.31	.08	4.0	—	.24
1960	**.50**	—	—	—	—	—	—	**.01**	**.03**	**1.34**	—	—	—	—	—
	1.80	—	—	—	—	—	—	.04	.11	4.80	—	—	—	—	—
1961	**.34**	**.59**	—	—	—	—	—	**.01**	**.04**	**3.11**	—	—	**1.40**	—	—
	1.20	2.10	—	—	—	—	—	.05	.14	11.10	—	—	5.00	—	—
1962	**.64**	**1.23**	—	—	—	—	**0**	**.01**	**.06**	**1.18**	—	—	**1.96**	—	—
	2.30	4.40	—	—	—	—	0	.04	.23	4.20	—	—	7.00	—	—
1963	**.39**	**.57**	**.01**	**0**	—	—	**0**	**.02**	**.04**	**1.78**	**.18**	**.13**	**1.7**	—	**.10**
	1.40	2.03	.05	0	—	—	0	.06	.13	6.37	.63	.48	6.0	—	.36
1964	**4.2**	**1.15**	**.04**	**0**	—	—	**0**	**.02**	**.07**	—	—	—	—	—	—
	1.51	4.12	.14	0	—	—	0	.08	.26	—	—	—	—	—	—
1965	**.38**	**.56**	**.02**	—	—	—	—	**.02**	**.08**	**1.34**	—	—	—	—	—
	1.35	2.00	.07	—	—	—	—	.09	.28	4.80	—	—	—	—	—
1966	**.36**	**.44**	**.01**	—	—	—	—	**.03**	**.06**	**1.96**	—	—	—	—	—
	1.28	1.56	.04	—	—	—	—	.09	.23	7.00	—	—	—	—	—
1967	—	—	—	—	—	—	—	—	**.76**	—	—	—	—	—	—
	—	—	—	—	—	—	—	—	2.72	—	—	—	—	—	—
1968	**.39**	—	—	**36.4**	—	—	—	**.01**	**.04**	**1.32**	—	—	—	—	—
	1.40	—	—	130.0	—	—	—	.02	.14	4.70	—	—	—	—	—
1969	**.46**	**.71**	**.02**	—	—	—	—	**.01**	**.05**	**1.76**	**.23**	**.08**	—	—	**.43**
	1.63	2.54	.06	—	—	—	—	.05	.16	6.27	.81	.27	—	—	1.52

(Continued)

MEATS

TABLE F-36 *(Continued)*

Item No.	Food Item Name	Approximate Measure	Weight	Moisture	Food Energy Calories	Protein	Fats	Carbo-hydrates	Fiber	Minerals (Macro) Calcium	Phos-phorus	Sodium	Mag-nesium	Potas-sium
			(g)	(%)	(kcal)	(g)	(g)	(g)	(g)	(mg)	(mg)	(mg)	(mg)	(mg)
	TURKEY (Continued)													
1970	[2]bologna	1 OZ	28	65.1	55.7	3.8	4.3	.3	–	23.5	36.7	245.8	3.9	55.7
			100	65.1	199.0	13.7	15.2	1.0	–	84.0	131.0	878.0	14.0	199.0
1971	[2]frankfurter	1 OZ	28	63.0	63.3	4.0	5.0	.4	–	29.7	37.5	399.3	–	50.1
			100	63.0	226.0	14.3	17.7	1.5	–	106.0	134.0	1426.0	–	179.0
1972	[2]pastrami	1 OZ	28	70.6	39.5	5.1	1.7	.5	–	2.5	56.0	292.6	3.9	72.8
			100	70.6	141.0	18.4	6.2	1.7	–	9.0	200.0	1045.0	14.0	260.0
1973	[2]patties, breaded or battered, fried	1 OZ	28	49.7	79.2	3.9	5.0	4.4	–	3.9	75.6	224.0	–	77.0
			100	49.7	283.0	14.0	18.0	15.7	–	14.0	270.0	800.0	–	275.0
1974	[2]salami	1 OZ	28	65.9	54.9	4.6	3.9	.2	–	5.6	29.7	281.1	4.2	68.3
			100	65.9	196.0	16.4	13.8	.6	–	20.0	106.0	1004.0	15.0	244.0
1975	[2]sticks, breaded or battered, fried	1 OZ	28	49.4	78.1	4.0	4.7	4.8	–	3.9	65.5	234.6	–	72.8
			100	49.4	279.0	14.2	16.9	17.0	–	14.0	234.0	838.0	–	260.0
1976	**TURTLE,** green, raw	1 OZ	28	78.5	24.9	5.5	.1	0	0	–	–	–	–	–
			100	78.5	89.0	19.8	.5	0	0	–	–	–	–	–
	VEAL													
1977	chop	1 OZ	28	54.6	75.3	7.6	4.7	0	0	3.4	69.4	22.4	5.6	140.0
			100	54.6	269.0	27.2	16.9	0	0	12.0	248.0	80.0	20.0	500.0
1978	chuck, medium fat, total edible, braised, 85% lean, 15% fat	1 OZ	28	58.5	65.8	7.8	3.6	0	0	3.4	42.3	22.4	–	140.0
			100	58.5	235.0	27.9	12.8	0	0	12.0	151.0	80.0	–	500.0
1979	cutlet, total edible, cooked	1 OZ	28	–	77.6	9.3	4.2	0	0	2.8	80.6	15.1	6.4	147.6
			100	–	277.0	33.2	15.0	0	0	10.0	288.0	54.0	23.0	527.0
1980	cutlet, breaded	1 OZ	28	–	89.3	9.6	4.2	2.9	0	0	80.6	15.1	6.4	147.6
			100	–	319.0	34.2	15.0	10.5	0	0	288.0	54.0	23.0	527.0
1981	flank, medium fat, total edible, stewed, 60% lean, 40% fat	1 OZ	28	43.8	109.2	6.5	9.0	0	0	3.1	32.8	22.4	–	140.0
			100	43.8	390.0	23.2	32.3	0	0	11.0	117.0	80.0	–	500.0
1982	foreshank, medium fat, total edible, stewed, 86% lean, 14% fat	1 OZ	28	60.1	60.5	8.0	2.9	0	0	3.4	43.1	22.4	–	140.0
			100	60.1	216.0	28.7	10.4	0	0	12.0	154.0	80.0	–	500.0
1983	loin, medium fat, total edible, broiled, 77% lean, 23% fat	1 OZ	28	58.9	65.5	7.4	3.8	0	0	3.1	63.0	22.4	–	140.0
			100	58.9	234.0	26.4	13.4	0	0	11.0	225.0	80.0	–	500.0
1984	plate, medium fat, stewed, 73% lean, 27% fat	1 OZ	28	52.1	84.8	7.3	5.9	0	0	3.4	38.6	22.4	–	140.0
			100	52.1	303.0	26.1	21.2	0	0	12.0	138.0	80.0	–	500.0
1985	rib, medium fat, roasted, 82% lean, 18% fat	1 OZ	28	54.6	75.3	7.6	4.7	0	0	3.4	69.4	22.4	5.6	140.0
			100	54.6	269.0	27.2	16.9	0	0	12.0	248.0	80.0	20.0	500.0
1986	round, w/rump, medium fat, total edible, broiled, 79% lean, 21% fat	1 OZ	30	60.4	64.8	8.1	3.3	0	0	3.3	69.3	24.0	–	150.0
			100	60.4	216.0	27.1	11.1	0	0	11.0	231.0	80.0	–	500.0
1987	[1]shoulder arm roast, slice, boneless, lean & fat, cooked	1 OZ	28	58.5	65.8	7.8	3.6	0	0	3.4	42.3	22.4	–	140.0
			100	58.5	235.0	27.9	12.8	0	0	12.0	151.0	80.0	–	500.0
	VENISON													
1988	lean only, raw	1 SLICE	40	74.0	50.4	8.4	1.6	0	0	4.0	99.6	28.0	11.6	134.4
			100	74.0	126.0	21.0	4.0	0	0	10.0	249.0	70.0	29.0	336.0
1989	lean only, cooked	1 OZ	28	–	48.7	8.4	1.7	0	–	2.5	95.5	–	9.2	–
			100	–	174.0	30.0	6.0	0	–	9.0	341.0	–	33.0	–
	MILK & PRODUCTS													
	BUTTER													
1990	regular, salted	1 TSP	5	15.9	35.9	0	4.1	0	0	1.2	1.2	41.3	.1	1.3
			100	15.9	717.0	.9	81.1	0.1	0	24.0	23.0	826.0	2.0	26.0
1991	salt free	1 TSP	5	–	35.8	0	4.1	0	0	1.0	.8	.4	.1	1.2
			100	–	716.0	.6	81.0	.4	0	20.0	16.0	8.0	2.0	23.0
1992	whipped	1 TBLS	9	16.0	64.0	.1	7.3	0	0	2.2	2.1	74.3	.2	2.3
			100	16.0	711.0	.9	81.0	.1	0	24.0	23.0	826.0	2.0	26.0
1993	oil or dehydrated butter	1 TSP	5	0.2	43.8	0	5.0	0	0	–	–	–	–	–
			100	0.2	876.0	.3	99.5	0	0	–	–	–	–	–
	BUTTERMILK													
1994	fluid, made from skim milk	1 CUP	244	90.5	87.8	8.4	2.1	11.7	0	295.2	217.2	256.2	26.8	368.4
			100	90.5	36.0	3.4	.9	4.8	0	121.0	89.0	105.0	11.0	151.0
1995	dried, made from skim milk	1 TBLS	7	2.8	27.1	2.4	.4	3.5	0	82.9	65.3	35.5	7.7	112.4
			100	2.8	387.0	34.3	5.8	50.0	0	1184.0	933.0	507.0	110.0	1606.0
	CARNATION INSTANT BREAKFAST MIX													
1996	chocolate	1 PKG	36	–	130.0	7.0	1.0	23.0	–	100.0	150.0	110.0	80.0	380.0
			100	–	361.1	19.4	2.8	63.9	–	277.8	416.7	305.6	222.2	1055.6
1997	vanilla	1 PKG	35	–	130.0	7.0	0	24.0	–	100.0	150.0	120.0	80.0	360.0
			100	–	371.4	20.0	0	68.6	–	285.7	428.6	342.9	228.6	1028.6
1998	vanilla, in whole milk	1 CUP	276	–	280.0	15.0	8.0	35.0	–	407.0	386.0	242.0	114.4	711.0
			100	–	101.4	5.4	2.9	12.7	–	147.5	139.9	87.7	41.4	257.6

Item No.	Minerals (Micro)			Fat-Soluble Vitamins			Water-Soluble Vitamins								
	Iron	Zinc	Copper	Vitamin A	Vitamin D	Vitamin E (Alpha Tocopherol)	Vitamin C	Thiamin	Ribo-flavin	Niacin	Panto-thenic Acid	Vit. B-6 (Pyri-doxine)	Folacin (Folic Acid)	Biotin	Vitamin B-12
	(mg)	(mg)	(mg)	(IU)	(IU)	(mg)	(mg)	(mg)	(mg)	(mg)	(mg)	(mg)	(mcg)	(mcg)	(mcg)
1970	**.43**	**.49**	**.01**	—	—	—	—	**.02**	**.05**	**.99**	—	—	—	—	—
	1.53	1.74	.03	—	—	—	—	.06	.17	3.53	—	—	—	—	—
1971	**.52**	—	—	—	—	—	—	**.01**	**.05**	**1.16**	—	—	—	—	—
	1.84	—	—	—	—	—	—	.04	.18	4.13	—	—	—	—	—
1972	**.47**	**.61**	**.02**	—	—	—	—	**.02**	**.07**	**.99**	—	—	—	—	—
	1.66	2.16	.05	—	—	—	—	.06	.25	3.5	—	—	—	—	—
1973	**.62**	—	—	—	—	—	—	**.03**	**.05**	**.64**	—	—	—	—	—
	2.20	—	—	—	—	—	—	.10	.19	2.30	—	—	—	—	—
1974	**.45**	**.51**	**.02**	—	—	—	—	**.02**	**.05**	**.99**	—	—	—	—	—
	1.61	1.81	.05	—	—	—	—	.07	.18	3.53	—	—	—	—	—
1975	**.62**	—	—	—	—	—	—	**.03**	**.05**	**.59**	—	—	—	—	—
	2.20	—	—	—	—	—	—	.10	.18	2.10	—	—	—	—	—
1976	—	—	—	—	—	—	—	—	—	—	—	—	—	—	—
	—	—	—	—	—	—	—	—	—	—	—	—	—	—	—
1977	**.95**	—	—	**0**	**0**	—	**0**	**.04**	**.09**	**2.18**	—	—	**1.40**	—	—
	3.40	—	—	0	0	—	0	.13	.31	7.80	—	—	5.00	—	—
1978	**.98**	—	—	**0**	**0**	—	**0**	**.03**	**.08**	**1.79**	—	—	**.84**	—	—
	3.50	(4.10)	—	0	0	—	0	.09	.29	6.40	—	—	3.00	—	—
1979	**1.18**	—	—	**0**	**0**	—	**0**	**.03**	**.09**	**1.79**	—	—	**.84**	**Trace**	**(.28)**
	4.20	—	—	0	0	—	0	.12	.32	6.40	—	—	3.00	Trace	(1.00)
1980	**1.18**	—	—	**0**	**0**	**.01**	**0**	**.03**	**.09**	**1.79**	—	—	—	—	—
	4.20	—	—	0	0	.05	0	.12	.32	6.40	—	—	—	—	—
1981	**.84**	—	—	**0**	**0**	—	**0**	**.01**	**.06**	**1.18**	—	—	**.84**	—	—
	3.00	—	—	0	0	—	0	.05	.22	4.20	—	—	3.00	—	—
1982	**1.01**	—	—	**0**	**0**	—	**0**	**.01**	**.07**	**1.40**	—	—	**.84**	—	—
	3.60	(4.10)	—	0	0	—	0	.05	.26	5.00	—	—	3.00	—	—
1983	**.90**	—	—	**0**	**0**	—	**0**	**.02**	**.07**	**1.51**	—	—	**.84**	—	—
	3.20	—	—	0	0	—	0	.07	.25	5.40	—	—	3.00	—	—
1984	**.92**	—	—	**0**	**0**	—	**0**	**.01**	**.07**	**1.29**	—	—	**.84**	—	—
	3.30	—	—	0	0	—	0	.05	.24	4.60	—	—	3.00	—	—
1985	**.95**	—	—	**0**	**0**	—	**0**	**.04**	**.09**	**2.18**	—	—	**.84**	—	—
	3.40	—	—	0	0	—	0	.13	.31	7.80	—	—	3.00	—	—
1986	**.96**	—	—	**0**	**0**	—	**0**	**.02**	**.08**	**1.62**	—	—	**.90**	—	—
	3.20	(4.10)	—	0	0	—	0	.07	.25	5.40	—	—	3.00	—	—
1987	**.98**	**1.00**	—	—	—	—	—	**.03**	**.08**	**1.79**	—	—	**.84**	—	—
	3.50	3.56	—	—	—	—	—	.09	.29	6.40	—	—	3.00	—	—
1988	**2.00**	—	—	**0**	—	—	**0**	**.09**	**.19**	**2.52**	—	—	—	—	—
	5.00	—	—	0	—	—	0	.23	.48	6.30	—	—	—	—	—
1989	**.62**	—	—	**0**	—	—	**0**	**.01**	**.07**	**1.26**	—	**.11**	—	—	**.40**
	2.20	—	—	0	—	—	0	.02	.24	4.50	—	.38	—	—	1.44
1990	**.01**	**0**	**.02**	**152.9**	**(.04)**	**.08**	**0**	**0**	**0**	**0**	**Trace**	**0**	**.15**	**Trace**	**Trace**
	.16	.05	.39	3058.0	(.76)	1.58	0	.01	.03	.04	Trace	0	3.00	Trace	Trace
1991	**0**	—	—	**165.0**	—	—	**0**	—	—	—	—	—	**.15**	—	—
	0	—	—	3300.0	—	—	0	—	—	—	—	—	3.00	—	—
1992	**.01**	**.01**	—	**275.2**	—	—	**0**	**0**	**0**	**0**	—	**0**	**.27**	—	—
	.16	.05	—	3058.0	—	—	0	.01	.03	.04	—	.03	3.00	—	—
1993	—	—	—	**187.5**	—	—	**0**	—	—	—	—	—	—	—	—
	—	—	—	3750.0	—	—	0	—	—	—	—	—	—	—	—
1994	**.12**	**.90**	**.01**	**80.5**	—	—	**2.44**	**.10**	**.44**	**.14**	**.75**	**.09**	**0**	**0**	**.54**
	.05	.37	0	33.0	—	—	1.00	.04	.18	.06	.31	.04	0	0	.22
1995	**.02**	**.28**	—	**15.4**	—	—	**.40**	**.03**	**.12**	**.06**	**.22**	**.02**	**3.29**	—	**.27**
	.30	4.02	—	220.0	—	—	5.66	.39	1.72	.90	3.20	.34	47.00	—	3.82
1996	**4.50**	**3.00**	**.50**	**1000.0**	**0**	**5.03**	**24.00**	**.30**	**.07**	**5.00**	**2.00**	**.40**	**100.00**	—	**600.00**
	12.50	8.33	1.39	2777.8	0	13.98	66.67	.83	.19	13.89	5.56	1.11	277.78	—	1666.67
1997	**4.50**	**5.00**	**.50**	**1000.0**	**0**	**5.03**	**24.00**	**.30**	**.07**	**5.00**	**2.00**	**.40**	**100.00**	—	**.60**
	12.86	14.29	1.43	2857.1	0	14.38	68.57	.86	.20	14.29	5.71	1.14	285.71	—	1.71
1998	**4.62**	**3.66**	**.52**	**1250.0**	**100.00**	**5.19**	**26.40**	**.40**	**.56**	**5.23**	**2.66**	**.51**	**100.00**	—	**1530.00**
	1.67	1.33	.19	452.9	36.23	1.88	9.57	.15	.20	1.90	.96	.19	36.23	—	554.35

(Continued)

TABLE F-36 *(Continued)*

Item No.	Food Item Name	Approximate Measure	Weight	Moisture	Food Energy Calories	Protein	Fats	Carbo-hydrates	Fiber	Minerals (Macro) Calcium	Phos-phorus	Sodium	Mag-nesium	Potas-sium
			(g)	(%)	(kcal)	(g)	(g)	(g)	(g)	(mg)	(mg)	(mg)	(mg)	(mg)
	CHEESE FOOD													
1999	**AMERICAN,** cold pack	1 OZ	**28**	**43.1**	**92.7**	**5.5**	**6.8**	**2.3**	**0**	**139.2**	**112.0**	**270.5**	**8.4**	**101.6**
			100	43.1	331.0	19.7	24.5	8.3	0	497.0	400.0	966.0	30.0	363.0
2000	**AMERICAN,** pasteurized process	1 OZ	**28**	**43.2**	**91.4**	**5.5**	**6.7**	**2.0**	**0**	**159.6**	**128.5**	**332.9**	**8.7**	**78.1**
			100	43.2	326.4	19.6	24.0	7.1	0	570.0	459.0	1189.0	31.0	279.0
2001	**SWISS,** pasteurized process	1 OZ	**28**	**43.7**	**90.4**	**6.1**	**6.8**	**1.3**	**0**	**202.4**	**147.3**	**434.6**	**7.8**	**79.5**
			100	43.7	323.0	21.9	24.1	4.5	0	723.0	526.0	1552.0	28.0	284.0
	CHEESE, NATURAL													
2002	BLUE	1 OZ	**28**	**42.4**	**100.5**	**6.0**	**8.0**	**.7**	**0**	**147.6**	**109.5**	**390.3**	**6.4**	**71.7**
			100	42.4	359.0	21.4	28.7	2.3	0	527.0	391.0	1393.9	22.7	256.0
2003	BRICK	1 OZ	**28**	**41.0**	**103.9**	**6.5**	**8.2**	**.8**	**0**	**188.7**	**127.4**	**156.7**	**6.7**	**37.5**
			100	41.0	371.0	23.2	29.4	2.8	0	674.0	455.0	559.7	24.0	133.8
2004	BRIE	1 OZ	**28**	**48.4**	**93.5**	**5.8**	**7.8**	**.1**	**0**	**51.5**	**52.6**	**176.1**	—	**42.6**
			100	48.4	334.0	20.8	27.7	.5	0	184.0	188.0	629.0	—	152.0
2005	CAMEMBERT	1 OZ	**28**	**51.5**	**83.8**	**5.5**	**6.8**	**.1**	**0**	**108.4**	**97.2**	**235.8**	**5.9**	**52.2**
			100	51.5	299.2	19.7	24.2	.5	0	387.2	347.0	842.0	21.1	186.5
2006	CARAWAY	1 OZ	**28**	**39.3**	**105.5**	**7.0**	**8.2**	**.9**	**0**	**188.3**	**137.0**	**193.2**	**5.9**	—
			100	39.3	376.6	25.1	29.1	3.1	0	672.3	489.3	689.9	21.1	—
2007	CHEDDAR	1 OZ	**28**	**38.5**	**112.7**	**7.0**	**9.3**	**.4**	**0**	**201.1**	**142.9**	**173.5**	**7.9**	**27.6**
			100	38.5	402.4	24.9	33.1	1.3	0	718.1	510.4	619.5	28.2	98.6
2008	CHESHIRE	1 OZ	**28**	**37.7**	**108.4**	**6.5**	**8.6**	**1.3**	**0**	**180.0**	**129.9**	**196.0**	**5.9**	**26.6**
			100	37.7	387.0	23.4	30.6	4.8	0	643.0	464.0	700.0	21.0	95.0
2009	COLBY	1 OZ	**28**	**38.2**	**110.3**	**6.7**	**9.0**	**.7**	**0**	**191.8**	**127.7**	**169.1**	**7.5**	**35.3**
			100	38.2	394.0	23.8	32.1	2.6	0	685.0	456.0	604.0	26.8	126.0
2010	**COTTAGE,** large or small curd, uncreamed, 2% fat	1 CUP	**226**	**79.3**	**203.4**	**31.1**	**4.4**	**8.2**	**0**	**153.7**	**339.0**	**917.6**	**13.6**	**217.0**
			100	79.3	90.0	13.7	1.9	3.6	0	68.0	150.0	406.0	6.0	96.0
2011	**COTTAGE,** large or small curd, creamed	1 CUP	**225**	**79.0**	**232.5**	**28.1**	**10.1**	**6.0**	**0**	**135.0**	**296.8**	**911.3**	**11.3**	**189.6**
			100	79.0	103.3	12.5	4.5	2.7	0	60.0	131.9	405.0	5.0	84.3
2012	**COTTAGE,** creamed, with fruit added	1 CUP	**226**	**72.1**	**280.2**	**22.4**	**7.7**	**30.1**	**0**	**108.5**	**235.0**	**915.3**	**9.0**	**151.4**
			100	72.1	124.0	9.9	3.4	13.3	0	48.0	104.0	405.0	4.0	67.0
2013	CREAM	1 TBLS	**15**	**54.2**	**52.4**	**1.1**	**5.1**	**.4**	**0**	**12.2**	**16.2**	**44.4**	**0.9**	**18.5**
			100	54.2	349.0	7.5	33.8	2.6	0	81.1	108.0	296.3	6.0	123.4
2014	EDAM	1 OZ	**28**	**41.6**	**100.0**	**6.8**	**7.8**	**.4**	**0**	**204.7**	**150.1**	**270.2**	**8.4**	**52.6**
			100	41.6	357.0	24.4	27.9	1.4	0	731.0	536.0	965.0	30.0	188.0
2015	FETA	1 OZ	**28**	**55.2**	**73.9**	**4.0**	**6.0**	**1.1**	**0**	**137.8**	**94.4**	**312.5**	**5.3**	**17.4**
			100	55.2	264.0	14.2	21.3	4.1	0	492.0	337.0	1116.0	19.0	62.0
2016	FONTINA	1 BAR	**28**	**37.9**	**108.9**	**7.2**	**8.7**	**.4**	**0**	**154.0**	—	—	**3.9**	—
			100	37.9	389.0	25.6	31.1	1.6	0	550.0	—	—	14.0	—
2017	GJETOST	1 OZ	**28**	**13.4**	**130.5**	**2.7**	**8.3**	**11.9**	**0**	**112.0**	**124.3**	**168.0**	—	—
			100	13.4	466.0	9.7	29.5	42.7	0	400.0	444.0	600.0	—	—
2018	[5]GOAT'S MILK	1 OZ	**28**	**65.1**	**48.4**	**4.5**	**2.9**	**1.0**	—	**86.8**	**40.9**	—	—	—
			100	65.1	173.0	16.0	10.3	3.7	—	310.0	146.0	—	—	—
2019	GOUDA	1 OZ	**28**	**41.5**	**99.7**	**7.0**	**7.7**	**.6**	**0**	**196.0**	**152.9**	**229.3**	**8.1**	**33.6**
			100	41.5	356.0	24.9	27.4	2.2	0	700.0	546.0	819.0	29.0	120.0
2020	GRUYERE	1 OZ	**28**	**33.2**	**115.6**	**8.3**	**9.0**	**.1**	**0**	**283.1**	**169.4**	**94.1**	—	**22.7**
			100	33.2	413.0	29.8	32.0	.3	0	1011.0	605.0	336.0	—	81.0
2021	LIMBURGER	1 OZ	**28**	**48.4**	**91.8**	**5.9**	**7.6**	**.1**	**0**	**139.2**	**110.0**	**224.0**	**5.9**	**35.8**
			100	48.4	328.0	21.2	27.2	.5	0	497.0	393.0	800.0	21.0	128.0
2022	MONTEREY	1 OZ	**28**	—	**104.7**	**6.9**	**8.5**	**.2**	**0**	**209.4**	**124.4**	**150.1**	**7.9**	**22.7**
			100	—	373.8	24.5	30.3	.7	0	747.7	444.4	536.1	28.2	81.1
2023	MOZZARELLA	1 OZ	**28**	**54.1**	**78.7**	**5.4**	**6.0**	**.6**	**0**	**144.8**	**103.9**	**104.4**	**5.3**	**18.8**
			100	—	281.0	19.4	21.6	2.2	0	517.0	371.0	373.0	19.0	67.0
2024	**MOZZARELLA,** low moisture	1 OZ	**28**	**48.4**	**89.0**	**6.0**	**6.9**	**.7**	**0**	**161.0**	**115.4**	**116.2**	**58.8**	**21.0**
			100	48.4	318.0	21.6	24.6	2.5	0	575.0	412.0	415.0	210.0	75.0
2025	**MOZZARELLA,** low moisture, part skim	1 OZ	**28**	**48.6**	**78.4**	**7.7**	**4.8**	**.9**	**0**	**204.7**	**146.7**	**147.8**	**7.3**	**2.7**
			100	48.6	280.0	27.5	17.1	3.1	0	731.0	524.0	528.0	25.9	9.5
2026	[1]**MOZZARELLA,** part skim	1 OZ	**28**	**53.8**	**71.1**	**6.8**	**4.5**	**.8**	**0**	**180.9**	**129.6**	**130.5**	**6.4**	**23.5**
			100	53.8	254.0	24.3	15.9	2.8	0	646.0	463.0	466.0	23.0	84.0
2027	MUENSTER	1 OZ	**28**	**42.5**	**103.0**	**6.5**	**8.4**	**.3**	**0**	**200.5**	**131.0**	**175.8**	**7.3**	**37.5**
			100	42.5	368.0	23.1	30.0	1.1	0	716.0	468.0	628.0	26.0	134.0
2028	NEUFCHATEL	1 OZ	**28**	**62.2**	**72.8**	**2.8**	**6.6**	**.8**	**0**	**21.0**	**38.1**	**111.7**	**2.1**	**31.9**
			100	62.2	260.0	10.0	23.4	2.9	0	75.0	136.0	399.0	7.6	114.0

Item No.	Minerals (Micro)			Fat-Soluble Vitamins			Water-Soluble Vitamins								
	Iron	Zinc	Copper	Vitamin A	Vitamin D	Vitamin E (Alpha Tocopherol)	Vitamin C	Thiamin	Ribo-flavin	Niacin	Panto-thenic Acid	Vit. B-6 (Pyri-doxine)	Folacin (Folic Acid)	Biotin	Vitamin B-12
	(mg)	(mg)	(mg)	(IU)	(IU)	(mg)	(mg)	(mg)	(mg)	(mg)	(mg)	(mg)	(mcg)	(mcg)	(mcg)
1999	**.22** .84	**.84** 3.01	— —	**197.4** 705.0	— —	— —	**0** 0	**.01** .03	**.13** .45	**.02** .07	**.27** .98	**.04** .14	**1.40** 5.00	— —	**.36** 1.28
2000	**.22** .80	**.84** 2.99	— —	**255.6** 913.0	— —	— —	**0** 0	**.01** .02	**.12** .44	**.04** .14	**.16** .56	— —	— —	— —	**.33** 1.18
2001	**.17** .60	**.99** 3.55	— —	**239.7** 856.0	— —	— —	**0** 0	**.0** .01	**.11** .40	**.03** .10	**.14** .50	— —	**.64** 2.30	— —	**.64** 2.30
2002	**.09** .31	**.74** 2.64	**(.03)** (.09)	**201.9** 721.0	**(.06)** (.23)	**(.20)** (.70)	**0** 0	**.01** .03	**.11** .38	**.28** .02	**.50** 1.80	**.05** .17	**10.19** 36.40	**.46** 1.64	**.34** 1.22
2003	**.12** .43	**.73** 2.61	— —	**303.2** 1083.0	— —	— —	**0** 0	**.0** .01	**.10** .35	**.03** .12	**.08** .29	**.02** .07	**5.69** 20.33	**.45** 1.59	**.35** 1.26
2004	**.14** .50	[6] **.62** 2.2	— —	**186.8** 667.0	— —	— —	— —	**.02** .07	**.15** .52	**.11** .38	**.19** .69	**.07** .24	**18.20** 65.00	— —	**.46** 1.65
2005	**.09** .33	**.67** 2.39	**(.02)** (.08)	**258.4** 923.0	**(.05)** (.18)	**(.17)** (.60)	**0** 0	**.01** .03	**.14** .49	**.18** .63	**.35** 1.25	**.06** .22	**17.42** 62.20	**1.60** 5.70	**.36** 1.30
2006	— —	**.88** 23.15	**.05** .17	**294.7** 1052.5	— —	— —	**0** 0	**.01** .03	**.13** .45	**.05** .18	**.05** .19	— —	**.08** .27	— —	**.08** .27
2007	**.19** .67	**1.12** 4.00	**.06** .22	**295.9** 1056.7	**(.07)** (.26)	**(.22)** (.80)	**0** 0	**.01** .03	**.11** .38	**.0** .01	**.12** .41	**.02** .07	**5.04** 18.00	**.48** 1.73	**.23** .83
2008	**.06** .21	— —	— —	**275.8** 985.0	— —	— —	**0** 0	**.01** .05	**.08** .29	— —	— —	— —	— —	— —	— —
2009	**.21** .76	**.86** 3.07	**.07** .24	**289.5** 1034.0	— —	— —	**0** 0	**.0** .02	**.11** .38	**.03** .09	**.06** .21	**.02** .08	— —	— —	**.23** .83
2010	**.36** .16	**.95** .42	— —	**158.2** 70.0	— —	— —	**0** 0	**.05** .02	**.42** .19	**.33** .14	**.55** .24	**.17** .08	**29.38** 13.00	— —	**1.61** .71
2011	**.32** .14	**.81** .36	**.05** .02	**366.4** 162.9	— —	— —	**0** 0	**.07** .03	**.37** .16	**.27** .12	**.50** .22	**.15** .07	**27.86** 12.38	**4.41** 1.96	**1.40** .62
2012	**.25** .11	**.66** .29	— —	**278.0** 123.0	— —	— —	**0** 0	**.04** .02	**.29** .13	**.23** .10	**.38** .17	**.12** .05	**22.60** 10.00	— —	**1.12** .49
2013	**.18** 1.20	**.08** .53	**Trace** (.04)	**214.3** 1428.4	**(.04)** (.28)	**(.15)** (1.00)	**0** 0	**0** .02	**.03** .20	**.02** .10	**.04** .27	**.01** .05	**1.95** 13.00	**.25** 1.64	**.06** .42
2014	**.12** .44	**1.05** 3.75	**(.01)** (.03)	**256.5** 916.0	**(.05)** (.18)	**(.22)** (.80)	**0** 0	**.01** .04	**.11** .39	**.02** .08	**.08** .28	**.02** .08	**4.48** 16.00	**.43** 1.52	**.43** 1.54
2015	**.18** .65	**.81** 2.88	**.04** .13	— —	— —	— —	**0** 0	— —	— —	— —	— —	— —	— —	— —	— —
2016	**.06** .23	**.98** 3.50	— —	**328.7** 1174.0	— —	— —	**0** 0	**.01** .02	**.06** .20	**.04** .15	— —	— —	— —	— —	— —
2017	— —	— —	— —	— —	— —	— —	**0** 0	— —	**.23** .81	— —	— —	— —	**1.40** 5.00	— —	— —
2018	**.22** .80	— —	— —	**37.3** 133.3	— —	— —	**0** 0	**0** .01	**.18** .63	**.06** .20	— —	— —	— —	— —	— —
2019	**.07** .24	**1.09** 3.90	— —	**180.3** 644.0	— —	— —	**0** 0	**.01** .03	**.09** .33	**.02** .06	**.10** .34	**.02** .08	**5.88** 21.00	— —	— —
2020	— —	— —	— —	**341.3** 1219.0	— —	— —	**0** 0	**.02** .06	**.08** .28	— —	**.16** .56	**.02** .08	**2.80** 10.00	— —	**.45** 1.60
2021	**.04** .13	**.59** 2.12	— —	**358.7** 1281.0	— —	— —	**0** 0	**.02** .08	**.14** .50	**.04** .16	**.33** 1.18	**.02** .09	**16.10** 57.50	**.63** 2.26	**.29** 1.04
2022	**.20** .71	**.84** 3.00	— —	**265.7** 948.8	— —	— —	**0** 0	— —	**.11** .39	— —	— —	— —	— —	— —	— —
2023	**.05** .18	**.62** 2.21	**.06** .23	**221.8** 792.0	— —	— —	**0** 0	**0** .02	**.07** .24	**.02** .08	**.02** .06	**.01** .06	**1.96** 7.00	**.45** 1.62	**.18** .65
2024	**.06** .20	**.69** 2.46	— —	**253.1** 904.0	— —	— —	**0** 0	**0** .02	**.08** .27	**.03** .09	**.02** .07	**.02** .06	**.20** .73	— —	**2.24** 8.00
2025	**.07** 2.25	— —	— —	**75.8** 628.0	— —	— —	**0** 0	**.01** .02	**.10** .34	**.03** .12	**.03** .09	**.02** .08	**2.80** 10.00	— —	**.26** .93
2026	**.06** .22	**.77** 2.76	— —	**163.5** 584.0	— —	— —	**0** 0	**.01** .02	**.09** .30	**.03** .11	**.02** .08	**.02** .07	**2.52** 9.00	— —	**.23** .82
2027	**.12** .41	**.79** 2.81	— —	**313.6** 1120.0	— —	— —	**0** 0	**0** .01	**.09** .32	**.03** .10	**.05** .19	**.02** .06	**3.39** 12.10	**.39** 1.39	**.41** 1.47
2028	**.08** .26	**.15** .52	**.04** .15	**317.5** 1134.0	— —	— —	**0** 0	**0** .02	**.06** .20	**.04** .13	**.16** .57	**.01** .04	**3.16** 11.30	**.54** 1.93	**.07** .26

(Continued)

TABLE F-36 *(Continued)*

Item No.	Food Item Name	Approximate Measure	Weight	Moisture	Food Energy Calories	Protein	Fats	Carbo-hydrates	Fiber	Minerals (Macro) Calcium	Phos-phorus	Sodium	Mag-nesium	Potas-sium
			(g)	(%)	(kcal)	(g)	(g)	(g)	(g)	(mg)	(mg)	(mg)	(mg)	(mg)
	CHEESE, NATURAL (Continued)													
2029	**PARMESAN,** hard	1 TBLS	**5**	**30.0**	**20.2**	**1.8**	**1.3**	**.1**	**0**	**59.2**	**34.7**	**80.1**	**2.2**	**4.6**
			100	30.0	403.0	35.8	26.5	2.9	0	1184.0	694.0	1602.0	44.0	92.0
2030	**PARMESAN,** grated	1 TBLS	**5**	**17.7**	**22.8**	**2.1**	**1.5**	**.2**	**0**	**68.8**	**40.4**	**93.1**	**2.6**	**5.4**
			100	17.7	456.0	41.6	30.0	3.7	0	1376.0	807.0	1862.0	51.0	107.0
2031	**PORT DU SALUT**	1 OZ	**28**	**45.5**	**98.6**	**6.7**	**7.9**	**.2**	**0**	**182.0**	**100.8**	**149.5**	—	—
			100	45.5	352.0	23.8	28.2	.6	0	650.0	360.0	534.0	—	—
2032	**PROVOLONE**	1 OZ	**28**	**41.3**	**98.3**	**7.4**	**7.3**	**.6**	**0**	**211.7**	**138.9**	**245.3**	**7.6**	**38.6**
			100	41.3	351.0	26.3	26.0	2.1	0	756.0	496.0	876.0	27.3	138.0
2033	**RICOTTA,** w/whole milk	1 OZ	**28**	**70.4**	**48.7**	**3.3**	**3.6**	**.9**	**0**	**58.0**	**44.2**	**23.5**	**3.2**	**29.4**
			100	70.4	174.0	11.7	13.0	3.0	0	207.0	158.0	84.0	11.3	105.0
2034	**RICOTTA,** part skim	1 OZ	**28**	**75.4**	**38.6**	**3.1**	**2.2**	**1.4**	**0**	**76.2**	**51.2**	**35.0**	**4.2**	**35.0**
			100	75.4	138.0	11.1	7.9	5.1	0	272.0	183.0	125.0	15.0	125.0
2035	**ROMANO**	1 OZ	**28**	**30.9**	**110.0**	**9.0**	**7.6**	**1.0**	**0**	**301.9**	**214.9**	**339.9**	—	—
			100	30.9	392.7	32.2	27.3	3.7	0	1078.1	767.6	1213.8	—	—
2036	**ROQUEFORT**	1 OZ	**28**	**40.0**	**103.3**	**6.0**	**8.6**	**.6**	**0**	**185.4**	**109.8**	**506.5**	**8.4**	**25.5**
			100	40.0	369.0	21.5	30.6	2.0	0	662.0	392.0	1809.0	30.0	91.0
2037	**SWISS,** domestic	1 OZ	**28**	**38.0**	**104.2**	**8.2**	**7.7**	**.9**	**0**	**269.1**	**169.4**	**72.8**	**10.1**	**31.1**
			100	38.0	372.0	29.2	27.6	3.4	0	961.0	605.0	260.0	35.9	111.0
2038	**TILSIT,** made w/whole milk	1 OZ	**28**	**42.9**	**95.2**	**6.8**	**7.3**	**.5**	**0**	**196.0**	**140.0**	**210.8**	**3.6**	**17.9**
			100	42.9	340.0	24.4	26.0	1.9	0	700.0	500.0	753.0	13.0	64.0
	CHEESE, PASTEURIZED PROCESS													
2039	American	1 OZ	**28**	**39.8**	**105.0**	**6.3**	**8.8**	**.4**	**0**	**172.5**	**208.6**	**400.4**	**6.2**	**45.4**
			100	39.8	375.0	22.5	31.3	1.6	0	616.0	745.0	1430.0	22.0	162.0
2040	American, w/pimiento	1 OZ	**28**	**40.0**	**103.9**	**6.4**	**8.5**	**.5**	**0**	**171.9**	**208.3**	**399.8**	**6.2**	**45.4**
			100	40.0	371.0	23.0	30.2	1.8	0	614.0	744.0	1428.0	22.0	162.0
2041	Swiss	1 OZ	**28**	**42.3**	**93.2**	**6.9**	**6.9**	**.6**	**0**	**216.2**	**213.4**	**383.6**	**8.1**	**60.5**
			100	42.3	333.0	24.7	24.5	2.1	0	772.0	762.0	1370.0	29.0	216.0
2042	**CHEESE SPREAD,** American pasteurized process	1 OZ	**28**	**48.6**	**80.6**	**4.5**	**5.9**	**2.4**	**0**	**158.2**	**199.4**	**376.6**	**8.1**	**67.2**
			100	48.6	288.0	16.0	21.2	8.7	0	565.0	712.0	1345.0	29.0	240.0
	CREAM													
2043	whipping, heavy, 37.6% fat	1 CUP	**240**	**56.6**	**828.0**	**5.3**	**90.7**	**6.7**	**0**	**156.0**	**141.6**	**76.8**	**16.8**	**180.0**
			100	56.6	345.0	2.2	37.8	2.8	0	65.0	59.0	32.0	7.0	75.0
2044	whipping, 31.3% fat	1 CUP	**240**	**73.8**	**468.0**	**5.2**	**46.3**	**7.1**	**0**	**165.6**	**160.8**	**86.4**	**16.8**	**244.8**
			100	73.8	195.0	2.2	19.3	3.0	0	69.0	67.0	36.0	7.0	102.0
2045	medium, 25% fat	1 TBLS	**15**	**68.5**	**36.6**	**.4**	**3.8**	**.5**	**0**	**13.5**	**10.7**	**5.6**	**1.2**	**17.1**
			100	68.5	244.0	2.5	25.0	3.5	0	90.0	71.0	37.0	8.0	114.0
2046	coffee or table, light, 20.6% fat	1 CUP	**240**	**77.9**	**511.2**	**6.4**	**49.4**	**8.8**	**0**	**228.0**	**199.2**	**103.2**	**21.6**	**283.2**
			100	77.9	213.0	2.6	20.6	3.7	0	95.0	83.0	43.0	9.0	118.0
2047	half-and-half	1 CUP	**240**	**79.7**	**312.0**	**7.1**	**28.1**	**10.3**	**0**	**259.2**	**228.0**	**110.4**	**21.6**	**309.6**
			100	79.7	130.0	3.0	11.7	4.3	0	108.0	95.0	46.0	9.0	129.0
2048	imitation, liquid, (frozen)	1 TBLS	**14**	**77.0**	**19.7**	**.2**	**1.4**	**1.6**	**0**	**1.3**	**9.0**	**11.1**	**0**	**26.6**
			100	77.0	141.0	1.2	10.0	11.3	0	9.0	64.0	79.0	0	190.0
2049	imitation, powdered	1 TSP	**3**	**2.0**	**16.1**	**.1**	**1.1**	**1.6**	**0**	**.7**	**12.7**	**5.4**	**.1**	**24.4**
			100	2.0	537.0	4.8	35.1	54.9	0	22.0	422.0	181.0	4.0	812.0
2050	substitute, coffee, rich, liquid	1 TBLS	**14**	**72.8**	**21.6**	**0**	**1.3**	**1.8**	**0**	**.3**	**5.3**	**5.6**	**.1**	**5.8**
			100	72.8	154.0	.3	9.4	12.8	0	2.0	38.0	40.0	1.0	40.0
2051	substitute, dried, w/cream, skim milk, lactose, & sodium hexametaphosphate	1 TSP	**3**	**0.9**	**15.3**	**.4**	**.8**	**1.6**	**0**	**14.9**	—	—	—	—
			100	1.4	508.0	8.5	26.7	61.3	0	82.0	—	17.3	—	—
2052	substitute, dried, w/cream, skim milk, lactose, calcium reduced	1 TSP	**3**	**1.4**	**15.2**	**.3**	**.8**	**1.8**	**0**	**2.9**	—	**17.3**	—	—
			100	1.4	508.0	8.5	26.7	61.3	0	82.0	—	575.0	—	—
2053	[1]whipped cream topping, pressurized	1 OZ	**28**	**61.3**	**72.0**	**.9**	**6.2**	**3.5**	**0**	**28.3**	**24.9**	**36.4**	**3.1**	**41.2**
			100	61.3	257.0	3.2	22.2	12.5	0	101.0	89.0	130.0	11.0	47.0
2054	**EGGNOG**	1 CUP	**254**	**74.4**	**342.9**	**9.7**	**19.0**	**3.4**	**0**	**330.2**	**281.9**	**132.1**	**45.7**	**419.1**
			100	74.4	135.0	3.8	7.5	1.4	0	130.0	111.0	52.0	18.0	165.0
	ICE CREAM													
2055	rich, 16% fat	1 CUP	**148**	**58.9**	**349.3**	**3.8**	**23.8**	**32.0**	**0**	**151.0**	**115.4**	**108.0**	**16.3**	**220.5**
			100	58.9	236.0	2.6	16.1	21.6	0	102.0	78.0	73.0	11.0	149.0
2056	regular, 12% fat	1 CUP	**135**	**62.1**	**282.2**	**5.4**	**16.6**	**27.8**	**0**	**166.1**	**133.7**	**54.0**	**18.9**	**151.2**
			100	62.1	209.0	4.0	12.3	20.6	0	123.0	99.0	40.0	14.0	112.0
2057	regular, 10% fat	1 CUP	**135**	**60.8**	**260.6**	**4.9**	**14.3**	**32.2**	**0**	**178.2**	**136.4**	**117.5**	**18.9**	**260.6**
			100	60.8	193.0	3.6	10.6	23.9	0	132.0	101.0	87.0	14.0	193.0
2058	cones	1 AVG	**12**	**8.9**	**45.2**	**1.2**	**.3**	**9.3**	**0**	**18.7**	**23.8**	**27.8**	—	**29.3**
			100	8.9	377.0	10.0	2.4	77.9	0	156.0	198.0	232.0	—	244.0

Item No.	Minerals (Micro)			Fat-Soluble Vitamins			Water-Soluble Vitamins								
	Iron	Zinc	Copper	Vitamin A	Vitamin D	Vitamin E (Alpha Tocopherol)	Vitamin C	Thiamin	Ribo-flavin	Niacin	Panto-thenic Acid	Vit. B-6 (Pyri-doxine)	Folacin (Folic Acid)	Biotin	Vitamin B-12
	(mg)	(mg)	(mg)	(IU)	(IU)	(mg)	(mg)	(mg)	(mg)	(mg)	(mg)	(mg)	(mcg)	(mcg)	(mcg)
2029	**.04**	**.14**	—	**30.2**	—	—	**0**	**0**	**.02**	**.02**	**.02**	**.01**	**.34**	**.10**	—
	.82	2.75	—	603.0	—	—	0	.04	.33	.29	.45	.09	6.80	1.71	—
2030	**.05**	**.16**	—	**35.1**	**(.01)**	**(.50)**	**0**	**0**	**.02**	**.02**	**.07**	**.01**	**.40**	**.10**	**(.42)**
	.95	3.19	—	701.0	(.27)	(.90)	0	.05	.39	.32	.53	.11	8.00	1.70	(1.50)
2031	—	—	—	**373.2**	—	—	**0**	—	**.07**	**.02**	**.06**	**.02**	**5.04**	—	**.42**
	—	—	—	1333.0	—	—	0	—	.24	.06	.21	.05	18.00	—	1.50
2032	**.15**	**.90**	—	**228.2**	—	—	**0**	**.01**	**.09**	**.04**	**.13**	**.02**	**2.91**	**.50**	**.41**
	.52	3.23	—	815.0	—	—	0	.02	.32	.16	.48	.07	10.40	1.79	1.46
2033	**.11**	**.33**	—	**137.2**	—	—	**0**	**0**	**.06**	**.03**	—	**.01**	—	—	**.10**
	.38	1.16	—	490.0	—	—	0	.01	.20	.10	—	.04	—	—	.34
2034	**.12**	**.38**	—	**121.0**	—	—	**0**	**.01**	**.05**	**.02**	—	**.01**	—	—	**.08**
	.44	1.34	—	432.0	—	—	0	.02	.19	.08	—	.02	—	—	.29
2035	—	—	—	**161.9**	—	—	**0**	—	**.11**	**.02**	—	—	**2.00**	—	—
	—	—	—	578.3	—	—	0	—	.38	.08	—	—	7.14	—	—
2036	**.14**	**.58**	—	**293.2**	—	—	**0**	**.01**	**.16**	**.21**	**.48**	**.03**	**13.72**	—	**.17**
	.50	2.08	—	1047.0	—	—	0	.04	.59	.73	1.73	.12	49.00	—	.62
2037	**.05**	**1.09**	**.03**	**236.6**	—	—	**0**	**.01**	**.11**	**.03**	**.12**	**.02**	**1.68**	**.26**	**.47**
	.17	3.90	.11	845.0	—	—	0	.02	.40	.09	.43	.08	6.00	.94	1.68
2038	**.06**	**.98**	—	**292.6**	—	—	**0**	**.02**	**.10**	**.06**	**.10**	—	—	—	**.59**
	.23	3.50	—	1045.0	—	—	0	.06	.36	.21	.35	—	—	—	2.10
2039	**.11**	**.84**	**(.14)**	**341.6**	**(.04)**	**.19**	**0**	**.01**	**.99**	**.02**	**.14**	**.02**	**2.18**	**23.07**	**.20**
	.39	2.99	(.50)	1220.0	(.15)	.67	0	.02	.53	.07	.48	.07	7.80	82.40	.70
2040	**.12**	**.83**	—	**353.4**	—	—	—	**.01**	**.10**	**.02**	**.14**	**.02**	**2.24**	**1.29**	**.20**
	.42	2.98	—	1262.0	—	—	—	.03	.35	.08	.49	.07	8.00	4.60	.70
2041	**.17**	**1.01**	—	**226.2**	—	—	**0**	**0**	**.08**	**.01**	**.07**	**.01**	—	—	**.34**
	.61	3.61	—	808.0	—	—	0	.01	.28	.04	.26	.04	—	—	1.20
2042	**.09**	**.73**	**(.03)**	**220.6**	**Trace**	—	**0**	**.01**	**.12**	**.04**	**.19**	**.03**	**1.96**	**23.07**	**.11**
	.33	2.59	(.09)	788.0	(.02)	—	0	.05	.43	.13	.69	.12	7.00	82.40	.40
2043	**.07**	**.55**	**(.38)**	**3528.0**	**240.00**	**1.92**	**1.39**	**.05**	**.26**	**.09**	**.61**	**.06**	**9.60**	**2.16**	**.43**
	.03	.23	(.16)	1470.0	100.00	.80	.58	.02	.11	.04	.26	.03	4.00	.90	.18
2044	**.07**	**.60**	—	**2704.8**	**120.00**	—	**1.46**	**.05**	**.29**	**.10**	**.62**	**.07**	**9.60**	—	**.48**
	.03	.25	—	1127.0	50.00	—	.61	.02	.12	.04	.26	.03	4.00	—	.20
2045	**.01**	**.04**	—	**141.3**	—	—	**.11**	**0**	**.02**	**.01**	**.04**	**.01**	**.30**	—	**.03**
	.04	.26	—	942.0	—	—	.71	.03	.14	.05	.27	.03	2.00	—	.22
2046	**.10**	**.65**	—	**1728.0**	—	—	**1.82**	**.07**	**.36**	**.14**	**.66**	**.08**	**4.80**	—	—
	.04	.27	—	720.0	—	—	.76	.03	.15	.06	.28	.03	2.00	—	—
2047	**.17**	**1.22**	—	**1041.6**	—	—	**2.06**	**.07**	**.36**	**.19**	**.69**	**.09**	**4.80**	—	**.79**
	.07	.51	—	434.0	—	—	.86	.03	.15	.08	.29	.04	2.00	—	.33
2048	**0**	**0**	—	**12.5**	—	—	**0**	**0**	**0**	**0**	**0**	**0**	**0**	—	**0**
	.03	.02	—	89.0	—	—	0	0	0	0	0	0	0	—	0
2049	**.04**	**.02**	—	**6.4**	—	—	**0**	**0**	**0.01**	**0**	**0**	**0**	**0**	—	**0**
	1.15	.51	—	213.0	—	—	0	0	0.17	0	0	0	0	—	0
2050	**.02**	**0**	**0**	**22.8**	—	—	**0**	**0**	**0**	**0**	—	—	—	—	—
	.15	.02	.02	163.0	—	—	0	0	0	0	—	—	—	—	—
2051	**.01**	**.02**	—	**15.6**	—	—	—	**0**	**.02**	**.01**	—	—	—	—	—
	.30	.51	—	520.0	—	—	—	.14	.71	.30	—	—	—	—	—
2052	**.01**	**.02**	—	**28.8**	—	—	—	**0**	**.04**	**0**	—	—	—	—	—
	.20	.51	—	960.0	—	—	—	.05	1.17	.10	—	—	—	—	—
2053	**.01**	**.10**	—	**255.6**	—	—	**0**	**.01**	**.02**	**.02**	**.09**	**.01**	—	—	**.08**
	.05	.37	—	913.0	—	—	0	.04	.07	.07	.30	.04	—	—	.29
2054	**.51**	**1.17**	—	**873.8**	**30.48**	—	**3.81**	**.13**	**.46**	**.27**	**1.06**	**.13**	**2.54**	—	**1.14**
	.20	.46	—	344.0	12.00	—	1.50	.05	.18	.11	.42	.05	1.00	—	.45
2055	**.10**	**1.21**	—	**896.9**	—	—	**.61**	**.04**	**.16**	**.12**	**.56**	**.05**	**2.96**	—	**.54**
	.07	.82	—	606.0	—	—	.41	.03	.11	.08	.38	.04	2.00	—	.36
2056	**.14**	**.68**	—	**702.0**	—	**.49**	**1.35**	**.05**	**.26**	**.14**	—	—	**2.70**	—	—
	.10	.50	—	520.0	—	.36	1.00	.04	.19	.10	—	—	2.00	—	—
2057	**.14**	**1.43**	**.01**	**550.8**	—	**.50**	**.72**	**.05**	**.28**	**.14**	**.66**	**.06**	**2.70**	—	**.64**
	.10	1.06	.01	408.0	—	.37	.53	.04	.21	.1C	.49	.05	2.00	—	.47
2058	**.05**	—	—	—	—	—	—	**.01**	**.03**	**.06**	—	—	—	—	—
	.40	—	—	—	—	—	—	.05	.21	.50	—	—	—	—	—

(Continued)

TABLE F-36 *(Continued)*

Item No.	Food Item Name	Approximate Measure	Weight	Moisture	Food Energy Calories	Protein	Fats	Carbo-hydrates	Fiber	Minerals (Macro) Calcium	Phos-phorus	Sodium	Mag-nesium	Potas-sium
			(g)	(%)	(kcal)	(g)	(g)	(g)	(g)	(mg)	(mg)	(mg)	(mg)	(mg)
	ICE CREAM (Continued)													
2059	cone, dipped in chocolate, small	**INDV**	**85**	–	**160.0**	**4.0**	**7.0**	**20.0**	–	–	–	–	–	–
			100	–	188.2	4.7	8.2	23.5	–	–	–	–	–	–
2060	cone, dipped in chocolate, large	**INDV**	**248**	–	**450.0**	**10.0**	**20.0**	**58.0**	–	–	–	–	–	–
			100	–	181.5	4.0	8.1	23.4	–	–	–	–	–	–
2061	bar, chocolate coated	**1 AVG**	**47**	**47.0**	**149.0**	**1.6**	**10.5**	**12.1**	–	**55.0**	**41.8**	**24.0**	**6.0**	**84.1**
			100	47.0	317.0	3.4	22.3	25.7	–	117.0	89.0	51.0	12.8	179.0
2062	chocolate	**1 CUP**	**133**	–	**295.3**	**5.0**	**16.0**	**32.8**	–	**186.2**	**167.6**	**74.5**	**17.7**	–
			100	–	222.0	3.8	12.0	24.7	–	140.0	126.0	56.0	13.3	–
2063	French custard, frozen	**1 CUP**	**133**	**63.2**	**256.7**	**6.0**	**14.4**	**27.7**	**0**	**194.2**	**153.0**	**83.8**	**10.3**	**240.7**
			100	63.2	193.0	4.5	10.8	20.8	0	146.0	115.0	63.0	7.8	181.0
2064	French vanilla, soft serve	**1 CUP**	**173**	**59.8**	**377.1**	**7.0**	**22.5**	**38.3**	**0**	**235.3**	**199.0**	**154.0**	**24.2**	**337.4**
			100	59.8	218.0	4.1	13.0	22.1	0	136.0	115.0	89.0	14.0	195.0
2065	strawberry	**1 CUP**	**133**	–	**250.0**	**4.3**	**12.0**	**31.3**	–	**146.3**	**123.7**	**58.5**	**19.2**	–
			100	–	188.0	3.2	9.0	23.6	–	110.0	93.0	44.0	14.4	–
2066	vanilla	**1 CUP**	**133**	**61.5**	**253.5**	**3.9**	**13.2**	**31.3**	**0**	**159.6**	**129.0**	**81.1**	**19.2**	–
			100	61.5	190.6	2.9	9.9	23.5	0	120.0	97.0	61.0	14.4	–
	ICE MILK													
2067	soft serve	**1 CUP**	**175**	**67.0**	**224.0**	**8.0**	**4.6**	**39.0**	**0**	**273.0**	**201.3**	**162.8**	**29.8**	**413.0**
			100	67.0	128.0	4.6	2.6	22.3	0	156.0	115.0	93.0	17.0	236.0
2068	hardened, strawberry	**1 CUP**	**133**	–	**196.8**	**6.3**	**6.8**	**32.7**	–	**238.1**	**178.2**	**94.4**	**17.7**	–
			100	–	148.0	4.8	5.1	24.6	–	179.0	134.0	71.0	13.3	–
2069	chocolate covered bar	**1 BAR**	**60**	–	**144.0**	**2.8**	**7.6**	**16.0**	–	**97.8**	**73.2**	**37.8**	**8.0**	–
			100	–	240.0	4.7	12.6	26.7	–	163.0	122.0	63.0	13.3	–
2070	**JUNKET,** fruit & vanilla flavors, made w/whole milk	**1 CUP**	**270**	–	**280.8**	**11.1**	**8.9**	**38.9**	**.3**	**321.3**	**237.6**	**129.6**	**38.0**	**364.5**
			100	–	104.0	4.1	3.3	14.4	.1	119.0	88.0	48.0	14.1	135.0
2071	fruit & vanilla flavors, made w/nonfat dehydrated milk	**1 CUP**	**232**	–	**176.3**	**9.7**	**.2**	**34.1**	**.2**	**285.4**	**211.1**	**116.0**	**32.6**	**317.8**
			100	–	76.0	4.2	.1	14.7	.1	123.0	91.0	50.0	14.1	137.0
2072	**[a]MILK, ASS,** fluid	**1 CUP**	**245**	**90.2**	**105.4**	**4.2**	**2.9**	**15.9**	**0**	**308.7**	**139.7**	–	–	–
			100	90.2	43.0	1.7	1.2	6.5	0	126.0	57.0	–	–	–
2073	**MILK, BUFFALO OR CARABAO**	**1 CUP**	**245**	**81.0**	**281.8**	**12.7**	**21.3**	**10.5**	**0**	**514.5**	**247.5**	–	–	–
			100	81.0	115.0	5.2	8.7	4.3	0	210.0	101.0	–	–	–
	MILK, COW'S													
2074	whole, 3.7% fat	**1 CUP**	**244**	**87.2**	**161.0**	**8.0**	**9.0**	**11.3**	**0**	**285.5**	**368.4**	**122.0**	**31.7**	**341.6**
			100	87.2	66.0	3.3	3.7	4.7	0	117.0	151.0	50.0	13.0	140.0
2075	whole, 3.3% fat	**1 CUP**	**244**	**87.8**	**148.8**	**8.0**	**8.2**	**11.4**	**0**	**290.4**	**217.2**	**122.0**	**31.7**	**370.9**
			100	87.8	61.0	3.3	3.3	4.7	0	119.0	89.0	50.0	13.0	152.0
2076	whole, 3.3% fat, multivitamin	**1 CUP**	**240**	**87.8**	**148.2**	**7.8**	**8.0**	**11.2**	**0**	**285.6**	**213.6**	**120.0**	**31.2**	**364.8**
			100	87.8	61.7	3.3	3.3	4.7	0	119.0	89.0	50.0	13.0	152.0
2077	whole, 3.25% fat	**1 CUP**	**244**	**87.8**	**150.0**	**8.0**	**8.2**	**11.4**	**0**	**291.0**	**228.0**	**120.0**	**33.0**	**370.0**
			100	87.8	61.5	3.3	3.3	4.7	0	119.3	93.4	49.2	13.5	151.6
2078	2% fat	**1 CUP**	**244**	**89.2**	**122.0**	**8.1**	**4.7**	**11.7**	**0**	**297.7**	**231.8**	**122.0**	**34.2**	**375.8**
			100	89.2	50.0	3.3	1.9	4.8	0	122.0	95.0	50.0	14.0	154.0
2079	1% fat	**1 CUP**	**244**	**90.1**	**102.5**	**8.0**	**2.6**	**11.7**	**0**	**300.1**	**234.2**	**122.0**	**34.2**	**380.6**
			100	90.1	42.0	3.3	1.0	4.8	0	123.0	96.0	50.0	14.0	156.0
2080	partially skimmed, w/1% nonfat milk solids added	**1 CUP**	**245**	**89.8**	**105.4**	**8.5**	**2.4**	**12.2**	**0**	**313.6**	**245.0**	**127.4**	**34.3**	**396.9**
			100	89.8	43.0	3.5	1.0	5.0	0	128.0	100.0	52.0	14.0	162.0
2081	partially skimmed, w/2% nonfat milk solids added	**1 CUP**	**245**	**87.0**	**125.0**	**8.5**	**4.9**	**12.2**	**0**	**313.6**	**245.0**	**127.4**	**34.3**	**396.9**
			100	87.0	51.0	3.5	2.0	5.0	0	128.0	100.0	52.0	14.0	162.0
2082	skim	**1 CUP**	**246**	**92.0**	**83.8**	**8.0**	**.4**	**11.9**	**0**	**302.6**	**250.9**	**127.9**	**27.1**	**408.4**
			100	92.0	34.1	3.3	.2	4.9	0	123.0	102.0	52.0	11.0	166.0
2083	skim, fortified	**1 CUP**	**246**	**89.4**	**91.0**	**9.7**	**.6**	**13.9**	**0**	**359.2**	**268.1**	**142.7**	**35.0**	**447.7**
			100	89.4	37.0	4.0	.3	5.7	0	146.0	109.0	58.0	14.2	182.0
2084	skim, w/nonfat milk solids added	**1 CUP**	**245**	**90.4**	**90.7**	**8.7**	**.6**	**12.3**	**0**	**316.1**	**254.8**	**129.9**	**36.8**	**419.0**
			100	90.4	37.0	3.6	.3	5.0	0	129.0	104.0	53.0	15.0	171.0
2085	fluid, filled, w/vegetable oil, lowfat	**1 CUP**	**240**	–	**110.4**	**7.9**	**3.6**	**11.3**	–	**307.2**	**232.8**	–	**31.2**	–
			100	–	46.0	3.3	1.5	4.7	–	128.0	97.0	–	13.0	–
2086	fluid, filled, with vegetable oil	**1 CUP**	**244**	–	**153.7**	**8.1**	**8.5**	**11.5**	–	**312.3**	**236.7**	–	**31.7**	–
			100	–	63.0	3.3	3.5	4.7	–	128.0	97.0	–	13.0	–
2087	low sodium	**1 CUP**	**244**	**88.2**	**148.8**	**7.6**	**8.4**	**10.9**	**0**	**246.4**	**209.8**	**4.9**	**12.2**	**617.3**
			100	88.2	61.0	3.1	3.5	4.5	0	101.0	86.0	2.0	5.0	253.0
2088	canned, condensed sweetened	**1 CUP**	**306**	**27.1**	**1003.7**	**24.8**	**26.6**	**166.2**	**0**	**868.0**	**790.5**	**396.8**	**76.5**	**1158.7**
			100	27.1	328.0	8.1	8.7	54.3	0	283.7	258.3	129.7	25.0	378.7

Item No.	Minerals (Micro)			Fat-Soluble Vitamins			Water-Soluble Vitamins								
	Iron	Zinc	Copper	Vitamin A	Vitamin D	Vitamin E (Alpha Tocopherol)	Vitamin C	Thiamin	Ribo-flavin	Niacin	Panto-thenic Acid	Vit. B-6 (Pyri-doxine)	Folacin (Folic Acid)	Biotin	Vitamin B-12
	(mg)	(mg)	(mg)	(IU)	(IU)	(mg)	(mg)	(mg)	(mg)	(mg)	(mg)	(mg)	(mcg)	(mcg)	(mcg)
2059	—	—	—	—	—	—	—	—	—	—	—	—	—	—	—
	—	—	—	—	—	—	—	—	—	—	—	—	—	—	—
2060	—	—	—	—	—	—	—	—	—	—	—	—	—	—	—
	—	—	—	—	—	—	—	—	—	—	—	—	—	—	—
2061	—	—	—	145.2	1.88	.03	—	.01	.07	.38	—	.02	—	—	.26
	—	—	—	309.0	4.00	.07	—	.02	.14	.80	—	.04	—	—	.55
2062	—	—	—	569.2	—	—	—	.06	.27	.15	—	—	—	—	—
	—	—	—	428.0	—	—	—	.05	.20	.11	—	—	—	—	—
2063	.13	—	.01	585.2	—	.49	1.33	.05	.28	.13	—	—	6.12	—	—
	.10	—	.01	440.0	—	.37	1.00	.04	.21	.10	—	—	4.60	—	—
2064	.43	1.99	—	794.1	—	—	.92	.08	.45	.18	1.07	.10	8.65	—	1.00
	.25	1.15	—	459.0	—	—	.53	.05	.26	.10	.62	.06	5.00	—	.58
2065	—	—	—	502.7	—	—	—	.05	.20	.15	—	—	—	—	—
	—	—	—	378.0	—	—	—	.04	.15	.11	—	—	—	—	—
2066	0	.67	.04	631.8	—	—	1.33	.05	.27	.13	—	—	2.66	—	—
	0	.50	.03	475.0	—	—	1.00	.04	.20	.10	—	—	2.00	—	—
2067	.18	.86	—	175.0	—	—	1.17	.09	.54	.19	1.03	.13	5.25	—	1.37
	.10	.49	—	100.0	—	—	.67	.05	.31	.11	.59	.08	3.00	—	.78
2068	—	.33	—	190.2	—	—	—	.08	.35	.15	—	—	—	—	—
	—	.25	—	143.0	—	—	—	.06	.26	.11	—	—	—	—	—
2069	—	—	—	82.8	—	—	—	.04	.14	.10	—	—	—	—	—
	—	—	—	138.0	—	—	—	.06	.23	.16	—	—	—	—	—
2070	.54	—	—	337.5	2.70	—	—	.08	.43	.68	—	.08	—	—	.54
	.20	—	—	125.0	1.00	—	—	.03	.16	.25	—	.03	—	—	.20
2071	.46	—	—	—	2.32	—	2.55	.09	.37	.58	—	.07	—	—	.70
	.20	—	—	—	1.00	—	1.10	.04	.16	.25	—	.03	—	—	.30
2072	.49	—	—	163.3	—	—	4.90	.05	.22	.25	—	—	—	—	—
	.20	—	—	66.7	—	—	2.00	.02	.09	.10	—	—	—	—	—
2073	.25	—	—	408.3	—	—	2.45	.10	.39	.25	—	—	—	—	—
	.10	—	—	166.7	—	—	1.00	.04	.16	.10	—	—	—	—	—
2074	.12	.98	.08	336.7	100.04	.15	3.59	.07	.42	.21	.76	.10	12.20	7.56	.87
	.05	.40	.03	138.0	41.00	.06	1.47	.03	.17	.08	.31	.04	5.00	3.10	.36
2075	.12	.78	.08	307.4	100.04	.15	2.29	.07	.42	.21	.77	.10	12.20	7.56	.87
	.05	.32	.03	126.0	41.00	06	.94	.03	.17	.08	.31	.04	5.00	3.10	.36
2076	.12	.77	.08	998.4	—	.14	18.72	2.40	.41	2.50	.75	.10	12.00	7.44	.86
	.05	.32	.03	416.0	—	.06	7.80	1.00	.17	1.04	.31	.04	5.00	3.10	.36
2077	.12	.98	—	307.0	100.04	.15	2.29	.09	.39	.21	.77	.10	12.20	7.56	.87
	.05	.40	—	125.8	41.00	.06	.94	.04	.16	.08	.31	.04	5.00	3.10	.36
2078	.12	.95	—	500.2	100.00	—	2.32	.10	.40	.21	.78	.11	12.20	—	.89
	.05	.39	—	205.0	40.98	—	.95	.04	.17	.09	.32	.04	5.00	—	.36
2079	.12	.95	—	500.2	100.00	—	2.37	.10	.41	.21	.79	.11	12.20	—	.90
	.05	.39	—	205.0	40.98	—	.97	.04	.17	.09	.32	.04	5.00	—	.37
2080	.12	.98	—	499.8	100.00	—	—	.10	.42	.22	.82	.11	12.25	—	.94
	.05	.40	—	204.0	40.82	—	—	.04	.17	.09	.34	.05	5.00	—	.38
2081	.12	.98	—	499.8	100.00	—	2.45	.10	.42	.25	.82	.11	12.25	—	.94
	.05	.40	—	204.0	40.82	—	1.00	.04	.17	.10	.34	.05	5.00	—	.38
2082	.10	.98	.07	501.8	100.41	Trace	2.46	.10	.34	.22	.81	.10	12.30	—	.98
	.04	.40	.03	204.0	40.82	Trace	1.00	.04	.14	.09	.33	.04	5.00	—	.40
2083	.15	.98	—	499.4	100.41	—	2.76	.11	.52	.25	.93	.12	14.76	—	1.05
	.06	.40	—	203.0	40.82	—	1.12	.05	.21	.10	.38	.05	6.00	—	.43
2084	.12	1.01	—	499.8	100.00	—	2.48	.10	.43	.22	.83	.11	12.25	—	.95
	.05	.41	—	204.0	40.82	—	1.01	.04	.18	.09	.34	.05	5.00	—	.39
2085	0	—	—	16.8	—	—	2.40	.07	.29	.24	—	.10	—	—	.82
	0	—	—	7.0	—	—	1.00	.03	.12	.10	—	.04	—	—	.34
2086	.24	—	—	1969.1	—	—	2.44	.07	.29	.24	—	.10	—	—	.83
	.10	—	—	807.0	—	—	1.00	.03	.12	.10	—	.04	—	—	.34
2087	—	—	—	317.2	—	—	—	.05	.26	.11	.74	.08	—	—	.88
	—	—	—	130.0	—	—	—	.02	.11	.04	.30	.03	—	—	.36
2088	.58	2.94	(.12)	1024.1	(.27)	.83	8.12	.25	1.30	.61	2.34	.16	34.68	(9.18)	1.36
	.19	.96	(.04)	334.7	(.09)	.27	2.65	.08	.42	.20	.77	.05	11.33	(3.00)	.44

(Continued)

TABLE F-36 *(Continued)*

Item No.	Food Item Name	Approximate Measure	Weight	Moisture	Food Energy Calories	Protein	Fats	Carbo-hydrates	Fiber	Minerals (Macro) Calcium	Phos-phorus	Sodium	Mag-nesium	Potas-sium
			(g)	(%)	(kcal)	(g)	(g)	(g)	(g)	(mg)	(mg)	(mg)	(mg)	(mg)
	MILK, COW'S (Continued)													
2089	canned, evaporated, filled, w/vegetable oil, diluted	1 CUP	256	—	174.1	8.7	9.7	12.8	—	332.8	258.6	—	30.7	—
			100	—	68.0	3.4	3.8	5.0	—	130.0	101.0	—	12.0	—
2090	canned, evaporated, filled, w/vegetable oil, undiluted	1 CUP	256	—	345.6	17.4	19.5	25.6	—	668.2	517.1	—	61.4	—
			100	—	135.0	6.8	7.6	10.0	—	261.0	202.0	—	24.0	—
2091	canned, evaporated, skimmed	1 CUP	256	79.4	199.7	19.3	.5	29.1	0	742.4	499.2	294.4	69.1	849.9
			100	79.4	78.0	7.6	.2	11.4	0	290.0	195.0	115.0	27.0	332.0
2092	canned, evaporated, unsweetened	1 CUP	256	74.0	343.0	17.4	19.4	25.7	0	668.2	517.1	271.4	61.4	775.7
			100	74.0	134.0	6.8	7.6	10.0	0	261.0	202.0	106.0	24.0	303.0
2093	dry, nonfat, calcium reduced	1 OZ	28	4.9	99.1	9.9	.1	14.5	0	78.4	283.1	638.4	16.8	190.4
			100	4.9	354.0	35.5	.2	51.8	0	280.0	1011.0	2280.0	60.0	680.0
2094	dry, skim, solids, instant	1 TBLS	4	4.0	14.1	1.4	0	2.1	0	49.2	39.4	22.0	4.7	68.2
			100	4.0	353.0	34.9	.8	51.6	0	1231.0	985.0	549.0	117.0	1705.0
2095	dry, skim, solids, regular	1 CUP	70	3.0	252.0	25.1	.6	36.6	0	915.6	711.2	372.4	65.8	1255.8
			100	3.0	360.0	35.9	.8	52.3	0	1308.0	1016.0	532.0	94.0	1794.0
2096	dry, whole, instant, high density	1 CUP	103	2.5	514.0	27.0	27.6	39.3	0	939.4	799.3	382.1	87.6	1369.9
			100	2.5	499.0	26.3	26.8	38.2	0	912.0	776.0	371.0	85.0	1330.0
2097	evaporated	1 CUP	256	73.8	350.7	17.9	20.2	24.8	0	645.1	524.8	302.1	64.0	775.7
			100	73.8	137.0	7.0	7.9	9.7	0	252.0	205.0	180.0	25.0	303.0
2098	'chocolate, whole	1 CUP	250	82.3	207.5	8.0	8.5	25.8	.2	280.0	250.0	150.0	32.5	417.5
			100	82.3	83.0	3.2	3.4	10.3	.1	112.0	100.0	60.0	13.0	167.0
2099	chocolate, 2% fat	1 CUP	250	82.8	175.0	8.3	5.0	27.3	.2	285.0	255.0	150.0	32.5	422.5
			100	82.8	70.0	3.3	2.0	10.9	.1	114.0	102.0	60.0	13.0	169.0
2100	chocolate, 1% fat	1 CUP	250	84.5	157.5	8.1	2.5	26.1	.2	287.5	257.5	152.5	32.5	425.0
			100	84.5	63.0	3.2	1.0	10.4	.1	115.0	103.0	61.0	13.0	170.0
2101	chocolate drink, w/whole milk, 3.3% fat	1 CUP	250	81.5	207.5	7.9	8.5	25.9	.2	277.5	250.0	150.0	32.5	417.5
			100	81.5	83.0	3.2	3.4	10.3	.1	111.0	100.0	60.0	13.0	167.0
2102	chocolate flavored, w/Ovaltine, skim milk	1 CUP	250	—	171.4	9.9	1.2	28.2	—	383.1	312.5	302.4	—	604.8
			100	—	68.5	4.0	.5	11.3	—	153.2	125.0	121.0	—	241.9
2103	chocolate, hot, homemade	1 CUP	250	80.5	250.0	8.3	12.5	26.0	.3	260.0	235.0	120.0	—	370.0
			100	80.5	100.0	3.3	5.0	10.4	.1	104.0	94.0	48.0	—	148.0
2104	chocolate, hot, w/cocoa, homemade	1 CUP	250	79.0	217.5	9.5	9.1	25.8	.3	295.0	282.5	127.5	55.0	480.0
			100	79.0	87.0	3.8	3.6	10.3	.1	118.0	113.0	51.0	22.0	192.0
2105	cocoa powder, w/nonfat, dehydrated milk	1 TBLS	7	1.9	25.1	1.3	.2	5.0	0	41.2	38.2	36.8	—	56.0
			100	1.9	359.0	18.6	2.9	70.8	.5	589.0	545.0	525.0	—	800.0
2106	malted, dry powder	1 TBLS	21	2.6	86.1	2.7	1.7	15.2	.1	55.9	79.8	92.4	19.5	159.2
			100	2.6	410.0	13.1	8.3	72.9	.6	266.0	380.0	440.0	93.0	758.0
2107	malted beverage	1 CUP	235	81.2	209.2	9.6	8.8	23.6	.1	317.3	286.7	190.4	47.0	470.0
			100	81.2	89.0	4.0	3.8	10.0	.1	135.0	122.0	81.0	20.0	200.0
2108	malted chocolate-flavor powder	1 TBLS	9	2.0	35.6	.6	.4	7.6	0	5.5	16.0	20.9	6.4	55.6
			100	2.0	396.0	6.5	4.5	84.9	.4	61.0	178.0	232.0	71.0	618.0
2109	malted chocolate-flavor beverage	1 CUP	265	81.2	233.2	9.4	9.1	29.2	.1	304.8	265.0	169.6	47.7	498.2
			100	81.2	88.0	3.5	3.4	11.0	0	115.0	100.0	64.0	18.0	188.0
2110	**MILK, GOAT**	1 CUP	244	87.5	163.5	8.1	9.8	11.2	0	314.8	258.6	122.0	34.2	497.8
			100	87.5	67.0	3.3	4.0	4.6	0	129.0	106.0	50.0	14.0	204.0
2111	**MILK, HUMAN,** U.S.A. samples	1 OZ	30	88.3	20.8	.3	1.3	2.1	0	9.9	4.2	4.8	.9	15.3
			100	88.3	69.1	1.0	4.4	6.9	0	33.0	14.0	16.0	2.9	51.0
2112	**MILK, INDIAN BUFFALO,** whole, fluid	1 CUP	244	83.4	236.7	9.2	16.8	12.6	0	412.4	285.5	126.9	75.6	434.3
			100	83.4	97.0	3.8	6.9	5.2	0	169.0	117.0	52.0	31.0	178.0
2113	**MILK, REINDEER**	1 CUP	244	64.1	571.0	26.4	47.8	10.0	0	619.8	483.1	383.1	—	388.0
			100	64.1	234.0	10.8	19.6	4.1	0	254.0	198.0	157.0	—	159.0
2114	**MILKSHAKE**	1 CUP	345	—	420.9	11.2	17.9	58.0	.3	362.3	320.9	—	—	—
			100	—	122.0	3.2	5.2	16.8	.1	105.0	93.0	—	—	—
2115	chocolate, thick	1 CUP	345	72.2	413.3	11.2	9.3	72.9	.9	455.4	434.7	383.0	55.2	772.8
			100	72.2	119.8	3.2	2.7	21.1	.2	132.0	126.0	111.0	16.0	224.0
2116	vanilla, thick	1 CUP	345	74.5	386.4	13.3	10.5	61.2	.2	503.7	396.8	327.8	41.4	631.4
			100	74.5	112.0	3.9	3.0	17.8	.1	146.0	115.0	95.0	12.0	183.0
2117	from soft-serve machine	1 AVG	226	73.0	287.0	11.0	9.0	38.4	—	379.7	391.0	—	—	—
			100	73.0	127.0	4.9	4.0	17.0	—	168.0	173.0	—	—	—
2118	**MILK, SHEEP,** whole, fluid	1 CUP	245	80.7	264.6	14.7	17.2	13.1	0	472.9	387.1	107.8	44.1	333.2
			100	80.7	108.0	6.0	7.0	5.4	0	193.0	158.0	44.0	18.0	136.0
2119	**MILNOT,** canned, fortified	1 CUP	240	—	288.0	16.8	14.4	23.2	—	571.2	441.6	655.2	55.2	—
			100	—	120.0	7.0	6.0	9.7	—	238.0	184.0	273.0	23.0	—

Item No.	Minerals (Micro)			Fat-Soluble Vitamins			Water-Soluble Vitamins								
	Iron	Zinc	Copper	Vitamin A	Vitamin D	Vitamin E (Alpha Tocopherol)	Vitamin C	Thiamin	Ribo-flavin	Niacin	Panto-thenic Acid	Vit. B-6 (Pyri-doxine)	Folacin (Folic Acid)	Biotin	Vitamin B-12
	(mg)	(mg)	(mg)	(IU)	(IU)	(mg)	(mg)	(mg)	(mg)	(mg)	(mg)	(mg)	(mcg)	(mcg)	(mcg)
2089	.26	–	–	501.8	–	–	2.56	.05	.41	.26	–	.05	–	–	.21
	.10	–	–	196.0	–	–	1.00	.02	.16	.10	–	.02	–	–	.08
2090	.51	–	–	1003.5	–	–	5.12	.13	.82	.51	–	.13	–	–	.41
	.20	–	–	392.0	–	–	2.00	.05	.32	.20	–	.05	–	–	.16
2091	.74	2.30	–	1003.5	225.73	–	3.17	.12	.79	.45	1.89	.14	23.04	–	.61
	.29	.90	–	392.0	88.18	–	1.24	.05	.31	.17	.74	.06	9.00	–	.24
2092	.49	1.97	.01	622.1	225.73	.33	4.81	.13	.82	.49	1.64	.13	20.48	11.520	.41
	.19	.77	0	243.0	88.18	.13	1.88	.05	.32	.19	.64	.05	8.00	4.500	.16
2093	–	–	–	2.2	–	–	–	.05	.46	.19	.93	.08	–	–	1.11
	–	–	–	8.0	–	–	–	.16	1.64	.67	3.31	.30	–	–	3.98
2094	.01	.18	0	94.8	–	.0	.22	.02	.07	.04	.14	.01	2.00	–	.16
	.31	4.41	.05	2370.0	–	.02	5.58	.41	1.78	.90	3.60	.35	50.00	–	3.99
2095	.22	2.86	.03	25.2	–	.01	4.90	.29	1.09	.63	2.52	.25	35.00	2.380	2.82
	.32	4.08	.04	36.0	–	.02	7.00	.42	1.55	.90	3.60	.36	50.00	3.400	4.03
2096	.52	3.44	–	949.7	–	–	8.90	.30	1.24	.67	2.47	.31	38.11	–	3.35
	.50	3.34	–	922.0	–	–	8.64	.30	1.21	.65	2.40	.30	37.00	–	3.25
2097	.26	(2.82)	(.10)	819.2	(.23)	.23	2.56	.10	.87	.51	1.64	.13	20.48	11.520	.41
	.10	(1.10)	(.04)	320.0	(.09)	.09	1.00	.04	.34	.20	.64	.05	8.00	4.500	.16
2098	.60	1.03	–	302.5	–	–	2.28	.75	.40	.33	.75	.10	12.50	–	8.3
	.24	.41	–	121.0	–	–	.91	.03	.16	.13	.30	.04	5.00	–	.33
2099	.50	1.03	–	500.0	100.00	–	2.50	.10	.40	.25	.75	.10	12.50	–	.85
	.20	.41	–	200.0	40.00	–	1.00	.04	.16	.10	.30	.04	5.00	–	.34
2100	.60	1.03	–	500.0	100.00	–	2.33	.10	.42	.32	.76	.10	12.50	–	.86
	.24	.41	–	200.0	40.00	–	.93	.04	.17	.13	.30	.04	5.00	–	.34
2101	.50	1.03	–	325.0	100.00	–	2.50	.08	.40	.25	.74	.10	12.50	–	.84
	.20	.41	–	130.0	40.00	–	1.00	.03	.16	.10	.30	.04	5.00	–	.33
2102	2.72	–	–	2268.1	181.45	–	29.24	.79	1.21	9.27	–	1.01	–	–	–
	1.09	–	–	907.3	72.58	–	11.69	.32	.48	3.71	–	.40	–	–	–
2103	.50	–	–	350.0	5.00	–	2.50	.08	.40	.25	–	–	–	–	–
	.20	–	–	140.0	2.00	–	1.00	.03	.16	.10	–	–	–	–	–
2104	1.00	1.23	–	317.5	–	–	2.50	.10	.45	.36	.81	.11	12.50	–	.87
	.40	.49	–	127.0	–	–	1.00	.04	.18	.15	.32	.04	5.00	–	.35
2105	.13	–	–	1.4	–	–	.21	.01	.05	.05	–	–	–	–	–
	1.80	–	–	20.0	–	–	3.00	.13	.73	.70	–	–	–	–	–
2106	.16	.21	–	67.8	–	–	0	.11	.14	1.07	–	.01	9.66	–	.16
	.76	.99	–	323.0	–	–	0	.53	.68	5.10	–	.04	46.00	–	.78
2107	.26	1.01	–	333.7	–	–	2.05	.14	.49	1.13	.68	.16	18.80	–	.92
	.11	.43	–	142.0	–	–	.87	.06	.21	.48	.29	.07	8.00	–	.39
2108	.16	.08	–	8.5	–	–	–	.02	.02	.19	–	.01	2.07	–	.02
	1.80	.89	–	94.0	–	–	–	.20	.18	2.06	–	.13	23.00	–	.22
2109	.50	1.11	–	326.0	–	–	2.31	.14	.43	.69	.77	.13	15.90	–	.92
	.19	.42	–	123.0	–	–	.87	.05	.16	.26	.29	.05	6.00	–	.35
2110	.12	.73	.17	451.4	4.88	–	3.15	.10	.34	.73	.78	.11	2.44	–	.16
	.05	.30	.07	185.0	2.00	–	1.29	.04	.14	.30	.32	.05	1.00	–	.07
2111	.01	.05	.02	72.0	(.07)	.05	1.50	0	.01	.06	.07	.00	1.50	.24	.01
	.02	.16	.05	240.0	(.03)	.15	5.00	.01	.04	.20	.22	.01	5.00	.80	.04
2112	.29	.54	–	434.3	–	–	5.49	.13	.33	.22	.47	.06	14.64	–	.89
	.12	.22	–	178.0	–	–	2.25	.05	.14	.09	.19	.02	6.00	–	.36
2113	.24	–	–	–	–	–	–	–	–	–	–	–	–	–	–
	.10	–	–	–	–	–	–	–	–	–	–	–	–	–	–
2114	1.04	–	–	686.6	3.45	–	4.14	.10	.55	.48	–	–	–	–	–
	.30	–	–	199.0	1.00	–	1.20	.03	.16	.14	–	–	–	–	–
2115	1.04	1.66	–	296.7	3.45	–	0	.10	.77	.48	1.25	.09	17.25	–	1.09
	.30	.48	–	86.0	1.00	–	0	.03	.22	.14	.36	.03	5.00	–	.32
2116	.35	–	–	393.3	–	–	–	.10	.67	.50	–	.15	24.15	–	1.79
	.10	–	–	114.0	–	–	–	.03	.20	.15	–	.04	7.00	–	.52
2117	–	–	–	–	–	–	–	–	–	–	–	–	–	–	–
	–	–	–	–	–	–	–	–	–	–	–	–	–	–	–
2118	.25	–	–	360.2	–	–	10.19	.16	.87	1.02	1.00	–	–	–	1.74
	.10	–	–	147.0	–	–	4.16	.07	.36	.42	.41	–	–	–	.71
2119	.21	–	.06	1166.4	232.80	–	2.64	.13	.91	.48	–	–	–	–	–
	.09	–	.03	486.0	97.00	–	1.10	.06	.38	.20	–	–	–	–	–

(Continued)

TABLE F-36 *(Continued)*

Item No.	Food Item Name	Approximate Measure	Weight	Moisture	Food Energy Calories	Protein	Fats	Carbo-hydrates	Fiber	Minerals (Macro) Calcium	Phos-phorus	Sodium	Mag-nesium	Potas-sium
			(g)	(%)	(kcal)	(g)	(g)	(g)	(g)	(mg)	(mg)	(mg)	(mg)	(mg)
2120	**RENNIN DESSERT,** chocolate, made from mix, w/milk	1 CUP	**255**	**77.9**	**260.1**	**8.7**	**9.7**	**36.0**	**.2**	**311.1**	**244.8**	**132.6**	–	**318.8**
			100	77.9	102.0	3.4	3.8	14.1	.1	122.0	96.0	52.0	–	125.0
2121	**SOUR CREAM**	1 TBLS	**12**	**71.0**	**25.7**	**.4**	**2.5**	**.5**	**0**	**14.0**	**10.2**	**6.4**	**1.2**	**17.3**
			100	71.0	214.0	3.1	21.0	4.3	0	116.0	85.0	53.0	10.0	144.0
2122	half & half, cultured	1 TBLS	**15**	**80.1**	**20.3**	**.4**	**1.8**	**.6**	**0**	**15.6**	**14.3**	**6.0**	**1.5**	**19.4**
			100	80.1	135.0	2.9	12.0	4.3	0	104.0	95.0	40.0	10.0	129.0
2123	imitation, made w/nonfat, dehydrated milk	1 CUP	**240**	**72.0**	**448.8**	**9.2**	**38.9**	**17.4**	–	**283.2**	–	–	–	–
			100	72.0	187.0	3.8	16.2	7.2	–	118.0	–	–	–	–
2124	imitation, nondairy, cultured	1 OZ	**28**	**71.6**	**58.2**	**.7**	**5.5**	**1.9**	**0**	**.6**	**12.3**	**28.6**	–	**44.8**
			100	71.6	208.0	2.4	19.5	6.6	0	2.0	44.0	102.0	–	160.0
	SOYBEAN MILK													
2125	fluid	1 CUP	**263**	**92.4**	**86.8**	**8.9**	**3.9**	**5.8**	**0**	**55.2**	**126.2**	–	–	–
			100	92.4	33.0	3.4	1.5	2.2	0	21.0	48.0	–	–	–
2126	liquid concentrate, sweetened	1 CUP	**305**	**74.4**	**384.3**	**14.6**	**22.3**	**37.5**	**.6**	**91.5**	**180.0**	**131.2**	–	**722.9**
			100	74.4	126.0	4.8	7.3	12.3	.2	30.0	59.0	43.0	–	237.0
2127	powder	1 OZ	**28**	**4.2**	**120.1**	**11.7**	**5.7**	**7.8**	**.1**	**77.0**	–	–	–	–
			100	4.2	429.0	41.8	20.3	28.0	.2	275.0	–	–	–	–
2128	powder, sweetened	1 OZ	**28**	**3.7**	**126.6**	**5.7**	**6.5**	**13.6**	**.1**	**32.2**	**79.8**	**.3**	–	**256.2**
			100	3.7	452.0	20.4	23.2	48.4	.5	115.0	285.0	1.0	–	915.0
	WHEY													
2129	sweet, fluid	1 TBLS	**14**	**93.1**	**3.6**	**.1**	**0**	**.7**	**0**	**7.1**	**6.4**	**7.6**	**1.1**	**22.5**
			100	93.1	26.0	.9	.3	5.1	0	51.0	46.0	54.0	8.0	161.0
2130	sweet, dry	1 TBLS	**8**	**3.2**	**28.3**	**1.0**	**.1**	**5.9**	**0**	**63.7**	**74.6**	**86.3**	**14.1**	**166.4**
			100	3.2	354.0	12.7	1.0	73.5	0	796.0	932.0	1079.0	176.0	2080.0
2131	acid, fluid	1 CUP	**246**	**93.4**	**59.0**	**1.9**	**.2**	**12.6**	**0**	**253.4**	**191.9**	**118.1**	**24.6**	**351.8**
			100	93.4	24.0	.8	.1	5.1	0	103.0	78.0	48.0	10.0	143.0
2132	acid, dry	1 CUP	**57**	**3.5**	**193.2**	**6.7**	**.3**	**41.9**	**0**	**1170.8**	**768.4**	**551.8**	**113.4**	**1304.2**
			100	3.5	339.0	11.7	.5	73.5	0	2054.0	1348.0	968.0	199.0	2288.0
	YOGURT													
2133	plain, made w/whole milk	1 CUP	**227**	**88.0**	**138.5**	**8.0**	**7.5**	**10.7**	**0**	**274.7**	**215.7**	**104.4**	**27.2**	**351.9**
			100	88.0	61.0	3.5	3.3	4.7	0	121.0	95.0	46.0	12.0	155.0
2134	plain, lowfat, made w/lowfat milk & nonfat milk solids	1 CUP	**227**	**85.1**	**143.0**	**12.0**	**3.6**	**15.9**	**0**	**415.4**	**326.9**	**158.9**	**38.6**	**531.2**
			100	85.1	63.0	5.3	1.6	7.0	0	183.0	144.0	70.0	17.0	234.0
2135	plain, skim milk, made w/skim milk & nonfat milk solids	1 CUP	**227**	**85.2**	**127.1**	**12.9**	**.5**	**17.5**	**0**	**451.7**	**354.1**	**172.5**	**43.1**	**578.9**
			100	85.2	56.0	5.7	.2	7.7	0	199.0	156.0	76.0	19.0	255.0
2136	fruit varieties, lowfat, w/nonfat milk solids	1 CUP	**227**	**74.5**	**231.5**	**10.0**	**2.3**	**43.4**	**0**	**345.0**	**270.1**	**131.7**	**34.1**	**440.4**
			100	74.5	102.0	4.4	1.1	19.1	0	152.0	119.0	58.0	15.0	194.0
	NUTS & SEEDS													
2137	**ALFALFA SEEDS**	1 OZ	**28**	**7.4**	**108.9**	**9.8**	**3.5**	–	**2.2**	**38.1**	–	–	–	–
			100	7.4	389.0	35.1	12.6	–	7.9	136.0	–	–	–	–
	ALMONDS													
2138	unshelled	1 CUP	**78**	–	**237.9**	**7.0**	**21.1**	**7.8**	–	**101.4**	**189.5**	**.8**	–	**275.3**
			100	–	305.0	9.0	27.0	10.0	–	130.0	243.0	1.0	–	353.0
2139	shelled	1 CUP	**142**	–	**906.0**	**26.4**	**76.5**	**27.7**	**20.3**	**332.3**	**715.7**	**4.3**	**(369.2)**	**1097.7**
			100	–	638.0	18.6	53.9	19.5	14.3	234.0	504.0	3.0	(260.0)	773.0
2140	unsalted	1 CUP	**142**	–	**910.2**	**27.0**	**76.5**	**28.4**	**4.3**	**355.0**	**681.6**	**4.3**	**383.4**	**979.8**
			100	–	641.0	19.0	53.9	20.0	3.0	250.0	480.0	3.0	270.0	690.0
2141	roasted & salted	1 CUP	**157**	**0.7**	**984.4**	**29.2**	**90.6**	**30.6**	**4.1**	**369.0**	**791.3**	**310.9**	–	**1213.6**
			100	0.7	627.0	18.6	57.7	19.5	2.6	235.0	504.0	198.0	–	773.0
2142	dried, shelled, whole	1 CUP	**142**	**4.7**	**849.2**	**26.4**	**77.0**	**27.7**	**3.7**	**332.3**	**715.7**	**5.7**	**415.6**	**1097.7**
			100	4.7	598.0	18.6	54.2	19.5	2.6	234.0	504.0	4.0	292.7	773.0
2143	dried, salted, unblanched	1 AVG	**1**	**4.7**	**6.2**	**.2**	**.6**	**.2**	**0**	**2.5**	**4.7**	**1.6**	**2.5**	**7.1**
			100	4.7	620.0	18.6	56.6	19.3	2.0	253.0	473.0	160.0	252.0	706.0
2144	meal, partially defatted	1 OZ	**28**	**7.2**	**114.2**	**11.1**	**5.1**	**8.1**	**.6**	**118.7**	**255.9**	**2.0**	–	**392.0**
			100	7.2	408.0	39.5	18.3	28.9	2.3	424.0	914.0	7.0	–	1400.0
2145	paste	1 OZ	**28**	–	**142.8**	**3.1**	**9.0**	**14.3**	–	**42.3**	**79.2**	**59.1**	–	**115.1**
			100	–	510.0	11.0	32.0	51.0	–	151.0	283.0	211.0	–	411.0
2146	**BEECHNUTS,** shelled	1 TBLS	**8**	**6.6**	**49.0**	**1.6**	**4.0**	**1.6**	**.3**	–	–	–	–	–
			100	6.6	612.0	19.4	50.3	20.3	3.7	–	–	–	–	–
	BRAZIL NUTS													
2147	shelled	1 CUP	**140**	**4.6**	**1001.0**	**20.0**	**95.5**	**15.3**	**4.3**	**260.4**	**970.2**	**1.4**	**444.5**	**1001.0**
			100	4.6	715.0	14.3	68.2	10.9	3.1	186.0	693.0	1.0	317.5	715.0

	Minerals (Micro)			Fat-Soluble Vitamins			Water-Soluble Vitamins								
Item No.	Iron	Zinc	Copper	Vitamin A	Vitamin D	Vitamin E (Alpha Tocopherol)	Vitamin C	Thiamin	Ribo-flavin	Niacin	Panto-thenic Acid	Vit. B-6 (Pyri-doxine)	Folacin (Folic Acid)	Biotin	Vitamin B-12
	(mg)	(mg)	(mg)	(IU)	(IU)	(mg)	(mg)	(mg)	(mg)	(mg)	(mg)	(mg)	(mcg)	(mcg)	(mcg)
2120	–	–	–	**357.0**	–	–	**2.55**	**.08**	**.38**	**.26**	–	–	–	–	–
	–	–	–	140.0	–	–	1.00	.03	.15	.10	–	–	–	–	–
2121	**.01**	**.03**	–	**94.8**	**.84**	–	**.10**	**0**	**.02**	**.01**	**.04**	**0**	**1.32**	–	**.04**
	.06	.27	–	790.0	7.00	–	.86	.03	.13	.07	.36	.02	11.00	–	.30
2122	**:01**	**.08**	–	**67.8**	–	–	**.13**	**.01**	**.02**	**.01**	**.05**	**0**	**1.65**	–	**.05**
	.07	.50	–	452.0	–	–	.86	.04	.15	.07	.36	.02	11.00	–	.30
2123	–	–	–	**9.6**	–	–	**.96**	**.05**	**.38**	**.19**	–	–	–	–	–
	–	–	–	4.0	–	–	.40	.02	.16	.08	–	–	–	–	–
2124	–	–	–	**0**	–	–	**0**	**0**	**0**	**0**	**0**	**0**	**0**	–	**0**
	–	–	–	0	–	–	0	0	0	0	0	0	0	–	0
2125	**2.10**	[6]**.53**	–	**105.2**	–	–	**0**	**.21**	**.08**	**.53**	–	–	–	–	–
	.80	.20	–	40.0	–	–	0	.08	.03	.20	–	–	–	–	–
2126	**2.44**	[6]**.61**	–	–	–	–	**0**	**.18**	**.09**	**.61**	–	–	–	–	–
	.80	.20	–	–	–	–	0	.06	.03	.20	–	–	–	–	–
2127	–	–	–	–	–	–	–	–	–	–	–	–	–	–	–
	–	–	–	–	–	–	–	–	–	–	–	–	–	–	–
2128	**1.40**	–	–	**5.6**	–	–	**0**	**.08**	**.07**	**.39**	**.22**	**.08**	–	–	**0**
	5.00	–	–	20.0	–	–	0	.30	.24	1.40	.80	.29	–	–	0
2129	**.01**	**.02**	–	**2.2**	–	–	**.01**	**0**	**.02**	**.01**	**.05**	**0**	**.14**	–	**.04**
	.06	.13	–	16.0	–	–	.10	.03	.14	.07	.38	.03	1.00	–	.28
2130	**.07**	**.16**	–	**4.0**	–	**0**	**.12**	**.04**	**.18**	**.10**	**.45**	**.04**	**.96**	–	**.16**
	.88	1.97	–	50.0	–	.03	1.49	.50	2.21	1.26	5.62	.58	12.00	–	2.00
2131	**.20**	**1.06**	–	**17.2**	–	–	**.15**	**.10**	**.34**	**.19**	**.94**	**.10**	**4.92**	–	**.44**
	.08	.43	–	7.0	–	–	.06	.04	.14	.08	.38	.04	2.00	–	.18
2132	**.71**	**3.60**	–	**33.1**	–	–	–	**.36**	**1.17**	**.66**	**3.21**	**.35**	**18.81**	–	**1.43**
	1.24	6.31	–	58.0	–	–	–	.62	2.06	1.16	5.63	.62	33.00	–	2.50
2133	**.11**	**1.34**	**(.09)**	**279.2**	**Trace**	**(.07)**	**1.20**	**.07**	**.32**	**.17**	**.88**	**.07**	**15.89**	–	**.84**
	.05	.59	(.04)	123.0	Trace	(.03)	.53	.03	.14	.08	.39	.03	7.00	–	.37
2134	**.18**	**2.02**	–	**149.8**	–	–	**1.82**	**.10**	**.49**	**.26**	**1.34**	**.11**	**22.70**	–	**1.28**
	.08	.89	–	66.0	–	–	.80	.04	.21	.11	.59	.05	11.00	–	.56
2135	**.20**	**2.20**	–	**15.9**	–	–	**1.98**	**.11**	**.53**	**.28**	**1.46**	**.12**	**27.24**	–	**1.39**
	.09	.97	–	7.0	–	–	.87	.05	.23	.12	.64	.05	12.00	–	.61
2136	**.16**	**1.68**	–	**104.4**	–	–	**1.50**	**.08**	**.40**	**.22**	**1.11**	**.09**	**20.40**	–	**1.06**
	.07	.74	–	46.0	–	–	.66	.04	.18	.10	.49	.04	9.00	–	.47
2137	**3.61**	**1.93**	–	–	–	**9.24**	**7.26**	**.30**	**.10**	**.50**	–	–	–	–	–
	12.90	6.90	–	–	–	33.00	26.00	1.08	.58	1.80	–	–	–	–	–
2138	**1.56**	[6]**.94**	–	**0**	**0**	–	–	**.01**	**.02**	**1.56**	–	–	–	–	–
	2.00	1.2	–	0	0	–	–	.01	.03	2.00	–	–	–	–	–
2139	**6.67**	**(4.42)**	**(.21)**	**0**	**0**	**21.30**	–	**.34**	**1.32**	**4.97**	**.82**	**.14**	**136.32**	**25.56**	**0**
	4.70	(3.11)	(.15)	0	0	15.00	–	.24	.93	3.50	.58	.10	96.00	18.00	0
2140	**5.68**	–	–	**0**	**0**	**21.30**	–	**.03**	**.10**	**(4.97)**	–	**.16**	**136.32**	–	–
	4.00	–	–	0	0	15.00	–	.02	.07	(3.50)	–	.11	96.00	–	–
2141	**7.38**	**4.02**	–	**0**	–	–	**0**	**.08**	**1.44**	**5.50**	**.39**	**.15**	–	–	**0**
	4.70	2.56	–	0	–	–	0	.05	.92	3.50	.25	.10	–	–	0
2142	**6.67**	**4.45**	**2.00**	**0**	–	**39.62**	–	**.34**	**1.31**	**4.97**	**.67**	**.14**	–	–	**0**
	4.70	3.13	1.41	0	–	27.90	–	.24	.92	3.50	.47	.10	–	–	0
2143	**0.05**	–	**0**	**0**	–	**0.28**	–	**0**	**.01**	**.05**	**.01**	**0**	–	–	**0**
	4.70	–	.14	0	–	27.90	–	.25	.66	4.60	.47	.10	–	–	0
2144	**2.38**	–	–	**0**	–	–	–	**.09**	**.47**	**1.76**	–	–	–	–	–
	8.50	–	–	0	–	–	–	.32	1.68	6.30	–	–	–	–	–
2145	**0.73**	–	–	**0**	–	–	–	**.01**	**.01**	**.78**	–	–	–	–	–
	2.60	–	–	0	–	–	–	.02	.04	2.80	–	–	–	–	–
2146	–	–	–	–	–	–	–	–	–	–	–	–	–	–	–
	–	–	–	–	–	–	–	–	–	–	–	–	–	–	–
2147	**4.76**	**5.92**	**3.34**	**0**	**0**	**9.10**	**14.00**	**1.34**	**.17**	**2.24**	**.32**	**.24**	**5.60**	–	**0**
	3.40	4.23	2.38	0	0	6.50	10.00	.96	.12	1.60	.23	.17	4.00	–	0

(Continued)

NUTS & SEEDS

TABLE F-36 *(Continued)*

Item No.	Food Item Name	Approximate Measure	Weight	Moisture	Food Energy Calories	Protein	Fats	Carbo-hydrates	Fiber	Minerals (Macro) Calcium	Phos-phorus	Sodium	Mag-nesium	Potas-sium
			(g)	(%)	(kcal)	(g)	(g)	(g)	(g)	(mg)	(mg)	(mg)	(mg)	(mg)
	BRAZIL NUTS (Continued)													
2148	salted	1 CUP	**140**	**4.6**	**1001.0**	**20.0**	**95.5**	**15.3**	**4.3**	**260.4**	**970.2**	**644.0**	**315.0**	**1001.0**
			100	4.6	715.0	14.3	68.2	10.9	3.1	186.0	693.0	460.0	225.0	715.0
2149	**BUTTERNUTS**	1 AVG	**3**	**3.8**	**18.9**	**.7**	**1.8**	**.3**	–	–	–	–	–	–
			100	3.8	629.0	23.7	61.2	8.4	–	–	–	–	–	–
	CASHEW NUTS													
2150	unsalted	1 CUP	**140**	**5.2**	**834.4**	**24.1**	**63.8**	**41.0**	**2.0**	**53.2**	**522.2**	**21.0**	**373.8**	**649.6**
			100	5.2	596.0	17.2	45.6	29.3	1.4	38.0	373.0	15.0	267.0	464.0
2151	salted	1 CUP	**140**	**5.2**	**834.4**	**24.1**	**63.8**	**41.0**	**2.0**	**53.2**	**522.2**	**280.0**	**373.8**	**649.6**
			100	5.2	596.0	17.2	45.6	29.3	1.4	38.0	373.0	200.0	267.0	464.0
	CHESTNUTS													
2152	fresh	1 CUP	**200**	**52.5**	**408.0**	**5.8**	**5.4**	**84.2**	**2.2**	**54.0**	**176.0**	**12.0**	**84.0**	**908.0**
			100	52.5	204.0	2.9	2.7	42.1	1.1	27.0	88.0	6.0	42.0	454.0
2153	dried	1 CUP	**100**	**8.4**	**377.0**	**6.7**	**4.1**	**78.6**	**2.5**	**52.0**	**162.0**	**12.0**	–	**875.0**
			100	8.4	377.0	6.7	4.1	78.6	2.5	52.0	162.0	12.0	–	875.0
2154	flour	1 CUP	**110**	**11.4**	**398.2**	**6.7**	**4.1**	**83.8**	**2.2**	**55.0**	**180.4**	**12.1**	–	**931.7**
			100	11.4	362.0	6.1	3.7	76.2	2.0	50.0	164.0	11.0	–	847.0
	COCONUTS													
2155	meat, fresh, grated	1 CUP	**130**	**50.9**	**481.9**	**4.4**	**46.2**	**12.2**	**5.2**	**16.9**	**123.5**	**29.9**	**59.8**	**332.8**
			100	50.9	370.7	3.4	35.5	9.4	4.0	13.0	95.0	23.0	46.0	256.0
2156	meat, dried, unsweetened	1 CUP	**130**	**3.5**	**860.6**	**9.4**	**84.4**	**29.9**	**5.1**	**33.8**	**243.1**	**68.9**	**117.0**	**764.4**
			100	3.5	662.0	7.2	64.9	23.0	3.9	26.0	187.0	53.0	90.0	588.0
2157	meat, dried, sweetened, shredded	1 CUP	**130**	**3.3**	**712.4**	**4.7**	**50.8**	**69.2**	**5.3**	**20.8**	**145.6**	**23.4**	**100.1**	**458.9**
			100	3.3	548.0	3.6	39.1	53.2	4.1	16.0	112.0	18.0	77.0	353.0
2158	milk	1 CUP	**244**	**65.7**	**614.9**	**7.8**	**60.8**	**12.7**	–	**39.0**	**244.0**	**129.3**	–	**463.6**
			100	65.7	252.0	3.2	24.9	5.2	–	16.0	100.0	53.0	–	190.0
2159	water	1 CUP	**244**	**94.2**	**53.7**	**.7**	**.5**	**11.5**	–	**48.8**	**31.7**	**61.0**	**68.3**	**358.7**
			100	94.2	22.0	.3	.2	4.7	–	20.0	13.0	25.0	28.0	147.0
2160	[5]**CUSHAW SEEDS**	1 OZ	**28**	**7.7**	**130.2**	**8.5**	**7.8**	**8.6**	**1.4**	**9.8**	**203.0**	–	–	–
			100	7.7	465.0	30.5	27.9	30.8	4.9	35.0	725.0	–	–	–
2161	**FILBERTS (HAZELNUTS),** whole, shelled	1 CUP	**135**	**5.8**	**945.0**	**17.1**	**87.3**	**22.5**	**4.1**	**282.2**	**455.0**	**2.7**	**234.8**	**950.4**
			100	5.8	700.0	12.7	64.7	16.7	3.0	209.0	337.0	2.0	173.9	704.0
2162	[3]**FLAXSEED,** dried	1 OZ	**28**	**6.3**	**139.4**	**5.0**	**9.4**	**10.4**	**2.5**	**75.9**	**129.4**	–	–	–
			100	6.3	498.0	18.0	34.0	37.2	8.8	271.0	462.0	–	–	–
	[4]**GINKGO SEEDS**													
2163	whole, dried, raw	1 OZ	**28**	**54.1**	**51.8**	**1.3**	**.4**	**10.7**	**.2**	**1.4**	**42.0**	**2.0**	–	**146.4**
			100	54.1	185.0	4.8	1.6	38.1	.6	5.0	150.0	7.0	–	523.0
2164	canned in water, solids	1 OZ	**28**	**68.6**	**35.3**	**.7**	**.3**	**7.6**	**0**	**2.8**	**13.4**	**89.6**	–	**44.8**
			100	68.6	126.0	2.4	.9	27.1	.1	10.0	48.0	320.0	–	160.0
2165	[6]**HAZELNUTS**	1 OZ	**28**	–	**106.5**	**2.1**	**10.1**	**1.9**	**1.1**	**1.23**	**65.0**	**.4**	**15.8**	**98.1**
			100	–	380.2	7.5	35.9	6.9	6.1	43.8	232.1	1.4	56.5	350.3
2166	**HICKORYNUTS**	1 SM	**1**	**3.3**	**6.7**	**.1**	**.7**	**.1**	**0**	–	**3.6**	–	**1.6**	–
			100	3.3	673.0	13.2	68.7	12.8	1.9	–	360.0	–	160.0	–
	MACADAMIA NUTS													
2167	shelled	1 CUP	**140**	**3.0**	**1086.4**	**10.9**	**106.0**	**22.3**	**3.5**	**67.2**	**225.4**	–	–	**369.6**
			100	3.0	776.0	7.8	75.7	15.9	2.5	48.0	161.0	–	–	264.0
2168	chocolate coated	INDV	**2**	–	–	–	–	–	–	–	–	**2.6**	–	**4.0**
			100	–	–	–	–	–	–	–	–	128.0	–	202.0
	MIXED NUTS													
2169	shelled	1 AVG	**2**	–	**12.5**	**.3**	**1.2**	**.4**	–	**1.9**	**8.9**	**.3**	–	**11.2**
			100	–	626.0	16.6	59.2	18.0	–	94.0	446.0	14.0	–	560.0
2170	dry roasted	1 OZ	**28**	**1.2**	**165.2**	**6.2**	**14.9**	**5.3**	**.3**	**19.6**	**121.8**	**145.6**	**63.0**	**196.0**
			100	1.2	590.0	22.1	53.1	19.1	.9	70.0	435.0	520.0	225.0	700.0
	PEANUTS													
2171	raw, w/skins	1 CUP	**150**	**5.6**	**846.0**	**39.0**	**71.3**	**27.9**	**3.6**	**103.5**	**601.5**	**7.5**	**237.5**	**1011.0**
			100	5.6	564.0	26.0	47.5	18.6	2.4	69.0	401.0	5.0	158.3	674.0
2172	raw, w/o skins	1 CUP	**142**	**5.4**	**806.6**	**37.3**	**70.6**	**25.0**	**2.7**	**83.8**	**580.8**	**7.1**	**292.5**	**957.1**
			100	5.4	568.0	26.3	49.7	17.6	1.9	59.0	409.0	5.0	206.0	674.0
2173	roasted, w/skins, whole	1 AVG	**3**	**1.8**	**17.5**	**.8**	**1.5**	**.6**	**.1**	**2.2**	**12.2**	**.2**	**5.3**	**21.0**
			100	1.8	582.0	26.2	48.7	20.6	2.7	72.0	407.0	5.0	175.0	701.0
2174	roasted, salted	1 CUP	**144**	**1.6**	**842.4**	**37.4**	**71.7**	**27.1**	**3.4**	**106.6**	**577.4**	**601.9**	**252.0**	**970.6**
			100	1.6	585.0	26.0	49.8	18.8	2.4	74.0	401.0	418.0	175.0	674.0

Item No.	Minerals (Micro)			Fat-Soluble Vitamins			Water-Soluble Vitamins								
	Iron	Zinc	Copper	Vitamin A	Vitamin D	Vitamin E (Alpha Tocopherol)	Vitamin C	Thiamin	Ribo-flavin	Niacin	Panto-thenic Acid	Vit. B-6 (Pyri-doxine)	Folacin (Folic Acid)	Biotin	Vitamin B-12
	(mg)	(mg)	(mg)	(IU)	(IU)	(mg)	(mg)	(mg)	(mg)	(mg)	(mg)	(mg)	(mcg)	(mcg)	(mcg)
2148	4.76	5.92	1.53	–	0	9.10	14.00	1.34	.17	2.24	.32	.24	5.60	–	0
	3.40	4.23	1.09	–	0	6.50	10.00	.96	.12	1.60	.23	.17	4.00	–	0
2149	0.20	–	–	–	–	–	–	–	–	–	–	–	–	–	–
	6.80	–	–	–	–	–	–	–	–	–	–	–	–	–	–
2150	5.32	6.13	–	140.0	–	–	–	.60	.35	2.52	1.82	–	–	–	0
	3.80	4.38	–	100.0	–	–	–	.43	.25	1.80	1.30	–	–	–	0
2151	5.32	6.13	–	140.0	–	–	–	.60	.35	2.52	1.82	–	–	–	0
	3.80	4.38	–	100.0	–	–	–	.43	.25	1.80	1.30	–	–	–	0
2152	3.40	[6]1.00	.12	160.0	0	1.00	12.00	.44	.44	1.20	.95	.66	–	(2.62)	0
	1.70	.50	.06	80.0	0	.50	6.00	.22	.22	.60	.47	.33	–	(1.31)	0
2153	3.30	–	–	–	–	–	–	.32	.38	1.20	–	–	–	–	–
	3.30	–	–	–	–	–	–	.32	.38	1.20	–	–	–	–	–
2154	3.52	–	–	–	–	–	–	.25	.41	1.10	–	–	–	–	–
	3.20	–	–	–	–	–	–	.23	.37	1.00	–	–	–	–	–
2155	2.21	.07	.03	0	0	.91	3.90	.07	.03	.65	.26	.06	31.20	–	0
	1.70	.05	.02	0	0	.70	3.00	.05	.02	.50	.20	.04	24.00	–	0
2156	4.29	[6]1.30	(.87)	0	0	–	0	.08	.05	.78	.26	–	–	–	–
	3.30	1.00	(.67)	0	0	–	0	.06	.04	.60	.20	–	–	–	–
2157	2.60	–	–	0	0	–	0	.05	.04	.52	.26	–	–	–	–
	2.00	–	–	0	0	–	0	.04	.03	.40	.20	–	–	–	–
2158	3.90	–	–	0	0	–	4.88	.07	–	1.95	–	–	–	–	–
	1.60	–	–	0	0	–	2.00	.03	–	.80	–	–	–	–	–
2159	0.73	–	–	0	–	–	4.88	–	–	.24	.12	.08	–	–	0
	0.30	–	–	0	–	–	2.00	–	–	.10	.05	.03	–	–	0
2160	.31	–	–	0	–	–	0	.04	.03	.62	–	–	–	–	–
	1.10	–	–	0	–	–	0	.13	.12	2.20	–	–	–	–	–
2161	4.59	3.29	1.73	144.5	0	28.35	4.05	.62	.73	1.21	1.55	.74	97.20	–	0
	3.40	2.44	1.28	107.0	0	21.00	3.00	.46	.54	.90	1.15	.55	72.00	–	0
2162	12.26	–	–	0	–	–	–	.05	.05	.39	–	–	–	–	–
	43.80	–	–	0	–	–	–	.17	.16	1.40	–	–	–	–	–
2163	.34	–	–	84.0	–	–	7.00	.07	.03	.78	–	–	–	–	–
	1.20	–	–	300.0	–	–	25.00	.24	.12	2.80	–	–	–	–	–
2164	.14	–	–	37.3	–	–	–	.02	.01	.03	–	–	–	–	–
	.50	–	–	133.3	–	–	–	.07	.02	.10	–	–	–	–	–
2165	.31	.67	.62	0	0	5.90	Trace	.12	–	.26	.32	.15	20.18	–	0
	1.11	2.38	.22	0	0	21.10	Trace	.42	–	.92	1.16	.55	72.03	–	0
2166	.02	–	–	–	0	–	0	.01	–	–	–	–	–	–	–
	2.40	–	–	–	0	–	0	.53	–	–	–	–	–	–	–
2167	2.80	2.39	–	0	0	–	0	.48	.15	1.82	–	–	–	–	–
	2.00	1.71	–	0	0	–	0	.34	.11	1.30	–	–	–	–	–
2168	–	–	–	–	–	–	–	–	–	–	–	–	–	–	–
	–	–	–	–	–	–	–	–	–	–	–	–	–	–	–
2169	.07	[6] .06	–	.4	0	–	–	.01	0	.08	–	–	–	–	–
	3.40	3.10	–	20.0	0	–	–	.59	.13	4.00	–	–	–	–	–
2170	1.04	1.06	–	–	–	–	–	.05	.06	1.32	–	–	23.24	–	–
	3.70	3.80	–	–	–	–	–	.18	.23	4.70	–	–	83.00	–	–
2171	3.15	4.86	1.18	24.0	0	14.55	0	1.71	.20	25.80	4.20	–	–	–	0
	2.10	3.24	.78	16.0	0	9.70	0	1.14	.13	17.20	2.80	–	–	–	0
2172	2.84	(4.12)	(.38)	0	0	13.77	0	1.41	.19	22.44	3.98	(.75)	(156.21)	–	0
	2.00	(2.90)	(.27)	0	0	9.70	0	.99	.13	15.80	2.80	(.53)	(110.01)	–	0
2173	.07	–	.01	10.8	0	.29	0	.01	0	.51	.06	.01	3.18	1.02	0
	2.20	–	.27	360.0	0	9.70	0	.32	.13	17.10	2.10	.40	106.00	34.00	0
2174	3.02	(4.26)	(.39)	0	0	13.97	0	.46	.19	24.77	3.02	.58	152.64	48.96	0
	2.10	(2.96)	(.27)	0	0	9.70	0	.32	.13	17.20	2.10	.40	106.00	34.00	0

(Continued)

TABLE F-36 *(Continued)*

Item No.	Food Item Name	Approximate Measure	Weight	Moisture	Food Energy Calories	Protein	Fats	Carbo-hydrates	Fiber	Minerals (Macro) Calcium	Phos-phorus	Sodium	Mag-nesium	Potas-sium
			(g)	(%)	(kcal)	(g)	(g)	(g)	(g)	(mg)	(mg)	(mg)	(mg)	(mg)
	PEANUTS (Continued)													
2175	blanched, salted	1 PKG	32	1.4	192.3	8.8	14.6	6.5	—	—	—	—	—	—
			100	1.4	601.0	27.5	45.5	20.2	—	—	—	—	—	—
2176	boiled	1 TBLS	15	36.4	56.4	2.3	4.7	2.2	.3	6.5	27.2	.6	—	69.3
			100	36.4	376.0	15.5	31.5	14.5	1.8	43.0	181.0	4.0	—	462.0
2177	redskin	1 PKG	35	2.2	212.8	9.8	16.8	5.7	—	—	—	—	—	—
			100	2.2	608.0	28.0	47.9	16.2	—	—	—	—	—	—
2178	butter, creamy	1 TBLS	32	—	190.0	8.1	16.2	5.4	(5.9)	(11.8)	105.7	190.0	(57.6)	200.0
			100	—	593.8	25.3	50.6	16.9	(7.6)	(36.8)	330.9	593.8	(180.3)	625.0
2179	butter, salted, w/small amount of added fat	1 TBLS	14	1.8	81.3	3.7	6.9	2.4	.3	8.8	57.0	85.0	24.2	93.3
			100	1.8	581.0	26.1	49.4	17.2	1.9	63.0	407.0	607.0	173.0	670.0
2180	butter, salted, sweetened, w/moderate amount of added fat	1 TBLS	15	1.7	88.4	3.8	7.6	2.8	.3	8.9	57.0	90.8	26.0	94.1
			100	1.7	589.0	25.2	50.6	18.8	1.8	59.0	380.0	605.0	173.0	627.0
2181	spread	1 TBLS	15	2.2	90.2	3.0	7.8	3.3	.2	7.5	48.3	89.6	—	79.5
			100	2.2	601.0	20.3	52.1	22.0	1.5	50.0	322.0	597.0	—	530.0
2182	**PECANS,** unsalted	1 CUP	108	3.4	797.7	10.2	77.1	15.8	2.5	78.8	312.1	11.9	118.8	651.2
			100	3.4	738.6	9.4	71.4	14.6	2.3	73.0	289.0	11.0	110.0	603.0
2183	**PILINUTS**	1 TBLS	7	6.3	45.2	.8	4.4	.6	.2	9.8	38.8	.2	—	34.2
			100	6.3	646.0	11.4	63.0	8.4	2.7	140.0	554.0	3.0	—	489.0
2184	**PINENUTS,** pignolias	1 TBLS	7	5.6	44.1	2.2	3.6	.8	.1	—	—	—	—	—
			100	5.6	630.0	31.1	51.0	11.6	.9	—	—	—	—	—
2185	**PINENUTS,** pinon	1 TBLS	7	3.1	41.5	.9	3.6	1.4	.1	.8	42.3	—	—	—
			100	3.1	593.0	13.0	51.0	20.5	1.1	12.0	604.0	—	—	—
2186	**PISTACHIO NUTS,** shelled	1 CUP	125	5.3	793.8	23.6	67.0	23.8	2.4	163.8	625.0	—	197.5	1215.0
			100	5.3	635.0	18.9	53.6	19.0	1.9	131.0	500.0	—	158.0	972.0
2187	[2]**POPPY SEEDS**	1 TSP	3	6.8	15.9	.5	1.3	.7	.2	43.4	25.4	.6	9.6	21.0
			100	6.8	533.0	18.0	44.7	23.7	6.2	1448.0	848.0	21.0	320.0	700.0
2188	[3]**PUMPKIN & SQUASH SEED KERNELS,** dry	1 CUP	140	—	774.2	40.6	65.4	21.0	2.7	71.4	1601.6	—	—	—
			100	—	553.0	29.0	46.7	15.0	1.9	51.0	1144.0	—	—	—
	SESAME SEEDS													
2189	dry, whole	1 OZ	28	5.4	157.6	5.2	13.7	6.0	1.8	324.8	172.5	16.8	50.7	203.0
			100	5.4	563.0	18.6	49.1	21.6	6.3	1160.0	616.0	60.0	181.0	725.0
2190	dry, decorticated	1 OZ	28	4.8	163.0	7.4	15.3	2.6	.8	36.7	217.3	11.2	97.1	114.0
			100	4.8	582.0	26.4	54.8	9.4	3.0	131.0	776.0	40.0	347.0	407.0
2191	**SUNFLOWER SEED KERNELS,** dry, hulled	1 CUP	145	4.8	812.0	33.4	68.6	28.9	5.5	174.0	1213.7	43.5	55.1	1334.0
			100	4.8	560.0	23.0	47.3	19.9	3.8	120.0	837.0	30.0	38.0	920.0
	WALNUTS													
2192	black, shelled, chopped	1 CUP	125	3.1	847.5	25.6	74.5	18.5	2.1	—	712.5	3.8	237.5	575.0
			100	3.1	678.0	20.5	59.6	14.8	1.7	—	570.0	3.0	190.0	460.0
2193	Persian or English, shelled, chopped	1 CUP	120	3.5	832.8	18.0	76.1	19.0	2.5	118.8	456.0	2.4	172.8	540.0
			100	3.5	694.0	15.0	63.4	15.8	2.1	99.0	380.0	2.0	144.0	450.0
	[4]**WATERMELON SEEDS**													
2194	whole, dried	1 OZ	28	4.6	150.1	6.4	11.5	7.7	.7	23.0	135.2	—	—	169.7
			100	4.6	536.0	22.7	41.2	27.5	2.5	82.0	483.0	—	—	606.0
2195	whole, pickled in soya sauce	1 OZ	28	4.4	142.2	7.3	9.9	8.4	.6	12.3	191.5	—	—	227.6
			100	4.4	508.0	26.1	35.3	30.1	2.0	44.0	684.0	—	—	813.0
2196	sugared	1 OZ	28	2.5	167.4	6.6	15.0	4.8	.5	19.3	226.8	—	—	380.5
			100	2.5	598.0	23.5	53.4	17.1	1.7	69.0	810.0	—	—	1359.0
	PICKLES & RELISHES													
2197	**CRANBERRY-ORANGE RELISH,** uncooked	1 TBLS	15	53.6	26.7	.1	.1	6.8	—	2.9	1.2	.2	—	10.8
			100	53.6	178.0	.4	.4	45.4	—	19.0	8.0	1.0	—	72.0
2198	**KIM CHEE** (vegetable pickle w/garlic, red pepper & ginger)	1 CUP	133	—	—	—	—	—	—	—	—	904.4	—	203.5
			100	—	—	—	—	—	—	—	—	680.0	—	153.0
	OLIVES													
2199	green	1 MED	5	78.2	5.8	.1	.6	.1	.1	3.1	.9	120.0	1.1	2.8
			100	78.2	116.0	1.4	12.7	1.3	1.3	61.0	17.0	2400.0	22.0	55.0
2200	ripe, Ascolano	1 MED	7	80.0	9.0	.1	1.0	.2	.1	5.9	1.1	56.9	—	2.4
			100	80.0	129.0	1.1	13.8	2.6	1.4	84.0	16.0	813.0	—	34.0
2201	ripe, Manzanilla	1 LG	6	80.0	7.7	.1	.8	.2	.1	5.0	1.0	48.8	—	2.0
			100	80.0	129.0	1.1	13.8	2.6	1.4	84.0	16.0	813.0	—	34.0

Item No.	Minerals (Micro) Iron	Zinc	Copper	Fat-Soluble Vitamins Vitamin A	Vitamin D	Vitamin E (Alpha Tocopherol)	Water-Soluble Vitamins Vitamin C	Thiamin	Ribo-flavin	Niacin	Panto-thenic Acid	Vit. B-6 (Pyri-doxine)	Folacin (Folic Acid)	Biotin	Vitamin B-12
	(mg)	(mg)	(mg)	(IU)	(IU)	(mg)	(mg)	(mg)	(mg)	(mg)	(mg)	(mg)	(mcg)	(mcg)	(mcg)
2175	**.63**	–	–	–	–	–	–	**.02**	**.05**	**1.11**	–	–	–	–	–
	1.98	–	–	–	–	–	–	.05	.15	3.46	–	–	–	–	–
2176	**.20**	–	–	–	–	–	**0**	**.07**	**.01**	**1.50**	**.31**	–	–	–	**0**
	1.30	–	–	–	–	–	0	.48	.08	10.00	2.05	–	–	–	0
2177	–	–	–	–	–	–	–	–	–	–	–	–	–	–	–
	–	–	–	–	–	–	–	–	–	–	–	–	–	–	–
2178	**(.67)**	**.93**	**(.22)**	**0**	**0**	**(1.49)**	**Trace**	**(.05)**	**(.04)**	**(4.89)**	**(.74)**	**(.16)**	**(16.90)**	–	**0**
	(2.09)	2.90	(.69)	0	0	(4.66)	Trace	(.16)	(.11)	(15.30)	(2.32)	(.51)	(52.90)	–	0
2179	**.28**	**.41**	**.09**	**0**	**0**	–	**0**	**.02**	**.02**	**2.20**	**.35**	**.05**	**11.06**	**5.46**	**0**
	2.00	2.90	.61	0	0	–	0	.13	.13	15.70	2.50	.33	79.00	39.00	0
2180	**.29**	**.44**	**.09**	**0**	**0**	–	**0**	**.02**	**.02**	**2.21**	**.38**	**.05**	**11.85**	**5.85**	**0**
	1.90	2.90	.61	0	0	–	0	.12	.12	14.70	2.50	.33	79.00	39.00	0
2181	**.23**	–	–	–	–	–	**0**	**.02**	**.02**	**1.86**	–	–	–	–	–
	1.50	–	–	–	–	–	0	.10	.10	12.40	–	–	–	–	–
2182	**2.59**	**4.43**	**1.19**	**140.4**	**0**	**1.30**	**2.16**	**.93**	**.14**	**.97**	**1.84**	**.20**	**25.92**	**29.16**	**0**
	2.40	4.10	1.10	130.0	0	1.20	2.00	.86	.13	.90	1.71	.18	24.00	27.00	0
2183	**.25**	–	–	**2.8**	–	–	–	**.06**	**.01**	**.04**	–	–	–	–	–
	3.40	–	–	40.0	–	–	–	.88	.09	.50	–	–	–	–	–
2184	–	[6] **.46**	–	–	–	–	–	**.04**	–	–	–	–	–	–	–
	–	6.50	–	–	–	–	–	.62	–	–	–	–	–	–	–
2185	**.36**	[6] **.46**	–	**2.1**	–	–	–	**.09**	**.02**	**.32**	–	–	–	–	–
	5.20	6.50	–	30.0	–	–	–	1.28	.23	4.50	–	–	–	–	–
2186	**9.13**	–	–	**287.5**	–	–	**0**	**.84**	–	**1.75**	–	–	**72.50**	–	–
	7.30	–	–	230.0	–	–	0	.67	–	1.40	–	–	58.00	–	–
2187	**.28**	**.31**	**.05**	**0**	–	**.05**	**.36**	**.03**	**.01**	**.03**	–	**.01**	–	–	**0**
	9.40	10.23	1.63	0	–	1.80	12.00	.95	.18	1.00	–	.45	–	–	0
2188	**15.68**	–	–	**98.0**	–	–	–	**.34**	**.27**	**3.36**	–	**.13**	–	–	**0**
	11.20	–	–	70.0	–	–	–	.24	.19	2.40	–	.09	–	–	0
2189	**2.94**	[6]**1.48**	–	**8.4**	–	–	**0**	**.27**	**.07**	**1.51**	–	–	–	–	**0**
	10.50	5.30	–	30.0	–	–	0	.98	.24	5.40	–	–	–	–	0
2190	**2.18**	**2.87**	–	**18.5**	–	–	**0**	**.20**	**.02**	**1.31**	**.19**	**.04**	–	–	**0**
	7.80	10.25	–	66.0	–	–	0	.72	.09	4.68	.68	.15	–	–	0
2191	**10.30**	**6.64**	**2.57**	**72.5**	–	**18.85**	–	**2.84**	**.33**	**7.83**	**2.03**	**1.81**	–	–	**0**
	7.10	4.58	1.77	50.0	–	13.00	–	1.96	.23	5.40	1.40	1.25	–	–	0
2192	**7.50**	–	–	**375.0**	–	–	–	**.28**	**.14**	**.88**	–	–	–	–	–
	6.00	–	–	300.0	–	–	–	.22	.11	.70	–	–	–	–	–
2193	**3.72**	**3.84**	**1.68**	**36.0**	–	**.48**	**2.40**	**.40**	**.16**	**1.08**	**1.08**	**.88**	**79.20**	**44.40**	**0**
	3.10	3.20	1.40	30.0	–	.40	2.00	.33	.13	.90	.90	.73	66.00	37.00	0
2194	**2.16**	–	–	**4.7**	–	–	–	**.06**	**.03**	**.73**	–	–	–	–	–
	7.70	–	–	16.7	–	–	–	.22	.10	2.60	–	–	–	–	–
2195	**2.21**	–	–	–	–	–	–	–	–	–	–	–	–	–	–
	7.90	–	–	–	–	–	–	–	–	–	–	–	–	–	–
2196	**3.33**	–	–	–	–	–	–	–	–	–	–	–	–	–	–
	11.90	–	–	–	–	–	–	–	–	–	–	–	–	–	–
2197	**.06**	–	–	**10.5**	–	–	**2.70**	**.01**	**0**	**.02**	–	–	–	–	–
	.40	–	–	70.0	–	–	18.00	.03	.02	.10	–	–	–	–	–
2198	–	–	–	–	–	–	–	–	–	–	–	–	–	–	–
	–	–	–	–	–	–	–	–	–	–	–	–	–	–	–
2199	**.08**	**0**	**.08**	**15.0**	**0**	–	–	–	–	–	**0**	–	**.05**	–	**0**
	1.60	.07	1.60	300.0	0	–	–	–	–	–	.02	–	1.00	–	0
2200	**.11**	–	–	**4.2**	–	–	–	–	–	–	**0**	**0**	**.07**	–	**0**
	1.60	–	–	60.0	–	–	–	–	–	–	.02	.01	1.00	–	0
2201	**.10**	–	–	**3.6**	–	–	–	–	–	–	**0**	**0**	**.06**	–	**0**
	1.60	–	–	60.0	–	–	–	–	–	–	.02	.01	1.00	–	0

(Continued)

TABLE F-36 *(Continued)*

Item No.	Food Item Name	Approximate Measure	Weight	Moisture	Food Energy Calories	Protein	Fats	Carbo-hydrates	Fiber	Minerals (Macro) Calcium	Phos-phorus	Sodium	Mag-nesium	Potas-sium
			(g)	(%)	(kcal)	(g)	(g)	(g)	(g)	(mg)	(mg)	(mg)	(mg)	(mg)
	OLIVES (Continued)													
2202	ripe, Mission	1 LG	5	73.0	9.2	.1	1.0	.2	.1	5.3	.9	37.5	—	1.4
			100	73.0	184.0	1.2	20.1	3.2	1.5	106.0	17.0	750.0	—	27.0
2203	ripe, salt cured, oil coated, Greek style	1 MED	2	43.8	6.8	0	.7	.2	.1	—	.6	65.8	—	—
			100	43.8	338.0	2.2	35.8	8.7	3.8	—	29.0	3288.0	—	—
2204	ripe, Sevillano	1 LG	10	84.4	9.3	.1	1.0	.3	.1	7.4	2.0	82.8	—	4.4
			100	84.4	93.0	1.1	9.5	2.7	1.2	74.0	20.0	828.0	—	44.0
	PICKLES													
2205	relish, sour	1 TBLS	15	93.0	2.9	.1	.1	.4	.2	4.4	3.0	—	—	—
			100	93.0	19.0	.7	.9	2.7	1.1	29.0	20.0	—	—	—
2206	relish, sweet	1 TBLS	15	63.0	20.7	.1	.1	5.1	.1	3.0	2.1	106.8	—	—
			100	63.0	138.0	.5	.6	34.0	.8	20.0	14.0	712.0	—	—
2207	chowchow, sour	1 OZ	28	87.6	8.1	.4	.4	1.1	.2	9.0	14.8	374.6	—	—
			100	87.6	29.0	1.4	1.3	4.1	.6	32.0	53.0	1338.0	—	—
2208	chowchow, sweet	1 OZ	28	68.9	32.5	.4	.3	7.6	.3	6.4	6.2	147.6	—	—
			100	68.9	116.0	1.5	.9	27.0	.9	23.0	22.0	527.0	—	—
2209	cucumber, dill	1 LG	135	93.3	14.9	.9	.3	3.0	.7	35.1	28.4	1927.8	16.2	270.0
			100	93.3	11.0	.7	.2	2.2	.5	26.0	21.0	1428.0	12.0	200.0
2210	cucumber, fresh, bread & butter	1 SLICE	8	78.7	5.8	.1	0	1.4	0	2.6	2.2	53.8	—	—
			100	78.7	73.0	.9	.2	17.9	.5	32.0	27.0	673.0	—	—
2211	cucumber, sour	1 LG	135	94.8	13.5	.7	.3	2.7	.7	23.0	20.3	1826.6	—	—
			100	94.8	10.0	.5	.2	2.0	.5	17.0	15.0	1353.0	—	—
2212	cucumber, sweet	1 LG	100	60.7	146.0	.7	.4	36.5	.5	12.0	16.0	—	1.0	—
			100	60.7	146.0	.7	.4	36.5	.5	12.0	16.0	—	1.0	—
	SANDWICH SPREAD													
2213	w/chopped pickle, regular	1 TBLS	20	45.4	75.8	.1	7.2	3.2	.1	3.0	4.0	125.2	—	18.4
			100	45.4	379.0	.7	36.2	15.9	.4	15.0	20.0	626.0	—	92.0
2214	w/chopped pickle, lo-cal (5 kcal/tsp)	1 TBLS	20	80.2	22.4	.2	1.8	1.6	.1	3.0	4.0	125.2	—	18.4
			100	80.2	112.0	1.0	9.0	8.0	.4	15.0	20.0	626.0	—	92.0
2215	no meat, unspecified oil	1 TBLS	15	40.8	58.4	.1	5.1	3.4	—	—	—	—	—	—
			100	40.8	389.0	.9	34.0	22.4	—	—	—	—	—	—
	SALADS													
2216	**CARROT & RAISIN**	1 OZ	28	59.9	22.1	.4	.5	4.5	.2	9.2	10.8	43.2	5.0	29.8
			100	59.9	78.9	1.4	1.7	16.2	.7	32.8	38.5	154.2	17.9	106.3
2217	**CHICKEN SALAD**	1 CUP	200	—	254.0	22.0	15.0	7.8	.6	44.0	136.0	—	—	—
			100	—	127.0	11.0	7.5	3.9	.3	22.0	68.0	—	—	—
2218	**COLESLAW**	1 OZ	28	89.8	7.7	.4	.1	1.6	.2	13.2	8.3	9.9	3.7	65.2
			100	89.8	27.6	1.3	.5	5.7	.8	47.1	29.7	35.3	13.1	232.8
2219	w/French dressing, homemade, w/corn oil	1 CUP	120	80.6	154.8	1.3	14.8	6.1	.8	50.4	30.0	157.2	—	236.4
			100	80.6	129.0	1.1	12.3	5.1	.7	42.0	25.0	131.0	—	197.0
2220	w/French dressing, commercial	1 CUP	120	82.6	114.0	1.4	8.8	9.1	.8	50.4	31.2	321.6	—	246.0
			100	82.6	95.0	1.2	7.3	7.6	.7	42.0	26.0	268.0	—	205.0
2221	w/mayonnaise	1 CUP	120	79.0	172.8	1.6	16.8	5.8	.8	52.8	34.8	144.0	—	238.8
			100	79.0	144.0	1.3	14.0	4.8	.7	44.0	29.0	120.0	—	199.0
2222	w/salad dressing	1 CUP	120	82.9	118.8	1.4	9.5	8.5	.8	51.6	33.6	148.8	14.4	230.4
			100	82.9	99.0	1.2	7.9	7.1	.7	43.0	28.0	124.0	12.0	192.0
2223	**CORNSALAD**, raw	1 CUP	140	92.8	29.4	2.8	.6	5.0	1.1	—	—	—	18.2	—
			100	92.8	21.0	2.0	.4	3.6	.8	—	—	—	13.0	—
2224	**COTTAGE CHEESE & VEGETABLES**	1 SERV	105	—	94.7	11.7	3.5	3.6	.2	85.8	134.7	202.1	—	125.5
			100	—	90.2	11.1	3.3	3.4	.2	81.7	128.3	192.4	—	119.5
2225	**CRAB SALAD**	1 OZ	28	—	40.6	3.3	2.4	1.4	—	10.6	36.1	—	7.3	—
			100	—	145.0	11.8	8.5	4.9	—	38.0	129.0	—	26.0	—
2226	**EGG SALAD**	1 CUP	240	—	780.2	21.4	75.6	3.6	—	108.0	324.0	—	21.6	—
			100	—	325.1	8.9	31.5	1.5	—	45.0	135.0	—	9.0	—
	GELATIN SALAD													
2227	w/chopped vegetables	1 SERV	164	—	114.8	2.2	5.7	15.1	.7	24.6	27.9	—	—	—
			100	—	70.0	1.3	3.5	9.2	.4	15.0	17.0	—	—	—
2228	w/fruit	1 SERV	188	—	139.1	2.1	5.6	21.6	.6	22.6	24.4	—	—	—
			100	—	74.0	1.1	3.0	11.5	.3	12.0	13.0	—	—	—
2229	w/fruit cocktail	1 OZ	28	81.6	15.9	2.1	.7	2.1	0	1.3	1.3	11.4	.9	17.2
			100	81.6	56.9	7.3	2.5	7.4	.1	4.8	4.5	40.6	3.1	61.5
2230	**LETTUCE & TOMATO**	1 SERV	100	—	19.0	1.4	0	4.5	.7	18.0	32.0	—	—	—
			100	—	19.0	1.4	0	4.5	.7	18.0	32.0	—	—	—
2231	**LOBSTER SALAD**	1 SERV	260	80.3	286.0	26.3	16.6	5.0	—	93.6	247.0	322.4	—	686.4
			100	80.3	110.0	10.1	6.4	2.3	—	36.0	95.0	124.0	—	264.0

Item No.	Minerals (Micro)			Fat-Soluble Vitamins			Water-Soluble Vitamins								
	Iron	Zinc	Copper	Vitamin A	Vitamin D	Vitamin E (Alpha Tocopherol)	Vitamin C	Thiamin	Ribo-flavin	Niacin	Panto-thenic Acid	Vit. B-6 (Pyri-doxine)	Folacin (Folic Acid)	Biotin	Vitamin B-12
	(mg)	(mg)	(mg)	(IU)	(IU)	(mg)	(mg)	(mg)	(mg)	(mg)	(mg)	(mg)	(mcg)	(mcg)	(mcg)
2202	**.09**	–	–	**3.5**	**0**	–	–	–	–	–	**0**	**0**	**.05**	–	**0**
	1.70	–	–	70.0	0	–	–	–	–	–	.02	.01	1.00	–	0
2203	–	–	–	–	–	–	–	–	–	–	–	–	–	–	–
	–	–	–	–	–	–	–	–	–	–	–	–	–	–	–
2204	**.16**	–	–	**6.0**	–	–	–	–	–	–	**0**	**0**	**.10**	–	**0**
	1.60	–	–	60.0	–	–	–	–	–	–	.02	.01	1.00	–	0
2205	**.17**	–	–	–	–	–	–	–	–	–	–	–	–	–	–
	1.10	–	–	–	–	–	–	–	–	–	–	–	–	–	–
2206	**.12**	–	–	–	–	–	–	–	–	–	–	–	–	–	–
	.80	–	–	–	–	–	–	–	–	–	–	–	–	–	–
2207	**.73**	–	–	–	–	–	–	–	–	–	–	**0**	–	–	**0**
	2.60	–	–	–	–	–	–	–	–	–	–	.01	–	–	0
2208	**.42**	–	–	–	–	–	–	–	–	–	–	**0**	–	–	**0**
	1.50	–	–	–	–	–	–	–	–	–	–	.01	–	–	0
2209	**1.35**	**.37**	–	**135.0**	**0**	–	**8.10**	–	**.03**	**.10**	–	**.01**	**1.35**	–	**0**
	1.00	.27	–	100.0	0	–	6.00	–	.02	.07	–	.01	1.00	–	0
2210	**.14**	–	–	**11.2**	–	–	**0.72**	–	**0**	–	–	**0**	–	–	**0**
	1.80	–	–	140.0	–	–	9.00	–	.03	–	–	.01	–	–	0
2211	**4.32**	–	–	**135.0**	–	–	**9.45**	–	**.03**	–	–	**.01**	–	–	**0**
	3.20	–	–	100.0	–	–	7.00	–	.02	–	–	.01	–	–	0
2212	**1.20**	**.14**	–	**90.0**	**0**	–	**6.00**	**0**	**.02**	–	–	**.01**	–	–	**0**
	1.20	.14	–	90.0	0	–	6.00	0	.02	–	–	.01	–	–	0
2213	**.14**	–	–	**56.0**	–	–	**1.20**	**0**	**.01**	–	–	–	–	–	–
	.70	–	–	280.0	–	–	6.00	.01	.03	–	–	–	–	–	–
2214	**.14**	–	–	**56.0**	–	–	**1.20**	**0**	**.01**	–	–	–	–	–	–
	.70	–	–	280.0	–	–	6.00	.01	.03	–	–	–	–	–	–
2215	–	–	–	–	–	–	–	–	–	–	–	–	–	–	–
	–	–	–	–	–	–	–	–	–	–	–	–	–	–	–
2216	**.24**	**.04**	–	**2068.8**	**.34**	–	**1.54**	**.02**	**.02**	**.14**	**.06**	**.03**	**1.51**	–	**0**
	.85	.16	–	7388.5	1.23	–	5.48	.06	.06	.51	.21	.10	5.40	–	.01
2217	**1.80**	–	–	**398.0**	–	–	**6.80**	**.06**	**.18**	**4.64**	–	–	–	–	–
	.90	–	–	199.0	–	–	3.40	.03	.09	2.32	–	–	–	–	–
2218	**.12**	**.04**	–	**318.9**	**0**	–	**10.98**	**.01**	**.01**	**.09**	**.07**	**.04**	**22.22**	–	**0**
	.42	.14	–	1139.0	0	–	39.20	.05	.05	.33	.23	.15	79.34	–	0
2219	**.48**	–	–	**132.0**	–	–	**34.80**	**.05**	**.05**	**.36**	–	–	–	–	–
	.40	–	–	110.0	–	–	29.00	.04	.04	.30	–	–	–	–	–
2220	**.48**	–	–	**132.0**	–	–	**34.80**	**.05**	**.05**	**.36**	–	–	–	–	–
	.40	–	–	110.0	–	–	29.00	.04	.04	.30	–	–	–	–	–
2221	**.48**	**.29**	–	**192.0**	–	–	**34.80**	**.06**	**.06**	**.36**	–	–	–	–	–
	.40	.24	–	160.0	–	–	29.00	.05	.05	.30	–	–	–	–	–
2222	**.48**	**.29**	–	**180.0**	–	–	**34.80**	**.06**	**.06**	**.36**	–	–	–	–	–
	.40	.24	–	150.0	–	–	29.00	.05	.05	.30	–	–	–	–	–
2223	–	–	–	–	–	–	–	–	–	–	–	–	–	–	–
	–	–	–	–	–	–	–	–	–	–	–	–	–	–	–
2224	**.37**	–	–	**878.0**	–	–	**5.68**	**.03**	**.22**	**2.35**	–	–	–	–	–
	.35	–	–	836.2	–	–	5.41	.03	.21	2.24	–	–	–	–	–
2225	**.17**	–	–	**27.2**	–	–	**.56**	**.02**	**.02**	**.36**	–	**.06**	–	–	**1.86**
	.60	–	–	97.0	–	–	2.00	.06	.06	1.30	–	.21	–	–	6.64
2226	**3.84**	–	–	**1077.6**	–	–	**0**	**.14**	**.50**	**0**	–	**1.92**	–	–	**2.28**
	1.60	–	–	449.0	–	–	0	.06	.21	0	–	.80	–	–	.95
2227	**.49**	–	–	**1976.2**	–	–	**7.87**	**.03**	**.07**	**.30**	–	–	–	–	–
	.30	–	–	1205.0	–	–	4.80	.02	.04	.18	–	–	–	–	–
2228	**.56**	–	–	**391.0**	–	–	**15.98**	**.04**	**.04**	**.28**	–	–	–	–	–
	.30	–	–	208.0	–	–	8.50	.02	.02	.15	–	–	–	–	–
2229	**.17**	**0**	–	**23.5**	**0**	–	**1.67**	**0**	**0**	**.04**	**0**	**0**	**0**	–	**0**
	.61	0	–	83.8	0	–	5.97	0	0	.15	0	.01	0	–	0
2230	**.70**	–	–	**1084.0**	–	–	**19.00**	**.06**	**.07**	**.50**	–	–	–	–	–
	.70	–	–	1084.0	–	–	19.00	.06	.07	.50	–	–	–	–	–
2231	**2.34**	–	–	–	–	–	**46.80**	**.23**	**.21**	–	–	–	–	–	–
	.90	–	–	–	–	–	18.00	.09	.08	–	–	–	–	–	–

(Continued)

TABLE F-36 *(Continued)*

Item No.	Food Item Name	Approximate Measure	Weight	Moisture	Food Energy Calories	Protein	Fats	Carbo-hydrates	Fiber	Minerals (Macro) Calcium	Phos-phorus	Sodium	Mag-nesium	Potas-sium
			(g)	(%)	(kcal)	(g)	(g)	(g)	(g)	(mg)	(mg)	(mg)	(mg)	(mg)
2232	**MACARONI SALAD,** w/onion & mayonnaise	1 CUP	**190**	—	**334.4**	**7.7**	**11.8**	**48.5**	**0**	**20.9**	**102.6**	—	—	—
			100	—	176.0	4.1	6.2	25.5	0	11.0	54.0	—	—	—
2233	**POTATO SALAD**	1 OZ	**28**	—	**42.3**	**.9**	**2.3**	**4.6**	**1.5**	**5.2**	**18.1**	**50.0**	**.9**	**37.3**
			100	—	150.9	3.3	8.4	16.4	5.4	18.7	64.7	178.7	3.2	133.2
2234	w/mayonnaise, French dressing, hard cooked egg, & seasoning	1 CUP	**250**	**72.4**	**362.5**	**7.5**	**23.0**	**33.5**	**1.0**	**47.5**	**157.5**	**1200.0**	—	**740.0**
			100	72.4	145.0	3.0	9.2	13.4	.4	19.0	63.0	480.0	—	296.0
2235	w/cooked salad dressing, seasonings, margarine	1 CUP	**250**	**76.0**	**247.5**	**6.8**	**7.0**	**40.8**	**1.0**	**80.0**	**160.0**	**1320.0**	—	**797.5**
			100	76.0	99.0	2.7	2.8	16.3	.4	32.0	64.0	528.0	—	319.0
2236	**TUNA SALAD,** w/celery, mayonnaise, pickle, onion, & egg	1 CUP	**200**	**69.8**	**340.0**	**29.2**	**21.0**	**7.0**	—	**40.0**	**284.0**	—	—	—
			100	69.8	170.0	14.6	10.5	3.5	—	20.0	142.0	—	—	—
2237	**TOSSED SALAD**	1 OZ	**28**	**85.2**	**3.9**	**.2**	**.3**	**.9**	**.1**	**5.7**	**6.0**	**4.9**	**2.9**	**50.5**
			100	85.2	13.9	.8	1.0	3.1	.5	20.5	21.4	17.6	10.4	180.4
2238	**WALDORF SALAD**	1 SERV	**108**	—	**137.3**	**.9**	**7.5**	**18.7**	**.9**	**18.8**	**32.7**	**77.5**	—	**200.2**
			100	—	127.1	.9	7.0	17.3	.8	17.4	30.3	71.8	—	185.3
	SALAD DRESSINGS													
	BLUE & ROQUEFORT													
2239	regular, w/salt	1 TBLS	**14**	**32.3**	**70.6**	**.7**	**7.3**	**1.0**	**0**	**11.3**	**10.4**	**153.2**	—	**5.2**
			100	32.3	504.0	4.8	52.3	7.4	.1	81.0	74.0	1094.0	—	37.0
2240	lo-fat (5 kcal/tsp), w/salt	1 TBLS	**14**	**83.7**	**10.6**	**.4**	**.8**	**.6**	**0**	**9.0**	**6.6**	**155.1**	—	**4.8**
			100	83.7	76.0	3.0	5.9	4.1	.1	64.0	47.0	1108.0	—	34.0
2241	lo-cal (1 kcal/tsp), w/salt	1 TBLS	**14**	**93.1**	**2.7**	**.2**	**.2**	**.2**	**0**	**4.9**	**3.4**	**158.8**	—	**4.1**
			100	93.1	19.0	1.4	1.1	1.4	.1	35.0	24.0	1134.0	—	29.0
2242	**COOKED,** homemade, w/margarine	1 TBLS	**14**	**69.2**	**21.8**	**.6**	**1.3**	**2.1**	**0**	**11.8**	**12.2**	**102.8**	—	**16.9**
			100	69.2	156.0	4.2	9.5	14.9	0	84.0	87.0	734.0	—	121.0
	FRENCH													
2243	homemade	1 TBLS	**14**	**24.2**	**88.3**	**0**	**9.8**	**.5**	**0**	**.8**	**.4**	**92.1**	**(1.9)**	**3.3**
			100	24.2	630.9	.1	70.2	3.4	.2	5.8	2.9	658.0	(13.3)	23.8
2244	regular	1 TBLS	**14**	**38.1**	**60.2**	**.1**	**5.7**	**2.5**	**.1**	**1.5**	**2.0**	**191.8**	**1.4**	**11.1**
			100	38.1	429.7	.6	41.0	17.5	.8	11.0	14.0	1370.0	10.0	79.0
2245	lo-cal (10 kcal/tsp), medium fat	1 TBLS	**14**	**78.8**	**21.8**	**.1**	**2.4**	**.2**	**0**	**1.5**	**2.0**	**110.2**	**1.4**	**11.1**
			100	78.8	156.0	.7	16.9	1.2	.3	11.0	14.0	787.0	10.0	79.0
2246	lo-cal (5 kcal/tsp), low fat	1 TBLS	**14**	**69.4**	**18.8**	**0**	**.8**	**3.0**	**0**	**1.5**	**2.0**	**110.2**	**1.4**	**11.1**
			100	69.4	134.2	.2	5.8	21.7	.3	11.0	14.0	787.0	10.0	79.0
2247	lo-cal (1 kcal/tsp), low fat	1 TBLS	**14**	**95.2**	**1.4**	**.1**	**0**	**.3**	**0**	**1.5**	**2.0**	**110.2**	**1.4**	**11.1**
			100	95.2	10.0	.4	.2	1.8	.3	11.0	14.0	787.0	10.0	79.0
	ITALIAN													
2248	regular	1 TBLS	**14**	**38.4**	**65.4**	**.1**	**6.8**	**1.4**	**0**	**1.4**	**.7**	**110.2**	—	**2.1**
			100	38.4	467.3	.7	48.3	10.2	.2	10.0	5.0	787.0	—	15.0
2249	lo-cal (2 kcal/tsp)	1 TBLS	**14**	**81.9**	**14.8**	**0**	**1.4**	**.7**	**0**	**.3**	**.7**	**110.2**	—	**2.1**
			100	81.9	105.4	.1	9.8	4.9	.3	2.0	5.0	787.0	—	15.0
	MAYONNAISE													
2250	regular	1 TBLS	**14**	**39.9**	**54.6**	**.1**	**4.7**	**3.3**	**0**	**2.0**	**3.6**	**99.5**	**.3**	**1.3**
			100	39.9	389.7	.9	33.4	23.9	0	14.0	26.0	710.8	2.0	9.0
2251	lo-cal (8 kcal/tsp)	1 TBLS	**14**	**80.7**	**19.0**	**.2**	**1.8**	**.7**	**.1**	**2.5**	**3.9**	**16.5**	—	**1.3**
			100	80.7	136.0	1.1	12.7	4.8	.5	18.0	28.0	118.0	—	9.0
2252	imitation milk & cream	1 TBLS	**15**	**79.6**	**14.5**	**.3**	**.8**	**1.7**	**0**	—	—	**75.6**	—	—
			100	79.6	96.6	2.1	5.1	11.1	0	—	—	504.0	—	—
2253	safflower & soybean	1 TBLS	**14**	**15.3**	**100.4**	**.2**	**11.1**	**.4**	**0**	**2.5**	**3.9**	**79.6**	—	**4.8**
			100	15.3	716.8	1.1	79.4	2.7	0	18.0	28.0	568.4	—	34.0
2254	soybean	1 TBLS	**14**	**15.3**	**100.4**	**.2**	**11.1**	**.4**	**0**	**2.5**	**3.9**	**79.6**	**.3**	**4.8**
			100	15.3	716.8	1.1	79.4	2.7	0	18.0	28.0	568.4	2.0	34.0
2255	imitation soybean	1 TBLS	**15**	**62.7**	**34.7**	**0**	**2.9**	**2.4**	**0**	—	—	**74.6**	—	—
			100	62.7	231.6	.3	19.2	16.0	0	—	—	497.0	—	—
2256	imitation soybean, w/o cholesterol	1 TBLS	**14**	**34.6**	**67.5**	**0**	**6.7**	**2.2**	**0**	—	—	**49.4**	—	—
			100	34.6	481.9	.1	47.7	15.8	0	—	—	353.0	—	—
	RUSSIAN													
2257	regular	1 TBLS	**14**	**34.5**	**69.2**	**.2**	**7.1**	**1.5**	**0**	**2.7**	**5.2**	**121.5**	—	**22.0**
			100	34.5	494.0	1.6	50.8	10.4	.3	19.0	37.0	868.0	—	157.0

Item No.	Minerals (Micro)			Fat-Soluble Vitamins			Water-Soluble Vitamins								
	Iron	Zinc	Copper	Vitamin A	Vitamin D	Vitamin E (Alpha Tocopherol)	Vitamin C	Thiamin	Ribo-flavin	Niacin	Panto-thenic Acid	Vit. B-6 (Pyri-doxine)	Folacin (Folic Acid)	Biotin	Vitamin B-12
	(mg)	(mg)	(mg)	(IU)	(IU)	(mg)	(mg)	(mg)	(mg)	(mg)	(mg)	(mg)	(mcg)	(mcg)	(mcg)
2232	.95 .50	— —	— —	39.9 21.0	— —	— —	2.09 1.10	.04 .02	.02 .01	.68 .36	— —	— —	— —	— —	— —
2233	.23 .81	.05 .17	— —	37.2 132.8	1.24 4.41	— —	4.67 16.68	.03 .10	.02 .06	.32 1.12	.06 .20	.01 .01	1.54 5.49	— —	.04 .13
2234	2.00 .80	— —	— —	450.0 180.0	— —	— —	27.50 11.00	.18 .07	.15 .06	2.25 .90	— —	— —	— —	— —	— —
2235	1.50 .60	— —	— —	350.0 140.0	— —	— —	27.50 11.00	.20 .08	.18 .07	2.75 1.10	— —	— —	— —	— —	— —
2236	2.60 1.30	— —	— —	580.0 290.0	— —	— —	2.00 1.00	.08 .04	.22 .11	10.00 5.00	— —	— —	— —	— —	— —
2237	.13 .46	.06 .21	— —	52.9 188.8	0 0	— —	1.49 5.31	.15 .53	.02 .05	.08 .29	.06 .20	.02 .06	4.93 17.60	— —	0 0
2238	.64 .59	— —	— —	133.9 124.0	— —	— —	3.34 3.09	.07 .06	.04 .04	.23 .21	— —	— —	— —	— —	— —
2239	.03 .20	— —	— —	29.4 210.0	— —	— —	.28 2.00	0 .01	.01 .10	.01 .10	— —	— —	— —	— —	— —
2240	.01 .10	— —	— —	23.8 170.0	— —	— —	.28 2.00	— —	.01 .07	.01 .10	— —	— —	— —	— —	— —
2241	.01 .10	— —	— —	11.2 80.0	— —	— —	.28 2.00	— —	0 .04	— —	— —	— —	— —	— —	— —
2242	.07 .50	.01 .11	— —	57.5 411.0	— —	— —	.08 .60	0 .06	.02 .15	.04 .25	— —	— —	— —	— —	— —
2243	.03 .20	— —	Trace (.01)	72.0 514.0	0 0	(.50) (3.80)	.08 .60	0 .01	0 .02	.02 .13	0 0	0 0	0 0	0 0	0 0
2244	.06 .40	.01 .08	— —	— —	— —	— —	— —	— —	— —	— —	— —	— —	— —	— —	— —
2245	.06 .40	.03 .18	.02 .12	— —	— —	— —	— —	— —	— —	— —	— —	— —	— —	— —	— —
2246	.06 .40	.03 .18	.02 .12	— —	— —	— —	— —	— —	— —	— —	— —	— —	— —	— —	— —
2247	.06 .40	.03 .18	.02 .12	— —	— —	— —	— —	— —	— —	— —	— —	— —	— —	— —	— —
2248	.03 .20	.02 .11	— —	— —	— —	— —	— —	— —	— —	— —	— —	— —	— —	— —	— —
2249	.03 .20	— —	— —	— —	— —	— —	— —	— —	— —	— —	— —	— —	— —	— —	— —
2250	.03 .20	— —	— —	30.8 220.0	— —	— —	— —	0 .01	0 .03	— —	— —	— —	.42 3.00	— —	— —
2251	.03 .20	— —	— —	30.8 220.0	— —	— —	— —	0 .01	0 .03	— —	— —	— —	— —	— —	— —
2252	— —	— —	— —	— —	— —	— —	— —	— —	— —	— —	— —	— —	— —	— —	— —
2253	.07 .50	.02 .12	— —	39.2 280.0	— —	— —	— —	— —	0 0	0 0	— —	— —	.42 3.00	— —	— —
2254	.07 .50	.02 .16	— —	39.2 280.0	— —	2.91 20.80	— —	0 0	0 0	— —	— —	— —	.42 3.00	— —	— —
2255	— —	.02 .11	— —	— —	— —	— —	— —	— —	— —	— —	— —	— —	— —	— —	— —
2256	— —	— —	— —	— —	— —	— —	— —	— —	— —	— —	— —	— —	— —	— —	— —
2257	.08 .60	.06 .43	— —	96.6 690.0	— —	— —	.84 6.00	0 .05	0 .05	.08 .60	— —	— —	— —	— —	— —

(Continued)

TABLE F-36 *(Continued)*

Item No.	Food Item Name	Approximate Measure	Weight	Moisture	Food Energy Calories	Protein	Fats	Carbo-hydrates	Fiber	Minerals (Macro) Calcium	Phos-phorus	Sodium	Mag-nesium	Potas-sium
			(g)	(%)	(kcal)	(g)	(g)	(g)	(g)	(mg)	(mg)	(mg)	(mg)	(mg)
	RUSSIAN (Continued)													
2258	low calorie	1 TBLS	16	65.0	22.6	.1	.6	4.4	0	3.0	5.9	138.9	—	25.1
			100	65.0	141.4	.5	4.0	27.6	.3	19.0	37.0	868.0	—	157.0
2259	**SESAME SEED**	1 TBLS	15	39.2	66.5	.5	6.8	1.3	.1	—	—	150.0	—	—
			100	39.2	443.1	3.1	45.2	8.6	.4	—	—	1000.0	—	—
	THOUSAND ISLAND													
2260	regular	1 TBLS	14	46.1	52.8	.1	5.0	2.1	.3	1.5	2.4	98.0	—	15.8
			100	46.1	377.3	.9	35.7	15.2	2.0	11.0	17.0	700.0	—	113.0
2261	lo-cal (10 kcal/tsp)	1 TBLS	14	68.3	22.2	.1	1.5	2.3	.2	1.5	2.4	140.0	—	15.8
			100	68.3	158.6	.8	10.7	16.2	1.2	11.0	17.0	1000.0	—	113.0
2262	**VINEGAR & OIL,** homemade	1 TBLS	16	47.4	71.8	0	8.0	.4	—	—	—	.1	—	1.2
			100	47.4	448.8	0	50.1	2.5	—	—	—	.5	—	7.5
	SNACK FOODS													
2263	**BACON RINDS**	1 SERV	28	—	144.5	19.2	7.7	0.0	—	1.4	0.0	217.4	—	—
			100	—	516.2	68.5	27.6	0.0	—	5.1	0.0	776.5	—	—
2264	**CHEESE PUFFS**	1 CUP	28	—	154.6	2.2	9.7	15.2	.1	15.7	19.9	170.0	3.9	64.1
			100	—	552.0	7.7	34.5	54.3	.5	56.0	71.0	607.0	14.0	229.0
2265	**CORN CHIPS**	1 CUP	40	—	220.4	2.7	14.8	20.9	.4	49.6	73.6	288.0	24.0	32.4
			100	—	551.0	6.8	37.0	52.3	1.0	124.0	184.0	720.0	60.0	81.0
	DIPS													
2266	**ENCHILADA**	1 OZ	28	—	36.7	1.9	1.5	3.8	.2	17.1	2.5	95.2	15.1	100.0
			100	—	131.0	6.9	5.4	13.6	.8	61.0	9.0	340.0	54.0	357.0
2267	**FRENCH ONION**	1 TBLS	15	—	24.2	.5	2.0	1.1	—	17.6	13.7	84.2	1.8	26.3
			100	—	161.0	3.3	13.0	7.6	—	117.0	91.0	561.0	12.0	175.0
2068	**BEAN**	1 CUP	240	—	285.6	13.9	12.0	31.2	5.3	82.8	184.8	744.0	64.8	352.8
			100	—	119.0	5.8	5.0	13.0	2.2	34.5	77.0	310.0	27.0	147.0
2269	**SOUR CREAM**	1 TBLS	15	—	28.1	.4	2.7	1.1	—	21.3	18.1	—	—	—
			100	—	187.5	2.5	17.9	7.5	—	141.7	120.8	—	—	—
2270	**GRANOLA CRUNCH**	1 PKG	32	—	154.8	4.0	7.0	19.0	—	—	—	—	—	—
			100	—	483.6	12.5	21.8	59.3	—	—	—	—	—	—
2271	**ONION RINGS**	1 SERV	28	—	131.8	.1	5.9	19.3	—	2.8	0	434.9	—	—
			100	—	470.8	.2	21.0	69.0	—	10.1	0	1553.1	—	—
	POPCORN													
2272	unpopped	1 CUP	170	9.8	615.4	20.2	8.0	122.6	3.6	17.0	448.8	5.1	—	—
			100	9.8	362.0	11.9	4.7	72.1	2.1	10.0	264.0	3.0	—	—
2273	popped, plain	1 CUP	6	4.0	23.2	.8	.3	4.6	.1	.7	16.9	0.2	—	—
			100	4.0	386.0	12.7	5.0	76.7	2.2	11.0	281.0	3.0	—	—
2274	popped, w/oil & salt	1 CUP	9	5.6	41.0	.9	2.0	5.3	.2	.7	19.4	174.6	—	—
			100	5.6	456.0	9.8	21.8	59.1	1.7	8.0	216.0	1940.0	—	—
2275	popped, sugar coated	1 CUP	35	4.0	134.1	2.1	1.2	29.9	.4	1.8	47.3	0.4	—	—
			100	4.0	383.0	6.1	3.5	85.4	1.1	5.0	135.0	1.0	—	—
	POTATO													
2276	**CHIPS**	1 CUP	20	1.8	113.6	1.1	8.0	10.0	.3	8.0	27.8	200.0	9.6	226.0
			100	1.8	568.0	5.3	39.8	50.0	1.6	40.0	139.0	1000.0	48.0	1130.0
2277	**STICKS**	1 CUP	40	1.5	217.6	2.6	14.6	20.3	.6	17.6	55.6	400.0	—	452.0
			100	1.5	544.0	6.4	36.4	50.8	1.5	44.0	139.0	1000.0	—	1130.0
2278	**PRETZELS**	1 AVG	13	4.5	50.7	1.3	.6	9.9	0	2.9	17.0	218.4	—	16.9
			100	4.5	390.0	9.8	4.5	75.9	.3	22.0	131.0	1680.0	—	130.0
2279	**PUMPKIN & SQUASH SEED KERNELS,** dry	1 CUP	140	4.4	774.2	40.6	65.4	21.0	2.7	71.4	1601.6	—	—	—
			100	4.4	553.0	29.0	46.7	15.0	1.9	51.0	1144.0	—	—	—
2280	**SUNFLOWER SEED KERNELS,** dry, hulled	1 CUP	145	4.8	812.0	33.4	68.6	28.9	5.5	174.0	1213.7	43.5	55.1	1334.0
			100	4.8	560.0	23.0	47.3	19.9	3.8	120.0	837.0	30.0	38.0	920.0
SOUPS & CHOWDERS														
	ASPARAGUS SOUP													
2281	cream of, made/= volume milk	1 CUP	240	85.2	165.6	6.5	8.2	17.2	.1	170.4	148.8	1056.0	—	225.6
			100	85.2	69.0	2.7	3.4	7.2	.1	71.0	62.0	440.0	—	94.0
2282	cream of, made/= volume water	1 CUP	240	91.7	84.0	2.3	3.4	11.3	.1	27.6	36.0	996.0	—	165.6
			100	91.7	35.0	1.0	1.4	4.7	.1	11.5	15.0	415.0	—	69.0
2283	cream of condensed, commercial	1 CUP	245	83.4	171.5	4.7	6.9	23.0	.2	56.4	76.0	2031.1	—	338.1
			100	83.4	70.0	1.9	2.8	9.4	.1	23.0	31.0	829.0	—	138.0

Item No.	Minerals (Micro)			Fat-Soluble Vitamins			Water-Soluble Vitamins								
	Iron	Zinc	Copper	Vitamin A	Vitamin D	Vitamin E (Alpha Tocopherol)	Vitamin C	Thiamin	Ribo-flavin	Niacin	Panto-thenic Acid	Vit. B-6 (Pyri-doxine)	Folacin (Folic Acid)	Biotin	Vitamin B-12
	(mg)	(mg)	(mg)	(IU)	(IU)	(mg)	(mg)	(mg)	(mg)	(mg)	(mg)	(mg)	(mcg)	(mcg)	(mcg)
2258	.10	–	–	–	–	–	–	–	–	–	–	–	–	–	–
	.60	–	–	–	–	–	–	–	–	–	–	–	–	–	–
2259	–	–	–	–	–	–	–	–	–	–	–	–	–	–	–
	–	–	–	–	–	–	–	–	–	–	–	–	–	–	–
2260	.08	.02	–	44.8	–	–	.42	0	0	.03	–	–	–	–	–
	.60	.14	–	320.0	–	–	3.00	.02	.03	.20	–	–	–	–	–
2261	.08	–	–	44.8	–	–	.42	0	0	.03	–	–	–	–	–
	.60	–	–	320.0	–	–	3.00	.02	.03	.20	–	–	–	–	–
2262	–	–	–	–	–	–	–	–	–	–	–	–	–	–	–
	–	–	–	–	–	–	–	–	–	–	–	–	–	–	–
2263	.00	–	–	0	–	–	0	.00	0	0	0	0	–	–	0
	.00	–	–	0	–	–	0	.00	0	0	0	0	–	–	0
2264	.10	.00	.04	71.1	–	3.86	0	.01	.03	.20	–	–	–	–	–
	.36	.00	.13	254.0	–	13.80	.0	.04	.11	0.71	–	–	–	–	–
2265	.44	.11	.18	141.2	–	2.41	0	.01	.03	0.42	–	0.08	–	–	–
	1.10	.28	.46	353.0	–	6.02	0	.03	.07	1.06	–	0.20	–	–	–
2266	.95	.01	.06	112.5	–	–	0	.01	.04	.28	–	–	–	–	–
	3.40	.04	.21	402.0	–	–	0	.03	.15	1.00	–	–	–	–	–
2267	.02	–	–	78.3	0.90	0.03	0	.01	.02	.13	–	.02	–	–	.08
	.10	–	–	522.0	6.00	0.20	0	.03	.16	.83	–	.10	–	–	.54
2268	3.31	.08	.43	547.2	–	0.82	0	.17	.14	1.68	–	–	–	–	–
	1.38	.04	.18	228.0	–	0.35	0	.07	.06	.70	–	–	–	–	–
2269	–	–	–	–	–	–	–	.03	–	–	–	–	–	–	–
	–	–	–	–	–	–	–	.21	–	–	–	–	–	–	–
2270	–	–	–	–	–	–	–	–	–	–	–	–	–	–	–
	–	–	–	–	–	–	–	–	–	–	–	–	–	–	–
2271	0	–	–	0	–	–	0	0	0	0	0	0	–	–	0
	0	–	–	0	–	–	0	0	0	0	0	0	–	–	0
2272	4.25	6.63	–	–	–	–	0	.66	.19	3.57	–	.58	–	–	0
	2.50	3.90	–	–	–	–	0	.39	.11	2.10	–	.34	–	–	0
2273	.16	.25	–	–	–	–	0	–	.01	.13	–	.01	–	–	0
	2.70	4.10	–	–	–	–	0	–	.12	2.20	–	.20	–	–	0
2274	.19	.27	–	–	–	–	0	.0	.01	.15	–	–	–	–	–
	2.10	3.00	–	–	–	–	0	.03	.09	1.70	–	–	–	–	–
2275	.46	–	–	–	–	–	0	–	.02	.39	–	–	–	–	–
	1.30	–	–	–	–	–	0	–	.06	1.10	–	–	–	–	–
2276	.36	.16	–	0	–	1.28	3.20	.04	.01	.96	–	.04	–	–	0
	1.80	.81	–	0	–	6.40	16.00	.21	.07	4.80	–	.18	–	–	0
2277	.72	–	–	–	–	–	16.00	.08	.03	1.92	–	–	–	–	–
	1.80	–	–	–	–	–	40.00	.21	.07	4.80	–	–	–	–	–
2278	.20	.14	–	0	–	.02	0	0	0	.09	.07	0	–	–	–
	1.50	1.08	–	0	–	.15	0	.02	.03	.70	.54	.02	–	–	–
2279	15.68	–	–	98.0	–	–	–	.34	.27	3.36	–	.13	–	–	0
	11.20	–	–	70.0	–	–	–	.24	.19	2.40	–	.09	–	–	0
2280	10.30	6.64	2.57	72.5	–	18.85	–	2.84	.33	7.83	2.03	1.81	–	–	0
	7.10	4.58	1.77	50.0	–	13.00	–	1.96	.23	5.40	1.40	1.25	–	–	0
2281	1.08	–	–	578.4	–	–	–	.17	.31	1.20	–	–	–	–	–
	.45	–	–	241.0	–	–	–	.07	.13	.50	–	–	–	–	–
2282	.96	–	–	386.4	–	–	–	.12	.11	1.08	–	–	–	–	–
	.40	–	–	161.0	–	–	–	.05	.05	.45	–	–	–	–	–
2283	1.96	–	–	786.5	–	–	–	.25	.22	2.21	–	–	–	–	–
	.80	–	–	321.0	–	–	–	.10	.09	.90	–	–	–	–	–

(Continued)

SOUPS & CHOWDERS

TABLE F-36 *(Continued)*

Item No.	Food Item Name	Approximate Measure	Weight	Moisture	Food Energy Calories	Protein	Fats	Carbo-hydrates	Fiber	Minerals (Macro) Calcium	Phos-phorus	Sodium	Mag-nesium	Potas-sium
			(g)	(%)	(kcal)	(g)	(g)	(g)	(g)	(mg)	(mg)	(mg)	(mg)	(mg)
	BEAN SOUP													
2284	w/bacon, diluted w/water	1 CUP	253	—	—	—	6.1	—	—	—	—	—	—	—
			100	—	—	—	2.4	—	—	—	—	—	—	—
2285	w/franks, diluted w/water	1 CUP	257	—	—	—	7.2	—	—	—	—	—	—	
			100	—	—	—	2.8	—	—	—	—	—	—	—
2286	w/ham	1 OZ	28	67.8	36.5	2.4	.8	5.1	.3	11.9	41.3	32.0	1.2	82.4
			100	67.8	130.3	8.7	2.8	18.3	1.2	42.6	147.4	114.4	4.2	294.4
2287	w/pork, condensed, canned	1 CUP	265	68.9	355.1	17.0	12.2	45.8	3.4	132.5	267.7	2135.9	—	837.4
			100	68.9	134.0	6.4	4.6	17.3	1.3	50.0	101.0	806.0	—	316.0
2288	w/pork, made w/ = volume water	1 CUP	250	84.4	167.5	8.0	5.8	21.8	1.5	62.5	127.5	1007.5	—	395.0
			100	84.4	67.0	3.2	2.3	8.7	.6	25.0	51.0	403.0	—	158.0
	BEEF SOUP													
2289	broth bouillon or consomme, condensed, canned	1 CUP	245	91.6	63.7	10.3	0	5.4	.2	—	63.7	1597.4	—	264.6
			100	91.6	26.0	4.2	0	2.2	.1	—	26.0	652.0	—	108.0
2290	broth boullion or consomme, canned, made w/ = volume water	1 CUP	240	95.8	31.2	5.0	0	2.6	—	—	31.2	782.4	—	129.6
			100	95.8	13.0	2.1	0	1.1	—	—	13.0	326.0	—	54.0
2291	w/noodle, condensed	1 CUP	246	86.4	140.6	7.8	6.2	14.3	—	15.1	98.4	1879.6	—	157.6
			100	86.4	57.1	3.2	2.5	5.8	—	6.1	40.0	764.1	—	64.1
2292	w/noodle, diluted w/water	1 CUP	241	93.2	67.3	3.8	3.1	7.0	—	7.0	48.2	920.8	—	77.3
			100	93.2	27.9	1.6	1.3	2.9	—	2.9	20.0	382.1	—	32.1
2293	w/noodle, dehydrated, made w/2 oz mix in 3 c water	1 CUP	240	93.1	67.2	2.4	1.2	11.5	—	9.6	26.4	420.0	—	40.8
			100	93.1	28.0	1.0	.5	4.8	—	4.0	11.0	175.0	—	17.0
2294	w/vegetables, condensed	1 CUP	251	—	—	—	3.8	—	—	—	—	—	—	—
			100	—	—	—	1.5	—	—	—	—	—	—	—
2295	w/vegetable,diluted w/water	1 CUP	244	—	—	—	2.0	—	—	—	—	—	—	—
			100	—	—	—	.8	—	—	—	—	—	—	—
2296	**BORSCHT**	1 AVG	250	—	80.0	.6	0	19.3	—	—	—	—	—	—
			100	—	32.0	.2	0	7.7	—	—	—	—	—	—
2297	**BOUILLON CUBES OR POWDER**	1 CUBE	4	4.0	(6.0)	(.6)	(.1)	(.6)	—	—	(8.0)	960.0	(2.0)	(15.0)
			100	4.0	(170.0)	(17.3)	(4.0)	(16.1)	—	—	(225.0)	24000.0	(50.0)	(403.0)
	CELERY SOUP, CREAM OF													
2298	condensed	1 CUP	246	84.6	180.4	3.4	10.8	18.2	—	98.4	74.3	1958.0	—	221.9
			100	84.6	73.3	1.4	4.4	7.4	—	40.0	30.2	795.9	—	90.2
2299	diluted w/milk	1 CUP	245	85.8	169.0	6.4	9.8	15.5	—	202.1	157.2	1060.6	—	295.0
			100	85.8	69.0	2.6	4.0	6.3	—	82.5	64.2	432.9	—	120.4
2300	diluted w/water	1 CUP	241	92.3	86.4	1.7	5.5	8.9	—	48.2	36.2	959.0	—	108.5
			100	92.3	35.8	.7	2.3	3.7	—	20.0	15.0	397.9	—	45.0
	CHICKEN SOUP													
2301	consomme, condensed	1 CUP	245	93.7	44.1	6.9	.2	3.7	—	24.5	144.6	1474.9	—	—
			100	93.7	18.0	2.8	.1	1.5	—	10.0	59.0	602.0	—	—
2302	consomme, made w/ = volume water	1 CUP	240	96.8	21.6	3.4	.5	1.9	—	12.0	72.0	722.4	—	—
			100	96.8	9.0	1.4	.2	0.8	—	5.0	30.0	301.0	—	—
2303	cream of, condensed	1 CUP	246	83.8	194.8	5.9	14.5	16.5	—	48.2	71.3	1990.1	(17.9)	162.7
			100	83.8	79.2	2.4	5.9	6.7	—	19.6	29.0	809.0	(7.3)	66.1
2304	cream of, diluted w/milk	1 CUP	245	85.4	182.7	7.4	11.5	14.5	—	175.6	155.2	1054.0	—	260.0
			100	85.4	74.6	3.0	4.7	5.9	—	71.7	63.3	430.2	—	106.1
2305	cream of, diluted w/water	1 CUP	241	91.9	94.4	2.9	7.2	7.9	—	24.1	34.1	974.0	—	79.3
			100	91.9	39.2	1.2	3.0	3.3	—	10.0	14.2	404.2	—	32.9
2306	cream of, condensed, commercial	1 CUP	245	81.7	232.8	4.9	17.2	15.4	.2	71.1	71.1	2045.8	—	166.6
			100	81.7	95.0	2.0	7.0	6.3	.1	29.0	29.0	835.0	—	68.0
2307	w/dumplings, condensed	1 CUP	246	—	—	—	11.8	—	—	—	—	—	—	—
			100	—	—	—	4.8	—	—	—	—	—	—	—
2308	w/dumplings, diluted w/water	1 CUP	241	—	—	—	6.0	—	—	—	—	—	—	— —
			100	—	—	—	2.5	—	—	—	—	—	—	—
2309	gumbo, condensed	1 CUP	246	87.6	113.5	6.4	2.7	15.0	—	39.2	51.2	1947.9	—	218.9
			100	87.6	46.1	2.6	1.1	6.1	—	15.9	20.8	791.8	—	89.0
2310	gumbo, diluted w/water	1 CUP	241	93.8	55.2	3.1	1.4	7.4	—	190.8	24.1	954.0	—	108.5
			100	93.8	22.9	1.3	.6	3.1	—	79.2	10.0	395.8	—	45.0
2311	gumbo, condensed, commercial	1 CUP	245	87.8	117.6	5.4	2.7	17.9	.5	46.6	49.0	1862.0	—	147.0
			100	87.8	48.0	2.2	1.1	7.3	.2	19.0	20.0	760.0	—	60.0
2312	noodle, condensed	1 CUP	246	86.6	130.5	6.9	1.5	16.3	—	17.1	74.3	2007.2	—	113.5
			100	86.6	53.1	2.8	.6	6.6	—	6.9	30.2	815.9	—	46.1
2313	noodle, diluted w/water	1 CUP	241	86.6	127.9	3.4	2.4	7.9	—	10.0	36.2	983.1	—	55.2
			100	86.6	53.1	1.4	1.0	3.3	—	4.2	15.0	407.9	—	22.9

Item No.	Minerals (Micro)			Fat-Soluble Vitamins			Water-Soluble Vitamins								
	Iron	Zinc	Copper	Vitamin A	Vitamin D	Vitamin E (Alpha Tocopherol)	Vitamin C	Thiamin	Ribo-flavin	Niacin	Panto-thenic Acid	Vit. B-6 (Pyri-doxine)	Folacin (Folic Acid)	Biotin	Vitamin B-12
	(mg)	(mg)	(mg)	(IU)	(IU)	(mg)	(mg)	(mg)	(mg)	(mg)	(mg)	(mg)	(mcg)	(mcg)	(mcg)
2284	–	–	–	–	–	–	–	–	–	–	–	–	–	–	–
	–	–	–	–	–	–	–	–	–	–	–	–	–	–	–
2285	–	–	–	–	–	–	–	–	–	–	–	–	–	–	–
	–	–	–	–	–	–	–	–	–	–	–	–	–	–	–
2286	**.61**	**.08**	–	**11.7**	**.66**	–	**1.23**	**.06**	**.03**	**.37**	**.04**	**0**	–	–	**.17**
	2.18	.29	–	41.9	2.35	–	4.40	.22	.09	1.33	.14	0	–	–	.69
2287	**4.77**	–	–	**1378.0**	–	–	**5.30**	**.29**	**.16**	**2.12**	–	–	**63.60**	–	–
	1.80	–	–	520.0	–	–	2.00	.11	.06	.80	–	–	24.00	–	–
2288	**2.25**	–	–	**650.0**	–	–	**2.50**	**.13**	**.08**	**1.00**	–	–	–	–	–
	.90	–	–	260.0	–	–	1.00	.05	.03	.40	–	–	–	–	–
2289	**.98**	–	–	–	–	–	–	–	**.05**	**.245**	–	–	–	–	–
	.40	–	–	–	–	–	–	–	.02	1.00	–	–	–	–	–
2290	**.48**	–	–	–	–	–	–	–	**.02**	**1.20**	–	–	**9.60**	– –	
	.20	–	–	–	–	–	–	–	.01	.50	–	–	4.00	–	–
2291	**1.71**	–	–	**120.5**	–	–	**2.01**	**.10**	**.12**	**2.21**	–	–	**9.84**	–	–
	.69	–	–	49.0	–	–	.82	.04	.05	.90	–	–	4.00	–	–
2292	**1.01**	**3.40**	–	**50.2**	–	–	**.00**	**.05**	**.07**	**1.01**	–	–	–	–	–
	.42	1.41	–	20.8	–	–	.00	.02	.03	.42	–	–	–	–	–
2293	**.48**	–	–	**24.0**	–	–	**.24**	**.10**	**.05**	**.72**	–	–	–	–	–
	.20	–	–	10.0	–	–	.10	.04	.02	.30	–	–	–	–	–
2294	–	–	–	–	–	–	–	–	–	–	–	–	–	–	–
	–	–	–	–	–	–	–	–	–	–	–	–	–	–	–
2295	–	–	–	–	–	–	–	–	–	–	–	–	–	–	–
	–	–	–	–	–	–	–	–	–	–	–	–	–	–	–
2296	–	–	–	–	–	–	–	–	–	–	–	–	–	–	–
	–	–	–	–	–	–	–	–	–	–	–	–	–	–	–
2297	**(.08)**	**(.01)**	–	–	**0**	–	**0**	**(.01)**	**(.01)**	**(.12)**	–	–	–	–	–
	(2.23)	(.21)	–	–	0	–	0	(.20)	(.24)	(3.30)	–	–	–	–	–
2298	**1.20**	–	–	**421.7**	–	–	**2.01**	**.02**	**.01**	**1.00**	–	–	–	–	–
	.49	–	–	171.4	–	–	.82	.01	0	.41	–	–	–	–	–
2299	**.72**	–	–	**398.1**	–	–	**2.00**	**.05**	**.27**	**.70**	–	–	–	–	–
	.29	–	–	162.5	–	–	.82	.02	.11	.29	–	–	–	–	–
2300	**.50**	–	–	**190.8**	–	–	**0**	**.02**	**.05**	**.0**	–	–	–	–	–
	.21	–	–	79.2	–	–	0	.01	.02	0	–	–	–	–	–
2301	**2.45**	–	–	–	–	–	–	–	–	–	–	–	**6.62**	–	–
	1.00	–	–	–	–	–	–	–	–	–	–	–	2.70	–	–
2302	**1.20**	–	–	–	–	–	–	–	–	–	–	–	–	–	–
	.50	–	–	–	–	–	–	–	–	–	–	–	–	–	–
2303	**1.00**	**(1.30)**	**(.07)**	**863.5**	**0**	–	**0**	**.02**	**.01**	**1.20**	–	–	**2.46**	–	**0**
	.40	(.53)	(.03)	351.0	0	–	0	.01	.00	.49	–	–	1.00	–	0
2304	**.51**	–	–	**610.0**	–	–	**2.00**	**.05**	**.27**	**.70**	–	–	–	–	–
	.21	–	–	249.0	–	–	.82	.02	.11	.29	–	–	–	–	–
2305	**.50**	–	–	**411.7**	–	–	**0**	**.02**	**.05**	**.50**	–	–	–	–	–
	.21	–	–	170.8	–	–	0	.01	.02	.21	–	–	–	–	–
2306	**1.23**	–	–	**1131.9**	–	–	**0**	**.03**	**.10**	**1.47**	–	–	**2.45**	–	–
	.50	–	–	462.0	–	–	0	.01	.04	.60	–	–	1.00	–	–
2307	–	–	–	–	–	–	–	–	–	–	–	–	–	–	–
	–	–	–	–	–	–	–	–	–	–	–	–	–	–	–
2308	–	–	–	–	–	–	–	–	–	–	–	–	–	–	–
	–	–	–	–	–	–	–	–	–	–	–	–	–	–	–
2309	**1.20**	–	–	**441.8**	–	–	**10.04**	**.05**	**.07**	**2.71**	–	–	–	–	–
	.49	–	–	179.6	–	–	4.08	.02	.03	1.10	–	–	–	–	–
2310	**.50**	–	–	**220.9**	–	–	**5.02**	**.02**	**.05**	**1.21**	–	–	–	–	–
	.21	–	–	91.7	–	–	2.08	.01	.02	.50	–	–	–	–	–
2311	**1.47**	–	–	**259.7**	–	–	**0**	**.05**	**.10**	**1.23**	–	–	–	–	–
	.60	–	–	106.0	–	–	0	.02	.04	.50	–	–	–	–	–
2312	**1.00**	–	–	**70.3**	–	–	**0**	**.02**	**.05**	**1.71**	–	–	**4.92**	–	–
	.41	–	–	28.6	–	–	0	.01	.02	.69	–	–	2.00	–	–
2313	**.50**	–	–	**50.2**	–	–	**0**	**.02**	**.02**	**.70**	–	–	–	–	–
	.21	–	–	20.8	–	–	0	.01	.01	.29	–	–	–	–	–

(Continued)

TABLE F-36 *(Continued)*

Item No.	Food Item Name	Approximate Measure	Weight	Moisture	Food Energy Calories	Protein	Fats	Carbo-hydrates	Fiber	Minerals (Macro) Calcium	Phos-phorus	Sodium	Mag-nesium	Potas-sium
			(g)	(%)	(kcal)	(g)	(g)	(g)	(g)	(mg)	(mg)	(mg)	(mg)	(mg)
	CHICKEN SOUP (Continued)													
2314	noodle, dehydrated, made w/2 oz mix in 4 c water	1 CUP	**240**	**94.7**	**52.8**	**1.9**	**1.4**	**7.7**	—	**7.2**	**19.2**	**578.4**	—	**19.2**
			100	94.7	22.0	.8	.6	3.2	—	3.0	8.0	241.0	—	8.0
2315	w/rice, condensed	1 CUP	**246**	**89.6**	**96.4**	**6.4**	**3.9**	**11.5**	—	**17.1**	**51.2**	**1879.6**	—	**201.8**
			100	89.6	39.2	2.6	1.6	4.7	—	6.9	20.8	764.1	—	82.0
2316	w/rice, diluted w/water	1 CUP	**241**	**94.8**	**48.2**	**3.1**	**1.9**	**5.8**	—	**7.0**	**24.1**	**920.8**	—	**98.4**
			100	94.8	20.0	1.3	.8	2.4	—	2.9	10.0	382.1	—	40.8
2317	w/rice, condensed, commercial	1 CUP	**245**	**87.5**	**105.4**	**6.4**	**2.5**	**14.0**	**.2**	**29.4**	**71.1**	**1641.5**	—	**110.3**
			100	87.5	43.0	2.6	1.0	5.7	.1	12.0	29.0	670.0	—	45.0
2318	w/rice, dehydrated, made w/1½ oz mix in 3 c water	1 CUP	**200**	**94.9**	**40.0**	**1.0**	**.8**	**7.0**	—	**6.0**	**8.0**	**518.0**	—	**8.0**
			100	94.9	20.0	.5	.4	3.5	—	3.0	4.0	259.0	—	4.0
2319	w/vegetable, condensed, commercial	1 CUP	**250**	**84.5**	**150.0**	**8.5**	**7.5**	**17.5**	**.8**	**35.0**	**85.0**	**1875.0**	—	**340.0**
			100	84.5	60.0	3.4	3.0	7.0	.3	14.0	34.0	750.0	—	136.0
	CLAM CHOWDER													
2320	Manhattan style, condensed	1 CUP	**251**	**83.7**	**165.7**	**4.5**	**4.5**	**25.1**	—	**73.3**	**95.4**	**1922.7**	—	**376.5**
			100	83.7	66.0	1.8	1.8	10.0	—	29.2	38.0	766.0	—	150.0
2321	Manhattan style, diluted w/water	1 CUP	**251**	**91.9**	**83.0**	**2.3**	**4.5**	**12.6**	—	**34.8**	**48.2**	**961.0**	—	**188.5**
			100	91.9	33.1	.9	1.8	5.0	—	13.9	19.2	382.9	—	75.1
2322	Manhattan style, w/tomatoes, condensed	1 CUP	**250**	**84.1**	**157.5**	**4.8**	**5.3**	**23.0**	**.8**	**47.5**	**82.5**	**2000.0**	**30.0**	**375.0**
			100	84.1	63.0	1.9	2.1	9.2	.3	19.0	33.0	800.0	12.0	150.0
2323	New England style, condensed & frozen, made w/milk	1 CUP	**250**	**78.5**	**270.0**	**9.3**	**16.0**	**22.3**	**.5**	**185.0**	**170.0**	**2175.0**	—	**462.5**
			100	78.5	108.0	3.7	6.4	8.9	.2	74.0	68.0	870.0	—	185.0
2324	chowder, New England style, condensed & frozen, made w/ = volume milk	1 CUP	**245**	**82.8**	**215.6**	**8.8**	**12.6**	**16.9**	**.2**	**235.2**	**198.5**	**1127.0**	—	**411.6**
			100	82.8	88.0	3.6	5.2	6.9	.1	96.0	81.0	460.0	—	168.0
2325	chowder, New England style, condensed & frozen, made w/ = volume water	1 CUP	**240**	**89.3**	**129.6**	**4.4**	**7.7**	**10.7**	**.2**	**88.8**	**81.6**	**1044.0**	—	**223.2**
			100	89.3	54.0	1.9	3.2	4.5	.1	37.0	34.0	435.0	—	93.0
2326	**[a]LENTIL**	1 CUP	**250**	—	**247.8**	**11.3**	**9.5**	**29.5**	**5.5**	**100.5**	**146.5**	**475.8**	**40.3**	**402.5**
			100	—	99.1	4.5	3.8	11.8	2.2	40.2	58.6	190.3	16.1	161.0
	MINESTRONE													
2327	condensed	1 CUP	**251**	**79.0**	**218.9**	**10.0**	**5.3**	**29.1**	—	**75.3**	**123.5**	**2041.1**	—	**640.6**
			100	79.0	87.2	4.0	2.1	11.6	—	30.0	49.2	813.2	—	255.2
2328	diluted w/water	1 CUP	**244**	**89.5**	**104.6**	**4.9**	**2.7**	**14.1**	—	**36.8**	**58.8**	**990.9**	—	**312.7**
			100	89.5	42.9	2.0	1.1	5.8	—	15.1	24.1	406.1	—	128.2
2329	condensed, commercial	1 CUP	**250**	**82.3**	**180.0**	**8.3**	**5.0**	**27.5**	**1.5**	**67.5**	**115.0**	**1862.5**	—	**637.5**
			100	82.3	72.0	3.3	2.0	11.0	.6	27.0	46.0	745.0	—	255.0
	MUSHROOM SOUP, CREAM OF													
2330	condensed	1 CUP	**246**	**79.3**	**273.1**	**4.7**	**18.9**	**20.7**	—	**83.3**	**105.4**	**1956.0**	—	**201.8**
			100	79.3	111.0	1.9	7.7	8.4	—	33.9	42.9	795.1	—	82.0
2331	condensed, diluted w/milk	1 CUP	**245**	**83.2**	**216.0**	**6.9**	**13.7**	**16.2**	—	**191.0**	**169.0**	**1039.0**	—	**279.0**
			100	83.2	88.2	2.8	5.6	6.6	—	78.0	69.0	424.1	—	113.9
2332	condensed, diluted w/water	1 CUP	**241**	—	**134.6**	**2.4**	**9.4**	**10.1**	—	**41.2**	**50.2**	**959.0**	**(10.4)**	**98.4**
			100	—	55.8	1.0	3.9	4.2	—	17.1	20.8	397.9	(4.3)	40.8
2333	condensed, commercial	1 CUP	**245**	**81.5**	**257.3**	**2.5**	**19.6**	**18.4**	**.2**	**53.9**	**76.0**	**1972.3**	—	**147.0**
			100	81.5	105.0	1.0	8.0	7.5	.1	22.0	31.0	805.0	—	60.0
	ONION SOUP													
2334	[1]canned, made w/ = volume water	1 CUP	**240**	**94.5**	**57.6**	**3.6**	**3.4**	**3.4**	**.2**	**9.6**	**38.4**	**1108.8**	—	**115.2**
			100	94.5	24.0	1.5	1.4	1.4	.1	4.0	16.0	462.0	—	48.0
2335	dehydrated, w/ 1½ oz mix in 4 c water	1 CUP	**240**	**95.8**	**36.0**	**1.4**	**1.2**	**5.5**	**.2**	**9.6**	**12.0**	**688.8**	—	**57.6**
			100	95.8	15.0	.6	.5	2.3	.1	4.0	5.0	287.0	—	24.0
	OYSTER STEW													
2336	homemade, 1 part oysters to 2 parts milk	1 CUP	**240**	**82.0**	**232.8**	**12.5**	**15.4**	**10.8**	—	**273.6**	**266.4**	**813.6**	—	**319.2**
			100	82.0	97.0	5.2	6.4	4.5	—	114.0	111.0	339.0	—	133.0
2337	homemade, 1 part oysters to 3 parts milk	1 CUP	**240**	**83.7**	**206.4**	**11.8**	**12.7**	**11.3**	—	**280.8**	**261.6**	**487.2**	—	**331.2**
			100	83.7	86.0	4.9	5.3	4.7	—	117.0	109.0	203.0	—	138.0
2338	condensed	1 CUP	**240**	—	—	—	**7.4**	—	—	—	—	—	—	—
			100	—	—	—	3.1	—	—	—	—	—	—	—
2339	condensed, diluted w/milk	1 CUP	**242**	—	—	—	**8.0**	—	—	—	—	—	—	—
			100	—	—	—	3.3	—	—	—	—	—	—	—
2340	condensed, diluted w/water	1 CUP	**238**	—	—	—	**3.8**	—	—	—	—	—	—	—
			100	—	—	—	1.6	—	—	—	—	—	—	—
2341	frozen, condensed	1 CUP	**240**	**79.8**	**244.8**	**11.0**	**15.1**	**16.6**	**.2**	**314.4**	**278.4**	**1632.0**	—	**492.0**
			100	79.8	102.0	4.6	6.3	6.9	.1	131.0	116.0	680.0	—	205.0

Item No.	Minerals (Micro)			Fat-Soluble Vitamins			Water-Soluble Vitamins								
	• Iron	Zinc	Copper	Vitamin A	Vitamin D	Vitamin E (Alpha Tocopherol)	Vitamin C	Thiamin	Ribo-flavin	Niacin	Panto-thenic Acid	Vit. B-6 (Pyri-doxine)	Folacin (Folic Acid)	Biotin	Vitamin B-12
	(mg)	(mg)	(mg)	(IU)	(IU)	(mg)	(mg)	(mg)	(mg)	(mg)	(mg)	(mg)	(mcg)	(mcg)	(mcg)
2314	**.24**	–	–	**48.0**	–	–	–	**.07**	**.05**	**.48**	–	–	–	–	–
	.10	–	–	20.0	–	–	–	.03	.02	.20	–	–	–	–	–
2315	**.70**	–	–	**321.3**	–	–	–	**0**	**.05**	**1.51**	–	–	–	–	–
	.29	–	–	130.6	–	–	–	0	.02	.61	–	–	–	–	–
2316	**.20**	–	–	**140.6**	–	–	–	**0**	**.02**	**.70**	–	–	–	–	–
	.08	–	–	58.3	–	–	–	0	.01	.29	–	–	–	–	–
2317	**.98**	–	–	**1369.6**	–	–	**0**	**.05**	**.07**	**2.70**	–	–	**2.45**	–	–
	.40	–	–	559.0	–	–	0	.02	.03	1.10	–	–	1.00	–	–
2318	–	–	–	**4.0**	–	–	–	–	**.08**	**.20**	–	–	–	–	–
	–	–	–	2.0	–	–	–	–	.04	.10	–	–	–	–	–
2319	**1.25**	–	–	**3880.0**	–	–	**1.00**	**.10**	**.13**	**2.75**	–	**.11**	–	–	–
	.50	–	–	1552.0	–	–	.40	.04	.05	1.10	–	.04	–	–	–
2320	**2.31**	–	–	**17.9**	–	–	–	**.05**	**.05**	**2.31**	–	–	–	–	–
	.92	–	–	7.1	–	–	–	.02	.02	.92	–	–	–	–	–
2321	**1.02**	–	–	**901.6**	–	–	–	**.02**	**.02**	**1.02**	–	–	–	–	–
	.41	–	–	359.2	–	–	–	.01	.01	.41	–	–	–	–	–
2322	**2.75**	–	–	**2157.5**	–	–	**8.25**	**.05**	**.05**	**2.25**	–	–	–	–	–
	1.10	–	–	863.0	–	–	3.30	.02	.02	.90	–	–	–	–	–
2323	**2.00**	–	–	**125.0**	–	–	–	**.08**	**.18**	**1.00**	–	–	–	–	–
	.80	–	–	50.0	–	–	–	.03	.07	.40	–	–	–	–	–
2324	**1.10**	–	–	**257.3**	–	–	–	**.09**	**.29**	**.61**	–	–	–	–	–
	.45	–	–	105.0	–	–	–	.04	.12	.25	–	–	–	–	–
2325	**.96**	–	–	**60.0**	–	–	–	**.04**	**.08**	**.48**	–	–	–	–	–
	.40	–	–	25.0	–	–	–	.02	.04	.20	–	–	–	–	–
2326	**3.03**	**1.30**	**.30**	**1824.3**	**.68**	–	**Trace**	**.20**	**.10**	**.73**	–	**.15**	–	–	**0**
	1.21	.52	.12	729.7	.27	–	Trace	.08	.04	.29	–	.06	–	–	0
2327	**1.81**	–	–	**4769.0**	–	–	–	**.15**	**.13**	**2.31**	–	–	**32.63**	–	–
	.72	–	–	1900.0	–	–	–	.06	.05	.92	–	–	13.00	–	–
2328	**1.00**	–	–	**2340.4**	–	–	–	**.07**	**.05**	**1.00**	–	–	–	–	–
	.41	–	–	959.2	–	–	–	.03	.02	.41	–	–	–	–	–
2329	**1.75**	–	–	**3437.5**	–	–	**3.25**	**.15**	**.13**	**2.25**	–	–	**32.50**	–	–
	.70	–	–	1375.0	–	–	1.30	.06	.05	.90	–	–	13.00	–	–
2330	**.70**	–	–	**150.6**	–	–	**0**	**.02**	**.25**	**1.51**	–	–	–	–	–
	.29	–	–	61.2	–	–	0	.01	.10	.61	–	–	–	–	–
2331	**.50**	–	–	**250.0**	–	–	**1.00**	**.05**	**.34**	**.70**	–	–	–	–	–
	.20	–	–	102.0	–	–	.41	.02	.14	.29	–	–	–	–	–
2332	**.50**	**.70**	**.10**	**70.3**	**0**	–	**0**	**.02**	**.12**	**.70**	–	**(.17)**	–	–	**0**
	.21	.29	.04	29.2	0	–	0	.01	.05	.29	–	(.07)	–	–	0
2333	**.98**	–	–	**0**	–	–	**0**	**.03**	**.20**	**1.47**	–	–	**5.64**	–	–
	.40	–	–	0	–	–	0	.01	.08	.60	–	–	2.30	–	–
2334	**.63**	–	**.12**	–	–	–	–	**.01**	**.02**	**.80**	–	–	–	–	–
	.25	–	.05	–	–	–	–	.01	.01	.35	–	–	–	–	–
2335	**.24**	**07**	–	**7.2**	–	–	**2.40**	–	–	–	–	–	–	–	–
	.10	.03	–	3.0	–	–	1.00	–	–	–	–	–	–	–	–
2336	**4.56**	–	–	**816.0**	–	–	–	**.14**	**.43**	**2.16**	–	–	–	–	–
	1.90	–	–	340.0	–	–	–	.06	.18	.90	–	–	–	–	–
2337	**3.36**	–	–	**672.0**	–	–	–	**.14**	**.43**	**1.68**	–	–	–	–	–
	1.40	–	–	280.0	–	–	–	.06	.18	.70	–	–	–	–	–
2338	–	–	–	–	–	–	–	–	–	–	–	–	–	–	–
	–	–	–	–	–	–	–	–	–	–	–	–	–	–	–
2339	–	–	–	–	–	–	–	–	–	–	–	–	–	–	–
	–	–	–	–	–	–	–	–	–	–	–	–	–	–	–
2340	–	–	–	–	–	–	–	–	–	–	–	–	–	–	–
	–	–	–	–	–	–	–	–	–	–	–	–	–	–	–
2341	**2.64**	–	–	**456.0**	–	–	–	**.14**	**.38**	**.72**	–	–	–	–	–
	1.10	–	–	190.0	–	–	–	.06	.16	.30	–	–	–	–	–

(Continued)

TABLE F-36 *(Continued)*

Item No.	Food Item Name	Approximate Measure	Weight	Moisture	Food Energy Calories	Protein	Fats	Carbo-hydrates	Fiber	Minerals (Macro) Calcium	Phos-phorus	Sodium	Mag-nesium	Potas-sium
			(g)	(%)	(kcal)	(g)	(g)	(g)	(g)	(mg)	(mg)	(mg)	(mg)	(mg)
	OYSTER STEW (Continued)													
2342	frozen, made w/milk	1 CUP	240	83.4	201.6	10.1	11.8	14.2	–	304.8	254.4	878.4	–	422.4
			100	83.4	84.0	4.2	4.9	5.9	–	127.0	106.0	366.0	–	176.0
2343	frozen, made w/water	1 CUP	240	89.9	122.4	5.5	7.7	8.2	–	158.4	139.2	816.0	–	244.8
			100	89.9	51.0	2.3	3.2	3.4	–	66.0	58.0	340.0	–	102.0
	PEA, GREEN, SOUP													
2344	condensed, commercial	1 CUP	255	68.9	318.8	17.6	4.1	47.2	1.3	48.5	244.8	1899.8	–	331.5
			100	68.9	125.0	6.9	1.6	18.5	.5	19.0	96.0	745.0	–	130.0
2345	condensed, made w/ = volume milk	1 CUP	200	79.9	170.0	8.4	5.2	23.4	.8	158.0	188.0	786.0	–	306.0
			100	79.9	85.0	4.2	2.6	11.7	.4	79.0	94.0	393.0	–	153.0
2346	dehydrated, made w/2 oz mix in 3 c water	1 CUP	200	86.7	100.0	6.2	1.2	16.8	.4	16.0	86.0	650.0	–	240.0
			100	86.7	50.0	3.1	.6	8.4	.2	8.0	43.0	325.0	–	120.0
2347	w/ham, condensed, commercial	1 CUP	250	71.8	272.5	19.0	5.8	36.3	3.5	60.0	255.0	1875.0	–	502.5
			100	71.8	109.0	7.6	2.3	14.5	1.4	24.0	102.0	750.0	–	201.0
2348	w/ham, made w/ = volume water	1 CUP	200	85.9	110.0	7.6	2.4	14.5	1.4	24.0	102.0	750.0	–	200.0
			100	85.9	55.0	3.8	1.2	7.3	.7	12.0	51.0	375.0	–	100.0
	PEA, SPLIT, SOUP													
2349	homemade	1 OZ	28	63.8	23.2	1.3	.5	3.0	.1	13.9	21.8	111.6	1.3	32.4
			100	63.8	82.9	4.5	1.9	10.8	.3	49.6	77.7	398.5	4.5	115.7
2350	condensed, made w/ = volume water	1 CUP	200	85.4	118.0	7.0	2.6	16.8	.4	24.0	122.0	768.0	–	220.0
			100	85.4	59.0	3.5	1.3	8.4	.2	12.0	61.0	384.0	–	110.0
2351	condensed, commercial	1 CUP	255	70.7	300.9	17.9	6.6	43.4	1.0	63.8	311.1	1955.9	33.2	561.0
			100	70.7	118.0	7.0	2.6	17.0	.4	25.0	122.0	767.0	13.0	220.0
	POTATO SOUP, CREAM OF													
2352	condensed	1 CUP	246	–	–	–	4.7	–	–	–	–	–	–	–
			100	–	–	–	1.9	–	–	–	–	–	–	–
2353	condensed, diluted w/milk	1 CUP	245	–	–	–	6.6	–	–	–	–	–	–	–
			100	–	–	–	2.7	–	–	–	–	–	–	–
2354	condensed, diluted w/water	1 CUP	241	–	–	–	2.4	–	–	–	–	–	–	–
			100	–	–	–	1.0	–	–	–	–	–	–	–
2355	frozen, condensed, commercial	1 CUP	250	79.8	225.0	6.8	10.8	25.3	.8	120.0	127.5	2360.0	–	462.5
			100	79.8	90.0	2.7	4.3	10.1	.3	48.0	51.0	944.0	–	185.0
	SCOTCH BROTH													
2356	condensed	1 CUP	251	–	–	–	5.5	–	–	–	–	–	–	–
			100	–	–	–	2.2	–	–	–	–	–	–	–
2357	condensed, diluted w/water	1 CUP	244	–	–	–	2.7	–	–	–	–	–	–	–
			100	–	–	–	1.1	–	–	–	–	–	–	–
	SHRIMP SOUP, CREAM OF													
2358	frozen, condensed, commercial	1 CUP	250	76.6	330.0	10.0	24.8	17.3	.8	77.5	100.0	2150.0	–	120.0
			100	76.6	132.0	4.0	9.9	6.9	.3	31.0	40.0	860.0	–	48.0
2359	frozen, condensed, made w/ = volume milk	1 CUP	250	81.8	250.0	9.5	17.3	14.8	.5	187.5	167.5	1137.5	–	247.5
			100	81.8	100.0	3.8	6.9	5.9	.2	75.0	67.0	455.0	–	99.0
2360	frozen, condensed, made w/ = volume water	1 CUP	240	88.3	158.4	4.8	11.9	8.3	.4	38.4	48.0	1032.0	–	57.6
			100	88.3	66.0	2.0	5.0	3.5	.2	16.0	20.0	430.0	–	24.0
	TOMATO SOUP													
2361	condensed	1 CUP	251	81.0	180.7	4.0	3.5	31.9	–	28.1	68.3	1987.9	–	471.9
			100	81.0	72.0	1.6	1.4	12.7	–	11.2	27.2	792.0	–	188.0
2362	diluted w/milk	1 CUP	248	84.0	171.6	6.4	6.2	22.3	–	166.7	153.8	1046.6	–	414.7
			100	84.0	69.2	2.6	2.5	9.0	–	67.2	62.0	422.0	–	167.2
2363	diluted w/water	1 CUP	244	90.5	87.6	2.0	2.0	15.6	–	14.9	33.9	966.0	–	229.1
			100	90.5	35.9	.8	.8	6.4	–	6.1	13.9	395.9	–	93.9
2364	w/rice	1 SERV	198	89.5	85.1	1.6	2.4	14.6	.2	15.8	25.7	655.4	–	257.4
			100	89.5	43.0	.8	1.2	7.4	.1	8.0	13.0	331.0	–	130.0
2365	w/vegetables & noodles, dehydrated, 2½ oz mix in 4 c water	1 CUP	245	92.5	66.2	1.5	1.5	12.5	.3	7.4	19.6	1046.2	–	29.4
			100	92.5	27.0	.6	.6	5.1	.1	3.0	8.0	427.0	–	12.0
	TURKEY & NOODLE SOUP													
2366	condensed	1 CUP	246	84.6	159.6	–	5.2	17.3	–	29.1	88.4	2046.3	–	157.6
			100	84.6	64.9	–	2.1	7.0	–	11.8	35.9	831.8	–	64.1
2367	condensed, diluted w/water	1 CUP	241	–	92.7	4.3	2.7	8.4	–	14.1	43.2	1002.2	–	77.3
			100	–	38.5	1.8	1.1	3.5	–	5.8	17.9	415.8	–	32.1

Item No.	Minerals (Micro): Iron	Zinc	Copper	Fat-Soluble Vitamins: Vitamin A	Vitamin D	Vitamin E (Alpha Tocopherol)	Water-Soluble Vitamins: Vitamin C	Thiamin	Ribo-flavin	Niacin	Panto-thenic Acid	Vit. B-6 (Pyri-doxine)	Folacin (Folic Acid)	Biotin	Vitamin B-12
	(mg)	(mg)	(mg)	(IU)	(IU)	(mg)	(mg)	(mg)	(mg)	(mg)	(mg)	(mg)	(mcg)	(mcg)	(mcg)
2342	**1.44**	–	–	**408.0**	–	–	–	**.12**	**.41**	**.48**	–	–	–	–	–
	.60	–	–	170.0	–	–	–	.05	.17	.20	–	–	–	–	–
2343	**1.44**	–	–	**240.0**	–	–	–	**.07**	**.19**	**.48**	–	–	–	–	–
	.60	–	–	100.0	–	–	–	.03	.08	.20	–	–	–	–	–
2344	**3.32**	–	–	**525.3**	–	–	**3.32**	**.20**	**.13**	**2.55**	–	–	**2.55**	–	–
	1.30	–	–	206.0	–	–	1.30	.08	.05	1.00	–	–	1.00	–	–
2345	**.80**	–	–	**420.0**	–	–	**8.00**	**.08**	**.22**	**1.00**	–	–	–	–	–
	.40	–	–	210.0	–	–	4.00	.04	.11	.50	–	–	–	–	–
2346	**1.60**	**.48**	–	**40.0**	–	–	–	**.12**	**.12**	**1.20**	–	–	–	–	–
	.80	.24	–	20.0	–	–	–	.06	.06	.60	–	–	–	–	–
2347	**4.00**	–	–	**462.5**	–	–	–	**.38**	**.15**	**2.50**	–	–	–	–	–
	1.60	–	–	185.0	–	–	–	.15	.06	1.00	–	–	–	–	–
2348	**1.60**	–	–	**186.0**	–	–	–	**.15**	**.06**	**1.00**	–	–	–	–	–
	.80	–	–	93.0	–	–	–	.08	.03	.50	–	–	–	–	–
2349	**.19**	**.03**	–	**40.6**	**3.83**	–	**.27**	**.01**	**.02**	**.15**	**.03**	**0**	**.47**	–	**.03**
	.67	.11	–	145.1	13.67	–	.96	.05	.08	.53	.11	.013	1.67	–	.12
2350	**1.20**	–	–	**360.0**	–	–	–	**.20**	**.12**	**1.20**	–	–	–	–	–
	.60	–	–	180.0	–	–	–	.10	.06	.60	–	–	–	–	–
2351	**2.81**	–	–	**918.0**	–	–	**2.55**	**.51**	**.31**	**2.81**	–	–	**5.10**	–	–
	1.10	–	–	360.0	–	–	1.00	.20	.12	1.10	–	–	2.00	–	–
2352	–	–	–	–	–	–	–	–	–	–	–	–	–	–	–
	–	–	–	–	–	–	–	–	–	–	–	–	–	–	–
2353	–	–	–	–	–	–	–	–	–	–	–	–	–	–	–
	–	–	–	–	–	–	–	–	–	–	–	–	–	–	–
2354	–	–	–	–	–	–	–	–	–	–	–	–	–	–	–
	–	–	–	–	–	–	–	–	–	–	–	–	–	–	–
2355	**1.75**	–	–	**862.5**	–	–	–	**.10**	**.13**	**1.00**	–	–	–	–	–
	.70	–	–	345.0	–	–	–	.04	.05	.40	–	–	–	–	–
2356	–	–	–	–	–	–	–	–	–	–	–	–	–	–	–
	–	–	–	–	–	–	–	–	–	–	–	–	–	–	–
2357	–	–	–	–	–	–	–	–	–	–	–	–	–	–	–
	–	–	–	–	–	–	–	–	–	–	–	–	–	–	–
2358	**1.00**	–	–	**232.5**	–	–	–	**.08**	**.13**	**.75**	–	–	–	–	–
	.40	–	–	93.0	–	–	–	.03	.05	.30	–	–	–	–	–
2359	**.63**	–	–	**317.5**	–	–	–	**.09**	**.28**	**.50**	–	–	–	–	–
	.25	–	–	127.0	–	–	–	.04	.11	.20	–	–	–	–	–
2360	**.48**	–	–	**112.8**	–	–	–	**.04**	**.06**	**.36**	–	–	–	–	–
	.20	–	–	47.0	–	–	–	.02	.03	.15	–	–	–	–	–
2361	**1.51**	–	–	**2038.1**	–	–	–	–	–	–	–	–	**30.12**	–	–
	.60	–	–	812.0	–	–	–	–	–	–	–	–	12.00	–	–
2362	**.79**	**1.07**	–	**1190.4**	–	–	**14.88**	**.10**	**.25**	**1.29**	–	–	–	–	–
	.32	.43	–	480.0	–	–	6.00	.04	.10	.52	–	–	–	–	–
2363	**.70**	**.17**	–	**995.9**	–	–	–	–	–	–	–	–	–	–	–
	.29	.07	–	408.2	–	–	–	–	–	–	–	–	–	–	–
2364	**.79**	–	–	**906.8**	–	–	–	**.06**	**.04**	**.89**	–	–	–	–	–
	.40	–	–	458.0	–	–	–	.03	.02	.45	–	–	–	–	–
2365	**.25**	–	–	**490.0**	–	–	**4.90**	**.05**	**.03**	**.49**	–	–	–	–	–
	.10	–	–	200.0	–	–	2.00	.02	.01	.20	–	–	–	–	–
2366	**1.21**	–	–	**391.6**	–	–	**0**	**.10**	**.10**	**2.51**	–	–	–	–	–
	.49	–	–	159.2	–	–	0	.04	.04	1.02	–	–	–	–	–
2367	**.70**	–	–	**190.8**	–	–	–	–	–	–	–	–	–	–	–
	.29	–	–	79.2	–	–	–	–	–	–	–	–	–	–	–

(Continued)

TABLE F-36 *(Continued)*

Item No.	Food Item Name	Approximate Measure	Weight	Moisture	Food Energy Calories	Protein	Fats	Carbo-hydrates	Fiber	Minerals (Macro) Calcium	Phos-phorus	Sodium	Mag-nesium	Potas-sium
			(g)	(%)	(kcal)	(g)	(g)	(g)	(g)	(mg)	(mg)	(mg)	(mg)	(mg)
	TURKEY AND NOODLE SOUP (Continued)													
2368	condensed, commercial	1 CUP	245	85.3	154.4	8.1	6.4	16.2	.2	19.6	93.1	1759.1	—	147.0
			100	85.3	63.0	3.3	2.6	6.6	.1	8.0	38.0	718.0	—	60.0
	VEGETABLE SOUP													
2369	condensed	1 CUP	251	83.4	160.6	5.5	4.3	27.6	—	40.2	80.3	1731.9	—	492.0
			100	83.4	64.0	2.2	1.7	11.0	—	16.0	32.0	690.0	—	196.0
2370	condensed, diluted w/water	1 CUP	244	91.7	77.7	2.7	2.2	13.4	—	19.9	38.8	841.6	—	239.0
			100	91.7	31.8	1.1	.9	5.5	—	8.1	15.9	344.9	—	98.0
2371	diluted w/water	1 CUP	244	91.9	77.7	5.1	3.4	9.6	—	12.0	48.8	1041.7	—	161.3
			100	91.9	31.8	2.1	1.4	3.9	—	4.9	20.0	426.9	—	66.1
2372	w/beef, condensed	1 CUP	251	83.8	163.7	10.5	7.0	19.9	—	25.1	98.4	2143.5	—	329.3
			100	83.8	65.2	4.2	2.8	7.9	—	10.0	39.2	854.0	—	131.2
2373	w/beef, condensed, commercial	1 CUP	250	84.0	165.0	10.0	5.8	17.5	.8	30.0	75.0	1887.5	30.0	375.0
			100	84.0	66.0	4.0	2.3	7.0	.3	12.0	30.0	755.0	12.0	150.0
2374	w/beef, frozen, commercial	1 CUP	250	82.7	170.0	13.5	5.8	16.5	1.3	52.5	157.5	2140.0	—	362.5
			100	82.7	68.0	5.4	2.3	6.6	.5	21.0	63.0	856.0	—	145.0
2375	w/beef broth, condensed, commercial	1 CUP	250	83.4	175.0	5.5	5.0	27.5	1.3	40.0	80.0	1600.0	30.0	425.0
			100	83.4	70.0	2.2	2.0	11.0	.5	16.0	32.0	640.0	12.0	170.0
2376	w/beef broth, condensed, made w/ = volume water	1 CUP	245	91.7	78.4	2.7	1.7	13.5	.7	19.6	39.2	845.3	24.5	240.1
			100	91.7	32.0	1.1	.7	5.5	.3	8.0	16.0	345.0	10.0	98.0
2377	vegetarian, condensed, commercial	1 CUP	250	83.8	155.0	4.5	3.8	26.0	1.3	37.5	75.0	1512.5	30.0	500.0
			100	83.8	62.0	1.8	1.5	10.4	.5	15.0	30.0	605.0	12.0	200.0
	VEGETABLES													
	ALFALFA SPROUTS													
2378	raw	1 CUP	38	88.3	15.6	1.9	.2	—	.6	10.6	—	—	—	—
			100	88.3	41.0	5.1	.6	—	1.7	28.0	—	—	—	—
2379	cooked	1 OZ	28	87.5	—	1.4	—	—	.4	7.9	—	—	—	—
			100	87.5	—	5.1	—	—	1.7	28.3	—	—	—	—
2380	[5]**AMARANTH LEAVES**	1 OZ	28	86.0	11.8	1.0	.2	2.1	.4	87.6	20.7	—	—	—
			100	86.0	42.0	3.7	.8	7.4	1.5	313.0	74.0	—	—	—
2381	[5]**ARROWROOT, BERMUDA**	1 OZ	28	57.2	44.0	.7	0	10.9	.5	5.6	6.7	—	—	—
			100	57.2	157.0	2.4	.1	39.0	1.9	20.0	24.0	—	—	—
2382	**ARTICHOKES,** Globe or French, boiled, drained	1 LG	100	90.2	38.0	3.4	.1	5.8	1.9	67.0	67.0	30.0	(27.2)	301.0
			100	90.2	38.0	3.4	.1	5.8	1.9	67.0	67.0	30.0	(27.2)	301.0
2383	**ARTICHOKES,** Jerusalem, raw	1 SM	25	79.8	10.3	.6	0	4.2	.2	3.5	19.5	—	2.8	—
			100	79.8	41.0	2.3	.1	16.7	.8	14.0	78.0	—	11.0	—
	ASPARAGUS													
2384	spears, raw	1 SPEAR	20	91.7	5.2	.5	0	1.0	.1	4.4	12.4	.4	4.0	55.6
			100	91.7	26.0	2.5	.2	5.0	.7	22.0	62.0	2.0	20.0	278.0
2385	fresh, cooked, drained	1 SPEAR	20	93.6	4.0	.4	0	.7	.1	4.2	10.0	.2	(2.9)	36.6
			100	93.6	20.0	2.2	.2	3.6	.7	21.0	50.0	1.0	(10.3)	183.0
2386	green, canned, drained solids	1 CUP	150	92.5	31.5	3.6	.6	5.1	1.2	28.5	79.5	354.0	—	249.0
			100	92.5	21.0	2.4	.4	3.4	.8	19.0	53.0	236.0	—	166.0
2387	green, canned, solids & liquid	1 SPEAR	22	93.6	4.0	.4	.1	.6	.1	4.0	9.5	51.9	4.4	36.5
			100	93.6	18.0	1.9	.3	2.9	.5	18.0	43.0	236.0	20.0	166.0
2388	green, canned, drained liquid	1 CUP	240	95.6	26.4	1.9	—	5.8	—	36.0	57.6	566.4	—	398.4
			100	95.6	11.0	.8	—	2.4	—	15.0	24.0	236.0	—	166.0
2389	green, canned, low sodium, drained solids	1 CUP	150	93.6	30.0	3.9	.5	4.7	1.1	28.5	79.5	4.5	—	249.0
			100	93.6	20.0	2.6	.3	3.1	.7	19.0	53.0	3.0	—	166.0
2390	green, canned, low sodium, solids & liquid	1 CUP	239	94.7	38.2	4.8	.5	6.5	1.2	43.0	102.8	7.2	47.8	396.7
			100	94.7	16.0	2.0	.2	2.7	.5	18.0	43.0	3.0	20.0	166.0
2391	green, canned, low sodium, drained liquid	1 CUP	240	96.8	21.6	1.9	—	4.8	—	36.0	57.6	7.2	—	398.4
			100	96.8	9.0	.8	—	2.0	—	15.0	24.0	3.0	—	166.0
2392	frozen, cuts & tips, boiled, drained	1 CUP	150	92.5	33.0	4.8	.3	5.3	1.2	33.0	96.0	1.5	21.0	330.0
			100	92.5	22.0	3.2	.2	3.5	.8	22.0	64.0	1.0	14.0	220.0
2393	frozen spears, boiled, drained	1 SPEAR	22	92.2	5.1	.7	0	.8	.2	4.8	14.7	.2	3.1	52.4
			100	92.2	23.0	3.2	.2	3.8	.8	22.0	67.0	1.0	14.0	238.0
2394	white, canned, drained solids	1 CUP	150	92.3	33.0	3.2	.8	5.4	1.2	24.0	61.5	354.0	—	210.0
			100	92.3	22.0	2.1	.5	3.6	.8	16.0	41.0	236.0	—	140.0
2395	white, canned, solids & liquid	1 SPEAR	22	93.3	4.0	.4	.1	.7	.1	3.3	7.3	51.9	4.4	30.8
			100	93.3	18.0	1.6	.3	3.3	.5	15.0	33.0	236.0	20.0	140.0
2396	white, canned, drained liquid	1 CUP	240	95.4	26.4	1.7	—	6.0	—	31.2	43.2	566.4	—	336.0
			100	95.4	11.0	.7	—	2.5	—	13.0	18.0	236.0	—	140.0
2397	white, canned, low sodium, drained solids	1 CUP	150	94.0	28.5	2.9	.3	5.3	1.1	24.0	61.5	6.0	—	210.0
			100	94.0	19.0	1.9	.2	3.5	.7	16.0	41.0	4.0	—	140.0

Item No.	Minerals (Micro)			Fat-Soluble Vitamins			Water-Soluble Vitamins								
	Iron	Zinc	Copper	Vitamin A	Vitamin D	Vitamin E (Alpha Tocopherol)	Vitamin C	Thiamin	Ribo-flavin	Niacin	Panto-thenic Acid	Vit. B-6 (Pyri-doxine)	Folacin (Folic Acid)	Biotin	Vitamin B-12
	(mg)	(mg)	(mg)	(IU)	(IU)	(mg)	(mg)	(mg)	(mg)	(mg)	(mg)	(mg)	(mcg)	(mcg)	(mcg)
2368	1.47 .60	— —	— —	294.0 120.0	— —	— —	— —	.15 .06	.15 .06	2.70 1.10	— —	— —	— —	— —	— —
2369	1.81 .72	— —	— —	6275.0 2500.0	— —	— —	— —	— —	— —	— —	— —	— —	— —	— —	— —
2370	.70 .29	1.81 .74	— —	3177.0 1302.0	— —	— —	— —	— —	— —	— —	— —	— —	— —	— —	— —
2371	.70 .28	3.86 1.58	— —	2689.0 1102.0	— —	— —	— —	.05 .02	.05 .02	1.00 .41	— —	— —	— —	— —	— —
2372	1.51 .60	— —	— —	5522.0 2200.0	— —	— —	— —	— —	— —	— —	— —	— —	— —	— —	— —
2373	1.75 .70	— —	.20 .08	2157.5 863.0	— —	— —	4.00 1.60	.08 .03	.10 .04	1.50 .60	— —	— —	— —	— —	— —
2374	2.00 .80	— —	— —	5680.0 2272.0	— —	— —	— —	.10 .04	.18 .07	3.75 1.50	— —	.18 .07	— —	— —	— —
2375	1.75 .70	— —	— —	2600.0 1040.0	— —	— —	5.00 2.00	.13 .05	.10 .04	2.50 1.00	— —	— —	— —	— —	— —
2376	.74 .30	— —	— —	3185.0 1300.0	— —	— —	— —	.05 .02	.03 .01	1.23 .50	— —	— —	— —	— —	— —
2377	2.00 .80	— —	— —	3037.5 1215.0	— —	— —	4.25 1.70	.13 .05	.08 .03	1.75 .70	— —	— —	22.50 9.00	— —	— —
2378	.53 1.40	.38 1.00	— —	— —	— —	— —	6.08 16.00	.05 .14	.08 .21	.61 1.60	— —	— —	— —	— —	— —
2379	.39 1.40	.28 1.00	— —	— —	— —	— —	3.08 11.00	.03 .12	.06 .20	.22 .80	— —	— —	— —	— —	— —
2380	1.56 5.60	— —	— —	448.0 1600.0	— —	— —	18.20 65.00	.01 .05	.07 .24	.34 1.20	— —	— —	— —	— —	— —
2381	.90 3.20	— —	— —	0 0	— —	— —	2.52 9.00	.02 .08	.01 .03	.20 .70	— —	— —	— —	— —	— —
2382	.90 .90	.35 .35	(.03) (.09)	90.0 90.0	0 0	— —	8.40 8.40	.07 .07	.03 .03	.93 .93	.21 .21	.07 .07	32.00 32.00	4.10 4.10	0 0
2383	.85 3.40	.02 .06	— —	5.0 20.0	— —	— —	1.00 4.00	.05 .20	.02 .06	.33 1.30	.02 .07	0 0	— —	— —	— —
2384	.20 1.00	[6]1.4 0.7	.01 .04	180.0 900.0	— —	— —	6.60 33.00	.04 .18	.04 .20	.30 1.50	.12 .62	.03 .15	21.8 109.0	.34 1.70	0 0
2385	.12 .60	(.09) (.31)	(.06) (.22)	180.0 900.0	— —	(.70) (2.50)	5.20 26.00	.03 .16	.04 .18	.28 1.40	(.04) (.14)	(.01) (.04)	(8.42) (30.06)	(.13) (.46)	0 0
2386	2.85 1.90	— —	— —	1200.0 800.0	— —	— —	22.50 15.00	.09 .06	.15 .10	1.20 .80	— —	— —	114.00 76.00	— —	— —
2387	.37 1.70	(.11) (.50)	(.03) (.14)	112.2 510.0	— —	— —	3.30 15.00	.01 .06	.02 .09	.18 .80	.04 .20	.01 .06	18.70 85.00	.37 1.70	0 0
2388	3.36 1.40	— —	— —	— —	— —	— —	36.00 15.00	.14 .06	.17 .07	1.92 .80	— —	— —	244.80 102.00	— —	— —
2389	2.85 1.90	— —	— —	1200.0 800.0	— —	— —	22.50 15.00	.09 .06	.15 .10	1.20 .80	— —	— —	114.00 76.00	— —	— —
2390	4.06 1.70	— —	— —	1218.9 510.0	— —	— —	35.85 15.00	.14 .06	.22 .09	1.91 .80	.47 .20	.13 .06	203.15 85.00	4.06 1.70	0 0
2391	3.36 1.40	— —	— —	— —	— —	— —	36.00 15.00	.14 .06	.17 .07	1.92 .80	— —	— —	244.80 102.00	— —	— —
2392	1.80 1.20	(.89) (.59)	(.21) (.14)	1275.0 850.0	— —	— —	34.50 23.00	.10 .07	.16 .10	1.50 1.00	— —	.03 .02	118.50 79.00	— —	— —
2393	.24 1.10	— —	— —	171.6 780.0	— —	— —	5.72 26.00	.01 .07	.02 .10	.24 1.10	.07 .32	0 .02	17.38 79.00	— —	— —
2394	1.50 1.00	— —	— —	120.0 80.0	— —	— —	22.50 15.00	.08 .05	.09 .06	1.05 .70	— —	— —	— —	— —	— —
2395	.20 .90	— —	— —	11.0 50.0	— —	— —	3.30 15.00	.01 .05	.01 .06	.15 .70	.03 .13	.01 .04	— —	— —	0 0
2396	1.69 .70	— —	— —	— —	— —	— —	36.00 15.00	.12 .05	.10 .04	1.68 .70	— —	— —	— —	— —	— —
2397	1.50 1.00	— —	— —	120.0 80.0	— —	— —	22.50 15.00	.08 .05	.09 .06	1.05 .70	— —	— —	— —	— —	— —

(Continued)

TABLE F-36 *(Continued)*

Item No.	Food Item Name	Approximate Measure	Weight	Moisture	Food Energy Calories	Protein	Fats	Carbo-hydrates	Fiber	Minerals (Macro)				
										Calcium	Phos-phorus	Sodium	Mag-nesium	Potas-sium
			(g)	(%)	(kcal)	(g)	(g)	(g)	(g)	(mg)	(mg)	(mg)	(mg)	(mg)
	ASPARAGUS (Continued)													
2398	white, canned, low sodium, drained solids & liquid	1 CUP	**150** 100	**95.0** 95.0	**24.0** 16.0	**2.1** 1.4	**.3** .2	**4.5** 3.0	**.8** .5	**22.5** 15.0	**49.5** 33.0	**6.0** 4.0	**30.0** 20.0	**210.0** 140.0
2399	white, canned, low sodium, drained liquid	1 CUP	**240** 100	**97.2** 97.2	**19.2** 8.0	**1.4** .6	— —	**4.3** 1.8	— —	**31.2** 13.0	**43.2** 18.0	**9.6** 4.0	— —	**336.0** 140.0
2400	[3]**BAMBARRA-GROUNDNUT**, whole seeds, dried	1 OZ	**28** 100	**10.3** 10.3	**102.8** 367.0	**5.3** 18.8	**1.7** 6.2	**17.2** 61.3	**1.3** 4.8	**17.4** 62.0	**77.3** 276.0	— —	— —	— —
	BAMBOO SHOOTS													
2401	raw	1 CUP	**133** 100	**91.0** 91.0	**35.9** 27.0	**3.5** 2.6	**.4** .3	**6.9** 5.2	**.9** .7	**17.3** 13.0	**78.5** 59.0	— —	— —	**706.9** 533.0
2402	canned	1 CUP	**133** 100	**95.0** 95.0	**21.3** 16.0	**1.9** 1.4	**.1** .1	**3.5** 2.6	**.9** .7	**13.7** 10.3	**20.1** 15.1	**5.1** 3.8	**6.0** 4.5	**128.1** 96.3
	BEANS, GREEN													
2403	raw	1 CUP	**100** 100	**90.1** 90.1	**32.0** 32.0	**1.9** 1.9	**.2** .2	**7.1** 7.1	**1.0** 1.0	**56.0** 56.0	**44.0** 44.0	**7.0** 7.0	**32.0** 32.0	**243.0** 243.0
2404	fresh, boiled in small amount of water, drained	1 CUP	**125** 100	**92.4** 92.4	**31.3** 25.0	**2.0** 1.6	**.2** .2	**6.8** 5.4	**1.3** 1.0	**62.5** 50.0	**46.3** 37.0	**5.0** 4.0	— —	**188.8** 151.0
2405	fresh, boiled in large amount of water, drained	1 CUP	**125** 100	**92.4** 92.4	**31.3** 25.0	**2.0** 1.6	**.2** .2	**6.8** 5.4	**1.3** 1.0	**62.5** 50.0	**46.3** 37.0	**5.0** 4.0	— —	**188.8** 151.0
2406	canned, drained solids	1 CUP	**220** 100	**91.9** 91.9	**52.8** 24.0	**3.1** 1.4	**.4** .2	**11.4** 5.2	**2.2** 1.0	**99.0** 45.0	**55.0** 25.0	**519.2** 236.0	**30.8** 14.0	**209.0** 95.0
2407	canned, solids & liquid	1 CUP	**239** 100	**93.5** 93.5	**43.0** 18.0	**2.4** 1.0	**.2** .1	**10.0** 4.2	**1.4** .6	**81.3** 34.0	**50.2** 21.0	**564.0** 236.0	**33.5** 14.0	**227.1** 95.0
2408	canned, drained liquid	1 CUP	**240** 100	**95.9** 95.9	**24.0** 10.0	**1.0** .4	**.2** .1	**5.8** 2.4	— —	**36.0** 15.0	**33.6** 14.0	**566.4** 236.0	— —	**228.0** 95.0
2409	canned, low sodium, drained solids	1 CUP	**220** 100	**93.2** 93.2	**48.4** 22.0	**3.3** 1.5	**.2** .1	**10.6** 4.8	**2.0** .9	**99.0** 45.0	**55.0** 25.0	**4.4** 2.0	**30.8** 14.0	**209.0** 95.0
2410	canned, low sodium, solids & liquid	1 CUP	**239** 100	**94.8** 94.8	**38.2** 16.0	**2.6** 1.1	**.2** .1	**8.6** 3.6	**1.4** .6	**81.3** 34.0	**50.2** 21.0	**4.8** 2.0	**33.5** 14.0	**227.1** 95.0
2411	canned, low sodium, drained liquid	1 CUP	**240** 100	**97.3** 97.3	**19.2** 8.0	**1.0** .4	**.2** .1	**4.3** 1.8	— —	**36.0** 15.0	**33.6** 14.0	**4.8** 2.0	— —	**228.0** 95.0
2412	frozen, cut, boiled, drained	1 CUP	**161** 100	**92.1** 92.1	**40.3** 25.0	**2.6** 1.6	**.2** .1	**9.2** 5.7	**1.6** 1.0	**64.4** 40.0	**51.5** 32.0	**1.6** 1.0	**32.2** 20.0	**244.7** 152.0
2413	frozen, french style, boiled, drained	1 CUP	**200** 100	**91.9** 91.9	**52.0** 26.0	**3.2** 1.6	**.2** .1	**12.0** 6.0	**2.2** 1.1	**76.0** 38.0	**60.0** 30.0	**4.0** 2.0	**40.0** 20.0	**272.0** 136.0
	BEANS, LIMA													
2414	immature seeds, boiled, drained	1 CUP	**184** 100	**71.1** 71.1	**204.2** 111.0	**14.0** 7.6	**.9** .5	**36.4** 19.8	**3.3** 1.8	**86.5** 47.0	**222.6** 121.0	**1.8** 1.0	— —	**776.5** 422.0
2415	canned, drained solids	1 CUP	**230** 100	**74.7** 74.7	**220.8** 96.0	**12.4** 5.4	**.7** .3	**42.1** 18.3	**4.1** 1.8	**64.4** 28.0	**161.0** 70.0	**542.8** 236.0	— —	**510.6** 222.0
2416	canned, solids & liquid	1 CUP	**184** 100	**80.8** 80.8	**130.6** 71.0	**7.5** 4.1	**.6** .3	**24.7** 13.4	**2.4** 1.3	**47.8** 26.0	**123.3** 67.0	**434.2** 236.0	**123.3** 67.0	**408.5** 222.0
2417	canned, drained liquid	1 CUP	**240** 100	**93.3** 93.3	**48.0** 20.0	**3.1** 1.3	— —	**9.4** 3.9	— —	**52.8** 22.0	**144.0** 60.0	**566.4** 236.0	— —	**532.8** 222.0
2418	canned, low sodium, drained solids	1 CUP	**230** 100	**75.6** 75.6	**218.5** 95.0	**13.3** 5.8	**.7** .3	**40.7** 17.7	**4.1** 1.8	**64.4** 28.0	**161.0** 70.0	**9.2** 4.0	— —	**510.6** 222.0
2419	canned, low sodium, solids & liquid	1 CUP	**184** 100	**81.7** 81.7	**128.8** 70.0	**8.1** 4.4	**.6** .3	**23.7** 12.9	**2.2** 1.2	**47.8** 26.0	**123.3** 67.0	**7.4** 4.0	**123.3** 67.0	**408.5** 222.0
2420	canned, low sodium, drained liquid	1 CUP	**240** 100	**94.4** 94.4	**45.6** 19.0	**3.4** 1.4	— —	**8.4** 3.5	— —	**52.8** 22.0	**144.0** 60.0	**9.6** 4.0	— —	**532.8** 222.0
2421	baby, frozen, boiled, drained	1 CUP	**173** 100	**68.8** 68.8	**204.1** 118.0	**12.8** 7.4	**.3** .2	**38.6** 22.3	**3.3** 1.9	**60.6** 35.0	**218.0** 126.0	**223.2** 129.0	**83.0** 48.0	**681.6** 394.0
2422	frozen, boiled, drained	1 CUP	**160** 100	**73.5** 73.5	**158.4** 99.0	**9.6** 6.0	**.2** .1	**30.6** 19.1	**2.6** 1.6	**32.0** 20.0	**144.0** 90.0	**161.6** 101.0	**76.8** 48.0	**681.6** 426.0
	BEANS, MUNG													
2423	sprouted seeds, uncooked	1 CUP	**210** 100	**85.9** 85.9	**111.3** 53.0	**9.0** 4.3	**.4** .2	**13.9** 6.6	**1.3** .6	**27.3** 13.0	**134.4** 64.0	**10.5** 5.0	**33.6** 16.0	**468.3** 223.0
2424	sprouted seeds, boiled, drained	1 CUP	**210** 100	**84.3** 84.3	**58.8** 28.0	**9.0** 4.3	**.4** .2	**10.9** 5.2	**1.5** .7	**27.5** 13.1	**100.8** 48.0	**8.4** 4.0	— —	**327.6** 156.0
2425	**BEANS,** refried, cooked in iron skillet	1 CUP	**210** 100	— —	— —	— —	— —	— —	— —	— —	— —	— —	— —	— —
2426	**BEAN SPROUTS**—See **BEANS, MUNG, LENTIL & SOYBEAN SPROUTS**													

Item No.	Minerals (Micro) Iron	Zinc	Copper	Fat-Soluble Vitamins Vitamin A	Vitamin D	Vitamin E (Alpha Tocopherol)	Water-Soluble Vitamins Vitamin C	Thiamin	Ribo-flavin	Niacin	Panto-thenic Acid	Vit. B-6 (Pyri-doxine)	Folacin (Folic Acid)	Biotin	Vitamin B-12
	(mg)	(mg)	(mg)	(IU)	(IU)	(mg)	(mg)	(mg)	(mg)	(mg)	(mg)	(mg)	(mcg)	(mcg)	(mcg)
2398	**1.35**	–	–	**75.0**	–	–	**22.50**	**.08**	**.09**	**1.05**	**.19**	**.05**	–	–	**0**
	.90	–	–	50.0	–	–	15.00	.05	.06	.70	.13	.04	–	–	0
2399	**1.68**	–	–	–	–	–	**36.00**	**.12**	**.10**	**1.68**	–	–	–	–	–
	.70	–	–	–	–	–	15.00	.05	.04	.70	–	–	–	–	–
2400	**3.42**	–	–	**9.3**	–	–	**0**	**.13**	**.04**	**.50**	–	–	–	–	–
	12.20	–	–	33.3	–	–	0	.47	.14	1.80	–	–	–	–	–
2401	**.67**	–	–	**26.6**	–	–	**5.32**	**.20**	**.09**	**.80**	–	–	–	–	–
	.50	–	–	20.0	–	–	4.00	.15	.07	.60	–	–	–	–	–
2402	**.36**	**.65**	**.01**	–	–	–	**1.33**	**.01**	**.03**	**.20**	–	–	–	–	–
	.27	.49	.01	–	–	–	1.00	.01	.02	.15	–	–	–	–	–
2403	**.80**	**.40**	–	**600.0**	–	–	**19.00**	**.08**	**.11**	**.50**	**.19**	**.08**	**27.50**	–	**0**
	.80	.40	–	600.0	–	–	19.00	.08	.11	.50	.19	.08	27.50	–	0
2404	**.75**	**.38**	–	**675.0**	–	–	**15.00**	**.09**	**.11**	**.63**	–	–	**50.00**	–	–
	.60	.30	–	540.0	–	–	12.00	.07	.09	.50	–	–	40.00	–	–
2405	**.75**	**.38**	–	**675.0**	–	–	**12.50**	**.08**	**.10**	**.38**	–	–	**50.00**	–	–
	.60	.30	–	540.0	–	–	10.00	.06	.08	.30	–	–	40.00	–	–
2406	**3.30**	**.66**	–	**1034.0**	–	–	**9.61**	**.07**	**.11**	**.66**	–	**.09**	**35.20**	–	**0**
	1.50	.30	–	470.0	–	–	4.37	.03	.05	.30	–	.04	16.00	–	0
2407	**2.87**	**.48**	**(.23)**	**693.1**	–	**.07**	**9.56**	**.07**	**.10**	**.72**	**.18**	**.10**	**43.02**	**3.11**	**0**
	1.20	.20	(.10)	290.0	–	.03	4.00	.03	.04	.30	.08	.04	18.00	1.30	0
2408	**2.16**	–	–	–	–	–	**9.60**	**.07**	**.07**	**.72**	–	–	**48.00**	–	–
	.90	–	–	–	–	–	4.00	.03	.03	.30	–	–	20.00	–	–
2409	**3.30**	–	–	**1034.0**	–	–	**8.80**	**.07**	**.11**	**.66**	–	**.09**	**29.92**	–	**0**
	1.50	–	–	470.0	–	–	4.00	.03	.05	.30	–	.04	13.60	–	0
2410	**2.87**	**.02**	–	**693.1**	–	–	**9.56**	**.07**	**.10**	**.72**	**.18**	**.10**	**32.50**	**3.11**	**0**
	1.20	.01	–	290.0	–	–	4.00	.03	.04	.30	.08	.04	13.60	1.30	0
2411	**2.16**	–	–	–	–	–	**9.60**	**.07**	**.07**	**.72**	–	–	**48.00**	–	–
	.90	–	–	–	–	–	4.00	.03	.03	.30	–	–	20.00	–	–
2412	**1.13**	**.34**	**(.10)**	**933.8**	–	**.18**	**8.05**	**.11**	**.15**	**.64**	**.21**	**.10**	–	–	–
	.70	.21	(.05)	580.0	–	.11	5.00	.07	.09	.40	.13	.06	–	–	–
2413	**1.80**	**.42**	**(.08)**	**1060.0**	–	–	**14.00**	**.12**	**.16**	**.60**	**.24**	**.09**	–	–	–
	.90	.21	(.05)	530.0	–	–	7.00	.06	.08	.30	.12	.04	–	–	–
2414	**4.60**	–	–	**515.2**	–	–	**31.28**	**.33**	**.18**	**2.39**	–	–	–	–	–
	2.50	–	–	280.0	–	–	17.00	.18	.10	1.30	–	–	–	–	–
2415	**5.52**	–	–	**437.0**	–	–	**13.80**	**.07**	**.12**	**1.15**	–	–	**89.70**	–	–
	2.40	–	–	190.0	–	–	6.00	.03	.05	.50	–	–	39.00	–	–
2416	**4.42**	**(1.20)**	**(.34)**	**239.2**	–	–	**12.88**	**.07**	**.07**	**.92**	**.24**	**.17**	**69.92**	–	**0**
	2.40	(.65)	(.19)	130.0	–	–	7.00	.04	.04	.50	.13	.09	38.00	–	0
2417	**5.52**	[6]**1.44**	–	–	–	–	**24.00**	**.10**	**.07**	**1.44**	–	–	**86.40**	–	–
	2.30	.60	–	–	–	–	10.00	.04	.03	.60	–	–	36.00	–	–
2418	**5.52**	–	–	**437.0**	–	–	**13.80**	**.07**	**.12**	**1.15**	–	–	**89.70**	–	–
	2.40	–	–	190.0	–	–	6.00	.03	.05	.50	–	–	39.00	–	–
2419	**4.42**	–	–	**239.2**	–	–	**12.88**	**.07**	**.07**	**.92**	**.24**	**.17**	**69.92**	–	**0**
	2.40	–	–	130.0	–	–	7.00	.04	.04	.50	.13	.09	38.00	–	0
2420	**5.52**	–	–	–	–	–	**24.00**	**.10**	**.07**	**1.44**	–	–	**86.40**	–	–
	2.30	–	–	–	–	–	10.00	.04	.03	.60	–	–	36.00	–	–
2421	**4.50**	–	–	**380.6**	–	–	**20.76**	**.16**	**.09**	**2.08**	**.31**	**.15**	**112.45**	–	–
	2.60	–	–	220.0	–	–	12.00	.09	.05	1.20	.18	.09	65.00	–	–
2422	**2.72**	**.77**	**.10**	**368.0**	–	–	**27.20**	**.11**	**.08**	**1.60**	**.29**	**.16**	**104.00**	–	–
	1.70	.48	.06	230.0	–	–	17.00	.07	.05	1.00	.18	.10	65.00	–	–
2423	**4.00**	**1.89**	–	**42.0**	–	–	**42.00**	**.29**	**.38**	**2.31**	–	–	–	–	–
	1.90	.90	–	20.0	–	–	20.00	.14	.18	1.10	–	–	–	–	–
2424	**4.00**	**1.89**	–	**42.0**	–	–	**33.60**	**.29**	**.38**	**2.52**	–	–	**21.00**	–	–
	1.90	.90	–	20.0	–	–	16.00	.14	.18	1.20	–	–	10.00	–	–
2425	**3.84**	–	–	–	–	–	–	–	–	–	–	–	–	–	–
	1.83	–	–	–	–	–	–	–	–	–	–	–	–	–	–
2426															

(Continued)

VEGETABLES

TABLE F-36 *(Continued)*

Item No.	Food Item Name	Approximate Measure	Weight	Moisture	Food Energy Calories	Protein	Fats	Carbo-hydrates	Fiber	Minerals (Macro)				
										Calcium	Phos-phorus	Sodium	Mag-nesium	Potas-sium
			(g)	(%)	(kcal)	(g)	(g)	(g)	(g)	(mg)	(mg)	(mg)	(mg)	(mg)
	BEANS, WAX													
2427	raw	1 CUP	100	91.4	27.0	1.7	.2	6.0	1.0	56.0	43.0	7.0	32.0	243.0
			100	91.4	27.0	1.7	.2	6.0	1.0	56.0	43.0	7.0	32.0	243.0
2428	boiled, drained	1 CUP	164	93.4	36.1	2.3	.3	7.5	1.6	82.0	60.7	4.9	—	247.6
			100	93.4	22.0	1.4	.2	4.6	1.0	50.0	37.0	3.0	—	151.0
2429	canned, drained solids	1 CUP	200	92.2	48.0	2.8	.6	10.4	1.8	90.0	50.0	472.0	—	190.0
			100	92.2	24.0	1.4	.3	5.2	.9	45.0	25.0	236.0	—	95.0
2430	canned, solids & liquid	1 CUP	250	93.7	47.5	2.5	.5	10.5	1.5	85.0	52.5	590.0	35.0	237.5
			100	93.7	19.0	1.0	.2	4.2	.6	34.0	21.0	236.0	14.0	95.0
2431	canned, drained liquid	1 CUP	240	96.1	26.4	1.0	.2	6.0	—	36.0	33.6	566.4	—	228.0
			100	96.1	11.0	.4	.1	2.5	—	15.0	14.0	236.0	—	95.0
2432	canned, low sodium, drained solids	1 CUP	200	93.6	42.0	2.4	.2	9.4	1.8	90.0	50.0	4.0	22.0	190.0
			100	93.6	21.0	1.2	.1	4.7	.9	45.0	25.0	2.0	11.0	95.0
2433	canned, low sodium, solids & liquid	1 CUP	250	95.2	37.5	2.3	.3	8.5	1.5	85.0	52.5	5.0	—	237.5
			100	95.2	15.0	.9	.1	3.4	.6	34.0	21.0	2.0	—	95.0
2434	canned, low sodium, drained liquid	1 CUP	240	97.7	16.8	1.0	.2	3.4	—	36.0	33.6	4.8	—	228.0
			100	97.7	7.0	.4	.1	1.4	—	15.0	14.0	2.0	—	95.0
2435	frozen, cut, boiled, drained	1 CUP	161	91.5	43.5	2.7	.2	10.0	1.8	56.4	49.9	1.6	33.8	264.0
			100	91.5	27.0	1.7	.1	6.2	1.1	35.0	31.0	1.0	21.0	164.0
	BEET GREENS													
2436	raw	1 CUP	33	90.9	7.9	.7	.1	1.5	.4	39.3	13.2	42.9	35.0	188.1
			100	90.9	24.0	2.2	.3	4.6	1.3	119.0	40.0	130.0	106.0	570.0
2437	boiled, drained	1 CUP	200	93.6	36.0	3.4	.4	6.6	2.2	198.0	50.0	152.0	—	664.0
			100	93.6	18.0	1.7	.2	3.3	1.1	99.0	25.0	76.0	—	332.0
	BEETS													
2438	raw	1 CUP	170	87.3	73.1	2.7	.2	16.8	1.4	27.2	56.1	102.0	42.5	569.5
			100	87.3	43.0	1.6	.1	9.9	.8	16.0	33.0	60.0	25.0	335.0
2439	fresh, boiled, drained	1 CUP	200	90.9	64.0	2.2	.2	14.4	1.6	28.0	46.0	86.0	—	416.0
			100	90.9	32.0	1.1	.1	7.2	.8	14.0	23.0	43.0	—	208.0
2440	canned, drained solids	1 CUP	166	89.3	61.4	1.7	.2	14.6	1.3	31.5	29.9	391.8	24.9	277.2
			100	89.3	37.0	1.0	.1	8.8	.8	19.0	18.0	236.0	15.0	167.0
2441	canned, solids & liquid	1 CUP	246	90.3	83.6	2.2	.2	19.4	1.2	34.4	41.8	580.6	36.9	410.8
			100	90.3	34.0	.9	.1	7.9	.5	14.0	17.0	236.0	15.0	167.0
2442	canned, drained liquid	1 CUP	240	92.2	62.4	1.9	—	14.9	—	12.0	36.0	566.4	—	400.8
			100	92.2	26.0	.8	—	6.2	—	5.0	15.0	236.0	—	167.0
2443	canned, low sodium, drained solids	1 CUP	166	89.8	61.4	1.5	.2	14.4	1.3	31.5	29.9	76.4	38.2	277.2
			100	89.8	37.0	.9	.1	8.7	.8	19.0	18.0	46.0	23.0	167.0
2444	canned, low sodium, solids & liquid	1 CUP	246	90.8	78.7	2.2	—	19.2	1.2	34.4	41.8	113.2	—	410.8
			100	90.8	32.0	.9	—	7.8	.5	14.0	17.0	46.0	—	167.0
2445	canned, low sodium, drained liquid	1 CUP	240	92.8	60.0	1.9	—	14.2	—	12.0	36.0	110.4	—	400.8
			100	92.8	25.0	.8	—	5.9	—	5.0	15.0	46.0	—	167.0
	BROADBEAN													
2446	[4]fried & salted	1 OZ	28	7.6	112.6	7.4	4.1	13.3	1.1	20.4	92.7	—	—	278.3
			100	7.6	402.0	26.4	14.8	47.4	3.8	73.0	331.0	—	—	994.0
	BROCCOLI													
2447	[2]spears, raw	1 CUP	155	89.1	49.6	5.6	.5	9.1	2.3	159.7	120.9	23.3	28.7	592.1
			100	89.1	32.0	3.6	.3	5.9	1.5	103.0	78.0	15.0	18.5	382.0
2448	spears, cooked	1 CUP	150	91.3	39.0	4.7	.5	6.8	2.4	132.0	93.0	15.0	—	400.5
			100	91.3	26.0	3.1	.3	4.5	1.5	88.0	62.0	10.0	—	267.0
2449	spears, frozen, cooked	1 CUP	188	91.4	48.9	5.8	.4	8.8	2.2	77.1	109.0	22.6	39.5	413.6
			100	91.4	26.0	3.1	.2	4.7	1.1	41.0	58.0	12.0	21.0	220.0
2450	chopped, frozen, cooked	1 CUP	188	91.6	48.9	5.5	.6	8.6	2.2	101.5	105.3	28.2	39.5	398.6
			100	91.6	26.0	2.9	.3	4.6	1.1	54.0	56.0	15.0	21.0	212.0
	BRUSSELS SPROUTS													
2451	raw	1 MED	10	85.2	4.5	.5	Trace	.8	.2	3.6	8.0	1.4	2.9	39.0
			100	85.2	45.0	4.9	.4	8.3	1.6	36.0	80.0	14.0	29.0	390.0
2452	cooked	1 CUP	150	88.2	54.0	6.3	.6	9.6	2.4	48.0	108.0	15.0	—	409.5
			100	88.2	36.0	4.2	.4	6.4	1.6	32.0	72.0	10.0	—	273.0
2453	frozen, cooked	1 CUP	150	89.3	49.5	4.8	.3	9.8	1.8	31.5	91.5	21.0	31.5	442.5
			100	89.3	33.0	3.2	.2	6.5	1.2	21.0	61.0	14.0	21.0	295.0

Item No.	Minerals (Micro)			Fat-Soluble Vitamins			Water-Soluble Vitamins								
	Iron	Zinc	Copper	Vitamin A	Vitamin D	Vitamin E (Alpha Tocopherol)	Vitamin C	Thiamin	Ribo-flavin	Niacin	Panto-thenic Acid	Vit. B-6 (Pyri-doxine)	Folacin (Folic Acid)	Biotin	Vitamin B-12
	(mg)	(mg)	(mg)	(IU)	(IU)	(mg)	(mg)	(mg)	(mg)	(mg)	(mg)	(mg)	(mcg)	(mcg)	(mcg)
2427	**.80**	–	–	**250.0**	–	–	**20.00**	**.08**	**.11**	**.50**	**.25**	–	**40.00**	–	**0**
	.80	–	–	250.0	–	–	20.00	.08	.11	.50	.25	–	40.00	–	0
2428	**.98**	–	–	**377.2**	–	–	**21.32**	**.12**	**.15**	**.82**	–	–	–	–	–
	.60	–	–	230.0	–	–	13.00	.07	.09	.50	–	–	–	–	–
2429	**3.00**	**.46**	–	**200.0**	–	–	**10.00**	**.06**	**.10**	**.60**	–	–	–	–	–
	1.50	.23	–	100.0	–	–	5.00	.03	.05	.30	–	–	–	–	–
2430	**3.00**	**.38**	**.16**	**150.0**	–	–	**12.50**	**.08**	**.10**	**.75**	–	**.11**	–	–	**0**
	1.20	.15	.07	60.0	–	–	5.00	.03	.04	.30	–	.04	–	–	0
2431	**2.16**	–	–	–	–	–	**12.00**	**.07**	**.07**	**.72**	–	–	–	–	–
	.90	–	–	–	–	–	5.00	.03	.03	.30	–	–	–	–	–
2432	**3.00**	**.56**	**.04**	**200.0**	–	–	**10.00**	**.06**	**.10**	**.60**	–	–	–	–	–
	1.50	.28	.02	100.0	–	–	5.00	.03	.05	.30	–	–	–	–	–
2433	**3.00**	–	–	**150.0**	–	–	**12.50**	**.08**	**.10**	**.75**	–	**.11**	–	–	**0**
	1.20	–	–	60.0	–	–	5.00	.03	.04	.30	–	.04	–	–	0
2434	**2.16**	–	–	–	–	–	**12.00**	**.07**	**.07**	**.72**	–	–	–	–	–
	.90	–	–	–	–	–	5.00	.03	.03	.30	–	–	–	–	–
2435	**1.13**	–	–	**161.0**	–	–	**9.66**	**.11**	**.13**	**.64**	–	–	–	–	–
	.70	.24	.06	100.0	–	–	6.00	.07	.08	.40	–	–	–	–	–
2436	**1.09**	**.01**	–	**2013.0**	–	–	**9.90**	**.03**	**.07**	**.13**	**.08**	**.03**	–	–	**0**
	3.30	.04	–	6100.0	–	–	30.00	.10	.22	.40	.25	.10	–	–	0
2437	**3.80**	–	–	**10200.0**	–	–	**30.00**	**.14**	**.30**	**.60**	–	–	–	–	–
	1.90	–	–	5100.0	–	–	15.00	.07	.15	.30	–	–	–	–	–
2438	**1.19**	**.09**	**.03**	**34.0**	–	–	**17.00**	**.05**	**.09**	**.68**	**.26**	**.09**	**22.95**	**3.23**	**0**
	.70	.05	.02	20.0	–	–	10.00	.03	.05	.40	.15	.06	13.50	1.90	0
2439	**1.00**	–	–	**40.0**	–	–	**12.00**	**.06**	**.08**	**.60**	–	–	**156.00**	–	**.17**
	.50	–	–	20.0	–	–	6.00	.03	.04	.30	–	–	78.00	–	.09
2440	**1.16**	–	–	**33.2**	–	–	**4.98**	**.02**	**.05**	**.17**	–	–	**39.84**	–	–
	.70	–	–	20.0	–	–	3.00	.01	.03	.10	–	–	24.00	–	–
2441	**1.48**	**.20**	**.22**	**24.6**	–	–	**7.38**	**.03**	**.05**	**.25**	**.25**	**.12**	**71.34**	–	**0**
	.60	.08	.09	10.0	–	–	3.00	.01	.02	.10	.10	.05	29.00	–	0
2442	**.96**	–	–	–	–	–	**7.20**	**.02**	**.05**	**.24**	–	–	**88.80**	–	–
	.40	–	–	–	–	–	3.00	.01	.02	.10	–	–	37.00	–	–
2443	**1.16**	**.60**	**.33**	**33.2**	–	–	**4.98**	**.02**	**.05**	**.17**	–	–	**39.84**	–	–
	.70	.30	.20	20.0	–	–	3.00	.01	.03	.10	–	–	24.00	–	–
2444	**1.48**	**.20**	–	**24.6**	–	–	**7.38**	**.03**	**.05**	**.25**	**.25**	**.12**	**71.34**	–	**0**
	.60	.08	–	10.0	–	–	3.00	.01	.02	.10	.10	.05	29.00	–	–
2445	**.95**	–	–	–	–	–	**7.20**	**.02**	**.05**	**.24**	–	–	**88.80**	–	–
	.40	–	–	–	–	–	3.00	.01	.02	.10	–	–	37.00	–	–
2446	**1.98**	–	–	**4.7**	–	–	**0**	**.03**	**.01**	**.76**	–	–	–	–	–
	7.10	–	–	16.7	–	–	0	.10	.05	1.00	–	–	–	–	–
2447	**1.71**	**1.01**	**1.24**	**3875.0**	**0**	**2.00**	**175.15**	**.26**	**.36**	**1.4**	**1.55**	**.33**	**201.50**	**.93**	**0**
	1.10	.65	.08	2500.0	0	1.30	113.00	.10	.23	.9	1.00	.21	130.00	.60	0
2448	**1.20**	**.23**	–	**3750.0**	–	–	**135.00**	**.14**	**.30**	**1.20**	–	–	**84.00**	**1.80**	–
	.80	.15	–	2500.0	–	–	90.00	.09	.20	.80	–	–	56.00	1.20	–
2449	**1.32**	**.28**	–	**3572.0**	–	–	**137.24**	**.11**	**.21**	**.94**	**.98**	**.20**	**112.80**	–	–
	.70	.15	–	1900.0	–	–	73.00	.06	.11	.50	.52	.10	60.00	–	–
2450	**1.32**	–	–	**4888.0**	–	–	**107.16**	**.11**	**.23**	**.94**	**.53**	**.14**	–	–	–
	.70	–	–	2600.0	–	–	57.00	.06	.12	.50	.28	.08	–	–	–
2451	**.15**	[6]**.05**	–	**55.0**	–	–	**10.20**	**.01**	**.02**	**.09**	**.07**	**.02**	**4.90**	–	**0**
	1.50	.50	–	550.0	–	–	102.00	.10	.16	.90	.72	.23	49.00	–	0
2452	**1.65**	**.54**	–	**780.0**	–	–	**130.50**	**.12**	**.21**	**1.20**	–	–	**54.00**	–	–
	1.10	.36	–	520.0	–	–	87.00	.08	.14	.80	–	–	36.00	–	–
2453	**1.20**	**.54**	**(.05)**	**855.0**	–	–	**121.50**	**.12**	**.15**	**.90**	–	–	**132.00**	–	–
	.80	.36	(.03)	570.0	–	–	81.00	.08	.10	.60	–	–	88.00	–	–

(Continued)

TABLE F-36 *(Continued)*

Item No.	Food Item Name	Approximate Measure	Weight	Moisture	Food Energy Calories	Protein	Fats	Carbo-hydrates	Fiber	Minerals (Macro) Calcium	Phos-phorus	Sodium	Mag-nesium	Potas-sium
			(g)	(%)	(kcal)	(g)	(g)	(g)	(g)	(mg)	(mg)	(mg)	(mg)	(mg)
	CABBAGE													
2454	common, raw	1 CUP	70	92.4	16.8	1.0	.1	3.8	.6	34.3	20.3	14.0	9.1	163.1
			100	92.4	24.0	1.4	.2	5.4	.8	49.0	29.0	20.0	13.0	233.0
2455	common, shredded, cooked in small amount of water	1 CUP	145	93.9	29.0	1.6	.3	6.2	1.2	63.8	29.0	20.3	–	236.4
			100	93.9	20.0	1.1	.2	4.3	.8	44.0	20.0	14.0	–	163.0
2456	common, wedges, cooked in large amount of water	1 CUP	145	94.3	26.1	1.5	.3	5.8	1.2	60.9	24.7	18.9	–	219.0
			100	94.3	18.0	1.0	.2	4.0	.8	42.0	17.0	13.0	–	151.0
2457	common, dehydrated, unsulfited	1 OZ	28	4.0	86.2	3.5	.5	20.6	.1	113.4	80.4	53.2	–	618.0
			100	4.0	308.0	12.4	1.7	73.7	.3	405.0	287.0	190.0	–	2207.0
2458	mustard, salted	1 PKG	200	–	–	–	–	–	–	–	–	3264.0	–	448.0
			100	–	–	–	–	–	–	–	–	1632.0	–	224.0
2459	red, raw	1 CUP	100	90.2	31.0	2.0	.2	6.9	1.0	42.0	35.0	26.0	17.0	268.0
			100	90.2	31.0	2.0	.2	6.9	1.0	42.0	35.0	26.0	17.0	268.0
2460	savoy, raw	1 CUP	50	92.0	12.0	1.2	.1	2.3	.4	33.5	27.0	11.0	14.0	134.5
			100	92.0	24.0	2.4	.2	4.6	.8	67.0	54.0	22.0	28.0	269.0
2461	spoon, raw	1 OZ	28	94.3	4.5	.4	.1	.8	.2	46.2	12.3	7.3	7.6	85.7
			100	94.3	16.0	1.6	.2	2.9	.6	165.0	44.0	26.0	27.0	306.0
2462	spoon, cooked	1 OZ	28	95.2	3.9	.4	.1	.7	.2	41.4	9.2	5.0	–	59.9
			100	95.2	14.0	1.4	.2	2.4	.6	148.0	33.0	18.0	–	214.0
2463	swamp, raw, shredded	1 CUP	60	89.7	17.4	1.8	.2	3.2	.7	43.8	30.6	–	–	90.0
			100	89.7	29.0	3.0	.3	5.4	1.1	73.0	51.0	–	–	150.0
2464	swamp, boiled, drained	1 CUP	100	92.7	21.0	2.2	.2	3.9	.9	55.0	32.0	–	–	88.0
			100	92.7	21.0	2.2	.2	3.9	.9	55.0	32.0	–	–	88.0
	CABBAGE, CHINESE													
2465	raw	1 CUP	44	95.0	6.2	.5	0	1.3	.3	18.9	17.6	10.1	6.2	111.3
			100	95.0	14.0	1.2	.1	3.0	.6	43.0	40.0	23.0	14.0	253.0
2466	[4]salted	1 PKG	183	86.8	71.4	2.0	.5	17.2	3.1	237.9	96.9	1647.0	–	344.0
			100	86.8	39.0	1.1	.3	9.4	1.7	130.0	53.0	900.0	–	188.0
2467	[5]**CALABASH GOURD**	1 OZ	28	92.2	7.3	.2	.1	1.8	.4	5.0	5.9	–	–	–
			100	92.2	26.0	.7	.2	6.3	1.5	18.0	21.0	–	–	–
	CARROTS													
2468	raw	1 LG	100	59.0	42.0	1.2	.2	9.7	1.0	37.0	36.0	47.0	18.5	341.0
			100	59.0	42.0	1.2	.2	9.7	1.0	37.0	36.0	47.0	18.5	341.0
2469	raw, boiled, drained	1 CUP	150	91.2	46.5	1.4	.3	10.7	1.5	49.5	46.5	49.5	9.3	333.0
			100	91.2	31.0	.9	.2	7.1	1.0	33.0	31.0	33.0	6.2	222.0
2470	canned, drained solids	1 CUP	150	91.2	45.0	1.2	.5	10.1	1.2	45.0	33.0	354.0	–	180.0
			100	91.2	30.0	.8	.3	6.7	.8	30.0	22.0	236.0	–	120.0
2471	canned, solids & liquid	1 CUP	250	91.8	70.0	1.5	.5	16.3	1.5	62.5	50.0	590.0	(30.0)	300.0
			100	91.8	28.0	.6	.2	6.5	.6	25.0	20.0	236.0	(12.0)	120.0
2472	canned, drained liquid	1 CUP	155	93.3	34.1	.6	0	8.5	–	21.7	23.2	365.8	–	186.0
			100	93.3	22.0	.4	0	5.5	–	14.0	15.0	236.0	–	120.0
2473	canned, low sodium, drained solids	1 CUP	150	93.0	37.5	1.2	.2	8.4	1.2	45.0	33.0	58.5	–	180.0
			100	93.0	25.0	.8	.1	5.6	.8	30.0	22.0	39.0	–	120.0
2474	canned, low sodium, solids & liquid	1 CUP	250	93.7	55.0	1.8	.3	12.5	1.5	62.5	50.0	97.5	–	300.0
			100	93.7	22.0	.7	.1	5.0	.6	25.0	20.0	39.0	–	120.0
2475	canned, low sodium, drained liquid	1 CUP	155	95.2	24.8	.6	0	6.2	–	21.7	23.3	60.5	–	186.0
			100	95.2	16.0	.4	0	4.0	–	14.0	15.0	39.0	–	120.0
2476	dehydrated	1 OZ	28	4.0	95.5	1.8	.4	22.7	2.6	71.7	65.5	75.0	20.4	544.3
			100	4.0	341.0	6.6	1.3	81.1	9.3	256.0	234.0	268.0	73.0	1944.0
	CASSAVA													
2477	[5]sweet, root, raw	1 OZ	28	65.2	37.0	.3	.1	9.2	.3	11.2	9.5	–	–	–
			100	65.2	132.0	1.0	.4	32.8	1.0	40.0	34.0	–	–	–
2478	[5]bitter, root, raw	1 OZ	28	60.6	41.4	.2	.1	10.5	.3	10.1	13.4	–	–	–
			100	60.6	148.0	.8	.3	37.4	1.0	36.0	48.0	–	–	–
2479	[5]bitter, flour	1 OZ	28	14.2	89.6	.5	.1	22.7	.5	41.4	29.1	–	–	–
			100	14.2	320.0	1.7	.5	81.0	1.8	148.0	104.0	–	–	–
	CAULIFLOWER													
2480	raw	1 CUP	100	91.0	27.0	2.7	.2	5.2	1.0	25.0	56.0	13.0	24.0	295.0
			100	91.0	27.0	2.7	.2	5.2	1.0	25.0	56.0	13.0	24.0	295.0
2481	boiled, drained	1 CUP	120	92.8	26.4	2.8	.2	4.9	1.2	25.2	50.4	10.8	(10.7)	247.2
			100	92.8	22.0	2.3	.2	4.1	1.0	21.0	42.0	9.0	(8.9)	206.0
2482	frozen	1 CUP	135	92.9	29.7	2.7	.3	5.8	1.1	25.7	56.7	14.9	17.6	303.8
			100	92.9	22.0	2.0	.2	4.3	.8	19.0	42.0	11.0	13.0	225.0
2483	frozen, boiled, drained	1 CUP	120	94.0	21.6	2.3	.2	4.0	1.0	20.4	45.6	12.0	15.6	248.4
			100	94.0	18.0	1.9	.2	3.3	.8	17.0	38.0	10.0	13.0	207.0

Item No.	Minerals (Micro) Iron	Zinc	Copper	Fat-Soluble Vitamins Vitamin A	Vitamin D	Vitamin E (Alpha Tocopherol)	Water-Soluble Vitamins Vitamin C	Thiamin	Ribo-flavin	Niacin	Panto-thenic Acid	Vit. B-6 (Pyri-doxine)	Folacin (Folic Acid)	Biotin	Vitamin B-12
	(mg)	(mg)	(mg)	(IU)	(IU)	(mg)	(mg)	(mg)	(mg)	(mg)	(mg)	(mg)	(mcg)	(mcg)	(mcg)
2454	**.28** .40	**.28** .40	**.04** .06	**91.0** 130.0	– –	**.04** .06	**32.90** 47.00	**.04** .05	**.04** .05	**.21** .30	**.14** .21	**.11** .16	**46.20** 66.00	**1.68** 2.40	**0** 0
2455	**.44** .30	**.58** .40	– –	**188.5** 130.0	– –	– –	**47.85** 33.00	**.06** .04	**.06** .04	**.44** .30	– –	– –	**26.10** 18.00	– –	– –
2456	**.44** .30	**.58** .40	– –	**174.0** 120.0	– –	– –	**34.80** 24.00	**.03** .02	**.03** .02	**.15** .10	– –	– –	**26.10** 18.00	– –	– –
2457	**1.09** 3.90	– –	– –	**364.0** 1300.0	– –	– –	**59.08** 211.00	**.13** .45	**.11** .40	**.84** 3.00	– –	– –	– –	– –	– –
2458	– –	– –	– –	– –	– –	– –	– –	– –	– –	– –	– –	– –	– –	– –	– –
2459	**.80** .80	**(.34)** (.34)	**(.10)** (.10)	**40.0** 40.0	**0** 0	**(.22)** (.22)	**61.00** 61.00	**.09** .09	**.06** .06	**.40** .40	**.32** .32	**.20** .20	**34.00** 34.00	**(.11)** (.11)	**0** 0
2460	**.45** .90	**(.16)** (.32)	**(.08)** (.08)	**100.0** 200.0	**0** 0	**(.12)** (.23)	**27.50** 55.00	**.03** .05	**.04** .08	**.15** .30	**(.11)** (.22)	**.10** .19	**(42.00)** (84.00)	**(.07)** (.14)	**0** 0
2461	**.22** .80	– –	– –	**868.0** 3100.0	– –	– –	**7.00** 25.00	**.01** .05	**.03** .10	**.22** .80	– –	– –	– –	– –	– –
2462	**.17** .60	– –	– –	**868.0** 3100.0	– –	– –	**4.20** 15.00	**.01** .04	**.02** .08	**.20** .70	– –	– –	– –	– –	– –
2463	**1.50** 2.50	– –	– –	**3780.0** 6300.0	– –	– –	**19.20** 32.00	**.04** .07	**.07** .12	**.42** .70	– –	– –	– –	– –	– –
2464	**1.50** 1.50	– –	– –	**5200.0** 5200.0	– –	– –	**16.00** 16.00	**.05** .05	**.08** .08	**.50** .50	– –	– –	– –	– –	– –
2465	**.26** .60	– –	– –	**66.0** 150.0	– –	– –	**11.00** 25.00	**.02** .05	**.02** .04	**.26** .60	– –	– –	**36.52** 83.00	– –	– –
2466	**2.0** 1.1	– –	– –	– –	– –	– –	**9.15** 5.00	**.15** .08	**.37** .20	**2.4** 1.3	– –	– –	– –	– –	– –
2467	**.14** .50	– –	– –	**0** 0	– –	– –	**5.32** 19.00	**.01** .04	**.00** .03	**.17** .60	– –	– –	– –	– –	– –
2468	**.70** .70	**.40** .40	**.01** .01	**11000.0** 11000.0	– –	**.45** .45	**0** 0	**.06** .06	**.05** .05	**.60** .60	**.28** .28	**.15** .15	**32.00** 32.00	**2.50** 2.50	**0** 0
2469	**.90** .60	**.45** .30	**(.12)** (.08)	**15750.0** 10500.0	**0** 0	**.17** .11	**9.00** 6.00	**.08** .05	**.08** .05	**.75** .50	**(.29)** (.19)	**(.15)** (.10)	**36.00** 24.00	**(.63)** (.42)	**0** 0
2470	**1.05** .70	**.45** .30	– –	**22500.0** 15000.0	– –	– –	**3.00** 2.00	**.03** .02	**.05** .03	**.60** .40	– –	– –	**9.00** 6.00	– –	– –
2471	**1.75** .70	**(.80)** (.32)	**.20** .09	**25000.0** 10000.0	– –	– –	**5.00** 2.00	**.05** .02	**.05** .02	**1.00** .40	**.33** .13	**.08** .03	**15.00** 6.00	**3.75** 1.50	**0** 0
2472	**1.24** .80	– –	– –	– –	– –	– –	**3.10** 2.00	**.03** .02	**.03** .02	**.62** .40	– –	– –	**9.3** 6.0	– –	– –
2473	**1.05** .70	– –	– –	**22500.0** 15000.0	– –	– –	**3.00** 2.00	**.03** .02	**.05** .03	**.60** .40	– –	– –	**9.00** 6.00	– –	– –
2474	**1.75** .70	– –	– –	**25000.0** 10000.0	– –	– –	**5.00** 2.00	**.05** .02	**.05** .02	**1.00** .40	**.33** .13	**.08** .03	**15.00** 6.00	**3.75** 1.50	**0** 0
2475	**1.24** .80	– –	– –	– –	– –	– –	**3.10** 2.00	**.03** .02	**.03** .02	**.62** .40	– –	– –	**9.30** 6.00	– –	– –
2476	**1.68** 6.00	– –	– –	**28000.0** 100000.0	– –	– –	**4.20** 15.00	**.09** .31	**.08** .30	**.84** 3.00	**.62** 2.20	– –	– –	– –	**0** 0
2477	**.39** 1.40	– –	– –	**0** 0	– –	– –	**2.80** 10.00	**.01** .05	**.01** .04	**.17** .60	– –	– –	– –	– –	– –
2478	**.31** 1.10	– –	– –	**1.4** 5.0	– –	– –	**11.20** 40.00	**.02** .06	**.01** .04	**.20** .70	– –	– –	– –	– –	– –
2479	**1.51** 5.40	– –	– –	**0** 0	– –	– –	**3.92** 14.00	**.02** .08	**.02** .07	**.45** 1.60	– –	– –	– –	– –	– –
2480	**1.10** 1.10	**.34** .34	**.14** .14	**60.0** 60.0	**0** 0	**(.22)** (.22)	**78.00** 78.00	**.11** .11	**.10** .10	**.70** .70	**1.00** 1.00	**.21** .21	**55.00** 55.00	**17.00** 17.00	**0** 0
2481	**.84** .70	**(.40)** (.33)	**(.04)** (.03)	**72.0** 60.0	**0** 0	**(.14)** (.12)	**66.00** 55.00	**.11** .09	**.10** .08	**.72** .60	**(.52)** (.43)	**(.16)** (.13)	**40.80** 34.00	**(1.26)** (1.05)	**0** 0
2482	**.81** .60	**.62** .46	**.02** .01	**40.5** 30.0	– –	– –	**75.60** 56.00	**.08** .05	**.08** .06	**.68** .50	**.73** .54	**.26** .19	**85.05** 63.00	– –	**0** 0
2483	**.60** .50	– –	– –	**36.0** 30.0	– –	– –	**49.20** 41.00	**.05** .04	**.06** .05	**.48** .40	**.54** .45	**.16** .13	**45.60** 38.00	– –	– –

(Continued)

TABLE F-36 *(Continued)*

Item No.	Food Item Name	Approximate Measure	Weight	Moisture	Food Energy Calories	Protein	Fats	Carbo-hydrates	Fiber	Minerals (Macro) Calcium	Phos-phorus	Sodium	Mag-nesium	Potas-sium
			(g)	(%)	(kcal)	(g)	(g)	(g)	(g)	(mg)	(mg)	(mg)	(mg)	(mg)
	CELERIAC ROOT													
2484	raw	1 AVG	4	88.4	1.6	.1	0	.3	0	1.7	4.6	4.0	.8	12.0
			100	88.4	40.0	1.8	.3	8.5	1.3	43.0	115.0	100.0	20.0	300.0
2485	[6]boiled	1 OZ	28	—	4.0	.4	Trace	.6	.6	13.2	19.7	7.9	3.3	112.0
			100	—	14.1	1.5	Trace	2.3	2.2	47.3	70.5	28.2	11.8	399.8
	CELERY													
2486	raw	1 SM	20	94.1	3.4	.2	0	.8	.1	7.8	5.6	25.2	1.7	68.2
			100	94.1	17.0	.9	.1	3.9	.6	39.0	28.0	126.0	8.7	341.0
2487	boiled, drained	1 CUP	125	95.3	17.5	1.0	.1	3.9	.8	38.8	27.5	110.0	3.9	298.8
			100	95.3	14.0	.8	.1	3.1	.6	31.0	22.0	88.0	3.1	239.0
2488	**[5]CHAYOTE**	1 OZ	28	90.8	8.7	.3	.1	2.2	.2	3.4	8.4	—	—	—
			100	90.8	31.0	.9	.2	7.7	.6	12.0	30.0	—	—	—
2489	fruit, raw	1 HALF	100	91.8	28.0	.6	.1	7.1	.7	13.0	26.0	5.0	14.0	102.0
			100	91.8	28.0	.6	.1	7.1	.7	13.0	26.0	5.0	14.0	102.0
2490	[5]root	1 OZ	28	79.0	22.1	.6	.1	4.5	.1	1.2	9.5	—	—	—
			100	79.0	79.0	2.0	.2	17.8	.4	7.0	34.0	—	—	—
2491	**CHERVIL,** raw	1 CUP	200	80.7	114.0	6.8	1.8	23.0	—	—	—	—	—	—
			100	80.7	57.0	3.4	.9	11.5	—	—	—	—	—	—
2492	**CHICORY GREENS,** raw	1 CUP	53	92.8	10.6	.8	.2	2.0	.4	45.6	21.2	9.5	6.9	222.6
			100	92.8	20.0	1.6	.3	3.8	.8	86.0	40.0	18.0	13.0	420.0
2493	**CHICORY,** Witloof, bleached head, raw	1 CUP	53	95.1	8.0	.5	.1	1.7	—	9.5	11.1	3.7	5.0	96.5
			100	95.1	15.0	1.0	.1	3.2	—	18.0	21.0	7.0	9.4	182.0
	CHIVES													
2494	raw	1TBLS	10	91.3	2.8	.2	Trace	.6	.1	6.9	4.4	—	2.4	25.0
			100	91.3	28.0	1.8	.3	5.8	1.1	69.0	44.0	—	24.1	250.0
2495	frozen	1 TBLS	10	—	3.0	.2	—	.6	.1	7.0	4.0	—	3.0	25.0
			100	—	30.0	2.0	—	6.0	1.0	70.0	40.0	—	30.0	250.0
	CHUFA													
2496	[5]flatsedge, tuber, raw	1 OZ	28	32.6	87.1	1.2	4.8	12.3	2.0	16.5	43.4	—	—	—
			100	32.6	311.0	4.4	17.2	43.9	7.1	59.0	155.0	—	—	—
2497	[5]flatsedge, tuber, dried	1 OZ	28	9.3	117.3	2.0	6.5	16.3	2.7	19.3	103.9	—	—	—
			100	9.3	419.0	7.2	23.1	58.2	9.6	69.0	371.0	—	—	—
2498	**COLLARDS** leaves w/stems, cooked in small amount of water	1 CUP	200	90.8	58.0	5.4	1.2	9.8	1.6	304.0	78.0	50.0	—	468.0
			100	90.8	29.0	2.7	.6	4.9	.8	152.0	39.0	25.0	—	234.0
2499	leaves w/o stems, cooked in small amount of water	1 CUP	200	89.6	66.0	7.2	1.4	10.2	2.0	376.0	104.0	—	—	524.0
			100	89.6	33.0	3.6	.7	5.1	1.0	188.0	52.0	—	—	262.0
2500	leaves w/o stems, cooked in large amount of water	1 CUP	200	90.2	62.0	6.8	1.4	9.6	2.0	354.0	96.0	—	—	486.0
			100	90.2	31.0	3.4	.7	4.8	1.0	177.0	48.0	—	—	243.0
2501	frozen, cooked	1 CUP	200	90.2	60.0	5.8	.8	11.2	2.0	352.0	102.0	32.0	70.0	472.0
			100	90.2	30.0	2.9	.4	5.6	1.0	176.0	51.0	16.0	35.0	236.0
	CORN, SWEET													
2502	raw, white & yellow	1 MED EAR	140	72.7	134.4	4.9	1.4	30.9	1.0	4.2	155.4	(1.4)	67.2	392.0
			100	72.7	96.0	3.5	1.0	22.1	.7	3.0	111.0	(1.0)	48.0	280.0
2503	fresh, white & yellow, cooked on cob	1 MED	140	74.1	127.4	4.6	1.4	29.4	1.0	4.2	124.6	(2.5)	(63.3)	274.4
			100	74.1	91.0	3.3	1.0	21.0	.7	3.0	89.0	(1.1)	(45.2)	196.0
2504	fresh, white & yellow, cut off cob before cooking	1 CUP	165	76.5	137.0	5.3	1.7	31.0	1.2	5.0	146.9	—	—	272.3
			100	76.5	83.0	3.2	1.0	18.8	.7	3.0	89.0	—	—	165.0
2505	canned, creamed, white & yellow, solids & liquid	1 CUP	166	76.3	136.1	3.5	1.0	33.2	.8	5.0	93.0	391.8	31.5	161.0
			100	76.3	82.0	2.1	.6	20.0	.5	3.0	56.0	236.0	19.0	97.0
2506	canned, creamed, low sodium, white & yellow, solids & liquid	1 CUP	166	77.3	136.1	4.3	1.8	30.7	.5	5.0	93.0	3.3	—	161.0
			100	77.3	82.0	2.6	1.1	18.5	.3	3.0	56.0	2.0	—	97.0
2507	canned, whole kernel, wet pack, white & yellow, drained solids	1 CUP	166	81.7	102.9	3.2	.9	24.7	1.3	5.8	68.1	350.1	21.2	195.5
			100	81.7	62.0	1.9	.5	14.9	.8	3.5	41.0	210.9	12.8	117.8
2508	canned, whole kernel, wet pack, white & yellow, solids & liquid	1 CUP	166	80.9	109.6	3.3	1.0	26.1	1.0	6.6	79.7	391.8	31.5	161.0
			100	80.9	66.0	2.0	.6	15.7	.6	4.0	48.0	236.0	19.0	97.0
2509	canned, whole kernel, wet pack, white & yellow, drained liquid	1 CUP	166	91.7	43.2	.8	—	11.5	—	5.0	74.7	391.8	—	161.0
			100	91.7	26.0	.5	—	6.9	—	3.0	45.0	236.0	—	97.0
2510	canned, whole kernel, low sodium, wet pack, white & yellow, drained solids	1 CUP	166	78.4	126.2	4.2	1.2	29.9	1.2	8.3	81.3	3.3	18.3	161.0
			100	78.4	76.0	2.5	.7	18.0	.7	5.0	49.0	2.0	11.0	97.0
2511	canned, whole kernel, low sodium, wet pack, white & yellow, solids & liquid	1 CUP	166	83.6	94.6	3.2	.8	22.6	.8	6.6	79.7	3.3	31.5	161.0
			100	83.6	57.0	1.9	.5	13.6	.5	4.0	48.0	2.0	19.0	97.0
2512	canned, whole kernel, low sodium, wet pack, white & yellow, drained liquid	1 CUP	166	94.8	28.2	.8	—	7.1	—	5.0	74.7	3.3	—	161.0
			100	94.8	17.0	.5	—	4.3	—	3.0	45.0	2.0	—	97.0

Item No.	Minerals (Micro) Iron	Zinc	Copper	Fat-Soluble Vitamins Vitamin A	Vitamin D	Vitamin E (Alpha Tocopherol)	Water-Soluble Vitamins Vitamin C	Thiamin	Ribo-flavin	Niacin	Panto-thenic Acid	Vit. B-6 (Pyri-doxine)	Folacin (Folic Acid)	Biotin	Vitamin B-12
	(mg)	(mg)	(mg)	(IU)	(IU)	(mg)	(mg)	(mg)	(mg)	(mg)	(mg)	(mg)	(mcg)	(mcg)	(mcg)
2484	.02 .60	— —	— —	— —	— —	— —	.32 8.00	Trace .05	Trace .06	.03 .70	— —	.01 .16	— —	— —	0 0
2485	.23 .81	— —	.03 .12	0 0	0 0	— —	1.2 4.3	— .04	.01 .04	.14 .51	— —	.03 .11	— —	— —	— —
2486	.06 .30	.01 .07	Trace .01	48.0 240.0	0 0	.09 .46	1.80 9.00	Trace .03	.01 .03	.06 .30	.09 .43	.01 .06	2.40 12.00	(.02) (.10)	0 0
2487	.25 .20	(.13) (.11)	.13 .10	287.5 230.0	0 0	.48 .38	7.50 6.00	.03 .02	.04 .03	.38 .30	(.37) (.30)	(.75) (.06)	(7.56) (6.05)	Trace Trace	0 0
2488	.16 .60	— —	— —	1.4 5.0	— —	— —	5.60 20.00	Trace .03	.01 .04	.11 .40	— —	— —	— —	— —	— —
2489	.50 .50	— —	— —	20.0 20.0	— —	— —	19.00 19.00	.03 .03	.03 .03	.40 .40	.48 .48	— —	— —	— —	0 0
2490	.22 .80	— —	— —	0 0	— —	— —	5.32 19.00	.01 .05	.01 .03	.25 .90	— —	— —	— —	— —	— —
2491	— —	— —	— —	— —	— —	— —	18.00 9.00	— —	— —	— —	— —	.05 .03	— —	— —	0 0
2492	.48 .90	(.11) (.22)	(.07) (.14)	2120.0 4000.0	0 0	— —	11.66 22.00	.03 .06	.05 .10	.27 .50	— —	.02 .05	27.56 52.00	— —	0 0
2493	.27 .50	— —	— —	— —	— —	— —	— —	— —	— —	— —	— —	.02 .05	— —	— —	0 0
2494	.17 1.70	— —	— —	580.0 5800.0	— —	— —	5.60 56.00	.01 .08	.01 .13	.05 .50	— —	.02 .18	— —	— —	0 0
2495	.20 2.00	— —	— —	580.0 5800.0	— —	— —	6.00 60.00	.01 .08	Trace .01	.10 1.00	— —	— —	— —	— —	0 0
2496	.67 2.40	— —	— —	— —	— —	— —	— —	.03 .09	— —	— —	— —	— —	— —	— —	0 0
2497	.92 3.30	— —	— —	— —	— —	— —	— —	.03 .12	— —	— —	— —	— —	— —	— —	0 0
2498	1.20 .60	— —	(.08) (.29)	10800.0 5400.0	— —	— —	92.00 46.00	.28 .14	.40 .20	2.40 1.20	— —	— —	— —	— —	0 0
2499	1.60 .80	— —	— —	15600.0 7800.0	— —	— —	152.00 76.00	.22 .11	.40 .20	2.40 1.20	— —	— —	— —	— —	0 0
2500	1.60 .80	— —	— —	15600.0 7800.0	— —	— —	102.00 51.00	.14 .07	.28 .14	2.20 1.10	— —	— —	— —	— —	0 0
2501	2.00 1.00	— —	— —	13600.0 6800.0	— —	— —	66.00 33.00	.12 .06	.28 .14	1.20 .60	— —	— —	100.00 50.00	— —	0 0
2502	.98 .70	.70 .50	.08 .06	560.0 400.0	0 0	.08 .06	16.80 12.00	.21 .15	.17 .12	2.38 1.70	.76 .54	.66 .47	46.20 33.00	8.40 6.00	0 0
2503	.84 .60	.56 .40	(.20) (.14)	560.0 400.0	0 0	(.73) (.52)	12.60 9.00	.17 .12	.14 .10	1.96 1.40	(.55) (.39)	(.24) (.17)	(46.35) (33.11)	— —	0 0
2504	.99 .60	.83 .50	— —	660.0 400.0	— —	— —	11.55 7.00	.18 .11	.17 .10	2.15 1.30	— —	— —	54.45 33.00	— —	0 0
2505	1.00 .60	.71 .43	.12 .07	547.8 330.0	— —	— —	8.30 5.00	.05 .03	.08 .05	1.66 1.00	— —	.33 .20	13.78 8.30	— —	0 0
2506	1.00 .60	— —	— —	448.2 270.0	— —	— —	8.30 5.00	.05 .03	.08 .05	1.66 1.00	— —	— —	— —	— —	—
2507	.37 .22	.66 .40	.04 .02	325.4 196.0	0 0	(.85) (.51)	8.30 5.00	.046 .028	.10 .06	1.71 1.03	.37 .22	.06 .04	59.76 36.00	— —	0 0
2508	.66 .40	.50 .30	(.15) (.09)	448.2 270.0	— —	.08 .05	8.30 5.00	.05 .03	.08 .05	1.49 .90	.37 .22	.33 .20	64.74 39.00	3.65 2.20	0 0
2509	.50 .30	— —	— —	— —	— —	— —	11.62 7.00	.05 .03	.07 .04	1.49 .90	— —	— —	84.66 51.00	— —	0 0
2510	.83 .50	— —	.02 .01	581.0 350.0	— —	— —	6.64 4.00	.05 .03	.08 .05	1.49 .90	.37 .22	.33 .20	59.76 36.00	— —	0 0
2511	.66 .40	— —	— —	448.2 270.0	— —	— —	8.30 5.00	.05 .03	.08 .05	1.49 .90	.37 .22	.33 .20	64.74 39.00	3.65 2.20	0 0
2512	.50 .30	— —	— —	— —	— —	— —	11.62 7.00	.05 .03	.07 .04	1.49 .90	— —	— —	84.66 51.00	— —	0 0

(Continued)

TABLE F-36 *(Continued)*

Item No.	Food Item Name	Approximate Measure	Weight	Moisture	Food Energy Calories	Protein	Fats	Carbo-hydrates	Fiber	Minerals (Macro) Calcium	Phos-phorus	Sodium	Mag-nesium	Potas-sium
			(g)	(%)	(kcal)	(g)	(g)	(g)	(g)	(mg)	(mg)	(mg)	(mg)	(mg)
	CORN, SWEET (Continued)													
2513	canned, whole kernel, vacuum pack, drained solids	1 CUP	166	76.4	134.5	0	1.2	31.9	—	8.0	90.5	439.1	32.0	260.8
			100	76.4	81.0	0	.7	19.2	—	4.8	54.5	264.5	19.3	157.1
2514	canned, whole kernel, vacuum pack, solids & liquid	1 CUP	166	75.5	137.8	4.2	.8	34.0	1.3	5.0	121.2	391.8	—	161.0
			100	75.5	83.0	2.5	.5	20.5	.8	3.0	73.0	236.0	—	97.0
2515	frozen, cooked on cob	1 MED	114	73.2	107.2	4.0	1.1	24.6	.8	3.4	109.4	1.1	38.8	263.3
			100	73.2	94.0	3.5	1.0	21.6	.7	3.0	96.0	1.0	34.0	231.0
2516	frozen, cut off cob before cooking	1 CUP	166	77.2	131.1	5.0	.8	31.2	.8	5.0	121.2	1.7	(29.9)	305.4
			100	77.2	79.0	3.0	.5	18.8	.5	3.0	73.0	1.0	(18.0)	184.0
	COW PEAS													
2517	young pods & seeds, cooked	1 CUP	160	89.5	54.4	4.2	.5	11.2	2.7	88.0	78.4	4.8	—	313.6
			100	89.5	34.0	2.6	.3	7.0	1.7	55.0	49.0	3.0	—	196.0
	CRESS, GARDEN													
2518	raw	1 AVG	2	89.4	0.6	.1	0	0.1	0	1.6	1.5	.3	—	12.1
			100	89.4	32.0	2.6	.7	5.5	1.1	81.0	76.0	14.0	—	606.0
2519	boiled, cooked in small amount of water a short time, drained	1 CUP	135	92.5	31.1	2.6	.8	5.1	1.2	82.4	64.8	10.8	—	476.6
			100	92.5	23.0	1.9	.6	3.8	.9	61.0	48.0	8.0	—	353.0
2520	boiled in large amount of water a long time, drained	1 CUP	135	92.9	29.7	2.4	.8	4.9	1.2	78.3	59.4	10.8	—	442.8
			100	92.9	22.0	1.8	.6	3.6	.9	58.0	44.0	8.0	—	328.0
	CUCUMBERS													
2521	raw, not pared	1 MED	100	95.1	15.0	.9	.1	3.4	.6	25.0	27.0	6.0	12.0	160.0
			100	95.1	15.0	.9	.1	3.4	.6	25.0	27.0	6.0	12.0	160.0
2522	raw, pared	1 MED	100	95.7	14.0	.6	.1	3.2	.3	17.0	18.0	6.0	10.0	160.0
			100	95.7	14.0	.6	.1	3.2	.3	17.0	18.0	6.0	10.0	160.0
	DANDELIONS													
2523	raw	1 CUP	142	85.6	63.9	3.8	1.0	13.1	2.3	265.5	93.7	107.9	51.1	563.7
			100	85.6	45.0	2.7	.7	9.2	1.6	187.0	66.0	76.0	36.0	397.0
2524	boiled, drained	1 CUP	200	89.8	66.0	4.0	1.2	12.8	2.6	280.0	84.0	88.0	—	464.0
			100	89.8	33.0	2.0	.6	6.4	1.3	140.0	42.0	44.0	—	232.0
	DOCK													
2525	raw	1 CUP	140	90.9	39.2	2.9	.4	7.8	1.1	92.4	57.4	7.0	—	473.2
			100	90.9	28.0	2.1	.3	5.6	.8	66.0	41.0	5.0	—	338.0
2526	boiled, drained	1 CUP	200	93.6	38.0	3.2	.4	7.8	1.4	110.0	52.0	6.0	—	396.0
			100	93.6	19.0	1.6	.2	3.9	.7	55.0	26.0	3.0	—	198.0
	EGGPLANT													
2527	raw	1 SLICE	30	92.4	7.5	.3	.1	1.7	.3	3.6	7.8	.6	24.6	64.2
			100	92.4	25.0	1.1	.2	5.6	.9	12.0	26.0	2.0	82.0	214.0
2528	boiled, drained	1 CUP	200	94.3	38.0	2.2	.4	8.2	1.8	22.0	42.0	2.0	—	300.0
			100	94.3	19.0	1.1	.2	4.1	.9	11.0	21.0	1.0	—	150.0
2529	**ENDIVE (CURLY & ESCAROLE),** raw	1 MED	7	93.1	1.4	.1	0	.3	.1	5.7	3.8	1.0	.7	20.6
			100	93.1	20.0	1.7	.1	4.1	.9	81.0	54.0	14.0	10.0	294.0
2530	**FENNEL,** common, leaves, raw	1 CUP	60	90.0	16.8	1.7	.2	3.1	.3	60.0	30.6	—	—	238.2
			100	90.0	28.0	2.8	.4	5.1	.5	100.0	51.0	—	—	397.0
2531	**GARLIC,** clove, raw	1 AVG	3	61.3	4.1	.2	0	.9	0	.9	6.1	.6	.7	15.9
			100	61.3	137.0	6.2	.2	30.8	1.5	29.0	202.0	19.0	23.0	529.0
2532	**GRITS,** hominy, cooked	1 CUP	242	87.1	123.4	2.9	.2	26.6	.2	2.4	24.2	—	7.3	26.6
			100	87.1	51.0	1.2	.1	11.0	.1	1.0	10.0	—	3.0	11.0
2533	**[5]GROUNDCHERRY TOMATO**	1 OZ	28	88.3	11.2	.4	.1	—	.5	2.8	9.5	—	—	—
			100	88.3	40.0	1.6	.5	—	1.7	10.0	34.0	—	—	—
	HORSERADISH													
2534	raw	1 TBLS	15	74.6	13.1	.5	0	3.0	.4	21.0	9.6	1.2	5.1	84.6
			100	74.6	87.0	3.2	.3	19.7	2.4	140.0	64.0	8.0	34.0	564.0
2535	prepared	1 TBLS	15	87.1	6.0	.2	Trace	1.4	Trace	9.0	5.0	14.0	—	44.0
			100	87.1	38.0	1.3	.2	9.6	.2	61.0	32.0	96.0	—	290.0
	HYACINTH BEANS													
2536	raw, young pods	1 CUP	90	88.8	31.5	2.5	.3	6.6	1.6	51.3	47.7	1.8	—	256.5
			100	88.8	35.0	2.8	.3	7.3	1.8	57.0	53.0	2.0	—	285.0
	KALE													
2537	boiled, drained, leaves w/stems	1 CUP	110	91.2	30.8	3.5	.8	4.4	1.2	147.4	50.6	47.3	—	243.1
			100	91.2	28.0	3.2	.7	4.0	1.1	134.0	46.0	43.0	—	221.0

Item No.	Minerals (Micro)			Fat-Soluble Vitamins			Water-Soluble Vitamins								
	Iron	Zinc	Copper	Vitamin A	Vitamin D	Vitamin E (Alpha Tocopherol)	Vitamin C	Thiamin	Ribo-flavin	Niacin	Panto-thenic Acid	Vit. B-6 (Pyri-doxine)	Folacin (Folic Acid)	Biotin	Vitamin B-12
	(mg)	(mg)	(mg)	(IU)	(IU)	(mg)	(mg)	(mg)	(mg)	(mg)	(mg)	(mg)	(mcg)	(mcg)	(mcg)
2513	.50	.61	.05	343.6	—	—	—	.08	.13	2.13	—	.09	59.76	—	0
	.30	.37	.03	207.0	—	—	—	.05	.08	1.28	—	.06	36.00	—	0
2514	.83	.66	—	581.0	—	—	8.30	.05	.10	1.83	—	—	64.74	3.65	0
	.50	.40	—	350.0	—	—	5.00	.03	.06	1.10	—	—	39.00	2.20	0
2515	.91	—	—	399.0	—	.22	7.98	.16	.09	1.94	—	—	41.04	—	0
	.80	—	—	350.0	—	.19	7.00	.14	.08	1.70	—	—	36.00	—	0
2516	1.33	(.06)	(.06)	581.0	—	.32	8.30	.15	.10	2.49	—	—	59.76	—	0
	.80	(.04)	(.04)	350.0	—	.19	5.00	.09	.06	1.50	—	—	36.00	—	0
2517	1.12	—	—	2240.0	—	—	27.20	.14	.14	1.28	—	—	—	—	0
	.70	—	—	1400.0	—	—	17.00	.09	.09	.80	—	—	—	—	0
2518	.03	—	—	186.0	—	—	1.48	Trace	.01	.02	—	.01	—	—	0
	1.30	—	—	9300.0	—	—	69.00	.08	.26	1.00	—	.25	—	—	0
2519	1.08	—	—	10395.0	—	—	45.90	.08	.22	1.08	—	—	—	—	0
	.80	—	—	7700.0	—	—	34.00	.06	.16	.80	—	—	—	—	0
2520	.95	—	—	9450.0	—	—	31.05	.05	.20	.95	—	—	—	—	0
	.70	—	—	7000.0	—	—	23.00	.04	.15	.70	—	—	—	—	0
2521	1.10	.12	.01	250.0	0	Trace	11.00	.03	.04	.20	.25	.04	(16.05)	(.41)	0
	1.10	.12	.01	250.0	0	Trace	11.00	.03	.04	.20	.25	.04	(16.05)	(.41)	0
2522	.30	[6].10	—	—	—	—	11.00	.03	.04	.20	—	.05	15.00	—	—
	.30	.10	—	—	—	—	11.00	.03	.04	.20	—	.05	15.00	—	—
2523	4.40	—	.21	19880.0	—	—	49.70	.27	.37	1.14	—	—	—	—	—
	3.10	—	.15	14000.0	—	—	35.00	.19	.26	.80	—	—	—	—	—
2524	3.60	—	—	23400.0	—	—	36.00	.26	.32	—	—	—	—	—	—
	1.80	—	—	11700.0	—	—	18.00	.13	.16	—	—	—	—	—	—
2525	2.24	—	—	18060.0	—	—	166.60	.13	.31	.70	—	—	—	—	—
	1.60	—	—	12900.0	—	—	119.00	.09	.22	.50	—	—	—	—	—
2526	1.80	—	—	21600.0	—	—	108.00	.12	.26	.80	—	—	—	—	—
	.90	—	—	10800.0	—	—	54.00	.06	.13	.40	—	—	—	—	—
2527	.21	—	0	3.0	—	—	1.50	.02	.02	.18	.07	.02	—	—	0
	.70	—	.01	10.0	—	—	5.00	.05	.05	.60	.22	.08	—	—	0
2528	1.20	—	—	20.0	—	—	6.00	.10	.08	1.00	—	—	32.00	—	—
	.60	—	—	10.0	—	—	3.00	.05	.04	.50	—	—	16.00	—	—
2529	.12	—	—	231.0	—	—	.70	.01	.01	.04	—	—	9.94	—	0
	1.70	—	—	3300.0	—	—	10.00	.07	.14	.50	—	—	142.00	—	0
2530	1.62	[6].30	—	2100.0	—	—	18.60	—	—	—	.15	.06	—	—	0
	2.70	.50	—	3500.0	—	—	31.00	—	—	—	.25	.10	—	—	0
2531	.05	.02	.01	—	—	—	.45	.01	Trace	.02	—	—	—	—	—
	1.50	.59	.32	—	—	—	15.00	.25	.08	.50	—	—	—	—	—
2532	.73	—	—	145.2	—	.75	0	.10	.07	.97	—	—	—	—	—
	.30	—	—	60.0	—	.31	0	.04	.03	.40	—	—	—	—	—
2533	.25	—	—	7.0	—	—	1.68	.03	.01	.67	—	—	—	—	—
	.90	—	—	25.0	—	—	6.00	.09	.04	2.40	—	—	—	—	—
2534	.21	—	(.02)	0	0	—	12.15	.01	Trace	(.08)	—	.02	—	—	0
	1.40	—	(.14)	0	0	—	81.00	.07	.03	(.52)	—	.16	—	—	0
2535	.10	—	—	—	—	—	—	—	—	—	—	—	—	—	—
	.90	—	—	—	—	—	—	—	—	—	—	—	—	—	—
2536	.90	—	—	522.0	—	—	18.00	.08	.10	.81	—	—	—	—	—
	1.00	—	—	580.0	—	—	20.00	.09	.11	.90	—	—	—	—	—
2537	1.32	[6].22	—	8140.0	—	—	68.20	—	—	—	—	—	—	—	—
	1.20	.20	—	7400.0	—	—	62.00	—	—	—	—	—	—	—	—

(Continued)

TABLE F-36 *(Continued)*

Item No.	Food Item Name	Approximate Measure	Weight	Moisture	Food Energy Calories	Protein	Fats	Carbo-hydrates	Fiber	Minerals (Macro) Calcium	Phos-phorus	Sodium	Mag-nesium	Potas-sium
			(g)	(%)	(kcal)	(g)	(g)	(g)	(g)	(mg)	(mg)	(mg)	(mg)	(mg)
	KALE (Continued)													
2538	boiled, drained, leaves w/o stems & midribs	**1 CUP**	**110**	**87.8**	**42.9**	**5.0**	**.8**	**6.7**	—	**205.7**	**63.8**	**47.3**	—	**243.1**
			100	87.8	39.0	4.5	.7	6.1	—	187.0	58.0	43.0	—	221.0
2539	frozen, boiled, drained	**1 CUP**	**110**	**90.5**	**34.1**	**3.3**	**.6**	**5.9**	**1.0**	**133.1**	**52.8**	**23.1**	**34.1**	**212.3**
			100	90.5	31.0	3.0	.5	5.4	.9	121.0	48.0	21.0	31.0	193.0
	KOHLRABI													
2540	raw	**1 CUP**	**149**	**90.3**	**43.2**	**3.0**	**.2**	**9.8**	**1.5**	**61.1**	**76.0**	**11.9**	**55.1**	**554.3**
			100	90.3	29.0	2.0	.1	6.6	1.0	41.0	51.0	8.0	37.0	372.0
2541	boiled, drained	**1 CUP**	**149**	**92.2**	**35.8**	**2.5**	**.1**	**7.9**	**1.5**	**49.2**	**61.1**	**8.9**	—	**387.4**
			100	92.2	24.0	1.7	.1	5.3	1.0	33.0	41.0	6.0	—	260.0
	LAMB'S-QUARTER													
2542	raw	**1 CUP**	**33**	**84.3**	**14.2**	**1.4**	**.3**	**2.4**	**.7**	**102.0**	**23.8**	—	—	—
			100	84.3	43.0	4.2	.8	7.3	2.1	309.0	72.0	—	—	—
2543	boiled, drained	**1 CUP**	**200**	**88.9**	**64.0**	**6.4**	**1.4**	**10.0**	**3.6**	**516.0**	**90.0**	—	—	—
			100	88.9	32.0	3.2	.7	5.0	1.8	258.0	45.0	—	—	—
2544	**LEEKS,** raw	**1 AVG**	**25**	**85.4**	**13.0**	**.6**	**.1**	**2.8**	**.3**	**13.0**	**12.5**	**1.3**	**5.8**	**86.8**
			100	85.4	52.0	2.2	.3	11.2	1.3	52.0	50.0	5.0	23.0	347.0
	LENTIL SPROUTS													
2545	raw	**1 CUP**	**130**	**72.7**	**135.2**	**10.9**	**.4**	—	**1.4**	**15.6**	—	—	—	—
			100	72.7	104.0	8.4	.3	—	1.1	12.0	—	—	—	—
2546	Cooked	**1 CUP**	**150**	**68.7**	—	**13.2**	—	—	**1.7**	**20.6**	—	—	—	—
			100	68.7	—	8.8	—	—	1.1	13.7	—	—	—	—
	LETTUCE													
2547	raw, Butterhead (Boston & Bibb)	**1 CUP**	**66**	**95.1**	**9.2**	**.8**	**.1**	**1.7**	**.3**	**23.1**	**17.2**	**5.9**	**10.6**	**174.2**
			100	95.1	14.0	1.2	.2	2.5	.5	35.0	26.0	9.0	16.0	264.0
2548	raw, Cos or Romaine (dark green & white Paris)	**1 CUP**	**66**	**94.0**	**11.9**	**.9**	**.2**	**2.3**	**.5**	**44.9**	**16.5**	**5.9**	**4.0**	**174.2**
			100	94.0	18.0	1.3	.3	3.5	.7	68.0	25.0	9.0	6.0	264.0
2549	raw, crisphead (Iceberg, New York, Great Lakes)	**1 CUP**	**60**	**95.5**	**7.8**	**.5**	**.1**	**1.7**	**.3**	**12.0**	**13.2**	**5.4**	**6.6**	**105.0**
			100	95.5	13.0	.9	.1	2.9	.5	20.0	22.0	9.0	11.0	175.0
2550	raw, looseleaf or bunching (Grand Rapids, Salad Bowl & Simpson)	**1 CUP**	**66**	**94.0**	**11.9**	**.9**	**.2**	**2.3**	**.5**	**44.9**	**16.5**	**5.9**	**9.9**	**174.2**
			100	94.0	18.0	1.3	.3	3.5	.7	68.0	25.0	9.0	15.0	264.0
2551	[5]**LUPINE,** South American	**1 OZ**	**28**	**7.7**	**114.0**	**12.4**	**4.6**	**7.9**	**2.0**	**25.2**	**152.6**	—	—	—
			100	7.7	407.0	44.3	16.5	28.2	7.1	90.0	545.0	—	—	—
2552	[3]**MALANGA TANNIA** leaves, raw	**1 OZ**	**28**	**89.9**	**9.5**	**.7**	**.3**	**1.5**	**.6**	**26.6**	**108.6**	—	—	—
			100	89.9	34.0	2.5	1.0	5.3	2.1	95.0	388.0	—	—	—
2553	[5]tubers	**1 OZ**	**28**	**65.9**	**37.0**	**.5**	**.1**	**8.7**	**.2**	**3.9**	**15.7**	—	—	—
			100	65.9	132.0	1.7	.3	30.9	.6	14.0	56.0	—	—	—
2554	**MANGOS** (sweet peppers)	**1 MED**	**40**	**93.4**	**8.8**	**.5**	**.1**	**1.9**	**.6**	**3.6**	**8.8**	**5.2**	**7.2**	**85.2**
			100	93.4	22.0	1.2	.2	4.8	1.4	9.0	22.0	13.0	18.0	213.0
2555	**MIXED VEGETABLES,** frozen, boiled, drained	**1 CUP**	**200**	**82.6**	**128.0**	**6.4**	**.6**	**26.8**	**2.4**	**50.0**	**126.0**	**106.0**	**48.0**	**382.0**
			100	82.6	64.0	3.2	.3	13.4	1.2	25.0	63.0	53.0	24.0	191.0
	MUSHROOMS													
2556	*Agaricus campestris*, raw	**1 SM**	**10**	**90.4**	**2.8**	**.2**	**0**	**.4**	**.1**	**.6**	**11.6**	**1.5**	**1.3**	**41.4**
			100	90.4	28.0	2.4	.3	4.4	.8	6.0	116.0	15.0	13.0	414.0
2557	other edible species, raw	**1 SM**	**10**	**89.1**	**3.5**	**.2**	**.1**	**.7**	**.1**	**1.3**	**9.7**	**1.0**	—	**37.5**
			100	89.1	35.0	1.9	.6	6.5	1.1	13.0	97.0	10.0	—	375.0
2558	*Agaricus campestris*, sauteed	**1 CUP**	**270**	—	**299.7**	**6.5**	**28.6**	**10.8**	**2.7**	**29.7**	**313.2**	—	—	—
			100	—	111.0	2.4	10.6	4.0	1.0	11.0	116.0	—	—	—
2559	*Agaricus campestris*, canned, drained	**1 CUP**	**270**	—	**51.3**	**3.3**	**.5**	**6.9**	—	**21.6**	**243.0**	—	**21.0**	—
			100	—	19.0	1.2	.2	2.6	—	8.0	90.0	—	7.8	—
2560	*Agaricus campestris*, canned, solids & liquid	**1 CUP**	**200**	**93.1**	**34.0**	**3.8**	**.2**	**4.8**	**1.2**	**12.0**	**136.0**	**800.0**	**16.0**	**394.0**
			100	93.1	17.0	1.9	.1	2.4	.6	6.0	68.0	400.0	8.0	197.0
	MUSTARD GREENS													
2561	raw	**1 CUP**	**33**	**89.5**	**10.2**	**1.0**	**.2**	**1.8**	**.4**	**60.4**	**16.5**	**10.6**	**8.9**	**124.4**
			100	89.5	31.0	3.0	.5	5.6	1.1	183.0	50.0	32.0	27.0	377.0
2562	boiled, drained	**1 CUP**	**200**	**92.6**	**46.0**	**4.4**	**.8**	**8.0**	**1.8**	**276.0**	**64.0**	**36.0**	—	**440.0**
			100	92.6	23.0	2.2	.4	4.0	.9	138.0	32.0	18.0	—	220.0
2563	frozen, boiled, drained	**1 CUP**	**200**	**93.8**	**40.0**	**4.4**	**.8**	**6.2**	**2.0**	**208.0**	**86.0**	**20.0**	**(30.0)**	**314.0**
			100	93.8	20.0	2.2	.4	3.1	1.0	104.0	43.0	10.0	(15.0)	157.0
	NOPAL CACTUS COCHINEAL													
2564	fruit	**1 OZ**	**28**	**83.4**	**16.8**	**.4**	**.1**	**4.1**	**.8**	**3.9**	**5.9**	—	—	—
			100	83.4	60.0	1.3	.3	14.7	2.8	14.0	21.0	—	—	—
2565	stem	**1 OZ**	**28**	**90.8**	**8.1**	**.4**	**0**	**1.9**	**1.9**	—	**4.8**	—	—	—
			100	90.8	29.0	1.3	.1	6.9	6.8	—	17.0	—	—	—

Item No.	Minerals (Micro)			Fat-Soluble Vitamins			Water-Soluble Vitamins								
	Iron	Zinc	Copper	Vitamin A	Vitamin D	Vitamin E (Alpha Tocopherol)	Vitamin C	Thiamin	Ribo-flavin	Niacin	Panto-thenic Acid	Vit. B-6 (Pyri-doxine)	Folacin (Folic Acid)	Biotin	Vitamin B-12
	(mg)	(mg)	(mg)	(IU)	(IU)	(mg)	(mg)	(mg)	(mg)	(mg)	(mg)	(mg)	(mcg)	(mcg)	(mcg)
2538	**1.76**	–	–	**9130.0**	–	–	**102.30**	**.11**	**.10**	**1.76**	–	–	–	–	–
	1.60	–	–	8300.0	–	–	93.00	.10	.18	1.60	–	–	–	–	–
2539	**1.10**	**(.19)**	**.05**	**9020.0**	–	–	**41.80**	**.07**	**.17**	**.77**	–	–	–	–	–
	1.00	(.17)	.05	8200.0	–	–	38.00	.06	.15	.70	–	–	–	–	–
2540	**.75**	**.34**	–	**29.8**	–	–	**98.34**	**.09**	**.06**	**.45**	**.25**	**.22**	–	–	**0**
	.50	.23	–	20.0	–	–	66.00	.06	.04	.30	.17	.15	–	–	0
2541	**.45**	–	–	**29.8**	–	–	**64.07**	**.09**	**.05**	**.30**	–	–	–	–	–
	.30	–	–	20.0	–	–	43.00	.06	.03	.20	–	–	–	–	–
2542	**.40**	–	–	**3828.0**	–	–	**26.40**	**.05**	**.15**	**.40**	–	–	**1.32**	–	–
	1.20	–	–	11600.0	–	–	80.00	.16	.44	1.20	–	–	4.00	–	–
2543	**1.40**	–	–	**19400.0**	–	–	**74.00**	**.20**	**.52**	**1.80**	–	–	**6.00**	–	–
	.70	–	–	9700.0	–	–	37.00	.10	.26	.90	–	–	3.00	–	–
2544	**.28**	**(.03)**	**(.03)**	**10.0**	**0**	**(.21)**	**4.25**	**.03**	**.02**	**.13**	**.03**	**.05**	–	**(.40)**	**0**
	1.10	(.12)	(.12)	40.0	0	(.84)	17.00	.11	.06	.50	.12	.20	–	(1.60)	0
2545	**3.90**	**1.95**	–	–	–	–	**31.20**	**.27**	**.12**	**1.43**	–	–	–	–	–
	3.00	1.50	–	–	–	–	24.00	.21	.09	1.10	–	–	–	–	–
2546	**4.65**	**2.40**	–	–	–	–	**36.00**	**.33**	**.14**	**1.80**	–	–	–	–	–
	3.10	1.60	–	–	–	–	24.00	.22	.09	1.20	–	–	–	–	–
2547	**1.32**	**.26**	**.02**	**640.2**	–	**.04**	**5.28**	**.04**	**.04**	**.20**	**.13**	**.04**	**13.86**	**2.05**	**0**
	2.00	.40	.04	970.0	–	.06	8.00	.06	.06	.30	.20	.06	21.00	3.10	0
2548	**.92**	**.26**	**.02**	**1254.0**	–	**.04**	**11.88**	**.03**	**.05**	**.26**	**.13**	**.04**	**118.14**	**2.05**	**0**
	1.40	.40	.04	1900.0	–	.06	18.00	.05	.08	.40	.20	.06	179.00	3.10	0
2549	**.30**	**.24**	**.02**	**198.0**	–	**.04**	**3.60**	**.04**	**.04**	**.18**	**.12**	**.03**	**22.20**	**1.86**	**0**
	.50	.40	.04	330.0	–	.06	6.00	.06	.06	.30	.20	.06	37.00	3.10	0
2550	**.92**	**.26**	**.02**	**1254.0**	–	**.04**	**11.88**	**.03**	**.05**	**.26**	**.13**	**.04**	**24.42**	**2.05**	**0**
	1.40	.40	.04	1900.0	–	.06	18.00	.05	.08	.40	.20	.06	37.00	3.10	0
2551	**1.76**	–	–	**0**	–	–	–	**.08**	**.14**	**.73**	–	–	–	–	–
	6.30	–	–	0	–	–	–	.28	.50	2.60	–	–	–	–	–
2552	**.56**	–	–	**1540.0**	–	–	**10.36**	–	–	–	–	–	–	–	–
	2.00	–	–	5500.0	–	–	37.00	–	–	–	–	–	–	–	–
2553	**.22**	–	–	**2.8**	–	–	**1.40**	**.04**	**.01**	**.20**	–	–	–	–	–
	.80	–	–	10.0	–	–	5.00	.13	.03	.70	–	–	–	–	–
2554	**.28**	**.01**	**.02**	**168.0**	–	–	**51.20**	**.03**	**.03**	**.20**	**.09**	**.10**	–	–	**0**
	.70	.03	.04	420.0	–	–	128.00	.08	.08	.50	.23	.26	–	–	0
2555	**2.60**	–	–	**9900.0**	–	–	**16.00**	**.24**	**.14**	**2.20**	**.52**	**.20**	–	–	–
	1.30	–	–	4950.0	–	–	8.00	.12	.07	1.10	.26	.10	–	–	–
2556	**.08**	**.01**	**.01**	**0**	**0**	**Trace**	**.30**	**.01**	**.06**	**.42**	**.22**	**.01**	**2.30**	**1.60**	**0**
	.80	.09	.07	0	0	Trace	3.00	.10	.46	4.20	2.20	.13	23.00	16.00	0
2557	**.14**	–	–	–	–	–	**.30**	**.01**	**.03**	**.68**	–	–	–	–	–
	1.40	–	–	–	–	–	3.00	.10	.33	6.80	–	–	–	–	–
2558	**2.70**	–	–	**653.4**	–	–	–	**.22**	**1.05**	–	–	–	–	–	–
	1.00	–	–	242.0	–	–	–	.08	.39	–	–	–	–	–	–
2559	**2.16**	**2.97**	–	**0**	–	–	–	**.05**	**.65**	**5.40**	–	–	**8.10**	–	–
	.80	1.10	–	0	–	–	–	.02	.24	2.00	–	–	3.00	–	–
2560	**1.00**	**2.20**	**.52**	–	–	–	**4.00**	**.04**	**.50**	**4.00**	**2.00**	**.12**	**6.00**	**14.60**	**0**
	.50	1.10	.26	–	–	–	2.00	.02	.25	2.00	1.00	.06	3.00	7.30	0
2561	**1.00**	–	–	**2310.0**	–	–	**32.00**	**.04**	**.07**	**.26**	**.07**	**.04**	**19.80**	–	**0**
	3.00	–	–	7000.0	–	–	97.00	.11	.22	.80	.21	.13	60.00	–	0
2562	**3.60**	–	–	**11600.0**	–	–	**96.00**	**.16**	**.28**	**1.20**	–	–	–	–	–
	1.80	–	–	5800.0	–	–	48.00	.08	.14	.60	–	–	–	–	–
2563	**3.00**	**(.46)**	**(.13)**	**12000.0**	–	–	**40.00**	**.06**	**.20**	**.80**	–	–	–	–	–
	1.50	(.23)	(.07)	6000.0	–	–	20.00	.03	.10	.40	–	–	–	–	–
2564	**.08**	–	–	**0**	–	–	**4.76**	**.00**	**.01**	**.14**	–	–	–	–	–
	.30	–	–	0	–	–	17.00	.01	.03	.50	–	–	–	–	–
2565	**.76**	–	–	**61.6**	–	–	**4.48**	**.01**	**.01**	**.11**	–	–	–	–	–
	2.70	–	–	220.0	–	–	16.00	.03	.04	.40	–	–	–	–	–

(Continued)

TABLE F-36 *(Continued)*

Item No.	Food Item Name	Approximate Measure	Weight	Moisture	Food Energy Calories	Protein	Fats	Carbo-hydrates	Fiber	Minerals (Macro) Calcium	Phos-phorus	Sodium	Mag-nesium	Potas-sium
			(g)	(%)	(kcal)	(g)	(g)	(g)	(g)	(mg)	(mg)	(mg)	(mg)	(mg)
	OKRA													
2566	raw	1 POD	12	88.9	4.3	.3	0	.9	.1	11.0	6.1	.4	4.9	29.9
			100	88.9	36.0	2.4	.3	7.6	1.0	92.0	51.0	3.0	41.0	249.0
2567	boiled, drained	1 POD	12	91.1	3.5	.2	0	.7	.1	11.0	4.9	.2	—	20.9
			100	91.1	29.0	2.0	.3	6.0	1.0	92.0	41.0	2.0	—	174.0
2568	frozen, cuts & pods	1 CUP	180	87.9	70.2	4.1	.2	16.2	1.8	169.2	91.8	3.6	95.4	394.2
			100	87.9	39.0	2.3	.1	9.0	1.0	94.0	51.0	2.0	53.0	219.0
2569	frozen, cuts & pods, boiled, drained	1 POD	13	88.3	4.9	.3	0	1.1	.1	12.2	5.6	.3	—	21.3
			100	88.3	38.0	2.2	.1	8.8	1.0	94.0	43.0	2.0	—	164.0
	ONIONS													
2570	young green, raw, bulb & entire top	1 MED	20	89.4	7.2	.3	0	1.6	.2	10.2	7.8	1.0	(2.2)	46.2
			100	89.4	36.0	1.5	.2	8.2	1.2	51.0	39.0	5.0	(11.1)	231.0
2571	young green, raw, bulb & white portion of top	1 AVG	8	87.6	3.6	.1	0	.8	.1	3.2	3.1	.4	2.0	18.5
			100	87.6	45.0	1.1	.2	10.5	1.0	40.0	39.0	5.0	25.0	231.0
2572	young green, raw, tops only (green portion)	1 CUP	100	91.8	27.0	1.6	.4	5.5	1.3	56.0	39.0	5.0	24.0	231.0
			100	91.8	27.0	1.6	.4	5.5	1.3	56.0	39.0	5.0	24.0	231.0
2573	mature (dry), white, raw	1 MED	100	89.1	38.0	1.5	.1	8.7	.6	27.0	36.0	10.0	12.0	157.0
			100	89.1	38.0	1.5	.1	8.7	.6	27.0	36.0	10.0	12.0	157.0
2574	mature (dry) white, boiled, drained	1 CUP	200	91.8	58.0	2.4	.2	13.0	1.2	48.0	58.0	14.0	(10.4)	220.0
			100	91.8	29.0	1.2	.1	6.5	.6	24.0	29.0	7.0	(5.2)	110.0
2575	mature (dry), white, dehydrated, flaked	1 CUP	64	4.0	224.0	5.6	.8	52.5	2.8	106.2	174.7	56.3	67.8	885.1
			100	4.0	350.0	8.7	1.3	82.1	4.4	166.0	273.0	88.0	106.0	1383.0
2576	mature (dry), yellow, raw	1 MED	110	89.1	41.8	1.5	.1	9.6	.7	29.7	39.6	11.0	13.2	172.7
			100	89.1	38.0	1.4	.1	8.7	.6	27.0	36.0	10.0	12.0	157.0
2577	mature (dry), yellow, boiled, drained	1 CUP	210	91.8	60.9	2.5	.2	13.7	1.2	50.4	60.9	14.7	—	231.0
			100	91.8	29.0	1.2	.1	6.5	.6	24.0	29.0	7.0	—	110.0
2578	mature (dry), yellow, dehydrated, flaked	1 TBLS	10	4.0	35.0	.9	.1	8.2	.4	16.6	27.3	8.8	10.6	138.3
			100	4.0	350.0	8.7	1.3	82.1	4.4	166.0	273.0	88.0	106.0	1383.0
2579	Welsh, raw	1 AVG	100	90.5	34.0	1.9	.4	6.5	1.0	18.0	49.0	—	—	—
			100	90.5	34.0	1.9	.4	6.5	1.0	18.0	49.0	—	—	—
2580	[a]**OXALIS,** oka, tubers, raw	1 OZ	28	83.8	17.6	.3	.2	3.9	.2	1.1	9.5	—	—	—
			100	83.8	63.0	1.0	.6	13.8	.8	4.0	34.0	—	—	—
	PARSLEY													
2581	raw	1 TBLS	4	85.1	1.8	.1	0	.3	.1	8.1	2.5	1.8	1.6	29.1
			100	85.1	44.0	3.6	.6	8.5	1.5	203.0	63.0	45.0	41.0	727.0
2582	dried	1 TSP	1	9.0	2.8	.2	0	.5	.1	14.7	3.5	4.5	2.5	38.1
			100	9.0	276.0	22.4	4.4	51.7	10.3	1468.0	351.0	452.0	249.0	3305.0
	PARSNIPS													
2583	raw	1 LG	200	79.1	152.0	3.4	1.0	35.0	4.0	100.0	154.0	24.0	91.2	1082.0
			100	79.1	76.0	1.7	.5	17.5	2.0	50.0	77.0	12.0	45.6	541.0
2584	boiled, drained	1 CUP	200	82.2	132.0	3.0	1.0	29.8	4.0	90.0	124.0	16.0	25.2	758.0
			100	82.2	66.0	1.5	.5	14.9	2.0	45.0	62.0	8.0	12.6	379.0
	PEAS													
2585	edible, podded, raw	1 CUP	133	83.3	70.5	4.5	.3	16.0	1.0	82.5	119.7	—	8.0	226.1
			100	83.3	53.0	3.4	.2	12.0	1.2	62.0	90.0	—	6.0	170.0
2586	edible, podded, boiled, drained	1 CUP	150	86.6	64.5	4.4	.3	14.3	1.8	84.0	114.0	—	—	178.5
			100	86.6	43.0	2.9	.2	9.5	1.2	56.0	76.0	—	—	119.0
2587	green, immature, raw	1 CUP	133	78.0	111.7	8.4	.5	19.2	2.7	34.6	154.3	2.7	46.6	420.3
			100	78.0	84.0	6.3	.4	14.4	2.0	26.0	116.0	2.0	35.0	316.0
2588	green, immature, sweet, boiled, drained	1 CUP	150	81.5	106.5	8.1	.6	18.2	3.0	34.5	148.5	1.5	(31.8)	294.0
			100	81.5	71.0	5.4	.4	12.1	2.0	23.0	99.0	1.0	(21.2)	196.0
2589	green immature sweet, canned, drained solids	1 CUP	160	79.0	128.0	7.4	.6	24.0	3.5	40.0	107.2	377.6	32.0	153.6
			100	79.0	80.0	4.6	.4	15.0	2.2	25.0	67.0	236.0	20.0	96.0
2590	green, immature, sweet, canned, solids & liquid	1 CUP	150	84.8	85.5	5.1	.5	15.6	2.1	28.5	87.0	354.0	30.0	144.0
			100	84.8	57.0	3.4	.3	10.4	1.4	19.0	58.0	236.0	20.0	96.0
2591	green, immature, sweet, canned, drained liquid	1 CUP	240	93.3	52.8	3.1	—	10.3	—	21.6	100.8	566.4	—	230.4
			100	93.3	22.0	1.3	—	4.3	—	9.0	42.0	236.0	—	96.0
2592	green, immature, sweet, low sodium, canned, drained, solids	1 CUP	160	81.8	115.2	7.0	.6	20.8	3.2	40.0	107.2	4.8	32.0	153.6
			100	81.8	72.0	4.4	.4	13.0	2.0	25.0	67.0	3.0	20.0	96.0
2593	green, immature, sweet, low sodium, canned, solids & liquids	1 CUP	150	87.8	70.5	5.0	.5	12.3	2.0	28.5	87.0	4.5	30.0	144.0
			100	87.8	47.0	3.3	.3	8.2	1.3	19.0	58.0	3.0	20.0	96.0

Item No.	Minerals (Micro)			Fat-Soluble Vitamins			Water-Soluble Vitamins								
	Iron	Zinc	Copper	Vitamin A	Vitamin D	Vitamin E (Alpha Tocopherol)	Vitamin C	Thiamin	Ribo-flavin	Niacin	Panto-thenic Acid	Vit. B-6 (Pyri-doxine)	Folacin (Folic Acid)	Biotin	Vitamin B-12
	(mg)	(mg)	(mg)	(IU)	(IU)	(mg)	(mg)	(mg)	(mg)	(mg)	(mg)	(mg)	(mcg)	(mcg)	(mcg)
2566	**.07** .60	[6]**.06** .50	**.02** .18	**62.4** 520.0	**0** 0	— —	**3.72** 31.00	**.02** .17	**.03** .21	**.12** 1.00	**.03** .26	**.01** .08	**10.56** 88.00	— —	**0** 0
2567	**.06** .50	[6]**.12** 1.0	— —	**58.8** 490.0	— —	— —	**2.40** 20.00	**.02** .13	**.02** .18	**.11** .90	— —	— —	**4.56** 38.00	— —	— —
2568	**1.08** .60	— —	— —	**864.0** 480.0	— —	— —	**28.80** 16.00	**.31** .17	**.38** .21	**1.80** 1.00	**.39** .22	**.08** .05	— —	— —	**0** 0
2569	**.07** .50	— —	— —	**62.4** 480.0	— —	— —	**1.56** 12.00	**.02** .14	**.02** .17	**.13** 1.00	**.03** .20	**.01** .05	— —	— —	— —
2570	**.20** 1.00	**.06** .30	**.01** .06	**400.0** 2000.0	— —	**Trace** Trace	**6.40** 32.00	**.01** .05	**.01** .05	**.08** .40	**.03** .14	**(.02)** (.11)	**(8.04)** (40.22)	**(.17)** (.93)	**0** 0
2571	**.05** .60	— —	— —	**26.4** 330.0	— —	— —	**2.00** 25.00	**.00** .05	**Trace** .04	**.03** .40	**.01** .14	— —	**2.88** 36.00	— —	**0** 0
2572	**2.20** 2.20	— —	— —	**4000.0** 4000.0	— —	— —	**51.00** 51.00	**.07** .07	**.10** .10	**.60** .60	— —	— —	**80.00** 80.00	— —	— —
2573	**.50** .50	**.30** .30	**.13** .13	**0** 0	**0** 0	**.21** .21	**10.00** 10.00	**.03** .03	**.04** .04	**.20** .20	**.13** .13	**.13** .13	**25.00** 25.00	**3.50** 3.50	**0** 0
2574	**.80** .40	**(.30)** (.15)	**(.16)** (.08)	**0** 0	**0** 0	**Trace** Trace	**14.00** 7.00	**.06** .03	**.06** .03	**.40** .20	**(.22)** (.11)	**(.14)** (.07)	**26.00** 13.00	**(1.32)** (.66)	**0** 0
2575	**1.86** 2.90	— —	— —	— —	— —	— —	**22.40** 35.00	**.16** .25	**.12** .18	**.90** 1.40	— —	— —	— —	— —	— —
2576	**.55** .50	**.33** .30	**.11** .10	**44.0** 40.0	— —	**.24** .22	**11.00** 10.00	**.03** .03	**.04** .04	**.22** .20	**.14** .13	**.14** .13	**27.50** 25.00	**3.85** 3.50	**0** 0
2577	**.84** .40	— —	— —	**84.0** 40.0	— —	— —	**14.70** 7.00	**.06** .03	**.06** .03	**.42** .20	— —	— —	**27.30** 13.00	— —	— —
2578	**.29** 2.90	— —	— —	**20.0** 200.0	— —	— —	**3.50** 35.00	**.03** .25	**.02** .18	**.14** 1.40	**.12** 1.20	**.05** .50	— —	— —	**0** 0
2579	— —	— —	— —	— —	— —	— —	**27.00** 27.00	**.05** .05	**.09** .09	**.40** .40	— —	— —	**66.00** 66.00	— —	— —
2580	**.22** .80	— —	— —	**0** 0	— —	— —	**10.36** 37.00	**.01** .05	**.02** .07	**.11** .40	— —	— —	— —	— —	— —
2581	**.25** 6.20	**.04** .92	**0** .02	**340.0** 8500.0	**0** 0	**(.07)** (1.84)	**6.88** 172.00	**Trace** .12	**.01** .26	**.05** 1.20	**.01** .30	**.01** .16	**4.64** 116.00	**.02** .42	**0** 0
2582	**.98** 97.86	**.05** 4.75	**.01** .64	**233.4** 23340.0	**0** 0	**(2.13)** (1.06)	**1.22** 122.04	**Trace** .17	**.01** 1.23	**.08** 7.90	**(.68)** (.34)	**.01** 1.00	**(61.20)** (30.60)	**Trace** Trace	**0** 0
2583	**1.40** .70	**(.22)** (.11)	**.20** .10	**60.0** 30.0	**0** 0	**(2.08)** (1.04)	**32.00** 16.00	**.16** .08	**.18** .09	**.40** .20	**1.20** .60	**.18** .09	**(134.22)** (67.11)	**(.26)** (.13)	**0** 0
2584	**1.20** .60	**(.22)** (.11)	**(.26)** (.13)	**60.0** 30.0	— —	— —	**20.00** 10.00	**.14** .07	**.16** .08	**.20** .10	— —	— —	— —	— —	— —
2585	**.93** .70	— —	— —	**904.4** 680.0	**0** 0	— —	**27.93** 21.00	**.37** .28	**.16** .12	— —	**1.09** .82	**.20** .15	**33.25** 25.00	**12.50** 9.40	— —
2586	**.75** .50	— —	— —	**915.0** 610.0	**0** 0	— —	**21.00** 14.00	**.33** .22	**.17** .11	— —	— —	— —	— —	— —	— —
2587	**2.53** 1.90	**1.20** .90	**.06** .05	**851.2** 640.0	**0** 0	**.13** .10	**35.91** 27.00	**.47** .35	**.19** .14	**3.86** 2.90	**1.00** .75	**.21** .16	**33.25** 25.00	**12.50** 9.40	**0** 0
2588	**2.70** 1.80	**1.05** .70	**(.24)** (.16)	**810.0** 540.0	**0** 0	**.83** .55	**30.00** 20.00	**.42** .28	**.17** .11	**3.45** 2.30	**(.50)** (.33)	**(.17)** (.11)	— —	**(.65)** (.43)	**0** 0
2589	**2.72** 1.70	**1.28** .80	— —	**1104.0** 690.0	— —	— —	**12.80** 8.00	**.18** .11	**.10** .06	**1.60** 1.00	— —	**.08** .05	**40.00** 25.00	— —	**0** 0
2590	**2.25** 1.50	— —	— —	**675.0** 450.0	— —	**.03** .02	**13.50** 9.00	**.17** .11	**.09** .06	**1.50** 1.00	**.23** .15	**.08** .05	**43.50** 29.00	**3.15** 2.10	**0** 0
2591	**2.64** 1.10	[6]**1.44** .60	— —	— —	— —	— —	**24.00** 10.00	**.29** .12	**.12** .05	**2.64** 1.10	— —	— —	**81.60** 34.00	— —	— —
2592	**2.72** 1.70	— —	— —	**1104.0** 690.0	— —	— —	**12.80** 8.00	**.18** .11	**.10** .06	**1.60** 1.00	— —	**.08** .05	**40.00** 25.00	— —	**0** 0
2593	**2.25** 1.50	— —	— —	**675.0** 450.0	— —	**.03** .02	**13.50** 9.00	**.17** .11	**.09** .06	**1.50** 1.00	**.23** .15	**.08** .05	**43.50** 29.00	**3.15** 2.10	**0** 0

(Continued)

TABLE F-36 *(Continued)*

Item No.	Food Item Name	Approximate Measure	Weight	Moisture	Food Energy Calories	Protein	Fats	Carbo-hydrates	Fiber	Minerals (Macro) Calcium	Phos-phorus	Sodium	Mag-nesium	Potas-sium
			(g)	(%)	(kcal)	(g)	(g)	(g)	(g)	(mg)	(mg)	(mg)	(mg)	(mg)
	PEAS (Continued)													
2594	green, immature, sweet, low sodium, canned, drained liquid	1 CUP	**240**	**94.9**	**43.2**	**3.1**	—	**8.2**	—	**21.6**	**100.8**	**7.2**	—	**230.4**
			100	94.9	18.0	1.3	—	3.4	—	9.0	42.0	3.0	—	96.0
2595	green, immature, frozen, boiled, drained	1 CUP	**160**	**82.1**	**108.8**	**8.2**	**.5**	**18.9**	**3.0**	**30.4**	**137.6**	**184.0**	**38.4**	**216.0**
			100	82.1	68.0	5.1	.3	11.8	1.9	19.0	86.0	115.0	24.0	135.0
2596	green, immature, Alaska, boiled, drained liquid	1 CUP	**240**	**92.3**	**62.4**	**3.1**	—	**12.5**	—	**24.0**	**115.2**	**566.4**	—	**230.4**
			100	92.3	26.0	1.3	—	5.2	—	10.0	48.0	236.0	—	96.0
2597	green, immature, Alaska, canned, drained solids	1 CUP	**160**	**77.0**	**140.8**	**7.5**	**.6**	**26.9**	**3.7**	**41.6**	**121.6**	**377.6**	**32.0**	**153.6**
			100	77.0	88.0	4.7	.4	16.8	2.3	26.0	76.0	236.0	20.0	96.0
2598	green, immature, Alaska, canned, solids & liquid	1 CUP	**150**	**82.6**	**99.0**	**5.3**	**.5**	**18.8**	**2.3**	**30.0**	**99.0**	**354.0**	**30.0**	**144.0**
			100	82.6	66.0	3.5	.3	12.5	1.5	20.0	66.0	236.0	20.0	96.0
2599	green, immature, Alaska, low sodium, canned, drained	1 CUP	**160**	**80.1**	**124.8**	**7.7**	**.6**	**22.9**	**3.2**	**41.6**	**121.6**	**4.8**	**32.0**	**153.6**
			100	80.1	78.0	4.8	.4	14.3	2.0	26.0	76.0	3.0	20.0	96.0
2600	green, immature, Alaska, low sodium, canned, solids & liquid	1 CUP	**150**	**85.9**	**82.5**	**5.4**	**.5**	**14.7**	**2.0**	**30.0**	**99.0**	**4.5**	—	**144.0**
			100	85.9	55.0	3.6	.3	9.8	1.3	20.0	66.0	3.0	—	96.0
2601	green, immature, Alaska, low sodium, canned, drained liquid	1 CUP	**240**	**94.1**	**52.8**	**3.4**	—	**9.8**	—	**24.0**	**115.2**	**7.2**	—	**230.4**
			100	94.1	22.0	1.4	—	4.1	—	10.0	48.0	3.0	—	96.0
2602	carrots, frozen, boiled, drained	1 CUP	**200**	**85.8**	**106.0**	**6.4**	**.6**	**20.2**	**3.0**	**50.0**	**114.0**	**168.0**	**37.8**	**314.0**
			100	85.8	53.0	3.2	.3	10.1	1.5	25.0	57.0	84.0	18.9	157.0
	PEPPERS, GREEN													
2603	immature, green, raw	1 MED	**40**	**93.4**	**8.8**	**.5**	**.1**	**1.9**	**.6**	**3.6**	**8.8**	**5.2**	**7.2**	**85.2**
			100	93.4	22.0	1.2	.2	4.8	1.4	9.0	22.0	13.0	18.0	213.0
2604	immature, green, boiled, drained	1 OZ	**28**	**94.7**	**5.0**	**.3**	**.1**	**1.1**	**.4**	**2.5**	**4.5**	**2.5**	**(2.8)**	**41.7**
			100	94.7	18.0	1.0	.2	3.8	1.4	9.0	16.0	9.0	(10.1)	149.0
2605	mature, red, raw	1 AVG	**90**	**90.7**	**27.9**	**1.3**	**.3**	**6.4**	**1.5**	**11.7**	**27.0**	**5.4**	**11.7**	**108.0**
			100	90.7	31.0	1.4	.3	7.1	1.7	13.0	30.0	6.0	13.0	120.0
	PEPPERS, HOT													
2606	chili, immature geeen pods, no seeds, raw	1 AVG	**74**	**88.8**	**27.4**	**1.0**	**.1**	**6.7**	**1.3**	**7.4**	**18.5**	**3.7**	**17.0**	**192.4**
			100	88.8	37.0	1.3	.2	9.1	1.8	10.0	25.0	5.0	23.0	260.0
2607	chili, immature green pods, no seeds, canned	1 AVG	**74**	**92.5**	**18.5**	**.7**	**.1**	**4.5**	**.8**	**5.2**	**12.6**	—	—	—
			100	92.5	25.0	.9	.1	6.1	1.2	7.0	17.0	—	—	—
2608	chili, mature, red, raw pods w/seeds	1 MED	**100**	**74.3**	**93.0**	**3.7**	**2.3**	**18.1**	**9.0**	**29.0**	**78.0**	**9.0**	**27.0**	**420.0**
			100	74.3	93.0	3.7	2.3	18.1	9.0	29.0	78.0	9.0	27.0	420.0
2609	chili, mature, red, raw pods, no seeds	1 OZ	**28**	**80.3**	**18.2**	**.6**	**.1**	**4.4**	**.6**	**4.5**	**13.7**	**7.0**	—	**157.9**
			100	80.3	65.0	2.3	.4	15.8	2.3	16.0	49.0	25.0	—	564.0
2910	chili, mature, red, dried pods	1 OZ	**28**	**12.6**	**89.9**	**3.6**	**2.5**	**16.7**	**1.7**	**36.4**	**67.2**	**104.4**	—	**336.3**
			100	12.6	321.0	12.9	9.1	59.8	6.2	130.0	240.0	373.0	—	1201.0
2611	**PIMIENTOS,** canned, solids & liquid	1 MED	**40**	**92.4**	**10.8**	**.4**	**.2**	**2.3**	**.2**	**2.8**	**6.8**	—	—	—
			100	92.4	27.0	.9	.5	5.8	.6	7.0	17.0	—	—	—
2612	[4]**PLANTAIN (COOKING BANANA),** boiled	1 OZ	**28**	**69.1**	**31.1**	**.2**	**.1**	**8.2**	**(1.8)**	**2.5**	**9.0**	**(1.2)**	**(9.6)**	**(92.6)**
			100	69.1	111.0	.8	.3	29.4	(6.5)	9.0	32.0	(4.2)	(34.3)	(330.6)
	POKEWEED (POKEBERRY, POKE)													
2613	shoots, raw	1 OZ	**28**	**91.6**	**6.4**	**.7**	**.1**	**1.0**	—	**14.8**	**12.3**	—	—	—
			100	91.6	23.0	2.6	.4	3.7	—	53.0	44.0	—	—	—
2614	shoots, boiled, drained	1 OZ	**28**	**92.9**	**5.6**	**.6**	**.1**	**.9**	—	**14.8**	**9.2**	—	—	—
			100	92.9	20.0	2.3	.4	3.1	—	53.0	33.0	—	—	—
	POTATOES													
2615	raw	1 AVG	**100**	**79.8**	**76.0**	**2.1**	**.1**	**17.1**	**.5**	**7.0**	**53.0**	**3.0**	**14.0**	**407.0**
			100	79.8	76.0	2.1	.1	17.1	.5	7.0	53.0	3.0	14.0	407.0
2616	baked in skin	1 MED	**100**	**75.1**	**93.0**	**2.6**	**.1**	**21.1**	**.6**	**9.0**	**65.0**	**4.0**	**(28.8)**	**503.0**
			100	75.1	93.0	2.6	.1	21.1	.6	9.0	65.0	4.0	(28.8)	503.0
2617	boiled in skin	1 MED	**100**	**79.8**	**76.0**	**2.1**	**.1**	**17.1**	**.5**	**7.0**	**53.0**	**3.0**	—	**407.0**
			100	79.8	76.0	2.1	.1	17.1	.5	7.0	53.0	3.0	—	407.0
2618	boiled, pared before cooking	1 MED	**100**	**82.8**	**65.0**	**1.9**	**.1**	**14.5**	**.5**	**6.0**	**42.0**	**2.0**	**(15.3)**	**285.0**
			100	82.8	65.0	1.9	.1	14.5	.5	6.0	42.0	2.0	(15.3)	285.0
2619	mashed, milk added	1 CUP	**200**	**82.8**	**130.0**	**4.2**	**1.4**	**26.0**	**.8**	**48.0**	**98.0**	**602.0**	—	**522.0**
			100	82.8	65.0	2.1	.7	13.0	.4	24.0	49.0	301.0	—	261.0
2620	mashed, milk & margarine added	1 CUP	**210**	**79.8**	**197.4**	**4.4**	**9.0**	**25.8**	**.8**	**50.4**	**100.8**	**695.1**	—	**525.0**
			100	79.8	94.0	2.1	4.3	12.3	.4	24.0	48.0	331.0	—	250.0
2621	scalloped & au gratin, w/o cheese, w/margarine	1 CUP	**245**	**76.7**	**254.8**	**7.4**	**9.6**	**36.0**	**.7**	**132.3**	**181.3**	**869.8**	—	**801.2**
			100	76.7	104.0	3.0	3.9	14.7	.3	54.0	74.0	355.0	—	327.0

Item No.	Minerals (Micro)			Fat-Soluble Vitamins			Water-Soluble Vitamins								
	Iron	Zinc	Copper	Vitamin A	Vitamin D	Vitamin E (Alpha Tocopherol)	Vitamin C	Thiamin	Ribo-flavin	Niacin	Panto-thenic Acid	Vit. B-6 (Pyri-doxine)	Folacin (Folic Acid)	Biotin	Vitamin B-12
	(mg)	(mg)	(mg)	(IU)	(IU)	(mg)	(mg)	(mg)	(mg)	(mg)	(mg)	(mg)	(mcg)	(mcg)	(mcg)
2594	**2.64**	—	—	—	—	—	**24.00**	**.29**	**.12**	**2.64**	—	—	**81.60**	—	—
	1.10	—	—	—	—	—	10.00	.12	.05	1.10	—	—	34.00	—	—
2595	**3.04**	**1.10**	**(.18)**	**960.0**	**0**	**.40**	**20.80**	**.43**	**.14**	**2.72**	**.45**	**.14**	**134.40**	**(.69)**	**0**
	1.90	.69	(.11)	600.0	0	.25	13.00	.27	.09	1.70	.28	.09	84.00	(.43)	0
2596	**3.12**	—	—	—	—	—	**24.00**	**.24**	**.10**	**2.40**	—	—	**81.60**	—	—
	1.30	—	—	—	—	—	10.00	.10	.04	1.00	—	—	34.00	—	—
2597	**3.04**	**1.28**	—	**1104.0**	—	—	**12.80**	**.14**	**.10**	**1.28**	—	—	**40.00**	—	—
	1.90	.80	—	690.0	—	—	8.00	.09	.06	.80	—	—	25.00	—	—
2598	**2.55**	**(.99)**	**(.18)**	**675.0**	—	**.03**	**13.50**	**.14**	**.08**	**1.35**	**.23**	**.08**	**43.50**	**3.15**	**0**
	1.70	(.66)	(.12)	450.0	—	.02	9.00	.09	.05	.90	.15	.05	29.00	2.10	0
2599	**3.04**	**2.08**	**.21**	**1104.0**	—	—	**12.80**	**.14**	**.10**	**1.28**	—	—	**40.00**	—	—
	1.90	1.30	.13	690.0	—	—	8.00	.09	.06	.80	—	—	25.00	—	—
2600	**2.55**	—	—	**675.0**	—	**.03**	**13.50**	**.14**	**.08**	**1.35**	**.23**	**.08**	**43.50**	**3.15**	**0**
	1.70	—	—	450.0	—	.02	9.00	.09	.05	.90	.15	.05	29.00	2.10	0
2601	**3.12**	—	—	—	—	—	**24.00**	**.24**	**.10**	**2.40**	—	—	**81.60**	—	—
	1.30	—	—	—	—	—	10.00	.10	.04	1.00	—	—	34.00	—	—
2602	**2.20**	—	—	**18600.0**	—	—	**16.00**	**.38**	**.14**	**2.60**	**.44**	**.18**	—	—	—
	1.10	—	—	9300.0	—	—	8.00	.19	.07	1.30	.22	.09	—	—	—
2603	**.28**	**.01**	**.02**	**168.0**	**0**	**(.32)**	**51.20**	**.032**	**.03**	**.20**	**.09**	**.10**	**7.60**	—	**0**
	.70	.03	.04	420.0	0	(.81)	128.00	.080	.08	.50	.23	.26	19.00	—	0
2604	**.14**	**(.06)**	**(.02)**	**117.6**	**0**	**(.23)**	**26.88**	**.017**	**.02**	**.14**	**(.05)**	**(.04)**	**(3.11)**	—	**0**
	.50	(.22)	(.06)	420.0	0	(.81)	96.00	.060	.07	.50	(.17)	(.15)	(11.11)	—	0
2605	**.54**	—	—	**4005.0**	—	—	**183.60**	**.072**	**.07**	**.45**	**.24**	—	—	—	**0**
	.60	—	—	4450.0	—	—	204.00	.080	.08	.50	.27	—	—	—	0
2606	**.52**	**.02**	—	**569.8**	—	—	**173.90**	**.07**	**.04**	**1.26**	**.51**	—	—	—	**0**
	.70	.02	—	770.0	—	—	235.00	.09	.06	1.70	.69	—	—	—	0
2607	**.37**	—	—	**451.4**	—	—	**50.32**	**.02**	**.04**	**.59**	—	—	—	—	—
	.50	—	—	610.0	—	—	68.00	.02	.05	.80	—	—	—	—	—
2608	**1.20**	**.02**	—	**21600.0**	—	—	**369.00**	**.220**	**.36**	**4.40**	**1.08**	—	**52.00**	—	**0**
	1.20	.02	—	21600.0	—	—	369.00	.220	.36	4.40	1.08	—	52.00	—	0
2609	**.39**	—	—	**6048.0**	—	—	**103.32**	**.028**	**.06**	**.81**	**.30**	—	**14.56**	—	**0**
	1.40	—	—	21600.0	—	—	369.00	.100	.20	2.90	1.08	—	52.00	—	0
2610	**2.18**	—	—	**21560.0**	—	—	**3.36**	**.064**	**.37**	**2.94**	—	—	—	—	—
	7.80	—	—	77000.0	—	—	12.00	.230	1.33	10.50	—	—	—	—	—
2611	**.60**	—	—	**920.0**	—	—	**38.00**	**.008**	**.02**	**.16**	**.07**	—	**2.40**	—	**0**
	1.50	—	—	2300.0	—	—	95.00	.020	.06	.40	.17	—	6.00	—	0
2612	**.34**	**(.06)**	**(.03)**	**161.0**	**0**	—	**3.36**	**.011**	**.02**	**.17**	**(.08)**	**(.09)**	**(5.11)**	—	**0**
	1.20	(.23)	(.12)	575.0	0	—	12.00	.040	.06	.60	(.27)	(.32)	(18.23)	—	0
2613	**.48**	—	—	**2436.0**	—	—	**38.08**	**.02**	**.09**	**.34**	—	—	—	—	—
	1.70	—	—	8700.0	—	—	136.00	.08	.33	1.20	—	—	—	—	—
2614	**.34**	—	—	**2436.0**	—	—	**22.96**	**.02**	**.07**	**.31**	—	—	—	—	—
	1.20	—	—	8700.0	—	—	82.00	.07	.25	1.10	—	—	—	—	—
2615	**.60**	**.30**	**.05**	**40.0**	**0**	**.05**	**20.00**	**.100**	**.04**	**1.50**	**.38**	**.25**	**19.00**	**(.13)**	**0**
	.60	.30	.05	40.0	0	.05	20.00	.100	.04	1.50	.38	.25	19.00	(.13)	0
2616	**.70**	**(.31)**	**(.18)**	**Trace**	**0**	**.03**	**20.00**	**.100**	**.04**	**1.70**	**(.22)**	**(.17)**	**(10.11)**	**Trace**	**0**
	.70	(.31)	(.18)	Trace	0	.03	20.00	.100	.04	1.70	(.22)	(.17)	(10.11)	Trace	0
2617	**.60**	**.30**	—	—	—	**.04**	**16.00**	**.090**	**.04**	**1.50**	—	—	—	—	—
	.60	.30	—	—	—	.04	16.00	.090	.04	1.50	—	—	—	—	—
2618	**.50**	**.30**	**(.12)**	**Trace**	**0**	**.04**	**16.00**	**.090**	**.03**	**1.20**	**(.21)**	**(.19)**	**(10.09)**	**Trace**	**0**
	.50	.30	(.12)	Trace	0	.04	16.00	.090	.03	1.20	(.21)	(.19)	(10.09)	Trace	0
2619	**.80**	—	—	**40.0**	—	—	**20.00**	**.160**	**.10**	**2.00**	—	—	**20.00**	—	—
	.40	—	—	20.0	—	—	10.00	.080	.05	1.00	—	—	10.00	—	—
2620	**.84**	—	—	**357.0**	—	—	**18.90**	**.168**	**.11**	**2.10**	—	—	**21.00**	—	—
	.40	—	—	170.0	—	—	9.00	.080	.05	1.00	—	—	10.00	—	—
2621	**.98**	—	—	**392.0**	—	—	**26.95**	**.147**	**.22**	**2.45**	—	—	—	—	—
	.40	—	—	160.0	—	—	11.00	.060	.09	1.00	—	—	—	—	—

(Continued)

TABLE F-36 *(Continued)*

Item No.	Food Item Name	Approximate Measure	Weight	Moisture	Food Energy Calories	Protein	Fats	Carbo-hydrates	Fiber	Minerals (Macro) Calcium	Phos-phorus	Sodium	Mag-nesium	Potas-sium
			(g)	(%)	(kcal)	(g)	(g)	(g)	(g)	(mg)	(mg)	(mg)	(mg)	(mg)
	POTATOES (Continued)													
2622	scalloped & au gratin, w/cheese, margarine	1 CUP	245	71.1	355.3	13.0	19.4	33.3	.7	311.2	298.9	1095.2	—	749.7
			100	71.1	145.0	5.3	7.9	13.6	.3	127.0	122.0	447.0	—	306.0
2623	hash browned after holding overnight	1 CUP	155	54.2	355.0	4.8	18.1	45.1	1.2	18.6	122.5	446.4	—	736.3
			100	54.2	229.0	3.1	11.7	29.1	.8	12.0	79.0	288.0	—	475.0
2624	fried from raw	1 CUP	170	46.9	455.6	6.8	24.1	55.4	1.7	25.5	171.7	379.1	—	1317.5
			100	46.9	268.0	4.0	14.2	32.6	1.0	15.0	101.0	223.0	—	775.0
2625	french fried in cottonseed oil	1 PIECE	5	44.7	13.7	.2	.7	1.8	.1	.8	5.6	.3	—	42.7
			100	44.7	274.0	4.3	13.2	36.0	1.0	15.0	111.0	6.0	—	853.0
2626	canned, solids & liquid	1 CUP	250	88.5	110.0	2.8	.5	24.5	.5	10.0	75.0	2.5	(35.0)	625.0
			100	88.5	44.0	1.1	.2	9.8	.2	4.0	30.0	1.0	(14.0)	250.0
2627	dehydrated, mashed flakes, w/milk, margarine	1 CUP	214	79.3	199.0	4.1	6.8	31.0	.6	66.3	100.6	494.3	—	612.0
			100	79.3	93.0	1.9	3.2	14.5	.3	31.0	47.0	231.0	—	286.0
2628	dehydrated, mashed flakes, w/milk, butter	1 SERV	113	—	80.0	2.4	1.0	15.0	—	35.0	58.0	239.3	—	285.0
			100	—	70.8	2.1	.9	13.3	—	31.0	51.3	215.0	—	252.2
2629	dehydrated, mashed granules, w/milk, margarine	1 CUP	200	78.6	192.0	4.0	7.2	28.8	.4	64.0	104.0	512.0	—	580.0
			100	78.6	96.0	2.0	3.6	14.4	.2	32.0	52.0	256.0	—	290.0
2630	frozen, diced for hash browning, cooked	1 CUP	155	56.1	347.2	3.1	17.8	45.0	1.1	27.9	77.5	463.5	—	438.7
			100	56.1	224.0	2.0	11.5	29.0	.7	18.0	50.0	299.0	—	283.0
2631	frozen, french fried, heated	1 CUP	200	52.9	440.0	7.2	16.8	67.4	1.4	18.0	172.0	8.0	—	1304.0
			100	52.9	220.0	3.6	8.4	33.7	.7	9.0	86.0	4.0	—	652.0
2632	frozen, mashed, heated	1 CUP	195	78.3	181.4	3.5	5.5	30.6	.8	48.8	81.9	700.1	—	419.3
			100	78.3	93.0	1.8	2.8	15.7	.4	25.0	42.0	359.0	—	215.0
	PRICKLY PEAR													
2633	raw, 1-½" diameter, 2-½" long	1 AVG	65	88.0	27.3	.3	.1	7.1	1.0	13.0	18.2	1.3	—	107.9
			100	88.0	42.0	.5	.1	10.9	1.6	20.0	28.0	2.0	—	166.0
2634	w/seeds, 1-½" diameter, 2-½" long	1 AVG	65	84.6	39.0	.9	.9	7.9	4.3	29.9	20.8	—	—	—
			100	84.6	60.0	1.4	1.4	12.1	6.6	46.0	32.0	—	—	—
2635	cholla, stem	1 OZ	28	88.8	10.4	.3	.1	2.5	.7	30.8	5.6	—	—	—
			100	88.8	37.0	1.1	.4	8.8	2.6	110.0	20.0	—	—	—
	PUMPKIN													
2636	[6]raw	1 OZ	28	—	4.2	.2	Trace	.9	.14	11.0	5.3	.3	—	86.3
			100	—	15.1	.7	Trace	3.3	.51	39.1	18.9	1.0	—	308.0
2637	canned	1 CUP	243	90.2	80.2	2.4	.7	19.2	3.2	60.8	89.9	12.2	63.2	537.0
			100	90.2	33.0	1.0	.3	7.9	1.3	25.0	37.0	5.0	26.0	221.0
2638	**PUMPKIN FLOWERS**	1 LG	30	94.8	4.8	.4	.1	.8	.2	14.1	25.8	—	—	—
			100	94.8	16.0	1.4	.3	2.7	.6	47.0	86.0	—	—	—
2639	[6]**PURSLANE,** common	1 OZ	28	91.2	7.3	.6	.1	1.4	.3	22.1	9.0	—	—	—
			100	91.2	26.0	2.0	.4	5.0	.9	79.0	32.0	—	—	—
	RADISHES													
2640	raw, common	1 SM	10	94.5	1.7	.1	0	.4	.1	3.0	3.1	1.8	1.5	32.2
			100	94.5	17.0	1.2	.1	3.6	.7	30.0	31.0	18.0	15.0	322.0
2641	raw, oriental	1 CUP	95	94.1	18.1	.9	.1	4.0	.7	33.3	24.7	—	—	171.0
			100	94.1	19.0	.9	.1	4.2	.7	35.0	26.0	—	—	180.0
	RUTABAGAS													
2642	raw	1 OZ	28	87.0	12.9	.3	0	3.0	.3	18.5	10.9	1.4	4.2	66.9
			100	87.0	46.0	1.1	.1	11.0	1.1	66.0	39.0	5.0	15.0	239.0
2643	boiled, drained	1 CUP	200	90.2	70.0	1.8	.2	16.4	2.2	118.0	62.0	8.0	—	334.0
			100	90.2	35.0	.9	.1	8.2	1.1	59.0	31.0	4.0	—	167.0
2644	[4]**SAGO PALM,** meal	1 OZ	28	13.1	100.0	.4	.1	24.1	.1	4.2	2.8	2.2	—	10.1
			100	13.1	357.0	1.4	.2	85.9	.2	15.0	10.0	8.0	—	36.0
	SALSIFY													
2645	raw	1 CUP	150	77.6	64.5	4.4	.9	27.0	2.7	70.5	99.0	30.0	34.5	570.0
			100	77.6	43.0	2.9	.6	18.0	1.8	47.0	66.0	20.0	23.0	380.0
2646	boiled, drained	1 CUP	150	81.0	61.5	3.9	.9	22.7	2.7	63.0	79.5	(13.2)	(21.5)	399.0
			100	81.0	41.0	2.6	.6	15.1	1.8	42.0	53.0	(8.8)	(14.3)	266.0
2647	**SAUERKRAUT,** canned, solids & liquid	1 CUP	150	92.8	27.0	1.5	.3	6.0	1.1	54.0	27.0	1120.5	(25.5)	210.0
			100	92.8	18.0	1.0	.2	4.0	.7	36.0	18.0	747.0	(17.0)	140.0
2648	**SHALLOT BULBS,** raw	1 OZ	28	79.8	20.2	.7	0	4.7	.2	10.4	16.8	3.4	—	93.5
			100	79.8	72.0	2.5	.1	16.8	.7	37.0	60.0	12.0	—	334.0
	SOYBEAN SPROUTS													
2649	**SPROUTED SEEDS,** raw	1 CUP	105	86.3	48.3	6.5	1.5	5.6	.8	50.4	70.4	—	—	—
			100	86.3	46.0	6.2	1.4	5.3	.8	48.0	67.0	—	—	—
2650	**SPROUTED SEEDS,** boiled & drained	1 CUP	125	89.0	47.5	6.6	1.8	4.6	1.0	53.8	62.5	—	—	—
			100	89.0	38.0	5.3	1.4	3.7	.8	43.0	50.0	—	—	—

Item No.	Minerals (Micro)			Fat-Soluble Vitamins			Water-Soluble Vitamins								
	Iron	Zinc	Copper	Vitamin A	Vitamin D	Vitamin E (Alpha Tocopherol)	Vitamin C	Thiamin	Ribo-flavin	Niacin	Panto-thenic Acid	Vit. B-6 (Pyri-doxine)	Folacin (Folic Acid)	Biotin	Vitamin B-12
	(mg)	(mg)	(mg)	(IU)	(IU)	(mg)	(mg)	(mg)	(mg)	(mg)	(mg)	(mg)	(mcg)	(mcg)	(mcg)
2622	**1.23**	–	–	**784.0**	–	–	**24.50**	**.147**	**.29**	**2.21**	–	–	–	–	–
	.50	–	–	320.0	–	–	10.00	.060	.12	.90	–	–	–	–	–
2623	**1.40**	–	–	–	–	–	**13.95**	**.124**	**.08**	**3.26**	–	–	**26.35**	–	–
	.90	–	–	–	–	–	9.00	.080	.05	2.10	–	–	17.00	–	–
2624	**1.87**	–	–	–	–	–	**32.30**	**.20**	**.12**	**4.76**	–	–	–	–	–
	1.10	–	–	–	–	–	19.00	.12	.07	2.80	–	–	–	–	–
2625	**.07**	–	–	–	–	–	**1.05**	**.01**	**Trace**	**.16**	–	–	**1.10**	–	–
	1.30	–	–	–	–	–	21.00	.13	.08	3.10	–	–	22.00	–	–
2626	**.75**	**(.98)**	**.18**	–	–	–	**32.50**	**.10**	**.05**	**1.50**	–	**.26**	–	–	**0**
	.30	(.39)	.07	–	–	–	13.00	.04	.02	.60	–	.10	–	–	0
2627	**.64**	–	–	**278.2**	–	–	**10.70**	**.09**	**.09**	**1.93**	–	–	–	–	–
	.30	–	–	130.0	–	–	5.00	.04	.04	.90	–	–	–	–	–
2628	**.30**	–	–	–	–	–	**3.80**	**0**	**.08**	**1.10**	–	**.25**	–	–	–
	.27	–	–	–	–	–	3.36	0	.07	.97	–	.22	–	–	–
2629	**1.00**	–	–	**220.0**	–	–	**6.00**	**.08**	**.10**	**1.40**	–	–	–	–	–
	.50	–	–	110.0	–	–	3.00	.04	.05	.70	–	–	–	–	–
2630	**1.86**	–	–	–	–	–	**12.40**	**.11**	**.03**	**1.55**	–	–	–	–	–
	1.20	–	–	–	–	–	8.00	.07	.02	1.00	–	–	–	–	–
2631	**3.60**	–	–	–	–	**.54**	**42.00**	**.28**	**.04**	**5.20**	**1.16**	**.41**	**50.00**	–	–
	1.80	–	–	–	–	.27	21.00	.14	.02	2.60	.58	.21	25.00	–	–
2632	**1.17**	–	–	**273.0**	–	–	**7.80**	**.12**	**.08**	**1.37**	–	–	–	–	–
	.60	–	–	140.0	–	–	4.00	.06	.04	.70	–	–	–	–	–
2633	**.20**	–	–	**39.0**	–	–	**14.30**	**.01**	**.02**	**.26**	–	–	–	–	–
	.30	–	–	60.0	–	–	22.00	.01	.03	.40	–	–	–	–	–
2634	**.78**	–	–	**6.5**	–	–	**14.30**	**.01**	**.02**	**.26**	–	–	–	–	–
	1.20	–	–	10.0	–	–	22.00	.02	.03	.40	–	–	–	–	–
2635	**.14**	–	–	**14.0**	–	–	**5.32**	**.01**	**.01**	**.06**	–	–	–	–	–
	.50	–	–	50.0	–	–	19.00	.04	.04	.20	–	–	–	–	–
2636	**.11**	**.06**	**.03**	**700.0**	**0**	**Trace**	**1.49**	**.01**	**.01**	**.01**	**.11**	**.02**	**3.87**	**.11**	**0**
	.41	.23	.09	2500.0	0	Trace	5.30	.04	.04	.03	.41	.06	13.80	.40	0
2637	**.97**	**.37**	**.27**	**66540.7**	–	–	**12.15**	**.07**	**.12**	**.97**	**(.97)**	**(.10)**	**36.45**	–	–
	.40	.15	.11	27383.0	–	–	5.00	.03	.05	.40	(.40)	(.04)	15.00	–	–
2638	**.30**	–	–	**60.0**	–	–	**5.40**	**.01**	**.03**	**.18**	–	–	–	–	–
	1.00	–	–	200.0	–	–	18.00	.02	.11	.60	–	–	–	–	–
2639	**1.00**	–	–	**210.0**	–	–	**6.44**	**.01**	**.03**	**.14**	–	–	–	–	–
	3.60	–	–	750.0	–	–	23.00	.02	.10	.50	–	–	–	–	–
2640	**.10**	**0**	**.01**	**1.0**	**0**	**0**	**2.60**	**Trace**	**Trace**	**.03**	**.02**	**.01**	**2.40**	–	**0**
	1.00	.02	.13	10.0	0	0	26.00	.03	.03	.30	.18	.08	24.00	–	0
2641	**.57**	–	–	**9.5**	–	–	**30.40**	**.03**	**.02**	**.38**	–	–	–	–	–
	.60	–	–	10.0	–	–	32.00	.03	.02	.40	–	–	–	–	–
2642	**.11**	–	–	**162.4**	–	–	**12.04**	**.02**	**.02**	**.31**	**.05**	**.03**	–	–	**0**
	.40	–	–	580.0	–	–	43.00	.07	.07	1.10	.16	.10	–	–	0
2643	**.60**	–	–	**1100.0**	–	–	**52.00**	**.12**	**.12**	**1.60**	–	–	**42.00**	–	–
	.30	–	–	550.0	–	–	26.00	.06	.06	.80	–	–	21.00	–	–
2644	**.39**	–	–	**0**	–	–	**0**	**Trace**	–	–	–	–	–	–	–
	1.40	–	–	0	–	–	0	.01	–	–	–	–	–	–	–
2645	**2.25**	–	–	**15.0**	–	–	**16.50**	**.06**	**.06**	**.45**	–	–	–	–	–
	1.50	–	–	10.0	–	–	11.00	.04	.04	.30	–	–	–	–	–
2646	**1.95**	–	**(.20)**	**15.0**	**0**	–	**10.50**	**.05**	**.06**	**.30**	–	–	–	–	**0**
	1.30	–	(.13)	10.0	0	–	7.00	.03	.04	.20	–	–	–	–	0
2647	**.75**	**1.22**	**(.14)**	**75.0**	–	–	**21.00**	**.05**	**.06**	**.30**	**.14**	**.20**	**15.00**	–	**0**
	.50	.81	(.10)	50.0	–	–	14.00	.03	.04	.20	.09	.13	10.00	–	0
2648	**.34**	–	–	–	–	–	**2.24**	**.02**	**.01**	**.06**	–	–	–	–	–
	1.20	–	–	–	–	–	8.00	.06	.02	.20	–	–	–	–	–
2649	**1.05**	–	–	**84.0**	–	–	**13.65**	**.24**	**.21**	**.84**	–	–	–	–	–
	1.00	–	–	80.0	–	–	13.00	.23	.20	.80	–	–	–	–	–
2650	**.88**	–	–	**100.0**	–	–	**5.00**	**.20**	**.19**	**.88**	–	–	–	–	–
	.70	–	–	80.0	–	–	4.00	.16	.15	.70	–	–	–	–	–

(Continued)

TABLE F-36 *(Continued)*

Item No.	Food Item Name	Approximate Measure	Weight	Moisture	Food Energy Calories	Protein	Fats	Carbo-hydrates	Fiber	Minerals (Macro) Calcium	Phos-phorus	Sodium	Mag-nesium	Potas-sium
			(g)	(%)	(kcal)	(g)	(g)	(g)	(g)	(mg)	(mg)	(mg)	(mg)	(mg)
	SPINACH													
2651	raw	1 CUP	**100**	**90.7**	**26.0**	**3.2**	**.3**	**4.3**	**.6**	**93.0**	**51.0**	**71.0**	**88.0**	**470.0**
			100	90.7	26.0	3.2	.3	4.3	.6	93.0	51.0	71.0	88.0	470.0
2652	boiled, drained	1 CUP	**180**	**92.0**	**41.4**	**5.4**	**.5**	**6.5**	**1.1**	**167.4**	**68.4**	**90.0**	**(106.2)**	**583.2**
			100	92.0	23.0	3.0	.3	3.6	.6	93.0	38.0	50.0	(59.1)	324.0
2653	canned, drained solids	1 CUP	**180**	**91.4**	**43.2**	**4.9**	**1.1**	**6.5**	**1.6**	**212.4**	**46.8**	**424.8**	**112.0**	**450.0**
			100	91.4	24.0	2.7	.6	3.6	.9	118.0	26.0	236.0	62.2	250.0
2654	canned, solids & liquid	1 CUP	**232**	**93.0**	**44.1**	**5.3**	**.9**	**7.0**	**1.6**	**197.2**	**60.3**	**547.5**	**146.2**	**580.0**
			100	93.0	19.0	2.3	.4	3.0	.7	85.0	26.0	236.0	63.0	250.0
2655	canned, drained liquid	1 CUP	**240**	**96.8**	**14.4**	**1.2**	**0**	**3.1**	—	**4.8**	**60.0**	**566.4**	—	**600.0**
			100	96.8	6.0	.5	0	1.3	—	2.0	25.0	236.0	—	250.0
2656	low sodium, canned, drained solids	1 CUP	**180**	**91.3**	**46.8**	**5.8**	**.9**	**7.2**	**1.8**	**212.4**	**46.8**	**57.6**	**112.0**	**450.0**
			100	91.3	26.0	3.2	.5	4.0	1.0	118.0	26.0	32.0	62.2	250.0
2657	canned, low sodium, solids & liquid	1 CUP	**232**	**92.8**	**48.7**	**5.3**	**.9**	**7.7**	**1.6**	**197.2**	**60.3**	**78.9**	**146.2**	**580.0**
			100	92.8	21.0	2.3	.4	3.3	.7	85.0	26.0	34.0	63.0	250.0
2658	low sodium, canned, drained liquid	1 CUP	**240**	**96.7**	**19.2**	**1.2**	**0**	**4.8**	—	**4.8**	**60.0**	**76.8**	—	**600.0**
			100	96.7	8.0	.5	0	2.0	—	2.0	25.0	32.0	—	250.0
2659	frozen, leaf	1 CUP	**80**	**91.3**	**20.0**	**2.4**	**.2**	**3.4**	**.6**	**84.0**	**36.0**	**42.4**	**83.2**	**308.0**
			100	91.3	25.0	3.0	.3	4.2	.8	105.0	45.0	53.0	104.0	385.0
2660	frozen, leaf, boiled, drained	1 CUP	**180**	**91.8**	**43.2**	**5.2**	**.5**	**7.0**	**1.4**	**189.0**	**79.2**	**88.2**	**117.0**	**651.6**
			100	91.8	24.0	2.9	.3	3.9	.8	105.0	44.0	49.0	65.0	362.0
2661	frozen, chopped	1 CUP	**140**	**91.6**	**33.6**	**4.3**	**.4**	**5.3**	**1.1**	**158.2**	**63.0**	**79.8**	**145.6**	**495.6**
			100	91.6	24.0	3.1	.3	3.8	.8	113.0	45.0	57.0	104.0	354.0
2662	frozen, chopped, boiled, drained	1 CUP	**180**	**91.9**	**41.4**	**5.4**	**.5**	**6.7**	**1.4**	**203.4**	**79.2**	**93.6**	**117.0**	**599.4**
			100	91.9	23.0	3.0	.3	3.7	.8	113.0	44.0	52.0	65.0	333.0
2663	New Zealand, raw	1 CUP	**33**	**92.6**	**6.3**	**.7**	**.1**	**1.0**	**.2**	**19.1**	**15.2**	**52.5**	**13.2**	**262.4**
			100	92.6	19.0	2.2	.3	3.1	.7	58.0	46.0	159.0	40.0	795.0
2664	New Zealand, boiled, drained	1 CUP	**180**	**94.8**	**23.4**	**3.1**	**.4**	**3.8**	**1.1**	**86.4**	**50.4**	**165.6**	—	**833.4**
			100	94.8	13.0	1.7	.2	2.1	.6	48.0	28.0	92.0	—	463.0
2665	**SPROUTS**—See **ALFALFA SPROUTS; BEANS, MUNG, SPROUTS; LENTIL SPROUTS**													
	SQUASH, SUMMER													
2666	all varieties, raw	1 CUP	**200**	**94.0**	**38.0**	**2.2**	**.2**	**8.4**	**1.2**	**56.0**	**58.0**	**2.0**	**32.0**	**404.0**
			100	94.0	19.0	1.1	.1	4.2	.6	28.0	29.0	1.0	16.0	202.0
2667	all varieties, boiled, drained	1 CUP	**210**	**95.5**	**29.4**	**1.9**	**.2**	**6.5**	**1.3**	**52.5**	**52.5**	**2.1**	**(25.2)**	**296.1**
			100	95.5	14.0	.9	.1	3.1	.6	25.0	25.0	1.0	(12.0)	141.0
2668	green zucchini & cocozelle, raw	1 CUP	**200**	**94.6**	**34.0**	**2.4**	**.2**	**7.2**	**1.2**	**56.0**	**58.0**	**2.0**	**47.4**	**404.0**
			100	94.6	17.0	1.2	.1	3.6	.6	28.0	29.0	1.0	23.7	202.0
2669	green zucchini & cocozelle, boiled, drained	1 CUP	**200**	**96.0**	**24.0**	**2.0**	**.2**	**5.0**	**1.2**	**50.0**	**50.0**	**2.0**	—	**282.0**
			100	96.0	12.0	1.0	.1	2.5	.6	25.0	25.0	1.0	—	141.0
2670	white & pale green scallop, raw	1 CUP	**200**	**93.3**	**42.0**	**1.8**	**.2**	**10.2**	**1.2**	**56.0**	**58.0**	**2.0**	**46.0**	**404.0**
			100	93.3	21.0	.9	.1	5.1	.6	28.0	29.0	1.0	23.0	202.0
2671	white & pale green scallop, boiled, drained	1 CUP	**200**	**95.0**	**32.0**	**1.4**	**.2**	**7.6**	**1.2**	**50.0**	**50.0**	**2.0**	—	**282.0**
			100	95.0	16.0	.7	.1	3.8	.6	25.0	25.0	1.0	—	141.0
2672	yellow, crook & straight neck, raw	1 CUP	**200**	**93.7**	**40.0**	**2.4**	**.4**	**8.6**	**1.2**	**56.0**	**58.0**	**2.0**	**44.0**	**404.0**
			100	93.7	20.0	1.2	.2	4.3	.6	28.0	29.0	1.0	22.0	202.0
2673	yellow crook & straight neck, boiled, drained	1 CUP	**200**	**95.3**	**30.0**	**2.0**	**.4**	**6.2**	**1.2**	**50.0**	**50.0**	**2.0**	—	**282.0**
			100	95.3	15.0	1.0	.2	3.1	.6	25.0	25.0	1.0	—	141.0
2674	yellow, crookneck, frozen, boiled, drained	1 CUP	**200**	**93.4**	**42.0**	**2.8**	**.2**	**9.4**	**1.2**	**28.0**	**64.0**	**6.0**	**32.0**	**334.0**
			100	93.4	21.0	1.4	.1	4.7	.6	14.0	32.0	3.0	16.0	167.0
2675	flowers	1 LG	**20**	**94.8**	**3.2**	**.3**	**.1**	**.5**	**.1**	**9.4**	**17.2**	—	—	—
			100	94.8	16.0	1.4	.3	2.7	.6	47.0	86.0	—	—	—
	SQUASH, WINTER													
2676	all varieties, raw	1 CUP	**200**	**85.1**	**100.0**	**2.8**	**.6**	**24.8**	**2.8**	**44.0**	**76.0**	**2.0**	**34.0**	**738.0**
			100	85.1	50.0	1.4	.3	12.4	1.4	22.0	38.0	1.0	17.0	369.0
2677	all varieties, baked	1 CUP	**205**	**81.4**	**129.2**	**3.7**	**.8**	**31.6**	**3.7**	**57.4**	**98.4**	**2.1**	—	**945.1**
			100	81.4	63.0	1.8	.4	15.4	1.8	28.0	48.0	1.0	—	461.0
2678	all varieties, boiled, mashed	1 CUP	**205**	**88.8**	**77.9**	**2.3**	**.6**	**18.9**	**2.9**	**41.0**	**65.6**	**2.1**	**(28.7)**	**528.9**
			100	88.8	38.0	1.1	.3	9.2	1.4	20.0	32.0	1.0	(14.0)	258.0
2679	acorn, raw	1 CUP	**200**	**86.3**	**88.0**	**3.0**	**.2**	**22.4**	**2.8**	**62.0**	**46.0**	**2.0**	**64.0**	**768.0**
			100	86.3	44.0	1.5	.1	11.2	1.4	31.0	23.0	1.0	32.0	384.0
2680	acorn, baked	1 CUP	**205**	**82.9**	**112.8**	**3.9**	**.2**	**28.7**	**3.7**	**80.0**	**59.5**	**2.1**	—	**984.0**
			100	82.9	55.0	1.9	.1	14.0	1.8	39.0	29.0	1.0	—	480.0

Item No.	Minerals (Micro) Iron	Zinc	Copper	Fat-Soluble Vitamins Vitamin A	Vitamin D	Vitamin E (Alpha Tocopherol)	Water-Soluble Vitamins Vitamin C	Thiamin	Ribo-flavin	Niacin	Panto-thenic Acid	Vit. B-6 (Pyri-doxine)	Folacin (Folic Acid)	Biotin	Vitamin B-12
	(mg)	(mg)	(mg)	(IU)	(IU)	(mg)	(mg)	(mg)	(mg)	(mg)	(mg)	(mg)	(mcg)	(mcg)	(mcg)
2651	**3.10**	**.80**	**.20**	**8100.0**	—	—	**51.00**	**.10**	**.20**	**.60**	**.30**	**.28**	**193.00**	**6.90**	**0**
	3.10	.80	.20	8100.0	—	—	51.00	.10	.20	.60	.30	.28	193.00	6.90	0
2652	**3.96**	**1.26**	**(.45)**	**14580.0**	**0**	**(3.62)**	**50.40**	**.13**	**.25**	**.90**	**(.40)**	**(.32)**	**163.80**	**(.20)**	**0**
	2.20	.70	(.25)	8100.0	0	(2.01)	28.00	.07	.14	.50	(.22)	(.18)	91.00	(.11)	0
2653	**4.68**	**1.44**	—	**14400.0**	—	—	**25.20**	**.04**	**.22**	**.54**	—	—	**73.80**	—	—
	2.60	.80	—	8000.0	—	—	14.00	.02	.12	.30	—	—	41.00	—	—
2654	**4.87**	**1.39**	—	**12760.0**	—	**.05**	**32.48**	**.05**	**.23**	**.70**	**.15**	**.16**	**134.56**	**5.34**	**0**
	2.10	.60	—	5500.0	—	.02	14.00	.02	.10	.30	.07	.07	58.00	2.30	0
2655	**2.16**	—	—	—	—	—	**33.60**	**.05**	**.17**	**.72**	—	—	**189.60**	—	—
	.90	—	—	—	—	—	14.00	.02	.07	.30	—	—	79.00	—	—
2656	**4.68**	—	—	**14400.0**	—	—	**25.20**	**.04**	**.22**	**.54**	—	—	**73.80**	—	—
	2.60	—	—	8000.0	—	—	14.00	.02	.12	.30	—	—	41.00	—	—
2657	**4.87**	**.30**	—	**12760.0**	—	**.05**	**32.48**	**.05**	**.23**	**.70**	**.15**	**.16**	**134.56**	**5.34**	**0**
	2.10	.13	—	5500.0	—	.02	14.00	.02	.10	.30	.07	.07	58.00	2.30	0
2658	**2.16**	—	—	—	—	—	**33.60**	**.03**	**.17**	**.72**	—	—	**189.60**	—	—
	.90	—	—	—	—	—	14.00	.02	.07	.30	—	—	79.00	—	—
2659	**2.00**	**.30**	**.07**	**6480.0**	—	—	**28.00**	**.08**	**.13**	**.40**	**.12**	**.12**	**122.40**	**5.52**	**0**
	2.50	.37	.08	8100.0	—	—	35.00	.10	.16	.50	.15	.15	153.00	6.90	0
2660	**4.50**	—	—	**14580.0**	—	—	**50.40**	**.14**	**.25**	**.90**	**.29**	**.15**	**194.40**	—	—
	2.50	—	—	8100.0	—	—	28.00	.08	.14	.50	.16	.08	108.00	—	—
2661	**2.94**	**.52**	**.12**	**11060.0**	—	—	**40.60**	**.13**	**.22**	**.70**	**.18**	**.27**	**214.20**	**9.66**	**0**
	2.10	.37	.08	7900.0	—	—	29.00	.09	.16	.50	.13	.19	153.00	6.90	0
2662	**3.78**	—	—	**14220.0**	—	—	**34.20**	**.13**	**.27**	**.72**	**.20**	**.36**	**194.40**	—	—
	2.10	—	—	7900.0	—	—	19.00	.07	.15	.40	.11	.20	108.00	—	—
2663	**.86**	**.26**	—	**1419.0**	—	—	**9.90**	**.01**	**.06**	**.20**	**.10**	—	—	—	**0**
	2.60	.80	—	4300.0	—	—	30.00	.04	.17	.60	.31	—	—	—	0
2664	**2.70**	**1.26**	—	**6480.0**	—	—	**25.20**	**.05**	**.18**	**.90**	—	—	—	—	—
	1.50	.70	—	3600.0	—	—	14.00	.03	.10	.50	—	—	—	—	—
2665															
2666	**.80**	—	—	**820.0**	—	—	**44.00**	**.10**	**.18**	**2.00**	**.72**	**.16**	**62.00**	—	**0**
	.40	—	—	410.0	—	—	22.00	.05	.09	1.00	.36	.08	31.00	—	0
2667	**.84**	**(.34)**	**(.11)**	**819.0**	—	—	**21.00**	**.11**	**.17**	**1.68**	**(.23)**	**(.12)**	**35.70**	—	—
	.40	(.16)	(.05)	390.0	—	—	10.00	.05	.08	.80	(.11)	(.06)	17.00	—	—
2668	**.80**	—	—	**640.0**	—	—	**38.00**	**.10**	**.18**	**2.00**	**.72**	**.16**	—	—	**0**
	.40	—	—	320.0	—	—	19.00	.05	.09	1.00	.36	.08	—	—	0
2669	**.80**	**.36**	—	**600.0**	—	—	**18.00**	**.10**	**.16**	**1.60**	—	—	—	—	—
	.40	.18	—	300.0	—	—	9.00	.05	.08	.80	—	—	—	—	—
2670	**.80**	—	—	**380.0**	—	—	**36.00**	**.10**	**.18**	**2.00**	**.72**	**.16**	**62.00**	—	**0**
	.40	—	—	190.0	—	—	18.00	.05	.09	1.00	.36	.08	31.00	—	0
2671	**.80**	—	—	**360.0**	—	—	**16.00**	**.10**	**.16**	**1.60**	—	—	—	—	—
	.40	—	—	180.0	—	—	8.00	.05	.08	.80	—	—	—	—	—
2672	**.80**	—	—	**920.0**	—	—	**50.00**	**.10**	**.18**	**2.00**	**.72**	**.16**	**62.00**	—	**0**
	.40	—	—	460.0	—	—	25.00	.05	.09	1.00	.36	.08	31.00	—	0
2673	**.80**	—	—	**880.0**	—	—	**22.00**	**.10**	**.16**	**1.60**	—	—	**34.00**	—	—
	.40	—	—	440.0	—	—	11.00	.05	.08	.80	—	—	17.00	—	—
2674	**1.40**	**.60**	**.28**	**280.0**	—	—	**16.00**	**.12**	**.08**	**.80**	—	—	**20.00**	—	—
	.70	.30	.14	140.0	—	—	8.00	.06	.04	.40	—	—	10.00	—	—
2675	**.20**	—	—	**40.0**	—	—	**3.60**	**Trace**	**.02**	**.12**	—	—	—	—	—
	1.00	—	—	200.0	—	—	18.00	.02	.11	.60	—	—	—	—	—
2676	**1.20**	—	—	**7400.0**	—	—	**26.00**	**.10**	**.22**	**1.20**	**.80**	**.31**	**34.00**	—	**0**
	.60	—	—	3700.0	—	—	13.00	.05	.11	.60	.40	.15	17.00	—	0
2677	**1.64**	—	—	**8405.0**	—	—	**26.65**	**.10**	**.26**	**1.44**	—	—	—	—	—
	.80	—	—	4100.0	—	—	13.00	.05	.13	.70	—	—	—	—	—
2678	**1.03**	**.35**	**(.10)**	**7175.0**	—	—	**16.40**	**.08**	**.20**	**.82**	**(.45)**	**(.23)**	—	—	—
	.50	.17	(.05)	3500.0	—	—	8.00	.04	.10	.40	(.22)	(.11)	—	—	—
2679	**1.80**	**.28**	**.24**	**2400.0**	—	—	**28.00**	**.10**	**.22**	**1.20**	**.80**	**.31**	**34.00**	—	**0**
	.90	.14	.12	1200.0	—	—	14.00	.05	.11	.60	.40	.15	17.00	—	0
2680	**2.26**	**.49**	**.16**	**2870.0**	—	—	**26.65**	**.10**	**.26**	**1.44**	—	—	—	—	—
	1.10	.24	.08	1400.0	—	—	13.00	.05	.13	.70	—	—	—	—	—

(Continued)

VEGETABLES

TABLE F-36 *(Continued)*

Item No.	Food Item Name	Approximate Measure	Weight	Moisture	Food Energy Calories	Protein	Fats	Carbo-hydrates	Fiber	Minerals (Macro) Calcium	Phos-phorus	Sodium	Mag-nesium	Potas-sium
			(g)	(%)	(kcal)	(g)	(g)	(g)	(g)	(mg)	(mg)	(mg)	(mg)	(mg)
	SQUASH, WINTER (Continued)													
2681	acorn, boiled, mashed	**1 CUP**	**205**	**89.7**	**69.7**	**2.5**	**.2**	**17.2**	**2.9**	**57.4**	**41.0**	**2.1**	—	**551.5**
			100	89.7	34.0	1.2	.1	8.4	1.4	28.0	20.0	1.0	—	269.0
2682	butternut, raw	**1 CUP**	**200**	**83.7**	**108.0**	**2.8**	**.2**	**28.0**	**2.8**	**64.0**	**116.0**	**2.0**	**33.0**	**974.0**
			100	83.7	54.0	1.4	.1	14.0	1.4	32.0	58.0	1.0	16.5	487.0
2683	butternut, baked	**1 CUP**	**205**	**79.6**	**139.4**	**3.7**	**.2**	**35.9**	**3.7**	**82.0**	**147.6**	**2.1**	—	**1248.5**
			100	79.6	68.0	1.8	.1	17.5	1.8	40.0	72.0	1.0	—	609.0
2684	butternut, boiled, mashed	**1 CUP**	**205**	**87.8**	**84.1**	**2.3**	**.2**	**21.3**	**2.9**	**59.5**	**100.5**	**2.1**	—	**699.1**
			100	87.8	41.0	1.1	.1	10.4	1.4	29.0	49.0	1.0	—	341.0
2685	hubbard, raw	**1 CUP**	**200**	**88.1**	**78.0**	**2.8**	**.6**	**18.8**	**2.8**	**38.0**	**62.0**	**2.0**	**38.0**	**434.0**
			100	88.1	39.0	1.4	.3	9.4	1.4	19.0	31.0	1.0	19.0	217.0
2686	hubbard, baked	**1 CUP**	**205**	**85.1**	**102.5**	**3.7**	**.8**	**24.0**	**3.7**	**49.2**	**80.0**	**2.1**	—	**555.6**
			100	85.1	50.0	1.8	.4	11.7	1.8	24.0	39.0	1.0	—	271.0
2687	hubbard, boiled, mashed	**1 CUP**	**205**	**91.1**	**61.5**	**2.3**	**.6**	**14.1**	**2.9**	**34.9**	**53.3**	**2.1**	—	**311.6**
			100	91.1	30.0	1.1	.3	6.9	1.4	17.0	26.0	1.0	—	152.0
	SUCCOTASH													
2688	(corn & lima beans)	**1 CUP**	**155**	**73.0**	**150.4**	**6.7**	**.6**	**33.3**	**1.4**	**21.7**	**136.0**	**69.8**	**54.3**	**423.2**
			100	73.0	97.0	4.3	.4	21.5	.9	14.0	89.0	45.0	35.0	273.0
2689	(corn & lima beans), frozen, boiled, drained	**1 CUP**	**200**	**74.1**	**186.0**	**8.4**	**.8**	**41.0**	**1.8**	**26.0**	**170.0**	**76.0**	—	**492.0**
			100	74.1	93.0	4.2	.4	20.5	.9	13.0	85.0	38.0	—	246.0
	SWEET POTATO													
2690	all varieties, raw	**1 SM**	**100**	**70.6**	**114.0**	**1.8**	**.4**	**26.3**	**.7**	**32.0**	**47.0**	**10.0**	**31.0**	**243.0**
			100	70.6	114.0	1.8	.4	26.3	.7	32.0	47.0	10.0	31.0	243.0
2691	baked in skin	**1 SM**	**100**	**63.7**	**141.0**	**2.1**	**.5**	**32.5**	**.9**	**40.0**	**58.0**	**12.0**	—	**300.0**
			100	63.7	141.0	2.1	.5	32.5	.9	40.0	58.0	12.0	—	300.0
2692	boiled in skin	**1 SM**	**100**	**70.6**	**114.0**	**1.7**	**.4**	**26.3**	**.7**	**32.0**	**47.0**	**10.0**	**(12.0)**	**243.0**
			100	70.6	114.0	1.7	.4	26.3	.7	32.0	47.0	10.0	(12.0)	243.0
2693	candied, w/butter	**1 SM**	**100**	**60.0**	**168.0**	**1.3**	**3.3**	**34.2**	**.6**	**37.0**	**43.0**	**42.0**	—	**190.0**
			100	60.0	168.0	1.3	3.3	34.2	.6	37.0	43.0	42.0	—	190.0
2694	canned, liquid pack, w/o sugar or salt, solids & liquid	**1 SM**	**100**	**88.0**	**46.0**	**.7**	**.1**	**10.8**	**.3**	**13.0**	**29.0**	**12.0**	—	**120.0**
			100	88.0	46.0	.7	.1	10.8	.3	13.0	29.0	12.0	—	120.0
2695	canned, liquid pack in syrup, solids & liquid	**1 SM**	**100**	**70.7**	**114.0**	**1.0**	**.2**	**27.5**	**.6**	**13.0**	**29.0**	**48.0**	**42.0**	**120.0**
			100	70.7	114.0	1.0	.2	27.5	.6	13.0	29.0	48.0	42.0	120.0
2696	canned, vacuum or solid pack	**1 SM**	**100**	**71.9**	**108.0**	**2.0**	**.2**	**24.9**	**1.0**	**25.0**	**41.0**	**48.0**	—	**200.0**
			100	71.9	108.0	2.0	.2	24.9	1.0	25.0	41.0	48.0	—	200.0
2697	canned, vacuum or solid pack, low sodium	**1 SM**	**100**	**71.9**	**108.0**	**2.0**	**.2**	**24.9**	**1.0**	**25.0**	**41.0**	**12.0**	—	**200.0**
			100	71.9	108.0	2.0	.2	24.9	1.0	25.0	41.0	12.0	—	200.0
2698	dehydrated flakes, dry	**1 OZ**	**28**	**2.8**	**106.1**	**1.2**	**.2**	**25.2**	**.9**	**16.8**	**22.4**	**50.7**	**28.0**	**157.4**
			100	2.8	379.0	4.2	.6	90.0	3.2	60.0	80.0	181.0	100.0	562.0
2699	dehydrated flakes, prepared w/water	**1 CUP**	**255**	**75.7**	**242.3**	**2.6**	**.3**	**57.6**	**2.0**	**38.3**	**51.0**	**114.8**	—	**357.0**
			100	75.7	95.0	1.0	.1	22.6	.8	15.0	20.0	45.0	—	140.0
	SWISS CHARD													
2700	raw	**1 CUP**	**166**	**91.1**	**41.5**	**4.0**	**.5**	**7.6**	**1.3**	**146.1**	**64.7**	**244.0**	**107.9**	**913.0**
			100	91.1	25.0	2.4	.3	4.6	.8	88.0	39.0	147.0	65.0	550.0
2701	boiled, drained	**1 CUP**	**166**	**93.7**	**29.9**	**3.0**	**.3**	**5.5**	**1.2**	**121.2**	**39.8**	**142.8**	—	**532.9**
			100	93.7	18.0	1.8	.2	3.3	.7	73.0	24.0	86.0	—	321.0
	TAPIOCA													
2702	dry	**1 TBLS**	**10**	**12.6**	**35.2**	**.1**	**0**	**8.6**	**Trace**	**1.0**	**1.8**	**.3**	**.3**	**1.8**
			100	12.6	352.0	.6	.2	86.4	.1	10.0	18.0	3.0	3.0	18.0
2703	minute	**1 TBLS**	**10**	—	**36.0**	**0**	**0**	**8.9**	**Trace**	**3.5**	**5.0**	**.8**	—	**1.8**
			100	—	360.0	.2	.1	89.1	.1	35.0	50.0	8.0	—	18.0
	TARO													
2704	corm, baked	**1 CUP**	**132**	—	—	—	—	—	—	—	—	**62.0**	—	**1360.9**
			100	—	—	—	—	—	—	—	—	47.0	—	1031.0
2705	[4]leaves, raw	**1 OZ**	**28**	**81.4**	**17.1**	**1.1**	**.3**	**3.3**	**.3**	**45.4**	**19.3**	**2.5**	—	**269.6**
			100	81.4	61.0	4.1	1.0	11.9	1.2	162.0	69.0	9.0	—	963.0
2706	[4]leaves, cooked	**1 OZ**	**28**	**85.7**	**13.4**	**.9**	**.2**	**2.8**	**.3**	**30.8**	**18.8**	—	—	—
			100	85.7	48.0	3.3	.6	9.9	.9	110.0	67.0	—	—	—
2707	[4]tubers, boiled	**1 OZ**	**28**	**67.3**	**34.7**	**.5**	**.1**	**7.8**	—	**13.4**	**13.4**	**3.1**	—	**139.0**
			100	67.3	124.0	1.9	.3	28.8	—	48.0	48.0	11.0	—	498.0
2708	[4]tubers, dried	**1 OZ**	**28**	**27.0**	**79.5**	**1.1**	**.1**	**18.7**	**.2**	**9.5**	**46.5**	—	—	—
			100	27.0	284.0	3.9	.5	66.7	.8	34.0	166.0	—	—	—

Item No.	Minerals (Micro) Iron	Zinc	Copper	Fat-Soluble Vitamins Vitamin A	Vitamin D	Vitamin E (Alpha Tocopherol)	Water-Soluble Vitamins Vitamin C	Thiamin	Ribo-flavin	Niacin	Panto-thenic Acid	Vit. B-6 (Pyri-doxine)	Folacin (Folic Acid)	Biotin	Vitamin B-12
	(mg)	(mg)	(mg)	(IU)	(IU)	(mg)	(mg)	(mg)	(mg)	(mg)	(mg)	(mg)	(mcg)	(mcg)	(mcg)
2681	**1.64**	–	–	**2255.0**	–	–	**16.40**	**.08**	**.20**	**.82**	–	–	–	–	–
	.80	–	–	1100.0	–	–	8.00	.04	.10	.40	–	–	–	–	–
2682	**1.60**	–	–	**11400.0**	–	–	**18.00**	**.10**	**.22**	**1.20**	**.80**	**.31**	**24.00**	–	**0**
	.80	–	–	5700.0	–	–	9.00	.05	.11	.60	.40	.15	12.00	–	0
2683	**2.05**	–	–	**13120.0**	–	–	**16.40**	**.10**	**.26**	**1.44**	–	–	–	–	–
	1.00	–	–	6400.0	–	–	8.00	.05	.13	.70	–	–	–	–	–
2684	**1.44**	–	–	**11070.0**	–	–	**10.25**	**.08**	**.20**	**.82**	–	–	–	–	–
	.70	–	–	5400.0	–	–	5.00	.04	.10	.40	–	–	–	–	–
2685	**1.20**	–	–	**8600.0**	–	–	**22.00**	**.10**	**.22**	**1.20**	**.80**	**.31**	**24.00**	–	**0**
	.60	–	–	4300.0	–	–	11.00	.05	.11	.60	.40	.15	12.00	–	0
2686	**1.64**	–	–	**9840.0**	–	–	**20.50**	**.10**	**.26**	**1.44**	–	–	–	–	–
	.80	–	–	4800.0	–	–	10.00	.05	.13	.70	–	–	–	–	–
2687	**1.03**	–	–	**8405.0**	–	–	**12.30**	**.08**	**.20**	**.82**	–	–	–	–	–
	.50	–	–	4100.0	–	–	6.00	.04	.10	.40	–	–	–	–	–
2688	**1.71**	**(.73)**	**(.10)**	**465.0**	–	–	**13.95**	**.17**	**.09**	**2.33**	**.69**	**.28**	–	–	**0**
	1.10	(.47)	.06	300.0	–	–	9.00	.11	.06	1.50	.44	.18	–	–	0
2689	**2.00**	–	–	**600.0**	–	–	**12.00**	**.18**	**.10**	**2.60**	–	–	–	–	–
	1.00	–	–	300.0	–	–	6.00	.09	.05	1.30	–	–	–	–	–
2690	**.70**	**.08**	**.15**	**8800.0**	**0**	**4.00**	**21.00**	**.10**	**.06**	**.60**	**.82**	**.22**	**12.00**	**4.30**	**0**
	.70	.08	.15	8800.0	0	4.00	21.00	.10	.06	.60	.82	.22	12.00	4.30	0
2691	**.90**	–	–	**8100.0**	–	–	**22.00**	**.09**	**.07**	**.70**	–	–	**18.00**	–	**0**
	.90	–	–	8100.0	–	–	22.00	.09	.07	.70	–	–	18.00	–	0
2692	**.70**	**.30**	**(.16)**	**7900.0**	**0**	**4.00**	**17.00**	**.09**	**.06**	**.60**	**(.65)**	**(.12)**	**18.00**	–	**0**
	.70	.30	(.16)	7900.0	0	4.00	17.00	.09	.06	.60	(.65)	(.12)	18.00	–	0
2693	**.90**	–	–	**6300.0**	–	–	**10.00**	**.06**	**.04**	**.40**	–	–	–	–	–
	.90	–	–	6300.0	–	–	10.00	.06	.04	.40	–	–	–	–	–
2694	**.70**	–	–	**5000.0**	–	–	**8.00**	**.03**	**.03**	**.60**	**.43**	**.07**	–	–	**0**
	.70	–	–	5000.0	–	–	8.00	.03	.03	.60	.43	.07	–	–	0
2695	**.70**	**.16**	**.06**	**5000.0**	–	–	**8.00**	**.03**	**.03**	**.60**	**.43**	**.07**	–	–	**0**
	.70	.16	.06	5000.0	–	–	8.00	.03	.03	.60	.43	.07	–	–	0
2696	**.80**	–	–	**7800.0**	–	–	**14.00**	**.05**	**.04**	**.60**	**.43**	**.07**	–	–	**0**
	.80	–	–	7800.0	–	–	14.00	.05	.04	.60	.43	.07	–	–	0
2697	**.80**	–	–	**7800.0**	–	–	**14.00**	**.050**	**.04**	**.60**	**.43**	**.07**	–	–	**0**
	.80	–	–	7800.0	–	–	14.00	.050	.04	.60	.43	.07	–	–	0
2698	**.62**	–	–	**13160.0**	–	–	**12.60**	**.027**	**.04**	**.36**	**.65**	–	–	–	**0**
	2.20	–	–	47000.0	–	–	45.00	.060	.13	1.30	2.33	–	–	–	0
2699	**1.53**	–	–	**30600.0**	–	–	**28.05**	**.051**	**.08**	**.77**	–	–	–	–	–
	.60	–	–	12000.0	–	–	11.00	.020	.03	.30	–	–	–	–	–
2700	**5.31**	–	–	**10790.0**	–	–	**53.12**	**.100**	**.28**	**.83**	**.29**	–	–	–	**0**
	3.20	–	–	6500.0	–	–	32.00	.060	.17	.50	.17	–	–	–	0
2701	**2.99**	–	–	**8964.0**	–	–	**26.56**	**.076**	**.18**	**.66**	–	–	–	–	–
	1.80	–	–	5400.0	–	–	16.00	.040	.11	.40	–	–	–	–	–
2702	**.04**	–	–	**0**	**0**	–	**0**	**0**	**0**	**0**	–	–	**.80**	–	–
	.40	–	–	0	0	–	0	0	0	0	–	–	8.00	–	–
2703	**.10**	–	–	**0**	–	–	–	**0**	**0**	**0**	–	–	–	–	–
	1.00	–	–	0	–	–	–	0	0	0	–	–	–	–	–
2704	–	–	–	–	–	–	–	–	–	–	–	–	–	–	–
	–	–	–	–	–	–	–	–	–	–	–	–	–	–	–
2705	**.28**	–	–	**2583.0**	–	–	**17.64**	**.04**	**.10**	**.42**	–	–	–	–	–
	1.00	–	–	9225.0	–	–	63.00	.13	.34	1.50	–	–	–	–	–
2706	**.22**	–	–	**2191.0**	–	–	**7.56**	**.03**	**.09**	**.28**	–	–	–	–	–
	.80	–	–	7825.0	–	–	27.00	.11	.32	1.00	–	–	–	–	–
2707	**.25**	–	–	**0**	–	–	**1.12**	**.02**	**.01**	**.17**	–	–	–	–	–
	.90	–	–	0	–	–	4.00	.08	.05	.60	–	–	–	–	–
2708	**1.57**	–	–	–	–	–	–	–	–	–	–	–	–	–	–
	5.60	–	–	–	–	–	–	–	–	–	–	–	–	–	–

(Continued)

TABLE F-36 *(Continued)*

Item No.	Food Item Name	Approximate Measure	Weight	Moisture	Food Energy Calories	Protein	Fats	Carbo-hydrates	Fiber	Minerals (Macro) Calcium	Phos-phorus	Sodium	Mag-nesium	Potas-sium
			(g)	(%)	(kcal)	(g)	(g)	(g)	(g)	(mg)	(mg)	(mg)	(mg)	(mg)
	TARO													
2709	[4]tubers poi. 2-finger. 17% solid	1 OZ	**28**	**83.0**	**18.8**	**.2**	**0**	**4.5**	—	**3.1**	**6.2**	—	—	—
			100	83.0	67.0	.6	.1	16.0	—	11.0	22.0	—	—	—
2710	[3]**TEPARY BEAN,** seeds. dried	1 OZ	**28**	**8.6**	**98.8**	**5.4**	**.3**	**19.0**	**1.3**	**31.4**	**86.8**	—	—	—
			100	8.6	353.0	19.3	1.2	67.8	4.8	112.0	310.0	—	—	—
	TOMATO													
2711	paste. canned	1 CUP	**249**	**75.0**	**204.2**	**8.5**	**1.0**	**46.3**	**2.2**	**67.2**	**174.3**	**1967.1**	**49.8**	**2211.1**
			100	75.0	82.0	3.4	.4	18.6	.9	27.0	70.0	790.0	20.0	888.0
2712	paste. low sodium. cannec	1 TBLS	**16**	**75.0**	**13.1**	**.5**	**.1**	**3.0**	**.1**	**4.3**	**11.2**	**6.1**	**3.2**	**142.1**
			100	75.0	82.0	3.4	.4	18.6	.9	27.0	70.0	38.0	20.0	888.0
2713	puree. canned	1 TBLS	**17**	**87.0**	**6.6**	**.3**	**0**	**1.5**	**.1**	**2.2**	**5.8**	**67.8**	**3.4**	**72.4**
			100	87.0	39.0	1.7	.2	8.9	.4	13.0	34.0	399.0	20.0	426.0
2714	puree. low sodium. canned	1 TBLS	**17**	**88.0**	**6.6**	**.3**	**0**	**1.5**	**.1**	**2.2**	**5.8**	**1.0**	**3.4**	**72.4**
			100	88.0	39.0	1.7	.2	8.9	.4	13.0	34.0	6.0	20.0	426.0
	TOMATOES													
2715	green. raw	1 SM	**100**	**93.0**	**24.0**	**1.0**	**.2**	**5.1**	**.5**	**13.0**	**27.0**	**3.0**	**14.0**	**244.0**
			100	93.0	24.0	1.0	.2	5.1	.5	13.0	27.0	3.0	14.0	244.0
2716	ripe. raw	1 SM	**100**	**93.5**	**22.0**	**1.0**	**.2**	**4.7**	**.5**	**13.0**	**27.0**	**3.0**	**17.7**	**244.0**
			100	93.5	22.0	1.0	.2	4.7	.5	13.0	27.0	3.0	17.7	244.0
2717	riped. boiled	1 CUP	**240**	**92.4**	**62.4**	**2.4**	**.5**	**13.2**	**1.4**	**36.0**	**76.8**	**9.6**	—	**688.8**
			100	92.4	26.0	1.0	.2	5.5	.6	15.0	32.0	4.0	—	287.0
2718	ripe. canned. solids & liquid	1 CUP	**240**	**93.7**	**50.4**	**2.4**	**.5**	**10.3**	**1.0**	**14.4**	**45.6**	**312.0**	**28.8**	**520.8**
			100	93.7	21.0	1.0	.2	4.3	.4	6.0	19.0	130.0	12.0	217.0
2719	ripe. low sodium. canned. solids & liquid	1 CUP	**240**	**94.1**	**48.0**	**2.4**	**.5**	**10.1**	**1.0**	**14.4**	**45.6**	**7.2**	**28.8**	**520.8**
			100	94.1	20.0	1.0	.2	4.2	.4	6.0	19.0	3.0	12.0	217.0
	TURNIP GREENS													
2720	raw	1 CUP	**200**	**90.3**	**56.0**	**5.8**	**.6**	**10.0**	**1.6**	**492.0**	**116.0**	**20.0**	**116.0**	**880.0**
			100	90.3	28.0	2.9	.3	5.0	.8	246.0	58.0	10.0	58.0	440.0
2721	boiled. in small amount of water. drained	1 CUP	**150**	**93.2**	**30.0**	**3.3**	**.3**	**5.4**	**1.1**	**276.0**	**55.5**	**(11.0)**	**(15.3)**	**(116.9)**
			100	93.2	20.0	2.2	.2	3.6	.7	184.0	37.0	(7.3)	(10.2)	(77.9)
2722	boiled in large amount of water. drained	1 CUP	**150**	**93.5**	**28.5**	**3.3**	**.3**	**5.0**	**1.1**	**261.0**	**51.0**	—	—	—
			100	93.5	19.0	2.2	.2	3.3	.7	174.0	34.0	—	—	—
2723	canned. solids & liquid	1 CUP	**200**	**93.7**	**36.0**	**3.0**	**.6**	**6.4**	**1.4**	**200.0**	**60.0**	**472.0**	—	**486.0**
			100	93.7	18.0	1.5	.3	3.2	.7	100.0	30.0	236.0	—	243.0
2724	frozen. boiled. drained	1 CUP	**150**	**92.7**	**34.5**	**3.8**	**.5**	**5.9**	**1.5**	**177.0**	**58.5**	**25.5**	**34.8**	**223.5**
			100	92.7	23.0	2.5	.3	3.9	1.0	118.0	39.0	17.0	23.2	149.0
	TURNIPS													
2725	raw	1 CUP	**132**	**91.5**	**39.6**	**1.5**	**.3**	**8.7**	**1.2**	**51.5**	**39.6**	**64.7**	**12.5**	**353.8**
			100	91.5	30.0	1.1	.2	6.6	.9	39.0	30.0	49.0	9.5	268.0
2726	boiled. drained	1 CUP	**150**	**93.6**	**34.5**	**1.2**	**.3**	**7.4**	**1.4**	**52.5**	**36.0**	**51.0**	—	**282.0**
			100	93.6	23.0	.8	.2	4.9	.9	35.0	24.0	34.0	—	188.0
2727	**VINESPINACH (BASELLA),** raw	1 CUP	**52**	**93.1**	**9.9**	**.9**	**.2**	**1.8**	**.4**	**56.7**	**27.0**	—	—	—
			100	93.1	19.0	1.8	.3	3.4	.7	109.0	52.0	—	—	—
2728	**WATER CHESTNUTS,** sliced. canned	1 PIECE	**6**	**87.8**	**2.7**	**.1**	**0**	**.6**	**Trace**	**.3**	**1.0**	**.3**	**.2**	**4.2**
			100	87.8	45.6	1.1	.1	10.3	.6	5.0	17.1	4.9	3.0	70.4
2729	**WATERCRESS,** leaves & stems. raw	1 PIECE	**1**	**93.3**	**.2**	**0**	**0**	**0**	**Trace**	**1.5**	**.5**	**.5**	**.2**	**2.8**
			100	93.3	19.0	2.2	.3	3.0	.7	151.0	54.0	52.0	20.0	282.0
	YAM													
2730	[6]raw	1 OZ	**28**	—	**36.7**	**.6**	**0**	**9.0**	**1.1**	**2.5**	**11.8**	—	**11.5**	**122.8**
			100	—	131.0	2.0	.2	32.3	4.0	9.1	42.0	—	41.2	438.5
2731	cooked in skin	1 CUP	**200**	—	**210.0**	**4.8**	**.4**	**48.2**	**1.8**	**8.0**	**100.0**	—	—	—
			100	—	105.0	2.4	.2	24.1	.9	4.0	50.0	—	—	—
2732	[4]**YAM LEAVES**	1 OZ	**28**	**86.0**	—	**.1**	**0**	—	—	**56.3**	—	—	—	—
			100	86.0	—	.4	0	—	—	201.0	—	—	—	—
	YAMBEAN, tuber													
2733	raw. cubed	1 CUP	**140**	**85.1**	**77.0**	**2.0**	**.3**	**17.9**	**1.0**	**21.0**	**25.2**	—	—	—
			100	85.1	55.0	1.4	.2	12.8	.7	15.0	18.0	—	—	—
2734	[4]cooked	1 OZ	**28**	**88.6**	**11.5**	**.2**	**0**	**2.9**	**.3**	**2.2**	**5.0**	—	—	—
			100	88.6	41.0	.8	0	10.2	1.2	8.0	18.0	—	—	—

* With the exception of the footnoted values, the data in this table were supplied by the Ohio State University Hospital Nutrient Data Base Catalogue, Columbus, Ohio. These values cannot be reproduced without their permission.
Trade and brand names are used only for information. The authors do not guarantee nor warrant the standard of any product mentioned; neither do they imply approval of any product to the exclusion of others which may also be similar.
[1]Data from the HVH-CWRU Nutrient Data Base developed by the Division of Nutrition, Highland View Hospital, and the Departments of Biometry and Nutrition, School of Medicine, Case Western Reserve University, Cleveland, Ohio.
[2]Data from *Composition of Foods*, Agriculture Handbook No. 8-5, USDA, 1979.

Item No.	Minerals (Micro)			Fat-Soluble Vitamins			Water-Soluble Vitamins								
	Iron	Zinc	Copper	Vitamin A	Vitamin D	Vitamin E (Alpha Tocopherol)	Vitamin C	Thiamin	Ribo-flavin	Niacin	Panto-thenic Acid	Vit. B-6 (Pyri-doxine)	Folacin (Folic Acid)	Biotin	Vitamin B-12
	(mg)	(mg)	(mg)	(IU)	(IU)	(mg)	(mg)	(mg)	(mg)	(mg)	(mg)	(mg)	(mcg)	(mcg)	(mcg)
2709	**.11**	—	—	**0**	—	—	**1.40**	**.01**	**.01**	**.08**	—	—	—	—	—
	.40	—	—	0	—	—	5.00	.04	.02	.30	—	—	—	—	—
2710	—	—	—	—	—	—	**0**	**.09**	**.03**	**.78**	—	—	—	—	—
	—	—	—	—	—	—	0	.33	.12	2.80	—	—	—	—	—
2711	**8.72**	**(1.99)**	**(1.47)**	**8217.0**	—	—	**122.01**	**.50**	**.30**	**7.72**	**1.10**	**.95**	—	—	**0**
	3.50	(.80)	(.59)	3300.0	—	—	49.00	.20	.12	3.10	.44	.38	—	—	0
2712	**.56**	—	—	**528.0**	—	—	**7.84**	**.03**	**.02**	**.50**	**.07**	**.06**	—	—	**0**
	3.50	—	—	3300.0	—	—	49.00	.20	.12	3.10	.44	.38	—	—	0
2713	**.29**	**(.04)**	**(.03)**	**272.0**	**0**	**(1.20)**	**5.61**	**.02**	**.01**	**.24**	**(.20)**	**.03**	**(24.0)**	**(1.36)**	**0**
	1.70	(.25)	(.20)	1600.0	0	(6.80)	33.00	.09	.05	1.40	(1.20)	.15	(141.0)	(7.98)	0
2714	**.29**	—	—	**272.0**	—	—	**5.61**	**.02**	**.01**	**.24**	—	**.03**	—	—	**0**
	1.70	—	—	1600.0	—	—	33.00	.09	.05	1.40	—	.15	—	—	0
2715	**.50**	—	—	**270.0**	—	—	**20.00**	**.06**	**.04**	**.50**	—	—	—	—	—
	.50	—	—	270.0	—	—	20.00	.06	.04	.50	—	—	—	—	—
2716	**.50**	**.20**	**.01**	**900.0**	**0**	**.40**	**23.00**	**.06**	**.04**	**.70**	**.33**	**.10**	**39.00**	**4.00**	**0**
	.50	.20	.01	900.0	0	.40	23.00	.06	.04	.70	.33	.10	39.00	4.00	0
2717	**1.44**	**.48**	—	**2400.0**	**0**	—	**57.60**	**.17**	**.12**	**1.92**	—	—	—	—	**0**
	.60	.20	—	1000.0	0	—	24.00	.07	.05	.80	—	—	—	—	0
2718	**1.20**	**.48**	**(.38)**	**2160.0**	**0**	**(2.71)**	**40.80**	**.12**	**.07**	**1.68**	**.55**	**.22**	**7.68**	**4.32**	**0**
	.50	.20	(.14)	900.0	0	(1.13)	17.00	.05	.03	.70	.23	.09	3.20	1.80	0
2719	**1.20**	**.10**	—	**2160.0**	**0**	—	**40.80**	**.12**	**.07**	**1.68**	**.55**	**.22**	**7.68**	**4.32**	**0**
	.50	.04	—	900.0	0	—	17.00	.05	.03	.70	.23	.09	3.20	1.80	0
2720	**3.60**	**3.82**	**.70**	**15200.0**	—	**4.48**	**278.00**	**.42**	**.78**	**1.60**	**.76**	**.53**	**190.00**	—	**0**
	1.80	1.91	.35	7600.0	—	2.24	139.00	.21	.39	.80	.38	.26	95.00	—	0
2721	**1.65**	**.62**	**.23**	**9450.0**	**0**	**(1.50)**	**103.50**	**.23**	**.36**	**.90**	**(.47)**	**(.23)**	**(165.32)**	**(.63)**	**0**
	1.10	.41	.15	6300.0	0	(1.00)	69.00	.15	.24	.60	(.31)	(.15)	(110.21)	(.42)	0
2722	**1.50**	**.62**	**.23**	**8550.0**	—	—	**70.50**	**.15**	**.35**	**.75**	—	—	—	—	**0**
	1.00	.41	.15	5700.0	—	—	47.00	.10	.23	.50	—	—	—	—	0
2723	**3.20**	—	—	**9400.0**	—	—	**38.00**	**.04**	**.18**	**1.20**	**.14**	—	**82.00**	—	**0**
	1.60	—	—	4700.0	—	—	19.00	.02	.09	.60	.07	—	41.00	—	0
2724	**2.40**	**.62**	**.23**	**10350.0**	—	—	**28.50**	**.08**	**.14**	**.60**	**.17**	**.13**	**58.50**	—	**0**
	1.60	.41	.15	6900.0	—	—	19.00	.05	.09	.40	.11	.09	39.00	—	0
2725	**.66**	**.49**	**.21**	**0**	**0**	**0**	**47.52**	**.05**	**.09**	**.79**	**.26**	**.12**	**26.40**	**(.16)**	**0**
	.50	.37	.16	0	0	0	36.00	.04	.07	.60	.20	.09	20.00	(.12)	0
2726	**.60**	**.42**	**.12**	**0**	**0**	**0**	**33.00**	**.06**	**.08**	**.45**	**(.23)**	**(.10)**	**19.50**	**Trace**	**0**
	.40	.28	.08	0	0	0	22.00	.04	.05	.30	(.15)	(.06)	13.00	Trace	0
2727	**.62**	—	—	**4160.0**	—	—	**53.04**	**.03**	—	**.26**	—	—	—	—	**0**
	1.20	—	—	8000.0	—	—	102.00	.05	—	.50	—	—	—	—	0
2728	**.03**	**.01**	**.0**	—	—	—	—	**.00**	**Trace**	**.01**	—	—	—	—	**0**
	.46	.17	.03	—	—	—	—	.01	.02	.14	—	—	—	—	0
2729	**.02**	**Trace**	**0**	**49.0**	**0**	**(.01)**	**.79**	**0**	**.00**	**.01**	**Trace**	**Trace**	—	**Trace**	**0**
	1.70	(.21)	.04	4900.0	0	(1.00)	79.00	.08	.16	.90	.31	.13	—	(.40)	0
2730	**.09**	**.11**	**.05**	**5.6**	**0**	—	**2.90**	**.03**	**.01**	**.18**	**.18**	—	—	—	**0**
	.33	.40	.17	20.0	0	—	10.30	.11	.04	.42	.64	—	—	—	0
2731	**1.20**	**.62**	**.44**	—	—	—	**18.00**	**.18**	**.08**	**1.20**	—	—	—	—	—
	.60	.31	.22	—	—	—	9.00	.09	.04	.60	—	—	—	—	—
2732	**.31**	—	—	**3042.7**	—	—	**8.68**	**.04**	—	—	—	—	—	—	—
	1.10	—	—	10866.7	—	—	31.00	.15	—	—	—	—	—	—	—
2733	**.84**	—	—	—	—	—	**28.00**	**.06**	**.04**	**.42**	—	—	—	—	—
	.60	—	—	—	—	—	20.00	.04	.03	.30	—	—	—	—	—
2734	**.11**	—	—	**0**	—	—	**2.52**	**.02**	**.02**	**.06**	—	—	—	—	—
	.40	—	—	0	—	—	9.00	.08	.06	.20	—	—	—	—	—

[3]Data from *Food Composition Table For Use in Africa*, FAO, United Nations, 1968.
[4]Data from *Food Composition Table for Use In East Asia*, FAO, United Nations, 1972.
[5]Data from *Food Composition Table For Use In Latin America*, Institute of Nutrition of Central America & Panama, and National Institutes of Health, USA, 1961.
[6]Footnote 6 and added values in parentheses are from reliable literature sources.

FOOD GROUPS

Contents *Page*

Four Food Groups—The Basic Four 990
Five Food Groups 990
Six Food Groups (Food Guide Pyramid) 990
Seven Food Groups 990
Other Food Groups 990
Meal Planning 991
Calorie Counting 993
Snacking 995
Strength and Weakness 995

The Food Groups represent an attempt to list foods together under headings which reflect their similarities as good sources of specific nutrients. Thus, the Food Groups are employed for planning and evaluating diets for nutritional adequacy.

Following World War II, the system of Seven Food Groups was used by most people in public health and nutrition in the United States. Gradually, the Seven Food Groups gave way to the Four Food Groups, to which a fifth group was added. With the introduction of the "Food Guide Pyramid" in 1992, six food groups are now preferred.

FOUR FOOD GROUPS—THE BASIC FOUR. This is the name applied to the daily food guide developed by Harvard's Department of Nutrition (presented at the 38th Annual Meeting of the American Dietetic Association in St. Louis, October 19, 1955, and published in the *Journal of the American Dietetic Association,* November 1955 [31:1103–1107, 1955]).

To ensure nutritional adequacy, use of the Four Food Groups calls for the daily diet to include definite amounts of foods from each of the Four Food Groups: (1) meats, poultry, fish and beans; (2) milk and cheeses; (3) vegetables and fruits; and (4) breads and cereals.

FIVE FOOD GROUPS. The Five Food Groups include the basic four groups plus a fifth group consisting of fats, sweets, and alcohols. Included in the fifth group are such foods as butter, margarine, mayonnaise, salad dressings, and other fats and oils, candy, sugar, jam, jellies, syrups, sweet toppings and other sweets, soft drinks and other highly sugared beverages, and alcoholic beverages. Also included are refined but unenriched breads, pastries, and flour products.

SIX FOOD GROUPS (FOOD GUIDE PYRAMID). For almost 50 years, the Four Food Groups were arranged on a wheel, which was hung in classrooms. In 1992, the U.S. Department of Agriculture released the *Food Guide Pyramid*, featuring Six Food Groups, designed to reflect the changing eating habits of consumers and to give the department's official recommendations of what is good for you. From a broad base, the design narrows progressively toward the top. In the pyramid, the hierarchy and daily servings of the six food groups are as follows: (1) 6 to 11 servings of bread, cereals, and pasta; (2) 3 to 5 servings of vegetables; (3) 2 to 4 servings of fruit; (4) 2 to 3 servings of milk, yogurt, and cheese; (5) 2 to 3 servings of meat, poultry, fish, dry beans, eggs, and nuts; and (6) at the apex of the pyramid and occupying the smallest space, the fats, oils, and sweets, along with the admonition to use sparingly. (See the Food Guide Pyramid.) Figs. F-54 to F-57a illustrate these Six Food Groups. Table F-37 is a summary of the Six Food Groups—The Food Guide Pyramid, listing the foods included, the amounts recommended, and the contribution to the diet.

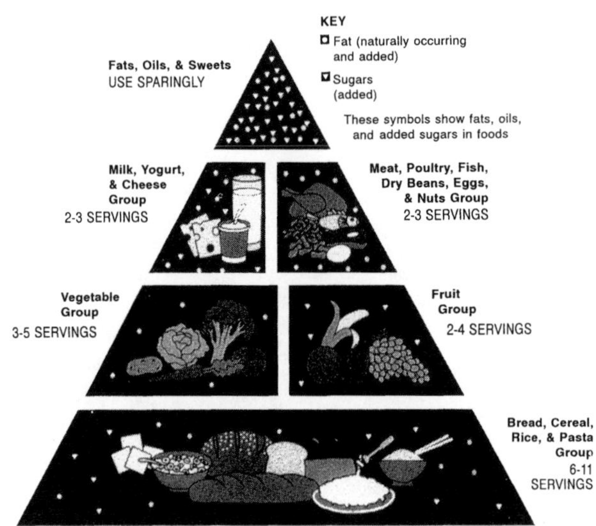

SEVEN FOOD GROUPS. Under the older Seven-Food-Groups' System, the following food groups were recognized: (1) meats, eggs, dried peas and beans; (2) milk, cheese, and ice cream; (3) potatoes, other vegetables, and other fruits; (4) green leafy and yellow vegetables; (5) citrus fruits, tomatoes, and raw cabbage; (6) breads, flour and cereals; and (7) butter and fortified margarine. From these seven groups the number of daily servings of each group were specified to provide a nutritionally adequate diet.

OTHER FOOD GROUPS. The number of food groups—four, five, six, or seven—is not sacred. Also, the groupings may vary in different countries depending upon food habits, economics, and dietary needs. Among the different countries, food grouping systems range from three groups to twelve groups.

The reasons behind the development of these food groups are: (1) the average adult may not use good judgment when it comes to choosing a well-balanced diet; (2) it would be an almost impossible task to calculate and balance a diet with respect to all nutrients involved, flavor, cost, and availability; and (3) the need for a simple but usable guideline for the daily diet.

Fig. F-54. Bread, cereal, rice, and pasta group.

Fig. F-55. Vegetable group.

Fig. F-55a. Fruit group.

Fig. F-56. Milk, yogurt, and cheese group.

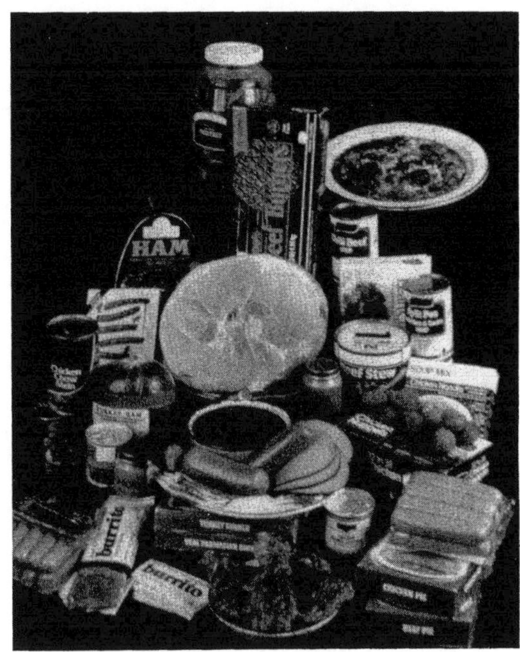

Fig. F-57. Meat, poultry, fish, eggs, dry beans and nuts group.

Fig. F-57a. Fats, oils, and sweets group.

TABLE F-37
THE SIX FOOD GROUPS/THE FOOD GUIDE PYRAMID

Group	Foods Included	Amount Recommended[1]	Contribution to the Diet	Comments
Bread, cereal, rice and pasta	All breads and cereals should be whole-grain or enriched products.	6-11 servings daily. 1 serving = 1 slice of bread 1 oz ready-to-eat cereal 1 small muffin 1 small piece of cake	Foods in this group furnish worthwile amounts of protein, iron, several of the B vitamins, and food energy. Whole-grain products also contribute magnesium, folacin, and fiber.	Includes all products made with whole-grains or enriched flour, or meal. Example: bread, biscuits, muffins, waffles, pancakes, cooked or ready-to-eat cereals, cornmeal, and pastas.
Vegetable	All vegetables, with emphasis on those that are valuable sources of vitamins A and C. (See Table V-5, p. 2163)	3-5 servings daily. 1 serving = 1 c leafy greens ½ c any other vegetable	Vegetables are valuable because of their minerals and vitamins. In addition, they contain fiber.	Should include one or two cruciferous vegetables (broccoli, collards, kale, mustard greens, parsley, and spinach).
Fruit	All fruits. One citrus fruit each day.	2-4 servings daily. 1 serving = 1 medium apple, banana, or orange ½ c other fruits 6 oz fruit juice	Fruits provide fiber, minerals, and vitamins.	All citrus fruits are high in vitamin C. Cantaloupe is high in vitamin A.
Milk, yogurt, cheese	All dairy products: milks, cheeses, ice creams, yogurts.	2-3 servings daily. 1 serving = 1 c milk 8 oz yogurt 1½ oz cheese	Milk and most milk products are our leading source of calcium. They also provide high-quality protein, riboflavin, vitamins A, B-6, and B-12, and vitamin D when fortified with this vitamin.	Milk used in cooked foods—creamed sauces, soups, puddings—can count towards quota for this group.
Meat, poultry, fish, dry beans, eggs, and nuts	Beef, veal, lamb, pork, liver, heart, kidney. Poultry. Fish and shellfish. Dry beans, peas, or lentils. Eggs. Nuts and nut butters.	2-3 servings daily. 1 serving = 2 oz cooked lean meat, poultry, or fish 1 c cooked beans 2 eggs 4 T peanut butter	Foods in this group are valued for their protein. They also provide iron, thiamin, riboflavin, niacin, and phosphorus.	Protein is necessary for growth and repair of body tissues—muscles, organs, blood, skin, and hair. Those of animal origin also supply vitamin B-12.
Fats, oils, and sweets	Butter, margarine, mayonnaise, salad dressing, jams, jellies, syrups, sweetened juices, and alcoholic beverages.	Use sparingly. Less than 30% of calories from fat.	The oils contribute the essential fatty acids, but otherwise, they contribute very little to the diet.	

[1]To convert to metric, see WEIGHTS AND MEASURES

MEAL PLANNING. The Food Groups provide a simple guide for planning the foundation for a day's meals. Choose at least the minimum number of servings from each food group, and select a variety of foods. Table F-38 demonstrates sample menus for breakfast, lunch, and dinner developed from the Food Groups.

TABLE F-38
SAMPLE MENUS DEVELOPED FROM THE FOOD GROUPS

Day	Breakfast	Lunch	Dinner
Day 1	Orange juice; cereal; poached eggs on toast; milk, coffee	Sandwich of ham, lettuce, mayonnaise on whole wheat bread; fresh fruit; cookie; milk, tea, or coffee	Meat loaf; baked potato; peas; lettuce salad; biscuit; carrot cake; milk, tea, or coffee
Day 2	One-half grapefruit; sausage; muffin without butter; milk, coffee	Vegetable soup; peanut butter sandwich; frozen fruit yogurt; milk, tea, or coffee	Pork chops; baked sweet potatoes; broccoli; relish tray; whole wheat roll; apple pie; coffee, tea, or milk

CALORIE COUNTING. Calorie counting is almost a national pastime for Americans. The energy (kcal), protein, mineral, and vitamin content of a large number of foods is given in Food Composition Table F-36, which can provide assistance when selecting food. Furthermore, to help count calories, foods and servings of the Food Groups can be classed as low, medium, and high calorie, as shown in Table F-39.

TABLE F-39
COUNTING CALORIES WITH THE SIX FOOD GROUPS[1]

	Food Group									
	Breads and Cereals		Vegetables		Fruits		Milk, Yogurt, and Cheeses		Meats, Poultry, Fish, Beans, Eggs, and Nuts	
General Caloric Level	Serving Size and Food[2]	Calories	Serving Size and Food[2]	Calories	Serving Size and Food2	Calories	Serving Size and Food[2]	Calories	Serving Size and Food[2]	Calories
		(kcal)		(kcal)		(kcal)		(kcal)		(kcal)
Low	2 medium biscuits, shredded wheat	44	1 c lettuce	9	1 raw apricot	19	1 c cultured buttermilk made from skim milk	90	½ c boiled navy beans	95
	1 slice of whole wheat bread	65	½ c cooked cabbage	15	1 large prune	24	½ c (single dip) ice milk	95	2 oz broiled chicken	95
	½ c steamed or boiled rice	85	½ c cooked spinach	20	1 raw fig	33	1 c milk, 1% fat	103	1 fried egg	101
	1 c plain cornflakes	95	½ c cooked broccoli	25	1 med. peach	38	1 oz Swiss or cheddar cheese	109	2½ oz broiled cod with butter or margarine	120
	½ c cooked noodles	100	½ c cooked Brussels sprouts	27	1 c boysenberries	50	½ c cottage cheese	110	2 oz lean roast beef	137
	1 c cooked Cream of Wheat cereal	103	½ c cooked beets	32	1 c watermelon	52	½ c (single dip) ice cream	125	2½ oz fried breaded ocean perch	160
	1 c oatmeal	123	½ c cooked okra	35	1 c fresh strawberries	56	1 c plain lowfat yogurt	145	1 c navy bean soup	170
Medium	1 corn muffin	125	½ c cooked green peas	57	1 c raw pineapple	77	1 c whole milk	160	½ fried chicken breast (2¾ oz) or 2 drumsticks (2½ oz)	160 to 180
	2 slices of whole wheat bread	130	½ c baked winter squash	65	1 c blackberries	84	1 c (two dips) ice milk	190	3 oz lean hamburger (without bun)	185
	1 c cooked spaghetti	155	½ c cooked parsnips	66	1 c apple	84	1 c vanilla flavored yogurt	195	3 oz fried beef liver	186
	1 c sugar-coated cornflakes	155	½ c cooked corn	69	1 c blueberries	87	2 c milk, 1% fat	206	3 oz regular hamburger (without bun)	235
	½ c fried rice without meat	185	½ c mashed potatoes	99	½ grapefruit	99	1 c cottage cheese	220	3 oz roasted cured ham	245
High	½ c rice pudding	235	½ c frozen cooked baby lima beans	102	1 c grapes	106	1 c yogurt with fruit or 2 dips frozen yogurt	225 to 244	1 c baked navy beans	310
	1 cup cooked rice polish	278	½ c au gratin potatoes	123	1 med. banana	112	1 c hot chocolate	250	3 oz Swiss steak	315
	3 average pancakes	312	½ c sauteed mushrooms	150	1 c cherries	140	1 c ice cream	250	2½ oz baked stuffed fish (½ c bread stuffing)	325
	2 waffles	420	1 sweet potato	170	1 c fruit cocktail	154	1 c vanilla milkshake	255	1 c chili con carne	335
	1 c plain granola	480	½ c hashed brown potatoes	170	1 sweetened baked apple	160	1 c cheese souffle	327	⅔ c beef stroganoff over noodles	525

[1]Calorie values were calculated from Food Composition Table F-36 of this book.
[2]To convert to metric, see WEIGHTS AND MEASURES.

Also, the following guidelines should be kept in mind when counting calories:

1. Cut down on high-fat foods such as margarine, butter, highly marbled or fatty meats, and fried foods. Salad dressings, cream sauces, gravies, and many whipped dessert toppings are also high in fat, and should be used sparingly.
2. Cut down on sugary foods such as candies, soft drinks and other sugar-sweetened beverages such as ades and punches, jelly, jam, syrups, honey, fruit canned in heavy syrup, pies, cakes, and pastries.
3. Cut down on or eliminate alcoholic drinks.
4. Cut down on portion sizes. Portions of some foods, such as meats, are hard to estimate. For example, a 3-oz (*84 g*) serving of cooked lean meat without bone is equivalent to a 3 in. by 5/8 in. (*7.5 cm by 1.6 cm*) hamburger pattie.
5. Use whole milk or whole-milk products (most cheeses and ice cream) sparingly. Lowfat and skim-milk products, such as ice milk and skim milk cheeses, provide fewer calories than their whole-milk counterparts.
6. Select cooking methods to help cut calories. Cook foods with little or no added fat and avoid deep-fat fried foods, which are high in calories because of the fat absorbed during cooking. For meat and poultry, trim off visible fat and either broil or roast on a rack. If braised or stewed, drain meat to remove fat. For fish, broil or bake. For vegetables, steam, bake, or boil. For an occasional change, stir-fry in a small amount of vegetable oil.
7. Be sure to count the nibbles and drinks enjoyed during social events and throughout the day as part of a day's calorie allotment.

Table F-40 demonstrates menus with three different levels of calories. These menus are based on the Food Guide Pyramid or Six Food Groups. Caloric levels are controlled by food choice and serving size.

TABLE F-40
SIX FOOD GROUP—FOOD GUIDE PYRAMID MENUS WITH DIFFERENT LEVELS OF CALORIES[1]

Meal	1,200 Calories	1,800 Calories	2,400 Calories
	(kcal)	(kcal)	(kcal)
Breakfast	Orange juice . . . ½ c Bran flakes with raisins . . . ½ c Lowfat (1%) milk . . . ½ c Whole-wheat toast . . . 1 slice Coffee or tea	Orange juice . . . ¾ c Bran flakes with raisins . . . ½ c Whole milk . . . ½ c Whole-wheat toast . . . 1 slice Jelly . . . 2 tsp Coffee or tea	Orange juice . . . 1 c Bran flakes with raisins . . . ½ c Whole milk . . . ½ c Whole-wheat toast . . . 1 slice Jelly . . . 1 Tbsp Coffee or tea
Lunch	Ham sandwich: Ham . . . 2 oz Cheese . . . ½ slice Lettuce . . . 1 leaf Tomato . . . ½ medium Enriched bread . . . 2 slices Apple . . . 1 medium Coffee or tea	Ham sandwich: Ham . . . 2 oz Cheese . . . 1 slice Lettuce . . . 1 leaf Tomato . . . ½ medium Enriched bread . . . 2 slices Salad dressing . . . 2 tsp Apple . . . 1 medium Coffee or tea	Ham sandwich: Ham . . . 2 oz Cheese . . . 1 slice Lettuce . . . 1 leaf Tomato . . . ½ medium Enriched bread . . . 2 slices Salad dressing . . . 2 tsp Apple . . . 1 medium Plain cookies . . . 4 Coffee or tea
Dinner	Beef roast . . . 3 oz Baked potato . . . 1 medium Broccoli . . . ½ c Skim milk . . . 1 c Oatmeal cookie, or . . . 1 Ginger snaps . . . 2	Beef roast . . . 4 oz Baked potato . . . 1 medium Broccoli . . . ½ c Roll . . . 1 Margarine . . . 1 tsp Lowfat (1%) milk . . . 1 c Angel food cake . . . 1/16 with strawberries . . . ½ c	Beef roast . . . 5 oz Baked potato . . . 1 medium Broccoli . . . ½ c Roll . . . 1 Margarine . . . 2 tsp Lowfat (2%) milk . . . 1 c Angel food cake . . . 1/12 with strawberries . . . ½ c and ice milk . . . 1/3 c
Snacks	Cucumber slices . . . 1 small Carrot sticks . . . 3 to 4 (2½ to 3 in. long sticks)	Fresh peach . . . 1 medium Celery sticks . . . 3 to 4	Fresh peach . . . 1 medium Fruit-flavored yogurt . . . 1 c Banana . . . 1 small

[1]To convert to metric, see WEIGHTS AND MEASURES.

SNACKING. Snacking can be good or bad depending upon what we snack on and how often. Of course, foods loaded with sugar and fat—calories—are not a good idea, mainly because these foods are also low in protein, vitamins, and minerals. In keeping with the objective of the Food Groups it just makes sense to include snacks from these same food groups; for example, an orange, an apple, peanut butter and crackers, a glass of juice, etc. However, the type of snack depends on age and activity. Small children often fail to consume the amount of food in regular meals that will add up to their nutritional needs. A slice of cheese, a wheat cracker, or a banana at various times helps supply the added energy they need. The active growing bodies of teenagers need nutritious snacks from the Food Groups which provide protein, vitamins, minerals, and energy. Senior citizens often have chewing or digestion problems which interfere with regular eating habits. Here again nutritious snacks are helpful.

Nutritious snacks are, however, no excuse for overindulgence. A calorie is still a calorie, and lots of little snacks add up. If more energy is consumed than used the excess is stored as fat.

Fig. F-58. Ideal snack foods, anytime of the day—served attractively. (Courtesy, USDA)

STRENGTH AND WEAKNESS. The major strength of the Food Groups is its simplicity, and therein lies its weakness, also. The Four Food Groups have been criticized and alternative plans for adaptations proposed, but for many people it is or can be an effective tool for balancing diets with foods instead of nutrients. It is a starting point for balanced nutritious diets.

(Also see CARBOHYDRATE[S]; FATS AND OTHER LIPIDS; GERONTOLOGY AND GERIATRIC NUTRITION, Table G-5 Major Food Groups for Senior Citizens; MINERAL[S]; PROTEIN[S]; and VITAMIN[S].)

FOOD INDUSTRY

In the colonial period consumers processed most foods in the home. In time, kitchen preparation techniques were extended and scaled up to furnish sufficient surplus products to be bartered or sold outside the household. Further enlargement of a business simply entailed building more and larger processing equipment—oil presses, baking ovens, or wine vats. Gradual improvements in design were made to increase yields and/or improve quality. This was the general pattern until the Industrial Revolution, at which time factories were greatly enlarged, much of the manual labor was replaced by machinery, and entirely new principles of processing and marketing were introduced, including canning, spray drying, and shipping by means of refrigerated railroad cars.

Today, the U.S. food industry is the largest and most important in the world. It comprises all business operations that are involved in producing raw food material, in processing it, and in distributing it through sales outlets. The entire complex of the industry includes farms, ranches, and feedlots; producers of raw materials, such as fertilizers, for agricultural use; water-supply systems; food-processing plants; manufacturers of packaging materials and food processing and transportation equipment; transportation systems; retail stores; and food-service operations such as restaurants, institutional feeding, and vending machine services.

FOOD MYTHS AND MISINFORMATION

Contents *Page*

The Medicine Man of Old 996
The Merchant of Menace 996
Definitions 996
Food Myth and Misinformation Dangers 997
Why are People so Gullible? 998
Food Myths and Misinformation Vs Facts 999

Without food, medicine as we know it today might never have been born, for until early in the 18th century health treatments were largely based on myths.

The *Papyrus Ebers,* written about 1550 B.C., on paper which the Egyptians made from the papyrus plant, describes scurvy and diabetes and gives a remarkable description of the practice of medicine in Egypt. Foods played many important roles at the time—as medicines, as beauty aids, and as remedies for household problems. Among the recommendations was one for making hair grow, by applying to the bald area either a mixture of honey, palm oil, and fruit juice, or the fat of several wild animals.

A thousand years later, Hippocrates (460 to 377 B.C.), the father of medicine and a contemporary of Socrates, made many wise observations about food and the science of nutrition; among them: Children produce more heat and need more food than adults; persons who are naturally very fat are apt to die earlier than those who are slender; people should exercise; and liver will cure night blindness. But the prescientific era of nutrition was

Fig. F-59. The medicine man of old, who, amid tom-toms and torchlights, plied his trade.

also characterized by a fascinating maze of myths, fallacies, fads, philosophies, taboos, bizarre superstitions, food cults, and religious precepts.

THE MEDICINE MAN OF OLD. In frontier America, the medicine man—the doc—plied his trade in a tent, as his show on the road moved from village to village. His potent snake oil and tiger's milk cure-all were always "smuggled out of the sacred tombs of ancient Egypt." Silk-hatted and suave, between band numbers, doc made his pitch:

> "Ladies and gentlemen! Boils and bunions, fevers and fits, gout and gas—these have plagued mankind since life began. But no more! No more for those of you, who for a mere pittance, avail yourselves of this marvelous cure—the guarded secret of health and long life."

Then the band played on while the ushers fanned down the aisles selling their treasures. Their shouts of "all sold out, doc" signaled product number two. Next, the doc gave his soap spiel—a product that would restore glow and youth to dry, harsh skin. Medicine number three was a linament, guaranteed to give relief from all aches and pains. It was a good show, and most of the products were quite harmless.

THE MERCHANT OF MENACE. The top hat and torchlights are gone, but the medicine man is still with us. Today, he's the self-styled scientist or health and nutrition expert, beating the drums for his potions and remedies at your doorstep, on lecture platforms, through the mail, in books, and on the radio and TV. He's the sophisticated salesman who bleats warnings against "that tired feeling"; "subclinical deficiencies"; "devitalized food," "hidden illness," and "aging before your time." Beware of this man! He may be a merchant of menace to you and your health. He has a product—usually an exotic pill, capsule, powder, tonic, special food, or food supplement—which will fortify your diet, increase appetite, build you up, cause you to lose weight, steady your nerves, strengthen your bones, enliven your blood, empty your bladder, roll back the stones from your dying kidneys, and regulate your bowels. Besides, he operates all over the world, from the witch doctors of darkest Africa to plush suites on Park Avenue.

DEFINITIONS. To understand the terms used in this section, several definitions follow:

• **Food myth**—The word *myth* comes from the Greek word *mythos*, meaning "story" or "fable." *A food myth is a traditional food story, belief, fable or legend, without proof or determinable basis of fact, which is accepted or used to justify one's own desires, interests, or practices.* To a large extent, food myths represent a reactionary response to advances in scientific agriculture and food technology.

• **Food misinformation**—To misinform is to give incorrect, untrue, or misleading information. *Food misinformation refers to a statement about food that is not in accord with the scientific facts.* Such false information may have arisen out of traditional fallacy, or it may represent belief in magic or folklore, or it may have been fabricated for sales purposes, or it may stem from ignorance.

• **Food fads**—*A food fad is a popular food pursuit or craze, without substantial basis, that is followed with zeal.* Most food fads are short-lived, but a few persist. Some are quite harmless, whereas others may adversely affect the health and welfare of their followers. Two examples of food fads follow:

1. In the early 1600s, coffee drinking became a fad when the brew was first introduced into Europe, and people met at coffeehouses for serious discussions. But the fad passed. Today, coffee drinking is no longer a fad, but a practice widely enjoyed.

2. In the early 1930s, one fad diet, known as the "Hay Diet," after its promoter Dr. William Howard Hay, stressed the incompatibility of proteins and carbohydrates, and that foods rich in each should be eaten in separate meals.

• **Food faddists**—*A food faddist is a person who takes up a food fad and follows it, usually ardently.* With missionary zeal, most food faddists try to convince others to eat the ingredients to which they attribute their vitality, which may include such things as molasses, wheat germ, yogurt, a "natural vitamin," soybean oil, etc. Generally speaking, food faddists focus on foods *per se,* not on the chemical agents in them—the nutrients—which are the actual physiologic agents of life and health.

• **Food quack (food charlatan)**—*A food quack is a pretender who claims to have skill, knowledge, and qualifications in foods, nutrition, and/or medicine, which he does not possess.* The quack is the modern counterpart of the medicine man of old. His motive is usually money.

The word *quack* is a shortened form of "quacksalver," a term invented centuries ago by the Dutch, to describe the pseudophysician or pseudoprofessor who sold worthless salves, magic elixirs, and cure-all tonics. The quacksalver made much noise, but had no real food, nutrition, and/or medical ability or knowledge. Like a barker, he proclaimed his wares in a patter that skeptical people came to compare to the quacking of a duck; hence, the derivation of the term *quack.*

The food quack, or charlatan, can usually be recognized by the following characteristics which are basic to his method of operation:

1. **He sells something.** He sells foods, supplements, a dietary plan, cooking utensils, cookbooks, magazines, a series of lectures, and/or books pertaining to foods, nutrition, and health. Often he makes a special introductory offer "for a limited time only," or on a "money back guarantee."

2. **He belittles normal foods.** This is the first sign of nutritional quackery.

3. **He appeals to the emotions rather than to the intellect.** He plays upon the fears and hopes of people. He either scares people about the dire consequences of failing to consume a certain food or product, or he makes exaggerated claims for curing certain diseases, or for enhancing beauty, vitality, and long life.

4. **He claims that he is persecuted by medical and nutrition groups.** Generally, he attacks the medical profession, nutritionists, federal regulatory agencies, American agriculture, and the food industry; and he creates distrust in science and technology.

5. **He speaks the language.** He glibly uses scientific terminology to impress people that he is learned in the profession. Often he quotes leading medical and nutritional authorities to support his claims, lifting their writings out of context.

6. **He usually has no academic preparation in college or in a professional school.** Nevertheless, in support of his self-styled expertise in nutrition, he makes claim to degrees, titles, membership in learned societies, and experience.

7. **His products are distributed through health food stores (which may be very good), by door-to-door sales, and/or through the mail.** Although there is nothing wrong with these outlets as such, the advertising and promotion are usually characterized by exaggerated claims, money-back guarantees, and testimonials.

8. **He possesses charisma.** He is usually a charming person, with a flair for words and a clever wit. Has anyone ever met a quack that they didn't like?

• **Food fallacy**—This refers to a false or mistaken idea or opinion with respect to food and nutrition. A fallacy generally arises through ignorance or through faulty interpretation of the science of food and nutrition as it is applied to the daily diet. For example, an erroneous claim is sometimes made that grapefruit possesses special enzymatic properties that enable it to break down excess fat. This claim may encourage obese persons to eat substantial quantities of grapefruit to the exclusion of other foods.

• **Food folklore (old wives' tales)**—*Folklore refers to the traditional food beliefs and customs held by a people and handed down from generation to generation—the old wives' tales.* For example, the belief that a pregnant woman's appetite for such things as pickles and clay indicate the body's need, and, therefore should be satisfied; or the influence that foods are believed to have on behavior, such as the bravery and courage instilled by eating flesh foods.

• **Folk medicine**—*Folk medicine is the medicine prescribed by nonprofessional people isolated from modern medical services.*

FOOD MYTH AND MISINFORMATION DANGERS. Food myths and misinformation abound! Most of them are charming, some are sheer nonsense, but others are hazardous. Some myths combine all three; for example, the claim that honey and vinegar will cure an amazing array of ills, ranging from arthritis to the common cold to digestive problems. It has been said that a couple of teaspoons (*10 ml*) of cider vinegar in a glass of water, taken at each meal, will cause the body to burn rather than store body fat. Thus, in easy steps, such pronouncements turn folk medicine into a hazardous practice.

But the following admonition tops the myth of honey and vinegar!

Amazingly, the "thinking man" has failed to grasp the terrifying significance of the term "in a pickle." Now, in a charming yarn attributed to a food faddist, all pickle eaters are warned that: "Pickles will kill you!" Then, the faddist backs up his pronouncement with the following facts:

Pickles are associated with declines of civilization—the ancient Egyptians ate them. Eating pickles breeds wars, causes auto accidents, and airline tragedies. Pickles are associated with all the major diseases of the body; and there is a positive relationship between crime waves and consumption of pickles. Then the faddist administers the coup de grace with the following alarming statistics:

1. 99.44% of all people who die from cancer have eaten pickles.
2. 100% of all soldiers killed in battle have eaten pickles.
3. 96.8% of all people involved in auto and air accidents ate pickles within 30 days preceding the accident.
4. 95.3% of all juvenile delinquents come from homes where pickles are served frequently.
5. 99% of all pickle eaters born prior to 1920 have wrinkled skin, false teeth, brittle bones, and/or failing eyesight—if the ills of eating pickles haven't already killed them. (If the above be true, I don't have long for this world. Pass me the pickles—please. M.E.E.)

Basically, food fads involve the following hazards, which concern all members of the nutrition and health professions:

1. **They may be a hazard to health.** In some cases, food faddism fosters malnutrition and other health problems. More importantly, self-diagnosis and self-treatment can be dangerous. By following such a course, a person with a serious illness may fail to secure proper medical care; falsely believing that the miraculous properties of this diet will cure their ailment. Many anxious patients with cancer, diabetes, or arthritis have been misled by quacks who fraudulently claimed to have a cure for these diseases, with the result that they postponed effective therapy until it was too late.

2. **They are usually costly.** Most foods and supplements used by faddists are expensive. Each year, Americans spend an estimated $2 billion on special foods, food supplements, and health lectures and literature. Today, conditions are ripe for "fast operators" to make a "quick buck." Many people have become health conscious. The elderly, the adolescent, the obese, the people whose living depends on their physical appearance, and the sickly are looking for a quick fix; and the charlatans happily accommodate them as they outsell reliable sources of nutrition information. As a result, many quacks, whose products and sales pitches are reminiscent of the medicine men of old, have developed a thriving business, pawning off a myriad of potions, cure-alls, and tonics.

3. **They stymie scientific progress.** The misinformation spread by charlatans hinders scientific progress. Besides, the superstitions that they perpetuate counteract sound nutrition and health teachings.

4. **They create distrust of the food team—the producer, the processor, and the marketer.** A positive program of education in the schools and among the adult population is needed to present the truth about foods and nutrition and to counter the distrust of the food team created by charlatans and faddists.

WHY ARE PEOPLE SO GULLIBLE? In human folly, facts have never stood in the way of myths and misinformation. Additional reasons for people being so gullible in foods and nutrition follow:

1. **Food is basic, and people's emotions become involved.** Aside from satisfying hunger, food means many things to many people—security, a soothing balm, a means of avoiding idleness, a symbol of success, a sensual appeal. Also, everyone has psychological reactions toward particular foods, determining likes and dislikes. Food quackery plays on these attitudes, plus the powerful incentives of human nature—the desire for health, and the fear of pain, disease, and death. *It is easier to believe a bizarre claim that reassures than a more scientific statement that offers little hope.*

2. **People are sadly uneducated about nutrition and diet.** Because food faddists and quacks talk a good line—they are messiahs and evangelists—it is often difficult for people to recognize what is myth and misinformation and what is fact. *With an informed public, the potential danger of anecdotal information replacing scientific findings can be lessened.*

Thus, there is need for more enlightened information, based on sound research. Additionally, there is need for continued vigilance by the government agencies entrusted to protect the public, especially the Food and Drug Administration, the Federal Trade Commission, and the U.S. Department of Agriculture. Also, several foundations, professional organizations, and trade associations are making monumental contributions through research and education. On a worldwide basis, the following divisions and agencies of the United Nations are engaged in food and nutrition work: Food and Agriculture Organization, World Health Organization, UNESCO, and UNICEF.

(Also see NUTRITION EDUCATION.)

3. **People long for the "good old days."** The phonies play upon the mistaken notion that our forefathers enjoyed an ideal food supply and good health. But, the truth is that in the "good old days" things weren't all that good; sickness, crippling diseases, and death rates were higher, and the life-span was shorter. Just before World War II, for example, rickets, beriberi, pellagara, and goiter—all extremely serious nutrition-deficiency diseases—were, according to the American Medical Association, "common diseases in the United States." Today, these maladies are no longer tabulated as causes of death in this country. Since World War II, more reliable information has been discovered about food production, food composition, and the functions of foods in conserving health than had been established in all previous history. Thus, instead of looking backward, people need to be informed and to look forward in the area of food and nutrition.

FOOD MYTHS AND MISINFORMATION VS FACTS. It is an intriguing test of one's own beliefs to see how they appear in light of scientific facts. There are so many myths and so much misinformation that almost everybody believes at least part of them. Some of these beliefs—most of them fallacious, along with the facts, are presented in Table F-41 Myths and Misinformation Vs Facts.

(Also see MEAT[S], section headed "Meat Myths"; and MILK AND MILK PRODUCTS, section headed "Milk Myths.")

TABLE F-41
MYTHS AND MISINFORMATION VS FACTS

About	Myth/Misinformation	Fact
Aches and pains	If you have aches and pains, more than likely you are suffering from a subclinical deficiency. (The word *subclinical* is a general, nonspecific term given to conditions for which there are no observable symptoms.)	Most people experience aches and pains at one time or another. Such symptoms may be caused by overwork, not pacing yourself, emotional stress, disease, lack of sleep, and/or poor nutrition. If such symptoms persist, see your physician.
Aluminum	Aluminum cooking utensils are dangerous to health.	Cooking in aluminum utensils is harmless. Aluminum is the second most abundant mineral in the soil, so it occurs naturally in many foods.
Antibiotics (Also see ANTIBIOTICS.)	Antibiotic residues in foods are harmful.	The Food and Drug Administration controls the amount of antibiotic residue which can be contained in foods derived from animals treated with or fed antibiotics. The direct addition of antibiotics to foods is not permitted in the U.S.
Apple (Also see APPLE.)	An apple a day keeps the doctor away.	Apples are good—and good for you. But they won't keep the doctor away.
Apple cider and honey	Apple cider and honey will cure all that ails you.	Neither of these foods has any therapeutic value. However, it's good to have faith in something. Any help from daily doses of cider and honey is psychological. There is nothing wrong with cider and honey provided the sugar intake need not be limited, but remember that honey is a concentrated sugar.
Beer (Also see BEERS AND BREWING; and BREAST FEEDING.)	Beer will increase lactation.	The lactating mother does not need beer, although it may be helpful in increasing total daily fluid intake. Also beer supplies liberal amounts of phosphorus, magnesium, riboflavin, vitamin B-6 and niacin.
Beets	Beets are a strong blood builder.	This superstition likely arose because beets, like blood, are red in color.
Breast milk (Also see BREAST FEEDING.)	Food flavors *do not* affect breast milk.	Food flavors *do* affect milk flavor, in all mammals. Numerous experiments with cows have shown that flavors enter the milk through both the respiratory and digestive systems. Feed flavors from such feeds as silage, wild onions, and skunk cabbage are detectable in the milk 20 minutes after the feed is consumed, and they are usually most pronounced at the end of 2 hours. Feed flavors that enter the milk through the respiratory system can usually be detected much sooner than those entering through the digestive system. So, it is logical to deduce that strong flavored foods, like cabbage and onions, will flavor breast milk. Since consumers want that cow's milk taste like milk, it is logical to assume that nursing infants want that breast milk taste like milk, and not like cabbage or onions.
Brewers' yeast	Brewers' yeast should be added to *all* diets.	It's in the nature of good insurance to add brewers' yeast, a rich supplemental source of B vitamins and unidentified factors, to an *average* diet. Further support to this statement is lent by the fact that lack of B vitamins is one of the forms of malnutrition that occur often throughout the world. But brewers' yeast need not be added to all diets.

(Continued)

TABLE F-41 (*Continued*)

About	Myth/Misinformation	Fact
Cannibalism	Cannibalism imparts the courage of the eaten one to the eater.	Cannibalism arose originally, according to legend, because man wanted to destroy the ghost of his killed enemy or relative, and acquire his courage. But it doesn't work that way. Many Western people would feel that they were committing cannibalism if they were to eat either dog or horse meat; they would likely experience a revulsion not unlike that from eating their own grandmother. Yet, dog is considered a delicacy by some Africans and Chinese, and horse meat is eaten with relish in Belgium and France.
Celery	Celery is a nerve tonic.	Celery is an excellent source of potassium, but this does not qualify it as a "nerve tonic," whatever the latter may be. Nerve tissue is built and repaired from amino acids contained in protein.
Citrus fruits	Citrus fruits such as oranges and lemons make the body acid; they produce an "acid stomach."	The hydrochloric acid, secreted by the parietal cells in gastric mucosal glands, forms the normally acid medium of the stomach; and this secretion is not affected by citrus fruits. The metabolism by the body of almost all fruits results in an alkaline, not an acid, residue.
Colds and fevers	Feed a cold, starve a fever.	Don't you believe it. This is not true.
Cookers, waterless	Waterless cookers retain food value in vegetables.	A small amount of boiling water, a tight lid, and a short cooking time help to save nutrients in vegetables. These factors are more important than the type of cooking vessel.
Cravings for unusual items during pregnancy (Also see PREGNANCY AND LACTATION NUTRITION, section headed "Cravings For Unusual Foods and Nonfood Items [Pica].")	Cravings for specific and unusual things like starch, clay, plaster, or chalk (known as pica) during pregnancy indicate needs; hence, consumption of these items should be permitted.	Eating of unusual foods (pica) should be discouraged, because (1) they are not nutritious, and (2) of the hazard of poisoning by contaminants. Recently, evidence has been found that some cases of pica may be associated with iron deficiency anemia.
Diet, reducing (Also see MODIFIED DIETS; and OBESITY.)	You must eat special foods if you wish to lose weight.	Your physician should prescribe any special diet that you may need. Personal experimenting and fad diets can be highly dangerous to your health. Successful weight control depends primarily on self-control of one's total calorie intake while maintaining a reasonable level of physical activity.
Diseases (Also see DISEASES; MALNUTRITION; and MALNUTRITION, PROTEIN-ENERGY.)	All diseases are the result of faulty diet. The proponents of this myth contend that certain chemical imbalances in the body are the cause of disease and that these imbalances are the direct result of improper or inadequate diet.	Good nutrition helps to maintain good health and prevent diseases. Also, a physician may prescribe diet therapy to treat some diseases. But diseases may be caused by inherited defects, microorganisms, or parasites. A large number of these causes have no relationship to nutrition. Some nutritional deficiency diseases, such as scurvy, rickets, pellagra, and kwashiorkor, are still a major health problem in the developing countries.
Eggs, fertile (Also see EGG[S], section headed "Nutritional Misconception.")	Fertile eggs (eggs from hens mated to roosters) are more nutritious and envigorating than nonfertile eggs.	There is no scientific proof for this statement. Furthermore, fertile eggs have a lower market value because danger of loss from the developing embryo.
Eggshell color (brown shelled eggs) (Also see EGG[S], section headed "Color.")	Eggs with brown shells are more natural and more nutritious than white-shelled eggs.	Eggshell color is determined by the breed of hens; and color of shell is not related to nutritional value. Brown shelled eggs are not superior to white shelled eggs, despite what many Bostonians think.

(*Continued*)

TABLE F-41 (*Continued*)

About	Myth/Misinformation	Fact
Eggs, raw	Raw eggs increase sexual potency.	Sorry, this is folklore for which there is no scientific support. Moreover, raw eggs contain the antivitamin, avidin, which binds biotin and makes it unavailable.
Fats and oils	Fats and oils should be eaten to lubricate the joints and help relieve arthritis.	Eating fats and oils will neither lubricate joints nor relieve arthritis.
Fertilizer, chemical	Chemical fertilizers are poisoning our soil.	Chemical fertilizers are not poisoning our soil. They, along with barnyard manure (organic fertilizer), are needed to keep our soil productive. Plants can't tell the difference between nutrients in chemical fertilizers and the same nutrients obtained from organic fertilizers.
Fish	Fish is a brain food; it will help brain tissue grow and increase intelligence.	Even though fish and brains are high in phosphorus, special foods do not build special tissues. No one food will help just the brain.
Food additives (Also see ADDITIVES.)	Food additives are dangerous.	Food additives are under the control of the Food and Drug Administration; hence, they are usually very safe. Many additives are useful for food preservation, for improvement of nutritional value, and for enhancement of quality. Remember that salt, sugar, baking soda, vinegar, vitamin C, and carotene are additives, as is sodium propionate.
Garlic (Also see GARLIC.)	Garlic is helpful in preventing heart disease and stroke.	It may be. Medical researchers in Great Britain, India, and the U.S., working independently, recently found that garlic contains adenosine, a substance that breaks down the blood-clot promoting protein called fibrin. This suggests that garlic may be helpful in preventing thrombotic types of heart disease and stroke. However, more research is needed.
Gelatin	Gelatin strengthens fingernails.	Although gelatin is a protein, it is a rather incomplete protein, notably deficient in the amino acids lysine and tryptophan. Of course, a deficiency of protein in the diet would tend to produce poor fingernails, but there are higher quality proteins than gelatin. There is no evidence that gelatin will improve the structure and toughness of fingernails.
Grapefruit	Grapefruit possesses special enzymatic properties that enable it to break down excess fat; hence, it should be included in a reducing diet.	Grapefruit is a fine citrus fruit, but it will not cause fat to go away. Worse yet, such false claims may encourage obese persons to eat substantial quantities of grapefruit to the exclusion of other foods.
Grape juice	Grape juice is a strong blood builder. Some food faddists recommend it as a cure for cancer.	This blood-building superstition likely arose because grape juice, like blood, is red in color. There is no experimental evidence that grape juice will cure cancer.
Honey (Also see HONEY.)	Honey has special nutritional and medicinal properties. Honey excites sexual desire.	One real advantage of honey is that, due to its fructose content, it is sweeter than table sugar—sucrose; hence, less of it is needed. Honey has no magical powers relative to either sexual desire or health.
Iron (Also see IRON.)	The use of cast-iron cookware increases the iron content of foods.	Some iron is obtained from cooking foods in iron pots, especially foods that have an acid reaction. However, it has not been established that the iron leached from iron cooking utensils is as available to the body as the naturally occurring iron in foods.
Lecithin (Also see LECITHIN.)	Lecithin should be added to the normal diet to help prevent heart attacks. The theory is that lecithin breaks up and disperses cholesterol in the blood so that it cannot become attached to the artery walls.	So far, research has not backed up this belief. Besides, the body synthesizes lecithin and it is found in a wide variety of foods.

(*Continued*)

TABLE F-41 (*Continued*)

About	Myth/Misinformation	Fact
Manure, barnyard (organic fertilizer)	Barnyard manure (natural organic fertilizer) is not only safer than chemical fertilizer, but it produces healthier crops.	Barnyard manure is not safer than chemical fertilizer; neither will it produce healthier crops. The average farm does not produce enough manure (organic fertilizer) to maintain the fertility of its soil; hence, the addition of chemical fertilizers is necessary. The major difference between chemical fertilizer and barnyard manure is that the latter contains organic matter, which almost all soils need, and which farmers cannot buy in a sack or tank. Organic fertilizers cannot be absorbed as such. They must be broken down by soil bacteria to produce the elements—nitrogen, phosphorus, and potassium. For this reason, they are not as quickly available as the nutrients in chemical fertilizers.
Meat	Meat will increase sexual potency.	This folklore likely had its origin in the primitive belief that eating certain foods gave the consumer the characteristics of the item eaten. It followed that meat (animal flesh) was supposed to arouse animal passions. Meat is not a "sex food." However, meat contributes to good health; and good health is requisite for sexual potency.
Meat, rare	Rare meats are more nutritious than those cooked medium rare or well done.	When cooked at moderate temperature, the same cut of meat has essentially the same nutrient content whether rare, medium, or well done. Excessively high heat, however, lessens the nutrient value.
Milk and fish	Milk and fish are a harmful combination, which should never be eaten together.	Both milk and fish are excellent foods, alone or together.
Milk, pasteurized	Pasteurized milk should be replaced by more nutritious raw milk.	Pasteurization destroys a small amount of vitamin C, but the quantity of vitamin C in milk is so small that it makes no significant contribution of this vitamin to the human diet. The control of microorganisms and the prevention of the spread of disease by pasteurization far outweigh any dietary loss.
Mineral supplementation (Also see NUTRITIONAL SUPPLEMENTS; and MINERAL[S], section headed "Food Processing.")	Mineral supplementation of the diet is not necessary.	Mineral supplementation is not necessary (1) if you eat a well-balanced diet based on the Four Food Groups, and (2) provided you eat *enough* of *proper* foods *regularly.* But these three key words—enough, proper, and regularly—are the catch! Much of the mineral content of grains is lost in milling, and most of the minerals present in raw sugar are removed when sugar is refined. Also, losses occur in cooking foods, especially when cooking in a lot of water then throwing away the cooking water. In certain instances, some nutrients are restored by enrichment after processing. More and more physicians and nutritionists recommend judicious mineral supplementation for buoyant good health.
Molasses, blackstrap (Also see MOLASSES.)	Blackstrap molasses is a rich source of iron. Some food faddists recommend it as a treatment for anemia.	Blackstrap molasses is a fair, but unreliable, source of iron. Moreover, there is less iron in molasses that is processed in stainless steel or aluminum vessels than in molasses processed in old-fashioned iron vats and pipes. Besides, molasses is eaten in such small amounts that it does not make an important contribution to the diet. *Blackstrap molasses is not recommended as a treatment for anemia.*
Nutmeg	Nutmeg is a good brain food.	No food as such builds any specific tissue. Rather, it is the specific nutrients in foods that are required for the building of specific tissue. Brain tissue is built and repaired from amino acids, which are found in many foods.
Olives	Olives increase sexual potency.	Sorry, this is folklore. But any food which contributes to a healthy body will enhance potency.

(Continued)

TABLE F-41 (*Continued*)

About	Myth/Misinformation	Fact
Onions (Also see ONIONS.)	Onions are effective in (1) prevention of colds, and (2) loosening of phlegm.	Europeans have long believed that onions have these medicinal properties. So far, there has been only limited support of this belief, consisting mainly of the finding that onions have a mild antibacterial effect. Copious use of onions may isolate one socially, and thereby ward off contagious diseases.
Oysters	Oyster increase sexual potency.	Sorry, this is folklore without basis in fact. Of course, any food that contributes to good health will enhance potency.
Organically grown foods (organic gardening, organic farming, natural foods, health foods) (Also see ORGANICALLY GROWN FOOD.)	Organically grown foods are superior. *Note well:* The designation "organically grown foods" generally refers to those produced without chemical fertilizers, pesticides, growth regulators, and livestock feed additives.	Before manure or compost can be used by plants, they must be broken down by soil bacteria into inorganic compounds—nitrogen, phosphorus, and potassium, which are the same elements found in chemical fertilizers. It follows that, whether grown on soils fertilized by chemical fertilizers or by manures, the foods are similar in nutritive value. *So, organically grown foods are not superior.* All foods are organic, and all edible foods, when selected for a balanced diet, are conducive to physiologic and psychological health regardless of whether they are organically grown (as defined) or conventionally grown.
Pesticides	Pesticides on food crops will poison you.	When pesticides on food crops leave a residue, Food and Drug Administration (FDA) and the Environmental Protection Agency (EPA) make sure the amount will be safe for consumers. The amount allowed, if any, is set at the lowest level that will accomplish the desired purpose, even though a larger amount might still be safe. Without pesticides to control insects and plant diseases, the food supply in many parts of the world would be so seriously threatened that the effects of food shortages and famine would far outweigh the possible hazards of pesticides.
Raw beef juice	Raw beef juice is a good nerve food.	Nerve tissue, including brain tissue, is built and repaired from amino acids contained in protein which is found in many foods.
Raw foods	Foods should be eaten in their natural state—mostly raw.	It is good to eat some foods raw. But others, such as meats and whole grains, are usually cooked to soften fiber, develop flavor, and promote digestibility.
Red wine	Red wine is a strong blood builder.	This superstition likely arose because red wine, like blood, is red in color.
Royal jelly	Royal jelly aids sexual rejuvenation. *Note well:* Royal jelly is the secretion of the honey bee that is fed to to the very young larvae and to all queen larvae.	That's an old wives' tale. Royal jelly should be left for the bees, for which it is intended.
Salt (Also see SALT.)	Sea salt is superior to salt taken from salt deposits in the earth.	Not really. Both sea salt and salt taken from deposits in the earth are sodium chloride (NaCl). Additionally, sea salt contains iodine, but so does iodized salt. Also, sea salt usually contains some other trace minerals, but so do most foods.
Sassafras	Sassafras tea will thin your blood. *Note well:* This folk medicine still persists in certain areas of the U.S.	Sassafras tea won't thin your blood, whatever that may mean. Moreover, the FDA now prohibits the sale of teas made from sassafras if they still contain its essential oil, safrole, a suspected cause of cancer.
Seawater	Seawater will cleanse the system and and lead to prolonged youth. *The reasoning:* All life once came from the sea, which means that our dietary needs are intimately linked with the ocean and its water.	Although ocean water does contain some minerals that are needed by the human body, these minerals are present in a variety of foods eaten in a normal diet. The money for seawater had best be kept by the consumer, rather than pocketed by the huckster.
Seaweed (kelp) (Also see SEAWEED.)	Seaweed (kelp) is highly prized as a protein and for its minerals and vitamins.	As a food, seaweed is low in protein, and the protein is of very low quality. Seaweed is used as a mineral-vitamin supplement. It is very rich in iodine.
Shellfish and ice cream or milk	A shellfish and ice cream or milk combination is bad for you.	Shellfish and ice cream or milk are excellent foods, alone or together.

(*Continued*)

TABLE F-41 (*Continued*)

About	Myth/Misinformation	Fact
Soil nutrients (Also see MINERAL[S], section headed "Mineral Composition of the Soil.")	Mineral deficiencies in soils may lead to reduced yields of plants, but not to plants containing subnormal levels of minerals.	There is a direct and most important relationship between the fertility of the soil and the composition of the plant. In human nutrition, the soil-plant relationship, confirmed by experiments, is especially noteworthy relative to iodine, phosphorus, and selenium.
Strawberries	Strawberry birthmarks on infants are caused by strawberries eaten by the mother during pregnancy.	That's an old wives' tale. No food can cause birthmarks on the developing fetus.
Sugar, raw (Also see SUGAR.)	Raw sugar is important in the diet because of its mineral content, especially iron.	Raw sugar contains a very small amount of iron, now that it is being processed in stainless steel or aluminum vessels. Besides, it is eaten in too small servings to be a source of iron. So, it should be eaten for reason of taste, rather than for any health properties.
Teflon-coated utensils	Cooking in Teflon-coated utensils is dangerous.	The Food and Drug Administration, which keeps strict control over all cooking ware, reports that there is no danger from normal kitchen use of Teflon-coated utensils, or from the overheating which may occur in the kitchen.
Tin cans	It is dangerous to refrigerate food in opened tin cans.	It is preferable to refrigerate food in covered cans since the cans are already clean. Today, tin cans have a protective lining which makes it safe to store foods in them after opening. Some acid foods may dissolve a little iron from the can, with the result that they taste like "tin", but this is not harmful or dangerous to health. Sauerkraut may turn dark, but it is still safe if properly refrigerated.
Tired, run-down feeling	If you have a tired, run-down feeling, chances are you're suffering from a subclinical deficiency. (The word *subclinical* is a general, nonspecific term given to conditions for which there are no observable symptoms.)	Lack of pep may be caused by overwork, not pacing yourself, emotional stress, disease, lack of sleep, and/or poor nutrition. If such symptoms persist, see your physician.
Tomatoes	Tomatoes clear the brain.	Tomatoes won't clear the brain, not even of "cobwebs" (nor will any other food).
Tomatoes and milk	A tomato and milk combination is bad for you.	Both tomatoes and milk are excellent foods, alone or together.
Tomato juice	Tomato juice is a strong blood builder.	This superstition likely arose because tomato juice, like blood, is red in color.
Vegetable juice, raw (liquefied vegetables)	Raw vegetable juice is better for you then vegetables themselves.	Raw vegetable juice has no magic because it is juice, *except* for someone with loose dentures. All vegetables are nutritious, regardless of the form in which eaten.
Vitamin supplementation (Also see NUTRITIONAL SUPPLEMENTS: and VITAMINS, section headed "Vitamin Supplementation of the Diet.")	Vitamin supplementation of the diet is not necessary.	Vitamin supplementation is not necessary (1) if you eat a well-balanced diet based on the Four Food Groups, and (2) provided you eat *enough* of *proper* foods *regularly*. But these three key words—enough, proper, and regularly—are the catch! Unfortunately, many of the vitamins naturally present in foods are lost by processing, cooking, and storing. For example, the losses in seven vitamins from milling wheat and making flour range from 50 to 86%. The fat-soluble vitamins (A, D, E, and K) are not easily lost by ordinary cooking, and they do not dissolve out in the cooking water. However, the water-soluble vitamins (B complex and C) are dissolved easily in cooking water; and some of these vitamins are destroyed by heating. For example, vegetables lose 57 to 77% of their vitamin B-6 and 56 to 78% of their pantothenic acid from canning. In some instances, some nutrients are restored by enrichment after processing. More and more physicians and nutritionists recommend judicious vitamin supplementation (1) during infancy, pregnancy, and breast feeding; and (2) for buoyant good health.
Vitamin, synthetic	Natural sources of vitamins are better for you than synthetic vitamins.	Vitamins are specific chemical compounds; and the human body can use them equally well whether they are synthesized by a chemist or by nature. A vitamin is a vitamin.

(*Continued*)

TABLE F-41 (*Continued*)

About	Myth/Misinformation	Fact
Water	Water is fattening.	Water may be held in the tissues (edema). But it has no energy value; hence, it cannot be converted to fat.
Wheat germ and wheat germ oil (Also see VITAMIN E.)	Wheat germ and wheat germ oil will increase the sexual potency and fertility of humans, and slow the aging process.	In humans, neither sexual potency nor fertility are increased by wheat germ or wheat germ oil, which are rich sources of vitamin E. The claim is made that wheat germ or wheat germ oil (vitamin E) slows the aging process by functioning as a natural inhibitor of the destruction of cells, and by protecting tissue from breaking down. But this is unproven.
Yogurt (Also see MILK AND MILK PRODUCTS, section headed "Cultured Milk and Other Products. • Yogurt.")	Yogurt is more nutritious than milk.	Yogurt is fermented milk. In the U.S., fermentation is usually accomplished by a one-to-one mixed culture of *Lactobacillus bulgaricus* and *Streptococcus thermophilus;* hence, yogurt is a practical means of introducing large numbers of these particular organisms into the digestive tract, which may have therapeutic value. However, the nutrients in yogurt are comparable to the milk products from which it is manufactured—whole milk, lowfat milk, or skim milk. The culture makes for a finer curd, so yogurt is easier to digest than milk. Also, persons with lactose intolerance can often tolerate yogurt, but not regular milk.

FOOD POISONING

Food poisoning is a term which describes a group of illnesses caused by different agents: bacteria, poisonous chemicals, and foods which are intrinsically poisonous, such as certain poisonous plants, certain varieties of mushrooms, rhubarb leaves, the green part of sprouting potatoes, certain clams and mussels, and the puffer fish from which the gland containing neurotoxin tetraodontoxin has not been removed.

America's food supply is the safest in the world. Nevertheless, cases of food poisoning continue to occur.

Prompt and up-to-date information on treatment of poison cases can be obtained by calling the Poison Control Center of the area. When the phone number is not known, (1) ask the operator for the Poison Control Center, or (2) call the U.S. Public Health Service at either Atlanta, Georgia, or Wenatchee, Washington.

In this book, food poisoning is categorized and discussed primarily under the following three sections, to which the reader is referred: BACTERIA IN FOOD; POISONOUS PLANTS; and POISONS.

FOODS, LOW ENERGY

These foods are generally filling, but not fattening. Thus, they contain small amounts of carbohydrates and fats, and great amounts of water and fiber. Many of the fresh fruits and vegetables fall into this category and are used in diets for weight reduction.

(Also see MODIFIED DIETS; and OBESITY, section headed "Selecting Low-Energy Foods.")

FOOD SCIENCE

The study of the physical and chemical characteristics of food is known as food science. Food scientists investigate the chemical, physical, and biological nature of food and apply the knowledge to processing, preserving, packaging, distributing, and storing an adequate, nutritious, wholesome, and economical food supply. About three-fifths of all scientists in food processing are engaged in research and development. Others work in quality control laboratories or in production or processing areas of food plants. Still others teach or do research in colleges and universities.

Food science is taught in more than 60 universities in the United States. Most of the colleges and universities that provide undergraduate food science programs also offer advanced degrees.

The professional organization for food scientists is the Institute of Food Technology.

(Also see INSTITUTE OF FOOD TECHNOLOGY.)

FOOD STAMP PROGRAM

Contents *Page*

History 1006
Food Stamp Provisions 1007
Criteria for Eligibility 1007
Financial Criteria 1007
Nonfinancial Criteria 1008
Determination of Allotment Values 1008
Items Which May Be Purchased With Food Stamps 1008
Causes for Disqualification from Participation 1008
Benefits 1009
Problems 1010
Future Prospects 1010

This program, which is administered by the U.S. Department of Agriculture, is the nation's primary means for providing food assistance to low-income Americans who meet certain eligibility requirements. Food stamps have been credited with helping to reduce infant deaths and malnutrition in areas of dire poverty. Additionally, food stamps have lessened the surplus food problem. However, there have been certain well publicized abuses that have tarnished its reputation. Furthermore, the program has required ever increasing expenditures by the federal government, as is shown in Fig. F-60. Recently, phasing out, or cutting, of the program has been considered because of the need to reduce federal expenditures and balance the budget. Therefore, the merits and demerits of food stamps are worthy of consideration.

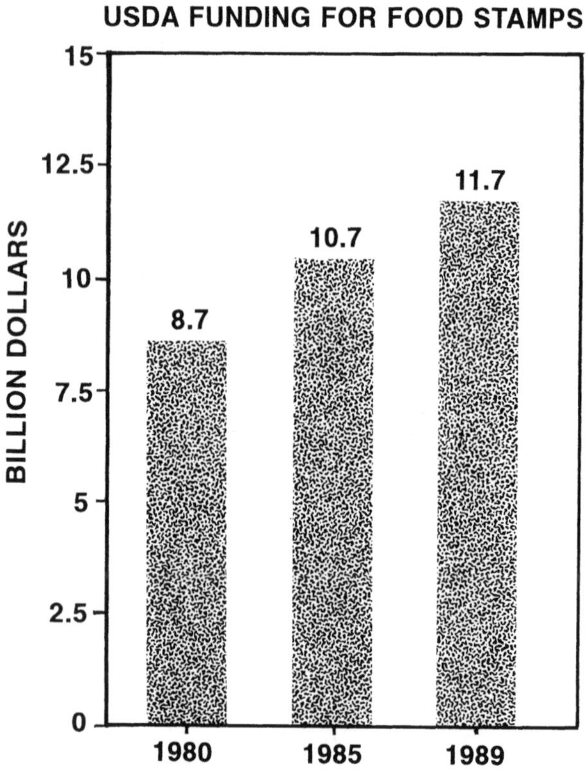

Fig. F-60. The mounting costs of the Food Stamp Program during the 1980s. (Based on data from *Statistical Abstract of the United States 1991*, p. 371, Table 611)

HISTORY. The first federal food stamp program was conducted on an experimental and limited basis between 1939 and 1943 for the purposes of (1) improving the diets of very needy people, and (2) raising the incomes of farmers by creating an increased demand for various foods. (Studies had shown that 14% of American families could afford to spend only about 5¢ per person per meal for food.) Eligibility for the program was limited to families on relief, WPA workers, and other needy persons certified by certain relief agencies. Participants purchased orange stamps in amounts roughly equal to their normal expenditures for food and received a bonus of one blue stamp for every two orange stamps that were purchased. The blue stamps could be used only for items that were declared surplus by the Secretary of Agriculture. Approximately four million people in selected areas of the country were served by the program until its termination during World War II, when employment rose sharply and food surpluses disappeared.

After World War II, farm production climbed to record levels and surpluses once again became a problem. However, the growth of the rest of the economy slowed down and there was a moderate increase in unemployment. Consequently, the Agricultural Act of 1949 made surplus food available to local governments for distribution to those in need. However, the surplus commodities consisted mainly of butter, cornmeal, flour, lard, nonfat dry milk, and rice. By 1960, it was evident that the diets of the impoverished people who subsisted on the donated commodities fell far short of providing good nutrition. Hence, in 1961 the surplus foods were augmented by other items such as canned pork and gravy, dried beans, dried egg, peanut butter, and rolled oats. Shortly thereafter, the number of families in the program grew from less than four million to almost six million. In addition, an experimental food stamp program was inaugurated in eight selected economically depressed areas of the country in which 133,400 people received the stamps. The participants in the program used their customary food expenditures to purchase stamps of higher monetary value, which could be used in local retail stores to purchase all types of foods, except certain ones that were imported. (The difference between the market value of the stamps and the amount paid for them represented the bonus value.)

By 1964, the program had grown to include 392,400 people from 43 areas. It is noteworthy that studies of the pilot projects showed that the recipients of the food stamps purchased increased amounts of meats, poultry, dairy products, and fruits and vegetables. The success of these projects in improving the diets of the poor led to the passage of the Food Stamp Act of 1964, which provided for a continuing, nationwide program. Participation was limited to those households whose income was a substantial limiting factor in the attainment of nutritionally adequate diets. Each state was given the responsibility of establishing standards of eligibility for food stamps, subject to the approval of the Secretary of Agriculture. Commodity distribution and food stamp programs were not to be operated in the same county, except in emergency situations.

The establishment of eligibility criteria by the states resulted in widely varying rules, low participation rates, inequitable allotments of stamps, and prohibitive purchase requirements in some food stamp areas. These problems led to major administrative modifications in late 1969 and legislative amendments in 1971. The amendments stipulated that the cost of a nutritionally adequate diet, and therefore the amount of the food stamp allotment, would be determined by the Secretary of Agriculture and adjusted annually to conform with changes in the price of food. Consequently, food stamp allotments were increased substantially above previous standards, especially among lower-income households. Furthermore, the amount required to purchase the full stamp allotment was lowered.

Other significant changes in the food stamp program were mandated by the Food and Agriculture Act of 1977, which (1) tightened eligibility standards to reduce costs, (2) eliminated the purchase requirement (effective in 1979, the stamps were provided free of charge) to encourage greater participation by the neediest of the recipients, and (3) extended the program's authorization. Subsequently, the Food Stamp Program has been amended and extended. As shown in Fig. F-61, the number of participants peaked at 21.1 million in 1980, then declined slightly during the decade of the 1980s.

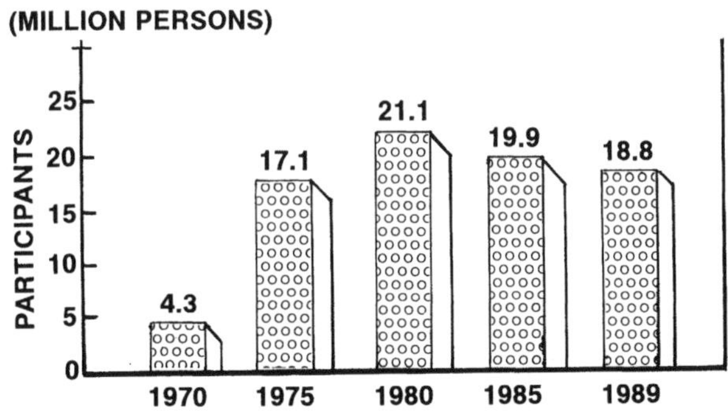

Fig. F-61. Participants in the Food Stamp Program from 1970 to 1989. (Based on data from *Statistical Abstract of the United States 1988*, p. 355, Table 591; *1991*, p. 370, Table 610)

FOOD STAMP PROVISIONS. The value of the stamps varies with income and represents a reasonable amount to spend for food. Stamps may be used as cash at participating and approved grocery stores. The Food Stamp Program is constantly being evaluated and redesigned to provide for more equitable use of funds. It is directed by the state agency responsible for administering other federally aided assistance programs.

It seems likely that current efforts to reduce federal spending and balance the budget will soon result in changes in some of the provisions of this act. Nevertheless, the major features of this law are noteworthy, since future discussions of new provisions may refer to the current ones that were modified.

Criteria for Eligibility. Participation in the Food Stamp Program is limited to households whose incomes and other financial resources, held singly or in joint ownership, are determined to be a substantial limiting factor in obtaining a nutritious diet. Uniform national standards of eligibility are established by the Secretary of Agriculture and must be uniformly applied by state agencies which conduct local programs.

FINANCIAL CRITERIA. The income standards of eligibility are based upon the nonfarm income poverty guidelines prescribed by the Office of Management and Budget and are adjusted annually in accordance with changes in the costs of food. Certain types of income and assessments are counted in determining eligibility, while others are not counted. Rules for assessing the income of applicants follow:

- **Countable sources of income**—The sources of income which are counted as contributing to the household income are wages; net earnings from self-employment; public assistance; retirement or disability benefits; veteran's, worker's, or unemployment compensation; old age, survivors', or strike benefits; support or alimony payments; scholarships, educational grants or loans, fellowships, or veteran's education benefits that are not used to pay tuition or mandatory fees; foster care payments for children or adults; and dividends, interest, and other payments.

- **Other countable assets**—The upper limit on household resources is based upon the assets which may be readily converted to cash. These resources include cash on hand, money in savings accounts, U.S. Savings Bonds, stocks and bonds, buildings and land other than those for the family home and lot, and a portion of the fair market values of nonbusiness vehicles.

- **Allowable deductions from counted income**—Income deductions include (1) 20% of earned income, (2) a standard deduction that is adjusted regularly to reflect changes in the cost of living, (3) shelter costs (includes rent, mortgage payments, and interest, utility payments, property taxes, and home insurance premiums) which take more than 50% of the household's income after all other deductions (the amount of this deduction cannot exceed the current maximum set by law), (4) costs of care for dependent children and disabled adults so that other members of the household can work, seek employment, or take a training course leading to a job.

• **Noncountable income**—Some of the kinds of income which are not counted in the determination of food stamp eligibility are nonmonetary income such as produce from a garden; earnings of a student under 18 who is attending school at least half-time; portions of educational loans, grants, scholarships, etc. which are used for tuition and mandatory fees; one-time lump sum payments from insurance refunds and inheritances; certain types of reimbursements; most types of loans; amounts paid for court-ordered support or alimony; the costs of self-employment, and other income so designated by current laws.

• **Noncountable assets**—the household resources which are not counted are a home and lot; the value of vehicles used mainly to produce income; life insurance policies; real estate that produces income consistent with its fair market value; and machinery and tools used to earn an income.

NONFINANCIAL CRITERIA. These rules define (1) the conditions under which a group of people living together and sharing meals constitutes an eligible household, (2) the citizenship or alien status of members, and (3) the work requirements of the members.

• **Definition of household**—A household is either (1) an individual living alone, or who, while living with others, customarily purchases food and prepares meals to eat at home separate and apart from the others, or else pays the others for his or her meals; or (2) a group of individuals who live together and customarily purchase food and prepare meals together to eat at home, or else live with others and pay them for their meals.

• **Residency**—Members of a household must live in the project area in which the application for participation is filed.

• **Citizenship and alien status**—In order to qualify for food stamps each member of a household must be a resident of the United States *and* a citizen of the United States or an alien meeting the conditions of one of the five categories stipulated in the Food Stamp Act.

• **Students**—People in this category cannot qualify for Food Stamp Program participation if they are properly claimed or could be properly claimed as tax dependents of ineligible households. Students who are not exempt under normal work registration rules must register for work during a school recess of more than 30 days. Nonexempt students must also register for 20 hours of work per week while school is in session unless they are already working 20 hours per week, earning an equivalent salary, participating in a federally financed work study program, or are the heads of households and supporting at least one other person.

• **Work registration requirements**—Persons who are otherwise eligible for food stamps, are physically and mentally fit, and between the ages of 18 and 60 must (1) register for work, (2) look for a job, (3) remain in a suitable job, and (4) accept a job paying either the state or federal minimum wage, whichever is higher, unless there is a good cause for refusing.

It is noteworthy that an experimental workfare program has been instituted in which unemployed food stamp recipients work off the value of the stamps by rendering services to certain cities, counties, and other governmental units.

Determination of Allotment Values. The values of the monthly allotments of food stamps issued to program participants are based upon amounts by which the current cost of the USDA Thrifty Food Plan exceeds 30% of their net income (after certain allowable expenses have been deducted). The data on food costs are revised in January of each year so that the allotments are brought in line with the changing costs of food.

Items Which May be Purchased with Food Stamps. The stamps may be used in authorized retail stores to purchase almost any type of food, *except* alcoholic beverages, foods that are eaten in the store, items that are ready to eat (such as barbequed chicken), and pet foods. However, eligible senior citizens may use them to purchase meals served in communal dining facilities (usually, those in senior citizen centers and certain restaurants), or to pay for authorized meal delivery services such as "Meals on Wheels." Participants may also buy seeds or plants to grow food that is used by the household.

It is noteworthy that special provisions of the law allow for a number of special uses of stamps by alcoholics and drug addicts who are regularly participating in rehabilitation programs, and that people living in remote areas of Alaska may use the coupons to buy hunting and fishing equipment that they can use to obtain food.

Causes for Disqualification from Participation. The failure of food stamp participants to meet the criteria for eligibility and/or to comply with other provisions of the Food Stamp Act may result in their disqualification for certain stipulated periods of time. For example, households who transfer resources knowingly for the purpose of qualifying for benefits are disqualified for up to 1 year. Other common reasons for disqualification are failure to comply with the work registration requirements, fraudulent use of authorization cards or food stamps, and failure to appear at hearings held by the local welfare agency that issues the stamps.

BENEFITS. The value of the food stamps has nearly tripled since 1975, as is shown in Fig. F-62. This increase in food assistance by the government has

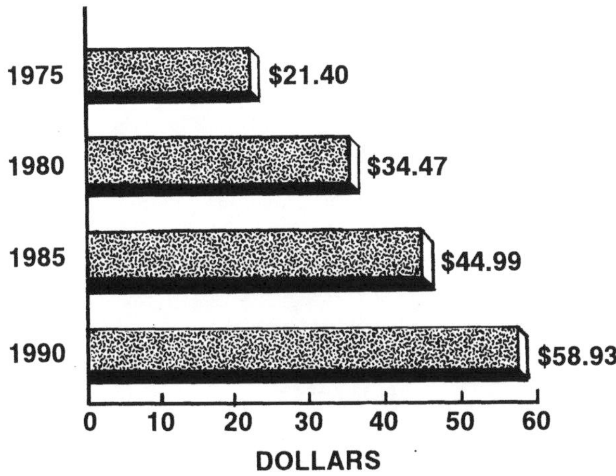

Fig. F-62. Average benefit per person has increased more rapidly than inflation. (*Agricultural Statistics*, 1991, USDA, p. 480)

prompted many people to ask whether there have been any significant benefits from this program. Although there is little hard data on precise cost-to-benefit ratios, some of the more pertinent recent observations follow:

• **Reduction in malnutrition**—In 1979, doctors who had conducted a Field Foundation survey of the urban and rural poor observed that although life in poverty was about as dismal as ever, there were far fewer grossly malnourished people than there were in 1969, thanks to the Food Stamp Program and other federal food assistance programs.[21]

• **It provides security**—Fig. F-63 shows that the Food Stamp Program provides security, as evidenced by the fact that when unemployment is high the Food Stamp participation is high.

[21]Vaden, A. G., "Child Nutrition Programs: Past, Present, Future," *The Professional Nutritionist*, Winter 1981, p. 10.

UNEMPLOYMENT RATE AND PARTICIPATION IN THE FOOD STAMP PROGRAM

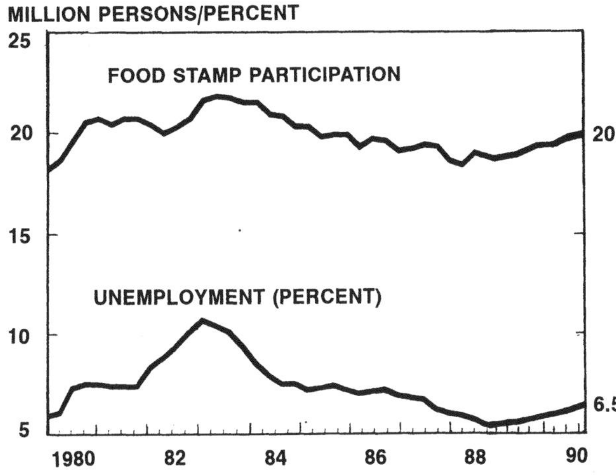

Fig. F-63. Unemployment rate and participation in the Food Stamp Program. (Courtesy, USDA)

• **Participants consume more animal products and vegetables**—Generally, the proportions of food stamp households with diets that meet the Recommended Dietary Allowances are the same or greater than propor-

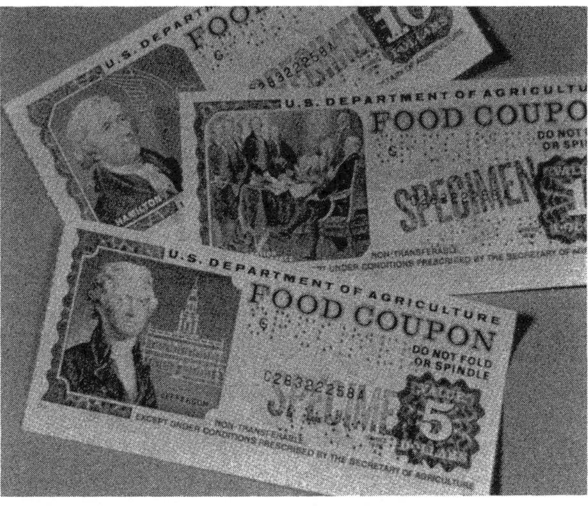

Fig. F-64. Food stamps. (Courtesy, USDA)

tions of households eligible for but not receiving food stamps. It appears that the recipients of food stamps used them to purchase greater quantities of meat, fish, poultry, fruits, and vegetables than people of similar economic status who do not receive the stamps.

• **Lifting people out of poverty**—The extra spending power represented by food stamps raises the incomes of about 30% of the participants who are below the "poverty line" to values which are above this level.[22]

• **Benefits to the overall economy**—A study concluded that the food manufacturing sector business receipts were higher by $589 million in 1972 and by $809 million in 1974 under the Food Stamp Program than they would have been without the program.[23] The study also estimated that between 56,000 and 77,000 new jobs were generated as a result of stimulation of the economy by the Food Stamp Program.

PROBLEMS. Some of the major longstanding problems of the Food Stamp Program follow:

• **Discouragement of needy applicants by bureaucratic procedures**—Some of the most needy people are poorly educated and find it very difficult to comprehend the procedures that are required for certification of eligibility for the program.

• **Loss of dignity of elderly participants**—Many of the elderly just cannot bring themselves to present the coupons in a local store where they may encounter the scorn of their neighbors. Therefore, they go without needed food even though their plight is little due to lack of prudence on their part, but rather to the erosion of the value of their social security and savings that has resulted from the long-term inflationary trend.

• **High costs of administering the program**—The procedures for establishing the eligibility of the applicants are complicated and time-consuming. Hence, they are expensive to execute. Furthermore, some of the regulations cause havoc at local levels, where the circumstances may be quite different from those envisioned by the writers of the regulations.

[22]*The Food Stamp Program: Income or Food Supplementation*, Congressional Budget Office, 1977, p. 31.

[23]Nelson, P. E., and J. Perrin, *Economic Effects of the U.S. Food Stamp Program Calendar Year 1972 and Fiscal Year 1974,* Agric. Econ. Report No. 331, USDA, 1976.

• **Theft and fraudulent uses of food stamps**—News stories and documentary programs have reported that there is considerable theft and fraudulent use of food stamps.

Sometimes, food stamps have been accepted by unscrupulous merchants from customers offering them as payment for unauthorized items such as alcoholic beverages and tobacco. There also appears to be criminal networks in some cities which specialize in the circulation of food stamps from liquor stores and other illegal traders in the coupons.

FUTURE PROSPECTS. It seems likely that any new legislation extending the Food Stamp Program will be required to trim the program costs, along with reducing the complexity and vulnerability to abuses.

FOOD TECHNOLOGY

Technology, in whatever field, is the application of science and the result of scientific research to the solution of practical problems. *Food technology is the application of science and engineering to the production, processing, packaging, distribution, preparation, and utilization of foods.* As defined by the Institute of Food Technologists, food technology is primarily based on the fundamentals of chemistry, physics, biology, and microbiology.

Man's technological progress started when he fashioned and used his tools for agriculture, weapons, and the crafts. Far into the age of the Industrial Revolution, all technological progress depended entirely on the exploitation of chance discovery and invention by enterprising men. In this manner, man developed the technology of food preservation by alcoholic and lactic acid fermentation; baked and pasta products from cereals; chilling and drying; confectionery and chocolate making; the conversion of milk to curd, butter, and cheese; cured meat and fish; drying; fat embedding; high salt and high sugar content; ice desserts; leaf fermentation of tea; milling of grain; oil processing; parboiling of cereals; sausage making; smoking; use of preservatives; and packaging. Most of these techniques can be traced back to the earliest days of recorded history. The 19th century contributed canning, margarine making, and other technology.

FOOL

A mashed fruit and cream with or without custard.

FORBIDDEN FRUIT

• The name given to the grapefruit (*Citrus paradisi*) when it was discovered growing on Barbados in 1750. It is believed that the grapefruit arose as either a mutation or a hybrid of the pummelo (*C. grandis*), which had been brought to Barbados in the previous century.

• The fruit of the Ceylonese tree (*Tabernaemontana dichotoma*) that has powerful narcotic seeds.

• An American liqueur of orange color made by grape brandy flavored with shaddock (a variety of citrus fruit).

• The fruit from the Tree of Knowledge in the Garden of Eden which Eve persuaded Adam to eat about 6,000 years ago.

FORMIMINOGLUTAMIC ACID (FIGLU)

An intermediary product of histidine metabolism. Since folic acid is necessary for its breakdown, the measurement of the urinary excretion of FIGLU may be used to determine folic acid status.

FORMULA

• Often a premixed powder or solution for feeding infants, or in some cases special diets for adults, is called a formula.

• A recipe or a prescription may be referred to as a formula.

• In chemical terms, a formula denotes the composition of a compound with symbols and numbers. The formula for table salt is NaC1—one sodium and one chloride atom—while water is H_2O—two hydrogen atoms and one oxygen atom.

• There are formulas which are mathematical expressions used to estimate some characteristic; for example, a formula for body surface area or a formula for a ponderal index, an index of slenderness.

FORTIFICATION

In the food industry, it is the addition of one or more nutrients—vitamins, minerals, amino acids, and/or protein concentrates—to a food thus raising its nutritive value. Enriched bread, and milk fortified with vitamin D, are probably the best well-known examples. Originally, the FDA differentiated between enrichment and fortification, but now the two terms are used interchangeably.

(Also see ENRICHMENT; FLOURS; RICE; and WHEAT.)

FRANGIPANE

• A dessert of almond cream flavored with frangipani (a perfume named after a French nobleman, which is derived from, or imitates, red jasmine flowers).

• A custard cream flavored with almonds and used as a tart filling.

FRANKFURTER (FRANK; WIENER; HOT DOG)

A sausage (made of beef, or of beef and pork, or of a mixture of meats and poultry) that is stuffed in casings or skinless, cured, and cooked. Although frankfurters (franks) and wieners (hot dogs) have merged their identity, it is noteworthy that, originally, wieners were a combination of veal and pork. Today, frankfurters are made of many kinds of meats and seasonings.

FRAPPE

• An iced and flavored semiliquid mixture served in glass.

• An afterdinner drink of liqueur served in a cocktail glass over shaved ice.

• A thick milk shake.

FREE FATTY ACIDS (FFA; UNESTERIFIED FATTY ACIDS, UFA; NONESTERIFIED FATTY ACIDS, NEFA)

A readily available energy source circulating in the blood, and resulting from the enzymatic liberation of fatty acids from triglycerides. Free fatty acids are transported loosely bound to the blood protein, albumin, in a concentration of about 10 mg/100 ml. However, fourfold increases are not uncommon. Liberation of free fatty acids is stimulated by the hormones epinephrine, glucagon, growth hormone, norepinephrine, and glucocorticoids. Insulin inhibits the release of free fatty acids.

(Also see FATS AND OTHER LIPIDS; FATTY ACIDS; ENDOCRINE GLANDS; METABOLISM; and TRIGLYCERIDES.)

FRENCH TOAST

A popular breakfast or supper dish made by dipping bread into beaten eggs and sauteing in a little hot fat. Any kind of bread can be used. French toast can be served with powdered sugar or a syrup, and bacon or sausage.

FROGS' LEGS

Frogs are a web-footed amphibian of which there are about 20 species. In the United States, there is a very large species called the *bull frog*. Only the legs are eaten.

FRUCTOSE

A hexose monosaccharide found abundantly in ripe fruits and honey. It is obtained, along with glucose, from sucrose hydrolysis, and is commonly known as fruit sugar.

(Also see CARBOHYDRATE[S]; and SUGAR.)

FRUIT(S)

Contents | *Page*

History 1013
Production 1015
Propagation and Growing 1016
Planting 1017
Pruning 1017
Cultivation, Fertilization, Irrigation, and Weed Control 1017
Pollination 1017
Thinning 1018
Harvesting 1018
Marketing of Fresh Fruits 1018
U.S. Department of Agriculture Grade Standards . 1019
Storage 1020
Processing 1020
Canning 1021
Drying 1022
Freezing 1024
Jams and Jellies 1025
Juices 1025
Federal Standards for Processed Fruit Products 1026
Recent Developments in Fruit Processing Technology 1026
Nutritive Values 1027
Effects of Processing and Preparation 1028
Fruits of the World 1029

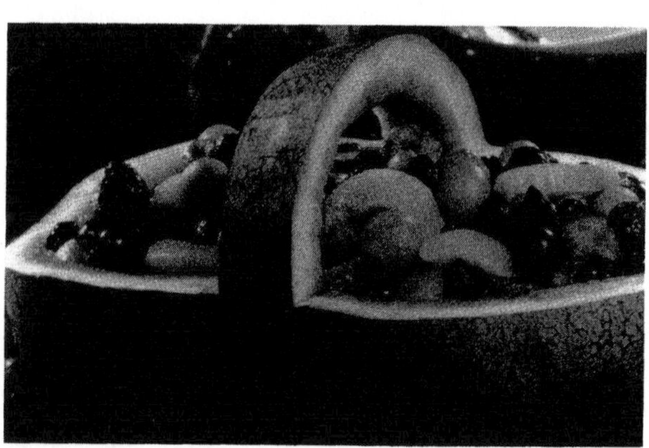

Fig. F-65. Fresh fruits. (Courtesy, Agriculture Canada, Ottawa, Ontario)

Botanically, a fruit is a ripened ovary of a female flower. This scientific definition covers both the succulent, fleshy items that lay persons regard as fruits, and the nuts, which are usually encased in hard shells. However, this article deals only with the soft, juicy types of fruits which most people serve raw at (1) breakfast, as accompaniments to the energy and protein foods, and (2) other meals, as appetizers and/or as dessert items. (Vegetables, in contrast to fruits, are usually cooked, and served with the main courses of the major meals other than breakfast. Hence, most people utilize squashes and tomatoes as vegetables rather than as fruits.)

There are various types of fruits, because the ovaries and seeds of the different flowers develop in different ways. Fig. F-66 shows the main types of fruits.

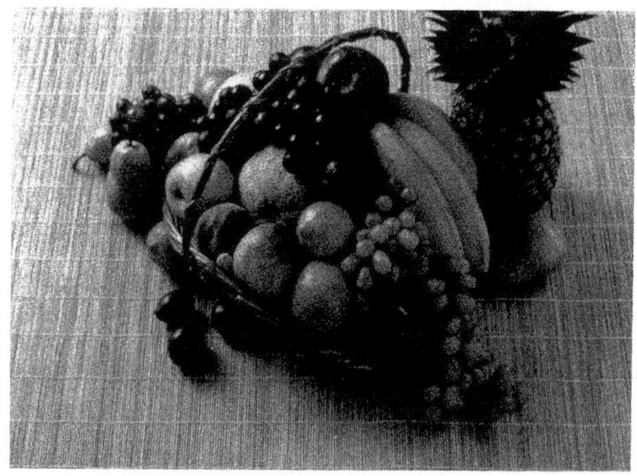

Fig. F-66. Major types of fruits. (Courtesy, USDA)

All fruits shown in Fig. F-66 develop from single or multiple ovaries, which are the bulbous bases of the female parts (pistils) of flowers. Sometimes the base of the whole flower (receptacle) also becomes a major part of the fruit. Descriptions of each type of fruit follow:

• **Aggregate fruit**—This type consists of many tiny seed-bearing fruits combined in a single mass, which develops from many ovaries of a single flower. In the case of the strawberry, the tiny fruits are embedded on an enlarged, fleshy receptacle.

• **Berry**—These fruits are each derived from a single ovary. They may contain one or more seeds. The banana is a berry which has lost its ability to develop seeds because growers have long propagated it vegetatively with the aim of getting rid of the seeds.

• **Drupe**—In this case the single-seeded stone fruit develops entirely from a single ovary.

• **False berry**—These many-seeded fruits result from the fusion of an ovary and a receptacle.

• **Hesperidium**—The citrus fruits are the most common examples of this type of fruit, which develops from a compound ovary into a many-seeded, multisectioned fruit enclosed in a tough, oily skin.

• **Multiple fruit**—The ovaries and receptacles from multiple flowers on a common base develop into these fruits. It is noteworthy that the pineapple belongs to this category, but that it is sometimes considered to be a false fruit because it contains all flower parts, rather than only ovaries and receptacles.

• **Pome**—This many-seeded type of fruit results from the fusion of an ovary and receptacle. The latter develops into the fleshy part of the fruit which is eaten, while the seed-containing ovary or core is usually discarded.

Many fruit-bearing plants have been so altered by domestication over the centuries that it is now difficult to identify their wild ancestors. Also, some can no longer grow without progagation by growers because they have lost the ability to propagate themselves. Hence, the dependent plants must be maintained continually, lest they be lost forever. Hopefully, the need for such maintenance will be duly recognized so that the long association between fruit crops and the human race might continue unabated. Some noteworthy details regarding these crops are given in the sections that follow, and in separate articles bearing the names of the individual fruits.

HISTORY. The use of fruits as food dates to the beginning of human existence. According to the Bible, Adam yielded to temptation in the Garden of Eden and was persuaded by Eve to eat of the forbidden fruit from the Tree of Knowledge about 6,000 years ago. From that date forward, the first couple and all their decendants knew but one creed: "In the sweat of thy brow shalt thou eat bread,..."[24]

[24]Genesis 3:19.

Early people apparently traveled in small bands and utilized freely the animals, birds, fishes, and fruits of the earth for many millennia. They did not sow; they only reaped the bounteous harvest that nature provided. The males usually hunted and fished to obtain foods rich in calories and protein, while the females gathered wild fruits and vegetables that supplied carbohydrates, fiber, minerals, vitamins, and water. It seems likely that the sweet, tart flavors of most fruits made them a welcome addition to the human diet. The more that fruits were eaten, the more that they grew, because the seeds discarded by man or excreted by animals and birds were often the beginnings of new wild stands which sprang up wherever the fruit eaters went. Also, man's use of fire to drive animals out of woodland cover led to the opening of new areas for the spread of the sun-loving, rapidly-growing berry plants. It is not known why some peoples stopped roaming and started to remain in certain areas about 20,000 years ago, and why after many generations of such residence, their descendants took up farming about 10,000 years ago.

The first farmers kept domesticated animals and grew grains, legumes, and/or starchy roots and tubers. In the beginning of this mode of existence, they continued to gather wild fruits, but soon they learned how to propagate their favorite items where they would be close at hand. Dates, figs, and grapes were cultivated by about 4000 B.C. in the Near East, and were made into wines. However, some of each crop was dried and kept for later use. Fig. F-67 shows where many of the important fruits were domesticated.

FROM WHENCE OUR FRUITS CAME

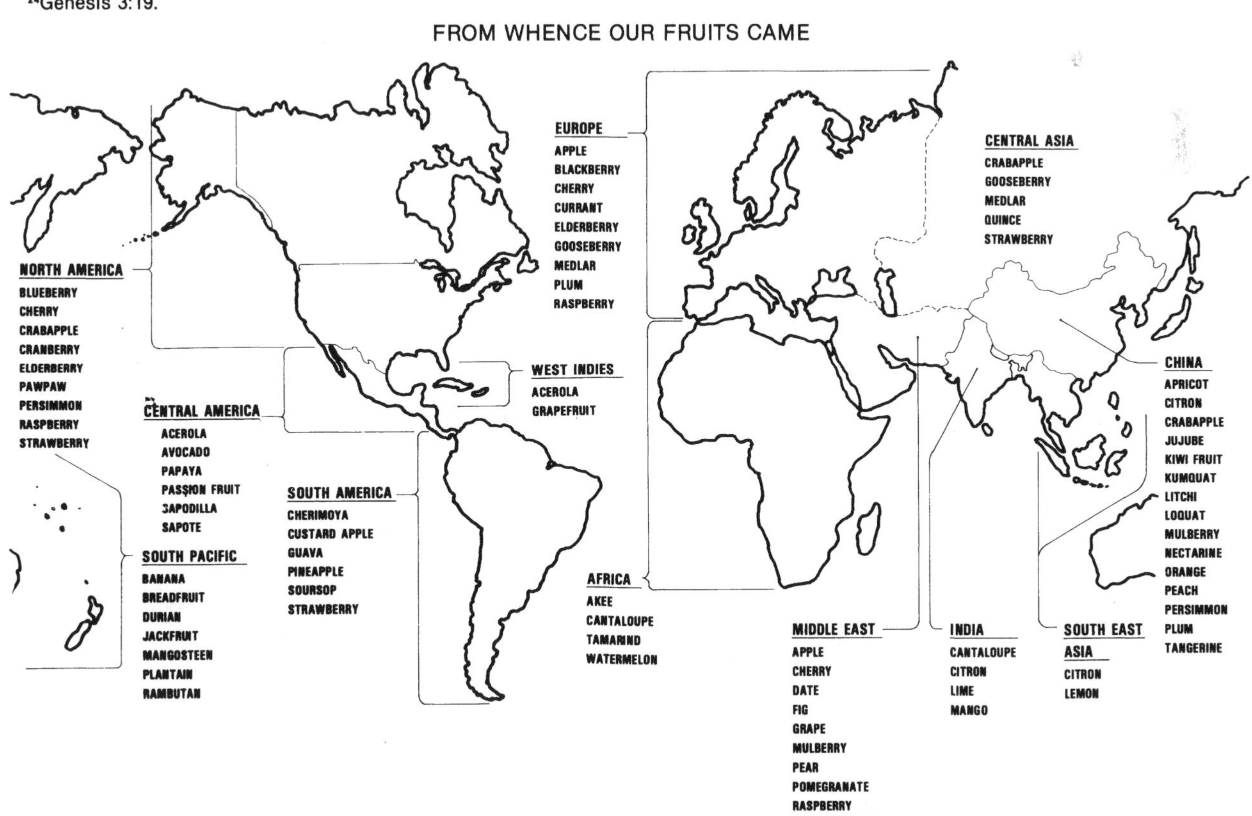

Fig. F-67. Places of origin and/or domestication of fruits.

Although the techniques used by the first fruit growers lacked many of the refinements of modern commercial horticulture, it is amazing how successful methods of propagation were developed many centuries before the basic principles were understood. For example, an ancient Babylonian monument depicts the placing of male date flowers in female trees to obtain a greater production of fruit. (The concept of the sex of flowers was not understood until the 17th century, A.D.) Similarly, the early grape growers propagated their vineyards by grafting. In some cases, the domesticated fruits lost the ability to grow without human assistance. For example, no seeds are formed by the banana, navel orange, and certain other fruits that are propagated by vegetative tissues from a parent plant.

Once fruit crops had become established in certain areas, they became candidates for dispersal to new places, by either intentional or accidental means. For example, the peach was cultivated in China by 2000 B.C., after which it was carried to ancient Persia, where it became an important crop. Much later (330 B.C.), Alexander the Great conquered the Persians and brought the peach to Europe. Other fruits traveled in both directions between the Mediterranean and China, by way of India. In time, each major agricultural area of the Old World grew certain fruit crops which were domesticated elsewhere. Also, some fruits were domesticated independently in two or more geographical regions since the wild fruits often grew over wide areas.

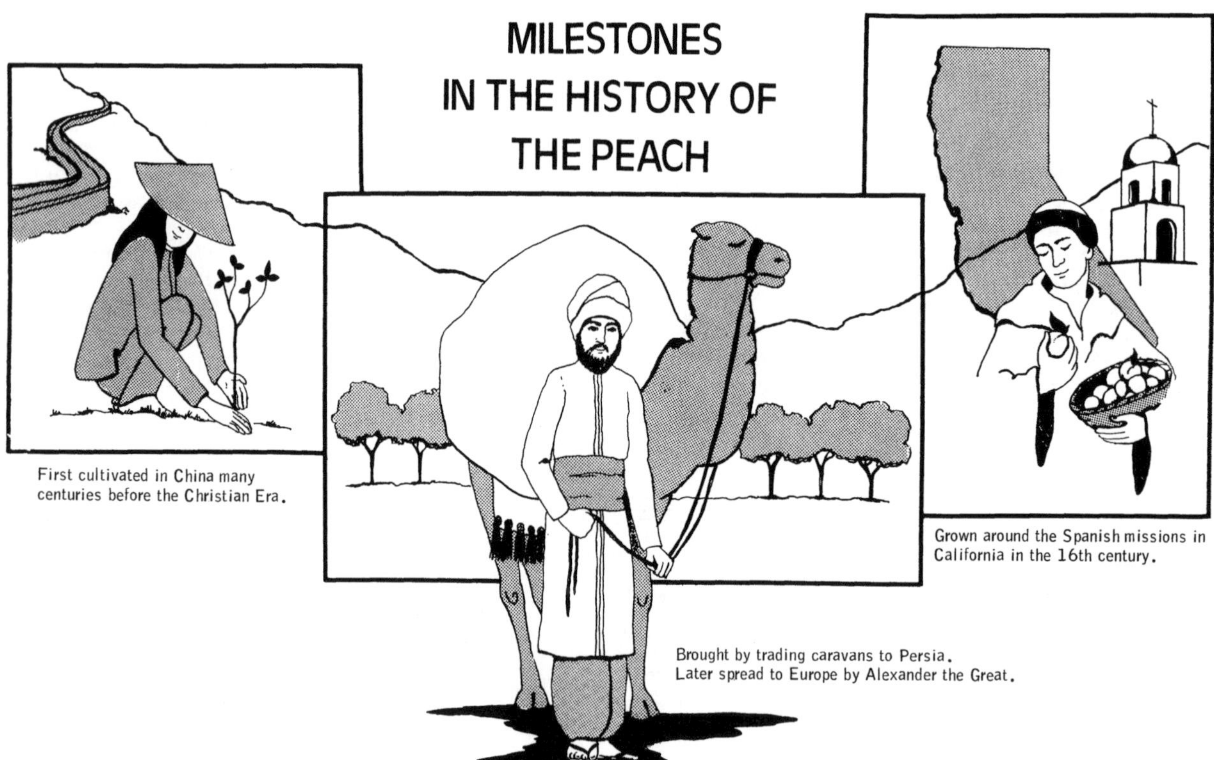

MILESTONES
IN THE HISTORY OF
THE PEACH
First cultivated in China many centuries before the Christian Era.
Brought by trading caravans to Persia.
Later spread to Europe by Alexander the Great.
Grown around the Spanish missions in California in the 16th century.

Fig. F-68. Many fruits are world travelers like the peach.

It is noteworthy, too, that many fruits prospered more in their new locations than in their native habitats. Some of the reasons for this development are: (1) the crop escapes some of its longtime competitors, diseases, and pests when it is moved to a new environment; (2) hidden genetic traits which were suppressed in the original habitat may express themselves in places where the suppressing factors are diminished or absent; and (3) the transplanted crop has new opportunities for hybridization with related species.

The many fruits which were known in Europe in the early Christian era were underutilized for centuries thereafter. Some historians believe that the limited consumption of fruits was due in part to (1) the Greek physician Galen (130 to 200 A.D.) who was a founder of medicine, because he taught that the longevity of his father was due to abstinence from fruits, and (2) a belief by many people that fresh, ripe fruits caused diarrhea. Perhaps, the latter belief arose because the ripening of fruits coincided with the warm season, when diarrhea might have resulted from food contamination and spoilage by rapidly-growing microorganisms. No doubt, lack of availability and high prices also contributed to the underutilization of fruit, and to the many cases of scurvy (a disease resulting from vitamin C deficiency) which occurred in northern European cities each year between late winter and early spring.

Scurvy also afflicted sailors on transoceanic voyages until the ships were stocked with fresh fruits and/or fresh vegetables. The use of limes by British ships led to their sailors being called "limeys." It is ironic that Columbus brought citrus fruits to the Caribbean in 1493, yet over 100 years later sailors on Spanish ships sailing to California were severely afflicted by scurvy. However, the Spanish settlers learned to cure their disease from Indians who used the fruits of cacti, such as the prickly pear.

The voyages of exploration and discovery by the Spanish, Portuguese, British, French, and Dutch in the 15th, 16th, 17th, and 18th centuries were the major means by which new fruits and vegetables were exchanged between the Old World and the New World. Towards the end of that period, botanical gardens were established which came to play major roles in the distribution of fruit crops around the world. One of the most renowned was the Royal Botanical Gardens at Kew, in England. One of the best known voyages was that of Captain Bligh of H.M.S. Bounty. However, Captain Bligh's first voyage, which left England in 1787 to bring breadfruit from Tahiti to the West Indies, was a failure; the crew mutinied after leaving Tahiti and cast the captain and 18 men adrift in a small boat. Nevertheless, Bligh eventually got back to England, from which he again sailed in 1791 on a much more successful 2-year voyage around the world. He visited the Cape of Good Hope, Tasmania, Tahiti, St. Helena, St. Vincent, and Jamaica in order to distribute for transplantation such fruits as apples, apricots, breadfruit, citron, figs, guava, mangoes, nectarines, pineapples, plums, pomegranates, quinces, and strawberries. Other useful plants were also distributed during this voyage, and some species were picked up for cultivation in the Kew gardens.

Efforts to cultivate potentially profitable fruit crops continues at the present time. Recently, the Chinese gooseberry (dubbed the "kiwi fruit") was introduced in the Central Valley of California after having been highly successful in Australia and New Zealand.

PRODUCTION. Fruit production has increased steadily throughout the 20th century because (1) the consumption of all types of fruit products has risen sharply in all parts of the world except eastern Asia and Africa;[25] and (2) the rapid growth of fruit processing now makes it possible for many people to have a wide variety of items the year around, instead of just a few months after harvesting.

However, the annual per capita consumption of fruit in the United States has decreased from 144 lb (*65 kg*) in 1910[26] to 92.3 lb (*42 kg*) in 1990 (Fig. F-71).

Data for fruit production in the world and in the United States are shown in Figs. F-69 and F-70.

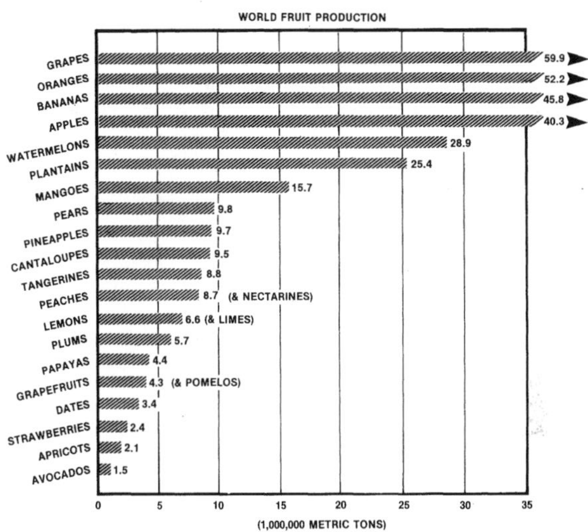

Fig. F-69. World production of the leading fruit crops. (Based on data from the *FAO Production Yearbook*, 1990, FAO/UN, Rome, Italy, Vol. 44)

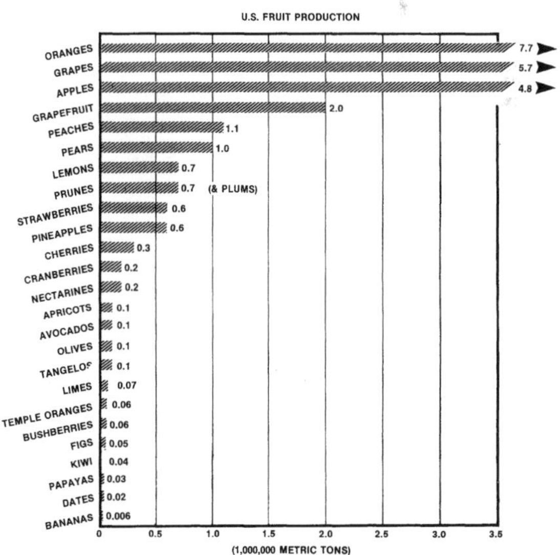

Fig. F-70. U.S. production of the leading fruit crops. (Based on data from *Agricultural Statistics*, 1991, USDA, p. 177, Table 254)

[25]Pekkarinen, M., "World Food Consumption Patterns," *Man, Food, and Nutrition*, edited by M. Recheigl, Jr., CRC Press, Inc., 1973, p. 25.

[26]Brewster, L., and M. F. Jacobson, *The Changing American Diet*, Center for Science in the Public Interest, Washington, D.C., 1978, p. 30, Fig. 13.

Fig. F-69 shows that grapes, oranges, bananas, and apples are by far the leading fruit crops of the world, followed by melons, plantains, and mangoes. It is noteworthy that most of these crops are produced in subtropical and tropical areas.

Fig. F-70 shows that citrus fruits, grapes, and apples are leading fruit crops in the United States. The citrus fruits require a subtropical climate. Hence, their production is limited mainly to California, Texas, and Florida. The great growth of the citrus industry after World War II was due in large part to advances in the technology for producing frozen juices from these fruits. (Fig. F-71 shows the trends in U.S. fruit consumption which have occurred during the past 3 decades.

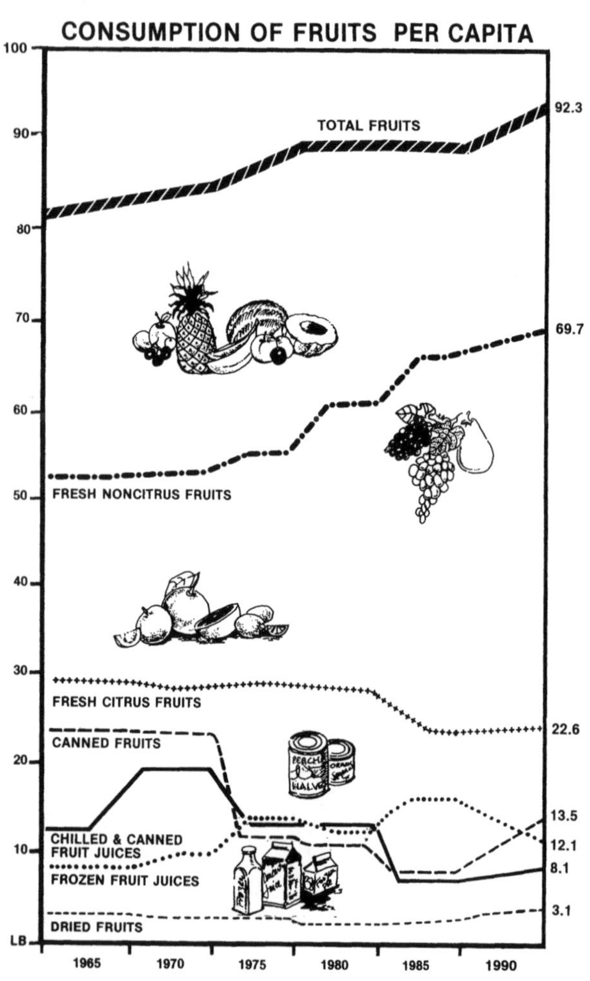

Fig. F-71. Recent trends in U.S. fruit consumption. (Based on data from *Agricultural Statistics*, 1978, USDA, pp. 253; 1988, p. 230; and 1991, p. 218.

Propagation And Growing. Fruit crops have been cultivated for about 6,000 years, during which time much has been learned about the factors which increase the fruit yields of domesticated plants by large amounts over those produced by uncultivated wild species. This body of knowledge constitutes the science of fruit production, which is called pomology. Therefore, the major pomological techniques are noteworthy.

The fruit-bearing parts of bushes, trees, and vines are rarely grown from seeds because (1) the new plant may differ in unpredictable ways from the parent plants, (2) a berry farm, an orchard, or a vineyard represents a considerable investment which does not yield profits unless genetic characteristics are controlled carefully, and (3) certain species of fruits do not produce viable seeds. Therefore, fruit growers almost always propagate fruit bushes, trees, and vines from certain vegetative tissues of the parent plants. *The use of plant parts other than seeds is called cloning*, because the new plants which are produced have the same genetic makeup as the parent plant, unless one or more mutations occur in the clones. (It is noteworthy that the navel orange, which first occurred as a mutation or sport on a normal tree, has been cloned ever since it was discovered.)

The main vegetative means by which fruit crops are propagated follow:

• **Budding**—This procedure consists of attaching a bud (which will eventually become a fruit-bearing branch) to an already rooted tree called the stock. Budding is used to propagate citrus species, stone fruits, and vines.

• **Cane tip rooting or layering**—In this method of propagation, new plants are developed by rooting stems while they are still attached to the parent plant. Blackberries and black raspberries will naturally form layers, whereas rootstocks for apple, cherry, and quince trees may be layered by mounding sawdust or soil around the lowest branches on the rootstocks.

• **Cuttings**—The fruit growers of ancient times most likely used this procedure, which consists of rooting detached pieces (hence, the term "cuttings") of roots or stems. Cuttings are used to grow blueberries, currants, figs, gooseberries, grapes, pomegranates, quinces, and other fruit-bearing species.

• **Grafting cuttings onto rootstocks**—This procedure utilizes a twig cut from a tree with the desired characteristics as the tissue which will ultimately bear fruit. The cutting (scion) is attached to the rootstock of the same species, or to a species that is closely related so that the tissues of both pieces grow together as a single plant. Rootstocks may be grown from seeds or rooted cuttings. Grafting is commonly used in the production of apples, citrus fruits, grapes, mangoes, peaches, pears, and similar fruits.

• **Offshoots or suckers**—Bananas, dates, and pineapples are propagated by detaching and planting the shoots that grow from the crowns (the parts of the stems just below the surface of the ground) of the parent plants.

• **Runners or stolons**—Some fruit-bearing plants, such as strawberries, have stems which spread across the surface of the soil and root themselves at their tips, when conditions are favorable. These stems are called runners or stolons. A new plant is formed wherever rooting occurs.

The propagation of fruit-bearing plants is just the beginning of the series of operations which must be performed by growers to obtain high yields of fruits. Therefore, the major details of these procedures are noteworthy.

PLANTING. Suitable land for growing fruit crops is becoming more expensive and harder to get with each passing day. Hence, growers space the young plants so as to maximize yields, while retaining sufficient open space to allow access for cultivation, fertilization, spraying, etc. There is a current trend towards the use of dwarf varieties of trees, which are easier to care for and to harvest than the older, full size trees. Furthermore, the trees in apple, cherry, pear, and plum orchards must be arranged so that cross pollination by bees and other insects can occur between related varieties of each species.

Fig. F-72. Irrigation of strawberries on a farm in Oxnard, California. (Courtesy, Union Pacific Railroad Company, Omaha, Nebr.)

PRUNING. The cutting back of fruit plants at the end of each bearing season yields benefits such as (1) the stimulation of more vigorous growth during the following season, (2) increasing the proportion of flowering and fruiting growth to vegetative growth, (3) ensuring the penetration of light to all leaves, and (4) allowing easy access for spraying and harvesting. Large, commercial growers often use special pruning machines in their orchards to cut large branches, and then employ laborers for the necessary handwork.

CULTIVATION, FERTILIZATION, IRRIGATION, AND WEED CONTROL. Fruit-bearing plants have high requirements for nutrients and water. These needs are met in different ways under different circumstances. For example, keeping the soil between the fruit plants free of other vegetation (clean cultivation) guarantees the maximum utilization of soil nutrients for fruit productivity, but this practice may promote soil erosion on hilly terrain. In the latter case, erosion is prevented by maintaining a sod in the orchard. However, the fruit trees may then require extra fertilizer and water to compensate for the amounts utilized by the sod.

It is often necessary to use irrigation on fruit crops, because rainfall may be inadequate when the water needs are greatest. Even in the humid tropics where rainfall exceeds 60 in. (*1,500 mm*) per year, large acreages of bananas are irrigated to maximize production. Furthermore, the usually high agricultural productivity of California, which supplies about half of all fruit produced in the United States, depends mainly upon irrigation, because only minimal amounts of rain fall during the peak growing season. Fig. F-72 shows a typical irrigation system.

POLLINATION. Although fruits such as the banana, certain citrus species, and the pineapple may develop satisfactorily from unpollinated flowers, most fruit-bearing plants require pollination to set fruit. For example, only one pollen grain from a male flower is needed to produce a peach, whereas about ten are needed for each apple, and over 100 for a melon. Each viable seed in a fruit usually develops as a result of the union of a sperm nucleus in a pollen grain with an ovule within the ovary of a female flower. Insects and the wind are usually the principal means by which pollination is accomplished in fruit crops. Fig. F-73 shows the pollination of watermelon flowers.

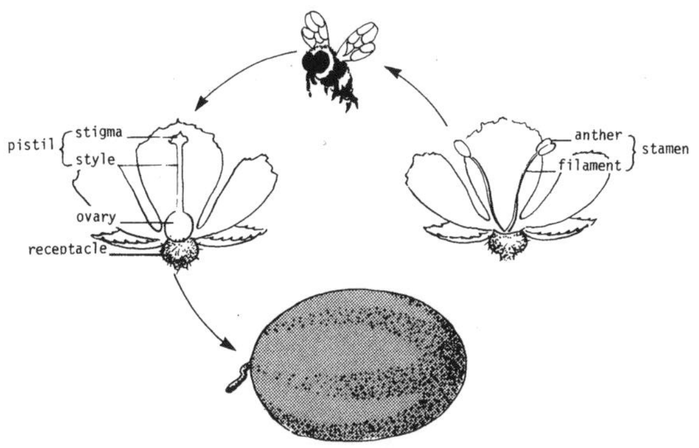

Fig. F-73. From flower to fruit. The production of a watermelon requires the transfer of pollen from the male flower to the female flower by a bee.

Bees and other insects are attracted to flowers by color, nectar, or odor. Fruit growers often arrange to have beekeepers place hives in their orchards for the purpose of pollination.

THINNING. The setting of a heavy load of fruit may tax a plant to the extent that it fails to bear during the following season. Also, the production of the maximum number of fruits is undesirable if each fruit is small and of poor quality. Therefore, excess flowers and/or fruit are removed (thinned) while fruiting is taking place. Then, the fruits that remain will put less strain on the plant, and they will be more likely to grow to an optimal size.

HARVESTING. The mechanization of fruit harvesting has been spurred by the ever growing cost and scarcity of farm labor. On the other hand, care must be taken to avoid injuries to the fruit which lower its market value. Hence, many commercial growers use a combination of machines and laborers to harvest their crops. A typical harvesting arrangement is shown in Fig. F-74.

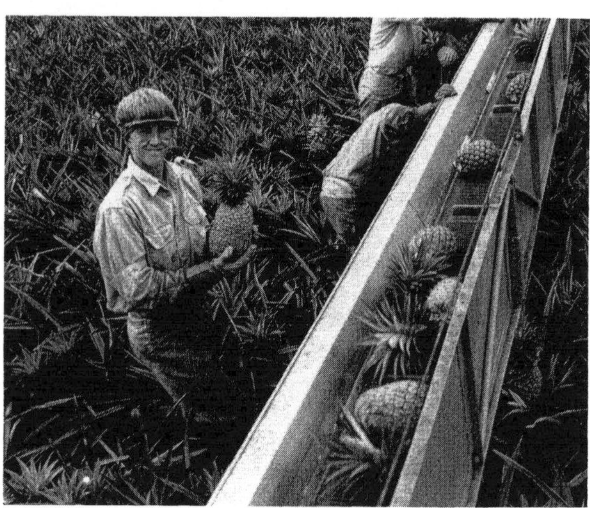

Fig. F-74. Harvesting pineapple in Hawaii. (Courtesy, Maui Pineapple Company, Ltd., Kahului, Maui, Hawaii)

The tree fruits are often harvested by mechanical tree shakers which bring the fruit down to moving belts which then convey it to collection bins. Other types of special equipment are utilized in the harvesting of cranberries and grapes. Each of these devices, when used properly, leaves the fruit in sound, wholesome condition, like the freshly harvested items shown in Fig. F-75.

Fig. F-75. Freshly harvested fruit. (Photo by J. C. Allen & Son, West Lafayette, Ind.)

MARKETING OF FRESH FRUITS. Fresh tree-ripened or vine-ripened fruits from the various exporting countries may be shipped across the world by air freight when the transportation cost is offset by high prices in the marketplaces. Hence, avocados, cherries, fresh figs, pineapples, and strawberries may be shipped from California to Europe in a day, or less. However, lower-priced items, such as bananas, are shipped by slower, less expensive means of transportation. Therefore, bananas are often harvested before they reach full ripeness so that they will be less susceptible to damage during shipment. Sometimes, the unripe fruits may be ripened artificially after arriving at their final destinations. A common artificial ripening procedure consists in exposing the fruit to ethylene gas in a sealed chamber. This means of ripening is more natural than it may appear, because bananas naturally produce ethylene and may be ripened at home in sealed plastic bags. However, refrigeration curtails the production of the gas. Fig. F-76 shows a large fruit counter in a food market.

Fig. F-76. An appealing display of fruit in a large food market. (Courtesy, California Tree Fruit Agreement, Sacramento, Calif.)

Growers who produce less-than-wholesale quantities of fruits may not be in a position to sell their product to wholesalers, grocery chains, or processors. Rather, they may have to utilize one or more of the alternative sale options which follow:

• **Roadside marketing**—This option presents an opportunity for the grower to sell at near retail prices and thereby get a larger share of the consumer's dollar than would be received from the various "middlemen." However, the operator of a roadside market must be prepared to bear the cost of promotion, merchandizing, quality control, and customer service. The best location for this type of market is on a heavily traveled highway near a large town or a city. Also, the fruit stand should be on (1) a straight stretch of road so that it may be seen from a sufficiently great distance to allow the automobiles to slow down, and (2) the side of the road leading to the nearest town or city, since city people on an excursion to the country are more likely to purchase fruit on the way home.

• **Pick-your-own fruit**—The harvesting of fruit in fields or orchards by customers has been growing in popularity because (1) the freshness of the fruit is assured, (2) prices of the commodity are usually lower, (3) the picker may select the desired size and quality of fruit, (4) labor cost for the producer is reduced greatly, and (5) the need for grading, packing, or storing is eliminated. Some popular pick-your-own crops are apples, cherries, cranberries, grapes, peaches, and strawberries.

Operators may need to give pickers-customers a little instruction in picking in order to avoid damage to plants or trees. Also, the possibility of financial loss due to liability for injuries to pickers should be minimized by (1) making certain that there are no harzards to pickers and their children, and (2) carrying sufficient accident and liability insurance.

• **Farmer markets**—Although the selling of fresh produce in farmers' retail markets declined with the development of refrigeration and rapid transportation to distant cities, there has recently been renewed interest in this form of marketing as a means of rejuvenating deteriorating areas of cities and towns. Also, more and more consumers are interested in purchasing "natural" foods. These markets vary from simply a place where growers can park their trucks to enclosed buildings with booths or stalls. Many farmers' markets charge a moderate rental or service fee to cover costs of facilities and operations.

Most markets operate seasonally, but some may be open all year. Usually, most sales are made on Fridays and Saturdays. Hence, insufficient parking for customers and traffic congestion may often accompany this type of operation. Nevertheless, neighboring businesses are likely to benefit from the increase in the number of shoppers in the area.

• **Grower cooperatives**—There is a growing trend among the producers of fruits and vegetables to join marketing cooperatives that help them to obtain maximum prices for their produce. Services by the "co-ops" range from simple pooling of the products supplied by the members to other forms of assistance such as harvesting, grading, packing, cooling, storage, processing, transporting, and providing means of marketing. The individual members usually own stock in the cooperative and they vote to determine operating procedures.

• **Food buying groups**—In the late 1960s, food buying clubs were formed by groups of people who wished to cut their food costs, and to avoid the traditional marketing channels. Many clubs were informal neighborhood organizations that (1) took orders and payment from members, (2) purchased produce and other items from farmers on the farm, or at farmers' markets, and (3) took the farm products to a place where members picked up their orders. Eventually, certain groups banded together to form cooperative buying units that entered into contracts with growers. These arrangements have benefitted both the small farmer and the consumer by eliminating some of the steps in the conventional marketing structure which tend to (1) reduce the prices paid to producers, and (2) raise the prices paid by consumers.

• **Sales to retail stores, restaurants, hospitals, and schools**—Usually, these purveyors of food are supplied by brokers or wholesalers who buy from growers, then arrange for storage, transportation, and delivery. However, enterprising small farmers may be able to sell their fresh fruit to one or more of these establishments, providing they can consistently provide the quantity, quality, and delivery guarantees that are required. Hence, the first contracts may be difficult to obtain unless the grower's reliability has already been established. The best items for this type of local marketing are cantaloupes, peaches, and strawberries which are at their highest quality when picked ripe and are likely to spoil if shipped for long distances.

• **Organic produce outlets**—The market for "organically-grown" produce has grown steadily, because an ever increasing number of people wish to purchase fruits grown without pesticides and other agricultural chemicals that they deem to be undesirable. However, many of the outlets for this type of produce may demand that the commodity sold as organic be certified by a regional organic farmers' association. Furthermore, the outlets often operate on a small scale, which means that they risk the loss of business due to the higher prices that must be charged in order to realize a profit.

U.S. Department Of Agriculture Grade Standards. Most fresh fruits are sold by wholesalers on the basis of USDA grades, although not many are identified by grade when sold by retailers. These voluntary grade standards provide (1) a common basis for wholesale trading, and (2) a means by which prices may be determined. One or more of the parties to a wholesale fruit transaction may request and pay for grading of the shipment by an inspector licensed by the USDA. The official USDA grades for fresh fruits are as follows:[27]

---

[27]*Uniform Grade Nomenclature Policy for United States Standards for Grades of Fresh Fruits, Vegetables, Nuts and Other Special Products,* USDA, 1976.

• **U.S. Fancy**—Premium quality; covers only the top quality range produced.

• **U.S. No. 1**—The chief trading grade; represents good average quality that is practical to pack under commercial conditions; covers the bulk of the quality range produced.

• **U.S. No. 2**—Intermediate quality between U.S. No. 1 and U.S. No. 3; noticeabley superior to U.S. No. 3.

• **U.S. No. 3**—The lowest merchantable quality practical to pack under normal conditions.

Most grading is visual. Internal as well as external quality of many products is examined. Models, color guides, and color photographs are available for graders to check samples for shape, degree of coloring, and degree of defects or damage. Products may be graded in the producing area; at assembly, packaging, or processing plants; at terminal markets; in a railcar, a receiver's warehouse, or a store.

Some of the other functions of the grade standards are: (1) to help establish loan values for produce in storage; (2) to assure that products purchased by the military services and various governmental agencies are of acceptable quality; (3) to provide purchasing specifications that might be used by restaurants, shipping lines, and other feeding establishments; and (4) to establish a basis for the trading of fruit crops on the futures market.

STORAGE. Freshly harvested ripe fruit is highly susceptible to spoilage. Hence, special measures must be taken if these crops are to be stored for longer than a week, because even in the United States, about 25% of the produce spoils.[28] The situation was much worse before the latter part of the 19th century, when cold storage facilities were first developed. Country people in the northern United States fared better than city dwellers because the basements of many rural homes contained fruit cellars, which were unheated, partitioned off sections. Also, the rural residents had outdoor sheds and special insulated pits for storing vegetables during the winter.

One of the first of the ice cooled storehouses was built in 1856 by the Reverend Benjamin Nyce, a preacher, teacher, and chemist of Decatur County, Illinois. Shortly thereafter, other fruit storage warehouses based upon similar principles were constructed in Chicago and New York City. By 1901, 600 cold storage facilities employing recently developed mechanical refrigeration units were in operation. Table F-42 gives some typical cold storage conditions for selected fruits.

TABLE F-42
RECOMMENDED STORAGE CONDITIONS AND STORAGE PERIOD FOR SELECTED FRUITS[1]

Commodity	Freezing Point		Place to Store	Storage Conditions		Length of Storage Period
				Temperature	Humidity	
	(°F)	*(°C)*		(°F)		
Apples	29.0	*−1.7*	Fruit storage cellar[2]	Near 32° *(0°C)* as possible	Moderately moist	Through fall and winter
Grapefruit	29.8	*−1.2*	Fruit storage cellar	Near 32° *(0°C)* as possible	Moderately moist	4-6 weeks
Grapes	28.1	*−2.2*	Fruit storage cellar	Near 32° *(0°C)* as possible	Moderately moist	1-2 months
Oranges	30.5	*−.8*	Fruit storage cellar	Near 32° *(0°C)* as possible	Moderately moist	4-6 weeks
Pears	29.2	*−1.6*	Fruit storage cellar	Near 32° *(0°C)* as possible	Moderately moist	6-8 weeks

[1]Adapted by the authors from Watt, J.A., *Storing Vegetables and Fruits at Home,* Purdue University Cooperative Extension Service, Bull. No. HO-125, p. 4.
[2]Or equivalent facility.

The storage times shown in Table F-42 are now extended greatly by a recently developed process called controlled atmosphere (CA) storage, which is combined with refrigeration to retard the normal aging process that occurs in fresh fruit. Even after picking, fruits are still living organisms which carry on respiration and other vital processes. The more rapid the respiration of the fruit, the faster it ages. Controlled atmosphere storage utilizes refrigerated, airtight rooms in which the carbon dioxide from the respiring fruit is allowed to accumulate to certain desired levels. The levels of oxygen and carbon dioxide in the storage rooms are monitored continually so that excesses of carbon dioxide may be removed, or extra oxygen may be added if needed. Over half of all the apples presently stored in the United States are kept in a controlled atmosphere.

PROCESSING. Until recent times, the most common methods of processing fruits were (1) using them to make wines, and (2) drying them in the sun. Now, various other procedures are used to extend the availability of fruits and fruit products over the entire year. Details of the major processes follow.

[28]Wu, M. T., and D. K. Salunkhe, "The Use of Certain Chemicals to Increase Nutritional Value and to Extend Quality in Economic Plants," *Storage, Processing, and Nutritional Quality of Fruits and Vegetables*, edited by D. K. Salunkhe, CRC Press, Inc., Cleveland, Ohio, 1974, p. 80.

Canning. In this process, fruits in metal or glass containers are heated to kill the microorganisms that cause spoilage. The more acid the products, the easier any spoilage organisms are destroyed by heat. Fortunately, most fruits are sufficiently acid to be processed safely at the temperature of boiling water in a boiling-water-bath canner. However, the proper procedures must be followed, for if spoilage organisms are not killed by adequate processing, they will continue to grow and reduce the acid content of the canned fruit, thus encouraging the growth of more dangerous organisms, such as *Clostridium botulinum*.

Fig. F-77. Checking the quality of canned pineapple prior to closing and sealing the can. (Courtesy, Maui Pineapple Company, Ltd., Kahului, Maui, Hawaii)

Table F-43 gives instructions for home canning some of the leading fruits.

TABLE F-43
CANNING CHART FOR SELECTED FRUITS[1]

Kind of Food	Preparation	Time (Minutes) to Process in Boiling Water Bath at 212°F *(100°C)*			
		Jars[2]		Cans	
		(Pt)	(Qt)	(No. 2)	(No. 3)
Apples	Pare, core, cut in pieces. Steam or boil in thin syrup or water 5 minutes. Pack hot; cover with hot liquid.	15	15	10	10
	Or make apple sauce, sweetened or unsweetened. Pack hot.	10	10	10	10
Berries (except strawberries)	If berries are firm, precook, adding just enough medium syrup or juice to prevent their sticking to pan. Pack hot; cover with hot liquid.	15	15	15	15
	If berries are soft, fill jars with raw fruit; cover with boiling syrup or juice.	20	20	15	15
Cherries	Can cherries with or without pits. Follow directions for berries.	20	20	20	20
Figs	Cover with boiling water; let stand for 5 minutes. Drain and rinse. Boil for 1 hour in medium syrup. Pack hot; cover with syrup.	20	20	15	15
Fruit juices	Berries, apple juice, grapes, or blends of fruits—remove any pits; crush the fruit. Heat gently to 170°F *(77°C)* (below simmering) until soft. Strain through a cloth bag. Add sugar if desired—about ½ to 1 cup *(120 to 240 ml)* sugar to 1 gal *(3.8 liter)* of juice. Heat to 170°F *(77°C)* again; pour into hot jars or bottles. Leave 1/8-in. (0.3 cm) headspace.	20	20	20	20
Peaches	Remove skins and pits. Precook juicy fruit slowly until tender to draw out juice. Pack hot or cold; cover with boiling juice or syrup. Heat in jar.	20	20	15	15
Pears	Peel, cut in halves, core. Otherwise same as for peaches.	20	20	20	20
Plums, prunes	Put up plums whole or in halves. Prick skin of each whole plum. Precook 3 to 5 minutes in juice or thin to medium syrup to sweeten. Pack hot; cover with boiling juice or syrup.	20	20	15	15
Strawberries	Stem berries and add ½ cup *(120 ml)* sugar to each quart of fruit. Bring slowly to boil. Remove from stove and let stand overnight. Bring to boil. Pack hot; cover with hot juice.	15	15	10	15

[1]Hindman, M.S., *Home Canning of Fruits and Vegetables*, HE Bull. 189, Clemson University Cooperative Extension Service, Clemson, S,C., 1970, pp. 16-17.
[2]One pt equals *470 ml* and one qt equals *950 ml*.

Drying. The drying of freshly harvested fruits in the sun was used as a means of preservation as long as 5 thousand years ago by the peoples of the Middle East. Then, the most common items to be dried were apricots, dates, figs, and grapes. Today, fruit drying is a major form of processing in California, which is the leading state for fruit production.

Dried fruits do not spoil readily because they do not contain sufficient water to support the growth of spoilage organisms. Drying requires a means of heating to evaporate the water present in the fruit, and some means of removing the water vapor. The simplest way to accomplish these objectives is to place the fruit in the sun. However, birds, insects, and other pests may eat or contaminate the fruit while drying is underway. Therefore, various types of drying equipment have been developed to insure protection against pests and various contaminants. Fig. F-78 shows the fundamental details of three common types of dryers.

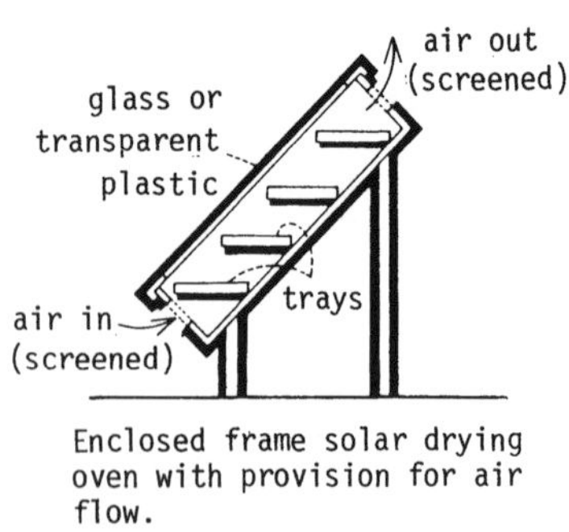

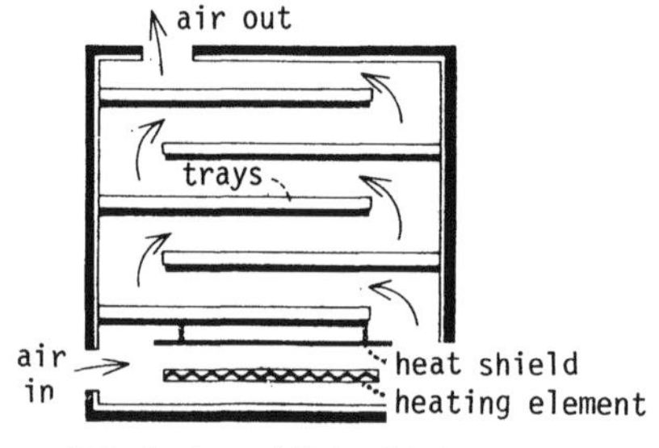

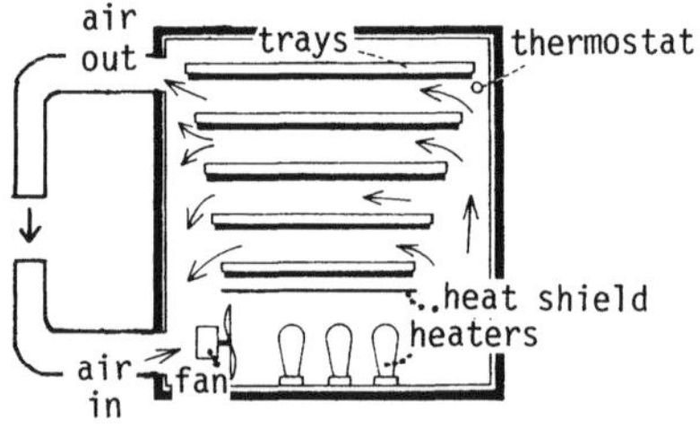

Fig. F-78. Drying cabinets commonly used for fruits and other fresh produce.

Fruits, like vegetables, require various treatments prior to drying because they contain enzymes that may cause color and flavor changes during drying unless enzyme activity is stopped. Table F-44 gives directions for preparing and drying selected fruits in a home oven.

TABLE F-44
DIRECTIONS FOR DRYING SELECTED FRUITS[1]

Fruit	Preparation	Pretreatment	Drying Procedure[2]	Dryness Test
Apples Mature, but not soft	Wash, pare, core, and cut in ¼-in. *(0.6 cm)* slices or rings. To prevent browning during preparation, treat cut portions with an ascorbic acid solution (2½ tsp *[12.5 ml]* of pure crystalline ascorbic acid in each 1 cup *[240 ml]* of cold water).	Blanch in steam 10 minutes; or Sulfur 30 to 60 minutes depending on size of pieces.	Set oven regulator at 150°F *(66°C)*. Arrange slices on trays spread not more than ½-in. *(1.3 cm)* deep—overlap rings. Drying time approximately 6-10 hours in controlled heat	Pliable, springy feel, creamy white, and moisture not visible when cut and squeezed
Apricots Fully ripened, but not soft	Wash. Do not peel. Cut in half and pit. Brush with ascorbic acid solution (1 tsp *[5 ml]* pure crystalline ascorbic acid in each cup *[240 ml]* of water) to prevent darkening.	Steam-blanch halves 5-10 minutes; or Sulfur 1-2 hours depending on size of pieces and ripeness of fruit. The riper the fruit, the more slowly it absorbs sulfur fumes; or Blanch in a hot syrup made of equal parts corn syrup and water or sugar and water.	Set oven regulator at 150°F *(66°C)*. Reduce to 140°F *(60°C)* when nearly dry. Arrange pit side up on drying trays. Average drying time for halves up to 14 hours in controlled heat	Pliable and leathery, and there is no moisture in center when cut and squeezed

Footnotes at end of table

(Continued)

TABLE F-44 *(Continued)*

Fruit	Preparation	Pretreatment	Drying Procedure[2]	Dryness Test
Berries	Wash. Leave whole, except cut strawberries in half.	No treatment necessary for soft berries. Steam firm berries ½-1 minute. Split tougher skins by dipping 15-30 seconds in boiling water, then in cold water. Drain thoroughly.	Heat oven to 140°F *(60°C)*. Place in layer of cheese cloth or nylon net on tray. Spread berries over cloth. Drying time approximately 5 hours	Berries are brittle and will rattle when shaken on the tray. No moisture visible when pressed between the fingers
Cherries	Wash. Remove stems and pits. If juicy, drain about 1 hour.	Split skins of whole cherries in boiling water 15-30 seconds, cool immediately. Drain. Syrup-blanch as for apricots.	Place trays in cold oven. Gradually heat to 140°F *(60°C)*. Reduce temperature near end of drying, if necessary. Drying time approximately 8 hours	Leathery, but sticky
Grapes (seedless)	Wash and clip into small bunches. Remove any blemished grapes. If grapes are large, pick from stem; leave whole.	Dip in boiling water 15-30 seconds to split skins; cool immediately. Drain.	Place trays in cold oven. Gradually heat oven to 150°F *(66°C)*. Reduce temperature near end of drying, if necessary. Drying time averages up to 8 hours	Pliable, and leathery
Peaches Ripe, freestone varieties preferred	Wash and peel. Cut in half, remove pit. Cut into quarters or slices as desired. To prevent browning during preparation, treat with ascorbic acid solution as for apricots.	Steam-blanch halves or quarters 15 to 20 minutes, slices 5 to 7 minutes. Sulfur 1-2 hours, depending upon size of pieces. If steam-blanched before sulfuring, cut sulfuring time about half.	Arrange on trays, pit side up to retain juices. Place trays in cold oven. Heat oven gradually to 150°F *(66°C)*. Turn halves over when visible juice disappears. Reduce temperature when nearly dry to prevent scorching. Drying time averages up to 15 hours for halves and about 6 hours for slices	Pliable, and leathery
Pears Ripe, Bartlett variety preferred	Pare, remove core and woody tissue. Cut into quarters, eighths, or thick slices. To prevent browning during preparation, treat with ascorbic acid as for apricots.	Steam-blanch 5 to 20 minutes depending upon size of pieces; or Sulfur as for peaches; or Syrup-blanch.	Arrange on trays and place in cold oven. Heat oven gradually to 150°F *(66°C)*. Reduce temperature when nearly dry to prevent scorching.	Leathery, springy feel
Plums	Wash, leave whole or cut in halves, remove pit.	Dip whole fruit in boiling water 30 seconds to split skins, cool and drain; or Steam-blanch halves 15 minutes, slices 5 minutes; or Sulfur whole fruit 2 hours, halves and slices 1 hour.	Dry like pears.	Pliable and leathery

[1]Miller, E.L., *Drying Foods,* Publication C-536, Kansas State University Cooperative Extension Service, Manhattan, Kans., 1976, pp. 14-19.
[2]Spread fruit in single layers on shallow wooden trays with slatted, perforated, or woven bottoms.

Freezing. The U.S. Department of Agriculture first conducted research on the freezing of fruits in 1904, although frozen strawberries were already being used in the production of ice cream. Within a few years, many commercial bakers began to use frozen berries and other fruits for making pies. The early frozen fruit packers learned by trial and error how to avoid deterioration of their product by careful preparation of the fruit before freezing. Table F-45 gives some guidelines for the preparation of various fruits that are to be placed in a home freezer.

TABLE F-45
FREEZING FRUIT GUIDE[1]

Fruit	Preparation	Type of Pack
Apples, slices Full-flavored, firm, crisp, ripe, but not mealy; free from bruises and decay	Wash, peel, quarter, and remove core. Slice medium apples into twelfths, large ones into sixteenths. *Raw:* Slice into solution 2-3 Tbsp salt to *1 gal water. Do not allow slices to* remain in solution more than 15-20 *minutes. Drain well. An ascorbic acid* solution instead of the salt solution can be used. *Steamed:* Slice, spread not more than ½-in. deep in steamer, steam 1½ minutes. Cool in ice water, drain.	Syrup pack is preferred if apples are to be used uncooked. *Unsweetened* *Sugar pack:* ½-1 cup sugar to 1 qt *apples.* *Syrup pack:* Slice apples directly into cold 30-40% syrup[2] to which ascorbic acid has been added. Leave head-space. It takes ½-⅔ cup of syrup for each pint package. *Unsweetened*
Blueberries, Elderberries, Huckleberries Ripe berries; full-flavored, tender skin, all about the same size	Sort, remove leaves, stems, inferior berries. Wash and drain. *Steamed:* If desired, steam 1 minute and cool. Pre-heating tenderizes skin and makes a better flavored product.	Syrup pack is preferred for berries to be served uncooked. *Syrup pack:* Pack berries into cold 40% syrup.[2] *Tray pack* *Unsweetened*
Cherries, Sour Bright red, tree-ripened	Discard soft, overripe or discolored cherries. Stem, wash, and pit.	Syrup pack is best for cherries to be served uncooked. *Syrup pack:* 60-65% syrup[2] *Sugar pack:* ¾ cup sugar to 1 qt fruit. To improve color retention mix ¼ tsp ascorbic acid with sugar.
Peaches Firm, ripe, well-colored, no green in skin	Sort, wash, pit, and peel, or immerse in boiling water for ½-1½ minutes to loosen skins. Cool in cold water. Gently rub off skins and pit, keeping under water. Slice or halve; drop immediately into syrup to prevent darkening or directly into prepared antidarkening solution.	For best quality, pack peaches in syrup or with sugar. *Syrup pack:* 40-50% syrup,[2] ½ cup per pint. Press fruit down, add syrup to cover. Add ½ tsp crystalline ascorbic acid to each quart of syrup for a better quality product. *Sugar pack:* ⅔ cup sugar to 1 qt fruit *Water-pack:* Cover with solution of 1 tsp ascorbic acid per quart of water.
Strawberries Firm, ripe, red, with slightly tart flavor and free from rot	Sort, wash, drain, remove hulls. Large berries are better sliced or crushed. Pureed.	Sugar and syrup packs make better quality frozen strawberries than berries without sweetening. *Syrup pack:* 50% syrup[2] *Sugar pack:* ¾ cup sugar to 1 qt fruit *Unsweetened:* Cover with water containing 1 tsp crystalline ascorbic acid to each quart of water. Seal and freeze. *Tray pack* *Sugar pack:* ⅔ cup sugar to 1 qt fruit

[1]VanZandt, D.P., *Preserving Foods: Freezing Fruits,* Fact Sheet 279, University of Maryland Cooperative Extension Service, College Park, Md., 1979, p. 3. To convert to metric, see WEIGHTS AND MEASURES.

[2]The various syrups are made up by mixing the specified amounts of sugar with 4 cups of water (in each case) as follows: 30%, 2 cups; 40%, 3 cups; 50%, 4¾ cups; and 60%, 7 cups.

Jams And Jellies. It appears that fruits preserved with sugar were used in Roman times, although the original reason for adding the sugar was probably for its sweetening effect. The high sugar content and acidity of fruit preserves prevent the growth of microorganisms. Eventually, cooks developed a variety of heavily sugared products that are similar in their semisolid consistency, but differ with respect to (1) the form(s) of fruit, (2) amounts of sugar and water, and (3) methods of preparation. Therefore, the features which distinguish the different items are noteworthy:

- **Butters**—Fruit pulp is cooked with sugar to yield a mixture which spreads easily. Apple butter is the most common of these products.

- **Conserves**—A mixture of fruits, with or without nuts and raisins, is used to make these items, which are more popular in European countries than in the United States.

- **Jams**—Pieces of a single type of fruit are used to prepare these easily spread mixtures.

- **Jellies**—Fruit juices are used to prepare these clear, firm gels which may sometimes require additional pectin to foster gelling.

- **Marmalades**—Citrus fruits and quinces, together with their pulp and peel or skin, are used to make clear, gelled products in which the fruit pieces are suspended.

- **Preserves**—Whole fruits or large pieces of fruit are packed in a syrup. Preserves resemble jams, except that they contain large pieces of fruits.

The gelling of the various products is due mainly to the substance pectin, which is present in varying amounts in most fruits. Acid is also needed to bring about gelling. If pectin is lacking, a commercial form of it may be added. Extra acid may be added in the form of citric acid or lemon juice. Most items are cooked to bring about the destruction of spoilage organisms. However; strong heating destroys considerable amounts of vitamins. Table F-46 gives directions for making uncooked jams from frozen fruits.

NOTE: These products must be kept in a refrigerator or a freezer because they are more susceptible to spoilage than cooked items.

TABLE F-46
DIRECTIONS FOR PREPARING UNCOOKED JAMS FROM FROZEN FRUITS AND JUICES[1,2]

Fruit	Powdered Pectin	Sugar	Lemon Juice	Water	Yield
	(pkg)	(cups)	(tsp)	(cups)	(glasses)
Peaches					
Commercial or home frozen with sugar—6 oz packs 3 pkgs finely chopped (6 cups)	2 pkgs citrus pectin[3]	7	2	3/4	12-14
Red raspberry					
Commercial or home frozen with sugar (4 1/2 cups)	1	5	—	3/4	9-10
Strawberries					
Commercial frozen, 16 oz packs 3 pkgs (6 cups)	2	7	—	1 1/2	12-14
Strawberries					
Fresh or frozen without sugar (2 cups)	1	4	—	1	6

[1]Adapted by the authors from Gibbs, J.C., *Freezing Fruits and Vegetables*, Bull. A-77, University of Arizona Cooperative Extension Service, Tucson, Ariz., 1973, p. 13.

[2]Mash or chop fruit as fine as possible. Stir in sugar; let stand until sugar is dissolved—about 20 minutes—stirring occasionally.

Boil pectin and water 1 minute, stirring constantly. Add fruit to pectin and water, and stir about 2 minutes.

Pour into glasses, cover loosely, and let stand until set (24 hours). Store in refrigerator.

[3]Citrus pectin recommended because it does not mask the peach flavor. Add lemon juice to peaches as soon as they are mashed.

Juices. There has been a great increase in the production of fruit juices since the end of World War II. The large volumes of these products have tended to keep prices down, so that they have increased much less than the prices for fresh fruits. As a result, shoppers have increased their purchases of juices, while decreasing their consumption of whole fruit.

The production of fruit juices involves (1) extraction of the juice by various types of crushing or squeezing processes, (2) removal of undesired fruit parts (pulp, rind, seeds, etc.) from the juice by passage through screens and/or other equipment, and (3) treatment of the refined juice by filtration, heat, or other means to prevent clouding, fermentation, and other types of deterioration during storage. However, the procedures vary considerably with the type of juice. Therefore, some noteworthy details regarding the production of the leading fruit juices follow:

- **Apple**—This juice was first sold as sweet cider in early America. Often, the cider fermented during storage, which was considered by some to be a good fortune.

Eventually, it was discovered that a higher quality product resulted from (1) the use of whole sound fruit; (2) removal of suspended solids by centrifugation, addition of settling material, and/or filtration; and (3) heating briefly to kill spoilage organisms (pasteurization).

• **Citrus**—Much of the market for fresh juice in bottles has been claimed by frozen citrus juice concentrates. The latter products are prepared from conventionally prepared citrus juices by concentration in a vacuum, packaging into cans, and rapid freezing.

• **Grape**—The production of the juice in the United States began in 1869 when it was used instead of wine in the communion services of certain churches. Usually, it is produced in ways similar to those used for other fruit juices, except that grapes are usually hot pressed to insure a desirable color in the juice.

• **Pineapple**—Many of the fruit residues and juices from the canning of the processed fruit may be used for juice production. Therefore, a considerable amount of blending of the different pressing fractions is required to obtain a uniform product. Also, finely pulverized pulp is often blended into the juice.

Federal Standards For Processed Fruit Products. All over the United States, fruit products such as apple butter and frozen cherry pie are essentially uniform with respect to ingredients and quality because food standards, established by the Federal Government set requirements which these products must meet if they move in interstate commerce. There are four major sets of standards which apply to various processed fruit products. These follow:

• **U.S. Department of Agriculture Grade Standards**—These standards, like those for fresh produce, are voluntary. They cover canned, frozen, and dried fruits; and related products such as preserves.

USDA provides official grading services, often in cooperation with State departments of agriculture, for a fee, to packers, processors, distributors, or others who wish official certification of the grade of a product. Also, the grade standards are often used by packers and processors as guidelines for their quality control.

• **Food and Drug Administration Standards of Identity**—These mandatory standards establish what a given food product *is*—for example, what a food must be to be labeled "catsup." They also provide for the use of optional ingredients, in addition to the mandatory ingredients, that may constitute the product.

Fruit products for which standards of identity have been formulated are canned fruits, canned fruit juices, fruit butters, jellies, preserves, and related products; and frozen cherry pie.

• **Food and Drug Administration Minimum Standards of Quality**—FDA standards of quality are mandatory, as opposed to USDA grade standards of quality, which are voluntary. They have been set for a number of canned fruits to supplement standards of identity, by specifying the minimum acceptable characteristics for such factors as tenderness, color, and freedom from defects.

• **Food and Drug Administration Standards of Fill of Container**—These mandatory standards tell the packer how full a container must be to avoid deception. They prevent the selling of air or water in place of food. For example, the amounts of headspace in cans and jars of fruit products are limited to those consistent with good manufacturing practices.

Recent Developments In Fruit Processing Technology. Food scientists have long been at work devising new means of preserving fruits and utilizing blemished, but otherwise wholesome, items. However, the most rapid advances in this field started after World War II, as a result of wartime breakthroughs in providing military personnel with the products to which they had grown accustomed back home. For example, the frozen citrus juice industry had a phenomenal growth rate during the 1950s. Lately, much of the research has been directed towards (1) using fruits in new ways, (2) developing products from edible fruit wastes, and (3) combining fruits with other foods. Therefore, some of the current developments are noteworthy:

• **Banana-soybean powder**—Many people in tropical countries use bananas as a staple food. As a result, their diets are often protein deficient unless supplemented with high-protein foods, which are not always readily available. One solution to this problem is the use of a mixture of powdered bananas and soybeans. Its production involves (1) preparation of a watery slurry of soybeans that have previously been soaked and blanched; (2) blending of fresh, ripe bananas into the soybean slurry, and sulfuring the mixture to prevent darkening; (3) drying the sulfured mixture on the surface of a revolving drum heated by steam; and (4) scraping the dried banana-soybean powder from the drum.

• **Bioflavonoids from citrus wastes**—The current interest in the reputed nutritional and therapeutic properties of these vitaminlike substances may make it profitable to extract them from citrus wastes. A typical process involves (1) pressing the pulp and peel residues from juice production to obtain citrus molasses, (2) extracting bioflavonoids from the molasses, and (3) purifying the bioflavonoid extracts.

• **Cherry pit flour**—This product, which consists of about a 50/50 mixture of finely ground cherry hulls and cherry kernels, has an almondlike flavor and contains about 10% protein and 10% fat. Hence, it may be used as a flavoring ingredient for baked goods, and various dessert items.

• **"Citrus milk"**—This term covers certain drinks made from citrus juices, cheese whey protein, starch, stabilizers, sweeteners, and flavorings. These beverages contain 3% protein, which is almost as much as that in milk. Also, the products may be marketed as dried powders which may be reconstituted with cold or hot water. Finally, the products have creamy textures and excellent flavors.

• **Firm pieces of canned fruit**—Most canned fruits are much softer than the original fresh, raw fruit because the canning process tends to cause the flow of the natural juices from the fruit to the fluid in the can. This problem can be alleviated by coating the fruit(s) to be canned with a carboxymethylcellulose gum that seals the fruit pores so that little juice escapes. Citric acid and/or other natural fruit acids may be added to the gum coating to stabilize it and to impart extra flavor.

• **Flaked dried fruits**—These products may be rehydratd much more rapidly than thicker pieces which usually require long periods of soaking and/or cooking to restore the original water content. The fruit flakes are produced by (1) grinding conventionally dried pieces of fruit, (2) softening the ground particles by heating, and (3) rolling the particles into flakes. They are particularly suitable as ingredients in breakfast cereals because the addition of milk to the cereal mixtures is sufficient to bring about rehydration of the flakes within a few minutes.

• **Instant applesauce**—This precooked product requires only the addition of cold water to make an applesauce that is ready to eat in from 3 to 5 minutes. It is made from cooked apple slices that have been made into tiny flakes. The instant applesauce is very convenient for backpackers and hikers, because only an ounce (*28 g*) or so of the flakes yields over ½ lb (*224 g*) of sauce when sufficient water is added.

• **Jams and jellies combined with butter or margarine**—Although fruit preserves may be mixed with table spreads at home, these mixtures are usually undesirable because they tend to separate. This tendency can be overcome by mixing the jam or jelly stock with butter or margarine prior to gelling. These mixtures are usually stabilized with pectin and sodium citrate. Salad oils such as corn, cottonseed, olive, peanut, safflower, soybean, and/or sunflower may be substituted for the butter or margarine.

• **Orange juice on a stick**—Unless homemade from fruit juices, popsicles are considered to be "junk food" because they usually supply only calories. Now, a much more nutritious product is available in the form of a frozen bar on a stick which is made from orange juice concentrate. The bar may be unsweetened, or sweetened.

• **Simulated fruit pieces**—A British concern has developed a line of pie fillings that contain simulated fruit pieces made from fruit purees, sugar, food starch, alginate gum, flavorings, vitamin C, and other ingredients. The simulated fruit pieces have a "skin" like ordinary fruit pieces, and they remain firm during the baking of pies.

• **Snacks made from grape waste**—A large part of the grape crop in the United States is used to make fruit juices, jelly, and wine. Usually, the grape pulp and skins are fed to livestock. However, it has been found that the grape wastes may serve as ingredients of baked goods, candies, and cereal-based snack foods. Also, the grape product may replace raisins and other dried fruits in the food products.

NUTRITIVE VALUES. Fruits are high in water content. However, most fruits are fair to excellent sources of calories (due to their sugars), fiber (poorly digested carbohydrate which stimulates movements of the digestive tract), various essential macrominerals and microminerals, vitamins, and vitaminlike factors. The nutrient composition of common fruits is given in Food Composition Table F-36.

Food Composition Table F-36 does not show the bioflavonoid content of the fruit items because this information is not readily available. Besides, at this time no conclusive evidence exists that bioflavonoids serve any useful role in human nutrition or in the prevention or treatment of disease in humans.

Bioflavonoids are vitaminlike substances which are reputed to have actions similar to those of vitamin C, such as the strengthening of capillary walls against breakage or leakage of fluid.[29] They may also act as antioxidants,[30] thereby sparing other nutrients that have similar functions or are prone to oxidation, such as selenium and vitamins A, C, and E.

It is noteworthy that bioflavonoids are members of a large family of compounds called flavonoids, which include (1) purple, blue, and red anthocyanin pigments in blackberries, blueberries, cherries, currants, gooseberries (only the ripened fruit), grapes, pomegranates, and apples with red skins; and (2) various colorless substances. It is not certain which of the flavonoids act as bioflavonoids. However, apricots, cantaloupes, cherries, citrus fruits, grapes, and papayas are believed to be among the richest sources of bioflavonoids.

---

[29]Kuhnau, J., "The Flavonoids. A Class of Semi-Essential Food Components: Their Role in Human Nutrition," *World Review of Nutrition and Dietetics*, Vol. 24, edited by G. H. Bourne, published by S. Karger, Basel, Switzerland, 1976, p. 175.

[30]Charley, H., "Fruits and Vegetables," *Food Theory and Applications*, edited by P. C. Paul and H. H. Palmer, John Wiley & Sons, Inc., New York, N.Y., 1972, p. 291.

Certain European researchers have found that the feeding of anthocyanins from fruits to human subjects resulted in beneficial effects such as inhibition of tumor growth, acceleration of the regeneration of visual purple (the pigment in the eyes which is involved in the visual process), and improvement of the adaptation of the eyes to darkness.

NOTE: Little or no research in these areas has been done in the United States. Hence, the authors make no claims regarding these reports, but merely present them here in order to stimulate further research.

(Also see BIOFLAVONOIDS.)

Effects Of Processing And Preparation. The maximum amounts of essential nutrients are supplied by raw, fresh fruits that have been harvested at just the right stages of maturity. However, it may not be advisable to use only fresh fruits, which are available only at certain times of the year. Therefore, the effects of the various processing and preparation procedures should be taken into account when planning menus that are to be nutritionally adequate.

NOTE: The information which follows is based on relatively few studies and may not be valid in all cases because (1) different fruits are affected in different ways; (2) the effects depend partly upon the condition of the fruits—whether underripe or fully ripe, fresh or taken from long storage, whole or in pieces, and untreated or treated; and (3) information is lacking on some of the effects of commonly used procedures.

• **Artificial ripening**—Most commercially grown apples, apricots, avocados, bananas, and pears are picked before full ripening in order to avoid excessive softness and oversusceptibility to damage and spoilage. (Ripening may be induced artificially just before marketing.) These fruits may have vitamin A and vitamin C contents equal to, or greater than, those of tree-ripened fruits, providing that they have been handled properly after picking.[31]

• **Canning**—The vitamin contents of eight commonly canned fruit products (apples, apricots, blueberries, cherries, orange juice concentrate, peaches, raspberries, and strawberries) were found to be considerably lower than those of their fresh counterparts.[32] This study showed that the average vitamin losses were as follows: vitamin A, 39%; vitamin B-1 (thiamin), 47%; vitamin B-2 (riboflavin), 57%; niacin, 42%; and vitamin C, 56%.

• **Chopping, dicing, grating, mashing, mincing, or slicing**—Any process which breaks many cells within fruits is likely to be responsible for significant losses of certain vitamins. The longer the broken pieces of fruit are held, the greater the losses. Therefore, vitamins may be conserved by (1) cutting up fruits just prior to serving, and (2) using a plastic knife for cutting, because metals speed up the destruction of vitamin C.

• **Drying**—Moderate amounts of carotene (provitamin A) and vitamin C may be destroyed in this process, unless the fruits are sulfured before drying. However, sulfuring destroys most of the thiamin content. Vitamin losses are greater during slow processes such as sun drying, than they are for more rapid processes like freeze drying.

• **Fermentation**—The production of wines from grapes and other fruits appears to render minerals like iron more available for absorption. Alcoholic iron tonics have long been used to treat anemia. Also, alcoholic fermentation by yeast may convert the inorganic chromium which is present in various fermentable materials to the glucose tolerance factor (GTF)—an organic form of chromium that acts with insulin to lower the blood sugar.

• **Freezing**—The vitamin contents of eight common frozen fruit products (apples, apricots, blueberries, cherries, orange juice concentrate, peaches, raspberries, and strawberries) were found to be lower than those of their fresh counterparts, but higher than those of canned fruit products.[33] This study showed that the average vitamin losses were as follows: vitamin A, 37%; vitamin B-1 (thiamin), 29%; vitamin B-2 (riboflavin), 17%; niacin, 16%; and vitamin C, 18%.

• **Paring**—Removal of the peels or skins of certain fruits may result in disproportionately high losses of nutrients because the outer layers are often richer in minerals and vitamins than the inner layers. Therefore, these measures, which are often performed mainly for cosmetic reasons, should be used only to the extent that is absolutely necessary.

• **Pureeing**—The preparation of purees by homogenizing fruits in water makes the iron in these items more available, because it breaks down the fibrous cell walls which enclose the nutrients.[34] However, the amount of homogenization should be kept to a minimum in order to avoid destruction of vitamins A and C, which occurs upon exposure of the fruit pulp to air.

• **Sulfuring**—This process, which consists of either exposing fruit to fumes from burning sulfur, or dipping them

---

[31]Krochta, J. M., and B. Feinberg, "Effects of Harvesting and Handling on Fruits and Vegetables," *Nutritional Evaluation of Food Processing*, 2nd ed., edited by R. S. Harris and E. Karmas, The Avi Publishing Company, Inc., Westport, Conn., 1975, pp. 102-106.

[32]Fennema, O., "Effects of Freeze-Preservation on Nutrients," *Nutritional Evaluation of Food Processing*, 2nd ed., edited by R. S. Harris and E. Karmas, The Avi Publishing Company, Inc., Westport, Conn., 1975, pp. 268, 269, Table 10.15.

[33]*Ibid.*

[34]Kakade, M. L., and I. E. Liener, "The Increased Availability of Nutrients from Plant Foodstuffs through Processing," *Man, Food, and Nutrition*, edited by M. Rechcigl, Jr., CRC Press, Inc., Cleveland, Ohio, 1973, p. 234.

in a sulfate solution, is used to prevent discoloration and spoilage. It is generally utilized before fruit is to be dried. Sulfuring helps to prevent the losses of both carotene (provitamin A) and vitamin C, but it destroys thiamin. Fortunately, fruits are utilized more as sources of the first two vitamins, than as sources of the last one.

• **Thawing**—The thawing of frozen fruits usually results in loss of nutrients in the juices. Freezing foods ruptures cell walls so that the cell contents escape readily during thawing. Although the thawed juices may be used, it may be dangerous to do so if there has been considerable time for the growth of microorganisms during thawing. Hence, it may be desirable to heat frozen fruit carefully, to speed up thawing so that the juices which escape may be used safely. This danger is minimal for fruits packed in sugar or syrup.

Microwave heating may be preferable to other ways of thawing frozen fruits because microwaves penetrate the interior of the product and bring about a rapid and uniform thawing, whereas other means of heating tend to cook the outer layers of the fruit, while the interior remains frozen.

FRUITS OF THE WORLD. About 70 or so species of fruit plants account for most of the world's fruit production. Noteworthy information on fruits is presented in Table F-47, Fruits of the World.

(Also see WILD EDIBLE PLANTS.)

TABLE F-47
FRUITS OF THE WORLD

Popular and Scientific Name(s); Origin and History	Importance; Principal Areas; Growing Conditions	Processing; Preparation; Uses	Calories; Nutritive Value[1]
Acerola (Barbados Cherry) *Malphigia glabra* **Origin and History:** Native to the West Indies, Mexico, and Central America.	**Importance:** The richest natural source of vitamin C **Principal Areas:** Puerto Rico and other islands in the Caribbean **Growing Conditions:** Subtropical or tropical climate	**Processing:** Most of the crop is processed into fruit products and nutritional supplements (mainly sources of vitamin C) **Preparation:** The fresh fruit, juice, and pulp are prepared as sources of vitamin C **Uses:** Ingredient of jams, jellies, juices, and vitamin C supplements	**Calories:** 28 kcal/100 g **Nutritive Value:** One of the richest known sources of vitamin C (1,300 mg/100 g); reputed to be excellent source of bioflavonoids
Akee *Blighia sapida* **Origin and History:** Native to West Africa. Brought to West Indies in the 18th century. The Latin name of the fruit is after Captain Bligh of H.M.S. Bounty.	**Importance:** A popular fruit in Jamaica, W.I. **Principal Areas:** West Africa and West Indies **Growing Conditions:** Tropical climate	**Processing:** Only the fresh ripe fruit is eaten **Preparation:** Serve raw or cooked **Uses:** Fruit or vegetable dish	Calories: 196 kcal/100g because of high fat content (20%) **Nutritive Value:** Moderately high in crude fiber (1.6%) and vitamin C (26 mg/100g); a good source of vitamin A (925 I.U./100 g) **Caution:** The unripe fruit is poisonous and should not be eaten. It contains hypoglycin A which lowers the blood sugar and induces vomiting.
Apple *Malus pumila* **Origin and History:** Originated in the Caucasus Mountains of western Asia and/or in eastern Europe. (Also see APPLE.)	**Importance:** One of the leading fruit crops in the world and in the U.S. The worldwide annual production is about 40 million metric tons. **Principal Areas:** U.S.S.R., China, U.S.A., Germany, and France **Growing Conditions:** Temperate climate	**Processing:** The better part of the crop is marketed as fresh fruit, while the rest is processed as canned, dried, or frozen apples, or as vinegar, cider, or juice **Preparation:** After washing, peeling, and coring, may be served fresh, or may be baked, stewed, dipped in batter and fried (apple fritters) **Uses:** Fresh fruit dishes, desserts, jams, jellies, juices, pies, salads, and syrups; and in the production of alcoholic beverages and vinegar	**Calories:** 56 kcal/100g **Nutritive Value:** Fair source of potassium and vitamin C. Dried apple slices have almost 5 times the solids and caloric content of fresh apples.

Footnote at end of table

(Continued)

TABLE F-47 *(Continued)*

Popular and Scientific Name(s); Origin and History	Importance; Principal Areas; Growing Conditions	Processing; Preparation; Uses	Calories; Nutritive Value[1]
Apricot *Prunus armeniaca* **Origin and History:** Native to China. (Also see APRICOT.)	**Importance:** Among the top 2 dozen fruit crops in the world and in the U.S.A. About 2.1 million metric tons are produced worldwide each year. **Principal Areas:** Turkey, Italy, U.S.S.R., Spain, and Greece **Growing Conditions:** Dry, temperate climate	**Processing:** A limited part of the crop is marketed fresh. The rest is canned, dried, frozen, pureed, or made into a fruit "leather" by drying the fruit pulp on a flat surface. **Preparation:** Wash, halve, and remove pit. Serve fresh. Dried fruit may be stewed. **Uses:** Fresh fruit dishes, desserts, jams, jellies, juices, pies, salads and syrups; and in the production of alcoholic beverages	**Calories:** 51 kcal/100 g **Nutritive Value:** An excellent source of vitamin A (2,700 I.U./100 g) and a fair source of vitamin C and potassium. Dried apricots have about 5 times the solids and caloric content of the fresh fruit.
Avocado (Alligator Pear) *Persea americana* **Origin and History:** Originated in Central America and was spread by Indians from Mexico to Peru by the time the Spanish explorers arrived in the Americas. (Also see AVOCADO.)	**Importance:** Among the top 2 dozen fruit crops in the world and in the U.S.A. About 1.5 million metric tons are produced worldwide each year. **Principal Areas:** Mexico, U.S.A., Dominican Republic, and Brazil **Growing Conditions:** Subtropical to tropical climate. Requires well-drained soil	**Processing:** Most of the crop is used fresh. A small part is extracted for avocado oil (an ingredient of certain cosmetics). **Preparation:** Wash, then serve in halves with catsup, to be scooped out with a spoon; or slice or mash and mix with other ingredients to make a salad dressing or a dip **Uses:** Fresh fruit dishes, dips, guacamole, salads, and salad dressing	**Calories:** 167 kcal/100 g because of high fat content (16.4%) **Nutritive Value:** Moderately high in fiber (1.6%) and potassium (290 mg/100 g); fair source of vitamin C; low in carbohydrates (6.3%); Florida avocados usually lower in fat (about 11%) than California varieties which are mainly Fuerte.
Banana *Musa paradisiaca* **Origin and History:** Appears to have originated on the Malay peninsula, about 4,000 years ago. Brought to Africa by Indonesian migrants (Also see BANANA.)	**Importance:** One of the world's leading fruit crops, with an annual production of about 45 million metric tons **Principal Areas:** India, Brazil, Philippines, Ecuador, and Indonesia **Growing Conditions:** Moist, tropical climate	**Processing:** Almost all of the crop which is to be consumed fresh is picked green (fortunately, this fruit ripens well after picking.) A small amount is dried into flakes. **Preparation:** Peel, then eat whole, sliced, mashed, baked, or dipped in batter and fried. **Uses:** Fresh fruit dishes, cakes, gelatin desserts, ice cream, pies, and salads	**Calories:** 72 kcal/100 g **Nutritive Value:** An excellent source of potassium (370 mg/100 g); fair source of vitamin C. Banana powder contains about 4 times as many solids and calories as the fresh fruit.
Blackberry (Brambles) *Rubus fructicosus* **Origin and History:** Long gathered from wild plants that were probably native to Europe. Not domesticated until the Middle Ages. (Also see BLACKBERRY.)	**Importance:** Blackberries grow throughout the world, with the exception of dry desert regions. **Principal Areas:** The largest production is in the Northern Hemisphere. Commercial production in the U.S.A. is largely limited to Oregon, Texas, Oklahoma, and Arkansas. **Growing Conditions:** Cool, temperate climate	**Processing:** Some of the crop is consumed fresh, while the rest is canned, frozen, or used to make other products. **Preparation:** Wash and stem, then serve fresh (alone or with cream), or make into various mixed dishes **Uses:** Fresh fruit dishes, cakes, gelatin desserts, ice cream, jams, jellies, juices, pies, salads, sherbets, and syrups; and in the production of alcoholic beverages	**Calories:** 58 kcal/100 g **Nutritive Value:** A good source of fiber (4.1%); fair source of potassium and vitamin C; rich source of bioflavonoids

Footnote at end of table

(Continued)

TABLE F-47 *(Continued)*

Popular and Scientific Name(s); Origin and History	Importance; Principal Areas; Growing Conditions	Processing; Preparation; Uses	Calories; Nutritive Value[1]
Blueberry *Vaccinium myrtilloides* **Origin and History:** Native to North America, where they have been cultivated since the early 19th century. (Also see BLUEBERRY.)	**Importance:** Among the top 2 dozen fruit crops in the U.S.A. The annual world-wide production is about 45 thousand metric tons. **Principal Areas:** U.S.A. and Canada. **Growing Conditions:** Cool, moist, temperate climate.	**Processing:** Sometimes, the best of the berries are picked by hand, and the rest are harvested mechanically. Most blueberries are sold fresh, canned, or frozen. **Preparation:** Wash, then serve raw (alone or with cream), baked in cakes and pies, or prepared in various other dishes. **Uses:** Fresh fruit dishes, baked products, gelatin desserts, ice cream, jams, jellies, pies, and syrups.	**Calories:** 62 kcal/100 g **Nutritive Value:** A good source of iron (1.0 mg/100 g), and bioflavonoids; and a fair source of potassium and vitamin C.
Breadfruit *Artocarpus altilis* and *Artocarpus communis* **Origin and History:** Native to South Pacific islands. Brought from Tahiti to the West Indies by Captain Bligh of H.M.S. Bounty (on which the well known mutiny occurred).	**Importance:** A staple food on certain islands in the Pacific and in the Caribbean. **Principal Areas:** South Pacific islands and West Indies. **Growing Conditions:** Tropical climate. Does best when rainfall is abundant.	**Processing:** The fresh fruit is usually cooked and eaten right after harvesting, without processing. **Preparation:** Baked, boiled, or fried. Also biscuits may be made from the fruit. **Uses:** Fruit or vegetable dish; ingredient of baked products and other mixed dishes.	**Calories:** 103 kcal/100 g **Nutritive Value:** Good source of crude fiber (1.2%), iron (1.2 mg/100 g), and vitamin C (29 mg/100 g).
Cantaloupe (Muskmelon) *Cucumis melo* **Origin and History:** Place of origin not definitely established; some authorities suggest Africa, others believe Asia (Persia) or India. (Also see CANTALOUPE.)	**Importance:** Among the top dozen fruit crops in the world with an annual production of 8.9 million metric tons. **Principal Areas:** China, U.S.A., and Spain. **Growing Conditions:** Warm, dry, temperate to subtropical climate.	**Processing:** Most of the crop is marketed fresh, but a small amount is frozen in forms such as melon balls. **Preparation:** Halved and served fresh; with lemon or ice cream; or made into melon balls. **Uses:** Fresh fruit dishes, and fruit salads.	**Calories:** 30 kcal/100 g **Nutritive Value:** An excellent source of vitamin A (3,400 I.U./100 g), and a good source of potassium (251 mg/100 g), vitamin C (33 mg/100 g), and bioflavonoids.
Cherimoya *Annona cherimolia* **Origin and History:** Native to the high valleys in the Andes mountains of Ecuador and Peru.	**Importance:** Little known outside of native habitat because it does not keep well in shipping. **Principal Areas:** Tropical areas of Central America and South America. **Growing Conditions:** Higher elevations in the tropics.	**Processing:** Most of the crop is consumed fresh, but some is made into processed fruit products. **Preparation:** The fresh fruit is usually served raw (chilled or unchilled). **Uses:** Desserts, fruit dishes, ice creams, fruit juices, and sherbet.	**Calories:** 94 kcal/100 g **Nutritive Value:** Moderately high in fiber (2.2%); a poor source of vitamin C.

Footnote at end of table

(Continued)

TABLE F-47 *(Continued)*

Popular and Scientific Name(s); Origin and History	Importance; Principal Areas; Growing Conditions	Processing; Preparation; Uses	Calories; Nutritive Value[1]
Cherry *Prunus avium* (sweet cherry) *Prunus cerasus* (sour cherry) **Origin and History:** Some kinds are native to North America; others may have originated in the areas between the Black and Caspian Seas. (Also see CHERRY.)	**Importance:** The tenth leading fruit crop in the U.S.A., with an annual production of 237,000 metric tons. **Principal Areas:** European countries and U.S.A. The leading U.S. sweet cherry producing states are: Washington, Oregon, California, and Michigan. Michigan produces more than 77% of the nation's sour cherries. **Growing Conditions:** Cool, temperate climate (sweet cherries require warmer weather than sour cherries).	**Processing:** Much of the crop is either marketed fresh or brined for manufacture of candied cherries or Maraschino cherries. (For the latter use, cherries are first bleached with sulfur dioxide to a white or pale yellow color, then dyed red). The remainder of the crop is canned, frozen, or made into juice or other products. **Preparation:** Wash and serve raw, or cook and use as a topping for various dishes. **Uses:** Fresh fruit dishes, cakes, gelatin desserts, ice creams, jams, jellies, juices, pies, sherbets, and syrups.	**Calories:** 58 kcal/100 g **Nutritive Value:** Sour cherries are an excellent source of vitamin A (1,000 I.U./100 g), potassium (191 mg/100 g), and bioflavonoids; and a fair source of vitamin C. Sweet cherries are a little higher in calories and carbohydrates, but they contain only about 1/9 as much vitamin A as sour red cherries.
Citron *Citrus medica* **Origin and History:** Originated in southeast Asia. Domesticated independently in both China and India. Brought to Near East and Mediterranean in ancient times. (Also see CITRON.)	**Importance:** Has limited commercial significance. **Principal Areas:** Italy, Greece, and Corsica. **Growing Conditions:** Subtropical or tropical climate. Requires a well-drained soil. May be killed by freezing temperatures.	**Processing:** Grown mostly for the rind, which is sliced and fermented in brine. Then, the rind is desalted and candied. **Preparation:** Use the candied rind in bakery products. **Uses:** Ingredients of baked goods and candies.	**Calories:** 314 kcal/100 g for candied citron **Nutritive Value:** Moderately high in fiber (1.4%).
Citrus Fruits: Citron; Grapefruit; Kumquat; Lemon; Lime; Mandarin Orange,Tangerine; Orange *Citrus spp* **Origin and History:** Somewhere in Southeast Asia between eastern India and the islands off the Pacific Coast; cultivated in China 4000 years ago; spread from China to Italy, Persia (Iran), Arabia, and Europe; introduced to New World by end of 16th century, large-scale breeding program in U.S. started in 1893. (Also see CITRUS FRUITS.)	**Importance:** Most citrus fruits rank among the top 20 fruits of the world and/or the U.S.A. **Principal Areas:** Brazil, Italy, Mexico, and U.S.A. **Growing Conditions:** Tropical or subtropical climate, with some being more tolerant of low temperatures than others.	**Processing:** Three fourths of the crop is processed, with even the peel and seeds utilized. Some of the edible products include: brandies, liqueurs, wines, frozen desserts, flavorings, fruit drinks, canned fruits, jams, jellies, marmalades, candied peel, and syrups. **Preparation:** Many eaten as fresh fruit or juice, but preparation depends upon the type of citrus fruit. **Uses:** Appetizers, desserts, fruit dishes, fruit salads, juices, gelatin dishes, jams, jellies, marmalades, and soft drinks; specific uses indicated under individual citrus fruits.	**Calories:** 20 to 50 kcal/100 g fresh, depending on fruit **Nutritive Value:** Excellent to good source of fiber, pectin, potassium, vitamin C, inositol, and bioflavonoids; juices not as nutritious as fresh whole fruit.

Footnote at end of table

(Continued)

TABLE F-47 *(Continued)*

Popular and Scientific Name(s); Origin and History	Importance; Principal Areas; Growing Conditions	Processing; Preparation; Uses	Calories; Nutritive Value[1]
Crabapple *Pyrus (malus), Ioensis* and *M. floribunda* **Origin and History:** Crabapple trees grow wild from Siberia (U.S.S.R.) to northern China and in North America. Most wild apples are crab apples. The ancestor of many cultivated varieties is the common wild apple of Europe and western Asia.	**Importance:** Not very important commercially. Certain wild species are valuable sources of genetic material which may be bred into commercial varieties of apples. **Principal Areas:** North America, Europe, and Asia **Growing Conditions:** Cool, temperate climate	**Processing:** Most of the limited production is made into jams, jellies and pickles. **Preparation:** Serve the processed fruit with other items. **Uses:** Appetizers, jams, jellies, pickles, relishes, and spiced fruit mixtures	**Calories:** 68 kcal/100 g **Nutritive Value:** A fair source of potassium and a poor source of vitamin C
Cranberry *Vaccinium macrocarpon* **Origin and History:** Native to Europe, North Asia, and North America (Also see CRANBERRY.)	**Importance:** One of the top 2 dozen fruit crops in the U.S.A., with an annual production of about 155,000 metric tons **Principal Areas:** Cool, temperate climate. Grows in low, acid bogs.	**Processing:** Most of the crop is harvested by special machines. Only about 1/4 of the crop is sold fresh. The rest is frozen and made into jellies, juices, and sauces. **Preparation:** Wash and stem, then may be ground raw for relish or jellied salads. Cook raw berries, then sweeten. **Uses:** Ingredient of jellies, juices, relishes, and sauces	**Calories:** 46 kcal/100 g for unsweetened raw cranberries **Nutritive Value:** A fair source of potassium and vitamin C; dehydrated cranberries contain about 8 times as much solids and calories as the fresh fruit.
Currant *Ribes nigrum* (black currants) *Ribes rubrum* (red currants) **Origin and History:** Currants are native to northern Europe, northern Asia, and northern United States. (Also see CURRANT.)	**Importance:** Among the top 2 dozen fruit crops in the world with an annual production of about 527 thousand metric tons **Principal Areas:** Germany, U.S.S.R., and Poland **Growing Conditions:** Cool, temperate climate	**Processing:** Marketed fresh or made into various products **Preparation:** Wash, stem, and serve raw (alone or with other fruits), cream, and/or ice cream). Baked in buns and cakes **Uses:** Baked goods, jams, jellies, juices, and syrups; and in the production of alcoholic beverages	**Calories:** 54 kcal/100 g for black European currants **Nutritive Value:** An excellent source of potassium (372 mg /100 g), vitamin C (200 mg/100 g), and bioflavonoids; moderately high in crude fiber (2.4%); red and white currants are considerably lower in potassium and vitamin C.
Custard Apple (Sweetsop) *Annona squamosa* **Origin and History:** Native to tropical America	**Importance:** Popularity is limited to the local area of production (too soft for shipping). **Principal Areas:** West Indies and South America **Growing Conditions:** Higher elevations in the tropics	**Processing:** Most of this crop is consumed fresh, but some is made into processed fruit products. **Preparation:** Serve the fresh fruit raw (chilled or unchilled) **Uses:** Dessert or fruit dish; and ingredient of ice creams, fruit juices, and sherbets	**Calories:** 101 kcal/100g **Nutritive Value:** A fair share of vitamin C

Footnote at end of table

(Continued)

TABLE F-47 (Continued)

Popular and Scientific Name(s); Origin and History	Importance; Principal Areas; Growing Conditions	Processing; Preparation; Uses	Calories; Nutritive Value[1]
Date *Phoenix dactylifera* **Origin and History:** Native to the deserts of southwestern Asia and the Middle East. Cultivated for over 8,000 years. (Also see DATE.)	**Importance:** One of the top 2 dozen fruit crops of the world, with an annual production of about 2.6 million metric tons **Principal Areas:** Egypt, Iraq, Iran, and Saudi Arabia **Growing Conditions:** Hot, dry, tropical climate	**Processing:** Large bunches of fruits are cut from the tree and lowered to the ground where the dates are shaken loose by machine. Most dates are dried to prevent spoilage. **Preparation:** Serve alone, or use in various dishes. **Uses:** Dessert or fruit dishes, baked goods, candies, ice creams, salads, and syrups; and in the production of alcoholic beverages	**Calories:** 271 kcal/100 g for dried dates **Nutritive Value:** Dried dates are rich in fiber (2.3%), iron (1.0 mg/100 g), potassium (666 mg/100 g), and niacin (2.2 mg/100 g); but contain little or no vitamin C.
Durian *Durio zibethinus* **Origin and History:** Native to the Malay peninsula and nearby areas. Has failed to thrive in other areas in which it has been cultivated.	**Importance:** Only the Malays and other southeastern Asians seem to appreciate the peculiar taste and odor of this fruit. **Principal Areas:** Malaysia and adjoining areas of southeast Asia	**Processing:** All of the fruit is consumed fresh near where it is grown. **Preparation:** Serve the fresh fruit raw. **Uses:** Dessert or fruit dish	**Calories:** 81 kcal/100 g **Nutritive Value:** Moderately high in fiber (1.6%); an excellent source of potassium (601 mg/100 g), and a good source of vitamin C (24 mg/100g)
Elderberry *Sambucus canadensis* Elderberries belong to the honeysuckle family. **Origin and History:** Various species are native to North America, Europe, and Asia.	**Importance:** Not very important commercially **Principal Areas:** North America, Europe, and Asia **Growing Conditions:** Cool, temperate climate	**Processing:** Much of the berries gathered from this wild shrub are used to make wine. **Preparation:** Serve the fresh fruit raw (alone or with cream). **Uses:** Dessert or fruit dish, making jellies, and in the production of brandies and wines	**Calories:** 72 kcal/100 g **Nutritive Value:** An excellent source of potassium (300 mg/100 g), and fiber (7.0%); a good source of vitamin C (36 mg/100 g), and a fair source of vitamin A
Fig *Ficus carica* **Origin and History:** Native to southwestern Asia and the eastern Mediterranean; Cultivated for thousands of years in the Near East. (Also see FIG.)	**Importance:** Among the top 2 dozen fruit crops of the U.S.A., with an annual production of about 40 thousand metric tons **Principal Areas:** Turkey, Greece, U.S.A., Portugal, and Spain **Growing Conditions:** Warm, semiarid areas throughout the world	**Processing:** Only small amounts of fresh figs are marketed for eating raw, canning, or making into jams. Most figs are dried. **Preparation:** Wash and serve fresh, canned, or dried—alone or with other items; or may be prepared in mixed dishes. **Uses:** Dessert or fruit dishes, baked goods, jams, relishes, salads, and syrups; and in the production of alcoholic beverages. Bakers use figs to make fig bars and fig newtons.	**Calories:** 80 kcal/100 g for fresh figs **Nutritive Value:** Fresh figs are a good source of potassium. Dried figs contain about 3½ times the solids and calories of the fresh fruit.

Footnote at end of table

(Continued)

TABLE F-47 *(Continued)*

Popular and Scientific Name(s); Origin and History	Importance; Principal Areas; Growing Conditions	Processing; Preparation; Uses	Calories; Nutritive Value[1]
Gooseberry *Ribes grossularia* **Origin and History:** Originated in northern Europe and Siberia. (Also see GOOSEBERRY.)	**Importance:** More popular in England than elsewhere. **Principal Areas:** England and other European countries; and North America. **Growing Conditions:** Cool, temperate climate.	**Processing:** Much of the crop is canned or frozen, and some is marketed fresh. **Preparation:** Served as fresh fruit or canned in a sugar syrup. **Uses:** Desserts, fruit dishes, jams, jellies, juices, and pies; and in the production of wines.	**Calories:** 39 kcal/100 g **Nutritive Value:** A good source of potassium (155 mg/100 g), fiber (1.9%), vitamin C (33 mg/100 g), and bioflavonoids.
Grape (Raisin) *Vitis vinifera* and others **Origin and History:** Believed to have originated in western Asia. Cultivated in the Near East since about 4000 B.C. (Also see GRAPE; and RAISIN.)	**Importance:** The world's leading fruit crop, with an annual production of about 60 million metric tons. **Principal Areas:** Italy, France, Spain, U.S.S.R., U.S.A., Turkey, and Argentina. **Growing Conditions:** Mild, temperate climate with long, sunny summers and mild winters. Requires a well-drained soil.	**Processing:** The bulk of the world grape crop is made into wine, and the rest is used either to make raisins (by drying in the sun or artificially) or juices, or to be sold as table (fresh) grapes. **Preparation:** Wash and serve fresh grapes alone or with other items. If not seedless, seeds need to be removed before using in mixed dishes. Raisins may be served alone or cooked in a variety of dishes. **Uses:** Desserts or fruit dishes, baked goods, gelatin desserts, ice cream, jellies, jams, juices, pies, salads, and syrups; and in the production of alcoholic beverages.	**Calories:** 69 kcal/100 g **Nutritive Value:** One of the richest fruit sources of chromium, which has recently been found to be an essential mineral; a good source of potassium (158 mg/100 g); dried grapes (raisins) contain about 4 times the solids and calories of the fresh fruit.
Grapefruit *Citrus paradisi* **Origin and History:** Developed in the West Indies around 1700 by crossing the sweet orange (*C. sinensis*) and the pummelo (*C. grandis*). (Also see GRAPEFRUIT.)	**Importance:** Among the top 2 dozen fruit crops of the world, with an annual production of about 4.3 million metric tons. **Principal Areas:** U.S.A. and Israel. **Growing Conditions:** Subtropical or tropical climate; may be killed by freezing temperatures.	**Processing:** About ½ of the crop is marketed fresh, and the rest is made into juice or canned in the form of segments. **Preparation:** Halve and serve fresh (alone or with other fruits and/or a little sugar), or broil briefly. **Uses:** Appetizer, fruit dishes, fruit salads, and juice.	**Calories:** 41 kcal/100 g **Nutritive Value:** A good source of potassium, vitamin C (38 mg/100 g), and bioflavonoids. Pink and red grapefruits contain moderate amounts of vitamin A (about 440 I.U./100 g).
Guava *Psidium guajava* **Origin and History:** Native to the American tropics. First cultivated by Inca Indians.	**Importance:** Interest in this crop is increasing steadily around the world. **Principal Areas:** Tropical areas of Asia, and the Americas. **Growing Conditions:** Subtropical or tropical climate.	**Processing:** The fruit may be marketed fresh, canned, frozen, or preserved in in other ways. **Preparation:** Serve the fresh fruit raw with a little sugar, bake in pies or tarts, or stew. **Uses:** Desserts, fruit dishes, "guava cheese" (a fruit pasta), jams, jellies, juices, "leather" (dried fruit puree), pies, and tarts.	**Calories:** 62 kcal/100 g **Nutritive Value:** High in fiber (5.6%); a good source of potassium (289 mg/100 g), and an excellent source of vitamin C (242 mg/100 g).

Footnote at end of table

(Continued)

TABLE F-47 *(Continued)*

Popular and Scientific Name(s); Origin and History	Importance; Principal Areas; Growing Conditions	Processing; Preparation; Uses	Calories; Nutritive Value[1]
Jackfruit *Artocarpus heterophyllus* **Origin and History:** Believed to have originated somewhere between Malaysia and India	**Importance:** Used mainly in areas of production. Most popular in Sri Lanka and India **Principal Areas:** Tropics of Asia and the Americas **Growing Conditions:** Tropical climate. Does best on well-drained, deep soils.	**Processing:** Little is processed, because most of the fresh fruits are either eaten raw or cooked in the areas where they are grown. **Preparation:** Serve the fresh fruit raw, or cook it briefly. **Uses:** Dessert or fruit dish; ingredient of fruit salads and pickles	**Calories:** 98 kcal/100 g **Nutritive Value:** High in potassium (407 mg/100 g); a poor source of vitamin C
Jujube (Chinese date) *Zizyphus jujuba* **Origin and History:** Native to China, where it has been cultivated for about 4,000 years	**Importance:** Minor fruit crop **Principal Areas:** China, India, Africa, U.S.A., and countries around the Mediterranean **Growing Conditions:** Dry, tropical climate	**Processing:** The fruits are sold fresh, candied, dried, or made into a juice **Preparation:** Eaten fresh, dried, or candied. **Uses:** Desserts, fresh fruits, beverages, cakes, and candies	**Calories:** 105 kcal/100 g **Nutritive Value:** A good source of potassium (269 mg/100 g); and vitamin C (69 mg/100 g). Dried jujubes contain about 2½ times the solids and calories of the fresh fruit.
Kiwi fruit (Chinese Gooseberry) *Actinidia chinensis* **Origin and History:** Originated in China. Brought to Australia, New Zealand, U.S.A., and Europe	**Importance:** The present limited cultivation of this crop appears to be increasing steadily **Principal Areas:** New Zealand and U.S.A **Growing Conditions:** Warm, temperate climate or subtropical climate	**Processing:** All of the crop is utilized as fresh fruit. **Preparation:** Eaten fresh or cooked. **Uses:** Desserts, fruit dish, jam, jelly, and pies.	Nutritional data is not readily available. Like the papaya, kiwi juice may be used as a meat tenderizer.
Kumquat *Fortunella japonica* and others **Origin and History:** Originated in China. Cultivated in China, Japan, Indochina, and Java for many centuries. Recently introduced into Australia, Europe, and Americas. (Also see KUMQUAT.)	**Importance:** Of much lesser importance than the citrus fruits, to which it is closely related **Principal Areas:** New Zealand, China, Japan, and Indochina **Growing Conditions:** Mild, temperate climate (more cold resistant than any of the citrus fruits)	**Processing:** This fruit may be utilized fresh, or it may be preserved by candying, pickling, and other means. **Preparation:** Serve the whole, unpeeled fruit raw, or use the preserved fruit raw, or use the preserved fruit in various dishes **Uses:** Desserts, fruit dishes, or garnish for roast duck; in candies, pickles, and sauces	**Calories:** 65 kcal/100 g **Nutritive Value:** High in fiber (3.7%); a good source of potassium (236 mg/100 g), vitamin A (600 I.U./100 g), and vitamin C (36 mg/100 g)

Footnote at end of table

(Continued)

TABLE F-47 *(Continued)*

Popular and Scientific Name(s); Origin and History	Importance; Principal Areas; Growing Conditions	Processing; Preparation; Uses	Calories; Nutritive Value[1]
Lemon *Citrus limon* **Origin and History:** Originated somewhere in southeastern Asia between India and southern China. (Also see LEMON.)	**Importance:** Lemons and limes are major fruit crops. Their combined worldwide production is about 7 million metric tons annually. **Principal Areas:** Italy, U.S.A., Mexico, India, and Spain **Growing Conditions:** Subtropical or tropical climate	**Processing:** Some of the fruit is marketed fresh, and the rest is made into juices and juice concentrates, and other products. **Preparation:** Squeeze the juice of a fresh lemon on salads, fish, meats, or fried foods; or utilize the fruit or juice in mixed dishes. **Uses:** Alcoholic drinks, cakes, candies, gelatin desserts, ice cream, ices, jams, jellies, marmalades, pies, sherbets, and soft drinks	**Calories:** 20 kcal/100 g **Nutritive Value:** A good source of vitamin C (77 mg/100 g) and bioflavonoids
Lime *Citrus aurantifolia* **Origin and History:** Originated somewhere in southeastern Asia between India and the East Indies. (Also see LIME.)	**Importance:** Lemons and limes are major fruit crops. Their combined worldwide production is about 5 million metric tons annually. **Principal Areas:** U.S.A., Mexico, Italy, Spain, and India **Growing Conditions:** Subtropical or tropical climate	**Processing:** About half of the crop is marketed fresh, and the rest is made into juice and other products. **Preparation:** The fruit or juice is used in various mixed dishes. **Uses:** Alcoholic drinks, cakes, candies, gelatin desserts, ice cream, ices, jams, jellies, marmalades, pies, sherbets, soft drinks, and syrups	**Calories:** 28 kcal/100 g **Nutritive Value:** Moderately high in vitamin C (37 mg/100 g); a fair source of potassium; and a good source of bioflavonoids
Litchi Nut *Litchi chinensis* **Origin and History:** Native to southern China. Introduced into tropical areas around the world.	**Importance:** Although not well known in the U.S.A., its annual production (hundreds of thousands of tons) exceeds that of some better known fruits. **Principal Areas:** China; India; South Africa; and Florida, U.S.A. **Growing Conditions:** Cool, dry subtropical climate	**Processing:** The fleshy part of the fruit is used fresh, canned in syrup, or dried (so that it resembles a large raisin). **Preparation:** Serve the fresh fruit raw. **Uses:** Desserts or fruit dishes (this fruit is highly esteemed by Chinese people living abroad)	**Calories:** 65 kcal/100 g **Nutritive Value:** A good source of potassium (170 mg/100 g) and vitamin C (50 mg/100 g). The dried fruit has about 3½ times the solids and calories of the fresh fruit.
Loquat (Japanese Medlar) *Eriobotrya japonica* **Origin and History:** Native to China and Japan	**Importance:** Minor, but popular, because it bears fruits very early in the year. **Principal Areas:** Japan, China, India, and Mediterranean countries. Also, it is grown commercially in California, U.S.A. **Growing Conditions:** Temperate climate; grows well at higher elevations in the tropics	**Processing:** Usually this fruit is eaten fresh, but it may also be candied, canned, or preserved in other ways. **Preparation:** Serve the fresh fruit raw or stewed, or add the preserved fruit to relishes and salads. **Uses:** Desserts, fruit dishes, candies, jams, jellies, relishes, and salads; and in the production of liqueurs	**Calories:** 48 kcal/100 g **Nutritive Value:** An excellent source of potassium (348 mg/100 g) and a good source of vitamin A (670 I.U./100 g)

Footnote at end of table

(Continued)

TABLE F-47 *(Continued)*

Popular and Scientific Name(s); Origin and History	Importance; Principal Areas; Growing Conditions	Processing; Preparation; Uses	Calories; Nutritive Value[1]
Mango *Mangifera indica* **Origin and History:** Native to the area extending from India to Burma. (Also see MANGO.)	**Importance:** Among the leading fruit crops of the world, with an annual production of about 15 million metric tons **Principal Areas:** India, Mexico, Pakistan, and Brazil **Growing Conditions:** Subtropical or tropical climate. Requires dry weather for flowering and fruiting.	**Processing:** Most of the mango crop is marketed fresh. However, small amounts are canned, or made into chutney, jam, and pickles. **Preparation:** Serve fresh (alone or in mixtures with other fruits). **Uses:** Desserts, fruit dishes, fruit salads, jams, chutney, and pickles	**Calories:** 66 kcal/100 g **Nutritive Value:** A good source of potassium (189 mg/100 g) and vitamin C (35 mg/100 g); and an excellent source of vitamin A (4,800 I.U./100 g)
Mangosteen *Garcinia mangostana* **Origin and History:** Native to Malaysia. (Also see MANGOSTEEN.)	**Importance:** Little known outside of native habitat **Principal Areas:** Malaysia **Growing Conditions:** Humid tropical climate (does not grow well outside of Malaysia)	**Processing:** All of this crop is used fresh in the vicinity of its production. Hence, none is processed. **Preparation:** Serve the fresh fruit raw **Uses:** Desserts or fruit dishes	**Calories:** 57 kcal/100 g **Nutritive Value:** High in crude fiber (5.0%); a fair source of potassium (135 mg/ 100 g); and a poor source of vitamin C
Medlar *Mespilus germanica* **Origin and History:** Native to Europe and central Asia. Grown in Europe in Roman times, when it was more popular than it is today.	**Importance:** Grown in some parts of Europe for its acid flavored, brown colored, applelike fruit **Principal Areas:** European countries **Growing Conditions:** Temperate climate. Although well adapted to cold weather, the fruit grown in mild climates is the most tasty.	**Processing:** Only the fruit grown in the warmer countries (such as Italy) ripens fully on the tree. Elsewhere, the fruit must be aged in a cool dry room to become palatable. Some of the crop is also preserved in various ways. **Preparation:** Only properly ripened or aged fruit should be served raw. **Uses:** Desserts, fruit dishes, jams, and jellies	Nutritional data is not available.
Mulberry, Black *Morus* spp. **Origin and History:** Native to the temperate regions of the Northern Hemisphere, and to China	**Importance:** Has limited commercial importance **Principal Areas:** China, Iraq, southern U.S.S.R., and various European countries **Growing Conditions:** Temperate climate	**Processing:** Most of this fruit is eaten fresh, but a small amount is used to make jams and wines. **Preparation:** Served as fresh fruit raw (alone or with cream). **Uses:** Desserts, fruit dishes, jams, and jellies; and in the production of wines (used mainly for its color)	**Calories:** 42 kcal/100 g **Nutritive Value:** An excellent source of iron (3.0 mg/ 100 g), a fair source of potassium (123 mg/100 g), and a good source of vitamin C (39 mg/100 g)

Footnote at end of table

(Continued)

TABLE F-47 *(Continued)*

Popular and Scientific Name(s); Origin and History	Importance; Principal Areas; Growing Conditions	Processing; Preparation; Uses	Calories; Nutritive Value[1]
Olive *Olea europaea* Olive is classed as a fruit, but used as a vegetable. See VEGETABLE(S), Table V-6 Vegetables of the World.			
Orange (Sour orange, Sweet orange) *Citrus aurantium* (sour orange) *Citrus sinensis* (sweet orange) **Origin and History:** Originated in southern China and in Indonesia (Also see ORANGE, SWEET; and SEVILLE ORANGE.)	**Importance:** The sweet orange is the leading fruit crop in the U.S.A and a major one of the world, with an annual worldwide production of more than 52 million metric tons **Principal Areas:** Brazil, U.S.A., China, Spain, Mexico, India, and Egypt	**Processing:** An ever increasing part of the crop ends up in the form of orange juice concentrate, but fresh oranges and fresh orange juice still have large shares of the market. **Preparation:** Peel and serve fresh orange halves or sections alone or with other fruit. **Uses:** Desserts, fruit dishes, alcoholic drinks, cakes, candies, gelatin desserts, ice cream, ices, jams, jellies, marmalades, sherbets, and soft drinks	**Calories:** 49 kcal/100 g **Nutritive Value:** A good source of potassium (200 mg/100 g), and an excellent source of vitamin C (50 mg/100 g), and bioflavonoids
Papaw *Asimina triloba* **Origin and History:** Native to the woodlands of southeastern U.S.A.	**Importance:** There is only a small scale commercial production in the eastern U.S.A. **Principal Areas:** Woodlands of eastern U.S.A. **Growing Conditions:** Temperate climate	**Processing:** Most of the fruit is eaten in the areas where it is produced, but some is converted to juice **Preparation:** Serve fresh fruit raw **Uses:** Desserts, fruit dishes, and in the production of juice	**Calories:** 85 kcal/100 g **Nutritive Value:** Over 5% protein
Papaya *Carica papaya* **Origin and History:** Native to Central America (Also see PAPAYA.)	**Importance:** One of the top 2 dozen fruit crops in the world with an annual production of about 4.4 million metric tons **Principal Areas:** Brazil, Mexico, and Thailand **Growing Conditions:** Tropical climate. Requires well-drained fertile soil and plenty of sunshine.	**Processing:** Much of the crop is utilized as a fresh fruit. Some is processed into juices and other products. **Preparation:** Wash, peel, slice fresh ripe fruit and serve with or without a little sugar. Boil unripe fruit as a vegetable. **Uses:** Fruit or vegetable dish, fruit salads, jams, jellies, juices, pickles, and sherbets	**Calories:** 39 kcal/100 g **Nutritive Value:** A good source of potassium (234 mg/100 g), an excellent source of vitamin A (1,750 I.U./100 g), vitamin C (56 mg/100 g), and bioflavonoids

Footnote at end of table

(Continued)

TABLE F-47 *(Continued)*

Popular and Scientific Name(s); Origin and History	Importance; Principal Areas; Growing Conditions	Processing; Preparation; Uses	Calories; Nutritive Value[1]
Passion Fruit (Granadilla) *Passiflora edulis* **Origin and History:** Native to Latin America. Spread throughout tropical areas of the world	**Importance:** The present limited commercial production of this fruit appears to be increasing steadily around the world. **Principal Areas:** Australia, Hawaii, South Africa, and tropical Americas **Growing Conditions:** Tropical highlands with moderate rainfall	**Processing:** Much of this crop is made into juice, candied fruit, and other products **Preparation:** People attempting to eat the fresh fruit will find that the seeds cannot be separated, and must be eaten with the fruit. **Uses:** Desserts, fruit dishes, fruit punch, candied fruit, jams, jellies, and juices	**Calories:** 34 kcal/100 g **Nutritive Value:** An excellent source of potassium, and a good source of iron (1.09 mg/100 g), vitamin A (16.9 I.U./100 g), and vitamin C (20 mg/100 g)
Peach and Nectarine *Prunus persica* **Origin and History:** Originated in China. Brought to India and Western Asia in ancient times. Alexander the Great introduced the fruit into Europe after he conquered the Persians. (Also see PEACH AND NECTARINE.)	**Importance:** Peaches and nectarines are major fruit crops. Their combined worldwide production is about 8.7 million metric tons annually. **Principal Areas:** Italy and U.S.A. **Growing Conditions:** Warm, temperate climate. Requires abundant sunshine	**Processing:** About half of the crop is marketed fresh, and the rest is processed. **Preparation:** Wash, peel, and slice fresh fruit and serve alone or with cream. There are many dessert recipes using peaches. **Uses:** Desserts, fruit dishes, baked goods, gelatin desserts, ice cream, jams, jellies, juices, pies, salads, sherbets; and in the production of alcoholic beverages	**Calories:** 38 kcal/100 g **Nutritive Value:** A good source of potassium (202 mg/100 g) and an excellent source of vitamin A (1,300 I.U./100 g). Dried peaches contain about 7 times the solids and calories of the fresh fruit.
Pear *Pyrus communis* **Origin and History:** The pear is indigenous to western Asia; and it has long been cultivated there and in Europe. (Also see PEAR.)	**Importance:** A major world fruit crop with an annual production of about 9.8 million metric tons **Principal Areas:** China, Italy, and U.S.A. **Growing Conditions:** Temperate climate (does not tolerate extreme heat or cold)	**Processing:** More than half of the crop is canned, while most of the rest is sold as fresh fruit. Only a very small amount is dried. **Preparation:** Wash, halve or quarter and remove core. Serve fresh or canned, alone or in mixed dishes. **Uses:** Desserts, fruit dishes, jams, jellies, juices, and salads; and in the production of alcoholic beverages	**Calories:** 61 kcal/100 g **Nutritive Value:** A fair source of potassium and vitamin C; dried pears contain about 4½ times the solids and calories of the fresh fruit.
Persimmon *Diospyrus virginiana* (American persimmon) *Diospyrus kaki* (Japanese persimmon [kaki]) **Origin and History:** The American persimmon is native to southeastern U.S.A. The Japanese persimmon is native to central and northern China.	**Importance:** A minor crop in the U.S.A., with an annual production of about 2 thousand metric tons; only the Japanese persimmon is grown commercially. **Principal Areas:** Subtropical regions of Asia, Europe, and U.S.A. **Growing Conditions:** Warm temperate to subtropical climate	**Processing:** The fruits are very astringent (puckery) unless they are fully ripened. Some persimmons are (1) dried and eaten like candy, or (2) frozen and eaten like popsicles. **Preparation:** Wash, remove skin layer, serve raw, cooked, or candied. **Uses:** Desserts, fruit dishes, jams, jellies, juices, and pies.	*Values for Japanese persimmon:* **Calories:** 77 kcal/100 g **Nutritive Value:** A good source of potassium (174 mg/100 g), and a fair source or vitamin C; Japanese or kaki persimmons are an excellent source of vitamin A (2,710 I.U./100 g), but the varieties native to the U.S.A. contain only small amounts of the vitamin.

Footnote at end of table

(Continued)

TABLE F-47 *(Continued)*

Popular and Scientific Name(s); Origin and History	Importance; Principal Areas; Growing Conditions	Processing; Preparation; Uses	Calories; Nutritive Value[1]
Pineapple *Ananas comosus* **Origin and History:** Native to South America, possibly originating in Brazil. Explorers took it to Europe where it was grown in hothouses. In the early 1900s, large plantations were established in Australia, the Azores, South Africa, and Hawaii. (Also see PINEAPPLE.) 	**Importance:** A major fruit crop of the world, ranking eighth among the top 20 fruits of the world, with a production of about 9.7 million metric tons. **Principal Areas:** Thailand, Philippines, China, Brazil, India, U.S.A, and Mexico. **Growing Conditions:** Semi-arid tropical climate.	**Processing:** Large amount consumed fresh, canned as sliced (rings), chunked, crushed, or juice. **Preparation:** Core and shell removed, sliced in rings or chunks. **Uses:** Served alone or used in pie, ice cream, puddings, baked goods, and salads or with meats; must be cooked or canned before using in gelatin.	**Calories:** 58 kcal/100 g, fresh **Nutritive Value:** Fair source of vitamin A (70 I.U./100 g) and vitamin C (17 mg/100 g).
Plantain *Musa paradisiaca* **Origin and History:** Appears to have originated on the Malay peninsula. (Also see PLANTAIN[S].) 	**Importance:** One of the leading fruit crops of the world, with an annual production of about 25.4 million metric tons. **Principal Areas:** Uganda, Colombia, Rwanda, Zaire, and Tanzania. **Growing Conditions:** Moist, tropical climate.	**Processing:** This type of banana usually has a higher starch content than ordinary eating bananas. Also, they are not as likely to be ripened after picking. **Preparation:** Baked, broiled, or fried. (The high starch content makes it necessary to cook them before eating.) **Uses:** Fruit or vegetable dish.	**Calories:** 119 kcal/100 g **Nutritive Value:** An excellent source of potassium (385 mg/100 g); fair source of vitamin C.
Plum and Prune *Prunus domestica* and others **Origin and History:** There are many cultivars; some originated in Europe, others in China and still others in America. (Also see PLUM AND PRUNE.) 	**Importance:** A major fruit with an annual worldwide production of about 5.7 million metric tons. **Principal Areas:** U.S.S.R., China, U.S.A., Romania, and Yugoslavia. **Growing Conditions:** Temperate climate.	**Processing:** Much of this crop is made into dried prunes, canned fruit, and various other products. **Preparation:** Served raw or stewed (usually dried plums). **Uses:** Desserts, fruit dishes, baked goods, candies, jams, jellies, juices, pickles, pies, and salads; and in the production of alcoholic beverages.	**Calories:** 66 kcal/100 g **Nutritive Value:** A source of potassium, but contains only negligible amounts of vitamin C; dried prunes contain about 4 times the solids and calories of fresh.
Pomegranate *Punica granatum* **Origin and History:** Originated in the Middle East. Long cultivated throughout the Mediterranean world. 	**Importance:** A minor fruit crop in the U.S.A., with an annual production of about 16 thousand metric tons. **Principal Areas:** Subtropical parts of North America, Europe, and Asia. **Growing Conditions:** Subtropical or tropical climate. Does best in areas with cool winters and hot, dry summers.	**Processing:** Marketed fresh or processed. **Preparation:** Cut open and remove the red sac-covered seeds. (The seeds and the white pulp are not eaten.) Serve as raw, fresh fruit. **Uses:** Desserts or fruit dishes; ingredient of alcoholic drinks (used in the form of grenadine syrup or as the fruit juice); jams, jellies, and syrups.	**Calories:** 63 kcal/100 g **Nutritive Value:** A good source of potassium (259 mg/100 g), but a poor source of vitamin C.

Footnote at end of table

(Continued)

TABLE F-47 *(Continued)*

Popular and Scientific Name(s); Origin and History	Importance; Principal Areas; Growing Conditions	Processing; Preparation; Uses	Calories; Nutritive Value[1]
Quince, Pineapple *Cydonia oblonga* **Origin and History:** Believed to have originated in western Asia. It is closely related to the apple and pear.	**Importance:** A minor fruit crop. **Principal Areas:** The temperate zones of Europe and Argentina. **Growing Conditions:** Mild, temperate climate.	**Processing:** The fresh fruit is quite acid and not very palatable. Hence, it is usually cooked or processed. **Preparation:** Stewed and sweetened. **Uses:** Desserts, fruit dishes, candies, jams, jellies, juices, and marmalades.	**Calories:** 57 kcal/100 g **Nutritive Value:** A good source of vitamin C and fiber (1.7%).
Rambutan *Nephelium lappaceum* **Origin and History:** Native to Malaysia. Seldom grown elsewhere.	**Importance:** Use is limited to native habitat and surrounding areas. **Principal Areas:** Malaysia **Growing Conditions:** Humid, tropical climate.	**Processing:** There is little or no processing because the fruit is consumed fresh in the areas near where it is grown. **Preparation:** Serve the fresh fruit raw or cooked. **Uses:** Desserts or fruit dishes.	**Calories:** 64 kcal/100 g **Nutritive Value:** Moderately high in crude fiber (1.1%); a good source of iron (1.9 mg/100 g) and vitamin C (53 mg/100 g).
Raspberry *Rubus idaeus* and others **Origin and History:** Various species of raspberry are native to eastern Asia, Europe and North America. (Also see RASPBERRY.)	**Importance:** Among the top 2 dozen fruit crops of the world, with an annual production of about 352 thousand metric tons. **Principal Areas:** U.S.S.R., Yugoslavia, Germany, Poland, U.S.A., and Hungary. **Growing Conditions:** Cool, temperate climate. Does best on light, neutral soils.	**Processing:** This crop is marketed as fresh fruit, canned, frozen, and processed in other ways. **Preparation:** Serve fresh fruit raw (alone or with cream), or in mixed dishes (such as shortcakes). **Uses:** Desserts, fruit dishes, baked goods, gelatin desserts, ice cream, ices, jams, jellies, pies, sherbets, soft drinks, and syrups; and in the production of liqueurs and wines.	**Calories:** 73 kcal/100g **Nutritive Value:** High in fiber (5.1%), a good source of potassium (199 mg/100 g) and bioflavonoids, and a fair source of vitamin C; red raspberries contain 3.0% fiber, 168 mg potassium/100 g, and are a good source of vitamin C (25 mg/100 g), and bioflavonoids.
Sapodilla *Manilkara zopota* **Origin and History:** Originated in Central America and Mexico.	**Importance:** Commercial cultivation throughout the tropics is increasing steadily. **Principal Areas:** Central America, West Indies, Mexico, Philippines, and Malaysia. **Growing Conditions:** Tropical climate.	**Processing:** None, because the fresh fruit is consumed in areas near to where it is produced. **Preparation:** Serve the fresh fruit raw. **Uses:** Desserts or fruit dishes.	**Calories:** 89 kcal/100 g **Nutritive Value:** A good source of fiber (1.4%) and potassium (193 mg/ 100 g); and a fair source of vitamin C.

Footnote at end of table

(Continued)

TABLE F-47 *(Continued)*

Popular and Scientific Name(s); Origin and History	Importance; Principal Areas; Growing Conditions	Processing; Preparation; Uses	Calories; Nutritive Value[1]
Sapote *Calocarpum sapota* (white sapote) *Calocarpum viride* (green sapote) **Origin and History:** Native to Central America and Mexico.	**Importance:** The present limited cultivation of this crop appears to be increasing steadily. **Principal Areas:** Central America, Mexico, and Caribbean Islands. **Growing Conditions:** Tropical climate.	**Processing:** Some of the crop is marketed fresh, while the rest is made into preserves. **Preparation:** Usually eaten fresh. **Uses:** Desserts, fruit dishes, jams, and jellies.	**Calories:** 125 kcal/100 g **Nutritive Value:** A good source of fiber (1.9%) and iron (1.0 mg/100 g), a fair source of vitamins A and C.
Soursop (Guanabana) *Annona muricata* **Origin and History:** Native to the American tropics.	**Importance:** Has a limited commercial production. **Principal Areas:** Central America and Peru. **Growing Conditions:** Tropical climate. Very sensitive to cold weather.	**Processing:** Much of the crop is used to make fruit drinks and ice cream, while the rest is marketed fresh. **Preparation:** Serve the fresh fruit raw with a little sugar. **Uses:** Desserts, fruit dishes, fruit drinks, and ice cream.	**Calories:** 65 kcal/100 g **Nutritive Value:** A good source of fiber (1.1%), and potassium (265 mg/100 g); a fair source of vitamin C.
Strawberry *Fragaria viginiana* and *Fragaria chiloensis* **Origin and History:** Some species are native to the temperate parts of the Old World, whereas others originated in North America and South America. (Also see STRAWBERRY.)	**Importance:** One of the top 2 dozen fruit crops of the world. About 2.4 million metric tons are produced annually. **Principal Areas:** U.S.A., Poland, Japan, Spain, and Mexico. **Growing Conditions:** Temperate climate. Requires a well-drained soil.	**Processing:** Fresh strawberries do not keep well, so much of the crop is frozen or made into jams and syrups. **Preparation:** Serve fresh fruit raw (alone or with cream), or in mixed dishes (such as shortcakes). **Uses:** Desserts, fruit dishes, baked goods, gelatin desserts, ice cream, ices, jams, jellies, pies, sherbets, soft drinks, and syrups; and in the production of liqueurs and wines.	**Calories:** 37 kcal/100 g **Nutritive Value:** A good source of iron (1.0 mg/100 g) and potassium (164 mg/ 100 g); an excellent source of vitamin C (59 mg/100 g).
Tamarind *Tamarindus indica* **Origin and History:** Originated in the savanna zone of tropical Africa. Brought to India in ancient times.	**Importance:** Of minor importance in the U.S.A. **Principal Areas:** India and other tropical areas throughout the world. **Growing Conditions:** Thrives in moderately dry tropical climates, but tolerates monsoon conditions if the soil is well drained.	**Processing:** This fruit is eaten fresh, and made into various products. **Preparation:** Serve the fresh fruit raw (alone or with a little sugar). **Uses:** Desserts, fruit dishes, candied fruit, chutneys, curries, fruit drinks, sauces, and sherbets.	**Calories:** 331 kcal/100 g **Nutritive Value:** Rich in fiber (5.1%), and potassium (781 mg/100 g); a poor source of vitamin C.

Footnote at end of table

TABLE F-47 *(Continued)*

Popular and Scientific Name(s); Origin and History	Importance; Principal Areas; Growing Conditions	Processing; Preparation; Uses	Calories; Nutritive Value[1]
Tangerine (Mandarin Orange; Tangelo; Tangor; Temple Orange) *Citrus reticulata* **Origin and History:** Originated in southern China. Cultivated in China and Japan for many centuries. (Also see TANGERINE.)	**Importance:** A major fruit crop of the world. About 8.8 million metric tons are produced annually. **Principal Areas:** Japan, Brazil, Korea, Rep., China, and U.S.A. **Growing Conditions:** Subtropical or tropical climate	**Processing:** Most of the crop is marketed fresh, but some is canned. **Preparation:** Wash, peel, and section, then serve fresh fruit raw. **Uses:** Desserts (including gelatin desserts) or fruit dishes	**Calories:** 46 kcal/100 g **Nutritive Value:** A good source of vitamin C (31 mg/100 g), and bioflavonoids; and a fair source of potassium and vitamin A
Tomato *Lycopersicon esculentam* Tomato is classed as a fruit, but used as a vegetable. See VEGETABLE(S), Table V-6 Vegetables of the World.			
Ugli Fruit *Citrus paradisi* (grapefruit) *Citrus reticulata* (tangerine) **Origin and History:** Naturally occurring hybrid of the mandarin orange, and the grapefruit or pummelo; recently developed hybrid resembles grapefruit	**Importance:** Present commercial production is limited. **Principal Areas:** Jamaica, W.I. **Growing Conditions:** Tropical climate	**Processing:** None, because all of the fruit is marketed fresh **Preparation:** Serve fresh (alone or with other fruits and/or a little sugar), or broil briefly **Uses:** Appetizer or fruit dishes	Nutritional data is not readily available.
Watermelon *Citrullus vulgaris* **Origin and History:** Native to subtropical and tropical regions of Africa. Cultivated in Egypt and throughout the Mediterranean in ancient times. (Also see WATERMELON.)	**Importance:** One of the leading fruit crops of the world, with an annual production of about 28.9 million metric tons **Principal Areas:** China, U.S.S.R., Turkey, Egypt, and U.S.A. **Growing Conditions:** Warm temperate, subtropical, or tropical climate. Requires well-drained, fertile soil. Thrives in hot, sunny weather.	**Processing:** Almost all of this crop is consumed fresh. Only small amounts are used to make processed fruit products. **Preparation:** Chill, slice, and serve with a fork or spoon; or remove seeds and make melon balls. **Uses:** Desserts, fruit dishes, ices, jams, and pickles (usually, only the white pulpy rind [not the green outer rind] is pickled)	**Calories:** 26 kcal/100 g **Nutritive Value:** A fair source of potassium and vitamin A, but a poor source of vitamin C

[1]Calories and nutritive values are for the fresh, raw fruit, unless stated otherwise, Canned, cooked, or dried fruits usually contain much less vitamin C than the fresh, raw fruits. For more information regarding the nutritive values of fruits the reader is referred to Food Composition Table F-36.

FRUIT BEVERAGES

These products range from pure fruit juices to highly diluted, artificially colored and flavored drinks that contain little or no fruit ingredient. Hence, the FDA has issued standards of identity for many types of fruit beverages so that consumers may be assured that certain types of products will contain specified levels of fruit ingredients; for example, "*cranberry juice cocktail*" must contain at least 25% cranberry juice.

(Also see the articles on the individual fruits for details regarding the various fruit beverages.)

FRUIT, CANNED

Many of the fruits that are grown in the United States are processed by canning, which ensures a supply of the item when fresh fruit is not readily available. Also, canning makes it possible to utilize bumper crops more efficiently since there is a limit to the demand for fresh fruit at harvest time.

Canned fruit is packed in a medium that may vary from plain water or fruit juice to syrups that range in density from "extra light" to "extra heavy." The denser the syrup, the higher its sugar content. Furthermore, the weight of solid fruit in the can or jar (which is called the "drained weight") may be as low as 50% of the net weight in the cases of grapefruit and whole plums to as high as 78% in the case of "solid pack" crushed pineapple. The FDA standards of identity for canned fruits specify minimum drained weights for certain products.

Canning may reduce the vitamin levels to about one-half of those in fresh fruits, but these fruits are still good sources of some of the vitamins. Minerals remain intact during processing, but some of these nutrients are likely to migrate from the fruit to the packing fluid. Canned fruits may be better tolerated by people with digestive disturbances because some of the fiber is broken down during canning.

(Also see FRUIT[S], section headed "Canning," and the articles on the individual fruits.)

FRUIT, DRIED

The fruits which are most commonly dried in the United States are apples, apricots, dates, figs, nectarines, peaches, plums, prunes, and seedless grapes. Drying reduces the water content of the fruit to about 24%, which is usually low enough to prevent spoilage. Slow drying of fruits in the sun may result in the destruction of much of the vitamin A and vitamin C contents, unless the fruits are sulfured before drying. However, sulfuring destroys thiamin.

Dried fruits are rich in calories, sugars, fiber, iron, potassium, and various other nutrients because the solids content is about 5 times that of the fresh fruits. Hence, dried fruit is generally a good buy when it does not cost more than 5 times the price per pound of the fresh fruit, since the former has no waste in the form of peel and seeds. Furthermore, the consumption of ample amounts of raw or stewed dried fruits may help to promote the regularity of bowel movements.

(Also see the nutrient compositions of fresh and dried fruits which are given in Food Composition Table F-36; and FRUIT[S], sections headed "Drying" and "Effects of Processing and Preparation.")

FRUIT SUGAR

Fructose is found in a free form in some fruits; hence, the name fruit sugar.

(Also see CARBOHYDRATE[S].)

FRUIT SYRUPS

Most of these products are fruit-flavored sugar syrups. For example, apple syrup is made by concentrating apple juice (by boiling or vacuum evaporation), then adding sugar until a syrup of the desired thickness (viscosity) is obtained. Syrups that have citrus fruit flavors may be made in much the same way, except that they may also contain citrus peel and/or essential oil extracted from the peel.

In California, fruit syrups are made from dried figs and dried raisins without added sugar by extracting the chopped fruit with boiling water, then concentrating the resulting juice. However, the fig syrup is rather expensive and is presently sold only in certain health food stores, whereas the raisin syrup is sold only to large food concerns.

Homemakers who would like to utilize other types of syruplike products that are made only from fruits might try thawed, unsweetened frozen fruit juice concentrates such as those made from apple, orange, and pineapple juices.

(Also see SYRUPS; and the articles dealing with the individual fruits.)

FRUMENTY

- A dessert made by boiling wheat in milk, flavored with sugar, spice, and raisins.

- A molded cereal dessert.

FUEL FACTOR

The average caloric value of each of three major nutrients is known as its *fuel factor*. One gram of carbohydrate yields 4 calories, 1 g of fat yields 9 calories, and 1 g of protein yields 4 calories.

FUNGI

Plants that contain no chlorophyll, flowers, or leaves, such as molds, mushrooms, toadstools, and yeasts. They may get their nourishment from either living or dead organic matter.

(Also see AFLATOXINS; BACTERIA IN FOOD, MOLDS; MUSHROOMS; and YEAST.)

FUNGICIDE

An agent that destroys fungi.

FUSTIC

A yellow dye derived from the wood of a tropical American tree.

PEACHES 'N SPICE 'N EVERYTHING NICE. Fresh California peaches are delicious spiced up in a fruit compote and stuffed with cottage cheese. (Courtesy, The California Peach Commodity Committee, Sacramento, Calif.)

GALACTOSE

A hexose sugar (monosaccharide) obtained along with glucose from lactose hydrolysis.

(Also see CARBOHYDRATE[S]; and SUGARS.)

GALACTOSEMIA

Literally means a galactose condition of the blood. Galactose builds up in the blood as a consequence of a genetic disorder in metabolism. The enzyme, galactose-1-phosphate uridyl transferase, necessary to change galactose to glucose, is missing; thus, galactose is not metabolized. In the infant, the source of galactose is lactose in the milk. Lactose is comprised of one glucose molecule and one galactose molecule linked together. Fig. G-1 illustrates the structure of glucose and galactose. Note how similar they are—only one subtle difference. Yet, the source of galactose must be removed from the infant's diet or serious complications result—vomiting, jaundice, growth failure, and mental retardation. The milk substitute usually used for infant feeding is Nutramigen, a complete protein hydrolysate that is free of galactose.

(Also see INBORN ERRORS OF METABOLISM.)

```
      H—C=O                 H—C=O
        |                     |
      H—C—O—H               H—C—O—H
        |                     |
  H—O—C—H               H—O—C—H
        |                     |
  H—O—C—H                 H—C—O—H
        |                     |
      H—C—O—H               H—C—O—H
        |                     |
      H—C—O—H               H—C—O—H
        |                     |
        H                     H

    Galactose             Glucose
```

Fig. G-1. The structure of galactose and glucose showing similarities and difference.

GALLATES

A class of organic compounds derived from gallic acid and used in the manufacture of writing ink, paper and dyes. In the food industry, propyl, octyl and dodecyl gallates are used as antioxidants, of which propyl gallate is the antioxidant of choice in the United States.

(Also see ADDITIVES; and PROPYL GALLATE.)

GALLBLADDER

The pear-shaped bag located under the right lobe of the liver. Its functions are the storage, concentration, and release of bile, a natural emulsifying agent. The presence of fats and other foods in the digestive tract stimulates the release of the hormone cholecystokinin-pancreozymin which in turn causes the contraction of the gallbladder, releasing bile into the duodenum of the small intestine. Bile is essential for the complete digestion of fats.

It is noteworthy that the following animals have no gallbladder: horse, deer, elk, moose, giraffe, camel, elephant, pigeon, dove, and rat; probably reflecting some relation to their normal diet.

(Also see DIGESTION AND ABSORPTION; and GALLSTONES.)

GALLBLADDER DISEASE

In general, it is a disorder in the normal structure and function of the gallbladder and bile ducts. Three types of gallbladder disorders exist: (1) cholecysititis or inflammation (infection) of the gallbladder; (2) cholelithiasis or the formation of gallstones; and (3) although uncommon, tumors or cancer.

Diet is extremely important to persons suffering from a gallbladder disease. The principal aim is to provide a fat-restricted diet, and to reduce the consumption of other foods such as onions, sauerkraut, and alcoholic beverages that may induce the recurrence of the symptoms—mild to severe pain, abdominal distention, nausea, and vomiting. Often persons suffering from gallbladder disorders are obese, and a reduction in total calories to achieve a weight loss is also necessary.

(Also see CHOLECYSITITIS; CHOLELITHIASIS; GALLSTONES; and MODIFIED DIETS.)

GALLON

A unit of liquid measure in the U.S. Customary System equivalent to 4 qt or 3.79 liter. The Imperial or British gallon equals 4.55 liter or 1.2 U.S. gallons.

GALLSTONES (CHOLELITHIASIS)

Contents Page

Symptoms 1048
Possible Causes 1048
Dietary 1049
Physiological 1049
Diagnosis 1050
Treatment 1050
Diet Therapy 1051

The gallbladder is a small pear-shaped sac which is attached to the system of bile ducts on the undersurface of the liver. It stores and concentrates (by the absorption of water) bile which the liver produces at a slow but steady rate. Bile is composed of bilirubin, bile acids—cholic acid, chenodeoxycholic acid, deoxycholic acid, and lithocholic acid—cholesterol, and lecithin. The bile acids are produced from cholesterol. Bile salts are the sodium and potassium salts of the bile acids combined with glycine or taurine. Gallstones—gravellike deposits—are formed when certain substances, which are normally dissolved in the bile, come out of solution as flakelike particles and clump together around a core which may be either tiny bits of sloughed-off lining tissue from the gallbladder, bile salts, calcium carbonate, bile pigments (bilirubin), bacteria, or, in rare cases, such parasites as roundworms (ascarids). Gallstones may vary in their composition from almost pure cholesterol to almost pure compounds of calcium and bile pigments.

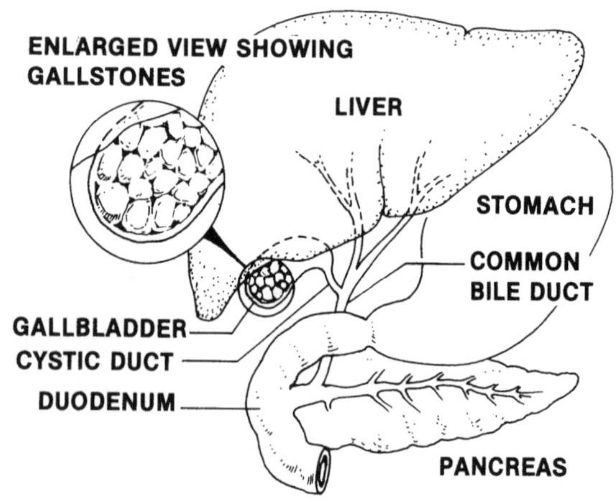

Fig. G-2. The location of the gallbladder with gallstones, the cystic duct and the common bile duct in relation to the liver, intestine, and pancreas.

However, only a few stones may cause problems because most of them are small enough to be easily carried through the bile ducts, and often stones remain in the gallbladder as silent stones. Trouble arises when the stones are too large to pass through the various biliary ducts. Large stones become wedged in the cystic duct, or in the common bile duct—choledocholithiasis—inducing a painful gallbladder attack.

• **Cholesterol gallstones**—This type of stone is by far the most common—about 85% of the cases in the United States. Cholesterol is a whitish, waxy, fatlike material (lipid) which is almost insoluble in most body fluids unless it is held in solution—emulsified—by means of special complexes. In bile, cholesterol is emulsified and solubilized by a combination of bile salts and lecithin (phosphatidyl-choline). Hence, cholesterol may drop out of the bile when there is a shortage of either of the two types of emulsifying agents.

The development of cholesterol gallstones is divided into three stages: (1) the chemical stage when the bile becomes supersaturated with cholesterol; (2) the crystallization stage when cholesterol begins to form microcrystals; and (3) the growth stage when microcrystals come together and form stones. These conditions may be induced by (1) too much absorption of water from the bile, (2) inadequate amounts of bile salts, and/or lecithin in the bile, (3) too much secretion of cholesterol in the bile, or (4) inflammation of the lining of the gallbladder.

• **Calcium-containing gallstones**—Normally, both the bile pigments and calcium stay in solution in the bile because (1) the liver usually combines the bile pigments with carbohydrate so as to form soluble, complex molecules, and (2) calcium is present in the bile as a soluble, charged particle or ion. However, certain bacteria, such as *E. coli*, are sometimes present in the gallbladder. Such bacteria may contain an enzyme which separates the carbohydrate from the bile pigments. Then, the pigments may react with calcium to form the insoluble substances that are found in gallstones.

SYMPTOMS. Generally, stones are undetected—silent—until they move and lodge in the bile duct. The hallmark of an acute gallbladder attack is a sharp pain which is felt on the right side of the body just under the rib cage, and sometimes traveling through to the back under the right shoulder blade. Frequently, the pain develops following a meal, when the gallbladder contracts and dumps its stored bile into the duodenum to promote fat digestion. (Pain caused by ulcer is relieved by eating.) Depending upon the location of the stones, jaundice, fever, chills and vomiting may also be symptoms.

(Also see DIGESTION AND ABSORPTION.)

POSSIBLE CAUSES. The exact causes of gallstones are not known. Many people believe that gallstones result from eating lots of fat and/or being fat. Also, there is an old saying that the people who are most likely to have gallstones are characterized by the "4 Fs"—fair, fat, female, and forty. Furthermore, a greater than average susceptibility to gallstones is noted in Jewish people, Pima Indians, and in people with such diseases as cirrhosis of the liver and diabetes. As with many disorders, there are no simple explanations. Various dietary and physiological factors seem to be contributory to stone-forming conditions.

The following discussion deals primarily with cholesterol containing gallstones, since they are the most common.

Dietary. Deficiencies or excesses of some nutrients may be involved in gallstone formation.

• **Cholesterol**—This substance is the major constituent of gallstones, so it would seem that eating too much of it may be one of the causes of the stones. However, many people who consume liberal amounts of cholesterol (more than 500 mg per day) have neither high blood cholesterol nor gallstones. Therefore, it appears that only certain people are susceptible to the buildup of cholesterol in their blood and/or bile. Obese individuals are more likely than lean individuals to be troubled with gallstones, but the reasons for this fact are uncertain. Perhaps one reason is that more cholesterol may be synthesized by the liver of the obese. Hence, their bile becomes loaded with cholesterol which may readily drop out of solution and form stones.

• **Fat**—Excessive consumption of fat over a period of years may lead to the formation of gallstones by stimulating the liver to synthesize greater-than-normal amounts of cholesterol. In particular, a high polyunsaturated to saturated fatty acid ratio in the diet has been suggested as promoting gallstone formation. In one study, patients who consumed large doses of polyunsaturated vegetable oils, for the purpose of lowering their blood cholesterol, demonstrated over twice the incidence of gallstones as similar patients whose dietary fat was more saturated.

• **Fiber**—Gallstones are rare in areas where considerable dietary fiber is eaten—rural Africa, but they are common where diets are low in fiber—the western world. Some studies have demonstrated that a high fiber diet lowers the levels of cholesterol and other fats. The means by which fiber affects cholesterol and lipid metabolism is unclear. Possibly fiber binds bile salts and increases the fecal excretion of cholesterol. This would place a twofold drain on cholesterol—fecal loss and more cholesterol required to synthesize more bile salts. Currently, though, the invovement of fiber is still speculative, and largely based upon epidemiological (studies of populations) evidence. Many factors other than fiber intake distinguish two populations of people.

(Also see FIBER.)

• **Lecithin**—This is an important constituent of bile which helps to keep cholesterol in solution. It is not generally considered to be an essential nutrient because it is thought that the body is able to synthesize the amounts it needs. However, some patients afflicted with gallstones have low levels of lecithin in their bile. Insufficient biliary lecithin may be due to dietary deficiencies of protein, vitamins, choline, and other nutrients which are involved in the synthesis of lecithin by the body.

(Also see CHOLINE; and FATS AND OTHER LIPIDS.)

• **Protein**—The digestion products of protein may protect against gallstone formation in the following ways:

1. The passage of partially digested proteins from the stomach to the small intestine helps to trigger the mechanism for the emptying of bile from the gallbladder. A static bile favors stone formation.
2. Certain amino acids—serine and methionine—from protein are the raw materials used by the body to make lecithin, which helps to prevent gallstones.

• **Vitamins**—Indirectly, a low intake of some vitamins may contribute to the formation of gallstones. This is demonstrated by the following salient points:

1. Laboratory animals that are deficient in vitamin A develop gallstones due to excessive shedding of the lining cells of the gallbladder, which act as nuclei for the formation of stones.
2. Various B vitamins, such as folacin, vitamin B-6, and vitamin B-12, are required for the production of lecithin by the body.
3. Animal studies suggest vitamin C is required for the conversion of cholesterol in the bile salts, which help to solubilize cholesterol.
4. Vitamin E protects red blood cells against a premature breakdown (hemolysis). Hence, it prevents the production of excessive amounts of bile pigments, which are products of red cell hemolysis. (One type of stone is composed of calcium and bile pigments.) The synthesis of lecithin requires polyunsaturated fatty acids (PUFA). Vitamin E protects these substances against peroxidation. It may also act along with vitamin A in protecting the gallbladder against excessive shedding of its lining cells.

Physiological. Some physiological factors also appear to predispose an individual to gallstone formation.

• **Bacteria and inflammation**—Under certain conditions, the small intestine may contain bacteria which convert the primary bile salts into secondary bile salts which are less effective emulsifying agents. Also, inflammation of the inside lining of the gallbladder resulting from a chronic low grade infection may change the absorptive characteristics of the gallbladder lining, possibly allowing excessive absorption of water, bile salts or lecithin—the substances necessary to keep cholesterol in solution. As a result, cholesterol begins to precipitate and the small crystals promote the formation of larger and larger crystals.

• **Diabetes**—It is well known that diabetics are more likely to be troubled with gallstones than are nondiabetics. The reasons for this unfortunate situation are that diabetes may often be accompanied by (1) accumulation of cholesterol and fat in the liver, and (2) the growth of intestinal bacteria which degrade bile salts. Furthermore, the administration of insulin increases the cholesterol saturation of bile—a condition which favors stone formation.

• **Disorders of the ileal region of the small intestine**—Gallstones may result form any disorder of the ileum in which the reabsorption of bile salts is drastically reduced so the liver cannot maintain the supply of these salts at the normal level. Some of these disorders are (1) celiac disease, which is an allergy to gluten—a protein found in such grains as barley, oats, rye, and wheat; (2) infectious diarrhea due to bacteria, viruses, or parasites; (3) chronic inflammations of the ileum like regional enteritis; and (4) surgical removal of part of the ileum.

(Also see MALABSORPTION SYNDROME.)

• **Hemolysis of red blood cells**—Excessive amounts of bile pigments—bilirubin—may accumulate in the liver and the bile when the rate of breakdown of red blood cells (hemolysis) is considerably more rapid than normal. Abnormally high rates of hemolysis may result from disorders such as anemia, drug toxicities, and nutritional deficiencies. High levels of bile pigments in the bile may increase the likelihood of gallstones.

• **Hypercholesterolemia**—Hereditary hypercholesterolemia—high blood cholesterol levels—appears to be related to the development of gallstones. Moreover, clofibrate, a drug used to lower the blood cholesterol of patients who are at risk of developing heart disease, increases the amount of cholesterol secreted by the liver in the bile, thereby increasing the likelihood that cholesterol stones may form. Still, benefits derived from its use may outweigh the risk of gallstones.

• **Liver disorders**—A logical place to look for abnormalities which may lead to gallstones is the liver—the organ where bile and its components are processed for secretion via the biliary tract. Gallstones occur about 2½ times more often in patients with cirrhosis of the liver—a degenerative disease in which functioning liver tissue is replaced by nonfunctioning fibrous tissue—than in noncirrhotic patients. The cause of cirrhosis is not always known, but it is often suspected to have resulted from one or more of such factors as alcoholism, infectious diseases, nutritional deficiencies, and toxic agents.

Another liver disorder which may be responsible for gallstones is the excessive accumulation of fat in the liver, a condition which is also known as *fatty liver.* However, the diagnosis of fatty liver is not made in humans as often as it might be, because it requires that a needle biopsy be taken. Hence, there are not enough data available to determine whether fatty liver and gallstones occur together. Nevertheless, various animal studies have shown that certain types of *stone-forming diets* produce the triple consequences of (1) elevated levels of fat in the liver, (2) high liver cholesterol, and (3) gallstones. Fatty liver may be caused by the same factors that cause cirrhosis. However, a recent study showed that there was a strong likelihood of the occurrence of fatty liver in people who (1) were markedly obese, and (2) had high blood levels of sugar, insulin, and/or triglycerides.[1]

• **Sex and hormones**—Women of child bearing age are much more likely than men to develop gallstones. Moreover, several pregnancies seems to predispose women to gallstones. Estrogens, the female hormones secreted by all normal women between the onset of puberty and the menopause, appear to increase the likelihood of stone formation. Women receiving supplemental estrogen demonstrate a higher than normal rate of gallbladder disease. However, not to be misleading, many men also develop gallstones.

DIAGNOSIS. Proper diagnosis of gallstones is essential so that (1) a potentially hazardous condition is corrected promptly, and (2) unnecessary surgery is avoided if the suspicion of gallstones proves to be unfounded.

The symptomatic pain suggests gallstones. However, these pains may be due to spasms or inflammation of the gallbladder with or without the occurrence of stones. In cases where diagnosis is uncertain, further diagnostic measures such as x rays or ultrasound are employed.

Ultrasound is not used as routinely but it offers the advantage of the absence of risk or discomfort. Sonarlike waves bounce off gallstones and other body structures forming their image on a screen. No dye is necessary.

TREATMENT. There is a very strong case for dissolving the stones, or for surgical removal of the gallbladder—cholecystectomy—when (1) large stones are detected by x rays, or other means, (2) gallbladder pain is unremitting, and (3) repeated gallbladder attacks occur.

While the surgery is major, life without a gallbladder generally does not require major changes in eating habits following recovery.

• **Solvent treatment for gallstones**—In 1989, Mayo Clinic researchers reported on a solvent treatment for gallstones that may eliminate the need for surgery in many cases. The treatment entails emptying the gallbladder of its liquid contents, then pumping a teaspoonful of the solvent, methyl tert-butyl ether, in and out of the gallbladder 4 to 6 times a minute. The solvent dissolved the gallstones in 71 of the 75 patients within 1 to 3 days.

The decision not to dissolve the stones or remove the gallbladder when stones have been positively demonstrated could cause complications.

• **Blockage of the bile duct(s)**—If the biliary tract is blocked by a gallstone(s), bile may back up into the liver where it may cause inflammation, or it may even accumulate in the blood so that the skin takes on a yellow tint—a condition which is known as *jaundice*. Sometimes excessive levels of bile pigments may also be detected in the urine.

• **Blockage of the pancreatic duct**—Occasionally, a gallstone may obstruct the pancreatic duct so that the pancreatic juices back up into the tissue of the gland, causing inflammation (pancreatitis). There may also be malabsorption of nutrients due to lack of pancreatic digestive juices in the small intestine.

• **Malabsorption syndromes**—Chronic obstruction of the biliary tract by gallstones—which prevent the passage

---

[1]Maruhama, Y., *et al.*, "Hepatic Steatosis and the Elevated Plasma Insulin Level in Patients with Endogenous Hypertriglyceridemia," *Metabolism*, Vol. 24, 1975, p. 653.

of bile into the small intestine—may result in great reductions in the absorption of certain nutrients. Hence, there may be signs of nutritional deficiency disorders such as (1) visual impairment and other problems which result from lack of vitamin A, (2) rickets in children or softening of the bones (osteomalacia) in adults due to malabsorption of calcium and vitamin D, (3) excessive breakdown (hemolysis) of red blood cells as a consequence of vitamin E deficiency, and (4) easy bruising or hemorrhaging from lack of vitamin K.

DIET THERAPY. Currently, no sure dietary method exists for preventing or treating gallstones. Possibly, a pill containing chenodeoxycholic acid—a bile acid—may be employed to dissolve small cholesterol gallstones. However, this requires ½ to 2½ years.

Prior to surgery, the diet is restricted in fat to reduce contractions of the gallbladder occurring via the action of the hormone cholecystokinin. Also, fat is often excluded from the diet during acute attacks thereby easing the pain. Following gallbladder removal, a fat-restricted diet may be eaten for several months for wound healing and comfort, but after this time most individuals can eat regular diets. Bile passes directly from the liver to the intestine during the digestive processes in individuals who have no gallbladder.

Table G-1 provides some general guidelines for restricting dietary fat, and foods which may cause gastric distress.

TABLE G-1
GUIDELINES FOR SELECTING LOW-FAT FOODS

Food Category[1]	Foods to Use	Foods to Avoid
Bakery products	Most kinds, except those with added fat.	Pastry, cakes, cookies, waffles, pancakes.
Beverages	Coffee, tea, carbonated soft drinks.	
Cereals and flours	All cooked or dry breakfast cereals.	Possibly macaroni, noodles, rice, spaghetti.
Desserts and sweets	Jam, jelly, sugar, sugar candies without nuts or chocolate; fruit whip; fruit pudding; gelatin (Jell-O).	Chocolate
Eggs and substitutes	Three per week.	Fried eggs.
Fats and oils	3 tsp *(15 ml)* butter or margarine per day.	Salad oil and cooking fats and oils.
Fish and seafoods	Lean fish baked, broiled or roasted without fat.	Fatty fish, fish canned in oil.
Flavorings and seasonings	In moderation: salt, pepper, spices, herbs.	Sometimes not tolerated: pepper; curries; meat sauces; excessive spices; vinegar.
Fruits	All kinds, if tolerated.	Avocado
Gravies and sauces		Cream sauces and gravies.
Juices	All kinds.	
Legumes and products		Dried, cooked peas and beans.
Meats	Lean beef, chicken, turkey, pork, veal, or lamb broiled, baked, or roasted without added fat.	Fatty meat of beef, lamb, or poultry; duck, goose, ham, bacon, sausage, and organ meats.
Milk and milk products	Skim milk or only 2 cups *(480 ml)* whole milk; cottage cheese.	Cream, ice cream, evaporated and condensed milk; all other cheese.
Nuts and seeds		All nuts and peanut butter.
Pickles and relishes		Many not tolerated.
Soups and chowders	Clear soups, bouillon or broth; no added fat, made with skim milk.	Cream soups.
Vegetables	All kinds, if tolerated; cooked without butter or cream.	Possibly omit the following if they cause distress: broccoli, Brussels sprouts, cabbage, cauliflower, cucumber, onion, peppers, radish, turnips.

[1]Food categories are those listed in the Food Composition Table F-36 of this book.

Fatty meats, gravies, oils, cream, lard, and desserts containing eggs, butter, cream, nuts, and avocados are avoided when attempting to lower fat intake.

When surgery is indicated in grossly obese individuals and it is not an emergency situation—prolonged symptoms indicative of impending gangrene and perforation—weight reduction is necessary. Therefore, the individual will be placed on a reducing diet since obesity presents an added risk for major surgery such as a cholecystectomy.

(Also see FATS AND OTHER LIPIDS; DIGESTION AND ASORPTION; MODIFIED DIETS; MALABSORPTION SYNDROME; METABOLISM; and OBESITY.)

GAME MEAT

Fig. G-3. Washington (third from left) relaxing with fellow hunters, following the kill of a deer. (Courtesy, General Services Administration, National Archives and Records Service, Washington, D.C.)

The King must hunt! The proper entertainment of sovereigns and envoys in ancient Egypt consisted of grand hunts, characterized by rich display. The Assyrian Kings were also eager to go down in history as "mighty hunters before the Lord." Following the hunt, there was usually a great feast, with much merriment.

Game meat refers to the flesh of wild animals or birds used for food.

Among the wild animals hunted for food are: bears, buffalo, deer, elk, hare, rabbits, roebuck, squirrels, and wild boar.

Among the wild birds hunted for food are: doves, ducks, geese, grouse, partridges, peacocks, pheasants, quails, turkeys, and woodcock.

GAME FOR GOURMETS. Game meats create romance for good eating. But, contrary to the legendary yarns, there are neither as many secrets relative to game preparation nor as many differences between preparing wild game and domestic animals and birds as most old-time hunters and guides would have us believe. Indeed, neither hunters nor cooks need be terrified about preparing wild game for the table.

The main difference between wild game and domestic animals and birds is that the wild ones are leaner and stronger flavored than a stall-fed ox or a caged bird—they have developed strong muscles from exercising. For this reason, the meat needs a blanket of fat before it is cooked—lard, fat bacon, or salt pork; and it may require tenderizing. Additionally, the following simple rules relative to preparing and cooking—most of which apply to all animals or birds, wild or domestic—should be observed:

1. Bleed all game, and pluck poultry, immediately. Game birds are easier to pluck while they are still warm. Do not wash game; have it dry when hung, and keep it dry. (Wipe with a dry or moist cloth if necessary.)
2. Limit hanging time in the open to a minimum; the period of time is dictated primarily by temperature, humidity, and the mutilated condition of the kill. While hanging, be sure to cover torn, shot-wounded areas to shut out flies.
3. Move the game to controlled refrigeration as soon as possible. Before refrigerating, draw (eviscerate). Age at 38° to 40°F (*3 to 4°C*) and at 70 to 90% humidity; dressed animals may be aged for 1 to 2 weeks, while dressed birds should be aged for only 2 to 4 days. (The length of aging time is determined primarily by the elapsed period between kill and refrigeration and the finish of the animal or bird.)
4. Skin and cut animal carcasses properly—cut across the grain, and separate the tender from the less tender cuts. Unless the hunter has time and experience, cutting, wrapping, and labeling had best be done on a custom basis by a person with the tools and expertise, such as the meat cutter at the local freezer locker.
5. Freeze at 0°F (*-17.8°C*) or lower. Ground meats may be frozen for 2 to 3 months; cuts may be frozen for 6 to 12 months. Birds may be frozen for 6 months or longer.
6. Remember that most game—except fat duck, goose, raccoon, or bear—is "dry" meat; so, it must be well lubricated while cooking in order to spark an exciting meal. This is accomplished by careful larding and frequent basting; otherwise, it will be tough and tasteless as it comes from the oven or grill.

Preparing and Cooking Game Animals. A tremendous amount of venison is eaten in this country. Buffalo (bison) is again becoming popular; the herds are no longer vanishing, now that some ranchers are raising them like cattle. The meat of buffalo is stronger than beef, but the more tender cuts are cooked in the same manner. Bear meat is rich, sweet, and delicious. It looks like beef, but it has the consistency of pork. Unless it has been frozen for a minimum of 2 weeks, bear meat should be cooked thoroughly, because of trichinosis.

(See TRICHINOSIS.)

The meat from older game animals should be tenderized (mechanically and/or marinated); and, regardless of the age of game animals, roasts should be larded and basted frequently, and steaks should be wrapped in bacon or salt pork.

Preparing and Cooking Game Birds. Wild duck is the most plentiful game that flies. Wild turkeys are scarce, and also shy, and therefore hard to shoot. Young geese are tender, but the old birds are tough. Grouse, quail, partridge, pheasant, woodcock, snipe, and doves are much more available. All of them have a rich aromatic flavor.

Wild fowl get a lot of exercise; hence, they are much leaner and drier than domestic birds. So, fat must be supplied when cooking. Young birds may be broiled, fried, or roasted. Older birds should be cooked with moisture—braised or stewed; also, they should be rubbed with butter or other shortening, or covered with bacon or fat.

GAMMON

- A ham or a side of cured pork.
- The lower portion of a side of bacon.

GARDEN HUCKLEBERRY (SUNBERRY; WONDERBERRY) *Solanum intrusum*

The fruit of a plant (of the family *Solanaceae*) that is believed to have originated in Africa. Garden huckleberries bear a close resemblance to the berries of the poisonous black nightshade, except that the former are a little larger.

Fig. G-4. The garden huckleberry, an edible berry which requires careful identification to distinguish it from its poisonous relative, the black nightshade.

Cultivated wonderberries are about ½ in. (*1.2 cm*) in diameter. The rather bland tasting fruit may be used in jams and pies.

GARLIC *Allium sativum*

The pungent cloves that make up the underground bulb of this plant are a well known seasoning agent which is relished greatly by some people and detested by others. Garlic, like the other onion vegetables (*Allium genus*), has long been assigned to the lily family (*Liliaceae*), although some botanists now place garlic and its close relatives in the *Alliaceae*. Fig. G-5 shows a typical garlic plant.

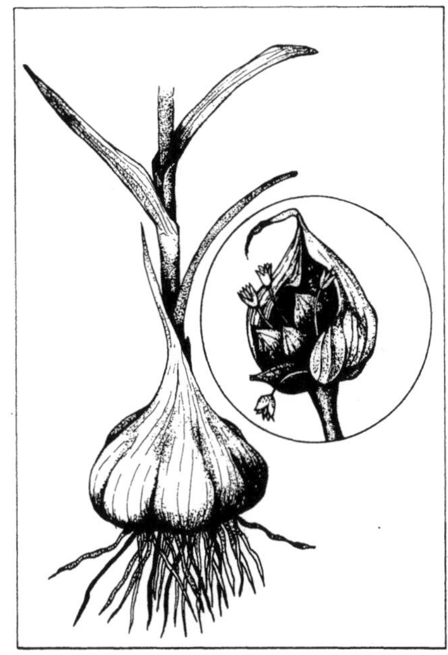

Fig. G-5. Garlic, an herb that has been used as a medicinal agent and a seasoning for many centuries. The flowers (shown in the circle) are often mixed with bulbils (tiny bulbs).

ORIGIN AND HISTORY. Wild varieties of the modern form of garlic are not found anywhere. Hence, it is believed that the garlic we use today has been derived from a wild ancestor that is native to Central Asia. Garlic has been cultivated for at least 5,000 years because ancient Egyptian writings attest to its importance as a crop as early as 3200 B.C. Also, the Greek historian Herodotus wrote that the laborers who built the Cheops pyramid around 2900 B.C. lived mainly on onions and garlic.

The use of garlic had apparently spread throughout a large region extending from the Mediterranean to China long before the Christian era. The Chinese sage Confucius, who lived from 551 to 479 B.C., considered it improper for one to reek of garlic. Although the bulbs were used medicinally in India by the 6 th century B.C., various ancient Indian writings described the eating of garlic and onions as loathsome practices. Later, the upper classes of ancient Greece showed a similar scorn for garlic. But, along the way, the use of garlic changed from scorn to chic. The first strike in history may have been called by Egyptian pyramid builders when their garlic rations were cut. Four garlic bulbs were found in King Tut's tomb, obviously intended to feed the pharaoh in afterlife. Roman gladiators ate garlic before combat; and Roman noblemen gave garlic to their laborers and soldiers. The ancient Israelites were very much attached to their garlic, in and out of bondage.

Garlic has continued to gain in popularity in recent times. For example, it became one of the major seasonings of the Chinese. In his writings, Marco Polo told how the poor people of Yunnan, China obtained liver from the slaughterhouses, then chopped it up, mixed the pieces in a garlic sauce, and ate them raw because they had little fuel for cooking. The Italians became known as garlic lovers, although it is mainly the Calabrians and Sicilians who use the seasoning in so many of their dishes. The northern Italians prefer various other spices.

PRODUCTION. About 2.9 million metric tons of garlic are produced worldwide each year. Fig. G-6 shows the amounts contributed by the ten top producers.

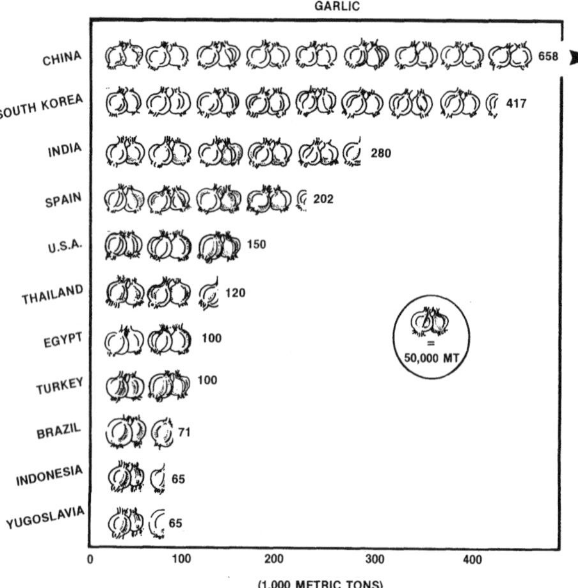

Fig. G-6. The leading garlic-producing countries of the world. (Based upon data from *FAO Production Yearbook*, 1990, FAO/UN, Rome, Italy, Vol. 44, p. 143, Table 59)

Most of the garlic produced in the United States comes from California.

Garlic is propagated by planting cloves or the tiny bulbs (bulbils) that develop in the flower head. The largest bulbs are obtained when garlic is planted in the late fall, provided that the ground does not freeze during the winter. The plant requires a fertile, moderately loose soil, since heavy clay soil causes the development of misshapen bulbs. California has a virtual monopoly on garlic production in the United States because the climate favors high yields.

The bulbs are harvested in June and July by loosening them with a cutter bar and pulling them from the ground. Then, they are usually dried in the sun for a week or more.

PROCESSING. Much of the U.S. crop is made into dehydrated garlic powder and garlic salt because these forms are more convenient for the food industry than raw garlic bulbs, which are susceptible to spoilage. Somewhat smaller amounts of garlic juice and garlic oil are also produced in the United States.

SELECTION AND PREPARATION. Young, plump cloves of garlic with the outer skin or sheath unbroken are the most desirable. Such cloves are usually contained in bulbs that are clean, compact, and "well cured," which means dry, but not soft and spongy, and with outer skin intact.

Soft or spongy garlic may have begun to sprout or be otherwise injured, and is undesirable. Split or broken skins of cloves may indicate sprouting. Decay may appear as mold, dry rot, or soft rot. Dry rot usually causes shrinking or shriveling and may progress downward from the top of the bulb until the whole bulb is powdery and useless.

Raw garlic which is chopped finely or crushed in a garlic press may be mixed with melted butter or margarine for spreading on toasted bread or rolls, or it may be added to relishes, salads, and salad dressing. Cooked garlic loses some of its pungency, yet it still imparts its characteristic taste to casseroles, meats, sauces, soups, and stews.

NUTRITIONAL VALUE. The nutrient composition of garlic is given in Food Composition Table F-36.

Some noteworthy observations regarding the nutrient composition of garlic follow:

1. Garlic does not usually contribute significant amounts of essential nutrients to human diets because only small amounts are consumed. However, the cooking of garlic in dishes such as casseroles, sauces, soups, and stews eliminates much of its pungency and allows much greater amounts to be consumed. If garlic were to be eaten as an ordinary vegetable, like boiled onions, it would be an excellent source of the essential mineral selenium (it is one of the richest vegetable sources of this nutrient); a good source of calories, protein, phosphorus, iron, and potassium; and a fair source of vitamin C.

2. The therapeutic uses of garlic are as old or older than the use of the vegetable for food. Across the ancient world from Rome to China, the bulbs were considered to be useful in treating deafness, dropsy, intestinal parasites, lack of appetite, leprosy, and respiratory disorders. However, the actual effectiveness of garlic falls short of these expectations.

3. The medicinal properties of raw garlic are being reported anew. It is claimed that its active ingredients, allicin and ajoene, (a) retard the growth of certain bacteria, (b) reduce artery-clogging cholesterol, (c) suppress the formation and growth of cancer cells, and (d) serve as a diuretic. But much research must be done before garlic can move from health-food stores and herbalists to pharmacies and physicians.

(Also see VEGETABLE[S], TABLE V-6 Vegetables of the World.)

GAS STORAGE

Primarily a method of storing fresh apples and pears in a controlled atmosphere where the normal 21% oxygen and 0.04% carbon dioxide content of the air is altered. These alterations double the storage life of the fruit. Conditions of storage vary for the different varieties. In general, the objective is to decrease the temperature and oxygen level and increase the carbon dioxide level. McIntosh apples store best at 37°F (3°C), 3% oxygen, 3% carbon dioxide (increased to 5% after one month), and 90% relative humidity. Apples will retain quality for more than 6 months. A high (60%) carbon dioxide concentration is also beneficial for storing eggs for long periods.

Due to the expense and difficulties involved in maintaining apples and pears in a controlled atmosphere, storage in heavy gauge polyethylene liners has proved to be a successful means of increasing the carbon dioxide to oxygen ratio, thereby extending the storage life by reducing the respiration rate similar to gas storage.

(Also see PRESERVATION OF FOOD.)

GASTRECTOMY

The surgical removal of part or all of the stomach.

GASTRIC

Pertaining to the stomach.

GASTRIC JUICE

A clear liquid secreted by the wall of the stomach. It contains hydrochloric acid and the enzymes rennin, pepsin, and gastric lipase.

(Also see DIGESTION AND ABSORPTION.)

GASTRIC MUCOSA

The membrane lining the stomach.

GASTRIC ULCER

An open sore occurring in the lower end of the esophagus or in the stomach. These occur when there is some breakdown in the protective mechanisms which normally prevent gastric secretions from ever touching the walls of the stomach. Hence, the wall of the stomach or esophagus erodes and an ulcer is formed. A common complaint of sufferers is pain in the upper abdomen 1 to 4 hours after eating. Foods or antacids relieve the pain.

(Also see ANTACIDS; HEARTBURN; MODIFIED DIETS; and ULCERS, PEPTIC.)

GASTRIN

A hormone secreted by the lining of the antral region of the stomach in response to the distension of the stomach, the presence of protein, and vagal nerve stimulation. In turn, gastrin stimulates the secretion of stomach acid (HCl), pepsin, and stomach motility.

(Also see DIGESTION AND ABSORPTION; and ENDOCRINE GLANDS.)

GASTRITIS

This refers to an inflammation of the stomach lining, a condition which most people have experienced, or will experience, during a lifetime. When it involves the intestines, it is called gastroenteritis. Gastritis has a variety of causes some of which are nonspecific. It occurs in three forms: acute, chronic, and toxic or corrosive. Causes, symptoms, diagnosis, and treatment of each form follows.

ACUTE GASTRITIS. This type of gastritis, which is experienced by most people, is probably the most common stomach ailment. It arises from a variety of causes, many of which can be summarized in one word—overindulgence.

- **Causes**—Alcohol, rich foods, spicy foods, coffee, strong tea, hot foods, and mechanically irritating foods (popcorn) in excess, can all result in acute gastritis. Some drugs (aspirin, sulfonamides, antibiotics, and quinine), and some viral and bacterial infections, cause acute gastritis. By far the most common drug causing acute gastritis is aspirin.

- **Symptoms and Diagnosis**—Loss of appetite, heartburn, pain in the upper abdomen, nausea, vomiting, cramps, and discomfort due to distention are all common symptoms. The intensity and number may vary, depending upon the type and amount of irritant. Symptoms pass rapidly and diagnosis is usually history.

- **Treatment**—Vomiting and/or diarrhea removes the irritant and often corrects acute gastritis. Other measures may be purely supportive. Small doses of a nonconstipating antacid, coating the stomach with half-and-half milk, and bed rest are recommended. In addition, a liquid of semisolid bland diet consisting of such items as unseasoned eggs, milk, clear soup, and light buttered toast should be eaten for 1 to 2 days following disappearance of the symptoms. Obviously, if the irritant which caused the attack can be identified it should be avoided, at least in excess.

CHRONIC GASTRITIS. This is a mild gastritis which recurs over weeks or months. It is difficult to diagnose, but chronic gastritis is more serious than acute gastritis, as it can signal other more pressing disorders.

• **Causes**—Chronic gastritis may be due to repeated exposure to an irritant, as described for acute gastritis, or due to emotional stress, or due to some unknown factors. Although the causes are difficult to sort out, it is often observed in cases of stomach cancer, pernicious anemia, gastric ulcer, gastric polyps, diabetes mellitus, adrenal or pituitary insufficiency, x-ray or surgical treatment of the stomach, and one unavoidable condition—normal aging.

• **Symptoms and diagnosis**—Symptoms occur intermittently and are very often overlooked by the sufferer. When they occur, they may consist of loss of appetite, constant mild nausea, recurrent indigestion, pain in the abdomen, a feeling of fullness. a bad taste in the mouth, anemia, and a general bodily discomfort.

Diagnosis is based on the patient's history, a biopsy, and gastroscopy. In addition, a series of x rays may be used to rule out other diseases such as tumors and ulcers.

• **Treatment**—Attempts should be made to eliminate the causative factors. However, until this is done the following recommendations are made: (1) avoid alcohol, caffeine, and smoking; (2) eat a bland diet; (3) eat small, but frequent, meals; (4) use a mild antacid or coating agent such as bismuth subnitrate or aluminum hydroxide gel. If symptoms persist, consult a physician.

TOXIC OR CORROSIVE GASTRITIS. This is probably the least common gastritis, but certainly the most dangerous. It often involves children.

• **Causes**—The victim, usually by accident, ingests a strong acid, caustic alkali, iodine, or solvent. In the average American home, these chemicals include the powerful cleaning detergents, drain openers, automatic dishwasher compounds, ammonia, paint thinner, turpentine, kerosine, and gasoline.

• **Symptoms and Diagnosis**—Severe stomach cramps occur, and these may cause collapse. Vomitus may contain blood. Some agents burn the inside of the mouth or leave a characteristic smell on the victim's breath. If the victim is unable to give information, perhaps a nearby container will identify the poison.

• **Treatment**—Toxic gastritis is an emergency that demands swift, calm action in order to minimize damage to the victim's esophagus and stomach, and quite possibly prevent death. Immediately give one or two glasses of milk or water to dilute the poison in the stomach. (In 1813, the French chemist Claude Bertrand reported that activated charcoal will absorb poisons like a sponge and prevent them from entering the bloodstream. Similar claims have persisted ever since, but experimental proof is lacking. So, the authors present this report without recommendation, with the hope that it will stimulate research work.) Do not induce vomiting. Vomiting causes further damage. A doctor, the Poison Control Center (ask the operator for the Poison Control Center, or call the U.S. Public Health Service at either Atlanta, Georgia or Wenatchee, Washington), and/or an ambulance should be contacted immediately. When the victim is transported to a doctor or hosp. emergency room, the container which held the poisc should also be taken.

Prevention is better than treatment. Keep household chemicals inaccessible to small children with big curiosities.

(Also see ANTACIDS; BACTERIA IN FOOD, Table B-5 Food infections (Bacterial); HEARTBURN; INDIGESTION; and POISONS.)

GASTROENTERITIS

Generally, a brief self-limiting inflammation of the stomach and intestines, which is very often caused by a foodborne bacterial infection, bacterial toxin, or viral influenza. However, some of the same agents causing gastritis may also cause gastroenteritis; for example, alcohol overindulgence, and drugs. Common symptoms are nausea, vomiting, cramps, and diarrhea. Treatment depends upon the causative factor, but a bland or liquid diet is recommended until symptoms pass. Some "over-the-counter" preparations may provide symptomatic relief. Bacterial infections may require an antibiotic prescribed by a doctor. Persistent, excessive vomiting and diarrhea are especially dangerous in infants and old people as dehydration can result.

(Also see BACTERIA IN FOODS, Table B-5 Food Infections [Bacterial], and Table B-6 Food Poisonings [Bacterial Toxins]; MODIFIED DIETS; DISEASES, Table D-13 Infectious and Parasitic Diseases Which May Be Transmitted by Contaminated Foods and Water; and GASTRITIS.)

GASTROINTESTINAL

Pertaining to the stomach and intestines.

GASTROINTESTINAL DISEASE

A very broad term suggesting any disorder in the structure or function of the digestive system. Often diets are modified for gastrointestinal tract disorders such as cholecystitis, cirrhosis of the liver, constipation, diarrhea, gastric carcinoma, gastritis, hiatal hernia, ileitis, malabsorption, and peptic ulcers, among others. On the other hand, some diets may be blamed for a specific gastrointestinal disorder, though the cause-and-effect nature of the diet in some cases is only suggestive and not demonstratable.

One or more of the following five diagnostic procedures commonly applied to gastrointestinal disorders are: (1) x-ray or fluoroscopic examinations; (2) gastroscopy; (3) gastric (stomach) content analysis; (4) peroral (by the mouth) suction biopsy; and/or (5) various absorption tests.

(Also see BACTERIA IN FOODS; DISEASES; INBORN ERRORS IN METABOLISM; MALABSORPTION SYNDROME; and MODIFIED DIETS.)

GASTROINTESTINAL TRACT

Often this is interchanged with the following terms: digestive tract, digestive system, and alimentary canal, but actually it refers to the stomach and intestines.

(Also see DIGESTIVE SYSTEM.)

GASTROSCOPY

A nonsurgical, diagnostic technique used to view the stomach lining. The instrument used is called a gastroscope, a telescopic system with a light source. The gastroscope is passed down the esophagus to the stomach. No anesthetic is required, but it does cause temporary discomfort.

(Also see GASTRITIS.)

GAVAGE

Introduction of material (as nutrients) into the stomach by means of a stomach tube.

(Also see TUBE FEEDING.)

GEL

A colloidal suspension which has solidified. It is the abbreviation for gelatinous.

GELATIN

A mixture of proteins not found in nature but derived from connective tissue (collagen) by hydrolytic action—boiling skin, tendons, ligaments and bones. Gelatin is digestible, but it is an incomplete protein. It lacks the essential amino acid tryptophan, and contains only small amounts of other essential amino acids. In its dry form, gelatin is colorless or slightly yellow, transparent, brittle, practically odorless, and tasteless. It is capable of swelling up and absorbing 5 to 10 times its weight of water to form a gel in solutions below 95° to 104°F (*35° to 40°C*). Gelatin has numerous food and nonfood uses. In foods, gelatin is used as a stabilizer, thickener and texturizer in such foods as confectionery, jellies, and ice cream. The FDA classifies gelatin as a GRAS (generally recognized as safe) additive. Nonfood uses consist of adhesives, capsules for medicinals, inks, plastic compounds, artificial silk, photographic plates and films, sizing of paper and textiles, plasma expander, and hemostasis.

(Also see ADDITIVES Table A-3; and PROTEIN[S].)

GELOMETER

An instrument used to measure jelly strength.

GENE

A unit of heredity arranged in a definite fashion on a chromosome.

GENETIC DISEASES

Harmful disorders in the normal structure and function of the body caused by an error in the genetic code. The disease is present at birth, though not always evident. It may manifest itself later in life due to some interaction with the environment. Genetic diseases occur (1) as a rare gene mutation; (2) as conditions that "run in the family" or hereditary disorders (dominant and recessive); or (3) as an accidental change in chromosome numbers (Down's Syndrome) or structure. A number of metabolic disorders have genetic origins, and once recognized can be treated by dietary alterations, for example, galactosemia and phenylketonuria (PKU).

Amniocentesis, obtaining fetal cells for genetic analysis, offers some hope for early detection of genetic diseases. However, only when we master genetic engineering will we be able to prevent genetic disorders. For now, the best treatment is early detection.

(Also see INBORN ERRORS OF METABOLISM.)

-GENIC

Suffix, meaning to produce or give rise to; e.g., ketogenic.

GENIPAPO *Genipa americana*

The fruit of a large tree (of the family *Rubiaceae*) that is native to eastern South America and the West Indies. Genipapo trees bear pear-shaped fruits that are about 3 in. (*7.5 cm*) in diameter and about 4 in. (*10 cm*) long. The granular pulp of the fruit contains many small seeds. The fruit is used mainly to make fruit drinks, marmalade, or wine.

GEOPHAGIA

The eating of earth, a practice which prevails among some of the population of Africa. "Edible earth," usually obtained from a particular spot, is commonly mixed with a bean or a green relish and consumed by the whole family. The cause is unclear. It may be a carry over of a habit acquired in childhood, from the exploratory putting of earth from the hut floor into the mouth; or it may be a means of satisfying hunger and malnutrition.

(Also see ALLOPRIOPHAGY; and PICA.)

GERBER TEST

This is a test for determining the fat content of milk. In principle, it is very similar to the commonly used

Babcock test. Sulfuric acid is mixed with milk, releasing the fat and leaving it free to rise. The Gerber bottle, in which this test is performed, has a thin graduated neck. Fat collects in the neck, where it is measured.

(Also see BABCOCK TEST.)

GERM

Embryo of a seed.

(Also see CEREAL GRAINS.)

GERONTOLOGY AND GERIATRIC NUTRITION (SENIOR CITIZENS AGE 65 AND OVER)

Contents Page

History 1059
Factors That May Accelerate Aging 1059
Changes That May Occur in the Body During Aging 1062
Nutritional Allowances 1068
Dietary Guidelines 1069
Menu Planning 1069
Saving Money in Food Purchasing and Preparation 1071
Dietary Modifications for Some Common Health Problems 1073

People are living longer. But they may not be staying younger. This article is concerned with the latter.

Many people are not as well informed about aging as they might be, considering that both the number and percentage of people 65 years of age or older are increasing steadily, as is shown in Fig. G-7.

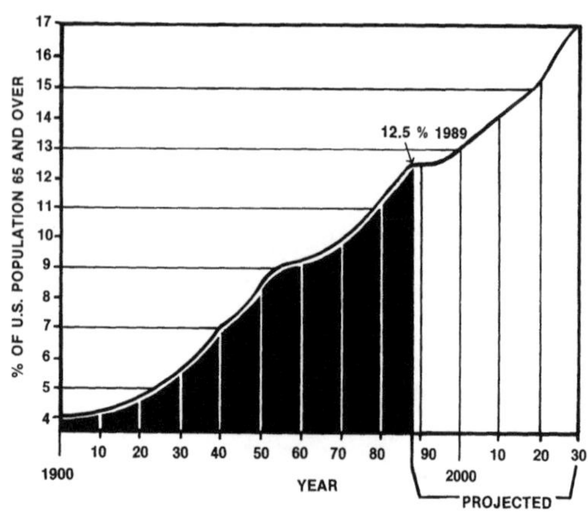

Fig. G-7. The increasing percentage of people aged 65 and over in the United States from 1900 to 1989, with projections for the period 1989 to 2030. (Source: *The Graying of Nations: Implications*, Hearing before the Special Committee on Aging of the U.S. Senate, First Session of the 95th Congress, November 10, 1977, U.S. Government Printing Office, 1978, p. 6, and *Statistical Abstract of the United States 1991*, p. 23)

It may be seen from Fig. G-7 that the percentage of senior citizens is expected to quadruple in the period which began in 1900 and will end in 2030, so that at the latter date this group of people will represent about one-sixth of all Americans.

Better to understand the nutritional needs and other characteristics of our older generation, it is first necessary to clarify some of the terms that are commonly applied to the group.

• **Aging**—Although this term is often applied to people of middle age (40 to 65 years old) and older, in the fullest biological sense it designates the sequences of mental and physical changes that begin at conception and end at death.

• **Elderly**—In common usage, this designation applies to people aged 65 and older. However, it is noteworthy that it is derived from the term "elder," which has long referred to persons with high rank or status in communities, families, and religious organizations. (In Biblical times, the eldest son inherited the authority and major means of the family when his father died.)

• **Geriatrics**—This term refers to the medical specialty which deals with the prevention and treatment of the diseases of aging. As such it is one of the areas of concentration within the much broader field of gerontology. Sometimes, the elderly are designated as "the geriatric set."

• **Gerontology**—This field of study covers many aspects of aging. Hence, it utilizes the methods of such disciplines as anthropology, biochemistry, biology, economics, history, medicine, physiology, psychology, and sociology. However, it should not be confused with geriatrics, which is limited to the medical aspects of aging.

• **Life expectancy**—Many people use this term incorrectly, in that they assume it to have a fixed value throughout a person's life. However, it refers to the number of years that a person may expect to live after having reached a certain age. (Life expectancy data are derived from the vital statistics that have been obtained from various groups of people and national populations.) It is noteworthy that the life expectancy at birth is usually less than that at age 65, because a newborn infant has yet to be exposed to the infectious diseases that threaten life, whereas the 65 year old has already survived many of these hazards.

• **Life-span**—This term usually designates the maximum or optimal amount of time that a person might live, provided that his or her health and life circumstances are quite favorable.

• **Longevity**—Depending upon the context in which it is used, this term means (1) a long life, or (2) the length of life.

• **Retirement age**—Retirement at age 65 is believed to have been first used by Chancellor Otto von Bismarck of Germany in 1889, when he set up the first nationwide old age pension plan. Since then, other nations and certain employers have established retirement ages which vary according to the work performed. For example, people who perform hard, laborious tasks may be allowed to retire as early as age 55, whereas, the usual retirement age for Roman Catholic clergy is 75.

• **Senility**—This condition, which is characterized by a marked loss of memory and mental deterioration, is *not* always irreversible as was once thought. Sometimes, the signs of senility are brought on by nutritional deficiencies and/or other correctable conditions such as dehydration, and a reduced flow of blood to the brain.

• **Senior citizens**—It is not known when this term was first applied to the elderly in America, but it appears to have been coined to give greater dignity to older, retired people. Furthermore, those who meet the requirement for this designation may be entitled to special benefits such as reductions in property taxes, lower fares for public transit, reduced meal prices at certain restaurants, lower price for hair cuts, lower dog license fees, and participation in programs at Senior Citizen Centers.

This article is concerned mainly with the conservation of the health, mental acuity, and physical capabilities of older people, rather than the extension of the life-span. The effects of nutrition on longevity are covered elsewhere in this book.

(Also see LIFE EXPECTANCY AND NUTRITION.)

HISTORY. Many ancient writings document man's longstanding search for ways to extend youthful health and vigor into old age. Both the Old and the New Testaments of the Bible tell of miraculous rejuvenations of women who conceived children long after their menses had ceased. Such was the case of Abraham's wife, Sarah (*Genesis* 18:10-14), and of John the Baptist's mother, Elizabeth (*Luke* 1:7-25). Furthermore, the ancient cultures of China, India, and Greece had their legends of the Fountain of Youth that was thought to be located somewhere along the course of a major river. Perhaps, this association was made because the rivers were responsible for bringing new life to the soils they watered.

In the early 1500s, the Spanish explorer Ponce de Leon came to the Americas to search for the Fountain of Youth that was believed to be present on an island in the Caribbean. He never found it, but in the course of his search he explored Puerto Rico and discovered Florida, where many older people now go to escape the ravages of winter and expose themselves to the healing power of the sun.

Although the search for more healthful places to live continues to the present day, increasing numbers of people have abandonded this type of quest to seek their rejuvenation through special therapies.

In 1869, Charles Edouard Broun-Sequard, who held the chair of physiology at the College de France, announced that he was 30 years younger after having injected himself three times with an extract of guinea pig testes. Shortly thereafter, doctors began to inject their patients with the extract, but the effects were short-lived, and interest in this therapy soon waned. It is now thought that the short-lived rejuvenating effects of Broun-Sequard's extracts resulted from the male hormone testosterone, which is produced in the testes. However, the liver promptly destroys excesses of the hormone in the blood, so repeated administrations are needed to stimulate vigor continuously.

The sex gland approach to rejuvenation was explored further in the first 3 decades of the 20th century when it was noted that some debilitated men were restored by (1) vasectomy (the operation causes the sperm and the hormone testosterone to be retained within the body), and (2) transplantation of animal testes in the abdominal layer of fat. In the 1940s, some people took pills made from dried "monkey glands" as tonics. Even today, some people take dried thyroid preparations to speed up the metabolic rate and permit a more liberal food intake without the prospect of becoming obese.

Interest in diet as a means of maintaining vigor into old age arose at the beginning of the 20th century, when the Russian scientist Metchnikoff advocated the consumption of liberal amounts of yogurt because he believed that this food was responsible for the unusual health and vigor of elderly Bulgarian peasants. The effects attributed to yogurt were believed to result from its content of *Lactobacillus bulgaricus* bacteria which supposedly prevented the formation of certain toxins in the intestine.

Shortly after the early interest in yogurt waned, the era of vitamins began. The dramatic responses that often followed the giving of minute amounts of these nutrients to deficient people led many people to believe that vitamins were potent metabolic stimulators. During and after World War II, military medical personnel administered injections of the vitamin B complex to their fellow servicemen who lined up at the dispensaries on the mornings after nighttime drinking bouts. However, it was ultimately shown by nutritional researchers that excesses of most vitamins have little beneficial effect once nutrient needs have been met.

Recently, a more holistic approach to the retardation of the degenerative changes during aging has emerged, as anthropologists and other social scientists have joined dentists, doctors, and nutritionists in studying various cultural groups in America and abroad to determine why some people age more slowly than others. These studies have focused upon such diverse groups as Mormons, Seventh-day Adventists, and Trappist Monks, and on the apparently long-lived and vigorous inhabitants of isolated mountainous areas such as (1) the valley of the Hunzas in northeast Pakistan, (2) Vilcabamba in the Andes of southern Ecuador, and (3) the Georgian region of the Soviet Union. In almost every case, the most healthy and long-lived people engaged in strenuous work throughout their lives and ate sparingly of reasonably good unrefined diets. However, there are still many unanswered questions regarding the precise physiological means by which above average good health in the elderly is achieved by diet and exercise. Hence, it is necessary to consider some of the known and postulated factors that may accelerate the aging of cells and tissues.

FACTORS THAT MAY ACCELERATE AGING.

The ever-growing percentage of elderly people in the United States and other developed countries has prompted concern over the ability of families, private groups, and governmental agencies to assist the needy seniors. Obviously, slowing of the crippling mental and physical deterioration that occurs in some, but not all, of the aged would help to increase the numbers of these people who could be assets rather than liabilities to society. However, it appears that a number of different factors contribute to the aging of the body tissues. Unfortunately, the physiological processes associated with the aging factors are not understood fully. Some of what is known is presented in Table G-2.

TABLE G-2
AGENTS AND PROCESSES THAT MAY CAUSE AGING

Factor	Means By Which Aging Is Accelerated	Relationship(s) to Diet and/or Hygiene
Accumulation of residues in cells and tissues	Substances such as amyloid,[1] asbestos, cholesterol, dust, fibrous matter, lipofuschin granules,[2] silica, trace metals, and uric acid may accumulate to the extent that the functioning of cells and tissues is impaired. Sometimes, the tissues are inflamed or irritated and fibrous matter or scar tissue is formed. Also, cancers may start at the sites of certain types of irritating deposits.	Nutritionally imbalanced diets may result in defects of metabolism that produce accumulation of certain residues. Inhalation of particle-laden air may result in deposits of the contaminants in the lung tissues. This may be prevented by the wearing of a respiratory mask.
Autoimmunity	The body develops antibodies that attack its own tissues. It is believed that autoimmune reactions may be responsible for some cases of Addison's disease,[3] arthritis, diabetes, kidney disease, lupus, myocarditis, opthalmia (damage to the eye), and thyroiditis. Sometimes, antibodies against tissues may be the result of mutations of the chromosomes in the antibody-producing cells.	Means for preventing autoimmunity by diet are uncertain. However, experimental animals fed high-protein diets are more prone to autoimmune disorders than those given low to moderate levels of dietary protein. Also, adequate nutrition may help to prevent the breakdown of tissues and the release of fragments into the blood, where autoimmune reactions may occur.
Cell loss from vital tissues	Certain types of cells may be lost more rapidly than they are replaced. Eventually, the tissues are left with too few functioning cells.	Good nutrition may minimize cell loss and maximize cell replacement in some, but not all, tissues.
Chromosome breakage	Radiation and some types of toxic agents cause damage to the chromosomes so that the genetic control of metabolism is altered. Fortunately, human chromosomes can be repaired.	Diets containing nucleic acids (DNA and RNA) or the nucleotide bases such as adenine (abundant in yeast) may help to speed up the repairing of chromosomes.
Collagen stiffening	The protein collagen, which is a major constituent of the blood vessel walls, bones, cartilages, skin, and tendons, gradually loses its elasticity and shrinks in size so that it interferes with the metabolism of cells. Also, collagen is often deposited in the form of fibrous tissue at the sites of inflammation and injury.	Experimental animals that were kept on low calorie diets for long periods of time after weaning had a much slower rate of collagen stiffening than those given more calorie-rich diets. However, the calorie-deprived animals had retarded growth and development.
Cross-linking of macromolecules	Large molecules such as nucleic acids and proteins may acquire cross-linkages that reduce functioning because of the increased rigidity which results. (DNA and collagen are highly susceptible to cross-linkage.)	It is believed that a heavy consumption of carbohydrates, fats, and/or proteins results in an increased production of aldehydes that form cross-linkages within macromolecules.
Free-radical damage	Polyunsaturated fatty acids that are present in the diet and in the cells are oxidized readily to highly reactive free radicals that damage the inner and outer membranes of cells and disrupt cellular functions. Free radical damage is similar to that caused by exposure to radiation.	The formation of free radicals is retarded by liberal amounts of dietary sulfur-containing amino acids, selenium, and vitamin E; and by avoiding an excessive consumption of polyunsaturated fatty acids.

Footnotes at end of table

(Continued)

TABLE G-2 *(Continued)*

Factor	Means By Which Aging Is Accelerated	Relationship(s) to Diet and/or Hygiene
Inborn programming of aging	The cells of the human body have limited lifespans. Hence, tissues age and die when the rates of cell death exceed the rates of replacement. Also, certain cells are not replaced after maturity.	Good nutrition and optimal hygienic practices may prevent the premature deterioration and death of the irreplaceable cells.
Loss of irreplaceable substances	Certain vital substances are produced only during the early stages of growth and development. However, there is a slow but continuous loss of these compounds during the normal functioning of the body throughout life. Rapid aging occurs when the cumulative losses reach critical levels.	Slowing the rates of growth and development of experimental animals by stringent dietary restriction has apparently slowed the metabolism and consequently the rate of loss of irreplaceable substances and aging. It is not certain whether similar results can be obtained in humans.
Mutations	Changes in the genetic material (chromosomes) of cells occur spontaneously and as a result of exposure to radiation and other injurious agents. These changes often produce abnormalities of cell functions. Generally, the rate of mutation increases with age.	The rate of mutations may be slowed somewhat by minimizing exposure to known mutagenic factors such as free radicals and radiation; and by consuming diets containing nucleic acids and other nutrients needed for the maintenance of healthy cells.
Nerve deterioration	The deterioration of nerve cells and/or of the various processes associated with nerve function results in the loss of control of vital functions. Disorders such as diabetes, multiple sclerosis, and certain nutritional deficiencies accelerate the rate of deterioration.	Nerves begin to develop before birth and require adequate nutrition throughout life. Furthermore, conditions in other tissues may ultimately be reflected in the nerves. Hence, a good diet plus the practice of sensible hygiene may prevent the premature deterioration of the nerves.
Radiation	Various types of radiation (neutrons, ultraviolet rays, x rays, etc.) cause damage to the chromosomes in cells that may result in abnormalities in cell function, and in some cases, cancer.	The effects of radiation have been counteracted in experimental animals by low-fat cereal diets that encourage the growth of lactobacilli in the intestine.
Reduced sensitivity of cells and tissues to hormones	The sensitivity of various tissues to hormones decreases as a result of (1) a reduction in the number of hormone-sensitive areas (receptor sites), and (2) an increase in the processes that counteract the hormonal effects. Hence, the body responds more slowly to stresses such as chilling, fasting, and the need for increased pumping of blood by the heart. The reduced responsiveness makes the body more vulnerable to various types of injury.	Injury to the body may be minimized by (1) eating a nutritionally balanced diet at regularly scheduled meals; (2) avoiding sudden changes in the rate of physical activity; and (3) taking care not to become chilled or overheated. However, moderate mental and physical challenges help to stimulate the functioning of the heart and the other vital organs.
Stress	Stresses that are sufficiently severe and/or prolonged to overtax the body's defense mechanisms may literally exhaust such glands as the adrenals and the pancreas; and/or they may cause the breakdown of tissues such as the stomach lining and the vertebrae.	Some of the destructive effects of stress may be counteracted by (1) a balanced diet that contains moderate amounts of calories, protein, minerals, and vitamins; and (2) sticking to a regular and sensible schedule of exercise and other activities.
Wear and tear on the tissues	Tissues that are subjected to chronic strains of various types may sustain injuries that are not repairable by the natural restorative processes. For example, many older people who have long been obese have a wearing down of their knee joints (osteoarthritis).	The "wearing out" of body tissues may be reduced by avoiding severe or prolonged overloads that may cause injury. Also, obese people should reduce their weight gradually to a more normal, healthy level.

[1]Amyloid deposits are carbohydrate-protein complexes occurring mainly in the tissues of the adrenal glands, digestive tract, kidneys, liver, lungs, muscles, skin, and spleen. They are believed to result from autoimmune reactions in which the tissues are attacked by antibodies. The amount of amyloid usually increases with age.

[2]Lipofuschin granules are yellow pigments (also called "age pigments") that accumulate mainly in the heart muscle and the brain. The pigments are the products of the destructive reactions between free radicals and various membranes within cells. These reactions may be minimized by adequate dietary amounts of protective nutrients such as the sulfur-containing amino acids, selenium, and vitamin E. However, diets rich in polyunsaturates promote the reactions.

[3]Addison's disease is an insufficiency of adrenal cortical secretion that may result from (a) diseases such as tuberculosis, (b) an autoimmune disorder, or (c) other causes.

CHANGES THAT MAY OCCUR IN THE BODY DURING AGING. Apparently, Americans are doing some of the things that counter the agents of deterioration discussed in the preceding section, since the death rate for older people from all causes has dropped sharply. Reductions in the death rates from the leading killer diseases are shown in Fig. G-8.

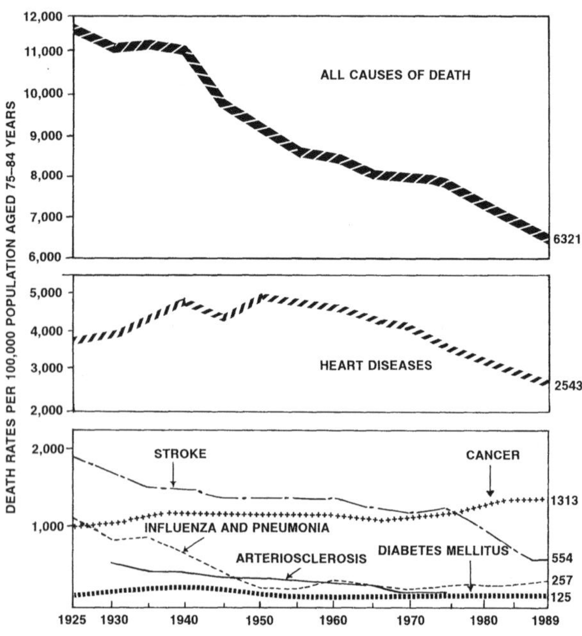

Fig. G-8. Death rates from the leading causes for persons aged 75 to 84 years, in the period from 1925 to 1973. (Based on data in *Health, United States 1975*, U.S. Dept. HEW, p. 553, Table CD. IV. 2, and *Statistical Abstract of the United States 1991*, p. 81, Table 118)

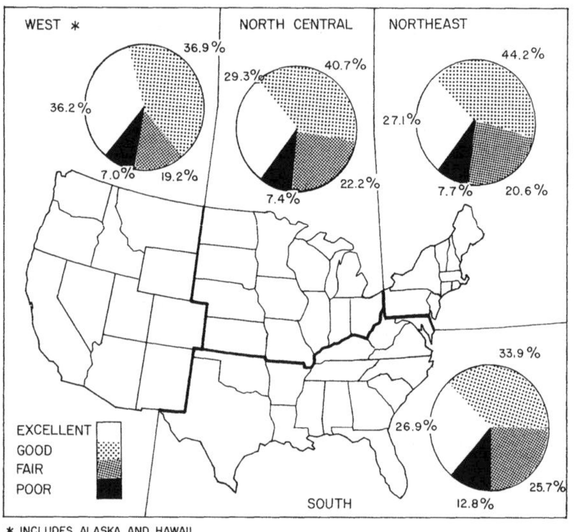

Fig. G-9. Self assessments of health in major regions of the United States. (Based on data in *Health, United States*, U.S. Dept. HEW, p. 551, Table CD. IV. 1)

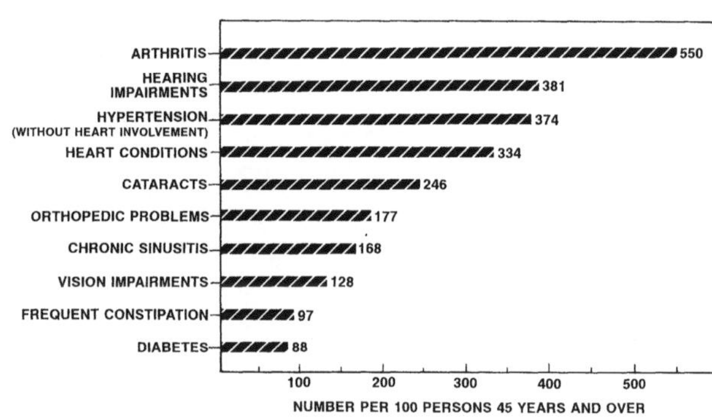

Fig. G-10. The top ten chronic conditions reported in health interviews by persons age 75 and older in 1988. (Based on data from *Statistical Abstract of the United States 1991*, p. 120, Table 195)

Furthermore, the majority of people aged 65 or over consider themselves to be in excellent or good health. However, self assessments of health vary somewhat between the different regions of the United States, as shown in Fig. G-9.

Nevertheless, older people are troubled with various chronic conditions such as those for which data is presented in Fig. G-10.

Many of the handicapping conditions shown in Fig. G-10 may be the result of one or more of the physiological changes which occur during aging. That is not to say that each type of degeneration occurs to an equal extent in every older person, since many of the elderly are healthier than some of their juniors. Rather, it should be understood that deterioration of body parts occurs at different rates in different people, depending upon their heredity, past medical history, diet, and type of hygiene practiced throughout life. The most common physiological changes which occur during aging are described in Table G-3.

TABLE G-3
CHANGES THAT MAY OCCUR IN THE BODY DURING AGING

Part or Function of the Body	Change(s)	Biological Significance	Comments
Basal metabolism	Decreases with aging.	The drop in basal metabolism (BMR) is due mainly to the reduced amounts of lean body tissue in older people.	Caloric needs of seniors may often be less than those of younger adults.
Body composition	The percent of lean body mass (mainly protein and water) decreases, and the percent of body fat often increases.	A significant amount of the lean body mass that is lost may come from the muscles if they are not used in physical activity. It is important to conserve muscle tissue because it serves as a source of protein during life-threatening emergencies.	Regular exercise and a good diet may help to maintain the lean body mass and prevent the accumulation of fat.
Body weight	Peak weight is usually attained by middle age, after which there is a gradual decrease in weight.	Excess weight which was acquired earlier in life is not lost readily by older people, although some people become much thinner as a result of long-term consumption of a low-calorie diet, illness, injury, or surgery.	It is not advisable for older people to try to lose weight rapidly because a drastic change in diet may be too stressful for the body. Furthermore, crash diets may cause the loss of considerable amounts of lean body tissue.
Circulatory system			
Heart	The heart rate and the volume of blood pumped per beat decrease gradually. There is also a loss in elasticity of the heart muscle fibers.	Sudden, strenuous exercise may overtax the hearts of older people who are unaccustomed to vigorous physical activity. However, moderate levels of sustained activity pose little threat to healthy people.	People who have long engaged in strenuous activities on a regular basis may have enlarged, but otherwise healthy, hearts that function well.
Blood	Some older people develop high blood levels of fatty substances (lipids) and/or mild to moderate degrees of anemia (lack of sufficient red blood cells). These abnormalities may result from hereditary factors, high fat diets, dietary deficiencies, and/or reduced absorption of iron, folacin, and/or vitamin B-12.	High levels of blood cholesterol and/or other lipids may result in an increased tendency to form atherosclerotic plaques and blood clots. Lack of sufficient red cells result in a reduction in the amount of oxygen delivered to the tissues. Pernicious anemia is the most dangerous of these conditions because an untreated deficiency of vitamin B-12 may lead to damage of the brain and nervous system.	The treatments of hyperlipidemias and anemias in older people are matters best left to a doctor, since those who have narrowed blood vessels and a tendency to form blood clots too readily may have reduced risks of heart attacks and strokes if they have subnormal amounts of blood cells.

Footnotes at end of table

(Continued)

TABLE G-3 *(Continued)*

Part or Function of the Body	Change(s)	Biological Significance	Comments
Circulatory system *(Continued)*			
Blood vessels	The walls of the blood vessels lose their flexibility, and even become somewhat hardened, (arteriosclerosis), thickened (a frequent complication of diabetes), and/or have atherosclerotic plaques. These changes occur earlier in some families than in others.	Rigid, partially blocked blood vessels may not be able to deliver sufficient blood to meet the needs of vital tissues such as the brain and the heart muscle. The blood-starved tissues may deteriorate and die.	It is believed that the regular participation in suitable physical activities helps to keep blood vessels open, and may even induce the formation of collateral circulations[1] in the brain and heart muscle.
Digestive system			
Teeth	Teeth may be lost due to cavities and/or periodontal disease (any of the various disorders of the gums and/or the underlying jaw bone).	Lack of teeth or suitable dentures may lead older people to select a poor diet that is comprised mainly of soft, starchy foods. Also, some foods may be swallowed in large pieces that are digested poorly.	Loss of teeth may be prevented by a good diet, adequate brushing and/or flossing, and regular visits to the dentist.
Tongue	Many older people have lost a significant percentage of their taste buds.	People with a poor sense of taste may add excessive amounts of salt and other seasonings to foods, or they may even develop a lack of enthusiasm for eating.	Heavy smokers lose more of their sense of taste than nonsmokers. Also, a deficiency of the essential mineral zinc may be responsible for loss of taste.
Stomach	Sometimes, people past age 40 develop an atrophy of the gastric cells that secrete acid or of those that produce the intrinsic factor for vitamin B-12 absorption.	Insufficient gastric juice may hamper the absorption of certain essential minerals, but the digestion and absorption of other nutrients is not affected. Lack of intrinsic factor may result in vitamin B-12 deficiency.	People who lack the intrinsic factor for vitamin B-12 absorption will most likely require periodic injections of the vitamin by a doctor. Sometimes, gastric cells are damaged by iron-deficiency anemia.
Large intestine	The movements of the large bowel may become sluggish, and chronic constipation may develop.	Constant straining during bowel movements may result in disorders such as diverticulosis (pouches on the wall of the bowel) and hemorrhoids.	Abundant quantities of water and dietary fiber help to correct many cases of chronic constipation.

Footnotes at end of table

(Continued)

TABLE G-3 *(Continued)*

Part of Function of the Body	Change(s)	Biological Significance	Comments
Endocrine glands			
Ovary	In postmenopausal women the number of ovarian follicles and the level of estrogen secretion is reduced greatly.	The marked decline in estrogen secretion after the menopause may result in (1) an accelerated rate of mineral loss from the bones, (2) thinner, less elastic skin, and (3) hot flashes. (These changes occur more slowly in obese women because estrogens are stored in body fat.)	Administration of estrogens to postmenopausal women carries some risks, such as promotion of breast and/or ovary cancer, and should be done only under the careful supervision of a doctor.
Pancreas	A decreased amount of insulin is released and the blood sugar remains elevated for a longer period after an oral dose of glucose is administered. However, the blood levels of insulin and glucose are usually normal after an overnight fast, except when diabetes is present.	Most diabetes diagnostic centers consider it to be normal for older people to have a higher level of blood sugar on a glucose tolerance test. However, it is not known whether the clearance of excess sugar from the blood is also delayed after a regular meal. A chronically elevated blood sugar may accelerate the deterioration of tissues.	Sometimes, increasing the carbohydrate content of the diet helps to increase the secretion of insulin from the pancreas and bring about the reduction of the blood sugar to a normal level. Regular meals and exercise also help to keep the blood sugar level stabilized.
Testes	Although the average rate of testosterone secretion is less in older men than in younger men, there is great variation in this function between different men of similar ages.	In certain cases, the lack of sufficient testosterone may be one of the causes of impotence and reduced vitality. However, this condition is not likely to be an important factor for most healthy older men. (There are many other reasons for impotence and chronic fatigue.)	There is little evidence to suggest that administration of testosterone would be beneficial to many older men. Furthermore, this treatment carries some risks, such as promotion of prostrate cancer, that need to be considered carefully.
Thyroid	The secretion of thyroid hormones is reduced somewhat in older people.	Most endocrinologists consider the lower level of thyroid hormone secretion to be indicative of reduced requirements for the hormones.	Except in the case of thyroid hormone deficiencies, little is gained by the taking of thyroid extract.
Hair and skin			
Hair	Both men and women get gray hair, and many men and some women may lose their hair and become bald.	The graying of hair and the development of baldness may occur in the healthiest of people. These conditions are abnormal only when they occur early in life.	Regular washing of the hair with a mild soap or shampoo helps to keep the scalp clean and prevent unhygienic conditions that cause premature loss of hair.

Footnotes at end of table

(Continued)

TABLE G-3 *(Continued)*

Part of Function of the Body	Change(s)	Biological Significance	Comments
Hair and skin *(Continued)*			
Skin	The skin becomes thinner, looser (due to loss of underlying fat), and wrinkled.	Aging of the skin is normal, except when it occurs prematurely. Some skin lesions are signs of nutritional deficiencies.	Nutritional deficiencies, exposure to strong sunlight, dust, and hot dry air speeds up the aging of the skin.
Immunity	Control of the immune system falters with aging, as evidenced by (1) weakening of the defenses against disease organisms, foreign matter, and abnormal cells and tissues; and (2) formation of antibodies that attack the body's own tissues (also known as autoimmunity).	Reduction of the defenses against potentially harmful agents means increased susceptibility to a wide variety of diseases, including cancer. Autoimmune antibodies may be at least partly responsible for some of the tissue deterioration of glands and organs in older people.	The immune systems of experimental animals were given an extended period of optimal functioning by feeding the animals diets low in calories and proteins from weaning to death. However, the growth and development of the animals was retarded significantly, and they were more susceptible to infectious diseases.
Kidneys	Older people have a reduced flow of blood through their kidneys and an increased tendency to pass some sugar in the urine. However, the kidneys are usually able to excrete wastes adequately, provided that no type of kidney disease is present.	The moderate reduction in the circulation of blood in the kidneys means that the reserve capacity of those organs is diminished. Therefore, other factors that reduce blood flow (dehydration, dehydrating diets, and lack of exercise) should be avoided in order not to stress the kidneys excessively.	Overloading of aging kidneys may be avoided by (1) drinking ample amounts of fluids (8 to 12 cups [*2 to 3 liter*] per day), (2) consuming diets that contain only moderate amounts of protein and mineral salts, and (3) exercising regularly to maintain a healthy circulation.
Lungs	Lung tissue loses some of its elasticity and the muscles used in breathing weaken with aging, so that there is a reduction in the amounts of air that can be inspired and expired. However, the rate of respiratory decline varies considerably among different peoples of similar ages.	At a given age, the number of years of life remaining can be predicted from a measurement of the amount of air that can be blown out of the lungs after a deep breath. (The measurement is called the "vital capacity.") Reduced respiratory functions limits the level of exercise that may be performed.	Smoking accelerates the decrease in vital capacity with aging, as do conditions such as bronchitis and emphysema. Vitamins A, C, E, and regular exercise may help to keep lung tissue healthy and maintain vital capacity.
Muscles and skeleton			
Bones	The thinning and weakening of the bones that occurs in older people usually results from loss of calcium. This loss is accelerated in postmenopausal women, and by poor diet, chronic stress, and lack of exercise.	Loss of bone in the jaw makes it more likely that teeth will be lost. Deterioration of the spinal column causes backaches and a shrinking of height. Weakening of the long bones makes them more susceptible to fractures. (Women are more vulnerable than men because their bones are smaller.[2])	Loss of calcium from bone may be retarded by (1) adequate calcium, phosphorus, and vitamin D in the diet; (2) consumption of only moderate intakes of protein and phosphorus,[3] and (3) regular exercise.

Footnotes at end of table

(Continued)

TABLE G-3 *(Continued)*

Part of Function of the Body	Change(s)	Biological Significance	Comments
Muscles and skeleton *(Continued)*			
Joints	As people grow older, their joints become increasingly susceptible to injury and the development of arthritis. (Archaeologists have found evidence of arthritis in the bones of people who lived thousands of years ago.)	Joints are usually the weakest link in the chain of bones, joints, ligaments, tendons, muscles, etc. Furthermore, joint pains deter people from engaging in necessary exercise. Overweight and stressful movements (such as rapid snaps of the arm in certain sports) increase the likelihood of injuries to joints.	Osteoarthritis may be avoided to some extent by keeping one's weight down. It is believed that some cases of rheumatoid arthritis result from allergies to certain foods. Hence, it may be worthwhile for arthritis sufferers to investigate this possibility.
Muscles	Muscle strength and endurance for sustained exercise is lost at variable rates during aging depending upon the level of physical activity that is performed on a regular basis.	It is noteworthy that muscles are the main reservoirs of amino acids and protein for times of need such as healing of burns, injuries, and surgery. Furthermore, lack of exercise results in wasting of muscles. (Physically active people have slower rates of muscle deterioration.)	The maintenance of healthy muscles throughout life requires a good diet and regular use of muscles. Crash dieting to lose weight rapidly may result in the loss of considerable protein and water from muscles.
Nervous system			
Brain	The nerve cells and mass of the brain decrease gradually, and varying amounts of age pigments (lipofuschin granules) are deposited. No new nerve cells can be formed in the adult brain.	Most of the aging of the brain is caused by factors outside of the tissue itself, such as (1) reduction in the supply of blood oxygen and/or nutrients to the brain, (2) poisoning of brain metabolism by ammonia, ketone bodies, and other metabolic products formed elsewhere in the body, and (3) injuries to the head.	Keeping a healthy brain involves (1) good nutrition and avoidance of toxic substances, (2) prevention of blood vessel disorders such as atherosclerosis or weakening of capillaries due to vitamin C deficiency, (3) controlling diseases that produce toxic metabolic products, and (4) protecting the head against injury.
Nerves	The speed with which nerve impulses are conducted decreases by only about 10% in healthy people during aging. Severe deterioration of the nervous system during aging is usually due to one or more diseases.	The nerve fibers, like the brain itself, are very sensitive to changes in the supply of oxygen nutrients, and other vital substances. Most nerve damage is probably due to (1) deprivation of essential substances, and (2) poisoning by metabolic products and other toxicants.	Various depressants (alcohol and certain drugs), nutrients (amino acids such as tryptophan and tyrosine, choline, and some of the vitamins) and stimulants (particularly caffeine) affect nerve functions.
Reflexes	Reflexes are generally slower in older people. However, extreme slowing may be a sign of a disease.	Certain activities, such as driving at high speeds, may be dangerous for older people.	Good health and long practice of certain activities may offset the effects of slowed reflexes.

Footnotes at end of table

(Continued)

TABLE G-3 (*Continued*)

Part of Function of the Body	Change(s)	Biological Significance	Comments
Sense organs			
Ears	The ability to hear and distinguish between some types of sounds declines variably in different people.	Loss of hearing may isolate an older person from the surrounding environment, unless a hearing aid is worn.	Some types of hearing loss are correctable by surgery and other means. Also, it can be compensated for by being very attentive when others are speaking.
Eyes	A large number of middle-aged and older people require glasses for reading since the usual tendency is to develop farsightedness (presbyopia). Also, the lens of the eye may develop occluded areas (cataracts).	The eyes are very susceptible to damage by rays from the sun, artificially produced ultra-violet light (such as that from fluorescent fixtures), nutritional deficiencies, nuclear radiation, abnormal blood levels of metabolic products such as sugars, and certain irritants.	Protection of the eyes from deterioration and injury requires (1) good nutrition, (2) wearing of protective goggles or other gear when exposed to hazardous conditions, and (3) prompt treatment of diabetes and other disorders.
Sleep	Many older people require less total sleep than younger people. Also, the elderly usually awaken more frequently and do not sleep as deeply.	Failure to obtain adequate amounts of sleep may result in poor performance of mental and physical tasks. However, an excessive need to sleep may be a sign of an emotional, nervous, or physical disorder. (Narcolepsy is an uncontrollable desire to sleep at frequent intervals.)	Restful sleep may be promoted by a regular rhythm of eating, mental and physical activity, and sleeping. Some people find that a glass of hot milk or a glass of wine before retiring helps them to sleep better. Also, it may help to abstain from drinking caffeine-containing beverages during late afternoon and evening.
Temperature regulation	The ability of the body to regulate its temperature within narrow limits appears to decline during aging.	Older people are highly susceptible to hypothermia (abnormal lowering of the body temperature when exposed to a cold environment) and are moderately susceptible to heat strokes.	All senior citizens should have adequate clothing and heating of their residences and workplaces. Some type of cooling system is advisable in hot, moist climates.

[1]A collateral circulation may develop from the tiny blood vessels which connect the major blood vessels. The collateral vessels enlarge and carry increased amounts of blood when the major vessels are occluded.

[2]Men and women lose about the same amounts of calcium from their bones during aging. However, the bone mass of women in early adulthood is only about 70% that of a male at the same age.

[3]Excessive amounts of dietary protein and phosphorus may reduce the utilization of calcium by the body, and increase the loss of the latter mineral in the urine.

It may be seen from Table G-3 that many, if not all, of the major degenerative processes that occur during aging may be slowed down by the appropriate dietary, hygienic, and medical measures. Furthermore, preventive measures are usually less stringent (and less expensive) than corrective measures. Therefore, it behooves each person to make every effort to apply what is known regarding preventive nutrition. Details are given in the sections that follow.

NUTRITIONAL ALLOWANCES. Recommended nutrient intakes for older adults are given in Table G-4. A more complete presentation of nutrient allowances for different age groups, including senior citizens, along with nutrient functions and best food sources of each nutrient, is given in this book in the section on Nutrients: Requirements, Allowances, Functions, Sources, and its accompanying Table N-6; hence, the reader is referred thereto.

TABLE G-4
RECOMMENDED DAILY NUTRIENT INTAKES FOR ADULTS AGED 51 AND OVER[1]

Category	Men	Women
Height .... in. (cm)	68 (173)	63 (160)
Weight .... lb (kg)	170 (77)	143 (65)
Energy .... kcal	1,530[2]	1,280[3]
Protein .... g	63	50
Macrominerals		
Calcium ....mg	800	800
Phosphorus ....mg	800	800
Sodium [4] ....mg	500	500
Chloride [4] ....mg	750	750
Magnesium ....mg	350	280
Potassium [4] ...... mg	2,000	2,000
Microminerals		
Chromium [4,5] ......mcg	50–200	50–200
Copper [4,5] ....mg	1.5–3.0	1.5–3.0
Fluoride [4,5] ....mg	1.5–4.0	1.5–4.0
Iodine ....mcg	150	150
Iron ....mg	10	10
Manganese [4,5] ....mg	2–5	2–5
Molybdenum [4,5] ....mcg	75–250	75–250
Selenium ....mcg	70	55
Zinc ....mg	15	12
Vitamins, fat-soluble		
Vitamin A ......mcg RE	1,000	800
Vitamin D ....mcg	5	5
Vitamin E ..... mg α TE	10	8
Vitamin K [4] ....mcg	80	65
Vitamins, water-soluble		
Biotin [4] ....mcg	30–100	30–100
Folate ....mcg	200	180
Niacin ....mg NE	15	13
Pantothenic acid [4] ...mg	4–7	4–7
Riboflavin ....mg	1.4	1.2
Thiamin ....mg	1.2	1.0
Vitamin B-6 ....mg	2.0	1.6
Vitamin B-12 ......mcg	2.0	2.0
Vitamin C ....mg	60	60

[1]*Recommended Dietary Allowances*, 10th ed., 1989, NRC–National Academy of Sciences.
[2]The recommended daily intake for men aged 76 and over is somewhat less as a result of reduced body size, REE (Resting Energy Equivalent), and activity.
[3]The recommended daily intake for women aged 76 and over is somewhat less as a result of reduced body size, REE, and activity.
[4]These figures are given in the form of ranges of recommended intakes because there is less information on which to base allowances.
[5]The upper values of the range should not be habitually exceeded, since the toxic levels of trace elements may be only several times usual intakes.

The recommended nutrient intakes shown in Table G-4 are *not* to be considered as absolute requirements, but rather as allowances which contain liberal safety margins for people with higher than normal requirements. Hence, they may be considered to represent the upper levels of intakes for use by dietitians and doctors in planning diets and therapies. The recommendations may also be used as a basis for choosing nutritional supplements when it is known that certain nutrients will not be supplied adequately by a diet consisting of ordinary foods because certain types of foods are not consumed regularly. For example, someone who does not eat calcium rich foods such as milk products or green leafy vegetables should take a supplement that provides about 800 mg of calcium per day.

DIETARY GUIDELINES. The first dietary guideline is: *Eat enough of the proper foods consistently.* To ensure nutritional adequacy, the elderly (and those responsible for their diets) are admonished to—

1. Read and follow the section on Nutrients: Requirements, Allowances, Functions, Sources, including Table N-6.

2. Eat a daily diet which includes definite amounts of foods from each of the Six Food Groups as detailed in the section on Food Groups, Table F-37.

But, lots of folks don't know what constitutes a good diet; and, worse yet, altogether too many people neglect or ignore the rules even if they know them. In either case, the net result is always the same; they shortchange themselves on the right amounts of good nutritious foods. As a result, more and more doctors and nutritionists are recommending judicious mineral and vitamin supplementation.

It is noteworthy, too, that those whose diets are restricted for the treatment of certain disorders may need to use supplements as sources of the nutrients that would ordinarily be provided by foods which are not allowed in their diets. For example, people who cannot drink milk or eat cheese may have to take a calcium supplement that provides this mineral.

Menu Planning. Nutritionists have found that it is easy to plan menus if the desired foods are put into the four food groups that are easily identified. It is noteworthy, however, that this procedure results in the best nutrition when a wide variety of foods are included in each group. A typical grouping of foods is given in Table G-5.

TABLE G-5
MAJOR FOOD GROUPS FOR SENIOR CITIZENS

Food Group	Amounts Recommended[1]	Typical Items
Meats, poultry, fish, dry beans, eggs, and nuts	Choose 2 or more servings every day. Count as a serving: 2 to 3 oz (without bone) of lean cooked meat, poultry, or fish. Equivalent in protein to 2 oz meat are 2 eggs; 1 cup cooked beans, dry peas, or lentils; 4 Tbsp peanut butter	Beef, veal, lamb, pork, organ meats such as liver, heart, or kidney; poultry and eggs; fish and shellfish; meat alternates—dry beans, dry peas, lentils, and peanut butter.
Milk, yogurt, and cheese	Use 2-3 cups of milk or the equivalent in a milk alternate, every day. *Note:* Calcium supplements in pill form are *not* the nutritional equivalent of milk and cheese because pills do not supply calories, protein, and the essential nutrients provided by the dairy foods.	Milk: fluid whole, skim, lowfat, evaporated, dry, or buttermilk. Milk alternates on the basis of calcium content are: Cheddar-type cheese, 1-in. cube = ½ cup milk Cream cheese, 2 Tbsp = 1 Tbsp milk Cottage cheese, ½ cup = ⅓ cup milk Ice cream, ½ cup = ⅓ cup milk Ice milk, ½ cup = ⅓ cup milk
Vegetables	Choose 3-5 servings daily including: One vegetable high in vitamin A, and one high in vitamin C. One serving equals: 1 cup of leafy greens, or ½ cup any other vegetable.	Broccoli, carrots, chard, collards, cress, kale, mango, persimmon, pumpkin, spinach, sweet potato, turnip greens, winter squash, or other dark green leaves are high in vitamin A. Asparagus, broccoli, Brussels sprouts, cabbage, cauliflower, collards, garden cress, green pepper, kale, kohlrabi, mustard greens, potato and sweet potato cooked in the jacket, rutabagas, spinach, sweet red pepper, tomato or tomato juice, turnip greens are good sources of vitamin C.
Fruits	Two to four servings of any fruit, including those that are valuable for vitamin C and vitamin A. Count as a serving: ½ cup *(120 ml)* fruit; or a portion as ordinarily served, such as 1 medium apple, banana, or orange, half a medium grapefruit or cantaloupe, or the juice of 1 lemon.	Apples, apricots, berries, cantaloupe, cherries, dates, figs, grapes, grapefruit, guava, lemon, mango, melons, orange, papaya, pear, persimmons, pineapple, plums, prunes, raisins, strawberries, watermelon.
Breads and cereals	Choose 6-11 servings daily. Count as a serving: 1 slice of bread; 1 oz ready-to-eat cereal; ½ to ¾ cup cooked cereal, cornmeal, grits, macaroni, noodles, rice, or spaghetti.	Breads, cooked cereals, ready-to-eat cereals, cornmeal, crackers, flour, grits, macaroni, spaghetti, noodles, rice, rolled oats, parboiled rice and wheat, quick breads, and other baked goods. (The most nutritious products are made from whole grain, or white flour that has been enriched or fortified.)

[1]One cup equals *240 ml*; 1 oz equals *28 g*; and 1 Tbsp equals *15 ml*.

(Also see FOOD GROUPS, Table F-37.)

A typical menu plan that utilizes the recommended servings of each of the major food groups is given in Table G-6.

TABLE G-6
A TYPICAL WEEKLY MENU PLAN FOR SENIOR CITIZENS[1]

Sunday	Monday	Tuesday	Wednesday	Thursday	Friday	Saturday
			Breakfast			
Grapefruit sections High-protein pancakes (part whole wheat) Honey or maple syrup	Cooked prunes Chicken livers Whole wheat English muffins	Tangerine Scrambled eggs Bran muffins	Mixed fruits French toast with maple syrup.	Applesauce Cheese omelet Cornbread	Canned or stewed figs Poached egg Mincemeat coffee cake (part whole wheat).	Orange slices Whole grain cereal w/milk Cinnamon rolls (part whole wheat)
			Lunch			
Barbecued pork chop Baked potato Spinach Ice cream-favorite flavor	Chicken a la king on part whole wheat baking powder biscuit. Asparagus Homemade chocolate pudding.	Glazed ham logs Scalloped potatoes Mixed vegetables Coconut cream pie	Meat loaf Augratin potatoes Broccoli Homemade butterscotch pudding with whipped cream or whipped evaporated milk.	Casserole of broccoli, ham, cream of chicken soup. Cinnamon roll Pumpkin pie	Macaroni and cheese Brussels sprouts Orange-grapefruit salad Peanut butter cookies	Lamb stew Rice pilaf (brown rice) Cole slaw Brownie pudding
			Supper			
Creamed dried beef on whole wheat toast Waldorf salad Chocolate chip cookies	Cream of pea soup Tuna sandwich (part or whole wheat bread) Relish plate Crushed pineapple or mixed fruit.	Hard-cooked eggs, peas, and white sauce Celery and peanut butter sliced salad Gingerbread	Salmon, tomato, green onion salad. Whole wheat roll Baked custard	Spinach salad w/hard-cooked eggs, bacon, croutons. Cheesecake	Fish chowder Carrot and raisin salad Herb bread Peach cobbler	Meat loaf Baked beans Steamed brown bread Apricots with custard sauce
			Evening Snack			
Cheese and whole grain crackers	Oatmeal-raisin cookies Milk or buttermilk	Graham cracker Milk or buttermilk	Granola w/milk	Cinnamon toast Milk	Yogurt w/fruit	Hot chocolate or carob

[1]Beverages are not included except in the evening snack, because of personal preferences. There are a wide variety of herbal teas, as well as the traditional tea and coffee from which to choose.

Saving Money In Food Purchasing and Preparation. The primary rule relative to money when purchasing food is that "the more processing that has gone into it, the more expensive." It is, therefore, less expensive to buy raw fruits and vegetables in season, than to purchase canned, frozen, or dried ones. Also, this applies to all baked goods; it is cheaper to bake your own breads, cakes, cookies, and desserts than to purchase them from the bakery. Further, it is less expensive to make your own soups, salad dressings, sauces, or ice cream, than it is to purchase these ready made. When buying (1) ready-to-cook or (2) cooked and ready-to-serve products you are paying for services, which, if you really want to save money, you can do yourself.

Fig. G-11. Not only do home-cooked meals save money, but there is considerable nostalgia and pride attached to preparing food for your family and friends. (Drawing by Dynamic Graphics, Inc.)

Other ways in which to save money in food purchasing and preparation include—

1. **Save and use coupons.** There are many coupons, either in the newspapers, in the mail, or in or on food packages. All of these can be saved and used to reduce the food bill.
2. **Purchase wholesale or in co-op outlets.** In some parts of the country, there are wholesale stores for canned and packaged products, where food can be purchased for less than in the supermarket.
3. **Watch for sales.** Above all, watch for the sales which are listed in the local newspaper or posted in the grocery store; and, before you shop, plan your menus around these items.
4. **Make quantity purchases at sale prices.** When you buy in quantity, be certain that you have planned how to use it. If perishable, and you have the facilities to store some of it, there shouldn't be any waste. But if you plan to eat it all before it spoils, make sure that you *will* be able to consume it within that time; and make certain that you won't become bored with it and throw out the remainder, which will cancel the money that you saved in the first place by buying it at the sale price. This is the danger in purchasing large quantities.
5. **Scrutinize the nonfood items in the monthly food bill.** When tallying the cost of food for the month, don't forget to deduct the nonfood items that are purchased at the grocery store. Although these items are not part of the cost of food, they are items which usually can be reduced considerably. You can get back to basics; for instance, purchase only soap, cleanser, laundry detergent, bleach, and vinegar (for windows). Armed with these and plenty of old rags made from worn out clothes, the whole house can be cleaned and shined, without the myriad of expensive cleaning solutions now on the market. Likewise, our grandmothers never had the luxury and expense of paper towels, paper napkins, or kleenex. The cost of these items can be eliminated completely.

There are numerous ways to save money, all of which will prove challenging to the thrifty shopper. A few tips for saving money in each of the four food groups follow:

- **Meats, poultry, fish, eggs, legumes, and nuts—**

1. Buy cuts of meat that are on sale. Inexpensive cuts, such as shank and heart, can be simmered, cubed, and jellied for an inexpensive meat dish.
2. With a little ingenuity, the cheaper cuts of meats, poultry, and fish can be made into tasty dishes.
3. Canned mackerel (usually the cheapest fish in the market) can be made into an inexpensive baked fish loaf, with the addition of oatmeal, milk, and egg. Or it can be used instead of salmon in other recipes.
4. Per unit of protein, eggs are one of the cheapest sources of protein.
5. Combine legumes, cereals, and/or nuts for an inexpensive high-quality protein meal. Most of the Spanish dishes take advantage of the combination of corn tortillas and refried beans. The Boston baked beans and steamed brown bread of New England, and black-eyed peas and rice in the deep south, are other examples of this combination.

- **Milk and cheeses—**

1. Nonfat dry milk, reconstituted, is less expensive than fluid fresh milk. Also, it has fewer calories than whole milk, it can be reconstituted in small amounts, and it takes little storage space. If the flavor becomes tiresome, it can be mixed with some whole milk to make a more acceptable drink. However, for cooking purposes this is not necessary. The powdered milk can be mixed with cocoa and sugar, then made into an instant chocolate drink by adding water.
2. The less expensive cheeses are just as nutritious as some of the more expensive ones.
3. Grated parmesan cheese is much more expensive than the brick parmesan. So, grate your own.

- **Vegetables and fruits—**

1. Be sure to watch the prices of fruits and vegetables and purchase when in season. Out of season fruits and vegetables can be very expensive.
2. If you have room around the house or apartment, or even some planter boxes on a patio, it will be advantageous, moneywise, to grow some of your own vegetables. Here again, you are performing some of the services that must be added to the price of the product at the store.
3. In the summertime, on the edge of most towns or in the farmers' market, fresh, homegrown vegetables usually can be purchased for less than in the supermarket.
4. In some areas of the country, growers will allow senior citizens and/or others to go into their fields and

harvest fruits and vegetables, usually for a minimum charge.

5. When you purchase a packaged amount of a vegetable—for instance, a head of cabbage—plan to use it in different ways. It may be served as cole slaw one day, as fried cabbage another day, as boiled cabbage a third day, and used to wrap around ground beef, covered with a sauce, and baked another day. Thus, the secret to success is to serve foods in different forms, and not the same way each meal.

• **Breads and cereals—**

1. As was mentioned earlier, it is less expensive to make all of your own breads and quick breads. Besides, the flavor is far superior. There is no finer dessert than a slice of homemade bread with butter or margarine, and homemade jam.

2. The recipes for bakery products are almost unlimited; thus, there is no reason for monotony in the meals.

3. There are many cereals besides wheat, which should be included in the diet. For instance, use oatmeal in scones, in meat loaves, and in cookies; use brown rice or millet with butter instead of potatoes or in dessert recipes; and add a little rice or barley to chicken or other kinds of soups. Use cornmeal in sauteeing fish, or to add to steamed brown bread. Whole barley can be used instead of potatoes.

(Also see CEREAL GRAINS.)

Perhaps last, but not least, in saving money the subject of leftovers should be considered. Some people throw out leftover food, but those who do so will never have enough money. Leftovers should be considered as treasures, to be used in a multitude of different ways, i.e., soups, casseroles, salads, or just mixed with another dish from the same food group. There are very few items that do not lend themselves to being used in this way. Don't forget that the more mixtures of foods that you eat, the closer you are coming to the golden rule of eating, which is: "the greater the variety of foods that you eat, the greater your chances of getting everything that your body requires for good health."

The recipe for English trifle was developed to use up stale cake. Fondue was developed years ago to use up stale bread. Stale bread can also be used for French toast, zwieback, croutons for soup, crumb topping for casseroles, fruit flummery, sauteed with seasoning for meat stuffings, or for breading chicken and chops. You will be able to think up many more ways to use stale bread.

There are unlimited ways to use up leftover foods. There is no reason to waste so much as a crumb of food; and don't forget, "a penny saved is a penny earned."

Dietary Modifications For Some Common Health Problems. Some of the health problems that affect senior citizens may be remedied by medical treatments prescribed by physicians, assisted by certain types of dietary modifications. However, people should *not* undertake to modify their diets without first consulting a doctor, since conditions other than the most evident problem(s) may require diagnosis and treatment before a dietary change is made. Therefore, the information given in Table G-7 is for the purpose of promoting a better understanding of some common dietary prescriptions.

(Also see NUTRIENTS: REQUIREMENTS, ALLOWANCES, FUNCTIONS, SOURCES.)

TABLE G-7
MODIFIED DIETS FOR SOME COMMON HEALTH PROBLEMS THAT AFFECT SENIOR CITIZENS[1]

Name and Purpose(s) of Diet	Indications for Use	Foods to Use	Foods to Avoid	Comments
Bland (to avoid irritation of inflamed or irritated tissues in the digestive tract by certain types of foods)	Colitis, diverticulitis, duodenal or gastric ulcer, esophagitis, gastritis, heartburn, and hiatus hernia.	Broiled meats, poultry, and fish; boiled eggs; milk drinks and mild cheeses; canned or cooked fruits and vegetables; breads and cereals made from refined flour or grains without hulls or seedcoats; butter, margarine, and mildly flavored mayonnaise and salad dressings; salt, and milk flavoring, herbs, and spices; weakly brewed coffee and tea, and herb tea.	Fried foods; highly seasoned items; aged or ripened cheeses; nuts; raw vegetables and fruits (except ripe bananas), and those with hulls, seeds, and skins; whole grain breads and cereals; and pungent spices.	A mineral and vitamin supplement may be needed if the diet is low in meats, poultry, fish, vegetables and fruits; and rich in refined breads and grains. Some types of soft drinks may be irritating. Some specialists are now recommending high-fiber diets for diverticulitis, with good results.

Footnote at end of table

(Continued)

TABLE G-7 *(Continued)*

Name and Purpose(s) of Diet	Indications for Use	Foods to Use	Foods to Avoid	Comments
High calorie (to promote a gain in weight and an efficient utilization of dietary protein for the healing of tissues)	Underweight, malabsorptive disorders, and convalescence after injury, starvation, surgery, and the treatment of cancer.	Meats, poultry, and fish broiled in a little butter, margarine, or salad dressing; eggs in creamy sauces; nuts as snacks; rich milk drinks such as malted milk; cooked vegetables with sauces; stewed dried fruits with honey; breads and cereals with butter, creamy milk products, honey, jam, margarine, and peanut butter; and liberal amounts of ice cream and other wholesome desserts (after other foods have been eaten).	Items that are low in calories and high in bulk such as lean meats, low fat milks and cheeses, raw vegetables and fruits, and whole grain cereal.	It may be easier for some people to eat this diet if it is taken in three small meals plus 2 to 3 daily snacks. Too much fat may be counterproductive in that it may spoil the appetite, interfere with digestion, and be harmful to health.
High fiber (to stimulate more regular bowel movements and/or slow or reduce the absorption of dietary cholesterol, fats, and sugars)	Chronic constipation, gallstones, and high blood levels of cholesterol, triglycerides, and/or sugar (glucose).	Substitute beans and other legumes for some of the meats, poultry, fish, and eggs normally eaten; low-fat milk products and cheeses; raw vegetables and fruits; stewed dried fruits; and whole grain breads and cereal products.	High-fat meats, poultry, fish, egg, nut, and milk products; canned or overcooked vegetables and fruits; and breads or cereal products made from highly refined flours and/or grains.	The fiber content of the diet should be increased gradually to avoid causing diarrhea and irritation of the digestive tract. Fatty foods tend to slow the movements of the digestive tract.
High protein (to increase the protein content of tissues that are subnormal in this respect)	Convalescence from injury, starvation, surgery, treatment of cancer, or other conditions that result in the wasting of lean body tissue.	Meats, poultry, fish, eggs, nuts, and legumes; milk products and cheeses; canned or cooked fruits and vegetables; and breads and cereal products fortified with eggs, milk, soy flour, and/or wheat germ.	Highly filling, low-protein foods such as fibrous and starchy vegetables and fruits that have a high water content, and bran cereals.	An adequate amount of calories must be consumed to ensure optimal utilization of dietary protein (carbohydrates are more effective than fats).

Footnote at end of table

(Continued)

TABLE G-7 *(Continued)*

Name and Purpose(s) of Diet	Indications for Use	Foods to Use	Foods to Avoid	Comments
Lactose restricted (to minimize the amount of milk sugar [lactose] that is consumed)	Lactose intolerance, due to the malabsorption of the milk sugar, lactose, caused by a decrease in, or absence of, the enzyme lactase.	Meats, poultry, fish, and egg products that contain *no* milk or cheese; vegetarian analogs of milk and cheese products; vegetables and fruits; and breads, cereals, desserts, and soups without milk.	Processed meats, poultry, and fish that contain milk products; except those that have been fermented; vegetables and fruits in cheese or cream sauces; breads and cereals with added milk; and snacks and desserts made with milk.	Small amounts of lactose taken with other foods may be tolerated by some people who cannot digest this sugar.
Low calorie (to reduce the dietary calories in order to bring about loss of body fat)	Adult-onset diabetes, high blood levels of cholesterol and/or triglycerides, high blood pressure, and obesity (excessive body fat content, *not* merely overweight).	Low-fat meat, poultry, fish, and egg products; milk, and cheese products made from skim milk; vegetables and fruits without high-calorie sauces and syrups; whole grain breads and cereal products containing only minimal amounts of fats and sugars.	Meat, poultry, fish, and egg products that are high in fat; full-fat milk and cheese products; vegetables and fruits with added fat-rich and/or sugar-rich sauces or syrups; breads and cereal products containing more than minimal amounts of fats and sugars; butter, margarine, oils, and other high-calorie dressings, sauces, and spreads; and desserts.	Bulky foods with high contents of water and fiber (legumes, vegetables, fruits, and cooked whole grains) may help to satisfy the appetite.
Low cholesterol and/or saturated fat (to reduce the accumulation of cholesterol and fat in the blood and in certain tissues)	Adult-onset diabetes, arteriosclerosis, atherosclerosis, high blood levels of cholesterol.	Low-fat meat, poultry, and fish products; egg whites; legumes; nuts; milk and cheese products made from skim milk; vegetables and fruits *without* cheese or cream sauces; whole grain breads and cereal products; margarine, mayonnaise, and salad dressings made with vegetable oils, but without egg yolks and animal fats.	High-fat meat, poultry, and fish products; egg yolks; legume dishes with meat fat; full-fat milks and cheeses; vegetables and fruits with cheese or cream sauces; breads and cereals with butter, high-fat milk products; or mayonnaise made with egg yolks; cakes and other desserts made from animal fats.	The benefits of these restrictions to older people are uncertain. However, some evidence suggests that the tendency to have strokes may be reduced somewhat. It may *not* be advisable to substitute polyunsaturated fats for saturated fats.

Footnote at end of table

(Continued)

TABLE G-7 *(Continued)*

Name and Purpose(s) of Diet	Indications for Use	Foods to Use	Foods to Avoid	Comments
Low residue (to avoid a laxative effect and distention or irritation of the digestive tract)	Colitis, distention of the digestive tract (indicated by feelings of excessive fullness or by protrusion of the abdomen), acute stage of duodenal or gastric ulcer, lack of appetite, and after surgery of the digestive tract or treatment of cancer by drugs and/or radiation.	Broiled meats, poultry, and fish; boiled eggs; 2 cups (*240 ml*) or the equivalent of milk and mild flavored cheese products; canned or cooked vegetables and fruits without hulls, seeds, or skins; breads and cereal products made from refined flour and/or grains; and ground or very finely chopped herbs and spices.	Fried meats, poultry, fish, and eggs; all beans, legumes, peas, and nuts; milk and cheese products that contain vegetables or fruits; all raw vegetables and fruits except those with minimal fiber content; and breads and cereal products made from whole grains.	Small, frequent meals are tolerated much better than a few large meals.
Low sodium (to restrict the dietary sodium)	Accumulation of excessive fluid in the tissues (edema), congestive heart failure, high blood pressure, and certain kidney disorders.	Meats, poultry, fish, eggs, legumes, nuts, vegetables, fruits, and cereals prepared without salt; milk products; breads and other baked goods leavened with yeast; and seasonings that do not contain salt.	Salted meats, poultry, fish, eggs, legumes, nuts, cheeses; canned vegetables and fruits that contain added salt; baked goods made with baking powder; sauces and soups containing salt or monosodium glutamate (MSG); and desserts, dressings, seasonings, and snack foods that contain salt or MSG.	Various salt substitutes are available. However, a doctor should be consulted before any of them are used. Small amounts of salt in yeast breads and similar products may be acceptable in some cases.

Footnote at end of table

(Continued)

Well balanced meals, with something from each of the Four Food Groups, is just as important at age 70 as at age 7. (Courtesy, National Live Stock & Meat Board, Chicago, Ill.)

TABLE G-7 *(Continued)*

Name and Purpose(s) of Diet	Indications for Use	Foods to Use	Foods to Avoid	Comments
Soft (to provide a diet that is easy to chew and swallow when these functions are weakened or diminished)	Convalescence after a severe illness, stroke, surgery, or treatment of cancer; and the patient finds it difficult to chew and swallow foods.	Tender meats, poultry, and fish that are baked, broiled, creamed, roasted or stewed; all milk products and cheeses (except those with strong flavors, whole seeds, spices); canned or cooked vegetables and fruits without hulls, membranes, seeds, or skins; breads and baked goods made from refined flour, but without fruits, nuts or seeds; refined cereal products; condiments and seasonings containing finely ground ingredients; and desserts such as custards, dessert gels, and ice creams which do *not* contain coarse fruits or nuts.	Meats, poultry, and fish that are (1) fried, (2) rich in bones or other hard to digest connective tissues, and (3) salted or smoked; all legumes, nuts, and seeds other than creamy peanut butter or highly refined soy products; all raw vegetables and fruits except avocados, bananas, and lettuce; fried potatoes; bread stuffing, fried doughs, chow mein noodles, wild rice, barley, and whole grain or bran cereals; and candied fruits, nut brittle; popcorn, relishes, and condiments that contain hulls, seeds, and skins.	Sometimes, soft diets are blenderized to make them easier to swallow; or, they may be made sufficiently liquid for tube feeding. Patients who require soft diets often lack sufficient digestive secretions to dilute the food that is consumed. Hence, ample fluids should be provided and excessive amounts of salts and sugars should be avoided because they may cause dehydration, diarrhea, nausea, and/or vomiting.

[1]Also see ELEMENTAL DIETS; HYPERALIMENTATION;LIQUID DIETS, TUBE FEEDING; and MODIFIED DIETS.

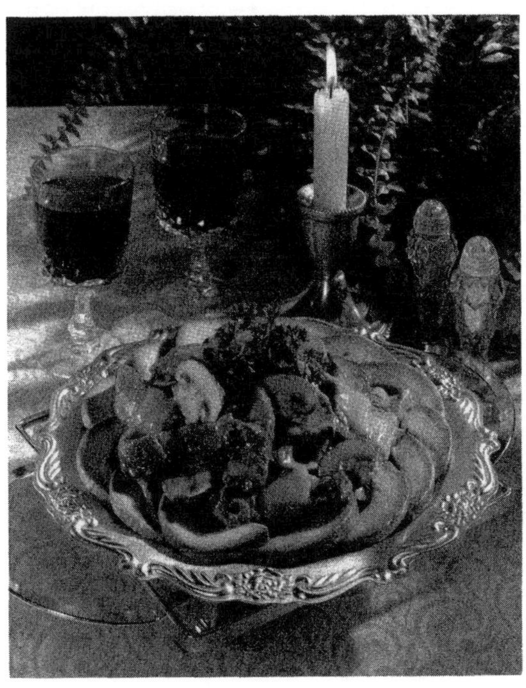

Simple but elegant meals can be accomplished easily during the senior citizen years when there is more time to prepare them and more time to savor them. (Courtesy, California Beef Council, Burlingame, Calif.)

GESTATION

The condition of carrying an unborn fetus. Pregnancy.

GHEE

A semifluid butter preparation from the milk of a buffalo, cow, sheep, or goat. It is nearly 100% milk fat and is used mostly in Asia and Africa.

(Also see MILK AND MILK PRODUCTS.)

GHERKIN *C. anguria*

Small cucumbers used for pickling are often called *gherkins*. But the true gherkin is another plant closely related to the cucumber, which bears many little spiny fruits shaped like olives, and which is cultivated chiefly in Brazil and the West Indies.

(Also see CUCUMBER AND GHERKIN.)

GIGOT

This is a French word for leg of lamb. The term was also applied to the leg-of-mutton sleeves that were popular in the days of Queen Elizabeth I of England.

GINGER BEER

This is not really a beer but an old time soft drink like root beer. Various recipes exist for making ginger beer, but in general it is made with water, dried gingerroot, sugar, lemon juice, and active yeast. Brief fermentation by the yeast serves to carbonate the drink—to make it fizzy.

(Also see SOFT DRINKS.)

GINGIVITIS

Inflammation of the gums.

GINKGO (MAIDENHAIR TREE) *Ginkgo biloba*

Often this tree is described as a "living fossil," because it so closely resembles its fossilized relatives. It was discovered in northern China, and it has now been transported to many areas of the world where it serves primarily as an ornamental. Ginkgoes are medium sized, deciduous trees growing up to 120 ft (*36 m*) high, shaped somewhat like a pine tree. The leaves are fan-shaped and fernlike with parallel veins. Males and females exist as separate trees. The fruit of the female resembles an apricot in size and color, but when these ripe fruits fall to the ground and begin to decompose, they emit an offensive odor. Also, the juice causes some people to develop a rash much like that caused by poison ivy. For this reason, male ginkgoes are more popular as ornamentals.

Despite the unpleasantness of the fruits, the nuts or seeds are good to eat, tasting like mild Swiss cheese. They may be purchased canned in water or they may be gathered in the autumn from the trees; however, this requires special preparation to remove the offensive, fleshy layer. After boiling and removing the shells, ginkgo seeds can be added to a variety of dishes such as duck or chicken. In oriental cooking, ginkgo seeds and chestnuts are often used together. Ginkgo seeds may also be roasted like any other nut or seed.

Ginkgo seeds are listed in Food Composition Table F-36. Overall, they are lower in protein and have fewer calories than most nuts, but they do contain significant amounts of many minerals and vitamins.

GLAND

An organ that produces and secretes a chemical substance in the body.

(Also see DIGESTION AND ABSORPTION; and ENDOCRINE GLANDS.)

GLIADIN

A protein classified as a prolamine, derived from the gluten of wheat, rye, and other grains. Gliadin and glutenin plus a liquid form the unique protein gluten.

(Also see GLUTEN; and PROTEIN[S].)

GLOBULINS

A type of protein which is abundant in nature. These proteins are slightly soluble in water, but solubility increases with the addition of salts. Globulins are rounded molecules; hence their name reflects their structure. Upon heating, globulins coagulate. Familiar examples include serum globulins (gamma globulin), muscle globulins, and numerous plant globulins.

(Also see PROTEIN[S].)

GLOMERULONEPHRITIS (NEPHRITIS)

In the kidney the functional unit is called a nephron—a system of tubules and a filter or glomerulus. Each kidney contains about one million of these nephrons. Thus, glomerulonephritis is a diseased condition, an inflammation, of the glomeruli of these functional units. The disease is serious. Glomerulonephritis can be either acute or chonic. It affects mainly children and adolescents, and it often follows a streptococcal infection. Since the disease disrupts the glomeruli—the points where filtration of the blood occurs—signs include albumin (protein) and blood in the urine, edema (water in the tissues), high blood pressure, and sodium retention. When edema, high blood pressure, and reduced urine excretion are noted, the primary dietary adjustments indicated are restriction of water and sodium intake. Sufficient calories in the form of carbohydrates and fat should be supplied in order to prevent the breakdown of the body's own proteins for energy. Protein sources should be rich in essential amino acids—meat, eggs, and milk.

(Also see HEMODIALYSIS; and MODIFIED DIETS.)

GLOSSITIS

An inflammation of the tongue.

GLUCAGON

A hormone secreted by the alpha cells of the islets of Langerhans in the pancreas. Its release is stimulated by a drop in blood glucose. When released, it acts to mobilize glucose from the liver, thereby raising the level of blood glucose.

(Also see DIABETES MELLITUS; ENDOCRINE GLANDS; and HYPOGLYCEMIA.)

GLUCOASCORBIC ACID ($C_7H_{10}O_7$)

A form of ascorbic acid (vitamin C; $C_7H_{10}O_7$) which has one minor chemical modification. It does not act like vitamin C. Rather, it is an antagonist to the vitamin and can be used experimentally to produce scurvy in animals.

(Also see VITAMIN C.)

GLUCOCORTICOIDS

A broad term which describes the specific hormones cortisol, corticosterone, and cortisone. These hormones are secreted by the cortex (outside layer) of the adrenal gland. Glucocorticoids stimulate gluconeogenesis, amino acid availability, protein synthesis, and fat mobilization (free fatty acids). They are necessary for the body to adapt to or resist stress.

(Also see ENDOCRINE GLANDS; and METABOLISM.)

GLUCONEOGENESIS

Formation of glucose from noncarbohydrate sources (amino acids and fats) when there is an insufficiency of dietary carbohydrates. The liver is the main site for gluconeogenesis. Amino acids are the primary precursors in gluconeogenesis. The process is essentially that of reverse glycolysis (glucose breakdown).

(Also see METABOLISM.)

GLUCOSE (DEXTROSE; GRAPE SUGAR; $C_6H_{12}O_6$)

Glucose is a monosaccharide—a carbohydrate—which serves as a chief source of fuel for the metabolic fire of life. Fig. G-12 shows the structure of glucose.

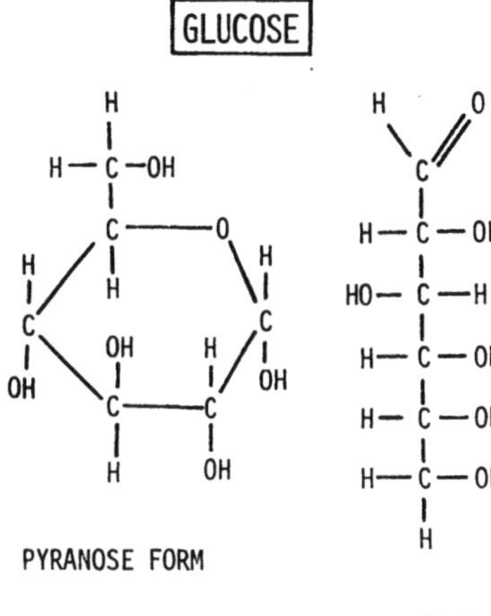

Fig. G-12. Structure of glucose.

Every tissue in the body is capable of removing glucose from the blood and utilizing it for the production of energy needed for body processes. Some tissues rely on the action of insulin for removal of glucose from the blood, while the brain, a large consumer of glucose, does not. All other digestible carbohydrates are eventually converted to glucose for transport in the blood and for utilization by the cells of the body. In the blood, the normal glucose—blood sugar—level ranges between 60 and 100 mg/100 ml, while excess glucose is stored as glycogen in muscles and the liver.

Plants manufacture glucose from carbon dioxide (CO_2) and water (H_2O) by the process of photosynthesis. Some glucose is found free in sweet fruits such as grapes, berries, and oranges, and some vegetables such as corn and carrots. However, much of the glucose manufactured by plants is converted to other carbohydrate forms. Cellulose and starch are long chains of glucose formed by plants. The disaccharides, sucrose and maltose, and even the nonplant disaccharide lactose, all contain glucose.

Commercially, glucose is employed in the manufacture of confections, in the wine industry, and in the canning industry. Since solutions of glucose rotate polarized light to the right, glucose is often called dextrose, especially in industry.

(Also see CARBOHYDRATE[S]; DEXTROSE; METABOLISM; and PHOTOSYNTHESIS.)

GLUCOSE METABOLISM

Body processes, hence life itself, depends upon a constant supply of energy. Much of this needed energy is derived from the catabolism—controlled combustion—of the sugar, glucose. In the cells of the body, this occurs via a series of enzymatic reactions which permit the orderly transfer of energy from glucose ($C_6H_{12}O_6$) to energy rich compounds, primarily adenosine triphosphate (ATP) as glucose is burned to carbon dioxide (CO_2) and water (H_2O). Then ATP becomes the driving force for those body processes requiring energy. Glucose is transported to the cells by the blood. The source of glucose in the blood is ultimately the diet. All digestible carbohydrate is eventually converted to glucose. However, since (1) eating is an intermittent process, (2) not all glucose is immediately converted to energy, and (3) blood glucose can be rapidly depleted, mechanisms exist for maintaining the level of blood glucose within the narrow limits of about 80 to 100 mg/100 ml. This ensures a constant supply of glucose (energy) to the cells of the body. Fig. G-13 illustrates the major routes of glucose metabolism, and the means whereby the glucose level of the blood is maintained.

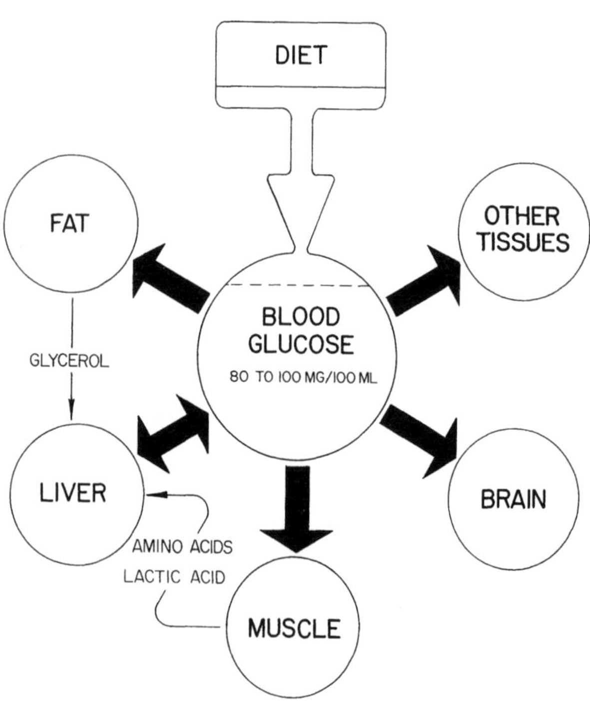

Fig. G-13. Glucose metabolism and the simplified relationships of the diet, tissues, and organs in maintaining a constant level of glucose in the blood.

Glucose may be converted to two storage forms of energy: (1) muscle and liver glycogen, and (2) fats. Also, the metabolism of glucose cannot be separated from the metabolism of fats and protein. Fats (glycerol) and protein (some amino acids) are potential sources of glucose, and glucose can be converted to glycerol, fatty acids, and certain amino acids. The liver plays a central role in maintaining blood glucose levels by converting (1) glycogen to glucose, (2) some amino acids to glucose, (3) glycerol from fat to glucose, and (4) lactic acid from the muscles to glucose. The brain, muscles, and other tissues consume glucose for the production of energy. Although the brain does not store energy, glucose is its major energy source. The overall direction that glucose metabolism takes depends on the needs of the body. Energy—maintaining the level of glucose in the blood—is a priority need.

Glucose metabolism is primarily directed by the hormones insulin, glucagon, epinephrine, and glucocorticoids. Entry of glucose into most of the cells requires the action of insulin. Glucagon and epinephrine mobilize glucose from liver glycogen. Glucocorticoids and glucagon promote the formation of glucose from protein and fats—gluconeogenesis.

Furthermore, adequate minerals and vitamins must be available for the proper metabolism of glucose. Minerals are cofactors with many of the enzymes involved, and with the B complex vitamins—thiamin, niacin, riboflavin, pantothenic acid, vitamin B-6, biotin, and folacin.

A well-known disorder of glucose metabolism—diabetes mellitus—is the fifth leading cause of death in the United States.

(Also see DIABETES MELLITUS; DIGESTION AND ABSORPTION; ENDOCRINE GLANDS; HYPOGLYCEMIA; METABOLISM, section headed "Carbohydrates"; MINERAL[S]; and VITAMIN[S].)

GLUCOSE OXIDASE

An enzyme produced by a number of fungi. It catalyzes the conversion of glucose plus oxygen to gluconic acid. The important feature of glucose oxidase is that it reacts with only glucose and no other hexoses—sugars. Glucose oxidase is commercially available; it is usually derived from *Aspergillus niger*. In the food industry, the most important application of glucose oxidase is for the removal of glucose from eggs before drying. This prevents the nonenzymatic browning reaction between glucose and proteins during storage, thereby stabilizing dried egg products. Glucose oxidase is also employed to remove oxygen from beverages, canned food products, dried or dehydrated foods, and mayonnaise. This minimizes flavor and color changes during storage.

GLUCOSE TM (GLUCOSE TRANSPORT MAXIMUM)

Glucose is an important substance which the body cannot afford to lose in the urine. Glucose TM is a measure of the ability of the kidneys to "salvage" glucose during the formation of urine. The letters "TM" stand for transport maximum. It is the maximum rate at which glucose can be transported from the filtrate formed in the kidney back into the blood. Healthy human kidneys can transport 320 mg of glucose per min. back into the blood. When this rate is exceeded glucose "spills over" into the urine (glucosuria). In individuals suffering from diabetes mellitus, blood glucose levels may reach 300 mg/100 ml, thus exceeding the transport maximum, and resulting in glucosuria. Healthy individuals excrete little, if any, glucose in the urine.

(Also see DIABETES MELLITUS; and RENAL GLUCOSURIA.)

GLUCOSE TOLERANCE FACTOR (GTF)

The complete identity of this hormonelike agent is not yet known, although it is certain that it contains the element chromium and the vitamin niacin, and perhaps amino acids such as glycine, glutamic acid and cysteine. Glucose tolerance factor is released into the blood—from perhaps the liver, kidneys or other tissues which store chromium—whenever there is a marked increase in the blood levels of sugar (glucose) and/or insulin. It, along with insulin, acts in making it easier for amino acids, fatty acids, and sugars to pass from the blood into the cells of the various tissues. It also promotes the metabolism of the nutrients within the cells. Much more insulin is required to accomplish these tasks when GTF is lacking; but GTF does not have any effect when insulin is absent.

(Also see CHROMIUM, section headed "Component of the Glucose Tolerance Factor.")

GLUCOSE TOLERANCE TEST (GTT)

A diagnostic procedure which determines the rate at which the cells of the body take up glucose (dextrose) from the blood. It is an index of the severity of diabetes and also of liver function and thyroid activity. Factors which reduce glucose tolerance—mainly insuffcient insulin—are those that slow the rate of glucose uptake by cells. Therefore, blood sugar (glucose) tends to rise above the levels considered normal under conditions of good health.

The test is administered in the morning, following an overnight fast. First, a sample of "fasting blood" is taken. Then, a test dose of glucose is given in the form of a water solution which the patient drinks. Blood samples are usually taken at hourly intervals for up to 6 hours. A diagnosis is made by comparing the hourly blood glucose levels of the patient to standard values. Depending upon the judgment of the physician and the laboratory methods used to measure glucose, a person is diagnosed as hypoglycemic, normal, or diabetic.

(Also see DIABETES MELLITUS, section headed "Glucose Tolerance Test"; and HYPOGLYCEMIA, section headed "Glucose Tolerance Test.")

GLUCOSURIA

The appearance of glucose (sugar) in the urine. Normally little if any glucose is present in the urine of healthy individuals. However, glucosuria may occur due to (1) a kidney defect; (2) a drug such as phlorhizin; (3) ingestion of a high-carbohydrate meal; (4) excessive stress; or (5) an infection. By far the most common cause of glucosuria is diabetes mellitus.

(Also see DIABETES MILLITUS; and RENAL GLUCOSURIA.)

GLUCURONIC ACID ($C_6H_{10}O_7$)

An acid derived from glucose. It occurs naturally in combination with a variety of chemicals in the urine. It is the body's way of detoxifying many substances, including those normally found in the body; for example, hormones, and those introduced into the body such as drugs and poisons. By combining these substances and glucuronic acid in the liver, they become water soluble and can be eliminated in the urine.

(Also see DISEASES, section headed "Detoxification of Poisonous Substances.")

GLUTAMATE, SODIUM (MONOSODIUM GLUTAMATE; MSG)

This chemical is better known as MSG or monosodium glutamate, the sodium salt of the amino acid glutamic acid (COOH $[CH_2]_2CH[NH_2]COON_2$). It is a widely used flavor enhancer, although it does not add any flavor of its own.

For centuries, Japanese cooks used a certain dried seaweed to flavor their soups and other foods. But it was not until the early part of the 20th century that they discovered that this seaweed contained MSG. In 1963, the FDA approved the use of MSG in foods. It is commercially produced (1) from the waste liquor of beet sugar refining. (2) from the hydrolysis of wheat or corn gluten, or (3) by organic synthesis. The worldwide consumption rate of MSG is now more than 150 million pounds (*67.5 million kg*) per year. It has, however, been associated with one minor disease—the Chinese Restaurant Syndrome.

(Also see ADDITIVES; and CHINESE RESTAURANT SYNDROME.)

GLUTAMIC ACID

One of the nonessential amino acids.

(Also see AMINO ACID[S].)

GLUTATHIONE

A tripeptide of cysteine, glutamic acid, and glycine, which can act as a hydrogen acceptor and hydrogen donor.

GLUTEN

A plant protein found mainly in wheat. Rye ranks as a poor second, followed by oats and barley. Corn and rice are low in gluten.

Gluten imparts the properties of elasticity and strength to flours. Actually, two proteins, gliadin and glutenin, form gluten when mixed with liquid. The elastic gluten, developed by kneading the dough, entraps the carbon dioxide, (1) formed during the fermentation of sugars and starches by yeast, (2) released by chemical leavening, or (3) beaten into the mixture. The result is the unique "rising" or expansion characteristics of wheat flour doughs.

Individuals suffering from celiac disease, sprue, or gluten allergy must avoid the numerous foods containing gluten, since it is often used as a food additive.

(Also see ADDITIVES; ALLERGIES, section headed "Wheat Allergy"; BREADS AND BAKING; CELIAC DISEASE; SPRUE; and WHEAT, section headed "Flour.")

GLUTEN-FREE DIET

Dietary sources of gluten—a protein abundantly present in wheat, rye, oats, and barley—must be eliminated from the diets of people who cannot tolerate it. This condition is characteristic of individuals suffering from celiac disease or sprue, and the condition which sometimes occurs after gastrointestinal surgery or severe diarrhea. Only products containing flours made from corn, rice, starchy vegetables like potatoes, or legumes such as soybeans and lima beans may be used. People who cannot tolerate gluten must use a minimum gluten diet for the rest of their lives.

(Also see ALLERGIES, section headed "Wheat Allergy"; CELIAC DISEASE; MODIFIED DIETS; and SPRUE.)

GLUTEN-FREE FOODS

Those foods manufactured without flour from wheat, rye, barley or oats, as these flours contain the protein gluten. Due to the versatility of wheat flour and the protein gluten, they are found in a large number of food products. Therefore, a person selecting gluten-free foods should exercise care, read the label and become familiar with the types of foods likely to contain wheat or gluten.

(Also see ALLERGIES, section headed "Wheat Allergy.")

GLYCERIDES

An ester of glycerol and fatty acids in which one or more of the hydroxyl groups of the glycerol have been replaced by acid radicals. Glycerides may contain one fatty acid (mono-), two fatty acids (di-), or three fatty acids (tri-).

(Also see FATTY ACIDS; FATS AND OTHER LIPIDS; and TRIGLYCERIDES.)

GLYCERIN (GLYCERINE)

The popular name for glycerol, an important biochemical and a common food additive.

(Also see ADDITIVES; and GLYCEROL.)

GLYCEROL ($C_3H_5[OH]_3$)

An alcohol containing 3 carbons and 3 hydroxy (OH) groups. it is a colorless, odorless, syrupy, sweet liquid. Glycerol is most commonly found in chemical combination with fats in compounds called triglycerides. Also, glycerol is a GRAS (generally recognized as safe) food additive employed to prevent drying, a humectant, as well as FDA approved for numerous other uses. Its popular name is glycerine.

(Also see ADDITIVES; FATS AND OTHER LIPIDS; and TRIGLYCERIDES.)

GLYCEROL-LACTO STEARATE (OLEATE OR PALMITATE)

A USDA approved food additive that is used as an agent to emulsify animal and vegetable fat.

(Also see ADDITIVES; and EMULSIFYING AGENTS.)

GLYCINE

One of the nonessential amino acids.

(Also see AMINO ACID[S].)

GLYCOCHOLIC ACID ($C_{26}H_{43}NO_6$)

The sodium salt of this acid is a normal constituent of bile. Thus, it is a natural emulsifying agent aiding in the digestion of fats. It is a GRAS (generally recognized as safe) food additive employed as an emulsifier.

(Also see ADDITIVES; DIGESTION AND ABSORPTION; and FATS AND OTHER LIPIDS.)

GLYCOGEN

The storage form of glucose within the liver and muscle cells. It is sometimes referred to as "animal starch" since it is similar to starch—both are composed of numerous glucose units. In a normal adult, there are about 108 g of glycogen in the liver and 245 g of glycogen in all the muscles combined. The storage of glucose as glycogen and the release of glucose from glycogen are hormonally controlled.

Generally, there are no dietary sources of glycogen since it is rapidly converted to pyruvic and lactic acid in the meat and liver of slaughtered animals. Only some seafoods—oysters, mussels, scallops, and clams—which are eaten virtually alive, contain small amounts of glycogen.

(Also see CARBOHYDRATE[S]; ENDOCRINE GLANDS; and METABOLISM.)

GLYCOGENESIS

Conversion of glucose to glycogen.

(Also see CARBOHYDRATE[S]; and METABOLISM.)

GLYCOGENIC

Of or pertaining to the formation of glycogen, the storage form of carbohydrate in humans and animals.

GLYCOGENOLYSIS

Conversion of glycogen to glucose.
(Also see CARBOHYDRATE[S]; and METABOLISM.)

GLYCOLYSIS

Conversion of carbohydrate to lactate or pyruvate by a series of enzymatic reactions.
(Also see CARBOHYDRATE[S]; and METABOLISM.)

GLYCOPROTEIN

Proteins containing less than 4% carbohydrate. This classification includes such proteins as egg albumin, serum albumins, and certain serum globulins.
(Also see PROTEIN[S].)

GLYCOSIDES

A group of naturally occurring plant toxicants, or poisons. (Not all glycosides are toxic; e.g., several of the common nonphotosynthetic plants pigments.) Specifically, they are various sugars attached to another chemical compound and are designated as glucoside (glucose), mannoside (mannose), galactocide (galactose), etc. Toxic glycosides include cyanogenetic glycosides that yield hydrocyanic (prussic) acid upon hydrolysis, goitrogenic substances that cause acute goiter, irritant oils such as mustard oil, coumarin glycosides, and steroid (cardiac and saponic) glycosides. Symptoms produced by these toxicants range from stomach irritation to cancer, to death. Some common sources of glycosides are choke cherry, peach and apricot pits, and wild black cherries. Two glycosides are familiar to most people: digitalis (digoxin), the heart stimulant; and Laetrile (a brand name), the questionable cancer cure.
(Also see POISONOUS PLANTS.)

GLYCOSURIA

An alternate spelling of glucosuria.
(Also see GLUCOSURIA.)

GLYCYRRHIZA (LICORICE)

A natural flavoring agent extracted from dried roots of *Glycyrrhiza glabra*. It is a GRAS (generally recognized as safe) food additive.
(Also see ADDITIVES.)

GOBLET CELLS

Secretory cells on the mucosal surface that produce mucus.

GOITER

Contents — *Page*

History of Goiter 1083
Causes of Goiter and Related Disorders 1084
Simple Iodine Deficiency Goiter 1084
Areas of Endemic Goiter 1085
Goitrogenic Agents 1085
Toxic Goiter 1085
Toxic Diffuse Goiter (Graves' Disease) 1085
Toxic Nodular Goiter 1085
Cretinism 1085
Myxedema 1086
Treatment and Prevention of Goiter and Related Disorders 1086
Protection Against the Effects of Radioactive Iodine 1087

Enlargement of the thyroid gland (which is located in front of the larynx at the base of the neck), known as goiter, is usually the result of dietary deficiencies of iodine. In some cases, however, it may be caused by such things as goitrogenic agents, inflammatory disorders, or tumors. Another type of goiter, called exophthalmic goiter (Graves' disease), is due to overactivity of the thyroid gland, which is usually—but not always—enlarged.

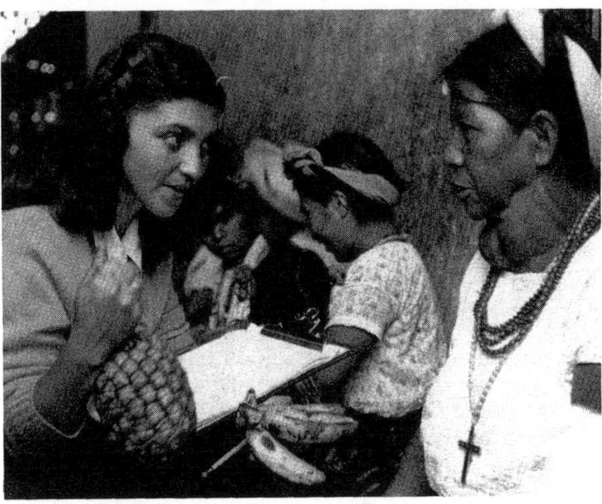

Fig. G-14. This villager suffering from goiter is shown talking to a nurse/dietitian at a local market in Guatemala. (Courtesy, WHO)

HISTORY OF GOITER. An early Chinese document dated about 3000 B.C. described the symptoms of goiter and blamed the disorder on the poor quality of drinking water, mountainous terrain, or emotional disturbances. The recommended cure consisted of the ingestion of seaweed and burnt sponge, which contained large amounts of iodine. It is known that the Chinese even administered dried thyroid glands of deer as a treatment for goiter. Today, physicians use similar dried glandular extracts from cattle, sheep, or swine.

The Ebers Papyrus of Egypt, which dates back to about 1500 B.C., described the surgical removal of the thyroid and the ingestion of salt obtained from a particular place in southern Egypt. Hippocrates believed that bad drinking water was the cause of goiter. Roman physicians noted that the thyroid gland in females seemed to increase during puberty, menstration, and pregnancy.

There is now evidence which supports this belief.

In the Middle Ages, the Germans thought that goiter was caused by strenuous work.

The name thyroid (after the Greek word for shield, *thyreos*) was given in a description of the gland by the English physician Wharton in 1656, who determined its location, size, and weight. Although there was much discussion at the time concerning the possible function of the gland, such an understanding required knowledge of the role of the mineral element iodine.

In 1811, the French chemist Courtois discovered iodine when he noted the violet fumes which rose from ashed seaweed heated with sulfuric acid during the cleaning of vats used to prepare saltpeter for the manufacture of gunpowder. Proof that iodine is an element was obtained several years later in the separate investigations of the French chemist Gay-Lussac and the English chemist Davy.

The first use of a purified form of iodine (potassium iodate) to treat goiter was by the English physician Prout in 1816. The Swiss physician Coindet, in 1820, recommended small doses of iodine in the treatment of goiter, since he observed that hyperthyroidism sometimes resulted from large doses.

From 1850 to 1876, the French chemist Chatin conducted investigations on the iodine content of air, water, soil, and plants from various parts of Europe. The data that he obtained led him to suggest that there was a relationship between goiter and low levels of the mineral in the environment. His ideas were rejected because of the limitations of his analytical techniques and the toxicity which had been observed to result from large doses of iodine.

During the last quarter of the 19th century, myxedema was described and thyroid extracts were used to treat patients after their thyroids had been surgically removed. Then, in 1895, the German physiologist Magnus-Levy demonstrated the regulation of the metabolic rate by the thyroid; and the German biochemist Baumann discovered that the gland was rich in iodine.

In 1907, Marine, an American medical scientist, studied thyroid disease and iodine deficiency in farm animals and fish. Then, in 1916 he began to apply his findings to the prevention of goiter in humans. His administration of sodium iodide to school girls in the fifth to twelfth grades in Ohio demonstrated that such treatment effectively prevented goiter.

In 1914, Kendall, a scientist at the Mayo Clinic in Minnesota, crystallized thyroxin, one of the thyroid hormones. The chemical structure of thyroxin was described in 1927 by Harrington and Barger, British university scientists. Harrington synthesized thyroxin, thereby paving the way for a more effective therapeutic agent than thyroid extracts which were variable in biological potency.

Chesney and his coworkers at John Hopkins Hospital demonstrated, in 1928, that goiter could be produced in rabbits by the feeding of cabbage, which was later found to contain goitrogenic agents. It was also found that the metabolic rate of the animals was lowered. The disorder was corrected by feeding Lugol's solution of iodine, but many animals developed hyperthyroidism and died, a demonstration of the effects of overzealous therapy.

Since then, the list of substances known to be goitrogenic has grown to include even bacteria in water. Hippocrates' theory was at least partially correct.

CAUSES OF GOITER AND RELATED DISORDERS. Although simple iodine deficiency goiter is the most common disorder of the thyroid gland, other factors may be responsible for abnormalities of this gland. Therefore, goiter and a group of related disorders are discussed in the sections that follow.

Simple Iodine Deficiency Goiter. The enlargement of the thyroid gland is usually the result of an attempt by the body to adjust to a deficient level of iodine in the diet. An iodine shortage slows the production of thyroid hormones since these hormones contain iodine. Reduced blood levels of thyroid hormones lead to increased secretion by the pituitary of a thyroid-stimulating hormone (TSH) which causes growth and metabolic activity in thyroid tissue. Secretion of TSH by the pituitary is believed to be under constant stimulation by the thyrotropin-releasing hormone (TRH) which flows directly to the pituitary from the hypothalamus. The flow of TRH is shut off when levels of thyroid hormones are adequate, but is operative when additional amounts of thyroid hormones are required, as in cold environments. Stimulation and enlargement of the thyroid results in a more efficient use of the limited supply of iodine for the synthesis of thyroid hormones. (See Fig. G-15.)

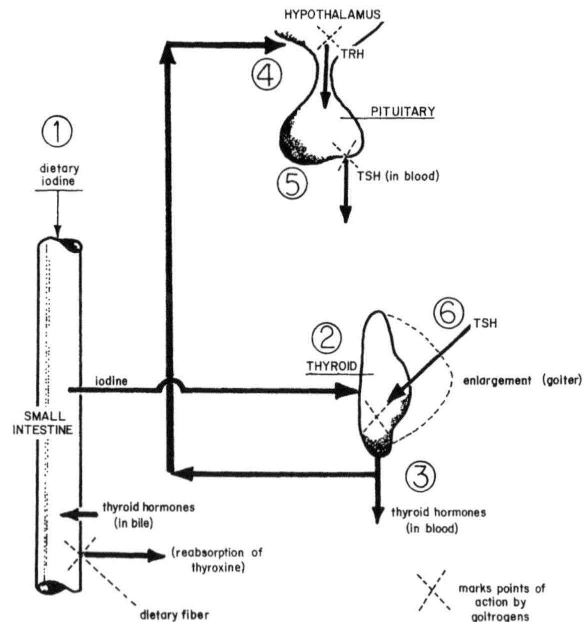

Factors In The Development Of A Goiter

1. Dietary deficiency of iodine and/or ingestion of goitrogens.
2. Reduced synthesis of hormones by the thyroid.
3. Deficient levels of thyroid hormones in blood.
4. Hypothalamus secretion of thyrotropin-releasing hormone TRH is controlled by blood levels of thyroid hormones.
5. TRH stimulates pituitary to release thyroid-stimulating hormone (TSH). Excessive release of TSH may be caused by goitrogens.
6. Enlargement of thyroid due to excessive TSH.

NOTE: Dietary fiber binds with thyroid hormones and prevents their reabsorption.

Fig. G-15. Common causes of goiter.

Goiter is more likely to develop during puberty and pregnancy, when there are greater than normal needs for the thyroid hormones. Also, it is more common in females than in males.

It has also been shown in animal studies that joint deficiencies of iodine and vitamin A result in a significant enlargement of the thyroid which ultimately leads to disordered functions of the gland since vitamin A is required for the normal growth and metabolism of the columnar epithelial cells of the thyroid.

AREAS OF ENDEMIC GOITER. There are regions where many persons have goiter (the disorder is said to be endemic when over 20% of adolescent females in an area are affected), due to iodine deficiency in foods produced in the area, or as a result of the ingestion of foods and water which contain goitrogenic agents. Geographical studies of the distribution of goitrous populations have shown that endemic goiter is likely to be found in inland areas where there is limited availability of iodine-containing seafoods, in mountainous areas, and where there has been extensive action of glaciers which results in topsoil of negligible iodine content. Examples of such regions are the western mountains and the Great Lakes and Pacific Northwest regions of the United States; the Andes Mountains of South America; the Alpine areas of Europe; the Himalayas of Asia; and the plains areas in Africa, Asia, and South America.

(Also see IODINE.)

GOITROGENIC AGENTS. A variety of substances produce goiter when they are ingested in food or water. Some sources of these agents are: (1) members of the cabbage family—such as cabbage, turnip, rutabaga, and kale (cooking may inactivate the goitrogenic factor); (2) milk from cows that have eaten plants containing goitrogens; (3) dietary excess of calcium or fluorine (the effect is enhanced by joint excesses of these minerals); (4) raw soybeans; (5) thiocyanate-containing drugs used to treat high blood pressure; (6) arsenic (arsenical ores are used in Alpine regions as a seasoning instead of onions and garlic); and (7) increased levels of indigestible residue (crude fiber) in the diet (unabsorbable fiber may bind the thyroxin secreted in the bile and prevent its reabsorption in the intestine).

Recent investigations have shown that some strains of *E. coli* bacteria in contaminated water supplies produce goitrogenic substances.

Iodine itself may be a goitrogen when it is ingested in large amounts. Excesses of iodine interfere with the synthesis of hormones by the thyroid and have been used clinically in the treatment of hyperthyroidism. Goiters are produced when iodine interference with thyroid function reduces hormone output to subnormal levels. Inhabitants of the coastal area of Japan may consume from 8 to 25 mg per day of iodine in seaweed (80 to 250 times the amount believed to prevent goiter in most people). Yet, about 6 to 12% of these inhabitants develop goiter, so it appears that other factors may contribute to susceptibility.[2]

Toxic Goiter. Hyperthyroidism (excessive secretion of thyroid hormones) is characterized by a goiter which is similar in appearance to a nontoxic goiter, except that there are other accompanying features which help examiners to distinguish between the two conditions. Thyroid enlargement and hypersecretion of hormones, in thyrotoxicosis, is usually accompanied by rapid heartbeat, nervousness, weight loss, fatigue, increased sweating, and sensitivity to heat. The two major types of toxic goiter are diffuse and nodular; both disorders are more common in females than in males.

TOXIC DIFFUSE GOITER (GRAVES' DISEASE). Most victims of this disorder have protruding eyeballs (exophthalmos). Graves' disease is believed to be caused by a long-acting thyroid stimulator (LATS) produced by lymphocytes (white blood cells which carry antibodies against tissues).

TOXIC NODULAR GOITER. Growth of thyroid nodules sometimes occurs without TSH simulation. In some cases, there may be escape of thyroid hormone production from the control of the pituitary; then, the uncontrolled nodules are described as having become autonomous. This condition is more likely to be found in older persons who may have cardiovascular disorders as a consequence, but have none of the other characteristics of hyperthyroidism.

Cretinism. Congenital deficiency of thyroid function, or cretinism, is characterized by dwarfism, mental retardation, deaf-mutism, and occasionally goiter, which is not present if the defect results in a complete absence of the gland. This condition is found to be more prevalent in areas of endemic goiter and is believed to result from iodine and thyroid hormone deficiencies in the pregnant mother because the developing fetus is dependent during the first 3 months of embryonic life on the maternal supply of thyroid hormones. Cretinism is more common among populations where there are many close kinship marriages. Although the Great Lakes region has long been recognized as an endemic goiter area, the rarity of cretinism has been attributed to the intermarriage of persons from diverse ethnic backgrounds and genetic stock. (See Fig. G-16.)

---

[2]Suzaki, H., *et al*, "Endemic Coast Goiter in Hokkaido, Japan," *Acta Endocrinologica*, Vol. 50, 1965, p. 161.

Fig. G-16. Cretins in central Africa. The three children in the foreground show the characteristic signs of lethargy, swollen cheeks (edema), and distended abdomens. (Courtesy, FAO, Rome, Italy)

Myxedema. Hypothyroidism (subnormal secretion of thyroid hormones), which is acquired anytime after birth, can develop into the severe clinical disorder of myxedema (named after the fluid accumulation apearing in the face and other areas of the body). Most cases of this disorder are found in females between ages 40 and 60, although it is found occasionally in young children. Although there seems to be some evidence that myxedema might be the end result of chronic lymphocytic thyroiditis (Hashimoto's disease), the cause of the chronic condition is not known for certain. It may begin with a slight enlargement of the thyroid. Signs of myxedema are nonpitting edema, impaired mental function (not as severe as in cretinism), increased susceptibility to chills due to a reduction in the basal metabolic rate, overweight, elevated blood cholesterol, constipation, poor circulation of the blood, coarsening of the skin and hair, loss of hair, hoarseness of voice, lack of muscle tone, partial deafness, and enlargement of the heart in severe cases.

TREATMENT AND PREVENTION OF GOITER AND RELATED DISORDERS. The longterm consequences of most of the chronic disorders of the thyroid gland may be disability or even death. Therefore, any sign of such a disorder should be investigated promptly, and treatment started as soon as the diagnosis is certain. Treatments for these disorders follow.

• **Simple iodine deficiency goiter**—This condition is first treated by the administration of thyroid hormone in doses large enough (0.1 to 0.3 mg of thyroxine per day) to block the secretion of thyroid-stimulating hormone. There may not be regression of the goiter if nodules have developed. After several weeks, the physician may conduct tests of thyroid function to make certain that the goiter does not contain autonomous nodules which could lead to hyperthyroidism. While therapeutic doses of potassium iodine (60 mg per day) may prevent further growth of a simple goiter, this agent is not as effective as thyroxine in obtaining regression of a thyroidal enlargement.

Prevention of goiter may sometimes be achieved in endemic areas of developing countries by the intramusclar injections of iodized oil once every 2 to 3 years. However, prevention of goiter is usually achieved in the developed countries by the use of iodized salt (3 g of which furnishes 228 mcg [micrograms] of iodine or slightly more than the *Recommended Dietary Allowance* [RDA] of 200 mcg per day for lactating women).[3] Most people use more than 3 g of salt per day.

Other dietary sources of iodine are seafoods, food products from animal sources (since all livestock receive iodine in their rations), fruits and vegetables grown on iodine-rich soils, and breads made from iodine-containing flours and/or iodates, which may be used as dough conditioners. However, extra iodine should be given with caution to persons living in iodine-poor environments since it has been frequently observed that persons in areas of endemic goiter are hypersusceptible to iodine and may even develop hyperthyroidism. This phenomenon is called the Jod-Basedow effect, after the man who first described the clinical signs of hyperthyroidism. Such an effect was recently observed in Tasmania where bread was fortified with 2 ppm of potassium iodate.[4] Iodate has been used as a dough conditioner in the commercial production of bread in the United States at levels as high as 12 ppm, resulting in 225 mcg of iodine per ounce of bread.[5] Some large commercial bakers, however, have discontinued the use of this additive.

• **Toxic goiter**—The most commonly used treatments for hyperthyroidism are antithyroid drugs which block the synthesis of thyroid hormones, surgical removal of part of the thyroid gland, and therapy with radioiodine which destroys the cells of the thyroid. It is sometimes necessary to balance these treatments with oral doses

---

[3]*Recommended Dietary Allowances*, 10th ed., NRC–National Academy of Science, 1989.

[4]Clements, F. W., *et al.*, "Goiter Prophylaxis by Addition of Potassium Iodate to Bread. Experience in Tasmania," *Lancet*, Vol. 1, p. 489.

[5]Pintauro, N. D., *Food Additives to Extend Shelf Life*, Noyes Data Corporations, Park Ridge, N.J., 1974, p. 174.

of thyroid hormones in order to prevent the effects of hypothyroidism. Administration of large excesses of iodine for the purpose of hormone synthesis is no longer used since the effect is only temporary.

• **Cretinism**—This condition requires lifelong provision of thyroid hormones, in the form of either the pure hormones or of extracts from the thyroid glands of animals such as cattle and sheep. Administration of iodine does not help since there is a lack of functioning tissue in the thyroid gland.

Early diagnosis and treatment of cretinism is important in order to minimize the amount of permanent disability and mental retardation. Some of the signs of the condition in newborn infants are: jaundice, feeding difficulties, respiratory problems, abdominal distention, edema, subnormal body temperature, delay in passing a stool (more than 20 hours after birth), lack of activity, birth weight over 9 lb (*4 kg*), and birth after more than 42 weeks of pregnancy.[6]

• **Myxedema**—This condition, like cretinism, requires lifelong provision of the thyroid hormones which cannot be made in the body. However, the prognosis for the disorder is usually better than that for cretinism, since it occurs at a later age when there has already been some normal growth and development.

PROTECTION AGAINST THE EFFECTS OF RADIOACTIVE IODINE. The radioisotope I-131 occurs in fallout from nuclear bomb tests and is very toxic to the thyroid, particularly in growing children. There have been suggestions that protection from such an effect might be obtained by high dietary levels of iodine (to prevent the rapid accumulation of radioiodine by iodine deficient thyroids). The isotope is likely to be concentrated in milk from cows grazed on pastures exposed to fallout of radioiodine from the atmosphere. Thus, another suggestion is to stockpile noncontaminated nonfat dry milk for use during periods of high atmospheric radioactivity.

(Also see IODINE.)

GOITROGENIC

Producing or tending to produce goiter.

(Also see GOITER.)

GOLDEN BERRY *Physalis peruviana*

Another name for the Cape Gooseberry, which is a husk-covered yellow berry of the family *Solanaceae* and is about ½ in. (*1 to 2 cm*) in diameter. It is native to the American tropics and is closely related to the Ground cherry (*P. pruinosa*) and the Strawberry tomato (*P. ixocarpo*). All of these fruits are relatives of the common tomato (*Lycopersicum esculentum*).

(Also see CAPE GOOSEBERRY; and STRAWBERRY TOMATO.)

GOOSEBERRY *Ribes spp*

This berry type of fruit grows on thorny bushes that are members of the family *Saxifragaceae*.

Fig. G-17. Gooseberries

Gooseberries are closely related to currants, except that they are larger and grow singly rather than in clusters. Also, the two types of plants do not crossbreed (hybridize) readily.

ORIGIN AND HISTORY. Various species of gooseberries grow wild in both the Old World and the New World. It seems likely that the fruit was gathered from wild plants in prehistoric times since there is evidence that the American gooseberry was used by the Cherokee Indians of the Carolinas.

The early history of gooseberry cultivation in Europe was linked closely to the conquests of other lands by the Normans (a group of Scandinavians who established themselves in Normandy, France) who brought the plants with them to France in the 10th century and most likely to England in the 11th century. Therefore, it is noteworthy that the 13th century records of King Edward I of England recorded the purchase of gooseberry plants from France. Two centuries later, the English had developed their own varieties of the fruit, but they did not grow them commercially until the 19th century. However, the initiation of commercial production was preceded by the organization of gooseberry clubs that held annual contests to identify the finest berries. Hundreds of varieties were developed or improved as a result of these contests.

European gooseberries were brought to America by the early colonists, but they did not fare very well in the new environment because of their susceptibility to an American type of mildew. Although the native American

[6]Schneider, J. M., "Congenital Hypothyroidism: A Newborn Diagnosis," *Perinatal Press*, September-October 1976, p. 4.

varieties of the fruit were resistant to the mildew, the berries were notably inferior to the larger and sweeter European berries. Furthermore, various federal and state regulations were enacted in the early 1900s to restrict the commercial production of gooseberries in the United States because the plants were found to be an intermediate host for the white pine blister rust which caused considerable pine tree losses around the end of the 19th century. In spite of these problems, American plant scientists are attempting to develop new varieties which are disease and pest resistant, because gooseberries are firm enough for mechanical harvesting and are therefore potentially profitable.

PRODUCTION. Statistics on the world production and the U.S. production of gooseberries are not readily available. Much of the world crop is grown in England and other countries in northern Europe, whereas the rather small U.S. crop is produced mainly in Oregon, Washington, and Michigan.

Gooseberries are usually grown in regions that have cool summers like those of northern Europe. Furthermore, they are among the most cold hardy of the commercial berry crops and are able to withstand temperatures as low as −40°F (*−40°C*). The plants are propagated by stem cuttings taken in the fall, or in the case of certain varieties, by mound layering. In the latter case, the main branches of the parent bushes are cut back drastically in the fall to stimulate the production of shoots. In the summer which follows soil is mounded up around the shoots to induce rooting. When the roots are well developed, the shoots are cut from the parent plants and grown in a nursery for a year or two. Well-rooted gooseberry cuttings are usually planted in the field in the fall or early spring.

A rich, well-drained, weed-free soil (preferably a clay loam) is required. Some type of nitrogenous fertilizer is required. Stable manure was long used for this purpose. Periodic shallow cultivation or mulching with straw is used to control the weeds. Shading of the plants during hot weather is also advisable, although large shade-providing plants may steal the soil nutrients needed by the gooseberries. Irrigation is required in dry areas.

Gooseberries are usually picked between the beginning of July and the middle of August. The berries turn from a rich green color to a whitish green when they ripen fully. It is noteworthy that some consumers expect gooseberries to be green and consider them to be over-ripened when the color has lightened. Furthermore, fruit which has not yet ripened completely makes the best jams, jellies, pies, and tarts, whereas the berries that are to be eaten raw are better when ripened fully. Picking is often done by hand with the picker stripping the berries onto a cloth laid under the bush. However, harvesting is also done by a special machine or by the application of a chemical (an abscission agent) that loosens the fruit so that it may be readily shaken from the bush.

PROCESSING. Some of the gooseberry crop is processed by canning in a sugar syrup or by freezing. Much of the frozen fruit is sold to bakeries and the manufacturers of jams and jellies.

SELECTION AND PREPARATION. Fresh gooseberries are not likely to be available, except in the vicinity of a few commercial plantings or when the bushes are grown by small scale gardeners. In the event that the fresh fruit is available, it is a rare treat when served with cream and sugar.

Canned gooseberries are sold in many supermarkets in the United States and are probably the form of fruit that is best known to most people. The berries retain a tart flavor even when canned in a sugar syrup. Hence, they serve well in pies, puddings, and tarts to which they contribute a tart flavor that is pleasing when adequate sweetener is used. The canned fruit is also good in compotes, fruit salads, and gelatin desserts. It also goes well as a topping for cottage cheese, ice cream, pancakes and waffles, and yogurt.

NUTRITIONAL VALUE. The nutrient compositions of various forms of gooseberries are given in Food Composition Table F-36.

Some noteworthy observations regarding the nutrient composition of gooseberries follow:

1. The raw fruit is fairly low in calories (39 kcal per 100 g) and carbohydrates (10%). It is a good source of fiber, potassium, and vitamin C, and a fair source of vitamin A.
2. Gooseberries canned in water have only about ⅔ the nutrient levels of the raw fruit, except that the vitamin C content is only ⅓ as much, due to destruction during canning.
3. Canned gooseberries packed in heavy or extra heavy syrup have about 2 or 3 times the caloric and carbohydrate contents of the raw fruit. However, the levels of the other nutrients are about the same as those in the canned fruit packed in water alone.

(Also see FRUIT[S], Table F-47 Fruits of the World.)

GOSSYPOL

The toxic yellow pigment contained in the glands of cottonseeds. Although there are no known cases of gossypol poisoning in humans, it is regarded as a potential toxicant based on animal experimentation. Also, it may cause discoloration of egg yolks during cold storage. Methods have been devised to extract the gossypol from cottonseeds. In the past, extraction was made at the expense of the amino acid content. However, new extraction techniques do not significantly lower the amino acid availability or the yield and quality of the oil. Also, glandless cottonseed, free of gossypol, is now available. As an inexpensive source of protein for humans, the future of cottonseed is bright.

(Also see COTTONSEED.)

GOURDS

• Sometimes the term *gourd* is used to refer to the entire gourd family. The gourd family or cucurbits (*Cucurbitaceae*) includes cucumbers, melons, squash, and pumpkins. Their fruits are usually large, fleshy, and have a thick rind. The plants are vines with broad leaves and trumpet shaped flowers.

• Gourds are also a large and varied group of plants which are trailing or climbing vines producing primarily an ornamental fruit, though a few are edible. they are similar to pumpkins and squash. Gourds occur in a variety of odd shapes, sizes and colors. Besides being used for decoration, the hard shells (rinds) are used for making cups, bowls, dippers, and cooking containers. The interior of the dishrag gourd may actually be used as a dishcloth or bath sponge. In India, the bottle gourds are used as sounding boxes for some musical instruments. For decorative purposes, gourds harvested when mature, and then varnished, keep for long periods.

(Also see MELON[S]; PUMPKINS; SQUASHES; and VEGETABLE[S].)

GOVERNMENT FOOD PROGRAMS

Contents *Page*

History **1089**
Details of the Major Programs **1089**
Benefits **1091**
Problems **1091**
Costs **1091**
Future Prospects **1092**

The major government food programs are those operated by the U.S. Department of Agriculture (USDA) for the purposes of (1) improving the nutritional status of infants, children, and low-income families; and (2) helping to support the income of farmers by providing mechanisms for utilizing surplus foods. The number of participants and the costs of most of the programs have grown rapidly. Recently, many of our elected officials and taxpayers have raised questions as to whether the benefits received have merited the great expenditures. Therefore, the aspects of major programs are noteworthy.

HISTORY. During the early 1930s, the depression led to widespread unemployment and great reduction in the food purchasing power of people in American cities. These unfortunate conditions led to surpluses of food and sharp drops in the incomes of farmers. The paradox of hungry people living in a country which had plenty of food fostered the idea of having the government purchase the surplus food from the farmers and distribute it to the poor. Hence, in 1933 the USDA and the Federal Emergency Relief Administration began the distribution of surplus pork, dairy products, and wheat to needy families.

The early efforts to dispose of farm surpluses led to establishment of the commodity distribution, school lunch, school milk, low-cost milk, and food stamp programs of the 1930s. People employed by the Works Progress Administration (WPA) provided much of the labor for operation of the programs. Some of the programs were discontinued in 1943 because World War II eliminated most of the unemployment together with the surplus since both people and food were needed to sustain the war effort. Thereupon, the USDA started a cash reimbursement to pay schools for food that was purchased locally for the school lunch program. Shortly thereafter, the National School Lunch Act of 1946 was enacted to establish the program on a permanent basis.

After World War II, there was a period of economic expansion that was accompanied by a high birthrate and general prosperity. Nevertheless, there were moderate recessions in the 1950s and farm surpluses reached new highs. So, the USDA reinstated its commodity distribution programs along the lines of those operated just before the outbreak of the war. However, it became evident during the early 1960s that groups of people living in certain areas, such as those in the Appalachian Mountains and Indians on reservations, were receiving poor diets because they depended entirely on the commodities. Shortly thereafter, public interest in families living in "pockets of poverty" became widespread, and new programs were instituted. For example, pilot food stamp programs were started in selected areas in 1961 and the Food Stamp Act was passed in 1964. A little later, the Child Nutrition Act of 1966 established the school breakfast program and added new features to the school lunch program.

The new programs of the 1960s apparently fell short of their goals since the Citizens Board of Inquiry into Hunger and Malnutrition in the United States reported in its 1968 study, *Hunger U.S.A.*, that (1) one-fifth of U.S. households had "poor" diets according to the nationwide survey conducted by the USDA in 1965, (2) 36% of low-income households subsisted on "poor" diets, and (3) people in 266 U.S. counties were living in such distressed conditions that a Presidential declaration called the counties "hunger areas." This report was publicized on a nationwide CBS television program on hunger in America. The national concern which was aroused led to expansion of the food assistance programs during the 1970s. Also, the Special Supplemental Food Program for Women, Infants, and Children, which is commonly known as the WIC program, was started on a pilot basis in 1972 and was expanded in the years that followed.

By the late 1970s, there were widespread concerns that (1) the costs of the food assistance programs would become unmanageable, (2) persons who were not needy were allowed to receive benefits, and (3) a permanent and growing class of dependent people was being fostered. Some of these concerns were addressed in the Food and Agriculture Act of 1977, which contained provisions designed to correct abuses of the food stamp program. Changes in other programs were achieved mainly by the issuance of new regulations.

DETAILS OF THE MAJOR PROGRAMS. Pertinent details of the major federal food assistance programs follow:

• **Food Stamp Program (FSP)**—Stamps that may be used to purchase foods are issued without charge to participants who meet certain criteria of eligibility. The criteria are based upon the net income left after certain costs for shelter, child care, and maintaining or seeking employment are deducted from countable contributions to income. Also, the value of the stamps issued is based upon the amount by which the current cost of the USDA Thrifty Food Plan exceeds 30% of the net income. Physically and mentally fit applicants between the ages of 18 and 60 years must also meet the work registration requirements in order to receive the stamps. In 1990, the Food Stamp Program cost $14.1 billion. Additional information is provided in the separate article on this program.

(Also see FOOD STAMP PROGRAM.)

• **School Lunch Program, National (NSLP)**—The USDA assists elementary and secondary schools in providing nutritious lunches through grants-in-aid, donated commodities, nonfood assistance, and technical guidance. Depending upon family income, pupils are served full-price, reduced-price, or free lunches. Each lunch must contain (1) a serving of cooked lean meat, poultry, fish, or a suitable meat alternative; (2) two or more servings of vegetables and/or fruits, one of which may be a fruit or vegetable juice; (3) one or more servings of bread and/or bread alternates such as cereals or pasta products; and (4) a serving of milk (preference is to be given to the serving of unflavored nonfat milk, skim milk, or cultured buttermilk). In 1990, 24,589,000 persons participated in the National School Lunch Program, which cost $3.2 billion. Additional information is provided in the separate article on this program.

(Also see pictures on p. 1108; and SCHOOL LUNCH PROGRAM.)

• **School Breakfast Program (SBP)**—Schools that serve breakfasts which meet USDA guidelines may be reimbursed at fixed rates for each full-price, reduced-price, and free breakfast that is served. The criteria of pupil eligibility for reduced-price or free breakfasts are the same as those for the corresponding types of school lunches. The guidelines require that each breakfast contain as a minimum the following food components in the amounts indicated: (1) ½ pint (*240 ml*) of fluid milk served as a beverage, on cereal, or used in part for each purpose; (2) ½ cup (*120 ml*) serving of fruit or vegetable or both, or full-strength fruit or vegetable juice; and (3) one slice of whole-grain or enriched bread, or an equivalent serving of corn bread, biscuits, rolls, muffins, etc., made of whole-grain or enriched meal or flour, or ¾ cup (*180 ml*) or 1 oz (*28 g*) (whichever is less) of whole-grain cereal or enriched or fortified cereal, or an equivalent quantity of any combination of these foods. It is further suggested that breakfasts served to children older than 1 year include as often as practicable meat or meat alternates such as a 1 oz serving of meat, poultry, or fish; or 1 oz of cheese; or 1 egg; or an equivalent quantity of any combination of any of these foods. In 1990, 4,235,000 persons participated in the School Breakfast Program, which cost $596 million.

• **Milk Program, Special (SMP)**—This program was established to provide partial reimbursement for milk served at low cost to children in schools where no facilities for serving regular school lunches existed. At the present time, it also promotes the serving of milk in addition to that included in school breakfasts or lunches by providing full or partial reimbursement for the cost of the extra milk. Full reimbursement is given to schools and nonprofit child care institutions for free milk that is supplied to needy children. The type of milk served may be unflavored or flavored whole milk, lowfat milk, skim milk, or cultured buttermilk. In 1990, the Special Milk Program cost $19 million.

• **Special Supplemental Food Program for Women, Infants, and Children, (WIC) Program**—Funds are provided to state health departments or comparable agencies for the purpose of supplying supplemental nutritious foods to low-income women, infants, and children who have been judged by health professionals to have special nutritional needs. Hence, the eligibility requirements were established by each of the states rather than by federal authority. Eligible participants receive supplemental food packages via (1) retail purchase with food vouchers, (2) home delivery, or (3) distribution by WIC clinics.

The contents of the packages vary according to the category of the recipient. At the present time, the categories and packages are as follows:

1. *Infants aged 0 to 3 months.* Iron-fortified infant formulas (may be concentrated, powdered, or ready-to-feed).

2. *Infants aged 4 to 12 months.* Iron-fortified infant formula, fruit or vegetable juice, and iron-fortified dry infant cereal.

3. *Children and women with special dietary needs.* Iron-fortified infant formula, fruit or vegetable juice, and certain types of hot or cold cereals.

4. *Children aged 1 to 5 years.* Unflavored whole milk, lowfat milk, skim milk, cultured buttermilk, evaporated whole milk, evaporated skim milk, dry whole milk, nonfat dry milk, and/or various types of cheeses; fruit or vegetable juice; certain types of hot or cold cereals; eggs; and peanut butter or mature dry beans or peas.

5. *Pregnant and breast-feeding women.* Same foods as No. 4 except that greater quantities of milk or cheeses are allotted.

6. *Nonbreast-feeding, postpartum women.* Same foods as No. 4 except that less fruit or vegetable juice is allotted, and no legumes are allotted.

It is noteworthy that participating agencies must use one-sixth of their administrative budgets for WIC on providing nutrition education, unless a waiver from this requirement is received. In 1990, the WIC Program cost $2.1 billion.

• **Child Care Food Program (CCFP)**—Cash and commodity assistance is provided for meal service for needy and nonneedy children in nonprofit day-care centers, as well as in family and group day-care centers. Meals may be provided to participating children free, at a reduced price, or at full price. The assistance covers one or more of the following types of meals: breakfast, lunch, supper, and supplemental food served between meals. However, the center shall not be reimbursed for supplemental food if it also participates in the special milk program. The federal guidelines for the foods served at the various meals are essentially the same as those for the School

Lunch and School Breakfast programs. Supper components are the same as those for lunch. Supplemental food must include (1) a serving of fluid milk, full-strength fruit or vegetable juice, a fruit or vegetable, or any combination of these foods; and (2) a serving of whole-grain or enriched bread, corn bread, biscuits, rolls, muffins, etc., or an equivalent quantity of any combination of these foods. In 1990, the Child Care Food Program cost $788 million.

• **Summer Food Service Program for Children (SFSPC)**—All meals served to children by local sponsors of this program are free. Participation is limited to sponsors in areas where at least one-third of the children would qualify for free or reduced-price meals under the National School Lunch and School Breakfast Programs or to institutions providing meals as part of an organized program for children enrolled in camps. Most of the meals are served during the summer months, although some are also served during other approved times. The guidelines for the foods served are the same as those for the other child feeding programs. In 1990, the Summer Food Service Program for children cost $162 million.

• **Commodity Supplemental Food Program (CSFP)**—This program provides surplus agricultural commodities to nutritionally vulnerable groups of needy people such as women, infants, children, and American Indians. However, the number of participants has dropped sharply in recent years as a result of (1) reduced availability of commodities, and (2) serving of the needy people by other federal food assistance programs such as the Food Stamp and WIC Programs.

• **Donation of Foods (commodities) to feeding programs**—This part of the commodity distribution program provides surplus agricultural commodities to schools which participate in the National School Lunch and School Breakfast Programs, child care institutions, nonprofit summer camps, eligible charitable institutions, state correctional institutions for minors, and nutrition programs for the elderly. The last item refers to projects conducted under Title VII of The Older Americans Act to assist in meeting the nutritional and social needs of persons aged sixty or older.

BENEFITS. Recent studies of the nutritional status of low income people in the United States and certain evaluations of the food assistance programs have yielded evidence that the most needy have benefited.

1. In 1979, the Field Foundation medical team in its report "Hunger in America: the Federal Response" credited the food programs with having been responsible for significant reductions in the numbers of grossly malnourished people in poverty stricken areas.

2. Generally, families in counties with the greatest poverty have received substantially more assistance than those in counties where average incomes are higher.

3. The school lunch program, which is now available to more than 90% of all elementary and secondary school children, has much higher student participation rates in the lower per-capita-income states than in the higher income states. It is noteworthy that schools considered "especially needy" (those in which 40% or more of the lunches served are free or reduced in price) may be reimbursed for up to 100% of the operating costs of the breakfast program.

4. The increase in retail food sales per capita is substantially greater in the counties which contain the largest numbers of needy and which receive the greatest amount of federal food assistance than in the counties where fewer people are assisted.

PROBLEMS. A long standing problem in all types of public assistance programs is that the less needy often find ways to obtain more than their just share, which tends to inflate costs and reduce the amounts available for the more needy. Furthermore, there is the possibility that food assistance programs will provide only extra food, and that poor dietary practices may remain unchanged; hence, only a small improvement in nutritional status is achieved. These problems have brought about the following noteworthy changes in the federal policies and regulations:

1. New regulations for the Food Stamp Program which were issued in 1980 require that students and other physically and mentally fit participants between the ages of 18 and 60 years register for employment at state employment agencies, actively seek a job, and maintain part-time or full-time employment.

2. The opportunities for the more affluent people to obtain surplus commodities are now limited greatly, because over 93% are now donated to schools, charitable institutions, and programs for the elderly; and, less than 7% is given to individuals.

3. In 1980, the USDA put into effect new rules for the school lunch program which restricted the sale of competitive foods of low nutritional value (carbonated drinks, water ices, chewing gum, candies, and candy-coated popcorn) until after the last lunch period.

4. A total of $20 million in nutrition education and training funds was awarded by the USDA in the 1980 fiscal year to the 50 states and 7 territories under federal jurisdiction. The money was for teaching children, teachers, and school food service personnel about nutrition.

COSTS. The numbers of participants and the costs for the USDA food assistance programs rose steadily from 1960 to 1989, as is shown in Table G-8.

It is noteworthy that the total cost of the USDA food assistance programs was more than $21.7 billion in 1990, with the Food Stamp and National School Lunch Programs accounting for 65% and 15%, respectively, of the total expenditures.

TABLE G–8
NUMBERS OF PARTICIPANTS AND COSTS OF FEDERAL FOOD PROGRAMS[1]

Program	1960	1970	1980	1989
Food Stamp:				
Participants (millons)	—	4.3	21.1	18.8
Federal cost (millions of dollars)	—	550	8,721	11,682
National School Lunch:				
Children participating (millions)	13.8	22.5	26.6	24.2
Federal cost[2] (millions of dollars)	94	300	2,279	3,006
School Breakfast:				
Children participating (millions)	—	0.5	3.6	3.8
Federal cost[2] (millions of dollars)	—	11	288	512
Special Milk:				
Quantity reimbursed (millions of ½ pt)	2,385	2,902	1,796	190
Federal cost (millions of dollars)	81	101	145	19
Women-Infant-Children:[3]				
Participants (millions)	—	0.1	2.0	4.4
Federal cost (millions of dollars)	—	8	603	1,553
Child Care Food:				
Children participating (millions)	—	0.1	0.7	1.4
Federal cost (millions of dollars)	—	6	210	612
Summer Food Service for Children:				
Children participating (millions)	—	0.2	1.9	1.7
Federal cost (millions of dollars)	—	2	106	132
Needy Family Commodity:				
Participants (millions)	3.9	3.8	0.1	0.1
Federal cost (millions of dollars)	59	282	24	52
Donation of Foods:				
Child nutrition (millions of dollars)	132	265	930	795
Charitable institutions (millions of dollars)	16	23	71	136
Emergency feeding (millions of dollars)	—	—	—	265

[1]Adapted by the authors from *Statistical Abstract of the United States 1980*, U.S. Dept. of Commerce, p. 133, Table 214; and *1991*, p. 370, Table 610.
[2]Excludes the cost of commodities donated to the schools, and State contributions to cost of meals.
[3]Covers only special supplemental food program and commodity supplemental food program.

FUTURE PROSPECTS. It seems likely that the policies and benefits of some or all of the federal food assistance programs will be revised in the near future in order to reduce federal expenditures and balance the budget. For example, the current regulations for the Food Stamp Program authorize a workfare pilot project in which unemployed participants aged 18 to 60 will work off the value of the stamps they receive. Some other possible changes are (1) curtailment of multiple benefits by families participating in two or more food assistance programs, and (2) strengthening of quality control, management evaluation, and planning to reduce errors in administration.

GOVERNOR'S PLUM (RAMONTCHI) *Flacourtia indica*

The fruit of a shrub (of the family *Flacourtiaceae*) that is native to the tropics of southern Asia. Governor's plums are purplish red or blackish berries that are about 1 in. (*2 to 3 cm*) in diameter. These fruits contain a red, juicy, seedy pulp that may be eaten fresh or made into jam or jelly. Governor's plums are moderately high in calories (94 kcal per 100 g) and carbohydrates (24%). They are a good source of fiber and potassium, and a poor source of vitamin C.

GRAHAM FLOUR

A whole wheat flour mix, named after Sylvester Graham, an advocate of dietary reform.

(Also see FLOUR[S]; and WHEAT.)

GRAINS

Seeds from cereal plants—members of the grass family, *Gramineae*.

(Also see CEREAL GRAINS.)

GRAM

• A metric unit of weight which equals 1000th of a kilogram or 0.035 oz. Four hundred fifty four grams equal 1 lb, and 1 oz equals about 28 g.

• The name used in the East Indies for any one of several leguminous plants, such as the chickpea and certain beans.

GRANADILLA, GIANT *Passiflora quadrangularis*

This species has the largest fruit of the more commonly known types of passion fruit (family *Passifloraceae*). They usually are from 8 to 12 in. (*20 to 30 cm*) long, and from 4 to 6 in. (*10 to 14 cm*) in diameter.

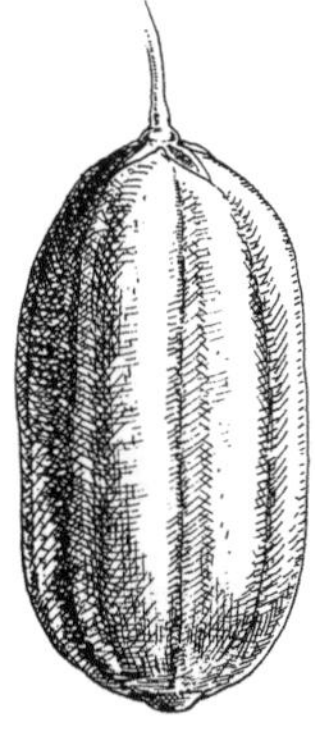

Fig. G-18. Giant granadilla.

The giant granadilla is native to the American tropics, but it differs from the purple passion fruit in that it grows better in the hot, humid, lowland than it does at higher altitudes.

The yellowish-green fruits have white, juicy seedy flesh that lacks the flavor of the purple passion fruit. Hence, they are used mainly to make beverages, ice cream, jam, jelly, and sherbet. The unripe fruits may be cooked as a vegetable.

The flesh of the ripe fruit has a high water content (94%) and is low in calories (20 kcal per 100 g) and carbohydrates (4%). It is a fair to good source of iron and vitamin C.

(Also see FRUIT[S], Table F-47 Fruits of the World—"Passion Fruit.")

GRAPE (RAISINS) *Vitis vinifera*

Contents — *Page*

Kinds of Grapes 1093
Origin and History 1094
Production 1094
Propagation and Growing 1095
Harvesting 1095
Processing 1095
Selection 1096
Preparation 1096
Nutritional Value 1097

Fig. G-19. A cluster of grapes of the Ribier variety. (Courtesy, California Table Grape Commission)

Grapes are the fruit of the woody vines of the genus *Vitis*. They are classified as berries and they grow in clusters of as few as 6 to as many as 300 berries. The color of grapes can be black, blue, golden, green, purple, red, or white, while the flavors range from fruity to spicy to muscat, depending upon the variety.

Grape vines climb by means of cylindrical-tapering tendrils and bear greenish flowers. Some plants bear perfect flowers, others bear only staminate flowers with rudimentary pistils.

Grapes are a versatile fruit. They are enjoyed fresh and processed into a number of popular products such as wines, raisins, juices, jellies, canned grapes, and frozen grapes. In both the Old and the New World, man enjoyed a long-time acquaintance with grapes.

KINDS OF GRAPES. There are three kinds of grapes: (1) European or Old World, (2) North American, and (3) hybrids.

1. **European or Old World.** The scientific name for Old World grapes is *Vitis vinifera*. About 95% of the grapes grown in the world are of this type. Among these are found the highest quality grapes for table, raisin, and wine use. Unfortunately, they are very susceptible to

diseases and some are damaged both by winter cold and by warm spells in the winter.

A few small plantings have been made outside the recognized West Coast vinifera areas but they are considered too risky to recommend for general commercial plantings.

2. **North American.** Before the colonists arrived, the Indians were already using several native grape species. Primarily two American species are important: (a) *Vitis labrusca,* commonly called American bunch grapes, and (b) *Vitis rotundifolia,*often called muscadine grapes. These are the hardy, disease-tolerant varieties. Most are derived in some degree from the wild American Fox grape *Vitis labrusca.* Well known examples are Concord, Delaware, and Niagara. Several modern varieties in this group are seedless and well suited for table use, but not for raisins.

Unlike the European type, the skins of the American types slip from the fruit pulp.

3. **Hybrids.** These include the so-called "French hybrids" developed in France over the past 80 years primarily for wine. They are more neutral in flavor than most American varieties, and a few are well adapted for table use.

The hybrids are crosses between European grape varieties and various native American species. They were selected for the fruit quality of their European ancestors and the disease and insect tolerance of their American ancestors.

Both traits vary widely, from nearly wild, with excellent resistance to pests, to high quality with modest resistance. As a group, hybrids are much easier to protect against the many pests than are the Old World varieties.

Ripening of hybrids differs widely; and there are varieties suited to most growing seasons. Also, in this group are several of the more recently introduced wine and table grapes originated in the United States.

ORIGIN AND HISTORY. Since prehistoric times, grapes have probably been available for man to eat. Fossilized grape leaf prints of grapes which grow in the wild in Europe, the Middle East, and northern India, have been found in France and Italy. Evidence suggests that grapes were domesticated in western Asia well before 5,000 B.C. Then, through the ages, man's familiarity with grapes continued. They are mentioned throughout Biblical writings, and tomb paintings dating back to 2,375 B.C. indicated that Egyptians cultivated grapes. In North America, the Indians were utilizing native grapes long before the colonists arrived.

Old World grapes were introduced into the New World soon after its discovery. Jesuit fathers took Spanish grapes to Mexico, and, about 1635, established a vineyard at what is now Socorro, New Mexico—the earliest planting in the Western United States. After California achieved statehood in 1850, widespread introduction of Old World varieties began. The dry central valleys of California were ideal for growing grapes under irrigation since the area was free from phylloxera (insects which attack the roots) and mildew.

The first American variety to achieve fame was the Concord. In 1852, E. W. Bull of Concord, Massachusetts, raised a seedling from fruit borne on a *Vitis labrusca* vine that had been growing near a Catawba in his garden. He named his new grape "Concord," a variety which became popular in the northeastern United States.

PRODUCTION. Worldwide, yearly production of grapes totals around 60 million metric tons, and almost 52% of this is produced in European countries. While the United States is a leading grape-producing country, it only produces around 8% of the world's grapes. The combined production of France, Italy, and Spain contributes 37% of the world's production and 72% of Europe's production. Grapes are the number one fruit crop of the world.

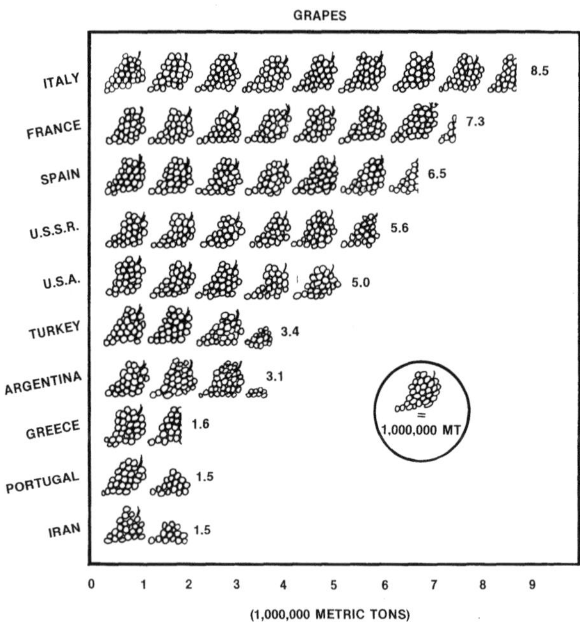

Fig. G-20. Leading grape-producing countries of the world. (Data from *FAO Production Yearbook*, 1990, FAO/UN, Rome, Italy, Vol. 44, p. 153, Table 65)

In the United States, California produces over 91% of the grapes, with Washington, New York, Pennsylvania, Michigan, Arizona, Ohio, Oregon, Arkansas, and Georgia contributing the remaining 9% in that order. The grape crop in California is worth about $1.5 billion.

American bunch grapes, primarily the Concord, are grown chiefly in the northeastern states and in Washington. Muscadine grapes are grown in the southeastern states. The best known variety of Muscadine grapes is the Scuppernong.

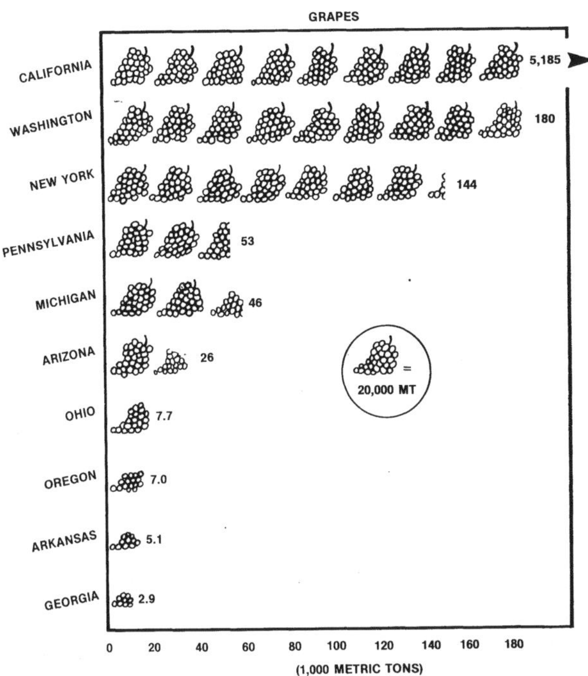

Fig. G-21. Leading grape-producing states. (Based on data from *Agricultural Statistics*, 1991, USDA, p. 197, Table 295)

Propagation and Growing. Grapes are propagated by cuttings. In the winter, branchlike growths called *canes* are cut from the vine. These canes are 12 to 18 in. (*30 to 46 cm*) long and contain three buds. Canes are buried in moist sand and stored in a cool place until early spring, when they are planted in nurseries leaving one bud above ground. By the following spring the cuttings have developed roots and they are then transplanted in a vineyard, about 400 to 600 vines to the acre. There they begin to develop into vines. As the grape plants grow they are trained onto trellises with proper tying and pruning. The trellis supports the plant and pruning controls production. Vines produce a partial crop the third and fourth years in the vineyard and a full crop thereafter. Even after reaching full production, vines are pruned every year during their dormant period. With proper care a vine may bear grapes every year for as long as 100 years.

The Old World or European type of grape vines grow best in climates that have warm to hot, dry summers and winters no colder than 10°F (*-12°C*). The North American kinds of grapes can withstand greater humidity and temperatures as cold as -10°F (*-23°C*).

With proper fertilization, weed control, irrigation, and pruning, vines produce 15 to 80 lb (*7 to 36 kg*) of grapes.

Harvesting. In California, European grapes are harvested from June to October about 60 to 120 days after blossoming. North American grapes are harvested during September and October, about 90 to 120 days after blossoming. Grapes may be harvested by hand or by machine.

PROCESSING. Harvested grapes reach consumers through several routes: table grapes, canned grapes, juice, jam, jelly, raisins, and wines. In recent years, the U.S. crop has been processed as follows: table grapes, 15%; canned, 0.7%; raisins, 31%; wines, 48%; and juice, jams, and jelly, 5.7%.

• **Table grapes**—Workers cut each cluster, and remove damaged berries. Then, the grapes are carefully packed in the fields, quickly cooled to 40°F (*4°C*), and treated with sulfur dioxide gas to slow decay. These packed, cooled grapes are shipped via refrigerated railroad cars or trucks to markets. *Caution:* The sulfur dioxide residue on table grapes must be kept below 10 ppm.

• **Canned grapes**—Almost all canned grapes are the Thompson Seedless or other seedless varieties. These grapes are mixed with peaches, pears, and pineapple as fruit cocktail or fruit salad.

• **Juice, jam and jelly**—The Concord grape is a popular variety for these uses. Juice may be single strength cans, cartons, or bottles of grape juice or drink, or juice may be concentrated and sold as a frozen concentrate.

• **Raisins**—Raisins are the sun-dried fruit of several varieties of grapes. These varieties have tender skin, rich flavor, and a high sugar content. The two most popular varieties are Thompson Seedless and the muscat, and the other two are the sultana and the Corinthian raisin, also known as currants. Muscats are larger than the seedless varieties—sultanas and Thompsons—and have their seed removed after drying.

Sundrying of grapes to produce raisins, takes place in the vineyard between the rows of grapevines. Ripe clusters of grapes are picked by hand and placed on the ground on paper trays. Then, depending upon the weather, 2 to 4 weeks are required to dry the grapes to raisins. Rain during this time is disastrous. Color of the fruit changes from a greenish-yellow to a brownish-purple and the moisture drops from around 75% to 16% or less. When the fruit is dry the paper trays are folded to form a package or "roll." The rolls are collected and transported to a collection site and then further processed.

At the processing plant, the raisins are cleaned, sorted, and fumigated. After washing and inspection, raisins are packaged and shipped.

Most raisins are sun-dried, but golden seedless raisins are fresh Thompson seedless grapes which are picked, washed and then placed on trays and exposed to the fumes of burning sulfur in a closed chamber. Following this treatment, the grapes are dehydrated in forced draft heating chambers. Their finished color is green to golden yellow to amber, since the sulfur dioxide treatment prevents the oxidative darkening that occurs with the sun-drying of fruit.

All U.S. raisins are produced in California, and the United States ranks as a leading producer of raisins, along with Turkey and Greece.

Fig. G-22. Grapes on trellises ready for harvest. (Photo by A. H. Ensminger)

• **Wines**—Varieties of the European grape are ideal for alcoholic fermentation. However, more than 5,000 varieties are believed to exist. Through years of selection, those varieties with higher sugar content and lower acid were selected. These are best for wine production. They contain 18 to 24% sugar with only 0.5 to 1.5% acidity. In the United States, about 10 to 15% of the wine production is from North American types such as Catawba, Concord, Delaware, Ives Seedling, Niagara, and the French hybrids. Worldwide, more than 85% of the grape crop is used for wine production.

In times past when labor was cheap, grapes for wine production were crushed by stomping on them with human feet. Besides the increased labor cost there are some aesthetic considerations, and now mechanical crushers are used to stem and crush grapes for fermentation.

The primary sugars present in grapes are fructose and glucose. During fermentation, yeasts convert these sugars to ethyl alcohol yielding wines of 13 to 15% alcohol. After crushing, the processing varies depending upon the type of wine to be produced; hence, production of wine is covered elsewhere. Wine types include white wine, rose (or pink) wine, red wine, or dessert wine.

(Also see WINE.)

SELECTION. Consumers have a choice when selecting table grapes—grapes for fresh consumption. These grapes are ripe when shipped to market. Grapes do not ripen after harvesting. When purchased, grapes should be well-colored, firmly attached to green, pliable stems, firm and wrinkle-free. Healthy bunches harbor no leaking berries. Also, the perfect grape is true to its varietal color. Green grapes are sweetest and best flavored when they are yellow-green; red varieties when all the berries are predominantly red; and blue-black varieties when grapes have a full, rich color.

After purchase, fresh grapes should be stored in the refrigerator where they will stay fresh for several days. Just before serving, the clusters are washed under a gentle spray of water and then drained or patted dry. Table grapes are best served slightly chilled to enhance their crisp texture and refreshing flavor.

PREPARATION. Since grapes are consumed in a number of forms, there are endless ways to prepare them. A few general suggestions follow.

• **Table grapes**—Aside from being a handy snack, table grapes also offer tantalizing color, flavor, and texture contrasts to all types of recipes. They can be enjoyed in a myriad of ways—in salads, pudding, pies, cakes, tarts, fruit cups, as meat accompaniments, condiments, and relishes, or just as is.

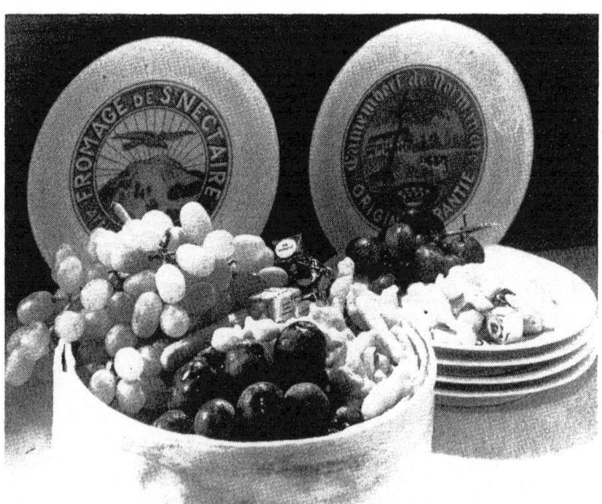

Fig. G-23. Grapes and cheese for a dramatic dessert. Mellow and delicate red grapes blend well with full-flavored Cheddars. The subtle, juicy, blue-black grapes are delightful with blue cheeses, while the sprightly green grapes team best with the light, milk cheeses such as Monterey Jack and Camembert. To enhance the flavor of both, grapes should be served chilled and cheeses served at room temperature. (Courtesy, California Table Grape Commission, Fresno, Calif.)

• **Canned grapes**—These are used chilled and directly from the can or they may be incorporated into fruit dishes, gelatin desserts, or baked goods.

• **Raisins**—Raisins are excellent as a snack when eaten alone or they may be used in puddings, cakes, candies, cookies, and bread. Chocolate covered raisins are sold as candy. The sultana raisins are used primarily in bakeries. Corinthian raisins are employed mainly to flavor bakery goods.

NUTRITIONAL VALUE. Grapes are about 80% water and they contain only about 70 Calories (kcal) per 100 g (about 3½ oz). The calories are derived primarily from the sugars which make grapes sweet. The sugar (carbohydrate) content of grapes averages 16%.

Raisins contain only 17% water; hence, many of the nutrients in grapes are more concentrated in raisins. Each 100 g of raisins contain a whopping 289 Calories (kcal), since the sugars (carbohydrate) comprise 77% of raisins. Additionally, raisins contain 2.8 mg of iron and 678 mg of potassium in each 100 g. Raisins have traditionally been promoted as good sources of iron, but other dried fruits are also good sources of iron.

Food Composition Table F-36 provides more complete nutritional data on fresh grapes, canned grapes, and raisins.

(Also see FRUIT[S], Table F-47 Fruits of the World; and RAISIN.)

GRAPEFRUIT *Citrus paradisi*

Contents *Page*

Origin and History 1097
Production 1097
Processing 1098
Selection 1098
Preparation 1098
Nutritional Value 1098

A large citrus fruit, popular as a breakfast or salad fruit. Grapefruit is misnamed, because it neither looks nor tastes like a grape. Like the other citrus fruits, it is a member of the rue family (*Rutaceae*).

Fig. G-24. The grapefruit, a regular feature of many American breakfasts. Mature grapefruit reach 3 to 6 in. (*8 to 15 cm*) in diameter.

ORIGIN AND HISTORY. The ancestry of the grapefruit is uncertain because it may have arisen as a bud sport (mutation) on some other type of citrus tree. It was first noticed in Barbados in 1750, when it was called the "small shaddock" because it bore a close resemblance to the pummelo or shaddock, which had been brought from Indonesia to Barbados in the preceding century by Captain Shaddock of the English East India Company. Some botanists believe that the grapefruit may have resulted from a natural crossing between the shaddock and another citrus species such as the sour orange or the sweet orange.

The name grapefruit was first used in Jamaica in 1814. Perhaps, certain trees bore large clusters of fruit that resembled giant clusters of grapes. It was first planted in Florida in the early 1800s, but it did not become an important commercial crop until 1880. All of the original varieties produced seedy fruit until a seedless sport (the first of the Marsh variety) was found growing in a Florida orchard in 1862. Later sports yielded (1) a seedy pink variety named Foster (1907), (2) the Thompson seedless pink variety (1913), and (3) the Ruby seedless red variety (1929).

PRODUCTION. The worldwide production of grapefruit averages about 4.3 million metric tons. At the present time, more than 41% of the world's grapefruit crop is produced in the United States, where it is the fruit most commonly served at breakfast. Florida is the leading grapefruit-producing state of the nation.

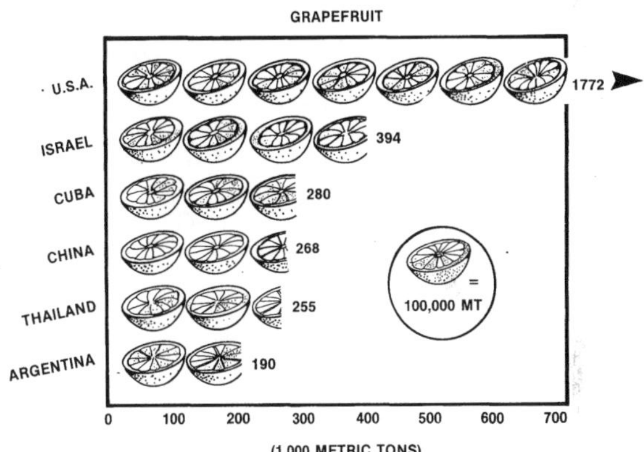

Fig. G-25. The leading grapefruit-producing countries of the world. (Based upon data from *FAO Production Yearbook*, 1990, FAO/UN, Rome, Italy, Vol. 44, p. 165, Table 72)

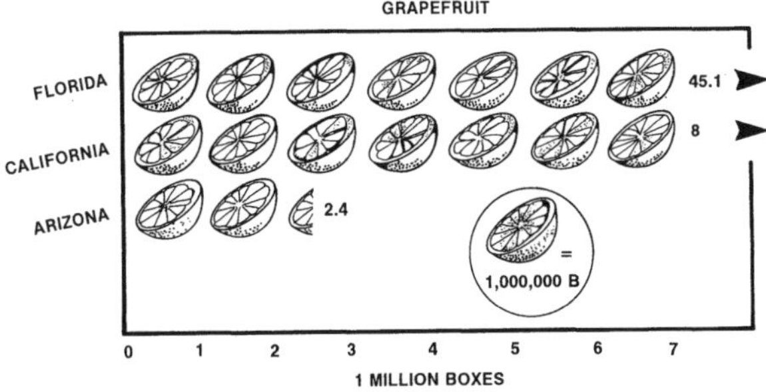

Fig. G-26. The leading grapefruit-producing states of the United States. (Based upon data from *Agricultural Statistics*, 1991, USDA, p. 189, Table 278)

The general details of citrus fruit growing are given elsewhere.

(Also see CITRUS FRUITS, section headed "Production.")

PROCESSING. About 60% of the U.S. grapefruit crop is processed. Frozen grapefruit juice concentrate and grapefruit juice each account for about 42% of all processed grapefruit products. The remainder of processed products is made up of (1) chilled grapefruit juice in glass bottles, plastic bottles, and paperboard cartons; (2) canned grapefruit sections; and (3) chilled grapefruit sections in glass or plastic containers.

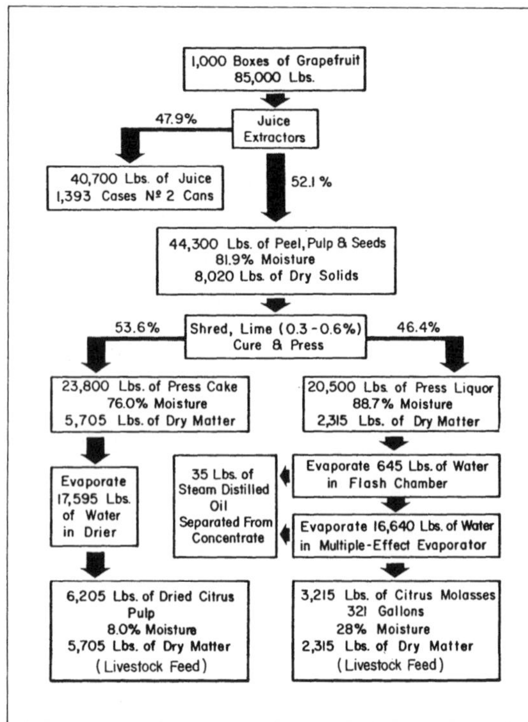

Fig. G-27. Utilization of grapefruit in the production of juice and animal feeds. (Adapted from R. Hendrickson, and J. W. Kesterson, *By-Products of Florida Citrus*, Bull. 698, University of Florida Ag. Exp. Sta., Gainesville)

Descriptions of these and other products are given elsewhere.

(Also see CITRUS FRUITS, section headed "Processing.")

SELECTION. Fresh grapefruits of good quality are firm, but springy to the touch; not soft, wilted, or flabby. They are well-shaped, and heavy for their size. Fruits heavy for their size are usually thin-skinned and contain more juice than those that have a coarse skin or are puffy or spongy.

Generally speaking, most of the defects found on the grapefruit in the markets (such as scale, scars, torn scratches, and discoloration) are minor in nature; they affect appearance only and not eating quality.

Decay is sometimes evident and should be avoided since it usually affects the flavor, making the taste flat and somewhat bitter. Decay sometimes appears as a soft, discolored area on the peel at the stem end or "button" of the fruit; or it may appear in the form of a water-soaked area, much of the natural yellow color within the area being lost, and the peel being so soft and tender that it breaks easily on pressure of the finger.

Sometimes a fruit is somewhat pointed at the stem end; it is likely to be thick-skinned particularly if the skin is rough, ridged, or wrinkled. Judgment in selecting this kind of fruit should be based on texture and weight for size.

Guidelines for the selection of processed grapefruit products are given elsewhere.

(Also see CITRUS FRUITS, section headed "Selection.")

PREPARATION. In the United States, grapefruit is one of the most popular fruits for breakfast. It is usually served as half of a grapefruit.

Fig. G-28. Grapefruit being processed in a packing plant. (Courtesy, USDA)

CAUTION: Some people are allergic to one or more constituents of citrus peel. When such allergies are suspected, neither the peel nor products containing it should be consumed, and fruit juice should be extracted gently to avoid squeezing the oil and other substances from the peel into the juice. It is noteworthy that some of the commercially prepared grapefruit juices and grapefruit drinks may contain peel and/or substances extracted from it, but that prepared citrus juice products for infants contain little or none of the peel constituents.

Suggestions for preparing dishes from fresh grapefruit and processed grapefruit are given elsewhere.

(Also see CITRUS FRUITS, section headed "Preparation.")

NUTRITIONAL VALUE. The nutrient compositions of various forms of grapefruit are given in Food Composition Table F-36.

Some noteworthy observations regarding the nutrient composition of grapefruit follow:

1. Fresh grapefruit is low in calories (only 41 kcal per half of an average fruit), but it is a good source of fiber, pectin, potassium, vitamin C, inositol, and bioflavonoids (vitaminlike substances), and a fair source of folic acid.

2. Frozen grapefruit juice concentrate is a rich source of the nutrients present in the fresh fruit.

3. Grapefruit drinks are low in all of the nutrients present in the fresh fruit, except for vitamin C, which is usually added during processing.

4. Citrus peel contains citral, an aldehyde that antagonizes the effects of vitamin A. Hence, people should make certain that their dietary supply of this vitamin is adequate before consuming large amounts of grapefruit peel.

(Also see BIOFLAVONOIDS; FRUIT[S], Table F-47 Fruits of the World; and VITAMIN C.)

GRAPEFRUIT NARINGIN

The flavonoid naringin is the principal component of grapefruit which contributes to the bitter taste of the fruit. Also, this substance, and similar ones, sometimes comes out of solution and forms unsightly crystals in canned citrus products. Hence, Japanese food technologists have developed an enzymatic process for breaking down the flavonoids to other less aesthetically objectionable substances.

(Also see BIOFLAVONOIDS.)

GRAPE JUICE

The fluid expressed from fully ripened grapes. Most of the grape juice produced in the United States is made from Concord and various other dark-colored varieties of grapes. Usually, the grapes are passed through a device which separates the skins, stems, and seeds from the pulp and juice by screening. Then, the pulp and juice are (1) preheated, (2) treated with enzymes that break up the pulp, and (3) pressed to extract the juice. The expressed juice is filtered, then heated briefly to pasteurize it so that it will keep well during storage.

The nutrient compositions of various forms of grape juice are given in Food Composition Table F-36.

Some noteworthy observations regarding the nutrient composition of grape juice follow:

1. Bottled or canned grape juice is moderately high in calories (66 kcal per 100 g) and carbohydrates (17%).

2. The juice is an excellent source of chromium (a mineral that is part of the glucose tolerance factor, which acts along with insulin to promote the utilization of sugar), a good source of potassium, and a poor source of vitamin C—unless supplemental vitamin C has been added during processing.

(Also see CHROMIUM; and GRAPE[S].)

GRAPE SUGAR

Glucose occurs in a free form in grapes. Thus, in times past it was commonly called grape sugar.

(Also see CARBOHYDRATE[S].)

GRAS (GENERALLY RECOGNIZED AS SAFE)

A designation of food additives that have been judged as safe by the Food and Drug Administration for human consumption by a panel of expert pharmacologists and toxicologists who consider available data, including experience of common use in food. The common use factor is often the major criterion on which judgment is based.

(Also see ADDITIVES.)

GRAYING OF HAIR

A deficiency of pantothenic acid produces graying of the hair in rats, monkeys, dogs, and foxes; usually, the color can be restored by adding pantothenic acid (one of the B vitamins) to the diet. In humans, no such relationship exists between pantothenic acid intake and graying of hair.

(Also see PANTOTHENIC ACID, section headed "HISTORY.")

GREEN BEANS

(See, BEAN, COMMON; and BEAN[S], Table B-10 Beans of the World.)

GREEN REVOLUTION

Contents *Page*

Historical Landmarks in the Improvement of the Leading Cereal Grains 1100
Wheat 1100
Rice 1101
Corn 1101
Benefits from the Green Revolution 1102
Shortcomings of the Green Revolution 1102
Future Needs and Prospects 1103

This term refers to certain recent breakthroughs in wheat and rice farming which have the potential to double or triple the supplies of these grains for the developing countries of the world. Gaud, who was the Director of United States Aid for International Development (USAID), coined the term in a speech he made in 1968. However, the father of the Green Revolution is considerd to be the American agricultural scientist *Norman Borlaug*, who was awarded the Nobel Peace Prize in 1970 for his breeding of new high-yielding varieties of wheat at the International Maize and Wheat Improvement Center in Mexico. A related development was the similar production of special varieties of rice by scientists at the International Rice Research Institute (IRRI) in the Philippines.

The new varieties of wheat and rice are no panacea for the problems that the world now faces in feeding its people. Nevertheless, they have provided food for many who might otherwise have had none, and their success has encouraged similar efforts to improve other crops which are vital to mankind. Therefore, it is noteworthy that the Green Revolution did not occur as an isolated event, but that it is merely the latest victory in the never-ending struggle to obtain better yields from the major cereal crops.

HISTORICAL LANDMARKS IN THE IMPROVEMENT OF THE LEADING CEREAL GRAINS. The selection and breeding of plants to suit man's needs began accidently about 10,000 years ago, when crops were first cultivated. Diversity among individuals of a species is a hallmark of all forms of life. Hence, there are always certain plants and animals which differ from most others of their species. For example, wild grain plants usually have seed heads which shatter and scatter upon ripening. This is nature's way of ensuring that a new crop is seeded each year. Man's intervention altered the ways by which certain cereals were propagated because the few nonscattering grains that remained on certain plants after ripening were those most likely to be harvested for food and for seed. These atypical plants thereby became the basis for man's domesticated cereal crops, which now cannot be propagated without human help.

Wheat. The origin of bread wheat is another example of unintentional cereal breeding by man. Wheat farming began in the arc of land in the Middle East known as the Fertile Crescent because it had the climate which favored the growth of the early types of wheat—mild, rainy winters and hot, dry summers. Year after year the early grain farmers planted their wheat farther and farther northward and eastward until their crops became accidentally hybridized with a wild grass that grew in northern Iran, Afghanistan, and to the east in Central Asia. The hybrid wheat that resulted was superior to its domesticated parent in such desirable characteristics as (1) loose-fitting husks which were readily removed by threshing, (2) tolerance to cold winters and rainy summers, and (3) an ample amount of the protein gluten—which is the sticky, elastic substance in bread flour that traps the bubbles of gas in leavened breads.

Over the centuries, people in various countries around the world selected and cultivated the types of wheat which were best adapted to the local growing conditions. However, it was not until 1866 that the Austrian monk Mendel first stated the principles governing the transmission of traits from generation to generation. Even then, 34 more years elapsed until his findings were confirmed by other biologists and announced to the world. Shortly thereafter, agricultural scientists engaged in numerous wheat-breeding projects in Canada and the United States. By the time of World War I, the new varieties of wheat which were grown in the Dakotas and Minnesota helped to stave off wartime food shortages. Hence, the stage was set for the work of Dr. Borlaug, who crossbred diverse strains of wheat which were truly global with respect to their places of origin. Fig. G-29 traces the origins of these strains.

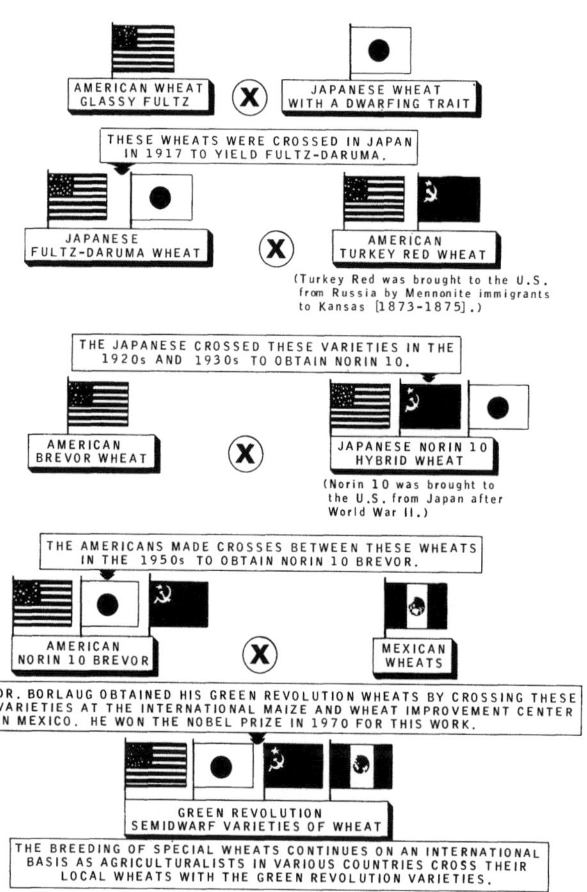

Fig. G-29. The origins of the varieties of wheat utilized by Dr. Borlaug in the development of his high-yielding varieties. (Based upon information in Wilkes, H. G., "The Green Revolution," *McGraw-Hill Encyclopedia of Food, Agriculture, & Nutrition*, edited by D. N. Lapedes, McGraw-Hill, Inc., New York, N.Y., 1977, p. 43)

Fig. G-29 shows that Dr. Borlaug's hybrid wheats were the results of crosses between varieties from the United States, Russia, Japan, and Mexico. This hybridization resulted in such favorable traits as (1) dwarfism, which prevents the plant from growing too tall and falling over (lodging) when extra water and fertilizer are applied; (2) an increase in the amount of grain per plant in response to increases in irrigation and fertilization; (3) resistance to diseases like rust; and (4) tolerance of a variety of

Fig. G-30. Harvesting wheat. Improved varieties and technology contributed richly to the green revolution. (Courtesy, Henry Fisher, Oakesdale, Wash.)

climates and day lengths. (Certain strains of cereal crops require a fixed number of daylight hours in order to flower and set seed.)

(Also see WHEAT.)

Rice. It seems likely that the first rice farming probably occurred in Thailand before 4000 B.C. when man selected a type with nonshattering seed heads for cultivation. All cultivated varieties, and most wild types, of rice belong to the species *Oryza sativa,* but the amount of variation within this species is very great. As in the case of the other cereals, there has also been some selection by man for varieties that are best suited to the climates and terrains of the different growing regions. For example, the *Japonica* types are adapted to temperate climates, whereas the *Indica* types do best in the tropics. There are also upland and lowland varieties of rice, but the yields of the latter are much higher. Therefore, most of the rice breeding efforts of the past 100 years or so have been directed towards the development of lowland varieties which might yield more grain with the application of fertilizer.

At the end of the 19th century the Japanese had begun to develop short-stemmed types of rice that responded to fertilizer by setting more grain rather than by growing tall and spindly. Tall and spindly plants are likely to fall over, or "lodge." Some of the "dwarf" strains of rice were grown by the Japanese on Taiwan just prior to World War II. All of these strains were of the *Japonica* type which is suited mainly to Japan, Korea, and Taiwan.

Fig. G-31. A close-up of rice. (Courtesy, USDA)

After World War II, the Chinese on Taiwan picked up where the Japanese had left off, and by 1956 had developed a dwarfed *Indica* variety of rice which (1) responded well to fertilization, and (2) was suited for growing in the tropics. Therefore, it is noteworthy that the Chinese rice geneticist Chang from Taiwan made a substantial contribution to the rice breeding program at the International Rice Research Institute (IRRI) in the Philippines. In 1966, the IRRI released the "miracle" IR-8 rice, which was a cross between a Chinese semidwarf strain and a tall, Indonesian *Indica* type. The yields of the new strain were more than triple those of the older types; but the new grains were chalky and had a strange taste. Since then, these defects have been overcome by further breeding to yield the IR-20 and the IR-22 types, which also have better resistance to diseases and pests than IR-8.

(Also see RICE.)

Corn. Although this grain is the leading cereal crop in the United States and the third leading one in the world, it is not considered to be part of the Green Revolution because (1) most of the production is fed to livestock, rather than to people; and (2) few of the developing countries outside of Latin America and Africa either produce or utilize it. However, this situation could change dramatically, as the demand for cereals continues to spiral upward, and the great breeding potential of this crop becomes more fully appreciated. Hence, the advances in corn farming which parallel those of wheat and rice production are presented here. It is noteworthy that Dr. Borlaug's work on wheat was performed at the center in Mexico which is also committed to the improvement of corn.

Tiny primitive ears of corn, which apparently were thousands of years old, have been found in caves in both Mexico and Peru. Neither type bore a close resemblance to modern-day types of corn. Therefore, it is suspected that at about 1500 B.C. the early types of corn became hybridized by wind pollination and developed markedly different characteristics because the Mexican Indians unknowingly planted their corn near stands of the wild grass teosinte —*Euchlaena mexicana*. Corn is readily cross-pollinated by the wind; so, many different varieties may emerge, unless measures are taken to control pollination.

Although the modern type of corn originated in Mexico, the first European explorers to reach the New World found it growing as far north as southern Canada, and as far south as Argentina and Chile. In fact, the settlers of the original 13 American colonies found that the Indians planted different types of corn in close proximity so that hybridization would occur and higher yields would result. The colonists copied the Indians' techniques, and soon many experiments in corn breeding were underway. Some of the hybridization was apparently accidental, such as in the case of the ancestor of the main variety now grown in the U.S. Corn Belt, which is believed to have resulted from an unintentional crossing of northern flint corn with southern dent corn, when the migrating settlers took their seeds with them to this area.

The elimination of many of the trial-and-error aspects of corn breeding came after the publication of Darwin's theories of evolution and heredity. He was the first to understand that the so-called "hybrid vigor" was produced by crossing unrelated varieties rather than by the act of crossing alone. Shortly thereafter, agricultural scientists in the United States applied Darwin's theories to the hybridization of corn, and subsequently discovered ways of rapidly developing desirable characteristics in this crop. Then, in 1917 a method of producing ample amounts of hybrid seeds was discovered by Jones, a researcher at the Connecticut Agricultural Experiment Station. It was called the "double cross" because it involved (1) the crossing of two pairs of inbred plants, and (2) crossing the descendents of the first cross. Twenty-five years of research were required to make the double cross profitable to seed growers. Fortunately, it was made available to farmers by the beginning of World War II, because the new hybrid corn increased the American yields by 20%, in spite of bad weather and labor shortages.

Today, 99% of the corn grown in the corn-producing sections of the United States comes from hybrid seed, which yield 8 to 10 bushels more per acre than the open pollinated varieties, or over one-half billion bushels more annually. However, it is still necessary to keep on breeding new hybrids in order to keep ahead of diseases and other pests which are undergoing continual evolution with respect to their abilities to attack this crop.

(Also see CORN.)

BENEFITS FROM THE GREEN REVOLUTION. Usually, much time and patience are required to replace older methods of agriculture with new ones that produce better results because (1) farmers may be highly experienced in certain traditional methods, but not very competent in carrying out new ones; (2) the margins of profit are very narrow for many farmers, so the taking of a large financial risk for a small or an uncertain return may not make good business sense; and (3) loans may not be available to some of the farmers who need them to purchase new supplies and equipment. When these obstacles to the modernization of farming practices are considered, it seems remarkable that the new varieties of grains were adopted so quickly by so many countries around the world. Some of the leading benefits which accompanied this rapid pace of agricultural innovation follow.

- **High yields of grain per acre**—The new, improved hybrid cereal crops may yield 2 to 3 times the grain per acre as the traditional varieties.

- **Modernization of agricultural practices**—Extra water, fertilizer, and pesticides are required in order to obtain the maximum yields from the special varieties of wheat, rice, and corn. Fortunately, the increase in yields from the new types is sufficiently high to justify the extra investment in modern farming practices. Hence, the Green Revolution has provided an incentive for the rapid modernization of agriculture around the world. It is noteworthy that, by the 1974-1975 crop year, the special hybrids had accounted for 38.4% of the wheat and 26% of the rice which were produced in the Near East and Asia.[7] Furthermore, almost all of the Mexican wheat crop was made up of the special wheats.

- **Additional crops may be grown each year**—Often, the adoption of the new hybrids has allowed the growing of an extra crop each year, because (1) the hybrids are ready for harvesting much sooner than the older types, so the next crop may be seeded at an earlier date; and (2) the new varieties are much less dependent upon a certain amount of daylight for flowering and setting seed, so they may be grown at various times during the year. When irrigation is available, the cereals may also be grown during the dry season, which has usually been the time when the land was allowed to lie fallow.

- **Conversion of grain-importing countries into grain exporting ones**—The Green Revolution has not only reduced the need for many of the developing countries to import grain, it also allowed certain ones—notably India and Pakistan-to build up their grain reserves in bountiful years, and sometimes, to be able to export part of these crops.

- **More jobs for farm workers**—Extra work is required to prepare the soil and to peform the other farming tasks which are necessary to ensure optimum yields from the new cereal crops. Also, the planting of additional crops, which is made possible by the shorter growing periods of the new grains, adds to the work requirements. Hence, the Green Revolution is likely to create more jobs for farmworkers, except when there is extensive mechanization of the agricultural operations.

SHORTCOMINGS OF THE GREEN REVOLUTION. Many criticisms have been leveled at the ways in which the Green Revolution has proceeded, because it appears to have intensified certain economic, nutritional, and social problems that have long existed in some of the developing countries. However, these problems existed before the agricultural innovations were

[7]Wilkes, H. G., "The Green Revolution," *McGraw-Hill Encyclopedia of Food, Agriculture & Nutrition*, edited by D. N. Lapedes, McGraw—Hill, Inc., New York, N.Y., 1977, p. 45.

adopted; and they can be solved only by the correction of such underlying causes as rapid population growth, socioeconomic inequities, and illiteracy. Nevertheless, it is worth noting some of the major shortcomings of the Green Revolution, so that more people may share in its benefits.

• **Special inputs of fertilizer, fuel, irrigation, mechanization, and pesticides are required to obtain optimal yields from the new grains**—It may be difficult for farmers in economically deprived areas to amass sufficient funds to pay for the inputs needed by the new cereal crops. Furthermore, supplies of fuel and chemical fertilizers may become scarce, as they did in 1974 during the global shortage of petroleum products. This crisis was a major factor in the reduction of India's wheat crop from a projected 33 million tons to the 25 million tons actually harvested.[8]

Professor Pimentel and his associates at Cornell University have calculated that if the rest of the world were to eat the same diet as Americans and were to have the same type of energy-consuming food production fueled solely by petroleum, the world supply of petroleum would run out in only 29 years.[9]

• **Large farmers may benefit much more than small farmers**—Although the size of the farm has no effect on the yields of the hybrid grains per acre, there are certain areas of the world where those who hold large amounts of land are more likely to receive loans and agricultural extension services than those who eke out a subsistence on small plots. Also, the more prosperous farmers may be in better positions to hold back some of their grain production from the market in years of high production and low prices, whereas the low income farmers must sell all of their crops in order to stay in business.

• **Increases in food supplies have slowed efforts to curb population growth**—It is no fault of the Green Revolution that many governments of densely populated countries slackened in their efforts to reduce fertility rates when greater quantities of grain become available. However, it is urgent that population growth in these countries be curtailed, because there are limits to the increases in agricultural productivity which may be expected in the future.

• **Malnutrition may be fostered with the displacement of legumes by grains**—The production of legumes had already started to decline in various parts of the world before the Green Revolution got underway. Even so, the availability of more profitable, high-yielding grains probably encouraged some of the recent reductions in the acerages allotted to beans, lentils, and peas. Therefore, it is noteworthy that growing children and pregnant women—who may be forced by necessity to eat mainly cereal products—need some source of supplemental protein like legumes to offset the amino acid deficiencies of the cereals. Usually, adult men and nonpregnant adult women can get by on the protein supplied by various grain products.

• **The new cereal crops are extra vulnerable to diseases, pests, droughts, and floods**—By 1970, over half of the total rice area in the Philippines was planted with the new, high-yielding varieties.[10] Unfortunately, their rice production in that year dropped sharply, because the new hybrids were attacked by a virus disease that reached epidemic levels.[11]

Another factor which may limit the benefits from the new rice is that short-stemmed plants cannot tolerate the degree of flooding which normally occurs in the fields of Southeast Asia during the growing season.

• **Utilization of a few high-yielding cereals around the world may result in the elimination of local varieties which have special traits**—Up to now, it has been possible to breed new varieties of cereal crops because strains with special traits have been available from various parts of the world. However, many of the special strains are likely to disappear if most of the world's farmland should be planted with only a few varieties of each species. For example, an epidemic of corn blight spread throughout the United States in 1970 and greatly reduced the yield of corn, because a certain virus had developed the ability to attack the variety which constituted most of the nation's crop. Fortunately, corn may be modified by crossbreeding within a much shorter time than the other cereals. Hence, U.S. agricultural scientists were able to develop a fungus-resistant variety in about a year by planting several off-season crops in Mexico.

FUTURE NEEDS AND PROSPECTS. Many agriculturalists believe that future increases in grain production will be less dramatic than those which have recently occurred under the Green Revolution. Hence, the provision of sufficient cereals for the peoples of the world will depend mainly upon reductions in human fertility rates in the developing countries, so that advances in agricultural production may be able to keep up with the demands for food. Also, our present varieties of grains are low in certain nutrients, so it will be necessary to (1) improve the nutritional qualities of the major cereal crops, and (2) produce sufficient amounts of the supplemental foods which supply the deficient nutrients in grains. Some approaches to these problems follow.

• **Breeding high-protein, high-lysine, and high-yielding cereals**—Agricultural scientists are currently attempting to breed varieties of cereals which have both high yields and high protein quality, since farmers are not utilizing fully the present types of high-lysine and high-protein grains because they have lower-than-average yields. The breakthrough that is most likely to occur in the near future is the development of a high-yielding, high-lysine corn because (1) corn may be crossbred much more readily than the other major grains, and (2) varieties which possess one or more of the desired traits are now under investigation by plant breeders.

• **Increasing the utilization of other important crops**—Although rice and wheat have long been the staple foods used by millions of people around the world, certain other crops such as potatoes, corn, and soybeans may yield

[8]Brown, L. R., and E. P. Eckholm, *By Bread Alone,* Praeger Publishers, New York, N.Y., 1974, p. 141.

[9]*Ibid,* p. 113.

[10]Wilkes, H. G., "The Green Revolution," *McGraw-Hill Encyclopedia of Food, Agriculture, and Nutrition,* edited by D. N. Lapedes, McGraw-Hill, Inc., New York, N.Y., 1977, p. 45, Table 2.

[11]Wade, N., "Green Revolution (II): Problems of Adopting a Western Technology," *Science,* Vol. 186, 1974, p. 1186.

more calories and/or protein per acre. Hence, the latter items should be more utilized. Fig. G-32 compares the yields of the five crops.

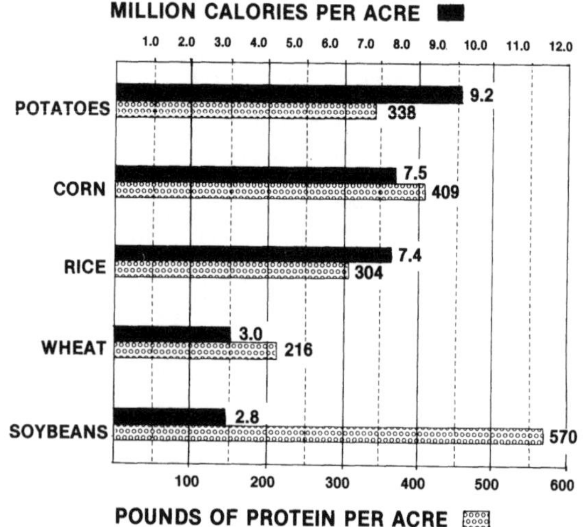

Fig. G-32. The yields of calories and protein from some leading food crops. (Values calculated from data in *A Hungry World: The Challenge to Agriculture*, University of California Food Task Force, pp. 37, 62, and 206, Tables 2.12, 3.7, and 6.9)

It may be seen from Fig. G-32 that potatoes outstrip by far the caloric yields of the leading grains, whereas soybeans, corn, and potatoes each provide more protein per acre than either rice or wheat. Unfortunately, the merits of the potato are not appreciated fully by much of the world, as evidenced by the fact that most of this crop is grown in the region of Europe which extends from Ireland to the Soviet Union.

Corn appears to be more highly regarded than potatoes, since the peoples in most of Latin America and in parts of Africa use corn as their major food. Even countries like France and Italy have recently begun to use more of this grain. It is noteworthy that the American poultry, meat, and dairy industries owe their success in part to corn because (1) the grain is the leading concentrate fed to swine, cattle and sheep in feedlots, and to dairy cattle; and (2) corn silage, which may be made from the leaves and stalks that might otherwise be wasted, is relished by and used extensively for both beef and dairy cattle.

Almost all of the efforts to increase man's supply of food have been directed towards improvement of the species which were domesticated thousands of years ago by primitive peoples. However, other lesser known plants might have even greater potential than the common crops used as staple foods around the world. For example, the native peoples of Latin America have long used the grainlike seeds of various species of *Amaranth* and *chenopodium,* each of which contain about 13% protein. Furthermore, it has been reported that cultivated strands of *Amaranth* yield a greater weight of seeds per acre than corn. At the present time the utilization of these plants is limited mainly to Latin America, although a few agriculturists in other parts of the world are studying their potential as alternative sources of food.

• **Shifting some of the land used for grains to legumes**—If population growth in the developing countries is curbed in the near future, and there is widespread utilization of high-yielding varieties of grains, it should then be possible to (1) produce the required amounts of cereals on less land than that which has been used up to now; and (2) shift some of the acreage now used for grain crops to legumes. Such a shift in the planting of crops would be desirable because (1) legumes contain from 2 to 3 times as much protein as grains; (2) the mixture of proteins present in combinations of these two types of foods is better utilized than that from either food alone; and (3) certain legumes, like peanuts and soybeans, convert nitrogen gas from the air to soil nitrogen. Also, the rotation of grain and legume crops on the same soil makes good sense because the former crops benefit from the nitrogen that was added to the soil by the latter crop.

There has been a shift of acreage from grains to soybeans in the United States, which took place in the 30-year period following World War II, because the substantial increases in the yields of corn and other grains have made it possible to produce almost double the amounts of cereals on only about three-fourths as much land. Table G-9 compares the production figures for grains and soybeans during World War II with the related data for 30 and 50 years later.

TABLE G-9
COMPARATIVE ACREAGE AND PRODUCTION OF SOME MAJOR GRAINS AND SOYBEANS[1]

Crop	Crop Production Data		
	1941-45	1971-75	1990
Corn (maize):			
Harvested acres (1,000) ...	91,941	63,125	66,952
Total production (1,000 tons)	103,589	191,044	230,059
Yield (tons per acre)	1.1	3.0	3.4
Wheat:			
Harvested acres (1,000) ...	56,396	56,819	69,353
Total production (1,000 tons)	29,297	52,787	85,527
Yield (tons per acre)	0.5	0.9	1.2
Sorghum:			
Harvested acres (1,000) ...	6,921	14,976	9,079
Total production (1,000 tons)	3,438	22,411	15,430
Yield (tons per acre)	0.5	1.5	1.6
Oats:			
Harvested acres (1,000) ...	39,300	14,100	5,940
Total production (1,000 tons)	20,272	11,232	5,714
Yield (tons per acre)	0.5	0.8	1.0
Rice:			
Harvested acres (1,000) ...	1,424	2,229	2,813
Total production (1,000 tons)	1,431	5,040	7,746
Yield (tons per acre)	1.0	2.3	2.8
Total for corn, wheat, sorghum, oats and rice:			
Harvested acres (1,000) ...	195,982	151,249	154,137
Total production (1,000 tons)	158,027	282,514	342,476
Soybeans:			
Harvested acres (1,000) ...	9,433	50,052	56,502
Total production (1,000 tons)	5,220	40,500	55,732
Yield (tons per acre)	0.6	0.8	1.0

[1]Adapted by the authors from Doering, O. C., III, "Current Production Systems for Grains and Oilseeds," *Plant and Animal Products in the U.S. Food System*, National Academy of Sciences, 1978, p. 41, Table 1; and *Agricultural Statistics*, 1991, USDA.

• **Coordinating grain production with the raising of livestock and fish**—In 1990, 67 million acres (*26.8 million ha*) of corn, yielding over 3.4 tons of grain per acre (*7.6 metric tons/ha*), were harvested in the United States.[12] Most of the crop residue, which amounted to the greater part of the corn plants, was left to rot in the fields. It is estimated that the equivalent of 184.2 million tons of feed were wasted, (2¾ tons/acre), enough to winter 93 million dry pregnant cows consuming an average of 33 lb (*15 kg*) of corn refuse per head per day over a 4-month period (3,960 lb/cow, or *1,796 kg/cow*). (Usually, cows are fed some type of supplement[s] along with the corn refuse; such as a low cost mixture of ammonia, molasses, and minerals.) Also, there are other residues from cereal crops, plus the by-products from the milling of grains, and the production of alcoholic grain beverages, which, if properly utilized, could be used to feed even larger numbers of livestock.

(Also see CEREAL GRAINS, "Feeds for Livestock"; and WORLD FOOD.)

Fish farming is considered to be one of the most promising means of producing food in the future. In this connection, it is noteworthy that the Chinese have long raised fish in their rice paddies. In fact, the middle and lower Yangtze Plain is known as the "land of fish and rice."[13] Recently, the Chinese have begun to coordinate the production of grain, hogs, and fish. The grain is fed to hogs in barns that are situated so that the pig manure may be flushed directly into adjacent ponds where fish are raised.[14]

Taking a cue from the Chinese, U. S. agriculturists have developed a plan for a triple crop rotation system on the Mississippi River delta land in the states of Alabama, Louisiana, and Mississippi which would involve the production of soybeans, rice, and fish.

• **Breeding of the less utilized grains**—Sorghum and millet are the most suitable cereal crops for areas that are too dry for wheat, rice, or corn. Futhermore, sorghum, which is nearly equal to corn in nutritive value, uses scarce supplies of water more efficiently than any of the other grains. Therefore, it is fortunate that two little known strains of sorghum that contain exceptionally high levels of both protein and lysine have been found recently in Ethiopia. Hopefully, there will soon be hybrids of sorghum and other less utilized grains which will be both high-yielding and of high-protein quality and quantity.

• **Collecting and storing diverse strains of cereal crops for future breeding stock**—The various diseases and pests which attack cereal crops are continually undergoing mutations that often give them new capabilities for overcoming the defenses of the plants. Hence, there is need to have some means of saving a large assortment of seeds and/or plants so that crossbreeding programs may be kept going to counter the ever changing threats to the survival of these crops.

Since World War II, collections of seeds have been kept in cold storage at various places around the world; the largest such facilities being those at the U.S. Department of Agriculture and at the Vavilov Institute in the Soviet Union. These facilities are not foolproof, as shown by the failure of the refrigeration units at a seed bank in Peru, where a major collection of corn seeds was lost.[15] Furthermore, it is not known for certain how long seeds may retain their viability during cold storage since some of those kept under dry conditions without refrigeration have lost their ability to germinate after about 2 years of such storage, whereas others have remained viable much longer. Therefore, some scientists have urged that natural preserves be maintained around the world where valuable varieties of cereal crops might be grown for generation after generation under field conditions. These preserves would also allow some natural adaptation of the plants to changing factors such as climate, diseases, and pests.

[12]*Agricultural Statistics 1991*, USDA, p. 31.

[13]Ensminger, M. E., and A. Ensminger, *China—the impossible dream*, published by M. E. and Audrey Ensminger, 1973, p. 100.

[14]*Ibid.*, p. 101.

GRIND

To reduce to small segments or a powder by impact, shearing, or attrition (as in a mill).

(Also see CEREAL GRAINS; and FLOUR.)

GRISKIN

This is an English term for a pork chop, especially the lean part of it. It is also used when referring to lean bacon—better known in this part of the world, as Canadian bacon.

GRITS

Coarsely ground grain from which the bran and germ have been removed, usually screened to uniform particle size.

(Also see CEREAL GRAINS.)

GROATS

Grain from which the hulls have been removed.

(Also see CEREAL GRAINS.)

GROSS ENERGY (GE)

The amount of heat, measured in calories, that is released when a substance is completely oxidized in a bomb calorimeter containing 25 to 30 atmospheres of oxygen. The gross energy of a food, feces, urine, tissue, or other material is determined by burning them in a bomb calorimeter.

(Also see CALORIMETRY.)

[15]Lappe, F. M., and J. Collins, *Food First*, Houghton Mifflin Company, Boston, Mass., 1977, p. 174.

GROUND BEEF (CHOPPED BEEF)

A product so labeled must be made with fresh and/or frozen beef, with or without seasoning, without the addition of fat as such, and shall contain no more than 30% fat.

GROUSE (CANADA SPRUCE GROUSE; PRAIRIE CHICKEN; AND RUFFED GROUSE)

This is an important game bird of the Northern Hemisphere. There are many species of grouse, all of which belong to the family, *Tetraonidae.* The ruffed grouse is genus *Bonasa,* species *B. umbellus.*

Grouse is a favorite with hunters of game birds, because its meat makes good eating. Young grouse is best roasted.

The prairie chicken (genus *Tympanuchus,* species *T. cupido*) is a member of the grouse family. It is an excellent game bird, esteemed for its delicate flesh.

GROWTH

The increase in size of bones, muscles, internal organs, and other parts of the body.

GRUEL

A food prepared by mixing ground ingredients with hot or cold water.

GUANINE

One of the nitrogenous bases in nucleic acids.

GUINEA CORN (SORGHUM)

The name given in West Africa to any of several grain sorghums, especially durra.

(Also see SORGHUM.)

GUINEA PIG

Domestic rodent of South American origin. Its flesh is edible, but in the United States it is used for laboratory experiments and as pets.

GUMS

As technically employed in industry, the term gum usually refers to polysaccharides (long chains of simple sugars) or their derivatives, which are dispersible in either hot or cold water to produce viscous mixtures. Gums also may include water-soluble derivatives of cellulose and modifications of other polysaccharides which in their natural form are insoluble. The original definition of gums applied only to the sticky gummy natural plant exudates.

Many of the plant gums originally used by man are still important today; for example, the dried exudates from trees and shrubs collected by hand in the hot, semiarid regions of the world. Also seed gums have been used for many centuries, and some ancient sources of gum include quince, psyllium, or flax seeds. Seaweeds provide another important source of gums.

Gums perform many food and nonfood functions. Since they are used in foods they are considered additives. In foods, gums serve as stabilizers and thickeners. They form viscous solutions which prevent aggregation of the small particles of the dispersed phase. In this way they aid in keeping solids dispersed in chocolate milk, air in whippping cream, and fats in salad dressing. Gum solutions also retard crystal growth in ice cream (ice crystals) and in confections (sugar crystals). Also, gums are employed to modify texture and increase moisture holding capacity; for example, as gelling agents in canned meats or fish, marshmallows, and jellied candies.

Nonfood uses of gums are extensive. Their thickening and stabilizing properties make them useful in water-base paints, printing inks, and drilling muds. Because of these properties they also are used in cosmetics and pharmaceuticals as emulsifiers or bases for ointments, greaseless creams, toothpastes, and lotions. The adhesive properties of gums make them useful in the production of cardboard, postage stamps, and as pill binders. A variety of other applications include the production of soluble surgery films, the sizing of textiles, cosmetics, beverage-clarifying, and plasma extenders.

Common or potentially useful gums, their source and composition are listed in Table G-10.

There are some completely synthetic chemical products of vinyl, acrylic, or ethylene oxide polymers which serve as gums. Some examples include polyvinylpyrrolidone, polyvinylalcohol, polyacrylic acid, and polyarylamide.

(Also see ADDITIVES, Table A-3 Common Food Additive; AGAR-AGAR; CARBOHYDRATE[S], section headed "Polysaccharides"; CARRAGEENAN; LEGUMES, Table L-2 Legumes of the World; and STARCH.)

TABLE G-10
SOURCES OF GUMS AND THEIR COMPONENT CARBOHYDRATES

Gum	Source	Carbohydrates (sugars) Present[1]
Seaweeds		
Agar	Red algae (*Gelidium group*)	Galactose, 3,6-anhydro-L-galactose plus sulfate acid ester groups.
Algin	Brown algae (*Macrocystis pyrifera*)	Mannuronic acid, guluronic acid.
Carrageenan	Red algae (*Chondrus crispus, Gigartina stellata*)	Galactose, 3,6-anhydro-D-galactose plus sulfate acid ester groups.
Facoidan	Brown algae (*Fucus,* and *Laminaria* groups)	Fucose plus sulfate acid ester groups.
Laminaran	Brown algae (*Laminaria* group)	Glucose and mannitol with branches.
Plant exudates		
Gum arabic	*Acadia* plants	Arabinose, galactose, rhamnose, glucuronic acid.
Ghatti	*Anogeissus latifolia*	Arabinose, xylose, galactose, mannose, glucuronic acid.
Karaya	*Sterculia urens*	Galactose, rhamnose, galacturonic acid.
Tragacanth	*Astragalus* plants	Galactose, xylose, glucuronic acid.
Plant extracts		
Pectin	Cell walls and intracellular spaces of all plants; commercial source, citrus waste.	Galacturonic acid, arabinose with galactose branches.
Larch gum	Western larch	Galactose, arabinose.
Ti	Tubers of *Cordyline terminalis*	Fructose, glucose.
Plant seeds		
Corn-hull gum	Corn seed coat.	Xylose, arabinose, galactose, glucuronic acid.
Guar	*Camposia teragonolobus* endosperm.	Mannose, galactose with branches.
Locust bean	Carob tree (*Ceratonia siliqua*) endosperm.	Mannose, galactose with branches.
Quince seed	*Cydonia vulgaris*	Arabinose, xylose, hexuronic acid, monomethyl hexuronic acid.
Psyllium seed	*Plantago* plants	Xylose, arabinose, galacturonic acid, rhamnose, galactose.
Flax seed	*Linum usitatissimum*	Galacturonic acid, xylose, rhamnose, arabinose, galactose, glucose.
Tamarind	*Tamarindus indica*	Glucose, galactose, xylose.
Wheat gum	Wheat	Xylose with arabinose branches.
Miscellaneous		
Cellulose derivatives	Plant cell walls, wood pulp, and cotton.	Glucose
Starch and derivatives	Cereal grains and tubers.	Glucose with some branch points.
Dextran	Bacterial action on sucrose.	Glucose
Xanthan	Bacterial action on glucose.	Glucose, mannose, glucuronic acid.
Chitin	Exoskeleton of insects.	Glucosamine

[1]The carbohydrates forming long chains which form gums are primarily sugars but some are carbohydrates or sugar derivatives such as guluronic acid, galacturonic acid, glucuronic acid and glucosamine.

Children participating in the School Lunch Program. (Courtesy, USDA)

HACKBERRY (SUGARBERRY) *Cettia occidentalis*

Purple fruit of a small tree (of the family *Ulmaceae*) that is native to North America. They were much used as food by the American Indians.

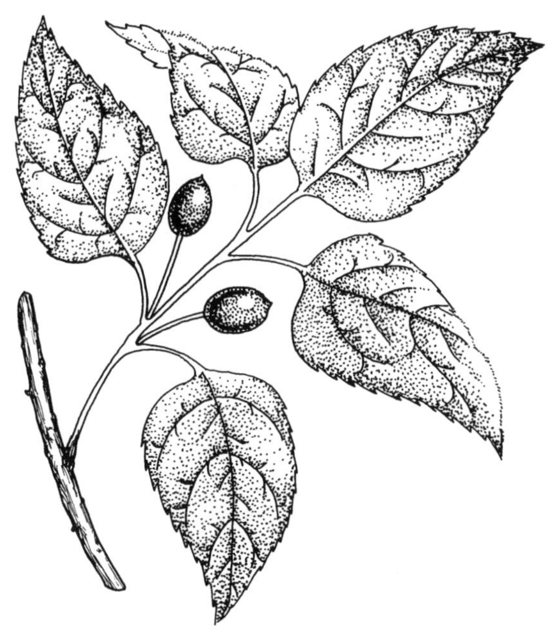

Fig. H-1. Hackberries.

The pulp of the berry is very sweet. It is moderately high in calories (72 kcal per 100 g) and carbohydrates (18%). Also, the fruit is a good to excellent source of iron and vitamin C.

HAFF DISEASE

This is a disease of humans characterized by the presence of hemoglobin in the urine (giving it a dark reddish color), muscular weakness, and pains in the limbs. Between World Wars I and II, it was rather prevalent among fishermen and their families around Haff in East Germany. Always on the day before an attack, fish (usually eel or burbot) had been eaten. The condition became known as Haff disease.

Haff disease appeared to be the human equivalent of Chastek paralysis, which affects mink and fox. Hence, it was concluded that it was caused by eating large quantities of inadequately cooked thiaminase-containing fish; with the enzyme thiaminase inactivating the thiamin molecule, resulting in a thiamin deficiency.

Some authorities now question that Haff disease is caused by thiaminase. Fortunately, whatever the cause, the malady is seldom reported nowadays.

HAGGIS

A favorite meat dish in Scotland, made of the heart, liver, and lungs of sheep or calves minced with suet, onions, oatmeal, and seasonings and boiled in the stomach of a sheep.

HAIR ANALYSIS

Minerals are deposited in hair as it grows; and, in theory, the hair reflects the mineral status of an individual at the time of hair growth. Scientific analytical methods, such as atomic absorption spectometry, neutron activation analysis, and x-ray fluorescence spectometry, are sensitive enough to detect the levels of minerals in hair samples. Because hair samples are easily and painlessly obtainable, and because hair samples are stable and store easily, there is considerable interest in the use of hair as a diagnostic tool for mineral deficiencies and/or toxicities. However, like many other diagnostic tests, hair analysis is only a tool to complement other tests and observations. It is not a panacea. For the reasons which follow, an individual's nutritional status cannot be assessed solely on the basis of a hair analysis:

1. The relationship between the concentration of a mineral or of a vitamin in the hair and that in other body tissues is unknown.
2. Laboratory results depend upon the proper treatment of the hair sample. Hair samples must be washed properly to remove outside contamination from environmental sources, but washing leaches some of the deposited minerals out of the hair.
3. Sampling procedures vary. Mineral levels at the ends of hair may be different than that near the scalp, or mineral levels from hair at the nape of the neck may be different than that from the hair on the top of the head.
4. Factors other than diet influence the mineral content of hair. Such factors include sex, age, and hair color. For example, red hair contains more iron than other hair colors.

Clearly, some scientific findings demonstrate that if dietary intakes of certain minerals (chromium, copper, iron, manganese, and selenium) are extremely low, if the diet includes toxic minerals (arsenic, cadmium, lead, and mercury), hair analysis can detect these changes. Furthermore, some disorders such as anemia, hepatitis, hyperthyrosis and nephrocalcinosis are reported to change the mineral levels in the hair.

Hair analysis is a growing industry, and some laboratories encourage individuals to submit hair samples to assess their nutritional status by multiple mineral hair analysis. Some of these laboratories then prescribe minerals and vitamins to correct the "metabolic imbalances," or deficiencies, discovered by the analysis. Not surprisingly, some of these laboratories also sell minerals and vitamins.

In summary, while hair analysis has been used to detect certain types of heavy metal poisoning (e.g., lead, arsenic, mercury) in populations, its value on an individual basis remains to be established. There are a number of limitations to hair analysis both in terms of analytical procedures and in interpretation of results. For example, the relationship between hair concentration of a trace element or of a vitamin and the concentration of other body tissues is unknown. Basically, hair analysis is of limited value for assessing mineral status and questionable for assessing vitamin status.

(Also see DEFICIENCY DISEASES, Table D-2 Minor Dietary Deficiency Disorders; HEALTH, sections headed "Signs of Good Health" and "Signs of Ill Health"; and MINERAL[S], section headed "Diagnosis of Mineral Deficiencies.")

HALF-LIFE

The time required to reduce a substance to ½ of its original amount, or to replace ½ of the original amount of a substance.

• **Physical**—In the field of radioactive isotopes, the half-life of radioactive carbon, carbon-14, is 5,568 years. In other words, if we start with 1 g of carbon-14, 5,568 years later only ½ g of carbon-14 will remain due to the radioactive decay. All radioactive isotopes have a specific half-life which may range from seconds to millions of years.

• **Biological**—In biology, chemicals in the body and tissues have half-lives. Follicle stimulating hormone has a half-life of 2 hours. That is, in 2 hours ½ of that secreted 2 hours earlier will have been eliminated from the circulation. One-half of the total protein in the body is replaced every 80 days. The radioactive element cesium-137 has a physical half-life of 28 years, while it has a biological half-life of about 140 days. That is, in 140 days, ½ of the initial cesium-137 would be excreted from the body.

(Also see RADIOACTIVE FALLOUT.)

HALIBUT LIVER OIL

Halibut liver oil is obtained from halibut livers. It is an excellent source of vitamins A and D and is frequently used as a primary material in making these vitamin products.

Although the livers of codfish are a good source of vitamins A and D, nature ordained that halibut livers be even richer for the following reason: Carotene is passed up through the food chain and concentrated as vitamin A in the livers of the higher level predators. Thus, codfish derive their vitamin A from the tiny vegetable plankton, upon which they feed, whereas halibut eat cod. It is noteworthy that at the top of the hierarchy is the polar bear, who eats the seal, who eats the halibut. Thus, the liver of the polar bear, which contains about 500,000 IU of vitamin A per gram, is toxic to man. A hungry explorer eating a 4-oz (*112 g*) serving of polar bear liver is in mortal danger, since he will have consumed 50,000,000 IU of vitamin A.

(Also see VITAMIN A.)

HALLUCINOGEN

A substance that induces hallucinations or false sensory perceptions.

HALVA

• A popular sweet in the lower Balkan area, made by crushing sesame seeds and mixing with honey or other syrup.

• In India, a halva is a pudding-like dessert that is always made from milk, but it is flavored with a variety of vegetables and fruits.

HAM

Cured and smoked hind leg of pork. Uncured hams should be labeled "fresh ham" or "pork ham."

HAMBURGER

This is the best known and most popular American food.

McDonald's, which features hamburgers, started with one fast-food restaurant at Des Plaines, Illinois, in 1955. Today, McDonald's competes with at least 15 other national chains, and the chains have more than 45,000 individual franchises. Different fast-food establishments feature different products, but the hamburger continues to be the mainstay of the industry.

The Code of Federal Regulation (CFR) defines *hamburger* as follows: chopped fresh and frozen beef, with or without added beef fat and/or seasonings. It shall not contain more than 30% fat, and it shall not contain added water, binders, or extenders. Beef cheek meat may be used up to 25% of the meat formulation. If hamburger contains extenders, it must be labeled as such.

(Also see CONVENIENCE AND FAST FOODS; and MEAT[S].)

HARD CIDER

Cider is usually the juice pressed out of apples. Before it is fermented, it may be referred to as sweet cider. However, after it is fermented and it contains alcohol, it is called hard cider. In some countries, the term cider alone refers to a fermented drink with an apple base. Fermenting cider is similar to the production of wine.

(Also see APPLE[S]; and WINE.)

HARDTACK (PILOT BREAD; SHIP'S BREAD)

A hard crackerlike bread which is similar to a soda cracker, but contains no salt. It was a staple on board the ships of old.

HARE

Wild rabbits with dark flesh, highly flavored, and excellent for eating. The mountain hare is more delicate than the plains hare.

HASLET

- A piece of meat roasted on a spit.

- The edible viscera (as the heart, liver, or kidneys) of pork.

- A braised dish made of edible viscera.

HAWS (HAWTHORNE) *Crataegus* spp.

The fruit of various species of hawthorne shrubs and trees (of the family *Rosaceae*) which grow wild throughout the New World and the Old World. "Haws" refers to the fruits, which resemble crab apples and are between ½ and 1 in. (*1 to 2.5 cm*) in diameter.

Fig. H-2. Haws, the fruit of wild hawthorne trees and shrubs.

Haws are usually made into jams and jellies or brewed in herb tea.

A typical hawthorne fruit (in this case *C. Pubescens* of Latin America) is moderately high in calories (89 kcal per 100 g) and carbohydrates (24%). It is a good source of fiber, iron, potassium, and vitamin C, and a fair source of vitamin A.

HAY DIET

The Hay Diet was first described by Dr. William Howard Hay in 1933 in his book, *Health Via Food*. Dr. Hay recommended that proteins and carbohydrates not be eaten at the same meal.

(Also see DIGESTION AND ABSORPTION.)

HEADCHEESE

A pork product made by skinning out a hog's head and removing the jaw bones, eyes, and ears; adding some hearts and tongues if desired; cooking the meat with the bones removed; grinding; adding enough broth to make a thick porridge; seasoning with salt, pepper, and marjoram to taste; and placing in crocks where the fat rises to the top. Usually, headcheese is eaten cold.

HEALTH

Contents *Page*

The Road to Good Health **1111**
Signs of Good Health **1113**
Signs of Ill Health **1113**
Rules of Good Health **1114**

Health is the state of complete physical, mental and social well-being, and not merely the absence of disease or infirmity. The word "health" comes from the Old English *haeth*, meaning the condition or state of being hail—safe or sound.

The custom of drinking to the health of people is very old, probably derived from the ancient religious rite of drinking to the gods. later, the Greeks drank to one another, and the Romans adopted the custom. By the beginning of the 17th century, health drinking had become a very ceremonious business in England; toasts were often drunk solemnly on bended knees. A Scots custom, still surviving, is to drink a toast with one foot on the table and the other on the chair. The French manner, when a health is drunk, is to bow to him that drank to you.

In 1992, the World Health Organization (WHO), the headquarters of which are located in Geneva, Switzerland, reported that at least 40% of the estimated 50 million annual deaths worldwide could be prevented by "improved health systems, drugs and vaccines, and a healthier life-style and education."

THE ROAD TO GOOD HEALTH. Sometimes a picture is worth a thousand words. No amount of persuasive narrative will bring home a point at times; thus, a briefing such as *The Road to Good Health* may help to "paint the picture." These are all well-known, documented facts to which everyone should adhere if they (1) aspire to be the "king of the castle," or (2) wish to have buoyant good health.

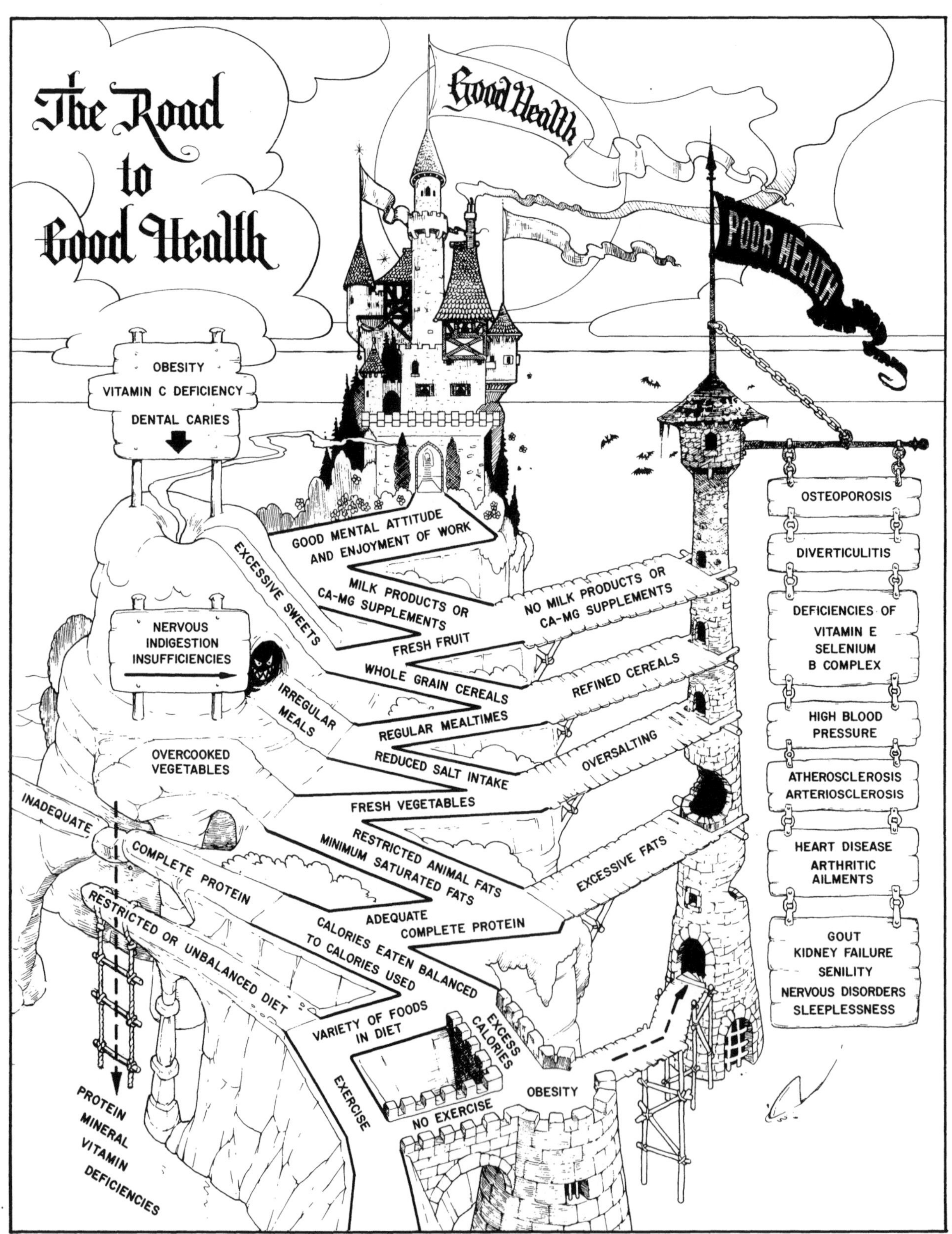

Fig. H-3. The road to good health.

SIGNS OF GOOD HEALTH. In order to know when disease strikes, people must first know the signs of good health, any departure from which constitutes a warning of trouble. Some of the signs of good health are:

1. **Seldom sick.** Healthy people are not susceptible to "every infection that comes along"; and they possess endurance and vigor.

2. **Good natured and full of life.** Healthy people are happy, contented, and full of "vim and vinegar."

Fig. H-4. Well fed, healthy, and happy. (Courtesy, USDA)

3. **Normal body temperature, pulse rate, and breathing rate.** For the average person, this means: mouth temperature, 98.6°F (37°C), and rectal about one degree higher; pulse, 72 times a minute; breathing varies with age: in the baby, it is about 45 times a minute; by age 6, it is down to about 25; between ages 15 and 25, it drops to about 18; and with advanced age, there is a tendency for it to increase.

Every family should have a thermometer. Before taking temperature, rinse the thermometer in cool water and shake it until the mercury falls below the 95°F mark. If an oral thermometer is used, it should be held under the tongue, with the mouth shut, for at least 3 minutes. After lubrication, a rectal thermometer should be inserted to the 98.6° line while the patient lies on his side; and it should remain in place for 3 to 5 minutes.

To take the pulse, place a finger on the radial artery at the wrist. The pulse is caused by the impact of the pressure of blood on the arteries as the heart beats.

Breathing rate can be determined by counting either the inhaling or exhaling of air.

4. **Average height for age, and average weight for height.** The body should not be undersized or poorly developed; and it should not be too thin (more than 10% underweight) or too fat and flabby (more than 10% overweight).

5. **Alertness.** Healthy people are alert; they are heads up.

6. **Good appetite.** In healthy people, the appetite is good and the food is eaten with relish; they're hungry when it's time to eat.

7. **Normal feces and urine.** The feces should be firm, and neither hard nor soft. The urine should be clear. Both the feces and urine should be passed without effort, and should be free from blood, mucus, or pus.

8. **Bright eyes and pink eye membranes.** In healthy people, the eyes are bright and clear; and the membranes—which can be seen when the lower lid is pulled down—are whitish pink in color and moist.

9. **Reddish pink tongue.** The tongue is reddish pink in color and not coated.

10. **Healthy gums and membranes of the mouth.** The gums should be firm, and the mucous membranes of the mouth should be reddish pink.

11. **Skin.** The skin is smooth, pliable, and elastic, and of a healthy color.

12. **Smooth, glossy hair.** The hair is lustrous (it has a gloss or shine) and firmly attached to the scalp (not easily pulled out).

13. **Firm, pink nails.** The nails are firm and pink, and not brittle or rigid.

SIGNS OF ILL HEALTH. Most sicknesses are ushered in by one or more departures from the signs of good health—by indicators that tell us that all is not well. If these signs exist and persist, you should see your doctor as soon as possible. Here are the most common signs of ill health:

1. **Loss of appetite; digestive disturbances.** Lack of appetite, nausea, indigestion, abdominal pain or cramps.

2. **Listless.** A tired run-down feeling—lack of energy.

3. **Abnormal feces.** Constipation, diarrhea, or passing blood.

4. **Abnormal urination or urine.** Difficulty in starting, stopping, dribbling, pain, or passing blood.

5. **Abnormal discharge from the nose, mouth, or eyes.** Pus or excess watering give reason for concern.

6. **Persistent cough or hoarseness.** When a cough or hoarseness lingers longer than normal, you should see a doctor.

7. **Lack of color of skin.** Paleness.

8. **Pale, dark red, or purple mucous membranes lining the eyes and gums.** Such signs give cause for concern.

9. **Sores on the skin, lips, or tongue.** If such sores fail to heal, there is reason for concern.

10. **Dull hair.** Lacking sheen; dry; and can be easily plucked.

11. **Nails.** Rigid, and brittle.

12. **Shortness of breath, or labored breathing.** This refers to shortness of breath when performing normal activities, and to labored breathing characterized by increased rate and depth.

13. **Chest pains.** Such pains may presage a heart attack.

14. **Persistent headaches.** The cause should be determined by your doctor.

15. **Prolonged aches in back, limbs, or joints.** See your doctor.

16. **Hard of hearing.** If hearing is worsening, see a specialist.

17. **Failing eyesight.** Dimming or fogging of vision should be checked by a specialist.

18. **Swelling of feet or ankles.** This may be serious. See your doctor.

19. **Sleeplessness.** Tendency to wake up during the night and difficulty in falling asleep again.

20. **Nervous, irritable, and depressed.** Such outward signs as crying spells, overwhelming sadness, worthlessness, and mental apathy may presage a nervous breakdown.

21. **Above normal body temperature.** If the temperature is more than 2° above normal.

22. **Sudden drop in weight or work.** Sudden loss of weight or work productivity should be checked.

23. **Malnutrition.** Malnutrition may result from a deficiency, excess, or imbalance of nutrients. This may include (a) undernutrition, which refers to a deficiency of calories and/or one or more essential nutrients; or (b) overnutrition, which is an excess of one or more nutrients and usually of calories.

24. **Noncontagious diseases.** Such noncontagious disease as allergies, anemia, arthritis, diabetes, diverticulosis, and heart disease are signs of ill health, all of which may be influenced by diet.

Special for women:

25. **Vaginal bleeding at unexpected times.**

26. **Lump on breast.** This may be cancerous.

RULES OF GOOD HEALTH. It is within easy reach of every individual to attain a good state of health and to retain it to a ripe old age by following the rules of good health. It is established and accepted medical opinion that good health does not imply merely the absence of disease, but presumes a state of physical, mental, and social well-being. Observance of the following rules will make for good health.

1. **Eat an adequate and balanced diet.** Proper food provides energy, builds new tissue, repairs worn out tissue, and keeps the body working well. A balanced diet contains adequate proportions of carbohydrates, fats, and proteins, along with the recommended daily allowances of the essential minerals and vitamins.

2. **Exercise regularly.** Exercise builds muscles, helps the blood circulate, enhances appetite, and helps the body use food properly.

3. **Relax regularly.** Relaxation makes a person feel happier and more comfortable, prevents tenseness and nervousness, and enhances both work and play.

4. **Get enough sleep.** Sleep lets the body rid itself of poisons, repair worn tissue, and grow properly.

5. **Get plenty of fresh air.** People must breathe air to live. Fresh air is desired because it is rich in oxygen (O_2) and low in carbon dioxide (CO_2).

6. **Take care of teeth.** Teeth cut, tear, and grind food to make it ready for digestion; hence, they are important for good health. Also, germs from decaying teeth may spread disease throughout the body. Eating a balanced diet and brushing the teeth after eating help prevent tooth decay, or cavities.

7. **Visit your doctor and dentist regularly.** Doctors and dentists guard your health. Many diseases develop slowly and do not cause pain in early stages. Often, doctors and dentists can find hidden signs of illness, and can take steps to correct disorders that may cause trouble. So, have your doctor and your dentist examine you at least once a year.

8. **Personal hygiene.** Personal cleanliness—the care of your body, including cleanliness of the skin, the body cavities, and the eliminative organs—improves appearance and helps keep a person well.

9. **Sanitize your immediate environment.** Keep your immediate environment sanitary, including your home and place of employment—be it office, farm, factory, or coal mine.

10. **Dress comfortably.** Loose, light clothing lets a person move freely, and allows air to reach the body.

11. **Keep good posture.** Posture refers to the carriage of the body in walking, standing, or sitting. Women in some cultures maintain an erect posture by carrying a basket or pot on the head; and in the western world, fashion models are taught to do so by walking with a book on the head.

In standing, the ideal posture is erect with the abdomen and chin drawn in and the shoulders square and high. In sitting, the body is erect and the head is poised in a line with the hip bones.

12. **Work and play safely.** Many accidents can be prevented by following safety rules and avoiding hazardous situations.

13. **Enjoy your work.** Fortunate is the person for whom work is fun and fun is work. (I have never felt that I worked a day in my life. M.E.E.) Pleasure is important for mental and social health.

14. **Be at peace with the world and all humanity.** Think positive, healthful thoughts; it can help make life more pleasant and healthful. Religion and philosophy help develop peace of mind.

15. **Practice restraint in taking drugs, medicines, and alcoholic beverages.** Take drugs and medications on your own only for the most minor conditions, and seek medical advice if there is no improvement within a day or two. Those who drink an alcoholic beverage should imbibe intelligently and in moderation.

16. **Follow the golden rule.** Find ways to help others, and think less about yourself. Cultivate wholehearted consideration of and genuine affection for your family, friends, and your fellowmen.

17. **Never grow old.** Keep on keeping on.

HEALTH FOODS

Contents Page

Historical Background and Current Trends 1115
Reasons for Popularity of Health Foods 1116
Commonly Considered Health Foods 1117

These are foods which are purported to improve health. The term "health food" encompasses both "natural" and "organic" foods, along with certain other foods. The two senior authors of this book patronize health food stores, especially for products that are not available in supermarkets.

Today, health food stores and health food sections in traditional supermarkets are doing a booming business—with retail sales of $3.3 billion in 1990. Their products may be extolled as "natural," "organically grown," "nothing artificial," "no preservatives," "no additives," "no chemicals," "anti-gray hair factors," or "dieters delight." Although most such claims are not supported by scientific evidence, it is difficult for the public to separate truth from fiction—particularly in regard to the use of the term *natural* for everything from whole grain flour to potato chips.

There is nothing wrong with the use of health foods. Most of them are beneficial. It is the exaggerated claims that some promoters make for them—the promise of better health through better eating—that needs to be questioned. However, it would be inaccurate to imply that all elements of the health food industry engage in questionable marketing practices. Also, not all items cost more; for example, at the time this section was written the senior author of this book checked the price of sesame seed at the local health food store and the local supermarket. A 16 oz (*454 g*) package of sesame seed in the health food store cost $1.35 vs $9.09 for the similar product in the supermarket.

Health food advocates are prone to make unfounded claims that a certain food or group of foods has special attributes or power which is out of proportion to its nutritive content. In fact, they often claim that particular foods can be used as preventive measures for health and even as curative measures for disease. Such claims can be downright dangerous, imperiling life and health by causing malnutrition, or by discouraging the consumer from seeking needed medical attention until it is too late. But the labels and promotions for fad foods or diets usually do not make any direct claims that can be shown to be false. Instead, they refer to a book, a pamphlet, a speech, or a magazine article that has praised the product. Thus, the indirect promotions receive the protection of the First Amendment.

Other areas upon which health food promoters capitalize are beauty, eternal youth, retention of sexual potency, and long life. Such promotion may give hope to the users and improve their health and well-being by meeting their emotional needs. There is nothing wrong with these desires.

Health foods can refer to natural and organic foods plus a wide range of other products. Such foods as alfalfa sprouts, blackstrap molasses, bone meal, ginseng, honey, kelp (seaweed), raw sugar, rose hips (seed pods of roses), seawater (bottled), sesame and sunflower seeds, wheat germ, whole grain flour, yeast, and yogurt have taken on a halo effect.

Science has found no magic or miracle foods. No one food is a preventive or cure-all for diseases. Foods are important because of their nutrient content; and no single food can serve as the source of all the essential nutrients. For example, wheat germ is considered a health food because it is a good source of vitamin E, the B vitamins, calories, essential fatty acids, protein, fiber, and most minerals. However, it lacks calcium, vitamin A, and vitamin C. So, the best way in which to ensure that nutritional requirements are met is to eat a variety of foods from each of the major food groups.

More and more health food stores are highly reputable. Some growers and distributors now supply affidavits or certificates of foods grown and handled according to *organic* or *natural* precepts. Also, many health food operators are very knowledgeable about nutrition, and take pride in sharing helpful health information with their customers. Health food stores are coming of age!

HISTORICAL BACKGROUND AND CURRENT TRENDS. Anthropologists, archaeologists, and other students of primitive peoples have found that beliefs in the superiority of certain foods were held by many ancient cultures. Some of the earliest concepts identified the organs of animals as beneficial to the analogous organs and functions in humans. For example, the American Indians and Eskimos considered brains, eyes, and glands of birds and animals as special foods, good for health and vitality. Later studies have shown that these items are especially rich in minerals and vitamins. And bird eggs and fish eggs have long been thought to be promoters of fertility and the bearing of healthy children, as evidenced by the practice (by natives in the South Sea islands) of feeding roe to prospective parents. Unfortunately, many useful and reasonably accurate observations were often clouded by cultural biases, superstitions, and taboos.

Another set of factors that has tended to complicate matters in man's search for a more healthful diet has been those associated with the settlement of large populations within the areas that could furnish only limited varieties of foods. This is best illustrated by the agricultural communities in which animal products were scarce and the people had to rely on only a few vegetable crops. Multiple nutritional deficiencies have often been found under these circumstances and it has often been surprising that the people have managed to propagate themselves. Any nutritious food that corrected even one of the severe deficiencies of these deprived people would have been regarded as a "health food." A contemporary example of this type is found among the natives in New Guinea who augment their predominantly sweet potato diets with plant ashes (rich in essential minerals) and small amounts of pork and other animal foods.

During the Middle Ages, some people felt that the more unpalatable the food or its source, the greater its curative properties. This philosophy culminated in the development of such delectables as ground coffee beans blended with fat.

The development of the modern scientific approach to nutrition began with the French chemist Lavoisier's experiments in metabolism during the latter part of the 18th century. Nevertheless, ideas regarding nutrient requirements were grossly oversimplified until the dawn of the Vitamin Era in the early part of the 20th century. (Around the end of the 19th century, agricultural chemists limited their analyses of food and feeds to carbohydrates, fats, proteins, fiber, a few minerals, and water. On that basis, they concluded that white corn was nutritionally equal to yellow corn. However, the farmers knew from experience that their animals did better on the latter grain, which was later found to be a much better source of vitamin A.

In 1773, Captain Cook fed sauerkraut to his sailors to prevent scurvy—the vitamin C deficiency disease; now, more than 200 years later, millions of Americans are taking vitamin C.

Meanwhile, the American health foods movement got underway in the early 1800s when the former Presbyterian preacher, the Reverend Sylvester Graham, toured the country giving lectures on the merits of eating coarse, whole grain breads and abstaining from fats and meat. He is immortalized by the graham cracker, which was named after him, and which since his death has undergone an evolution to a soft modern product made from a mixture of graham flour and white flour. (A 100% whole wheat graham cracker is still sold in some health food stores.) Also, Graham's ideas were promoted by Sister Ellen White, one of the early leaders of the Seventh-day Adventist church, and a founder of the Western Health Reform Institute that was established in Battle Creek, Michigan in 1866. The institute was later operated by Dr. John Kellogg, who, with his brother Will, developed a dry, ready-to-eat breakfast cereal that marked the birth of the cereal industry in Battle Creek. It is noteworthy that (1) one of the patients of Dr. Kellogg was C. W. Post, who also started a cereal company in Battle Creek and invented a cereal substitute for coffee; and (2) the vegetarian dietary practices of the Seventh-day Adventists, who abstain from animal flesh, were the main reason for the development of the first meat analogs to be used in the United States. (However, the idea for making meatlike dishes from vegetable protein foods did not originate in America, but in Asia, where such dishes have been prepared for many centuries.)

Health food stores became a notable part of the American scene in the 1920s and 1930s when shrewd businessmen capitalized on the news of the newly discovered vitamins and the foods from which they were obtained. An early mecca for these establishments was Hollywood, where actors and actresses were influenced by the ideas of the German immigrant Gaylord Hauser, who advocated five "wonder foods"—blackstrap molasses, brewers' yeast, powdered skim milk, yogurt, and wheat germ. In the eastern part of the United States, an interest in health, natural, and organically grown foods was sparked by the writings of Louis Bromfield, and later by Jerome Rodale, who moved from the dingy Lower East Side of New York City to the green meadows of Emmaus, Pennsylvania and started his own organic farming and gardening movement. (Rodale's entry into the field was financed by the successful electrical fixtures business which he operated with his brother. The continued success of the Rodale enterprises led to the founding of *Prevention* magazine in 1950.)

In its early days, the industry had its most ardent supporters among adults who were striving to retain or regain their youthful beauty, virility, and/or vitality. In the 1960s and 1970s, the health food industry received a very large boost from the ecology movement, when many young people, including the hippies, started to patronize health and natural food stores. Even the more conventional sector of the food industry has sought to capitalize on the evergrowing demand for "natural" foods, as evidenced by new products such as (1) many new types of yogurts; (2) natural whole-grain cereals with little or no added sweetener; (3) the substitution of carob powder for cocoa in many products (however, a major reason for this action was the high price of cocoa); (4) meat analogs made from soy protein; (5) natural coloring agents made from beets, marigolds, and turmeric; and (6) natural orange flavor, consisting of orange oil artificially extracted from natural oranges.

REASONS FOR POPULARITY OF HEALTH FOODS. The public has been introduced to health foods through news media, magazines, books, and self-styled nutrition experts. The reasons for changing to health foods are numerous, among them, those that follow:

- **Affluence**—In today's affluent society, most people are assured of their basic needs for food and security. Consequently, they can devote both time and money to a wider range of interests, including health foods.

- **Arthritic treatment**—A study of a large population of arthritic patients revealed that half of them had followed dietary treatments at different times.

- **Athletic prowess**—Much nutritional fallacy goes with athletics. There isn't a health food (including royal jelly) known to man that is not advocated somewhere by a coach or athletic director to the boys or girls under his care.

- **Emotions**—For others, health foods serve as a symbol in meeting emotional needs. For them, food is a symbol of acceptance, friendliness, and socialization. So, to get "with it," health foods are used.

- **Environment**—Others are concerned about the effect of chemical fertilizers and pesticides on the environment. Their concern causes them to buy health foods because they feel that less chemical fertilizer and pesticides have been used in their production.

- **Illness**—Some believe that particular foods have curative powers for certain diseases. At the same time they seek modern medical treatment, many persons look for something beyond what scientific medicine offers. This search is for a miracle drug, a secret formula, a special herb, or a wonder food.

• **Lifetime customer and advocate**—Once they are hooked, they usually become lifetime customers. It is virtually impossible to convince a true believer that their favorite health foods do not possess the purported magic. So, once they have concluded that health foods are their salvation from nutritional sins and the panacea and preventive for the world's mental and physical ills, they usually remain lifetime health food customers and advocates.

• **Nature**—According to some health enthusiasts, nature cannot be improved upon; hence, it follows that natural, organic, or health foods are the best.

• **Nutrition "experts"**—Millions of people pose as self-styled nutrition "experts." Some of these are content to ply their trade on themselves and their families. Others give their advice free to all who will listen. Still others are the modern counterpart of the medicine men of old—they sell their "expertise." Television, newspapers and magazines, best-selling books—all feed us information and advice that is a mixture of scientific fact, half-truths, and just plain nonsense.

(See FOOD MYTHS AND MISINFORMATION, section headed "The Medicine Man of Old.")

• **Nutritious and safe foods**—Some people are demanding health foods because they believe that they are getting more nutritious and safer foods. They associate their appearance and well-being with what they eat.

• **Obesity therapy**—Consumers are inundated with quick weight-loss plans; from megavitamin therapy to herbalism, from fasting to the wondrous grapefruit diet billed as follows:

"Lose 10 lb (*4.5 kg*) in 10 days
on grapefruit diet."

Books abound, assuring the weight watcher that he or she can lose weight while eating all he wants.

• **Personal attention**—Most health food store operators get to know their customers, much like the pop and mom corner grocery of old. By contrast, supermarkets seldom impart a personal touch.

• **Profit motive**—Many health food manufacturers and quack "doctors" have exploited the beliefs of some consumers at considerable profit to themselves.

• **Rebellion**—For some, health foods serve as a symbol against the establishment. Their reasoning: The food industry is linked with the establishment; therefore, foods normally found in supermarkets are undesirable.

• **Summary**—The movement toward a more primitive lifestyle has created a market as well as many gross untruths about health foods. Most health foods are very good—and good for you. For example, bread made with stoneground whole wheat flour may be enjoyed by all. Also, health food stores serve useful purposes by (1) focusing attention on certain foods that are more nutritious than others, (2) providing some items that are not found in other stores, and (3) stimulating beneficial innovations that might not otherwise occur in the food industry.

Food fads and fallacies! How do you fight them? What do you do, for example, when you discover that one of your friends or acquaintances is following a dangerously unbalanced diet prescribed by the newest media guru? Or when you are asked a tough question about a best-selling book that recommends hair analysis and a host of cure-alls? The answer: Until accurate and practical nutrition information is provided to all people, beginning with their early experiences and education in the family and in the school, large numbers of people will lack sufficient knowledge to sift through the barrage of nutrition misinformation being circulated so widely today.

NOTE WELL: Some people have the impression that health food promotors—manufacturers; health food store operators; authors and publishers of books, pamphlets, leaflets, magazines, or newspapers; and health lecturers to live audiences, or over radio or TV—couldn't make the claims that they do if these claims were not proven and true. Nothing could be further from the truth. *The government can only regulate claims made on product labels and in product advertising.* The Food and Drug Administration (FDA) enforces proper labeling; and both the U.S. Postal Service and the Federal Trade Commission are empowered to halt false advertising claims. Also, the American Medical Association maintains a Bureau of Investigation, which monitors the facts where there appears to be promotion and distribution of products that may be harmful to individuals. However, under the First Amendment (constitutional guarantees of a free press and free speech), purveyors of nutrition misinformation can (1) write, publish, and distribute books and other literature, or (2) lecture, proclaiming miraculous, but totally unfounded, benefits from foods.

COMMONLY CONSIDERED HEALTH FOODS. The selling of certain items by health food stores and not by supermarkets is as much a matter of economics and marketing as it is a matter of providing items reputed to be of high nutritional value. (Both types of food stores sell a range of products that are designed to suit the preferences of their customers, and which differ widely in their nutritional values.) Furthermore, many health food stores carry more tablet forms of nutritional supplements than actual food items. Finally, quite a few products are merely combinations and/or variations of certain basic foods. Table H-1 is a summary of commonly considered health foods.

(Also see FOOD MYTHS AND MISINFORMATION; NATURAL FOODS; ORGANICALLY GROWN FOOD; PESTICIDES; POISONS; PROCESSED FOODS, HIGHLY; VEGETARIAN DIETS; and ZEN MACROBIOTIC DIET.)

TABLE H-1
COMMONLY CONSIDERED HEALTH FOODS

Name of Food	Type of Product(s)	Uses	Reported Values and Claims	Comments
Acerola (Barbados cherry)	Tablets made from powdered, dried acerola berries.	Vitamin C supplement.	Rich in vitamin C and bioflavonoids.	Acerola has the highest ascorbic acid (vitamin C) content of any known food.
Acidophilus	Bottled suspensions of *Lactobacillus acidophilus* bacteria, and powder and tablets made from the dried, but viable bacteria.	To promote the growth of this bacteria in the intestine.	The lactic acid produced by the bacteria may aid in the absorption of calcium and other minerals, while other substances it produces prevent the growth of less desirable bacteria.	The conditions that favor the growth of this bacteria are not known for certain.
Acidophilus milk	Fluid milk to which a culture of *Lactobacillus acidophilus* has been added.	To increase the *L. acidophilus* component of the gut microflora, thereby helping to maintain the balance of microorganisms in the intestinal tract. To regenerate intestinal flora after antibiotic treatment or other conditions that upset the microfloral balance in the gut.	Same as that for acidophilus, plus those values for milk.	The older types of products had a sour taste, but a sweet acidophilus milk is now available.
Adzuki bean	A small dark red dried bean that is popular in the Orient.	Cooked and eaten with brown rice, or ground into a powder that is used to make a vegetarian milk substitute.	Good source of calories, carbohydrates, proteins, minerals (particularly phosphorus and iron), and the B vitamins.	Sweetened fillings and pastes may be made from the mashed cooked beans.
Agar-Agar (Japanese gelatin) (Also see AGAR-(AGAR.)	A firm gumlike substance extracted from seaweed.	Emulsifier, gelling agent, and thickener in foods.	The calories and carbohydrates are not available, but iron, iodine, and other minerals may be utilized.	Sometimes used by vegetarians as a substitute for gelatin.
Alfalfa flour (powder)	A green, grassy-flavored flour made from the dried leaves of the alfalfa plant.	In baked goods, gravies, soups, and stews.	An excellent source of proteins, calcium and trace minerals, carotene, vitamins E and K, and the unidentified factors. Used by some health food enthusiasts for treatment of diabetes.	Only small amounts should be used until one becomes accustomed to the color and flavor. There is no proof that alfalfa powder is effective as a treatment for diabetes.
Alfalfa seeds	Tiny seeds of the alfalfa plant.	Making a herb tea, and sprouting to obtain a raw salad vegetable.	The unsprouted seeds are *not* usually consumed.	The tea is reputed to ease aches and pains in the joints, but this is without experimental proof.

(Continued)

TABLE H–1 *(Continued)*

Name of Food	Type of Product(s)	Uses	Reported Values and Claims	Comments
Alfalfa sprouts	Sprouts made from alfalfa seeds.	As a raw salad vegetable, sandwich filling, and in soups.	Similar to alfalfa flour; rich in vitamin C.	Fresh sprouts keep for only a few days in a refrigerator. There are cheaper sources of vitamin C.
Alfalfa tablets	Tablets made from alfalfa powder	Nutritional supplement.	Same as alfalfa flour.	Some people are allergic to alfalfa powder.
Aloe vera	A gel made from the leaves of the aloe vera plant, a tropical and semitropical plant of the lily family, resembling a cactus.	Healing of burns, cuts, and ulcers.	Uncertain.	Should *not* be taken internally unless indicated on the product tablet.
Apricot kernels (Also see LAETRILE.)	The kernels from apricot pits.	As a source of the highly controversial substance laetrile (also known as amygdalin.)	Marketed as a dietary supplement as well as a drug for treatment of cancer, hepatitis, heart disease, alcoholism, allergies, and other conditions. Laetrile is not a nutrient.	Laetrile has no known value for human. Anyone who consumes these kernels in excess runs the risk of cyanide poisoning.
Arrowroot starch	The powdered starch from the root of a tropical herb.	Ingredients of biscuits, gravies, sauces, and soups for infants and invalids.	Provides mainly calories and is easily digested.	Arrowroot cookies have long been given to teething infants.
Artichoke noodles	A pasta product made from the flour of Jerusalem artichoke tubers.	Dietetic pasta products for people who need to limit their consumption of calories and carbohydrates.	The major starchy constituent of the tubers is inulin, a carbohydrate that is not utilized by the body.	The actual calorie and carbohydrate contents of this product are rarely given on the label.
Avocado oil	Vegetable oil extracted from avocados.	In salad oils and cosmetics.	Source of essential fatty acids.	This oil is quite expensive.
Baking mixes, dietetic	Mixes for preparing baked goods that are suitable for people who have conditions that warrant the use of modified diets.	Preparation of special dietetic types of biscuits, breads, cakes, cookies, muffins, pancakes, rolls, etc.	Varies according to the type of mixture. May be deficient in the nutrients that are present in the ingredients which were replaced by substitutes.	Also see BREADS AND BAKING, Table B-18 "Special Bakery Products."
Baking powder, low sodium	Chemical leavening agent in which potassium salts replace sodium salts.	Preparation of low sodium "quick" breads.	Rich in potassium, but contains little or no sodium.	Recipes may have to be modified to compensate for bitter taste of potassium salts.
Banana flakes	Flakes made from dried bananas.	Adding calories, flavor, and thickening to drinks and foods.	Rich in calories, magnesium and potassium; easily digested.	May be mixed in cereals, fruit drinks, or milk for infants or invalids.
Barbados molasses	Molasses residue from sugar refining.	As a flavoring in milk drinks, an ingredient or a spread for baked goods, and a laxative.	It is a variation of blackstrap molasses. Used by some enthusiasts as a general health food, and in the treatment of anemia.	Has a more pleasant flavor than blackstrap molasses, but rest of comments made for blackstrap molasses are applicable to Barbados molasses.

(Continued)

TABLE H–1 *(Continued)*

Name of Food	Type of Product(s)	Uses	Reported Values and Claims	Comments
Barley flour	Flour milled from barley grains at various stages of refinement.	Preparation of baked goods for people with allergies to wheat.	Similar to wheat flours of equivalent degrees of refinement.	Contains gluten and is *not* useful for gluten-free diets.
Blackstrap molasses	Molasses residue from sugar refining.	As a flavoring in milk drinks, an ingredient or spread for baked goods, and a laxative.	Blackstrap molasses is purported to be a health-imparting food and a rich source of iron; it is used by advocates as a cure for anemia and rheumatism.	There is no experimental proof that molasses improves health or cures anemia or rheumatism. Molasses is a fair, but unreliable source of iron. Molasses is eaten in such small amounts that it does not make an important contribution to the diet.
Bone meal	Tablets or powder made from ground sterilized bones of inspected slaughter animals.	Nutritional supplement that provides calcium, phosphorus, and other minerals.	Rich source of calcium, phosphorus and other minerals.	The consumption of excessive amounts may reduce the absorption of certain essential trace minerals.
Bran	The seed coat of wheat and other grains. Available as flakes or in tablets.	May be added to cereals and baked goods to increase the fiber content.	Low in calories, but rich in fiber, protein, minerals, and B vitamins. Slightly laxative.	Should be eaten with plenty of fluid to prevent irritation of the g.i. tract.
Branch-chain amino acids	Individual amino acids in tablet or powder form.	As a dietary supplement—an alternative to anabolic steroids. To augment the protein in the regular diet.	Superior athletic performance.	Protein overloading may increase the amount of water the body loses and alter kidney function, and cause diarrhea and abdominal cramps. More creditable research on this product is needed.
Brewers' yeast (Also see YEAST.)	Dried food yeast derived from the brewing of beer and ale (cannot be used for leavening). Available in flake, powder, or tablet form.	Used primarily as a source of B vitamins and unidentified factors that may be taken with water or added to baked goods, drinks, puddings, soups, etc.	Rich source of B vitamins and unidentified factors. Also, an excellent source of protein of good quality.	The high nucleic acid content makes it unsuitable for people with gout or high blood levels of uric acid.
Buckwheat flour	Flour from dehulled buckwheat.	Preparation of buckwheat pancakes and blintzes (rolled thin pancakes filled or served with fish, fruit, meat, and/or sour cream).	Similar to that of wheat flour, except that amino acid patterns of the 2 flours are complementary. (Hence, mixtures of the flours are more nutritious than either one alone.)	This flour has a rich hearty flavor and produces baked goods that are a little heavier and moister than wheat flour items.
Buckwheat groats (kasha)	Dehulled and lightly milled pieces of buckwheat grains.	Cooked and served as a cereal, side dish, or stuffing for poultry.	Similar to that of buckwheat flour.	The groats add flavor when mixed with more bland cereal products.

(Continued)

TABLE H–1 *(Continued)*

Name of Food	Type of Product(s)	Uses	Reported Values and Claims	Comments
Bulgar wheat	Grains of wheat that have been soaked, cooked, dried, lightly milled, and cracked.	The cooked grain is used as a cereal, stuffing for meats and poultry or a substitute for rice.	Similar to that of whole wheat grains.	Cooks more quickly than whole wheat grains.
L-carnitine	This nonprotein amino acid is synthesized from two essential amino acids in the liver and kidneys.	As a dietary supplement—an alternative to anabolic steroids. Weight reduction. Endurance in athletic events.	Some endurance athletes feel that taking carnitine supplements helps them go the distance.	L-carnitine may have a sparing action of glycogen and improve fatty acid metabolism. More creditable research on this product is needed.
Carob powder (Also see CAROB.)	Chocolate colored, sweet powder made from deseeded carob pods.	A chocolate-like ingredient of cakes, candies, drinks, and puddings.	Lower in calories and fats than most chocolate products. Carob powder is low in sodium and high in potassium; hence, desirable for diets which restrict sodium and encourage potassium, such as those used in the treatment of congestive heart failure and hypertension.	
Carrageenan (Irish moss) (Also see CARRAGEENAN.)	A vegetable gum extracted from the seaweed Irish moss.	Emulsifier, gelling agent, and thickener for various types of food products.	Low in calories (most of the carbohydrate content is not utilizable), but rich in calcium, iodine, and other minerals.	Sometimes used by vegetarians as a substitute for gelatin.
Cheeses, raw milk	Mild-flavored cheeses made from certified raw (unpasteurized) milk.	Same as for cheeses made from pasteurized milk.	Similar to counterparts made from pasteurized milk.	Some people prefer not to use products made from pasteurized milk.
Chia seeds	Seeds from a wild sage that grows in Mexico and in southwestern U.S.	May be added as a thickener to cereals, doughs for baking, sauces, spreads, etc.	Data on the nutrient composition is not readily available.	These seeds become jellylike when soaked in a liquid.
Chickpeas (Also see LEGUMES.)	Salted roasted chickpeas or uncooked dried chickpeas.	Roasted peas—as a snack like peanuts. Boiled peas—as most other dried legumes.	Similar to that of most legumes.	Canned cooked chickpeas are sold in many American supermarkets.
Chicory root powder	Powder made from roasted roots of chickory.	As an extender or a substitute for coffee, which it resembles in flavor.	The amounts that are generally used (1 tsp or 5 *g*) are too small to contribute much in the way of nutrients.	The practice of mixing chicory with coffee was brought to New Orleans by French settlers.
Chlorophyll	Green pigment extracted from the leaves of plants that is sold in liquid and tablet forms.	As a breath deodorizer and a "cleanser" of the digestive tract. For digestive disorders.	The amounts ingested (1 tsp or 5 *g*) are too small to contribute much in the way of nutrients. It is reputed to alleviate ulcers, but experimental proof is lacking.	May be obtained more economically by eating liberal amounts of fresh green leaves from edible plants.

(Continued)

TABLE H–1 *(Continued)*

Name of Food	Type of Product(s)	Uses	Reported Values and Claims	Comments
Chromium picolinate	It is a natural compound of chromium and picolinic acid. Picolinic acid is a natural chelating agent that occurs normally in the human body to facilitate the assimilation of minerals.	As a dietary supplement—an alternative to anabolic steroids. Weight reduction and muscle building.	It is purported to speed fat loss while building muscle.	It works by enhancing insulin-dependent mechanisms that control hunger, stimulates metabolism, and conserves body protein. Chromium picolinate is one of the most widely used dietary supplements. More creditable research on this product is needed.
Coffee substitutes	Powdered, toasted cereal preparations that may also contain chicory, figs, molasses and/or other ingredients.	For beverage use by people who may wish to avoid consuming caffeine.	The amounts consumed (1 tsp or *5 g*) are too small to contribute much in the way of nutrients.	These products are now sold in many American supermarkets.
Cold-pressed oils	Vegetable oils that have been extracted from seeds heated to 200°F to 300°F (*94°C to 148°C*). (The term "cold-pressing" is a misnomer).	By people who expect the oils to have higher nutritive values than those extracted by the more conventional process.	May be a little higher in vitamin E and tocopherol content than oils extracted by higher temperature processes.	Some of the vitamin E and related tocopherol content may be destroyed even by so-called "cold-pressing."
Corn, flaked	Flakes produced by rolling whole kernels of corn.	Whole grain cooked cereal.	Higher in protein, fiber, and certain trace minerals than ready-to-eat cornflakes.	Much more filling than the more commonly used breakfast cereal.
Corn flour, whole ground	Flour produced by milling whole kernels of corn.	Baked goods; breading for fish, meats, and poultry; soups; and stews.	Higher in calories, corn oil, and vitamin E than degerminated corn flours.	Should be kept refrigerated to prevent the development of rancidity.
Cornmeal, whole	Meal produced from whole kernels of corn.	Baked goods; cooked breakfast cereal; cornmeal mush, etc.	Higher in calories, corn oil, and vitamin E than degerminated cornmeal.	Yellow cornmeal is richer in vitamin A than white cornmeal.
Cottonseed flour	Flour made from ground cottonseed from which the oil has been extracted.	Mixed with cereal flours that are used to prepare baked products.	Richer in protein, calcium, phosphorus, iron and B vitamins than most cereal flours. (Also, the protein complements that of the grain flours.)	About 7% cottonseed flour may be used in a mixture with wheat flour.
Couscous	A semolina made from finely ground hard durum wheat.	As a side dish or an addition to soups and stews.	About the same as that for other products made from partly milled wheat.	Couscous is a staple food in North Africa and the Middle East.
Date syrup	A syrup made from ground fresh dates.	An ingredient or a topping for hot breads and desserts.	Provides mainly calories and carbohydrates.	Produces moist baked goods when used in place of some of the sugar.

(Continued)

TABLE H–1 *(Continued)*

Name of Food	Type of Product(s)	Uses	Reported Values and Claims	Comments
Dolomite	Tablets or powdered, made from ground dolomitic limestone.	Nutritional supplement that provides calcium, magnesium, and other minerals.	Rich source of calcium, magnesium, and other minerals. May also be used as an antacid.	The consumption of excessive amounts may reduce the absorption of certain trace minerals.
Eggs, fertile	Fertile eggs, laid by hens mated to roosters.	Same as nonfertile eggs.	Advocates claim that fertile eggs are more nutritious and invigorating than nonfertile eggs.	Fertile eggs are no more nourishing than nonfertile eggs. Besides, they spoil more quickly due to the developing embryo.
Eggs, raw	Uncooked eggs.	Eaten raw, usually with salt and pepper added.	Faddists claim that raw eggs increase potency.	Sorry, this is folklore. Moreover, raw eggs contain the antivitamin, avidin, which binds biotin and makes it unavailable.
Egg replacers	Egg-free mixtures made from various hypoallergenic vegetable gums and starches.	Baked preparations for people who cannot eat eggs (as in certain allergies.).	Provide mainly calories and carbohydrates.	These products are designed to provide binding for batter and doughs.
Egg white, dried	A powder prepared by drying separated egg white.	In baked goods and other foods prepared for people who wish to avoid egg yolks.	Moderately high in calories and very rich in high-quality protein.	1 lb (*454 g*) of dried egg white is equivalent to about 14 cups (*3,360 ml*) of liquid egg whites.
Fig syrup	Syrup made by evaporating a water extract of ground dried figs.	A flavoring for milk drinks, an ingredient or topping for baked items, and as a laxative.	Provides mainly calories and carbohydrates.	Whole dried figs are a more effective, but more irritating laxative.
Flaxseed	Whole or ground seeds from the flax plant.	Additions to cereals and porridges, for making a soothing tea, and as a laxative.	High in polyunsaturated fatty acids, protein, and fiber.	The seeds contain a mucilage that is extracted by steeping in hot water. (Women once used the mucilage for setting their hair.)
Fructose	Sugar obtained from certain fruits or from the chemical conversion of dextrose (glucose).	As a substitute for table sugar, because it is sweeter and less is needed.	Same caloric and carbohydrate values as most other pure sugars.	There is little evidence to support the use of fructose in treating blood sugar disorders.
Gamma oryzanol and other plant sterols	The chemical structure of this product is similar to cholesterol.	As a dietary supplement—and alternative to anabolic steroids. Users hope to obtain an increase in lean muscle mass.	Plant sterols are purported to increase testosterone production and stimulate human growth hormone release.	Gamma oryzanol is produced from rice bran oil. Currently, no published scientific evidence shows any anabolic effect from plant sterols or oryzanol. More creditable research on this product is needed.

(Continued)

TABLE H-1 (*Continued*)

Name of Food	Type of Product(s)	Uses	Reported Values and Claims	Comments
Garlic	Pills	Medicinal purposes.	It may lower blood cholesterol levels.	Livens up the taste of some foods; helps reduce salt consumption.
Ginger root	Fresh root of the ginger plant.	Seasoning for various dishes, and for making a spicy tea that is reputed by some to help open a clogged nose.	The amounts consumed (1 tsp or 5 *g*) are too small to contribute much in the way of nutrients.	The flavor of fresh ginger root is much more distinctive than that of powdered dried ginger.
Ginseng	Root of the ginseng plant which is sold in capsules, extract, instant powder, paste, tea (sometimes made from leaves), and whole root.	Claimed to be a general tonic for digestive troubles, impotence, and overall lack of vitality.	The amounts consumed (1 tsp or 5 *g*) are too small to contribute much in the way of nutrients.	Ginseng may raise the blood pressure of certain susceptible people.
Gluten flour	A high-protein highly-elastic flour obtained by washing much of the starch from ordinary wheat flour.	Preparation of low-carbohydrate, high-protein breads and in baking mixtures with other flours or meals of low gluten content.	Compared to ordinary wheat flour, it is lower in carbohydrate, higher in protein, and lower in minerals and vitamins.	Gluten flour strengthens and lightens doughs made with barley, corn, oat, potato, rice, rye, or soy flours.
Goat's milk	Fresh or evaporated goat's milk.	For people with an intolerance to cow's milk.	About the same as cow's milk.	Some infants digest goat's milk more readily than cow's milk.
Graham flour	Whole wheat flour made from winter wheat.	Baking of whole grain breads.	Similar to that of whole wheat flour.	Available in some supermarkets.
Granola cereal	A ready-to-eat toasted cereal mixture containing rolled oats, one or more sweeteners, oil, raisins, nuts, and/or flavorings.	Breakfast cereal or ingredient of bars, breads, muffins, pancakes, etc. Topping for other foods such as desserts.	Usually higher in calories and sometimes higher in fiber and certain minerals, but otherwise not much better than fortified breakfast cereals.	Granola mixtures vary considerably in their compositions and may contain several different sweeteners.
Guarana	Tablets or a bitter chocolatelike powder, made from the seeds of a Brazilian climbing shrub, that are mixed into drinks.	As a stimulant (see comments).	The amounts used are too small to make any significant nutritional contributions.	Guarana contains 3 to 5 times as much caffeine as coffee or tea.
Honey	Produced by bees from the flower nectar of 1 or more plants such as alfalfa, buckwheat, clover, orange, sage, and tupelo.	Sweetener for baked goods, desserts, drinks, etc. Spread for breads, muffins, pancakes, etc.	Due to its fructose content, honey is sweeter than table sugar; hence less of it is needed. Some cultures claim that honey excites sexual desire. Unfounded medical claims have been ascribed to honey.	Honey has no magical powers relative to either sexual desire or health. Breads and cakes that contain honey keep moist longer than those sweetened with sugar.

(*Continued*)

TABLE H-1 (*Continued*)

Name of Food	Type of Product(s)	Uses	Reported Values and Claims	Comments
Honey and Vinegar	A mixture of honey and vinegar.	As a cure-all.	Honey and vinegar have been purported to cure an amazing array of ills, ranging from arthritis to the common cold and digestive disturbances.	Don't you believe it! The presence of honey makes the swallowing of vinegar more tolerable.
Kefir	Milk product similar to yogurt but produced by a double fermentation (bacterial and yeast).	Beverage, dessert, dressing, and ingredient of baked products.	The lactic acid content may increase the absorption of calcium and other minerals.	Kefir may be made by using granules.
Kelp	A seaweed that is rich in iodine. Available in powder and tablet form.	As an iodine supplement to prevent a deficiency of the mineral.	Very rich source of iodine.	Consumption should be limited to amounts which provide the RDA for iodine. (Too much iodine may be toxic). Kelp may contain high levels of arsenic, as do many products from the sea.
Lactobacillus	Bacteria (such as *L. acidophilus*) which ferment milk sugar (lactose) to lactic acid. Available as dried powder, liquid suspensions, and tablets.	Preparation of fermented milks. Colonization of the intestines by lactobacilli and elimination of undesirable intestinal microorganisms.	Dietary lactic acid may promote mineral absorption and the counteraction of ammonia accumulation in the intestine. (Excessive ammonia is toxic.)	There is still considerable uncertainty regarding the diet and other conditions which promote the colonization of the intestine by *Lactobacilli*.
Lecithin	Fatty substance that is obtained from egg yolks, unrefined soybean oil and other oils.	Emulsifying agent (helps to keep constituents of mixtures from separating out). Certain therapeutic applications.	Some enthusiasts claim that added lecithin will prevent or cure arthritis, gallstones, heart disease, nervous disorders, and skin disorders. Recent reports suggest that one of the constituents of lecithin, choline, may counter senile deterioration of the brain.	Research has not backed up these claims. Supplements of lecithin contain little choline. Besides, more research is needed on senile deterioration of the brain.
Licorice root	Extracted from the root of a herb and used in the production of certain candies and drinks.	Chips from the root are used for brewing teas and making other home remedies.	Little nutritional value. May sooth minor colds and sore throats.	Licorice has a laxative effect for some people.
Liver powder	Powder or tablets made from defatted or undefatted dried liver.	Nutritional supplement for undernourished people.	Excellent source of protein, nucleic acids, phosphorus, iron, copper, and B vitamins.	Defatted liver powder is much more palatable than undefatted forms.

(*Continued*)

TABLE H-1 (*Continued*)

Name of Food	Type of Product(s)	Uses	Reported Values and Claims	Comments
Maple sugar	Sugar obtained by evaporating maple syrup.	Ingredient of candies and other sweets.	Provides mainly calories and sugar.	Some products contain mixtures of maple and other sugars.
Maple syrup	The syrup obtained by boiling down the sap from maple trees.	Ingredient of baked goods, candies, pies, etc. Spread or topping for pancakes, waffles, etc.	Provides mainly calories and sugar.	Some products are only "maple-flavored" syrups, or blends of maple and other syrups.
Meat analogs	Vegetarian imitations of meat products that are made from cereal and legume derivatives such as gluten and soy protein.	Extenders or substitutes for fish, meats, and poultry. Vegetarian diets.	Values vary acording to the ingredients. Usually the protein values approximate those of meat, but the levels of calories and other nutrients may be quite different.	Some products are much higher in sodium than fish, meats, or poultry.
Milk, raw	Unpasteurized milk, straight from the udder to the consumer.	For people who do not wish to use pasteurized milk.	Some supporters claim that raw milk is more nutritious than pasteurized milk---that pasteurization destroys vitamin C.	The amount of vitamin C in raw milk is so small that it makes no significant contribution of this vitamin to the diet. Raw milk, even if certified, (produced under rigid sanitary conditions prescribed by the American Association of Medical Commissioners) can be a source of undulant fever (brucellosis) and tuberculosis---two dreaded diseases.
Milk substitutes	Milklike fluids that are usually made from soy derivatives and a sweetener.	Vegetarians who consume no animal foods. People who have an intolerance to milk.	Usually, amino acids, minerals, and vitamins are added to make these products nutritionally similar to milk.	Now available in pharmacies and supermarkets.
Millet	A tiny grain that is much more commonly used for food in Africa, the Middle East, and India than in the U.S. (It is used mainly for birdseed in America.)	Cooked breakfast cereal, side dish, or topping for other dishes. The ground grain may be added to flour mixtures used for breads, muffins, rolls, etc.	Millet is generally superior to wheat, rice, and corn in protein quality. However, like other grains, it is low or lacking in calcium, and in vitamins A, D, C, and B-12.	Millet has a bland taste that is improved by browning or toasting it slightly before cooking.

(*Continued*)

TABLE H-1 (*Continued*)

Name of Food	Type of Product(s)	Uses	Reported Values and Claims	Comments
Miso	A fermented food paste made from soybeans.	Ingredients of dips, dressings, salads, sandwich fillings, soups, and stews.	Varies considerably according to the way in which it is prepared. (However, it is a good source of calories and protein.)	Made with sea salt; hence, it has a high sodium content.
Mung bean sprouts	Small beans from India and other tropical countries that are used for sprouting.	Mung bean sprouts are good in salads, sandwiches, soups, and stir-fried dishes.	The sprouts are low in calories, but a good source of protein, fiber, and vitamin C.	The unsprouted beans are rarely used in the U.S.
Mushrooms, dried	Dried pieces of various species of mushrooms.	Salads, soups, and stews.	Varies according to the species. Proteins complement those of cereal grains.	The best flavored dried mushrooms are the wild species from Europe.
Oat flour	Flour made by milling dehulled and debranned oats.	Alone or with other cereal flours in baked products.	Slightly higher in protein than the other common cereal flours.	Oat flour produces a heavier and moister baked product than wheat flour.
Oats, steel cut	Oat grains cut into granules.	Cooked cereal, puddings, and soups.	Same as oat flour.	Also known as Scottish oats.
Papain	A protein-digesting enzyme extracted from papaya that is available in tablet and powder forms.	Meat tenderizer. Digestive aid for people lacking digestive secretions.	The amounts generally used are too small to contribute much in the way of nutrients.	Papain has also been used to heal festering sores by digesting rotten tissues.
Peanut butter, natural	Unhydrogenated peanut butter in which the oil separates and floats to the top of the jar.	Same as for hydrogenated peanut butter, except that in this case the oil may be skimmed off and used separately.	Similar to that for hydrogenated peanut butter, except that the natural oil is richer in essential fatty acids than the hydrogenated oil.	Natural peanut butter is now sold in many supermarkets.
Peanut flour	More of a meal than a true flour. It is made from finely ground peanuts.	In baked products (when mixed with wheat flour).	The proteins of peanut flour complement those of wheat flour.	The calorie and fat contents are variable. Flours may be low, medium, or high fat.
Pectin	A dry powder or a water solution of a carbohydrate gum used to make gels.	Preparation of jellies and other types of gels. In an antidiarrheal remedy.	Little or no nutritional value, but used medicinally to lower blood cholesterol.	Citrus pectin appears to be much more effective than apple pectin in lowering cholesterol.
Pollen	Male reproductive cells of flowering plants that are sold as granules or in tablets.	A nutritional supplement (according to some who promote its use).	Uncertain. However, it is said to be useful in desensitizing allergic people. Also, it is reported to be good for general health and vitality.	The claims for beneficial effects of pollen have *not* been confirmed in scientific tests.

(Continued)

TABLE H-1 (*Continued*)

Name of Food	Type of Product(s)	Uses	Reported Values and Claims	Comments
Potato flour	Flour made from cooked, dried potatoes.	In baked goods, and as a thickener for gravies.	Good source of energy. Dried potato flour contains 8% protein, which is about the same as corn and rice; and potatoes supply lysine, the amino acid which is lacking in cereal grains. Potato flour may be substituted for wheat flour for those allergic to the latter.	Can be used to make instant mashed potatoes.
Protein supplements	Candylike bars, liquids, powders, and tablets made from high-protein derivatives of eggs, milk, and soybeans.	For protein-deficient people, muscle builders, and patients recovering from burns, cancer, injuries, and surgery.	High protein intakes during periods of great stress may help to prevent the loss of proteins from the body tissues.	An excessively high protein intake may lead to a large urinary loss of calcium and other undesirable effects.
Psyllium seeds	Ground refined or unrefined seeds of *Plantago ovata* in granule or powder form.	As a source of dietary fiber and a laxative.	Little or no nutritive value, but binds bile salts and lowers the blood cholesterol.	People with easily irritated bowels should use only the refined psyllium products.
Pumpkin seeds	Dried, salted or unsalted, toasted or raw seeds of pumpkins or squashes.	Snack food. Nutritional supplement.	Rich source of calories, fats, proteins, minerals, and vitamins.	The prehistoric Indians of the Americas grew pumpkins and squashes for the seeds.
Rice, brown	Grains of rice with the bran layer intact.	Cooked cereals, desserts, ingredient of entrees, soups, and stews.	A moderately good source of calories and protein, and a good source of minerals and B vitamins.	Asians have consumed brown rice for thousands of years.
Rice flours	Flours produced by milling grains of raw or parboiled brown, glutinous, or white rice.	Cookies, infant cereals, low-gluten baked goods. Thickener for sauces, soups, and stews.	The same as that of the type of rice used to make the flour.	Most infants, invalids, and sufferers from allergies tolerate rice flour well.
Rice, glutinous (sweet rice)	Grains or flour of a variety of rice that becomes sticky (glutinous) when cooked.	Desserts and fillings for pastries.	About the same as ordinary rice.	Available in Chinese markets and some food stores.
Rice polish	Tiny flakes of the inner bran layers of the rice grain that are obtained by milling. Also available as an extract or syrup.	Ingredient of infant cereals, and special cakes (to which it imparts tenderness and moistness). Nutritional supplement.	Easily digested, low in fiber, rich in thiamin, and high in niacin.	This item is a well documented "health food" where thiamin is needed because it was first used to cure beriberi in Asia.

(Continued)

TABLE H-1 (*Continued*)

Name of Food	Type of Product(s)	Uses	Reported Values and Claims	Comments
Rose hips	The bulb-shaped fleshy fruits that remain after the rose petals fall off. Available in dried pieces, powders, syrups, and tablets.	Preparation of fruit soups (a European dish), jams, jellies, syrups, and teas.	A very rich source of vitamin C and bioflavonoids.	Used as a source of vitamin C in Great Britain during World War II.
Royal jelly	Substance produced by worker bees for feeding the queen. (Available in capsules).	Nutritional supplement. Ingredient of certain expensive cosmetics.	Very rich in certain B vitamins such as pantothenic acid. Some claim that royal jelly aids sexual rejuvenation.	An expensive item that has been promoted with many extravagant claims. The claim of sexual rejunenation is an old wives' tale. Royal jelly should be left for the bees, for which it is intended.
Rye flours	Flours that vary from light to dark that are milled from rye grains at various stages of refinement.	Alone or in mixtures with other flours in baked products.	Dark rye flours are less refined and more nutritious than light rye flours.	Rye flours are low in gluten and produce heavier, moister breads than wheat flours.
Rye grain	Whole, cut, or rolled grains.	Cooked cereal, special breads for people allergic to wheat.	Similar to wheat, but a little higher in proteins, fiber, minerals, and B vitamins.	Rye has long been an important cereal in Northern Europe, where wheat grows poorly.
Safflower oil	Oil from the seeds of a hardy, thistlelike plant.	Same as other vegetable oils---frying, salads, shortenings, etc.	Rich in essential polyunsaturated fatty acids.	Excessive consumption of vegetable oils may promote the formation of gallstones.
Salt-free foods	Items made without added salt.	For people who are on low-sodium diets.	About the same as the salted counterparts.	Some items may contain sources of sodium other than salt (Hence, labels should be read carefully.)
Sea salt	Salt produced by the evaporation of sea water.	Seasoning and source of certain trace minerals.	Contains more trace minerals than common salt.	The form of salt used by ancient peoples.
Seawater	Water from the sea.	As drinking water.	Good source of iodine and fluorine.	Seawater is abundant and cheap. Another great advantage, shared by few manufacturers: The raw material is also the finished product!
Sesame seeds	Tiny hulled or unhulled seeds of the sesame plant.	Baked products, candies (such as halvah), desserts, meat and poultry dishes, and salads.	Good source of calories, essential fatty acids, proteins, minerals, and B vitamins.	Taste best when lightly browned or toasted.

(Continued)

TABLE H-1 (*Continued*)

Name of Food	Type of Product(s)	Uses	Reported Values and Claims	Comments
Sesame tahini	A creamy spread made by blending or grinding sesame seeds.	Alone or mixed with various sauces or other spreads.	About the same as that for sesame seeds, unless mixed with other items.	Should be kept refrigerated to prevent the development of rancidity.
Sorghum syrup	The syrup produced by evaporation of the sap expressed from sorghum stalks.	Same as other syrups and sweeteners. Sometimes, it is fermented to produce alcohol.	Rich in calories and sugar, and a good source of minerals.	Has a milder flavor than either blackstrap or Barbados molasses.
Soy flour	Flours made from heat treated soybean meals of varying fat contents.	Protein supplement, extender in meat products, and ingredient of baked products.	Caloric content depends upon fat content. Excellent source of protein, minerals, and vitamins.	Full-fat, medium-fat, low-fat, and "extra-lecithin" soy flours are available.
Soy milk powder	Finely ground soy powder of low to medium fat content.	Preparation of soy "milks" and "cheeses," and fermented products.	Same as those of a soy flour having the same fat content.	Long used by the Chinese for the preparation of a milk substitute.
Soy protein powder	A concentrated source of protein (85 to 98%) produced by removal of most of the carbohydrates from a soybean flour or meal.	Protein supplement. Preparation of extenders or substitutes for fish, meat, milk, poultry, and seafood.	Very rich in protein. The levels of the other nutrients depend upon the extraction processes used.	The consumption of excessive amounts of protein may promote high urinary losses of calcium and other undesirable effects.
Spirullina	Powder or tablets made from a dried blue-green algae grown in Mexico and Africa.	Food and nutritional supplement, sold at very high prices in the U.S.	Good source of protein, fiber, minerals, and vitamins. (One of the few wholly vegetarian foods that contains vitamin B-12.)	Used for centuries by the African tribes around Lake Niger. Recently promoted as an aid to weight reduction.
Sprouts	Sprouted (germinated) seeds prepared by soaking them in water in the dark.	In salads, soups, stews, and stir-fried dishes.	Low in calories, but a fair to good source of proteins, fiber, minerals, and vitamins.	The exposure of newly emerged sprouts to sunlight develops the chlorophyll content.
Sugar, raw (Brown sugar)	A mixture of sugar crystals coated with a film of molasses.	Same as for white sugar, except that the taste of raw sugar is a little more appealing.	Closer in composition and nutritional value to white sugar than to molasses. Raw sugar is promoted on the basis of its mineral content, especially iron.	Raw sugar contains little iron now that it is processed in stainless steel and aluminum vessels. Besides, it is eaten in small servings. An adult woman would have to eat 2 1/2 cups (*600 ml*) of brown sugar daily in order to meet her requirement for iron. So, brown sugar should be eaten for reason of taste, rather than for any health properties.

(Continued)

TABLE H-1 (*Continued*)

Name of Food	Type of Product(s)	Uses	Reported Values and Claims	Comments
Tamari	A soy sauce made by water extraction of a fermented mixture of soybeans, grains, water, and salt.	As a flavoring for Oriental dishes such as chop suey, chow mein, sprouts, and stir-fried mixtures.	High in sodium content. Hence, only small amounts should be used, making the nutritional contribution small.	Some products contain added monosodium glutamate.
Tapioca	Starchy globules produced by heating a paste made from cassava root.	Foods for infants and invalids, and starchy puddings.	Provides mainly calories and starch. Easily digested by infants and invalids.	In Latin America, the name tapioca is given to various cassava products.
Tapioca flour	Finely ground cassava root starch.	Same as those for other edible starches.	Same as for tapioca.	Hard to find in the U.S.
Textured vegetable product (TVP)	Extracted soy protein that has been made into granules or spun into chewy pieces that have a texture like cooked animal tissues.	Extender or substitute for fish, meats, poultry, or seafood.	Moderately high in calories and high in protein. (The levels of the other nutrients vary with the type of product.)	Some products have a rather insipid taste and a dry crumbly texture. (They should be mixed with other more tasty and juicy ingredients.)
Tofu	A mild-flavored, cheeselike type of soybean curd.	As a cheese extender or substitute; and in desserts, salads, soups, and vegetarian dishes.	High in protein and the B vitamins.	Now sold in the produce section of many supermarkets.
Torula yeast (Also see YEAST.)	Species of yeast grown specially for food use. (Available in powder or tablet form.)	Ingredient of fortified fruit or milk drinks, gravies, sauces, etc.; and nutritional supplement.	Contains from 50 to 62% protein, and is a good source of minerals and B vitamins.	More palatable than brewers' yeast. Some brands contain added nutrients.
Vegetable salt	Sea salt flavored with powdered dried vegetables.	Seasoning for various dishes.	Provides mainly salt and some trace minerals.	Excessive use of any form of salt may be detrimental to good health.
Vegetarian gelatin	Any of several vegetable gelling agents such as agar-agar and carrageenan.	In place of gelatin in the preparation of gelled desserts, salads, and soups.	Much of the carbohydrate in these items is not utilized but they furnish some essential minerals.	Used mainly by vegetarians who do not use animal gelatin.
Wheat, cracked	Cracked grains of wheat.	Baked products, cooked cereals, desserts, filler in meat and poultry dishes, salads, soups, and stews.	Good source of calories, carbohydrates, proteins, fiber, minerals, and B vitamins.	Cooks more rapidly than whole grains of wheat.
Wheat germ	The embryo of the wheat kernel.	As a ready-to-eat breakfast cereal; or in baked goods; breading for fish, meats, or poultry; casseroles; cooked cereals; desserts; salads; soups; and stews.	Good source of calories, essential fatty acids, proteins, fiber, minerals, vitamin E, and the B vitamins.	Should be kept refrigerated to prevent the development of rancidity. Light toasting improves the flavor.

(*Continued*)

TABLE H-1 (*Continued*)

Name of Food	Type of Product(s)	Uses	Reported Values and Claims	Comments
Wheat germ flour	Finely ground wheat germ.	Preparation of baked products.	Same as for wheat germ.	Should be stored in a refrigerator.
Wheat germ oil	Oil extracted from wheat germ.	Salad oil, and nutritional supplement.	Primarily as a rich vitamin E supplement.	Heating (as in frying) may destroy some of the vitamin E.
Whey powder	A powder obtained by drying the whey from cheese production.	In baked goods, desserts, fruit or milk drinks, and soups; and a nutritional supplement.	Rich in milk sugar (lactose), protein; and a fair to good source of calcium, phosphorus, and B vitamins.	Much of the whey produced in the U.S. goes to waste because of lack of demand.
Whole wheat berries	Whole grains of wheat.	Cereal, desserts, puddings, and stuffings.	Good source of calories, carbohydrates, proteins, fiber, minerals, and B vitamins. Also a fair source of vitamin E.	The long cooking time may be shortened by using a pressure cooker.
Yeast, Instant (Also see YEAST.)	Powdered yeast treated to make it blend readily with various liquids.	Nutritionally fortified drinks for undernourished people.	Rich source of proteins, nucleic acids, and B vitamins.	Powdered yeast has long been used as a protein and vitamin supplement.
Yeast, Nutritional (Also see YEAST.)	Powdered or tableted yeast grown specially for use as a food.	Nutritional supplement or tonic for undernourished or debilitated people.	Very rich in proteins, nucleic acids, essential minerals, and B vitamins.	Has a milder flavor and a higher protein content than brewers' yeast.
Yogurt	Mildly fermented milk product (milk sugar converted to lactic acid by bacteria).	Dessert, entree, salad dressing, or snack. May be added to baked goods, in which it reacts with baking soda to leaven them.	Similar to milk, except that *Lactobacilli* (if alive) prevent the growth of certain undesirable microorganisms in the intestines.	A dairy food that has had a recent rapid growth in popularity. Although yogurt does not have any magical health-imparting properties, it is a nutritious food.

HEARTBURN

This condition is characterized by a burning and painful sensation under the sternum (breastbone), sometimes accompanied by a sour regurgitation (bringing up of gas and small bits of food from the stomach). The most common response to these symptoms is to take antacids to relieve what is thought to be hyperacidity. It is better, however, not to rely solely on these remedies, but to try to discover the cause(s) of these symptoms.

CAUSES OF HEARTBURN. Although the pain experienced in heartburn is sometimes due to a peptic ulcer, in many cases there is some type of irritation or spasm of the esophagus. Two of the most common esophageal disorders follow.

• **Esophageal reflux (esophagitis)**—This type of irritation, also known as esophagitis, results from regurgitation of stomach acid into the esophagus. Such a situation may result from any of the following: (1) pressure on the stomach, resulting in its upward displacement; (2) the posture of the victim (it happens more readily in a prone position); (3) relaxation of the lower esophageal sphincter (where the esophagus and stomach join) due to nervousness; (4) weakness of the sphincter; or (5) stomach surgery. Hyperacidity increases the amount of irritation.

• **Hiatus hernia**—Many persons must avoid conditions which force the upper part of the stomach up through the opening (hiatus) in the diaphragm; among them, pressure on the stomach by tight clothing or adjoining tissue (as in obesity), lying down when the stomach is full, distention of the stomach by large meals, chronic coughing or sneezing, and straining at stool. Esophageal reflux is more likely to occur along with hiatus hernia when the shape of the flap valve between the esophagus and stomach is distorted.

(Also see HIATUS HERNIA.)

TREATMENT AND PREVENTION OF HEARTBURN. It is wise to take prompt action in treating and preventing heartburn since chronic irritation of the esophagus may lead to the development of ulcers and formation of scar tissue which may narrow the opening of the tube and make swallowing difficult. Also, some people develop a pattern of habitual regurgitation and hiatal herniation, which may persist for many years without serious complications, and which they regularly treat with large doses of antacids and a restricted diet. Some suggestions for more rational treatment of this condition are as follows:

1. After strenuous exercise, first rest for a while, then eat (it appears that fatigue is frequently accompanied by digestive problems).

2. Avoid taking large volumes of food and drink at a single meal. Persons who have chronic heartburn might consider eating 5 to 6 small meals or snacks per day instead of 2 or 3 large meals. It may also be wise not to drink much fluid with meals, but to take fluids between meals.

3. Develop a relaxed frame of mind at meals and avoid bolting of food. Sometimes a glass of wine before or with a meal is helpful.

4. Chilling of the body immediately before or after meals should be avoided. Likewise, excessively hot or cold food and beverages should be allowed to reach more moderate temperatures before their ingestion.

5. Limit the amount of foods and drinks known to cause irritation or stimulation of acid secretion, such as strong alcoholic beverages (like whiskey), coffee and tea, acid foods, and spices like pepper, salt, and monosodium glutamate.

6. Prevent pressure on the abdomen during and after meals by loosening clothing and sitting upright or slightly reclined (not hunched over). Reduction of obesity has helped some persons.

7. Treat continued burning with small doses of mild antacids. Liquids are preferable to tablets, as the latter may not fully disintegrate until they reach the stomach. Chewing and swallowing tablets with water may help to soothe the esophagus.

8. Plan meals so as to avoid eating just prior to lying down. Elevating the head of the bed and/or lying on the left side also help since food may remain in the stomach for several hours after eating.

NOTE: Sufferers from chronic heartburn should consult a physician for diagnosis of the cause(s). They might then inquire as to the advisability of using antacids. (Indiscreet use of these compounds may result in other digestive disorders such as malabsorption.)

(Also see ANTACIDS; DISEASES; HIATUS HERNIA; and ULCERS, PEPTIC.)

HEART DISEASE

Contents *Page*

Heart and Blood Vessel Disorders 1134
Angina Pectoris 1134
Atherosclerosis 1134
Coronary Occlusion 1135
Heart Failure 1135
Myocardial Infarction 1135
Irregular Heartbeats 1135
Factors Affecting Cardiovascular Diseases 1135
Ethnic Groups 1135
The Framingham Study of Risk Factors 1137
Dietary Components and Heart Disease 1139
Alcoholic Beverages 1140
Caffeine-Containing Beverages 1141
Carbohydrates 1141
Energy 1141
Fatty Substances 1141
Fiber 1143
Macrominerals 1143
Microminerals (Trace Elements) 1144
Vitamins and Vitaminlike Factors 1145
Other Risk Factors Associated with Heart Disease 1147
Atherosclerosis and Hardening of Arteries 1147
Allergic Reactions 1148
Blood Abnormalities Other Than High Cholesterol 1148
Body Builds 1149
Chronic Constipation 1149
Climate and Weather 1149
Congenital Defects 1149
Crash Dieting 1149
Drugs 1149
Emotional Stresses 1151
Glandular or Hormonal Disorders 1151
Heavy Meals 1152
Heredity 1152
Injuries to the Blood Vessels and/or the Heart 1152
Kidney Diseases 1152
Living in Densely Populated Areas 1152
Medical Treatments 1153
Overly Aggressive (Type A) Behavior 1153
Physical Stresses 1153
Soft Drinking Water 1153
Stickiness of Blood Platelets 1153
Toxic Substances 1154
Signs and Symptoms of Heart Disease 1155
Easily Recognizable Abnormalities 1155
Detection of Cardiovascular Disorders by Physical Examination 1156
Special Diagnostic Procedures 1156
Treatment and Prevention of Heart Disease 1157
Emergency Care for Victims of Heart Attacks 1157
Cardiac Arrest 1157
Heart Attack Without Cardiac Arrest 1158
Angina Pectoris 1158
Heart Failure 1158
Long-Term Treatments 1158
Biofeedback 1158
Changing of Behavior 1159
Dietary Measures 1159
Donations of Blood 1161
Drugs 1162
Exercise 1163
Surgery 1164
Summary 1164

This term refers to a wide variety of disorders, ranging from congenital defects in the valves of the heart to degeneration of the heart muscle. However, the mass media and the public tend to identify heart disease with what is medically known as *coronary heart disease*—the various conditions that are associated with blockage of the coronary artery, which delivers blood to the heart muscle. Coronary heart disease is responsible for about 80% of the one million deaths from diseases of the heart and blood vessels (cardiovascular diseases) which occur in the United States each year.[1] Another name for coronary heart disease is *ischemic heart disease*. (The term "ischemia" means a drastic reduction in the flow of blood to a tissue.

Lack of blood in the heart muscle may lead to a *heart attack*—an event usually characterized by the sudden onset of severe chest pain which may spread to nearby points of the body, although some people may suffer so-called "silent heart attacks," which are not noticed at the time they occur, but which leave evidence of damage to the heart muscle. Some characteristic features of a heart attack are depicted in Fig. H-5.

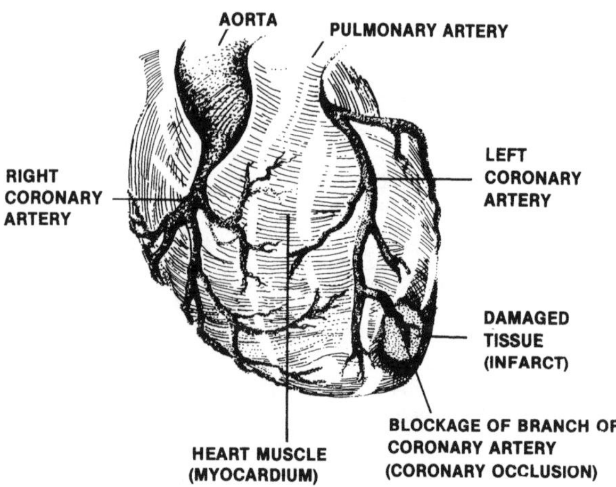

Fig. H-5. Some characteristic features of a heart attack. Interference with the flow of blood to the heart muscle is usually accompanied by such symptoms as (1) severe pain in the chest and surrounding areas, and (2) shortness of breath.

The three potentially modifiable heart disease risk factors of greatest concern are: cigarette smoking, high blood pressure, and high serum cholesterol.

[1]DeBakey, M., and A. Gotto, *The Living Heart*, David McKay Company, Inc., New York, N.Y., 1977, p. 219.

Details of the major cardiovascular disorders are given in the section that follows.

HEART AND BLOOD VESSEL DISORDERS.

Many of the ideas about heart disease which have been popularized by the mass media are limited to the dramatic aspects of blood clots, cardiac arrest, clogging of blood vessels with plaques, excruciating pain, and sudden death. However, these aspects represent only the later stages of the disease. Detection and treatment of the early stages may forestall the later stages for a long time. Hence, it is worthwhile to consider the various aspects of the cardiovascular disorders that follow.

Angina Pectoris. This is the Latin term for an intense pain in the chest. It usually results from an insufficient flow of blood to the heart muscle for meeting the demands which are imposed. Often, the pain occurs during emotional excitement or exercise, and subsides when the stress factor is removed. However, similar pain felt in the chest may also originate from the upper part of the digestive tract in such disorders as gallstones, hiatal hernia, ulcers of the stomach or duodenum; or when the small intestine is distended with fluid, food, or gas.

Lying down may aggravate the pain from a diseased heart because (1) more blood than usual flows into the chambers of the heart, with the result that greater muscular force is needed to expel it; or (2) a full stomach may press up against the heart, cutting off some of its blood supply. Therefore, the fastest relief from angina may usually be obtained by remaining still in either a sitting or a standing position.

Angina may be a warning of an impending heart attack, so it should lead one to obtain prompt diagnosis and treatment.

Atherosclerosis.

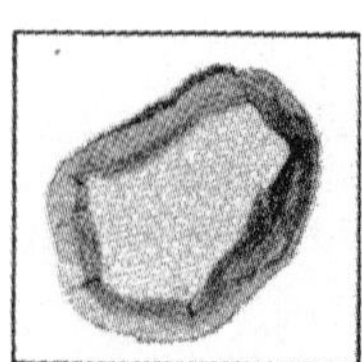

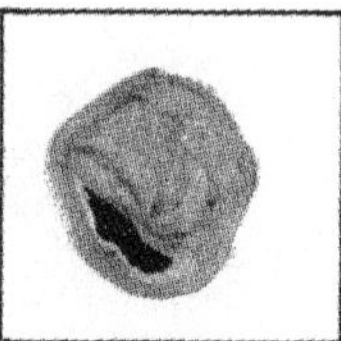

NORMAL ARTERY — FATTY DEPOSITS IN VESSEL WALL — PLUGGED ARTERY WITH FATTY DEPOSITS AND CLOT

ACTUAL PHOTOGRAPH OF THE DEVELOPMENT OF ATHEROSCLEROSIS

Fig. H-6. The development of atherosclerosis in an artery. (Reprinted with permission, American Heart Association, Dallas, Tex.)

Although this degenerative disease, which is characterized by the buildup of abnormal patches (plaques) on the walls of arteries, is thought to be the main cause of heart attacks, autopsies sometimes indicate that there were other causes of fatal attacks. Furthermore, the plugging up of certain branches of the coronary artery by atherosclerotic plaques is sometimes partly compensated for by enlargement of the smaller arteries nearby so that just enough blood gets to the heart muscle. This compensatory effect is commonly called "the development of a collateral circulation."

Coronary Occlusion (Coronary Thrombosis). If a coronary artery is partially blocked by a blood clot, it is commonly called a coronary thrombosis. If the blood clot completely blocks the blood flow, it is called a coronary occlusion.

Occlusion, or blocking, of the coronary circulation is usually caused by extensive atherosclerosis and/or a blood clot lodged in an artery. Occlusion is likely to occur in the coronary system before it does in other major arteries because, during the heartbeat, there is a moment when the blood literally "stands still" in these arteries. This condition occurs because each contraction of the heart muscle squeezes the coronary arteries and temporarily stops the forward flow of blood in them. The flow of blood in these arteries resumes when the muscle relaxes between contractions. Slowing of the movement of blood favors the development of such obstructions as atherosclerotic deposits and clots, when other conditions are also favorable for these developments.

Heart Failure. The heart may fail to pump sufficient blood to various tissues when there are conditions such as those which follow:

1. The heart muscle is weakened by chronic overwork, infectious disease, nutrient deficiencies, or a reduction in its blood supply.
2. There is damage to the heart valves which regulate the flow of blood.
3. The transmission of the heartbeat through the heart muscle is abnormal.
4. The beating of the heart muscle fibers is not coordinated in a regular rhythm.
5. The demands on the heart to pump blood are stepped up by such factors as emotional excitement, high blood pressure, strenuous exercise, an expansion of the volume of blood in circulation resulting from the enlargement (dilation) of small blood vessels, or a speeding up of metabolism such as occurs during a fever, or when the thyroid gland is overactive.
6. There is an increased resistance to the flow of blood because the openings in the blood vessels are narrowed or obstructed, the blood itself is thickened, or there is an accumulation of excessive fluid in various tissues.

Myocardial Infarction. When the coronary artery or one of its branches is obstructed (occluded), tissue may be damaged if there are areas of the heart muscle which do not receive sufficient blood. These damaged areas are called *infarcts*, and the coronary occlusion is simply called a "coronary." Usually, a myocardial infarction is accompanied by severe pain which does not subside upon resting, although a few people might have "silent" attacks where there is no noticeable pain. The presence of infarcts may often be confirmed by an electrocardiogram or by elevations in the blood levels of enzymes which leak out of the damaged heart muscle.

Irregular Heartbeats. Normally, the beating of the individual fibers in the heart muscle is coordinated so that the muscle contractions are regular and smooth. However, the normal rhythm of the heartbeat may be lost when the heart is sorely taxed and there may be a rapid and irregular twitching of the individual muscle fibers, which is called *fibrillation*. When this occurs in the muscle around the large chambers (ventricles) of the heart, the pumping action may stop, producing what is commonly called "cardiac arrest." Death usually occurs within a few minutes, unless the fibrillation is corrected. Some people who have survived longer periods of fibrillation have had severe brain damage due to lack of blood. Disorders of heart rhythm are believed to be the main cause of sudden death due to cardiovascular disease in the United States.

FACTORS AFFECTING CARDIOVASCULAR DISEASES. Right after World War II, many doctors in the developed countries around the world reported that there had been significant declines in the number of cases of diabetes and heart disease during the war, compared to the prewar period. It seemed logical to suspect that rationing and/or shortages of foods might have had something to do with the decline because the disease rates rose when more food became available after the war ended. For more than four decades since then, there have been countless studies of both animals and humans which were aimed at identifying the specific dietary and other factors responsible for cardiovascular diseases. Details follow.

Ethnic Groups. The first large-scale studies found that the countries with high rates of the disease were characterized by diets rich in butter, cheese, cream, and meat, and in highly refined sugars. However, these diets and the corresponding rates of heart disease were also associated with other factors which characterize the developed countries, such as industrialization, mechanization of production, processing of foods, large numbers of obese people, and sedentary life-styles. Other researchers are quick to point out that certain peoples who eat plenty of animals fats have exceptionally low rates of heart disease. Fig. H-7 depicts three typical peoples:

Fig. H-7. Three long-lived peoples:
a. Eskimos who live in or near the Arctic regions of North America.
b. Georgians from the Caucasus Mountains of the Soviet Union.
c. Masai herdsmen who live on the high plains of East Africa.

The Masai, a tribe of nomadic herdsmen in East Africa, have low death rates from cardiovascular diseases, even though their diet is rich in fat from blood, meat, and milk. (They regularly draw blood from the jugular veins of their cattle.)[2] Before one jumps to conclusions regarding their diet, it should be noted that these people are quite different from those in developed countries in that (1) they are much more slender (the average man is 5 ft 8 in [*1.7 m*] tall, but only weighs about 135 lb [*61 kg*]; (2) their bodies compensate for diets high in cholesterol by cutting down on the internal production of this substance (this compensation is *not* as well developed in most other peoples); (3) their life-style is obviously more vigorous

[2]Mann, G. V., *et al.*, "Cardiovascular Disease in the Masai," *Journal of Atherosclerosis Research*, Vol. 4, 1964, p. 289.

since they eat about 3,000 Calories (kcal) per day, yet they remain slender; and (4) there is no attempt to practice sanitary measures, so there are widespread parasitic infestations and infectious diseases such as malaria and tuberculosis.[3]

It is noteworthy that Dr. Barnes, an American doctor who has studied 70,000 reports of autopsies which were made between 1930 and 1970, found that when death rates from tuberculosis rose within such groups as Europeans and native Africans, the number of fatal heart attacks dropped, and vice versa.[4] Dr. Barnes and his colleagues suggest that the marked decline in deaths from heart disease in war-torn countries in World War II might have been due to the sharp rise in tuberculosis fatalities, rather than to the reduced consumption of cholesterol-containing foods.

Another group of native Africans, the Bantu of South Africa, who were originally farmers, but who now often work in the cities or in the mines, once had a low rate of heart disease and a high rate of tuberculosis. However, they are now succumbing in increasing numbers to cardiovascular diseases as the death rate from TB continues to decline.[5]

It is well known that the people in the Georgian region of the Soviet Union are among the most long-lived in the world, yet they consume plenty of dairy products, meats, and sweets—foods which are often blamed for the high rates of heart attacks in other societies.[6] These people have heart attacks like other people, but it seems that the attacks are mostly of the "silent' type which do not handicap them because their daily exercise is so great that their heart muscles are much stronger than those of urbanized peoples. Furthermore, the Georgians eat large amounts of fresh vegetables; hence, their total diet may be very good.

Finally, it has often been said that the Eskimos who still live under primitive conditions are an outstanding contradiction of the theory that animal fats cause heart disease, since their diet consists mainly of animal flesh and fish. Also, it is well known that when these people move to settlements near trading posts they adopt new eating patterns and develop such degenerative diseases as atherosclerosis, diabetes, and gallstones.[7] These conditions are rarely found in Eskimos who live in their traditional manner, so it seems that their recent adoption of diets containing large quantities of sugar may be the main disease-causing factor. This assumption is evidenced by (1) the rampant tooth decay in settlement Eskimos, and (2) their abnormally high blood sugar levels when they are given a test dose of glucose. These responses are closer to normal when the Eskimos are fed a meat meal about 90 minutes before administration of the tests. However, the largely carnivorous diets of the nomadic Eskimos are much higher in polyunsaturated fats than those of other societies that depend upon hunting animals for food. The reasons for this are (1) their major foods are fish and fish-eating animals such as polar bears, seals, and walruses; and (2) the fat laid down by animals in cold environments is more polyunsaturated than that of animals living in warmer climates.

Findings from studies such as those cited above neither prove nor disprove that dietary fat is a cause of heart disease. Rather, they show that it is likely that this type of disease is the end result of many different dietary and nondietary conditions, which may not be obvious to those investigating only a few such conditions.

Now, with the aid of electronic computers, it is possible to study many different factors at the same time; the only limitations being the amounts of energy, money, and time available for data collection. Consequently, some recent investigations have provided new insights concerning various risk factors for cardiovascular diseases. Details follow.

The Framingham Study Of Risk Factors. This well-known study was conducted from 1949 to 1969 on over 5,000 men and women, aged 30 to 62, in Framingham, Massachusetts, which is about 20 miles southwest of Boston. The subjects were free of clinical signs of coronary heart disease at the beginning of the study, and they were examined once every other year for signs of the disorders which often accompany this disease. Data obtained in the study were used to estimate the risks for the development of heart disease by men and women who had certain characteristics.

It is noteworthy that, although high blood cholesterol was found to add to the risk of heart disease, no direct relationships were found to exist between the amounts and types of foods consumed by the study participants and their blood levels of cholesterol. However, it was found that blood cholesterol steadily increased in the men who gained weight while the study was underway.[8]

Graphical representations of the effects of some of the important risk factors for middle-aged men and women are given in Fig. H-8.

The graphs for each combination of the risk factors shown in Fig. H-8 were plotted separately for normal, moderately elevated, and high levels of both blood cholesterol and blood pressure because increases in these factors were consistently associated with increased risks of heart disease in the people who participated in the Framingham Study.

[3]Biss, K., *et al.*, "Some Unique Biologic Characteristics of the Masai of East Africa," *The New England Journal of Medicine*, Vol. 284, 1973, p. 694.

[4]Barnes, B. O., M. Ratzenhofer, and R. Gisi, "The Role of Natural Consequences in the Changing Death Patterns," *Journal of the American Geriatrics Society*, Vol. 22, 1974, p. 176.

[5]Laurie, W., J. D. Woods, and G. Roach, "Coronary Heart Disease in the South African Bantu," *American Journal of Cardiology*, Vol. 5, 1960, p. 48.

[6]Leaf, A., "Search for the Oldest People," *National Geographic*, Vol. 143, 1973, p. 93.

[7]Schaefer, O., "When the Eskimo Comes to Town," *Nutrition Today*, Vol. 6, November/December 1971, pp. 8-15.

[8]Kannel, W. B., "The Disease of Living," *Nutrition Today*, Vol. 6, No. 3, 1971, pp. 2-11.

Risk Factors for the Development of Coronary Heart Disease (CHD) by 45-Year Old Man [1]

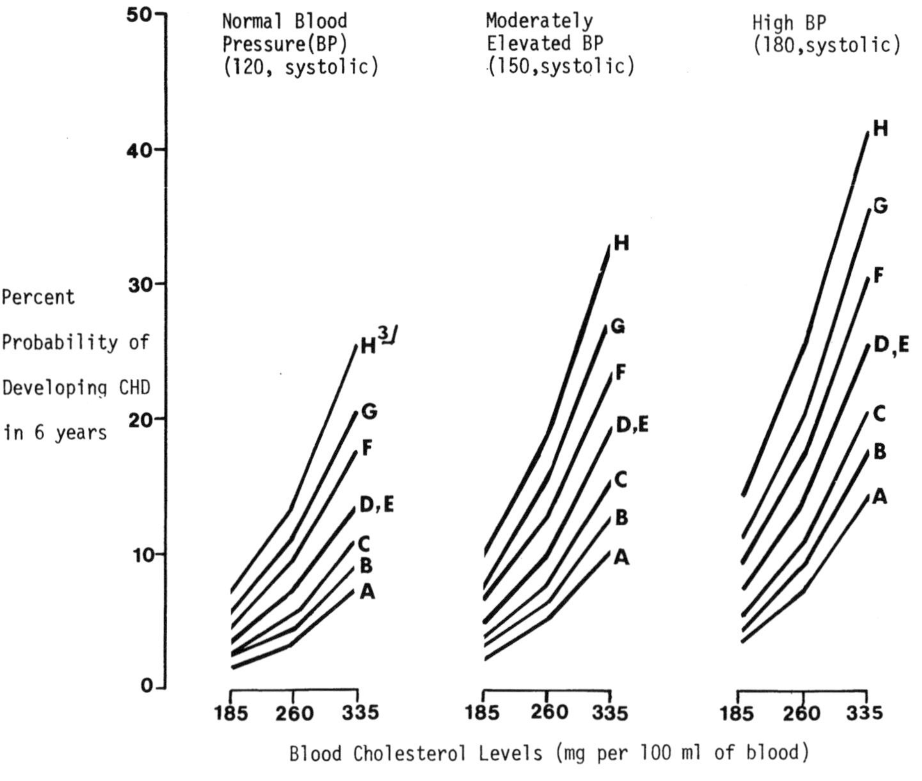

Risk Factors for the Development of Coronary Heart Disease (CHD) by 45-Year Old Woman [2]

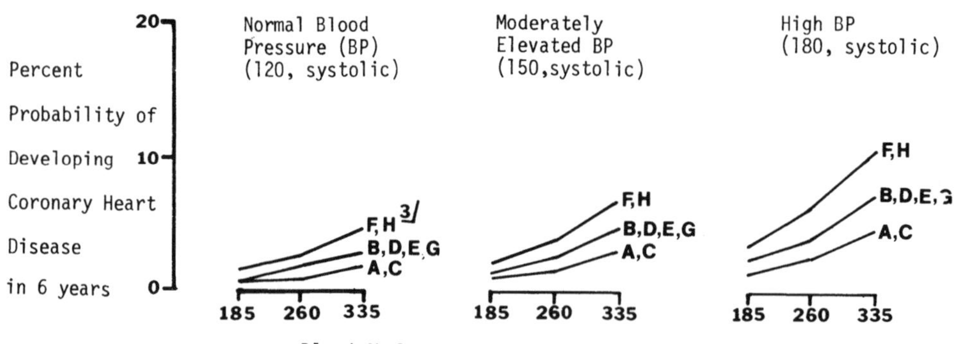

[1] Data for plotting the graphs was obtained from Coronary Risk Handbook, American Heart Association, New York, N.Y., 1973, pp. 12 and 13.

[2] Ibid., pp. 22 and 23.

[3] Capital letter designates risk factor classification as follows:

A. Nonsmoker with normal electrocardiogram (ECG) and normal glucose tolerance (GT)
B. Nonsmoker with normal ECG and abnormal GT
C. Smoker with normal ECG and normal GT
D. Smoker with normal ECG and abnormal GT
E. Nonsmoker with abnormal ECG and normal GT
F. Nonsmoker with abnormal ECG and abnormal GT
G. Smoker with abnormal ECG and normal GT
H. Smoker with abnormal ECG and abnormal GT

Fig. H-8. The combined effects of risk factors for coronary heart disease. (Data for plotting the graphs was obtained from *Coronary Risk Handbook*, American Heart Association, New York, N.Y.)

Discussions of individual and combined risk factors follow:

• **Abnormal electrocardiograms (ECG) and/or enlargement of the heart**—While abnormalities in an ECG may not always indicate heart disease, the data from this study showed that ECG abnormalities usually indicated a high risk of heart disease.

Although signs of heart enlargement may often be found in athletes who regularly participate in competitive events requiring exceptional endurance, this enlargement might also indicate chronic strain of the heart muscle due to obesity, poor nutrition, rapid heart rate, reduction in the supply of blood resulting from partial coronary occlusion, or weakness of the muscle itself.

• **Elevated blood cholesterol**—This condition alone contributed only a fraction of the risk compared to that which existed when it was accompanied by such other conditions as an abnormal ECG, high blood sugar, high blood pressure, and moderate to heavy smoking. However, even though the risk due to high blood cholesterol alone was small, a blood value of 335 mg/100 ml of blood conferred about three times the risk of a value of 185.

• **Glucose intolerance**—The investigators in the Framingham Study defined this abnormalitiy as one or more of the following: a blood sugar value of 120 or more, diabetes, or sugar in the urine. This condition, whether alone or accompanied by other risk factors, was always associated with a higher risk of heart disease. For example, the risk of angina pectoris or sudden death for obese diabetics was 2½ times as great as that for obese nondiabetics, and four times as great as that for nonobese nondiabetics.[9]

• **High blood pressure**—The chances of developing heart disease markedly increased as the blood pressure rose. For example, when other factors were the same, people with a systolic blood pressure (the highest reading) of 180 had about twice the risk as those who had a blood pressure of 120.

• **Lack of exercise**—Although this factor is not shown on Fig. H-8, it was found that active persons had only from ⅓ to ½ the risk of angina pectoris, myocardial infarction, or sudden death, as sedentary persons.[10]

[9]Kannel, W. B., G. Pearson, and P. M. McNamara, "Obesity as a Force of Morbidity and Mortality," *Adolescent Nutrition and Growth,* F. P. Heald, Editor, Appleton-Century-Crofts, New York, N.Y., 1969, p. 60.

[10]Kannel, W. B., "The Disease of Living," *Nutrition Today*, Vol. 6, 1971, p. 5.

• **Obesity**—People who were 20% or more heavier than their ideal weights had about double the risk of angina pectoris, triple the risk of sudden death in a heart attack, and about an equal risk of myocardial infarction as persons whose weights were ideal.[11]

• **Sex of subject**—When all of the other risk factors were similar, 45-year-old men were about three times as likely to develop heart disease as 45-year-old women.

• **Smoking**—Fig. H-8 shows that the risk of heart disease for smokers was consistently about 1½ times as great as that for nonsmokers, when the other risk factors are the same for both groups.

Like many other similar investigations, the Framingham Study had its limitations, and the participants did not necessarily represent a cross section of the people in the United States; so, it would not be wise to make sweeping generalizations concerning the applicability of its findings. Nevertheless, it showed that the risk of cardiovascular diseases increased with each added risk factor, which suggested that preventative measures should be directed towards systematic treatment of those factors that may be managed.

However, it was surprising that about ⅔ of the study participants who died from heart attacks died outside of hospitals. Furthermore, over ½ of the deaths were sudden; that is, they occurred within an hour of the first signs of a heart attack. Finally, about ⅓ of the people who died suddenly had shown no previous signs of heart disease. It seems that only some of the factors which might predispose people to this disease are commonly considered, while others may be overlooked. Therefore, discussions of many factors follow.

Dietary Components and Heart Disease. The American Heart Association has long asserted that most people in the United States might reduce their risk of developing cardiovascular diseases by (1) reducing their consumption of dietary sources of saturated fats and cholesterol such as beef, pork, eggs, butter, cheeses, and ice cream; and (2) substituting polyunsaturated vegetable oils for saturated animal fats. On the other hand, the opponents of these recommendations have argued that there has been a modest decline in heart disease mortality, even though the per capita consumption of beef and pork has recently reached an all-time high. Fig. H-9 shows by means of a bar graph how food consumption patterns in the United Stated differed between 1960 and 1977, while Table H-2 summarizes data on overall nutrient consumption, amounts of dietary energy, tobacco use, and the death rates from heart disease during these years.

[11]*Ibid.*

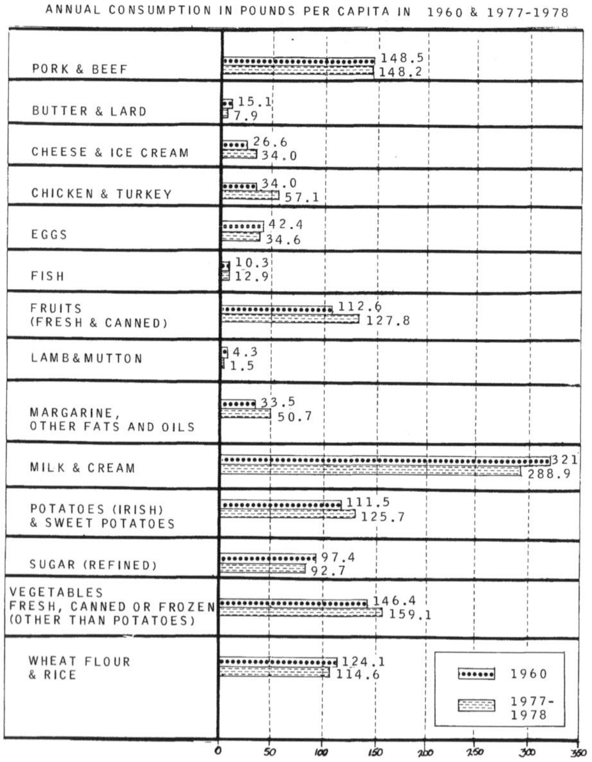

ANNUAL CONSUMPTION IN POUNDS PER CAPITA IN 1960 & 1977-1978

	1960	1977-1978
PORK & BEEF	148.5	148.2
BUTTER & LARD	15.1	7.9
CHEESE & ICE CREAM	26.6	34.0
CHICKEN & TURKEY	34.0	57.1
EGGS	42.4	34.6
FISH	10.3	12.9
FRUITS (FRESH & CANNED)	112.6	127.8
LAMB & MUTTON	4.3	1.5
MARGARINE, OTHER FATS AND OILS	33.5	50.7
MILK & CREAM	321	288.9
POTATOES (IRISH) & SWEET POTATOES	111.5	125.7
SUGAR (REFINED)	97.4	92.7
VEGETABLES FRESH, CANNED OR FROZEN (OTHER THAN POTATOES)	146.4	159.1
WHEAT FLOUR & RICE	124.1	114.6

0 50 100 150 200 250 300 350

Fig. H-9. Consumption of major food commodities in 1960 and in 1977-78. The graph was plotted from data given in *National Food Review*, Economics, Statistics, and Cooperative Service, USDA, April 1978, p. 53.

TABLE H-2
MAJOR DIETARY COMPONENTS, TOBACCO USE, AND MORTALITY FROM HEART DISEASE IN THE YEARS 1960 AND 1974[1]

	1960	1974	Change From 1960-1974
			(%)
Major dietary components:[2]			
Protein, per person per day ...... (g)	95	99	+4
Fat, per person per day ...... (g)	143	153	+7
Carbohydrate, per per person per day ...... (g)	375	376	0
Energy, per person per day ... (kcal)	3,130	3,240	+4
Consumption of tobacco products, per person per year .... (lb)	11.8	9.4	-20
Death rate from cardiovascular diseases, deaths per 100,000 population	515.1	478.2	-7

[1]*Statistical Abstract of the United States 1976*, pp. 65, 94, and 778.
[2]Nutrients available for civilian consumption. No deductions were made for loss or waste during preparation of meals, or for use as pet food.

It seems unlikely that the decline in heart disease mortality between 1960 and 1974 was due mainly to changes in diet because (1) the 20% decrease in the use of tobacco products might well have been responsible; (2) Table H-2 shows that the dietary energy consumption increased by 100 Calories (kcal) per day, which is not likely to be beneficial to a population which was overfed to begin with; and (3) the average weights of both men and women of all ages increased between 1960 and 1974.[12] Furthermore, such overall statistics tell little about individual consumption of the various items or about the amount of wastage; so one cannot make a case for any "cause-and-effect" relationships. Therefore, some of the current, but controversial, ideas about individual foods and nutrients merit consideration.

ALCOHOLIC BEVERAGES. There have been many debates on the merits and demerits of the use of various alcoholic beverages by persons suffering from heart disease, since some doctors have long believed that a shot of whiskey eases the pain of angina pectoris. However, a 3-day study of 12 men who had coronary artery disease, and who suffered from angina pectoris showed that both 2 oz (*60 ml*) and 5 oz (*150 ml*) respectively, of vodka reduced the length of time that the men could engage in exercise.[13] It might be that vodka, which is more purified than other alcoholic beverages, lacks certain nonalcoholic, but beneficial, components which may be present in such drinks as beer, wine, and whiskey. This possibility is suggested by the research studies of both American and European scientists who found that solutions of pure alcohol had different effects than various wines.[14] Other evidence for beneficial effects of the moderate use of alcoholic beverages was obtained in a study of the incidence of heart disease in over 7,000 Japanese men living in Hawaii.[15] The men who drank the equivalent in alcohol content of a bottle of beer or of two shots of whiskey per day had only about one-half the rate of cardiovascular disorders as did men who abstained from these beverages. There did not appear to have been any additional benefits from the consumption of greater amounts of alcohol. Therefore, the amounts which were beneficial may be a good guide to the upper limits of daily alcohol consumption since it is well known that excesses may lead to serious disorders of the heart, including degeneration of the heart muscle and irregularity of heartbeats.

(Also see ALCOHOLIC BEVERAGES; BEERS AND BREWING; DISTILLED LIQUOR; and WINE.)

[12]"Land of the Fat," *Time*, January 2, 1978, p. 53.

[13]"Ethanol and Angina Pectoris," *Nutrition Reviews*, Vol. 34, 1976, p. 300.

[14]Lucia, S. P., *Wine and Your Well-Being*, Popular Library, New York, 1971, pp. 67-77.

[15]Yano, K., G. G. Rhoads, and A. Kagan, "Coffee, Alcohol and Risk of Coronary Heart Disease Among Japanese Men Living in Hawaii," *The New England Journal of Medicine*, Vol. 297, 1977, p. 405.

CAFFEINE-CONTAINING BEVERAGES. Although the findings of certain studies which were conducted in the early 1970s suggested that heavy coffee drinking was a causative factor in heart disease, it was later found that these studies did not take into account the fact that the heavy coffee drinkers were often heavy smokers. Nevertheless, people known to have heart disease or irregularities of the heartbeat should abstain from caffeine-containing beverages because they may aggravate these conditions. These beverages include coffee, cola beverages and certain other types of soft drinks, and tea.

(Also see CAFFEINE; COCOA AND CHOCOLATE; COFFEE; SOFT DRINKS; and TEA.)

CARBOHYDRATES. Interest in the possibility that dietary carbohydrates might somehow be involved in heart disease was stimulated by the observation that diabetes and other disorders of carbohydrate metabolism were frequently accompanied by various signs of heart disease. The most common of these disorders is high blood levels of triglycerides (hypertriglyceridemia) due to poor utilization of carbohydrates and fats. Many research studies on both animals and humans have produced evidence that high blood triglycerides may be associated with rapid clotting of the blood and angina pectoris.[16]

It seems that the disorder(s) which result(s) in high triglycerides and poor carbohydrate utilization may be at least partly corrected in some people by the administration of glucose tolerance factor (GTF), a complex substance containing chromium, which increases the effectiveness of insulin.[17] The chronic consumption of table sugar (pure sucrose), white flour, and other highly refined sources of carbohydrates may deplete the body's stores of chromium because (1) the carbohydrates in the foods cause increased excretion of chromium, and (2) refining of these foods removes their naturally occurring chromium.

(Also see CARBOHYDRATE[S]; CHROMIUM; DIABETES MELLITUS; GLUCOSE TOLERANCE FACTOR.)

ENERGY. Consumption of more food energy (kilocalories) than is needed eventually leads to obesity and an increased risk of heart disease. (The findings of the Framingham Study regarding obesity were discussed earlier in this article.) Many studies have shown that the blood fats (cholesterol and triglycerides) which are believed to be involved in atherosclerosis rise in obesity and fall when the excess weight is lost. Also, it seems that cholesterol is synthesized, then stored in the body fat to a much greater extent in obese people compared to those who are lean.[18]

[16]Scheig, R., "Diseases of Lipid Metabolism," in *Duncan's Diseases of Metabolism,* 7th ed., edited by P. K. Bondy and L. E. Rosenberg, W. B. Saunders Company, Philadelphia, Pa., 1974, p. 383.

[17]Doisy, R. J., *et al.,* "Chromium Metabolism in Man and Biochemical Effects," *Trace Elements in Human Health and Disease,* Vol. II, edited by A. S. Prasad and D. Oberleas, Academic Press, New York, N.Y., 1976, pp. 91-96.

[18]Scheibman, P. H., "Diet and Plasma Lipids," in *Diet and Atherosclerosis,* edited by C. Sirtori, G. Ricci, and S. Gorini, Plenum Press, New York, N.Y., 1975, p. 159.

FATTY SUBSTANCES. The relationships between the various fatty substances in the diet, the blood levels of the substances, and the formation of clots and plaques in the major arteries are complex and not well understood. Nevertheless, some of the current ideas on the effects of the various fatty substances found in most diets are noteworthy.

Cholesterol. It has been found that the feeding of extra dietary cholesterol fails to produce high blood cholesterol in the Masai people in Africa because of an adjustment in their body's rate of cholesterol synthesis. However, many studies of other peoples have found a direct relationship between the amounts of cholesterol in the diet and the levels of blood cholesterol.[19] The results obtained in such studies appear to depend upon such factors as (1) the ages, sex, health status, and heredity of the subjects; (2) length of the study; (3) types of foods which accompany dietary sources of cholesterol; and (4) the time(s) of day when blood samples are collected, because blood cholesterol rises and falls in a regular daily rhythm.

For example, in a recent study it was found that, although the eating of two eggs a day for 3 to 4 weeks produced significant increases in the blood cholesterol levels of young and middle-aged men, the values dropped back to normal after another 2 weeks on these diets.[20] Hence, it appears that people differ considerably in their response to dietary cholesterol, so it might be foolish to tell everyone to cut down on eggs if only a limited number of people have a tendency to develop high blood cholesterol from eating them. Furthermore, other factors such as dietary energy, amount and type of fats, and fiber appear to affect the utilization of dietary cholesterol. However, the American Heart Association recommends a maximum intake of 300 mg of cholesterol per day.

Recent evidence suggests that total blood cholesterol is a poor indicator of a heart attack. Rather, the level of the lipoproteins which carry cholesterol in the blood, specifically high-density lipoproteins (HDL) and low-density lipoproteins (LDL) are better indicators. HDL has a protective role since it transports cholesterol back to the liver, while LDL seems to deposit cholesterol in cells, including blood vessels. Also, evidence indicates that high HDL decreases the risk of heart attack, while high LDL increases the risk.

(Also see CHOLESTEROL; and HYPERLIPOPROTEINEMIAS.)

Essential fatty acids. Certain polyunsaturated fatty acids (PUFA)—essential fatty acids (EFA)—are required in small amounts because they are not synthesized in the body. The most important of the essential fatty acids is linoleic acid because it has several vital functions in the body, one of which is to serve as raw material for the

[19]Mattson, F. H., B. A. Erickson, and A. M. Kligman, "Effects of Dietary Cholesterol on Serum Cholesterol in Man," *The American Journal of Clinical Nutrition,* Vol. 25, 1972, p. 589.

[20]Slater, G., *et al.,* "Plasma Cholesterol and Triglycerides in Men with Added Eggs in the Diet," *Nutrition Reports International,* Vol. 14, 1976, p. 249.

synthesis of other compounds. Although linoleic acid and other polyunsaturated fatty acids may lower blood cholesterol and retard the clotting of blood, it does not appear that the various cardiovascular disorders are due to deficiencies of these fatty acids, since they are abundantly supplied by most vegetable oils, fats from pork and poultry, and, in smaller amounts, by fats from beef and lamb.

Fats: Polyunsaturated Vs Saturated. The current interest in this subject started in the 1950s when it was discovered that diets rich in vegetable oil sources of polyunsaturated fatty acids lowered the blood cholesterol, while certain saturated fatty acids raised it. However, almost three decades later there is still controversy regarding the importance of these effects since it is not certain where the cholesterol goes when the blood level is lowered. It may merely be shifted from the blood to the tissues.[21] Furthermore, there is the possibility that the effects of dietary fats on blood cholesterol may not be as important as their effects on the heart muscle or on the clotting of blood.

It seems that under certain conditions the feeding of extra dietary fat may somehow injure the heart muscle (myocardium). Furthermore, autopsies have shown that a heart attack may sometimes occur without much blockage of the coronary artery. Therefore, it seems likely that damage to the heart muscle might sometimes be the cause rather than the result of a heart attack. This possibility is also suggested by the results of animal experiments which have yielded evidence of direct damage to the heart by dietary fat. In one such study, only one-fifth of a group of rats given a synthetic stress hormone (fluorocortisol) plus sodium phosphate developed heart lesions, similar to those found in man, and died. However, fatal heart lesions were produced in almost all of the rats given extra fat (chicken fat, corn oil, olive oil, peanut oil, or pork fat) in addition to the hormone and phosphate.[22] The type of fat which was added to the diet did not seem to make much difference in the severity of the heart lesions, which, oddly enough, were *not* accompanied by noticeable atherosclerosis.

In another much larger study, similar damage to the heart muscle was found where the various groups of rats were not stressed, but were fed different fats over their life-spans of approximately 2 years.[23] The findings in this investigation differed from those in the one previously cited in that each of the vegetable oils (coconut, corn, cottonseed, olive, or soybean) produced lesions in more rats than any of the animal fats (beef fat, butter, chicken fat, and lard). It is noteworthy that the experimental diets were supplemented with vitamin E, but not selenium. Both substances protect against tissue damage due to toxic oxidation products which may be formed from vegetable oils rich in polyunsaturated fatty acids.[24]

Another important effect of certain dietary fats is their slowing or speeding of blood clotting. Although drugs which slow the clotting of blood have long been used to treat people suffering from heart trouble, little has been done to prevent blood clots by dietary measures. This is surprising since it has been known for almost 20 years that diets high in stearic acid (which is abundant in such foods as beef, butter, cocoa butter, and lamb) tend to induce more rapid clotting of the blood and somehow act to reduce the clot-dissolving activity in the blood. On the other hand, linoleic acid (which is the major component of corn oil, cottonseed oil, safflower oil, and soybean oil) counteracts the clot-promoting effects of stearic acid, and acts to increase the clot-dissolving activity.

However, not all vegetable oils have the same effects, since arachidonic acid, a polyunsaturated fatty acid found mainly in peanut oil, promotes clotting. Furthermore, it is well known that both coconut oil and peanut oil induce atherosclerosis in monkeys.

It is not certain that the increased clotting tendency produced by certain fatty acids is in itself a major cause of heart attacks since other factors such as calcium, fatty acids released from fatty tissues, hormones, rate of blood flow, various clot-promoting proteins, and vitamin K are also involved in blood clotting. While it might be desirable to avoid large quantities of the fats from beef, dairy products, and lamb at individual meals, one should be aware that even larger quantities of similar saturated fats may be synthesized in the body from dietary carbohydrate, particularly when there is a greater energy intake than expenditure, or when there is a tendency towards diabetes. Also, large amounts of saturated fatty acids may be released from the body fat stores during periods of emotional stress.

According to the American Heart Association, up to 30% of the total calorie intake may be from fat, but no more than one-third of this should be saturated fats. Sources of polyunsaturated fats are: canola (rapeseed), corn, cottonseed, olive, peanut, safflower, sesame, soybean, sunflower, and walnut oils.

(Also see FATS AND OTHER LIPIDS, section headed "Fatty Acids"; OILS, VEGETABLE, section headed "Antinutritional and/or Toxic Factors"; and TRANS FATTY ACIDS.)

Lecithin and Other Phospholipids. These compounds act as bonds between fats and proteins in cell membranes, and in the special lipoproteins which carry cholesterol and fats (triglycerides) in the blood. They also promote the absorption of fatty substances in the intestine. For example, it is believed that the high rate of absorption of cholesterol from eggs is due in part to the lecithin which is present in the egg yolk.

The interest in lecithin as a protective factor against heart disease arose from the finding that high ratios of cholesterol to phospholipids in the blood were correlated with the severity of atherosclerosis.[25] Furthermore, it ap-

[21]Grundy, S., and E. H. Ahrens, Jr., "The Effects of Unsaturated Dietary Fats on Absorption, Excretion, Synthesis, and Distribution of Cholesterol in Man," *Journal of Clinical Investigation*, Vol. 49, 1970, p. 1135.

[22]Selye, H., "On Just Being Sick," *Nutrition Today*, Vol. 5, Spring 1970, p. 6.

[23]Kaunitz, H., and R. E. Johnson, "Influence of Dietary Fats on Disease and Longevity," *Proceedings of the Ninth International Congress of Nutrition (Mexico, 1972)*, Vol. 1, edited by A. Chavez, H. Bourges, and S. Basta, published by S. Karger, Basel, Switzerland, 1975, p. 369, Table III.

[24]"The Biological Effects of Polyunsaturated Fatty Acids," *Dairy Council Digest*, Vol. 46, 175, p. 31.

[25]Harper, H. A., *Review of Physiological Chemistry*, 15th ed., Lange Medical Publications, Los Altos, California, 1975, p. 320.

pears that the phospholipids which are present in the platelets may play an important role in the clotting of blood.[26]

It is thought that the body synthesizes the phospholipids that it requires as long as there is a sufficient body supply of methionine, magnesium, protein, polyunsaturated fatty acids, and vitamin B-6 (pyridoxine). However, this may not be the case when the feeding of diets rich in polyunsaturated fats causes a rise in the cholesterol content of bile, which may lead to gallstones. It has been reported that the feeding of lecithin raised the levels of phospholipid in the bile of patients afflicted with gallstones.[27]

(Also see LECITHIN; and PHOSPHOLIPIDS.)

FIBER. Certain types of indigestible dietary carbohydrates lower blood fats because they bind with bile salts and other fatty substances in the intestine and cause their excretion in the stool. Cholesterol within the body is lowered because additional amounts are converted to bile salts to replace those lost in the stool. It is also noteworthy that diets high in fiber may counteract the effects of fatty foods which raise blood cholesterol. For example, in an experiment in India, each male participant ate about 5½ oz (*154 g*) of butter per day, which resulted in an average cholesterol level of 206 mg per 100 ml of blood.[28] Then, the addition of a little over ½ lb (*224 g*) of chick peas (known in India as Bengal gram) per day was substituted for the cereals in their diets, and, after 20 weeks, their average blood cholesterol dropped to 160, which is considerably below that of American men. For example, the average cholesterol level of 494 healthy men of Modesto, California, whose ages ranged from 25 to 79, was about 200.[29] The findings of these and other studies suggest that lack of sufficient fiber to counteract the effects of dietary cholesterol may be one of the factors which contributes to the high blood levels of cholesterol in the developed countries where the food sources of carbohydrates contain only traces of fiber.

(Also see FIBER.)

MACROMINERALS. Most of the mineral elements that are required in relatively large amounts (macrominerals) are supplied by diets containing a moderate variety of plant and animal foods. However, certain types of diets, which are low in such nutritious foods as dairy products, fish, fresh vegetables, lean meats,and unrefined grains, may furnish only marginal amounts of the essential macrominerals which, in addition to other functions, may protect against heart disease. Also, excesses of certain elements may aggravate deficiencies of others. Discussions of macrominerals associated with the functioning of the heart and blood vessels follow.

(Also see MINERAL[S].)

Calcium. Normally, the blood level of this element is maintained even when there is a severe dietary deficiency because it is released from bone when the amount in the blood drops too low. However, a British study showed that the death rates from heart disease decreased as the intake of calcium increased.[30] The average daily intake of calcium in England and Wales is about 1,000 mg. Districts where the average daily consumption equaled or exceeded this amount had the lowest rates of cardiovascular diseases.

On the other hand, elevated blood levels of calcium (hypercalcemia) may promote irregular heartbeats, toxicity of cardiac drugs, and/or deposits of the mineral in the arteries and kidneys. Hypercalcemia is *not* usually due to dietary calcium, but to such factors as an excess of vitamin D or a deficiency of magnesium.

(Also see CALCIUM; and MINERAL[S].)

Magnesium. This element appears to provide at least partial protection against such cardiovascular disorders as arteriosclerosis, excessive clotting of the blood, high blood pressure, nonrhythmic heartbeats, and metabolic abnormalities in the heart muscle.[31] Magnesium also appears to protect arteries against the calcium deposits which occur with aging (arteriosclerosis).[32]

Although some nutritionists claim that the diets consumed by most Americans contain sufficient magnesium, others doubt whether this is true because (1) there has been a steady decline in the consumption of such sources of the element as green, leafy vegetables, legumes, and whole grains; and (2) the conditions of acidosis, alcoholism, chronic use of diuretic drugs, diabetes, and diarrhea tend to increase the loss of magnesium from the body.

(Also see MAGNESIUM; and MINERAL[S].)

Potassium. Generally, a deficiency of potassium sufficient to impair the functioning of the heart does *not* result from a dietary shortage of the element since many foods are good sources. However, the supply of the mineral in the cells of the heart muscle may be low due to such factors as (1) excessive losses in the digestive tract (as in diarrhea and vomiting), in sweat (particularly during hot and humid weather), or in urine (as a result of acidosis, diuretics, and various stress factors); and/or

[26]Luscher, E. F., "The Effects of Lipids and Fatty Acids on Blood Coagulation and Platelets in Relation to Thrombosis," *Diet and Atherosclerosis*, edited by C. Sirtori, G. Ricci, and S. Gorini, Plenum Press, New York, N.Y., 1975, p. 107.

[27]Tompkins, R. K., *et al.*, "Elevation of Phospholipid Concentrations in Human Bile by Feeding Lecithin," (Abstract), *Federation Proceedings*, Vol. 27, 1968, p. 573.

[28]Mathur, K. S., M. A. Khan, and R. D. Sharma, "Hypocholesterolemic Effect of Bengal Gram: A Long-Term Study in Man," *British Medical Journal*, Vol. 1, 1968, p. 30.

[29]Wood, P. D. S., *et al.*, "Prevalence of Plasma Lipoprotein Abnormalities in a Free-Living Population of the Central Valley, California," *Circulation*, Vol. 45, 1972, p. 114.

[30]Knox, E. C., "Ischaemic Heart Disease Mortality and Dietary Intake of Calcium," *Lancet*, Vol. i, 1973, p. 1465.

[31]Seelig, M. S., and H. A. Heggtveit, "Magnesium Interrelationships in Ischemic Heart Disease: A Review," *The American Journal of Clinical Nutrition*, Vol. 27, 1974, p. 59.

[32]Szelenyi, I., "Magnesium and Its Significance in Cardiovascular and Gastro-Intestinal Disorders," *World Review of Nutrition and Dietetics*, Vol. 17, edited by G. H. Bourne, published by S. Karger, Basel, Switzerland, 1973, p. 195.

(2) lack of sufficient magnesium for the process which pumps potassium into the cells. Furthermore, depletion of potassium from the heart muscle renders it more susceptible to nonrhythmic beating, particularly when digitalis is used to treat heart failure.

An excess of potassium is dangerous and may cause cardiac arrest when kidney function is impaired because, when the latter condition exists, the body cannot get rid of an overload. It has long been common practice to advise people taking diuretics to eat extra amounts of fruits and vegetables which are rich in potassium, but this advice may not always apply because some of the new diuretic drugs do *not* cause excessive loss of potassium.

(Also see POTASSIUM; and MINERAL[S].)

Sodium. This element is vital to life, so it is fortunate that the human body has remarkable ways of conserving it when dietary intakes are low. However, the ability to conserve short supplies, or to excrete excesses in the urine, varies considerably among different people, and the processes also vary according to the condition of health. For example, some of the people who consume moderate to large excesses of sodium may fail to excrete completely such excesses and, therefore, suffer such consequences as buildup of excessive fluid in the blood and tissues, damage to the kidneys, heart failure, and/or high blood pressure. There is evidence that susceptibility to these sodium-induced disorders may be a trait that runs in certain families.

At the other extreme are people who lose excessive amounts of sodium in their urine, due to such conditions as Addison's disease or kidney diseases. This may lead to excessive loss of water from the body, dangerous declines in the pressure and volume of the blood, and failure of the heart and blood vessels to deliver sufficient blood to the tissues. These circumstances may lead to the complete cutoff of blood flow to the kidneys, a condition which might cause kidney damage.

It is noteworthy that certain vegetarians may consume only small amounts of sodium because their diets are based mainly upon unprocessed, unsalted plant foods which contain only traces of the mineral. Animal foods and commercially processed plant foods usually contain sufficient sodium so that the adding of extra salt is not necessary. However, normally healthy people on low intakes of salt may suffer from sodium deficiency if they are exposed to hot environments, or if they engage in strenuous activities that provoke heavy losses of sweat.

(Also see SODIUM; and MINERAL[S].)

MICROMINERALS (Trace Elements). The importance of these essential elements may often be overlooked in investigations of the causes of heart disease, because only small quantities are required and deficiencies may develop only after long periods on poor diets. However, a variety of these elements is necessary for (1) various functions of the cardiovascular system, and (2) the regulation of metabolic processes which directly or indirectly affect the heart and blood vessels. Furthermore, they are often the nutrients most likely to be removed during the processing of such foods as whole grains, or to be rendered unavailable because they are bound by naturally occurring food constituents like oxalates and phytates, or by food additives like ethylenediaminetetraacetic acid (EDTA). (EDTA is added to foods so as to bind metals like copper and zinc which may cause discoloration of the products.) Therefore, a discussion of some trace elements and their functions follows.

Chromium and the Glucose Tolerance Factor. It was recently found that chromium in the form of glucose tolerance factor (GTF) may prevent disorders of carbohydrate metabolism which could lead to heart disease. Evidence of a direct relationship between chromium deficiency and cardiovascular disorders in the United States consists of the findings that (1) chromium could not be detected in the aortas of people dying from heart disease, but it was found in those of people who died from accidents; (2) the levels of chromium in the hearts and aortas of people in the United States were only a fraction of the levels found in these tissues from people in most other parts of the world.[33]

(Also see CHROMIUM; GLUCOSE TOLERANCE FACTOR; and MINERAL[S].)

Copper. This mineral element is required for the production of red blood cells, the formation of connective tissue, and as a component of various enzymes. When rapidly growing young animals are given diets deficient in copper, they become susceptible to rupture of such major blood vessels as the aorta, enlargement of the heart, and degeneration of the heart muscle.[34] These findings suggest that similar disorders in man might be due in part to copper deficiencies during periods of rapid growth. It is noteworthy that high dietary levels of the minerals iron, molybdenum, and zinc may interfere with the utilization of copper.

(Also see COPPER; and MINERAL[S].)

Iodine. The sole function of iodine in the human body is as a component of the hormones secreted by the thyroid gland. A dietary deficiency of iodine may lead to low levels of these hormones (hypothroidism), a condition which is often accompanied by high blood cholesterol, and in some cases, atherosclerosis. Studies in Finland have shown that, although the dietary levels of fat are similar throughout the country, there seem to be more cases of cardiovascular diseases in the areas where the rates of goiter (an enlargement of the thyroid due to iodine deficiency) are also high.[35]

(Also see IODINE; and MINERAL[S].)

Iron. It is well known that severe deficiencies of iron may cause the heart to race in order to pump enough oxygen-poor blood to the tissues. (The blood becomes oxygen-poor due to lack of red cells which require iron

[33]Schroeder, H. A., "The Role of Chromium in Mammalian Nutrition," *The American Journal of Clinical Nutrition,* Vol. 21, 1968, p. 230.

[34]"Brain and Myocardial Lesions in Copper-Deficient Young Rats," *Nutrition Reviews, Vol. 33, 1975, p. 306.*

[35]Davidson, S., R. Passmore, and J. F. Brock, *Human Nutrition and Dietetics,* 5th ed., The Williams and Wilkins Company, Baltimore, Md., 1972, p. 323.

for their production.) Although these deficiencies are nowhere near as common in the developed countries as in the developing countries where iron-deficiency anemia may be a leading cause of disability or death, certain conditions such as heavy menstrual flow, hemorrhage, or pregnancy may lead to their occurrence in the developed countries.

(Also see IRON; and MINERAL[S].)

Manganese. This element is required for blood clotting, carbohydrate metabolism, and as a component of various enzymes. Studies have shown that animals deficient in manganese have abnormal metabolism of glucose (a simple sugar), a defect which is corrected by manganese supplementation.[36]

(Also see MANGANESE; and MINERAL[S].)

Selenium. It has been reported that the region of the United States where the crops have the highest content of this mineral (roughly, an area lying between the Mississippi River and the Rocky Mountains)[37] have the lowest incidence of deaths from heart disease.[38]

Other evidence of the protective effects of selenium against heart disease consists of (1) its action as part of the enzyme which breaks down toxic peroxides that may damage the heart muscle, particularly those formed from polyunsaturated fats;[39] and (2) its counteraction of the toxicity of cadmium, a common environmental pollutant which causes kidney abnormalities leading to high blood pressure.[40]

Although many nutritionists maintain that the feeding of selenium supplements to meat animals minimizes the possibility of human deficiencies, cooperative experiments in both a low-selenium area (Michigan) and a high-selenium area (South Dakota) showed that, while inorganic selenium prevents the deficiency in swine, organically bound selenium in feed grains was much more effective in raising the level of the element in the muscle meat.[41]

(Also see SELENIUM; and MINERAL[S].)

Silicon. This mineral element, which is a major constituent of such inorganic materials as glass, quartz, and sand, was recently shown to be essential for the formation and the healing of various connective tissues.[42]

A deficiency of this mineral might be a factor in the development of cardiovascular diseases, because the silicon content of human aortas decreases with aging and with the development of atherosclerosis.

Although silicon is one of the most abundant chemical elements on the face of the earth, its inorganic compounds are poorly utilized by man and the higher animals. Its more readily utilized organic forms are generally present in the connective tissues of animals and in the fibrous tissues of plants, two forms of food which are often spurned by people who prefer soft, highly refined items. For example, most of the silicon is removed during the production of table sugar from sugar cane, and in the milling of whole wheat to make flour.

(Also see SILICON; and MINERAL[S].)

Zinc. Diabeticlike disorders of metabolism have been observed in zinc-deficient animals, but the basis for this abnormality is not understood.

Better known is the promotion of wound healing by zinc. This property appears to have been responsible for the marked improvement in the circulations of patients suffering from severe atherosclerosis.[43]

Finally, studies over the past 10 years or so have produced strong evidence that the counteraction of the toxic effects of cadmium by zinc may help to prevent various forms of cardiovascular diseases. For example, a study in North Carolina of the associations between cardiovascular disorders and the trace element levels in the organs of deceased people showed that low ratios of zinc to cadmium in the kidneys were associated with both atherosclerosis and high blood pressure.[44]

(Also see ZINC; and MINERAL[S].)

VITAMINS AND VITAMINLIKE FACTORS. Certain vitamins may be involved in processes associated with the cardiovascular system. Also, excesses of some vitamins may result in deficiencies of others. There is also evidence that the massive doses which may be used to treat deficiencies and other disorders may do more harm than good. Details follow.

Inositol. This vitaminlike substance is synthesized within the body. Nevertheless, under certain circumstances it may be like a vitamin in that a dietary supply is required for optimal health.

---

[36]Everson, G. J., and R. E. Shrader, "Abnormal Glucose Tolerance in Manganese Deficient Guinea Pigs," *The Journal of Nutrition,* Vol. 94, 1968, p. 89.

[37]Hodgson, J. F., W. H. Allaway, and R. B. Lockman, "Regional Plant Chemistry as a Reflection of Environment," *Trace Element Geochemistry in Health and Disease,* edited by J. Freedman, The Geological Society of America, Inc., Boulder, Colo., 1975, p. 61, Fig. 2.

[38]Sauer, H. I., and F. R. Brand, "Geographic Patterns in the Risk of Dying," *Trace Element Geochemistry in Health and Disease,* edited by J. Freedman, The Geological Society of America, Inc., Boulder, Colo., 1975, p. 137.

[39]Hoekstra, W. G., "Biochemical role of Selenium," *Trace Element Metabolism in Animals-2,* edited by W. G. Hoekstra, *et al.,* University Park Press, Baltimore, Md., 1974, p. 61.

[40]Underwood, E. J., *Trace Elements in Human and Animal Nutrition,* 4th ed., Academic Press, New York, N.Y., 1977, p. 253.

[41]Ullrey, D. E., "The Selenium-Deficiency Problem in Animal Agriculture," *Trace Element Metabolism in Animals-2,* edited by W. G. Hoekstra, *et al.,* University Park Press, Baltimore, Md., 1974, p. 61.

[42]Carlisle, E. M., "Silicon," *Nutrition Reviews,* Vol. 33, 1975, p. 257.

[43]Pories, W. J., W. H. Strain, and C. G. Rob, "Zinc Deficiency in Delayed Healing and Chronic Disease," *Trace Element Geochemistry in Health and Disease,* edited by J. Freedman, The Geological Society of America, Inc., Boulder, Colo., 1975, p. 90.

[44]Voors, A. W., M. S. Shuman, and P. N. Gallagher, "Atherosclerosis and Hypertension in Relation to Some Trace Elements in Tissues," *World Review of Nutrition and Dietetics,* edited by G. H. Bourne, published by S. Karger, Basel, Switzerland, 1975, Vol. 20, p. 299.

Like choline, which is part of the compounds which are commonly called lecithins, inositol is an important component of phospholipids which help to stabilize blood cholesterol and prevent its deposition on the walls of arteries. Sometimes, the administration of inositol has helped to reduce accumulations of fatty substances (lipids) in the blood and the liver. Hence, it is designated as an agent which aids in the utilization of lipids (a lipotropic factor). However, it appears that the actions of inositol are closely tied to those of such other nutrients as choline, essential fatty acids, phospholipids, niacin, and vitamin B-6 (pyridoxine). Finally, it is noteworthy that the heart muscle contains high levels of inositol, which suggests that it has an important cardiac function.

The body's requirements for inositol may exceed the rate of its synthesis when (1) the urinary excretion of this nutrient is high, such as in diabetes, or when there is copious urination resulting from the consumption of alcoholic beverages; or (2) there is an accumulation of fatty substances in the blood and/or the liver due to bad diets or various metabolic disorders.

(Also see INOSITOL; and VITAMIN[S].)

Vitamin A. Although various animal studies have shown that supplemental doses of this vitamin may help to heal atherosclerotic lesions and to lower blood cholesterol, similar studies in humans have yielded disappointing results. Hence, it is *not* certain at this time whether Vitamin A supplements are beneficial for people with various cardiovascular disorders.

However, it is well established that vitamin A is involved in (1) the laying down of the connective tissue which lines the large blood vessels; (2) synthesis of various stress hormones; and (3) counteracting the effects of vitamin K in blood clotting (when large amounts of vitamin A are consumed). Therefore, it appears that adequate levels of vitamin A may help in maintaining the health of the cardiovascular system, but that excesses of the vitamin might be harmful.

(Also see VITAMIN A; and VITAMIN[S].)

Thiamin (Vitamin B-1). It is well known that people suffering from severe thiamin deficiency, or beriberi, may die suddenly from heart failure. However, this severe deficiency is not often seen in the developed countries, except in certain alcoholics who eat poor diets. Also, hospital patients given only intravenous glucose (sugar) solutions without supplementary thiamin and other vitamins may develop severe thiamin deficiencies when such treatment lasts for prolonged periods of time (1 to 2 weeks).

(Also see THIAMIN; VITAMIN[S]; and VITAMIN B-COMPLEX.)

Niacin (Nicotinic Acid; Nicotinamide). Massive doses of this vitamin have been used to reduce elevated levels of blood fats. This practice may be dangerous because there is now evidence that it reduces the energy stores in the heart muscle.[45] (One of the effects of massive doses of niacin is the blocking of the release of free fatty acids from the fatty tissues, so the heart is forced to rely on its own stores of fat and glycogen.) However, the doses which have such effects are many times the amounts present in most diets.

(Also see NIACIN; VITAMIN[S]; and VITAMIN B-COMPLEX.)

Vitamin B-6 (Pyridoxine). Various studies conducted during the past 20 years or so have shown that monkeys fed a low-fat, vitamin B-6-deficient diet develop lesions of their blood vessels which are very similar to the atherosclerotic lesions in man. Hence, it has recently been suggested that perhaps some cases of atherosclerosis might be due to lack of this vitamin and/or excessive dietary amounts of the amino acid methionine, which requires pyridoxine for its metabolism.[46] Furthermore, the high-protein foods which are rich in cholesterol, such as dairy products, eggs, and meat, are also rich in methionine. Since the dietary requirement for pyridoxine depends to a large extent on the amount of protein eaten, the people who regularly consume large excesses of such foods might not receive sufficient amounts of this vitamin.

(Also see VITAMIN B-6; VITAMIN [S]; VITAMIN B-COMPLEX; and ATHEROSCLEROSIS.)

Vitamin B-12 (Cobalamin). Deficiencies of this vitamin are most likely to occur either among strict vegetarians who eat no animal foods, or among people over 40 who have various stomach disorders. Hence, some doctors inject large doses of this vitamin into people who show signs of the deficiency since it is generally thought that these doses may be beneficial, but never harmful. However, it has recently been found that massive doses have caused blood clots in certain people.[47] The explanation for this effect was that severe deficiency of the vitamin resulted in pernicious anemia, a condition where the formation of blood cells was impaired, and the time required for the blood to clot was prolonged. Repeated injections of large doses of the vitamin over short periods of time (1,000 mcg daily for 1 or 2 weeks) caused overcorrection of the abnormalities, so that the blood clotted more rapidly than normal.

(Also see VITAMIN B-12; VITAMIN[S]; and VITAMIN B-COMPLEX.)

Vitamin C and the Bioflavonoids. Some recent reports have presented evidence of the beneficial effects of vitamin C (ascorbic acid) on heart disease, while others state that the vitamin failed to produce such effects.

Perhaps one reason for these contradictory reports is that many investigators have relied upon the reduction of blood cholesterol as an indicator of a beneficial effect in man, whereas in animal studies the arteries are usually examined for atherosclerotic lesions. For example, a series of studies of the effects of a mild lack of dietary vitamin C in guinea pigs (which, like man, need dietary ascorbic acid) have shown that a chronic defi-

[45]"Niacin and Myocardial Metabolism," *Nutrition Reviews,* Vol. 31, 1973, p. 80.

[46]"Study Questions Cholesterol Role," *Journal of the American Medical Association,* Vol. 212, 1970, p. 257.

[47]"A Qualitative Platelet Defect in Severe Vitamin B-12 Deficiency," *Nutrition Reviews,* Vol. 32, 1974, p. 202.

ciency led to (1) elevations in blood and liver cholesterol; (2) a reduced rate of conversion of cholesterol to bile salts (which may favor the buildup of cholesterol); and (3) atherosclerotic lesions, even when no cholesterol was given in the diet.[48]

Another reason for the variable results of vitamin C supplementation might be that certain people may need less of the vitamin than those who are exposed to chronic stresses and smoking. During stress, the ascorbic acid content of the adrenal glands drops markedly, while the synthesis of cholesterol in these tissues is greatly increased.

Finally many of the natural sources of vitamin C also contain substances called *bioflavonoids* which appear to have effects like those of the vitamin (such as strengthening capillary walls against breakage or leakage of fluid), and which may protect it against destruction.[49] Hence, the effects of a given amount of dietary vitamin C may vary according to the amounts of bioflavonoids which are present in the diet.

(Also see VITAMIN C; BIOFLAVONOIDS; and VITAMIN[S].)

Vitamin D. It is well known that overdoses of this vitamin may promote the deposition of calcium in such soft tissues as the kidneys. Recently, it has been suggested that overdoses might also be responsible for various cardiovascular disorders. It seems that high intakes of the vitamin increase the requirement for magnesium, and that some of the damage might be prevented by equally high intakes of this mineral.

(Also see VITAMIN D; and VITAMIN[S].)

Vitamin E. This vitamin has long been extolled as beneficial for both the prevention and treatment of heart disease. Certainly, it is necessary to consume greater amounts of vitamin E when extra polyunsaturated fats are taken, or when the dietary levels of selenium are low. (Vitamin E prevents the formation of toxic peroxides from polyunsaturated fats, while selenium is part of the enzyme which breaks down the peroxides once they are formed.)

Although the so-called anticlotting effect of large doses of vitamin E has often been described, the basis for the effect is not well understood. A recent report has attributed it to interference with the action of vitamin K, a factor which is required for the clotting of blood.[50] This report came from doctors who noted bleeding in the skin of a man treated with an anticlotting drug, who also took, without their advice, a daily dose of up to 1,200 IU of vitamin E per day. However, the anticlotting drug which was used also interfered with vitamin K activity, so it is not known whether vitamin E alone might have produced such a drastic effect. Nevertheless, this observation suggests that self-treatment with large doses of vitamin E might be dangerous, particularly for people on anticlotting drugs, or for those who have high blood pressure.

(Also see VITAMIN E; and VITAMIN[S].)

Vitamin K. Many people may not have to worry about whether they get sufficient amounts of this vitamin, if they regularly eat green leafy vegetables (the best of the food sources of the vitamin) or carry within their intestines bacteria such as *E. coli* which synthesize the vitamin (except when the bacteria are killed by medicines that are taken orally for various infectious diseases).

The traditional anticlotting therapy for victims of heart disease consists of drugs which block the action of vitamin K. However, people so treated have sometimes died from massive hemorrhages in the heart muscle, particularly in the days before the need for frequent checks on blood clotting times was recognized. A recent report suggests that the mineral manganese also must be present in adequate amounts in order for vitamin K to have its full effect on blood clotting.[51] Hence, people who are deficient in manganese may be extra susceptible to the effects of anticoagulant drugs.

(Also see VITAMIN[S]; and VITAMIN K.)

Other Risk Factors Associated with Heart Disease. The consequences of poor diet may be aggravated by various other factors. For example, a person with an inherited tendency to have high blood levels of various fats might be much more affected by a high-cholesterol, high-fat diet than a person who has no hereditary abnormality. Likewise, people with certain temperaments may be more susceptible to emotional stress factors. Hence, it is worth considering some of the other risk factors that may be involved in cardiovascular diseases. Details follow.

ATHEROSCLEROSIS AND HARDENING OF THE ARTERIES. These degenerative disorders are often associated with heart disease because even a partial blockage of the blood vessels leading to the heart is likely to lead to a dangerous reduction in the flow of blood to the heart muscle. Also, a clot is more likely to lodge in a blood vessel lined with such irregular projections as plaques than in a healthy one with a smooth lining.

Too often it is assumed that the control of blood cholesterol ensures that the degeneration of the arteries will be delayed until old age. However, arterial degeneration may occur in people with normal blood cholesterol. In fact, a growing number of investigators believe that first there must be some type of abnormality in the lining of the artery before cholesterol and/or other blood-borne materials are deposited in plaques.

One of the newest theories suggests that atherosclerotic plaques may stem from mutations in the

[48]Ginter, L., "Vitamin C in Lipid Metabolism and Atherosclerosis," *Vitamin C,* edited by G. G. Birch and K. J. Parker, Applied Science Publishers, Ltd., Essex, England, 1974, p. 179.

[49]Kuhnau, J., "The Flavonoids. A Class of Semi-Essential Food Components: Their Role in Human Nutrition," *World Review of Nutrition and Dietetics,* Vol. 24, edited by G. H. Bourne, published by S. Karger, Basel, Switzerland, 1976, p. 175.

[50]"Hypervitaminosis E and Coagulation," *Nutrition Reviews,* Vol. 33, 1975, p. 269.

[51]Dolsy, E. A., Jr., "Effects of Deficiency in Manganese Upon Plasma Levels of Clotting Proteins and Cholesterol in Man," *Trace Element Metabolism in Animals-2,* edited by W. G. Hoekstra, *et al.,* University Park Press, Baltimore, Md., 1974, p. 668.

cells lining the arteries.[52] This concept is based upon the observation that the early stages of atherosclerotic plaques in humans do not contain much fatty material, but consist mainly of mutated cells similar to those which line the arteries. It was also found that some samples of human blood contained mutagen-causing substances which had been produced within the body from cholesterol and from substances present in cigarette smoke. Hence, the evidence still points to the dangers of excess cholesterol and of smoking.

(Also see HYPERLIPOPROTEINEMIAS.)

ALLERGIC REACTIONS. Severe allergic reactions may injure artery linings so that atherosclerosis results. For example, when bovine serum albumin (a protein from the blood of cattle) is injected into baboons that have elevated blood cholesterol, severe atherosclerosis develops in their aortas and coronary arteries.[53] However, these injections fail to produce this effect when the baboons have normal cholesterol levels.

Another way in which allergies may cause cardiovascular problems is by the large amounts of histamine released in allergic reactions. Histamine causes blood vessels to enlarge (dilate) so that a larger volume of blood must be pumped to maintain the blood pressure. An extreme example is anaphylactic shock where the heart may not be able to pump sufficient blood to fill the dilated arteries and veins. Animal studies have shown that allergic reactions may be more severe when the blood sugar is low.

(Also see ALLERGIES; and HYPOGLYCEMIA.)

BLOOD ABNORMALITIES OTHER THAN HIGH CHOLESTEROL. Various chronic conditions such as diabetes, excessive numbers of red cells (polycythemia), gout, and obesity may be accompanied by blood abnormalities which increase the risk of developing cardiovascular diseases. In some cases, the blood abnormality may be diagnosed when there are no signs of any disease. Hence, it would be wrong to assume that these characteristics of the blood invariably indicate one or more systemic disorders. Details follow.

• **Acidosis and/or ketosis**—These conditions, which are characterized by elevated levels of acids and/or ketones, may accompany diabetes and kidney disease, or they may be the results of crash dieting or starvation. The abnormalities may not disturb normal heart functions, but they may be dangerous when there is already some extra stress on the heart muscle.

(Also see ACID-BASE BALANCE; and DIABETES MELLITUS.)

• **Elevated blood sugar**—This condition may be an indication of diabetes, or of other factors which predispose people to heart disease. Furthermore, an excess of sugar may also lead to thickening of the blood, high blood pressure, and an increased likelihood of clots.[54]

• **Elevated free fatty acids**—Free fatty acids are released into the blood by various fatty tissues under such conditions as emotional and physical stress, fasting, a heavy consumption of alcohol, low blood sugar, obesity, and diabetes. Normally, the heart muscle may use free fatty acids in its energy metabolism; but when it is overtaxed by various stress factors, excesses of these fats in the blood might cause irregularity in the rhythm of the heartbeat which may lead to cardiac arrest and sudden death.

• **Excessive blood volume**—Failure of the kidney to excrete excess sodium may lead to the retention of excessive water, an expanded volume of blood, and high blood pressure.

• **High triglycerides**—Like cholesterol, these fatty substances are suspected of hastening the development of atherosclerosis. They may originate from excess dietary carbohydrates and/or fats; or from the free fatty acids liberated by various fatty tissues.

• **High uric acid levels**—At one time it was thought that this condition, which is associated with heart disease and gout, was the result of eating too much meat. Now, the evidence points to its causes as being (1) an inherited tendency to overproduce uric acid; and (2) chronic acidosis and/or ketosis which interferes with the excretion of uric acid in the urine.

(Also see GOUT.)

• **Imbalances between various ions (electrolyte imbalances)** —Many of the vital processes in the body require the correct proportions of various charged particles (ions). The ions which appear to be most important for proper functioning of the heart are bicarbonate, calcium, hydrogen (acid ion), hydroxyl (alkali ion), magnesium, potassium, and sodium. These ions are also called electrolytes because they form solutions which carry electric currents, a property which is very important for the transmission of the electrical impulses which trigger the heartbeat in the fibers of the heart muscle. Too little or too much of such ions may be the result of a poor diet and/or processes within the body such as absorption, excretion in the sweat or urine, or the release of minerals from bone. It is believed that electrolyte imbalances are a major cause of irregular heartbeats, cardiac arrest, and sudden death due to heart attacks.

• **Thickening of the blood**—This condition is characterized by an increased resistance to the flow of blood through the small blood vessels. It is dangerous because it may lead to blood clots and/or high blood pressure. The

[52]Benditt, E. P., "The Origin of Atherosclerosis," *Scientific American,* Vol. 236, 1977, p. 74.

[53]Howard, A. N., "Cardiovascular Disease and Dietary Experiments in Animals," *Proceedings of the Ninth International Congress of Nutrition,* Vol. 1, edited by A. Chavez and S. Karger, Basel, Switzerland, 1975, p. 343.

[54]Anderson, J. W., "Metabolic Abnormalities Contributing to Diabetic Complications. 1. Glucose Metabolism in Insulin—Insensitive Pathways," *The American Journal of Clinical Nutrition,* Vol. 28, 1975, p. 273.

thickening of the blood may be due to (1) an excessive number of red blood cells (polycythemia); and/or (2) an increased ratio of dissolved solids to water such as occurs in dehydration or diabetes. Sedentary, middle-aged, obese men are most likely to have thickening of the blood, although it may occur in some women who use birth control pills.

BODY BUILDS. Cardiovascular disease is more likely to be found in people with heavy muscular builds (mesomorphs) and in those with large, rounded abdomens whose arms and legs are short in proportion to the length of their trunks (endomorphs), than in people who are tall and slender with disproportionately long arms and legs (ectomorphs).[55] An approximate estimate of heaviness or slenderness of body-build may be obtained by calculating the ponderal index (height in in. $\div \sqrt[3]{\text{weight in lb}}$). Thus, a person who is 67 in (*170 cm*) tall and who weighs 165 lb (*75 kg*) has an index of 12.2 ($67 \div \sqrt[3]{165}$) Values of the ponderal index which are less than 12.0 suggest a high risk for heart disease, whereas values greater than 13.0 usually mean a low risk. (The average adult male Masai has a ponderal index of 13.3.)

(Also see OBESITY.)

CHRONIC CONSTIPATION. Straining at stool may generate pressures in the chest cavity, abdomen, and/or in the leg veins. Such stresses may sometimes lead to distended veins in the rectum (hemorrhoids), varicose veins in the legs, and dislodging of clots within blood vessels. Hence, preexisting cardiovascular disorders may be aggravated by chronic constipation. Occasionally victims of heart attacks and strokes have been found to have collapsed in the bathroom.

CLIMATE AND WEATHER. The body reacts to the humidity and temperature of the outside environment by means of (1) shifts in the distribution of fluid between the blood and various tissues; (2) constriction or dilation of the blood vessels on or near the surface of the body; and (3) slowing or speeding of the heart rate. Hence, living in certain climates or exposure to unpleasant weather may tax the cardiovascular system. The effects of such weather conditions follow:

- **Cold weather**—Chilling of the skin causes a constriction of the blood vessels, and may provoke a rapid rise in blood pressure, which is called the "cold pressor effect." Hence, cold weather may bring on attacks of angina pectoris in susceptible persons.

- **Dry weather**—People who have circulatory troubles, which lead to fluid accumulation in tissues, often feel better when the humidity is low because their tissues may "dry out" somewhat. Persons who lose too much water (that is, they become dehydrated) may develop irregular heartbeats, or even cardiac arrest. Furthermore, the cardiovascular system might be taxed by the chronic coughing and wheezing which sometimes results from the breathing of very dry air.

- **Humid weather**—The combination of heat and humidity, which commonly occurs in the summer in the southeastern United States and throughout the year in the Carribean and Hawaii, may make it advisable for people with very weak hearts to spend most of their time in air-conditioned buildings. Humid environments may stress the heart because (1) extra fluid is retained in the body tissues, which swell and press against nearby blood vessels so that there is an increased resistance to the flow of blood; and (2) the heart has to pump more blood to the surface of the body for cooling because the evaporation of sweat is reduced by the humidity.

- **Warm weather**—Overheating of the body causes the blood vessels in the skin to enlarge (dilate) as part of the body's attempt to maximize the loss of heat from its surface. Also, the heart rate is stepped up and the flow of blood to the skin is greatly increased. The circumstances may tax the heart because it works harder, yet it receives a smaller proportion of the circulating blood for its sustenance. The brain may be similarly shortchanged of its normal supply of blood so that fainting results.

However, warm environments with low humidity are tolerated better than those which are more humid, because there is a greater cooling of the body by the evaporation of perspiration when the air is dry.

CONGENITAL DEFECTS. Abnormalities in the heart and its connecting blood vessels may originate during fetal development, or after birth. The fetal circulation of blood differs from that in newborns in that (1) blood is oxygenated in the placenta rather than in the lungs; and (2) blood flows through an opening (*foramen ovale*) between the right and left sides of the heart, and through the *ductus arteriosus* between the pulmonary artery and the aorta. Normally, both the foramen ovale and the ductus arteriosus close soon after birth, when the baby takes its first few breaths of air. Occasionally, the fetal openings fail to close at birth and cause problems in the functioning of the heart. Usually, surgical correction of these defects is done sometime in the first few years of life.

Other common congenital abnormalities are defects in the valves of the heart, and narrowing (coarctation) of the aorta, the main artery which carries blood from the heart to the rest of the body. These defects are not always discovered, so they may never be corrected. The causes of such defects are not known.

CRASH DIETING. Diets which produce rapid loss of weight may also cause acidosis and/or ketosis, dehydration, electrolyte imbalances, nutritional deficiencies, or even permanent damage to the heart muscle which may be fatal.

(Also see OBESITY.)

DRUGS. Many of the commonly used drugs may have either direct or indirect effects on the cardiovascular system. It might be wise in some cases to reduce the need for these drugs by a more conscientious adherence to a good diet and/or other hygienic measures such as avoidance of sources of infectious organisms and unnecessary stresses, getting enough exercise and sleep, keeping calm, and selecting alternative methods of birth control when advisable. Therefore, it is worth noting the

[55]Mayer, J., *Human Nutrition*, Charles C. Thomas, Publisher, Springfield, Ill., 1972, p. 265.

effects of some of the drugs which are used most often and for the longest periods of time:

• **Antacids**—Many of the common remedies in this category may reduce the absorption of the essential minerals calcium, chromium, copper, iron, magnesium, manganese, and zinc. Hence, chronic use of large doses of antacids may lead to deficiencies of nutrients which may make the cardiovascular system more susceptible to certain disorders.

(Also see ANTACIDS.)

• **Antibiotics**—These agents may kill the bacteria in the digestive tract which synthesize vitamin K, a factor that is needed for clotting of the blood. It is important to consider this possibility, particularly when the patient may be exposed to anticlotting agents such as aspirin, drugs which slow clotting, and vitamin E.

(Also see ANTIBIOTICS.)

• **Antidepressants**—This category includes a group of substances called monoamine oxidase inhibitors (MAOI) which block the body's destructive metabolism of such stimulatory (biogenic) amines as adrenaline, dopamine, and serotonin. (These amines are synthesized within the body and are found mainly in the brain and nerves. They appear to be associated with states of emotional and nervous excitement.) Hence, people who take these drugs feel less depressed because their biogenic amines remain active for longer periods of time. However, they may be very prone to develop high blood pressure when they consume such foods as aged or ripened cheeses, Chianti wine, herring, and certain yeast products. These foods all contain amines which may act like the biogenic amines, and raise the blood pressure, particularly when the body's normal process for breaking down these compounds is impaired by MAOI drugs.

• **Birth control pills**—These drugs, also known as oral contraceptives or simply as "the pill," may be hazardous for women who have a tendency to develop blood clots or diabetes. The failure to identify in advance those who are extra susceptible to such side effects leads to higher rates of certain cardiovascular disorders in users of the pill compared to nonusers. For example, a British study found that women aged 30 to 39 who used oral contraceptives were 2.7 times as likely to have a heart attack as nonusers, whereas the users aged 40 to 44 had 5.7 times the risk of nonusers.[56]

It may be in order to recommend the use of nutritional supplements by users of oral contraceptives, since these drugs alter the metabolism of such nutrients as calcium, folic acid (one of the B-complex vitamins), magnesium, phosphorus, vitamin A, vitamin B-1 (thiamin), vitamin B-2 (riboflavin), vitamin B-6 (pyridoxine), vitamin B-12, vitamin C, and vitamin E.[57]

• **Cortisone**—This antiinflammatory hormone should be used with great caution in the treatment of people who are susceptible to blood clots, diabeticlike utilization of carbohydrates, and high blood pressure. Long-term use might also weaken the heart muscle, since the hormone promotes the breakdown of protein in various tissues.

• **Digitalis**—Although the main use of this drug is to slow and strengthen a fluttering heartbeat in people who have heart trouble, it also may be used in combination with thyroid hormones in pills used by people trying to lose weight. (Digitalis counteracts the racing of the heart provoked by thyroid hormones.)

People who are deficient in either magnesium or potassium are extra susceptible to the toxic effects of digitalis (irregular heartbeats and faulty transmission of the heartbeat impulse through the heart muscle).

• **Insulin and oral antidiabetic drugs**—There is evidence that excesses of the hormone insulin, whether injected as a drug or naturally secreted by the pancreas, may contribute to the development of atherosclerosis.[58] Likewise the oral diabetic drugs which are currently in use act to increase the amount of insulin secreted by the pancreas, so they also might contribute to cardiovascular diseases. However, it was noted earlier in this article that sometimes diabetics can get along on less insulin if they are given supplemental chromium in the form of glucose tolerance factor, and they stick to their dietary prescription.

(Also see CHROMIUM; DIABETES MELLITUS; and GLUCOSE TOLERANCE FACTOR.)

• **Laxatives**—Frequent use of these agents over long periods of time may promote excessive losses of mineral salts, vitamins, and water in the stool. Hence, the user may end up in a weakened condition which overtaxes the heart. For example, the chronic loss of potassium may sometimes be the underlying cause of irregular heartbeats.

• **Pep pills (or "uppers")**—This term usually refers to the *amphetamines*, which are generally taken to obtain stimulation or to get a "lift." They also depress the appetite, so they are sometimes prescribed for people who are trying to lose weight. Pep pills may cause excessive excitability, so they should *not* be used by those who have conditions such as advanced arteriosclerosis, coronary artery disease, high blood pressure, or irregular heartbeats.

• **Rainbow pills for weight control**—These multicolored pills may contain mixtures of amphetamines, barbiturates, digitalis, diuretics, laxatives, and thyroid hormones. The use of these combinations requires close medical supervision, since it seems that they might have been responsible for the deaths of six women in Oregon.[59]

[56]Mann, J. I., *et al.*, "Myocardial Infarction in Young Women with Special Reference to Oral Contraceptive Practice," *British Medical Journal*, Vol. 2, 1975, p. 241.

[57]"Diet-Drug Interactions," *Dairy Council Digest*, Vol. 48, 1977, p. 7.

[58]Stout, R. W., and J. Vallance-Owen, "Insulin and Atheroma," *The Lancet*, Vol. ii, 1969, p. 1078.

[59]Henry, R. C., "The Fatal Interaction," *Nutrition Today*, Vol. 3, March 1968, p. 18.

The autopsies on the women showed signs like those which accompany congestive heart failure. However, these women might have eaten poorly as a result of the appetite depressants (amphetamines) in these pills.

• **Thyroid hormones**—There is a great temptation for certain doctors to prescribe these hormones for people who are struggling to lose weight, although this use may consitute an unwarranted risk when the patient has a normally functioning thyroid gland. A common side effect of supplemental doses of these hormones is a speeding up of the heart rate. Hence, the treatment may be dangerous for people with cardiovascular disorders.

• **Water Pills**—These substances, which are also known as diuretics, are designed to help people get rid of an excessive buildup of sodium and/or water. Unfortunately, they may be given to those who are trying to lose weight, since the loss of water is the fastest (but not the safest) way to accomplish this goal. The loss of large amounts of water may be accompanied by excessive losses of magnesium and potassium, minerals which are essential for optimal functioning of the heart and blood vessels. Subnormal levels of these minerals in the heart muscle might render it more susceptible to irregular beating and the toxic effects of such drugs as digitalis.

EMOTIONAL STRESSES. Most people have had acquaintances, friends, or relatives who suffered heart attacks which were apparently brought on by emotional stresses such as angry arguments, severe business losses, divorce, bereavement, or other catastrophes. There is mounting evidence that severe psychological stresses may cause marked abnormalities in the heartbeat rhythm (called arrhythmias or ventricular fibrillation), even when the cardiovascular system appears to be healthy. A tentative explanation for these effects is that (1) the emotional disturbances lead to the outpouring of adrenal hormones which stimulate the release of abundant quantities of free fatty acids from fatty tissues, and (2) the resulting elevation in the blood level of fatty acids may produce abnormal heart rhythms.[60] Other harmful effects of such stresses are a more rapid clotting of the blood, speeding up of the heart rate, and a wrinkling of the lining of arteries so that atherosclerotic degeneration may occur more readily.

Furthermore, it seems that certain people, who are said to be "field-dependent," are more likely to react emotionally to external factors than others who rely more on an internal sense of judgment. The latter type of people are more immune to external stress factors and are said to be "field-independent.[61]

Also, situations which have uncertain or unfamiliar aspects, like association with new co-workers, are more likely to be stressful than those that are familiar. Along these lines, a pair of researchers in psychology has developed a Social Readjustment Rating Scale which lists major life events or changes according to the degree that may produce stresses.[62] For example, some of the most stressful events are death of spouse, divorce, marital separation, jail term, death of a close family member, and personal injuries or illness. People who experience several of these events in the same year may be very prone to develop some type of heart trouble.

Finally, various social factors may be responsible for chronic stress. It is noteworthy that people in lower social classes, such as certain minority groups, are much more susceptible to stress due to disorganization of their families than are the socially prominent.

GLANDULAR OR HORMONAL DISORDERS. Abnormalities in the patterns of hormonal secretion may be due to (1) defects in the tissues which secrete the hormones, (2) nutritionally imbalanced diets, or (3) various types of emotional and physical stresses. Therefore, it is worth considering those hormones which have the most important effects on the cardiovascular system:

• **Aldosterone and renin**—These adrenal cortical and kidney hormones act to raise the blood pressure and to increase the amounts of sodium and water retained by the body. Hence, they may be the causes of abnormally high blood pressures in certain susceptible people, particularly under the circumstances of emotional and physical stresses, high levels of salt in the diet, and major surgery. Another undesirable effect of aldosterone is an excessive loss of potassium in the urine, which may lead to irregular heartbeats and a slowing of the release of insulin from the pancreas.

• **Insulin**—This hormone is required for the optimal utilization of carbohydrates, fats, and proteins by the heart muscle. Hence, some diabetics may have poor functioning of the heart when it is deficient. On the other hand, it was mentioned earlier in this article that excessive secretion of insulin may speed the atherosclerotic deterioration of arteries. This oversecretion may occur after large meals, particularly in people who are obese. Various types of stresses may also induce abnormalities of insulin secretion.

• **Stress hormones**—Conditions such as chilling, emotional distress, irregular patterns of eating and sleeping, low blood sugar, and strenuous exercise induce the release of stress hormones that enable the body to cope with these challenges. The major hormones in this category are adrenocorticotropic hormone (ACTH) a growth hormone from the pituitary; epinephrine, norepinephrine, and the steroid hormones from the adrenals; and glucagon from the pancreas. Depending upon the condition, some or all of those hormones may be secreted and act together to produce such effects as (1) elevations in blood pressure and in the blood fats and sugar; (2) a more rapid and stronger beating of the heart; (3) the resistance of fatty tissues to the action of insulin, similar to that which occurs in adult onset diabetes; (4) an increased urinary excretion of magnesium and potassium, but a retention of sodium and water in the

[60] Kurien, V. A., and M. F. Oliver, "A Metabolic Cause for Arrythmias During Acute Myocardial Hypoxia," *The Lancet,* Vol. i, 1970, p. 813.

[61] Cleghorn, J. M., G. Peterfy, E. J. Pinter, "Psychophysiology of Lipid Mobilization," *Progress in Human Nutrition,* Vol. 1, edited by S. Margen, The Avi Publishing Co., Inc., Westport, Conn., 1971, p. 169.

[62] Holmes, T. H., and R. H. Rahe, "The Social Readjustment Rating Scale," *Journal of Psychosomatic Research*, Vol. 11, 1967, p. 213.

body, which are effects that may lead to heart failure; (5) a more rapid clotting of the blood; and (6) the wrinkling of the inner lining of blood vessels, a condition that favors the attachment of clots. Each of these hormonal resonses to stress may be lifesaving in certain crises, but they may also bring on or aggravate various cardiovascular disorders when the oversecretion of stress hormones occurs on a regular basis.

• **Thyroid hormones**—People who suffer from deficiencies of these hormones may be obese and have high blood levels of cholesterol plus an increased susceptibility to atherosclerosis. On the other hand, excesses of the hormones foster slenderness and keep the blood cholesterol low, but they are also likely to induce a racing of the heart that may be dangerous. The hormone deficiencies might be due to insufficient dietary iodine, or to goitrogenic substances in foods and water which interfere with the utilization of dietary iodine. While only a few people have excessive secretion of thyroid hormones, others with normal thyroid function sometimes obtain extra hormones on a doctor's prescription for the purpose of weight reduction. However, taking extra thyroid hormones when they are not needed is both foolish and hazardous, because the surplus hormones suppress the secretion of thyroid-stimulating hormones from the pituitary, and the thyroid glands slow their activity. Hence, the practice produces "lazy thyroids," which may remain subnormal when the taking of extra hormones is discontinued.

(Also see ENDOCRINE GLANDS.)

HEAVY MEALS. Loading the stomach with food and drink may have adverse effects on the cardiovascular system because (1) a distended stomach might press up against the heart and constrict its blood flow; (2) fats and sugars may rapidly build up to dangerous levels in the blood; (3) the clotting of blood is likely to occur as a result of high blood levels of fats and sugar; and (4) a few large meals per day usually result in a greater laying down of body fat than more frequent, smaller meals.

HEREDITY. It has been suggested that small groups of people who have inherited tendencies to diet-induced heart diseases may account for a disproportionately large number of cases of these diseases in the countries where the diets are high in cholesterol and other fatty substances. For example, over half of a group of 50-year-old men with hereditary susceptibility to high blood cholesterol had heart disease, whereas only about 5% of the men at age 50 in the Framingham Study (cited earlier in this article) had this disease.

INJURIES TO THE BLOOD VESSELS AND/OR THE HEART. The cardiovascular system may fail to function well under stress when it has sustained various types of injuries. Therefore, it is worth noting the types of injuries which may mar the performance of the blood vessels and the heart.

Atherosclerotic plaques and clots are most likely to form at points of injury to the linings of blood vessels since healthy, unscathed lining tissues are resistant to the buildup of such materials. Some of the major factors leading to damage of vascular linings are (1) deficiencies of such nutrients as copper, manganese, vitamin A, vitamin C, and zinc, because they are needed for the metabolism of these tissues; (2) diabetes, which causes abnormal thickening of blood vessel linings; (3) high blood pressure, because the force exerted against the lining cells is increased; and (4) inflammation resulting from such conditions as allergies, Buerger's disease which occurs mainly in smokers, and infections.

A hard working heart is more susceptible to failure, irregular beating, or even rupture (a condition which is usually fatal) if it has been weakened by (1) chronic overwork due to excessive thyroid hormones, iron-deficiency anemia, or to high blood pressure; (2) deficiencies of such nutrients as calcium, chromium, iron, magnesium, potassium, selenium, thiamin (vitamin B-1), vitamin C, or vitamin E; (3) diabetes, since insulin inadequacy impedes the uptake of blood-borne nutrients by the heart muscle; and (4) infectious diseases like pneumonia, rheumatic fever, syphilis, and tuberculosis—which may injure the inner lining of the heart (endocardium), the heart muscle (myocardium), the sac which surrounds the heart (pericardium), or the heart valves.

KIDNEY DISEASES. The kidneys help to regulate the composition and volume of the blood which flows through the cardiovascular system. Hence, many medical scientists classify disorders of these systems together under the designation of *cardiovascular-renal (kidney) diseases.*

For example, the retention of excessive amounts of sodium and water by the kidneys often leads to high blood pressure (hypertension) which may damage the blood vessels and the heart. The hypertensive actions of the kidney might result from such conditions as (1) excessive consumption of salt and/or other forms of sodium, like monosodium glutamate; (2) an inherited tendency towards oversecretion of renin from the kidneys or of aldosterone from the adrenals, since both of these hormones raise the blood pressure; or (3) damage to the kidneys by dehydration, deficiencies of nutrients (particularly potassium and zinc), infectious diseases, or by toxic substances like cadmium (an environmental polutant).

Another renal disorder which affects the cardiovascular system is kidney failure. This condition is characterized by (1) a great reduction in the production of urine, and (2) by a retention within the blood of many substances which would normally be excreted. Some of the retained wastes are toxic to the cardiovascular system. For example, elevated blood levels of magnesium and/or potassium may cause the heart to stop beating. In some cases kidney failure results from heart failure because the latter condition may deprive the kidneys of an adequate supply of blood.

(Also see KIDNEY DISEASES; and DEFICIENCY DISEASES.)

LIVING IN DENSELY POPULATED AREAS. It has often been stated that the incidence of cardiovascular diseases in American cities and suburban areas is higher than the rate in rural areas. For example, some of the data supporting this conclusion were obtained in a study which was conducted in Indiana.[63] The investigators found that above-average rates of heart disease were associated with such conditions as (1) a high percentage of people living in multifamily dwellings, (2) a large number of people per square mile, and (3) a high percentage of people who were in white-collar occupations.

The findings which are applicable to the United States do not necessarily apply to the other countries of the world since New York City has a very high death rate for heart disease, whereas Tokyo, the world's largest city, has one of the lowest death rates from this disease. Lest it be thought that the difference in disease rates between the United States and Japan might be due to such factors as diet and race, let it be noted that a study of some 4,000 Japanese men living in the San Francisco area showed that those who became aggressive and competitive like Americans were five times as likely to suffer heart attacks as the men who retained behavior more characteristic of the Japanese.[64] (The study was conducted over a 10-year period by a team headed by D. M. Marmot from the University of California at Berkeley.)

MEDICAL TREATMENTS. Various types of medical treatments may have harmful effects on the cardiovascular system unless certain precautions are taken. Details of some of the more common measures follow:

- **Anesthesia**—The most frequently occurring cardiovascular effects of anesthetics are (1) altered responses of the heart muscle to nervous stimulation, (2) a lowering or raising of blood pressure, and (3) irregular heartbeats. Fortunately, most well-equipped hospitals have electronic systems which are used to monitor heart actions during anesthesia and surgery.

- **Dialysis of patients with kidney failure**—People with poorly functioning kidneys often need to have either hemodialysis (this procedure requires a special kidney machine) or peritoneal dialysis to remove accumulated waste products from their blood. The latter procedure not only removes wastes, but also considerable amounts of amino acids and water-soluble vitamins. Loss of these nutrients may weaken the heart. Hence, the appropriate nutritional supplements are needed in order to prevent deficiencies.

- **Intravenous administration of glucose**—Long-term use of this therapy without nutrient supplementation may provoke deficiency diseases like beriberi and kwashiorkor, both of which may cause severe damage to the heart. Many recent medical articles have stressed the need for supplemental minerals, protein, and vitamins during this treatment.

- **Tube feeding of liquid formula diets**—Care has to be taken during this type of therapy to avoid the complications of dehydration, diarrhea, and/or vomiting due to the rapid administration of concentrated mixtures (those which contain high proportions of dissolved materials to water).

OVERLY AGGRESSIVE (TYPE A) BEHAVIOR. Type A behavior is characterized by traits such as (1) an obsession with doing all things according to a tight schedule, (2) a setting of goals which are difficult to achieve, and (3) an overly angry and aggressive reaction to various frustrations.

Although a certain amount of aggressive behavior may be prerequisite for success in highly industrialized countries like the United States, some people exhibit these traits to a much greater degree than others. When all other factors such as diet, smoking, and weight are the same, the people with habitual Type A behavior are much more prone to cardiovascular diseases.

PHYSICAL STRESSES. Physical stresses such as allergies, burns, dehydration, excessively cold or hot environments, extreme fatigue, fractured bones, head injuries, infections, loud noises, surgery, or toxic agents may provoke the oversecretion of adrenal hormones; a condition which might have harmful effects on the cardiovascular system when the stresses are severe and/or prolonged.

SOFT DRINKING WATER. Statistical studies in various countries around the world have shown that the harder the drinking water, the lower the incidence of cardiovascular disease. These findings have led some scientists to suggest that the extra calcium and magnesium which is usually present in hard water somehow protects against heart disease. However, in Japan, where all of the drinking water supplies are soft, the death rates from stroke are higher when the acidity of the water is higher. An explanation linking these observations has been provided by the late Dr. Schroeder, director of the Trace Element Laboratory in Vermont.[65] His studies showed that (1) the amounts of dissolved cadmium in drinking water are likely to be higher when the water is either more acid or softer than average, and (2) cadmium which is present in food and drinking water is absorbed into the blood and deposited in the kidneys where it may cause high blood pressure.

STICKINESS OF BLOOD PLATELETS. Under certain conditions the platelets in the blood (small cell-like bodies which participate in blood clotting) become sticky and adhere to each other or to the walls of blood vessels. Furthermore, it is suspected that certain atherosclerotic lesions might originate from platelet clumps. Therefore, it is worth noting the factors which may affect the stickiness of platelets.

Factors which promote platelet clumping are (1) the increased secretion of adrenal hormones during various stress conditions; (2) meals high in beef or lamb fat; (3) coconut and peanut oils; and (4) table sugar (sucrose) in certain susceptible persons.

---

[63]Klusman, R. W., and H. I. Sauer, "Some Possible Relationships of Water and Soil Chemistry to Cardiovasular diseases in Indiana," *Environmental Geochemistry in Health and Disease*, edited by H. L. Cannon and H. C. Hopps, The Geological Society of America, Boulder, Colo., 1971, p. 38.

[64]"Culture and Coronaries," *Time*, August 18, 1975, pp. 54-55.

[65]Schroeder, H. A., *The Trace Elements and Man*, The Devin-Adair Company, Old Greenwich, Conn., 1975, pp. 105-113.

The best known substances which inhibit the clumping of platelets are (1) aspirin; and (2) corn oil, cottonseed oil, safflower oil, and soybean oil.

Also, it was recently discovered that garlic and onions contain unidentified anticlumping factors.[66]

TOXIC SUBSTANCES. Substances which have toxic effects on the cardiovascular system are sometimes present in air, foods, or water. The number of people affected by these substances is not known, but it seems likely that the incidence of toxicity might be higher in crowded and industrialized urban area where CHD rates are higher than in rural areas. Also, the amount of exposure to toxicants might depend upon various environments at home and at work. For example, workers engaged in sanitation work in New York City have double the rate of heart disease and other groups of comparable workers in the city. It is suspected chronic exposure to carbon monoxide in the exhaust fumes from garbage trucks might be partially responsible for the higher rates of CHD in these people, who otherwise might be expected to have a lower rate of disease because of the great amount of physical activity required in their work. Details concerning other toxicants follow:

• **Cadmium**—This toxic metal, which has been discussed earlier in this article, may sometimes be the cause of high blood pressure because it interferes with the vital functions of zinc in the kidneys. However, it is not known how many people might be so affected, since (1) the amounts of cadmium which get into the bodies of different people depend upon the levels of the metal in their air, food, and water; (2) drinking water which is hard apparently reduces the absorption of cadmium by the body; and (3) the toxic effects of cadmium in the body are more severe when the blood and tissue levels of zinc are low.

• **Cobalt**—Although this metal is a constituent of vitamin B-12, it has no vital function for man apart from the vitamin, and in some cases it may have toxic effects. In the mid-1960s, there were outbreaks of a mysterious type of heart disease among certain heavy beer drinkers in Canada and the United States. Cobalt had been used by various brewers to prevent excessive foaming of their beers. It seems that the combination of alcohol and cobalt had toxic effects which greatly surpassed those that might have been expected from the amounts of cobalt that were ingested. Higher doses of cobalt have been safely used to stimulate the production of red blood cells in anemic people. However, excesses of this agent may cause overproduction of red cells and thickening of the blood, such as occurs in the disease polycythemia.

• **Honey from poisonous plants**—Occasionally people are poisoned by honey from bees that have collected nectar from such plants as azalea, mountain laurel, oleander, and rhododendron. The honey is poisonous because these plants contain substances which are chemically simialar to digitalis and which may induce irregbular heartbeats.

(Also see MEDICINAL PLANTS; POISONS; and POISONOUS PLANTS.)

[66] Bordi, A., and H. C. Bansai, "Essential Oil of Garlic in Prevention of Atherosclerosis," *The Lancet*, Vol. ii, 1973, p. 1491.

• **Lung Irritants**—Inflammation of the lungs by such irritants as asbestos fibers, automobile exhaust fumes, coal dust, cotton fibers, molds, pollens, soot, and tobacco smoke may lead to chronic diseases like emphysema where breathing is obstructed and there is insufficient oxygenation of blood. These conditions increase the workload of the heart and may even contribute to its failure.

• **Overheated fats**—Certain cooking practices may cause the chemical breakdown of fats and oils into toxic substances such as (1) agents which may cause atherosclerosis, and (2) peroxides which may damage cells in the heart muscle. Some of the evidence comes from an animal nutrition experiment in which it was found that the experimental groups fed heated corn oil developed more atherosclerosis than the control groups fed unheated corn oil.[67] Therefore, it is appropriate herein to review the various ways in which fats may be converted to toxicants by cooking. Details follow:

1. The fats most susceptible to breakdown during heating are polyunsaturated oils derived from corn, cottonseed, safflower, and soybeans, whereas the ones least likely to deteriorate are the saturated fats from beef and lamb.
2. Oils which have been used for deep fat frying over prolonged periods of time may become dark-colored like automobile crankcase oil, a change which is likely to be accompanied by the formation of toxic substances. Also, heating such oils until they start smoking hastens their deterioration.
3. Even fats from beef and lamb may be degraded when cooking temperatures are sufficiently high. When the meats are broiled on a grate over charcoal, some of the melted meat fat may drip into the hot coals where it is converted to toxic substances which are then carried upward in the smoke to coat the bottom of the meat.

(Also see FATS AND OTHER LIPIDS, section headed "Fats and Health Problems".)

• **Pesticides**—Certain insecticide sprays may cause slowing of the heart and other disorders when they enter the blood after inhalation into the lungs.

• **Poisonous potatoes**—Irish potatoes which are green or sprouted may contain toxic levels of solanine, an alkaloidal substance which slows the heartbeat and the rate of breathing.

• **Tobacco smoke**—Cigarette smoking will eventually kill one in five people in the industrialized countries, or 250 million people—which is equal to the entire population of the United States, according to a report that appeared in the May 23, 1992 issue of *The Lancet*, a British medical journal.

[67] Kritchevsky, D., and S. A. Tepper, "Cholesterol Vehicle in Experimental Atherosclerosis. IX. Comparison of Heated Corn Oil and Heated Olive Oil," *Journal of Atherosclerosis Research*, Vol. 7, 1969, p. 647.

SIGNS AND SYMPTOMS OF HEART DISEASE.

Some of the characteristic features of heart disease may be so obvious that even lay people can recognize them. However, other more subtle signs may require special laboratory tests in order that experienced physicians may be certain of their diagnoses. Furthermore, the earlier that heart disease is detected, the better the outlook for the patient. Hence, it is worth noting the various ways in which heart disease may be diagnosed.

Easily Recognizable Abnormalities. These conditions, which are readily recognizable by lay people, are often associated with major disorders of the heart although they may sometimes result from other types of medical problems. Therefore, it is wise to consider the possibility of heart disease whenever there are problems like those which follow:

• **Pain in the chest (angina pectoris)**—This is the best known and most characteristic feature of coronary heart disease. However, the pain may not be sharp, rather, it may be a dull ache or it may feel like a heavy weight pressing on the heart. Sometimes angina is also felt in such neighboring areas of the body as the upper abdomen, back, shoulders, arms, elbows, wrists, neck, or mouth. The pain is usually brought on by one or more of the factors commonly designated as the four E's—exercise, emotional agitation, eating, or exposure to cold. Often angina is relieved by resting in a standing or sitting position, but it may not go away when the victim is lying down, because in the latter position there is a greater filling of the heart with blood which results in a greater pumping load for the heart muscle.

• **Shortness of breath**—Breathlessness after mild exertion may indicate problems of the heart and/or the lungs. Also, people with heart disease may awaken an hour or two after retiring with breathlessness due to the greater workload on their hearts when they are lying down. Hence, they may need to rest with their head and back raised by several pillows.

• **Fainting**—Sudden loss of consciousness, in the absence of trauma, is usually due to an inadequate flow of blood to the brain. Occasionally, it may be due to a "stroke." This condition occurs in many types of cardiovascular disorders since the problem may originate in the heart itself and/or in the major blood vessels which carry blood to the brain. For example, vigorous exercise in a warm environment or other factors which cause overheating may lead to fainting because, under these conditions, a high proportion of the blood pumped by the heart is shunted to the skin for the dissipation of the excess heat. Also, partial blockage of the carotid artery in the neck by atherosclerotic deposits may be accompanied by frequent spells of dizziness or fainting. Finally, certain types of emotional reactions or factors which stimulate the nerves of the digestive tract may cause a marked slowing of the heart rate.

• **Fluid accumulation in various tissues**—This disorder, which is also called *edema*, may often be due to an impairment in the pumping action of one or both of the main chambers (ventricles) of the heart. Failure of the right side of the heart leads to fluid accumulation in the legs, whereas left-sided heart failure leads to fluid accumulation (congestion) in the lungs. Fluid buildup in the abdomen (ascites) occurs in the advanced stage of heart failure, but it may also occur as a result of a liver disorder like cirrhosis.

• **Enlargement of the veins in the neck**—The veins in the neck and upper part of the chest may be unusually prominent when there is some type of interference with the flow of blood through the right side of the heart.

• **Chronic tiredness**—A person may feel unusually weary before a heart attack, although there are many major and minor disorders which may be responsible for this condition. Some typical heart-related causes of chronic tiredness are (1) slowing of the heart rate produced by various emotional states or nervous stimuli; (2) heart failure; (3) a drastic reduction in the volume of blood from such causes as hemorrhage or dehydration; and (4) a weakening of the heart muscle due to nutritional deficiencies and/or imbalances between such mineral elements as calcium, magnesium, sodium, and potassium. Severe anemia might also be responsible for this condition.

• **Palpitations**—This term refers to the feeling that one's heart is fluttering or throbbing. However, some people chronically complain of palpitations although no abnormalities are detected by their doctors, whereas others may not even notice gross abnormalities. Nevertheless, it often pays to have the patient describe his or her palpitations as accurately as possible, since they may only occur in certain circumstances; yet they may have serious consequences when they do occur.

• **Coughing up blood**—Most people associate this sign with tuberculosis. However, it might be due to the fluid accumulation in the lungs which accompanies failure of the left side of the heart.

• **Bluish coloring of the skin**—This sign, which is also called *cyanosis*, is due to a lack of sufficient oxygen in the tissues because of a disorder of the heart and/or the lungs. However, it is difficult to recognize mild degrees of this condition, so it is not very useful as an early diagnostic sign of disease.

• **Arterial blockage in the legs**—Some of the common symptoms of this condition are lameness, numbness, or pain in such parts of the legs as the buttocks, hips, thighs, lower legs, or the feet. It is also called *intermittent claudication* because the symptoms often appear during walking, and disappear during resting. A common sign of this condition is a draining of color from the skin of the leg when it is elevated above the rest of the body. Obvious artery disease in the legs may be a sign that advanced atherosclerosis is present in other parts of the body. Furthermore, the problems in the legs may sometimes be due to blockage of blood vessels by blood clots which might become dislodged, and travel to the heart, lungs, or brain. A similar condition in the arms may be indicated by pain in the muscles of the arm which is brought on by such activities as painting, washing windows, typing, or writing.

• **Wheezing**—This sign may indicate that the lungs are congested with fluid, a condition which may result from heart failure. Hence, this condition may be called *cardiac asthma*. Heart failure may also bring on or aggravate *bronchial asthma*, which may be mistakenly thought to be due to an allergy.

Detection Of Cardiovascular Disorders By Physical Examination. Although it is usually preferable for a doctor to conduct a physical examination for heart problems, emergency situations may require that the examination be made by a nurse, paramedic, or even an untrained lay person. Hence, brief descriptions of some common examination procedures follow:

• **Taking of the pulse**—The usual procedure is for the examiner to place his (her) index finger over the radial artery (which may be found on the underside of the wrist, just below the base of the thumb) and to count the heart rate in beats per minute. (Usually, the rate is counted for only 15 to 30 seconds, and the count obtained is multiplied by 4 or by 2.) Sometimes, it may be easier to feel the pulsations in either the right or left carotid artery in the neck. (These arteries run from just above the upper points of the collarbone to just below the back edges of the lower bone of the jaw.) Normal pulse rates range from 68 to 85 beats per minute. However, the heartbeat increases from 8 to 10 beats per minute for each degree Fahrenheit above normal body temperature, so that a markedly rapid pulse may be expected to accompany a high fever. Contrariwise, certain highly trained athletes may have pulse rates as low as 40. Doctors and nurses may often be able to identify various heart problems by noting whether the pulse is slow or fast, weak or strong, and regular or irregular.

• **Listening to the sounds of the heart**—The doctor or nurse listens to the heart by placing the bell of the stethoscope on the chest over the heart. Various types of unusual vibrations, which are commonly called *heart murmurs*, may indicate the presence of certain abnormalities. However, having a murmur does *not* necessarily mean that one has a type of heart disease, since many different circumstances, such as anemia and athletic endurance training, may cause murmurs which disappear when the circumstances change. It is noteworthy that children may be hampered in their normal development if they are wrongly diagnosed as having heart disease on the basis of a single examination in which a murmur is detected. Two doctors have shown that the rate of illness was greater in a group of children who had been falsely thought to have had heart disease than in a similar group who actually had this type of disease.[68]

• **Measurement of blood pressure**—A device for the taking of the blood pressure (sphygmomanometer) is sold in many drugstores. However, learning how to use this instrument correctly may take some practice in identifying the sounds which accompany the *systolic* blood pressure (the highest reading) and the *diastolic* blood pressure (the lowest reading). The normal blood pressure readings for adults up to 45 years of age range from 110 to 140 systolic blood pressure, and from 65 to 95 diastolic blood pressure. After age 45, the upper limit of normal systolic pressure is 100 + years of age up to a maximum reading of 160. Sometimes, the use of a blood pressure cuff which is too narrow results in readings which are erroneously high. Hence, special wide cuffs should be used to take the blood pressures of markedly obese people.

• **Examination of the eyes**—Certain types of heart disease may cause bulging, or even pulsation of the eyeballs. Also, a doctor using an ophtalmoscope may examine the blood vessels in the interior of the eyes and look for evidence of atherosclerosis, diabetes, and high blood pressure.

• **Exercise testing of heart function**—The simplest type of these tests consists of having the patient step up and down on a standard wooden step (or on a stool of similar height) at a given rate for a fixed period of time. More sophisticated testing equipment consists of such items as special exercise bicycles and treadmills. Heart function is usually assessed both before and after exercise in terms of pulse rate, blood pressure, and the time required for the pulse to return to normal after exercise. Symptoms of fatigue, breathlessness, and angina may reveal hidden weaknesses of the cardiovascular system.

A few people have suffered heart attacks during exercise testing, so great care has to be used in estimating the ability of the patient to withstand such stress. Hence, careful observations must be made of the patient during testing so that the test(s) may be stopped at the first sign(s) of trouble. Many of the laboratories which regularly perform exercise tests have on hand electric defibrillators (devices for inducing more regular heartbeats), heart drugs, and supplies for intravenous administration of fluids. Also, their staffs are trained in resuscitation procedures.

Special Diagnostic Procedures. Some of the characteristic features of heart disease are ambiguous in that they may sometimes be due to other causes. For example, severe pains in the chest may be due to a flare-up of gallstones, rather than a heart attack, Hence, the diagnostic procedures which follow are noteworthy:

• **Electrocardiogram (ECG)**—This term refers to a representation of the rise and fall of electrical charges in various parts of the heart muscle. It may be traced on a piece of paper or it may be displayed on the screen of an oscilloscope. Usually, an abnormal ECG is associated with such conditions as myocardial infarctions, nonrhythmic heartbeats, and other disorders where there is an abnormal flow of electrical charges in the heart muscle. However, certain types of heart attacks may not be detected in an ECG.

• **Analysis of blood for certain enzymes**—Elevations in the blood levels of certain enzymes indicate their leakage due to damage to the heart muscle, which often occurs in a heart attack.

[68]Bergman, A. B., and S. J. Stamm, "The Morbidity of Cardiac Nondisease in School Children," *New England Journal of Medicine*, Vol. 276, 1966, p. 1008.

• **X-rays of the heart**—These pictures show the shape of the heart and of the major blood vessels connected to it. Hence, abnormal bulges or enlargement of the heart or its attached blood vessels might be spotted in the x rays before other signs of the trouble are noted.

• **Analysis of blood for certain ions (electrically charged particles or electrolytes)**—It was mentioned earlier in this article that imbalances between sodium, potassium, calcium, magnesium, and chloride ions may lead to irregularities in the heartbeat. Hence, low or high blood levels of these ions may be indicative of similar conditions in the heart muscle. Also, certain types of electroylte imbalances may be responsible for some of the abnormalities which show up in electrocardiograms.

• **Measurement of hemoglobin in the blood**—Anemia, or lack of sufficient hemoglobin in the blood, may be the cause of a racing heart, or even heart murmurs. This condition is most likely to be found in young growing children, adolescents, pregnant women, and the aged.

On the other hand, an abnormally high blood level of hemoglobin may indicate polycythemia, a condition characterized by an excess of red blood cells, which is usually found in sedentary, middle-aged men. This abnormality may make the blood clot too readily.

• **Insertion of a catheter into the heart**—Sometimes, it may be necessary for a doctor to pass a long, slender, flexible tubing, called a *catheter*, up through a vein in an arm or leg into the heart. This is done in order to obtain direct measurements of (1) the amount of oxygen in the blood right after it comes from the lungs, (2) location(s) of lesions in the heart muscle, (3) effectiveness of the heart valves in regulating the flood of blood between the various parts of the heart and the attached blood vessels, and (4) the pressures which develop in various areas of the heart during the pumping of blood.

• **Analysis of blood for various fatty substances (lipids)**—High levels of blood fats are most likely to be associated with severe atherosclerosis in people who come from families with histories of such conditions. In others, high blood fats may not be inherited, but might be the consequences of (1) emotional or physical stress factors, or (2) the long-term consumption of a diet high in fats, sugars, or calories.

TREATMENT AND PREVENTION OF HEART DISEASE.

Many people with severe heart disease have lived full lives for long periods after the onset of their medical problems, thanks to well-designed therapies. Therefore, brief descriptions of a variety of helpful measures follow.

Emergency Care For Victims Of Heart Attacks.

While it is beyond the scope of this article to provide comprehensive descriptions of the various emergency measures for heart attacks, it is nevertheless appropriate to consider briefly what actions might be taken in certain critical situations, so that concerned people may take steps to prepare themselves for dealing with such emergencies. Hence, descriptions of both life-threatening emergencies and recommended countermeasures follow. (It is noteworthy that in many communities there are special courses in which lay persons are taught how to assist heart attack victims through severe crises.)

CARDIAC ARREST. Stoppage of the pumping of blood by the heart is usually followed promptly by cessation of breathing. These are the most critical conditions which may accompany a heart attack because cells in the brain begin to die within a few minutes after the delivery of blood and oxygen is halted. Hence, *immediate* actions must be taken in order to (1) get the blood back into circulation, and (2) force air into the lungs.

Whether or not the heart is beating may be determined by feeling for a pulse in either one of the carotid arteries in the neck, or in the left or right femoral artery which runs along the inside of the thigh. The pulse in the wrist may be very difficult to detect when the heartbeat is weak. If a pulse appears to be absent in the patient, one should call for help and immediately begin the resuscitation procedure illustrated in Fig. H-10.

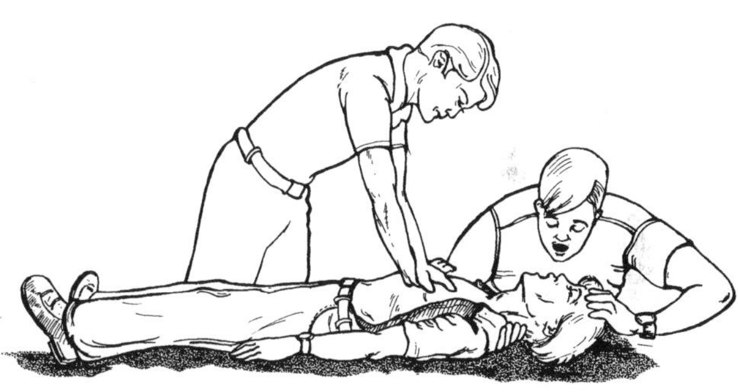

Fig. H-10. Revival of a person whose heart has stopped. this procedure is called *cardiopulmonary resuscitation (CPR)*. It involves (1) rhythmically pressing down the lower part of the breastbone against the heart 60 to 80 times per minute so as to pump blood to the tissues, and (2) mouth-to-mouth breathing 16 to 20 times per minute so as to force oxygenated air into the lungs of the victim. When resuscitation is attempted by only one person, every 15 chest compressions should be alternated with two respirations.

The procedure shown in Fig. H-10 should be continued until (1) the heart action and breathing of the victim resumes and remains steady, (2) help arrives with a defibrillator and a respirator, (3) at least 60 minutes have elapsed after cardiac arrest, when neither events (1) or (2) have occurred. Although *brain damage* may occur within a few minutes after cardiac arrest, the rhythmic compression of the chest and mouth-to-mouth breathing by the person(s) attempting resuscitation may provide a flow of blood to the brain which is sufficient to prevent such damage.

Another means of treating cardiac arrest is shown in Fig. H-11.

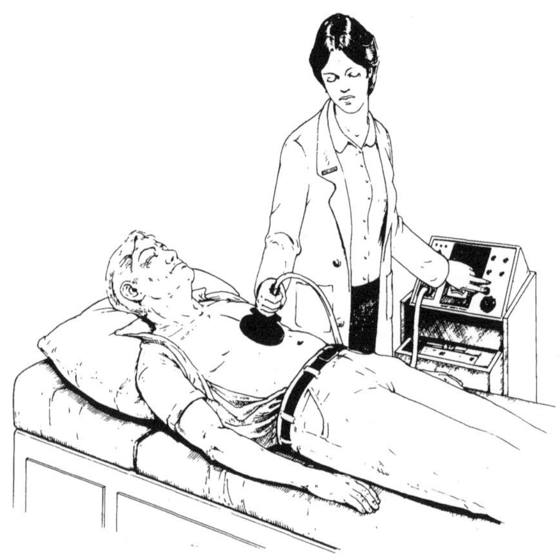

Fig. H-11. Using a defibrillator on a person in cardiac arrest. An electrical shock is applied to the chest of the victim in an attempt to induce the heart to resume its beating.

In addition to these emergency measures, doctors, nurses, or paramedics may inject one or more doses of adrenalin, atropine, digitalis, lidocaine, quinidine, or sodium bicarbonate so as to obtain the desired rhythm, speed, and force of the heartbeat.

If the attempts at reviving the patient have been successful, he or she is usually placed in the coronary care unit (CCU) of a hospital until it is fairly certain that the crisis has passed.

HEART ATTACK WITHOUT CARDIAC ARREST. A person might have the persistent chest pain of a heart attack, yet his or her heart rate and the blood pressure may remain normal. However, there is always the danger that the heartbeat will become highly irregular, or stop altogether, if the attack is not promptly and properly treated. Hence, some recommendations for handling this kind of emergency follow:

1. Send for assistance—a doctor, an ambulance, or a rescue squad. In some communities, paramedic teams may work with the firemen and police.
2. Make the patient as comfortable as possible by the loosening of tight clothing, helping him or her settle into a relaxed position on a soft surface, minimizing discomfort due to chilling or overheating, and generally going about things in a calm way so as to avoid exciting the patient.
3. Observe the patient carefully so that you might notice any turns for the worse, such as loss of consciousness and cardiac arrest. Be prepared to carry out promptly the revival measures outlined in the preceding section if such should be necessary.

ANGINA PECTORIS. This condition is characterized by pain similar to that which occurs in a heart attack, only it differs in that it is usually relieved in a few minutes by resting. However, it is noteworthy that the long-term survival rate for people with angina is only about as good as that for people who have had heart attacks. The outlook is disappointing because heart attacks may often strike those afflicted by one of the three dangerous types of angina which follow:

- **The first occurrence of angina pectoris**—It is difficult to predict the outcome when the patient has had no prior experience with this condition.

- **Angina which is not readily relieved by resting and nitroglycerine**—Pain which lasts after 15 to 20 minutes of resting and a moderate dose of nitroglycerine may indicate the onset of a heart attack.

- **Steadily worsening angina**—Sometimes, ever higher doses of nitroglycerine are needed to relieve the pain, the episodes of pain last longer, and they occur more frequently.

A doctor should be called for any of the three conditions described above.

- **Stable angina**—This refers to the type of angina which is predictably brought on by certain circumstances, and which is readily relieved by resting and/or moderate doses of nitroglycerine. The outlook for survival is much better in the case of stable angina than it is for any of the three dangerous types of the condition.

HEART FAILURE. The characteristics which differentiate heart failure from a heart attack are that the former (1) usually develops more slowly; (2) may occur without pain; (3) is often accompanied by difficulty in breathing, weakness and chronic fatigue, fainting and/or dizziness, a rapid and feeble pulse, lack of appetite, swelling and tenderness of the liver, and fluid accumulation in the legs, lungs, and/or the abdomen.

However, the early stages of heart failure may be difficult for physicians to diagnose. Hence, the lay person is not likely to be in a position to make such a judgment. Therefore, anyone who suspects that a person they know has been stricken with heart failure should call a doctor. While awaiting medical assistance, someone should remain with the patient and be ready to carry out cardiopulmonary resuscitation, if necessary.

Long-Term Treatments. There is good news for those who suffer from heart disease! Since 1975, there has been a reduction in deaths from heart disease in the United States. By following proper treatment, a heart patient can add many years to his or her life, and add life to the years. The sections that follow tell how this can be achieved.

BIOFEEDBACK. This term refers to various techniques by which a limited amount of voluntary control is gained over certain body functions that were hitherto thought to be strictly automatic and therefore not subject to such control. It may be necessary for certain excitable people to learn how to exert control over their nerves so that they may prevent steep rises in blood pressure and/or racing of the heart when they become excited. Overreaction to emotional excitement, if habitual, could offset the benefits which accrue from the treatment of their disease. However, they may be taught how partially to control their blood pressure and heart rate by biofeedback, thereby countering the threat of various cardiovascular disorders.

Learning these techniques requires special exercises which are too detailed to describe here. However, briefly stated, the basic principles are: (1) the subject must have some means of measuring blood pressure, breathing, and heartbeat; (2) efforts are made to change the measurements by concentration, conjuring up various images, movements of certain muscles, and relaxation; and (3) the subject learns which of the various types of action were successful in altering certain functions. Hence, the *feedback* is the conscious *input* of the subject which alters the *output* of his or her body. For example, he or she watches for the tightening of the neck or shoulder muscles. Then, relaxation techniques are employed to loosen up the muscles.

Details of biofeedback may be obtained from recently published books on this subject, which are available in many public libraries.

(Also see BOOKS.)

CHANGING OF BEHAVIOR. The effects of this type of therapy may augment or complement those obtained from biofeedback in that a patient learns how to avoid or minimize behaviors which may be self-destructive.

An important type of desirable behavior change is the development of certain attitudes which lessen stressful emotional reactions to common frustrations such as dealing with aggravatingly slow-paced people or driving in traffic jams. Also, the patient learns how to plan ahead so as to minimize the occurrence of stress-producing situations. Some suggestions for these behavior changes follow:

1. Try to anticipate slow traffic in bad weather, and allow extra time to get to work or to meet other commitments.
2. Drop unproductive, time-consuming activities such as participation in social affairs which fail to provide any tangible personal benefits.
3. Put in extra working hours, when necessary, on a regular basis so as to avoid "burning the midnight oil" on several consecutive nights prior to deadlines.
4. Beg off optional engagements which require staying up late on nights before working days.
5. Avoid excessive striving to win every argument, contest, or game engaged in with other people when the outcomes are not important.
6. Practice biofeedback, meditation, prayer, and/or other means of obtaining emotional peace without drinking, overeating, or taking drugs.

DIETARY MEASURES. It is widely believed that many people might forestall the onset of cardiovascular disease by their adoption of more healthful diets. However, these diets should be planned on an individual basis so as to take into account the variable factors such as age, health status, heredity, level of activity, and sex. Not all people will need or benefit from drastic dietary changes. Hence, it is worth noting the recommendations which follow.

Modified Diets for People Who May Be at Risk. People who might be expected to have more than an average chance of developing heart disease are those who (1) have elevated blood levels of cholesterol (values over 240 mg per 100 ml of blood) and/or triglycerides (values over 150 mg per 100 ml of blood); (2) come from families with histories of premature heart disease; (3) show signs of angina pectoris or poor circulation of blood to various parts of the body; (4) have diabetic tendencies; or (5) who are obese (overweight by 20% or more). These people need to consult a physician in order to obtain the proper diagnosis and advice as to the dietary measures which might be most suitable for them. Brief descriptions of some modified diets follow:

- **Low-cholesterol diets**—People with high blood cholesterol are put on this type of diet because it is believed that lowering the cholesterol level might minimize the chances of its deposition in atherosclerotic plaques. Generally, eggs are restricted to one every other day, for a total of not more than four per week. Also, it is usually recommended that the intake of saturated animal fats be at least partially curtailed, and that they be replaced by margarines and vegetable oils which are rich in polyunsaturated fatty acids (PUFA).

NOTE: The body's need for essential fatty acids may be met by 1 to 2 teaspoons (*5 to 10 ml*) of polyunsaturated oils per day. Large amounts of such oils may have harmful effects because they are easily oxidized to toxic substances unless there is sufficient dietary selenium and/or vitamin E to counteract such oxidation. Furthermore, diets which are high in fat may not be sufficiently filling when the total energy intake is restricted. Hence, it may not be wise to replace completely the deleted animal fats with vegetable oils. It might be better to fill the gap with such fiber-containing starchy foods as beans and whole grain products.

- **Low-energy diets**—Reduction in excess weight frequently leads to lowering of blood cholesterol and triglycerides, and correction of diabetic tendencies. These diets are made up of foods low in carbohydrates and fats, but rich in protein, minerals, and vitamins. Some typical items are lean meats, whole grain breads and cereals, green leafy vegetables, and low-fat dairy products.

- **Low-fat diets**—Low-fat diets are usually prescribed for people who have hereditary tendencies for high levels of blood fats. These diets are usually bulkier than ordinary diets because foods low in fat usually contain plenty of carbohydrates and water. Generally, the restricted foods are animal fats, whole milk, eggs, butter, rich cheeses, fatty meats, ice cream, margarine, and vegetable oils.

- **Low-salt (low sodium) diets**—The restriction of dietary sodium may be necessary for people who are susceptible to the accumulation of fluid in their tissues, heart failure, and high blood pressure. Thus, in addition to salt, this calls for restricting other sources of sodium such as baking soda (sodium bicarbonate) and monosodium glutamate. Unfortunately, it is often difficult to determine the amounts of sodium which are present in the various types of foods. The best source of this information is a dietitian who has had experience in planning low-salt diets.

• **Low-sugar diets**—These diets are often prescribed for people who have high blood levels of triglycerides. They are based upon drastic restricton of simple sugars, fruits, and sweets, but they allow for a moderate intake of breads, potatoes, and other starchy foods which are low in sugars.

• **Vegetarian diets**—There has long been interest in the practice of vegetarianism for religious reasons and for countering various health problems, although only a few physicians have recommended this practice. However, the current concern over heart disease has led to recent research into the nutritional merits and demerits of vegetarian diets. For example, a study of certain participants in the Framingham Study showed that vegetarians had blood levels of cholesterol and triglycerides which were about one-third lower than those in meat eaters.[69]

Some demerits of very strict vegetarian diets (those which eliminate meats, fish, dairy products, and eggs) are (1) they are likely to lead to a deficiency of vitamin B-12 unless a supplement containing this nutrient is taken, (2) the forms in which the trace elements and vitamins are present in plant foods are often less well utilized than those in animal products, and (3) it may sometimes be difficult for people to eat the quantities and varieties of vegetarian foods which are needed to provide adequate nutrition.

The diets discussed above are most likely to be effective if (1) the food is distributed between three or more meals; (2) foods rich in energy, fats, and/or sugars are eaten prior to periods of physical activity, rather than before taking it easy or retiring for the night; and (3) moderate to vigorous exercise is engaged in on a regular basis (daily, or at least 4 days per week).

Please note that no specific diets are detailed here, rather, the reader might obtain guidance in following a diet which has been prescribed by his or her doctor, by referring to the examples given elsewhere in this book, and to the information presented in the section which follows.

(Also see DEFICIENCY DISEASES; DIABETES MELLITUS; HYPERLIPOPROTEINEMIAS; HYPOGLYCEMIA; MODIFIED DIETS; OBESITY; SALT DIET, LOW; and VEGETARIAN DIETS.)

Selection of Foods for Modified Diets. Diets which are modified in order to avoid excessive intakes of cholesterol, energy, fats, and/or sugar may lead to nutritional deficiencies unless special efforts are made to select foods which have low levels of the restricted nutrients, but which have moderate to high levels of protein, minerals, and vitamins. These foods are generally called "protective foods" because the nutrients they contain help to maintain the health and strength of tissues against the onslaught of aging and various other stress factors. For example, a sound heart muscle might better withstand the effects of a "coronary" than one which has been weakened by deficiencies of various essential minerals and vitamins.

It would be desirable to rely as much as possible on obtaining nutrients from foods, and to resort to nutritional supplements only when certain nutrient-rich foods are not liked by the patient, or cannot be eaten because of medical problems. In order to plan diets containing optimal amounts of the required nutrients, it is necessary to (1) eat a wide variety of foods from both animal and plant sources, and (2) select minimally refined foods whenever possible. This means that a typical diet might contain such items as those which follow:

• **Alcoholic beverages**—A glass of beer or wine before retiring may help one to get a good night's sleep. Also, the undistilled beverages (beer and wine) might contain such vital substances as chromium in the form of the glucose tolerance factor. However, beverages containing alcohol should *not* be taken by people with tendencies towards alcoholism, high blood levels of triglycerides, or irregulr heartbeats.

(Also see ALCOHOLIC BEVERAGES; BEERS AND BREWING; and WINE.)

• **Beans and peas, eggs, fish, meats, nuts, oilseeds, and poultry**—Usually, 4 to 6 oz (*112 to 168 g*) of high-protein foods from this group, when taken with suitable quantities of breads or cereals and dairy products, provides a liberal daily allotment of protein. However, eggs, beef, pork, and lamb should be limited to one serving per day (1 egg, or 2 to 3 oz [*56 to 84 g*] of meat) or less for people with high blood cholesterol. The nuts and oilseeds (pumpkin, soybean, and sunflower seeds) are rich in polyunsaturated fat; but they are also rich in calories. Hence, people who are required to cut down on both cholesterol and fats may have to obtain much of their protein from beans and peas, fish, very lean cuts of meat, and the breast meat of poultry.

(Also see BEAN[S]; EGG[S]; FISH AND SEAFOOD[S]; MEAT[S]; NUTS; OILSEED; POULTRY; and PULSES.)

• **Breads and cereal products**—The low-cholesterol, low-fat diets published by the American Heart Association allow 12 bread exchanges (each exchange is equivalent to one slice of bread or an equivalent amount of another high-carbohydrate food) for women on a 2,000 Calorie (kcal) per day diet, or 15 exchanges for men on a 2,400 Calorie (kcal) diet.[70] It would be desirable if as much as possible of this daily allowance were to be whole-grain products because (1) they contain more fiber and are more filling than the highly milled (white flour) products, and (2) whole grains are good sources of essential minerals and vitamins.

(Also see BREADS AND BAKING; and CEREAL GRAINS.)

[69]Sacks, F. M., *et al.*, "Plasma Lipids and Lipoproteins in Vegetarians and Controls," *New England Journal of Medicine*, Vol. 292, 1975, p. 1148.

[70]*A Maximal Approach to the Dietary Treatment of the Hyperlipidemias Diet A: The Low Cholesterol (100 mg) Moderately Low Fat Diet, American Heart Association*, New York, N.Y., 1973, pp. 12, 14.

• **Cocoa, coffee, and tea**—These beverages contain caffeine and/or similar types of stimulants which are usually harmless when taken in moderate quantities (not more than 5 cups [*1.2 liter*] per day). However, the stimulants may cause irregularity of the heartbeat in susceptible people, and may also keep some people awake if taken just before retiring. Those who are so affected by these items might switch to (1) carob mixtures which resemble cocoa poducts, (2) decaffeinated or imitation (toasted cereal) coffee, or (3) herb teas.

(Also see CAFFEINE; CAROB; COCOA AND CHOCOLATE; COFFEE; HERBS; and TEA.)

• **Dairy products**—People with high blood lipids (hyperlipoproteinemias) and/or cardiovascular disease are usually advised to eat low-fat or nonfat types of cheeses and milk, and to avoid cream, creamy cheeses, ice cream, and whole milk. However, it is very important for everyone (except those with milk allergies) to eat *every* day the calcium equivalent of 3 cups (*720 ml*) of milk (800 mg of calcium) so as to obtain sufficient amounts of this mineral to forestall the bone loss which often accompanies aging.

(Also see MILK AND MILK PRODUCTS.)

• **Fruits and Vegetables**—These nutritious items fit well in diets with restrictions on cholesterol, energy, and fats. Hence, alomost all dieters may take liberal amounts of these items.

It is recommended that the minimum intake of fruits and vegetables be four servings of 2 to 4 oz (*56 to 112 g*) each, and that at least one item be rich in vitamin A and another be rich in vitamin C (such as a citrus fruit). There is *no limit* to the amounts which may be taken of very low-calorie raw vegetables such as chicory, Chinese cabbage, endive, escarole, lettuce, parsley, radishes, and watercress.

(Also see FRUIT[S]; and VEGETABLE[S].)

• **Herbs and spices**—Many of these items are rich in trace minerals and vitamins. They are also a boon to people on salt-restricted diets, because, unlike salt, they are more likely to stimulate the kidneys to excrete rather than to retain water. Furthermore, it was noted earlier in this article that garlic and onions contain substances which counteract the tendency for blood to clot more rapidly after a meal rich in animal fat.

CAUTION: It may be dangerous to try certain unfamiliar herbs and wild plants since they may contain strong, druglike substances that affect the heart. For example, extracts from such plants as foxglove, lily of the valley, oleander, and hawthorn have long been used as medicines to regulate the heartbeat. Dangerous toxicity may result from overdoses. Even the familiar red peppers contain a highly irritating substance (capsaicin) which is suspected of causing inflammation of the lining of blood vessels when taken in excessive quantities.

(Also see HERB; POISONOUS PLANTS; SPICES; and WILD EDIBLE PLANTS.)

• **Sweeteners**—The problem with most of these items is that they provide energy, but little else in the way of nutrients. Furthermore, it may be difficult to achieve control of the blood sugar in people with diabetic tendencies when more than a minimal proportion of the dietary energy is supplied by highly refined simple sugars. However, it appears that certain crude forms of sugar such as the unprocessed cane juice and dark molasses contain chromium, a trace mineral which is suspected of having a protective effect on the heart and blood vessels. Unfortunately, chromium and the other trace minerals that are removed from sugar during its refining are concentrated in the dark molasses which is usually given to livestock. Hence, it might be a good idea for more people to use molasses rather than white sugar for sweetening of their foods.

(Also see CARBOHYDRATE[S]; CHROMIUM; GLUCOSE TOLERANCE FACTOR; and SUGAR.)

• **Vegetable oils and margarines**—In the early 1970s, when there were great expectations that liberal amounts of polyunsaturated fatty acids might prevent atherosclerosis, certain health professionals recommened that 2 tsp (*10 ml*) of vegetable oil or margarine be consumed for each ounce (*28 g*) of meat. This recommendation, if followed, might have led to heavy meat eaters getting more than half of their dietary energy from fat. Since then, a number of research reports have suggested that overconsumption of hydrogenated fats (present in most margarines, salad oils, and cooking oils) might be hazardous to health. Also, the current trend is to recommend the replacement of dietary fat by starch and other complex carbohydrates. Therefore, 3 to 4 Tbsp (*45 to 60 ml*) per day of oil and/or margarine might be considered to be plenty for most people.

A related matter of concern is whether most Americans consume enough vitamin E for protection against the toxic peroxides which may be formed from the polyunsaturated fats present in the oils. Fresh, unheated oils from cottonseed and safflower are much richer in vitamin E than those derived from corn and soybeans. It was noted earlier that the prolonged use of polyunsaturated vegetable oils for frying foods at high temperatures hastens the destruction of vitamin E and promotes the formation of toxic substances.

Most doctors and dietitians recommend that their patients select soft margarines which contain liquid vegetable oils because these products contain more polyunsaturated fats than those containing mainly hardened (hydrogenated) oils.

(Also see FATS AND OTHER LIPIDS, section headed "Fatty Acids"; OILS, VEGETABLE, section headed "Antinutritional and/or Toxic Factors"; and TRANS FATTY ACIDS.)

DONATIONS OF BLOOD. In the not too distant past, the application of leeches to the skin of patients and bloodletting (the medical term is phlebotamy) were standard medical practices. Even today, doctors may periodically withdraw blood from people who have too many red cells (the condition called polycythemia) because an abnormally thickened blood clots too readily. Hence, it might be a good idea for those who have a high red cell count to make donations of their blood after first receiving assurance from their physicians that it is safe for them to do so.

DRUGS. It may be tempting for people who have high blood levels of lipids and/or the earlier signs of cardiovascular disease to make only half-hearted attempts to adhere to stringent diets when they know that there are drugs which may be used to treat their disorders. However, more and more doctors are recommending that their patients make every effort to control their conditions by dieting, exercise, and healthy living habits because there are many dangers associated with most of the drugs which are commonly used. Nevertheless, there may be no choice other than drugs for people with hard-to-manage disorders. Hence, brief discussions are provided so that the reader might become more aware of some of the problems which may often be avoided, or at least minimized, by close communication with one's doctor and careful following of his or her instructions. Details follow.

Interactions Between the Diet and the Actions of Drugs. One reason for the variations in the effects of drugs on different people is that there are numerous interactions between nutritional factors and the responses of the body to the drugs. Some typical interactions follow:

1. The destructive metabolism of drugs by the liver is speeded up by (a) increasing the level of dietary protein, and (b) taking vitamin E supplements. Contrariwise, nutritional deficiencies render people more susceptible to the toxic effects of many drugs.
2. Erratic patterns of eating such as fasting one day and feasting the next may cause fluctuations in the rate of drug metabolism, thereby making it difficult to arrive at the proper dosage for drugs.
3. Chronic consumption of liberal amounts of alcohol and of the barbiturate type of sleeping pills stimulates the activity of the drug-metabolizing enzymes in the liver so that larger doses of drugs may be required to achieve the desired results.
4. Crash diets may weaken the body so that it becomes more susceptible to the toxic effects of drugs.
5. It is often important to heed carefully how a drug is to be taken in relation to meals. That is, it may work best when it is taken (a) before meals, (b) after meals, or (c) between meals.
6. Diseases of the liver and of the kidneys are usually accompanied by slowing of the breakdown and excretion of drugs. Hence, the drug effects may be more pronounced, or even toxic.
7. Cocoa, coffee, cola beverages, and tea contain caffeine and related substances which tend to stimulate the heart to work harder. Hence, large quantities of these beverages may either augment or counteract the effects of certain cardiac drugs.

Side Effects of Some Commonly Used Drugs. It may often be necessary to use strong drugs in the treatment of certain cadiovascular problems because the healthy human body has built-in mechanisms which oppose drastic changes in the heart rate, diameters of blood vessels, and the amounts of fluid retained in the blood and tissues. The amount of natural resistance to these changes varies among different people so that what is a safe and effective dose of a drug for some may be either ineffective or excessive and toxic for others. Therefore, it is worth noting that some of the commonly used cardiovascular drugs may have side effects such as those which follow:

• **Anticlotting (anticoagulant) drugs**—These drugs, which are also known as blood thinners, must be carefully monitored by the frequent measurement of blood clotting time because certain investigators have found an increased incidence of hemorrhage and rupture of the heart in patients given these drugs.[71] It is estimated that about 25,000 people die yearly from cardiac rupture. The most frequently used anticoagulants are the coumarin type, which block the effects of vitamin K on blood clotting. Hence, the effects of these drugs might be increased to dangerous levels by other factors which interfere with blood clotting, among which are (1) administration of antibiotics that kill the microorganisms which synthesize vitamin K; (2) the taking of large supplemental doses of vitamin E that antagonize the effects of vitamin K; (3) ingestion of the drug cholestyramine which reduces the absorption of vitamin K and similar fat-soluble substances; and (4) diets that contain little or no green, leafy vegetables, which are the best food sources of vitamin K.

Another commonly used cardiovascular drug is heparin, which is both an anticoagulant and a blood-lipid-lowering compound. However, it is derived from animal tissues and may provoke such allergic reactions as asthma, histamine release, hives, runny nose, and tearing of the eyes. Also, large doses have caused osteoporosis because it binds with calcium in the blood. Finally, it may allow the prolongation of bleeding and prevent the healing of ulcers and other lesions of the digestive tract.

• **Cholestyramine**—This substance binds the bile salts, which are secreted with the bile from the liver into the intestine, prevents their reabsorption, and carries them out of the body with the stool. The body's supply of cholesterol is diminished by the action of cholestyramine because more of the sterol has to be converted to bile salts in order to replace those which were withdrawn from the body.

It is probably wise to administer this drug at different times than those at which various fat-soluble drugs and vitamins are taken, since it is uncertain to what extent the absorption of the latter may be affected by the former.

• **Clofibrate**—People with high blood chlosterol are given this drug because it slows the rate at which the sterol is synthesized by the body. It is often necessary to reduce the dosage of anticoagulant drugs when clofibrate is given in order to prevent bleeding complications.

• **Digitalis or digitoxin**—These compounds stabilize the force and rhythm of the heartbeat in people who have heart failure or racing and/or irregularity of the heartbeat. They also pose some dangers in that the margins between doses which are effective and those which are toxic are small and sometimes difficult to determine. Furthermore, patients who are deficient in potassium or magnesium are likely to be highly susceptible to toxic

[71]Bates, R.J., *et al.*, "Cardiac Rupture—Challenge in Diagnosis and Management," *The American Journal of Cardiology*, Vol. 40, 1977, p. 433.

effects. Some early signs of toxicity are lack of appetite, nausea, and/or vomiting. It is noteworthy that the injection of calcium salts may bring on the toxic effects of these drugs.

• **Diuretics**—There is now a wide range of substances—including alcoholic beverages, coffee, and tea—which help to flush excess water from the body. Hence, the precautions to be taken depend upon the type(s) of drug(s) used and on the health status of the patient. Nevertheless, it is important to note that certain types of diuretics also flush potassium out of the body so that extra amounts must be provided in the diet. However, the indiscriminate taking of potassium salts may be very dangerous for people who have kidney disease. Therefore, the instructions of the doctor and/or the dietitian should be carefully followed in this regard.

• **Nitroglycerin**—It is generally believed that this drug relieves the pain of angina pectoris because it increases the flow of blood to critical areas of the heart muscle. A few people who are extra sensitive to nitroglycerin may have such side effects as a subnormal blood pressure, dizziness, flushing due to dilation of blood vessels in the skin, and weakness resulting from a reduction in the flow of blood to the brain. The consumption of alcoholic beverages increases the likelihood of these side effects.

Sometimes, patients develop a tolerance to nitroglycerin so that larger doses are needed to achieve the same effect. For this reason, people should not take more than is needed to relieve the pain, even though this drug is considered to be one of the safest for this use.

• **Papaverine hydrochloride**—This drug relaxes the heart muscle and the smooth muscle which controls the diameter of the arteries. Hence, its action results in an increased flow of blood to the heart muscle and to such other tissues as the brain. Although papaverine hydrochloride is quite safe for most people, a few experience lack of appetite, nausea, vomiting, constipation, dizziness, sweating, sleepiness, and headache.

• **Propanolol hydrochloride**—People who have narrowing of the aorta, racing of the heart, irregular heartbeats, and/or excessive secretion of the hormone adrenaline are given this drug because it reduces the sensitivity of the heart to the stimulatory effects of the hormone. However, it may be dangerous to administer propanolol hydrochloride when (1) the heart rate is excessively slow, (2) there is a heart failure, or (3) when other relaxant drugs have been administered; because the heart might then lack sufficient stimulation to do its work.

This drug also makes other tissues less sensitive to adrenalin, which may lead to life-threatening conditions in such uses as (1) *hypoglycemia*, where it prevents the appearance of adrenalin-induced warning signs; (2) *bronchial asthma* and other allergic disorders, because adrenalin opens up the bronchial tubes so that more air may be taken into the lungs; and (3) *mental depression*, because adrenalin is also a psychic stimulant.

• **Quinidine sulfate**—This drug is used to restore and maintain the regularity of the heartbeat. Also, it is often used right after the revival of people who have suffered cardiac arrest.

Some of the side effects of quinidine sulfate are dizziness, blurred vision, ringing in the ears, trembling, nausea, vomiting, and diarrhea. The digestive disturbances may be minimized by taking the drug with food.

• **Reserpine**—People who are overly excitable, and prone to racing of the heart and high blood pressure may benefit from this long-used drug. Reserpine slows the beating of the heart and brings down the blood pressure by depleting the tissues of adrenalin and other stimulatory substances. Hence, it must be used with caution to avoid such side effects as an excessively slow heart rate, a subnormal blood pressure, extreme mental depression (which may sometimes verge on the suicidal state), flushing, impotence, stuffy nose, lack of appetite, nausea, vomiting, diarrhea, and aggravation of such disorders as ulcers and colitis.

EXERCISE. Recently, there have been many reports describing how people with heart trouble so serious that they might in the past have been consigned to lives of semiinvalidism, have been given new leases on life by supervised programs of strenuous exercise. The reasons for such dramatic recoveries are still unclear, although it has long been suspected that exercise-induced enlargement of the small arteries of the heart, which are sometimes called the collateral circulation, has been a factor in this regard.

Now that it is known that recovered heart patients may safely tolerate a moderate amount of exercise stress, more and more health scientists are measuring various cardiovascular functions before and after periods of moderate to intensive physical training. One study showed that 6 weeks of training on an exercise bicycle, twice a day, 5 days per week enabled six middle-aged people who had suffered attacks to more than double the time that they could pedal vigorously before the onset of chest pain.[72] These patients were tested again 8 to 12 months later, and it was found that all of them had maintained their improved tolerance for exercise. However, one patient had suffered a nonfatal heart attack 3 months after the training period, which shows that there is still a lot to be learned on this subject. For example, electrocardiograms which were taken of the participants in similar training programs have shown that damage to the heart muscle often remains for months or years, even though the ability to engage in strenuous activity may be greatly improved.

Furthermore, a few people have even had heart attacks while participating in exercise tests by physicians. Therefore, it is not wise for a customarily sedentary person to engage in vigorous exercise without first trying less strenuous activities. Of all such activities, walking at a moderate pace is probably the safest. It is heartening to note that a few people who recovered from heart attacks were able to build themselves up to the

[72] Redwood, D. R., D. R. Rosing, and S. E. Epstein, "Circulatory and Symptomatic Effects of Physical Training in Patients with Coronary-Artery Disease and Angina Pectoris," *The New England Journal of Medicine*, Vol. 286, 1972, p. 959.

point where they were able to complete marathons. At any rate, the patient should consult a doctor who will make some type of evaluation of cardiovascular fitness before specifying the exercises which are most suitable.

NOTE: It may be dangerous for people who have certain cardiovascular problems to perform isometric types of exercise (those which cause both the buildup of tension and the constriction of blood vessels in muscles) because these activities tend to raise the blood pressure. For example, pushups, knee bends, weight lifting, and similar activities have isometric components, whereas such activities as running and walking are basically nonisometric in nature.

SURGERY. There have been remarkable advances in the surgical treatment of cardiovascular diseases since World War II. Hence, it is now possible, with certain limitations, to replace worn out or diseased parts of the heart and blood vessels with either those from human or animal donors, or with artificial analogs of human organs. Brief descriptions of some common types of surgery follow:

• **Bypass graft(s) of a piece of leg vein onto a coronary artery**—In this well-known operation, which is frequently called a *coronary bypass*, a piece of a leg vein is grafted onto a blocked coronary artery. The piece of vein is attached to the artery so that it allows the blood to flow from the heart side of the occluded artery, around the obstruction, and back into the artery on the other side of the occlusion. However, a few patients may develop atherosclerotic blockage of the vein graft within a year or so. The reason for this development is not known for certain, although it is suspected to be due to whatever factors might have been responsible for the original coronary occlusion.

• **Grafts of dacron patches of tubes onto blood vessels**—This type of surgery uses dacron, rather than living tissue, to repair diseased blood vessels. For example, a dacron patch may be sewn into the wall of a blood vessel which has been surgically opened, because such a procedure allows for a larger opening in the vein or artery than might be obtained by merely stitching the cut edges together. Also, a knitted dacron tube may be used in lieu of a vein graft in a coronary bypass operation.

• **Heart transplantation**—In this procedure, the patient is connected to a heart-lung machine, then, his or her diseased heart is removed and replaced with a healthy heart from a donor. However, few medical centers now conduct this surgery because (1) there may be a rejection of the transplanted heart by the patient's body; and (2) treatments which decrease the likelihood of the transplant's rejection also render the patient more susceptible to infectious diseases.

• **Implantation of an artificial pacemaker**—Sometimes, it is necessary to provide artificial electrical stimulation of the heart for a patient who has what is commonly called a "heart block." A heart block is a pathological condition in which there is some type of interference with the natural pacemaker action in the heart.

Certain types of heart blocks may be counteracted by an artificial electronic pacemaker which is surgically implanted in the body. The usual procedure is for the surgeon to insert the pacemaker's electrodes, which are connected by wires to the main body of the device, up through a vein and into the heart. Then, the main body of the pacemaker which contains the power supply and the electronic impulse generator is placed under the skin on the chest. It is usually necessary to replace the power source every 5 to 7 years.

Patients who are fitted with artificial pacemakers should discuss with their physicians the precautions which should be taken to avoid electronic interference from microwave ovens, x-ray machines, and various types of electronic equipment.

• **Replacement of diseased heart valves with prosthetic ball valves**—The special flaps of tissue which serve as valves for the heart may become damaged or diseased, and require replacement. Most surgeons now use prosthetic valves as replacement parts rather than human or animal tissues because the artificial devices are better tolerated by the body. Hence, the type of valve which is implanted in the heart is likely to be a wear-resistant ball or disc that moves back and forth in a small wire cage in response to the ebb and flow of blood from the heart.

• **Surgical removal of atherosclerotic plaques (endarterectomy)**—This operation consists of the surgical removal of localized plaque from a blood vessel. A Dacron patch may be sewn onto the blood vessel to close the incision because the procedure produces a larger opening in the vessel than that of stitching the cut edges together.

SUMMARY. The thought of suffering from some form of cardiovascular disease has led many people to become overly anxious about their hearts. Yet, it may be seen from this article that there has been a steady advance in the ability of physicians to detect and treat all stages and types of heart disease. However, people who wish to minimize their problems in this regard might consider the remarks which follow:

1. The earlier that heart disease is detected and treated, the safer the treatment(s), and the better the outcome.
2. Regular exercise and restraint in eating are about the best preventative measures for people whose hearts are still healthy.
3. Whole foods, which are richer in fiber, minerals, and vitamins than their refined counterparts, may provide the nutrition that the heart needs to withstand the emotional and physical stresses of a full life.

HEAT EXHAUSTION

The result of overtaxing the body's mechanisms for keeping cool, and for maintaining balance between body water and various mineral salts.

• **Water depletion heat exhaustion**—Inadequate water replacement during physical labor or exercise may result in water depletion heat exhaustion. It is characterized by thirst, fatigue, giddiness, fever, and decreased urine output (oliguria).

- **Salt depletion heat exhaustion**—During hard physical labor or excercise both sodium and water are lost in the sweat, especially in unacclimatized persons. Water may be replenished by drinking; however, a sodium deficit still remains. Inadequate sodium replacement leads to heat exhaustion characterized by fatigue, nausea, vomiting, and giddiness.

Water loss usually exceeds sodium loss. Hence, free access to water is essential whenever one expects significant sweat losses. Free access to water is especially important when additional sodium (salt) is provided.

(Also see DISEASES, section headed "Injuries from Physical Agents"; SALT; and SODIUM.)

HEAT LABILE

Unstable to heat.

HEAT OF COMBUSTION

The heat of combustion or gross energy of a material is determined in a bomb calorimeter by the following procedure: An electric wire is attached to the material being tested, so that it can be ignited by remote control; 2,000 g of water are poured around the bomb; 25 to 30 atmospheres of oxygen are added to the bomb; the material is ignited; the heat given off from the burned materials warms the water; and a thermometer registers the change in temperature of the water. For example, if 1 g of material is burned and the temperature of the water is raised 1 °C, 2,000 Calories (kcal) are given off. Hence, the material contains 2,000 Calories (kcal) per gram. This value is known as the gross energy (GE) content of the material.

(Also see CALORIC (ENERGY) EXPENDITURE.)

HEIGHT INCREMENTS

Human growth is composed of two periods of rapid growth: (1) the first year of life, and (2) during puberty or adolescence. As determined by height increments, the human grows an average of 9 to 10 in. (*23 to 25 cm*) the first year of life, following which growth slows to 2 to 2¼ in. (*5.1 to 5.6 cm*) per year. Then, at puberty there is a growth spurt lasting about 4 years. This growth spurt usually begins earlier in females. During this time, height increments increase to about 2½ to 3 in. (*6.4 to 7.6 cm*) per year. In total, boys may grow 8 in. (*20 cm*) and gain 40 lb (*18 kg*) during this time. Girls grow slightly less during their growth spurt. Following this, the yearly height increments decline rapidly and by 17 or 18 years of age only slight height increments are noted. Understandably, during these growth spurts of adolecence, additional food requirements are indicated and usually met via increased food intake at mealtimes and between meal snacks. Little actual information exists as to the nutritional requirements during this time.

HEM-, HEMA-, HEMO-

Prefixes referring to blood.

HEMAGGLUTININS (LECTINS OR PHYTOAGGLUTININS)

Naturally-occurring plant toxins capable of causing red blood cell agglutination—clumping—are proteins called hemagglutinins. These are found in soybeans and other legumes. Adequate cooking destroys hemagglutinins. Experimentally, these toxins induce weight loss and inhibit the growth of rats.

(Also see POISONOUS PLANTS.)

HEMATOCRIT

An instrument for separating the solid elements (largely red cells) of the blood from the plasma.

HEMATOPOIESIS (HEMOPOIESIS; HEMOPOIETIC)

The formation of blood or of blood cells within the body.

HEMATURIA

The presence of blood or blood cells in the urine.

HEMERALOPIA

A defect of vision characterized by reduced visual capacity in bright lights—also called day blindness.

(Also see NIGHT BLINDNESS.)

HEMIPLEGIA

The most common type of paralysis. It often involves the loss of strength in the arm, the leg and sometimes the face on the same side of the body. In order of importance, hemiplegia may be caused by vascular diseases of the brain, trauma (injury), brain tumors, brain abcess, encephalitis, demyelination disease, and syphilis. Due to reduced activity, and the difficulties encountered in preparing and eating food, some dietary modification may be necessary.

(Also see MODIFIED DIETS.)

HEMOCHROMATOSIS

An iron overload in the body due to an inherited defect in iron metabolism or a high intake of iron in the diet. The clinical symptoms may include hyperpigmentation of the skin, deposition of iron in the liver and pancreas, cirrhosis of the liver, diabetes, and heart failure.

(Also see INBORN ERRORS OF METABOLISM; and IRON, section headed "Interrelationships.")

HEMODIALYSIS

Those individuals whose kidneys are completely nonfunctional—normal function continues when as little as two-thirds of one kidney remains—must have the waste products of metabolism removed from their blood. Without this, the concentrations of urea, uric acid, and creatinine increase. Also, acidosis or hyperkalemia (high blood potassium) and edema may develop. If not controlled death results in 8 to 14 days. Hemodialysis, using an artificial kidney, provides a means of removing the waste products from the blood.

The blood leaves the body via an artery in the arm and flows through a dialyzer where a semipermeable cellophane membrane separates the blood and a dialyzing fluid. Metabolic waste products pass into the dialyzing fluid and are carried away. Blood proteins and red blood cells are too large to pass through the cellophane membrane. The "refreshed" blood is warmed to body temperature and then returned to the body via a vein in the arm. These connections, "hookups", in the arm are usually permanent since dialysis must be repeated 2 or 3 times per week. Each treatment requires 4 to 12 hours depending upon the type of artificial kidney used. Blood is leaving and entering the body on a continuous basis until the process is complete. The treatments are expensive, but have saved many lives and allowed individuals to enjoy quite normal lives.

Persons using dialysis require restriction of sodium, potassium and water intake. Protein intake is also closely controlled and will vary from 30 to 60 g daily. The aim is to provide about ¾ of the protein allowance as protein of high biological value. Furthermore, to prevent high blood urea and potassium levels, it is especially important that adequate energy be supplied to prevent the catabolism of body protein. Dialysis patients also require vitamin supplementation in addition to the dietary controls.

(Also see KIDNEY DISEASES; METABOLISM; and MODIFIED DIETS.)

HEMOGLOBIN

The oxygen-carrying, red-pigmented protein of the red corpuscles.

(Also see ANEMIA; ERYTHROCYTES; and HEMOGLOBIN SYNTHESIS.)

HEMOGLOBIN SYNTHESIS

The red oxygen-carrying pigment of the blood, hemoglobin, is comprised of two substances. The first substance, the red color, is the heme which is manufactured in the developing red blood cells from acetic acid, the amino acid glycine, and iron. Heme is rather an odd-shaped octagon with a hole in the center wherein the iron is found. Once the heme is manufactured, it combines with the second substance, globin, a protein classified as a globulin; forming hemoglobin. About 7 g of hemoglobin are formed every day. It constitutes 90% of the dry weight of the red blood cell. Iron is essential for the synthesis of hemoglobin. Copper, vitamin C, vitamin B-6 (pyridoxine), folic acid and vitamin B-12 are also required. Each day, adult males normally lose about 1,000 mcg of iron; adult females may lose 1,500 mcg per day due to menstrual losses. If a deficiency of iron occurs, and is not corrected, iron deficiency anemia results. Liver, red meats, and dietary supplements are rich sources of iron.

(Also see ANEMIA; and IRON.)

HEMOLYSIS

Destruction of red blood cells.

HEMOLYTIC

Causing the separation of hemoglobin from the red blood cells.

HEMORRHAGE

Copious loss of blood through bleeding.

HEMORRHOIDS (PILES)

They are dilated and engorged veins of the anus and rectum which may be located either externally or internally. The veins in the area of the rectum and anus become stretched, dilated, and then knotted as the muscle fibers in the walls of the veins break down. Blood returning through these veins to the heart is slowed down and pooled thus distending a portion of the vein causing it to protrude and be covered by only the lining of the rectum (internal) or by the skin around the anus (external). Hemorrhoids are often noted during the following conditions: constipation, overweight, pregnancy, diarrhea, chronic liver disease, tumors, or incomplete evacuation of the feces. Likely, there is also a hereditary factor involved. The first signs of hemorrhoids is severe rectal pain and/or rectal bleeding. Most hemorrhoids respond to conservative therapy such as hot sitz baths, dietary changes to soften the stool and remove irritating roughages, and the proper use of local medication.

(Also see CONSTIPATION.)

HEMOSIDERIN

The insoluble iron oxide-protein compound in which iron is stored in the liver and spleen if the amount of iron in the blood exceeds the storage capacity of ferritin. Such accumulation of excess iron occurs in diseases that are accompanied by rapid destruction of red blood cells (malaria, hemolytic anemia).

HEPARIN

A mucopolysaccharide that prevents clotting of blood by preventing the formation of fibrin.

HEPATITIS

The name hepatitis means *liver inflammation.* It can be caused by viruses, drugs, and chemicals (including alcohol), to name the most common offenders. In the minds of most people, however, it means a viral infection.

Type A hepatitis (previously called *infectious hepatitis*) is spread via feces contamination of food, eating utensils, toys, clothing, etc. It is mainly a disease of children and young adults.

Persons in contact with a hepatitis patient or travelers to countries where sanitation is poor, should receive gamma globulin shots, which confer protection that lasts 3 to 4 months.

(Also see DISEASES.)

HEPATOMEGALY

Enlargement of the liver.

HERB

The word *herb* is usually applied to a plant or plant part valued for its savory, aromatic, or medicinal qualities.

Some herbs are used in cooking—to flavor foods. Others give scent to perfumes; and still others are used for medicines. Some herbs, like balm and sage, are valued for their leaves. Saffron is picked for its buds and flowers. Fennel seeds are valued in relishes and seasoning. Ginseng is valued for its aromatic roots.

Some people grow herbs in their gardens. The plants do well with little care. When they are grown, the leaves, stems, or seeds are harvested and dried. Then, they are generally pounded to a fine powder, placed in airtight containers, and stored for later use.

Although herbs have little nutrient value, they make food tasty and more flavorful. Cooking with herbs is a culinary art, which adds great interest to menus.

(Also see FLAVORINGS AND SEASONINGS.)

HEXAMIC ACID (CYCLAMIC ACID)

An artificial sweetener whose calcium or sodium salt, cyclamate, was historically used in many foods. It is about 30 times as sweet as refined cane sugar. In 1970 the FDA banned all cyclamate-containing products. However, cyclamates may now be used under medical supervision.

(Also see ADDITIVES; CYCLAMATES; and SWEETENING AGENTS.)

HEXOSEMONOPHOSPHATE SHUNT (PENTOSE SHUNT)

An alternate mechanism for the breakdown of glucose. The function of this pathway is twofold: (1) the generation of energy through the use of NADPH, and (2) the production of pentoses—five carbon sugars—especially ribose which can be used in the synthesis of nucleic acids—DNA and RNA.

(Also see METABOLISM.)

HEXOSES (6-CARBON SUGARS)

Glucose, fructose, galactose, and mannose are the principal dietary hexoses. All are sugars with a "backbone" of six (hexa) carbon atoms. In nature, only two—fructose and glucose—occur in free form. Galactose, together with glucose, forms the disaccharide, lactose, or milk sugar. Mannose is found in the polysaccharide mannan. Glucose is probably the most important sugar in the metabolism of carbohydrates since both anaerobic and aerobic catabolism of carbohydrates are highly sensitive to levels of this sugar.

(Also see DIGESTION AND ABSORPTION ; and METABOLISM.)

HIATUS HERNIA

Many persons have, under certain circumstances, a slight protrusion of the upper part of the stomach through the opening (or hiatus) in the diaphragm. Normally, the esophagus passes down through this opening and is attached to the stomach just below the hiatus. This disorder sometimes causes *heartburn* and sour regurgitation. These conditions may be wrongly attributed to heart disease or peptic ulcer. (It is difficult to diagnose the origin of pain in the region under the breastbone.) It is rare for more than a small portion of the stomach to protrude up into the chest cavity. Also, the amount of protrusion is usually variable with the hernia sliding up and down in reponse to the size of the hiatus, the amount of pressure on the stomach, and the length of the esophagus. (The esophagus may be shortened due to irritation or distention.) Serious consequences may result when there is chronic esophagitis caused by reflux of stomach acid (hiatus hernia makes this more likely to happen as it alters the action of the sphincter between the esophagus and the stomach), or under rare circumstances, where a large portion of the stomach protrudes into the chest cavity. (See Fig. H-12.)

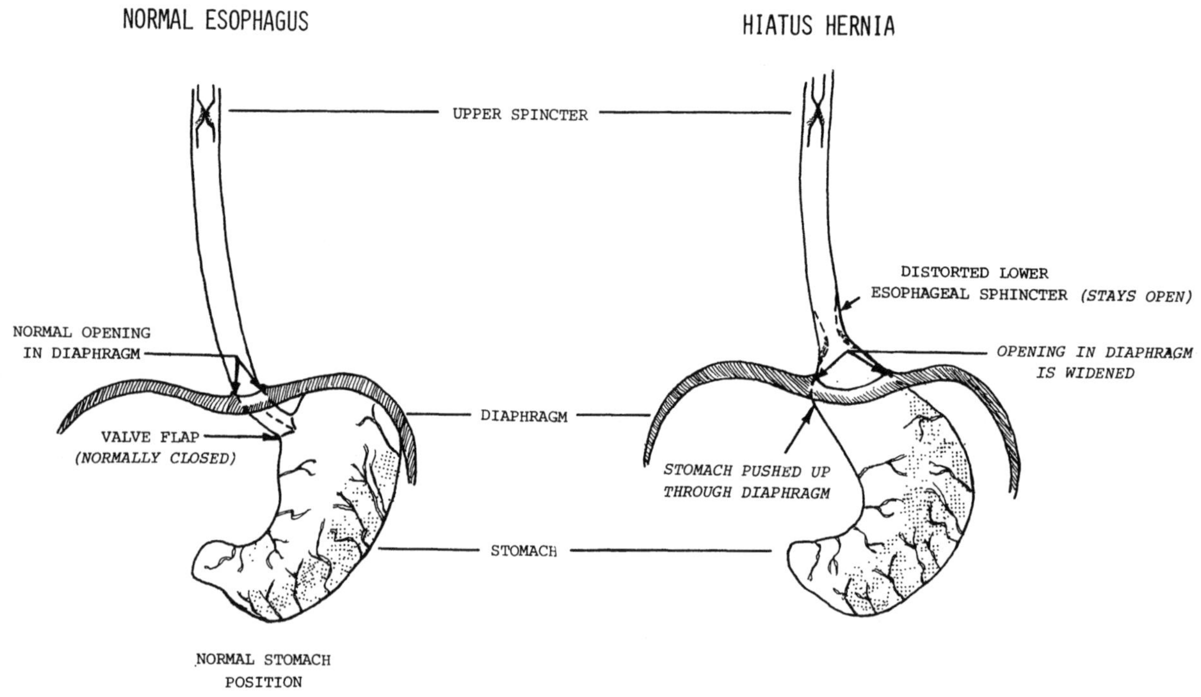

Fig. H-12. Displacement of the esophagus and stomach in hiatus hernia.

CAUSES OF HIATUS HERNIA. Some people appear to be more susceptible to this disorder than others. As in other hernias, an increased intra-abdominal pressure is the most common cause of herniation. Ways of avoiding such pressure are discussed in the section that follows.

TREATMENT AND PREVENTION OF HIATUS HERNIA. Although surgical reduction of the tendency for herniation might appear to be the best treatment, most physicians are reluctant to advise such surgery, except when the hernia is very large. Therefore, the most common approach to the problem is to advise patients how to avoid large increases in intra-abdominal pressure. Preventive measures involve (1) reducing obesity; (2) avoiding straining at the stool, distention of the stomach with food and drink, chronic coughing and sneezing, and lying down soon after meals (the stomach is held in a lower position by gravity when the patient is in an upright position); and (3) sleeping with the head high and drinking water when there are signs of regurgitation coming on.

(Also see HEARTBURN.)

HIGH BLOOD PRESSURE (HYPERTENSION)

Contents *Page*

Symptoms **1169**
Dangers **1169**
Types **1169**
Secondary Hypertension **1169**
Primary or Essential Hypertension **1170**
Possible Causes **1170**
Treatment **1170**
Prevention **1171**

About 25% of Americans have some degree of high blood pressure. If not corrected, it is a risk factor for both heart disease and strokes. Each year about 300,000 people die from diseases linked to high blood pressure.

The force which the blood exerts against the wall of the blood vessels—blood pressure—is measured at two readings. The higher reading, which is called systolic blood pressure, occurs when the heart exerts its maximum force of contraction (systole). The lower reading, which is called the diastolic blood pressure, occurs when the heart rests between contractions (diastole). Normal blood pressure values for an adult between the ages of 18 and 45 center around readings of 120 systolic and 80 diastolic—when written as 120/80 and spoken as "120 over 80." High blood pressure, or at least the upper limits of normal is considered as 160/95. However, the range for high blood pressure is from 140/90 to 165/95, and severe cases of high blood pressure may have readings as high as 200/115. Aside from the high extremes, a diagnosis of high blood pressure is based upon an individual's medical history, and the judgment of the physician.

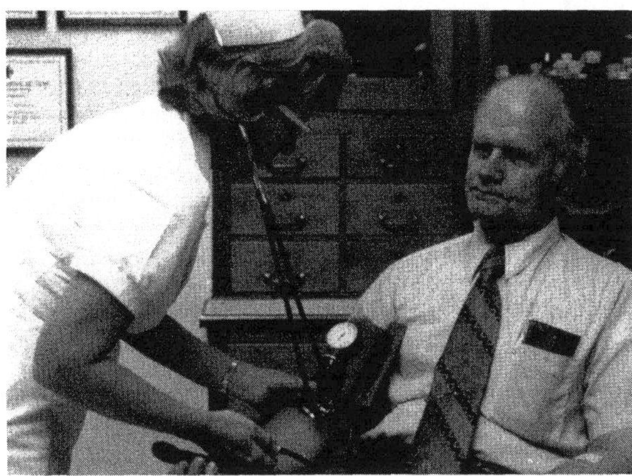

Fig. H-13. Taking blood pressure with a sphygmomanometer. The gauge is calibrated in millimeters of mercury. A stethoscope is placed over an artery in the arm to detect sounds characteristic of the systolic reading and the diastolic reading as the pressure is released from the cuff surrounding the arm.

The numbers used to express blood pressure are pressures in millimeters of mercury—the height that blood flow in an artery pushes a column of mercury in a glass tube—determined by an instrument known as a manometer. Normally, blood pressure is taken in the large artery of the upper arm. An inflatable pressure cuff connected to a mercury manometer (a sphygmomanometer) is wrapped around the upper arm. The pressure cuff is inflated until the wall of the artery collapses and blood flow ceases. Then a stethoscope is held below the inflated cuff, just over the artery. Next, pressure is gradually released from the cuff. Just at the instant when blood is heard to spurt through the artery—a tapping sound—the reading on the manometer is noted. This reading is the *systolic* pressure, since blood is being forced through the artery only when the heart contracts. Gradually more pressure is released from the cuff and the spurting sounds or vibration are still noted through the stethoscope. The reading noted on the manometer the moment the spurting sounds disappear is the *diastolic* pressure.

The diagnosis of high blood pressure should be based upon at least two separate blood pressure readings taken under relaxed surroundings with the same equipment, and determined by the same individual. High blood pressure is a pattern of consistently elevated blood pressure. However, the initial blood pressure check of some individuals may be so high that further verifiction is unnecessary.

Blood pressure varies from day to day and even minute to minute. Emotional excitement, exercise, relaxation or sleep all alter blood pressure. Of course, blood pressure varies from individual to individual. No one is typically normal.

SYMPTOMS. Many diseases have warning signals indicating things are not quite right in the body. Thus, measures to cure the disease can be initiated before serious damage. However, high blood pressure is often referred to as a "silent" disease. Individuals are not aware of the dangerous changes caused by high blood pressure until the damage is done—sometimes too late. About one out of every five indivduals will be affected by high blood pressure during their lifetime.

DANGERS. Untreated high blood pressure may, over a period of time, affect vital areas of the body, particularly the heart, brain, and kidneys.

High blood pressure causes the heart to pump harder than normal which in turn causes the heart to get larger but pump less effectively. Since the heart pumps less effectively, it dilates causing the signs and symptoms of heart failure to appear. High blood pressure may also speed up the process of atherosclerosis—a contributor to heart attacks.

High blood pressure increases the risk of a stroke due to the chance of accelerated blockage of an artery feeding the brain or hemorrhage (rupture) of a brain blood vessel. Both result in the destruction of brain tissue.

Aside from the brain, the other nervous tissue involved is the eye. High blood pressure may cause the tiny blood vessels in the back of the eyes to undergo atherosclerotic changes (narrowing), or hemorrhaging, resulting in blurred vision and even blindness. These small blood vessels of the eye can be examined directly by a physician, providing a means of following the effects of high blood pressure.

Oftentimes the blood vessels of the kidneys are the hardest hit by high blood pressure. Narrowing or rupturing of the small blood vessels of the kidneys eventually leads to the inability of the kidneys to perform their task of removing wastes from the blood—kidney failure.

(Also see HEART DISEASE.)

TYPES. High blood pressure is classified as (1) secondary hypertension, or (2) primary or essential hypertension. Most individuals suffer from primary or essential hypertension.

Secondary Hypertension. This is high blood pressure due to a clearly definable cause, primarily kidney disorders, endocrine malfunctions, pregnancy, and "the pill."

Diseases of the kidneys such as kidney infections, narrowing and hardening (sclerosis) of the small blood vessels of the kidney, and inflammation may all elevate blood pressure. The kidneys control blood pressure via their ability to conserve or excrete sodium and water, and their control over the formation of angiotensin—a powerful constrictor of blood vessels.

Occasionally the cortex of the adrenal gland, an endocrine gland, malfunctions and begins secreting excessive amounts of the hormone aldosterone. This excessive aldosterone causes the kidneys to retain sodium (salt) and water, thereby increasing the blood volume and, in turn, increasing the blood pressure above normal. Also, the medulla of the adrenal gland may develop a tumor which secretes large quantities of the hormones epinephrine and norepinephrine. These two hormones promote intense constriction of the blood vessels, resulting in high blood pressure.

High blood pressure may also be a complication of pregnancy-toxemia. A woman experiencing high blood pressure during pregnancy stands a greater chance of suffering from high blood pressure in the future.

"The pill"—oral contraceptives—evokes high blood pressure in some individuals predisposed to high blood pressure, or those who gain excessive weight during their menstrual cycle.

In secondary hypertension, the underlying cause is dealt with first, thereby eliminating the high blood pressure.

Primary or Essential Hypertension. Unfortunately, relatively few individuals demonstrate a clear-cut cause of hypertension. Rather, about 90% of the people suffering from high blood pressure show no single, easily identifiable cause. This type of high blood pressure is called primary, or essential, or sometimes idiopathic hypertension.

POSSIBLE CAUSES. It is likely that primary hypertension is due to a variety of causes—some interrelated. The following list contains some of the most probable causes:

1. **Age.** When hypertension begins early in life it becomes progressively worse with age. Furthermore, slightly elevated blood pressure is noted for people older than 45 years of age.
2. **Emotions.** A temporary rise in blood pressure may be elicited by a number of psychological factors such as fear and anger. However, it is still uncertain whether the daily emotional strains of life can account for permanent rises in blood pressure. Still, some individuals react to daily life as if it were a series of emergencies. Furthermore, some studies on college women, U.S. Air Force officers, and insurance company employees used the following terms to describe high blood pressure individuals: "irritable, "anxious," "restless," "tense," "dominant," "assertive," "overcontrolling," "emotionally more responsive," "guarded," and "apprehensive." An interesting observation, which may or may not be related, is that hypertension occurs more frequently in people with low incomes and low educational levels. For now, the role of emotions in the development of high blood pressure is theoretical and needs research.
3. **Endocrine glands.** Obviously the endocrine glands, specifically the adrenal glands, can alter blood pressure, but the stimulus affecting secretion of the endocrine glands is not specific. It could be a psychological stimulus, such as emotions, or a physical stimulus, such as some dietary factor.
4. **Heredity.** Hypertension often runs in families. Moreover, American blacks are twice as likely as whites to have high blood pressure.
5. **Salt and diet.** Epidemiologic studies (studies of populations) have demonstrated a positive correlation between salt and hypertension. The highest intakes of salt are found in northern Japan (28 g/day) where about 38% of the population is hypertensive. In contrast, Alaskan natives rarely add salt to food (4 g/day), and they rarely have hypertension. Other studies which ranked inviduals according to reported salt intake, or by the amount of sodium excreted in the urine, indicate that elevated blood pressure levels are more prevalent among those who have high salt intakes. In addition, prolonged feeding of diets high in salt to experimental animals induces hypertension.

   A sodium-restricted diet, along with drug therapy, is an accepted treatment to lower blood pressure. Whether or not a high level of sodium consumption causes the development of hypertension is controversial, based on studies of the salt intake and prevalence of hypertension among various populations. It is doubtful that excess salt (sodium chloride) is the sole cause of high blood pressure. Rather, the effects of excessive dietary sodium on blood pressure may be determined by an inherited predisposition and enhanced by low levels of potassium consumption, which are typical of the current American diet pattern. Moreover, the hidden salt (sodium) in processed foods of the typical American diet may further contribute to the problem for susceptible individuals.

   Cadmium, a widely used industrial metal, may get into food and water. Regions which have high levels of cadmium in the environment are likely to have greater than normal incidences of high blood pressure because cadmium causes damage to the kidneys. Fortunately, certain dietary minerals, such as selenium and zinc, may counteract the effects of cadmium.

   (Also see CADMIUM; MINERAL[S]; POTASSIUM; SALT; and SODIUM.)
6. **Sex.** High blood pressure is more frequent up to age 55 in men than women, but after 55 the reverse is true—more women than men. Still, men are more likely to develop complications.
7. **Smoking.** Cigarette smoking is the single most preventable cause of death, and heavy smoking is implicated in the development of high blood pressure. A component of cigarettes, nicotine, is known to raise blood pressure.
8. **Weight.** Obesity aggravates hypertension, but the exact relationship is unknown. As weight increases so does the prevalence and incidence of hypertension. The greater the weight gain, the greater the rise in blood pressure. Data also indicate that when people with high blood pressure lose weight their blood pressure generally decreases. Sometimes, in cases of mild hypertension, weight loss serves as an effective treatment. Nevertheless, some hypertensive individuals are thin.

TREATMENT. Traditional treatment of high blood pressure depends upon dietary modifications, drugs, and life-style changes.

• **Dietary modifications**—Rigid, long-term sodium (salt) restriction can be successfully used to control, at least in part, high blood pressure. It is not difficult to outline a sodium restricted diet, but it is difficult to follow. Low sodium foods are just not very palatable, and the normal

American diet contains numerous sources of sodium. An initial step in reducing sodium intake is to give up the salt shaker at the table and to avoid those foods which obviously contain large amounts of added salt. Sodium restricted diets may be mild (2,000 to 4,500 mg/day), moderate (1,000 mg/day), or strict (500 mg/day), depending upon the need. With modern antihypertensive drugs, a severe dietary restriction is not necessary. However, the effectiveness of the antihypertensive drugs is enhanced by a moderate restriction of sodium.

Sodium and potassium are closely interrelated. Although sodium intake may be the most important dietary determinant of blood pressure, variations in the sodium:potassium ratio in the diet affect blood pressure under certain circumstances. In rats with high blood pressure due to a high intake of sodium, blood pressure may be lowered to a more normal level by increasing the potassium intake and lowering the sodium intake. It seems that a 1:1 ratio may be somewhat protective against the blood pressure-elevating effects of a given level of sodium. This ratio can be achieved by increasing the intake of potassium, or lowering the intake of sodium, or both. Good dietary sources of potassium include meats, dried fruit, green leafy vegetables, Lima beans, soybeans, winter squash, and wheat bran or wheat germ.

Researchers in both Finland and Denmark report that, for people who do not eat fish regularly, fish oil may be helpful.

• **Drugs**—Individuals with blood pressure levels above 160/100 generally require drugs to lower their blood pressure. These drugs can be classified into three major groups: (1) diuretics which promote increased urinary excretion of sodium; (2) sympatholytics which block nerves controlling blood vessels, and allow blood vessels to dilate thus decreasing the pressure; and (3) vasodilators which act directly on blood vessels causing them to widen. The long-term use of diuretics leads to a potassium deficiency. Hence, dietary potassium supplementation is required.

• **Life-style changes**—These may be different for every individual suffering from hypertension. Life-style changes which appear to be beneficial include weight reduction, cessation of smoking, and regular exercise. Possibly an evaluation of the day-to-day events—coping with the stress of living—may point out the need for engaging in some relaxing activity. Biofeedback and meditation offer a means of relaxation and regaining control for some individuals. Perhaps just working out some personal problems will help.

PREVENTION. Most Americans who suffer from high blood pressure suffer from primary or essential hypertension—a condition without specific causes. Hence, preventative measures are difficult to designate. However, some suggestions can be made based on the current knowledge.

• **Diet**—Average Americans consume about 10 g of salt each day. This amount translates into 4,000 mg of sodium daily—considerably more than the upper limit of 500 mg currently recommended by the National Research Council. Furthermore, under ordinary circumstances the body requires only 115 mg of sodium daily and under conditions of profuse sweating only 780 mg daily. So, while no direct evidence links salt intake to development of hypertension, there are no known benefits to the healthy person of excessive salt consumption. Moreover, it is reasonable to assume that a voluntary lowered salt intake will reduce the risk of developing high blood pressure in the 10 to 30% of all Americans born with a genetic predisposition to hypertension. In addition, a 1:1 ratio of sodium and potassium intake may be somewhat protective.

• **Exercise and weight control**—Many Americans suffer from too many calories and too little exercise. Since obesity is implicated with high blood pressure, the benefit of maintaining an ideal weight appears obvious. A regular exercise program is also beneficial in maintaining and achieving weight control. Moreover, exercise relaxes and promotes a sense of well-being, self-reliance, decreased anxiety, and relief from mild depression which may also be preventative.

• **Smoking**—Give it up. Cigarette smoking is the principal preventable cause of chronic disease and death in the United States.

• **Periodic blood pressure checks**—Hypertension causes permanent damage to the heart, brain, and kidneys without any warning symptoms. Hence, periodic screening is an invaluable preventative measure, so treatment can be started before damage occurs. Many government, voluntary health agencies, community leaders, medical societies and health care providers have been active in screening efforts. Adults should have a screening exam for high blood pressure at least every 5 years, and every 2 or 3 years if over the age of 40, or if there is a family history of high blood pressure.

It is encouraging to note that death rates attributable to high blood pressure have been reduced by as much as 35% in the last 10 years due largely to earlier detection and effective treatment.

(Also see ENDOCRINE GLANDS; HEART DISEASE; MODIFIED DIETS; OBESITY; POTASSIUM; SALT; SODIUM; STRESS; and WATER AND ELECTROLYTES.)

HIGH DENSITY LIPOPROTEINS (HDL)

Also called alpha-lipoproteins.

(Also see ALPHA-LIPOPROTEINS; and LIPOPROTEINS.)

HIGH-NUTRIENT DENSITY

A high proportion of a specific nutrient to total kilocalories in a given amount of food.

HIGH RATIO FLOUR

Flour of very fine and uniform particle size, treated with chlorine to reduce the gluten strength, used for making cakes. It is possible to add up to 140 parts of sugar to 100 parts of this flour. The product can be classed as a "weak" flour which does not make good bread, but does make excellent cakes.

(Also see FLOUR, Table F-25 Major Flours.)

HIGH TEMPERATURE-SHORT TIME METHOD (HTST)

A rapid method of pasteurization employed by the food industry.
(Also see FLASH-PASTEURIZATION.)

HIOCHIC ACID ($C_6H_{12}O_4$)

Originally isolated from Japanese sake, it was found to promote the growth of *Lactobacillus acidophilus*. Later, however, hiochic acid proved to be mevalonic acid.
(Also see MEVALONIC ACID.)

HISTIDINE

One of the essential amino acids.
(Also see AMINO ACID[S].)

HISTIDINEMIA

A condition of the blood wherein the levels of the amino acid histidine are abnormally high. It is attributed to a genetic disorder in the metabolism of histidine—lack of the enzyme histidine ammonia lyase. Histidinemia causes mental retardation and slurred, inarticulate speech. Treatment consists of a low intake of histidine-containing proteins. The effectiveness of the treatment is, however, questionable.
(Also see INBORN ERRORS OF METABOLISM.)

HISTOLOGY

Microscopic anatomy; pertaining to minute structure, composition, and function of tissues.

HISTORY OF NUTRITION

In nutrition, more than in any other science, history is most important. In the final analysis, food and people are not only inseparable from history—they're part of it.

Fig. H-14. The biotic pyramid.

Without them, there would be no history—no mankind.

For thousands of years, the quest for food has shaped the course of history. It dictated population growth, urban expansion, migration, and settlement of new lands. It profoundly influenced economic, social, and political theory. It promoted early sailing and the discovery of new worlds, widened the horizons of commerce, caused wars of dominion, and played no small role in the creation of empires.

Food has played an important role in many things. In religion, it helped to define the separateness of one creed from another by means of dietary taboos. In science, the prehistoric cook's discoveries of the effect of heat applied to raw materials laid the foundations on which much of early chemistry was based. In technology, the waterwheel, first used in the milling of grain, achieved immense industrial importance. In wars, battles were sometimes postponed until the harvest had been gathered in, and the well-fed armies usually defeated the hungry ones.

Although the ancient Greek philosophers were interested in science, logical reasoning—rather than experimentation—was the Greek way. Hippocrates (460 to 377 B.C.), known as the "father of medicine," was the first great physician to indicate an interest in nutrition. Among his famous quotes were "Children produce more heat and need more food than adults"; and "Persons who are naturally very fat are apt to die earlier than those who are slender."

The naturalistic, or prescientific, era of nutrition was characterized by a fascinating maze of philosophy, taboos, bizarre superstitions, and religious precepts. During this period, little of truly scientific nature was accomplished in the field of nutrition. Then came the successful merger of the art and the science of food and nutrition, ushering in a new era. It stimulated increased research. Many maladies which had long plagued mankind were traced to dietary deficiencies, imbalances, and toxicities.

Obviously, all of the significant milestones in a subject as broad as nutrition cannot be presented in this section because of space limitations. Obviously, too, the milestones were a progressive series of developments, rather than unrelated individual discoveries. All researchers contributed to the kaleidscope of knowledge that we have today, whether it was to prove or disprove a theory, or to add just one more piece to the puzzle. Neither can the dates given always be pinpointed because some research projects spanned several years; others were completed on a certain date; but not reported or published until later; and still others led to the wrong conclusion, which, with further study, was changed.

The important thing in reviewing this history of nutrition is to recognize that it must be causing Americans to do something right, for we are living longer and healthier than ever before.

Winston Churchill once said, "The farther back you can look, the farther forward you are likely to see." So, it is hoped that this return to the past will be of interest and value to today's nutritionists, despite their preoccupation with the "here and now" and the "where to go from here."

A chronology of the history of nutrition is given in Table H-3.

TABLE H-3
THE HISTORY OF NUTRITION

Date	Nutritional Milestones
2600 B.C.	Beriberi, which affects the nervous system, was known to the Chinese as early as 2600 B.C. The word *beriberi* means *I cannot*, referring to the fact that persons with the disease cannot move easily. But the cause of beriberi in man remained elusive for centuries.
2000	Arsenicum or arsenic (As_2O_3), the sublimate from copper and iron ores, was discovered.
1550	Scurvy and diabetes were described by the Egyptians on medical papyrus rolls (man's first writing paper, made by the Egyptians from the papyrus plant as early as 2400 B.C.), discovered in Thebes by George Moritz Ebers, a German Egyptologist and novelist, who, in 1874, edited it in a romantic historical novel on medicine which he titled *Papyrus Ebers.*
1000	The Chinese, in their herbal medicine, used a concoction rich in vitamin A as a remedy for night blindness; and cod-liver oil was used in treating or preventing rickets long before anything was known about the cause of the disease.
1100 to 500	In the Old Testament, which was written over a long period, reference is made to scurvy.
460 to 364	Hippocrates, the Father of Medicine and a contemporary of Socrates, believed that the loss of body weight in fasting was due to the loss of "insensible perspiration" from the skin and to a loss of heat which he conceived to consist of a fine material. Also, he advocated liver as a cure for night blindness, and described the symptoms of scurvy.
47	Scribonius Largus insisted on the importance of diet.
25	Celsus, in the first Latin treatise on medicine, classified foodstuffs and emphasized their role in maintaining health.
200 A.D.	Arataeus, the Cappadocian, described diabetes, diptheria, and sprue.
1100	The chemists and apothecaries learned how to attach a cooled condenser when they were distilling various materials. Along in the 1500s, this opened the way for the preparation of ether, formic acid, acetic acid, alcohol, ammonia water, succinic acid, benzoic acid, and the chemical analysis of foods.
1250	Elemental arsenic was first prepared by Albertus Magnus. Jean Sire de Joinville, the French chronicler, accompanied Louis IX of France to Cyprus and Egypt. In 1309, he completed in final form the *History of Saint Louis*, an account of the Crusade, in which he told of a disease "which attacked the mouth and legs" (obviously scurvy).
1497	Vasco de Gama, a Portuguese navigator, sailed around the Cape of Good Hope to establish the first European trading colony on the coast of India. One hundred of his crew of 160 men perished of scurvy on the voyage.
1535	In Canada, during the winter, Jacques Cartier, the daring explorer who laid claim to Canada for France, recorded in his log that the lives of many of his men dying of scurvy were saved when they learned from the Indians that a brew made from the tips of pine or spruce trees cured and prevented the malady. (It is now known that the brew contained vitamin C.)
1542	Andrew Borde, in his book, *A Dyetary of Health*, described factors affecting the health of man and the part played by diet.
1564	Ronsseus, a Dutch physician, was the first to recommend oranges as the antiscorbutic for sailors. Unfortunately, it was to be over 200 years before this addition to the diet of sailors became mandatory.
1601	Sea Captain James Lancaster recorded that lemon juice protected against scurvy. Thus, some of the lucky sailing ships were freed from the scourge of scurvy.
1615	Fabricious Bartholetti first described a milk sugar. However, Von Haller claimed that it had been known long before that time in India.
1621 to 1679	Robert Boyle and John Mayow, two English scientists, both observed that animals and burning candles used some component from the air, but they did not know what it was.
1624 to 1689	Thomas Sydenham, the "father of modern clinical medicine," described diabetes.

(Continued)

TABLE H-3 (*Continued*)

Date	Nutritional Milestones
About 1650	Leeuwenhoek pioneered in using the microscope to describe the appearance of starch granules.
1669	Hennig Brand procured phosphorus from urine.
1700s	Several scientists were involved in trying to solve the mysteries of body proteins, including George Fordyce, Albrecht Von Haller, P.J. Macquer, and Denis Papin.
1713	Iron was demonstrated to be present in the ash of blood.
1730	A Spanish physician, Gaspar Casal, was the first to describe pellagra, which appeared soon after the introduction of corn (maize) into Europe.
1742	I.B. Beccari recorded that, if we consider only our bodies, it must be true that we are composed of the same substances which serve as our food.
1745	I.B. Beccari was the first to describe gluten, which was produced when starch was washed out of such cereal grains as wheat or barley.
1746	Menghini found that the iron content of blood could be increased.
1747	Andreas Marggraf, a German pharmacist, wrote of isolating glucose from raisins.
1747 to 1753	James Lind, a British naval surgeon, and the first great experimental nutritionist, tested six remedies on 12 sailors who had scurvy, and found that oranges and lemons were curative. His classical studies, the results of which were published in *A Treatise of Scurvy*, 1753, are generally credited as being the first experiments to show that an essential food element can prevent a deficiency disease. But another 50 years elapsed before the British Navy required rations of lemons or limes on sailing vessels, which resulted in the sailors being nicknamed "limeys."
1750	Out of this misty background, the world would see the beginning of modern chemistry develop before the end of the 18th century.
1771	Francesco Frapoli, an Italian physician, named the disease pellagra (*pelle*, for skin; and *agra*, for sour).
1772	Daniel Rutherford, following some experimental work, described what is now called nitrogen gas.
1768 to 1775	On two historic voyages, each of three years' duration, from 1768 to 1771 and from 1772 to 1775, British Captain James Cook, avoided scurvy---hitherto the scourge of long sea voyages. He had his ship stocked with concentrated slabs of thick brown vegetable soup and barrels of sauerkraut. Of the sauerkraut he said: "It is not only a wholesome vegetable food, but, in my judgment, highly antiscorbutic, and spoils not by keeping." In addition, he sent seamen ashore at every port visited to gather all sorts of fresh fruits and green vegetables (including grasses), which the crew prepared, served, and ate. As a result, not one of the crew died from scurvy.
1773	H. M. Rouelle discovered urea and hippuric acid in urine.
1774	Priestley discovered "dephlogisticated air," later renamed oxygen.
1776	Cavendish, a nephew of the third Duke of Devonshire, discovered hydrogen.
1777	Crawford was the first individual to publish experiments on animal calorimetry.
1783	The great French chemist, Antoine Laurent Lavoisier, the "Father of Nutrition," published the results of his experiments on respiratory metabolism. He used guinea pigs to measure body heat, oxygen consumed, and carbon dioxide expired and concluded that the human body was like a little furnace using food to produce heat and energy. Further, he was able to show that heat production in the animal body is directly related to oxygen consumption. Lavoisier compared animal heat to that produced by a lamp or candle.
1790	Spallanzani, Berzelius, W. F. Edwards, and G. Lussac were all pioneers in the study of the relationship between oxygen, carbon dioxide, nitrogen, respiration, and the blood.
1791	Dr. George Fordyce of England was the first to use a control group in an experiment with chickens. He found that they required a calcareous (calcium) supplement in order to produce healthy eggs, that did not break easily in the nest.
1797	Dr. John Rollo advocated a meat diet for diabetes. W. Cruickshanks, who was a chemist in the British artillery, made the first attempt to study quantitatively the products which resulted from the destructive distillation of sugar.

(Continued)

TABLE H-3 (*Continued*)

Date	Nutritional Milestones
1800s	Throughout the 1800s, a host of researchers worked on the source of muscular energy, energy metabolism studies, calorie values of foods, determination of respiratory quotients of different foods, specific dynamic action of foods, and basal metabolic rate. They contributed richly to the study of nutrition. Working in laboratories throughout Europe and the United States, these illustrious scientists included W. Allen, Barral, M. Berthelot, F. Bidder, T. L. W. Bischoff, J. B. Boussingault, C. T. Coathupe, Despretz, Dulong, J. B. A. Dumas, A. Fick, E. Frankland, J. H. Gilbert, J. B. Lawes, Nicholas Lunin, F. Magendie, A. Magnus-Levy, Pepys, Lyon Playfair, H. V. Regnault, J. Reiset, Charles Richet, Max Rubner, C. Schmidt, E. Smith, T. Thomson, C. Voit, Max von Pettenkofer, Justus von Liebig, F. J. Wislicenus, and Nathan Zuntz. J. W. Dobereiner, a professor of chemistry at Jena, Germany, did some brilliant work. He first noted the relationship between the atomic weights of calcium, strontium, and barium; he recognized platinum acting as a catalyst; he developed a preparation of furfurol and aldehyde ammonia; and, he invented a quick process to make vinegar.  Fig. H-15. Although the microscope was developed about 1590 by Zacharias Janssen, a Dutch spectacle maker, few improvements were made to it until the early 1800s, when the interest in metabolism and anatomical fine structure ballooned.
1802	Valentin Rose discovered inulin.
1803	J. R. Young of Pennsylvania proved that gastric digestion involved a chemical process.
1804	J. L. Proust of France, prepared, among other things, dextrose (glucose) in its pure form from grape juice.
1806	Vauquelin and Fourcroy made extensive and profound observations on the essentials of an adequate diet. Besides many other findings, they discovered that beans contained something more than starch.
1807	Sodium, potassium, calcium, magnesium, sulfur, and boron were all isolated by Humphrey Davy.
1810	Cystic oxide (cystine) was isolated from urinary calculus by Wollaston. This heralded the discovery of the first amino acid.
1811	Gay-Lussac and Thenard made the first accurate analysis of sugar. Berthollet determined the quantitative differences between proteins.
1811	Various researchers established that the sugar of grapes, of honey, in the urine of diabetics, and from the acid hydrolysis of starch and cellulose were all identical.
1811 to 1820s	The French chemist, Chevreul, described many of the components and properties of fats. At the same time, the French naturalist, Braconnot, used acid hydrolysis to convert cellulose-rich plant materials to glucose and the proteins to their constituent amino acids.
1812	A Russian chemist named Kirchhoff found that, when starch was boiled with dilute sulfuric acid, it was converted into a sugar which was identical to grape sugar (dextrose).

(Continued)

TABLE H-3 (*Continued*)

Date	Nutritional Milestones
1814	Chevreul, a French professor, showed that fats were composed of fatty acids and glycerol. Also, he contributed many experimental observations to the meager knowledge of fats. He named *stearin, elain, margarine* (Greek for *mother of pearl*), *oleic acid*, and *caproic acid*.
1815	J. J. Colin, a professor at the St. Cyr military academy, and Gaulthier de Claubry, a professor of chemistry at the Paris School of Pharmacy, discovered that iodine would react with starch to form a blue color. That same year, Strohmeyer, at Gottingen, described the same reaction.
c. 1818	Berzelius worked out the modern symbols for elements.
1819	M. H. Braconnot discovered that, by using Kirchhoff's procedure, many things such as sawdust, straw, linen rags, and the bark of trees, also yielded dextrose. Proust discovered leucine.
1820	Braconnot discovered how to prepare an amino acid from a protein by acid hydrolysis, and isolated both glycine and leucine.
1821	John Gorham of Harvard University did the first chemical analysis of an American food---he analyzed Indian corn, or maize.
1822	J. B. Boussingault, a Frenchman, while traveling in South America, observed that the villagers who used salt containing iodine were free from goiter, but that those who used plain salt were afflicted.
1824	Cod-liver oil, long known as a folk medicine, was found to be important in the treatment of rickets. But the remedy lost favor with the medical profession because physicians could not explain its action. Hydrochloric acid was proven by Wm. Prout to be the free acid in gastric juice.
1825	From 1825 into the early 1900s, a host of scientists worked on what they called *albuminous substances*, now known as *proteins*. Much research was done on the digestion, absorption, and synthesis of proteins in the body, and gradually some of the mysteries were unwound.
1826	F. V. Raspail was the first to describe the swelling of starch granules when heated in water.
1827	Wm. Prout classified food constituents as saccharine (now called carbohydrates), fats, and protein. Wholer isolated aluminum.
1832	Creatine was isolated by Chevreul.
1833	Anselm Payen, an industrial chemistry professor in Paris, and J. F. Persoz, a pharmacy professor in Strasbourg, produced a malt extract that converted starch to sugar. They proceeded to add alcohol and separated it into two products which they called diastase and amidin. William Beaumont (1785 to 1853) became an outstanding pioneer in the study of gastric digestion with his observations of Alexis St. Martin, who had a permanent opening into his stomach because of a gunshot wound.
1834	J. Fritsche correctly described the structure of starch granules. Eberle showed that digestion could take place without the stomach.
1835	J. B. A. Dumas refined the method of determining carbon and hydrogen in organic substances by mixing the food with copper oxide. Beriberi was described in a treatise by John Grant Molcolmson. Pepsin action and properties were described by Schwann.
1836	Jean Baptiste Boussingault (1802 to 1887) was perhaps the founder of scientific agriculture because he studied it both in the field and in the laboratory. Among other things, he demonstrated that some plants took up nitrogen from the air and added it to the soil. He made studies on how much of a certain food was digested and absorbed. He did work on the mineral requirement for bone growth. He developed the first method to distinguish between the nitrogen of urea and of the ammonium salts in urine. He analyzed a curative salt from South America and proved that it was iodine that cured goiter. He proved that ordinary salt, potassium, calcium, and phosphate were all indispensable nutrients. He was the first to use animals and birds along with the techniques of the chemical laboratory for his experiments. He was a clear thinker and an enthusiastic, industrious worker. He made many significant contributions to the history of nutrition.
1838	G. J. Mulder, a Dutch chemist, did some work on albumin, fibrin, casein, and other substances similar in reaction and concluded that they all contained the same radical which he called *protein*. The French chemist, Jean Baptiste Andre Dumas, named the main end product of the digestion of disaccharides and starch, *glucose*. Berzelius, the noted Swedish chemist, observed that the iron in hemoglobin made it possible for the blood to absorb oxygen.

(Continued)

TABLE H-3 (*Continued*)

Date	Nutritional Milestones
1840s	The German chemist, von Liebig, who is considered to be the "father of organic chemistry," and his coworkers improved the methods of food analysis and suggested the designations of *carbohydrate, fat,* and *protein* for the major organic nutrients.
1842	Rutin was first prepared by a German pharmacist-chemist, who obtained it from garden rue, *Ruta graveolens*; hence, the name *rutin.*
1843	J. B. Boussingault performed calcium-balance studies with animals and showed the importance of calcium. A controversy developed between scientists over whether the body could convert carbohydrates to fat. Von Liebig showed that some animals produced more fat than they ingested, thus, he was convinced that the body could take carbohydrates and make fat. Finally, Boussingault and Persoz, who were originally convinced that he was wrong, proved, with repeated experiments, that von Liebig *was* correct.
1844	C. Schmidt, a student of von Liebig, showed the presence of sugar in the blood.
1844 to 1846	N. T. Gobley isolated a substance from egg yolk, which he called *lecithin* (from the Greek, *lekithos*, meaning *egg yolk*). This substance contained glycerol, fatty acids, phosphorus and nitrogen.
1846	Justus von Liebig produced and named the amino acid, tyrosine. Subsequently, F. Bopp did extensive studies on it in his laboratory.
1849	Strecker, a German chemist, isolated a compound from hog bile, to which he subsequently (in 1862) applied the name *choline* (after *chole*, the Greek word for bile).
1850s	Henneberg and Stohmann, of Germany, developed the system of proximate analysis, which is still used in the analysis of foods today.
1854	Berthelot established the chemical nature of fats, but up to this period none of the scientists had suggested that the fats eaten were in any way changed when they became fat in the body.
1855	C. G. Lehmann studied all the evidence which miscroscopic examination of the intestinal tract revealed and came to the conclusion that fats did, indeed, undergo changes before absorption into the body. Thus, the nutritional puzzle was gradually being unwound.
1856	C. Bernard discovered the formation and storage of glycogen in the liver. Pasteur began his experiments that marked the beginning of the study of bacteria, and their role in the scheme of nutrition and health.
1866	The English chemist, Edward Frankland, measured the heat content of various foods that he burned in a crude calorimeter. Studies of calorie expenditure of people and animals had been measured before, but little had been done to determine the energy content of foods.
1866 to 1867	Baeger and Wurtz, working independently, determined the correct structure of choline and carried out the first synthesis of it. But the compound did not attract the attention of nutrition investigators at the time.
1867	Nicotinic acid was first discovered and named when Huber, a German chemist, prepared it from the nicotine of tobacco, but it lay idle on the shelf for another 70 years. No one suspicioned that it was a cure for pellagra.
1869	DNA (deoxyribonucleic acid) was discovered by F. Miescher.
1871	J. B. A. Dumas was the first to question a purified diet of protein, carbohydrate, and fat. During the siege of Paris by the Germans, they fed infants an imitation milk with disastrous results, but his paper on this subject, giving his observations, went completely unnoticed.
1873	Van Lent was the first to conclude that the type of diet had something to do with the origin of beriberi. By reducing the ration of rice in the diet of the sailors in the Dutch navy, he was able to eradicate beriberi almost entirely.
1879	The existence of a yellow-green fluorescent pigment in milk whey was observed by A. W. Blyth. Subsequently, other workers found this pigment in such widely varying sources as liver, heart, and egg white. This pigment was called *flavin.* But, at the time, the biological significance of the pigment was not understood.
1880	Jungfleisch and Lefranc crystallized inulin.
1881	E. Schulze and Barbiere discovered phenylalanine, an indispensable amino acid. F. Miescher prepared nucleic acid from the nuclei of the heads of salmon sperm. Nicholas Lunin, while a student of Von Bunge at the University of Dorpat, had come to the conclusion that certain foods, such as milk, contain small quantities of unknown substances essential to life. He discovered that a purified diet of protein, fat, and carbohydrate would not maintain life.

(Continued)

TABLE H-3 (*Continued*)

Date	Nutritional Milestones
1882	Kanehiro Takaki, Director-General of the Japanese Navy, cured beriberi in sailors of the Japanese Navy by giving them less rice and more meat and milk, but he concluded that the higher protein content of the diet was responsible for the cure.
1883	Mills and Snodgrass proposed a "bromine number" for unsaturated fats. J. Kjeldahl developed the method of boiling substances with sulfuric acid and a catalyst, which converted the nitrogen of organic substances into ammonia. This method was so simple, quick, and accurate that it is still used today.
1884	Hubl developed a method for determining the iodine number of fats. The German physiologist, Rubner, utilizing data from dietary studies of Voit, calculated the heat of combustion of carbohydrates, fats, and proteins. He corrected for the incomplete metabolism of protein in the body by subtracting the heat of combustion of the urinary product, urea.
1886	E. Schulze and E. Steiger isolated arginine, another of the indispensable amino acids.
1888	Gaertner discovered the bacterium, *Bacillus enteritidis*, during an outbreak of meat poisoning. This gave great impetus to the study of intestinal flora.
1889	E. Drechsel isolated lysine, another essential amino acid.
1890	Palm, an English physician, observed that where sunshine was abundant, rickets was rare, but where the sun seldom shone, rickets was common.
1890s	Atwater, the American nutritionist, who had worked for a while with Voit and Rubner, refined the caloric conversion factors of Rubner by correcting them for the percent losses that occurred in digestion. This required many feeding trials with animals and people in which both the foods consumed and the excretory products were analyzed.
1891	C. Voit, a pupil of von Liebig, was one of the most successful investigators of metabolism. For many years, his laboratory in Munich was a Mecca for students. One of his studies showed that glucose, levulose, cane sugar, and maltose were all converted into glycogen in the liver, but that lactose and galactose were not of much value.
1896	A. Kossel, Heidelberg, and S. G. Hedin independently discovered histidine. Wilbur Atwater, who is considered the "father of American nutrition," published the first extensive table of food values. Atwater approximated that on a typical American diet each gram of carbohydrate, fat, and protein would yield 4, 9, and 4 Calories (kcal), respectively.
1897	Christiaan Eijkman, a Dutch physician assigned to a prison hospital in the East Indies, observed beriberi among the inmates and sought the answer through experiments with chickens. To save money, he fed the birds scraps, mostly polished rice, from the patients' meals. The chickens developed a bad nerve ailment, which resulted in paralysis. Later, the director of the hospital withdrew his permission to use the scraps, so Dr. Eijkman had to buy unmilled rice for the chickens. All of them promptly recovered. This led to the first experimental work that proved a nutritional deficiency. Dr. Eijkman observed that the chickens' ailment was similar to beriberi, but he erroneously attributed it to (1) too much starch from the rice, and (2) the presence of a nerve poison in the endosperm of rice, which the outer layers of unmilled rice neutralized.
1899	K. A. H. Morner isolated cystine from a protein hydrolysate.
1900	F. G. Hopkins, one of the most advanced thinkers in nutrition, and Cole, isolated and described tryptophan, an essential amino acid. Emil Fischer, the German carbohydrate chemist, established the structure of glucose, fructose, galactose, and mannose.
1901	Emil Fischer reported the discovery of the essential amino acid, valine. Dr. G. Grijns continued the work of Eijkman on beriberi, but he correctly interpreted the results as a deficiency of an essential ingredient. E. Wildiers expressed the belief that yeast required for its nutrition an organic substance which he called bios.
1900s	Pellagra reached epidemic proportions in southern United States, where the diet was based primarily on corn, which is extremely low in both available niacin and in tryptophan. In 1915, 10,000 people died from the disease; and in 1917 to 1918, there were 200,000 cases of pellagra in this country. In 1914, the U.S. Public Health Service sent a team under the direction of Dr. Joseph Goldberger, a physician-researcher, to study the cause of, and hopefully find a cure for, pellagra. In his studies, Goldberger proved that the disease was caused by a dietary deficiency, and not an infection or toxin.
1903	Ehrlich discovered the indispensable amino acid, isoleucine.

(Continued)

TABLE H-3 (*Continued*)

Date	Nutritional Milestones
1905	Carnitine was first isolated from meat extract in 1905, but its structure was not established until 1972. C. A. Pekelharing, a German professor, observed that milk had some unknown substance in minute quantity that would maintain life, but his discovery went largely unnoticed.
1906	Sir Frederick G. Hopkins, of Cambridge University, England, the most advanced thinker in nutrition of his time, determined by careful experiments that rats became sick and died on a purified diet; but that the addition to the diet of either (1) a small amount of milk, or (2) an alcoholic extract of milk solids or of certain dried vegetables enabled the animals to live and grow; however, the ash of either milk or vegetables was ineffective. Thus, Hopkins showed that the essential unknowns which existed in certain foods were organic in nature rather than inorganic (minerals). He called these substances *accessory food factors.*
1907	E. Metchnikoff recommended in his book, *The Prolongation of Life*, that *Bacillus bulgaricus*, found in sour milk, would safeguard the body against premature aging due to the putrefactive bacteria that had been observed in a meat-only diet. E. V. McCollum, one of the century's nutritional giants, fed purified diets and tried to make them more palatable, because he thought that the problem was unpalatability, but he only proved once again that a purified diet would not maintain life. Holst and Frolich, of Norway, produced scurvy experimentally in guinea pigs by feeding them a diet deficient in foods containing ascorbic acid.
1909	T. B. Osborne and L. B. Mendel demonstrated by the use of protein-free whey, that there was considerable difference between proteins, that some were complete and some were incomplete. Their work established the quantitative biological values for proteins.
1911	Casimir Funk, a Polish biochemist at the Lister Institute, London, England, isolated crystals with B complex activity. He coined the name *vitamin* from the Latin *vita* or life, plus *amine*, and applied it to the antiberiberi substance. Later, it was shown that these substances did not contain nitrogen, so eventually the "e" was dropped.
1912	Funk isolated nicotinic acid from rice polishings while attempting to isolate the antiberiberi vitamin; and, that same year, Suzuki, in Japan, isolated nicotinic acid from rice bran. But both researchers lost interest in the acid when they found it ineffective in curing beriberi. So nicotinic acid was again left on the shelf unobserved.
1913	Acceptance of the vitamin theory was not immediate, however. Working independently, McCollum and Davis at the University of Wisconsin, and Osborne and Mendel at the Connecticut Experiment Station, demonstrated the presence of an essential dietary substance in fatty foods, such as butterfat and egg yolk (Wisconsin), and cod-liver oil (Connecticut). These researchers believed that only one factor, which they called fat-soluble A, was needed to supplement the purified diets. Michaelis and Menten isolated cholorphyll.
1914	Dr. Goldberger visited an orphanage in Jackson, Mississippi, and noted that 68 of the 211 children suffered from pellagra, but that none of the employees in the institution had ever contracted the disease. This caused him to doubt the prevailing theory that pellagra was a communicable disease. He concluded that the absence of pellagra in the better-fed employees was due to the presence of meat and milk in their diets; additionally, he found that some of the orphans who were free of the disease were able to pilfer some milk and meat from the limited supply.
1915	Dr. Goldberger continued his research with prisoners from one of the penitentiaries. Out of the 12 volunteers, five developed pellagra, which he cured by the addition of yeast to the diet.
1916	Dr. Elmer V. McCollum of the University of Wisconsin designated the concentrate that cured beriberi as *water-soluble B*, thinking that it was one factor only. Bateman reported that raw egg white had a detrimental effect on rats, but that the raw egg white was rendered innocuous by heat. Most medical men refused to accept Goldberger's findings for curing pellagra, so Goldberger and his faithful disciple, Dr. G. A. Wheeler, made themselves the first subjects of a series of experiments designed to prove to the doubters that pellagra could not be contagious. In April, both of them swabbed their throats with secretions obtained from the nose and throat of pellagra patients, but they did not contract the disease. In addition, a group of 21 men and one woman took capsules containing blood, feces, and urine from pellagra patients, but none contracted the disease.
1917	E. V. McCollum showed that xerophthalmia, or night blindness, is due to lack of fat-soluble A.
1918	Sir Edward Mellanby of England, working with puppies, demonstrated that rickets was a nutritional deficiency disease. He cured them with cod-liver oil, but incorrectly attributed the cure to the newly discovered fat-soluble A.

(Continued)

TABLE H-3 (*Continued*)

Date	Nutritional Milestones
1919	R. J. Williams discovered the nutritive significance of pantothenic acid in the proliferation and fermentative activity of yeast cells.
1921	Banting and Best, physiologists, at the University of Toronto, obtained insulin from the pancreas of dogs. The insulin which they obtained cured the diabetes of depancreatized dogs. Following many tests on dogs, the hormone was administered to a male diabetic human, who experienced a remarkable recovery. Banting was awarded the 1923 Nobel Prize for medicine for the discovery, but Best was not even considered for the award because he held only an undergraduate degree at the time. However, Banting shared the money with him.
1922	McCollum, now at Johns Hopkins University, found that, after destruction of all the vitamin A in cod-liver oil (oxidation, by passing heated air through cod-liver oil), it still retained its ricket-preventing potency. This proved the existence of a second fat-soluble vitamin, carried in liver oils and certain other fats, which he called the *calcium-depositing vitamin.* It is of interest to note that, though McCollum discovered the existence of vitamin D, he did not call it by this name until after this designation was in common use by others. Evans and Bishop of the University of California, discovered that a fat-soluble dietary factor (then called "factor X") in lettuce and wheat germ was essential for successful reproduction in rats. For many years, what subsequently became known as vitamin E was known as the "antisterility vitamin," because of its effect on the fertility of rats. Mueller discovered another one of the essential amino acids, methionine.
1924	Sure, of the University of Arkansas, named the antisterility factor found in wheat germ and lettuce, *vitamin E.* The mystery of how sunlight could prevent rickets was partially solved. Dr. Harry Steenbock and Dr. A. Hess, working independently, showed that the antirachitic activity could be produced in foods and in animals by ultraviolet light. The process, known as the Steenbock Irradiation Process was patented by Steenbock, with the royalties assigned to the Wisconsin Alumni Research Foundation of the University of Wisconsin.
1925	E. B. Hart, who together with Stephen Babcock, helped to make the University of Wisconsin world famous, reported that white rats developed anemia if fed a milk diet, and that the anemia was not cured by iron salts. However, the ash of lettuce, which contained copper, was an effective cure. Mere traces of copper added to pure iron salts were found to be adequate for the cure of iron-deficiency anemia. G. H. Whipple showed that liver was a great benefit in blood regeneration of dogs rendered anemic by bleeding. Goldberger and Tanner classified common foods on the basis of their effectiveness in preventing or curing pellagra. They found that yeast rated abundant; lean meat and milk were good; and peas, beans, and other vegetables were fair.
1926	Minot and Murphy of the Harvard Medical School reported that feeding large amounts of raw liver restored the normal level of red blood cells in cases of pernicious anemia. For this discovery, they shared a Nobel Prize with G. H. Whipple. B. C. P. Jansen and W. P. Donath, in Holland, isolated the antiberiberi vitamin, which, at first, was called *aneurin* because of its specific action on the nervous system. Goldberger and Wheeler showed that pellagra in man and blacktongue in dogs were similar. They confirmed it by curing the blacktongue when they fed yeast to the dogs. They designated this preventive and curative factor as *P-P* (pellagra-preventive); others designated it *vitamin G* for Goldberger. Sumner isolated an enzyme and proved it to be a protein. Jansen and Donath isolated vitamin B-1 (thiamin) from rice polishings. It took over 30 years of research to isolate this vitamin!
1927	Boas in England reported that feeding egg white to rats produced a dermatitis.
1928	It became evident that what had been called vitamin B was not a single vitamin. Numerous investigators found that a growth-promoting substance remained after heat had destroyed the beriberi-preventive factor (thiamin) in yeast. This unknown substance was called vitamin G in the United States and vitamin B-2 by British scientists. At the time, it was thought to be only one vitamin; later, it was found that the heat-stable fraction was composed of several vitamins. A Hungarian scientist, Albert Szent-Gyorgy, working in Hopkins' laboratory at Cambridge University, England, isolated a substance from the ox adrenal glands, oranges, and cabbage leaves, which he called *hexuronic acid* (eventually to be called *ascorbic acid*); but he did not test it for antiscorbutic effect. Dr. George Oswald Burr, University of California, Berkeley, published a paper in which he proved that fats were essential in the diet of rats. Subsequently (in 1929), at the University of Minnesota (to which he transferred), he and Mildred M. (Mrs.) Burr found that the essential fat was linoleic acid.

(Continued)

TABLE H-3 (*Continued*)

Date	Nutritional Milestones
1929	Professor Carl Peter Hendrik Dam, biochemist at the University of Copenhagen, Denmark, observed that certain experimental diets produced fatal hemorrhages in chicks. Bleeding could be prevented by giving a variety of foodstuffs, especially alfalfa (lucerne) and fishmeal. Further, it was found that the active principle in these materials could be extracted with ether; thus, a new fat-soluble factor (later called vitamin K) was discovered. Sir Frederick G. Hopkins shared the Nobel Prize in Medicine with Christiaan Eijkman for their work with the vitamins. W. B. Castle of Harvard showed that pernicious anemia could be controlled by feeding patients beef incubated in normal gastric juice, although neither the beef nor the gastric juice alone was effective. Thus, he found that there is an "extrinsic factor" in food, and an "intrinsic factor" in normal gastric secretion; which, given together, caused red blood cell formation in pernicious anemia.
1930	William C. Rose and co-workers of the University of Illinois used rats to study the essentiality of each amino acid. Later (1935 to 1955), Rose conducted studies on humans to determine how much of each amino acid is needed. These experiments proved to be brilliant additions to the field of nutrition.
1931	While working in a maternity hospital in Bombay, India, Dr. Lucy Wills described a macrocytic anemia in pregnant women which improved when they were given extracts of yeast. Subsequent work showed that the factor in yeast was a group of chemically related vitamins with folic acid activity which was given the generic name *folacin.* P. Karrer, a Swiss researcher, determined the chemical formula for vitamin A---the first vitamin to have its chemical structure determined. For this, and for his work with riboflavin, he received the Nobel Prize.
1932	Dr. C. H. Best, codiscoverer of insulin, made the discovery that choline was important to nutrition. Further research proved that choline has a definite relationship in the metabolism of fats. Best and his co-workers at the University of Toronto reported that choline prevented fatty livers in rats fed high fat diets. Windaus of Germany and Askew of England isolated crystals of pure vitamin D_2 (ergocalciferol) from irradiated ergosterol. Charles Glen King, a graduate of Washington State College, and W. A. Waugh, working at the University of Pittsburgh, isolated a crystalline material from lemon juice that possessed antiscorbutic activity in guinea pigs. This marked the discovery and isolation of vitamin C, a deficiency of which caused the centuries-old scourge of scurvy.
1933	R. J. Williams fractionated a compound from yeast and called it pantothenic acid. Vitamin C was synthesized by Reichstein, a Swiss scientist. Northrup isolated the two enzymes in crystalline form---pepsin and trypsin. Allison and co-workers isolated a nitrogen-fixing bacteria in legume nodules, which they named *coenzyme R.* Kuhn, working at the University of Heidelberg, isolated pure riboflavin from milk, and demonstrated its similarity to that which Warburg and Christian had isolated.
1934	Dr. Paul Gyorgy found a cure for the severe dermatitis which resulted when Goldberger and Lillie tried to produce pellagra. But the curative compound of the extract was neither thiamin, niacin, nor riboflavin, but a substance which he called *vitamin B-6.*
1935	Kuhn at the University of Heidelberg, and the Swiss researcher, P. Karrer, both accomplished the synthesis of riboflavin. Karrer named it *riboflavin*, because it was found to have a pentose side chain---ribitol (similar to the sugar, ribose)---attached to a flavinlike compound. Dam named the antibleeding factor "koagulation vitamin" (the Danish word for *coagulation*), which was later shortened to *vitamin K* (from the first letter of Koagulation). W. C. Rose of the University of Illinois discovered threonine (1935 to 1936), thereby (1) paving the way for carefully controlled studies with diets made of synthetic amino acids, and (2) making it possible to learn the dispensability and indispensability of the various amino acids for different species of animals and the role of amino acid imbalance in nutrition. Most of this work was done with rats. But the doctoral students conducting these investigations confirmed the results on themselves; they tested each amino acid at different levels and determined the amount needed for optimum utilization of dietary protein. Von Euler, Albers, and Schlenck studied the preparation of cozymase, the coenzyme which is necessary for the alcoholic fermentation of glucose by apozymase, shown later to be diphosphopyridine nucleotide (DPN). On hydrolysis, cozymase yielded nicotinic acid. This was the first evidence that nicotinic acid formed a part of the structure of an enzyme, and placed it among the organic compounds of great importance in biological chemistry.

(Continued)

TABLE H-3 (*Continued*)

Date	Nutritional Milestones
1936	Szent-Gyorgy, who won a Nobel Prize in Medicine for his work with vitamin C, isolated a material from citrus rind that he called citrin, which he showed consisted of a mixture of flavonoids. His initial test of the new substance with scorbutic guinea pigs seemed to indicate that, in combination with ascorbic acid, it was effective in strengthening the body's smallest blood vessels, the capillaries, in addition to curing scurvy. In 1938, Szent-Gyorgy reported that subsequent tests failed to confirm the results of his earlier experiments. Brockman of Germany isolated crystals of pure vitamin D_3 (cholecalciferol) from tuna liver oil. Evans and co-workers isolated crystalline vitamin E from wheat germ oil and named it *tocopherol*, from the Greek words *tokos* (offspring) and *pherein* (to bear), meaning to *bear offspring*. Robert R. Williams, an American, determined the structure of and synthesized the antiberiberi vitamin, which he named *thiamine* because it contains sulfur (from *thio*, meaning *sulfur-containing*) and an amine group. Subsequently, the "e" was dropped. Samuel Lepkovsky, Jukes, and Krause, at the University of California, isolated vitamin B-6, or pyridoxine. Almost simultaneously, Keresztesy and Stevens, and Gyorgy reported its isolation. And in 1938, Kuhn and Wendt in Germany, and Ichiba and Michi in Japan, independently isolated vitamin B-6. Kogl and Tonnis of Germany isolated a crystalline substance from boiled yolks of duck eggs, which they called *biotin*, since they believed it to be identical to the *bios* factor needed for yeast growth. The two German scientists, Warburg and Christian, showed that nicotinamide was an essential component of the hydrogen transport system in the form of nicotinamide adenine dinucleotide (NAD).
1937	Paul Gyorgy found that a substance, which he named *vitamin H*, would prevent the pathological condition that resulted from feeding rats and chicks raw egg white. Later, he demonstrated that vitamin H and biotin were one and the same thing. Dr. Conrad Elvehjem, who spent most of his career at the University of Wisconsin, where he eventually became President, discovered that niacin (as either nicotinic acid or nicotinic acid amide, isolated from liver) cured blacktongue in dogs. Shortly thereafter, it was established that niacin was effective in the prevention and treatment of pellagra, and that it was essential to man, monkeys, pigs, chickens, and other species. Dr. Elvehjem also did much work on the role of copper and iron in nutrition, and amino acid interrelationships.
1938	P. Karrer and his associates succeeded in synthesizing alpha-tocopherol, and in proving its biological activity. Five different laboratories, working independently, isolated vitamin B-6 (pyridoxine) in crystalline form, but credit for obtaining the first crystals is generally given to Lepkovsky of the University of California. The study of vitamins accelerated as new discoveries were made. Three of the B complex---thiamin, riboflavin, and niacin, were now available in crystalline form.
1939	Dam and Karrer isolated vitamin K in pure form; and, that same year, Almquist and Klose synthesized vitamin K. In 1943, Dam shared the Nobel Prize in physiology and medicine with Edward Doisy who had studied the chemical properties of vitamin K. Williams isolated pantothenic acid from liver. A group of scientists from Merck & Co. discovered the structure of vitamin B-6 (pyridoxine), and S. A. Harris and K. Folkers synthesized the vitamin.
1940	D. W. Woolley, one of the famous Wisconsin group, isolated inositol from liver. Sure and Gyorgy-Goldblatt, working independently, reported that choline is essential for growth of rats, thereby indicating its vitamin nature. Gyorgy and associates obtained conclusive experimental evidence that coenzyme R, biotin, and vitamin H were the same substance. Pantothenic acid was synthesized by R. J. Williams and two other laboratories, all working independently. Thus, many scientists were studying the previous research papers and investigations into the vitamins were at the peak. D. W. Woolley, at the University of Wisconsin, demonstrated that inositol could prevent alopecia (patchy-hair, or baldness) in mice. E. Nielson, Oleson, and Elvehjem reported the concentration and fractionation of para-aminobenzoic acid.
1941	The name *folic acid*, the forerunner of the term *folacin*, was suggested by Mitchell, Snell, and Williams, of Texas, for the bacterial growth factor that was found in spinach and known to be widely distributed in green leafy plants. The word *folic* is derived from the Latin word *folium*, meaning *foliage* or *leaf*. Link and co-workers at the University of Wisconsin discovered dicoumarol---an antagonist or antimetabolite of vitamin K.

(Continued)

TABLE H-3 (*Continued*)

Date	Nutritional Milestones
1942	V. du Vigneaud and associates, at Cornell, suggested the correct structural formula for biotin based on a study of its degradation products. Snell discovered the vitamin B-6 activity of two closely related substances in natural products, which he named pyridoxal and pyridoxamine.
1943	Harris and his co-workers of Merck and Company synthesized biotin. Recommended Dietary Allowances (RDA) were first published in 1943---the work of the then newly established (1940) Food and Nutrition Board of the National Research Council-National Academy of Sciences whose first assignment was to determine dietary standards for people of different ages. The Recommended Dietary Allowances are revised at approximately 5-year intervals, as additional data become available.
1945	Angier and co-workers, isolated and synthesized folacin (folic acid). That same year, Dr. Tom Spies showed that folic acid was effective in the treatment of megaloblastic anemia of pregnancy and of tropical sprue. Willard Krehl and his associates at the University of Wisconsin finally solved the mystery of pellagra when they discovered that tryptophan is a precursor of niacin, thereby explaining two things: (1) why milk, which is low in niacin but high in tryptophan, will prevent or cure pellagra; and (2) why, in earlier concepts, protein deficiency was often related to pellagra.
1946	Lipman and co-workers showed that coenzyme A is essential for acetylation reactions in the body.
1947	Isler, working in Switzerland, synthesized vitamin A.
1948	Rickes and co-workers of Merck & Co., New Jersey, and Smith and Parker of England, isolated from liver concentrate, a crystalline, red pigment, which they called *vitamin B-12*. R. West of Columbia University, New York, showed that injections of vitamin B-12 induced a dramatic beneficial response in patients with pernicious anemia.
1950	Lipmann and co-workers showed that pantothenic acid was a constituent of coenzyme A.
1952	R. B. Woodward of Harvard accomplished the first total synthesis of a form of vitamin D (in this case vitamin D_3). He was awarded the Nobel Prize in Chemistry in 1965 for this and other similar achievements.
1953	Molybdenum (Mo) was found in the essential enzyme, xanthine dehydrogenase. Subsequently, it was found to be a component of three different enzyme systems which are involved in the metabolism of carbohydrates, fats, proteins, sulfur-containing amino acids, nucleic acids (DNA and RNA), and iron.
1955	The structure of vitamin B-12, cyanocobalamin, was determined by Dorothy Hodgkin and co-workers, at Oxford. In 1964, Hodgkin was awarded the Nobel Prize for her work. Woodward's group at Harvard synthesized vitamin B-12, using a very complicated and expensive procedure. Fortunately, soon thereafter, it was found that it could be produced from cultures of certain bacteria and fungi grown in large tanks containing special media.
1957	Selenium was discovered to be an essential element by Klaus Schwarz, a German medical researcher, who at the time was a visiting scientist at the National Institute of Health (NIH). In 1951, Schwarz reported that the American type of brewers' yeast contained an unidentified "Factor 3" which apparently acted along with Vitamin E and sulfur-containing amino acids in protecting the liver against damage due to certain types of diets. In 1957, he and co-worker, C. M. Foltz, reported that Factor 3 contained selenium; thereby making the discovery that selenium is an essential element. Other studies by the group at NIH, and by Scott and his co-workers in the Department of Poultry Husbandry at Cornell, showed that the addition of selenium salts to the diets of chicks prevented certain disorders which resulted from Vitamin E deficiencies. Finally, in 1973, Rotruck and co-workers at the University of Wisconsin reported that selenium acted as a co-factor for a recently-discovered enzyme (glutathione peroxidase) which breaks down toxic peroxides, most of which are formed from the oxidation of polyunsaturated fats. Thus, the link between selenium and Vitamin E was established.
1959	Chromium was found to be an essential element by W. Mertz and K. Schwarz, German medical scientists working at the U.S. National Institute of Health. They (1) made the important discovery that the feeding of chromium salts were necessary in the diet for normal metabolism of blood glucose in the rat, and (2) announced the specificity of chromium as the glucose tolerance factor (GTF).
1966	A food labeling law, known as "The Fair Packaging and Labeling Act," was enacted in 1966 and took effect in July, 1967. It initiated regulations requiring fuller and more prominent information on the labels of packaged foods. The regulations also involve special diet foods, with particular reference to vitamin and mineral supplementation and low-calorie foods; the dietary food regulations became effective in 1969.

(Continued)

TABLE H-3 *(Continued)*

Date	Nutritional Milestones
1967	Dr. George Wald of Harvard University was awarded the Nobel Prize in Medicine for clarifying the role of vitamin A in vision, which, to that time, had been an extremely complicated picture.
1969	Prostaglandins ushered in a new era of research. Although a group of lipid compounds—20-carbon fatty acids with various functional groups and configurations—were first extracted from human semen and sheep seminal vesicles in the early 1930s, and named prostaglandins in 1935, little attention was paid to their physiological action at the time. In 1969, Pharris and co-workers of the Upjohn Company reported that prostaglandin F_2a would regress the corpus luteum in rats, following which prostaglandin research increased manyfold. Prostaglandins are a class of hormonelike compounds, made in various tissues of the body from arachidonic acid (and other derivatives of linoleic acid), which are important in the regulation of such diverse reactions as gastric secretion, pancreatic functions, release of pituitary hormones, smooth muscle metabolism, and control of blood pressure. Several prostaglandins have been identified.
1970	Dr. Norman Borlaug, an American agricultural scientist employed by the Rockefeller Foundation and known as the *Father of the Green Revolution*, was awarded the Nobel Peace Prize for breeding new high-yielding, short-strawed varieties of wheat. A related development was the development of special varieties of rice by scientists at the International Rice Research Institute in the Philippines.
1977	Gene splicing, also known as recombinant DNA technique, ushered in a new era of genetic engineering, when scientists at the University of California–San Francisco reported a major breakthrough as a result of altering genes—turning ordinary bacteria into factories capable of producing insulin, a valuable hormone essential to the survival of diabetics. But this was only the beginning! Building upon the present knowledge and understanding of the nucleic acids DNA and RNA, the *Biotechnology Era* arrived.
1990	Americans ate more broilers and turkeys than ever before because of (1) lower prices in relation to other meats, and (2) new and further processed products. Poultry production became specialized to the point that most producers contracted with large integrated companies to produce eggs, broilers, and turkeys. The companies supplied the chicks/poults, feed, litter, fuel, and medication, and the contract growers provided housing and labor for the care of birds.

THE SEARCH GOES ON. Today, we know that the chemical substances found in food materials can be used, and are necessary, for life itself. We know that the nutrient chemicals—more than 40 of them, including amino acids, minerals, and vitamins—reach the body cells and tissues and are utilized for growth, development, physical activity, reproduction, lactation, maintenance of health, and recovery from illness or injury.

Like other sciences, modern human nutrition does not stand alone. It draws heavily on the basic findings of chemistry, biochemistry, physics, microbiology, physiology, bacteriology, medicine, food technology, agricultural science, genetics, mathematics, endocrinology, behavior, cellular biology, and, most recently, genetic engineering.

Although the chasm between awareness and application is becoming smaller, there is still much to be learned about foods and nutrition. We need to know more about human nutritional requirements; we need to know more about food composition and availability; we need to know more about diet, disease causation, and food safety; we need to know more about food consumption and nutritional status.

We need to be able to distinguish between myths and facts. We are told that Americans are living happily against the odds. We are warned about cholesterol in our eggs, about nitrate in our bacon, and about caffeine in our coffee—all of which has taken the fun out of eating. But there is no escape! If we go on bread and water, just as soon as we begin to enjoy it, someone will tell us that it's bad for us.

So, despite the advances that typify nutrition, there are many unknowns. The search goes on!

HIVES

Small swellings under or within the skin. They appear suddenly over large portions of the body and can be caused by food allergies.

(Also see ALLERGIES.)

HOGGET (HOGGETT)

Term applied to a male lamb from weaning time until shorn.

HOGS VS PIGS

Terms relating to apparent age of swine. Animals weighing more than approximately 120 lb(*54 kg*) are called hogs, whereas those weighing less than 120 lb (*54 kg*) are called pigs.

HOME ECONOMISTS

Specialists in home economics. Home economists work to improve products, services, and practices that affect the comfort and well-being of the family. Some specialize in specific areas, such as child development and family relations, clothing and textiles, consumer economics, foods and nutrition, home furnishings and equipment, home management, and housing. Others have broad knowledge of the whole professional field.

Home economists are engaged in the following types of work: (1) teaching in colleges, high schools, and adult education programs; (2) research for the federal government, state agricultural experiment stations, colleges, and private organizations; (3) Cooperative Extension Service; (4) private business firms and trade associations; (5) social welfare programs conducted by federal, state, and local governments; (6) dietitians for hospitals, schools, group food services, and restaurants; and (7) health services.

About 350 U.S. colleges and universities offer a bachelor's degree in home economics.

The professional organization for home economists is: American Home Economics Association, 2010 Massachusetts Ave. N.W., Washington, D.C. 20036.

Fig. H-16. Bees on a honeycomb. (Photo by J. C. Allen & Son, West Lafayette, Ind.)

HOMEOSTASIS

A state of equilibrium of the body's internal environment; for example, the lungs provide oxygen as it is required by the cells; and the kidneys, through the processes of reabsorption and elimination, help to maintain normal blood composition.

HOMOGENIZATION

A process which creates small uniform particles in a mixture. Fat globules in milk tend to clump together, separate and rise to the surface forming cream. By forcing milk through small openings under high pressure these fat globules can be broken down into small uniform particles which will remain evenly distributed throughout the milk and will not form cream. Milk so treated is called homogenized milk. Many substances are homogenized when a uniformity of particle size is desired.

(Also see MILK AND MILK PRODUCTS.)

HOMOGENIZED MILK

Milk in which the fat globules and other solids have been broken up, made uniform, and distributed evenly.

(Also see MILK AND MILK PRODUCTS.)

HONEY

Contents *Page*

Origin and History 1185
Production 1186
Processing 1187
By-Products 1187
Nutritional Value and Uses 1187

Essentially, honey is an invert sugar—a mixture of glucose and fructose—dissolved in 14 to 20% water with minor amounts of organic acids, along with traces of minerals and vitamins. Honey is derived from the nectar of flowering plants which the honeybee collects. Nectar consists primarily of 10 to 50% sucrose, glucose, and fructose, and 50 to 90% water. Each worker bee has a special pouch, called a honey bag, inside its body. Here the bee stores the nectar that it collects. After collection, the bees (1) supply an invertase enzyme which converts the sucrose to glucose and fructose, (2) store it in honeycombs, and (3) reduce the water content. The color, flavor, and proportion of the sugars vary with the source of nectar. In the United States, most honey is produced from alfalfa and clover. These honeys are light colored and delicately flavored. Tupelo honey, from southern United States, contains about twice as much fructose and glucose; hence, it seldom granulates. Nectar from the flowers of goldenrod and aster yield a dark honey. Other common honeys include buckwheat from eastern United States, orange-blossom honey, and sage honey from California.

ORIGIN AND HISTORY. Honey originated with honeybees. They make honey to use as their food, and man exploits this trait. Early man probably discovered, quite by accident, the sweet-tasting substance in honeycombs from a hollow tree, log, or cave. In Africa, native villagers use a bird, the honey guide, to help them locate nests of the African honeybee so they can rob the honey. Before cane and beet sugar, honey was the only sweetener known to mankind. It was once considered a mark of great wealth. The Bible tells us that the Israelites were promised "a land flowing with milk and honey" (Exodus 3:8).

Beekeeping is an ancient art. It probably goes back to the days of the early Egyptians, who used honey in medicine and embalming, as well as for food, some 2,000 to 5,000 years ago. The honeybee is native to Europe and

Africa, but the domesticated honey bee has been distributed all over the world by man. In most places they have flourished.

One of the early types of hives—the straw skep—is the traditional symbol of beekeeping. It was inefficient. The only way to remove the honey was to kill the bees and cut the combs out. Early settlers often used hollow sections of the gum tree or tupelo for hives. An American, L. E. Langstroth, started scientific beekeeping in 1851. Langstroth discovered the key dimensions—bee space—and designed the movable frame hive. This discovery increased honey production since it allowed beekeepers easily to manipulate a colony, harvest honey, or examine a brood nest, since bees no longer cemented parts of the hive in place with bee glue.

Other discoveries followed which increased production. In 1857, it was discovered that bees could be forced into building a straight comb in a wooden frame by giving them a piece of wax on which the cells were already embossed—a foundation. Then, in 1865, another discovery made the operation more efficient; by using an extractor—a centrifugal force machine—it was found that honey could be removed from the combs without damaging the comb. Thus, combs could be returned intact to the hive, saving the bees the time and energy required to build more combs. For bees to make a wax comb, approximately 16 lb (*7 kg*) of honey must be consumed to make 1 lb (*0.45 kg*) of wax. The modern smoker with bellows and a nozzle was developed in 1873. This modern smoker effectively directs smoke where it is needed. Smoke causes bees to engorge with honey and become gentle, and without smoke, manipulation of the frames in the hive is not possible. By 1880, a commodity that had once been scarce, was becoming abundant. Today, there are about 300,000 beekeepers in the United States, and beekeeping is done for fun and profit.

PRODUCTION. The honeybees do all of the work and man reaps the benefits. Considering that a good colony—a peak population of 30,000 to 60,000 workers in early summer—can store as much as 2.2 lb (*1 kg*) per day, the work of the honeybee is truly an amazing feat. Production of honey requires hives, honeybee colonies, flower sources, efficient management of the hives and the harvesting and separation of the honey.

Most beehives used in the industry today are the wooden ten-frame Langstroth hives. The removable frames contain the combs. The frames of beeswax comb foundation are the backbone of the colony. Commercial honey producers feel that they need to own 500 to 2,000 colonies, while those who make a hobby of beekeeping may only keep two colonies.

A colony of bees in an artificial hive contains about 10,000 to 12,000 worker bees (about 3 lb, or *1.4 kg*) and a queen in midwinter. During early summer, the same colony would have a peak population of 30,000 to 60,000 workers, a queen, and possibly 3,000 drones, which are the males. Drones only exist for the purpose of mating with the queen, and once this is accomplished they are deprived of food and driven from the hive. The size and composition of the colony population is governed by the number of eggs reared each season, and the availability of pollen ultimately controls the number of young reared. Without large quantities of pollen only a small number of young are reared. Bees store honey and pollen—their food source of energy and protein—separate from the brood-rearing area which contains eggs, larvae, and pupae. This factor makes beekeeping possible.

Successful, commercial beekeepers live in areas where nectar-producing plants abound. Bees search the area surrounding a beehive, and once they locate a nectar source, scout bees return to the hive and communicate the location to workers of the hive by a dance—described by the great Austrian bee watcher Karl von Frisch.

Fig. H-17. Honey and honey wheat bread. (Courtesy, California Honey Advisory Board, Whittier, Calif.)

In 1990, 16.6 million gallons of honey were produced in the United States. Of this amount, almost half was produced by six states: California, Florida, North Dakota, South Dakota, Minnesota, and Texas. Every state except Alaska produces some honey. It is considered uneconomical to keep bees in Alaska because an excess of 100 lb (*45 kg*) of honey per colony is required by the bees, leaving little to be harvested. In the United States, clover and alfalfa provide an abundant source of nectar for bees. Also, the orange blossoms of California and Florida provide a high quality and quantity of nectar. Furthermore, in a given area bees forage nectar from a variety of flowering trees, bushes, and ground plants.

When bees return to a hive with nectar, it contains 50 to 90% water, 10 to 50% sugar (mainly sucrose), and 1 to 4% aromatic substances, coloring and minerals. To transform the nectar into honey the bees do two things: (1) add an invertase enzyme which converts the sucrose to fructose and glucose, and (2) reduce the moisture content to 14 to 19%. Bees store their honey in the tiny, six-sided containers called cells, of the honeycomb.

Efficient management of honeybees involves: (1) proper location of the hives near nectar-producing plants;

(2) adequate storage of pollen and honey by bees for winter usage; (3) prevention of swarming; and (4) proper harvesting of the honey.

Harvesting the honey requires critical timing on the part of the beekeeper. Honey is harvested (1) when necessary to obtain more empty combs; (2) after a major production peak is completed; (3) when bees are actively foraging for nectar, to prevent stealing; and (4) before the onset of cool weather. Only surplus honey is harvested. A generous supply is left in the hive as bee food.

In farming areas, pesticides are a major risk to honey bee production. Other risks include disease such as American Foulbrood, animals—especially bears and skunks—vandalism, and fire.

PROCESSING. Originally honey was sold in the comb. Now, commercial producers remove the honey from the comb in a honey extractor. This machine whirls the honeycombs around thus forcing the honey out and leaving the comb intact for reuse by the bees, or for sale as beeswax. Commercial honey is heated to destroy yeasts, filtered, bottled, and sold. Raw honey may be strained, bottled, and sold. Honey with its low pH (3.5 to 4.0) and high osmotic pressure, due to the high sugar content, is inhospitable to microorganisms.

Fig. H-18. Honey contained in a honeycomb from a wooden Langstroth hive. (Photo by J. C. Allen & Son, West Lafayette, Ind.)

Honey must be kept sealed as it absorbs moisture from the air and may ferment if yeasts are present. Most honeys granulate during storage, but this can be reversed by immersion of the honey container in hot water. All honeys become darker and change flavor during prolonged heating and/or storage at room temperature. Long term storage—years—is best in the freezer. Honey that is sold as creamed honey is actually honey that has formed crystals.

BY-PRODUCTS. Honey production results in two major by-products: (1) beeswax, and (2) plant pollination. In some instances the honey actually becomes a by-product of the pollination.

1. **Beeswax.** Bees secrete the yellow wax to build their honeycombs of six-sided (hexagonal) cells. About 4 million lb (*1.8 million kg*) of beeswax are produced annually in the United States. In addition, the United States imports beeswax from Canada, Dominican Republic, Mexico, Haiti, Brazil, Chile, Egypt, and Ethiopia. Throughout history, beeswax has been used as a polish, in the preparation of molds for metal castings, for waterproofing and in many other ways. It was civilization's first plastic. Today, beeswax is used in the manufacture of wax paper, candles, modeling of artificial fruit and flowers, in process engraving, shoe polish; and in the preparation of ointments and plasters.

2. **Pollination.** Bees have hairy bodies; so, after wallowing around in flowers, they become thoroughly dusted with pollen. Then, in the process of moving from flower to flower, they bring about efficient cross-pollination. Approximately one-third of all the food we eat depends on honeybee pollination; and in the United States, honeybees provide about 80% of the insect pollination. The cross-pollinating ability of bees is so important that each year thousands of colonies are rented by crop growers. These bees are moved into groves, orchards, and fields, usually at night when all bees are in the hive, just when the plants begin blooming. Generally hives are placed about two per acre. In the agricultural economy, the real importance of bees is cross-pollination. Honey is a by-product for the beekeeper.

Fig. H-19. Beekeeper moving a group of hives. Note the wooden Langstroth hives. (Photo by J. C. Allen & Son, West Lafayette, Ind.)

NUTRITIONAL VALUE AND USES. Honey is pleasing to the senses—especially taste—so pleasing that some cultures have considered it an aphrodisiac (a product that excites sexual desire). Also, honey has acquired a special reputation as a nutritional food or medicine. Unfortunately, neither reputation is deserved. As Table H-4 shows, honey supplies substantial energy, since it is 75 to 80% fructose and glucose. But there are only traces of other nutrients. Despite its unimpressive chemical composition, honey continues to be appreciated by all who (1) can afford it and (2) enjoy its pleasant taste.

TABLE H-4
NUTRITIONAL VALUE OF HONEY[1]

Nutrient[2]	Unit	Amount	
		Per 100g	Per Tbsp (20 g)
Energy	kcal	304.00	61.00
Protein	g	.30	.10
Carbohydrates	g	82.30	16.50
Calcium	mg	5.00	1.00
Phosphorus	mg	6.00	1.20
Sodium	mg	5.00	1.00
Magnesium	mg	3.00	.60
Potassium	mg	51.00	10.20
Iron	mg	.50	.10
Zinc	mg	.10	.02
Copper	mg	.20	.04
Vitamin C	mg	1.00	.20
Riboflavin	mg	.04	.008
Niacin	mg	.30	.06
Pantothenic acid	mg	.20	.04
Pyridoxine	mg	.02	.004
Folic acid	mcg	3.00	.60

[1]Values are from the Food Composition Table F-36 of this book.
[2]Nutrients not listed are not present.

The other major constituent of honey is water, about 14 to 20%. Hence, very little is left over for essential nutrients. Indeed, numerous other food sources provide more of the essential nutrients. One real advantage in using honey is that, due to the fructose content, it is sweeter than table sugar—sucrose—and less is needed. As a sweetener, honey is used on breads, cereals, and desserts. About half the honey produced is used this way and the other half is used in the baking industry.

Because honey absorbs and retains moisture (it is hygroscopic), it is used by the baking industry to keep bread or cakes moist and fresh. Furthermore, the fructose in honey improves the browning quality. These same benefits of honey can be enjoyed in home baking. When honey is substituted for sugar, it is usually necessary to reduce the amount of liquid in the recipe by ¼ cup for each cup (*240 ml*) of honey. Many recipe books for cooking with honey are available. Honey may also be used in canned or frozen fruits, in jams and jellies, and in drinks. One popular drink in Viking and Elizabethan times was mead—fermented honey.

Honey has enjoyed some application in medicine for thousands of years. Any number of unfounded medical properties have been ascribed to it; among them, relief from nasal and bronchial pneumonia. And some really wild claims have been made for honey combined with vinegar; without doubt, the sweetness of the honey makes swallowing the vinegar tolerable, but not curative. Nowadays, honey is employed in some medicinal compounds just to cover up harsh bitter flavors to prevent granulation.

It may not be wise to feed honey to infants under 1 year of age, since infant botulism—botulism resulting from the production of toxins after the ingestion of *Colostridium botulinum*—may result from the ingestion of raw agricultural products. Honey has been implicated as

Fig. H-20. Honey over hot cereal. (Courtesy, California Honey Advisory Board, Whittier, Calif.)

a source in a very few cases. This type of botulism does not occur in older children and adults.[73]

(Also see CARBOHYDRATE[S]; DIGESTION AND ABSORPTION; METABOLISM; and SWEETENING AGENTS.)

HORMONE

The word hormone comes from the Greek word *hormon,* which means "to spur on," "to set in motion," or "to excite to action." These phrases are all very descriptive of a hormone. Released by the endocrine glands, hormones are "chemical messengers" which travel via the blood to specific organs or tissues and direct such processes as growth, reproduction, metabolism and behavior. In the blood they exist in extremely small quantities—millionths and billionths of a gram. Yet, their effects upon the body are profound, as demonstrated by diseased conditions where there is an over or under secretion of some hormone.

NOTE WELL: Hormone therapy for whatever reason, and whether oral or by injection, should always be under the direction of a physician.

(Also see ENDOCRINE GLANDS.)

[73]*The Harvard Medical School Letter*, November, 1978, p. 5.

HORSEMEAT

This refers to the flesh of the horse, which is used as a human food in most European countries, the U.S.S.R., and Japan. The leading horsemeat consumers are Belgium, Luxembourg, and France. Elsewhere in the Old World, horseflesh has been avoided by most Hindus, Buddhists, Jews, Moslems, and Christians, all of whom have followed closely the precepts of their religion or their cultural prejudices. Additionally, the notion that eating horseflesh is disreputable has persisted for a variety of reasons, not the least of which is man's attachment to the horse as a good friend and stout companion.

A variety of recipes for beef can be used in the preparation of horsemeat.

HORSERADISH *Armoracia rusticana*

This pungent root vegetable, which is one of the five bitter herbs of the Jewish Passover, is a member of the mustard family (*Cruciferae*). The grated root is used chiefly as a condiment or table relish to promote appetite. Formerly it was held in high repute as an antiscorbutic. Fig. H-21 shows a typical horseradish plant.

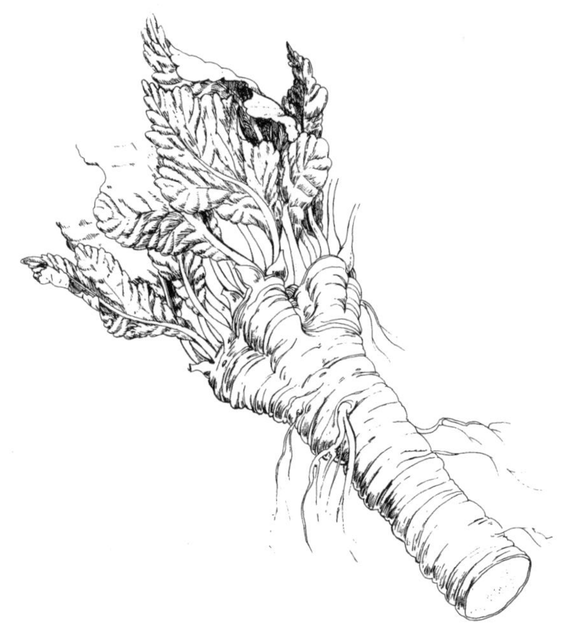

Fig. H-21. Horseradish, a strong-flavored root vegetable and seasoning agent.

ORIGIN AND HISTORY. Horseradish has been cultivated in the Near East for over 2,000 years. It has long been used as one of the bitter herbs for the Passover ritual which is described in Exodus 12:8. However, modern scholars are in disagreement as to when it was first used as a seasoning, since many of the ancient botanical descriptions are imprecise and subject to misinterpretation. Although the exact place of its first cultivation is uncertain, it is well established that in recent times its culinary use spread from eastern Europe to the British Isles, and from there across the Atlantic to North America.

PRODUCTION. There is little large-scale commercial production of horseradish in the United States, except for certain areas in the states of Illinois, Missouri, and Wisconsin. River bottom soils are ideal for growing horseradish. Elsewhere, the crop is grown on a small scale. Hence, statistics on the size of the crop are not available.

The seeds of horseradish are sterile. Hence, the plant is propagated by root cuttings (usually the slender branches that have been trimmed from the main root) which are placed in well cultivated soil in the early spring. Often, the soil has to be fertilized with nitrogen, phosphorus, and potassium to ensure that the roots will grow to an optimal size. Also, the young growing roots may be removed from the soil after the plants have grown to a height of 12 in. (*30 cm*) or more above the ground, so that branch roots may be trimmed from the main root. The removal of the branches ensures that the main root receives maximum nourishment. Then, the trimmed roots are replaced in the soil. This process may be repeated one or more times during the growing season.

Usually, the horseradish roots are dug from the ground in late fall or early winter, although they will usually survive the winter if left in the ground. After harvesting, the tops and side roots are removed from the main roots prior to sending the latter to market or to a processor. The yields of marketable roots range from 3,000 lb to 6,000 lb per acre (*3,360 to 6720 kg/ha*).

PROCESSING. Statistics showing the amounts of this root which are marketed fresh, and which are processed are not readily available. However, it seems likely that only a few American consumers make homemade preparations from the fresh root. Hence, it appears that most of this crop is utilized in commercial products that have been processed to give them long-lasting quality and consumer appeal.

Freshly-grated horseradish cannot be stored for long without further processing because the rupturing of the root cells during grating releases enzymes that activate the substances which contribute the strong odor and flavor. Furthermore, these pungent components are volatile and may escape with the passage of time. Hence, the enzyme action must be stopped by (1) the addition of an acid such as vinegar, or (2) drying. Also, jars of grated horseradish preparations should be refrigerated after opening so that the loss of volatile flavor components is minimized.

Some typical horseradish products sold in grocery stores, and served in restaurants are (1) grated fresh horseradish mixed in vinegar, tomato sauce, and/or mayonnaise; and (2) powdered dried horseradish.

SELECTION AND PREPARATION. Large, straight, undamaged roots are the best buy for those who wish to make their own horseradish preparations. Furthermore, the fresh root should be grated and mixed with other ingredients just prior to use. Some of the most common ways of preparing and serving grated horseradish follow:

1. Preparations of the grated root in vinegar are traditionally served with roast beef, flank steak, corned beef, and pastrami.

2. The applesauce commonly served with roast pork may be spiced by a little horseradish.
3. "Russell Chrain" is a mixture of horseradish and fermented beet juice, which is one of the five bitter herbs served during the Jewish Passover celebration.
4. Many restaurants serve oysters, clams, and shrimp with a special tomato sauce that contains the grated root.
5. Salad dressings may be made from sweet cream or sour cream mixtures that are flavored with horseradish.

NUTRITIONAL VALUE. Horseradish is a good source of potassium and vitamin C, but it is not likely to serve as a significant source of these or any other essential nutrients because the irritating nature of its flavor components limits the amounts which may be consumed. However, it may make a valuable nutritional contribution by stimulating the appetite and enhancing the enjoyment of food.

HOSPITAL DIET

Depending upon the needs of the patient it may be a regular, light, soft, or full liquid diet. Regular diets are for those persons who are ambulatory and require no therapeutic diet. Light diets are generally given during the early stages of convalescent periods, and differ from regular diets mainly in their mode of preparation. Rich pastries, fried foods, and most raw vegetables and fruit are avoided. The soft diet is an intermediate step between the light and liquid diet. Some hospitals, however, may omit the light diet. A soft diet consists mainly of liquids and semisolid foods; it is required in some gastrointestinal tract conditions, debilitated patients, and acute infections. Soft diets are low in residue and easily digested. They contain little or no spices and condiments, and are more restrictive in fruits, meats, and vegetables. The liquid diet is usually given postoperative, or to patients acutely ill with an infection, with gastrointestinal tract disturbances, or with myocardial infarction. It contains nothing but liquids such as fruit and vegetable juices, broth, milk, eggnogs (with powdered high-protein supplements added), junket, ice cream, etc.
(Also see MODIFIED DIETS.)

HOTHOUSE LAMBS

Considered by epicureans as the most delectable of the lamb age groups. They are very young lambs—usually less than 3 months of age at slaughter—which are born and marketed out of season. Such milk-fat lambs are usually marketed during the period from Christmas to the Easter holidays at weights ranging from 30 to 60 lb (*14 to 27 kg*). Hothouse lambs may consist of ewe, wether, or ram lambs.
(Also see LAMB.)

HOT POT

This dish, which originated in Lancashire, England, became popular as a national dish. It consists of layering sliced potatoes, lamb chops (trimmed), lamb kidneys, onions, and mushrooms in a baking dish. Water is added, and it is baked covered in the oven. Sometimes meats other than lamb, are used.

HUCKLEBERRY *Gaylussacia* spp.

The fruit of a bush that grows wild throughout the United States and southern Canada.

Fig. H-22. Huckleberry.

Huckleberries resemble blueberries, except that they are usually smaller, contain larger seeds, and are almost always gathered from wild bushes rather than from cultivated plants. They may be eaten fresh, or used to make jam, jelly, juice, pies, tarts, or wine.
(Also see BLUEBERRY[S].)

HULLS

Outer covering of grain or other seed, especially when dry.

HUMBLE PIE

Originally, this was a meat pie made from the inferior parts of the carcass, usually eaten by the huntsman and the servants. Ironically, these variety meats were superior, nutritionally, to the more desired cuts.

HUMECTANT

A term for additives which prevent foods from drying out. Humectants keep marshmallows and shredded coconut soft. Glycerine (glycerol) and sorbitol are utilized for this purpose.
(Also see ADDITIVES.)

HUMULONE

One of the two resins of hops which imparts the traditional odor and bitter taste to beer.
(Also see BEERS AND BREWING.)

HUNGARIAN PARTRIDGE (CHUKAR; EUROPEAN PARTRIDGE; OR GRAY PARTRIDGE).

True partridges belong to the partridge family, *Phasianiidae*. There are about 150 kinds of these birds, the best known of which are: (1) the Hungarian (or gray partridge; European partridge), native to Europe, northern Africa, and Western Asia, which is genus *Perdix,* species *P. perdix*); and (2) the Chukar, native to Asia and Europe, which is *Alectoris chukar.*

The Hungarian, or gray partridge, was introduced into western United States as a game bird early in the 20th century. Today, most of these birds are found on the Canadian plains and in the North Central and Northwestern United States.

The chukar, which was also introduced into the United States, is an important game bird in northwestern United States.

HUNGER

Hunger is a physiological desire for food following a period of fasting. Appetite, on the other hand, is a learned or habitual response, which arises with the customary intervals of eating and may be influenced by numerous external and internal phenomena. Satiety is the opposite of hunger—a feeling of complete fulfillment of the desire for food.

(Also see APPETITE.)

HUNGER, WORLD

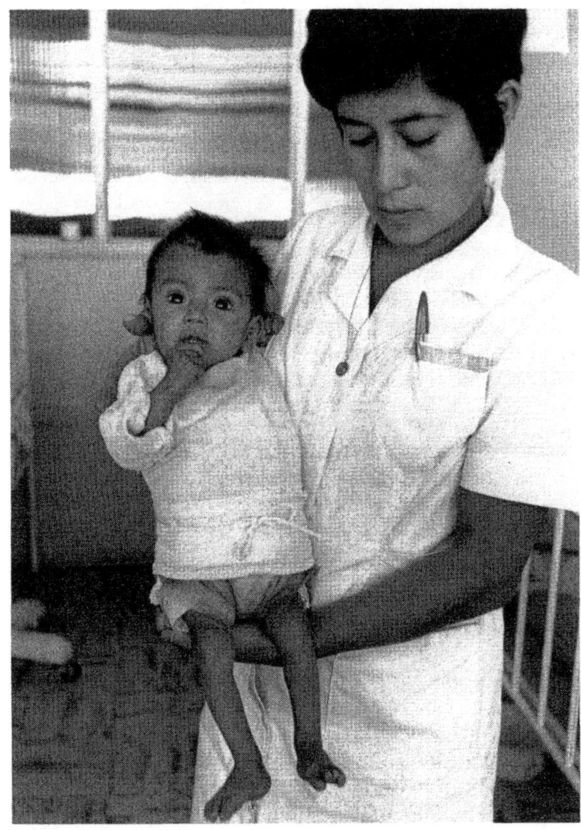

Fig. H-23. Hunger is a tragedy for millions of children. This 6-month-old baby, suffering from malnutrition, was treated in a Chilean hospital. (Courtesy, FAO, Rome, Italy)

Many knowledgeable people subscribe to the view that the world is on the brink of being subjected to food shortages of unprecedented scale and urgency—a world broken by unshared bread.

The troubled 21st century will open with 1.3 billion malnourished people, up from one-half billion in 1986. Starvation will claim increasing numbers, and many of the surviving babies will grow up physically and mentally retarded. More disturbing yet, all nations and all people must realize that a hungry man is impelled only by his hunger and the right to survive, and that there can be no peace on an empty stomach.

The best way in which to lessen hunger and malnutrition in the developing countries is by a massive infusion of science, technology, and education—by self-help programs—so that they can produce more of their own food.

Improved genetics, along with improved feeding and management, have made for more meat, milk, and eggs in the United States, as shown in Fig. H-24. And similar achievements have been made in crops.

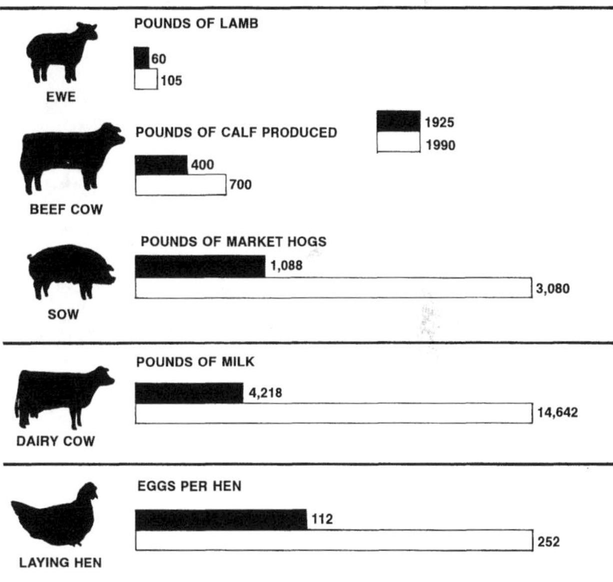

Fig. H-24. Improved efficiency in U.S. meat and milk animals from 1925 to 1990. The application of science and technology made the difference. *Sources:* Milk and egg statistics from USDA. Ewe, beef cow, and sow statistics estimated by Dr. M. E. Ensminger.

The science and technology that have been so successful in the United States may be adapted to and applied by developing countries.

(Also see MALNUTRITION, PROTEIN-ENERGY; POPULATION, WORLD; PROTEIN, WORLD PER CAPITA; and WORLD FOOD)

HUSKS

Leaves enveloping an ear of corn (maize).

HYACINTH BEAN (LABLAB BEAN) *Dolichos lablab*

This legume, which is little known in the developed countries, is an important crop in India and other dry tropical areas, because of its resistance to drought. Fig. H-25 shows this legume.

Fig. H-25. Hyacinth bean. (Courtesy, Minnesota Agricultural Experiment Station, University of Minnesota)

ORIGIN AND HISTORY. It is believed that this plant originated in India, where it has been cultivated for thousands of years. Over the centuries, explorers and traders took the hyacinth bean to other dry tropical areas in Africa, Asia, Central America, and the West Indies.

PRODUCTION. The agricultural practices used in the production of this legume vary, with the field variety having different requirements than the garden variety.

• **Field variety**—This type of hyacinth bean is grown mainly for dry mature seeds and livestock fodder. Often, it is intercropped with millet or another cereal and given little care because of its tolerance to poor soil (which must be well drained) and its drought resistance. The yield of dry beans per acre ranges from 400 lb to 1,300 lb (*448 to 1,456 kg/ha).*

• **Garden variety**—Gardeners in India give this crop ample manure and water, and provide supports for the climbing vines. The immature pods and seeds are harvested continuously.

It is noteworthy that hyacinth beans are utilized as green feed for livestock in some parts of the world, and that as much as 10,000 lb to 20,000 lb (*11,200 to 22,400 kg/ha*) of plant material per acre may be obtained. Also, because of its purple flowers and attractive purplish-red seed pods, it is sometimes grown for ornamental purposes.

PROCESSING. In India, the mature beans are dried and split for utilization in ways similar to those for split peas.

SELECTION AND PREPARATION. The immature pods and seeds may be utilized as a green vegetable in much the same way as snap beans.

The dried, whole mature seeds require a long cooking period. Hence, many people in India use the split seeds in ways similar to split peas, because firewood and other fuels are often scarce.

CAUTION: The seeds of the hyacinth bean, like those of the common bean, contain (1) inhibitors of trypsin, which may reduce the amount of dietary protein that is digested fully; and (2) red blood cell clumping agents (hemagglutinins). However, these potentially harmful substances may be rendered ineffective by soaking the beans overnight, then cooking them thoroughly. In the event that the beans are to be ground into flour, they should be soaked and treated prior to grinding.

NUTRITIONAL VALUE. The nutritive composition of hyacinth beans is given in Food Composition Table F-36.

Some noteworthy observations regarding the nutritive value of hyacinth beans follow:

1. The immature pods and seeds are low in calories and other nutrients because of their high water content. However, they are a fair to good source of vitamins A and C.
2. It appears that the dried mature seeds of the hyacinth bean are very similar in composition to the seeds of the kidney bean. Therefore, it might be assumed that the nutritive values of the cooked dry beans of the two species are similar, and that one cup (*240 ml*) of the cooked mature hyacinth beans contain about 14 g of protein (the amount in 2 oz [*56 g*] of lean meat) and 218 kcal (about twice the calories furnished by the equivalent amount of meat).
3. The dried mature seeds contain only about one-sixth as much calcium as phosphorus, whereas the immature pods and seeds contain approximately equal amounts of the two elements. (It is suspected that low dietary calcium to phosphorus ratios sometimes result in poor utilization of calcium by the body.) Hence, foods higher in calcium and lower in phosphorus, such as dairy products and green vegetables, should also be consumed.
4. Both the mature and the immature beans provide ample amounts of iron per calorie.

Protein Quantity and Quality. Some of the people in Africa, India, and Southeast Asia utilize the dried mature beans as a staple food. Hence, certain facts concerning the bean protein are noteworthy.

1. A 3½ oz (*100 g*) portion (about ½ cup) of the cooked beans furnishes about 118 Calories (kcal) and about 7 g of protein, the amount provided by 1¾ oz (*49 g*) of cottage cheese (*53 kcal*) or 1 large egg (*82 kcal*).
2. The protein quality of the dried beans is inferior to that of most animal proteins, because the bean protein is deficient in the sulfur-containing amino acids methionine and cystine.
3. Cooked dried beans are from 2 to 4 times as rich in protein as cooked cereal grains such as corn and rice.
4. Combinations of the beans and cereals have higher quality protein then either food alone because their amino acid patterns are complementary. That is, the cereals supply extra amounts of the amino acids lacking in the beans, and vice versa.

(Also see BEAN[S], Table B-10 Beans of the World; and LEGUMES, Table I-2 Legumes of the World.)

HYDRAULIC PROCESS

A process for the mechanical extraction of oil from seeds, involving the use of a hydraulic press. Sometimes referred to as the old process.

(Also see OILS, VEGETABLE.)

HYDROCHLORIC ACID (HCL)

Formed by the chemical combination of hydrogen (H) and chlorine (Cl), it is important in the digestion of proteins. Hydrochloric acid secreted by the cells of the stomach breaks down some proteins, but, more importantly, it provides an environment conducive to the activity of the enzyme pepsin.

(Also see DIGESTION AND ABSORPTION; and HYPERCHLORHYDRIA.)

HYDROCOOLING

To avoid loss of produce through decay and to preserve freshness, many fruits and vegetables are substantially cooled shortly after harvesting. In the past, packing produce with ice when loading for transport was the major practice. Now, many fruits and vegetables are cooled by spraying them continuously with cool water while they are in wholesale containers—hydrocooling. The water used is recirculated. It often contains salt to lower the freezing point and a fungicide or bacteriocide. There are portable hydrocoolers which cool fruits and vegetables as they are picked and before loading them into refrigerated trucks or railroad cars.

(Also see VACUUM COOLING.)

HYDROGENATION

The chemical addition of hydrogen to any unsaturated compound. However, the term is usually applied to the process for hardening vegetable oils.

(Also see OILS, VEGETABLE.)

HYDROLYSATE

The product of hydrolysis; for example, protein hydrolysate is a mixture of the constituent amino acids when the protein molecule is split by acids, alkalies, or enzymes.

HYDROLYSIS

The splitting of a substance into the smaller units by its chemical reaction with water.

(Also see DIGESTION AND ABSORPTION; and ENZYME.)

HYDROPONICS

Hydroponics is the growing of plants in chambers under conditions of controlled temperature and humidity with their roots immersed in an aqueous solution containing the essential mineral nutrient salts, instead of in soil. This means that food plants may be produced with water and chemicals, without dirt.

HYDROSTATIC PRESSURE

The pressure exerted by a liquid on the surface of the walls that contain it, which is equal on all containing walls. In the body, hydrostatic pressure usually refers to the blood pressure; together with the plasma proteins, it maintains fluid circulation and volume in the blood vessels.

HYDROSTATIC STERILIZER

A type of continuous pressure cooker which is open to the atmosphere at the inlet and outlet. Steam pressure in the sterilizing chamber is balanced by tall columns of hot water in the inlet and cold water in the outlet. If the steam in the sterilizing chamber is 260°F (*127°C*), to balance the pressure developed the water columns would have to be 46 feet (*14 meters*) high at the outlet and the inlet. The hydrostatic sterilizer is well suited for large scale production of canned goods. Canned goods on a conveyor transverse down the inlet column into the sterilizing chamber and then, after adequate time, leave via the outlet column.

HYDROXYLATION

The introduction of a hydroxy group (-OH) into an organic compound (a carbon-containing compound).

HYDROXYPROLINE

One of the nonessential amino acids.

(Also see AMINO ACID[S].)

HYGROSCOPICITY

The degree of tendency to absorb and retain moisture. For example, nonfat dry milk (NDM) is hygroscopic in that it tends to absorb moisture.

HYPER-

A perfix meaning over, above, or beyond.

HYPERALIMENTATION, INTRAVENOUS

The term *hyperalimentation* alone literally means the provision of nutrients at levels considerably higher than those received in a normal diet, although in medical circles it is commonly used as a synonym for total parenteral nutrition (TPN). The latter procedure involves the provision of all nutrient requirements by intravenous means. However, the term hyperalimentation might also be used to indicate feeding by means of orally administered liquid diets which provide high levels of one or more nutrients. Furthermore, some patients on TPN are normal weight or overweight and will require only normal or somewhat subnormal levels of calories and nutrients such as carbohydrates and fats. Therefore, confusion about the nature of the various nutritional procedures may be avoided by using the terms *intravenous* and *parenteral* to denote feeding through the veins. (These subjects are covered in detail elsewhere in this book.)

(Also see INTRAVENOUS [PARENTERAL] NUTRITION, SUPPLEMENTARY; and TOTAL PARENTERAL [INTRAVENOUS] NUTRITION.)

HYPERCALCEMIA

An excessive quantity of calcium in the blood, above the normal level of 11 mg per 100 ml. Hypercalcemia may be caused by some cancers (especially lung cancer), overactivity of the parathyroid glands, chronic ingestion of large doses of vitamin D, and alkali and milk therapy for peptic ulcers. Affected persons demonstrate a loss of appetite, vomiting, flabby muscles, and possibly kidney stones.

(Also see CALCIUM, section headed "Calcium Related Diseases"; and ENDOCRINE GLANDS.)

HYPERCALCIURIA

Excess calcium in the urine, as in hyperparathyroidism.

HYPERCHLORHYDRIA (HYPERACIDITY)

A term used to denote excess acidity of the gastric juice, most often hydrochloric acid, in the stomach. It is usually accompanied by burning pain after meals, heartburn, and acid indigestion. It is a common symptom of emotional upsets, duodenal ulcers, or inflammation of the gallbladder and is associated with hunger pains.

The distress may be relieved by antacids, including calcium carbonate, aluminum hydroxide gel, milk of magnesia, magnesium carbonate, and magnesium trisilicate.

(Also see ANTACIDS; GASTRITIS; HEARTBURN; INDIGESTION; and ULCERS, PEPTIC.)

HYPERCHOLESTEREMIA (HYPERCHOLESTEROLEMIA)

Excess of cholesterol in the blood.

HYPEREMIA

An excess of blood in any part of the body.

HYPERESTHESIA

Excess sensitivity to touch or pain.

HYPERGLYCEMIA

Excess of sugar in the blood.

HYPERKALEMIA

Excess potassium in the blood. Hyperkalemia is a serious complication of kidney failure, severe dehydration, or shock; it causes the heart to dilate, and the heart rate is slowed by weakened conditions.

HYPERLIPOPROTEINEMIAS (HYPERLIPIDEMIAS)

The term hyperlipoproteinemias refers to a group of disorders characterized by excessive concentrations of one or more of the lipoproteins in the blood.

Fats or lipids, which are not soluble in a water medium such as blood, are transported in the blood in the form of water-soluble, fat-protein complexes—lipoproteins. All of these lipoproteins contain phospholipids, triglycerides, and cholesterol, but in varying amounts as shown in Table H-5.

TABLE H-5
PERCENT COMPOSITION OF LIPOPROTEINS OF THE BLOOD

Lipoprotein	Abbreviation	Chemical Makeup			
		Protein	Phospholipids	Triglycerides	Cholesterol
		←———(%)———→			
Chylomicrons		0.5-2.5	3-15	79-95	2-12
Very low density (prebeta-lipoproteins)	VLDL	2-13	10-25	50-80	9-24
Low density (beta-lipoproteins)	LDL	20-25	20-22	10	43-45
High density (alpha-lipoproteins)	HDL	45-55	24-30	5-8	17-18

The lipoproteins are classified as chylomicrons, very low density lipoproteins (VLDL), low density lipoproteins (LDL), or high density lipoproteins (HDL)—according to their density, from least dense to most dense. They are also named prebeta-lipoproteins (VLDL), beta-lipoproteins (LDL), and alpha-lipoproteins (HDL), based upon their migration on paper subjected to an electrical field—electrophoresis (see Table H-5).

Levels of lipoproteins in the blood are controlled by diet, a number of hormones (anterior pituitary hormones, adrenal hormones, insulin, and thyroxine), age, weight change, emotions and stress, drugs, and illness. Abnormalities of blood lipid levels are best determined by measurement of the lipoproteins rather than the analyses of the blood lipid fractions alone. Thus disorders in blood lipid levels are classified according to the concentrations of the lipoproteins. The term hyperlipoproteinemias is used to described a group of disorders characterized by excessive concentrations of one or more of the lipoproteins in the blood. Table H-6 provides a simple classification, the characteristic features and treatment, primarily dietary, of the hyperlipoproteinemias.

TABLE H-6
THE HYPERLIPOPROTEINEMIAS

Type and Name	Characteristic Features	Treatment	Comments
TYPE I Hyperchylomicronemia (normal or elevated cholesterol with markedly elevated triglyceride)	Elevated blood levels of chylomicrons; yellow papules with reddish base develop over skin and mucous membranes (xanthomas); enlarged spleen; enlarged liver; abdominal pain; possible pancreatitis.	Low fat intake (25 to 36 g/day); possible use of medium chain triglycerides; high energy, high protein diet; no alcohol.	Occurrence rare; often due to inherited deficiency or decreased activity of the enzyme lipoprotein lipase; generally appears in childhood; the disease dysgammaglobulinemia, diabetes and lupus erythematosus also cause elevation of chylomicrons.
TYPE II A Hyperbeta-lipoproteinemia or hypercholesterolemia (increased LDL)	Increased blood levels of beta-lipoprotein (LDL) and cholesterol (300 to 600 mg/100 ml); yellow lipid deposits in skin, tendons, and cornea; accelerated atherosclerosis.	Saturated fat and cholesterol limited sharply; cholesterol restricted to less than 300 mg/day; polyunsaturated fats increased; use of the drug cholestyramine.	A common hereditary disorder; detected in early childhood in severe cases; secondary disorders which may be confused with Type II A include dietary excesses of cholesterol, obstructive liver disease, hypothyroidism, nephrosis, porphyria, and some tumors.
TYPE II B Hypercholesterolemia with hyperglyceridemia or combined hyperlipidemia (increased LDL and VLDL)	Elevated blood levels of both beta-lipoprotein (LDL) and prebeta-lipoprotein (VLDL); cholesterol and triglyceride blood levels also increased. Yellow or orange deposits of lipid on the skin (xanthomas); accelerated atherosclerosis.	Low cholesterol (less than 300 mg/day); weight reduction if necessary; decreased use of saturated fats and increased use of polyunsaturated fats; limited intake of sucrose, concentrated sweets and alcohol.	Usually not detected until adulthood; possibly inherited; quite a common disorder; possibility of confusing with myxedema or dysgammaglobulinemia.

(Continued)

TABLE H-6 *(Continued)*

Type and Name	Characteristic Features	Treatment	Comments
TYPE III Broad-beta or floating beta (increased ILDL)	Elevated blood levels of an abnormal prebeta-lipoprotein (VLDL), cholesterol, and triglycerides; tendon xanthomas; buttock, knee, and elbow xanthomas; creases in palm of hands show as orange-yellow lines; accelerated atherosclerosis of coronary and peripheral blood vessels; heart disease.	Low cholesterol intake (less than 300 mg/day); achieve ideal weight; no concentrated sweets; reduce total fat consumption to 30% or less of total calorie intake; use of polyunsaturated fats and restriction of saturated fats; drug of choice, clofibrate.	Occurrence relatively rare; no absolute diagnostic test; an inherited disorder; appears after 30 to 45 years of age; abnormal prebeta-lipoprotein also called intermediate density lipoprotein (IDL); patients respond rapidly to dietary and drug therapy.
TYPE IV Hyperpre-beta-lipoproteinemia (increased VLDL)	High blood levels of pre-beta-lipoprotein (VLDL) and triglycerides; cholesterol normal or elevated; accelerated heart disease; glucose intolerance.	Maintenance of ideal weight; controlled carbohydrate intake, about 45% of energy; moderate cholesterol restriction to 300 mg; polyunsaturated fats preferred; no concentrated sweets. Drugs of choice: clofibrate or nicotinic acid.	Common occurrence; possibly inherited; obesity and alcohol aggravate; usually appears after age 20; frequently observed in women age 20 to 50 using oral contraceptives; diabetes, glycogen storage diseases, nephrotic syndrome, pregnancy, stress and alcoholism can secondarily cause Type IV.
TYPE V Mixed hyperlipidemia (increased chylomicrons and VLDL)	Cholesterol may be elevated or normal. Elevated chylomicrons, prebeta-lipoproteins (VLDL), and triglycerides (as high as 1,000 to 6,000 mg/100 ml); eruptive xanthomas (orange-yellow lipid deposits); abdominal pain; lipid in retina of eye; enlarged liver and spleen.	Fat restricted to 30% of energy; use of polyunsaturated fats; high protein diet; maintenance of ideal weight; reduce cholesterol intake to 300 mg daily; no concentrated sweets; possible use of nicotinic acid; no alcohol.	Clinical features similar to Type I; manifested in early adulthood; may be inherited as a dominant trait; may be a consequence of diabetic acidosis, alcoholism or nephrosis.

The specific causes of the hyperlipoproteinemias are not completely understood. In a broad sense, hyperlipoproteinemias occur for two reasons: (1) inadequate removal of the lipoprotein from the blood, and (2) increased production of lipoproteins. In turn this may be caused by secondary diseases such as diabetes, liver disease, and hypothyroidism. On the other hand, hyperlipoproteinemias may be the primary disorder, but their cause is not obvious. Many of the primary hyperlipoproteinemias indicate a genetic origin. They "run in families."

Elevated lipoproteins are a major health concern due to their association with accelerated atherosclerosis—a basic pathologic process in coronary heart disease. Fortunately, the hyperlipoproteinemias generally respond to diet therapy and within a few weeks lipoproteins may be lowered. Both the American Heart Association and the National Heart and Lung Institute have published detailed diets for controlling the hyperlipoproteinemias. Overall, the therapeutic diet is palatable, selected from ordinary foods, and no more expensive than the usual diet.

If a personal medical history, family history, and physical examination suggest the existence of a hyperlipoproteinemia, the next step is to measure blood levels of cholesterol and triglycerides. These measurements are made on blood drawn after an overnight fast during a steady state period in an individual's habits and environment. Elevated levels of cholesterol and/or triglycerides, or the presence of a "creamy" layer on the top of plasma after overnight refrigeration indicates to a physician the type of hyperlipoproteinemia. The actual lipoproteins—chylomicrons, VLDL, LDL or HDL—which are involved may need to be determined by ultracentrifugal fractionation or electrophoretic analysis. However, both of these laboratory techniques are too time-consuming and complex for routine blood analyses.

(Also see DIGESTION AND ABSORPTION; ENDOCRINE GLANDS; FATS AND OTHER LIPIDS; HEART DISEASE; INBORN ERRORS OF METABOLISM; and METABOLISM.)

HYPERPHOSPHATEMIA

Excess phosphorus in the blood serum, which may result when the kidneys do not excrete phosphorus adequately or from hypoparathyroidism, which causes an insufficient secretion of parathyroid hormone. When serum phosphorus rises, serum calcium falls, causing tetany.

HYPERPLASIA

Abnormal increase in number of normal cells.

HYPERTENSION

The medical term for high blood pressure. It is not a disease as such; rather, it is a sign of a variety of diseases differing in nature and significance. High blood pressure is indicated by a persistent elevation above normal limits. The normal: 120-150 systolic pressure (when the heart contracts), and 80-100 diastolic pressure (when the heart is at rest). Blood pressure is a measure of the pressure of the blood against the walls of the blood vessels produced by the beating of the heart.

(Also see HEART DISEASE; HIGH BLOOD PRESSURE.)

HYPERTHYROIDISM

Overactivity of the thyroid gland.

(Also see GOITER.)

HYPERTONIC DEHYDRATION

Loss of water from cells as a result of hypertonicity (excess solutes [dissolved substances]; hence, greater osmotic pressure) of the surrounding extracellular fluid.

HYPERTRIGLYCERIDEMIA

Increased levels of triglycerides in the blood.

HYPERTROPHIED

Having increased in size beyond the normal growth.

HYPERURICEMIA

Excess of uric acid in the blood—a characteristic of gout.

HYPERVITAMINOSIS (VITAMIN OVERDOSES)

This refers to an excessive intake of certain vitamins. the best known vitamin toxicities are those which result from vitamin A (known as hypervitaminosis A) and vitamin D (known as hypervitaminosis D), because (1) these vitamins are fat-soluble, (2) small amounts of them have strong effects, and (3) they tend to accumulate in the liver. Toxic effects do not occur so readily with vitamins E and K, which are also fat-soluble, unless high potency supplements are taken. Excesses of water-soluble vitamins (vitamin C and the vitamin B complex) are not stored in the body to any great extent, so toxicities from food sources of these nutrients are rare. However, people who take very large doses (megadoses) of any vitamin in the form of supplements should be under the supervision of a physician or nutritionist.

(Also see DISEASES, section headed "Essential Nutrients"; and VITAMINS.)

HYPO-

Prefix meaning lack, or a deficiency.

HYPOALBUMINEMIA

Abnormally low albumin level of the blood.

HYPOCALCEMIA

A subnormal concentration of ionic calcium in blood resulting in convulsions, as in tetany. The most common cause of this problem in man is lack of parathormone, a secretion of the parathyroid glands.

(Also see CALCIUM; and ENDOCRINE GLANDS.)

HYPOCHLOREMIC ALKALOSIS

An abnormal decrease of chlorides in the blood, with bicarbonate replacing the chloride ions, may lead to hypochloremic alkalosis. This condition may follow excessive loss of gastric secretion (hydrochloric acid) with the accompanying loss of chlorides. Excess vomiting may cause hypochloremic alkalosis unless the chloride is replaced promptly.

HYPOCHLORHYDRIA

Decreased secretion of hydrochloric acid by the cells of the stomach.

HYPOCHROMIC

An abnormal decrease in the hemoglobin content of the red blood cells, characterized by below normal color.

HYPOCUPREMIA

A deficiency in blood copper, which may be caused by urinary loss of ceruloplasmin (the copper binding protein of the plasma) in nephrosis (degeneration of the kidneys) or by malabsorption of copper in sprue.

HYPOGEUSIA

A diminished (or blunted) sense of taste.

HYPOGLYCEMIA (LOW BLOOD SUGAR)

Contents *Page*

Factors Which Raise or Lower the Blood Sugar in Healthy People 1198
Control of Blood Sugar After Meals 1198
Control of Blood Sugar During Fasting or Starvation 1199
Major Causes of Hypoglycemia 1200
Other Disorders Which May Result from Hypoglycemia 1202
Diagnosis of Hypoglycemia 1204
Signs and Symptoms 1204
Glucose Tolerance Test (GTT) 1204
Conditions Which May Mimic Hypoglycemia 1204
Treatment and Prevention of Hypoglycemia 1205
Emergency Treatment 1205
Diet Therapy 1205
Practical Hints for Controlling Blood Sugar 1206
Nondietary Measures 1208

This condition results when the level of glucose, the main sugar found in the blood, is below normal. It may be accompanied by such irritating signs as anxiety, mental confusion, rapid pulse, and tiredness which is greater than might be expected under the circumstances. The diverse complaints accompanying hypoglycemia are due to the sensitivity of the brain and the nerves to deprivation of glucose, which is the major source of energy for these tissues. Some physicians believe the condition to be uncommon, while others suspect it to be regularly present in a fairly large group of people. Those doctors who subscribe to the latter thinking have, at various times, suggested that low blood sugar may be associated with such diverse disorders as alcoholism, excess coffee, allergies, behavior problems, brain damage in infants, depression, diabetes, drug addiction, fatigue, high blood pressure, impotency, obesity, ulcers, and underachievement in school. Therefore, a review of what is known about this controversial subject is important. Pertinent facts follow.

FACTORS WHICH RAISE OR LOWER THE BLOOD SUGAR IN HEALTHY PEOPLE. Primitive man often faced starvation, practicularly in those areas of the world where harsh environments limited the supply of animals and plants which might be used for food. At other times, he gorged himself on food when it was bountiful and likely to spoil. Through the years, some populations, such as the Eskimos, have subsisted almost entirely on fish and meat, which contain only negligible amounts of carbohydrates. In stark contrast, many Asian peoples traditionally consume diets in which carbohydrates supply over 80% of the energy. The amazing part of it is that some of the descendants of these diverse peoples have been able to adapt to radically different diets which have been necessitated by immigration, food shortages, and cultural changes. The means by which people adapt to irregularities in the supply of dietary carbohydrates, and yet maintain normal functioning of their brain and nerves, is explored in the next two sections.

Control Of Blood Sugar After Meals. The factors which affect the blood sugar after a meal are outlined in Fig. H-26.

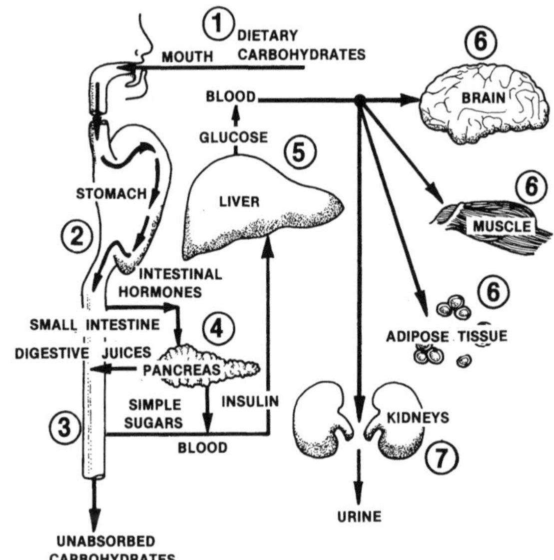

FACTORS WHICH AFFECT THE BLOOD SUGAR AFTER A MEAL

1. **Size and nutrient composition of the meal**
2. **Rate at which the stomach empties into the small intestine**
3. **Efficiency of digestion and absorption**
4. **Rate of insulin secretion by the pancreas**
5. **Metabolism of sugars and insulin in the liver**
6. **Utilization of glucose by such tissues as brain, muscle, and adipose tissue**
7. **Spilling of sugar from the blood into the urine**

Fig. H-26. Regulation of blood sugar after meals.

A discussion of the schematic outline follows:

1. **Nutrient content of the meal.** Foods rich in carbohydrates produce rapid, but often only temporary, rises in the blood sugar which are frequently followed by sharp declines, depending upon the size of the meal, the rate of nutrient absorption, and the response of the pancreas. Proteins contribute to a steady level of blood sugar because some of them are converted to sugar when needed. Fats slow both the absorption and the rate of

utilization of sugars by the various tissues. Hence, moderate amounts of dietary proteins and fats help to counteract the ups and downs in the blood sugar caused by carbohydrates.

2. **The rate of gastric emptying.** Meals which are large and/or fatty slow the rate at which the stomach empties its contents into the small intestine, whereas chilled foods tend to accelerate this process. Nervous people may have rapid emptying of the stomach. Rapid emptying of the stomach may result in speedy carbohydrate absorption and a surge in the blood sugar level, followed by a sharp drop due to overstimulation of insulin secretion by the pancreas.

3. **Efficiency of digestion and absorption.** Some people do not fully digest the complex dietary carbohydrates to readily absorbable sugars, but excrete the former in their stools. Such malabsorption may give rise to bacterial fermentation, gas, and diarrhea. For example, many people around the world lack the digestive enzyme lactase, which splits unabsorbable milk sugar (lactose) into simple sugars (galactose and glucose) that are absorbed. Thus, administration of milk sugar to such persons fails to produce a rise in blood sugar; instead, it may cause intestinal cramps and loose stools. Excessively rapid movement of the digestive tract, such as is found in very nervous people, may produce similar effects.

4. **Rate of insulin secretion by the pancreas.** The entry of food into the small intestine triggers the intestine to release hormones into the blood, which stimulate the pancreatic secretion of both digestive juices and the hormone insulin. The release of insulin is also stimulated by the rise in blood sugar following the absorption of simple sugars. A rapid and copious release of insulin produces a quick drop in blood sugar because the main function of this hormone is to promote the passage of sugars, fatty acids, and amino acids from the blood into the cells of the body.

5. **Metabolism of sugars and insulin in the liver.** The liver controls the amounts of sugars and insulin which reach the other body tissues since these substances must first travel through the veins which connect (a) the intestine and the liver, and (b) the pancreas and the liver. Thus, the sugars which circulate to the other tissues are those remaining after the liver has used portions of them for energy metabolism and the synthesis of fat and glycogen (the complex carbohydrate which is stored in the liver). Other simple sugars, such as fructose (fruit sugar) and galactose (derived from milk sugar), may be converted into glucose by the liver. Also, some of the insulin from the pancreas may be degraded by the liver, which has enzymes (insulinases) for this purpose.

6. **Utilization of blood sugar by various tissues.** The blood sugar level drops as glucose passes into cells throughout the body. This process requires insulin; hence, it is curtailed when this hormone is lacking or is ineffective. (Blood levels of insulin are subnormal in starvation and juvenile diabetes, whereas they are normal, but ineffective, in stress situations, in obesity, and in adult onset diabetes.) Muscle tissue uses part of the glucose for energy, part for the synthesis of glycogen (like that stored in the liver), and the remainder for the synthesis of fat. Fatty (adipose) tissue uses some glucose for energy, and the rest is converted into fat. The brain and nerves *cannot* store energy as either fat or glycogen, hence they require a steady supply of glucose from the blood. Utilizaton of glucose by the satiety center in the brain results in a reduction of the appetite in normal, healthy people. Thus, the body has a built-in mechanism which provides partial protection against overeating. However, there are various circumstances under which this mechanism is ineffective.

7. **Spilling of sugar into the urine.** If, for some reason, the blood glucose remains elevated after a meal rich in carbohydrates, some of the sugar may spill over into the urine. Healthy people may occasionally have sugar in their urine, but a regular occurrence of this condition suggests the presence of diabetes or another abnormality.

Control Of Blood Sugar During Fasting Or Starvation. Many people are able to withstand considerable deprivation of food without experiencing signs of abnormalities in their mental and nervous functions. Others, not so fortunate, may have symptoms of low blood sugar when they go too long between meals. The major factors which help to maintain the blood sugar during fasting or starvation are outlined in Fig. H-27.

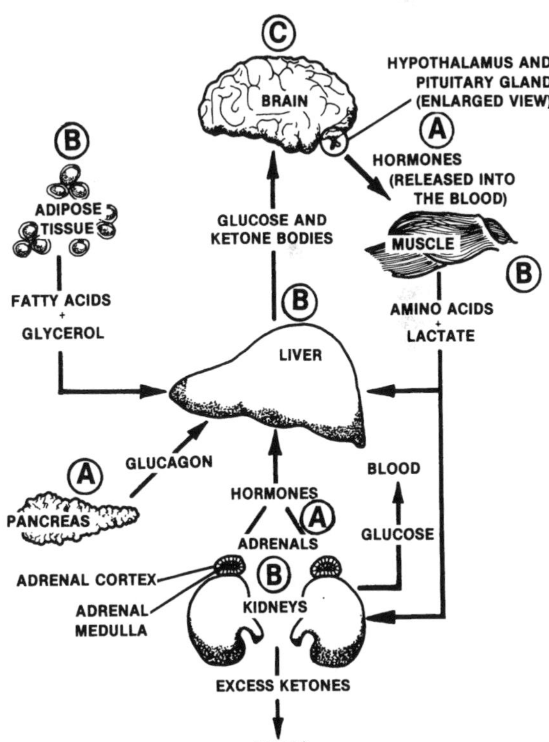

Fig. H-27. Regulation of blood sugar during fasting or starvation.

A discussion of the outline follows:

A. **Release of hormones.** A drop in blood sugar to a subnormal level triggers the release of hormones which counteract the hypoglycemia. The pancreas releases the hormone glucagon, the pituitary releases the adrenocorticotropic hormone (ACTH) and the growth hormone, and the cores (medullas) of the adrenals release adrenalin (epinephrine). ACTH from the pituitary stimulates the outer layers (cortices) of the adrenals to release cortisol.

B. **Effect of hormones on various tissues.** Although all of the hormones described in the preceding paragraph serve to raise the blood sugar, they act in different ways on the various tissues. A knowledge of these hormonal effects helps in understanding the causes, treatment, and prevention of low blood sugar. Therefore, discussions of these effects follow.

Glucagon from the pancreas promotes the processes in the liver which (1) break down glycogen to glucose and (2) convert amino acids to glucose. The glucose produced in these processes is then released into the blood.

Adrenalin from the adrenal medulla triggers the breakdown of glycogen to glucose in *both* the liver and the muscle. However, the glucose in muscles cannot be released into the blood, but must first be metabolized (during exercise) to lactic acid, which is then released into the blood. The blood carries lactic acid to the liver where it is converted to glycogen and/or glucose. Thus, the initial reduction of liver glycogen is followed by its replenishment at the expense of muscle glycogen when a person engages in moderate exercise. Another effect of adrenalin is the mobilization of free fatty acids and glycerol from stored fats (triglycerides). Finally, adrenalin delays the secretion of insulin, which would normally be secreted by the pancreas in response to an elevation in blood sugar. Stress factors which stimulate adrenalin secretion, such as low blood sugar, chilling, fear, and rage, tend to produce elevated levels of glucose and fatty acids, which the body uses as sources of energy for coping with the stresses.

ACTH from the pituitary acts to counteract the effects of low blood sugar by (1) stimulating the adrenal glands to release the hormones that convert amino acids and glycogen to glucose, and (2) promoting the breakdown of stored fats (in the form of triglycerides) into fatty acids and glycerol, which are released into the blood. Glycerol may then be converted to glucose in the liver, while elevations of fatty acids in the blood level help to keep the blood sugar up by depressing the utilization of blood glucose by various tissues.

Growth hormone from the pituitary, like ACTH, promotes the mobilization of fatty acids and glycerol from stored fat. It also speeds the release of glucose from the liver, and it antagonizes the effects of insulin by blocking the utilization of glucose in muscle.

The adrenal cortical hormone cortisol is usually secreted according to a day-to-day rhythmic pattern (circadian rhythm). Blood levels of the hormone usually peak during the early hours of the morning, and drop to low levels after meals. Cortisol and glucagon from the pancreas promote the processes which convert protein into blood glucose. The conversion of protein to blood sugar involves (1) the breakdown of muscle protein to amino acids, (2) transport of the amino acids through the blood to the liver, (3) conversion of amino acids to glucose within the liver and kidneys, and (4) release of glucose from the liver to the blood. Cortisol is usually the last of the hormones to respond to the low blood sugar, which is fortunate because protein is one of the most essential substances in the body.

C. **Reduced activity to the satiety center in the brain.** When plenty of glucose is available to the brain, the satiety center puts a brake on the appetite for food. However, a subnormal level of blood glucose results in decreased activity of the satiety center and an increased appetite for food.

MAJOR CAUSES OF HYPOGLYCEMIA. Low blood sugar is often nature's way of telling a person that he or she needs to eat some food; for if the blood sugar could be maintained at a normal level during fasting, one might go without eating and starve to death. Fortunately, there are also other factors, such as force of habit and rumblings of the stomach, which provoke us to eat. However, even persons who are well fed and apparently healthy may sometimes suffer from bouts of hypoglycemia. Furthermore, it would not be sensible to advise all persons with a tendency for low blood sugar to rely solely on eating when they have symptoms of this condition, since some might become obese, which could lead to even worse problems. Last, but not least, there is always the possibility that hypoglycemia might be due to a disorder requiring medical treatment. Therefore, it is important to note the causes of low blood sugar which follow:

(Also see OBESITY.)

• **Adrenal cortical exhaustion, or chronic insufficiency of the glands**—Lack of the adrenal cortical hormones slows the conversion of protein to glucose during fasting or starvation. Exhaustion of the adrenal cortical glands may occur after a person has been subjected to prolonged or severe stresses like exposure to extreme cold, febrile diseases, starvation, surgery, and tuberculosis. Chronic adrenal cortical insufficiency may be due to an unexplained atrophy of the adrenal cortices.

(Also see ADRENAL CORTEX INSUFFICIENCY.)

• **Alcoholism**—Chronic alcoholism blocks the conversion of lactate (from the breakdown of muscle glycogen) to glucose in the liver.

(Also see ALCOHOLISM.)

• **Caffeine consumption**—Caffeine-containing beverages, such as coffee, may give a temporary lift in energy and sharpen mental acuity, but some people have a let down feeling about an hour or two later. The reason for the lift is that caffeine stimulates the central nervous system and promotes the breakdown of glycogen to glucose in the liver, which raises the blood sugar level. However, the elevation in blood glucose may be short lived in some people, because their pancreas reacts by secreting insulin. Hence, they may feel let down due to a drop in their blood sugar.

(Also see CAFFEINE.)

• **Children born to diabetic mothers**—An unborn child in the womb of a diabetic mother is exposed to a high level of glucose which may overstimulate the growth of the insulin-secreting cells of the pancreas. Often, the oversecretion of insulin in the newborn lasts only a short time after birth. However, sometimes it persists through infancy and childhood when it may be aggravated or perpetuated by the feeding of too much food.

(Also see DIABETES MELLITUS.)

• **Diabetes, adult onset type**—In contrast to juvenile diabetics, who lack sufficient insulin to keep their blood sugar down to normal, people with adult onset diabetes (this order may also occur in some children) apparently secrete normal to excessive amounts of insulin. The latter type often experience great swings between high and low levels of blood sugar because their secretion of insulin is delayed, and they may also have resistance to insulin.

(Also see DIABETES MELLITUS.)

• **Dietary factors**—When meals are few and far apart, there is more time for the blood sugar to drop to low levels between meals. Also, foods high in carbohydrates may cause rapid elevation of the blood sugar after meals, a condition which may cause the pancreas to secrete excessive amounts of insulin. In some people, high-protein diets may also overstimulate the secretion of insulin.

• **Drugs**—First and foremost of the substances producing hypoglycemia are the agents used to reduce blood sugar levels in diabetics (insulin and oral drugs), because it is sometimes difficult to determine the proper dosages of these medications. Other drugs which may lower the blood sugar in susceptible people are tranquilizers—like chlorpromazine, cardiac medicines—like propranolol (which blocks the effects of adrenalin), and overdoses of aspirin.

• **Emotional stress**—Anxiety and nervousness may cause hypoglycemia by excessively speeding up the motions of the digestive tract, by triggering a rapid release of insulin, and/or by depletion of stored carbohydrate (glycogen) due to oversecretion of adrenalin.

• **Exercise**—Prolonged or strenuous exercise may deplete both liver and muscle of their glycogen stores and produce hypoglycemia. However, special types of athletic training coupled with the dietary maneuvers may increase the glycogen stores in muscle by as much as 50%.

(Also see GLYCOGEN; and MUSCULAR WORK.)

• **Fasting or starvation**—Some people can barely tolerate an overnight fast; they may be awakened in the early hours of the morning by stomach pains due to low blood sugar. Others may lack the ability to convert proteins to glucose rapidly enough to maintain their blood sugar levels.

• **Gastrointestinal problems**—Rapid emptying of the stomach into the small intestine following a meal rich in sugars but low in fat (the latter nutrient tends to slow the emptying of the stomach) may lead to hypoglycemia. Such a situation is much more likely to occur in people who have had their stomachs removed, because the food then passes directly from the esophagus to the small intestine. (Also see DUMPING SYNDROME.) However, some highly nervous people may have overactive digestive tracts. Also, hypoglycemia may result from poor digestion and/or absorption of carbohydrates.

• **Glucagon insufficiency**—It is suspected that some people do not secrete enough glucagon when their blood sugar is low because they have fewer than the normal number of pancreatic alpha cells which secrete this hormone. Evaluation of glucagon secretion is complex because it has recently been found that the small intestine secretes a hormone which is very similar to glucagon in its chemical properties and in its effects on tissues.

(Also see ENDOCRINE GLANDS.)

• **Inborn errors of metabolism**—Among the types of genetic disorders which may give rise to hypoglycemia are: (1) lack of some of the enzymes involved in the conversion of amino acids to glucose, (2) inability to break glycogen down to glucose (glycogen-storage diseases), and (3) inability to utilize fructose (fructose intolerance). These disorders are rare in persons of mixed genetic background; they are generally limited to populations in which there have been many close kinship marriages, or inbreeding.

• **Insulin excess**—This situation was first observed in diabetics given too much insulin. Later, it was found that the chronic eating of a high-carbohydrate diet may cause enlargement (hypertrophy) of the pancreatic beta cells which secrete insulin. These cells may then produce and excrete excessive amounts of insulin.

• **Kidney disease**—Impairment of the various kidney functions may result in hypoglycemia due to (1) failure in the conversion of amino acids to glucose, or (2) loss of blood glucose in the urine.

• **Liver disease**—The main ways in which diseases of this organ produce hypoglycemia are (1) interference with the metabolism of glycogen, and (2) failure in the conversion of amino acids to glucose.

• **Obesity**—More than 80% of adult onset diabetics are overweight. Cells which are swollen with fat resist the actions of insulin; hence, obese people may have the paradoxical condition of high blood sugar levels and high blood levels of insulin occurring simultaneously. However, their pancreas may continue to pump out insulin until the blood sugar starts to drop. Often the excess of insulin produces a drastic drop in the blood sugar level. It is also believed that chronic oversecretion of insulin may sometimes be the cause of obesity. Hence, it may be difficult to know which came first—the oversecretion of insulin, or the obesity.

(Also see DIABETES MELLITUS; and OBESITY.)

• **Pituitary insufficiency**—Deficiencies of either ACTH (which blocks the effects of insulin and stimulates the release of hormones from the adrenal cortices), or of the growth hormone (like ACTH, it may also counter the effects of insulin) may result in hypoglycemia. Such deficiencies are most likely to be found in persons suffering from severe protein deficiencies and other forms of malnutrition.

• **Thyroid deficiency**—As in the case of pituitary insufficiency, malnutrition can produce a thyroid deficiency in people who would otherwise have normal glandular functions. Lack of thyroid hormones may contribute to hypoglycemia due to (1) reducing the absorption of sugars from the intestines (these hormones usually promote such absorption), and (2) slowing the breakdown of glycogen to glucose (thyroid hormones accelerate such breakdown).

• **Tumors**—Tumors of the pancreas may produce great excesses of insulin. Also, tumors of other tissues may release into the blood substances which have insulinlike effects, or they may utilize glucose so rapidly for their growth that they cause a drop in the blood sugar.

• **Women after childbirth**—Pregnant women usually have greater-than-normal levels of blood glucose due to hormonal and other metabolic factors unique to this situation. The elevation of blood levels of sugar and other nutrients allows the fetus to meet its needs better. After pregnancy, the nervous system may have to adjust to a lower level of blood sugar than that to which it had become accustomed during the preceding 9 months. It is noteworthy that the first diagnosis of hypoglycemia in a woman may be made after a pregnancy. This prompts the question: Are postpregnancy bouts of depression in any way related to low blood sugar?

OTHER DISORDERS WHICH MAY RESULT FROM HYPOGLYCEMIA. Just about everyone may have a mild hypoglycemic episode from time to time. Perfect regulation of the blood sugar in the face of all challenges is more than might be expected of a normal, healthy body. Experienced athletes and other people who strive hard to reach their goals learn that temporary weaknesses due to low blood sugar soon pass. However, severe forms of acute or chronic hypoglycemia pose some dangers, since upsetting the normal processes of the body may, along with other factors, produce harmful effects. Some possible effects of abnormally low levels of blood sugar follow:

• **Accident proneness**—Such special senses of the body as sight and hearing depend upon good nutrition of the nerves which feed sensory data into the brain. The failure of a person to react swiftly to a hazardous situation may mean involvement in an accident which might otherwise have been prevented. Low blood sugar has been shown to produce alterations in the electrical impulses generated within the brain and to slow response time.

• **Alcoholism**—From time to time, it has been suggested that chronic hypoglycemics may seek refuge in an alcoholic stupor in order to escape the irritations which accompany their condition. The nervous symptoms of hypoglycemia are due to the outpouring of adrenalin as the body attempts to bring the blood sugar up to normal. Alcohol depresses the activity of the central nervous system, thereby counteracting the effects of adrenalin. However, there is no proof that hypoglycemia is a major cause of alcoholism, only the reverse—that chronic alcoholism may produce hypoglycemia.

• **Brain damage in infants**—Hypoglycemia frequently occurs in newborn infants who are premature and/or who are smaller than normal. Failure to treat this condition may result in defects of brain development.[74]

• **Depression**—Some people become depressed when their blood sugar is low, although there are often other causes of this emotional problem. However, persons seeking to get rid of depression should make certain that the possibility of hypoglycemia has been ruled out before they undertake a long and expensive course of psychiatric treatment.

(Also see the section of this article—"Conditions Which May Mimic Hypoglycemia.")

• **Diabetes**—In some people, chronic hypoglycemia is due to constant overactivity of the insulin-secreting cells of the pancreas. These cells may become "burned out" and diabetes may ensue. Adult onset diabetes might result from an abnormal pattern of endocrine secretion which is found in some obese people who have long been overeating in an attempt to alleviate the annoying symptoms of low blood sugar.

• **Drug addiction**—As in the case of alcoholism, some investigators have suggested that chronic hypoglycemics may be tempted to use psychoactive drugs to stabilize their emotional ups and downs. After all, many businessmen rely on coffee to keep them going during the day, and on alcohol or sleeping pills to calm themselves down at night.

• **Extra sensitivity to allergens**—Exhausted people seem more prone to develop allergies. Now, there is evidence to support the idea that low blood sugar heightens allergic responses. It has been shown that more histamine is secreted in response to an injected allergen when the blood sugar is low than when it is normal or high.[75] Histamine is the culprit responsible for many of the irritating features of allergic responses.

• **Fatigue**—Lack of glucose for the nervous system produces feelings of tiredness and lack of drive. Many people, however, are exceptionally strong willed and are able to continue their activities in spite of such feelings. When there is a prolonged deprivation of glucose, such as that which occurs after long periods of fasting or starvation, the brain and nerves begin to use ketone bodies (derived from the mobilization of stored fats) in lieu of glucose.

• **Glycogen depletion**—Liver glycogen may drop very low when this stored carbohydrate is drawn upon to meet the

---

[74]Chase, H. P., *et al.*, "Hypoglycemia and Brain Development." *Pediatrics*, Vol. 52, 1973, p. 513.

[75]Adamkiewicz, V. W., and P. J. Sacra, "Histamine and Sugars," *Federation of American Societies for Experimental Biology Proceedings*, Vol. 26, 1967, p. 209.

needs for glucose. The liver is extra susceptible to damage by various stresses and toxic factors when it has a low content of glycogen.

• **High blood pressure (hypertension)**—Oversecretion of epinephrine (Adrenalin) has been implicated as one of the factors responsible for hypertension. Renin, a hormone secreted by the kidney, also acts to raise the blood pressure. Recently, it was found that hypoglycemia triggered the secretion of both adrenalin and renin, and that theophylline (a substance which is similar to caffeine, and is found in tea) enhanced the overall response.[76]

• **Hunger for sweets**—A person with chronic hypoglycemia soon learns that sugary beverages and foods bring about the quickest relief from the aggravating symptoms of the disorder. However, the practice of using sugar as a remedy may do more harm than good, if the pancreas is stimulated to produce and release even more insulin.

• **Irritability**—Anyone who has frequent outpourings of adrenalin is likely to be irritable, and sometimes even antagonistic. It is suspected that the Qolla—a very aggressive group of people living in the border area between Bolivia and Peru, noted for high rates of murder and rape—owe their ferocity to severe hypoglycemia.[77]

• **Ketosis**—A shortage of carbohydrates forces the body to draw upon its fat stores for energy. Ketosis may result when energy metabolism is based upon a very high proportion of fat and a low proportion of carbohydrates. Such a condition is usually prevented by the conversion of protein to glucose when the need arises, but this process may not work well for some people. Children appear to be more likely to develop ketosis than adults.

• **Liver damage**—The depletion of glycogen from the liver in order to meet the demands for blood glucose leaves the organ very vulnerable to damage by noxious substances. This condition, plus the flood of fat which is mobilized from the fatty tissues by the various antihypoglycemic hormones (ACTH, adrenalin, cortisol, glucagon, and growth hormone), is a deadly combination which sorely taxes the defenses of the liver.

• **Neurotic behavior**—Anyone who is subject to the ups and downs of mood which accompany swings in blood sugar may develop abnormal behavior patterns in order to avoid anguish and humiliation. For example, a salesperson may not feel up to meeting with potential customers when his or her blood sugar is low. Sometimes, business may be conducted during a lunch which is well lubricated with alcoholic drinks. Others may find that getting angry at the least provocation helps to dispel weakness due to hypoglycemia since the adrenalin released during anger helps to raise the blood sugar. Perhaps psychiatrists should test more of their patients for hypoglycemia.

• **Obesity**—A person whose hypoglycemia is due to excessive secretion of insulin may get fat, since insulin (1) promotes the synthesis of fat from carbohydrates, and (2) blocks the mobilization of fat from fatty tissues. Furthermore, the symptoms of hypoglycemia may drive some people to eat in order to obtain relief.

• **Oversusceptibility to toxic agents**—The liver is the body's main defense against toxic agents which get into the body via the food that is eaten or the air which is breathed into the lungs. Hence, depletion of liver glycogen by hypoglycemia reduces the body's ability to detoxify poisons.

• **Slackening of mental or physical performance**—It has long been thought that the poor morning performance of both schoolchildren and adult workers is due mainly to either (1) skipping breakfast, or (2) consuming high-carbohydrate diets, which produce only a short-lived rise in blood sugar. The effects of breakfast on work performance were investigated in a series of well-known studies conducted by Iowa State University.[78]

The researchers found that work performance was highest when an adequate breakfast was eaten; next best when coffee and rolls were consumed at a mid-morning break, after breakfast had been skipped; and lowest when there was neither breakfast nor a mid-morning snack. Other studies have shown that small, frequent meals (eating 5 times a day) generally result in better work performance than eating fewer, but larger, meals (such as 2 or 3 per day). Thus, the provision of milk or coffee breaks may be a wise move on the part of a school or an employer seeking to prevent slacking off of work due to hypoglycemia.

• **Sleeplessness**—This grievance usually besets its victim in the wee hours of the morning, when the liver glycogen which has been built up from supper becomes depleted. In most persons, the hypoglycemia which ensues is offset by the secretion of adrenal cortical hormones which promote the conversion of protein to glucose. However, a few people may awake with a start, and have hypoglycemic symptoms such as a rapid pulse, irritability, stomachache, or nervousness. These complaints are generally due to the outpouring of adrenalin which is ineffective in raising the blood sugar because the liver's supply of glycogen has been exhausted.

• **Ulcers**—Hypoglycemia and peptic ulcers occur together in some people. It seems that low blood sugar stimulates the secretion of acid and digestive juice by the stomach. Normally, the lining of the stomach is protected by an acid-resistant barrier of mucus. However, adrenal cortical hormones, which are secreted in response to hypoglycemia and/or emotional stresses, may set up conditions favorable for ulcer formation

[76]Lowder, S. C., M. G. Frozen, and G. W. Liddle, "Effect of Insulin-Induced Hypoglycemia upon Plasma Renin Activity in Man," *Journal of Clinical Endocrinology and Metabolism*, Vol. 41, 1975, p. 97.

[77]Trotter, R., "Aggression: A Way of Life for the Qolla," *Science News*, Vol. 103, 1973, p. 76.

[78]Tuttle, W. W., and E. Herbert, "Work Capacity with No Breakfast and a Mid-Morning Break," *Journal of the American Dietetic Association*, Vol. 37, 1960, p. 137.

because they reduce the amount of mucus secreted by the stomach.

DIAGNOSIS OF HYPOGLYCEMIA. A doctor should be relied upon to decide whether or not one has a tendency towards an abnormally low blood sugar level because a laboratory test on the blood is needed in order to be certain of the diagnosis. However, everyone should be able to recognize the major features of the condition so as to know when a physician should be consulted. Therefore, signs and symptoms of hypoglycemia follow.

Signs And Symptoms. When hypoglycemia develops within a short period of time (such as from the oversecretion of insulin), the symptoms, which are mainly due to the adrenalin response, include rapid pulse, trembling, hunger, and a slight amount of mental confusion. However, hypoglycemia which has developed over many hours (such as when meals are skipped) is characterized by signs that the brain has undergone glucose deprivation, including headache, depression, blurred vision, incoherent speech, considerable mental confusion, and, occasionally, coma or convulsions. None of the above signs and symptoms can be attributed solely to hypoglycemia. There are similar syndromes in drunkenness and in certain nutritional deficiencies. Therefore, a person suspected of having a tendency to low blood sugar should be given a glucose tolerance test.

Glucose Tolerance Test (GTT). The term *glucose tolerance* literally means the rate at which the cells of the body take up glucose from the blood. Factors which reduce glucose tolerance are those that slow the rate of glucose uptake by cells. Therefore, such factors tend to raise the blood sugar above the level considered normal under conditions of good health.

The test is administered in the morning, following an overnight fast. Sometimes, the doctor or dietitian tells the person who is to be tested to eat a diet moderately high in carbohydrates (250 to 300 g) for each of the 3 days preceding the test. The purpose of this so-called "carbohydrate-loading" is to stimulate the pancreas to its full capacity for secreting insulin. First, a sample of "fasting blood" is taken. Then, a test dose of glucose is given in the form of a water solution which the patient drinks. Blood samples are usually taken at hourly intervals for up to 6 hours. A diagnosis is made by comparing the hourly blood glucose levels of the patient to standard values. Depending upon the judgment of the physician and the laboratory methods used to measure glucose, a person is diagnosed as hypoglycemic when his or her blood sugar falls to values below 60 mg of glucose per 100 ml of blooa. Fig. H-28 shows normal, hypoglycemic, and diabetic blood sugar curves. Explanations of these curves follow:

The normal blood sugar curve shows a moderate rise as the test dose of glucose is absorbed from the intestine and enters the blood. Soon after absorption, insulin secreted by the pancreas starts to promote the utilization of glucose by various tissues, thereby causing the blood sugar to drop back to the fasting level. A fasting blood sugar of around 80 mg per 100 ml of blood is considered to be normal.

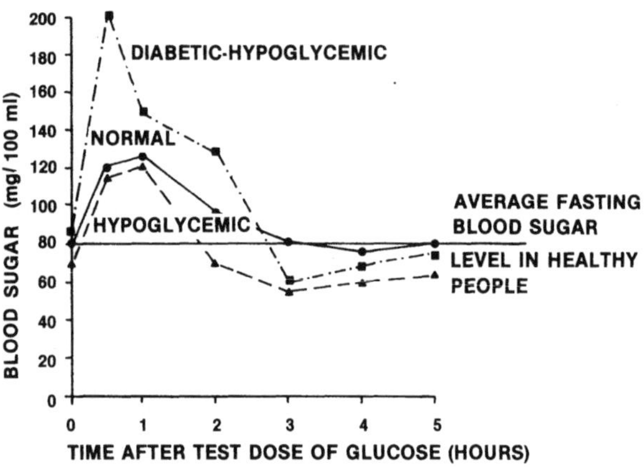

Fig. H-28. Glucose tolerance curves.

A person with hypoglycemia may have a fasting blood sugar slightly below normal (70 instead of 80). However, the early response of such a person to a test dose of glucose will be very close to normal for the first hour or so of the test. Then, there is likely to be an abnormally sharp drop in the blood sugar due to an oversecretion of insulin. After about 3 hours, the blood glucose level of the hypoglycemic is likely to be below the normal fasting level. It may then start back up towards the normal level after the 4th or 5th hour. People whose blood sugar curve indicates hypoglycemia may be totally free of symptoms. For this reason, some doctors and health scientists think that a lowered blood sugar should not be considered hypoglycemic until it drops to 40. On the other hand, some people start to have symptoms while their blood sugar is on its way down, but is above 60.

Finally, many people with adult onset diabetes may have a blood sugar curve which typifies what at one time was called "dysinsulinism," in which the insulin response is delayed and the blood sugar soars to a diabetic level. At this point, there is a heavy secretion of insulin which causes the blood sugar to plummet to a hypoglycemic level. Hence hypoglycemia is sometimes a prelude to, or an accompaniment of, diabetes.

(Also see DIABETES MELLITUS.)

Conditions Which May Mimic Hypoglycemia. Many of the signs and symptoms of low blood sugar are similar to those which occur in other, unrelated disorders. Some of these conditions follow:

- **Reduced flow of blood to the brain and other tissues**—Symptoms of dizziness, followed by jumpiness and rapid pulse, are more often due to problems related to the circulation of blood than to hypoglycemia. There are various regions throughout the body, such as the carotid artery in the neck, which delivers blood to the brain, that are sensitive to drops in blood pressure and/or the oxygen content of the blood. These conditions often provoke a nervous response which triggers the release of epinephrine (Adrenalin). Epinephrine (Adrenalin) and a similar hormone, norephinephrine, act to speed the heart rate and to raise the blood pressure.

Factors responsible for a slowing of the circulation are lack of exercise, partial clogging of blood vessels by clots and/or fatty deposits, pooling of the blood in the digestive system after eating, weakened pumping action of the heart, or postural hypotension (the term used to denote a drop in blood pressure following a sudden change from a supine to an upright position, such as when a person jumps up from bed in the morning).

• **Depression**—Often, feelings of apathy, boredom, or despair are attributed to low blood sugar. More than likely, these dismal feelings are due to biochemical factors, emotional problems, lack of stimulation, or poor mental outlook. For example, bouts of depression appear to increase with aging, and may be the result of heightened activity of the enzyme monoamine oxidase (MAO) which breaks down adrenalin and other stimulatory substances (amines) into biologically inactive compounds. Also, it is noteworthy that very depressed people, such as suicidal types, have low brain levels of serotonin, a nerve transmitter derived from the amino acid tryptophan, present in varying amounts in protein foods. Brain levels of serotonin may be affected by diet, so it soon may be possible to devise an antidepression diet.

• **Nutritional deficiencies**—Just as a hypoglycemic condition may sometimes mimic a drunken condition, certain nutritional deficiencies have features like those of both alcoholism and low blood sugar. Mental and nervous disorders may be due to lack of thiamin (vitamin B-1), niacin, vitamin B-6 (pyridoxine), and vitamin B-12.

Rapid heartbeat is not only due to the adrenalin response to hypoglycemia, but also occurs in iron-deficiency anemia where the heart strives to compensate for oxygen-poor blood by pumping greater amounts of blood to the deprived tissues.

• **Alkalosis**—Like hypoglycemia, this condition is accompanied by such symptoms as feelings of lightheadedness and tiredness, nausea, trembling, and sweating. There may be a mild alkalosis for a short time after meals, due to the so-called "alkaline tide," which is a mild elevation in the alkalinity of the blood resulting from the secretion of acid in the stomach. Another cause of this condition is the exaggeration of such actions as coughing, deep breathing, laughing, and yawning due to force of habit or nervousness. Such actions cause excessive loss of carbon dioxide from the lungs. Since carbon dioxide contributes acidity to the blood, its loss produces alkalinity.

TREATMENT AND PREVENTION OF HYPOGLYCEMIA. It is important to treat severe hypoglycemia promptly, so as to avoid permanent damage to the brain and nervous system. After treatment has been given, measures should be taken to prevent such episodes in the future, since the victim might have another attack when alone or at night while sleeping. It makes one wonder how many times hypoglycemia might have been a contributing factor to deaths occurring in the early hours of the morning. Such thoughts give point and purpose to the recommendations which follow:

Emergency Treatment. Every person known to have a tendency to diabetes or hypoglycemia should always have a source of sugar at hand, such as hard candy, fruit juice, or just plain packets of sugar. These items will pull a person out of the severe stage of hypoglycemia, but they may also render the victim more susceptible to a subsequent attack. Therefore, long-term measures to prevent future attacks need to be instituted at the first opportunity.

Diet Therapy. The standard treatment for low blood sugar was formerly a low carbohydrate, high-fat diet. However, many of those so treated developed adult onset diabetes. It is conjectured that the hypoglycemia of these patients was actually the first sign of diabetes. If that were the case, a high-fat diet did more harm than good, since it is well known that fat blocks the actions of insulin. This effect of fat was formerly thought to be desirable for hypoglycemics, because the therapy was aimed at countering what was thought to be the result of excessive insulin secretion. Recent research has shown that replacement of dietary carbohydrate with either fat or protein has undesirable effects on patients who develop low blood sugar after eating.[79] Therefore, the current dietary treatment for hypoglycemia is based upon the following considerations:

1. Insulin secretion should be regulated by avoiding excessive stimulation of the pancreas by either large meals or too much readily available carbohydrate.

2. Six small meals daily are preferable to three large ones. Reducing the intervals between meals from about 5 hours on a three-meal pattern to about 3 hours on a six-meal pattern helps to keep the blood sugar from dropping too low.

3. The dietary pattern should be essentially the same as that for adult onset diabetes.

4. Hard-to-manage cases may require nondietary treatments, rather than diets which are extremely low in carbohydrates.

Typical diets used for hypoglycemia are given in Table H-7.

[79]Anderson, J. W., and R. H. Herman, "Effect of Carbohydrate Restriction on Glucose Tolerance of Normal Men and Reactive Hypoglycemic Patients," *The American Journal of Clinical Nutrition*, Vol. 28, 1975, p. 748.

TABLE H-7
TYPICAL MENUS FOR CONTROL OF MILD HYPOGLYCEMIA (LOW BLOOD SUGAR)

Types of Food Exchanges[1] (used for each meal or snack)	Number of Exchanges per Meal (based upon daily energy requirements) Calories (kcal) per Day					
	1,200	1,500	1,800	2,100	2,400	2,700
Breakfast:						
Bread (or cereals, crackers, pancakes, etc.)	1	1	2	2	2	2
Fat (bacon, butter, margarine, etc.)[2]	1	1	2	3	3	3
Fruit (or equivalent in juice)	1	1	1	1	1	1
Lean meat (cheese, eggs, fish, fowl, meats, and/or legumes)	1	1	1	1	1	1
Optional beverage: bouillon or clear broth, caffeine-free coffee, club soda, herb tea, or water	Any amount, but *no* alcohol or sugar.[3]					
Mid-morning snack:						
Bread	1	1	1	1	1	2
Fat	1	1	0	2	2	2
Fruit (or juice)	0	0	0	0	0	1
Milk, nonfat (if whole milk is used, deduct two servings of fat from the day's allowance)	1	1	1	1	1	1
Optional beverage (see Breakfast)	Any amount, but *no* alcohol or sugar.					
Lunch:						
Bread	1	1	2	2	2	3
Fat	1	2	2	3	4	4
Fruit (or juice)	0	0	0	1	1	1
Meat (see Breakfast)	2	2	2	3	3	3
Optional beverage (see Breakfast)	Any amount, but *no* alcohol or sugar.					
Raw vegetable, or salad w/o dressing	1	1	1	1	1	1
Mid-afternoon snack:						
Bread	0	1	1	1	1	2
Fat	0	1	0	2	2	2
Fruit (or juice)	0	0	0	0	0	1
Milk, nonfat	1	1	1	1	1	1
Optional beverage (see Breakfast)	Any amount, but *no* alcohol or sugar.					
Supper:						
Bread	1	2	2	2	3	3
Fat	2	3	3	3	5	5
Fruit (or juice)	0	1	1	1	2	1
Meat (see Breakfast)	3	3	3	3	3	3
Optional beverage (see Breakfast)	Any amount, but *no* alcohol or sugar.					
Raw vegetable, or salad w/o dressing	1	1	1	1	1	1
Nonstarchy vegetable, cooked	1	1	1	1	1	1
Late evening snack:						
Bread	0	0	0	1	1	2
Fat	0	1	1	1	2	2
Fruit (or juice)	1	1	1	1	1	1
Milk, nonfat	1	1	1	1	1	1
Optional beverage (see Breakfast)	Any amount, but *no* alcohol or sugar.					

[1]Exchanges are portions of foods which have been grouped together because they contain similar proportions of carbohydrates, fats, proteins, and calories. Portion sizes have been calculated for the foods within each group so that they may be substituted for one another because their nutritive values are approximately equal.

(Also see MODIFIED DIETS, section headed "Food Exchanges.")

[2]The total day's allowance for fat plus the distribution of fat at meals and snacks may have to be adjusted so as to compensate for foods which contain considerably more fat than the amount typical of their exchange groups. For example, two fat exchanges should be deducted from the amounts allowed each time that whole milk is used instead of the nonfat type, or one fat exchange deducted to compensate for the use of 2% milk or plain yogurt in lieu of nonfat milk. Similarly, one fat exchange should be deducted for each exchange of high-fat meat used instead of lean meat, or 1/2 fat exchange deducted for each exchange of medium-fat meat. Also, many doctors recommend that the items used as fat exchanges be selected from sources of polyunsaturated fats, such as salad oils and soft margarines.

[3]The optional beverage might be sweetened with some fruit juice taken from the day's fruit allowance, or the coffee might be lightened with some of the milk which is allowed. Plain club soda may contribute to a mild alkalosis in people who are susceptible to this condition after meals. Therefore, they should either omit it altogether, or add an acid fruit juice (taken from their fruit allowance) to it so as to improve its flavor and neutralize the alkali.

Further information relative to each of the exchange groups listed in Table H-7 follows:

• **Meat**—The high-protein foods comprising these exchanges include cheese, eggs, fish and seafood, fowl, meats, and legumes. Each lean meat exchange contains 7 g of protein, 3 g of fat, and 55 Calories (kcal).

• **Milk**—This group contains the various forms of milk. (Other dairy products like butter are grouped with the fat exchanges, while the cheeses are grouped with the meat exchanges.) The standard measure is the equivalent of an 8-oz (*240 ml*) cup of skim milk, which contains 12 g of carbohydrate, 8 g of protein, and 80 Calories (kcal).

• **Vegetable, raw**—Some of the members of this group are salad ingredients like chicory, endive, escarole, lettuce, parsley, radishes, and watercress, which contribute negligible amounts of available carbohydrates, fats, proteins, and calories when they are eaten raw.

• **Vegetable, cooked nonstarchy**—Vegetables such as beets, carrots, kale, okra, onions, pumpkins, rutabagas, and squash contain about 5 g of carbohydrate, 2 g of protein, and 25 Calories (kcal) per ½ cup serving (approximately 100 g).

• **Fruit**—Only fruits make up this group. Each exchange, such as a small apple, furnishes 10 g of carbohydrates and 40 Calories (kcal).

• **Bread**—These are high-carbohydrate, low-protein items. In addition to breads, it includes cereals, crackers, flour, and starchy vegetables. Each item contains 15 g of carbohydrates, 2 g of proteins, and 70 Calories (kcal).

• **Fat**—This group of high-fat foods includes such items as bacon, butter, cream, lard, margarine, mayonnaise, nuts, and olives. Each serving contains the equivalent of a teaspoon of fat (5 g of fat and 45 Calories [kcal]).

(Also see MODIFIED DIETS, section headed "Food Exchanges.")

Here is how to use Table H-7: First, determine the desirable weight for (1) a male by allowing 106 lb (*48 kg*) for the first 5 ft (*152 cm*) of height and 6 lb (*2.7 kg*) for each additional inch (*2.5 cm*) of height over 5 ft, or (2) a female by allowing 100 lb (*45 kg*) for the first 5 ft of height and 5 lb (*2.3 kg*) for each additional inch (*2.5 cm*). Then, multiply the desirable weight by 15 to obtain the caloric allowance. Second, select foods from each of the exchanges.

For example, a lady who is 5 ft 4 in. (*163 cm*) tall and of medium build should weigh about 120 lb (*54 kg*) (4 × 5 + 100 = 120). If she is not very active, but occasionally goes for a walk, her energy needs are 1,800 Calories (kcal) per day (120 × 15 – 1,800). She may use the menu plan for 1,800 Calories (kcal) per day, if her weight is what it should be (120 lb). However, if she is overweight and wishes to lose from 1 to 2 lb (*0.45 to 0.9 kg*) per week, then she must limit her diet to around 1,200 Calories (kcal) per day. Each pound per week of weight loss requires about 500 Calories (kcal) less per day. The same principle applies to energy needed for gaining weight. Some typical meals and snacks for a 1,200 Calorie (kcal) daily plan follow:

Meal	Menu		Energy (Calories [kcal])
Breakfast . . .	Ham, lean	1 slice . .	55
	Whole wheat toast . .	1 slice . .	70
	Butter or margarine . .	1 tsp	45
	Grapefruit	½	40
	Decaffeinated coffee.	1 cup, no cream or sugar . . .	0
Mid-morning snack	Corn muffin.	1 (2-in. diameter)	115
	Nonfat milk.	1 cup or 8-oz glass. . . .	80
Lunch	Hamburger, lean	2 oz.	110
	Hamburger bun	. . . .	70
	Lettuce and dill pickle slices	no limit. .	0
	French dressing	1 Tbsp . .	45
	Bouillon	1 cup . . .	0
Mid-afternoon snack	Nonfat milk.	1 cup . . .	80
Supper	Roast chicken	3 slices .	165
	Mashed potato	½ cup . .	70
	Butter or margarine . .	1 tsp	45
	Cooked beets	½ cup . .	25
	Salad made with Chinese cabbage, parsley, herbs, vinegar, and oil (1 tsp)	1 cup of salad. . . .	45
	Decaffeinated coffee.	1 cup, no cream or sugar . . .	0
Late-evening snack	Apple	1 (2-in. diameter)	40
	Nonfat milk.	1 cup . . .	80
		Total.	1,180

Practical hints for Controlling Blood Sugar. People who have a tendency towards low blood sugar may often spare themselves from the agonies of this condition if they regularly eat a good diet and practice certain other health habits. Some beneficial practices follow:

1. Eat regularly, well-balanced meals.
2. Avoid highly refined starches and sugars, especially between meals.
3. Avoid caffeine.
4. Exercise and avoid stress.

Nondietary Measures. It may be necessary for a physician to resort to nondietary treatments to control hypoglycemia in some people. Discussions of some typical measures follow:

• **Adrenal cortical extract**—Some recent books and articles for the lay public have promoted injections of adrenal cortical extract as a remedy for low blood sugar. The American Medical Association is strongly opposed to such treatments on the grounds that (1) they are nowhere near as effective as injections of pure, steroid hormones; and (2) many people who have been given these injections do not need extra adrenal cortical hormones. A hypoglycemic glucose tolerance does *not* demonstrate the lack of sufficient adrenal cortical secretion. Special tests of the glands' function are necessary for such a diagnosis.

• **Antispasmodic (anticholinergic) drugs**—These drugs block some of the nerve impulses to the stomach and intestine, thereby slowing the rate of emptying of the stomach, and reducing the secretion of intestinal hormones which trigger the release of insulin. Such therapy has helped some highly nervous people who literally had jittery digestive systems.

• **Psychotherapy**—Although these treatments may be expensive and time-consuming, they may in the long run be well worth the cost, since continual emotional distress may ruin one's health. However, a thorough physical examination is often helpful in determining the extent to which hypoglycemia is mental and/or physical.

• **Steroid hormone treatments**—Sometimes, tests of adrenal cortical function reveal an insufficiency in the secretion of these glands. Then, it is necessary for the physician to inject regularly the hormones which are lacking.

• **Surgery**—Surgical removal of part of the pancreas may be necessary if it contains a tumor which is secreting a large excess of insulin.

HYPOGLYCEMIC AGENTS

These agents are capable of lowering the sugar (glucose) content of the blood. The hormone insulin from the pancreas is the natural hypoglycemic agent. Some sulfa-type drugs (sulfonylureas) are used as hypoglycemic agents for some forms of diabetes. However, the action of these drugs is to stimulate the pancreas to release insulin.

(Also see DIABETES MELLITUS, section headed "Treatment and Prevention of Diabetes"; ENDOCRINE GLANDS; and HYPOGLYCEMIA.)

HYPOKALEMIA

Low potassium in the blood. Hypokalemia is a serious complication of severe diarrhea, in which large amounts of potassium are lost in intestinal secretions. Replacement therapy should involve added potassium.

HYPOMAGNESEMIA

An abnormally low level of magnesium in the blood. (Also see MAGNESIUM.)

HYPOPROTEINEMIA

An abnormally small amount of total protein in the blood plasma.

HYPOTHALAMUS

A group of nuclei at the base of the brain, includes (1) centers of appetite control and (2) cells that produce antidiuretic hormone.

HYPOTHYROIDISM

Under activity of the thyroid gland, of which there are two forms: (1) cretinism, a congenital defect, and (2) myxedema, subnormal secretion occurring anytime after birth.

(Also see ENDOCRINE GLANDS; GOITER, sections headed "Cretinism," "Myxedema," and "Prevention of Goiter and Related Disorders.")

HYPOTONE DEHYDRATION

Increase of water in cells (cellular edema) and decrease of extracellular fluid, as a result of hypotonicity (decreased solutes [dissolved substances]; hence, diminished osmotic pressure) of the extracellular fluid surrounding the cell.

HYPOVITAMINOSIS (Avitaminosis)

A deficiency of one or more vitamins. Often the term is used in combination with a specific vitamin to identify the vitamin involved; for example, hypovitaminosis A.

HYSSOP

A herb whose natural camphoric characteristics are extracted and used to flavor liqueurs such as Benedictine and Chartreuse. It is a GRAS additive.

Printed and bound by CPI Group (UK) Ltd, Croydon, CR0 4YY
27/10/2024
01779863-0001